American College of Veterinary Internal Medicine (ACVIM) Forum 2015

Indianapolis, Indiana, USA
3-6 June 2015

ISBN: 978-1-5108-0409-8

Printed from e-media with permission by:
Curran Associates, Inc.
57 Morehouse Lane
Red Hook, NY 12571

CURRAN ASSOCIATES INC. proceedings.com

Some format issues inherent in the e-media version may also appear in this print version.

Printed by Curran Associates, Inc. (2015)

Address correspondence to:
AMERICAN COLLEGE OF VETERINARY INTERNAL MEDICINE (ACVIM)
1997 WADSWORTH BLVD.
LAKEWOOD, CO 80214-5293
www.ACVIM.org

Additional copies of this publication are available from:

Curran Associates, Inc.
57 Morehouse Lane
Red Hook, NY 12571 USA
Phone: 845-758-0400
Fax: 845-758-2634
Email: curran@proceedings.com
Web: www.proceedings.com

TABLE OF CONTENTS

RESEARCH REPORTS-CARDIOLOGY

RESEARCH REPORTS-NEUROLOGY

RESEARCH REPORTS-ONCOLOGY

RESEARCH REPORTS-SMALL ANIMAL INTERNAL MEDICINE

RESEARCH REPORTS-EQUINE

RESEARCH REPORTS – FOOD ANIMAL

PRE-FORUM CARDIOLOGY

AMERICAN ASSOCIATION OF VETERINARY NUTRITION/AMERICAN COLLEGE OF VETERINARY NUTRITION SMALL ANIMAL

AMERICAN COLLEGE OF VETERINARY CLINICAL PHARMACOLOGY/AMERICAN ACADEMY OF VETERINARY PHARMACOLOGY AND THERAPEUTICS

AMERICAN SOCIETY FOR VETERINARY CLINICAL PATHOLOGY

AMERICAN SOCIETY OF VETERINARY NEPHROLOGY & UROLOGY

BUSINESS MANAGEMENT

CARDIOLOGY CLINICAL WORKSHOP

CARDIOLOGY COMPREHENSIVE REVIEW

ANIMAL PARASITE COUNCIL

GASTROENTEROLOGY SOCIETY

EQUINE CLINICAL WORKSHOP

EQUINE COMPREHENSIVE REVIEW

EQUINE POST-GRADUATE COURSE

FOOD ANIMAL CLINICAL WORKSHOP

FOOD ANIMAL COMPREHENSIVE REVIEW

FOOD ANIMAL POST-GRADUATE COURSE

FOUNDATION RESEARCH REPORT

INTERNATNIOAL SOCIETY FOR COMPANION ANIMAL INFECTIOUS DISEASES

LIVER STUDY GROUP

NEUROLOGY CLINICAL WORKSHOP

NEUROLOGY COMPREHENSIVE REVIEW

ONCOLOGY CLINICAL WORKSHOP

ONCOLOGY COMPREHENSIVE REVIEW

ONE MEDICINE

SMALL ANIMAL INTERNAL MEDICINE CLINICAL WORKSHOP

SMALL ANIMAL INTERNAL MEDICINE COMPREHENSIVE REVIEW

SMALL ANIMAL INTERNAL MEDICINE POST-GRADUATE COURSE

SOCIETY FOR COMPARATIVE ENDOCRINOLOGY

VETERINARY COMPARATIVE RESPIRATORY SOCIETY (LARGE ANIMAL) COMPREHENSIVE REVIEW

VETERINARY COMPARATIVE RESPIRATORY SOCIETY (SMALL ANIMAL) COMPREHENSIVE REVIEW

VETERINARY EMERGENCY AND CRITICAL CARE SOCIETY

TECHNICIAN PROGRAM

CARDIOLOGY

CANINE MITRAL VALVE INTERSTITIAL CELL GROWTH IS IMPROVED BY CANINE WHARTON'S JELLY MESENCHYMAL STEM CELL CONDITIONED MEDIA

Vicky Yang, Dawn Meola, Sarah Crain, Kristen Thane, Airiel Davis, Andrew Hoffman

Tufts University Cummings School of Veterinary Medicine, North Grafto*n, Massachusetts, USA*

Myxomatous mitral valve disease (MMVD) is the most common acquired cardiac disease in dogs and the most common cause of congestive heart failure in dogs. Chronic valvular disease will lead to valvular insufficiency, cardiac enlargement, and sometimes contractile dysfunction in the later stages of the disease. As the disease progresses, the amount of valvular regurgitation increases, eventually leading to volume overload, pulmonary venous congestion, and finally congestive heart failure with pulmonary edema. The common histologic findings seen in myxomatous valves include disarray of the collagen and elastin fibers as well as disruption of the microstructure within the valves. This disarray and disorganization is partly a result of phenotypic transdifferentiation of the valvular interstitial cells (VICs) from quiescent fibroblastic cells to myofibroblastic-like cells, which follows the classical paradigm of fibrogenesis. As the transformation takes place, the VIC density decreases, and these cells then express alpha smooth muscle actin (α-SMA) instead of vimentin. Similarly, TGFβ stimulation of VICs *in vitro* results in fibroblastic (vimentin^{high+}, αSMA^{low+}) to myofibrolastic (vimentin^{low+}, αSMA^{high+}) transition. Given that mesenchymal stem cells (MSC) can exert anti-fibrotic effects, we investigated the effects of conditioned media (CM) derived from canine Wharton's Jelly MSCs (WJ-MSC) on the growth potential of canine VICs and its ability to counter the effects of TGFβ. VICs were isolated and cultured from normal and diseased valves. Immunohistochemistry and real-time (q-) PCR were used to evaluate the expression of αSMA, vimentin, elastin, and collagen in VICs. CM was collected from WJ-MSC culture, and CM exosomes were isolated by either ultrafiltration or ultracentrifugation. Cell growth and replicative capacity were evaluated using MTT assay and colony forming units (CFU). Internalization of WJ-MSC exosomes by VICs was imaged with membrane or RNA staining. qPCR showed that diseased valve VICs had increased myofibroblastic phenotype and decreased cell growth by MTT and CFU. Culturing of VICs of either phenotype with WJ-MSC CM or exosomes alone resulted in increased number of viable cells, and depletion of exosomes had the opposite effect. Exosomal membrane and RNA labelling confirmed exosome uptake into VIC cytoplasm and nucleus, although RNA containing exosomes were internalized by <20% of VICs, implying selective endocytosis of RNA containing exosomes or they are less numerous. Increase in responsiveness to TGFβ1 stimulation when VICs were cultured in CM was noted. WJ-MSC CM or its exosomes improve cell growth for VICs isolated from both normal and diseased canine mitral valves and may exert protective effects on VICs to slow the progression of canine MMVD.

CARDIAC BIOMARKERS AND SEMI-DOMINANCE IN A GENETIC MODEL OF FELINE HYPERTROPHIC CARDIOMYOPATHY

Joshua Stern[1], Kun-Ho Song[1], Eric Ontiveros[1], Samantha Harris[2]

[1]University of California Davis; School of Veterinary Medicine; Department of Medicine & Epidemiology, Daivs, CA, USA, [2]University of Arizona; College of Medicine; Department of Cellular and Molecular Medicine, Tucson, AZ, USA

A purpose-bred colony of mixed breed cats harbors a mutation associated with development of hypertrophic cardiomyopathy (HCM) in Maine coons. HCM in Maine coon cats carrying the same A31P mutation in the cardiac regulatory protein myosin binding protein C is a disease with age-related and incomplete penetrance. Here we hypothesized that cardiac biomarkers will reveal a semi-dominant inheritance pattern in cats with the A31P mutation and may identify cats at risk for HCM development earlier than echocardiography.

In 27 mixed-breed cats, A31P genotype status, echocardiography, serum 2nd-generation NTproBNP and plasma high-sensitivity cardiac troponin-I (CTNI) were analyzed. No cats were echocardiographically diagnosed with HCM in this study. 12 homozygous (HO), 9 heterozygous (HET) and 6 wild type (WT) cats were included. The median ages of WT, HET, and HO cats were 27, 31 and 15 months, respectively. The median and IQR for each biomarker was calculated for each group and compared for significant differences ($P<0.05$) with a Mann-Whitney test. CTNI was not different between groups. NTproBNP values were as follows: WT 15(12,16.75); HET 20(16.5,27.0); HO 49(32,131) and were significantly different for each genotype.

In contrast to previous studies of A31P mutant cats, significant NTproBNP elevation was present in the absence of echocardiographic diagnosis of HCM. NTproBNP elevation follows a semi-dominance model where heterozygous cats are less severely elevated than homozygous cats. This matches previous observations in clinical characterization of disease. In context of incomplete penetrance, NTproBNP may represent a tool for early identification of HCM affected cats.

SAFETY AND BIOCOMPATIBILITY OF THE MITREX® EPICARDIAL ANNULOPLASTY DEVICE IN A CHRONIC MODEL

Jeffrey Solomon[1], Thomas Fogarty[2], Evan Anderson[1], Pierluca Lombardi[3]

[1]*Infiniti Medical, Menlo Park, California, USA,* [2]*Fogarty Institute for Innovation, Mountain View, California, USA,* [3]*Maquet Cardiovascular, Wayne, New Jersey, USA*

This study evaluated the safety of the myocardial compression required to perform epicardial annuloplasty and the biocompatibility of the Mitrex® device.

Ten swine (seven test and three control) were used. The Mitrex® device was placed in all subjects such that the septo-lateral dimension of the mitral valve was reduced by 15-35%. Echocardiography and angiography were performed pre implant, post implant and at term. Test devices were secured in the test group and removed from the animals in the control group. Necropsy was performed at 180 days. Hearts were pressure fixed and analyzed.

Test devices were placed without incident. Coronary flow, ejection fraction, left ventricular wall motion and mitral valve anteroposterior dimension were normal post implantation and at term.

There were no remarkable postoperative events and all subjects survived to term with the exception of one test animal that was euthanized due to a non device related complication (refractory pleural effusion).

Devices were well tolerated causing only minimal to mild fibrosis and chronic inflammation. No significant changes were observed in the myocardium except for muscle fiber atrophy near the tip of the anterior arm. There appeared to be ample tissue over the tip and no danger of perforation in all but one subject.

No meaningful changes were noted in cardiac shape, ventricular wall thickness, chamber size, heart valves, and blood vessels.

Myocardial compression necessary to perform epicardial annuloplasty was well tolerated. The Mitrex® device was safe and biocompatible.

ECHOCARDIOGRAPHIC ASSESSMENT OF LEFT ATRIAL SIZE IN HORSES: DIAMETER OR AREA, DOES IT MATTER?

Colin Schwarzwald, Iris Huesler

Vetsuisse Faculty, University of Zurich, Zurich, Switzerland

Echocardiographic assessment of left atrial (LA) size is routinely done in horses. A variety of conventional (linear) and novel (area-based) indices of LA size have been described, but their clinical use is not standardized and novel indices are poorly established. The goals of this study were to define reference intervals for indices of LA size in horses and to provide prove of concept for the use of area-based indices of LA size in this species. The agreement between conventional linear measurements and novel area-based indices of LA size was assessed in a population of healthy horses and horses with valvular regurgitation.

Forty healthy horses (11 ± 4 years, 547 ± 84 kg) and 112 horses with mitral and/or aortic insufficiency (15 ± 6 years, 554 ± 76 kg) were included in this study. Echocardiographic examination was performed by a single operator (CCS) using a standardized protocol (GE Vivid 7 Dimension). The maximum LA diameter measured in a right-parasternal long axis view (LAD_{max}), the maximum LA diameter measured in a left-parasternal long axis view ($LAD_{llx\text{-}max}$), the maximum LA area measured in a right-parasternal long axis view (LAA_{max}), and the maximum LA area measured in a right-parasternal short axis view ($LA_{sx}A_{max}$) were measured by the same operator. Measurements were allometrically scaled to a body weight of 500 kg. Reference intervals were calculated based on a subpopulation of 31 healthy Warmblood horses (Reference Value Advisor). Method agreement was investigated using linear regression, calculation of weighted Kappa (κ) and Bland-Altman analyses. The level of significance was 0.05.

For this study population of healthy and diseased horses, LAD_{max} (reference interval 10.5-13.2 cm) was 0.9 (-0.9 to +2.7) cm [mean bias (limits of agreement)] smaller than $LAD_{llx\text{-}max}$ (reference interval 11.8-14.0 cm); they were significantly related ($p < 0.01$, $r^2 = 0.45$) and in moderate agreement ($\kappa = 0.59$) for classification of LA dimensions in reduced, normal, or increased size. Of the horses with a LAD_{max} within normal limits, 10% (12/118) had a $LAD_{llx\text{-}max}$ above normal. Conversely, 6% (7/110) of horses with $LAD_{llx\text{-}max}$ within normal limits had a LAD_{max} above normal. The LAA_{max} (reference interval 82-103 cm^2) was 18 (-9 to +44) cm^2 smaller than $LA_{sx}A_{max}$ (reference interval 84-134 cm^2); they were significantly related ($p < 0.01$, $r^2 = 0.51$). LAD_{max} was significantly associated with LAA_{max} ($p < 0.01$, $r^2 = 0.80$) and $LA_{sx}A_{max}$ ($p < 0.01$, $r^2 = 0.50$), respectively. However, out of all horses that had a LAD_{max} within normal limits, 9% (11/121) were diagnosed with an enlarged LA based on LAA_{max} ($\kappa = 0.76$) and 3% (4/121) had an enlarged LA based on $LA_{sx}A_{max}$ ($\kappa = 0.48$). Of the horses with both LAD_{max} and $LAD_{llx\text{-}max}$ within normal limits, 7% (7/104) had an enlarged LA based on LAA_{max} and 3% (3/104) had an enlarged LA based on $LA_{sx}A_{max}$, respectively. Of the horses with both LAA_{max} and $LA_{sx}A_{max}$ within normal limits, 2% (2/94) had an enlarged LAD_{max} and 6% (6/94) had an enlarged $LAD_{llx\text{-}max}$. Method agreement between LAA_{max} and $LA_{sx}A_{max}$ was fair ($\kappa = 0.29$). Of the horses with a LAA_{max} within normal limits, 4% (4/110) had a $LA_{sx}A_{max}$ above normal. Conversely, 17% (23/132) of horses with $LA_{sx}A_{max}$ within normal limits had a LAA_{max} above normal.

In conclusion, linear measurements and area-based measurements of LA size are not always in good agreement. Therefore, LA enlargement should not be diagnosed based on a single uni-dimensional measurement. Instead, multiple measurements of LA size, including linear and area-based variables, should be obtained to complement subjective assessment of LA dimensions in clinical patients.

EVALUATION OF POINT-OF-CARE LUNG ULTRASOUND (VETBLUE PROTOCOL) FOR THE DIAGNOSIS OF CARDIOGENIC PULMONARY EDEMA IN DOGS AND CATS WITH ACUTE DYSPNEA

Jessica Ward[1], Greg Lisciandro[2], Sandra Tou[1], Bruce Keene[1], Teresa DeFrancesco[1]

[1]NC State University College of Veterinary Medicine, Raleigh, NC, USA, [2]Hill Country Veterinary Specialists & FASTVetTM, San Antonio, TX, USA

Point-of-care lung ultrasound (LUS) is an emerging imaging technique that can suggest the presence of cardiogenic pulmonary edema (CHF) by identifying ultrasound artifacts (B-lines) caused by interstitial or alveolar fluid. This study was designed to determine the accuracy of a protocolized LUS technique for diagnosing CHF in dyspneic dogs and cats.

Seventy-six dogs and 24 cats were enrolled for evaluation of acute dyspnea. Exclusion criteria included trauma, pleural effusion, and the lack of a thoracic radiograph within 6 hours of LUS. Patients underwent LUS, quantifying the presence of B-lines at 4 sites on each hemithorax. An individual site was scored as positive if >3 B-lines were observed. LUS with ≥2 positive sites on each hemithorax was considered positive for CHF. Medical records were then evaluated for final diagnosis.

Patients with CHF had a higher number and different distribution of positive LUS sites compared to patients with noncardiac disease. Sensitivity and specificity of LUS for diagnosing CHF was 84% and 72%, respectively. When considering cats only, sensitivity and specificity of LUS was 87% and 89%, respectively. Diagnostic accuracy of LUS was similar to thoracic radiographs. LUS tended to misdiagnose CHF in cases of diffuse interstitial/alveolar disease, such as ARDS. Inter-observer variability for quantification of B-lines was low (kappa statistic >0.85).

In conclusion, LUS was useful in predicting CHF as the cause of dyspnea, particularly in cats. Considering the utility and rapidity of this technique, point-of-care LUS should be considered as a diagnostic tool for dyspneic veterinary patients.

USE OF NT-proBNP AND cTnI CONCENTRATIONS TO DETECT HEART DISEASE IN WHIPPETS

Rebecca Stepien[1], Virginia Luis Fuentes[2], Heidi Kellihan[1]

[1]University of Wisconsin School of Veterinary Medicine, Madison, WI, USA, [2]Royal Veterinary College, London, UK

Systolic heart murmurs in whippets may be due to myxomatous mitral valve disease (MMVD) or may be unassociated with any structural heart disease (HD). Echocardiography (ECHO) may be required to differentiate MMVD from functional murmurs in screening programs but is not always available. In this pilot study, we compared the ability of NT-proBNP (BNP) and cTnI concentrations versus auscultation to predict ECHO-documented HD in a cohort of outwardly healthy whippets screened at a national show.

One hundred thirty five dogs underwent auscultation, ECHO and blood sampling for BNP and cTnI analysis. ECHO results were reviewed to identify the presence of HD based on presence of any of the following: mitral valve prolapse, mitral regurgitation, left atrial/ventricular dilation and presence of ventricular premature complexes during ECHO exam, and categorized as ECHO negative (ECHO−) or positive (ECHO+) for evidence of HD. Blood test results were analyzed ($p<0.05$) for all dogs as a group (n=135), and for dogs with heart murmurs (n=73). [BNP] is expressed as pmol/L; [cTnI] as nmol/mL. ECHO+ prevalence in this population was 47/135 (35%). 73/135 (54%) dogs had systolic murmurs (MUR), 41/73 (30%) left basilar (LB) and 32/73 (24%) left apical (LA). ECHO+ prevalence in MUR dogs was 55% (40/73).

Median [BNP] was higher in dogs with LA (634[range 169−2303], p<0.0001) versus no murmur (304 [112−1048]) or LB (279[164−804]). [BNP] was higher in ECHO+ dogs (574[169−2303], n=47) versus ECHO− (297[112−988], n=88, p<0.0001). In MUR dogs, [BNP] was significantly higher in ECHO+ dogs (578[169−2303], n=40) versus ECHO− (364[164−804], n=33, p=0.004).

Median [cTnI] was higher in LA dogs (0.07[0.02−0.26], p<0.0001) versus those with no murmur (0.03[0.01−0.25]) or LB (0.03[0.01−0.15]). [cTnI] was higher in ECHO+ dogs (0.06[0.01−0.26], n=47) versus ECHO− (0.03[0.01−0.25], n=88, p<0.0001). In MUR dogs, [cTnI] was higher in ECHO+ dogs (0.07[0.02−0.26]) versus ECHO− (0.03[0.01−0.07], p<0.0001). ROC analysis was used to predict diagnostic cut-off values for [BNP] and [cTnI]. Test characteristics are reported below:

All dogs n=135 (ECHO+ disease prevalence: 35%)	Se	Sp	PPV	NPV	AUC	P value
• [BNP] > 575 pmol/L	46	88	69	73	0.70	<0.0001
• [cTnI] > 0.065 nmol/mL	42	98	91	74	0.83	<0.0001
• Presence of any MUR	85	63	55	89	NA	<0.0001
Dogs with MUR n=73 (ECHO+ disease prevalence: 55%)						
• [BNP] > 575 pmol/L	53	88	83	43	0.70	=0.004
• [cTnI] > 0.065 nmol/mL	53	97	95	63	0.87	<0.0001
• Presence of LA murmur	81	96	98	73	NA	<0.0001

The results of this pilot study suggest that in the absence of auscultation, both [BNP] and [cTnI] may be useful to distinguish between whippets with HD from those without HD. If murmur status is known, the PPV of both tests is improved.

STENT ANGIOPLASTY FOR TREATMENT OF BALLOON RESISTANT CANINE VALVULAR PULMONIC STENOSIS

Simon Swift[1], Ivan Sosa[1], Amara Estrada[1], Ashley Jones[1], Curt Fudge[2]

[1]College of Veterinary Medicine, Gainesville, Florida, USA, [2]Congenital Heart Center, Gainesville, Florida, USA

Canine valvular pulmonic stenosis has been categorized based on the degree of leaflet thickening, leaflet fusion and pulmonary artery hypoplasia. Dogs with mild and moderate disease have normal life expectancies but those with severe disease die prematurely. Balloon dilation is often successful in severe cases with thin fused leaflets, resulting in decreased clinical signs and increased life expectancy. Dogs with thickened, dysplastic valves and pulmonary artery hypoplasia show limited response to balloon dilation. Two case reports describe the use of stents to treat obstruction to pulmonary blood flow in dogs. In one, 2 dogs with supravalvular stenosis were treated successfully with stents. In the second, 2 dogs with severe dysplastic pulmonary valve stenosis initially did well, but stenosis recurred within 6 months.

We describe the use of bare metal stents to treat severe dysplastic pulmonary valve stenosis in 3 dogs. All dogs initially had a single stent implanted with a good reduction in pressure gradients. This has been maintained long term in 2 dogs. One dog suffered a stent fracture resulting in stent embolization. That dog subsequently underwent repeat cardiac catheterization with four additional stents placed in the right ventricular outflow tract to treat severe, dynamic subvalvular pulmonic stenosis. All dogs continue to receive atenolol and clopidogrel.

Our findings suggest that in selected patients, stenting of severe dysplastic pulmonary valve stenosis is a viable option and can provide long term relief of the obstruction.

ONLINE SURVEY TO ASSESS INTER- AND INTRA-OBSERVER AGREEMENT ON ECHOCARDIOGRAPHIC CLASSIFICATION OF CARDIOMYOPATHY IN CATS

Lois Wilkie[1], Virginia Luis Fuentes[1], Mark Rishniw[2]

[1]The Royal Veterinary College, Hatfield, Hertfordshire, UK, [2]Cornell University, Ithaca, New York, USA

Aim: to investigate inter- and intra-observer agreement on echocardiographic classification of cardiomyopathy in cats.

An online survey was devised with video loops and still images of echocardiographic recordings compiled from 21 cats with a range of cardiac phenotypes. Signalment and a brief clinical history were provided. The survey was distributed via the Veterinary Information Network (VIN) to members of a cardiology list serve. A list of 13 possible diagnoses was provided and participants were asked to indicate which described each case most accurately. Selection of multiple diagnoses was not allowed and respondents were asked to include a brief justification for their choice. Intra-observer agreement was assessed by repeating 4 of the cases. Fleiss kappa (κ) was calculated for all participants with complete responses to determine inter and intra-observer agreement. In order to assess whether agreement was greater among board-certified cardiologists, agreement was also calculated using only responses from ACVIM or ECVIM diplomates.

The survey was attempted by 86 participants routinely involved in echocardiographic assessment of cats with myocardial disease. Participants' clinical experience in veterinary cardiology ranged from 6 months to 42 years; 49% were ACVIM diplomates and 25% were ECVIM diplomates. No case was scored with perfect inter or intra-observer agreement. Overall, inter-observer agreement ranged from κ0.17 - 0.96 depending on the case, and intra-observer agreement κ0.33-0.67. Among ACVIM/ECVIM diplomates, inter-observer agreement was κ0.18-0.95 and intra-observer agreement κ0.29-0.67.

The observed range of inter-observer Fleiss kappa values demonstrates that classification of cats with cardiomyopathy was inconsistent between participants, suggesting that different criteria may be in use. Intra-observer agreement was also poor, suggesting that irrespective of the diagnostic criteria used, they are inconsistently applied. Agreement might be improved by standardizing the echocardiographic criteria used to diagnose different cardiomyopathy phenotypes.

FELINE HYPERSOMATOTROPISM IS A NATURALLY OCCURRING, REVERSIBLE CAUSE OF MYOCARDIAL REMODELING

Kieran Borgeat[2], Stijn Niessen[1], Christopher Scudder[1], Ruth Gostelow[1], Sophie Keyte[1], Julia Sargent[1], Patrick Kenny[1], Yaiza Forcada[1], David Church[1], Virginia Luis Fuentes[1], David Connolly[1]

[1]Royal Veterinary College, London, UK, [2H]ighcroft Veterinary Referrals, Bristol, UK

In humans, hypersomatotropism caused by a functional pituitary mass is recognized as a cause of increased left ventricular (LV) mass, and cardiovascular complications are a major cause of morbidity and mortality. We hypothesized that feline hypersomatotropism was associated with increased LV wall thickness and left atrial (LA) dilation, compared to control groups of non-hypersomatotropic diabetic and age-matched healthy cats.

Cats with confirmed hypersomatotropism (IGF-1>1000ng/ml and pituitary mass; n=57) were prospectively recruited, as were two control groups: diabetic cats (IGF-1<1000ng/ml; n=27) and healthy cats with no history of diabetes or cardiovascular disease (n=41). Cats with other endocrinopathies were excluded. Echocardiography was performed in all cases and studies were measured by one trained operator.

Normally distributed continuous variables were compared using a one-way ANOVA. Non-normally distributed variables were compared using a Kruskal-Wallis test. Categorical data were compared using Chi-square tests. Paired data were compared using Wilcoxon's signed rank test. Significance was p<0.05, with correction for pairwise comparisons.

There was no age difference between groups (p=0.243). Cats with hypersomatotropism had a greater maximum LV wall thickness (6.7mm, 4.1-10.1mm) than diabetic (5.2mm, 4.0-9.1mm; p<0.001) or normal cats (5.3mm, 3.9-6.5mm; p<0.001). LA diameter was greater in cats with hypersomatotropism (17mm, 14.1-29.5mm) than in diabetic (15.3mm, 10.1-21.3mm; p<0.001) and healthy cats (15.8mm, 11.2-21.5mm; p<0.001). Aortic insufficiency was more common in both cats with hypersomatotropism and diabetes than in normal cats (p<0.001). After hypophysectomy (n=8), echocardiographic changes were mostly reversible (Figure 1).

Hypersomatotropism is a naturally occurring, reversible cause of LV hypertrophy and LA dilation in cats.

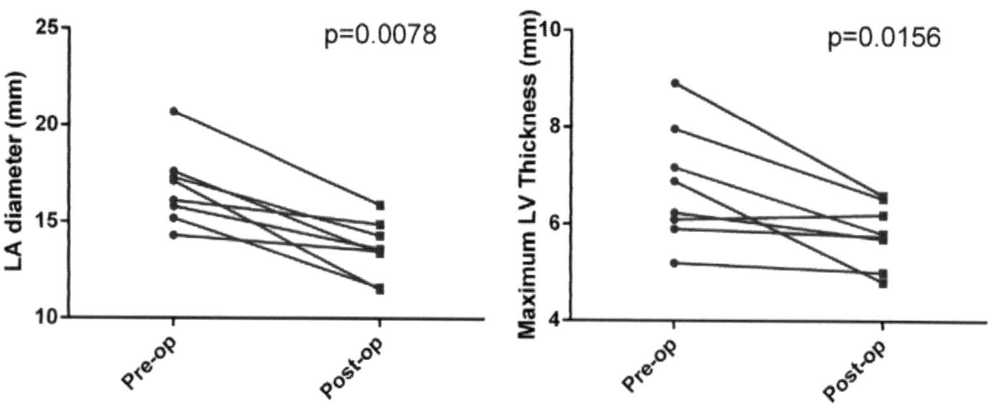

DIAGNOSTIC VALUE OF RIGHT PULMONARY ARTERY DISTENSIBILITY INDEX IN DOGS WITH PULMONARY HYPERTENSION

Lance Visser, Minu Im, Joshua Stern

University of California Davis; Department of Medicine & Epidemiology; School of Veterinary Medicine, Davis, CA, USA

We sought to determine the value of right pulmonary artery distensibility index (RPADi) for the prediction of Doppler-derived estimates of systolic pulmonary artery pressure (sPAP) in dogs with pulmonary hypertension (PH) compared to other echocardiographic indices of PH.

Dogs with tricuspid regurgitation (TR) permitting Doppler-derived estimates of sPAP were prospectively recruited and grouped into control (n=20; sPAP <30 mmHg) and dogs with mild (n=12; sPAP 30-50 mmHg), moderate (n=12; sPAP 50-75 mmHg), and severe (n=14; sPAP >75 mmHg) PH. Indices of PH quantified were RPADi (percent change in diameter of the PA from systole to diastole), pulmonary artery-to-aortic diameter (PA:Ao), acceleration time to peak PA flow velocity (AT), and AT-to-ejection time of PA flow (AT:ET). Associations between indices of PH and sPAP, right ventricular fractional area change (FAC), age, gender, heart rate, and body weight were performed using linear regression. Receiver operating characteristic analysis was performed to determine the optimal cutoff values for the indices of PH in the prediction of moderately increased sPAP (>50 mmHg).

RPADi (r=-0.89) showed the strongest correlation to sPAP followed by PA:Ao (r=0.74), AT and AT:ET (both r=-0.69), and FAC (r=-0.52). AT weakly (r=-0.42) correlated with heart rate. No other significant correlations were identified. Cutoffs to predict moderate PH were defined for RPADi (29%; sensitivity [Sn] 81%; specificity [Sp] 94%), AT:ET (0.30; Sn 65%; Sp 100%), AT (54 ms; Sn 81%; Sp 83%), and PA:Ao (1.04; Sn 100%; Sp 67%).

RPADi may be useful and predictive of PH in dogs if TR is absent.

EVALUATION OF SYSTEMIC MICRORNA ADMINISTRATION FOR IMMUNOMODULATION IN THE TREATMENT OF CANINE GLIOMA

C. Elizabeth Boudreau[1], Nasser Yaghi[2], Padmanabh Chivukula[3], Xiaoyang Ling[2], Brian Porter[1], Gwendolyn Levine[1], Joseph Payne[3], Amy Heimberger[2], Jonathan Levine[1]

[1]Texas A&M University, College of Veterinary Medicine and Biomedical Sciences, College Station, TX, USA, [2]University of Texas MD Anderson Cancer Center, Houston, TX, USA, [3]Arcturus Therapeutics, San Diego, CA, USA

The signal transducer and activator of transcription 3 (STAT3) pathway is a key regulator of tumorigenesis and tumor-mediated immune suppression. MicroRNA (miR)-124 has been shown to be an inhibitor of STAT3 signaling. In rodent models of glioma, systemic administration of nanoparticle-encapsulated miR-124 (designated LUNAR-301) results in down-regulation of STAT3, up-regulation of anti-tumor immune effector responses, tumor regression, and enhanced survival. Additionally, miR-124 can reverse immune suppression and T-cell anergy in immune cells from human glioblastoma patients.

We investigated STAT3 and miR-124 expression levels in canine astrocytomas (n=28) and normal brain (n=6) to determine if these tumors share key biologic features with human and rodent gliomas. We found that miR-124 expression is absent and STAT3 is increased in canine gliomas relative to normal brain. The level of STAT3 expression and the number of glioma-infiltrating CD3+ T cells directly correlated with astrocytic grade. Next we investigated the safety and pharmacokinetics of LUNAR-301 in healthy dogs (n=5). No significant adverse events or toxicity was noted during single-dose escalation or during sustained dosing. LUNAR-301 induced *in vivo* up-regulation of immune effector responses as detected by flow cytometry. Pharmacokinetic studies confirmed delivery of miR-124 to the immune compartment during intravenous administration of LUNAR-301. Cumulatively, these data indicate that STAT3 is an operational therapeutic target in canine high-grade gliomas and that systemic administration of LUNAR-301 is safe and feasible. We are initiating a phase I/II clinical trial in client-owned canines with spontaneously arising high-grade gliomas to ascertain if LUNAR-301 can induce radiographic regression and prolong survival.

ONSCREEN-GUIDED BRAIN TUMOR RESECTION THROUGH REGISTRATION OF A VARIABLE-SUCTION TISSUE RESECTION DEVICE WITH A NEURONAVIGATION SYSTEM

Rebecca Packer[1], Stephanie Engel[2]

[1]Colorado State University, College of Veterinary Medicine and Biomedical Sciences, Fort Collins, CO, USA, [2]Colorado State University, College of Veterinary Medicine and Biomedical Sciences, Fort Collins, CO, USA

This study describes a surgical technique in which a variable-suction tissue resection device is linked to a neuronavigation system for direct onscreen guidance of brain tumor excision. This technique is considered the first step towards minimally-invasive, guided neurosurgery to excise otherwise inaccessible deep intraaxial tumors. Goals of this study were 1) improved access to deep or poorly accessible masses, 2) reduced surgical trauma by using targeted approaches and improved resection tools.

A retrospective evaluation of 7 dogs and 1 cat that underwent brain tumor excision using the NICO® Myriad (NICO Corporation) and Brainsight neuronavigation system (Rogue Research) was performed. The patient and resection instrument were registered to the neuronavigation system. Surgery was guided by real-time onscreen visualization of the resection instrument position relative to the pre-operative MRI images, cross-referenced with direct visualization where possible. In 7 of 8 cases degree of resection was evaluated on post-operative MRI: gross-total resection (GTR, no residual tumor evident), near-total resection (NTR, <100% but ≥90% resection), or sub-total resection (STR, <90% resection). Of these 7 cases, 2 achieved GTR (1 equivocally), 3 achieved NTR (99%, 91%, 90%), 2 achieved STR (35%, 26%).

Neuronavigation-guided resection of cortical and subcortical brain tumors using the NICO® Myriad was feasible, and study goals were achieved in part. Some refinements are required. Further studies will adapt this technique to minimally-invasive port-based surgical approaches for deep brain tumor resection (BrainPath™), to improve visibility and minimize effects of brain shift. The impact on clinical outcome for minimally-invasive approaches must also be determined.

A SIMPLIFIED METHOD OF WALKING TRACK ANALYSIS TO ASSESS LOCOMOTION IN DOGS WITH ACUTE SPINAL CORD INJURY

Rachel Song, Maureen Oldach, Ronaldo da Costa, Sarah Moore

The Ohio State University, Columbus, OH, USA

The utility of a "finger painting" technique for walking track analysis was evaluated in 20 control dogs and 29 dogs with acute thoracolumbar spinal cord injury (SCI) caused by spontaneous intervertebral disc extrusion (IVDE) . Stride length (SL), base-of-support (BS) and the co-variance (COV) for both parameters were measure in all four limbs at three separate time points in normal dogs and on days 3, 10 and 30 following laminectomy in dogs with SCI. SL and BS were compared between control and SCI-affected dogs at each time point during recovery. $P< 0.05$ was considered a statistically significant difference.

Mean SL (cm) for control dogs was 43.69, 43.68, 43.43, 43.54 for the left thoracic limb (TL), right TL, left pelvic limb (PL), and right PL respectively. COV-SL was 0.13 in each of four limbs in control dogs. Mean SL of all four limbs was significantly shorter in SCI-affected dogs at day 3, 10 and 30 compared to normal dogs. The mean difference in SL between normal and SCI-affects dogs declined significantly over time with recovery from injury. BS-TL (cm) was significantly wider in SCI-affected dogs compared to controls at days 3 and 30 after surgery. BS-PL (cm) did not differ between groups at any time-point. These findings support the utility of this simplified method of walking track analysis to compare differences in pelvic limb SL between normal and SCI-affected dogs, and to assess changes in SL as a marker of recovery after SCI.

ATTEMPTS TO BREED OUT CHIARI-LIKE MALFORMATION: IS A CROSS THE ANSWER?

Susan Knowler[1], Henny van den Berg[1], Eric Noorman[2], Roberto La Ragione[1], Clare Rusbridge[1]

[1]*University of Surrey, Guildford, UK,* [2]*Dierenkliniek den heuvel, Best, The Netherlands*

Chiari-like malformation (CM) and syringomyelia (SM) are complex inherited disorders observed most commonly in toy breed dogs and can cause a significant loss in quality of life though pain and disability. Brachycephalicism is a risk factor and it has been suggested that crossbreeding with a different breed/s then backcrossing may produce individuals free of disease. This two and a half year project took advantage of a cross between a mesaticephalic normal Australian terrier and CM affected Griffon Bruxellois (GB) with subsequent backcrossing to a GB to investigate the inheritance and phenotype of these conditions and a means of reducing the incidence of CM and risk of SM.

The study cohort comprised 2 control dogs (CM affected Australian terrier and unrelated GB without CM) and a single of family of 29 dogs, 12 of which were used in 8 different mating combinations resulting in 19 progeny. T1-weighted sagittal DICOM images were analyzed for traits (2 angles, 2 lines and a "best fit" circle diameter) shown previously to have the greatest significance for CM in the GB. The quantitative findings in this study revealed these traits to be significant for CM-affectedness. Furthermore the external phenotypes showed that by outcrossing breed types and selecting appropriate conformation characteristics in the F1 generation, it is possible to regain the GB breed standard in the F2 generation and reduce the degree of CM. However this is dependent on careful selection of conformation and screening for CM and SM at one year of age. The 4 dogs affected with SM in the study all exhibited reduced caudal skull development compared to their relatives.

We showed that traits on MR images were useful to distinguish the phenotype and these exhibited segregation and may be additive towards the severity of CM. It suggests such traits might be useful to quantify the condition and the risk to SM. We propose that grading of CM takes account of quantitative traits that can be used in Estimated Breeding Values (EBV) to assist breeders with their mate selections. Such a system will have to be verified to ensure appropriateness for all breeds at risk.

CYTOGENETIC ALTERATIONS DEFINE TUMOR SUBTYPES AND KEY PATHWAYS IN CANINE PRIMARY GLIOMAS

Peter Dickinson, Daniel York, Robert Higgins, Richard Lecouteur, Danika Bannasch

Uc Davis School Of Veterinary Medicine, Davis, CA, USA

Defining commonly occurring cytogenetic abnormalities in canine glioma is essential to an understanding of gliomagenesis, and the definition of appropriate therapeutic strategies. We used Illumina 173K HD SNP arrays to determine copy number alterations and allelic imbalances in 38 histologically confirmed spontaneous canine gliomas (11 astrocytomas, 23 oligodendrogliomas, 4 mixed oligoastrocytomas and the J3T cell line). Copy number calls and allelic events were determined by matched paired analysis using BioDiscovery Nexus Copy Number™ software.

Hierarchical clustering by aberration profile revealed two major groups with 5/6 glioblastomas and 2/3 grade II astrocytomas comprising one group. Comparative analysis showed loss of the *INK4A* locus (*CDKN2A*/*CDKN2B* tumor suppressor genes) was significantly associated with glioblastomas. Querying for aberrant events in specified genes showed common losses/loss of heterozygosity of tumor suppressor genes and amplification of oncogenes in the TP53, RB1 and PI3K/AKT/RAS pathways. Specifically, loss of *TP53*, *CDKN2A*/P14, *CDKN1A*/P21 and gain of *MDM4* was present in the TP53 pathway; loss of *RB1*, *CDKN2A*/P16, *CDKN2B*/P15, and *CDKN2C*/P18 was present in the RB1 pathway and loss of *PTEN* and gain of *PDGFRA* and *FGFR1* was present in the PI3K/AKT/RAS pathways. Evaluation of oligodendrogliomas for deletions syntenic to the commonly described 1p loss in human oligodendrogliomas revealed a syntenic loss in CFA 5 in 36% of canine tumors.

These preliminary data suggest that canine gliomas have pathway alterations in common with human tumor counterparts, however further definition of the pathways in greater detail is warranted to inform appropriate choice of targeted therapies and interpretation of future therapeutic trials

METRONOMIC CHLORAMBUCIL CHEMOTHERAPY FOR CANINE GLIOMAS: A PHASE I/II CLINICAL TRIAL

R. Timothy Bentley[1], Aaron Cohen-Gadol[2], David Jones[2], Deborah Knapp[1]

[1]Purdue Veterinary Medicine, West Lafayette, IN, USA, [2]Indiana University, Indianapolis, IN, USA

Gliomas in dogs are typically fatal and limited information is available to guide therapy. Spontaneous brain tumors in client-owned dogs have been suggested as a translational model for the development of human therapies. Chlorambucil is reported to display negligible penetration of the blood-brain-barrier in normal animals, but penetration of the blood-brain tumor-barrier has not been investigated. The purpose of this translational study was to perform a safety and pharmacokinetic study of metronomic chlorambucil chemotherapy in dogs with gliomas.

Four client-owned dogs with high-grade gliomas began receiving metronomic chlorambucil (4mg/m^2 PO q 24hr) and underwent surgical resection. The chlorambucil concentrations of serum, cerebrospinal fluid and surgical brain tumor specimens were analyzed. Patients additionally received CCNU post-operatively and were monitored with monthly examinations and laboratory work. Repeat magnetic resonance imaging was performed. One patient suffered a post-operative cerebrovascular accident and two are currently in complete remission. A fourth was euthanized after 5 months and brain tumor was harvested for repeat chlorambucil concentration analysis.

No increase in seizures was observed and no cytopenias or other major chemotherapy complications were detected. Brain tumor chlorambucil concentrations ranged from below the limit of quantification (0.024 ng/specimen) to well in excess of the serum concentration of the same patient, indicating very variable penetration of the blood-brain tumor-barrier. Chlorambucil could not generally be detected in the cerebrospinal fluid.

Metronomic chlorambucil chemotherapy is well tolerated in dogs with glioma and may penetrate the blood-brain tumor-barrier in individual cases. There is preliminary evidence of activity.

CANINE PERIPHERAL NERVE SHEATH TUMORS: CLINICAL ASPECTS, MAGNETIC RESONANCE IMAGING FINDINGS AND COMPARISON OF PALLIATION, SURGERY AND STEREOTACTIC RADIOTHERAPY

Mario Dolera, Luca Malfassi, Simone Pavesi, Massimo Sala, Giovanni Mazza, Silvia Marcarini, Nancy Carrara, Sara Finesso

La Cittadina Fondazione Studi e Ricerche Veterinarie, Romanengo, Italy

No updates for canine peripheral nerve sheaths tumor (PNST) appeared in recent literature. The aim of this study was to evaluate the correlation between clinical aspects and MRI findings of tumors involving a major peripheral nerve, plexus or root and to determine the survival time in dogs treated with palliation, surgery or stereotactic radiotherapy (SRT).

Records of dogs with PNST evaluated from 2000 to 2014 were reviewed to determine signalment, duration of clinical signs, neurological examination, MRI features, treatment option (palliation, surgery, stereotactic hypo fractionated radiotherapy). Time to first event, survival times and statistical differences across categories were calculated by the Kaplan-Meier product limit method and log-rank test.

Forty-seven dogs (median age 9 years, male:female ratio 1.76) were included, with Labrador retrieveroverrepresented (17%). Roots lesions were the most frequent (46.8%), with C5-T1, V nerve and left side more involved (25.5%, 19.1% and 61.7%). Presenting sings were lameness, paresis and pain. Mean duration of clinical signs was 90 days. MRI findings comprises increased diameter, hyper intense and contrast enhancing nerve roots (57.1%), plexus or peripheral nerve (42.9%), focal hypomiotropy and muscle hyper intensity (73%). The time to first event was 30 days after surgery and 240 days after SRT. Overall mean survival was 97, 144 and 371 days with palliation, surgery and SRT.

A predilection for Labrador retriever is observed. Comparing our results with published data, SRT seem to promise better results than palliation or surgery and warrant further evaluation.

PERIPHEAL GLYCAEMIA IN DOGS WITH LIMB THROMBOSIS: A PROSPECTIVE STUDY

Mario Dolera, Luca Malfassi, Roberto Vailati Facchini, Sara Finesso, Giovanni Mazza

La Cittadina Fondazione Studi e Ricerche Veterinarie, Romanengo, Italy

The aim of this study was to document the pheripheral glycaemia variations in hypoperfused limbs of patients affected by Magnetic Resonance Imaging (MRI)-confirmed arterial thrombosis.

Eleven dogs were recruited. Inclusion criteria were a clinical examination supportive of limb hypoperfusion and availability of blood cell count, biochemical profile and urine analyses. Two blood samples were sampled, one from the affected limb and one from a healty limb. Plasmatic glycaemia was measured using an automated glucose analyser. All the patients underwent a total body MRI that provided the final diagnosis.

The thrombus was located: in the abdominal aorta (7/11), in the subclavian artery (1/11), in the axillary artery (1/11), in the iliac arteries (2/11). Of the total abdominal aortic thrombosis, 3/7 involved also the internal iliac arteries, 2/7 the external ones and 2/7 both. The extent of the thrombosis was classified as grade 1 when the greatest portion of the thrombus did not reach half of the vessel lumen (1/11); grade 2 when the greatest portion of the thrombus was between 1/2 and 2/3 of the vessel lumen (7/11); grade 3 when the thrombus exceded 2/3 of the lumen (3/11). A substantial decrease in pheripheral glycaemia values was found in sampling arising from the affected limbs. Comparing affected limbs values with healthy limbs measurements from the same patient, the reduction was found from 17.65% to 34.41%. Accounting only the grade 3 scored patients, the percentage of reduction was found up to the 28.34%.

SURGICAL STABILIZATION OF CANINE LUMBOSACRAL SPINE WITH STOP-SCREWS AND ILIAC WINGS SCREWS

Mario Dolera, Luca Malfassi, Simone Pavesi, Massimo Sala, Giovanni Mazza, Silvia Marcarini, Nancy Carrara

La Cittadina Fondazione Studi e Ricerche Veterinarie, Romanengo, Italy

Surgical stabilization of canine lumbosacral spine can be challenging. The aim of this research was to evaluate two surgical techniques to achieve lumbosacral stabilization in dogs either with normal or transitional vertebrae.

Lumbosacral instability and degenerative stenosis were evaluated by dynamic Computed Tomography (CT) and Magnetic Resonance Imaging (MRI). In dogs with normal vertebrae two 4.5 mm screws were bicortically inserted in S1 with the heads behind the caudal articular process of L7 to prevent the extension of the lumbosacral joint; if ventral listhesis of S1 was evident, the head of the screws were augmented by methyl methacrylate. In dogs with transitional vertebrae, two 4.5 mm screws were inserted in the iliac wings, two 3.5 mm screws were inserted in the spinous process of L6 and L7; the emerging screws were embedded in methyl methacrylate after flexion of the lombosacral spine. In cases of residual radicular compression, dorsal laminectomy and partial discectomy were accomplished. Serial clinical and imaging follow up examinations were performed.

Twenty-two large breed dogs were enrolled. In 14 dogs stop-screws (in 4 augmented) and in 8 dogs iliac wings screws were inserted. 2 dogs required additional decompression. During a mean follow up of 36 months, clinical examination and imaging reveals amelioration of presenting complaints and reduction of radicular compression, with no surgical complications.

Stop-screws and iliac wings technique are effective methods to obtain stabilization and indirect decompression of the lumbosacral joint. Comparing with other described surgical procedures, our obtained results are better but with lesser complications.

FRAMELESS STETEOTAXY FOR VERTEBRAL IMPLANT GUIDANCE IN DOGS

Fred Wininger[1], Nicholas Archambault[2], Stephen Frey[3]

[1]*Veterinary Specialty Services, St. Louis, Missouri, USA,* [2]*University of Missouri, Columbia, Missouri, USA,* [3]*Rogue Research, Montreal, Quebec, Canada*

Multiple pathologies of the dog spine require stabilization with vertebral implants. Optimal implant corridors are described and vary between specific vertebrae throughout the spine. Current placement guidelines use general surface landmarks and proposed angles without opportunity to visualize vital structures at depth. Preferred corridors are often inaccessible because of soft tissue impediment. Potential complications of vertebral implant placement include neuro-vascular compromise or insufficient bone purchase for required biomechanical loads. Trans-articular implants present the added challenge of aligning a vertebral motion units and countering biomechanical stresses. Recent reports suggest the frequency of spinal canal violation by vertebral implants is high. Frameless stereotaxy for brain neuronavigation has been described in veterinary literature and validation studies show a high level of accuracy. These techniques utilize high resolution cross sectional imaging to create 3D data sets. Identification of fiducial markers visible on 3D models (image space) and on the subject intraoperatively (anatomic space) permits fusion for co-localizing guidance. A position sensor is used to track, in real-time, reflective instruments affixed to the skull and surgical tools within a virtual space. The surgeon can identify any point within that virtual space for tissue sampling or treatment. The purpose of this study was to develop a technique for implant placements in the dog spine using image-guided intervention. Unlike the brain, which is encased within the solid calvarium and once registered with the navigation software, all structures can be reliably identified relative to the origin of the coordinate system; There are two major obstacles when considering neuronavigation of the vertebral column. First, multiple vertebral motion units require individual registration. The vertebral column is made of multiple articulating bones, making a single bone registration inapplicable for the remaining vertebrae. A novel subject tracker was created for attachment to single vertebra for individual registration. The second obstacle in adapting frameless neuronavigation to the spine is the assignment of fiducial markers. Current brain protocols require pre-imaging procedures for placement of artificial fiducial makers rigidly affixed to the skull or teeth. Pre-imaging surgery and affixing fiducial arrays is logistically difficult in spinal applications. For this reason, identification of natural boney fiducial markers identifiable in both image and anatomic space is preferable. Evaluated landmarks included spinous processes, accessory processes and cranial costovertebral joints. Initially a cadaveric study validating the technique in the thoracolumbar spine was performed. Qualitative measures of canal compromise and bone purchase in addition to quantitative measures of trajectory and depth were evaluated. Safety and efficacy findings of zero canal violations, statistically accurate placement, and mostly cross-body implants warranted clinical use. Completed clinical cases included three thoracolumbar fractures with vertebral body screws, two atlantoaxial stabilizations and three lumbosacral stabilizations with transarticular screws. Stereotactic guidance enabled the surgeon to visualize cross-sectional implantation in real time. All cases had post-operative imaging with adequate placement of implants and satisfactory clinical outcomes. The described adaptation provides a means of safe and effective placement of vertebral implants. Multiple trajectories could be created with the computerized guidance. Subjectively less soft tissue dissection was necessary and surgical time was minimally extended.

PLA BEAD TREATMENT OF REFRACTORY EPILEPSY IN DOGS

Roger Clemmons

University of Florida, Gainesville, FL, USA

Gold-bead implants have been used to help treat epilepsy, but they interfere with modern imaging techniques like MRI examinations. Polylactic acid (PLA) is a biodegradable, implantable plastic that appears to offer similar effects to gold material, but degrades over time and is invisible to modern imaging. The purpose of this study was to investigate whether PLA beads (discs 1mm by 0.5 mm) could be useful in assisting in the control of refractory epilepsy in dogs when placed in specific acupuncture points thought to affect seizures.

This is a descriptive study of a case series which was done as a non-blinded, preliminary study. Ten patients presenting to the University of Florida Neurology Service who were experiencing frequent seizures that were refractory to more than 2 anticonvulsant medications were enrolled in the study. All patients had systemic evaluations to demonstrate no significant findings, MRI evaluations to demonstrate no significant changes in the central nervous system, and CSF analysis revealing no abnormalities. Many patients experienced cluster seizures requiring emergency interventions to control the seizures. No further adjustments to medications were made during the study, although some anticonvulsants could be reduced once the PLA bead treatment was given. An EEG was performed in dexmedetomidine-sedated dogs (10-15 µg/kg) before and after PLA bead placement and analyzed by a statistical package (NeuroStat package of NeuroGuide by Applied Neuroscience, Inc., Seminole, FL). PLA beads were placed in transpositional and classical acupuncture points by a certified veterinary acupuncturist (Chi Institute, Reddick, FL). The injection sites were wiped with ethyl alcohol gel (Purell Advanced Hand Sanitizer, Gojo Industries, Akron, OH) and the PLA beads inserted using a modified 16 ga needle and syringe. Points used were GV20, Long hui, Nao shu, An shen, Tian ping, HT/PC7, LI4, LIV3, Bai hui, CV15 and the auricular point, Shen men. Owners were asked to keep calendars of seizure events for comparison of pre- and post-treatment seizure frequencies. Data was compared by pair-t test and ANOVA using a statistical program (SPSS ver 14.0)

Of the ten patients enrolled in the study, 9 were deemed to have satisfactory results by the owners. These patients showed a reduction of seizures by greater than 50%, moreover, the seizure characteristics changed so that seizures were more likely to be singular rather than multiple. ($p < 0.5$) The patient who did not respond was euthanized by the owner because of seizures that occurred 2 weeks after bead placement. We later recognized that this is a period of time when patients are most likely to seizure and it was speculated that this was a consequence of the healing from the implantation procedure before the bead took over the protection. One patient who had severe clusters every 13 days prior to bead implantation remained seizure free for 4 years (using re-implantation of beads every 6 months). From preliminary investigations, we found that the bead implantation lasted about 8 months on average, so patients were re-implanted every 6 months, based upon owner's desires. EEG examinations of patients prior to implantation showed slow-wave, high-amplitude activity with occasional epileptiform discharges and spike activity. The most frequent wave form was an enhanced alpha wave pattern. Following bead implantation there was a significant reduction (48 $\pm$ 13%) in amplitude of EEG activity. ($p < 0.05$)

PLA bead placement in specific points around the head and in other acupuncture points which have been found to reduce seizures appears to significantly reduce seizure activity in refractory epileptic patients. There is also evidence of an acute reduction in the electrical activity of the cortex based upon changes in the EEG following bead implantation. This may represent a promising new method to help control refractory seizures in dogs.

THE SPECTRUM OF INHERITED MYOPATHIES IN YOUNG LABRADOR RETRIEVERS

G. Diane Shelton

University of California San Diego, La Jolla, CA, USA

Over the past 10 years, several inherited myopathies have been reported in young Labrador Retrievers. While some have distinct clinical phenotypes and can be easily recognized, others require in depth examinations including state of the art evaluation of muscle biopsies and DNA testing. This report describes the current status of inherited myopathies in the Labrador Retriever breed with a guide to clinical recognition.

An early onset of severe and generalized muscle weakness occurring in the **first several weeks** of life has been described in the congenital myasthenic syndrome associated with an autosomal recessive mutation in *COLQ* (2014), which encodes the collagenous tail of acetylcholinesterase. Weakness in this disorder is fatigable and anticholinesterase drugs exacerbate the weakness. Severe generalized and progressive weakness has also been described in the X-linked myotubular myopathy associated with a mutation in the lipid phosphatase *MTM1* (2010). In both of these disorders, creatine kinase (CK) activity is normal or only minimally elevated. Another potentially severe myopathy occurring in this age group is X-linked dystrophin deficient muscular dystrophy (XLMD). A key finding is markedly and persistently elevated CK activity. This finding, along with a dystrophic phenotype in muscle biopsies, helps to distinguish this severe myopathy from other congenital myopathies.

Other less severe myopathies can occur at **several months** of age including autosomal recessive centronuclear myopathy associated with a mutation in *PTPLA* (2005). The clinical presentation can appear similar to that of x-linked myotubular myopathy, however, is not as rapidly progressive. Affected dogs can live well into adulthood. The CK activity is normal or only mildly elevated and a muscle biopsy shows a myopathic phenotype with numerous central nuclei. A very mild form of XLMD has recently been identified in several related male Labrador Retrievers with markedly elevated CK activity detected at the time of neuter. No clinical evidence of weakness or muscle atrophy is observed. Muscle biopsies show a dystrophic phenotype and dystrophin is not detectable by western blot. What is protecting these dogs from the devastating effects of dystrophin deficiency is yet to be determined. A myopathy with progressive gait abnormalities and multiple joint deformities was recently identified associated with sarcolemmal specific collagen VI deficiency (2014), similar to that of human Ulrich's congenital muscular dystrophy. The mutation has been confirmed in *COL6A3* but not yet published. Although not a myopathy, the syndrome of exercise-induced collapse has been characterized and a mutation in the CNS specific *DNM1* gene described (2008). Dogs with this syndrome are clinically normal in appearance and activity, which helps to separate this disorder from all other characterized inherited myopathies.

In summary, many inherited myopathies have been described in young Labrador Retriever dogs and clinicians need to be aware of the specific phenotypes. Knowledge of these myopathic disorders and recognition of the myopathic phenotype should enable a more precise clinical diagnosis and guide breeders to eliminate these disorders from their lines.

A P.P248LFS3* MUTATION IN *RAB3GAP1* IN LARYNGEAL PARALYSIS AND POLYNEUROPATHY WITH OCULAR ABNORMALITIES AND SPONGIFORM ENCEPHALOPATHY IN BLACK RUSSIAN TERRIERS

Dennis O'brien[1], Tendai Mhlanga-Mutangadura[1], Gayle Johnson[1], Diane Shelton[2], Jeremy Shomper[1], Nicolas Granger[3], Jeremy Taylor[1], Robert Schnabel[1], Gary Johnson[1]

[1]University of Missouri, Columbia, MO, USA, [2]University of California San Diego, San Diego, CA, USA, [3]University of Bristol, Langford, North Somerset, UK

Juvenile onset laryngeal paralysis/polyneuropathy has been reported in Black Russian Terriers (BRT). The purpose of this study was to further characterize the phenotype and identify the mutation responsible.

Three BRT with laryngeal paralysis and weakness examined at the University of Missouri (MU) underwent electrodiagnostic testing. Medical records of 13 other BRT were reviewed and histopathology examined on 9 dogs. Whole genome sequencing (WGS) was performed on one affected BRT and rare homozygous variants ascertained by comparison to 111 other canid WGS.

All affected BRT presented at 3 months of age with laryngeal paresis/paralysis. The 3 MU cases had miotic pupils, cataracts, microphthalmia, and/or persistent pupillary membranes. Neurologic examination showed sensory and motor deficits. Electromyograms showed prolonged insertional activity, fibrillations and complex repetitive discharges. Tibial nerve studies showed moderately reduce conduction velocity and compound motor unit action potential amplitudes. Muscle and nerve pathology was most prominent in the larynx, and consisted of a neurogenic pattern of muscle fiber atrophy, decreased density of large myelinated fibers, and inappropriately thin myelin sheaths. Intraneuronal vacuolation was found in the cerebellar roof nuclei and Purkinje cells.

WGS identified a unique homozygous single base pair deletion: *RAB3GAP1:c.743del*. The produced frame shift predicts a premature stop codon and a truncated gene product missing 730 C-terminal amino acids. DNA was available from 8 other affected BRT which were all homozygous for the variant, while 48 normal BRT were either heterozygous or homozygous for the wild-type allele. Mutations in *RAB3GAP1* cause a similar phenotype in humans, Warburg micro syndrome.

PAROXYSMAL NON-KINESOGENIC DYSKINESIA IN SOFT COATED WHEATEN TERRIERS IS ASSOCIATED WITH A MISSENSE MUTATION IN *PIGN* AND RESPONDS TO ACETAZOLAMIDE THERAPY

Dennis O'Brien[1], Ana Kolicheski[1], Rebecca Packer[2], Stephanie Thomovsky[3], Jeremy Taylor[1], Robert Schnabel[1], Jason Berg[4], Laura Vasquez[5], Gary Johnson[1]

[1]University of Missouri, Columbia, MO, USA, [2]Colorado State University, Fort Collins, CO, USA, [3]Purdue University, West Lafayette, IN, USA, [4]Animal Specialty Center, Yonkers, NY, USA, [5]Gulf Coast Veterinary Neurology & Seurosurgery, Houston, TX, USA

The purpose of this study was to characterize a hereditary, paroxysmal, non-kinesogenic dyskinesia in Soft Coated Wheaten Terriers (SCWT) and identify the responsible mutation.

Medical records and videos were reviewed for 20 SCWT or SCWT/poodle crosses that presented for a movement disorder. Whole genome sequences (WGS) were performed on 2 affected SCWT and rare homozygous variants ascertained by comparison to 111 other canid WGS.

Affected dogs had episodes of dystonia and/or rapid, irregular involuntary movements of the limbs that typically began between 8 months to 3 years of age. The episodes lasted a few minutes up to several hours and became more frequent and severe over time. They occurred when the dogs were awake, but most dogs showed no clear trigger. Benzodiazepines, antiepileptic drugs and muscle relaxants showed limited benefit. Five of 8 dogs treated with acetazolamide improved and 3 had complete resolution of the dyskinesia. Histopathology of two dogs showed no pathology.

WGS identified a homozygous *PIGN:c.398C>T* variant in both affected SCWT. All 12 other affected SCWT and crosses were homozygous for the variant, while 762 normal SCWT were either heterozygous or homozygous for the wild-type allele. Three-hundred-eighty-eight randomly selected poodles and 119 dogs of other breeds were all homozygous for the wild-type allele.

PIGN codes for phosphatidylinositol glycans-N, an enzyme involved in synthesis of glycosyl-phosphatidylinositol (GPI). GPI anchors numerous proteins to lipid rafts in the cell membrane including carbonic anhydrase 4. Mutations in *PIGN* in humans cause a syndrome characterized by congenital abnormalities, neonatal hypotonia, seizures and choreoathetosis.

EVALUATION OF SECONDARY NEURONAL INJURY AFTER INTRACEREBRAL HEMORRHAGE IN DOGS BY PERFUSION-WEIGHTED MRI, PROTON MR SPECTROSCOPY AND HISTOPATHOLOGY

Sun-Hyung Cho[1], Dae-Gi An[1], Junyoung Park[2], Chulhyun Lee[2], Dong-In Jung[3], Dongwoo Chang[1], Ji-Houn Kang[1], Mhan-Pyo Yang[1], Byeong-Teck Kang[1]

[1]*Chungbuk National University, Cheongju, Republic of Korea,* [2]*Korea Basic Science Institute, Ochang, Republic of Korea,* [3]*Gyeongsang National University, Jinju, Republic of Korea*

Intracerebral hemorrhage (ICH) is a subtype of stroke with high mortality and disability. The effects of ICH upon brain tissue are biphasic, including primary and secondary injury. Although hematoma volume is the main factor that influences ICH outcome, secondary brain injury itself results in severe neurological deficits. Therefore, the purpose of this study was to investigate the mechanisms of secondary neuronal injury by evaluating cerebral blood flow (CBF), metabolites and pathological changes in and around the intracerebral hematoma.

ICH was induced in 5 healthy laboratory beagle dogs by injecting 500 U of bacterial collagenase from *Clostridium histolyticum*, which was delivered into the parietal lobe over 5 min with a micro infusion pump. Perfusion-weighted magnetic resonance imaging (PW-MRI) and proton MR spectroscopy (MRS) were performed serially at 6 different time points by using a 3T MR system: before and at 12 h, 2.5, 4.5, 9.5 and 23.5 days after ICH. Prussian blue and immunohistochemical staining were performed to demonstrate iron, glial fibrillary acidic protein (GFAP), tumor necrosis factor-α (TNF-α), interleukin-6 (IL-6) and IL-10 in and around the intracerebral hematoma.

In hematoma, cerebral blood volume (CBV) was significantly decreased at 2.5 and 4.5 days after ICH. Additionally, significant decrease of CBF was observed at 4.5 days after ICH. In perihematoma, there were no significant changes of CBV and CBF except for the decrease of CBV at 4.5 days after ICH. Proton MRS revealed no abnormality of the levels of *N*-acetyl-asparate and lactate in perihematoma. A significant loss of GFAP immunoreactivity was noted in hematoma, whereas iron deposition and expressions of TNF-α, IL-6 and IL-10 were significantly increased. However no pathological and immunohistochemical changes were observed in perihematoma.

In conclusion, decreased CBF, iron overload and inflammation may contribute to ICH-induced injury and could provide therapeutic targets in ICH of dogs.

SERUM MELATONIN LEVELS IN NORMAL DOGS AND DOGS WITH SEIZURES

Stephanie Thomovsky[1], Annie Chen[2], David Deavila[2], Alecia Kiszonas[2]

[1]Purdue University, W. Lafayette, IN, USA, [2]Washington State University, Pullman, WA, USA

Epilepsy or recurrent seizures is reported to be the most common neurologic condition in dogs. 20-30% of epileptic dogs are considered to be pharmacoresistent to one medication. Recent lab animal literature, in addition to studies in human literature, show the hormone melatonin, to have significant anti-convulsant effects. Epileptic humans actually have significantly lower inter-ictal serum melatonin levels compared to unaffected people. The central hypothesis of our study is that serum levels of melatonin will be lower in dogs with seizures than in normal dogs. Our aim will be to compare mean serum melatonin levels in normal dogs versus dogs with seizures.

Sixty two dogs were enrolled in the study: 29 normal dogs and 33 dogs with seizures. Sampling was done during the daylight hours at three separate time points (8:00 hours, 12:00 hours and 16:00 hours). Melatonin levels were measured in pg/mL using a radioimmunosassay.

Mean melatonin values at 8:00, 12:00 and 16:00 hours in normal dogs were 0.65 pg/mL, 0.55 pg/mL and 0.57 pg/mL, respectively. Values for dogs with seizures were 0.64 pg/mL, 0.56 pg/mL and 0.54 pg/mL at the same time points. No statistical significance existed between mean serum melatonin values in either group at any of the three time points.

In conclusion, there were no notable differences in daytime serum melatonin values in normal dogs versus dogs with seizures.

ANTINEOPLASTIC EFFECTS OF AURANOFIN IN CANINE LYMPHOMA CELLS

Douglas Thamm, Barbara Rose, Travis Laver, Alex Pyuen

Colorado State University, Fort Collins, CO, USA

Lymphoma is a serious condition for which there remain unmet medical needs in humans and dogs. The gold complex auranofin has been utilized as a human therapeutic, primarily as an antirheumatic agent, and some limited information about its use in dogs has been reported. Antiproliferative and pro-apoptotic activity has been observed in a variety of human tumor-derived cell lines, including carcinomas of the breast, head and neck, ovary, lung, and a variety of hematopoietic tumors, including lymphoma. Putative antitumor mechanisms include inhibition of NF-κB and STAT3 signalling, and induction of reactive oxygen species via thioredoxin reductase (TrxR1) inhibition. The dog is a well-established model for spontaneous lymphoma in humans, owing to striking similarity in biology and gene expression. The goal of the current study was to investigate the antineoplastic effects of auranofin in a panel of canine lymphoma-derived cell lines, and to identify potential biomarkers of drug activity for future translational studies.

Four canine lymphoma -derived cell lines were incubated in varying concentrations of auranofin +/- bortezomib, elesclomol, or the conventional antineoplastic agents doxorubicin (DOX), CCNU and vincristine (VCR) for 24-72 hours. Relative viable cell number was assessed using MTS. Cell cycle distribution and apoptosis induction were evaluated using flow cytometry with propidium iodide staining and Annexin V / PI staining respectively. Induction of ROS was assessed using the redox-sensitive fluorescent dye CM-H_2DCFDA. TrxR activity was assessed using a commercial colorimetric kit (Cayman). Changes in STAT3 and NF-κB phosphorylation status were assessed via western analysis.

Auranofin induced dose-dependent antiproliferative effects in all canine lymphoma cell lines, with 50% inhibitory concentrations (IC_{50}s) between 0.1 and 1 uM. These are similar to IC_{50}s observed in human hematopoietic tumor cells, and well within clinically achievable serum concentrations. Dose- and time-dependent apoptosis induction was observed, and additive to synergistic antiproliferative/propaptotic activity was observed with bortezomib and elesclomol, but not DOX, CCNU or VCR. Dose- and time-dependent induction of ROS and inhibition of TRxR activity was observed following auranofin treatment, and antiproliferative activity could be partially blocked with the free-radical scavenger N-acetylcysteine. There was no change in STAT3 of p65 phosphorylation following auranofin treatment.

In conclusion, antitumor activity is observed with auranofin treatment *in vitro* in canine lymphoma cells, at pharmacologically achievable concentrations. This is similar to activity observed in human lymphoma and leukemia. Cooperative activity was observed with the targeted agents bortezomib and elesclomol. These data support clinical evaluation of auranofin in canine lymphoma. Measurement of ROS accumulation and/or TRxR activity may be useful pharmacodynamic markers of drug activity.

RABACFOSADINE AND PREDNISONE: EFFICACY OF A Q21 DAY ADMINISTRATION SCHEDULE IN CANINE LYMPHOMA

Douglas Thamm[1], Michelle Morges[1], Craig Clifford[2], Kristine Burgess[3], Corey Saba[4], David Vail[5], Cheryl London[6]

[1]*Colorado State University, Fort Collins, CO, USA,* [2]*Hope Veterinary Specialists, Malvern, PA, USA,* [3]*Tufts University, North Grafton, MA, USA,* [4]*University of Georgia, Athens, GA, USA,* [5]*University of Wisconsin - Madison, Madison, WI, USA,* [6]*The Ohio State University, Columbus, OH, USA*

Lymphoma is one of the most common canine cancers. While current therapies induce remission in most naïve dogs with lymphoma, drug-resistant relapse is common and there is a distinct need for novel agents. The acyclic nucleotide phosphonate 9-(2-phosphonylmethoxyethyl)-guanine (PMEG) forms an active phosphorylated metabolite, PMEGpp, in cells and causes cytotoxicity in dividing cells due to inhibition of DNA polymerases α, δ, and ε; however, PMEG's use as an anticancer agent is limited by poor cellular permeability and nonspecific toxicity. Rabacfosadine (VDC-1101/GS-9219/TANOVEA™), a novel double prodrug of PMEG, was designed to preferentially target lymphoid cells with significantly reduced systemic toxicity. Rabacfosadine has been administered on a variety of dosing schedules to dogs with lymphoma. Objective responses were noted in 100% of chemotherapy-naïve dogs and 60% of refractory dogs, with a median remission duration of 128 days. Given the ease of administration and equivalent activity with a q21-day administration schedule, we sought to generate additional data regarding efficacy of this regimen through completion of a prospective clinical trial. Additionally, we explored the potential of concomitant low-dose prednisone to mitigate previously observed cutaneous and pulmonary adverse effects.

Dogs with cytologically or histologically confirmed lymphoma were treated with rabacfosadine (0.82 mg/kg free base, as a 30-minute IV infusion once every 21 days). Dogs received concomitant prednisone (1 mg/kg PO QOD) throughout treatment. Dogs experiencing a complete response (CR) received 5 doses of rabacfosadine, followed by monthly rechecks. Complete clinicopathological assessment and clinical assessment of remission and adverse effects (AEs) were performed every 21 days. Response was assessed according to published VCOG criteria and AEs according to the VCOG-CTCAE v1.1.

74 dogs were prospectively enrolled. 63 were evaluable for response assessment and 73 were evaluable for progression free interval (PFI) assessment. While 13% of evaluable dogs were treatment-naïve and 29% had received a single line of previous treatment, the majority of dogs (59%) had received 2 or more lines of previous therapy. 50 evaluable dogs had B cell lymphoma and 13 had T cell lymphoma. The overall response rate (ORR) was 57% (25% CR, 32% PR). The ORR was 64% and 23% for B cell and T cell respectively. Degree of pre-treatment impacted response rate: the ORR was 88% in naïve dogs, 56% in dogs treated 2nd-line, and 51% in 3rd-line and beyond. The median PFI was 112 days for dogs experiencing a CR and 42 days for a PR (overall median PFI 41 days). Degree of pre-treatment significantly impacted PFI (164 d, 84 d and 32 d for naïve, 2nd-line and >2nd line respectively).

The majority of AEs were mild and self-limiting: gastrointestinal (GI) and hematologic AEs were most common. Grade 3 AEs included liver enzyme elevation (4), lethargy (4), GI (2) and urinary (1). 3 dogs experienced grade 4 hematologic toxicity, and 2 developed severe hemorrhagic gastroenteritis leading to euthanasia, several weeks after the first rabacfosadine treatment and therefore of uncertain attribution. 4 dogs experienced grade 1 dermatologic AEs, and 1 dog developed grade 1 pulmonary fibrosis.

In conclusion, rabacfosadine was generally well tolerated and had substantial antitumor activity in dogs with both treatment-naïve and refractory lymphoma when administered on a q21-day schedule. Response rates and PFI observed in this study are comparable to historical data with rabacfosadine when degree of pre-treatment is accounted for. There was a reduction in both the frequency and severity of AEs relative to historical data; however, it is not clear whether this is a result of less frequent dosing, concurrent prednisone, or lower cumulative rabacfosadine exposures in this heavily pre-treated population. Further studies are warranted to explore rabacfosadine at higher doses.

DEFINITIVE HIGH-DOSE HYPO-FRACTIONATED TOTAL PELVIC IRRADIATION WITH SIMULTANEOUS BOOST IN CANINE URINARY TRANSITIONAL CELL CARCINOMA: A FEASIBILITY STUDY AND FIRST CLINICAL EXPERIENCES

Mario Dolera, Luca Malfassi, Massimo Sala, Giovanni Mazza, Simone Pavesi, Silvia Marcarini, Nancy Carrara, Sara Finesso

La Cittadina Fondazione Studi e Ricerche Veterinarie, Romanengo, Italy

The lower urinary tract transitional cell carcinoma (CCT) poses challenge in order to the appropriate radiotherapy (RT) regimen. Organs at risk (OARs) within the irradiation field (ureters, rectum) are the limiting factors in dose escalation. The primary aim of this study was to evaluate the technical feasibility of high-dose hypo-fractionated dynamic IMRT in non-resectable lower urinary CCT affected dogs. The secondary goal was to evaluate the toxicity and the efficacy of the RT regimens.

Three dogs with lower urinary tract CCT were treated with definitive high-dose hypo-fractionated RT with volumetric modulates arc therapy (VMAT) technique. The volume treatment definition include the gross tumor (GTV), the PTV1 (GTV+3mm), limphatics (PTV2), the entire bladder, prostate in males and uretra (PTV3), the entire pelvis except the rectal volume (PTV4). Dose prescriptions were 40 Gy to PTV1, 38 Gy to PTV2, 34 Gy to the PTV3, 30 Gy to the PTV4, in 6 fractions on alternate days. A piroxicam was subministered to all dogs. Serial clinical and CT/MRI examination were performed. Disease control and toxicity effects were evaluated according to RECIST and VRTOG criteria.

Three T2N0M0 urinary tract CCT were treated. Prescription goals were obtained in all three cases with V95%>95% and V107%>2%. During follow-up (mean 6 months) one partial response and two complete responses were obtained. Two grade I cystitis developed. Non rectal toxicity was recognized.

The initial experiences with the RT regimen adopted indicate a feasibility and effectiveness in lower urinary CCT. Longer follow-up and larger treatment series are needed.

DEFINITIVE HIGH-DOSE HYPO-FRACTIONATED STEREOTACTIC BRAIN-SPARING IRRADIATION OF STAGE IV CANINE NASAL TUMORS: A FEASIBILITY STUDY AND FIRST CLINICAL EXPERIENCES

Luca Malfassi, Mario Dolera, Simone Pavesi, Massimo Sala, Giovanni Mazza, Silvia Marcarini, Nancy Carrara, Sara Finesso

La Cittadina Fondazione Studi e Ricerche Veterinarie, Romanengo, Italy

The prognosis for canine nasal tumours with intracranial extension is poor with an expected survival of 1 month with palliation and 6.7 months with irradiation. However, studies regarding stage IV nasal tumours treated with brain-sparing irradiation techniques are lacking. The aim of this prospective study was to evaluate feasibility and efficacy of definitive intent stereotactic radiotherapy in dogs with nasal tumours with massive intracranial extension.

Seven dogs with stage IV nasal tumours were treated with high-dose hypo-fractionated stereotactic radiotherapy with VMAT technique. Dose prescriptions were 32-36 Gy in four consecutive-day fractions to the gross tumour and 30 Gy to limphatics. Adjuvant treatment included carboplatin. Serial clinical and CT/MRI examination were performed. Disease control and toxicity effects were evaluated according to RECIST and VRTOG criteria. Median survival time (MST) was evaluated using Kaplan-Meier curves.

Six carcinoma and 1 sarcoma were treated. Prescription goals were obtained in four cases with V95%>95% and V107%>2% whereas in 3 dogs V95%=86-90% was accepted to limit maximum brain punctual dose <27 Gy. Two partial response and 5 complete responses were obtained. MST was 9 months. One grade II late brain radiotoxicity and two brain ascending infections were observed. Relapse pathways involves diffuse meningeal and sphenoid invasion.

The initial experiences with the RT regimen adopted indicate a feasibility and effectiveness in modified stage IV nasal tumours. The relapse pathways observed suggest to evaluate alternative adjuvant treatment in dogs treated with stereotactic radiotherapy.

INVESTIGATION OF THE FOXM1 TRANSCRIPTIONAL PATHWAY IN CANINE LYMPHOMA

Jackie Wypij, Sara Goldschmidt, Holly Pondenis

University of Illinois at Urbana-Champaign, Urbana, IL, USA

Forkhead box M1 (FOXM1) is a transcription factor regulating cell cycle progression, angiogenesis, and apoptosis. Increased FOXM1 is observed in human lymphoma/leukemia and is recognized as both a prognostic factor and a therapeutic target. Thiostrepton, an antibiotic used in commercial veterinary topical products, is a putative FOXM1 inhibitor. The purpose of this study was to 1) evaluate the expression of FOXM1 pathway proteins in canine lymphoma cell lines and 2) evaluate the anti-proliferative effects of the FOXM1 inhibitor thiostrepton on canine lymphoma in vitro.

Four canine lymphoma cell lines (17-71, GL-1, CL-1, and OSW) were used. Western blot was performed to assess protein expression of FOXM1 and FOXM1 pathway proteins cyclin B, survivin, and Cdc25. 17-71 canine lymphoma cells were treated with thiostrepton (0-100μM) for 24 hours. Cell viability (cell proliferation and cell death) were assessed with a commercial MTS assay and Trypan Blue exclusion.

Canine lymphoma cell lines express FOXM1 and FOXM1 pathway proteins. Thiostrepton decreases cell viability in a dose-dependent manner ($p<0.05$).

In vitro results support the FOXM1 pathway as a putative target in canine lymphoma. The FOXM1 inhibitor thiostrepton has anti-cancer activity in vitro. These results warrant further evaluation of thiostrepton *in vitro* for canine lymphoma. This research may lead to validation of drug repurposing of thiostrepton-containing topical compounds for treatment of canine lymphoma, which may be particularly relevant in cutaneous lymphoma.

TUBULIN-MODULATING EFFECTS OF THE ANTHELMINTIC MEBENDAZOLE IN FELINE CANCER

Jackie Wypij

University of Illinois at Urbana-Champaign, Urbana, IL, USA

Limited effective treatment options are available for feline cancer patients. One mechanism for new drug development is drug repositioning of low-toxicity pharmaceuticals with inherent anti-cancer activity. Benzimidazole anti-parasitics such as mebendazole (MBZ) are putative novel mitotic spindle inhibitors. The purpose of this study was to 1) evaluate the potential synergism between MBZ and a traditional mitotic spindle inhibitor *in vitro,* 2) evaluate MBZ effect *in vitro* on tubulin polymerization and 3) evaluate serum tubulin polymerization activity in healthy cats treated with MBZ.

Two feline cancer cell lines (SCCF1, oral squamous cell carcinoma and K12, feline mammary carcinoma) were incubated with MBZ at 0-100µM and vinorelbine at various concentrations for 24 hours and cell proliferation was assessed with a commercial MTS assay. Synergism was computed utilizing a combination index. Two healthy adult cats were treated with 22mg/kg MBZ daily. A commercial assay was used to assess direct tubulin polymerization activity of MBZ as well as the tubulin polymerization of treated cat sera.

MBZ decreased cell proliferation ($p<0.05$) and appears synergistic with vinorelbine. MBZ decreased the rate and total polymerization of tubulin *in vitro.* There was measurable decreased tubulin polymerization activity in serum of the treated cats.

MBZ demonstrates *in vitro* anti-neoplastic effects in feline cancer cell lines. MBZ was well tolerated *in vivo* and resulted in measurable anti-tubulin activity in treated cats. Given the known low toxicity, MBZ may be a potential candidate for future clinical trials in pet cats with cancer.

TREATMENT OF CANINE B-CELL LYMPHOMA WITH CHEMOTHERAPY AND A CANINE ANTI-CD20 MONOCLONAL ANTIBODY: A PROSPECTIVE DOUBLE-BLIND, RANDOMIZED, PLACEBO-CONTROLLED STUDY

G.K. Ogilvie[1], D.R. Proulx[1], L VanHorn[1], G Archer[1], B Monreal[1], E Moe[1], L Pope[1], G Hansen[2], B Alkuzweny[2], W.M. FitzPatrick[1]

[1]Angel Care Cancer Center, California Veterinary Specialists, Carlsbad, CA, USA, [2]Aratana Therapeutics, Inc., Kansas City, KS, USA

Management of human B-cell lymphoproliferative disorders, including diffuse large B-cell lymphoma (DLBCL), includes a monoclonal antibody that targets the B-cell antigen CD20. The aggressive form of DLBCL has many similarities to the most common form of lymphoma in dogs. Addition of the species-specific antibody rituximab to chemotherapy for the treatment of DLBCL in humans is associated with little increase in toxicity, yet enhanced efficacy. This study evaluates a monoclonal antibody that binds specifically with the canine B-cell antigen CD20 (canine Mab), used with chemotherapy.

Dogs were enrolled in a prospective, randomized, blinded, placebo-controlled study, treated with one 4-week cycle of L-CHOP chemotherapy and then randomized to either receive treatment with the canine Mab or placebo. Once remission was lost, all dogs received one dose of doxorubicin followed by treatment with the canine Mab. Efficacy was assessed by evaluation of measurable lymph nodes, cytology and clinical status. Safety was based on Veterinary Cooperative Oncology Group's common terminology criteria (VCOG-CTCAE v1.1).

Dogs (n=27 enrolled, 24 randomized) all had measurable, cytologically-confirmed and flow cytometrically immunophenotyped DLBCL. The median progression-free and overall survival times in the canine Mab arm of the study were 167 and 325 days, respectively, compared to 93.5 and 177 days for the placebo arm. Adverse events were restricted to the L-CHOP cycle.

The canine Mab was well-tolerated and, in conjunction with L-CHOP, caused significant increases in survival in dogs with DLBCL.

PRECLINICAL COMPARISON OF THREE INDENOISOQUINOLINES CANDIDATES IN TUMOR-BEARING DOGS

Amy LeBlanc[1], Jenna Burton[2], Chand Khanna[1], Christina Mazcko[1], Susan Lana[3], Kristen Weishar[3], Cheryl London[4], William Kissberth[4], Erika Krick[5], David Vail[6], Michael Childress[7], Jeffrey Bryan[8], Kimberly Selting[8], Carolyn Henry[8], Jennifer Wilcox[8], Sandra Axiak-Betchel[8], Lisa Barber[9], Jiuping Ji[10], Joesph Covey[10], Thomas Pfister[11]

[1]Comparative Oncology Program, Center for Cancer Research, National Cancer Institute, Bethesda, MD, USA, [2]UC Davis School of Veterinary Medicine, Davis, CA, USA, [3]Colorado State University, College of Veterinary Medicine,, Fort Collins, CO, USA, [4]Veterinary Biosciences, The Ohio State University, Columbus, OH, USA, [5]School of Veterinary Medicine, University of Pennsylvania, Philadelphia, Pennsylvania, USA, [6]School of Veterinary Medicine, University of Wisconsin, Madison, Wisconsin, USA, [7]Purdue University College of Veterinary Medicine, West Lafayette, Indiana, USA, [8]Department of Veterinary Medicine and Surgery, College of Veterinary Medicine, University of Missouri-Columbia, Columbia, MO, USA, [9]Department of Clinical Sciences, Cummings School of Veterinary Medicine at Tufts University, North Grafton, Massachusetts, USA, [10]Division of Cancer Treatment and Diagnosis, National Cancer Institute,, Bethesda, MD, USA, [11]Laboratory of Human Toxicology and Pharmacology, Applied/Developmental Research Support Directorate, SAIC-Frederick, Inc., Frederick National Laboratory for Cancer Research, Frederick, MD, USA

Indenoisoquinolones act as topoisomerase 1 inhibitors and are known to be active anticancer agents in both humans and dogs. A clinical trial in tumor-bearing dogs with lymphoma was conducted through the NCI Comparative Oncology Trials Consortium to define the safety, pharmacokinetics (PK) and pharmacodynamic modulation of three related novel indenoisoquinolones.

Sixty-eight dogs with lymphoma were enrolled in the dose escalation phase. Dogs were randomized to receive 5 consecutive intravenous daily doses of one agent. Each agent was escalated independently with 3 dogs/cohort. Serial pre- and post-treatment tumor biopsies, tumor aspirates, and bone marrow aspirates were collected. Two 24-hour PK curves were completed after the first and last dose of drug. Tumor and plasma PK, TOP1 levels, and gH2AX induction data were collected along with clinical toxicity and response data.

The maximum tolerated dose (MTD) was defined for two of three indenoisoquinolines (NSC725776 17.5mg/m2, NSC706744 100mg/m2). An MTD was not reached for NSC743400. Toxicity profiles for all agents included expected gastrointestinal, hematologic and constitutional events. Objective responses were seen across all agents with a dose-response relationship evident for NSC725776 and NSC734400. The overall response rates for evaluable dogs were 30% (7/23), 39% (3/23), and 78% (14/18) for NSC725776, NSC743400, and NSC706744, respectively. Correlative relationships between clinical response, pharmacokinetic profiles and pharmacodynamic markers were established for all 3 agents.

All 3 novel agents studied herein are both tolerable and effective in tumor-bearing dogs. Study of NSC706744 at 100mg/m2 in dogs with lymphoma and measurable soft tissue sarcoma is ongoing.

Combination AT-014 (ADXS31-164) *Lm*-LLO immunotherapy plus palliative radiation delays tumor progression and prolongs overall survival in canine osteosarcoma.

Nicola Mason[1], Josephine Gnanandarajah[1], Ana Caceres[1], Lilian Duda[1], Kim Agnello[1], Cara Blake[1], Falon Grey[1], Julie Engiles[1], Anu Wallecha[2], Yvonne Paterson[1]

[1]*University of Pennsylvania, Philadelphia, PA, USA,* [2]*Advaxis Inc, Princeton, USA*

Radiation therapy (RT) induces immunogenic tumor cell death and promotes anti-tumor immunity that can result in regression of metastatic lesions, a phenomenon known as the abscopal effect. We hypothesized that combining RT with a recombinant HER2/neu expressing *Lm*-LLO immunotherapy (AT-014, ADXS31-164) could promote anti-tumor immunity, delay primary tumor progression and prevent metastatic disease in dogs with osteosarcoma.

Ten dogs with histopathologically confirmed, treatment naïve, appendicular osteosarcoma were enrolled to this pilot study. All dogs received 16 Gy of RT in 2 fractions on consecutive days, followed by AT-014 intravenously every 3 weeks for 8 administrations. Radiographs were performed at weeks 0, 10, 22 and every 2 months thereafter to assess primary tumor progression and development of pulmonary metastases. At these times dogs were evaluated for toxicity, lameness and quality of life (QOL) and PBMCs were collected to evaluate HER2/neu specific T cell responses. The primary endpoint was time to progression (TTP); secondary end points were safety and overall survival.

Repeat AT-014 administrations were well tolerated. Lameness and QOL improved in 7 dogs. At present, 6 dogs are still alive. After 8 doses of AT-014, 5 dogs showed minimal radiographic evidence of primary tumor progression. 4 dogs developed pulmonary metastatic disease and 4 suffered pathological fractures. Median TTP is 204 days and median survival time is 285 days. Immunological assessment is pending. Results compare favorably to historical reports of RT alone and warrant performance of a randomized, placebo controlled clinical trial.

CHARACTERIZATION OF THE RELATIONSHIP BETWEEN BODY CONDITION SCORE AND SURVIVAL IN DOGS WITH LYMPHOMA AND OSTEOSARCOMA

Frank Romano[1], Cailin Heinze[1], Lisa Barber[1], Joel Mason[2], Lisa Freeman[1]

[1]*Cummings School of Veterinary Medicine, North Grafton, MA, USA,* [2]*USDA Human Nutrition Research Center at Tufts University, Boston, USA*

In humans, obesity exacerbates many types of cancer by increasing the risk of cancer development, tumor aggressiveness, recurrence, and the risk of cancer death. However, this relationship in dogs has not been thoroughly evaluated. The aim of this study was to determine whether body condition score (BCS) at the time of lymphoma (LSA) or osteosarcoma (OSA) diagnosis in dogs was predictive of survival time (ST). We hypothesized that an overweight BCS at the time of cancer diagnosis would be associated with a shorter ST. Medical records of dogs diagnosed and treated for LSA or OSA between 2000 and 2010 were reviewed. Data on signalment, body weight, BCS, treatment, and survival were collected. Dogs were grouped by BCS (underweight, normal, and overweight) and STs were compared using Kaplan-Meier survival analysis. A total of 325 dogs (LSA = 270 and OSA = 55) were included in the study. Overall, 5.5% of dogs were underweight, 54.2% were normal weight, and 40.3% were overweight. On univariate analysis, ST was significantly shorter for underweight dogs with LSA ($P = 0.017$), but not for OSA ($P = 0.518$). Dogs with LSA that gained > 10% body weight after diagnosis had a longer ST ($P = 0.003$). On multivariate analysis of all dogs, change in body weight ($P < 0.001$) and anemia ($P = 0.026$), but not BCS group, were significantly associated with ST. In conclusion, although no relationship between BCS and ST was detected, changes in body weight after diagnosis were associated with ST.

TREATMENT OF FELINE HYPERSOMATOTROPISM - EFFICACY, MORBIDITY AND MORTALITY OF HYPOPHYSECTOMY

Patrick Kenny, Christopher Scudder, Sophie Keyte, James Swann, Robert Fowkes, David Church, Yaiza Forcada, Stijn Niessen

Royal Veterinary College, London, UK

Hypersomatotropism (HS) is an important cause of diabetes mellitus (DM) in cats. Though excision of the somatotrophinoma is generally recommended in humans with HS, surgical therapy has so far been infrequently reported in cats. We aim to describe the efficacy of hypophysectomy as a treatment for HS and the morbidity and mortality encountered in a cohort of cats.

Hypophysectomy has been offered to owners who presented diabetic cats with confirmed HS (insulin-like growth factor-1 [IGF-1]>1000ng/mL, pituitary mass) to the Royal Veterinary College since 2012. All cats were operated on by one neurosurgeon. Hypophysectomy was performed by manual extirpation using fine surgical tools via a trans-oral trans-sphenoidal approach. All cats were rigidly positioned in a commercially available surgical head frame, and the approach to the pituitary fossa guided by referencing bony landmarks to computed tomographic scans. Cats received intensive peri- and post-operative monitoring of electrolytes, glucose and blood pressure, and were initially administered conjunctival desmopressin (DDAVP), intravenous infusions of insulin and hydrocortisone, before being transitioned to conjunctival DDAVP, oral hydrocortisone and levothyroxine and subcutaneous glargine insulin.

In total, 21 diabetic cats with HS underwent hypophysectomy from April 2012 – Oct 2014 (median, range; age: 10.3years, 5.4-14.8; pituitary height: 6.0mm, 4.0-10.6; IGF-1: 1833ng/mL, 1138->2000, fructosamine: 574µmol/L, 339-1076). Other than mild pelvic limb weakness, no cat displayed overt neurological deficits prior to surgery.

Three (14%) cats died post-operatively. Two cats did not recover from anaesthesia and were euthanized within 24hrs; one cat developed septic meningitis and was euthanized 17 days post-operatively.

All surviving cats (n=18) saw a reduction of serum IGF-1, 16 cats (89%) showed IGF-1 normalization (Median post-op serum IGF-1: 38ng/mL, 15-1955; Wilcoxon Signed Rank Test, $p<0.001$).

Fourteen of the 18 surviving cats (78%) achieved diabetic remission; the remaining 4 achieved superior glycaemic control with lower insulin dosages (median fructosamine pre- and post-operatively: 692µmol/L and 547µmol/L respectively; median insulin dose pre- and post-operatively: 20.5units/kg/day and 3.5units/kg/day respectively).

Congestive heart failure was encountered as transient problem in 4/19 cats that recovered from the surgery, all 4 cases occurred prior to implementing a reduction in volume of intravenous fluid delivered as part of the post-operative protocol. Two cats developed paresis of the left orbicularis oculi muscle, which resolved in the surviving cat. Cardiac arrest occurred in 1 cat post-operatively at time of jugular catheter placement, which was successfully revived and made an uneventful recovery. One cat developed a left pelvic limb monoparesis, which improved but did not resolve. Palatal wound breakdown was not encountered.

Hypophysectomy as a treatment for feline HS/DM resulted in a high incidence of diabetic remission and resolution of HS.

DURATION OF FASTING BUT NOT DIURNAL VARIATION AFFECTS THE RESPONSE TO GLUCAGON IN HEALTHY CATS

Chen Gilor, Rebecca Glock, Shir Gilor

The Ohio State University, Columbus, OH, USA

The role of glucagon disturbances in diabetes is increasingly recognized. Glucagon stimulation tests have been described in healthy and diabetic cats previously but information is lacking on the response of healthy cats to glucagon under specific conditions. The aim of this study was to assess the effect of diurnal variation and duration of fasting on the response to glucagon in healthy cats.

Five healthy cats were used in this repeated-measures study. Cats were free-fed regularly at 0730h and 1730h for 30 minutes. Glucagon stimulation tests (20mcg/kg, IM) were performed on each cat 3 times, 2 weeks apart: At 1900h after a 25h fast (PM25), at 0900h after a 25h fast (AM25), and at 0900h after a 15h fast (AM15). Glucose and insulin concentrations were measured at -15, 0, 15, 25, 35, 45 and 60 minutes post stimulation. Baseline (mean of -15 and zero minutes) and peak concentrations were compared using the Friedman test ($P<0.05$ considered significant).

Baseline glucose and insulin did not differ significantly between treatment groups. Peak glucose concentrations occurred at 15 minutes and were significantly higher ($P=0.0085$) at AM15 (mean±SD = 185.2±43.0mg/dL) but did not differ between AM25 (144.4±10.5mg/dL) or PM25 (128.0±18.4mg/dL). Similarly, peak insulin concentrations occurred at 15 minutes and were significantly higher ($P=0.04$) at AM15 (1911±1153 ng/L) but did not differ between AM25 (739±452ng/L) or PM25 (549±366ng/L).

In conclusion, prolonged fasting (25h) significantly blunted the glycemic response to glucagon compared to shorter fasting (15h) but diurnal variation had no significant effect on glucose or insulin responses.

A feline-specific anti-nerve growth factor antibody is safe and effective for the alleviation of inflammatory pain in cats

David Gearing[1], Patrick Gearing[1], Elena Virtue[1], Duncan Lascelles[2], Alexander Drew[1]

[1]Nexvet Biopharma, Melbourne, VIC, Australia, [2]North Carolina State University, Raleigh, NC, USA

Options for the management of chronic pain in cats are extremely limited - in particular, the use of NSAIDs is limited by the potential for significant side effects. A safe and effective analgesic for cats is highly desirable.

Neutralizing monoclonal antibodies (mAb) against Nerve Growth Factor (NGF) are analgesic in rodent models and in humans with osteoarthritis. Using a novel technique for inter-species conversion of antibodies based on expressed cDNA sequence analysis (PETizing) we have recently described a canine-specific anti-NGF mAb (NV-01) that alleviates pain due to osteoarthritis in dogs. Using the same approach, we now describe the design and development of a novel therapeutic feline-specific anti-NGF mAb (NV-02) for the alleviation of pain in cats.

The PETized amino acid sequences of the NV-02 mAb heavy and light chains were converted to cDNA sequences by complete oligonucleotide-based chemical synthesis, cloned into a mammalian cell expression vector and expressed in Chinese Hamster Ovary (CHO) cells in chemically-defined animal-component free media. The NV-02 mAb was purified to homogeneity by affinity chromatography and sterile filtration. Purified NV-02 mAb demonstrated very high affinity and potency for neutralizing NGF and did not bind complement.

In pilot studies, no adverse events were observed following injection of NV-02 mAb i.v. or s.c. in four cats. The NV-02 mAb had an elimination half-life of approximately one week and did not induce neutralizing antibodies following repeated injection. Further studies, using the kaolin model of inflammatory pain in the cat, demonstrated that NV-02 mAb was effective at reducing signs of lameness due to inflammation in a dose-dependent manner ($P<0.05$).

Together, these data suggest that NV-02 mAb has the potential to be a safe and effective therapeutic analgesic in the cat. Studies are now in progress in cats with degenerative joint disease.

SPIKE GENE MUTATIONS IN FELINE CORONAVIRUS AND THEIR CORRELATION TO FELINE INFECTIOUS PERITONITIS

Christian Leutenegger[1], Nancy Sanders[1], Jane Robertson[1], Peter Rottier[2]

[1]*IDEXX Laboratories, Inc., West Sacramento, USA,* [2]*Utrecht University, Utrecht, The Netherlands*

The feline coronaviruses (FCoV) occur as 2 pathotypes with an enigmatic, even controversial, relationship: the low virulence or nonvirulent feline enteric coronavirus (FECV) and the highly lethal feline infectious peritonitis virus (FIPV). Recently described spike gene mutations within the putative fusion region were described to correlate with the mutated form of FCoV (FIPV) leading to the clinical presentation of feline infectious peritonitis (FIP). In this presentation, the development and validation of an allelic discrimination real-time PCR typing test which can identify each mutation separately will be described.

The diagnostic sensitivity and specificity will be reported from a set of 203 European clinical samples acquired from either FIP confirmed cats or from healthy cats that previously tested FCoV positive. Of these archived samples, 187 FCoV positive samples were included into the validation. From these, 13 samples did not pass quality control and 17 had virus levels that were below the limit of detection of the PCR assay. Of the remaining 157 samples, 156 were typed correctly with an accuracy of 99.4%. One FIP characterized sample was typed FECV (diagnostic sensitivity 98.6%) while all of the healthy cats were typed FECV (100% diagnostic specificity).

To confirm that these spike gene mutations are not unique to European cats with FIP, additional validation studies from US and Japanese samples were conducted. The US clinical study included 68 cases, 46 from FIP suspicious cases and 22 with non-FIP compatible disease. The FIPV RealPCR biotyping assay was able to accurately differentiate between the FIP or non-FIP (FECV) etiologies ($p<0.0001$) and did not biotype cats with confirmed non-FIP disease as FIPV, confirming the high diagnostic specificity of the molecular test.

EVALUATION OF WEIGHT LOSS OVER TIME IN CATS WITH CHRONIC KIDNEY DISEASE

Lisa Freeman[1], Marie-Paul Lachaud[2], Sean Matthews[3], Linda Rhodes[4], Bill Zollers[4]

[1]Tufts Cummings School of Veterinary Medicine, North Grafton, MA, USA, [2]Aratana Therapeutics, Inc, Paris, France, [3]Statistical Consultant, Dublin, Ireland, [4]Aratana Therapeutics, Inc, Kansas City, KS, USA

Thin body condition, weight loss, and muscle loss are common in cats with chronic kidney disease (CKD). However, the time course and progression of weight loss before and after diagnosis have not been thoroughly evaluated. Therefore, the purpose of this retrospective analysis was to describe the weight loss experienced by cats with CKD before and after diagnosis. Cats with CKD from 6 US veterinary practices for which International Renal Interest Group (IRIS) Stage was available were eligible. Only those with age, date of CKD diagnosis, and weight measurements available in the 3 years before and after diagnosis were included in the analysis. A total of 569 cats, with a mean age at diagnosis of 14.5±2.8 yrs, were evaluated (55.5% spayed females and 44.5% castrated males). Cats were categorized at diagnosis as IRIS Stage 1 [n=34 (6%)], Stage 2 [n=345 (61%)], Stage 3 [n=141 (25%)], and Stage 4 [n=49 (9%)]. Median body weight at diagnosis was 4.2 kg (range, 1.6-9.9 kg). Cats lost a median of 8.9% of body weight in the 12 months before diagnosis, but weight loss began as early as 3 years before diagnosis and accelerated after diagnosis of CKD. Cats below median body weight (4.2 kg) at diagnosis had a significantly shorter survival time compared to cats ≥4.2 kg at diagnosis ($P<0.0001$). Weight loss can be detected in cats before diagnosis of CKD, accelerates after diagnosis, and is associated with survival so careful monitoring could benefit feline health.

PROTEINURIA IN MINIATURE SCHNAUZER DOGS WITH AND WITHOUT HYPERTRIGLYCERIDEMIA

Eva Furrow[1], Valerie Parker[2], Jordan Jaeger[4], Susan Murdoch[3], John Brunzell[3]

[1]University of Minnesota, College of Veterinary Medicine, Saint Paul, MN, USA, [2]The Ohio State University, College of Veterinary Medicine, Columbus, OH, USA, [3]University of Washington, School of Medicine, Seattle, WA, USA, [4]Carolina Veterinary Specialists, Charlotte, NC, USA

Laboratory rodents with spontaneous hyperlipidemia develop glomerular injury and proteinuria. Idiopathic hypertriglyceridemia is common in Miniature Schnauzers, and a previous report demonstrated a possible link to proteinuria in the breed. The objective of this study was to evaluate the relationship between hypertriglyceridemia and proteinuria in Miniature Schnauzers.

Twenty-seven Miniature Schnauzer dogs were enrolled in the study. Fasting serum triglycerides and urine protein-to-creatinine ratios (UPC) were determined. The relationship between these two variables was analyzed by the chi-square test and with simple linear regression as well as multiple regression including age, sex, and body condition score as covariates.

Nine dogs had normal serum triglyceride concentrations (median = 68 mg/dL, range = 14-81 mg/dL), and 18 dogs had hypertriglyceridemia (median = 303 mg/dL, range = 87 – 2,089 mg/dL). There was a strong positive correlation between triglyceride concentration and UPC ($r = 0.75$, $p < 0.001$). Ten of 18 dogs (56%) with hypertriglyceridemia had proteinuria (UPC ≥ 0.5) compared to none of the dogs with normal triglyceride concentrations ($p = 0.009$). Proteinuria was most severe in dogs with triglyceride concentrations >400 mg/dL; 5/6 dogs (83%) with triglyceride concentrations above this level had a UPC ≥2.0. None of the dogs were azotemic or hypoalbuminemic. Endocrine disease was not diagnosed in any dog.

In conclusion, significant proteinuria is present in dogs with idiopathic hypertriglyceridemia and may be due to lipid-induced glomerular injury. Longitudinal studies and evaluation of renal pathology are warranted to further investigate these findings.

PREVALENCE OF CANINE INFECTIOUS RESPIRATORY DISEASE IN DOGS IN CHICAGO OUTBREAK (MARCH-APRIL 2015)

Jill A. Richardson[1], Amy Glaser[2], Nyssa Reine-Salz[1], Edward J. Dubovi[2]

[1]*Merck Animal Health, Madison, New Jersey, USA,* [2]New York State Veterinary Diagnostic Laboratory, Cornell University, Ithaca, New York, USA

Canine infectious respiratory disease complex (CIRDC) is a common disease complex caused by many different viruses and bacteria, including *Bordetella bronchiseptica*, *Mycoplasma cynos*, adenovirus type 2, distemper, influenza A virus, parainfluenza virus, pneumovirus and respiratory coronavirus. In March 2015, veterinarians in the Chicago area noted an increase in incidence of signs of canine infectious respiratory disease in dogs. Nasal and pharyngeal swabs from dogs showing clinical signs were submitted to the Cornell University Animal Health Diagnostic Center (AHDC). A canine respiratory polymerase chain reaction (PCR) screening panel was utilized which allows identification of the following CIRDC pathogens: *B. bronchiseptica*, *Mycoplasma cynos*, adenovirus type 2, distemper, influenza A, parainfluenza virus, pneumovirus and respiratory coronavirus.

On March 16, 2015, a diagnostic sampling program was initiated which tested over 350 samples from dogs with respiratory signs using this screening panel as of April 23, 2015. Samples were collected by the reporting clinics and shipped based on the Cornell AHDC specifications.

Of the over 350 screened sick dogs, 198 dogs tested positive for canine influenza using a PCR assay. None of the tested dogs were confirmed to have the H3N8 influenza strain. Further testing by Cornell and the University of Wisconsin identified the strain as Canine Influenza H3N2. The H3N2 influenza virus is of avian origin and was first isolated from clinically ill dogs in China in 2006 and South Korea in 2007. Canine H3N2 influenza virus has been associated with severe respiratory signs and other clinical signs such as fever, reduced body weight, and interstitial pneumonia. In addition, 24 dogs tested positive for parainfluenza virus, nine dogs tested positive for *B. bronchiseptica*, three dogs tested positive for adenovirus type 2, 27 dogs tested positive for pneumovirus and 29 dogs tested positive for respiratory coronavirus.

Records were reviewed to determine the vaccination status of the positive dogs. None of the 198 dogs that tested positive for the H3N2 influenza virus were vaccinated against the H3N8 influenza virus. Twenty-two of the 24 dogs that tested positive for parainfluenza virus had received an injectable distemper, adenovirus type 2, parvovirus and parainfluenza virus combination vaccine, 13 had received a monovalent *B. bronchiseptica* vaccine and five had been vaccinated with an intranasal *B. bronchiseptica* and parainfluenza combination product. Of the five dogs that received the *B. bronchiseptica* combination vaccine, timing of vaccination suggests that the virus sampled may have been of vaccine origin.

The information gathered from this testing program represents the first time that the H3N2 influenza virus has been isolated from dogs in the United States. A point of origin could not be determined. In addition, data collected supports the role of parainfluenza virus as a major preventable pathogen in CIRDC and route of vaccination should be considered in vaccination protocols.

NEUTROPHIL EXTRACELLULAR TRAP FORMATION IS INCREASED IN DOGS WITH IMMUNE MEDIATED HEMOLYTIC ANEMIA

Stephanie Smith, Maureen McMichael, Ron Achiel, Katrina Jung

University of Illinois, Urbana, IL, USA

It has recently been discovered that stimulated neutrophils can release chromatin fibers decorated with enzymes to form extracellular traps (NETs). Formation of NETs (NETosis) is associated with histone hypercitrullination. NETs are part of the innate immune response, but also associated with thrombosis, organ damage, and autoimmunity.

In this prospective observational study, EDTA plasma was collected from 30 apparently healthy dogs, and 18 dogs with IMHA. Plasma DNA concentration was measured by diluting samples 10 to 160 fold in phosphate-buffered saline/0.1% albumin, and adding 1 mM SytoxGreen. Fluorescence was recorded (Ex485/Em538) and corrected for autofluorescence using diluted samples without SytoxGreen. DNA concentrations were calculated based on a standard curve. Nucleosomes (DNA-histone complexes) were quantified by ELISA and normalized to canine pooled normal plasma (defined as 1). The presence of citrullinated histone H3 in plasma was confirmed using western blot. Results were compared between healthy and IMHA dogs by Wilcoxon rank sum test.

Plasma from dogs with IMHA (median 3.1 ug/mL, range 0-27.1) contained significantly ($p<0.001$) higher concentrations of free DNA than normal plasma (median 0.7, range 0.2-1.2). IMHA plasma also contained significantly ($p<0.001$) higher concentrations of nucleosomes (median 6.8, range 1.4-103) as compared to normal plasma (median 0.9, range 0.4-3.1). Plasma from 16 of 18 (89%) dogs with IMHA contained citrullinated histone H3, but this marker was identified in only 4 of 30 (13%) of healthy dogs.

Excessive NETosis appears to be a feature of canine IMHA, and may contribute to the prothrombotic state in this disease.

FELINE *T. FOETUS* CYTOTOXICITY CAN BE INHIBITED BY SELECTIVE, SMALL-MOLECULE CYSTEINE PROTEASE INHIBITORS

Katie Tolbert, Emily Gould, Mabre Brand

University of Tennessee, Knoxville, TN, USA

Tritrichomonas foetus (Tf) is a mucosal protozoan parasite that infects the feline distal ileum and proximal colon resulting in chronic diarrhea. Tf has a worldwide distribution with no consistently effective drugs to treat the infection. Cysteine proteases (Cp) have recently been demonstrated to promote adhesion-dependent cytotoxicity of feline Tf to the intestinal epithelium. These results support further investigation of Cp inhibitors as potential therapeutic targets for ameliorating the pathological effects of feline trichomonosis. To reduce host toxicity, anti-protease drugs must be tailored to inhibit only those proteases produced by pathogens that are responsible for cytopathic effects and that have redundant mechanisms of production or function with other host proteases. Thus, the aim of this study was to investigate currently available small-molecule Cp inhibitors for their feasibility in targeting specific regions of feline Tf Cp activity and ameliorating Tf-induced cytopathogenicity.

The effect of small molecule Cp inhibitors on specific Cp activities of protein extracts obtained from 4 feline Tf isolates were identified by means of SDS-PAGE *In gel* zymography in the presence or absence of 0.1-1.0 mM calpain inhibitor, antipain, cystatin, leupeptin, chymostatin, WRR, and K11777. The effect of inhibition of specific Cp activity on Tf cytopathogenicity was determined using feline Tf that were allowed to adhere to monolayers of porcine intestinal epithelial cells (IPEC-J2) in co-culture. The cytopathogenic effect of Tf following inhibition of specific Cp activity was evaluated by light microscopy, crystal violet spectrophotometric analysis and immunoblotting Tf-infected IPEC-J2 cells for the M30 antigen of cleaved cytokeratin 18 as a marker of apoptosis. A minimum of 8 replicates were performed for each cytotoxicity experiment. Data were analyzed using Systat software ($p<0.05$).

Patterns of gel zymography, demonstrated by 4 different feline isolates of Tf, were similar and revealed the ability of the vinyl sulfone Cp inhibitors (K11777, WRR) and cystatin to target specific zones of feline Tf Cp activity. These inhibitors had no effect on Tf growth and significantly inhibited cytotoxicity towards the intestinal epithelium in all 4 feline Tf isolates tested ($p<0.001$).

These studies establish that currently available small-molecule inhibitors of cysteine proteases are capable of inhibiting specific regions of feline Tf cysteine protease activity and ameliorating Tf-induced cytopathogenicity to the intestinal epithelium in vitro. The results of these studies provide strong evidence-based justification for identification of the specific cysteine proteases inhibited. Moreover, examination of the effect of specific cysteine protease inhibitors on amelioration of clinical signs in cats naturally infected with Tf is warranted.

OUTCOME FOLLOWING URETERAL STENT PLACEMENT IN DOGS FOR BENIGN URETERAL OBSTRUCTIONS: 44 DOGS (57 URETERS) 2009-2013

Philippa Pavia[1], Allyson Berent[1], Chick Weisse[1], Demetrius Bagley[2]

[1]The Animal Medical Center, New York, NY, USA, [2]Thomas Jefferson University, Philadelphia, PA, USA

Ureteral obstructions are a serious clinical problem in dogs and traditional surgery is associated with a high complication rate. Ureteral stenting provides a minimally-invasive option for immediate decompression, and is considered the standard-of-care in human medicine.

To describe the use of double-pigtail ureteral stents in dogs for benign ureteral obstructions hypothesizing it is a safe and effective treatment option.

44 dogs(57 ureters).

The diagnosis of benign ureteral obstruction was made via ultrasonography, radiography, and ureteropyelography. Ureteral stents were placed endoscopically and/or surgically, with fluoroscopic-guidance. The medical records were reviewed for pre-, intra, and postoperative data and outcome.

44 dogs(57 ureters) underwent stent placement for ureterolithiasis (48/57[84%]), stricture (5/57[9%]), or both (4[7%]). Endoscopic/fluoroscopic or surgical/fluoroscopic techniques were successful in 45/55(82%) and 12/12 ureters, respectively. Pre-operative azotemia was present in 52%(23/44) (median 2 mg/dL) and 37%(16/43) remained azotemic (median 1.3) after decompression. Urinary tract infections were present in 59%(26/44) pre-operatively, and 58%(25/43) post-operatively, 76%(19/25) clearing with treatment. The perioperative mortality rate was 2% (1/44), and was not procedure-related. Three major complications occurred during stent placement. Major peri-operative, short-term, and long-term complications requiring further intervention occurred in 0%, 4%, and 14%, respectively. Median follow-up time was 1158 days (range, 3-> 1,555), with 30/44 alive at last follow-up.

Findings suggest that ureteral stents are a safe and effective minimally invasive short and long-term treatment option for benign ureteral obstructions in dogs. Complications were typically minor, but may necessitate stent exchange or use of an alternative device.

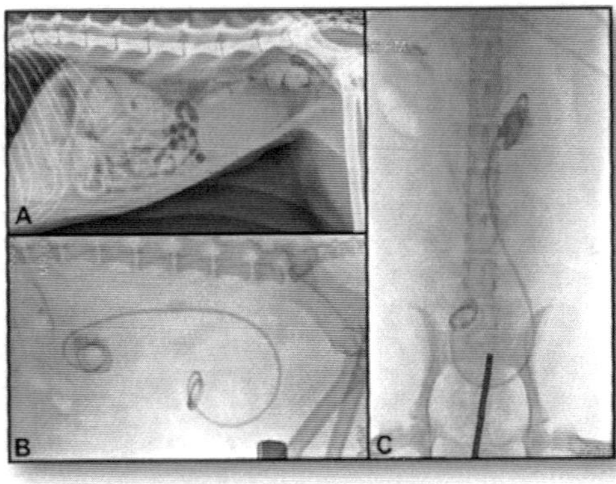

Table 1: Complications *= requiring surgical conversion #= requiring stent exchange or SUB

Complications	Procedural	Post-operative (<7 days)	Short term (<30 days)	Long term (>30 days)
Guidewire obstruction	#1 (2%)	0	0	0
Ureteral perforation	1(2%) 1*(2%)	0	0	0
Ureteral avulsion	*1 (2%)	0	0	0
Contrast extravasation	1 (2%)	0	0	0
Migration	0	0	2 (4%)	1 (2%) 2# (4%)
Occlusion	0	0	0	5# (9%)
Proliferative tissue at UVJ	0	0	0	3 (5%)
Ureteritis	0	0	0	3 (5%)
Encrustation	0	0	0	1 (2%)
Hematuria	0	9 (20%)	3 (7%)	3 (5%)
Hernia	0	0	1 (2%)	2 (4%)
Incontinence	0	0	1(2%)	0

PROGNOSTIC VALUE OF SYMMETRIC DIMETHYLARGININE (SDMA) TO CREATININE RATIO IN DOGS AND CATS WITH CHRONIC KIDNEY DISEASE (CKD)

Mahalakshmi Yerramilli[1], Murthy Yerramilli[1], Edward Obare[1], Dennis Jewell[2], Jean Hall[3]

[1]IDEXX Laboratories Inc, Westbrook, ME, USA, [2]Hill's Pet Nutrition, Inc, Topeka, KS, USA, [3]Oregon State University, Corvallis, OR, USA

SDMA is one of the dimethylated derivatives of arginine and is released into cytoplasm after proteolysis. As circulating SDMA is mostly eliminated by the kidneys, plasma concentrations are affected by changes in GFR. We have previously shown that increased serum SDMA concentrations were observed in cats and dogs with reduced renal function. We have also demonstrated that serum SDMA allows for earlier detection of chronic kidney disease (CKD) in cats and dogs when compared with serum creatinine. In general, serum SDMA and creatinine concentrations correlate with each other ($R^2 \sim 0.7$-0.9). However in some CKD patients, SDMA concentrations increased disproportionately compared with serum creatinine. This discordance is seen in both cats and dogs, but the prevalence is much higher in cats. This observed discordance in CKD patients can be quantified in terms of a SDMA to creatinine ratio; the higher the ratio the greater the discordance. Although the typical SDMA to creatinine ratio is < 10 in the majority of animals with CKD, we have observed that the larger the SDMA to creatinine ratio (>10) in CKD patients, the greater the chance of mortality. It should be noted that the ratio only applies to CKD animals; dogs and cats with creatinine and/or SDMA concentrations within normal limits can have ratios > 10. We have previously reported that the upper normal reference limit for SDMA in dogs and cats is < 14 µg/dL.

For example, in a retrospective study consisting of 21 CKD cats, 15/16 that had a SDMA to creatinine ratio of >10 were deceased within 1 year, whereas 5/5 CKD cats that had a normal ratio of SDMA to creatinine (<10) were still alive at that time. Similar trends were observed in another study with dogs. Kaplan-Meier survival curves were generated for both these cats and dogs based on SDMA concentrations greater than and less than 14 µg/dL. Cats with SDMA <14 µg/dL survived approximately 1.6 times longer than cats with SDMA >14 µg/dL. Dogs with SDMA <14 µg/dL survived approximately 2.6 times longer compared with dogs with SDMA >14 µg/dL. All cats and dogs were maintained with high quality care, including optimal nutrition and veterinary health care with opportunities for socialization and play time with caretakers and with daily opportunities to exercise, and play with toys.

Elevated SDMA concentrations in general, and SDMA to creatinine ratios in particular, for feline and canine CKD patients could potentially provide prognostic value. These results warrant further study.

THE CONVERGENCE OF EPIGENETICS AND CD8$^+$ T CELL DYSFUNCTION IN THE FELINE IMMUNODEFICIENCY VIRUS MODEL OF VIRAL PERSISTENCE

Jonathan Fogle, Yan Wang

North Carolina State University College of Veterinary Medicine, Raleigh, NC, USA

Generation of mature, fully-functional CD8$^+$ lymphocytes is dependent upon epigenetic changes such as DNA demethylation and histone acetylation at the promoter regions of cytokines essential for CD8$^+$ T cell function and differentiation. Conversely, dysfunctional CD8$^+$ T cells exhibit altered patterns of methylation and acetylation, and failure of complete differentiation, suggesting that there are early transcriptional events contributing to the induction of CD8$^+$ T cell dysfunction. For many persistent viral infections such as Herpes Simplex Virus (HSV), Hepatitis C Virus (HCV), Rotavirus (RV), Human Immunodeficiency Virus (HIV), and Feline Immunodeficiency Virus (FIV), virus specific CD8$^+$ T cell activation is followed by the induction of CD8$^+$ T cell dysfunction, which is characterized by lack of interleukin-2 (IL2) production and poor proliferation following T cell receptor (TCR) stimulation. Loss of IL2 production signals the onset of a progressive decline in cytokine production and antiviral function. Using the feline immunodeficiency model (FIV) for lentiviral persistence, we have clearly demonstrated that virus-activated CD4$^+$CD25$^+$ T regulatory (Treg) cells provide a strong inhibitory signal and induce cell cycle arrest and down-regulation of IL2 in *activated* CD8$^+$ lymphocyte targets early and progressively during the course of infection. Further, we have shown that Treg cells are activated by FIV infection and induce the expression of the repressive transcription factor Foxp3 in CD8$^+$ T cell targets following Treg cell / CD8$^+$ coculture. Finally, we have recently demonstrated that Foxp3 can bind the IL2 promoter in "bulk" (virus non-specific) CD8$^+$ T cells and that blocking epigenetic rearrangements prevents Foxp3 binding to the IL2 promoter in virus non-specific CD8$^+$ T cells.

Recent findings from our laboratory suggest that early epigenetic changes (histone acetylation and DNA demethylation), while essential for antiviral function, *render the CD8$^+$ T cell highly permissive to Treg-induced, Foxp3-mediated repression of IL2 transcription.* Our hypothesis is that activated Treg cells exploit virus-induced epigenetic changes in CD8$^+$ T cells to cause dysfunction. The purpose of these investigations was to demonstrate that Foxp3 binds the IL2 promoter in virus specific CD8$^+$ T cells. Using chromatin immunoprecipitation (ChIP) and bisulfite reduction of DNA, we have assessed histone acetylation and DNA demethylation at the IL2 promoter region in virus specific CD8$^+$ T cells from FIV$^+$ cats. Briefly, virus specific T cells were identified by ex-vivo proliferation in response to inactivated virus. These cells were then purified and co-cultured with autologous Treg cells. In concert with our previous work in virus non-specific CD8$^+$ T cells, our data suggests that following coculture with autologous Treg cells, virus specific CD8$^+$ T cells exhibit increased Foxp3 mRNA and decreased IL2 mRNA by RT-qPCR (n=2). More importantly, our most recent ChIP studies demonstrate that Foxp3 binds the IL2 promoter in virus specific CD8$^+$ T cells from FIV$^+$ cats following Treg cell coculture (n=2). Collectively, these results suggest that it will be possible to block epigenetic rearrangements in virus specific CD8$^+$ T cells thereby blocking Foxp3 binding to the IL2 promoter. These mechanistic studies will provide a new avenue of investigation for restoring antiviral CD8$^+$ T cell function during the course of persistent viral infection.

PHARMACOKINETICS OF VORICONAZOLE IN HEALTHY CATS

Polina Vishkautsan[1], George Thompson[2], Mark Papich[3], Jane Sykes[1]

[1]UC Davis VMTH, Davis CA, USA, [2]UC Davis School of Medicine, Davis CA, USA, [3]North Carolina Unversity, Raleigh, NC, USA

Voriconazole is a potent azole antifungal drug but is contraindicated in cats because of life-threatening adverse effects. We sought to determine the pharmacokinetics of voriconazole in healthy cats after oral and IV administration so that a safe dose might be established. Six cats were administered 1 mg/kg of voriconazole IV. Plasma voriconazole concentrations were measured at multiple time points for 24 hours after administration, using high performance liquid chromatography. Subsequently, voriconazole suspension was administered to 3 groups of 2 cats at 4, 5 and 6 mg/kg PO. Plasma concentrations were measured at multiple time points for 24 hours after administration. The pharmacokinetics of tablet and suspension preparations were also compared using a range of doses from 4.0 to 4.7 mg/kg. Plasma measurements were performed over a 72-hour period after administration.

Voriconazole half-life after IV administration was approximately 20 hours. After oral administration, maximal plasma concentrations were reached in 30-60 min. A dose of 4 mg/kg resulted in optimal plasma drug concentrations (1-4 mcg/mL). The predicted half-life after oral administration was 80-90 hours. Adverse effects included hypersalivation coincident with oral suspension administration (not tablets) and miosis. Miosis was associated with peak plasma voriconazole levels > 2.5 mcg/mL and persisted up to 48 hours.

Voriconazole has excellent oral bioavailability in cats. An oral dose of 4 mg/kg q48h may maintain adequate antifungal plasma concentrations, but multi-dose administration studies are required. Miosis was a significant adverse effect and occurred at plasma concentrations at the high end of the target range recommended for humans.

PAIN MANAGEMENT AND END OF LIFE CARE: RESULTS OF A NATIONAL, CROSS-SECTIONAL, SURVEY OF SMALL ANIMAL OWNERS

Roschelle Heuberger[1], Michael Petty[2], Page Burnia[1], Jill Prior[1]

[1]Central Michigan University, Mt. Pleasant, MI, USA, [2]Arbor Pointe Veterinary Hospital, Canton, MI, USA, [3]Central Michigan University, Mt. Pleasant, MI, USA, [4]Central Michigan University, Mt. Pleasant, MI, USA

The objective of this study was to describe knowledge, attitudes and beliefs (KAB) of a national sample of owners with regard to End of Life (EOL) care for their pets. The hypothesis included: paradigms for EOL decision-making are few and dependent on factors such as owner education. EOL decisions are influenced by many variables, requiring large samples (n=945) for data segmentation. Additional objectives included: describing EOL and Quality of Life (QOL) awareness, beliefs regarding Pain Management (PM), and recognition of terms such as "In home care" for pets by owners.

This study was "exempt category #2 anonymous survey" by Federal Statute on 2/12/14 by Institutional Review Board. Computerized links opened with informed consent. Instrumentation was developed through modifications of validated surveys of KAB in EOL. Veterinarians and specialty associations with known involvement in EOL and QOL were consulted, followed by owner focus groups and piloting. Designed in Survey Monkey™, the data were password protected. Recruitment through social media was performed (2014-2015) and data were processed in SPSS v.21.

Respondent age was μ=45.4±13 years, μ=2±1.2 pets/household, pet age μ=3±1.7 years and pet losses/lifetime μ=10±1.2. Only 4% were first time pet owners, and only 14% had pet health insurance. Respondents were primarily: Female (92%) Caucasian (94%) Married (55%) Religiously Unaffiliated (41%), Education >HS (75%), Non-Healthcare Occupation (75%). Familiarity with "hospice" or "in home EOL care for pets" was < 33%, QOL rating scales was 20%, but willingness to use QOL scales in the future was >70%. In terms of PM in EOL care, owners were likely or very likely to use oral medication in pill form (80%), liquid (82%), injection (70%), acupuncture (55%), acupressure/massage (64%), physical therapy (67%), hydrotherapy (61%) surgery (62%) nerve block (41%) corticosteroid injections (52%), opt for euthanasia (15%). Other (5%) included Chiropractic, transdermal patch, laser, herbals, naturopathy, aromatherapy, prayer etc. Thirty-five percent of owners reported fiscal concerns as influential in continued EOL care. If pain could not be controlled almost 80% of owners said they would euthanize.

Further data segmentation is warranted to discern differences among owners with regard to KAB in EOL, PM, QOL and euthanasia. These findings will provide a basis for educational outreach and protocol development for owners of ill or geriatric animals.

DISCOVERY AND CLINICAL EFFECTIVENESS OF A COMPOSITION THAT PROMOTES HAIR GROWTH (PATENT PENDING)

Gen-Ichiro Soma[1], Hiroshi Okawa[2]

[1]BioMedical Research Group Inc., Tokyo, Japan, [2]Scarecrow Inc., Tokyo, Japan

To overcome alopecia related to a variety of diseases in pets, we have identified materials that can promote hair growth, particularly in bald spots, leading to the development of products that can be helpful in the field of food and medicine.

We developed a composition that promotes hair growth in animals, which can be used in numerous formulations, including drugs, quasi-drugs, dressings, adhesive bandages, supplements, foods, snacks, water, hair care supplies, dental supplies, clothes, shoes, and socks. This formulation contains lipopolysaccharide (LPS) as an active ingredient. LPS tablets and pine bark polyphenol tablets were orally administered to dogs, cats, rabbits, and hamsters with alopecia in a study conducted in Japan, with the cooperation of practitioners. The study involved no restriction on food intake and medication, and prohibited the concomitant use of other supplements. The animals were divided into 2 groups, those that received LPS tablets alone and those that received LPS tablets and pine bark polyphenol tablets. The duration of the study was approximately 30 days. Hair growth was noted in 10 of 24 animals (41.6%) administered LPS tablets alone and all 13 animals (100%) administered LPS tablets and pine bark polyphenol tablets.

A CLINICAL AND NECROPSY EVALUATION OF ACUTE LUNG INJURY (nALI) AND ACUTE RESPIRATORY DISTRESS SYNDROME (nARDS) IN NEONATAL FOALS

Daniela Bedenice, Samuel Jennings, Mauricio Solano, Mary Rose Paradis

Cummings School of Veterinary Medicine at Tufts University, North Grafton, MA, USA

The study purpose was to evaluate the clinical and post-mortem characteristics of neonatal foals fulfilling the criteria of neonatal ALI (nALI) and ARDS (nARDS). A retrospective cohort analysis of 786 foals (1990-2014) admitted to a university hospital under 4-weeks old was performed. Diagnostic criteria for nALI and nARDS were applied to a preexistent database, including all patients with lateral thoracic radiographs and arterial blood gas results obtained on room air within 24-hours of each other. Neonates born following c-section, induction or with a diagnosis of prematurity were excluded. The diagnostic PaO_2/FiO_2 threshold was adjusted for age as previously described (Wilkins-2007). Thoracic radiographs were evaluated using a pre-existent scoring system. Histopathologic abnormalities were graded according to the distribution and frequency (0-3) of edema/congestion/hypercellularity, fibrin exudation, hyaline membranes, type-2 pneumocyte hyperplasia, and interstitial fibrosis to establish a diagnosis of diffuse alveolar damage (DAD). All data were presented descriptively.

Thirty-four of 786 (4.3%) admitted foals fulfilled the clinical and PaO_2/FiO_2 threshold criteria for nALI (24/786) and nARDS (10/786). Of these, 10/24 (42%) and 4/10 (40%) survived, respectively. Histopathologic review of lung tissue was available for 10 affected foals, and showed severe, diffuse edema or congestion in all patients. Hyaline membranes, and thus DAD, were observed in 7/10 (70%), being rare in 4/7 (57%). Interstitial fibrosis was not identified. Associated pulmonary diagnoses in foals with DAD included meconium aspiration (n=2), EHV-1 infection (n=2), and suppurative bronchopneumonia (n=1).

In conclusion, histopathology showed exudative rather than proliferative or fibrotic DAD in foals with nALI or nARDS.

METABOLIC RESPONSES TO A STANDARDIZED FIELD EXERCISE TEST IN ARABIAN ENDURANCE HORSES WITH A HISTORY OF EXERTIONAL RHABDOMYOLYSIS

Erica McKenzie[1], Lauren Eyrich[1], Mark Payton[2], Stephanie Valberg[3]

[1]Oregon State University, Corvallis, OR, USA, [2]Oklahoma State University, Stillwater, OK, USA, [3]University of Minnesota, Saint Paul, MN, USA

At least 4% of Arabian endurance horses suffer from exertional rhabdomyolysis (ER) of unknown etiology. This study compared muscle histopathology and metabolic responses to a field exercise test between endurance Arabians with and without previous ER (ER: n=10, age 15 ± 6 yrs; control: n=9, 13 ± 6 yrs). No horses possessed the *GYS1* mutation.

After 24-48 hours stall rest, paired ER and control horses, fitted with a telemetric ECG, performed 47 minutes of standardized intervals of walk and trot, and one interval of canter. Blood samples were obtained before, immediately after and 3 hours post-exercise, and plasma was frozen immediately in liquid nitrogen. Percutaneous gluteus medius muscle samples frozen in liquid nitrogen were obtained before and 3 hours post-exercise for glycogen analysis. Pre-exercise muscle samples were also fixed in formalin for histopathologic analysis. Fisher's exact test, t-test, and ANOVA were performed ($P<0.05$).

No horses displayed clinical signs of ER and log-transformed serum creatine kinase activity 3 hours post-exercise was not significantly different between ER and control Arabians. Muscle glycogen, heart rate, PCV, and plasma total protein, glucose, lactate and electrolyte concentrations did not differ between exercising ER and control horses. ER horses had more central nuclei in mature myofibers, and higher myopathic scores compared with control horses.

Arabian endurance horses with a history of ER have similar metabolic responses to a submaximal exercise test as healthy Arabians, in spite of histopathologic indicators of chronic ER. Thus, Arabian ER does not appear to be associated with a consistent metabolic myopathy.

LONGITUDINAL EFFECT OF A MULT-STRAIN PROBIOTIC ON THE MICROBIOTA OF NEONATAL FOALS

Angelika Schoster[1], Henry Staempfli[2], Miranda Abrahams[2], Mohammad Jalali[3], Luca Guardabassi[4], Scott Weese[3]

[1]University of Zurich, Vetsuisse faculty, Clinic for Equine Internal Medicine, Zurich, Switzerland, [2]University of Guelph, Ontario Veterinary College, Departement of Clinical Studies, Guelph, Ontario, Canada, [3]University of Guelph, Ontario Veterinary College, Departement of Pathobiology, Guelph, Ontario, Canada, [4]University of Copenhagen, Faculty of Health and Medical Sciences, Departement of veterinary Disease Biology, Frederiksberg, Copenhagen, Denmark

It is suggested that probiotics modify the intestinal microbiota but scientific proof is lacking. The objective of this study was to evaluate the effect of a probiotic containing strains selected for anti-clostridial activity, on the intestinal microbiota of neonatal foals.

14 healthy foals were randomly allocated to receive a probiotic or placebo formulation for 3 weeks. The probiotic formulation contained two *Lactobacillus rhamnosus* strains, two *L. plantarum* strains and *B. animalis lactis*. Fecal samples were collected at 2, 4 and 6 weeks of age. Mothur analysis of next generation sequencing data was used to compare data between treatment groups.

There was no effect of treatment group on alpha diversity indices, Chao richness, Simpson's diversity and Shannon evenness (all $p>0.37$). There were no significant differences in the relative abundances of any phyla, classes, or predominant genera, including *Bifidobacterium* (p= 0.37, 0.18 and 0.90 at 2, 4 and 6 wks of age) and *Lactobacillus* (p=0.61, 0.16 and 0.52 at 2, 4 and 6 wks of age). Several differences between treatemnt groups were noted using LEfSe. Most notably at week 6, an unclassified Lactobacillaceae (LDA score 2.1, *P*=0.016) was enriched in the probiotic group. There was no significant difference in beta diversity between groups, as assessed by unifrac applied to Jaccard and Yue&Clayton trees (all $p>0.30$).

Probiotic treatment had limited effects on the composition of the microbiome. Further studies including larger numbers of foals have to be performed to assess whether this difference is significant and can be used to therapeutically modify the microbiome of foals.

SKELETAL MUSCLE GENE EXPRESSION PROFILE IN TYPE 1 POLYSACCHARIDE STORAGE MYOPATHY

Raffaella Teixeira, James Mickelson, Stephanie Valberg, Molly McCue

University of Minnesota, St Paul, MN, USA

Type 1 Polysaccharide Storage Myopathy (PSSM1) is an inheritable neuromuscular disorder caused by a gain of function in skeletal muscle glycogen synthase that leads to excessive glycogen and abnormal polysaccharide accumulation in muscle fibers. PSSM1 horses utilize glycogen, but develop exercise intolerance and rhabdomyolysis during sub-maximal exercise. The link between excessive glycogen, abnormal polysaccharide and rhabdomyolysis is unknown. Controlled regular exercise and a low starch/high fat diet improve PSSM1clinical signs. *We hypothesized that 1) excess glycogen and/or improper regulation of glycogen synthase results in altered gene expression in skeletal muscle energy metabolism pathways leading to rhabdomyolysis in PSSM1; and 2) that clinical improvement with daily exercise is due to normalization of gene expression in PSSM1 horses.* Gluteal muscle biopsies were collected from PSSM1 cases and controls on a high starch diet prior to training (t1), before (t2) and immediately after exercise (t3) following a standardized 3-week training protocol. RNA was isolated and sequenced at 40-million reads/sample; reads were mapped to EquCab2 and *de novo* assembly was performed to identify novel transcripts. Differential gene expression was measured by Edge-R. 195, 294 and 739 genes were differentially expressed between cases and controls at t1, t2 and t3, respectively. The largest differences in expression were in cases between t1-t3 (4822 genes) and t2-t3 (2474 genes). Genes involved in glycogen metabolism, glycolysis and mitochondrial electron transport were differentially expressed between cases and controls. Understanding altered energy regulation in PSSM1 should lead to new therapies and new insights in metabolic myopathies and muscle glycogen regulation.

BLOOD AND CEREBROSPINAL FLUID ALPHA-TOCOPHEROL AND SELENIUM CONCENTRATIONS IN NEONATAL FOALS WITH NEUROAXONAL DYSTROPHY

Carrie Finno[1], Krista Estell[1], Laramie Winfield[1], Aaron Rendahl[2], Jamie Textor[3], Danika Bannasch[1], Birgit Puschner[1]

[1]University of California, Davis, Davis, CA, USA, [2]University of Minnesota, St. Paul, MN, USA, [3]Total Performance Equine, Martnez, CA, USA

Equine neuroaxonal dystrophy/equine degenerative myeloencephalopathy (NAD/EDM) is a neurodegenerative disorder affecting genetically predisposed foals maintained on an α-tocopherol (α-TP) deficient diet during the first year of life. We hypothesized that (1) the administration of E-Se® at 4-days of age would have no significant effect on serum or cerebrospinal fluid (CSF) [α-TP] in healthy foals and (2) serum and CSF [α-TP], but not [Se], would be significantly decreased in NAD/EDM-affected foals during the first year of life. Fourteen Quarter horse foals were included in the study; ten healthy foals supplemented with 0.02 ml/kg E-Se® (n=5) or saline (n=5) at 4d of age and four unsupplemented NAD/EDM-affected foals. Complete neurologic examinations, blood and CSF collections were performed before (4d of age) and after supplementation at 10, 30, 60, 120, 180, 240 and 360d of age. At 540d, NAD/EDM affected foals and one unsupplemented healthy foal were euthanized and full necropsies performed. A significant decrease in blood and CSF [α-TP] and [Se] was found during the first year of life in all foals, with the most significant changes in serum [α-TP] occurring from 4-150d. Dam [α-TP] and [Se] significantly impacted foal concentrations through 10d of age. An injection of E-Se® did not significantly increase CSF [Se], blood or CSF [α-TP] in healthy foals. NAD/EDM-affected foals had significantly lower CSF [α-TP]. Despite all 14 foals remaining deficient in α-TP, only the four genetically predisposed foals developed NAD/EDM.

RELATIONSHIP BETWEEN TRANSCRANIAL MAGENTIC MOTOR EVOKED POTENTIAL LATENCIES AND AXON LOSS IN EQUINE CERVICAL VERTEBRAL STENOTIC MYELOPATHY

Cody Alcott[2], Nicholas Jeffery[1], David Wong[1], Stephen Reed[1], Brett Sponseller[1], Hilary Hu[1], Kate Hepworth-Warren[1], Elizabeth Whitley[1]

[1]Iowa State University College of Veterinary Medicine, Ames, IA, USA, [2]Veterinary Specialty Center of Tucson, Tucson, AZ, USA

Current diagnostic strategies for equine cervical vertebral stenotic myelopathy (CVSM) are limited to identification of structural changes within the vertebral canal and inferring spinal cord dysfunction. Delayed latencies of transcranial magnetic motor evoked potentials (TMMEPs) have been associated with spinal cord dysfunction in horses, although histopathologic confirmation and correlation with axon loss is lacking. The purpose of this study was to investigate the relationship between TMMEP latency and axonal loss in horses diagnosed with CVSM. The hypothesis is that cervical spinal cord axonal loss will correlate with TMMEP latency.

TMMEP latencies from 20 horses were recorded bilaterally from the extensor carpi radialis and cranialis tibialis muscles. Ten horses had clinical signs consistent with CVSM (CVSM group) while the ten remaining horses presented for non-neurologic disorders (control group). All horses had lateral cervical radiographs performed prior to humane euthanasia. Cerebrospinal fluid from all horses tested negative for *Sarcocystis neurona* (SAG2,4/3 ELISA). Postmortem harvesting of cervical spinal cords in all horses was via dorsal laminectomy (C1-T2), in-situ formalin fixation, transverse sectioning (cranial, dorsal and caudal to each intervertebral disc space). Tissue blocks were stained with luxol fast blue and an H&E counterstain. Digital images of the spinal cord transverse sections (20X) were acquired and pixel density counts were determined (Adobe Photoshop©) to determine axon density at each of the following locations within each section; dorsal lateral funiculus (DLF), dorsal funiculus (DF), ventral funiculus (VF) and lateral funiculus (LF). TMMEP latencies of the extensor carpi radialis and cranialis tibialis were compared between control and affected horses (unpaired t test). The ratio of axon densities at each site between CVSM and control horses (termed the CCPR) was plotted against the TMMEP latency of the cranialis tibialis muscle and examined for correlation. Individual funiculus axon density from the most severely affected transverse sections were plotted against TMMEP latency of the cranialis tibilais muscles to investigate the effect of location within the spinal cord.

TMMEP latency showed a significant difference between groups for the cranialis tibialis muscle exclusively. Cranialis tibialis TMMEP latency linearly correlated with CCPR. Loss of axons was most severe in the ventromedial funiculi of CVSM horses.

These data confirm that horses affected by CVSM have prolonged TMMEP latencies and that TMMEP latency correlates with severity of axon loss. Both of these findings indicate that TMMEP might be useful to identify and quantify severity of spinal cord damage in clinical CVSM. The severity of ventral funicular changes is a unique finding and might suggest that this area is more vulnerable to injury in horses with CVSM than previously recognized.

NOVEL DETERMINANT CONFERS MACROLIDE, LINCOSAMIDES, AND STREPTOGRAMIN B RESISTANCE IN *RHODOCOCCUS EQUI*

Steeve Giguere[1], Elisa Anastasi[2], Londa Berghaus[1], Mary Hondalus[1], Jennifer Willingham[1], Sonsiray Alvarez[2], Noah Cohen[3], Marilyn Roberts[4], Jose Vazquez-Boland[2]

[1]University of Georgia, Athens, GA, USA, [2]University of Edinburgh, Edinburgh, UK, [3]Texas A&M University, College Station, TX, USA, [4]University of Washington, Seattle, WA, USA

The incidence of macrolide and rifampin resistance in *R. equi* isolated from foals has increased considerably in recent years. The objective of this study was to identify the molecular mechanism of emerging macrolide resistance in *R. equi* and to determine its transferability.

Macrolide-resistant (n=62) and macrolide susceptible (n=62) clinical isolates of *R. equi* from foals in the USA were studied. Whole genome sequencing of a sample of 18 macrolide-resistant and 6 macrolide-susceptible *R. equi* was performed. PCR was used to screen for the presence of the resistance determinant in the other isolates. Mating experiments were performed to document transfer of the determinant.

The genomes of resistant isolates were virtually identical whereas there was marked chromosomal variability among susceptible isolates, suggesting expansion of a resistant clone. A novel *erm* gene, *erm*(46) was identified only in resistant isolates. There was perfect association between macrolide resistance and presence of *erm*(46) as detected by PCR in 124 isolates of *R. equi*. Expression of *erm*(46) in a macrolide-susceptible strain of *R. equi* induced high level resistance to macrolides, lincosamides, and streptogramins B, but not to other classes of antimicrobial agents. Transfer of *erm*(46) from resistant to susceptible strains of *R. equi* was confirmed and occurred at a transfer frequency of up to 2×10^{-3}.

This is the first molecular characterization of macrolide, lincosamides and streptogramins B resistance in *R. equi*. Resistance is caused by a novel *erm* gene, *erm*(46), which is transferrable likely by conjugation.

DETERMINING OPTIMAL SAMPLING SITE FOR *STREPTOCOCCUS EQUI* SUBSP *EQUI* CARRIERS USING A LOOP-MEDIATED ISOTHERMAL PCR ASSAY

Ashley Boyle[1], Darko Stefanovski[1], Shelley Rankin[2]

[1]*University of Pennsylvania Clinical Studies New Bolton Center, Kennett Square, PA, USA,* [2]*University of Pennsylvania School of Veterinary Medicine Pathobiology, Philadelphia, PA, USA*

We hypothesized that samples obtained from the guttural pouch would be more sensitive than samples obtained from the nasopharynx to identify carriers of *Streptococcus equi* (*S. equi)* and that a Loop-mediated isothermal (LAMP) PCR assay that targeted the *eqbE* gene would be more sensitive than a realtime PCR assay that targeted the *seel* gene.

Three samples were collected from each horse: nasopharyngeal flocked nylon swab (NPFS), nasopharyngeal wash (NPW), and endoscopically-guided guttural pouch lavage (GPL). *eqbE* LAMP assay was performed on NPFS, NPW and GPL samples. The GPL sample was split into 3 aliquots; *S. equi* culture, *eqbE* LAMP and *seel* realtime PCR. Logistic regression and area under the receiver-operator curve (ROC) were performed using STATA 13. *P*-values ≤0.05 were considered significant.

A total of 123 samples were obtained from 40 horses (41 NPFS, 38 NPW, 44 GPL). 1/41 NPFS, 6/38 NPW and 24/44 GPL samples were positive by *eqbE* LAMP. 18/26 GPL samples were positive with *seel* PCR. *S. equi* was isolated from 4/44 GPL samples. GPL was the best sample to detect carriers compared to NPFS (OR 48, P<0.001) and NPW (OR 6.4, P=0.001). When *eqbE* LAMP results were compared to an endoscopically abnormal guttural pouch, sensitivity was 74.07% and specificity was 76.47%: LAMP ROC=0.75, *seel* PCR ROC=0.78. Sensitivity and specificity of *eqbE* LAMP was 83.3% and 65.4% compared to the *seel* PCR: ROC=0.70.

Our study demonstrates that GPL should be used to detect *S. equi* carriers and that *eqbE* LAMP assay was comparable to *seel* PCR.

PRO-RESOLUTION MECHANISMS OF INFLAMMATION IN EQUINE RECURRENT AIRWAY OBSTRUCTION: TAMOXIFEN AS A NEW THERAPEUTIC OPTION

Benjamin Uberti, Natalia Morales, Constanza Borlone, Claudio Henríquez, José Sarmiento, Gabriel Morán

Universidad Austral De Chile, Valdivia, Región De Los Ríos, Chile

Equine recurrent airway obstruction (RAO) can severely limit athletic function and quality of life of adult horses. Inhalation of a variety of environmental allergens induces chronic non-infectious respiratory inflammation, leading to hyper-responsiveness of the lower airways, increased production of mucus and airway remodeling. Swelling and narrowing of the airways leads to decreased lung function and limited gas exchange. There is evidence that many cells and structures of the lung play an important role in the pathophysiology of the disease: the contribution of bronchial epithelial cells, vascular endothelium, lymphoid and myeloid cells has been well documented. The interaction of an allergen with resident innate immune cells induces the arrival of more inflammatory cells, via cytokines and chemokines. Once the inflammatory reaction has fulfilled its task, pro-resolutive mechanisms take the stage: the shut-down of chemokine signaling leads to the abrogation of inflammatory cell influx (mainly neutrophils). Neutrophils initiate their apoptotic program and release molecules that attract macrophages, which in turn perform neutrophil clearance. The phagocytosis of apoptotic neutrophils eventually reprograms macrophages towards an anti-inflammatory phenotype. The successful completion of the pro-resolutive mechanism leads to the restoration of tissue homeostasis. Our preliminary results demonstrate that tamoxifen, a selective estrogen receptor modulator used as treatment in all stages of estrogen-positive human breast cancer, has a marked effect on equine neutrophil function and apoptosis.

In vitro results showed that tamoxifen significantly inhibited neutrophil chemotaxis in horses and humans in response to IL-8, from 1 µM onwards ($p<0.05$, one-way ANOVA). Additionally, tamoxifen significantly inhibited production of reactive oxygen species from 2.5 µM onwards. Early neutrophil apoptosis was induced in approximately 90% of cells after 30 minutes of incubation with tamoxifen at 5 µM (Sarmiento, Perez, Morales *et al*, 2013). This induction of apoptosis coincided with *in vivo* results, in which neutrophils obtained from tamoxifen-treated animals showed a marked increase of early apoptosis.

IV and PO pharmacokinetic analysis of tamoxifen in horses is currently under way. A proof-of-concept trial was performed recently, in which 4 stalled horses with acute exacerbation of RAO were treated with tamoxifen (100 mg PO q 24 hours for 7 days). This dose was extrapolated from human and murine pharmacokinetic data. Horses received no other treatment and underwent no management changes. There was a marked decrease in bronchoalveolar lavage neutrophil count after treatment (day 0=28%±6; day 8=4%±7). Clinical parameters (spirometry, respiratory rate and pattern) improved from 3 days of treatment onwards. Tracheal mucus content did not change significantly over the course of treatment. These results suggest that tamoxifen has an effect on neutrophil survival in the pulmonary environment, and warrant further investigation. Recruitment of additional subjects with acute exacerbation of RAO is ongoing.

THE IMPACT OF PROLONGED HYPERINSULINAEMIA ON THE DOWNSTREAM INSULIN SIGNALING PATHWAY IN EQUINE STRIATED MUSCLE AND DIGITAL LAMELLAE

Allison Campolo[1], Lauren Keith[1], Melody de Laat[2], Christopher Pollitt[2], Veronique Lacombe[1]

[1]*Oklahoma State University, Stillwater, USA,* [2]*University of Queensland, Brisbane, Australia*

Equine metabolic syndrome, an increasingly recognized disease, is characterized by obesity and insulin resistance. Although hyperinsulinemia predisposes horses to developing laminitis, the exact relationship between these conditions remains undetermined. We hypothesized that gene expression of proteins involved in the downstream insulin signaling and glucose transport pathways will be altered during prolonged hyperinsulinemia-induced laminitis.

Standardbred horses were treated with a prolonged euglycaemic, hyperinsulinaemic clamp (p-EHC) for 48 hours or a balanced electrolyte solution. Gene expression of key proteins involved in the insulin signaling pathway was evaluated in archived striated muscle and lamellar samples using real-time reverse transcription PCR with primers selected for insulin receptor substrate-1 (IRS-1), AKT-2, and glycogen synthase kinase beta (GSK-3β). Gene expression of the basal glucose transporter 1 (GLUT-1) and the insulin-sensitive GLUT-4 were evaluated using RT-PCR.

The p-EHC group became markedly hyperinsulinaemic while maintaining normoglycemia and developed clinical laminitis. There was no significant difference in gene expression of IRS-1, AKT-2, GSK-3β, GLUT-1 and GLUT-4 in skeletal muscle and lamellar tissue between groups. In contrast, there was a significant upregulation of AKT-2, GSK-3β, GLUT-1 and GLUT-4 expression in cardiac tissue of the p-EHC group compared to controls ($P < 0.05$).

These data suggested that the prolonged, euglycaemia hyperinsulinaemic clamp induced an increase in insulin sensitivity in the heart, as well as a transcriptional activation of glucose transport. In addition, the lack of downregulation of the insulin signaling pathway in both the skeletal muscle and digital lamellae supports that insulin resistance is not required for the onset of laminitis.

TISSUE MICRODIALYSIS STUDIES OF EQUINE LAMELLAR ENERGY METABOLISM AND MICROVASCULAR BLOOD FLOW

Carlos E Medina-Torres[1], Christopher C Pollitt[1], Claire Underwood[1], E Matias Castro-Olivera[1], Mark P Hodson[2], Dean W Richardson[3], Andrew W van Eps[1]

[1]Australian Equine Laminitis Research Unit, School of Veterinary Science, The University of Queensland, Gatton, QLD, Australia, [2]Metabolomics Australia, Queensland Node, Australian Institute for Bioengineering and Nanotechnology, The University of Queensland, Brisbane, QLD, Australia, [3]New Bolton Center, Department of Clinical Studies, School of Veterinary Medicine, University of Pennsylvania, Kennett Square, PA, USA

Disruption of cellular energy pathways may play a role in the different forms of laminitis. The aim of this work was to characterize the normal lamellar bioenergetic profile and establish if changes in metabolite patterns and microvascular perfusion occur as a result of modifications in limb load cycling activity and during the development of experimentally induced sepsis-associated laminitis.

First, a microdialysis technique for serial measurement of lamellar tissue energy metabolites (glucose, lactate, pyruvate, urea and glycerol) over a 24 h period was developed, and the lamellar energy profile of 14 clinically normal horses characterized. Subsequently, 9 horses instrumented with lamellar and skin microdialysis probes were subjected to sequential interventions designed to modify tissue perfusion (4 interventions) and limb load cycling activity and weight bearing (4 interventions). Urea (20 mmol/L) was added to the perfusion fluid to investigate if urea clearance (UC; a microdialysis-based method used to assess local blood flow) could identify lamellar perfusion changes. Energy metabolite concentrations, UC, and the standard indices of energy metabolism lactate to glucose (L:G) and lactate to pyruvate (L:P) ratios were determined during each intervention. Then, changes in lamellar bioenergetic composition and perfusion during the development of experimentally induced, sepsis-associated laminitis were investigated in 12 horses. Treatment (n=6) and control (n=6) horses received oligofructose (OF) and water via nasogastric tube. Lamellar metabolite composition, L:G, L:P and UC were determined bihourly and compared. Lastly, using lamellar and skin dialysate and tissue sections from healthy horses (n=7), and lamellar dialysate from OFT (n=4) and CON (n=4) horses, targeted metabolomic analysis (i.e. 44 central carbon metabolites: CCM) was performed using liquid chromatography coupled with tandem mass spectrometry (LC-MS/MS). The differences between lamellae and skin in healthy horses and the differences between OFT and CON horses were investigated.

In the intervention studies, UC alone and in combination with fluctuations in energy metabolite patterns detected profound and mild changes in lamellar perfusion, respectively. Significant changes in lamellar energy metabolite composition and UC were observed when limb load cycling activity was modified: increases in glucose and UC, compatible with increased lamellar perfusion, were observed when limb activity increased. In the OF study, glucose decreased significantly in OFT lamellar dialysate (OFT-L), and was consistently lower in OFT than in CON horses. Lactate was higher in OFT skin (OFT-S) compared with OFT-L. Pyruvate decreased significantly in OFT-L and was significantly less than CON-L. L:G and L:P increased significantly in OFT-L and OFT-S, but not CON-L. Urea decreased significantly (increased UC) in OFT-L and remained stable in CON-L. Hyperaemia rather than ischaemia together with some bioenergetic disturbances occurred during the developmental phase of OF-induced laminitis. Though perfusion increased, lamellar glucose was markedly reduced, which could have compromised lamellar basal epithelial cell metabolism despite the lack of definitive evidence of lamellar bioenergetic failure. Metabolomic analysis of lamellar and skin dialysate showed no difference in metabolite composition in these two tissues in healthy horses, and the metabolome of dialysate and tissue samples was also found to be similar. A difference in CCM between OFT and CON horses was observed. Metabolomic analysis of lamellar CCM was capable of differentiating OFT from CON horses. Lamellar malate, pyruvate, aconitate and glycolate, were identified as the source of differentiation between OFT and CON groups.

These studies demonstrated that lamellar microdialysis and UC can be used to assess lamellar bioenergetic and perfusion changes. Microdialysis UC suggests an increase in lamellar perfusion during the development of sepsis-associated laminitis. The findings also suggest that lamellar perfusion and energy balance may be related to limb load cycling and are particularly affected by ambulation, supporting the hypothesis that reduced limb load cycling, but not increased weight bearing, may be associated with reduced lamellar perfusion.

INCRETINS POTENTIATE THE INSULIN RESPONSE TO ORAL GLUCOSE IN INSULIN-RESISTANT PONIES

Melody de Laat, Jessica van Haeften, Martin Sillence

Queensland University of Technology, Brisbane, Queensland, Australia

Equine insulin secretion varies between individuals and is affected by diet and metabolic status, with dysregulation resulting in hyperinsulinaemia, a significant laminitis risk. Gastrointestinal hormones may exacerbate insulin responses to feeding. This study investigated whether the incretin hormones, glucagon-like peptide-1 (GLP-1) and glucose-dependent insulinotropic peptide (GIP), augment insulin secretion following oral glucose, compared with intravenous glucose infusion.

Ponies (n=9) received D-glucose (0.75mg/kg) in a low glycaemic ration (0.8%BW). Assays were validated for insulin, glucose and incretin analysis of blood samples collected before and every 30 minutes for 6 hours after feeding. Seven days later the experiment was replicated however the D-glucose was administered intravenously at a variable rate matched to individual meal consumption (<75min).

Ponies were stratified for insulin sensitivity status using basal values/proxies. Although intravenous glucose elicited a larger area under the curve for insulin ($AUC_{insulin}$), the insulin response to oral glucose was larger ($p<0.05$) and bimodal in insulin-resistant (compared to insulin-sensitive) ponies and correlated well with $AUC_{glucose}$. Importantly, insulin resistance did not affect insulin responses to intravenous glucose, whereas with oral glucose the $AUC_{insulin}$:$AUC_{glucose}$ was higher ($p<0.05$) in insulin-resistant ponies. Secretion of GLP-1_{active} and GIP, but not GLP-1_{total}, was higher ($p<0.05$) following oral glucose (compared to intravenous) and GLP-1_{active} was positively correlated ($p<0.05$) with insulin. Further, oral glucose stimulated a larger ($p<0.05$) $AUC_{GLP\text{-}1active}$ in insulin-resistant ponies.

Thus, the larger insulin response to oral glucose in insulin-resistant ponies is likely due in part to incretin action. In particular, GLP-1_{active} may be a key factor in equine insulin dysregulation.

USE OF DAILY DICLAZURIL PELLETED TOP DRESS FOR THE PREVENTION OF *SARCOCYSTIS NEURONA* INFECTION IN FOALS

Nicola Pusterla, Andrea Packham, Sarah Mackie, Philip Kass, Laszlo Hunyadi, Patricia Conrad

SVM, UC Davis, Davis, CA, USA

Therapeutic treatment strategies for the prevention of *Sarcocystis neurona* infection in horses have been empirical. A pelleted top dress 1.56% diclazuril anti-protozoal drug recently introduced to the equine market and labelled for the treatment of equine protozoal myeloencephalitis (EPM) has the potential to be used for the prevention of EPM due to its convenient formulation. A low-dose of diclazuril given at 0.5 mg/kg body weight has shown to reach plasma and cerebrospinal fluid concentrations at steady-state in excess of the minimal concentration known to be inhibitory to *S. neurona* merozoite production in cell culture. The purpose of this study was to evaluate the temporal serological response against *S. neurona* in foals receiving daily diclazuril and in untreated herdmates.

Thirty-three foals from a farm with a high exposure rate to *S. neurona* were randomly assigned to either an untreated control group or a diclazuril treated group. Treatment consisted of the administration of 0.5 mg/kg body weight of diclazuril pelleted top dress starting at 4 weeks of age until the foals were 12 months of age. Whole blood was collected from every dam and foal 24 hours post-foaling and monthly thereafter from every study foal for the duration of the study. The blood was tested for IgG against *S. neurona* using the indirect fluorescent antibody test.

Following ingestion of colostral antibodies to *S. neurona*, there was a steady and continuous decline in seroprevalence and antibody titers to *S. neurona* until the foals from both groups reached weaning age. Thereafter, untreated foals showed an increase in monthly seroprevalence ranging from 53 to 88%. Diclazuril treated foals showed significant lower monthly seroprevalences ranging from 6 to 29%. At the end of the study, 88% and 6% of untreated and treated foals, respectively, tested seropositive to *S. neurona*.

In conclusion, daily supplementation of diclazuril pelleted top dress at 0.5 mg/kg body weight demonstrated a significant reduction in seroconversion to *S. neurona* between diclazuril treated and control foals up to 12 months of age. To the author's knowledge this is the first report determining dose and duration of treatment for the prevention of seroconversion in foals originating from a farm with high infection rate to *S. neurona*. From an economic standpoint, supplementation of diclazuril at a low-dose during early exposure and high-risk periods may benefit the horse industry by reducing costs associated with EPM.

MEDICAL ALTERNATIVES TO CONVENTIONAL CYCLOOXYGENASE INHIBITORS FOR TREATMENT OF ACUTE FOOT PAIN IN A REVERSIBLE LAMENESS MODEL IN HORSES

Jonathan Foreman, Catherine Foreman, Benjamin Bergstom

University of Illinois, Urbana, IL, USA

In the treatment of acute equine foot pain such as laminitis, conventional treatments include the pan-cyclooxygenase (COX) inhibitors phenylbutazone (PBZ) and flunixin meglumine (FM). These pan-COX inhibitors (conventional NSAIDs) are known to cause side effects such as gastrointestinal ulceration and renal disease. Veterinarians treating laminitis often search for efficacious alternatives to conventional NSAIDs. The hypothesis in this series of 4 experiments was that non-pan-COX-inhibitor therapies would be shown to be similar to conventional NSAIDs for the treatment of acute equine foot pain in a reversible model of lameness in horses.

8 horses were shod with an adjustable heart bar shoe on one foot. Horses were monitored by a blinded investigator for 12- or 24-hrs post-medication. Variables included HR and Lameness Score with results compared by RM ANOVA and Tukey's test with significance set at $P<0.05$.

Firocoxib (0.09 mg/kg IV) was no different than saline and both were less effective than PBZ (4.4 mg/kg IV) ($P<0.05$). A firocoxib loading dose (0.27 mg/kg IV) was similar in efficacy to PBZ (4.4 mg/kg IV). Acetaminophen (20 mg/kg PO) and oral FM (1.1 mg/kg PO) administered separately were comparable to one another and both were better than saline ($P<0.05$). Acetaminophen (20 mg/kg PO BID) combined with firocoxib (0.27 mg/kg SID) was similar to PBZ (4.4 mg/kg IV BID) and both regimes were better than saline ($P<0.05$) in a 24-hr study.

Intravenous firocoxib (loading dose 0.27 mg/kg IV SID) and oral acetaminophen showed promise in alleviating acute equine foot pain.

INITIAL CHARACTERIZATION OF THE TRACHEAL MICOBIOMES IN HEALTHY HORSES AND HORSES WITH HEAVES

Julia Montgomery, Katharina Lohmann, Bonnie Chaban, Lisa Johnson, Scott Dos Santos, Jordan Steedman, Janet E. Hill

Western College of Veterinary Medicine, University of Saskatchewan, Saskatoon, SK, Canada

Microbial communities colonizing the airways may play a role in the development of chronic inflammatory airway diseases. To investigate whether this may be true for equine heaves, we characterized the tracheal microbiomes of three healthy horses and five horses with historical, clinical and/or bronchoalveolar lavage cytological findings consistent with heaves. On two occasions, approximately two weeks apart, tracheal aspirate (TA) samples were collected transendoscopically using a two-stage sampling catheter. Total genomic DNA was extracted from one aliquot and DNA integrity verified by PCR amplification of the horse mitochondrial *cox*1 gene. Total bacterial load was estimated by quantitative PCR targeting the 16S rRNA gene, and PCR amplicon libraries based on the universal 60 kDa chaperonin (cpn60) gene were prepared using established protocols and sequenced using the 454 GS Junior pyrosequencing platform. Sequences were assembled into operational taxonomic units (OTU) and compared to the cpnDB reference database (www.cpndb.ca) to identify the closest reference sequence (nearest neighbor) for each OTU. Phylum assignment of OTU sequences was based on the taxonomic lineage of the nearest cpnDB reference sequence. Alpha and beta diversity were calculated using QIIME, and hierarchical clustering of microbiome profiles was performed in R. A second aliquot of each sample underwent routine microbiologic analysis using aerobic culture on blood and McConkey agar plates, followed by biochemical testing and identification. Up to 12 colonies per microbiologic culture plate were amplified using cpn60 universal primer PCR, sequenced and compared to the microbiome library from the corresponding horse.

A total of 15 TA samples were available for analysis. Of the five horses with heaves, two had active disease at the time of sampling and three were considered to be in remission. Total bacterial load (number of 16S rRNA copies per ml of TA) appeared higher in horses with heaves although considerable variation existed within groups and between samplings. A total of 94,800 sequence reads were analyzed (median 1,488 per sample; range 137-35,926). Consistent with the 16S rRNA copy number, microbiomes of horses with heaves appeared to have higher species richness than those of healthy horses. Microbiome profiles of horses in both clinical categories were dominated by Firmicutes, Actinobacteria and Proteobacteria. OTU sequences corresponded to 329 nearest neighbor reference sequences. OTU with similarity to *Pseudomonas fluorescens*, *Rothia* spp., *Pantoea agglomerans*, and *Streptococcus* spp. were the most prevalent among all samples and were frequently the most abundant in individual microbiomes. Both phylum level profiles and hierarchical clustering of OTU level profiles based on Bray-Curtis dissimilarity showed that microbiome profiles of individual horses were largely similar between the two samplings, and that microbiomes did not cluster by clinical category. Sequence data were obtained from 51 bacterial colonies, which represented 35 unique sequences. Only 8/35 of these sequences were >99% identical to OTU in the relevant microbiome libraries. Species identified by both culture and microbiome sequencing included only Firmicutes and represented only 6.7% of the sequence data generated from all horses, consistent with the more comprehensive view of the microbiome provided by the culture-independent approach.

In conclusion, this pilot study suggests that horses with heaves may have a higher bacterial load with greater species richness, but their microbiome profiles are not distinguishable from those of healthy horses at the phylum or OTU level. For most of the horses, tracheal microbiomes appeared stable during the time frame of the study. Microbiome analysis at the genetic level enhanced the assessment of microbial communities compared to conventional culture methods. Work is ongoing to confirm these initial findings in a larger group of horses.

FOOD ANIMAL RESEARCH REPORTS

CHARACTERIZATION OF THE PATHOGENICITY AND SHEDDING OF BOVINE PARAINFLUENZA 3 GENOTYPES

Benjamin Newcomer[1], John Neill[2], Binu Velayudhan[3], Paul Walz[1]

[1]Auburn University, Auburn, AL, USA, [2]National Animal Disease Center, Ames, IA, USA, [3]Texas A&M Veterinary Medical Diagnostic Laboratory, Amarillo, TX, USA

Bovine parainfluenza-3 virus (PI3) is a widespread respiratory pathogen of cattle capable of causing disease as a unique pathogen but more commonly predisposing cattle to secondary bacterial infection and development of pneumonia. Non-A genotypes in the United States (US) have only recently been reported but little is known of their pathogenicity or shedding compared to PI3-A. Therefore, the purpose of this pilot study was to compare the pathogenicity and shedding of clinical US isolates of all three genotypes (A, B, C). Sixteen weaned Jersey bull calves seronegative to PI3 were randomly assigned to one of four groups; calves were challenged intranasally with approximately 10^7 $CCID_{50}$ of one of the PI3 genotypes or served as negative controls. The calves were maintained in isolation rooms and daily nasal swab collection and observation for signs of disease by a blinded investigator were performed throughout the 14-day study. Mild clinical signs were observed similarly in all challenged groups and included serous nasal discharge and coughing. Onset of nasal shedding differed between groups with all PI3-A challenged calves shedding beginning on Day 4, 3 of 4 PI3-B challenged calves on Day 5, and PI3-C challenged calves on Day 6. Shedding of PI3-C lasted 5 days; shedding of the other viruses extended 7-8 days. This study demonstrates a difference in shedding dynamics between PI3 genotypes.

PERCUTANEOUS TUBE CYSTOSTOMY USING RUTNER SUPRAPUBIC CATHETER FOR TREATMENT OF OBSTRUCTIVE UROLITHIASIS IN FIVE GOATS

Katharine Simpson[1], Robert Streeter[2], Sarah Depenbrock[1], Caitlin Jablonski[1], Joseph Lozier[1]

[1]The Ohio State University, Columbus, OH, USA, [2]Oklahoma State University, Stillwater, OK, USA

Obstructive urolithiasis in male small ruminants is a common metabolic disease that often requires surgical intervention. Temporary or permanent urinary diversion is indicated if urethral process amputation and/or retrograde urethral hydropulsion are unsuccessful in restoring urethral patency. Surgical cystostomy tube placement is often considered the treatment of choice to allow for urinary diversion and ultimately the reestablishment of urethral patency. However, the costs associated with general anesthesia, surgery and post-operative care can be substantial. Advantages of percutaneous tube cystostomy rather than surgical tube cystostomy placement include shortened procedure time, cost-effectiveness, and provision of rapid urinary diversion in systemically compromised patients. Complications of general anesthesia are avoided as the procedure can be performed utilizing local anesthesia with or without sedation. We describe use of a commercially available suprapubic catheter[1] for percutaneous tube cystostomy in conjunction with urinary acidification as a primary treatment modality in cases of obstructive urolithiasis in goats with suspected or confirmed struvite urolithiasis.

Five goats (four Pygmy, one French-Alpine) were diagnosed with obstructive urolithiasis. Retrograde urethral catheterization was unsuccessful in relieving the obstructions in four of the goats. The other goat was considered to be emergent and a poor anesthetic candidate. A Rutner suprapubic catheter[1] was percutaneously inserted into the urinary bladder utilizing ultrasound guidance in all five goats to provide urinary diversion. Patency of the urethra was reestablished in 4 of the 5 goats. Mean and median cost associated with percutaneous cystostomy was $677.86 and $701.54 respectively, compared to $1406.05 (mean) and $1393.14 (median) associated with surgical tube cystostomy during the same year. Complications noted post-cystostomy included obstruction of the catheter (n=1), transient uroabdomen (1), and urethral rupture (1).

Based on these outcomes, we suggest that percutaneous tube cystostomy using a commercially available Rutner suprapubic catheter[1] (16 Fr, 4 mL balloon) in conjunction with urinary acidification may be a useful non-invasive and cost-effective alternative to surgical interventions for treatment of obstructive urolithiasis; particularly in cases with uroliths amenable to dissolution. Additional research is indicated to fully elucidate post-procedural complications and long-term prognosis.

[1]Rutner Suprapubic Balloon Catheter Set, Cook® Medical, 16 French, 4 mL balloon

PHARMACOKINETICS OF AN EXTENDED RELEASE FORMULATION OF EPRINOMECTIN IN HEALTHY ADULT ALPACAS AND ITS USE IN ALPACAS CONFIRMED WITH MANGE

John Pollock[1], Daniela Bedenice[1], Mark G. Papich[2]

[1]*Cummings School of Veterinary Medicine at Tufts University, Grafton, MA, USA,* [2]*2North Carolina State University, Raleigh, NC, USA*

The study purpose was to determine the pharmacokinetics (PK) and clinical effects of extended release 5% eprinomectin (Longrange®, Merial) following subcutaneous (SC) injection in healthy (n=6) and mange-infected (n=4) adult alpacas. High-performance liquid chromatography was performed on plasma samples obtained at regular intervals for 161-days following a single 5mg/kg injection SC in healthy alpacas, and for 5-days following each dose (3 treatments, 2 months apart) in mange affected animals. Clinical monitoring included biweekly hematology and physical examination; monthly weight and fecal egg counts (FEC-double centrifugation). Skin scrapings and biopsies were performed pre- and post-treatment at two comparable sites in alpacas with mange. Four alpacas served as healthy controls. Pharmacokinetic analysis utilized a non-compartmental model (WinNonlin-6.0). Results were compared between time-points using repeated-measures ANOVA and paired samples t-test.

Eprinomectin plasma concentrations showed a biphasic peak (Cmax-1 and 2) in all animals:

	Cmax-1 (ng/mL)	Tmax-1 (days)	Cmax-2	Tmax-2	AUC infinity (day*ng/mL)	T1/2 terminal (days)
Mean (SD)	5.72 (3.25)	3.88 (5.2)	6.06 (2.47)	77 (12.52)	456.6 (154.5)	15.53 (3.16)

Eprinomectin plasma concentrations remained above 1.27±0.96 ng/mL for up to 120 days. Hematocrit (35.8 vs. 31.3%, $P<0.003$) and albumin (3.5 vs. 2.8 g/dL, $P<0.006$) reduced significantly over 6 months in multi-dose animals, while FEC did not differ between groups. Self-limiting injection-site reactions occurred in 9/10 animals. Pre- and post-treatment skin biopsies showed reduced hyperkeratosis, but increased fibrosis, with 1/4 alpacas remaining positive on skin scraping for mange.

In conclusion, alpacas require a higher eprinomectin dose (5mg/kg SC) than cattle, to reach comparable plasma concentrations.

DEVELOPMENT OF A ZINC IMPLANT-BASED MODEL FOR UROLITHIASIS IN GOATS

Meredyth Jones, Allen Roussel, W. Shawn Ramsey

Texas A&M University, College Station, Texas, USA

The study of urolithiasis in small ruminants is hampered by the lack of a convenient and reliable model that would allow the quantitative assessment of the calculogenic potential of rations. The objective of this study was to develop a nonsurgical experimental model of urolithiasis in goats.

Ten, 1-year old Boer-cross doelings were fed a calculogenic diet designed to promote struvite urolithiasis. Zinc washers, 6-13 mm in diameter, were used with some washers tied or wrapped with chromic catgut or polyester fiber suture. Goats were anesthetized and the implants were placed into the lumen of the urinary bladder retrograde through the urethra. Different combinations of washer size and suture material and pattern were used in 4 separate experiments. At the end of each experiment, 28-118 days after placement, the implants were retrieved via the urethra. Two implants required removal via cystotomy due to the large amount of crystalline precipitation.

Across experiments, 25-67% of implants were spontaneously passed. All implants that remained in the bladders until the end of the experiment had grossly visible calculus accumulation. Discs with suture material accumulated 1.5-46 mg/day compared to 0.03-1.9 mg/day for those without suture material. Urolith analysis revealed that the predominant material accumulated was either struvite or amorphous magnesium calcium phosphate.

This study demonstrates that uroliths can be created consistently in a nonsurgical model in goats. This model should be useful in the study of rations and to evaluate interventions that may reduce or prevent the formation of urinary calculi in small ruminants.

Efficacy of different multivalent MLV vaccines administered to calves possessing maternally-derived immunity and subsequently challenged with BVDV or BVDV and BoHV-1

Manuel F. Chamorro[1], Paul H. Walz[1], Thomas Passler[1], Kay Ridell[1], Benjamin Newcomer[1], Roberto Palomares[2]

[1]Auburn University, Auburn, Alabama, USA, [2]University of Georgia, Athens, Georgia, USA

The objective of these studies was to evaluate efficacy of different commercially-available multivalent MLV vaccines to prevent clinical disease, viremia, and virus shedding in early weaned beef calves challenged with BVDV or BVDV and BoHV-1. In the first study, 48-male beef calves were early weaned at a median age of 72.2 days and assigned to 1 of 4 treatment groups: control (A), vaccine B, Vaccine C, and Vaccine D. Forty-five days after vaccination calves were challenged intranasally with virulent BVDV 2. In the second study, 54-male beef calves were early weaned (median age 93.5 days) and assigned to 1 of 5 treatment groups: control (A), vaccine B, vaccine C, vaccine D, and vaccine E. Forty-five days after vaccination calves were simultaneously exposed to 6 cattle PI with BVDV and 8 calves acutely infected with BoHV-1. Samples were collected in all calves for virus isolation and virus neutralization analysis before vaccination and before and after virus challenge. Calves vaccinated with vaccines B and C in the first study and calves vaccinated with vaccines B, C, and E in the second study had greater BVDV-antibody responses and a lower proportion of viremia and BVDV shedding compared with controls. Vaccination of early weaned beef calves resulted in increased BVDV antibody responses and reduced viremia and BVDV shedding. Differences in vaccine efficacy to prevent BVDV viremia and shedding were observed.

COMPARISON OF JOHNE'S DISEASE PREVALENCE ON ORGANIC AND CONVENTIONAL DAIRY FARMS IN PENNSYLVANIA

Marie-Eve Fecteau, Helen Aceto, Terry Fyock, Raymond Sweeney

University of Pennsylvania, School of Veterinary Medicine, Kennett Square, PA, USA

Johne's disease (JD) affects approximately 70% of all US dairies, and can be a cause of great economic loss to dairy producers. To qualify for the label "organic", certain restrictions on management practices may predispose for the transmission of JD on the farm. The objectives were: 1- to compare JD prevalence between Pennsylvania organic and conventional dairy farms, and 2- to identify risk factors associated with differences in JD prevalence. A JD milk ELISA was performed on individual milk samples from each lactating cows in the study herds. Information regarding management practices was collected during a farm visit. Overall herd prevalence, and within-herd prevalence were compared between groups. Logistic regression was used to identify risk factors associated with differences in JD prevalence between groups. A total of 2,739 cows from 50 herds (25 organic and 25 conventional) were included in the study. Median herd size was 58 (range 20-114 cows) for conventional farms, and 39 (range 20-211 cows) for organic farms. Of all the farms included in the study, 27/50 (54%) were positive, with 15/25 (60%) positive conventional farms and 12/25 (48%) positive organic farms. From the conventional farms, 25/1,506 (2%) cows were positive, compared with 28/1,233 (2%) cows from organic herds. After adjusting for herd size, there was no significant difference in herd prevalence (P=0.56), or within-herd prevalence (P=0.41) between conventional and organic farms. Risk factors are currently being analyzed. Preliminary results indicate that there is no difference in JD prevalence between Pennsylvania organic and conventional dairy farms.

A RANDOMIZED CLINICAL TRIAL EVALUATING METABOLISM OF COLOSTRAL AND PLASMA DERIVED IMMUNOGLOBULIN G IN JERSEY BULL CALVES

Kelly Pipkin, Jill Hagey, Maire Rayburn, Munashe Chigerwe

UC Davis, CA, USA

The objective of this study was to determine the rate of catabolism of colostral derived IgG administered by oroesophageal tubing compared to IV administered plasma IgG.

A randomized clinical trial was performed. Thirty newborn Jersey calves were enrolled. Fifteen were fed colostrum (CL group) and 15 were given bovine plasma IV (PL group). Calves in the CL group were fed 3 L of colostrum once, by oroesophageal tubing. Calves in the PL group were given plasma IV at a dosage of 34 ml/kg. Serum and fecal samples were collected at 0 h, 6 h, 12 h, 48 h, 5 d and 7 d. Serum and fecal IgG concentrations were determined by radial immunodiffusion.

Calves in the CL group maintained serum IgG concentrations consistent with adequate transfer of immunity (≥ 1000 mg/dL) throughout the study period. Calves in the PL group achieved median IgG concentrations of ≥ 1000 mg/dL at 6 h but the concentrations were < 1000 mg/dL by 12 h. Calves in the PL group were 5 times more likely to experience mortality compared to the CL group (hazard ratio = 5.01). Fecal IgG concentrations were not different between the 2 groups during the first 48 h ($P > 0.05$). Catabolism of plasma derived IgG occurs rapidly during the first 12 h after transfusion. Fecal excretion did not explain the fate of the plasma derived IgG.

PLASMA LIPIDOMIC PROFILE IN COWS DURING THE TRANSITION PERIOD AND IN COWS WITH FATTY LIVER.

Christian Gerspach[1], Hanspeter Naegeli[1], Maude Gubler[1], Claudine Bieli[1], Endre Laczko[2], Maja Rütten[1], Sandro Imhasly[1]

[1]Vetsuisse Faculty, University of Zurich, Zurich, Switzerland, [2]Functional Genomics Center Zurich, Zurich, Switzerland

The transition time around parturition and early lactation involves critical physiologic changes in dairy cows. An excessive demand for nutrients due to the increased performance required for milk production results in a negative energy balance. One major adjustment consists in the rapid mobilization of energy sources from tissue depots in the form of non-esterified fatty acids. Cows poorly adapting to NEB are at high risk of developing disease. Many transition period diseases, including fatty liver, occur in a subclinical form, affecting milk production and reproductive performance of dairy cows. Fatty liver is interrelated with other production diseases. A noninvasive and accurate test would be helpful to diagnose fatty liver in cattle.

The purpose of a large-scale screening approach was to provide a hypothesis for the development of a novel diagnostic test based on lipidome profiles that is less invasive and more accurate.

We compared the plasma lipidome of diseased dairy cows using liquid chromatography coupled to quadrupole time-of-flight mass spectrometry. A study on 63 cows revealed 20 masses, which could distinguish between healthy cows and cows with different stages of disease.

In a study on 12 clinical healthy cows, during their transition period, 30 masses could be analyzed by MS/MS as potential biomarkers for determination of changes within the lipidome relative to calving. The main lipid groups detected, were triacylglycerides, phosphatidylcholines, and lysophosphatidylcholines.

BAYESIAN ESTIMATION OF THE ACCURACY OF CLINICAL EXAMINATION AND SYSTEMATIC THORACIC ULTRASONOGRAPHY FOR THE DIAGNOSIS OF BOVINE RESPIRATORY DISEASE IN PRE-WEANED DAIRY CALVES

Sébastien Buczinski[1], Teresa Ollivett[2], Nandini Dendukuri[3]

[1]*Faculté de médecine vétérinaire, Université de Montréal, St-Hyacinthe, QC, Canada,* [2]*Department of Medical Sciences, University of Wisconsin-Madison, Madison, WI, USA,* [3]*Department of medicine, Department of epidemiology, Biostatistics and Occupational Health, McGill University, Montréal, QC, Canada*

There is no gold standard method for the diagnosis of bovine respiratory disease (BRD) complex in Holstein pre-weaned dairy calves. Systematic thoracic ultrasonography (TUS) has been used as a proxy for BRD that can be performed in the field by veterinarians, but cannot be directly used by producers. The TUS examination focuses on lung consolidation, a common finding in bronchopneumonia. The Wisconsin calf respiratory scoring chart (CRSC) is a simpler alternative that systematically assesses rectal temperature, cough, ear position, nasal and ocular discharge and attributes a score to each of these items as well as the decision based on the total score obtained (i.e. do nothing, monitor, or treat). However, the accuracy of CRSC is still unknown.

Our objective was to estimate the accuracy of CRSC in two different populations, while adjusting for the lack of a gold-standard to define BRD status.

Two cross sectional study populations with a high BRD prevalence (n = 106 pre-weaned Holstein calves) and an average BRD prevalence (n = 85 pre-weaned Holstein calves) from North America were studied. All calves were simultaneously assessed using CRSC (cutoff ≥ 5) and TUS (cutoff ≥1 cm of lung consolidation). When using TUS as a gold standard CRSC Se were 20.0% and 55.4% and Sp were 58.0% and 100.0% respectively. Bayesian latent class models with conditional dependence were used with informative priors for BRD prevalence in both settings and TUS accuracy (sensitivity (Se_{TUS}) and specificity (Sp_{TUS})) and non-informative priors for CRSC accuracy (Se_{CRSC}/ Sp_{CRSC}) .

The Se_{CRSC} (95% credible interval (CI)) and Sp_{CRSC} were 62.4% (47.9-75.8) and 74.1% (64.9-82.8). The Se_{TUS} was 79.4% (66.4-90.9) and Sp_{TUS} 93.9% (88.0-97.6). Sensitivity analysis revealed the results were robust to prior specification.

Despite their imperfect accuracy both tools are helpful for BRD management. Improvement of the accuracy of BRD detection is a key step to decrease its negative impact as well as overuse of antimicrobials for false positive cases.

IMPACT OF MILK FEEDING LEVELS AND HOUSING ON THE INCIDENCE OF RESPIRATORY DISEASE AND SUBSEQUENT PRODUCTIVITY OF YOUNG DAIRY CALVES

Maria Prado, John Wilkerson, Peter Krawczel, Chris Boyer, Arnold Saxton

University of Tennessee, Knoxville, TN, USA

Dairy calf pneumonia (DCP) continues to be a highly prevalent condition affecting calves during the pre-weaning and/or post-weaning periods. However, its impact extends beyond the actual disease episode, having a negative effect on subsequent productivity and survivability of replacement stock. The ability to identify calves during the initial stages of respiratory disease development by monitoring activity and feeding behavior would allow for earlier disease intervention and thus potentially decrease the effects that DCP has later on the calf's life. Therefore, the objective of this study was to determine the impact of different milk feeding levels (standard or high) and housing (individual or group) on respiratory disease development and subsequent productivity of young dairy calves.

Female Holstein calves (n=215) from a commercial dairy farm were used in this study. Calves were randomly assigned to one of three treatment groups: (1) individual hutch with standard milk (4 L/day), (2) individual hutch with high milk (8 L/day), and (3) paired hutches (2 calves) with standard milk in a randomized block design. Social interaction/activity levels as well as feeding behavior were measured by fitting calves with collars containing sensors during the pre-weaning period. Calves were screened daily for clinical signs of disease. DCP was confirmed by PCR or serology. Body weights were collected from pre-weaned and weaned calves. Data collected were analyzed to identify behavior changes in social interaction and/or feeding associated with the development of respiratory disease. Serum samples were collected for determining passive transfer at 48 hours after birth. Farm records are being used to evaluate the effects of management during the pre-weaning phase on the long-term health and productivity of these calves by a retrospective analysis.

The automated sensor system successfully recorded activity levels of calves in the different treatment groups. Total activity levels were correlated with milk consumption showing significant increases during the time preceding the feedings. In addition, we established a unique time series signature from the acceleration data (activity levels) that corresponds with normal specific behaviors such as sleeping, standing, walking and/or bottle feeding. We used these data to identify any deviations from the norm. Overall, respiratory disease levels at this farm were low. All calves developed diarrhea when moved to the hutches. Calves in the high milk group gained an average of 1.28 Lbs/day versus 1.06 Lbs/day in the control group. Average daily gain was comparable between calves in the high milk group and the paired group (1.25 Lbs.). Overall, the calves fed high milk levels gained the most weight and the incidence of respiratory disease was low (0.93%, 2/215 of all enrolled calves). Milk yield, health and reproduction data will be presented.

EDUCATING THE EDUCATORS

Simon Dennis, BVtM, MVM, PGCert(Vet Ed), MRCVS, DECVIM-CA (Cardiology)
New York, NY

INTRODUCTION

"*Some people talk in their sleep. Lecturers talk while other people sleep*" (quote attributed to Albert Camus)

All board-certified cardiologists and cardiology residents are both educators and students. Each day that we learn something new we are students. Each day that we help someone learn something new we are educators. Teaching a student something new does not guarantee that it will be learnt effectively. A common example of this is the use of traditional lectures to disseminate information. Traditional lectures involve the passive transfer of information to a (hopefully) receptive audience. A wealth of literature exists supporting that students only retain a minority of information that is presented to them in a lecture and there is little impact on 'deep' learning (Long & Lock 2010). This is at odds with the perceptions of adult educators that information transfer is the most important purpose of lectures (Sutherland & Badger 2004). Nevertheless, traditional lectures remain one of the most common teaching methods in veterinary education, partly because they are an efficient method of delivering information in a content-rich curriculum. In our world in which the amount of biomedical information increases substantially each year, the challenge of delivering large amounts of information in a way that encourages students to retain it is highly relevant for veterinary education. An effective educator needs more than just knowledge of the subject that they teach. An effective educator needs to know how to learn.

PEDAGOGY AND ANDRAGOGY

Pedagogy is characterized by 'teacher-centered', dependent learning (teachers transmit information in a passive manner to a student) in which motivation to learn is the result of external factors (primarily to meet the demands of assessment). Andragogy is characterized by 'learner-centered', independent learning (students are active participants in their learning) in which the motivation to learn is the result of internal factors (interest in the subject, self-esteem, career progression) (Knowles 1998). Pedagogical learning promotes **superficial learning** in which information is memorized for the short term and often without context (subject-centered learning). It is regurgitated when required, typically for assessment (learning for the test). Andragogical learning promotes **deep learning** in which information is learnt by understanding and constructing on prior experiences (life-centered learning). This leads to long-term recall of information (learning for life). While both learning styles are appropriate to reach the goal of passing the requirement of veterinary licensing, the goal of creating life-long independent learners with an interest in continuing education is best served by the promotion of andragogy. Unfortunately there are features of most veterinary curricula that encourage pedagogy over andragogy. These include student perception of information overload, the reliance on didactic teaching (particularly in the preclinical years), and the importance of assessment (Parkinson & St. George 2003). These prior educational experiences of students entering clinical rotations are important for any educator to consider. If material continues to be taught in a pedagogical environment or in a manner not aligned with students' aspirations, the most likely result is superficial learning. Creating an andragogical learning environment and aligning the learning objectives with what students desire to learn are effective ways to encourage retention of knowledge in the long term and the skills to be effective life-long learners. A summary of the features of a teaching environment that promotes andragogical learning in medical education is presented as follows:

1. Learners should feel "safe and free to express themselves without judgment or ridicule".
2. Learners should be encouraged to formulate their own learning objectives, with guidance from educators to ensure that they are relevant and achievable.
3. Learners should be encouraged to identify their individual needs and learning strategies to meet their objectives.
4. Learners should be encouraged to identify learning resources, with guidance from educators.
5. Leaners should be helped to accomplish their learning objectives.
6. Learners should be involved in evaluating their own learning.

This summary is based on the description of the principles of andragogical learning by Kaufmann & Mann (2010).

DESIGNING LEARNING OBJECTIVES AND CONSTRUCTIVE ALIGNMENT

The 6 principles of andragogical learning listed above require students to have explicit knowledge of the learning objectives of the course. While it is essential that the learning objectives of a cardiology rotation be aligned with the assessment for licensing, in order to promote andragogical learning it is important to know what students prioritize for learning on their clinical cardiology rotation. A survey of students undertaking the cardiology rotation at the University of Pennsylvania between 2013 and 2014 revealed that the 5 most common learning objectives are:

Learning objective	% of student rotation groups requesting this learning objective
Auscultation & heart murmurs	100
Cardiac drugs & managing heart disease	94
ECGs & antiarrhythmic therapies	56
Echocardiography	50
Emergency treatment of cardiac disease	31

Constructive alignment of a curriculum is the process by which what we want students to learn, how students are taught and how students are assessed are in an aligned system. In this system, the assessment, teaching and learning are all driven by the learning objectives. The educator who founded the term, John Biggs, describes it as follows: "*It is a fully criterion-referenced system, where the objectives define what we should be teaching; how we should be teaching it; and how we could know how well students have learned it. In aligned teaching, there is maximum consistency throughout the system. The curriculum is stated in the form of clear objectives, which state the level of understanding required rather than simply listing the topics to be covered. The teaching methods chosen are those that are likely to realize those objectives; you get students to do the things that the objectives nominate. Finally, the assessment tasks address the objectives, so that you can test to see if the students have learned what the objectives state they should be learning. All components in the system address the same agenda and support each other. The students are "entrapped" in this web of consistency, optimizing the likelihood that they will engage the appropriate learning activities.*" (Biggs 1999) This process, which is also referred to as "blueprinting" has been applied extensively in the design of medical education programs and assessment (van der Vleuten and Schuwirth 2005).

ADOPTING FLEXIBLE TEACHING STYLES

Students have different learning styles. A study into 150 veterinary students at North Carolina State University (Neel & others 2010) reported that learning styles of individual students vary notably with regards to how individuals prefer to perceive information, process information and progress in the understanding or a subject. About 80% of veterinary students prefer to perceive information in a 'sensory' manner (facts, practical, sensations) than an 'intuitive' manner (theories and models), and of the sensory information that is presented, most students prefer visual and kinesthetic learning styles to written and auditory learning. However, only two-thirds preferred a 'sequential' approach to understanding information rather than a 'global' approach, and the proportions of students who preferred to process information in an 'active' manner rather than 'reflective' manner were fairly even. The challenge for veterinary educators is to develop teaching sessions and learning resources allow all students to learn in the most effective manner. This requires flexible teaching styles.

Similarly, the transition from pedagogical to andragogical learning requires further flexibility in the teaching style of the educator. Grow (1991) describes how 'dependent' learners require guidance to become 'self-directed' learners. A dependent learner responds well the teaching of the 'expert/authority' teacher who teaches in a purely instructional and pedagogical manner. Self-directed learners respond best to a 'delegator' teacher who guides their learning with ideas and encouragement. It is important to recognize that dependent learners often resent the delegator teacher for not teaching them adequately, and self-directed learners often resent the expert/authority teacher for attempting to control their learning. While usually unintentional, both mismatches in teaching and learning styles are severe enough to damage the effective educator-student relationship and prevent effective learning. Matches and mismatches in teaching and learning styles, as proposed by Grow (1991) are listed below:

Learning stage	Learner style	Teaching style match	Teaching style mismatch
Stage 1	Dependent	Expert/authority	Delegator, Facilitator
Stage 2	Interested	Motivator	Delegator
Stage 3	Involved	Facilitator	Expert/authority
Stage 4	Self-directed	Delegator	Expert/authority, Motivator

Transitioning a student from a stage 1 dependent learner to a stage 4 self-directed learner requires an educator to recognize the learning style of the student, match with the appropriate teaching style, use positive feedback to build the student's confidence in their knowledge, and encourage them to set their own goals. As the student progresses up a learning stage, the process can be repeated with the gradual replacement of teaching with goal setting. Eventually teaching is no longer required and the student becomes a fully autonomous learner.

POSITIVISM AND CONSTRUCTIVISM

The preclinical years of veterinary curricula require the learning of a large number of facts. This is facilitated by a positivist approach to learning. In positivism, the epistemology is objectivism and the ontology is realism (facts are known and independent of the observer) (Illing 2010). For example, students learn that the renal site of furosemide action is the ascending loop of Henle. Similarly, clinical decisions based on multiple randomized, controlled trial or meta-analyses are conducive to learning by positivism. In veterinary medicine, there is an absence of such high level evidence to guide clinical decisions. For example, even among veterinary cardiologists there is no single accepted way of managing feline hypertrophic cardiomyopathy; instead decisions are based on theoretical benefits and anecdote (Rishniw & Pion 2011). The challenge of taking an undergraduate student trained in the positivist approach and creating a clinician able to deal with the uncertainties and subjectivism of clinical medicine has been the subject of discussion in medical education literature (Brass 2009; Fincher and others 2009). It has been argued that the positivistic view is often inadequate for clinical medicine in which "medical decisions are seldom based on science alone, but rely instead on a complex amalgam of factual knowledge, personal experience, anecdote and empathy, played out against a background of professionalism and underpinned by a sense of care and compassion" (Kneebone 2002). In constructivism, the epistemology is subjectivism and the ontology is relativism (multiple possible realities exist and the observer may influence how a fact is viewed). The constructivist approach to learning may therefore be appropriate for developing the skills of a veterinary clinician. Clinical rotations are an opportunity to encourage the constructivist approach to learning as a complement to facts learnt during preclinical years.

THE LEARNING CYCLE FOR DEVELOPING SKILLS

The development of mastery in a skill or knowledge involves a transition that every person goes through (Peyton 1998). In the first stage a student is not competent but is not aware of what they need to do to improve (unconscious incompetence). With learning a student understands what they need to do to become competent while remaining incompetent (conscious incompetence). When the student initially develops competence they are aware of the processes they went through to develop the state of competence (conscious competence). A person at this stage is often good at teaching competence in the skill or knowledge to incompetent students, as they are aware of the steps they themselves have just gone through. After time, a consciously competent person will be able to undertake the skill or knowledge without thinking about how they developed the competency (unconscious competence). Such individuals are usually experts in the skill or knowledge. As a result of losing the awareness of the processes they went through to develop competency, such unconsciously competent individuals may be less effective teachers than consciously competent individuals. Unconscious competence is relatively common in subject specialists in medical disciplines. For example, because the knowledge of the pharmacology of loop diuretics is second nature to a boarded cardiologist, such a person in an educator role may forget what it was like to be an unconsciously competent student and therefore be less effective in teaching this competency to the student that a person who has just learnt the pharmacology of loop diuretics. Peer-to-peer teaching, such as of later year students to early year students, represents the teaching of a consciously competent individual to an unconsciously competent one. For educators that are unconsciously competent, the challenge is deconstruct the skill or knowledge in order to regain the knowledge of a consciously competent individual. Reflecting on how students learn and refining teaching sessions so that the subject is taught in a deconstructed manner are methods that can improve an unconsciously competent educator's teaching.

THE IMPORTANCE OF ROLE MODELING

What educators believe that they teach and what students learn are not necessarily aligned. The 'intended curriculum' is what an educator intends to deliver to students during a teaching session. The 'delivered curriculum' is what an educator actually delivers during a teaching session. The delivered curriculum may not be the same as the intended curriculum. For example, an educator may intend to teach the pharmacology of loop diuretics to a student group, but end up teaching the clinical application of diuretics instead. This may be more common in teaching sessions that have more flexibility of structure, such as small group teaching and rounds. Most educators will be aware when the intended and delivered curricula do not match. The 'received curriculum' is what students actually learn during the teaching session. This is the end result of all teaching. The received curriculum may not be the same as the intended or delivered curricula. The aspects of the received curriculum that an educator is not aware that they are delivering are the 'hidden curriculum' (Portelli, 1993). The hidden curriculum can involve course content that is directly relevant to the learning objectives of the teaching session. It may also represent teaching and learning styles. More worryingly, it can represent educator behaviours such as interactions with colleagues, clients and students, empathy towards patients, appearance, and time-management. For example, if an educator behaves in a disrespectful manner about a colleague who refers a case, the hidden curriculum that the student receives may be that this type of behavior is acceptable and is copied by the student, or that the student loses respect for the educator and becomes reluctant to refer. Recent studies into the hidden curriculum in medical and veterinary education show that it can have a powerful impact on student learning; and one that is not always positive (Mossop & Cobb 2013). All educators are role models.

The impact of the 'hidden curriculum' and the educator as a role model may be easily overlooked as these are more related to humanities than scientific learning. The qualities of an effective clinical supervisor have been defined as instruction, empathy, support, tolerance, flexibility, feedback, encouragement and respect (Kilminster & Jolly 2000). The Association of American Veterinary Medical Colleges has defined the prime characteristics of veterinarians as compassion, expertise, humaneness, judgment, care and understanding (Willis & others 2007). If our goal is to create veterinarians, cardiologists and educators that have these qualities, then we in turn need to role model the same level of professionalism, respect, and empathy every single day.

REFERENCES

Biggs J. (1999) What the Student Does: teaching for enhanced learning. *Higher Education Research and Development* **18**(1), 57-75.

Brass EP. (2009) Basic Biomedical Sciences and the Future of Medical Education: Implications for Internal Medicine. *Journal of General Internal Medicine* **24**(11), 1251-1254.

Fincher RE, Wallach PM, Richardson S. (2009) Basic Science Right, Not Basic Science Lite: Medical Education at a Crossroad. *Journal of General Internal Medicine* **24**(11), 1255-1258.

Grow G. (1991) Teaching learners to be self directed. *Adult Education Quarterly* **41**, 125-149.

Kaufman DM, Mann KV. (2010) Teaching and learning in medical education: how theory can inform practice. *In*: Understanding Medical Education: evidence, Theory & Practice. *Ed*: T. Swanwick, Wiley-Blackwell, Chichester, UK.

Kilminster SM, Jolly BC. (2000) Effective supervision in clinical practice settings: a literature review. *Medical Education* **34**, 827-840.

Kneebone R. (2002) Total internal reflection: an essay on paradigms. *Medical Education* **36**, 514-518.

Knowles MS, Holton EF, Swanson RA. (1998) The Adult Learner. 5th ed. Gulf, Houston, TX.

Long A, Lock B. (2010) Lectures and large groups. *In*: Understanding Medical Education: Evidence, Theory & Practice. *Ed*: T. Swanwick, Wiley-Blackwell, Chichester, UK. Pp. 141.

Mossop LH, Cobb K. (2013) Teaching and assessing veterinary professionalism. *Journal of Veterinary Medical Education* **40**(3), 223-232.

Neel JA, Grindem CB. (2010) Learning-Style Profiles of 150 Veterinary Medical Students. *Journal of Veterinary Medical Education* **37**(4), 347-352.

Parkinson TJ, St George AM. (2003) Are the Concepts of Andragogy and Pedagogy Relevant to Veterinary Undergraduate Teaching? *Journal of Veterinary Medical Education* **30**(3), 227-233.

Peyton JWR. (1998) Teaching and Learning in Medical Practice. Manticore Europe Ltd. Pp. 174-177.

Portelli JP. (1993) Exposing the hidden curriculum. *Journal of Curriculum Studies* **25**(4), 343-358.

Rishniw M, Pion PD. (2011) Is treatment of feline hypertrophic cardiomyopathy based in science of faith? A survey of cardiologists and a literature search. *Journal of Feline Medicine and Surgery* **13**, 487-497.

Sutherland P, Badger R. (2004) Lecturers' Perceptions of Lectures. *Journal of Further and Higher Education* **28**(3), 277-289.

van der Vleuten CP, Schuwirth LW. (2005) Assessing professional competence: from methods to programmes. *Med Educ* **39**(3): 309-317.

Willis NG, Monroe FA, Potworowski JA, & others. (2007) Envisioning the Future of Veterinary Medical Education: The Association of American Veterinary Medical Colleges Foresight Project, Final Report. *Journal of Veterinary Medical Education* **34**(1), 1-41.

Pericardial Disease: Constrictive pericarditis secondary to coccidioidomycosis

Whit M. Church, DVM, DACVIM (Cardiology)
Gilbert, AZ

Restrictive/constrictive physiology is characterized by decreased ventricular compliance or pericardial constraint, with an associated rise in ventricular filling pressures. It can result from constrictive pericarditis, restrictive cardiomyopathy, and any condition that results in rapid dilation of one or both ventricles.[1] The classic constrictive physiologic findings on invasive pressure measurements during cardiac catheterization include: a steep y descent with an abrupt rise and plateau after early diastolic filling in all 4 chambers, elevated atrial pressures during atrial contraction and ventricular systole, large atrial reversal waves with reduced S and D waves on pulmonary venous flows, high filling pressures of both ventricles and atria, and diastolic pressure equilibration of all four chambers.[1] All of these characteristics can also be associated with restrictive cardiomyopathy.[1]

Many cases of constrictive pericarditis can be addressed surgically, but making the recommendation for a thoracotomy with pericardectomy takes confidence. We must often rely on the information obtained from invasive pressure measurements and often echocardiography alone. The main characteristic for recognizing constrictive pericardial disease during invasive pressure measurements is to observe the presence of significant respiratory variation and ventricular interdependence.[2] In constrictive pericarditis, respiratory variation in ventricular filling and ventricular interdependence are much more prominent than in restrictive myocardial disease. This means that during inspiration, a patient with constrictive pericarditis will have increased filling of the right ventricle at the expense of left ventricular filling.[1,4] This results from the combination of mechanical constraint of the ventricles and a shared interventricular septum. During expiration the opposite phenomenon occurs with decreased right ventricular filling and increased left ventricular filling. In Figure 1, there are elevated right atrial pressures with steep a and v waves. These waves, combined with the prominent x and y descents, give the atrial pressure trace this characteristic M or W configuration. During inspiration, we can see the increase in right atrial pressure. The atrial premature beat produces a pause that allows visualization of the characteristic dip and plateau (square root sign), which indicates all ventricular filling occurs early in diastole. The plateau portion or "h" wave, shows the elevated diastolic pressures of the right atrium and ventricle. The diastolic plateau is often obscured due to an elevated heart rate. The respiratory variation in right atrial pressures suggests constrictive pericarditis, but simultaneous left sided pressures were not obtained, so ventricular interdependence was not demonstrated.

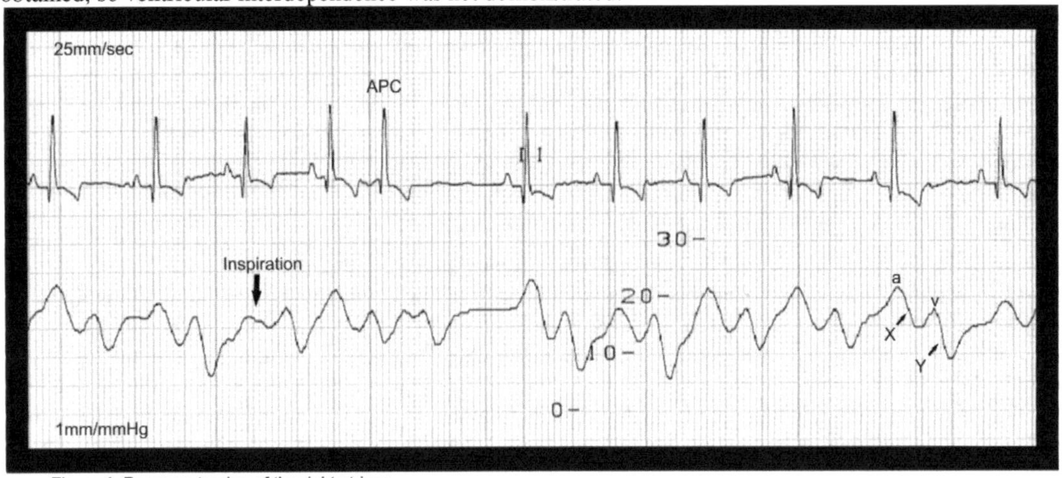

Figure 1. Pressure tracing of the right atrium.

The echocardiographic characteristics described to help identify constrictive pericarditis on M-mode or 2D include: thickened pericardium (Pericardial thickness of $\geq$ 3mm), lack of independent movement between the epicardium and pericardium (Adhesion), abnormal ventricular septal motion (Septal bounce), flattening of the LV posterior wall during diastole, respiratory variation in the ventricular size, dilated caudal vena cava, caudal vena caval plethora, ascites, pleural effusion, edematous thickening of the gallbladder wall, and dilated hepatic veins. However, these findings are inconsistently found in patients with constrictive physiology. I have found that mild thickening of the pericardium can be missed, but lack of independent movement of the pericardium and epicardium (adhesion) is more easily identified. A septal bounce is typically specific, but not always found in my patients. Distention of the vena cava and hepatic veins is almost always present to some degree. Vena caval plethora is often not reliably visualized at the right atrial-caudal vena caval junction, so observation of the lack of a 50% reduction in diameter after deep inspiration is often visualized on the abdominal side of the diaphragmatic caval foramen.

Doppler echocardiography may be a better identifier of constrictive physiology than the 2D imaging findings. These include a restrictive mitral inflow pattern (E/A ratio >1.5), 25% or greater variation in the mitral inflow E wave velocity with respiration, increased diastolic flow reversal and decreased diastolic forward flow during expiration in the hepatic veins. Normal mitral annular E' velocities, and normal E'/A' ratio are other variables that can indicate constrictive pericarditis.

Respiratory variation in transmitral and transtricuspid flow is due to the ventricular interdependence that results from a constricting pericardium. This results in improved filling of the right ventricle during inspiration, with subsequent hindrance of left ventricular filling. The opposite of this occurring during expiration. However, the expected variation in mitral inflow E wave by >25% is often absent due to the frequency of elevated left atrial pressure in constrictive disease. This is caused by the severe diastolic dysfunction or mitral regurgitation in the case of infectious pericarditis. If the atrial pressure is elevated, the mitral valve opens during a very steep portion of the left ventricular pressure curve that is resistant to the effects of respiration. Thus, many patients with constrictive pericarditis do not manifest this finding.

Pulse wave Doppler of the hepatic vein does seem to be a very reliable indicator for elevated right atrial pressures regardless of the cause, but if respiratory variation is pronounced, constrictive physiology should be considered. The diastolic flow reversal is pronounced immediately following expiration unlike restrictive disease which has the reversal wave more pronounced following inspiration.

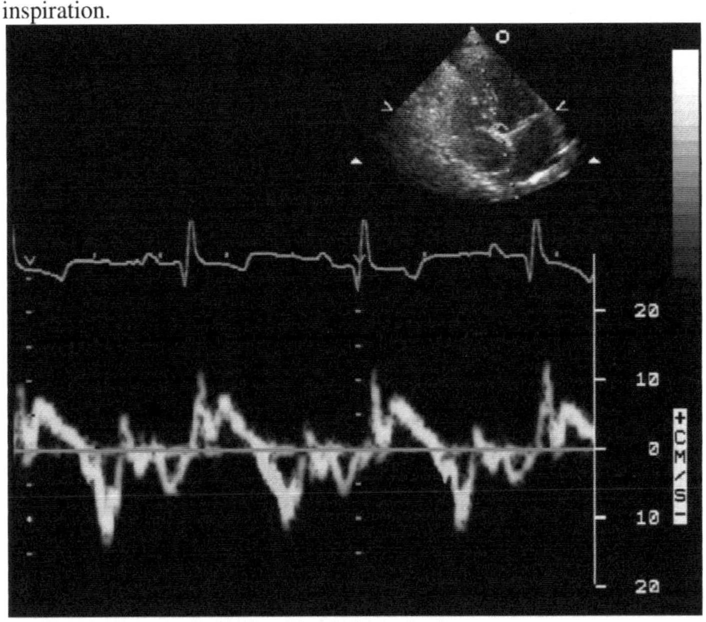

Figure 2. Tissue doppler imaging of the mitral valve annulus.

The mitral annular velocity tracing in Figure 2 shows normal E' and A' velocities and E'/A'ratio in a patient with constrictive epicarditis after pericardectomy. An E' value of <8 cm/s in either the septal or lateral wall is considered abnormal, and the E'/A' ratio should be >1.[2,4] This would seem inappropriate for a disease that often presents with a restrictive filling pattern on mitral valve inflow velocities. However, it is thought that pericardial constriction limits the circumferential expansion of the heart, and may exaggerate the longitudinal motion of the mitral annulus.[3] Thus, E' velocity is preserved or even augmented. This is unlike restrictive cardiomyopathy in which E' velocity is significantly reduced due to loss of the mechano-elastic properties of the myocardium.[2,3,4] This difference in E' wave velocity response, can significantly improve our ability to distinguish between constrictive pericarditis and restrictive cardiomyopathy. This becomes even more important when a patient continues to have signs of right heart failure post-pericardectomy.

An interesting finding in constrictive pericarditis secondary to coccidioidomycosis is the effect on the mitral valve. Mitral regurgitation is found in a very high percentage of these patients resulting in moderate dilation of the left atrium, often with the size right atrium being normal or mildly enlarged. I have assumed this to be a rheumatic-type valve lesion, which sometimes resolves with antifungal treatment, and on rare occasions progressives to severe valvular regurgitation with secondary left-sided congestive heart failure.

REFERENCES

1. Lorell BH, Grossman W. Profiles in constrictive pericarditis, restrictive cardiomyopathy, and cardiac tamponade. In Baim DS, ed Grossman's Cardiac Catheterization, Angiography, and Intervention. Philadelphia, Lippincott WW. 2000: 829-838.
2. Sengupta PP, Mohan JC, Mehta V, Arora R, Pandian NG, Khandheria BK. Accuracy and Pitfalls of Early Diastolic Motion of the Mitral Annulus for Diagnosing Constrictive Pericarditis by Tissue Doppler Imaging. Am J Cardiol 2004;93:886–890.
3.Ha JW, Oh JK, Ling LH, Nishimura RA, Seward JB, Tajik AJ. Annulus paradoxus: transmitral flow velocity to mitral annular velocity ratio is inversely proportional to pulmonary capillary wedge pressure in patients with constrictive pericarditis. Circulation 2001;104:976 –978.
4. Ha JW, Oh JK, Ommen SR, Ling LH, Tajik AJ. Diagnostic value of mitral annular velocity for constrictive pericarditis in the absence of respiratory variation in mitral inflow velocity. J Am Soc Echocardiography. 2002 Dec;15(12):1468-71.

THORACOSCOPIC PROCEDURES FOR PERICARDIAL DISEASE

Keith Richter, DVM, DACVIM (SAIM)
San Diego, CA

INTRODUCTION

There are several diagnostic options to evaluate dogs and cats with pericardial disease, including radiographs, ultrasonography, CT scan, thoracoscopy, and exploratory thoracotomy.[1] Advances in thoracoscopy have occurred due to improvement in video technology and in manufacture of instruments that allow a versatile array of procedures similar to those performed in open thoracotomy. Thoracoscopy has many advantages, allowing direct visualization of the thoracic cavity, including pulmonary parenchyma, mediastinum, pleura, pericardium, heart, and major blood vessels. In many cases, the view obtained during thoracoscopy is superior to that obtained during open thoracotomy due to the ability to get the tip of the scope near thoracic structures, magnification, and ability to get the scope in places that are inaccessible to a surgeon during open thoracotomy. In particular, thoracoscopy can be extremely useful in treating pericardial disease, including the creation of pericardial windows and subtotal pericardiectomy. Over 300 of these procedures have been performed at the author's hospital, with a high success rate and minimal complications. Most of the cases have had heart base masses, with lesser numbers of idiopathic pericardial effusion, right atrial masses, or mesotheliomas. A partial pericardiectomy gives permanent relief of tamponade and allows excellent palliation of these patients, usually resulting in survival times equal to that of an open surgical pericardiectomy. However, one study reported a longer survival time and disease free interval in dogs with non-neoplastic pericardial effusion with open thoracotomy compared to thoracoscopy.[2] The longer survival time and disease free interval was not different between the two methods in dogs with neoplastic disease.[2] The author has also performed several successful subtotal pericardiectomy procedures in dogs with constrictive pericarditis. A thorough knowledge of the anatomy, appropriate preoperative assessment, and experience with equipment are all necessary to complete a successful procedure. Thoracoscopy also results in decreased postoperative pain, fewer wound complications, and more rapid return to function compared to open thoracotomy.[3]

INSTRUMENTATION

Most of the equipment used in abdominal laparoscopy can be used for thoracoscopy. A rigid scope ("telescope") is used to visualize the thoracic viscera. The author prefers a 5 or 10 mm diameter scope. Smaller scopes have a smaller image with less field of view. In addition, a greater light intensity is needed for smaller scopes. A 5 mm scope is suitable to examine any dog or cat. Scopes are also available in various degrees of field of view, including zero-degree (direct forward viewing) up to 70°. The zero-degree angle is easier to use, and generally preferred for most procedures. The author occasionally uses a 30° scope to view structures to the side of the tip, and through rotation, to expand the field of view. Operating scopes have a 5 or 6 mm channel, with an eyepiece extending from the proximal end at a 45-90° angle. These scopes allow introduction of instruments through the same puncture as the scope. Some operators believe them to be technically easier because the tip of the instrument is always in the field of view, without the need for "triangulation" to locate the tip of the instrument when they are introduced through accessory or secondary puncture sites. Operating scopes are generally slightly longer, so instruments must be long enough to protrude through the channel. The disadvantage of operating scopes is the limited ability to manipulate instruments passing through the channel. Scopes and instruments are introduced with a trocar/cannula system. Unlike laparoscopic cannulas, thoracoscopic cannulas are shorter to minimize trauma to thoracic viscera and allow protrusion of instruments closer to the thoracic wall, thus permitting work on the surface of the heart or lungs, or other structures in the near field. Accessory instruments include biopsy forceps ("spoon/clamshell" style), grasping forceps (straight, curved, serrated, toothed, etc.), scissors, suction devices, retractors, suturing and stapling devices, blunt probes, and cautery devices (Ligasure©).

ANESTHETIC MANAGEMENT

Preoperative evaluation should consider metabolic and cardiopulmonary status. If there is significant pleural or pericardial effusion, withdrawal of the fluid is often helpful to stabilize the patient and simplify the procedure. In addition, ascites (often present with pericardial effusion) must be removed, otherwise it will flow into the thoracic cavity after the procedure through the rent in the diaphragm created for insertion of the scope (see below). The procedure is usually performed in dorsal recumbency, and standard tracheal intubation is used for two-lung ventilation.[4,5] Once the thorax is entered, positive pressure ventilation must be started.

APPROACHES INTO THE THORAX

Several approaches to the thorax can be used to introduce the scope, including a lateral intercostal, transdiaphragmatic paraxiphoid, or thoracic inlet approach. The transdiaphragmatic paraxiphoid approach is used for most pericardial procedures.[3-6] It is performed with the animal in dorsal recumbency, with the puncture site for the trocar/cannula assembly just lateral to the xiphoid cartilage. The trocar/cannula assembly is directed forward at a 45° angle, with the actual puncture occurring during expiration immediately following several assisted breaths. This approach permits examination of the heart/pericardium, pleura, lung, diaphragm, and cranial and caudal mediastinum. The main limitation of this approach is examination of dorsal structures in the thorax, including the hilus of the lung and dorsal mediastinal structures. The lateral approach is only used when evaluating dorsal structures in the thorax such as when pericardiectomy is combined with thoracic duct ligation in patients with idiopathic chylothorax.

TECHNIQUE

Once the scope is introduced into the thorax, the video camera is attached to allow visualization of the thoracic viscera. With the transdiaphragmatic paraxiphoid approach, the ventral mediastinum often hangs down to obscure the view. This is removed with a combination of sharp and blunt dissection with a ligating device (Ligasure©) or scissors attached to the cutting setting of an electrocautery unit. With any of the approaches, carbon dioxide insufflation is occasionally helpful to expand the pleural space and improve visualization. Care must be taken to prevent instillation of too much gas, which could cause respiratory compromise. A drop in oxygen saturation determined by pulse oximetry usually detects this. Once a clear view of thoracic structures is obtained, the scope is slowly moved to completely evaluate all surfaces visible. Blunt probes and retractors can be introduced at secondary puncture sites to facilitate examination. The technique for introducing accessory instruments involves careful planning of where the tip of the instrument is desired. The puncture site is determined by estimating the course of the instrument, and locating the puncture site by visualizing with the scope a finger jabbing the intercostal space at the desired accessory entry site (usually at the level of the heart base to the middle of the heart). The trocar/cannula assembly is then introduced under direct visualization to prevent injury to thoracic viscera. The visual examination of the thorax is completed, and pericardiectomy then performed.

Often the pericardium is covered by pericardial fat making identification of the actual pericardium difficult. A combination of sharp and blunt dissection using scissors and forceps may be necessary to identify the pericardium. Once the pericardium is identified, it is grasped with forceps. Often this must be very forceful, especially if the pericardium is taught from fluid or is thickened. A forceps with teeth often make grasping the pericardium easier. In addition, the ability to lock the forceps makes holding onto the pericardium easier and less tedious. If the pericardium can not be grasped after multiple attempts, a scissors with a sharp point can be used to punch a hole in the pericardium. This maneuver requires extensive operator experience to avoid puncture of the epicardium, especially if pericardial-epicardial adhesions are present (as is often the case with constrictive pericardial disease). Once the pericardium is grasped with forceps, the scissors introduced from the opposite side of the thorax are used to slowly cut a hole in the pericardium. Suction applied to a suction probe is often helpful to remove residual pericardial fluid and keep the field of view as clear as possible. As large a hole as possible is cut. This often requires alternating the position of the forceps and scissors to complete the cut, and removing portions of the pericardium in multiple attempts. The dorsal aspect of the cut is limited by the position of the phrenic nerve. The author has inadvertently cut the phrenic nerve on one side on two different occasions. No clinical problems were seen in these patients. Each time a piece of pericardium is cut free, it is removed from the thorax using forceps. This is accomplished by removing the forceps and cannula as a unit. Once a small portion of the pericardium exits the thorax, it can be grasped with hemostats to get the most secure grip possible. The remainder of the pericardium is pulled out of the thorax and submitted for histopathologic examination. Sometimes a definitive diagnosis of neoplasia can be diagnosed in this manner, even when idiopathic pericarditis is suspected (most commonly mesothelioma). When the final cut is completed, the "window" is inspected to determine its margins. Usually approximately one-third to one-half of the epicardium can be visualized. Additionally, if a cardiac-associated mass can be visualized, the author has biopsied these with "spoon" or cup forceps (5 mm diameter) with very few complications. Though the potential for a pericardial window to spontaneously close has been suspected,[2] the author has not had closure of a "window" in any case.

Once it has been determined that there is no significant residual hemorrhage or air leakage, the accessory instruments and cannulas are removed. The puncture sites are sutured, usually with single sutures in the muscle, subcutis, and skin. The scope is then withdrawn and a no. 14 to no. 18 French red rubber catheter is inserted through the cannula into the thorax. The cannula is removed and a purse-string suture placed around the red rubber catheter. Gas/air is manually evacuated with a 60-cc syringe until negative pressure is obtained. Local anesthetic is instilled into the pleural space to provide postoperative analgesia. A total of 1.5 to 2.0 mg/kg each of lidocaine and bupivacaine is instilled. The tube is then gradually removed with the syringe attached to provide negative pressure. Once the tube exits the thorax, the purse-string suture is tightened. A single suture is then placed in the subcutis and skin to complete the procedure. Most animals are discharged the following day.

Potential complications include inadvertent lung trauma with a trocar or instrument, extra puncture sites from initially entering at the wrong location, uncontrolled bleeding from vascular injury or following cardiac-associated mass biopsy, bleeding in a patient with a coagulopathy, inadvertent cutting of a phrenic nerve, ascites secondary to pericardial effusion escaping into the thorax through the diaphragmatic hole created (for this reason ascites should be completely drained prior to the procedure), excessive postoperative pain, and portal site metastasis of mesothelioma.[7]

OTHER PROCEDURES PERFORMED IN CONJUNCTION WITH THORACOSCOPIC PERICARDIECTOMY

Other procedures can be performed in conjunction with pericardiectomy. As mentioned above, cardiac-associated masses can usually be biopsied. It is imperative that the mass be distinguished from normal heart base structures such as the aortic root, right atrium, or right atrial appendage. Despite the risk of significant bleeding, this has rarely occurred in the author's experience.

In dogs with idiopathic chylothorax, thoracic duct ligation can be performed thoracoscopically, and usually in conjunction with pericardiectomy.[8] The thoracic duct ligation is performed through a lateral intercostal approach. Once this portion of the procedure is completed, the animal is repositioned in dorsal recumbency for the pericardiectomy as described above. In most of these cases, the pericardium is normal or only mildly thickened. This makes the pericardiectomy much easier than in patients that present with pericardial effusion. Right atrial mass resection using a laparoscopic stapling device can also be performed following pericardiectomy.[9,10] Case selection is very important for this procedure as this is generally only feasible for pedunculated masses or those at the tip of the right atrial appendage (as opposed to sessile tumors involving the atrial or ventricular wall).

REFERENCES

1. Scollan KF, et al. **J Vet Intern Med** 2015; 29:79. 2. Case JB, et al. **JAVMA** 2013; 242:493. 3. Walsh PJ, et al. **Vet Surg** 1999; 28:472. 4. Jackson J, et al. **J Vet Intern Med** 1999; 13:529. 5. Dupre GP, et al. **Vet Surg** 2001; 30:21. 6. Atencia S, et al. **J Sm Anim Pract** 2013; 54:564. 7. Brisson BA, et al. **JAVMA** 2006; 229:980. 8. Allman DA, et al. **Vet Surg** 2010; 39:21. 9. Ployart S, et al. **JAVMA** 2013; 242:237. 10. Crumbaker DM, et al. **JAVMA** 2010; 237:551.

ACVIM CARDIOLOGY CANDIDATE FUNCTION: UNDERSTANDING CORE STATISTICAL METHODS USED IN VETERINARY LITERATURE

Mark A. Oyama, DVM, DACVIM (Cardiology)
Philadelphia PA

INTRODUCTION

Consumers of scientific literature must assess the validity of results based on careful assessment of the reported materials and methods. It is standard for investigators to report the statistical methods that were used in the study, however, this information is often passed over by the reader, who assumes that the proper methods were employed. This may or may not be the case. Some journals employ a statistical editor whose job it is to ensure that proper methodology and reporting is performed. Some journals leave evaluation of these methods to the associate editor or to the individual reviewers. Thus, quality control of statistics is highly variable from study to study and from journal to journal. Mark Twain famously stated "There are three kinds of lies: lies, damned lies, and statistics." Consumers of scientific information that understand commonly used core statistical methods can better evaluate what they read and make determinations as to the validity of reported results. This review will cover some basic statistical premises including evaluation of normality, multiple comparison testing, linear regression and use of Bland-Altman plots, and survival curves.

EVALUATION OF NORMAL DISTRIBUTION

The distribution of continuous or categorical data is often evaluated prior to performing statistical evaluation. Graphical tests to visually inspect the distribution of data include frequency distributions, box plots, probability-probability (P-P) plots, and quantile-quantile (Q-Q) plots. Often, graphical evaluation can be used to "eyeball" data and determine if an assumption of normality seems reasonable. Statistical tests to determine if data adheres to a "normal" distribution include Shapiro-Wilk, D'Agostino-Pearson, Kolmogorov-Smirnov, and the Chi-square goodness of fit test. Each of these tests compares the cumulative probability of the variable against the cumulative probability of a particular distribution (such as a normal distribution). A series of scores are calculated and compared against the scores that would arise if data with the same mean and standard distribution was normally distributed. The null hypothesis is that the data is normally distributed, and P values <0.05 cause one to reject the null and assume a non-normal distribution. Data that is normally distributed are evaluated using parametric statistical techniques such as Student t-tests, while data that is not normally distributed can be arithmetically transformed to achieve a normal distribution or evaluated using non-parametric techniques such as Wilcoxon rank tests.

MULTIPLE COMPARISON TESTING

Many medical studies perform statistical evaluation of multiple hypotheses simultaneously. An example is a study that hypothesizes that any of 15 different echocardiographic variables will be associated with the outcome in question, for instance, presence of congestive heart failure. Interpretation of a single hypothesis and therefore single P value is straightforward. If the P value is <0.05, one concludes that the probability that you would detect a difference in your group outcome at least as large as that which was observed in your study by random chance is <0.05. However, if you are evaluating multiple hypotheses at once, the likelihood of finding a result by chance increases as function of the number of hypotheses under inspection ($100\% \times [1-0.95^n]$) where n=number of hypotheses tested. If 13 different echo parameters are tested at once, the probability of at least 1 of those parameters to meet the P<0.05 threshold by chance is 49%! To circumvent this pitfall, the P value required to meet your definition of "statistically significant" is increased. A common correction is the Bonferroni correction which sets the definition of statistical significance as 0.05/n. Thus, in an experiment with 15 echo parameters, a P<0.0033 is needed to meet the definition of statistical significance. In analysis that utilize ANOVA, Tukey Dunnett or Holm-Sidak tests can be used to correct for multiple comparisons, and in analysis using non-parametric mean rank tests, the Dunn test can be employed. For analysis that calculate very large numbers of t-tests, such as genome wide association tests or gene chip analysis, false discovery (FDR) analysis is used to avoid false positive results.

LINEAR REGRESSION AND EVALUATION OF SYSTEMATIC BIAS

Linear regression is a common analytic technique to assess the association between two (usually continuous) variables. Linear regression evaluates association and not causation. The most common technique to fit a regression line is the least-squares method which attempts to reduce the sum of the squares of the vertical distance each point has to the fitted line. Linear regression assumes 4 important features regarding the data you are attempting to fit: 1) the expected relationship between the dependent and independent variable is a straight line function, 2) any errors in the two variables are independent of each other, 3) there is constant variance of the errors over time, prediction, or any independent variable, and 4) the errors are normally distributed. These assumptions should be tested rather than assumed when attempting to perform linear regression. Many statistical software packages will provide output as to these assumptions as part of the linear regression procedure. Two common measures of the linear association are the r statistic (correlation coefficient) and r^2 statistic (coefficient of determination). Both values are indices of the strength of the association between the variables. In the case of the r statistic, the direction of the association is also known based on whether the coefficient is positive or negative. The r^2 value is interpreted as the amount of total variation in the dependent variable is attributable to the independent variable. Thus a linear relationship with r^2=0.88 indicates that 88% of the total variation is y can be explained by the linear relationship with x.

Linear regression can be used to compare two techniques or methods that each purport to measure the same thing. An example would be two different commercial BUN analyzers each designed to measure serum BUN. Linear regression is used to compare results of paired samples analyzed by each of the machines. If the level of agreement was 100% between the two machines (for instance paired sample results were 5,5; 18,18; 35,35; 47,47) the r^2 value is 1.0, and one assumes that the results from both machines are in complete agreement. However, an r^2 value of 1.0 could also be achieved if the results of one machine were consistently 2x higher than the other machine (5,10; 18,36; 35,70; 47,94). Thus, knowing the r^2 value does not provide all the necessary information. Options for further evaluating the level of agreement and evidence of systematic bias include inspection of the slope of the regression equation or constructing Bland-Altman (BA) plots. BA plots graph the difference between paired results on the y axis against the mean of paired results on the x axis. 95% confidence intervals are calculated and if these intervals fall outside a difference of 0, a systemic bias exists. BA plots can be visually examined and are easy to interpret. Several examples will be shown during this presentation.

BASIC KAPLAN-MEIER SURVIVAL CURVE ANALYSIS

Time to event data can be analyzed using either life table or Kaplan Meier (KM) analysis. KM analysis is routinely used to analyze effect of treatment on survival or some other clinically meaningful outcome. KM analysis plots probability of survival or outcome against time. The KM estimates that are used to plot this curve reflect the probability of the outcome at a given time while also taking into account the probabilities of all the times that proceeded the given time. An important concept of KM analysis is what to do with individuals that have not yet reached the endpoint at the time the study ended or those individuals that were lost to follow up at some point during the study. These "censored" data are incorporated into the KM estimate up until the last known time they were in the study. KM curves of two different populations, for instance those receiving a drug vs. those receiving placebo are compared using a chi-square based test called the log-rank test and $P<0.05$ indicate that the survival functions of the two groups are significantly different.

REFERENCES

Rich JT et al. A practical guide to understanding Kaplan-Meier curves. Otolaryngol Head Neck Surg 2010;143:331-6.
Hou W and Carden D. Statistics for the nonstatistician: part II. South Med J 2012;105:131-5.
Ghasemi A and Zahediasl. Normality tests for statistical analysis: a guide for non-statisticians. Int J Endo Metab 2012;10:486-9.
Bland JM and Altman DG. Statistical methods for assessing agreement between two methods of clinical measurement. Lancet 1986;8476:307-10.

Systematic Reviews and Meta-analysis

Jan Sargeant DVM, PhD[1] and Annette O'Connor, BVSc, MVSC, DVSC, FANZCV[2] and
[1]Guelph, Ontario, Canada, [2]Ames, IA

INTRODUCTION

Systematic reviews are an increasingly common method of research synthesis in veterinary science. Meta-analysis is a statistical approach to combining data from multiple primary studies that can be used in conjunction with a systematic review. The goal of a systematic review is to address a very specific question. A further goal of a systematic review is to be transparent about identifying potential sources of bias in the primary research informing the review conclusion and the potential biases in the review methodology. This proceedings paper will briefly discuss the rationale for employing the systematic review methodology and discuss the explicit steps of a systematic review. The unique aspects of a systematic review are: 1) the framing of the review scope in an explicit question, 2) development of a protocol describing proposed conduct of the review, 3) the conduct of an extensive search designed to reduce the impact of retrieval bias, 4) assessment of the risk of bias of primary studies informing the review, 5) extraction of the explicit outcome from each relevant study and synthesis with meta-analysis if possible, and 6) assessment of the risk of bias in the review. These elements however are not unique to a systematic review; rather, it is the combination of these elements into one review that distinguishes the systematic review as such.

How does a systematic review differ from other research synthesis approaches?

Systematic reviews differ from other research synthesis approaches in that they focus on a particular question. Many reviews aim to be a broad overview of a topic. For example a review may aim to describe treatment options for a disease, the epidemiology of a disease or the mechanisms of disease pathogenesis. Such reviews often provide a useful overview or general concepts in an area and are sometimes referred to as narrative reviews. Alternatively, systematic reviews aim to address a particular question. The distinction is that the systematic review question can be envisaged as a primary study itself. For example, while it is not likely that an author would conduct a primary study that aims to describe the epidemiology of *Clostridium difficile* in foods and animals, it is likely an author would conduct a study to determine the association between *Clostridium difficile* and diarrhea in dogs. Similarly, while it is not possible to envision a primary study that would address treatment options and diagnosis of canine visceral leishmaniasis , it is possible to design a study that would calculate the sensitivity and specificity of a new assay for detection compared to the currently available assays . For the *Clostridium difficile* and canine visceral leishmaniasis examples, the latter studies presented have a testable hypothesis. It is easy to envision that the study could have been conducted by multiple researchers and those findings from multiple studies summarized by a systematic review with a meta-analysis.

What questions are suitable for systematic review?

Because of the emphasis of systematic reviews on questions with a narrow focus that have a testable hypothesis, there are four main styles of question that systematic reviews aim to address: 1) questions about effect of interventions, 2) questions about etiology, 3) questions about diagnostic test characteristics (sensitivity and specificity), and 4) questions about the incidence or prevalence of a characteristic such as a disease . This focus on specific questions also means that systematic reviews do not lend themselves to questions about lists i.e. what are all the tests available for diagnosis of canine visceral leishmaniasis. Such questions do require a review of the literature but do not use the systematic review methodology, as there is no testable hypothesis.

The steps of systemic reviews

In the following sections, we briefly outline the steps in a systematic review; more detail about these steps is available in other publications

Pre-Step: Assemble a review team and develop a systematic review protocol

The development of an explicit protocol is a critical aspect of a systematic review. Frequently, narrative reviews do not start with an explicit question, but rather a theme or objective. This means that the focus of a narrative review can evolve over time dependent upon findings identified during the conduct of the review. As systematic reviews are focused it is necessary to define that scope at the start of the review. To define the scope, a review team needs to be assembled with the expertise needed to ensure the review is comprehensive and addresses sources of bias. Once the review team is selected, the approach to each step of the review can be designed and documented in the protocol. After development, the protocol should be peer-reviewed prior to the conduct of the review. Peer review should be conducted by content experts and systematic review experts. It is now possible to have review protocols peer-reviewed in veterinary science . The process for developing the protocol and registration should be included in the final report of the review .

Step 1: Define the review question

Systematic reviews should begin by defining explicitly the components of the scope based on the review question. For reviews about interventions the format to create the review question is summarized by the acronym PICO(S), which stands for P=Population, I=Intervention, C=Comparator, O=Outcome, and, optionally, S=Study design. For reviews about etiology, the format to create the review question is summarized by the acronym PECO(S): P=Population, E-Exposure, C=Comparator, O=Outcome, and, optionally, S=Study design . For reviews about diagnostic test accuracy (DTA) the format uses the acronym PIT: P = Population, I = Index test(s), and T=Target condition or disease . For reviews about disease burden (prevalence or incidence) the format uses the acronym PO: P=Population and O=Outcome . By defining each of these components, it will be clear to the end user what studies are relevant to the review question. This is the same as defining the eligibility criteria i.e. what is relevant is eligible.

Step 2: Conduct an extensive search for studies.

The aim of conducting an extensive search is to ensure as many relevant results as possible are included in the review. The rationale is to reduce the bias associated with the accessibility of studies based on their outcome, sometimes called retrieval bias. It may be necessary to consult with an information scientist familiar with systematic reviews to design a comprehensive search . The search should be designed based on some or all of the concepts included in the review question (i.e. the PICO(S), PECO(S), PO(S) or PIT(S) components). It is not always necessary to include all components in the search. For example, frequently the outcomes are not included in the search for reviews of interventions, as these may not be explicitly reported in the abstracts of all studies. The aim is to design a search that will capture as many relevant studies as possible (high sensitivity) . More details about performing the search are available . Approaches to reporting the search for a systematic review are described in the PRISMA statement .

Step 3: Selecting relevant studies from the results of the search

Once the citations have been retrieved it is necessary to evaluate them and identify those relevant to the review. This step is called screening because the aim is to screen out non-relevant studies. This study selection process usually has two steps, a preliminary rapid step followed by a closer evaluation of potentially relevant studies. The 1st level of screening is usually based on the title and abstract, and the aim is to determine if a paper is likely to be relevant. Often a 2nd step is included that is based on the full text and is usually sufficient to determine if the paper is relevant.

Step 4: Collecting data from relevant studies

Once papers that are relevant to the review have been identified, the next step in a systematic review is the extraction of data from the relevant studies. The explicit extraction of data from relevant studies is one of the key features that differentiates systematic reviews from narrative reviews. Where a narrative review might explain that several papers found "a protective association", a systematic review would publish the estimated effect reported for each study (provided it was published in the original study). Traditionally, the data extracted for each paper is 1) the study population characteristics including sources of heterogeneity such as breed, age etc., 2) the intervention or exposure or assay information if applicable, and 3) the outcome. .

Step 5: Assess the *risk of bias* in relevant studies

Another key feature of systematic reviews is the assessment of the risk of bias for the studies included in the review. The rationale for assessing and reporting the risk of bias is that it enables end users to understand the potential and direction of biases that occur. For PICO(S) and PECO(S) questions, the reporting guidelines checklists such as the CONSORT statement or the REFLECT statement should not be employed as a risk of bias tool. These reporting guidelines provide a yes or no answer about the presence of absence of design features; however, this does not equate to risk of bias. The Cochrane Handbook of Systematic Reviews for Interventions risk of bias tool is best employed with little modification for interventions . This tool should can likely be applied to animal health systematic reviews with little or no modification. One possible modification that is required includes accounting for the impact of non-independence of populations in the analysis.

For diagnostic test evaluation, the STARD statement is a reporting guideline that can be used to learn more about sources of bias in diagnostic tests . The QUADAS statement is designed as a quality assessment tool of diagnostic test assessments.

Step 6: Synthesize the results

Having extracted the data about the results, study characteristics, and potential sources of contextual or methodological heterogeneity, the next step is to synthesize the results. If the data are amenable to quantitative analysis, a meta-analysis can be conducted, although a meta-analysis is not always a component of a systematic review. Meta-analysis involves assessing heterogeneity and calculating the summary effect measure if appropriate. For a meta-analysis to be conducted, there needs to be a reasonable assumption that studies are measuring the same metric. Figure 1 provides an example of a forest plot, an approach to presenting the results of a meta-analysis. Frequently in animal health, animal welfare and food safety it is not possible to combine the results in a meta-analysis. This often occurs because of poor reporting or differences in metrics used to measure the outcomes. In this situation a quantitative synthesis will not occur and the presentation of the results and discussion takes the place of this step.

Step 7: Presenting the results

The presentation of the results of the review should include the following components:

- The results of search and study selection
- Summary information about the characteristics of the studies identified as relevant to the review (including those that could not be included in a meta-analysis.)
- Risk of bias assessment for the individual studies
- Outcomes reported by the relevant studies
- Meta-analysis results including subgroup analysis and/or meta-regression if conducted
- Risk of bias across the studies

Step 8: Interpreting the results and discussion

The conclusions about the results of the review and the potential biases of the review should be discussed. The PRIMSA guidelines for reporting systematic reviews suggest the following aspects be covered: summarize the evidence, discuss the limitations of the review and provide conclusions. The discussion of the impact should include the expected direction of bias rather than mere mention that such biases could occur. The PRISMA statement and its accompanying explanation and elaboration document should be strictly adhered to when reporting all the steps of the systematic review.

Figure 1: A forest plot of hypothetical data

Study	Experimental Events	Experimental Total	Control Events	Control Total	OR	95%-CI	W(fixed)	W(random)
1	25	55	35	50	0.36	[0.16; 0.80]	9.2%	7.5%
2	35	60	42	59	0.57	[0.26; 1.21]	8.1%	8.3%
3	25	59	35	60	0.53	[0.25; 1.09]	9.2%	9.0%
4	34	49	38	50	0.72	[0.29; 1.74]	5.3%	6.2%
5	45	67	42	67	1.22	[0.60; 2.48]	6.4%	9.4%
6	33	56	44	60	0.52	[0.24; 1.14]	8.0%	7.9%
7	32	50	38	51	0.61	[0.26; 1.43]	6.2%	6.7%
8	25	60	39	60	0.38	[0.18; 0.80]	10.5%	8.8%
9	18	52	24	53	0.64	[0.29; 1.40]	7.2%	7.8%
10	22	51	24	53	0.92	[0.42; 1.99]	6.2%	8.0%
11	20	50	36	50	0.26	[0.11; 0.60]	10.0%	6.9%
12	25	50	35	50	0.43	[0.19; 0.97]	8.1%	7.2%
13	34	50	38	50	0.67	[0.28; 1.62]	5.6%	6.3%
Fixed effect model		**709**		**713**	**0.56**	**[0.45; 0.70]**	**100%**	**--**
Random effects model					**0.56**	**[0.45; 0.71]**	**--**	**100%**

Heterogeneity: I-squared=5.2%, tau-squared=0.0089, p=0.3946

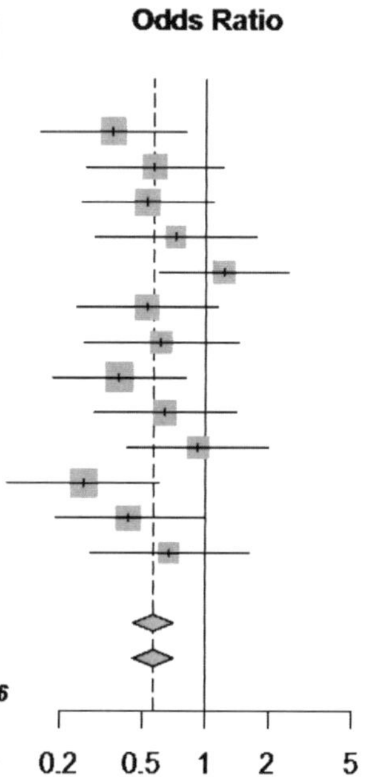

REFERENCES

Rodriguez-Palacios, A., et al., **Anim Health Res Rev,** 2013, 14 (1): Wetterwik, K. J., et al., **Acta Vet Scand,** 2013, 55 ; Noli, C., et al., **Vet J,** 2014, 202 (3):Laurenti, M. D., et al., **Vet Parasitol,** 2014, 205 (3-4):; O'Connor, A. M., et al., **Animal Health Research Reviews,** 2014, 15 (01):3; Sargeant, J. M., et al., **Zoonoses Public Health,** 2014, 61 Suppl 1 2; O'Connor, A. M., et al., **Zoonoses Public Health,** 2014, 61 Suppl 1 52; EFSA, **EFSA Journal,** 2010, 8 (6):90 O'Connor, A. M., et al., **Animal Health Research Reviews,** 2014, 15 (01):1; Khangura, S., et al., **Syst Rev,** 2012, 1 (1):Deeks, J. J., **BMJ,** 2001, 323 (7305):O'Connor, A. M., et al., **Zoonoses Public Health,** 2014, 61 Suppl 1 28; Grindlay, D. J., et al., **J Vet Med Educ,** 2012, 39 (4):; Moher, D., et al., **J Clin Epidemiol,** 2009, 62 (10):1006; Khan, K. S., et al., **Evid Based Med,** 2012, ; Schulz, K. F., et al., **Obstet Gynecol,** 2010, 115 (5):1063; O'Connor, A. M., et al., **Zoonoses Public Health,** 2010, 57 (2):95; Sargeant, J. M., et al., **Zoonoses Public Health,** 2010, 57 (2):105; Bossuyt, P. M., et al., **Clin Chem,** 2003, 49 (1):1; Christopher, M. M., **Vet Clin Pathol,** 2007, 36 (1):6; Liberati, A., et al., **J Clin Epidemiol,** 2009, 62 (10):1

Clinical Trial Design and Reporting

Annette O'Connor, BVSc, MVSC, DVSC, FANZCV[1] and Jan Sargeant DVM, PhD[2]
[1]Ames, IA [2]Guelph, Ontario, Canada

INTRODUCTION

Clinical trials are a unique study design used to assess the efficacy of interventions. Although clinical trials can be used to assess any intervention, in clinical veterinary medicine trials are most frequently conducted to assess either preventive interventions or therapeutic interventions. In this paper we briefly introduce clinical trials as a design, key features, rationale for conduct, analyzing and reporting clinical trials. There are numerous resources for the design, conduct, analysis and reporting of trials.

The aim of clinical trials

Clinical trials aim to assess comparative efficacy in interventions. In veterinary clinical medicine, there are generally two types of interventions; preventive and therapeutic. For preventive interventions, the aim is to assess if the intervention can prevent the development of an outcome. Clinical trials of preventive interventions such as vaccines enroll a population that is disease free at the time of enrollment but at risk of developing the outcome. The study population is watched over time for the development of the outcome. For therapeutic interventions, the aim is still to assess if the intervention affects the occurrence of an outcome. However, in a trial of a therapeutic intervention, the diseased individuals are enrolled and the outcome is either an adverse event or a positive event such as death, remission or recovery

For all clinical trials it is necessary to decide what metric will be used to compare the groups. The outcome measured in clinical trials can be continuous such as weight, categorical such as disease occurrence or rates such as survival. For categorical outcomes, it is important to clarify the event of interest because the event of interest may be defined as a positive outcome (survival or remission) or a negative outcome (death). In clinical trials of preventive interventions, it is most common to define the event as the occurrence of a negative outcome (or adverse event). The working hypothesis for most trials of preventive interventions would be that the "new" intervention would protect against the event and therefore reduce the occurrence of the adverse event. For a therapeutic trial, it is more common to describe the event as a positive outcome (or a favorable event). The working hypothesis for most trials of therapeutic interventions would be that the "new" intervention would increase the occurrence of the favorable event.

What are the features of a clinical trial?

Clinical trials share many design features with other designs; it is the combination of features that makes a clinical trial.

- Clinical trials compare an outcome across two or more groups (often referred to as arms) and
- The groups receive different interventions which may / may not include a placebo group and
- The units of concern are randomly allocated to the groups (with or without randomization restrictions) and
- blinded assessment of the outcome is used where ever possible

in addition

- For therapeutic trials, the study population develops the disease naturally (not a disease model) or
- For preventive trials, the study population is exposed to the disease naturally (no disease challenge)

Why use clinical trials to assess the efficacy of interventions?

Clinical trials compare interventions by evaluating differences in-group characteristics such as group means, proportions or average survival times. Other study designs such as observational studies and experiments can also be used to compare interventions. However clinical trials have a unique role because they are often more externally valid that experiments and more internally valid than observational studies, and therefore provide the best evidence for the efficacy of interventions.

Impact of external validity and internal validity when comparing interventions?

External validity is a concept that relates to how realistically the study population refers to the target population. Veterinarians would like data from studies that assess the intervention in patients that they expect to be presented with in practice. Experiments often have poor external validity because of the study population and the disease model used. In experiments, the study population is often an experimental animal. For example, experiments may use purpose-bred beagles or colostrum derived piglets, which lack the variety of of breeds, weights and concurrent diseases experience that veterinarians see in practice. More importantly, the approach to reproducing the disease or exposure used in experiments may not realistically represent the disease observed in practice. For example, an experiment that models bovine respiratory disease by injecting high doses of *Mannheimia hemolytica* into the trachea of 2-month-old colostrum derived calves is likely a poor model of bovine respiratory disease in modern feedlots. Internal validity refers to the absence of errors in the comparison of the interventions made within the study. Internal validity is affected by random error and systematic biases; confounding, information bias and selection bias. The internal validity of experiments is often impacted by random error, as experiments used many design tools to control for systematic errors. Despite few issues with internal validity, the lack of external validity implies it is difficult to expect the intervention to behave in the "real" world as it did in the experiment.

Alternatively, observational studies often have strong external validity because they are often conducted on patients in practices using records obtained from practice. The major issue with observational studies is internal validity, which is discussed below. It is beyond the scope of the article to discuss all aspects of bias, but the major issue with using observational studies to assess the efficacy of interventions is confounding by indication. Often animals received an intervention because they have a characteristic, and if that characteristic is associated with the outcome, it will not be possible to attribute any differences in group outcome to the intervention. For example, if equine sarcoids can be treated by cryofreezing or surgery, but surgery is reserved for cases that are located in readily accessible places such as the flank, then an observational study that compared successful remission of sarcoids and found surgery more effective than cryofreezing is confounded by location. It is unclear if cryofreezing had been used on flank lesions, that the same remission rate would have occurred. Clinical trials use randomization to address confounding by indication.

Clinical trials are designed to address many issues that affect experiments and observational studies, but here we have only room to discuss two: external validity and confounding. With respect to external validity, clinical trials lie between experiments and observational studies. The extent to which trial populations reflect practice populations is affected by exclusion criteria. Some clinical trials use very restrictive criteria for enrollment in the trial, other trials have very few eligibility criteria and the study populations of a trial are very reflective of target populations. For example, trials of bone cancer therapy in dogs may restrict the population to those without any renal disease or cardiac disease, and these concurrent diseases may occur quite commonly with bone cancer patients in practice. In this situation it is difficult to know if the results of a trial of bone cancer therapy in the restrictive dog population apply to the "real" population. Alternatively, trials of the effect of antibiotics in beef cattle in feedlots often have very few restrictive criteria and generally reflect the cattle that practitioners would expect to treat.

The other major issue discussed in confounding, i.e., the uneven distribution of a third characteristic that is associated with the outcome across the intervention groups. Clinical trial control this issue by ensuring that random allocation methods are used to allocated animals to the treatment groups. The advantage of random allocation is that it controls both none and unknown confounders provide the sample size is sufficient. There are many approaches to randomization and restricted randomization can be used to increase the power of a study, and may be very helpful in trials that are likely to be small. In summary, the main reason to prefer information about the effect of interventions obtain from clinical trials is that "on average" well executed clinical trials have better external validity than experiments and better internal validity than observational studies and combined these factors suggest that clinical trials provides the best evidence of the effect of an intervention. .

Assessing the intervention – calculating the effect size for continuous outcomes

The effect size is the most important outcome of the trial; it is the measure of the comparison. The effect size is a summary of the comparison and describes by how much the interventions differ for the outcome. For continuous outcomes, the effect size is usually calculated as the difference in the average of the groups. If the intervention has no effect then the difference in the averages would be zero (note some differences will occur due to random sampling error). The effect size is always reported with a measure of precision such as a standard error or a confidence interval. For example, consider a 2-arm clinical trial was conducted that compared the weight gain of pigs that received a vaccine compared to those that received the placebo. The investigators measure weight gain in each pig, and then calculate the average weight of pigs in each group. The vaccinated group has 35 pigs with a mean weight gain of 680 grams per day with a standard deviation of 20. The unvaccinated group has 35 pigs with a mean weight gain of 668 grams per day with a standard deviation of 21. For such a trial the effect size is 12 grams i.e., 680-668. The 95% confidence interval for the mean difference is 2.22 to 21.8. This wide confidence interval suggest that we know the effect size with uncertainty, which is not surprising given the variation and the small group sizes. It is also possible to test if the effect size is likely to come from a population with mean zero (i.e the vaccine has no effect). The t-test p value for this test is 0.016 and we reject the null hypothesis that the mean difference is zero. Knowing the magnitude of the effect size and its precision is more informative than the p value and hence the effect size and measure of precision should always be reported.

Assessing the intervention – the effect size for categorical outcomes

For trials that have categorical outcomes, the effect size is calculated as a ratio of proportions rather than the difference in proportions. If the intervention has no effect, then the proportion will be the same in each group (some differences will occur due to random sampling error) and the ratio will be zero. The rationale for using the ratio of proportions rather than the difference in proportions is based on statistical properties. For example, consider a 2-arm clinical trial was conducted that compared post surgery for dogs that received surgery versus dogs that received hydrotherapy. The investigators measure the outcome returns to full function by 3 months post surgery, and then calculate the proportion of dogs that meet this definition in the two groups. The surgery group has 30 dogs and 10 meet the definition of recovery by 3 months (10/30). The hydrotherapy group has 20 pigs and 7 meet the definition of recovery by 3 months (7/30). The ratio of these proportions is 1.28, and the 95% confidence interval is 0.5-2.9. This effect size (1.28) suggests that recovery was higher in the surgery group, however we know this number with great uncertainty, and the number 1 which suggest no effect is included in the confidence interval, as are numbers less than one. Therefore overall we can't reach a conclusion about this intervention. The p value for these data using uncorrected 2 sided Yates chi-square test is 0.7486, suggesting there is no reason to reject the null hypothesis that the ratio is 1.

Reporting clinical trials.

Clinical trials represent a risk to participants, if one treatment is better than the other the owner and animal risk not receiving that treatment. Instead of participating in a trial the owner of an animal can choose to not participate, and direct the practitioner to give the patient the best treatment in the practitioner opinion. By participating in a trial owners of animals are choosing to contribute to the knowledge of the community rather than selecting an option entirely in their self interest. Such consent, to participate (at some even minimal risk) means that investigators who conduct clinical trials have ethical responsibility with respect to the results of clinical trials. As trials are designed to contribute to the overall level of knowledge, principle investigators must fulfill their responsibility to make the results of the trial available in a manner that enables complete and comprehensive understanding of the trial. Fortunately, several guidelines are available to increase the potential for comprehensive reporting .

Table 1: Reflect statement checklist items for reporting interventions studies

Paper Section And topic	Item	Descriptor of REFLECT statement item
Title & Abstract	1	How study units were allocated to interventions (eg, "random allocation," "randomized," or "randomly assigned"). Clearly state whether the outcome was the result of natural exposure or was the result of a deliberate agent challenge.
Introduction Background	2	Scientific background and explanation of rationale.
Methods Participants	3	Eligibility criteria for owner/managers and study units at each level of the organizational structure, and the settings and locations where the data were collected.
Interventions	4	Precise details of the interventions intended for each group, the level at which the intervention was allocated, and how and when interventions were actually administered.
	4b	Precise details of the agent and the challenge model, if a challenge study design was used.
Objectives	5	Specific objectives and hypotheses. Clearly state primary and secondary objectives (if applicable).
Outcomes	6	Clearly defined primary and secondary outcome measures and the levels at which they were measured, and, when applicable, any methods used to enhance the quality of measurements (eg, multiple observations, training of assessors).
Sample size	7	How sample size was determined and, when applicable, explanation of any interim analyses and stopping rules. Sample-size considerations should include sample-size determinations at each level of the organizational structure and the assumptions used to account for any non-independence among groups or individuals within a group.
Randomization -- Sequence generation	8	Method used to generate the random allocation sequence at the relevant level of the organizational structure, including details of any restrictions (eg, blocking, stratification)
Randomization -- Allocation concealment	9	Method used to implement the random allocation sequence at the relevant level of the organizational structure, (eg, numbered containers), clarifying whether the sequence was concealed until interventions were assigned.
Randomization -- Implementation	10	Who generated the allocation sequence, who enrolled study units, and who assigned study units to their groups at the relevant level of the organizational structure.
Blinding (masking)	11	Whether or not those administering the interventions, caregivers and those assessing the outcomes were blinded to group assignment. If done, how the success of blinding was evaluated. Provide justification for not using blinding if it was not used.
Statistical methods	12	Statistical methods used to compare groups for all outcome(s); Clearly state the level of statistical analysis and methods used to account for the organizational structure, where applicable; methods for additional analyses, such as subgroup analyses and adjusted analyses.
Results Study flow	13	Flow of study units through each stage for each level of the organization structure of the study (a diagram is strongly recommended). Specifically, for each group, report the numbers of study units randomly assigned, receiving intended treatment, completing the study protocol, and analyzed for the primary outcome. Describe protocol deviations from study as planned, together with reasons.
Recruitment	14	Dates defining the periods of recruitment and follow-up.
Baseline data	15	Baseline demographic and clinical characteristics of each group, explicitly providing information for each relevant level of the organizational structure. Data should be reported in such a way that secondary analysis, such as risk assessment, is possible.
Numbers analyzed	16	Number of study units (denominator) in each group included in each analysis and whether the analysis was by "intention-to-treat." State the results in absolute numbers when feasible (eg, 10/20, not 50%).
Outcomes and estimation	17	For each primary and secondary outcome, a summary of results for each group, accounting for each relevant level of the organizational structure, and the estimated effect size and its precision (e.g., 95% confidence interval)
Ancillary analyses	18	Address multiplicity by reporting any other analyses performed, including subgroup analyses and adjusted analyses, indicating those pre-specified and those exploratory.
Adverse events	19	All important adverse events or side effects in each intervention group.
Discussion Interpretation	20	Interpretation of the results, taking into account study hypotheses, sources of potential bias or imprecision, and the dangers associated with multiplicity of analyses and outcomes. Where relevant, a discussion of herd immunity should be included. If applicable, a discussion of the relevance of the disease challenge should be included.
Generalizability	21	Generalizability (external validity) of the trial findings.
Overall evidence	22	General interpretation of the results in the context of current evidence.

REFERENCES

Friedmen, L. M., et al., **Fundamentals of Clinical Trials, Fourth Edition,** 2010, O'Connor, A. M., et al., **Zoonoses Public Health,** 2010, 57 (2):95; Sargeant, J. M., et al., **Zoonoses Public Health,** 2010, 57 (2):105; Schulz, K. F., et al., **Obstet Gynecol,** 2010, 115 (5):1063; Moher, D., et al., **J Clin Epidemiol,** 2010

MENTORING THE MENTORS

Michelle Henry Barton DVM, PhD, DACVIM (LAIM)
Athens, GA

INTRODUCTION

Supervising Diplomates, residency training facilities, and the American College of Internal Medicine (ACVIM), invest a considerable amount of time, effort, and resources in training candidates to become board certified specialists in internal medicine. Likewise, the ACVIM has a vested interest in the success of our candidates, as they will direct the future of our College. In the spirit of welcoming assistance, the specialty of large animal internal medicine has been the first specialty under the umbrella of the ACVIM to have an ombudsman. Appointed in the fall of 2012, the ombudsman's main role is to serve as a facilitator and an arbiter for candidates engaged in residency training and for those who have finished their training, but have not completed the steps necessary to obtain Diplomate status. The word ombudsman has a Swedish origin, meaning "trusty manager." A key message is that the ombudsman is not a "spy!" The ombudsman is to serve as a communication link between candidates, irrespective of their home training facility, and the ACVIM. With no such previous link, the ACVIM found that it could not "fix" or address issues that they were unaware existed.

Several keys steps followed in attempt to open up lines of communication. A candidate listserv was created on ACVIM360, of which the ombudsman has access to all posts. This enables the ombudsman to screen questions or identify concerns posted by individuals. It also serves as a key method by which the ACVIM office, President of the Specialty of LAIM, and the ombudsman can post reminders, updates, or notices of any kind to the candidates. On the ACVIM's homepage for candidates, the purpose of the ombudsman is listed and includes contact information. This particular resource has been very helpful as numerous individual candidates have been able to directly contact the ombudsman with more personal questions or concerns that might perhaps involve their individual training program. It is important for Diplomates to appreciate that all attempts are made to facilitate while attempting to enable the candidate to maintain anonymity with respect to their training facility or direct mentors. To date, no individual supervising Diplomates have been revealed, either by a candidate or by the ombudsman. The roles that the ombudsman serves is to provide counsel on general governing rules providing by the ACVIM (such as the General Information Guide), suggestions on potential paths of action or solutions, or to provide and facilitate contact with individuals that might be better suited to assist. It is not uncommon for a candidate to be unsure of what components of a training program the ACVIM would have "jurisdiction" over and what parts would be under the jurisdiction of the policies of the sponsoring program. Two additional resources that have been provided at the ACVIM Forum for candidates have been open group "resident chat sessions" with the ombudsman (2013 and 2014) and a mentorship workshop for LAIM candidates facilitated by Drs. Julie Carey and Nicki Wise at the 2014 Forum. All of these resources have provided valuable feedback from candidates. This feedback has not only highlighted common concerns among candidates, it has opened the door of communication on how some of the most pressing concerns could be addressed. To engage a larger population of residents and candidates, in the fall of 2014, the ACVIM helped organize an online survey including 32 multiple-choice questions using a likert scale and two open-ended questions. All residents actively enrolled in a registered ACVIM LAIM program and candidates or Diplomates that completed their program within the last three years were solicited via email to participate. Candidates and residents were also solicited to participate via the ACVIM360 candidate listserv and Diplomate business listserv. Supervising Diplomates were also solicited via the ACVIM360 business listserv to encourage their residents to complete the survey. As incentive for participation, all that completed the survey were enrolled in a drawing for free registration at the 2015 ACVIM Forum. The questions were provided via Survey Monkey, which was monitored by an ACVIM staff member. Anonymity was maintained, unless a candidate elected to reveal the location of their program.

The main purpose of this session is not to provide you with the ombudsman's view on what is good mentorship. Although I have mentored candidates for over 25 years, I do not have any particular expertise on mentorship. Rather, I will share with the audience the responses that we collectively received from the survey on mentorship. In other words, the mentees will reveal to us how "good" of a job we are doing with mentorship and providing training. It is anticipated that our strengths and weaknesses will be revealed and what our candidates are seeking in mentorship in their programs will be made known.

Survey Results

Participation in the survey was excellent! Of approximately 180 candidates or recent Diplomates, 119 (66%) completed the survey. Fifty-four of the responses were from individuals that finished their program in the last three years (52% board certified) and 65 were from candidates still within a residency program (34% year one, 28% year 2, and 38% year three). From all respondents, 44% were equine emphasis, 16% food animal and 40% mixed large animal. Anticipating that there may be considerable variation in the responses to some of questions on mentorship, candidates were asked to anonymously rate individual mentors in their program. For those questions, the mentors were simply listed as A, B, C, D, etc. The responses were therefore tallied against the total number of mentors being evaluated (for a total of between 381 to 388 mentor evaluations). An asterisk indicates these questions.

The expectations of my training and responsibilities as a resident were made clear and I could comfortably ask for clarification if I was unsure. Strongly agree: 22% Agree: 54% Neutral: 6% Disagree: 14% Strongly disagree: 3%

The experience and training that I had during my residency program met my expectations as described by the VIRMP or the advertisement that the program was given.
Strongly agree: 20% Agree: 48% Neutral: 14% Disagree: 14% Strongly disagree: 3%

At least one of my mentors was aware of the ACVIM training requirements and served as a solid reference for all residents in the program. Strongly agree: 36% Agree: 49% Neutral: 5% Disagree: 8% Strongly disagree: 1%

The caseload and intensity or complexity of the caseload are adequate for training internal medicine.
Strongly agree: 28% Agree: 41% Neutral: 15% Disagree: 13% Strongly disagree: 2%

Case discussion rounds without students are routinely scheduled and attended by mentors.
Strongly agree: 29% Agree: 34% Neutral: 8% Disagree: 13% Strongly disagree: 13%

Discussion on cases that is inclusive, from pathophysiology, to clinical signs, to current literature findings, and case management, routinely occurs with my mentor.(out of a total of 381)*
Strongly agree: 146 (38%) Agree: 106 (28%) Neutral: 62 (16%) Disagree: 47 (12%) Strongly disagree: 25 (6%)

*Week**day** case rounds are routinely attended by my mentor.*(out of a total of 387)*
Strongly agree: 211 (54%) Agree: 107 (28%) Neutral: 36 (9%) Disagree: 32 (8%) Strongly disagree: 1 (0%)

*Week**end** case rounds are routinely attended by my mentor.*(out of a total of 381)*
Strongly agree: 112 (29%) Agree: 113 (30%) Neutral: 64 (17%) Disagree: 63 (16%) Strongly disagree: 39 (10%)

When performing procedures, my mentor is routinely willing to be present until I am confident with the skill.(out of a total of 382)*
Strongly agree: 196 (51%) Agree: 136 (36%) Neutral: 30 (8%) Disagree: 15 (4%) Strongly disagree: 4 (1%)

I feel confident with my mentor's clinical abilities.(out of a total of 385)*
Strongly agree: 231 (60%) Agree: 106 (28%) Neutral: 30 (8%) Disagree: 29 (8%) Strongly disagree: 9 (2%)

Topical review sessions, discussions, or classes are available for board review and are helpful.
Strongly agree: 26% Agree: 28% Neutral: 13% Disagree: 16% Strongly disagree: 14%

Journal Club is routinely scheduled to meet the GIG requirements of 80 sessions in a 3 year residency.
Strongly agree: 57% Agree: 35% Neutral: 3% Disagree: 2% Strongly disagree: 0%

Journal Club is routinely attended by my mentor.(out of 387 total)*
Strongly agree: 147 (38%) Agree: 106 (30%) Neutral: 36 (9%) Disagree: 70 (18%) Strongly disagree: 25 (6%)

I am provided instruction or adequate mentorship in scientific writing.
Strongly agree: 18% Agree: 36% Neutral: 20% Disagree: 19% Strongly disagree: 5%

I am given sufficient time away from clinic duty for ACVIM General and Specialty Exam preparation.
Strongly agree: 21% Agree: 43% Neutral: 18% Disagree: 12% Strongly disagree: 6%

I was routinely provided with verbal or written feedback on my performance, strengths and weaknesses.
Strongly agree: 25% Agree: 47% Neutral: 16% Disagree: 9% Strongly disagree: 2%

I feel that I could go to or call my mentor at any time to ask for help on a case or to ask for personal advice.(out of a total of 386)*
Strongly agree: 199 (51%) Agree: 94 (24%) Neutral: 39 (10%) Disagree: 35 (9%) Strongly disagree: 19 (5%)

I do not feel ridiculed by the actions or words of my mentor.(out of a total of 386)*
Strongly agree: 185 (52%) Agree: 127 (33%) Neutral: 24 (6%) Disagree: 30 (8%) Strongly disagree: 20 (5%)

In my program, I had access to a professional that could help deal with stressful situations, fatigue, or other difficult situations and I felt comfortable accessing these resources. (A similar question was asked with respect to mentors and the responses were similar.)
Strongly agree: 13% Agree: 25% Neutral: 25% Disagree: 26% Strongly disagree: 10%

I can tell that my mentor truly cares about my personal as well as my professional development.(out of a total of 388)*
Strongly agree: 177 (46%) Agree: 113 (29%) Neutral: 48 (12%) Disagree: 28 (7%) Strongly disagree: 22 (6%)

There is generally the right balance between good mentorship with appropriate support and the opportunity to be independent.(384)*
Strongly agree: 124 (32%) Agree: 146 (38%) Neutral: 55 (14%) Disagree: 47 (12%) Strongly disagree: 11 (3%)

Considering the overall quality of the caseload, training experience, and mentorship, I would recommend my program to others seeking internal medicine training.

Strongly agree: 28% Agree: 42% Neutral: 15% Disagree: 10% Strongly disagree: 6%

Open Ended Survey Questions

Candidates were also given an opportunity to provide open-ended feedback with two questions. There was a wide variety of responses to these two questions, but here is a sampling.

First open-ended question: "**Given the opportunity to provide constructive feedback to my mentors, I would advise them to**:"

- *"To be that mentor that their residents will think back on frequently, want to emulate in their future dealings with interns/residents/students and call for help with a tricky case or situation, rather than that boss that you make awkward small talk with at future meetings."*
- *"Please realize how influential you are and the great power and responsibility you have to be a positive influence in the lives of the next generation of ACVIM Diplomates."*
- *"Understand that they are mentors; this is a valuable and life changing position."*
- *"Praise more, critique less."*
- *"I would advise mentors to remember that it can sometimes be difficult to take the initiative to ask for constructive feedback or to help with a problem if you are in a junior role."*
- *"Although I felt ready to handle any emergency when it walked through the door, I didn't feel at all ready to discuss the underlying disease mechanisms, etc., related to the disease processes and treatment options." "Provide some structured time for pathophysiologic discussions with regards to cases." "I would have liked to be challenged more from you on a daily basis with questions regarding pathophysiology, etc."*
- *"Mentors need to be the advocate for the resident when administrators do not understand the unique requirements for residency programs."*
- *"Care more about their residents."*
- There were several comments from residents that they would like more help with preparation for the exams or case report selection.

The second open-ended questions was" "**Given the opportunity to provide constructive feedback to the ACVIM regarding my program I would say:"**

- *"ACVIM should promote more mobility of the resident through different hospitals/programs."*
- *"Costs for attending annual conference beyond the budget of most residents."*
- *"ACVIM should intervene to limit the number of candidates."*
- *"Making a ridiculously difficult exam at the END of a program that covers only minutia and is not a reflection of one's true internal medicine abilities is not a fair practice."*
- *"More oversight to ensure that universities/practices understand the requirements for residency programs."*

There were several comments on case report grading and help on case reports. There were also several comments on low caseload, providing mentorship to mentors, and to look at institutional performance records in terms of success of candidates becoming Diplomates.

Summary

Firstly, I would like to thank all of our candidates or recently board certified Diplomates for providing input on types of questions to include and for taking the time out of your busy schedules to participate in the survey. In addition to this presentation, I will provide the full survey results (minus institutional identification) on the ACVIM website for both candidates and Diplomates. I also plan to provide the survey results to key members of our specialty, including the LAIM President and Executive Committee, and the chairs of the Residency Training (RTC) and Credentials Committee. From my interactions with candidates and from the results of this survey, overall, we are doing a good job, but we can do better! The message I am hearing is that residents want their mentors to care, to be engaged in all aspects of their training, to provide constructive feedback, and to help them know how to navigate through the process of becoming a specialist. From the survey responses, we need to provide clear expectations of candidates in training, both mentors and mentees need to be abreast of the credentialing requirements, case rounds and journal club should be attended by mentors, candidates need help in preparing case reports and examinations, and we need to provide methods by which candidates within individual programs can have access to trained professionals if help is needed with conflict resolution or stress.

Some of these areas are already being addressed. The Residency Training and Credentials Committees have extensively revised the requirements for annual program renewal, including information on caseload, evaluations, and how much time the candidate has been under supervision in the clinic. These committees will also be providing simple annual credential checklists for candidates and mentors alike. Input is being collected from candidates on ACVIM360 on case reports concerns. I would like to charge the specialty to provide an outline of the details that should be included in VIRMP descriptions of programs and to encourage all seeking residencies to contact current or former candidates in programs to gain insight in the ranking process. Through annual exit surveys, perhaps similar to the one included here, we can follow our progress on these areas and other areas of weakness. We should consider making an outcome assessment database by program that would be accessible to the RTC and perhaps to candidates seeking programs. Although this concept might make us uneasy, performance of academic mentors frequently includes student teaching or service evaluations. Why would we not look at similar assessments for accreditation for training programs for residents?

INTRACRANIAL EEG - DRUG EFFECTS, SEIZURE DETECTION AND PREDICTION IN DOGS AND PEOPLE

Edward (Ned) E. Patterson DVM, PhD, DACVIM (SAIM)
St. Paul, Minnesota

INTRODUCTION

Seizure detection and prediction (forecasting) has the potential to create new therapeutic strategies for epilepsy, e.g. providing patient warnings and delivering preemptive therapy to stop seizures as soon as possible (detection) and/or before the seizure occurs via forecasting. Progress on seizure forecasting, however, has been hindered by lack of sufficient data to rigorously evaluate the hypothesis that epileptic seizures are preceded by measurable physiological changes, and are not simply random events. EEG features for prediction currently being evaluated include but are not limited to: spectral power in band (PIB), inter-ictal spikes, and high frequency oscillations.

A trans-disciplinary team of veterinarians, pharmacologists, neurologists, and biomedical engineers have investigated seizure detection and forecasting in 5 dogs with naturally occurring focal epilepsy implanted with a device recording continuous intracranial EEG (iEEG). The device was manufactured by NeuroVista Inc., though the company is now defunct with intellectual property rights residing with Cyberonics Inc. The iEEG spectral power in six frequency bands: delta (0.1-4 Hz), theta (4-8 Hz), alpha (8-12 Hz), beta (12-30 Hz), low-gamma (30-70 Hz), and high-gamma (70-180 Hz), were used as electrophysiological features[1] for seizure detection[2] and seizure prediction[3] in dogs.

SEIZURE DETECTION IN DOGS

Five dogs with naturally occurring epilepsy were implanted with the iEEG device that consists of 16 subdural EEG leads that wirelessly transmit continuous intracranial EEG (iEEG) to an external (page sized) device embedded with a seizure detection and/or prediction algorithms. The dogs were in a room equipped with 24 hour a day video monitoring. An automated alert by text message was sent if prolonged (> 5 mins) or repetitive seizures (> 3 seizures in 4 hours) occurred, and a rescue therapy protocol with 0.5mg/kg diazepam IV and/or 6mg/kg IV phenobarbital was implemented. The performance of the alert system was evaluated over 8 weeks. 3 of the 5 dogs experienced spontaneous unprovoked partial seizures that secondarily generalized and lasted 1–2 minutes. Three repetitive seizure episodes meeting alert criteria occurred in 2 of the 3 dogs. On each occasion, the alert successfully messaged the on call veterinarian who confirmed the seizure activity via remote video-monitoring, and rescue medication was administered. [2]

The seizure detection algorithm was based on a beta wavelength (12-30 Hz), spectral power in band (PIB) features. The performance of the algorithm was demonstrated to have high sensitivity (99 -100%) and specificity (91%). The average latency of detection based on expert visual review of seizure onset was 6.75 seconds. Rescue medication was administered within 15 min (range 2–15 min) of an alert. Seizure activity was aborted in each instance as determined by visual observation of motor seizures and confirmed by cessation of electrical seizure activity on iEEG. There were no additional seizures between the alert and drug administration. Phenobarbital concentrations at 30 min were 9.95 and 10.6 μg/mL. Diazepam concentrations were 0.8 and 1.1 μg/mL at 30 min. On one occasion, a dog had another seizure 1.5 h following diazepam administration. At the time of the additional seizure, the diazepam plasma concentration was approximately 0.35 μg/mL. IV Phenobarbital was then administered and the seizure activity was aborted. [2]

These results in 2 epileptic dogs with cluster seizures demonstrated the feasibility of a caregiver alert for prolonged or repetitive seizures, and enabling rescue medications to be delivered in a timely manner. Future (less invasive?) EEG based alert systems may improve the management of both canine and human epilepsy by alerting caregivers of seizures, enabling early intervention and the improvement of outcomes and quality of life of patients and caregivers.

SEIZURE PREDICTION (FORECASTING)

In People

NeuroVista Inc., along with collaborators concurrently and independently enrolled human patients with drug resistant, non–surgically resectable, epilepsy at three centers in Melbourne, Australia, from 2010-2011 and surgically implanted the exact same iEEG device in 15 people. [4] There was data collection phase, during which an algorithm for identification of periods of high, moderate, and low seizure likelihood was individually established for each patient. If the algorithm met performance criteria (ie, sensitivity of high-likelihood warnings greater than 65% and performance better than expected through chance prediction), there was an advisory phase and the patients received information about seizure likelihood. The device met advising criteria in 11 of the 15 patients with high likelihood estimated sensitivities ranging from 65% to 100%. Three patients' algorithms did not meet performance criteria and one patient required device removal because of an adverse event before sufficient training data was acquired.[4] This human study showed that iEEG monitoring and forecasting is feasible in human patients with drug-resistant epilepsy, though larger replication studies are needed.

In Dogs

Two of the 5 dogs implanted with the iEEG device had infrequent seizures that were insufficient for seizure forecasting. In the 3 of the implanted dogs, a total of 125 spontaneous seizures were detected spanning 6.5 to 15 months. In each of the 16 iEEG channels, the iEEG record was partitioned into non-overlapping 1-minute blocks, and the resulting power spectrum (0.1–200 Hz) divided into the 6 aforementioned frequency bands: Within each frequency band, the power was summed over band frequencies to produce a power-in-band (PIB) feature resulting in 96 PIB values (16 channels × 6 bands) Logistic regression classifiers were trained to discriminate labeled pre-ictal and inter-ictal blocks. Each classifier used 10 of 96 available PIB features individually selected for each dog during the training process via forward selection. For training purposes, blocks between 90 minutes preceding a seizure and the seizure itself were given a pre-ictal label, and all other blocks were labeled as inter-ictal.[3] The seizure forecasting algorithm triggered a seizure warning when the classifier output exceeded a defined seizure risk threshold. The threshold was chosen adaptively so that total time in warning approximately matched a predefined target time in warning. Once triggered, the warning persisted for a defined period of 90 minutes.

These canine seizure forecasting algorithms performed significantly better than a Poisson-model chance predictor constrained to have the same time in warning for a range of total warning times for all 3 dogs when considering all seizures with a majority of seizure prediction sensitivities ranging from 73.3 to 90.4%. When considering a subset of seizures separated by at least 4 hours (lead seizures), the performance remained better than chance for a subset of algorithm parameters,[3] but needs significant improvement as the overall sensitivity was low for these lead seizures. The results of the this prediction study demonstrated that seizures in naturally occurring canine epilepsy are not randomly occurring events, and highlights the feasibility of long-term seizure forecasting using iEEG monitoring.

American Epilepsy Society Seizure Prediction Challenge Competition

https://www.kaggle.com/c/seizure-prediction - The competition was sponsored by sponsored by the National Institutes of Health (NINDS), the Epilepsy Foundation, and the American Epilepsy Society. The iEEG data from the 3 dogs for seizure forecasting was made publically available for 2 months. In addition, datasets from human patients with epilepsy undergoing intracranial EEG monitoring to identify a region of brain that can be resected to prevent future seizures was also included, but this period is usually limited to a week or less.

The challenge was to distinguish between ten minute long data clips covering an hour prior to a seizure (pre-ictal), and ten minute iEEG clips of inter-ictal activity with a low false positive rate. For the contest only lead seizures, defined here as seizures occurring four hours or more after another seizure, were included in the training and testing data sets. Participants utilized the NIH-sponsored International Epilepsy Electrophysiology portal (http://ieeg.org) to review and download annotated inter-ictal and pre-ictal data from human patients and the dogs 504 teams consisting of 653 participants submitted 17,780 prediction algorithms. The winning team's http://blog.kaggle.com/2015/01/26/american-epilepsy-society-seizure-prediction-challenge/ best algorithm predicted 82% of lead seizures which was a substantial improvement over the original published results for lead seizures in these same dogs.[3] The results of this contest show the power or large datasets being made publically available and crowd-sourcing enabling the substantial improvement of complex algorithms.

REFERENCES

[1]Davis KA, Sturges BK, Vite CH, Ruedebusch V, Worrell G, Gardner AB, Leyde K, Sheffield WD, Litt B. A novel implanted device to wirelessly record and analyze continuous intracranial canine EEG (2011). Epilepsy Research 96: pp. 116-122

[2]Coles LD, Patterson EE, Sheffield WD, Mavoori J, Higgins J, Michael B, Leyde K, Cloyd JC, Litt B, Vite C, Worrell GA. (2013) Feasibility of a caregiver alert system in canine epilepsy. Epilepsy Res. Oct;106(3):456-60

[3]Howbert JJ, Patterson EE, Stead MS, Vasoli V, Crepeau D, Vite CH, Sturges B, Ruedebusch V, Mavoori J, Leyde K, W, Sheffield WD, Litt B, Worrell GA (2014). Forecasting seizures in dogs with naturally occurring epilepsy. PLOS One. Jan 8;9(1):e81920

[4]Cook MJ, O'Brien TJ, Berkovic SF, Murphy M, Morokoff A, Fabinyi G, D'Souza W, Yerra R, Archer J, Litewka L, Hosking S, Lightfoot P, Ruedebusch V, Sheffield WD, Snyder D, Leyde K, Himes D. (2013) Prediction of seizure likelihood with a long-term, implanted seizure advisory system in patients with drug-resistant epilepsy: a first-in-man-study. Lancet Neurology Jun;12(6):563-71

ACVIM NEUROLOGY RESIDENT NEUROPATHOLOGY REVIEW

Joe N. Kornegay, DVM, PhD, ACVIM (Neurology)
College Station, TX

(Modified from notes first presented at a resident review session at the 2009 ACVIM Forum & Canadian Veterinary Medical Association Convention, Montreal, Quebec, Canada and again at the 2012 ACVIM Forum, New Orleans, Louisiana. Additional case discussions were added in 2012).

Neuropathology is the study of reactions of nervous tissue to disease processes. Most diseases affecting the nervous system are characterized by a predictable set of lesions. The acuity of onset, progression, and location of these lesions determines the clinical signs of the disease and their course. For sake of this Neurology Resident Review session, we will first cover cells found in the nervous system and then review common microscopic lesions associated with neurologic diseases. This session will focus particularly on diseases causing necrosis/malacia. A series of cases are discussed at the end to further illustrate characteristic lesions.

Image 1

Image 1 is a photomicrograph of normal canine cerebral cortical gray matter. Cells present include neurons and glial cells. The homogeneous pink background consists of the processes of these cells and is referred to as neuropil. The glial cells include astrocytes, oligodendrocytes, and microglia. Neurons, astrocytes, and oligodendrocytes develop from neuroectoderm; neurons from neuroblasts and astrocytes and oligodendrocytes from spongioblasts. Microglia develop from mesoderm. The larger cells in **Image 1** are neurons. What are their morphologic characteristics? Well, they are typically bigger than other cells, with larger nuclei and cell bodies. The size of neurons actually varies from the relatively small granular cells of the cerebellum to large cells such as cerebellar Purkinje cells, ventral horn cells of the spinal cord, and the pyramidal cells of the cerebrum. What else serves to distinguish neurons from other cells of the central nervous system (CNS)? Neurons usually have prominent nucleoli, while those of other CNS cells are only faintly visible. In addition, on H&E stains, only neuronal cytoplasm contains basophilic granular material referred to as Nissl substance (ribosomes). Do you see the neurons in **Image 1**?

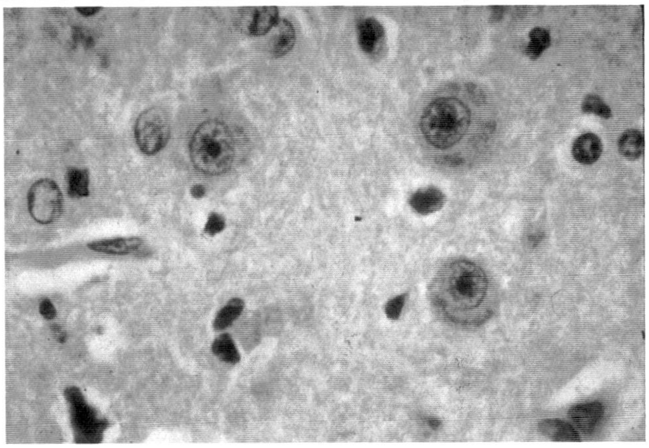

Image 2

The diagram in **Image 2** is from *Veterinary Neuroanatomy and Clinical Neurology* by Alexander de Lahunta. It depicts the layered arrangement of neurons in the cerebral cortex. Six layers of neurons are distinguishable in primate brains from lateral to medial: molecular, external granular, external pyramidal, internal granular, internal pyramidal, and multiform. These layers are much less prominent in subprimates. Why do we attach significance to this laminated arrangement of cerebrocortical neurons? It seems that certain layers are more susceptible to injury than others. For instance, layers three and five are more susceptible to hypoxia. When a single layer of neurons is necrotic, the histologic lesion is termed laminar cerebrocortical necrosis. We don't encounter this lesion very often in veterinary medicine. However, it is not unusual to see multiple layer involvement, in which case the term pseudolaminar cerebrocortical necrosis applies.

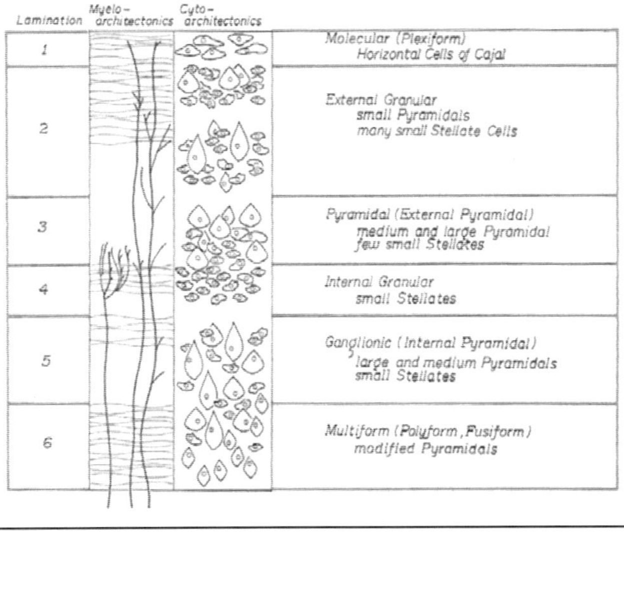

Image 3

Causes of pseudolaminar cerebrocortical necrosis encountered in veterinary medicine include cerebral hypoxia, systemic hypoglycemia as occurs with insulin secreting tumors, polioencephalomalacia (necrosis of cerebral gray matter) due to thiamine deficiency and lead poisoning in ruminants, and salt poisoning in pigs. Note in **Image 3** that the outer rim of gray matter stains considerably more intensely than the gray matter adjacent to the white matter. The inner neurons are necrotic and there is vacuolation of the neuropil.

Can you think of any factors that might predispose certain neurons to injury while others are unaffected? What about their metabolic demands? Perhaps, some neurons have higher requirements for oxygen and substrates such as glucose. Make sense?

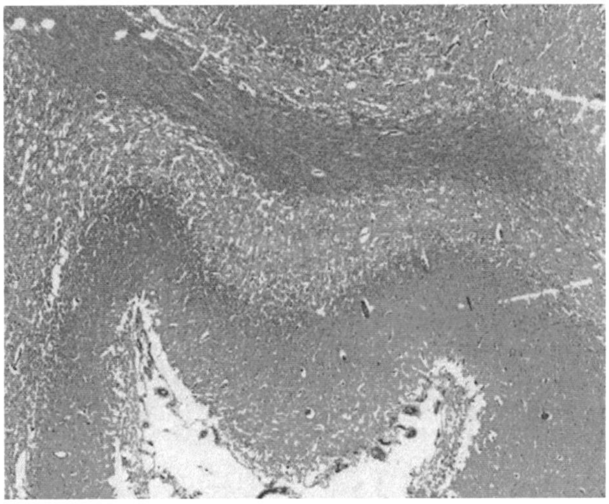

Image 4

We've seen some normal neurons. Another one is seen here in **Image 4**. Look at the large cell to the left. It has Nissl substance in its cytoplasm and a prominent nucleolus. You also can see a process extending from it at the bottom (probably its axon) and less obvious processes on either side toward the top (probably dendrites). What about the large red cell towards the middle of the photomicrograph? This is a necrotic neuron. The nucleus is displaced eccentrically and is pyknotic; also note that its cytoplasm is shrunken and intensely eosinophilic. This is referred to as an ischemic neuron because these changes are seen following cerebral ischemia. Similar neuronal changes occur subsequent to cerebral hypoxia and metabolic insults, including hypoglycemia. You can also see an expanded space (vacuole) to the left of the ischemic neuron and an additional space to the right. Spaces can also be seen around the normal neuron to the left. It's difficult to be absolutely sure of the origin of these spaces. Open spaces or vacuoles surrounding cells and vessels often occur due to swollen astrocyte processes associated with cytotoxic brain edema. This histopathologic lesion frequently accompanies ischemic neuronal cell change caused by metabolic diseases. In this form of edema, the fluid accumulates intracellularly rather than in the extracellular space, as is the case with vasogenic edema. The principal cause of cytotoxic edema is hypoxia-induced failure of the sodium pump. Intracellular accumulation of sodium necessitates imbibition of water to maintain osmotic equilibrium. Neurons, glia and endothelial cells are all affected. However, the associated cell swelling is most noticeable in astrocytes.

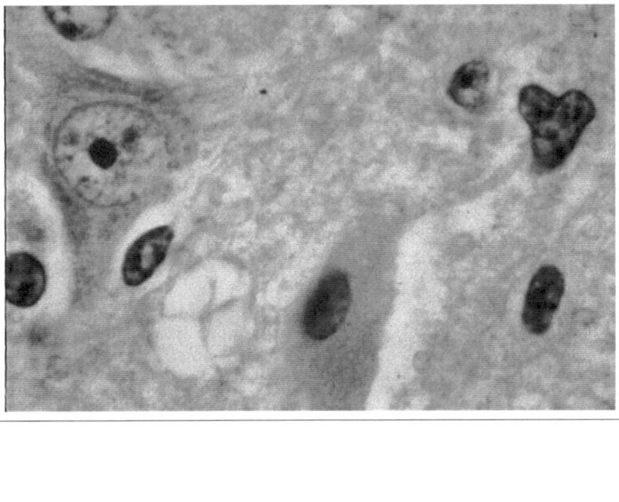

Images 5-7

These photomicrographs are sequentially higher powers from the brain of a dog with hypoglycemia due to an islet cell tumor. Note that there is prominent cerebrocortical necrosis and associated vacuolation due the combined effects of neuronal cell loss and cytotoxic edema. Interestingly, while there is neuronal cell loss, overall cell numbers are increased, especially at the periphery of the cortex (**Image 5, top**). This increased cellularity has occurred due to the combined effects of proliferating vessels (neovascularization) and astrocytes and microglia reacting to the neuronal necrosis.

What happens to necrotic neurons? As you would expect, phagocytes remove them. Many of the necrotic neurons in this slide are associated with cells that have small, irregularly shaped hyperchromatic nuclei (**Image 6, middle**). These cells are microglia, the CNS phagocyte. The process of neuronal necrosis and subsequent phagocytosis is called neuronophagia. The aggregate of cells is referred to as a glial nodule or glial star. Glial nodules also may be seen unassociated with degenerating neurons. This is particularly true of viral encephalitis.

There is also an increase in astrocytes in this brain. Astrocytes have round nuclei, no appreciable cytoplasm, and indistinct nucleoli. When there is brain disease, astrocytes usually become reactive and undergo both hyperplasia and hypertrophy. This process is called gliosis. Several astrocytes are seen among the necrotic neurons (**Image 7, bottom**). Because they are reacting to the ongoing disease, the nuclei are somewhat larger than normal and the nucleoli are more prominent. Clustering of astrocyte nuclei is typically seen in gliosis (**see, in particular, the two adjacent nuclei towards the center of Image 7**). Binucleate forms also occur. Cytoplasm may become more evident. This is particularly true of so-called gemistocytic astrocytes (gemistocytes), as discussed further below.

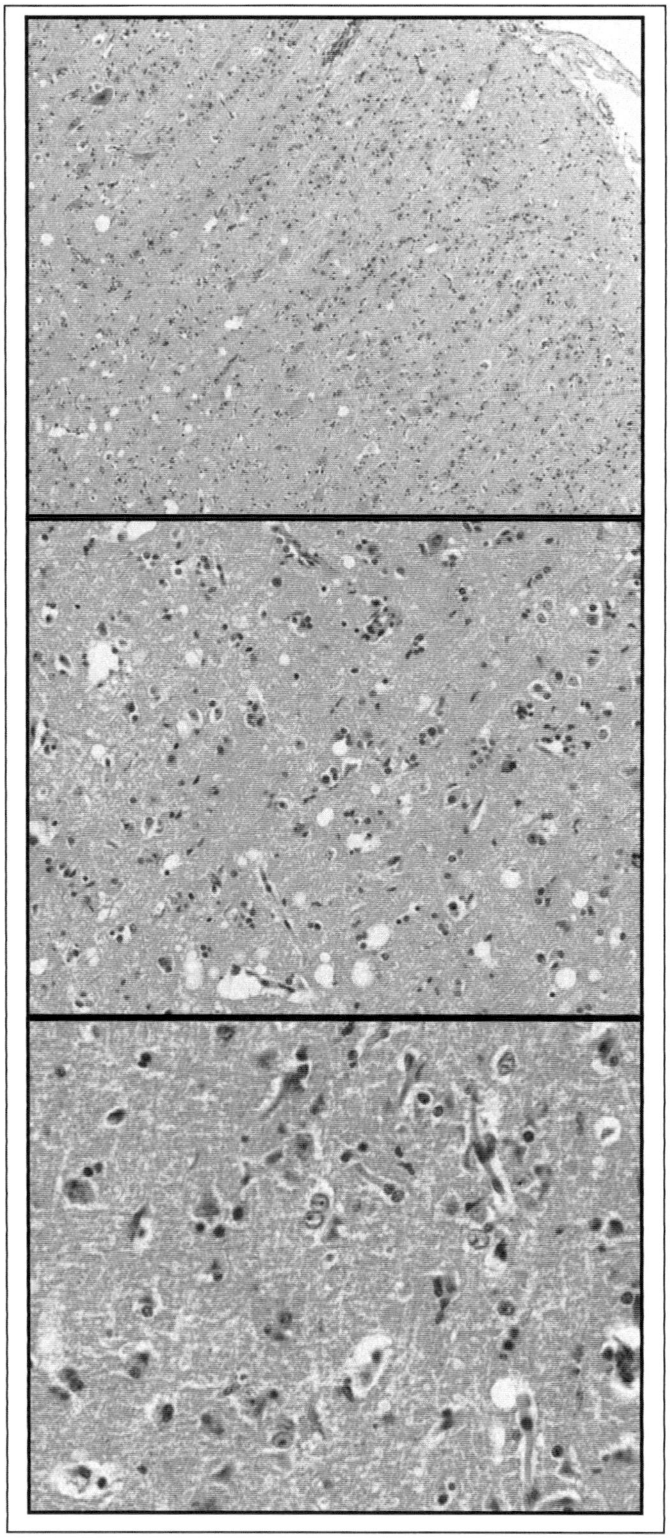

Images 8 and 9

The photomicrographs on the next page are from different cases and illustrate typical features of astrocytes. **Image 8 (left)** is from a horse with cerebral hypoxia. Note the ischemic neuron at the center of the picture and several normal neurons with prominent nucleoli and Nissl substance in their cytoplasm. The larger more vesicular nuclei in this picture are those of reactive astrocytes. You can also appreciate faint eosinophilic cytoplasm in some of the cells.

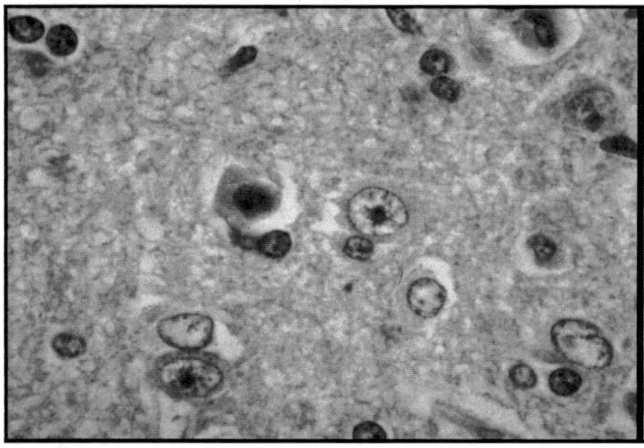

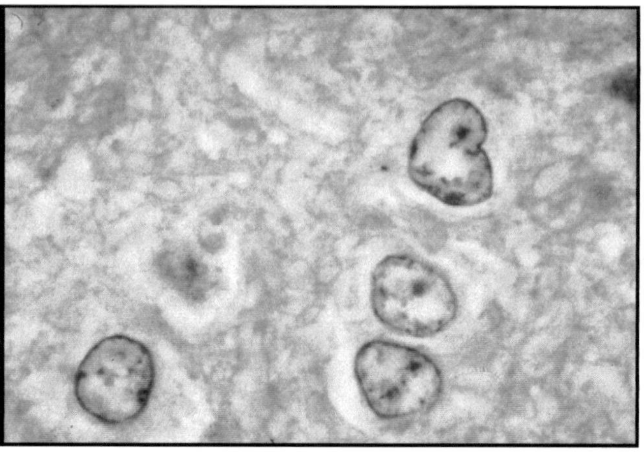

The four nuclei in **Image 9 (right)** are also those of reactive astrocytes. The nuclei are larger and more vesicular than those of normal astrocytes. The one at the upper right appears to be dividing. However, note that their nucleoli are relatively small and there is no Nissl substance. These features allow them to be distinguished from neurons. In fact, the limited cytoplasm of these cells is faintly eosinophilic rather than basophilic. As briefly mentioned above and discussed further below, some reactive astrocytes have considerable eosinophilic cytoplasm. Such cells are called gemistocytic astrocytes.

Images 10-12

With thiamine deficiency and hypoglycemia, there is a generalized metabolic insult that leads to widespread neuronal changes. In contrast, localized cerebral (or spinal cord) ischemia affects a more limited area of nervous tissue. Neuronal necrosis and edema occur acutely and brain cavitation/atrophy is seen chronically. The nature of the lesion will vary depending on whether the underlying lesion is primarily ischemic or hemorrhagic. **Image 10** is from the brain of a cat with ischemic encephalopathy. The deep gray matter within the gyrus at the bottom of the photomicrograph is rarefied. On higher power, most neurons had undergone necrosis and were becoming indistinct histologically. This gyrus is progressing toward cavitation and eventual atrophy. Neurons in adjacent gyri showed less pronounced involvement compatible with ischemic cell change.

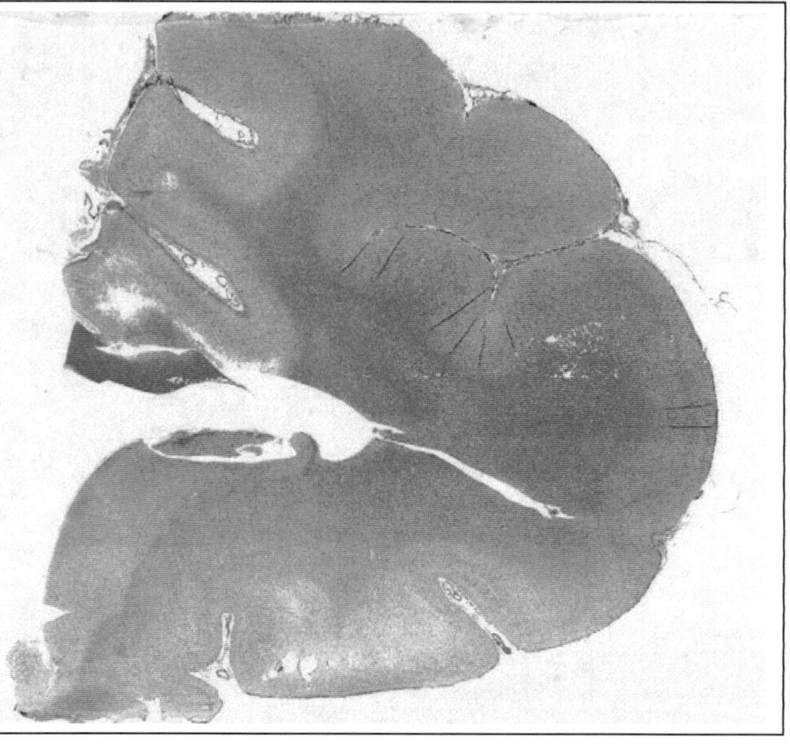

Two additional classical histologic changes are evident in the white matter of the adjacent corona radiata (**see Images 11 and 12 on the next page**). Note the many cells with open faced to somewhat hyperchromatic nuclei and appreciable eosinophilic cytoplasm typical of gemistocytic astrocytes in **Image 11 (left)**. In general, as depicted here, astrocytes function somewhat similar to fibroblasts elsewhere in the body. Their processes tend to wall off areas of necrosis. In this way, gliosis is analogous to fibrosis.

The white matter of the corona radiata is also markedly vacuolated in **Image 12 (right)**. These vacuoles likely reflect the combined effects of demyelination and brain edema.

When neuronal cell bodies become necrotic, there is concomitant degeneration of the axon and its myelin sheath. The process whereby axons and myelin sheaths degenerate distal to the point of neuronal injury is termed Wallerian degeneration. This is the most common form of demyelination in animals. In the case illustrated here, the axon and myelin sheath have degenerated because of the underlying neuronal necrosis. As you might imagine, the axon is dependent on the neuronal cell body for "nutrients" that would

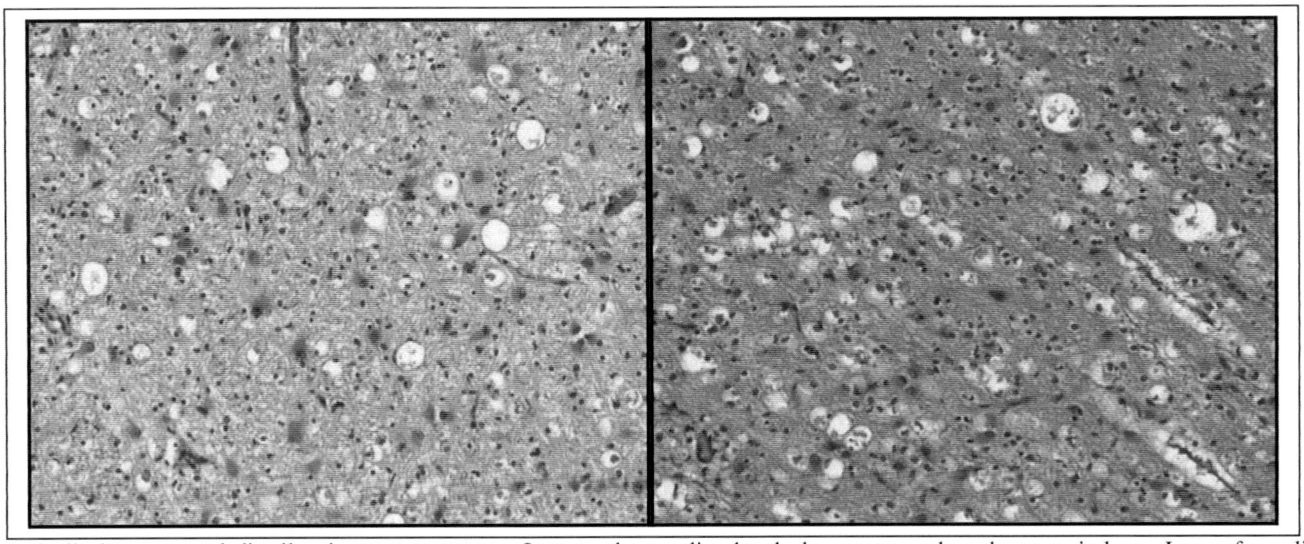

typically be conveyed distally via axon transport. In turn, the myelin sheath degenerates when the axon is lost. Loss of myelin secondary to neuronal cell body or axonal injury is termed secondary demyelination.

Wallerian deneneration is particularly marked after chronic spinal cord compression. Histologic lesions begin within 30 hours of axonal severance and gradually progress over a period of six months. Axons and their myelin sheaths are fragmented into a series of ovoid segments referred to as ellipsoids. The axonal fragments and myelin debris gradually are removed by macrophages, leaving residual vacuoles throughout the white matter.

With compressive spinal cord lesions, Wallerian degeneration leads to a characteristic pattern of spinal cord vacuolation. Remember that axons and their myelin sheaths degenerate distal to the point of the lesion. Also, recall that most sensory (ascending) spinal cord pathways are in the dorsal and dorsolateral columns; in contrast, motor (descending) pathways are in the ventral and ventrolateral columns. This characteristic pattern distinguishes Wallerian degeneration from that seen with diseases such as degenerative myelopathy and Afghan myelopathy.

Image 13

The diagram in **Image 13** on the next page illustrates Wallerian degeneration that has occurred subsequent to a compressive lesion at T5. The red mass, potentially a tumor, is situated ventrolaterally on the left side of the vertebral canal. The gray circles in the spinal cord are intended to reflect the characteristic pattern of vacuolation that occurs both at the site of the lesion due to local axonal injury and cranially and caudally subsequent to Wallerian degeneration. As depicted by the arrows at T5, the resultant compression occurs radially across the full cross section of the spinal cord. Thus, while the vacuoles at T5 are concentrated somewhat immediately adjacent to the mass, you also see loss of axons with resultant vacuolation throughout the cord here. The effects of Wallerian degeneration are illustrated cranial (C4) and caudal (L3) to the mass. Descending motor and ascending sensory fibers are shown, with their cell bodies located in the cerebral cortex and a dorsal root ganglion, respectively. Cranial to the lesion, the ascending sensory fibers are now detached from the cell body and degenerate, resulting in vacuoles primarily in the dorsal funiculi and dorsolateral portions of the lateral funiculi. In contrast, caudal to the lesion, the motor fibers, which are primarily in the ventral funiculi and inner aspects of the lateral funiculi, degenerate, leading to vacuoles there.

Wallerian Degeneration

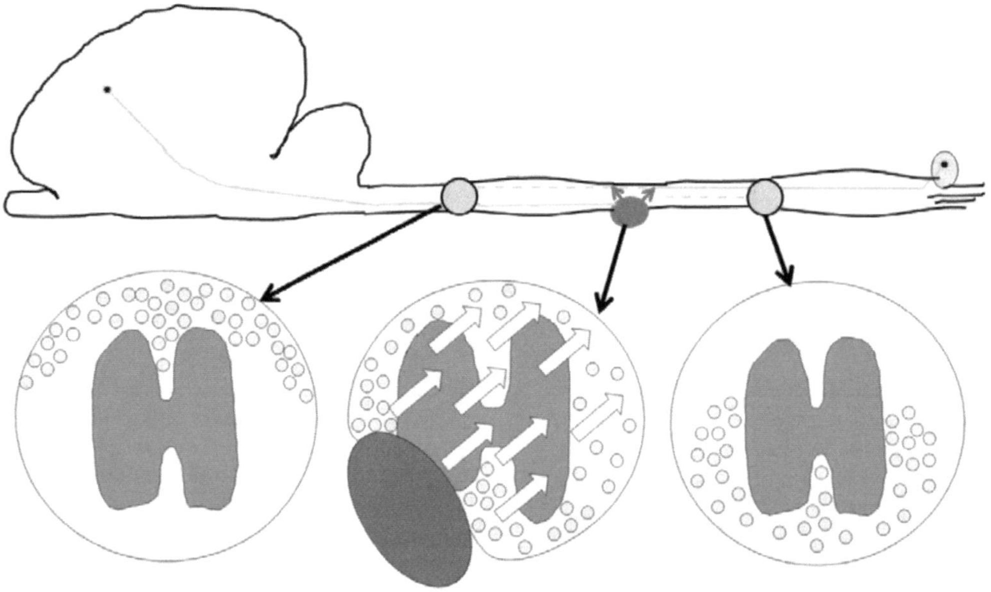

Before we leave the concept of demyelination, let's clear up one more thing. If secondary demyelination refers to myelin loss due to initial axonal injury or separation from the neuron, what do you suppose occurs with primary demyelination? That's correct; it's when myelin loss occurs without antecedent axonal injury. The most common cause in animals is canine distemper virus infection. In this disease, myelin is stripped from axons by invading lymphocytes as a result of a poorly defined, immune-mediated process. You've heard a lot about another primary demyelinating disease that affects people, multiple sclerosis.

Image 14

In **Images 10-12** from the cat with ischemic encephalopathy, we saw histologic evidence of a subacute ischemic infarct (stroke). A stroke is caused by either acute cerebral infarction or intracerebral hemorrhage. Such lesions are common in human beings because of their high incidence of cerebral atherosclerosis but occur relatively infrequently in domestic animals. One potential cause in dogs is hypothyroidism (see further below). **Image 14** illustrates an example of a chronic cerebral infarct in a dog. Do you see the lesion? Compare the left and right sides of the brain. Note the atrophy of the left thalamus and parietal cortex and the consequent enlargement of the left lateral ventricle. On which side of the body would the neurologic deficits be in this animal? Do you know the term used to refer to passive dilation of the ventricular system due to loss of adjacent neural tissue? Yes, hydrocephalus ex vacuo.

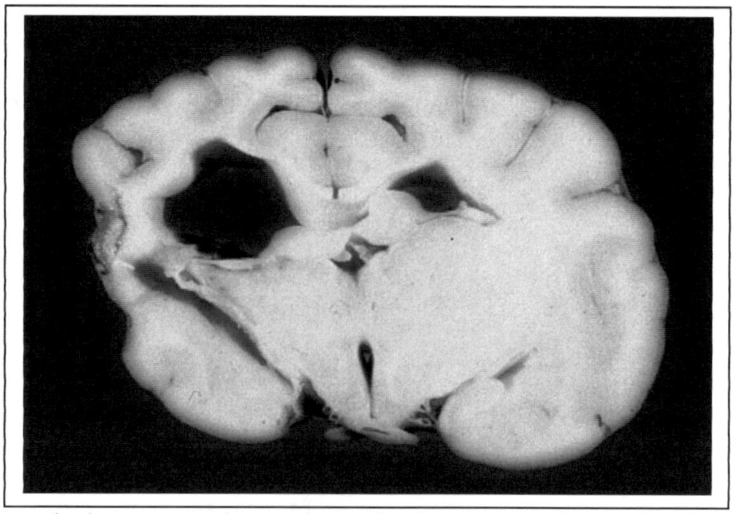

Images 15 and 16

Now, have a look at **Image 15 (top)**. You can clearly see the area of infarction on the left side of the brain corresponding to the gross photograph in the previous image. There is only a slender rim of tissue in the area of the parietal cortex at the site of the infarct. At higher power, there was little, if any, evidence of neural tissue there. Neurons of the adjacent gyri appeared normal. However, there were reactive astrocytes immediately adjacent to the infarct, acting to wall off the area of prior necrosis.

Even at relatively low power in **Image 16 (bottom)**, two additional characteristic lesions of cerebral infarction, cholesterol clefts and macrophages, can be seen. The cholesterol clefts look like ice crystals. They occur because of the high content of lipids in myelin and represent areas where there once was cholesterol that washed out in processing. The macrophages have phagocytized both red blood cells and lipid breakdown products of degenerating cell membranes/myelin. Remember that macrophages in the brain may originate from either microglia or blood borne monocytes. Microglia, normally, have rod-shaped or elliptical hyperchromatic nuclei such as those seen earlier associated with neuronophagia. (Note, there is one cell that can be easily confused with the microglia. That's the endothelial cell of cerebral vessels. Of course, it will usually be associated with vessel lumens and red blood cells). Microglia that become phagocytic often have small, round, hyperchromatic nuclei and appreciable foamy cytoplasm. These cells have been called compound granular corpuscles, foamy macrophages, and "gitter" cells. The vacuolated appearance results because of phagocytosis of myelin lipids. Cells seen at the bottom of **Image 16** have a brownish-yellow hue, most likely reflecting phagocytosis of the pigment hemosiderin from degenerating red blood cells. Such cells are sometimes called hemosiderophages (macrophages that have phagocytized hemosiderin).

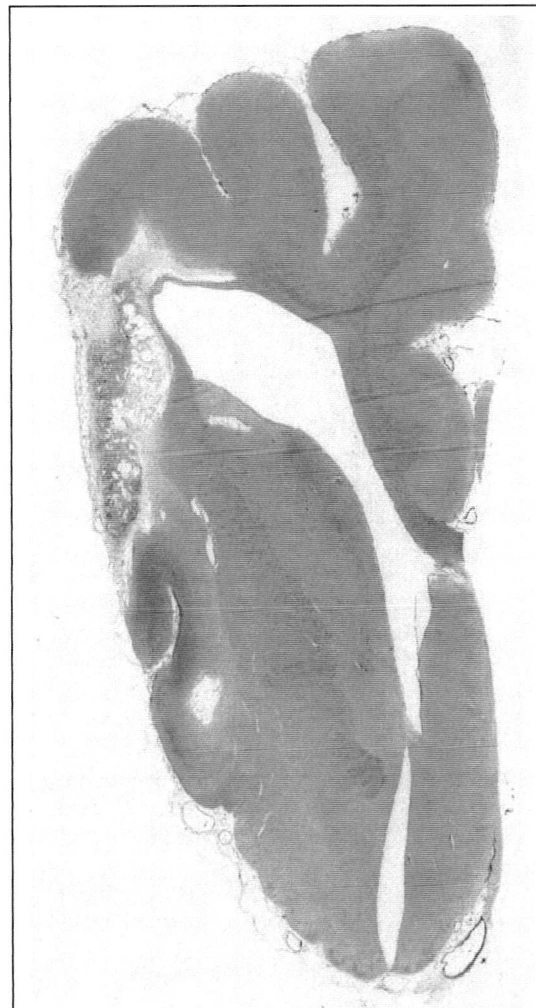

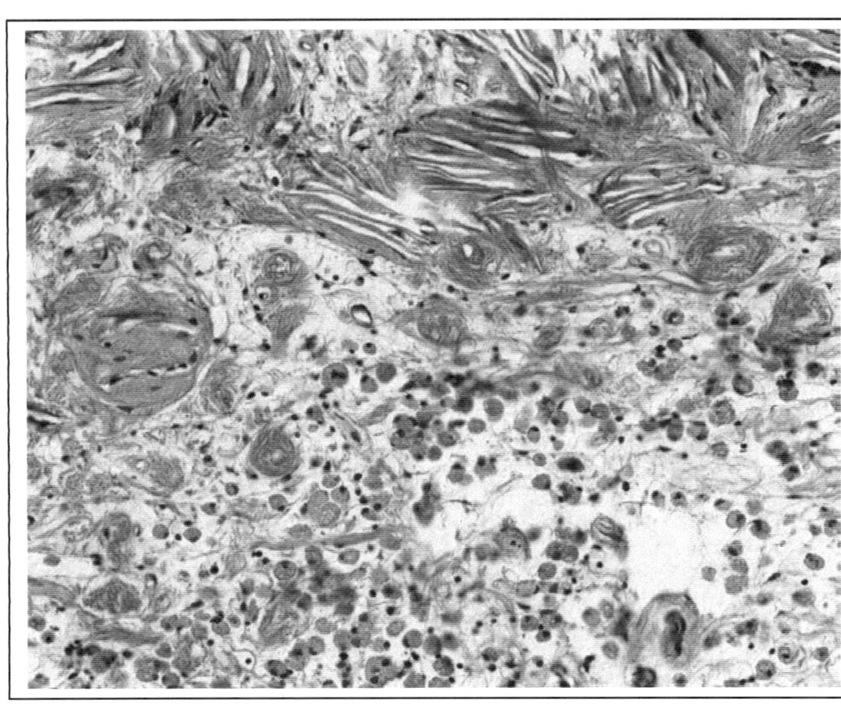

Images 17-19

Let's move on to another example of nervous tissue infarction. Take a look at **Image 17 (top)** prepared from the spinal cord of a 6-year-old, male Siberian husky dog with an acute onset of asymmetric weakness. This is a transverse section of spinal cord. First, get your bearings. The cleft at the bottom of the section is the ventral median fissure dividing the two ventral funiculi. And, at the top of the section, notice the less prominent dorsal median sulcus, which separates the dorsal funiculi. A large pale spherical area, representing necrosis, affects the left dorsal and lateral funiculi. Also, at this lower power, you can see that the ventral spinal artery is partially occluded by a thrombus. At higher powers, you would have been able to see that the spinal cord in this area was progressing towards liquefactive necrosis and cavitation, with little remaining recognizable neural tissue.

Image 18 (middle) is a photomicrograph taken at the interface between the infarct on the right and more normal white matter on the left. Numerous "gitter cells" are present in the infarct. The adjacent white matter is vacuolated. Some vacuoles contain fragmented axons and macrophages in keeping with the ellipsoids described earlier. Interestingly, there are relatively few reactive astrocytes at this acute stage of the lesion. On the left side, numerous round eosinophilic structures are evident. These are axonal spheroids. Recall, earlier, we mentioned that Wallerian degeneration occurs due to disruption of nutrition from the neuronal cell body. Axoplasmic flow is disrupted and axoplasm accumulates in the proximal fragment, resulting in the axonal spheroids seen here.

Now have a look at a higher power view of the ventral spinal artery, which is occluded by a large thrombus containing hemorrhage, cholesterol clefts, and fat droplets (**see Image 19 - bottom**). The vessel wall, itself, also contained vacuoles. Based on examination of other organs at necropsy and with histology, the dog was determined to have hypothyroidism due to thyroiditis. Hypothyroidism is one of the few causes of atherosclerosis in dogs. (*Slide courtesy of Dr. Peter Moisan, North Carolina Veterinary Diagnostic [Rollins] Laboratory*)

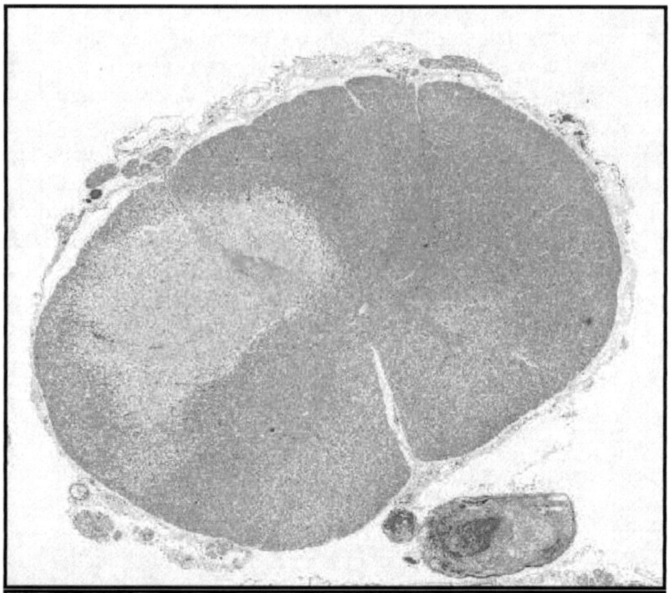

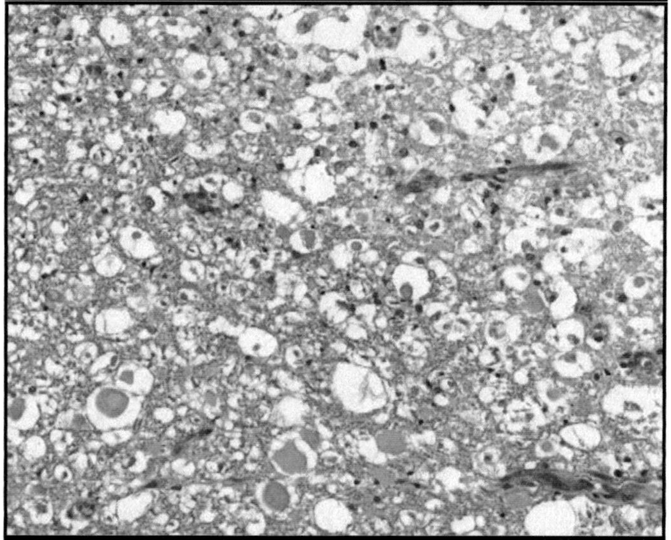

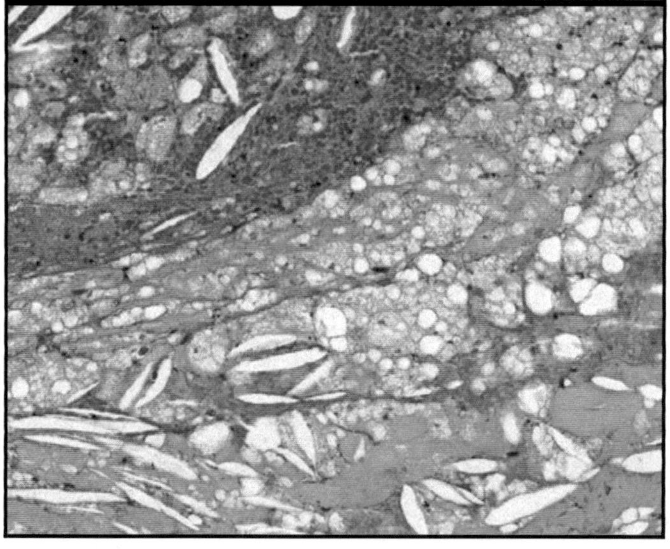

Image 20

Let's circle back around to the general topic of brain necrosis. At the start, we illustrated a case of polioencephalomalacia (cerebrocortical necrosis) due to thiamine deficiency in a calf. Of course, as the name polioencephalomalacia suggests, this lesion/disease principally involves gray matter. We've also discussed vascular-based diseases that affect both white and gray matter. Now, we'll consider a group of diseases that primarily involve white matter.

Image 20 is a photograph of brain from a horse with a disease named leukoencephalomalacia. As you can see, the name is appropriate, because there is selective white matter involvement. This disease is seen in association with feeding of moldy corn and is thought to be due to the fungus Fusarium moniliforme. The pathogenesis is poorly understood. Could it be that a toxin produced by the fungus alters CNS myelin in some way?

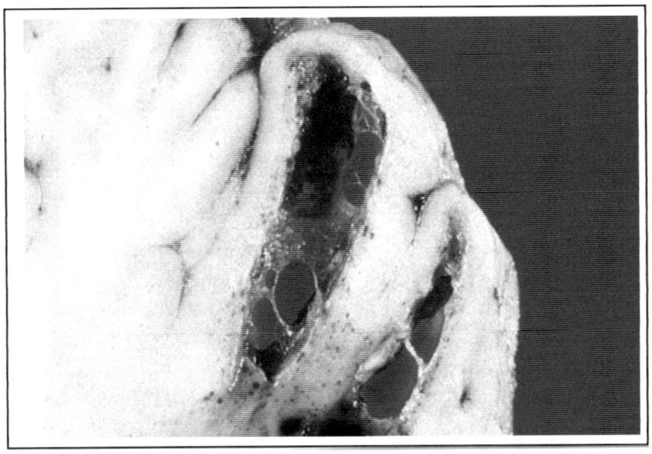

Images 21 and 22

Images 21 and 22 are from a 5-month-old Labrador retriever with signs suggesting forebrain dysfunction. Upon histologic evaluation of the brain, lesions were confined to the cerebrum. At low power in **Image 21**, the corona radiata stains palely. As you move to a higher power in **Image 22**, you can see that the corona radiata is vacuolated; more severely affected areas had undergone necrosis and approached cavitation. The normal architecture was lost and minimal neural tissue remained. In these areas, large foamy macrophages predominated. Scattered axonal spheroids were also present.

Moving peripherally from these severely affected areas, there were many reactive astrocytes, including gemistocytes. Vessels were prominent. The adjacent more normal areas of the corona radiata contain vacuoles consistent with edema and/or demyelination (**see Image 22**). A littermate and another related dog were affected, suggesting a heritable basis. The strict cerebral white matter involvement and cavitating nature of these lesions are consistent with changes seen with other leukodystrophies in dogs and humans, such as globoid cell leukodystrophy and Canavan's disease. Underlying enzymatic defects in lipid metabolism can lead to myelin instability and necrosis. (See Neer TM and Kornegay JN: J Vet Intern Med 9:100-104, 1995)

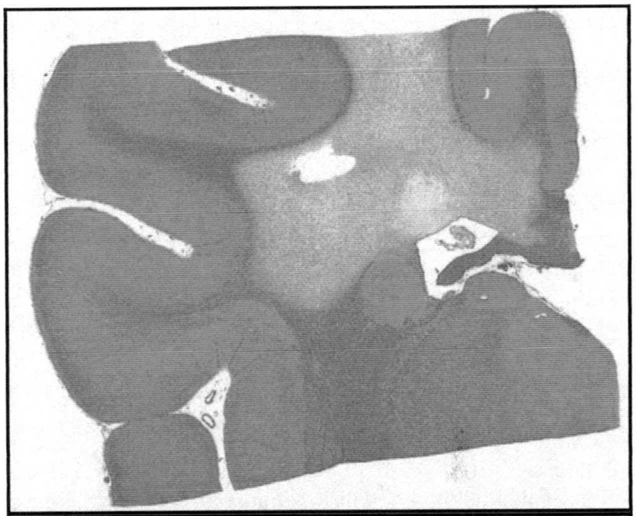

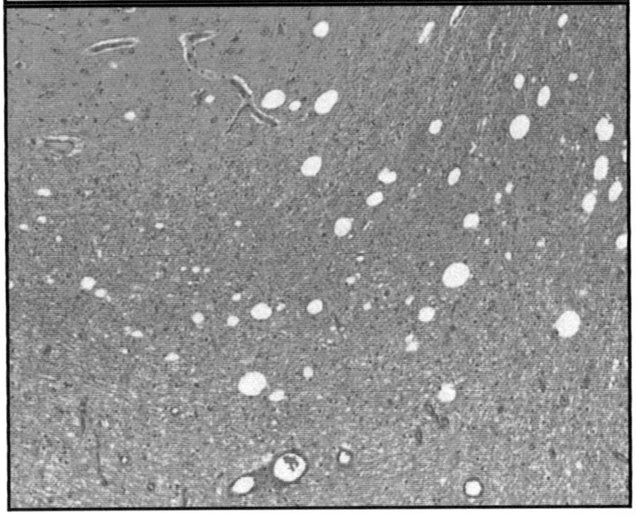

Images 23 and 24

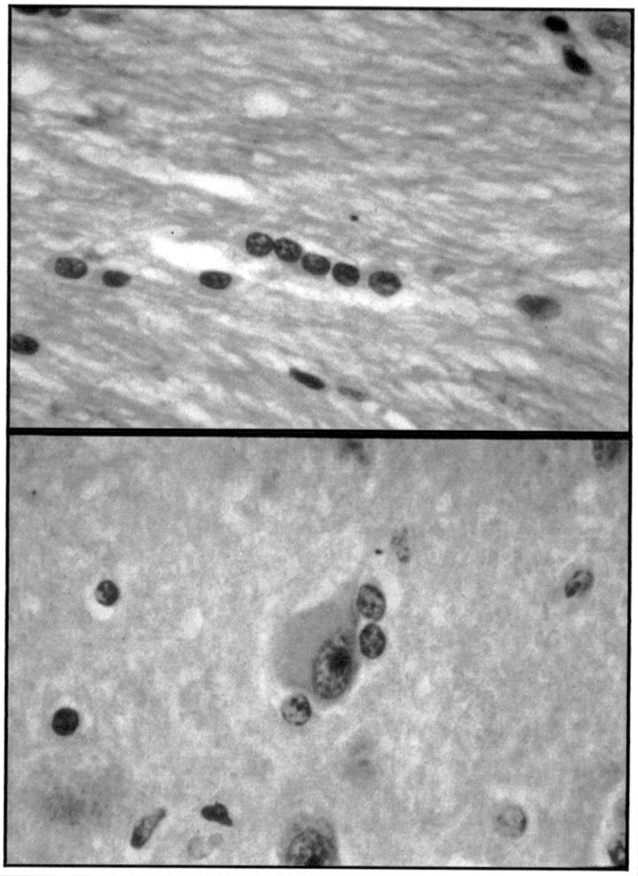

In discussing disorders of myelin, we should round out our overview of the histology of the nervous system by covering one more cell, the oligodendrocyte. These cells have small, round, hyperchromatic nuclei and no appreciable cytoplasm. Their nuclei are smaller and more hyperchromatic than those of astrocytes and round as compared to the rod or elliptically shaped nuclei of microglia. Another factor that serves to distinguish oligodendrocytes is their location certain areas of the nervous system. They often are aligned in rows parallel to axons in the white matter and in small groups around neuronal cell bodies in the gray matter.

The small, round, hyperchromatic nuclei arranged longitudinally in the white matter in **Image 23 (top)** are those of oligodendrocytes. Oligodendrocytes are the dominant cell in white matter, owing to their role in myelination of CNS axons. This function is analogous to that of the Schwann cell of the peripheral nervous system. However, while each Schwann cell is responsible for the internodal segment of only one axon, oligodendrocytes may myelinate many axons in the CNS. The lack of one-to-one precision between the oligodendrocyte and axon in CNS myelination may be a factor in the failure of remyelination subsequent to CNS injury. This contrasts with the peripheral nervous system where remyelination is more complete.

The tendency for oligodendrocytes to cluster around neurons, as demonstrated in **Image 24 (bottom)**, probably relates to myelination of axons as they leave the cell body. This has been called "satellitosis" and was once thought to indicate neuronal disease. You should be aware that this is normal. How would you distinguish this from neuronophagia?

Images 25 and 26

Armed with this added information on the role of oligodendrocytes in myelination, let's look at another disorder that primarily affects white matter. **Image 25** contains a section of cerebral cortex from a young West Highland white terrier with a progressive degenerative disease termed globoid cell leukodystrophy or Krabbe's disease. The brain section is stained with H&E/LFB. Myelin stains blue with LFB. Notice that the white matter of the corona radiata is undergoing necrosis with cavitation. Globoid cell leukodystrophy is a form of lysosomal storage disease caused by a deficiency of the enzyme galactosylceramidase (galactocerebrosidase β-glactosidase). Other breeds of dogs, including in particular the Cairn terrier, are affected. Galactosylceramide is a key component of the myelin sheath and paradoxically does not accumulate in affected individuals, presumably because oligodendrocytes are rapidly lost in the early stages of the disease. CNS myelin degenerates and there is an associated infiltrate of large macrophages (globoid cells) that phagocytize the affected myelin. There were groups of these globoid cells in areas of necrosis. Many of them had phagocytized myelin, which stains blue with LFB, as seen on higher power in **Image 26**. In areas of the corona radiata that were better preserved, there were vacuoles consistent with edema and/or demyelination. In addition, numerous gemistocytic astrocytes were present.

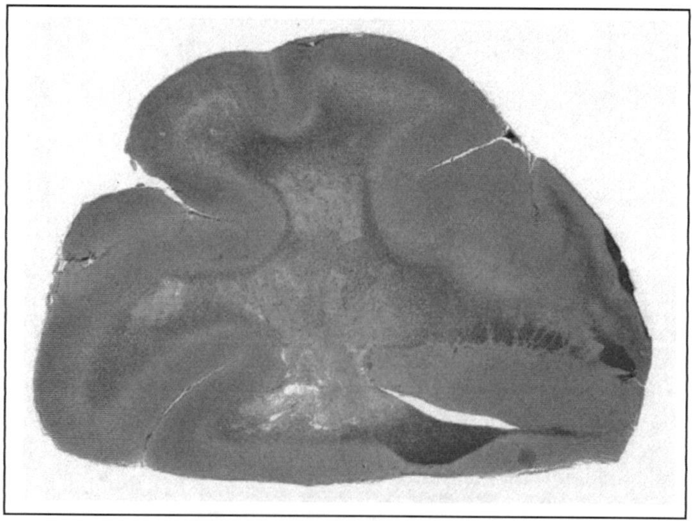

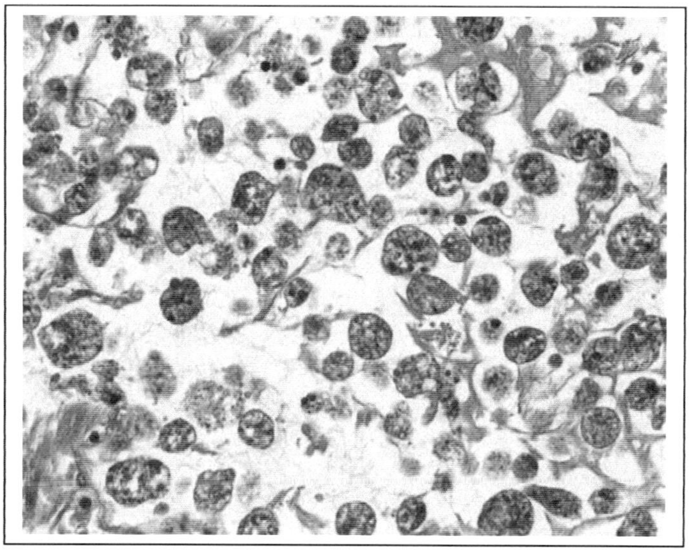

ACVIM NEUROPATHOLOGY SESSION - ADDITIONAL CASES
(Modified from a Presentation Made at the 2012 ACVIM Forum, New Orleans, Louisiana s)

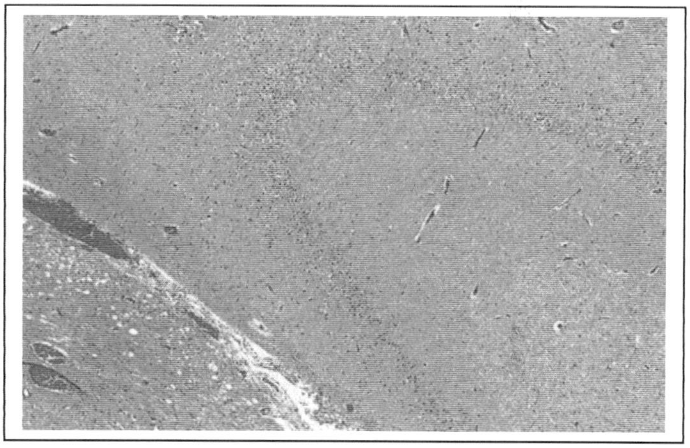

Images 27-29 are from a young, unvaccinated dog with signs of progressive multifocal disease. **Image 27 (top)** was taken from the cerebrum at the level of the midbrain. The relevant lesions are present in the hippocampus (to the right) and a section of optic tract (on the left). In the hippocampus, there are numerous necrotic neurons, with characteristic pyknotic nuclei and eosinophilic cytoplasm **(Image 28, bottom left)**. This pattern of cerebrocortical necrosis has been seen with canine distemper virus infection. Mechanisms responsible for the necrosis are unclear but do not appear to involve direct viral cytopathic effects. The adjacent optic tract in **Image 29 (bottom right)** contains vacuoles, consistent with demyelination, also typical of canine distemper. In this same image, note characteristic intranuclear inclusion bodies in astrocytes (green circles). (*Slide courtesy of Dr. Peter Moisan, North Carolina Veterinary Diagnostic [Rollins] Laboratory*).

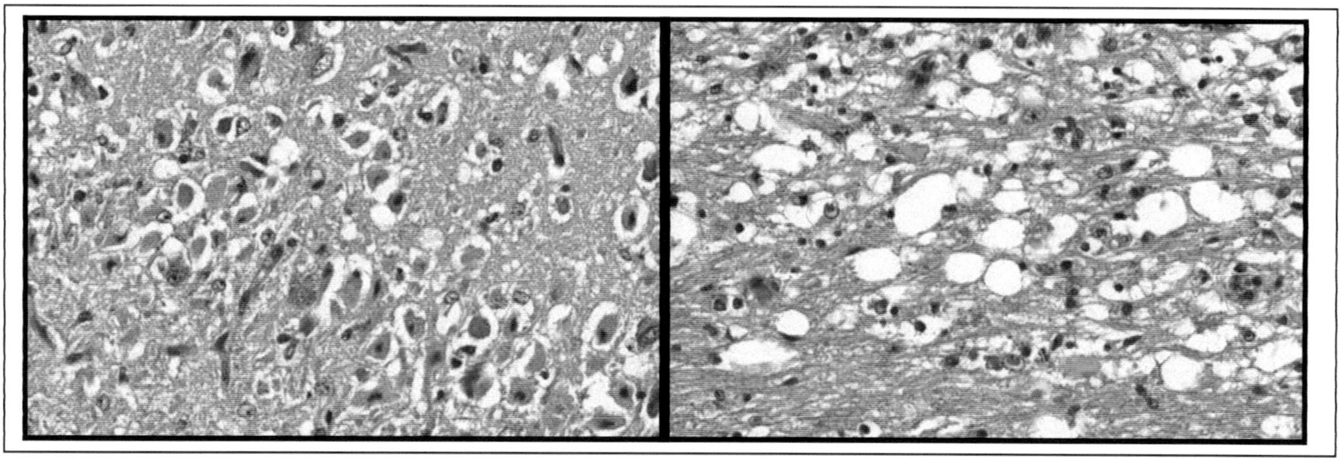

A young pug dog with seizures and other signs of forebrain disease is illustrated in **Images 30-32**. The clinical syndrome was typical of a breed-associated disease termed *pug dog encephalitis*. The section in **Image 30 (left)** is at the level of the caudate nucleus and cingulate gyrus (outlined by the rectangle). Meningoencephalitis, with greater involvement of white matter, is seen in **Image 31 (right, top)**. Also, note in **Image 31**, laminar vacuolation (green arrows) in the adjacent gray matter, which at higher power in **Image**

32 (right, bottom) is associated with ischemic neuronal cell change. This neuronal injury has been reported in other cases of *pug dog encephalitis* and may result from the metabolic insult of severe seizures.

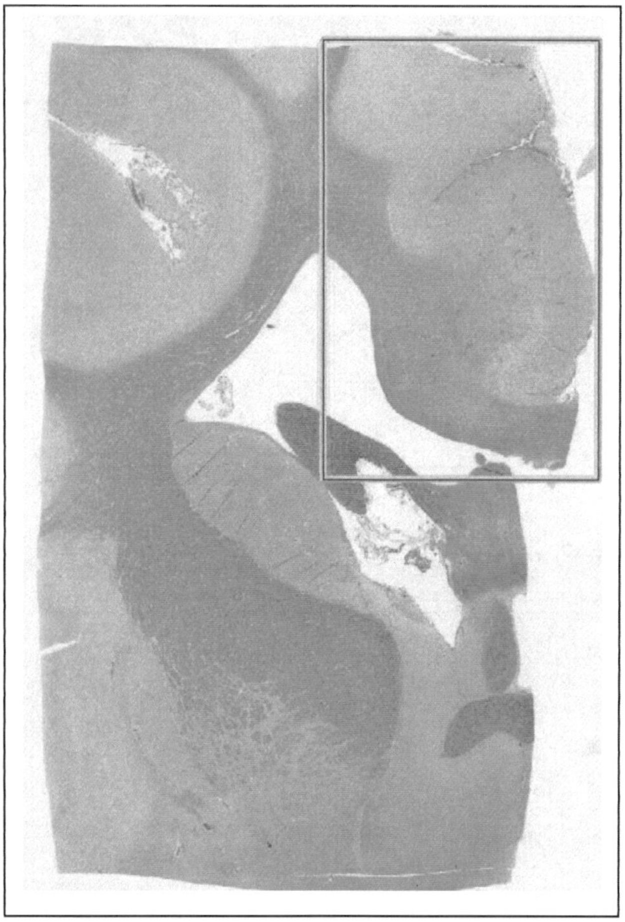

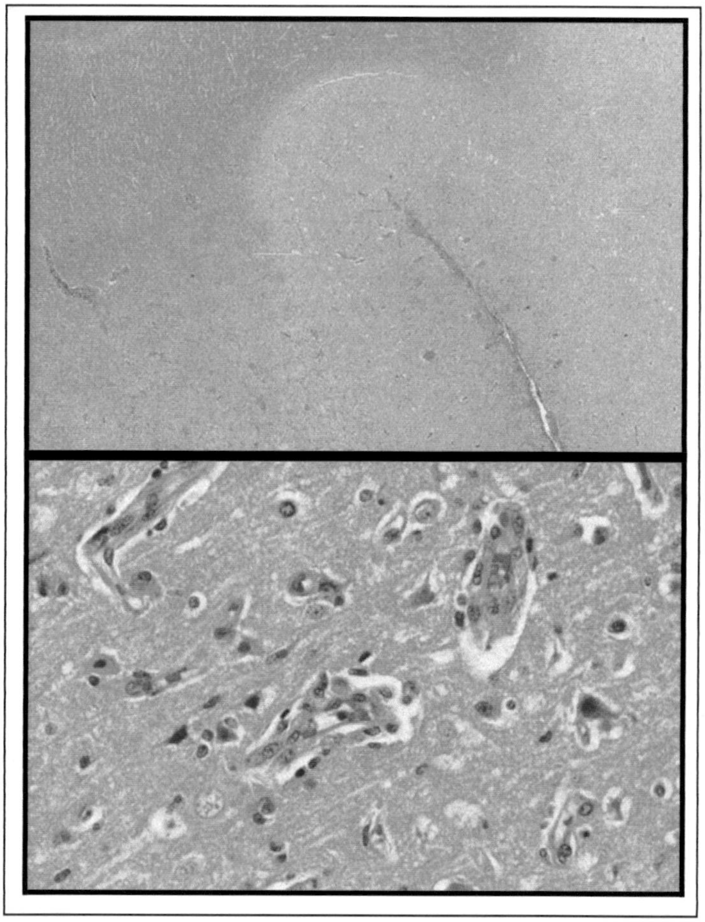

Images 33-35 are from a goat with signs of central vestibular disease. **Image 33** is at the level of the medulla oblongata (obex of the fourth ventricle). Note the multiple, variably sized cell aggregates, which on higher power in **Image 34 (next page, left)**, contained degenerate neutrophils (microabscesses). Several additional lesions are seen in **Image 35 (next page, right)**. The eosinophilic structures among the neutrophils are axonal spheroids. These spheroids occurred due to compression and infiltration of the adjacent white matter, further evidenced by the accompanying vacuoles that resulted from demyelination and edema. These changes are typical of listeriosis. (*Slide courtesy of Dr. Peter Moisan, North Carolina Veterinary Diagnostic [Rollins] Laboratory*).

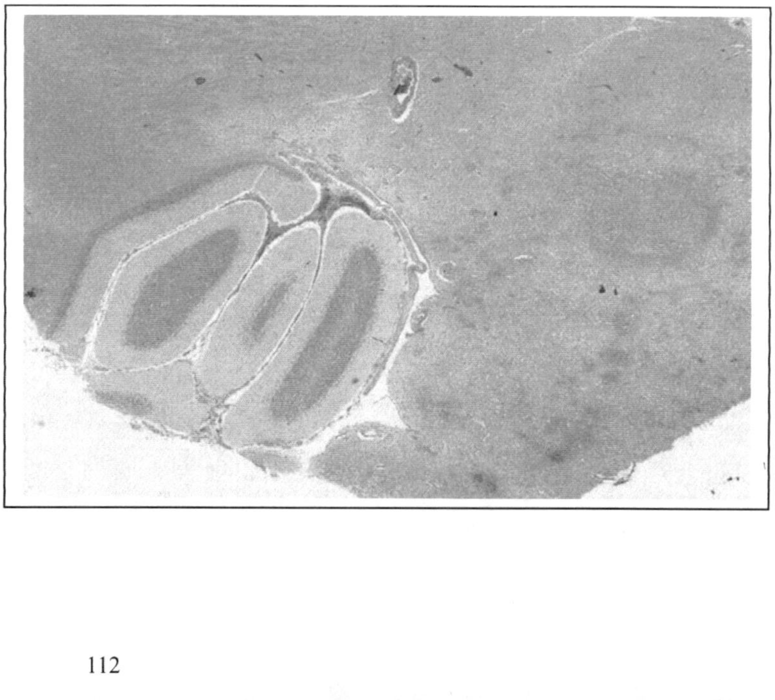

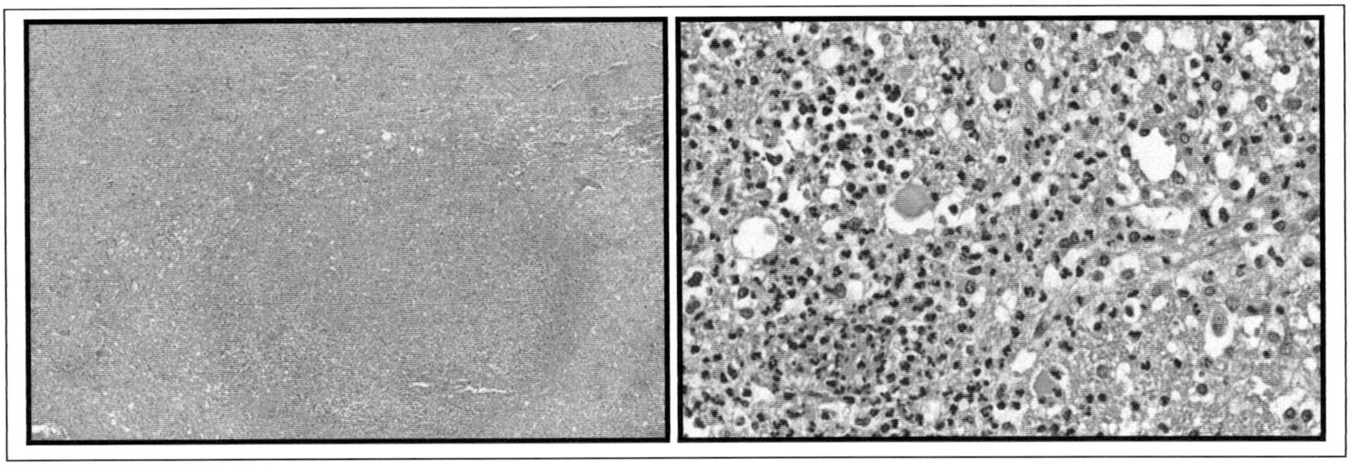

An intriguing syndrome of cerebral infarction is illustrated in **Images 36 38** taken from the cerebrum of a cat with acute lateralizing signs of forebrain disease. Termed ischemic encephalopathy, the pathogenesis of this condition has long been debated, with variable underlying potential pathogenetic mechanisms involved. Recently, in some cases, the syndrome has been attributed to aberrant cuterebra larva migration. **Image 36 (top)** shows a cyst containing a cuterebra larva. Note the vacuoles adjacent to the cyst, which is caused by ischemic neuronal necrosis and cell loss (**Image 37, bottom left**). These changes are associated with fibrinoid necrosis of vessel walls (**Image 38, bottom right**). (*Side courtesy of Dr. Elizabeth Howerth, College of Veterinary Medicine, University of Georgia*).

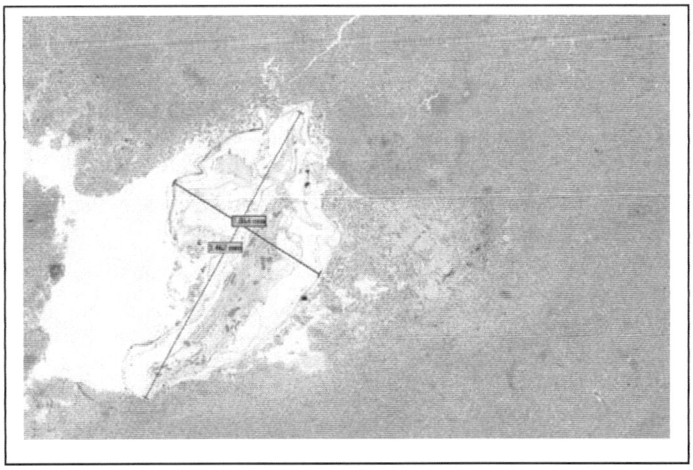

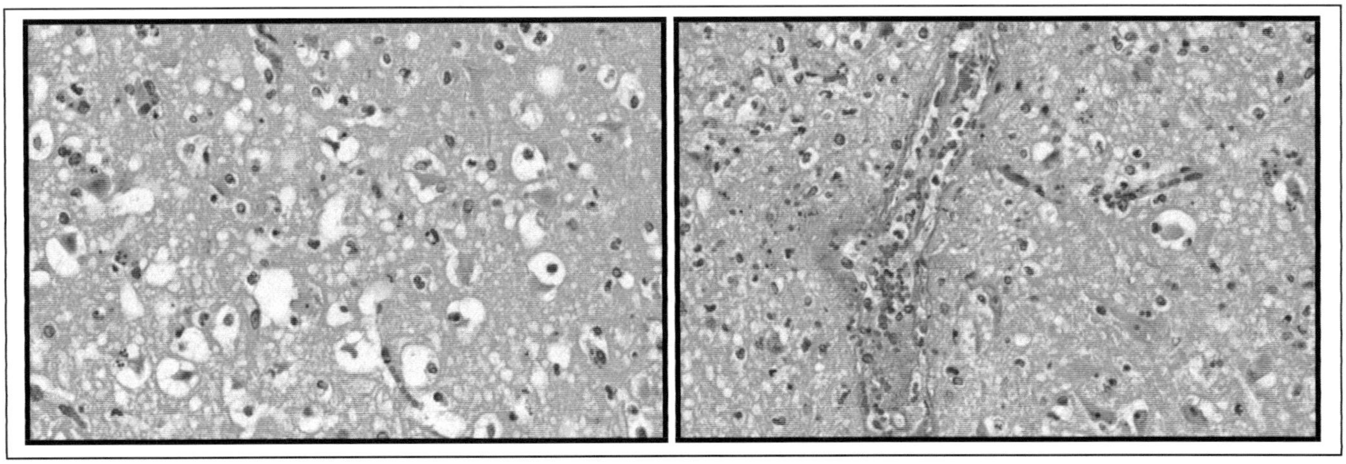

WHAT DO WE CALL IT: PARESIS, ATAXIA AND OTHER NEUROLOGIST'S PERSPECTIVES

Rodney S. Bagley DVM, DACVIM (Neurology and Small Animal Internal Medicine)
Ames, IA

Through examination and assessment of functions controlled by the nervous system, clues to the presence and location of disease within the nervous system are found. Observing the animal is paramount in determining the presence, location, and extent of any disease process. Many clinically observable neurologic abnormalities are indicative of dysfunction of specific functional components within the nervous system. Clinical signs resulting from dysfunction of the nervous system, however, are rarely pathognomonic for a specific disease process.

Unfortunately, "observation" and more importantly interpretation of the observation is "subjective" (adjective existing in the mind; belonging to the thinking subject rather than to the object of thought (opposed to objective).- http://dictionary.reference.com/browse/subjective?s=t)

Of critical importance is the actual "label" the observer places ("thinks") based on what they see, hear, feel, and in some cases smell. The label, for example, may be "head tilt" or some other medically learned description. This is an important step in clinical diagnosis as an inappropriate "label" will influence subsequent steps in the diagnostic process, and influence subsequent medical communication.

Ideally, the terminology associated with a clinical observation would be standardized. From a scientific standpoint, when patient evaluation and assessment is a component of clinical research, patient assessments are frequently unquestioned. Medical terminology such as "ataxia", "paresis", or "lame" are commonly applied and recorded descriptors for patient abnormalities and are associated with certain clinical "connotations". When, however, the medical definitions for these terms are assessed, it is readily apparent that these definitions are less than "specific". This situation, therefore, has the potential for introduction of cognitive errors into the clinical assessment process and for "connotation" or "subjective interpretation" to become individualized operational definitions.

In an effort to bring clarity to this situation, it would ideal if there could be clinical observational and operational "facts" which would have the characteristics of being universally agreed upon as clinical abnormality descriptors to counterbalance the inherently subjective aspects of clinical abnormality assessment. The following, then, are the goals of this session:

1. Identify terminology associated with dog gait assessment where there is not universal agreement on operational usage
2. Determine the range of current operational definitions
3. Establish a plan for resolution of any discrepancies among board-certified neurologists

Specific Aims

- To collect survey data on the operational literal definition and associated clinical connotation of medical terminology used to describe neurologic signs by expert clinicians (board certified veterinary neurologists)
- To determine from a range of options the most consistent terminology used to describe clinical signs in dogs

Background

Gait is commonly defined as a regularly repeating series of leg movements during walking or running. The nervous system controls the actions of the muscles associated connective tissue important for this function. Bones, joints, and ligaments provide a mechanical support system for movements. Disease of any or all of these components can result in a gait abnormality. Normal walking is produced by the recruitment of stepping reflexes, which alternate between the extensor and flexor muscles.

All board-certified neurologists are human beings. Therefore, they are susceptible to the vast array of cognitive biases present in human beings. These cognitive biases, however, can result in inaccurate medical descriptions of patient abnormality. Medical description inaccuracies have a number of negative consequences in both clinical practice and clinical science. Some of the most common sources of potential for mislabeling are associated with terms such as "ataxia", "hypermetria", "paresis", and "lameness".

Ataxia

Ataxia - Function: *noun*: an inability to coordinate voluntary muscular movements that is symptomatic of some nervous disorders
http://www.merriam-webster.com/medlineplus/ataxia
Definition of *ATAXIA*: an inability to coordinate voluntary muscular movements that is symptomatic of some central nervous system disorders and injuries and not due to muscle weakness —called also *incoordination*
http://www.merriam-webster.com/dictionary/ataxia

Noun -ataxia (*countable and uncountable, plural* ataxias) - (pathology) Lack of coordination while performing voluntary movements, which may appear to be clumsiness, inaccuracy, or instability.
http://en.wiktionary.org/wiki/ataxia

Ataxia is a failure of muscular coordination or an irregularity of muscle action. In some situations, it is defined as lack of an axis, implying an inability to travel along a straight line. Ataxia is usually associated with a "falling" or "swaying" of the body from side to side. It is also commonly present in association with other forms of abnormal limb movements such as dysmetria.

Ataxia results from a variety of anatomic lesions within the nervous system, most commonly of the cerebellum, vestibular system, and spinal cord sensory pathways. Ataxia, without motor dysfunction, usually implies cerebellar or cerebellar pathway disease. Sensory ataxia due to loss of joint position sense is often made worse by blindfolding the animal, effectively preventing visual compensatory mechanisms.

Hypermetria
Function: *noun***:** a condition of cerebellar dysfunction in which voluntary muscular movements tend to result in the movement of bodily parts (as the arm and hand) beyond the intended goal
http://www.merriam-webster.com/medlineplus/Hypermetria

Definition of *HYPERMETRIA*: a condition of cerebellar dysfunction in which voluntary muscular movements tend to result in the movement of bodily parts (as the arm and hand) beyond the intended goal—compare hypometria
http://www.merriam-webster.com/medical/hypermetria
Noun- hypermetria

1. dysmetria in which the patient tends to overshoot the intended target

http://en.wiktionary.org/wiki/hypermetria

Dysmetria is improper estimation of distance during muscular activity. Dysmetria includes both *hypo-* and *hypermetria.* With hypermetria, voluntary muscular movement overreaches the intended goal and is essentially a failure to limit the degree of flexion. With hypometria, voluntary movement falls short of the intended goal. Hypermetria is more commonly recognized than hypometria. Both of these abnormalities are most often associated with lesions of the cerebellum or cerebellar pathways.

Paresis
Function: *noun*
Inflected Form(s): *plural* pa·re·ses \-ˌsēz\
1: slight or partial paralysis 2: general paresis
http://www.merriam-webster.com/medlineplus/paresis

Definition of *PARESIS*-
: slight or partial paralysis
: general paresis
http://www.merriam-webster.com/medical/paresis

Noun -paresis (*plural* pareses) – 1) A paralysis which is incomplete or which occurs in isolated areas, 2) Inflammation of the brain as a cause of dementia or paralysis.
http://en.wiktionary.org/wiki/paresis

Paresis is derived from the Greek word for relaxation and implies weakness of motor function without complete paralysis (some voluntary motion remains). *Paralysis* implies loss of motor function from either a neurologic or muscular abnormality. The term "paresis," however, has different connotations including limb dysfunction from various causes. In many clinical descriptions, paresis and conscious proprioceptive abnormalities are used interchangeably, with the term "paresis" implying a conscious proprioceptive abnormality.

Lame (ness)
Function: *adjective*
Inflected Form(s): lam·er; lam·est
: having a body part and especially a limb so disabled as to impair freedom of movement : physically disabled
—lame·ly *adverb*
—lame·ness *noun*
http://www.merriam-webster.com/medlineplus/lameness

Definition of *LAME*
: having a body part and especially a limb so disabled as to impair freedom of movement : physically disabled
http://www.merriam-webster.com/medical/lameness

Noun- lameness (*countable and uncountable, plural* lamenesses) -A impediment to walking due to the feet or legs.
http://en.wiktionary.org/wiki/lameness

The aim of this session is two-fold. First the current state of medical descriptions will be discussed, followed by audience feedback opportunities. The outcome is to record audience perspectives on currently used medical terminology often used to describe abnormalities of gait in dogs.

Materials and Methods (intended)
This study will be approved by the institution review board of Iowa State University. All participation will be voluntary. Appropriate study explanations and consent will be obtained prior to participation. Board certified veterinary neurologists attending

the premiere international scientific meeting of this specialty group will be invited to participate in this study. While actual numbers of participants can only be determined at the time of the study, it is anticipated that the participant numbers will be greater than 50. At present, there are 269 board certified neurology diplomates in the American College of Veterinary Internal Medicine subspecialty of Neurology.

The study will be performed through audience participation at the American College of Veterinary Internal Medicine Meeting between June 3 and 6 2014 in Indianapolis Indiana, USA.

The facilitator will project currently published medical definitions of the following terms:

- Ataxia
- Paresis
- Hypermetria
- Lameness

Diplomates will view the presentation which will consist of clinically descriptive words and associated medical dictionary or standard textbook literal definitions. Diplomates will then rate the accuracy in their professional opinion of the definitions presented based on their operational definitions. Diplomates will then view videos of affected dog patients during voluntary movement (gaiting) and subsequently record the medical terminology they would enter into the patient medical record to describe the their assessment from a list of appropriate terminology.

Finally, Diplomates will have an opportunity to record their operational clinical definitions for these abnormalities.

Data will be recorded through responses in Turningpoint™ and recorded with clicker technology.

Data assessment will be performed by an independent assessment individual (DS).

Following recording of all responses, the responses will be shared with the audience. There will be an opportunity for audience discussion and feedback.

REFERENCES

Bagley, R. S. 2005. Fundamentals of Clinical Neurology, Blackwell Publishing.
Brown, C. M. 1986. Dog Locomotion and Gait Analysis. Hoflin Publishing.
Dewey C.W. 2008. A Practical Guide to Canine and Feline Neurology, Wiley Blackwell.

GENE-ENVIRONMENT INTERACTIONS AND CANCER RISK IN DOGS

Lauren A. Trepanier, DVM, PhD, DACVIM, DACVCP
Madison, WI

TUMORS ASSOCIATED WITH ENVIRONMENTAL EXPOSURES

A number of human cancers have strong associations with environmental toxins. The individual risk of developing cancer, however, is modified by genetic variability, including genes in pathways such as carcinogen bioactivation, xenobiotic detoxification, and DNA repair. Human cancers with well-established environmental links include lung cancer from tobacco smoke, colon cancer from heterocyclic amines in grilled meats, esophageal cancer and alcoholism, and hepatocellular carcinoma and aflatoxin exposure. Non-Hodgkin lymphoma (NHL) and transitional cell carcinoma (TCC) of the bladder are also associated with environmental toxins in human patients, and these tumors are common in dogs.

Lymphoma

Exposure to pesticides is a risk factor for NHL in humans.[1,2] In particular, the use of phenoxy herbicides, including the commonly used 2,4-D, has been linked in multiple studies with risk for NHL.[3-7] 2,4-D is found in agricultural fields, golf courses, parks, treated lawns, and even indoors in household carpet dust of nearby homes.[8,9] 2,4-D itself is not highly mutagenic; however, other components of commercial 2,4-D formulations may be important in cancer development. In dogs, lymphoma has also been associated with exposure to pesticides,[10] and two publications found a significant association between lymphoma and lawn treatment with 2-4-D.[11,12] These data were refuted by an industry-funded task force, however, and remain controversial.[13,14] Dogs do absorb 2,4-D from treated lawns,[15] and excrete this compound in urine as glucuronide, sulfate, and other conjugates.[16] 2,4-D excretion involves active renal tubular transport, which is apparently less efficient in dogs compared to humans and rodents.[17] The elimination half-life of 2,4-D in dogs is quite long (over 4 days);[16] this may make dogs more susceptible to adverse effects from standard 2,4-D applications. This risk may be modulated further by pharmacogenetic differences in renal transporters or conjugation pathways among individual dogs or dog breeds, and this deserves further evaluation.

In addition to herbicide exposure, NHL and other lymphoid cancers have been associated with exposure to benzene,[18-21] an air pollutant found in cigarette smoke, vehicle exhaust, and industrial solvents. The risk of NHL is also increased with exposure to methylene chloride,[22,23] an organic solvent used in food, pharmaceutical, and electronics industries. In the household, methylene chloride is found in degreasers, paint strippers, and aerosol spray propellants (www.epa.gov). Both benzene and methylene chloride are found as pollutants in landfills and near manufacturing sites in most states in the U.S. (http://www.atsdr.cdc.gov/). In dogs, lymphoma has been associated with living in industrial areas and in proximity to polluted sites.[24,25] Further, household storage of paints and solvents was associated with an earlier age of onset of lymphoma compared to non-exposed dogs.[25] Overall, these findings suggest a possible causal relationship between certain environmental chemicals and canine lymphoma. However, it is unclear which chemicals are important and what other factors may modulate risk after exposure.

Bladder TCC

Transitional cell carcinoma of the bladder is strongly associated with smoking in humans,[26] with up to 50% of risk attributed to tobacco use in men. The tobacco constituent 4-aminobiphenyl is a known bladder carcinogen,[27] and DNA adducts from this chemical are found in the urothelium of smokers.[28] 4-aminobiphenyl also leads to bladder cancer experimentally in dogs.[29] Although one study did not find an association between second-hand smoke and bladder TCC in dogs,[30] this area deserves further exploration. Exposure to halogenated byproducts of chlorinated water is also a bladder cancer risk factor in humans,[31] although a recent study in 100 dogs with bladder tumors did not find an association.[32]

In both humans and dogs, bladder cancer rates map to regions with high levels of manufacturing.[33] The industrial solvent benzidine, in particular, is an occupational risk for bladder cancer in people, and is an experimental bladder carcinogen in dogs.[29] Benzidine is an aromatic amine previously used in dye manufacturing, and is still found in some polluted sites in the U.S.[34] It is not known whether environmental exposure to benzidine or similar aromatic amines plays a role in naturally occurring bladder cancer in pet dogs. Other exposures that have been linked to bladder TCC in dogs include flea and tick dips,[30] and access to a lawn or garden treated with herbicides in the year prior to diagnosis.[35] Exposure to phenoxy herbicides, to include 2,4-D, was implicated in particular (OR 4.42, 95% CI, 1.74-11.19).[35]

GENETIC VARIABILITY AND CANCER RISK

Many carcinogens are mutagenic via reactive metabolites that bind to DNA, and therefore require bioactivation to initiate carcinogenesis. Bioactivation, detoxification, and DNA repair pathways can each modulate individual responses to the same carcinogen exposure. Understanding individual variability in these pathways can strengthen associations between environmental pollutants and cancer outcomes. There are numerous data in humans that can inform our understanding of risk exposures in dogs.

Carcinogen bioactivation

The CYP1A family is a predominant group of cytochrome P450's involved in bioactivation of carcinogens. CYP1A2, which is expressed in the liver, bioactivates the bladder carcinogen 4-aminobiphenyl, the lung carcinogen benzo[a]pyrene (found in cigarette smoke and car exhaust), and the colon carcinogen PhIP (found in meats cooked at high temperatures).[36] In addition, CYP1A2 expression is induced by many of these substrates. Rapid activity of CYP1A2 has been linked to colorectal cancer, squamous cell carcinoma of the lung, and bladder cancer, particularly in association with smoking.[37-39] CYP1A2 also bioactivates aflatoxin, and low activity alleles appear to protect against hepatocellular carcinoma in susceptible human patients.[40]

In dogs, CYP1A2 is known to be polymorphic. Four to 17% of colony beagles are homozygous for a nonsense mutation (1117C>T; Arg373*) in canine *CYP1A2* that leads to a truncated and inactive protein.[41-44] This homozygous defect has also been found in Irish Wolfhounds, Whippets, Dalmations, Australian Shepherds, and pet beagles.[45] In another survey, the defective homozygous *CYP1A2* genotype was not found in any of 105 boxers, 84 Rottweilers, 110 German shepherds, and 87 mixed breed dogs in Brazil.[46] Absence of CYP1A2 activity in some breeds may be protective against carcinogen bioactivation, and should be considered in risk assessment for carcinomas of the bladder, lung, colon, and liver in dogs.

Carcinogen detoxification

In contrast to oxidation reactions mediated by cytochrome P450s, conjugation by Phase II enzymes usually leads to non-toxic metabolites. *N*-acetyltransferases (NAT) conjugate reactive compounds to an acetyl group; this can neutralize the generation of free radicals. Both *N*-acetyltransferase enzymes, NAT1 and NAT2, are important in the detoxification of carcinogens such 4-aminobiphenyl, found in inhaled and second hand tobacco smoke. About 50% of Americans have polymorphisms that lead to decreased or absent NAT enzyme activity ("slow acetylators"), and this is a risk factor for bladder cancer in smokers.[47] The dog genome lacks *NAT1* and *NAT2* genes,[48] so dogs cannot *N*-acetylate carcinogens. This could possibly contribute to the risk of bladder TCC in dogs as a species.

In some instances conjugates can undergo cleavage to release reactive electrophiles that can damage DNA. NATs bioactivate the dietary carcinogens MeIQx and PhIP, leading to the formation of DNA adducts in the colon and other tissues. A high activity NAT1 allele (*NAT1*10*) appears to contribute to the risk of colorectal cancer, in humans[49] as well as cancers of the lung,[50] stomach,[51,52] and prostate.[53,54] Since dogs lack *NAT* genes, this may be protective against colon carcinogens that are bioactivated by this pathway.

Glutathione-S-transferases (GSTs) are Phase II biotransformation enzymes that conjugate xenobiotics and endogenous hydroperoxides to glutathione; this typically neutralizies these reactive molecules that could otherwise damage DNA.[55] There are multiple families of human GSTs, to include GST-mu (GSTM), GST-theta (GSTT), GST-pi (GSTP), and others. These families contain various isoforms, and exhibit considerable inter-individual variability in expression and activity. Both GSTM1 and GSTT1 show a "null" genotype in humans, where the respective genes are partially missing at the genomic level, and expression and activity are absent.[56] These *GSTT1* and *GSTM1* null genotypes are present in about 20 and 50% of Americans, respectively. In addition, single nucleotide polymorphisms in GSTs can lead to impaired enzyme expression or catalytic activity.

Toxic GST substrates include butadiene, DDT, lindane, alachlor, aflatoxin, and benzo[a]pyrene, as well as methylene chloride, a chlorinated hydrocarbon that is linked to lymphoma in humans.[57] Null GST genotypes in humans are associated with DNA adducts from aflatoxin, 4-aminobiphenyl, and benzo[a]pyrene; mutations in the oncogenes p53, Ki-ras, and HPRT; chromosomal damage, and increased urine mutagenicity.[58,59] Further, defective GST alleles are over-represented in patients with various solid tumors,[59-61] and with hematopoietic malignancies.[62-65] For example, the null *GSTT1* genotype has been associated with Hodgkin and non-Hodgkin lymphoma in multiple human studies,[65-69] and the *GSTM1* null genotype is a risk factor for bladder TCC.[70]

In dogs, GST-theta hepatic activities vary more than 45-fold,[71] with 12% of beagles lacking GST-theta activity and protein expression.[71,72] We hypothesized that, as in humans, defective or null *GSTT1* genotypes would be present in dogs, and would be associated with an increased risk of lymphoma. We resequenced *GSTT1* in 93 dogs with lymphoma and 86 age-matched healthy controls, and found 27 *GSTT1* polymorphisms. One splice site variant, I2+28 G>A, was significantly over-represented in dogs with lymphoma (OR 6.26, 95% CI, 1.77-22.2). The I2+28 AA genotype was found in 18.3% of LSA dogs but in only 3.5% of healthy dogs ($P = 0.002$). This SNP was predicted to affect splicing of *GSTT1* mRNA; functional characterization is underway. We also found a variant haplotype in the 3'untranslated region (3'UTR) of canine *GSTT1* which was observed with a minor allele frequency (MAF) of 0.129 in dogs with LSA, compared to healthy dogs (MAF 0.048; $P = 0.009$). This variant leads to a 50% reduction in gene expression (Craft and Trepanier, manuscript in preparation), which may lead to impaired carcinogen detoxification in dogs. We are currently evaluating these variants in the setting of canine bladder cancer risk, in collaboration with Drs. Deborah Knapp, Larry Glickman, and Alison Motsinger-Reif.

DNA repair

DNA repair pathways combat DNA damage from both exogenous and endogenous reactive molecules. DNA that is oxidized, alkylated, dimerized, or adducted can be repaired by base excision repair or nucleotide excision repair.[73] Therefore, variants in DNA repair genes also impart cancer risk, especially in combination with exposure to environmental mutagens. For example, polymorphisms in the base excision repair genes *OGG1*, *XRCC1*, and *APE1* have been associated with lymphoma and tumors of the GI tract, lung, bladder, and colon in humans.[74] *BRCA2* (the "breast cancer gene") functions to repair DNA strand breaks, and polymorphisms in this gene show a strong association with familial breast cancer risk in women.

In dogs, *BRCA2* mutations have also been documented,[75] and a variant *BRCA2* gene locus was associated with a 4-fold increase in risk for mammary cancer in unspayed dogs.[76] Springer spaniels were over-represented. In Golden retrievers, lymphocytes from dogs with untreated lymphoma showed increased susceptibility to exogenous *in vitro* DNA damage compared to healthy age-matched Golden retrievers or mixed breeds.[77] This suggests that a subset of Golden retrievers may be deficient in one or more DNA repair mechanisms, and this remains to be explored.

FUTURE DIRECTIONS

Exposures to pesticides and environmental pollutants are known risk factors for many tumor types in humans including lymphoma, bladder TCC, primary lung tumors, hepatocellular carcinoma, leukemia, and gliomas. Many of these cancers affect specific dog breeds, to include Golden retrievers (lymphoma),[78] Scottish terriers (bladder TCC),[79] German Shepherds (leukemia),[80] and boxers (gliomas),[81,82] which suggests possible gene-environment interactions in dogs. There is enormous potential for us to better understand the molecular epidemiology of environmental cancer risk in dogs, and this can be achieved with multi-center collaborations among veterinary specialists in oncology, pharmacogenetics, and epidemiology.

Selected gene-environment interactions for cancer risk in humans: relevance for dogs

TUMOR	ENVIRONMENTAL RISK IN HUMANS	ENVIRONMENTAL RISK IN DOGS	MODULATING GENES IN HUMANS AND DOGS
Non-Hodgkin lymphoma	Phenoxy herbicides, including 2,4-D[3-7] Benzene (pollutant) Methylene chloride (food, pharmaceutical, and electronics manufacturing pollutant)	Pesticides[1,2] 2,4-D[11,12] Living in proximity to industrial or polluted sites.[24,25] Household storage of paints and solvents.[83]	Increased risk in humans: Low activity *GSTM1* and *GSTT1* genotypes[62,65-69] Increased risk in dogs: *GSTT1* polymorphisms[84] Impaired DNA repair (Some Golden retrievers)[77]
Bladder TCC	Tobacco smoke Benzidine (pollutant from dye manufacturing)	Negative for tobacco smoke in one study[30] Flea and tick dips[30] Lawn herbicides, including 2,4-D[35]	Increased risk in humans: Slow *NAT2* alleles in smokers[47] GST null variants[70] Risk in dogs: All dogs lack *NAT1* and *NAT2* genes[48] Study on GST variants underway
Primary lung tumors	Smoking Second hand smoke Radon	Associated with accumulation of particulates (anthracosis), which may reflect inhaled pollutants[85] Second hand smoke (in short-nosed breeds only)[86]	Increased risk in humans: High activity CYP1A1 and CYP1A2[37,87]
Nasal adenocarcinoma	Rare in humans	Indoor exposure to coal or kerosene heaters, and second hand smoke, in dolichocephalic breeds[88,89]	Not evaluated in humans due to low incidence
Hepatocellular carcinoma	Aflatoxin ingestion (contaminated grain and corn)	Not evaluated	Decreased risk in humans: Low activity CYP1A2 alleles[40] Note: Null activity *CYP1A2* genotype found in some Irish Wolfhounds, Whippets, Dalmations, Australian Shepherds, and Beagles[45]
Leukemia	Benzene exposure (cigarette smoke, vehicle exhaust, and industrial pollution) 1,3 butadiene;[90] found in air and water near rubber and plastics plants	Not evaluated	Increased risk in humans: Low activity GST genotypes[63,64]
Gliomas	Chlorpyrifos exposure[90,91]	Not evaluated	Increased risk in humans: Low activity GSTP alleles[92-94]
Colorectal adenocarcinoma	Dietary carcinogens (PhIP and other heterocyclic amines) found in grilled meats	Not evaluated	Increased risk in humans: Rapid CYP1A2 alleles, in association with smoking and red meat intake[38,95]

REFERENCES

1.Fritschi L, et al. *Am J Epidemiol* 2005;162:849-857. 2.Pahwa M, et al. *Int J Cancer* 2012;131:2650-2659. 3.Hardell L, et al. *Br J Cancer* 1981;43:169-176. 4.Hoar SK, et al. *JAMA* 1986;256:1141-1147. 5.Persson B, et al. *Br J Ind Med* 1989;46:516-520. 6.Mills PK, et al. *Cancer Causes Control* 2005;16:823-830. 7.Miligi L, et al. *Ann N Y Acad Sci* 2006;1076:366-377. 8.Colt JS, et al. *J Expo Anal Environ Epidemiol* 2004;14:74-83. 9.Ward MH, et al. *Environ Health Perspect* 2006;114:893-897. 10.Takashima-Uebelhoer BB, et al. *Environ Res* 2012;112:171-176. 11.Hayes HM, et al. *Environ Res* 1995;70:119-125. 12.Hayes HM, et al. *J Natl Cancer Inst* 1991;83:1226-1231. 13.Carlo GL, et al. *Regul Toxicol Pharmacol* 1992;16:245-252. 14.Kaneene JB, Miller R. *Vet Hum Toxicol* 1999;41:164-170. 15.Reynolds PM, et al. *Cancer Epidemiol Biomarkers Prev* 1994;3:233-237. 16.van Ravenzwaay B, et al. *Xenobiotica* 2003;33:805-821. 17.USDA. 2,4-D: Human health and ecological risk assessment. Arlington, VA: USDA Forest Service, Department of Agriculture, 2006. 18.Hayes RB, et al. *J Natl Cancer Inst* 1997;89:1065-1071. 19.O'Connor SR, et al. *J Pathol* 1999;189:448-453. 20.Cocco P, et al. *Occup Environ Med* 2010;67:341-347. 21.Vlaanderen J, et al. *Environ Health Perspect* 2011;119:159-167. 22.Barry KH, et al. *Am J Epidemiol* 2011;173:404-413. 23.Cooper GS, et al. *Int J Environ Res Public Health* 2011;8:3380-3398. 24.Pastor M, et al. *J Vet Intern Med* 2009;23:301-310. 25.Gavazza A, et al. *J Vet Intern Med* 2001;15:190-195. 26.Marcus PM, et al. *Pharmacogenetics* 2000;10:115-122. 27.Poirier MC, et al. *Carcinogenesis* 1995;16:2917-2921. 28.Malaveille C, et al. *Carcinogenesis* 1996;17:2193-2200. 29.Deichmann WB, Radomski JL. *J Natl Cancer Inst* 1969;43:263-269. 30.Glickman LT, et al. *J Toxicol Environ Health* 1989;28:407-414. 31.Villanueva CM, et al. *Am J Epidemiol* 2007;165:148-156. 32.Backer LC, et al. *J Am Vet Med Assoc* 2008;232:1663-1668. 33.Hayes HM, Jr., et al. *Am J Epidemiol* 1981;114:229-233. 34.Brown SC, et al. *Am J Ind Med* 2011;54:300-306. 35.Glickman LT, et al. *J Am Vet Med Assoc* 2004;224:1290-1297. 36.Zhou SF, et al. *Drug Metab Rev* 2010;42:268-354. 37.Singh AP, et al. *Cancer Biomark* 2010;8:351-359. 38.Bae SY, et al. *Cancer Sci* 2006;97:774-779. 39.Tao L, et al. *Int J Cancer* 2012;130:1174-1183. 40.Chen X, et al. *Pharmacogenet Genomics* 2006;16:219-227. 41.Mise M, et al. *Pharmacogenetics* 2004;14:769-773. 42.Mise M, et al. *Drug Metab Dispos* 2004;32:240-245. 43.Tenmizu D, et al. *Xenobiotica* 2004;34:835-846. 44.Whiterock VJ, et al. *Drug Metab Lett* 2007;1:163-165. 45.Aretz JS, Geyer J. *J Vet Pharmacol Ther* 2011;34:98-100. 46.Scherr MC, et al. *J Vet Pharmacol Ther* 2011;34:184-186. 47.Marcus PM, et al. *Cancer Epidemiol Biomarkers Prev* 2000;9:461-467. 48.Trepanier LA, et al. *Biochem Pharmacol* 1997;54:73-80. 49.Lilla C, et al. *Cancer Epidemiol Biomarkers Prev* 2006;15:99-107. 50.Abdel-Rahman SZ, et al. *Mutat Res* 1998;398:43-54. 51.Boissy RJ, et al. *Int J Cancer* 2000;87:507-511. 52.Katoh T, et al. *Int J Cancer* 2000;85:46-49. 53.Fukutome K, et al. *Cancer Lett* 1999;136:83-87. 54.Hein DW. *Mutat Res* 2002;506-507:65-77. 55.Blair IA. *Curr Drug Metab* 2006;7:853-872. 56.Hayes JD, et al. *Annu Rev Pharmacol Toxicol* 2005;45:51-88. 57.Seidler A, et al. *J Occup Med Toxicol* 2007;2:2. 58.Musak L, et al. *Mutat Res* 2008;641:36-42. 59.Rebbeck TR. *Cancer Epidemiol Biomarkers Prev* 1997;6:733-743. 60.Strange RC, et al. *Arch Toxicol Suppl* 1998;20:419-428. 61.Sorensen M, et al. *Int J Cancer* 2004;110:219-224. 62.Soucek P, et al. *Int Arch Occup Environ Health* 2002;75 Suppl:S86-92. 63.Krajinovic M, et al. *Pharmacogenetics* 2002;12:655-658. 64.Ye Z, Song H. *Eur J Cancer* 2005;41:980-989. 65.Al-Dayel F, et al. *Leuk Lymphoma* 2008;49:122-129. 66.Kerridge I, et al. *Br J Haematol* 2002;118:477-481. 67.Hohaus S, et al. *Clin Cancer Res* 2003;9:3435-3440. 68.Ruiz-Cosano J, et al. *Mol Carcinog* 2011. 69.Abdel Rahman HA, et al. *J Cancer Res Clin Oncol* 2012;138:1363-1368. 70.Simic T, et al. *Nat Rev Urol* 2009;6:281-289. 71.Watanabe T, et al. *Arch Toxicol* 2004;78:218-225. 72.Watanabe T, et al. *Arch Toxicol* 2006;80:250-257. 73.Grosse N, et al. *BMC cancer* 2014;14:203. 74.Karahalil B, et al. *Hum Exp Toxicol* 2012;31:981-1005. 75.Yoshikawa Y, et al. *J Vet Med Sci* 2005;67:1013-1017. 76.Rivera P, et al. *Cancer research* 2009;69:8770-8774. 77.Thamm DH, et al. *PLoS One* 2013;8:e69192. 78.Wilkerson MJ, et al. *Vet Immunol Immunopathol* 2005;106:179-196. 79.Childress MO, et al. *J Am Vet Med Assoc* 2011;239:350-356. 80.Tasca S, et al. *Vet Clin Pathol* 2009;38:2-12. 81.Lurie DM, et al. *Vet Comp Oncol* 2004;2:171-175. 82.Stoica G, et al. *Vet Pathol* 2011;48:266-275. 83.Gavazza A, et al. *J Vet Intern Med* 2001;15:190-195. 84.Ginn J, et al. *Vet Comp Oncol* 2014;12:227-236. 85.Bettini G, et al. *Vet J* 2010;186:364-369. 86.Reif JS, et al. *Am J Epidemiol* 1992;135:234-239. 87.Kawajiri K, et al. *Princess Takamatsu Symp* 1990;21:55-61. 88.Bukowski JA, et al. *J Toxicol Environ Health A* 1998;54:579-591. 89.Reif JS, et al. *Am J Epidemiol* 1998;147:488-492. 90.Clapp RW, et al. *Rev Environ Health* 2008;23:1-37. 91.Karunanayake CP, et al. *J Agromedicine* 2012;17:30-39. 92.De Roos AJ, et al. *Cancer Epidemiol Biomarkers Prev* 2003;12:14-22. 93.Custodio AC, et al. *Genet Mol Res* 2010;9:2328-2334. 94.Ezer R, et al. *J Neurooncol* 2002;59:123-134. 95.Badawi AF, et al. *Prog Clin Biol Res* 1996;395:109-140.

BARTONELLOSIS AND CANCER: IS THERE A LINK?

Edward B. Breitschwerdt, DVM, DACVIM (Small Animal Internal Medicine)
Raleigh, NC

INTRODUCTION

Bartonella species are fastidious Gram-negative bacteria that are highly adapted to a mammalian reservoir host and within which the bacteria usually cause a long-lasting intra-erythrocytic bacteremia.[1-3] As an increasing number of animal reservoir hosts have been identified for various *Bartonella* species. These intra-erythrocytic, endotheliotropic, zoonotic pathogens have become of particular importance to veterinarians and due to their zoonotic potential, physicians. Among numerous other examples, *Bartonella henselae* has co-evolved with cats, *Bartonella vinsonii* subsp. *berkhoffii* has co-evolved with dogs and wild canines, and *Bartonella bovis* has co-evolved with cattle. Importantly, the list of reservoir-adapted *Bartonella* species, including a large number of rodent species that might serve as "pocket pets", continues to grow exponentially, as new *Bartonella* spp. are discovered.[2-4] Prior to 1990, there were only two named *Bartonella* species, whereas there are now at least 32 named and numerous unnamed or candidatus species, based upon deposited GenBank sequences or preliminary reports, respectively.

As recently reviewed,[4] an increasing number of *Bartonella* species (currently 14 species) have been identified as zoonotic pathogens; transmitted by animal bites, scratches, arthropods, and even by needle sticks to veterinary workers. Considering the diversity of newly discovered *Bartonella* species and subspecies, the large number and ecologically diverse animal reservoir hosts and the evolving spectrum of arthropod vectors that can transmit these bacteria among animals and humans, the cumulative diagnostic challenges posed by *Bartonella* transmission in nature are presumably much more complex than is currently appreciated by diagnosticians, vector biologists, ecologists, physicians or veterinarians. Historically the term bartonellosis was attributed to infections with *Bartonella bacilliformis,* transmitted by sandflies in the Peruvian Andes, whereas bartonellosis currently encompasses a more inclusive medical use of this term and now includes infections caused by any *Bartonella* sp., anywhere in the world. Potentially, because *Bartonella* spp. can infect erythrocytes, endothelial cells, pericytes, CD34+ progenitor cells, and various macrophage-type cells (including microglial cells, dendritic cells and circulating monocytes) *in vitro*, the clinical and pathological manifestations of bartonellosis appear to be very diverse in both sick animals and human patients.

INFECTIOIN, INFLAMMATION, AND CANCER

Within the past 25 years, considerable research has been conducted on the oncogenic properties of infectious agents such as bacteria, viruses, mycoplasma, and protozoa. Currently, infectious agents are accepted as causes or co-factors in nearly 20% of human cancers worldwide.[5] A majority of these infectious agents are viruses, such as Epstein Barr virus, human papillomaviruses, and Kaposi's sarcoma-associated herpesvirus, which have direct oncogenic properties by integration of viral genomes into host cells or by secretion of gene products into healthy cells to create tumor cells.[5] Although other infectious agents, such as bacteria, may lack the inherent oncogenic properties of their viral counterparts, these organisms can indirectly promote cancer development through persistent replication, inflammation, and chronic tissue damage.[6] *Helicobacter pylori*, for example, colonizes and replicates within the gastric mucosa and induces a chronic pro-inflammatory response, which promotes cancer risk and oncogenesis of gastric cancer through functional changes associated with epithelial cell proliferation and apoptosis.[6,7] Compared to viruses, the potential role of bacteria in cancer causation is less well studied.

BARTONELLA SPECIES AND THE POSTULATE OF COMPARATIVE INFECTIOUS DISEASES CAUSATION

Recently, we emphasized the comparative medical importance and pathogenic relevance of combining clinical, microbiological, and pathological observations across animal species in order to elucidate the capacity of a specific microorganism (or combination of microorganisms in the context of co-infections) as a cause of similar disease manifestations or identical pathology in more than one mammalian species.[8] Based upon recent research observations derived from studies involving the genus *Bartonella*, we proposed that the following postulate be added to Koch's postulates so as to assist in the establishment of disease causation by stealth pathogens: **Causation can be established, if the same infectious agent (or combination of agents) are isolated or organism-specific DNA sequences are amplified from a naturally-occurring pathological entity found in at least three different mammalian species.** Although the selection of a minimum of three different mammalian species is arbitrary, the strength of this type of association would be supported by the fact that random detection of DNA of a specific organism in an uncommon pathological lesion obtained from different animals at different time points would have a low probability of being an unrelated occurrence. We suggested that the new postulate be referred to as the "**Koch's postulate of comparative infectious disease causation**".[8] It is possible that the application of this postulate could be used in conjunction with high throughput sequencing to investigate the role of specific or combinations of infectious agents as a cause of cancer across animal species.

BARTONELLA INFECTION AND VASOPROLIFERATIVE PATHOLOGY

In North America prior to the 1990s, *Bartonella* species were not known to infect animals or human beings. If it were not for the epidemic of acquired immunodeficiency syndrome (AIDS) and the recognition of two prototypical vasoproliferative lesions in this population of immunocompromised patients, it is likely that much of the contemporary literature relative to the genus *Bartonella* as pathogens of increasing microbiological relevance in human and veterinary medicine would not exist. Early in the 1990s, *Bartonella quintana* and *Bartonella henselae* were implicated as the cause of bacillary angiomatosis (BA), a vasoproliferative lesion of the skin, which subsequently was described most often in immunocompromised patients infected with the human immunodeficiency virus (HIV).[9] Subsequently, BA was reported for the first time in an iatrogenically-immunosuppressed dog infected with *Bartonella vinsonii* subsp. *berkhoffii.*[10] This observation was to become of comparative microbiological relevance in the context of disease pathogenesis,[8] as subsequent studies documented the ability of *B. vinsonii* subsp. *berkhoffii* to induce the production of vascular endothelial growth factor (VEGF),[11] as previously described for *B. quintana* and *B. henselae*.[12-14] Immunosuppression apparently contributed to the subsequent development of BA in this dog, as has been documented for HIV-infected human BA patients. Also, on a comparative pathological/infectious disease basis, *B. henselae* is the only member of the genus that has been associated with peliosis hepatis (PH; a vasoproliferative lesion in the liver characterized by the presence of irregular blood-filled spaces within the parenchyma) in human AIDS patients or in dogs.[12,13] Thus, on a comparative infectious disease causation basis, DNA of the same *Bartonella* species has been amplified from dogs and people with the same clinically suspected and histopathologically confirmed vasoproliferative lesions (PH and BA).[9,12]

As described below, angiogenesis or angiogenic factors play an important and potentially central role in a variety of pathological processes including cancer. A substantial body of literature now supports the ability of *B. henselae, B. quintana, B. bacilliformis* and *B. vinsonii* subsp. *berkhoffii* to induce the production of VEGF and resultant proliferation of endothelial cells.[11,14-18] As discussed above, infection with a *Bartonella* spp. may be far more ubiquitous in animals and humans than has been historically realized.[1-4] Thus, these 'stealth', intravascular bacteria could activate an 'angiogenic switch', secondary to tumor hypoxia within the stroma[16] and concurrent, secondary inflammation resulting in HIF-induced bacterial signaling, so as to ensure increased vascularization and exponential tumor growth. In the context of naturally-occurring cancers, increased expression of VEGF has been documented in association with several vascular tumors including hemangiosarcoma in dogs[19] and hemangiopericytoma and epithelioid hemangioendothelioma (EHE) in human patients[20,21] In 2009, *B. vinsonii* subsp. *berkhoffii* genotype II was isolated from the blood of a boy with hepatic EHE and from a dog with a hemangiopericytoma.[22] Subsequently, infection with *B. henselae* and *B. koehlerae* was documented in patients with EHE from Australia and England[23] and DNA of *B. henselae, B. vinsonii* subp. *berkhoffii,* or both organisms was amplified and sequenced from hemangiopericytomas in dogs, a horse, and a red wolf.[11] Thus, intravascular infection with one or more *Bartonella* species has been documented in EHE patients from three continents and in hemangiopericytomas from three animal species, suggesting that persistent intravascular infection with these bacteria might contribute to the development of these biologically unusual and uncommon vascular tumors in both immunocompetent people and animals, respectively. On a comparative microbiological and pathological basis, *B. henselae, B. koehlerae* or *B. vinsonii* subsp. *berkhoffii* DNA was PCR amplified and sequenced[11] from formalin-fixed and paraffin wax-embedded tissues from cats[24] and a steer[25] with systemic reactive angiomatosis. This rare neoplastic-like disease is characterized pathologically by the proliferation of vessels throughout numerous tissues within the animal's body. Collectively, the above comparative microbiological and pathological findings served as the basis for a retrospective study designed to determine if *Bartonella* spp. might play a role in the pathogenesis of canine hemangiosarcoma (a malignant tumor of vascular endothelial cells that generally arises in either the spleen or the right atrium of the heart), one of the most common cancers affecting dogs.[26] Using surgically obtained formalin-fixed and paraffin-wax embedded tissue biopsy samples from fifty dogs with splenic hemangiosarcoma, we found a statistically higher prevalence of *Bartonella* spp. DNA (26%), as compared to DNA from hemotropic *Mycoplasma* spp. (2%) or *Babesia* spp. (6%). In addition, *Bartonella* spp. DNA was statistically more prevalent in the spleens of dogs with hemangiosarcoma (26%), as compared to the spleens from dogs with lymphoid nodular hyperplasia (10%).[26] Organism-specific PCR did not amplify DNA of *Babesia, Bartonella* or hemotropic *Mycoplamsa* spp. from the spleens of specific-pathogen free research dogs. Although causation cannot be established by a geographically limited, retrospective study, the comparative pathological evidence outlined above implicates a potential role for *Bartonella* species in immunocompetent animals and people, including EHE in human patients, hemangiopericytoma, and hemangiosarcoma in dogs and systemic reactive angiomatosis in cats. If the association between infection with a *Bartonella* spp. and hemangiosarcoma is confirmed as casual, the development of a vaccine to prevent *Bartonella* sp. infections might decrease or eliminate one of the more common and serious malignancies affecting dogs, analogous to the papillomavirus vaccine that was recently introduced to decrease the prevalence of cervical cancer in woman.[27]

BARTONELLA SPECIES AND LYMPHOMA

Bartonella spp. are biologically capable of invading numerous host cells, within which the bacteria modify cellular functions by injecting peptides and potentially transporting bacterial DNA into the host cell.[16,28,29] In addition, several *Bartonella* sp., including *B. henselae,* contain bacteriophages[30] that might also facilitate DNA translocation events. Previously, we compared the molecular prevalence of *Bartonella* species DNA in the blood and lymph nodes of healthy Golden retrievers and Golden retrievers with lymphoma.[31] No difference in molecular prevalence was found among the two groups; however, it should be pointed out that these dogs were tested at a single point in time and that the overall *Bartonella* sp. prevalence in both groups was unexpectedly high (18%). Comparative studies of naturally-occurring lymphoma in animals and humans are warranted and could potentially generate evidence that supports a comparative infectious disease causation.[8] Although not well studied in human patients, dogs experimentally-infected with *Bartonella vinsonii* subsp. *berkhoffii* became immunosuppressed.[32] Potentially, DNA translocation events in conjunction with infection-induced immune suppression could be additive cofactors in the development of lymphoma in a subset of dogs.

CONCLUSION

As 75% of emerging infectious diseases are zoonoses, many of which are vector-transmitted by an arthropod, a One Health approach to bartonellosis and other zoonotic infections is needed to properly address animal health, public health and environmental factors that influence the distribution and transmission of these bacteria. The One Health concept encourages a spirit of cooperation among animal, environmental, and human health professionals and promotes developing integrated solutions for complex problems that impact the health of animals, humans and the planet. Importantly, substantial research is needed to define the medical importance of the genus *Bartonella* as a cause of animal and human illnesses and whether these bacteria might play a role in cancer.

REFERENCES

[1]Breitschwerdt EB, et al **Clin Microbiol Rev** 2000;13:428. [2]Chomel BB, et al **Ann NY Acad Sci** 2003;990:1. [3]Breitschwerdt EB, et al **J Vet Emerg Crit Care** 2010;20:8. [4]Breitschwerdt EB. **ILAR J** 2014;55:46. [5]Pagano JS, et al **Seminars in Cancer Biology** 2014;14:453. [6]Coussens LM et al **Nature** 2002;420: 860. [7]Lax AJ et al **Trends Microbiol** 2002;10: 293. [8]Breitschwerdt EB, et al **J Comp Pathol** 2013;1:1. [9]Koehler JE et al **Clin Infect Dis** 1993;17(4):612. [10]Yager JA, et al **Vet Dermatol** 2010;21:420. [11]Beerlage C, et al **Med Microbiol Immunol** 2012;201(3):319. [12]Koehler J, et al **N Engl J Med** 1997 Dec 25;337(26):1876-83. [13]Kitchell BE, et al **J Am Vet Med Assoc** 2000;216: 519. [14]Kempf VA, et al **Cell Microbiol** 2001;3:623. [15]Resto-Ruiz SI, et al **Infect Immun** 2002;70:4564. [16]Kempf VA, et al **Circulation** 2005;111:1054. [17]Berrich M, et al PLoS One 2011;6(5):e20204. [18]Dehio C. **Nat Rev Microbiol** 2005;3:621. [19]Yonemaru K, et al **Vet Pathol** 2006;43(6):971. [20]Emamaullee JA, et al **Liver Transpl** 2010;16:191. [21]Hatva E, et al **Am J Pathol** 1996;148:763. [22]Breitschwerdt EB, et al **J Clin Microbiol** 2009;47(6):1957. [23]Mascarelli PE, et al **J Clin Microbiol** 2011;49:4006. [24]Fuji RN, et al **Vet Pathol** 2005;42:608. [25]Breshears MA, et al. **Vet Pathol** 2008;45:645. [26]Varanat M, et al **J Vet Intern Med** 2011;25:1284. [27]Jenkins M, et al **Int Rev Immunol** 2012;31:3. [28]Pulliainen AT, et al **FEMS Microbiol Rev** 2012;36:563. [29]Ben-Tekaya H, et al **Cold Spring Harb Perspect Med** 2013;3(8):pii: a010231. [30]Maggi RG, et al. **J Mol Microbiol Biotechnol** 2005;9:44. [31]Duncan AW, et al **J Vet Intern Med** 2008;22:88. [32]Pappalardo BL, et al. **Vet Immunol Immunopathol** 2001;83:125.

BIOMARKERS FOR TARGETED ONCOLOGY DRUGS

Anthony J. Mutsaers, DVM, PhD, DACVIM (Oncology)
Guelph, ON, Canada

INTRODUCTION

There are a variety of definitions and use of the term "biomarker" in the fields of medicine and oncology. In general, this term refers to a broad subcategory of medical signs, which can be measured accurately and reproducibly.[1] As such, it is clear that biomarkers are utilized in patient management on a daily basis. In 1998, the National Institutes of Health Biomarkers Definitions Working Group defined biomarkers as "a characteristic that is objectively measured and evaluated as an indicator of normal biological processes, pathogenic processes, or pharmacologic responses to a therapeutic intervention."[2] This presentation is focused specifically on how tissue, imaging, blood and urine-based biomarkers are being investigated and may be applied in the context of targeted oncology drug application. Therefore, biomarkers associated with cancer detection, (e.g. early detection) and monitoring are not covered in detail, but may still be discussed in the context of targeted drug selection and monitoring.

Targeted Oncology Drugs

All drugs in oncology are considered targeted to a certain degree. Cytotoxic chemotherapy drugs have tissue selectivity based on their preferential targeting of rapidly dividing cells and indeed, a number of patient biomarkers have been investigated with regard to patient selection for proper chemotherapy choice, or perhaps more accurately, chemotherapy avoidance or dose reduction. The most common veterinary example would be testing for patient MDR1 status,[3] with consequent dose reduction in agents impacted by this drug resistance mechanism, such as doxorubicin and vincristine, among others. In human medicine, testing for certain genetic polymorphisms has reached the FDA approval stage for predicting side effects to drugs such as irintotecan, through UGT1A1 genotyping. The most common targeted oncology drugs currently being investigated for biomarker application in both human and veterinary oncology are small molecule and antibody-based kinase inhibitors.

DEFINING THE PROPER PATIENTS TO TREAT

There has been much interest in identifying patients who can accurately be predicted to respond to a specific targeted drug. Examples include kit mutation and canine mast cell tumors (MCT), HER2 overexpression and use of the monoclonal antibody trastuzumab in breast cancer, EGFR mutation and response to the small molecule inhibitor erlotinib or gefitinib, and lack of prediction of cetuximab monoclonal antibody response from EGFR expression in colorectal cancer. In this final example, the use of ras mutation as a biomarker actually predicts resistance to cetuximab treatment in this same patient population. Finally, for cancers in which a particular drug target is highly expressed or mutated, such as kit in gastrointestinal stromal tumors (GIST), or BCR-Abl in chronic myelogenous leukemia, drugs such as imatinib are highly effective in the vast majority of patients and these markers are utilized as an integral component of the diagnostic process. Therefore, expression of the drug target may not be considered a predictive biomarker in these particular patient populations because of its high frequency. However, these biomarkers may still have utility in identifying the very rare patient that may not respond to drug therapy due to a lack of target expression.

Predictive vs. Prognostic Biomarkers

These terms may often be confused in the context of interpretation of targeted drug clinical trial results. In the case of HER2 overexpression, this protein is both a prognostic indicator of poor outcome in breast cancer patients, as well as a robust predictor of response to trastuzumab treatment. In the case of canine mast cell tumors, c-kit mutation is a weaker predictor of response to kinase inhibitor therapy with drugs such as toceranib, as a subset of c-kit wild type tumors will also respond to the drug. In addition, both kit mutation and protein localization have been shown to impact prognosis of canine mast cell tumors, independent of how they may cases be treated.

MONITORING TREATMENT RESPONSE

RECIST and VCOG criteria have been useful guidelines for objective measurement of tumor response. Unfortunately there is a potential lack of objective response with certain categories of targeted drug use. The nature of blood vessel targeting with anti-angiogenic drugs is an example where it has been argued that sustained stable disease may be the most likely and desirable end point. Coincident with an appreciation of potentially valuable tumor response criteria beyond RECIST has been the desire to prove target modulation as a biomarker of targeted drug application. Importantly, there is a clear difference between documentation that a targeted drug has been optimally applied versus documentation of the subsequent tumor response.

(Functional) Imaging Studies

Imaging has always played a vital part of assessing the objective response to cancer therapy, and never more so than with targeted drug application. However, with the emergence of functional imaging modalities, such as positron emission tomography (PET), dynamic contrast-enhanced MRI or CT, and nuclear scintigraphy, pharmacodynamic assessment of specific target modulation is also achievable through assessment of tumor receptors or antigens. Finally, In addition to a specific drug target, many more generalized aspects of the tumor microenvironment can be assessed with functional imaging, such as vascularity, metabolism, hypoxia, bone turnover, and proliferative status, among others.

PHARMACODYNAMIC BIOMARKERS – OPTIMAL DRUG APPLICATION

The field of targeted therapy represented a departure from the well-known therapeutic concept in oncology that pertains to the maximum tolerated dose (MTD) of cytotoxic chemotherapy drugs. Targeted agents are generally given at the optimal biologic dose (OBD), often with minimal break periods and over prolonged periods of time. Conceptually, this approach is similar to the use of conventional chemotherapy agents in "metronomic" protocols.

To aid the optimization of monoclonal antibody therapies that may lack overt dose-limiting toxicities, we demonstrated a dose-dependent increase in circulating growth factors in the plasma that bind to the receptor being targeted. For example, monoclonal antibodies targeting vascular endothelial growth factor receptor-2 (VEGFR2) lead to dose dependent increases in circulating VEGF that plateaus at the optimal biologic dose.[4] This finding that VEGF could be used as a pharmacodynamic biomarker for the OBD of VEGFR2 antibodies was also demonstrated for EGFR-directed antibody therapy. In the case of EGFR, ligands, such as TGF-alpha became elevated in the plasma of patients being treated with the monoclonal antibody cetuximab.[5] The use of antibody therapeutics has historically been irrelevant to veterinary oncology, given the species-related difficulties inherent in the use of human monoclonal antibody drugs in veterinary patients. However, recent developments within this drug class mean these pharmacodynamic biomarker studies may have relevance as new antibody-based therapies make their way into veterinary oncology.

The targeted drug class that has the most current application to veterinary oncology are the small molecule kinase inhibitors, such as toceranib and masitinib. Very similar drugs used in human oncology, such as sunitinib, have also demonstrated dose-dependent changes in circulating cytokines, as was shown previously with antibody drugs. In the case of kinase inhibitors, growth factors considered "on target" such as VEGF and stem cell factor (SCF) were elevated during sunitinib therapy.[6] Interestingly however, certain "off target" growth factors are also modulated by treatment. The biologic consequences of these growth factor changes in the circulation are currently unknown, but pre-clinical evidence has suggested that they may play a role in altering the tumor microenvironment and possibly create a pre-metastatic niche in susceptible tissues. Finally, preliminary studies performed in vitro and with mouse models, suggest that the EGFR-targeted small molecule inhibitor erlotinib may also modulate certain off-target cytokines, such as granulocyte colony stimulating factor (G-CSF), with as-yet unknown consequences.

Veterinary Examples

Important examples already exist that illustrate the clinical application of these biomarker concepts using the drug toceranib in veterinary oncology. The first example concerns the OBD. The FDA-approved dose of toceranib, which was based on conventional clinical trial establishment of the MTD, is 3.25 mg/kg EOD.[7] Through clinical experience, it was suggested that a lower dose, which was better tolerated by patients, would still retain sufficient efficacy, and perhaps actually be the OBD. Follow up pharmacokinetic studies supported a dose of approximately 2.5 mg/kg EOD or MWF as optimal, and interestingly, this dose was indeed sufficient to show an increase in the pharmacodynamic biomarker VEGF in circulation,[8] as was previously demonstrated with sunitinib. Finally, preliminary data supports the possibility that again, like sunitinib, use of toceranib at the OBD results in elevation of certain cytokines considered off-target for this drug.

The Future

Where must we go from here to improve our understanding of the optimal use of biomarkers to guide targeted drug applications in oncology? As genomic and other "omic" approaches become more fully integrated into clinical decision making, the use of appropriate biomarkers for proper drug selection will become commonplace. The addition of functional imaging studies to serially monitor for treatment response and relapse in a non-invasive fashion will improve our ability to make treatment changes as drug resistance develops during tumor progression. Of course, many challenges remain in the validation of biomarkers and overcoming the complicated nature of tumor heterogeneity in the course of proper targeted drug selection.

REFERENCES

[1]Stirmbu K and Tavel JA. Curr Opin HIV AIDS 2010;5(6):463-466. [2]Biomarkers Definition Working Group. Clin Pharmocol Therapeutics 2001;69:89-95. [3]Mealey KL. J Vet Pharmacol Ther 2004;27(5):257-264. [4]Bocci, et al. Cancer Res 2004;64(18):6616-25. [5]Mutsaers AJ, et al. Clin Cancer Res 2009;15(7):2397-405. [6]Ebos JML et al, Proc Natl Acad Sci USA 2007;104(43):17069-74. [7]London CA, et al. Clin Cancer Res 2009;15(11):3856-65. [8]Bernabe LF, et al. BMC Vet Res 2013;9:190.

MICROBIOME AND HUMAN CRC: LESSONS LEARNED FROM ANIMAL MODELS

Christian Jobin, PhD
Gainesville, FL

INTRODUCTION

Despite the increasing implementation of colonoscopy screening and advances in chemotherapeutic and biological agent-based therapies, colorectal cancer (CRC) remains one of the most common and deadliest malignancies in the United States [1]. CRC includes hereditary, sporadic, and colitis-associated CRC (CAC). Cancer genome studies have linked many somatic mutations to CRC [2-4], but the etiology of CRC can only be partially attributed to genetics [2,5]. This suggests the involvement of environmental factors in disease pathogenesis. A prominent risk factor for CRC is duration and severity of inflammatory bowel diseases (IBDs), which significantly enhances CRC risk by ~60% compared to healthy controls [2,6]. Another key environmental factor in the development of CRC is the large microbial community inhabiting the gastrointestinal tract, particularly the colon [7-9]. Humans are populated with trillions of microbes at various body sites, with the intestine representing the major reservoir (~10^{14}). These intestinal microbiota consist of more than 10^6 bacterial genes [10,11] with a metabolic capacity matching that of the liver. Through the action of various microbial structural components, microbial gene products and metabolites, this microbiota plays essential roles in host homeostasis, regulating host immunity, gut barrier function and metabolic activity [8,12-14]. Although the intestinal microbial community exists in a state of homeostasis called eubiosis, changes in richness, diversity and stability of the gut bacterial ecosystem, a state referred to as microbial dysbiosis, are commonly associated with intestinal pathologies such as IBD and CRC [2,8]. Environmental (e.g: diet, stress, inflammation) and genetic factors can lead to microbial dysbiosis and associated pathologies. Numerous laboratories have compared the microbiome of patients at various states (or stages) of CRC with that of healthy controls and found evidence of microbial dysbiosis [8,15,16]. Among the various microbial candidates contributing to dysbiosis, *Fusobacterium sp*, especially *Fusobacterium nucleatum,* stood out as one of the most reproducible and robust bacterium found in CRC patients [17]. In addition, a higher prevalence of *Escherichia*, *Klebsiella*, *Enterococcus*, *Streptococcus*, and *Peptostreptococcus* was observed in the luminal compartment of CRC patients compared to controls in two separate Chinese cohorts as measured by 16S rDNA deep sequencing analysis [18,19]. Moreover, culture-based studies have validated the genomic approach and shown increased prevalence of adherent-invasive *Escherichia coli* (AIEC) in IBD and CRC patients compared to controls [14,20-22].

FUNCTIONAL IMPACT OF BACTERIA IN CRC USING PRE-CLINICAL MODELS

Albeit clinically important, microbiome studies performed with human intestinal samples (tissue or fecal materials) are essentially descriptive and correlative experimental approaches. Although these studies are essential for the identification of potential microbial candidates associated with CRC, animal studies are essential to identify mechanisms responsible for microbial dysbiosis and establishing cause/effect. A series of experiments performed using microbes identified from microbiome studies have greatly contributed to our understanding of the interplay between bacteria and intestinal pathologies. For example, a pair of papers demonstrated that *F. nucleatum* could promote CRC in a model of tumor transplantation or in $Apc^{min/+}$ mice, which carry a mutation in one allele of the tumor suppressor gene adenomatous polyposis coli (*Apc*) and are genetically predisposed to developing intestinal tumors [23,24]. The mechanism responsible for carcinogenesis is not fully understood but may involve increased cell proliferation through activating the WNT signaling [24] and/or induction of cancer-promoting immune responses [23]. As mentioned above, *Escherichia coli* has been associated with development of both IBD and CRC and functional studies point to an important role of this bacterium in CRC pathogenesis [25]. For example, mono-association (gnotobiotic technology) of the non-pathogenic *E. coli* mouse strain NC101 enhanced colonic tumor development in azoxymethane (AOM)-treated interleukin-10 knockout ($Il\text{-}10^{-/-}$) mice, a mouse model of colitis-associated CRC [2]. In addition, the human CRC mucosa-associated *E. coli* strain 11G5 promoted intestinal neoplastic changes when introduced in $Apc^{min/+}$ mice [26]. Another human CRC *E. coli* isolate CCR20, increased tumor formation in AOM/dextran sulfate sodium (DSS)-treated mice [7]. Therefore, *E. coli* represents a group of microorganisms that are implicated in the pathogenesis of both IBD and CRC. Subsequent experiments revealed that these *E. coli* (NC101 and 11G5) produce the genotoxin colibactin, which is encoded by a multi-enzymatic machinery present in the polyketide synthase (*pks*) island [26,27]. Colibactin-producing *E. coli* induce a double strand DNA break (DSBs) in various cells *in vitro* and *in vivo*, including intestinal epithelial cells [2,28-30].

FROM MICROBIOME ANALYSIS TO MICROBIAL GENES: DISSECTING MECHANISMS OF ACTION

Microbiome analysis and gnotobiotic technology are powerful means to establish functional impact of microorganisms in intestinal pathologies. However, deeper mechanistic investigations where microbial genes are genetically manipulated are necessary to generate therapeutic targets. Such investigations have been carried with *E. coli* carrying the polyketide synthase (*pks)* island. Deletion of the *pks* gene from *E. coli* prevents development of adenocarcinoma in a preclinical model [2]. Furthermore, higher prevalence of *E. coli pks* positive strains are found in patients with CRC compared to non-CRC controls [2,9]. Therefore, *E. coli* represents a group of microorganisms that are implicated in the pathogenesis of both IBD and CRC. *E. coli* is a bacterium which colonizes massively the mammalian gut within a few days after birth and becomes the predominant facultative anaerobic bacterium in the adult microbiota. The majority of *E. coli* strains can be assigned to seven phylogroups (A, B1, B2, C, D, E, F) [31]. Recent studies showed that the prevalence of isolates of the B2 group is increasing in human microbiota of the Western population [32,33], with strains reported to have enhanced capacity to colonize the human gut [33]. There is a concomitant increase in populations from industrialized countries of immune-mediated diseases along with a shift in *E. coli* commensal population from the declining phylogenetic group A to the newly dominant phylogenetic group B2, encompassing colibactin-producing strains carrying the genotoxin polyketide synthase (*pks*) island [34]. The functional implication of this *E.coli* phylogenic shift in human population is unknown.

Since its discovery in 2006 by the Oswald group [29], numerous laboratories have been working on purifying the final natural *pks* product, named colibactin. Unfortunately, isolation and characterization of colibactin has remained elusive. Colibactin is the product of a multi-enzymatic assembly line involving polyketide synthase and non-ribosomal peptide synthase. *E.coli* producing colibactin has been shown to induce DNA double strand breaks (DSBs) in vitro and in vivo, in various mammalian cells, including intestinal epithelial cells [2,28-30]. The event leading to colibactin production and mammalian host response is virtually unknown. Moreover, factors influencing colibactin production and activity are largely undefined. Using 16SrRNA analysis (Illumina sequence), we recently observed that inflammation enhances *E. coli* resilience and *pks* activity in *Il10*$^{-/-}$ mice [10]. We then used microbial RNA-sequence technique to study *E. coli* NC101 transcriptomic response to inflammation/cancer in *Il10*$^{-/-}$ mice. To investigate the effect of inflammation in *E. coli* NC101 gene expression, germ-free *Il10*$^{-/-}$ mice were mono-associated with *E. coli* NC101 for 2 weeks (mild inflammation), 12 weeks (chronic inflammation) and 20 weeks (inflammation and CRC) and microbial RNA was isolated from stool samples and transcript expression levels determined by RNA-sequence using Illumina HiSeq2000. Principal Component Analysis showed that *E. coli* NC101 gene expression profile clustered differently at 2 weeks, 12 weeks and 20 weeks in *Il10*$^{-/-}$ mice (Figure 5). A number of genes were differentially expressed over time (~1000 genes) and mapping these genes to KEGG pathways using Pathview revealed the majority belong to general metabolic pathways. We next evaluated the effect of cancer on *E. coli* transcriptome by comparing gene expression between *Il10*$^{-/-}$ mice (colitis) vs. AOM/*Il10*$^{-/-}$ mice (colitis/CRC). We found that the differential expression of 66 *E. coli* genes is driven by cancer status at 20 weeks post-colonization. Importantly, five genes present in the *pks* island were significantly induced during development of inflammation and CRC, including the colibactin-initiating enzyme *clbN*, colibactin-editing enzymes *clbL* and *clbG*, and colibactin-transport protein *clbM gene* [10]. Disruption of inflammatory signaling by crossing *Il10*$^{-/-}$ mice to *Rag2*$^{-/-}$ mice prevented *E. coli pks*-induced colitis-associated CRC, without altering *E. coli* abundance [10]. Therefore, the carcinogenic potential of *pks*/colibactin is highly regulated by environmental factors such as inflammation. Moreover, these findings suggest that microbial activity and not abundance per se may be the cause of CRC.

Since *clbM* is among the five pks genes influenced by host inflammation/cancer, we elected to study this gene in more detail. Bioinformatics analysis shows that *clbM* potentially encodes for a sodium/drug antiporter, suggesting that the island is equipped with a gene allowing the transport of colibactin. We have preliminary results demonstrating that ClbM is a 12-transmembrane antiporter, proposed to export colibactin across the inner membrane. To specifically dissect the role of *clbM* in *E.coli* mediated genotoxic effect, we deleted the gene using a gene targeted system. *clbM* deletion did not affect bacterial growth, or attachment/invasion to IEC-6 cells but reduced *E.coli* NC101-induced DNA damage and γH2AX phosphorylation. Moreover, the *pks*-derived metabolite *N*-myristoyl-D-asparagine was present in the supernatant of NC101 but undetectable in NC101*ΔclbM* culture. Therefore, *clbM* is essential for transport and activity of the *E. coli* genotoxin colibactin and targeting this transporter using small molecule inhibitors may represent a novel therapeutic option to prevent *E. coli*-induced colorectal cancer.

OVERVIEW AND PERSPECTIVE.

The field of microbiome research has literally exploded in the past 5 years and the scientific community has recognized the key role the microbiota play in health and diseases. Amongst the various pathologies associated with microbial dysbiosis, CRC has been the subject of intense investigation.

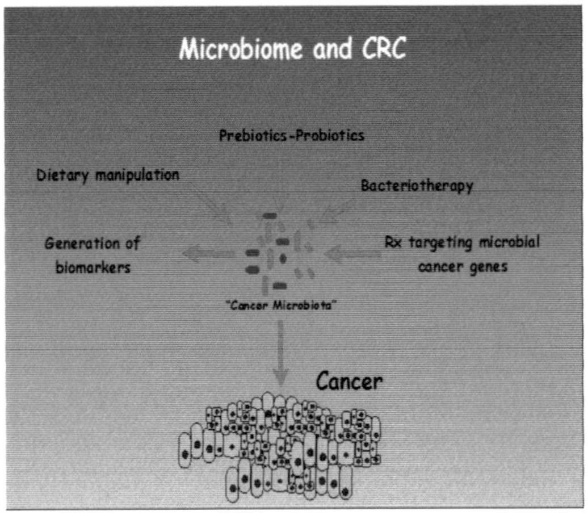

Fig:1 Mining the microbiome for therapeutic purpose. Cancer microbiome could be manipulated through the use of probiotics-prebiotics, bacteriotherapy, drug targeting microbial genes and by dietary modification. In addition, the microbiome could be mined for microbial non-invasive gene biomarkers.

It is clear from this body of work that microbial dysbiosis is present in subjects with CRC and that clinical isolates from these patients have the potential to promote carcinogenesis in preclinical models. In depth investigation has identified microbial genes as causative factors of CRC and these findings reinforce the notion that the microbiome may be mined for therapeutic purposes. For example, recent publications have demonstrated the potential use of microbes as clinical biomarkers for CRC[35,36]. Manipulating microbial composition through either targeted therapy (e.g bacteriophages) or wide spectrum approaches (diet, bacteriotherapy, probiotics) are possible means by which a "cancer-associated biota" could be restored to a healthy status. A sophisticated approach using small molecule inhibitors of microbial genes causing CRC is another example of how the microbiota could be therapeutically targeted. A similar approach has been used to alleviate the toxicity of a chemotherapeutic drug [37] or to prevent the tumorigenic effect of colibactin [38]. In summary, the field of cancer microbiome is in full expansion and it is likely that the next decade will bring novel therapeutic strategies for this pathology.

REFERENCES

1. Siegel R, Desantis C, Jemal A. Colorectal cancer statistics, 2014. CA Cancer J Clin 2014;64:104–117.
2. Arthur JC, Perez-Chanona E, Muhlbauer M, et al. Science 2012;338:120–123.
3. Esteban-Jurado C, Garre P, Vila M, et al. WJG 2014;20:1961–1971.
4. Watson IR, Takahashi K, Futreal PA, et al. Nat Rev Genet 2013;14:703–718.

5. Rustgi AK. Genes & development 2007;21:2525–2538.
6. Herrinton LJ, Liu L, Levin TR, et al. Gastroenterology 2012;143:382–389.
7. Cougnoux A, Dalmasso G, Martinez R, et al. Gut 2014.
8. Schwabe RF, Jobin C. Nature reviews 2013;13:800–812.
9. Buc E, Dubois D, Sauvanet P, et al. PLoS One 2013;8:e56964.
10. Arthur JC, Gharaibeh RZ, Mühlbauer M, et al. Nat Commun 2014;5:4724.
11. Consortium THMP. Nature 2012;486:207–214.
12. Martin P, Marcq I, Magistro G, et al. Nassif X, ed. PLoS pathogens 2013;9:e1003437.
13. Clemente JC, Ursell LK, Parfrey LW, et al. Cell 2012;148:1258–1270.
14. Sartor RB. Gastroenterology 2008;134:577–594.
15. Jobin C. Médecine sciences : M/S. June 2013:582–585.
16. Jobin C. Cancer Discov 2013;3:384–387.
17. Allen-Vercoe E, Jobin C. Immunol Lett 2014.
18. Wu N, Yang X, Zhang R, et al. Microb Ecol 2013;66:462–470.
19. Wang T, Cai G, Qiu Y, et al. ISME J 2011;6:320–329.
20. Martin HM, Campbell BJ, Hart CA, et al. Gastroenterology 2004;127:80–93.
21. Darfeuille-Michaud A, Neut C, Barnich N, et al. Gastroenterology 1998;115:1405–1413.
22. Swidsinski A, Khilkin M, Kerjaschki D, et al. Gastroenterology 1998;115:281–286.
23. Kostic AD, Chun E, Robertson L, et al. Cell Host and Microbe 2013;14:207–215.
24. Rubinstein MR, Wang X, Liu W, et al. Cell Host and Microbe 2013;14:195–206.
25. Arthur JC, Jobin C. Gut Microbes 2013;4:253–258.
26. Bonnet M, Buc E, Sauvanet P, et al. Clin Cancer Res 2014;20:859–867.
27. Nowrouzian FL, Oswald E. Microbial pathogenesis 2012;53:180–182.
28. Cuevas-Ramos G, Petit CR, Marcq I, et al. Proceedings of the National Academy of Sciences of the United States of America 2010;107:11537–11542.
29. Nougayrede JP, Homburg S, Taieb F, et al. Science 2006;313:848–851.
30. Secher T, Samba-Louaka A, Oswald E, et al. PLoS One 2013;8:e77157.
31. Clermont O, Christenson JK, Denamur E, et al. Environ Microbiol Rep 2013;5:58–65.
32. Escobar-Páramo P, Grenet K, Le Menac'h A, et al. Applied and environmental microbiology 2004;70:5698–5700.
33. Nowrouzian FL, Adlerberth I, Wold AE. Microbes and infection / Institut Pasteur 2006;8:834–840.
34. Tenaillon O, Skurnik D, Picard B, et al. Nature reviews 2010;8:207–217.
35. Zeller G, Tap J, Voigt AY, et al. Mol Syst Biol 2014;10:766.
36. Zackular JP, Rogers MAM, Ruffin MT, et al. Cancer Prevention Research 2014;7:1112–1121.
37. Wallace BD, Wang H, Lane KT, et al. Science 2010;330:831–835.
38. Cougnoux A, Delmas J, Gibold L, et al. Gut 2015.

PHARYNGEAL AND ESOPHAGEAL DYSPHAGIAS IN DOGS: FROM BENCH TO BEDSIDE

Stanley L. Marks BVSc, PhD, DACVIM (Small Animal Internal Medicine, Oncology), DACVN
G. Diane Shelton DVM, PhD, DACVIM (Small Animal Internal Medicine)
Davis, CA and La Jolla, CA

THE BEDSIDE: RECOGNITION OF THE CLINICAL PHENOTYPE OF DYSPHAGIA

The normal swallowing reflex is a four-stage process, characterized by the oral preparatory phase, oral phase, pharyngeal phase, and esophageal phase. Pharyngeal and esophageal dysphagias in dogs can be a diagnostic challenge and occur when the elaborate mechanism of bolus transit from the pharyngeal cavity into the esophagus or along the length of the esophagus becomes compromised. A history of repetitive swallowing, gagging, and retching associated with meals, nasal regurgitation with meals, swallow-related coughing, falling of food from the mouth during swallowing, and recurrent pneumonia should cause the clinician to suspect pharyngeal dysphagia. The assessment of dysphagia encompasses multiple dimensions that include (1) review of the signalment, (2) review of medication history and inquiry regarding recent anesthesia, (3) physical examination (prefeeding assessment), (4) neurologic examination, (5) clinical feeding and swallowing evaluation, and (6) laboratory and other testing to provide a basic data set for neurologic evaluation, including imaging studies and endoscopic evaluation of swallowing. Age and breed associations with dysphagia have been well documented in dogs. Breeds that have a hereditary predisposition or a high incidence of pharyngeal dysphagia include the golden retriever (pharyngeal weakness), cocker and springer spaniels (cricopharyngeal dysphagia), Bouvier des Flandres and cavalier King Charles spaniel (muscular dystrophy), and boxer (inflammatory myopathy).

Physical and Neurologic Examination

Physical examination of the dysphagic animal must include careful examination of the oropharynx using sedation or anesthesia if necessary to help rule out morphologic abnormalities such as dental disease, foreign bodies, cleft palate, glossal abnormalities, and oropharyngeal tumors. The pharynx and neck should be palpated carefully for masses, asymmetry, or pain. The chest should be auscultated carefully for evidence of aspiration pneumonia. Evaluation of cranial nerves should be performed, including assessment of tongue and jaw tone, and abduction of the arytenoid cartilages with inspiration. A complete physical and neurologic examination may identify clinical signs supporting a generalized neuromuscular disorder, including muscle atrophy, stiffness, or decreased or absent spinal reflexes. The gag reflex should be evaluated by placing a finger in the pharynx; however, the presence or absence of a gag reflex does not correlate with the efficacy of the pharyngeal swallow nor the adequacy of deglutitive airway protection.

Observation of Eating and Drinking

Careful observation of the dysphagic animal while it is eating (kibble and canned food) and drinking in the hospital cannot be overemphasized, and such observation helps to localize the problem to the oral cavity, pharynx, or esophagus. Dogs with an abnormal oral phase of swallowing typically have difficulty with prehension or aboral transport of a bolus to the tongue base, and these disorders often can be diagnosed by watching the animal eat. Dysphagias affecting the pharyngeal phase of swallowing can be more challenging to diagnose and often present with nonspecific signs such as gagging, retching, and the necessity for multiple swallowing attempts before a bolus is moved successfully into the proximal esophagus. These patients have abnormal transport of bolus from the oropharynx to the hypopharynx or from the hypopharynx to the proximal esophagus. Dogs with esophageal strictures or esophagitis can exhibit evidence of odynophagia and regurgitation seconds to minutes following ingestion of the bolus.

THE BENCH: DIAGNOSTIC WORKUP

Routine Laboratory Screening

Routine laboratory screening including a CBC, serum chemistry profile, and urinalysis can aid in the identification of a number of common systemic diseases that may result in neuromuscular weakness.[1] Specific abnormalities that may be identified on routine screening include anemia, hypo- and hyperglycemia, hypo- and hyperkalemia, hypo- and hypernatremia, hypo- and hypercalcemia, hypophosphatemia, and hypomagnesemia. Additional testing procedures that should be part of baseline testing in any dog with clinical neuromuscular disease includes measurement of serum creatine kinase (CK) activity, cardiac troponin I, plasma lactate concentration, and determination of thyroid status. Serum CK activity should be on every neuromuscular minimum database, and most importantly on pre-neuter blood evaluations in young dogs. Creatine kinase activity may be normal in the presence of muscle disease, and muscle disease should not be ruled out based on normal CK activity. Marked or persistent increases of CK may be indicative of a congenital or inherited muscle disease even though the animal may be clinically asymptomatic at the time. The most marked increases in serum CK activity (>20,000 IU/L) are associated with necrotizing myopathies or muscular dystrophies. Generalized inflammatory myopathies usually show moderate increases in CK activity (2,000-20,000 IU/L), while the CK activity in focal inflammatory myopathies such as masticatory muscle myositis (MMM), endocrine myopathies, neuropathies, and other congenital muscle diseases are normal or only mildly increased (0-2000 IU/L).

Cervical and Thoracic Radiography

The pharynx of healthy animals is evident on radiographs because it is air filled. The size of the air-filled space can be decreased by local inflammation or neoplasia, laryngeal edema, or elongation of the soft palate. Pharyngeal size also can appear increased with dysfunction of the pharynx or upper esophageal sphincter, chronic respiratory (inspiratory) disease, and chronic severe megaesophagus. The normal esophagus is not visible on survey radiographs. An exception occurs following aerophagia due to excitement, nausea, dyspnea, or anesthesia.

Videofluoroscopic Swallow Study

Contrast videofluoroscopy involves real-time capture of images of the animal as it is swallowing liquid barium or barium-soaked kibble and is one of the most important procedures for assessing the functional integrity of the swallow reflex. Videofluoroscopy is used to determine the normal sequence of events that make up a swallow and to measure the timing of these events in relation to one another. Additionally, the movement of certain anatomic structures is measured in relation to a fixed point to assess function further. Swallowing events that occur out of sequence, at inappropriate times, or with reduced vigor can cause significant morbidity. One problem with videofluoroscopy is that animal positioning is not standardized in veterinary medicine. Alterations in body position (sternal versus lateral recumbency) do not appear to affect measurements of pharyngeal constriction ratio or the timing of swallowing in healthy dogs; however, cervical esophageal transit is significantly delayed when dogs are imaged in lateral recumbency.[2]

The fluoroscopic swallow study typically involves assessment of five swallows each of 5 to 10 ml of liquid barium (60% weight per volume) followed by five swallows of canned food mixed with barium and finally 5 swallows of kibble soaked in barium. The timing of the swallow can be determined easily when the swallow video is viewed frame by frame, with each frame representing 1/30th of a second in the National Television System Committee (NTSC) system, the analog television system used in the United States. The frame in which the epiglottis is observed to close over the larynx is considered as the starting point for all time measurements, and frames are counted until the observation of maximal contraction of the pharynx, opening of the proximal esophageal sphincter (PES), and closing of the PES. The swallow is considered completed when the epiglottis is observed to reopen, which usually takes five or six frames in healthy dogs. More recently, a contrast videofluoroscopy method for quantifying pharyngeal contractility in the dog has been described.[3] The pharyngeal constriction ratio is calculated by dividing the pharyngeal area at maximum contraction by the pharyngeal area at rest. As pharyngeal contractility diminishes, the ratio approaches 1.0. This simple procedure provides important information regarding the strength of pharyngeal contraction in dysphagic dogs.

Laryngoscopy, Pharyngoscopy, and Esophagoscopy

Thorough laryngeal examination is important in all animals with pharyngeal and esophageal dysphagia to rule out laryngeal paralysis associated with a polyneuropathy.[4] Geriatric, large-breed dogs can experience a progressive generalized neuropathy, with associated pharyngeal weakness, pharyngeal dysphagia, and esophageal dysmotility.[5] Pharyngoscopy and esophagoscopy provide anatomic information about the structures involved in the oropharynx and esophagus, but both procedures are of limited diagnostic utility for evaluating functional disorders in anesthetized animals. This is an important limitation of these diagnostic procedures, particularly in animals that are dysphagic secondary to dynamic disorders such as cricopharyngeal dysphagia or esophageal dysmotility. Esophagoscopy is helpful for diagnosing esophagitis, esophageal strictures (that can be missed on barium swallow studies), and hiatal hernias.

Miscellaneous Laboratory Screening

The acetylcholine receptor antibody test should be performed in all cases of acquired dysphagia.[6] This test is not useful for congenital dysphagia as an immune-basis is unlikely. The gold standard for the diagnosis of acquired MG remains the demonstration of serum AChR antibodies against native AChR by immunoprecipitation radioimmunoassay. This assay involves precipitation of serum IgG and IgM antibodies that bind to solubilized AChR complexed with a high-affinity peptide agonist, ^{125}I-labeled α-bungarotoxin. The precipitate's γ-emission reflects the amount of AChR bound to immunoglobulin. The assay is specific, sensitive and documents an autoimmune response against muscle AChRs. Although there is some cross-reactivity in AChR recognition of antibodies among species, the assay is relatively species specific and a canine specific assay system should be used. Antibody titers in dogs are in general lower than in humans, and low-titer positives may be missed if human AChR is used as antigen. A positive AChR antibody titer, however, is not predictive of the degree of weakness. Within an individual, AChR antibody levels correlate with the disease severity, but antibody levels between patients are highly variable and do not correlate well with severity.

Electrodiagnostic Testing

Electrodiagnostic evaluation, including electromyography and measurement of motor and sensory nerve conduction velocities, does not provide a specific diagnosis in most cases but can supply important information as to the severity, distribution, and character of a myopathic or neuropathic disease process and assist in selecting the optimal anatomic site for biopsy. Electrodiagnostic testing also should include evaluation of the pharyngeal muscles and tongue. The health status of the animal must be taken into consideration because the lengthy procedure is performed under general anesthesia.

Esophageal Manometry

Esophageal manometry measures pressure within the esophageal lumen and sphincters, and provides an assessment of the neuromuscular activity that dictates function in health and disease. Manometric techniques have improved in a step-wise fashion from a single pressure channel to the development of high-resolution manometry (HRM) with up to 36 pressure sensors. Advances in computer processing allow pressure data to be presented in real time as a compact, visually intuitive "spatiotemporal plot" of esophageal pressure activity. This spatiotemporal plot provides objective measurements of the forces that drive food and fluid from the pharynx to the stomach. This diagnostic modality has been shown to be feasible in fully awake dogs and provides a sensitive functional assessment of the PES, esophagus, and distal esophageal sphincter (DES).[7]

Esophageal pH/Impedance Testing

Esophageal pH/impedance testing is a useful diagnostic tool that is used to diagnose acid and non-acid reflux in animals with suspected gastroesophageal reflux (GER), unexplained esophagitis, or hiatal hernias. The technology of esophageal pH testing has advanced tremendously in recent years, and clinicians have several choices when selecting esophageal pH probes. The catheter-free Bravo pH Monitoring System from Medtronic is the first catheter-free system used to measure esophageal pH in human patients and dogs that are suspected of having GER, and is revolutionizing the way esophageal pH testing is done, because it is allows people and animals to maintain their regular diet and activities during pH testing. The Bravo system is an alternative to the traditional pH trans-nasal pH catheter that can cause patient discomfort, and is easily dislodged by dogs and cats if the animal is not closely monitored. The main disadvantage of the Bravo system is that one can only record esophageal pH, and the system does not utilize impedance technology that allows one to measure both acid and non-acid reflux. Esophageal pH testing has been extensively utilized in awake[8] and anesthetized dogs in an effort to identify risk factors for GER and assess the effects of prokinetic agents on GER.[9]

Tissue Biopsies

Muscle, and in some cases peripheral nerve biopsies, should be collected early in the course of diagnostic evaluation of a dog with dysphagia suspected to be caused by a neuromuscular disease.[10] A delay in collection of muscle biopsies can result in extensive muscle damage, fiber loss, and fibrosis that may be irreversible diminishing the chances for successful treatment. Muscle and nerve biopsies should be evaluated by laboratories with expertise in neuromuscular diseases. Muscle biopsies should be collected by an open-biopsy technique and evaluated in frozen sections, using a standard panel of histochemical stains and reactions including fiber typing. The usefulness of evaluating muscle and nerve in only paraffin sections is limited. Ultrastructural analysis is necessary for definition of structural abnormalities in selected congenital myopathies and is useful for peripheral nerve diseases. Results of tissue biopsies should guide further diagnostic testing focusing on laboratory tests that are relevant to the specific disease group, or identify a specific disease and therapy if available. In addition to the diagnosis, results of muscle biopsies can also help determine a prognosis in many cases, as some neuromuscular diseases are treatable and others are not.

BACK TO THE BEDSIDE: TREATMENT

Disorders of the Proximal Esophageal Sphincter: Cricopharyngeal Dysphagia

Cricopharyngeal dysphagia is a swallowing disorder of the PES characterized by either cricopharyngeal dyssynchrony (functional) or cricopharyngeal achalasia (structural). *Cricopharyngeal dyssynchrony* is essentially a pump problem in which the weak pharyngeal muscles are unable to propel the bolus through the PES. Early evidence in the authors' laboratories points toward a neuropathy in these dogs. On videofluoroscopy, there is evidence of incoordination between the contraction of the dorsal cranial and middle pharyngeal contractor muscles (hyopharyngeus, pterygopharyngeus, and palatopharyngeus muscles) and opening of the PES (cricopharyngeus and thyropharyngeus muscles). A comprehensive workup should be completed in an effort to find a treatable cause of the suspected neuropathy (complete blood count and serum chemistry panel, AChR antibody titer, CK measurement, muscle and nerve biopsy). The prognosis for these dogs is less favorable compared to dogs with cricopharyngeal achalasia. An effort should be made to identify the optimal consistency of food and water that these dogs will tolerate (by adding commercial food thickeners such as Thick-It), although these animals will ultimately succumb to repeated bouts of aspiration pneumonia and malnutrition. Enteral feeding via a gastrostomy tube is a viable alternative in these animals; however, silent aspiration and pneumonia can occur despite the use of enteral feeding devices. *Cricopharyngeal achalasia* is the inability of the cricopharyngeus muscle to open during the cricopharyngeal phase of swallowing. The exact underlying cause has not been determined, although the disorder can be reproduced by transection of the pharyngeal branch of cranial nerve X. Cricopharyngeal dysphagia and achalasia are diagnosed via contrast videofluoroscopy. Cricopharyngeal achalasia has been well documented in miniature dachshunds and a variety of other toy breeds in the author's laboratory, and all dogs had marked hypertrophy of the cricopharyngeus muscle (cricopharyngeal bar) causing severe obstruction to propulsion of the bolus through the PES.

In dogs with cricopharyngeal dysphagia a comprehensive workup must be undertaken before surgical intervention to ensure that systemic disorders (myopathies, polyneuropathies) are ruled out and aspiration pneumonia is managed properly. A fluoroscopic swallow study must be performed in dogs suspected of having cricopharyngeal dysphagia to assess pharyngeal function before surgical intervention. Dogs that are diagnosed with underlying neuropathies or myopathies are managed conservatively with alterations of feeding practice or the use of low-profile gastrostomy devices if specific management of the underlying neuropathy or myopathy is not possible. Definitive treatment of cricopharyngeal achalasia involves surgical myotomy or myectomy of the cricopharyngeal muscle. In veterinary medicine, the standard surgical approach for myotomy or myectomy has remained constant over the years, and the cricopharyngeal and thyropharyngeal muscles are approached either by a standard ventral midline approach with 180-degree rotation of the larynx on its longitudinal axis or by a lateral approach with 90-degree rotation of the larynx.

Megaesophagus

Idiopathic megaesophagus is the most common type of megaesophagus in the dog. The syndrome may be manifested either in puppies at the time of weaning or in adulthood. The *etiology* of idiopathic megaesophagus is unknown. The congenital form of the disease may be due to a delay in maturation of the esophageal neuromuscular system; a theory that explains why young dogs may improve with careful feeding management. Idiopathic megaesophagus has been shown to be inherited in the wire-haired fox terrier and the miniature schnauzer. A breed predisposition also exists for the German Shepherd, Great Dane and Irish Setter. The site and pathogenesis of the lesion in idiopathic megaesophagus is unknown. Suggested hypotheses include abnormalities of the afferent limb of the reflex arc (receptors, neurons) or of the swallowing center in the CNS. Acquired megaesophagus may result from a large number of systemic diseases including, autoimmune myasthenia gravis, SLE, polymyositis, polymyopathies, dermatomyositis, polyneuropathies, dysautonomia, botulism, distemper, neoplasia, brain stem disease, lead and thallium toxicity, Addison's disease, hypothyroidism, pituitary dwarfism, and thymoma. Many obstructive esophageal diseases can also lead to megaesophagus if they are of sufficiently chronic duration.

Repeat AChR antibody testing is important in dogs 3-4 weeks following a first negative AChR antibody test, particularly if the original titer is 0.4-0.6 nmol/l (borderline). Measurement of the CK activity is critical. Additional diagnostic procedures that can be performed based on the animal's signalment, history, and neurological examination include an EMG, nerve conduction velocities, and muscle biopsies. Medical management of idiopathic generalized megaesophagus involves modification of feeding practices. Treatment of the underlying cause (secondary megaesophagus) is of paramount importance. Dogs with megaesophagus generally tolerate a liquid or semi-liquid gruel better than solid food. Feeding from an elevated position allows gravity to help move the liquid into the stomach. If possible the animal should be held in a vertical position for 5 - 10 minutes after eating. This can often be accomplished with the use of a Bailey chair or similar device. Multiple feedings rather than one large single meal may also help minimize food accumulation in the esophagus. We have successfully placed low-profile-gastrostomy tubes for feeding in many dogs with idiopathic megaesophagus in an effort to minimize aspiration pneumonia. The frequency of aspiration pneumonia has been markedly reduced in comparison to oral feeding and this therapeutic modality should be considered when a client is willing to dedicate the time to tube maintenance and feeding. The prognosis for dogs with megaesophagus is very variable depending upon the underlying etiology, the degree of dysfunction and the systemic status of the dog.

Esophageal Strictures

Esophageal strictures are a relatively common problem in dogs and can be caused by benign and malignant conditions, although the latter are relatively uncommon in dogs. The most common cause of esophageal stricture formation is *gastroesophageal reflux in association with general anesthesia.* This phenomenon has been reported to occur in up to 65% of cases of esophageal stricture, with a median onset of clinical signs occurring 7.5 days post-anesthesia. The incidence of gastroesophageal reflux (GER) in dogs during anesthesia varies from 16-55%, and occurs secondary to a decrease in DES pressure. Reduction of DES pressure occurs secondary to a variety of anesthetic agents, including atropine, morphine, acepromazine, thiopentol, xylazine, and isoflurane. The second most common cause of esophageal stricture formation is from *esophagitis induced by administration of doxycycline or clindamycin.* The proposed mechanism of tablet-induced esophagitis and stricture formation is from tablet retention in the esophagus due to poor esophageal clearance with a "dry" swallow. Other important causes of esophagitis and secondary stricture formation include chronic vomiting of acid contents from the stomach, foreign body ingestion, and swallowing of caustic substances. Esophagitis is associated with a weakening of the DES pressure that can result in further reflux of gastric contents and increased damage to the esophageal mucosa. Damage to the muscularis layer of the esophagus is often associated with fibroblastic proliferation and contraction leading to stricture formation. The clinical signs associated with severe esophagitis and/or esophageal stricture formation include odynophagia (painful swallowing), dysphagia, increased salivation, regurgitation, anorexia, coughing (secondary to aspiration pneumonia), and weight loss. These signs are often insidious at the onset and are often missed by owners, but are progressive as the esophageal lumen gets progressively narrowed.

Mechanical dilation of the stricture is best accomplished using balloon dilation or bougienage. The theoretical advantage of balloon dilation is that the forces applied to the stricture are a radial stretch, in contrast to the longitudinal forces applied with the rigid bougienage instrument. However, a retrospective case series in 20 dogs and 8 cats with benign esophageal strictures that underwent bougienage treatment suggested that this procedure was safe and effective for most dogs and cats with benign esophageal strictures, with outcomes similar to balloon dilation.[11] The administration of triamcinolone into the stricture site using a four-quadrant approach before the balloon dilation procedure has been associated with a reduced rate of restricture formation[12] We generally inject approximately 2.5 mg triamcinolone into each of the quadrants using a Wang needle (or similar transbronchial needle) that can be threaded down the biopsy channel of the endoscope. The steroid is generally used for the first 2-3 dilation procedures. Topical mitomycin C has also been shown to be beneficial for preventing restricture formation[13] Clinicians generally apply 5 mg of mitomycin using a soaked gauze sponge that is placed endoscopically at the stricture site for approximately 5 minutes. The site is then rinsed with 60 mL of water following the removal of the sponge.

Intraluminal stents are being used with increasing frequency in veterinary medicine for patients that have failed balloon dilation or for patients with recurrent stricture formation. Stents are available both covered (polypropylene) and uncovered. The covering helps prevent the ingrowth of tissue within the stent. Available stent materials include Nitinol (nickel plus titanium), Elgiloy (cobalt, nickel, plus chromium), stainless steel, polyester plastic/silicone, or a biodegradable material such as PDS. The selection of a particular stent is based upon the characteristics of the stricture such as its location and length, and the need for removal of the stent. Once the stent is deployed, it must be anchored in place or it will rapidly migrate into the stomach. The stent can be secured in place using a suturing device (GI Stitch, Pare Surgical) that can be used through a double channel endoscope.

Esophagitis

Medical management of esophagitis involves a combination of proton pump inhibitors such as omeprazole, sucralfate suspension, and a potent prokinetic agent such as cisapride to minimize further gastric reflux and facilitate gastric emptying. Proton pump inhibitors can initially be administered q 12 hrs for 3-4 days to decrease acid production more rapidly, before continuing with once daily administration. H2-receptor antagonists can also be administered at night to help minimize nocturnal acid breakthrough.

Gastroesophageal Reflux Disease

Gastroesophageal reflux disease (GERD) is a common chronic disorder in the Western world. The basic cause of GERD has been well characterized - the fundamental defect is a loss of integrity of the gastroesophageal barrier. What is less clear is the most appropriate means of addressing this reflux. In dogs, most cases of GERD occur secondary to hiatal hernia which is commonly recognized in brachycephalic breeds; however, the author (SLM) has documented cases of severe flaccidity of the lower esophageal sphincter in the absence of a hiatal hernia. Medical therapy aims at decreasing acidity by suppressing proton secretion and has been well established in people. Antacids do little to prevent the reflux of gastric contents, and antireflux surgery is usually warranted. In people, antireflux surgery is commonly performed laparoscopically, and is aimed at reinforcing and repairing the defective barrier through plication of the gastric fundus.[14] A left-sided gastropexy, esophageal plication, and esophagopexy (fixation of the esophagus to the diaphragmatic crus) via celiotomy is most commonly performed in dogs with a favorable outcome in most animals.

REFERENCES

[1]Shelton GD **Vet Clin Pathol** 2010;39:278-295; [2]Bonadio C et al. **J Vet Int Med** 2009;23:801-5; [3]Pollard RE et al. **Vet Radiol Ultrasound** 2007;48:221-6; [4]VanHaesebrouck AE et al. **J Neurol Sci** 2008;275:100-105; [5]Stanley B et al. **Vet Surg** 2010;39:139-149; [6] Shelton GD et al. **J Vet Int Med** 1990:4:281-284; [7]Kempf J et al. **Am J Vet Res** 2013;74:895-900; [8]Kook PH. **J Vet Int Med** 2014;28:1716-23; [9]Zacuto AC et al. **J Vet Int Med** 2012;26:518-525; [10] Ryckman LR et al. **J Am Vet Med Assoc** 2005;226:519-523; [11]Bissett SA et al. **J Am Vet Med Assoc** 2009;235(7):844-850; [12]Orive-Calzada A et al. **Surg Laparosc Endosc Percutan Tech** 2012;22(6):518-522; [13]El-Asmar et al. **J Pediatr Surg** 2013;48(7):1621-1627; [14]Peters JH et al. **Ann Surg** 1998;228:40-50.

TRACHEOBRONCHOSCOPY IN THE DIAGNOSIS AND MANAGEMENT OF SMALL ANIMAL RESPIRATORY DISEASES

Carol Reinero, DVM, DACVIM, PhD
Columbia, MO

OVERVIEW

Tracheobronchoscopy is an important diagnostic tool for a variety of respiratory diseases in dogs and cats, and ultimately can help guide selection of appropriate therapy and provide a prognosis for the patient. It allows for direct visual inspection of the interior of the tracheobronchial tree during the phases of inhalation and exhalation, which may highlight dynamic changes in airway caliber not well appreciated by other modalities. It is used most commonly to determine the type, extent, and severity of lesions within the tracheobronchial tree (e.g., narrowing or collapse of the trachea or bronchi), abnormal orientation of airways (e.g., lung lobe torsion), masses, abnormal appearance to the mucosa, luminal secretions, hemorrhage, foreign material, bronchomalacia, or bronchiectasis.[1-3] Targeted sampling of regions of the lung, based on imaging or gross visual inspection, is possible. By instilling and aspirating sterile saline, it allows for the collection of cells and epithelial lining fluid of the airways for analysis. Brushes can also be used for collection of superficial cells for cytology or for culture. With specialized fine-needle aspiration/biopsy needles that fit down the biopsy channel, it can allow for aspiration of enlarged hilar lymph nodes and other masses. Tracheobronchoscopy is useful for the non-surgical removal of foreign bodies, and using a snare technique, can be used to remove obstructing tracheal masses.[4] This lecture will focus on equipment required, special anesthetic considerations, examination technique, indications, contraindications, complications, and special procedures (e.g., bronchoalveolar lavage, brush cytology, fine-needle aspiration, biopsy, foreign body retrieval, etc.).

ENDOSCOPES AND ANCILLARY EQUIPMENT

A detailed discussion of the types of endoscopes used for tracheobronchoscopy is beyond the scope of this lecture, but a few general comments may be made. Flexible fiberoptic and videoscopes are commonly used as with gastroduodenoscopy and colonoscopy; however, there are important differences. Bronchoscopes are generally of a shorter working length (50-100cm) and have a smaller outer diameter (usually in the 3.5-6mm range). There is no house suction or water supplied; rather there is a single channel that can be used to pass instruments (e.g., cytologic brush catheter, foreign body retrieval instrument, mucosal "pinch" biopsy forcep, shielded needles for fine-needle aspiration), or for attachment of a syringe filled with sterile saline for bronchoalveolar lavage fluid collection. Additionally, this port may be used to provide supplemental oxygen, especially when the larger airways are being examined (note: supplemental oxygen should not be used when the scope is tightly wedged in an airway to prevent pneumothorax). Many scopes have two-way deflection; this is generally not an obstacle to orientation in any direction as the scope can be turned by using gentle simultaneous rotation at the hand piece and tip while placed within the airways. An ideal endoscope would be long enough to reach the lower airways of a large dog, sufficiently small to fit in the small airways of cats or puppies, and have a large enough biopsy channel to fit the major instruments of interest (at least a biopsy forcep and tracheobronchial aspiration needle). Unfortunately, there is no single perfect endoscope that fits all of these criteria. By necessity, endoscopes with small outer diameters that fit in the airways of cats and small dogs have smaller biopsy channels and tend to be shorter. Some practices will buy a small bronchoscope to use in cats and small to medium sized dogs, and may adapt a small gastric endoscope for large breed dogs. If the latter is done, it is absolutely essential that appropriate disinfection be performed between endoscopies of the airways and gastrointestinal tract. Cleaning and disinfection are also beyond the scope of this lecture, but should closely follow manufacturer's instructions to prevent hospital-acquired infections.
Patients should be placed in sternal recumbency. Depending on the size of the animal, the scope may either be placed directly in the airway, or through an endotracheal tube (ETT). A modified ETT which has been shortened to decrease the dead space through which the scope must pass can be used in medium to large breed dogs.

A double lumen Y adaptor may be placed to facilitate gas exchange (oxygen and carbon dioxide) through one port, and allow passage of the scope through the other port. By using a modified ETT, inhalational anesthesia can be employed. If the scope is introduced directly into the airways, injectable anesthesia (commonly proprofol) must be used. In smaller dogs and in cats, oxygen is generally administered via a red rubber feeding tube and the scope is advanced alongside the tube/catheter. Alternatively a catheter attached to a jet ventilator could be used. The endoscope can be used to assess appropriate placement of the tube supplying oxygen. It is not uncommon for the oxygen tube to be introduced too far, or to be inadvertently advanced as the endoscope drags it down the airways; thus, it is essential to make sure the catheter lies within the trachea and is not fed down one of the mainstem bronchi. Alternatively, as mentioned above, supplemental oxygen may be provided through the biopsy channel of the endoscope. The author premedicates cats with an injectable bronchodilator prior to performing bronchoscopy and bronchoalveolar lavage to minimize airflow limitation induced by this procedure.

Other equipment should include a spay hook and dental mirror (to allow for a pharyngeal/retropharyngeal exam prior to endoscopic examination), a mouth gag (to prevent inadvertent biting of the scope; avoid spring loaded mouth gags in cats[5]), and warmed sterile saline (for collection of bronchoalveolar lavage fluid). Use of warmed sterile saline is especially important in cats that tend to have more hyperreactive airways. The endoscope can also be used to retroflex above the soft palate for a choanal exam, but this should be done AFTER tracheobronchoscopy to minimize inadvertent innoculation of bacteria from the caudal nasopharynx into the airways. A choanal exam is particularly useful in patients with concurrent lower airway disease and nasal disease, especially if post-nasal drip or gastroesophageal reflux disease is thought to be contributing to tracheobronchitis or recurrent pneumonia. Miscellaneous equipment, depending on the underlying disease in the patient, may include biopsy forceps and foreign body retrieval instruments, tracheobronchial aspiration needles, aspiration catheters, cassettes, and a small container of formalin (for biopsy specimens).

EXAMINATION TECHNIQUE

Endoscopic examination should begin with a thorough evaluation of the oral, retropharyngeal, pharyngeal and laryngeal regions. If laryngeal paralysis is suspected based on stridor or voice change, an appropriate anesthetic drug with minimal interference with laryngeal function should be used (e.g., propofol is a reasonable choice). Dopram should be administered intravenously while one individual examines the arytenoids for movement and a second individual calls out each time the animal inspires. If the examiner is not aware of the cycle of respiration while monitoring the arytenoids, it is possible they may incorrectly identify exhalation-induced opening of the arytenoids as adequate movement (ie, laryngeal paralysis may be missed).

Knowledge of normal tracheobronchial anatomy is critical and will be reviewed in this lecture. Prior to starting endoscopic examination, a review of cervical and thoracic radiographs and CT should be performed to correlate site of disease noted by imaging, with the anatomic location to be identified with endoscopy. A systematic approach should be taken with each endoscopic examination. The trachea is examined first, with particular attention paid to evidence of collapse (dynamic or static, location, severity), masses or foreign bodies, exudate and/or mucus, and the appearance of mucosa and vessels. It is helpful to pause at the carina and ensure proper orientation before a lower airway examination. Of note, with the patient in sternal recumbency, the right hemithorax will be visualized to the operator's left and the left hemithorax will be visualized to the operator's right. Static and dynamic collapse of the major bronchi should be evaluated for prior to advancing the scope. The author will generally take a photo at the carina prior to and after examination of the lower airways so that changes in diameter of the airways can be assessed. It is critical to keep the endoscope centered over the airways to allow smooth advancement. If the operator is having a difficult time centering the endoscope at the carina, the position of the head and neck must be assessed, and extended or straightened as needed. The lower airways are subsequently evaluated, using the bronchial "road map" which will help in the systematic approach to each major airway and its branches. The author starts at the bronchus leading to the right cranial lung lobe, then progressing to the right middle, accessory and right caudal lobes; then moves to the left side starting at the bronchus leading to the cranial portion of the left cranial lung lobe, then progressing to the caudal portion of the left cranial lung lobe then the left caudal lung lobe. As needed, reorientation can be performed by moving back to the carina and positioning the scope so that the dorsal tracheal membrane is located dorsally. The same criteria applied to evaluation of the trachea are applied to the lower airways. It is also important to pay particular attention to areas of exudate, as foreign material may be hidden behind purulent exudate. In patients with known or suspected infection in one or a few lobes, examination of those lobes should be performed last to prevent inadvertent inoculation of infectious material to healthy lobes.

A choanal examination (retroflex above the soft palate to view the back side of the nasal passages) may also be indicated as post-nasal drip may be a reason for cough or predispose to recurrent pneumonia from aspiration. Other lesions such as polpyps/abscesses/granulomas, nasopharyngeal stenosis, nasopharyngeal turbinates, foreign material, evidence of inflammation or hemorrhage may be noted. The author performs a choanal examination after tracheobronchoscopy so as to not contaminate the scope with oropharyngeal secretions.

INDICATIONS, CONTRAINDICATIONS AND COMPLICATIONS

The indications for tracheobronchoscopy include, but are not limited to, evaluation for collapsing trachea or mainstem bronchi, bronchomalacia, strictures, masses, hilar lymphadenopathy, infectious or inflammatory processes, trauma, neoplasia, foreign bodies, and bronchiectasis. The contraindications for tracheobronchoscopy include a significant anesthetic risk (eg, concurrent cardiac disease or other serious underlying illness), hypoxemia unresponsive to oxygen therapy, severe hypercapnia, and uncontrolled bleeding diatheses.

As with any invasive procedure involving the respiratory tract, there are complications associated with tracheobronchoscopy. It is not surprising that hypoxemia and hypercapnia can result from respiratory compromise. Care must be taken with cats as they have narrowed airways and a propensity towards airway hyperresponsiveness. Pneumothorax, while uncommon, can also occur. Most often it results from preexisting bullae that rupture with mechanical ventilation. Infection and hemorrhage have been reported as well. The operator must always be prepared to deal with emergent situations. As previously mentioned, cats in particular frequently benefit from pre-medication with a bronchodilator. Supplemental oxygen before and after endoscopic examination is also helpful; dogs or cats with severe respiratory disease can decompensate for up to 24 hrs, so the author will not discharge these patients to the owner's care the day of the procedure. In patients at high risk for pneumothorax, equipment to tap the chest should be readily available. If there is suspicion for disease that might benefit from surgical intervention, a surgeon should be on stand-by in case surgery becomes emergently needed.

SPECIALIZED PROCEDURES

Tracheobronchoscopy is not just used for direct visual examination of the interior of the tracheobronchial tree—there are a number of diagnostic procedures that can be performed using the endoscope. Bronchoalveolar lavage and tracheal or bronchial mucosal brush biopsies can be obtained for cytologic examination and culture. There is no consensus on volume of sterile saline to be instilled for lavage, with some individuals using a single aliquot (small or large volume), some using multiple aliquots, and some basing the volume on body weight (i.e., dosing on a ml/kg basis). The author standardly uses 20 ml aliquots in all but the smallest of patients (ie patients under 3 kg). The presence of surfactant as a foamy layer on top of the fluid indicates a deep lavage. It is critical to wedge the scope with an airway open in the center of the field of view. If the tip of the scope is wedged more tangentially against mucosa, there is a good chance the mucosa will be sucked down over the channel and prevent good retrieval. Saline can be instilled as quickly as feasible, and gentle suction (either manual or with a suction trap) applied. The tip of the scope can be subtly moved up and down to improve retrieval and moved back incrementally. An adequate yield is in the 50-70% range of what was instilled. Very small volumes (e.g. 1-5 ml, depending on the volume capacity of the channel) may not penetrate deep enough in the tissues to get representative samples of the epithelial lining fluid. This is in large part why the author prefers a larger volume aliquot of 20 mls. Depending on the yield, patient clinical status after lavage, and presence/absence of surfactant, subsequent aliquot(s) may be indicated. Additionally, if there are different "types" of lesions in different locations of the respiratory tract, targeted lavages can be performed with the aliquots kept separate. An example of this might be a patient with suspected chronic bronchitis (i.e. diffuse airway hyperemia, mucosal thickening, mucus accumulation, etc.) with a focal purulent exudate compatible with bacterial or aspiration pneumonia in a solitary lobe.

Specialized biopsy instruments can be passed down the biopsy channel to obtain biopsies of masses or of the mucosa. The biopsies obtained are usually very small and frequently demonstrate crush artifact, and may be better used as cytologic specimens (i.e., the tissue can be rolled or smeared on a slide) or for culture. If tracheal polyps are noted, pinch biopsy forceps can be used to undermine the base of the polyp. A snare can subsequently be used to debulk the remainder of the polyp. Specialized aspiration needles may be useful for sampling enlarged hilar lymph nodes; the endoscopic approach is faster and more direct than CT-guided fine needle aspiration of hilar nodes.

Tracheobronchoscopy can be also used therapeutically. If there are foreign bodies that are visualized within the airways, they can be removed using bronchoscopic basket retrieval instruments, snares, or forceps. Tracheal stents may be placed with endoscopic guidance. Debulking of luminal masses may be attempted using snares. Intracorporeal nebulization of chemotherapeutics may be administered to target an endobronchial neoplasm.

REFERENCES

1. Bottero E et al. *J Vet Intern Med* 2013;27:840-6.
2. Cannon MS et al. *Vet Radiol Ultrasound* 2013;54:351-7.
3. Johns S et al. *J Am Anim Hosp Assoc* 2014;50:273-7.
4. Queen EV et al. *J Vet Intern Med* 2010;24:990-3.
5. Stiles J et al. *Vet J* 2012;193;367-73.

UROENDOSCOPY

Larry G. Adams, DVM, PhD, DACVIM (SAIM); Jodi L. Westropp, DVM, PhD, DACVIM (SAIM)
West Lafayette, IN; Davis, CA

INTRODUCTION

Uroendoscopy (or transurethral cystoscopy) is an important component in the diagnostic evaluation of dogs and cats with recurrent or persistent lower urinary tract disease. In veterinary medicine, uroendoscopy has mainly been utilized for visualization and biopsy of lesions in the bladder and urethra. Uroendoscopy is also the basis of more advanced endourology techniques such as laser lithotripsy and ureteral stent placement.

EQUIPMENT FOR UROENDOSCOPY

Several rigid and flexible endoscopes designed for use in humans may be used for uroendoscopy in dogs and cats. The size of the patient's urethra dictates the size of cystoscope. Rigid cystoscopes used for diagnostic cystoscopy in female dogs and cats typically have a 30° angle view. For female dogs weighing approximately 5 to 30 kg, 2.7 mm diameter and 18 cm long rigid pediatric cystoscopes with a 14 French sheath are utilized for transurethral cystoscopy. For female dogs smaller than 5 kg and female cats, 1.9 cm diameter and 18 cm long pediatric cystoscopes are small enough to pass through the urethra in most animals. For larger female dogs, 4.0 mm diameter and 30 cm long cystoscopes are required for access to the cranial aspect of the urinary bladder. These scopes are used with a 17 or 19 French sheath. The working channels range from 3 French in flexible ureteroscopes and smaller rigid cystoscopes to 5 French in the larger rigid cystoscopes. Instruments that are commonly used through the working channel include urologic guide wires, biopsy forceps, graspers, stone baskets, and laser fibers. Because male dogs have smaller diameter and longer urethral length, 2.5 to 2.8 mm (7.5 to 8.5 French) diameter and 70 cm long flexible ureteroscopes may be utilized for transurethral cystoscopy. These ureteroscopes are fragile and require replacement or repair after a variable number of procedures.

Flexible tipped urologic guide wires are available in a variety of sizes and designs for endourologic procedures. Guide wires may be placed past urethral tears or obstructions under visual or fluoroscopic guidance during uroendoscopy, and then an open-ended indwelling catheter can be passed over the guide wire to bypass the tear or obstruction. Guide wires are also required for placement of ureteral and urethral stents. Hydrophilic guide wires (Weasel Wire, Infiniti Medical, Melno Park, CA; BiWire, Cook Urologic, Spencer, IN) are most effective for initial passage up the ureter and urethra. Once initial access is established, the hydrophilic wire may be exchanged for a stiff shaft working wire by passage of an open-ended ureteral catheter over the initial guide wire followed by wire exchange through the catheter.

Uroendoscopy techniques involving the upper urinary tract also require c-arm fluoroscopy in addition to the cystoscopy equipment. This approach is required for retrograde placement of urologic guide wires up the ureter for advanced endourology procedures such as ureteral stent placement, retrograde ureteropyelography, lavage of the renal pelvis for pyonephrosis, ureteroscopy, and sclerotherapy of idiopathic hematuria.[1,2] These advanced endourology procedures require specialized training.

ROUTINE CYSTOSCOPIC EXAMINATION OF THE LOWER URINARY TRACT

Prior to cystoscopy the hair around the vulva in female dogs and around the prepuce in male dogs is clipped and the skin surrounding the area is cleaned using standard presurgical cleaning. The prepuce or vaginal vestibule is flushed with dilute chlorhexidine or dilute povidone iodine solution. Dogs and cats may be positioned in dorsal or lateral recumbency with the vulva positioned at the end of the table.

For cystoscopic procedures, aseptic principles are observed including wearing sterile gloves and surgical gowns. Cystoscopes and all related equipment should be sterile and flushed generously with isotonic fluids before the procedure.

Routine cystoscopy should follow a standard protocol to inspect all aspects of the lower urinary tract before focusing on any abnormalities and before any procedures that may induce hemorrhage such as biopsy of the bladder wall. The cystoscope is passed proximally up the urethra into the urinary bladder under visual guidance with fluid distention. In females, the cystoscope should be passed retrograde into the vestibule and into the urethra without entering the body of the vagina to avoid contamination of the scope with normal vaginal flora. Once the cystoscope is in the urinary bladder, the urine is drained and the bladder is refilled with sterile, warm isotonic solution. This process is repeated as necessary to provide clear visualization of the bladder lumen. The irrigation solution is allowed to flow by gravity with the fluids positioned < 80 cm above the patient to avoid over-distension that may result in bleeding with resultant impaired visualization of the urinary tract.

Normal bladder mucosa allows for easy visualization of submucosal blood vessels. Edema or inflammation of the bladder wall makes the vessels more difficult to visualize and may completely obscure the smaller vessels. Normal ureteral orifices are c-shaped with the opening pointing medially. In cats, the ureteral orifices are more caudally positioned than in dogs and may appear to be in the proximal urethra. The bladder mucosa should be visually inspected throughout the entire urinary bladder. If a 30° view cystoscope is utilized, the scope should be rotated 180 degrees along the long axis to view the dependent portion of the bladder to inspect for uroliths, masses, or other lesions.

DISEASE CONDITIONS and ENDOUROLOGIC PROCEDURES

Idiopathic cystitis in cats

Idiopathic cystitis in cats is an exclusion diagnosis that is made on the basis of demonstrating lower urinary tract disease in the absence of UTI, urolithiasis, neoplasia or other causes. Cystoscopy is not routinely recommended and is usually only performed in refractory cases of suspected FIC to be certain no other lesions are found that could contribute to the cat's clinical signs.

Localization of hematuria

Using cystoscopy allows the source of gross hematuria to be definitively localized to the genital tract, urethra, bladder, and left or right kidney and ureter. Bleeding lesions in the lower urinary tract can be confirmed and sometimes treated depending on the extent and location of the lesion. Determination of upper tract hematuria can be determined by either direct visualization of gross hematuria from the ureteral orifice or by retrograde catheterization of each ureter. Recently endoscopic-guided sclerotherapy has been used as a renal sparing treatment for idiopathic renal hematuria.[1] Dogs with idiopathic renal hematuria may have bleeding from both kidneys at the time of initial diagnosis or at a later date, so renal sparing techniques are preferred over nephrectomy.

Diagnostic evaluation of recurrent urinary tract infection

Uroendoscopy is also useful for diagnostic evaluation of dogs with recurrent UTI to evaluate for anatomic defects and to obtain biopsies for cytology, histopathology, and microbiologic culture. Cultures of mucosal biopsies obtained via cystoscopy may detect organisms not isolated by simultaneous urine culture. Although one recent study failed to show a benefit of cultures of bladder mucosal biopsies, cultures (aerobic and mycoplasma) of bladder mucosa do have utility in some individual cases with negative urine cultures. Lesions present in the lower urinary tract from dogs with recurrent UTI should be also biopsied for culture and histopathology.

Masses or polyps

Masses or polyps in the urinary bladder may be visualized and biopsied during cystoscopy. Transitional cell carcinoma (TCC), the most common neoplasia of the lower urinary tract in dogs and cats and has a variable cystoscopic appearance. Most TCC masses have papillary frond-like projections from the surface of the mass. Polypoid cystitis in dogs and is a benign polyp lesion that often occurs secondary to urolithiasis and/or chronic urinary tract infection.[5] Benign polyps tend to have a smoother surface and have shorter projections than TCC. However, definitive differentiation of transitional cell carcinoma from polypoid cystitis requires histopathologic evaluation of biopsies. To increase the size of biopsies, placement of the biopsy forceps or stone baskets across the base of pedunculated or papillary lesions yields larger biopsy samples from the lesions.[6] Some polyps or solitary tumors with narrow stalks can be removed using biopsy instruments, electrocautery snares or by diode or holmium: YAG laser.

Removal of Uroliths

Stone baskets may be used to remove uroliths from the lower urinary tract provided the uroliths are smaller than the dilated urethra. Stone baskets are passed through the working channel of the cystoscope, the basket is opened, and once the urolith is engaged in the basket, the basket is closed around the urolith and positioned at the end of the cystoscope. Then the cystoscope and basket containing the urolith are slowly withdrawn through the urethra with direct visualization to ensure there is no bulging of the urothelium around the urolith edges, which may indicate the urolith is becoming wedged in the urethra. This technique may be safely used to remove some uroliths that will not readily pass during voiding urohydropropulsion. If the uroliths are too large for basket extraction, they may be fragmented into smaller passable fragments by laser lithotripsy.[7,8] Nonsurgical removal of uroliths by basket extraction or laser lithotripsy is easier, faster, and more efficacious in female dogs than male dogs.[7-8] Most female dogs are good candidates for these procedures whereas male dogs must be carefully evaluated to determine if this approach is appropriate. Male dogs smaller than approximately 5 kg are often too small enough to permit easy passage of the flexible ureteroscope through the urethra. The number and location of uroliths is also an important consideration. Male dogs with no more than 2-3 lower tract uroliths (<0.7mm) that require fragmentation can usually be successfully managed with laser lithotripsy. Laser lithotripsy is highly effective for 2-4 uroliths in the urethra provided the cystoscope can be easily passed proximally through the urethra. Urethroliths may contribute to urethral strictures and proliferative tissue in the urethra, which narrows the urethral lumen and may prevent passage of the cystoscope through the urethra.

Urinary incontinence

Uroendoscopy is a valuable diagnostic procedure for evaluation of dogs with urinary incontinence. The precise location of the ureteral openings can be readily identified by uroendoscopy in female dogs and this technique is equal to or superior to imaging techniques for diagnosis of ectopic ureters.[9] Cystoscopic-guided laser ablation of intramural ectopic ureters is possible to avoid open surgical correction.[10,11] Urethral bulking agents may be injected into the urethra via uroendoscopy for treatment of refractory urethral incompetence.[12]

REFERENCES

1. Berent AC, et al. **J Am Vet Med Assoc**. 2013;242(11):1556-1563. 2. Berent AC. J **Vet Emerg Critical Care** 2011;21(2):86-103. 3. Cannizzo KL, et al. **Vet Clin North Am Small Anim Pract** 2001;31(4):789-807. 4. Westropp JL, Buffington CAT.. In: Ettinger SJ, Feldman EC, eds. *Textbook of veterinary internal medicine: diseases of the dog and cat*. Vol 7. St. Louis: Saunders Elsevier; 2010:2069-2086. 5. Martinez I, et al. **J Vet Intern Med** 2003;17(4):499-509. 6. Childress MO, et al. **J Am Vet Med Assoc** 2011;239(3):350-356. 7. Adams LG, et al. **J Am Vet Med Assoc** 2008;232:1680-1687. 8. Grant DC, et al. **J Vet Intern Med** 2007;21:600-600. 9. Cannizzo KL, et al. **J Am Vet Med Assoc** 2003;223(4):475-481. 10. Berent AC, et al. **J Am Vet Med Assoc** 2012;240:716-725. 11. Berent AC, et al. **J Am Vet Med Assoc** 2008;232:1026-1034. 12. Byron JK, et al. **J Vet Intern Med** 2011;25(5):980-984.

PHEOCHROMOCYTOMAS AND ADRENAL CARCINOMAS: DIAGNOSIS, SUPPORTIVE CARE, AND DEFINITIVE TREATMENT

Sandra Bechtel, DVM, DACVIM (Oncology) and Anne Corrigan, DVM, DACVIM (SAIM)
Columbia, MO and Grenada, West Indies

INTRODUCTION

The increasing availability of advanced imaging has increased the number of adrenal masses found in dogs and cats, including benign, malignant, and functional masses. In dogs, adrenocortical tumors are the most common adrenal tumors found in necropsy studies with adenomas more common than carcinomas, followed by adrenal medullary tumors (pheochromocytomas) and metastatic lesions to the adrenal glands. In cats, tumors metastatic to the adrenal glands are more common than adrenocortical tumors and pheochromocytomas.[1]

The adrenal gland is structurally divided into the cortex and the medulla, and the cortex had 3 functionally distinct layers that release different hormones under the control of separate stimulatory mechanisms. Mineralocorticoid release occurs from the zona glomerulosa under the control of the renin-angiotensin-aldosterone system, serum potassium and ACTH. Glucocorticoids are released from both the zonae fasciculata and reticularis, primarily under the control of ACTH. The medulla acts as a modified post ganglionic sympathetic neuron and releases epinephrine (EPI) and norepinephrine (NE); canines release primarily the former and felines release about 50/50 of each.[2]

ADRENAL CORTICAL CARCINOMA

Benign adenomas are more common than carcinomas and are non-invasive, while carcinomas are often fixed and invasive. Functioning adrenal tumors, either adenomas or adenocarcinomas, secrete cortisol independent of pituitary control; secretion may intermittent and random.

Diagnosis of adrenal dependent hyperadrenocorticism (ADH)

Common diagnostics for ADH include screening tests, specific diagnostic tests for hyperadrenocorticism and differentiating tests to distinguish ADH from pituitary dependant (PDH). Imaging is required for diagnosis as well as for preparing for treatment and to evaluate for caval invasion or metastasis. One screening test is the urine cortisol:creatinine ratio (UCCR). The diagnosis of hyperadrenocorticism can be achieved with ACTH stimulation tests or low-dose dexamethasone suppression test (LDDST), however both of these tests can have both false positive and false negative results. Differentiation tests include endogenous ACTH levels (which can be difficult to interpret) and advanced imaging; abdominal ultrasound, or abdominal CT or MRI. The LDDST can often act as a differentiating test but only for identifying PDH. You must use additional imaging to confirm ADH.[3]

Clinical presentation

Dogs tend to be older with a median age of 11 years and clinical signs are consistent with hyperadrenocorticism. These signs include polyuria, polydipsia, polyphagia, weakness, lethargy, dermatologic changes, abdominal enlargement, and panting.[4-5]

Differentiation of adrenal adenoma versus carcinoma

Adrenal adenomas tend to be small (less than 2cm), well encapsulated or at least partially encapsulated, and well differentiated on histopathology. In contrast, carcinomas tend to be greater than 2cm in diameter with peripheral fibrosis and invasion into the capsule. On histopathology, hemorrhage, necrosis, and a high Ki67 proliferation index are associated with carcinomas. Up to 20% of carcinomas invade the phrenicoabdominal vein with extension into the renal vein and caudal vena cava. Adrenal carcinomas may also metastasize with the liver and lungs as the most commonly reported sites. Other sites reported include kidney, ovary, mesenteric lymph node, peritoneal cavity, and thyroid gland. The metastatic rate reported ranges from 14-50%.[4-5]

Treatment

Adrenelectomy is the treatment of choice for adrenal tumors; with adrenal carcinomas a median survival time of 230-778 days is reported. Potential complications of surgery are cardiovascular events and hemorrhage; dogs are also at risk for adrenocortical insufficiency, pulmonary thromboembolism, pancreatitis, renal failure, and wound dehiscence. Perioperative mortality rates range from 9-60%. Post-operatively, ACTH stimulation testing within 24 to 48 hours should be considered, in addition to monitoring of glucose and electrolytes. Heparin and glucocorticoids may be indicated in addition to mineralocorticoid supplementation in some dogs.[5]

Medical therapy for functional adrenal tumors include mitotane, although much higher doses are required than when used for PDH and relapses are common (up to 60% of dogs) and untoward side effects are also common in > 50% of dogs, not including dogs who develop mineralocorticoid deficiencies. The median survival time reported is 16.4 months. Trilostane can also be used for medical management of HAC, where the quality of life improves and clinical signs decrease however the size of the adrenal tumor may remain the same.[3, 6]

Prognosis

The median survival time in dogs with adrenocortical carcinomas treated with surgery alone is 778 days; in dogs that survive the peri-and post-operative period and are discharged from the hospital, the median survival time is reported as 992 days in one study and 17.5 months in another.

Conn's syndrome

Conn's syndrome (primary aldosteronism) is an aldosterone secreting adrenocortical tumor that is increasingly recognized in cats. Both adrenal adenomas and carcinomas have been reported to cause this syndrome, in addition unilateral and bilateral adrenal lesions are reported. The most common clinical sign is muscle weakness resulting from hypokalemia; additional findings are arterial hypertension and, uncommonly, hypernatremia. A normal plasma aldosterone level in the face of hypokalemia is supportive of but not diagnostic for Conn's syndrome and an oral fludrocortisone suppression test or aldosterone: renin ratio may be more useful for definitive diagnosis. Adrenalectomy is the treatment of choice with unilateral disease and outcomes reported are good, even if invasion into vena cava is present. Medical management with potassium supplementation, antihypertenisive drugs, and spironolactone (aldosterone antagonist) can be used for non-surgical patients.[7]

PHEOCHROMOCYTOMA

Pheochromocytomas are functional tumors of the adrenal medulla derived from chromaffin cells. Normal chromaffin cells are stimulated by sympathetic nerve fibers to secrete hormones into the blood stream – epinephrine, norepinephrine, and dopamine. In contrast, pheochromocytomas are not innervated, so catecholamine release is not initiated by neural impulses and their release is variable and unpredictable. Pheochromocytomas are uncommon in dogs (0.01-0.1% of all canine tumors) and rare in cats (3 case reports).

Catecholamine synthesis, degradation, and physiologic effects

Catecholamines are derived from tyrosine and are processed through dopa and dopamine into NE which may undergo methylation into EPI. The are released into the blood where they exert a variety of effects through adrenergic receptors, NE acting mostly on alpha receptors and EPI acting equally on both alpha and beta receptors. Tissues can respond differently based on the types of receptors present on their cell membranes, in general NE has greater effects on the vasculature and EPI will have greater effects on the heart and on the metabolic rate. Catecholamines have a very short half life and are metabolized by catechol-*O*-methyl transferase (COMT) in the liver one to a few minutes after release.[2, 8] Pheochromocytomas have been reported to produce their own COMT and therefore the metabolites and catecholamines can be highly elevated in animals with this tumor and these metabolites may be used for assessment. [9]

Signalment, behavior, and clinical signs

In dogs, the median age of diagnosis is 11 years with no sex predilection. Most often, pheochromocytomas are unilateral (approximately 5% are bilateral), slow growing, vascular, and malignant. Vascular invasion is reported in up to 82% of cases and metastasis in up to 40% to the liver, spleen, lung, regional lymph nodes, bone, and central nervous system.

Clinical signs are often vague and episodic in nature due to the intermittent release of catecholamines. The most common clinical signs associated with pheochromocytoma are weakness and collapse. Additional signs are tachypnea, anxiety, restlessness, exercise intolerance, decreased appetite, weight loss, polyuria/ polydipsia, and seizures. Acute collapse may also occur due to tumor rupture and hemorrhage.[10-11]

Diagnosis

Physical examination may be normal, or findings non-specific: tachypnea, panting, tachycardia, weakness, pallor, arrhythmias, and hypertension may be noted; 5% of dogs will have a palpable abdominal mass and 10% will have findings consistent with a space occupying mass in the abdomen (ascites or peripheral edema). No consistent findings are reported for complete blood count, chemistry panel, or urinalysis. Hypertension is documented in 40-50% of dogs with pheochromocytoma. Vascular smooth muscle has alpha-1 and beta-2 receptors, with alpha-1 receptors predominating. Therefore, the result of catecholamine release is vasoconstriction and hypertension. Hypertension should be suspected with a systolic blood pressure of greater than 160mmHg. Note that a normal blood pressure does not rule out a diagnosis of pheochromocytoma, as catecholamine release is episodic in nature. Ocular examination may reveal mydriasis, retinal hemorrhage, or, less commonly, retinal detachment. Neurologic examination may reveal spinal pain, rear limb ataxia, or head tilt.[10-11]

Thoracic radiographs may reveal cardiomegaly due to concentric hypertrophy, distension of the caudal vena cava from tumor thrombus, and, in 5-10% of cases, evidence of pulmonary metastasis. Abdominal ultrasound detects an adrenal mass in 85% of dogs with pheochromocytoma; adrenal asymmetry may be detected if the adrenal mass is not large or invasive. Vascular invasion may also be detected into the caudal vena cava or the phrenicoabdominal vein in addition to tumor thrombus. Advanced imaging using CT or MRI provides more imaging detail in regards to the presence and location of the adrenal mass, the size, shape, and architecture of the adrenal glands, mass invasion, and metastasis to other abdominal organs. Advanced imaging is especially helpful for surgical planning.[12] Scintigraphy using ^{131}I or ^{123}I labelled meta-iodobenzylguanidine may be used to image pheochromocytomas in people and has been reported in the dog. Meta-iodobenzylguanidine has a similar structure to norepinephrine and the mechanism of cell uptake is the same, concentrating in the neurosecretoy storage granules of chromaffin cells. P-[18F] fluorobenzylguanidine is a norepinephrine analogue and PET imaging agent used successfully to image 2 dogs with pheochromocytoma.[13-14]

Measurement of plasma or urine catecholamine metabolites are superior to measurement of the catecholamine in human medicine due to the short half-life of catecholamines in circulation and the sporadic and unpredictable release of catecholamines from the pheochromocytoma. Membrane bound catechol-O-methyltransferase (COMT) on the tumor cells is responsible for the metabolism of tumor released catecholamines – since catecholamines are only intermittently released by the tumor, the continuous release of the metabolites metanephrine and/ or normetanephrine leads to more consistent test results. In fact, one study found plasma metanephrine to have moderate sensitivity with excellent sensitivity, and normetanephrine to have excellent sensitivity and specificity, in differentiating dogs with pheochromocytoma from healthy dogs, dogs with non adrenal disease, and dogs with adrenal cortical tumors. Urine normetanephrine: creatinine ratio may also be helpful in the diagnosis of pheochromocytoma. False positive results for both serum and urine metanephrine and normetanephrine measurements can be obtained if the dog is on phenoxybenzamine.[15-16]

Definitive diagnosis requires histopathology and immunohistochemistry. Chromogranin A will differentiate between pheochromocytoma and adrenal carcinoma, as chromagranin A is present the in secretory granules of endocrine cells; pheochromocytomas will also stain with synaptophysin, a membrane component of synaptic vesicles in neurons and neuroendocrine cells.[10-11]

Treatment and prognosis:

Adrenalectomy is the treatment of choice for pheochromocytoma in both the dog and cat. Prior to consideration of treatment for a suspected pheochromocytoma, phenoxybenzamine, a non-selective alpha-1 and -2 receptor blocker, is recommended to decrease blood pressure, decrease ventricular arrhythmias, and support blood volume expansion. A study in dogs found that pre-treatment with phenoxybenzamine reduced the overall mortality rate; treated dogs were six times more likely to survive than dogs that did not receive phenoxybenzamine as a pre-treatment (mortality rate 13% versus 48% in untreated dogs). Potential complications of adrenalectomy include hypertension, hypotension, cardiac arrhythmias, and hemorrhage.[10-11, 17]

Overall prognosis following surgery is a median survival time of 1 year, with some dogs surviving 2 and 3 years post-operatively. Actual metastatic rates are difficult to determine as, histologically, the appearance of pheochromocytomas are variable and not predictive of malignancy. Dogs with tumors larger than 5cm, evidence of metastasis, or thrombosis have a worse prognosis than dogs without these factors. Adjuvant chemotherapy has not been studied in dogs and is not thought to be helpful in people; radiation therapy has also not been studied.[10-11]

Medical management is indicated for dogs in which surgical resection is not a viable treatment option. Phenoxybenzamine is recommended long term for medical management of catecholamine release. Beta blockers, such as propranolol or atenolol, may be considered for severe tachycardia to prevent unopposed alpha stimulation and uncontrolled hypertension.

One dog with a non-resectable pheochromocytoma has been successfully treated with ^{131}I meta-iodobenzylguanidine; this dog had stable disease on CT scan 6 weeks post treatment. Progressive disease was noted 5.5 months later and a second treatment was initiated. The dog was euthanized following the 2nd treatment due to gastrointestinal ulcers, but this therapy may hold promise in the future for non-resectable disease.[13]

INCIDENTALOMA

With the increased availability of advanced imaging, incidental adrenal masses are found more frequently. When an incidental adrenal mass is found, determining whether or not the mass is functional is an important diagnostic step. Blood pressure, fundic exam, and endocrine testing to rule out functional tumors should be considered. Also consider staging for other tumor types since metastatic lesions to adrenal glands are common, especially in cats. Adrenalectomy should be considered for functional tumors, locally invasive tumors, or those larger than 2.5cm. If the mass is determined to be nonfunctional and is less than 2cm with no invasion, repeating an abdominal ultrasound regularly is recommended.

REFERENCES

[1]Labelle P et al. Vet Path 2005; 42: 52-8; [2]Kemppainen R and Behrend E. Vet Clin Small Ani 1997; 27(2),173-186; [3] Melian et al in Ettinger and Feldman eds., Textbook of Veterinary Internal Medicine 7th ed.1816-1840 ; [4]Labelle P et al Vet Path 2004; 41: 490; [5]Anderson CR et al, J Am Anim Hosp Assoc 2001; 37: 93-7; [6] Eastwood et al, J of Small Ani Practice 2003 Mar, 44(3) 126-31; [7]Lo AJ et al. J Vet Intern Med. 2014; 28: 137-43; [8] Guyton and Hall Textbook of Medical Physiology 10th ed. 2000, 697-704; [9] Gostelow et al. J Vet Intern Med 2013; 27:83-90; [10]Barthez PY et al. J Vet Intern Med 1997; 11: 272-79; [11]Gilson SD et al. J Vet Intern Med 1994; 8: 228-32; [12]Rosenstein DS Vet Rad and US 2000; 41: 499-506; [13]Bommarito DA et al J Am Anim Hosp Assoc 2011; 47: e188-94; [14]Berry CR et al. Vet Rad US 2002; 43: 183-6; [15]Gostelow et al. J Vet Intern Med 2013; 27: 83-90; [16]Quante S et al. J Vet Intern Med 2010; 24: 1093-7; [17]Herrera MA et al. J Vet Intern Med 2008; 22: 1333-9.

PARANEOPLASTIC SYNDROMES: DIAGNOSIS, SUPPORTIVE CARE, AND DEFINITIVE TREATMENT

Sandra Bechtel, DVM, ACVIM (Oncology) and Anne Corrigan, DVM, MS, ACVIM (SAIM)
Grenada, West Indies and Columbia, MO

INTRODUCTION

Paraneoplastic syndromes are clinical disorders caused by malignant neoplasia that are not directly caused by the physical effects of the tumor or its metastases. Instead, these syndromes are caused by tumor production of substances that directly or indirectly cause distant clinical signs, depletion of normal substances, or the host response to the tumor leads to an effect at a distant site or immune cross-reactivity. Examples are tumor production of hormones, enzymes, fetal proteins, or cytokines, metabolism of steroids by tumor cells, or tumor stimulation of antibody production. The progression and resolution of paraneoplastic syndromes may correlate with progression and resolution of the primary tumor, although those of immunologic or neurologic etiology are unpredictable and may not resolve with tumor treatment. Furthermore, a paraneoplastic syndrome may be recognized prior to the malignancy itself.

CANCER CACHEXIA

Recent research in human medicine has redefined and reclassified cancer cachexia. Currently, cancer cachexia is defined as a multifactorial syndrome characterized by an ongoing loss of skeletal muscle mass, with or without loss of fat mass, that cannot be fully reversed by conventional nutritional support, leading to functional impairment[1-3]. The major hallmarks of cachexia are anorexia or hyporexia, fatigue, metabolic and endocrinologic alterations, and loss of lean body mass. Cachexia is initiated by a combination of negative protein and energy balance caused by a combination of reduced caloric intake and abnormal metabolism. Cachexia leads to a decline in quality of life, decreased tolerance and response to cancer therapy, and increased morbidity and mortality. In fact, in people, it is estimated that 50% of cancer patients die with the presence of cachexia, and up to 20% of deaths are due to cachexia[1-3]. The incidence of cancer cachexia in dogs and cats is not known, although it is thought to be less than that seen in people. In one study investigating body condition score (BCS) and weight loss in oncology patients, 4% of dogs met the criteria for cachexia in that study based on a BCS of < 3, however, more than 1/3 experienced a greater than 5% weight loss and 35% had muscle wasting present[4]. In a study of cats with cancer, 91% exhibited muscle loss[5]. Difficulty detecting preclinical cachexia (pre-cachexia), referral bias and the option of euthanasia for veterinary patients likely has an effect on studies of the incidence of cancer cachexia in veterinary medicine. Furthermore, in dogs without clinical evidence of cachexia, alterations in protein and carbohydrate metabolism are documented in dogs with lymphoma, osteosarcoma, and nonhematopoietic malignancies[6-8]; alterations in lipoprotein profiles in dogs with lymphoma are also evident[9]. Pre-cachexia in people is defined as the following: underlying chronic disease, unintentional loss of less than or equal to 5% of body weight in 6 months, chronic or recurring systemic inflammation, and anorexia or anorexia related signs. The definition of cachexia in people is defined as weight loss of 5% or greater and 3 of the following: decreased muscle strength, fatigue, anorexia, low fat free mass index, and abnormal biochemistry results (increased inflammatory markers, anemia, or hypoalbuminemia) [1-3]. Pre-cachexia is likely under recognized in veterinary cancer patients.

Anorexia or hyporexia is likely multifactorial – enhanced inflammatory cytokine responses can result in hypothalamic arcuate nucleus suppression of appetite, gastrointestinal cancer may lead to difficulty eating, changes or loss in smell and taste can cause hyporexia, and cancer therapy can contribute to cachexia loss or alteration in taste and smell[1-3, 10]. The metabolic changes associated with cancer cachexia are multifactorial and multifaceted. Metabolic changes, tumor factors, host-tumor interactions, and host response factors all play a role in the development of cachexia. Cytokines involved in enhancing proteolysis are tumor necrosis factor-alpha, interleukin (IL)-1beta, and IL-6; these cytokines are produced by the tumor cells themselves or by the host inflammatory response. These cytokines can interact with appetite regulation through the melanocortin system or the neuropeptide Y system leading to decreased appetite, increased basal metabolic rate, and decreased lean body mass[10].

The end result of cachexia is a state of active inflammation resulting from both tumor and host factors leading to a catabolic state. Early recognition and intervention are critical in the treatment of cachexia. Management of diet and appetite in the early stages may contribute to the overall well-being and outcome of dogs and cats with cachexia. Underlying concurrent diseases should be managed appropriately (diseases causing chronic pain, dental disease), diet should be evaluated and modified, and appetite stimulants may halt the progression. Light exercise has been helpful in the treatment of people with cancer cachexia and omega-3 fatty acid supplementation may decrease the inflammation contributing to cachexia[10]. Research is ongoing in veterinary and human medicine to support cancer patients and reverse cachexia, with the goal of improving quality of life and outcomes of cancer therapy.

HYPERCALCEMIA

Approximately 2/3 of dogs and 1/3 of cats with hypercalcemia have cancer as the underlying cause. In dogs, lymphoma is the most common cause of hypercalcemia; other common neoplastic cases are apocrine gland anal sac adenocarcinoma, thyroid carcinoma, and multiple myeloma. Other reported tumors causing hypercalcemia are bone tumors, thymoma, squamous cell carcinoma, mammary gland adenocarcinoma, melanoma, primary lung carcinoma, chronic lymphocytic leukemia, renal angiomyxoma, and parathyroid tumors. In cats, the most common neoplastic causes of hypercalcemia are lymphoma and squamous cell carcinoma[11].

Causes of hypercalcemia of malignancy include ectopic parathyroid hormone (PTH), pararthyroid hormone related peptide (PTHrp), extensive and multifocal bone lysis (osteosarcoma, multiple myeloma, tumor metastasis to bone), prostaglandins (PGE1/2), interleukin (IL)-1beta (osteoclast activating factor), transforming growth factor (TGF)-beta, and receptor activator or nuclear factor kappa-B (RANKL).

PTH is produced ectopically with parathyroid adenomas and adenocarcinomas. The clinical presentation of both adenomas and carcinomas is similar, and surgical excision is the treatment of choice for both[12]. PTHrp is an embryonal peptide similar to PTH, and functions as a normal endocrine hormone for calcium regulation in the fetus and normal paracrine factor in fetal and adult tissues. In hypercalcemia of malignancy, the tumor produces PTHrp, which has a N-terminal region that binds to and stimulates PTH receptors with an affinity equal to that of PTH. The result is osteoclastic bone resorption and increased renal resorption of calcium. Unlike PTH, PTHrp cannot stimulate renal formation of vitamin D to increase calcium absorption from the gastrointestinal tract. Anal sac adenocarcinoma and lymphoma in the dog both result in hypercalcemia through the production of PTHrp. An additional mechanism of hypercalcemia in lymphoma is the conversion of vitamin D to its active form through 1-alpha hydroxylase, which some lymphocytes contain. The treatment of choice for hypercalcemia of malignancy is to treat the underlying cause, however, supportive care may be indicated in many cases prior to definitive therapy. Supportive care measures to treat emergent hypercalcemia include saline diuresis, furosemide (only in the fully hydrated patient), prednisone (may interfere with diagnosis of lymphoma), and bisphosphonates[11-14].

HYPOGLYCEMIA

The most common cause of paraneoplastic hypoglycemia in the dog is insulinoma. The tumor cells secrete insulin independent of glucose, and hyperinsulinemia results in hypoglycemia through suppression of glucose release and production. Signs of neuroglycopenia are weakness, ataxia, collapse, disorientation, behavioral changes, and seizures. Catecholamine release stimulated by hypoglycemia may lead to muscle tremors, shaking, anxiety, and hunger. Clinical signs are usually intermittent and precipitated by fasting, exercise, excitement, and eating. Other tumor types that cause hypoglycemia are extra pancreatic neoplasia resulting in paraneoplastic hypoglycemia (for example, hepatoma, hepatocellular carcinoma leiomyoma, leiomyosarcoma, lymphoma, hemangiosarcoma, oral melanoma, multiple myeloma). Nonislet cell cancers can induce hypoglycemia through increased tumor utilization of glucose, decreased hepatic glycogenolysis or gluconeogenesis, secretion of insulin or insulin like growth factors-1 and 2, upregulation of insulin receptors, increased insulin binding by M proteins with multiple myeloma, or increased production of somatomedins. Treatment is aimed at supportive care followed by definitive tumor therapy.[15]

HYPERTROPHIC OSTEOPATHY

Hypertrophic osteopathy (HO) is characterized by painful periosteal proliferation and soft tissue swelling along the diaphysis of long bones which may be bilateral or involve all 4 limbs. The most common neoplastic cause in dogs and cats is pulmonary neoplasia, either metastatic or primary. Other non-pulmonary tumors reported to result in HO in dogs are urinary bladder rhabdomyosarcoma, esophageal tumors, malignant Sertoli cell tumor, renal transitional cell carcinoma, and nephroblastoma; in cats, renal adenoma, papillary carcinoma, and adrenocortical carcinoma have been reported to cause HO. Non-neoplastic disease can also result in HO including heartworm disease, heart disease, esophageal granuloma, congenital megaesophagus, and infectious or inflammatory lung disease.[16]

Although the etiology is not well defined, pathophysiologic changes are increased peripheral blood flow, proliferation of vascular connective tissue, and periosteal proliferation that may be the result of neurologic stimulation. Vascular endothelial growth factor, platelet derived growth factor, increased circulating vasodilators, and increased growth hormones may also be involved. Clinical signs of HO are lameness (sometimes shifting leg lameness), reluctance to ambulate, and heat and swelling of the distal limbs. Diagnosis is based on radiographs of affected limbs, which reveal a characteristic periosteal proliferation forming horizontal to the long axis of the bone; HO tends to start in the distal limb and progress proximally. Further investigation with thoracic radiographs is indicated to investigate for pulmonary masses. The definitive treatment of HO, if indicated, is excision of the pulmonary mass. In cases where this is not an option, non-steroidal anti-inflammatory drugs and additional pain control measures are indicated. Treatments such as vagotomy and bisphosphonate have not been well described in dogs.[16]

CUTANEOUS PARANEOPLASTIC SYNDROMES

Feline paraneoplastic alopecia

Feline paraneoplastic alopecia, reported with pancreatic carcinoma and biliary carcinoma, is characterized by progressive, bilateral symmetrical alopecia of the ventrum and medial limbs that may progress to the head. The skin is thin and glistening but not fragile and hair easily epilates from non-alopecic areas. Foot pads may also be involved and appear dry and crusted or erythematous and moist. Systemic clinical signs are related to the primary tumor and include anorexia, weight loss, vomiting, and diarrhea. Most cats with pancreatic carcinoma have advanced stage disease at the time of diagnosis, and successful treatment of metastatic pancreatic cancer in cats is not reported. One case report of a cat with a single pancreatic carcinoma surgically resected did have resolution of the paraneoplastic alopecia.[17-18]

Feline exfoliative dermatitis

Exfoliative dermatitis is characterized by diffuse nonpruritic cutaneous erythema with exfoliation and scaling of the head and pinnae. As scaling increases, lesions progress to other areas and alopecia develops. Crusts, ulcerations, and brown waxy debris accumulation occurs. Pruritis occurs in some cases which may or may not be associated with underlying infection. Thymomas cause this syndrome and thoracic radiographs should be performed if exfoliative dermatitis is suspected; surgical excision is the treatment of choice for thymomas; external beam radiation therapy has been utilized successfully in non-resectable canine thymomas and can be considered in feline thymomas that are not resectable. Underlying infection, in particular *Malassezia* spp, should be investigated for and treated if present.[17-18]

Nodular dermatofibrosis

Nodular dermatofibrosis is characterized by multiple, firm, well-circumscribed cutaneous or subcutaneous nodules primarily in the extremities that can progress to diffuse distribution. This particular syndrome is associated with a mutation mapped to exon 7 on the Birt-Hogg-Dube locus on chromosome 5 in dogs. This syndrome includes bilateral renal cystadenoma or renal cystadenocarcinoma and, in intact females, an increased risk for uterine leiomyomas. This is most commonly seen in German shepherd dogs, and has also been reported in one Golden retriever and one German shepherd cross. As the renal tumors are bilateral, nephrectomy is not curative and does not extend survival times. Treatment is palliative in nature with excision of ulcerated or infected masses as indicated. Nodular dermatofibrosis is a slowly progressive disease, however, and the mean time from observation of cutaneous lesions to time of death is 2.5 years.[18]

Superficial necrolytic dermatitis

The most common presenting clinical sign of glucagonoma in dogs is superficial necrolytic dermatitis, which is also known as diabetic dermatopathy, hepatocutaneous syndrome, or necrolytic migratory erythema. Superficial necrolytic dermatitis (SND) is comprised of hyperkeratosis, crusting, ulceration and erosion of the footpads, mucocutaneous junctions, external genitalia, distal extremities, pressure points, and ventral abdomen. A diagnosis of SND can be determined on histopathology (other causes of SND include advanced liver disease, phenobarbital, and in cats pancreatic carcinoma) and the treatment of choice for glucagonoma is surgical resection. Superficial necrolytic dermatitis may improve after surgery or medical therapy; and other reported treatments for SND are amino acid infusions, oral protein supplementation using protein powder of egg yolks, zinc, and essential fatty acids.[19]

HEMATOLOGIC PARANEOPLASTIC SYNDROMES

Hypergammaglobulinemia

Hypergammaglobulinemia is most commonly associated with multiple myelomas due to excessive production of immunoglobulins from a monoclonal population of plasma cells or lymphocytes. Clinical signs are those of hyperviscosity syndrome, (lethargy, ataxia, hypoxia, cardiac disease, seizures, coma), hypoxia, bleeding, and ocular changes (retinal hemorrhage, detachment). Lymphoma and leukemia can also cause a paraneoplastic hypergammaglobulinemia.

Anemia

A mild to moderate normocytic, normochromic, nonregenerative anemia is commonly seen in veterinary cancer patients. This is most consistent with anemia of chronic disease, caused by three concurrent mechanisms: shortened red blood cell (RBC) survival (associated with IL-1 and oxidant damage to RBC membrane and binding of immunoglobulin molecules which accelerate removal), impaired iron mobilization or utilization (altered ferritin production, alterations in transferrin receptors leading to increased iron storage and decreased availability), and impaired RBC production (erythroid cells become refractory to increased erythropoietin (Epo) due to effects of inflammatory cytokines on precursors in addition to decreased Epo production). Immune mediated hemolytic anemia (IMHA) is a highly regenerative anemia that is the result of the immune mechanisms triggering the early destruction of RBCs by binding of immunoglobulins, complement, or both to the RBC membrane. Lymphoma and leukemia are most commonly associated with IMHA. The treatment of choice is supportive care, immunosuppressive therapy, and treatment of the underlying neoplasia.[20]

Erythrocytosis

Although uncommon, erythrocytosis has been reported in dogs and cats with cancer. Renal tumors, lymphoma, lung tumors, liver tumors, cecal leiomyosarcomas, nasal fibrosarcomas, and transmissible venereal tumors have all been reported in association with erythrocytosis. It is caused by tumor production of erythropoietin, renal hypoxia caused by the tumor leading to increased Epo production, or increased production of hypoxia inducible factors.

Neutrophilic leukocytosis

Paraneoplastic neutrophilic leukocytosis is characterized by a marked mature neutrophilia not attributed to inflammatory disease, infectious disease, or primary leukemia. This syndrome is reported in canine rectal adenomatous polyps, renal tubular adenocarcinoma, metastatic fibrosarcoma, and in cats with lymphoma, dermal tubular adenocarcinoma, mast cell tumor, oral squamous cell carcinoma, and various other sarcomas and carcinomas. The mechanism is reported as tumor production of G-CSF and GM-CSF by the primary tumor and the treatment of choice is treatment of the primary tumor.[21-22]

Thrombocytopenia

Thrombocytopenia as a paraneoplastic syndrome is caused by antiplatelet or antimegakaryocytic antibodies. Immunosuppressive therapy and treatment of the primary tumor are the treatments indicated.[23]

REFERENCES

[1]Muscaritoli M et al. Crit Rev Oncol/Hematol 2014; doi: 10.1016/j.critrevonc.2014.10.014; [2]Evans WJ et al. Clin Nut 2008; 27: 793-99; [3]Donohoe CL et al. Gastroenterol Res Pract 2011 doi: 10.1155/2011/601434; [4]Michel et al. J Vet Intern Med 2004; 18: 692-5; [5]Baez JL. J Fel Med Surg 2007; 9: 411-417; [6]Mazzaferro EM et al. AJVR 2001; 62: 1234-9; [7]Vail DM J Vet Intern Med 1990; 4: 8-11; [8]Ogilvie GK et al. AJVR 1997; 58: 277-81; [9]Ogilvie GK et al. J Vet Intern Med 1994; 8; 62-6l; [10]Freeman LM J Vet Intern Med 2012; 26: 3-17; [11]Bergman J Topics in Compan An Med 2012; 27: 156-158; [12]Sawyer ES. Vet Comp Oncol 2011; 10: 57-64; [13]Eliot J. J Small Anim Prac 1991; 32: 564-7; [14]Savary KC et al. J Vet Intern Med 2000; 14: 184-9; [15]Goutal CM et al. J Am Anim Hosp Assoc 2012; 48: 151-163; [16]Withers SS et al. Vet Comp Oncol 2013; doi: 10.1111/vco.12026; [17]Outerbridge CA. Vet Clin North Am Small Anim Pract 2013; 43: 135-52; [18]Turek MM. Vet Derm 2003; 14: 279-296; [19]Mizuno T et al. Vet Dermatol 2009; 20: 72-9; [20]Childress Vet Clin North Am Small Anim Pract 2012; 42: 123-55; [21]Sharkey LC et al. J Vet Intern Med 1996; 1: 405-8; [22]Lucroy MD J Am Vet Med Assoc 2001; 218: 736-9; [23]Botsch et al. Vet Record 2009; 164: 647-51.

THE WHITE COAT EFFECT IN SMALL ANIMAL MEDICINE

Anthony P. Carr, DMV, DACVIM
Saskatoon, Saskatchewan, Canada

Measuring blood pressure in small animals can be a challenge. Fear and stress in clinic can lead to elevations on blood pressure termed white coat effect (WCE) that do not truly reflect the individual's usual blood pressure. This elevation can make the patient appear hypertensive at which time it could be called white coat hypertension (WCH). It is however also possible to see a drop in blood pressure in response to stress which can lead to a misdiagnosis of normotension which is termed masked hypertension. As a result, if only in clinic values are used it is possible to make diagnostic mistakes and thereby potential therapeutic errors. This is commonly seen in humans as well and is usually addressed either with ambulatory blood pressure (ABP) measurement or with home blood pressure (HBP) measurements. ABP has been found to more closely correlate with end-organ damage and cardiovascular events than in-clinic blood pressure measurements. Home blood pressure was found to more accurately predict left ventricular hypertrophy than either ABP or office blood pressure, provided that an adequate number of readings were taken.

Whether WCH poses a risk is the subject of some debate in humans. It certainly will make the accurate diagnosis of blood pressure impossible. In humans, generally blood pressure decreases with sequential visits to the doctor's office. It is common for white coat hypertensives to go on to develop hypertension, though overall the risk of cardiovascular problems does not appear to be increased by having WCH. The reasons for the WCE are unclear. In humans it has been shown that increased office BP, female gender, increased daytime blood pressure variability, increasing age, and smoking were major factors determining the severity of WCE. Labeling also was a major factor, meaning that people that were diagnosed with hypertension had more marked WCE, probably related to increased anxiety.

White coat effect has been documented to occur in dogs and cats. In one study 14 dogs were tested in clinic as well as at home. At home all dogs had blood pressures that were considered normal whereas in clinic 5 of 14 had systolic blood pressures >165 mmHg or diastolics greater than 95 mmHg. In one part of another study, 10 clinically normal dogs were measured in clinic and at home. This did not show a statistically significant difference in blood pressure values obtained. In another study dogs from 3 different breeds were subjected to various measurement scenarios. Standard protocol was to have the veterinarian measure with the owner restraining, then the owner measured without veterinarian in the exam room, then the veterinarian measured without the owner, and then the veterinarian measured without the owner and used a squeaky toy to cause increased alertness in the pet. This found that at least in Labrador Retrievers having the veterinarian measure without the owner increased systolic and diastolic blood pressure in comparison to having the owner in the exam room. In Dachshunds and Cavalier King Charles Spaniels there was no difference between the various scenarios. This group also had the owners measure at home with the same scenarios and no difference was found to in-clinic readings as a group. In a study with retired racing greyhounds significantly higher blood pressure values were found in hospital than at home, independent of whether the in-home pressure was taken by the owner or the veterinarian.

The only study on WCH in cats that has been published looked at cats that had been implanted for telemetric blood pressure measurement. These cats were exposed to a simulated veterinary clinic visit and physical examination. The study compromised 13 cats, with 7 having partial renal insufficiency. Blood pressure increased on average 17 mm/Hg systolic during the visit. Heart rate increases did not correlate to blood pressure changes. It was also found that repeated visits did not decrease the magnitude of the white coat elevation. The cats with renal insufficiency had more pronounced elevations in blood pressure than the normal cats. The range of changes in blood pressure was wide from +75 to -27 mmHg difference from normal. The lowering of blood pressure under stress is especially bothersome as it is counter-intuitive. This is analogous to masked hypertension in humans, where hypertension is not found in-clinic but can be documented with HBP or ABP.

Telemetry is not a realistic approach to clinical patients. We have been gathering HBP measurement in our patient population with the owners obtaining the measurements. Generally HBP has been a challenge in small animals. Doppler technology is unsuited for widespread owner use. Oscillometric blood pressure devices are better suited, however there is considerable variation in the readings obtained and it is not possible to determine which readings are significantly impacted by artifact. Use of the HDO (S + B medVET) allows previously obtained oscillometric tracings to be downloaded and evaluated for artifact. This makes it possible to discard any values that are based on artifact or are obtained with other issues (low amplitude, inappropriate deflation speed of cuff, etc.). It has been our experience that the tail is a superior site for blood pressure measurements especially in cats. This makes this technique very easy for owners to carry out.

If we detect hypertension in clinic (even if borderline) we will frequently send the HDO home with the owner. The owners are instructed on how to take readings using the tail. The cuff is adjusted so that it can be slipped up the tail till it is snuggly seated. The owners are instructed to have 3 to 4 blood pressure reading sessions, obtaining at least 5 readings per session. The unit is returned and the oscillometric curves are downloaded for evaluation on a computer. Only values that are free of artifact are used to assess blood pressure.

The initial pilot data from 8 cats showed significantly different blood pressure measurements for HBP vs. clinic blood pressure, both systolic and diastolic (Carr AP, Egner B. In home blood pressure monitoring, a pilot study. ECVIM abstract 2010). Heart rate was not statistically different. Whereas in clinic 7 had systolic hypertension, HBP showed only 3 were affected. Diastolic pressures were even more susceptible to change with 5 cats in clinic being hypertensive but only one at home. We have continued to expand our numbers and have to date not had an owner fail at getting readings.

Similar pilot data obtained from dogs was presented at the ACVIM Forum in 2012. In this study 10 dogs had blood pressure measured in clinic and at home. All dogs had systolic hypertension in clinic, but only 5 were hypertensive at home. In clinic, 7 were categorized as having severe risk of TOD, yet at home only 1 had severe risk of TOD. Diastolic hypertension was present in 5 dogs in clinic but only 1 at home. When analyzed as a group, systolic pressure, mean pressure, and heart were significantly different at home vs. in-clinic, diastolic was not. Two dogs had higher pressures at home than in-clinic.

There does appear to be a disparity between some of the studies showing a difference and others not. However it does appear that healthy animals may be less at risk of WCE than animals that have an underlying disease. This was supported by the study in telemetry cats where the cats that had induced renal insufficiency were more prone to WCE. Based upon the evidence available at this time it would certainly appear prudent to consider in-home blood pressure measurements in dogs and cats. This should not be restricted to animals that are hypertensive in-clinic since masked hypertension has been documented in dogs and a cat as well.

REFERENCES

Cohen DL, Townsend RR. How significant is white coat hypertension? J Clin Hypertension 2010; 12: 625-626., Shimbo D, Pickering TG, Spruill TM, et al. Relative utility of home, ambulatory, and office blood pressures in the prediction of end-organ damage. Am J Hypertension 2007; 20:476-482., Kallet AJ, Cowgill LD, Kass PH. Comparison of blood pressure measurements obtained in dogs by use of indirect oscillometry in a veterinary clinic versus at home. J Am Vet Med Assoc 1997;210:651-654., Remillard RL, Ross JN, Eddy JB. Variance of indirect blood pressure measurements and prevalence of hypertension in clinically normal dogs. Am J Vet Res 1991; 52:561-565., Belew AM, Barlett T, Brown SA. Evaluation of the white-coat effect in cats. J Vet Intern Med 1999;13:134-142., Sprull TM, Pickering TG, Schwartz JE, et al. The impact of perceived hypertension status on anxiety and the white coat effect. Ann Behav Med 2007; 34:1-9,, Manios ED, Koroboki EA, Tsivgoulis GK, et al., Am J Hypertens 2008;21:153-158., Hoeglund K, Hanas S, Carnabuci C, et al. Blood pressure, heart rate, and urinary catecholamines in healthy dogs subjected to different clinical settings. J Vet Intern Med 2012;26:1300-1308., Marino CL, Cober RE, Iazbik MC, Couto CG. White-coat effect on systemic blood pressure in retired racing greyhounds. J Vet Intern Med 2011;25:861-865.

HYPERKALEMIA AND THE ECG: A TALE OF TWO DISEASES

Anthony P. Carr, DMV, DACVIM
Saskatoon, Saskatchewan, Canada

Hyperkalemia is occasionally encountered in veterinary medicine. The most profound effects of hyperkalemia are on the cardiac system. These include decreased myocardial excitability, increased myocardial refractory period, and slowed conduction; all effects which can cause critical cardiac disturbances. Severe hyperkalemia is a potentially life-threatening complication in a variety of conditions. Because of time it takes to get potassium concentration determined, early diagnosis of severe hyperkalemia can be made using an ECG. What is often underappreciated is that potassium is not the sole determinant of ECG changes seen in hyperkalemia.

"Classic" electrocardiographic findings in hyperkalemic patients include bradycardia, peaked T waves, decreased R and P wave amplitudes, prolonged QRS intervals, prolonged PR intervals, flat or invisible P waves, and deviation of the ST segment. The expected progression of electrocardiographic changes associated with hyperkalemia is well described in human medical literature. In general, the first change to occur is the development of a peaked T-wave. Next the P-R interval becomes prolonged as does the QRS duration. After this, the P-wave becomes smaller or absent due to lack of atrial depolarization. The sinus node still fires, leading to a sino-ventricular rhythm. This can then progress to a sine wave type configuration followed by ventricular fibrillation. Other findings include complete heart block and ventricular arrhythmias. What is essential to note is that the relationship between potassium levels and electrocardiographic changes may vary between individual human patients. There is also the realization that the ECG is not always a reliable marker of hyperkalemia. Normal ECG's, ECG's with minimal abnormalities, and those with "atypical" results can be obtained in human patients with severe hyperkalemia.

There have been few studies on cats with urethral obstruction, with regard assessing and quantifying elevated serum potassium levels in conjunction with other parameters such as ionized calcium and acid-based balance. Most of our information to date on electrocardiographic manifestations of hyperkalemia is based on human patients. We have carried out a study looking at ECGs in blocked cats and looking at correlations to blood gas parameters. In our pilot study, we were able to find statistically significant correlation between ionized calcium values and the following; P wave amplitude and duration, T wave amplitude, heart rate, and duration of QT, P-R, and QRS intervals. Significant correlation was also found between serum potassium concentration and amplitude of T, heart rate, and duration of PR and QRS intervals. A significant association was found between serum sodium: potassium ratio and heart rate, and duration of PR and QRS interval.

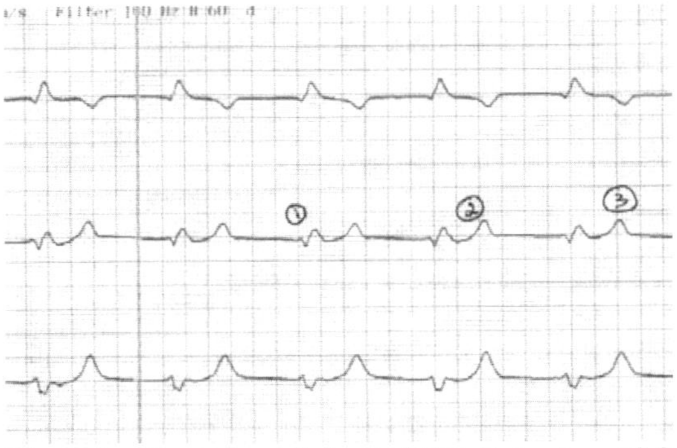

Figure 1. Classic hyperkalemic ECG, Potassium 10.8 mmol/L, ionized calcium 0.58 mmol/L

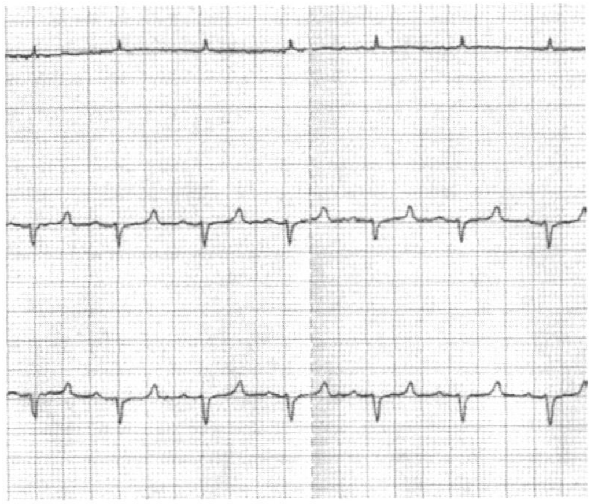

Figure 2. Intermediate ECG, Potassium 8.76 mmol/L, ionized calcium 0.99 mmol/L

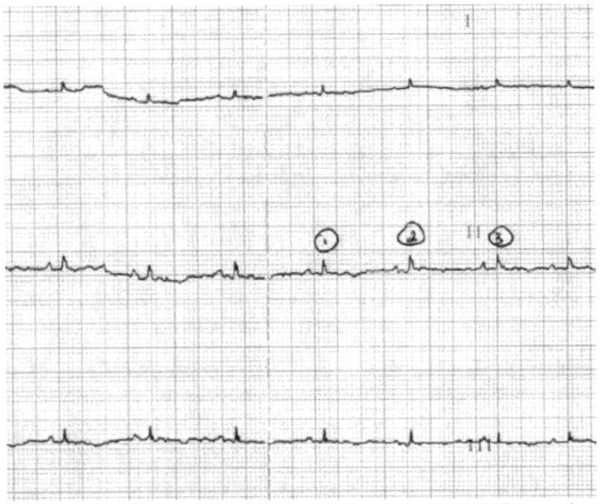

Figure 3. "Normal ECG" in hyperkalemic cat, Potassium 8.32 mmol/L, ionized calcium 1.28 mmol/L

In dogs, the most common cause of significant hyperkalemia is hypoadrenocorticism. What is an interesting contrast between hypoadrenocorticism and urethral obstruction is that in cats with UO hypocalcemia is common whereas with hypoadrenocorticism, normal to elevated ionized calcium is usually encountered. In addition, there is also usually a concurrent hyponatremia that could exacerbate hyperkalemia induced ECG changes. In experimental models on dogs with bupivacaine infusions, both hyponatremia and hyperkalemia prolonged QRS duration and ventricular conduction time. There have been some studies that have documented the ECG changes in dogs with hypoadrenocorticism. Out of 122 dogs with hypoadrenocorticism that had an ECG performed, almost half did not have a P-wave, almost 30% were bradycardic, 20% had prolonged QRS duration and almost 8% had a tall T-wave.

We have also looked at data from dogs with hypoadrenocorticism. P-waves are still present in all dogs whose potassium is less than 7.3 mmol/L. In one dog that had a potassium concentration of 10.3 mmol/L the P-wave was still present. This dog also had a marked ionized hypercalcemia. In 11 dogs with a potassium concentration of greater than 7.3 mmol/L, 3 dogs still had visible P-waves. All 3 of these dogs had ionized calcium concentrations greater than 3 mmol/L. The contrast between these two diseases emphasizes the importance of calcium concentrations to ECG changes expected in hyperkalemic patients. In addition it is clear that patients can have near normal ECGs and yet be suffering from severe hyperkalemia, this is especially true in dogs with hypoadrenocorticism.

REFERENCES

Mattu A, Brady WJ, Robinson DA. Electrocardiographic manifestations of hyperkalemia. American Journal of Emergency medicine 2000; 18(6): 721-729., Yu AS. Atypical electrocardiographic changes in severe hyperkalemia. American Journal of cardiology 1996; 77: 906-908., Dittrich KL, Walls RM. Hyperkalemia: ECG manifestations and clinical considerations. American Journal of Emergency medicine 1986; 4:449-455., Martinez-Vea A, Bardaji A, Garcia C, et al. Severe hyperkalemia with minimal electrocardiographic manifestations: A report of 7 cases. Journal of electrocardiology 1999; 32: 45-49., Martinez-Lopez JI, ECG of the month. Harbinger of evil. Hyperkalemia. J La State Med Soc 1997; 149: 103-104., Parks J. Electrocardiographic abnormalities from serum electrolyte imbalance due to feline urethral obstruction. Journal of the American Animal Hospital Association. 1975; 11(1): 102-106., Schaer M. Hyperkalemia in cats with urethral obstruction. Veterinary Clinics of North America. 1977; 7(2): 407-414., Schaer M. The use of regular insulin in the treatment of hyperkalemia in cats with urethral obstruction. Journal of the American Animal Hospital Association. 1975; 11(1): 106-109., Finco DR. Induced feline urethral obstruction: Response of hyperkalemia to relief of obstruction and administration of parenteral electrolyte solution. Journal of the American Animal Hospital Association. 1976; 12 (2): 198-202., Tilley LP. Essentials of Canine and Feline Electrocardiology, 3rd ed. Philadelphia, PA: Lea & Febiger; 1992., Peterson ME, Kintzer PP, Kass PH Pretreatment clinical and laboratory findings in dogs with hypoadrenocorticism: 225 cases (1979-1993). J Am Vet Med Assoc 1996;208:85-91.

DEVELOPING IMMUNOLOGIC CLASSIFICATION SCHEMES FOR SEPSIS

Amy E. DeClue, DVM, MS, DACVIM (SAIM)
Columbia, MO

Sepsis is a syndrome of immune dysregulation and subsequent globalized tissue damage secondary to infection. For years sepsis has been viewed as a pro-inflammatory phenotype, yet clinical trials aimed at blunting this pro-inflammatory shift have been largely unsuccessful. The realization that sepsis is not simply inflammation but a complex, variable syndrome that encompasses aspects of both exuberant inflammation and immunosuppression has alerted researchers to the need for better classification systems for sepsis. The concept that sepsis is, well, sepsis, has gone by the wayside. There is a strong need to develop mechanisms to identify the type of immunodysfunction, either hyper-inflammatory, hypo-inflammatory or a mix, so that targeted therapies aimed at the specific immunopathology can be instituted. Currently, there are no rigorously tested immunologic classification systems for sepsis in veterinary medicine. The goal of this talk is to start the brainstorming process of how we, as clinician scientists, can work together to tease out a rational approach to developing a classification system.

Classically sepsis has been considered an imbalance between hyper-inflammatory and hypo-inflammatory responses. The hyper-inflammatory response has largely been blamed for the morbidity and mortality associated with sepsis. The hyper-inflammatory response (i.e., SIRS) is characterized by activation of the innate immune system and production of a cytokine storm. This leads to cellular recruitment, tissue damage, and the classical manifestations of sepsis. However, animals with sepsis can also develop a hypo-inflammatory response to infection called the compensatory anti-inflammatory response syndrome (CARS). The compensatory anti-inflammatory response is characterized by immunoparalysis or lack of immune response to infection. Millions of dollars and years of time have been spent trying to identify treatments to dampen the hyper-inflammatory response during sepsis with minimal success. Recently, the importance of the anti-inflammatory aspects of the immune response to sepsis has been realized.

One of the hypotheses for why many clinical trials aimed at restoring immune homeostasis have failed is that all patients have been treated as if they have SIRS (i.e., hyper-inflammatory response). Administration of an immunosuppressive therapy to a patient with CARS would be unlikely to be helpful. Likewise, administration of a pro-inflammatory stimulant to a patient with SIRS would be unlikely to be helpful. By grouping patients with SIRS and CARS together in clinical trials as if they had the same pathology, treatment benefits may have been missed. However, it isn't quite that simple. There are permutations of both SIRS and CARS and many animals likely have distinct features of each. Developing an immunologic classification system will increase the likelihood that immunomodulatory therapy will result in improved outcome by allowing targeted selection of patients most likely to benefit from that therapy based on individualized assessment.

The first step in developing a classification scheme for sepsis is to develop a better understanding of the immunopathogenesis of naturally developing sepsis in dogs and cats. This is an enormous challenge since there are multiple factors that influence the immunopathogenesis of sepsis. The manifestations of sepsis vary based on both pathogen and host factors. Pathogen factors include the type of pathogen, load, virulence and site of inoculation while the host factors include genetic factors, anatomic regions of infection, and co-morbid diseases. Additionally, treatment interventions alter the host immune response. Thus, a successful classification system will need to allow for variation in all of these factors, yet still allow for appropriate selection of interventions. Based on work in people and in animal models of sepsis, there are several key aspects of the immune response that should be considered as classification systems are tested. These include:

- Circulating inflammatory mediators: Evaluating panels of circulating inflammatory mediators could lead to a better understanding of the underlying immunopathology by not only suggesting the current immunopathogenesis but by helping to predict dynamic changes in the immune system. For example, Th1 cells and Th17 cells cytokines such as TNF, IL-2, IL-12, IFN-γ, and leukotrienes or IL-17A, respectively, which typically drives a hyper-inflammatory phenotype. Conversely, a hypo-inflammatory phenotype is observed when there is a Th2 response resulting in production of cytokines such as IL-4, IL-5, IL-9, IL-10 and IL-13. Differentiation of monocytes into the M1 phenotype (hyper-inflammatory) is through IFN-γ signaling while the M2 phenotype (hypo-inflammatory) develops in the presence of IL-4 and IL-13. Further, interferon-γ activates phagocytes and promotes intracellular killing thus a deficiency in IFN-γ may lead to ineffective phagocytosis. Classical cytokines associated with sepsis have also been used to differentiate hypo- and hyper-inflammatory response. Serum TNF:IL-10 ratio has proven helpful in identifying people with a low ratio indicating a hypo-inflammatory response and a high ratio indicating a hyper-inflammatory response.
- Leukocyte cytokine production: Stimulated leukocyte cytokine production capacity gives insight into leukocyte function and the predominant cytokine pathways being activated. The degree of cytokine production along with the relative ratio of pro- vs anti-inflammatory (e.g., TNF:IL-10) cytokines produced help differentiate animals that have a hypo- versus hyper-inflammatory response.
- Phagocytic function: Professional phagocytes recognize, engulf and kill pathogens by generating reactive oxygen species, proteases and uninhabitable pH. Phagocytosis is a key element for pathogen removal. Suppression of either recognition, phagocytosis of the pathogen or pathogen killing mechanisms can lead to deadly infection. Conversely, overactive phagocytes can mediate tissue damage through expulsion of oxidative burst products. Reduction in phagocytosis or respiratory burst function would suggest a hypo-inflammatory response while upregulation of spontaneous respiratory burst suggests a hyper-inflammatory response.
- Cell surface receptor expression: Alterations in cell surface receptor expression can lead to either a hyper- or hypo-inflammatory response. Down regulation of HLA-DR expression reduces antigen presentation and T cell activation consistent with a hypo-inflammatory shift. Cell surface pattern recognition receptor (e.g., TLR, NOD) expression upregulation suggests a hyper-inflammatory shift and promotes cytokine production by increasing sensitivity to small

concentrations of PAMPs. Conversely, down-regulation of cell surface pattern recognition receptor expression is associated with reduced response to PAMPs, reduced cytokine production and a hypo-inflammatory state.

- Th1/Th2: The Th1 phenotype is important during sepsis. Cytotoxic-killer T cells play an important role in pathogen clearance and Th1 cytokines promote a hyper-inflammatory state. Conversely, the Th2 phenotype leads to a hypo-responsive immune system.
- T regulatory cells: T regulatory cells play an important role in tolerance and have immunosuppressive effects. Expansion of the T regulatory cell population during sepsis is associated with immunodysfunction and the development of opportunistic infections.
- Apoptosis of immune cells: During sepsis neutrophils may become immortal. Activation of NF-kappa B leads to attenuation of caspase-mediated apoptotic pathways. Neutrophil immortality has been associated with tissue damage during sepsis. The opposite situation happens with lymphocytes. Upregulation of lymphocyte apoptosis is a major mechanism of immunosuppression during sepsis and many people with sepsis will have near total lymphocyte depletion of the spleen, Peyer's patches, bone marrow and thymus. CD4+ T cells, CD8+ T cells and B cells are most commonly affected.

This list encompasses only a few immunologic parameters that could be evaluated as possible markers for immunodysfunction during sepsis. Future studies should select a panel of these immunologic markers/function tests and then evaluate their association with outcome. There are likely to be unique species variations and thus evaluation in the species of interest is necessary. Ultimately classification schemes may be developed that will allow identification of key immunodysfunctions. For example, a dog might be classified as having exuberant stimulated cytokine production, high plasma TNF:IL-10 ratio and upregulated spontaneous respiratory burst production. Based on that information the dog might be treated with a drug aimed at calming a hyper-inflammatory immune system. Conversely, a dog might have blunted stimulated cytokine production, low plasma TNF:IL-10 ratio and down regulation of phagocytosis which would be treated with an immunostimulant. Perhaps most importantly, dogs with mixed immunodysfunction could be identified and treatment specifically targeted. A dog with normal cytokine production but a reduction in phagocytic function could be managed with a drug or cytokine developed to upregulate phagocytosis with little to no additional immunomodulatory effects.

There are many obvious challenges associated with developing immunologic classification systems for sepsis and many we have not yet discovered. One key obstacle is identification of optimal outcome measures to test the classification system. Initially, the goals of immunologic investigations may be to learn more about the immunopathogenesis of sepsis in dogs and cats. Outcome measures in this situation could simply be survival or the identification of secondary or opportunistic infections. However, both of these outcome measures are challenging in veterinary medicine. Often financial, emotional and ethical concerns influence the medical decision to continue treatment, which induces bias into survival statistics. Some animals that might have survived with continued aggressive care could be placed into a non-survival group skewing our perception of the relative importance of the measured parameter. Identification of secondary or opportunistic infections might also present as a challenge in that repeated culture or other testing for microorganisms is expensive and negative test results often fail to rule out disease. Despite these limitations, assessing outcome and development of secondary infections are good starting points for evaluation of immunoclassification systems. Eventually, as we move towards targeted immunotherapy for sepsis, classification will allow for identification of targeted interventions in animals with that specific immunodysfunction. Ultimately the value of classifying the nature of the immunodysfunction will be measured by our success in utilizing targeted therapies.

Sepsis is a rapidly progressive syndrome and thus one of the other key challenges to developing an immunologic classification system is the need for rapid resulting. Ideally, key immunologic features that could be tested at the bed side should be utilized. Currently, that limits testing to predominately identification of inflammatory mediators and cell surface markers, although bedside methodologies for dogs and cats would need to be developed from current technology. Platforms for cell function testing and would require extensive development. However, it is possible that surrogate markers for some cellular functions could be identified. Thus, as classification systems are investigated, the goal of a bedside test should be maintained and association between cellular function and more easily measured surrogate clinical or immunologic parameters considered.

REFERENCES

Aziz M, Jacob A, Yang WL, et al. Current trends in inflammatory and immunomodulatory mediators in sepsis. Journal of leukocyte biology 2013;93:329-342. King EG, Bauza GJ, Mella JR, et al. Pathophysiologic mechanisms in septic shock. Laboratory investigation; a journal of technical methods and pathology 2014;94:4-12. Schulte W, Bernhagen J, Bucala R. Cytokines in sepsis: potent immunoregulators and potential therapeutic targets--an updated view. Mediators of inflammation 2013;2013:165974. Hotchkiss RS, Monneret G, Payen D. Sepsis-induced immunosuppression: from cellular dysfunctions to immunotherapy. Nature reviews Immunology 2013;13:862-874. Hotchkiss RS, Monneret G, Payen D. Immunosuppression in sepsis: a novel understanding of the disorder and a new therapeutic approach. The Lancet infectious diseases 2013;13:260-268. Unsinger J, Kazama H, McDonough JS, et al. Sepsis-induced apoptosis leads to active suppression of delayed-type hypersensitivity by CD8+ regulatory T cells through a TRAIL-dependent mechanism. Journal of immunology 2010;184:6766-6772. Yu D, Kim J, Park C, et al. Serial Changes of CD4+CD25+FoxP3+ Regulatory T Cell in Canine Model of Sepsis Induced by Endotoxin. The Journal of veterinary medical science / the Japanese Society of Veterinary Science 2014;76:777-780. DeClue AE, Williams KJ, Sharp C, et al. Systemic response to low-dose endotoxin infusion in cats. Veterinary immunology and immunopathology 2009;132:167-174. DeClue AE, Sharp CR, Harmon M. Plasma inflammatory mediator concentrations at ICU admission in dogs with naturally developing sepsis. Journal of veterinary internal medicine / American College of Veterinary Internal Medicine 2012;26:624-630. DeClue AE, Delgado C, Chang CH, et al. Clinical and immunologic assessment of sepsis and the systemic inflammatory response syndrome in cats. Journal of the American Veterinary Medical Association 2011;238:890-897. Deitschel SJ, Kerl ME, Chang CH, et al. Age-associated changes to pathogen-associated molecular pattern-induced inflammatory mediator production in dogs. Journal of veterinary emergency and critical care 2010;20:494-502. Odunayo A, Dodam JR, Kerl ME, et al. Immunomodulatory effects of opioids. Journal of veterinary emergency and critical care 2010;20:376-385.Osterbur K, Mann FA, Kuroki K, et al. Multiple Organ Dysfunction Syndrome in Humans and Animals. Journal of veterinary internal medicine / American College of Veterinary Internal Medicine 2014. Venet F, Lukaszewicz AC, Payen D, et al. Monitoring the immune response in sepsis: a rational approach to administration of immunoadjuvant therapies. Current opinion in immunology 2013;25:477-483. DeClue AE, Nickell J, Chang CH, et al. Upregulation of proinflammatory cytokine production in response to bacterial pathogen-associated molecular patterns in dogs with diabetes mellitus undergoing insulin therapy. Journal of diabetes science and technology 2012;6:496-502. Fowler BL, Axiak

SM, DeClue AE. Blunted pathogen-associated molecular pattern motif induced TNF, IL-6 and IL-10 production from whole blood in dogs with lymphoma. Veterinary immunology and immunopathology 2011;144:167-171.

LET'S TALK: TEACHING EFFECTIVE COMMUNICATION SKILLS

Erika L. Krick, VMD, DACVIM (Oncology)
Philadelphia, PA

INTRODUCTION

The ability to communicate effectively with clients, co-workers, colleagues, etc., is an essential tool for anyone in the medical and veterinary field. A common misconception regarding communication skills is that they are innate and thus cannot be learned—either someone is a good communicator or not, and that cannot be changed because it is intimately linked with the individual's personality. In fact, communication skills can be taught, learned, and practiced similar to any other clinical skill, such as performing a thorough physical examination or a surgical procedure.

One aspect of effective communication is the exchange of information between two parties as opposed to the majority of information given from one party to another. Characteristics of different relationships and roles for the physician and patient have been described.[1,2] Traditional relationships between practitioners and patients/clients have been primarily doctor-centered, meaning that the doctor determines what will be discussed and gives information to the patient/client without expecting or encouraging feedback or questions from him or her. This role is often referred to as the "paternalistic" or "guardian" role for the physician. In contrast, in patient-centered care, the focus of the interaction is the patient—his or her questions, concerns, and opinions, and the physician is acting primarily as a consultant. This role is often referred to as the "consumerist" or "consultant" role. In between these extremes lies relationship-centered care, in which the focus is establishing a joint partnership between the physician and patient/client. This role is often referred to as the "advisor" role for the physician. Each of these roles is associated with a different type of communication pattern between the physician and patient. A biomedical communication pattern is often used in doctor-centered care. With this pattern the doctor is verbally dominant, the majority of information discussed relates to the medical condition at hand, and little patient input occurs. The consumerist pattern, most often occurring in patient-centered care, is characterized by high patient input and control of the conversation with the physician responding to the patient's many questions. A biopsychosocial or psychosocial pattern is characterized by mutual exchange of information and is associated with relationship-centered care. A recent observational study of communication patterns used by physicians found that the majority used the biomedical (65%) and biopsychosocial (28%) patterns, while a small percentage (8%) used the consumerist pattern.[2] The use of relationship-centered care in the human medical field has been shown to result in increased patient and physician satisfaction, improved patient health outcomes, and reduction in malpractice complaints.[2-5]

As both the medical and veterinary fields have gradually moved to an emphasis on relationship-centered care versus doctor-centered care, patterns of communication used by practitioners, the frequency and impact of using (or lack thereof) effective communication skills, and the outcome of teaching students to use these skills have been evaluated. The majority of research has been performed in the human medical profession; however, multiple studies regarding teaching and using communication skills in the veterinary field have recently been published. In this session we will discuss communication patterns used by veterinarians, client expectations regarding communication, and the outcome of using effective communication skills for veterinarians. Once we've established why proper utilization of these skills is beneficial to the practice of veterinary medicine, we will discuss methods for teaching these skills and the outcome of these methods. We will finish with a brief story to describe how the communication curriculum at Penn Vet has developed over the last few years as a concrete example of one way to build such a program.

COMMUNICATION IN VETERINARY MEDICINE

Several observational studies have documented common communication patterns used by companion animal veterinarians and evaluated variables that may alter these patterns, such as gender of the veterinarian and client or the type of visit (routine wellness visit or sick patient visit).[6-8] Most of the data comes from a series of appointments that were videotaped and coded according to a published quantitative method for categorizing medical conversations into a set of discrete variables.[2,6-8,14] The dataset includes 300 videotaped appointments (6 per veterinarian, 3 wellness and 3 sick visits per veterinarian) which were coded according to the method described above. The results provide an interesting snapshot into the communication patterns and skills used by veterinarians and a good starting point for comparing these results to client expectations and for the development of outcome variables by which to measure the effect of communication training.

For all of the visits, the percentage of time the veterinarian spent on specific parts of the appointment (data gathering/history taking, informing and educating the client about diagnostic tests, diagnosis, and treatment, building a relationship with the client, and promoting partnering with the client) was measured.[6] Veterinarians spent the majority of the time on client education (48%), followed by relationship building (30%), then data gathering (9%) and partnering (2%). A similar breakdown was found when the client's questions and statements were evaluated—the majority of time was spent giving information to the vet (46%) and relationship building (44%), followed by question asking (4%) and partnering (3%). The use of specific communication skills of open-ended questions and empathy statements was uncommon. Veterinarian asked an average of 13 closed-ended questions and 2 open-ended questions per appointment, and in 25% of appointments no open-ended questions were asked. Verbal expressions of empathy were rare, occurring in 7% of appointments. One could argue that there were few opportunities presented for empathy statements, but given that 50% of the appointments were health problem-related, that seems unlikely. Additionally, in one study of human patients, 60% of encounters between patients and physicians contained at least one opportunity for the physician to express empathy, and it is likely that the percentage is at least that high in veterinary appointments.[9]

The authors then applied the communication patterns observed in human physician and patient encounters to the veterinarians and appointments in the study with the purpose of documenting the use of these patterns in veterinary medicine and determine the factors associated with the use of one pattern versus another.[7] None of the veterinarians used the consumerist pattern. The majority

(58%) of appointments followed the biomedical pattern, while 42% followed the biolifestyle-social pattern. This term was used instead of biopsychosocial or psychosocial because it seemed more appropriate to veterinary patients. The patterns used by individual veterinarians were not consistent—almost all of the veterinarians used each of the 2 styles in at least one appointment, and veterinarians were classified by pattern according to the one they used in at least 4 of the 6 appointments. The biomedical was most commonly used by veterinarians (46%), while 38% used mixed patterns (3 appointments biomedical and 3 biolifestyle-social) and 16% employed a biolifestyle-social pattern in the majority of appointments. Communication pattern was most consistent for the appointment type, with the biomedical pattern used significantly more frequently in sick visits and the biolifestyle-social pattern used significantly more frequently in wellness visits. Interestingly, visits in which the biomedical pattern was used were significantly longer than visits in which the biolifestyle-social pattern was used (12 versus 10.4 minutes, $p = 0.02$). One could conclude that that difference was due the confounding effect of longer sick versus wellness visits; however, no significant difference in appointment duration between sick and wellness visits was found. High relationship-centered care scores were significantly more likely in wellness versus sick appointments as well (57% versus 11%, $p< 0.01$), and the median relationship-centered care score was higher when the biolifestyle-social versus the biomedical pattern was used (1.1 versus 0.4, $p < 0.01$). Overall, the authors concluded that veterinarians use 2 different communication patterns with clients, and the variables associated with which pattern is used include the type of appointment (wellness or sick), the duration of the visit, and the relationship-centered care score.

When specific comparisons were made between wellness and sick visits, several differences in communication patterns and skills were noted.[8] For example, the number of client education statements and statements of concern was significantly higher in sick visits as was the likelihood of the veterinarian to ask about the client's understanding. Relationship building statements were significantly more likely to occur, and veterinarians were more likely to ask about the client's opinion in wellness visits. The number of data gathering questions and the percentage of time spent on data gathering was not significantly different between sick and wellness visits.

It is then interesting to compare these findings to reported expectations from clients. When clients who were presenting their dogs to a tertiary care oncology service were interviewed, their priority for the content they expected from the veterinarian was that the information given would be the truth or the facts.[10] The amount of information expected varied, but consistently clients preferred to be given accurate factual information. A second study described client expectations regarding how that information was communicated.[11] The most frequently described expectations by clients included delivery of information in an understandable manner free of technical jargon and in multiple formats (verbally and written, possibly diagrams if indicated) with positivity, empathy, compassion, and a non-judgmental attitude. They also valued an unrushed environment and staff that took time to answer questions and repeat information as needed, and consistency with which staff members they interacted. Another study of focus groups of clients found similar results, specifically that clients wanted information presented up front, to be given information in multiple forms, to be listened to and to partner with the veterinarian.[12] Interestingly, these priorities are almost identical to those of human patients, who expect to receive information in a clear and understandable manner with compassion, empathy, and patience from a service provider with whom they are familiar.[12] Human patients also expect to be listened to, without judgment, and to have a part in decision making.

Additional studies have examined the outcome of communication skills and specific aspects of communication used by veterinarians. One study evaluated adherence of clients to recommendations regarding dental and surgical procedures.[13] Interactions with higher relationship-centered care scores were associated with increased client adherence to recommendations. In addition, clients who were satisfied at the end of the appointment had higher adherence, as did clients who received a clear recommendation from the veterinarian. Given the expectations of many clients for information to be delivered clearly and upfront, it is not surprising that a clear recommendation increased adherence. Another study (using the same 50 veterinarians and 300 appointments described above) found that veterinarian satisfaction with visits was associated with communication that increased relationship building with clients.[14] Increased veterinarian satisfaction was also higher when the veterinarian was working with a client with whom he or she already had a relationship.

TEACHING COMMUNICATION SKILLS

Both theoretical and experiential learning with adequate opportunities to practice and give and receive feedback are essential to learning many clinical skills, and communication skills are no exception. Didactic lectures, small group discussions, observations of videotaped or live interactions between practitioners and patients/clients as well as interactions with simulated clients have all been used in client communication curricula in both medical and veterinary medical education. Similar to comparing the use of effective communication skills on physician, patient, and client outcomes, it is important to establish ways to assess the effectiveness and outcome of teaching these skills. The validity of using standardized clients and trained faculty members as assessors of student communication skills has been documented.[15] Web-based training has been shown to be effective in teaching the theory and practice of communication skills to medical students to achieve core competencies but has not yet been evaluated in veterinary medical education.[16] A systematic review of publications that describe the use of web-based methods for teaching communication skills found that both interactive web-based modules and the use of online virtual patients were described.[16] The authors concluded that the results of the studies supported the use of web-based virtual patients for teaching history taking skills and to improve learners' knowledge of communication skills and clinical reasoning. As might be expected, web-based learning did not effectively teach or elicit verbal and non-verbal expressions of empathy, and encounters with virtual patients resulted in lower relationship-building responses.

How do these results compare to communication skills taught in small group settings with a mediator? One randomized prospective study of a cohort of veterinary students attempts to answer that question.[17] The goal of the study was to determine if small group or web-based learning had a greater effect on teaching specific communication skills used in an interaction with a client. 96 consenting students in a veterinary school, all of whom had some initial communication skills education as part of their veterinary school curriculum, were randomized to receive no additional training (control group), participate in a 2 hour small group communication skills training session that included working with a simulated client, or complete a 2 hour web-based learning experience that was both informative and interactive. In order to assess the effect of the intervention, all participating students underwent an Objective Structured Clinical Examination (OSCE) before and after the intervention. The OSCE consisted of 4 stations that each included a simulated client and a rater who scored the communication skills demonstrated by the student. There was no significant difference among the groups in the pre-intervention OSCE scores, and the post-intervention scores were significantly higher for all 3 groups. Specifically, the small group teaching group had a significantly greater change in OSCE scores pre and post intervention compared to the other 2 groups, and the difference in pre and post scores were not different between the web-based and the control groups. A potential confounding variable is the presence of a facilitator in the small group learning and a lack of such a facilitator in the self-directed web-based learning; however, the authors' recommendation that a combination of web-based and small group learning be used in communication curricula seems reasonable.

BUILDING A CURRICULUM

Now that we have established the importance and outcome of using effective communication skills in veterinary medicine, learned about client expectations from us regarding communication, and described some effective teaching strategies, how do we use that information to develop a communication curriculum that results in effective leaning of communication skills? How do we assess the effectiveness of our curriculum, and how can we apply these teaching strategies to veterinarians and other members of the team in a veterinary practice, such as nurses, technicians, and receptionists?

Certainly not all of these questions have been answered; however, as more veterinary schools include communication training in their curricula, more research regarding effective teaching strategies and the effects of such strategies will emerge. That information can then possibly be extrapolated and applied to communication training for practicing veterinarians, including general practitioners, interns and residents, and specialists. One case study describes the effect of communication training in a private veterinary practice.[18] In this study, the 4 veterinarians in the practice underwent a 12 month (one 8 hour session per month) communication training intervention that occurred at the practice. Interactions between the veterinarians and clients were videotaped before and after the intervention (6 per veterinarian before and 6 per veterinarian after), and the videotapes were analyzed according to the methods described earlier in these proceedings. Overall, the veterinarians' communication after the intervention was more relationship centered. Specifically, veterinarians asked significantly more lifestyle-social questions and used more partnership and positive rapport building communication. Interestingly, during encounters after the intervention, clients were significantly more likely to express statements of emotion and give lifestyle-social information. Although this is a small study of just one practice, the results are intriguing and suggest that communication training is feasible for practicing veterinarians and results in changes in communication style. It would be interesting to perform a similar study with more practices and veterinarians and have longer follow up to determine long-term effects of such an intervention.

Over the past several years the communication curriculum at Penn Vet has been expanded to include lecture and discussion hours dedicated to different aspects of communication skills as well as laboratory sessions during which the students are put into small groups, each with a trained facilitator, and each student has a coached interaction with a standardized client. Overall the student review of these laboratories are positive with most students finding them to be a valuable learning opportunity and many requesting that more opportunities be built into the curriculum. This past spring we held similar workshops for our house officers and some faculty and staff.

CONCLUSION/FUTURE DIRECTIONS

As relationship-centered care becomes more common in veterinary medicine due to client and veterinarian expectations, the teaching and practicing of communication skills for students will become increasingly more essential. The body of literature describing the effects of communication skills training will continue to grow, and as it does, additional metrics for assessing the effectiveness of such education will be more clearly identified. Communication skills training cannot occur in a vacuum, however, as it takes repeated practice for learners to become comfortable with using them as well as good modeling by senior clinicians and mentors. Future goals for our program include the incorporation of communication skills training for all faculty, staff, and hospital employees so that we all become models for each other and for students and house officers. As more literature continues to emerge regarding the benefits of using effective communication skills when working with clients and other members of the veterinary team and evaluation of the methods and impact of communication training in veterinary medicine, client communication will more easily be recognized as an essential clinical skill.

REFERENCES

[1]Roter D. **Patient Educ and Counsel** 2000;39:5. [2]Roter DL et al. **JAMA** 1997;277:350. [3]DiMatteo et al. **Health Psychol** 1993;12:93. [4]Stewart MA. **Can Med Assoc J** 1995;152:1423. [5]Roter DL. **J Health Care Law Policy** 2006;9:304. [6]Shaw JR et al. **J Am Vet Med Assoc** 2004;225:222. [7]Shaw JR et al. **J Am Vet Med Assoc** 2006;228:714. [8]Shaw JR et al. **J Am Vet Med Assoc** 2008;233:1576. [9]Bylund CL et al. **Patient Educ Counsel** 2002;48:207. [10]Stoewen DL et al **J Am Vet Med Assoc** 2014;245:773. [11]Stoewen DL et al. **J Am Vet Med Assoc** 2014;245:785. [12]Coe JB et al. **J Am Vet Med Assoc** 2008;233:1072. [13]Kanji et al. **J Am Vet Med** Assoc 2012;240:427. [14]Shaw JR et al **J Am Vet Assoc** 2012;240:832. [15]Artemiou et al. **Vet Rec** 2014; doi: 10.1136/vr.102633 [16]Artemiou et al. **J Vet Med Educ** 2014;41:44. [17]Artemiou et al. **J Vet Med Educ** 2013;40:242. [18]Shaw JR et al. **Patient Educ Counsel** 2010;80:337.

FLIPPING THE CLASSROOM FOR SMALL GROUP LEARNING

Jo R. Smith, MA, VetMB, PhD, DAVIM (SAIM)
Athens, GA

INTRODUCTION

"Flipping the classroom" reverses the traditional way of lecture-based teaching. Students acquire basic content outside of the classroom and then work together in class on application-oriented activities[1]. In-class time is "re-purposed" for inquiry, application, and assessment to better meet the needs of the individual learners. Students can pace their own learning process when studying course material outside of class using readings, pre-recorded video lectures, or research assignments.

During class, instructors can facilitate the learning process by helping students work through problems individually or in groups. This approach seeks to preserve the value of lecture (expertise and custom delivery), while using class time for active learning strategies. The goal in flipping a class is to encourage deeper, richer active learning experiences for students while the instructor is present to coach and guide them. Emphasis is on higher-order thinking skills and application to complex problems.

This technique supports Fink's idea of starting with the end product in mind when designing a course[2]. Rather than using a content-centered approach, he suggests a learning-centered approach. Specific learning goals are defined, then teaching and learning activities, as well as feedback and assessment techniques, are designed and integrated to support these goals.

Furthermore, this approach encourages students to learn how to study more effectively rather than rely on the expertise of the teacher, i.e. finding resources, making sense of them and then applying that knowledge. Rather than focus on content, it concentrates on the process and helps develop the skill set required for life-long learning.

Transforming a course takes time (typically much more than we anticipate!) and commitment. It may be helpful to repurpose reliable high-quality materials already available on the internet. Start with a single class and focus on what and how students are learning at that scale, and then consider applying this technique to additional lessons. Flipping should be iterative – reflect on what works well and what needs to be modified - as you implement this technique.

FIGURE OUT WHERE "FLIPPING" MAKES THE MOST SENSE FOR YOUR COURSE[3,4]

These questions may help you to identify a good starting point – whether your course is designed around learning objectives/outcomes or learning units. Consider *why* you teach *what* you teach.

- What do I want my students to know and be able to do as a result of completing this sequence of the course?
- How does what I want them to know or do fit into the bigger picture of the unit and course?
- What topics within the course have students struggled with l?
- What common misconceptions are there in your field?
- Which topics would be better served if students were given the opportunities during class to apply their knowledge and skills actively?
- What learning outcomes need to be introduced or refined to target higher order use of knowledge and skills?

LOOK FOR IN-CLASS ACTIVITIES REQUIRING STUDENTS TO APPLY WHAT THEY ARE LEARNING

The key point of flipping the classroom is to utilize class time effectively to encourage active learning strategies. Previous teaching experience may help you anticipate where trouble spots will be.

- What "homework" questions could be tackled during class?
- What activities have you developed that are currently rushed through due to time constraints?
- What activity could be designed that would appropriately challenge students to apply concepts and engage them in the types of thinking required in your field?

Different instructional and assessment strategies can provide students with feedback about what they do and do not know; provide instructors with real-time data that informs the instruction that needs to occur in class; and encourages students to engage during class by holding them accountable for their preparation.

Peer instruction – students prepare before class and provide instructors with feedback about areas they found confusing or difficult. Use of discussion board software may facilitate students posting questions prior to class and help signpost topics that need addressing in class. Conceptual questions that elicit and confront misconceptions can then be resolved by mini-lectures and peer discussion.

Team-based learning – students prepare before class and undertake a short quiz at the start of the class to verify they have learned the content. These quizzes are taken on an individual basis first, and then as a team. Teams get immediate feedback on their performance and instructors can address gaps in understanding via mini-lectures. Suitable technologies for both individual and group-based assessments include classroom response systems (AKA "clickers"); "scratch and reveal" multiple-choice answer forms (Instant Feedback Assessment Technique, IF-AT)[5,6], and software such as Learning Catalytics[7]. Teams then engage in structured application activities that conclude with teams simultaneously making a specific choice.

Case-based learning – students work as individuals or in groups to analyse problems(with guidance from the instructor) to evaluate a suggested course of action or devise one of their own.

Process-oriented guided inquiry learning – (AKA problem-based learning) students ask questions, find information, generate conclusions and then apply these concepts to new situations. These activities are structured to develop process skills, critical thinking, problem-solving and collaboration.

IDENTIFY THE CONTENT STUDENTS WILL ENGAGE TO PREPRARE FOR CLASS

Once you have a clear idea for how students will be asked to apply their knowledge and skills during class, begin considering what students will need to read or view in advance of class. Online video or audio content is often used with the flipped classroom model and includes podcasts, videos, on-line micro-lectures, simulations or demonstrations, (e.g. those found on You Tube or Ted.Ed). However, you can also flip a class by repurposing traditional reading materials such as textbooks chapters and primary literature or review articles. While lecture capture technology can permit students to search within slides, bookmark, annotate and collaborate, it is suggested that multiple short videos are more effective than one 50 minute lecture.

- What essential content do students need to acquire before class?
- What existing resources would supply students with the information needed?
- What could be included by producing your own videos in 3-5 segments with each lasting 3-5 minutes?

For this technique to succeed, it is essential that students are held accountable for doing pre-class preparatory work. Techniques include:

Build learning modules – for students to complete as homework, e.g. a combination of text, video and assessment. Integrating assessments into their learning experience ensures that students are required to test their understanding of the content immediately after it is presented to them. This will allow them to self-identify areas at which they excel and others that may need more study.

Just-in-time-teaching approach – students answer 2-4 conceptual questions targeted at common misconceptions prior to class. These results may be used to inform instruction during class.

Reflective questions – students can be encouraged to pause and consider the answer to each question and write down the answer in their notes (e.g. what concepts did you find most challenging?)

Require students to pose questions about the content – technologies such as the discussion platform in your classroom management system or different forms of social media can be utilized.

PREPARE STUDENTS FOR THE UNIQUE ROLES EVERYONE WILL HAVE DURING CLASS

Be prepared for push-back from the students. Sitting passively through a didactic lecture can require little cognitive effort of a student whereas active learning strategies take much more energy. Students may be unfamiliar with the need to prepare adequately for class time and may resent being held accountable. Conversely, the instructor needs to ensure that the extra-curricular time required to study prior to class is reasonable given the students' concurrent curricular demands. Some self-directed study sessions may have to be timetabled within "standard" curricular hours.

- What expectations and procedures need to be communicated to students about how they prepare before, and engage during, class?
- What additional tools or techniques would help you in your role as a "cognitive" coach?
- How much extra-curricular time is available for students to devote to this course given their concurrent study demands?

REFERENCES

1. http://www.washington.edu/teaching/teaching-resources/engaging-students-in-learning/flipping-the-classroom/
2. Fink, LD. Creating significant learning experiences: an integrated approach to designing college courses. 1st ed, 2003. Published by Jossey-Boss, John Wiley & Sons, Inc., San Francisco, CA.
3. http://ctl.utexas.edu/sites/default/files/utflipquickstartguide112114.pdf
4. http://ctl.utexas.edu/teaching/flipping-a-class/how
5. http://www.epsteineducation.com/home/about/default.aspx
6. http://www.uc.edu/content/dam/uc/cetl/docs/IF-ATinstructions.pdf
7. https://learningcatalytics.com/

EDUCATING THE MILLENNIAL VETERINARY STUDENT

Cynthia R. Ward, VMD, PhD, DACVIM (Small Animal Internal Medicine)
Athens, Georgia

Who are the millennials? They are generally defined as people born between 1982-2001(1). They have experienced significant tragedies such as 9/11 and school violence that have shaped societal interaction.

The millennials are digital natives. They have grown up in a world full of digital media including cell phones, video games, and the Internet. Their immersion and comfort with this technology affects their learning styles in ways not familiar with many educators who are not of this generation. Because of their comfort level with technology, computers and the Internet are part of their lives and not just tools they use to reach an endpoint. They do not need to depend on educators for information. They assume that online source materials are community property and should not be restricted.

Millennials' approach to learning is trial and error, similar to the strategies that they have used to be successful at gaming. They feel that doing something is more important than learning something; therefore, results and actions are more desirable than factual accumulation. They tend to push buttons and do not read manuals.

Millennials have had baby boomers as parents. This social group tended to have children later in life and have allowed their children to have input into family decisions and education options. Millennials grew up with role-playing games that allowed a change in ending based upon decisions made during the games. This results in the creation of a generation of negotiators who will apply this technique to parents, law officers, as well as their teachers. Some educators may perceive this approach as argumentative.

Millennials have grown up in a 24/7 world. They expect immediate response and results to demands. They are not comfortable with delays. They are expert multi-taskers and are used to balancing input from various sources at once. It can be hard for millennials to focus on one thought process as many students are more comfortable listening to lecture materials surfing the Internet. This creates intolerance for delay, even in educational concepts. It can be difficult for educators to keep up with new technology with which to engage students in learning.

Millennials have also been introduced to consumerism at a young age. They may see education as a product in which they should be sought for input even though their educators may feel that they have no expertise to determine appropriate and effective educational materials.

These students are a challenge for teachers, and we must consider changes to traditional information delivery if we are to teach these students effectively.

REFERENCES

Strauss and Howe, Generations: the history of america's future 1584-2069, Mangold, K. 2007. Education a new generation: teaching baby boomer faculty about millennial students, Nurse Educator 32, 21-23, Frand J, Hagner P, Bates, T, Natalicio D, Smith K, Gunderson M, Jackson G, Larsen R, Balestri D, Katz R. 2000. The information-age mindset, Educause Review 35, 5.

THE CONNECTION BETWEEN DOG HEALTH AND OWNERSHIP ON HUMAN WELL-BEING

Mark A. Oyama, DVM, DACVIM (Cardiology)
Philadelphia PA

Dog ownership is associated with physical health benefits such as increased exercise, reduced blood pressure, reduction of stress hormones, and prevention of illness and facilitation of recovery. In addition to these physical health benefits, psychological health benefits including, companionship, reduced loneliness, improved social contact with others, and increased self-esteem have also been associated with dog ownership. These effects on overall well-being and quality of life (QoL) are a critical component of human health and deserving of study. The impact of dog ownership on human QoL has been previously reported and most, but not all, reports indicate that dog ownership is associated with improved QoL. Measurement of QoL associated with dog ownership is hindered by the lack of QoL instruments that are specifically validated for this type of research. Existing QoL instruments, including two that are widely used, the WHOQOL-BREF and SF-36, are not particularly robust in detecting changes in QoL associated with pet ownership. Previous studies have used a variety of different QoL instruments and the lack of a single accepted QoL tool makes comparisons across studies difficult. Finally, the potential for dog ownership to have negative impact on their owner's QoL has been underappreciated. Negative aspects, including increased caretaking responsibility, financial burden, and stress associated with the dog becoming ill, are unlikely to be accounted for in existing QoL instruments. Thus, development of a balanced and robust QoL instrument specifically designed to assess the many potential effects of dog ownership on human QoL is needed.

If such an instrument was available, studies investigating the effect of dog ownership on QoL are still difficult to design and execute. One commonly cited study reported the effect of new pet (dog or cat) ownership on owner physical and psychological health status over a 10 month period. The results indicated improved QoL in dog owners but were limited by the general nature of the QoL instrument (General Health Questionnaire-30) used in the study. Moreover, because new dog ownership could not be investigated using a conventional randomized double-blinded study design, changes in the QoL of new dog owners were not compared against a true control group. This hurdle in study design can be overcome using special types of statistical methods. Increasingly, observational studies that lack a natural control group use techniques to estimate the likelihood that any individual will receive the intervention being studied (in this case, choosing to acquire a dog). When analysis is conditioned on this variable, which is called a propensity score, baseline variables are made similar between the two treatment groups (i.e., dog owners vs. non-dog owners) and differences in outcome due to the treatment can be calculated using standard univariable and multivariable techniques. Thus, two important roadblocks to better understanding the effect of dog ownership on human QoL are lack of a specific measurement instrument and difficulties in study design. The presenter and other are interested in measuring the potential beneficial and negative effects of dog ownership on QoL by developing a specifically validated QoL instrument, the *DOG Owner qUality of life SurvEy (DOGhOUSE)*.

REFERENCES

Friedman E, Son H. The human-companion animal bond: how humans benefit. Vet Clin Small Anim 2009;39:293-326.

Levine GN, Allen K, Braun LT, et al. Pet ownership and cardiovascular risk: a scientific statement from the American Heart Association. Circ 2013;127:2353-2363.

Serpell JA. Beneficial effects of pet ownership on some aspects of human health and behavior. J Royal Soc Med 1991;84:717-720.

IMPACT, MONITORING, AND MANAGEMENT OF HYPERKETONEMIA IN TRANSITION DAIRY COWS

Daryl Nydam, DVM, PhD, Paula Ospina, DVM, PhD, Jessica McArt, DVM, PhD, Tom Overton, PhD
Ithaca, NY

INTRODUCTION

As energy demands surpass dry matter intake during the transition period, elevated non-esterified fatty acids (NEFA) and B-hydroxybutyrate (BHBA) concentrations can be used as markers of the extent of negative energy balance both at the individual cow (Duffield et al., 2009) and herd level (Oetzel, 2004). Previous studies have identified an objective level where NEFA and/or BHB are excessive and cause detrimental effects on health (Ospina et al., 2010a), reproduction and milk production (Ospina et al., 2010b). Recently, the herd alarm level, i.e., the proportion of sampled animals with metabolite concentrations above the threshold which was associated with detrimental herd level outcomes, was determined (Ospina. et al., 2010c).

The objectives are to review the: 1) association between the NEFA and BHB critical thresholds and the risk of disease and the effect on milk production and reproductive performance; 2) the proportion and prevalence of sampled cows with elevated metabolites resulting in herd-level effects i.e., the herd-alarm level; 3) use of pooled versus individual samples to estimate the herd-alarm level in free-stall, TMR-fed herds; 4) suggest a herd-testing strategy.

MATERIALS AND METHODS

A convenience sample of approximately 100 farms in the Northeast of the USA was selected to participate in a prospective cohort study. All farms consented to participate, and this study was approved by the Cornell University Institutional Animal Care and Use Committee. To be included in the study a herd must have: 1) had greater than 250 milking cows, 2) free-stall housing, 3) fed a total mixed ration (TMR), and 4) participated in DHIA and/or use Dairy Comp 305 (Valley Ag. Software, 2009).

All farms received a standardized consent form, a survey, and case definitions for diseases of interest. The survey collected information on: farm demographics, feeding times in relation to blood collection, voluntary waiting period, and ovulation synchronization protocols. Farm personnel were instructed to document any incident cases of the diseases of interest: displaced abomasusm (DA), clinical ketosis (CK), and metritis (MET) and/or retained placenta (RP).

Farms were visited once, and during the farm visit, two cohorts of animals were selected: those 14-2 days pre-partum and those 3-14 days post-partum. Within each cohort, convenience samples of 15 apparently healthy animals were evaluated. The evaluation included simultaneous blood collection and body condition scoring (BCS) (Ferguson et al., 1994). Guidelines for blood collection and sample handling were based on previous studies (Stokol and Nydam, 2006). Briefly, a plain evacuated red-top tube was used to collect 10 ml of blood from the coccygeal vein or artery. The sera from the pre-partum cohort were analyzed for NEFA and hemolysis. The sera from animals sampled post-partum were analyzed for NEFA, BHB, and hemolysis. For animals sampled, the incidence of the diseases of interest within 30 DIM, time to pregnancy within 70 days post voluntary waiting period and Mature Equivalent 305 (ME 305) milk at 120 DIM were recorded.

DATA ANALYSIS

Summary:

The analytical approach was done in four stages. 1) The first stage was to identify critical NEFA and BHB concentration thresholds that were predictive of disease and this was done in three steps. The first step identified significant risk factors with a multivariable model where the outcome was the development of any combination of the diseases of interest (DA or CK or MET and/or RP). The second step analyzed the continuous significant predictors from the multivariable model with receiver operator characteristic (ROC) curves to identify critical thresholds for prediction of individual diseases (e.g. DA) and any combination of the diseases. Once the range of critical thresholds predictive of disease was identified, the covariates were treated as categorical variables within this range. The third and final step analyzed the magnitudes of the associations between these categorical predictors with disease, reproduction and production were evaluated. 2) The second stage used the NEFA and BHB critical thresholds identified as predictors of disease and evaluated the association between these predictors and milk and reproductive performance. 3) The third stage identified the proportion of animals with metabolite concentrations above the disease thresholds that were predictive of herd level outcomes, i.e. herd-alarm level. 4) The final stage compared the interpretation of pooled versus individual samples when determining the herd-alarm level status.

Stage I: Critical thresholds for disease

Evaluation of significant risk factors:

The metabolites, NEFA and/or BHB were the main risk factors and at this level of analysis were treated as continuous predictors. Parity, season, BCS, time of blood collection, and all biologically plausible 2-way interactions were evaluated as covariates in the model. They were modeled with PROC GENMOD using a Poisson distribution, a log link function, p-scale option for over-dispersion, and an exchangeable correlation matrix (Spiegelman and Hertzmark, 2005). This statistical method allows for clustering of cows within herds (i.e. including herd as a random effect) while adjusting for continuous or categorical covariates. There was no adjustment for varying time spans (offset term) because the length of the time interval at risk was the same for every individual in the sample (Allison, 2007).

ROC curves:

The continuous, significant risk factors identified in the multivariable model were evaluated using ROC curves to determine the critical threshold for predicting disease. The point on the ROC curve that has the highest combined sensitivity and specificity was considered the critical threshold. Interpretation of this critical threshold depends on the area under the curve (AUC), such that if the AUC >0.7 the test is considered accurate (Swets, 1988).

Effect on disease risk, reproduction, and production:

Disease risk - Once the critical thresholds for prediction of disease (DA, CK, MET/RP, or any combination) were identified with ROC analysis, the covariates were dichotomized at the critical threshold. The risk of disease, given these categorical covariates, was evaluated with PROC Genmod, using a poisson distribution, log link function, p-scale option for over-dispersion, and an exchangeable correlation matrix

Stage II: Reproduction and milk production

Effect on reproduction - The effect of elevated NEFA and/or BHB concentrations on reproductive performance was evaluated with time-to-event analysis (PROC Phreg). Cox proportional hazard models (Cox, 1972) were analyzed accounting for clustering of cows within herds. The covariates were: BCS, parity, and ME 305 milk at 120 DIM. ME 305 data was dichotomized based on the median production of the pre- or post-partum group. Animals culled before the end of voluntary waiting period were excluded from the analysis and those not pregnant by the end of the follow-up period were right censored. The proportional hazards assumption was checked statistically by evaluating time dependent covariates and non-informative censoring was evaluated with sensitivity analysis (Allison and SAS Institute, 1995). The categorical metabolite value selected from within the range of critical threshold predictive of disease that resulted in the smallest chance of committing a type I error was kept in the final model.

Effect on milk production - The effect of elevated NEFA and/or BHB concentrations on ME305 milk were evaluated with mixed effects models with herd as a random effect. The covariates were: BCS, season, and when applicable both parity, and the interaction between parity and the metabolite level. In all models, the metabolites, NEFA and BHB were dichotomized and evaluated within the range of values identified as critical thresholds for prediction of disease. The categorical metabolite value that resulted in the smallest chance of committing a type I error was kept in the final model.

Stage III: Herd-alarm level

The outcomes of interest were: herd-level pregnancy rate (PR), incidence of DA and clinical ketosis, and average 120 DIM ME 305 milk production for animals sampled. Pregnancy rate was averaged over two 21 d periods post individual-herd voluntary waiting period. Statistical analyses were performed using SAS version 9.1. Mixed models were evaluated using PROC mixed with herd as a random effect. The range of critical NEFA and BHB levels was based on previous results of concentrations predictive of cow-level health, reproductive and production effects. The effect of the proportion of sampled animals with elevated metabolites on the herd level outcomes of interest was evaluated.

Stage IV: Pooling versus individual samples

For each herd, NEFA and BHB were measured in 12 individual cow serum samples and in a pooled sample, derived from equal volumes (200 μL) of each serum sample. The interpretation of herd-alarm level status was based on the herd alarm level research and was compared between the proportion of individual animal samples and the pooled samples.

Two methods were used for this comparison: 1) the proportion of animals above the threshold was used as the reference test and the sensitivity and specificity of the pooled concentration was evaluated 2) the Kappa statistic was evaluated to test the agreement beyond chance between the proportion and the pooled concentration. All analyses were performed using SAS v. 9.1 (SAS Inst., Inc., Cary, NC).

RESULTS

Stage I and II: disease, reproduction and milk production

Study population - Of the 104 herds, 4 were excluded from the study due to missing data. 2758 cows from the remaining 100 herds were included in the study and of these cows, 1440 were sampled pre-partum (35 % heifers and 65% cows) and 1318 were sampled post-partum (37% heifers and 63% cows). The number of milking cows per herd averaged 840.

Multivariable analysis - In the three multivariable models, the metabolites were the only significant predictors of any of the diseases of interest: pre-partum NEFA (p=0.028), post-partum NEFA (p=0.0005) and when BHB was the only main predictor in the model (p=0.005) it was also the only significant predictor. No other covariate or interaction term in any of the three multivariable models had a p-value <0.1.

ROC- critical thresholds for prediction of disease - The critical thresholds identified with ROC analysis are summarized in Table 1 with their AUC values. Briefly the NEFA critical thresholds for predicting any of the diseases of interest in the pre- and post-partum cohort were 0.29 and 0.57 mEq/L, respectively and the BHB critical threshold was 10 mg/dL. Figure 1 is a graphical representation of an ROC curve with DA as the outcome and concentrations of post-partum NEFA as the test.

Risk of disease - The risk of disease based on NEFA and BHB concentrations greater than or equal to the critical thresholds are also summarized in Table 1. For example, experiencing elevated metabolite levels post-partum increased the risk of developing a DA by up to 10 times, and elevated levels of post-partum NEFA contributed the greatest risk of disease development.

Effect on reproduction - Table 2 summarizes the results of elevated metabolite levels on reproduction with estimates for metabolites and significant covariates reported. Animals with elevated metabolite levels (within the range identified as predictors of disease) took longer to get pregnant; the hazard ratio for pregnancy within 70 days post-voluntary waiting period decreased. Figure 2 is a graphical representation of a Kaplan-Meier curve, where animals with elevated pre-partum NEFA levels took longer to get pregnant.

Effects on production - The results of elevated metabolite concentrations are reported separately for heifers and cows sampled post-partum because the effect of the elevated metabolite thresholds on ME305 milk was different between these two groups. Generally, elevated metabolite levels predicted a decrease of several hundred kilograms of ME 305 milk; however, in heifers sampled post-partum elevated metabolites levels predicted an increase in ME 305 milk. The results of this analysis are summarized in Table 3 with metabolite results and significant covariates reported.

Stage III: Herd-alarm level

Pre-partum sampled cohort:

If more than 15% of the animals had NEFA concentrations ≥0.27mEq/L, pregnancy rate (PR) decreased by 1.2% ($p=0.006$), ME 305 milk from sampled animals decreased by 282 kg ($p=0.002$), and the percent of diseases increased by 3.6% ($p=0.006$).

Post-partum sampled cohort:

If more than 15% of the animals had post-partum NEFA concentrations ≥ 0.70mEq/L: percent of disease increased by 1.7% ($p=0.04$); in cows ME 305 milk decreased by 593 kgs ($p < 0.0001$). If more than 15 % of animals had post-partum NEFA concentration ≥ 0.60 mEq/L, then PR decreased by 0.9% ($P=0.05$), and in heifers ME 305 milk decreased by 288 kg ($p = 0.07$).
If more than 15% of the animals had BHB concentrations ≥12 mg/dL: PR decreased by 0.8% ($p=0.03$), percent of disease increased by 1.8% ($p=0.03$). If more than 15% of cows had BHB ≥ 10mg/dL then ME 305 milk decreased by 358 kgs and if more than 20% of heifers had BHB ≥ 12 mg/dL then ME 305 milk decreased by 534 kgs.

Prevalence:

75% of herds had >15% of sampled animals with pre-partum NEFA levels ≥0.30 mEq/L and 65% of herds had >15% of sampled animals with post-partum NEFA levels ≥0.70mEq/L, lastly 40% of herds had >10% of animals sampled with BHB levels ≥ 12mg/dL.

Stage IV: Pooled versus individual samples

Based on the herd-alarm level study, herds were defined as being above the herd alarm level if the proportion of animals with metabolite concentration above the threshold was more than 15 % (≥ 2 animals from 12). The metabolite thresholds were 0.3mEq/L for pre-partum NEFA, 0.6 mEq/L for post-partum NEFA, and 12 mg/dL for BHBA. Sensitivity of the pooled metabolite concentration as a single test was 58% (95% Confidence Intervals: 28 to 85) for pre-partum NEFA; 55% (95% CI 24 to 83) for post-partum NEFA; and 36% (95% CI 11 to 69) for BHBA. Specificity of the pooled sample was 100% (95 % CI: 17 to 100, 19 to 100, and 19 to 100) for pre-partum NEFA, post-partum NEFA and BHBA, respectively. The Kappa statistics were: 0.2 (95% CI: -0.1 to 0.5), 0.3 (95% CI: -0.07 to 0.6), and 0.2 (95% CI: -0.07 to 0.4) for pre-partum NEFA, post-partum NEFA and BHBA, respectively. The McNemar's p-values for the Kappa statistics were 0.02, 0.02, and 0.008, respectively.

SIGNIFICANCE

These studies demonstrated that excessive negative energy balance (as measured by NEFA and BHB concentrations) in the transition period are strong predictors of clinical disease, and negative reproductive and productive performance in cattle from free-stall, TMR-fed dairies averaging 840 milking cows. The magnitude of the association between elevated NEFA and/or BHB and diseases of interest, measured by risk ratios, was large (range: 1.75-9.7). The effects of elevated metabolite levels on reproduction decreased the relative risk of pregnancy within 70 days post-voluntary waiting on average by 20 %, with parity as the only other significant covariate (cows took longer to get pregnant than heifers). Milk production, showed mixed results, and although further investigation about homeorhesis in heifers is warranted, there was strong evidence of significant ME305 milk loss in cows sampled post-partum and all animals in the pre-partum cohort.

The NEFA herd alarm level was 15 - 20 % and the BHB herd alarm level was 10% at the defined critical thresholds. These data demonstrate that the proportion of sampled animals with NEFA and BHB concentrations above critical levels can have detrimental effects on herd-level disease status, reproduction and production. Additionally, the prevalence information indicates that at least 66% of the herds sampled can improve the energy status of their transition animals.
When evaluating herds to determine opportunities in transition cow management, pooled samples have low sensitivity and there is only poor to fair agreement between this and the proportion of animals above the threshold test. Although pooled samples may seem desirable due to lower laboratory cost, the low sensitivity and lack of agreement between pooled samples and the proportion can lead to an increased number of false negatives. An increased number of false negatives may result in missed opportunities for herds looking to improve transition cow negative energy balance.

Management programs focused on minimizing the risk of these diseases and minimizing negative effects of decreased reproductive and productive performance may consider the following as general guidelines for monitoring NEFA and BHB concentrations in individuals: NEFA concentrations ≥ 0.3 mEq/L for cattle 14-2 d pre-partum; and NEFA concentrations ≥0.6 mEq/L and BHB ≥10 mg/dL for those 3-14 d post-partum; and no more than approximately 15% (2/12) cows sampled with metabolite concentrations above the defined critical thresholds.

Testing strategies

An initial sample for prevalence estimation should be performed following sample size guidelines, e.g., test at least 20 at risk animals who are subjectively healthy and 3 to 14 DIM. Any animal with a BHB concentration ≥ 1.2 mmol/L should receive 300 mL of propylene glycol for 5 days.

1. If 15% or fewer of the animals sampled (3 to 14 DIM) have a BHB concentration ≥ 1.2 mmol/L, the recommendation is to sample again every other week to monitor herd-level prevalence of elevated BHB. More frequent sampling may be indicated when there are significant changes in diet formulation, management, or environment.

2. If > 15 to 40% of the animals sampled (3 to 14 DIM) have a BHB concentration ≥ 1.2 mmol/L, the recommendation is to test all animals that are 3 to 9 DIM twice weekly (e.g. Tuesday and Friday) and treat all positive cows. This more frequent testing scheme is warranted in order to identify and treat most of the cows with elevated BHB, thus reducing their risk of negative downstream outcomes. If at least two consecutive prevalence tests independently result in fewer than 15% of the animals testing positive then one could consider stopping this testing and treating protocol and monitor the prevalence every two weeks as described above.

3. If > 40% of the animals sampled (3 to 14 DIM) have a BHB concentration ≥ 1.2 mmol/L the recommendation is to start treating all fresh cows with PG at 3 DIM for five days. This treatment scheme will help reduce the negative effects of elevated BHB concentrations in herds with a very high prevalence. Recheck the prevalence in two weeks to determine the next course of action, e.g. you can stop treating all cows and move to the test and treat positive cow scheme or remain in the treat all cows protocol. This monitoring scheme should continue until at least two consecutive prevalence tests independently result in fewer than 40% of the animals testing positive.

RECOMMENDATIONS

As summarized, several studies have reported that there is an association between elevated NEFA and BHB and negative downstream outcomes. There are also several concentration thresholds above which negative outcomes are more likely; the chosen NEFA and BHB cut-point used will depend on the level of sensitivity and specificity that is most important in the situation. Currently, there are only two studies which report herd-alarm levels. These herd-level cut-points should be applied in the same way as individual cut-points were evaluated: if false negatives are more important, the cut-point with the highest sensitivity (i.e., lower herd alarm level) should be used for evaluation; conversely, if false positives are more important, then the cut-point with the highest specificity (i.e., higher alarm level) should be chosen (Tables 3 and 4).

METABOLITE TESTING

NEFA vs. BHB - Concentrations of BHB and NEFA are not well correlated when sampled on the same day, i.e., they are not both elevated in the same animal at the same time. In addition, reports that evaluated both NEFA and BHB more often than not found an association between NEFA and negative downstream outcomes when the association was not present with BHB. Generally, NEFA was found to be a better predictor of negative downstream outcomes; however, BHB is easy and inexpensive to measure accurately cow-side if the appropriate test is chosen. Therefore, BHB should be used as the primary monitoring tool, but when there is evidence that the herd is experiencing issues with transition cows without significant elevation in BHB, NEFA should be evaluated.

Sample size - When evaluating prevalence, it is recommended to sample at least 20 cows at risk; a larger sample size will result in a more precise estimate.

Sampling time frame - The first time cows are sampled on a farm, the sampling window should include cows between 3 to 14 DIM. However, when twice per week sampling is warranted, this window is decreased to cows between 3 to 9 DIM.

Sample frequency, testing and treatment strategies - Figure 3 provides an overview of the sampling guidelines based on initial prevalence of elevated BHB concentrations (Ospina et al., 2013).

TAKE HOME MESSAGE

In conjunction with BHB or NEFA monitoring, it is necessary to continuously evaluate other herd-level factors which increase the risk of excessive negative energy balance, such as overcrowding, lack of heat abatement, excessive pen moves, excessive energy in dry cow diets, and unbalanced dietary protein. The goal is to prevent excessive NEB and use these testing and monitoring guidelines to evaluate the effectiveness of your program or make changes based on objective measures of the success of your dairy.

REFERENCES

Allison PD 2007. Logistic Regression using the SAS System: Theory and Application. Bauman DE and WB Currie. **J Dairy Sci** (1980) 63:1514. Cheong SH et al. **J Dairy Sci** (2011), 94: 762. Cox D. **J Royal Stat Soc Series B** (1972) 34:187. Duffield TF et al. **J Dairy Sci** (2009) 92:571. Ferguson JD et al. **J. Dairy Sci** (1994) 77:2695. Hayirli A et al. **J Dairy Sci** (2002) 85:3430. Herdt TH **Vet Clin North Am Food Anim Pract** (2000)16:215. McArt JAA et al. **J Dairy Sci** (2011), 94: 6011. McArt JAA et al. **J Dairy Sci** (2012), 95: 2505. McArt JAA et al. **J Dairy Sci** (2012), 95: 5056. McArt JAA et al. **J Dairy Sci** (2013), 96: 198. Oetzel GR. **Vet Clin North Am Food Anim Pract** (2004) 20:651. Ospina PA et al. **J Dairy Sci** (2010a) 93:546. Ospina PA et al. **J Dairy Sci** (2010b) 93:1596. Ospina PA et al. **J Dairy Sci** (2010c) 93: 3595. Ospina PA et al. **Vet Clin N Am.** (2013) Vol 29; 387. Schoonjans F (2008)broekstraat 52, 9030 Mariakerke, Belgium.9.5.2.0:. Spiegelman D and Hertzmark E. **Am J Epidemiol** (2005).162:199. Stokol T. and Nydam DV. **J Vet Diagn Invest** (2006) 18:466. Swets JA. **Science** (1988) 240:1285-1293. Valley Ag. Software. 2009. K St. Tulare California 93274. Dairy comp 305.

Table 1. Receiver operator characteristic (ROC) curve determination of critical NEFA (mEq/L) and BHB (mg/dL) thresholds as predictors of disease and risk ratios of disease based on these critical thresholds.

Animals Sampled Pre-partum					
Disease	Critical pre-partum NEFA[1]	AUC[2]	Risk Ratio	95 % CI[3]	P-value
DA	0.27	0.6	2.0	1.1 – 3.7	0.03
CK	0.26	0.6	1.8	1.2 – 2.5	0.001
Met and/or RP	0.37	0.6	2.2	1.6 – 3.0	< 0.0001
Any of the three	0.29	0.6	1.8	1.4 – 2.2	< 0.0001
Animals Sampled Post-Partum					
Disease	Critical post-partum NEFA[1]	AUC[2]	Risk Ratio	95 % CI[3]	P-value
DA	0.72	0.8	9.7	4.2 – 22	<0.0001
CK	0.57	0.7	5.0	2.3 – 11	<0.0001
Met	0.36	0.6	17	2.0 – 134	0.008
Any of the three	0.57	0.7	4.4	2.6 – 7.3	< 0.0001
Disease	Critical BHB[1]	AUC[2]	Risk Ratio	95 % CI[3]	P-value
DA	10	0.8	6.9	3.7 – 12.9	<0.0001
CK	10	0.7	4.9	3.2 – 7.3	<0.0001
Met	7	0.6	2.3	1.1 – 5.1	0.037
Any of the three	10	0.7	4.4	3.1 – 6.3	<0.0001

[1] Highest combined specificity and sensitivity, [2] Area under the curve, [3] Risk ratio confidence interval

Table 2. Cox proportional hazard model of the effect of NEFA (mEq/L), and/or BHB (mg/dL), covariates, and animals clustered within herds on days to conception after voluntary waiting period.

Sampled population	Variable	Hazard	P-value
Pre-Partum Cohort	NEFA ≥ 0.27	0.81	0.01
	Parity	0.73	0.001
Post-Partum Cohort	NEFA 0.72	0.84	0.05
	BHB ≥ 10	0.93	0.4
	Parity	0.81	0.01
Post-Partum Cohort	BHB ≥ 10	0.87	0.1
	Parity	0.80	0.01

Table 3. Mixed models for the effect of NEFA (mEq/L), and/or BHB (mg/dL), covariates, and herd as a random effect on milk production measured by 120 DIM ME 305 (kg).

Sampled Population	Variable	Difference in ME milk yield (kg)	P-value
Pre-Partum	NEFA ≥ 0.33	-683	0.001
	Parity	-556	0.01
Post-partum- heifers	NEFA ≥ 0.57	+488	0.02
	BHB ≥ 10	-143	0.5
Post-partum- heifers	BHB ≥ 9	+ 403	0.04
Post-partum- cows	NEFA ≥ 0.72	-647	0.001
	BHB 10	-165	0.4
Post-partum- cows	BHB 10	-393	0.04

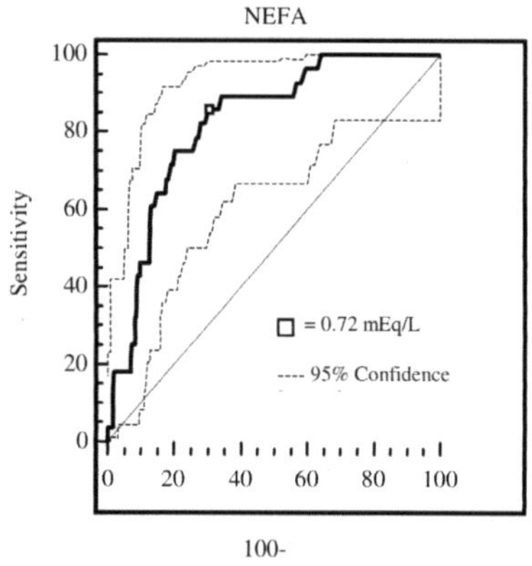

Figure 1. ROC curve determination of critical threshold (upper most left hand corner) for NEFA concentrations predicting DA in animals sampled post-partum.

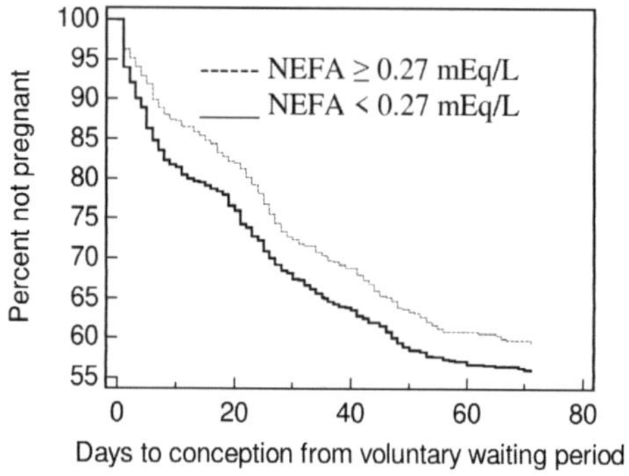

Figure 2. Kaplan-Meier curves of time to pregnancy of cows and heifers with NEFA ≥ 0.27 mEq/L or < 0.27 mEq/L measured in serum 14-2

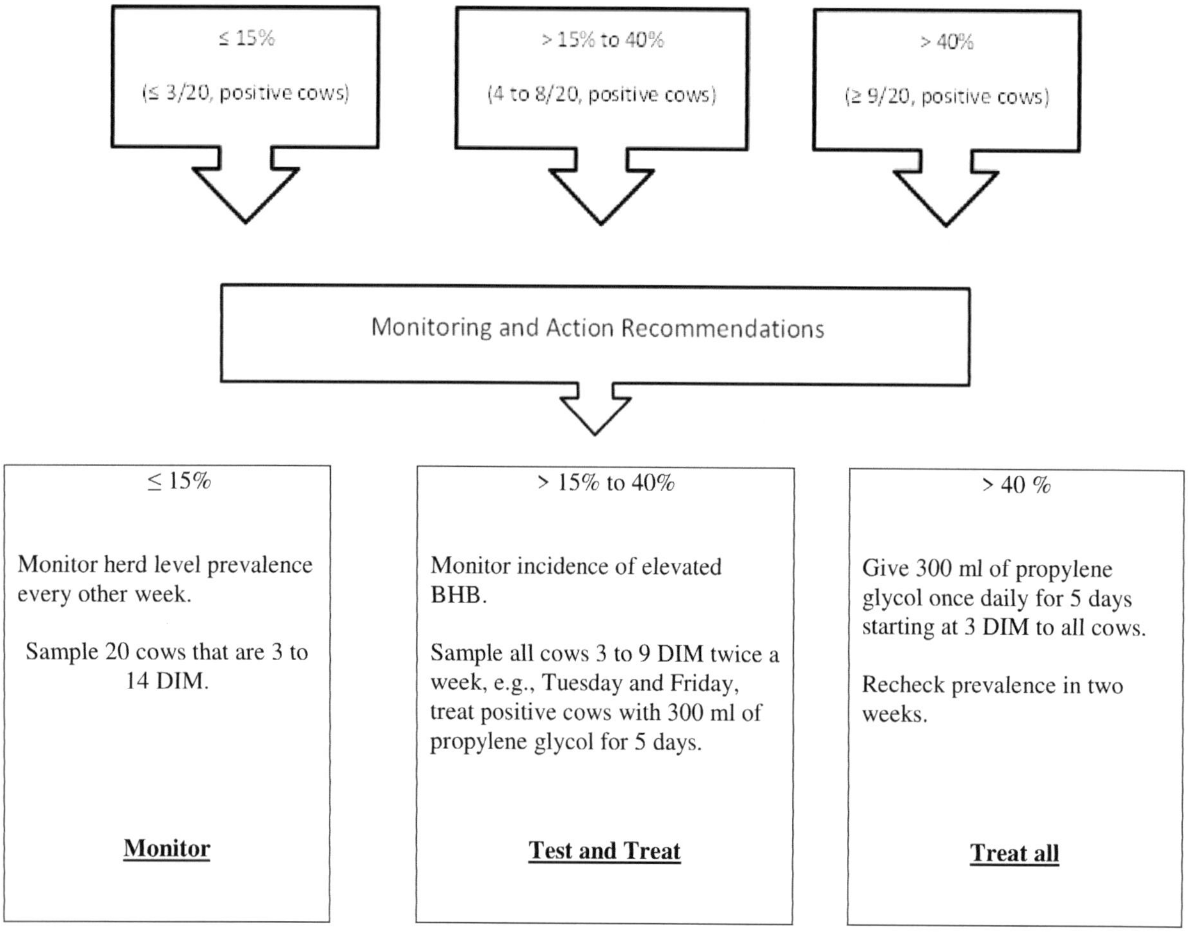

Figure 3. Monitoring recommendations for β-hydroxybutyrate (BHB) testing based on herd-level prevalence of elevated BHB

DRY COW ENERGY: EFFECTS ON NEGATIVE ENERGY BALANCE, PRODUCTION, MUSCLE AND ADIPOSE TISSUE IN PERIPARTUM DAIRY CATTLE

Sabine Mann, DVM, DECBHM, Tom Overton PhD, Joseph Wakshlag DVM, PhD, DACVN, DACVSMR, and Daryl Nydam, DVM, PhD
Ithaca, NY

INTRODUCTION

A successful transition of dairy cows from the dry period into lactation is reflected by preventing disease and ensuring the profitability of dairy cows (Drackley, 1999). Research efforts have concentrated on both pre- and postpartum management strategies to identify nutritional, housing and preventive procedures that can be employed to accomplish this goal. Nutritional strategies for the dry cow have varied during the last decades with several different ones being used on dairy farms today. Because dry matter intakes (DMI) invariantly drop before calving, some authors have proposed to increase the energy content of the prepartum ration in order to improve production and health in early lactation (Grummer, 1995; McNamara et al., 2003) and it was shown that DMI is increased in the prepartum period when higher energy diets are fed. On the other hand, research and experience in the field have shown beneficial effects to feeding cows a controlled energy diet prepartum (Drehmann, 2000; Douglas et al., 2006; Janovick et al., 2011; Vickers et al., 2013). A difference was shown for the effect of diets differing in energy being fed either during the far-off or close-up period and the authors concluded that overfeeding in the far-off period had more severe negative consequences than overfeeding in the close-up period (Dann et al., 2006). The objective was to evaluate the effect of each dry cow ration on postpartum negative energy balance, ketone body production and productivity as well as differences in pancreatic hormone concentrations involved in glucose balance.

Adipose and muscle tissue represent major tissues with insulin-dependent glucose uptake. Research has focused mainly on the role of adipose tissue in peripartum dairy cows, specifically because of its role in potentially altering insulin sensitivity due to production of different adipokines, such as tumor necrosis factor alpha (TNFα) (Ronti et al., 2006). Excessive adipose tissue mobilization in postpartum dairy cows is a key factor in the development of metabolic disorders (Vernon, 2005). However, muscle tissue plays a big role in the adaptation to early lactation in dairy cows also (Bell, 1995; Kuhla et al., 2011). Therefore in this study, we also aimed at describing the changes in adipose and muscle tissue mass, as well as adipokine expression.

EXPERIMENTAL APPROACH

In this randomized block design, 84 Holstein cows from the Teaching and Research Dairy herd were blocked by expected calving date at enrollment (57 days before expected parturition) and assigned to one of three dry cow dietary energy levels. Cows in group **C** received a controlled energy diet, formulated to supply 100% of energy requirements during the dry period. Cows in group **H** received a diet supplying approximately 150% of energy requirements. Cows in group I received the controlled energy diet for the first four weeks after dry off and were switched to a diet formulated to supply 125% of requirements until the day of calving. All treatments were formulated to contain 300 mg/day of monensin. After calving, all cows were fed the same fresh cow diet **(F)** until the end of the study at 42 DIM. The ingredient composition of major components of the dry and fresh cow diets is described in Table 1. All diets were fed ad libitum and adjusted to allow for a minimum of 5% refusals. Individual intakes were recorded daily and dry matter intake (DMI) determined from weekly dry matters of all TMRs. Weekly blood samples were taken from all cows before the morning feeding. Sampling frequency was increased to three times per week from three weeks before to three weeks after calving. Concentrations of ß-hydroxybutyrate (BHBA) in whole blood were measured cow-side using the Xtra Precision handheld device (Abbott Laboratories) and concentrations of NEFA in serum were measured using a commercial kit (WAKO Life Sciences). Milk yield was recorded for every milking; samples were collected weekly and submitted for analysis of components to Dairy One (Ithaca, NY). The analyzed composition of the diets is partially described in Table 2.

Energy balance was estimated in CNCPS (v. 6.1) for each week and individual by entering the weekly average of the animal's body weight and dry matter intake and either days carried calf for dry period estimates or milk yield as well as percent milk fat, protein and lactose for the postpartum period. Energy balance was expressed as percentage of requirements. On days 28 and 10 before expected calving as well as on days 4 and 21 after calving, baseline concentrations of insulin and glucagon were determined via radioimmunoassay (Millipore) four hours after removal of orts. On the same days as well as on the day of enrollment and at the end of the study, ultrasonography was used to determine the thickness of the subcutaneous fat depot in the thurl area as well as the diameter of the M. longissimus dorsi at the height of the 4th lumbar vertebra. Quantitative real-time PCR was used to describe the expression of target genes in adipose tissue.

Chi-squared tests were generated with PROC FREQ of SAS (SAS 9.3, SAS Institute Inc., Cary, NC) for differences in episodes of BHBA $\geq$1.2 mmol/L as well as treatment episodes for clinical ketosis. Repeated measures ANOVA was performed for the outcomes DMI, energy balance, BHBA and NEFA concentrations, milk and ECM yield using PROC MIXED in SAS. Five covariance structures were tested for each variable analyzed (simple, compound symmetry, autoregressive order 1, Toeplitz and unstructured). The covariance structure with the smallest Aikaike information criterion was chosen. Fixed effects were treatment group and parity with the REPEATED statement for the time variable. Data were analyzed separately for pre- and postpartum. When results of the ANOVA analysis yielded a p-value of $\leq$ 0.05, Tukey´s posthoc test was used for comparison of means across all groups to control experimentwise error rate.

EFFECT OF DRY PERIOD ENERGY LEVEL ON ENERGY METABOLITES, DRY MATTER INTAKE AND MILK YIELD

Concentrations of BHBA during the postpartum period were highest in group H (0.63 ± 0.06, 0.77 ± 0.06 and 0.85 ± 0.06 in group C, I and H, respectively, P=0.05). Four cows in group I and five cows in group H were treated for clinical ketosis, while none of the cows in group C required treatment (p=0.07). Of all possible tests in the first 21 DIM, when blood was tested three times per week for each cow, there were 13,32 and 31 positive episodes in group C, I and H, respectively (p=0.007). Concentrations of NEFA were highest in group H postpartum (659 ± 36.4, 665 ± 36.7 and 796 ± 39.5 μEq/L in the C, I and H group, respectively, *P*=0.02). Dry matter intake prepartum was different among the groups with the highest intakes in group H and the lowest intakes in group C. No differences in DMI or milk yield were measured postpartum. Numerical differences were found for % of milk fat and ECM yield with highest fat % and ECM in group H and lowest in group C. Energy balance estimations produced with CNCPS v.6.1 are presented in Figure 2.

EFFECT OF DRY PERIOD ENERGY LEVEL ON INSULIN AND GLUCAGON

Insulin concentration was lowest in group H at day 4 postpartum while the glucagon concentration was highest in this group at the same timepoint. Calculation of a molar insulin: glucagon ratio as an indication of negative energy balance showed that the ratio decreased substantially in all cows in the immediate postpartum period, but was lowest in cows in group H.

EFFECT OF DRY PERIOD ENERGY LEVEL ON MUSCLE AND ADIPOSE TISSUE

Cows in groups I and H showed numerically higher diameters of subcutaneous adipose tissue immediately prepartum, and a greater rate of decrease in the diameter in the postpartum period. Together with higher concentrations of NEFA in these groups, we hypothesize that it reflects a greater degree of lipolysis in cows being fed higher energy diets compared to those fed a controlled energy diet. This hypothesis was further supported by the finding that gene expression of fatty acid synthase, an enzyme involved in catalyzing the synthesis of fatty acids was increased in cows in group H. In parallel to changes in adipose tissue, muscle diameter was also numerically larger in the higher energy diet groups, indicating that accretion of muscle mass was increased compared to cows being fed the controlled energy diet. Cows in all groups showed a decrease in muscle diameter in the postpartum period. No difference was found in the gene expression of TNFα between a subset of cows in the controlled and high energy group, respectively.

IMPLICATIONS AND CONCLUSIONS

Data from this study shows that cows fed a controlled energy diet throughout the whole dry period mobilized less adipose tissue as reflected by lower concentrations of NEFA and also had lower concentration of BHBA postpartum while milk production and postpartum DMI intake were not affected. Higher milk fat % in the group fed a high energy diet prepartum is likely due to a greater amount of NEFA being used for milk fat synthesis. Lower concentrations of insulin and higher concentrations of glucagon at 4 DIM reflects the greater degree of negative energy balance and decreased inhibition of insulin on lipolysis explains the higher concentration of NEFA in cows being fed a higher energy diet prepartum.

Cows fed an intermediate energy level in a step-up system had the same milk yield and postpartum DMI as cows in the other two groups. When taking into account concentrations of NEFA and BHBA postpartum, as well as treatment for clinical ketosis and episodes of subclinical ketosis, cows fed an intermediate energy level in a step-up system prepartum showed some of the same effects of more pronounced negative energy balance as cows being fed a high energy diet for the whole duration of the dry period, despite the fact that had received the same controlled energy diet during the far-off period.

Although sample size in this study was not adequate to compare health events, epidemiological data linking high BHBA and NEFA concentrations to an increased risk for several periparturient disorders, including displacement of the abomasum, metritis, reduced reproductive success as well as decrease in milk production and removal from the herd (Ospina et al., 2010a; b; Chapinal et al., 2011; Chapinal et al., 2012; McArt et al., 2012; McArt et al., 2013; Suthar et al., 2013) is available. The conclusion of the authors is therefore that a controlled energy diet fed throughout the whole dry period showed clear advantages in preventing excessive negative energy balance that could lead to downstream disease without affecting milk yield.

REFERENCES

Bell AW. **J Anim Sci.** (1995)73:2804. Chapinal NM et al. **J Dairy Sci.** (2011)94:4897. Chapinal NS et al. **J Dairy Sci.** (2012) 95:5676. Dann HM et al. **J Dairy Sci.** (2006) 89:3563. Douglas GN et al. **J Dairy Sci.** (2006) 89:2141. Drackley J K. **J Dairy Sci.** (1999)82:2259. Drehmann PC. **Hoard´s Dairyman.** (2000) 145:641. Grummer RR. **J Anim Sci.** (1995) 73:2820. Janovick NA et al. **J Dairy Sci** (2011) 94:1385. Kuhla BG et al. **J Proteome Res.** (2011) 10:4252. McArt JA et al. **J Dairy Sci.** (2012) 95:2505. McArt JA et al. **Vet J.** (2013)198:560. McNamara SF et al. **J Dairy Sci.** (2003) 86:2397. Ospina PA et al. **J Dairy Sci.** (2010a) 93:1596. Ospina PA et al. **J Dairy Sci.** (2010b) 93:3595. Ronti TG et al. **Clin Endocrinol** (Oxf). (2006) 64:355-365. Suthar VS et al. **J Dairy Sci.** (2013) 96:2925-2938. Vernon, R. G.. **J Dairy Res.** (2005) 72:460. Vickers LA et al. **J Anim Sci.** (2013)91:886.

Table 1. Partial ingredient composition of diets (% of total dry matter).

Ingredient	Prepartum			Postpartum
	Controlled	Intermediate	High	
Corn silage	28.45	42.2	55.9	44.19
Wheat Straw	35.56	24	12.42	-
Hay Crop Silage	-	-	-	10.72
Hay Switchgrass	-	-	-	2.73
Amino Plus®	10.54	6.03	1.52	4.53
Canola Meal	6.82	7.86	8.92	9.82
Citrus Pulp	3.67	4.38	5.10	-
Corn Germ Meal	-	-	-	3.83
Corn ground	2.64	3.633	4.64	4.73
Distillers Grain	4.52	3.038	1.56	7.24
Soybean Hulls	2.42	3.678	4.95	5.97
Rumensin 90®	0.01	0.01	0.01	0.01
Soy Chlor®	1.71	1.258	0.82	-

Table 2. Analyzed composition of diets, presented as average of eleven monthly composites ± SD.

Component	Prepartum			Postpartum
	Controlled	Intermediate	High	
Dry matter (DM) %	56.4 ± 3.8	49.8 ± 3.1	44.7 ± 3.9	46.8 ± 3.1
CP % of DM	14.2 ± 1.6	13.9 ± 0.7	12.5 ± 0.9	17.8 ± 0.9
NDF % of DM	48.4 ± 5.0	42.2 ± 4.5	41.0 ± 4.2	35.4 ± 2.3
ADF % of DM	30.1 ± 4.2	28.5 ± 3.2	26.6 ± 3.3	21.3 ± 4.4
Starch % of DM	15.0 ± 2.5	20.1 ± 3.6	23.7 ± 2.6	21.2 ± 2.3
DCAD meq/100g	-0.76 ± 1.65	-0.39 ± 3.32	-0.53 ± 2.24	-

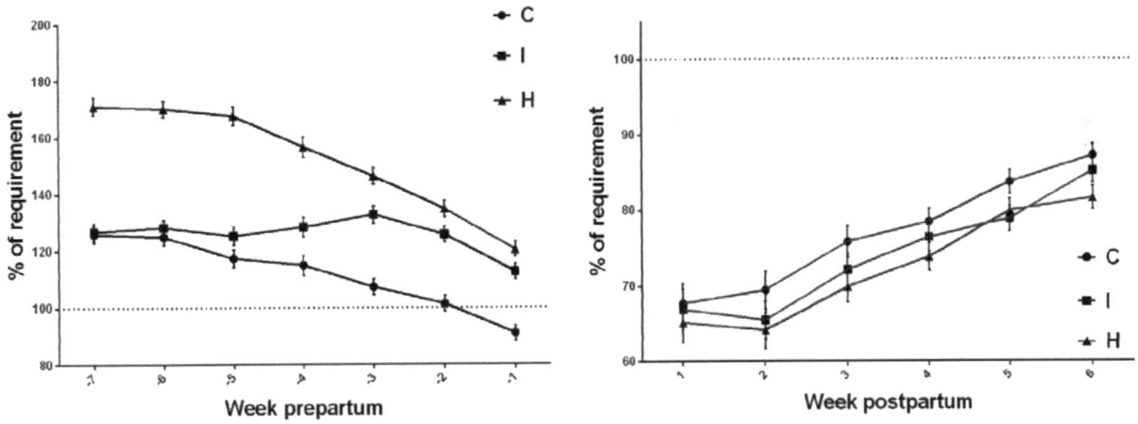

Figure 2. Energy balance estimation expressed as % of requirement using CNCPS v. 6.1 (C=controlled energy diet, I= intermediate energy diet, H= high energy diet).

LONG TERM FOLLOW-UP OF AGED CATS FED DIFFERENT SODIUM CONTENT DIETS

Brice S. Reynolds, DVM, PhD
Toulouse, FRANCE

INTRODUCTION

Sodium levels in cat food have been a controversial issue in veterinary science. The main concerns relate to the safety of chronic high dietary sodium intake. They originate mostly from still debated human health counterpart emphasized by associated public health prevention and educational campaigns. Indeed, increased dietary sodium intake in humans is held responsible for increased cardiovascular and renal risks and sodium restriction is also recommended to prevent calcium urolithiasis and osteoporosis in patients at risk for these conditions. Until not long ago, the implications of high-sodium content diets, a nutritional strategy proposed in the management of urolithiasis, for cat health remained disputable. However, recent scientific reports have shed new light on the debate.

BACKGROUND

Knowledge of some key points on sodium physiology, balance, dietary requirements and main nutritional practices is a prerequisite to a good understanding of the issue of sodium content in feline nutrition.

Body functions and homeostasis of sodium

Sodium is one of the major electrolytes of extra-cellular fluid (ECF) and is generally restricted to this body fluid compartment. Thus, ECF volume varies in line with total body sodium content.[1] There is very little physiologic regulation of gastrointestinal intake of sodium. Therefore, in normal conditions, kidneys are the cornerstone of sodium homeostasis and basically match urinary excretion of sodium to dietary sodium intake. The kidneys play a central role in regulation of ECF volume through sodium and water excretion. Sodium homeostasis is mainly driven by 2 opposite forces: pressure natriuresis and activation of the renin-angiotensin-aldosterone system (RAAS). Both are interrelated to blood pressure regulation which also involves brain, heart and vasculature under the complex influence of neural, hormonal (systemic) and intrinsic (local) factors.

Sodium dietary requirements in cats

Sodium requirement were determined to be 10.6 mg/kg body-weight (BW) or 0.15 g Na/Mcal of metabolizable energy (ME) in adult cats, based on plasma aldosterone concentration.[2] Accordingly, the National Research Council (NRC)[3] and European Pet Food Industry Federation (FEDIAF)[4] established recommended sodium minimum contents of 0.17 and 0.19 g Na/Mcal ME, respectively, in food for adult cats. Sodium requirement of healthy older cats is not likely to be different from those of young adults,[5] and mineral absorption in general does not seem to decline with increasing age.[6]

Signs of sodium deficiency have been observed only in kittens (12-15 weeks of age) fed 1.6 mg Na/kg BW/day and were anorexia, impaired growth, polyuria, polydipsia, hemoconcentration, reduced urinary specific gravity and elevated plasma aldosterone concentrations.[7]

Conversely, healthy adult cats appear to tolerate dietary sodium levels far above their needs as long as drinking water is provided ad libitum. Accordingly, a safe upper limit for sodium content in adult cat food has not been determined to date and is reported to be >3.8 g Na/Mcal ME,[3] based on anecdotal reports.[8]

Sodium content in feline commercial diets

The range of sodium content in most popular prescription commercial cat foods in Europe, including diets claiming reduced or increased sodium content (e.g. intended for cats with chronic kidney disease or urolithiasis), is 0.4-3.4 g Na/Mcal ME.[9]

Sodium restriction has been historically advocated for cats in some disease states (mainly cardiovascular and kidney diseases). This was essentially based on studies performed in other species or derived from human medicine recommendations. No study has confirmed to date the benefit of such dietary intervention in cats.

On the other hand, sodium supplementation has gained attention from some pet food manufacturers as a means to stimulate water consumption and enhance diuresis. Urine dilution is indeed recommended as part of treatment or prevention strategies for feline lower urinary tract disease (FLUTD). In that context, expected and potential adverse effects of sodium supplementation in cats have been more thoroughly addressed in the past years.

EVIDENCES ON THE EFFECTS OF DIETARY SODIUM INTAKE FROM MISCELLANEOUS STUDIES

Several studies of varying durations performed in cats of various age and characteristics have addressed the benefits and potential risks of increased dietary sodium intake in that species.

Effects on urine composition

Increasing water intake promotes increased urine flow and dilution. Both are considered beneficial in the management of FLUTD. This can be achieved notably through canned feeding, but cats or owners may be reluctant to transition from a dry to an exclusively canned food diet.[10] Several studies have confirmed that increasing the sodium content of a dry diet was a reliable mean to increase water intake as well as urine volume and/or dilution.[11-14] This may be of particular interest in the management of urolithiasis which is a common condition in cats.

Actually, complete dissolution of naturally occurring struvite uroliths was obtained within 10-42 days in 9 cats fed a commercially available dry diet containing 3.4 g Na/Mcal ME and inducing a relative supersaturation index (RSSI) for struvite <1.[15]

In the same time, higher sodium intake has been associated with higher calciuria in cats.[14, 16] This has raised concerns on possible increased risk of Calcium Oxalate (CaOx) stone formation in cats fed high sodium content diets. However, several studies have shown no change or a decrease in CaOx RSSI, as an indicator of urine crystallization potential, when dietary sodium intake was increased, presumed to be a result of concomitant sodium induced dilution effect on calcium, oxalate or both urine solutes concentrations.[12, 13, 16-18]

Effects on blood pressure

No correlation was found in several studies of various (up to 6 months) durations between dietary sodium levels and blood pressure in healthy cats of various ages[14, 17, 19] and in cats with experimentally reduced renal function.[20] There is no evidence that blood pressure or systemic hypertension are salt sensitive in cats. Accordingly, the current recommendation for hypertensive cats is to avoid high dietary salt intake without making a specific effort to restrict it.[21] Insight into physiopathogenic mechanisms of hypertension does rather suggest that, in this species, increased vascular tone may be of particular importance and relative hyperaldosteronism could be one contributor.[1]

Effects on cardiac structure/function

One study has examined the effects of sodium dietary intake on cardiac functional and anatomical parameters assessed by conventional echocardiography in 30 cats with various physiologic status: 10 healthy adults (2-6 years), 10 obese (>30% body fat), 10 aged (>10 years) and 6 cats with serum creatinine concentration >1.5 mg/dL.[14] In that study, cats were successively fed for 3 months with a diet containing either 0.9 or 2.9 g Na/ Mcal ME in a cross-over experimental design. The only difference observed was a slightly higher shortening fraction (50 vs 47%) measured at the end of the high-sodium period. All assessed parameters remained within the normal ranges for healthy cats.

Effects on kidney function

In the aforementioned study[14], higher slopes of initial versus final serum creatinine, urea nitrogen and phosphate concentrations were observed for the high-sodium period. This led to the claim that the highest sodium intake was associated with a progressive decline in renal function. However, in all other studies performed to date in healthy cats and in cats with experimentally reduced renal mass, no renal adverse effect of high dietary sodium intake could be detected.[17, 20, 22] In one study, 24 middle-aged and older cats, including a subset of 9 cats with serum creatinine concentration >1.5 mg/dL, were assessed over 6 months in a parallel group experimental design. One group was fed a diet containing 2.9 g Na/Mcal ME and the other one a diet containing 1.4 Na/Mcal ME. Serum creatinine, urea nitrogen and phosphate concentrations as well as urinary protein creatinine ratio (UPC) were unaffected by sodium intake throughout the study period.[17] Moreover, when renal function was directly assessed through glomerular filtration rate (GFR) measurement, in healthy cats fed different levels of sodium, it was not negatively altered by a sodium content of 3.2 vs 0.5 g Na/Mcal ME for 4 to 6 months.[22] Similarly, when diets containing 0.7, 1.4 and 2.7 g Na/Mcal ME were fed for 1 week, the GFR was significantly higher with the highest sodium intake in healthy cats and no difference in GFR according to the diet was observed in cats with experimentally reduced renal mass. In addition, RAAS activation and kaliuresis were observed with the lowest sodium intake and suppressed by sodium supplementation.[20]

Effects on bone mass

No difference according to dietary sodium intake in bone mineral content or bone mineral density assessed by dual-energy X-ray absorptiometry was found in 2 studies of 3 and 6 months durations.[14, 17]

LONG TERM FOLLOW-UP OF AGED CATS FED 2 DIFFERENT SODIUM CONTENT DIETS

A 2-year prospective randomized, blinded, and controlled study in aged cats was recently performed to overcome the discrepancy in previous conclusions regarding renal and cardiovascular safety of high dietary sodium intake in cats. Results of this study were the object of 2 separate reports.[23, 24]

Materials and methods

Twenty healthy cats from a research colony, 10 males and 10 females, aged 10.1±2.4 years in mean, were enrolled in the study. Cats were declared healthy according to medical history, physical examination, BP measurement (Doppler method), routine urine (including UPC) and blood (including total thyroxin) analyses, kidney ultrasonography, standard echocardiography and conventional Doppler examination at baseline. In addition, cats were subjected to additional assessments including cardiac Tissue Doppler Imaging (TDI) and GFR measurement (Plasma exogenous creatinine clearance test).

To ensure that the cats in each diet group were well matched with regard to renal function and cardiac structure/function (i.e. to control for the potential confounding factor of cardiac functional alteration on renal function) cats were first stratified in 3 different subsets according to TDI examination results. Cats were then ranked according to their GFR and paired in each subset separately. In each pair of cats, the first was randomly allocated to one diet group and the second was assigned to the other diet group.

Cats were fed either a high-sodium maintenance dry expanded diet (HSD) with a sodium content of 3.1 g/Mcal ME or a control diet (CD) of exactly the same composition except for the level of sodium (1.0 g/Mcal), according to the group, and monitored over 2 years (i.e. all baseline measurements were repeatedly performed over this period).

Results

Sixteen cats completed the 2 year study. Among the 64 variables of interest assessed, the only 4 statistically affected by dietary sodium intake were plasma aldosterone and urinary sodium/creatinine ratio, respectively higher and lower in the CD group all over the study period, urinary specific gravity and the radial early on late diastolic velocity ratio assessed in the endocardium of the left ventricular free wall, lower in the HSD group at 3 and 12 months, respectively.

Conclusion

This prospective study confirms, over a longer period and in older cats, likely to be at higher cardiovascular and renal risk than younger healthy cats, prevailing evidence from previous reports indicating that increased sodium intake does not appear to induce deleterious effects on blood pressure and renal function.

In addition, it is the first in which the long-term effects of dietary sodium intake on cardiac structure/function were thoroughly assessed using, standard 2-dimensional and M-mode transthoracic echocardiography, conventional Doppler examination, as well as two-dimensional color tissue Doppler imaging, demonstrating that cats do not seem to be sensitive to the detrimental cardiovascular effects of high dietary sodium intake described in other species.

Further studies are needed to confirm this lack of sodium-sensitivity in larger and other populations of cats, and to understand the underlying mechanisms of this potential species specificity.

REFERENCES

[1]Syme HM. **Vet Clin Small Anim** 2011;41:63. [2]Yu S, Morris JG. **J Nutr** 1999;129:419. [3]NRC. 2006 **Requirements of Dogs and Cats**; 145. [4]FEDIAF. 2014 **Nutritional Guidelines for Complete and Complementary Pet Food for Cats and Dogs**. [5]Hand MS. 2010 **Small Animal Clinical Nutrition**, 5th Edition. [6]Teshima E et al. **J Am Physiol Anim Nutr** 2010;94:e251. [7]Yu S, Morris JG. **J Nutr** 1997;127:494. [8]Burger I. **Pedigree Digest** 1979;6:10. [9]Nguyen et al., **J Anim Physiol Anim Nutr**. Under Review. [10]Kirk CA. ACVIM forum, 2002. [11]Burger IH et al. 1980 **Nutrition of the cat and dog**; 145. [12]Hawthorne AJ, Markwell PJ. **J Nutr** 2004;134:2128S. [13]Xu H et al. **J Vet Intern Med** 2006;20:738 [abstract]. [14]Kirk CA et al. **Vet Ther** 2006;7:333. [15]Houston DM et al. **Br J Nutr** 2011;106:S90. [16]Biourge V et al. **J Vet Intern Med** 2001;15:301 [abstract]. [17]Xu H et al. **J Feline Med Surg** 2009;11:435. [18]Paßlack N et al. **Vet J** 2014:201;401. [19]Luckschander N et al. **J Vet Intern Med** 2004;18:463. [20]Buranakarl C et al. **Am J Vet Res** 2004;65:620. [21]Brown S et al. **J Vet Intern Med** 2007;21:542. [22]Cowgill LD et al. **J Vet Intern Med** 2007;21:600 [abstract]. [23]Reynolds BS et al. **J Vet Intern Med** 2013;27:507. [24]Chetboul V et al. PLoS One. 2014 Jun 18;9(6):e97862.

DOG DIGESTIVE SENSITIVITY ACCORDING TO SIZE: A SUMMARY OF 16-YEAR RESEARCH

Mickaël, P. Weber, PhD
Aimargues, France

INTRODUCTION

From the Chihuahua under 1 kg of body weight (BW) up to giant breed dogs, over 90 kg BW, dogs come in all sizes andshapes. If morphologic differences are obvious, canine breeds also present physiological differences that may influence their digestion process. Different authors[1,2] have indeed pointed out the fact that fed the same diet, large and giant breed dogs (LB-dogs) present an increased frequency of soft feces than do small breed dogs (SMA-dogs). This synthesis first aims at identifying the reasons that could explain this difference in digestive sensitivity. In a second part, specific nutritional recommendations for LB-dogs will be proposed, in order to improve their fecal quality.

PART 1 - INFLUENCE OF BODY SIZE ON DOG DIGESTIVE FUNCTION

Different hypotheses can be proposed to explain the high digestive sensitivity observed in LB-dogs. A majority of these hypotheses have been screened in different successive studies performed at the National Veterinary School of Nantes (France), between 1998 and 2014. Four breeds of six healthy female dogs were used in this work. SMA-dogs were represented by Miniature Poodles (MP) and Standard Schnauzers (SS) whereas Giant Schnauzers (GS) and Great Danes (GD) represented LB-dogs. All those dogs were fed the same dry expanded diet in the same conditions of environment.

Influence of body size on stool quality

Results obtained confirmed the predisposition of LB-dogs to poorer fecal consistency and higher fecal water content than SMA-dogs. A strong positive correlation was observed between moisture content and the dogs' weight ($r=0.89$; $p<0.0001$).[3]

Influence of body size on digestive efficiency

It was first supposed that LB-dogs digestive efficiency could be lower than in SMA-dogs: effectively, if intestinal absorption of food in reduced in the small intestine, it induces a larger flow of by-products entering the colon and, due to high fermentative activity and osmotic gradient, a larger quantity of water is left in the feces. The global digestive capacity of SMA and LB-dogs was evaluated by measuring the total apparent digestibility of the dry diet tested.[3] Their intestinal absorption capacity was assessed by measuring the urinary excretion of two sugars (D-xylose and 3-O-methyl-D-glucose) actively absorbed by the small intestine.[4]

Contrary to our expectation, results showed that digestibility of dry matter, organic matter, crude protein and gross energy was higher for LB-dogs than for SMA-dogs. Furthermore, ratio of urinary xylose to 3-O-methyl-D-glucose was similar in all four breeds, suggesting that transport of these sugars was comparable among SMA and LB-dogs. These results suggest that lower fecal quality observed in healthy LB-dogs does not appear to be related to lower nutrient utilization.

Influence of body size on upper gastrointestinal transit time

Many studies have reported a strong relationship between gastrointestinal transit time and poor fecal consistency in human and animals.[5,6] It could also be hypothesized that differences in gastrointestinal transit time exist between SMA and LB-dogs. By decreasing time for digestion and absorption, a short gastric emptying time and/or small intestinal transit time could potentially increase the flux of fluids and nutrients entering the colon, which could affect colonic function and consequently stool quality.

Half gastric emptying time (GET 50), predictive total gastric emptying time (pTGET), small intestinal transit time (SITT) and oro-cecal transit time (OCTT) of food have been evaluated in SMA and LB-dogs by using radiopaque markers or sulfasalazine mixed with food.[7,8] However, no correlation of body size or BW with GET50, pTGET, SITT or OCTT was found. There results indicated a lack of relationship in healthy dogs between upper gastrointestinal tract transit time and fecal parameters.

Influence of body size on colonic and total transit time

The efficiency of minerals and water absorption is dependent, to a large extent, on colonic motility. A short colonic transit time may affect fecal quality by decreasing time for fluid and mineral absorption in colon. A reduction in large intestinal transit time (LITT) has already been reported in non-healthy dogs with loose and watery feces, inducing a decreased capacity for electrolyte and water absorption.[9]

Total transit time (TTT) determined by plastic pellets showed a positive correlation between body size and mean TTT.[10,11] Results also revealed significantly longer LITT in LB-dogs with times varying from 9.1±1.1 hours in MP to 39.4±1.6 hours for GS. Given the times observed, mean LITT accounted 39% of mean TTT for SMA-dogs and 70% for LB-dogs. These results suggest that time for water absorption in the large intestine is not the limiting factor behind the looser fecal consistency in LB-dogs, compared with SMA-dogs. Their prolonged LITT could however result in higher fermentative activity, affecting the stool quality.

Influence of body size on colonic fermentative activity

High fermentative activity results in an increase in bacterial biomass and considerable production of metabolites due to fermentation, essentially short-chain fatty acids (SCFA) and lactic acid. Longer transit time of contents in the large intestine promotes fermentation[12] and the impact of colonic fermentation on fecal quality has been reported.[6]

To check whether fermentative activity is really higher in LB-dogs, colonic fermentative activity of SMA and LB-dogs was indirectly assessed by quantifying fermentative by-products in their stools[13]. Lactic acid and total short-chain fatty acid (SCFA) concentrations were significantly higher in LB-dogs. Although these molecules are strongly absorbed from the colon, many studies have associated increased fecal concentrations of those acids with high fermentative activity in dogs.[14,15] The large quantity of organic acids produced could thus exceed the absorption capacity of the colonic mucosa, thereby leading to their accumulation in the colonic lumen. These compounds, with their high osmotic power, would then increase intra-luminal osmotic pressure, thereby leading to high water secretion in the colon. This high fermentative activity in LB-dogs could explain why their apparent digestibility values are so high. Increased bacterial fermentation overestimates the real nutrient digestibility.

Influence of body size on colon size

Anatomical differences between canine breeds are not often evoked in the literature. However, previous work has pointed out important differences in the respective proportions of the different part of the digestive tract. Post-mortem examination of dogs has revealed obvious differences concerning the volume of the large intestine.[16] When GD dogs are compared with Miniature Spitz, the respective surface and volume of the colon were respectively 15 and 47 higher in GD. Barium radiographs performed in the four breeds of dogs of the study have confirmed these important variations in the colorectal proportions. This larger colon size could explain the longer LITT observed in LB-dogs.

Conclusion – Part 1

It seems that the production of soft stools in LB-dogs might be explained by both anatomic and physiologic differences, influencing the water absorption process and/or colonic fermentation. LB-dogs present a highly developed large intestine. These characteristics, associated with a longer LITT, suggest a greater fermentative activity in LB-dogs. This hypothesis is confirmed by the more important production of lactic acid and SCFA in the feces of LB-dogs. All together, these elements could be a possible cause of their poor fecal quality observed. This effect would be reinforced by the fact that strong intestinal permeability and reduced sodium absorption have been clearly shown in LB-dogs.[4,13,17] These phenomena are obviously linked to the fluid absorption process.

PART 2 – NUTRITIONAL RECOMMENDATIONS FOR LARGE BREED-DOGS

If intense colonic fermentation is one of the keys to explain why LB-dogs present a poorer fecal quality compared to smaller ones, it is necessary to pay special attention to the composition and the formulation of their diet. A nutritional strategy to improve the quality of feces in LB-dogs could effectively consist of reducing the quantity of undigested nutrients reaching the colon, and so, minimizing fermentation. This second part screens the nutrients that can influence colonic fermentation and consequently, stool quality. It also presents nutritional recommendations, which could improve fecal quality.

Influence of protein sources and contents

Proteins constitute a large proportion of a carnivorous diet and their ileal digestibility strongly varies according to the quality of their source. Undigested proteins arriving in the large intestine generate major reactions of colonic putrefactions, which can affect digestive tolerance, as well as thedog's health.

Nery et al.[18,19] studied the effect of protein sources and contents on nutrient digestibility, and fecal characteristics in SMA and LB-dogs (German Shepherds): four dry-extruded, nutritionally complete diets, varying in protein sources and concentrations were tested. Poultry meal (PM) and wheat gluten (WG) were used as protein sources: PM is the most common protein source used in dog food and WG is one of the most digestible sources of proteins available (85% and 99% *in vitro* ileal digestibility level respectively). Two protein concentrations were also tested: low [LP], 22% and high [HP], 39% CP, on a DM basis. Both these levels are commonly observed in commercial maintenance diets.

Compared with PM, WG was superior in terms of fecal consistency and fecal moisture in all dogs. The lowest fecal moisture content was observed for WG diets. Dry matter and crude protein digestibility were significantly greater for WG than for PM diets, considering all groups of dogs ($p<0.001$). This beneficial effect of WG on fecal quality is likely due to the reduced flow of undigested proteins entering the colon, resulting in lower protein fermentation. Protein concentration in diet also affected fecal score and moisture. The PM-HP diet resulted in the greatest fecal moisture content, regardless of dog size. On the contrary, the LP diets led to improved fecal scores. Given the results obtained, an increased protein intake can increase the quantity of substrate available for colonic fermentation, and consequently lead to a poorer fecal consistency. This additionally indicates that, within the range observed in commercially available diets for normal healthy dogs, protein concentration has significant effect on fecal quality. Overall, during diet formulation, it is therefore relevant to consider not only protein quality, but also protein quantity to improve fecal quality.

Influence of starch forms and sources

In extruded canine diets, starch represents between 20 and 50% of the formula. It represents one of the main sources of energy for animals and is necessary to get the specific honeycomb structure of kibbles. However, its digestibility can strongly influence stool quality.

A previous study evaluated the influence of starch forms (purified starches *vs.* flours) and sources (corn *vs.* rice *vs.* wheat) in SMA-dogs (Miniature Schnauzers) and LB-dogs (German Shepherds) on nutrient digestibility, faecal characteristics and fermentation products.[20] When using flours in dog food, the influence of the source was profoundand resulted in large variations in stool quality. Maize and rice appear to give the best tolerance, whereas wheat should be used in moderate quantity in LB-dogs. Purified starches led to stools with better consistency, lower moisture and lower concentration of fermentative end-products whatever their source. Purified starch could be particularly interesting in LB-dogs. Its use is however not recommended in animals with constipation signs or animals predisposed to hard and dry feces, especially in small dogs.

Influence of fiber sources

Dietary fiber remains undigested along the small intestine and, therefore, it arrives virtually intact in the large intestine where it is either heavily fermented by the colonic microflora (fermentable fiber), or very slightly decomposed (non-fermentable fiber). Considering the different characteristics of fermentable and non-fermentable fiber, an optimal ratio in dog's food appears essential to maintain a good colonic ecosystem and a good quality of the stools. Due to the high fermentative activity of LB-dogs, this fermentable and non-fermentable fiber ratio has to be different in diets designed for SMA and LB-dogs.

In LB-dogs, stool quality is improved with the incorporation of non-fermentable fiber (i.e. cellulose). Fermentable fiber (*i.e.*: beet pulp and especially fructo-oligosaccharides) must however be strictly limited. On the contrary, SMA-dogs are tolerant toward the inclusion of fermentable fiber. but cellulose in excess can induce constipation. For these dogs, fiber issued from cereals is a more suitable source of non-fermentable fiber than the use of cellulose.

CONCLUSION

This synthesis confirms that LB-dogs present digestive characteristics that must be taken into account in their daily diet. Prolonged colonic transit and large colon size compared with smaller dogs are elements that promote colonic fermentation and soft stools. When choosing a particular diet in order to improve their stool quality, several aspects of the formulation must be taken into account. The general goal is to avoid any ingredient that could increase the level of fermentable undigested residues and, in fine, exacerbate colonic fermentation.

REFERENCES

[1]Zentek J, et al..**J Small Anim Pract** 1995;36:354. [2]Meyer H, et al. **Zentralbl Veterinarmed** 1999;46:155. [3]Weber M, et al. **J Anim Physiol Animal Nutr** 2003;87:1. [4]Weber M, et al. **Am J Vet Res** 2002;63:1323. [5]Jian R, et al..**Gut** 1984;25:728. [6]Lewis L, et al. **J Nutr** 1994;124:2716S. [7]Weber M, et al.. **Am J Vet Res** 2002;63:677. [8]Weber M, et al. **Am J Vet Res** 2003;64:1105. [9]Rolfe V, et al. **J Small Anim Pract** 2002;43:290.[10]Hernot D, et al. **J. Anim Physiol Anim Nutr** 2005;89:189. [11]Hernot D, et al. **Am J Vet Res** 2006;67:342. [12]MacFarlane S, et al. **Proc Nutr Soc** 2003;62:67. [13]Weber M, et al. **J Anim Physiol Anim Nutr** 2004;88:356. [14]Schünemann V, et al. **Adv Anim Physiol Anim Nutr** 1989;19:44. [15]Zentek J, **J Anim Physiol Anim Nutr** 1996; 76:36. [16]Hernot D, et al. **European College of Veterinary Internal Medecine** 2003 (Uppsala, Sueden). [17]Hernot D, et al. **J Anim Physiol Anim Nutr** 2009;93:703. [18]Nery J, et al. **J Anim Sci** 2010;88:159. [19]Nery J, et al. **J Anim Sci** 2012:90:2570. [20]Goudez R, et al. **The American Academy of Veterinary Nutrition** 2012 (New Orleans, US).

MASTITIS: EVIDENCE-BASED TREATMENT

Sarah A. Wagner, DVM, PhD, DACVCP
Fargo, ND

INTRODUCTION

Veterinarians are often consulted about mastitis treatment decisions for individual animals or on a herd basis. The purpose of this article and presentation is to review evidence for antimicrobial treatment of clinical mastitis in dairy cows with drugs that are approved for such use in the United States. Evidence for the treatment of dairy cows with antimicrobial drugs at the time of drying off is beyond the scope of this presentation. A comprehensive review of evidence for the treatment of both lactating and dry cows with intramammary antimicrobial drugs approved for use in the United States has recently been published. (Royster and Wagner, 2015)

CASE DEFINITIONS, CURES, AND CAUSES OF MASTITIS

Evaluations and comparisons of treatment efficacy in mastitis are complicated by the variability in clinical manifestation of the disease and the different treatment outcomes that are desirable and measurable. Mastitis may decrease milk production and milk quality even in the absence of clinical signs (Halasa et al, 2007); such cases of mastitis are referred to as subclinical and can be diagnosed by the presence of elevated cell numbers (Somatic Cell Count, SCC) in the milk. Clinical mastitis is characterized by the presence of abnormal milk with or without observable inflammation of the udder or systemic clinical signs. Comparison and evaluation of clinical trials of mastitis may be challenging as there is no standard definition of a case or a cure for mastitis. A case of mastitis may be enrolled based on clinical signs, elevated SCC, positive microbial culture of milk (once or more than once), or some combination of these signs. Likewise, treatment success may be defined as improvement in SCC, improvement of clinical signs, and/or negative milk culture, once or more than once, or some combination of these. Some studies enroll only one affected quarter per cow, while others enroll multiple quarters from a single cow, yet evaluate the treatment effects on each quarter independently.

Clinical or subclinical mastitis is seldom due to causes other than bacterial infection. Most pathogens that commonly cause mastitis may be detected using relatively simple methods for microbial culture of milk from affected cows; however, pathogens may shed variably and it is common for ≥ 25% of milk samples from cows with clinical mastitis to yield no growth on microbial culture. (Ruegg, 2012) In addition, mastitis may be caused by *Mycoplasma* species organisms, which require specialized growth media and incubation conditions for successful culture.

COW FACTORS IN THERAPEUTIC DECISION-MAKING

Prior to initiating treatment of a case of mastitis it is also important to consider factors that may influence the prognosis and whether the cow's value to the herd justifies the expense of mastitis treatment. Factors include:

- Somatic Cell Count
- Production Level
- Reproductive Status
- Chronicity of condition
- Number of quarters affected
- Causative pathogen (if known)
- Age of cow
- Presence of other illnesses or issues

Treatment of older cow without high production, efficient reproduction, and otherwise good health, may not be economically justifiable. If a cow is close to the end of her lactation, it may benefit prognosis and decrease cost to wait and treat the cow at drying off. Some dairies will discontinue milking a single quarter if it has a chronic or recurrent case of mastitis, saving the money that would be spent on repeated treatments and avoiding the loss in milk quality that results from adding high SCC milk to the herd bulk milk tank. Finally, it is important to consider that chronicity or recurrence of mastitis and concurrent disease may decrease the likelihood of successful treatment, and that some pathogens are poorly responsive to treatment with antimicrobial drugs.

TREATMENT OF CLINICAL MASTITIS: EVIDENCE FOR INTRAMAMMARY (IMM) DRUGS APPROVED IN THE UNITED STATES

The clinical trials presented here do not necessarily include ALL clinical trials published for each drug. Unless it is clearly stated that no published trials are unmentioned, some trials may not be mentioned here. Due to space constraints, only the trials that are most applicable, in the author's judgment, are presented here. Clinical trials performed as part of the drug approval process are not described here, as it can be safely assumed that such trials demonstrated some benefit to treatment.

Penicillins: Amoxicillin, Hetacillin, and Cloxacillin

The spectrum of activity of amoxicillin, barring the presence of antimicrobial drug-resistant populations, is considered to include Gram-positive bacteria (that do not produce beta-lactamase) and the Enterobacteriaceae. A recent literature search did not discover any prospective clinical trials treating naturally-occurring clinical mastitis with IMM amoxicillin published within the last 20 years. Older studies had limited statistical power.

Hetacillin is rapidly transformed to ampicillin after administration and is expected to have the same spectrum of activity as amoxicillin. As with amoxicillin, there is a lack of published research about the efficacy of IMM hetacillin for the treatment of mastitis.

Cloxacillin is a penicillinase-resistant penicillin with reduced activity against Gram-negative pathogens when compared to amoxicillin and ampicillin. Published prospective clinical trial results for IMM treatment of naturally-occurring mastitis with cloxacillin could not be found.

Cephalosporins: Cephapirin, Ceftiofur

Cephapirin is a 1st generation cephalosporin, with a spectrum of activity that includes the Enterobacteriaceae and most gram-positive bacteria. It is considered to be effective against Gram-positive bacteria that produce beta-lactamase. Cephapirin was included in a clinical trial designed to determine the non-inferiority of the cephapirin to ceftiofur, a 3rd generation cephalosporin, for the treatment of mild to moderate clinical mastitis (Schukken et al, 2013). Cephapirin was administered IMM twice at a 12 hour interval, while ceftiofur was administered IMM 5 times at 24 hour intervals. For mastitis caused by gram-positive pathogens, cephapirin had comparable efficacy to ceftiofur for the resolution of clinical signs (65 % and 62% clinical cures, respectively), and for microbiological cure (68 % and 67%, respectively) but for Gram-negative pathogens, ceftiofur was found to have a higher microbiological cure rate than cephapirin (79% for ceftiofur versus, 50% for cephapirin). There were no untreated controls in that study.

An earlier study (Rosenberg et al, 2002) that did include untreated controls found that microbiological cures for subclinical mastitis were higher for cephaprin than for untreated controls (53% for cephapirin versus 30% for untreated controls). Based on the data from that study, one would have to treat 4.3 cows to produce one additional cure beyond the number of cures observed to occur spontaneously.

Ceftiofur, a 3rd generation cephalosporin, is considered to have more activity against Gram-negative pathogens than 1st generation cephalosporins such as cephapirin, but less activity against Gram-positive bacterial species. It is the only IMM antimicrobial drug for which the FDA-approved indication includes a Gram neagative pathogen (*Escherichia coli*) as well as Gram-positive pathogens. Ceftiofur has an unusually flexible approved dosing regimen; the approved dose of 125 mg may be administered at 24 hour intervals between 2 and 8 times. A study comparing untreated controls to cows treated using 2-, 5- and 8- day regimens was published in 2004 (Oliver et al), but this study evaluated subclinical cases of mastitis, not clinical cases. In that study, all ceftiofur treated cows had significantly higher rates of microbiological cure than untreated controls (39%, 54%, and 66% for 2-, 5-, and 8 day treated cows, 11% for untreated controls), and cows treated with an 8 day regimen had a significantly higher microbiological cure rate that cows treated for 2 days. Based on these data, one would have to treat more than 3 cows in order to obtain an additional cure when comparing 2 doses to no treatment or 8 doses to 2 doses ((3.6 and 3.7, respectively), and about 2 additional cases (1.8) to achieve an additional cure when comparing 8 doses to the untreated controls. How these outcomes relate outcomes that might be expected when treating clinical cases is not known.

Macrolide: Pirlimycin

Pirlimycin is expected to be active against Gram-positive pathogens. Like ceftiofur, the product has a flexible approved dosing regimen. One tube containing 50 mg of pirlimycin may be administered IMM every 24 hours for 2 to 8 total doses. Results of clinical trials of pirlimycin treatment of mastitis have been published in the last 15 years; these trials were focused on the treatment of subclinical infections and enrolled more than one affected quarter per cow.

CONCLUSIONS

Published evidence about the efficacy of IMM drugs for the treatment of clinical mastitis in dairy cows, based on prospective, randomized studies involving naturally-occurring disease, is scant, especially when compared with the rate of publication of clinical trial outcomes for bovine respiratory disease. Clinical trials for the treatment of naturally-occurring clinical mastitis are expensive and time-consuming, and complicated by the lack of standard definitions of mastitis cases and cures and the wide variety of pathogens that may cause the disease.

The absence of substantial high-quality published evidence about antimicrobial drug treatment for clinical mastitis leaves a void for those who desire to practice evidence-based veterinary medicine. The best approach may be to create one's own evidence by keeping records of treatment outcomes on the farm and selecting the antimicrobial drugs that have the highest rate of cure among comparable cases of mastitis. In addition, microbial culture of milk from cases of mastitis can help with selection of drugs with appropriate spectra of activity, and may reduce the frequency of treating cases in which treatment is unnecessary or futile. Research published within the last 5 years has presented trials evaluating selective, culture-based mastitis treatment, in which IMM drug treatment was reserved for mastitis with Gram-positive or mixed infection confirmed by milk culture (Lago et al, 2011). Outcomes of culture-based selective treatment were found to be to be similar to outcomes of when all clinical cases of mastitis were treated using an IMM antimicrobial drug.

An excellent source of information about the characteristics of various mastitis pathogens, along with much general information about the control of mastitis on dairy farms, is the website of the NMC (formerly National Mastitis Council) at www.nmconline.org.

REFERENCES

Royster, E. and Wagner, S. Vet Clin North Am: Food An Prac, in press, March, 2015. Halasa, T., et al, Vet Quarterly 2007; 29(1): 18. Ruegg, P.L. Vet Clin North Am: Food An Prac 2012;28(2):149. Schukken Y.H., et al, J Dairy Sci 2013;96(10):6763.Rosenberg J.B., et al, Vet Ther 2002;3(4):381. Oliver S.P., et al J Dairy Sci 2004;87(8):2393. Lago, A. et al, J Dairy Sci, 2011;94(9):4441.

FELINE DRUG METABOLISM AND DISPOSITION: DIFFERENCES BETWEEN CATS AND DOGS AND OTHER SPECIES

Michael H. Court, BVSc, PhD, DACVA
Pullman, WA

Although it is widely appreciated that cats respond differently to certain drugs when compared with other companion animal species, the causes of these differences are poorly understood. This lecture will critically evaluate evidence for altered drug effects in cats, focusing on pharmacokinetic differences between cats, dogs and humans, and the molecular mechanisms underlying these differences.

A review of published pharmacokinetic studies[1] indicate that acetaminophen, propofol, carprofen, and acetylsalicylic acid (aspirin) are cleared significantly more slowly in cats versus dogs and humans (Figure 1). All of these drugs are metabolized by conjugation. Cats lack the major phenol UDP-glucuronosyltransferase (UGT) enzymes[2,3], including UGT1A6 and UGT1A9, that glucuronidate acetaminophen and propofol (Figure 2a). Deficient glucuronidation may also explain slower carprofen clearance, although there is no direct evidence for this. Table 1 lists drugs with direct evidence that they are glucuronidated more slowly, or with similar efficiency compared with most other mammalian species. Although pharmacology textbooks commonly state that sensitivity of cats to the effects of aspirin (and related salicylates) is a consequence of poor glucuronidation by cats, there is no evidence to support this. In fact, the main pathway for elimination of aspirin (and other salicylates) in most species is glycine conjugation of the salicylate to form salicylurate that is excreted in urine[4]. As shown in Figure 2b, cats readily excrete salicylate glucuronides in the urine but poorly eliminate salicylurate suggesting a deficiency in glycine conjugation[5]. Cats are also deficient in several other conjugation enzymes, including N-acetyltransferase (NAT) 2 and thiopurine methyltransferase (TMPT)[6,7]. NAT2 deficiency may be the reason cats are more prone to developing methemoglobinemia rather than hepatotoxicity from acetaminophen[8]. TMPT deficiency may predispose cats to azathioprine toxicity. No evidence was found for slower elimination of drugs cleared by oxidation or unchanged into urine or bile. Piroxicam, an oxidized drug, was cleared much more rapidly in cats than humans and dogs, although the mechanism for this difference is unclear.

In conclusion, cats are deficient in a number of drug conjugation pathways that can lead to relatively slow elimination of certain drugs, and the need for dose adjustment or alternative therapies to avoid serious adverse effects. More work is needed to better understand drug metabolism and disposition differences in cats, thereby enabling more rational prescribing of existing medications, and the development of safer drugs for this species.

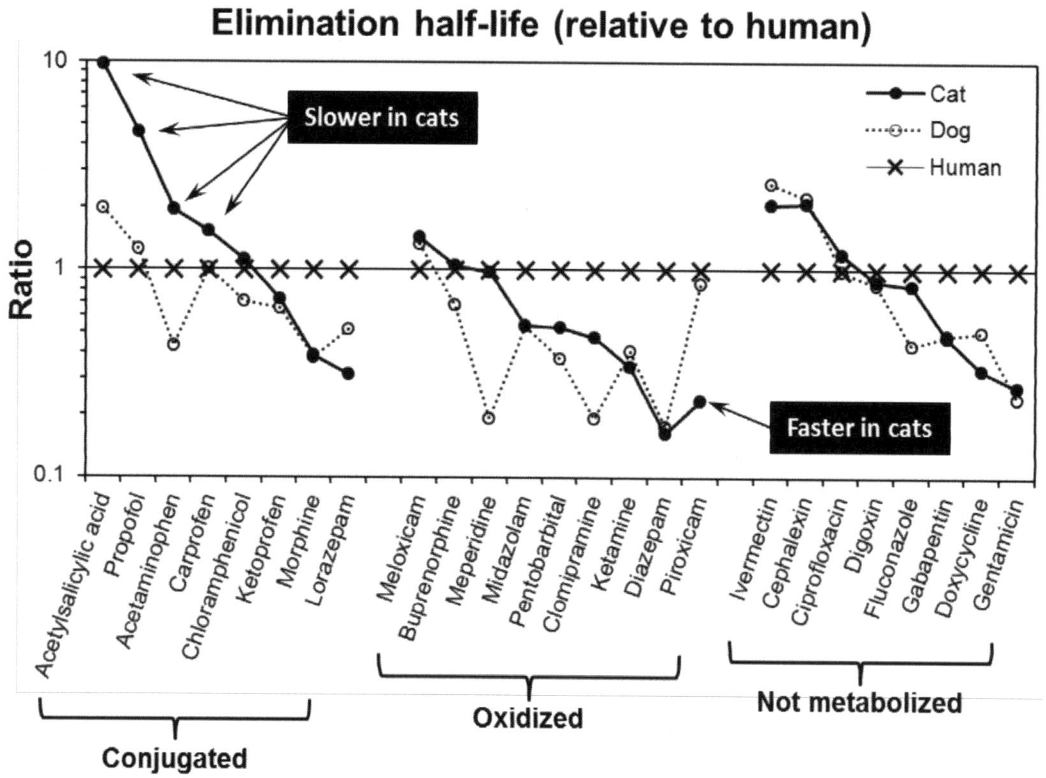

Figure 1. Evidence for differences in drug elimination rates between cats, dogs, and humans. Shown is a comparison of published elimination half-life values in cats (filled circle), dogs (open circle) and humans (X symbol) for representative drugs that are primarily eliminated by conjugation (glucuronidation, sulfation, and glycination), oxidation (CYP enzymes) or primarily excreted unchanged into urine and/or bile. All values are expressed as a ratio of the human value. Details of studies providing these data are provided elsewhere[1].

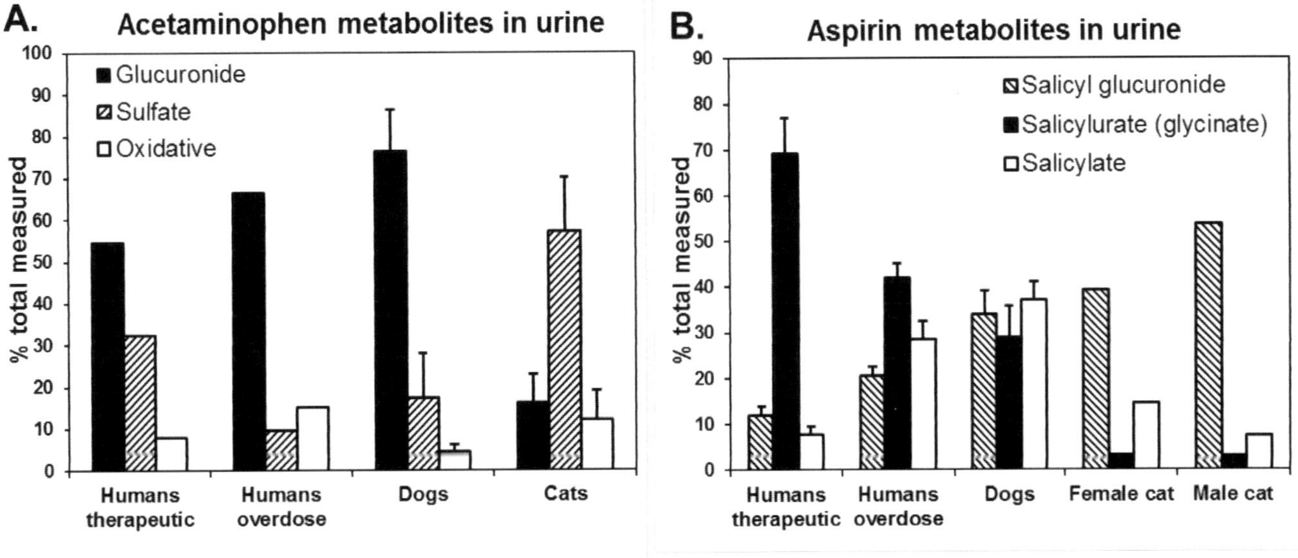

Figure 2. Comparison of acetaminophen and aspirin metabolite concentrations (expressed as a percentage of total) in cats, dogs and humans. A. Cats are sensitive to the toxic effects of acetaminophen, in part because they glucuronidate acetaminophen less efficiently than humans or dogs. B. In contrast, cats can readily glucuronidate salicylate, but they poorly conjugate salicylate with glycine (forming salicylurate). Details regarding these data are provided elsewhere[1].

Table 1. Drugs with direct evidence that they are glucuronidated more slowly (left column) or with similar efficiency (right column) in cats when compared to other mammalian species. Details of data used to support these conclusions are provided elsewhere[1].

Drugs glucuronidated slowly in cats	Drugs glucuronidated efficiently in cats
Acetaminophen	Flurbiprofen
Chloramphenicol	Ibuprofen
Clofibrate	Lorazepam
Morphine	Phenolphthalein
Orbifloxacin	Pirprofen
Valproate	Salicylate
	Telmisartan

REFERENCES

[1]Court MH. Vet Clin North Am Small Anim Pract. 2013 Sep;43(5):1039-54.
[2]Court MH, et al. Pharmacogenetics 2000;10(4):355.
[3]Court MH, et al. Biochem Pharmacol 1997;53(7):1041.
[4]Chen Y, et al Pharmacogenet Genomics 2007;17(8):571-579.
[5]Davis LE, et al. Am J Vet Res 1972;33(6):1253-1262.
[6]Trepanier LA, et al. Pharmacogenetics 1998;8(2):169-179.
[7]Salavaggione OE, et al. J Pharmacol Exp Ther 2004;308(2):617-626.
[8]McConkey SE, et al. J Vet Pharmacol Ther 2009;32(6):585-595.

EVIDENCE-BASED TREATMENTS FOR FELINE GASTROINTESTINAL DISEASE

Jody L. Gookin, DVM, PhD, DACVIM (SAIM)
Raleigh, NC

INTRODUCTION

Gastrointestinal disease is common in the cat. Despite this, there are surprisingly few studies that provide scientific evidence to support the efficacy of treatments commonly chosen for medical management of these patients. In this presentation we will discuss key mainstays of therapy in cats with chronic enteropathy for which published evidence for efficacy or non-efficacy can be found.

"Food-Responsive" Gastrointestinal Disease

A considerable number of diets are currently marketed for use in treatment of inflammatory bowel disease in cats (Table 1). Very few studies have been published that examine the impact of diet on clinical signs in cats with gastrointestinal disease. Despite this, there is compelling evidence that a large percentage of cats with chronic enteropathies will respond to dietary intervention alone. The first, and perhaps most compelling, of these studies is that of Guilford *et al* published in 2001[1]. In this study, 55 cats with "chronic idiopathic gastrointestinal problems", that underwent an extensive diagnostic effort to rule out underlying causes of gastrointestinal disease, were exclusively fed a chicken or venison-based novel protein diet (chosen based on historical lack of prior exposure) for a duration of 4 weeks. Clinical signs of gastrointestinal disease resolved in 49% (27/55) of the cats in the study. Moreover, clinical signs resolved in most cats within 2-3 days suggesting that an elimination trial as short as 1 week is likely sufficient to determine if cats have diet-responsive gastrointestinal disease. In approximately half of the food-responsive cats, a true "food allergy" was demonstrated based on recrudescence of signs when challenged with their original diet and resolution of signs when re-introduced to the elimination diet. The most common allergens identified were beef, wheat and corn gluten. No clinical or histological features were identified as able to distinguish between cats that did or did not respond to the novel protein dietary trial. In a more recent study by Jergens *et al*, 26% (6/23) of cats with chronic enteropathy responded to treatment with an intact protein elimination diet fed exclusively for a minimum of 14 days[2]. Compared to the use of novel protein diets, there has so far been no studies published examining the effectiveness of hydrolyzed protein diets in the treatment of cats with gastrointestinal disease.

Diets marketed for the treatment of IBD in cats: listed in alphabetical order by company	Canned	Dry
Hill's Prescription Diet		
• z/d ULTRA*	✓	
• z/d® Low Sensitivity*		✓
• d/d® Skin Support (Duck and Venison varieties)	✓	✓
• i/d® Gastrointestinal Health	✓	✓
Iams Veterinary Formula		
• Skin and Coat Plus Response™ LB	✓	
• Intestinal Plus Low-Residue™	✓	✓
Purina Veterinary Diet		
• Hypoallergenic® HA*		✓
Royal Canin Veterinary Diet®		
• Gastrointestinal™ High Energy	✓	✓
• Gastrointestinal™ Moderate Calorie	✓	✓
• Hypoallergenic Hydrolyzed Protein Adult HP*		✓
• Selected Protein Adult PD	✓	✓
• Selected Protein Adult PR	✓	✓
• Selected Protein Adult PV	✓	✓

*Hydrolyzed protein

Table prepared by Dr. Jessica Harris, Nutrition Service, North Carolina State University

While the high response rate of cats to a novel protein diet is compelling other factors about the diet are clearly also influential. This would make sense based on the multifactorial potential causes of chronic enteropathy in cats and a multitude of influences of diet on gastrointestinal homeostasis, the microbiota, and immune responses. In a study by Laflamme *et al*, a diet formulated to simply incorporate factors thought to be of value in the management of diarrhea in cats (high protein, low carbohydrate, and moderate fat) was effective in resolving diarrhea in 7/15 (46.7%) cats[3]. Additional studies by Laflamme *et al* demonstrate that a highly digestible diet is effective regardless of the fat content[4]. In these studies, 55 pet cats with chronic diarrhea were fed a highly digestible diet with either low (23.8% calories) or high fat (45.1% calories) content using a double-blinded crossover design. There were no significant differences between treatment groups, with 78.2% of cats responding (36% with return of normal stools) beginning in the first week of treatment with maximal responses by the end of 3 weeks. The role of dietary carbohydrates in treatment of cats with diarrhea is less well studied. At least one study has demonstrated that cats with inflammatory bowel disease have evidence of carbohydrate malabsorption but appear to be highly tolerant to any adverse effect of carbohydrate source or amount on severity of diarrhea[5].

Cobalamin (B12) Supplementation

Cobalamin deficiency is recognized as common in cats with chronic enteropathy. This deficiency is largely attributed to malabsorption of cobalamin-intrinsic factor complexes by the ileum. In 2001, Simpson *et al* reported a 61.3% prevalence of low cobalamin in 80 cats with clinical signs suggestive of chronic enteropathy[6]. More recently, Kook *et al* (2012) published a large case series of 261 cats with gastrointestinal disease in which a low cobalamin was diagnosed in 41.4% of cats[7]. Clinical experience suggests that cats with low cobalamin are often less responsive to therapy if they are not concurrently treated with cobalamin[8]. Evidence in support of the clinical benefit of cobalamin supplementation include significant decreases in serum methylmalonic acid (an indicator of cellular cobalamin availability) and owner-reported improvement in signs of gastrointestinal disease and weight gain in cats supplemented with parenteral cobalamin for 4 weeks[9]. Cats appear to have rapid cobalamin turnover compared to humans. It was demonstrated in 2 cats with inflammatory bowel disease (IBD) and low cobalamin that parenteral (SQ) cobalamin has a much shorter half-life (5 days) than in healthy cats (2-weeks)[6]. Successful treatment of IBD in these cats was associated with a substantial increase in the half-life of cobalamin. The exact dietary requirement of cobalamin in cats is unknown. Current recommendations[8,10] are to treat hypocobalaminemic cats with 250 μg of cobalamin subcutaneously once a week for 6 weeks, then once after a further 30 days and then to recheck serum cobalamin 30 days after that dose. If cobalamin in within the reference range then once monthly doses are recommended. If cobalamin is below the reference range then treatment should be reinstituted at a minimum of once every 2 weeks (and in the case of severely low cobalamin once a week should be considered). It is expected that cats requirements for cobalamin will diminish with increasing resolution of their underlying gastrointestinal disease Cats responding to primary treatment of their underlying intestinal disease will usually have cobalamin concentrations above the reference range at the 30 day time point and discontinuation of treatment can then be considered.

"Steroid-Responsive" Gastrointestinal Disease

Corticosteroids are well-accepted as highly effective drugs for treatment of feline inflammatory bowel disease[2]. Compared to dogs, cats are generally considered to require higher doses and to be less susceptible to the side effects of steroids than dogs. Species differences are apt to account for some of these observations. For example, cats have been demonstrated to have approximately half the density of glucocorticoid receptors in the skin and liver as compared to dogs and cat glucocorticoid receptors have a lower binding affinity[11]. Cats reportedly also have better clinical responses to use of the metabolically active drug prednisolone compared to the pro-drug prednisone. Prednisone and prednisolone can be interconverted to one another in the liver by 11β-hydroxysteroid dehydrogenase-1. Recent studies in cats have shown that the area-under-the-curve for prednisolone is significantly greater (approximately 4-fold) after administration of oral prednisolone than after oral prednisone[12,13]. The reason for lower bioavailability of prednisone in cats remains unknown but may be due to decreased absorption, increased metabolism, or more rapid elimination. These studies also suggested that prednisolone dosing should be based on lean or ideal body weight. Increased serum concentrations of prednisolone result when over-conditioned cats are dosed based on actual body weight. This increase in metabolically active serum drug concentration is speculated to increase the susceptibility of over-conditioned cats to steroid side effects such as diabetes mellitus and hepatic lipidosis[13].

No studies have examined the use of dexamethasone or budesonide in cats for treatment of inflammatory bowel disease. Parenteral (SQ) use of dexamethasone is advocated by some clinicians as a more effective initial route for treatment of patients with moderate to severe intestinal malabsorption or underlying cardiac disease (the latter due to lack of dexamethasone mineralocorticoid activity). Budesonide is sometimes chosen for cats with concurrent diseases such as diabetes mellitus or cardiac disease. The rationale for use of budesonide is based on a presumption of extensive enterohepatic recirculation that attenuates systemic drug absorption and hypothalamic-pituitary-adrenal axis suppression; however this has not been examined in cats.

Nitroimidazoles

The nitroimidazoles, most notably metronidazole, are commonly used in cats for treatment of diarrheal disease. Metronidazole is reported to be effective in reducing anaerobic bacteria in the duodenum of cats[14] and is effective for the treatment of *Giardia* spp. infections[15]. Use of metronidazole in cats can be challenging due to its bitter taste and tendency to cause inappetence. Studies have shown that metronidazole benzoate (a metronidazole ester of benzoic acid) is much more palatable to cats[15]. Metronidazole benzoate needs to be compounded and is only ~60% metronidazole by weight. Suggested dosing of metronidazole benzoate for treatment of *Giardia* is 25 mg/kg PO BID (equivalent to 15 mg/kg PO BID of metronidazole) for 7 days[15]. While benzoate can be toxic to cats, the amount of benzoate is considered to be safe at this dosage. It is worth noting that metronidazole has been shown in cats (and in people) to induce DNA damage in circulating peripheral blood mononuclear cells both in vitro and in vivo in as few as 7 days of treatment. The DNA damage appears reversible as it is no longer observed 7 days after drug withdrawal[16].

Ronidazole remains the only drug with evidence for an ability to eradicate *Tritrichomonas foetus* infection as demonstrated in experimentally infected cats[17]. Based on pharmacokinetic studies, the recommended dose of ronidazole in cats is 30 mg/kg orally once a day[18]. Treatment failure and resistance of *T. foetus* to ronidazole are now well documented[19]. Until recently, the efficacy of ronidazole for treatment of naturally infected cats under field conditions was unknown. In a retrospective study by Xenoulis *et al*, outcome data was reported for 45/79 cats diagnosed with *T. foetus* and undergoing treatment with ronidazole at a dose of 30 mg/kg once a day[20]. A complete or close-to-complete resolution of diarrhea was reported for 64% (29/45) of cats and a partial or no improvement or relapse of diarrhea in 36% (16/45) of cats. Among the cats not responding, 31% (5/16) were treated with less than the recommended dose or duration of ronidazole. Follow up testing for persistence of underlying *T. foetus* infection in these cats was not reported. A study by Holliday *et al*, reported 100% resolution or dramatic reduction in clinical signs of diarrhea in 24 cats with *T. foetus* treated with ronidazole at 30 mg/kg PO BID[21]. Upon completion of treatment, all cats were negative for *T. foetus* by culture of a rectal swab however positive results were observed in the only cat tested by means of PCR. Bell *et al*, reported 100% resolution of diarrhea in 12 cats with *T. foetus* treated with ronidazole at 30 mg/kg PO BID[22]. After treatment, 9 cats tested negative for *T. foetus* by means of PCR (7 cats) or fecal smear (2 cats). Based on these reports, a conservative estimate is that 60% of cats treated with ronidazole will have close-to-complete resolution of clinical signs. What remains unclear is to what extent ronidazole has eradicated versus merely suppressed the infection below the limit of detection. Toxic effects of ronidazole including lethargy, inappetence, trembling, and ataxia have been reported in 4 cats[23]. Until recently the prevalence of these side effects among cats undergoing treatment with ronidazole was unknown. Xenoulis *et al* report discontinuation of ronidazole in 4/79 (5%) of cats due to development of anorexia or neurological signs[20].

Orally-Administered Gastric Acid Suppressants

With the exception of studies obtained from cats with surgically placed gastric fistulas, there has been no published evidence for the efficacy of commonly used gastric acid suppressants in cats until recently. Recent studies by Parkinson *et al*[24] evaluated the effectiveness of famotidine and 2 different formulations of omeprazole for control of gastric pH in normal cats as measured using a novel endoscopically-placed pH measuring device (Bravo pH). Omeprazole was utilized as either the equine oral paste (Gastroguard) mixed in cod liver oil or as fractionated enteric-coated omeprazole tablets. This group demonstrated that omeprazole formulations provide superior acid suppression compared to famotidine (each at a dose of 1 mg/kg PO BID). Moreover, only omeprazole formulations administered to cats approached goals established for treatment of people with gastroduodenal ulcers and gastroesophageal reflux disease, demonstrated to be optimal at an intragastric pH of ≥3 and ≥4 for approximately 75% and 67% of the day, respectively. Even more recent studies by Sutalo *et al*[25], evaluated the effectiveness of ranitidine (2 mg/kg PO BID) and enteric-coated granules (i.e. removed from a capsule) of omeprazole at either 1 mg/kg PO SID or 1 mg/kg PO BID. Using the Bravo pH measuring device they determined that ranitidine and omeprazole SID were no different than placebo for control of intragastric pH. However omeprazole BID achieve an intragastric pH of ≥3 and ≥4 for approximately 67% and 55% of the day, respectively.

Ponazuril for Coccidiosis

Sulfadimethoxine (Albon) is the only drug approved for treatment of coccidiosis in dogs and cats. This drug is coccidiostatic and requires a prolonged duration of administration. Recently the use of ponazuril (toltrazuril sulfone) has gained popularity for treatment of coccidiosis in dogs and cats. Ponazuril is available in the U.S. in paste form (Marquis paste®, Bayer Animal Health) as a treatment for *Sarcocystis neurona* in horses. A recent study examined the efficacy of treatment with ponazuril paste at each of three dosages in shelter-housed dogs and cats with confirmed coccidiosis[26]. Dogs and cats treated with 50 mg/kg q 24 h for 3 days showed the greatest reduction in oocyst counts. The treatment protocol was associated with a 92.2% clearance of infection in dogs and 87.5% in cats. Animals with high pre-treatment oocyst counts were more likely to remain infected, many of which cleared the infection with a second treatment course. Due to prolonged infectivity of *Cystoisospora* spp. oocysts, environmental decontamination requires contaminated surfaces to be left in contact with 10% ammonia for at least 10 min or steam cleaned. Bathing infected animals may reduce oocyst contamination of the haircoat.

For dilution of Marquis paste, add 20 mL of water to 10 ml (=10 mg) of Marquis paste (15% ponazuril) to achieve a final solution of approximately 50 mg/ml. Mixture does not need to be refrigerated. Shake well before dosing; dispose of after 30 days. Alternatively ponazuril can be commercially compounded.

REFERENCES

[1]Guilford WG, et al. J Vet Intern Med 2001;15:7-13. [2]Jergens AE, et al. **J Vet Intern Med** 2010;24:1027-1033. [3]Laflamme DP, et al. **J Feline Med Surg** 2012;14:669-677. [4]Laflamme DP, et al. **J Vet Intern Med** 2011;25:230-235. [5]Ugarte C, et al. **J Nutr** 2004;134:2068S-2071S. [6]Simpson KW et al. **J Vet Intern Med** 2001;15:26-32. [7]Kook PH et al. **Schweiz Arch Tierheilkd** 2012;154:479-486. [8]Ruaux CG. **The Vet J** 2013;196:145-152. [9]Ruaux CG et al. **J Vet Intern Med** 2005;19:155-160. [10]http://vetmed.tamu.edu/gilab/research/cobalamin-information#dosing. [11]van den Broek AHM, et al. **Res Vet Sci** 1992;52:312-315. [12]Graham-Mize CA and Rosser EJ. **Vet Dermatol** 2004;15:9 (abstract). [13]Center SA, et al. **Res Vet Sci** 2013;95:225-230. [14]Johnston KL, et al. **Am J Vet Res** 2000;61:1106-1112. [15]Scorza AV, et al. **J Feline Med Surg** 2004;6:157-160. [16]Sekis I et al. **J Feline Med Surg** 2009;11:60-68. [17]Gookin JL et al. **Am J Vet Res** 2001;62:1690-1697. [18]LeVine DN et al. **J Feline Med Surg** 2011;13:244-250. [19]Gookin JL, et al. **J Vet Intern Med** 2010;24:1003-1007. [20]Xenoulis et al. **J Feline Med Surg** 2013;15:1098-1103. [21]Holliday et al. **J Feline Med Surg** 2009;11:131-134. [22]Bell et al. **J Feline Med Surg** 2010;12:889-898. [23]Rosado et al. **J Vet Intern Med** 2007;21:328-331. [24]Parkinson S, et al **J Vet Intern Med** 2014; DOI.1111/jvim.12493. [25]Sutalo S et al. **J Vet Intern Med** 2015 in press. [26]Lister AL, et al. **Vet Parasitol** 2014;202:319-325.

MANAGEMENT OF FELINE INFLAMMATORY AIRWAY DISEASES

Carol Reinero, DVM, DACVIM, PhD
Columbia, MO

OVERVIEW

Inflammatory airway diseases of the cat include asthma, chronic bronchitis, parasitic bronchitis, and secondary bacterial bronchitis; they are among the most common respiratory disorders in this species. This lecture will primarily focus on non-infectious inflammatory disorders in the cat, with brief mention of treatment of the other infectious disorders. While there can be suggestions that one type of disorder might be more likely than another, all airway diseases to a certain extent can have some overlapping clinical signs and thoracic radiographic features. Discrimination of these disorders requires consideration of signalment, historical signs, physical examination findings, thoracic radiography, testing for infectious causes, analysis of lavage fluid cytology and response to therapeutic trials. Although there are some similarities in the management strategy (e.g., relevant parasitic control in at risk environments and environmental modulation to reduce inhaled irritants), there are also some specific treatments.

THERAPEUTIC APPROACH TO INFECTIOUS INFLAMMATORY FELINE AIRWAY DISEASES

Aelurostrongylus abstrusus, heartworm associated respiratory disease (HARD) and secondary bacterial infections can cause inflammatory airway lesions. There has recent experimental evidence that *Toxocara cati* also causes airway-oriented lesions, [1] although the clinical significance of these lesions in pet cats is unknown. Treatment considerations should include fenbendazole or other anthelmintic, an appropriate year-round heartworm preventative in endemic areas (regardless of whether the cat is indoor only), and an antibiotic ideally based on culture and sensitivity that passes the blood bronchus barrier (e.g., doxycycline), respectively. Cats with infectious inflammatory airway disease also benefit from glucocorticoid therapy while infection is active. Chronic infections may predispose to chronic bronchitis, which is not a curable disease, but can be managed with lifelong therapy (see below).

OVERVIEW OF THERAPEUTIC APPROACH TO NON-INFECTIOUS INFLAMMATORY FELINE AIRWAY DISEASES

For decades, the cornerstone of management for feline asthma and chronic bronchitis has consisted of environmental modulation, glucocorticoids and bronchodilators. Environmental modulation consists of avoiding or minimizing inhalation of irritants or in asthmatic cats, allergens to which a cat has been sensitized. Irritants might include smoke (tobacco smoke, smoke from wood stoves or fireplaces, aerosols, powders, dusty cat litter, etc) and can trigger non-specific airway hyperresponsiveness (AHR). HEPA filters are useful when cats are indoors. Glucocorticoids are critical for both asthma and chronic bronchitis as they are anti-inflammatory. Bronchodilators are most important for asthmatic cats, especially those with wheeze or episodic respiratory distress; they may not be necessary for asthmatic cats that have cough as the sole clinical manifestation or in cats with chronic bronchitis.

GLUCOCORTICOIDS

Glucocorticoids are the mainstay of therapy for inflammatory airway disease as unchecked inflammation can predispose to airway hyperresponsiveness and remodeling. While there are many routes of administration including oral, inhaled, injectable and transdermal, not all are good therapeutic options because of adverse effects (e.g., injectable repositol forms) or lack of efficacy (transdermal). Additionally, there are many types of glucocorticoid preparations, most of which have not been studied specifically in cats. The vast majority of the literature on use of glucocorticoids for feline lower airway disease has been in experimental models.

The author prefers to start treatment of asthma and chronic bronchitis with oral prednisolone, which has higher bioavailability than oral prednisone. [2] If oral prednisolone is not effective, oral dexamethasone has been advocated; however, it has greater diabetogenic effects than prednisolone.[3] Inhaled glucocorticoids have all but replaced oral glucocorticoids in humans with asthma as the drug is delivered directly to the site of disease and systemic effects are minimized. Inhaled glucocorticoids in cats have minimal systemic immune effects but can suppress the hypothalamic-pituitary-adrenal axis. [4] Inhaled flunisolide and fluticasone can effectively suppress airway inflammation in cats. [5,6] There is a plateau effect where high doses do not achieve a larger benefit,[7] and the exact dose for each type of inhalant steroid for pet cats is not known. Experimentally, very high short term doses of fluticasone have been shown to be as effective as prednisolone in reducing airway inflammation.[8] Transdermal administration of prednisone or prednisolone (2 mg/kg BID X 3 weeks) does not lead to detectable blood concentrations of prednisolone and thus should not be used therapeutically.

BRONCHODILATORS

Bronchodilators are a critical component of therapy for cats with evidence of acute bronchospasm (e.g., cats having "asthma attacks" or status asthmaticus). They are perhaps not as critical for cats who only have cough as the chief complaint when glucocorticoids are given to suppress airway inflammation. It is the author's opinion that bronchodilators should not be given as monotherapy to asthmatic cats or cats with chronic bronchitis, as these are primarily inflammatory diseases and inflammation must be addressed to halt further damage to the airways. Bronchodilators have adverse effects such as hyperexcitability, systemic hypertension and tachycardia; additionally, owner compliance may be poor with oral medications given 2-3 times daily. Thus unless cats have clinically relevant airflow limitation, the emphasis should be placed on glucocorticoids to manage inflammation.

While short acting beta 2 agonists (SABA) are the most important lifesaving therapy for a cat with status asthmaticus, there is evidence that chronic inhaled use may paradoxically exacerbate airway inflammation and airway hyperresponsiveness. In an experimental model of feline asthma, inhaled racemic albuterol which is a 1:1 mixture of R- and S-albuterol given twice daily for 2 weeks led to significant increases in airway eosinophilia; the R-isomer alone did not. [9] Thus, the author recommends racemic albuterol to be used predominantly as a rescue medication at home and ideally no more than twice weekly. If sustained use of a SABA is needed, inhaled levalbuterol, a commercially available form of R-albuterol, or oral bronchodilators (SABA or methylxanthines) may be administered. Inhaled long acting beta 2 agonists (LABA) are thought to have similar paradoxical reactions, although these effects have not been specifically studied in cats. Overall, LABA (salmeterol) is thought to have weak but more sustained bronchodilatory effects in healthy cats[10] but does not appear to be an effective bronchodilator in asthmatic cats[11]. If given the choice of an injectable or inhaled SABA for an asthmatic cat in crisis, the author will preferentially use an injectable medication as with severe bronchospasm inhaled medications may not adequately reach the target airways. [12]

Other medications have been investigated in experimental asthma models for potential bronchodilatory and anti-inflammatory properties. These include nebulized lidocaine, anticholinergics, cysteinyl leukotriene antagonists, anti-serotonergics and anti-histaminics. Nebulized 2% lidocaine 2mg/kg TID for 2 weeks had no anti-inflammatory effects but significantly reduced AHR compared with placebo. [13] The anticholinergic ipratropium has synergistic bronchodilatory effects with albuterol. [12] There has been no evidence to suggest that the other aforementioned classes of drugs including zafirlukast, cyproheptadine or cetirizine beneficially reduce AHR or airway inflammation. [6,14]

OTHER ASTHMA THERAPEUTICS

There have been a variety of other miscellaneous treatments which have been assessed for use in feline asthma using experimental models. Dietary omega-3 polyunsaturated fatty acids and luteolin (an antioxidant flavinoid) administered to experimentally asthmatic cats for 4 weeks showed no effect on airway inflammation, but did show a decrease in airway reactivity. [15] Importantly, although this prophylactic treatment holds promise to diminish airway hyperresponsiveness, because it does not blunt eosinophilic airway inflammation (which ultimately contributes to worsening airway hyperresponsiveness and airway remodeling), it should not be given as monotherapy to treat feline asthma. Inhaled N-acetylcysteine, a mucolytic and antioxidant medication, was also tested in experimental feline asthma and was found to increase airway resistance putting into question its safety in cats with preexisting airway disease.[16] No studies to date have evaluated the effects of N-acetylcysteine administered either orally or by inhalation on airway inflammation or mucus quality in asthmatic cats.

It would seem rational to design therapy which could block the effects of the allergic asthmatic cascade. Modulation of T lymphocyte activity, cytokine inhibition, antibody blockade (especially IgE), small molecule inhibition (e.g., tyrosine kinases), and other immunomodulators are thought to be rational targets. Not all of these have been evaluated in cats, but represent future areas of research.

Since allergic asthma is thought to be driven by a T helper 2 lymphocytic process, modulating Th2 lymphocyte activity would seem to hold promise for therapy. The treatments which have been tested which function by altering T cell activity include cyclosporine, allergen-specific immunotherapy and possible tyrosine kinase inhibitors and/or stem cells. In experimental feline asthma, cyclosporine did not inhibit the early phase response to allergen challenge (mediated in large part by mast cells), but it was effective at blunting airway hyperresponsiveness to acetylcholine and airway remodeling. [17,18] Because of the need for therapeutic monitoring and the potential side effects of cyclosporine, it is NOT advocated for routine management of feline asthma, but may be considered in severe or refractory cases. Other potential therapies being evaluated for the treatment of feline asthma include allergen-specific immunotherapy ("allergy shots"). Allergen-specific immunotherapy is most commonly administered to humans with allergic rhinitis, but has also been used in the treatment of asthma. There is evidence that an abbreviated form of conventional immunotherapy, called rush immunotherapy, may dampen eosinophilic inflammation in experimental feline asthma.[19] Addition of an adjuvant (CpG immunostimulatory sequences which polarize the immune response towards a T helper 1 and away from a T helper 2 response) demonstrated good safety and efficacy. [20] Comparing the subcutaneous route of injection with topical mucosal delivery of allergen-specific rush immunotherapy was also performed in experimental feline asthma. [21] The latter route of administration more closely mimics the natural route of exposure to aeroallergens. Both routes of administration of allergen-specific immunotherapy decreased eosinophilic airway inflammation and were generally associated with minimal side effects; either could be used. However, the subcutaneous protocol demonstrated more consistent resolution of scored clinical signs after allergen challenge by aerosol. Concurrent use of oral glucorticoids during immunotherapy diminishes its efficacy; however, inhaled steroids do not alter beneficial effects of immunotherapy. [22] The major difference between allergen-specific immunotherapy and other forms of treatment (eg, steroids and bronchodilators) is that immunotherapy has the potential for cure of the disease by re-training the immune system to be tolerant to allergens. This has not yet been tested in pet cats with asthma and there are differences in the protocol used in this model versus conventional immunotherapy—a clinical trial in pet cats still needs to be performed.

Inhibition of tyrosine kinases, a group of proteins which regulate cell survival, growth and differentiation, has more recently been of interest for treatment of asthmatic patients. Tyrosine kinase inhibitors are small molecule inhibitors which block the ATP binding sites of kinases. In asthma, the c-KIT receptor has been associated with proliferation and degranulation of mast cells and eosinophils in humans and mice and seems to be a logical target for therapeutic intervention. Masitinib had beneficial effects on the asthmatic phenotype; however, adverse effects were common and limiting. [23] Toxicity of these small molecule inhibitors is a concern and the side effects must be carefully weighed against the potential benefits on airway inflammation and airflow limitation.

Stem cell therapy has been advocated for use in a number of chronic lung disorders, including asthma. Allogeneic adipose-derived mesenchymal stem cells have been assessed in experimental feline asthma. These are different from commercial autologous stem cells which are really "stromovascular fraction" that likely have few stem cells. While with the protocol used there was no reduction of airway inflammation or AHR, airway remodeling assessed by computed tomographic changes was dampened compared with placebo.[24] Additional study of stem cell protocols is warranted.

REFERENCES

1. Dillon AR, Tillson DM, Hathcock J, et al. Lung histopathology, radiography, high-resolution computed tomography, and bronchio-alveolar lavage cytology are altered by Toxocara cati infection in cats and is independent of development of adult intestinal parasites. *Vet Parasitol* 2013;193:413-426.

2. Center SA, Randolph JF, Warner KL, et al. Influence of body condition on plasma prednisolone and prednisone concentrations in clinically healthy cats after single oral dose administration. *Res Vet Sci* 2013;95:225-230.

3. Lowe AD, Graves TK, Campbell KL, et al. A pilot study comparing the diabetogenic effects of dexamethasone and prednisolone in cats. *J Am Anim Hosp Assoc* 2009;45:215-224.

4. Reinero CR, Brownlee L, Decile KC, et al. Inhaled flunisolide suppresses the hypothalamic-pituitary-adrenocortical axis, but has minimal systemic immune effects in healthy cats. *J Vet Intern Med* 2006;20:57-64.

5. Kirschvink N, Leemans J, Delvauz F, et al. Inhaled fluticasone reduces bronchial responsiveness and airway inflammation in cats with mild chronic bronchitis. *J Fel Med Surg* 2005;8:45-54.

6. Reinero C, Decile K, Byerly J, et al. Effects of drug treatment on inflammation and hyperreactivity of airways and on immune variables in cats with experimentally induced asthma. *Am J Vet Res* 2005;66:1121-1127.

7. Cohn LA, DeClue AE, Cohen RL, et al. Effects of fluticasone propionate dosage in an experimental model of feline asthma. *J Feline Med Surg* 2010;12:91-96.

8. Leemans J, Kirschvink N, Clercx C, et al. Effect of short-term oral and inhaled corticosteroids on airway inflammation and responsiveness in a feline acute asthma model. *Vet J* 2012;192:41-48.

9. Reinero CR, Delgado C, Spinka C, et al. Enantiomer-specific effects of albuterol on airway inflammation in healthy and asthmatic cats. *Int Arch Allergy Immunol* 2009;150:43-50.

10. Leemans J, Kirschvink N, Bernaerts F, et al. A pilot study comparing the antispasmodic effects of inhaled salmeterol, salbutamol and ipratropium bromide using different aerosol devices on muscarinic bronchoconstriction in healthy cats. *Vet J* 2009;180:236-245.

11. Leemans J, Kirschvink N, Bernaerts F, et al. Salmeterol or doxycycline do not inhibit acute bronchospasm and airway inflammation in cats with experimentally-induced asthma. *Vet J* 2012;192:49-56.

12. Leemans J, Kirschvink K, Clercx C, et al. Functional response to inhaled salbutamol and/or ipratropium bromide in Ascaris suum-sensitised cats with allergen-induced bronchospasms. *Vet J* 2010;186:76-83.

13. Nafe LA, Guntur VP, Dodam JR, et al. Nebulized lidocaine blunts airway hyper-responsiveness in experimental feline asthma. *J Feline Med Surg* 2013.

14. Schooley EK, McGee Turner JB, Jiji RD, et al. Effects of cyproheptadine and cetirizine on eosinophilic airway inflammation in cats with experimentally induced asthma. *Am J Vet Res* 2007;68:1265-1271.

15. Leemans J, Cambier C, Chandler T, et al. Prophylactic effects of omega-3 polyunsaturated fatty acids and luteolin on airway hyperresponsiveness and inflammation in cats with experimentally-induced asthma. *Vet J* 2010;184:111-114.

16. Reinero CR, Lee-Fowler TM, Dodam JR, et al. Endotracheal nebulization of N-acetylcysteine increases airway resistance in cats with experimental asthma. *J Feline Med Surg* 2011;13:69-73.

17. Mitchell R, Cozzi P, Ndukwu I, et al. Differential effects of cyclosporine A after acute antigen challenge in sensitized cats in vivo and ex vivo. *Br J Pharmacol* 1998;123:1198-1204.

18. Padrid P, Cozzi P, Leff A. Cyclosporine A inhibits airway reactivity and remodeling after chronic antigen challenge in cats. *Am J Repir Crit Care Med* 1996;154:1812-1818.

19. Reinero CR, Byerly JR, Berghaus RD, et al. Rush immunotherapy in an experimental model of feline allergic asthma. *Vet Immunol Immunopathol* 2006;110:141-153.

20. Reinero CR, Cohn LA, Delgado C, et al. Adjuvanted rush immunotherapy using CpG oligodeoxynucleotides in experimental feline allergic asthma. *Vet Immunol Immunopathol* 2008;121:241-250.

21. Lee-Fowler T, Cohn L, DeClue A, et al. Evaluation of subcutaneous versus mucosal (intranasal) allergen-specific rush immunotherapy in experimental feline asthma. *Vet Immunol Immunopathol* 2009;129:49-56.

22. Chang C, DeClue A, Cohn L, et al. Oral glucocorticoids may diminish efficacy of allergen-specific immunotherapy in experimental feline asthma. *30th Annual ACVIM Forum*. New Orleans, LA, 2012.

23. Lee-Fowler T, Guntur V, Dodam J, et al. The tyrosine kinase inhibitor masitinib blunts airway inflammation and improves associated lung mechanics in a feline model of chronic allergic asthma *Int Arch Allergy Immunol* 2012;158:369-374.

24. Trzil JE, Masseau I, Webb TL, et al. Long-term evaluation of mesenchymal stem cell therapy in a feline model of chronic allergic asthma. *Clin Exp Allergy* 2014;44:1546-1557.

EVIDENCE-BASED TREATMENT OF CHRONIC RENAL DISEASE IN CATS

Jessica Quimby, DVM, PhD, DACVIM
Fort Collins, CO

WHY EVIDENCE-BASED TREATMENT?

Establishing a solid relationship with the pet owner and tailoring recommended therapies to each individual case is an important part of successful management of feline chronic kidney disease (CKD). The treatment plan will fail if the owner does not understand why the medication is important, if they cannot administer it to the cat, and if follow-up visits and phone calls are not made to troubleshoot medication administration. It may be necessary to prioritize medications based on their medical necessity, the likely benefit to the patient, and the owner's ability to give the medication. The level of evidence available to support the clinical efficacy of the treatment should therefore be considered when determining medication priority.

LEVELS OF EVIDENCE

Grade 1: Randomized, controlled clinical study
Grade 2: Studies with models or simulations, historical control clinical study
Grade 3: Non-randomized clinical study, cohort or case-controlled clinical study, dramatic uncontrolled experiments
Grade 4: Opinions, expert committees, pathophysiologic justification, descriptive or comparative studies

NUTRITION AND DIET IN CKD

Several studies have documented the therapeutic value of specially formulated diet in the management of CKD (Grade 1), including amelioration of renal secondary hyperparathyroidism, decreased incidence of uremic crisis and increased survival.[1-3] Renal diet has also recently been demonstrated to decrease FGF-23 levels, a marker that is positively correlated with disease stage, and is thought to be a mediator of renal secondary hyperparathyroidism.[1] These diets typically contain conservative amounts of high quality protein, adequate non-protein calories, and are restricted in phosphorus. The failure of the patient to eat the diet negates the benefit of dietary management, and therefore a key therapeutic target for these patients is the maintenance of appetite and food intake. Poor body condition is associated with decreased prognosis in several species and has a negative effect on immune function, wound healing and strength.[4] Additionally, poor appetite is perceived by owners as a significant quality of life concern and anorexia in companion animals can cause emotional distress to owners.[5] Cats with CKD often suffer from poor appetite, and although renal diets contain adequate dietary protein, the animal will be protein deficient if not eating its caloric requirement. Management of long term hyporexia is important. Addressing complications of decreased kidney function (hydration, hypokalemia, anemia, etc.) that have the potential to affect appetite is important. If cats cannot be enticed to eat an appropriate amount of renal diet, then complete nutrition needs to be obtained via other sources. As cats are particularly sensitive to environmental factors, suggestions for food acceptance and appetite enhancement include: slow transition to renal diet over 10-14 days, quiet environment with exclusive access without interference from bothersome household members, novel food type in cases of possible aversion, similar food type in case of food exclusivity, warming the food particularly if olfaction is an issue, alternatively chilling the food if aromas appear to result in nausea, social interaction while eating, and feeding small frequent meals as premature satiety is associated with many disease states.

Does "Uremic Gastritis" Exist in Cats?

Cats with CKD have been shown to have elevated concentrations of gastrin that increase with the severity of renal failure,[6] but the relationship between gastrin, gastric acid secretion, and gastric pathology has not been fully described. In humans and dogs, gastrin is excreted by the kidneys, and it is hypothesized that as renal function declines, hypergastrinemia develops, resulting in gastric hyperacidity.[6] Cats that have gastrin-secreting tumors with levels of hypergastrinemia similar to those found in cats with CKD have significant gastric pathology; however no study has shown this to be the case in cats with CKD.[7] Even in humans, the development of hyperacidity in association with CKD appears to be inconsistent, and may be related to the presence of *Helicobacter spp.* infection.[8] A recent study evaluated the type and prevalence of histopathologic lesions in the stomach of cats with CKD.[9] Cats with CKD appear more likely to have gastric fibrosis and mineralization, rather than uremic gastropathy lesions previously described in dogs and humans (uremic gastritis, ulceration, vascular injury, edema). Therefore, the administration of gastric protectants such as sucralfate may not be justified, unless obvious clinical evidence of gastrointestinal hemorrhage such as melena or hematemesis is appreciated. The exact role of hypergastrinemia in contributing to gastric hyperacidity and/or gastric lesions in cats with CKD is still unclear. Thus there is very little available evidence on which to base recommendations for the use of acid-reducing medications such as H_2 blockers, proton pump inhibitors in cats with uremia. Limiting gastric acidity with the use of H_2 blockers or proton pump inhibitors such as famotidine or omeprazole respectively anecdotally appears to palliate inappetence in some CKD patients (Grade 4), however as previously mentioned both the degree of hyperacidity present in CKD and the efficacy of these medications for management of cats with CKD remains unproven. Recent studies of the effect of omeprazole on the gastric pH in normal cats indicates that it is superior to famotidine in its ability to inhibit acid production at 1 mg/kg twice daily (Grade 1).[10]

Anti-nausea Therapy

Gastrointestinal symptoms in CKD cats may not necessarily be the result of gastric lesions, but perhaps the consequence of circulating uremic toxins interacting with the chemoreceptor trigger zone in the brain. Medical management of gastrointestinal symptoms with anti-emetic and anti-nausea drugs may therefore be more appropriate (Grade 4). Several anti-nausea therapies are available. These include maropitant, ondansetron and dolasetron. These drugs work at nausea center in brain as well as in gut and can be given as an injection. Ondansteron has been documented to be helpful in human patients suffering from uremia.[11] However recent pharmacokinetic studies in cats have demonstrated that oral bioavailability of ondansetron (0.5 mg/kg) is poor in cats (~35%) and the half -life is very short (approximately 1 hour) making it a q 8hr medication.[12] Subcutaneous ondansetron has a slightly longer half-life of 3 hours. In addition to its appetite-stimulating properties, mirtazapine demonstrates anti-nausea properties as it acts at the 5HT3 receptor similarly to ondansetron and dolasetron (Grade 4).[13, 14] A recent study assessed the efficacy of Cerenia for management of chronic vomiting and inappetence associated with feline CKD (Grade 1).[15] When given orally (4 mg) daily for two weeks, Cerenia was demonstrated to palliate vomiting associated with chronic kidney disease, however did not appear to significantly improve appetite or result in weight gain in cats with Stage II and III CKD within the timeframe of the study. A pharmacokinetic and toxicity study in cats indicated that longer-term usage appears safe.[16]

Appetite Stimulant Therapy

Cyproheptadine (Grade 4) has been used for some time as an appetite stimulant and has anecdotal efficacy in many patients, however its efficacy has never been scientifically evaluated. Twice daily administration is necessary in many cases and this can prove a challenge for owners, particularly long-term. Sedation is a common side effect. Mirtazapine has become more commonly used and recent placebo-controlled, double-blinded crossover clinical trial was performed to evaluate the effects of mirtazapine on body weight, appetite, and vomiting in cats with CKD (Grade 1).[14] Mirtazapine is an effective appetite stimulant in cats with CKD and resulted in significantly increased appetite and weight. Mirtazapine also appears to have anti-emetic properties and resulted in significantly decreased vomiting in cats with CKD. Hyperactivity and vocalization are side effects and are generally dose-dependent; smaller, more frequent doses are recommended (1.87 mg PO q 48 hours) (Grade 1).

SYSTEMIC HYPERTENSION

Systemic hypertension appears to be common in cats with CKD (20-65%), but the exact pathophysiologic relationship is unknown. Systemic hypertension can have other deleterious effects such as retinal hemorrhage and detachment, neurologic and cardiac impairment. Elderly cats, particularly those with renal impairment should be routinely screened. Unless blood pressure is >200 mm Hg and/or evidence of target organ damage is seen, such as retinal hemorrhage, blood pressure should be rechecked on three occasions to rule out white coat hypertension. Azotemic cats with blood pressure persistently >160 mm Hg are candidates for treatment. Amlodipine is documented to be an effective treatment for hypertension in the cat and is the drug of choice at 0.625-1.25 mg/cat daily (Grade 2).[17, 18]

PROTEINURIA

Intraglomerular and systemic hypertension, tubular dysfunction and diseases of the glomerular apparatus can lead to proteinuria. The renal tubular epithelial cells are responsible for processing this protein and this increased their metabolic load and oxidative stress. Proteinuria is associated with poorer prognosis in several species including cats.[19] Several studies looked at efficacy of ACE inhibitors (benazepril) for treatment of proteinuria in cats and it was documented that benazepril decreases proteinuria in cats with CKD (Grade 1), but results were only suggestive of long-term benefit, not conclusive.[20] Nonetheless it is generally recommended that cats with renal proteinuria with a urine protein creatinine ratio > 0.4 on at least two repeated measurements should be treated (0.25-0.5 mg/kg benazepril orally daily). Careful monitoring is required because ACE inhibitors may result in decreased GFR and subsequent worsening of renal function. Telmisartan is an angiotensin receptor blocker that has recently become available in Europe for treatment of proteinuria in cats. A placebo-controlled randomized study regarding the short and long-term efficacy of the drug is not currently available.

HYDRATION

As the kidney becomes disease, the ability to concentrate the urine is lost. Although it has not been systematically assessed in a clinical trial, adequately maintaining hydration by giving subcutaneous fluids anecdotally appears to substantially help quality of life, improve appetite and activity (Grade 4). It can be a very helpful tool for owners in management of disease, but may not be necessary for every patient. The best candidates for SQ fluid therapy (75-150 ml SQ every 1-3 days) are those cats that appear to gain clinical benefit from management of hydration, are prone to secondary complications of chronic dehydration such as constipation, and do not suffer quality of life concerns from the procedure. If possible, supplementation with free water (orally or with a feeding tube) is preferred to avoid the sodium load that comes with the electrolyte solutions available for subcutaneous use. Feeding canned food instead of dry, or adding water to food is another way to potentially increased water consumption. Paying special attention to water sources in the house – fresh, accessible, water fountains etc., is also key.

ANEMIA

The kidney is responsible for producing the hormone erythropoietin which stimulates the bone marrow to produce red blood cells. As kidney disease progresses, the hormone levels decrease and anemia can result. Chronic low-grade gastrointestinal hemorrhage may also contribute to anemia and iron deficiency may be noted. The deficient hormone can be supplemented with artificial products such as recombinant human erythropoietin Darbepoetin alpha. Darbepoetin is a longer-acting form of erythropoietin and is thought to have less association with anti- erythropoietin antibodies than the previously used Epogen (Grade 3).[21] Darbepoetin is currently the product of choice, but expense can limit its use for some owners. Iron supplementation is recommended when therapy is initiated (iron dextran 50 mg/cat IM, Grade 4). Recommended starting dose is 1 mcg/kg SQ once weekly until the low end of the normal PCV range is reached, then injections are decreased to once every 2-3 weeks as needed to maintain PCV (Grade 3).

HYPOKALEMIA

Inadequate dietary intake, increased urinary loss and activation of the renin-angiotensin-aldosterone system are all thought to contribute to hypokalemia. Hypokalemia is associated with development and worsening of CKD in humans and appears to exacerbate damage to tubular epithelial cells.[22] Serum potassium levels are not representative of systemic tissue potassium levels and cats with low normal serum potassium may actually be systemically depleted.[23] Potassium is vital for normal muscle function, GI motility and supplementation has been demonstrated to correct hypokalemic myopathy (Grade 3).[24] Some clinicians recommend supplementation even when serum potassium is in the low normal range[11], with a goal of maintaining serum potassium levels above 4 mg/dL, but the effect on the progression of CKD has not been evaluated.

HYPERPHOSPHATEMIA

The kidneys are the main route of phosphorus excretion and as kidney function declines, hyperphosphatemia develops. Hyperphosphatemia is detrimental because it contributes to renal secondary hyperparathyroidism, tissue mineralization and progression of CKD. Hyperphosphatemia has been identified as a predictor of progression in feline CKD, thus controlling phosphorus intake through diet and phosphate binders is important part of CKD management (Grade 2).[19, 25] If a renal diet is initiated and phosphorus is still elevated after 4-6 weeks, then a phosphate binder is recommended with the goal of keeping serum phosphorus in the low normal range.[26] These medications bind the phosphorus in the food and make it less bioavailable so it is critical that the owners understand the medication should be given with each meal. Several different phosphate binders are available including aluminum hydroxide (30-100 mg/kg/day), calcium carbonate (90-150 mg/kg/day), and lanthanum carbonate (30 mg/kg/day),[26] As phosphate binders are dosed to effect – with a target phosphorus range for each IRIS stage as outlined by the IRIS recommendations for CKD therapy (www.iris-kidney.com) – these ranges are just a starting point. However the main problem with administration is palatability and this often limits the amount that can be reasonably administered. Serum calcium should be monitored when calcium-containing phosphate binders are used.

CALCITRIOL

Renal production of calcitriol, the active form of Vitamin D, is decreased in CKD and hyperphosphatemia further inhibits its synthesis. Calcitriol is an important inhibitor of parathyroid hormone and low levels leads to renal secondary hyperparathyroidism which has multiple deleterious consequences. In dogs and people, use of calcitriol has been shown to be beneficial (Grade 1 in dogs),[26] but a study performed in cats was not supportive.[27] Nonetheless, controversy remains and many clinicians recommend supplementation at 2 ng/kg orally daily on an empty stomach (Grade 4). Careful monitoring of calcium and PTH levels is required as hypercalcemia can result from over-supplementation, particularly if given with food. If hypercalcemia results, the daily dose can be doubled and administered every other day.[26]

ANTI-OXIDANT THERAPY

Several processes in the diseased kidney potentially result in oxidative stress including hypoxia, proteinuria, and inflammation. Oxidative stress documented in CKD cats[13] and potentially contributes to the progression of disease. Vitamins E, C and B-carotene have been shown to decrease oxidative stress in CKD cats (Grade II) however the effect of mitigating oxidative stress on the progression of CKD has not been evaluated in cats.

REFERENCES

[1]Geddes RF, et al. **J Vet Intern Med** 2013; 27: 1354-61. [2]Ross SJ, et al. **J Am Vet Med Assoc** 2006; 229: 949-57. [3]Elliott J, et al. **J Small Anim Pract** 2000; 41: 235-42. [4]Chan DL. **J Feline Med Surg** 2009; 11: 925-33. [5]Reynolds CA, et al. **J Vet Intern Med** 2010; 24: 1421-6. [6]Goldstein RE, et al. **J Am Vet Med Assoc** 1998; 213: 826-8. [7]Liptak JM, et al. **J Feline Med Surg** 2002; 4: 27-42. [8]El Ghonaimy E, et al. **Nephron** 1985; 39: 86-94. [9]McLeland SM, et al. **J Vet Intern Med** 2014; 28:827-837. [10]Parkinson S, et al. **J Vet Intern Med** 2015; 29: 104-12. [11]Ljutic D, et al. **Kidney Blood Press Res** 2002; 25: 61-4. [12]Quimby JM, et al. **J Vet Pharmacol Ther** 2013; 37: 348-353. [13]Riechelmann RP, et al. **Am J Hosp Palliat Care** 2010; 27: 106-10. [14]Quimby JM, et al. **Vet J** 2013; 197: 651-5. [15]Quimby JM, et al. **J Feline Med Surg** 2014. [16]Hickman MA, et al. **J Vet Pharmacol Ther** 2008; 31: 220-9. [17]Jepson RE, et al. **J Vet Intern Med** 2007; 21: 402-9. [18]Jepson RE, et al. **J Vet Intern Med** 2014; 28: 144-53. [19]Chakrabarti S, et al. **J Vet Intern Med** 2012; 26: 275-81. [20]King JN, et al. **J Vet Intern Med** 2006; 20: 1054-64. [21]Chalhoub S, et al. **J Vet Intern Med** 2012; 26: 363-9. [22]Wang HH, et al. **PloS one** 2013; 8: e67140. [23]Theisen SK, et al. **J Vet Intern Med** 1997; 11: 212-7. [24]Dow SW, et al. **J Am Vet Med Assoc** 1987; 191: 1563-8. [25]Kidder AC, et al. **J Feline Med Surg** 2009; 11: 913-24. [26]Polzin DJ. **J Vet Emerg Crit Care** 2013; 23: 205-15. [27]Hostutler RA, et al. **J Vet Intern Med** 2006; 20: 1307-13.

ADVANCED DIAGNOSTICS FOR CANINE LYMPHOPROLIFERATIVE DISEASES

Davis Seelig, DVM, PhD, DACVP
St. Paul, MN

INTRODUCTION

The lymphoproliferative diseases (LPDs) are a biologically heterogeneous group of neoplasms that result from the clonal expansion of lymphocytes at varied and unique stages of development. In the dog, lymphoma is the most common LPD and traditionally it has been diagnosed using tools that have emphasized the morphologic characteristics of the suspect cellular population, namely cytology and/or histopathology. However, a series of recent reports have expanded our understanding of the heterogeneity of canine LPD and the scientific, clinical, and therapeutic importance of going beyond the diagnosis of "lymphoma" or "leukemia." Chief among are studies illustrating the applicability of a canine-adapted version of the World Health Organization (WHO) classification scheme. Work by Valli *et al*, Ponce *et al*, and Vezzali *et al* revealed that are at least 18 subtypes of canine LPD, many of which demonstrate unique clinical profiles including aggressive and indolent forms of the disease.[1,2,3] In addition to demonstrating the biologic heterogeneity of canine LPD, these studies illustrate that the traditional approach to their diagnosis is insufficient in its ability to capture their scope. As such, there is increasing evidence that moving beyond traditional diagnostics into advanced diagnostics expands our ability to identify unique forms of canine LPD and offer unique prognostic and therapeutic guidance, all in a minimally-invasive manner and often without the need for biopsy.

Practically, advanced diagnostics for LPD offer complementary information that extends the traditional morphologic diagnosis of "lymphoma" by providing cell lineage information (i.e. immunocytochemistry [ICC], immunohistochemistry [IHC] and flow cytometry [FC]) and molecular evidence of lymphoid clonality (i.e. polymerase chain reaction for antigen receptor rearrangement [PARR]). While this information has academic significance, by increasing our understanding of the biology of canine LPD and informing the design of scientific and clinical studies, it also provides useful prognostic and therapeutic information. However, while there are some situations in which advanced diagnostics can replace a traditional diagnostic (i.e. flow cytometry in the diagnosis of leukemia and certain forms of lymphoma), it is currently recommended advanced diagnostics be used as additions to, not replacements for, traditional diagnostics.

TRADITIONAL APPROACHES IN LYMPHOMA DIAGNOSIS

Traditionally, the diagnosis of canine lymphoma has relied upon a combination of fine-needle aspiration cytology (FNAC) with or without biopsy with histopathology. While the utility of both is well-recognized, it is increasingly apparent that the diagnosis of "lymphoma" is generic and fails to fully capture the biologic and clinical spectrum of this heterogeneous disease. As our recognition of the number of subtypes of canine lymphoma has expanded, by necessity so has the diagnostic approach.

Cytology

The low cost, limited invasiveness, low barrier to entry, and the ease of interpretation has made the examination of cytologic preparations, including peripheral blood films, fluid samples, and fine-needle aspiration (FNA) samples, the most common tool in the diagnosis of canine lymphoma. As the most common form of canine lymphoma is a lymph node-effacing disease composed of large and atypical cells (i.e. diffuse large B-cell lymphoma, DLBCL), most cases of canine lymphoma can usually be made by cytology alone.[4] However FNA cytology has well-recognized shortcomings, including its limited utility in: 1) small- or medium-sized cell lymphoma, 2) emergent lymphoma, 3) differentiating lymphoma from robust reactive hyperplasia, and 4) forms of lymphoma that do no affect the entire lymph node (i.e. nodular forms). Moreover, while there are some variably-sized descriptive reports on the subject, it is important to recognize that the role of FNAC in the classification of canine lymphoma using the WHO system or in the prediction of disease behavior remains to be defined.[3] However it is important to note that FNAC samples are suitable for follow-up advanced diagnostics, including ICC and PARR.

Histopathology

Incisional or excisional biopsy with histopathology has long been considered the gold standard for the diagnosis of canine lymphoma. Histopathology allows for the complete evaluation of tissue architecture, which can overcome many of the limitations of cytology. Moreover, the current canine-adapted WHO system relies upon histopathology for complete classification of nodal lymphomas, namely by stratifying them according to the distribution of the neoplastic lymphoid cell into diffuse and nodular forms. While biopsy certainly offers tremendous advantages for diagnostic and research purposes, its inherent invasiveness, cost, and prolonged turnaround time increasingly make it a less-attractive tool for many forms of LPD. Additionally, as the complexity of the canine LPD classification systems increase, a pathologist trained in the application of such systems is likely to be increasingly vital. However, there is increasing evidence that the combined application of minimally-invasive traditional and advanced diagnostics (i.e. FNAC and FC) may be sufficient for both the diagnosis and classification of certain forms of canine lymphoma and may supplant the need for biopsy.

ADVANCED DIAGNOSTICS IN LYMPHOPROLIFERATIVE DISEASE

The advanced diagnostics are assays that examine non-morphologic characteristics of a neoplastic lymphoid population in order to diagnose, classify, and prognose lymphoproliferative disease.

Immunocytochemistry (ICC) & Immunohistochemistry (IHC)

ICC and IHC are antibody-based staining protocols that are applied to cytology or histology samples, respectively, in order to classify cells in specific lineages. In canine lymphoproliferative disease, both IHC and ICC are most commonly used as to classify cells as either T or B-cell in origin. For the immunophenotyping of lymphoma, it is recommended that 3-4 unstained smears of adequate cellularity be submitted to the laboratory. In general, these slides will be stained with an anti-B-cell antibody (Pax5, CD79a, or CD21) and a T-cell antibody (CD3) with the remaining slide(s) used as negative control.

Flow Cytometry

Flow cytometry (FC) is an analytic method in which unique characteristics of cells in a fluid suspension are revealed. Using a series of lasers positioned at differing angles, cells are stratified according to their size and nuclear morphology. However, much of the utility of FC in the diagnosis and classification of canine LPD incorporates antigen-specific fluorescent antibodies to provide advanced immunophenotypic characterization of the neoplastic cell population. Using panels of cell lineage-specific antibodies (Table 1), FC is able to detect, quantify, and objectively characterize neoplastic lymphoid populations.

Table 1. Commonly Used Antibodies in the Canine LPD Flow Cytometry

Antigen	Cell specificity
CD3	T cells
CD5	T cells
CD4	Helper T cells and canine neutrophils
CD8	Cytotoxic T cells
CD79a or b	Mature B cells
CD21	Mature B cells
CD22	Mature B cells
CD45	Pan leukocyte
CD14	Monocytes
CD18	Pan leukocyte (+/- lymphocytes)
MHC II	Antigen presenting cells + mature lymphocytes
CD34	Hematopoietic progenitor cell

A further advantage of FC is that samples for analysis can be obtained using minimally-invasive methods, including body fluids and solid tissue aspirates. Bloody fluid samples, including peripheral blood and bone marrow, should be collected in an ethlyenediaminetetracetic acid (EDTA) anti-coagulant. Other fluid samples, including cerebrospinal, pleural, and peritoneal fluids, should be collected in either an EDTA anti-coagulant or serum tube. Solid tissue aspirates, including lymph node, spleen, and liver aspirates, should be placed in a tube containing either tissue culture medium (e.g. RPMI-1640) or 0.9% saline supplemented with serum (e.g. 10% fetal calf or patient serum). To ensure cell viability, samples should be analyzed within 48 hours of collection (ref).

FC and prognosis in canine LPD

In canine LPD, FC is primarily used to provide complementary immunophenotyping information once diagnosis of lymphoma has been made, including many cases diagnosed by FNAC. The utility of this approach, particularly using lymph node aspirates, is well recognized.[5] Moreover, recent work by Joetzke *et al* has confirmed the suitability of both liver and spleen aspirates for the detection of extranodal lymphoma in dogs.[6] Beyond categorization of lymphoid neoplasia as either T or B cell origin, a number of studies have demonstrated the prognostic utility of FC (Table 2).

Table 2. Prognostic Information Provided by Flow Cytometry of Canine LPDs

Neoplastic cell feature	Prognostic impact
Cell size and MHC II status in B-cell lymphoma	Large cells and low MHC class II expression are associated with a decreased survival.[7]
CD45 status in T-cell lymphoma	Loss of CD45 expression in T-cell lymphoma is associated with an indolent clinical course.[8]
CD4 expression in T-cell lymphoma	CD4 / CD45 positive T-cell lymphomas are associated with an aggressive clinical course.[9]
CD34 expression in lymphocytosis	CD34-positive lymphocytosis is associated with a very poor clinical course.[10]
Cell number in T-cell lymphocytosis	Dogs with > 30,000 CD8-positive cells/μL had a worse prognosis than dogs with fewer than <30,000 cells/.[10]
Cell size in B-cell lymphocytosis	Dogs with large CD21-positive cells had a worse prognosis than dogs with small cells.[10]

In a study evaluating dogs with newly-diagnosed nodal B-cell lymphoma, Rao *et al* illustrated the prognostic utility of cell size and MHC II expression in that dogs with lymphoma composed large cells or cells with low MHC class II expression had median survival times significantly shorter than their counterparts with neoplasms composed of larger cells or cells expression higher levels of

MHC II expression.[7] Two subsequent complementary publications have provided similar data in T-cell lymphoma. Avery *et al* showed that CD4 and CD45-positive lymphoma was associated with an aggressive clinical course whereas Seelig *et al* demonstrated that CD45-negative lymphoma was associated with a biologically indolent clinical course.[8,9] In addition to lymphoma, a number of studies have confirmed the prognostic utility of FC in canine leukemia. Comazzi *et al* found that T-cell chronic lymphocytic leukemia (CLL) had a better prognosis than B-cell CLL.[11] This finding was expanded by Williams *et al*, who demonstrated the importance of cell number and size in canine CLL, namely the association of cell size and cell number in B- and T-cell CLL respectively and the poor clinical outcome in dogs with acute (CD34-positive) leukemia.[10]

Can FC be used to diagnose canine LPD?

While there is an increasing body of literature demonstrating the utility of FC as a tool for lineage assignment of neoplastic cells within the bone marrow, peripheral blood, liver, spleen, and lymph node of dogs with lymphoma, there have been no objective studies evaluating its ability to definitively diagnose canine LPD and in particular, lymphoma. This knowledge gap largely results from the lack of published objective criteria upon which a diagnosis of lymphoma or leukemia can be definitely made using flow cytometry. As such, the true sensitivity and specificity of FC for the diagnosis of canine lymphoma is unknown. Despite this, there is general acceptance that FC can be used to diagnose lymphoma when an unusually high percentage of a single cell population is present or when phenotypically-aberrant cells are detected.

Work by Williams *et al*, which compared FC to FNAC in the diagnosis and monitoring of dogs with peripheral lymphadenopathy and lymphoma, reported that only 70% of dogs could be diagnosed by FC, although this low sensitivity likely resulted from, in part, the limited antibody panel used at the time.[12] It remains to be seen if the combination of FNAC and immunophenotyping by FC will replace histopathology in both the diagnosis and sub-classification of canine lymphoma. Long term follow-up studies comparing FC, FNAC, and histopathology are needed to address this question.

PCR for Antigen Receptor Rearrangement (PARR)

The PARR assay is a test of cellular clonality that determines whether a lymphoid population in a given sample is derived from a single precursor (i.e. is a clonal population) or not. In general, the presence of a clonal population supports a neoplastic process whereas a polyclonal population suggests a reactive (i.e. non-neoplastic) disorder. However, not all clonal lymphoid populations are neoplastic nor are all polyclonal expansions non-neoplastic, but the overall specificity of the assay is considered high.[13] The most notable example of a non-neoplastic clonal lymphoid population can be found in dogs with *Ehrlichia canis* infection.

Clinically, the PARR assay is most useful in the evaluation of samples in which the initial diagnosis is equivocal for the presence of a LPD. This is most commonly observed on FNAC samples with an ambiguous diagnosis of possible or probable lymphoma. The PARR assay can be performed on any sample in which sufficient amounts of high quality DNA can be harvested, including previously-stained cytology samples. Approximately 50,000 cells are needed for the assay.[14] In addition to diagnostic information, the PARR assay can provide limited immunophenotype data (i.e. a T- vs B-cell expansion). However owing to sensitivity and specificity concerns resultant from incomplete primer coverage, clonality in non-lymphoid malignancies, and poor performance compared to FC and IHC, PARR should be not be used as a tool for immunophenotyping.[15] Moreover, in light of the complexity of the assay and the varied protocols used by differing laboratories, it is important submitting clinicians understand sensitivity and specificity of their particular laboratory.

In addition to providing initial diagnostic and immunophenotyping data, PARR and FC can be used for the monitoring of canine patients with LPD. Work by Aresu *et al* demonstrated the utility of PARR, in concert with FC, in the detection of minimum residual disease in dogs with treated DLBCL.[16]

REFERENCES

1. Valli V, et al Vet Path 2011;48(1):198
2. Vezzali E, et al Vet Comp Oncol 2009;8(1):38
3. Ponce F, et al Vet Path 2010;47(3):414
4. Teske E, et al Vet Q 1996;18(3):112
5. Sozmen M, et al J Vet Diagn Invest 2005;17(4):323
6. Joetzke A, et al AJVR 2012;73(6):884
7. Rao S, et al J Vet Intern Med 2011;25(5):1097
8. Seelig D, et al J Vet Intern Med 2014;28(3):878
9. Avery P, et al J Vet Intern Med 2014;28(2):538
10. Williams, et al J Vet Intern Med 2008;22(3):596
11. Comazzi, et al J Vet Intern Med 2011;25(1):100
12. William L, et al JAVMA 2005;226(4):562
13. Burnett R, et al Vet Pathol 2003;40(1):32
14. Burkhard M, et al Vet Clin NA Small An Prac 2013;43(6):1331
15. Thalheim L, et al J Vet Intern Med 2013;27(6)1509
16. Aresu L, et al Vet J 2014;200(2)318

CLASSIFICATION OF CANINE LYMPHOPROLIFERATIVE DISEASES

Davis Seelig, DVM, PhD, DACVP
St. Paul, MN

INTRODUCTION

The lymphoproliferative diseases (LPDs) are a biologically heterogeneous group of neoplasms that result from the clonal expansion of lymphocytes at varied and unique stages of development. The heterogeneity of this disease is reflected clinically by its highly variable prognosis. According to the most recent World Health Organization (WHO) classification system, which segregates diseases according to morphologic, genetic, immunophenotypic, and clinical characteristics, there are more than 70 forms of the human disease, including mature and precursor forms of both B- and T-cell lymphoma and leukemia as well as histiocytic and dendritic cell neoplasia.[1]

Lymphoma is the most common canine LPD and an expanding body of work suggests substantial cytologic and histopathologic overlap with human non-Hodgkin's lymphoma. In light of that work, and an increasing appreciation of a similar degree of immunophenotypic and clinical heterogeneity in canine lymphoma, there have successful efforts to adapt the WHO for the classification of canine LPD. Resultant from this work are three studies that have demonstrated the reproducibility of this system for the successful classification of lymphoma in dogs, including one that demonstrated that a large group of veterinary pathologists without specific expertise in hematopathology could reach consensus on a series of over 200 lymphomas.[2-4]

LYMPHOMA CLASSIFICATION

Canine lymphoma has historically been classified using a number of different schemes that were originally developed for humans, including those emphasizing cell type and architectural pattern (the Working Formula) or cytomorphology (the Kiel classification scheme). Based upon these initial efforts it was clear that the vast majority of canine lymphomas (>80%) are composed of a rapidly proliferating, high-grade population of large cells with a diffuse, lymph node-effacing, architecture. Coincident with these advancing efforts in classification was the development of lineage-specific antibodies that enabled the immunophenotyping of lymphoma as either T- or B-cell origin. With this came the recognition that dogs with T-cell lymphomas tend to have a lower therapeutic response rate and shorter survival time than dogs with B-cell disease.[5]

Most recently, there has been successful adoption of a canine-adapted version of the human WHO classification scheme. The classification of lymphoid neoplasms using this scheme incorporates clinical presentation, histopathologic, immunophenotypic, and disease progression information. In the WHO scheme, canine lymphomas are histopathologically divided by growth pattern into diffuse (i.e. neoplasms that efface the affected tissue) or nodular (i.e. neoplasms that are multifocal in the affected tissue) variants. Next, the neoplastic cells are characterized according to their relation to residual lymphoid tissue and by size, nuclear morphology, and the number of mitoses per high power field. Finally, in concert with T- vs B-cell immunophenotyping and knowledge of their organ distribution, the disease is classified as a distinct entity.[2] This immunophenotypic information often results from immunohistochemistry on biopsy specimens, although there is increasing utilization of immunocytochemistry, flow cytometry (FC), and the PCR for antigen receptor rearrangement (PARR) clonality assay. Although it is likely that the complete spectrum of canine lymphoma has yet to be realized, in part owing to a lack of genetic and broad immunophenotypic information, there is general consensus that there are at least 17 variants of canine lymphoma (Table 1).

Table 1. Common Canine LPDs According to a WHO Classification Scheme[2-4]

B-cell	T-cell
Lymphoblastic Lymphoma	Lymphoblastic lymphoma
Diffuse large B cell lymphoma	Peripheral T cell lymphoma – not otherwise specified
Chronic lymphocytic leukemia / Small cell lymphoma	Chronic lymphocytic leukemia / Small cell lymphoma
Mantle cell lymphoma	T-zone lymphoma
Marginal zone lymphoma	Cutaneous T-cell lymphoma (mycosis fungoides)
Follicular lymphoma	Hepatosplenic T-cell lymphoma
Plasma cell neoplasia	T-cell Large granular lymphoma / leukemia
Burkitt's-like lymphoma	Intestinal T-cell lymphoma
T-cell-rich B-cell lymphoma	

RELEVANCE OF CLASSIFICATION

The goal of any canine LPD classification system is to provide the end user (clinician, pet owner, or scientist) with neoplasm-specific data that can inform expected response to therapy, duration of remission, and survival. Moreover, it should be capable of reproducibly segregating patients into unique categories so as to inform the practice of robust basic and clinical research. Additionally, the classification system should be broad enough to capture the spectrum of known diseases, yet flexible enough to incorporate new entities as they are identified. Finally, it needs to be sufficiently reproducible amongst those that are charged with applying it (i.e. pathologists). If any of these criteria are not addressed, a particular classification scheme is unlikely to be successful.

The shortcomings of many of the earliest canine LPD classification schemes resulted from an overreliance on only 1 or 2 pieces of information, including cellular morphology, architectural distribution, or the T vs B immunophenotype. Recent studies, including those that confirm the inability of routine cytology or histology to predict immunophenotype and those that illustrate the therapeutic implications of lymphoma classification, emphasize the importance of moving beyond the isolated "T or B cell" or "high-grade or low-grade" paradigms. (Flournel-Fleury & Ponce – T-cell disease). Included are studies that have identified uniquely indolent and aggressive forms of the disease (Table 2).

Table 2. The Clinical Impact of the Classification of Canine Lymphoma

Lymphoma subtype	Clinical Significance	Reference
Indolent lymphomas	Median lymphoma specific survival time of 52 months.[6]	Flood-Knapik
T-zone lymphoma	Median survival time of 21 - 33 months.[6,7]	Flood-Knapik / Seelig
CD4+ Peripheral T-cell Lymphoma and Lymphoblastic Lymphoma	Median survival time of 5.3 months.[8]	Avery
Marginal zone lymphoma	Median lymphoma specific survival time of 21.2 months.[6,9]	Flood Knapik

Indolent canine lymphoma

The most clinically impactful rationale for the classification of canine lymphoma is provided by recent work describing unique forms of indolent lymphoma. In contrast to most forms of canine lymphoma, which are biologically-aggressive and show relatively short survival times, the indolent lymphomas are slowly progressive and, in at least one case, demonstrate increased responsiveness to less aggressive chemotherapy.[6,10,11] There are four recognized variants of canine indolent lymphoma, including 3 B-cell diseases (marginal zone lymphoma [MZL], mantle cell lymphoma [MCL], and follicular lymphoma [FL]) and 1 T-cell disease (T zone lymphoma [TZL]). In 3 publications, the indolent lymphomas represent between 12-29% of all canine lymphoma.[6,10,11] The prognostic uniqueness of these diseases has been demonstrated in work by Flood-Knapik *et al*, which reported a median lymphoma specific survival time of 52 months.[6] As with other forms of canine LPD, it is currently recommended that indolent lymphoma be diagnosed using a combination of histopathology and immunophenotyping. However, according to one report, there is strong evidence that at least one form of these diseases (TZL) can be diagnosed using FC alone.

Marginal Zone Lymphoma

Marginal zone lymphoma is a form of indolent B-cell lymphoma that results from the clonal proliferation of a unique population of lymphocytes that reside within the marginal zone of the lymphoid follicle. In the dog, MZL has been shown to be of primary spleen or lymph node origin. The incidence and prevalence of canine MZL is unknown, in part owing to the small number of reports describing it and in the reported difficultly in distinguishing MZL from benign marginal zone hyperplasia.[9] However in 2 publications, it is reported to be the most common B-cell variant of canine indolent lymphoma.[6,10] While there is limited outcome information in canine MZL, Flood-Knapik *et al* report a median survival time of 21.2 months.[6] A more recent report by O'Brien *et al* describing the clinical behavior of canine splenic MZL reports a median survival time of 383 days in dogs undergoing splenectomy and 1,153 days in dogs without clinical signs referable to MZL.[9]

T-Zone Lymphoma

In humans, T-zone lymphoma (TZL) is a morphologic variant of peripheral T-cell lymphoma not otherwise specified (PTCL-NOS) characterized by the clonal expansion of T lymphocytes that manifest a unique architectural and cytomorphologic pattern. Similar clonality and histopathologic findings are found in canine TZL. As one of the indolent lymphomas, TZL is characterized by a prolonged clinical course. In contrast to the poor prognosis for dogs with a generic diagnosis of T-cell lymphoma, which have a reported median survival time of 159 days, dogs with T-zone lymphoma have a median survival time of 21 to 33 months.[6,7,12] Moreover, although the number of patients in the study was small, Flood-Knapik *et al* reported no difference in survival between dogs with TZL treated with a less aggressive chemotherapeutic regime (chlorambucil and prednisone) as compared to dogs treated with L-CHOP (L-asparaginase, cyclophosphamide, hydroxydaunorubicin, vincristine, and prednisone).[6] While the true incidence and prevalence of canine T-zone lymphoma is unknown, 10% of the submissions from dogs with suspected LPD to a leading US veterinary immunodiagnostics laboratory are reported to have an LPD with a TZL immunophenotype. Two additional publications similarly indicate that TZL is a relatively common canine LPD, comprising between 15.5 and 62% of all canine indolent lymphomas, which themselves comprise up to 29% of all canine lymphomas.[6,10]

ALTERNATE APPROACHES TO CLASSIFICATION

As the primary goal of canine LPD classification is to provide both pet owner and clinician with the most clinically-actionable information (i.e. tumor subtype, patient prognosis, and therapeutic options) in the most cost-effective and minimally invasive manner, it is increasingly attractive to consider flow cytometry the most efficient diagnostic option. A similar approach, which combines FNAC with multiparameter FC, is well-established in human medicine and can offer an 85% sensitivity and 100% specificity for both the diagnosis and classification of Non-Hodgkin's lymphoma.[13] While there are certainly limitations which prevent immediate and widespread adoption of this approach in canine medicine, chief of which is the limited scope of specific reagents, there are at least 2 examples where FC may be capable of replacing biopsy with histopathology as tool for the diagnosis and classification of canine lymphoma.

A recent report by Avery *et al* reports the utility of FC in the detection and classification of the most common form of canine T-cell lymphoma, namely CD4+ T-cell lymphoma.[8] In this report, FC on peripheral lymph node aspirates stratified canine CD4+ T-cell lymphoma into 2 groups, namely those that are CD45-positive with low class II MHC expression and those that are CD45-negative with high class II MHC expression. Using follow-up biopsy with histopathology and immunohistochemistry as the gold standard, the authors demonstrate that the CD45-positive variant represents 2 WHO subtypes, namely PTCL-NOS and lymphoblastic lymphoma. Seelig *et al* extend this work in a companion report in which they show that the CD45-negative variant is, in fact, canine TZL.[7] In addition to demonstrating the diagnosis and classification potential of FC, these reports illustrate the prognostic prowess of FC by reporting median survival times of 5.3 months and 33 months for the CD45-positive and CD45-negative groups, respectively. However, it is important to note FC alone is currently insufficient to classify many of the remaining forms of canine LPD, including all B-cell lymphomas.

REFERENCES

1. Swerdlow S, et al IARC Press, 2008
2. Valli V, et al Vet Path 2011;48(1):198
3. Vezzali E, et al Vet Comp Oncol 2009;8(1):38
4. Ponce F, et al Vet Path 2010;47(3):414
5. Ruslander D, et al In Vivo 1997;11(2):169
6. Flood-Knapik, et al Vet Comp Oncol 2013;11(4):272
7. Seelig D, et al J Vet Intern Med 2014;28(3):878
8. Avery P, et al J Vet Intern Med 2014;28(2):538
9. O'Brien D, et al J Vet Intern Med 2013;27(4):949
10. Valli V, et al Vet Path 2013;50(5):738
11. Aresu L, et al Vet J 2014;200(2)318
12. Vail D, et al in Small Animal Clinical Oncology 5th ed, Elsevier Saunder 2013
13. Bangerter, et al Acta Cytol 2007;51(3):390

VALIDATION OF TISSUE BANK SPECIMENS

Amy L. MacNeill, DVM, PhD, DACVP
Fort Collins, CO

INTRODUCTION

In the 1940's, the first organized collections of human tissues were established for medical and research use. The need for access to tissue banks for medical treatments continued to grow and, during the 1960's, small tissue banks were maintained in hospitals. As the demand for viable tissues became higher, laws regarding organ donation and accreditation of tissue banks became necessary to ensure quality of the tissues and safety of the patients. More recently, safety issues for the people collecting and processing the tissues have been addressed.[1]

The most organized tissue banks in veterinary medicine are blood banks. Easy access to blood products from healthy animals that are routinely screened for infectious diseases has occurred over the past decade. Veterinary diagnostic laboratories keep banks of formalin-fixed paraffin-embedded sections of tissues that are useful for retrospective histologic studies, but may not be suitable for many types of molecular analysis. Banks of fresh and fresh-frozen tissues suitable for molecular analyses are much less common and are typically small and maintained in individual research laboratories.

Similarly, there are few biorepositories available for cell culture specimens. It is well known that many research laboratories inadvertently utilized cells in culture that were contaminated by other cells or by infectious agents , which negates a significant amount of previously published cell culture data. Even large companies that maintain collections of cells may not screen the samples to ensure quality and accurate labeling. It is up to the researcher to ensure the quality and identification of tissues and cells used in experiments.

HUMAN TISSUES

The ethics of tissue banking are a prominent concern for tissue banks that maintain products derived from human patients. Novels such as *The Immortal Life of Henrietta Lacks* by Rebecca Skloot[2] have introduced these issues to the general public, inciting the argument that tissue donors should benefit when samples become products used in research. Proper consent forms are now used to inform patients of the need for tissues and how the samples may be used medically or for research purposes. The issue of which biological therapeutics should be overseen by which regulatory agencies is an ongoing legal discussion. The safety and effectiveness of biological products remain the primary concerns of the United States Federal Drug Administration and Health Canada.[3]

Scientists have identified short falls of the current system of tissue banking and oversight and have offered suggestions for improvement. Many of these issues are relevant for veterinary medicine. To make improvements, adequate funding is needed to maintain biorepositories and provide oversight of tissue banks. This is a confounding issue that may be overcome by establishing specific foundations that promote tissue banking.[4]

Key recommendations for improved tissue banking include:[4-6]

- Development of international ethical rules and standards of tissue banking
    - Informed, voluntary donation only
    - Open access of materials to the scientific community
    - Tissue banks act as "custodians" of specimens not "owners"
    - Biorepositories should be non-profit institutions
- Verification of tissue type and quality
    - Pathologists are critical for identification and characterization of samples[7-9]
    - Close interaction between the patient, surgeon, and pathologist is necessary
- Improved and integrated information technology systems
    - Protected data
    - Confidential data
    - Genetic testing shared with patients by genetic counselors if requested
- Enforceable regulatory oversight to ensure quality and safety
    - Credible quality management systems
    - Appropriate standard operating procedures in place
        - Maintain sample integrity
        - Biosafety for tissue bank personnel
    - Regular auditing of tissue bank procedures

VETERINARY TISSUES

Veterinary blood banks typically have standard operating procedures in place to ensure the safety of their products. Cats are often tested for feline leukemia, feline immune deficiency virus, toxoplasmosis, feline infectious peritonitis, and hemobartonella. Dogs are screened for pathogens including brucellosis, *Borrelia burgodorferi*, *Dirofilaria immitus*, and *Ehrlichia canis*. Donors are vaccinated regularly and are maintained in closed colonies to minimize exposure to pathogens. Ideally, physical examinations, complete blood counts, and serum chemistry profiles are routinely performed to monitor the overall health of the animals. Still, veterinarians must pay close attention to the quality of the sample they administer to patients. Blood products are easily damaged by improper sample handling.

Tissue samples for transplantation tend to be collected immediately prior to use. Often, owners agree to adopt the animal that is chosen as a donor. Recipients of tissues may continue to have serious medical issues following transplantation that the owner must be prepared for. Very few veterinary institutions offer tissue transplantation operations. Those that do perform transplants should have protocols in place that include extensive owner education about the procedure and necessary long-term medical maintenance of their pets' health.

TISSUE SECTIONS

Microscopic review of tissue samples used in molecular studies must be performed to ensure the tissue section being analyzed is an adequate, viable, representative sample. Pathologists have the necessary training to perform these critical reviews. Studies that would benefit greatly by having samples reviewed by a pathologist include gene expression studies, any methods that utilize tissue microarrays, and samples submitted for spectral imaging analysis.[9]

CELL CULTURES

Researchers frequently use cell lines and primary cell cultures to identify biological differences between healthy and diseased tissues and to screen novel therapeutics. The characteristics of the cell type used can drastically alter the outcome of these experiments. The use of cells cultured from veterinary species is vital to understand disease in individual species and gain an initial understanding of the effects of chemical compounds in cells of different species. Cultured primary cells and cell lines from various species are available through large tissue banks, including the American Type Culture Collection (www.atcc.org) and the Pfizer-Canine Comparative Oncology and Genomics Consortium Biospecimen Repository, and through individual research laboratories.

It is critical for researchers to know what cell type they are evaluating and ensure that the cells they use are not contaminated or infected with unknown products or organisms. Recurrent screening of cells for mycoplasmas and viruses is an important quality control procedure that laboratories should perform. This is easily accomplished by periodically submitting a sample of the cell culture for PCR analysis to detect possible pathogens.

In 2010, the World Health Organization (WHO) updated its *Recommendations for the evaluation of animal cell cultures as substrates for the manufacture of biological medicinal products and for the characterization of cell banks* which are available online at http://www.who.int/biologicals/Cell_Substrates_clean_version_18_April.pdf. These recommendations are particularly useful to researchers who are isolating, propagating, or utilizing cells in culture. Observation of Good Cell Culture Practices is stressed throughout the WHO recommendations. Key items of these practices include adequate documentation of cellular identity, origin, phenotypic and genotypic characteristics, and verification that cells are free of contamination by other cells or microbes. Culture conditions should be fully documented including specific culture reagents used during isolation and passage. Types of tests that aid in verification of cell type include karyology, short tandem repeat analysis, detection of multiple single nucleotide polymorphisms, major histocompatibility typing, isoenzyme testing, and DNA barcoding. Advantages and disadvantages of molecular tests to identify cells are indicated in the **Table**. Note that in primary cell cultures, a mixture of cell types may be present which can cause variability in the identity profile of the cells. Additionally, the WHO recommends that cell banks periodically evaluate stability and viability of the cells that they are providing to scientists. The WHO maintains reference cell banks that may be used for vaccine development, among other things. They provide information about the origin, materials used to create the seed stock, published scientific characterizations, and biological safety of the cells.

Table: Tests to verify the identity of cultured cells (modified from the 2010 WHO recommendations[10]).

Test	Advantages	Disadvantages
Karyotyping	Can identify the species of origin	Not specific for the individual animal Requires expertise
Isoenzyme analysis	Can identify the species of origin	Not specific for the individual animal No longer the preferred test
Cytochrome C oxidase I testing	Can identify the species of origin	Not specific for the individual animal
Short tandem repeat analysis	Rapid Specific for the individual animal	Not readily available for most species
Evaluation of multiple single nucleotide polymorphisms	Specific for the individual animal	Not commonly performed
Major histocompatibility class II typing	Somewhat specific for the individual animal	Not available for many species
Mitochondrial DNA barcoding	Can identify the species of origin	Not specific for the individual animal
Exon primed intron crossing PCR	Can identify the species of origin Can be used for all species	Not commonly performed (yet)

REFERENCES

1. Pedraza JM, et al. Cell and Tissue Banking 2012;13:15-25
2. Skloot, R. Crown Publishers 2010.
3. Farrugia A. Cell and Tissue Banking 2006;7:325-335
4. Ravid R. Cell and Tissue Banking 2008;9:121-137
5. Manyalich M, et al. Transplantation Proceedings 2009;41:2035-2043
6. Pedraza JM, Herson MR. Cell and Tissue Banking 2012;13:103-117
7. Leyland-Jones BR, et al. Journal of Clinical Oncology 2008;26(34):5638-5644
8. Bevilacqua G, et al. Virchows Archiv 2010;456:449-454
9. Hostetter G, et al. Veterinary Pathology 2014;51(1):270-280
10. World Health Organization. Recommendations for the evaluation of animal cell cultures as substrates for the manufacture of biological medicinal products and for the characterization of cell banks 2010

HISTIOCYTIC NEOPLASMS: MARKERS AND TREATMENT OPTIONS

Amy L. MacNeill, DVM, PhD, DACVP
Fort Collins, CO

INTRODUCTION

Diseases of histiocytes include both inflammatory and neoplastic conditions. Advances in the understanding of the basic immunology of histiocytic cells drive the continuous revision of theories about the underlying pathogenesis of histiocytic diseases. This has led to some confusing terminology. Specifically, the term malignant histiocytosis is no longer favored; the preferred term is disseminated histiocytic sarcoma. Also, malignant fibrous histiocytosis is inaccurate because, using immunochemical techniques, these lesions can be reclassified as specific subtypes of sarcoma. Likewise, it is now known that lesions diagnosed as fibrohistiocytic nodules actually include many different disease processes. The currently recognized histiocytic diseases are histiocytic inflammation, reactive histiocytoses, canine cutaneous histiocytoma, and histiocytic sarcomas.

HISTIOCYTE DIFFERENTIATION

Histiocytic cells include monocyte, macrophage, and dendritic cell lineages. These cells are derived from CD34$^+$ hematopoietic stem cells. When exposed to the appropriate microenvironment, stem cells differentiate into macrophage/dendritic cells precursors (MDPs) that express cell surface molecules CX_3CR1, CD115, and CD135.[1] Stimulation of MDPs by cytokines in the fetal liver and in adult hematopoietic tissues drive the development of several subtypes of monocytes, macrophages, and dendritic cells.

Monocytes and macrophages

Histiocytic precursor cells exposed to granulocyte, monocyte-colony stimulating factor (GM-CSF) and monocyte-CSF, tend to differentiate into monocytes. Monocytes are histiocytes that circulate in the peripheral blood. They express CD14, CD18, major histocompatibility complex class II (MHC II), and various other cell surface markers. Monocytes can be induced to differentiate into macrophages and dendritic cells in cell culture.[2] *In vivo*, if inflammatory cytokines [interleukin-12 (IL-12) and interferon-γ (IFNγ)] are present when the monocyte leaves the vasculature, the monocyte may differentiate into an active inflammatory macrophage, which is microbicidal, pro-inflammatory, and phagocytic.[3] If other cytokines (IL-4 and IL-13) are present, monocytes tend to differentiate into classic antigen presenting cells (APCs) that stimulate the adaptive immune response. Macrophages traffic to the tissues and are important in inducing inflammation (M1 macrophages) and tissue repair (M2 macrophages). Macrophage subtypes that reside in tissues can be identified by expression of β_2-integrins on the cell surface. For example, Kupffer cells of the liver, microglia, and type A synoviocytes are CD11b/CD18$^+$; whereas splenic, bone marrow and lymphoid macrophages express CD11d/CD18.[4]

Conventional dendritic cells

Exposure of MDPs to combinations of GM-CSF, IL-4, tumor necrosis factor-α (TNFα), and transforming growth factor-β1 (TGFβ1), induces the development of common dendritic cell precursors (CDPs). CDPs may then differentiate into plasmacytoid dendritic cells (which mediate inflammation) or into pre-dendritic cells. Pre-dendritic cells develop into conventional dendritic cells, which are potent APCs.[5] Conventional dendritic cells, including Langerhans cells and interstitial dendritic cells, are the cell types involved in many of the known histiocytic neoplasms. Langerhans cells develop from pre-dendritic cells of the fetal liver.[5] They are epidermal dendritic cells that are derived following exposure to TFβ1, a cytokine expressed by epidermal cells. The TFβ1-stimulated cells down regulate CD14 and upregulate CCR6, a chemokine receptor that triggers migration of the cells to the epidermis. As they localize to the epidermis, Langerhans cells increase expression of E-cadherin and CD1a. Dendritic cell differentiation from pre-dendritic cells and monocytes is driven by GM-CSF and IL-4. Interstitial dendritic cells can be found perivascularly in all tissues except for the brain, although they are present in the meninges and choroid plexus.[6] Subtypes of interstitial dendritic cells include interdigitating dendritic cells in lymph nodes and dermal dendritic cells. Many interstitial dendritic cells in the tissues migrate to lymph nodes following antigenic stimulation.

Morphology

Histiocyte subtypes are morphologically similar to each other and tend to be 15-20 μm in diameter with an ovoid to bean-shaped nucleus and blue-grey cytoplasm with small vacuoles and occasional granules. Morphologic changes occur in inflammatory macrophages which can be difficult to differentiate from neoplastic changes. When activated, macrophages can be up to 80 μm in diameter and can be swollen and filled with phagocytosed debris. They can have an epithelioid morphology, making sheets of macrophages difficult to differentiate from neoplastic epithelial cells. Additionally, macrophages can fuse together and become multinucleated giant cells. In order to differentiate histiocyte subtypes, immunologic evaluation of cell surface molecules and cytochemical detection of intracellular proteins is necessary. Cell surface markers vary between species and should be validated for each new species they are used in. The **Table** lists several of the markers that have been reported in domestic animals.[4,7-12] Note that not all markers listed have been validated in every species (for example, antibodies that detect feline CD11 molecules are not available). Currently available histiocyte markers can differentiate macrophages from subtypes of dendritic cells, but they are not good at differentiating inflammatory macrophages from neoplastic histiocytes. This confounds the pathologic diagnosis of histiocytic disease in many cases.

HISTIOCYTIC INFLAMMATION

Macrophages regulate inflammatory responses and clean up damaged cellular and microbial debris. They are critical components of the immune response and wound healing. Histiocytic or granulomatous inflammation is associated with fungal infections, atypical bacterial infections, foreign body reactions, and chronic inflammation.

Histiocytic ulcerative colitis

Histiocytic ulcerative colitis is characterized by severe colonic inflammation predominated by macrophages. The disease has been reported in both dogs and cats, but is most common in Boxers. Clinical signs include large bowel diarrhea and marked weight loss that often lead to death. Initially, this disease was defined as an idiopathic immune-mediated disease, but it is now believed to be an atypical response to bacterial overgrowth. The presence of invasive *Escherichia coli* within the colonic mucosa and the positive response to antibiotic therapy support this etiology.[13]

REACTIVE HISTIOCYTOSIS

In veterinary medicine, reactive histiocytoses are currently classified as non-neoplastic diseases. Underlying pathology seen with reactive histiocytoses includes Langerhans cell invasion of vasculature causing lymphohistiocytic vasculitis.[4]

Cutaneous and systemic histiocytosis

Both cutaneous histiocytosis and systemic histiocytosis have been documented in dogs and cats; cutaneous histiocytosis is more commonly seen.[4] Animals with cutaneous histiocytosis present with multiple skin lesions that may wax and wane. Spontaneous regression of the cutaneous lesions has been reported, but long-term immunosuppressive treatment is often needed. More than half of the dogs diagnosed with reactive histiocytoses respond to immunosuppressive treatments.[14] However, the overall prognosis is guarded due to ulceration of the lesions and disease recurrence. Cutaneous T cell lymphoma is an important differential diagnosis.

Cutaneous Langerhans cell histiocytosis

Cutaneous Langerhans cell histiocytosis is a rare condition of dogs that present with multiple skin lesions. It is most commonly diagnosed in the Shar pei breed. Delayed regression of the lesions has been documented. In other cases, lesions have metastasized to the lymph nodes and have led to death. Lesions may respond to treatment with lomustine (CCNU) but they typically recur.[4] In 57% of cutaneous Langerhans cell histiocytosis cases in humans, a point mutation in the gene encoding BRAF was found that alters the amino acid sequence of the protein and supports classification of this disease as a neoplastic condition.[5]

Pulmonary Langerhans cell histiocytosis

Feline pulmonary Langerhans cell histiocytosis is a fatal disease that has been diagnosed in low numbers of cats (typically 10-15 year-old) with respiratory distress. The disease causes a bronchointerstitial pattern with a miliary to nodular appearance on thoracic radiographs. Extrapulmonary spread is seen. Histologically, Birbeck granules can be identified in the neoplastic histiocytes. Birbeck granules are formed by intracellular aggregation of langerin (CD207), a C-type lectin found in Langerhans cells. In many species, langerin has Ca^{2+} binding sites that interact with mannose. These binding sites appear altered in dogs and may explain why Birbeck granules are not observed in canine histiocytic diseses.[4]

CUTANEOUS HISTIOCYTOMA

Cutaneous histiocytoma is a benign neoplasm of dogs. Typically, a cutaneous lesion is detected which regresses over time. Lesions are often round, raised, hairless, and mildly erythemic. Cells have a benign histiocytic appearance with abundant rounded lightly basophilic cytoplasm and an ovoid to bean-shaped nucleus. Occasional mitotic figures are often observed. A lymphocyte infiltrate occurs as the tumor begins to regress. This infiltrate can become the predominant cell type and may complicate the diagnosis of histocytoma. Treatment is not necessary unless the lesion fails to regress within a few months of diagnosis.

HISTIOCYTIC SARCOMA

Diagnosis of histiocytic sarcoma is suspected in cases where there are marked characteristics of malignancy in individualized, rounded cells with abundant pale cytoplasm, pleomorphic nuclei, and prominent nucleoli. Cells may exhibit phagocytic activity, vacuolation, and frequent multinucleation. Differential diagnoses may include anaplastic carcinomas, amelanotic melanoma, and other sarcomas. Differentiation between these tumor types can be made with immunochemical and cytochemical staining. Histiocytic cells should be positive for ionized calcium-binding adaptor molecule 1 (Iba1), lysozyme, vimentin, MHC II, CD11b, CD18, and CD204.[15] A recent publication demonstrated that mRNA expression of MHC IIα, CD11b, CD11c, and CD86 was significantly higher in biopsies from histiocytic sarcomas as compared to inflammatory diseases and other tumors.[15] Additionally, ferritin, fibrinogen, and C-reactive protein may be useful biomarkers for canine histiocytic sarcoma.[16-18]

Histiocytic sarcoma is seen frequently in Bernese mountain dogs and flat coated retrievers and is the cause of death in approximately 14% of these dogs.[19] In these breeds, the genetic inheritance is not fully recessive, but is oligogenetic, with a few key genes interacting to affect disease. Recurrent deletions in chromosomes 2, 11, 16, 22 and 31 have been reported. Defects associated with these deletions include loss of the tumor suppressor genes CDKN2A/B, RB1, and PTEN.[8,20]

Case studies of localized, systemic and hemophagocytic histiocytic sarcoma in dogs are often evaluated together, even though the disease subtypes have distinctly different clinical signs. Dogs between 1 and 15 years of age have been affected.[21] Response rates to CCNU range from 29-55%.[14,21,22] Similar response rates were seen with nimustine (ACNU) therapy. No response was observed following treatment with L-asparaginase or doxyrubicin.[21] Overall survival rates with treatment range from 2-468 days and 1 year survival times range from 0-30% have been reported.[14] Dogs with anemia consistently have decreased median survival times. Thrombocytopenia and hypoalbuminemia also were negatively associated with prognosis.[23]

Localized histiocytic sarcoma

These tumors arise from a single site but often metastasize to draining lymph nodes. Cutaneous lesions on the limbs are most frequently reported. Articular histiocytic sarcoma tends to be associated with mixed inflammation and may be difficult to distinguish from chronic inflammation or synovial cell sarcoma. The two neoplastic differentials can be differentiated using immunochemistry.[4] Subdural histiocytic sarcoma has been reported as a localized lesion the cerebrum and spinal cord, but more diffuse lesions can be observed.[24] There also are several case reports of ocular histiocytic sarcoma. The median survival time of splenic histiocytic sarcoma ranges from 1 month to 1 year.[14] Aggressive surgical removal of localized histiocytic sarcoma lesions is recommended.

Disseminated histiocytic sarcoma

Disseminated histiocytic sarcoma was previously termed malignant histiocytosis. Disseminated histiocytic sarcoma is a disease of neoplastic interstitial dendritic cells that affects lymphoid tissues, bone marrow, and liver.[25] Common clinical signs include anorexia, weight loss, lethargy, and dyspnea.[26] Treatment of systemic histiocytic sarcoma is difficult as the neoplasm responds somewhat poorly to chemotherapy. Treatment with CCNU has been reported most frequently.[21,23,27]

Hemophagocytic histiocytic sarcoma

Hemaphagocytic histiocytic sarcoma has been documented in dogs, cats, cattle, and horses.[28-30] The disease is caused by neoplastic macrophages which originate in the spleen or bone marrow. Clinical signs include splenomegaly and jaundice. Animals are often anemic, thrombocytopenic, hypoalbuminemic, hypocholesterolemic, and hyperbilirubinemic. The prognosis is grave with a reported median survival time of 4 weeks in dogs. This disease can be differentiated from immune mediated hemolytic anemia by identification of the neoplastic histiocytes.[4,31] Neoplastic cells tend to be large and individualized. They have abundant cytoplasm, pleomorphic and multiple nuclei, and prominent irregular nucleoli. Erythrophagocytosis is a predominant feature of this neoplasm.

Feline progressive histiocytosis

Feline progressive histiocytosis is caused by neoplastic dermal dendritic cells. Cats from 7-17 years-old have been diagnosed with the disease. Single or multiple lesions may be seen and are often found on the head or distal limbs. These lesions inevitably progress and spread systemically. The disease course is prolonged, but full regression has not been documented.[4,32]

Table: Identification of histiocytic cell subtypes. The most frequently utilized markers are in bold font.

Histiocytic cell	Associated diseases	Key cellular markers present	Cellular markers with variable expression	Key cellular markers absent
Monocytes	Acute myelomonocytic and monocytic leukemias	CD14, CD18, MHC II		CD3, CD4, CD8, CD79a
Macrophages	Hemophagocytic histiocytic sarcoma	CD11b/CD18, CD11c/CD18, **CD11d/CD18**, CD14, CD15, CD43, CD45, CD45RO, CD68, CD163, CD204, Mac387, MHC I, MHC II, Iba1, lysozyme, vimentin	**CD1a**	CD3, CD4, CD8, CD30, CD79a, Melan-A, cytokeratin, myeloperoxidase
Langerhans cells (LC)	Cutaneous LC histiocytosis Pulmonary LC histiocytosis Histiocytoma	**CD1a**, CD11a/CD18, **CD11c/CD18**, CD44, CD45, CD207, MHC I, MHC II, ICAM-1, Iba1, vimentin	CD11b/CD18, CD54, CD90, **E-cadherin**, lysozyme	CD3, **CD4**, CD8, CD30, CD45RA, CD79a, Melan-A, cytokeratin, myeloperoxidase
Reactive interstitial dendritic cells	Cutaneous histiocytosis Systemic histiocytosis	**CD1a**, **CD4**, **CD11c/CD18**, **CD90**, CD204, **MHC II**, Iba1, vimentin	Lysozyme	CD3, CD8, CD30, CD79a, **E-cadherin,** Melan-A, cytokeratin, myeloperoxidase
Neoplastic interstitial dendritic cells	Feline progressive histiocytosis Histiocytic sarcoma Dendritic cell leukemia	**CD1a**, **CD11c/CD18**, MHC I, MHC II, Iba1, vimentin	CD90, CD80, CD86, CD204, E-cadherin, lysozyme	CD3, **CD4**, CD8, CD30, CD79a, Melan-A, cytokeratin, myeloperoxidase

REFERENCES

1. Auffray C, et al. **Journal of Experimental Medicine** 2009;206(3):595-606
2. Strauss-Ayali D, et al. **Journal of Leukocyte Biology** 2007;82(2):244-252
3. Ferrante CJ, Leibovich SJ. **Advances in Wound Care** 2012;1(1):10-16
4. Moore PF. **Veterinary Pathology** 2014;51(1):167-184.
5. Berres ML, et al. **Advances in Immunology** 2013;120:127-161.
6. D'Agostino PM, et al. **Acta Neuropathologica** 2012;124(2):599-614.
7. Takahashi M, Nakamura S. **Journal of Clinical and Experimental Hematopathology** 2013;53(1):1-8.
8. Dobson JM. **ISRN Veterinary Science** 2012;2013:1-23.
9. Affolter VK, Moore PF. **American Journal of Dermatopathology** 2000;22(1):40-48.
10. Moore PF. **Veterinary Pathology** 1986;23(6):757-762.
11. Moore PF. **Veterinary Pathology** 1986;23(6):763-769.
12. Kato Y, et al. **Veterinary Clinical Pathology** 2014;43(4):589-593.
13. Mansfield CS, et al. **Journal of Veterinary Internal Medicine** 2009;23(5):964-969.
14. Coomer AR, Liptak JM. **Compendium: Continuing Education for Veterinarians** 2008;30(4):202-204.
15. Yamazaki H, et al. **Journal of Veterinary Internal Medicine** 2014;28:204-210.
16. Nielson LN, et al. **Veterinary and Comparative Oncology** 2011;10(2):124-134.
17. Friedrichs KR, et al. **Journal of Veterinary Internal Medicine** 2010;24:904-911.

18. Nielsen LN, et al. **Veterinary Journal** 2013;198(2):424-428.
19. Erich SA et al. **Veterinary Journal** 2013;198(3):678-683.
20. Hedan B, et al. **BMC Cancer** 2011;11:1-14.
21. Takahashi M, et al. **Journal of Veterinary Medical Science** 2014;76(5):661-666.
22. Rassnick KM, et al. **Journal of Veterinary Internal Medicine** 2010;24(6):1528-1531.
23. Skorupski KA, et al. **Journal of Veterinary Internal Medicine** 2007
24. Ide T, et al. **Journal of Veterinary Diagnostic Investigation** 2011;23(1):127-132.
25. Affolter VK, Moore PF. **Veterinary Pathology** 2002;39(1):74-83.
26. Schmidt ML, et al. **Veterinary Quarterly** 1993;15(3):117-120.
27. Marina E, et al. **Veterinary Dermatology** 2014;25(6):559-562.
28. Matsuda K, et al. **Veterinary Pathology** 2010;47(2):339-342.
29. Paciello O, et al. **Research in Veterinary Science** 2013;94:648-650.
30. Anjiki T, et al. **Journal of Veterinary Medical Science** 2000;62(12):1235-1240.
31. Moore PF, et al. **Veterinary Pathology** 2006;43(5):632-645.
32. Affolter VK, Moore PF. **Veterinary Pathology** 2006;43(5):646-655.

DISCUSSION OF CHALLENGING CASES OF FELINE HEMATOLOGY

John W. Harvey, DVM, PhD, DACVP
Gainesville, FL

OVERVIEW

A brief history and physical findings will be presented for five feline cases in which the acquisition and interpretation hematologic data was essential in reaching the disease diagnosis. Results from CBCs are provided for each case. The significance of abnormalities identified in each CBC will be discussed in context with history and other hematology findings. The disease present will be characterized to the extent possible from the CBC alone, and the need for additional information and/or additional diagnostic tests (bone marrow evaluation, serologic assays, etc.) will be identified. Occasionally, a diagnosis can be made from careful examination of a stained blood film, such as when an infection agent is identified, but usually additional diagnostic tests are required to reach a diagnosis. After all available information is considered, a final diagnosis will be rendered for each case.

CASE 1

History and Physical Findings

This 6-year-old male mixed breed cat presented with a 2-week history of weight loss. even though the animal is still eating. Physical examination revealed a thin cat with increased respiratory rate and slight dehydration. Peripheral lymphadenopathy (bilateral prescapular and inguinal) was observed and a large mid-abdominal mass was palpated.

Hematology Findings

Parameter	Patient	Reference Interval
HCT (%)	33L	34-51
MCV (fL)	50	42-52
MCHC (%)	34H	30-33
RDW (%)	18H	13-16
Ict Index (units)	<5	<5
TPP (g/dL)	7.6	6.2-8.0
Fibrinogen (mg/dL)	100	100-300
Platelets ($x10^3/\mu L$)	390	160-502
Total WBC ($x10^3/\mu L$)	16.4H	5.4-15.4
Bands	0.2	0-0.3
Neutrophils	13.3H	2.3-9.8
Lymphocytes	0.7 L	0.9-5.5
Monocytes	1.8H	0-0.8
Eosinophils	0.4	0-1.8
Basophils	0	0-0.2

Erythrocyte morphology: 1+ elliptocytes

CASE 2

History and Physical Findings

A four-year-old castrated male mixed-breed cat was presented with a history of pustules and crusting on the skin of the ears and feet. Coughing was present that was attributed to an allergic tracheobronchitis. The cat was treated for 3 weeks with prednisone (glucocorticoid) and 16 days with griseofulvin (fungistatic drug used to treat dermatophyte infections). Anorexia, diarrhea and fever began 3 days ago and griseofulvin therapy was stopped 2 days ago. Scaling and crusting of footpads was observed during the physical examination and inspiratory crackles were heard on auscultation of the chest. The rectal temperature was 106°F.

Hematology Findings

Parameter	Patient	Reference Interval
HCT (%)	38	34-51
MCV (fL)	47	42-52
MCHC (%)	31	30-33
Ict Index (units)	<5	<5
TPP (g/dL)	8.9H	6.2-8.0
Fibrinogen (mg/dL)	500H	100-300
Platelets (x10^3/µL)	Normal	160-502
Total WBC (x10^3/µL)	1.2 L	5.4-15.4
Bands	0	0-0.3
Neutrophils	0.1 L	2.3-9.8
Lymphocytes	0.9	0.9-5.5
Monocytes	0	0-0.8
Eosinophils	0.2	0-1.8
Basophils	0	0-0.2

Erythrocyte morphology: Marked rouleaux with 1+ anisocytosis.

CASE 3

History and Physical Findings

An 11-year-old male DSH cat was presented with lethargy and anorexia for one day. It was an indoor cat but had escaped 3 days earlier. Physical examination revealed dehydration, weight loss, prominent 3rd eyelids and a grade II/VI left side heart murmur. The rectal temperature was 105°F.

Hematology Findings

Parameter	Patient	Reference Range
HCT (%)	29 L	34-51
MCV (fL)	44	42-52
MCHC (%)	33	30-33
Icterus Index (units)	<5	<5
Fibrinogen (mg/dL)	500H	100-300
TPP (g/dl)	7.1	6.2-8.0
Platelets (x10^3/µL)	Normal	300-800
Total leukocytes (x10^3/µL)	2.7 L	5.4-15.4
Bands	0	0-0.3
Neutrophils	2.5	2.3-9.8
Lymphocytes	0.1 L	0.9-5.5
Monocytes	0.1	0-0.8

Eosinophils	<0.1	0-1.8
Basophils	0	0-0.2

Erythrocyte morphology: Prominent rouleaux, Neutrophil morphology: 1+ toxicity in most cells

CASE 4
History and Physical Findings

A three-year-old male castrated DSH cat was presented to a practicing veterinarian for a non-healing skin lesion on the face. Physical examination also revealed an ulcer on the roof of the mouth. The skin lesion was surgically removed and the histopathologic diagnosis was eosinophilic granuloma. A CBC done at the time revealed at PCV of 30%, slight monocytosis, and slight eosinophilia. A FeLV test was negative. The cat was returned to the practicing veterinarian 25 days later with a complaint of anorexia for two days. Mucous membranes appeared pale and the rectal temperature was 102 F. Hematology findings from the second visit are given below.

Hematology Findings

Parameter	Patient	Reference Range
HCT (%)	12 L	34-51
Icterus Index (units)	<5	<5
Platelets (x10^3/μL)	normal	160-502
Corrected total leukocytes (x10^3/μL)	13.2	5.4-15.4
Bands	0.3	0-0.3
Neutrophils	3.1	2.3-9.8
Lymphocytes	8.1H	0.9-5.5
Monocytes	1.4H	0-0.8
Eosinophils	0.3	0-1.8
Basophils	0	0-0.2
Nucleated RBC (x10^3/μL)	46.8H	0

Erythrocyte morphology: normal for mature erythrocytes, but some nucleated erythrocytes exhibited were immature and some exhibited dysplastic changes. Low numbers of *M. haemofelis* were also present.

CASE 5
History and Physical Examination

A two-year-old neutered female cat was presented with a history of decreased appetite for one week. She became icteric and began vomiting 4 days ago. The animal was apprehensive, had bilateral ocular discharge and had slight hepatomegaly. The rectal temperature was normal.

Hematology Findings

Parameter	Patient	Reference Range
HCT (%)	42	34-51
MCV (fL)	45	42-52
MCHC (%)	35H	30-33
Icterus index (units)	50H	<5
Total plasma protein (g/dL)	7.0	6.2-8.0
Fibrinogen (mg/dL)	100	100-300

Platelets ($x10^3/\mu L$)	305	160-502
Total leukocytes ($x10^3/\mu L$)	12.9	5.4-15.4
Bands	0.1	0-0.3
Neutrophils	6.3	2.3-9.8
Lymphocytes	2.1	0.9-5.5
Monocytes	1.5	0-0.8
Eosinophils	2.9H	0-1.8
Basophils	0	0-0.2

Erythrocyte morphology revealed 3+ echinocytosis and 1+ acanthocytosis. Small Heinz bodies were present in 15% of the erythrocytes.

REFERENCES

Harvey JW. Veterinary Hematology. A Diagnostic Guide and Color Atlas. Elsevier/Saunders, St. Louis, MO, 2012.

DISCUSSION OF CHALLENGING CASES OF CANINE HEMATOLOGY

John W. Harvey, DVM, PhD, DACVP
Gainesville, FL

OVERVIEW

A brief history and physical findings will be presented for five canine cases in which the acquisition and interpretation hematologic data was essential in reaching the disease diagnosis. Results from CBCs are provided for each case. The significance of abnormalities identified in each CBC will be discussed in context with history and other hematology findings. The disease present will be characterized to the extent possible from the CBC alone, and the need for additional information and/or additional diagnostic tests (bone marrow evaluation, serologic assays, etc.) will be identified. Occasionally, a diagnosis can be made from careful examination of a stained blood film, such as when an infection agent is identified, but usually additional diagnostic tests are required to reach a diagnosis. After all available information is considered, a final diagnosis will be rendered for each case.

CASE 1

History and Physical Findings

Regenerative anemia and thrombocytopenia were first recognized five weeks ago in an eight-year-old male mixed-breed dog. A presumptive diagnosis of immune-mediated hemolytic anemia and immune-mediated thrombocytopenia was made by the referring veterinarian. The dog was treated with prednisone, danazol, cyclosporine, and doxycycline with little improvement. A bone marrow aspirate revealed megakaryocytic hyperplasia and erythrocytic hyperplasia 3 days ago. IV vincristine was given after examining the bone marrow. Anorexia and vomiting began one day ago. Physical examination revealed pale, icteric mucous membranes and red blood in the stool.

Hematology Findings

Parameter	Patient	Reference Interval
HCT (%)	20 L	40-56
MCV (fL)	83 H	64-74
MCHC (%)	30 L	33-38
RDW (%)	22 H	11-14
Reticulocytes ($x10^3/\mu L$)	7 L	8-65
Icterus index (units)	15 H	<5
Total plasma protein (g/dL)	9.1 H	6.0-8.0
Fibrinogen (mg/dL)	600 H	100-400
Platelets ($x10^3/\mu L$)	42 L	134-396
Total leukocytes ($x10^3/\mu L$)	2.7 L	5.0-13.0
Bands	<0.1	0-0.3
Neutrophils	1.4 L	2.7-8.9
Lymphocytes	0.8 L	0.9-3.4
Monocytes	0.4	0.1-0.8
Eosinophils	0.1	0.1-1.3
Basophils	0	0-0.1

Erythrocyte morphology: 4+ anisocytosis, 1+ codocytes, rare NRBC

CASE 2

History and Physical Examination

A six-month-old male miniature poodle was castrated one week ago and treated postoperatively with antibiotics. He exhibited anorexia, lethargy, and red urine beginning four days ago and vomited yesterday. He was given IV fluid and referred to the UFVH. He was depressed, had bilateral nasal discharge, was icteric, had red urine, and had increased lung sounds. SQ hemorrhage was present at the castration site.

Hematology Findings

Parameter	Patient	Reference Interval
HCT (%)	11 L	40-56
MCV (fl)	64	64-74
MCHC (%)	35	33-38
RDW (%)	28 H	11-14
Retics (x10^3/μl)	295 H	8-65
Ict Index (units)	Hemol H	<5
TPP (g/dl)	8.0	6.0-8.0
Fibrinogen (mg/dl)	800 H	100-400
Platelets (x10^3/μl)	388	134-396
Total WBC (x10^3/μl)	79.0 H	5.0-13.0
Metas	0.8 H	0
Bands	20.5 H	0-0.3
Neutrophils	36.3 H	2.7-8.9
Lymphocytes	2.4	0.9-3.4
Monocytes	19.0 H	0.1-0.8
Eosinophils	0	0.1-1.3
Basophils	0	0-0.1
NRBC (x10^3/μl)	4.0 H	0

Erythrocyte Morphology: 2+ polychromasia, anisocytosis, and poikilocytosis

CASE 3

History and Physical Examination

A nine-year-old castrated male golden retriever had lethargy for six months and episodic bilateral epistaxis for five months. Bilateral hip dysplasia was diagnosed previously. He was treated with tetracycline and a vitamin/mineral supplement (Lixotinic) when the PCV was 14%. The dog has lived in Tallahassee, Florida for last year. On physical examination, the dog was thin and lethargic, with slight pallor of the mucous membranes. Bilaterally enlarged popliteal and prescapular lymph nodes were noted.

Hematology Findings

Parameter	Patient	Reference Interval
HCT (%)	21 L	40-56
MCV (fL)	56 L	64-74
MCHC (%)	34	33-38
Icterus index (units)	<5	<5
Total plasma protein (g/dL)	10.0 H	6.0-8.0
Fibrinogen (mg/dL)	300	100-400
Platelets (x10^3/μL)	148	134-396
Total leukocytes (x10^3/μL)	18.7 H	5.0-13.0

Bands	0	0-0.3
Neutrophils	16.3 H	2.7-8.9
Lymphocytes	0.7L	0.9-3.4
Monocytes	1.5 H	0.1-0.8
Eosinophils	0.2	0.1-1.3
Basophils	0	0-0.1

Erythrocyte morphology: Slight anisocytosis

CASE 4
History and Physical Examination

A six-year-old Doberman was presented with a history of lethargy and multiple SQ masses. Diarrhea with black-colored feces and red urine had recently been recognized. Physical examination revealed pale mucous membranes, an enlarged submandibular lymph node and multiple small SQ masses. The abdomen appeared to contain abdominal fluid. The temperature was 105°F. Excessive hemorrhage occurred following phlebotomy.

Hematology Findings

Parameter	Patient	Reference Interval
PCV (%)	21 L	40-56
MCV (fL)	63 L	64-74
MCHC (%)	35	33-38
RDW (%)	16	11-14
Retics ($x10^3/\mu L$)	ND	8-65
Ict Index (units)	15 H	<5
TPP (g/dl)	5.0 L	6.0-8.0
Fibrinogen (mg/dl)	<100 L	100-400
Platelets ($x10^3/\mu L$)	6 L	134-396
Total WBC ($x10^3/\mu L$)	30.0 H	5.0-13.0
Bands	5.4 H	0-0.3
Neutrophils	22.1 H	2.7-8.9
Lymphocytes	0.9	0.9-3.4
Monocytes	1.3 H	0.1-0.8
Eosinophils	0	0.1-1.3
Basophils	0	0-0.1
PT (sec)	24 H	8 control
APTT (sec)	28 H	11 control
FDP	neg 1/5 pos 1/20 H pos 1/100 H	neg

Erythrocyte morphology: 1+ poikilocytosis (acanthocytes) and polychromasia and 2+ anisocytosis.

CASE 5

History and Physical Examination

A seven-year-old male golden retriever dog was presented to a referring veterinarian 10 days ago for intermittent limping, depression, and poor appetite. A clinical chemistry panel revealed hypercalcemia and azotemia. The dog was referred to the UFVH for diagnostic evaluation. He appeared normal on physical examination.

Hematology Findings

Parameter	Patient	Reference Interval
HCT (%)	44	40-56
MCV (fL)	68	64-74
MCHC (%)	34	33-38
Icterus index (units)	<5	<5
Total plasma protein (g/dL)	6.0	6.0-8.0
Fibrinogen (mg/dL)	100	100-400
Platelets (x10^3/µL)	65 L	134-396
Total leukocytes (x10^3/µL)	1.9 L	5.0-13.0
Bands	0	0-0.3
Neutrophils	0.8 L	2.7-8.9
Lymphocytes	0.9	0.9-3.4
Monocytes	0.2	0.1-0.8
Eosinophils	0	0.1-1.3
Basophils	0	0-0.1

Erythrocyte morphology: normal

REFERENCES

Harvey JW. Veterinary Hematology. A Diagnostic Guide and Color Atlas. Elsevier/Saunders, St. Louis, MO, 2012.

DIAGNOSIS OF INFECTIOUS AGENTS ON BLOOD FILMS – CHALLENGING CASES

Jennifer Johns, DVM, PhD, DACVP (Clinical Pathology)
Stanford, CA

BLOOD FILMS:

The preparation of high-quality blood films (smears) for evaluation and diagnosis of potential infectious agents relies on the use of good technique in blood collection and smear preparation. In most small animal patients, blood smears will be prepared using blood that has been collected into EDTA anticoagulant. If more convenient, the blood smear can be made from blood before mixing it in anticoagulant, but any delay can result in clotting of the sample. Use of heparin anticoagulant for hematology samples is discouraged as it can drastically impair staining of the smear.

Stains and stain care:

In-house stain sets quickly become contaminated when used for cytology samples such as ear swab smears and skin scrapes. The best solution is to maintain a separate stain set for blood smears only, or for blood smears and "clean" cytology samples (e.g. organ aspirates and body fluid smears) to avoid contamination problems. Bacterial contamination of stains can cause confusion when evaluating a stained smear for evidence of infection. In addition, stains should be filtered and/or changed regularly to avoid precipitate buildup, as precipitate can also impede smear evaluation for infectious agents, and to keep them fresh as quick stains lose potency over time/use. In our experience, glass coplin jars are superior to plastic for cleaning and removal of stain precipitate; scrub out stain jars completely and wash and dry them well before refilling them.

Sources of contamination and confusion in blood smears:

Blood smears are less prone to contamination than cytology smears (see below discussion of contaminants in tissue cytology smears), but several contaminants including glove powder and superficial epithelial cells (if smears/slides are handled by non-gloved hands) can potentially be seen.

Common sources of confusion in blood smears when evaluating for infectious agents are pyknotic cells and cellular debris. Additionally, when evaluating for hemotropic mycoplasmas, the presence of Howell-Jolly bodies in RBCs can be a source of confusion. Howell-Jolly bodies are generally larger than most *Mycoplasma* seen on the surface of red blood cells (RBCs); they vary in size more dramatically and are generally seen singly in RBCs, in contrast to the chains of bacteria that these *Mycoplasma* spp. can form. Basophilic stippling and siderotic inclusions in RBCs can also cause potential confusion with hemotrophic mycoplasmas, but both are fairly rare findings in blood films from small animal patients. Other rare blood smear findings that can potentially be confused with microorganisms include staining cytoplasmic granules due to lysosomal storage disorders (see images in [1]). Artifacts that can cause confusion include drying artifact (refractile crescent shapes seen in cells; this is due to staining the smear before it is completely dry), and stain precipitate.

Evaluation of smears for potential infectious agents:

In some cases, characteristic hematologic abnormalities will be seen in conjunction with infectious agents in blood smears, strengthening the likelihood of the diagnosis. For example, hemotropic mycoplasmas may induce hemolytic anemia that can become regenerative in several days' time; agglutination and/or spherocytes may be noted in the blood smear. Some bacterial agents that can be seen in blood may cause a left shift, neutrophil cytoplasmic toxicity, and/or neutrophilia or neutropenia depending on the duration and severity of infection. However, acute or peracute bacteremia may not have associated leukocyte abnormalities. Some bacteria in family Anaplasmataceae (*Ehrlichia* and *Anaplasma* species) frequently induce thrombocytopenia (generally moderate) during acute host infection; the absence of thrombocytopenia would lessen the likelihood of diagnosis when identification of the organisms is questionable.

In cases where infectious agents of leukocytes are suspected, preparation of a buffy coat smear should be considered to increase the sensitivity of the cytology review. In one recent study, cytology of a buffy coat smear proved to be more sensitive than cytology of a routine blood smear or bone marrow smear for detection of *Hepatozoon* canis.[2] Other diagnostic techniques not discussed here, including serology and PCR, are often utilized for diagnosis of infectious disease in blood when definitive cytologic evidence of infectious agents is lacking and/or when complete speciation is required. For some infectious diseases, a combination of cytology of blood smears and tissue aspiration smears yields greater sensitivity than when each is performed alone. In one study, the combined sensitivity of buffy coat smears and lymph node aspiration smears for cytologic detection of *Ehrlichia canis* was superior to the individual sensitivity of the two specimen types.[3]

REFERENCES

1. Harvey JW. 2012. Veterinary Hematology – a diagnostic guide and color atlas;
2. Otranto D, et al. Diagnosis of Hepatozoon canis in young dogs by cytology and PCR. Parasit Vectors. 2011 Apr 13;4:55;
3. Mylonakis ME, et al. **Vet Microbiol.** 2003 Feb 2;91(2-3):197-204.

DIAGNOSIS OF INFECTIOUS AGENTS BY TISSUE CYTOLOGY – CHALLENGING CASES

Jennifer Johns, DVM, PhD, DACVP (Clinical Pathology)
Stanford, CA

TISSUE CYTOLOGY:

Collection technique is often selected based on operator preference/experience and specific circumstances, e.g. ultrasound-guided collection. Fine needle aspiration of most tissues and organs can be performed either with or without suction applied via syringe. A 12-cc syringe is a useful general-purpose size for FNA[1]. If suction is not used, it is convenient to aspirate a few cc's of air into the syringe prior to aspiration, to facilitate expelling the sample onto the slide. Careful preparation of the smear following FNA is critical. If smears are insufficiently smeared, the material may be too thick to evaluate. Conversely, fragile cells can easily be ruptured during smearing. If cell disruption is a recurrent problem, one useful technique is to use a coverslip to smear the material onto the slide. The coverslip can be stained and read in-house, and the slide can be submitted to a referral laboratory if indicated. Referral laboratories generally do not accept coverslip smear submissions for cytology.

Other techniques include swabbing a lesion then rolling the swab onto a slide; scraping a lesion then smearing the material on a slide; and making imprints/impressions by dabbing a biopsy tissue/ gently onto the slide, or alternately blotting the slide directly on the cutaneous lesion (see Table 1.1 in [1]). Fluids are generally handled in a standard fashion: a direct smear is first made, then (if possible) automated cell counts are obtained and concentrated smears are prepared based on the nucleated cell count.
Collection technique may be optimized by use of endoscopy for direct visualization in some sites, e.g. nasal passages. One study found a marked improvement in sensitivity of cytology for detection of fungal hyphae when nasal samples were collected using rhinoscopy for visualization of lesions.[2]

Stains and stain care:

See adjacent section on stain care for blood films. Maintenance of a separate stain set for "clean" smears is highly recommended. Regular filtration and replacement of stains is necessary to avoid buildup of precipitate and to preserve the function of the stains, as quick stains can lose potency over time.

Sources of contamination and confusion in tissue cytology smears:

Cytology smears are prone to contamination with a variety of agents. As with blood smears, glove powder and superficial epithelial cells can be found in smears due to handling of the smears or the slides prior to smearing. Other common contaminants can be found under certain collection conditions: for example, ultrasound gel can be seen in smears from ultrasound-guided fine needle aspirates (FNAs), and lubricant may be seen when used during sample collection. Stain precipitate is a common contaminant in all smears when stains are inadequately maintained. One type of contaminant that is most common in swabs and imprint cytology samples are superficial bacteria on the skin/fur of the patient.

Common sources of confusion in cytology smears when evaluating for infectious agents are phagocytosed debris and other material within leukocytes, karyorrhexic nuclei within pyknotic cells (can be mistaken for intracellular bacterial cocci), capillary fragments and streaming nuclear material (can be mistaken for fungal hyphae).

Evaluation of smears for potential infectious agents:

The presence of characteristic cellular changes usually strengthens the likelihood of a diagnosis of infectious disease when identification of the organism itself is questionable or difficult. A primarily inflammatory cell population is generally seen in cytology of an infectious process, with the predominance of specific cell type(s) dependent on the infectious agent present.[1,3] The presence of karyolysis or nuclear "degeneration" within neutrophils is additionally suggestive of bacterial infection, particularly with gram-negative bacteria.[2] In many situations, the presence of extracellular bacteria alone, particularly without an accompanying inflammatory response, should not be considered diagnostic for bacterial infection as surface bacterial contamination may be the source of the organisms. One potentially confusing situation is evaluation of abdominal fluid for evidence of bacterial infection due to intestinal perforation, as accidental aspiration of the bowel during fluid collection can result in inadvertent collection of intestinal contents. Evidence of intracellular bacteria and inflammation are very helpful in such situations to confirm true infection, although a peracute intestinal rupture will potentially lack these features.

Examples of typical inflammatory populations are: a pure or predominantly neutrophilic inflammation (also referred to as "suppurative" or "purulent" inflammation) consistent with many bacterial infections; pyogranulomatous inflammation with neutrophils, macrophages and (usually) multinucleated giant cells, consistent with fungal infection, infection with filamentous bacteria, or Mycobacterial infection. Note that non-infectious differentials exist for some inflammatory responses, e.g. pyogranulomatous inflammation can also be seen in the context of a foreign body response.

If infection is suspected but organisms are not found on evaluation of routinely stained cytology smears, special stains may be useful. Gram staining is easily performed in the clinic, and is helpful for identification of gram-positive bacteria. Other special stains including methenamine silver (GMS), periodic acid-Schiff (PAS), and acid-fast staining may be utilized in-house, or alternately a reference laboratory can perform these staining procedures. Other diagnostic modalities not covered here, e.g. bacterial culture and PCR, are often utilized when clinical signs and/or cytologic evidence of inflammation are consistent with infection but organisms are not seen on cytology. One recent study showed that evidence of inflammation was present in all bronchoalveolar lavage specimens from dogs with culture-confirmed lower respiratory infection, while organisms were seen cytologically in roughly 74% of the specimens.[3]

Site-specific and organism-specific issues with cytology for infectious agents:

The specific site chosen for sampling can be very important in detecting a suspected infectious agent. It is always important to consider the usual cytologic findings for the site being sampled. Potential "normal" findings include resident flora, e.g. mixed bacterial flora with *Simonsiella* in the oral cavity, and low numbers of mixed bacterial flora and yeast in the external ear canal. Also, resident cell population(s), e.g. lymphocytes with scattered histiocytes and lower numbers of granulocytes in a lymph node, should be known for cytologic evaluation of a site.

The expected sensitivity of a sample for a specific infectious agent should be considered. One sampling site may be preferable to others in detecting a microorganism, so be aware of relative sensitivity and consider sampling multiple sites. If organisms are only found rarely within cytologic samples, multiple aspiration samples with preparation of numerous slides will increase the sensitivity of cytologic detection. One example is diagnosis of *Coccidiodes immitis*, as organisms are often rare in cytologic preparations. It is also important to remember the relative size of the infectious agent of interest; e.g. *Coccidiodes* spherules can approach 100 μm in diameter. In this situation, scanning all cytology smears at 10x objective prior to starting a detailed search for smaller structures using an oil objective is a reasonable approach.

REFERENCES

[1]Valenciano AC and Cowell RL. 2013. Cowell and Tyler's diagnostic cytology and hematology of the dog and cat, 4th ed.; [2]De Lorenzi D. **J Small Anim Pract.** 2006 Jun;47(6):316-9; [3]Raskin RE. and Meyer D. 2009. Canine and feline cytology: a color atlas and interpretation guide, 2nd ed.; [4]Johnson LR., et al. **J Vet Intern Med.** 2013 Mar-Apr;27(2):259-67.

PHARMACOLOGIC CONSIDERATIONS IN KIDNEY DISEASE

Mark G. Papich, DVM, MS, DACVCP
Raleigh, North Carolina

INTRODUCTION

The kidneys receive a greater percentage of the cardiac output (20%) and have a higher metabolic requirement than most other organs. The kidneys are also a major route of drug excretion, which exposes the kidney tubules to high concentrations of drugs and drug metabolites. Drug-induced kidney disease is a concern when administering many drugs. Drug induced kidney injury is classified according to the area affected: 1) glomerular nephritis, 2) tubular necrosis, 3) interstitial nephritis, and 4) obstructive nephropathy, and whether it is chronic or acute.

DRUGS THAT PRESENT A RISK TO THE KIDNEYS

There are many drugs that may injure the kidneys. Included among those used in veterinary medicine are: aminoglycosides, tetracyclines, amphotericin B, cyclophosphamide, non-steroidal anti-inflammatory drugs (NSAIDs), cisplatin (Platinol), cyclosporine, sulfonamides, and radiographic contrast agents. Some of these are discussed in more detail below.

AMINOGLYCOSIDES:

Aminoglycoside antibiotics (eg, gentamicin, amikacin, tobramycin, kanamycin) have been the drugs most studied with respect to drug-induced nephrotoxicity. Aminoglycosides preferentially concentrate in the cells of the proximal tubules, reaching the tubular cells from the luminal side. Uptake into the proximal tubular cells is saturable; therefore, high tubular drug concentration is not necessarily more toxic than a lower one. However, persistent concentrations from multiple doses will increase the risk of kidney injury. When aminoglycosides enter the proximal tubular cells, they disrupt electrolyte and water regulation in cell membranes and accumulate within lysosomes (myeloid bodies). Proximal tubular dysfunction is the result of impairment of organelle function.

The risk factors that contribute to aminoglycoside nephrotoxicity include: a) prolonged aminoglycoside therapy: > 7 to 10 days, b) electrolyte depletion (K^+, Na^+), c) volume depletion, and d) pre-existing renal disease. One review of predisposing factors (Brown et al, 1985) found that most small animal cases of gentamicin-induced nephrotoxicity had received the drug for an average of 6.8 days, the dosages were greater than 2 to 4 mg/kg every 6 to 8 hours, and the diagnosis was based largely on azotemia and rule-outs of other causes. Most patients had normal renal function before gentamicin was administered. However, animals with previous renal function impairment are at a greater risk for toxicity (Frazier, et al, 1988).

The diagnosis of aminoglycoside nephrotoxicity is made on the basis of a history of drug administration and clinicopathological signs of acute kidney disease (polyuria, azotemia, and enzymuria). The management of acute kidney disease entails supportive care. With large accidental overdoses, other drugs may be administered (eg, injectable penicillin drugs) to bind aminoglycosides in the plasma, but this will not be effective when kidney injury is caused by doses in the usual range. In many patients, aminoglycoside injury can be reversed if the problem is recognized early enough. To support the patient one should employ fluid diuresis to decrease azotemia and uremic signs and maintain electrolyte balance. However, fluid diuresis will not increase the clearance of aminoglycosides from the body; it will only produce more dilute urine.

To prevent kidney injury from aminoglycosides, the most important preventative measure is to ensure that the patient has healthy renal function prior to drug administration. Since it is unlikely that glomerular filtration rates (GFR) – or a surrogate measure of GFR such as creatinine clearance – will be measured prior to drug administration, the most consistent parameter that will predict a risk of nephrotoxicity is the ability to concentrate urine and the serum creatinine level. If these parameters indicate reduced renal function, aminoglycoside therapy should be done cautiously, if at all.

Alternatively, therapeutic drug monitoring can be used to predict aminoglycoside drug clearance and measurement of drug concentrations can predict drug-induced toxicosis. This can be determined by obtaining at least two plasma samples (preferably three or four) at least one hour after drug administration. If drug elimination is slow, reduced clearance should be suspected, which increases the risk of kidney injury.

The current accepted method of administration for aminoglycosides is once per day injections. This has reduced the risk of nephrotoxicosis from aminoglycosides because it reduces the exposure to the kidneys. Common dosages used are: gentamicin 10-15 mg/kg for dogs, and 5-8 mg/kg for cats; amikacin 15-30 mg/kg for dogs, and 10-14 mg/kg for cats. (All doses listed are IV, SC, or IM). Since drug uptake into the proximal tubular cells is independent of the concentration in the tubule, a high dose given once a day is less likely to injure the kidneys than smaller doses given more frequently. Some recommendations have suggested that the choice of specific drug affects the risk of kidney injury, but there is no established ranking of differences in toxicity between gentamicin, tobramycin, and amikacin (Hanberger, et al, 2013).

NONSTEROIDAL ANTI-INFLAMMATORY DRUGS (NSAIDs)

The kidney is an active site for prostaglandin production. Prostaglandins play a role in tubular function as well as vascular tone. Healthy animals are somewhat immune from adverse effects from NSAID (Lobetti & Joubert, 2000). The NSAIDs are not inherently toxic to the kidneys. However, if there is renal compromise (eg, dehydration, tubular dysfunction, electrolyte depletion, anesthesia, etc), the kidney is dependent on cyclooxygenase 1 and 2 (COX-1 and COX-2) for prostaglandin synthesis to autoregulate water metabolism, tubular function, and renal blood flow (Gambaro & Perazella, 2003). Kidney injury associated with NSAIDs is characterized by decreased renal perfusion, sodium and fluid retention, and decreased tubular function. In people, pain in the kidney area has been recorded. NSAID-associated kidney injury is associated with inhibition of renal prostaglandins, which are involved in the maintenance of renal hemodynamics and sodium and fluid balance. NSAIDs inhibit prostaglandin synthesis and renal ischemia can result (Gunson et al, 1983; Rubin 1986; Clive and Stoff, 1984). The combination of anesthesia and NSAIDs may increase the susceptibility to kidney injury. In most cases, NSAIDs at recommended doses are safe to animals with healthy kidneys. However, animals with kidney disease are more at risk for dehydration, which can increase the likelihood of NSAID-induced nephropathy.

One should not assume that NSAIDs that are more specific for the COX-2 enzyme are safer for the kidneys. Both COX-1 and COX-2 enzymes are involved in kidney blood flow regulation and tubular function. Some of the prostaglandins that play an important role in salt and water regulation and hemodynamics in the kidney are synthesized by COX-2 enzymes (Rossat et al 1999). Constitutive COX-2 is found in various sections of the kidney and administration of drugs that are selective for COX-2, may adversely affect the kidney during conditions in which the kidney is stressed because of dehydration, decreased perfusion, or disease. Administration of a specific COX-2 inhibitor to salt-depleted people decreased renal blood flow, glomerular filtration rate, and electrolyte excretion (Rossat et al 1999).

NSAIDs in Dogs

Of the currently-available NSAIDs used in dogs, carprofen and meloxicam's effect on renal function has been the most extensively studied. Because these drugs are used in perioperative situations in an injectable formulation, investigations were performed to determine if there was any evidence of renal toxicity, particularly during anesthesia. Based on these studies, the risk of kidney injury from these medications appears to be low. In one study, carprofen, ketorolac and ketoprofen were examined in healthy dogs undergoing surgery, but without IV fluid administration. There were minor increases in renal tubular epithelial cells on urine sediment, but carprofen had no adverse effects on renal function (Lobetti et al, 2000). Some ketorolac and ketoprofen-treated dogs had transient azotemia. In other studies, administration of carprofen to anesthetized healthy dogs had no adverse effects on renal function (Ko et al 2000; Böstrom et al 2002; Bergmann et al, 2005; Crandell et al, 2004; Frendin et al, 2006). Meloxicam did not produce adverse renal effects in healthy dogs after short-term administration, with and without pimobendan (Fusellier et al, 2008). In healthy dogs anesthetized and treated with acepromazine to produce hypotension, preanesthetic administration of meloxicam did not produce any altered renal function (Boström, et al, 2006). Healthy dogs administered meloxicam prior to anesthesia and electrical nociceptive stimulus did not have decreased renal function associated with treatment (Crandell et al, 2004). Adverse effects on the kidney from deracoxib administration to dogs were reported by the manufacturer. At high doses, there is a dose-dependent effect on renal tubules. It is well-tolerated in most dogs up to 10 mg/kg for 6 months, but there is a potential for a dose-dependent renal tubular degeneration/regeneration at doses of 6 mg/kg or higher (clinically approved dose for long-term treatment is 1-2 mg/kg per day). Long-term administration of carprofen, etodolac, flunixin, ketoprofen, or meloxicam to dogs did not induce any evidence of renal injury as measured by urinalysis and serum biochemistry (Luna et al, 2007).

The common design in many of the studies cited above was that dogs were healthy, generally young, and NSAID doses were from a single dose and within the recommended range. Deviations from this design, use of higher doses, longer treatment, or administration to clinical patients with other problems could produce different results. Ultimately, the pre-operative use of a NSAID is justified as long as systemic blood pressure (and therefore renal blood flow) is maintained within normal physiologic ranges.

NSAIDs in Cats

Because it is often aged cats that are treated with NSAIDs, often for osteoarthritis, veterinarians and pet owners are concerned that these drugs may injure feline kidneys. Acute effects are a concern in any cat that is dehydrated or has renal compromise. The long-term effects also have attracted attention. One of the reasons for concern is a black box warning issued by the U.S. FDA in 2010 that stated that, "Repeated use of meloxicam in cats has been associated with acute renal failure and death. Do not administer additional injectable or oral meloxicam to cats". This prompted investigations because chronic kidney disease (CKD) is common in the population of cats treated with long-term NSAIDs. Follow-up studies have helped to clarify the effects of meloxicam in cats. The study by Surdyk et al (2013) showed that in cats with reduced renal function, but euvolemic, NSAIDs did not affect kidney function. In their study, neither meloxicam nor acetylsalicylic acid had a measurable effect on urinary clearance of exogenously administered creatinine, serum creatinine concentration, or the protein to creatinine ratio. They concluded that kidney function of euvolemic cats with normal or reduced renal function is not dependent on cyclooxygenase function.

Meloxicam is licensed for chronic use in cats in several countries, and is the drug recommended for chronic administration, by consensus, from a group of feline medicine experts in 2010 (Sparkes, et al, 2010). Three important studies have concluded that chronic administration of meloxicam to cats can be safe. Oral meloxicam was safe and palatable for long-term treatment of osteoarthritis in cats when given with food at a dose of 0.01-0.03 mg/kg/day (Gunew et al, 2008). Gowan et al, (2011) concluded in a study with aged cats with CKD that long-term therapy with meloxicam (0.02 mg/kg daily) can be administered safely. These cats tolerated meloxicam with daily administration despite their kidney disease. And finally, in a follow-up study, Gowan et al, (2012) concluded that long-term management with oral meloxicam in cats did not reduce the lifespan of cats with pre-existent stable CKD. Many cats actually had a better outcome associated with meloxicam treatment. These studies with meloxicam have helped resolve some concerns about the use of NSAIDs, at least for meloxicam, when administered to aged cats, and to cats with CKD. It should be noted that the doses prescribed in these studies are low compared to the approved label dose, and in fact lower than the dose recommended in the review by Sparkes et al (2010). Studies have also been completed on the newest NSAID approved for cats, robenacoxib. The cats in the safety studies were all healthy cats, but the data shows that robenacoxib was tolerated well even for a treatment that was longer than the label indication (King et al, 2012).

ANGIOTENSIN CONVERTING ENZYME INHIBITORS (ACE-INHIBITORS)

Azotemia has been associated with administration of ACE-inhibitors in some patients. However, in most patients that have been monitored, ACE-inhibitors produce little risk to the kidneys (Atkins et al 2002). In this study of 139 dogs with heart disease (mitral regurgitation) that received either enalapril or placebo, the authors concluded that, administration of an ACE-inhibitor for up to 2 years did not have any demonstrable adverse effects on renal function in dogs. Nevertheless, ACE-inhibitors carry a risk of decreased renal perfusion. Use of diuretics should be limited and renal parameters should be monitored in treated animals periodically. Pre-existing kidney disease and dehydration will increase this risk. In people, there is concern that, in some patients, the combination of an ACE-inhibitor and an NSAID may increase the risk of kidney injury (Loboz & Shenfield 2005). Such an association has not been shown in dogs or cats. Only one study examining this combination has been published for dogs (Fusellier et al, 2005). It that study, the authors concluded that tepoxalin (an NSAID no longer available for dogs) did not alter renal function in healthy Beagle dogs receiving an ACE inhibitor. Such an effect of other NSAID combinations has not been adequately studied in veterinary medicine to make conclusions.

A comprehensive review of the use of ACE-inhibitors in patients with renal disease was published by Lefebvre & Toutain (2004). As reviewed by these authors, ACE-inhibitors can be beneficial in chronic kidney disease because they relieve glomerular pressure, slow progression of lesions, and decrease proteinuria. In the same article, and in other papers (Lefebvre et al 2006), the authors documented the changes in disposition of ACE-inhibitors in animals with renal impairment. In dogs and cats enalaprilat (metabolite of enalapril) is predominantly cleared by the kidneys. In dogs, ramiprilat, and benazeprilat (metabolites of ramipril and benazepril) are cleared by both liver and kidney. Subsequently, these authors demonstrated ACE-inhibitor changes (both pharmacokinetic and pharmacodynamic) caused by renal impairment for enalaprat in dogs but not ramiprilat and benazeprilat. In cats with reduction in GFR, there was no difference in disposition measured for benazeprilat.

MANAGING PATIENTS WITH KIDNEY DISEASE

For drugs that are cleared by the kidneys, drug dose rates should be altered when there are documented decreases in creatinine clearance or elevations in serum creatinine. For a drug that relies on the kidneys for drug clearance (Cl_R) a loss in renal function will proportionately decrease drug excretion. That is, a 75% loss in renal function results in a 75% loss in renal drug clearance. Dosage adjustments can be made from estimates in the loss of renal function. The most exact method of assessing renal function is to measure creatinine clearance (Cl_{CR}), which is an estimate of GFR. The decrease in the patient's Cl_{CR} compared to a healthy animal can be used to either adjust the dose interval or the dose as follows:

Constant interval, decreased dose method:

$$\text{Dose}_{\text{Kidney Disease}} = \text{Dose}_{\text{Normal}}\ x\ (\text{Patient's } Cl_{CR} / \text{Normal } Cl_{CR})$$

Or, *Constant dose, increased interval method (τ = dose interval):*

$$\tau_{\text{Kidney Disease}} = \tau_{\text{Normal}}\ x\ (\text{Normal } Cl_{CR} / \text{Patient's } Cl_{CR}).$$

Creatinine clearance is rarely performed clinically in animals. Alternatively, a less precise method is to make a dose adjustment based on serum creatinine:

Constant interval, decreased dose method:

$$\text{Dose}_{\text{Kidney Disease}} = \text{Dose}_{\text{Normal}}\ x\ (1/\text{serum creatinine})$$

Or, *Constant dose, increased interval method (τ = dose interval):*

$$\tau_{\text{Kidney Disease}} = \tau_{\text{Normal}} / (1/\text{serum creatinine})$$

In this example, the patient's creatinine should be in units of mg/100 mL. Because serum creatinine is relatively insensitive to changes in creatinine clearance (Finco, et al 1995), this method is only a rough approximation and assumes that serum creatinine in healthy animals is 1.0 mg/100 (range 0.5 – 1.5 mg/dL). The method of relying on serum creatinine is less reliable when creatinine levels are high (eg, above 4 mg/dL). An estimate of dose adjustment using serum creatinine is more risky in geriatric compared to younger animals because serum creatinine is affected by animal's muscle mass. A geriatric animal with low muscle mass and poor renal function may have a falsely low serum creatinine.

DRUG SELECTION FOR PATIENTS WITH KIDNEY DISEASE (Antibiotic Example)

In patients with renal disease, drug selection is challenging. Avoid potentially nephrotoxic drugs that may worsen the existing condition. Drugs that are highly metabolized by the liver will be less likely to accumulate, although one should be aware that drug metabolites may rely on the kidneys for excretion. These metabolites can accumulate and result in toxicity.

With every drug administration, the clinician should consider whether or not adjustments in dose or dose interval should be considered. With respect to antibiotic therapy (one of the most common therapies used in these patients), avoid tetracyclines, and aminoglycosides. However, doxycycline is safe because it is also eliminated by non-renal routes. Clearance of minocycline in dogs and cats appears to be renal, but there is not enough data to determine conclusively. The drugs that rely on metabolism for their elimination can be used cautiously. This includes macrolides (erythromycin, azithromycin) and clindamycin. Avoid metronidazole because the most common adverse effect is CNS toxicity and this may be difficult to distinguish from problems caused by uremia.

For other drugs there is always a concern that, in a patient with reduced renal function there may be decreased renal clearance and a need to adjust dosages. For example, some fluoroquinolones accumulate in human patients with renal compromise (Fillastre, et al 1990), but others such as ciprofloxacin do not. When marbofloxacin was administered to dogs with renal impairment, there was no significant effect on drug disposition and no need for dose modification. (Lefebvre et al 1998) Renal failure had only a minimal and biologically irrelevant effect on marbofloxacin disposition in the dog. Apparently, other routes of drug clearance can play a role in elimination when renal function is compromised.

All beta-lactam antibiotics carry a risk of accumulation in patients with impaired renal function. This includes penicillins, ampicillin, amoxicillin, and cephalosporins. The most common adverse effects from accumulation are CNS toxicity (eg, tremors, seizures) or gastrointestinal (vomiting). If a beta-lactam drug is considered reduce dose frequency if any signs of toxicity are observed. For treatment of gram-negative infections, extended-spectrum cephalosporins, cefotaxime, cefpodoxime axetil, or ceftazidime can be a good choice because of their activity. All except cefpodoxime must be injected. If there is persistent azotemia the frequency of administration should be decreased in proportion to the decrease in renal function as described above. The effect of kidney disease on clearance of cefovecin in dogs and cats has not been investigated. The signs that may indicate accumulation of beta-lactam antibiotics in patients with kidney disease are gastrointestinal problems (vomiting, nausea), or neurological signs, such as seizures. (Chow et al, 2004; Wallace 1997). For serious resistant infections, the carbapenems can be considered. Of those available, meropenem (Merrem) is preferred over imipenem. It is not associated with neurotoxicity that is seen with imipenem accumulation.

REFERENCES AND ADDITIONAL READING

Atkins Atkins CE, Brown WA, Coats JR, Crawford MA, DeFrancesco TC, Edwards J, Fox PR, Keene BW, Lehmkuhl L, Luethy M, Meurs K, Petrie JP, Pipers F, Rosenthal S, Sidley JA, Straus J. Effects of long-term administration of enalapril on clinical indicators of renal function in dogs with compensated mitral regurgitation. J Am Vet Med Assoc. 2002 221(5):654-658.

Bergmann HML, Nolte IJA, Kramer S. Effects of preoperative administration of carprofen on renal function and hemostasis in dogs undergoing surgery for fracture repair. Am J Vet Res 66: 1356-1363, 2005.

Boström IM, Nyman GC, Lord PF, et al. Effects of carprofen on renal function and results of serum biochemical and hematologic analyses in anesthetized dogs that had low blood pressure during anesthesia. Am J Vet Res 63: 712-721, 2002.

Brown SA, Barsanti JA, Crowell WA: Gentamicin-associated acute renal failure in the dog. J Am Vet Med Assoc 186: 686-690, 1985.

Chow KM, Szeto CC, Hui A C-F, Li P K-T. Mechanisms of antibiotic neurotoxicity in renal failure. International Journal of Antimicrobial Agents 23: 213-217, 2004.

Clive DM, Stoff JS: Renal syndromes associated with nonsteroidal anti-inflammatory drugs. New Engl J Med 310: 563-571, 1984.

Crandell DE, Mathews KA, Dyson DH. Effect of meloxicam and carprofen on renal function when administered to healthy dogs prior to anesthesia and painful stimulation. Am J Vet Res.;65(10):1384-1390, 2004.

Frendin JH, Boström IM, Kampa N, Eksell P, Häggström JU, Nyman GC. Effects of carprofen on renal function during medetomidine-propofol-isoflurane anesthesia in dogs. : Am J Vet Res; 67(12):1967-1973, 2006.

Fillastre JP, Leroy A, Moulin B, et al. Pharmacokinetics of quinolones in renal insufficiency. J Antimicrob Chemother 26 (suppl B): 51-60, 1990.

Finco DR, Brown SA, Vaden SL, & Ferguson DC. Relationship between plasma creatinine concentration and glomerular filtration rate in dogs. J Vet Pharmacol Therap 18: 418-421, 1995.

Frazier DL, Aucoin DP, Riviere JE: Gentamicin pharmacokinetics and nephrotoxicity in naturally-acquired and experimentally-induced diseases in dogs. J Am Vet Med Assoc 192: 57-63, 1988.

Fusellier M, Desfontis JC, Madec S, Gautier F, Marescaux L, Debailleul M, Gogny M. Effect of tepoxalin on renal function in healthy dogs receiving an angiotensin-converting enzyme inhibitor. J Vet Pharmacol Ther. 2005 28(6):581-586.

Gambaro G, Perazella MA. Adverse renal effects of anti-inflammatory agents: Evaluation of selective and nonselective cyclo-oxygenase inhibitors. Journal of Internal Medicine 253: 643-652, 2003.

Gowan, R. A., Lingard, A. E., Johnston, L., Stansen, W., Brown, S. A., & Malik, R. (2011). Retrospective case–control study of the effects of long-term dosing with meloxicam on renal function in aged cats with degenerative joint disease. Journal of Feline Medicine & Surgery, 13(10), 752-761.

Gowan, R. A., Baral, R. M., Lingard, A. E., Catt, M. J., Stansen, W., Johnston, L., & Malik, R. (2012). A retrospective analysis of the effects of meloxicam on the longevity of aged cats with and without overt chronic kidney disease. Journal of feline medicine and surgery, 14(12), 876-881.

Gunew, M. N., Menrath, V. H., & Marshall, R. D. (2008). Long-term safety, efficacy and palatability of oral meloxicam at 0.01–0.03 mg/kg for treatment of osteoarthritic pain in cats. Journal of feline medicine and surgery, 10(3), 235-241.

Gunson DE: Renal papillary necrosis in horses. J Am Vet Med Assoc 182: 263-266, 1983.

Hanberger, H., Edlund, C., Furebring, M., G. Giske, C., Melhus, Å., Nilsson, L. E., ... & Eliasson, E. (2013). Rational use of aminoglycosides-Review and recommendations by the Swedish Reference Group for Antibiotics (SRGA). Scandinavian journal of infectious diseases, 45(3), 161-175.

Hiemenz and Walsh, 1996

King, J. N., Hotz, R., Reagan, E. L., Roth, D. R., Seewald, W., & Lees, P. (2012). Safety of oral robenacoxib in the cat. Journal of Veterinary Pharmacology and Therapeutics, 35(3), 290-300.

Ko JCH, Miyabiyashi T, Mandsager RE, Heaton-Jones TG, Mauragis DF: Renal effects of carprofen administered to healthy dogs anesthetized with propofol and isoflurane. Am J Vet Med Assoc 217: 346-349, 2000.

Lefebvre HP, Schneider M, Dupouy V, et al. Effect of experimental renal impairment on disposition of marbofloxacin and its metabolites in the dog. J Vet Pharmacol Therap 21: 453-461, 1998.

Lefebvre HP, Jeunesse E, Laroute V, Toutain PL. Pharmacokinetic and pharmacodynamic parameters of ramipril and ramiprilat in healthy dogs and dogs with reduced glomerular filtration rate. J Vet Intern Med. 2006 20(3):499-507.

Lefebvre HP, Toutain PL. Angiotensin-converting enzyme inhibitors in the therapy of renal diseases. J Vet Pharmacol Ther. 2004 27(5):265-281.

Lobetti RG, Joubert KE: Effect of administration of nonsteroidal anti-inflammatory drugs before surgery on renal function in clinically normal dogs. Am J Vet Res 61: 1501-1506, 2000.

Loboz KK, Shenfield GM. Drug combinations and impaired renal function – the triple whammy. British Journal of Clinical Pharmacology 59: 239-243-, 2005.

Luna SPL, Basílio AC, Steagall PVM, Machado LP, Moutinho FQ, Takahira RK, and Brandão CVS. Evaluation of adverse effects of long-term oral administration of carprofen, etodolac, flunixin meglumine, ketoprofen, and meloxicam. Am J Vet Res 68: 258-264, 2007.

Rossat J, Maillard M, Nussberger JU, et al: Renal effects of selective cyclooxygenase-2 inhibition in normotensive salt-depleted subjects. Clin Pharmacol Ther 66: 76-84, 1999.

Rubin SI: Nonsteroidal anti-inflammatory drugs, prostaglandins, and the kidney. J Am Vet Med Assoc 188: 1065-1068, 1986.

Sparkes, A. H., Heiene, R., Lascelles, B. D. X., Malik, R., Real, L., Robertson, S., Scherk M & Taylor, P. (2010). ISFM and AAFP consensus guidelines: long-term use of NSAIDs in cats. Journal of feline medicine and surgery, 12(7), 521-538.

Surdyk, K. K., Brown, C. A., & Brown, S. A. (2013). Evaluation of glomerular filtration rate in cats with reduced renal mass and administered meloxicam and acetylsalicylic acid. American journal of veterinary research, 74(4), 648-651.

Wallace KL. Antibiotic-induced convulsions. Critical Care Clinics 13: 741-761, 1997.

MANAGING ANEMIA IN CHRONIC KIDNEY DISEASE

Cathy Langston, DVM, DACVIM (SAIM)
Columbus, OH

CAUSES OF ANEMIA

Anemia results in reduced oxygen delivery to all organs and a subsequent decline in cell metabolism.[1,2] Anemia triggers numerous adaptive response mechanisms due to tissue hypoxia, some of which may be detrimental in the long term.[3,4] Anemia of renal disease is multifactorial in its pathogenesis. The lack of erythropoietin (EPO) is the main cause of anemia in CKD patients. Additional factors include anemia of inflammation, other causes of suppressed erythropoiesis, the influence of uremic toxins on red blood cell survival, and gastrointestinal hemorrhage or other sources of blood loss.

Blood Loss

Uremic bleeding may be due to retention of small substances that affect platelet function, vascular damage caused by uremic toxins, changes in platelet responsiveness and function, changes in platelet-endothelial interactions, increased levels of nitric oxide (a potent platelet antagonist) in uremic states, and dysregulation of coagulation factors responsible for normal platelet function.[5-7] Significant blood loss can occur through the gastrointestinal tract in people with CKD in the absence of identifiable discrete bleeding ulcers.[8,9] Studies of the prevalence of upper gastrointestinal ulceration in humans with CKD yield conflicting results, although recent evidence suggests that CKD patients do not have a higher prevalence of ulceration compared to patients with normal kidney function.[9-12] Although gastrin concentration in the blood is elevated in people and cats with CKD, hyperacidity and gastric ulceration are variably present.[13-15] Gastric pathology is common in dogs and cats with kidney disease, but ulceration is uncommon.[16,17] In one study of very low birth weight infants, blood sampling accounted for loss of approximately 2 ml/kg/day blood loss over the first 28 days of life.[18] Whether a similar degree of sampling is performed in hospitalized dogs and cats is unknown.

Impaired Erythropoiesis

Several uremic inhibitors of erythropoiesis have been identified, but their relative contribution to anemia is controversial. It is thought that these substances affect erythrogenesis at various stages within the bone marrow.[19,20] Renal secondary hyperparathyroidism can cause EPO hyporesponsiveness and may have a direct toxic effect on erythroid progenitor cells, and also an indirect effect by inducing bone marrow fibrosis.[21] Deficiency of B vitamins may also impair erythrogenesis. Common therapies used for CKD may impair RBC production. Angiotensin-converting enzyme inhibitors or angiotensin receptor blockers are suspected to decrease angiotensin II-induced EPO release, as well as prevent pluripotent hemopoietic stem cell recruitment.[21,22] Aluminum toxicity, from aluminum containing phosphate binders, can interfere with iron metabolism, leading to a microcytic anemia that is not responsive to iron administration.[23-26]

Erythropoietin. Erythropoietin is the hematopoietic growth factor responsible for stimulating erythrogenesis in response to anemia. The majority of EPO is produced in the peritubular interstitial cells of the inner renal cortex and outer renal medulla; however, the liver produces 10 to 15 per cent of the total EPO.[27] EPO has a circulating half-life of 6 to 10 hours. The main trigger for the synthesis of EPO is renal hypoxemia, which increases production of hypoxia-inducible factor (HIF). HIF stimulates transcription of the EPO gene and is inactivated by HIF-alfa prolyl and asparaginyl hydroxylases. HIF-alfa hydroxylases require oxygen and iron to function.[28]

EPO stimulates proliferation and maturation of erythrocytes in the bone marrow directly by binding specific growth factor receptors on the red blood cell precursors. EPO has the greatest effect on the colony-forming unit erythroid cell (CFU-E), although it has some effect also on the differentiation of the burst-forming unit erythroid cell (BFU-E) and erythroblasts.[29] In the absence of EPO, these progenitor cells undergo apoptosis, whereas they differentiate and proliferate in the presence of EPO.[30] It takes approximately 5 days from stimulation by EPO for the release of reticulocytes into the circulation.[31] In addition to its effects on red blood cell differentiation, EPO induces synthesis of hemoglobin and red blood cell membrane proteins. It facilitates release of reticulocytes into the blood by lowering the barrier between the marrow and the blood.[32]

When renal function is diminished, renal oxygen consumption decreases. The increased oxygen tension in the renal microenvironment stimulates HIF-prolyl hydroxylase, which inactivates HIF, thus decreasing EPO gene transcription. EPO levels are decreased in patients with CKD relative to patients with a similar degree of anemia from other causes such as blood loss or hemolysis.

Iron. Orally ingested iron is absorbed into the duodenal enterocytes via the apical protein divalent metal transporter (DMT1), and enters the circulation via ferroportin on the basolateral side. Ferroportin is the principal cellular efflux channel for iron and is present in enterocytes, macrophages, and hepatocytes. Hepcidin is considered to be the central regulator of systemic iron homeostasis. The biological actions of hepcidin are mediated by its binding to and internalization of ferroportin. Hepcidin is suppressed by normal erythropoietic activity and iron deficiency, allowing duodenal absorption of iron and release of iron for hemoglobin production. Hepcidin production is increased in response to inflammation and iron loading. Hepcidin traps iron in the enterocytes by enhancing degradation of ferroportin, and it prevents release of iron stored in macrophages and hepatocytes as well.[33,34] The decrease in enteral absorption and the sequestration of iron in macrophages eventually leads to anemia by decreasing the availability of iron for hemoglobin production.

Anemia of inflammation contributes to anemia of renal disease, and CKD itself is known to be associated with a chronic pro-inflammatory state.[35] Cytokines such as interferon-α, -β, and –γ, as well as tumor necrosis factor-α, interleukin-1, and interleukin-6 are produced in inflammatory states and create a relative iron deficiency by induction of hepcidin, an acute phase protein.[35,36] In addition to induction of hepcidin production, the aforementioned inflammatory cytokines can negatively affect red blood cell survival and directly decrease responsiveness to erythropoietin, all contributing to worsening anemia.[35,37]

Hepcidin concentrations are increased in human CKD patients, even in those without significant inflammation, in part due to decreased renal clearance.[34,36] Although inadequate EPO is considered to be the major cause of the anemia of CKD, iron deficiency is present in 25-38% of anemic people with CKD and is the leading cause of failure to achieve target hematocrit in people on ESA therapy.[21,22] Serum iron concentrations in cats with CKD have been reported to be lower[38] or not different[39] from healthy cats, whereas median serum iron was below the reference range in dogs with anemia of CKD.[40] Although an absolute iron deficiency may not be present, there may be a functional iron deficiency, in which total body store of iron is not depleted, but the iron is unavailable for hematopoiesis. In that situation, serum iron and transferrin saturation are expected to be low, but serum ferritin would be high, as ferritin is a positive acute phase reactant. Changes in the complete blood count (low mean cell volume, low mean cell hemoglobin content) are not sensitive indicators of iron deficiency.[41]

TREATMENT OF ANEMIA

The most immediate way to treat anemia is with blood transfusion. Transfusion is indicated when there is acute blood loss or when a patient has severe clinical signs of anemia. Disadvantages of blood transfusions include the possibility of immune reactions and donor-recipient incompatibility, the limited availability of blood products, the reduced lifespan of infused blood products in a uremic patient, the associated costs of blood product transfusion, and the lack of long term effectiveness of these products.[42]

Erythropoiesis stimulating agents (ESA) are recombinant forms of erythropoietin or other substances that can increase erythropoiesis. Canine EPO shares 81.3% homology with the amino acid sequence of human EPO, while homology in cats is 83.3%.[43,44] Thus, human recombinant products can be effective in animals. Epoetin-alfa (e.g., Epogen®, Procrit®) was the first product available. Darbepoetin is a newer product that has a prolonged half-life compared to epoetin and thus can be given only once a week, instead of three times a week. Survival for cats responding to darbepoetin therapy was a median of 238 days.[45] There was a significant difference in responder cat survival compared to non-responders, with responder cats having a longer survival. Therefore, it is quite possible that darbepoetin therapy, by improving anemia, increases survival time for animals with renal disease. However, a controlled, randomized study would be necessary to prove this hypothesis. It is known that correcting anemia can improve quality of life in humans and animals.

The use of ESAs has been associated with a number of complications, such as iron deficiency, hypertension, arthralgia, fever, seizures, polycythemia, and pure red cell aplasia (PRCA). PRCA is one of the most serious side effects, and the incidence of clinically apparent PRCA appears to be decreased with darbepoetin use (approximately 10% of dogs and cats) when compared to epoetin administration (approximately 20-25% of dogs and cats).

Lack of available iron for erythropoiesis is recognized as the leading cause of treatment failure with ESA in humans.[46-48] Although iron should be administered to all dogs and cats receiving ESA therapy, it may not be fully utilized. Cats with CKD who responded to treatment with feline erythropoietin had a decrease in mean cell volume serum iron concentration and percentage transferrin saturation despite receiving oral iron supplementation.[49] The decrease in circulating iron parameters during ESA therapy may represent utilization of iron for RBC production in excess of intake or in excess of available iron stores, inadequate intestinal iron absorption despite oral iron supplementation, or functional iron deficiency due to sequestration of iron in storage sites such as the bone marrow with an inability to utilize that iron due to alterations in hepcidin metabolism.[40,49] Evaluation of serum iron parameters is recommended at the start of ESA therapy and monthly thereafter.

Several other drugs are in development to treat anemia of CKD. Peginesitide (Omontys, Hematide) is a small peptide with no homology to EPO that never-the-less stimulates the EPO receptor. Unfortunately, limited use in dogs and cats suggests that it will not be effective in those species. Drugs that inactivate prolyl hydroxylase would decrease degradation of HIF, and thus EPO gene transcription would be enhanced. One such drug is in a Phase 3 trial in people. The fact that these drugs can increase native EPO production suggests that the traditional explanation that too few peritubular cells exist to produce EPO is an oversimplification. These drugs can be given orally, and because they induce native EPO production, they eliminate the risk of anti-EPO antibody formation (i.e., PRCA). However, prolyl hydroxylase inhibition may impact other hypoxia-sensitive genes, such as those involved in angiogenesis. Another target to increase EPO gene transcription is to inhibit an inhibitor of EPO gene transcription, GATA. Because the excess hepcidin associated with inflammatory diseases limits iron utilization, inhibitors of hepcidin production or action have been hypothesized as mechanisms to treat anemia. Some potential inhibitors have been identified and are under development.

REFERENCES

[1]Brazier J, et al **Surgery** 1974;75(4):508. [2]Cain SM. **J Appl Physiol** 1977;42(2):228. [3]Fishbane SB, et al **Brenner & Rector's The Kidney**. 8 ed. 2008:1733. [4]Wilson HE, et al **J Feline Med Surg** 2010;12(12):904. [5]Linthorst GE, et al **J Am Soc Nephrol** 2010;21(5):753. [6]Boccardo P, et al **Semin Thromb Hemost** 2004;30(5):579. [7]Boyle JM, et al **Am J Med** 1983;75(3):409. [8]Rosenblatt SG, et al **Am J Kidney Dis** 1982;1(4):232. [9]Karagiannis S, et al **World J Gastroenterol** 2006;12(32):5182. [10]Stolic RV, et al **Am J Med Sci** 2008;336(1):39. [11]Abu Farsakh NA, et al **Nephrol Dial Transplant** 1996;11(5):847. [12]Shirazian S, et al **Nat Rev Nephrol** 2010;6(8):480. [13]Goldstein RE, et al **J Am Vet Med Assoc** 1998;213(6):826. [14]Ciccotosto GD, et al **J Clin Endocrinol Metab** 1996;81(9):3231. [15]Kes P. **Acta Med Croatica** 1992;46(1):47. [16]Peters RM, et al **J Vet Intern Med** 2005;19(3):315. [17]McLeland SM, et al. **J Vet Intern Med** 2014;28(3):827. [18]Obladen et al, **Eur J Pediatr** (1988) 147:399. [19]Brunati C, et al **Am J Nephrol** 1992;12(1-2):9. [20]Radtke HW, et al **J Clin Invest** 1981;67(6):1623. [21]Kwack C, et al **Semin Dial** 2006;19(2):146. [22]Kanbay M, et al **Blood Purif** 2010;29(1):1. [23]Segev G, et al **J Vet Intern Med** 2008;22(6):1432. [24]Priyadarshi A, et al **Semin Dial** 2006;19(4):273. [25]Jeffery EH, et al **J Toxicol Environ Health** 1996;48(6):649. [26]Cannata-Andia JB, et al **Nephrol Dial Transplant** 2002;17 Suppl 2:9. [27]Car B. **Schalm's veterinary hematology**, 2000, p. 105. [28]Macdougall IC. **Clin J Am Soc Nephrol** 2008;3:200. [29]Giger U. **Compend Contin Educ Pract Vet** 1992;14:25. [30]Fisher JW. **Exp Biol Med** 2003;228:1. [31]Duncan R, et al **Veterinary Laboratory Medicine** 1986, p. 4. [32]Bunn H. **Yale J Biol Med** 1990;63:381. [33]Nemeth E, et al **J Clin Invest** 2004;113(9):1271. [34]Young B, et al. **Clin J Am Soc Nephrol** 2009;4(8):1384. [35]Stenvinkel P. **Nephrol Dial Transplant** 2001;16 Suppl 7:36. [36]Nemeth E, et al **Acta Haematol** 2009;122(2-3):78. [37]Weiss G, et al. **N Engl J Med** 2005;352:1011. [38]Javard R, et al **J Vet Intern Med** 2014; 28(3):1073. [39]Gest J, et al **J Vet Intern Med** 2014; 28(3):1072. [40]Randolph JF, et al **J Vet Intern Med** 2004;18(1):81. [41]Steinberg JD, et al **Vet Clin Pathol** 2005;34(1):23. [42]Cowgill LD. **Semin Vet Med Surg (Small Anim)** 1992;7(3):175. [43]MacLeod JN. **Proc ACVIM** 2001. p. 578. [44]MacLeod JN, et al **Am J Vet Res** 1998;59(9):1144. [45]Chalhoub S, et al **J Vet Intern Med** 2012;26(2):363. [46]Macdougall IC. **Kidney Int** 1999;55(Suppl 69):S61. [47]Schiesser D, et al. **Nephrol Dial Transplant** 2006;21(10):2841. [48]Sjaastad OV, et al. **Acta Vet Scand** 1996;37:133. [49]Randolph JF, et al **Am J Vet Res** 2004;65(10):1355.

CHARACTERIZING URINARY BLADDER EICOSANOID METABOLISM IN HEALTHY CATS AND CATS WITH IDIOPATHIC CYSTITIS (FIC)

John M. Kruger DVM, PhD, DACVIM; Patrick J. Venta PhD; Matti Kiupel DVM, PHD; Scott D. Fitzgerald DVM, PhD; Lisa M. Singer DVM; Maciej Parys DVM; Lorraine M. Sordillo PhD
East Lansing, Michigan

INTRODUCTION

Feline idiopathic cystitis (FIC) is an inflammatory disorder that affects a quarter to a half million cats annually in North America. It is the most common urinary disorder of young to middle-aged cats and is characterized by frequent, bloody, and painful urinations, house-soiling, and in some cases, life-threatening urinary obstruction. The clinical and morphologic features of FIC are strikingly similar to those of an idiopathic disorder of people called interstitial cystitis. Although clinical and light microscopic features of FIC and interstitial cystitis are well characterized, underlying causative factors in the pathogenesis of both disorders remain obscure. Recent studies, however, suggest that human interstitial cystitis and FIC may be associated with unique changes in urinary bladder eicosanoid expression.[6,12,16,18]

EICOSANOID BIOSYNTHESIS AND FUNCTIONS

Eicosanoids compromise a class of bioactive lipid mediators formed from the metabolism of cell membrane-derived polyunsaturated fatty acids [linoleic, arachadonic, eicosapentaenoic (EPA) and docosahexanoic (DHA) acids] by cyclooxygenase (COX), lipoxygenase (LOX), cytochrome P450, and nonenzymatic pathways. (Figure 1). Eicosanoid biosynthetic enzymes are expressed epithelial cells, endothelial cells, reticulocytes, monocytes, and macrophages. Oxidation of polyunsaturated fatty acids through either enzymatic or non-enzymatic pathways results in the production of prostaglandins (PG), thromboxanes (TX), leukotrienes (LT), lipoxins (LX), protectins, resolvins and maresins.[3,17,19]

Eicosanoids are involved in numerous essential homeostatic physiologic functions and pathologic processes including inflammation, autoimmunity, allergies, and carcinogenesis. In healthy urinary bladders, prostaglandin E_2 (PGE_2) is synthesized in urothelium and detrusor smooth muscle and has shown to have important physiologic roles in bladder afferent signaling, smooth muscle contractility, urothelial tight junction integrity, glycosaminoglycan synthesis, cytoprotection, and repair (proliferation, migration, differentiation, angiogenesis).[4,7,11] Eicosanoid production is also influenced by other physiologic factors such as tissue type, age, reproductive status, and dietary or pharmacologic interventions.[15,19]

Altered COX and LOX expression and changes in local eicosanoid production has been associated with pathologic bladder disorders including carcinoma,[1,13] overactive bladder syndrome,[8] urinary obstruction,[14] urinary tract infection,[5] drug-induced cystitis,[9] interstitial cystitis,[2] and stress.[23] In general, the COX and LOX arachadonic acid metabolites (e.g. prostaglandins, thromboxanes, and leukotienes) are considered potent mediators of inflammation, cell proliferation, and altered immune function. However, it is now apparent that certain LOX pathway-generated metabolites have dual anti-inflammatory and pro-resolving functions (Figure 1). Lipoxins are lipid mediators derived from arachadonic acid via LOX pathways and have been shown to block granulocyte chemotaxis and trans-epithelial migration, stimulate non-phlogistic clearance of apoptotic granulocytes by macrophages, down regulate epithelial cell secretion of pro-inflammatory cytokines, reduce organ fibrosis, and reduce inflammation-induced pain.[17,21] Resolvins, protectins and maresins are additional LOX-derived metabolites of omega-3 fatty acids EPA and DHA that have similar anti-inflammatory, pro-resolving properties.[17,21] Thus, products of the COX and LOX pathways are key regulators of both acute and chronic inflammation that may either enhance or resolve the inflammatory response depending on the context of their expression in tissues.[17,19]

Studies characterizing urinary bladder eicosanoid expression in healthy cats and cats with various bladder disorders have not been reported. Due to the multiple biologic actions and complex interactions of various eicosanoids, understanding of the role these lipid mediators in feline health and disease requires a comprehensive integrated systems biology approach involving genomic, proteomic, and lipidomic analyses of eicosanoid expression.[3,19]

URINARY BLADDER EICOSANOID EXPRESSION IN HEALTHY CATS

Urinary bladder tissue concentrations of 23 targeted COX and LOX eicosanoid metabolites were quantified by liquid chromatography-mass spectroscopy (LC/MS) in specimens obtained from 12 healthy male and female cats.[10] Both pro- and anti-inflammatory lipid mediators, and mediators associated with lipid peroxidation and oxidative stress were readily detected and quantified in urinary bladder tissues. In addition, formalin-fixed paraffin-embedded bladder specimens were immunohistochemically stained for COX-1, COX-2, 5-LOX, and PGDH.[18] Strong COX-1, 5-LOX, and PGDH immunoreactivity was observed in the urothelium of specimens from normal cats; no appreciable urothelial COX-2 immunoreactivity was observed.

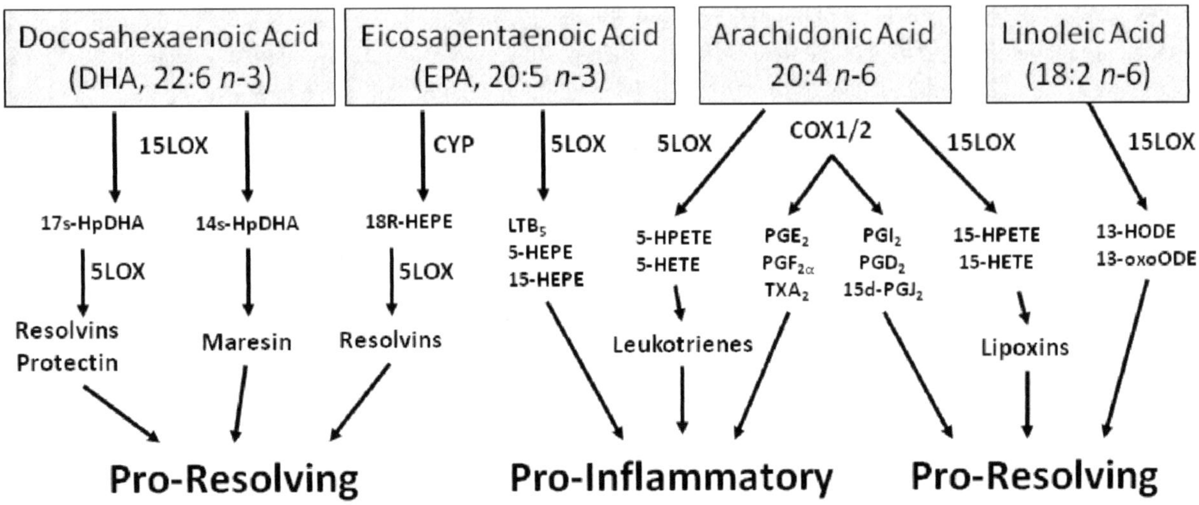

Figure 1. Enzymatic eicosanoid biosynthetic pathways illustrating production of selected lipid mediators from cell membrane phospholids. COX= cyclooxygenase; CYP= cytochrome –P450 complex; HEPE=hydroeicosapentaenoic acid; HETE= hydroxyeicosatetraenoic acid; HPETE= hydroperoxyeicosatetraenoic acid; HODE= hydroxyoctadecadienoic acid; oxoODE= oxooctadecadienoic acid; PG= prostaglandin; TX= thromboxane. Adapted from Sordillo LM, et al. *Anim Health Res Rev* 2009;10:53.

The structures of genes encoding 9 key eicosanoid metabolizing enzymes 5-LOX (*ALOX5*), FLAP (*ALOX5AP*), 1-LOX-1 (*ALOX15*), 15-LOX-2 (*ALOX15B*), 12R-LOX (*ALOX12B*), PGDH (*HPGD*), mPGES (*PTGES*), cPGES (*PTGES3*), and COX-1 (*PTGS1*) were also determined.[19,22] Except for 15-LOX1, the predicted amino acid sequences of feline gene orthologues shared considerable sequence identity with other species. Interestingly, the feline 15-LOX1 gene appears to have become an unprocessed pseudogene. Presence of intact exons in 4 other carnivores and representative species for 7 other mammalian Orders suggest that loss of this gene in non-feline species is rare. Since 15-LOX metabolites were detectable in feline tissues, we hypothesize that 15-LOX-2 or other LOX isoforms have assumed the functions of 15-LOX-1 found in other species.

URINARY BLADDER EICOSANOID EXPRESSION IN CATS WITH FIC

Tissue lipid profiles and relative gene expression where evaluated in a pilot study of 2 symptomatic cats with a history of chronic or frequently recurrent FIC. In one cat, COX-2 gene expression was significantly down-regulated in bladder mucosa and submucosa compared to healthy controls. In the other cat, COX-2 was significantly up-regulated and was associated with a substantial increase in tissue PGE_2 concentrations. Increased COX-2 expression in this cat was localized to inflammatory cell infiltrates associated with urothelial ulceration. Increased PGDH and 12-LOXB gene expression was observed in both cats. In general, concentrations of proinflammatory lipid mediators in urinary bladders from affected cats were similar to healthy controls. However, concentrations of anti-inflammatory/pro-resolving lipid mediators tended to be lower in both affected cats compared to controls.

CONCLUSION

In conclusion, these are the first studies characterizing the feline urinary bladder microenvironment expression of key eicosanoid biosynthetic enzymes and their downstream lipid mediator products. Results of these studies will provide the basis for future genomic, proteinomic, and lipidomic studies investigating the role of eicosanoids in the pathogenesis of FIC. Manipulation of eicosanoid biosynthesis through nutrition-based or pharmacologic interventions may offer attractive targets for developing more consistently effective means of treating cats with FIC.

REFERENCES

[1]Bommer NX, et al. *J Feline Med Surg* 2012;14:527. [2]Bouchelouche K, et al. *J Urol* 2001;166:2121 .[3]Buczynski MW, et al. *J Lipid Res* 2009;50:1015. [4]Bultitude MI, et al. *Br J Urol* 1976; 48: 631. [5]Farkas A, et al. *J Urol* 1980;124:455. [6]Khan OA, et al. *J Vet Intern Med* 2011;25:724. [7]Khan MA, et al. *Prostaglandins Leukot Essent Fatty Acids.* 1998;59:415. [8]Kim JC, et al. *J Urol* 2006;175:1773. [9]Klinger MB,et al. *Am J Physiol Regul Integr Comp Physiol* 2007; 293: R677. [10]Kruger JM, et al. *J Vet Intern Med* 2013;27:740. [11]McCaffery GP, et al. *Am J Physiol Renal Physiol* 2008;295:F507. [12]McHowat J, et al.*Fed Am Soc Exp Biol J* 2010;24:999. [13]Mohammed SI, et al. *Cancer Res* 1999;59:5647. [14]Park JM, etal. *Am J Physiol Renal Physiol* 1999;445:F129. [15]Poulsen RC, et al. Am J Hematol 2008;83:437. [16]Rastogi P, et al. *Am J Physiol Renal Physiol* 2008;294: F1129. [17]Serhan CN, et al. *Nature Rev Immun* 2008;8:349. [18]Singer LM, et al. *J Vet Intern Med* 2013;27:740. [19]Sordillo LM, et al. *Anim Health Res Rev* 2009;10:53. [20]Snyder CL, et al.*J Vet Intern Med* 2013;27:735. [21]Uddin M, et al.. *Prog Lipid Res* 2011; 50:75. [22]Venta PJ, et al. *Proceedings of the 2014International Conference on Feline Health,* Davis CA, 2014. [23]Yamamoto K, et al. *Int J Urol* 2012;19:155.

URINARY TRACT INFECTIONS IN HUMANS AND COMPANION ANIMALS

J. M. Blondeau, BSc, BSc(Hons), MSc(Micro), PhD(Med Micro)
Saskatoon, SK, Canada

INTRODUCTION

Humans

Urinary tract infection or acute cystitis is one of the most common reasons that patients seek physician advice.[1] This infectious disease affects primarily women and more rarely men, however in later years, urinary tract infections increase in men in association with surgical procedures, instrumentation and catheterization. Diabetes, being uncircumcised and having an enlarged prostrate are also risk factors for UTI.[2, 3] Despite this, UTI's remain primarily a female disease with up to 50% of women having at least one infection by age 30[4, 5] and recurrent infections are common with upward of 25% of women experiencing a second infection within 6 months.[6] Simple acute cystitis is an infection of the bladder and is different than upper tract infection or pyelonephritis or kidney infection.

Acute cystitis is most often characterized by painful urination, urinary frequency and/or urgency in urination (Table 1). In contrast, pyelonephritis also may include flank pain, fever and nausea.[1, 3]

Risk factors for UTI include a shorter urethra in female patients facilitating ascending infection, family history and/or sexual intercourse. For pyelonephritis, bladder infection may precipitate upper tract involvement or infection may result from haematogenous spread.

In elderly patients, typical symptoms of UTI may be absent and replaced by symptoms of incontinence, changes in mental status or fatigue or in more advanced cases sepsis with or without fever.[7] In younger patients (children) fever may be the only symptom and in others symptoms of incontinence may be the presenting symptoms [8]; however, in older children symptoms of dysuria, urinary frequency, hematuria, abdominal pain or new daytime incontinence may be present.

The impact of UTIs should not be underestimated. Fox *et al* indicated the average episode of cystitis is associated with 6.1 days of symptoms, 2-4 days of restricted activity, 1.2 days of not working/attending school and 0.4 days in bed.[9] While such similar information may be difficult to determine in companion animals, clearly social and financial impacts likely occur to pet owners related to therapy costs and possibly lost productivity due to absenteeism from work/school or altered plans while attending to a sick pet.

Companion Animals

Like humans, female animals have a higher rate of infection than do males and the risk of infection increases with weakened immunity, older age, cancer, dental disease, diabetes, obesity, poor grooming, incontinence, bladder stones and co-incidental infection (i.e. feline leukemia or immunodeficiency viruses). UTI's occur more commonly in dogs than in cats and some animals may not be obviously symptomatic while other may exhibit a number of symptoms as summarized in Table 1.

Untreated infections may lead to complications such as lower urinary tract dysfunction, prostatitis, infertility, urolithiasis, sepsis, pyelonephritis, scarring and renal failure. Stone formation is a further complication. Prostatitis as a complication presents therapeutic challenges and bacteria may serve to seed to genitourinary tract – in some instances facilitating abscess formation. Recurrent infections in companion animals may relate to:

- Improper antibiotic
- Wrong dose
- Emergency of drug resistant pathogen
- Failure to eliminate predisposing cause
- Altered normal host defense (i.e. viable bacterial in struvite uroliths)

EPIDEMIOLOGY

In the USA, UTI are the most frequent bacterial infection of women occurring most frequently between the ages of 16-35 years with some 60% of women having infection over their lifetime. Recurrent infections are not uncommon and may affect as many as 50% of women within a year. Pyelonephritis occurs 20-30 times less frequently than cystitis but is a common cause of hospital acquired infection. In children, 7% of those between 2-24 months and 8% of those between 2-19 years were diagnosed with a UTI.[10]

Asymptomatic bacteriuria increases with age from some 2-<10% in women of child-bearing age, 2-10% of pregnant women and up to ~50% in elderly women in long care facilities. In men over 75 years of age, asymptomatic bacteriuria ranges between 7-10% [11]; however, asymptomatic bacteriuria rates may be even higher when considering patients with indwelling catheter, spinal cord injury and residence in long term care facilities.

UTIs account for ~7-8 million office visits per year in the USA, ~1 million emergency department visits and ~100,000 hospitalizations with an estimated direct cost of treatments to be ~1.6 billion dollars. Additionally, estimates on days of restricted activities including reduced productivity and absenteeism from the workplace are not insignificant as summarized above.

In dogs, more than 70% of UTIs are caused by a single bacterial species.[12] *E. coli* continues to be the single most common pathogen associated with acute and recurrent UTI in dogs and cats causing 37-45% of cases [13], however, infections are less common overall in cats, but increase in prevalence with older age.[14] It is estimated that 14% of all dogs develop a UTI during their lifetime [15] with infection most commonly in female spayed dogs and male castrated dogs. In intact male dogs, lower UTI is thought to extend to the prostate. UTI infections are less common in intact female and male dogs.[13] A summary of factors to consider in UTIs in companion animals is shown in Figure 1 with symptom related disease shown in Figure 2.

Gram-positive organisms – *Staphylococcus* spp., *Streptococcus* spp. and *Enterococcus* spp. account for some 25-30% of naturally occurring UTIs. Methicillin-resistant strains of *Staphylococcus aureus* have been reported from dogs.[16] Polymicrobial infection may account for up to 30% of infections. Non-bacterial causes of UTI (i.e. yeast and fungus) are seen only rarely and almost always in immunocompromised patients. Bailiff *et al* reported on 5 dogs and 2 cats with *Corynebacterium urealyticum* associated UTI.[17] Pre-existing urinary tract disorders were common in infected animals and antimicrobial therapy along with surgical debridement of bladder plaques was required in some animals. This study serves to identify that uncommon pathogens might be causative in some hosts and be challenging for optimal therapy. The finding of uncommon pathogens might be a clue to investigate for urinary tract abnormalities.

PATHOGENESIS

Bacteria typically enter the bladder via the urethra. Colonizing bacteria from the urogenital and perianal anatomy are introduced into the urethra and ascend to the bladder. The shorter urethra in female patients increases the risk of bacteria reaching the bladder and is likely the primary reason why UTI infections are more common in female versus male patients. UTIs arise from organisms that colonize the urogenital tracts with bladder infection following ascending organisms in the urethra and ascending organisms from the bladder to kidney might be the most common route leading to pyelonephritis.

Escherichia coli remains the most common bacteria species associated with UTI accounting for up to 80% of uncomplicated acute cystitis cases.[2] Pili present on the surface of the bacteria facilitate the attachment of the organism to the bladder wall thereby initiating an inflammatory response and the symptoms associated with UTI. Attached bacteria are not easily eliminated by voiding, however, frequent voiding serves to flush or eliminate bacteria from the bladder.

Depending on the age of the patient, organisms such as *Staphylococcus saprophyticus* may account for up to 10-15% of cases with *Klebsiella* species, *Enterobacter* species, *Proteus* species, *Pseudomonas* species and *Enterococcus* species being less frequent causes.[18] In elderly patients polymicrobial infections may be seen in up to 33% of cases and Enterobacteriaceae predominate in children. Certain underlying conditions such as diabetes may also influence etiology.

A summary of bacterial pathogens associated with urinary tract infections in humans and companion animals are summarized in Table 2. Table 3 summarizes host defenses protecting against urinary tract infection.

Diagnosis

Clinical diagnosis can be made on presenting signs and symptoms, and in humans laboratory diagnosis is made based on urinalysis and culture of urine for the presence of bacteria. Urine for culture should be collected before initiating antimicrobial therapy. Collected urines should be transported to the laboratory without delay (<4 hours) or be refrigerated or collected into a collection device containing preservative (i.e. boric acid). McCarter *et al* suggested that urines should be processed within 2 hours of collection or be refrigerated.[19]

In human patients with simple cystitis, a midstream urine is the specimen of choice. Other specimens from more complicated patients could include in-out catheter collection or surgically collected urine. In neonates and young children, bag urine specimens may be collected. For patients with suspected pyelonephritis including fever, blood culture should also be collected. In the lab, urine should be cultured using a calibrated loop on media to recover both Gram-positive and –negative bacteria. Bacterial counts of $\geq$100,000 colony forming units (cfu)/ml (or $\geq 10^8$ cfu/L) are considered significant, however, some patients may have low colony count bacteriuria. One difficulty in culturing urine in suspected UTI cases relates to a single versus multiple bacterial species. When a single bacterial species is recovered (i.e. *E. coli*) and in sufficient colony counts, then the microbiology diagnosis seems straight forward. As well, when a known pathogen (*E. coli*) is recovered on culture plates along with viridans streptococci and/or coagulase-negative staphylococci (i.e. *Staphylococcus epidermidis*) the reporting seems straight forward. However, when 2 or 3 known uropathogens are recovered on culture plates (eg. *E. coli*, *Klebsiella* spp., *Proteus* spp. or *Enterococcus* spp., etc) then reporting becomes more challenging. Consultation between the clinical microbiologist and the patient's most responsible physician is necessary.

Patients with anatomic or functional abnormalities may have multiple bacterial species. Common Gram-negative organisms include *E. coli*, *Proteus* spp., *Klebsiella* spp., *Pseudomonas* spp. and *Enterobacter* spp. *Pasteurella* spp. are more commonly associated with UTI in cats than in dogs.[20]

In animals, midstream urine collections are more difficult than in humans with a higher probability of contamination during collection. Three methods for consideration for urine collection include:

1. Cystocentesis – sampling of urine directly from bladder by needle aspiration. Proper skin preparation is necessary to reduce contamination.
2. Catheterization – catheter insertion to bladder via urethra. Aseptic technique is necessary to reduce contamination during collection.
3. Midstream urine – high probability for contamination in dogs – not recommended. May occasionally be useful in cats with severe dysuria.

In dogs and cats, quantitative cultures should be performed and for cat $\geq 10^3$ cfu/ml and in dogs $\geq 10^4$ cfu/ml should be considered significant.

Factors to consider in the diagnosis of UTI in companion animals is summarized in Figure 3.

Antimicrobial Resistance/Susceptibility

Organisms recovered from urine specimens need to be tested for susceptibility/resistance to antimicrobial agents – particularly those used for UTI therapy. Susceptibility testing is always required where the susceptibility or resistance to antimicrobial agents is unpredictable. For urinary pathogens, susceptibility is not predictable.

Increasing antimicrobial resistance with pathogens associated with UTI is a major concern. In humans, antimicrobials typically used for treating acute uncomplicated UTIs include nitrofurantoin, beta-lactams, trimethoprim/sulfamethoxazole (TMP/SMX) and fluoroquinolones. Focusing on *E. coli* strains, data from 2001 in Canada indicated resistance to ampicillin ranged from 23-41% and to TMP/SMX ranged from 8.4-19.2%; quinolone resistance was <2%. Karlowsky *et al* reported on antimicrobial resistance in Canada for UTI pathogens collected between 2007-2009.[21] For 1581 *E. coli* strains analyzed, 4% of strains were not susceptible to nitrofurantoins, approximately 20% resistant to ciprofloxacin and more than 22% were resistant to TMP/SMX. Approximately 4% of strains were not susceptible to amoxicillin-clavulanate.

A more recent concern is with UTIs caused by Gram-negative bacilli that produce extended spectrum beta-lactamase (ESBL). ESBLs hydrolyze extended spectrum cephalosporins including 3rd and 4th generation agents such as ceftriaxone, cefotaxime, ceftazidime and cefepime. For bacterial strains producing ESBLs, few antimicrobial agents may be available for therapy. In Canada from 2008-2011, ESBL producing E. coli, K. pneumoniae, K. oxytoca and P. mirabilis increased from 3.8% to 11.6% (p=0.001). The highest increase was seen for E. coli where the prevalence of ESBL producing strains rose from 4-8% to 14% over the 4 years studied.22 Additionally, ESBL producing bacteria may also be co-resistant to TMP/SMX, fluoroquinolones and aminoglycosides.[23, 24] In some cases, a carbapenem (imipenem, meropenem, ertapenem) may be the most suitable option. A further alarming trend are with bacterial strains producing carbapenemases, thereby conferring resistance to all carbapenem antibiotics.[25] ESBL producing organisms have been described in bacteria recovered from humans [26] and animals.[27-29]

Treatment

Guidelines for the therapy of UTIs in humans[30] and companion animals[31] have been published. Therapeutic guidelines have been published for the treatment of uncomplicated, complicated and asymptomatic bacteriuria in humans. For uncomplicated infection, drugs such as fosfomycin, nitrofurantoin and trimethoprim/sulfamethoxazole have been recommended as first line agents whereas fluoroquinolones, extended-spectrum cephalosporins and amoxicillin/clavulanic acid have been recommended as 2nd or 3rd line agents. Table 4 shows antimicrobial treatment recommendations for UTIs in humans and companion animals.

REFERENCES

[1]Mazzulli T. **Canadian Journal of Urology** 2012; 19: 42-8.[2]Nicolle LE. **Urol Clin North Am** 2008; 35: 1-12.[3]Lane DR, et al. **Emerg Med Clin North Am** 2011; 29: 539-52.[4]Schappert SM. **Vital Health Stat 13** 1997: 1-38.[5]Foxman B, et al. **Infect Dis Clin North Am** 2003; 17: 227-41.[6]Foxman B. **Dis Mon** 2003; 49: 53-70.[7]Beveridge LA, et al. **Clin Interv Aging** 2011; 6: 173-80.[8]Robinson JL, et al. **Paediatr Child Health** 2014; 19: 315-25.[9]Foxman B, et al. **Ann Epidemiol** 2000; 10: 509-15.[10]Shaikh N, et al. **Pediatr Infect Dis J** 2008; 27: 302-8.[11]Nicolle LE, et al. **Clin Infect Dis** 2005; 40: 643-54.[12]Ling GV, et al. **J Vet Intern Med** 2001; 15: 341-7.[13]Cohn LA, et al. **J Vet Diagn Invest** 2003; 15: 338-43.[14]Bartges JW. Diseases of the urinary bladder. In: Birchard SJ, Sherding RG, editors. Saunders Manual of Small Animal Practice. Philadelphia, PA: WB Saunders; 1999: 943-57.[15]Ling GV. **J Am Vet Med Assoc** 1984; 185: 1162-4.[16]Watts JR, et al. **J Small Anim Pract** 1996; 37: 54-60.[17]Bailiff NL, et al. **J Am Vet Med Assoc** 2005; 226: 1676-80.[18]Ronald A. **Dis Mon** 2003; 49: 71-82.[19]McCarter YS, et al. **Cumitech 2C** 2009: 1-25.[20]Wooley RE, et al. **J Clin Microbiol** 1976; 4: 326-9.[21]Karlowsky JA, et al. **Antimicrob Agents Chemother** 2011; 55: 3169-75.[22]Poutanen SM, et al. **Canadian Association for Clinical Microbiology and Infectious Diseases (CACMID)**; 2013.[23]Pitout JD, et al. **Lancet Infect Dis** 2008; 8: 159-66.[24]Prakash V, et al. **Antimicrob Agents Chemother** 2009; 53: 1278-80.[25]Gupta N, et al. **Clin Infect Dis** 2011; 53: 60-7.[26]Pallett A, et al. **J Antimicrob Chemother** 2010; 65: iii25-33.[27]Nobrega DB, et al. **J Infect Dev Ctries** 2014; J Infect Dev Ctries: 954-60.[28]O'Keefe A, et al. **Antimicrob Agents Chemother** 2010; 54: 3489-92.[29]Wu G, et al. **PLoS One** 2013; 8: e75392.[30]Nicolle LE. **Drugs Aging** 2001; 18: 243-54.[31]Weese JS, et al. **Vet Med Int** 2011: 1-9.[32]Mazzulli T. **Canadian Journal of Urology** 2001; 8 2-5.[33]Damron DJ, et al. **J Clin Microbiol** 1986; 24: 400-4.[34]Waites KB, et al. **Urology** 1991; 38: 589-96.[35]Nicolle LE, et al. **Antimicrob Agents Chemother** 1994; 38: 1368-73.[36]Cox CE, et al. **Clin Infect Dis** 1995; 21: 86-92.

Figure 1: Factors to consider with UTIs in companion animals.

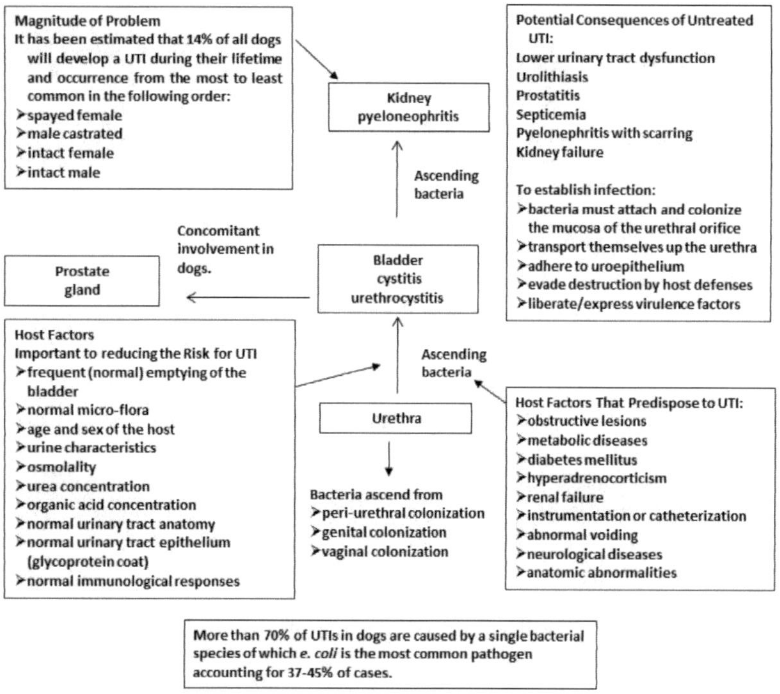

Figure 2: Symptoms/complications associated with UTI in companion animals.

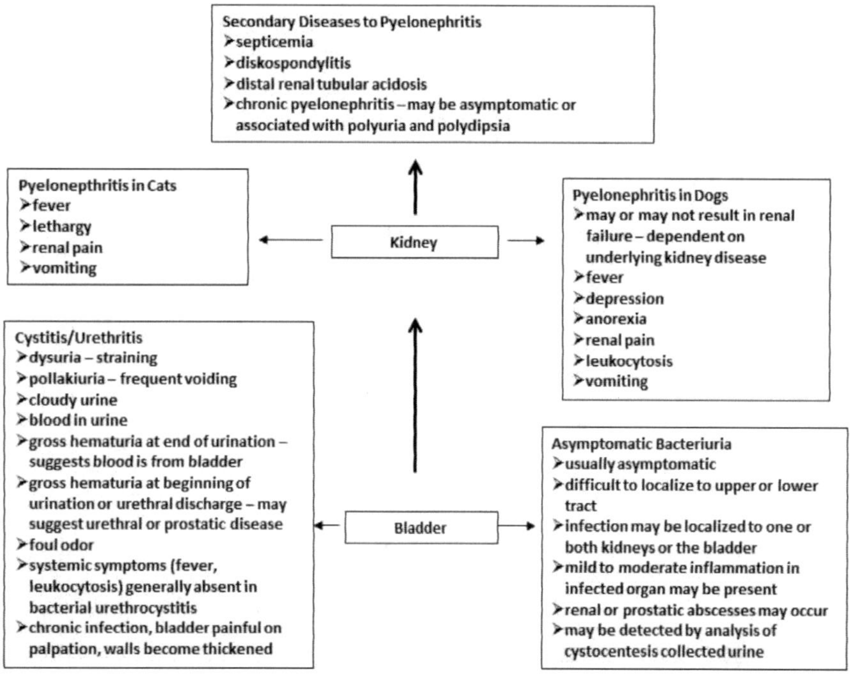

Figure 3: Factors to consider for the diagnosis of UTIs in companion animals.

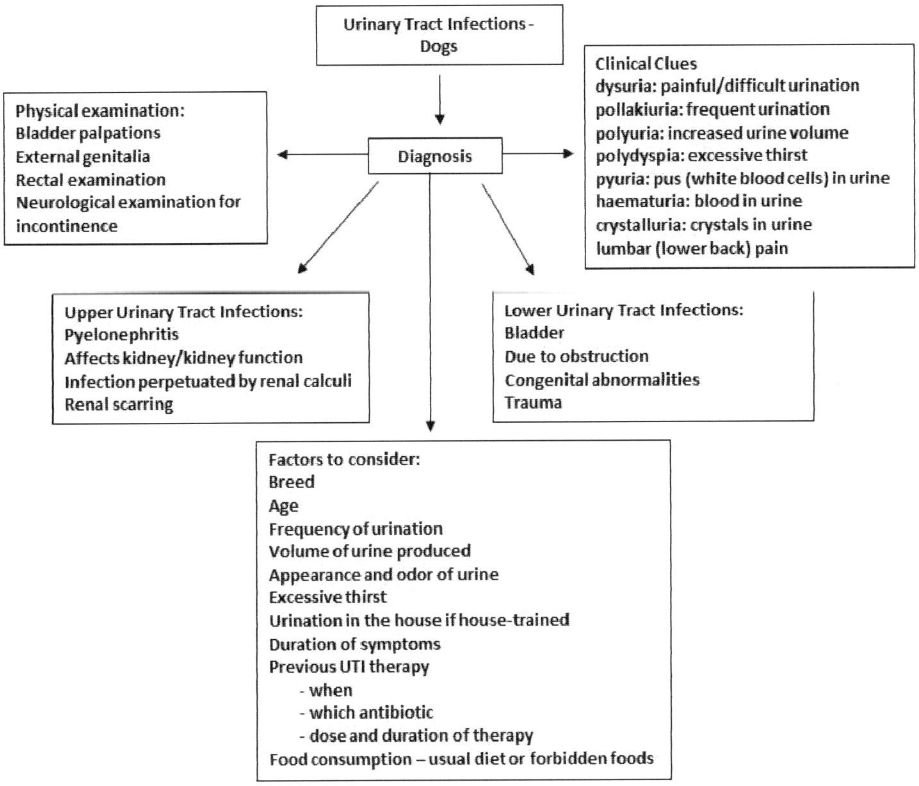

Table 1: Symptoms associated with urinary tract infections.

Humans			**Companion Animals**	
Children	**Adults**	**Elderly**	**Dogs**	**Cats**
• Blood in urine • Cloudy urine • Foul/strong odour • Frequency • Malaise • Pain/burning on urination • Pressure/pain in lower back/pelvis • Wetting problems in toilet trained children	Women • Burning with urination • Increased frequency of urination/small volumes of urine passed • Bloody urine • Cloudy urine • Discoloured urine (tea or dark color) • Foul smell • Pelvic pain in women Rectal pain in man	• Acute irritative symptoms - Dysuria - Frequency - Supra-pubic discomfort - Incontinence - Urgency • Invasive infection - Pyelonephritis - Bacteremia • Other - Epididymo-orchitis - Urethritis - Periurethral abscess • **Possible Symptomatic Infection Institutionalized Elderly** - Fever without localizing findings - Hematuria - Clinical deterioration - Malodourous urine • Confusion/delirium like state • Agitation • Hallucinations • Behavioral changes • Poor motor skills • Dizziness • Falling	• Small volumes during urination • Straining • Pain • Loss of control in house • Increased thirst • Blood in urine • Decreased appetite • Odorous fur • Lethargy • Loss of activity • Limping • Difficulty urinating • Cloudy/foul smelling urine • Frequent urination • incontinence	• Difficult/painful urination • Blood in urine • Abnormal/frequent urination • Urination in inappropriate places • Blocked urine flow • Thickened/firm bladder wall

Table 2: Frequency (%) of uropathogens recovered from companion animals[12, 20] and humans[32].

Organism	**Companion Animals**		**Humans**	
	Dogs	**Cats**	**Uncomplicated**	**Complicated**[33-36]
E. coli	33-45	31	70-90	9-55
Staphylococcus spp./epidermidis	1.5-12		5-20*	1
Klebsiella pneumoniae/spp.	4-11	15	4	3-16
Proteus mirabilis/spp.	11-29	13	2	7-15
Others including *Mycoplasma* spp.	3-10	5		
Streptococcus spp.	5-8	13		1
Enterobacter spp.	3-5			3-5
Pseudomonas aeruginosa/spp.	2-3			2-14
Pasteurella multocida		25		
Staphylococcus aureus	17	19		0.5-2
Citrobacter spp.				5-6
Serratia spp.				7
Enterococcus spp.				6-11
Morganella morganii				3-11
Providencia spp.				6-18

**Staphylococcus saprophyticus*

Table 3: Host defences protecting against urinary tract infections.

Humans	Companion Animals
• Frequent (normal) emptying of the bladder • Normal microflora • Age and sex of the host • Urine characteristics • Osmolality • Urea concentration • Organic acid concentration • Normal urinary tract anatomy • Normal urinary tract epithelium (glycoprotein coat) • Normal immunological responses	• Normal micturition • Normal anatomy • Intact mucosal defense • Surface glycosaminoglycans • Cell exfoliation • Normal microflora • Local antibody production • Epithelial cell antimicrobial properties • Antimicrobial properties of urine • Hyperosmolality concentration • Tamm-Horsfall mucoprotein • Systemic immunocompetence

Table 4: Antimicrobial agents recommended as first line agents for treatment of urinary tract infections.

Human[2]	Companion Animals[31]
• Amoxicillin – 500 mg tid 7 days • Amoxicillin/clavulanic acid – 500 mg tid 7 days • Cephalexin – 250-500 mg qid 7 days • Cefuroxime axetil – 500 mg bid 7 days • Cefixime – 400 mgs od 7 days • Cefpodoxime proxetil – 100 mg bid 3 days • Norfloxacin – 400 mg bid 3 days • Ciprofloxacin – 250-500 mg bid 3 days; extended release – 500 mg od 3 days • Ofloxacin – 200 mg 3 days • Levofloxacin – 250-500 mg bid 3 days • Nitrofurantoin: macrocrystal 50-100 mg qid 7 days; monohydrate macrocrystals 100 mg bid 7 days • Pivmecillinam – 400 mg bid 3 days or 200 mg bid 7 days • Fosfomycin trometamol – 3 g single dose • Trimethoprim/sulfamethoxazole – 160/800 mg bid 3 days • Trimethoprim – 200 mg bid 3 days	• Amoxicillin – 11-15 mg/kg PO q8H • Amikacin – 15-30 mg/kg IV/IM/SC q24h (dogs); 10-14 mg/kg IV/IM/SC q24h (cats) • Amoxicillin/clavulanate – 12.5-25 mg/kg PO q8h (dose based on combination of amoxicillin + clavulanate) • Ampicillin • Cephalexin, cefadroxil – 12-25 mg PO q12h • Cefovecin – 8 mg/kg single SC injection. Can be repeated once after 7-14 days • Cefpodoxime proxetil – 5 to 10 mg/kg q24 PO • Ceftiofur – 2 mg/kg q12-24h SC • Chloramphenicol – 40-50 mg/kg PO q8h (dogs); 12.5-20 mg/kg PO q12h (cats) • Ciprofloxacin – 30 mg/kg PO q24h • Doxycycline – 3-5 mg/kg PO q12h • Enrofloxacin – 5 mg/kg PO q24h (cats); 10-20 mg/kg q24h (dogs) • Imipenem-cilastatin – 5 mg/kg IV/IM q6-8h • Marbofloxacin – 2.7-5.5 mg/kg PO q24 h • Meropenem – 8.5 mg/kg SC/IV q12 (SC) or 8 (IV) h • Nitrofurantoin – 4.4-5 mg/kg PO q8h • Orbifloxacin: Tablets – 2.5-7.5 mg/kg PO q24h; oral: Suspension – 7.5 mg/kg PO q24h (cats) or 2.5-7.5 mg/kg PO q24h (dogs) • Trimethoprim-sulfadiazine – 15 mg/kg PO q12h (note – dosing is based on total trimethoprim + sulfadiazine concentration

DECISION MAKING IN URINARY TRACT INFECTION AND BACTERIURIA

J Scott Weese, DVM, DVSc, DACVIM
Guelph, Ontario, Canada

INTRODUCTION

Urinary tract disease is a common and important problem in veterinary medicine, accounting for a large percentage of veterinary visits. Urinary tract disease can be divided into 'infectious' and 'non-infectious'; however, there can be overlap between those groups, along with debate about what constitutes 'disease'. Often, the presence of bacteria in urine (bacteriuria) is used as an indication of the presence of disease, but it is increasingly clear that bacteriuria does not always indicate the presence of disease or the potential for future disease. As such, consideration of what constitutes 'disease' and what needs to be treated is necessary to optimize patient care, reduce treatment costs and adverse effects, and reduce antimicrobial resistance selection pressure. The continued emergence of multidrug resistant bacteria and the limited treatment options for some bacteria isolated from urine have heightened awareness of bacteriuria vs infection in human medicine and increasing awareness in veterinary medicine.

BACTERIURIA VS URINARY TRACT INFECTION

It is important to understand some key definitions. *Bacteriuria* is defined as the presence of bacteria in urine, with *asymptomatic bacteriuria* (ASB or ABU) indicating bacteria in the absence of clinical signs. However, a more detailed definition is often used in humans to help differentiate ASB from contamination. Infectious Disease Society of America (IDSA) guidelines define bacteriuria in asymptomatic women as the presence of $\geq 10^5$ CFU/ml of the same bacterium in two consecutive clean voided samples.[1] For men, bacteriuria is defined as a single clean-catch voided sample with one bacterial species present at $\geq 10^5$CFU/ml. In animals, ASB is not an appropriate term because of the inability of animals to report symptoms. Therefore, *subclinical bacteriuria* (SB) is more appropriate. Objective definitions of bacteriuria are not available for animals. While CFU/ml cut-offs similar have been used in animals, they are lacking in objective support. Whether CFU/ml counts are relevant in animals, where collection methods differ (cystocentesis vs free-flow samples) and where collection of clean free-flow samples is difficult, is unclear. In contrast, urinary tract infection (UTI) indicates the presence of disease in association with bacteriuria.

How common is bacteriuria?

Broad studies have been performed in humans, with wide ranges in prevalences depending on the study population. Examples include 40% prevalence of ASB in diabetic women,[2] 17% in uncatheterized hospital inpatients[3] and 12% in people undergoing total hip or knee arthroplasty.[4] Rates in low risk individuals can also sometimes be high, such as a 9% ASB rate in healthy controls in one study.[5]

Corresponding data from animals (at least from large studies and diverse study populations) are largely lacking. A recent study of healthy female dogs identified bacteriuria in 9/101 (8.9%).[6] In another study, bacteriuria (reported as 'asymptomatic bacterial UTI') was identified in 9% of obese and 31% of morbidly obese dogs.[7] A study of dogs undergoing orthopedic surgery reported SB in 2.1% (3/80).[8] Higher rates can be found in specific populations. For example, 38% of dogs that underwent surgery for thoracolumbar intervertebral disk extrusion developed a UTI (based on positive culture) within 3 months of surgery.[9] Sixty percent of these UTIS were classified as 'occult', presumably indicating the presence of SB, not UTI. Thirty percent (26/87) dogs treated with cyclosporine had at least one positive urine culture over an undefined retrospective study period.[10] A point prevalence study of the same dogs, considering their first reported urine cultures, identified bacteriuria in 15% of cyclosporine-treated dogs and 3% of untreated controls. It was unclear whether any of these dogs had clinical UTIs, but there were no reports of abnormal clinical signs in the medical records for any dog, so it is likely that most or all represented SB. Similarly, 42% of dogs with diabetes mellitus or hyperadrenocorticism had "UTI"s (defined as positive culture), with <5% of those having clinical abnormalities.[11] Thus, while studies are limited and tend to be small, there is enough evidence to suggest that SB is common in some dog (and presumably cat) populations.

MANAGEMENT OF UTI AND BACTERIURIA

Humans

It is clear that UTI (disease) should be treated. Numerous large and well-controlled studies have evaluated different treatment approaches, and comprehensive treatment guidelines are available to maximize clinical response, minimize recurrence and optimize antimicrobial use.[12]

It is also clear that ASB should not be treated in most situations,[1] yet compliance with this recommendation is often poor.[3,13,14] Significant efforts have been underway to reduce treatment of ASB in humans because various studies have identified no beneficial effect to patients. While the bacterium presence in the bladder may be eliminated during treatment, treatment of ASB does not reduce the risk of UTI or other complications (e.g. post-operative infection).[2,4,15] Lack of efficacy combined with the potential for massive use of antimicrobials with such a common condition highlight the importance of controlling ASB treatment as part of antimicrobial stewardship programs. Treatment of ASB is largely reserved for pregnant women (as ASB has been associated with pyelonephritis in the mother as well as low birth weight and pre-term delivery)[1] and individuals undergoing urologic procedures that involve comprising the urinary tract mucosa (e.g. transurethral resection of the prostate).[1] Beyond that, there is little evidence of efficacy of treatment. A recent systemic review identified a slight increased risk of UTI in people with ASB, but also lack of a beneficial effect of treatment on risk of subsequent infection.[16] Therefore, even if there is a risk posed by ASB, elimination of bacteriuria may not be a viable mechanism to reduce that risk, in large part because of the inability to address the underlying problem and the likelihood of recurrence of bacteriuria shortly after cessation of antimicrobials. One recent study has even indicated that treatment of ASB might be harmful. That study of sexually active women with ASB reported a significantly higher UTI recurrence rate in treated individuals,[15] perhaps because less virulent strains are eliminated by treatment, creating an environment amenable to colonization (and subsequent infection) by more virulent counterparts. While pyuria may be present with both ASB and UTI, pyuria in the absence of clinical abnormalities is not an indication to treat.[1]

Animals

Veterinary data pertaining to management of SB are very sparse; however, a recent study of healthy female dogs reported that no dogs with SB developed a clinical UTI during a 3 month follow-up period.[6] While small (with only 9 dogs with SB), this study provides some support to the recommendation that dogs with SB not be treated with antimicrobials. Ultimately, larger studies are required but given this study, the apparent high SB rate in some dog populations (without clear evidence of increased UTI incidence) and abundant human data, it is reasonable to support a recommendation not to treat most animals with SB.[17]

Anecdotally, SB is often treated in some animal populations. For example, while treatment of diabetic human patients with ASB has been shown to have no impact[2] and is not recommended,[1] some veterinary clinicians screen and treat diabetic patients routinely. While evidence is lacking either way, it is difficult to make a plausible story as to why a diabetic animal would be at greater risk of complications from SB than a human. This cannot be taken to say that treatment of SB in animals is poor practice, since there is an absence of evidence for either approach; however, it should be taken as a reason to consider whether screening and treatment are necessary and to prompt proper study. The presence of SB perhaps most importantly indicates the need for investigation of causes (e.g. endocrinopathy, immunosuppressive disorder, anatomical abnormalities, conformation) that might also predispose to UTI and be manageable.

As with any clinical decision, a variety of factors must be considered when determining whether treatable UTI is present. One is the method of sample collection, with little confidence provided by a positive free-flow culture.[17] Whether CFU counts provide valuable information is currently unclear. It is reasonable to consider both the CFU/ml and the number and nature (e.g. typical urovirulence) of the bacterial isolate, but decision-making based on free-flow samples will be questionable, regardless of the CFU count. Clinical signs are the differentiating factor between UTI and SB, so careful history-taking and physical examination are important. Cytology is typically given greater consideration in animals compared to humans when interpreting the clinical relevance of bacteriuria,[17] in part because of the problem that veterinary patients cannot as effectively convey urinary tract discomfort and because some veterinary patients at increased risk of disease cannot adequately indicate the presence of disease (e.g. animals with hindlimb paresis/paralysis that do not urinate unaided). However, there can be poor agreement between pyuria and bacteriuria in dogs[8,11] and pyuria can have various non-infectious causes. The distinction between SB and UTI can therefore difficult in some cases.

Discussions regarding when to treat SB are closely tied to questions about when to culture urine. In humans, there are efforts underway to reduce urine culture submissions. This may seem counter-intuitive given the emphasis on submitting proper diagnostic specimens for other diseases. However, a general guideline in humans is "*If treatment of bacteriuria is not beneficial, screening of asymptomatic populations to identify bacteriuria is not indicated*".[1] In fact, the concept of '*knowing less leads to doing less*'[18] with 'less' being a good thing, has been supported in human medicine. While culture is certainly indicated when there is evidence of disease, it is difficult to support screening of individuals in the absence of disease or a plausible reason why treatment of a positive culture would be necessary. Better data are needed to define when treatment of SB in veterinary patients (if ever) is required, to facilitate better screening recommendations, but avoiding submission of culture specimens when there is no indication to treat a positive culture is a useful step.

KNOWLEDGE GAPS

Profound knowledge gaps exist for diagnosis and management of bacteriuria (and UTI) in animals. While abundant study has been performed to characterize bacteriuria and UTI in humans, with large studies of diverse patient populations, veterinary data are comparatively very limited. There is limited information about the prevalence of bacteriuria in different populations, something that hampers assessment of the relevance of results in an individual patient. There has been limited study of the clinical implications of bacteriuria in veterinary patients. There has also been no study of treatment of bacteriuria, including whether it can be successful in the short- and long-term, whether it reduces the likelihood of subsequent UTI or other complications or whether it increases or decreases the risk of patient morbidity (e.g. antimicrobial complications, antimicrobial resistance).

Advances in genomics might facilitate interpretation of bacteriuria and identification of specific situations where treatment is or is not likely to be beneficial. As more information is known about urovirulence factors, characterization of bacteria from urine could play a role in decision-making. A recent (and relatively small) study of *E. coli* from children with UTI or ASB reported a significant difference in virulence gene profile between the groups.[19] Whether this will lead to evidence that there is a clinical benefit of treating ASB caused by strains with selected virulence factors is debatable but not beyond the realm of possibility.

Ultimately, better research is required to provide veterinary-specific guidance on optimal management of urinary tract disease. In the absence of such guidance, veterinary clinicians should look to their human medical counterparts and, while recognizing that humans are different than veterinary species, consider the approach that is taken to UTI and bacteriuria in human patients.

REFERENCES

[1] Nicolle LE, et al. ***Clin Infect Dis*** 2005;40:643-654. [2] Harding GKM, et al. ***N Engl J Med*** 2002;347:1576-1583. [3] Leis JA, et al. ***Clin Infect Dis*** 2014;58:980-983. [4] Sousa R, et al. ***Clin Infect Dis*** 2014. [5] Georgiadou SP, et al. ***Clin Infect Disease*** 2014. [6] Wan SY, et al. ***Journal of the American Veterinary Medical Association*** 2014;245:106-112. [7]Lusby AL, et al. Prevalence of asymptomatic bacterial urinary tract infections in morbidly obese dogs. *ACVIM Forum*. Denver, CO, 2011. [8] McGhie JA, et al. ***Australian Veterinary Journal*** 2014;92:33-37. [9] Olby NJ, et al. ***J Vet Int Med*** 2010;24:1106-1111. [10] Peterson AL, et al. ***Veterinary Dermatology*** 2012;23:201-e243. [11] Forrester SD, et al. ***J Vet Intern Med*** 1999;13:557-560. [12] Warren J, et al. ***Clin Infect Dis*** 1999;29:745-758. [13] Drekonja DM, et al. ***Infect Control Hosp Epidemiol*** 2014;35:574-576. [14] Phillips CD, et al. ***BMC Geriatrics*** 2012;12:73. [15] Cai T, et al. ***Clin Infect Dis*** 2012;55:771-777. [16] Dull

RB, et al. ***Pharmacotherapy*** 2014;34:941-960. [17] Weese JS, et al. ***Vet Med Int*** 2011;4:1-9. [18] Naik AD, et al. ***Clin Infect Dis*** 2014;58:984-985. [19] Yun KW, et al. ***J Microbiol Immunol Infect*** 2014;47:455-461.

DEVELOPING A SUCCESSFUL SOCIAL MEDIA MARKETING STRATEGY

Eric Garcia
Tampa, Florida

1. Why should a veterinary hospital use social media?

The real question should be, "Why wouldn't veterinary practices use social media?"

Social media is the form of communication that gives your message the largest reach, one that includes not only your clients, but your prospects, too.

What happens when you mail out postcards or give your business card to people who don't need your services right away?

We all like to think they save the information, but the reality is that when they do need you, that card is nowhere to be found. So they go to the Internet, and you can only hope they find you there.

If your practice has a social media (which you also listed on that business card), the prospective client can follow it or subscribe to its feed, and that puts your practice name in front of them several times a week.

This can help stimulate reminders about refilling heartworm and flea medications, coming in for annual vaccines and other things. Seeing your name is a reminder of the services you offer, and when they need those services, they can easily find you by logging into social media.

Used correctly, social media can actually increase compliance with pet owners.

2. What's the best way to use social media to stimulate business?

Most consultants will tell you not to use social media as a place to sell. I agree—the minute you start to sell to clients in "their" space, which is where they want to avoid advertising, they'll unsubscribe from your feed.

This was what businesses did when social media first took off, and they learned the repercussions quickly. But you can market services to clients through social media if you do it correctly.

I always tell people there are three parts to using social media successfully:

- A mention of a service you want to market
- Some sort of fun and engaging fact
- The most important thing: the social part

Here's an example: "Wow, we just received Penny's DNA results back from the Mars Veterinary Lab, and it turns out she's 40% Yorkie and 60% Maltese. What do you think your mixed breed is made of?"

You can easily see the service we're trying to market, but we made it cute. People look at the photo of Penny, think about her breed and socialize back with you. And then, they might call you to ask more about DNA testing, maybe schedule an appointment outside of a routine visit or share it on their neighbor's page or their own wall. And what's attached to that posting? The practice name and page, so now everyone can see the services you offer.

The average Facebook user has 140 friends. Say your practice has 100 followers, and three of those people share one of your posts. One hundred of your followers, plus 140 friends of each of those users equals a reach of 520 people! The possibilities are endless.

3. How do you ensure that social media messages are on-point and appropriate?

The best way to do that is make sure that a few people on your team go through training. Visit a conference and spend a little extra for the all-day social media lab, or hire a consultant to come in for one-on-one training. You may have a technician or receptionist who knows how to use social media, but knowing how to use it and how to communicate properly and effectively with clients are two totally different things.

I don't suggest outsourcing social media, by the way. When you outsource social media, it becomes an argument about return on investment. If you spend a few hundred dollars a month having another company communicate on your behalf, at some point you're going to start questioning the value of that communication.

It also makes no sense to outsource social media because the whole point is to honestly engage in conversation with your clients. How does it look if your clients find out you've hired an outside company to do that? It becomes fake social media—it's a lie.

4. How often should a veterinary practice post to social media? Which channels are best to focus on?

If you're just getting into social media, I suggest two postings per week. It's a realistic number. Start small and work your way up. If you've been doing this for a while, ramp up to three to five postings per week—not too few, but not overwhelming.

As far as the social channels:

Facebook: Use it. Most users log on multiple times a day.

Twitter: Use this, too. There are many users and you can easily link to Facebook if you don't have the number of followers to engage in true Twitter talk.

Pinterest: This is still too new to be effective for veterinary practices. If you're an advanced social media user and want to dabble in this, go for it. If you don't have the time or are new to this, then hold off. Don't rush to be part of every new channel—you'll burn out.

Google + : This is also still very new, plus there are many spammers. Here's how I gauge whether this channel is worth it or not: If Dr. Marty Becker has 881 people in his Google + circles and he's on top news channels and travels the nation to spread the good veterinary word. What would this look like for a veterinary practice? One distinct difference between Google+ and Facebook is Facebook appeals to every demographic with strong numbers. Google + can't say that yet. However, Google itself is an unstoppable force and you do not want to miss out on the opportunity to be part of the world's largest search engine's social network. I suggest mastering Facebook first then tap into Google+.

5. What's my 30 minute a week strategy for social media?

Make sure you preplan your marketing. This should include what your focus is (such as dental care, laser therapy, senior care). Pinpoint the dates you want to post and put them on a calendar. Finally, plan your resources for each topic: Visit a few websites in advance and copy/paste those links into a Word document. This way, you'll know when you will post, what you will post, and where to get a good reference to share when you do.

Most practices stop posting on social media because they wind up sitting there, trying to think of something productive to say. They waste time and lose sight of effective postings.

DEVELOPING A SUCCESSFUL SOCIAL MEDIA MARKETING STRATEGY

Eric Garcia
Tampa, Florida

Our reputation precedes us when we enter a new environment, though now the World Wide Web is usually the messenger, for better or worse. Whether it's a new job, a meeting, a fundraiser, a date, or something in-between, there are more ways than ever for those around us to do their homework, *on us,* before meeting face to face.

That's right, a simple Google keyword search can bring up anything from your business website, to your Facebook page. While it may seem like you have no control over what people find out about you online, the truth is that you have more of a say in the matter than you think.

Google has a remarkably comprehensive system in place for coordinating user search results and delivering them in a timely, deliberate and meaningful fashion. What might seem like random results at first glance is actually a highly methodical process unfolding in real-time. "Crawlers" (know singularly as Googlebot) continuously sort through billions of web pages from each corner of the globe, algorithmically analyzing, assessing, ranking, and indexing web pages and sorting through each applicable result.

So what does this mean for you? It means that when you're looking for new clothes online and nonchalantly typing in "Summer Outfits Sale"; your results aren't random. In fact, they're far from it. Google, which accounts for approximately 70% of search engine market share, is calculating your keywords and weighing-out millions of results before delivering them to you, all in about 0.5 seconds flat. The key takeaway here is this: business owners can, and frankly need to, monitor and enhance their business's search results in order to maintain and leverage their Online Reputation.

For example, when searching online for a local veterinary practice, customer reviews will gravitate toward the top of the screen, and instantly display the name, website, address, phone number, and here's the kicker, one to five star average rating, of nearby practices. In a Google Search (by far the most common kind of online perusal) this one to five score is the combined average rating of the "Google Reviews" that have been left for your practice. These scores can dramatically influence the amount of new clients that find, and ultimately choose your business. This is why business owners must safeguard and even encourage the establishment of a positive Online Reputation in order to fully optimize their practice.

Ok, so Keyword Searches and Google Reviews are all important factors to consider when managing my Online Reputation. But what can I do about it and how can I use these tools to enhance my practice? Well, we're happy you asked.

***

Quick Tips: Improve Your Online Reputation

(1) First, you've got to search for your business online. If you've never looked, you might be a bit intimidated at what you might find. But you can do it, dive right in. We recommend that you search on Google, Bing, Yahoo and Yelp, and then "Claim Your Listing" once you've found the right result. Verify that this truly is your business, and presto, you've claimed your online listing.

(2) Next, setup your online alerts. By using both google.com/alerts and mention.net you'll have two powerhouse services monitoring the Internet for results about your business. Now you'll be notified by email whenever a new result or review is available. Convenient? We think so.

(3) Now, bookmark your top online listings in a folder within your browser. This will save you time when you wish to check back on your results, and lets you easily review your listings at least several times a month.

(4) Here's the fun part; engage with positive reviews! The average veterinary practice receives one review per month. You can reply to a positive review online, call and thank the client, or get creative and send a personalized Thank You card. Engage how you see fit, but definitely reach back to those who've been impressed with your practice. This little bit of effort can most definitely go a long way when it comes to encouraging positive reviews and ensuring client retention.

(5) Finally, go the extra mile. You can engage directly with happy clients at checkout by encouraging them to leave a positive review for your practice. If they're onboard, send them a friendly reminder email at the end of the week, with a direct link to help them leave a Google Review.

***

It's really that simple to enhance your online presence and manage your online reputation. Doing so could mean the difference between countless new customers, and a haphazard web-presence. By pro-actively managing your Online Reputation, you'll stay ahead of the curb, reach more clients and ensure your veterinary practice maximizes its resources. In the digital age, you may not be able to control everything that's published online, but you can certainly nudge it in the right direction.

HOW WE CAN MAXIMIZE THE VALUE OF ACVIM SPECIALISTS

Lucas Pantaleon, DVM, MS, DACVIM (LAIM), MBA
Versailles, KY

INTRODUCTION

Strategy is not figuring out what the competition is doing and trying to do the same thing, this is to compete to be the best but it is not to compete on strategy. This type of competition (zero sum) will not create a sustainable competitive advantage. [1] Competition focuses more on meeting customer needs than on demolishing rivals. [2]

Strategy could be defined as the underlying variability among firms that lead to a significant difference in their performance. [3] This means deliberately choosing a set of activities to deliver a unique mix of value that is making a set of choices that no one else has made. [1] Strategy explains how an organization, faced with competition, will achieve superior performance. [2] Veterinary Specialists must think strategically in a way that will deliver value to its clients, in this way the perception of specialists within the veterinary community will be improved, thus benefiting everyone.

The company's strategically valuable resources and capabilities are what give a competitive advantage, which is enabling the enterprise to perform activities better and cheaper than the competition. [4] Superior performance is based on developing a competitively distinct set of resources and capabilities and deploying them in a well-conceived strategy.[4] The objective is to compete to be unique, this concept is all about value and how to create it, and to command a price premium and or operate at lower cost than rivals. [2,5] Competing to be the best feeds on imitation while competing to be unique thrives on innovation. [2]

It can't be forgotten that best practices (operational effectiveness) must be followed and that it is the foundation of competitive advantage.[5] A company must be operationally effective, continually embracing best practices and standards of care. However this is rarely a source of sustained competitive advantage, since the competition will institute those practices and catch up. [5]

RESOURCES AND CAPABILITIES

Resources can be physical assets (i.e. prime location), financial assets (i.e. cash), capital (i.e. equipment, buildings) or intangible assets (i.e. knowledge, reputation, human resources). Capabilities are defined as a bundle of resources that work together (i.e. integrated practice units, IPU's). [4] IPU's are defined as a dedicated team made up both of clinical and non-clinical staff that provide the full care cycle for the patient's condition.[6] It is important to keep in mind that the interplay of market forces is what determines the value of resources, thus evaluating resources in isolation is futile. [4]

The Resource Based View (RBV) is an important concept in management since it explains how the company's resources drive its performance in a dynamic competitive environment. It combines the analysis of the internal environment (resources and capabilities) with the analysis of the external environment's competitive forces (what the market demands and what competitors offer). [4]

RBV sees companies as very different collections of physical and intangible resources and capabilities, no two companies are alike. [4] Thus a company will thrive if it has the best and most appropriate collection of resources and capabilities for its business and strategy. [4]

It is important to keep in mind that managers need to look internally, within the company, and externally, market forces, at the same time. Interestingly most of the best resources are often intangible (culture, knowledge, leadership), thus sometimes overlooked. [4] Veterinary specialists have a unique resource, which is knowledge and the ability to treat complex medical cases, thus this key resource could be where part of our competitive advantage resides.

Even companies that are lucky enough to have unique resources and capabilities cannot lower their guard from competitors. Furthermore valuable resources need to be joined with other resources and embedded in a set of functional policies and activities that makes the company unique in the market. [4]

Managers should look forward and think about the next round of competition. That is continually invest and upgrade their resources, however good they are today, and leverage them with effective strategies into attractive industries or segments within an industry in which they can contribute to a competitive advantage. [4] Failure to do this will lead to value erosion from time and competition since resources depreciate, thus effective strategy requires constant understanding of the market, the competition and a continual investment to maintain and build valuable resources. [4]

COMPETITIVE ADVANTAGE

Competitive advantage is the goal of strategic thinking. It results from a combination of effective market positioning and defense against the competition in order to create superior value. [7,8] For competitive advantage to be sustainable a good offense (effective market positioning) and a strong defense (isolating mechanisms, that is customer retention and preventing imitation) must be present at the same time. [7]

Having a real competitive advantage means that when compared to the competition, you operate at a lower cost, command a premium price, or both. [5,8] Thus strategy must be directly linked to the company's financial performance.[8]

Table 1: relationship between price, cost and profitability

	Competitive advantage	
	Profitable company	**Non profitable company**
Relative price	higher	lower
Relative cost	lower	higher
	Profit = market price – product cost	

A company can sustain a premium price only by offering a unique and valuable product or service to its customers (differentiation). Raising the value to buyers increases their Willingness to Pay (WTP), which is the price the customers is willing to pay for a product or service. [8] Two important concepts about value; the customer and not the seller determines value and value is not price. [7,9]

Buyer's value can be tangible, that is quantifiable in economic terms. Here a manufacturer would pay more for a machine that compared to a lower priced one will save in labor costs that exceed the higher price. [8] Veterinarians are willing to pay more for a digital x-ray machine, since the speed, efficiency and labor savings offset the higher price.

At the consumer level the WTP is likely to be more intangible, for example based on trust to a brand or status. An example would be consumers WTP large price premiums for a carton of organic eggs, due to the concern for factory farms. [8] Strong trust amongst some of our loyal clients is what commands their WTP for our specialty veterinary services. However, it appears that the value for specialty services has been eroded in the veterinary community, thus clients are less willing to pay for more expensive expert opinions and are migrating to less expensive services that offer less value.

Relative cost is also an important part of the profitability equation, and managing to produce at a lower cost than the competition is important to remain profitable. [8] To be able to achieve this goal finding more efficient ways to create, produce, deliver, sell and support a product or service are needed. [8] Cost advantages could come from lowering operating cost or using capital more efficiently. [8]

ECONOMIC CONTRIBUTION

Economic contribution is defined as the difference between value and cost. The firm's resources and capabilities act as value and cost drivers. Value drives increase value, thus WTP. On the other hand, cost drives lower cost. A firm's market position should ideally have the highest economic contribution, thus have the highest value (WTP) at the lowest cost. Value and cost drivers lead to a superior economic contribution, in turn creating a superior market positioning and greater competitive advantage.

Example of resources and capabilities acting as value drivers are: quality, service, technology, reputation, geography, etc. Example of cost drivers are organizational practices, scope economies, economies of scale, learning curve, vertical integration, etc. There are two components to economic contribution, buyer's surplus and supplier's profits

Key to achieve the highest economic contribution is to create the highest buyer's surplus by increasing value to the customer (WTP). On the other end of the equation profits should be maximized by achieving the lower product costs possible. The following diagram depicts these relationships. [7]

Figure 1: Economic contribution diagram.

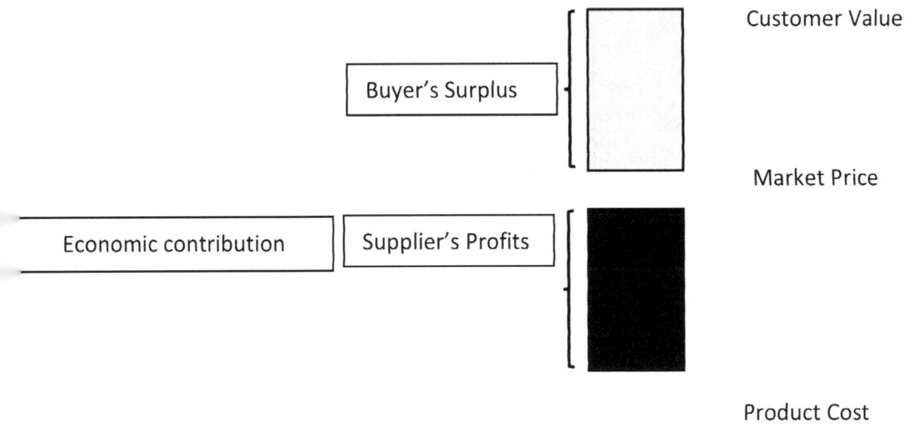

THE VALUE AGENDA

One of the new strategies proposed to transform health care is based on maximizing the value for patients, value defined as health outcomes achieved per dollar spent.[6,9] The key aspects of such strategy are:

- Move from a system organized around what doctors do towards a patient centered approach organized to meet the needs of patients.

- Move from a volume and profitability of services provided (doctors' visits, hospitalizations, procedures, and diagnostics) to a patient outcome achieved.
- Move from a fragmented system in which every local provider offers a full range of services, to a system in which services for a particular medical condition are centered in health delivery organizations and in the right location to deliver high value of care.

Hospitals that implemented these changes are already seeing results with regards to improved outcomes and efficiency, and growth in the market share. [6] Value is ultimately determined by the way medicine is practiced, thus the implementation of some or all of these strategic changes must come from within the organization and involve all stakeholders. [6]

Today increasing profits in human and veterinary medicine depend on increasing the volume of services, and not delivering good results (value). In health care, and arguably veterinary medicine, the focus has been on growing volumes and maintaining margins. The goal for success should be based on increasing value, since failure to do this would lead to failure. [6]

The value agenda has six components which are independent and mutually reinforcing. That means that implementing some of these components will have a positive effect, also leaving room for improvement by widening its implementation. The components of the value agenda are: [6]

- Organize into IPU's
- Measure outcomes and cost for every patient
- Move to bundled payments for care cycles
- Integrate care delivery across separate facilities
- Expand excellent services across geography

One point that could be adapted to veterinary medicine is the organization into IPU's, which is to organize around patients, customers (owners and referral veterinarians) and their needs. This would require a departure from the rigid structure to organize around the patient's medical condition. [6] In an IPU the team works towards the common goal of maximizing the patient's overall outcomes as efficiently as possible. [6] The basis for the team is that they are experts in the patient's condition, the onsite and remote team members know and trust one another and work in a coordinate manner to minimize wasted time and resources. Team members meet often to review data on their own performance, in order to improve care. They do this by establishing new protocols or more efficient ways to engage with customers. [6] Furthermore accountability for value is shared amongst team members. [9]

One of the key aspects of providing better care is that costs are reduced, since productivity increases, rather than depending on more fee for service visits to drive revenue from unnecessary or duplicated tests and care. [6] Hospitals adopting this system operate in a more efficient manner and are able to increase the number of patients under the same space with the same number of personnel, thus leading to a rise in revenue. [6]

The IPU model not only is applicable to hospitals or specialty practices but also to primary care. Appling this concept to primary care is more challenging since at this level the approach is more general, concerned with the circumstances and needs of the patient. There is a wide range of patients managed at this level, from healthy young to the sick elderly.[9] Value improvement is more difficult due to the patient heterogeneity and very difficult to measure outcomes. IPU's should be multidisciplinary teams organized to serve groups of patients with similar primary and preventive care needs. [6,10] Segmentation is based on similarities of care needed which reflects patient's condition and the severity of those conditions. [10] Furthermore there is an opportunity for specialists to be part of this model by providing support and advice to the primary care team, thus adding value.[10] Value here will lead to a closer follow up with less complications, improved outcome and happier owner.

A very important step in improving health care is to meticulously measure value, which is outcome and cost. In management rapid improvement requires measuring results. Despite this the human side and I would argue that the animal side, fail for the most part to track outcomes or costs by medical condition for individual patients. [6] Interestingly the only measure of quality are outcomes that matter to the patients, and in our case owners. When those outcomes are measured, collected and reported, this creates incentives to improve and adopt best practices, thus improving outcomes. [6] Outcomes should be measured by medical condition (diabetes, Cushing's), and not by specialty (internal medicine, surgery) or intervention (eye examination, ultrasound). [6] The outcomes that matter to patients for a particular medical condition fall into three categories: [6,9]

- Health status achieved or retained. Patients care about survival and also about their functional status.
- Process of recovery. Related to the nature of the care and recovery.
- Sustainability of health. For example: how long an implant would last?

Measuring the full set of outcomes that matter to the patient or owners is paramount to better meet patient's and owner's needs. This also is in health care a powerful vehicle to lower costs. [6] For example Cleveland Clinic, a pioneer on The Value Agenda, publishes several different "outcome books" such as the one from their Respiratory Institute. [11] It would be ideal if human and veterinary providers across the world would measure outcomes by condition, this would allow comparison and stimulate improvement.

CONCLUSION

Creating a strategy by which an internist can increase value to clients, thus giving them greater surplus would be paramount for regaining lost market share. It is important to keep in mind that valuable services and procedures need to be performed in a way that will not erode profitability. However the balancing act is to decrease cost without eroding value or compromising care. That is since value is outcome relative to cost, cost reduction disregarding outcome is dangerous and self-defeating, leading to false savings and jeopardizing care.[9]

The question is how we can implement this knowledge into practice and how to translate human health care concept to veterinary medicine.

Improving value within the internal medicine specialty will not only benefit veterinary internist, but other specialties and generalists, and most importantly our animal (and human) patients. This will be fundamental in creating a sustainable veterinary medicine industry. As in health care, shifting the thinking from volume to value could be a key to this challenge. [9]

REFERENCES

1. Magretta J. Creating Value: The core. In: J M, ed. Understanding Michael Porter. The Essential Guide to Competition and Strategy Boston, MA: Harvard Business Review Press; 2012:95-120.
2. Magretta J. Competition: The Right Mind Set. In: Magretta J, ed. Understanding Michael Porter. The essential Guide to Competition and Strategy Boston, MA: Harvard Business Review Press; 2012:19 - 33.
3. Moliterno T. What is Strategy? In, University of Massachussets Organization and Strategy MBA Course ed. Isenberg School of Management University of Massachusetts Amherst 2013.
4. Collis D. Competing on Resources. Harvard business review 2008:2-12.
5. Porter M, Heppelmann, JE. How Smart Cinnected Products are Transforming Competition. Harvard business review 2014;92:65-88.
6. Porter M, Lee, TH. The Strategy That Will Fix Health Care. Harvard business review 2013:1-19.
7. Moliterno T. Competitive Advantage. In. Isenberg School of Management University of Massachusetts Amherst 2013.
8. Magretta J. Competitive Advantage: The Value Chain and Your P&L. In: Magretta J, ed. Understanding Michael Porter. The essential Guide to Competition and Strategy Boston MA: Harvard Business Review Press; 2012:63-90.
9. Porter ME. What is value in health care? The New England journal of medicine 2010;363:2477-2481.
10. Porter ME, Pabo EA, Lee TH. Redesigning primary care: a strategic vision to improve value by organizing around patients' needs. Health affairs 2013;32:516-525.
11. Hatipoglu U. Respiratory Institute Outcomes. In: Outcomes 2013 10th ed. Cleveland, OH: The Cleveland Clinic Foundation 2013:1-39.

DIAGNOSTIC TESTING AS A TOOL TO ENHANCE THE REFERRAL RELATIONSHIP

Louise S. Dunn, Saundra E Willis DVM DACVIM (Small Animal Internal Medicine)
Pfafftown, NC

INTRODUCTION

Diagnostic testing is an important step in the Problem Oriented Medical Record (POMR), taking us from the problem list and associated differential list to the diagnosis. Diagnostic tests are the tools we use to diagnose the disease condition causing the patient's clinical signs such that appropriate and effective treatment can be instituted. Diagnostic testing should be offered to every owner in their healthcare or treatment plan. The decision to do testing, which ones in what order, etc. are based on the communication between client and primary care veterinarian based on the many diverse needs of the client. The management side of the equation is how we deliver this diagnostic testing, how we develop our own in-clinic testing with associated quality assurance, quality control, how we price our diagnostic testing, and how do we develop communication roles and protocols for our staff.

COMMUNICATION THROUGH THE MEDICINE BEHIND DIAGNOSTIC TESTING

How do specialists use diagnostic testing to optimize the relationship between themselves and the referring primary care veterinarians? Generally, the primary care veterinarian will perform initial diagnostic steps such as general blood panels and survey radiographs. Often it is this work that is inconclusive and suggests that the problem requires the additional expertise and/or procedures available to the specialist. The decision for where additional diagnostics such as endoscopy, ultrasound, cytology or biopsy should be done depends on factors including skill and expertise of the primary care veterinarian, needs of the pet and wishes of the owner. Availability of the specialist may be a significant factor as some specialists are located miles away from the primary care veterinarian. These procedures should not be done twice. Diagnostic blood testing whose results would not be available at the time of referral might best be left for the specialist to order.

Diagnostic tests done at the referral center can serve an important role in the partnership between the primary care veterinarian, specialist and pet owner throughout the ongoing care of the patient. They are used by both the primary care veterinarian and specialist. Both parties can review them with the owner, and the owner in fact often carries the results between the specialty care clinic and referring veterinary hospital. Diagnostic tests are used to inform and educate not only the pet owner, but they are a vehicle by which the specialist can enhance the diagnostic skills and knowledge of the primary care veterinarian. It is important that this is done diplomatically. The primary care veterinarian-specialist partnership goes beyond referral of cases. There are so many ways the specialist can enhance the knowledge and expertise of their referring primary care veterinarians by a quick consult by phone or email. Diagnostic tests document the medical work up of both primary care veterinarian and specialist and show the steps by the primary care veterinarian or specialist. Because some testing can be done in clinic and also at a veterinary diagnostic laboratory such as a complete blood count (CBC), chemistry and urinalysis, diagnostic testing can stimulate discussion on quality and accuracy of testing. There are times when diagnostic work needs to be repeated, for example in-clinic CBC findings confirmed with a CBC and slide review by a pathologist at a veterinary diagnostic laboratory. Biopsy results often reflect the knowledge and experience of the pathologist and a need may arise for a second opinion. Two heads are often better than one on complicated cases.

The relationship based on diagnostic testing is there…but often times it is a weak link. This more often arises when diagnostic testing is thought to be incomplete and or even unnecessary. It is these weak links in your chain that can be causing poor delivery of patient care and loss of revenue for business growth! To deal with these weak lines, it will be necessary to drill down into both the medicine and the management of your emergency/referral practice. More specifically, to look at the service of diagnostic testing and how your emergency/referral practice can improve upon the management and the medicine of diagnostic testing.

COMMUNICATION THROUGH MANAGEMENT OF DIAGNOSTIC TESTING

Manage for Success

An important step in offering exceptional patient care is to manage the services. Examine your diagnostic testing services. Most likely, you offer a plethora of diagnostic tests to enhance the care of your emergency/referral patients. However, do you "manage" these services for optimal performance and delivery? Each member of your team has an important role in executing effective, consistent diagnostic testing. In order to ensure optimal performance, you must manage protocols, practice culture and communication.

Manage Protocols

Create protocols when appropriate for diagnostic testing and communication to both the client and referring veterinarian. Establish safety, maintenance and calibration protocols for all diagnostic equipment. Maintain a system of medical record keeping that is consistent and thorough. Craft concise specialized forms and templates for documenting testing results and communicating with the client and primary care veterinarian. Provide checklists for your entire hospital team.

Manage your Culture

Educate the team on both the medical importance and business aspects of diagnostic testing. All staff needs to be on the same page and promote the same message. Staff needs to understand the value of diagnostic testing in diagnosing disease quickly and efficiently such that effective and not simply supportive treatment can be given to the pet. Provide continuing education for the team to increase their skills, knowledge and confidence. It may help for staff to see the benefits of diagnostic testing for their own pet for early detection of disease screening. Contact your veterinary diagnostic laboratory for pricing options for staff wellness programs. Some laboratories also offer routine discounts for staff as well as veterinarians on blood work for their own pets. Boost compliance among team members and clients with sound medical needs recommendations and consistent pricing and payment structures. Maintain open discussions with your team regarding any barriers to optimal diagnostic testing procedures, record documentation or communication with the primary care veterinarian and/or client. Designate a Patient Care Coordinator as the "go to" person for the client and/or primary care veterinarian to have questions answered, schedule appointments and coordinate communication with the doctor in your practice.

Manage Communication

Outline for pet owners the process of the medical workup from the history and physical exam findings, to creation of the problem list, to a few of the differential diagnoses and the initial diagnostic testing to rule-in or out these diseases. Ask pet owners about their concerns related to their pet's medical condition including quality of life. Discuss the value of diagnostic testing and how the results of each test will affect management of the pet's disease and may lead to further testing.

Inform and update regarding all costs. Explain the treatment or healthcare plan and associated costs. The word estimate better defines the sale of tangibles products like cars and water heaters. Treatment/healthcare plan puts the emphasis back on the pet. Set up recheck appointments and/or phone contacts to monitor the pet's condition. Will a staff member make the phone contact or the veterinarian? Communicate among your team members regarding client concerns. Communicate with the primary care veterinarian in a timely manner soon after the client leaves the emergency clinic/referral hospital. Establish protocols for ongoing communication regarding a long standing case.

It's All about Communication

As you scan through the three important points to manage, you probably notice that communication plays a role in all of them. It is important to have a clear understanding of communication types and the communication roles of everyone (primary care veterinarian, the specialist, team members and clients) involved in the treatment of the patient. To have a truly collaborative partnership, take some time to understand communication.

Communication can be one-way or two-way. In one-way communication, you are persuading or manipulating the audience to influence their action and behavior. In two-way communication, you are negotiating, demonstrating mutual understanding and respect, balancing interest to influence action and behavior. Two-way communication has the potential to affect not only the audience, but also to affect the speaker. In two-way communication, the organization and the audience are changed for the better because of the mutual exchange of ideas and needs.

Think about your communication technique. Do you dole out protocols without ever having open dialogue with other doctors or team members? That would be one-way communication. Here is the protocol – do exactly how I told you. Never mind the fact that the protocol ignores the needs of the primary care veterinarian or that the team runs into documentation problems. One-way communication will miss out on the input of others.

Is your communication with the pet owner one-way or two-way? Are you the 'specialist with all the answers' so you work at persuading the pet owner to do as you say? Or are you part of a partnership working with the primary care veterinarian and the pet owner in communication? Participating in two-way communication enables you to learn about the wants and needs of others involved in the care of the patient.

Two way communications is integral to the specialist who wants to know what the primary care veterinarian wants, what the client wants and what the emergency/referral practice needs. Some times the needs and wants are similar to all parties but oftentimes there is give and take. The overriding concern needs to be the pet and the owner.

Just what does the primary care veterinarian want from your emergency/referral practice? They want optimum care for their client and their pet which often includes an initial phone consultation with the specialist before the client is referred, management of the case at the time of referral including consultation on follow up care, prompt referral reports, avoiding unnecessary tests or procedures on the patient and giving clients a fair and affordable option for treating their pets. The specialist should address the initial problem for which the patient was referred even if another problem, perhaps more significant is identified. It would create a sense of healthcare teamwork between the specialist and primary care veterinarian if at the time of the referral appointment, a quick call was made to the primary care veterinarian. This is particularly important when the specialist has identified a significant health program for which the pet was not referred for and/or they feel that work up of the case should take a significant departure from that indicated by the primary care veterinarian. For example, the ill older dog is referred for an ultrasound examination of the abdomen when additional history and physical examination at the referral center does not indicate a need for ultrasound.

What does the client want? They want to understand the medical problems facing their pet, to be involved in the care and feel confident in their choices for medical care and to have a plan for follow up care, even lab work, back with their primary care veterinarian. It may surprise some people to realize that a two-way communication style involves the pet owner in more of the treatment and thus improves client compliance more than the one-way lecture.

Moreover, what about you, the specialist in the emergency/referral practice? What do you want? Information helpful for case management, confidence in client and primary care veterinarian compliance with follow up, assurance of financial payment and ongoing status updates. Do you get any of that with the scribbled notes faxed over from the office of the primary care veterinarian or the comment "do whatever it takes Doc" from the client? One-way communication is not very helpful for your needs either. Take your "needs" a step further and consider your communication with your team and with the primary care veterinarian. How do you prefer to receive the referring records? Do you prefer a summary of the case or the actual day by day medical records in addition to all laboratory work, radiographs, biopsies, etc? It is expected that you would prefer copies of all laboratory work even that which is "normal". Normal is relative and subtleties within the normal ranges, trends between successive normal results, may reveal important clues to the pet's illness. A discussion of these subtleties on a particular case is an excellent opportunity to educate the primary care veterinarian if done diplomatically. Is communication within the walls of your practice one-way or two-way? Look back at how you manage your protocols, your culture and your team's communication training.

The AAHA Referral Guidelines[1] is an excellent resource for companion animal practices addressing all matters of communication between the specialist and primary care veterinarian.

Diagnostic Testing and Your Circle of Care

Diagnostic testing are services provided every day in your practice. How these "services" perform can have a positive impact on all aspects of your veterinary practice, from patient care to staff morale, from client compliance to practice profitability. Delivering optimal diagnostic testing can be easily overlooked in your day-to-operations if you fail to manage the business side of these services. Namely, these include implementing consistent protocols, identifying cultural barriers and strengthening communications. After all, without proper management of diagnostic testing services you risk missing returns (ROI) from all the investments in testing equipment.

As stated in the beginning, offering diagnostic services is a two-prong approach. The first is management and the second in the medicine. Your team plays a vital role in delivering effective diagnostic testing for every patient, every client, every time. Diagnostic testing, although considered a medical offering, is a value-added service that benefits both patient and practice. The next step is to recognize and enhance the medical opportunities for performing diagnostic tests.

REFERENCES

1. Brogdon R et al 2013 AAHA Referral Guidelines www.aaha.org

RESPECTFUL CAT HANDLING VS. CAT WRANGLING: FROM THE CAT'S POINT OF VIEW

Margie Scherk, DVM, DABVP (Feline Practice)
Vancouver, BC, Canada

INTRODUCTION

In many clinics, some veterinarians and other team members do not enjoy working with cats because they believe that cats are unpredictable and feel anxious about getting hurt. By understanding why cats feel that they need to defend themselves, by learning to identify the warning cues, managing the interactions in a positive manner, and making relatively minor changes to what the cat is exposed to, this fear can be reduced.

The basis for working cooperatively with cats is being empathic to their nature and behaviors and trying to imagine what *their* experience is like. Cats are a species with a social structure different from ours. We need to look at cats differently, slow down and adjust our interactions. Minor modifications to the physical facility help reduce the strangeness and threats that cats experience in the veterinary clinic.

The goal of these two presentations is to look at how to change the experience for cats thereby removing some of the obstacles to routine feline veterinary care. This benefits cats and their human companions with the resulting side effect of clinic growth.

WHY CATS RESPOND THE WAY THEY DO

In the wild, the number of feral cats living together depends on the availability of resources. These are food, water, privacy and safety, toileting areas, and availability of sexual partners. Mice and small birds are single portions; they are not large enough to be shared. After weaning, cats are responsible for feeding themselves. The resource density determines the number of cats living in a given area. In order to reduce conflict and the potential for physical harm associated with fighting, cats have developed an impressive repertoire of signals to maintain distance and protect resources within their territory. This results in little competition and a social structure that does not require sharing or taking turns. Stress is minimal unless resources become scarce. Aggressive communication signals developed in order to keep distance between individuals and to prevent contact with outsiders. Cats need to avoid physical injury in order to be able to hunt and protect themself. When resources are plentiful, a colony will develop consisting of related female cats with their young, who they jointly defend and nurse. Males are relegated to the periphery and vie for breeding privileges; only one mature tom usually lives with the group.

Many of the behaviours cats show in a clinic situation stem from the fact that while they are predators of mice and small birds, they are prey relative to almost all other larger animals, including larger birds. When they feel threatened, they rely on "fight or flight" and will try to escape situations that they view as dangerous. When they can't flee, they fight (self-defense) or freeze. From the perspective of a cat, humans are, (and what we do is), dangerous. As a result, we see frightened and defensive cats every day. Cats try to avoid physical confrontation through by using intimidating sounds and postures. This small creature feels more threatened than we do; it is important to refrain from becoming frightened ourselves.

Reading and understanding the cues and signals that cats use is important to detecting incipient fear. This allows us to respond respectfully as well as redirect the progression of an emotion and reshape experiences. We can learn to avoid using signals that are hostile (e.g., scruffing, making shushing/hissing sounds, looking into their faces) when we know how cats communicate.

FELINE SIGNALING: READING THEIR CUES

Tactile Sense

Touch is very important to cats. They rub against each other (allorubbing), against us, and against inanimate objects. Whether a full-body rub or rubbing a flank, tail, cheek or other body part, rubbing is believed to be an affiliative behavior seen between members of the same social group, feline or human. Rubbing is not only tactile, but is also a means of depositing the colony (family) scent. Cats often rub against us; unfortunately, we often misinterpret it as a request to be fed.

Allogrooming (mutual grooming) may precede a playful attack, follow a stressful interaction, and appear to be conciliatory or may simply be grooming. Kneading and treading occurs in adults either as a kitten-regressive behavior or as a component of sexual interaction.

The neck bite/scruffing is used by cats in three contexts: for transportation of young kittens, for restraint during copulation, and for dominance in a fight. Our use of scruffing fits most closely with the last and does not promote shaping safe, respectful cooperation. (See AAFP and ISFM Feline Friendly Handling Guidelines.)

Olfactory Cues

The role of smell and scent in feline communication is something we human beings are ill-equipped to appreciate. It has been estimated that the size of the olfactory epithelium in cats can be up to 20 cm^2, whereas humans have only 2 to 4 cm^2 of olfactory epithelium. While olfactory signals may be left by several methods, the one that is most problematic for people is urine spraying. This is a potent and important method of communication that we fail to appreciate. Other forms of olfactory messaging are cheek marking an object or individual, scratching to leave scent from glands below the footpads, and midden, (i.e., leaving a deposit of feces uncovered in a strategic place). All of these have several advantages over visual cues. The message persists over time and in the absence of the sender, allowing for remote communication without the potential for conflict that direct interaction risks. This is especially useful at night and in areas with poor visibility. These signals help cats spread out over space as well as time-share territory. The disadvantage of this form of communication is that the sender cannot change the message once it has been deposited; it cannot be altered or removed and no adjustments can be made in response to the recipient's reaction. So, urine marking in the home is an attempt to signal to the other cats when "I was 'here'" and to establish a routine so that the cats can keep a distance by time-sharing the same space without needing to come into conflict. Every time we remove the urine, we interfere with this communication!

Because of our less well-developed olfactory sense, we fail to "read" the signals a patient may be giving us and are unable to fathom the overwhelming olfactory messages from previous patients and substances used in the hospital that the clinic experience presents to cats.

Visual Cues: Body Language (Posture, Face, Tail)

Body language and facial expression are extremely effective at maintaining or increasing distance between individuals potentially competing for resources. This requires having an unobstructed view, adequate ambient light, and, unlike olfactory cues, that the two individuals are in the same space at the same time. Body posture cues the big picture of emotional state but facial expression (eyes, ears, whiskers, mouth, visibility of teeth) provides the finer details and changes more rapidly. In a clinic setting, for us to appreciate the mental/emotional state of an individual, to avoid provoking them and getting hurt, it is extremely important to watch and interpret facial changes.

As a species that generally leads a solitary existence, survival depends on speed, stealth, self-reliance, and outsmarting others. As a consequence, cats may "bluff". When they act aggressively, they are generally hiding fear; "stoicism" hides vulnerability; subtle changes in behavior mask pain or significant illness. Body postures communicate confidence and physical prowess that may not be present. Keeping a threat at a distance may eliminate the need for a physical confrontation. The arched back "Halloween cat" typifies this façade of confidence. Making oneself smaller, on the other hand, to minimize threat and evade attention is portrayed by a crouch and withdrawal. In these postures, the weight remains on all four paws so that flight or chase remains possible. A cat feeling less fearful does not need to be on his or her feet. However, an extremely fearful threatened cat will roll exposing his or her abdomen with all four feet ready for self-defense. This cat may be screaming while showing all of its weapons (nails and teeth).

Cats have extremely mobile ears. When the ears are forward, a cat is listening and is generally relaxed or alert but not emotionally aroused. Turned laterally, flat "airplane ears" indicate that the cat is more fearful or feels threatened. When ears are back and tight to the head, the cat is feeling very threatened and frightened. This cat will have a partially or fully open mouth and be hissing, spitting, yowling, or screaming. Cats will protect themselves if we fail to reduce the level of perceived threat. Ears turned back but erect indicates the most reactive and aggressive state. In this case, the mouth will be closed and the cat will be emitting a low growl with or without swallowing. This is the cat to be apprehensive of.

Vocalization

This form of communication requires the direct presence of the recipient. It has the benefit of being easy to adapt from moment to moment. As with other signals, cats have a well-developed repertoire of sounds to convey a need or wish to increase the distance between individuals. The sounds made for encouraging socialization are a trill/chirrup, purr, puffing, prusten, chatter, miaow, and sexual calling. The cat that is open-mouth screaming is highly aroused but is probably less aggressive than the cat that is close-mouthed growl/wah-wah/mowling.

Cats use a combination of these different signals in any situation. We have to learn to look for all of them and interpret them together.

REFERENCES

Dards JL. **Appl Anim Ethol** 1983; 10: 133-153. Warner R. **J Wildlife Manage** 1985; 49: 340. The domestic cat: The Biology Of Its Behaviour. 2nd ed. Turner DC, Bateson P (eds.). Cambridge, U.K.: Cambridge University Press, 2000. Crowell-Davis SL et al. **J Feline Med Surg** 2004:6:19. Gourkow N. Vancouver, BCSPCA 2004.www.spca.bc.ca/welfare/professional-resources/catsense/. Gourkow N et al. **Anim Welfare** 2006;15:371.Horn JA, et al. **J Wildlife Manage** 2011; 75: 1177.Rodan I. In Little S ed. The Cat Clinical Medicine and Management 2013: 2. Bradshaw J. Cat Sense. Philadelphia, PA; Basic Books, 2013.

RESPECTFUL CAT HANDLING VS. CAT WRANGLING: PUTTING PURRSPECTIVE INTO YOUR PRACTICE

Margie Scherk, DVM, DABVP (Feline Practice)
Vancouver, BC, Canada

INTRODUCTION

Making the clinic environment more "feline friendly" requires imagining how a cat perceives it. The exercise becomes one of identifying potential threats and removing or reducing their significance.

Reducing Threats In The Hospital Setting

It is important to reduce exposure to true predators (dogs, people, other cats) and to other *perceived* threats. Visual barriers in the seating/waiting area help to prevent cats from seeing dogs. Covering the carriers with a towel will also help so that cats don't see each other. Using chairs or ledges, keep kennels off the floor. If possible, have a separate cat-only waiting area. Reserve at least one examination room only for cats in order to reduce the smells of predators and to be able to furnish it with necessary items for examining cats as well as to strive to achieve cat comfort.

Looking over our clinic/hospital environment, what can we do to reduce the stress and threat level of the physical and social environment? What things or events assault the five senses of a cat? How can we make positive changes to these? *Table 1* shows a chart that can be completed by the clinic team. For example: Scary smells include alcohol, disinfectants, odours of other carried animals; this can be remediated by wiping the area to which alcohol had sparingly been applied with a damp cloth and using venipuncture sites far from the nose (medial saphenous) when possible. Disinfectant should be allowed to evaporate before a cat is placed/replaced in a kennel. Carry and examine all patients in their own, fresh towel rather than have their smells embed themselves in your clothing.

Table 1. Chart For Evaluating Perceived Threats To Cats In Hospital Setting

Sense	Threat	Reduce threat by
Smell		
Hearing		
Sight		
Taste		
Touch		

Handling (Examination, Hospitalization, Diagnostics, And Treatments)

The goal is to handle our patients respectfully and provide an appeasing environment to build positive, long-term relationships. This is achieved by reducing threat and, thus, the cat's need to react defensively. Avoid doing things in a way that use threatening feline body language or tone. The aggressive cat is upright, stiff-legged, and large; sit down to examine cats. Never stare a frightened cat in the face: examine cats from behind and, other than for ophthalmic evaluation, avoid direct frontal facial viewing. Using a sideways glance with hooded eyelids indicates a desire to cooperate. A slow blink is a reassuring signal to a cat similar to a human smile.

The aggressive cat growls and uses low tones; use light, upper register tones, perhaps chirruping as cats do when they are relaxed with conspecifics. Shushing a cat to try to calm her as we might a child is the equivalent to hissing at her. Short repetitive sounds should be avoided, since these may resemble spitting rhythms. Purrs, chuffing, trills, and chirrups are welcoming sounds.

When cats feel secure and safe, even just able to hide their faces in an elbow or a towel, they allow most procedures. Try to keep all four of their paws on the floor and avoid changing their body position as much as possible. A comprehensive examination, blood and urine collection, body temperature and blood pressure evaluation can all be done without changing the cat's position. Examine her in the base of her own carrier if the lid can be removed. Don't hang a cat's forelimbs over the edge of a table for jugular venipuncture. For the frightened individual, additional lack of support under the paws is not reassuring.

Reaching into a kennel to pick up a patient blocks the light; to the cat you appear as a looming, frightening stranger (smells, sounds, visual input). Instead, approach the opening of a kennel from the side so that some light still enters. Do not block every chance for escape; if the possibility to have some control over her environment and situation exists, she will be much more cooperative. Because cats rely on flight and fight for survival and are not reliant on others, when it comes to restraint, the mantra holds true: *Less is more*! Cats inherently resist intimate handling and restraint. By restraining them, we take away their sense of control and cause them to react. It is very easy to condition negative emotional responses. Scruffing is strongly discouraged as it is an act of dominance that cats may resent. Similarly, stretching is an inappropriate, disrespectful and unnecessary way to apply restraint. Every future experience builds on the previous negative (or positive) experience. Cat bags, masks, and gloves all carry the scents of similarly terrified patients plus other sundry smells (anal gland secretion, pus, blood, halitosis, etc.) A towel is all that is needed to wrap a cat in, in order to protect the handler. Remember, a cat would rather flee than attack.

Train all staff in respectful cat interactions and handling. An excellent and comprehensive resource is the American Association of Feline Practitioners (AAFP) and International Society of Feline Medicine (ISFM)'s Feline Friendly Handling Guidelines, downloadable at: www.isfm.net/wellcat/UK/FFHG.pdf. It is well worth reviewing and refining cat examination techniques as a clinic team for a consistent approach, the goal being to make them less threatening. Because value is "perceived worth" and because every visit is a valuable opportunity to educate the client, talk to the client and the cat throughout the entire procedure. Source and provide feline friendly medications, being sure to follow up one or more times with the client to find out how the patient is doing and if the client needs a refresher course on how to administer the medications. Be sure to send home an exam report with home care instructions for the client to refer to. Schedule recheck appointments or the next wellness visit before the client leaves the practice. The AAFP has created the Cat Friendly Practice program through which any interested clinic can raise its cat care IQ. (catfriendlypractice.catvets.com)

Meeting Environmental Needs Improves Health

Recently, it has been recognized that emotional well-being is highly dependent on meeting the environmental needs of cats. These include those relating to the indoor and outdoor physical environment, as well as a cat's social interactions, human and otherwise. In the AAFP and ISFM Feline Environmental Needs Guidelines, five pillars are described that form the basis of a healthy feline environment (Ellis, 2013). These pillars are:

1. A safe space
2. Multiple and separated resource stations (food, water, toileting areas, scratching areas, play areas, perches, resting and sleeping areas)
3. Opportunity for play and expression of predatory behaviors
4. Positive, consistent and predictable interactions with humans
5. An environment that respects the importance of a cat's sense of smell

When these are not met, cats become stressed to varying degrees. Some may express illness (such as inflammatory bowel disease, lower urinary tract inflammation), while others will manifest their distress through inappropriate elimination.

Other Considerations

As cats age, they tolerate less time in the clinic. Siamese cats are especially prone to becoming depressed. Three days may be as long as a cat can stand the anxieties and indignities of hospitalization, even with daily visits from the owner. Consider capping intravenous catheters and send patients home, having them return for outpatient care. Even for in-hospital care, capping catheters off overnight (administering the overnight dose via the subcutaneous route) allows greater ease of movement, avoids alarms, which keeps patients awake. In either case, administer the overnight fluid volume subcutaneously.

Because cats "see" the world in overlapping clouds of smells, we should strive to provide familiar smells and reduce foreign, medicinal smells. Client-worn shirts or toys from home are helpful in cages. Feline facial pheromone may help to reduce stress. Because cats' sense of hearing is tuned more finely than ours, a quiet and reassuring environment is desirable. Cats should not be exposed to the sounds of predators, namely barking dog, but strange machines (faxes, printers, phones, dishwashers, centrifuges, etc.) should also be remdiated. Reducing noises should be addressed when using certain induction agents as some enhance hearing (e.g., ketamine).

Avoid changing a cat's diet during hospitalization as is likely to result in inappetence and possibly the development of an aversion. If a change in diet is required for therapeutic reasons, try to make that change gradually in the safety of the home territory.

Taking a thorough history is especially important given cats' tendency to hide illness. Listening carefully to clients and their concerns is extremely important. Often clients detect changes that represent real problems. This is probably more common than the client who is blissfully unaware of significant health problems. By asking open-ended questions, we elicit a more detailed history than using only specific questions. For example, asking, "Have you noticed any changes in the contents of the litter box?" will probably evoke a yes or no answer. Whereas: "What does his stool look like?" followed by: "Would you describe it as hard pellets, moist logs, cowpie, or colored water? What colour is it? When did you first notice this?" will probably provide more useful answers. "Is there anything else?" is a very effective question.

Schedule a recheck appointment to evaluate the effect of any medical or nutritional therapy. Reassessing important variables (e.g., body weight, body condition score, previously abnormal laboratory results) and updating the patient history allows us to provide better care for our feline patients. Care of the client is essential to providing complete patient care. It is only through listening to, educating, and working with the client that we are able to offer the very best veterinary care.

Examples of practical applications

1. If a cat is uncooperative, a comprehensive physical examination can usually be done using only a towel as a protective barrier. Facing the cat away from you is less threatening for her. Confining the cat between your legs as you sit on the floor provides adequate persistent firm restraint that is reassuring rather than frightening.
2. Swaddling a cat's forelimbs and torso may help with blood and urine collection, placing the cat in lateral recumbency for cystocentesis and using the medial saphenous vein. This vein is also a superb choice for catheter placement and administration of intravenous medications. If the cat is allowed to have her front end in a sternal position while the back end is in lateral recumbency, she may struggle less.
3. Allow the client to be with the kitty as much as, and whenever, possible.
4. Recognize that a *persistently* elevated systolic value above 160 or 170 mm Hg probably represents true hypertension rather than the stress response. If in doubt, repeat the value later during the visit.
5. Feliway™ (Ceva Animal Health), a synthetic analog of a feline facial pheromone, may have a calming effect on cats. Spray (or wipe) it into kennels and carriers and even on your clothing before handling an anxious cat. Let the substance evaporate for a few minutes before placing the cat into the sprayed space. Feliway diffusers plugged into treatment and hospitalization areas as well as reception and consultation rooms can help patients relax.
6. Elevated blood glucose and glucosuria may be a result of persistent stress. A diagnosis of diabetes, therefore, should be confirmed by finding an elevated serum fructosamine.

REFERENCES

Buffington CAT. **J Am Vet Med Assoc** 2002; 220 (7): 994. Ellis SL, et al. **J Feline Med Surg** 2009; 11: 901. Rodan I, et al. **J Feline Med Surg** 2011;13:364. Stella JL, et al. **J Am Vet Med Assoc** 2011; 238 (1) 67. Ellis SL, et al. **J Feline Med Surg** 2013 15: 219.

IMPROVING CLIENT COMPLIANCE IN THE FELINE PRACTICE

Margie Scherk, DVM, DABVP (Feline Practice)
Vancouver, BC, Canada

INTRODUCTION

While the number of cats kept as companions in North American homes continues to increase, the number of feline visits to clinics has been declining since 2001. Based on the AVMA's 2007 pet ownership and demographics survey, there were 13% more cats than dogs, yet cats failed to receive the same degree of veterinary attention. In small-animal practices, dogs represented 59% of office visits, cats only 39%. The 2011 Bayer Brakke study further noted three client-driven factors that limited the number of feline visits.

1. Inadequate understanding of the need for regular preventive health visits other than for vaccination.
2. Resistance to bringing a cat to the clinic because of the distress caused by putting a cat into a carrier and making the trip to the clinic.
3. The cost of veterinary care, in particular the frequency and size of price increases. (The state of the economy was a separate, external factor.)

In November 2012, Bayer conducted an online survey of 401 veterinary practice owners across the USA. The Bayer Veterinary Care Usage Study III: Feline Findings noted that 78% of veterinarians believed that better care for cats represented one of the most significant, missed opportunities for the profession. Yet, while 70% of those veterinarians were familiar with the earlier Bayer-Brakke studies, and while most recognized that cat owners consider a clinic visit to be stressful for themselves and their cats, nearly one-third of practices did not have staff trained on how to make visits less stressful for clients. Additionally, relatively few practices had adopted exam rooms used only for cats (35%), cat-only waiting areas that are physically and visually separated from dogs (18%), and cat-only days and appointment hours (11%). The study found that 46% of the surveyed clinics had recently started taking specific steps to increase visits among current feline patients, attract more cat-owning clients, and make their practices more "cat friendly".

Clearly there is a need to grow compliance among cat owners. However, part of the lack of awareness (at best) or reluctance (at worst) for making simple, inexpensive changes in attitude and facility is that many veterinarians and veterinary staff members prefer, or feel more comfortable, working with dogs than cats. The survey also identified that veterinarians find it easier to diagnose dog cases.

IMPROVING CLIENT COMPLIANCE

The verb "comply" means to act in accordance with a wish or command (Oxford), to conform, submit, or adapt (as to a regulation or to another's wishes) as required or requested (Miriam-Webster). For clients to comply with our recommendations, they have to fully understand, be willing and able to perform the actions that we are recommending. We need to enroll them so that they believe in the importance of these actions. But explanations aren't enough: on-going *caring* communication are needed to enhance client compliance.

Many clients believe that cats are self-sufficient, have very few needs, and are low maintenance pets. They don't understand that cats live as solitary hunters because their prey are too small to share. This means that they lack the resources of a supportive society. To avoid showing vulnerability they hide illness well. Additionally, cats are prey to larger birds and other species. This is is critical in understanding why cats so easily feel self-defensive and how to work with cats. Educating potential and existing clients what these subtle signs of sickness are is a huge opportunity for increasing compliance. All veterinary team members also have to recognize that any admission of illness by a cat may signal a problem that has been going on for longer than recognized. Following are descriptions of signs that clients can be taught to look via direct emails, newsletters, the clinic website, and social media.

Subtle Signs Of Sickness (adapted from http://www.haveweseenyourcatlately.com/Home.html)

1. *Inappropriate Elimination*: Regardless of how "deliberate" it may seem to be, when a cat is avoiding or not using the litter box, they are trying to tell you something. The message may be that they are in physical discomfort or psychological distress. Physical problems include inflammation of the bladder or bowel, arthritis, hyperthyroidism, diabetes, dementia. Psychological distress may be due to social disturbance, anxiety due to other animals, children or adults, boredom, or a lack of opportunities to perform the full, natural repertoire of cat behaviours.
2. *Changes In Interaction*: Changes in how a cat interacts with people, other animals or his/her environment may indicate pain or distress.
3. *Changes In Activity*: A decrease in energy may be abrupt or gradual. The latter is often attributed to "just getting older", however, a healthy individual does not inherently "slow down" due to increasing age. A cause for such changes should be investigated. Dehydration, pain - from anything, including arthritis-, and hypokalemia are some of the problems that should be evaluated. The reverse is also true: an increase in energy in a previously normal cat may be an indicator of incipient illness, most notably, hyperthyroidism or hypertension.
4. *Changes In Sleeping Habits*: This refers to pattern of sleeping (times of the day and night) as well as postures and locations. A cat with pain or with dementia may either sleep for longer or for shorter periods than previously. With FIV infection, the latter may occur. Night-time yowling suggests a decline in vision or hearing, hypertension, hyperthyroidism, pain or dementia.
5. *Changes In Food And Water Consumption*: This refers to quantity, preferences, as well as changes in behaviours associated with these activities (i.e., where, how often, attitude, amount at each instance, body posture, etc.).
6. *Unexplained Weight Loss Or Gain*: As gratifying as it is to see rapid weight loss in a previously obese patient, even for those on appropriate dietary regimes, dramatic changes are neither desirable nor the norm. Oral pain may cause inappetence. Gradual weight and muscle loss may be related to ageing but should be monitored and investigated. Weight

gain is most often from excess calories but could also be due to abdominal or thoracic fluid accumulation. Helpful tools include repeated assessment of body weight percentage weight change, body and muscle condition scoring.

7. *Changes In Coat And Grooming*: Excessive grooming may be caused by skin irritation (e.g., allergy, fleas, dryness), neuropathy, or be psychogenic (a way to reduce stress by releasing endorphins). A decrease in grooming is often associated with pain, often arthritic or oro-dental. Hairballs may be a sign of dermatologic or psychogenic problems, altered digestive motility or pain.
8. *Signs Of Stress*: Along with aforementioned inappropriate elimination and overgrooming, signs of distress include hiding, chewing on non-food items, a flicking tail, ears placed further back than normal, unprovoked attacks.
9. *Changes In Vocalization*: These may be a change in tone, pitch, or urgency and frequency of vocalizing. See above regarding night-time yowling.
10. *Bad Breath Or Smelly Coat*: Numerous oral and dental conditions result in halitosis; periodontal disease is extremely common in cats. The odour from sialoadenitis, infected ulcers, tumours, abscesses and anal gland secretions may be spread onto the coat via grooming.

However, even recognizing that their cat has a problem may not be enough to motivate the client to bring their cat in to the veterinarian. Screening to proactively identify disease early and to provide solid medicine can be an even harder "sell" because most people do not like bringing their cats in to the clinic. Many cat owners would rather provide care at home or even skip any form of consultation unless there is "something serious going on"! The second significant opportunity to improve the lives of our patients and be of help to our clients is therefore enabling less stressful trips to the clinic.

Getting Cats To The Clinic

It is no fun bringing a cat to a veterinary clinic (for the owner or the cat)! All veterinary team members should be trained in coaching clients how to make the trip less stressful, from the experience at home, during transit, and once they arrive at the clinic. This conversation begins when the client calls to make an appointment or at the first visit with their cat. The American Association of Feline Practitioners (AAFP) has a free downloadable client handout entitled: Getting Your Cat to the Veterinarian (catvets.com/uploads/PDF/2011FelineFriendlyClientHandout.pdf). Clicker training can be used to help create positive associations with the carrier. Catalyst Council (www.catalystcouncil.org) has created excellent videos that clinic teams and clients can watch to learn how to accomplish this.

The frightening experience begins at home. Imagine the scenario from the cat's point of view: *The carrier comes out, your caregiver is nervous, she chases you around and tries to force you into the carrier. You resist and may resort to self-defense. There are new or residual smells of human sweat, fear, maybe even blood. You may feel so anxious that you soil yourself! Eventually you are in the carrier. Everyone is exhausted. Then you are moved into a "car" that makes you move without any intention or action on your part. You may be a bit nauseous; certainly you are scared. You cry out repeatedly. You may vomit or soil yourself. Then the "car" stops and you get carried on a noisy and unfamiliar street and into a place with overwhelming smells and sounds! Help! And you are already aroused and anxious....look out!*

We can reduce the stressors the cat encounters, or, in the case of a new cat, prevent them from occurring by teaching or habituating the cat to associate positive experiences with the carrier, the car, and even the clinic. By leaving the carrier out (or using a Hide-Perch-Go box/carrier) so that the cat sees it as non-threatening and enters it for meals, treats or other rewards, we reduce the initial tension and fight. Taking the cat on short car rides unassociated with the clinic helps to recondition the cat's negative associations with the clinic. Finally, taking the kitty to the clinic to be fussed over or only to get a treat will help teach the cat that the clinic isn't necessarily a horrible place.

Taking The Household Pet Inventory

While there are a lot of cats who never receive veterinary care, there are a lot of cats living with existing clients we never see. We don't even know that they exist! If the cat is well or if the client has had a really bad experience in the past with a cat (or anticipates "bad behavior" from a cat), they are unlikely to voluntarily bring them in for preventive care. By asking whether they have any other cats or pets when they any patient we can identify the un-served animals.

IMPROVING THE CLINIC EXPERIENCE

From the client's point of view: *It wasn't fun to bring her, she isn't happy about being in the clinic and it isn't fun watching her be "manhandled"*. Once at the clinic, already stressed and frightened, it is extremely important to minimize or eliminating any further perceptions of threats. This requires that we imagine or try to see the clinic from the cat's point of view. The second and third of these presentations will speak in depth to these matters.

Making the environment more "feline friendly" can be as simple as having visual barriers in the seating/waiting area to prevent cats from seeing dogs. Covering the carriers with a towel will also help so that cats don't see each other. If possible, have separate cat-only waiting area. Restrict at least one examination room solely for cats to reduce the smells of predators and to be able to furnish it items needed for cat examinations and comfort.

Train all staff in respectful cat handling. An excellent and comprehensive resource is the AAFP and International Society of Feline Medicine (ISFM)'s Feline Friendly Handling Guidelines, downloadable at: www.isfm.net/wellcat/UK/FFHG.pdf. It is well worth reviewing and refining cat examination techniques with the goal of making them less threatening. Because value is "perceived worth" and because every visit is a valuable opportunity to educate the client, talk to the client and the cat throughout procedures. Source and provide feline friendly medications, being sure to follow up one or more times with the client to find out how the patient is doing and if the client needs a refresher course on how to administer the medications. Be sure to send home an exam report with home care instructions for the client to refer to. Schedule recheck or the next preventive healthcare appointments before the client leaves the practice.

The AAFP has created the Cat Friendly Practice program through which any interested clinic can raise its cat care IQ. (catfriendlypractice.catvets.com)

Facilitating Finances

The Bayer study showed that clients want costs spread out over time. Fear of large bills is another significant factor preventing owners from bringing their cats to the clinic. Many practices have wellness plans. Directing clients toward pet health insurance that covers *both* preventive and accident/illness before it is needed is sound medical advice. This can save lives that might otherwise be lost because of hesitation to seek care or a decision to euthanize a pet out of financial concerns.

FACILITATING COMPLIANCE AT HOME

Having a library of YouTube links or making your own clinic "how-to" videos is extremely helpful. YouTube videos made by lay people may have the advantage of being more convincing rather than those by healthcare professionals. Find ones that you and your staff think are best. Ask clients which ones they found and liked. There are many good links. Examples of useful illustrative clips to have on hand include how to:

- Give your cat a pill (see below)
- Give subcutaneous fluids: www.youtube.com/watch?v=OLOVw35w4Ns
- Administer insulin: http://www.youtube.com/watch?v=XeZgKLfIJn4
- Measure blood glucose: www.veterinarypartner.com/Content.plx?A=605
- Use an inhaler for asthma medications: www.youtube.com/watch?v=INF1W8uaPEA
- Feed with a feeding tube: contact the author at hypurr@aol.com
- Change a KittyKollar (video) and Living with an E-tube (handout): www.kittykollar.com

You might also want to include syringe/assisted feeding. Cat caregivers like to show their skills and help others. Compile a selection of reading materials on the internet that you have vetted and feel comfortable with to guide those clients who want to learn more about their companion's medical condition. Superb client specific books are available from www.vetprofessional.com. They include: Caring For A Cat With Hyperthyroidism, Caring For A Cat With Chronic Kidney Disease, Caring For A Blind Cat among others.

Cornell University has a Series Of Videos On A Number Of Procedures And Diseases At Www.Partnersah.Vet.Cornell.Edu. They Include: Brushing Your Cat's Teeth, Giving Your Cat A Pill Or Capsule, Giving Your Cat Liquid Medication, Taking Your Cat's Temperature, Trimming Your Cat's Nails. Other free videos include: Caring For Your Diabetic Cat, Gastrointestinal Diseases In Cats, Cat Owner's Guide To Kidney Disease, Managing Destructive Scratching Behaviour In Cats and A Pet Owner's Guide To Cancer.

The ISFM has an excellent owner website: http://www.icatcare.org/advice. It includes handouts on medical conditions, general cat care, and videos. Websites specific for conditions include: chronic kidney disease: www.felinecrf.org and diabetes: www.felinediabetes.com, www.petdiabetes.com.

Follow-Up

In addition to reviewing discharge appointments of discussing future care, it is helpful to discuss reintroducing the patient to his/her home environment and housemates. Scheduling the next appointment before the client leaves, be it for a recheck evaluation, or for their semiannual or annual visit, encourages continuation of care. Especially useful in a first year of life program, it helps to lay the foundation for a culture of lifelong preventive healthcare.

Even with a motivated and educated client, the likelihood of a successful outcome depends on that client having a relationship built on trust, communication and knowing that the clinic team cares. Follow-up phone calls are a very effective investment of time.

SUMMARY

Educating clients about the subtle signs of illness, reducing stressful travel and clinic visits, following up with personalized care, improves compliance. Taking a household pet inventory allows us to discover an un-served pet population. This gives us the opportunity to:

- Provide preventive healthcare
- Detect disease early when we can prevent or alleviate suffering and save expense
- Protect life and enhance welfare
- Build trust with our clients
- Increase clinic visits.

REFERENCES

AVMA. US pet ownership and demographics sourcebook. Schaumburg, Ill: AVMA, 2007. Lue TW, et al. **J Am Vet Med Assoc** 2008; 232: 531. Shepherd AJ. **J Am Vet Med Assoc** 2008; 232:695. Volk JO, et al. **J Am Vet Med Assoc** 2011;238:1275. Volk JO, et al. **J Am Vet Med Assoc** 2011;239:1311. Loftus L. **Vet Nurs J**. 2012; 27:294.

HOME CARE AND END OF LIFE ISSUES

Margie Scherk, DVM, DABVP (Feline Practice)
Vancouver, BC, Canada

"The death we fear most is dying in pain, unnoticed, and isolated from loved ones. Concern about such an undignified and difficult death has engendered the debate over authorizing voluntary active euthanasia and physician-assisted suicide. Death is fundamental to the nature of being human." (Anon.)

INTRODUCTION

What person does not have an emotional response to thoughts of death and dying? We are aware of our mortality and to that of those around us. No different from our clients, we fear the pain and suffering of friends and companions, human or non-human. Nor do we cherish the probabilities of deteriorating health or abilities, long-term nursing care or terminal illness.

Normal aging begins at the moment of conception. It involves cell differentiation and maturation and, at some point, the progressive loss of functional capacity characteristic of senescence culminating in death. This occurs at an organismal level as well as at a cellular level. The former may be affected by genetics, social environment, nutrition, and the occurrence of age-related diseases. Cellular aging, on the other hand, includes progressive accumulation of sublethal injury (e.g., from free radical damage), resulting in either cell death or diminished capacity of the cell to repair itself.

Home care and end of life considerations encompass matters of age-related and age-appropriate illnesses. In cats, these include chronic kidney diseases progressing to renal failure, hyperthyroidism, diabetes mellitus, degenerative joint disorders, neoplasia and chronic digestive disorders including inflammatory bowel disease, pancreatitis and cholangiohepatitis. By addressing organ and cell function, we can often impact well-being or quality of life (QOL).

One approach to end of life is to match the level of dysfunction (cell and organism) and types of care to these final stages in life.

1. **Support function**
2. **Alleviate discomfort, optimize comfort**
3. **Preparing for an ending; dying with dignity**
4. **Caring for the caregivers**

1. SUPPORT FUNCTION OF CELLS AND ORGANISM

Hydration is of utmost importance and should be included in any home care program. Most clients are able to give fluids subcutaneously if the care team believes in the importance of the therapy and that the client is capable. Fluids help make the cat feel better. Cellular dehydration causes headaches, sluggishness, inappetence, and lethargy. Dehydrated cells reclaim water from the urinary and gastrointestinal tracts relating in constipation. Dehydrated cells don't function optimally, can't transport toxins or nutrients well, aren't well oxygenated, suffer further damage and lethal changes.

Whether fluids are given intravenously or subcutaneously (SC), the volume needed is determined in the same manner, namely deficit (as a percentage of hydrated body weight in kg), plus maintenance (60 ml/kg ideal weight/day) plus on-going losses sustained by diarrhea or vomiting. If this volume is too large for SC administration at a single time, it may be divided into multiple treatments during the day. To maintain hydration, 60 ml/kg ideal weight are given SC once a day. Warming the fluids to body temperature, using Teflon coated needles and giving a treat makes the experience more pleasant for many cats.

Nutrition is essential to provide calories from protein and fat, anti-oxidants, fiber and other micronutrients. Carbohydrates are less essential for cats as obligate carnivores, however, they can be a good source of energy as well. The goal is to have a cat eat approximately 50 kcal/kg ideal weight/day on his/her own. Often illness interferes with meeting this goal, so we have to assist.

Pharmacological agents such as cyproheptadine (1 mg/cat PO q12h) or mirtazapine (2 mg/cat PO q48h) can be used. Discontinue either agent should it be ineffective after 2-3 days. Monitor both the calories as well as the amount of protein that patient needs to eat.

Pain may result in inappetance. There may be oral pain from periodontal disease, from a resorptive lesion, an ulcer or from a mass. Dental health should be optimized wherever possible. Musculoskeletal pain makes crouching or bending the neck uncomfortable. Shapes, sizes and placement of bowls may help. Declining senses may result in lack of awareness of food (hearing, vision, smell). Warming food to a freshly killed prey's temperature increases its palatability. Small, frequent meals may suit the older patient better than two-three meals a day. The texture of the food may play an important role. Should the cat like them, canned foods are preferable as they contain significantly more water. When recommending a prescription diet for a specific ailment, if the patient dislikes a prescription diet and is not ingesting adequate nutrients, especially with terminal illness, it is not advisable to persist with that diet. It is more important *that* they eat, than *what* they eat.

Feeding tubes save lives. They make administration of nutrients and medications less stressful for the client and for the patient. Naso-esophageal (n-e) tubes can be used short-term and require a liquid diet. Human enteral diets are too low in protein for long-term use and have a high osmolarity resulting in diarrhea. Oral syringe feeding can be performed with minimal stress if several tips are considered. Face the cat away from you; small volume syringes are preferable as administration of more than one ml at a time is larger than the oral capacity of the kitty; place the tip of the syringe at the back of the mouth to make it harder for the cat to spit out the food; room or body temperature food is less unpleasant. A cat's stomach can hold up to 100 ml in health, so starting with 6 ml and increasing in 6 ml increments to 48 ml total per feeding is realistic with most cats.

Large bore tubes are preferable to n-e tubes for periods longer than one-two days because a wider variety of diets can be used. The easiest one for home care is an esophagostomy tube. It requires only a brief anaesthetic and doesn't have post-op risks of peritonitis that a gastrotomy tube potentially does. G-tubes must be aspirated before infusing food to determine residual gastric volume. Both large tubes must be flushed with water following feeding to prevent clogging. These tubes allow the use of Hill's Prescription Diet a/d (1.3 kcal/ml), Royal Canin Recovery (1.2 kcal/ml) or Iams Veterinary Formula Maximum-Calorie (2.1 kcal/ml). If the client wants to dilute the diet for ease of syringing, they should use CliniCare or Rebound to prevent loss of caloric density. These are complete and balanced for cats and have 1 kcal/ml.

Mobility often declines in older or ill cats. While estimates vary regarding the frequency of joint disorders, it is accepted that older cats have a greater incidence of joint problems that are clinically under-diagnosed, either because cats aren't an exercised species or because they hide their discomfort well. Specific questions about movement may uncover that stiffness or discomfort is present. "Have you noticed: a change in how he jumps/climbs up? down? walks?" Care of the older cat must include attention to mobility.

Manifestation of problems with movement include constipation, urinating/defecating outside of the litter box, falling when jumping onto or off of the bed, hesitation climbing stairs, inability to crouch to eat resulting in weight loss. Regular nail trimming helps to maintain proper joint relationships. Ramps and steps onto favourite sleeping/perching spots benefit wellbeing. Warm, soft, padded sleeping places for stiff, painful, possibly bony joints should be considered. Adding a litter box so that a cat doesn't have to walk as far may reduce "accidents" as well as encourage regular voiding and defecation. Similarly, it helps to lower the rim of the litter box and ensure that the opening into the box is large enough. Scoop the box several times a day and make sure that the litter is neither too deep nor too sparse.

2. ALLEVIATE DISCOMFORT, OPTIMIZE COMFORT

Optimizing cell function may require medications and supplements. Compliance is best achieved through helping the client understand the cat's illness and how the medication helps. There are numerous handouts and good blogs/websites available on diabetes, chronic kidney diseases, hyperthyroidism and lymphoma, cancer terminology, treatment, and making decisions.

A stiff, painful or weak cat may neglect their coat care. Grooming may also fall by the wayside in a cat that feels ill or has dementia. Cats with cognitive dysfunction have similar brain changes to humans with dementia. Diligent, patient grooming may be necessary to help keep a cat's coat clean and healthy. Muscle massage may be welcomed. Warm, soft padded places to lie are helpful.

A common complaint is night-time yowling. Differentials include: loss of special senses (hearing and vision), hypertension, hyperthyroid agitation, pain, and cognitive dysfunction. The first may be discernable and alleviated by calling the cat, so that he/she is able to locate where you are. Nightlights may help a cat with visual impairment. Hypertension and hyperthyroidism are readily diagnosed and medically controlled. Pain may difficult to assess. *Often the best way to determine if pain is present is to administer pain relief and see if behaviour normalizes.* Cognitive dysfunction is a diagnosis made once the others have been excluded.

The Veterinarian's Oath states that: ***"I solemnly swear to use my scientific knowledge and skills for… the relief of animal suffering…"*** Above all else, clients have the right to expect that our focus will be on alleviating (and preventing) pain. Fortunately, over the past decade, cats have received more attention than they did previously in this regard, so that we now have a slightly wider menu to choose our analgesic protocols from. Use of a multi-modal approach is often preferable in order to minimize the potential adverse effects of a single agent by using lower doses of several agents. Often a narcotic (such as buprenorphine, hydromorphone, butorphanol or fentanyl) is combined with a non-steroidal anti-inflammatory (NSAID) like meloxicam, robenacoxib or aspirin or, for single dose, ketoprofen or cartrophen. Topical and local analgesia may also be provided either with EMLA crème or a local block or acupuncture. Gabapentin may be useful for pain and anxiety. Corticosteroids should not be combined with NSAIDs but can be used in conjunction with the other drugs.

Because of concern about possible effects on renal function, the general recommendation is to avoid NSAIDs in patients with renal disease. However, at this stage of life, as long as the client has been fully informed about the potential (not assured) risk, quality of life without arthritic pain may well be preferable to a painful, risk-free existence. The clinical signs of arthritis include weight loss, anorexia, depression, urinating/defecating outside the litter box, poor grooming, and lameness. In two separate studies, Gowan and Gunew showed no impact on renal function in older cats with or without chronic kidney disease receiving meloxicam (0.03-0.05 mg/kg lean, hydrated weight/day PO) for six months. Additionally, nutraceuticals and chondroprotectants, such as glucosamine and chondroitin sulfate may help in the management of degenerative joint disease. A good source of information is the Veterinary Academy of Pain Management at: http://www.cvmbs.colostate.edu/ivapm/ and IVAS - The International Veterinary Acupuncture Society at: http://www.ivas.org/

Many patients in the later stages of life require numerous medications. The number and frequency of administration is a source of stress for both the patient and the client. Whenever possible, the importance of the particular agents should be prioritized to ensure that the most important ones are given diligently. If the less critical ones cannot be avoided perhaps they can be administered by a different, less psychologically invasive route. Feeding tubes allow most oral medications to be given without restrictive handling. Clients are often comfortable giving injections subcutaneously of agents that can be administered in this fashion.

3. PREPARING FOR AN ENDING: DYING WITH DIGNITY

As we have the wonderful opportunity to know our patients from cradle to grave, we often know their people for a long time. As a feline companion reaches middle to late years, it may be a good idea to encourage clients to give consideration to introducing another companion, not only for themselves, but also for other critters who will be left alone. Should they want a kitten, a pair is better for the older cat to avoid some of the indignities of kitten behaviour! If they are considering adopting another single cat, then a young adult who has learned his/her manners is more suitable.

Probably the most difficult thing clients fear is wondering how they will know "when it is the right time". Our aim has to be to not take time away from an individual's life or from the time that that cat and their people can spend together and balance that with trying to avoid going beyond the point that the kitty wishes to be alive.

I encourage the human caregivers to step inside their cat's skin and try to imagine what he/she is experiencing. Some people are very clear in their assessment of how kitty is doing, others less so. Using a scale of one through ten, with ten being the best day of their life and one being equivalent to agony and hopelessness, most of us live at a six or seven. If clients use this as a means to score a given day or part of a day, it provides a measure of objectivity that is an emotional buffer from the roller coaster of emotions they are living in. (Or, for those who are hiding or numb, it adds content to their *cat's* experience.) When the scores are mostly twos and threes, it is time to consider helping kitty pass on. This tool can be helpful for a person who is afraid that they allowed their last companion to suffer too long, but it also encourages the person who can't let their friend go, to see that it is inhumane to continue treatment.

Advice from Rollin is to ask a client to write a list, as long as possible, of the activities that their cat enjoys doing while the cat is still well and to place this list in a drawer and in the medical record. As the cat declines, they can review this list and see what changes have occurred. This may help as a "reality check", to gauge progression and provide reassurance that the decision to euthanize is appropriate.

Because this is such a difficult period, people like to feel that they have some control should their cat's condition deteriorate quickly. Make sure that they have appropriate phone numbers and know where the emergency facility is. Give them a photocopy of the most recent medical record and lab work to keep by their car keys in case they need to go to the emergency clinic. Clients often forget the names of medications and doses that they are giving when they are upset.

Let people know what euthanasia entails preemtively. Tom Ewing wrote in the June 2006 issue of CatWatch newsletter: "The term *euthanasia* is derived from two Greek words — *eu*, which means good, and *thanatos*, which means death. In a report published in 2001, the American Veterinary Medical Association (AVMA) defined this "good death" as follows: "Euthanasia is the act of inducing humane death in an animal. It is our responsibility as veterinarians and human beings to ensure that if an animal's life is to be taken, it is done with the highest degree of respect, and with an emphasis on making the death as painless and distress-free as possible."

Reassure the client that the dose of barbiturate is painless. We place kitty on a nice thick towel on the client's lap and tell them that cats generally keep their eyes open and that because muscles relax, the cat may empty his/her bladder or bowels. I also let them know that some cats may still make breathing movements as the body shuts down. While IV administration (direct or via catheter) is the most common route for euthanasia, unless a cat is agonal, I prefer to administer euthanasia solution intraperitoneally (IP), just caudal to a kidney. This avoids restraint and the accompanying fear for the patient. Additionally, the transition from life to death is less sudden: it may take 2 minutes or 20 minutes. The cat becomes drowsy, falls asleep and then becomes anaesthetized. As soon as the cat is anaesthetized, should the client be wanting to "finish it", a vein can be accessed for an additional dose without any distress for the patient or the observer. In my experience, clients who have witnessed IV euthanasia, prefer the more gradual passing with the IP route. The time of waiting gives them a good opportunity to remember and cry and laugh. This helps me know that they are working through their grieving normally and are going to be okay emotionally.

4. CARING FOR THE CAREGIVERS

Most people are able to cope with a loss if they know that it is imminent and if they have a support network. Too often, in our modern life, the veterinary team may be the only support the client has. This is especially sad when friends and family don't appreciate the attachment the person has for their cat. Along with sending a personal card, it is usually appreciated greatly when we call the person after a few days. If there is any concern about the client's emotional security or that they are suicidal, enlist help from the human health care system.

It helps people to know that they may go through a whole range of emotions, from grief to guilt to anger to uncertainty and emptiness. This is normal and healthy. I counsel them that: "It is even normal for these feelings to overwhelm you days or weeks after the death of a beloved companion. This reflects the unconscious mind working through things and letting go a little bit at a time. What is NOT normal or healthy is when you get stuck in one emotion." There are also instances in which the death of a cat companion is a reminder of unfinished grieving for another person or pet. There are numerous wonderful grieving support networks via veterinary school telephone hotlines: www.vet.cornell.edu/public/petloss/other.htm. Some may prefer the electronic support group found at the Rainbow Bridge: www.petloss.com.

As mentioned in the previous section, if clients haven't already adopted a new friend, they should give thought to that. Everyone recognizes that the new cat isn't a replacement for the one who has died, but by rescuing another pet, the newcomer receives a wonderful home and a heart to fill to ease grief and bring joy. When the time is right, that new cat will show up.

Finally there is **the cost of caring**. We, the care team, experience dying and recovering from it approximately five times as often as our human health care equivalents. "Compassion fatigue goes beyond just normal burnout. Compassion fatigue is a type of physical, emotional, and spiritual exhaustion that comes with frequent exposure to death and having to offer support to clients in highly emotional situations over long periods of time."(Durrance) Some of the issues that have been associated with compassion fatigue in veterinary medicine include:

- Difficulty accepting that the patient's physical problems cannot always be controlled
- Frustration at having invested large amounts of energy in caring for a patient who then dies, taking this investment with them
- Disappointment if expectations for patients to die a "good death"- however this may be defined- are not met
- Difficulty ending a life you once saved
- Difficulty establishing realistic boundaries and expectations on veterinary care

- Caring for an animal more than the owner does
- Guilt arising from a cat's death

Without risking the tragedy of arms-length detachment, it *is* possible to take care of yourself. Suggestions for dealing with and protecting your self proactively from compassion fatigue include:

- Allowing yourself to be human
- Acknowledging and honouring your own grief and emotions
- Embracing your personal life away from work
- Allowing time to debrief and support other team members
- Saying your own private "good-bye" to your patients
- Believing in your ability to provide comfort and love to your patients
- Re-defining death: not as a failure but as an inevitable part of the life cycle
- After euthanasia, knowing that the individual is no longer suffering
- Thinking of euthanasia as a gift that owners want and appreciate
- With euthanasia, realizing that you are providing a loving and caring time for your patients and clients

REFERENCES

Cellular injury and cellular death In Cotran RS, Kumar V, Robbins SL (eds): Robbins: Pathologic basis of disease, W B Saunders, Philadelphia, 5th ed, 1994, 32. Robertson SA. **Vet Clin N Am-Small** 2008; 38: 1267. Lascelles BDX, et al. **J Feline Med Surg** 2010; 12 (3): 200. Sparkes AH et al. **J Feline Med Surg** 2010; 12 (7): 521. Gunn-Moore DA, et al. **J Feline Med Surg** 2006; 8(4): 234. Landsberg GM, et al. **J Feline Med Surg** 2010; 12 (11): 837. Gunew MN, et al. J Feline Med Surg 2008;10(3):235. Gowan R, et al. J Feline Med Surg 2011; 13 (10): 752.McMillan FD. **J Am Vet Med Assoc** 2000; 216:1904. McMillan FD. **J Am Vet Med Assoc** 2001; 219:1204. Tannenbaum J. Veterinary Ethics: animal welfare, client relations, competition and collegiality. 2nd ed. St. Louis: Mosby, 1995. Thayer V, et al. **J Feline Med Surg** 2010; 12 (9): 728. Adams RE, et al. **Am J Orthopsychiatry** 2006; 76(1): 103. Durrance D: Compassion Fatigue. Proceedings of the 2005 Western Veterinary Conference. Rollin BE. **J Feline Med Surg** 2007; 9:326. Thayer V, et al. **J Feline Med Surg** 2010; 12 (9): 728. Shaw JR, et al. **Vet Clin N Am-Small** 2007; 37 (1): 95. Rank MG, et al. **Best Practices in Mental Health** 2009; 5 (2): 40. Hewson C. **Vet Nurs J** 2014; 29 (12): 388.

CAN WE STANDARDIZE LEFT ATRIAL MEASUREMENTS?

Mark Rishniw, BVSc, MS, PhD, DACVIM (Cardiology)
Ithaca, NY

INTRODUCTION

Left atrial size is one of the most routine measurements obtained when performing echocardiography. Numerous studies have detailed the clinical importance of estimating left atrial size, showing it to be a predictor of outcome in patients with left heart disease.[1-6] Several studies have detailed methods of measuring the left atrium in dogs and cats.[7, 8] It would seem reasonable that consistent intra-observer and inter-observer measurements are required if patients from different populations (or the same population) are to be compared. However, current data suggest that investigators vary considerably in their measurements of the left atrium.

WHICH VIEW IS BEST?

The earliest estimates of left atrial size in human patients were obtained from M-mode imaging, with the M-mode cursor transecting the aorta and the left atrium directly beneath it. This allowed investigators to calculate the left-atrial-to-aortic ratio (LA:Ao). Using these guidelines veterinary cardiologists began to measure the left atrium, but, with the advent of two-dimensional echocardiography, realized that the rotational displacement of the left atrium in dogs and many cats resulted in the M-mode cursor transecting the left auricle (left atrial appendage), rather than the body of the left atrium. Thus, the initial estimates of left atrial size could be more accurately described as estimates of left auricular dimensions.

With evolution imaging technology, two-dimensional measurements of the left atrium and aorta became commonplace. These allowed more accurate cursor placement without loss of spatio-temporal resolution. The first published studies examining multiple methods of estimating left atrial size in healthy dogs appeared in 2002.[7, 8] These described the approach used to measure the left atrium, including the timing of the measurement, and the positioning of the cursors. In both cases, the left atrium was measured from the right parasternal short axis view at the onset of ventricular diastole, just prior to opening of the mitral valve. This point in the cardiac cycle allowed the observer to measure the left atrium at its largest. The two groups of investigators used different axes to measure the cross-sectional area or diameter of the aorta against which they compared the left atrium. Interestingly, the reference intervals that these groups defined for normal dogs differed substantially. Rishniw and Erb, examining dogs of various breeds and sizes, found that the upper limit for LA:Ao was 1.6, while Hansson and colleagues, examining Cavalier King Charles Spaniels, defined the upper limit as 1.3. Initially, these differences were conjectured to be the result of 2 factors: (A) subject selection (Cavalier King Charles Spaniels vs various breeds) and (B) different axes for the measurement of the aortic dimension. However, a subsequent study found no difference between measurements of the aorta along either axis.[9] Additionally, unpublished data from the 2 groups involved in the original studies found that their methods of imaging the left atrium from the right parasternal short axis view were similar and interchangeable, and the points to and from which they measured the left atrium in these short-axis images were similar. This suggested that systematic measurement errors (differences) could not account for the discrepancies observed in the reference intervals. Thus, the remaining, untested hypothesis is that breed differences or sample subject population variation most likely account for the differences in reference intervals.

Rishniw and Erb additionally determined the LA:Ao for left atrial measurements obtained from a right parasternal long axis view.[7] Reproducibility of the left atrial measurement in this view appeared better than in the short axis view, and the left atrial dimension was larger, with reference intervals that were not interchangeable between the two views. This measurement was essentially identical to that proposed by LeBobinnec, where the left atrium was measured from the right parasternal short axis view in a line extending from the interatrial septum to the left-most border of the left atrium (at the bottom of the left atrial image as viewed on the screen) (unpublished). Reference intervals proposed by LeBobinnec were similar to those proposed by Rishniw using the right parasternal long-axis view.

Subsequently, other investigators proposed alternative measurement techniques: one group examined the longitudinal and transverse left atrial dimensions obtained from the left apical four-chamber view (unpublished); another used the right parasternal long axis view, but measured the left atrial dimension at the level of the mitral annulus.[10] In all these cases, reference intervals differed.

All these methods had inherent measurement errors: those obtained in long-axis did not have the aorta against which to ratio the left atrial measurement, and did not have a predictable way of determining that the maximal dimension of the left atrium was being imaged. Those obtained in short-axis had potential issues of obliquity through the left atrium (if the transducer had been positioned too apically), and in many dogs, one of the pulmonary venous confluences obscured the posterior wall of the left atrium in exactly the location to which the left atrial measurement point was to extend. The apical view is not routinely used by less experienced clinicians, and, as with the long-axis view, the maximal dimensions of the left atrium could not be guaranteed.

Rishniw examined the inter-observer variability in left atrial measurements (using the right parasternal short axis view) (unpublished). Unsurprisingly, the starting point of the measurement was relatively consistent (near the commissure of non-coronary and left coronary cusp of the aorta). However, the end-point, along the caudal border of the left atrium varied considerably between observers, despite the original studies defining the measurement line to be used. This variability increased considerably from image to image, being worst for severe enlargement and images where the pulmonary venous confluence obscured the caudal atrial border. The variability was sufficient to impact the classification of patients by up to 2 categories of atrial enlargement.

However, what was lacking in all these studies was any means of determining if one method more precisely determined degrees of left atrial enlargement in dogs with cardiac disease. In no study was an independent reference standard (e.g. MRI) used to determine accuracy. Furthermore, none of the studies compared all the methods to determine intra- and inter-observer measurement precision.

From these studies, it is apparent that no method is inherently "better" than any other. Consequently, it might be better to ask "what method is most commonly used"?

WHICH VIEW IS MOST COMMONLY USED?

From examination of the literature, and informal polling of cardiologists, it appears that the most commonly used method is the right parasternal short axis view, although some clinicians and investigators also use the method proposed by LeBobinnec, or the right parasternal long axis method proposed by Rishniw and Erb. This is in part because it is easily obtained, and provides an opportunity to measure the aorta in the same image as the left atrium.

WHAT ABOUT CATS?

Only one study has examined left atrial measurement techniques in healthy cats.[11] These investigators examined cats using the same approach as that of Rishniw and Erb. They determined that healthy cats had LA:Ao < 1.4. However, subsequently, cardiologists have suggested that since cats vary relatively little in body size, obtaining a bodyweight-independent measure of left atrial enlargement might be unnecessary, and could introduce an additional potential error from the aortic measurement. Consequently, some cardiologists have suggested that the absolute left atrial measurement might be sufficient to define left atrial enlargement. Since most cats have an aortic dimension close to 1cm (10mm), the LA:Ao essentially converts to an absolute left atrial measurement (i.e. LA:Ao of 1.4 is simply a left atrial dimension of 14mm).

DOES THE AORTIC MEASUREMENT MATTER?

The aorta viewed from the right parasternal short axis view has three axes of symmetry. As mentioned earlier, Rishniw and Hansson measured the aorta along two different, but symmetrical planes. Consequently, it is not surprising that the diameters measured by either of these methods were interchangeable.[9] However, discussions amongst cardiologists suggested that the point in the cardiac cycle at which the aorta was measured from this view differed. This was in part due to the relatively poor image resolution at the point of aortic valve closure (defined as the time-point used by both Rishniw and Hansson). Consequently, other cardiologists measured the aorta just prior to mitral valve closure (at the onset of the R-wave) or during atrial depolarization (at the peak of the p-wave). Since the aorta should be unchanging during diastole, this should not matter. Dickson and colleagues examined the aortic diameter at these 3 time-points in 163 dogs and 40 cats (unpublished). Somewhat surprisingly, the measurements differed, with the early diastolic measurement being consistently the largest, and the early systolic measurement being consistently the smallest. By comparing these measurements either to a fixed left atrial measurement, or by comparing the LA:Ao obtained at each of these time-points, they found that the changes in aortic dimension were substantial enough to frequently affect the classification of patients with different degrees of heart disease. In other words, the same patient, measured by these different methods could be classified as "normal" by one method, and "mildly diseased" or even "moderately diseased" by another. The cause of these discrepancies in aortic measurements were conjectured to be due to basilar displacement of the aorta during diastole, thereby shifting the plane being transected by the transducer beam.

Therefore, either the timing of aortic measurements should be agreed upon, or different reference intervals and severity thresholds should be established for each method of aortic measurement.

HOW DO CARDIOLOGISTS DEFINE LEFT ATRIAL ENLARGEMENT?

Most cardiologists translate a quantitative LA:Ao measurement into a categorical ordinal estimate of severity. This is either a quadripartite scale ("Normal", "Mild", "Moderate" and "Severe")[12] or a sexipartite scale (same as quadripartite, but with "Mild-Moderate" and "Moderate-Severe" interspaced between the measures of enlargement). In humans, specific thresholds have been developed to define the degree of pathology (using the quadripartite scale). However, no such thresholds exist in veterinary medicine. Brown proposed thresholds for a sexipartite scale from M-mode left atrial measurements which was based on the number of standard deviations from a Normal distribution of healthy animals (unpublished). Rishniw conducted a study (unpublished) that asked cardiologists to assign a series of images of the left atrium (right parasternal short-axis view) to one of 4 categories (normal, mild, moderate or severe). Based on responses from 80 cardiologists and cardiology residents, 75% of responses where "Normal" was selected had a calculated LA:Ao < 1.55; 50% of responses where "Mild" was selected had a calculated LA:Ao between 1.6 and 1.9; 50% of responses where "Moderate" was selected had a calculated LA:Ao between 1.9 and 2.2 and 75% of responses where "Severe" was selected had a calculated LA:Ao >2.3. From these data, it appears that at least a majority of cardiologists agree on what constitutes "Normal", "Mild", "Moderate" and "Severe" enlargement. Intraobserver agreement in that study was high (>80%), and the estimate of severity was mostly unaffected by knowing the actual LA:Ao for a specific image. This, therefore, serves as a starting point for establishing a consensus for defining left atrial size in cardiac disease in dogs.

Ljungvall and colleagues subsequently proposed an even more stringent approach to defining LA enlargement for dogs with mitral valve disease.[12] Rather than accepting just the LA:Ao, these investigators used a "2-out-of-3" approach. They considered the LA enlarged if either the LA:Ao & wLA (using thresholds proposed by),[13] the LA:Ao & wLVIDd, or the LA:Ao & wLVIDd were enlarged. This reduced the probability of a "false positive" classification of a patient.

A similar scaling was developed for cats, using different thresholds.[14]

WHY DOES IT MATTER?

The left atrium is very commonly measured, and its size is a relatively simple measure of the severity of many cardiac diseases. It is an independent predictor of outcome. Furthermore, many clinical trials of common cardiac diseases in dogs and cats require a certain degree of severity to be present in a potential enrollee – severity that is often defined as LA:Ao above a critical value.

While intraobserver variability is relatively small, interobserver variability can be profound. Consequently, sample populations defined by LA:Ao obtained by different investigators could be objectively different from each other, and consequently not comparable. Outcomes associated with severity could appear to differ because the way that severity was defined differed.

WHAT CAN WE DO?

Since potentially more accurate technologies (3D echocardiography, MRI) are not going to be readily and cheaply available to most cardiologists in the foreseeable future, we are likely resigned to estimating severity of left heart disease using two-dimensional techniques. Consequently, identifying the optimal imaging plane would be the first place to begin. Based on the speaker's observations, the right parasternal short axis view appears to be favored by most cardiologists, both in day-to-day dealings with clients and in clinical studies. However, more specific and detailed imaging instructions are required so that different clinicians obtain images sufficiently similar to each other to be interchangeable. The issue of inconsistency of measurement should be addressed, possibly by finding and defining atrial echocardiographic landmarks that are consistently identifiable in most patients. This will reduce inter-observer measurement errors. It is possible that new and different reference intervals and severity thresholds will need to be defined. Timing of the image measurement with respect to the cardiac cycle should be standardized, or phase-specific intervals and thresholds should be defined. Investigators reporting estimates of left atrial size in studies should provide detailed specifics about the method(s) used to determine left atrial size, so that readers can make valid comparisons between studies. Finally, in the absence of independent estimates of severity, threshold values should be determined first by consensus, and subsequently, by longitudinal studies examining clinical outcomes (such as time to onset of clinical signs or survival).

REFERENCES

1. Lord PF, et al. **J Vet Intern Med** 2011;25(6):1312. 2. Hezzell MJ, et al. **J Vet Intern Med** 2012;26(2):302. 3. Ferasin L, et al. **J Vet Intern Med** 2013;27(2):286. 4. Ljungvall I, et al. **Am J Vet Res** 2009;70(5):604. 5. Hezzell MJ, et al. **J Vet Cardiol** 2012;14(1):269. 6. Guglielmini C, et al. **Vet J** 2000;160(3):235. 7. Rishniw M, et al. **J Vet Intern Med** 2000;14(4):429. 8. Hansson K, et al. **Veterinary Radiology and Ultrasound** 2002;43(6):568. 9. Georgiev R, et al. **J Vet Cardiol** 2013;15(2):131. 10. Hetyey C, et al. **Acta Vet Hung** 2005;53(1):23. 11. Abbott JA, et al. **J Vet Intern Med** 2006;20(1):111. 12. Ljungvall I, et al. **J Small Anim Pract** 2014;55(11):545. 13. Brown DJ, et al. **J Vet Intern Med** 2005;19(4):542. 14. Ljungvall I, et al. **J Feline Med Surg** 2014;16(4):281.

HEARTWORM RESISTANCE TO MACROCYCLIC LACTONES: DISASTER, MYTH OR WAKE-UP CALL?

Clarke Atkins, DVM, DACVIM (SAIM, Cardiology)
Raleigh, NC

Prophylaxis

Prevention of HWI is an obvious and attainable goal for the veterinary profession. Prevention failure results from ignorance on the part of owners as to the presence or potential severity of HWI, lack of owner compliance, or from inadequate instruction on preventative measures by the attending veterinarian.[1,32-34] Studies of owner compliance have revealed that approximately 55% of dog owners that use veterinary care purchase heartworm preventative, and enough medication is dispensed only to meet the needs of approximately 56% of those dogs. Hence the proportion of "cared for" dogs in the population that receive adequate heartworm prophylaxis is less than one third.[33] If one takes into consideration doses purchased but not administered and dogs that are never taken to a veterinarian, the percentage of protected dogs falls drastically. This was emphasized in North Carolina in 1999, when Hurricane Floyd caused extensive flooding and disruption in the poorest part of the state. Of dogs rescued from the floodwaters, 67% were infected with heartworms (personal communication, Dr..Kelli Ferris, North Carolina State University, 2003). In addition, evidence suggests that the veterinary profession is failing in its education of clients. New and colleagues,[35] upon questioning veterinary clients purchasing macrolide preventatives, found that 38% did not realize that their prescribed drug's spectrum was broader than solely preventing HWI.

Macrocyclic Lactone (Macrolide) Antibiotics

The introduction of the macrocyclic lactone endectocides (macrolides), ivermectin (Heartgard®, Iverhart ®, TriHeart®), ivermectin with pyrantel pamoate (Heartgard® Plus, Iverhart Plus®, TriHeart® Plus), ivermectin with pyranel pamoate and praziquantel (Iverhart Max®), milbemycin oxime (Interceptor®), milbemycin with lufenuron (Sentinal®) and with spinosad (Trifexis®), selamectin (Revolution™), and moxidectin (ProHeart®, ProHeart® 6), and moxidectin with imidicoprid (Advantage/Multi™) has provided the veterinary profession with highly effective, incredibly safe heartworm preventatives in a variety of formulations and with a variety of spectra. These agents, because they interrupt larval development (L3 and L4) during the first 2 months after infection, have a large temporal window of efficacy and are administered monthly. These products have enjoyed great efficacy, "virtually 100%", when used as directed. Recently, a single isolate (MP3) from north-eastern Georgia has shown restistance/tolerance to some macrocyclic lactones, when administered <u>once</u> 30 days after heavy experimental challenge.

All are safe in collies when used as directed at preventive dosages. They each have microfilaricidal efficacy and render female heartworms sterile. Hence microfilarial tests for HWI cannot be reliably used in dogs receiving these products. Prophylaxis should be commenced no later than 6 to 8 weeks of age in endemic areas or as soon thereafter as climatic conditions dictate.[19,49] Macrolides should be administered precisely as indicated by the manufacturer. If accidental lapses of more than 10 weeks occur, the preventative should be reinstituted at recommended doses and maintained for at least12 consecutive months.[25] In the event of a lapse in preventative administration during a time of known exposure risk, an antigen test should be performed 7-8 months after the last possible exposure to determine if infection has occurred. It is recommended by the AHS and by CAPC that these agents be used year-around in all areas of the U.S.

<u>Off-Label Use</u>: The macrolides are effective microfilaricides, with varying microfilarial kill rates, but microfilariae, in reduced numbers, are often found in the circulation for months after treatment has begun. Some but not all macrolides have adulticidal activity if used continuously for prolonged periods.[39-41]

Macrocyclic Lactone Resistance/Lack of Effectiveness

In 2005, the FDA-CVM reported an increase in the reports of LOEs (Lack of Effectiveness) for macrocyclic lactones and required that such agents no longer be labelled as "perfect" in terms of efficacy. This failure of complete rapid microfilarial clearing, coupled with concern in the Mississippi River delta region (areas of LA, AR, MS, TN), has caused concern that resistance to this class of drugs may be developing.[41a] The proof of this is small, but taken together, the data argue that a small percentage of microfilariae, isolated from dogs in this region have characteristics suggesting tolerance to the drug group. A joint consensus of the AHS and CAPC stated the following (excerpts). "There is evidence in some HW populations for genetic variations that are associated with decreased in vitro susceptibility to the macrocyclic lactones. Whether the observed genetic variations constitute heritable resistance is being investigated. Most credible reports of LOE that are not attributable to compliance failure are geographically limited at this time. The extent of the problem is obscured by demonstrated lack of owner and DVM compliance, possible changes in environmental/vector factors, and more effective antigen testing. The potential for resistance is *not* a reason to abandon use of approved preventive products."

The concern relative to the presence of circulating microfilariae in dogs that are started and maintained on monthly preventives is that they could be a source of propagation of microfilariae that are preselected for resistance to macrocyclic lactones. In 2005, Prichard wrote "Consideration of the proportion of the *D. immitis* population in refugia, the life cycle stage targeted, and the anthelmintic dosages used suggest that it is unlikely that significant avermectin/milbemycin [macrolides] resistance will be selected in *D. immitis* with current treatment strategies."[118] However, this belief was based upon the assumption that people were using the preventives as per label instructions, not using them as adulticides and microfilarial suppressants. The prudent approach is simply to administer the products as approved by the FDA: as preventives that should be given to microfilaria-negative dogs. This means that the "soft" or "slow" kill approach to adulticidal therapy should be avoided. Likewise, one could argue against the use of macrocyclic lactones in microfilaria-positive dogs prior to beginning them on adulticidal therapy (ie, a method advocated by this author; see below) as 10% to 20% of these dogs will have circulating microfilariae for months after they start this regimen – microfilariae that have seen a macrocyclic lactone. If this approach is utilized, the clinician must ensure that microfilaria are eradicated in the first months of macrolide therapy. All current heartworm preventives belong to the same class of molecule, the macrocyclic lactones, and thus, we need to be very prudent in our long-term stewardship of these drugs.

Very recently (July, 2013), data presented at the AAVP Meeting demonstrated resistance in certain strains of *D. immitis*. There were 4 strains studied and all the current molecules were less than perfect in preventing HWI in dogs receiving them for prophylaxis. Since *all* molecules performed at a lower level than when approved by the FDA, it was concluded that this was a class effect. It has been shown that not all formulations of the same molecule perform identically, however. Furthermore, HW from "suspect-resistant Delta strains" were shown to drift toward resistance (by increasing homozygosity for the G-gp gene) when macrocyclic lactone administration was provided to HW-infected dogs when the dosage was gradually uptitrated over time. This substantiates the fear that "slow-" or "soft-kill" adulticidal therapy with macrocyclic lactones will indeed create problems with resistance and that this approach should be avoided, if at all possible.

After scrutinizing the data, this author feels that resistance has occurred and is possibly due to inappropriate use of macrocyclic lactones. However, the problem in the Mississippi is not "due to" this resistance and not all of the least susceptible strains emanate from the Delta, nor are they similar genetically to those strains from the Delta. It appears to me (as a "non-geneticist") that there have been spontaneously resistant *D. immitis* strains develop without apparent macrocyclic lactone pressure and some have developed with that pressure, presumably by our use of the soft-kill methodology. Other factors that may (likely) have contributed include poor compliance; LOE (lack of effectiveness) over-reporting; increased mosquito and increased infected mosquito pressure on dogs, possibly due to increased hurricane incidence in last decade in MS Delta region; increased wild animal (coyote) reservoirs; lack of focus of owners and DVMs on preventive practices (complacency); FDA-CVM rule change, dictating that all LOE be reported to the FDA-CVM as an adverse reaction; economic downturn since 2007; as well as, improved and increased heartworm antigen testing. Evidence that the Delta crisis is not purely resistance lies in some logistical findings, such as the fact that the concern has never spread beyond the original Delta boundaries (lines drawn between Jonesboro, AR, Memphis, TN, Gulfport, MS, and Lake Charles, LA) and the fact that the LOE reports are falling since mid-2009, and that Delta DVMs indicate that the problem is abating or has abated. This is not the characteristic behavior of resistance.

REFERENCES

Available upon request.

CANINE HEARTWORM DISEASE – WHAT HAS CHANGED?

Clarke Atkins, DVM, DACVIM (SAIM, Cardiology)
Raleigh, NC

Wolbachia

In 1975, organisms were observed with the electron microscope in intra-uterine microfilariae and developing embryos of *Dirofilaria immitis* and *Brugia pahangi*. Research on these organisms then stalled until 1995 when it was shown with molecular methods that the organisms in *Dirofilaria immitis* were related to the *Rickettsia*-like *Wolbachia* endosymbionts of arthropods. In 1999, it was shown that tetracycline had negative effects on the embryogenesis of *D. immits*, and that long-term treatment of cattle infected with *Onchocerca ochengi* using tetracycline resulted in the death of the adult filarial worms, apparently associated with the death of their endosymbiotic bacteria. Long-term tetracycline or doxycycline chemotherapy does have negative effects on the larvae and adults of human parasitic filarioid nematodes, *Brugia* and *Onchocerca*; however, the required long-term and regular administration of these compounds appears to provide no significant advantage over periodic treatment with microfilaricidal ivermectin for control through the prevention of transmission.

Doxycycline and *Wolbachia*

The benefits of doxycycline result from its ability to remove or reduce the burden of *Wolbachia*, a rickettsial organism that exists in a symbiotic relationship with heartworms (and other filarids), occupying the reproductive tract and lateral chords of the host parasite. *Wolbachia* are necessary for the parasite (in this case, *Dirofilaria immitis*) to develop, thrive, reproduce, and maintain infectivity.

Doxycycline has been used to (presumably) rid the parasite of *Wolbachia* organisms; therefore, *D immitis* organisms do not thrive, may deteriorate and die, and have reduced reproductive potential, which helps manage HWD in infected dogs and reduces potential for infection in other dogs. There has been concern raised as to the potential for resistance of bacteria to doxycycline being developed if it is used in all cases. This, in my opinion, is worth considering, as doxycycline is widely used and valuable to our profession (and physicians) in the treatment and prevention of a variety of infectious processes. On the other hand, it is widely used for other diseases and the concerns for resistance generally occur when drugs are used at sub-optimal dosages and durations of therapy. Neither of these will result from the current recommendations for HWD.

Potential and realized benefits derived from anti-*Wolbachia* therapy include:

1. **Reduce Ability of Parasite to Reproduce**
   It has been shown that the *Wolbachia* organism is suppressed (killed) by doxycycline and the resulting, negative effects on the heartworm reproductive system renders the parasite infertile or less fertile (temporarily?), with reduced microfilarial numbers.[1,2,3]

2. **Reduce Infectivity**
   In doxycycline-treated dogs, even if microfilariae are produced and ingested in a mosquito's blood meal, the resultant L3 are incapable of producing infection, reducing the spread of HWD.[1,2,3]

3. **Potentiate Slow- or Soft-kill Efficacy**
   Most agree that *Wolbachia* is an obligatory symbiont for *D immitis*, which makes us hope that *Wolbachia* eradication with antibiotics would result in the nematode's demise. Unfortunately, prolonged doxycycline therapy does not kill heartworms because they are not sufficiently bound to their bacterial symbionts.[4]
   However, two studies[3,5] have demonstrated that doxycycline shortens the time until worm death when administered chronically with ivermectin/pyrantel at preventive dosages but with a decreased dosing interval:

4. **Reduce Microfilarial Burdens More Effectively, Safely, & Rapidly**
   In the transplanted worm model mentioned in Study 1, it was shown that a combination of weekly ivermectin (6 mcg/kg) and daily doxycycline (10 mg/kg Q 24 H) eliminated microfilariae over 8 to 12 weeks.[3]

   This elimination is relatively fast, but not so quick that therapy results in the adverse, shock-like reactions seen with rapid destruction of large numbers of microfilariae. In addition, subacute removal of microfilariae lessens the chance of macrocyclic lactone resistance, especially when the practitioner is forced to use the slow-kill method due to owner finances or unavailability of adulticide (ie, melarsomine).

5. **Reduce Lung Reaction to Worm Death (Spontaneous & Post-adulticide)**
   It has been shown that the combination of weekly ivermectin (6 mcg/kg) and daily doxycycline (10 mg/kg Q 24 H) significantly reduced lung lesions after melarsomine therapy.[3,6]

6. **Kill Developing Larvae**
   Recently, McCall, et al, demonstrated that, while doxycycline (even with ivermectin) does not have rapid adulticidal efficacy, doxycycline monotherapy does stop the progression of infective larvae to adulthood when administered for the first 30 days of infection at 10 mg/kg Q 24 H.[7]

- If the 30 days of administration begin on day 40 of infection, however, the effect is partially lost, with 2% of L3 reaching adulthood.
- If the larvae reach 65 days before doxycycline is begun, only 52% reach adulthood.[7]

Therefore, in addition to reducing adverse effects from heartworm death, doxycycline begun on the day of diagnosis will help close the potential *seasonal window of continuous infection*, which means that, during certain times of year when exposure is continuous (warmest months), the host may have developing larvae of all stages. However, younger—tissue-stage larvae become less vulnerable with development, maturing to a point where (1) macrocyclic lactone therapy is ineffective and (2) they are too young for elimination with melarsomine. This treatment failure, potentially alleviated with pre-adulticide doxycycline, can result in the perception of product failure (preventive or adulticide).

Doxycycline Dosing

Currently the best data we have argues that the dosage, if tolerated, is **10 mg/kg Q 12 H for 30 days**, administered prior to adulticidal therapy (Figure 1). If this dose is not tolerated, it can be reduced to 5 mg/kg Q 12 H.[7] I advocate a second month's delay in adulticidal therapy to allow the parasite to deteriorate maximally and, thereby, further reduce the pulmonary reaction to worm death. Benefits include:

- Prevention of maturation of recently acquired infection (tissue phases)
- Reduced pulmonary reaction to dying worms
- More rapid and complete eradication of microfilariae (potentially reducing risk of heartworm resistance to macrocyclic lactones)
- Enhancement of vermicidal efficacy of ivermectin if using slow-kill method.

Adulticidal therapy

An important breakthrough in the management of heartworm infection (HWI) is the adulticide melarsomine (Immiticide®), an organoarsenical superior in safety and efficacy to thiacetarsemide.[17] This product, which is administered twice, at 2.5 mg/kg q24h, has a mean retention time 5 times longer than thiacetarsemide and its metabolites are free in the plasma, on which HW feed.[18] In a study of 382 dogs with HWI receiving melarsomine, none required cessation of therapy due to hepato-renal toxicity, as compared to 15-30% with thiacetarsemide, the agent previously used.[18]

"Split-Dose" Adulticidal Therapy

With 2 doses, the efficacy is over 90% (FDA pivotal study) with the useful flexibility of a 50% worm kill with 1 dose. This then allows "split-dose" protocol to be utilized in severely afflicted individuals or in those in which pulmonary thromboembolism (PTE) is a concern. This method allows destruction of only one-half the worms initially (1 IM injection of 2.5 mg/kg), thereby lessening the chance for embolic complications. This single dosage is followed by a 2 dose regimen in 1-3 months, if clinical conditions permit. While the manufacturer recommends this protocol (Figure 1) for severely affected dogs, the author employs it in **all cases** unless there is financial constraint or underlying concern for arsenical toxicity (for example, preexistent severe renal or hepatic disease).[18a] One disadvantage to the "split-dose" method, in addition to the expense, is the need for 2 months' exercise restriction.

In 55 dogs, with severe heartworm disease (HWD) and treated in this 3-dose manner, 96% had a good or very good outcome with >98% negative for antigenemia 90 days post-therapy.[18] Although symptomatic and even fatal PTE can result from treatment with melarsomine, no case of severe PTE was seen in the 382 dogs of this series.[19] Of the 55 severely affected dogs, 31% had "mild or moderate PTE"; no fatalities resulted. The most common sign was fever, cough, and anorexia 5-7 days post-treatment. This was associated with mild perivascular caudal lobar pulmonary radiographic densities and subsided spontaneously or after corticosteroid therapy.

The most common complication to melarsomine therapy is the local inflammatory reaction at the injection site. This can be minimized by following the manufacturer's directions explicitly (change needles before injecting, choose deep IM site with care, put pressure on site after injection, and alternate sites). In addition, corticosteroids (e.g. dexamethasone) or NSAIDs can be given at the time melarsomine is administered to lessen the reaction.

Standard 2-Dose Melarsomine Method

There are clear advantages and disadvantages to each method of adulticidal therapy. The major drawbacks to the 2 methodologies outlined above are the cost (often \$1000 to >\$1500, and even more, if there are complications) of "split-dose" protocol and the risk of resistance, poor compliance, and ongoing disease progression over 1-3 years with the "soft-kill" approach. Perhaps the solution is the "back to the future" approach of using 2 injections, 24 hours apart, the protocol used by the majority of practitioners when melarsomine was brought to market. The advantages include lower cost, rapid clearing of the parasites with an approximate 90% of worms killed (note: this is not % of dogs cleared, which is as low as 70%). This is not, in the authors' preferred approach when there are no financial considerations. However, when the choice is between doing nothing at all and "soft-kill" adulticidal therapy, the standard 2-dose protocol is likely superior in young, otherwise healthy dogs with HWI. Furthermore, the diagnostic workup considered to be "good medicine" for the well-to-do client, can be avoided for the less affluent owner of a young, otherwise healthy dog. This was demonstrated in the 4 years following Hurricane Fran (1996). Over 350 HW-infected dogs, owned by "low-income" NC families, were treated, using the standard 2-dose method without pre-treament imaging or laboratory workup (Personal communication, K. Ferris, NCSU). Of these, 2 older dogs died, giving a mortality rate of only 0.5%, certainly an acceptable number under these circumstances. It is imperative, when "all" the worms are killed with two injections of melarsomine, that exercise be ***strictly*** curtailed for at least 30 days.

This approach is not without disadvantages, as compared to the split-dose regimen, including a lower kill rate, inability to observe the dogs' responses to a single dose and to react accordingly, and the fact that if the macrolide is started at the time of diagnosis, as advised by the author, one has to wrestle with concerns of microfilaremic dogs receiving macrolides (resistance). That said, I believe that this approach is probably ideal for the vast majority of dogs adopted from shelters, an estimated 75-90% of dogs seen in general practice (varying based on socioeconomic of practice clientele) and perhaps 10% of dogs in referral practices. The point is that this approach is superior to "soft-kill" in the vast majority of patients and is a little-considered option. Lastly, if 30 days' doxycycline administration is employed, concerns about microfilariae are largely removed. If this treatment is not possible due to financial constraints, high-dose ivermectin (50 ug/kg) or preventive-dose milbemycin or moxidectin/imidacloprid may be used as microfilaricidal agents <u>and</u> preventives (ivermectin dosage returns to preventive dosage [6 ug/kg], after first month).

"Soft-" or "Slow- Kill" Adulticidal Therapy

It is now known that certain macrolides have adulticidal properties.[6,16,20] Ivermectin, when administered for 31 months continuously has nearly 100% efficacy in young heartworm infections.[20] It has been shown, however, that lung and pulmonary vascular manifestations of HWD still result when ivermectin "prophylaxis" is begun 5.5 and 6.5 months post-infection and continued for 1 year.[21] Selamectin, when administered continously for 18 months killed approximately 40% of transplanted worms.[16] Sustained release moxidectin also appears to have some adulticidal efficacy.[12] Recent data suggests that an aggressive macrolide protocol (ivermectin, given at 6 ug/kg weekly instead of monthly), coupled with a complex regimen of doxycycline (10 mg/kg/day) will hasten worm destruction, with worm eradication with approximately 9 months' therapy.[21a] Furthermore, microfilariae are eradicated more quickly in this manner. This has caused many to invoke the use of doxycycline routinely in the management of heartworm infection in dogs, with a current protocol calling for ivermectin at preventive dosages given monthly (some advocate administering it every 2 weeks for the first 6 months), coupled with doxycycline for 30 days at 10 mg/kg/day. The American Heartworm Society advocates, when melarsomine is unavailable, to use doxycycline at this dosage one month on, two months off, etc, until the patient reverts to an antigen-negative status. While there may be a role for this therapeutic strategy (Slow Kill) in cases in which patient age, financial constraints or concurrent medical problems prohibit melarsomine therapy, the current recommendations are that **marcolides <u>not</u> be adapted as the primary adulticidal approach.**

Surgery

Surgical removal of HW can minimize PTE, as compared to pharmacologic adulticides, such as melarsomine.[22,23] This procedure, however, requires specialized training and instrumentation, including fluoroscopic imaging capabilities. Nevertheless, it remains an alternative for the management of high risk patients.

Ancillary therapy

As mentioned above, aggressive macrolide therapy (ivermectin, given at 6 ug/kg weekly instead of monthly), coupled with a complex regimen of doxycycline (10 mg/kg/day) hastens worm destruction and quickly eradicates microfilariae. This has resulted in increasing use of doxycycline in the management of HWI in dogs.

Cage rest is an important aspect of the management of HWD after adulticidal therapy, after PTE, or during therapy of heart failure. This can often be best, or only, accomplished in the veterinary clinic. If financial constraints preclude this, crating at home and/or tranquilization are useful alternatives.

Microfilaricidal therapy

Microfilaricidal therapy is traditionally instituted 4-6 weeks after adulticide administration. The macrolides offer a new and effective alternative to levamisole and dithiazanine.[5,5a] Microfilariae are efficiently and rapidly cleared with ivermectin at 50 ug/kg (approximately 8 times preventative dose) or milbemycin at 500 ug/kg (preventative dose), although this represents an extra-label use of these drugs.[5,5a] The off-label use of livestock formulations of ivermectin is discouraged because of the possibility of dosing errors and resultant toxicity. Adverse reactions, the severity of which is likely related to microfilarial numbers, were observed in 6% of 126 dogs receiving ivermectin at the microfilaricidal dose (50 ug/kg).[30] Signs included shock, depression, hypothermia, and vomiting. With fluid and corticosteroid (dexamethasone at 2-4 mg/kg IV) therapy, all dogs recovered within 12 hours. One fatality was observed 4 days after microfilaricidal therapy. Similar findings and frequency have been reported with milbemycin at the preventative dosage.[31] Dogs so treated should be hospitalized and carefully observed for the day. Dogs <16 kg, harboring >10,000 microfilaria per ml blood, are more apt to suffer adverse reactions.[32] Benadryl (2 mg/kg IM) and dexamethasone (0.25 mg/kg IV) can be administered prophylactically to prevent adverse reactions to microfilaricidal doses of macrolides. A 90% microfilaricidal success rate can be expected with ivermectin[25], while milbemycin at 500 mcg/kg cleared 6/8 (75%) dogs which had received adulticide therapy and did not harbor male and female adults; microfilarial numbers were reduced by 99% on the day after after treatment.[31] A slower microfilarial kill rate can also be achieved with ivermectin, moxidectin, moxidectin-imidacloprid, and selamectin at preventative doses.[1-4] The combination of imidacloprid and moxidectin has recently received FDA approval as a safe and effective microfilaricide. The time-honored approach to ridding the patient of microfilariae involves macrolide therapy (50 mcg/kg for ivermectin or 500 mcg/kg milbemycin) instituted 3-6 weeks after adulticide.[5,5a] In 2-3 weeks, a second microfilaria concentration test is performed and, if negative, preventative started. If still positive, the treatment is repeated or alternatively, chemoprophylaxis begun (assuming that no adverse reaction occurred on the initial treatment).[5,5a] Persistent antigenemia (after 6-7 months) indicates continued patent infection. This author chooses an alternative approach (Figure 1[18a]), beginning the administration of a macrolide preventative at the time of diagnosis, often days to weeks prior to adulticidal therapy. With the "slow microfilaricides" (ivermectin, moxidectin, or selamectin), there is little chance of an adverse reaction, but the owner is warned and advised to administer the medication on a day when he/she will be at home. If Milbemycin (a superior microfilarial agent) is used, it is administered in the hospital and/or preceded by administration of dexamethasone and benadryl, as described above. If this approach is used, the dog must be rendered microfilariae-free by 1-3 months post-diagnosis. Recent evidence demonstrates that concurrent usage of a macrolide and doxycycline reduces microfilarial numbers more rapidly, rendering dogs negative in less than 3 months. It is imperative that dogs on macrocyclic lactones be rendered microfilaria-free.

Figure 1. The author's preferred approach to adulticidal therapy in virtually *all* (severely affected or not) dogs infected with heartworms includes 3 doses of melarsomine **in which finances do not play a role. If finances do play a role, the author prefers the standard 2-dose protocol with doxycycline** for 1 month, followed by another month of preventive only, and then 2 doses of melarsomine (2.5 mg/kg) daily for 2 injections. Macrolide prophylaxis is begun at the time of diagnosis, if not already in use. *If microfilaremic, care should be taken to prevent or observe and treat adverse reactions, based on microfilarial numbers and macrolide used. It is imperative to rid patients receiving macrolides of microfilariae to reduce the chance of resistance.

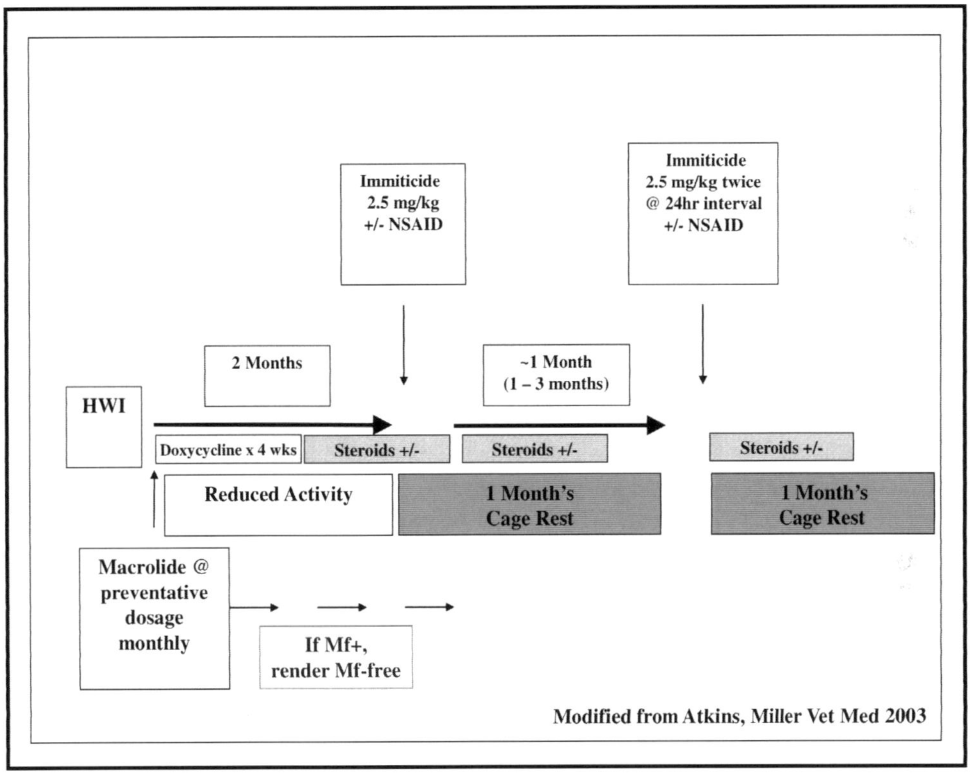

REFERENCES

Available upon request.

FELINE HEARTWORM DISEASE: BASICS TO BREAKTHROUGHS

Clarke Atkins, DVM, DACVIM (SAIM, Cardiology)
Raleigh, NC

Though recognized since 1921[1], heartworm infection (HWI) in the cat has received increasing interest over the last decade. Reasons for this are multiple, with an increasing diagnostic armamentarium, an increasingly aware and demanding public, and the development of safe, effective, and broad-spectrum preventative drugs. With this comes the responsibility of understanding the pathogenesis, clinical signs, diagnostic process, treatment, and most importantly, preventing HWI in cats.

Etiology

The domestic cat, though an atypical host, can be parasitized by *Dirofilaria immitis* (HW) with resultant heartworm disease (HWD). The clinical manifestations of the disease are different and more severe in this species, but the infection rate is only 5-20% of that of the dog.[1] Experimental infection of the cat is more difficult than in the dog; <25% of L3 reach adulthood. This resistance is also reflected in natural infections, in which feline heartworm burdens are usually less than 6, and typically only 1-3 worms.[1] Other indications of the cat's inherent resistance to this parasite are a shortened period of worm patency, high frequency of amicrofilaremia or low microfilaria counts, and shortened life span of adult heartworms (2-3 years).[2] Additionally, while some species of mosquito may feed on cats, most prefer the dog and, for a cat to become infected, the mosquito must first have fed on a dog. Nevertheless, studies have shown a prevalence as high as 10-14% in shelter cats[1,1a] and a study performed at NCSU revealed HWD in 9% of cats presented with cardiorespiratory signs.[3] Furthermore, antibody testing showed 26% of 100 of these cats to have been exposed to HW.[3] Similar to dogs, some studies have shown the male to be at higher risk for HWI than the female. Aberrant worm migration appears to be a greater problem in cats than in dogs.

Pathogenesis

The pathological, clinicopathological, and clinical response to infection with *D. immiits* in cats is not well understood. The pulmonary arterial response to adult heartworms is more severe than that of the dog, although pulmonary hypertension has infrequently been reported. Dillon demonstrated pulmonary enlargement within one week of transplantation of adults, suggesting an intense host-parasite interaction.[4] A severe myointimal and eosinophilic response produces pulmonary vascular narrowing and tortuosity, thrombosis, and possibly hypertension.[5] Because the feline pulmonary artery tree is smaller than that of the dog and has less collateral circulation, embolization, even with small numbers of worms, produces disastrous results with infarction and even death. Although uncommon, cor pulmonale and right heart failure can be associated with chronic feline HWD and is manifested by pleural effusion (hydro- or chylothorax) and/or ascites. The lung per se also is insulted by HWI, with eosinophilic infiltrates in the lung parenchyma (pneumonitis), pulmonary vasculature, and air spaces. The pulmonary vessels may leak plasma producing pulmonary edema, which has been considered by some to represent acute respiratory distress syndrome (ARDS). If the cat survives this initial insult, type II cells proliferate, replacing damaged type I cells, potentially impairing O_2 diffusion.[4] The end result is diminished pulmonary function, hypoxemia, dyspnea, and cough.

Acute or sudden death is typically associated with worm death and fulminant pulmonary failure, possibly associated with pulmonary embolism. Recent research suggests, however, an immune-mediated reaction to HW antigens in the feline shock organ (lung).[a] Fatal respiratory failure probably results when HW antigen is released, producing bronchioloar and bronchial constriction, pulmonary congestion, superficial pulmonary hemorrhage, and periarterial hemorrhage.[a]

"HARD" It has been known since 1996 that cats exposed to heartworms but which reject maturation develop radiographic lesions.[13] Recently, studies of both natural and experimental infections have confirmed this finding.[4a,4b] In natural infections, pulmonary arterial lesions (myointimal proliferation and thrombotic obliteration) have been demonstrated in HW-free, antibody-positive cats [4a] and airway, pulmonary artery and pulmonary interstitial lesions have been demonstrated in cats heavily, experimentally-infected cats in which pharmacological abortion of the infection at the early L_5 stage was performed.[4b] This combination of data from experimental and natural infections indicates that when cats abort infections at the early 5th stage, radiographic and histological changes develop in the lungs, likely producing the most commonly recognized signs in cats, cough, dyspnea and wheezing. The American Heartworm Society has utilized this information in a campaign for heatworm awareness, labeling it "HARD" or "**H**eartworm-**A**ssociated **R**espiratory **D**isease".[b] The fact that larvae that never fully mature can cause disease is important to our understanding of this syndrome. We do not know if the resultant pathology is persistent whether it can result in fatality, and if it explains signs later in the disease course. It also causes confusion as to the exact terminology for HWI as a cat may fall outside the standard concept of "exposed and uninfected" as opposed to "infected". This has led me to invoke a new term for the HWI infection in cats that is aborted at the young L_5 stage: "Pulmonary Larval Dirofilariasis" meaning a HW larval infection that produces disease without full larval maturation.

In recent years, interest has developed in the symbiotic bacterium, *Wolbachia*, found within filarid parasites in all stages.[4c] This bacterium is essential for filarial reproduction and well-being. It has been hypothesized that antigens from these bacteria are proinflammatory, contributing to the HWD, particularly upon the death of the adult HW and that, possibly, treatment of bacteria with tetracycline might be a strategy in the treatment adult dirofilariasis.[4d] Currently, there are no data to support this interesting hypothesis.

Clinical Signs

Cats with HWI may be asymptomatic and, when present, clinical manifestations may be either peracute/acute or chronic.[3,4,6-8] Acute or peracute presentation is usually due to worm death, embolization or aberrant migration and signs variably include salivation, tachycardia, shock, dyspnea, hemoptysis, vomiting and diarrhea, syncope, dementia, ataxia, circling, head tilt, blindness, seizures, and death.

More commonly, the onset of signs is less acute (chronic form). Reported historical findings in chronic feline HWD include anorexia, weight loss, lethargy, exercise intolerance, signs of right heart failure (pleural effusion; rare), cough, dyspnea, and vomiting. We have found dyspnea and cough to be a relatively consistent findings and, when present, should cause suspicion of HWD in endemic areas.[8]

In a report of 50 natural cases of feline HWI seen at North Carolina State University, presenting signs were most commonly related to the respiratory system (32 cats; 64%), with dyspnea (24 cats; 48%) being most often noted, followed by cough (19 cats; 38%), and wheezing.[8] Vomiting was reported in 17 (38%) cats and was noted frequently in 8 (16%). Five (10%) heartworm-infected cats were reported to have exhibited vomiting without concurrent respiratory signs and vomiting was a presenting sign in 7 (14%). Neurological signs (including collapse or syncope, which were described in 5 [10%]) were reported in 7 (14%) cats. Five (10%) of the cats were dead at the time of presentation. Murmurs were infrequently noted in cats that did not have concurrent heart disease, independent of heartworm infection. Heart failure was present in 1 cat but this cat had concurrent hypertrophic cardiomyopathy. Heartworm infection was considered to be an incidental finding in 14 (28%) of these cats.

Physical examination is often unrewarding although a murmur, gallop, and/or diminished or adventitial lung sounds may be audible. In addition, cats may be thin and/or dyspneic. If heart failure is present, jugular venous distension, dyspnea, and rarely ascites are typically detected.

Diagnosis

The diagnosis of HWI/HWD in cats poses a unique and problematic set of issues.[4] First, the clinical signs are often quite different from those seen in the dog In addition, the overall incidence in cats is low, so suspicion is lessened; eosinophilia is transient or absent; electrocardiographic findings are minimal; and most cats are amicrofilaremic.

Immunodiagnostic methods are also imperfect in cats because of the low worm burdens (1-7, mean = 3)[1] and hence, antigenic load. In a recent study, ELISA antigen tests were positive on sera from 36-93% of 31 cats harboring 1-7 female HW, with sensitivity increasing as female worm burden increased.[9] Cats with male worm(s) were not detected as positive. Therefore, false negative tests occur frequently, depending on test used, maturity and gender of worms, and worm burden. All tests were, however, virtually 100% specific. It is important to realize that infection with signs may be present prior to the presence of detectable antigen (from gravid adult females); McCall reports that, in natural infections, the antigen test detects less than 50% of cases.[10] Snyder and colleagues present differing results from cats natural infections in which blood was obtained as long as 2 hours post-euthanasia. In this study the antigen test was found to be more sensitive than previous reports (74%).[11] Recently an antigen test "for cats" (IDEXX's SNAP[R] Feline Heartworm Antigen Test) has been marketed. This is an adaptation of the canine test with a reported increase in sensitivity of 15% over conventional antigen tests.

Though not specific for **mature** infections, heartworm antibody tests are of use in the detection of **exposure** to (and partial development of) heartworms. These tests are, therefore, useful to determine cats at risk for HWI and to determine the potential for HWI in antigen-negative cats, in which HWI is a consideration. This author uses a negative antibody test to "rule out" HWI in cats with HWD-compatible signs. It must, nevertheless, be kept in mind that 14% of cats with a proven natural infection were shown to be the antibody-negative.[8] This "false-negative" number reached 50% in smaller study of 10 cats.[1a] Furthermore, although infrequent (2%), cats may be antibody-negative and antigen-positive, leading some to suggest that the 2 tests be run in tandem when HWI is suspected.[11a] Nevertheless, for routine screening, the antibody test is preferred. There is now an "in clinic" feline heartworm antibody test available (HESKA™ SOLO STEP™ FH).

Thoracic radiographs have been suggested as a screening test for HWI in cats. However, Schafer and Barry showed that the most sensitive radiographic criterion (left caudal pulmonary artery greater than 1.6 times the 9th rib at the ninth intercostal space) was only detected 53% of cases.[12] Furthermore, even though most cats with clinical signs have some radiographic abnormality, the findings are often not specific to HWD. In addition, a study by Selcer, et al. demonstrated that radiographic findings were often transient and that radiographic abnormalities were found in cats which ultimately resisted maturation of HW and were negative on post-mortem (ie "false positive").[13] Radiographic findings include enlarged caudal pulmonary arteries, often with ill-defined margins, pulmonary parenchymal changes include focal or diffuse infiltrates (interstitial, broncho-interstitial, or even alveolar), perivascular density, and occasionally, atelectasis. Pulmonary hyperinflation may also be evident. Pulmonary angiography has also been utilized to demonstrate radioluscent linear intravascular "foreign bodies", as well as enlarged, tortuous, and blunted pulmonary arteries with loss of normal patterns of aborization.

Echocardiography, in our experience, is more sensitive in cats than in dogs.[3,14] Typically, a "double-lined echodensity" is evident in the main pulmonary artery, one of its' branches, the right ventricle, or occasionally at the right atrioventricular junction. We found HW echocardiographically in 78% of 9 naturally occurring cases[3] as did Selcer in 16 experimentally-infected cats.[13] It is important to realize that heartworms most often inhabit the main pulmonary artery or its branches, which require some expertise and an index of suspicion from the sonographer. Heartworm infection can be missed by ultrasound when worms are immature (hence, smaller) or when they have died and compacted into the more distal pulmonary arteries.

Prevention

The question arises as to whether heartworm prophylaxis is warranted for cats because they are not the natural host and because the incidence is low. Necropsy studies of feline HWI in the Southeast have yielded a prevalence of 2.5 to 14% with a median of 7%.[1] When considering the question of institution of prophylaxis, it is worth considering that this prevalence approximates, or even exceeds that of FeLV and FIV infections in comparable populatons.[15] A 1998 nationwide antibody survey of over 2000 largely asymptomatic cats revealed an exposure prevalence of nearly 12%.[16] While outdoor cats are at greater risk for HWI[16], it is noteworthy that, based on owners' information, nearly one-third of cats diagnosed with HWD at NCSU were housed solely indoors.[8] Lastly, the consequences of feline HWD are potentially dire, with no clear therapeutic solutions. Therefore, I advocate preventative therapy in cats in endemic areas. There are now three drugs with FDA approval and which are marketed for use in cats. Ivermectin is provided in a chewable formulation, milbemycin as a flavored tablet and selamectin, a broad-spectrum parasiticide, comes in a topical formulation. Additionally, a new broad-spectrum, topical preventive (moxidectin and imdacloprid) has been used in Australia and Canada and is now approved in this country. The spectrum, as well as the formulation of these products varies, and hence the clients' individual needs are easily met in most cases (Table 1). The risk of an adverse reaction to dying microfilariae is small because of the microfilarial "slow-kill" property of most macrolide preventatives (with milbemycin being the exception) and because most cats are amicrofilaremic or have low microfilarial burdens. Nevertheless, precaution should be taken (i.e., in-clinic or home-observed administration after first macrolide dose) in known microfilaremic cats.

Treatment

The use of arsencial-adulticides is problematic. Thiacetarsemide, if available, poses risks even in normal cats. Turner and colleagues reported death due to pulmonary edema and respiratory failure in 3 of 14 normal cats given of thiacetarsemide (2.2 mg/kg twice over 24 hours).[17] Dillon could not confirm this acute pulmonary reaction in 12 normal cats receiving thiacetarsemide, but 1 cat did die after the final injection.[18] More importantly, a significant, though unquantified, percentage of cats with HWI develop pulmonary thromboembolism after adulticidal therapy.[4,6,7] This occurs several days to a week after therapy and is often fatal. In 50 cats with HWI, seen at North Carolina State University, 11 received thiacetarsemide. There was no significant difference in survival between those receiving thiacetarsemide and those receiving symptomatic therapy.[8]

Data on melarsomine in experimental (transplanted) HWI in cats are limited and contradictory. Although there is an abstract report in which 1 injection (2.5 mg/kg; ½ the recommended canine dosage) of melarsomine was used in experimentally-infected cats without treatment related mortality, the worm burdens after treatment were not significantly different than those found in untreated control cats.[19] Diarrhea and heart murmurs were frequently noted in treated cats. A second abstract report, using either the standard canine protocol (2.5 mg/kg twice over 24 hours) or the "split-dosage" (1 injection, followed by 2 injections, 24 hours apart, in 1 month), gave more favorable results.[20] The standard treatment and split-dosage regimens resulted in 79% and 86% reduction in worm burdens, respectively and there were no adverse reactions. Although promising, these unpublished data need to be interpreted with caution as the transplanted worms were young (<8 months-old and more susceptible), the cats may not have had time to develop antibodies to HW antigens, thereby reducing the risk of anaphylaxis, and the control cats experienced a 53% worm mortality (average worm burden was reduced by 53% by the act of transplantation). Additionally, the clinical experience in naturally-infected cats has been generally unfavorable, with an unacceptable mortality. Because of the inherent risk, lack of clear benefit, and the short life expectancy of heartworms in this species, this author does not advocate adulticidal therapy in cats. Surgical removal of heartworms has been successful and is attractive because it minimizes the risk of thromboemboli. The mortality seen in the only published case series was, unfortunately, unacceptable (2 of 5 cats).[21] This procedure holds promise for the future, however.

Cats with HWI should be placed on a monthly preventative and short-term corticosteroid therapy (prednisone at 1-2 mg/kg q48h-tid) used to manage respiratory signs. If signs recur, alternate day steroid therapy (at the lowest dosage that controls signs) can be continued indefinitely. For respiratory emergencies, oxygen, corticosteroids (dexamethasone at 1 mg/kg IV or IM or prednisolone sodium succinate at 50-100 mg IV/cat) and bronchodilators (aminophylline at 6.6 mg/kg IM q12h, theophylline sustained release at 10 mg/kg PO or terbutaline at 0.01 mg/kg SC) may be employed. Bronchodilators have logic, based on the ability of agents, such as the xanthines (aminophylline and theophylline), to improve function of fatigued respiratory muscles. In addition, the finding of hyperinflation of lung fields may indicate bronchoconstriction, a condition for which bronchodilation would be indicated. Nevertheless, this author does not routinely utilize bronchodilators in feline HWD.

Recently, doxycycline (10 mg/kg/day for 30 days) has been used to clear *Wolbachia* from a heartworm-infected dog with proteinuria and antibodies against *Wolbachia,* in hopes of reducing proinflammatory mediators (e.g. interleukin 8), prior to adulticidal therapy.[21a] While logical, there are no published studies to indicate that the use of doxycline in should become routine practice in the management of HWI.

The use of aspirin has been questioned as vascular changes associated with HWI consume platelets, increasing their turnover rate and effectually diminishing the antithrombotic effects of the drug. Conventional doses of aspirin did not prevent angiographically-detected vascular lesions.[22] Dosages of aspirin necessary to produce even limited histological benefit approached the toxic range. However, because therapeutic options are limited; because at conventional doses (80 mg PO q72h), aspirin is generally harmless, inexpensive, and convenient; and because the quoted studies were based on relatively insensitive estimates of platelet function and pulmonary arterial disease (thereby possibly missing subtle benefits), the author continues to advocate aspirin for cats with HWI. Aspirin is **not** prescribed with concurrent corticosteroid therapy.

Since the vast majority of cats are amicrofilaremic, microfilaricidal therapy is unnecessary in this species. Management of other signs of HWD in cats is largely symptomatic.

Prognosis

In the aforementioned study of 50 cats with natural heartworm infection, at least 12 cats died of causes other than heartworm disease. Seven of these and 2 living cats were considered to have survived heartworm disease (lived ≥1000 days).[8] The median survival for all heartworm-infected, cats living beyond the day of diagnosis, was 1460 days (4 years; range 2-4015 days), while the median survival of all cats (n=48 with adequate follow-up) was 540 days (1.5 years; range 0-4015 days). Survival of 11 cats treated with sodium caparsolate (mean 1669 days) was not significantly different from that of the 30 managed without adulticide (mean 1107 days). Likewise, youth (≤ 3 years of age), presence of dyspnea, cough, ELISA-positivity for heartworm antigen, presence of echocardiographically-identifiably worms, or gender of the cat did not appear to affect survival.[8] The effect of HWI on survival has been compared to that of other cardiovascular diseases.[23] Overall, the prognosis for HWI in cats is comparable to that of hypertrophic cardiomyopathy, the most benign of primary feline heart diseases.

Table 1. Spectra comparisons of "macrolide' heartworm preventives currently in use in cats.

	Heartworms	Hookworms	Roundworms	Fleas	Ear Mites	
Ivermectin	+	+				chewable tablet
Milbemycin*	+	+	+			flavored tablet
Selamectin	+	+	+	+	+	topical
Imidacloprid/ Moxidectin	+	+	+	+	+	topical

*Not currently marketed for cats

REFERENCES

Available upon request.

RENOCARDIOVASCULAR FUNCTION: BEYOND RAAS

Geoff Culshaw, BVMS, DVC (RCVS Specialist in Veterinary Cardiology), MRCVS
Edinburgh, UK

INTRODUCTION

In human cardiology, the rise of renal disease is a major cause for concern. Chronic kidney disease (CKD) has reached epidemic proportions worldwide, with approximately 11% of people in the western world having some degree of renal dysfunction.[1] Terminal renal failure, however, is not the most common cause of death in these patients. Most patients with CKD die from cardiovascular disease.[2]

This link between cardiovascular disease and renal disease is typified in cardiorenal syndrome in which disease or dysfunction in the heart or kidneys results in disease or dysfunction in the other organ, in other words, the dysfunction is bidirectional.[3] As veterinary cardiologists, we are used to managing patients with congestive heart failure that have concurrent renal dysfunction which we usually ascribe to a combination of poor cardiac output, altered renal haemodyamics brought about by our therapy and, in unlucky patients, pre-existing renal disease. The corollary is that renal dysfunction enhances pre-existing cardiac dysfunction through RAAS by increasing preload and afterload. Cardiorenal syndrome though is also commonly observed in people with chronic systemic disorders such as obesity or diabetes, themselves worldwide epidemics and major contributors to the increase in CKD. There is strong evidence that the unifying link between these disease states (commonly grouped under the term metabolic syndrome) is vascular dysfunction that leads to hypertension. Hypertension is the single biggest contributor to global disease burden[4] and is an independent predictor of deteriorating renal function[5] and cardiovascular risk[6]. The kidneys are where vascular function and regulation of blood pressure integrate. Further understanding of this dynamic relationship is useful to cardiologists assessing the impact of our cardiac therapy on serum biochemistry and the impact of renal and systemic disorders on vascular function and cardiac disease.

EXTRINISIC AND INTRINSIC RENAL REGULATION

The kidneys are unique in that they are the only organ in the body to have 2 separate capillary networks in series to one another- the glomerulus and the *vasa recta*. These networks control excretion of waste products and regulate electrolyte, water and acid/base balance. Although highly effective, in many ways they are far from efficient, consisting of a relatively non-selective filter followed by a selective reabsorption system. This arrangement is energy dependent and means that the kidney requires a high proportion of cardiac output (approximately 20-25%) and is susceptible to the effects of poor perfusion and hypoxia (although it routinely operates within hypoxic conditions).

Renal regulation at the level of the glomerulus

In its vascular series circuit, the kidney prioritises glomerular filtration over perfusion of the *vasa recta*. Glomerular filtration operates independently over a range of systemic arterial blood pressures, termed autoregulation. Downstream of the glomerulus, the tubular network's *vasa recta* are perfused by blood devoid of filtrate which gradually replenishes itself selectively and non-selectively of water, ions and small molecules as it supplies the medulla, while still performing an additional excretory function, particularly of larger organic molecules that are too large to pass through the glomerular filter.

Glomerular filtration rate (GFR) is tightly regulated by variations in tone of the afferent and efferent arterioles that supply and remove the arterial blood of the glomerulus. Afferent arterioles have an inherent ability to dilate and constrict according to blood flow (myogenic response), which is modified by sensory cells in the distal nephron (*macula densa*) that are sensitive to the level of filtration (especially sodium ions). This is termed tubuloglomerular feedback (TGF). TGF represents a balance between the activities of 3 key signalling molecules released from the *macula densa*. As the concentration of sodium within the distal nephron increases (suggesting GFR is too high), release of dilatory PGE_2 wanes and ATP release increases, promoting afferent arterioconstriction and lowering of GFR. Further increases in sodium concentration within the distal nephron result in paradoxical release of dilatory nitric oxide. This may serve to ensure that the *vasa recta* achieves adequate blood flow without increasing GFR by dilating efferent as well as the afferent arteriole. TGF introduces the concept of ATP (and other purinergic molecules) as major players in cellular signalling rather than simply an energy store. Further evidence for this comes from the high density of mitochondria within *macula densa* cells despite the cell being relatively inactive metabolically. In fact, the importance of ATP, ADP and AMP signalling within the kidney and the plethora of different purinergic receptor subtypes is very much *in vogue* in human renal physiology.

Renal regulation via filtrate flow

The efferent arterioles are sensitive to a range of vasoactive agents such as angiotensin II. Because GFR is prioritised over tubular perfusion, tubular function and health are affected for better or worse by endogenous or exogenous manipulation of afferent or efferent arteriolar tone.

Downstream of the glomerulus, renal control of blood pressure and sodium and water balance is tightly regulated both extrinsically and intrinsically via extremely complex neurohumoral and paracrine/autocrine interactions. While extrinsic regulation, unsurprisingly, comes from the renal arterial and neural (particularly sympathetic) input, intrinsic regulation is now known to include two-way vascular/tubular interaction and even tubular/filtrate communication. The importance of the circulatory renin-angiotensin-aldosterone system (RAAS) has been redefined in recent years as an extrinsic regulator of intrinsic RAAS activated within the proximal convoluted tubule. This then leads to a cascade of RAAS-mediated events as angiotensin II secreted into the tubule is carried along by the filtrate along the length of the nephron binding to receptors as it goes and modifying activity of different sodium channels to increase sodium (and water) reabsorption. Another example of the filtrate acting as a vehicle for intrinsic renal regulation involves urinary exosomes. These lipid bilayer membrane bound microvesicles bud off from renal tubular cells along the length of the nephron and encapsulate soluble proteins, mRNA and miRNA. Originally thought of an additional component of renal excretion in which tubular cells could eliminate unwanted intracellular material, it is now believed that urinary exosomes may act as shuttles of important proteins and genetic material allowing manipulation of cellular activity distally and downstream by the proximal and upstream tubular cells. Thus urinary exosomes and their proteome may be an extremely useful therapeutic target and biomarker.

Additional renal intrinsic factors and their significance will be outlined below (Fig. 1).

Predicting the impact of diuresis on renal function

Dysregulation of blood pressure is associated with impairment of the acute pressure natriuresis response[7], an important homeostatic function of the kidney influencing the set point of blood pressure.[8] After ingestion of salt and water, plasma volume expands and blood pressure rises slightly. GFR is autoregulated by efferent arteriolar dilation and blood flow to the tubules and *vasa recta* is enhanced. Post efferent flow is shunted from the renal cortex to the medulla.[9] Enhanced blood flow increases medullary interstitial hydrostatic pressure inhibiting sodium reabsorption from the ascending loop of Henle and promoting natriuresis and diuresis. This mechanism (and hence regulation of blood pressure) is clearly susceptible to reduced cardiac output, volume overload cardiac disease and the therapy to treat it.

Normally administration of excessive diuretic leads to pre-renal azotaemia typified by elevated serum urea and normal or only slightly raised creatinine. Both urea and creatinine are filtered through the glomerulus but only urea is reabsorbed, principally because it is a smaller molecule. The main sites of reabsorption are the proximal convoluted tubule and the inner medullary collecting duct. Urea follows water, so where blood volume is excessively lowered by diuresis and a compensatory anti-diuretic hormone (ADH) response is initiated, there is enhanced urea and water reabsorption from the renal filtrate, raising serum urea.[8] However, this rise in urea may be dampened by the effect of diuresis. Increased renal filtrate flow in the distal nephron and, particularly, reduced water reabsorption will increase urea excretion. Furthermore, if perfusion of the *vasa recta* is compromised by the lowered blood volume and preferential maintenance of GFR, urea reabsorption will be reduced according to the Fick principle. If efferent tone is reduced by adding in an ACE inhibitor with a compensatory reduction in afferent arteriolar tone to maintain GFR, tubular blood flow will be increased. This will increase reabsorption of urea according to the Fick principle but, equally, increase urea excretion through activation of pressure natriuresis. The result is the potential for marked inter and intra-patient variability in renal serum biochemistry. The effect on a single sample taken at one moment in time will depend on the fluid status of the patient and the distribution of fluid between different body compartments, the timing of diuresis and ACE inhibition and their pharmacokinetics in an individual, access to water and food along with appetite and thirst, environmental temperature and many other variables such as the rate of urea and creatinine production. This would explain why patients can have widely fluctuating renal parameters, marked changes in renal parameters for only modest changes in medication and the poor correlation in many patients between renal parameters and well-being. Thus serum biochemistry in cardiac patients receiving medication should be interpreted with caution.

We need more accurate biomarkers of GFR and tubular flow in order to improve our understanding and therapeutic management of CHF and optimise renal function for as long as possible. An example of these would be creatinine clearance testing or measurement of inulin and para-amino hippurate (PAH) clearance. These are relatively cumbersome and time-consuming techniques that do not necessarily lend themselves to our veterinary patients. Unfortunately, alternative biomarkers or imaging techniques that accurately measure GFR and effective renal plasma flow are currently too costly for routine use in the clinic.

An alternative to RAAS

The kidney clearly plays a crucial role in the pathogenesis of CHF, mediating the increase in blood volume that increases cardiac preload as a compensatory measure for cardiac dysfunction. Many effects ascribed to the mineralocorticoid aldosterone can be performed by cortisol, which has the same affinity for mineralocorticoid receptor but is usually at 1000 times the level of aldosterone in plasma. Mineralocorticoid receptor is protected from cortisol's influences by the membrane bound enzyme 11 beta hydroxysteroid dehydrogenase 2 (11βHSD2) which converts cortisol to inactive cortisone.[10] Changes in the expression of 11βHSD2 have resulted in phenotypes similar to those displayed in metabolic syndrome and heterozygous and homozygous knockout mouse models develop sodium and water retention and salt sensitive hypertension.[11] We are currently investigating whether changes in cortisol metabolism at the tissue level, via changes in 11βHSD2 activity, occur in CHF which can be counteracted by mineralocorticoid receptor inhibition. As well as offering the potential of a novel therapeutic target in the form of 11βHSD2, confirmation would lead naturally into a more widespread investigation of cortisol's influences in the systemic vasculature and the influence of states of dysregulated metabolism such as diabetes and other endocrinopathies, and obesity (of increasing significance in the veterinary patient population) in which cortisol is thought to play a major role in people

RENAL AND CARDIOVASCULAR DISEASE

Pressure natriuresis is key to regulation of blood pressure. It has been proposed that impaired pressure natriuresis leads to higher blood pressure in people during the night.[7] This loss of nocturnal dipping predicts deterioration in renal function.[12] The vascular dysfunction that underlies impaired pressure natriuresis is currently under investigation. Two key intrinsic paracrine/autocrine pathways are good candidates that have profound effects on renal tubular and vascular function as well as being ubiquitously expressed throughout the cardiovascular system. These are the endothelin and purinergic systems.

Renal endothelin signalling

Renal endothelin-1 (ET-1) is predominantly expressed in the inner medullary collecting duct. Probably released in response to renal filtrate flow and sodium concentration [13,14], it acts on neighbouring endothelin B receptors inhibiting the epithelial sodium channel, ENaC, and the action of ADH. Overspill ET-1 may interact with vasculature in a negative feedback loop on pressure natriuresis via constrictive pericytes around the *vasa recta*.[15]

Renal purinergic signalling

The purinergic system (ATP and ADP) has multiple regulatory roles along the entire length of the nephron, including TGF and control of urinary pH. There is now strong evidence that ATP released form the medullary thick ascending limb buffers the effects of vasoconstricting agents such as angiotensin II on pericytes of the *vasa recta*.[16] Such a buffering effect enhances diuresis. Where the signalling is dysfunctional, there is increased susceptibility to hypertension and loss of circadian rhythm of blood pressure. ATP and ADP are also produced by fibroblasts and endothelial cells in response to mechanical forces and act via ERK-dependent pathways. Purinergic pro-fibrotic responses are regulated by ectonucleoside tri/diphosphate hydrolases (ENTPD), with the eventual production of adenosine, which is anti-fibrotic through its activation on adenosine (A2) receptors.

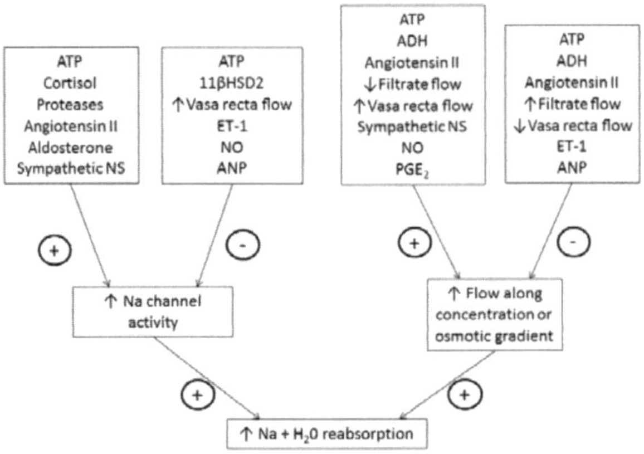

Figure 1. Intrinsic and extrinsic regulatory factors of sodium and water reabsorption. This is not a comprehensive list but illustrates the complexity of sodium regulation and hence regulation of blood pressure. Note that some factors appear in more than one box and often elicit contrary effects.

Common pathways in the kidney and the mitral valve

Systemic vascular dysfunction and mediation of fibrosis may be directly relevant to the pathogenesis of myxomatous mitral valve disease (MMVD). The mitral valve apparatus is an endothelial lined structure consisting of fibroblast-type cells (valvular interstitial cells, VICs), collagen, mucopolysaccharide, cardiac muscle and neuronal fibres.[17] We have previously shown that MMVD is characterised by dyscollagenesis, VIC phenotypic changes and endothelial cell damage and loss.[18-21] The series of events that lead to matrix degeneration are not known but are presumed to be driven by VIC-dependent catabolic processes and ineffective collagen production and maturation. Unlike myocardial fibroblasts, VICs do not respond to valve injury by initiating fibrosis. We have demonstrated VIC proliferation and accumulation at the valve edge close to damaged endothelium but also have transcriptomic and immunohistochemical data to suggest that endothelial-to-mesenchymal transition is contributing to this cell population. We also suspect that endothelial cell-VIC signalling contributes to disease pathogenesis. Investigation of endothelin and purinergic signalling within MMVD is ongoing, using techniques acquired from studying these systems in the kidney. Whether MMVD might represent part of a systemic endothelial disorder that also impacts on renal function and exerts systemic effects via loss of regulation of blood pressure remains to be determined, but this "unifying theory" of endothelial dysfunction is an attractive one.

BLOOD PRESSURE AS A BIOMARKER OF RENAL AND VASCULAR FUNCTION

Arterial blood pressure results from left ventricular outflow against a high resistance systemic vascular circuit, and can be considered as an index of the body's ability to maintain blood flow to vital organs. In general, it is regulated in the short term by changes in vascular (particularly arteriolar) vascular tone and in the long term by changes in blood volume via renal regulation of sodium balance, which in turn is dependent upon GFR, and intrinsic and extrinsic regulation of sodium channel activity including renal tubulovascular interactions. In people, hypertension is the single largest contributor to global disease burden and it independently predicts morbidity and mortality in many cardiovascular and renal disease states. There have been major advances in the measurement of blood pressure in small animals in general practice but there are still significant limitations. Circadian variation in blood pressure, particularly nocturnal dipping, which appears to be very important in people has not been fully investigated in healthy pet dogs in the home environment let alone in dogs in disease states. We are currently investigating the feasibility of invasive measurement of blood pressure in pet dogs using radiotelemetry implants placed into a peripheral artery. There remain significant practical and ethical obstacles and the role of blood pressure as a biomarker and as a risk factor for morbidity will not be fully determined until these are overcome.

REFERENCES

[1]Hallan SI, et al **J Am Soc Nephrol** 2006;17:2275. [2]Helset E et al **Mediators Inflamm** 1993;2:417. [3]Ronco C et al **J Am Coll Cardiol** 2008;52:1527. [4]Lim SS et al **Lancet** 2012;380:2224. [5]Mogensen CE et al **N Engl J Med** 1984;311:89. [6] UK Prospective Diabetes Study Group. **BMJ** 1998;317:703. [7]Fukuda M et al **Kidney Int** 2004;65:621. [8]Guyton AC et al **Textbook of Medical Physiology**. 11th edn 2006. [9]Roman RJ at al **Hypertension** 1988;12:168. [10]Stewart PM et al **Vitam Horm** 1999;57:249. [11]Walker BR et al **Clin Endocrinol (Oxf)** 1993;39:221. [12]Lurbe E et al **N Engl J Med** 2002;347:797. [13]Lyon-Roberts B et al **Am J Physiol Renal Physiol** 2011;300:F650. [14]Pandit MM et al **Am J Physiol Renal Physiol** 2012;302:F1325. [15]Silldorff EP et al **J Clin Invest** 1995;95:2734. [16]O'Connor PM et al **Hypertension** 2012;60:965. [17]Culshaw GJ et al **Am J Vet Res** 2010;71:194. [18]Black A et al **Am J Vet Res** 2005;66:1408. [19]Corcoran BM et al **Am J Vet Res** 2004;65:198. [20]Han RI et al **Vet J** 2013;197:388. [21]Han RI et al **Vet J** 2013;197:502.

CARDIORENAL SYNDROME: A NEW APPROACH TO AN OLD PROBLEM

Joao S. Orvalho, DVM, DACVIM (Cardiology), Larry D. Cowgill, DVM, PhD, DACVIM (SAIM)
San Diego, CA

INTRODUCTION

With development of water and salt preserving mechanisms in animals, a relationship between the cardiovascular system and kidneys was established. Only recently, it has become recognized increasingly by cardiologists and nephrologists that there are important bidirectional functional and pathological interactions between these systems. These systems responsible for maintaining hemodynamic balance, when maladaptive, become mutual and interactive participants in the pathophysiology of cardiovascular and renal disease. The clinical consequences of these interactions have gained attention and have prompted further definition, classification, and understanding of the relationship, and are the bases for the clinical entity termed Cardiorenal Syndrome (CRS) in human medicine. [1-4] Cardiorenal Syndrome has not been well characterized in veterinary medicine, but a recent attempt has been made to define a consensus for cardiovascular-renal disorders (CvRD) of the dog and cat.

A DEFINITION

The definition of CRS includes a variety of acute or chronic conditions in which the primary failing organ: the heart, the kidney, or both (due to a systemic condition), promotes dysfunction and/or failure of the other organ system.[4]

CLASSIFICATION

Five subtypes of CRS have been suggested based on the organ affected by the primary insult and subsequent time frame of development. [3-5]

Table 1. Human and Veterinary Classification of Cardiorenal Syndrome

Type of Cardiorenal Syndrome (Human Classification)	**CvRD Consensus**	**Brief definition**	**Conditions**
Type 1 - Acute cardiorenal syndrome	$CvRD_H$ unstable	Acute impairment of the cardiac function leading to acute kidney injury (AKI)	Acute heart failure Cardiogenic shock
Type 2 - Chronic cardiorenal syndrome	$CvRD_H$ stable	Chronic cardiovascular disease causing progressive chronic kidney disease (CKD)	Chronic heart failure “Congestive kidney failure”
Type 3 – Acute renocardiac syndrome	$CvRD_K$ unstable	Acute primary worsening of kidney function that leads to cardiac dysfunction.	Acute kidney injury Hyperkalemia, uremia
Type 4 – Chronic renocardiac syndrome	$CvRD_K$ stable	Primary chronic kidney disease that contributes to cardiac dysfunction	Chronic glomerular disease, anemia, syst. hypertension
Type 5 – Secondary cardiorenal syndrome	$CvRD_O$	Cardiac and renal dysfunction secondary to an acute or chronic systemic condition	Diabetes mellitus Sepsis

Type 1 CRS - Acute Cardiorenal Syndrome is characterized by a rapid impairment of cardiac function leading to acute kidney injury. $CvRD_H$ as classified by the veterinary consensus statement, consists of numerous and complex mechanisms by which acute heart failure or acutely decompensated chronic heart failure leads to acute kidney injury (AKI).[6] The onset of AKI may result from reduced renal perfusion secondary to decreased left systolic function[7], neuroendocrine and sympathetic systems activation, and passive renal congestion. The congestive state may also induce decreased diuretic responsiveness, which may lead to excessive use of diuretics and further kidney injury.[8,9] In CRS Type 1 and $CvRD_H$, the early recognition of AKI remains a challenge due to the lack of early stage biomarkers. Serum creatinine (sCr) and blood urea nitrogen (BUN) have been the classic markers for AKI, but when the concentrations of these markers are detectably elevated the injury process is well established, and it is often too late to protect the kidney or prevent progressive damage. The discovery and validation of novel urinary biomarkers such as neutrophil gelatinase-associated lipocalin (NGAL), cystatin c, symmetrical dimethylarginine (SDMA) and others, may allow an earlier recognition of AKI and CRS.[10,11]

Type 2 CRS - Chronic Cardiorenal Syndrome is caused by chronic cardiovascular disease leading to progressive chronic kidney disease (CKD).[3-5] Chronic heart failure is likely to lead to persistently reduced renal perfusion, chronic renal congestion (“congestive nephropathy”), and neurohormonal changes associated with chronic sympathetic stimulation (production of epinephrine, angiotensin, endothelin, and release of natriuretic peptides and nitric oxide).[13,14] Congestive heart failure (CHF) therapy using diuretics and renin-angiotensin aldosterone system (RAAS) blockers can cause drug-induced hypovolemia and hypotension.[6] In humans, the prevalence of renal dysfunction in CHF patients is high and constitutes an independent predictor of outcomes and mortality; therefore it is of the upmost importance to preserve the renal function in these patients.[15]

Type 3 CRS – Acute Renocardiac Syndrome is characterized by a primary acute worsening of kidney function that leads to acute cardiac dysfunction. AKI can affect cardiac function through multiple mechanisms, such as fluid overload, electrolyte disturbances, neurohormonal activation, and myocardial depressant factors, potentially contributing to the development of arrhythmias, pericarditis and acute heart failure.[16,17] Diagnosis of AKI in patients concurrently treated for heart failure may force clinicians to reduce the dose or discontinue heart failure medications, further decompensating the cardiovascular system in order to prevent additional kidney injury.

Type 4 CRS– Chronic Renocardiac Syndrome consists of primary chronic kidney disease that contributes to cardiac dysfunction. Decreased systolic function, left ventricular hypertrophy, systemic hypertension, and a high output state (secondary to anemia) are some of the potential long-term cardiac sequelae of CKD.[5,12] The medical management of CKD and concurrent CHF is not as problematic as the acute forms of these conditions, but most CHF patients are likely undertreated, due to concerns of further worsening renal function and creating a vicious cycle of bidirectional damage secondary to the specific pathophysiology and pharmacotherapy of both conditions.

Type 5 CRS – Secondary Cardiorenal Syndrome is characterized by cardiac and renal dysfunction secondary to an acute or chronic systemic condition. Sepsis is the most common acute condition that affects both the heart and the kidney.[5,18,19] Diabetes mellitus and hyperadrenocorticism are typical chronic diseases in dogs that have a similar effect on the urinary and cardiovascular systems.[20-22]

EVALUATION OF CARDIOVASCULAR-RENAL DISORDERS IN THE DOG

The aim of the study is to use existing clinical instruments for the assessment of cardiovascular and kidney disease in addition to novel biomarkers with the potential to document concurrent or induced specific short-term and long-term injury to these respective organ systems in response to the onset of dysfunction of either organ or in response to its therapeutic management.

HYPOTHESIS

In dogs like in people, intrinsic dysfunction of the cardiovascular system and/or its management promotes secondary injury and dysfunction of the kidney ($CvRD_H$). In addition, intrinsic dysfunction of the kidney and/or its management promotes secondary injury and dysfunction of the cardiovascular system ($CvRD_K$).

METHODOLOGY

Dogs presented to the University of California Veterinary Medical Center-San Diego (UCVMC-SD) for the initial evaluation of stable or unstable cardiovascular or kidney disease will be screened for enrolment in the study. The dogs of either sex, between 6 months and 12 years of age will be categorized accordingly to the ACVIM cardiac disease severity classification (Table 2), IRIS AKI grading (Table 3), and IRIS CKD staging (Table 4) classification.[11,31-33] The patients will be enrolled prospectively for further study. Dogs presenting for interventional cardiovascular procedures will also be enrolled prospectively for further study.

Table 2. ACVIM cardiac disease classification.

ACVIM Classification	Class A	Class B_1	Class B_2	Class C_1	Class C_2	Class D_1	Class D_2
Brief Definition	Patients at risk	Asymptomatic No cardiomegaly	Asymptomatic Cardiomegaly	Heart failure Hospitalized	Heart failure At home	Refractory HF Hospitalized	Refractory HF At home

Table 3. IRIS acute kidney injury (AKI) grading criteria

AKI Grade	Serum creatinine	Clinical description
Grade I	< 1.6 (mg/dL) (<140 µmol/l)	**Non Azotemic AKI:** - Documented AKI: (Historical, clinical, laboratory, or imaging evidence of acute kidney injury, clinical oliguria/anuria, volume responsiveness) and/or - Progressive non azotemic increase in serum creatinine; ≥0.3 mg/dL (≥ 26.4 µmol/l) within 48 hours - Measured oliguria (<1 ml/kg/hr) or anuria over 6 hrs
Grade II	1.7 – 2.5 mg/dL (141-220 µmol/l)	**Mild AKI:** Documented AKI and static or progressive azotemia Progressive azotemic increase in serum creatinine; ≥0.3 mg/dL (≥ 26.4 µmol/l) within 48 hours), or volume responsiveness - Measured oliguria (<1 ml/kg/hr) or anuria over 6 hrs
Grade III	2.6 – 5.0 mg/dL (221-439µmol/l)	**Moderate to Severe AKI:** Documented AKI and increasing severities of azotemia and functional renal failure
Grade IV	5.1 – 10.0 mg/dL (440-880 µmol/l)	
Grade V	> 10.0 mg/dL (>880 µmol/l)	

Courtesy IRIS Board

Biomarkers

Since there are no diagnostic tests available to predict worsening renal function (WRF). Renal dysfunction is inferred from measurements of serum creatinine or calculated GFR.

The ideal marker of renal injury in heart failure should be a sensitive and specific early marker renal injury as well as disease severity, morbidity and mortality. It should detect changes in renal impairment and renal function and allow guidance of therapy.

Some of the currently described markers of glomerular filtration are BUN, creatinine, cystatin c, and SDMA. To evaluate glomerular damage, proteinuria and microalbuminuria can be used; and for tubular injury, urinary NGAL and cystatin c show promise.

Cardiac biomarkers such as N-terminal pro-B-type natriuretic peptide (NT-proBNP), which reflects ventricular wall stress and correlates loosely with left ventricular filling pressures, and cardiac troponin-I (cTnI), a well-established indicator of cardiac disease or injury, may help guide therapy.[26-30] Identification of novel tubular biomarkers of early kidney injury, such as NGAL, clusterin and cystatin c, have promise for the recognition of CRS and management of both conditions.[5,10,11] Therefore, NT-proBNP, cTnI, NGAL, SDMA, clusterin, and cystatin c were measured at each data point.

The preliminary results of the study will be presented and discussed during the session.

Table 4. IRIS chronic kidney disease (CKD) classification

CKD Stage	Serum creatinine (mg/dl)	Comments
Stage I	<1.4 Dogs <1.6 Cats	**Non-azotemic** Some other renal abnormality present, e.g: inadequate urinary concentrating ability without identifiable nonrenal cause; abnormal renal palpation and/or abnormal renal imaging findings; proteinuria of renal origin; abnormal renal biopsy results; increasing blood creatinine concentrations in samples collected serially
Stage II	1.4-2.0 Dogs 1.6-2.8 Cats	**Mild renal azotemia** (lower end of the range lies within reference ranges for many laboratories but the insensitivity of creatinine as a screening test means that animals with creatinine values close to the upper reference limit often have excretory failure) Clinical signs usually mild or absent
Stage III	2.1-5.0 Dogs 2.9-5.0 Cats	**Moderate renal azotemia** Many extrarenal clinical signs may be present
Stage IV	>5.0 Dogs and Cats	**Severe renal azotemia** Many extrarenal clinical signs are usually present

Courtesy, J. Elliott and A.D.J. Watson, and the IRIS Board

MANAGEMENT OF THE CARDIORENAL SYNDROME PATIENT

In patients with cardiovascular disease the renal hemodynamics are affected and considered the main contributor to secondary kidney disease.[13,23] A low cardiac output stage triggers neurohormonal activation through the sympathetic nervous system (SNS) and the RAAS. This process leads to initial preservation of the glomerular filtration rate (GFR), but subsequently decreased renal perfusion and secondary hypoxic tubular injury promotes loss of nephrons. These systems also promote water and sodium retention, elevating central venous pressure and organ congestion ('congestive nephropathy'). Renal congestion in turn also stimulates the SNS and RAAS (and likely increases in renal parenchymal pressure contributing to progressive reductions of GFR and tubular fluid flow.[13, 23-25]

Research should focus on clinically useful knowledge in order to understand the hemodynamic drivers of cardiorenal interactions and how renal function should influence therapeutic decisions. In human medicine, low forward flow or decreased cardiac output (CO) do not appear to be consistently associated with WRF, and central venous pressure (CVP) is not a good way to access congestion induced renal dysfunction. The best marker for congestion induced renal dysfunction is intra-abdominal pressure,[13] which may represent a useful diagnostic and therapeutic target, but additional research is necessary to validate this marker. WRF is also strongly related with BP changes and a reduction in blood pressure appears to be the strongest hemodynamic driver for worsening of renal function.[34]

Abnormal kidney function may lead to metabolic and functional disturbances in multiple organs, which include the cardiovascular system. These processes may directly affect the heart, such as systemic hypertension causing decreased systolic function, left ventricular hypertrophy and diastolic dysfunction, or indirectly by reducing the production of erythropoietin leading to anemia.[35-37]

Cardiologists treat a significant number of patients with AKI and CKD, but there is no established protocol for the treatment or monitoring of cardiovascular disorders in these patients. Patients with kidney dysfunction may receive suboptimal treatment for concurrent cardiovascular conditions, even though they may benefit from the standard therapies. This may account in part for the worsening prognosis attributed to patients with renal disease. In human medicine, early referral to a nephrologist is associated with a delayed progression of CKD. CRS patients may benefit from co-management by cardiologists and nephrologists.[38]

Angiotensin converting enzyme-inhibitors and angiotensin receptor blockers (ARBs) are beneficial in cardiovascular and renal diseases, but patients with renal dysfunction are less likely to receive this type of drugs due to the concern of worsening renal function.[39-42] A better understanding of the relative risk of using these and other drugs may be very important in CRS patients. Worsening renal function that occurs in the context of a therapeutic maneuver may not represent an adverse prognosis. The patient should be treated and if a mild to moderate increase in serum creatinine occurs, that should be expected and may be tolerated-within limits.

Mineralocorticoid receptor antagonists (aldosterone blockers) have the potential for renal and cardiac protection, therefore the use of spironolactone in this subset of patients may be beneficial.[31] Loop diuretics may have conflicting effects on the renal function. By reducing renal congestion they may improve GFR and delay the progression of CKD, but on the other hand, excessive doses of diuretics may also decrease renal perfusion and consequently GFR. The combination of loop diuretics and thiazide diuretics has a synergistic effect that may cause excessive volume depletion and electrolyte disturbances, therefore they should be used with caution in the CRS patient. Pimobendan improves the systolic function, which may increase GFR. Pimobendan does not enhance or suppress furosemide-induced RAAS activation.[40,43] Digoxin and other drugs with predominant renal excretion may require closer monitoring and potential reduction of the dose. Omega-3 fatty acids are a recommended oral supplement that has been used as an antioxidant and appetite stimulant in patients with heart and kidney disease.[22-35]

CONCLUSIONS AND FUTURE DIRECTIONS

An accurate appreciation of the kidney and the cardiovascular system and their interactions may have practical clinical implications. A multidisciplinary evaluation including the expertise of cardiologists and nephrologists may be the most appropriate approach for the patient at risk for CRS.[44] The outcome of CRS patients is likely to improve with the increasing awareness and ability to identify and understand the pathophysiological characteristics of cardiorenal syndrome.

REFERENCES

[1]Schrier RW. Nat Clin Pract Nephrol 2007;3:637. [2]Berl T, Henrich W. Clin J Am Soc Nephrol 2006;1:8 –18. [3]Ronco C. Int J Artif Organs 2008;31:1–2. [4]Berl T, Henrich W.. Clin J Am Soc Nephrol 2006;1:8 –18. [5]Ronco, C. et al.. J Am Coll Cardio. 2008; 52:1527–39 [6]Liang KV, Williams AW,et al. Crit Care Med 2008;36:S75– 88. [7]Jose P, Skali H, Anavekar N, et al. Am Soc Nephrol 2006;17:2886 –91. [8]Ellison DH. Semin Nephrol 1999;19:581–97. [9]Almeshari K, Ahlstrom NG, Capraro FE, Wilcox CS. J Am Soc Nephrol 1993;3:1878–83. [10]De Loor, J. et al. J Vet Intern Med 2013;27:998–1010. [11]Segev, G.,et al. J Vet Intern Med 2013;27:1362–1367. [12]Graziani, G. et al. Heart Fail Rev. 2013. [13]Mullens et al. J Am Coll Cardiol. 2009, 53:589-96. [14]Jessup, M. Costanzo,M. 2009;7:597–9. [15]Hillege HL, Nitsch D, Pfeffer MA, et al. Circulation 2006;113:671– 8. [16]Blake P, et al. ASAIO J.1996;42:M911–5. [17]Meyer TW, Hostetter TH. Uremia. N Engl J Med 2007;357:1316–25. [18]Langhorn, R. et al. J. Vet. Intern. Med. Am. Coll. Vet. Intern. Med. 2013; 27:895–903. [19]RJ et al. J Immunol 2002;168:5817–23. [20]Waldum, B. Os, I. Cardiology 2013;126:175–186 [21]Atkins, C. E. Vet. Clin. North Am. Small Anim. Pract. 1991. 21, 1035–1080. [22]Polzin, D. J. Vet. Clin. North Am. Small Anim. Pract. 2011; 41, 15–30 [23]Damman K,et al. Eur J Heart Fail 2007; 9: 872–878. [24]Ruiz-Ortega, M. et al. Kidney Int. Suppl. 2002; S12–22 [25]Remuzzi, G., Perico, N.et al. Kidney Int. Suppl. 2002; S57–65 [26]Sharkey, L. C. et al. J. Am. Vet. Med. Assoc. 2009; 234:767–770. [27]Häggström, J. et al. J. Vet. Cardiol. Off. J. Eur. Soc. Vet. Cardiol. 2000; 2, 7–16 [28]Serres, F. et al. J. Vet. Cardiol. Off. J. Eur. Soc. Vet. Cardiol. 2009;.11:103–121 [29]Schmidt, M. K. et al. J. Vet. Cardiol. Off. J. Eur. Soc. Vet. Cardiol. 2009.; 11 Suppl 1, S81–86 [30]Hezzell, M. J. et al. J. Vet. Intern. Med. Am. Coll. Vet. Intern. Med. 2012; 26:302–311 [31]Atkins C et al. J Vet Intern Med 2009.,23:1142-50. [32]Ross, S.. Vet. Clin. North Am. Small Anim. Pract. 2011; 41:1–14 [33]http://www.iris-kidney.com. [34]Dupont et al., Eur J Heart Fail 2013;15:433.[35]Brown, S. et al. J. Vet. Intern. Med. Am. Coll. Vet. Intern. Med. 2007; 21:542–558. [36]Polzin, D. J. J. Vet. Emerg. Crit. Care San Antonio Tex 2001. 2013; 23:205–215. [37]Harison, E., Langston, C., Palma, D. et al.Am. Coll. Vet. Intern. Med. 2012; 26:1093–1098. [38]Campbell GA, Bolton WK:. Adv Chronic Kidney Dis 2011; 18: 420–427. [39]BENCH Study Group.J. Vet. Cardiol. Off. J. Eur. Soc. Vet. Cardiol.1999; 1:7–18. [40]Häggström, J. et al. the QUEST study. J. Vet. Intern. Med. Am. Coll. Vet. Intern. Med. 2008; 22, 1124–1135. [41]Ettinger, S. J. et al. (LIVE) Study Group. J. Am. Vet. Med. Assoc. 1998; 213:1573–1577. [42]The IMPROVE Study Group. J. Vet. Intern. Med. Am. Coll. Vet. Intern. Med. 1995; 9:234–242. [43]Lantis, A. C., Atkins, C. E. et al. Am. J. Vet. Res. 2011; 72:1646–1651. [44]Bock, J. S. Gottlieb, S. S. Circulation 2010; 121, 2592–2600

CARDIOVASCULAR PATHOLOGY

Philip R. Fox, DVM, MSc, DACVIM (Cardiology), DECVIM (Cardiology), DACVECC
New York, NY

Introduction

Clinical diagnosis of cardiopulmonary disease has improved during the past decade, facilitated by commercial cardiac biomarker tests, improved schemes for disease classification and staging, and remarkable enhancements in diagnostic imaging. Nevertheless, the need to identify and confirm diseases suspected from ante mortem evaluation that increasingly depends upon surrogate markers, is increasingly relevant- to validate results from noninvasive assessments, to improve our understanding of how cardiovascular aging masks or simulates pathothologic changes, and to learn how disease progresses, how cardiac structures remodel, and to better define variations of normal. In today's rapid advancements led by molecular and genetic medicine, reliance on 'normal' echocardiographic examination findings may not assure absence of cardiac disease, as has been identified by Doppler echocardiographic techniques. Indeed, changes affecting desmosomes, gap junctions, ion channels and other conditions may represent the primary or secondary causes that lead to cardiac injury, morbidity, and mortality.

Verify what you are looking for and interpret what you see

A thorough and effective gross pathologic necropsy examination enables the clinician to recognize, describe, and interpret lesions that relate to ante mortem findings and often, cause of death. In the majority of cases, gross pathologic findings directly relate to echocardiographic and radiographic changes, particularly conditions causing substantial volume or pressure overload in both congenital and acquired heart disease. In some circumstances routine necropsy examination and related histopathology will not be sufficiently distinctive or specific enough to permit cause-effect relationships to be clarified. More complete pathologic findings can often be obtained by examining tissues using special histopathologic stains, or techniques such as PCR to detect infectious etiologies, EM, or immunohistochemistry.

Principles, practice, techniques, and interpretation

Selection of the optimal approach to assess cardiovascular and respiratory structures is best guided by history, physical examination, and diagnostic modalities whenever possible. This is also particularly important for deciding appropriate selection of tissues- particularly when the opportunity exists to utilize echocardiographic findings as landmarks to direct sections intended for microscopic evaluation.

Materials and supplies

Basic equipment and supplies include a variety of scalpel blade sizes and eye blades, small and large sharp scissors, tweezers, mosquito clamps, impervious cutting board, sponges, gloves, safety glasses, apron, red-rubber feeding tubes of different diameters and large syringes to insert into vessels and flush out blood in the heart chambers, different sized screw top, stackable containers, packets or containers for fresh tissue to snap freeze, water resistant labels and water resistant marking pens, labels that actually stick to wet containers, millimeter ruler, 10% buffered formalin, liquid nitrogen and RNA*later* storage reagent (ideally), good lighting, a scale that can measure to 0.1gm, and a photographic setup.

Approaches for gross cardiac evaluation

Selecting the optimal necropsy approach should be guided by history, physical examination, radiographic and or echocardiographic findings. Removing the left chest wall permits one to view the thoracic contents in situ, approximating the right lateral thoracic radiograph. A ventral sternectomy can be useful when wishing to approach the thorax in an anatomical setting approximating the ventral dorsal or dorsal ventral radiographic view. Prior to removing the heart and lungs, one observes all thoracic structures including the presence (and measured quantity) of plural and pericardial effusion.

Removing the heart and lungs

If the chest wall is not removed, the heart and lungs can be extracted through an intercostal or sub costal, trans-diaphragmatic incision. If effusion is present, extract and record the volume and appearance. All relevant attachments (esophagus, trachea, anterior and posterior vena cava) are severed. The descending Ao is cut. This releases the pluck which is removed from the chest cavity. Trim off pericardium; dissect off trachea, esophagus and each lung.

Preparing the heart

Flush the heart to remove clotted blood. This is facilitated by making a 3-4cm incision in the dorsal aspect of the LA and RA. Then, catheterize the Ao and PA with a red rubber feeding tubes and flush out clotted blood using cold water. Note- you may wish to modify or skip this step if you are looking to detect fragile or friable lesions such as thrombi or vegetations.

Gross inspection

Visually examine the pericardium and all tissue surfaces including mediastinum and pleural space. Record depressions, elevated regions, color changes, ecchymosis, petechiae, nodules, masses, and any abnormal surface features. Note lesions and related great vessels Excise the pericardium. Examine the epicardium, great vessels, cardiac chambers and record gross changes and relative chamber size. (can use normal, 1+, 2+. 3+. 4+ scheme to semi-quantify size of LA/Lau; RA/Rau; RV; LV; great vessels). To assess cardiac mass, measure the heart wt: body wt and heart wt: brain wt.

Fixation

Use 10% buffered formalin with 20:1 volume:tissue. Section within 6-12 hrs.

Cardiac sectioning

When possible, guide sectioning using echocardiography defined tomographic planes or other clinical imaging. Section options: longitudinally (Apical-4C, or optimized for LV inflow and LV outflow); cross section ('bread loaf') approaches; Inflow/outflow technique (used by pathologists to dissect the heart along lines of blood flow, but destroys tomographic/anatomic relationships); window method- selective applications.

A. Gross cardiac description Cogent and clinically useful gross description includes a basic and systematic approach designed to identify and describe pathologic lesions in findings.

1. Semi-quantify external gross changes. Can designate using a scheme such as: normal, 1+, 2+, 3+, 4+ scheme to estimate size of LA/Lau; RA/Rau; RV; LV; great vessels). To assess cardiac mass, measure the heart wt: body wt and heart wt: brain wt.
2. Section heart when fixed (usually 6-12 hours in formalin will permit sectioning and facilitate further fixation.
    a. Assess chamber dilation as above.
    b. Measure thickness of IVS, LVW, RVW at level of chordae as per echo landmark. Also measure anything or any segment that seems abnormally thickened or thin.
3. Valves- describe lesions
    a. MV- use Whitney grading system used to describe MMVD
    b. Examine all components of MV apparatus
    c. Identify and describe LVOT
    d. Record any mid or apical LV lesions, abnormal moderator band networds, thrombi, or any abnormality.
    e. Evaluate papillary muscles
    f. Tricuspid valve- describe as above.
    g. Examine semilunar valves
4. Detect
    a. endocardial jet lesions or LV scar
    b. Infarctions
    c. Regional thinning
    d. Pericardial disease
    e. Masses, clots, nodules
5. Lungs
    a. Excise pulmonary arteries to examine for PTE
    b. Describe lung features (floats, sinks, nodular, spongy, etc)

Describe:

B. Distribution of Lesions (i.e., the spatial arrangement in the tissue or structure)

1. random
2. symmetrical (some amount of organization) or asymmetrical
3. focal
4. multifocal (> 1 lesion)
5. coalescing (> 1 lesion and appearing to fuse or combine)
6. miliary (numerous tiny foci)
7. segmental
8. diffuse (lesion effects ever extent of the organ or tissue)
9. chronicity (subacute [eg,hemorrhage], acute, chronic [e.g., scar])

C. Size (measure)- For multiple lesion, sizes may be uniform or nonuniform

D. Extent (percentage of organ or structure with the lesion)

E. Severity (minimal, mild, moderate, marked, severe- i.e., 5 pt scale)

F. Color

G. Demarcation (whether and how much the lesion is defined or distinct from adjacent tissue)

H. Contour (elevated or depressed vs. surrounding tissues)

I. Shape (geometric configuration)

J. Assess odor if present

Fluids

1. serous – clear
2. serosanguineous - clear but blood tinged
3. sero-fibrinous or fibrinous
4. chylous- milky white
5. purulent- opaque, may be thickish

Descriptive reports

Morphologic description- describes characteristic findings or principle lesions. e.g.

- Diffuse granulomatous pericarditis
- Severe left auricular thrombus
- Focal segmental LVW thinning
- Multifocal, coalescing pulmonary nodules
- Moderate mitral and tricuspid leaflet endocardiosis
- Dirofilariasis

Histopathologic Description

There is no single most correct method to describe microscopic findings but the following guidelines may be useful.

1. List the tissue first
2. Develop consistent descriptive pattern
    a. To avoid missing lesions, develop a consistent manner by incorporating principal morphologic components
    b. When describing heart tissue – epicardium, myocardium, endocardium, vessels, matrix
    c. When describing lung tissue - pleura, vasculature, airways, alveolar septa valvular space
    d. When describing vasculature- lumen and walls (with associated layers)
3. Describe patterns resulting from the majority of the slide
4. Descriptive terminology
5. location (e.g., mid mural, subendothelial, endocardial)
6. Distribution (e.g., focal, multifocal)
7. Size of lesions
8. Describe cell types in order prevalence (e.g., plasma sites in neutrophils surrounded by lesser numbers of macro fudges and fibroblasts)
9. Quantify architecture (e.g., small, few, abundant, moderate, sheets of, etc.)a descriptive interpretations were possible (e.g. fibrinous vasculitis, alveolar hemorrhage, etc.)
10. Etiologic agent location, etiology (e.g., bacilli, photo, etc.), inclusion bodies)
11. Morphologic diagnosis
12. List- the site and location (e.g., left ventricular outflow tract, epicardium, valve, etc.
13. Describe the lesion (e.g. valvulitis, myocarditis, pericarditis, myxomatous valve degeneration)
14. Describe the distribution (e.g., focal, extensive, mild interstitial fibrosis versus replacement fibrosis, etc.)
15. Describe the severity

REFERENCES

Liu SK et al. In: Fox PR, Sisson DD, Moïse NS (eds). A Textbook of Canine and Feline Cardiology. Clinical Practice and Scientific Principles. 2nd Ed. WB Saunders Co, Philadelphia, 1999, pp 817-844

Sheaff M et al. Postmortem technique handbook. 2nd Ed, Springer, 2013

Bryce CH. J Forensic Sci Criminol 1(1);2013

Collins K. Special autopsy dissections. 2010. ISBN: 978-0-930304-97-3

SCREENING FOR SUBAORTIC STENOSIS IN BOXERS IN ITALY

Serena Crosara, DVM, PhD, DECVIM (Cardiology), Cecilia Quintavalla, DVM, PhD, David Chiavegato, DVM, PhD, Claudio Bussadori, DVM, DM, PhD, DECVIM (Cardiology)
Parma, Italy

INTRODUCTION

Fixed subaortic stenosis (SAS) is by far the most common form of aortic stenosis in the dog[1]. SAS is one of the most common congenital heart diseases in Boxer dogs[2-11]. The Boxer breed is widespread throughout Italy, accounting for one of the largest canine populations. Boxer dogs breeding is regulated by Boxer Clubs. The Boxer Club of Italy (BCI) has drown up the guidelines for a "Selection" process of dogs with several attitudinal and morphological standards. The BCI breeders members are encouraged to only use the dogs who have passed the "Selection" for the reproduction. From 1992 to 1995 a survey on 500 Boxer dogs, randomly chosen among training dogs at regional BCI centers, showed a high prevalence (17,8%) of congenital heart disease (CHD). SAS accounted for the 90% of cases[6]. The results of that survey encouraged the BCI to adopt a cardiovascular screening program aimed to monitor the prevalence of CHD in Boxer breed. The BCI established that the cardiovascular screening is mandatory for the admission to the "Selection" process in purpose to reduce the prevalence or CHD in Boxer dogs in Italy.

THE SCREENING PROGRAMME IN ITALY

The screening program is ongoing since 1999. Collected data have been recorded over digital files and submitted to a central register, handled by the BCI. Since 2012 the operators can directly submit data online on the BCI website. The central register is open for the BCI breeders members only. The owners are then allowed to make the cardiac status of their dogs public. Since 2009 the screening program is regulated by the "Fondazione Salute Animale (FSA)" through the "Osservatorio delle Malattie Cardiovascolari del Cane e del Gatto" (Animal Health Foundation through the Observatory on Cardiovascular Disease of Dogs and Cats).

Operators

Screening data have been collected by either board-certified cardiologists or operators with experience in canine echocardiography. Operators have been selected through a practical exam after a practical training with a board-certified cardiologist. During the exam different operators examined the same Boxer and their findings were compared with those of the board-certified trainer (inter-observer variability). In addition, each operator performed three measurements on the same dog during the same day (intra-observer variability). The criteria used to allow an operator to become an "official breed screener" were: (1) Complete agreement with cardiac auscultation findings between operator and trainer; (2) Less than 10 % difference in M-mode and Doppler measurements between operator and trainer; (3) Less than 5 % difference in B-mode measurements of annular diameters between operator and trainer, (4) Complete agreement with diagnosis between operator and trainer; and (5) Less than 10 % difference in M-mode, B-mode and Doppler measurements in repeated measurements on the same dog. At the time of writing 43 cardiologists are included in the list of breed screeners.

The screening procedures

The screening is performed on purebreed Boxer dogs with implanted microchip and a pedigree registered to the "Registro Italiano delle Origini (ROI)" ("Italian Origin Register") of the "Ente Nazionale della Cinofilia Italiano" ("Italian Dogs National Association"). Screening includes patient history, physical examination and transthoracic echocardiography (M-mode, 2-D-mode and echo-Doppler) on both breeder- and client-owned Boxers. The same operator has to accomplish physical examination and transthoracic echocardiography on the same dog. Animal care and handling are in compliance with regulations for animal welfare. The owners sign an informed consent before the examination. Dogs are non-sedated for echocardiographic examination.

Physical examination

Cardiac auscultation is performed in a quiet room while the dog is standing and gently restrained. The presence or absence of a cardiac murmur is recorded and the murmur severity graded on a scale of 1-6, with a grade 1 murmur the softest and a grade 6 the loudest. Respiratory rate, heart rate, and capillary refill time are recorded in the database as well.

Echocardiography

Two-dimensional, M-mode, spectral and color-flow Doppler transthoracic echocardiography (TTE) is performed with a machine equipped with multi-frequency phased array probes: 2.5–3.5 and 3.5–5.0 MHz. Dogs are restrained consecutively in right and left lateral recumbence on a table with an opening that allows transducer manipulation and examination from beneath the animal. TTE is performed according to standards from right parasternal, subcostal, left cranial parasternal, and left apical parasternal windows[12], with concurrent continuous electrocardiographic monitoring. High-quality clips of standard echocardiographic views and high-quality M-mode, PW- and CW-Doppler images of at least 6 cardiac cycles are acquired and stored using the echo-machine software for off-line measurements. Echo-Doppler examination is used to help detect congenital or acquired heart disease. With reference to left ventricular outflow obstructions, 2D imaging, Doppler measurements, and morphologic diagnosis were in compliance with established guidelines for echocardiographic studies of suspected SAS and pulmonic stenosis[13]. In particular the parameters contemplated for SAS screening are: presence or absence of overt lesion of the left ventricular outflow, diameter of the aortic annulus, aortic peak velocity and gradient. From the right and left parasternal long-axis views, the left ventricular outflow tract (LVOT) is scrutinized for evidence of lesions by 2-D echocardiography and for the presence or absence of turbulence by color-flow Doppler. In case of detection of overt lesions in the LVOT, SAS is classified as type 1 to 3 (Table 1).

According to Doppler aortic peak gradient, SAS is graded as mild, moderate or severe[13]. Aortic annulus diameter is measured at the hinge points of the aortic leaflets in early systole from the right and left parasternal long axis view of the left ventricular outflow (**Figure 1**). To not underestimate the annulus diameter, the view is optimized so that the non-coronary and the right coronary cusps appear equal in length and the apposition point in diastole is centrally located in the aortic root. For the purpose of the screening, the largest measurement of the annulus is reported in the patient's record.

Aortic peak velocity is acquired by CW-Doppler and PW-Doppler (sample volume located at the aortic annulus level) from the subcostal window[15,16]. The average CW-Doppler peak aortic velocity is measured over three cardiac cycles on periods of regular sinus rhythm. In case of marked sinus arrhythmia, five cardiac cycles are averaged. If aortic peak velocity falls in a "grey zone" (2.2-2.6 m/sec), it is measured once again after up to 15 minutes to give the dog time to acclimate to the environment and to physical restraint, in order to correctly diagnose and/or grade aortic stenosis. In fact, the effect of emotional stress on Doppler-derived AoPV should be taken into account as it could affect the grading of disease severity in SAS-affected dogs or lead to a misdiagnosis of mild cases in which overt lesions are not evident on TTE[17]. In case of absence of overt lesions in the LVOT and aortic peak velocity in the grey zone, the diameter of the annulus is critical to judge the patient as a small annulus is a risk factor for future development of lesion in the LVOT with time, both in humans and dogs[18,19,21]. In particular, dogs with an annulus ≤ 16 mm are tested again at 2 years of age before a definitive diagnosis. Dogs with an annulus ≥ 18 mm are considered free of the disease. Dogs with an annulus > 16 mm but < 18 mm are rechecked after 6 months. Equivocal cases are referred to the President of the Committee of the Observatory on Cardiovascular Disease of Dogs and Cats (CB) for a second opinion and transesophageal echocardiography in case.

Table 1. Classification of SAS according to Pyle-Patterson-Chako applied to echocardiography[14]

Type 1 SAS	Variable number of small (1-2 mm), slightly raised nodules on the endocardial surface immediately below the aortic valve; sometimes on the ventral surface of the aortic valve cusps
Type 2 SAS	Narrow ridge of thickened endocardium extended partially around the LVOT
Type 3 SAS	Fibrous band, ridge or collar completely encircling the LVOT (tunnel lesion)

Figure 1A: Aortic annulus measurement from the right parasternal long axis left ventricular outflow view. **1B:** Aortic annulus measurement from the left parasternal long axis left ventricular outflow view (inverted image).

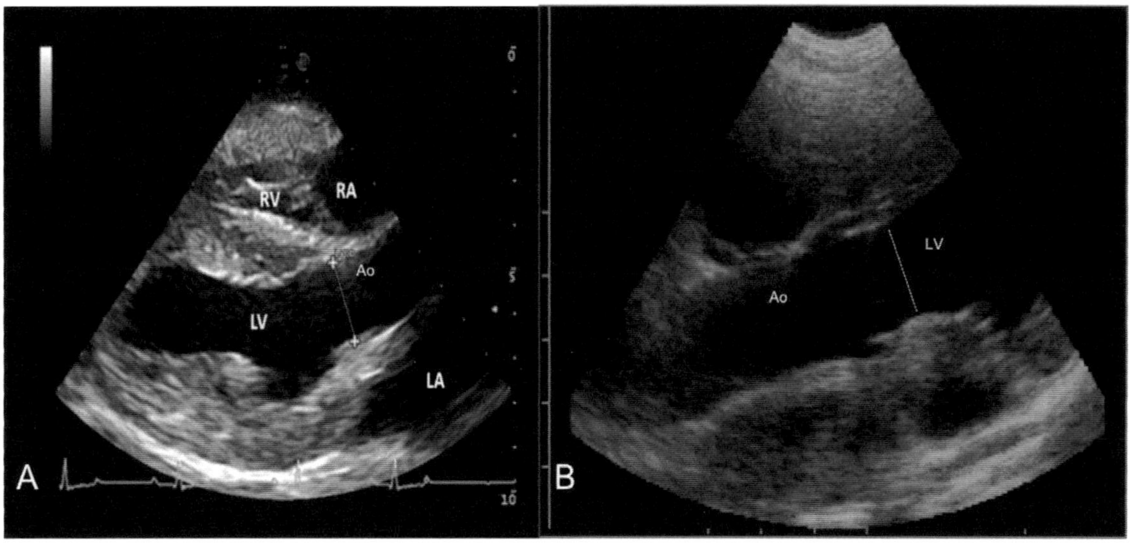

CONCLUSIONS

By now, the screening has been conducting on a large population of Boxers (more than 3000), with a homogeneous distributions for age, body weight, and gender. According to our findings the prevalence of SAS in Boxers in Italy has been gradually decreased in the last fifteen years, even if SAS is still the most frequent congenital heart disease in this breed. In particular, we observed a reduction of the severity of SAS and a reduction of cases with SAS associated with other CHD (especially pulmonic stenosis). We are aware that this population is not representative of the whole Boxer population in Italy since dogs not intended for breeding or belonging to owners not subscribed to BCI were not involved in the screening program. However, this represents a large scale screening which is representative of the breeding Boxers population, indeed this screening is mandatory for any Boxer to be included in the list of recommended breeding animals.

Soft left basilar heart murmurs (1–3/6) were found in a high percentage of dogs without overt signs of left or right ventricular outflow obstruction or other congenital or acquired heart disease. Grade 1 to 2 cardiac murmurs were the most frequent finding. This agrees with reports of other Authors in different Countries[5,8,20]. However a large part of Boxers with SAS had 1–3/6 grade heart murmur. Even if heart murmur distribution significantly differs between healthy and affected dogs, there is an obvious overlap when heart murmur loudness is used as a screening parameter. Thus, auscultation would not be a reliable criterion to discriminate between affected and unaffected dogs if used on its own in screening programs. The difficulties inherent in cardiac auscultation in Boxers and the greater number of operators involved in this breed screening, even if preventively trained and tested, may explain this discrepancy, although it reinforces the subjectivity and unreliability of cardiac auscultation in screening for heart disease in Boxers. In conclusion, the Italian breed screening for left ventricular outflow obstruction appears to have reduced the prevalence of SAS in Boxers.

REFERENCES

[1]McDonald KA. **Vet Clin N Am - Small** 2006;36:503, [2]Patterson DF **Circ Res** 1968;23:171. [3]Patterson DF. **J Small Anim Pract** 1989;30:153. [4]Buchanan JW. **Canine Pract** 1993;18:12. [5]Heiene R, et al. **Vet Rec** 2000;147:152. [6]Bussadori CB, et al. **J Vet Cardiol** 2001;3:7. [7]Chetboul V, et al. **J Vet Med A** 2006;53:346. [8]Linde A, Koch J. **J Vet Cardiol** 2006;8:79. [9]Bussadori C, et al. **Vet J** 2009;181:187. [10]Hopfner R, et al. **Schweiz Arch Tierh** 2010;152:319. [11]Oliveira P, et al. **J Vet Intern Med** 2011;25:477. [12]Thomas WP, et al. **J Vet Intern Med** 1993;7: 247. [13]Bussadori C et al. **J Vet Card** 2000;2:17. [14]Pyle RL, et al. **Am Heart J** 1976;92:324. [15]Abbott JA, et al. **Vet Radiol Ultrasoun**, 2003;44:695. [16]Riesen SC, et al. **Vet Radiol Ultrasoun**, 2007;48:570. [17]Pradelli D, et al. **J Vet Inter Med**, 2014; 28:1724. [18]Javard R, et al. **J Am Vet Med Assoc** 2014;245:1367. [19]Sigfússon G, et al. **J Am Coll Cardiol** 1997;30:255. [20]Koplitz SL, et al. JAVMA 2003;222:770. [21]Menegazzo L., et al. **J Anim Sci** 2012;90:419.

AN UPDATE ON CARDIAC DISEASE IN THE IRISH WOLFHOUND: THE NORTH AMERICAN EXPERIENCE

William D. Tyrrell, Jr.[1], DVM, DACVIM (Cardiology); Jonathan Abbott, DVM, DACVIM (Cardiology); Henry Green, DVM, DACVIM (Cardiology); Steven Rosenthal, DVM, DACVIM (Cardiology); Mariellen Dentino, MD; Frances Abrams, PhD.
[1]Leesburg, VA

INTRODUCTION

Dilated cardiomyopathy (DCM) and atrial fibrillation (AF) have been reported to be common in the Irish Wolfhound (IWH). Several investigators have found a high prevalence of DCM and AF in the IWH throughout Europe, specifically in northwestern Europe as well as Great Britain.[1-4] An initial study in Great Britain described the prevalence of arrhythmias detected in the IWH at 22.2% and AF being the most common rhythm abnormality recorded in 10.5% of dogs examined.[1] Another study evaluated the prevalence of DCM based on a retrospective review of 500 IWH that had presented to the investigator's hospital. Abnormalities were found in 41.8% of IWH presenting to the hospital with DCM diagnosed in 24.2% of which nearly 88% were in concurrent AF.[2] In another study in western Europe, 232 IWH were sequentially examined via physical examination, echocardiography, and electrocardiography. 65.5% were deemed normal, 12.1% had DCM, 4.7% had left atrial enlargement, 4.3% had right ventricular enlargement, 8.2% were in AF with normal appearing echocardiograms and 5.2% were in sinus rhythm but had intraventricular conduction disturbances.[3]

Dr. Vollmar's extensive research in western Europe has looked at the results of serial evaluations of over 1,000 IWH. In this population, DCM and/or AF were diagnosed in approximately one third of these dogs. Males were overrepresented with disease in her studies. Mean age at time of diagnosis was approximately four years of age. Patients in this study population were commonly reported to develop pleural effusion with some pulmonary edema as a manifestation of congestive heart failure.[4] The character of the pleural effusion was often chylous.

Over the past ten years, we sought to investigate the incidence, prevalence, and clinical characteristics of heart disease affecting the IWH in North America and determine whether or not this mirrored findings of the European population of IWHs. Three main studies (The IWH Lifetime Cardiac Study, Incidence of Atrial Fibrillation in the IWH from 2000-2007 and 2007-2014, and Echocardiographic vs. Electrocardiographic Diagnosis of Cardiac Disease in the IWH) have taken place over the past 15 years. The purpose of this lecture will be to summarize the findings of these three investigations and to provide further insight into cardiac disease in the IWH.

INCIDENCE OF ATRIAL FIBRILLATION FROM 2000-2007 AND 2007-2014

Evaluation of apparently healthy IWHs at dogs shows began in the mid 1990's and demonstrated that some IWH had unsuspected atrial fibrillation. Some of these dogs, on subsequent echocardiographic examination showed no evidence of structural cardiac disease. Lone atrial fibrillation did not appear to affect the life expectancy in many cases.[5] In April 2000, a large epidemiological study, with initial enrollment of over 1,000 dogs, was implemented to evaluate all-cause mortality and lifetime effects of heart disease in the IWH. As part of this study, annual electrocardiographic (ECG) testing was performed at many large shows and gatherings of the breed. The study was closed in 2005 when enrollment reached 1,200 IWH, however, ECG testing continued and in 2006, echocardiograms were added to the testing program. The initial testing period of 2000-2006 resulted in early detection of heart disease in IWHs which might otherwise have been asymptomatic. Many of those dogs were eliminated from breeding programs, although dogs with later age onset often had already been bred before symptoms or findings emerged. This study compares the prevalence of atrial fibrillation for two seven year time periods. While many IWHs do live past the age of seven, repeated studies have shown the average life span is approximately seven years. The two seven year periods were each intended to encompass two generations of IWH.

Testing was done by a veterinarian, medical doctor or licensed technician. Six lead ECGs were run on all dogs. Behavior of the dog was noted. The length of tracings varied and was often dependent on patient cooperation and stress levels. Tracings were then sent to an ACVIM board certified cardiologist for interpretation if none was present when the ECG was run. Dogs were sorted into those with atrial fibrillation and those in sinus rhythm. Other rhythm disturbances were also noted and included 1st degree atrioventricular block, ventricular premature contractions (VPCs), atrial premature contractions (APCs), right bundle branch block (RBBB), left anterior fascicular block (LAFB), and other intraventricular conduction disturbances. Many dogs were tested annually and for multiple years. In those cases, individual tests were treated as separate data points. Data recorded included patient identifying information, heart rate on auscultation, sex, and age.

Overall summary results of the two periods of 2000-2006 and 2007-2014 are tabulated below in Figures 1A and 1B. Further statistical analysis will be discussed during the lecture. The difference in prevalence of AF during the two periods is small (11.48% v 12.01%) but there is some indication that AF first became evident in older IWH during the second period. One explanation for this observation is the culling of previously affected AF IWH from the breeding population, however, the mean age in the testing population from 2007-2014 did increase.

01/01/2000-12/31/2006

Figure 1A	AF		Sinus	
Age (months)	n	%	n	%
0-11	0	0	360	100
12-23	15	3.03	480	96.97
24-35	31	5.98	487	94.02
36-47	41	9.51	390	90.49
48-59	52	13.79	325	86.21
60-71	54	23.08	180	76.92
72-83	54	23.08	180	76.92
84-95	46	29.87	108	70.13
>96	57	28.79	141	71.21
UNK DOB	1	5.88	16	94.12
Total	351		2707	
% Total	11.48		88.52	

01/01/2007-10/25/2014

Figure 1B	AF		Sinus	
Age (months)	n	%	n	%
0-11	0	0	7	100
12-23	0	0	57	100
24-35	7	3.02	225	96.98
36-47	13	6.95	174	93.05
48-59	13	8.9	133	91.1
60-71	20	14.71	116	85.29
72-83	17	13.82	106	86.18
84-95	19	17.59	89	82.41
>96	52	28.26	132	71.74
UNK DOB	2	18.18	9	81.82
Total	143		1048	
% Total	12.01		87.99	

	AF		Sinus	
Sex	n	%	N	n%
M	178	10.43	1528	89.57
F	173	12.87	1171	87.13
UNK	0	0	8	100
Total	351		2707	

	AF		Sinus	
Sex	n	%	n	%
M	77	11.05	620	88.95
F	65	13.29	424	86.71
UNK	1	20.00	4	80.00
Total	143		1048	

LIFETIME CARDIAC STUDY (LCS): PRELIMINARY RESULTS

The IWH Club of America (IWCA) began their Lifetime Cardiac Study in 2000. The purpose of this investigation was to both determine all-cause mortality within the IWH breed in North America and evaluate for underlying cardiac arrhythmias and/or abnormalities on their ECG. The study's goal was to run an annual ECG and obtain questionnaire information on each enrolled IWH. Hounds were recruited at national specialty shows, local fun matches, and regional breed gatherings in order to achieve as much of a diverse population as possible. The population of IWHs varied from breeding stock to older neutered males/females as well as rescue IWH with unknown genetic backgrounds. The survey was a questionnaire to each IWH owner and was used to collect data regarding diet, exercise, supplements, medications, relatives affected by cardiac disease, and overall general health issues.

Total enrollment in the LCS was 1,259 IWHs. Verbal information regarding cause and/or date of death was ultimately available on 825 IWHs (Table 2), however, there were only 326 hounds where a minimum of three annual ECGs and questionnaires were present in the data.

Table 2: All-Cause Mortality in IWH (IWCA Lifetime Cardiac Study)

Cause of Death (M&F)	n	%	Mean Age at Death (mos.)
Hemangiosarcoma	17	2.06	94.71
Chondrosarcoma	2	0.24	52
Lymphoma	38	4.61	74.96
Osteosarcoma	167	20.24	86.09
Neoplasia-other	74	8.97	95.76
Neoplasia-ALL	*298*	*36.12*	*88.09*
Cardiac	126	15.27	89.28
Sudden death	21	2.55	85.4
Post op hemorrhage	7	0.85	78.31
Bloat and/or GDV	48	5.82	82.27
Megaesophagus	9	1.09	103.67
Pneumonia	59	7.15	91.83
Trauma (HBC, GSW, etc.)	99	12.00	76.7
Pelvic limb paralysis	81	9.82	109.95
Epilepsy/uncontrolled seizures	1	0.12	32
Renal	20	2.42	90.1
Unknown	56	6.79	90.28
TOTAL	825		88.85/7.40 years

All ECGs were interpreted by an ACVIM board certified cardiologist. Cardiac rhythm was determined to be sinus or atrial fibrillation. Other abnormalities were noted to include APCs, VPCs, RBBB, LAFB, first degree atrioventricular block and other intraventricular conduction disturbances. IWH diagnosed with single APCs did trend toward development of AF. Twenty-four IWH had documented APCs and were followed for at least two consecutive years with 14/24 (58.33%) developing AF. Significant ventricular ectopy was noted in several IWH. Many of these hounds were followed via Holter monitoring and were found to have aggressive ventricular ectopy that was determined to likely benefit from anti-arrhythmic therapy. Ultimately, 21 of these IWHs suffered sudden death of which was likely attributable to their ventricular arrhythmias. In conclusion, the main cause of mortality in the IWH was neoplasia at 36.12% with osteosarcoma representing 20.24%. Cardiac disease was the second leading cause of death at 15.27%. All deaths considered to be related to cardiac disease were documented via echocardiography and/or radiography demonstrating congestive heart failure. IWHs with sudden cardiac death were not included in this mortality percentage. Interestingly, another epidemiologic study conducted by Gretchen Bernardi for the IWCA from 1966-1986 in the United States identified a 15.1% mortality rate for cardiac disease and 33.9% for all neoplasia (21% osteosarcoma).[6]

ECHOCARDIOGRAPHIC VS. ELECTROCARDIOGRAPHIC DIAGNOSIS OF CARDIAC DISEASE IN THE IWH

Echocardiographic (ECHO) and ECG examinations were performed on IWHs between 2006 and 2014. Seven hundred twenty-six ECHO and ECG studies were performed on 576 separate IWHs (346 female/230 male). Many of these hounds were followed for years and some were examined via ECHO/ECG only once. Age of the IWHs at the time of the initial examination ranged from 1-11.43 years (median 3.14 years) and body weight ranged from 43.1-103.9 kg (mean 64.6kg). The purpose of this study was to evaluate associations between ECG and ECHO findings and to determine if an ECG could serve as a stand-alone diagnostic test for IWH DCM.

Echocardiographic reference values for the IWH have previously been published.[7] Left ventricular end-diastolic and end-systolic dimensions were compared to body weight specific reference intervals from previously published normative data based on the assumption that breed-specific normative data is likely superior to intervals derived from studies of breed-diverse samples. DCM was defined as a dilated left ventricle with concurrent myocardial dysfunction; left ventricular dilation and myocardial dysfunction were identified when the end-diastolic and end-systolic left ventricular dimension exceeded the upper limits of the respective reference intervals.

Left ventricular dilation was evident in 99/576 (17.19%) examined IWHs; more females (73/346, 21.10%) than males (26/230, 11.30%) were affected ($p < 0.0022$). Myocardial dysfunction was found in 48/576 (8.33%) IWHs of which 29/346 (8.38%) were female and 19/230 (8.26%) were male. Ventricular tachyarrhythmias were recorded in 17/576 (2.95%) with 11/346 (3.18%) females and 6/230 (2.61%) males being affected. AF was present in 51/576 (8.85%) IWH with 33/346 (9.54%) females and 18/230 (7.83%) males affected by AF.

In this population of IWHs, the sensitivity of AF as a diagnostic marker of DCM was 40.5% and for myocardial dysfunction was 37.5%. Past research has suggested that the presence of AF predicts future development of DCM in the IWH.[4] AF had a 58.49% sensitivity for identifying left atrial enlargement defined by the left atrial:aortic ratio (LA:Ao) via M-mode. However, all IWH ECHOs in this study were performed in standing position and alignment for M-mode LA:Ao measurements was difficult at times especially if the left atrium was enlarged. At the time of writing, we are still calculating the sensitivity of AF for the identification of 2D left atrial dimensions and LA:Ao from 2D dimensions. The primary author's opinion is that AF will have a will have a very high sensitivity for 2D left atrial size/dilation once these statistics are calculated. This will be presented in the lecture.

Lastly, the primary author's description of IWH type DCM does vary from what is typically called canine DCM. On ECHO, most IWHs diagnosed with IWH type DCM do not have markedly reduced left ventricular systolic performance evaluated by ejection-phase indices such as fractional shortening, nor do they develop the magnitude of left ventricular dilation that is often observed in the Doberman Pinscher and other breeds affected by DCM. Most IWHs initially diagnosed with AF have mild to moderate enlargement of left atria with +/- minimal left ventricular chamber dilation and preserved left ventricular myocardial function. Mild to moderate mitral valve regurgitation is usually present which is likely secondary to annular dilation, however, one could argue all observed changes are secondary to progressive mitral valve disease. It is the primary author's experience that the majority of IWHs do not develop significant myocardial dysfunction or left ventricular dilation but do develop progressively severe left atrial dilation resulting in left sided congestive heart failure manifested as pulmonary edema. These IWH type DCM patients will often resemble a small breed dog with progressive myxomatous mitral valve disease in regard to left atrial and left ventricular size and function.

REFERENCES

1. Brownlie SE: An electrocardiographic survey of cardiac rhythm in Irish wolfhounds. *Vet Rec 1991: 129:470-471.* 2. Vollmar AC: The prevalence of cardiomyopathy in the Irish Wolfhound: A clinical study of 500 dogs. *J Am Anim Hosp Assoc 2000; 36: 125-132.* 3. Vollmar AC, Fox PR. Clinical, echocardiographic and ECG findings in 232 sequentially examined Irish wolfhounds. *J Vet Int Med 15; 279, 2001.* 4. Vollmar, AC et al: Proceed. 22nd ACVIM Forum 2004; 92. 5. Harpster, NK: Cardiac arrhythmias in the Irish wolfhound: preliminary study. *Proceed. 12th ACVIM Forum 1994; 1:17-22.*
6. http://www.wolfhoundweb.com/breedinfo/health/longevity/bernardi.html. 7. Vollmar AC: Echocardiographic measurements in the Irish wolfhound: reference values for the breed. *J Am Anim Hosp Assoc 1999;35:271-7.*

ANESTHESIA FOR THE CARDIAC PATIENT UNDERGOING NON-CARDIAC PROCEDURES

Khursheed Mama, DVM, DACVAA
Fort Collins, CO

INTRODUCTION

As other veterinary specialties continue to advance, it is increasingly common that patients with cardiac disease(s) need anesthesia for elective and emergent procedures. This clinically focused manuscript and accompanying presentation includes information on two common cardiac diseases encountered in small animals, namely regurgitant mitral valve disease in dogs and restrictive/hypertrophic cardiac disease in cats. Broad based suggestions for management of cardiac patients in the peri-anesthesia period are also included and may be referred to for this and the subsequent presentation titled 'Anesthesia for the Cardiac Patient Undergoing (Surgical) Attenuation of their Disease'.

PREOPERATIVE ASSESSMENT AND PREPARATION

The history and physical examination should be directed at assessing the overall condition of the patient. For example is the patient thin and lethargic, does he/she have a history of coughing, exercise intolerance or syncope? Do you notice ascites or limb edema? Do you auscultate a murmur or abnormal rhythm? Are pulse deficits present, extremities unusually cool to the touch, etc.?

Diagnostic procedures should include 'routine' blood work such as a CBC, serum chemistry profile and urinalysis. Examples of information these tests may reveal include polycythemia as might be associated with right to left shunting of blood or an elevation of BUN/Cr which may be indicative of poor renal perfusion. Radiographs to evaluate the size and shape of the cardiac silhouette and pulmonary vasculature and an ECG to identify arrhythmias and conduction disturbances provide additional useful information that may be used during anesthesia management. Cardiac ultrasound, arterial blood gas assessment, etc. may also be part of the diagnostic plan and will help document the source and severity of cardiac pathology.

If diagnostics reveal that the animal has significant cardiovascular compromise or heart failure and the procedure is elective, the recommendation is to postpone the anesthesia and stabilize the cardiac condition. In animals where surgery cannot be postponed, their condition should be optimized (with the aid of invasive monitoring if necessary) prior to anesthetic induction. Things to consider (and these will vary with different cardiac disease) include ensuring adequate preload while also preventing excessive preload, treating pulmonary edema, managing arrhythmias, etc.

ANESTHESIA PLAN

Broad based suggestions for managing animals with cardiac disease are included.

Premedication

Drugs may be administered IM or SC with the intent of reducing patient anxiety and stress during restraint and catheter placement. Premedication also facilitates a reduction in anesthetic induction drug doses. However, excessive sedation which could lead to hypoventilation (and in certain circumstances hypoxemia) in room air breathing animals should be avoided. IV premedication is acceptable in patients with a pre-placed catheter; drug and dose modification may be needed. Drugs that exhibit cardiovascular safety and provide analgesia are ideal premedicants in these patients; opioids are a good option for both dogs and cats (although cats may not sedate with these drugs). Use anticholinergics as necessary to maintain heart rate (cardiac output) within the desired range (this will vary with the disease).

Anesthesia Induction and Maintenance

This should be performed with regard for the specific disease and the general knowledge of the effects of the selected drugs on the cardiovascular system. Familiarity with the effects of different anesthetic agents on the cardiovascular system and the ability to detect adverse effects and intervene if necessary are important. Intravenous opioids (e.g., fentanyl) and benzodiazepines (e.g., midazolam) in the dog and etomidate and benzodiazepines in the cat are among the safest techniques for inducing anesthesia in patients with cardiovascular disease but not necessary or desirable in all circumstances due to additional drug related considerations. Anticholinergics may be necessary to offset the bradycardia commonly seen with opioids following IV induction doses.
Balanced anesthetic techniques (opioid infusion with an inhalant are commonly used for maintenance of anesthesia in compromised patients. In some circumstances a muscle relaxant may be added (in which case ventilation is essential).

Pre and intraoperative monitoring and support

In our practice environment we generally pre-place the Doppler and ECG in animals with moderate to significant cardiovascular compromise so blood pressure and rhythm changes may be monitored throughout induction of anesthesia. In rare cases additional support equipment such as pacing leads or pads may be placed. Preoxygenation is routine. As intubation conditions may not be ideal with the aforementioned induction drug combinations, it is advisable to have a small amount of lidocaine (patient size dependent) to facilitate this; the local anesthetic is placed directly on the arytenoids and intubation attempted 30 sec to a minute later. A pulse oximeter, capnograph and thermistor are routinely placed following anesthesia induction. An arterial line to facilitate direct blood pressure monitoring and sampling of blood for assessment of blood gases and electrolytes may be placed if warranted following anesthesia induction. Judicious use (so as not to compromise venous return and cardiac output) of positive pressure ventilation of these patients is generally suggested as maintaining normocapnia and normal pH can be helpful when treating cardiac based hypotension. Additional management is focused on maintaining myocardial performance and avoiding decompensation of the patient with stable disease or providing additional support (in the form of anti-arrhythmic medications, inotropes, etc. as appropriate) to decompensated patients. The manner in which this is done will vary with the type of cardiac disease, severity of the same and knowledge of therapeutic modalities the patient is receiving at presentation. We will address these more specifically for each of the diseases but the basic concepts are the same as for any patient requiring anesthesia aka to maintain tissue perfusion; appropriate preload, inotropic support, heart rate and rhythm, and afterload must all be factored in when managing these patients.

Anesthesia Recovery

This period should be as stress free as possible. Make sure the patient is able to maintain his/her airway and breathing before extubation. Oxygen supplementation during the recovery period may be warranted. Additional support and monitoring should be maintained until patient is stable and doesn't require this.

MITRAL VALVE INCOMPENTENCE

This is common in middle aged or older small (and medium) breed dogs. An incompetent mitral valve permits regurgitation of blood from the left ventricle into the left atrium during ventricular contraction. While fairly large regurgitant flows can initially be accommodated with little change in atrial pressure, a decrease in forward left ventricular stroke volume occurs early in the disease. The body responds to the resulting reduction in perfusion (renal/renin-angiotensin-aldosterone) and sodium and water are retained to increase blood volume. The increased blood volume stretches the myocardium of the left ventricle which grows larger increasing end diastolic volume (eccentric hypertrophy) and allowing the ventricle to pump a greater volume of blood to compensate for the valvular leak. The condition typically worsens over time with the mitral orifice enlarging, the regurgitation worsening, and the ventricle growing bigger. As the disease progresses (or in acute MR) left ventricular end-diastolic pressure increases and may lead to increased pulmonary venous pressure and pulmonary edema. Bronchial compression and premature atrial contractions or atrial fibrillation may also be seen due to the increase in atrial size. Left ventricular myocardial failure is seen with severe disease.

Patients with early asymptomatic mitral regurgitation tolerate anesthesia well and special management is usually not necessary other than to avoid excessive fluid administration and maintain heart rate in the normal range. Symptomatic patients often receive diuretics (e.g., furosemide), vasodilators or angiotensin converting enzyme inhibitors (e.g., benazepril) and may also receive drugs that stimulate contractility (e.g., pimobendan). These help slow the progression of the disease and minimize the consequence to the patient but also reflect increasing severity. While it has been the author's preference not to discontinue medications prior to anesthesia, others recommend discontinuation of the ace inhibitor because of the potential to exacerbate (cause) hypotension. Bradycardia or an increase in systemic vascular resistance will increase the duration of ventricular ejection and the fraction of regurgitant flow. Significant acute increases in regurgitant flow can precipitate heart failure. Hence during anesthesia, bradycardia should be avoided and heart rate maintained in the high normal range for that animal. Inotropic drug therapy and afterload reduction also help reduce the fraction of regurgitant flow. Patients in severe heart failure secondary to mitral insufficiency are at greater anesthetic risk and should have elective surgery postponed until their condition is controlled and stabilized with medical management.

Dogs are frequently premedicated with an opioid (e.g., oxymorphone 0.05 – 0.1 mg/kg IM or SC or 0.01 – 0.03 mg/kg IV). Benzodiazepines (e.g. midazolam 0.1 – 0.2 mg/kg, IV, SC or IM) may be used in addition in older or sicker dogs where paradoxical excitation is less likely to be observed. Acepromazine decreases afterload and in early stages of the disease may be a useful adjunct if the dog requires additional tranquilization. However it also has the potential to decrease preload (and subsequently contractility) which can be challenging to manage during anesthesia. Alpha-2 agonist drugs while excellent sedatives are not recommended as they increase systemic vascular resistance and decrease heart rate which in turn increase regurgitant flow and decrease forward stroke volume. Depending on choice of premedication and subsequent anesthesia medications, an anticholinergic (e.g., atropine 0.02-0.03 mg/kg IM or SC) might be added to prevent bradycardia (e.g., with use of IV opioids) and maintain heart rate in a high normal range; tachycardia should be avoided as this can negatively affect ventricular filling.

For an animal with mild to moderate mitral insufficiency, judicious use of any intravenous induction agent should suffice. Hence, propofol and alfaxalone are acceptable for dogs with mild disease and may be used judiciously in patients with moderate disease. Ketamine is a useful alternative in patients with moderate disease and may be used in combination with a benzodiazepine or low dose of propofol. Anesthesia induction for patients with advanced disease is safely accomplished using drugs with a low likelihood of or manageable cardiovascular depression, such as a combination of an opioid (e.g., fentanyl 10 microgram/kg) or etomidate (1 – 2 mg/kg) with a benzodiazepine (e.g., midazolam 0.1 – 0.2 mg/kg); bradycardia can be offset with use of anticholinergics as mentioned previously.

Maintenance with inhalation anesthetics is appropriate for dogs with mild to moderate disease with supplemental analgesia provided as needed. In patients with moderate to severe disease, an opioid based, balanced technique allows for a reduction in the dose of inhalation agent and in turn their dose dependent myocardial depressant effects. Respiratory depression is however likely so if mechanical ventilation is not possible, alternative approaches using anesthetic sparing infusions of ketamine and lidocaine may be considered.

Intraoperatively it is important to maintain adequate preload while avoiding excessive fluid administration. The author prefers to use crystalloid fluids (vs. colloids) as inadvertent volume overload is more manageable. Recall that CVP is an inaccurate reflection of left ventricular filling pressures in these animals and should not be used to guide therapy. If hypotension is noted, often treating a low heart rate with an anticholinergic will improve blood pressure by enhancing forward stroke volume. Alternatively of in addition drugs that increase contractility such as dobutamine (2 - 7 microgram/kg/min, IV) may be used. Ephedrine (0.05 – 0.1 mg/kg, IV) and dopamine (2 – 5 microgram/kg/min, IV) may also be used to provide cardiovascular support but have mixed cardiovascular effects.

Dysrhythmias may be observed in the peri-anesthetic period and in addition to addressing common causes such as hypoxia, hypercapnia, a light plane of anesthesia, etc. treatment may be necessary if there is a risk for the dysrhythmias to progress (e.g., ventricular ectopy to tachycardia or fibrillation) or tissue perfusion is adversely affected. Lidocaine (1 – 2 mg/kg, IV as bolus followed if needed by a CRI of 75 – 100 microgram/kg/min) is a good first line choice for treatment of dogs with ventricular dysrhythmias and has the added benefit of having synergistic effects with other anesthetic agents. For an in depth review and additional options for treatment of rhythm abnormalities, the reader is referred to other sources.

FELINE HYPERTROPHIC CARDIOMYOPATHY

This is a complex cardiovascular disease characterized by concentric hypertrophy and accompanying left ventricular diastolic dysfunction. Both the ventricular free wall and ventricular septum may be symmetrically or asymmetrically thickened and cause a narrowing of the left ventricular outflow tract. Papillary muscle thickening is also common. Even in basal conditions the hypertrophic myocardium has vastly increased oxygen requirements. Additionally, intramural coronary arteries are abnormally small and narrowed. Increased left ventricular diastolic pressure further compresses these vessels limiting blood supply. Mitral regurgitation and (dynamic) outflow tract obstruction from the mitral valve leaflets coming into contact with the septal wall add complexity to management of these animals during anesthesia. Slow heart rates (prolonged mitral leaflet-septal contact time), increased contractility, decreased preload and afterload (e.g. vasodilation) are all thought to contribute to the dynamic nature of the obstruction which can contribute to the myocardial supply and demand imbalance.

Hypertrophic cardiomyopathy should be a differential in any cat with a systolic murmur, a gallop rhythm, cardiomegaly, arterial thromboembolism, or congestive left heart failure. Contributing or frequently associated diseases include hyperthyroidism, aortic stenosis, acromegaly and systemic hypertension. The disease can affect cats over a broad range of ages with males more commonly affected than females. There is a genetic form recognized in Maine Coon cats. Yet many cats with hypertrophic cardiomyopathy are asymptomatic and often the diagnosis is made after auscultation of a murmur or gallop rhythm, or radiographic evidence of cardiomegaly during examination of the cat for other reasons. HCM may be a cause of 'sudden death' in otherwise apparently healthy cats undergoing anesthesia. Sinus tachycardia and/or ventricular ectopy are present in many cats with HCM. Exercise intolerance when reported, reflects the inability of the heart to increase cardiac output with exertion due to outflow obstruction and impaired diastolic filling. The symptomatic hypertrophic cardiomyopathy animal often presents with pulmonary edema resulting from transmission of elevated end diastolic left ventricular pressures to the pulmonary vasculature. Tachypnea as opposed to coughing may be noted with pulmonary edema in cats.

Because of the multi-faceted aspects of this disease syndrome, cats are highly susceptible to myocardial ischemia during anesthesia. Any increase in myocardial oxygen demand such as increased heart rate or contractility may cause decompensation and so should be avoided. The poorly compliant left ventricle depends heavily on adequate intravascular volume and synchronous atrial contraction for adequate filling so correction of any volume deficits and maintenance of a normal sinus rhythm are of extreme importance in HCM cats and especially so when ventricular outflow tract obstruction is a feature of the disease.

Anesthetic protocols for HCM typically include pre-anesthetic administration of opioids (methadone 0.3 mg/kg or oxymorphone 0.05 mg/kg IM or SC) in an attempt to minimize sympathetic stimulation from anxiety, excitement, or pain and facilitate catheter placement. IV opioids should be used cautiously in cats as they may lead to excitement with inappropriate dosing. Anticholinergics should be used judiciously and generally only if needed to treat bradycardia. Induction of anesthesia should be as stress free as possible taking special care to avoid hypoxemia. The author has a preference for using a combination of etomidate (0.5 – 1.5 mg/kg) and diazepam or midazolam (0.1 – 0.3 mg/kg) for anesthesia induction as these drugs tend to minimally affect cardiovascular function and other side effects are generally manageable. Ketamine and tiletamine/zolazepam (Telazol®) both increase myocardial contractility and heart rate resulting in increased myocardial oxygen consumption in the face of limited supply and so should not be used. Propofol causes vasodilation and has been reported to impair function of the myocardium. Occasionally a cat with HCM proves challenging to handle and while not ideal, modern inhaled anesthetics (following premedication if possible) can be used to facilitate restraint. Alpha-2 agonist use is reported for sedation of cats with HCM and low doses for premedication may be warranted in this circumstance. Intramuscular alfaxalone administration may also be considered as an alternative. In the animal with significant disease, anesthesia is best maintained with a balanced technique using an opioid (e.g., fentanyl) in combination with low doses of inhaled anesthetics.

Intraoperative monitoring should focus on monitoring heart rate and rhythm (ECG), non-invasive blood pressure, body temperature, oxygen saturation and ventilation. If the cat has pulmonary edema, or other evidence of heart failure, or if it is undergoing a lengthy or invasive procedure an arterial catheter should be placed to facilitate direct pressure measurements and sampling of blood for gases and electrolytes. Fluids must be administered with care. Due to poor left ventricular compliance, left sided pressures rise quickly with excessive fluid administration and can quickly result in pulmonary edema. On the other hand left ventricular filling is very dependent on adequate volume so fluid deficits and intraoperative losses must be promptly replaced. Because of the heavy reliance of the stiff left ventricle upon synchronous atrial contraction for filling and the deleterious effect of tachycardia on myocardial oxygen demand/supply ratios, intraoperative arrhythmias should be treated aggressively. Hypotension that is not due to hypovolemia is best managed by increasing systemic vascular resistance using an alpha adrenergic agonist, such as phenylephrine. Increasing contractility (using inotropes) is controversial in HCM as it has the potential to increase myocardial work and oxygen demand in the face of decreased supply.

REFERENCES

[1]Amara Estrada. Plumbs Therapeutics Brief. 2015; 2 (1), 24.[2] Nattel S et al. In Opie LH (ed). Drugs for the Heart, 8th edition. Saunders Elsevier, Philadelphia PA, 2013;272. [3]Wiese AJ et al. AJVR 2012; 73(6):908. [4]Kittleson MD et al. Circulation 1999:3172. [5]Abbott JA Vet Clin NA SA practice 2010; 40(4):485. [6]Sargent J. J Vet Cardiology 2015 (epub-January). [7]Lamont LA et al. JAVMA 2002; 221(9):1276.

ANESTHESIA FOR THE CARDIAC PATIENT UNDERGOING ATTENUATION OF THEIR DISEASE

Khursheed Mama, DVM, DACVAA
Fort Collins, CO

INTRODUCTION

This clinically based manuscript and accompanying presentation will focus on anesthesia management of dogs and to a lesser degree cats with cardiac disease where attenuation of the disease is the primary reason for anesthesia. Attempted attenuation may be surgical or interventional in nature. While the diseases are diverse in scope, the principles of anesthesia preparation and management are similar. For a brief review, the reader is referred to the lecture notes for the previous session titled 'Anesthesia for the cardiac patient undergoing non-cardiac procedures'. Individuals managing anesthesia in these animals should also have a basic knowledge of the procedure (site/approach, need for anticoagulation, specific medications or diagnostic modalities [e.g., echocardiography] that might be employed during the procedure, etc.) and equipment (e.g., pacing or defibrillation equipment). The approach might range from external defibrillation to balloon catheterization using fluoroscopy to use of cardiopulmonary bypass depending on the disease process and practice circumstance necessitating diverse management skills, flexibility and a team approach. As with any procedure, clients should be made aware of related costs and potential risks to their animal. This manuscript will review highlights of peri-anesthesia management for selected cardiac diseases where attenuation / correction may be more commonly attempted.

PATENT DUCTUS ARTERIOSUS (PDA)

Patent ductus arteriosus is one of the most common congenital heart defects in dogs (rare in cats) and typically diagnosed by a continuous murmur ausculted over the left cranial thorax during routine physical examination. The disease can however have different clinical manifestations and an animal may also present at the other extreme with signs of left heart failure and associated varying degrees of pulmonary edema (tachypnea, dyspnea, coughing). The palpated pulse pressure is wide, often with prominent or bounding ('water hammer') pulses. This is the result of systolic blood pressure being increased because of increased stroke volume delivered to the aorta and diastolic blood pressure being very low because of rapid runoff of blood through the ductus during diastole. Radiographically there is evidence of left heart enlargement, which may be marked. A 'bulge' in the descending aorta approximating the origin of the ductus may be seen on the DV view. An increase in the size of the pulmonary vasculature may be present. Pulmonary edema will be present in dogs in heart failure. An echocardiogram may also be used to confirm the diagnosis of a PDA and is a useful tool in documenting severity of cardiac changes associated with the same.

Right to left shunting through a PDA can occur and is usually a later manifestation (but may accompany early severe forms) of the disease. It typically occurs because over time increased pulmonary artery pressure combined with increased pulmonary blood flow creates a pathological response in the pulmonary arteries. The pulmonary vascular pathology progresses until ultimately pulmonary vascular resistance (PVR) exceeds systemic vascular resistance (SVR) and a right to left shunt is created. Clinical diagnosis may be made by the observation of differential cyanosis (i.e., oral mucous membranes remain well oxygenated and are pink, but caudal mucous membranes may have a bluish tinge reflecting poor oxygenation). This occurs because deoxygenated blood enters the aorta distal to the brachiocephalic trunk and left subclavian arteries (supplying head and forelimbs). Decreased renal oxygenation stimulates increased release of erythropoietin, resulting in increased erythropoiesis and measurable polycythemia. The increase in red cell mass improves oxygen delivery, with a maximum benefit achieved at a PCV of ~55-65%. Prognosis for animals with a R to L shunt is less favorable.

Correction of a PDA is possible by either surgical or interventional occlusion of the ductus with the majority of procedures performed electively in young puppies in stable condition. While these animals may have evidence of volume overload they are not in failure. With this clinical picture a variety of anesthetic techniques are suitable; additional consideration should be given to size and age related concerns such as hypothermia and hypoglycemia. As most puppies sedate well with opioids, it is common in our practice to administer an opioid and anticholinergic (as needed) for premedication. Both as a result of the disease and age, maintenance of a normal to slightly elevated heart rate can be helpful in maintaining systemic perfusion. In uncomplicated disease premedication may be followed by an induction and maintenance agent of the clinicians choosing. In dogs with significant cardiac changes, opioid and benzodiazepine based protocols minimize additional cardiovascular depression. Fluids should be administered with care taken not to overload these animals. In animals with severe disease including a right to left shunt it is important to keep in mind decreases in systemic vascular resistance as are possible with acidemia, hypercapnia and hypoxemia or as caused by certain drugs will worsen this condition. Increased pulmonary vascular resistance as with hypoxemia should also be avoided. Routine anesthesia monitoring including ECG, non-invasive blood pressure, oxygen saturation, capnography and body temperature are suggested. Direct blood pressure monitoring is useful to assess labile patients and can also help confirm occlusion of the correct structure. It is typical to see at least a transient increase in diastolic pressure when the ductus is occluded.

Surgical correction of a PDA is performed via a lateral thoracotomy. Bleeding is minimal unless a vessel is accidentally torn, then it can become catastrophic if not addressed. Additionally as the ductus can resemble the pulmonary artery or the aorta in size either can be inadvertently occluded instead of the ductus. Direct continuous blood pressure monitoring will aid in rapidly identifying this and preventing the surgeon from ligating it permanently. If a surgical approach is chosen, analgesia and optimization of oxygenation and ventilation should be addressed. Transvenous approaches to correction of PDA have become increasingly routine. Although less invasive than a thoracotomy, serious complications have occurred during these procedures including perforation of major blood vessels with catheters and guide wires and dropped or dislodged occlusion devices that may embolize other vessels in the pulmonary or systemic circulation. Dysrhythmias (most typically ventricular in origin) may be observed with either approach due to cardiac manipulation and the anesthetist should be ready to treat these; lidocaine is usually effective.

PULMONIC STENOSIS

Pulmonic stenosis is the result of valvular, subvalvular, or supravalvular resistance to flow in the right ventricular outflow tract. As a result of this resistance, the normally thin walled compliant right ventricle is modified to thick walled, poorly compliant chamber. Additionally with severe pulmonic stenosis right ventricular systolic pressures can exceed aortic pressures impairing coronary flow (whereas in mild to moderate disease right ventricular coronary perfusion is minimally affected). Aberrant coronary vasculature has also been reported with this disease syndrome and may be relevant to clinical management. Echocardiography is typically used to characterize the severity of the disease. Trans-esophageal echocardiography may provide added benefit when assessing the coronary vasculature; angiography prior to balloon valvuloplasty may also be used.

Pulmonic stenosis may be attenuated by surgery (often under a brief period of 'inflow' occlusion) or more commonly by balloon valvuloplasty using a transvascular approach. During anesthesia it is important to maintain preload as the noncompliant ventricle requires increased filling pressures to achieve the same end diastolic volume versus more compliant chambers. Trying to maintain atrial synergy and adequate time for diastolic filling are also important and may be negatively influenced not only by reduced preload, but also by tachycardia and supraventricular arrhythmias. While tachycardia will reduce time for diastolic filling, bradycardia will also reduce cardiac output since stroke volume is relatively fixed (due to the obstruction). Hence the best balance between maximizing end diastolic volume and cardiac output is achieved by maintaining a normal heart rate. It is also important to maintain afterload as profound hypotension can result from vasodilation given the fixed stroke volume. Diastolic hypotension is especially concerning as this will reduce coronary perfusion. Treatment of hypotension can be challenging as increasing contractility (or heart rate) increases myocardial oxygen demand and in the face of a fixed output may not cause an increase in stroke output. In the authors experience (in dogs managed with opioid based anesthetic protocols) dopamine has been used safely and effectively as part of the strategy for managing hypotension; additional use of pressors (e.g., phenylephrine) has not been needed.

An anesthetic drug plan that uses cardiovascularly sparing drugs and considers the physiology of the disease when treating complications such as hypotension during anesthesia will provide the best chance for a positive outcome. Since ventricular dysrhythmias are common during cardiac manipulation, lidocaine (which also has an inhaled anesthetic sparing effect) may be used as a precautionary measure or for treatment as is the anesthetist's preference. Development of bundle branch block which may be observed during catheterization should not be confused with a ventricular rhythm. In addition to routine monitoring as has been previously mentioned, direct arterial blood pressure monitoring is highly recommended despite the challenges encountered in these animals with fixed stroke volume. This also facilitates sampling of blood for PCV, TP, blood gas, electrolyte, blood glucose, lactate and acid base analysis which can aid in patient management.

ATRIAL FIBRILLATION

Atrial Fibrillation is not uncommonly diagnosed in small animals. Often this diagnosis is made in older pets with underlying heart disease indicating decompensation and disease progression. However this dysrhythmia can also occur in the absence of cardiac disease most commonly in 'giant' breed dogs. The loss of synchronous atrial contraction results in a decrease in cardiac output and may manifest clinically as exercise intolerance or weakness even in dogs without primary cardiac disease. It is postulated that the variation in ventricular rate which is a common feature of this disease puts animals at a risk for developing myocardial hypoxia as coronary perfusion and oxygen demand are out of balance. This contributes to development or worsening of myocardial dysfunction.

Auscultation typically reveals an irregularly irregular rhythm associated with variable pulse quality. Heart rate can be quite variable but is often significantly elevated in animals with clinical symptoms. The ECG additionally reveals the absence of p waves and presence of f or fibrillation waves. As with auscultation, the ventricular rate noted on the ECG is variable. Further evaluation may include holter monitoring, laboratory tests, radiographs and/or an echocardiographic evaluation. The latter can be especially helpful in determining contributing causes and cardiac function. Thromboembolic disease is a common sequelae to atrial fibrillation in people and may be observed in cats occasionally necessitating anticoagulant or clot resolution therapy.

Anesthesia may be required in these animals to facilitate unrelated emergent or elective procedures or in an attempt to restore normal sinus rhythm with electrical cardioversion or ablation of foci that trigger atrial fibrillation. While there are differences in patient management for these two different scenarios basic tenants apply. For animals with primary cardiac disease knowledge of medications (and their recent administration) being used to manage the disease are important as these might impact anesthesia management. Animals may also be receiving medications to control heart rate related to their atrial fibrillation. Examples of drugs used to decrease ventricular rate include calcium channel blockers (e.g., diltiazem) or beta blockers (e.g., sotalol). These can contribute to hypotension during anesthesia and may also influence the effectiveness of inotropes when used at traditional doses. The anesthetist may also choose to use these to manage poor cardiac output if related to high ventricular rates (and poor diastolic filling) during anesthesia. A drug that has become increasingly used in dogs with atrial fibrillation scheduled for cardioversion is amiodarone. In addition to anti-arrhythmic properties, it is also reported to maintain coronary blood flow and reports suggest that normal sinus rhythm is better maintained following cardioversion in dogs pretreated with this drug. In our hospital dogs have typically received oral amiodarone for 1 – 2 weeks prior to scheduled cardioversion; the original IV formulation resulted in marked anaphylactic reactions in dogs as a result of the co-solvent polysorbate 80. The availability of a cyclodextrin preparation without these reported side effects for IV use may allow for more acute administration during the peri-anesthetic period.

For animals specifically presented for cardioversion, preparation should include assessment of serum electrolytes, specifically potassium and magnesium. If values are low, these should be normalized prior to anesthesia as it is suggested that this can improve the success of conversion to a normal sinus rhythm. When possible dogs are pre-instrumented (ECG [button electrodes], Doppler/NIBP and on some occasions an arterial catheter) following premedication and preoxygenated prior to anesthesia induction. The thorax is clipped and external pads (if being used instead of paddles) for administration of the electric current are pre-placed. While anesthesia may be maintained with balanced techniques (opioid +inhalation agent), given the short nature of these procedures and potential influence on conduction, many prefer a combination of injectable drugs including an opioid (e.g., fentanyl), amnesic drug with muscle relaxant properties (e.g., midazolam) and hypnotic agent (e.g., etomidate). During cardioversion, it is important that no individual comes in contact with the machine, table or dog and that the defibrillator is synchronized to deliver the 'shock' at the appropriate phase of the cardiac cycle (or ventricular fibrillation may result). Following conversion the animal may be gradually recovered or kept anesthetized for a brief period to ensure the rhythm does not revert.

PACEMAKER IMPLANTATION

While cardiac pacemakers may be placed for short term management during the peri-anesthesia period (e.g., atrial standstill following CPB, high second degree AV block noted after anesthesia induction), using percutaneous, surgical epicardial or transvenous lead placement or transthoracic pacing (with patches), it is more common that cardiac pacemaker implantation is utilized as a long term treatment. In veterinary medicine the two most common reasons for placement are third-degree atrioventricular block (AVB) and sick-sinus syndrome (SSS). The former is characterized by a complete dissociation of conduction between the atria and ventricles and a resulting default generally to a ventricular rate of approximately 30 – 50 beats per minute in dogs and 70-90 beats per minute in cats. Sometimes owners report syncopal episodes that may be mistaken for seizures; weakness or exercise intolerance are also symptoms of the disease. This diagnosis is confirmed electrocardiographically with non-conducted p waves appearing at one rate and wide ventricular 'escape' complexes appearing at a different and slower rate as mentioned. Secondary cardiac disease may or not be a feature. Systolic blood pressure tends to be high whereas diastolic and consequently mean arterial blood pressure tends to be low. Sick-sinus syndrome is more challenging to diagnose as animals may show normal ECG activity and behavior interspersed with cardiac changes and symptoms which are associated with the lack of impulse generation from the sinoatrial node resulting asystole. Typically neither condition is responsive to anticholinergic therapy; beta agonists may work to increase heart rate in some animals. In the author's experience SSS has proved to be more unpredictable and therefore more challenging to manage during anesthesia and it is prudent to have a temporary pacing system available or in place for anesthesia in the event of a sinus arrest.

In addition to knowledge of the disease and any underlying cardiac pathology, the anesthetist should be aware of the type of pacemaker (e.g., single or dual chamber, triggered or inhibited, etc.) being placed and whether this will be performed using a surgical (typically trans-diaphragmatic approach with epicardial leads) or transvenous approach (typically jugular). The pacemaker system consists of the pulse generator and pacing lead(s) and are universally described by a 5 letter code. While new generators have a life expectancy of 8-10 years, circumstances may be such that this is shorter when these devices are acquired for veterinary use and hence occasionally require replacement.

Anesthesia management of these animals should be planned after a complete assessment of their clinical presentation and level of compromise. While use of inhaled anesthetic agents has been questioned due to potential effects on conduction, they are commonly used as part of a balanced anesthesia technique in our hospital. As mentioned previously diastolic and mean arterial pressures tend to be low despite high normal systolic blood pressure. This is thought to be the consequence of a lengthy period of diastolic run off associated with the low heart rate. If temporary pacing is not available and since anticholinergics are ineffective, beta agonists may be used to try increase heart rate. Both dopamine and dobutamine have been used (at doses of 5-10 µg/kg/min) with mixed success. Other agonists such as isoproterenol are not usually recommended as vasodilation is a common feature associated with their use and heart rate elevation is generally not sufficient in these animals to compensate. While ventricular dysrhythmias may be observed during lead placement, lidocaine should not be used until pacing capture is certain for fear this anti-arrhythmic drug may suppress the ventricular escape beats keeping the animal alive. One final feature the anesthetist should be aware of is the phenomenon of 'overdrive suppression' as this can be unnerving if unexpected. Often following placement of the pacemaker and pacing capture, if the pacemaker is disconnected for any reason (as for example tunneling of the wires to the generator), the heart will typically stop beating (resulting in a no cardiac output state) for a few seconds before reverting to its escape rhythm until paced again.

SUMMARY

Due to the many different forms of cardiac disease and constant advances in technology and knowledge it is essential that a team approach is utilized to ensure a successful outcome for animals requiring anesthesia either for attenuation/treatment of their disease or unrelated problem. While additional procedures such as valvular replacements or repairs are performed in dogs, these are highly specialized, require use of cardiopulmonary bypass and are not yet routine in clinical practice.

REFERENCES

[1]Bright JM. J Vet Cardiology 2005; 7:85, [2]Johnson MS. JVIM 2004:18 (5):656. [3]Tobias AH. Vet Clin NA-SA Pract.2010; 40 (4):581. [4]Saunders AB. JVIM 2006; 20 (4):921. [5]Bright JM. JVIM 2008; 22 (1):114. [6]Pedro B, J Small An Pract 2002;53 (1):19. [7]Cushing DJ. Europ J of Pharmacol 2009:167

CARDIAC MANIFESTATIONS OF CHAGAS DISEASE IN DOGS

Ashley B. Saunders, DVM, DACVIM (Cardiology), Sonya G. Gordon, DVM, DVSc, DACVIM (Cardiology)
College Station, TX

INTRODUCTION

Trypanosoma cruzi is a protozoal organism transmitted by *Triatoma* insect vectors from the *Reduviidae* family, also known as cone-nose or kissing bugs. The *T. cruzi* organism is transmitted to the host from a bite from an infected bug or ingestion of an infected bug. Transplacental transmission and laboratory accidents have also been reported as methods of transmission. The metacyclic trypomastigotes enter and invade host cells. They become amastigotes in the myocardium, resulting in various degrees of myocarditis and fibrosis that manifests clinically as arrhythmias, cardiomyopathy, heart failure and sudden death in dogs. The acute stage of the disease is described as intense inflammation that often goes undetected or can be associated with fever, lethargy, lymphadenopathy and arrhythmias.[1] The latent/indeterminate/chronic stage is described as myocardial fibrosis and clinical signs, if they develop, are related to arrhythmias, cardiomyopathy (dilated cardiomyopathy phenotype), heart failure, and sudden death. All dogs are at risk of infection, especially those from or that have traveled to areas where the insect vector has been identified. Dogs housed outdoors may be at increased risk. The kissing bug vectors typically feed at night and are attracted to light. The disease is endemic in Latin America, and is present in predominately southern states in the US.

CLINICAL PRESENTATION AND DIAGNOSIS

Characteristic clinical problems include arrhythmias (ventricular, sinus node dysfunction, atrioventricular block), cardiomegaly, ventricular dysfunction and heart failure. Physical examination findings support heart disease (typically auscultable arrhythmias or clinical signs of heart failure) but are not specific for Chagas disease. Abnormal findings could include tachycardia, arrhythmias, abnormal pulse rhythm and quality, tachypnea or dyspnea, murmur and ascites. Lymphadenopathy has been documented as well.

The index of suspicion increases for dogs from an endemic location or when abnormalities characteristic of the disease are identified in an atypical breed (i.e. a non-Doberman Pinscher with arrhythmias and DCM phenotype). The authors have also diagnosed Doberman Pinschers with DCM and a positive Chagas titer. Consider testing when: 1. The dog originated, has traveled to, or is currently living in an area where Chagas disease has been reported including most of the US, although the disease is more common in southern states; 2. Dogs with arrhythmias, dilated cardiomyopathy phenotype, and congestive heart failure (Doberman or atypical breed); 3. Dog whose mother, littermate, or housemate (dog or cat) has been diagnosed with Chagas disease.

Diagnostic tests

In general, echocardiography and ECG are used to characterize the heart disease but are not specific for diagnosing Chagas disease. Additional tests may be useful for assessing the patient but will not specifically diagnose Chagas disease (i.e. thoracic radiographs, blood pressure, CBC, biochemistries, cardiac troponin I).

Specific tests for Chagas disease:

- Buffy coat smear to identify trypomastigote organisms in the acute stage of the disease.
- Fine needle aspirate and cytology of lymph nodes may identify amastigotes if lymphadenopathy is present.
- Serology including serum antibody titer (immunofluorescent antibody or ELISA). The Chagas screening test for *Trypanosoma cruzi* is simply for "screening" purposes and does not include a serologic titer. Serology that includes a titer is useful. A single negative titer does not rule out the disease and a negative titer does not necessarily indicate the dog has cleared the infection. A titer may be negative in the acute stage of the disease, and cross-reaction with *Leishmania* is possible. If the initial test is negative, a titer can be repeated if Chagas disease remains of clinical concern.
- Polymerase chain reaction (PCR) assay of blood or tissue is available in a few, select laboratories.

MANAGEMENT AND PROGNOSIS

Current management is directed at controlling the cardiac effects of the disease, rather than targeting the organism itself. In symptomatic dogs, management is based on clinical signs at that time. For ventricular arrhythmias, antiarrhythmic therapy options include lidocaine, sotalol and mexilitine. For atrioventricular block or sinus node dysfunction resulting in bradycardia, weakness, or collapse, pacemaker implantation may be indicated.[2] Standard supportive therapy for congestive heart failure is instituted as indicated and would include oxygen supplementation if dyspneic, pimobendan, ACE Inhibitor, diuretic therapy with furosemide/spironolactone, and abdominocentesis if ascites is present and compromising breathing or patient comfort.

Arrhythmias and heart failure typically require long-term management for the remainder of the life of the dog. Even in the asymptomatic patient, routine recheck evaluations are advised to monitor for development of clinical signs of heart disease. Anti-trypanosomal medications used to specifically treat the parasitic organism (ie. benznidazole, nifurtimox) do not currently have consistently proven efficacy for curing the disease and are not readily available in North America. Veterinarians may contact the Centers for Disease Control and Prevention in Atlanta, GA (www.cdc.gov/parasites/chagas) for the most up-to-date information regarding current recommendations as well as use and availability of anti-protozoal drug therapy for both dogs and humans.

Not all dogs with Chagas disease develop clinical signs but those that do can have disease progression, and require repeat evaluation. Dogs with acute signs, congestive heart failure, or that were infected at a young age (<6 months of age) typically have a worse prognosis. Disease progression is common despite symptomatic therapy and could include diminishing response to antiarrhythmic medication or progression of heart failure. New arrhythmias may develop and require alterations in treatment. Sudden death suspected to be related to arrhythmias is possible.

CONTROL AND PREVENTION

If a dog is diagnosed with Chagas disease, the other dogs in the household or litter should be tested for the infection as well. Affected female dogs should not be used for breeding due to the increased risk of transplacental transmission. Insect control is an important component of managing the disease. The insects are attracted to light and often feed at night. Dogs housed outdoors are more likely to encounter the insect vectors. Animal reservoirs in North America include wood rats, raccoons, opossums, and armadillos. Eliminate woodpiles, wood rat nesting sites and other brushy areas that can serve as breeding areas for the insect vectors.

REFERENCES

1. Barr SC, et al. Trypanosomiasis. In: Sykes J, ed. Canine and Feline Infectious Diseases. Philadelphia, PA: Saunders Elsevier; 2013. p. 760-770.
2. Saunders AB, et al. J Vet Intern Med 2013;27(4):890.

NEW INSIGHTS INTO THE COMMONALITIES BETWEEN HUMAN AND CANINE MYXOMATOUS MITRAL VALVE DISEASE

Kathryn H. Driesbaugh, PhD and Mark A. Oyama, DVM, DACVIM (Cardiology)
Philadelphia PA

Myxomatous mitral valve disease (MMVD) is a common cardiac disease in both human and canine patients. Many of MMVD's clinical, gross, histological, and molecular traits are similar between these two species, and the study of MMVD in dogs informs medical professionals about MMVD in humans and vice versa. This presentation is designed to highlight the particularly intriguing commonalities and differences between human and canine MMVD, with an eye towards how collaborative investigations between medical and veterinary researchers can help advance our understanding and treatment of MMVD.

TERMINOLOGY AND CLASSIFICATION OF MITRAL REGURGITATION

Common diseases of the mitral valve in human patients include the following: mitral stenosis, mitral regurgitation (MR), and mitral valve prolapse (MVP). MR can be caused by primary (usually degenerative) conditions or be secondary to other conditions such as ischemic heart disease or idiopathic dilated cardiomyopathy. In dogs, the vast majority of clinically significant MR is caused by primary myxomatous disease. In humans, MVP is defined by echocardiographic evidence of superior displacement of the valve leaflet tissue >2mm above the plane of the mitral valve annulus in the long axis view. A similar definition for MVP in the dog exists. However, whereas MVP in humans is typically regarded as a separate disease entity with a natural history that is often different from that of primary MR, MVP in the dog is typically regarded as part of the clinical findings associated with MMVD. In the presenters' view, these differences make direct comparison of MVP between the 2 species problematic.

PATHOPHYSIOLOGICAL MECHANISMS OF MMVD

In recent years, much attention has been paid to the pathological mechanisms that underlie MMVD. Important pathways include those involving oxidative stress, transforming growth factor-beta, matrix metalloproteinases and their inhibitors, and serotonin. Many of these pathways are closely related, and our understanding of the exact sequence of events during the development and progression of MMVD in dogs and humans remains opaque. The presenters and others have investigated the role of oxidative stress and serotonin in human and canine MMVD. Increasingly, serotonin and the 2B serotonin receptor subtype have been implicated in pathogenesis of MMVD in both species.

TREATMENT OF MMVD

MMVD in human patients is considered a surgical disease insofar as mitral valve replacement or repair is the treatment of choice in patients with moderate to severe disease (Figure, adapted from Braunwald *et al*). Thus, data involving efficacy or safety of chronic medical therapy is lacking in human patients. MMVD in dogs is primarily a medical disease that is treated with diuretics, ACEI, and positive inotropes. Despite advances in medical therapy, the prognosis of dogs with congestive heart failure due to MMVD remains very poor. Treatment of MMVD in dogs with earlier asymptomatic stages of disease with medications such as ACEI is controversial. No data regarding treatment of dogs at risk for development of MMVD are available. Advances in our understanding of the pathophysiological mechanisms of disease make clinical trials in earlier stages of MMVD possible.

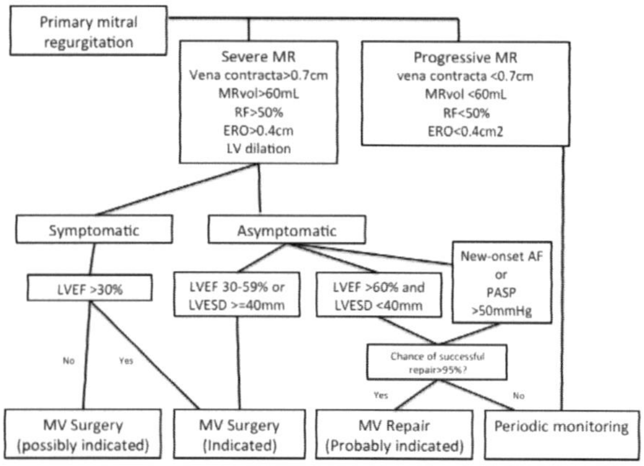

REFERENCES

Pedersen HD and Haggstrom J. Mitral valve prolapse in the dog: a model of mitral valve prolapse in man. Cardiovasc Res 2000;47:234-43.
Borgarelli M and Buchanan JW. Historical review, epidemiology, and natural history of degenerative mitral valve disease. J Vet Cardiol 2012;14:93-101.
Dal-Bianco JP et al. Basic mechanisms of mitral regurgitation. Can J Cardiol 2014;30:971-81.

REAL TIME 3D-ECHOCARDIOGRAPHIC MORPHOLOGIC AND QUANTITATIVE ASSESSMENT OF MITRAL VALVE IN NORMAL DOGS AND DOGS WITH MYXOMATOUS MITRAL VALVE DEGENERATION

M. Borgarelli, DMV, PhD, DECVIM-CA (Cardiology)
Blacksburg, VA

The mitral valve (MV) has a complex three-dimensional structure that is difficult to investigate with conventional techniques, such as two-dimensional echocardiography or angiography. Particularly, the valve's annulus is a hyperbolic paraboloid (commonly defined as "saddle-shape"), characteristically conserved across a number of mammalian species. This peculiar geometry has been demonstrated to confer a mechanical advantage and to be important in reducing leaflet stress and maintaining valve competency in humans. Myxomatous mitral valve disease (MMVD) is the most common acquired cardiovascular disease in dogs. Diagnosis of MMVD is commonly based on the presence of a left apical cardiac murmur and identification of MV leaflet thickening and mitral valve regurgitation by means of conventional 2D (2DE) and Doppler echocardiography. However, because of the complexity of MV apparatus and of changes occurring with MMVD, a comprehensive appreciation of mitral valve lesions and morphological changes using 2D echocardiographic presents several limits. Recently, real-time three-dimensional echocardiography (RT-3DE) has become available and it has been demonstrated to provide a better non-invasive assessment of MV in vivo. Real-time 3DE acquires full volume datasets of the MV that, when analyzed using appropriate post-processing, allows us to see MV in multiple planes. In human beings, RT-3DE has become a standard technique to assess patients with mitral valve prolapse and MMVD and the technique has been reported being superior to 2DE to correctly identify and localize MV prolapse. Moreover, in this species, changes of MV morphology detected by RT-3DE have been associated with an increased risk of disease progression: in particular, a flattening of the annular saddle-shape, was associated with leaflet billowing and increased frequency of chordal rupture. Finally, in humans, RT-3DE allows surgeons to predict the complexity of mitral valve repair. In veterinary medicine, RT-3DE has been recently introduced and it has been demonstrated to provide accurate information about left atrial enlargement. Moreover, RT-3DE assessment of the left ventricle in dogs with MMVD suggested that left ventricular remodeling is mainly due to changes that occur in the middle segments. Analysis of the MV by RT-3DE requires image acquisition from both right parasternal long axis and left apical views. During acquisition, the MV should be positioned in the middle of the ideal 3D pyramidal volume and should include a portion of left atrium, left ventricle and the aorta. Every attempt should be made to reduce stitching artifacts due to excessive thoracic motion. Moreover, in order to correctly post-process the datasets, a simultaneous recording of a clean ECG trace is required. Multi-planar assessment of the acquired images can be used to identify the presence of mitral valve prolapse or a flailing leaflet, and which scallop of the MV is affected. Acquired full volume images can be post-processed using dedicated software that allows construction of a three-dimensional color coded valvular model (parametric map). This analysis not only permits the correct identification of which mitral valve scallops are affected, but also provides quantitative data about MV morphology such as mitral valve annulus area, area of mitral valve leaflets, measure of mitral valve annulus planarity and tenting height (distance from mitral valve annulus to valve leaflets coaptation point), tenting volumes and tenting areas.

Our group has tested the feasibility of RT-3DE to assess the canine MV over the past two-years. In this period of time, we acquired and analyzed data from both healthy dogs, and dogs affected by MMVD. We have reported that RT-3DE is a feasible technique for analyzing canine mitral valves in clinical settings, but that there is a learning curve, with the percentage of analyzable datasets increasing over time. We also reported that the MV of healthy dogs is elliptical, with a smaller antero-posterior diameter compared to the anterolateral-posteromedial one, and has a saddle-shape, described by an annulus height to commissural width ratio (AHCWR) of 21%, about the same reported in healthy humans. In our studies, dogs affected by MMVD instead have a significantly more circular annulus, and a reduced AHCWR (that is, a more flat annulus). In addition, the valve's tenting height decreases significantly with the severity of the disease. The annulus height is already significantly reduced in pre-clinical affected dogs, compared with normal dogs. Accordingly, our data suggest that dogs with MMVD exhibit changes in MV annular and leaflet morphology, and that these changes are associated with the severity of the disease. However, these data do not provide any information about the predictive value of morphological changes to identify dogs with a higher risk of progression. Therefore, we are now planning to perform a longitudinal study to assess the effects of changes in MV morphology on the progression of MMVD in dogs.

REFERENCES

1. Levine RA, Handschumacher MD, Sanfilippo AJ, Hagege AA, Harrigan P, Marshall JE, Weyman AE. Three-dimensional echocardiographic reconstruction of the mitral valve, with implications for the diagnosis of mitral valve prolapse. Circulation 1989;80:589–598
2. La Canna G, Arendar I, Maisano F, et al. Real-time three-dimensional transesophageal echocardiography for assessment of mitral valve functional anatomy in patients with prolapse-related regurgitation. The American journal of cardiology 2011;107:1365-1374.
3. Tsang W, Lang RM. Three-dimensional echocardiography is essential for intraoperative assessment of mitral regurgitation. Circulation 2013;128:643-652; discussion 652.
4. Tidholm A, Bodegard-Westling A, Hoglund K, et al. Comparisons of 2- and 3-dimensional echocardiographic methods for estimation of left atrial size in dogs with and without myxomatous mitral valve disease. Journal of veterinary internal medicine / American College of Veterinary Internal Medicine 2011;25:1320-1327.
5. Ljungvall I, Hoglund K, Carnabuci C, et al. Assessment of global and regional left ventricular volume and shape by real-time 3-dimensional echocardiography in dogs with myxomatous mitral valve disease. Journal of veterinary internal medicine / American College of Veterinary Internal Medicine 2011;25:1036-1043.
6. Tsang W, Weinert L, Sugeng L, et al. The value of three-dimensional echocardiography derived mitral valve parametric maps and the role of experience in the diagnosis of pathology. Journal of the American Society of Echocardiography : official publication of the American Society of Echocardiography 2011;24:860-867.
7. Grewal J, Suri R, Mankad S, et al. Mitral annular dynamics in myxomatous valve disease: new insights with real-time 3-dimensional echocardiography. Circulation 2010;121:1423–1431

8. Moustafa SE, Mookadam F, Alharthi M, et al. Mitral annular geometry in normal and myxomatous mitral valves: three-dimensional transesophageal echocardiographic quantification. J Heart Valve Dis 2012;21:299–310.
9. Lee AP-W, Hsiung MC, Salgo IS, et al. Quantitative analysis of mitral valve morphology in mitral valve prolapse with real-time 3-dimensional echocardiography: importance of annular saddle shape in the pathogenesis of mitral regurgitation. Circulation 2013;127:832–841
10. Menciotti G, Borgarelli M, Wesselowski S, Abbott J. Quantitative evaluation of canine mitral valve in dogs using three-dimensional echocardiography. Res. Commun. 24th ECVIM-CA Congr. Mainz, Germany: Wiley Online Library; 2015. p. 436–437.

FORECASTING OF HEARTWORM PREVALENCE AND THE IMPACT OF PREVENTION

Dwight D. Bowman, MS, PhD, DACVM (Honorary, Parasitology)
Ithaca, NY

The Companion Animal Parasite Council (CAPC) has been collecting data on the number of antigen positive tests for heartworm from around the nation based upon test results collected by IDEXX Laboratories, Inc. and ANTECH Diagnostics. These data are collected monthly and formatted into interactive maps that appear on the CAPC website – www.capcvet.org. The amount of data that is being collected has become a very large data set. Currently available for viewing are data from the years 2011, 2012, 2013, and 2014 which map 4.8, 4.2, 5.7, and 7.2 million data points/year, respectively. There is now data being collected in all of the states of the United States, and Puerto Rico, with some states having the majority of positive tests, e.g., in 2014 Texas represented 19.75% and Florida represented 8.95% of all positive tests performed, and some states are represented with only a very few tests being performed, e.g., in 2014 Alaska had only 1,241 tests, Wyoming had only 2,358 tests and Montana had only 2,995 tests performed. The maps are designed to provide a cross section between the scientific data included and a display that is interpretable by clients as well as veterinarians; thus, it appears on the maps as a gradation of risk ranked as Low, Medium, or High based on the prevalence of reported cases that year. It must be remembered, that the maps show the prevalence of infection in dogs visiting veterinarians, and thus, the prevalence is less than would be expected in dogs that are strays or collected in surveys performed at shelters. Also, there is no clinical history of any sort collected on the sampled dogs; therefore, it is not possible to say that a dog that is positive in Montana was infected within Montana or whether there may be more than one test performed on the same dog in Texas. However, because the data that is collected does represent data on circulating antigen, the results are not representing exposure but the presence of circulating antigen in the blood that can be interpreted as infection. The maps overall do not typically present any significant surprises when viewed, but states collecting few data points (e.g., Alaska and Wyoming) may appear as of High Risk because veterinarians in these states are using the tests as confirmatory diagnostics for dogs with signs of infection rather than as part of annual screening. Often such states with low-volume samples change risk status year-to-year as would be expected of populations where only a few dogs are tested annually.

Figure 1 CAPC Heartworm Prevalence 2014 (www.capcvet.org)

CAPC is working hard to begin to tease information out of the map data. Maps can be spatially smoothed for the purposes of minimizing the influence of states with only minimal data being collected. This was done for the data collected in 2011 when the a group of parasitologists, veterinarians, and entomologists met to determine what factors might be most important in trying to associate the maps with various trends (Figure 2).[1] In these maps it can be seen that many of the small disparities tend to disappear as the data is entered by county, but areas still appear, e.g., Nevada in Figure 2, that seem a bit out of synch with what would be expected by most familiar with the typical prevalence of heartworms in given areas. The question then becomes more interesting, i.e., is this an accident of mapping the collected data or does it represent a trend that will persist over time. These are some of the questions that can then be asked with the maps that are generated. This is the first inkling of what may be possible with the maps to examine the potential spread of heartworm to new areas or potentially contraction of infection status with the intervention of additional preventive (or mosquito control) protection of dogs. However, as the data continues to be collected there are now three larger questions that can be asked that are of direct concern to practitioners around the nation. First, what is the chance that a dog walking into a practice in a given area is going to test positive for heartworm? Can we predict trends for a given area as to how many dogs may be coming into a practice next year in a given month based on various environmental factors, i.e., can we suggest annual trends compared to various externally measured factors? And most importantly, will it be possible to examine the effects of intervention on reducing the amount of heartworm in an area, i.e., what percentage of increased heartworm protection of a dog population might be required to cause a 10% reduction in observed cases in a given state or county?

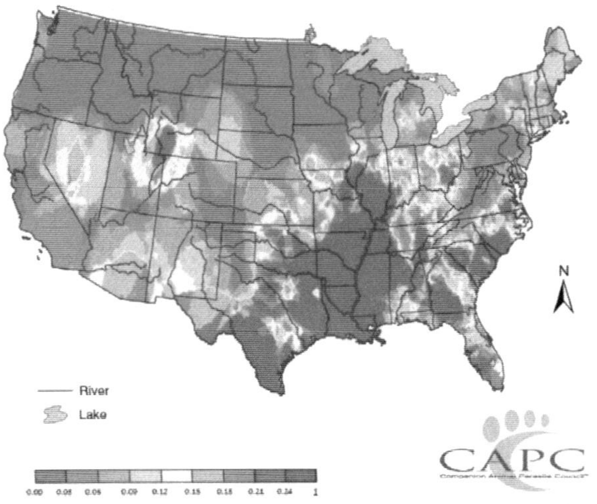

Figure 2 Smoothed Heartworm Prevalence Data Collected in 2011[1]

Along these lines, the next work that was done by the mapping team was to examine what could be learned by including the different identified factors in maps that dealt with heartworm infection and parasite-vector ecology, i.e., by including in the mapping the effects of potential changes in factors and their effects over the course of a year. This work was presented in a paper dealing with the utilization of the identified factors and the development of modeling methods for including these factors in the production of more accurate maps.[2] This work was based on the data collected in 2011 and 2012 and represented 9.5 million or so tests (Figure 3). The raw data was represented simply by counties and then, it was smoothed using a method called "head-banging" which is a method similar to what occurs when you have pins in a board that are used to produce an image of a hand that is placed on the board that smooths the trends between the different parts of the hand to produce a smoothed image of the hand – the more pins, the more "rough" the image. Thus, the amount of smoothing depends on the size of the individual points that are sampled. Simple smoothing by this method produced the map that is seen in Figure 4. This map takes into account none of the factors that were identified as important – i.e., climate conditions (annual temperature, precipitation, and relative humidity), socio-economic conditions (population density, household income), local topography (surface water and forestation coverage, elevation), and vector presence (several mosquito species). This map simply smoothes the data presented in Figure 3, and it can be seen that there is increased sensitivity added by the presence of the larger numbers of samples present on the map, and again, the representation is similar to what would be expected by those familiar with where heartworm prevalence is expected to be higher or lower in the United States. However, this map does not take into account the fact that the defined factors may play a significant role in where heartworm prevalence is higher in different areas in the country.

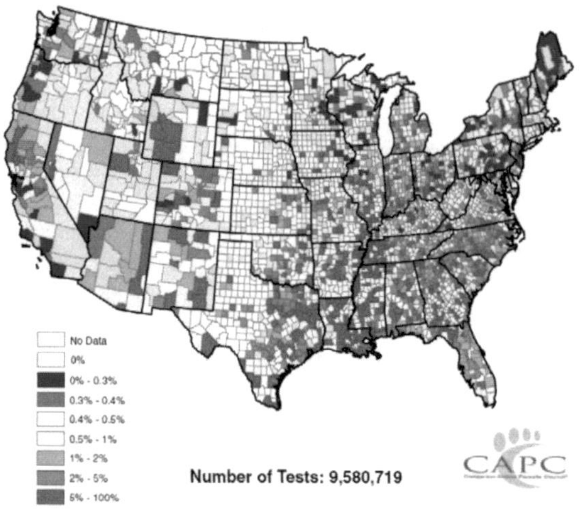

Figure 3 Raw Heartworm Prevalence Data from 2011 and 2012[2]

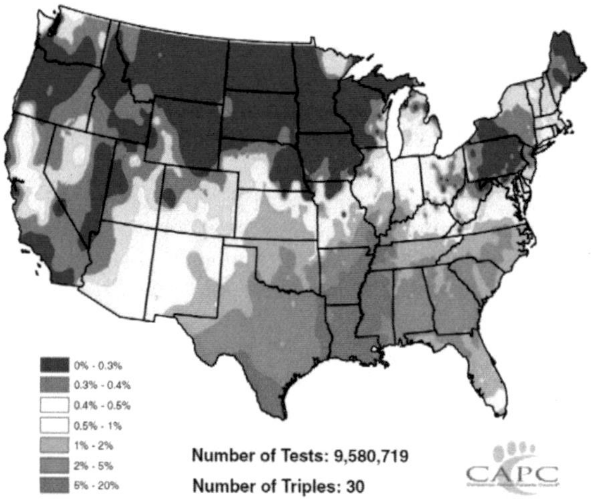

Figure 4 Smoothed Heartworm Prevalence Data from 2011 and 2012[2]

The next step was to add these different factors into the maps. This was done by the utilization of logistic regression where a model was produced that incorporated the different factors described above into an equation that defines each point based on the inclusion of the factors and the different magnitudes of their impact on the final location of prevalence on the maps (Figure 5). Thus, it then produces a method by which each point on the map is determined mathematically with a sum of the prevalence based on the different factors being built into the maps. When this is done, it provides a map that will allow for the inclusion of changes in these factors, i.e., when there is an increase in temperature noted in a given area or an increase in the median household income in an area. Thus, it becomes possible to look at what will happen when there are such changes that might occur, such as, an especially cold winter, warm summer, or another economic downturn as occurred in 2008. It can be seen that this map is not a perfect representation of the map seen in Figure 4, but it is a pretty good representation of the map, and it needs to be remembered that this map is showing the chance that a dog walking into a practice will be positive based on the factors that were determined from the original map but with the data being a mathematical summary of prevalence combined with the various factors that play a significant role in the determining the prevalence of a positive heartworm test within a given region. The map seems to predict a slightly higher risk of infection in the upper western states of the United States, but now it would be possible to combine the effects of hot dry weather in the western United States on the potential for a decrease in the number of positive cases in the west or the effects of a drought in the southeastern United States. These maps are no longer locked directly into what has been observed directly relative to prevalence, but they can now take into account changes that will change heartworm prevalence in a given area. Combining these factors allows then for a means of determining what might be if conditions change. This is the basis for where the maps can have a significant input by being able to forecast what might be. These can have major implications for defining how prevention might be affected by various temporal changes in the factors included in the mapping, and how intervention might change prevalence in a given area.

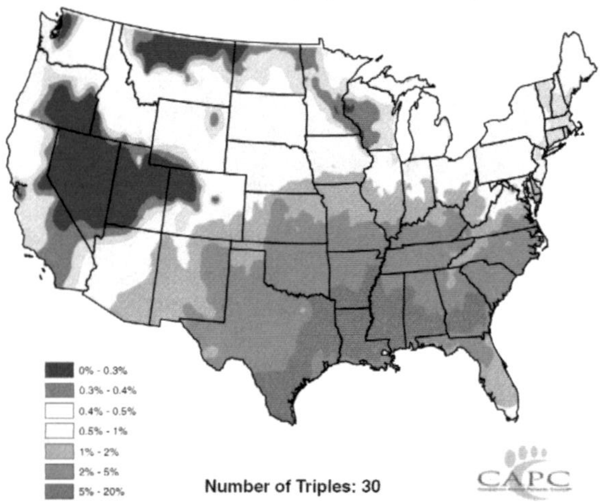

Figure 5 Predicting Heartworm Prevalence with Significant Factors[2]

Forecasting is a word that worries many when it is discussed in terms of disease mapping. People worry about the value of the information and the fact that the information may be more a supposition than a guess. However, most of us have become used to various forecasts based on similar calculations, examples would be weather and economic trends for a given area. We are aware that such methods provide information, but our concern always remains that it is difficult to decide to what extent they are representing what will really happen. We have learned that often the forecasts for weather may be generally close to some sort of accurate description of what is to come, but we also are aware that they are sometimes incorrect, and sometimes very incorrect. However, things have developed to the point, where there is an expectation that weather predictions, at least in the near term, are often fairly close to the mark. Now, it is possible to begin to forecast what might be and to compare it to the data on the basis of what actually takes place. The Companion Animal Parasite Council has begun an effort to try such forms of forecasting. It is just in its formative stages, but things are progressing and might lead to some interesting results in years to come.

First, there is the potential to simply utilize the factors and past prevalence rates to predict what trends might be for the coming year or coming months. The CAPC statisticians have forecast what was to be seen in 2013 based on the predicted upcoming temperatures for the year and the map generated from the data collected in 2012 and 2013. Basically, the prediction was too low in the southeastern United States and too high in the central and northwestern United States (Figure 6). However, overall the trend was fairly good. The large white areas in Figure 6 show that there were large areas of the United States where there was a very good fit between the predicted and the observed. It is believed that with the collection of a larger and larger baseline sample, improved mapping methods that are under development, and the identification of additional factors of influence that the maps will just get better at predicting the percentage of dogs that enter a practice testing positive for heartworm infections.

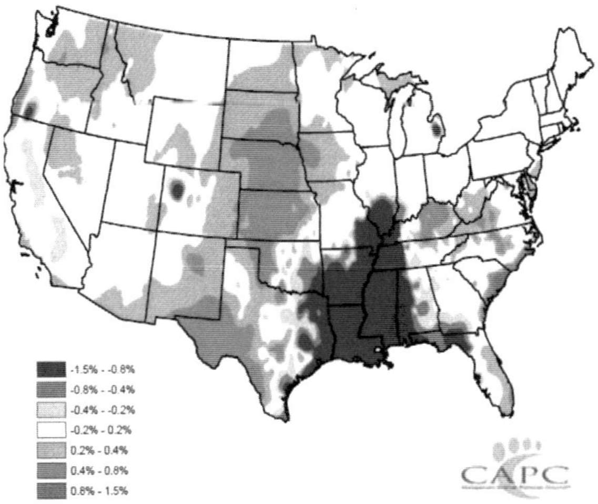

Figure 6 2013 Heartworm Prevalence - Predicted versus Observed

The most important thing that the maps can ultimately do is to assess the impacts of intervention. In 2014 an important paper was published out of the University of Pennsylvania that showed the effects of the introduction of broad-spectrum heartworm preventatives and flea products on the prevalence of internal parasites in dogs entering and having fecal examinations within the University's clinic. The effect was dramatic and significant.[3] Basically, the declines seen in roundworms, hookworms, whipworms, and tapeworms were all associated with the year of introduction of preventive products (Interceptor® (milbemycin oxime), (c) Heartgard Plus® (ivermectin + pyrantel pamoate), (d) Program® (lufenuron), and (e) Frontline® (fipronil)) that had effects on these parasites either directly for the hookworms, roundworms, and whipworms, or indirectly in the case of the tapeworm *Dipylidium caninum* as the fleas came under the excellent control provided by the products when they entered the marketplace. The significance of this paper is that it very carefully illustrates that these products and their utilization are having the desired effect. They are reducing the presence of these parasites and hence any associated morbidity or mortality within the pets for which these products are prescribed. Ultimately, it is hoped that the CAPC maps can do the same on a larger scale relative to heartworm and the other parasites and tick transmitted diseases for which data is being collected and which is currently being mapped.

In 1989, before the introduction of monthly heartworm preventives, a paper was published from Florida, that showed that the population of heartworms as determined by necropsy of 876 shelter dogs (89.2% from northeastern Florida and 16.8% from central Florida) that 520 (59.4%) had heartworms in their heart, lungs, pulmonary arteries, or venae cavae.[4] When one looks at similar counties today in Florida on the CAPC maps, the prevalence in dogs seeing veterinarians is 1.3%. Also, remarkably, when one looks at shelter dogs in Florida from an area similar to that examined in 1989, the current prevalence by antigen testing was 14.6%,[5] which raises the possibility that the excellent protection provided to pet dogs may be great enough that it is even having an impact on dogs that are not seeing veterinarians within the state of Florida. CAPC has begun work looking at the association between prevalence rates of different parasitic diseases across the United States and how they relate to drug sales by treatment category by state. Ultimately, it is hoped that it will be possible to model the effects of intervention with questions being posed such as: What would a 2% increase in heartworm preventives leaving each practice have on the prevalence of heartworm within a given state? Or What increase in heartworm preventives leaving each practice would be required to reduce heartworm positive tests in Texas by 1% in the coming year? Such predictions, even if close, will provide targets for intervention programs with measureable outcomes. This will markedly change the discussions around the value of prevention in any given area.

REFERENCES

[1]Brown et al. **Parasites and Vectors** 2012; 5 245; [2]Wang D et al. **Parasites and Vectors** 2014; 7 264; [3]Gates MC, Nolan TJ **Veterinary Parasitology** 2014; 204 (3-4): 265; [4]Courtney CH, Zeng QY **Proceedings of the American Heartworm Society, Heartworm Symposium** '89 1989; 1; [5]Tzipory N et al. **Veterinary Parasitology** 2010; 171: 136

FORECASTING OF THE PREVALENCE OF CANINE TICK TRANSMITTED DISEASES AND THE IMPACT OF PARASITE CONTROL AND PREVENTION

Dwight D. Bowman, MS, PhD, DACVM (Honorary, Parasitology)
Ithaca, NY

The Companion Animal Parasite Council (CAPC) has been collecting data on the number of dogs with detectable antibodies against *Borrelia burgdorferi*, *Anaplasma* spp, and *Ehrlichia* spp. These data are collected monthly and formatted into interactive maps that appear on the CAPC website – www.capcvet.org (Figure 1). The amount of data that is being collected has become a very

Figure 1 CAPC Canine 2014 Seroprevalence Maps for *Borrelia burgdorferi*, *Anaplasma* spp, and *Ehrlichia* spp. (www.capcvet.org)

large data set. Currently available for viewing are data from the years 2011, 2012, 2013, and 2014 which map for antibodies to each agent test results for about 1.3, 1.1, 2.1, and 3.1 million data points/year, respectively. There is now data being collected in all of the states of the United States, and Puerto Rico, with some states where the agents are highly prevalent performing the majority of tests, e.g., in 2014 New York had 0.23 million tests for positive Lyme serology and 12.54% of the sera testing positive in the United States, and Pennsylvania had 0.21 million Lyme serology tests with a total of 14.73% of the total positive sera within the United States. Some locals, as would be expected have very few tests done each year, Puerto Rico is not always represented year to year, Alaska only had 19 dogs tested form antibody to *B. burgdorferi* tested in 2014, and Montana only tested 85 dogs for Ehrlichia antibodies this past year (2 were positive). The maps are designed to provide a cross section between the scientific data included and a display that is interpretable by clients as well as veterinarians; thus, it appears on the maps as a gradation of risk ranked as Low, Medium, or High based on the prevalence of reported cases that year. It must be remembered, that the maps show the prevalence of positive serologic tests in dogs visiting veterinarians. Also, there is no clinical history of any sort collected on the sampled dogs; therefore, it is not possible to say that a dog that is positive in Montana was infected within Montana or whether there may be more than one test performed on the same dog in Pennsylvania. The data that is collected is a map of seroprevalence, and since there is no clinical associated with the reported test result (positive or negative), the maps are not really showing disease incidence but rather seroprevalence. And unfortunately, since there is no tracing of animals from year to year, there will not be a method to assess the number of new cases each year. Overall, the maps overall do not typically present any significant surprises when viewed, but states collecting few data points (e.g., Alaska and Montana) may appear as of High Risk because veterinarians in these states are using the tests as confirmatory diagnostics for dogs with disease signs rather than as part of annual screening. Often such states with low-volume samples change risk status year-to-year as would be expected of populations where only a few dogs are tested annually.

The prevalence of dogs with positive Lyme serology is undoubtably expanding. When the author arrive in Ithaca, NY, there were no cases of Lyme present in people or animals. Now infections occur in both. If one looks at the CAPC maps from 2011 to 2014 (Figure 2), one can see that the prevalence has gone from 9.67% to 12.54%. However, of more interest perhaps is that all counties that are collecting data have moved from into the high risk level (except Schuyler County which tested 1 dog in each of 2013 and 2014). To those who live in this area, there is the shared opinion that Lyme has now spread across the state, and this data just verifies what has been observed from other cases. Also, there is data that is pretty convincing that when the prevalence of positive serology in dogs is greater than 5% that human cases begin to occur;[1] – so, now, basically, all of New York state is an at risk area for human exposure. There is general consensus that all the tick transmitted diseases are increasing in prevalence for a number of reasons, and unfortunately, there does not appear to be any significant method that has been developed to extend their growth and spread. Thus, it can only be expected that unless there is some phenomenal breakthrough in environmental tick control, the areas of tick disease transmission will continue to expand for years into the future until they perhaps come upon some form of geographical boundary that will hold them back for a period of years.

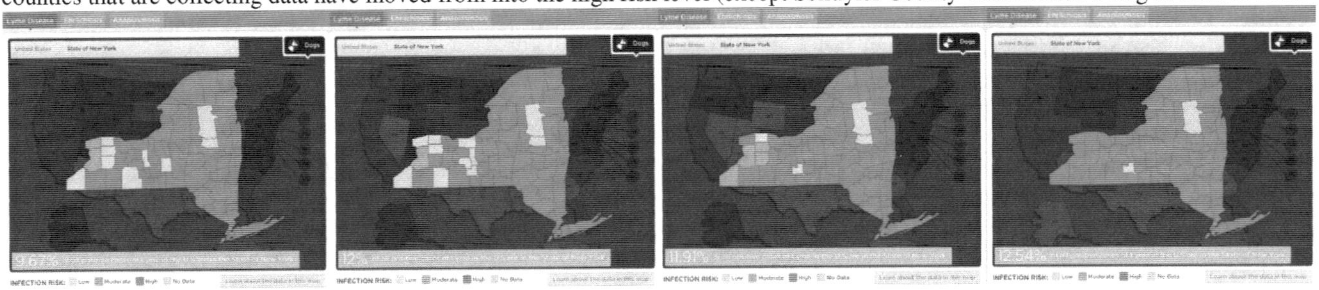

Figure 2 Seroprevalence of Lyme in New York 2011 through 2014 (www.capcvet.org)

As the Companion Animal Parasite Council has done for heartworm, it is also working to perform maps and forecasts around the different tick-transmitted disease for which data has been collected. The group held a meeting to identify factors for inclusion in tick-transmitted agent forecasting maps.[2] The factors that were chosen for the *Ixodes* transmitted agents were forest cover, relative humidity, annual precipitation, human population density, deer/vehicle collisions, topography, temperature, proximity of forest to highways, human case distribution, distribution or abundance of the ticks, household income, and the forest fragmentation index. For *Ehrlichia* the factors were maximum, minimum, and average temperature, precipitation, remote light reflectance transmission data, vegetation cover, reservoir densities, human population, median income, and seasonal precipitation. Thus, maps have been generated utilizing much of this data; however, it has yet to been carefully gone over relative to significance testing as to the value of the different factors compared to each other as has occurred in the case of canine heartworm.

The work is now progressing with *Anaplasma* along a similar path to what has occurred with heartworms.[3] After developing the list of factors for inclusion, the model had to be built and the factors tested as to their statistical usefulness in the model. Thus, the various factors have been built into models and the significant factors have been selected. Now, maps have been prepared with and without controls for known areas of endemicity; i.e., areas where the disease is already known to occur. This has the advantage that strange and unusual locations of disease prevalence outside of known endemic areas are not allowed to occur, but this has the potential to hide information such as where people might be at the time the disease is diagnosed or when the disease moves into a new area. If one looks in northwest Montana in Figure 3, there is a lot of *Anaplasma* there, and it seems that it is probably due to summer residents from other areas or few data points being collected. Similarly, the hot area surrounded in blue in Arizona might represent winter residents in Arizona who move there for half the year from areas where *Anaplasma* is endemic. After the *Anaplasma* model is fully stress tested, the model will be verified and reexamined with the seroprevalence of *Borrelia burgdorferi* that will involve the inclusion of the same or very similar factors.

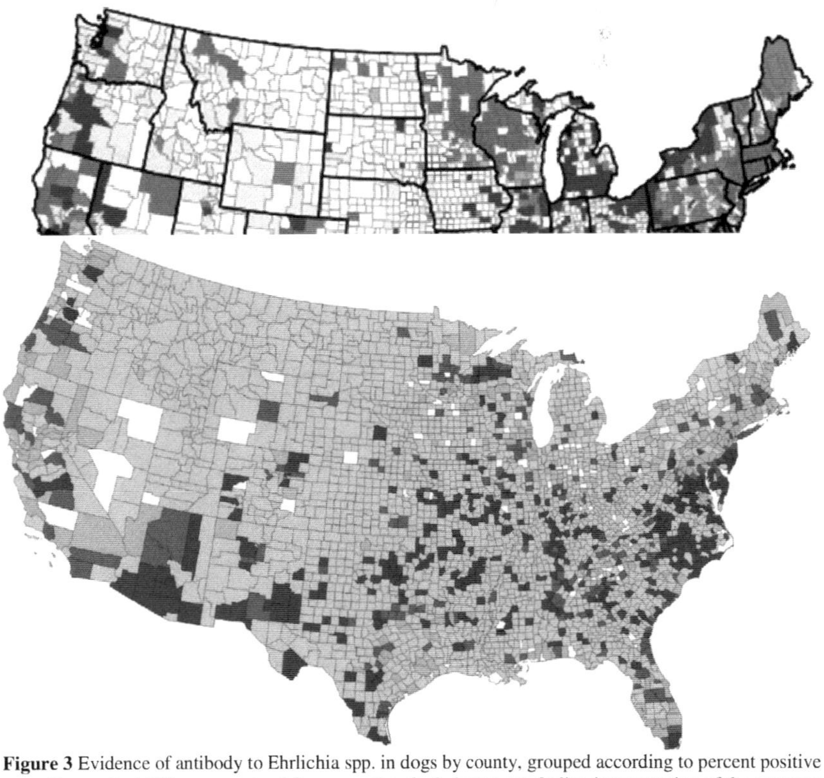

Figure 3 Evidence of antibody to Ehrlichia spp. in dogs by county, grouped according to percent positive tests. No results (<30) were received from counties shaded gray, precluding interpretation of the presence of antibody in dogs from these areas. Counties depicted in white had no dogs reported as positive (0%). Remaining counties were coded as follows: 0.1-0.5% (light purple), 0.5-1.0% (purple), 1.1-2.0% (dark purple), and > 2.0% (very dark purple).[5]

Ehrlichia spp. are probably the group of organisms that are putting more dogs at risk around a greater area of the United States than the other two agents. The reason is that it is a large species complex about which not enough is known. Prior work showed that there was a form of *Ehrlichia* sp. other than *E. canis*, *E. chaffeensis*, or *E. ewingii*, called *E. muris*, that caused the presence of *Ehrlichia* in unexpected regions of the upper midwest.[4] More recently similar work has shown a greater than expected percent positive *Ehrlichia* test results identified in Clark and King Counties in Washington, where results from more than 2,000 dogs were available.[5] It seems that there may be another species of *Ehrlichia* that is getting into dogs in a new geographic region. What is amazing is that when one looks at the United States there is very little that is *Ehrlichia* free – the geographic distribution is larger for *Ehrlichia* spp than either *Borrelia* or *Anaplasma*.

Work is now beginning to determine how best to measure the impact of tick preventives and vaccines on the seroprevalence of these diseases in dogs. This is a complicated problem because unlike heartworm where the test is for antigen, these tests are for antibody that may take months to clear after treatment. There are "preventives." In the case of *B. burgdorferi*, it is possible to see protection in OspA vaccinated dogs as monitored with a test detecting antibody to the C6 protein.[6] Tick protection has also been shown to protect against these infections in a number of studies with different products in the laboratory,[7] however, success in the field appears to be, as logic would suggest, associated with owner compliance in the routine application of product.[8] It needs to be remembered that the assays measure only whether the dogs have antibodies to one of these three agents. There is no information on whether the dog has been sampled multiple times, is on tick prevention, vaccinated, or previously infected and cleared. Significant tick protection could result in lowered prevalence rates, and it is also possible that some vaccines might also provide significant protection against seroconversion.

CAPC is working towards being able to look at product sales in relation to disease reduction. The goal is to utilize data on tick protection products along with vaccine sales data. The tick preventive information will be organized such that it will not include information as to details about the route of application or method of prevention. Thus, the maps when used to monitor results of sales, they will simply be able to produce fairly direct results, such as: a 5% increase in tick prevention product sales in a given state will result in a 1% reduction in tick-transmitted diseases within this area. However, they could be used predictively with statements such as: if we vaccinate 25% more of the dogs in this region we should be able to virtually eliminate *E. canis* from our state in two or three years. The goal of the maps is not to track disease spread, or to identify new hotspots. The reason to have the maps, is to make the prevalence in all areas equal to zero, or at least close to zero (Figure 5). People need to understand that there may be a risk, but simply knowing there is a risk provides no incentive for change if you feel incapable of modifying the outcome. And with the excellent products now in the veterinary pharmacopoeia, minimal improvements in product delivery are liable to have very large positive benefits in reducing the risk of pets developing disease from these tick-transmitted agents.

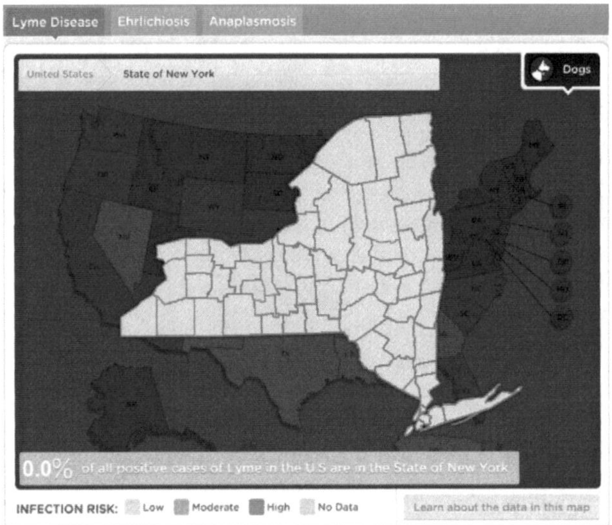

Figure 5 NY after treatment has reduced Lyme seroprevalence to ZERO

REFERENCES

[1]Mead P et al. **Emerging Infectious Diseases** 2011; 17(9): 1720; [2]Stich RW et al. **Parasites and Vectors** 2014; 7: 417; [3]McMahan CC et al. **Parasites and Vectors** 2015 – submitted; [4]Bowman DD et al. **Veterinary Parasitology** 2009; 160: 138; [5]Little SE et al. **Parasites and Vectors** 7 257; [6]Eschner AK, Mugnai K. **Parasites and Vectors** 2015: 8: 92; [7]Banks SD et al. **Medical and Veterinary Entomology** 2014; 28 (S1): 14; [8]Leshnick M et al. **Parasites and Vectors** 6: 62

ANTIEMETIC THERAPY IN HUMAN PATIENTS

Eli D. Ehrenpreis, MD
Chicago, IL

Introduction

The symptoms of nausea and vomiting are especially difficult for the human patient. To illustrate this point, Josh Billings a 19th-century American humorist stated that "one of the best temporary cures for pride and affection is seasickness; a man who wants to vomit never puts on airs". I am sure that all of us attending my lecture can appreciate the degree of disability that a person experiences when they suffer from the symptoms of nausea and vomiting.

Nausea and vomiting are commonly experienced symptoms in human patients. These symptoms may initially be managed in an outpatient setting by primary care physicians. Nausea and vomiting are also commonly encountered in emergency room patients. Patients with chronic issues related to nausea and vomiting may ultimately be referred for specialized care. The purpose of this lecture is to describe some of the mechanisms that result in nausea and vomiting, discuss common causes of nausea and vomiting in the human patient, to review the special case of chemotherapy induced nausea and vomiting, and to review anti-medic agents that are commonly prescribed.

Pathophysiology of nausea and vomiting

Research has indicated that the sensation of nausea arises within the autonomic nervous system. As shown in the diagram below, nausea can originate from a variety of processes. Visceral stimulation, resulting from gastrointestinal disease is a common cause of nausea. Gastrointestinal disorders such as those affecting the esophagus, including gastroesophageal reflux disease, gastric conditions such as peptic ulcer disease and exposure to toxic substances, small intestinal mechanical obstruction or inflammation, pancreatic diseases, and gallbladder disorders such as acute cholecystitis often are associated with nausea and vomiting. Cerebral disorders, including thromboembolic events, migraine headache, and neoplastic disorders are also commonly associated with nausea and vomiting. Stimulation of the vestibular apparatus as seen in patients with motion sickness or alcohol ingestion are additional common sources of these symptoms. Emotional stress and pain are other causes of nausea and vomiting. The neurotransmitters utilized to transmit information that produces nausea and vomiting to the central nervous system vary according to the type of stimulation occurring. Dopamine, serotonin, histamine, and acetylcholine neurotransmitters are all involved.

Vomiting occurs when the aforementioned processes result in stimulation of the vomiting center located in the dorsolateral reticular formation of the medulla. The medullary vomiting center is stimulated by afferent neurons coming from the chemotactic trigger zone (CTZ). Other stimulation comes from the cerebral cortex, the limbic system, visceral stimuli from within the gastrointestinal tract and the vestibular apparatus of the inner ear, (See figure 1).

Figure 1-Stimulation of the vomiting center of the medulla occurs via stimulation from a variety of sources and neurotransmitters, including serotonin, histamine, dopamine, and acetylcholine

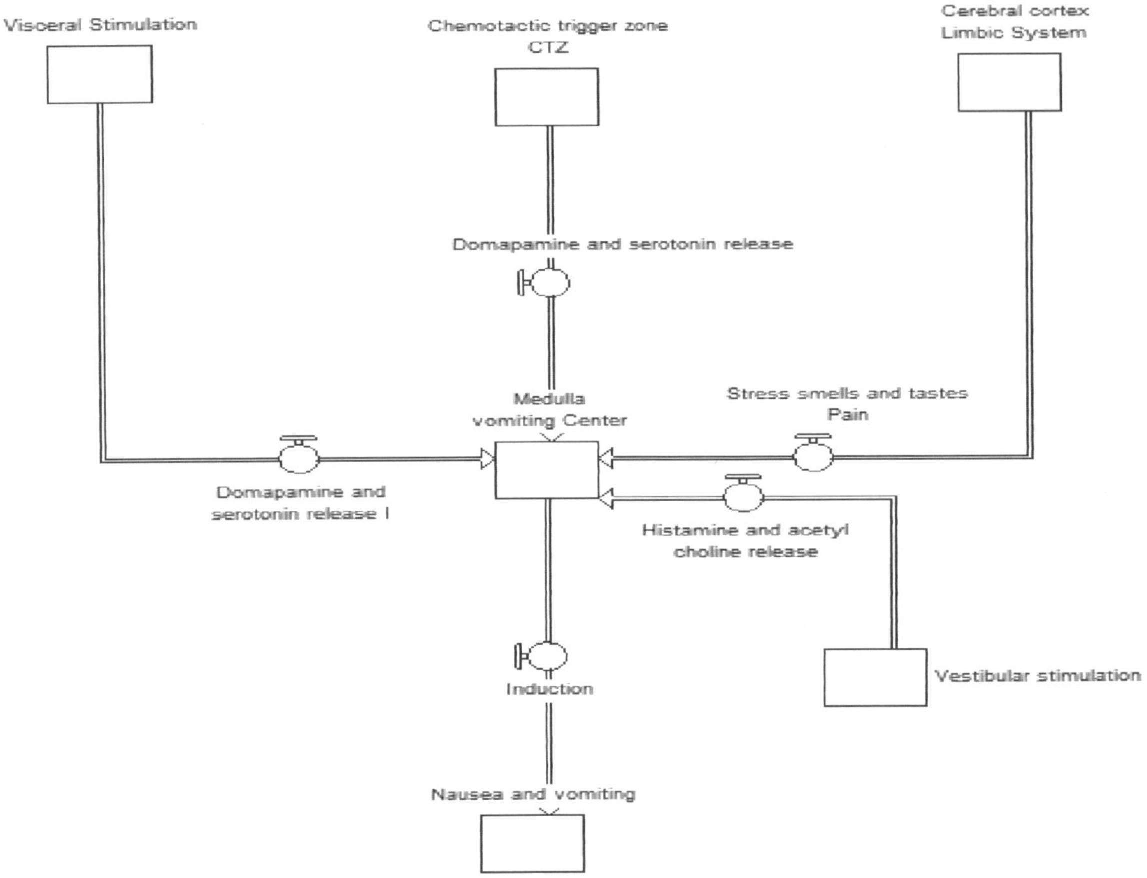

Common causes of nausea and vomiting in humans

Nausea and vomiting has been divided in some classification systems as acute, subacute, or chronic. The etiology of nausea and vomiting determines the acuity of symptoms with conditions such as gastroenteritis presenting acutely, while chronic nausea occurs in patients with primary gastrointestinal disorders such as gastroparesis.

Table 1-Common causes of nausea and vomiting

- Gastroenteritis
- Post-operative nausea and vomiting
- Chemotherapy-induced nausea and vomiting
- Pregnancy-induced nausea and vomiting
- Central nervous system conditions
    - Migraine
    - Vestibular stimulation (motion sickness)
- Drugs
    - Opioids, Nonsteroidal anti-inflammatory agents (NSAIDs)

Antiemetic medications

A variety of antiemetic regimens are administered for the prevention and treatment of nausea and vomiting. The rationale for antiemetic therapy is modulation of neurochemical factors that promote vomiting. As previously described, peripheral neuroreceptors and the CTZ possess a variety of receptors, including those for serotonin, histamine (H_1 and H_2), dopamine, and acetylcholine. Receptors for a variety of endogenous neurotransmitters, including opioid and GABA, are also present. Antiemetics generally work by competitively blocking these receptors. The overall effect of these drugs is to inhibit stimulation of peripheral nerves at the CTZ and to control input into the vomiting center.

The following is a list of antiemetic agents used in humans that will be discussed during the lecture.

Table 2- Antiemetic agents used for the prevention and treatment of nausea and vomiting

- •Phenothiazines
  - •Prochlorperazine
- •Butyrophenones
  - •Droperidol and haloperidol
- •Dopamine$_2$ Antagonists
  - •Metoclopramide
- •5-HT$_3$ Receptor Antagonists
  - •Ondansetron
  - •Granisetron
  - •Dolasetron
  - •Palonosetron
- •Substance P antagonists (NK$_1$ Receptor Antagonists)
- •Histamine receptor blockers
- •Corticosteroids
- •Cannabanoids
- •Benzodiazepines
  - •Lorazepam
- •Olanzapine
- •Other Pharmacologic Agents
- •Ginger

Nausea and vomiting in cancer patients

Patients with cancer represent complex challenges as for the management of nausea and vomiting. These patients are especially at risk due to emetogetic properties of many chemotherapeutic agents. They may also suffer from direct central nervous system or gastrointestinal tract involvement of their tumor. In addition, anxious patients are at higher risk for CNS side effects of chemotherapy. Anticipatory nausea (occurring prior to administration of chemotherapy) is a common problem, requiring prophylactic antiemetic therapy in these patients.

REFERENCES

1) Camilleri M, Prather CM. In: Sleisenger and Fordtran's Gastrointestinal Disease, 6th ed, Feldman M, Scharschmidt BF, Sleisenger MH (Eds), WB Saunders, Philadelphia 1998. p.572.
2) American Gastroenterological Association medical position statement: nausea and vomiting. American Gastroenterological Association. Gastroenterology. 2001;120(1):261.
3) Andrews PK, Hawthorn J. The neurophysiology of vomiting. Bailleires Clin Gastroenterol 1988,2(1):141-68.
4) Flake CA, Scalley RD, Bailey AG. Practical Selection of Antiemetics. *Am Fam Physician.* 2004 Mar 1;69(5):1169-1174.
5) Aaprol, M. Molassiotis A, et al. The effect of guideline-consistent antiemetic therapy on chemotherapy-induced nausea and vomiting (CINV): the Pan European Emesis Registry (PEER)Ann Oncol (2012) doi: 10.1093/annonc/mds021 First published online: March 6, 2012
6) Feyer, P. Jordon, K. Update and new trends in antiemetic therapy: the continuing need for novel therapies Ann Oncol (2011) 22 (1): 30-38. doi: 10.1093/annonc/mdq600 First published online: October 14, 2010
7) Madjunkova S, Maltepe C, et al. Patterns of antiemetic use among American women with nausea and vomiting of pregnancy..Obstetrics and Gynecology [2014, 123 Suppl 1:155S]

HOW I USE ANTIEMETICS ON THE CLINIC FLOOR

David C. Twedt, DVM, DACVIM (Small Animal Internal Medicine)
Fort Collins, CO

INTRODUCTION

Nausea and vomiting are common clinical conditions and anti-emetics play an important part in the management of many of these cases.[1] Antiemetics are indicated when vomiting is severe, resulting in dehydration, electrolyte loss and acid base disturbances, and for the prevention of motion sickness. Additionally, antiemetics should be used for prevention of aspiration pneumonia from vomiting associated with radiation or chemotherapy, management of nausea resulting in inadequate nutritional intake and to enhance patient comfort.

HOW ANTIEMETICS WORK

Antiemetics work either peripherally, centrally or both.[2] Understanding the basic receptors involved in vomiting provides the rational for selection of an appropriate antiemetic drug.[3] Activation of the emetic center (EC) occurs either through a humoral pathway initiated via blood-borne substances or by activation through various peripheral neural pathways leading to the EC. Neural stimulation of the EC arises through either afferent vagal, sympathetic, vestibular or cerebrocortical pathways. Activation of peripheral receptors found throughout the body can stimulate these neural pathways. Most GI associated afferents arise through vagal innervation. The EC is also stimulated indirectly via a humoral pathway by activating the chemoreceptor trigger zone (CRTZ) located in the area postrema at the base of the fourth ventricle, an area devoid of a blood brain barrier, that allows exposure to chemical stimuli found in the circulation. Blood-borne substances stimulating the CRTZ include certain drugs, uremic toxins, chemicals, osmolar and acid-base disorders as well as a number of other metabolic derangements. There is also evidence in the dog that vestibular stimulation passes through the CRTZ before activating the EC but not in the cat. Motion sickness, inflammation of the labyrinth or lesions in the cerebellum can all result in vomiting via this pathway.

VOMITING RECEPTORS

The CRTZ (the area of the postrema of the brain) contain dopamine (D_2), histamine (H_1), muscarinic cholinergic (M_1), alpha adrenergic (α_2), serotonin ($5HT_3$), enkephalin ($ENK_{\mu,\check{e}}$) and neurokinin 1 (NK_1) receptors that are activated by blood borne substances.[2] Presently only $5HT_1$, NK_1 and α_2 have been documented in the regulation of emesis at the level of the EC.[2] Vestibular activation or motion sickness associated with vomiting is associated with blocking M_1, histamine (H_1) or NK_1 receptors. The antagonists of histamine and acetylcholine have proved useful in treating emesis associated with motion sickness however antihistamines are not very effective in cats. There are a number of different mechanisms by which stimuli within the gastrointestinal tract (such as inflammation, stretching or distention) activate the emetic center.[2] The $5HT_3$ and NK_1 receptors appear to play a pivotal role in the process of vomiting being abundant on vagal afferent neurons, other neurons and smooth muscle within the abdominal cavity and presynaptically on vagal afferent terminals in the medulla.[1] The NK_1 receptors are the newest class of receptors found in high concentration in the EC, CRTZ and in peripheral afferent nerves. The inflammatory neurotransmitter substance P binds to NK_1 receptors initiating that stimulus. The role of receptors within the cerebral cortex has been implicated in humans by using opioids, benzodiazepines and cannabinoids for controlling chemotherapy-associated nausea and vomiting with $ENK_{\mu,\check{e}}$ receptors being implicated. Specific receptors in the cerebral cortex have not yet been established in dogs and cats.

ANTIEMETIC DRUGS

The following table lists some of the more common clinically used antiemetics in small animal veterinary medicine. Domperidone is not yet available in the US. The reader should refer to other references for a more extensive list of antiemetics, mechanism of action, doses and side effects.[3]

DRUG	RECEPTOR(s)	SITE OF ACTION	SIDE EFFECTS
Maropitant	NK_1	EC, CRTZ, VA, V	Injection stinging
Dolasetron	$5HT_3$	EC, CRTZ, VA	Sedation
Ondansetron	$5HT_3$	EC, CRTZ, VA	Sedation
Metoclopramide	D_2, $5HT_3$	CRTZ	Extrapyramidal signs
Domperidone	D_2	CRTZ	Extrapyramidal signs
Prochlorperizine	D_2, α_2, M_1	EC, CRTZ	Hypotension, sedation
Chlorpromazine	D_2, α_2, M_1	EC, CRTZ	Hypotension, sedation
Diphenhydramine	H_1	CRTZ, V	Sedation
Aminopentamide	M_1	EC, CRTZ, V	GI hypomotility
Butorphanol	$ENK_{\mu,\check{e}}$	CRTZ	Sedation
Mirtazapine	$5HT_3$	EC, CRTZ, VA	Appetite stimulation Serotonin effects

ANTIEMETICS I USE IN THE CLINIC

Of the antiemetics listed in the table the predominant ones I use in the clinic are metoclopramide, ondansetron and maropitant. For the last 9 years maropitant has been my number one choice for several reasons. First, it is FDA approved for dogs and cats, is broad-spectrum, proven clinical efficacy and there is substantial research on its role in control of both nausea and visceral pain. Mirtazapine, an appetite stimulant, may also be a clinically applicable antiemetic. The following is an overview of these antiemetics.

Metoclopramide

Metoclopramide (Reglan™) has antiemetic effects through several mechanisms. At low doses it inhibits dopamine (D_2) receptors and at higher doses inhibits serotonin ($5HT_3$) receptors in the CRTZ.[1] Metoclopramide is useful for the control of CRTZ mediated vomiting from such things as chemotherapy, azotemia, or endotoxemia. Cats are reported to have few if any CRTZ dopamine receptors and consequently a poor antiemetic choice. Metoclopramide is rapidly metabolized through the liver and the best results are observed when given at continuous rate infusion. At high doses it causes CNS excitement from dopamine antagonism. It is reported to have prokinetic effects but they are weak. A recent study reports metoclopramide to have no effect on distal esophageal sphincter pressure in the dog. Metoclopramide should be avoided in epileptics or in animals receiving other drugs that are likely to cause extrapyramidal reactions and should not be used in conjunction with phenothiazine tranquilizers due to an additive effect. I have on occasion used metoclopramide in combination with a second antiemetic with severe uncontrollable vomiting such as in dogs with parvovirus, pancreatitis or patients treated with chemotherapy.

Ondansetron

The serotonin antagonists ondansetron (Zofran™) or dolasetron (Anzemet™) were developed as antiemetics primarily for the control of vomiting and nausea in humans receiving chemotherapy. They are effective at blocking 5HT3 receptors found peripherally, in the CRTZ and EC but not very effective in relieving motion sickness.[1] Ondansetron can be given PO or IV and is frequently given to animals vomiting from parvovirus, pancreatitis or azotemia. It is however not approved for use in dogs and cats. We recently performed a blinded study comparing ondansetron to maropitant in parvovirus and believe maropitant may be superior (see below).

Maropitant

Maropitant (Cerenia™) is the first FDA approved antiemetic for the control of acute vomiting and motion sickness in the dog and cat. Numerous reports find it to be an excellent broad-spectrum antiemetic.[3,4] Maropitant is approved for dogs over 8 weeks of age and for a total of five consecutive days of therapy at the dose of 1 mg/kg SQ or 2 mg/kg PO q 24h. I will however use it for a longer duration of therapy if required. Pharmacokinetic studies have shown a high margin of safety given SQ, PO or IV and given for up to 30 days.[5,6] It does sting in some patients when given SQ and some recommend refrigeration or diluting the drug in saline before administration with mixed results. We will usually administer maropitant IV if the patient has an IV catheter. It should be given slowly IV over a minute or two as rapidly administration can cause transient hypotension. Because the drug requires hepatic metabolism it should be used with caution in patients that have hepatic dysfunction. I tend to decrease the dose by approximately 50% in cases having advanced hepatic disease. It should be noted however there are yet no reported pharmacokinetic studies in patients with hepatic compromise. For prevention of motion sickness maropitant is given at a dose of 8 mg/kg PO for two days. I have found however some dogs may respond to a lower dose for motion sickness.

Our experience finds maropitant very effective in preventing vomiting and nausea associated with chemotherapy, in management of parvovirus cases and many other causes of vomiting such as pancreatitis.[1,7] We compared maropitant to ondansetron in the therapy of parvovirus. We observed no difference between the drugs but found the maropitant group gained weight during hospitalization while the ondansetron group lost weight ($p<0.05$). Although the sample size was small ($n=22$) our impression was that the dogs of the maropitant group may have had less nausea and ate slightly more than the dogs of the ondansetron group. In that study however, a few dogs that failed their antiemetic therapy (3/11 in maropitant, 5/11 in ondansetron group) were subsequently treated with combined maropitant and ondansetron therapy.

Another clinical observation was that many patients appeared more comfortable when given maropitant and this led us to investigate maropitant's role in blocking visceral pain. Visceral pain pathways contain many NK_1 receptors and our hypothesis was that maropitant likely also blocks those pain receptors. Consequently, we developed a visceral pain model by stretching the ovarian ligament in anesthetized dogs and cats.[8,9] Maropitant significantly lowered the anesthetic MAC for pain perception during ovarian ligament stretch compared to placebo. Others have shown maropitant decreased nausea or vomiting when given with preanesthetic narcotics.[10,11] We also compared preanesthetic effects of morphine vs. maropitant in an OHE model. We found significant difference during surgery with lower heart rate, blood pressure and isoflurane concentration in the maropitant group and found that dogs pretreated with maropitant had a significantly lower postop pain score at the time of extubation and ate significantly more during the three-hour postop recovery period than dogs that were not pretreated with maropitant. Our clinical impression is that these dogs had a more stable plane of anesthesia and also seemed to have less nausea and were more confortable. We will administer maropitant as a preanesthetic in patients where pre- or post-op vomiting is a concern. Other studies have shown improved return to appetite following anesthesia when given maropitant. In another study we found maropitant to have no effect on GI motility when compared to a saline placebo using Smartpill™ technology. There are numerous anecdotal reports of maropitant for the treatment of many other conditions (dermatologic, rhinitis, respiratory) but they lack critical evaluation with convincing data. However, maropitant has been shown in mice to reduce the itching in experimental ulcerative dermatitis; studies are however lacking in dogs and cats.[12]

Maropitant also is my choice of an antiemetic for cats as well. The recommended dose is 1 mg/kg IV, SQ or PO.[13] Cats should be over 16 weeks of age and the same dose is used for prevention of motion sickness. In a CSU feline chronic kidney disease (CKD) study the investigators found maropitant vs. placebo given orally for two weeks significantly reduced ($p<0.01$) vomiting episodes and may be helpful in nutritional management of cats with CKD.[14] We are currently investigating the effectiveness of maropitant in the management of visceral pain associated with feline interstitial cystitis. Others have suggested that maropitant may reduce inflammation and bronchospasm in patients with asthma however, in an experimental feline asthma model using maropitant was ineffective.[15]

Mirtazapine

Another serotonin antagonist sometimes used in cats is mirtazapine (Remaron™). This drug works primarily as a $5HT_3$ antagonist, reducing vomiting and stimulating appetite. It is likely if other serotonin antagonists (i.e., ondansetron) are given in conjunction with mirtazapine they would be poorly effective. In a CSU placebo controlled crossover study in cats with CKD mirtazapine treated cats had a significantly increased appetite ($p=0.02$) and decreased vomiting ($p=0.04$).[16] Our clinical impression is that mirtazapine has antiemetic effects in cats as an added benefit when using it as an appetite stimulant but that maropitant is still our clear choice of an antiemetic for control of vomiting.

REFERENCES

[1]Elwood C, et.al. **J Small Anim Pract** 2010;51(1):4, [2]Boothe DM. **Vet Clin North Am Small Anim Pract** 1999;29(2):343, [3]Benchaoui HA, et.al. **J Vet Pharmacol Ther** 2007; 30(4):336, [4]Benchaoui HA, et.al. Vet Rec. 2007; 29;161(13):444, [5]Lesman SP, et.al. **J Vet Pharmacol Ther** 2013;36(5):462, [6]Ramsey DS, et.al. **J Vet Pharmacol Ther** 2008;31(6):538, [7]Mason SL, et.al. **J Small Anim Pract**. 2014;55(8):391, [8]Boscan P, et.al. **Am J Vet Res** 2011;72(12):1576, [9]Niyom S, et.al. **Vet Anaesth Analg** 2013;40(4):425, [10]Hay KBL. **Vet Anaesth Analg** 2013;40(1):28, [11]Claude AK, et.al. **J Vet Intern Med** 2014;28(5):1414, [12]Williams-Fritze MJ, et.al. **J Am Assoc Lab Anim Sci** 2011;50(2):221, [13]Hickman MA, et.al. **J Vet Pharmacol Ther** 2008;31(3):220, [14]Quimby JM, et. al. **J Feline Med Surg** 2014; Epub, [15]Graham A, et.al. **ACVIM Forum** (abstract) 2013; Seattle, [16]Quimby JM, et.al. **Vet J** 2013;197:651.

RATIONAL SELECTION AND USE OF ACID SUPPRESSANT MEDICATIONS

Katie Tolbert, DVM, PhD, DACVIM (SAIM)
Knoxville, TN

INTRODUCTION

The pathophysiology of acid-related diseases is complex, but the end result is a compromised mucosal barrier and, often, gastric acid, bile acids and/or pepsin-induced tissue injury. Such diseases are prevalent in humans and, left untreated, lead to increased morbidity and poor quality of life.[1] Development of gastric acid suppressants, including histamine-2 receptor antagonists (H2RAs) and proton pump inhibitors (PPIs), in the late 20th century transformed the medical management of acid-related diseases and gastrointestinal (GI) bleeding in humans. Gastric acid secretion is triggered by hormonal and neural stimulation of parietal cells, the acid producing cells of the stomach. Acid suppressant drugs, which take aim at these physiological targets on the parietal cell surface, have largely supplanted antacids (acid-neutralizing drugs) as the drugs of choice for acid-related diseases. H2RAs are competitive inhibitors for the interaction of histamine with its receptor but differ in potency (famotidine > ranitidine > cimetidine hydrochloride). Peak concentration of H2RAs occurs within hours after oral administration. In contrast, PPIs (e.g. omeprazole, pantoprazole, esomeprazole) form disulfide bonds with active H+-K+-ATPase enzymes and are thus inactivators. De novo H+-K+-ATPase enzymes are subsequently recruited and require further inactivation; thus delaying the peak effect (approximately 4 days) with PPIs.[2-4] Pharmacokinetic and pharmacodynamic differences exist between PPI drugs, which might affect clinical response depending on the individual patient and disease. However, most clinicians use drugs within the PPI class interchangeably.

ACID SUPPRESSANTS IN HUMAN MEDICINE

Indications and guidelines for use of acid suppressants in human adults

Acid suppressants are recommended as a first line medical treatment for acid-related diseases including gastroduodenal ulceration, gastroesophageal reflux disease, and Zollinger-Ellison syndrome. Treatment objectives for acid-related diseases are to promote healing of damaged mucosa, provide symptomatic relief, and prevent recurrence. Clinical trials in humans have demonstrated that healing of mucosal injury and reduction of gastrointestinal hemorrhage is dependent on achieving sustained and maximal gastric acid suppression. Clinical goals established for people with acid-related diseases are centered on three key parameters, which include: the degree and the duration of acid suppression in a given day and the length of acid suppression in weeks.[5] For example, the goals for healing of gastroduodenal ulceration and gastroesophageal reflux disease are to maintain the intragastric pH at or above 3 and 4 for at least 75% and 67% of the day, respectively.[5] For this reason, acid suppressant drugs are the treatment of choice for the medical management of acid-induced injury and many gastric and non-gastric causes of GI bleeding in people. PPIs target the final common pathway of acid production and are therefore significantly more effective than H2RAs in increasing gastric pH and preventing and healing acid-related tissue injury. Additionally, pharmacological tolerance and/or exaggerated first dose response may develop with prolonged H2RA use. Thus, PPIs are often chosen over H2RAs when chronic acid suppressant therapy is anticipated.[5] However, after oral administration, H2RAs are rapidly absorbed in the small intestine and, unlike PPIs, may provide immediate relief of symptoms (e.g., heartburn) when taken on an as needed basis. H2RAs are also effective against basal acid secretion and therefore may have efficacy against nocturnal gastric acidity.

Evidence for adverse effects of acid suppressants in human adults

Short-term acid suppressant therapy is considered to be relatively safe in human adults. However, because acid suppressants frequently provide symptomatic relief and are available over-the-counter, patients often take acid suppressants chronically without consultation and/or follow up with a gastroenterologist.[6] Long-term use of acid suppressants in the absence of routine monitoring by a clinician can lead to adverse effects. The side effect profile of acid suppressants is under debate as studies have shown inclonclusive results. More controlled, prospective trials are needed to critically evaluate the effects of chronic acid suppressant therapy. However, several studies in humans suggest that chronic acid suppressant administration might result in severe adverse effects, including community acquired pneumonia and *Clostridium difficile*-associated diarrhea, bacterial/fungal overgrowth, thrombocytopenia, vitamin B12 deficiency (PPI only), hypomagnesemia (PPI only), altered bone metabolism and pathologic fractures (PPI only).[7-11] In addition, undesired drug interactions with drugs metabolized by cytochrome p450 are also possible as a result of the inhibitory effects of many of the PPIs as well as cimetidine on cytochrome p450 enzymes. pH-dependent absorption of drugs can also be altered by co-administration of acid suppressants. Rebound acid hypersecretion in patients receiving chronic acid suppression, as a result of hypergastrinemia and increased enterochromaffin-like cell mass, can develop following rapid drug cessation. Therefore, slow tapering of acid suppressants is recommended for human patients receiving prolonged acid suppressant therapy.

ACID SUPPRESSANTS IN VETERINARY MEDICINE

Indications and guidelines for use of acid suppressants in dogs and cats

In contrast to humans, there are no guidelines for the use of acid suppressants in dogs and cats. Moreover, there are limited studies investigating the use of acid suppressants in dogs and cats with acid-related disorders or diseases in which gastric hyperacidity is thought to be a sequela. This lack of guidelines often results in inadequate treatment for acid-related disorders or use of acid suppressants for conditions in which they have not been proven to be beneficial. Acid suppression is considered to be standard of care for dogs and cats with presumptive acid-related disorders including gastroduodenal ulceration, reflux esophagitis and severe stress-related mucosal bleeding. Omeprazole is more effective than famotidine in raising canine[12-13] and feline[14] intragastric pH and in preventing exercise-induced gastritis in dogs.[15] Intragastric pH studies have not been performed in dogs and cats receiving chronic H2RA treatment. Nevertheless, as observed in humans, recent studies suggest that prolonged famotidine administration might result in tolerance and/or an exaggerated first dose response in dogs.[16-17] For this reason and because of their superior efficacy, PPIs are widely considered the best choice for prevention or treatment of acid-induced tissue injury in dogs and cats.

Evidence for use of acid suppressants for acid-related disorders in dogs and cats

Gastroduodenal ulceration: The most common causes of gastroduodenal ulceration in the dog and cat are NSAIDS with or without concurrent steroid administration (D & C), hepatic disease (D), and pancreatic and gastroduodenal neoplasia (adenocarcinoma, lymphoma, gastrinoma, mastocytosis) (D & C).[18-19] Spinal cord injury and critical illness are also infrequent causes of ulceration.[18,20] Omeprazole has been documented to control clinical signs of gastrinoma in cats and dogs.[19,21,22] Studies in dogs suggest that omeprazole dosed at 1mg/kg twice daily is more effective than once-daily dosing for raising intragastric pH and healing gastroduodenal ulceration.[12-13,20,23]

Reflux esophagitis: Gastroesophageal reflux disease secondary to LES incompetence is a common disorder in people but is not a commonly recognized problem in dogs and cats.[24] However, vomiting and anesthetic-induced reflux, as well as esophageal leiomyoma at the LES, are well-documented causes of acid-induced esophagitis and esophageal stricture in dogs and cats.[25-28] Esomeprazole, in combination with cisapride, dosed 12-18 hours and 1-1.5 hours before induction was an effective treatment for prevention of anesthetic-induced reflux in dogs.[29] Short-term administration of acid suppressants is recommended for dogs and cats at risk for anesthetic-induced reflux (e.g., patients with delayed gastric motility) and vomiting-induced esophageal injury. Acid suppressants should be discontinued following anesthetic recovery or successful treatment of the underlying cause of vomiting.

Stress-gastritis: Stress-related mucosal disease (SRMD) represents a continuum from stress-related superficial erosions to life-threatening bleeding ulcers. Critically ill human patients are at an increased risk for developing SRMD secondary to several risk factors including breakdown of the gastric mucosal barrier, ischemia and reperfusion injury, splanchnic hypoperfusion and delayed gastric motility. SRMD has a reported incidence as high as 60-100% in critically ill human patients.[30] Studies investigating the incidence of SRMD in critically ill dogs and cats are lacking. Until such studies are performed, the author administers acid suppressants to critically ill dogs and cats solely when continuous vomiting or regurgitation is observed and/or compromise of the gastric mucosal barrier (e.g., hematemesis, melena, clinicopathologic evidence for bacterial translocation) is suspected. When acid suppression is warranted, co-administration of IV famotidine and pantoprazole does not appear to be more effective than pantoprazole alone even in the first few days of therapy.[31]

Evidence for use of acid suppressants for extra-gastrointestinal disease in dogs and cats

Renal disease: Gastric ulceration and *H. pylori* infection are common complications of end-stage renal disease in human patients.[32] Urea is a substrate for the urease-producing bacterium. Therefore, acid suppression in combination with other medications for the eradication of *H. pylori* as well as treatment of gastric ulceration are often recommended for human patients with advanced renal disease.[33] *Helicobacter* is not recognized as a co-morbidity in dogs and cats with renal disease. Moreover, gastric mineralization, gastric gland atrophy, and hypergastrinemia, but not ulcerative or erosive gastropathy, are typically documented in dogs and cats with advanced renal disease.[34-36] Intragastric pH studies have not been performed in companion animals with acute kidney injury or chronic kidney disease. Therefore, the contribution of hypergastrinemia and, if present, gastric hyperacidity, to the development of gastric mineralization has not been determined. Until such studies are performed, the author administers acid suppressants solely to dogs and cats with renal disease who are anorexic or when concern for GI bleeding (e.g. melena, severe iron deficiency anemia) or vomiting-induced esophagitis exists.

Hepatic disease: A recent study demonstrated no difference in serum gastrin between dogs with liver disease and healthy controls; however, continuous intragastric pH monitoring and histopathologic examination of the gastrointestinal tract were not performed.[37] Liver disease is reportedly one of the most common predisposing factors for ulceration in dogs with the duodenum being a common site of ulceration.[18] The contribution of gastric hyperacidity versus compromised hemodynamics to GI bleeding and the efficacy of acid suppressants as a preventative for ulcerative disease in cats and dogs with liver disease requires further study. However, acid suppression may be considered for feline and canine patients with advanced liver disease at risk for ulceration. Cimetidine, however, should be avoided due to an association with acute liver injury in people.

Pancreatitis: The use of acid suppressant therapy for patients with pancreatitis is controversial. Although there are scattered case reports describing a causal association of acid suppressants with acute pancreatitis in humans, larger studies have shown discordant results.[38-39] PPIs have also had mixed results in experimental rodent model and *in vitro* studies. A handful of reports suggest PPIs are anti-inflammatory and reduce pancreatic secretions whilst others demonstrate no effect or suggest a pro-inflammatory effect of PPIs.[40-42] Pantoprazole possesses free radical scavenger activity and ameliorates inflammation in rodent models of pancreatitis[42]; however, a recent placebo-controlled study failed to demonstrate a benefit to pantoprazole administration in human patients with acute pancreatitis.[43] Acid suppression is warranted for companion animals with pancreatitis and persistent vomiting to prevent esophageal mucosal injury. However, the beneficial effect of acid suppression in dogs and cats with pancreatitis but without persistent vomiting is unknown. Until such studies are performed, judicious use of acid suppressants in dogs and cats with pancreatitis is recommended.

Thrombocytopenia: There is little evidence to support the use of acid suppressants to control thrombocytopenia induced-GI hemorrhage in human patients likely because platelet transfusions are more readily available. Additionally, there are reports of PPI-induced thrombocytopenia in humans, therefore PPI therapy may be contraindicated in these patients. However, available case reports describe the successful treatment of gastric bleeding using octreotide and omeprazole in a myelodysplastic patient with thrombocytopenia and prednisone and omeprazole in an idiopathic thrombocytopenia purpura patient.[44, 45] Maintaining an intragastric pH > 6.8 is required to promote platelet aggregation and clot formation in patients with GI bleeding.[46] Studies performed in healthy dogs and cats using omeprazole, pantoprazole, or famotidine at currently recommended doses did not maintain this degree of acid suppression for a prolonged period.[13-14,31] More studies are necessary to determine the efficacy and optimal dose of acid suppressants for the treatment of thrombocytopenic-induced GI bleeding in dogs and cats.

Evidence for adverse effects of acid suppressants in dogs and cats

Few side effects, with the exception of self-limiting diarrhea and a transient alteration of GI microbiota, have been described with short-term acid suppressant administration in dogs.[13,47, 48] Esomeprazole, omeprazole and cimetidine are cytochrome P450 (CYP) inhibitors. Therefore, careful monitoring is recommend if these drugs are used concurrently with other drugs metabolized by CYP. Sub-therapeutic doses of omeprazole given daily for 7 years resulted in no detectable effects in beagle dogs.[48] However, no studies have evaluated the effects of chronic administration of therapeutic doses of acid suppressants in dogs and cats. These studies are needed to determine if dogs and cats are at risk for the same untoward effects observed in human patients receiving chronic acid suppressant therapy. Until such studies are performed, judicious use of acid suppressant therapy in dogs and cats along with careful monitoring when chronic acid suppression is warranted are recommended.

REFERENCES

1. Majumdar, S.R., et al., **Am J Gastroenterol** 2003;98(11): 2409.
2. Kromer, W., et al., **Pharmacology** 2000;60(4):179.
3. Larsson, H., et al., **Gastroenterology** 1983;85(4):900.
4. Wallmark, B., et al., **J Biol Chem** 1985;260(25):13681.
5. Huang, J.Q., et al., **Best Pract Res Clin Gastroenterol** 2001;15(3):355.
6. Shi, C.W., et al., **Am J Gastroenterol** 2004;99(4):606.
7. Lam, J.R., et al., **JAMA** 2013;310(22):2435.
8. Johnson, D.A., et al., **Clin Gastroenterol Hepatol** 2013;11(5):458.
9. Tleyjeh, I.M., et al., **PLoS One** 2013;8(3):e56498.
10. Binnetoglu, E., et al., **Platelets** 2015;26(1):10.
11. Compoginis, J.M., et al., **Am Surg** 2011;77(12):1580.
12. Bersenas, A.M., et al., **Am J Vet Res** 2005;66(3): 425.
13. Tolbert, K., et al., **J Vet Intern Med** 2011;25(1):47.
14. Parkinson, S., et al., **J Vet Intern Med** 2014.
15. Williamson, K.K., et al., **J Vet Intern Med** 2010;24(2):285.
16. Mordecai, A., et al., **J Vet Intern Med** 2011;25(6):1248.
17. Parente, N.L., et al., **J Vet Intern Med** 2014;28(5):1465.
18. Stanton, M.E., et al., **J Vet Intern Med** 1989;3(4):238.
19. Liptak, J.M., et al.,. **J Feline Med Surg** 2002;4(1): 27.
20. Neiger, R., et al., **J Vet Intern Med** 2000;14(1): 33.
21. Diroff, J.S., et al., **J Vet Intern Med** 2006;20(5):1245.
22. Brooks, D., et al., **J Vet Intern Med** 1997;11(6):379.
23. Jenkins, C.C., et al., **Am J Vet Res** 1991;52(5):658.
24. Kook, P.H., et al., **J Vet Intern Med** 2014;28(6):1716.
25. Wilson, D.V., et al., **Am J Vet Res** 2005;66(3):386.
26. Bissett, S.A., et al., **J Am Vet Med Assoc** 2009;235(7):844.
27. Lux, C.N., et al., **J Am Vet Med Assoc** 2012;240(9):1100.
28. Kook, P.H., et al., **Schweiz Arch Tierheilkd** 2009;151(10):497.
29. Zacuto, A.C., et al., **J Vet Intern Med** 2012;26(3):518.
30. Choung, R.S., et al., **Curr Mol Med** 2008;8(4):253.
31. Tolbert, K., et al., **J Vet Intern Med** 2015;in press.
32. Bunchorntavakul, C., et al., **J Med Assoc Thai** 2014;97 Suppl 11:S62.
33. Sotoudehmanesh, R., et al., **Endoscopy** 2003;35(6):502.
34. McLeland, S.M., et al., **J Vet Intern Med** 2014;28(3):827.
35. Peters, R.M., et al.,. **J Vet Intern Med** 2005;19(3):315.
36. Cianciolo, R.E., et al., **J Am Vet Med Assoc** 2008;233(5):729.
37. Mazaki-Tovi, M., et al., **Vet Rec** 2012;171(1):19.
38. Eland, I.A., et al., **Br J Clin Pharmacol** 2000;49(5):473.
38. Sundstrom, A., et al., **Pharmacoepidemiol Drug Saf** 2006;15(3):141.
40. Burdan, F., et al., **J Physiol Pharmacol** 2000;51(3):463.
41. Cai, J., et al., **In Vitro Cell Dev Biol Anim** 2007;43(1):25.
42. Hackert, T., et al., **Life Sci** 2010;87(17-18):551.
43. Yoo, J.H., et al., **Korean J Gastroenterol** 2012;60(6):362.
44. Gurzo, M., et al., **Orv Hetil** 1994;135(15):805.
45. Dang, S., et al., **Gastrointest Endosc** 2009;70(1):167.
46. Green, F.W., Jr., et al., **Gastroenterology** 1978;74(1):38.
47. Garcia-Mazcorro, J.F., et al., **FEMS Microbiol Ecol** 2012;80(3):624.

48. Safholm, C., et al., **Digestion** 1994;55(3):139.
49. Dossin, O., et al., **J Vet Intern Med** 2011; 25(3): 688.

MOTILITY MODIFYING DRUGS

Eli D. Ehrenpreis MD
Chicago, IL

Introduction

Functional bowel disorders remain some of the most complex conditions treated by internists and gastroenterologists, and are characterized by abnormalities in motility as well as abnormal sensory signals from the gastrointestinal tract to the central nervous system. Normal digestion and absorption of nutrients requires a highly coordinated system to change macronutrients into simpler building blocks and utilizes complex mechanisms to assimilate fat, protein, carbohydrates, vitamins and minerals coming from our diet. Abnormal motility is defined as an interruption in the coordinated movements of the organs of the gastrointestinal tract, and complicates these processes. Difficulty in patient management with functional disorders stems from poor understanding of the pathophysiology of these conditions, limited testing methods available to document abnormalities in these patients and strong psychological overlay that often complicates the conditions. Treatments are often inadequate due to lack of high-quality studies demonstrating efficacy, high placebo response rate in clinical trials and drug side effects that may be worse than the condition themselves. Like psychiatric disorders, accurate diagnoses of functional gut disorders requires adherence to guidelines that outline diagnostic criteria. In spite of these considerations, up to 50% of the patients seen by gastroenterologists have functional gastrointestinal disorders. Some recent advances in the management of patients with functional gastrointestinal disorders have occurred. In this monograph, we will review some of the common functional gut disorders, the role of motility in the development of these conditions and treatments that are currently used in these patients.

Common Motility Disorders in the Human Patient

Several gastrointestinal organs are affected by motility disorders. In the esophagus, achalasia has been the best characterized. Achalasia is a disorder defined by spasm of the lower esophageal sphincter and poor peristalsis of the lower esophagus. Gastroparesis, or delayed gastric emptying, occurs as an idiopathic disorder or as a secondary effect of other medical conditions such as diabetes mellitus. Gastroparesis can also occur as a paraneoplastic condition. Small intestinal pseudo-obstruction is a vexing problem due to altered motility. It is characterized by abdominal pain, bloating, nausea and vomiting and symptoms of small intestinal bacterial overgrowth. Irritable bowel syndrome, commonly occurring in the US, involves abnormalities of sensory and motor function in the small and large intestine. Constipation can be an idiopathic disorder or may be due to diminished motility in the colon or abnormalities of the pelvic floor. Disordered colonic motility can occur secondary to autoimmune conditions such as scleroderma, endocrinopathies, including hypothyroidism, and neurologic diseases such as Parkinson's disease. Medications are a frequent culprit in patients that develop chronic constipation. Antihypertensive drugs, antidiarrheals, anti-seizure medications, antidepressants, antipsychotics and pain relieving medications all may produce drug induced constipation.
Diarrhea can occur from a large number of gastrointestinal conditions, food intolerances or medication side effects.

Normal motility in the gastrointestinal tract

Nerve stimulation of the gastrointestinal tract occurs via extrinsic pathways from the sympathetic and parasympathetic systems or via intrinsic innervation from the myenteric plexus. The myenteric pathway consists of a large group of neurons extending between longitudinal and circular muscle layers of the gastrointestinal tract and is the primary intrinsic source of nerve impulses that influence gastrointestinal motility. The submucosal plexus, on the other hand, primarily functions in the intrinsic modification of mucosal secretion and absorption.

Motor activity, occurring in a random fashion, produces mixing of foods in the stomach. Peristaltic waves of circular muscle contractions in the small intestine moves partially digested food along the intestines. Peristalsis is triggered by a gastric pacemaker that transmits slow waves with a frequency of 3 cycles per minute. These waves of electrical impulses stimulate muscular contractions in the antrum of the stomach and empty food through the pylorus into the small intestine. The migrating motor complex, also known as the housekeeping wave occurs every 90 minutes and sweeps through the entire small intestine to the terminal ileum. Other factors involved in gastrointestinal motility include extrinsic release of acetylcholine and epinephrine, and a variety of enteric hormones, including motilin and vasoactive intestinal peptide (VIP).

Epidemiology of Irritable Bowel Syndrome and Chronic Constipation

Symptoms of irritable bowel syndrome occur in up to 20% of the US population. A recent study suggests that the worldwide prevalence of irritable bowel syndrome is approximately 11%. Determining the prevalence of constipation is difficult as there are a variety of definitions for this complaint. For example, patients often define constipation as straining or difficulty with bowel movements. Physicians tend to refer to constipation as having fewer than three bowel movements per week. According to Rome II - an expert panel, constipation is defined as: "a symptom-based disorder defined as unsatisfactory defecation and is characterized by infrequent stools, difficult stool passage, or both. Difficult stool passage includes straining, a sense of difficulty passing stool, incomplete evacuation, hard/lumpy stools, prolonged time to stool, or need for manual maneuvers to pass stool". Using these criteria, approximately 14% of the world's population has chronic constipation. Chronic constipation is more common in patients from a low socioeconomic background, female gender and the elderly.

Treatment Options for Intestinal Motility Disorders

A. **Prokinetic Agents**

Because a number of intestinal motility disorders involve abnormalities of the migrating motor complex and diminished contractions of gastrointestinal organs, use of agents that stimulate intestinal motility would appear to be an attractive treatment for these conditions. Unfortunately, development of these agents and their use in clinical practice has been hindered by serious side effects associated with these drugs. As a consequence, a number of promising prokinetic agents have been released on the market and subsequently withdrawn by the FDA. Others, including metoclopramide (Reglan), have received Blackbox Warnings that have limited their use in the majority of patients. Prokinetic agents that have historically been used in the treatment of patients with motility disorders include the following:

Cisapride - withdrawn due to severe cardiac side effects (Torsades de points)

Domperidone - never FDA approved due to prolongation of the QT interval and risk of drug interactions

Metoclopramide - Black box warning for the risk of tardive dyskinesia, a permanent neurologic disorder

Alosetron – withdrawn from the market due to cardiac side effects

B. Antimotiltiy Agents

Drugs that slow intestinal motility are primarily used as antidiarrheals. Examples of these are shown below:

Loperamide (Imodium) - an opioid agonist that does not cross the blood brain barrier

Diphenoxylate with atropine (Lomotil) - an opioid agonist combined with an anti-cholinergic agent. Anticholinergic side effects are a concern with this drug

Other opioids are used to treat severe diarrhea. These include paregoric, tincture of opium and codeine.

Octreotide, a somatostatin analog (Sandostatin) - this has been used for patients with severe diarrhea. Somatostatin decreases intestinal secretions and slows motility in the small intestine and gallbladder. It also reduces pancreatic enzyme secretion.

Newer Approaches to the Treatment of Irritable Bowel Syndrome and Chronic Constipation

Polyethylene glycol, lubiprostone, linaclotide and rifaxamin will be discussed in the lecture.

Table 1. Summary of results of monograph on interventions for IBS (reference #21)

Statement	No. of trials	No. of patients	RR symptoms (95% CI)	NNT (95% CI)	Recommendation	Quality of evidence
Specialized diets may improve symptoms in individual IBS patients.	3	230	NA	NA	Weak	Very low
Fiber provides overall symptom relief in IBS.	14	906	0.86 (0.80–0.94)	10 (6–33)	Weak	Moderate
Psyllium, but not bran, provides overall symptom relief in IBS (data presented for psyllium).	7	499	0.83 (0.73–0.94)	7 (4–25)	Weak	Moderate
There is insufficient evidence to recommend prebiotics or synbiotics in IBS.	2	198	NA	NA	Weak	Very low
Taken as a whole, probiotics improve global symptoms, bloating, and flatulence in IBS.	23	2,575	0.79 (0.70–0.89)	7 (4–12.5)	Weak	Low
Rifaximin is effective in reducing total IBS symptoms and bloating in IBS-D.	5	1,805	0.84 (0.78–0.90)	9 (6–12.5)	Weak	Moderate
Certain antispasmodics provide symptomatic short-term relief in IBS.	23	2,154	0.69 (0.59–0.81)	5 (4–9)	Weak	Low
Peppermint oil is superior to placebo in improving IBS symptoms.	5	482	0.51 (0.33–0.79)	3 (2–4)	Weak	Moderate
There is insufficient evidence to recommend loperamide for use in IBS.	2	42	0.44 (0.14–1.42)	NA	Strong	Very low
As a class, antidepressants are effective in symptom relief in IBS.	17	1,084	0.67 (0.58–0.77)	4 (3–6)	Weak	High
A variety of psychological interventions are effective in improving IBS symptoms.	32	2,189	0.68 (0.61–0.76)	4 (3–5)	Weak	Very low
Alosteron is effective in females with IBS-D.	8	4,987	0.79 (0.69–0.90)	8 (5–17)	Weak	Moderate
Mixed 5-HT_4 agonists/5-HT_3 antagonists are not more effective than placebo at	9	2,905	0.96 (0.83–1.11)	NA	Strong	Low

Statement	No. of trials	No. of patients	RR symptoms (95% CI)	NNT (95% CI)	Recommendation	Quality of evidence
improving symptoms of IBS-C.						
Linaclotide is superior to placebo for the treatment of IBS-C.	3	2,028	0.80 (0.75–0.85)	6 (5–8)	Strong	High
Lubiprostone is superior to placebo for the treatment of IBS-C.	3	1,366	0.91 (0.87–0.95)	12.5 (8–25)	Strong	Moderate
There is no evidence that polyethylene glycol improves overall symptoms and pain in patients with IBS.	2	166	NA	NA	Weak	Very low

CI, confidence interval; 5-HT_3, serotonin subtype 3; 5-HT_4, serotonin subtype 4; IBS, irritable bowel syndrome; IBS-C, IBS with constipation; IBS-D, IBS with diarrhea; NA, not available; NNT, number needed to treat; RR, relative risk.

REFERENCES

1) Longstreth GF, Thompson WG, Chey WD et al. Functional bowel disorders. Gastroenterology 2006;**130**:1480–1491.
2) Quigley EM, Abdel-Hamid H, Barbara G et al. A global perspective on irritable bowel syndrome: a consensus statement of the World Gastroenterology Organisation Summit Task Force on Irritable Bowel Syndrome. J Clin Gastroenterol 2012;**46**:356–366.
3) Suares NC, Ford AC. Prevalence of, and risk factors for, chronic idiopathic constipation in the community: systematic review and meta-analysis. Am J Gastroenterol 2011;**106**:1582–1591.
4) Koloski NA, Jones M, Wai R et al. Impact of persistent constipation on health-related quality of life and mortality in older community-dwelling women. Am J Gastroenterol 2013;**108**:1152–1158.
5) Brandt LJ, Bjorkman D, Fennerty MB et al. Systematic review on the management of irritable bowel syndrome in North America. Am J Gastroenterol 2002;**97** (11 Suppl): S7–S26
6) Masand PS, Pae C-U, Krulewicz S et al. A double-blind, randomized, placebo-controlled trial of paroxetine controlled-release in irritable bowel syndrome. Psychosomatics 2009;**50**:78–86.
7) Gershon MD. Review article: serotonin receptors and transporters -- roles in normal and abnormal gastrointestinal motility. Aliment Pharmacol Ther 2004;**20** (Suppl 7): 3–14
8) Rao S, Meduri K, Tuteja A et al. Effects of lubiprostone on gastrointestinal symptoms, bowel function, and bowel satisfaction in patients with methanogenic flora and chronic constipation. Am J Gastroenterol 2012;**107** (Suppl 1s): S712–S713
9) Lembo AJ, Schneier HA, Shiff SJ et al. Two randomized trials of linaclotide for chronic constipation. N Engl J Med 2011;**365**:527–536
10) Chey WD, Lembo AJ, Lavins BJ et al. Linaclotide for irritable bowel syndrome with constipation: a 26-week, randomized, double-blind, placebo-controlled trial to evaluate efficacy and safety. Am J Gastroenterol 2012; **107: 1702-1712**
11) Corazziari E, Badiali D, Bazzocchi G et al. Long term efficacy, safety, and tolerability of low daily doses of isosmotic polyethylene glycol electrolyte balanced solution (PMF-100) in the treatment of functional chronic constipation. Gut 2000;**46**:522–526
12) Ashraf W, Park F, Lof J et al. Effects of psyllium therapy on stool characteristics, colon transit and anorectal function in chronic idiopathic constipation. Aliment Pharmacol Ther 1995;**9**:639–647.
13) Drossman DA, Chey WD, Johanson JF et al. Clinical trial: lubiprostone in patients with constipation-associated irritable bowel syndrome - results of two randomized, placebo-controlled studies. Aliment Pharmacol Ther 2009;**29**:329–341
14) Vahedi H, Merat S, Momtahen S et al. Clinical trial: the effect of amitriptyline in patients with diarrhea-predominant irritable bowel syndrome. Aliment Pharmacol Ther 2008;**27**:678–684.
15) Merat S, Khalili S, Mostajabi P et al. The effect of enteric-coated, delayed-release peppermint oil on irritable bowel syndrome. Dig Dis Sci 2010;**55**:1385–1390.
16) Hovdenak N. Loperamide treatment of the irritable bowel syndrome. Scand J Gastroenterol 1987;**130**:81–84
17) Pimentel M, Park S, Mirocha J et al. The effect of a nonabsorbed oral antibiotic (rifaximin) on the symptoms of the irritable bowel syndrome: a randomized trial. Ann Intern Med 2006;**145**:557–563.
18) Pimentel M, Lembo A, Chey WD et al. Rifaximin therapy for patients with irritable bowel syndrome without constipation. N Engl J Med 2011;364:22–32.
19) Vazquez Roque MI, Camilleri M, Smyrk T et al. A controlled trial of gluten-free diet in patients with irritable bowel syndrome-diarrhea: effects on bowel frequency and intestinal function. Gastroenterology 2013;144**:**903–911.
20) Quigley EM. What we have learned about colonic motility: normal and disturbed. Current Opinion in Gastroenterology 2010, 26:53–60
21) Ford AC, Moayyedi P, LemboAH, et al. American College of Gastroenterology Monograph on the Management of Irritable Bowel Syndrome and Chronic Idiopathic Constipation. Am J Gastroenterol 2014; 109:S2–S26.

HOW I USE MOTILITY MODIFIERS ON THE CLINIC FLOOR

Frédéric P. Gaschen, DMV, Dr.habil., DACVIM (SAIM), DECVIM-CA (IM)
Baton Rouge, LA

INTRODUCTION

Motility is often the "missing link" in the complex equation of gastrointestinal (GI) health in dogs and cats. The prevalence of GI motility disorders in small animals cannot be precisely documented because obtaining a definitive diagnosis is often difficult, sometimes even impossible. However, it is thought that these disorders are of significant clinical importance. Proper diagnosis and treatment of any underlying disease that might affect GI motility is an essential premise.

Therapy of functional, non-obstructive disorders of gastric motility is based on two main pillars: dietary modification and judicious use of prokinetic drugs. Dietary modifications are an essential part of the management of GI motility disorders, and are based on our knowledge of GI physiology. First, gastric emptying of liquid food is faster than that of solid foods. Also, diets with high caloric density tend to remain longer in the stomach. In addition, gastric emptying of fat is slower than that of proteins, which is slower than that of carbohydrates. Consequently, feeding liquid or semi-liquid diet of low caloric density low in fat and protein should maximize gastric emptying. Finally, increased meal frequency and decreasing meal size are also useful.

PARALYTIC ILEUS

It is the most common indication for the use of GI prokinetics in the author's practice.

There is evidence that dogs have decreased gastric emptying after general anesthesia and laparotomy.[1-3] Early experimental studies performed on anesthetized cats demonstrated that mechanical stimulation of the small and large intestine, intestinal surgery, and peritoneal inflammation all induce gastric inhibition. Probable pathways included sympathetic nerves, vagus nerve, and spinobulbar ascending pathways as well as local vago-vagal reflexes.[4] Four major pathways have been identified for postoperative ileus (POI) in animal models and in people: (1) neurogenic pathway with stimulation of inhibitory neural pathways associated with surgical stress; (2) inflammatory pathway with stimulation of macrophages and neutrophils upon bowel manipulation, and release of pro-inflammatory mediators that reduce GI motility; (3) hormonal pathway with release of corticotrophin-releasing factor and stimulation of pro-inflammatory cytokines in the bowel. Finally, POI may be compounded by (4) pharmacologic interventions, in particular by the use of exogenous opioids as analgesics and their general inhibitory effect of GI motility.

In addition to POI, motility disturbances are frequently observed in association with GI inflammation. There is preliminary data suggesting delayed GI transit times occur in dogs and cats with pyloric diseases, pancreatitis and uremic syndrome.[1] Acute pancreatitis, peritonitis, sepsis, intestinal ischemia, uremic syndrome, spinal cord injuries, electrolyte imbalances are processes that have the potential to decrease GI motility.

Clinical signs include nausea, vomiting, anorexia, cranial abdominal discomfort, abdominal distension, and bloated abdomen in association with signs of the primary underlying disease. Gut sounds are usually absent on abdominal auscultation. Typically, abdominal radiographs show moderately distended bowels loops filled with gas and liquid content, and abdominal ultrasound reveals distended and hypomotile stomach and intestine.

It is the authors conviction that timely recognition and appropriate management of paralytic ileus is an absolutely essential part of the management of diseases such as acute pancreatitis or severe gastroenteritis or enterocolitis

GASTROINTESTINAL PROKINETICS

Prokinetic drugs should not be administered before a GI obstruction has been definitely ruled out. The available published evidence on the effect of various prokinetic drugs in healthy dogs and cats is tenuous, while it is almost nonexistent for dogs and cats with GI dysmotility. Therefore, collective clinical experience plays an important role in most recommendations for drug usage and dose (see table 1)

Serotonergic drugs

The drugs act on pre-synaptic 5-hydroxytryptamine (5-HT) receptors of different types. Beside its antiemetic effect, metoclopramide acts on 5-HT_3 (antagonist) and 5-HT_4 receptors (agonist). 5HT_4 receptors are located on enteric cholinergic neurons that innervate intestinal smooth muscle. Activation results in release of acetylcholine in the synaptic cleft and excitation of the postsynaptic cell membrane. These effects ultimately stimulate contraction of smooth muscle cells of the stomach and intestine. Metoclopramide was shown to shorten the time to restoration of GI motility after surgery in dogs.[5] In the author's clinical experience IV administration using a constant-rate infusion of 1-2 mg/kg/day is the most efficient option in dogs and cats with paralytic ileus. Cisapride is a serotonergic drug that was withdrawn from the pharmaceutical market due to its cross-reactivity with serotonergic receptors on the myocardium and the risk of lethal cardiac re-entry arrhythmias. It is a substrate of CYP3A, a cytochrome P450 enzyme in the liver. Unlike what is observed in people, cisapride does not cause prolongation of the QT interval and polymorphic ventricular arrhythmia in cats unless administered at doses 20 times higher than currently recommended.[6] It is currently available in the USA as a generic substance from compounding pharmacies. The principal mode of action of cisapride is its ability to bind to 5-HT_4 receptors, and to stimulate smooth muscle contractions. Prucalopride is a new highly selective 5-HT_4 agonist that has been recently approved for use in people in many countries but not in the USA. It has prokinetic properties on the canine gastric antrum, small intestine and colon.[7] Prucalopride induces defecation in cats, but there is no published evidence of its effects on the various segments of the feline GI tract.[8] Mosapride is a selective 5HT_4 agonist that is approved for use in dogs in Japan (Pronamid®). It has also been reported to be helpful in constipated cats.

Motilin agonists

Erythromycin is a macrolide antibiotic. At reduced doses, it exerts gastrokinetic effects similar to those of motilin, the hormone that triggers the onset of migratory motor complexes type III, a motility pattern responsible for cleaning the canine stomach and small intestine during the interdigestive phase.[9] The administration of erythromycin to dogs stimulates gastric emptying without any attention to particle size. Cats do not display well-organized interdigestive motility patterns and the actions of motilin feline GI motility are not well defined. Therefore, the prokinetic effect of erythromycin in cats is unknown.

Acetylcholinesterase inhibitors

Beside their effect as H_2-receptor antagonists, ranitidine and nizatidine are also prokinetics due to their effect as acetylcholinesterase (AChE) inhibitors, which increase the concentration of acetylcholine in the synaptic cleft between postganglionic myenteric neurons and smooth muscle cells of the stomach and intestine.

Novel prokinetics for use in dogs and cats

Mirtazapine is an inhibitor of central presynaptic α_2-adrenergic receptors (resulting in increased central release of noradrenaline) and a potent inhibitor of 5-HT_2 and 5-HT_3 receptors. The drug is often used as an appetite stimulant and antinausea drug. A recent study showed that mirtazapine at a very high dose of 45 mg/dog administered orally 1.5 h before a meal accelerated gastric emptying and colonic transit in healthy beagles.[10] It is possible that part of the positive effects of the drug in nauseous patients may be due to improved digestive motility. The recommended posology of mirtazapine as an appetite stimulant is 0.6 mg/kg q24h (max. 30 mg/dog) in dogs, and 1.88 mg/cat q48h in cats with chronic kidney disease.

Yohimbine (Yobine®) is an α_2-adrenergic receptor antagonist used in reversal of xylazine sedation in dogs. Intravenous doses of 0.5 to 3 mg/kg have been shown to induce giant migrating contractions in the colon of healthy dogs.[11] This property may be interesting in the treatment of postoperative ileus. However, the package insert mentions that doses of 0.55 mg/kg; may occasionally induce "brief seizures" and muscle tremors, therefore caution is recommended. It is not known if other α_2-adrenergic receptor antagonists such as atipamezole exert similar direct prokinetic activities.

Capsaicine is the major pungent ingredient of hot peppers. It has been shown to stimulate the release of peptides, including the prokinetic molecule calcitonin gene-related peptide (CGRP). In healthy dogs, intragastric capsaicin induced contractions in the antrum, small intestine, and colon within 15 minutes. Once in the small bowel, the compound had inhibitory effects on GI motility. Finally, intracolonic application of 5 to 10 mg capsaicin stimulated colonic motility and defecation.[12]

Lidocaine: a constant-rate infusion (CRI) of lidocaine (initially 0.025 mg/kg/min IV after loading bolus of 0.5 mg/kg IV) during any surgical intervention was thought to decrease the severity of POI through the drug's antinociceptive, antihyperalgesic and anti-inflammatory properties. However, the current consensus is that lidocaine CRI does not have any direct effect on GI motility.

Table 1: Mode and site of action and posology of commonly used prokinetics in small animals

Name	Mode of action	Site of action	Dose (for dogs and cats unless specified)
Metoclopramide	Serotonergic (5-HT4 receptors)	Pyloric antrum Duodenum (?)	0,2-0,5 mg/kg q8h, PO, SC CRI: 1-2 mg/kg/24h
Cisapride	Serotonergic (principally 5-HT4 receptors)	LES Pyloric antrum Small intestine Colon	0.1-0.5 mg/kg PO, q8-12h
Mosapride	Serotonergic (5-HT4 receptor-specific)	Pyloric antrum	0.25-2 mg/kg PO q12-24h (D). Only available in Japan.
Prucalopride	Serotonergic (5-HT4-receptor-specific)	Pyloric antrum Small intestine (?), Colon	0.01-0.6 mg/kg PO q12-24h. Only available outside USA
Erythromycin	Motilin analog	Pyloric antrum Small intestine Colon	0.5-1.0 mg/kg PO, IV q8h
Ranitidine	Acetyl-cholinesterase inhibitor (stimulation of M3 receptors)	Pyloric antrum (Small intestine, colon)	1-2 mg/kg q12h, PO or slowly IV (after dilution)
Nizatidine	Same as ranitidine	Same as raniditine	2.5-5 mg/kg q24h, PO

LES: lower esophageal sphincter, D: dog

REFERENCES

1.Allan FJ, et al. Journal of Veterinary Internal Medicine. 1994:151.
2.Schmitz S, et al. Tierarztliche Praxis Ausgabe K, Kleintiere/Heimtiere. 2013;41(2):87-92.
3.Boscan P, et al. Veterinary anaesthesia and analgesia. 2014;41(1):73-81.
4.Glise H, et al. Scandinavian Journal of Gastroenterology. 1980;15(7):769-74.
5.Graves GM, et al. Veterinary surgery : VS. 1989;18(1):27-33.
6.Kii Y, et al. Pharmacology & toxicology. 2001;89(2):96-103.
7.Wong BS, et al. Clin Exp Gastroenterol. 2010;3:49-56.
8.Briejer MR, et al. Gastroenterology. 1997;112:A705.
9.Ogawa A, et al. Am J Physiol Regul Integr Comp Physiol. 2012;302(2):R233-41.
10.Yin J, et al. Am J Physiol Gastrointest Liver Physiol. 2014
11.Nagao M, et al. Digestive Diseases and Sciences. 2007;52(9):2204-10.
12.Hayashi K, et al. Surgery. 2010;147(6):789-97.

ANTI-INFLAMMATORIES AND IMMUNE MODULATORS: TREATMENT OPTIONS FOR INFLAMMATORY BOWEL DISEASE

Eli D. Ehrenpreis. MD
Chicago, IL

Introduction

Inflammatory Bowel Disease is a term for two conditions, Crohn's disease and ulcerative colitis. Inflammatory bowel disease results from altered regulation of the immune system, causing inflammation within the gastrointestinal tract. Crohn's disease, also known as regional enteritis, is characterized by normal and affected areas of the bowel. Crohn's disease can affect the entire gastrointestinal tract from the mouth to the anus. Crohn's disease is also a transmural disorder that affects all layers of the bowel wall. In contrast, ulcerative colitis is an inflammatory disorder that is specific to the colon only. Ulcerative colitis is a mucosal disease that only affects the inner most layer of the colon. It is a disease that begins in the rectum and extends proximally in the colon. Thus some patients have ulcerative colitis only affecting the rectum, others the rectum and sigmoid colon and some have involvement of the entire colon. Ulcerative colitis is not characterized by skip areas of normal tissue, as seen in patients with Crohn's disease.

Pathophysiology

Although incompletely understood, it is generally accepted that inflammatory bowel disease occurs as an inappropriate response to an environmental stimulus in patients having a genetic susceptibility to these conditions. The putative cause of the initial inflammatory response that ultimately leads to autoimmunity is not fully understood. However, various bacteria and other intestinal microorganisms clearly play a role in the development of these disorders. The most fully described environmental factors that interact with the development and severity of inflammatory bowel disease is cigarette smoking and the use of nonsteroidal anti-inflammatory agents (NSAIDs). For example, cigarette smokers are more likely to develop Crohn's disease, present with more severe manifestations of Crohn's disease, and require more immunosuppressant agents to manage their condition. Cigarette smokers also are more likely to undergo surgery for Crohn's disease and have postsurgical relapses of the disease that occur more quickly and are more severe than non-smokers. There are a variety of theories as to why the association between cigarette smoking and Crohn's disease occurs, including the possibility that localized ischemia due to decreased blood flow from smoking triggers intestinal damage. Of interest, patients with ulcerative colitis are more likely to be non-smokers or former smokers than current smokers. This observation has resulted in a few small clinical trials of nicotine patches for the treatment of ulcerative colitis. Nonsteroidal anti-inflammatory agents have been shown to trigger the development of Crohn's disease, to increase the likelihood of symptomatic flare-ups of inflammatory bowel disease and worsening of ongoing inflammatory bowel disease. A number of genes have been identified that are more common in patients with Crohn's disease and ulcerative colitis than the general population. The first one discovered was a mutation in the NOD2/CARD15 gene, which was found to be associated with developing Crohn's disease. Up to 20% of IBD patients in North America and Europe may have a mutation in the NOD2/CARD15 gene. Many studies have also shown alterations in the gut microbiota in patients with inflammatory bowel disease. Usage of antibiotics or the drug isotretinoin (Accutane), a treatment for severe acne, have also been suggested as risk factors for the development of ulcerative colitis.

Epidemiology

A population-based study conducted from 1940 to 2011 in Olmsted County, Minnesota, examined the incidence of IBD. The total number of new cases of Crohn's disease diagnosed each year (incidence) was 10.7 per 100,000 people, or approximately 33,000 new cases per year. The total number of new cases of ulcerative colitis diagnosed each year was 12.2 per 100,000 people, or approximately 38,000 new cases per year. Approximately 1.6 million Americans currently have Crohn's disease or ulcerative colitis. As many as 70,000 new cases of IBD are diagnosed in the United States each year. Of interest, the highest rates of IBD are assumed to be in developed countries, and the lowest are considered to be in developing regions; colder-climate regions and urban areas have a greater rate of IBD than those of warmer climates and rural areas.

Clinical Manifestations

Two forms of symptoms of inflammatory bowel disease are seen in patients with these conditions. The first of these are a direct result of local inflammation within the gastrointestinal tract. Common gastrointestinal symptoms include diarrhea, abdominal pain, urgency and rectal bleeding. Systemic symptoms that are due to either complications of the disorder or indirect effects include fever, weight loss, night sweats, growth failure or fatigue (Table 1).

Table 1-Symptoms of inflammatory bowel disease

Symptoms related to inflammation of the GI tract:	General symptoms that may also be associated with IBD:
Diarrhea	Fever
Abdominal pain	Fatigue
Urgent need to move bowels	Loss of appetite
Rectal bleeding	Weight loss
Sensation of incomplete evacuation	Loss of normal menstrual cycle
	Night sweats

Complications of Inflammatory Bowel Disease
Ulcerative colitis

Severe diarrhea
Significant blood loss
Intractable abdominal pain
Toxic megacolon
Intestinal perforation

Crohn's disease

Intestinal stricture
Intra-abdominal, perianal or perirectal abscesses
Nutrient malabsorption
Vitamin deficiency (B12, vitamin D)
Severe bleeding

Other organ systems affected by inflammatory bowel disease include the liver (hepatitis, sclerosing cholangitis), the joints, skin, and the eyes. Kidney stones and osteoporosis are also more common in patients with inflammatory bowel disease.

Table 2. Extraintestinal Complications of IBD in Swiss Patients

Complication	Crohn Disease	Ulcerative Colitis
Arthritis	33%	4%
Aphthous stomatitis	10%	4%
Uveitis	6%	3%
Erythema nodosum	6%	3%
Ankylosing spondylitis	6%	2%
Psoriasis	2%	1%
Pyoderma gangrenosum	2%	2%
Primary sclerosing cholangitis	1%	4%

Source: Vavricka SR, Brun L, Ballabeni P, et al. Frequency and risk factors for extraintestinal manifestations in the Swiss inflammatory bowel disease cohort. Am J Gastroenterol. 2011; 106:110-9.

Treatments for Inflammatory Bowel Disease
(These will be discussed in detail in the lecture)

A variety of treatment options are available for the medical management of patients with Inflammatory Bowel Disease. Since there are a large number of inflammatory mediators that have been associated with these conditions, latest research on pharmacotherapy has focused on development of antibodies against these mediators and/or methods to stop their production. Therapy for Crohn disease is generally less effective than that for ulcerative colitis. In general, treatment involves a stepwise approach during which drugs with the fewest side effects or immunosuppression are started first. In patients with advanced or aggressive disease that are deemed to be at high risk for complications, data suggests that early therapy with more potent immunomodulators and biologic therapy will reduce the likelihood of complications in these patients.

5-Aminosalicylic Acid

5-aminosalicylic acid (5-ASA) derivatives are topical gastrointestinal anti-inflammatory agents that are available in oral and intrarectal forms (suppositories and enemas). They are approved for the acute treatment of flares of mild to moderate ulcerative colitis. Less effectiveness has been demonstrated for Crohn's disease. Oral 5-ASAs are administered in a variety of delayed release forms to prevent oral absorption. %-ASAs on the market include:

Sulfasalazine (Azulfidine)
Mesalamine (Asacol HD, Pentasa, Canasa, Rowasa, Lialda, Apriso, Delzicol)
Balsalazide (Colazal)
Olsalazine (Dipentum)

Antibiotics

Metronidazole (Flagyl) and ciprofloxacin (Cipro) have some efficacy in treating inflammatory bowel disease. Rifaximin (Xifaxan) has also being studied.

Corticosteroids

Corticosteroids are commonly used for treatment of acute inflammatory bowel disease (IBD), especially in those patients with severe disease. Numerous acute and chronic side effects occur with corticosteroid therapy. In general, corticosteroids are not used as maintenance therapy for inflammatory bowel disease. Budesonide, a steroid with high first pass metabolism is available in a delayed release (Entocort) for the treatment of ileal Crohn's disease and ulcerative colitis (Uceris).

Immunosuppressants

Immunosuppressant agents are administered in patients requiring long-term or high-dose steroids as steroid-sparing agents and for patients with fistulizing Crohn's disease. These include:

Azathioprine (Imuran)
6-Mercaptopurine (Purinethol)
Methotrexate
Cyclosporine (Sandimmune, Neoral)

TNF Inhibitors

These are monoclonal antibodies directed against tumor necrosis factor alpha (TNFα). They function to decrease the production and activity of endogenous TNF. TNFα is a proinflammatory cytokine that enhances interleukin production, activates neutrophils and enhances leukocyte migration. Infliximab was the first of this class for use in inflammatory bowel disease. Monoclonal antibodies for tumor necrosis factor alpha (TNFα) promote mucosal healing, heal perianal and enterocutaneous fistulas, and induce clinical remission and mucosal healing. A variety of infectious complications have been described with the use of these drugs. These include disseminated fungal infections, tuberculosis and a variety of opportunistic infections. Lymphomas, including hepatosplenic T cell lymphoma, a universally fatal disease, and other cancers have also been described with the use of inhibitors of TNFα. These agents include:

Infliximab (Remicade)
Adalimumab (Humira)
Certolizumab pegol (Cimzia)
Golimumab (Simponi)

Alpha 4 Integrin Inhibitors

Integrins are adhesion molecules that bind to endothelial receptors, contributing to the enhancement of the immune response and chronic inflammation. Two of these are available on the market for the treatment of Inflammatory Bowel Disease:

Natalizumab (Tysabri)
Vedolizumab (Entyvio)

REFERENCES

1. Vavricka SR, Brun L, Ballabeni P, et al. Frequency and risk factors for extraintestinal manifestations in the Swiss inflammatory bowel disease cohort. Am J Gastroenterol. 2011;106:110-9.
2. http://www.ccfa.org/assets/pdfs/updatedibdfactbook.pdf
3. Rodriguez LAG , Gonzalez-Perez A , Johansson S *et al.* Risk factors forinfl ammatory disease in the general population . Aliment Pharmacol Th er 2005 ; 22 : 309 – 15
4. Mahid SS , Minor KS , Soto RE *et al.* Smoking and infl ammatory bowel disease: a meta-analysis . Mayo Clinic Proc 2006 ; 81 : 1462 – 71 .
5. Timmer A , McDonald JWD , MacDonald JK . Azathioprine and 6-mercaptopurinefor maintenance of remission in ulcerative colitis . Cochrane Database Syst Rev 2007 , Issue 1. Art. No.: CD000478. DOI: 10.1002/14651858.CD000478.pub2 .
6. Sandborn WJ , Sutherland LR , Pearson D *et al.* Azathioprine or 6-mercaptopurinefor induction of remission in Crohn ' s disease . Cochrane DatabaseSyst Rev 1998 , Issue 3. Art. No.: CD000545. DOI: 10.1002/14651858.CD000545
7. Sutherland LR , MacDonald JK . Oral 5-aminosalicylic acid for induction ofremission in ulcerative colitis . Cochrane Database Syst Rev 2006 , Issue 2. Art. No.: CD000543. DOI: 10.1002/14651858.CD000543.pub2 .
8. Akobeng AK , Gardener E . Oral 5-aminosalicylic acid for maintenance ofmedically-induced remission in Crohn ' s disease . Cochrane Database Syst Rev 2005 , Issue 1. Art. No.: CD003715. DOI: 10.1002/14651858.CD003715.pub2 .
9. Sutherland LR , MacDonald JK . Oral 5-aminosalicylic acid for maintenance of remission in ulcerative colitis . Cochrane Database Syst Rev 2006 , Issue 2. Art. No.: CD000544. DOI: 10.1002/14651858.CD000544.pub2 .
10. Benchimol EI , Seow CH , Steinhart AH *et al.* Traditional corticosteroids for induction of remission in Crohn ' s disease . Cochrane Database Syst Rev 2008 , Issue 2. Art. No.: CD006792. DOI: 10.1002/14651858.CD006792 pub2 .
11. Behm BW , Bickston SJ . Tumor necrosis factor-alpha antibody for maintenance of remission in Crohn ' s disease . Cochrane Database Syst Rev 2008 , Issue Art. No.: CD006893. DOI: 10.1002/14651858.CD006893 .
12. Lawson MM , Th omas AG , Akobeng AK . Tumour necrosis factor alpha blocking agents for induction of remission in ulcerative colitis . Cochrane Database Syst Rev 2006 , Issue 3. Art. No.: CD005112. DOI: 10.1002/14651858.CD005112.pub2 .
13. Talley NJ, Abreu , MT , Achkar JP, et al. An Evidence-Based Systematic Review on Medical Therapies for Infl ammatory Bowel Disease. Am J Gastroenterol 2011; 106:S2 – S25.

HOW I USE ANTI-INFLAMMATORIES ON THE CLINIC FLOOR

Albert E. Jergens, DVM, PhD, DACVIM (Small Animal Internal Medicine)
Ames, IA

INTRODUCTION

Idiopathic inflammatory bowel disease (IBD) is a common cause of chronic gastrointestinal (GI) disease in dogs. Medical therapy for induction of clinical remission is largely empirical and consists of dietary management and use of nonspecific anti-inflammatory drugs, with corticosteroids providing the most consistent benefit. Accumulating evidence in human IBD and experimental colitis in animal models suggests that the resident microbiota contribute to the pathogenesis of chronic immunologically-mediated intestinal inflammation. Dogs with IBD have distinctly different duodenal microbial communities compared to healthy dogs. Additionally, canine histiocytic ulcerative colitis now is recognized to be associated with adherent and invasive *E. coli*, and affected dogs frequently respond to fluoroquinolone antimicrobial therapy. Similarly, GI signs in cats with IBD have been linked to altered microbial composition with subsets of intestinal bacteria and histopathologic inflammation. Thus, there appears to be a sound rationale for evaluating canine IBD treatment protocols that target enteric bacteria and the harmful immune responses that they might evoke.

Therapeutic intervention of IBD is aimed at counteracting inflammation and dysbiosis and combating nutritional (cobalamin) deficiencies. Practical drug treatment recommendations are determined by disease severity and clinical course (e.g., presence of mild IBD versus more serious gut inflammation leading to PLE), the involved segments of the alimentary tract, clinician experience, size of the patient impacting drug costs, and the potential side-effects associated with use of some medications.

PRACTICAL GUIDELINES FOR DRUG TREATMENT OF HUMAN IBD

Similar to companion animal IBD, current guidelines for IBD therapy are distinguished by the course, place, and severity of the clinical disease. Recent discussions in human IBD are concerned with recommendations for a step-down or step-up therapy. In Crohn's disease (CD), a mild active form, a moderate or severe active form or a severe chronic active form can be differentiated. Patients can also be defined as steroid-dependent or steroid-refractory and endoscopic evidence of mucosal healing, as well as the segments of the gut that are involved (SI vs. colon), are salient. In ulcerative colitis (UC), differentiation is much easier since only the colon is involved; thus, treatment is determined by the clinical disease course and the involved segments of the colon (left-sided disease vs. pancolitis) as well as the previous response to steroids. Corticosteroids are effective for the induction of clinical response and remission in moderate-to-severe CD and in left-sided UC. Induction of remission typically averages 7-14 days and then dosages are tapered. Conventional steroids have not been shown to be efficacious in maintenance treatment of HIBD due to side effects. As the clinical course of 36% of human IBD patients is defined as steroid-refractory, immunomodulators are recommended for the treatment of chronic active IBD. Only limited trial data exists on the efficacy of immunosuppressants in subsets of CD patients and evidence-based data is missing on the use of azathioprine for the prevention of post-operative recurrence of CD. Potential immunomodulators for use include azathioprine/6-mercaptopurine, methotrexate, mycophenolate mofetil, cyclosporine, and the biologics (e.g., Infliximab).

IMMUNOSUPPRESSIVE DRUG CATEGORIES AND GENERAL MECHANSISMS OF ACTION

Glucocorticoids represent the most common form of immunosuppressive drug therapy for immune-mediated GI disorders, especially canine and feline IBD. The mechanisms underlying the immunosuppressive effects of glucocorticoids are complex, and thought to involve drug interaction with the regulatory factor, NFκ-B, to suppress the transcription of genes that encode proteins with pro-inflammatory and immune-stimulating activity. **Prednisone/prednisolone** are used most commonly with prednisolone being preferred by some clinicians for use in cats. Other forms of glucocorticoids administered include **dexamethasone** (D), **methylprednisolone acetate** (MPA), and **budesonide.** Both D and MPA may be administered parenterally and are recommended for use in animals with severe mucosal disruption where oral medications may be inadequately absorbed. **Budesonide** has an extensive first pass hepatic transformation and may be associated with fewer systemic effects as compared to other oral glucocorticoids. Budesonide is rapidly absorbed and metabolized in dogs with IBD. Separate studies indicate and that there is adequate therapeutic response and no adverse effects when budesonide is used short-term for canine IBD.

Calcineurin inhibitors are potent immunosuppressive drugs that include **cyclosporine** (CYC) and **tacrolimus**. Inhibition of calcineurin causes decreased production of key cytokines (i.e., IL-2) and modulate both B- and T-lymphocyte responses. Cyclosporine is available in several different formulations, which may impact the specific GI absorption and hepatic metabolism between formulations. Atopica is the only specifically licensed formulation for use in dogs and cats. There is some evidence-based data that (CYC) is useful in the management of steroid-refractory canine IBD. 1% topical tacrolimus has been used successfully in the management of dogs with perianal fistulas.

The **antimetabolite drug** used most commonly for immunosuppression in dogs with IBD is **azathioprine** (AZA). The active moiety of AZA is 6-mercaptopurine, which acts to block DNA synthesis and mRNA transcription causing impaired B- and T- cell immune responses. While AZA is a mainstay drug for maintenance of remission in human IBD, its use in treatment of canine/feline IBD is much less commonly reported. Potential concerns with AZA include (1) an apparent "lag effect" of 2-3 weeks before desired clinical benefits are seen and (2) anecdotal reports of bone marrow, hepatic, and/or pancreatic toxicities. This drug should be used very cautiously in cats.

Alkylating agents and other immunosuppressive agents may be of value in dogs with severe enteropathy, including PLE, and in cats where differentiation of severe IBD vs. alimentary LSA is problematic. Alkylating agents such as **cyclophosphamide** and **chlorambucil** (used most often) block DNA synthesis and inhibit both B- and T-cell host responses. Chlorambucil is often used as a rescue drug for steroid-refractory IBD, in canine PLE, and when well-differentiated feline LSA cannot be conclusively excluded. Still **other potential drugs** for immunosuppressive use but with considerably less clinical experience and anecdotal observations include **mycophenolate mofetil** and **leflunomide**.

THERAPEUTIC APPROACHES TO CANINE AND FELINE IBD

Most clinicians prefer a sequential step-up approach for lymphocytic-plasmacytic IBD (most common), using diet → antibiotics → steroids → various immunosuppressive drugs. The optimal drug or drug combination as well as duration of therapy for induction and maintenance of remission of clinical signs have not been determined and are based on therapeutic trials.

Canine IBD

Different clinical trials and case series describing immunosuppressive therapy for dogs with IBD will attest to the efficacy of glucocorticoids in inducing short-term (weeks to months) clinical remission.

- One standard therapy with glucocorticoids uses **prednisone** at a starting dose of 1-2 mg/kg PO daily for 2 weeks as induction therapy in small and medium sized dogs. Dosage reductions at 25% q 2 weeks are then recommended if the desired clinical response is noted – 50% reduction in baseline CIBDAI and/or maintenance of serum albumin > 2g/dL.
- Dogs > than 25 kg body weight are best treated with prednisone on the basis of body surface area (BSA) expressed as M^2. Higher steroid doses in these dogs will often cause adverse effects including PU/PD, polyphagia, and excessive panting. Guidelines for prednisone therapy using BSA are empirical and published guidelines are unavailable.
- Use **budesonide** (1-3 mg/day) in diabetic dogs or those with a history of adverse reactions to prednisone/prednisolone. Consider budesonide as a first-choice glucocorticoid in large breed dogs (> 25 kg) to avoid adverse effects that are commonly observed when prednisone is used at traditional immunosuppressive (1-2 mg/kg/day) doses. Budesonide is also useful as a maintenance agent in dogs requiring long-term immunosuppression due to severe intestinal disease or recurring clinical signs (flares). In my experience, there is a 10-14 day "lag time" before clinical response to budesonide is realized.
- **Azathioprine** may be used as a maintenance agent in canine IBD – its use in human IBD is well recognized; however, there are surprisingly few studies in the veterinary literature attesting to its efficacy in the treatment of canine IBD. **Chlorambucil** is an alternative maintenance agent to be considered and is the adjunct treatment of choice along with glucocorticoids in canine PLE.
- **Cyclosporine** may be used as a first line immunosuppressive in dogs with severe IBD; however, drug costs may limit its use in mid-sized and larger dogs. It is also reported to be efficacious in the therapy of steroid-refractory IBD based on a single published study.
- Monitor for bone marrow suppression when potent immunosuppressive drugs are administered.
- When using glucocorticoids alone or in combination with other immunosuppressive drugs, I favor tapering these agents first, then followed by cessation of the other immunosuppressive agent. Relatively few dogs will require long-term (months) drug therapy and it is always my goal to maintain remission (if possible) using dietary therapy alone.

Feline IBD

Different clinical trials and case series describing immunosuppressive therapy for cats with IBD will attest to the efficacy of glucocorticoids in inducing short-term (weeks to months) clinical remission.

- One standard therapy with glucocorticoids uses **prednisolone** at a starting dose of 1-2 mg/kg PO daily for 2 weeks as induction therapy. Dosage reductions at 25% q 2 weeks are then recommended if the desired clinical response is noted.
- **Glucocorticoids** should be used with caution in cats to avoid aggravation of cardiac disease and promotion of diabetes mellitus.
- **Chlorambucil** may be used with steroids in cats having severe IBD or when biopsy results cannot conclusively discriminate between IBD versus alimentary LSA.

SUGGESTED READINGS

Craven M, Simpson JW, Ridyard AE, et al. Canine inflammatory bowel disease: retrospective analysis of diagnosis and outcome in 80 cases (1995-2002). J Small Anim Pract 2004; 45:336-342.

Jergens AE, Schreiner CA, Frank DE, *et al.* A scoring index for disease activity in canine inflammatory bowel disease. J Vet Intern Med 2003; 17:291-297.

Allenspach K, Wieland B, Grone A, et al. Chronic enteropathies in dogs: evaluation of risk factors for negative outcome. J Vet Intern Med 2007; 21:700-708.

Allenspach K, Rufenacht S, Sauter S, et al. Pharmacokinetics and clinical efficacy of cyclosporine treatment of dogs with steroid-refractory inflammatory bowel disease. J Vet Intern Med 2006; 20:239-244.

Dye TL, Diehl KJ, Wheeler SL, et al. Randomized, controlled trial of budesonide and prednisone for the treatment of idiopathic inflammatory bowel disease in dogs. J Vet Intern Med. 2013; 27:1385-91.

Simpson K, Jergens A. Pitfalls and progress in the diagnosis and management of canine inflammatory bowel disease. The Veterinary clinics of North America Small animal practice 2011;41:381-398.

Trepanier L. Idiopathic inflammatory bowel disease in cats. Rational treatment selection. Journal of feline medicine and surgery 2009;11:32-38.

SERUM AMYLOID A

Bonnie Barr, VMD, DACVIM
Lexington, KY

The acute phase response is a nonspecific inflammatory response that occurs immediately after any process that leads to tissue damage. A major step in the acute phase response is a change in the concentrations of acute phase proteins. Acute phase proteins are defined as those proteins whose plasma concentration increase or decrease by at least 25% during inflammatory disorders. The change in acute phase proteins is due to cytokines produced during the inflammatory response. The liver is the primary producer of acute phase proteins. Conditions that lead to changes in plasma acute phase proteins include bacterial and viral infection, parasite infestation, trauma, surgery, ischemic necrosis, burns and neoplasia.

Serum amyloid A is an acute phase protein that is normally found in the plasma in low or undetectable levels in healthy individuals. In response to inflammation, serum amyloid A rapidly increases to more than 10 fold; often 100 or 1000 fold. Once appropriate treatment has been initiated and inflammation has resolved the plasma levels decrease to a normal value. The complete sequence of equine serum amyloid A has been published and 3 isoforms have been identified. Evidence exists of extra hepatic synthesis in the digestive tract, airways, mammary glands and synovial tissue. Serum amyloid A is a nonspecific marker of inflammation thus cannot be used for etiologic diagnosis.

Several studies have suggested that serum amyloid A is a more sensitive indicator of equine inflammation than classical markers of inflammation such as white blood cell count and fibrinogen. In a recent report clinically abnormal horses had a significantly higher increase in serum amyloid A, whereas only a minimal increase in white blood cell count was noted. Diseases diagnosed in these horses included bacterial and viral pneumonia, bacterial cholangiohepatitis, meningitis, enterocolitis, colic, pulmonary and musculoskeletal abscesses and septic tenosynovitis. Fibrinogen, which is a commonly monitored acute phase protein, is normally found in high amounts in the plasma of healthy horses and only increases by 1-10 times in response to inflammation. The time to peak response of fibrinogen with infection and inflammation is longer than serum amyloid A. Other factors including consumptive coagulopathies or increased vascular permeability that are commonly observed in critically ill horses and can falsely lower fibrinogen make interpretation challenging.

Several methods for measuring equine serum amyloid A have been developed including ELISA, single radial immunodiffusion, electroimmunoassay, slide reversed passive latex agglutination test and latex agglutination immunoturbidimetric assay. Initially many of the systems were not commercially available and lacked validation in horses. More recently an immunoturbidimetric assay developed for humans has been validated in horses providing a rapid and reliable method to monitor serum amyloid A. Smaller stall size methods have been developed and show promise in providing a rapid and accurate means to monitor serum amyloid A.

Bacterial infections have been reported to give rise to a high plasma serum amyloid A concentration in both adults and foals. In a recent report comparing serum amyloid A plasma concentrations in clinically normal and clinically abnormal horses those with bacterial infection had a higher plasma serum amyloid. This was in contrast to white blood cell count and fibrinogen, which were only minimally increased in these cases. Elevated levels were also noted in horses with viral infections and horses experimentally challenged with Equine Herpes Virus-1 showed a marked but transient rise in the value.

Rhodococcus equi can illicit increased serum amyloid A concentrations in foals and because of this response the test has been evaluated as an early screening tool for R equi pneumonia. One study obtained serum samples from foals at two time points one to two weeks apart and from foals with clinical signs of pneumonia and normal foals. The results indicated that at the specific time points there was no significant difference between serum amyloid A values in foals with Rhodococcus equi pneumonia and those without. A smaller scale study with more frequent sampling produced similar results. Therefore in this situation serum amyloid A does not appear to be beneficial as a screening tool. Neonatal foals with infection were found to have a significantly higher value when compared to those with trauma or non-infectious disorders such as prematurity or neonatal encephalopathy.

Experimentally induced non-septic arthritis results in an elevation of plasma serum amyloid A. It has been shown that serum amyloid A can also be measured in synovial samples. Both serum and synovial samples have been found to have increased levels of serum amyloid A after naturally and experimentally occurring joint diseases. Those with bacterial infections and infectious arthritis were found to have the highest levels. Once the disorder was treated appropriately the values returned to normal. Repeated sampling did not alter the synovial value.

Serum amyloid A has been researched as a marker to determine if a horse is fit for training. A study evaluating serum amyloid A values in endurance horses showed that most horses that did not finish had elevated serum amyloid A values prior to the race. It is postulated that the increase in this value was due to underlying subclinical disease, overtraining or slight injuries that may worsen with exertion and lead to elimination from the ride. Since a large number of health issues that impact training can trigger an inflammatory response, serum amyloid A may indicate a subclinical problem in horses that are performing poorly. A pilot study in racehorses suggested a strong correlation with increased serum amyloid A and poor performance. Detection of subclinical problems would be useful information to trainers and veterinarians, thereby providing a management tool for adjustments to training intensity that could potentially decrease the risk of horses breaking down. Interestingly, in a recent paper comparing normal horses to abnormal horses admitted to an equine veterinary hospital, two clinically normal horses with mildly elevated serum amyloid A ultimately developed clinical signs of infection within 24 hours.

Another area in which serum amyloid A may prove to be useful is placental pathology. In the normal pregnant mare serum amyloid A values remain low but show a significant increase within 12 hours of parturition. In experimentally induced placentitis the plasma serum amyloid A value increases and remains high until abortion or successful treatment with delivery of a live foal.

Serum amyloid A has been reported to be a useful monitoring tool for response to treatment. Numerous reports have documented a rapid decrease in serum amyloid A with appropriate treatment. A good example of this is bacterial pneumonia patients in which the initial serum amyloid A value is significantly high but with appropriate treatment the value rapidly declines. Serum amyloid A is degraded by the liver and its plasma half-life is very short. Thus once the inflammatory response decreases the value should return to normal.

The acute phase response is non-specific and is induced by many diverse infective, inflammatory and traumatic stimuli. Serum amyloid A measurement can therefore never be diagnostic by itself but the results can provide valuable information in clinical practice. A particularly high concentration is most likely to occur in bacterial infections and serial studies indicate that effective therapy is associated with a rapid fall in serum amyloid A to normal.

REFERENCES

Gaby, C, et al. NJM 2006;340(6):448. Belgrave R, et.al. JAVMA 2013;243(1):113. Jacobsen S, et al. EVE 2007;19(1):38. Jacobsen S, et al. Vet J 2006;172(2):315. Jacobsen S, et al AJVR 2006;67(10):1738. Cywinska, A, et al. EVJ 2010;42(38):23. Paltrinieri, S, et al Vet J 2008;176:393. Coutinho da Silva, M, et.al. EVJ 2012;45:619.

HOW TO USE L-LACTATE CLINICALLY

Pamela A. Wilkins, DVM, MS, PhD, DACVIM (LAIM), DACVECC
Champaign-Urbana, IL

INTRODUCTION

Abbreviations:	
LAC	L-lactate
[LAC]	L-lactate concentration
$[LAC]_{Admit}$	Admission [LAC]
[LAC]Δ	Change in [LAC] with time
LAC_{Area}	Area under [LAC] *vs.* time curve

Blood lactate (LAC) concentration ([LAC]), a metabolite normally found in blood, was recognized for its role as a biomarker to predict morbidity and mortality many years ago. Blood [LAC], reflecting both production and elimination/metabolism (sometimes termed 'clearance') of LAC, is readily measured using inexpensive hand held lactate meters and is commonly monitored clinically. Hyperlactatemia can be classified as Type A (due to inadequate oxygen delivery or increased oxygen demand) or Type B (due to inadequate oxygen utilization, decreased clearance, congenital errors of metabolism, drugs/toxins or other causes).[1] Serial [LAC] measurements may be more reliable for evaluating response to therapy and predicting outcome than [LAC] measurement at a single time point.[2-6] The rate or degree of [LAC] decrease has been associated with survival in septic human patients and in critically ill foals and horses, supporting the concept that serial lactate measurements may be more reliable predictors of survival.[2-6] Numerous studies, mostly retrospective, have evaluated the role of [LAC] in sick neonatal foals and have consistently shown increased [LAC], in addition to either apparently impaired LAC removal or increased production, or both, in non-survivors.[3,7-9] Prospective serial evaluation of [LAC] in adult equine emergencies showed similar results.[2]

L-LACTATE USE IN THE EQUINE NEONATE

A recent prospective multicenter study[10,11] evaluating the relationship between [LAC] at admission ($[LAC]_{Admit}$) and short term outcome to hospital discharge in ~ 580 foals confirmed the findings of several smaller equine studies,[3,7-9] and was consistent with findings of several human retrospective and prospective studies,[4,5,12-15] in that $[LAC]_{Admit}$ is associated with outcome. Overall, there was an increased risk of non-survival of 14% (OR = 1.14) for each 1 mmol/l increase in $[LAC]_{Admit}$. Uniquely, and perhaps more importantly, that multicenter prospective study suggested that the utility of $[LAC]_{Admit}$ as a prognostic indicator may be different and vary by clinicians' major diagnosis category in the ill equine neonate. The clinicians' major diagnoses of 'sepsis' (OR = 1.21), 'unspecified enterocolitis'(OR = 1.28), 'unspecified colic'(OR = 1.21), 'trauma'(OR = 1.45), 'immune related (not FPT)' (OR = 1.58) and 'respiratory'(OR = 2.82) each carried a statistically significant and clinically relevant increased risk of non-survival for each 1 mmol/l increase in [LAC] at admission, despite having very similar [LAC] to each other and to other diagnoses.[10] In fact, the similarity of [LAC] at admission across the clinicians' major diagnoses was of interest, particularly when combined with the observation that differences between major clinical diagnoses related to $[LAC]_{Admit}$ were only clearly distinguished by the OR analysis. This is in alignment with the clinical observation that sick foals consistently present with increased [LAC], perhaps of different causes, and that use of [LAC] at admission only correctly classifies ~75-80% of critically ill foals. The relationship of clinical diagnosis and survival associated with [LAC] at admission was lost, except for foals with sepsis, when assessed as either serial [LAC] measurement over time or as the change in [LAC] over time, although persistent hyperlactatemia was associated with non-survival across all foals. Further study is indicated as, despite the size of the study and its prospective multicenter nature, some diagnostic categories contained insufficient patient numbers for meaningful analysis.

There is an age-dependent decrease in [LAC] over the first few days of life that may confound the evaluation of the various LAC measures as tools for evaluating prognosis and disease severity; prematurity may then be an important confounder in L-lactate studies in this population.[16-18] The fetus utilizes L-lactate as an important energy source and failure to adapt to post-fetal energy metabolism may be a feature of some disease processes.[17,18] Another potential confounder in the previous large prospective study[10,11] is that no effort was made to ensure that all [LAC] measuring devices were similar. We have unpublished data from a related study in adult horses that shows differences between [LAC] measuring devices and, while internally consistent, measuring techniques are not necessarily concordant. This suggests that appropriate ranges for 'normal' values should be obtained by each location for the type of measuring device in use and also suggests that there should be no hard and fast cut-points for L-lactate measures. Rather, any published cut-points should be considered as guidelines that need local refinement depending on both local conditions and type of L-lactate measuring device used.

Prior to the previously mentioned multicenter study of sick equine neonates[11], serial [LAC] had not been evaluated prospectively as a prognostic indicator in a large population. One objective of that study was to investigate serial [LAC] and [LAC]Δ over the first 120 hour time period of hospitalization at approximate 24 hour intervals. An increased or positive value for [LAC]Δ indicates a decrease in [LAC], while a smaller or negative value indicates either minimal change or an increase in [LAC] between the sampling periods. When considering all foals, logistic regression analyses of outcome for [LAC]Δ over all time intervals showed that [LAC]Δ in non-surviving foals did not itself decrease over time, indicating persistent hyperlactatemia, while survivors showed significant positive (decreasing [LAC]) [LAC]Δ for admission and 24-Hours ($[LAC]_{A-24}$) and all other time periods ($p < 0.001$). Additionally, logistic regression analysis showed that the odds of survival decreased for each 1 mmol/l [LAC] increase at all testing points for all sick foals, independent of major final diagnoses. Septic foals showed significantly greater [LAC] at all testing points compared with non-septic foals ($p < 0.001$). [LAC]Δ in septic foals was significantly more positive (suggesting better clearance) at Days 1 and 3 ($p < 0.01$) compared to other time periods, while in non-septic foals [LAC]Δ was significantly positive on Day 1 compared to all other time periods ($p < 0.001$). Together these data demonstrate that the admission to 24-Hour period, Day 1, is quite dynamic regarding LAC and worthy of more detailed study.

The recent studies emphasizing the prognostic value of LAC associated measures such as LAC_{Admit} and [LAC]Δ suggest that both duration and severity of hyperlactatemia are related to mortality and disease severity in critically ill patients.[5,10-13,15,19] [LAC] at admission represents only a patient's initial status, not change over time. [LAC]Δ only demonstrates duration of hyperlactatemia. Lactate area (LAC_{Area}) is defined as the sum of the areas under the curve of serial (every 6 hour) [LAC] determinations during the first 24 hours following each patient's admission using the trapezoidal rule. This newer concept reflects both the severity and duration of hyperlactatemia in critically ill patients. A recent study of human pediatric sepsis patients using LAC_{Area} -in addition to [LAC]Δ and $[LAC]_{Admit}$- identified only LAC_{Area} as significantly correlated with 28-day mortality.[19] LAC_{Area}, while not yet critically studied in critically ill foals, has been evaluated in healthy foals. The normal values of LAC_{Area} in equine neonates are now known to be age-dependent and different between Day 1 and 3 of post-gestational life, reaching adult values by Day 7.[20] In clinically affected foals LAC_{Area} is likely to be related to both the persistence and the severity of increased [LAC] during the early course of treatment; in other words it is influenced by the exposure time to hyperlactatemia as well as the actual [LAC]. Preliminary data suggest that LAC_{Area} will be a useful L-lactate indicator for use in ill neonatal foals.[20,21]

Currently, [LAC] at admission appears most useful as a prognostic indicator when applied to septic neonatal foals, neonatal foals with gastrointestinal disturbances such as diarrhea and colic, neonatal foals suffering trauma, neonatal foals with diseases such as neonatal isoerythrolysis, and in neonatal foals with significant respiratory comprise. Neonatal foals with sepsis have a persistent association with both [LAC] at admission and change in [LAC] over time that may be clinically useful. LAC_{Area} may prove a useful indicator of both survival and disease severity, potentially proving useful in guiding therapy. Following [LAC] as a guide to resuscitation therapy in neonatal foals is appealing, but not yet critically investigated.

L-LACTATE USE IN THE ADULT HORSE

A retrospective study examining adult horses with large colon volvulus found that mean [LAC] at admission is significantly larger in non-survivors compared with survivors.[22] In that study mean [LAC] decreased in both groups 24 hours postoperatively; however, while [LAC] returned to normal in survivors it remained significantly increased in nonsurvivors.[21] A more recent large single location prospective study[2] demonstrated that admission [LAC] in a general adult equine emergent population is significantly larger in non-survivors, in agreement with earlier reports [22-24] although there was considerable overlap between survivors and non-survivors. [LAC] appears to decrease rapidly toward the normal range in non-survivors, most likely in response to fluid resuscitation and restoration of intravascular volume, but then remains significantly increased above normal at all sampling points and may even increase later in hospitalization. The author has noted that it is not unusual for [LAC] to increase in the few hours following admission with fluid resuscitation and during exploratory laparotomy. While these observed increases are common, they are generally transient and have been ascribed to perfusion of previously hypoxic or ischemic tissues and release of 'trapped' LAC, and are not necessarily associated with survival.

Hyperlactatemia secondary to hypovolemia and tissue hypoxia is expected to respond rapidly to volume replacement and, if necessary, inotrope or vasopressor therapy. Hyperlactatemia secondary to an inflammatory process is less likely to respond or is unlikely to respond completely to therapies primarily aimed at improving O_2 delivery, including restoration of circulating blood volume and peripheral circulation. This suggests that persistently increased [LAC] in severely ill patients, even if only mildly increased above the normal range, might result from persistent and severe inflammation. The persistence of hyperlactatemia unrelated to O_2 delivery is potentially an important contributor to poorer outcome in these patients, primarily associated with problems of large intestinal strangulation and colitis. Increased [LAC] later in hospitalization, even to a small degree, should prompt thorough reassessment of the patient and re-evaluation of treatment as it is significantly associated with increased odds of non-survival. The author has recognized this later increase in [LAC] as a harbinger of significant complications ranging from incipient colitis to colonic/cecal infarction, although it may also be an indicator in insufficient vascular volume requiring additional fluid replacement therapy that is fully related to the primary clinical problem.

Another study[6] investigated the possible association of serial lactate measurement and outcome in horses with acute colitis, suggested in the first study of horses with emergency admissions to a referral hospital.[2] That study examined the association between serial measurements of [LAC] during the first 24 hours of hospitalization with outcome to discharge in horses with acute colitis. The investigators reported that [LAC] at both 4–8 and 24 hours after admission and the % reduction of [LAC] from baseline were the only clinical or clinicopathologic parameters examined that independently predicted survival to discharge in their group of horses with colitis.

Both blood and peritoneal [LAC] have been studied as diagnostic and prognostic indicators in equine colic.[25-27] Peritoneal and blood [LAC] are lower in healthy horses compared to horses with colic. Horses with intestinal strangulating obstructions have much higher peritoneal [LAC] compared to horses with non-strangulating obstructions and predictive [LAC] to differentiate non-strangulating from strangulating obstructions in blood and peritoneal fluid have been reported.[25,27] Increased peritoneal fluid [LAC] over time is known to be a strong predictor of strangulating obstructions,[27] while an increase of 1 mmol/L of [LAC] in plasma or peritoneal fluid is associated with an increased risk of need for abdominal surgery, intestinal resection, development of post-operative ileus, and fatal outcome in horses with colic.[25]

Clearance of exogenously administered L-lactate, defined as the volume of blood from which the substance of interest has been removed per unit time, has been reported for normal horses and is currently being investigated in a clinical trial as a tool for monitoring and prognosis development, in addition to being used to describe L-lactate kinetics, in horses with acute gastrointestinal diseases.[28,29] Preliminary results of the clinical trial suggest that exogenously administered L-lactate kinetics are different in horses with acute gastrointestinal disease particularly in horses with medical and or surgical problems.[29] This study may support the concept that continued [LAC] concentration monitoring is most useful for monitoring and prognostication in adult horses with colitis, or other inflammatory conditions.

The utility of using [LAC] concentration monitoring as a guide to initial resuscitation has not been critically investigated in adult horses. Based on one study most horses presenting with emergent conditions appear to respond favorably to initial fluid resuscitation, ameliorating the effects of Type A hyperlactatemia. It is probable that during the later stages of treatment, when Type B hyperlactatemia may predominate, serial [LAC] monitoring will prove most useful. An investigation of the application of $[LAC]_{Area}$ in adult horses with gastrointestinal disease associated emergency admission is ongoing, re-evaluating a subset of data previously published (unpublished data). Preliminary results show that $[LAC]_{Area}$ is larger in non-survivors (unpublished data).

LACTATE ANALYZERS

The type of lactate analyzer used is important and the reader is cautioned that published cut-points are only appropriate to use only if determined under similar conditions using an analyzer identical to that which the cut-point was determined.[29] Generally, at admission, the differences between abnormal and normal blood or peritoneal [LAC] are not subtle and strict application of cut-points is generally not required clinically, particularly when taken as a part of the whole evaluation of the patient. Lactate point-of-care (POC) analyzers are generally internally consistent but not necessarily concordant with each other.[29,30] However, despite minor differences, most seem to produce similar results, identify changes with similar magnitude and appear clinically useful. Measurements of [LAC] on different devices are NOT directly comparable, repeated determinations used for monitoring and prognostication should always be made using the same device. It is not uncommon to see clinicians comparing blood [LAC] reported from a blood gas device with peritoneal [LAC] measure on a different POC device. This may result in failure to recognize important differences in these measurements or finding a difference where none really exists, potentially confounding or delaying appropriate interventions. Pick a device and use it consistently.

CONCLUSIONS

It is important that the clinician recognize that uses of [LAC] are not directly applicable between species and that significant differences exist in response and utility of these measures between species, diseases and age groups. Extrapolation of findings in adult horses should not be done regarding foals and *visa versa*. Reported resuscitation end-points for humans are not directly applicable in our species of interest and require detailed examination is needed in both adult horses and foals.

Neonatal Foals

L-lactate concentrations measured at admission in neonatal foals can provide guidance regarding survival prognostication but is more useful in some diseases than others, primarily sepsis, gastrointestinal disturbances, respiratory problems, trauma and NI. Repeated [LAC] determination after admission is most useful for prognosis in septic foals, but has application in all neonatal foals. Persistent hyperlactatemia is an indicator of poor outcome in all sick neonatal foals, apparently regardless of diagnosis, but requires further investigation. $[LAC]_{Area}$ may be more useful when determined over the first 24 hours of hospitalization but may require adjustment based on age at admission. Further study, particularly related to differences in [LAC] parameters by diagnosis or over time as change in [LAC] and $[LAC]_{Area}$, is indicated.

Adult horses

L-lactate concentration determination appears most useful as a prognostic indicator in horses with gastrointestinal disease particularly if measured at admission or if persistent. If peritoneal can be obtained measurement of peritoneal [LAC] and blood [LAC] may provide evidence supporting the presence of compromised/ischemic bowel when peritoneal [LAC] is increased above blood [LAC] and both should be determined if at all possible. Further investigation of [LAC], change in [LAC] with time as response to treatment and for prognostication purposes, and $[LAC]_{Area}$ for similar purposes, are indicated. Future [LAC] studies in adult horses should evaluate these measures by type of disease. The use of [LAC] as an indicator of response to treatment or as an end-point for resuscitation deserves investigation.

Lactate analyzers

Most commercially available L-lactate POC (handheld, stallside) devices are useful and internally consistent. However, the results from one are not necessarily concordant with another and all sequential measurements used for monitoring or prognostication should be performed on the same analyzer.

REFERENCES

[1]Cohen RD, Woods HF. **In: Clin Biochem Aspects Lactic Acidosis**, 1976 p. 77-91. [2]Tennent-Brown, et al. **J Vet Intern Med**. 2010 24:198-205. [3]Wotman K, et al. J Vet Intern Med. 2009 23, 598-605. [4]Levraut J, et al. **Crit Care Med.** 2003 31:705-710. [5]Nguyen BH, et al. **Crit Care Med.** 2004 32:1637-1642. [6]Hashimoto-Hill S, et al. **J Vet Intern Med**. 2011;25(6):1414-1419. [7]Castagnetti C, et al. **Theriogenology** 2010; 73:343-357. [8]Corley KTT, et al. **Eq vet J**. 2005; 37:53-59. [9]Henderson ISF, et al. **J Vet Emerg Crit Care** 2008 18:496-502. [10]Borchers A et al. **Eq vet J.** 2012 Suppl;(41):57-63. [11]Borchers A et al. **Eq vet J**. 2013;45(S45):4-7. [12]Mikkelsen ME et al. **Crit Care Med**. 2009 37:1670-1677. [13]Howell MD, et al. **Intensive Care Med.** 2007 33:1892-1899. [14]Jansen TC, et al. **J Trauma**. 2009 66:377-385. [15]Trzeciak S, et al. **Intensive Care Med**. 2007 33:970-977. [16]Magdesian KG. **J Vet Emerg Crit Care** 2003;13(2):174.(abstract) [17]Père MC. **Reprod Nutr Dev.** 2003 43(1):1-15. [18]Pirrone A, et al. **Theriogenology** 2012; 78(6):1182–1189. [19]Young AK, et al. **Intensive Care Med**. 2013; 39:1818-1823. [20]Sheahan B et al. **J Vet Emerg Crit Care**, 2014;24(S1):S24. (abstract) [21]Wilkins PA et al. **J Vet Intern Med.** 2015. (accepted not yet in press) [22]Johnston K, et al. **Vet Surg** 2007 36:563-567. [23]Moore JN, et al. **Eq vet J** 1976;8:49–54. [24]Furr MO, et al. **Vet Surg** 1995;24:97–101. [25]Latson KM, et al. **Eq vet J** 2005;37:342–346. [26]Delesalle C, et al. **J Vet Intern Med** 2007;21:293–301. [27]Peloso JG, Cohen ND. **J Am Vet Med Assoc** 2012;240:1208–1217. [28]DePedro P et al. **J Vet Emerg Crit Care** 2012;22(5):564-72. [29]VanderWerf K, et al. **J Vet Emerg Crit Care** 2014;24(S1):S2.(abstract) [30]Tennent-Brown BS, et al. **J Vet Intern Med** 2007;21:1090–1098 [31]Nieto JE, et al. **Vet Surg.** 2014 Early View Jun 24. doi: 10.1111/j.1532-950X.2014.12231.x. [Epub ahead of print]

ANALGESIA FOR EQUINE VISCERAL PAIN

L. Chris Sanchez, DVM, PhD, DACVIM
Gainesville, FL

OVERVIEW AND PAIN SCORING

Colic (along with lameness) is one of the most clinically and economically important medical problems facing horses and their owners. Pain is a critical component of colic, regardless of cause, and its alleviation is often critical to a successful outcome. A limited number of analgesics are available for use in horses and many of these options are associated with significant untoward effects. This talk will focus on methods used for recognition of visceral pain in horses and available options for analgesic drug use.

In order to be useful, a pain scoring system should include clearly defined assessment criteria, be suitable for all observers, be simple and quick to use, be sensitive, have identified strengths and weaknesses, and be validated. Possible deficiencies include bias, inter- and intra-observer variability. A lack of agreement between observers is one of the flaws of simple scoring systems such as the visual analog scale (VAS), in which pain is scored on a numerical scale, when used in humans.

One of the most useful clinical applications of integrated pain scoring systems is the assessment of interventions (i.e. assess pain score, apply treatment, then re-assess pain score at various intervals). When critically assessing scoring systems, the investigator should control for signalment (age, breed, sex), observer (veterinarian, student, owner/trainer), procedure, and other effects (food withdrawal, anesthesia, management, etc.). Continuous video assessment allows for quantification of either time budgets (locations within the stall, ear position, head position, eating, lying down, etc.) or events (vocalizing, stomping feet, shifting weight, etc.). Time budget and event analyses has been performed on horses following arthroscopy.[1] A numerical rating system was also used to assess the effect of a butorphanol constant rate infusion on physiologic and outcome variables in horses following colic surgery.[2] All horses received flunixin following surgery. Interestingly, horses receiving butorphanol had improved behavior scores in the 24 hours following surgery, decreased cortisol concentration, lost less weight, and were discharged from the hospital faster than horses receiving a placebo infusion. Facial pain scoring systems have recently been proposed for use in horses.[3,4] Similar scoring systems have been used in humans and rodents and clearly show promise for use in horses as well.

Objective measures such as vital signs, plasma cortisol concentration, and force plate analyses alleviate the subjective nature of assessment. But, vital signs and cortisol are affected by a variety of factors in addition to pain, including hydration status, perfusion, sepsis and/or endotoxemia, fear, and anxiety; thus, they are not specifically useful indicators of pain.

ANALGESIC OPTIONS

A complete discussion of all available analgesic agents for use in horses is beyond the scope of these proceedings. A brief description of relevant data by class will be presented.

Opioids

All opioids have the potential for adverse effects in horses, including increased locomotion, excitement, and decreased gastrointestinal motility. Butorphanol has been the most widely used opioid in horses. Intravenous use can be associated with excitement, ataxia, and increased locomotion when used alone. Intramuscular administration had decreased systemic availability (37%). Butorphanol constant rate infusion (CRI) minimized behavioral and gastrointestinal adverse effects, relative to single injection in normal horses and decreased weight loss, improved recovery characteristics and earlier discharge from the hospital when administered for 24 hours after colic surgery, relative to control horses.[2] Typical recommended dosage for a constant rate infusion is an 18 µg/kg IV bolus then 10-23 µg/kg/hr.

Morphine has been used with varying success in horses. Several reports have documented positive results with peri-operative morphine use, typically in combination with other agents.[5,6,7] Though morphine has been shown to decrease gastrointestinal transit time in healthy horses [8], morphine did not appear to increase the risk of colic in one report of horses undergoing MRI or non-abdominal surgery [9], whereas morphine use was associated with an increased risk of colic following orthopedic surgery in another. Buprenorphine (0.005-0.01 mg/kg IV) appears to have a good safety and efficacy profile in horses, with the advantage of a longer (8-12 hour) duration of action.[10]

Transdermal fentanyl patches would theoretically provide a fantastic route of opioid administration in horses. Unfortunately, uptake from the patches is highly variable and extremely high plasma concentrations (associated with agitation in some horses) are needed for MAC reduction[11] or visceral or somatic anti-nociception in healthy horses.[12]

Tramadol is an analog of codeine, but it has less abuse potential and fewer cardiorespiratory side-effects than drugs classified as opioids. Tramadol has a short half-life and very low oral bioavailability (~3%) in horses and did not significantly alter nociceptive withdrawal latency.[13] In a clinical setting, tramadol alone failed to provide pain relief in horses with naturally occurring laminitis but did appear to help when used in combination with subanesthetic doses of ketamine.[14]

Alpha$_2$-adrenoceptor agonists

Alpha$_2$-adrenergic agonists are frequently used for both sedation and short-term analgesia. These drugs are not ideal for prolonged analgesic therapy, as they cause an immediate and profound decrease in gastrointestinal motility, amongst other cardiovascular effects, and have a relatively short duration of action. Importantly, one should note that sedative effects may require lower dosages and/or last longer than analgesic effects with the majority of drugs in this class. Alpha$_2$-adrenergic agonists provide dose-dependent visceral and somatic anti-nociception of varying duration, as well as an opioid-sparing effect. Thus, a combination of an alpha$_2$-adrenergic agonist and opioid provide a variety of commonly used multimodal analgesic protocols.

Sodium channel blockers

Lidocaine is an aminoamide local anesthetic which prevents propagation of action potentials by binding to voltage-gated sodium channels. Lidocaine, administered as an intravenous infusion, is commonly used in horses for its potential analgesic, prokinetic and anti-inflammatory properties. Clinical signs of toxicity in conscious horses include skeletal muscle tremors, altered visual function, anxiety, ataxia, and collapse. Intravenous lidocaine infusion obtunded the EEG response to castration in anesthetized ponies[15] and increased somatic nociceptive threshold in healthy horses[16], thus lending support to its use as a visceral and somatic analgesic agent. It is important to note that drug accumulation can be noted in a clinical setting after prolonged administration and lower infusion rates should be used in horses receiving highly protein bound drugs. Typical recommended dosage is 0.4-1.2 mg/kg/hr as a constant rate infusion.

N-methyl-D-aspartate antagonists

Ketamine is a non-competitive N-methyl-D-aspartate receptor antagonist and can modulate central sensitization and exert an anti-hyperalgesic effect at sub-anesthetic doses. Antinociception has not been demonstrated in healthy horses receiving infusions of ketamine [17] but in clinical settings there seems to be some beneficial effects.[14,18] In laminitic horses, addition of ketamine resulted resulted in decreased blood pressure, decreased forelimb offloading frequency and increased forelimb load, relative to tramadol alone.[14] Typical recommended dosage is a 1.3 mg/kg IV bolus over 15 minutes then 3 mg/kg/hr (75 ml/hr of 2% for a 500kg horse) as a constant rate infusion.

Antispasmodic medications

N-butylscopolammonium bromide (NBB) has both anticholinergic and antispasmodic properties and is labeled for the treatment of spasmodic colic. NBB administration also decreases rectal tone, facilitating rectal examination, [19] and may be useful in horses with esophageal obstruction.

Multimodal therapy

In severely painful horses, combination of ketamine with lidocaine and/or butorphanol could potentially provide additional analgesia, relative to infusion of a single drug. In healthy horses, butorphanol-containing combinations resulted in delayed total gastrointestinal time and reduced fecal output,[20] thus one should pay careful attention to fecal output when using said combinations.

Available options for systemic analgesia in horses

Dosages are summarized from the available literature and personal communication. These amounts must be titrated to the individual animal and situation.

Class	**Drug**	**Dosage (mg/kg unless noted)**	**Route**	**Duration (h)**	**Comments**
NSAIDs	Flunixin	0.25-1.1	IV, PO	8-24	Avoid IM; avoid max. dose more than 2x/day
	Ketoprofen	2.2	IV	24	
	Phenylbutazone	2.2-4.4	IV, PO	12-24	
	Carprofen	0.7	IV, PO	24	
	Firocoxib	0.1	PO, IV	24	
	Etodolac	2.3	IV, PO	12	10-fold reduction from original dose
α-2 Agonists	Xylazine	0.2-1.1	IV, IM	prn	Sedation outlasts analgesia
	Detomidine	0.005-0.04	IV, IM	prn	Sedation outlasts analgesia
	Romifidine	0.04-0.12	IV	prn	Analgesia may be insufficient for many procedures
	Medetomidine	0.004-0.01	IV	prn	
Opioids	Butorphanol	0.01-0.05	IV	3-4	Can combine w/ α-2; good alone IM in foals
		0.04-0.1	IM	4-6	
	Butorphanol	18 μg/kg then 10-23 μg/kg/hr	IV	CRI	
	Morphine	0.12-0.66	IV	4-6	Combine w/ α-2
	Buprenorphine	0.005-0.01	IV	8-12	
Misc.	Ketamine	0.4-1.2 mg/kg/hr	IV	CRI	
	Lidocaine	1.3 mg/kg bolus then 3 mg/kg/hr	IV	CRI	75 ml/hr for 500kg horse (2%)
	Buscopan	0.3	Slow IV	Once	Tachycardia; less for palpation

REFERENCES

1. Price, J., *et al.* **Veterinary Anaesthesia and Analgesia** 2003; 30, 124. 2. Sellon, D.C., *et al.* **J Vet Intern.Med.** 2004; 18, 555. 3. Gleerup, K.B., *et al.* **Vet Anaesth Analg** 2015; 42, 103. 4. Dalla Costa, E., *et al.* **PLoS One** 2014; 9, e92281. 5. Love, E.J., *et al.* **Vet.Anaesth.Analg.** 2006; 33, 179. 6. Mircica, E., *et al.* **Vet Anaesth.Analg** 2003; 30, 147. 7. Clark, L., *et al.* **Vet Anaesth.Analg.** 2008; 35, 22. 8. Boscan, P., *et al.* **Am J Vet Res** 2006; 67, 998. 9. Andersen, M.S., *et al.* **Equine Vet J.** 2006; 38, 368. 10. Love, E.J., *et al.* **Vet Anaesth Analg** 2014. 11. Thomasy, S.M., *et al.* **Br.J Anaesth.** 2006; 97, 232. 12. Sanchez, L.C., *et al.* **J Vet Intern.Med** 2007; 21, 1067. 13. Dhanjal, J.K., *et al.* **Vet Anaesth Analg** 2009; 36, 581. 14. Guedes, A.G., *et al.* **Am J Vet Res** 2012; 73, 610. 15. Murrell, J.C., *et al.* **Vet Anaesth.Analg.** 2005; 32, 212. 16. Robertson, S.A., *et al.* **Equine Vet J** 2005; 37, 122. 17. Fielding, C.L., *et al.* **Am J Vet Res** 2006; 67, 1484.
18. Wagner, A.E., *et al.* **J Am Vet Med Assoc** 2011; 238, 1629. 19. Luo, T., *et al.* **Vet Ther.** 2006; 7, 243. 20. Elfenbein, J.R., *et al.* **BMC Vet Res** 2014; 10 Suppl 1, S6.

EQUINE NEUROAXONAL DYSTROPHY-GENETIC FACTORS

Carrie J. Finno, DVM, PhD, DACVIM

INTRODUCTION

Neuroaxonal dystrophy (NAD) is a bilaterally symmetric neuroaxonal degeneration of selected nuclei and axonal processes in the central nervous system (CNS). Equine degenerative myeloencephalopathy (EDM) is a neurologic disease reported in young horses of various breeds, and is considered as a pathologically more advanced form of equine NAD. In Quarter Horses (QHs) with histopathologic lesions consistent with EDM, we have demonstrated that there also are lesions consistent with NAD.[1] Therefore, the term NAD encompasses the primary disease process, with EDM considered a more severe variant of NAD. The term neuroaxonal dystrophy/ equine degenerative myeloencephalopathy (NAD/EDM) has been used to describe features that have been associated with both disease variants.

Neurologic abnormalities with NAD/EDM consist of symmetric ataxia, conscious proprioceptive deficits, and weakness that typically develop during the 1st year of life.[1-2] To date, the etiology of NAD/EDM remains unknown. Based on previous research, there appears to be a strong heritable component to the disease.[1-3] In addition, there has been an association of NAD/EDM with vitamin E deficiency[1,4-6] and supplementation with vitamin E has decreased the prevalence of EDM in genetically predisposed horses.[1,4] Overall, there is very strong evidence that NAD/EDM is an inherited disorder and that, in susceptible families, dietary vitamin E during early development modifies the severity of the phenotype.

GENETIC INVESTIGATION

Inheritance

NAD/EDM occurs in Standardbreds, Paso Finos, Quarter horses, Mongolian horses, Appaloosas, Haflingers, Arabians, Morgans, Lusitanos, Thoroughbreds, Paint horses, Tennessee Walking Horses, Norwegian Fjord Horses, Welsh Pony, Pony of the Americas and various mixed breeds. Although clinical signs of NAD/EDM are similar across breeds, it is not known whether all breeds share the same etiologic basis. Both a breeding trial and clusters of cases of related horses provide strong evidence for a genetic basis of NAD/EDM.[2,4,7] An autosomal dominant mode of inheritance with variable expression has been suggested in Morgan horses.[2] One study found that foals from dams that had an NAD/EDM-affected foal were 25x more likely to develop NAD/EDM than foals from other dams further supporting a genetic basis.[8] Although the clinical disease appears similar across breeds, there is currently no evidence that the underlying etiology is shared.

A more recent study demonstrated high heritability (0.7) of NAD/EDM in a family of Quarter horses (QHs) maintained in a high-risk (i.e. low dietary vitamin E) environment.[3] This study excluded an X-linked or fully penetrant autosomal dominant mode of inheritance for NAD/EDM. Although an autosomal recessive mode of inheritance could not be fully excluded in this study, it was concluded that NAD/EDM was most likely inherited as an incompletely penetrant or polygenic trait. Therefore, a horse may have the potential genetic mutation for NAD/EDM but not display a consistent moderate to severe ataxia. This is consistent with clinical findings demonstrating that if a genetically-susceptible horse receives enough dietary vitamin E during the first year of life, they are unlikely to develop a moderate to severe ataxia.[1,4] More recent breeding crosses that we have performed support an incompletely penetrant autosomal dominant mode of inheritance (unpublished results).

Candidate Gene Evaluation

Comparatively, both the clinical and histologic findings in horses with NAD/EDM resemble ataxia with vitamin E deficiency (AVED) in humans. AVED is caused by various mutations in the α-tocopherol transfer protein gene (*TTPA*), which encodes for the protein responsible for vitamin E transport in the liver and incorporation of α-tocopherol, a main component of vitamin E, into very low-density liposomes for transport throughout the body.[9,10] Human patients with AVED develop neurologic abnormalities similar to those seen in horses with NAD/EDM and pathologic findings in AVED are similar to those observed in NAD/EDM, including neuroaxonal dystrophy within the gracile and cuneate nuclei.[11]

TTPA was evaluated as a candidate gene for NAD/EDM by sequencing of coding and regulatory regions within the gene and performing an association analysis using eight identified variants within *TTPA* and 71 surrounding single nucleotide polymorphisms (SNPs).[3] There was no significant association of any *TTPA*-associated variant with the NAD/EDM phenotype and RT-PCR for *TTPA* demonstrated no evidence of alternative splicing of *TTPA*. It was concluded that variants for *TTPA* were not responsible for NAD/EDM in the family of QHs evaluated.[3]

Genome-wide Association (GWA)

Based on comparative knowledge of the clinical and histological presentation of NAD/EDM, there are no additional strong candidate genes to evaluate and either a linkage or association study is required to further investigate the etiology. The first-generation equine single nucleotide polymorphism (SNP) array contains 54,000 SNP markers. This tool was used to genotype NAD/EDM-affected and control horses across 54,000 markers and subsequently perform association testing to determine if the genotype at any of the markers was significantly associated with the NAD/EDM phenotype. Using this first-generation equine SNP array, a GWA study was performed on two populations: (1) 37 clinically affected NAD/EDM QHs and 62 clinically unaffected QHs, and (2) a subset of the previous population that included only the 10 affected NAD/EDM QHs in which the disease was confirmed at post-mortem plus the same 62 clinically phenotyped unaffected QHs.[12] Candidate genes were prioritized within the region of significant association for further investigation. A significant GWA in this study was not achieved, although a suggestive association was uncovered when only the most stringently phenotyped NAD-affected horses (n = 10) were included (chromosome 8:62130605 and 62134644 [log(1/P) = 5.56]).[12] Three candidate genes (*PIK3C3, RIT2*, and *SYT4*) associated with synaptic function were located within this genomic region. The potential genetic roles of *PIK3C3, RIT2*, and *SYT4* in NAD/EDM were excluded through sequencing, association testing of uncovered variants and quantitative RT-PCR. It was concluded that variants in *PIK3C3, RIT2,* and *SYT4* were not responsible for equine neuroaxonal dystrophy in this family of QHs.[12]

With the recent development of the third generation equine SNP array, containing 670,000 SNP markers, mapping efforts are currently underway to pursue candidate regions of genetic association for NAD/EDM using this higher density tool.

Next-generation sequencing

Studies using next-generation sequencing techniques, both in RNA (RNA-Seq) and DNA (DNA-Seq), are currently being performed to identify genes that may be dysregulated in NAD/EDM and search for associated genetic variants.

VITAMIN E SUPPLEMENTATION

In equine families where previous NAD/EDM cases have been diagnosed, supplementation above the baseline 2007 NRC requirement is advised to pregnant broodmares and foals during the first two years of life. This is based on the findings that supplementation appears to decrease the overall severity of neurologic defects in genetically susceptible foals.[1,4] It is important to realize, however, that the foal may still be mildly affected with neurologic deficits due to NAD/EDM DESPITE adequate vitamin E supplementation. Breeding changes are required to completely prevent new NAD/EDM cases.

In genetically-susceptible populations, we are currently recommending that pregnant broodmares maintain a serum α-tocopherol (i.e. the primary antioxidant isoform of vitamin E) > 3 μg/mL. If the mare is determined to be deficient, the most effective method of supplementation is to provide 5,000 IU/500kg horse/day of RRR-α-tocopherol (in a liquid formulation) to pregnant mares and reassess their serum α-tocopherol concentrations after one week. The dose of supplementation can be tapered to maintain a stable serum α-tocopherol concentration > 3 μg/mL. Foals appear to obtain most of their α-tocopherol from colostrum (unpublished results). At birth, it is prudent to check foal serum α-tocopherol concentrations at 3-4 days of age and again at 15 and 30 days of age. If the serum concentrations decrease below 2 μg/mL, supplementation of the foal is warranted with 5-6.6 IU/kg/day of RRR-α-tocopherol (in a liquid formulation). It is strongly advised to monitor serum α-tocopherol concentrations carefully in genetically-susceptible foals often throughout the first two years of life.

REFERENCES

1. Aleman M, et al. **J Am Vet Med Assoc** 2011;239(6):823-33.
2. Beech J, et al. **Am J Vet Res** 1987;48:109-113
3. Finno C, et al.. **J Vet Intern Med** 2013;27(1):177-85.
4. Mayhew IG, et al. **Cornell Vet** 1978;68 Suppl 6:1-207.
5. Blythe LL, et al. **J Am Vet Med Assoc** 1991;198:1005-1013.
6. Gandini G, et al. **Equine Vet J** 2004;36:83-85.
7. Baumgartner W, et al. **J Comp Pathol** 1990;103:114-119.
8. Dill SG, et al. **Am J Vet Res** 1990;51:1300-1305.
9. Ouahchi K, et al. **Nat Genet** 1995;9:141-145.
10. Gotoda T, et al. **N Engl J Med** 1995;333:1313-1318.
11. Yokota T, et al. **J Neurol Neurosurg Psychiatry** 2000;68:521-525.
12. Finno C, et al. **Vet J** 2014;202(3):543-9.

VITAMIN E

Carrie J. Finno, DVM, PhD, DACVIM; Stephanie J. Valberg, DVM, PhD, DACVIM

INTRODUCTION

Vitamin E (vitE) is an essential fat-soluble vitamin that is required for healthy neuromuscular function. VitE is actually a general term used to describe a family of eight naturally occurring closely related compounds.[1] The family consists of two subgroups: tocopherols (saturated) and tocotrienols (unsaturated). Within each subgroup, there are four individual isoforms (α, β, γ, and δ) (Table 1). Alpha-tocopherol (α-TP), specifically the natural RRR stereoisomer (RRR-α-TP; bolded in Table 1), is the most bioavailable and bioactive form in animal tissues due to preferential uptake by the liver.[2] The major dietary source of vitE in horses is forage, optimally providing approximately 2,000 IU/day of natural α-TP to grazing horses.[3] However, the amount of fresh pasture available to horses is markedly decreasing, especially with recent drought conditions,[4,5] and the amount of α-TP in hay declines markedly with processing and storage.[6] Commercial feeds usually provide α-TP in the form of synthetic vitamin E, a less bioavailable formulation than natural α-TP containing of a mixture of eight isomers (dl-α-tocopherol [all-rac-α-tocopherol]), of which only one is identical to the natural isomer. Thus, many horses receive far less α-TP than the dietary requirements of 1-2 IU/kg body weight, set by the 2007 National Research Council.[7]

Table 1: Components of vitamin E

Tocopherols	*Tocotrienols*
α-tocopherol 8 stereoisomers (**RRR**, SRR, RRS, RSS, RSR, SSR, RSS, SSS)	α-tocotrienol
β-tocopherol	β-tocotrienol
γ-tocopherol	γ-tocotrienol
δ-tocopherol	δ-tocotrienol

VITAMIN E DEFICIENCY

A deficiency of whole body vitE is typically assessed by measuring serum concentrations of one component of vitE, α-TP, with normal equine serum concentrations >2 μg/mL.[2] Whether or not a deficiency of vitE has an impact on health depends upon individual genetic factors, the temporal occurrence of deficiency during development and the duration of deficiency (Table 2). If genetically susceptible foals are vitE deficient during the first year of life they may develop neuroaxonal dystrophy/equine degenerative myeloencephalopathy (NAD/EDM), a neurodegenerative disease characterized by a general proprioceptive ataxia. We have found that foals with NAD/EDM have very low cerebrospinal fluid (CSF) α-TP concentration (unpublished results). Older horses that have been vitE deficient for ≥18 months[8] can develop equine motor neuron disease (EMND) resulting in clinical signs of weakness and neurogenic muscle atrophy.[9] Some adult horses, potentially those with shorter duration vitE deficiency, have a vitE deficient myopathy(VEM) and present with muscle weakness, low muscle α-TP concentrations, mitochondrial alterations in skeletal muscle but no evidence of neurogenic atrophy.[10]

Table 2: Equine neuromuscular diseases associated with vitE deficiencies

	EMND	NAD/EDM[12,13]	VEM[10]
Age of onset	Adult; median age 10 years[8]	6-12 months	Adult
Clinical signs	Muscle wasting and fasciculation, abnormally low head carriage, prolonged recumbency[9]	Symmetric ataxia, abnormal stance, proprioceptive deficits	Muscle wasting and fasciculation, abnormally low head carriage
Effect of treatment with vitamin E	40% of cases improve, 40% stabilize but remain disfigured, 20% progress[11]	Supplementation will not improve affected cases but can prevent NAD/EDM in susceptible horses	Affected horses can recover within 3 months

VITAMIN E SUPPLEMENTATION FOR DEFICIENT STATES

The type and formulation of vitE supplements has a major impact on their bioavailability. Natural vitE, RRR-α-TP, is more bioavailable than synthetic vitE and therefore many equine supplements now strongly market the "natural" (i.e. RRR) form. The natural formulation is available as an esterified form (α-TP acetate) to prolong shelf life (i.e. powder or pellet) or water dispersible (i.e. liquid) formulation with higher biopotency (Table 3).

Table 3: Forms of vitamin E supplementation available

Type of Supplement	Form of vitE in supplement	Bioavailability	Cost
Natural vitE (liquid)	RRR-α-TP	HIGH	HIGH
Natural vitE (powder or pellet)	RRR-α-tocopheryl acetate	INTERMEDIATE	INTERMEDIATE
Synthetic vitE (powder or pellet)	All-rac α-tocopheryl acetate (all stereoisomers) OR dl-α-tocopheryl acetate (RRR and SRR-stereoisomers)	LOW	LOW

Alpha-TP acetate may have less bioavailability because the ester has to be removed and the α-TP made water-dispersible by the action of bile salts (micellization). In contrast to the α-TP acetate, the liquid formulation leads to a more rapid increase (within 24 hours) in serum α-TP concentrations in horses.[14] Both formulations are useful for equine practice and selection of the appropriate formulation will depend upon whether or not vitE supplements are being used to prevent deficiency, to treat a deficiency or to treat a disease associated with vitE deficiency. Thus the essential starting point is assessment of serum α-TP concentrations. For maintenance requirements or for vitE deficient horses without evidence of neuromuscular disease, α-TP acetate is an economical alternative for supplementation. Serum α-TP concentrations may take weeks to achieve normal concentrations with acetate formulations, however, and periodic evaluation of serum α-TP after approximately 4 weeks is recommended to ensure horses achieve normal serum concentrations. If an adequate response to supplementation does not occur, the dose or formulation of α-TP should be altered.

Dosage: Most studies of vitE supplementation have used healthy horses and assessed serum α-TP rather than neuromuscular tissue concentrations. The absorption and metabolism of vitamin E in healthy horses may differ from that of horses with diseases associated with vitamin E deficiency. As a result it is essential to measure serum α-TP concentrations, monitor α-TP concentrations and adjust the dose in response to supplementation. Recommended dosages for vitE in the literature vary widely from 2,000 to 20,000 IU/day[15-17] and it is also very important to note that, at the time these higher recommendations were made, the type and formulation of vitE was not considered. With the ability to rapidly increase serum α-TP with new vitE products, such high doses may be unnecessary. The most rapid rise in serum α-TP concentrations can be obtained by providing natural-source water-dispersible forms of α-TP at 10 IU/kg body weight. This formulation is 5-6 times more bioavailable than synthetic α-TP acetate, and a 5000-IU dose/horse more than doubles serum vitamin E levels within 12 hr.[18] Research into healthy horses with marginal or deficient serum α-TP concentrations found that 14 days of water dispersible natural α-TP supplementation at 10,000 IU/ horse/day significantly increased CSF α-TP.[19] VitE has also been used or to treat horses suffering from neurologic disease of ANY cause, however, the use of vitE for these conditions is empirical based on a belief that it could be neuroprotective in disorders not related to a deficiency of vitE. We know nothing about vitE metabolism in horses, the appropriate dosage or its actual efficacy in preventing or treating neurologic conditions other than EMND, NAD/EDM and VEM. Supplementation with vitamin E may alter drug metabolism and disposition since, in humans, the same cytochrome isoforms that metabolize vitamin E metabolize >50% of therapeutic drugs.[20] The NRC has set the upper safe diet concentration at 20 IU/kg BW based on biopotency of synthetic α-tocopherol (10,000 IU/500 kg horse). Above this level, coagulopathy and impaired bone mineralization have been reported.[7]

Veterinarians should be aware that when horses are switched from the same dose of natural water dispersible α-TP to the powdered actetate formulation, a precipitous drop in serum α-TP can occur. Based on our research, we propose that the best and most economical approach to supplementing deficient horses over time would be to overlap treatment with the water dispersible and acetate formulations, and gradually withdraw the water dispersible α-TP, leading to a better steady state of serum α-TP and higher tissue α-TP concentrations of continued supplementation (Figure 1).

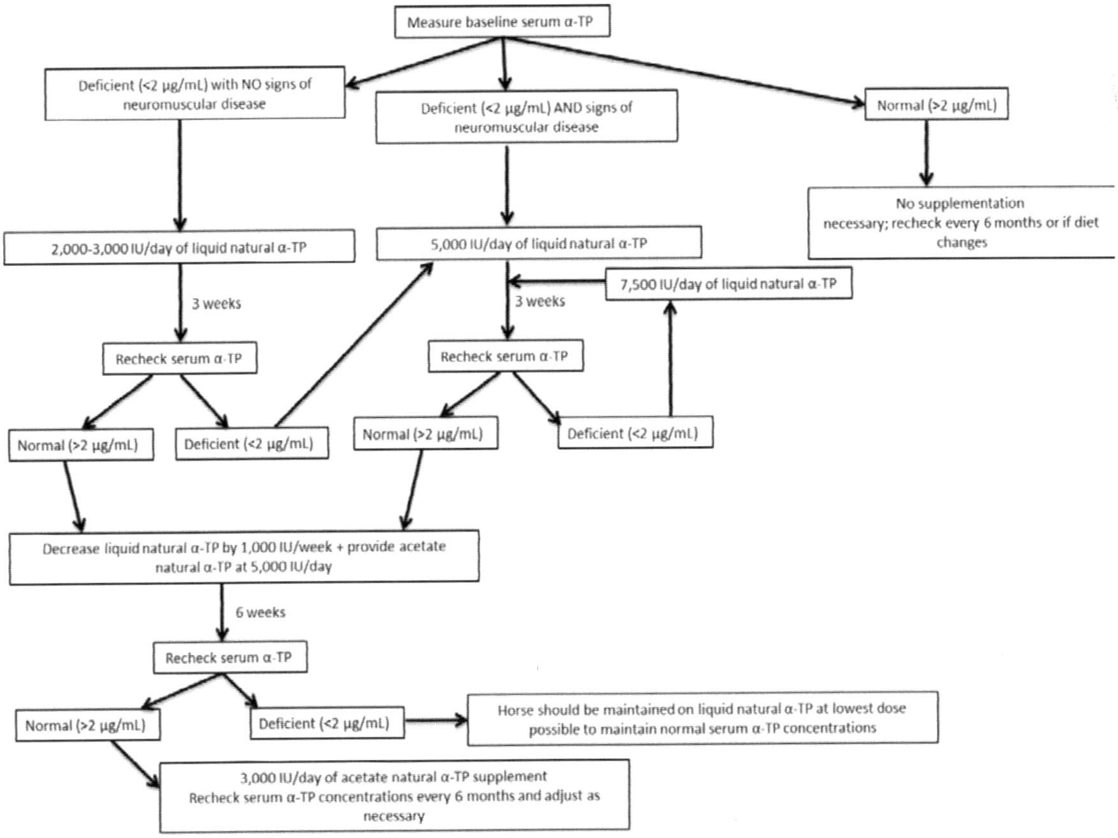

Figure 1: Proposed protocol for supplementing α-TP-deficient horses

CONCLUSIONS

Previous recommendations for dosages of vitE have not taken into consideration the highly variable bioavailability of vitE supplements or differences in vitE absorption between individuals. Serum α-TP concentrations should be assessed prior to and during supplementation to assure that serum concentrations achieve normal levels. In horses with vitE deficiencies and clinical signs of neuromuscular disease, liquid natural formulations of α-TP are the best means to rapidly increase serum and tissue levels. Powdered natural forms of α-TP may be a more cost effective means of meeting daily requirements in horses with normal α-TP status.

REFERENCES

[1]Schneider C. **Mol Nutr Food Res** 2005;49:7-30.
[2]Finno CJ, et.al **J Vet Intern Med** 2012;26:1251-1266.
[3]Craig AM, et.al **38th Annual Am Assoc Equine Pract Proc** 1992;579-580.
[4]Rice D. **USA Today** 2012; http://usatoday30.usatoday.com/weather/drought/story/2012-04-11/mild-winter-expands-usa-drought/54225018/1
[5]**U.S. Drought Monitor** 2014; http://droughtmonitor.unl.edu/
[6]McDowell LR. In: McDowell LR,ed. **Vitamins in Animal Nutrition** 1989;93-131.
[7]NRC **National Research Council** 2007.
[8]Mohammed HO, et.al. **Acta Vet Scand** 2007;49:17.
[9]Divers T, et.al **Equine Vet J** 1994;26:409-415.
[10]Bedford H, et al. **J Am Vet Med Assoc** 2013;242:1127-1137.
[11]Divers T, et.al **Equine Vet Educ** 2001;13:63-67.
[12]Aleman M, et.al **J Am Vet Med Assoc** 2011;239:823-833.
[13]Mayhew IG, et.al. **J Vet Intern Med** 1987;1:45-50.
[14]Pagan JD, et al. **Pferdeheilkunde** 2005;21:101-1-2
[15]Orsini JA, et.al In:Orsini JA, Divers TJ,ed. **Manual of Equine Emergencies** 2003.
[16]Bertone JJ. In:Berton JJ,ed. **Equine Clinical Pharmacology** 2004.
[17]Stewart RH, et al. In:Reed SM, Bayly WM, Sellon DC,ed. **Equine Internal Medicine** 2004;594-598.
[18]Pagan JD, et.al **Nordic Feed Sci Conf Proc** 2010;112-115.
[19]Pusterla N, et.al **Vet Rec** 2010;166:366-368.
[20]Kliewer SA, et.al **Endocr Rev** 2002;23:587-702

THE GUT MICROBIOTA: CAN WE PREDICT AND PREVENT GASTROINTESTINAL DISEASE?

J. Scott Weese, DVM, DVSc, DACVIM
Guelph, Ontario, Canada

INTRODUCTION

The "germ theory" was a landmark concept in medical history that formed the basis for revolutionary thoughts about the origins, treatment, and prevention of disease. Yet, this may have led to an over-simplification of the host–pathogen relationship and a focus on "one pathogen–one disease." While this is certainly true for many pathogens and diseases, the relationship between the host and microbes is complex, and it is clear that the one pathogen–one disease model fails to account for all that happens in nature. Much of this failure revolves around the lack of consideration of the body's large, complex, and poorly understood microbiota and how it influences health and disease.

THE COMMENSAL MICROBIOTA

The body harbors a vast microbial population consisting of an array of commensal bacteria, viruses, fungi, and parasites. This microbial population (the microbiota) outnumbers the hosts' own cells by a factor of ~10. The sum of the genetic components of the microbiota (the microbiome) exceeds the genetic material of the host by a factor of ~100 (or more), and this has been referred to as our 'second genome'.[1] Yet, the host typically escapes unscathed from this constant microbial and antigenic exposure. While medicine focuses on the harmful aspects of microbial encounters, the microbiota is critical for normal development and maintenance of health of the host, and the absence of the commensal microbiome is largely incompatible with life. At the same time, an inability of the host to develop adequate immunological tolerance of the accompanying massive immunological onslaught, interact "normally," and to prevent the opportunistic pathogens amongst the microbiota (a rather large component) from causing disease is at the root of myriad infectious and inflammatory conditions, with increasing evidence of a role in other conditions such as allergy, metabolic disease, neoplasia, and obesity.[2–4]

Every body site that communicates with the external world will have a microbiota, the composition of which can vary both between and within sites (e.g., the skin microbiota is different from the respiratory tract microbiota, and the composition of the skin or respiratory microbiotas differ at different body sites). Understanding the composition of microbiotas, their function, how they vary between and within individuals normally, and how they are impacted by disease and various endogenous and exogenous factors is an important area of study.

THE GASTROINTESTINAL MICROBIOTA

While all external sites have microbiotas, and these probably all play roles in the pathogenesis and protection from disease, most of the focus has involved the intestinal tract. This is reasonable because the intestinal tract harbors one of the most dense microbial populations on the planet,[5] and alterations in this microbiota have been associated with disease of various species for some time. In humans, it has been estimated that there are 10^{12}–10^{14} bacteria per gram of feces,[5,6] with estimates of the number of different bacterial species ranging from 1000 to many thousands. There is no reason to believe this would be different in other animal species. Estimates of the numbers of different species have steadily increased with advances in sequencing and bioinformatics technologies, and the upper limit is yet to be determined. The actual composition of the microbiota is somewhat of a moving target, with general aspects understood but with details being refined as technology advances.

Assessment of the microbiota

It is only with very recent advances in technology that microbiotas are starting to be understood. Traditionally, assessment of the bacterial microbiota relied on bacterial culture. While providing some useful information, culture has severe limitations because of the inability of a large percentage of microorganisms to grow under conventional culture conditions and the corresponding underestimation of bacterial richness (number of different species present). Ready growth by a small number of bacteria leads to overestimation of the relative abundance of those organisms, with corresponding underestimation of diversity. Even if most or all organisms were culturable, culture is completely impractical for description of a population with billions of organisms per gram and 1000 or more different species. Molecular methods can detect organisms that are not culturable using conventional methods, as well as detect unknown microorganisms, and offer a distinct advance to culture. Early methods such as denaturing gradient gel electrophoresis (DGGE) provided insights into microbiota variation,[7] but it had limited ability to determine components that account for variation or to provide specific information about the overall microbial composition. Quantitative PCR can be used to quantify components of the microflora,[7] but this method is not useful for large-scale and broad studies because each PCR reaction can only be used to assess one specific organism or group, and target organisms must be known to create appropriate primer sets. Group-specific oligonucleotide probes have been used for quantitative analysis of the microbiota,[8] but only provide information about limited, targeted organisms. Fluorescent in situ hybridization (FISH) can provide detailed information about specific organisms or groups and can be very important in some situations, but is not able to provide broad microbiota assessment and has had little application in horses. Ready availability of sequencing then led to major advances. In particular, recent advances in next generation sequencing, first with the 454-sequence platform then through Illumina® sequencing, have allowed for sequencing of complex microbial populations at unprecedented sequence depths. Paired with advances in bioinformatics, assessment of complex microbial populations can now be cost-effective and routine, generating massive amounts of data.

The "normal" microbiota

While powerful laboratory and analytical tools are available, care must be taken to ensure that there is truly a proper understanding of the "normal" microbiota through investigation of adequate numbers of animals, study of animals of different breeds and under different management, and evaluation of the spectrum of microbiota compositions (metacommunities) that may occur in healthy individuals. Care must also be taken to ensure that proper testing and analytical tools are used, as massive datasets like this can be markedly mismanaged and manipulated, resulting in erroneous interpretations.

While what truly constitutes the "normal" microbiota is probably not completely understood (and likely varies with many factors such as diet, management, age, and myriad other factors), general conclusions can be made. In most, if not all, mammalian species, a few bacteria phyla predominate, and while thousands of bacterial species can be present, a small number account for the bulk of the microbiota. In horses, the phyla Firmicutes, Bacteroidetes, Proteobacteria, Actinobacteria, and Verrucomicrobia dominate.[9–15] Differences in proportions of these phyla vary between studies, likely as a result of different study populations and different methodologies; however, Firmicutes tend to dominate. This phylum of gram-positive bacteria (including Clostridia, one of the dominant classes, and Clostridiales, a prominent order) is receiving increasing attention as its role in health becomes apparent. This can run somewhat counter to typical (superficial) perceptions of the microbiota, particularly the "bad" nature of Clostridia and the desire to limit their populations. Despite the dominance of this one phylum, a large number of different phyla can be identified, with estimates of a few thousand distinct bacterial species being present per horse.[15]

The composition of the microbiota is not static over the horse's life. The foal is presumably born with a sterile gastrointestinal tract, yet it rapidly becomes colonized after birth and pronounced richness and diversity are evident within 24 h of life.[16]

It's not just about bacteria

While often overlooked and poorly studied, the microbiota also consists of Archaea, viruses, fungi, and parasites. These are less well understood, in large part because of limited investigation. Little is known about Archaea, members of a Kingdom of single-celled prokaryotes previously classified as bacteria, with limited study of both the composition of the Archaeal microbiota and its role in health and disease. It is likely that Archaea play very minor roles and may be present largely as a simple reflection of environmental exposures. The viral microbiota is likely vast, with a combination of bacteriophages and mammalian viruses. Assessment of the viral microbiota is more complicated and costly compared with the relatively straightforward bacterial assessment. There has been no study of this component, as well as of the fungal microbiota, in horses.

Limitations in microbiota assessment

While new technologies have revolutionized our understanding of microbiotas, there are some potentially important limitations. 16S rRNA gene-based studies (the dominant approach to bacterial assessment) are dependent on adequate reference databases. Further, assessment of the composition of the microbiota does not necessarily indicate its function. Indeed, there is evidence of profound functional redundancy (the ability of genetically diverse populations to achieve the same end goal) in the human microbiota.[17] Thus, while understanding of composition and changes in composition are useful, microbial community structure does not always clearly indicate its function. Functional aspects can be investigated through "shotgun" sequencing approaches, whereby sequencing is performed on whole genome DNA fragments, not 16S rRNA gene PCR products. This eliminates potential PCR bias and provides more information about functional genes; however, it requires much larger sequence depth (and corresponding cost) and is hampered by limitations in databases. Further, presence of genes or gene fragments does not indicate that they are being expressed. Transcriptomics, sequencing of cDNA created from RNA from a sample, can provide a better understanding of the function of the sample (at that time), but it is more technically challenging and much more costly because of the required sequence depth. Limitations in databases (i.e., understanding the function of the transcribed gene) also exist, but it can be assumed that continued advances in technology will make these techniques more accessible. Ultimately, combinations of many different "-omics" approaches (e.g., metagenomics, transcriptomics, proteomics…) will be used to more accurately assess the microbiota and its function.

THE GASTROINTESINTAL MICROBIOTA IN HEALTH AND DISEASE

Comprehensive overview of the association of the gastrointestinal microbiota with normal development and its role in disease processes of horses and other species is beyond the scope of these proceedings. In horses, understanding of the composition of the microbiota and changes associated with different diseases is increasing but still rather superficial. It should not be surprising that an altered fecal microbiota is present in horses with colitis.[9] Significant differences in various phyla and lower taxonomic orders, along with changes in community membership (species present) and structure (species present and their relative abundances) were reported. Further, diarrheic horses had significantly lower relative abundances of some key families that are increasingly associated with "gut health" in other species, particularly Lachnospiraceae and Ruminococcaceae. The increase in Proteobacteria (as a whole or specific members of that phylum) in disease is consistent with studies in other species whereby an increase in Proteobacteria and a decrease in other major phyla, particularly Firmicutes, is an indicator of "dysbiosis." Changes in the fecal microbiota have also been identified in horses with carbohydrate overload-associated laminitis[11] and potentially with chronic laminitis.[14] Antimicrobials can modify the community membership, structure, and other ecological parameters in clinically normal horses, with different impacts with different antimicrobials.[15] Recovery of the microbiota to a state similar to pre-treatment occurred relatively soon after cessation of antimicrobials; however, some changes were still evident 25 d after cessation of treatment (the last sampling date that was tested).

Associations between the microbiota and health or disease must be interpreted for what they are — associations — not necessarily causation. Differentiating cause from effect is difficult and requires more vigorous study. This can be difficult, particularly when there may be different influences on both the disease process and microbiota, including differences in diet.[18] As a result of the various potential influencing factors and the degree of normal inter-horse variation, large scale or longitudinal studies (or both) may be required to help identify causation. For example, decreases in Lachnospiraceae in horses with gastrointestinal disease suggest that this family might play an important role in gut health in the horse. Yet what does this really tell us? More informative was the subsequent study of post-partum mares that identified significant increases in Proteobacteria and depletion of Lachnospiraceae prior to the onset of large colon volvulus compared with controls that did not go on to develop colic.[19] While not necessarily proof of causation, prospective studies can provide stronger support for the clinical relevance of observed changes, albeit with much greater effort in study execution. For many sporadic conditions, this may be impractical (i.e., massive numbers of animals would need to be enrolled to result in an adequate population of diseased animals for most conditions).

CONCLUSIONS

It is clear that the bacterial microbiota of the intestinal tract plays an important role in development, immune regulation, and maintenance of health. It is also evident that discernable changes in the microbiota occur with certain diseases, and that some of these changes may be identifiable before the onset of clinical disease. Therefore, it is likely that targeted approaches could be used to identify horses at increased risk of developing diseases such as large colon volvulus or perhaps a broader range of colic types. Technological advances will probably make rapid microbiota analysis feasible in the near future, increasing the chance that this could be a clinically relevant surveillance tool for high-risk groups. Yet, there is currently inadequate information about what changes are relevant and how those vary across different horse populations. Broader study of the microbiota as a predictor of different diseases and in different horse populations is required to be able to develop the evidence needed to interpret microbiota assessment results, since it is clear that *technology* currently outpaces *knowledge*.

REFERENCES

[1] Zhao L. ***Nature*** 2010;465:879–880. [2] Schwiertz A, et al. ***Obesity*** 2010;18:190–195. [3] Sanz Y, et al. ***Proc Nutr Soc*** 2010;69:434–441. [4] Bisgaard H, et al. ***J Allergy Clin Immunol*** 2011;128:646–652.e641–645. [5] Whitman WB, et al. ***Proc Natl Acad Sci USA*** 1998;95:6578–6583. [6] Furrie E. ***Gut*** 2006;55:141–143. [7] Grønvold A-MR, et al. ***Veterinary Microbiology*** 2010. [8] Daly K, et al. ***FEMS Microbiology Ecology*** 2003;44:243–252. [9] Costa MC, et al. ***PLoS ONE*** 2012;7:e41484. [10] Costa MC, et al. *American College of Veterinary Internal Medicine Forum*. Nashville, TN, USA, 2014. [11] Moreau MM, et al. ***Veterinary Microbiology*** 2013. [12] O' Donnell MM, et al. ***Lett Appl Microbiol*** 2013;57:492–501. [13] Proudman CJ, et al. ***Equine Veterinary Journal*** 2014;doi:10.1111/evj.12324. [14] Steelman SM, et al. ***BMC Vet Res*** 2012;8:231. [15] Costa MC, et al. ***BMC Veterinary Research*** 2015;11:19. [16] Costa MC, et al. *American College of Veterinary Internal Medicine Forum*. Seattle, WA, 2013. [17] Consortium HMP. ***Nature*** 2012;486:207–214. [18] Dougal K, et al. ***PLoS ONE*** 2014;9:e87424. [19] Weese JS, et al. ***Equine Veterinary Journal*** 2014; doi: 10.1111/evj.12361

INFLAMMATORY BOWEL DISEASE IN THE HORSE

Benjamin W. Sykes, BSc, BVMS, MS, DACVIM, DECEIM, MBA
Upper Orara, NSW, Australia

INTRODUCTION

Classic inflammatory bowel disease (IBD) is an uncommon disease condition. However, another, less specific, variation of the disease that presents as chronic weight loss and/or low grade diarrhea is becoming increasingly recognized[1]. This talk consists of four main sections covering:

- Classic IBC
- Nonspecific IBD (NS-IBD)
- The diagnostic approach and utility of different tests
- Prognostication of horses with a diagnosis (presumptive or confirmed) of IBD

Considering the above four key points, the talk will conclude with a summarized approach to the diagnosis and management of IBD in the horse.

CLASSIC INFLAMATTORY BOWEL DISEASE (IBD)

As mentioned above, classic IBD is a relatively rare diagnosis. Five classic manifestations of disease have been described including eosinophilic enteritis (EE), multicentric eosinophilic epitheliotrophic disease (MEED), granulomatous enteritis (GE), lymphocytic/plasmacytic enteritis (LPE), and lymphosarcoma (LS)[2]. The prognoses for EE, MEED, and LS are all reported to be poor[2]. However, in the author's experience, EE and MEED are extremely rare while LS is slightly more common, representing a handful of cases in the author's 10 years of referral practice experience. Further, such cases tend to present to referral centers with severe manifestations of disease with the diagnosis often made during exploratory laparotomy. Considering this, these manifestations will not be discussed further.

In contrast, GE and LPE are relatively common and, in the author's opinion, often under diagnosed. A major contributory factor to this is that obtaining a definitive diagnosis typically requires full-thickness intestinal biopsies. However, the obtainment of such biopsies, even by laparoscopy, is invasive and expensive. Instead, a presumed diagnosis can be made clinically on the basis of a variety of diagnostic tests. The focus of this talk is on this nonspecific manifestations of IBD, which include GE and LPE, in cases in which the diagnosis is made on the basis of minimally invasive testing.

NONSPECIFIC INFLAMMATORY BOWEL DISEASE (NS-IBD)

As discussed above, awareness of NS-IBD as a disease entity appears to be increasing[1,3]. Furthermore, long-term survival rates of 50%[3] to 65%[1] have been reported suggesting that successful treatment of affected cases is possible. This is in direct contrast to EE, MEED, and LS, all of which carry a poor prognosis[2]. Horses with NS-IBD most commonly present with clinical signs consistent with low-grade, chronic gastrointestinal disease such as recurrent colic and/or weight loss. The initial diagnostic evaluation for each of these presentations may vary somewhat, but in the author's opinion it can largely be standardized to include a complete history to ensure that inadequate or inappropriate nutrition is not a likely cause, physical examination, examination *per* rectum, dental examination (although evidence supporting the link between dental abnormalities and an increased risk of colic is, at best, weak) and routine hematology/biochemistry, all of which can be performed readily in the field/first opinion setting. Fecal analysis for parasites may be beneficial but the reliability of the single fecal egg count in predicting the worm burden of a horse is questionable at best. Further, in the author's experience, the majority of horses presenting with chronic gastrointestinal disease have typically been repeatedly dewormed, although not always appropriately. Abdominocentesis has been advocated by some authors[3,4] and is justified, although in the author's experience it is low yield especially if the hematology and biochemistry are within normal limits. As such, it is typically reserved for second-round testing.

First-round tests may result in a diagnosis with specific therapy indicated. If a specific diagnosis is not present then second-round testing is usually indicated. An exception to that statement is if hypoproteinemia, hypoalbuminemia, and/or hypoglobulinemia is/are identified on routing hematology/biochemistry in the absence of a clear cause (i.e., severe liver or kidney disease). In this case, third-round diagnostics may be considered sooner rather than later as the diseases screened for in second-round testing, with the possible exception of sand enteropathy, are unlikely to cause hypoproteinemia, hypoalbuminemia and/or hypoglobulinemia. Instead, the presence of one (or more) of these findings is highly suggestive of IBD.

Second-round tests for both weight loss and recurrent colic include gastroscopy, abdominocentesis (if not performed as part of the initial evaluation), and, where geographically relevant, radiography for sand or enteroliths. In the author's opinion, gastroscopy is indicated at this stage regardless of the horse's usage as the risk for glandular ulceration appears largely independent of the intensity of exercise/management. However, if significant hypoproteinemia, hypoalbuminemia, and/or hypoglobulinemia are present, gastroscopy may not be necessary as gastric ulceration is unlikely to cause any of these parameters to significantly change. If second-round tests fail to identify a cause, third-round testing is indicated.

At this point of the clinical approach, the diagnosis of NS-IBD is largely one of exclusion. In the author's opinion, the purpose of the third-round of testing is to try to identify supportive evidence of intestinal inflammation and/or nonspecific dysfunction that may be relatable to inflammation (i.e., malabsorption) and/or to provide information as a baseline for monitoring the response to treatment. The author's approach to third-round testing is to perform them as a complete set as the information gained from each individual test can be difficult to interpret while the complete set of tests often yields a more consistent picture (much like a jig-saw puzzle). Third-round testing consists of functional (xylose/glucose absorption testing), structural (ultrasound), and histological (rectal ± duodenal mucosal biopsies) evaluation of the gastrointestinal tract. An abnormal finding in any/all of these tests is/are supportive of NS-IBD at this stage of the investigative process.

DIAGNOSTIC AND PROGNOSTIC UTILITY OF THE TESTS

Approaching a case of weight loss and/or recurrent colic in a systematic, staged manner is medically logical and all of the testing described is minimally invasive and safe. However, such an approach is expensive and the return on investment/risk should be considered for each test. All of the first-round tests proposed are readily performed in the field and, in the author's opinion, necessary as a minimum database before further evaluation should be performed.

The return on investment for second-round testing is variable and dependent on a number of factors. As mentioned above, in the author's opinion, gastroscopy is indicated to rule out equine gastric ulcer syndrome, regardless of the intensity of management of the horse, unless significant hypoproteinemia, hypoalbuminemia, and/or hypoglobulinemia is/are present. Further, gastroscopy is indicated to rule out gastric impaction, which may present with similar clinical signs[5]. Lastly, the collection of a duodenal biopsy for histopathology can be performed at the same time although the value of this is questionable (see below). As discussed above, the value of abdominal radiography is highly region-specific.

Within the third-round tests, ultrasound is the simplest test to perform and one with which most clinicians are comfortable. Transabdominal ultrasound yielded an abnormal finding in 44% of horses in one study[3] and 50% of horses in another[1]. However, the non-specific finding of intestinal thickening was not associated with prognosis in the latter study suggesting that, although potentially diagnostically useful, non-specific ultrasound findings should be interpreted with caution. Malabsorption is a common finding in this population with 45%[3] and 70%[1] determined to be abnormal. Caution should be exercised in interpreting abnormal glucose absorption tests as the percentage of affected horses did not vary between survivors and non-survivors in one study[1]. However, in the same study, high mean peak xylose concentrations were present in survivors[1] and, in the author's experience, "flat line" absorption tests are often associated with a poor outcome. Histopathology of duodenal biopsies is often advocated but little is published on their diagnostic accuracy. In one study, abnormal findings were not present in any of the nine samples submitted for examination[3] and this is consistent with the author's experience that their diagnostic value is low. In contrast, rectal mucosal biopsies, despite their limitations, appear to be useful. In one study, 89% of horses in which rectal mucosal biopsy was performed had abnormal findings, although the procedure was only performed in selected cases[3]. In another prospective study, less than 50% of horses examined had abnormal histopathology of the rectal mucosa but, when changes were present, they were considered significant and correlated with subsequent post-mortem examination in non-survirors[1]. This is consistent with the initial investigations into the diagnostic utility of histopathology of the rectal mucosa performed by Lindberg *et al.* in which it was considered diagnostic in 40% of cases of IBD[6].

Together, this suggests that the findings of the third-round tests are often non-specific and of little prognostic utility. As such, the return on investment of each of the tests is questionable and, despite being advocated above as medically logical, the use of these tests may not be justifiable in all cases in terms of yield on investment. This is particularly true for abdominal ultrasound and absorption tests given that the tests require specialized equipment (ultrasound) or are time-consuming to conduct (glucose/xylose absorption tests). Of the third-rounds testing proposed, histopathology of the rectal mucosa appears most useful because, although the test lacks sensitivity, abnormal findings appear to correlate well with disease state and the test can readily be performed in the field.

Interestingly, the four factors that appear to have greatest prognostic value in NS-IBD are the presence of hypoproteinemia, hypoalbuminemia, and/or hypoglobulinemia[1,3], and the response to treatment with anthelmentics and corticosteroids[1]. In a prospective study evaluating the long-term outcome of horses with NS-IBD, the three-year survival rate of horses that responded to larvicidal deworming (fenbendazole 10 mg/kg PO SID for 5 days) and prednisolone (1mg/kg PO SID for 3 weeks) was 80%, which was significantly higher than the survival rate of those that did not respond to treatment (20%: $p = 0.031$)[1]. In that study, "response to treatment" was based on clinical response and improvement in third-round test parameters. As such, it may be that third-round diagnostics are more useful in monitoring the response to therapy than as predictors of outcome in their own right.

SIMPLIFIED APPROACH TO IBD DIAGNOSIS AND MANAGEMENT

Following on from the above discussion, the author now approaches this in a less regimented manner than the strict three-tier approach outlined above. First-round diagnostics are performed on all cases as a minimum database. However, from there the case management is heavily dictated by the owners' resources and expectations. If no specific findings are present on the first-round diagnostics, second-round diagnostics are pursued. However, if hypoproteinemia, hypoalbuminemia, and/or hypoglobulinemia are present, three weeks of anthelmintic and corticosteroid therapy are performed. If no response is seen, second-round diagnostics are performed. If no specific findings are present on second-round diagnostics, third-round diagnostics are performed (if owners' wishes and finances allow) understanding that their value is as much for collecting information to monitor the response to treatment as it is for getting a specific diagnosis in most cases. Alternatively, if it has not already been performed, and the owners' wishes and/or finances dictate, a trial therapy of three weeks of anthelmintic and corticosteroid therapy can be performed. The advantage of this approach is that it can be entirely field-based and is of relatively low cost. If no response to treatment is seen, third-round diagnostics are often performed attempting to establish a definitive diagnosis for euthanasia, although the poor response to treatment itself would suggest that the prognosis is poor.

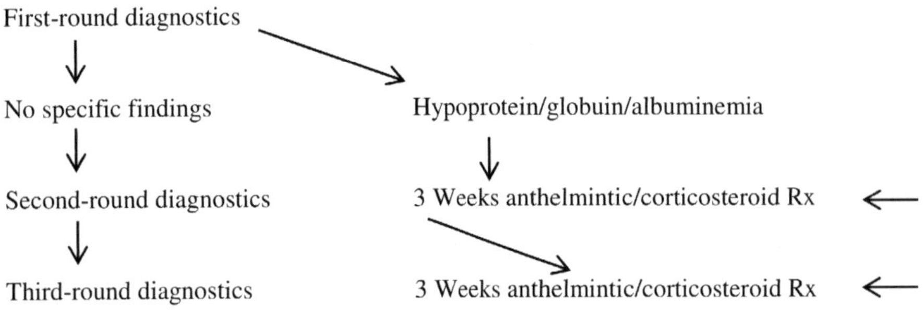

REFERENCES

1. Kaikkonen, R. *et al.* Diagnostic evaluation and short-term outcome as indicators of long-term prognosis in horses with findings suggestive of inflammatory bowel disease treated with corticosteroids and anthelmentics. *Acta Vet. Scand.* **Early View,** 1–12 (2014). 2. Kalck, K. a. Inflammatory Bowel Disease in Horses. *Vet. Clin. North Am. - Equine Pract.* **25,** 303–315 (2009). 3. Metcalfe, L. V. a, More, S. J., Duggan, V. & Katz, L. M. A retrospective study of horses investigated for weight loss despite a good appetite (2002–2011). *Equine Vet. J.* **45,** 340–345 (2013). 4. Tamzali, Y. Chronic weight loss syndrome in the horse: A 60 case retrospective study. *Equine Vet. Educ.* **18,** 289–296 (2006). 5. Vainio, K., Sykes, B. W. & Blikslager, A. T. Primary gastric impaction in horses: A retrospective study of 20 cases (2005–2008). *Equine Vet. Educ.* **23,** 186–190 (2011). 6. Lindberg, R., Nygren, A. & Persson, S. G. B. Rectal biopsy diagnosis in horses with clinical signs of intestinal disorders: A retrospective study of 116 cases. *Equine Vet. J.* **28,** 275–284 (1996).

NUTRITIONAL OPTIONS FOR THE CRITICALLY ILL EQUINE

Richard Hepburn, BVSc, MS, CertEM (IntMed), DACVIM (LAIM), MRCVS
Gloucestershire, UK

INTRODUCTION

Nutritional deficit in the form of starvation ultimately leads to death, hence the main goal of artificial nutrition is improving outcome in critical care. It is considered a treatment provided to anorectic patients which improves energy and protein intake, but with concerns that it is unphysiologic and may produce complications.[1] Current nutritional guidelines in man recommend the use of low calorie enteral nutrition in critical care cases, rapid reintroduction of food after surgery and avoidance of early introduction of parenteral nutrition; rather than the historic wait for motility and give increased calories parenterally approach. Despite these recommendations many human clinicians "preserve these sacred cows of medical dogma" and starve for prolonged periods and then aggressively hyperalimentate.[28]

There is limited information regarding the potential benefits and complications of different strategies for nutritional support of the critically ill adult horse, indeed nutrition is often overlooked in clinical practice and seen as unduly complicated. As many of the nutrient recommendations for critically ill horses are based upon these out-dated human guidelines we need to update and simplify our practice. This should be followed by systematic critical appraisal of the effect of timing, composition and method of delivery of nutrition in the commonest equine critical care patient, those with acute medical and surgical gastrointestinal disturbance.

CURRENT NUTRITIONAL RECOMMENDATIONS IN MAN

The hope of caloric and protein support is to minimize muscle atrophy, however the mechanisms underlying this are more complex than mere caloric deprivation,[2] which is borne out by the observation that aggressive nutritional support results mainly in fat gain.[3] Artificial nutrition may also suppress autophagy, which is essential for removal of the toxic protein aggregates and damaged organelles that develop with systemic inflammation, suggesting that too aggressive a nutritional plan may increase tissue damage.[4]

Human observational trials have suggested an association between caloric deficit and negative outcome, however this is predominantly a concern in obese patients.[5] Similar studies performed with early initiation enteral nutrition are conflicting, with some showing lower mortality and others increased complication rates and duration of hospitalization.[1] One potential explanation is the illness severity, with more severely ill patients benefiting less.[6] Randomized control trials however have shown that permissive enteral underfeeding (defined as 60% resting caloric goal) significantly reduced mortality in patients remaining in the ICU for at least 48hours.[7] Trophic feeding involves provision of even fewer calories enterally (25% caloric goal), and compared to 80% caloric goal feeding results in fewer GI disturbances and less alteration in blood glucose-insulin levels.[8] These studies suggest that in the human ICU early enteral calorie restriction is not deleterious, and that early trophic enteral nutrition is at least as good as full enteral nutrition. The early addition of parenteral nutrition (PN) to cases where it is not possible to reach enteral goals has a significant *negative* effect on outcome (defined by mortality, rate of infection and duration of hospitalization), compared withholding PN for 7 days.[9] The historic use of caloric multipliers that *increased* estimated caloric needs has been replaced by a general recommendation of initial *underfeeding*.[1] When PN is required for <10 days in man, the current recommendation is that it should be low dose.[21]

The balance between the time at which nutrient restriction becomes harmful and the point where the benefits of artificial nutrition outweigh the risks is unknown, and likely to vary by patient and by presentation.[1] In order to better understand which patients may benefit from artificial nutrition a nutritional risk score has been validated, which is calculated from nutritional status (itself determined from measures of feed intake, body mass index and estimates of recent weight loss), severity of disease (an indicator of stress metabolism and increased nutritional requirements) and an age adjuster.[10] Selecting nutritional support only for patients at increased nutritional risk has been shown to lower the complication rate.[11]

There is now extensive clinical data that prolonged periods of *nil per os* after GI surgery whilst awaiting a return of bowel sounds/defecation worsens outcome.[15] The rationale for bowel rest post surgery is based upon a concern that GI secretions and feed will accumulate causing reflux, pain, anastomotic leak or wound dehiscence. Twelve prospective randomized controlled trials however have shown no benefit to the traditional process of placing a nasogastric tube for post operative aspiration/decompression, waiting for evidence of motility then slow mechanical progression from liquid through to a solid diet.[28] Clinical trial data actually shows that enteral feeding started immediately after surgery *improves* anastomotic healing and reduces dehiscence, and delaying feeding for even a few hours may promote ileus.[16,17] When considering specific formulations of artificial nutrition, the use of whole protein instead of peptides is recommended otherwise there is insufficient data to recommend either high fat/low CHO *vs.* low fat/high CHO, high protein *vs.* low protein, or the inclusion of fibre.[21]

Nutritional pharmacology describes nutritional supplementation with key pharmaconutrients that could improve outcome.[15] Several clinical studies have shown that enteral arginine supplementation helps to restore T-lymphocyte numbers and function following major surgery in non-septic patients, reducing postoperative infection risk,[18] however this supplementation is not a blanket recommendation as in patients with severe sepsis it *worsens* SIRS and *increases* the risk of mortality.[19] Glutamine is the most abundant free amino acid in the body and rapidly becomes depleted with sepsis and surgery.[15] It is a metabolic substrate for enterocytes, supporting barrier function, and immune cells where it acts as a messenger cellular recovery pathways. A glutamine dose of 0.5g/kg/d can reduce mortality in patients receiving parenteral nutrition, with the greatest benefit in truly critically ill patients,[20] but *not* in patients in unstable shock or MODS. The current clinical guidelines do not recommend supplementation of either enteral nutrition or PN with arginine or glutamine, except in cases of severe burns.[21]

Refeeding syndrome describes the potentially fatal fluid and electrolyte disturbances that can occur in malnourished patients receiving both enteral and parenteral nutrition.[25] The hallmark is hypophosphatemia, but thiamine deficiency, hypokalemia, hypomagnesemia can also occur. The incidence of refeeding syndrome is 0.43%, with malnourished patients (e.g. anorexia nervosa) or those unfed for >10d that subsequently receiving hyperalimentation most affected. It is recommended that patients at risk be re-fed slowly starting with low levels of energy, and with the addition of thiamine and multivitamin supplements. Any fluid or electrolyte disturbances created by refeeding should be concurrently corrected.

BASIC CALORIE AND PROTEIN REQUIREMENTS OF THE HEALTHY ADULT HORSE

Nutrient requirements for horses are estimated from digestible energy (DE), which represents the stored energy in a feed that is digested and not lost in feces.[12] The daily maintenance energy requirement describes the caloric need of a normally active horse (non-working, at pasture) and is calculated differently for horses depending upon body weight: *BWT<600kg*: DE(Mcal/d) = (BWT in kg x 0.03) + 1.4; *BWT>600kg*: DE(Mcal/d) = 1.82 + (0.0383 x BWT in kg) – [0.000015x(BWT in kg)^2]. In most cases this DE requirement can be met if the horse eats between 1.5% and 2.5% BWT/d as hay.[14] The daily resting DE requirement is that required for a stabled, idle horse and is approximately 70% of maintenance DE. In all horses the resting can be calculated: Resting DE (Mcal/d)= (BWT in kg x 0.021)+0.975. The crude protein requirement of healthy adult horses can be calculated from DE: CP(g/d) = 40 x DE(Mcal/d). For a 500kg horse this represents a resting caloric requirement of 11.5Mcal/d and 656g protein.

Assessing nutritional status is most accurately accomplished with walk on digital scales. Weight estimations made from girth and body length measurements should not be used to monitor body weight changes in hospitalized horses due to their large standard error.[12] Body condition score (BCS) can also be used to estimate potential nutritional risk (underweight, obese).[13,14] The utility of monitoring changes in body condition score in hospitalized horses is unknown.

THE EFFECTS OF FEED DEPRIVATION IN HORSES

Fasted horses deplete stored liver glycogen within 36 hours, increasing lipolysis and fat utilization leading to an increase in plasma concentrations of glycerol and non-esterified fatty acids which plateaus after three days.[14] Some mobilized fat is re-esterified in the liver and transported to the peripheral tissues, resulting in elevated serum triglyceride concentrations. Pregnancy, lactation, obesity and severe illness emphasize the effect of fasting on plasma triglyceride concentrations, this effect is most severe in obese ponies. Prolonged malnutrition will eventually lead to emaciation and place the horse at risk of refeeding syndrome.[24]

Extensive small intestinal resection (>60-80%) leads to rapid and substantial weight loss in ponies, clinical experience suggests that difficulty in maintaining condition also occurs when ≥50% of the small intestine is removed or when malabsorption is present.[14] Horses with extensive large colon resection lose weight post-operatively but typically regain lost condition within 6 months, although chronic vitamin-mineral supplementation is required as the majority of mineral absorption occurs in the large colon.[26] These cases also have increased fecal water and increased voluntary hay intake.[27]

Horses in good body condition score that are not pregnant or lactating can probably tolerate up to 4 days of starvation, although the effect of this on development of ileus is unknown.[14] Immediate nutritional support is indicated for thin horses (BCS≤3) or those that have recently sustained ≥10% body weight loss. Emaciated horses and those nutrient deprived for more than 5-7 days should have nutrition introduced *gradually* to minimize the chance of refeeding syndrome: starting at 50% resting DE, the amount of feed given should be gradually increased over a 7-10d period.[24] Over-conditioned horses and ponies (BCS≥7) may be poorly tolerant of even short periods of starvation, being at risk of hypertriglyceridemia or hyperlipaemia.[22] This population may require early nutritional support and should be monitored closely.[22]

NUTRITIONAL SUPPORT IN EQUINE ACUTE GASTROINTESTINAL DISEASE

There are no controlled studies in adult horses of the relationship between the level of nutritional support and clinical endpoints such as mortality, duration of hospitalization or perioperative complication rate. Similarly the caloric and protein requirements of horses after colic surgery or other critical illness have not been determined, and so the resting DE is generally considered the initial recommendation when formulating a nutritional plan.[12] Historically the use of caloric multipliers, based upon then current human data, was suggested when calculating the energy requirement of ill horses.[14] As there is no equine evidence supporting their use, and as they have been superseded by underfeeding in man, this author does not recommend them. Similarly increased protein intake in sick horses was also advised, but as current data does not support high protein nutrition in human critical illness this author also does not recommend it.[14,21] Given the relatively short duration of equine critical illness maybe we should revise our nutrition plans to follow human permissive under-nutrition, aiming for an initial caloric goal of 60% resting DE (6.9Mcal/d for a 500kg horse).

The current advice regarding what horses should eat with specific intestinal disorders seems appropriate.[12] Surgical large colon disorders should be given small amounts of low bulk feed (alfalfa pellets, complete senior pellets) every 1-2 hours. The addition of high quality grass hay may reduce post-operative diarrhea risk.[28] Cases of acute typhlocolitis often experience SIRS and partial or complete anorexia.[12] Low bulk, low soluble CHO, highly digestible diets (complete senior pellets, alfalfa) are recommended, again with the addition of grass hay and small amounts of fresh cut grass. Diets high in soluble CHOs (grains, lush grass) should be minimized to limit colonic gas distension and osmotic diarrhea.[12] Small intestinal cases should be given small volumes of highly digestible feeds such as alfalfa leaves, beet pulp, complete senior pellets and rice bran, with the addition of small amounts of fresh grass.[12, 14] Essentially the recommendations for large and small intestinal diets are identical, with the exception of hay feeding. It should therefore be possible to formulate a basic diet appropriate for both voluntary enteral feeding and assisted enteral feeding. Using the excellent calculations given in a review article it is possible to create a trophic feeding diet and a permissive underfeeding diet for a 500kg horse, using a commercial complete pelleted feed (Equine Senior, Purina Mills, 2.6Mcal per kg) and corn oil (0.68Mcal per 100ml), with 100ml vitamin E added to prevent lip peroxidation:[12]

Trophic nutrition (25% resting DE = 2.9Mcal/d) = 1.1kg Equine Senior *or* 0.9kg Equine Senior + 100ml corn oil.
Permissive underfeeding (60% resting DE = 6.9Mcal) = 2.7kg Equine Senior *or* 2.4kg of Equine Senior + 100ml corn oil

Voluntary enteral nutrition is obviously dependent upon the horse being willing and able to eat, if appetite is poor it may be stimulated by grazing in hand for short periods (15-20 minutes). Assisted enteral nutrition can be administered via an indwelling nasogastric tube, which is easy if one is already placed to assess reflux, or via a narrow gauge open ended feeding tube placed into the stomach (rather than the distal esophagus).[12] The use of equine specific narrow gauge enteral tubes may also permit concurrent voluntary feeding,[29] this author recommends endoscopic assistance to aid placement in dysphagic cases. The enteral diet should be blended to a thin slurry for easy administration administration, and addition of brewers yeast as a source of multiple B vitamins is advised in horses in low body condition.[12]

Critical evaluation of gastrointestinal function in the context of the expected clinical course is considered good practice when formulating a nutritional plan, [23] yet there is no objective equine evidence that specific conditions require the withholding of enteral nutrition. It makes clinical sense in duodenitis-proximal jejunitis given the physical limitation to ingesta transit. Otherwise post-operatively feed has traditionally been withheld until after a period of assessment of gastric reflux and demonstration of motility, then water is given in increasing quantity before small, frequent feeds of highly digestible feeds are started.[12] In man this practice is outdated and may promote ileus, so should we know follow their lead and immediately start enteral underfeeding to *prevent* ileus?[16,17,28] It would be interesting to place the stomach tube as normal, and if no reflux was obtained, give the trophic diet shown above split into 4 feeds over the first 24 hours, increasing the quantity over the next 48 hours to the permissive underfeeding level before transitioning to voluntary intake at resting DE over the next 48 hours. Rapid reintroduction of frequent small amounts of feed is already advised in horses recovering from large intestinal surgery.[17] The potential for a standard nutritional plan for all post-operative colic cases would greatly simplify clinical practice. A similar approach for typhlocolitis may require shorter term assisted enteral nutrition, as anorexia resolves, but a more gradual transition from voluntary permissive underfeeding to resting DE levels, at a rate determined by the speed of diarrhea resolution.

Parenteral nutrition should be viewed as a temporary measure, whilst the disease which is limiting enteral nutrition is treated.[12] In normal horses total parenteral nutrition administered for 10 days led to maintenance of 94% BWT and was well tolerated.[32] The classic indication for PN is intolerance to enteral feeding which is seen as reflux, ileus and abdominal distension.[12] Probably the most appropriate use of sole parenteral nutrition is duodenitis-proximal jejunitis, where the volume of reflux prevents enteral administration and prokinetics are often ineffective.[31] It is worth considering the possibility that the traditional combination of prolonged starvation and elevated early caloric goals are contributing to enteral intolerance.[1] For instance excessive enteral feeding in typhlocolitis could lead to nutrient delivery to the diseased large intestine, whereas trophic or permissive levels may not. The enteral intolerance in typhlocolitis could also arise from adynamic ileus caused by the primary disease,[12] and so it is important to identify the origin of the intolerance, potentially with ultrasonographic assessment of motility.[30] When intolerance occurs in man current guidelines recommend the use of a prokinetic, rather than changing to PN.[21] The use of prokinetics to improve tolerance to enteral nutrition has not been reported.

There is minimal information regarding the benefit of PN in horses recovering from GI disease. Whilst improved indices of nutritional status have been reported, there is no information regarding outcome (mortality, duration of hospitalization, cost-benefit).[33] Complications of PN are of three types: catheter-related, septic and metabolic, with particular care taken to avoid hyperglycemia in SIRS cases.[12,21] There is little data on the frequency of equine complications: phlebitis was reported to occur in 4/79 horses receiving PN,[31] diarrhea occurred in 2/4 normal horses on longterm PN,[32] and hyperglycemia can develop warranting a reduction in delivery rate.[31]

Short term PN using dextrose only solutions is often advocated in horses to provide approximately caloric needs for up to 48 hours,[12,14] this practice is strongly *not* recommended in man when the GI route is intact.[21] If the GI route is unusable then low dose PN should be given for the first 10 days. As parenteral nutrition in adult horses is often prohibitively expensive, this reduction in caloric target to the permissive underfeeding level would be beneficial. There are no reports of combined enteral and parenteral nutrition techniques in horses, nor any detailing the effect of timing of PN introduction. This likely reflects the relatively short duration of nutritional support that sick horses require compared to human critical care patients.

REFERENCES

[1]Schetz M et al. **Critical Care**. 2013;17:302-9, [2]Llano-Diez M et al. **BMC Genomics** 2011;12:602, [3]Streat S et al. **J Trauma** 1987;27:262-6, [4]Cuervo A, Macian F **Mol Aspects Med** 2012;33:2-13, [5]Alberda C et al **Intensive Care Med** 2009;35:1728-37, [6]Huang H et al. **Nutr J** 2012;11:30, [7]Arabi Y et al **Am J Clin Nutr** 2011;93:569-77, [8]Rice T et al **JAMA** 2012;307:795-803, [9]Caser M et al. **N Engl J Med** 2011:365:506-517, [10]Kondrup J et al **Clin Nutr** 2003;22:321-36, [11]Jie B et al **Nutrition** 2012;28:1022-7, [12]Magdesian K **Vet Clin Equine** 2003;19:617-44, [13]Burkholder W **JAVMA** 2000;217:650-4, [14]Geor R **World Eq Vet Rev** 2000;5: 12-14, [15]Wischmeyer P **Curr Op Anesthiol** 2011;24:381-8, [16]Lewis S **BMJ** 2001;323:773-6, [17]Andersen H **Cochrane Database Syst Rev** 2006;4:CD004080, [18]Drover J **J Am Coll Surg** 2011;212:385, [19]Heyland D **Intensive Care Med** 2003;29:669-71, [20]Weitzel L **Crit Care Clin** 2010;26:515-25, [21] http://www.criticalcarenutrition.com, [22] Dunkel B **EVJ** 2003;35;590-5, [23] Dunkel B **Vet Clin Equine** 2004;20:107-26, [24] Witham C **JAVMA** 1998;212:691-6, [25] Mehanna H **BMJ** 2008;336:1495-8, [26]Bertone A **AJVR** 1989;50:253-8, [27]Bertone A **AJVR** 1989;50:1628-32, [28]Cohen **JAVMA** 1996;209:810-3, [28] Franklin G **J Parent Ent Nutr** 2011;35:337-42, [29]Lopes M **EVE** 2003;15:107-112, [30]Mitchell C **Can Vet J** 2005;46:134-40, [31]Lopes M **EVJ** 2002;34:250-7, [32]Hansen T **AJVR** 1988;49:122-4, [33]Durham A **EVJ** 2004;36:390-6,

FLUID THERAPY AND GI DISEASE

Dr Gayle D Hallowell, MA, VetMB, PhD, CertVA, DACVIM, DACVECC, PFHEA, MRCVS, Nottingham, UK
Dr Lucas Pantaleon, DVM, MS ,DACVIM (LAIM), MBA, Lexington, KY

INTRODUCTION – INTRAVENOUS FLUID THERAPY

In humans and animals fluid therapy is a core element for the resuscitation of critically ill patients, since fluid therapy is essential to decrease mortality and accelerate recovery[1-4]. In equids fluid therapy is a fundamental component of the treatment for diseases, such as SIRS, sepsis, colic, hypovolemia and shock[5,6]. The purpose of fluid therapy are to restore effective blood volume, correct hypotension, improve cardiac output, normalize tissue oxygenation and correct electrolyte and acid base abnormalities[1,7,8]. The ultimate aim is to prevent organ dysfunction[8]. Thus fluid therapy is the mainstay in the treatment of SIRS and sepsis induced hypovolemia[7,9].

Sepsis is a complication of critical illness with high degree of mortality. In horses with gastrointestinal disease SIRS is a common complication, and as in man it carries a high mortality rate. In adult horses SIRS commonly originates secondary to gastrointestinal disease (i.e. colitis), due to bacterial and endotoxin translocation through the gut wall[10]. The most current Surviving Sepsis Campaign guidelines recommend crystalloids over hydroxyethyl starches (HES) as the preferred fluid for resuscitation[11,12]. However the preferred fluid therapy remains a matter of debate in humans and veterinary medicine and the optimal fluid for resuscitation for a given clinical situation remains divisive[2,4,13-15]. The most commonly used type of intravenous fluid therapy for resuscitation in humans and equids are polyionic isotonic crystalloid fluids, Lactated Ringers and 0.9 % sodium chloride respectable[8,16,17].

Intravenous fluid classification

Intravascular fluids can be classified as crystalloids and colloids. The crystalloid family includes isotonic and hypertonic solutions, which are also categorized into non-buffered (i.e. isotonic saline) and buffered solutions (i.e. Ringers lactate, acetate)[7]. Another way of classifying crystalloids is based on use for resuscitation or maintenance fluid therapy regimes. Maintenance crystalloids contain a higher potassium concentration and lower sodium and chloride concentrations compared to resuscitation crystalloids. Hypertonic saline solution is a hyperosmotic crystalloid with high sodium content (7.2 to 7.5%). Colloids solutions include hypo-oncotic (i.e. gelatin, 4% and 5% albumin) and hyper-oncotic (i.e. dextran, hydroxyethyl starches, 25% albumin)[7]. Furthermore colloids can be of two types: natural (plasma and whole blood) or synthetics [hydroxyethyl starch solutions (Hetastarch and Pentastarch), dextran and polymerized hemoglobin][1].

Intravenous fluid therapy effects and side effects

Intravenous fluids expand the intravascular space. However depending on the fluid type, leakage into the extravascular space varies in speed and degree. Crystalloids fluids counteract the movement into the extravascular space due to the osmotic pressure exerted by its solutes. Whilst colloids create oncotic pressure gradients to keep fluids within the intravascular space[7]. Thus in theory the blood volume expansion may be proportional to the solute tonicity or oncotic power[7]. Colloid fluids remain in the intravascular space longer than crystalloids, thus less fluids are needed to achieve similar hemodynamic effects, as shown by the recent CHEST trial[17]. Colloids are used for rapid and long lasting circulatory stabilization, although data supporting this practice is lacking[4,9].

The major complications of fluid resuscitation are pulmonary and interstitial edema[8]. There are concerns with the use of colloids with regards to immune effects in critical illness, acute kidney injury, coagulopathy, increased risk of death and higher costs[7,9,18]. On the other hand the administration of large volumes of 0.9% sodium chloride has been associated with hyperchloremic metabolic acidosis due to the increase in plasma chloride and decrease in the strong ion difference[19,20]. Furthermore hyperchloremia may cause renal vasoconstriction and decrease glomerular filtration rate, leading to acute kidney injury and higher mortality[21].

The recent open labeled, randomized CRISTAL trial, was designed to assess the effect of colloids on mortality compared with crystalloids in a wide range of critically ill septic patients with hypotension and lactic acidosis[7]. Patients were randomized into two groups: control (mainly isotonic saline) and experimental (mainly HES). An important restriction was that HES were not allowed to be administered at a dose larger than 30 ml/kg per day[7]. No difference in mortality at 28 days was observed between groups. Surprisingly at 90 days, there were less patients that died in the colloid group[7]. These findings suggested lack of harm, however more studies need to be done in order to prove or refute this finding[7,22]. Some argue that these findings are against other work that provided high quality evidence on increased mortality and renal injury, thus calling for further scrutiny of the CRISTAL trial results[23].

In humans large doses of HES 200/0.5 (70 ml/kg) lead to renal failure[18]. However, as stated previously, it appears that lower doses of HES (30 ml/kg) show no difference in renal insult when compared with crystalloids[7].

Third generation HES solutions with a lower molecular weight (130 KD) and degree of substitution (0.4) have replaced older formulations[9]. The evidence for using these newer solutions is scant in patients with sepsis and its use has been discouraged[9]. The 6S was a blinded randomized clinical trial comparing the use of a third generation HES, with crystalloids in septic patients[9]. HES was administered at a maximal daily dose of 33 ml/kg with a total median volume received of 44 ml/kg. In this study more patients in the HES group had impaired renal function and higher mortality at 90 days. When HES 200/0.5 and HES 130/0.4 were used the absolute risk of death at 90 days increased by approximately 8% in both trials[9,18]. The CHEST trial found no difference in mortality at 90 days when compared patients treated with 6% HES (130/0.4; Voluven) or 0.9% sodium chloride[17]. However it did show increases in renal injury; maximal doses of HES allowed were 50 ml/kg/day[17].

As previously stated the debate on fluid therapy, specifically HES and 90 day mortality remains to date[22]. With only one study, the 6S trial showed a small increase in 90 day mortality in the HES group [9]. Another two clinical trials showed no effect at 90 days[17,18] and the CRISTAL trial showed lower mortality[7,22]. The inconsistency of studies over HES effects on mortality and renal injury are likely to be due to a combination of factors including variation in patient population, types and doses of HES administered, safety profile of the comparator fluid and research misconduct or author bias[4].

Intravenous fluid therapy in GI disease

Colic is a major cause of morbidity and mortality in the horse and any delay in performing surgery would negatively affect outcome, particularly in strangulated lesions[24,25]. Resuscitation of painful horses is a challenge and the volume of fluids that could safely be administered before anesthesia is limited. The ease of transport, speed of administration and rapid correction of hemodynamics of colloids or colloids combined with hypertonic crystalloids could be especially useful for resuscitating horses with severe abdominal pain both, in the field prior to referral and preceding anesthesia. Therefore it can be hypothesized that initial resuscitation and cardiovascular stabilization with these type of fluids, followed with isotonic crystalloids, could have a positive effect on critically ill horses with abdominal disease.

The use of colloids in equine medicine as an initial resuscitation fluid at doses not higher than 10 ml/kg are unlikely to have the long term adverse effects observed in humans treated with large doses of HES for several days.

INTRODUCTION – ENTERAL FLUID THERAPY

For the gastro-intestinal patient with SIRS, intravenous fluid therapy will be the mainstay for replacing circulating volume. However enteral fluids when they can be used has many advantages not least from a financial point of view, particularly in the management of impactions, oral supplementation of electrolytes, provision of nutrition and treatment of dehydration when present without hypovolemia. Interestingly NICE guidelines[26], based on meta-analyses of the literature recommend that justification is required as to why intravenous fluids are chosen over enteral fluids and that the former if used should be stopped as soon as practical.

Water absorption and blood flow

Water absorption from the gastro-intestinal tract in the normal horse primarily occurs in the large intestine, and more specifically the large colon. In an adult horse a volume of up to 100L of fluid and associated secretions is absorbed during the course of the day[27-29]. In the hypovolemic horse, in order to protect the vital organs, blood flow is diverted from the gastro-intestinal tract. Once blood flow is reduced, so too are gastro-intestinal motility and absorption. In addition, obtunded, hypovolemic horses have a reduced thirst drive. It is for this reason that using oral fluid therapy in hypovolemic animals is unsuccessful at best and detrimental in certain scenarios.

Non-parenteral fluid administration

Enteral fluids can be delivered in the form of water from a bucket, via an indwelling narrow nasogastric tube (continuous administration of oral fluids), via a conventional nasogastric tube (intermittent fluid administration) and when data is extrapolated from other species instilled per rectum.

Fluids provided in the stable can simply be plain water or a second bucket containing either electrolytes or sweetening agents (fruit juice or molasses) to encourage water intake. Some horses will select oral fluids containing electrolytes when they have specific derangements, whereas others will not. Fluids with electrolyte supplementation should be hypotonic or at most isotonic, or will not be consumed.

Intermittent fluid administration via nasogastric tube allows the clinician to have complete control over the amount and timing of fluids administrated. The stomach tube can be left in and capped to prevent aerophagia or removed after each fluid administration.

Continuous enteral fluid therapy has the advantage that intermittent gastric distention causing abdominal discomfort is prevented when compared with intermittent nasogastric fluid administration. This is relatively easily undertaken using either commercially available enteral fluid therapy kits or using equine nasogastric feeding tubes. As these tubes are thin, it is prudent to be sure of their position in the esophagus, which may require endoscopy or radiography. They need to be appropriately secured and a muzzle will prevent them being rubbed out. Clean, but non-sterile, coiled fluid administration sets and intravenous fluid bags can be used to hold the water or electrolyte solutions.

A maximum intermittent rate of enteral fluid would be 6-8L/500Kg horse at no less than 30 minute intervals. Checking for nasogastric reflux prior to the next dose is essential. The maximum rate usually possible for continuous administration would be 5L/500Kg/hr and maximum rates should be 10L/500kg/hr.

Proctoclysis

There are various descriptions in the literature of the use of fluids per rectum in a variety of veterinary species and in man[30-32]. They appear to be an effective way of providing maintenance requirements or provision of enough fluids to allow other vascular access to be obtained. They have been reported to be used for over 100 years in man on the battlefield[33]. They are commonly reported for use in elephants[34] and have been partially used successfully by one of the authors (GH). A recent letter has been published reporting their successful use in the horse[30] for the management of suspected botulism as well as other anecdotal reports. In humans, the maximum amount of fluid that can be absorbed in a 70Kg adult is 150ml/hr with a maximum installation rate of 500ml/hr via a foley catheter[33]. Similar fluid instillation rates proportionate to body size have been used in elephants and horses and significantly higher rates experimentally in rabbits[31]. The suggestion from one study is that absorption of fluids per rectum will still occur in the face of hypovolemic shock[31].

Fluids used enterally

These include water, hypo or isotonic fluids supplemented with electrolytes (e.g. sodium chloride and potassium chloride) and also hypertonic magnesium sulphate. Hypotonic fluids given enterally are likely to be absorbed from the gastro-intestinal tract, whereas isotonic fluids are more likely to remain the gastro-intestinal lumen.

Magnesium sulphate is used enterally as a cathartic agent as an initial treatment for large colonic impactions. Initial use of 0.5g/kg is recommended diluted in water. Doses up to 1g/kg can be used, but can result in hypermagnesemia. Experimentally magnesium sulfate is not as effective in increasing colon water content when compared with a balanced electrolyte solution but does increase the water content of feces in the small colon[35].

Balanced electrolyte solutions can be made from sodium chloride and combined sodium and potassium chloride (LoSalt®) with approximately 5g of each per 1L of water. If only using sodium chloride, add a maximum of 9g per 1L of water.

In horses that have severe hypokalemia, often secondary to the administration of resuscitation fluids and food being withheld, potassium chloride can be used orally (0.1-0.2g/kg/day) in addition to intravenous fluid supplementation.

Large colonic impactions

An experimental[36] and clinical study[37] have shown that correction of large colonic impactions is more effective using enteral than parenteral fluids unless significant hemodilution occurs. The reported parenteral fluid rates required to hydrate impactions is not only expensive but results in significant hemodynamic changes[36]. In addition, another experimental study proposed that the main effect of a selection of oral fluids that were compared was activation of the gastrocolic reflex following filling of the stomach[35]. In clinical patients with colonic impactions, resolution rates were fastest in those that received five doses of isotonic electrolyte solutions every 30 minutes or hourly when compared to those that received fluids every 2 hours or intravenously. Complication rates were similar between groups, but severity of abdominal pain was greatest in those treated every 30 minutes[37].

Diarrhea in other species

There are few studies in any species comparing the effects of enteral versus parenteral fluids for the treatment of diarrhea. There has been one systematic review[38] comparing these treatments for management of acute gastroenteritis caused by E. coli in children where the evidence is felt to be very poor. The data suggests that enteral fluids reduce hospital stay, but there was no difference between the two treatments regarding duration of diarrhea and weight gain. The overall conclusions were that intravenous fluids should be used if the patient was hypovolemic, but that enteral fluids would likely have some efficacy in mild to moderate hypovolemia.

Products not recommended

Mineral oil can be used as a marker of gastrointestinal transit (18 hours to anus if transit time is normal). For impactions, it works its way around without penetrating it as well as hindering water penetrating. Based on this observation, increased cost over water and the fatality if this product ends up in a horse's lungs it cannot be recommended for treatment of pelvic flexure impactions.

Dioctyl sodium sulphosuccinate is a detergent that should penetrate impacted fecal material by affecting surface tension, thus allowing water to enter the feces. Care should be taken as a 3-fold overdose is fatal[39] and also increases absorption of mineral oil so should not be administered with it. There is no benefit of this over water based on an experimental study[35].

Sodium sulphate is an even more potent cathartic than magnesium sulphate, but consistently causes hypernatremia and hypocalcemia, so is not recommended[35].

Oral glucose in fluids does not provide sufficient nutrition to be valuable in horses and glucose and glycine containing fluids in an experimental diarrhea model resulted in incomplete fluid absorption[40,41].

SUMMARY

In summary, enteral fluids can be beneficial for treatment of large colonic impactions and electrolyte supplementation, but are not appropriate for hypovolemic animals or those with gastro-intestinal ileus. The use of fluids per rectum may be of value in patients where there are challenges with enteral fluid administration or financial constraints regarding the administration of intravenous fluids. Further work to evaluate how much fluid can safely be administered and what absorption rates can be expected would be beneficial.

REFERENCES

1.McFarlane D. *Compendium* 1999;21:867-877; 2.Choi PT, et al. *Crit Care Med* 1999;27:200-210; 3.Khandelwal P, et al. *Crit Care* 2002;6:286-288; 4.Zarychanski R, et al. *JAMA* 2013;309:678-688; 5.Jones PA et al *Am J Vet Res* 1997;58:541-548; 6.Seahorn JL, Seahorn TL. *Vet Clin North Am Equine Pract* 2003;19:665-679; 7. Annane D et al. *JAMA* 2013;310:1809-1817; 8. Hollenberg SM et al. *Crit Care Med* 2004;32:1928-1948; 9. Perner A, et al. *N Engl J Med* 2012;367:124-134; 10.Roy MF. *Vet Clin North Am Equine Pract* 2004;20:41-61; 11.Dellinger RP et al. *Crit Care Med* 2013;41:580-637; 12.Dellinger RP, et al. *Intensive Care Med* 2013;39:165-228; 13.Bedenice D. *Vet Clin North Am Equine Pract* 2007;23:293-316; 14.Marik PE, Iglesias J. *Crit Care Med* 2000;28:2652-2654; 15.Perel P, Roberts I. *Cochrane Database Syst Rev* 2012;6:CD000567; 16.Shuster R et al *J Am Vet Med Assoc* 1997;210:87-92; 17.Myburgh JA et al. *N Engl J Med* 2012;367:1901-1911; 18. Brunkhorst FM et al. *N Engl J Med* 2008;358:125-139; 19.Kellum JA. *Crit Care Med* 2002;30:300-305; 20.Story DA. *Crit Care Resusc* 2004;6:188-192; 21.Yunos NM et al. *JAMA* 2012;308:1566-1572; 22.Annane D, Chevret S. *JAMA* 2014;311:1069-1070; 23.Bellomo R, et al *JAMA* 2014;311:1067-1068; 24. van der Linden MA et al , *J Vet Intern Med* 2003;17:343-348; 25.Hallowell GD, Corley KT. *J Vet Intern Med* 2006;20:980-986; 26.https://www.nice.org.uk/guidance/cg174/chapter/introduction,; 27. Argenzio RA. F*Cornell Vet* 1975;65:303-330; 28. Argenzio RAet al. *Am J Physiol* 1974;226:1035-1042; 29.Argenzio RA, Stevens CE. *Am J Physiol* 1975;228:1224-1230; 30.Gardiner M. A*Vet Rec* 2013;172:430; 31.Girisgin AS et al. *Emerg Med J* 2006;23:862-864; 32.Grocott MP et al *Wilderness Environ Med* 2005;16:209-211; 33.Tremayne V. *Nurs Stand* 2009;24:46-48; 34. Sanchez CR et al. *J Zoo Wildl Med* 2004;35:397-399; 35. Lopes MA, et al. *Am J Vet Res* 2004;65:695-704; 36. Lopes MA, et al. *Equine Vet J* 2002;34:505-509; 37.Hallowell GD. *Equine Vet J* 2008;40:411-413; 38.http://bestpractice.bmj.com/bestpractice/monograph/1105/treatment/evidence/intervention/0314/0/sr-0314-i2.html; 39.Moffatt RE et al. *Can J Comp Med* 1975;39:434-441; 40.Ecke P, Hodgson et al *Vet J* 1998;155:161-170; 41.Ecke P et al. *Vet J* 1998;155:149-159.

EQUINE TRANSFUSION MEDICINE

Rose D. Nolen-Walston, DVM, DACVIM (LAIM)
Kennett Square, PA
Joy Tomlinson, DVM, DACVIM (LAIM)
Ithaca, NY

INTRODUCTION

Equine transfusion medicine is a field in infancy compared to human and small animal practice. Due to the large volumes of blood required and the smaller numbers of horses compared to humans and companion animals, commercial equine blood banking has not been developed, though a variety of equine plasma products are available for a host of indications. Allogenic whole blood transfusion is an accepted method of resuscitation in cases of anemia in horses where oxygen delivery to the tissues is below acceptable limits. Almost all blood is collected from a donor immediately before use, and is given as whole blood rather than packed red cells (RBCs) simply because processing the blood involves specialized equipment and expertise. Washed RBCs are ideally used in cases of Neonatal Isoerythrolysis, where maternal RBCs are a guaranteed safe choice if the antibody-containing plasma is removed. Plasma is often purchased frozen from commercial sources. Aliquots obtained from horses with high levels of gamma-globulins are marketed for foals with failure of passive transfer, whereas plasma from horses hyper-immunized against specific pathogens can be used for anti-endotoxin (J5;E. coli cell wall), anti-botulism, anti-Rhodococcus, anti-West Nile Virus, and other therapeutic uses. For replacement of coagulation factors, plasma must be "fresh frozen" (FFP) within 8 hours of collection. Platelets are obtained either via plateletpheresis or separated from fresh blood using centrifugation. They cannot be refrigerated as it inactivates thrombocyte function, so must be used within 3 days of collection. The majority of this presentation will focus on whole blood use and storage.

DONOR SELECTION

Most blood donors in equine medicine are either horses maintained for this purpose by larger clinics, or barn mates of the potential recipient who are press-ganged into service when an emergency transfusion is necessary in the field. For the "professional" blood donor, several characteristics are useful. Ideally the donor should be Qa and Aa antigen and antibody negative (this is as close to a universal donor as we have) as these blood types are the most frequently implicated in transfusion reactions. Usually Standardbreds and Quarter horses are most likely to meet this description. Prospective donors can be blood typed by a commercial laboratory as part of their assessment process. As most equine blood transfusions aim to use blood from just one donor, the ideal animal is large and of quiet temperament, a gelding or maiden mare, and with a normal PCV >35%. These horses should be fully vaccinated, have appropriate anthelmintic usage, and be routinely tested for local blood borne disease (especially Equine Infectious Anemia). Most horses can be collected every 3 weeks or so. Blood is collected aseptically through a 14g or even 10g catheter ideally placed retrograde in the neck, if speed is of the essence. It is safe to take up to 15-18 ml/kg (approximately 8-10L for a 500kg horse), and some clinicians choose to replace with 10-20L of crystalloid. Whole blood can be collected into acid citrate dextrose (ACD), sodium citrate, or citrate-phosphate-dextrose with adenine (CPDA-1) The dextrose and adenine allow the RBCs to maintain ATP during storage, and the citrate is an anti-coagulant. These can be in premade 450ml bags designed for human use, or large glass bottles, though collection into vacuum bottle damages the cells considerably.

CROSS-MATCHING AND BLOOD-TYPING

Equine blood types are described using both blood group and factor. Horses have seven blood groups: A, C, D, K, P, Q, and U, and each group can have multiple factors, of which 30 exist in total. For example, horses with blood group A can have factors a, b, c, d, e, f, g. These factors and types can combine to result in ~400,000 possible blood types. Serendipitously, though, horses rarely have significant alloantibodies against non-self blood types.[1] Anti-RBC antibodies are typically acquired by exposure (usually pregnancy), and Qa and Aa are associated with most transfusion reactions. Anti-Ca antibodies can naturally occur, and cause milder reactions. As horses age, they seem to develop more alloantibodies, possibly due to cross-reaction with ingested protein exposure from their diet (Dr. Sean Owens, personal communication). Nonetheless, ideally blood-type and cross-match compatibility is used to determine an appropriate donor. A point-of-care equine blood-typing method has been developed[2] but is not widely used due to lack of a source of anti-Aa and anti-Ca antibodies. Therefore, in most emergency situations in the field or in smaller clinics, blood-typing is unavailable. Cross-matching is more widely available in equine referral hospitals and is commonly relied on to minimize the chance of transfusion reaction although there is disagreement in the literature about the reliability of this.[3,4] Cross-matching equine blood generally includes both major (donor erythrocytes washed and incubated with recipient serum) and minor (donor serum incubated with washed recipient erythrocytes) cross-match, and is performed using the same methods as in small animal and human medicine. Positives are determined by assessing both agglutination, and unlike other species, hemolysin-mediated hemolysis should be evaluated after adding complement to the reaction mixture. This procedure is requires technical training, and is usually only performed in larger hospitals or labs.

Blood samples banked for up to 1 month are typically used to perform pre-transfusion testing in humans and small animals, but a study using equine blood samples refrigerated for 0-4 weeks found that the reproducibility of cross-match compatibility was poor using stored blood. This was especially true in tests using stored erythrocytes (major antigen cross-matches), with significant differences from baseline at all weeks. Most of these differences were positive, indicating poorer compatibility. This decreased apparent compatibility after storage makes exclusion of compatible donors more likely than mistaken administration of incompatible blood, but when possible, fresh samples should be collected from potential donors before cross-matching equine blood.[5]

TRANSFUSION TRIGGERS AND TECHNIQUE

The normal blood volume for an adult horse is about 8% of bodyweight (80ml/kg). Clinical signs of blood loss are not usually seen until 15% of total volume is lost, with fatal exsanguination occurring after an acute loss of >40-50% of blood. With acute blood loss (for example, uterine artery rupture) PCV may stay normal for several hours until interstitial fluid is recruited into the intravascular space to restore volume. Often the first hematologic sign of hemorrhage is a rapidly dropping plasma total solids, which often precedes the drop in PCV. Clinical signs usually occur when PCV <12-15% for peracute loss, but may get as low as 9-10% with chronic anemia before signs are seen. Blood deficit can be calculated using an intuitive formula:

(desired PCV - current PCV)/desired PCV x 0.08 x bodyweight (kg)

A further handy guide is that assuming donor blood is ~ PCV 40%, every 2.2 ml of whole blood/kg body weight will increase the PCV by 1%. Indications for RBC transfusion include anemia with clinical signs such as tachycardia (HR >60 bpm), colic (presumed gastrointestinal hypoxia), dullness, depression, or anorexia. Laboratory parameters that may trigger transfusion include blood lactate level of 4 mmol/L or more after volume resuscitation, oxygen extraction ratio 50% or greater[1] (this is suggested by a low PvO_2, indicating maximal tissue oxygen extraction).

Blood should be administered using a blood transfer set which includes a filter to remove clots; this should be changed every 4 liters (8 bags). The transfusion should be given slowly for the first 15 minutes, with a physical examination and vital parameters obtained every 5 minutes for first 30 minutes, then every 15 minutes until transfusion is complete. Throughout transfusion, the horse should be monitored for signs of reaction including increased rectal temperature, piloerection, pruritis, sweating, tachycardia, colic, or diarrhea. Most transfusion reactions can be managed by reducing the rate or discontinuing the transfusion, if possible. Milder reactions can be treated with NSAIDS, diphenhydramine (use with caution, generally IM is safer), or epinephrine if the reaction is severe (0.01 ml/kg; 1-5ml of 1:1,000 epinephrine IV or SQ in a 450kg horse). Though the first transfusion can often be given without compatibility testing, within a week the recipient will mount an immune response against the foreign antigen.[8] All transfusions given >3-5days after the initial blood should be preceded by a full cross-match.

TRANSFUSION REACTIONS

Transfusion reactions may occur secondary to leukocytes, platelets, proteins, or poorly described red blood cell antigens that are not detected in current compatibility testing, and have been reported in 16% of blood transfusions in horses.[6] In one study, 2/4 horses that received allogenic blood transfusions of cross-match compatible blood had anaphylactic transfusion reactions, while 2/2 receiving cross-match incompatible blood did not have reactions.[4] Some reports suggest that cross-matching in horses does not accurately predict the risk of transfusion reaction,[3,4] while another retrospective study reported that 4/6 horses with transfusion reactions had incompatibilities on cross-match.[4] There has even been the suggestion that cross-match results are so fallible that testing should be replaced entirely by a 50 ml test transfusion.[7] Unpublished data by the authors show that cross-match incompatibility with cross-match greater than +1 agglutination or hemolysis was significantly associated with febrile transfusion reaction in recipients. The degree of reaction was mild and did not prevent completion of the transfusion, despite 6/8 recipient horses receiving blood that showed 4+ agglutination and strongly positive hemolysin testing on pre-transfusion cross-match. Mild fever (range 101.3 – 103.1°F) occurred in all but one horses given cross-match incompatible blood, and one horse developed moderate colic signs 25 minutes into the transfusion which resolved with 4.4 mg/kg phenylbutazone IV. Fever was first noted between 45 minutes and 25 hours after completion of the transfusion and lasted up to 2 days. Additional signs of transfusion reaction were urticaria or pruritis, tachycardia of 48-60 beats per minute, tachypnea, sweating and muscle fasciculations , transient murmur, arrhythmia, and mild icterus. No horse developed anaphylactic signs nor evidence of severe hemolysis (icterus, pigmenturia) within the first 5 days after transfusion. This agrees with prior literature in that cross-match testing does not appear to be predictive of life-threatening transfusion reaction.[3,4,7] However, the major limitation of this conclusion is that the majority of recipient horses used in this study had anti-Ca antibodies which are known to be clinically less significant than anti-Aa or anti-Qa antibodies; reactions in horses with these antibodies might be considerably more severe. Interestingly, in the 10 pairs of horses used for this study, no horse with known additional instances of blood-type incompatibility (6 horses, 10 instances) developed new anti-RBC antibodies following transfusion of cross-match incompatible blood.

TRANSFUSED RED BLOOD CELL LIFESPAN

As newer technologies are used, our knowledge of the lifespan of the transfused RBC has changed. The lifespan of the equine RBC in the body is ~150 days. Looking at transfused blood, the original studies used 59Fe radiolabeling and found allogenic RBC survival times of 2-6 days.[4] Newer studies using biotin labelling show the half-life of autologous re-transfused blood at 45 days,[9] and that of well-matched allogenic blood stored for 24 hours at 20 days.[10] Cross-matching has been reported not to predict the survival of allogenic RBCs measured by 59Fe-labeling,[4] but the accuracy of the radiolabeling method has been called into question. The difference between methods could come from two possible sources. Firstly, the radiolabel could have decayed at faster than expected rates while biotin labels were more stable. Secondly, the radiolabeled study used only cross-matching to determine compatibility, but the biotin labeled study used blood-type as well, suggesting that the survival times could have truly been different if the blood-types were poorly matched in the radiolabeled studies. Recent unpublished work by the authors looking at transfusion with poorly matched blood showed that cross-match incompatibility was significantly associated with decreased survival time ($P < .001$). The half-life of transfused incompatible (cross-match > 1+) allogeneic equine RBC was 4.7 days versus 33.5 days for compatible pairings. These data suggest that unmatched blood has limited clinical utility beyond immediate patient stabilization, but that with the exception of mild fever, the transfusion reactions were not severe.

BLOOD BANKING

To the best of the authors' knowledge, there are no commercial sources for whole blood or RBCs. However, clinics may store whole blood for their own use, either for autologous transfusion (for example, in a patient scheduled for sinus surgery where high blood losses may be expected) or allogenic blood that was collected and then not used. Blood should be stored carefully in a refrigerator preferably dedicated to blood storage, or at the very least one that is opened infrequently to reduce temperature fluctuations.[1] Temperatures below 2°C result in hemolysis, and those above 6°C are permissive of bacterial growth. Over time, RBCs hemolyze and potassium and lactate concentrations climb. Equine blood stored for 28 days in CPDA-1 has a half-life of 29 days when transfused autologously, but inapparent storage lesions may degrade the RBC and result in reduced oxygen carrying capacity and increased transfusion reactions.[9] At standard storage temperatures of 2–6°C, within 3 days the RBC Na^+/K^+ pump becomes inactive and potassium is eluted from the cell and replaced with sodium. Extra-cellular potassium concentration can increase to 50 mEq/l within the stored blood.[11] After the RBCs are transfused back into the body, intracellular sodium concentration normalizes within 24 h though potassium is not recovered until day 4. Additionally, biochemical changes that occur during storage include falls in 2,3-diphosphoglycerate, intracellular RBC adenosine triphosphate, and nitric oxide which reduce RBC deformability and end-organ tissue oxygenation.[12]

REFERENCES

[1]Mudge MC. **Vet Clin North Am Equine Pract**. 2014;30(2):427, [2]Owens SD, et al. **Vet Clin Pathol** 2008;37(1):49; [3]Tocci LJ, et al. **J Vet Emerg Crit Care (San Antonio)** 2009;19(1):66; [4]Kallfelz FA, et al **Am J Vet Res** 1978;39(4):617; [5]Harris M, et al. J **Vet Intern Med** 2012;26(3):662; [6]Hurcombe SD, et al. **J Am Vet Med Assoc** 2007;231(2):267; [7]Durham AE. **Eq Vet Edu** 1996;8(1):8; [8]Wong PL et al**. Am J Vet Res** 1986;47(12):2566; [9]Owens SD, et al. **Am J Vet Res** 2010;71(8):960; [10]Mudge MC, et al. **Vet Clin Pathol** 2012;41(1):56; [11]van de Watering L. **Vox Sang**. 2011 Jan;100(1):36; [12]Kor DJ, et al. **Bosn J Basic Med Sci.** 2009;9(Suppl 1):21;

PPID AND INSULIN RESISTANCE IN THE HORSE: NOT JUST FEET AND FAT

Peter R. Morresey, BVSc, DACT, DACVIM (LAIM)
Lexington, KY

INTRODUCTION

Laminitis and obesity are well known effects of endocrine and metabolic dysfunction in the horse. However, information from the horse and other species suggests multiple organ systems may be affected by the induced chronic pro-inflammatory state. In addition to the well-described regional adiposity and lamellar changes, reproductive, renal, hepatic, and other manifestations may occur.

PATHOBIOLOGY OF DISEASE: AN UPDATE

Pituitary pars intermedia dysfunction (PPID, Cushing's syndrome) is a common condition of aged horses and ponies in which inhibition of the intermediate lobe of the pituitary is lost due to degeneration of the hypothalamic dopaminergic neurons. During the early stages of this syndrome cellular hyperplasia of the intermediate lobe results in increasing levels of pro-opiomelanocortin (POMC), a precursor molecule cleaved into many peptides including α-melanocyte stimulating hormone (α-MSH), adrenocorticotropic hormone (ACTH), β-endorphin and corticotropin-like intermediate lobe peptide (CLIP). As the syndrome progresses, the pars intermedia enlarges until a singular intermediate lobe adenoma develops that compresses the adjacent pituitary and hypothalamic structures.

Considerable debate exists regarding the etiology and pathogenesis of metabolic syndrome as no single unifying mechanism has as yet been elucidated.[1] A complex interaction between genetics, hormonal status and nutrition is most likely. Metabolic syndrome describes a set of clinical findings in human patients including insulin resistance, hyperinsulinemia, dyslipidemia, hypertension and atherosclerosis, however obesity, while common, is not a consistent finding. Inflammatory mediators such as TNF-α have been shown to mediate insulin resistance.[2] Adipose tissue is not simply a store of excess energy, but is rather an organ of diverse functions which plays a pivotal role in development of metabolic syndrome.[3] Adipokines, biologically active secretions of adipose tissue, may play a role in the pathogenesis of insulin resistance as some have been shown to have effects on insulin sensitivity and signaling.[4] Plasma concentrations of some proinflammatory adipokines including IL-6, TNF-α, C-reactive protein, and resistin are elevated in metabolic syndrome, whereas the levels of anti-inflammatory adipokines such as adiponectin are reduced.[5, 6] It is the resulting hyperinsulinemic, proinflammatory state that is thought to drive the laminitis seen in affected horses. However, some studies do not support the occurrence of a chronic proinflammatory state in insulin resistant horses;[7, 8] while others find an association between cytokine levels, obesity and insulin sensitivity consistent with human reports of insulin resistance and metabolic syndrome.[9]

In the horse, breed-related differences in the response by insulin to oral and intravenous glucose have been demonstrated.[10] The concept of metabolic syndrome in the horse while relatively new is widely researched and reviewed.[11] Linkage between the mechanism of PPID and metabolic syndrome has been suggested. However, a recent study concluded PPID does not always alter insulin sensitivity in aged horses.[12]

LAMINITIS UPDATE

The innate immune system is important in the pathophysiology of laminitis, although the role in hyperinsulinemic laminitis has not been fully defined. Although LPS is the primary ligand for toll-like receptor 4 (TLR4), signaling can be activated by free fatty acids.[13, 14] This provides a link between metabolism and inflammation, with TLR4 commonly located in insulin sensitive tissue.[15-17] As some reports have shown infiltration of inflammatory leukocytes is minimal during insulin-induced laminitis episodes.[18, 19] Inflammation may in fact be a sign, not a cause, of laminitis, and may have a role in the repair and remodeling phase.[20] The normal response to insulin of digital arterial and venous rings is relaxation, however *in vitro* studies have shown the onset of insulin resistance alters responses of digital vessels to contraction.[21] Endothelial dysfunction results from hyperinsulinemia, insulin resistance or hyperglycemia. Altered insulin dynamics are often present in obese insulin-resistance horses or those with PPID, with those predisposed to laminitis having higher insulin than those not developing laminitis.[22, 23]

EFFECTS ON THE HYPOTHALAMIC-PITUITARY-GONADAL AXIS

In many species metabolic and endocrinological changes have been proven to have significant effects on female reproductive processes, with ovarian activity, establishment of pregnancy, pregnancy maintenance to term and the overall health of the fetus impaired. Metabolic syndrome affects the insulin-like growth factor system, this being crucial to follicle selection and dominance in the mare, leading to disturbances of ovarian function.[24, 25] Obese mares with reduced insulin sensitivity have been shown to have prolonged interovulatory and luteal phases.[26] Insulin has a major role in the regulation of ovarian steroidogenesis, follicular development and granulosa cell proliferation.[27] Insulin resistance has been recognized as the major factor related to polycystic ovary syndrome. Compensatory hyperinsulinemia stimulates ovarian androgen production and the peripheral aromatization of androgens to estrogens, thus altering gonadotropin secretion and therefore follicular development.[28] While effects on pregnancy of metabolic syndrome in the mare are not proven, gestational diabetes, abortion and fetal compromise have been reported in humans.[29, 30] Interestingly, a reduction is seen in first trimester spontaneous abortion in women with metformin usage[31], and no evidence of teratogenic effects on the fetus with this approach has been reported.[32, 33]

Effects on the reproductive competence of the male are also known to occur. As in the mare, reproductive effects of metabolic syndrome have not been clearly defined, although there is a wealth of comparative information available. It has been shown that there is an interaction between metabolic syndrome and testicular function that is bidirectional in nature. Obesity can affect spermatogenesis, and alterations in the hypothalamic-pituitary testicular axis occur. Resulting endocrine function changes in the testes have been shown to impact systemic metabolic parameters.[34]

ENDOTHELIAL DYSFUNCTION

Insulin resistance is an important but inconsistent component of metabolic syndrome.[35] Insulin stimulates endothelial cell production and release of nitric oxide (NO), causing vasodilation[36]. The vasodilating action is attenuated in insulin-resistant states.[37] Insulin impairs endothelial function by inducing oxidative stress[38] and stimulating release of endothelin-1.[39] During hyperinsulinemia, decreased NO production, increased production of reactive oxygen species, decreased activity of vasodilating agents (e.g. prostacyclin), and increased activity of vasoconstrictive substances (e.g. endothelin-1 and angiotensin II) contribute to endothelial dysfuntion.[36] Endothelial dysfunction in skeletal muscle may decrease blood flow contributing to insulin resistance, as NO dependent increases in blood flow to skeletal muscle are responsible for substantial glucose uptake in response to insulin stimulation.[36, 37] Furthermore, in metabolic syndrome lower circulating adiponectin, and higher resistin and leptin levels are present.[40]

RENAL EFFECTS

Metabolic syndrome in humans is considered a risk factor for microalbuminuria and chronic kidney disease.[41] Microalbuminuria may be a marker of generalized endothelial dysfunction.[42] Insulin resistance is associated with hypertension in humans, leading to systemic and glomerular capillary pressure increases. Increased glomerular filtration pressure predisposes to renal injury, which includes glomerulosclerosis. Salt and water conservation expands blood volume. Insulin resistance is associated with end-stage renal disease[43], and lower insulin sensitivity is present in non-diabetic individuals suffering mild-to-moderate kidney disease.[44] Severity of insulin resistance is correlated with elevated glomerular filtration rate in some studies.[45, 46]

HEPATIC EFFECTS

In human insulin-resistant patients, hepatic free fatty acid flux is elevated, triglyceride synthesis and storage are increased, and excess triglyceride is secreted as very low density lipoprotein.[47] Adiponectin acts directly on hepatocytes inhibiting fatty acid synthesis and uptake while promoting fatty acid oxidation. Low adiponectin levels are associated with increased concentrations of aspartate aminotransferase and gamma-glutamyltransferase, indicating adiponectin may be involved in maintaining hepatic health.[48] Adiponectin enhances hepatocyte insulin sensitivity.[49] Low adiponectin concurrent with high TNF-α and increased hepatic exposure to free fatty acids precipitates hepatic steatosis and profound hepatic insulin resistance.[50] Increased production of numerous pro-inflammatory cytokines and oxidative hepatic changes may over time precipitate cirrhotic changes.[51]

OTHER MANIFESTATIONS

Immune-mediated and inflammatory conditions share common mediator pathways. Co-morbidities with insulin resistance identified in human medicine include dermatological conditions (psoriasis, acanthosis nigricans)[52-54] and coagulopathies.[55, 56]

REFERENCES

[1] Ferrannini E. J Clin Endocrinol Metab 2007;92:396-398. [2] Hotamisligil GS, et al. Science 1993;259:87-91. [3] Burns TA, et al. J Vet Intern Med 2010;24:932-939. [4] Kershaw EE, et al. J Clin Endocrinol Metab 2004;89:2548-2556. [5] Wisse BE. Journal of the American Society of Nephrology 2004;15:2792-2800. [6] Dandona P, et al. Circulation 2005;111:1448-1454. [7] Carter RA, et al. Equine Vet J 2009;41:171-178. [8] Holbrook TC, et al. Vet Immunol Immunopathol 2012;145:283-289. [9] Tanti JF, et al. Front Endocrinol (Lausanne) 2012;3:181. [10] Bamford NJ, et al. Domestic Animal Endocrinology 2014;47:101-107. [11] Frank N, et al. J Vet Intern Med 2010;24:467-475. [12] Mastro LM, et al. Domestic Animal Endocrinology 2015;50:14-25. [13] Devaraj S, et al. Diabetologia 2009;52:1665-1668. [14] Dasu MR, et al. Diabetes Care 2010;33:861-868. [15] Könner AC, et al. Trends in Endocrinology & Metabolism 2011;22:16-23. [16] Reyna SM, et al. Diabetes 2008;57:2595-2602. [17] Waller AP, et al. Vet Immunol Immunopathol 2012;149:208-215. [18] de Laat MA, et al. Journal of Comparative Pathology 2011;145:399-409. [19] de Laat MA, et al. The Veterinary Journal 2013;195:305-312. [20] de Laat MA, et al. Vet Immunol Immunopathol 2014;157:78-86. [21] Venugopal C, et al. The FASEB Journal 2014;28. [22] Bailey SR, et al. J Am Vet Med Assoc 2007;231:1365-1373. [23] Borer KE, et al. Journal of Animal Science 2012;90:3003-3011. [24] Beg MA, et al. Reproduction 2006;132:365-377. [25] Ginther OJ, et al. Biology of Reproduction 2004;70:1374-1379. [26] Vick MM, et al. Reprod Fertil Dev 2006;18:609-617. [27] Willis D, et al. Journal of Clinical Endocrinology & Metabolism 1996;81:302-309. [28] Diamanti-Kandarakis E, et al. Obes Rev 2001;2:231-238. [29] Lanzone A, et al. Hum Reprod 1996;11:2382-2386. [30] Glueck CJ, et al. Hum Reprod 2002;17:2858-2864. [31] Glueck CJ, et al. Fertil Steril 2001;75:46-52. [32] Coetzee EJ, et al. S Afr Med J 1984;65:635-637. [33] Coetzee EJ, et al. S Afr Med J 1980;58:795-802. [34] Corona G, et al. The Journal of Sexual Medicine 2011;8:272-283. [35] Reaven GM. Diabetes 1988;37:1595-1607. [36] Kim J, et al. Circulation 2006;113:1888-1904. [37] Steinberg HO, et al. Journal of Clinical Investigation 1994;94:1172-1179. [38] Arcaro G, et al. Circulation 2002;105:576-582. [39] Cardillo C, et al. Circulation 1999;100:820-825. [40] Hivert MF, et al. The Journal of Clinical Endocrinology & Metabolism 2008;93:3165-3172. [41] Chen J, et al. Annals of Internal Medicine 2004;140:167-174. [42] Deckert T, et al. Diabetologia 1989;32:219-226. [43] Mccalebeb ML, et al. Endocr Res 1985;11:113-125. [44] Dzúrik R, et al. Nephron 1995;69:281-285. [45] Dengel DR, et al. Hypertension 1996;28:127-132. [46] Fliser D, et al. Kidney Int 1998;53:1343-1347. [47] Lewis GF. Diabetes Care 1996;19:390-393. [48] López-Bermejo A, et al. Clinical Endocrinology 2004;60:256-263. [49] Stefan N, et al. Hormone and Metabolic Research 2002;34:469-474. [50] Yamauchi T, et al. Nat Med 2001;7:941-946. [51] Diehl AM, et al. Gut 2005;54:303-306. [52] Boehncke S, et al. British Journal of Dermatology 2007;157:1249-1251. [53] Davidovici BB, et al. J Invest Dermatol 2010;130:1785-1796. [54] Kahn CR, et al. N Engl J Med 1976;294:739-745. [55] Schneider DJ. Coronary Artery Disease 2005;16:473-476. [56] Juhan-Vague I, et al. Pathophysiology of Haemostasis and Thrombosis 2002;32:269-273.

CO_2 METABOLISM IN HEALTH AND DISEASE

Modest Vengust, DVM, DVSc, PhD, DACVIM (LAIM), DACVSMR
Ljubljana, Slovenia
George Heigenhauser, BA (hons), MSc, PhD, FACSM
Hamilton, Ontario, Canada

INTRODUCTION

The oxidation of carbohydrates (CHO) and non-esterified fatty acids (NEFA) provide most energy by which ATP is generated, and are, therefore, the main source of CO_2 in respiring organisms. Oxidation of CHO produces 1 mol of CO_2 per 6 mol of ATP, whereas the oxidation of NEFA yields 1 mol of CO_2 for 8 mol of ATP[1].

The total amount of CO_2 (TCO_2) in blood is the sum of bicarbonate (HCO_3^-), the dissolved CO_2, and CO_2 bound to hemoglobin. The majority of CO_2 in all compartments is HCO_3^- (> 90%). The contribution of HCO_3^- to CO_2 transport requires rapid conversion of CO_2 into HCO_3^- and H^+. Carbonic anhydrases catalyzes the reversible reaction involving the hydration/dehydration of CO_2 ($H_2O + CO_2 \xleftrightarrow{CA} H_2CO_3 \leftrightarrow H^+ + HCO_3^- \leftrightarrow H^+ + CO_3^{2-}$) and increases the rate of the Jacobs Stewart cycle and chloride exchange (Hamburger shift) several folds to optimize CO_2 elimination by facilitating the transformation of CO_2 to HCO_3^- in peripheral tissues and HCO_3^- back to molecular CO_2 across the lung. The half time for completion of these reactions is ~1 sec. Carbonic anhydrase inhibition slows the hydration/dehydration of CO^2 with consequent hypercapnia (increased TCO_2) and acidosis, which has important implications for muscle metabolism and the development of fatigue[1,2,3].

Carbon dioxide metabolism is closely linked to pulmonary transvascular fluid dynamics in horses and possibly other athletic mammalian species. Any alteration in the Jacobs-Stewart cycle may affect transvascular fluid shifts in the pulmonary vasculature and the diffusion distance between the pulmonary alveoli and pulmonary capillaries; a phenomena that can be independent of Starling forces.[4,5]

CO_2 METABOLISM DURING EXERCISE

During exercise ATP turnover increases CO_2 production. Factors limiting CO_2 equilibration across the lung during exercise are the short pulmonary capillary transit time, inadequate alveolar ventilation, and a long half time for the Hamburger shift. In a horse increased ventilation cannot offset hypercapnia during sub- and maximal exercise[1,2,6,7].

Interpretation of the muscles energy metabolism during submaximal and maximal exercise is still bound to a historical concept of the anaerobic threshold. Anaerobic threshold is the point at which lactate (Lac^-) begins to accumulate after Lac^- production exceeds the capacity for its clearance. Lactic acid is considered to be exclusively responsible for the acidosis during exercise. Lactate production was attributed to inadequate O_2 supply to the working muscle (anaerobic metabolism). In the anaerobic threshold concept lactic acidosis is buffered by HCO_3^-, which yields water and CO_2 ($2H^+ + 2HCO_3^- \rightarrow 2CO_2 + 2H_2O$). Carbon dioxide is then removed with increased rate of ventilation. Lactate production and buffering, and final CO_2 formation, were considered stochiometricaly equivalent[6].

It is important to realise that Lac- production is not due to anaerobic metabolism but due to a metabolic mismatch between the rate of glycolysis and the rate at which pyruvate can enter mitochondria through the rate limiting pyruvate dehydrogenase[7]. The advent of the Peter Stewart's physico-chemical approach to acid-base chemistry firmly established that the concept of Lac-accumulation, HCO_3^- buffering, and CO_2 elimination has a poor quantitative relationship. Peter Stewart's physico-chemical approach to acid-base chemistry quantifies the relative contributions of three independent variables to changes in dependent variables [hydrogen ion (H^+), bicarbonate (HCO_3^-)] in aqueous solutions. These independent variables are: strong ion difference (SID), weak electrolyte concentrations (A_{tot}), and partial carbon dioxide pressure (PCO_2). Changes in H^+ can be achieved only by changing one or more of these three independent variables. The system is constrained by the following fundamental laws of the chemistry of aqueous solutions: conservation of mass, electro-neutrality, and equilibrium constraints on dissociation reactions. Increased partial pressure of CO_2 will increase [H^+] (decrease in pH), regardless of the behavior of other independent variables[9,10,11].

Concept of bicarbonate buffering acidosis (Lac^- acidosis) has prompted treatment with bicarbonate alkalinising solutions (sodium hydrogen carbonate = $NaHCO_3$) in clinical settings and in athletic horses (milkshaking). The ingested HCO_3^- becomes part of the open CO_2 system and is rapidly normalized with ventilation. Therefore, it is not HCO_3^- but the cation Na^+, which through the increase in SID reduces [H^+]; delivering Na^+ into the body by other means (sodium citrate, sodium Lac-) will have similar effect to NaHCO3[1,11].

The plasma or serum concentration of TCO_2 provides a clinically useful screening test for the detection of the prerace administration of alkalinising agents to racehorses. Strict and unified standards for sample collection, manipulation, storage and analysis should be adhered to to prevent erroneous TCO_2 results. Blood samples collected for the purpose of detection of TCO_2 should be centrifuged immediately and analysed as soon as possible, or the results will underestimate TCO_2 values. A serum [TCO_2] threshold of 36 to 37 mmol/l, depending on the racing jurisdiction, is used to identify horses administered a prohibited alkalinising agent. A wide range of normal TCO_2 (23–38 mmol/l) has been determined in racehorses. TCO_2 is naturally affected by exercise, feeding practices, electrolyte administration and diurnal variation; however, these factors are unlikely to cause an increase to or beyond the value of 36 mmol/l. Waller et al. (2010)[12] proposed that commercial pelleted feed, as opposed to oats or grain, were responsible for the lower [Cl^-], which increased TCO_2 levels close to, or in excess of, the 37 mmol/l testing threshold. This was not confirmed in the study by Balaskonis et al. (2010)[13]. It is more likely that a combination of dietary, environmental and handling practices, which elevate plasma [SID] (subclinical dehydration, furosemide treatment), lower [A_{tot}] (high grain diet) and lower the concentration of unmeasured weak acids are the primary contributors to elevated TCO_2 in horses not receiving $NaHCO_3$[12,14,15].

HYPERCAPNIA UNRELATED TO EXERCISE

Circulation and the state of alveolar ventilation influence CO_2 production and output. Impaired circulation causes local hypercapnia due to intracellular hypoxia and increased Lac- production, and general hypercapnia due to the delay in CO_2 reaching the lung. Although it is generally accepted that hypoxia causes increase in Lac^- and CO_2 production, the apparent Michaelis–Menten constant for oxygen (Km′) is still several orders of magnitude higher than the level at which mitochondrial respiration would be supressed. Only a severe circulatory collapse would probably decrease Km′ to a level where mitochondria would not be able to sustain oxidation. Most plausible mechanism for increased Lac^- and CO_2 production in circulatory impairment is the increased intracellular concentration of ADP because of the increased ATP hydrolysis. ADP then stimulated glycolysis and Lac^- production[16,17].

Most often, however, hypercapnia arises from hypoventilation = reduced alveolar ventilation. Reduced alveolar ventilation is caused by 1) decreased ventilatory drive (as a result of central nervous system pathologies, compensatory metabolic alkalosis, or depressant medications) and 2) worsening respiratory mechanics and/or severely abnormal gas exchange (as a result of non-neuromuscular chest wall disease, immature lungs, or interstitial lung disease). Hypercapnia develops due to the lag in removal or inability to remove CO_2 with ventilation. Hypercapnia in such instances is traditionally considered as not beneficial. However, it has been shown that partial carbon dioxide pressure (PCO_2) above the traditional safe range has beneficial effects on the physiology of the respiratory, cardiovascular, and nervous system. It is well documented in humans that permissive hypercapnic acidosis is generally well-tolerated by patients and is more beneficial to their wellbeing than normocapnia. Even though adverse effects of hypercapnia have been reported, especially in patients with central nervous system pathology with increased intracranial pressure and/or chronic infection, critical care clinicians often artificially increase PCO_2 to take advantage of its positive effects on compromised tissues, especially in neonates. This is referred to as therapeutic hypercapnia. Clinically hypercapnia is commonly seen and therapeutically utilised as a consequence of a low tidal volume ventilation strategy for acute respiratory distress syndrome. Therapeutic effect of hypercapnia seems to be independent of changes in extracellular $[H^+]$. Carbon dioxide, similar to O_2, can also alter gene transcription, where some pathways are thought to have beneficial effect on chronic inflammatory disease; a therapeutic phenomenon finding more implications in medicine in recent years[18,19,20,21].

Hypercapnic respiratory acidosis is common in critically ill newborn foals. The safety levels of PCO_2 and data on the effectiveness of different CO_2 levels are not available for newborn foals or adult horses. Permitting arterial PCO_2 ($PaCO_2$) to remain slightly above the traditional safe range may be beneficial for pulmonary development and function, lung tissue injury arising from inflammation, and systemic circulation, as well as having neuroprotective properties. These authors subjectively consider the level of permissive hypercapnic acidosis in newborn foals to mean maximum $PaCO_2$ and pH of not more than 60 mmHg and not less than 7.25, respectively. Beneficial effects diminish or the effect of hypercapnic acidosis may become harmful when $PaCO_2$ rises above 60 mmHg and approaches 70 mmHg, especially when $PaCO_2$ increases acutely or fluctuates rapidly. In animals with suspected hypoxic-ischemic encephalopathy or any other intracranial pathologies and/or in foals showing signs of colic, attempts should be made to keep $PaCO_2$ closer to 50 mmHg[19,22,23,24].

Acute, fluctuating or profound metabolic acidosis has a negative effects on cardiovascular function, facilitates cardiac arrhythmias, stimulates inflammation, and suppresses immune response. Sodium bicarbonate administration for treatment of lactic acidosis and ketoacidosis does not reduce morbidity or mortality, or improve cellular function. One of the reasons is the exacerbation of intracellular acidosis, especially in patients with reduced ability to remove CO_2 with ventilation. In patients with reduced alveolar ventilation the infused HCO_3^- becomes part of the closed system and increases TCO_2. Carbon dioxide diffuses across membranes causing intracellular acidosis, whereas Na^+ is contained to plasma where it causes alkalosis. Reduction in ionized Ca^{2+}, and production of hyperosmolality are other potential negative effects of $NaHCO_3$ treatment. Volume correction with polyconic fluid treatment will most often have a better alkalising effect in patients with metabolic acidosis than treatment with $NaHCO_3$[11,25].

Horses are prone to reduction in cardiovascular function and alveolar hypoventilation when recumbent, which is mostly evident during general anaesthesia. In contrast, recumbence in unprovoked newborn foals does not affect $PaCO_2$, but will reduce partial pressure of oxygen (PaO_2) (2). Physically struggling newborn foals, e.g., during arterial blood collection will show temporary decrease in both $PaCO_2$ and PaO_2[19].

HYPOCAPNIA

Arterial partial pressure of CO2 should not decrease below 30 mmHg. Hypocapnia can worsen ventilation–perfusion matching and gas exchange in the lung, has a negative effect on central nervous and cardiovascular system. Physiologically horses develop hypocapnia when introduced to high altitude[26]. Pathologic hypocapnia usually occurs as a consequence of hypoxemia (and hyperventilation) in pulmonary disorders, and due to fever, pain (colic), sepsis, acidosis, central nervous disorder (anxiety, infection, neoplasia) or drug induced (salicylates, methylxanthines, β-adrenergic agonists, progesterone)[27]. A pronounced postexercise hypocapnia has been reported in horses diagnosed with equine asthma syndrome and exercise-induced pulmonary hemorrhage, which most likely was the result of alveolar hyperventilation due to a more severe exhaustion compared to normal horses[28].

Ponies develop hypocapnia during exercise unlike race horses where hypercapnia occurs during sub- and maximal exercise as discussed above. It seems that ponies for their level of exercise capacity develop less pulmonary diffusion and transient time limitations[29].

CONCLUSION

It is important to understand CO_2 metabolism in horses, which can have a profound effect on the performance, or on the physiology of different tissues in healthy and diseased animals. The concept of HCO_3^- buffering of lactic acidosis during exercise with production of H_2O and CO_2 is outdated. Identification of CO_2 related changes, including those related to $NaHCO_3$ treatment, is best facilitated by use of the physicochemical approach to acid-base status, which uses dependent and independent variables to describe acid-base state.

The ability of CO_2 to alter gene transcription, where some pathways may cause a desirable therapeutic outcome is expected to have a significant implication in patient care. Evidence regarding the effects of hypercapnia and hypocapnia in horses is still weak/anecdotal. More research work is needed in this area. The evidence seems to be insufficient even in human medicine where therapeutic strategies for managing PaCO2 still vary significantly among medical specialists.

REFERENCES

[1]Jones NL. Appl Physiol Nutr Metab 2008; 33(4):641, [2]Jones NL & Heigenhauser GJ. Clin Sci (Lond) 1996; 90(5):323, [3]Vengust M. et al. Equine Vet J Suppl 2010;(38):40, [4]Vengust M et al. J Physiol 2006; 570(Pt 2): 397, [5]Vengust M et al. J Physiol 2013; 591(Pt 18):4499, [6]Brooks GA. Med Sci Sports Exerc 1991; 23(8):895, [7]McDonough P. et al. J Appl Physiol 2002;92(4):1499, [8]Parolin ML. et al. Am J Physiol Endocrinol Metab 2000;278(3):E522, [9]Lindinger MI et al. Am J Physiol 2005;289:R891, [10]Spriet LL, Exercise Metabolism, 2nd edition, Human Kinetics, 2006, [11]Stewart PA. Can. J. Physiol. Pharmacol 1983;61:1444, [12]Waller AP. et al. Equine Vet J Suppl 2010;(38):592, [13]Balaskonis V. et al. Comparative Exercise Physiology 2010; 7(1):43, [14]Tinkler SH. et al. Equine Vet J Suppl. 2012;43:57, [15]Kline K. et al. Journal of Equine Veterinary Science 2006; 26(7): 317, [16]Mole PA et al. Am J Physiol 1999;277:R173, [17]Shulman RG. Exerc Sport Sci Rev 2005;33(4):157, [18]Brown LK. Chest 2008;134(1):4, [19]Vengust M. Can J Vet Res 2012;76(4):275, [20]Rogovik A. & Goldman R. Emerg Med Clin North Am. 2008;26(4):941, [21]Taylor CT & Cummins EP. J Physiol 2011;589(Pt 4):797, [22]Palmer JE. Vet Clin North Am Equine Pract 2005;21(2):457, [23]Giguère S. et al. Am J Vet Res 2007;68(12):1407. [24]Giguère S. et al. J Vet Intern Med 2008;22(2):401, [25] Irvine RO & Dow J. Metabolism 1966;15(11):1011, [26]Greene HM. et al. Equine Vet J Suppl 1999;(30):71, [27]Curley G. et al. Crit Care 2010;14(2):220, [28]Couëtil LL & Denicola DB. Equine Vet J Suppl 1999;(30):77, [29]Katz LM. et al. Equine Vet J Suppl 1999;(30):49.

QUINIDINE CARDIOVERSIONS: LESSONS FROM 35 YEARS OF CLINICAL EXPERIENCE

Virginia B. Reef, DVM, DACVIM, DACVSMR, Assoc ECVDI
Kennett Square, PA

INTRODUCTION

Atrial fibrillation (AF) is the most common arrhythmia affecting performance in horses. Its effect on horses performing rigorous work usually precludes their being able to compete successfully. Cardioversion to normal sinus rhythm (NSR) is necessary for affected horses to return to their prior level of athletic competition.

Paroxysmal AF is common, with conversion to NSR usually occurring within the first 24 -48 hours.[2-4] Cardioversion to NSR is recommended for those horses with persistent AF with no evidence of underlying cardiac disease ("lone" AF), in those with echocardiographic evidence of structural heart disease, in those with only mild or mild to moderate left atrial enlargement and in those horses with excessively rapid ventricular response rates, aberrant conduction or ventricular arrhythmias (VA) that are present during exercise or secondary to other sympathetic nervous system stimulation.[5] Cardioversion of AF is not recommended for horses with severe valvular regurgitation, advanced cardiomyopathy or for those with congestive heart failure (CHF). Cardioversion of horses with AF to NSR using quinidine has been described for over 60 years. Early reports described using digitalis tincture prior to the administration of quinidine sulfate (QS) to horses for conversion of AF.[1]

DIAGNOSTIC WORK-UP

A complete history should be obtained including any performance abnormalities, suspected AF duration (either based on auscultatory findings of an irregularly irregular rhythm or length of time since the horse's last successful performance) and medication history (in particular, furosemide, and bicarbonate or thyroid hormone supplementation). A thorough physical examination should be performed with careful auscultation for the presence of any significant cardiac murmurs. A complete echocardiogram should be performed looking for evidence of structural heart disease, particularly atrial enlargement (usually left) and valvular regurgitation (usually mitral or, less frequently, aortic). Decreased systolic function as evidenced by a decreased fractional shortening may be detected in horses with AF.[6] Fractional shortenings of $\geq$ 24% have been reported in horses with AF that subsequently return to normal following cardioversion.[6] A complete blood count, fibrinogen and chemistry screen is usually normal but a fractional excretion of potassium is indicated in high performance horses, particularly when receiving sodium bicarbonate or furosemide. If there is a history of supplementation with Thyro-L®, T3 and T4 concentrations should be measured. Cardiac troponin I (cTnI) concentrations are usually within normal limits but should be measured if myocardial injury is suspected.

QUINDINE CARDIOVERSION

All horses being treated for AF should be monitored with ***continuous electrocardiography*** so that detection and treatment of rapid ventricular response rates, aberrant conduction or VA that could be life-threatening can occur.[5] Administration of digoxin should be considered prior to QS treatment in horses that have a labile heart rate (easily excitable with rapid sustained increases in rate) or echocardiographic evidence of systolic dysfunction.[7] More recently successful cardioversion of horses with rapid AF in which transvenous electrical cardioversion (TVEC) was not possible has been accomplished with a combination of sotalol and quinidine. The mild ß-blocking effects of the sotalol helped to control the ventricular response rate and the sotalol's antiarrhythmic properties may have had a synergistic effect with quinidine resulting in successful cardioversion.

Although fasting horses prior to treatment with QS has been recommended by some authors, we routinely allow horses free choice access to hay and water during treatment (as long as there is no indwelling NG tube). There was no significant difference between fed and fasted horses by the time of the 4th dose of quinidine (different dosing regimen) in one study.[8]

Quinidine Gluconate (QG) Treatment (Acute Onset AF)

QG administered IV is effective in the treatment of acute onset AF but is most efficacious in horses with no echocardiographic evidence of structural heart disease ("lone" AF) and AF of short duration ($\leq$ 7 days).[9] Although cardioversion of horses with "lone" AF and a duration of AF > 2 weeks are possible with QG, this regimen is less likely to be successful than QS. QG administered IV at 1.0 to 1.5 mg/kg over one minute, repeated every 10-15 minutes until cardioversion was achieved resulted in 75% of horses converting to NSR.[9] The 3 horses in this study that did not convert had a suspected duration of AF of approximately 3 months. Two of these horses were subsequently treated successfully with QS.

We routinely administer QG at 0.5-2.2 mg/kg intravenously (IV) every 10 minutes to effect, not to exceed 12 mg/kg. Practically speaking QG comes in 800 mg vials which, for the average 450 - 550 kg horse, equals 1.78 – 1.45 mg/kg IV, respectively, every 10 minutes until a total of 12 mg/kg have been administered. Cardioversion to NSR may occur during the QG treatment or within the next 24 hours. If cardioversion has not occurred by the following morning and the horse is not exhibiting signs of quinidine toxicity, QS treatment can safely begin.

Quinidine Sulfate Treatment

QS administered via nasogastric (NG) intubation is indicated for horses with "lone" AF of > 2 weeks duration and horses with echocardiographic evidence of structural heart disease (mild or mild to moderate valvular regurgitation or mild left atrial enlargement) if there are no contraindications for QS treatment. QS should be administered at a dose of 22 mg/kg per NG tube every 2 hours until the horse converts to NSR, the QRS duration exceeds 25% of its pretreatment value, significant adverse reactions or signs of quinidine toxicity are detected or the horse has received 4 - 6 treatments of 22 mg/kg every (q) 2 hours. If quinidine plasma concentrations can be obtained they should be obtained 1 hour after the 4th 22 mg/kg treatment is administered to determine if additional quinidine treatments q 2 hours are indicated. ***If the plasma quinidine concentration exceeds 4 µg/ml at this time, QS administration should be prolonged to q 6 hours***. Many horses will convert to NSR during this 6 hour period.

Most horses will not tolerate more than 4 doses of QS administered q 2 hours without showing significant adverse effects or signs of quinidine toxicity! If measuring plasma quinidine concentrations is not possible, the quinidine treatments should be prolonged to every 6 hours after four 22 mg/kg doses administered q 2 hours. Further adjustments in the dosing schedule can be made as needed recognizing that the half-life of quinidine after oral administration has been determined to be 6.65 +/- 3 hours (in a small number of horses).[10] If the horse has not converted by day 2 and is tolerating the treatment well, digoxin can be administered at 0.011 mg/kg orally (or in NG tube with QS) twice daily for the second day. Peak plasma digoxin concentrations should be obtained on day 3 before administering additional digoxin along with the QS, due to the elevation of plasma digoxin concentrations that occur with concurrent quinidine administration (plasma digoxin concentrations almost double).[11]

Critical to the successful administration of QS to horses with AF that have longer duration of AF or underlying structural heart disease is ***patience*** while ***carefully monitoring the horse for significant adverse reactions or signs of quinidine toxicity***, ***continuously monitoring the electrocardiogram (ECG)*** and monitoring plasma quinidine concentrations. Plasma quinidine concentrations should ideally be maintained between 3-5 µg/ml while on the q 6 hour QS regimen. Most horses will tolerate 3 consecutive days of QS treatment, if needed. Although successful cardioversion of horses has occurred with longer duration of treatment, the chances of success are significantly reduced. TVEC should be considered in these cases, if the horses are a candidate for TVEC. Once cardioversion is achieved, further treatment with quinidine should be discontinued. A plasma quinidine concentration to determine the concentration of quinidine in the plasma at the time of cardioversion which may be helpful if AF recurs.

Adverse Reactions to Quinidine/Quinidine toxicity

Side effects of quinidine treatment are more likely in horses with longer duration of AF and in horses with underlying heart disease.[12] Adverse or toxic reactions to quinidine include depression, paraphimosis, urticaria/wheals, nasal mucosal swelling/ upper respiratory tract (URT) obstruction, laminitis, ataxia, bizarre behavior, convulsions, flatulence, oral ulcerations, diarrhea, colic, pulmonary edema, tachycardia, supraventricular or VA, torsades de pointes, prolongation of the QRS duration, hypotension, CHF and sudden death. While this list of adverse or toxic reactions is intimidating, most of these complications are manageable. Colic and QRS prolongation are associated with the horse receiving higher doses of quinidine while URT obstruction and neurological signs are signs associated with quinidine toxicity and should prompt immediate discontinuation of quinidine administration.[7] Many horses that experience quinidine toxicity do not convert during that treatment attempt. The rapid quinidine-induced supraventricular tachycardia is an idiosyncratic reaction to quinidine and, although not associated with quinidine toxicity, should prompt immediate intervention if sustained at high heart rates > 100 bpm. The detection of large number of VA should also prompt discontinuation of quinidine administration. Antiarrhythmic drugs should be administered if arrhythmias persist after discontinuation of quinidine or if arrhythmias are life threatening. Sudden cardiac death is most likely caused by VA deteriorating to ventricular fibrillation.

TREATMENT OF QUINDINE-INDUCED ARRHYTHMIAS

1. **Don't panic!**
2. Determine if arrhythmia is supraventricular or ventricular.
3. Obtain blood pressure, if possible.
4. Discontinue quinidine administration.

If arrhythmia is supraventricular

If rate is sustained > 100 bpm administer digoxin (0.0022 mg/kg IV) or 0.0011 mg/kg PO or diltiazem (125mg/kg IV slowly as a bolus, monitoring blood pressure).

If rate is sustained > 150 bpm or pressures are poor, administer digoxin (0.0022 mg/kg IV) or 0.0011 mg/kg PO or diltiazem (125mg/kg IV slowly as a bolus, monitoring blood pressure). Can repeat if necessary. Also administer $NaHCO_3$ (1 mEq/kg IV).

If rate still high or pressures are poor

1. Administer propranolol (0.03 mg/kg IV) to slow rate (do not begin not until diltiazem has been metabolized).
2. Administer phenylephrine (1 µg/kg/min) to effect up to a total dose of 0.01 mg/kg to improve blood pressure (will also slow heart rate).
3. Administer verapamil (0.025-0.05 mg/kg IV every 30 minutes; can repeat up to 0.2 mg/kg total dose).

If arrhythmia is ventricular

1. Administer MgSO4 (1-2.5 g/450 kg/minute IV to effect or up to 25 gms/450 kg). Preferable to administer rapidly IV but can bolus if necessary. ***Treatment of choice for quinidine-induced torsades de pointes (wide QRS tachycardia).***
2. Administer lidocaine hydrochloride (20-50 µg/kg/min or 0.25-0.5 mg/kg very slowly IV; can repeat in 5- 10 minutes).
3. Administer procainamide (1 mg/kg/min to maximum of 20 mg/kg).
4. Administer propafenone (0.5-1 mg/kg IV in 5% dextrose slowly IV over 5 to 8 minutes).
5. Administer bretylium (3-5 mg/kg IV; can repeat up to 10 mg/kg total dose).

PROGNOSIS

There is no relationship between the method of cardioversion and success of cardioversion or recurrence of AF. The majority of horses (82 - 89%) with AF, including numerous horses with structural heart disease, convert to NSR with the administration of quinidine.[7,12,13] Horses with shorter duration of AF and "lone" AF are more likely to convert to NSR.[12] The duration of AF and the presence of atrial enlargement or mitral regurgitation (MR) are important factors that increase the likelihood of horses experiencing a recurrence of AF.[12,14]

CONCLUSION

QG and QS remain viable choices for successful cardioversion of AF when used with continuous ECG monitoring and monitoring of therapeutic quinidine plasma concentrations. In the absence of the ability to monitor plasma quinidine concentrations, quinidine should be administered judiciously to avoid quinidine toxicity.

REFERENCES

1. Kroneman J. Treatment of atrial fibrillation in the horse with digitalis tincture and quinidine sulfate. Tijdschr Diergeneesk 1966;91:223-229.
2. Slack J, Boston RC, Soma LR, et al. Occurrence of cardiac arrhythmias in Standardbred racehorses. Equine Veterinary Journal 2014:n/a-n/a.
3. Holmes JR, Henigan M, Williams RB, et al. Paroxysmal atrial fibrillation in racehorses. Equine Vet J 1986;18:37-42.
4. Amada A, Kurita H. Five cases of paroxysmal atrial fibrillation in the racehorse. Experimental Reports of Equine Health Laboratory 1975;12:89-100.
5. Reef VB, Bonagura J, Buhl R, et al. Recommendations for Management of Equine Athletes with Cardiovascular Abnormalities. Journal of Veterinary Internal Medicine 2014;28:749-761.
6. Marr CM, Reef VB, Reimer JM, et al. An echocardiographic study of atrial fibrillation in horses: before and after conversion to sinus rhythm. J Vet Intern Med 1995;9:336-340.
7. Reef VB, Reimer JM, Spencer PA. Treatment of atrial fibrillation in horses: new perspectives. J Vet Intern Med 1995;9:57-67.
8. Bouckaert S, Voorspoels J, Vandenbossche G, et al. Effect of drug formulation and feeding on the pharmacokinetics of orally administered quinidine in the horse. J Vet Pharmacol Ther 1994;17:275-278.
9. Muir WW, 3rd, Reed SM, McGuirk SM. Treatment of atrial fibrillation in horses by intravenous administration of quinidine. J Am Vet Med Assoc 1990;197:1607-1610.
10. McGuirk SM, Muir WW, Sams RA. Pharmacokinetic analysis of intravenously and orally administered quinidine in horses. Am J Vet Res 1981;42:938-942.
11. Parraga ME, Kittleson MD, Drake CM. Quinidine administration increases steady state serum digoxin concentration in horses. Equine Vet J Suppl 1995:114-119.
12. Reef VB, Levitan CW, Spencer PA. Factors affecting prognosis and conversion in equine atrial fibrillation. J Vet Intern Med 1988;2:1-6.
13. Deem DA, Fregin GF. Atrial fibrillation in horses: a review of 106 clinical cases, with consideration of prevalence, clinical signs, and prognosis. J Am Vet Med Assoc 1982;180:261-265.
14. De Clercq D, Decloedt A, Sys SU, et al. Atrial Fibrillation Cycle Length and Atrial Size in Horses with and without Recurrence of Atrial Fibrillation after Electrical Cardioversion. Journal of Veterinary Internal Medicine 2014;28:624-629.

EQUINE BIG DATA – YES WE CAN!

Marco Duz, MedVet, MVM(Res), DECEIM, MRCVS
Sutton Bonington, United Kingdom

INTRODUCTION

In the computer era the use of practice management software systems (PMSS) has helped many veterinary practices to improve management of accounting, drug dispensary as well as client and patient electronic medical records (EMR). The amount of data digitally stored in PMSS is vast and is suitable for research to investigate relevant clinical topics to ultimately improve clinical practice. "Big data" is usually defined as a dataset that is too large to be evaluated by traditional manual processes and certainly this term applies to multicentre EMR datasets including hundreds of thousands of animals, which would be better representative of the overall equine population. It is intuitive how such a large amount of information is not easily evaluated manually and would require automated analysis tools.

A wide variety of PMSS are available today and veterinary practices have the opportunity to choose the PMSS that best meets their specific needs. This variability creates a great challenge to research as these systems often store data of variable quality in different formats. Nevertheless, the number of projects developed to collect data from veterinary practice to look at large EMR datasets both retrospectively as well as prospectively is on the increase[1-7]. Much of the data has been obtained from first opinion veterinary practice for disease surveillance but also to evaluate disease prevalence, risk factors and survival.

Studies that collect data prospectively, mostly focused on small animal practice, require commitment by participating practices to produce good quality data. In the United Kingdom two main systems have been implemented. The first, SAVSNET (Small Animal Veterinary Surveillance Network), is a point and click application that allows a reasonably quick classification by syndrome/body system, while detailed information is obtained only in a randomised subsample of cases[4]. The second approach is based on the use of a predetermined coding system, VeNom (Veterinary Nomenclature), which allows the precise classification of each syndrome or diagnosis. The VeNom coding system contains several hundreds of possible codes and requires clinicians to commit to learn this large coding system in order to provide accurate/complete data. Further, the use of a restricted coding system might limit freedom of expression and should ideally be combined with free-text mining techniques.

Natural language processing (NLP) is an area of computer science dedicated to study the interactions between computers and human natural language. This discipline started to develop in the late 1960s and focuses on enabling computers to derive meaning from human natural language[8]. Text mining is a component of NLP and aims to find or create a structure from free-text to predispose it for further automated analytic processes[9-11]. The final aim is providing structure to an otherwise fundamentally unstructured ensemble of words and sentences[9]. This is achieved by classification of sentences by the combination of words that form those sentences, according to pre-determined user-defined categories. The classification of these free-text EMR allows evaluation of the relationships between those categories[12-13]. Free-text mining techniques has the advantage that they do not require any co-operation by clinicians other than fill the EMR for each of their patient, as they are be normally required to do in daily practice. On the other hand, data analysis becomes more laborious and technically complicated as data is less structured.

Techniques for mining free-text EMR have been described in the veterinary literature[13-14] and also in equine science[5-7] using a purpose-built commercially available software package: SimStat/WordStat by Provalis Research. An early study described the reasons for retirement in a population of 3727 horses from the Hong-Kong Jockey Club[7], whilst another investigated prevalence and risk factors of cervical vertebral stenotic myelopathy in a 1289 foals born at a thoroughbred breeding farm[5]. The study by Anholt and colleagues[14] illustrates its use in small animal practice in Canada to study the frequency of antimicrobial usage in pets with diarrhoea, demonstrating the potential of this methodology also for disease surveillance. While these studies describe briefly the technique used to extract the information of interest from their respective free-text EMR, none actually evaluated in detail sensitivity and specificity of the methodology. Technique validation is extremely important as for any assay or analysis in research and this also applies to text mining of EMR. A recent study, dedicated to the validation of the use of WordStat for text mining veterinary EMR against manual classification[13]. This study describes in details how analysis can be done following software manufacturer's recommendations using functions and syntax inbuilt to the Software. In that study, the dataset was imported as a comma separated value file format and included 25,000 records from 12 first opinion small animal practices in Canada[13]. The dataset included a column for the anonymised identification number of each animal, patient species, breed, gender and date of birth plus a rough indication of the owner post-code area and also a column with the free-text clinical notes (including chief complaint, history, physical examination findings, procedures performed and relative results) added by the attending clinicians. The dataset did not include any standardised diagnostic coding or fixed vocabulary. The analysis was performed to automatically characterise the text contained within the free-text notes and resulted in high sensitivity (87%) and excellent specificity (99%). These authors elected to sacrifice the identification of some true positive in order to maximise specificity. This study demonstrated the usefulness of WordStat to examine large EMR datasets, but while specificity was excellent, sensitivity could be improved further with a substantial change in the methodology to identify false positive cases so that no significant amount of true positive cases has to be misclassified.

MATERIALS AND METHODS

Data

A "big dataset" of EMR from equine practice from North America was obtained from a PMSS company, which had permission from some of its client practices to share their anonymised data for research purposes.

Method validation

The free-text EMRs were analysed using WordStat by Provalis Research, with a technique modified from that of Anholt and colleagues (2014). All subsequent data manipulation and analysis was performed using R v3.0.0 (R Development Core Team). Briefly, the software provides a term frequency table including a list of all words present in the dataset, which is used to identify all terms that might identify a specific group (eg. "condition A") that are used to create an "inclusion dictionary". This process maximises analysis sensitivity by including all misspellings and abbreviations. Once the inclusion dictionary has been created, cases (rows of data) including the terms in the "condition A" group are identified by their row number. The obtained "inclusive condition A" dataset is subsequently mined to identify combinations of terms that identify false positive cases (eg. "the horse did not have condition A, but had condition B"), which are subsequently added in an "exclusion dictionary". This process is also extremely important as it determines analysis specificity. The cases identified from the exclusion dictionary are then removed from the "inclusive condition A" dataset using R and the result is a dataset that includes only "true condition A" cases. This analysis process has undergone a validation process, comparing its results against manual classification in a sample of lifelong EMRs of 300 animals. The automated identification of conditions with expected mid-high (NSAIDs administration and colic) and low prevalence (renal failure and right dorsal colitis) was compared to manual classification.

Caseload of equine veterinary practice in North America

Following validation the technique has been used to describe the nature of first opinion equine practice in North America. All dataset entries were classified as belonging to one of 19 categories (bandaging, chemotherapy, cardiology, dentistry, dermatology, endocrinology, gastroenterology, haematology, hepatology, muscle disease, neurology, ophthalmology, orthopaedics, pre-purchase examinations, reproductive medicine, respiratory medicine, urology, vaccinations and wounds). Prevalence estimates and 95% confidence intervals were also calculated.

Equine Colic in North America

All horses that suffered at least one episode of abdominal discomfort were included in the colic group while horses with a medical record of >1-year duration and no diagnosis of colic were grouped as "control". The sub-group of horses in the colic group that underwent an exploratory laparotomy were identified and designated as the "laparotomy" group. The earliest episode of colic for each affected horse was identified. The prevalence of colic and laparotomy was determined stratified by year. Median survival was calculated for control, colic and laparotomy groups. The number of laparotomies as a proportion of the number of colic cases per year was compared using chi-square analysis.

NSAIDs toxicity in North America

The database was searched for records reporting the clinical sign diarrhoea and for NSAIDs, antimicrobials, corticosteroids, anthelmintic, laxative and intravenous fluid administration. Single and multiple conditional logistic regression was used to evaluate the relationship between diarrhoea and drug administration in the 1 to 2 weeks preceding the development of clinical signs. The best fitting logistic regression model was selected and explanatory variables' significance was set for $p<0.05$.

These studies were performed with the approval of the Ethics and Welfare Committee of the University of Glasgow, UK.

RESULTS

Data

The dataset included data from 9 veterinary groups including a total of 27 branches distributed throughout 16 states between the U.S.A. and Canada. The overall dataset consisted in a spreadsheet of 50 columns by 11,699,875 rows stored in the ".csv" format. The dataset included all the EMR of 312,634 equids in the system of participating practices. The mean number of rows of data for a single horse was 37.4 (range: 1 to 9263 rows of data). Data was collected by 309 veterinarians between 1994 and 2013. Each patient had a unique, anonymous, numeric identification, which allowed tracking through time. Part of the data was imported from other previous PMSS and might have resulted with some of the dates of data entry in the record being wrong (eg. horse vaccinated on 01/01/1900), only cases with a valid date were included. This resulted in a total subpopulation of 225,777 animals.

Method validation

The randomly selected 300 horses yielded a total of 17,561 rows of data, which were then classified manually (~100 man-hours) and automatically. Errors (39/0.22%) were more common with manual classification than with the automated process (12/0.07%). However, assuming that manual classification represented "gold-standard" the sensitivity and specificity of automated analysis with WordStat were both greater than 99.7% for all four diagnosis/treatment categories.

Caseload of equine veterinary practice in North America

The nature of the caseload of equine practice has not been described before on a large population of equids including hundreds of thousands of animals. The most prevalent examinations were orthopaedic (35.5% of cases), followed by dental (20.1%), dermatological (17.5%) and of the alimentary tract (10.1%). Prevalence of reproductive, respiratory, ophthalmological, neurological, muscular and endocrine conditions ranged between 8.5% and 3.4%. Cardiologic, hepatic and urinary examinations were the least prevalent (1.1-0.5%). Haematopoetic disorders had a high prevalence (28.8%) as the dictionary included generic terms such as leukopaenia, neutrophilia etc. that often related to primary conditions in other categories, but suggesting that testing had been performed. Each year 5.9% of horses suffered a wound/laceration requiring veterinary care, while 9.0% of horses were likely to require some form of bandage application. Vaccinations were recorded with reference to only 35% of horses in each year. Pre-purchase examinations accounted for 2.5% of all records in each year.

Equine colic in North America

Medical records of >1-year duration were extracted for 71,186 horses out of the 225,777 horses in the dataset that had a correct date of data entry. A total of 21,386 horses suffered at least one episode of colic. The mean yearly prevalence of colic between 2007 and 2012 was 5.2% (range: 4.5-5.7%) of all cases under veterinary care in each year. Among all colic cases, 1799 underwent laparotomy (8.5%). Between 2007-2012, the mean proportion of colic cases undergoing laparotomy was 8.9% (n=1575 of 17738 colic episodes). The proportion of colic cases undergoing exploratory laparotomy was significantly greater in 2008 (8.7%), 2009 (9.6%), 2010 (10.5%), 2011 (11.4%) and 2012 (12.2%) than in 2007 (6.0%) ($p < 0.001$). A significantly greater proportion of colic cases underwent exploratory laparotomy in 2011 (11.4%) and 2012 (12.2%) than in 2008 (8.7%) and 2009 (9.6%) ($p<0.05$). Colic-to-death interval was available in 4019 horses of the colic group, of which 52.1% (n=2055) died within the first 24 hours following the initial veterinary evaluation of abdominal pain. Excluding horses that died within 24 hours, the median survival was 54 days (range: 1 day to 7.6 years). One year after the first colic episode 72.5% of 10,588 horses for which follow-up data was available were still alive. For each of the following five years survival was >90%. For non-survivors in the laparotomy group, the median survival was two days (eight days after excluding those dead within 24 hours). Horses in the colic group were significantly more likely to die during any of the five years follow-up than the control population ($p<0.001$).

NSAIDs usage in North America

A total of 5618 cases of diarrhoea were identified and prevalence (±95%CIs) was determined (2.5%; 95%CIs:2.4-2.6). The results of multiple conditional logistic regression suggested that horses receiving flunixin meglumine in the preceding 7 days were 2.5 times more likely to develop diarrhoea (95%CIs:2.1-3.0; $p<0.001$), those receiving phenylbutazone in the preceding 2 weeks were 0.7 times as likely to develop diarrhoea (95%CIs:0.5-0.8; $p<0.001$), those receiving antimicrobials in the preceding week were 1.7 times more likely to develop diarrhoea (95%CIs:1.5-2.0; $p<0.001$), those receiving corticosteroids in the preceding 2 weeks were 0.36 times as likely to develop diarrhoea (95%CIs:0.28-0.46; $p<0.001$), those receiving anthelmintic drugs in the preceding 2 weeks were 0.6 times as likely to develop diarrhoea (95%CIs:0.5-0.8; $p<0.001$), those receiving a laxative treatment in the preceding 48 hours were 4.1 times more likely to develop diarrhoea (95%CIs:3.0-5.6; $p<0.001$) and those receiving fluids intravenously in the preceding 2 days were 2.1 times more likely to develop diarrhoea (95%CIs:1.7-2.6; $p<0.001$). There was no significant relationship between administration of firocoxib ($p=0.73$) and practice ($p>0.7$) and the development of diarrhoea.

DISCUSSION

These studies illustrate how automated analysis can assist clinical research on datasets that are too large to be evaluated manually. The most time consuming part of the process is dictionary definition and evaluation to ensure that all positive terms are detected with a comprehensive inclusion dictionary. Equally important and time consuming is the effort to ensure that all false positive terms are also identified and removed with a detailed exclusion dictionary. This process requires a reasonable amount of time and effort as well as technical expertise but its convenience is proportional to the size of the dataset to be analysed.

The studies here discussed show just a small example of the possible applications for this type of research. Describing the caseload of veterinary practice provides an objective view of how veterinary practice and business is evolving, but also could guide academic institutions to adapt the undergraduate curriculum emphasizing disciplines most relevant to first opinion practice and ensuring that new graduates possess the most relevant day-one competences. The study of colic also shows how data can be used to estimate disease prevalence or for survival analysis, but also to evaluate the effect that an economic recession might have had on practice's caseload and revenue. This is in line with the concept of applying evidence-based principles not only to veterinary practice but also to improve practice management and business strategy. There should be a direct interest by veterinary practices to provide data of good quality in order to confidentially compare to the national/regional average. This could allow practices to identify areas where they might consider changing their clinical practice or to direct future investments for business expansion.

Finally the study on diarrhoea also represents an example of how the data that could be used to evaluate the relationship between drug administration and side effects on a large population, accounting for a large number of confounding factors, without the limitation of the study becoming underpowered.

In conclusion, "Yes, we can big data" in equine practice and we should! Fundamentally, this paper has the ambitious goal to raise awareness and interest among equine veterinarians. Not only this type of research is now doable but also has the potential to help improve clinical practice as well as practices' business and revenue. While using clinical data retrospectively without client consent is an ethical hot topic in smaller studies where results might also refer to a specific individual, data is also anonymous with larger datasets and results do not focus on single patients. One might also argue that choosing not to participate to large multicentre studies of this kind could prevent research advancement and new knowledge discovery.

This paper hopes to stimulate a discussion to convince first opinion as well as referral private practices and academic institutions of the need to share EMR data to create a national and international database of anonymous EMR that could be accessible by any participating research group or institution. The final goal is to produce research that is more universal than simply applicable to a specific environment/population. The basic principle is very simple: "together we are stronger", together we can obtain more and better data and we should all participate in this collective effort to promote the advancement of equine veterinary science.

REFERENCES

1. O'Neill D, et al. In: **Proceedings: 13th ISVEE**; 2012:59.
2. Mattin M, et al. **Vet Rec** 2014;174:349–349.

3. Radford AD, et al. **Vet Rec** 2011;169:310–310.
4. Jones PH, et al. **Vet J** 2014; 201(3):412-418.
5. Oswald J, et al. **Vet Rec** 2010;166:82–83.
6. Lam K, et al. **Vet Rec** 2007;161:547–551.
7. Lam K, et al. **Eq Vet J** 2007;39(2):143–148.
8. Sager N. **Adv Comput** 1967;8:153–188.
9. Heinze DT, et al. **Proc AMIA Symp** 2001:254–258.
10. Mooney RJ, Bunescu R. **ACM SIGKDD Explor Newsl** 2005;7:3–10.
11. Meystre SM, et al. **Yearb Med Inform** 2008;128–144.
12. Erstad TL, **J Healthc Inf Manag** 2003;17:51–57.
13. Anholt RM, et al. **Prev Vet Med** 2014;113:417–422.
14. Anholt RM, et al. **PLoS ONE** 2014;9(7):e103190.

OVERVIEW OF DRUG COMPOUNDING

Patricia M. Dowling, DVM, MSc, DACVIM (Large Animal Internal Medicine), DACVCP
Saskatoon, SK, Canada

INTRODUCTION

Compounding of veterinary drugs is both necessary and beneficial for the treatment of animals. In October 1994, the Animal Drug Use Clarification Act of 1994 (S 340) was passed as a federal law to codify extralabel drug use (ELDU) in animals by veterinarians. AMDUCA amended the Federal Food, Drug, and Cosmetic Act to legalize ELDU under a valid veterinary-client – patient relationship (VCPR) and other specific conditions according to Food and Drug Administration (FDA) regulations, creating the professional flexibility that veterinarians need to adequately treat animals. Under AMDUCA, ELDU is limited to drug treatment when the health of the animal(s) is threatened or suffering or death may result from failure to treat. Therefore, ELDU should be the exception and not the rule in prescribing and AMDUCA does not encourage ELDU as a way to circumvent the drug approval process. AMDUCA requires that only FDA approved human or veterinary drugs be used can be used in an extralabel manner. The use of compounded drugs is permitted under AMDUCA only when there is not a suitable approved product and compounded drug products can be made only from approved human or animal drug products. Compounded drugs are not the same as generic drugs. Generic drugs are FDA-approved and have demonstrated bioequivalence to the "pioneer brand name" drug product. An approved drug product (brand name or generic) can be identified by the NADA (new animal drug approval number) on its label or by cross-checking with the FDA Green Book of Approved Animal Drug Products. For specific information on compounding of veterinary drugs, see the FDA's policy guidelines in Sec. 608.400 - Compounding of Drugs for Use in Animals. Unlike regulations that establish "laws", guidance documents are not binding on the FDA and enforcement is therefore exercised at the FDA's discretion.

In Canada, approved drugs can be identified by a "Drug Identification Number". Unlike the United States, where pharmaceutical compounding is federally codified, veterinary compounding in Canada is guided by Health Canada policy, provincial veterinary and pharmacy regulations, and guidelines from the National Association of Pharmacy Regulatory Authorities (NAPRA) and the Canadian Veterinary Medical Association (CVMA).

COMPOUNDING DONE RIGHT

The first step in valid veterinary compounding is the establishment of a valid veterinarian-client-patient relationship. The veterinarian must establish that there is no approved animal or human drug that, when used on label or extralabel according to AMDUCA, can be used to treat the diagnosed condition and that the health of the patient is threatened, or suffering or death may result from failure to treat with the compounded product. The compounding must be performed by a licensed pharmacist upon the prescription of a veterinarian or by a veterinarian if allowed by their state's pharmacy law. The compounded product must be safe and effective and the compounding operation must be consistent with providing small quantities of product for very specific patient needs.

Preparations can only be compounded from FDA-approved animal or human drugs. If an approved veterinary drug can be used for the compounding, it is not permissible to compound from an approved human drug. The FDA regulations do not permit compounding from bulk (active pharmaceutical ingredient, API) chemicals. Veterinarians may occasionally face situations where they need to treat conditions for which there are no FDA-approved products is available for compounding (e.g., pergolide for horses before Prascend® was approved, cisapride for cats). Although the FDA considers compounding from APIs a violation of the Federal Food, Drug and Cosmetic Act, it acknowledges the need for such compounding within certain areas of veterinary practice and exercises enforcement discretion in such situations.

COMPOUNDING DONE WRONG

When drugs are compounded without adherence to current good compounding practices, veterinary patients may be harmed, the public health may be endangered and the prescribing veterinarian may be liable. Because compounded drugs are typically used in small numbers of patients, if things go wrong it usually does not draw much attention. But recent compounding mishaps were significant enough to draw international attention. In 2009, twenty-one Venezuelan polo horses died acutely in Florida after Franck's Pharmacy, a veterinary compounding pharmacy, incorrectly prepared a vitamin-and-mineral cocktail that was injected into the horses prior to their polo match. Due to a mathematical error (the easily misplaced decimal point) in the concentration of selenium, the solution was too potent and caused fatal selenium toxicity. In October 2012, an outbreak of fungal meningitis, paraspinal/spinal infections and peripheral joint infections occurred in the United States. The US Centers for Disease Control and Prevention traced the outbreak to contamination with a black mold called *Exserohilum rostratum* in three lots of preservative-free methylprednisolone acetate solution used for steroid injections. The FDA-approved methylprednisolone acetate (Depo-Medrol®) was sold by Pfizer and two generics companies, but since the compounder's version did not contain preservatives, it sidestepped the regulatory process with tragic results. Doses from these three lots had been distributed to 75 medical facilities in 23 states, and administered to approximately 14,000 patients. As of May, 2013, 741 people have had infections from the contaminated product and 51 people have died.

Compounding Versus Manufacturing

Unless conditions set forth in 21 CFR 530.13(b) are met, the compounding of a new animal drug from an approved human or animal drug results in an adulterated new animal drug. Traditional compounding was limited to a pharmacist or a physician meeting the therapeutic needs of a specific patient. Once disconnected from individual patients, compounding essentially becomes drug manufacturing. Often compounded products are marketed as cheaper alternatives to the approved drug. All compounding guidelines state that cost is not a defensible reason for prescribing a compounded drug. But in actuality, price is the main reason for using compounded drug products when approved products are readily available. The FDA has taken action against some pharmacies for large scale compounding from bulk APIs and aggressive marketing of such products.[1]

Problematic Packaging

Guidelines state that a compounded product should not be made to look like an approved product. Yet I frequently see packaging that looks very much like an approved veterinary product and "compounded" is often not indicated on the box or the bottle. In Canada, an approved human or veterinary drug product has a DIN (drug identification number). One compounding pharmacy puts a "CIN" number on their products, but it has no meaning other than to the compounder. I suppose it means "Canadian Identification Number", but to me it means it's a sin to sell this stuff!

Problematic Product

Compounding is legitimate if an approved product is available but the appropriate method for dosing or the drug concentration is not suitable and a practical alternative does not exist. One of the most frequent examples that I see violating this requirement is compounded oral liquid phenylbutazone. Considering that there are approved tablet, paste and flavoured granule formations, it is very difficult to make the case that a liquid formulation is necessary to save equine lives!

There are numerous publications documenting the poor quality of compounded drug products.[2,3] Many products contain less than the stated drug concentration, but some have been to contain more drug than stated.[4,5] So there is a much greater chance of under dosing (leading to therapeutic failure) or overdosing (leading to toxicity) than with an approved drug product. In performance horses, there is the additional issue of poor quality compounded drugs resulting in "positive" tests during competition, resulting in loss of purse or prize money, suspensions from competition and the public humiliation of being identified as violating competition drug rules.

Problematic Prescription Labeling

State and provincial pharmacy regulations are explicit regarding the information required on a veterinary prescription. I've seen numerous instanced of mislabelling on compounded products and sometimes they have had no prescription labelling at all. The expiration date of an approved product should be based on known stability data, but stability testing is rarely done by compounding pharmacies for their products. If no stability data exists, products compounded for re-dispensing should have a "beyond use date" identical to the last date the treatment will be administered, as per the duration of the prescription, with a maximum dating of 180 days. A number of publications have demonstrated problems with stability of compounded veterinary drug products such as pergolide and doxcycline.[6,7]

PRESCRIBER BEWARE

If you choose to prescribe compounded drugs, you need to counsel your client regarding potential adverse reactions, including therapeutic failure, and the potential for unintended human or animal exposure to the drug. You are obligated to inform your client that the compounded preparation has not been evaluated by the FDA (or Health Canada) for potency, purity, stability, efficacy or safety, and client consent should be obtained in writing.

As a practicing veterinarian, if you choose to prescribe or dispense a compounded drug, you are responsible for its quality, safety, efficacy and potency and for following pharmacy regulations for proper labeling. You should report suspected adverse events including therapeutic failure and quality defects involving compounded drugs to the compounding pharmacist, the state or provincial Board of Pharmacy and the FDA Center for Veterinary Medicine or Health Canada's Veterinary Drug Directorate. Should an adverse drug reaction, toxicity or lack of therapeutic effect occur from a poorly prepared or labeled compounded product, you are responsible for the consequences, which can include client complaints and disciplinary action by your state or provincial veterinary medical regulators. Pharmacies are not required to carry product liability insurance, so you should be sure that your liability insurance will cover you. This may be difficult as there are no "registered label indications" on which an insurance company can base their decision on whether or not a product was used in accordance with best practices. If the manufacture of the compounded product violates federal regulations (e.g. compounded from bulk APIs), your liability insurance will not protect you. So maybe that approved drug product isn't so expensive after all!

COMPOUNDING GUIDELINES

American Veterinary Medical Association Compounding Policy (https://www.avma.org/KB/Policies/Pages/Compounding.aspx)
FDA Green Book of Approved Animal Drug Products (www.fda.gov/AnimalVeterinary/Products/ApprovedAnimalDrugProducts/UCM042847)
Compounding of Drugs for Use in Animals (www.fda.gov/downloads/ICECI/ComplianceManuals/CompliancePolicyGuidanceManual/UCM200461.pdf)
USP Compounding Standards & Resources (http://www.usp.org/usp-healthcare-professionals/compounding)
Health Canada: Policy on Manufacturing and Compounding Drug Products in Canada (http://www.hc-sc.gc.ca/dhp-mps/compli-conform/gmp-bpf/docs/pol_0051-eng.php)
CVMA Guidelines for the Legitimate Use of Compounded Drugs in Veterinary Practice (http://www.canadianveterinarians.net/programs/national-issues.aspx#.UaeFTdj8_kc)
NAPRA Guidelines to Pharmacy Compounding (http://napra.ca/pages/Practice_Resources/guidelines_to_pharmacy_compounding.aspx)

REFERENCES

[1]Kuehn BM. **J Am Vet Med Assoc** 2004;225:993-994.
[2]Merritt AM, et. al. **Equine Vet J** 2003;35:691-695
[3]Nieto JE, et. al. **J Am Vet Med Assoc** 2002;221:1139–1143.
[4]Stanley S, et.al. **Proceedings, 49th Annual AAEP Convention** 2003;238-242.
[5]Davis JL, et. al. **J Am Vet Med Assoc** 2009;234:385-389.
[6]Papich MG, et. al. **J Am Vet Med Assoc** 2013;242:1674-1678.
[7]Thompson JA, et. al. **J Am Vet Med Assoc** 2011;239:842-849.

AVMA TASK FORCE ON COMPOUNDING

K. Gary Magdesian, DVM, DACVIM, DACVECC, DACVCP
Davis, CA

In 2014 the AVMA selected eight volunteer veterinarians to serve on a task force to advocate for federal legislation addressing veterinary compounding needs. The goal of the task force is to lead a national dialogue among veterinarians, pharmacists, drug manufacturers, and the FDA. The end goal is to ensure safe and effective veterinary compounding.

Members of the task force include: Drs. Lori Thompson (chair), Fred Gingrich, Kurt Sladky, Dawn Boothe, Colleen Currigan, Charles Lemme, Wanda West, and Gary Magdesian. These members represent a wide range of veterinary medicine sectors, including small animal, farm animal, equine, zoo and wildlife, and clinical pharmacology.

Task: Evaluate the need for legislation on veterinary compounding. Develop such legislative proposal regarding compounding pursuant to AVMA policies as guidance.

There are currently > 12,000 compounding pharmacies. More than 20% of the animal drug industry in 2009 consisted of compounded preparations, and this figure has likely increased. Some of the recognized problems with veterinary compounding include compounding from bulk (raw pharmaceutical ingredients), anticipatory compounding, variable quality, contamination, super-potency and sub-potency of compounded preparations, and mimicking of FDA-approved products. The FDA and federal courts have held that the federal drug laws prohibit compounding from bulk chemicals, because these compounds constitute unapproved new animal drugs. Veterinarians cannot guarantee the potency, purity, safety, or stability of these unapproved bulk substances. The legality of maintaining compounded medications for office dispensing and administration ("office stock") is another issue that requires resolution, as these are obtained without an individual patient prescription and are currently regarded as illegal.
Unapproved, marked drugs, such as potassium bromide (KBroVet) and atropine sulfate injection, are another area needing oversight. These drugs are manufactured, not compounded, and meet good manufacturing practices, but are problematic in terms of regulation because they are unapproved drugs.

Last year two lots of compounded pyrimethamine/toltrazuril preparations caused illness in 10 horses, with four of these having died or been euthanized, as a result of pyrimethamine overdose. In 2009 21 polo horses died as a result of a selenium overdose from a compounded preparation substitute for Biodyl, a product containing vitamins and minerals. In 2012 there was an outbreak of fungal meningitis in 751 human patients, with 64 deaths from fungal contamination of compounded methylprednisolone used for epidural injections. As a result of this outbreak, the Drug Quality and Security Act (DQSA) was passed by Congress and signed by the President in 2013, but this does not address veterinary compounding.[1] The DQSA addresses compounding for human medicine, but did not change the laws governing veterinary compounding.[1] In January of 2014, Congress asked the Government Accountability Officer to prepare a report on the compounding of drugs for use in animals.

The AVMA currently has 3 policies regarding veterinary compounding, last updated and approved in June 2013:[2-4]

1. *Veterinary Compounding*[2]: This policy addresses the need for veterinary compounding and recommends that veterinarians be allowed to keep sufficient quantities of compounded medications for urgent/emergency use. It also recommends clear labeling and oversight, such as United States Pharmacopoeia (USP) standards.
   This policy recommends the following actions in order to minimize risks of adverse events associated with compounded preparations:[2]

   i. The decision to use a compounded preparation should be veterinarian driven within a veterinarian-client-patient relationship, rather than pharmacist driven. The veterinarian should make that decision utilizing evidence-based medicine whenever available.
   ii. Compounding should be implemented in compliance with the Animal Medicinal Drug Use Clarification Act (AMDUCA) as well as the FDA Compliance Policy Guide 608.400 ("Compounding of Drugs for Use in Animals"). Use of compounded preparations in food animals may have food safety concerns that preclude their use unless information exits about withdrawal times to assure avoidance of violative drug residues.
   iii. Use of compounded preparation should be limited to:
      a. Individual patients for which no other method or route of drug delivery is practical
      b. Drugs for which safety, efficacy, and stability have been demonstrated in the compounded form in the target species, or
      c. Disease conditions for which a quantifiable response to therapy can be monitored or when drug concentration can be measured.
   iv. Use of a compounded preparation should be accompanied by the same precautions followed when using an approved drug. These include counseling of the client regarding potential adverse reactions, including therapeutic failure. Attention should be given to the potential for unintended human or animal exposure to the drug. Clients should also be informed that the compounded preparation has not been evaluated by the FDA for potency, purity, stability, efficacy or safety, and client consent for its use should be obtained.
      a. Veterinarians should report suspected adverse events including therapeutic failure as well as quality defects involving the compounded preparation to the compounding pharmacist, the State Board of Pharmacy and the FDA Center for Veterinary Medicine. Pharmacists should instruct pet owners to contact both the prescribing veterinarian and pharmacist immediately in the event of an adverse event, including therapeutic failure, and quality defects.

b. Veterinarians should comply with all aspects of the federal extra label drug use regulations. This includes record-keeping and labeling requirements. The labeled compounded preparation should also indicate that it is not FDA approved.

This policy also asserts that veterinarians should be able to legally maintain sufficient quantities of compounded preparations in their office ("office use") for urgent or emergency situations. It also recommends that the compounding processes follow the guidelines of the USP monographs for good compounding practices.[2]

2. *Compounding from Unapproved (Bulk) Substances in Food Animals*[3]: This policy addresses the need for compounded medications for certain uses in food animals. These uses are specifically outlined in the policy as limited to poison antidotes and compounds used for euthanasia or depopulation that are not approved or commercially available. If adequate scientific information is not available to determine a withdrawal time, the compound cannot be used in a food-producing animal or the treated animal cannot enter the food supply. It also emphasizes that a Veterinarian-Client-Patient Relationship (VCRP) is required before compounding preparations can be prescribed.[3]

3. *Compounding from Unapproved (Bulk) Substances in Non-Food Animals*[4]: While the FDA prohibits compounding from raw ingredients, this policy recognizes the medical necessity of compounded medications in certain situations. It also emphasizes that a VCPR is required. The AVMA policy states that there are 3 general circumstances in which compounding from bulk substances be medically necessary:[4]

i. The approved product is no longer commercially available.

ii. The needed compounded preparation cannot be made from the approved product (e.g. for ophthalmic use).

iii. There is no approved product from which to compound the needed preparation (e.g. cisapride for feline megacolon).

In addition, there should be some expectation of safety and effectiveness of the compounded preparation made from bulk ingredients. There should be some means to monitor the animal for positive or negative results, and there should be no other viable choice to adequately treat the animal.

REFERENCES

[1] Public Law 113-54, Drug Quality and Security Act. 113th Congress. Nov 27, 2013. [2]Veterinary Compounding. American Veterinary Medical Association, 2015. https://www.avma.org/KB/Policies/Pages/Compounding.aspx. [3]Compounding from Unapproved (Bulk) Substances in Food Animals. American Veterinary Medical Association, 2015. https://www.avma.org/KB/Policies/Pages/Compounding-from-Unapproved-Bulk-Substances-in-Food-Animals.aspx. [4]Compounding from Unapproved (Bulk) Substances in Non-Food Animals. American Veterinary Medical Association, 2015. https://www.avma.org/KB/Policies/Pages/Compounding-from-Unapproved-Bulk-Substances-in-Non-Food-Animals.aspx

USP STANDARDS, PHARMACY LAW AND THE DVM

Wilson Gwin, RPh
West Lafayette, IN

Benefits and Risks of Compounded Preparations

The vast physiological differences between species treated in veterinary medicine coupled with differences within species create the necessity for a widely diverse array of dosing options. Historically, this need has not, and in reality, cannot be met by the drug manufacturing community. This leads to gaps in treatment options. Simple economics prohibit the ability for commercial drug manufacturers to deliver such a demand. Pharmaceutical compounding offers a clinician resources to fill in the gaps. Pharmaceutical compounding predates large scale manufacturing by centuries, even millennia. Egyptian artifacts display complex descriptions as to how certain remedies were to be prepared. From the earliest, preparers understood the need for reproducibility, or standardization to assure quality of the final product, therefore these "compounders" generated documents.

Compounding allows a clinician to choose dosing forms that are custom designed for their patient's need. Compounding allows for treatment options that may be temporarily, or permanently unavailable from federally licensed manufacturers. Compounding offers a clinician the opportunity to optimize care for their patient. However, compounding is not the panacea for pharmaceutical utilization. When done incorrectly, or inappropriately, compounding can result in harm to the patient. We've all seen the headlines of Polo Ponies killed, and fungal contamination of compounded products intended for spinal injection. While there is credible argument that the type of compounding that led to these tragedies is not the type of compounding that was intended by our laws, this is beyond the scope of this program but the reality is that there is compounding as defined under current law occurring daily that fails to achieve the goal of benefitting the patient. There are various reasons for this. Remember the Egyptians? They recorded guidelines, recipes, or formulas to assure a standard of reproducibility. The knowledge and ability to prepare a suspension is not enough. The compounder must understand that stability of the final product as well as the bio-availability of the final product along with the accuracy of potency are vital. The compounder must also know the quality of the starting "ingredients". The compounder must understand the chemistry behind a formulation. How then, does a compounder know what is right? How does the clinician know they can trust that their patient will receive the treatment intended? There are tools in place to assure that the compounder and the clinician are equipped with resources to deliver the best standard of care to our patients.

USP Standards

The United States Pharmacopeia (USP) is an organization that dates back to the 1820 when the USP was founded. In 1906, the Federal Food and Drugs Act passed, officially recognizing USP and National Formulary (NF) standards for strength, quality, and purity. In 1938, the Federal Food, Drug, and Cosmetic Act passed. The FDA now recognized USP and NF standards for strength, quality, purity, packaging, and labeling and enforced those standards. From the early days of our country, the USP has grown to be global in nature, with facilities in China, Switzerland, Ghana, Brazil and India. More than 140 countries adopt USP standards and many countries are incorporating USP standards into law much like the United States. The United States Federal Government has clearly mandated that USP standards establish criteria by which pharmacy practice must adhere to. Most state Boards of Pharmacy incorporate USP standards automatically into their State Pharmacy Practice Laws. USP standards are established for drugs, biologics, compounded preparations and nomenclature. The purpose of USP standards is to establish parameters by which products are measured and compared in order to protect everyone, including non-human patients, as it is very important to note that none of USP's standards define patients as "human-only". The USP makes possible the level of safety and care enjoyed by our modern society. Quoting from the USP website, the USP mission is, "To improve global health through public standards and related programs that help ensure the quality, safety, and benefit of medicines and foods." A wealth of information can be found at www.usp.org. The USP publishes a compendia named The United States Pharmacopeia-National Formulary, (USP-NF). The USP-NF is composed of several components. The USP contains monographs for drug products, substances, and preparation. Monographs for excipients are found in the NF. A monograph contains information that establishes standards for comparison. Within the USP are General Chapters that address various topics, including consistent preparation and quality control, minimizing risk, personnel training, detailed documentation, engineering controls, quality assurance and testing. USP also publishes a compendium specific to compounding, the *USP Compounding Compendium* offers compounding practitioners convenient access to all compounding-related general chapters from the United States Pharmacopeia–National Formulary (USP–NF), that are enforceable under Federal law in the United States. It is important to note, that the Food, Drug and Cosmetic Act considers USP General Chapters numbered below <1000> to be legally enforceable standards, and General Chapters numbered greater than <1000> to be informational in nature. Within USP's compounding standards, it is within 3 specific chapters numbered below <1000> that we find critical information that particularly impacts veterinary medicine today.

Pharmacy Law

Chapters 795, 797 and 800 provide direction for compounders. Specifically, chapter 795 addresses non-sterile compounding, such as topical ointments, capsules, suspensions, solutions, etc. Chapter 797 addresses sterile compounding such as compounded ophthalmic preps, injectable compounds and IV preparations. Chapter 800 address compounding with hazardous chemicals. In the introduction to chapter <800>, veterinarians and veterinary technicians are included in the list of healthcare professionals who may be exposed to hazardous drugs when preparing compounds. The intent of Chapter 800 is to equip handlers and preparers of hazardous drugs, with knowledge to minimize risks associated with these products. When one remembers that USP standards are recognized as standards by which enforcement of federal law is based, it is clear that the veterinarian understands and utilizes this resource.

The current definition of compounding provided by the FDA is, "compounding is a practice in which a licensed pharmacist, a licensed physician, or, in the case of an outsourcing facility, a person under the supervision of a licensed pharmacist, combines, mixes, or alters ingredients of a drug to create a medication tailored to the needs of an individual patient." As you can see, this does not mention veterinarians or animals as individual patients. However, the FDA Center for Veterinary Medicine provides the following guidance that addresses veterinary compounding. CPG Sec. 608.400 Compounding of Drugs for Use in Animals is a section of the Federal Code that establishes guidance to the staff of the FDA, veterinarians and pharmacists. Guidance documents are meant to describe the FDA's current thinking on enforcement of the law. Additionally, the Animal Medicinal Drug Use Clarification Act of 1994 (AMDUCA) permits veterinarians to prescribe extra label uses of certain approved new animal drugs and approved human drugs for animals under certain conditions. While some chapters of the USP do not address veterinary compounding specifically, it is clear that the FDA includes veterinary compounding in enforcement of the federal law. In November, 2013, the Drug Quality and Security Act was signed into law. This law was a direct result of tragedies associated with compounding. This law created within the FDCA, section 503B under which a compounder can become an "outsourcing facility". The law defines an outsourcing facility as "a facility at one geographic location or address that is engaged in the compounding of sterile drugs; has elected to register as an outsourcing facility; and complies with all of the requirements of section 503B". 503B does not include veterinary pharmacy, however, this is creating much discussion and debate and it would not be surprising if changes occur in the future that incorporate veterinary compounding. Veterinary compounding is crucial in providing the highest level of care to veterinary patients. In summation, the Federal Code provides laws and guidelines that enable veterinarians and pharmacists to compound medication for veterinary patients. The USP is designated by Federal Code as the source of standards that are the foundation of compounding.

Information to equip the DVM

The veterinarian and the compounding pharmacist, when considering treatment options for veterinary patients, must consider the following. Is there a Federally Approved product available? If not, should a compounded alternative be prepared? Is the compound stable, bioavailable, safe and effective? What is the shelf life or beyond use date (BUD)? Will this work and how can that be determined? These questions must be asked, and there must be communicative links between the prescriber and the compounder. Treatment failures associated with compounding include improper composition of active ingredients, lack of stability, quality of preparation, contamination, potency, vehicles, preservatives, dyes, flavorings, and even sweeteners. Veterinarians are taught little about the science behind compounding. Pharmacists are taught very little about veterinary pharmacotherapy, if any. While changes need to me made in curriculum design that takes years, even after recognizing the need. Where can the DVM turn today? One option is the USP. As discussed, the USP prepares monographs. A monograph describes standards for finished products, specific ingredients, or chemicals, their definition, packaging, storage requirements, labeling as well as testing parameters to validate quality, purity and strength. Compounded preparation monographs are documents that provide formulas, specific preparation direction, packaging and storage, stability and beyond use dates (BUD), and assay information for quality assurance. In 2010, there were only 5 official monographs for veterinary compounded preparations. Since 2010, 16 additional veterinary compounded preparation monographs were added. Currently, 14 more are pending. Obviously, there needs to be more monographs. Suggestions for future monographs can be submitted to the USP and the AVMA, the Society of Veterinary Hospital Pharmacists (SVHP) and the International Academy of Compounding Pharmacists (IACP) are three sources for suggestions. The USP is an atlas to be consulted during the journey, offering a roadmap to follow. There are journals that are dedicated to compounding and are excellent resources. Veterinary teaching hospital pharmacists specialize in veterinary pharmacy and are leaders addressing these issues. There are reputable companies that specialize in assisting healthcare professionals in analyzing and enabling compliance to these standards. Utilizing these resources will assist the veterinarian in the mission of providing the best care for the patient.

Initiating Change

Opportunities exist for all healthcare providers in veterinary medicine and pharmacy to participate in initiating change and improvement of veterinary patient care. The AVMA conducted a webinar on compounding in Feb, 2014. The USP conducted the 2nd Veterinary Stakeholders Meeting in Feb, 2014. There are compounding forums at national symposia of ACVIM, SVHP, IACP and NABP (National Association of Boards of Pharmacy). Veterinary pharmacy electives are available online and at some colleges of pharmacy, (there needs to be more). Peer-reviewed publications exist and there's always a need for more articles. USP Stimuli Articles are utilized to encourage discussion and input into potential changes and additions to the USP-NF. Participation in these opportunities assists in the development of policies and guidelines. Participation in these opportunities establishes communication that is vital to ultimately serving the needs of the patient.

Bottom Line-The Patient

Establishing relationships with pharmacists who study and utilize standards established by the USP is paramount to veterinary medicine. Communicating with your pharmacist and asking how they comply with these standards should comprise early conversations in developing these relationships. One aspect of USP Chapters 795 and 797 that hasn't been discussed is that compounding pharmacies are required to have quality assurance testing and records to validate quality, particularly in areas of content, potency and sterility. Remember the USP mission. A compounding pharmacy should be eager to share this information. Simply asking "can you make this?" cannot comprise the extent of communication between the veterinarian and the pharmacist. An important component of the federal code that addresses compounding for veterinary patients resides in AMDUCA and the extralabel drug use regulation does not permit animal drug compounding from active pharmaceutical ingredients (API) or commonly referred to as "bulk drugs". Compounding with API's is permitted in compounded preparations for human patients. This inconsistency is the topic of much debate and discussion. Veterinary and pharmacy associations as well as industry must pool resources to address this issue.

Other current challenges include the need for product monographs to be donated to the USP to expand the umbrella of safety for our patients. More monographs addressing veterinary compounded products are needed to standardize patient care. There is much that needs done to continue to raise the standard of care for our veterinary patients. Working together, all associations and organizations can raise the standard. Associations and organizations are comprised of individuals with something in common, a desire for providing the best possible patient care, and that is a strong foundation to build upon.

REFERENCES

Code of Federal Regulations-Title 21-Food and Drugs http://www.accessdata.fda.gov/scripts/cdrh/cfdocs/cfCFR/CFRSearch.cfm accessed Feb 2015 United States Pharmacopeia Chapter 795 http://www.usp.org/search/site/795 accessed Feb 2015 United States Pharmacopeia Chapter 797 http://www.usp.org/search/site/797 accessed Feb 2015 United States Pharmacopeia Chapter 800 http://www.usp.org/usp-nf/notices/general-chapter-hazardous-drugs-handling-healthcare-settings accessed Feb 2015 FDA Safety Med Alert http://www.fda.gov/Safety/MedWatch/SafetyInformation/SafetyAlertsforHumanMedicalProducts/ucm322849.htm accessed Mar 2015 Polo Tragedy http://www.ftboa.com/fl-equine-pub/the-florida-horse/743-polo-tragedy-the-inside-story accessed Mar 2015 FDA Drug Quality and Security Act http://www.fda.gov/Drugs/DrugSafety/DrugIntegrityandSupplyChainSecurity/DrugSupplyChainSecurityAct/default.htm accessed Feb 2015

ALTERED PLASMA CORTISOL BINDING DYNAMICS DURING CRITICAL ILLNESS IN HORSES AND FOALS

Kelsey A. Hart, DVM PhD DACVIM (LAIM), Melanie A. Fratto, Natalie A. Norton, MS
Athens, GA

DEFINITIONS AND INTRODUCTION

Clinical and clinicopathologic evidence of systemic inflammation -- such as hyperemic mucous membranes, fever, tachycardia, tachypnea, leukopenia or leukocytosis – are frequent findings in critically ill horses and foals, and horses with these signs can often be classified as having the Systemic Inflammatory Response Syndrome (SIRS).[1] In adult horses, SIRS most commonly develops in association with gastrointestinal disease (colic, colitis), which in neonatal foals, local or systemic bacterial infection most often triggers SIRS.[1,2] SIRS that occurs in conjunction with infection, such as in the newborn foal or in bacterial colitis in horses, is defined as sepsis.[1] Septic shock develops when sepsis results in organ dysfunction and evidence of acute circulatory failure (e.g. persistent hypotension or hyperlactatemia).[1]

SIRS, sepsis, and septic shock result in activation of both the immune and endocrine systems, to respectively help the patient eliminate the pathogen and cope with the physiologic stresses of illness respectively. The endocrine response to SIRS/sepsis is primarily mediated by the hypothalamic-pituitary-adrenal (HPA) axis, and culminates with the synthesis and secretion of the steroid hormone cortisol by the adrenal cortices. Cortisol is vital for the stress response to severe illness, and plays an essential role in the regulation of immune and inflammatory responses by inhibiting leukocyte migration into tissues, stabilizing lysosomal membranes, and decreasing pro-inflammatory cytokine production.[3] *These anti-inflammatory effects of cortisol are critical in sepsis, as an unregulated systemic inflammatory response can be as dangerous as the inciting infection, culminating in multi-organ failure or death.*[4-16]

In people, HPA axis dysfunction that results in cortisol insufficiency – termed critical illness-related corticosteroid insufficiency or CIRCI – occurs frequently in sepsis and is associated with decreased survival rates.[17,18] Several studies have recently demonstrated evidence for CIRCI in up to 50% of septic foals, and found significant increases in incidence of septic shock, multi-organ failure and mortality in septic foals with CIRCI as compared to septic foals with appropriate cortisol responses.[19-22] CIRCI also appears to occur and be similarly associated with disease severity in adult horses with SIRS and critical illness,[23] though specific diagnostic criteria for CIRCI in foals or horses are not fully defined or universally accepted.

Despite the high incidence of CIRCI, our current understanding of its pathogenesis in equine – or human – SIRS/sepsis is very limited. In some cases, the adrenal glands are damaged or suppressed and fail to produce sufficient cortisol, but in the majority of CIRCI cases, plasma cortisol concentrations are increased compared to resting levels in healthy unstressed controls but fail to generate appropriate anti-inflammatory effects.[17,19,24,25] Evidence in people and horses suggests that CIRCI may result from resistance to cortisol's activity in peripheral tissues,[17,26-30] but mechanisms resulting in this so-called "peripheral cortisol resistance" in sepsis are not yet defined in any species.

Recent evidence in several species suggests that changes in the relative availability of free versus protein-bound cortisol during sepsis may be integral to the development of peripheral cortisol resistance in immune and inflammatory cells, resulting in CIRCI and exaggerated pro-inflammatory responses that can decrease survival in sepsis. Specifically, cortisol bound to its carrier protein cortisol binding globulin (CBG) – rather than free, unbound cortisol as previously believed – appears to play a critical role in mediating cortisol's anti-inflammatory effects in local and systemic inflammation.[31,32] The role of CBG in mediating cortisol's anti-inflammatory effects in people and rodent models, and our current understanding of plasma cortisol binding dynamics during health and illness is horses and foals, are discussed below.

THE IMPORTANCE OF CORTISOL-BINDING GLOBULIN

As mentioned above, alterations in cortisol's plasma protein binding dynamics resulting in changes in the relative availability of free versus protein-bound cortisol appear to play an important role in mediating cortisol's effects in peripheral tissues, particularly at sites of inflammation.[31,32] Cortisol is lipophilic, and thus in healthy adults circulates in plasma with $\geq$ 90% bound to plasma proteins – predominantly to CBG.[24] Because cortisol-receptors are cytosolic, the free fraction is generally presumed to be the biologically active portion that is available to diffuse into cells and interact with these intracellular receptors. Once bound to a free cortisol molecule, these cytoplasmic receptors dimerize and enter the nucleus, where they ultimate interact directly with DNA at hormone response elements and up- or down-regulate transcription of specific steroid-responsive target genes.

Protein-bound cortisol is generally thought of as providing a circulating cortisol reservoir, to deliver cortisol to target tissues and protect circulating cortisol from renal and hepatic elimination. However, recent work suggests that CBG-bound cortisol may play a much more direct role in mediating cortisol's action in target tissues. *In vitro* studies in people and mice have shown that CBG facilitates cortisol delivery to sites of active inflammation via elastase-mediated mechanisms – specifically, cleavage of CBG by neutrophil elastase results in irreversible release of cortisol directly at sites of local inflammation.[31,32] In addition, increased body temperature – certainly common in sepsis – and certain bacterial proteases produced by *Pseudomonas* spp., *Burkholderia* spp., and *Staphylococcal* spp. all have been shown to reduce CBG's binding affinity for cortisol.[33,34] This could be beneficial if it promotes the release of cortisol to modulate inflammation specifically at infection sites, but in widespread or systemic infections such as sepsis/septic shock, fever or bacteria-mediated decreases in CBG binding affinity, coupled with decreased CBG production in these conditions, could ultimately result in loss of the circulating cortisol reservoir and failure of cortisol to localize to sites of inflammation.

CBG and/or CBG-bound cortisol may also have direct effects on cellular activity via as-yet-undefined cell surface mechanism. High affinity binding sites for CBG on plasma membranes have been described,[35] and the CBG-cortisol complex has been shown to directly increase intracellular cyclic AMP concentrations by binding to these sites,[36] though specific CBG receptors have not been identified. Further, CBG-deficient mice – which have *only* free cortisol – exhibit *increased* pro-inflammatory responses and increased mortality in experimental sepsis, compared to wild-type mice with both free and CBG-bound cortisol.[37] Finally, in human burn patients, increased serum IL-6 concentration is associated with low serum CBG concentrations and high serum free cortisol fraction.[38] In sum, this evidence suggests that CBG and/or CBG-bound cortisol play an integral role in delivering cortisol specifically to sites of inflammation and directly mediating cortisol's local and systemic anti-inflammatory effects.

CBG is a negative acute phase protein, and thus hepatic production of CBG is down-regulated in acute inflammation.[38-40] In 30 adult patients with septic shock, CBG concentrations at ICU admission were half of concentrations measured in healthy controls, and remained below levels in healthy individuals for at least the first 7 days of treatment. A decrease in CBG in acute SIRS/sepsis has been thought to promote cortisol's anti-inflammatory activity by increasing free cortisol availability during the acute inflammatory response.[39-41] However, in severe illness such as septic shock, if this decrease in CBG is extreme or persists too long, it could permit excessive or too rapid elimination of circulating free cortisol, as free cortisol is preferentially eliminated over protein-bound cortisol.[42] When rapid cortisol excretion is combined with direct HPA axis suppression by pro-inflammatory cytokines, as is well documented in sepsis,[17,26-28,43,44] relative cortisol insufficiency could actually result, despite higher free cortisol availability. Finally, if modulation of neutrophil and other innate inflammatory responses is mediated in part or predominantly by CBG-bound cortisol, excessive or dangerous degrees of systemic inflammation could occur with persistent CBG deficiency in septic patients, similar to what is seen in CBG-deficient mice in an experimental rodent sepsis model.[37]

To date, our understanding of plasma cortisol binding dynamics in SIRS/sepsis in horses and foals has been limited, and straightforward methodology for direct measurement of equine free cortisol or CBG was not available until recently. Our laboratory has recently optimized and validated an ultrafiltration radioligand binding assay for measurement of free cortisol in equine samples, as well a commercially available ELISA assay for measurement of equine CBG, to permit further characterization of plasma cortisol binding dynamics during health and illness in horses and foals.

PLASMA CORTISOL BINDING DYNAMICS IN HEALTHY HORSES AND FOALS

Several studies have demonstrated healthy neonatal foals have substantially greater free cortisol than adult horses during the first week of life.[24,45,46] Specifically, healthy foals have up to 60% free cortisol at birth, which decreases steadily to approximately 20% free at 5-7 days of age, in comparison to approximately 10% free cortisol in healthy adult horses.[24,46] Further, comparison of the pharmacokinetics of hydrocortisone (synthetic cortisol) in healthy horses and adult foals demonstrated decreased peak plasma concentrations and significantly greater volume of distribution at steady state and total body clearance rate in 2-3 day old foals as compared to adult horses, consistent with decreased plasma cortisol binding capacity in foals.[47] Serial measurement of free cortisol in healthy foals from birth to 12 weeks of age demonstrated that foals have comparable serum free cortisol fraction as adult horses by 2 weeks of age, but their total cortisol concentration is significantly lower than adult horses from 12 hours to at least 12 weeks of age.[48] Breed (horse vs. pony) or induction of parturition with oxytocin did not significantly impact total or free cortisol in neonatal foals, suggesting that differences in HPA axis activity between horses and ponies detected in adults are not present at birth.[48] Together, these findings strongly suggest that plasma cortisol binding capacity is substantially decreased in healthy foals for at least the first 3 months of life, most likely due to a decrease in CBG.

Our laboratory has recently validated a commercially available equine CBG ELISA (MyBioSource, San Diego, CA, USA) and utilized this methodology to quantify serum CBG concentrations in healthy foals from birth to 8 weeks of age and in healthy adult horses. The results of this study demonstrated that while there was no significant effect of foal age on CBG concentration in healthy foals, healthy foals at all ages had significantly – approximately 25% – lower serum CBG concentrations than adult horses.[49]

In sum, these data demonstrate that healthy neonatal foals have higher free cortisol than adult horses during the first 7-14 days of life,[24] which increases the amount of biologically-available cortisol to enter cells and bind cytoplasmic steroid receptors but results in more rapid clearance of circulating cortisol. By two weeks of age, healthy foals have comparable serum free cortisol fraction to adult horses, but significantly lower total cortisol concentration until at least 3 months of age.[48] This is due to a reduction in protein-bound cortisol during this time period, due to a 25% decrease in serum CBG concentrations in healthy foals compared to adult horses during the first several months of life.[49] These findings are consistent with one previous report that employed different methodology to demonstrate lower serum CBG concentrations in neonatal foals compared to adult horses.[46]

PLASMA CORTISOL BINDING DYNAMICS IN CRITICALLY ILL HORSES AND FOALs

In theory, this increased free cortisol in neonatal foals during the first two weeks of life should make them more effective at controlling inflammatory responses in sepsis, but in practice – and in a number of clinical studies – septic foals actually appear to exhibit more robust or altered systemic inflammatory responses with increases in certain pro-inflammatory cytokines.[5,50,51] Because of the preferential excretion of free over bound cortisol,[42] this increased free cortisol results in more rapid clearance of plasma cortisol in neonatal foals[47] and could contribute to these exaggerated inflammatory responses by causing cortisol elimination before it can produce adequate anti-inflammatory effects in septic foals. However, if CBG-bound cortisol is important for regulating some of cortisol's anti-inflammatory effects in foal leukocytes as in the human and rodent studies discussed above, this disparity between increased circulating free cortisol but clinical evidence of cortisol insufficiency in some septic neonatal foals could be explained by further decreases in CBG availability in septic foals. In septic people, CBG concentrations are decreased due to negative acute phase effects,[39,40] but measurement of CBG concentrations in critically ill horses or foals has not been reported to date.

Our laboratory recently conducted a preliminary study to determine the effect of illness on CBG concentrations in horses and foals. Archived serum samples collected at hospital admission from 26 hospitalized foals $\leq$ 7 days of age with a diagnosis of sepsis based on a positive blood culture and/or sepsis score $\geq$ 11 at hospital admission, and from 58 adult horses with gastrointestinal disease were analyzed, and compared with age-matched healthy foals and horses from the above study.[49] Sick animals were further stratified by severity of illness and survival after evaluation of medical records. Mean serum CBG concentration was significantly lower in septic foals (56.4 ± 18.5 mg/L) than in healthy foals (78.0 ± 7.6 mg/L), but significantly higher in horses with GI disease (169.4 ± 65.2 mg/L) than in healthy horses (109.0 ± 23.5). There was no difference in serum CBG concentrations between survivors and non-survivors in horses or foals. Similarly, serum CBG concentration was not different among horses with strangulating, non-strangulating, or inflammatory intestinal lesions, nor between foals with and without shock or multi-organ dysfunction syndrome.

CONCLUSIONS AND FUTURE DIRECTIONS

These results provide further evidence for substantial CBG deficiency in septic foals, as compared to healthy foals and adult horses. Serum CBG concentration was decreased by 25% in septic foals compared to healthy age-matched foals, by 50% compared to healthy adult horses, and by 75% compared to adult horses with GI disease. This is best explained by age- and illness-mediated suppression of CBG production in septic neonatal foals, as described in other species.[40,41,52,53] Our data also demonstrated a significant increase rather than the anticipated decrease in serum CBG in adult horses with GI disease as compared to healthy horses. In animal models of acute environmental stress (i.e. novel environment or restraint stress), massive and immediate hepatic release of CBG occurs resulting in peak CBG concentrations by 30 minutes after the stressor.[54] If similar CBG response to the stresses of illness, shipping, and hospital admission occurs in adult horses, this could explain the increased serum CBG concentration we found in sick adult horses at hospital admission; neonatal foals may lack this CBG reserve due to previously described hepatic immaturity.[55] Further, the group of adult horses with GI disease included any horse presenting for colic, so this heterogeneous group might not accurately reflect inflammatory mediated changes in CBG specifically in sick adult horses with SIRS/sepsis.

Age- and illness-mediated CBG-deficiency in foals results in decreased circulating CBG-bound cortisol, to the greatest degree in septic foals. If CBG-bound cortisol is critical for mediating some of cortisol's anti-inflammatory effects in foals as in other species,[31,37] this CBG-deficiency could contribute to the exaggerated and potentially detrimental pro-inflammatory responses seen in some septic foals.[5,50,51] Additionally, in a small preliminary study on archived samples, serum CBG was not predictive of disease severity or survival. However, further prospective study is needed to confirm these findings in larger numbers of septic foals, to investigate the effects of CBG-bound cortisol on equine leukocyte function, and to determine relationships between CBG availability and systemic inflammation during acute and prolonged illness in horses and foals.

REFERENCES

[1]Moore J, Vandenplas M. **Vet Clin North Am Equine Pract** 2014;30:337-351; [2]Hart KA, Mackay R. In: Smith B, ed. **Large Animal Internal Medicine, 5th ed.** St. Louis, MO: Elsevier, Inc. (Mosby); 2015:682-695; [3]Bailey S. **Vet Clin North Am Equine Pract** 2010;26:277-285; [4]Gibson S, et.al. **Emerg Med Clin North Am** 2005;23:909-929; [5]Gold J, et.al. **J Vet Intern Med** 2007;21:482-488; [6]Harris M, et.al. **J Pediatr** 2005;147:462-368; [7]Kozyra E, et.al. **Ann Pharmacother** 2005;39:691-698; [8]Kumar A, et al. **Indian J Crit Care Med** 2009 13:74-78; [9]Oppert M, et.al. **Crit Care Med** 2005;33:2457-2464; [10]Wang H, Ma S. **Am J Emerg Med** 2008;26:711-775; [11]Casey L, et.al. **Ann Intern Med** 1993;119:771-778; [12]Haddad J, et.al. **J Neuroimmunol** 2002;133:1-19; [13]Marti L, et.al. **Gerontology** 2007;53:239-244; [14]Ng P, et.al. **Arch Dis Child Fetal Neonatal Ed** 2003;88:F209-F213; [15]Pinsky M, et al. **Chest** 1993;103:565-575; [16]Shapiro N, et.al. **Ann Emerg Med** 2006;48:583-590; [17]Marik P. **Chest** 2009;135:181-193; [18]Marik P, et.al. **Crit Care Med** 2008;36:1937-1949; Hart K, et.al. **J Vet Intern Med** 2009;23:901-912; [20]Gold J, et al. **J Vet Intern Med** 2007;21:791-796; [21]Hurcombe S, et al. **J Vet Intern Med** 2008;22:639-647; [22]Wong D, et.al. **J Am Vet Med Assoc** 2009;234:126-132; [23]Stewart A, et.al. Identification of critical illness related corticosteroid insufficiency in critically ill horses. Research Abstract, Annual Forum of the American College of Veterinary Internal Medicine, Denver, CO 2011; [24]Hart K, et.al. **J Vet Intern Med** 2011;25:345-355; [25]Hart K, et.al. **Pediatr Res** 2011;70:72-77; [26]Cohen J, Venkatesh B. **Crit Care Resusc** 2009;11:287-289; [27]Liu D, et.al. **Crit Care Med** 2002;30:623-627; [28]Liu L, et al. **Am Rev Respir Dis** 1993;148:878-881; [29]Meduri G, et.al. **Neuroimmunodulation** 2005;12:321-338; [30]Hoffman CJ, et al. **J Vet Intern Med** 2015;29:626-635; [31]Hammond G, et al. **J Clin Endocrin Metab** 1990;71:34-39; [32]Lin HY, et al. **Mol Cell Endocrinol** 2010;316:3-12; [33]Simard M, et al. **Endocrinology** 2014;155:2900-2908; [34]Westphal U. **Mol Cell Biochem** 1983;55:145-157; [35]Strel'chyonok OA, Avvakumov GV. **J Steroid Biochem Mol Biol** 1991;40:795-803; [36]Nakhla AM, et al. **Biochem Biophys Res Commun** 1988;153:1012-1018; [37]Petersen H, et al. **Mol Cell Biol** 2006;26:7326-7245; [38]Bernier J, et al. **Crit Care Med** 1998;26:452-460; [39]Beishuizen A, et al. **Intensive Care Med** 2001;27:1584-1591; [40]Hamrahian A, et al. **N Engl J Med** 2004;350:1629-1638; [41]Ho J, et al. **J Clin Endocrin Metab** 2006;91:105-114; [42]Bright G. **J Clin Endocrin Metab** 1995;80:770-775; [43]Aubry J, et al. Endocrinology 1997;138:1621-1626; [44]Qahwash I, et al. **Endocrine** 2002;18:13-20; [45]Cudd TA, et al. **J Endocrinol** 1995;144:271-283; [46]Irvine C, Alexander S. **J Reprod Fert Suppl** 1987;35:19-24; [47]Hart K, et al. **Am J Vet Res** 2012;73:68-75; [48]Hart K, et al. Effects of age, breed, and parturition induction on serum free cortisol fraction in neonatal foals. Research Abstract, Dorothy Havemeyer Neonatal Septicemia Workshop, Saugerties, NY 2013; [49]Fratto M, et al. Cortisol binding globulins in septic foals, healthy foals and healthy horses. Research Abstract, ACVIM Annual Forum, Indianapolis, IN 2015; [50]Burton A, et al. **Vet Immunol Immunopathol** 2009;132:122-128; [51]Pusterla N, et al. **Am J Vet Res** 2006;67:1045-1049; [52]Rokicki W, et al. **Biol Neonate** 1990;57:21-29; [53]Stahl F, et al. **Endokrinologie** 1979;74:243-246; [54]Qian X, et al. **Endocrinology** 2011;152:3738-3748; [55]Barton M, LeRoy B. **J Vet Intern Med** 2007;21:508-513.

DIARRHEA IN YOUNG FOALS

Ashley Whitehead, DVM, BSc, DVSc, DACVIM (LAIM)
Calgary, AB, Canada

INTRODUCTION

Diarrhea is one of the most common infectious disease conditions requiring veterinary care of young foals [1]. Although most episodes of diarrhea are not life threatening, a substantial subset requires intensive veterinary care at considerable cost to the owners. Recently, two studies have investigated the infectious etiological agents that may be responsible for foal diarrhea. The first study in central Kentucky found that there is a high prevalence of co-infections in foals sick with gastrointestinal disease; rotavirus and coronavirus were most prevalent in these diseased animals [2]. In Florida, a retrospective study of hospitalized diarrheic foals identified rotavirus, *Clostridium perfringens* and *Salmonella* spp. as the most frequently identified causative agents [3]. An understanding of the primary infectious agents, of co-infection and transmission patterns in cases of diarrhea in young foals is essential to develop intervention strategies to reduce the prevalence of disease.

ROTAVIRUS

Equine rotavirus is one of the most diagnosed agents associated with diarrhea in foals [2-7]. While the diarrhea caused by equine rotavirus is usually mild, outbreaks are not uncommon on farms. Villous blunting and atrophy result in malabsoption and subsequent diarrhea[6, 8]. This loss of brush border leads to a reduction of disaccharidase enzymes, such as lactase and therefore results in increased lactose concentration in the aboral intestinal tract and leading to a secondary osmotic diarrhea which may persist past the resolution of the viral disease as seen in other species including humans [8,9]. Rotavirus is generally a self-limiting disease though the resultant dehydration can lead severe complications including death in a small percentage of foals[6, 10]. Neonatal foals appear to be more susceptible to severe disease and complications.

CORONAVIRUS

Equine coronavirus is frequently identified in the feces of both healthy and diarrheic foals, however, the true pathogenicity of the virus in foals is unknown[2, 3, 11, 12]. Co-infection with other infectious agents is common in diarrheic foals and it has been speculated that equine coronavirus may predispose a foal to secondary pathogens that then result in diarrhea[2]. Recent descriptions of cases and outbreaks of equine coronavirus in adults have highlighted the paucity of scientific knowledge of the pathogenesis, transmission and viral dynamics of coronavirus in equine patients[13-15]. To date there has been one study investigating an adult experimental model of coronavirus disease, [16] however a foal experimental model has not yet been established.

CLOSTRIDIUM DIFFICILE

While *Clostridium difficile* can be identified in low levels in some healthy foals,[17, 18] the presence of Clostridium toxin A and B are essential to the pathogenesis of diarrheic disease. Major risk factors such as previous antibiotic administration and hospitalization identified in horses and people do not appear to be important in the pathogenesis of disease in young foals[19, 20]. High mortality rates can occur in neonatal foals though variables such as inoculation with vegetative cells verses spores, dose of inoculation and immunocompetence of the foal are likely important in the severity of disease[19]. Co-infection with toxigenic *C. difficile* and *C. perfringens* has been reported in severe cases of diarrhea in young foals resulting in death[21]. The clinical signs and pathological findings are usually indistinguishable between *C. difficile*, *C. perfringens* and salmonellosis in the neonatal foal and testing of feces or pathology samples are essential to distinguishing between these conditions.

CLOSTRIDIUM PERFRINGENS

The role of *C. perfringens* in foals with neonatal enterocolitis has been investigated worldwide[5, 22,-26]. While toxigenic *C. perfringens* strains may cause diarrhea of varying severity in young foals, *C. perfringens* type A has been identified in the feces of non-diarrheic foals[25,27, 28]. Severe cases of *C. perfringens* associated diarrhea in foals present with a hemorrhagic, necrotizing enteritis involving especially the jejunum[23,29]. This clinical presentation is most commonly associated with type C-associated infections, [29] however, certain cases of *C. perfringens* type A associated infections have considerable clinical and pathological similarities especially when CPE and beta2 toxins are produced[30, 31]. Recently Dr. Prescott (University of Guelph) has identified a novel *C. perfringens* toxin, which is strongly associated with severe necrotizing enterocolitis in foals and hemorrhagic gastroenteritis in dogs.

Necrotizing enterocolitis associated with *C. perfringens* is well recognized by Alberta practitioners. In 2014, a pilot study was undertaken in Alberta and of 15 cases of foal diarrhea seen in a referral hospital, thirteen of the foals had *Clostridium perfringens* cultured from their feces, two of which had the novel toxin identified by Dr. Prescott. The identification of this novel toxin provides a new area of investigation into the pathogenesis of *C. perfringens* diarrhea in foals and may help determine the causative agent for some of the undiagnosed necrotizing enterocolitis cases.

Table 1. Clinical features of selected infectious agents causing diarrhea in young foals

Agent	Age	Clinical Signs	Incubation	Transmission	Shedding	Vaccine
Rotavirus (RVA)	3 days to 5 months (most are less than 30 days)	Anorexia, depression, diarrhea, dehydration, pyrexia.	12-24 hours[32]	Feco-oral	Before onset of diarrhea and for up to 12 days post-resolution of the diarrhea[33]	Killed vaccine (Serogroup A, subtypes P12, G3)
Equine Coronavirus (ECoV)	Generally, less than two weeks	Anorexia, depression, diarrhea, dehydration, pyrexia and leukopenia.	Unknown	Feco-oral	3-19 days (adults)	No
Clostridium difficile	Any aged foal including less than 24 hours old	Variable from low grade diarrhea to severe diarrhea with ileus. Anorexia, depression, diarrhea, dehydration, pyrexia. Enterotoxemia with associated clinical signs.	< 24 hours[32]	Feco-oral	Unknown but likely 3 to 7 days after clinical resolution[32]	Experimental toxoid vaccine (alpha, beta2 and CPE toxins)
Clostridium perfringens	Any aged foal including less than 24 hours old	Colic with or without diarrhea. Ileus may be present. Anorexia, depression, diarrhea, dehydration, pyrexia. Enterotoxemia with associated clinical signs.	< 24 hours[32]	Feco-oral	Unknown but likely 3 to 7 days after clinical resolution[32]	No
Cryptosporidium	4-21 days (most are 1-2 weeks)	Greenish, yellow malodourous feces with or without mucus. Variable systemic signs. Severity of clinical signs may be related to immunocompetance.	3-7 days[32]	Feco-oral	10 to 40+ days experimentally[34]	No
Salmonella spp.	12 hours to six months	Severe diarrhea that may contain casts and blood. Fever, lethargy, anorexia. Secondary septicemia.	12+ hours (dependent on dose, virulence, etc)	Feco-oral Airborne transmission also possible	Highly variable, days to extended periods. Shedding may start by 1.3 days post inoculation[35]	Experimental inactivated bacterin vaccine

Table 2. Pathological findings and diagnostic testing recommendations of selected infectious agents causing diarrhea in young foals

Agent	Gross Pathology Findings	Histopathology Findings	Diagnostic Testing
Rotavirus	Undifferentiated diarrhea	Destruction of the villous tips. Edema, mononuclear cell infiltration, vacuolation and virus particles visible on electron microscopy within the epithelial cytoplasm.[6,8]	Gold standard: Electron microscopy Real time-PCR, best if performed in conjunction with ELISA Recommend three negative tests to confirm negative
Equine Coronavirus	Undifferentiated diarrhea	Villous atrophy and inflammation in the lamina propria, mild mononuclear infiltrate in the cytoplasm of the affected enterocytes. May have crypt epithelial hyperplasia.[36]	Rule out other causes of diarrhea. Fecal testing with real time RT-PCR. Electron microscopy available but not as sensitive. May be detected in nasal secretions (adults).[16]
Clostridium difficile	Extensive mucosal damage with intestinal inflammation. Younger foals tend to have more small intestinal lesions with less frequent lesions in the colon and cecum than older foals. Extensive hemorrhagic necrotizing damage. Luminal fluid is frequently hemorrhagic. Pseudomembranous lesions may be noted.[20]	Necrotizing or necrohemorrhagic enteritis with mucosal or submucosal thrombosis. Bacteria are noted on the intestinal surface and within the affected mucosal cells. Mural edema and mild to moderate neutrophilic infiltrations.[20, 29]	Gold standard: cytotoxin assay Tox A/B II ELISA has been validated in horses Screening: PCR for toxins in feces
Clostridium perfringens	Segmental or diffuse hemorrhage and thickening of the intestinal mucosa. Mucosal lesions include ulcers and necrotic damage. Pseudomembranous lesions may be noted. Luminal fluid is frequently hemorrhagic.[29]		CPE ELISA (not yet validated for horses) PCR from culture or feces for CPE, beta2 toxins (alpha toxin is of limited value)
Cryptosporidium	Usually limited to signs of inflammation of the aboral small intestine.	Hyperplasia of the crypts of Lieberkühn. Villous atrophy of variable severity. Presence of significant numbers of cryptosporidium among the microvilli of the ileum.[36]	Exclusion of other causes of diarrhea Fecal testing using sugar floatation, acid fast staining, immunofluorescence, PCR or ELISA
Salmonella spp.	Few gross lesions are noted in some foals. Other may have a fibrinous enterocolitis. Lesions may be hemorrhagic with significant mucosal damage. Signs of septicemia. Mesenteric lymph node enlargement with occasional hemorrhage and edema.[36]	Variable. Hemorrhage, necrosis and diphtheritic membranes present along the mucosa of the small intestines. Thrombosis in the proprial blood vessels. Significant number of neutrophils within the lumen of the bowel.[36]	Culture of blood and/or feces. Multiple cultures are required. Real time PCR for screening followed by culture.

Table 3. Environmental persistence and disinfection protocols for infectious agents causing diarrhea in young foals.
Note: Always follow the manufacture's directions for concentration, contact time, use in presence of organic material and optimum environmental temperature as these parameters may be different for each organism

Agent	Persistence in the environment	Disinfection protocols
Rotavirus	9 months in feces at room temperature, one hour at 60°C.[8]	**Hands:** Chlorhexidine is not effective alone. Povodine iodine or alcohol based products are recommended. **Environment:** In presence of organic matter - glutaraldehyde In absence of organic material - Chlorine and iodine based disinfectants or benzalkonium chloride.[10, 37]
Equine Coronavirus	Unknown	Sodium hypochlorite, ethyl alcohol, povodine iodine, phenols, accelerated hydrogen peroxide, peroxide based disinfectants.[32]
Clostridium difficile	Spores last years in the environment and are resistant to UV light, low and high temperatures and desiccation.[32]	**Hands:** Alcohol based products are ineffective. Appropriate hand washing measures. **Environment:** Bleach, accelerated hydrogen peroxides.[32] **Zoonotic risk**
Clostridium perfringens		
Cryptosporidium	Several months in cool moist environments. Oocytes are inactivated at temperatures less than 0°C and greater than 65°C	**Hands:** Alcohol based products are ineffective. Appropriate hand washing measures. **Environment:** Cryptosporidium is extremely resistant to chemical disinfectants. Removal of fecal material is essential to environmental management. When using pressure sprayers to clean infected areas, be aware that oocytes may become entrapped in aerosolized water droplets. Ammonium hydroxide, hydrogen peroxide, chlorine dioxide, 10% formalin, 5% ammonia.[32] **Zoonotic risk**
Salmonella spp.	Months to years in fecal material and under adverse condition	Susceptible to most chemical disinfectants. Organic material must be removed prior to disinfection and porous surfaces may be inadequately disinfected. **Zoonotic risk**

REFERENCES

[1]Wohlfender FD, et al **Equine Vet J** 2009; 41(2):179. [2]Slovis NM, et al **Equine Vet J** 2014; 46:311. [3]Frederick J, et al **J Vet Intern Med** 2009; 23:1254. [4]Woode GN, et al **Infect and Immun** 1976; 14:804. [5]Netherwood T, et al **Epidemiol Infect** 1996; 117:375. [6]Conner ME, et al **Am J Vet Res** 1980; 41:1699. [7]Matthijnssens J, et al **Vet Microbiol** 2015; doi: 10.1016/j.vetmic.2015.01.011. [8]Woode GN, et al **J Am Vet Med Assoc** 1978; 173:522. [9]Hyams JS, et al **J Pediatr** 1981; 99(6):916. [10]Bailey KE, et al **J Vet Microbiol** 2013; 167:135. [11]Bass EP, et al **The Lancet** 1975:822. [12]Guy JS, et al **J Clin Microbiol** 2000; 38:4523. [13]Oue Y, et al **Vet microbial** 2011;150:41. [14]Pusterla N, et al **Vet Microbiol** 2013; 162:228. [15]Fielding CL, et al **J Vet Intern Med** 2015; 29:307. [16]Nemoto M, et al **Arch Virol** 2014;159:3329. [17]Baverud V, et al **Equine Vet J** 2003;35(5):465. [18]Abrahams M, et al **ACVIM Forum Proceedings** 2014. [19]Arroyo LG, et al **J Vet Intern Med** 2004;18(5):734. [20]Diab SS, et al **Vet Pathol** 2013; 50:1028. [21]Uzal FA, et al **J Vet Microbiol** 2012; 156:395. [22]Kanoe M, et al **Microbios** 1990; 64:153. [23]East LM, et al **Am Vet Med Assoc** 1998; 212:1751. [24]Netherwood T, et al **Epidemiol Infect** 1998; 120:193. [25]Silva ROS, et al **Equine Vet J** 2013; 45(6):671. [26]Gohari IM, et al **Can J Vet Res** 2014; 78:1. [27]Netherwood T, et al **Vet J** 1998;155, 289. [28]Tillotson K, et al **J Am Vet Med Assoc** 2002; 220:342. [29]Diab SS, et al **Vet Pathol** 2012; 49:255. [30]Timoney JF, et al **Vet Rec** 2005; 157:810. [31]Hazlett MJ, et al **J Vet Diagn Invest** 2011; 23:373. [32]Sellon DC, Long MT **Equine Infectious Diseases** 2nd Edition; 2014. [33]Dwyer RM, et al **Proc Am Assoc Equine Pract** 1990; 36:335. [34] Xiao L, et al **Equine Vet J** 1994; 26(1):9. [35] Aceto H, et al **J Am Vet Med Assoc** 2011; 238 (10);1333. [36]Maxie, MG (editor) **Pathology of Domestic Animals**, Vol 2; 2007. [37]Nemoto M, et al **J Vet Med Sci** 2014; 76:106.

CURRENT TOPICS IN NEONATAL GASTROINTESTINAL SONOGRAPHY

Cristobal Navas de Solis LV, PhD, DACVIM (LAIM), Jonathan E Palmer VMD, DACVIM (LAIM)
Bern, Switzerland, Kennett Square, PA

INTRODUCTION

Gastrointestinal disease is one of the most common problems neonatal foals face during the first weeks of life. The most frequently encountered problem in this age group is contagious gastroenteritis which has been well described. Less well understood and less common is the gastrointestinal (GI) dysfunction which often accompanies neonatal encephalopathy (also referred to as hypoxic ischemic encephalopathy, asphyxia syndrome) and neonatal nephropathy during the first days of a foal's life. This triad of neurologic, renal and GI syndromes often with signs of sepsis is the most common complaint of foals less than 48 hours old admitted to our Neonatal Intensive Care Unit. The GI component has a wide range of clinical signs and severity form very mild dysmotility to severe necrotizing disease. The necrotizing disease we see has some similarities (and some differences) to the disease well recognized in human infants called Necrotizing Enterocolitis (NEC). This talk will describe our experience with two recent studies which have brought attention to the occurrence of necrotizing gastrointestinal (GI) disease and asymptomatic intussusceptions in neonatal foals.

The relevance and association of clinical and sonographic findings in neonates with gastrointestinal disease will be discussed from the perspective of a sonographist and a neonatologist. During this session we will not attempt to review the technique, normal or abnormal findings in abdominal sonograms in foals. However, we want to point out 2 technical notes. The sonographic examination of the neonate with abdominal disease can be done with the foal standing, in lateral or sternal recumbency. It is important to realize that lesions tend to localize in the most dependent part of the abdomen and that these lesions can be mobile. Therefore it is important to scan carefully the gravity dependent area. This implies to place the probe under the foal to scan the ventral abdomen if the foal is sternal, to scan the left side if the foal is in left lateral recumbency and the right side if the foal is in right lateral recumbency. The author finds microconvex probes to be the most useful probes. These probes typically have a frequency of 5-11MHz. The frequency varies depending on the equipment used. To assess carefully the echogenicity of the wall, the wall layering or to be precise when measuring the thickness high frequency linear probes (usually 8-15 MHz) are required. Soaking the hair with alcohol is, in most foals, enough to obtain adequate images for 'fluid checks' or to detect severe abnormalities. To obtain detailed images clipping may be necessary. Clipping and using gel reduces the scanning time although it increases the preparation time.

Abdominal ultrasonography has been used for years as a valuable diagnostic aid and has become routine in the assessment of foals with critical gastrointestinal disease, however information about its usefulness, accuracy and limitations is not available. Abdominal ultrasound examinations are performed commonly by non-radiologists in human and veterinary intensive care units. In human critical care units, the time from presentation to operative care, length of hospitalization, complications and cost decrease when emergency abdominal sonograms are performed, potentially due to a more rapid diagnosis[1]

INTESTINAL WALL LAYERING AND COLOR DOPPLER

The characterization of the layering of the GI wall has received little attention in equine medicine. There are specific case reports that have shown that this has potential increase the precision of diagnosticians.[2-3] Sonographic to pathologic correlation has been demonstrated for several disease processes including pediatric colitis[4] and GI emergencies.[5,6] In these cases the careful assessment of the echogenicity and layering patterns can give information about potential specific diagnoses often helping patient management. Of particular interest to the diagnosis of GI disease in human neonates was the study by Baud et al. In this study different patterns of abnormal intestinal echogenicity, based on the loss/maintenance of stratification and folding helped in obtaining a precise diagnosis. In our opinion the wide use of high quality portable or hospital based ultrasound machines makes possible the assessment of intestinal wall layering. We performed a study[7] to define the normal layering pattern of the GI of normal neonates. Equine neonates showed distinct layering of the stomach and large intestine, and indistinct layering of the small intestine. The reason why the layering of the stomach and large intestine was distinct, but not in the small intestine, is uncertain. It is known that rapid changes in the intestinal wall occur in the first days of life, including marked changes in total and relative thickness of the different histologic layers with the muscularis layer changing quantitatively the most in this period. We believe that the evaluation of the changes in layering patterns could help clinicians be more precise about diagnosis in neonates with GI disease and hopefully help in the management of these cases. Multicenter studies are needed to prove this and to investigate the association of different sonographic patterns with specific diagnoses in foals.

Sonographic evaluation of GI blood supply using color flow Doppler is well correlated with histology and outcome in human neonates with GI disease and allows the differentiation of focal from diffuse necrosis in the setting of NEC.[8-10] Increased flow is reportedly obvious to the untrained observer. Color Doppler sonography was found to be more accurate than clinical examination and plain abdominal radiography in the prediction of necrosis in human neonates with NEC.[10] The usefulness of this technique in equine medicine had not been evaluated. In our prospective study of normal neonates diagnostic color Doppler signals of mural blood flow could not be obtained. It is possible that different ultrasound equipment, different settings, sedating foals to decrease motion artifacts, or the use of this technique in foals with abnormal patterns might help in obtaining diagnostic color flow Doppler readings. The authors believe that this modality has potential in the evaluation of GI in horses but further work is needed before it this potential can be defined.

NECROTIZING ENTEROCOLITIS (NEC) AND PNEUMATOSIS INTESTINALIS (PI)

The recognition in our hospital population of hyperechoic echoes of gas within the gastrointestinal wall (pneumatosis intestinalis, PI) of foals hospitalized with GI or critical illness led us to retrospectively investigate this phenomenon within the context of all sonographic examinations performed in foals with GI disease. The presence of PI is considered virtually pathognomonic for

NEC in human neonates with a compatible clinical picture.[11-13] NEC is a disease process of the GI tract of human neonates that is marked by inflammation and bacterial invasion of the bowel wall. NEC occurs frequently in human newborn intensive care units,[14-15] primarily in premature neonates[16-17] and had been described in foals in isolated case reports.[18-21] We have the clinical impression, and several texts have also suggested, that this may be an under recognized syndrome.[22-25] Based on our retrospective study we concluded that: 1-PI was diagnosed frequently in severe cases. 2-Foals with PI have a worse prognosis than foals without PI. 3- Variables associated with the presence of necrotizing GI disease were: prematurity, blood in the feces, gastric reflux, abdominal distension, abnormal colon echogenicity and leucopenia.

Several similarities exist between human NEC and GI disease in equine neonates with a necrotizing component resulting in gas in the wall of the intestinal tract. We chose to describe this group of foals as having a necrotizing component of the GI and avoid call this process NEC as the pathogenesis may differ. This nomenclature has been used in these proceedings and referring to NEC we refer to the GI disease in human babies.

It is important to realize that PI is not a diagnosis or a clinical problem *per se*, but a pathological or imaging finding that is a consequence of the underlying disease.[12] PI has been attributed to at least 58 causes in man[8] but it is virtually pathognomonic of NEC in human neonates with compatible clinical presentation.[13] Therefore, and as for all imaging findings, PI has to be interpreted in in the context of the clinical picture. The process of gas production and accumulation within the intestinal wall is poorly understood. The composition of the gas in human neonates is variable but it is often 30% hydrogen.[11] Three possibilities have been proposed as the source of the gas: (1) intraluminal GI gas that migrates due to injury, (2) bacterial production of gas due to bacterial invasion, and (3) pulmonary gas. The pattern or extent of PI does not correlate with the severity of the signs or the severity of the underlying diseases in human neonates and the presence of intramural gas can even precede the presence of any clinical signs.[28]

Epidemiology

An understanding of the epidemiological aspects of NEC may help to place into context our clinical observations of necrotizing gastrointestinal disease in foals. The precise etiology of NEC, which is an ischemic and inflammatory necrosis of bowel,[29] is unknown but three aspects are felt to be major risk factors for NEC to occur in human neonates: ischaemic/hypoxic mucosal injury, feeding and the presence of bacteria. Prematurity is the only independent determinant of human NEC. 90–95% of cases occur in infants born before 36 weeks of gestational age and the incidence varies inversely with birth weight and gestational age. Those most susceptible appear to be infants weighing less than 1000 g at birth and under 28 weeks' gestation.[16] Less than 10% of NEC cases occur in term infants.[17] This close correlation with prematurity does not occur in foals in our experience as almost half the foals (46.2%) we identified were not premature. However, in our retrospective study, there was a difference (P = 0.03, OR 5.1) in the presence of necrotizing disease between premature foals with GI disease (53.8% of premature foals) and term (18.75%) foals with GI disease. In infants the clinical signs usually start between day 3-10 but NEC can be diagnosed up 90 days of age.[11] In our clinical experience delayed onset beyond the first week does not occur in foals. Approximately 25–33% of all infants diagnoses with NEC die and 27–63% require surgical intervention.[16] Only 33% of our cases with necrotizing disease survived but this figure includes fatalities because of economic concerns as well as humane considerations and so direct comparison might be misleading. 17/85 of foals that had abdominal sonograms due to signs of GI disease were diagnoses as having necrotizing GI disease. This frequency seems subjectively more than what is seen in other veterinary hospitals. This could be a reflection of regional characteristics, the type of cases that present to different hospitals or the specific concern of clinicians about these particular problems. Foals with PI have a worse prognosis than foals without PI. However, it is important to realize that many foals in which PI was diagnosed survived and foals with PI were likely to have more severe disease which introduces the confounders of financial limitations and humane perceptions. The survival rate of foals for which PI was observed sonographically was 37% vs. 72% in foals in which PI was not observed. When comparing the disease seen in foals to that seen in infants, the equine disease may be more similar to NEC seen in full term infants which in general has an earlier onset, shorter course and is less severe (less frequent perforations) and has been speculated to have a different pathogenesis.[30-32]

Clinical presentation

Clinical manifestations of NEC in infants can be specific to the GI tract (such as feeding intolerance, vomit, blood in the feces, abdominal distention, painful or discolored abdominal wall) and others can be vague: apnea, hypoxemia, bradycardia, lethargy, temperature instability, blood pressure instability, erytema or edema. In many cases NEC may be indistinguishable from sepsis.[11,15] Laboratory values of babies with NEC are often the reflection of infection, activated coagulation and fluid retention. Leukopenia or leukocytosis, anemia, thrombocytopenia, hypo or hyperglycemia, electrolyte abnormalities and metabolic acidosis are common in this population. Bacteremia is reported in approximately 35% of cases of human NEC.[16] Variables associated with the presence of necrotizing GI disease in foals are: prematurity, blood in the feces, gastric reflux, abdominal distension, abnormal colon echogenicity and leucopenia showing some parallelism between NEC and necrotizing GI disease in equine neonates. Bacteremia was observed in 55% of foals with necrotizing GI disease. This was not statistically different to the group of foals with GI disease without a necrotizing component in which 49% were bacteremic. In general the clinical picture seen in foals with necrotizing GI disease is consistent with sepsis perhaps reflecting that the final common pathway of intestinal injury is from the activation of the inflammatory cascade as has been speculated in infants.[33-35]

Diagnosis

The diagnosis of NEC can be challenging and it seems that it is often made based on clinical and imaging findings. Classically radiographs showing PI, a thickened intestinal wall, free air in the abdomen (pneumoperitoneum), and portal venous gas were the basis of the diagnosis. The imaging findings (radiography, ultrasonography and computed tomography) have been described.[28] Radiographic PI has been described in equine neonates.[18] A recent review in infants described 11 sonographic features of NEC[8]: 1. Increased intestinal echogenicity reflecting wall edema, inflammation or hemorrhage, 2. Intestinal thickening (2.7 mm or more), 3. Intestinal thinning (1 mm or less), 4. Increased intestinal perfusion, 5. Absent bowel perfusion, 6. PI, 7.Portal venous gas, 8. Free gas, 9. Anechoic free intraperitoneal fluid, 10. Free intraperitoneal fluid with echoes, and 11. Focal fluid collections. In our group of foals the presence of PI was part of the definition and apart from this finding only the presence of colonic thickening was different between the group with and the group without necrotizing disease. PI was most frequently imaged as scattered hyperechoic foci within the intestinal wall, although large clusters of hyperechoic foci within the bowel wall were also imaged in some foals. The intramural gas was imaged more frequently in the small intestine (58%) than in the large intestine (32%) and infrequently (11%) in both. The presence of increased peritoneal fluid was somewhat common (25%) but the presence of focal accumulations were rare (5%). The presence of hypo or dysmotility is also frequent in equine neonates with necrotizing GI disease. In our retrospective study only sonography was used. Comparisons with radiography would be interesting. Our clinical impression is that sonograms tend to be more informative when evaluating foals with GI disease. When compared to CT, sonograms have the advantage of being stall side and less expensive. We defined necrotizing disease as the presence of signs of GI disease (colic, diarrhea, gastric reflux or abdominal distension) and sonographic evidence of PI or pathologic presence of necrosis. In foals there are no reliable clinical signs that correlate with the presence of necrotizing GI lesions. Even dramatic signs such as hemorrhagic or coffee ground reflux or hemorrhagic or melanotic diarrhea may occur without any gross or histologic lesions. The major sequela in foals surviving with necrotizing lesions is segmental strictures which can be found in the small intestine, large colon or small colon. These strictures respond well to surgical resection.

Treatment

Medical treatment of NEC in infants typically consists of GI rest and decompression, broad spectrum antibiotic therapy, supportive care to manage electrolyte or acid base imbalances and parenteral nutrition.[15] Although full thickness necrosis and secondary peritonitis may occur in affected foals, unlike infants, perforations are not seen in our experience. We have had anecdotal success treating foals at risk of or recovering from necrotizing disease with trophic feeding initially using fresh or frozen colostrum (even in foals up to 20 days old) or fresh mare's milk and with judicious gradual volume increases dictated by feeding tolerance. The prevention of NEC is a complex matter. Avoiding preterm birth, judicious use of antenatal steroids in preterm deliveries, breast-milk feedings (vs. formula) and trophic feedings have been suggested as potential strategies to reduce the incidence of NEC in human neonates. The prophylactic use of oral antibiotics or the use of probiotics have received attention but its general use has not been adopted for reasons such as the emergence of antibacterial-resistant organisms or the questionable efficacy of probiotics.[15] It's been hypothesized that NEC can be a contagious disease and in some reports clusters of cases have occurred. *Klebsiella, E. coli, Enterobacter, Salmonella, Pseudomona, Clostridia, Staphylococci*, coronavirus, rotavirus and enterovirus have been implicated as potential causative organisms.[11,16] There are reports associating *C. difficile*,[20] *C. perfringens*[21] and *Rhodococcus equi*[19] with necrotizing GI disease in foals. We feel that the necrotizing lesions in these reports are a direct result of bacterial pathogenic properties whereas the necrotizing disease we are reporting occurs as a primary problem which may then predispose the foal to secondary pathogenic colonization. In the group of foals that we reported blood cultures, PCRs for detection of *Salmonella* in feces and fecal testing for the detection of clostridial toxins were performed frequently and a positive result was not associated with a necrotizing component to the gastrointestinal disease. In fact the percentage of foals positive to clostridial toxin detection analysis was lower (not statistically significantly higher) in foals with necrotizing GI disease.

Asymptomatic intussusceptions

As a part of our investigations in equine gastrointestinal disease we designed a prospective study in which normal neonatal standardbreds were scanned. Incidentally asymptomatic intussusceptions were found in 10/18 neonates. Finding jejuno-jejunal intussusceptions in these normal foals was unexpected, as asymptomatic intussusceptions had not been previously reported in horses. Jejuno-jejunal intussusceptions are the most common type of intussusception in foals with abdominal pain, but intussusceptions are reported to uniformly require surgical correction in horses.[36-38] Dysrhythmic peristaltic activity has been proposed as the cause of intussusception in horses and people. Asymptomatic small bowel intussusceptions occur in humans and dogs and can resolve spontaneously (up to 20% in people[39] and 8% in dogs[40]) but as stated above this had not previously reported in horses. Sonographic signs that have been associated with transient, nonclinical small intestinal intussusceptions in small animals and infants are the absence of identifiable intestinal lesions, normal wall thickness, length of less than 3.5 cm, normal undilated proximal bowel, normal vascularity on color Doppler, respected layering, and the intussusception being compressible.[40,41] We did not recognize pathologic lead points, increased wall thickness, or altered echogenicity in any of the asymptomatic intussusceptions we observed. Compressibility, length or mesenteric, and mural blood flow in the area of the intussusceptions were not evaluated. The wall thickness of the intussuscipiens was 0.1–0.2 mm thinner than the lower end of the reference range for normal jejunal wall thickness in 4 cases and the average wall thickness of the intussuscipiens was less than the average jejunal wall thickness. Clinical intussusceptions in humans[42] are characterized by an increased thickness of both intussusceptum and intussuscipiens, and the marginal thinning observed in foals with asymptomatic intussusceptions is of uncertain relevance. It is possible that asymptomatic intussusceptions are a normal occurrence in foals of this age range because of initial development of gastrointestinal function and motility. This unexpected finding helped us realize that the finding of intussusceptions in equine neonates should not be considered an absolute indication for surgery and that the clinical, clinicopathologic and rest of imaging findings should be used to decide the course of therapy. Recheck sonograms may be useful in this subset of foals.

REFERENCES

1. Levy JA et al. *Pediatrics 2008* ;121:1404-1412. 2. Navas de Solis et al. Vet Radiol Ultrasound. 2014;0:1-4. 3. Navas de Solis et al. Equine Vet J Suppl. 12;(41):64-8. 4. Baud C et al. Eur Radiol;2004:14:2105-2119. 5. Puylaert JB *Eur. Radiol.*2001 ; 11:1867-1877. 6. Pastore GE et al. *J Am. anim. Hosp. Ass .2007;*43:264-269. 7. Abraham M et al. J Vet Int Med. 2014;28:1580-1586. 8. Silva CT et al. Pediatr Radiol 2007;3:274-282. 9. Faingold R et al. Radiology 2005;235:587-594. 10. Epelman M et al. Radiographics 2007;27:285-305. 11. Walsh MC et al. *Pediatr. Clin. North Am.*1986;33:179-201. 12. St. Peter SD et al. Arch. Surg. 2003;138:68-75. 13. Kim WY et al. Invest. Radiol. 2007;42:312-318. 14. Babcock DS. J. Ultrasound Med. 2002;21:887-899. 15. Thompson AM et al. Drugs. 2008;68:1227-1238. 16. Lee JS et al. Semin. Neonatol. 2003; 8:449-459. 17. Obladen M. Neonatology 2009;96:203-210. 18. Cudd T et al. Comp. cont. Educ. Pract. 1987;9:88-89. 19. Yager JA Vet. Microbiol. 1987;14:225-32. 20. Jones RL et al. J. Am. vet. med. Ass. 1988;193:76-79. 21.East LM et al. J. Am. vet. med. Ass. 1998;11:1751-1756. 22. Slovis NM Clin. Tech. equine Pract. 2003;2:79-86. 23. Magdesian KG Vet. Clin. N. Am: Equine Pract. 2005;21:295-312. 24. Vaala WE In: Large Animal Internal Medicine, 4th edn. Mosby Elsevier, St Louis. 2009:306-315. 25. Wilkins PA In: Equine Internal Medicine, 3rd edn. Saunders, St Louis. 2005:1311-1363. 26. King S et al. Pediatr. Radiol. 1992;22: 577-578. 27. Gurland et al. J Pediatr. Surg. 1998;33:1309-1311. 28. Shawn D et al. Arch Surg. 2003;138:68-75. 29. Caplan M In Fetal and Neonatal Physiology. 4th ed. Philadelphia: Elsevier Saunders; 2011:1256-1261. 30. Wiswell TE et al. Am J Dis Child. 1988;142:532-535. 31. Martinez-Tallo N et al. Biol Neonate. 1997;71:292-298. 32. Bolisetty K et al. J Paediatr Child Health. 2001;37:413-414. 33. Hsueh W et al. Acta Paediatr Suppl. 1994;396:11-17. 33. 34. Caplan MS et al. Clin Perinatol. 1994;21:235-246. 35. Caplan MS et al. Curr Opin Pediatr. 2001;13:111-115. 36. Nelson BB et al. Compend Contin Educ Vet 2012;34:E1–E5. 37. Bernard WV et al. J Am Vet Med Assoc 1989;194:395–397. 38. Gift LJ et al. J Am Vet Med Assoc 1993;202:110–112. 39. Lindor RA et al. J Emerg Med 2012;43:1–6. 40. Patsikas MN et al. J Am Anim Hosp Assoc 2008;44:41–47. 41. Mateen et al. Ultrasound findings andclinical significance. Abdom Imaging 2006;31:410–416. 42. Kim JH. Korean J Radiol 2004;5:178–184.

NEONATAL PHYSIOLOGY AND THE STRESSED NEONATE

Jon Palmer, VMD, DACVIM (LAIM)
Kennett Square, PA

INTRODUCTION

Neonates are not miniature adults. Their physiology is unpredictably different from adults. The neonatal period is the physiologic transition period from the aquatic, self-contained fetal life with its full dependence on maternal nutrition, waste handling and environmental homeostasis to independent extrauterine life where birth and breaking the umbilical cord literally means the neonate must adapt its physiology to provide all the necessities of independent life. The term "neonate" is derived from the Latin natus (to be born) and refers to a newborn during the first weeks of life during the physiologic transition. Although strictly speaking it should encompass the entire period until the transition is complete for all organ systems, by convenience it is usually defined as the first 3 to 4 weeks of life in most domestic species.

FLUID PHYSIOLOGY

Fluid dynamics in the fetus and neonate is much different than that of the adult and these differences have significant implications in responses to fluid therapy. These differences are due to unique characteristics of the interstitium, lymph flow and endothelial permeability. The interstitium is a heterogeneous space which dynamically controls its fluid content much like the vascular space but by poorly understood mechanisms. The interstitial space of the fetus and neonate is easily expanded relative to that of the adult with compliance in the ovine fetus roughly 10 times that observed in the adult. The volume of fluid within the lymphatic system is surprisingly small and averages only 1 mL/kg body weight in adult dogs and may be similar in the fetus and neonate. In fetal sheep, thoracic duct lymph flow averages 0.25 mL/minute/kg which is substantially higher (about 5x) than those in the adult. As with most physiologic phenomena the neonate appears to fall between these 2 extremes. The lymph flow from subcutaneous tissue in anesthetized puppies is approximately twice that observed in adult dogs per kg body weight. Furthermore, lymph flow from the lungs is higher in anesthetized newborn lambs and puppies than in adults. The local, as well as whole body lymph flow rates are significantly greater in the neonatal period than later in life. The increased lymphatic flow during the neonatal period is probably indicative of the elevated interstitial volume in the neonate in relation to the adult and the higher capillary permeability. The capillary filtration rate of fluid in fetal lambs is approximately 5x that of the adult and the filtration rate for proteins about 15x higher. This increased filtration rate has been thought to be primarily due to poor precapillary tone resulting in higher capillary hydrostatic pressure. The role of the glycocalyx, the major endothelial barrier, has not been investigated. The neonate, depending on the maturity at birth, may have very similar endothelial filtration characteristics as the fetus in many vascular beds.[1,2]

During the last few days of gestation fetal blood pressure increases significantly (by 20% in fetal lambs) although fetal blood pressures are still much lower than adult blood pressure. Part of this increase in blood pressure is thought to be transmitted to the capillaries because of poor precapillary tone resulting in increased transcapillary filtration of fluid and protein. Further increases in blood pressure during labor and birth is thought to result in a decrease in plasma and blood volume (18% and 12% respectively in fetal sheep) resulting in an increased hematocrit secondary to these fluid shifts. Other reasons for the fluid shift include direct compression or transformational change of the fetus leading to increased venous pressure and increases in vasoactive hormones including arginine vasopressin, norepinephrine, cortisol, and atrial natriuretic factor.[1]

As the result of these fluid shifts drawing from the fetal fluids or maternal circulation and accumulating in the fetal interstitium all neonates are born fluid overloaded to one degree or another. In the normal neonate the rate of loss of this fluid varies. Foals hold on to this extra fluid and slowly lose it over weeks. Many other species may lose 10-15% of their body weight in the first days after birth. [1]

CONSEQUENCES OF NEONATAL FLUID PHYSIOLOGY

Response to Hemorrhage

Perinatal blood loss of the fetus or neonate can occur after rupture of umbilical vessels, during premature placental separation, by bleeding of the fetus into the maternal circulation (fetomaternal transfusion) or into a twin fetus (fetofetal transfusion), or by internal bleeding, for example, long bone fractures or intestinal hemorrhage secondary to necrotizing enterocolitis. In adult dogs, cats, and sheep, 24 to 48 hours is required for blood volume to return to normal after a 30% loss of blood (without therapeutic intervention). This restoration of volume occurs as plasma volume returns to or rises above normal, whereas RBC volume remains reduced. The time required for full volume restoration after hemorrhage in the fetus or neonate is shorter than in the adult. Fetal sheep restore twice the volume than adults within 30 minutes after rapid hemorrhage. The ovine fetus restores its blood volume to normal within 3 to 4 hours after a 30% hemorrhage which is in one tenth the time required in the adult. This rapid restoration is mediated by a translocation of fluid and protein from the interstitial space into the vascular space. Neonatal kittens and rabbits are also better able to tolerate blood loss than adults, as more blood per kg must be removed before arterial pressure decrease. This was attributed to a more rapid mobilization of interstitial fluid in the young animals in the first week of life. Neonatal lambs also rapidly restore their blood volume to normal after hemorrhage.[1]

Responses to Volume Loading

Rapid intravascular infusions of isotonic solutions expand plasma volume of fetal lambs by only 6 to 7% of the infused volume after 30 to 60 minutes because of rapid transfer of the infused fluid into the interstitial spaces. When adults receive IV fluids average intravascular retention is 20% to 50% of the infused volumes after 30 to 60 minutes. The poor intravascular retention of crystalloid during fetal life and the early neonatal period is due to the high interstitial compliance and the high capillary filtration coefficient which permits very rapid fluid shifts.[1,3]

Once fluid overloaded the neonate retains the fluid much longer than the adult which is thought to be largely because of the reduced intravascular retention of fluid in the neonate. After volume loading, normal adults (dogs, sheep) will excrete the entire volume load through their kidneys over a period of several hours as plasma renin activity and plasma concentrations of arginine vasopressin and atrial natriuretic factor all change in a direction appropriate for elevating urine flow. Puppies and lambs and likely other neonates will retain the same fluid load for 24 to 36 hours probably because the volume load escapes the intravascular before plasma volume sensors can recruit these hormones to exert their diuretic response. Thus the neonatal urine flow rapidly returns to normal after rapid vascular volume expansion.[1]

Both fetal and newborn sheep undergo an increase in left thoracic duct lymph flow up to 3.5 times normal after vascular volume expansion augmented by angiotensin II. Because of the dynamics of this increased lymph flow even small increases in central venous pressure as will occur in response to rapid administration of intravenous fluids will result in a dramatic decrease in the lymphatic flow rate making the neonate receiving intravenous fluid therapy particularly susceptible to the development of edema.[1]

Responses to Hypoxia

Hypoxia in the fetal lamb results in elevation of capillary pressure within the fetal body. There are no changes in fetal arterial or venous pressure if the hypoxia is mild because of vasodilation in selected organs but if the hypoxia is moderate to severe there will be an increases in arterial and venous pressures. This change in capillary pressure due to hypoxia will result in a greater transfer of fluids into the interstitial space leading to increased fluid overload. So although all neonates at birth and just after are fluid overloaded, the fetus or neonate suffering from hypoxia/asphyxia has a greater degree of fluid overload often resulting in the somewhat unique situation of having a hypovolemic individual who is fluid overloaded.[1]

RENAL PHYSIOLOGY

There is marked variation in the degree of renal maturation at birth between species.[4] In normal lambs, foals and calves nephrogenesis is complete by birth and GFR reaches adult levels in days. In the dog nephrogenesis continues for at least 2 weeks after birth.[5-6] In puppies serum creatinine levels and BUN concentrations are lower than in the adult dogs. Foals, calves, lambs, kids and crias also have BUN levels lower than adults and these levels stay low as long as the neonate has a positive energy balance. Creatinine levels in foals, calves, kids and crias is much higher than adult levels at birth. If the neonate is normal these levels will decrease to below adult levels (generally below 1.1 mg/dl and as low as 0.4 mg/dl depending on the species) within 48 hours of birth. If there has been fetal distress the birth creatinine levels in these species at birth may be as high as 20-30 times normal. This increase is not a reflection of renal disease but rather an indication of fluid shifts from the creatinine rich fetal fluids to the neonate as outlined above in the section on fluid physiology (the fetal foal's allantoic fluid Cr is 120-160 mg/dl and anionic fluid Cr 8 to 12 mg/dl). Serum phosphorous concentrations are typically increased in neonates because of special renal handling of that ion.[11]

In general in all species fetal and early neonatal arterial blood pressure is lower than that of adults but this is balanced by lower vascular resistance insuring perfusion. In the fetus about 3-5% of cardiac output perfuses the kidneys. At birth this percentage rapidly increases to 15% due to a modest increase in arterial pressure and although renal vascular resistance also increases it does so to a lesser extent relative to other vascular beds resulting in the kidneys capturing a greater percent of cardiac output. The speed in which the arterial pressure and vascular resistance reach adult levels vary between species and is closely associated with changes in the baroreceptor response and set point and autonomic maturation. The range of blood pressure within which renal blood flow is maintained in the immature kidney is the normal range for that age. The "autoregulatory range" increases as blood pressure increases to adulthood levels.[13] In the foal, calf and lamb the GRF becomes adult-like independent of the increase in blood pressure. In puppies renal blood flow is directly correlated with arterial pressure and does not seem to be altered by inhibition of angiotensin until approximately 6 weeks of age.[6] The balancing of perfusing blood pressure and vascular resistance is vital for proper renal function during the neonatal period. At least in large animal clinical neonatology the most common cause of renal dysfunction is Neonatal Vasogenic Nephropathy where this balance is not maintained because of abnormal levels of vasoactive substances and sympathetic tone. Prostaglandins are vital in counterbalancing endogenous vasoconstrictors by promoting afferent arteriolar vasodilation. Unlike the adult, high prostaglandin activity is physiologically necessary to maintain sufficient perfusion of the newborn kidney. Nonsteroidal anti-inflammatory drug therapy has a much greater potential for producing renal toxicity in neonates than in adults. NSAIDs reduce GFR and RBF so are associated with Neonatal Vasogenic Nephropathy, oliguria and fluid overload. Both nonselective and selective cyclooxygenase inhibitors likely share the same adverse effect so their use is not advised in the neonate.[6] Other major differences in the neonate's response to vasoactive agents from adults is beyond the scope of this talk. It is interesting to note that in the neonatal rabbit, decreases in body temperature as little as 2 C are able to induce renal vasoconstriction and decrease GFR.[14] Hypothermic neonates are at risk. In normal neonates change in environmental temperature may play an important role in the sympathoexcitatory response seen at birth. This response occurs before a decrease in core temperature and is reversible with rewarming, suggesting it is mediated by sensory input from cutaneous cold-sensitive thermoreceptors rather than a response to a change in core temperature.

There is considerable variation in the number of nephrons in normal subjects, and there is a linear relationship between the number of glomeruli and birth weight at least in humans.[2] Despite a significant increase in glomerular volume, both normal and compensatory renal growth are primarily caused by an increase in proximal tubular mass. Intrauterine growth restriction can result in fewer nephrons than in normal neonates. The timing of intrauterine growth restriction may determine whether nephron numbers are reduced. Perinatal asphyxia or severe circulatory disturbances can cause irreversible nephron loss. Exposure of the fetus to maternal administration of prostaglandin inhibitors, glucocorticoids, aminoglycosides, beta lactam antibiotics, or cyclosporine may impair nephrogenesis.

In general tubular function is poorly developed in most species at birth. Even in the species where GFR is normal just after birth, the tubules are short and the carrier density less than adult levels resulting in limited tubular function. In puppies the urine specific gravity is limited (1.006 to 1.017) and initially it is normal to find protein, glucose, and various amino acids in the urine. In large animal neonates urine specific gravity has a broad range (1.001 to > 1.035) within 24 hours. On the usual milk diet the urine specific gravity is generally < 1.004 but in foals the first urine, which is delayed approximately 12 hours, usually has a urine specific gravity > 1.035. Although blood and protein may appear in urine during the first 48 hours of life generally the urinalysis is similar to the adult. Acid base control develops postnatally in all species.[4]

It is interesting to note that although fractional Na absorption in the proximal tubules is not fully developed at birth enhanced absorption in the distal tubule insures Na conservation allowing for the positive Na balance needed for growth despite the fact that a fresh milk diet contains very small amounts of Na. This is achieved by up-regulation of several Na carrier systems. For example in the face of a Na load total nephron fractional Na reabsorption was 0.91 for adult dogs and 0.98 for the puppy ($p < .01$). However, proximal tubule Na fractional reabsorption was greater in the adult dog (0.64) than in the puppy (0.48, $p < .01$), whereas distal nephron Na fractional reabsorption was much greater in the puppy (0.51) than in the adult dog (0.26, $p < .01$).[16] The distal tubule handling of Na seems to be less responsive to Na loads, continuing maximum reabsorption in some neonates rapidly leading to Na overload. Because the neonatal kidney function predisposes to Na overload and the neonatal fluid physiology predispose to prolong fluid overload and (at least in puppies) there may be limited urine dilution, fluid therapy should be administered with great care.[17] Sodium and fluid overload is a common iatrogenic problem in all neonatal species.

CARDIOVASCULAR SYSTEM

At birth, the cardiovascular system changes dramatically with arterial blood pressure, heart rate, and cardiac output increasing, and blood flow distribution undergoing significant regional changes. Cardiac output in the newborn lamb is four times greater than in the adult.[18] The neonate initially retains the fetal low-resistance–high-flow system.[6] To maintain peripheral perfusion, the neonate must maintain a higher heart rate, cardiac output, plasma volume, and central venous pressure as compared with the adult. In the studies of normal blood pressure in large animal neonates restraint artifacts have confounded the results. However clinical experience suggest that most large animal neonates make a rapid transition from the fetal low vascular resistance circulatory system to a physiologic state similar to adults. But a few retain the low vascular resistance (low blood pressure) circulatory system and maintain excellent perfusion. These patients may be retaining the fetal baroreceptor set point having delayed transition to the higher vascular resistance/blood pressure physiology of the adult. Critically ill neonates are more likely to retain the low vascular resistance and low baroreceptor set point. In these patients blood pressure cannot be used as the only surrogate for perfusion and absolute blood pressure numbers make poor and sometimes dangerous therapeutic goals.

Resetting the arterial baroreflex is defined as a change in the relationship between arterial pressure and heart rate or between pressure and sympathetic and parasympathetic nerve activities.[20] The sensitivity of the baroreflex changes with maturation, and shifts (resets) toward higher pressures. This shift occurs during fetal life, is present immediately after birth, and continues with postnatal maturation, paralleling the naturally occurring increase in blood pressure. In all species the baroreceptor reflexes exhibit depressed sensitivity at birth, with a gradual postnatal maturation to adult levels.[18] Studies evaluating heart rate response to hypoxia (decrease in inspired oxygen from 21% to 17% fraction of inspired oxygen) in a variety of newborn unanesthetized mammalian species failed to detect an alteration in heart rate.[24] We recognize a population of critically ill neonates which seem to retain the fetal baroreceptor set point and thus have what appears to be an inappropriate bradycardia in response to hypoxemia with low blood pressures. They appear to be retaining the fetal response to hypoxemia. This autonomic dysregulation may be transient but while present requires careful medical management.

Functional closure of the ductus arteriosus is thought occur in almost half of neonates by 24 hours and in 90% by 48 hours. Until anatomic closure which occurs within weeks, the neonate possesses a unique tool to deal with pulmonary hypertension.[25] Pulmonary hypertension may occur secondary to hypoxemia or, more commonly in my clinical experience, secondary to sepsis. In the neonate because of the right to left shunt, left sided cardiac output will be maintained eliminating the ischemia and increasing the likelihood of survival.

RESPIRATORY AND GASTROINTESTINAL SYSTEMS

Space limitations prevent discussions of the transition of these important organ systems. See http:\\nicuvet.com.

REFERENCES

1. Bauer K, et.al. Chapter 137 In: Polin RA, et.al. Fetal and Neonatal Physiology. 4th ed. Philadelphia: Elsevier Saunders; 2011, p. 1436. 2. Chevalier RL. Chapter 135 In: Polin RA, et.al. Fetal and Neonatal Physiology. 4th ed. Philadelphia: Elsevier Saunders; 2011, p.1436. 3. Brace RA. Am J Obstet Gynecol 1983;147:777. 4. Zoetis T, Hurtt ME. Birth Defects Res B Dev Reprod Toxicol 2003;68:111. 5. Evan AP, et.al. Anat Rec 1979;194:187. 6. Grundy, SA. Vet Clin Small Anim 2006;36:443. 7. Olbing H, et.al. J Clin Invest 1973;52:2885. 8. Hay DA, et.al. Anat Rec 1979;195:273. 9. Kleinman LI, Banks RO. Pressure natriuresis during saline expansion in newborn and adult dogs. Am J Physiol 1984;246:F828–834. 10. Kleinman LI, et.al. J Physiol 1974;239:225. 11. Chevalier RL, et.al. Chapter 124 In: Polin RA, et.al. Fetal and Neonatal Physiology. 4th ed. Philadelphia: Elsevier Saunders; 2011, p. 1317. 12. Rudolph MA, et.al. Circ Res. 1979;26:289. 13. Chevalier RL, et.al. Am J Physiol. 1983;244:F483. 14. Toth-Heyn P, et.al. Pediatr Nephrol 2000;14:227. 15. Macintire, DK. Vet Clin Small Anim 2008;38:621. 16. Banks RO. Proc Soc Exp Biol Med. 1983;173:231. 17. Feld LG, et.al. Chapter 128 In: Polin RA, et.al. Fetal and Neonatal Physiology. 4th ed. Philadelphia: Elsevier Saunders; 2011, p.1349. 18. Hew KW, et.al. Birth Defects Research (Part B) 2003;68:309. 19. Adelman RD, et.al. Dev Pharmacol Ther 1985;8:396. 20. Segar JL. Chapter 70 In: Polin RA, et.al. Fetal and Neonatal Physiology. 4th ed. Philadelphia: Elsevier Saunders; 2011, p.789. 21. Mace SE, Pediatr Res 1983;17:491. 22. Fox MW. In: Fox MW, et.al. Canine pediatrics. Springfield (MO): Charles C Thomas; 1966. pp. 22. 23. Swann HG, et.al. Surg Gynecol Obstet 1954;99:5. 24. Finley JP, et.al. Can J

Physiol Pharmacol 1985;64:122. 25. Clyman RI. Chapter 73 In: Polin RA, et.al. Fetal and Neonatal Physiology. 4th ed. Philadelphia: Elsevier Saunders; 2011, p.821 26. Zoetis T, et.al. Birth Defects Research (Part B) 2003;68:121. 27. Fox MW. J Anim Behav 1964;12:301. 28. Burrin DG. Chapter108 In: Polin RA, et.al. Fetal and Neonatal Physiology. 4th ed. Philadelphia: Elsevier Saunders; 2011, p.1182. 29. Holsapple MP, et.al. Birth Defects Res B 2003;68:321. 30. Felsberg PJ. Hum Exp Toxicol 2002;21:487.

WHEN TREATING DOWN COWS BECOMES AN OPPORTUNITY

Gilles Fecteau, DMV, DACVIM and André Desrochers, DMV, MS, DACVS, DECBHM
St-Hyacinthe, QC, Canada

INTRODUCTION

Adult recumbent bovine remains a challenge for veterinary practitioners, especially if the resources available on a farm are limited. To make an accurate diagnosis and establish optimal therapy remains the primary objective. Determining the prognosis of each case is next to impossible without a definitive diagnosis, which may only come after necropsy in some cases. Over the last 20 years, systems to lift down cows have advanced and diversified. In particular, flotation tanks designed to "float" cows have become popular due to their effectiveness, their ability to avoid secondary problems, facilitate examination, and provide comfort to the cow.

CASE DEFINITION AND CLASSIFICATION

The list of conditions possibly causing an adult cow to become recumbent are numerous however, they can be summarised in a few broad categories: electrolyte imbalance (calcium, potassium, magnesium and phosphorus), musculoskeletal injuries (coxofemoral luxation, gastrocnemius rupture), systemic inflammatory illness (acute metritis and mastitis) and nerve paresis (sciatic, obturator or central origin (lymphoma)). Several terms are used – down cow, downer cow, alert downer, etc – and definitions vary depending on author, which complicates comparisons between published studies. In our hospital we generally use "down cow" to describe a cow that cannot exit the trailer that transported them to the hospital. Once examined, non-ambulatory cows are then subdivided into two categories: 1) those in which a primary cause had been identified and is considered reversible (eg. hypocalcemia) and 2) those in which there is no clear explanation for the incapacity to stand.

WHAT IS HAPPENING WHILE THE COW IS DOWN!

No matter why the animal is down, complications will develop with time. The chances of success decrease as the number of hours increases.

Importance of recumbency duration

In 1969, Fenwick showed that the incidence of problems related to recumbency following milk fever increased with time elapsed between onset of recumbency and treatment with calcium. For recumbency lasting under six hours before the first treatment, the rate of down cows is negligible; for seven to twelve hours, he estimated the rate of complications to 26%; for twelve to eighteen hours, it is 32%; and for more than eighteen hours, it reaches 38% (1).

Origin of muscle necrosis

Cox was first to hypothesise that ischemia due to prolonged pressure on a limb during recumbency was involve in down cows (2). With an increase of pressure on the external soft tissue, vascular permeability increases, producing an extracellular accumulation of fluid and therefore swelling. If the fascia that surrounds the muscle prevents distension, the internal pressure will increase dramatically, even beyond the external pressure that initiated the problem. Tissue anoxia leads to cell damage and inflammation, which in turn induces an increase in pressure within the tissue (compartment syndrome).

Degree of involvement in the development of neurological complications

In 1982, Cox et al. investigated the consequences of compression in down cow syndrome (2). Fifteen adult cows were kept in sternal recumbency on a rubber mat for six, nine, or twelve hours with the right hind limb under the body. Afterward, half of the cows were able to stand up while eight remained down. Among the latter, the right limb was stiff and swollen. During necropsy (10-14 days after), the most obvious injury was the sciatic nerve in the caudal region of the proximal femur. The nerve was discolored and surrounded by a deposit of connective tissue. Muscle damage was limited to the same caudo-proximal region of the thigh: the semitendinosus was the most affected (very pale and necrotic) as well as the gluteal muscle on the greater trochanter. Regular measurements of CK on all animals showed values within normal limits three and six hours after anesthesia and similar levels for all the animals after twelve hours. The differences between the two groups (the cows able to stand and those not able) appeared only twenty-four hours after anesthesia. The levels of CK in the animals that stood peaked at 24 hours and began to drop, while the levels from down animals continued to increase and peaked at 48 hours. Six days later, both groups found comparable levels of CK. The comparable levels during the first day show that the two groups of animals had similar muscle injuries, meaning only the sciatic nerve lesions found during autopsies could explain the trend toward prolonged recumbency in some cows.

The real question

The question now arises as to why some animals develop lesions in the sciatic nerve and others do not. One of the hypotheses is that very slight differences in leg position can have a decisive influence on the severity of lesions. The nerve branches pass through a small anatomic region caudal to the greater trochanter. A compression point located in this area will affect the sciatic nerve directly or the muscle groups surrounding the muscle, or a combination of the two. With the results of Cox in mind, it is clear that the positioning and comfort of the animal is decisive in her ability to recover.

FIVE IMPORTANT STEPS IN THE SYSTEMATIC APPROACH

1) History

The cow's history is very important in the diagnostic process. In particular, you should collect the following information.
Age: cows over 5 years old are more likely to suffer from hypocalcemia and lymphoma.
Parturition: dystocia and forced extraction lead to inflammation of the pelvic inlet. The most common peripheral nerve problems are obturator and sciatic paralysis.
Trauma or a fall: recumbent cows that are not in a postpartum period have often suffered trauma. The precise diagnosis is sometimes difficult, especially when proximal structures are involved.
Sequence of events since recumbency: a detailed history of what has happened since recumbency is very important. Did the animal stood after the initial treatment? Was it found in an awkward position and never stood after? Did anyone see the animal slip and fall?
Treatment and response: a detailed list of administered drugs and their posology (dose, frequency, and routes of administration) is essential in the treatment of a case (calcium, antibiotics, anti-inflammatories). To the best knowledge of the producer, have any previous or established treatments helped? Was the cow able to stand after calcium therapy?
BLV status: individual and herd

2) Physical examination

A physical examination of a cow that is recumbent requires extra effort. You are often working by yourself in a small space while surrounded by a mixture of wet litter, faeces, urine and colostrum. Nonetheless, time and energy should be directed toward this first and important task.
Acute systemic inflammatory conditions: rule out the frequent inflammatory conditions that can occur during the puerperal period: acute metritis and mastitis (check all 4 quarters, even those difficult to reach since under the cow).
Musculoskeletal problem: examine all 4 limbs, which necessitate turning the cow to see both sides. Look for pain, heat, redness, swelling, angularity, crepitus, fracture, and evident muscle tears.
Neurological deficit: examination of skin sensitivity may be useful in locating the nerve structures involved. Remember that in most cases several nerve branches are affected, complicating the clinical aspect of a problem. For example, the sciatic and obturator nerves are often injured during difficult parturition.
Electrolyte imbalance: clinical signs associated with electrolyte imbalance should be evaluated carefully however laboratory tests are often necessary to state with certainty that the electrolyte concentrations are within normal range. In most cases, previous treatments will mask the clinical signs of classic milk fever and marginal residual hypocalcemia cannot be identified without a laboratory analysis.

3) Laboratory analyses

Laboratory analyses are very useful in the diagnostic work up of down cows, in particular in the assessment of electrolyte imbalances. This is a crucial step because electrolyte imbalances must be resolved before you should encourage a cow to stand.
Serum biochemistry profile. Electrolytes, renal and hepatic function and integrity are important. Also, AST and CK provide insight on the extent of the muscle damage and help determine prognosis (3).
Complete blood count (CBC). The CBC may reveal acute inflammatory response (mastitis and metritis associated with endotoxemia), severe anemia (abomasal ulcer) or persistent lymphocytosis (BLV).

4) Ancillary tests

Ultrasound: Use this method to assess the integrity of a muscle or tendon or for more information about a mass detected during the physical examination. Coxofemoral joint could also be assessed by this technique (4).
Analysis of the cerebrospinal fluid: This additional test is useful in diagnosing problems with the spinal cord (severe trauma, vertebral body abscess and lymphoma).
Radiographic examination of the pelvis: radiographs are particularly useful in confirming a diagnosis of fracture or luxation of the hip (4).

5) Assess the producer's attitude

The producer's attitude greatly influences the chances of success. If your perception is that he feels the battle is already lost, your effort could not replace his dedication. Remember that your contribution to the therapy is circumscribed by the duration of your visit, and the overall quality of care the animal receives rests on the shoulders of the producer (while you are not there). If he is not convinced that the chances are favorable, will he make the extra effort needed? It is very clear from our experience that every little details that is done to care for a down cow can significantly influence prognosis.

MAINTAINING YOUR CHANCES FOR A FAVOURABLE OUTCOME

Minimize trauma

Transport must be undertaken with caution. Always position the tail well to reduce the risk of trauma. The pelvic limbs must be tied together (hobbles) to prevent abduction that can result in severe muscle tears and even dislocated hips. Moving an 800 kgs animal is not routine work. Multiple things can go wrong including the security of the people surrounding the animal.

Prevention of injuries associated with recumbency

The consequences of recumbency must be prevented at all costs by moving the affected animal to a floor other than concrete and keeping it in a NORMAL anatomical position. Ideally, the cow should have a large stall – where she can turn around and move without difficulty – with a dirt floor covered with a thick layer of straw. If she is unable to change sides herself, the cow has to be rolled over regularly (ideally every 4-6 hours) to minimize the pressure on the muscles and nerve structures. After she moves, pay special attention to the limb kept under her body. Make sure the limb bearing most of the weight of the animal is in a normal sleeping position and not stretched under her.

Helping the cow to stand

The following is an evaluation of the advantages and disadvantages of the different assistive technologies available to lift down cows.

Physical assistance: You can best help a cow get up by supporting the base of the tail, which acts as an anchor. Take the time to properly position the limbs beforehand. It is also a good idea to change the side a cow is on just before giving her a boost in case the limb on which she spent the night is a bit weak.

Slings: Several models exist and each offers its own advantage, but they all have a major drawback: the animal squeezed in a sling is just not comfortable. In addition, if the cow is unable to stand, the pressure in some areas quickly leads to complications (wounds, swelling).

Hip clamps: This technique is outdated and we strongly discourage its use.

Inflatable air bag: Much more useful in vehicles than on the farm.

Flotation tanks/Aqualift: The advantage of this system is that it facilitates standing with little inconvenient. Once the animal is standing, it is easy to examine the pelvis and limb position and to note any swelling, deformities of the proximal limbs, and if the weight is equal on all 4 legs. Flotation tanks are limited by the fact that they require access to a large amount of water (ideally warm) (5).

Prevention of intramammary infections

You must continue milking down cow and ensuring that the udder is only in contact with clean, dry bedding. In fact, using a teat dip is a good idea. A word of caution: using wood shavings as a litter can promote Klebsiella mastitis.

USEFUL TREATMENTS FOR RECUMBENT COWS

Fluid replacement therapy

A recumbent animal is often dehydrated and fluid replacement should be considered. Oral or intravenous fluid therapy will help restore basic functions. The fluid should contain calcium and potassium unless blood results show that these electrolytes are within normal range. Many dairy cows in the immediate post partum period are challenge by some degree of lipomobilisation. The presence of ketones should be measured and glucose added to the fluid therapy if necessary.

Pain control and anti-inflammatory drugs

The administration of analgesic improves the comfort of the animal and helps maintain appetite (clinical signs of abomasal ulcers should be monitored when prolonged use). Whether it is to reduce inflammation around the nerve structures or to decrease the acute inflammatory response associated with endotoxemia, the use of anti-inflammatory drugs in the treatment of recumbent cows is no longer a debate. What is open for discussion is how to determine which one is more useful, or even what family should be used – steroidal or nonsteroidal? We tend to use steroidal when nerve structures are involved and non steroidal when acute inflammatory response is diagnosed.

Antibiotic therapy

This will be indicated if an infectious process is identified by clinical or laboratory examination. Be aware that many recumbent cows develop clinical signs of pneumonia (fever, tachypnea, increased respiratory sounds, and sometimes a cough). The majority of cases appear to benefit from antibiotic treatment.

CONCLUSION

The causes of recumbency are diverse and each situation is unique. A methodical, rigorous, and systematic approach is the only way to increase the chances of success. Down cow syndrome remains a frustrating problem for the veterinarian practitioner. New ways to assist in lifting the cows could lead to better evaluation, more accurate diagnosis, and ultimately improved accuracy of prognosis.

REFERENCES

1:Fenwick, D.C. (1969). *Australian Veterinary Journal* **45**:111. 2: Cox, V. S., C. J. McGrath, et al. (1982) *American Journal of Veterinary Research* **43**(1): 26-31. 3: Shpigel, N. Y., Y. Avidar, et al. (2003) *The Veterinary Record* **152**(25): 773-776. 4: Marchionatti E, Fecteau G, Desrochers A. Vet Clin North Am Food Anim Pract. 2014 Mar;30(1):247-64. 5: Smith, B. P., J. Angelos et al. (1997) *Proceedings 30th American Associon of Bovine Practioners Conference*: 43-45.

PHYSICAL EXAMINATION OF DOWN COWS REGARDING ORTHOPEDIC CONDITIONS AND PROGNOSTICATION

André Desrochers, DMV, MS, DACVS, DECBHM and Gilles Fecteau, DMV, DACVIM
St-Hyacinthe, QC, Canada

In peripartum, hypocalcemia and other electrolytes imbalance are often the primary causes preventing a cow to stand. If not responding to the electrolytes therapy, she will be called a down cow. Many things can go wrong on a heavy animal trying to rise and stand on a slippery concrete floor. Prompt and thorough examination are essential to decide if the animal has a fatal condition, can be treated conservatively or would benefit of a floating tank.

ASSESSMENT OF A DOWNER COW

The physical examination is performed methodically keeping in mind 4 common causes of recumbency: Acute systemic inflammatory conditions, Musculoskeletal problems, Neurological injuries and Electrolyte imbalance. In large animals, it is common practice to perform the physical examination by side, right and left. The same principle applies to down cows. However it should be divided in sternal and lateral recumbency for each side. The animal must be in lateral recumbency for the musculoskeletal examination and to evaluate the mammary gland. The exam will start with whichever side the cow is lying with sternal and lateral recumbency. Then the animal will be switch on the other side to repeat the same process.

MUSCULOSKELETAL EXAMINATION

Pelvic limb

The thigh and gluteal muscles are palpated for fibrosis, swelling, hardness and pain. Skin sensitivity is evaluated with a hemostatic forceps by pinching the skin starting from the coronary band up to the stifle. With the limb slightly elevated, the patellar reflex is evaluated by knocking on ligament insertion at the tibial crest. Then the limb is palpated and manipulated from the claws to the hip. Each joints and bones are stressed to elicit pain; the joints are flexed and extended, long bones are firmly palpated. The integrity of the gastrocnemius and the peroneus tertius is evaluated by flexing or extending the tarsus and stifle. Being part of the reciprocal apparatus, the stifle and the tarsus are flexing or extending in harmony. With a ruptured gastrocnemius, the hock can be flexed while the stifle partially extended. With a ruptured peroneus tertius, the hock is extended while the stifle is partially flexed.

The stifle and the hock are evaluated for articular fracture and collateral ligament rupture. Lateral and medial movement of the joint will help to determine joint instability and abnormal cranio-caudal limb angulation. Although cruciate ligament injury is possible, it is the authors' opinion that it will not prevent a cow to stand. Condylar or supracondylar femoral fracture are not uncommon in adult cattle. It is diagnosed by abducting and adducting the distal limb. Crepitation can be felt at the stifle level and the limb can be abnormally abducted.

The coxofemoral joint is first assessed by the relative position of the greater trochanter to that of the tuber coxae and the tuber ischii. The normal position of the greater trochanter is ventral to both of these bony prominences and imaginary lines drawn between them will create a "triangle" with a shorter side between the greater trochanter and the tuber ischia. Positioning of the greater trochanter in-line with the tuber coxae and tuber ischii suggests dorsal luxation of the coxofemoral joint. If a ventral luxation is present, the greater trochanter cannot be seen or palpated as usual because of its abnormal medial positioning. The hip is further manipulated by forcefully abducting and adducting it. A subluxated hip can be ventrally luxated by excessive abduction which is diagnostic. Palpation and location of the femoral head per rectum can also be performed if limb manipulation is inconclusive.

Thoracic limb

With the cow in lateral recumbency as previously stated, the larger muscles are palpated for fibrosis, swelling and pain. Skin sensitivity is evaluated from the coronary band up to the elbow. Triceps reflex is assessed with the Frick speculum hitting the tendon at its insertion on the olecranon. Then each joints and bones are stressed to elicit pain with pressure, flexion and extension.

Vertebral column

Trauma to the vertebral column is suspected if there is bilateral paresis or paralysis of the limbs or if the animal is in pain for unknown reason. Needless to say that the vertebral column is difficult to evaluate on a down cow. The lateral aspect of the cervical vertebra can be palpated. Neck is carefully flexed in different direction to elicit reaction from the animal. An S shape neck is often observed on hypokalemic animal from muscle weakness. The spinous process and epaxial muscles of the thoracic and lumbar vertebrae are palpated for abnormal deviation, swelling or pain. Crepitation can be felt over luxated or fractured area.

Other areas of concern

The inadequate use of hip lifter can seriously damage the oblique and other surrounding muscles of the tuber leading to massive irreversible muscle damage keeping the cow down even if the primary problem was successfully treated. Ribs can be fractured from a fall or during careless manipulation of the down cow. The affected animals will be in pain, tachypneic with a very superficial respiratory pattern. Ribs are individually palpated and pain reaction observed. Affected area can be ultrasound to confirm fracture. We have seen lungs' trauma from fractured ribs. The tail is often neglected when manipulating or moving the animal. The sensitivity, motility, color are evaluated upon admission of the animal. A necrotic tail can be at the origin of ascending neurological infection.

NON FATAL MUSCULOSKELETAL CONDITIONS

Muscular damage

Muscular damages are of two origins: a specific torn muscle or myopathy or muscle degeneration and necrosis from prolong recumbency. All downer cows have some degree of muscle necrosis. It is difficult to give a prognosis based on the location of the affected muscles. Usually the prognosis is based on the treatment's response. The affected animals must be on a deep bedding, inflammation and pain controlled with NSAID's, and judicious use of the floating tank. Although difficult to objectively assess, acute muscle degeneration and necrosis preclude the animal to stand in the tank. It is very painful and it is the authors' opinion that those animals will benefit to simply rest in a well bedded stall than trying to stand in the tank.

Peroneus tertius rupture

The peroneus tertius in cattle has muscle fibers compared to the horse. It originates from the extensor fossa of the distal femur along with the digital extensor and inserts in the area of the tarso-metatarsal joint. It will rupture after and hyperextension of the tarsus while the stifle remains flexed. This situation is often encountered when the animal slides with the limb backward. Its rupture rendered hock flexion more difficult and the animal might fumble having difficulty to clear the ground with its foot. Based on ultrasound finding, the rupture is more a muscle tear than as straight tendon rupture precluding any surgical attempt to suture it. It should not prevent the animal to stand neither to walk. Other than stall rest for weeks, there is no specific treatment for this condition.

Brachial, Radial nerve paralysis

While attempting to get up and stand, the cow might slide and over abduct a thoracic limb. The more dramatic consequence will be radial nerve damage. Prolong lateral decubitus on a hard surface can also be at the origin of the damage. Typically the elbow will drop and the carpus remains partially flexed. Animals with ruptured triceps, humeral or olecranon fracture will show a similar posture. Affected animals have difficulty to get up and stand lacking strength to keep their balance. Physical examination rules out fracture. Skin pricking or pinching as well as triceps reflex confirm the diagnosis of radial nerve paralysis. Splinting at the caudal aspect of the forelimb, from the foot up to the elbow might help the animal to stand. If there is a neurapraxia, strength will come back in less than 3 weeks. If paresis is still present after that, nerve damage is important and a prognosis is difficult to establish.

FATAL MUSCULOSKELETAL CONDITIONS

Needless to say that the pain and weakness engendered by some conditions will prevent the cow to get up even in a floating tank. Even though repair of a luxation, or fracture is surgically successful, the chance of survival are slim. The amount of pressure or stress on the repair is tremendous on a debilitated cow trying to rise and stand. Postoperatively, those animals will have to be floated otherwise prolonged decubitus will worsen the prognosis. Floating a cow in a tank with a cast, a splint or simply with a fresh surgical wound seems feasible with appropriate measures (sticky surgical drape on incisions, plastic bag over a cast) but often unrewarding. Although it is feasible, a grave prognosis with thorough explanation of the situation must be given to the owner.

Coxofemoral luxation

As reported in the literature, most cows affected with ventral luxation are presented down (68%-83%) compared to dorsal luxation (6%-43%). The origin of this ventral luxation is most likely secondary to an extreme abduction because of a slippery floor or obturator nerve paresis or paralysis. Although clinical examination is often diagnostic, a ventrodorsal radiographic image of the pelvis will confirm the luxation but not its exact location. Acetabular, femoral head fractures will worsen an already grave prognosis. Therefore downer cows with ventral luxation are most often immediately euthanized. If reduction should be attempted, it must be within 48 hours. Close reduction of a ventral luxation is extremely difficult compared to a dorsal luxation. An open approach is possible but also extremely difficult. The cartilage, capsular and adjacent muscles are extensively damage preventing efficient surgical stabilization and increasing postoperative relapse. According to one study, the prognosis factors for successful outcome were: standing at admission, less than 3 years old, body weight less than 400 kg and duration of less than 12 hours. Those criteria are rarely met in downer cows.

Ruptured gastrocnemius

The rupture is usually at the tendinous-muscular junction. To keep a cow down, the rupture must be complete. The condition and the repair carry a grave prognosis and the animal is rapidly euthanized.

Femoral fracture

This is a fatal condition in adult because of the severity of the fracture. On a down cow, the femoral fracture is comminuted, condylar, multiple with extensive soft tissue damage. The cow is unable to stand because of the pain and the lack of support from the shattered femur. Abnormal angulation and crepitation upon manipulation of the limb will confirm the diagnosis.

Vertebral trauma and lymphoma

Vertebral trauma often results in spinal cord injury. Affected animals are in pain and limbs can be paralyzed depending of the level of the trauma. The vertebral column is palpated and the clinician should look for abnormal tumefaction, pain, deviation of the spinous process and paniculus reflex. Careful examination of the skin sensitivity and reflexes of the limbs might help to localize the affected area. Special attention must be in the thoracolumbar region. Lateral radiographic views of the vertebral column can be performed to confirm the diagnosis. Cerebrospinal fluid tap at the level of the lumbosacral area can confirm the diagnosis of trauma or lymphoma depending of the severity and the location of the disease.

CONCLUSION

A thorough examination of the musculoskeletal system on a down cow will help the clinician to recommend an appropriate treatment or euthanasia. In many cases it might prevent unnecessary hauling and suffering of a cow with a fatal disease.

REFERENCES

Jubb TF et. al. Prognostic factors for recovery from coxo-femoral dislocation in cattle. **Aust Vet J** 1989; 66: 354.

Tulleners EP, et. al. Coxofemoal luxations in cattle: 22 cases (1980-1985). **J Am Vet Med Assoc** 1987; 191: 569.

Marchionatti E, et. al. Traumatic conditions of the coxofemoral joint: luxation, femoral head-neck fracture, acetabular fracture. **Vet Clin North Am Food Anim Pract** 2014;30(1):247.

Burton AJ et al. Prognostic indicators for nonambulatory cattle treated by use of a flotation tank system in a referral hospital: 51 cases (1997-2008). **J Am Vet Med Assoc** 2009;234:1177.

Green AL et al. Factors associated with occurrence and recovery of nonambulatory dairy cows in the United States. **J Dairy Sci** 2008;91:2275.

USING POPULATION DATA FOR SYSTEMATIC ANALYSIS OF HEALTH INTERVENTIONS IN DAIRIES

Gerald Poppy, DVM, MBA
Cedar Rapids, IA

As dairies consolidate in size, populations of lactating cows in a single location of 5,000 or even 10,000 cows are becoming more common-place. This consolidation and increase dairy size has led to the rapidly changing role of the Veterinarian in US dairies. Traditionally, we as veterinarians are trained in individual animal care. The best examples of individual animal care are surgeries for displaced abomasums (DA), C-sections, or even palpating individual cows for pregnancy. It is the role of the management teams for larger dairies, which now includes the veterinarian, to look for answers to systematic questions as opposed to dealing with questions at the individual cow level. What are the herd level questions that veterinarians should be involved in? Herd level questions that have an economic outcome could be:

- Should the farm be doing DA's on cows or culling them?
- What is the evidence for or risk of treating mastitis cows with antibiotics?
- Should we feed yeast culture for decreasing salmonella in calves?
- What antibiotic is superior for treating BRD?
- Should we implement herd level treatment in our fresh cows for subclinical ketosis?

A useful approach to large herd interactions is to view the dairy herd as an organism. Is the dairy sick or healthy? Dave Galligan, DVM, MBA once stated that "low milk production is a disease of dairy cows". What are the tools we use to assess the health of the dairy, much like we use the thermometer or stethoscope on the individual cow? Some of these tools are herd records such as Dairy Comp 305 or DHI+ or PCDart. Herd records alone fail to adequately asses the health of the dairy. From the perspective of the dairy ownership, the healthy dairy enterprise is one that is returning an adequate profit for the amount of risk it is assuming. It might be better for the dairy investor to invest in treasury notes for instance instead of in the dairy.

When we begin to look at profit on the dairy enterprise, there are so many variable that it becomes overwhelming. Breaking down profit into manageable or actionable areas is the first important step. What goes into the profit of a dairy farm? Profit which is revenue – costs on the dairy farm comes from milk production and culling of unproductive cows (revenue) and the costs are expenses from feed and reproduction and labor. The raising of replacements for the herd and the associated costs are a bigger question than we want to get into in this paper. Breaking the system further, the milk production for revenue is made up of the number of animals milked times the amount of milk per cow. Amount of milk per cow is determined by how many cows are at what stage of lactation, number of cows in each parity and the quality of feeding program, along with some environmental effects.

We can see that profit is at the top of the chain events. You could look at it as a portfolio of possible combinations of events. Each portion of these events can be assessed for their impact on the final profit of the dairy as a marginal decision. Marginal revenue is the amount of revenue a firm makes when it sells one more unit of output. A marginal profit is the amount of marginal revenue minus marginal cost. The ability to analyze the marginal decision is a key tool helping dairies make daily management changes that have the most impact. Veterinarians are in a unique position to help and drive decisions based on marginality as opposed to utilizing average costing.

The opposite of average costing is marginal analysis. This is where only the next unit input is analyzed for its potential for profit. The fixed system is ignored as it isn't relevant to the analysis.

An example of a marginal decision that is fun to think through for veterinarians is Fetrow's marginal pig example (J. Fetrow, personal correspondence; Eicker et al. 2006). The example goes as follows:

There is a feeder pig operation with the following economics.

Number of pigs	1	1000
Cost of pig	$ 40	$ 40,000
Starter Feed	$ 15	$ 15,000
Labor	$ 5	$ 5,000
Vaccine	$ 3	$ 3,000
Total Cost	$ 63	$ 63,000
Sale per pig	$ 60	$ 60,000
Net profit per pig	$ -3	$ -3,000

Figure 1. Partial budget of weaned pig enterprise.

It is clear this producer is having problems. He is raising 1,000 baby pigs and in doing so he is losing $3,000 dollars on this enterprise. A veterinarian is doing a walk through on the farm and notices a baby pig with a hernia. He offers to fix the baby pig hernia for the cost of $20. There is a decision to make. Should the producer spend $20 for the hernia repair even though they are already losing an average of $3 pig or should the baby pig be euthanized so there is no more cost? The second option would save the money spent on starter feed, labor and vaccinations.

Starter Feed	$0
Labor	$0
Vaccine	$0

Figure 2. Cost of raising the euthanized baby pig

Let's look at the partial budget for the decision to do the surgery repair (Figure 3).

Repair Hernia	$	20
Starter Feed	$	15
Labor	$	5
Vaccine	$	3
Total Cost	$	43
Sale per pig	$	60
Net profit per pig	$	17

Figure 3. Partial budget of raising the baby pig after the hernia repair.

As you can see in this example, fixing the hernia repair turns this baby pig into a magic "marginal" pig and would result in $17 more income for this farm. The reason for this profit is because the cost of $40 per pig was already paid for prior to finding the hernia. This cost is a fixed cost. In this scenario, the producer would also only sell 999 pigs instead of the 1000 had he elected to euthanize the pig. The final partial budgets comparing the two options are below.

	Euthanize		Repair	
Number of pigs		999		1,000
Cost of Pigs	$	40,000	$	40,000
Repair Hernia	$	-	$	20
Starter Feed	$	14,985	$	15,000
Labor	$	4,995	$	5,000
Vaccine	$	2,997	$	3,000
Total Cost	$	62,977	$	63,020
Sale of pigs	$	59,940	$	60,000
Net profit per pig	$	(3,037)	$	(3,020)
Advantage			$	17

Figure 4. Partial budget of pig enterprise with or without doing the hernia repair.

The first thing we can see from this example is that partial budgets are a useful tool to analyze decisions. Second, it is important to know what is excluded from the partial budget to focus on the real question at hand.

In this example, we evaluated the decision to do a hernia surgery with the assumption of a 100% success rate for the surgery. One prime use for partial budgets is evaluating health interventions such as vaccinating the herd, putting new feed additives into the herd, changing milking frequency, or purchasing new equipment. This will help to determine if the new intervention is going to make a difference and what is the "risk" surrounding the true response for the new intervention.

We have all studied statistics and the experimental process. Often the meaning of the outcomes of the statistical analysis from a decision making process are overlooked. We can assist the producer in making these decisions and helping them understand the impact of trials and variability. If we have a product that is evaluated in a recent random control trial that shows an improvement of 3 lb of milk with a P-value of 0.05, what does this tell us? Does this mean that 95% of the farms that feed this product will get an additional 3 lb of milk and 5% will not get an additional 3 lb of milk? This is often the common thought on what this means. However, this actually only tells us there is good evidence the true response is not 0 lb. We need to move away from evaluating products and interventions in trials by using P values and begin to use confidence intervals. We use P values as a method for limiting our Type I error in the design of our hypothesis testing. Not making a Type I error, we think the results are different than 0 when they really are not, is important for our clients and we do not want to mistakenly recommend the wrong intervention for the dairy, but It is important to understand that in this process there are in fact 2 errors that can be made in recommending the intervention to the dairy. In biology we look at only the biological response. When we are dealing with the dairy enterprise, it is the biological response times the value of the biological response minus the cost of the intervention. For this purpose we the partial budget as in the pig example above to evaluate these changes.

We all have experience in dealing with the Type I error (hypothesis testing for $P < 0.05$ or another set value). There are actually 4 possible outcomes that are made in each decision to use or not use an intervention (Figure 5; Williamson, 1975; Galligan,

1991). For example. Let's say a new product is introduced that will decrease subclinical ketosis SCK in fresh cows. As a veterinary advisor you look at the data and evaluate the trials. You look at a partial budget for the cost of the product and the impact it would have on the herd. The first question; is the new intervention profitable or above breakeven for the dairy. The firm selling the product claims it is. You decide not to use the product because you think it doesn't have enough data. Maybe there is only a couple trials and they both came back at $P=0.09$ and when you add this to the cost you decide it is too risky. This might have been the correct decision, you avoided a Type I error. On the other hand, the product might have done just as they claimed and you are now responsible for the dairy to continue to lose 4% to culling that is attributable to ketosis complications in the dairy. (On this dairy of 10,000 cows that is about $800,000 per year). You made a Type II error. You should have used the product. The other outcome; is you decided the product looked good and there is a lot of labor involved in drenching all the cows in the fresh pen so you recommend using the product. The product works as proposed and you made the correct decision. On the other hand the product doesn't actually work and you made a Type I error. (The product costs $0.40 per cow per day and is fed only to the fresh pen for 15 days. You lost the dairy $60,000).

Type I and II Errors

Actual Results

Decision Results	Above Breakeven (PROFITABLE)	Below Breakeven (UNPROFITABLE)
USE PRODUCT	CORRECT	INCORRECT TYPE 1 ERROR
NOT USE PRODUCT	INCORRECT TYPE 2 ERROR	CORRECT

Figure 5. Four possible decision outcomes associated with binary decisions regarding the use of profitable or unprofitable products. (Galligan et al. 1991)

For dairy management, we do not really care if the product biologically works at $P < 0.05$. We care if the biological change along with the economics of the change is greater than the break-even cost (the cost of the intervention plus other variable costs). This is a much more difficult question than it appears. We can see in this example the cost of the errors might be important in the decision. I might be more willing to take the risk of a Type I error if I realized the cost of the Type II error was 10X the Type I error. This is why the economics of the partial budget are important. Another important consideration is that when we only look at the biological change by not implementing the process, you only forgo the cost of the product. When you look at the biological times the economic change there is many possible outcomes that could have occurred. For the Type I error you use the product but it is below break-even, but it might be just a bit below break-even. Suppose you got half the benefit but it wasn't above breakeven. What if you got ¼ of the benefit? What if it prevented the death of 30 cows instead of all 400 that died due to ketosis related problems? You still would have made a Type I error but now it only cost $100. The Type II error, you should have used the product but you didn't, has even greater potential cost. In the case of the product that decreased SBK, suppose you also got a milk response. You cannot realize this improvement because it was never in the ration. Galligan (1991) following on an earlier paper by Williamson (1975) built a brilliant model for calculating out the cost of making the wrong decision (Figure 6)

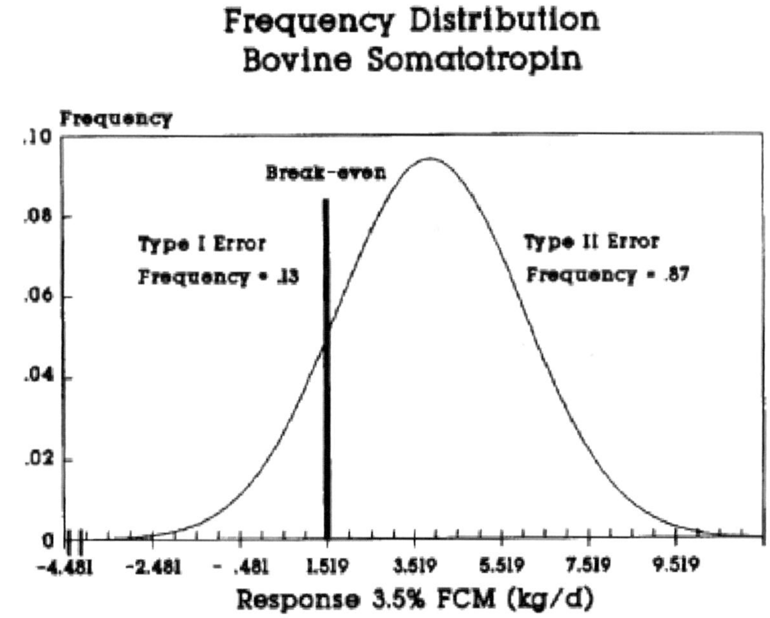

Figure 6. Frequency distribution and break-even production level for bST milk production responses. (Galligan 1991)

This example looks at a decision to use bST in a dairy herd. We can see that the mean and the standard deviation of the response create a bell shaped frequency distribution. There are an infinite number of possible responses in the frequency distribution, some of which are above the break-even response an some that are below. We know if you don't use the product there is 0 response. If you use the product and it is below break-even, there are many possible responses. Most of them close to the break-even. The mean of the Type II error side, you didn't use the product but should have, has a mean close to 5 kg/d, but again the response might have been anywhere from break-even to over 9 kg/d.

Historically, as scientists, we really wanted to avoid Type I error. I thought the medicine was safe when in fact it will kill me. Death makes avoiding the Type I error very important. However, from a dairy management standpoint, it is important to determine if more money can be made by using or not using an intervention. The conservative stance is to not use anything new because the current system is working well enough. On the other hand, investing pennies could be beneficial and result in additional profit. This would be worth the risk.

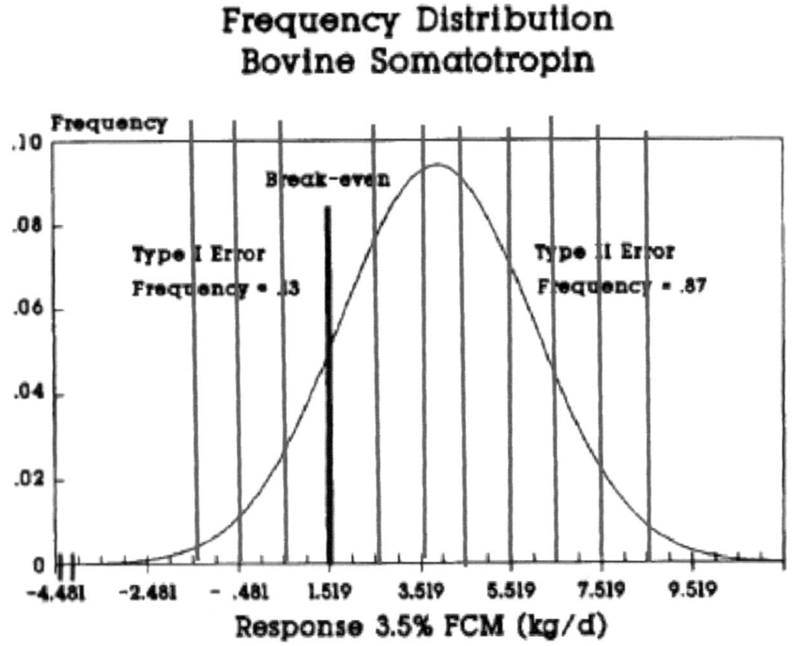

Figure 7. Frequency distribution and break-even production level for bST milk production responses with the Type II frequency divided into discrete responses. (Galligan 1991)

The cost of making the wrong decision is dependent on what the response might have been had the producer used the product. If we now take and divide up the frequency distribution into small segments, we can use this to calculate the cost of making a Type II error vs. making a Type I error (Figure 7). We can calculate the cost by multiplying the revenue (expected value of the milk) with each of the expected possible responses times the frequency of the occurrence of that response. In Figure 7, we can see that there is a possibility of getting as much as 9.5 kg of response, but the frequency is very low. It is more likely to get a response of 4 kg. One can keep dividing the intervals smaller and smaller to become more accurate. However, this makes the calculation more complicated. In this bST example, Dr. Galligan calculated the cost of making the Type I error to be $0.13 per day per cow, and the cost of a Type II error to be $0.87 per cow/day (using cost and revenue data from 1991). We can see that the farmer might actually be willing to use a much riskier intervention when the cost of error is high and there is little downside to not using a product (low Type I). Too often as scientists we insist on $P = 0.05$. However, this may not be in the best interest of the dairy.

This is obviously not an easy analysis to whip up to a dairy farm each time you visit or are presented with a decision. Understanding the variation surround the outcome of each intervention would be necessary. Some of this information we could get from journals articles or from an "expert opinion". If we look back at our marginal pig partial budget, we can should now ask what the outcome would be if we were not 100% successful in our surgery. We know that the "marginal pig" most likely will not eat exactly $15 worth of feed and probably will not sell for exactly $60. As we analyze the farm and move away from one pig, we can see that the pig might have been contributing to the cull or death loss on the farm. This has some frequency distribution, which we may or may not be able to improve based on some risk or probability of occurrence. In this case, there is a different way to calculate the cost of the Type I and Type II errors. This is accomplished by building the partial budgets using stochastic analysis. In a stochastic analysis, similar to the frequency distribution above, we can just ask the computer to keep running simulation instead of doing calculus. This simulation will run over and over again with different numbers in each of the frequency distribution based on the expected frequency. If you run this simulation 1,000, or even 10,000 times in all combinations, you will see a risk profile of the possible outcomes. Software such as ModelRisk, @Risk or Crystal Ball are tools for this type of analysis.

In conclusions, making a decision of benefit to the farm begins with understanding the biology of the process that might be impacting the farm. It is important to evaluate possible proposed interventions without the distractions the average cost per pig for example. The decision to implement an intervention or not has consequences for the cost of making an error. We often overlook what might be the more expensive error, the Type II because of our conservative science training and the mistaken use of P values for evaluating production questions.

REFERENCES

Eicker S., J. Fetrow, S. Stewart. Marginal Thinking: Making money on a dairy farm. Proceedings. Western Dairy Seminar Advances in Dairy Technology 2006, 137. Galligan D. T., W. Chalupla, C.F. Ramberg Jr. Application of type I and II errors in dairy farm management decision making. 1991 J. Dairy Sci. 74:902. Poppy, G. D., A. R. Rabiee, I. J. Lean, W. K. Sanchez, K. L. Dorton, P. S. Morley. 2012. A meta-analysis of the effects of feeding yeast culture produced by anaerobic fermentation of Saccharomyces cerevisiae on milk production of lactating dairy cows. J. Dairy Sci. 95:6027. Williamson, D. B., The use of Decision Analysis to Evaluate the Economic Effects of Heat Mount Detectors in Two Dairy Herds. 1975 Australian Veterinary Journal, Vol 51 March:114

OBJECTIVE EVALUATION OF HEALTH INTERVENTIONS IN DAIRY POPULATIONS

Gerald Poppy, DVM, MBA
Cedar Rapids, IA

The key to making good decisions is to have good information available to use in the decision making process. However, it is to also important to have good models that are useful for helping handle the complexity of the questions. The questions we have with economics that are tied to biological questions on large dairies are not different from the questions on 50 or 60 cow dairies. The key questions on large or small dairies are:

- Are we getting good milk production?
- What should we feed the cows to maximize or optimize the conversion of feed into milk?
- What are the best health processes that should be in place to have a healthy productive herd?
- What health treatment should I use for a particular disease risk?

The difference today when working with a consolidated dairy enterprise is our advice may impact 5,000, 10,000 or maybe even more cows. Making an economic error, either Type I or Type II, (Galligan et al., 1991) has a large impact on the system. The question we have to ask ourselves as we suggest interventions, or potentially suggesting to forego an intervention, in a dairy system is "Do I really know everything I need to know about this intervention or do I just think I know?". How strong is the evidence for what I am recommending? Often we find that we have an opinion about a lot of things, but did not stop to evaluate how strong the evidence behind this assumption is.

We have all seen charts for the hierarchy of evidence. At the bottom of the list is Opinions, antidotal and then expert opinion. Often because we are on farms and observing animals, we tend to use a lot of anecdotal evidence from observing the effects. These may be correct under some instances but rarely can we use this information to understand interactions and variances. Above these are case series and case reports. These are observational reports, while they may have good information, they lack negative controls and control of confounding. Above the case reports are the observational studies. . Observational studies (the domain of epidemiologists) provide fairly strong evidence. These studies have the benefit of observing changes within a real exposure setting but lack the ability to control for unknown confounding by the use of randomization. Random Control trials (RCT) are the best evidence from a single study standpoint. These studies need to be blindly randomized and controlled for known confounding factors. These studies should test the correct population to have validity. Also, these studies should be thoroughly evaluated by the practitioner prior to validating the trial information. Often, we find that the power of the randomized controlled trial fails to provide for minimizing the Type II error. The Type II error is when we fail to find enough evidence to say the new mean is different than 0, but there actually is a difference (the outcome is false). This could be the variability of effect of the intervention or it could be the failure to have a large enough sample size. Utilizing the confidence interval to help us understand about the how certain we are of the mean is an important part of understanding the value the RCT plays in our ability to make inference about the economic value of the finding.

What happens when we have serval RCT trials that don't all agree on the value of an intervention? It is not unusual to have some papers report a positive response to an intervention and others a negative response, or some are significant and others fail to reject the null hypothesis? How does one make a decision. We may mentality weight the last few papers we read. We may know one of the authors or the University affiliated with a couple of the papers. We tend to give them more validity and make a decision. Rarely does a consultant feel they have read all the papers, or evaluated all the papers in the field. This is where the use of the systematic reviews and meta-analysis of the reviews is important and why these sit at the top of the evidence hierarchy. Systematic reviews look at the cumulative data surrounding a specific intervention and summarize the information utilizing meta-analytic methodology. By reviewing an earlier meta-analysis, we can see the impact of this method.

Streptokinase is used in human medicine as an intravenous thrombolytic therapy for acute infarction. If you were a clinician in 1959, this was not one of the tools you would have used. Between 1959 and 1988, there were 33 random controlled trials using between 20 and 1700 patients. Six of the 33 trials showed a significant improvement on mortality and 27 showed no significant change. Based on these results, the question that would arise would be if you had a heart attack would you want to be given streptokinase with only 6 out of 33 studies showing a significant benefit. In 1991 a meta-analysis (Lau et al., 1992) of these RCT showed there was a 21% decrease in death (95% CI = 0.72 - 0.87). By plotting the studies in a forest plot and mathematically combining the studies weighted by their variance, we get a clearer picture of the true impact of the intervention.

The use of meta-analysis also has other values. One of these is looking at the heterogeneity of effect. Analysis for heterogeneity is important in meta-analysis, because it tests the amount of variance within the group of studies compared to the within study variation. The chi-square test of Q is a test of the null hypothesis that all the studies share a common effect size. The I2 statistic is the ratio of the between study variation or true heterogeneity to the total variance across the observed effect estimates. A high I2 suggests the difference between individual study outcomes is greater (or more variable) than expected. Excess variation may indicate more than one outcome is being measured and may not be appropriate to combine the studies for an average effect (Higgins et al. 2003, Borenstein, 2009b). For example, suppose an intervention is fed but the meta-analysis has a high I^2. This might mean the effect of the intervention is different among subgroups within the population included in the study (i.e. mid-lactation and early lactation cows). Meta-regression or sub-group analysis would then be appropriate to try and identify where the variation is coming from. This would not be possible when only doing a narrative review of a few RCT papers.

When evaluating interventions in the dairy, we are not actually interested only in the biological outcome of the intervention. We are also interested in the financial outcome of the intervention. An intervention for milk production is going to impact not only the fluid milk, but also might impact the fat test, the solids test, the somatic cell count. All of these impacts must be calculated against the current milk price that is being paid, which could be called Dollar Corrected Milk (DCM) (Poppy, 1996). Along with the change in DCM from a revenue side, one also needs to look at the change in cost, in which a change in the cost of feed consumed to produce the additional milk is usually the most important and significant. All these variables and the variance of each of these outcome must be accounted for to measure the risk of being greater or less than breakeven or making a reasonable projection that will translate into how many dollars to project for future planning.

One method for accomplishing this is to combine the use of partial budgets and meta-analysis with stochastic analysis. Stochastic analysis is one where with the help of modern computing power we can utilize separate distributions of numbers for each variable within a partial budget and combine them into a projected theoretical distribution of outcomes. This is contrasted to a deterministic formula. In a deterministic formula for instance 71 pounds of milk X 100 cows X $20.00 per CWT of milk would provide the revenue to the dairy of $1,420 per day of income. If the cows were eating 52 lbs. of dry matter intake (DMI) of a ration that cost $0.12 per lb. it would cost them $624 per day to feed their herd leaving them with $796 of operating income per day. The trouble is we know that none of these figures are true. The herd probably is never going to produce exactly 71 lbs. per day. It might product 70.5, or 72.2. Some days there might be some bad weather and it only produces 65 lbs. of milk, although that doesn't happen often.

Here is a partial budget (deterministic) looking at the input of Diamond V Yeast Culture on milk production (Figure 1). Partial budgets of this type help frame the question for the producer. This question is: What is the difference in milk that is going to be produced as well as the component data? This helps put the information more in the context of a dollar corrected milk standpoint. You can adjust various inputs up and down to develop some amount of sensitivity range. Do we think the feed costs are going to continue at $0.12/lb? Is milk price going to continue to be $19.60 as in this example? A stochastic analysis can analyze all these various questions simultaneously and provide a view of how risky all the variances are when combined. All we need to know is the chances or frequency distributions for all of these events. When evaluating a health or production intervention outside of the risk for milk price and feed costs, one can utilize the literature for this information.

Early Lactation	Control	Diamond V
Milk Yield, lb	71	74.02
Fat %	3.40%	3.43%
lb of fat	2.414	2.54
Protein %	3.00%	2.95%
lb of Protein	2.13	2.19
OS %	5.70%	5.70%
lb of OS	4.05	4.22
DMI, lb	52	53.4
$ / DM lb	0.12	
DM, $/cow/day	6.24	6.40
DV/Cost/Cow		$ 0.05
$ / cow / day	$ 13.92	$ 14.44
$ / cwt	$ 19.60	$ 19.50
IOFC Advantage, $	$ 7.68	$ 7.98
DV Milk Value Advantage		$0.304

Figure 1. Partial Budget of Diamond V yeast culture impact on milk production, (deterministic). Milk price, $19.60, feed cost $0.12 /lb. dry matter, 52 lb. of dry matter intake/cow/day. Milk response from meta-analysis (Poppy et al. 2012), milk 3.02 lb/d, protein yield 0.056 lb/d, fat yield 0.126 lb/d.

From the meta-analysis on yeast culture (Poppy et al., 2012), we have the information we need to build a stochastic partial budget for the expect response and understand what the risk is for being above break-even for a decision to feed or not feed Diamond V Yeast Culture. From the meta-analysis, the mean milk response in early lactation is 3.03 lb/day with a variance of 0.31 and SE of 0.82 lb/d. The fat yield data has a mean of 0.126 with a variance of 0.002 lb/d and a SE of 0.062. The protein yield has a mean of 0.056 with a variance of 0.001 and SE of 0.027 lb/d. The dry matter intake in early lactation is 1.36 lb/d with a variance of 0.095 and SE of 0.458. We know from statistics that if we have the mean, the SE tells us how sure are we of its exact location on the number line. The meta-analysis is made up of individual studies, so the square root of the variance tells us our certainty about the next study in sample. We are going to use a normal distribution around the mean. The stochastic analysis then samples (5,000 times) from SE distribution for the mean, taking more samples near the mean and less further to the tails of the distribution. Then it uses this distribution and again samples it by the square root of the variance. The analysis does the same for each of the other components in the partial budget and multiplies out the values. The data is then plotted out in a histogram (Figure 2).

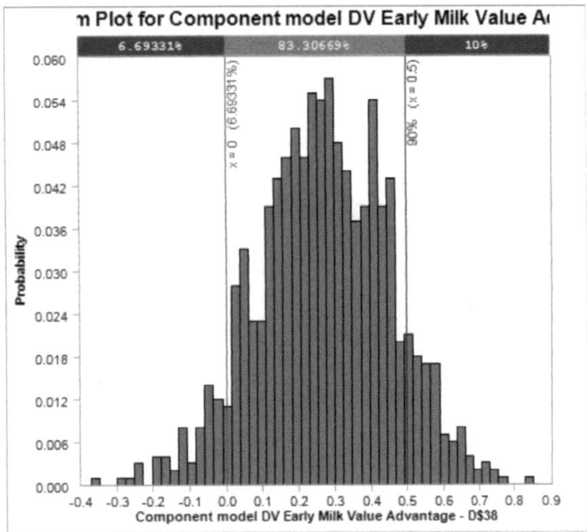

Figure 2. Histogram of the response for Diamond V yeast culture in early lactation (Poppy et al. 2012). There is a 6.69% chance the true response is below 0 and 93.1% chance it is greater than $0. There is an 83.3% chance it is between 0 and $.50 /cow/d. (ModelRisk, Vosesofterware.com)

We can see from the histogram that it looks like the mean response is around $0.25 /cow/day. It also tells us that there is a high probability the true response is greater than 0. In fact, it says that there is about a 6.9 % probability the true response is less than $0.00 so we know there is a 1 in 6.9% chance or 93.1% chance, the value is above $0.00. We can also see that we have an 83.3% probability we will be between $0.00 and $0.50 per day for each cow we feed the DV yeast culture to in early lactation. We can use stochastic software to easily see and evaluate information from interventions that have meta-analysis information. Diamond V yeast culture in mid late lactation has a different histogram (Figure 3).

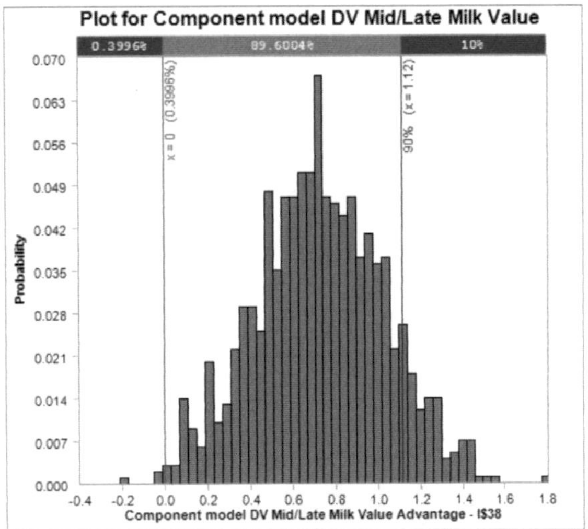

Figure 3. Histogram of the response for Diamond V yeast culture in mid to late lactation (Poppy et al. 2012). There is a 0.39% chance the true response is below 0 and 99.6% chance it is greater than $0. There is an 89.6% chance it is between 0 and $1.10 /cow/d. (ModelRisk, Vosesofterware.com)

We can see the mean response in mid to late lactation for Diamond V Yeast Culture is closer to $0.65 /cow per day. The early lactation period would be considered "less risky (Figure 4). Even though the mid/late lactation mean is greater than early lactation, we can see the "spread" is greater in mid/late lactation. Because of this spread, we are less certain where the true mean is.

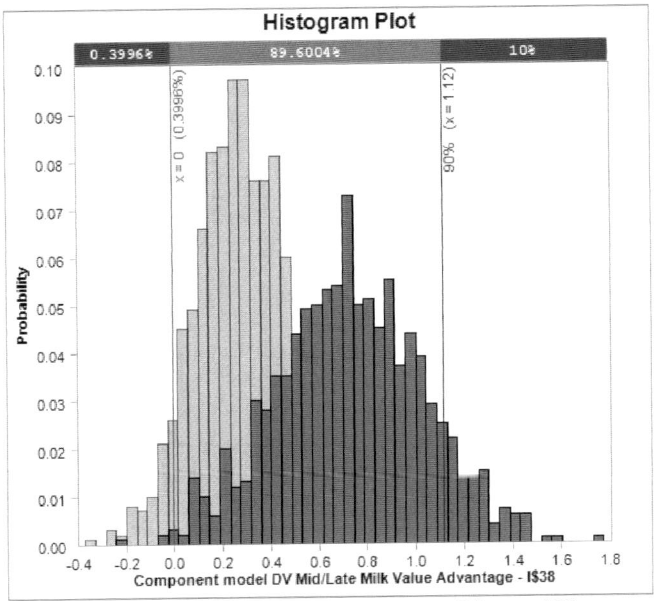

Figure 4. Histograms of the milk response to Diamond V yeast culture in early lactation (pink graph) and mid/late lactation (blue graph). The late lactation response is wider therefore more risky than the lower early lactation response.

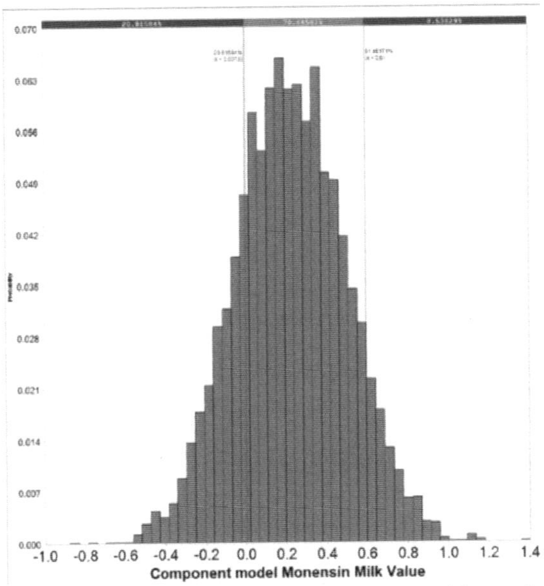

Figure 5. Histogram of the milk response to Monensin (Duffield 2008). The mean response is close to \$0.20/cow/day. The probability the true economic response is less than \$0.00 is 20.81% and greater than \$0.00 is 79.1%.

For example from a meta-analysis in 2008, we can extract the data for Monensin's milk and component response (Figure 5). This shows a mean of around \$0.30 per/cow/day and 79.1% probability the true response is greater than breakeven.
It is possible to do a stochastic model without using meta-analysis data. To do this, it is necessary to derive all the variance and distributions for each of the components in the partial budget. Overton in 2006 examined the use of various Presynch-CoSynch strategies by using this approach (Overton, 2006) using the program @Risk. In addition, the authors utilized the stochastic model to calculate Type I and Type II costs.

In conclusion, as veterinarians, the role we have with producers is to understand both the quality of the scientific data and weigh the value of the evidence and the variance that are known about the intervention. The use of partial budgets by a deterministic model help the producer understand both the marginal revenue and the marginal costs. Using meta-analysis information to further look at the variance and modeling the intervention by the use of stochastic programs can further enhance the producers understanding of the risk for using or not using an intervention.

REFERENCES

Borenstein, M., L. Hedges, J. Higgins, and H. Rothstein. 2009b. Identifying and Quantifying Hetrogeneity. Pages 105-133 in Introduction to Meta-Analysis. John Wiley & Sons, Ltd., West Sussex, UK. Duffield, T.F., A.R Rabiee, I.J.Lean. A Meta-Analysis for the Impact of Monensin in Lactating Dairy Cattle. Part 2. Production Effects. 2008. J. Dairy Sci. 91:1347. Galligan D. T., W. Chalupla, C.F. Ramberg Jr. Application of type I and II errors in dairy farm management decision making. 1991 J. Dairy Sci. 74:902. Higgins, J., S.G. Thompson, J.J. Deeks, D.G. Altman. 2003. Measuring inconsistnecy in meta-analysis. BNJ. 327. Lau J., Antman,

E. M.,J. Jimeniz-Silva, B Kupelnick, F. Mosteller, T. Chalmers. 1992. Cumulative Meta-Analysis of therapeutic trials for myocardial infarction. N. Engl. J. Med 327 No 4:248. Overton, Improving Decision Making by Consideration of Type I and Type II Error Costs. Proc AM Assoc Bov Pract 39:190. Poppy, GD. 1996. A research Study to Determine the Effects on Marginal Income Over Feed Cost Due to an Intervention of Diamond V XP Yeast in Dairy Cows. MBA Thesis, University of Phoenix. Poppy, G. D., A. R. Rabiee, I. J. Lean, W. K. Sanchez, K. L. Dorton, P. S. Morley. 2012. A meta-analysis of the effects of feeding yeast culture produced by anaerobic fermentation of Saccharomyces cerevisiae on milk production of lactating dairy cows. J. Dairy Sci. 95:6027.

KEY CONSIDERATIONS WHEN EXAMINING THE COST OF DISEASE

Michael W. Overton, DVM, MPVM
Athens, GA

INTRODUCTION

Understanding and communicating the economic consequence of common periparturient diseases can aid veterinary consultants in their efforts to improve the profitability of dairies. If the cost of infectious and metabolic disease issues can be quantified, greater justification for allocation of resources for prevention and treatment can be made based upon the expected cost/benefit analysis. This paper briefly reviews a few issues and precautions that should be considered when attempting to estimate via partial budgeting the cost of disease that occurs during the periparturient period.

The Vital 90™ Days begins approximately 60 days prior to calving and continues through the first 30 days of lactation. During this transition period, dairy cows experience a series of biological and physiological alterations that are usually accompanied by large changes in feed intake, dramatic shifts in hormonal profiles, and major fluxes in hepatic demands and function. The resulting negative energy and negative protein balance as well as immune suppression often lead to a multitude of metabolic and infectious problems including, but not limited to, retained fetal membranes, ketosis, metritis, displaced abomasum and mastitis.

These postparturient disease issues all impact the welfare and productivity of dairy cattle and the profitability of the dairy to varying degrees. Partial budgeting is a widely used approach for quantifying the economic costs and benefits of a planned intervention such as the adoption of a new technology or for consideration of an alternative outcome such as a change in disease incidence. It focuses only on the marginal income and marginal expenses that result as a consequence of a specific alternative and completely ignores any fixed cost. In evaluating the economic value of some change, partial budgets quantify the net impact of the difference between A) any increase in income or reduction in costs (together, these represent increases in marginal revenue) and B) any increase in costs or reduction in income (together, these represent increases in marginal cost). A net impact (A – B) that is positive indicates that the change under consideration will yield an improvement in farm income while a negative result indicates that the change will yield a reduction in farm income.

ISSUES AND PRECAUTIONS FOR CONSIDERATION

Both direct and indirect impacts of disease must be considered

The financial impact, i.e., the consequence cost, of a disease refers to the total direct and indirect disease costs that are realized as a result of experiencing the disorder under consideration. Direct disease costs include any cost associated with diagnostics and therapeutics for clinical cases treated, discarded milk during treatment and during any required withdrawal period, veterinary services, labor, and death losses that are directly associated with specific periparturient disease issues. Some people may refer to direct disease costs as explicit costs since these often seem more tangible and typically occur at or very near to the time of disease diagnosis. Indirect disease, however, are usually more implicit in nature and often represent the "lost opportunity cost" of disease. The contributors to indirect disease costs include any predicted future milk production losses, future culling losses, on-going diagnostic or monitoring costs, future reproductive losses, and in some cases, other diseases that are attributable to the initial disease in question. By summing the total direct and indirect disease costs, the total consequence cost is determined.

The value of milk NOT produced subsequent to disease must be considered

Diseases that occur during The Vital 90 Days often cause significant milk production losses. As in the case of culling, milk production losses can be measured in terms of immediate impact and in long term impact. Immediately, there is acute milk loss due to systemic and local impacts of inflammatory changes as well as a marketing requirement to discard milk during and following therapy for a prescribed period of time due to pharmaceutical contamination of milk. Diseases such as metritis and mastitis that occur during the postparturient period are also likely associated with longer term milk loss either via negative impacts to the lactation curve that are never fully recovered or via long term damage to milk secretory cells. Also, early lactation disease such as mastitis is expected to have a greater negative impact on milk production than a similar case that occurs later in lactation due to the differences in days remaining in the current lactation.

Often, there is a failure to consider the value of the milk that is not produced. This economic loss is typically considered a lost opportunity cost. When considering the lost opportunity cost of milk that is not produced, one cannot simply take the total projected loss over the remaining lactation and multiply that by the anticipated milk price. First, subsequent culling risk must be considered. For example, a mature cow that experiences a severe gram negative case of mastitis in early lactation may be predicted to lose 669 lbs of milk in the subsequent 50 days of lactation and even more beyond that.[1] However, if the cow is culled prior to completing the current lactation, any remaining unrealized predicted milk loss must be removed from the cost estimate. Second, one must only consider the marginal value of the milk not produced. According to the NRC, a cow must eat enough feed to provide 0.32 Mcal NE_L per lb of marginal milk produced if that milk contains 3.8% milk fat, 3.1% protein, and 4.8% lactose.[2] If the current feed cost is $0.13 per lb of ration dry matter and contains 0.78 Mcal NE_L per lb, the marginal feed cost required to support the production of a pound of milk is $0.055. Therefore, if milk has a value of $0.20 per lb, the value of 669 lbs of milk not produced in the first 50 days following the gram negative mastitis case is ($0.20*669) - ($0.055*669) or $133.80 - $36.80 = $97 and NOT $133.80, after accounting for the marginal feed required to produce the marginal milk.

Only consider the attributable consequences

One of the challenges to understanding and estimating the economic impact of disease consequences during The Vital 90 Days is the apparent association between diseases. The appearance of one disease challenge is often associated with another downstream disease issue. For example, cows with hyperketonemia (BHBA > 1200 μmol/L) have a greater risk of experiencing a displaced abomasum (DA).[3,4] This is not to say that one disease or issue causes another, but rather that there are often associations that exist such that when one condition is observed, there is a higher likelihood of seeing other related conditions. In general, cows that experience greater immune suppression during the periparturient period are more likely to experience retained fetal membranes, metritis, and mastitis; whereas cows that experience more severe negative energy balance are more likely to experience ketosis, displaced abomasum, and ovarian dysfunction.

While the literature is clear regarding the associations between diseases, actually quantifying the economic impact can be a bit of a challenge. Consider the following statement: "cows that have hyperketonemia in early lactation have an 8.5X higher risk of developing an LDA". Upon hearing this statement, many people would intuitively guess that if the whole herd incidence of DA in a herd was 5%, the risk of DA in cows with hyperketonemia would be 8.5*5% = 42.5%. However, this conclusion is incorrect because it applies the relative risk of 8.5 across the entire population instead of applying only to those non-hyperketonemic cows. Instead, we must consider the attributable risk within the exposed cows in order to determine how many cases are truly attributable to hyperketonemia.

To illustrate, I have created a hypothetical 2X2 table showing the expected relationship between hyperketonemia and DA as shown in Table 1. In this example, cows that have ketosis (hyperketonemia) have an 8.5X higher risk of also having a DA than cows that do not have ketosis (12.5%/1.47% = 8.5%). To estimate the attributable risk, we use the following equation: ($Risk_K$ - $Risk_N$)/ $Risk_K$ = (12.5% - 1.47%)/12.5% = 88%. In other words, 88% of the cases of DA in the cows with hyperketonemia can be attributed to the hyperketonemia (0.88*40 = 35 cases) and we would have expected only 5 cases of DA in this same set of cows had hyperketonemia not been present. Therefore, we must account for the 35 cases of DA in the cows with hyperketonemia when estimating the economic cost of hyperketonemia. If we assumed that a DA case costs $750 accounting for all of the direct and indirect costs, the contribution of DA to the cost of an average case of hyperketonemia = ($750*35)/320 or $82 per case of hyperketonemia in this herd's scenario.

Table 1. A 2X2 table showing the stratification of cows by disease (DA and Ketosis) within a herd over one year.

	DA	No DA	Total
Ketosis	40	280	320
No Ketosis	10	670	680
	50	950	1000

Ketosis incidence =		**32%**
$Risk_K$ =	**40 / 320 =**	**12.5%**
$Risk_N$ =	**10 / 680 =**	**1.47%**
DA incidence =		**5%**

Care should be taken to carefully consider and value attributable culling issues

Cows that experience one or more metabolic or infectious challenges during the Vital 90 Days usually experience higher premature culling (sold for slaughter) and mortality risks. The reason for the higher risk being associated with this periparturient time is multi-fold. First, due to a large decline in feed intake and the hepatic challenge of increased demand for gluconeogenesis, cows are more likely to experience metabolic stress in addition to infectious challenge. Secondly, a significant proportion of cows that experience an adverse health issue actually suffer more than one such event further increasing the negative impact on performance and increasing the culling risk. Finally, cows that experience negative health consequences in early lactation often have carryover effects that impact future milk production, reproductive performance, and even future disease risk [3,5-7]. As a consequence, the short term culling and mortality risk is higher and there is carryover impact on future culling risk. For both culling and mortality, care should be taken to estimate the attributable risk, as shown previously, in order to more carefully estimate and assign the true economic impact to the disease under consideration.

Once the number of attributable cases of premature culling and mortality have been determined for a disease, a value representing the magnitude of the loss must be estimated. For the purposes of estimating the current market value of a cow, a straight line depreciation approach starting with a purchase price (or fresh heifer value) of $2,200 to a cull value of $1,264 and using 7 years of potential productive life with typically observed parity-specific culling risks was used to determine a remaining market value at the end of each lactation. Then the average market value ($1,917 in this case) was simply the weighted average remaining market value for all mature cows. In the case of culled cows that were sold, the value lost was calculated as the difference between the average market value of either $2200 for first lactation or $1917 for mature cows and the market cow value which was determined by the product of predicted body weight*market cow price. In the case of mortality cost, the full current market value of either $2200 or $1917 was assigned as the loss, depending upon parity group under consideration.

One cannot simply sum the total cost of each disease to estimate the total impact

When calculating disease costs, two different estimates can be created. The first one is the component cost, and the second is the total cost. The component cost of hyperketonemia includes only costs directly attributable to this condition on milk production, treatment, culling, mortality and reproduction. The total cost refers to the sum of the component cost plus any other costs from other diseases that may be attributable to hyperketonemia. In other words, when discussing the cost of hyperketonemia and considering the economic impact of some management intervention specifically aimed at reducing the risk of hyperketonemia, we would use the total cost. However, when tallying the cumulative impact of a variety of diseases on the profitability of a farm, we need to only add the component disease costs since each disease will be considered individually. If we add the total cost of hyperketonemia to the total cost of metritis, we overvalue the economic impact of the diseases since DA costs have been attributed to each one. Consider the cost of hyperketonemia as presented in McArt et al and shown in Table 2.[8] The authors estimated the component cost of hyperketonemia to be $38 (direct cost) + $79 (indirect cost) = $117. By adding in the DA and metritis costs attributable to hyperketonemia, the total cost grows to $289. If discussing the potential impact of some management change that lowers the risk of hyperketonemia, it is appropriate to use the total cost of $289. However, if this management change impacts not only hyperketonemia but also the risk of metritis, DA, and retained fetal membranes, then the more appropriate approach would be to use the component cost of each disease including $117 for hyperketonemia.

Table 2. Direct and Indirect Cost of Hyperketonemia per Case (Values are shown to the nearest whole dollar. Totals do not appear to match due to rounding issues.)

Direct Costs		**Indirect Costs**	
Diagnostics	$1	Future milk losses	$30
Therapeutics	$3	Future culling losses	$9
Discarded milk (tx & w/d)	$0	Future repro losses	$40
Veterinary service	$0	**Indirect Cost (per case)**	**$79**
Labor	$2		
Death loss	$31	DA costs attributable to hyperketonemia	**$76**
Direct Cost (per case)	**$38**	Metritis costs attributable to hyperketonemia	**$95**
		Total Cost of Hyperketonemia (per case)	**$289**

Consider accuracy and consistency of disease recording

One challenge facing the dairy industry is the inconsistency of disease definition, detection and recording. When evaluating disease records, care must be taken to understand just how the farm defines and detects each of the commonly observed periparturient diseases. The owner and veterinarian may agree on a definition and approach used but the herdsman might have a completely different understanding. If the definitions used are drastically different, or if the detection pressure applied to identify sick cows is inconsistent, the records may convey a different message than reality. A great example is ketosis. On farm A, the only recorded cases of ketosis are those where the cow is very clearly clinically affected and treatment is applied. On farm B, there is an active screening approach using a blood-based BHBA test to examine every cow two or three times during early lactation. In these contrasting scenarios, farm A may report an average incidence of 5% where farm B reports an incidence of 26%. In reality, these two farms may be very similar, but the numbers convey a very different message. The cost of hyperketonemia as describe earlier would apply very well for farm B but would drastically underestimate the financial impact of hyperketonemia in farm A.

CONCLUSIONS

Because of the negative impacts of metabolic and infectious challenges that occur during the Vital 90 Days, dairy producers and influencers typically spend considerable time, money and effort to mitigate these negative impacts and consequences. However, there is little guidance available to determine the correct economic impact of each periparturient disease. While it is possible to quantify the cost of various disease issues using partial budgeting approaches, extreme care should be exercised to create the most correct estimates possible. Using these estimates, consultants are better positioned to appropriately motivate management change and to allocate resources for prevention and treatment based upon an accurate cost/benefit analysis.

REFERENCES

1. Schukken YH, Hertl J, Bar D, et al. Effects of repeated gram-positive and gram-negative clinical mastitis episodes on milk yield loss in Holstein dairy cows. J Dairy Sci 2009;92:3091-3105. 2. NRC. Nutrient requirements of dairy cattle, 7th ed. In. Washington: National Academy Press; 2001. 3. Duffield TF, Lissemore KD, McBride BW, et al. Impact of hyperketonemia in early lactation dairy cows on health and production. J Dairy Sci 2009;92:571-580. 4. McArt JA, Nydam DV, Oetzel GR. Epidemiology of subclinical ketosis in early lactation dairy cattle. J Dairy Sci 2012;95:5056-5066. 5. Santos J, Bisinotto R, Ribeiro E, et al. Role of animal health on reproduction of dairy cows. In: 2013 Dairy Cattle Reproduction Council Conference, Indianapolis, IN 2013;32-48. 6. Wilson DJ, González RN, Hertl J, et al. Effect of Clinical Mastitis on the Lactation Curve: A Mixed Model Estimation Using Daily Milk Weights. J Dairy Sci 2004;87:2073-2084. 7. Overton M, Fetrow J. Economics of postpartum uterine health. In: Proc. of the Dairy Cattle Reprod. Council Conv. , Omaha, Nebraska 2008;39-43. 8. McArt JAA, Nydam DV, Overton MW. Hyperketonemia in early lactation dairy cattle: A deterministic estimate of component and total cost per case. Journal of Dairy Science 2015;98:2043-2054.

QUANTIFICATION OF THE COST OF DISEASE VIA THE ECONOMIC ASSESSMENT TOOL

Michael W. Overton, DVM, MPVM
Athens, GA

INTRODUCTION

The Vital 90™ Days begins approximately 60 days prior to calving and extends through 30 days postpartum. During this time, dairy cows experience major biological and physiological transitions that include mammary development and initiation of lactation (heifers), cessation of lactation and mammary gland involution followed by re-initiation of lactation (cows), and major fetal growth and parturition (both). Accompanying these various changes are large drops in feed intake, dramatic shifts in hormonal profiles, depression of immune status, and major fluxes in hepatic demands and function. The resulting negative energy balance, negative protein balance and immune suppression often lead to a multitude of metabolic and infectious problems including, but not limited to, retained fetal membranes, ketosis, metritis, displaced abomasum and mastitis. Cows that experience these events experience higher culling and mortality risks; survivors of these insults usually have carry over effects of diminished milk production and reproduction well into lactation. As a consequence, dairy producers and influencers typically spend considerable time, money and effort to mitigate the negative impacts and consequences of these metabolic and infectious challenges that occur during The Vital 90 Days, but few have carefully considered the magnitude of cost incurred by each cow that calves. The purpose of the Economic Assessment Tool is to facilitate the estimation and demonstration of the wide variety of costs incurred during this critical period of time and to present a new metric for evaluating transition dairy cow performance, The Vital 90 Days Cost per Calving.

MATERIALS AND METHODS

The Economic Assessment Tool was created to evaluate the various costs that are incurred during The Vital 90 Days which includes the far off (-60 to -21d prepartum), close up (-21d to calving), and maternity-fresh (calving to 30 days in milk) periods. The tool is part accounting worksheet and part economic model. It does not suggest or predict management needs, but instead tabulates the various costs currently being incurred during The Vital 90 Days. Users of the tool are first asked to input general herd parameters which include milk price, feed price, labor cost, culling risk, distribution of the herd by parity group (first calving vs. second plus calving), market cow price, replacement heifer price, and other herd-specific inputs that are used in the economic tool. Next, users select the specific preventive procedures used in each time period by parity group from a predefined pick list, or they may enter additional options if not listed. Preventive procedures considered during The Vital 90 Days include confirmation of pregnancy, routine hoof care, vaccinations, parasiticides, feed additives, dry cow therapeutics, and other supplements. Representative prices for products in each area are provided in the tool, but may be customized by the user, as can the dose and duration of treatment. The total investment costs for The Vital 90 Days are determined by summing the far off, close up and maternity-fresh preventive costs.

After completing the preventive cost inputs, users are asked to input the herd-specific incidence of periparturient disease issues by parity group. Diseases evaluated include retained placenta, milk fever, hyperketonemia, clinical ketosis, mastitis, metritis, displaced abomasum, pneumonia, ovarian dysfunction, digestive or diarrhea issues and dystocia. For each disease issue, incidence risk refers to the estimated proportion of cows calving that experience the periparturient issue in question one or more times within the first 30 days of lactation. Hyperketonemia refers to the occurrence of one or more ketone tests of 1200 μmol/L or greater. Clinical ketosis refers to the proportion of cows calving that are treated for ketosis. Mastitis is stratified into mild, moderate and severe cases based on clinical signs and treatment strategy. Metritis is divided into routine and severe cases based upon presence or absence of systemic involvement while dystocia is split into moderate (DHIA calving score of 2 or 3 out of 5) or severe (DHIA calving score of 4 or 5 out of 5). Ovarian dysfunction refers to either cystic ovarian disease or anestrous/annovulatory cows that are individually identified and treated.

In order to estimate the cost of each disease issue, users must specify the most commonly used treatment protocol for each disease under consideration. Again, the tool contains a pick list of some of the most commonly used options for treatment of each disease but is customizable to allow for the careful quantification of the cost of each protocol including labor and milk withdrawal costs. In each protocol, users can specify if veterinarians are used in the treatment protocols for direct diagnostic and therapeutic management, if on-farm labor manages this aspect, or if both are involved. In the background, the tool estimates the total labor cost using predetermined times for each treatment activity.

Once all inputs are completed, the user must hit the "calculate" button. When this occurs, all costs are tabulated and then presented in a variety of ways. The new metric that is calculated, The Vital 90 Days Cost per Calving, is comprised of two major types of cost. The first cost for consideration is called the total investment cost which is the sum of the far off, close up, and maternity-fresh preventive costs. The second summary cost presented is the consequence cost. Despite typical preventive efforts, cows experience a variety of transition-related disorders. Consequence cost refers to the direct disease costs and indirect disease costs that are realized as a result of experiencing these disorders. Direct disease costs include any cost associated with diagnostics and therapeutics for clinical cases treated, discarded milk during treatment and during any required withdrawal period, veterinary services, labor, and death losses that are directly associated with specific periparturient disease issues. Indirect disease costs differ in that they include future milk production losses, future culling losses, on-going diagnostics or monitoring costs, and future reproductive losses attributable to the disease in question. By summing the total direct and indirect disease costs, the total consequence cost is determined. These consequence costs may be viewed as stratified into direct and indirect disease costs or by parity group.

Once all of the required inputs have been made and the calculations have been completed, the user will see the estimated Vital 90 Days Cost Per Calving metric as well as the total investment cost and consequence cost. On the main presentation page, the user can also see some of the more important general herd parameters and parity-specific disease incidences that were previously entered. At this point, there are several options for additional investigation:

1. By clicking on the Cost per Calving dial, the user can visualize a stratification of all costs by parity, disease category, and stage. Click "Back" in the upper right to return to the main output page.
2. By clicking on the Investment Costs dial, the user can visualize a stratification of all Investment costs by parity group and by stage of The Vital 90 Days. Click "Back" in the upper right to return to the main output page.
3. By clicking on the consequence costs dial, the user can see the stratification into direct and indirect disease costs.
   a. By clicking on direct disease costs, the user can see these costs stratified on a "Cost per Calving" or "Cost per Case" basis. On this page, the user can move one slider bar at a time to investigate the impact of changing one disease incidence on the direct disease costs for the farm. Remember, many of the transition diseases impact the risk and cost of other diseases in The Vital 90 Days. For example, reducing the risk of hyperketonemia results in a reduction in the expected incidence of retained fetal membranes, metritis, and DA. After investigating the impact of a change in one disease incidence, click the "Reset" button in the upper right to reset incidence and cost to the original calculated levels. Repeat as desired. When finished, click "Back" in the upper right to return to the consequence costs page.
   b. By clicking on indirect disease costs, the user can see the indirect disease costs broken out on a "Cost per Calving" or "Cost per Case" basis. On this page, the user can move one slider bar to investigate the impact of changing one disease incidence on the indirect disease costs for the farm as previously described. After investigating the impact of a change in one disease incidence, click the "Reset" button in the upper right to reset incidence and cost to the original calculated levels. Repeat as desired. When finished, click "Back" in the upper right to return to the consequence costs page.
4. To investigate the impact of a change in one of the general herd parameters or to visualize the impact of a change in disease incidence on both direct and indirect disease costs, the user can move the slider bars on the summary page. Note: multiple general herd parameter sliders can be moved without hitting reset, but only one disease incidence slider can be moved before resetting the inputs. For example, a user could lower milk price, raise replacement heifer prices, increase feed price and lower the expected incidence of mastitis and examine the impact but in order to change another disease incidence, the tool must be reset before proceeding.

In the estimation of the cost of a specific disease, extreme care has been taken to try and avoid double counting costs. The Economic Assessment Tool takes a very unique approach in this regard. When multifactorial diseases are being evaluated, care is taken to estimate the number of cases of culls sold for market beef, culls removed due to mortality, and other diseases that are considered attributable to the specific disease in question. Two disease cost estimates can be calculated for the most important multifactorial diseases that occur very early in lactation and that are linked to other costly outcomes further downstream (clinical hypocalcemia, retained placenta, hyperketonemia, and metritis). The component cost of a disease includes only costs directly attributable to this condition on milk production, treatment, culling, mortality and reproduction without consideration of further downstream impacts on other diseases. The total cost refers to the sum of the component cost plus any other costs from other diseases that may be attributable to the original disease under consideration. For example, in the case of hyperketonemia, there is a strong association between the presence of elevated ketones in the blood in the early postparturient period and a multitude of adverse health events including, but not limited to, an increased risk of culling, death, metritis, displaced abomasum, and future reproductive challenges. If the total cost of hyperketonemia was added to the total cost of metritis and the total cost of displaced abomasa, we would grossly overestimate the cost of hyperketonemia and the predicted number of cases of metritis and displaced abomasa would greatly exceed the herd's actual number of cases. However, if we added the component costs of each disease, the grand total should more accurately reflect the total impact on the herd. When increasing or decreasing a specific disease incidence, algebraic equations are used to estimate the change in culling and in attributable (associated) diseases. Due to the complexity of this approach, once the herd's cost per calving is estimated, "what if" scenarios for the multifactorial disease issues are performed individually and a reset is performed prior to the next scenario.

CONCLUSION

The Vital 90 Days consists of a series of physiologic changes in the life of a dairy cow and is of critical importance to the economic performance of any dairy. Many disease issues can be attributed to this time period. Wise investment into the proper preventive management opportunities may result in an increase in the total investment cost but should yield a reduction in the total consequence cost and positively influence profitability. The Economic Assessment Tool and the Vital 90™ Days Cost per Calving metric should help producers and consultants better understand how much is currently being spent during this critical time period, and over time, help make better decisions to improve profitability.

ECONOMIC IMPACT OF SCHMALLENBERG VIRUS INFECTION IN SWISS DAIRY HERDS

Mireille Meylan, PhD, DACVIM, DECBHM; Marianne Wüthrich, med.vet.; Gertraud Schüpbach, PhD, DECVPH
Berne, Switzerland

INTRODUCTION

In the late summer and fall of 2011, a new disease entity manifested by high fever, diarrhea and reduced milk yield was observed in adult cattle in Northern Germany and in the Netherlands. The causative agent was identified as a virus of the *Bunyaviridae* family in November 2011, and was named Schmallenberg Virus (SBV) after the German town where the samples it was first isolated from had been taken from cattle with fever and diarrhea.[1] Starting in December 2011, an increased frequency of malformations was noticed in lambs and calves, and SBV was isolated from malformed fetuses.[2,3,4]

The Schmallenberg Virus

The SBV is an arthropod-borne RNA virus belonging to the genus *Orthobunyavirus*, Simbu serogroup. Transmission of SBV among susceptible ruminants occurs via biting midges of the *Culicoides* species.[1,5,6,7] The best known representatives of the Simbu serogroup in Asia, Africa and Oceania include the Akabane, Aino, and Shamonda viruses which are also transmitted mainly by *Culicoides* midges and can be pathogenic in cattle. The SBV was the first virus of the Simbu serogroup isolated in Europe.[1] How this virus developed and arrived to Northern Europe remains unclear to date. It has been shown that the virus can persist through cold winters although activity of infected vectors may not be detectable.[6]

After infection of naive animals, clinical signs of the disease appear after an incubation time of 2 to 5 days, but the acute phase often remains unnoticed in adult animals. The phase of viremia is short in cattle and sheep (3-4 days, maximal 6 days), although SBV genome can be detected in lymphoreticular tissue for extended periods of time.[7,8] Therefore, analyses to demonstrate the presence of the virus itself in the blood of infected animals are often unsuccessful. Neutralizing antibodies were found 2-3 weeks after experimental infection in cattle, but the duration of protective immunity remains unknown to date.[8]

The disease is mostly relevant in cattle, sheep and goats, but antibodies to SBV have also been detected throughout Europe in bison, moose, alpacas, buffalos, fallow deer, roe deer and red deer, as well as in a dog in Sweden.[8]

Clinical signs of SBV infection

The clinical symptoms that have been associated with acute SBV infection include fever, diarrhea, reduced milk yield and abortion in adult domestic ruminants (cows, sheep and goats), and malformation of the fetuses in all 3 species. In contrast to cattle, acute symptoms have rarely been observed in adult small ruminants, and malformations were the main manifestation of the disease in sheep and goats. It is however possible that this difference is rather due to less intensive observation of the animals than to true differences in the clinical signs of disease.[1,5,9,10]

The most common malformations associated with intrauterine SBV infection have been described as "arthrogryposis-hydranencephaly syndrome" and can include arthrogryposis, torticollis, head deformations (e.g. shortened lower jaws), and deformations of the spine. Anhydrencephaly, porencephaly, hydrocephalus, cerebral and cerebellar hypoplasia, and micromyelia are common malformations of the central nervous system. Malformed fetuses are generally delivered at term. Normal and malformed fetuses have been observed in the same litter of goat kids and lambs.[2,3,10,11] The critical phase for fetal malformations after infection with the closely related Akabane virus ranges between the third and sixth month of gestation in cattle, and between 30 and 50 days in sheep. Despite the lack of specific data, it is believed that the critical periods are similar for SBV.[8,12] Malformation rates were reported to be relatively low with less than 1% in cattle herds and less than 3% in sheep herds.[12] The fact that the abortion and malformation rates were distinctly lower than the observed seroconversion rate of pregnant dams was interpreted as indicating a low vertical transmission rate.[13]

Epidemiology and impact of SBV infection

After the first descriptions of cases of SBV infection in Northern Germany and the Netherlands in the late summer of 2011, the infection spread to the neighboring countries and had reached Belgium, Luxemburg, France, Italy, Spain, and the South of the United Kingdom by April 2012. It continued to spread rapidly during the following year, reaching Scandinavia and South-Eastern Europe in 2013. Seroconversion rates over 90% in cattle and up to 30% in small ruminants were observed in the affected areas.[8,9] Nevertheless, the impact of SBV on adult cattle health and productivity was estimated to be limited in terms of milk production and reproductive performance, and negligible in terms of mortality. The impact of intra-uterine infections was also considered to be limited with an incidence of 1-4% in cattle and 3-7% in sheep, and only 0.34% of the cattle herds and 5.15% of the sheep herds reporting more than one case of malformations. These figures are likely underestimated due to under-reporting after confirmation of the first cases at the beginning of the outbreak of the disease. The possibly main economic impact of SBV, due to the international trade restrictions placed by some non-European countries, particularly for live animals and semen, has not been quantified to date.[5,12,13,14] The high rate of seroconversion is expected to provide protective immunity to the infected ruminant populations after first infection, but its duration is unknown to date. Nevertheless, the impact of the presence of SBV in Europe is expected to decrease in the next years as the ruminant population is no longer naive.[8]

Measures to control SBV infection in Europe

In analogy with the outbreak of Bluetongue virus serotype 8 in Europe in 2006, the potential of vaccination to protect cattle and small ruminants from SBV infection was explored. An inactivated trivalent vaccine against Akabane virus, Aino virus and the teratogenic reovirus Chuzan available in Japan did not prevent experimental infection with SBV.[15] In contrast, a specific SBV inactivated vaccine was shown to be able to prevent viremia after challenge infection, and two commercial vaccines are currently available in some European countries. It is however not known whether these vaccines will be used in practice given the high seroconversion rates following natural spread of the virus.[5,8]

Further measures to limit the damages caused by the virus would be e.g. to minimize the number of pregnant sheep in the fall, during maximal activity of the insect vectors. Repellents and insecticides may also be used to reduce the infection pressure on susceptible animals, however the results of one study in Germany indicated that this measure did not protect the animals from infection with SBV.[8] Restrictions on animal movements are not expected to have any effect on the transmission of disease given the rapid spread via arthropod vectors observed in the first phase of the outbreak.[7]

SCHMALLENBERG VIRUS INFECTION IN SWISS DAIRY HERDS

Swiss herd managers and veterinarians had been actively informed about the new disease entity observed in Northern Europe and asked to be attentive to signs of the disease such as fever with diarrhea, drop in milk yield, abortions, malformations and stillbirths in early 2012. The first case of SBV infection in Swiss cattle was confirmed in July 2012 in Western Switzerland, and by December of the same year 99.5% of the Swiss dairy herds were found to be SBV positive by bulk tank milk serology. Thus, the infection had spread rapidly eastward, at an estimated velocity rate of 0.903 km/day. Since all animals tested for SBV were found to be seropositive after October 2012, the mandatory diagnostic investigations ordered for suspected cases by the official veterinary services were discontinued.[16,17]

A case-control study was started in the fall of 2012 in order to describe the clinical signs associated with acute SBV infection in adult cattle and assess the economic impact of the disease in Swiss dairy herds. Herds where SBV infection had been confirmed by PCR or serology based on clinical signs suggestive of the disease were recruited (case herds), and control herds were matched to the case herds by size, breed, production level, and geographical location. A total of 176 dairy herds, were visited and data regarding animal health and performance (milk yield and reproduction) were compiled for the period preceding the appearance of SBV in Switzerland (July 2011 to June 2012), the period of spread of the virus (July to December 2012), and the following period (January to December 2013). Of the 141 herds for which adequate data could be collected, 77 had observed acute signs of the disease during the summer and fall 2012, the other 84 (control) herds had not. Seropositive animals were present in all herds, except for 2 control herds. Between-herd comparisons were performed between herds with and without clinical signs of acute SBV infection, and within-herd comparisons were done for the 3 consecutive time periods described above.

Clinical signs of acute SBV infection in Swiss dairy herds

The mean morbidity rate (percentage of cows with clinical signs of the disease) in case herds was 14.8% (range: 2.5%-100%). The most commonly observed clinical sign of acute SBV infection at the individual cow level was a drop in milk yield (87.9%), followed by diarrhea (83.5%), decreased appetite (82.8%), and prolonged periods of recumbency (68.9%). Fever was only mentioned in 60.7% of the cases, possibly because the temperature of diseased animals was not always measured. Of the affected animals, 70% recovered completely within 14 days.

Animal losses in the case herds (cows and calves) ranged from 1 animal (56.1% of the herds) to 6 animals (3.9%). The main reasons for animal losses were abortion (45.5%), stillbirth (24.7%), and early culling of the cows (22.3%). Malformations were only rarely reported (median in all herds = 0%, IQR = 0-1.2).

According to indications of the managers of the case herds, the number of services per conception was significantly higher in the year after acute SBV infection than the year before (median: 2.0 vs. 1.8), as was the observed abortion rate (6.5% vs. 3.7%) and the rate of stillbirths (5.7% vs. 4.3%). The percentage of cows treated for retained membranes was also slightly but significantly higher in 2013 than in 2011-2012 (11.8%, IQR = 7.1-17.5 vs. 11.8%, IQR = 7.8-18.8). Review of treatment records in the case herds did not reveal significant differences in the frequency of therapies for dystocia, retained membranes, or metritis before and after the apparition of SBV in Switzerland.

Economic impact of acute infection with SBV in Swiss dairy herds

The mean costs of treatment for the acute phase of the disease were estimated to 149 US$ per affected cow and 372 US$ per herd. Based on detailed information available for a limited number of cows, the costs due to reduced milk yield were estimated to 188 US$ per herd. For both parameters, a large variability was observed among herds. In 30% of the case herds, the managers reported that one or several cows had shown a persistent reduction in milk yield after acute infection, which could last up to 180 to over 250 days and corresponded to estimated losses of 225-2850 US$ per herd.

Over 50% of the herd managers estimated that the main costs associated with SBV infection were due to decreased milk yield, alone or in combination with other causes related to the disease. Another 26% of the farmers named the loss of calves and cows as the main loss, and only 7.8% estimated reduced reproductive efficacy to be the main cause of losses associated with SBV infection. The farmers estimated the total costs of SBV infection for their operation to a mean of 5400 US$ (range: 325-39,000).

DISCUSSION AND CONCLUSION

The present results show that the course of SBV infection in Swiss dairy herds was mostly subclinical, however a small number of herds were affected more severely and the consequences of infection lasted for several months. The most common clinical sign during the acute phase was not fever, diarrhea or abortion, but a marked drop in milk yield. Abortions and fetal malformations were reported less frequently than expected, as several practicing veterinarians had reported an increased frequency of dystocia due to malformed fetuses (mainly arthrogryposis).

Although the impact of SBV infection on cattle health and productivity, including reproductive efficiency, was generally limited in Swiss dairy herds, large differences were observed among individual herds and some operations suffered severe losses distributed on several months after acute infection. These observations are in accordance with those of others in Europe.[13,18]

Infection with SBV is likely to become endemic in affected regions. Protecting antibodies have been shown to be produced after clinical and subclinical acute infection with SBV and to persist for at least 2 years. How long this protective effect remains efficient is unknown to date and will only become evident in the next years.[8]

REFERENCES

[1]Hoffmann B. et al. **Emerg. Infect. Dis.** 2012: 18: 469. [2]van den Brom R. et al. **Tijdschr Diergeneeskd.** 2012: 137: 106. [3]Herder V. et al. **Vet. Pathol.** 2012: 49: 588. [4]Bilk S. et al. **Vet. Microbiol.** 2012: 159: 236. [5]Doceul V. et al. **Vet. Res.** 2013: 44: 31. [6]Goffredo M. et al. **Prev. Vet. Med.** 2013: 111: 230. [7]Gubbins S. et al. **Prev. Vet. Med.** 2014: 116: 380. [8]Wernicke K. et al. **Prev. Vet. Med.** 2014: 116: 423. [9]Beer M. et al. **Epidemiol. Infect.** 2013 : 141: 1. [10]Ganter M. et al. **Sm. Rum. Res.** 2014: 118: 63. [11]Hahn K. et al. **Emerg. Infect. Dis.** 2013: 19: 155. [12]Afonso A. et al. **Prev. Vet. Med.** 2014: 116: 391. [13]Veldhuis A.M.B. et al. **Vet. Microbiol.** 2014 : 168 : 281. [14]Tarlinton R. et al. **Vet. J** 2012: 194: 10. [15] Hechinger S. et al. **Vet. Res.** 2013: 44: 114. [16]Schorer M. et al. **Schweiz. Arch. Tierheilkd.** 2012: 154: 543. [17]Balmer S. et al. **Prev. Vet. Med.** 2014: 116: 370. [18]Martinelle L. et al. **Transbound. Emerg, Dis.** 2014: 61: 285.

ACUTE PUERPERAL METRITIS: RISK FACTORS AND CURRENT CONCEPTS ON PREVENTION

Brent Credille, DVM, PhD, DACVIM
Athens, GA

INTRODUCTION

Dairy cattle are susceptible to numerous metabolic and infectious disorders in the transition period. Acute puerperal metritis (APM), clinically defined as the presence of a fetid, watery, reddish-brown uterine discharge, an enlarged, flaccid uterus, and overt signs of systemic illness that may include fever, dehydration, and depression is one of the most commonly encountered infectious peri-parturient diseases in modern dairy practice. Histologically, affected cattle demonstrate leukocytic infiltration of all uterine layers, uterine edema, myometrial degeneration, and mucosal congestion.[1] The median lactational risk for APM in North American dairy cattle has historically been approximately 10%.[2] However, more recent work has shown the incidence to be much higher with most modern dairy farms reporting a range of 18-35%.[3] While the ultimate outcomes of this disease are impaired reproductive performance, reduced milk production, increased culling risk and, in severe cases, death, the most significant consequence of APM is the economic losses absorbed by dairy producers. Current estimates suggest that each case of APM costs producers $350-400.[4] Therefore, in a herd of 1,000 cattle with a median lactation incidence of APM approaching 20%, costs for this one disease alone can exceed $70,000 per year.[4] When applied to the United States dairy industry as a whole, APM represents a loss of $350,000,000-650,000,000 annually.[3,4]

EFFECTS OF ACUTE PUERPERAL METRITIS ON DAIRY PRODUCTION

The goals of reproductive management within a dairy herd are to ensure cows are inseminated and conceive within a timely manner after calving. Improved reproductive performance will lead to more revenue from increased milk production due to cows averaging fewer days in milk, increased revenues from the sale of excess calves, and increased value of cull cows.[5] While APM can certainly contribute to mortality on dairy farms, the mortality risk for affected cattle is typically less than 2%.[6] Thus, it's clear that the economic losses resulting from APM come from some other sources. Indeed, cattle diagnosed with APM typically have more days open, a lower 21-day pregnancy rate, and lower milk production.[4,5] Overton and Fetrow developed an economic model to evaluate the costs of APM within a single dairy herd in California. In this herd, cattle with APM were open on average 16 days longer and had a 21-day pregnancy rate 3-6 percentage points lower than that of healthy cattle.[4] In addition, culling risk for cattle with APM was 4.2% greater than the culling risk in other groups. These same animals produced almost 5 lbs less milk per day over the first 120 days of lactation.[4] Of note is that each cow in this study was treated with standard therapies (antimicrobials, anti-inflammatories, fluid therapy, etc.), a factor that suggests that APM still has significant costs despite treatment. A separate study demonstrated that cattle affected by APM produce 3.7 kg less milk at first test and 259 kg less milk over the course of a single lactation than unaffected herdmates.[7] Interestingly, cattle diagnosed with both placental retention and APM produce over 1,000 kg less milk per lactation than healthy herdmates, suggesting an additive effect of the two disorders.[7] This same study also showed that cattle with APM had a first service pregnancy risk of 27.2% as compared to 35.3% in unaffected cattle.[7] In contrast to the work done by Overton and Fetrow, however, this study found no relationship between APM and culling risk.[7] Other work studied the effects of APM on milk production, reproductive performance and the efficacy of ceftiofur therapy in Argentinian cattle.[8] These data showed that cattle diagnosed with APM produced less milk (2,367 kg vs 2,647 kg) by 90 days in milk (DIM) and had a lower risk for pregnancy by 100 DIM than unaffected herdmates (adjusted odds ratio = 0.19).[8] Therapy with ceftiofur was associated with an increased pregnancy risk at insemination (adjusted odds ratio = 2.7) but not risk of clinical cure or milk yield.[8] Lastly, a study that evaluated the effects of APM on fertility of grazing dairy cattle found that cattle with APM had significantly lower pregnancy per artificial insemination (P/AI) than healthy herdmates (52.4% vs 63.9%).[9] In addition, cattle with APM were less likely to be cycling by 49 days post-partum.[9]

Mechanisms of Decreased Fertility in Cattle with Acute Puerperal Metritis

It is well established that cattle with APM experience decreased fertility when compared to cattle that remain healthy. The mechanisms for this are quite complex and still under investigation. From the work that has been done, it is likely that the causes of infertility are multifaceted and involve effects acting both locally and systemically. For example, cows with APM have slower growth of dominant follicles, reduced risk of ovulation, and lower blood concentrations of both estradiol and progesterone compared to healthy cattle.[10] Furthermore, in cattle that do ovulate, the corpora lutea formed after ovulation are typically smaller and secrete less progesterone, a factor that may impair embryonic survival should fertilization occur.[10] Infusion of LPS into the systemic circulation reduces the secretion of GnRH from the hypothalamus and the secretin of LH from the pituitary gland, factors that may be responsible for the reduced risk of ovulation in cattle with APM.[11] Bromfield and Sheldon showed that, when bovine ovarian cortex cells are exposed to LPS, the primordial follicle pool is significantly reduced and this reduction in primordial follicles is associated with increased follicle atresia.[12] In a separate study, Bromfield and Sheldon demonstrated that the reduction in primordial follicle cells is likely due to arrested meiotic progression, a factor shown to be clearly related to LPS and the accumulation of inflammatory mediators within the granulosa cell.[13] One group exposed bovine ovarian theca and granulosa cells to both LPS and TNF-α and evaluated steroid secretion *in vitro* and risk of ovulation *in vivo*.[13] They found that LPS significantly reduces granulosa cells estradiol secretion and TNF-α decreases the production of androstenedione and estradiol by both theca and granulosa cells. In addition, fewer animals ovulated following intrauterine infusion of either LPS or TNF-α.[13] Further investigations into this phenomenon by this same group showed that the reduced secretion of estradiol by granulosa cells is primarily mediated by a reduction in aromatase activity and not cell death.[13]

In spite of the detrimental effects of APM on ovarian function, some affected animals do ovulate and, under certain circumstances, the ovulated oocyte can be fertilized. However, cattle with APM show increased rates of embryonic death and pregnancy loss when compared to healthy controls.[14] For example, when exposed to LPS prior to fertilization *in vitro*, bovine oocytes are less likely to develop to the blastocyst stage.[14] When co-cultured with neutrophils, *in vitro* produced bovine embryos display few, if any, developmental issues. However, when embryos are exposed to a cell-free uterine lavage solution obtained from cattle with experimentally induced endometritis, development is significantly impaired.[14] Thus, it is likely that cytokines and other inflammatory mediators cause the adverse effects on the developing embryos. When evaluated more closely, embryos obtained from animals with uterine inflammation have fewer trophoectoderm cells and, because trophoectoderm cells are the source of interferon tau, the signal for the maintenance of pregnancy is lost.[14] It is important to note that, while most studies have utilized LPS or *E. coli* to evaluate the effects of bacterial infections on uterine health, gram positive infections are no less detrimental. In fact, recent work has found significant associations between the presence of certain gram positive pathogens and uterine disease in post-partum dairy cattle.[15-17]

MEDICAL THERAPIES AVAILABLE FOR MANAGING ACUTE PUEPERAL METRITIS

While APM is multifactorial and successful treatment may require multiple different therapeutic modalities, antimicrobials are the mainstay of therapy for cattle diagnosed with clinical disease. In fact, APM is the most common reason for parenteral antimicrobial use in dairy cattle. Currently, two antimicrobials in three different formulations are labeled for systemic use in cattle with APM in the United States: oxytetracycline dihydrate, ceftiofur hydrochloride, and ceftiofur crystalline free acid. Oxytetracycline dihydrate, sold under the trade name Liquamycin LA 200®, is the oldest and least expensive of the available compounds still in use. It is labeled for therapy of acute metritis caused by susceptible strains of *Staphylococcus* spp. and *Streptococcus* spp. given SQ at a dose of 11 mg/kg once daily. However, because of the prolonged withdrawals (28 days meat and 4 days milk) and lack of data demonstrating efficacy in clinical trials, this product has fallen out of favor. Intrauterine infusions of tetracycline-based products have been closely evaluated in recent years. For example, a study comparing an intrauterine infusion of 6 grams of oxytetracycline to parenteral procaine penicillin G and ceftiofur sodium found no differences in the reduction in rectal temperature over the first 5 days after diagnosis or daily milk yields for 12 days following initiation of treatment.[18] In addition, Goshen and Shpigel evaluated the effect of infusion of 5 grams of chlortetracycline into the uterus twice weekly for 2 weeks.[19] Cattle receiving this treatment produced 1,438 lbs more milk over the course of lactation and conceived 29 days sooner than untreated controls.[19] Furthermore, conception risk in treated cattle was 42.5% compared to 38.3% in clinically normal animals and 18% in untreated controls.[19] In addition, at current market prices, treated cattle would yield a net return to the producer of approximately $74.70/head, despite the prolonged milk withdrawal required with this therapeutic regimen (> 21 days). Nevertheless, this treatment modality is not approved in the United States, making this protocol impractical for routine clinical use.[20]

Ceftiofur, a 3rd generation cephalosporin, has become the gold-standard therapeutic agent for cattle with APM. Two preparations of ceftiofur, ceftiofur hydrochloride (Excenel®) and ceftiofur crystalline free acid (Excede®), are labeled for use in cattle with APM and have the advantages of short slaughter (3 and 13 days, respectively) and no milk withdrawal. In addition, the effects of ceftiofur on clinical cure and, to some extent, reproductive performance, have been thoroughly evaluated. Work has demonstrated that ceftiofur hydrochloride, when given at a dose of 2.2 mg/kg IM once daily for 5 days, is effective in reducing rectal temperature and improving the character of uterine discharge in cattle with APM.[21] In a study that compared the efficacy and economic efficiency of treating cattle diagnosed with APM with either ceftiofur hydrochloride or a combination of intrauterine and systemic antimicrobials, cattle treated with ceftiofur had rates of clinical cure and reproductive performance similar to cattle in the other groups.[22] Furthermore, economic analysis revealed that costs for the group treated with only ceftiofur were less, primarily as a result of reduced milk withholding periods.[22] A German group compared the clinical efficacy of ceftiofur hydrochloride to oxytetracycline in cattle with APM. Probability of clinical cure at day 7 was numerically higher in the ceftiofur group (64.8%) when compared to oxytetracycline treated cattle (58.1%), however this difference was not statistically significant.[23] Between days 2-5 after treatment, ceftiofur treated cattle had significantly lower rectal temperatures than cattle treated with oxytetracycline.[23] Unfortunately, comparisons of milk production and reproductive performance between groups were not evaluated in this study. Ceftiofur crystalline free acid (CCFA) is labeled for treating APM and is given subcutaneously at the base of ear at a dose of 6.6 mg/kg every 72 hours for two doses. One study evaluated the effect of two doses of CCFA on clinical cure risk in cows with APM and found that, when compared to a saline treated control, cows treated with CCFA are more likely to cure (74.3% vs 55.3%, respectively).[24] A second study evaluated the effects of CCFA on clinical cure risk, milk yield, and reproductive performance and found that CCFA had no influence on clinical cure or milk yield but did increase risk of pregnancy at insemination (adjusted odds ratio = 2.69).[8]

Recent work evaluating the potential utility of ampicillin trihydrate as a therapeutic agent in dairy cattle with APM has been done. In this study, ampicillin trihydrate was compared to ceftiofur HCl and the results showed that cattle treated with ampicillin had a faster rate of clinical cure based on vaginal discharge score on day 5 after diagnosis than cattle treated with ceftiofur (37.1% vs 25.2%, respectively).[25] In addition, cattle treated with ampicillin had a significantly lower risk of developing purulent vaginal discharge (PVD) than cows treated with ceftiofur (57.7% vs 67.8%. respectively).[25] There was no difference in the proportion of cattle that developed cytologic evidence of endometritis. In addition, pregnancy per AI (P/AI) was similar between treatments.[25] While the results of this study suggest that ampicillin is a potentially useful therapeutic agent for cattle with APM, its use in cattle with this disease is considered extra-label drug use (ELDU) under the Animal Medicinal Use Clarification Act (AMDUCA) and therefore must be justified by the prescribing veterinarian.

RISK FACTORS FOR ACUTE PUERPERAL METRITIS

Significant factors contributing to APM include subclinical hypocalcemia and negative energy balance. In fact, it has been shown that the population attributable risk for APM due to subclinical hypocalcemia is 91%.[26] Other work found that cattle experiencing dystocia, retained fetal membranes, and elevated pre-partum serum non-esterified fatty acid (NEFA) concentrations are 2.58 times more likely to develop APM than cattle without these issues.[8] Studies performed in Canada yielded similar results and found that cows with elevated pre-partum serum NEFA levels, cows that experienced dystocia, and cows with placental retention are more likely to develop APM.[27] From this work and work done by others, it appears that APM and other infectious disorders occurring in the immediate post-parturient period are mediated by a number of different factors. Of these, the metabolic status of the cow in the period prior to parturition seems to play the most significant role. Indeed, hypocalcemia and elevations in plasma NEFA concentration suppress the immune system and contribute to placental retention and impaired bacterial clearance.[28-30] One study evaluated the role of prepartum dry matter intake as a risk factor for APM and found that cows that went on to develop APM consume significantly less feed beginning 2 weeks prior to parturition than healthy cattle. In this study, for every 10-minute decrease in average daily feeding time, the odds of a cow developing APM increased by 1.72.[30] In addition, for every 1 kg decrease in dry matter feed intake, the odds of a cow developing APM increased 3 times.[30] Cows later diagnosed with APM also had fewer aggressive interactions at the feed bunk than cattle that remained healthy.[30] Lastly, more recent studies have shown that lactation number, locomotion score, dystocia, and previous lactation length play a role in the development of post-partum disease. This work found that cows that are older, had greater degrees of lameness, problems at calving, or longer lactation lengths have an increased risk of being treated for at least one post-partum disease.[31] Clearly, the development of APM is preceded by changes occurring weeks before the disease is diagnosed. Most of these changes seem to result from social interactions and alterations in feed intake that lead to negative energy balance and disruptions in calcium homeostasis.

PREVENTION OF ACUTE PUERPERAL METRITIS IN MODERN DAIRY HERDS

It is clear that APM is an expensive disease with significant impacts on dairy herds. While reasonably effective therapies are available, the disease is still costly and, therefore, prevention becomes the key factor for reducing the impact of APM on the welfare of affected cattle and profitability of dairy operations as a whole. When developing strategies to prevent disease, it is important to evaluate the factors thought to place animals at higher risk for a specific condition. The risk factors for the development of APM are well understood and are outlined above. To summarize here, important factors that predispose cattle to developing APM include the following:

1. Dystocia
2. Reduced pre-partum feed intake and negative energy balance
3. Hypocalcemia
4. Retained placenta

By focusing on these risk factors, the occurrence of APM in a herd can be reduced to acceptable levels and its impact on an individual cow can be minimized. Therefore, this discussion on preventing APM will focus on 5 key areas of dairy transition management. These 5 key areas are as follows:

1. Controlling the length of the dry period
2. Minimize dry matter intake (DMI) depression prepartum
3. Minimize the risk and effects of dystocia
4. Minimize the risk and effects of hypocalcemia
5. Maximize DMI postpartum

Controlling dry period length

Prior to calving, dairy cattle are given a period of time during which they are not milked. This period, commonly referred to as the "dry period", gives the mammary gland time to involute and the cow some time to prepare for the metabolic challenges of the coming lactation. The length of the dry period is often 45-60 days in length. Nevertheless, in certain situations, cattle may calve while still lactating, giving them effectively no dry period or, alternatively, cattle may have a dry period that is excessively long. Recent work has shown that cattle with dry periods of > 70 days in length are approximately 2 times more likely to have elevations in blood ketone bodies than cattle with shorter dry periods.[32] The reasons for this likely lie in the fact that cattle fed ad libitum during the dry period overconsume energy relative to their absolute requirements. This places cattle at greater risk of insulin resistance and subsequent problem with lipid metabolism.[32]

Minimizing dry matter intake depression pre-partum

On average, DMI will decrease 25-52% during the final 14 days of gestation and this decrease in DMI is a normal phenomenon in all cattle.[33] Generally, the decrease in DMI and resultant negative energy balance is more severe for heifers than cows because heifers are still growing. While the exact cause of the DMI depression is unknown, fetal growth and hormonal changes have been implicated, with most believing that hormonal changes play a larger role.[33] Minimizing the effects of decreased prepartum feed intake involves removing obstacles and stressors that limit feed availability. Ensuring appropriate stocking density in housing areas is absolutely essential. Cows should have adequate access to both feed and resting areas alike. In addition, minimizing the effects of heat stress by providing fresh water, shade, fans, and soakers will help promote feed intake. Reducing the frequency with which pens are changed and grouping animals of a similar signalment together can be beneficial. Lastly, preventing cows from becoming overconditioned is essential as excessively conditioned cattle will experience a greater drop in intake and suffer from more problems with energy balance than cattle in more moderate condition.

Minimizing the effects of dystocia on cattle health

Dystocia increases the risk of a cow developing APM by causing trauma to the reproductive tract and increasing the degree of bacterial contamination within the uterine lumen. When developing strategies to manage dystocia there are three factors to evaluate: calf factors, human factors, and cow factors. Calf factors include size relative to the dam and presentation, with relative calf size being a far more common cause of dystocia than malpresentation. Human factors include timing of intervention (too early vs too late) and the manner in which obstetrical procedures are performed (cleanliness, lubrication, etc). Cow factors include pelvic size, body condition, and periparturient calcium status. Management interventions that may help reduce the impact of dystocia on the development of APM include sire selection, employee training, heifer growth, and nutrition. Utilization of artificial insemination and choosing bulls with favorable birthweight EPDs can be of tremendous benefit to dairies. In addition, because cattle giving birth to bull calves have a greater risk of dystocia, the use of sexed semen might serve as an important management tool by providing a greater proportion of heifer calves in animals at greatest risk of birthing problems. Training employees as to when intervention is necessary and how obstetrical procedures should be performed is a necessity. Growing heifers to the appropriate frame score and providing adequate nutrients can increase animal size and make the dam's pelvis larger relative to the calf. The use of selection tools such as pelvic measurements and reproductive tract scoring are used with some frequency in beef cattle. Preliminary investigations into their utility in dairy cattle have been favorable and might serve as a useful tool in the future.

Minimize the risk and effects of hypocalcemia

The onset of lactation represents a tremendous drain on the dairy cow's calcium stores. There are approximately 11 g of calcium in the body's extracellular fluid with roughly 3.5 g of calcium in the serum. In early lactation, 20-30 g of Ca will be partitioned into milk daily. To compensate for this sudden increase in calcium need, calcium is drawn from bone via osteoclast recruitment and activation and removal from bone fluid. Nevertheless, some animals are unable to compensate for this as this need for calcium represents a demand that sometimes overwhelms the body's homeostatic mechanisms. While hypocalcemia is often associated with the classic clinical "milk fever" cases, subclinical hypocalcemia is far more common and has much more significant consequences. The most important of these consequences is impaired immune function, a phenomenon that leads to an increased risk of retained placenta and an inability to control the uterine contamination that normally occurs after calving.

Because the most significant effects of hypocalcemia result from subclinical manifestations of the disorder, prevention is a more effective strategy than treatment. Prevention might involve alteration of the diet of periparturient cattle by selecting forages low in potassium. Restricting prepartum calcium intake has also been described. The most effective means of preventing hypocalcemia in post-partum cattle is through feeding a negative DCAD diet in the periparturient period. Other strategies employed include the oral administration of calcium containing solutions or pastes to cattle after parturition.

Maximize dry matter intake post-partum

The goal of maximizing fresh cow feed intake is directed at a more rapid return to a positive energy balance. Greater levels of intake lead to less risk of ketosis, higher first test milk, higher peak milk levels, less weight loss, and a reduced time to first ovulation. Similar to the strategies employed to minimize the decrease in feed intake pre-partum, maximizing post-partum feed intake involves removing stressors that may hold cows back. Reducing the occurrence of fresh cow diseases, improving cow comfort, optimizing stocking density, and the utilization of high quality feedstuffs are critical control points that should be evaluated.

OTHER METHODS OF PREVENTION

Metaphylaxis

Administration of antimicrobials to cattle at high risk of developing uterine disease (HRUD, dystocia, twins, abortion, retained fetal membranes) has been evaluated. In one two-part study, cattle at HRUD were given CCFA at the label dose or saline within 24 hours of an abnormal calving event. In the first part of the study, 20.2% of cattle given CCFA developed metritis by day 14 after calving as compared to 36.8% of controls.[34] In the second part of the study, 28.7% of cattle given CCFA developed metritis by day 14 as compared to 43.5% of controls.[34] In addition, rectal temperatures of cattle given CCFA were lower for the first 6 days of lactation than controls. Another study found that, when cattle at HRUD are given CCFA within 24 hours of calving, there was no association of treatment with probability of metritis overall.[35] However, it was found that administration of CCFA decreased the risk of metritis in cattle without retained fetal membranes and cattle of parity >2.[35] In addition, administration of CCFA decreased the risk of diagnosis of purulent vaginal discharge in cattle at HRUD.[35] While these results seem promising, administration of CCFA to cattle at HRUD to prevent APM is considered extra-label drug use.

Prophylactic vaginal lavage

One study evaluating the effect of a prophylactic vaginal lavage containing 30 ml of 10% povidone-iodine in 2L of 7.2% hypertonic saline on the risk of metritis in post-partum dairy cattle found no benefit to the health or productivity of treated animals.[36]

Vaccination

Recent work evaluating the efficacy of a vaccine containing virulence factors of the organisms most commonly associated with APM in cattle has been promising. Cattle vaccinated subcutaneously with this product were significantly less likely to be diagnosed with APM than either controls or cattle vaccinated via other routes.[37] Nevertheless, this product is not commercially available and more data are needed before evidence-based recommedations can be made regarding its use.

Recombinant bovine granulocyte-colony stimulating factor (rbG-CSF)

rbG-CSF has been evaluated as a means of improving neutrophil numbers and function in the immediate periparturient period. In the most recent studies, rbG-CSF, when given as a SQ injection twice prior to calving, significantly increased both circulating numbers and myeloperoxidase activity of neutrophils.[38] Older studies showed increases in both of these parameters as well as improved phagocytosis and cytotoxicity.[39] In addition, rbG-CSF has been found to increase the number and alter the relative proportions of peripheral blood mononuclear cells (PBMCs) in periparturient cattle.[39] Nevertheless, this product is not commercially available and more data are needed before evidence-based recommendations can be made regarding its use.

CONCLUSIONS

Acute puerperal metritis is a common disorder with significant consequences on cattle production and welfare. While treatments are available, their ability to minimize the untoward consequences of APM on production are limited. Therefore, effective management of APM is targeted at prevention. Changes occurring before, during, and after calving place cattle at greater risk of APM. By managing these risk factors the occurrence of APM in dairy herds can be minimized and cattle productivity can be optimized. Technologies such as vaccines and rbG-CSF will be available in the future and could serve as valuable means of disease prevention moving forward.

REFERENCES

1. Sheldon IM, et al. Theriogenology 2006;65:1516.2. Kelton DF, et al. J Dairy Sci 1998;81:2502.3. Sheldon IM, et al. Biol Reprod 2009;81:1025.4. Overton MW, et al. Economics of Post-Partum Uterine Health. Dairy Cattle Reproductive Council Convention 2008;39.5. Overton MW. Periparturient Management for Post-Parturient Success. North American Veterinary Conference 2008;30.6. Gardner IA, et al. Preventive Veterinary Medicine 1990;8:157.7. Dubuc J, et al. J Dairy Sci 2011;94:1339.8. Giuliodori MJ, et al. J Dairy Sci 2013;96:3621.9. Ribeiro ES, et al. J Dairy Sci 2013;96:5682.10. Sheldon IM, et al. Reprod Domest Anim 2009;44 Suppl 3:1.11. Williams EJ, et al. Am J Reprod Immunol 2008;60:462.12. Bromfield JJ, et al. Biol Reprod 2013;88:98.13. Bromfield JJ, et al. Endocrinology 2011;152:5029.14. Gilbert RO. Reprod Fertil Dev 2011;24:252.15. Wagener K, et al. Vet J 2014;202:527.16. Wagener K, et al. Vet Microbiol 2015;175:286.17. Prunner I, et al. Theriogenology 2014;82:1316.18. Smith BI, et al. J Dairy Sci 1998;81:1555.19. Goshen T, et al. Theriogenology 2006;66:2210.20. Galvao KN, et al. Identifying and Treating Uterine Disease in Dairy Cows. In: Extension UoF, ed. IFAS Extension: 2013.21. Chenault JR, et al. J Am Vet Med Assoc 2004;224:1634.22. Drillich M, et al. J Dairy Sci 2001;84:2010.23. Schmitt E, et al. American Association of Bovine Practitioners 2001;199.24. McLaughlin CL, et al. J Dairy Sci 2012;95:4363.25. Lima FS, et al. J Dairy Sci 2014;97:5401.26. Martinez N, et al. J Dairy Sci 2012;95:7158.27. Dubuc J, et al. J Dairy Sci 2010;93:5764.28. Hammon DS, et al. Vet Immunol Immunopathol 2006;113:21.29. Kimura K, et al. J Dairy Sci 2006;89:2588.30. Huzzey JM, et al. J Dairy Sci 2007;90:3220.31. Vergara CF, et al. J Dairy Sci 2014;97:4127.32. Rollin E, et al. J Dairy Sci 2010;93:978.33. Grummer RR, et al. Vet Clin North Am Food Anim Pract 2004;20:447.34. McLaughlin CL, et al. Theriogenology 2013;79:725.35. Dubuc J, et al. J Dairy Sci 2011;94:1325.36. Smith I, et al. American Association of Bovine Practitioners 2010;187.37. Machado VS, et al. PLoS One 2014;9:e91734.38. Kimura K, et al. J Dairy Sci 2014;97:4842.39. Harp JA, et al. Vet Immunol Immunopathol 1991;28:29.

Q FEVER IN RUMINANTS: AN ANIMAL AND HUMAN HEALTH ISSUE

Raphaël Guatteo, DVM, PhD, DECBHM
Nantes, France

INTRODUCTION

Q fever is a zoonotic disease caused by an obligatorily intracellular bacterium, *Coxiella burnetii (Cb)*, well known to produce a dense, small and highly resistant spore-like form. This disease, described for the first time among abattoir workers in Australia is now recognised as being endemic worldwide except in New-Zealand. Although asymptomatic in more than 60% of cases, Q fever may manifest in humans either as an acute or chronic disease. While acute disease is mainly a flu-like syndrome, with a self-limited febrile illness associated with headaches and myalgia, severe complications such as pneumonia or hepatitis can occur. In chronic disease, endocarditis, especially in patients with previous valvulopathy, is the most frequent manifestation and *Coxiella burnetii* is also frequently involved in chronic fatigue syndrome. Pregnant women are particularly susceptible to *Coxiella burnetii* infections which could lead to premature delivery or abortions.

In livestock, *Cb* infection would lead mainly to reproductive disorders like abortions, premature delivery, stillbirth, weak calf syndrome (APSW complex). Some authors suggest also the implication of *Cb* in metritis and infertility, with associated economic impact for the herd. Therefore, *Cb* infection is both a human and animal health issue. The aim of this presentation is mainly to (i) describe the impact of *Cb* infection in humans, (ii) to summarize the evidence regarding the implication of *Cb* infection in reproductive disorders in ruminants and lastly (iii) to summarize the evidence regarding the effectiveness of different measures to control *Cb* infection in ruminants.

Q FEVER: A ZOONOSIS

Clinical signs in humans

One of the characteristic features of Q fever is its clinical variability and clinical picture, often asymptomatic, which complicates the diagnosis. Thus, in humans, the prevalence of Q fever is not accurately known. As an example, in France, tests searching for the involvement of *Coxiella burnetii* in events suggestive of Q fever suggest an annual incidence estimate of the infection between 0.1 and 1 per 1,000 inhabitants, a figure that confirms data stating the infection rate in 0.5 per 1,000 inhabitants (Raoult *et al.*, 2005). Under this assumption, the number of cases per year would be set by extrapolation between 6,000 and 60,000 cases, which is equivalent to between 2,400 and 24,000 asymptomatic cases and between 120 and 1,200 hospitalizations.

Primary infection occurs after the contact between a non-immune individual and *Cb.* It has been estimated that this primary infection can be either asymptomatic (60 % of the cases) or symptomatic, thus leading to acute Q fever (Raoult *et al.*, 2005). The most frequent clinical picture is a flu-like syndrome regressing spontaneously without treatment (Raoult, 2005). Humans become infected by inhaling contaminated aerosols. The role of consuming contaminated raw milk in considered as null to negligible.

Among human Q fever acute cases, 4 % required hospitalization due to the severity of the clinical signs (Dupuis *et al.,* 1985) (Figure 1). These signs were variable: hepatitis, pneumonia or meningoencephalitis were the most frequent. In some people, *Cb* can multiply despite the immune response occurring after primary infection. Thus, when the immune system is unable to control the infection, a chronic form of the infection develops (Maurin and Raoult, 1999). Chronic forms occur mainly in subjects with predisposing factors: patients suffering from valvular heart disease, immunocompromised individuals or pregnant women. As it can be deduced, Q fever is commonly underdiagnosed. The size of inoculums, the geographic distribution, the moment of infection as well as individual factors seem to influence the incubation period and the clinical picture. When human cases occurred and epidemiological investigations were carried out, the studies confirm the major role domestic ruminants have as the main reservoir for human infection.

Figure 1. Possible evolutions of *Coxiella burnetii* infection in humans (according to Dupuis et al., 1985)

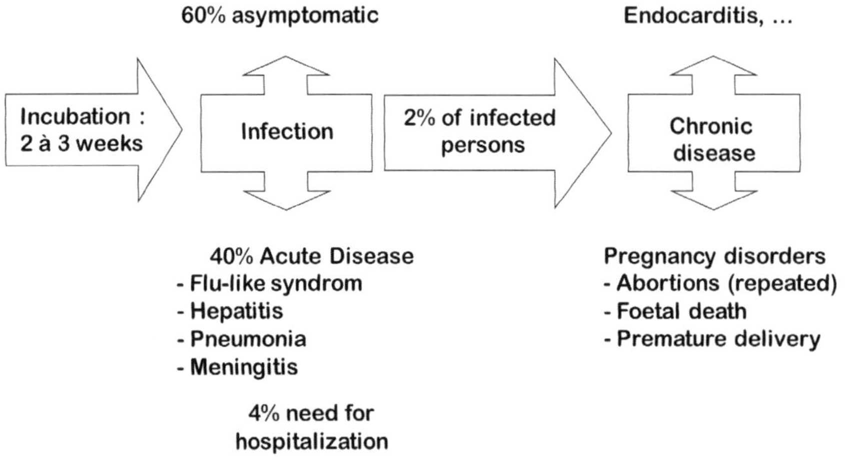

The Dutch outbreak

Since 2007, the Netherlands has been facing an important human Q fever epidemic. At the end of 2009, 2,357 new cases of human Q fever were reported: 60 % were asymptomatic and approximately 20 % of cases required hospitalization, most often due to pneumonia processes (source: www.eurosurveillance.org). Patients were around 50 years old and 60 % of them were men. Since the beginning of the epidemic, the distribution of human cases has progressively expanded; in all cases, infected individuals lived in the proximity of small ruminant farms-mainly goat farms-which had experienced outbreaks of abortions in the past. As a result, in February 2009, several mandatory health protection measures were implemented in Dutch sheep and goat farms, together with a systematic milk tank screening of *Cb* using PCR. This screening was carried out irrespective of the presence of clinical signs in the farm. Between April and November 2009, almost 250,000 small ruminants received a compulsory vaccination (phase I vaccine namely Coxevac®). Initially, milk tanks were analyzed once a month, but in December 2009 they were screened every fifteen days. In April 18th 2010, 81 goat farms and 2 sheep farms (out of a total of 360 goat farms and 40 sheep farms with more than 50 animals) were found infected with *Cb*. In July 2010, the 91st infected farm was detected. Within these infected farms, pregnant females were slaughtered and non-pregnant females were removed from reproduction. On the contrary, animals coming from small farms and/or from school farms were not affected by these measures (especially under the pressure of public opinion about school farms issue).

Thus, nearly 60,000 animals were euthanized (and a total compensation cost of more than 20 million euros was established). Costs related to the implementation of hygienic measures have been estimated at around 11 million euros, whereas the vaccination campaign to nearly 8 million Euros in 2010. Since June 1st 2010, the frequency of screening had been reduced to a single analysis of milk tank every 2 months.

The recent report of the EFSA (European Food Safety Agency), together with a pool of experts in this domain, point out the following reasons to explain this unusual Q fever outbreak in the Netherlands:

- Intensification of goat rearing. The number of goat farms doubled in the last 10 years.
- the number of goats tripled
- Increase in the livestock (herds between 1,000 and 7,500 individuals)
- Rearing method Open buildings (increasing the risk of bacteria aerial dissemination)
- Accumulated bedding (increasing the risk of aerial dissemination when spreading the manure)
- Farm localization High concentration of goat farms close to areas with high human population density
- Climatic conditions, particularly the dryness during the last three years, which might facilitate the aerial dissemination of *Cb*.

Up to June 21th 2010, 421 cases of Q fever in humans (including 5 deaths) were reported in the Netherlands, against almost 1,800 cases in 2009. So, a significant reduction of cases has been noted and the threat of an epidemic peak observed for the last 2 years has disappeared. A study conducted in infected patients following the Dutch outbreak reported a fatigue syndrome in 52 % of cases, more than 6 months after the initial diagnosis. The syndrome was still present in 25 % of them a year later and high levels of antibodies anti-phase I and II up to 3 months after the onset of the disease, followed by a slow and progressive decline, were detected. Between December 2009 and July 2010, 92 goat farms were found infected. The last 2 farms were detected in July 2010. All the flocks of more of 50 sheep or goats were vaccinated with phase I vaccine; this vaccination was still mandatory in 2011. Currently, the general prohibition of breeding non-pregnant females has been suspended..

This episode highlights the zoonotic aspect of Q fever, which can be potentially serious for human health, and points out the damaging effects on the ruminant sector. Ruminants being the main source for human infection, any measure aiming to decrease or prevent Cb shedding by infected ruminants would lead to decrease the risk of transmission of *Cb* from ruminants to humans and between ruminants too and would also benefit to the dairy industry.

Q FEVER IN RUMINANTS

The theory ...

While the pathogenesis of *Cb* infection is well known in humans, few data are available about the precise and detailed pathogenesis of *Cb* infection in ruminants. Given the current knowledge of *Cb* infection in humans, the suspected pathogenesis regarding uterine infection with *Cb* in pregnant ruminants is displayed in figure 2.

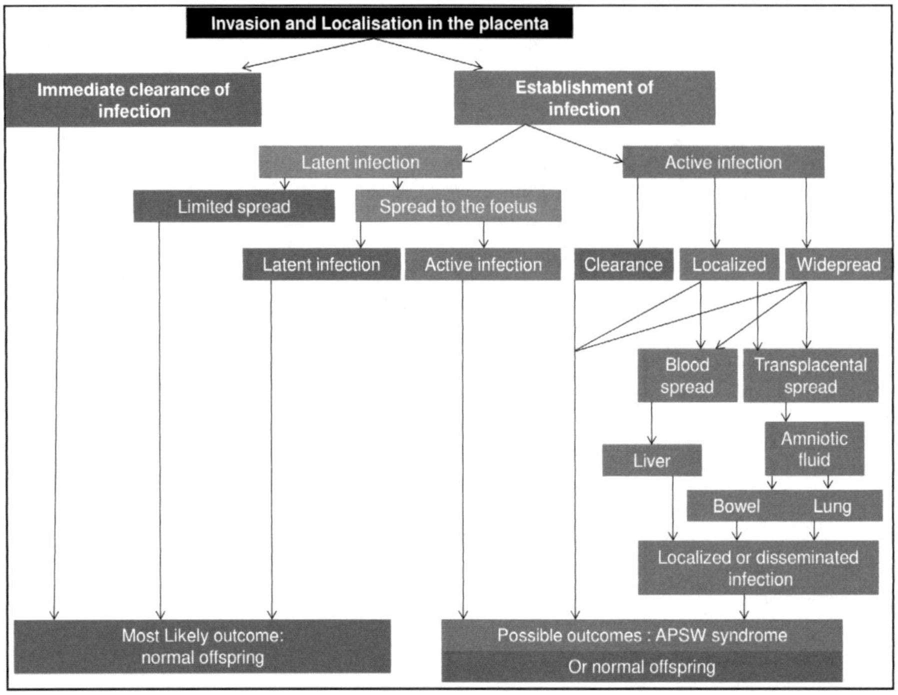

Figure 2. Suspected pathogenesis of uterine infection with *Coxiella burnetii* in ruminants (adapted from Agerholm, 2013) (ASPW: Abortions, Stillbirth, Premature delivery, Weak calf syndrom)

The evidence …

Clinical manifestations of Q fever are less referenced in ruminants than in humans. In fact, infection due to *Cb* appears often to be subclinical (Rodolakis, 2006). The most common clinical signs observed are related to reproductive disorders such as abortions, stillbirths or birth of weak offspring. Q fever seems also associated with endometritis and retained placentas (Arricau-Bouvery and Rodolakis, 2005). However, it is still unknown if this relationship is due to the bacterium itself or it is the abortion which predisposes to retained placentas which, in turn, predisposes to metritis. Thus, Q fever should be included in infertility differential diagnoses (EFSA, 2010). The diagnosis of abortions due to Q fever relies on positive PCR (on placenta, vaginal swab or stomach content of the aborted calf) combined with positive serology (using blood ELISA) on a subsample of cows with reproductive disorders (at least 50% of seropositive animals) (EFSA, 2010).

Other manifestations are also attributed to Q fever. Some studies report that experimentally infected ewes suffered from fever or respiratory disorders as Plommet et al in cattle (Plommet et al., 1973). These signs have been observed during the acute phase of the disease and followed by reproductive disorders (stillbirths, birth of non-viable lambs). In goats, Q fever would be also responsible for pneumonia. *Cb* has been sometimes blamed for inducing subclinical mastitis and clinical mastitis in dairy cattle. However, these studies cannot differentiate between *Cb* as the trigger for mastitis and the simple detection of *Cb* shedding in cattle with high concentrations of somatic cells in milk.It has been clearly shown through experimental and/or field studies (table 1) that Q fever is responsible for :

Table 1. Scientific evidence regarding the implication of *Coxiella burnetii* in different reproductive disorders in Ruminants (+ when evidence, ? when still unknown or need for research)

Species	ASPW Complex	Endometritis	Fertility	Retained
Cattle	+	?	+	+
Goat	+	?	?	+
Sheep	+	?	?	?

TREATMENT AND PREVENTION OF Q FEVER IN CATTLE

Non-medical measures

The non-medical control actions putatively feasible within infected herds would aim to reduce or prevent the exposure of susceptible animals (and humans) to contaminated aerosols. To reach this goal, several actions can be put forward. However, these measures are not specific to Q fever and the efficacy of each to reduce the infectious pressure or the *Cb* transmission between cows and between herds remains unknown:

- Disinfection of the calving box after each calving

- Avoiding putative contact in pasture between with other ruminants from other herds
- Nowadays, no standardized and efficient protocol for bedding or manure treatment is available. However, spreading manure on pasture when the wind blows should be avoided
- Prevention of tick infestation or regular treatments against tick could be a relevant measure in risky areas.

All together, these control measures could limit the risk of transmission among animals and from animals to humans. However, owing to the existence of apparently healthy (i.e. non-aborted) ruminants shedding *Cb* in various specimens, as well as the supposed high infectivity of *Cb:*

- The non- medical control actions are not sufficient and medical control actions are necessary.
- The exposed workers (especially farmers, veterinarians and abattoir workers) must be informed about the risks and clinical signs relative to human Q fever, in order to allow for early detection of the disease.

Medical measures

Several studies were conducted in goats, sheep and cattle to assess the effectiveness of vaccination and/or antibiotics to prevent *Cb* shedding or the clinical signs due to Q fever infection (Guatteo et al., 2008; Taurel et al., 2014 for instance). The table 2 summarizes the evidence about the effectiveness of vaccination (using phase I vaccine) and antibiotics to control Q fever in Ruminants.

Table 2. Scientific evidence regarding the effectiveness of vaccination and/or antibiotics to manage Q fever in ruminants (Guatteo, 2013)

Outcome	Vaccination using phase I vaccine (Coxevac®)	Oxytetracycline once at drying off (20mg/kg)	Oxytetracyclines once at calving (20mg/kg)
Abortions	Reduction (evidence in goats and cattle)	Unknown	No effect expected
Retained Placenta	unknown	Reduction of occurrence (evidence in cattle)	No effect
Fertility	Diminution of the risk of return ot service (evidence for nulliparous cattle)	No effect	Slight reduction or risk of return to service (evidence for seropositive cows)
Shedding	Prevention of shedding (evidence at calving for nulliparous cattle and goats and sheep) Reduction of amount of Cb shed at calving (evidence for dairy cows and goats)	Prevention of shedding in vaginal mucus at calving (evidence in cattle)	No effect

IN PRACTICE

Finally, altogether, the results of the different studies support the following scheme in herds clinically infected by *Cb* (Figure 3).

Figure 3. Proposal of control scheme for Q fever in Ruminants (Guatteo, 2013)

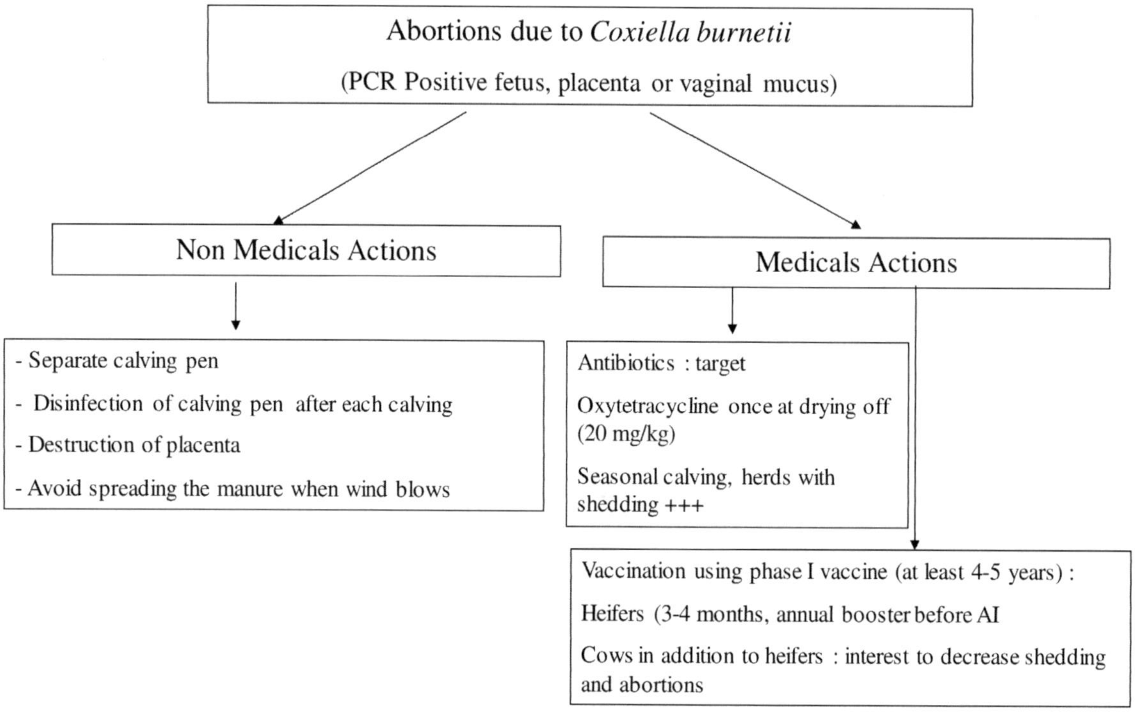

REFERENCES

[1]Raoult D, et al. Lancet Infect Dis 2005; 5:219-226. [2]Maurin M. et al. Clin. Microbiol Rev 1999 ;12:518-553.[3]Dupuis G, et al. Schweiz. Med. Wochenschr 2005 ;115:814-818. [4]EFSA. 2010. http://www.efsa.europa.eu/fr/supporting/doc/48e.pdf. 48p.[5]Agerholm JS. Acta Vet Scand 2013;18 :55:13. [6]Rodolakis, A. Small Rum. Res 2006;62:121-124. [7]Arricau-Bouvery N, et al. Vet Res 2005;36, 327-349. [8]Plommet M, et al. Ann Rech Vet 1973;4:325-346. [9]Guatteo R, et al. Vaccine 2008;26(34):4320-8. [10]Taurel AF, et al. Comp Immunol Microbiol Infect Dis. 2014;37(1):1-9. [11]Guatteo R. Servet Publications. 2013. 134p

ANTIMCROBIAL USE AND RESISTANCE FROM A EUROPEAN PERSPECTIVE

Raphaël Guatteo, DVM, PhD, DECBHM
Nantes, France

INTRODUCTION

To allow the production of milk and or/meat by healthy livestock and to ensure the welfare of animals as well to control the spread and the consequences of bacterial diseases , sick animals have to be treated in a responsible manner. However, in that context, the use of antimicrobials may enhance the development and/or the spread of antimicrobial resistance (AMR). In human medicine, antimicrobial resistance is a growing and global public health concern as the loss of effective antimicrobials is reducing the ability to protect the persons from infectious diseases. The use of antimicrobials by veterinarians, especially in livestock, is a hot topic in Europe due to the fact that during the last decade the evidence of an association between the use of antimicrobials in livestock and the presence of resistant bacteria was reported. Moreover the same families of antimicrobials are available and used in both veterinary and human medicine. As a consequence, since 2007, several European countries (Denmark, The Netherlands, France, and Switzerland for instance) have implemented national plans with the aim to reduce and rationalize the use of antimicrobials in animals. The common basis, regardless the country, was to monitor and reduce the use of antimicrobials, in order to achieve a decline in antimicrobial resistance in animals. The aim of this presentation is (i) to describe the current attitudes of European veterinarians in terms of antimicrobials prescription, (ii) to describe the trends in terms of sales of antimicrobials in European countries and (iii) to report the current proposed measures or perspectives to minimize antimicrobial resistance arising from the use of antibiotics in veterinary medicines and especially in livestock and cattle medicine.

WHAT ARE THE ANTIMICROBIALS USED MOST COMMONLY TO TREAT CATTLE IN EUROPE

The Heads of Medicines Agencies and the Federation of Veterinarians of Europe made a survey in 2013 to gain an insight into European prescribing of antimicrobials for animals. The goal was in particular to highlight the diseases for which antibiotics are most commonly prescribed and which different classes, including human critically important antibiotics (CIAs). The survey was conducted and completed by 3004 veterinary practitioners from 25 European countries (deBriyne et al., 2014) (Table 1). Older antimicrobials (eg, penicillins, tetracyclines) are cited most frequently to treat the main food producing species. The therapeutic area 'others' concerned perioperative (including caesarean-section), sepsis, infection, peritonitis, traumatic reticulitis, wounds, abscess, urinary, meningitis, nephritis, eye disease and abortion. In calves, diarrhea and respiratory covered 84 per cent of the indicators mentioned. In the total population of cattle (including calves), the relative mentions of CIAs to non-CIAs was 26 per cent compared to 74 per cent.

Clear 'preferences' between countries can be observed between antibiotic classes (Table 2). There were differences between countries in the antibiotics veterinarians said they use. Penicillins are the antibiotic class most frequently mentioned in all countries particularly in Sweden where this class by far predominates the others (84 per cent). The use of CIAs was specified more frequently for Germany (42 per cent v 26 per cent average), while no use was cited for Sweden.

Table 1. Antimicrobials mentioned and frequency with which they were mentioned for the top five indications where antibiotics are said to be prescribed for cattle and calves by veterinary practitioner (deBriyne et al., 2014)

Therapeutic area	Percentage mentioned		Percentage critically important antibiotics (CIAs) vs percentage other antibiotics	Frequency of citation of the different classes of antibiotics
	Adult cattle	Calves		
Mastitis	40%	-	CIAs: 22% Non-CIAs: 78%	Penicillins 41% 1st and 2nd generation cephalosporins 12% 3rd and 4th generation cephalosporin 11% Macrolides 6%
Respiratory disease	22%	27%	CIAs: 45% (44% calves) Non-CIAs: 55% (56% calves)	Macrolides 27% (28% calves) Phenicols 22% (19% calves) Tetracyclines 19% (18% calves) Fluoroquinolones 13% (12% calves) Pénicillins 7%
Diarrhea	14%	57%	CIAs: 29% (26% calves) Non-CIAs: 71% (74% calves)	Polymyxins 40% (44% calves) Fluoroquinolones 20% (18% calves) Pénicillins 13% (13% calves) Aminoglycosides 9% 3rd and 4th generation cephalosporin 8% (8% calves)
Uterine	8%	-	CIAs: 21% Non-CIAs: 79%	Penicillins 37% 3rd and 4th generation cephalosporin 18% Aminoglycosides 16% Tetracyclines 16% 1st and 2nd generation cephalosporins 7%
Locomotion	8%	-	CIAs: 31% Non-CIAs: 69%	Penicillins 33% Tetracyclines 24% 3rd and 4th generation cephalosporin 22% Macrolides 9%

				Aminoglycosides 9%
Other	8%	16%	CIAs: 13% (10%) Non-CIAs: 87% (90%)	Penicillins 60% (55% calves) Aminoglycosides 12% (18%) 3rd and 4th generation cephalosporin 8% Lincosamides 7% (12 % calves) Tetracyclines 4%

Table 2. Overview of antibiotics cited most frequently as being prescribes for use in cattle in Europe (deBriyne et al., 2014)

Antibiotics classes (*CIAs in italic*)	**Total all countries**	**Belgium**	**France**	**Germany**	**Spain**	**Sweden**	**UK**
3rd and 4th generation cephalosporins	10	19	9	23	11	0	6
Macrolides	9	9	10	5	10	0	13
Fluoroquinolones	7	11	6	14	13	0	1
1st and 2nd generation cephalosporins	6	3	7	3	3	0	12
Aminoglycosides	9	8	12	7	8	6	14
Lincosamides	1	3	1	2	1	0	0
Penicillins	34	28	25	31	29	84	32
Phenicol	5	5	4	5	8	0	10
Polymixin	8	7	12	1	2	0	0
Sulphonamide	2	2	3	3	3	2	2
Tertracyclines	8	5	10	4	12	7	10
Other	1	<1	1	2	0	1	0

TRENDS IN THE SALES OF VETERINARY ANTIMICROBIALS AGENTS IN EUORPENA COUNTIRES IN 2005-2009

The European Medicines Agency is the European Union body responsible for coordinating the existing scientific resources put at its disposal by Member States for the evaluation, supervision and pharmacovigilance of medicinal products. The European Surveillance of Veterinary Antimicrobial Consumption (ESVAC) project collects information on how antimicrobial medicines are used in animals across the European Union (EU). This type of information is essential to identify possible risk factors that could lead to the development and spread of antimicrobial resistance in animals. (for more details, **http://www.ema.europa.eu**). The amounts of veterinary antimicrobial agents sold in the different countries are, among others, linked to the animal demographics in each country, which may vary over time. In this report, the annual sales figures in each country were divided by the estimated weight at treatment of livestock and of slaughtered animals in the corresponding year. In the ESVAC report, the population correction unit (PCU) is used as the term for the estimated weight; the PCU is purely a technical unit of measurement, used only to estimate temporal trends in individual countries and across countries. Therefore in ESVAC reports 1 PCU = 1 kg of different categories of livestock and slaughtered animals. The distribution of estimated PCU over the period 2005-2009 is displayed in table 3. The level of PCU was quite stable among this period. Since 2007-2008 several countries decided due these figures to implement national plan (more or less compulsory) to reduce and rationalize the use of antimicrobials).

Table 3. Estimated PCU (in 1,000 tons) of the animal population, by country during the period 2005-2009 (ESVAC report)

Country	**2005**	**2006**	**2007**	**2008**	**2009**
Czech Republic	888	887	875	840	771
Denmark	2443	2543	2537	2523	2447
Finland	562	558	554	541	524
France	7801	7666	7789	7707	7539
Netherlands	3170	3214	3288	3133	3109
Norway	436	437	429	438	440
Sweden	837	828	819	819	825
Switzerland	-	729	731	734	743
United Kingdom	6142	6190	6202	6018	5925
Total	22280	23052	23224	22754	22322

WHAT ABOUT ANTIMICROBIAL RESISTANCE IN CATTLE: THE FRENCH EXAMPLE

In France, for several years, all the data concerning antimicrobial resistance are gathered annually by RESAPATH (Réseau d'épidémiosurveillance de l'antibiorésistance des bactéries pathogènes animals: network for the surveillance of antimicrobial resistance of pathogens). An example of this monitoring is displayed in table 4. Such information is used in practice by the veterinarian to adapt their treatment.

Table 4. Antimicrobial susceptibility pattern of a range of bacteria tested in France in 2013 (non-exhaustive data)

Antibiotic	*E Coli* neonatal diarrhea		*E coli* from mastitis		*Salmonella typhimurium*		*Mannheimia haemolytica*	
	n	% S	n	% S	n	% S	n	% S
Amoxicilline	2226	15	566	75	149	23	117	82
Amox Ac. Clanulanic	2718	42	663	83	156	53	122	98
Céfalexine	2008	74	544	87	134	100	92	97
Ceftiofur	2647	92	524	98	156	100	117	98
Cefquinome	2634	85	584	99	140	100	121	97
Streptomycine	1332	13	341	72	84	23	33	24
Tétracycline	2284	21	619	81	151	19	117	65
Florfenicol	2044	76	383	97	133	51	121	996
Enrofloxacine	2494	75	549	97	151	98	118	91
Marbofloxacine	2423	79	578	97	143	100	116	94
Sulfamides	330	19	96	84	40	12	112	94

A GENERAL DECREASE IN SALE OF VETERINARY ANTIMICROBIALS SINCE 2007

Following the implementation of national schemes in the different countries, an overall decrease of the sales of antimicrobials was observed. The sales of antibiotics for use in animals fell overall by 15% between 2010 and 2012 in Europe, according to a report published today by the European Medicines Agency (EMA). Of the 20 European countries that provided data over this timespan, 18 observed decreases ranging from 0.4% to 49% (Figure 1 for instance in the Netherlands). The use of critically important antibiotics decreased as well as the use of oral antibiotics.

According to Member States, different factors may have contributed to the decline:

- national programs and campaigns on the responsible use of antimicrobials
- restrictions on the use of certain antimicrobial
- increased awareness of the threat of antimicrobial resistance
- reduction targets for the use of antimicrobials in animal production in certain Member States (for instance 50% in Netherlands and 25% France over a 5 years period)
- fluctuations in size and types of animal populations.

Figure 1. Sales figures for antibiotics licensed for therapeutic use in animals (kg x 1000) in the Netherlands from 1999 up to 2012 (Mevius et al., 2014). Beginning of the national plan in 2007

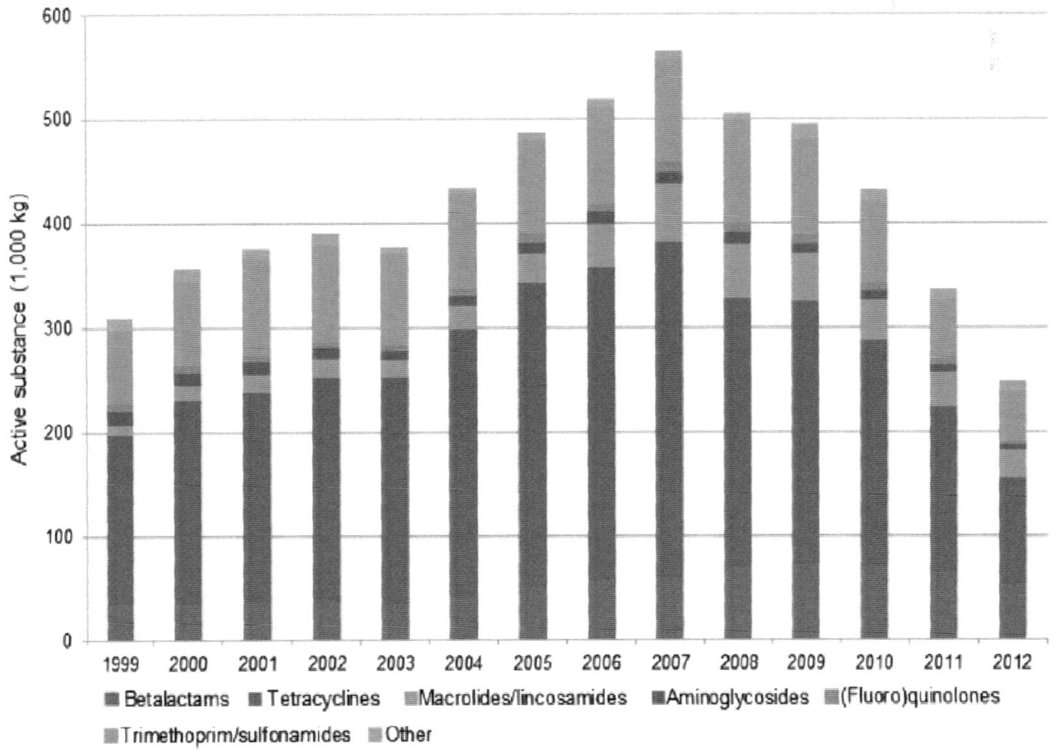

Figure 2. Sales figures for antibiotics licensed for therapeutic use in animals (kg x 1000) in France from 1999 up to 2013 (Anses report)

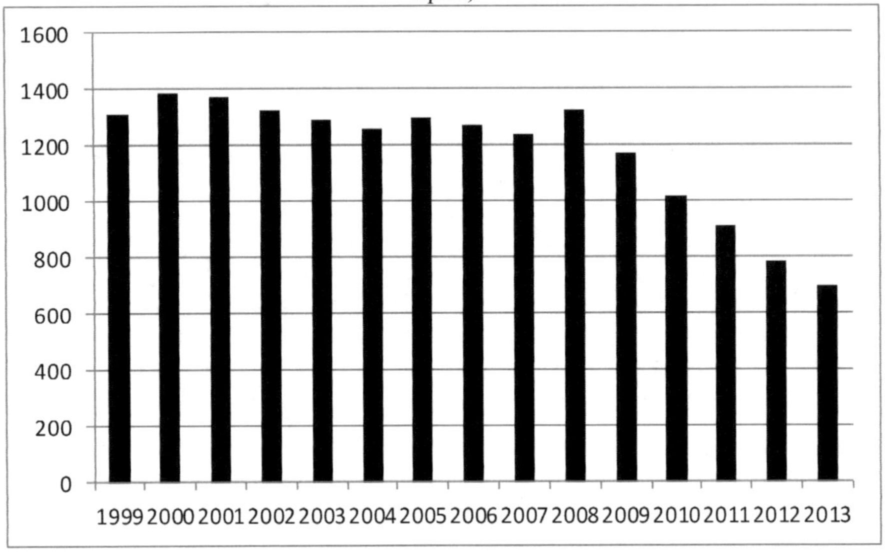

PERSPECTIVE FOR MITIGATING RISKS DUE TO THE USE OF ANTIBIOTICS IN ANIMALS

Currently, European experts propose measures to minimize antimicrobial resistance arising from the use of antibiotics in veterinary medicines:

- Reinforcing the existing risk assessment of new veterinary antimicrobial substances by identifying the main potential risks to public health at a very early stage of authorization;
- Monitoring the development of resistance in specific bacteria according to plans agreed between regulators and companies when a new antimicrobial substance is first approved in veterinary medicine;
- Introducing legal tools to restrict the use of antibiotics in animals if a significant risk to public health is identified;
- Monitoring changes in the use of antimicrobials in animals as part of the Agency's European Surveillance of Veterinary Antimicrobial Consumption (ESVAC) project to measure the impact of the actions implemented; in particular, fluoroquinolones and cephalosporins should be monitored as these two classes of antibiotics are critically important for human medicine.

A number of recommendations relate to the off-label use of veterinary medicines such as the systematic prevention using antimicrobials (in veal calves for instance).

The experts also propose to classify antimicrobials from the WHO CIA list in three different categories:

- Category 1 as antimicrobials used in veterinary medicine where the risk for public health is estimated as low or limited,
- Category 2 as antimicrobials used in veterinary medicine where the risk for public health is estimated higher. Category 2 includes those antimicrobial classes listed as CIAs by WHO for which the risk to public health from veterinary use is only considered acceptable provided that specific restrictions are placed on their use (i.e. fluoroquinolones and systemically administered (parenteral and oral), 3rd - and 4th generation cephalosporins). These reserved antimicrobials should be used only when there are no alternative antimicrobials authorized for the respective target species and indication.
- Category 3 as antimicrobials not approved for use in veterinary medicine.

The recommendations propose the introduction of flexible tools as part of the on-going review of the veterinary legislation to allow banning or limiting the off-label use in animals of certain antimicrobials authorized only in human medicine following a specific risk assessment of the potential risks arising from such off-label use. It is also suggests that prescribers should keep records of all off-label use and that authorities should be encouraged to collect data on off-label use of antibiotics used in animals. The advice will act as valuable input into the discussions that have now started in the European Council and the European Parliament on the proposal for revision of the legislation covering veterinary medicines.

In terms of research, the current perspectives concern

- Promotion of good farming practices and animal husbandry.
- The off label use of antimicrobials in animals;
- The extent of metaphylactic use of orally administered AMs and the impact of this practice on the development and persistence of resistance in the gut microflora of the animals;
- Pathways of dissemination of antimicrobial-resistant bacteria from animals to food;
- Methods for the quantification of the spread of resistance genes from commensals to pathogens in foods and the environment;
- Methodologies to evaluate the potential economic consequences and impact on both human and animal health and welfare that would result from the introduction of new risk management measures

- Appropriate strengths and pharmaceutical forms of those antimicrobials identified with a lower risk should be available and authorized for veterinary use in all EU countries. Antimicrobials should be marketed with the adequate pack size, according to the required posology for animal treatment
- For cattle, the main areas of research concern the early detection of sick animals for individual treatments, the determination of metaphylactic rules and the proper implementation of selective treatment strategies.

REFERENCES

[1]Cogliani C, et al. Microbe 2011 ; 6:274-279. [2]De Briyne N, et al. Vet Rec 2014;175(13):325. [3]Graveland Het al. . Int. J. Med. Microbiol 2011;301:630–634. [4]Hendriksen RS, et al. Acta Vet. Scand 2008; 50: 28. [5]MEvius D et al., J Verbr Lebensm 2014;9:177-181.

PETTING ZOOS: THE GOOD, THE BAD, AND THE BIOHAZARDOUS

Maureen E.C. Anderson, DVM, DVSc, PhD, DACVIM (LAIM)
Guelph, Ontario, Canada

INTRODUCTION

Animals in public exhibits such as petting zoos have the potential to transmit pathogens to a large number of people in the community. Numerous outbreaks of disease associated with petting zoos have been documented,[1] but it is likely that many instances of transmission, particularly those resulting in sporadic gastrointestinal illness, go undetected.[2-4] Careful attention to animal health and infection control at these venues is therefore important to public health.

THE GOOD AND THE BAD

Petting zoos and other animal exhibits facilitate contact of the public with live animals as part of agricultural fairs and other seasonal or permanent attractions. They are entertaining, but also serve an important educational function for individuals of all ages regarding animals and animal husbandry, and they help to increase compassion for animals of different kinds.[1,5] The risk of infectious diseases associated with human-animal contact at public animal exhibits varies according to the animals and individuals involved, as well as the degree of contact that occurs. Risks are higher for young children, with whom petting zoos and similar events are often particularly popular.[1,6-10] Risks are also higher among individuals who are immunocompromised and the elderly,[1,6,10] all of whom may also potentially visit such exhibits, as well as among pregnant women, who were observed touching animals at 41% (14/34) of petting zoos investigated during one study.[11]

Numerous disease outbreaks have been associated with animal contact at public venues. The most commonly reported pathogen in these outbreaks is *Escherichia coli* O157:H7, but others include *Salmonella*, *Cryptosporidium*, *Giardia*, *Coxiella burnetii* (Q fever) and dermatophytes (ringworm).[1,9,12-18] There is also potential risk of transmission of other zoonotic pathogens including *Campylobacter,* methicillin-resistant *Staphylococcus aureus* (MRSA) and rabies.[1,19-21] Most recently, transmission of influenza from swine to humans at agricultural fairs has also been recognized.[22-24] While outbreaks receive the greatest attention, there may be a significant level of morbidity from sporadic infections that are not reported to public health officials or linked to petting zoo exposure. Overall, there is limited information regarding the disease burden associated with petting zoos.

SIZING UP THE PROBLEM

Based on a general population survey of FoodNet sites in the US, it was estimated that 3.88% of individuals visited a petting zoo or farm in the previous week, and in the UK there are thought to be up to 10 million recreational visitors to open farms and similar venues annually, which indicates how popular these types of activities are.[25,26] More specific demographic information regarding people who attend or do not attend public animal exhibits is lacking. It has been postulated that such events are actually primarily a draw for urban dwellers who otherwise may not have contact with agricultural species, but attendance would also be affected by proximity to the venues which may be more often located in or near smaller, more rural communities.

There are very few studies examining pathogen prevalence in animals used in public displays/exhibits that have been undertaken outside of outbreak investigations. Because *E. coli* O157 is the pathogen most commonly associated with outbreaks occurring at petting zoos or similar events, most of the studies on the animals used in these venues focus only on this pathogen, and there is far less information regarding carriage of other pathogens such as *Salmonella*, *Campylobacter*, *Cryptosporidium* and *Giardia* in these animals (although *Cryptosporidium* would be the second-most investigated pathogen).

There is a paucity of information available with regard to quantitative or semi-quantitative measurement of pathogen shedding by animals used in public displays, and contamination of the environment at such venues. This is in part due to the diagnostic testing methods use (e.g. enrichment culture, fecal antigen tests), but with the increasing use of molecular detection techniques such as PCR there needs to be a push to develop quantitative tests as well (i.e. quantitative PCR, semi-quantitative antigen tests). There is virtually no information available regarding coat or oral carriage of common petting zoo pathogens in animals with public contact. Although it is expected that such carriage would be very low, it could still be a significant source of pathogens due to the increased degree of contact people have with the coat and mouth of animals. However, a Dutch study describing a quantitative microbiological risk assessment (QMRA) model for *Campylobacter* in petting zoos was recently published, that more specifically examined the risk of fecal contamination on high-touch surfaces (including parts of the animals) and hand-to-mouth transmission.[27] The model could potentially also be applied to examine the risk of direct transmission of other fecal pathogens from contact with animals and their environments.

MAKING THINGS BETTER

Several organizations have published recommendations for management of live animal exhibits such as petting zoos, intended to help decrease the risk of pathogen transmission from animals to the public.[1,6,10,12,28] Some of the key recommendations typically include avoiding use of potentially dangerous or clinically ill animals, taking special precautions to prevent contact between high-risk individuals (e.g. children less than five years of age) and high-risk animals (e.g. young ruminants, birds, reptiles, amphibians), physical layout considerations to control visitor contact with animals and overall flow through the exhibit, exclusion of food and drink for human consumption from the area, and visitor education with particular emphasis on hand hygiene and avoiding high-risk (e.g. hand-to-mouth) behaviours. While these are sound guidelines based on general infection control principles, there has been no objective assessment of the efficacy of their use at public venues, and implementation of these recommendations is still poor in many facilities.[11,20,29] Studies to help pinpoint those factors with the greatest potential to decrease pathogen transmission risk are needed.

According to the National Association of State Public Health Veterinarians, "*the recommendation to wash hands is the most important prevention step for reducing the risk for disease transmission associated with animals in public settings.*"[1] Failure to perform hand hygiene has been identified as a significant risk factor for disease in numerous outbreaks associated with animal exhibits.[7,12,13,18,30] Previous outbreaks have involved both visitors who did and did not report direct contact with animals, and outbreak strains of *E. coli* and *Salmonella* have been found on numerous surfaces at petting zoos and in similar environments.[7,13,18,30-32] Thus it is recommended that all individuals who enter an animal exhibit perform hand hygiene before leaving the area, regardless of whether they had direct contact with an animal.[1,16,18] However, hand hygiene compliance at petting zoos and other public exhibits has been reported to be poor, with mean compliance ranging from 31-58%.[5,11,29,33] One study with an overall compliance of 58% found higher compliance among visitors who had direct contact with an animal or pen (71%) compared to those who did not (35%).[33] Other studies have found that the presence of a hand hygiene station on an exit route, the presence of hand hygiene reminder signs, availability of running water and the presence of a staff member within or at the exit of the animal area were all positively associated with hand hygiene compliance.[11,29] Anderson and Weese[33] found that interventions involving active reminders from personnel to perform hand hygiene were the most effective at increasing hand hygiene compliance. Active interventions may be more likely to be noticed, more difficult to ignore or may be taken more seriously. Based on this, having designated personnel to supervise the hand hygiene area at an animal exhibit and provide verbal reminders to visitors to perform hand hygiene could be considered an important safety measure, just as having personnel to supervise animal areas is considered important for the safety of visitors (and animals).[33] Other important characteristics for public animal exhibits include a layout that facilitates one-way flow of traffic.[1,10,33] Having a well-designed petting zoo that follows recommended guidelines can improve hand hygiene compliance, yet there is still room for improvement even in facilities designed and operated according to current recommendations. Identification of specific interventions that are effective for increasing hand hygiene compliance would be beneficial.

HAND SANITIZER VS HAND WASHING

In a video observation study of a relatively small petting zoo at a University open house, hand sanitizer was used by 198/583 (34%) visitors instead of washing their hands at the sink (i.e. using running water).[33] Line-ups for the sink may have affected this proportion, as a small number of visitors were noted to leave the line-up after standing in it for some time. Ensuring that an adequate number of hand hygiene stations (accessible to children and adults) and sufficient product are available for the number of visitors expected at a given venue is critical to optimizing hand hygiene compliance. Alcohol-based hand sanitizers have been shown to be equally effective to soap-and-water hand washing for reducing hand contamination with coliform bacteria (such as *E. coli* 0157:H7) in a livestock-show setting,[34] and for reducing overall bacterial load on the hands following contact with horses.[35] Waterless hand sanitizers are often more convenient to use in barns and temporary animal exhibits where access to sinks may be limited. Although some studies have reported that hand sanitizer availability reduced the risk of disease associated with animal contact, other studies have shown no protective effect.[1,9,13,18] It is also important to recognize that alcohol-based hand sanitizers are ineffective against some pathogens that may be encountered in a petting zoo environment (e.g. *Cryptosporidium*, clostridial spores), and their effectiveness may be significantly impaired when hands are visibly soiled. In these instances, hand washing with soap and water to physically remove gross contamination and alcohol-resistant pathogens is recommended.[1,6,10] If hand sanitizer is used at the exit from animal areas because soap and running water is not readily available, it is still recommended that visitors wash their hands with soap and water as soon as possible thereafter.[1]

CHILDREN VS ADULTS

At least two studies have found that children had better compliance with hand hygiene than adults at petting zoos.[5,33] This may be due to increased awareness of the need for hand hygiene among school-aged children if there is growing use and emphasis on hand hygiene in schools, or due to parents instructing their children to perform hand hygiene after visiting the exhibit but failing to do so themselves. While healthy adults are generally at lower risk of infection than children, it is still important for parents and guardians to perform hand hygiene, as indirect transmission may result from contamination of their own hands followed by close contact with their children. Adults have also commonly been involved in outbreaks of disease associated with animal exhibits.[30,31,36] Anderson and Weese[33] found that very young children (approximately 2-5 years of age) were not significantly more likely to perform hand hygiene than adults. The reason for this may in part be due to the inability of children of this age to access or use the hand hygiene stations without assistance from an adult. A negative association was found between hand hygiene and being a baby.[33] Infants less than two years of age are less likely to benefit from briefly seeing and contacting animals in a petting zoo compared to older children who are much more cognizant of and responsive to the experience. The cost-benefit for a baby in terms of the risk of infectious disease transmission from either direct contact with the petting zoo environment or animals, or indirect contact via the person accompanying the baby, is less easily justified.

OTHER KNOWLEDGE GAPS

There are a number of areas in which additional information would be useful for assessing the risk (and by extension ways to mitigate the risk) of pathogen transmission to the general public from animals in public exhibits. For example, the species and numbers of animals used in petting zoos and public displays that may have direct or indirect contact with the public is not consistently provided in published studies. Weese et al.[11] reported the most common species at 36 petting zoos in Ontario, Canada included goats, sheep and adult or juvenile cattle at over 50% of exhibits, as well as baby chicks (17% of exhibits) and neonatal calves (11% of exhibits) which are considered high-risk animals.[1,10] Pritchard et al.[37] investigated 31 open farms in England/Wales, and found cattle (age not specified), sheep, goats, pigs, horses and small pet mammals were present on over 50% of facilities. One farm also displayed lizards and amphibians, another high-risk animal group.[1,10,37]

The source of animals used in these exhibits is also frequently unclear. If display animals are in fact being sourced from production groups (e.g. calves, chicks, lambs) then the prevalence of enteropathogens from the later could be used to estimate the prevalence in the former (at least when the animals are initially acquired). If however these animals are obtained from other sources (i.e. calves of "pet" cows or backyard/private animals) then these estimates are not as applicable.

Characterization of the proportion of enteropathogens shed by animals that are potentially zoonotic is also important, particularly for parasites such as *Giardia* and *Cryptosporidium.* This kind of characterization is becoming more common as researchers' ability to characterize species/genotypes/assemblages improves. Inclusion of this kind of testing in future studies will greatly increase the ability to estimate the potential risk to public health posed by animal carriers of these pathogens.

CONCLUSION

Despite the availability of guidelines to help reduce zoonotic disease risks associated with live animal exhibits, implementation of these recommendations is still poor in many facilities.[11,20] The risk of infectious disease transmission between animals and people in any setting can never be completely eliminated. However, by increasing adherence to management and safety recommendations and increasing hand hygiene compliance among visitors the risk can be reduced. This will allow these exhibits to continue to play their role in educating the public about animals and animal husbandry and promoting compassion towards different animal species.

REFERENCES

[1]National Association of State Public Health Veterinarians Animal Contact Compendium Committee 2013. **J Am Vet Med Assoc** 2013;243:1270–1288. [2]Taylor LH, et al. **Philos Trans R Soc Lond, B, Biol Sci** 2001;356:983–989. [3]Majowicz SE, et al. **Can J Public Health** 2005;96:178–181. [4]Cleaveland S, et al. **Curr Top Microbiol Immunol** 2007;315:85–111. [5]McMillian M, et al. **J Am Vet Med Assoc** 2007;231:1036–1038. [6]Ontario Ministry of Health Long-Term Care. Recommendations to prevent disease and injury associated with petting zoos in Ontario. Queen's Printer for Ontario 2009:1–28. [7]CDC. **Morb Mortal Wkly Rep** 2001;50:293–297. [8]CDC. http://www.cdc.gov/healthypets/. Accessed Jan-2015. [9]Goode B, et al. **Arch Pediatr Adolesc Med** 2009;163:42–48. [10]Erdozain G, et al. **Zoonoses Public Health** 2014 (epub ahead of print). [11]Weese JS, et al. **Clin Infect Dis** 2007;45:10–15. [12]Bender JB, et al. **J Am Vet Med Assoc** 2004;224:1105–1109. [13]Friedman CR, et al. **J Pediatr** 1998;132:802–807. [14]Hullinger GA, et al. **Vet Dermatol** 1999;10:73–76. [15]LeJeune JT, et al. **J Am Vet Med Assoc** 2004;224:1440–1445. [16]Pickering LK, et al. **Pediatrics** 2008;122:876–886. [17]Sayers GM, et al. **Commun Dis Rep CDR Rev**1996;6:R140–4. [18]Steinmuller N, et al. **Clin Infect Dis** 2006;43:1596–1602. [19]Friedman CR, et al. **Clin Infect Dis** 2004;38 Suppl 3:S285–96. [20]Heuvelink AE, et al. **Epidemiol Infect** 2007;135:1174–1183. [21]Dressler AE, et al. **Vet Rec** 2012;170:495. [22]Wells DL, et al. **J Am Med Assoc** 1991;265:478–481. [23]Killian ML, et al. **Zoonoses Public Health** 2013;60(3):196-201. [24]Wong KK, et al. **Emerg Infect Dis** 2012;18:1937–1944. [25]Dawson A, et al. **Commun Dis Rep CDR Rev**1995;5:R81–6. [26]CDC. Foodborne Active Surveillance Network (FoodNet) Population Survey Atlas of Exposures. 2011:1–29. [27]Evers EG, et al. **Risk Anal** 2014;34:1618–1638. [28]Environmental Health Directorate, Western Australia. Petting Zoo Guidelines. 2007:1–18. [29]Erdozain G, et al. **Zoonoses Public Health** 2013;60(4):304-310. [30]Crump JA, et al. **N Engl J Med** 2002;347:555–560. [31]CDC. **Morb Mortal Wkly Rep** 2005;54:1277–1280. [32]Varma JK, et al. **J Am Med Assoc** 2003;290:2709–2712. [33]Anderson MEC, et al. **Epidemiol Infect** 2012;140:182–190. [34]Davis MA, et al. **Epidemiol Infect** 2006;134:1024–1028. [35]Traub-Dargatz JL, et al. **Can Vet J** 2006;47:671–676. [36]David S, et al. **Canada Commun Dis Rep** 2004;30:1–7. [37]Pritchard GC, et al. **Vet Rec** 2009;164:545–549.

UTERINE DISEASES IN POST-PARTUM DAIRY CATTLE

Brent Credille, DVM, PhD, DACVIM
Athens, GA

INTRODUCTION

Diseases of the reproductive tract are a common cause of impaired productivity in dairy cattle. Studies have shown that the majority of cattle will have some degree of contamination of the uterus with bacteria after calving. While most cattle clear this contamination, some will be diagnosed with inflammatory disorders that affect fertility. Recent estimates suggest that up to 50% of all cattle that calve will develop at least one uterine health disorder that compromises reproductive performance.[1] The effects of the various uterine health disorders on individual animal production and the relative frequency with which these diseases occur make them costly for both single herds and the dairy industry as a whole. The primary drivers for the high incidence of post-partum reproductive disorders are primarily attributable to altered immune function and disruptions in metabolic health in the periparturient period.[1,2] In addition, recent work has shown that the presence of certain bacteria or strains of bacteria may increase the risk of disease through the expression of virulence factors or other mechanisms that have not been described.[3-5]

DISEASE DEFINITIONS

Historically, the definitions used to characterize the various uterine health disorders varied from person to person and lab to lab. Because of this variation, early studies that aimed to characterize the pathophysiology or untoward consequences of uterine diseases in dairy cattle are difficult to evaluate. As a result, consensus definitions were developed to help researchers and clinicians communicate in a more standardized way. Acute puerperal metritis (APM) is defined as an acute systemic illness due to infection of the uterus within the first 21 days post-partum. Affected animals will have an abnormally enlarged and flaccid uterus that contains a fetid, watery, reddish-brown discharge.[6] In addition, animals with APM will have a fever (>103F) and other systemic signs of illness (decreased milk yield, obtundation, anorexia).[6] Clinical metritis is defined as an animal that is not systemically ill but has an abnormally enlarged uterus that contains purulent discharge within 21 days of calving.[6] Purulent vaginal discharge (PVD), a disease formerly known as clinical endometritis, is characterized by the presence of purulent or mucopurulent exudate in the vagina 21 or more days post-partum.[6,7] Subclinical endometritis is defined as endometrial inflammation in the absence of purulent material.[6] Pyometra is characterized by the presence of purulent or mucopurulent material in the uterine lumen, palpable uterine distension, and an active corpus luteum.[6]

EFFECTS OF UTERINE DISEASES ON DAIRY PRODUCTION

In a study by Overton and Fetrow, cattle with APM were open on average 16 days longer and had a 21-day pregnancy rate 3-6 percentage points lower than that in healthy cattle.[8] In addition, culling risk for cattle with APM was 4.2% greater than the culling risk in other groups.[8] These same animals produced almost 5 lbs less milk per day over the first 120 days of lactation.[8] Of note is that each cow in this study was treated with standard therapies (antimicrobials, anti-inflammatories, fluids, etc), a factor that suggests that APM still has significant costs despite treatment. A separate study demonstrated that cattle affected by APM produce 3.7 kg less milk at first test and 259 kg less milk over the course of a single lactation than unaffected herdmates.[9] Interestingly, cattle diagnosed with both placental retention and APM produce over 1,000 kg less milk than healthy herdmates, suggesting an additive effect of the two disorders.[9] This same study also showed that cattle with APM had a first service pregnancy risk of 27.2% as compared to 35.3% in unaffected cattle. Neither PVD nor subclinical endometritis have any effect on milk production.[9] Nevertheless, both diseases have significant impacts on reproductive performance. Early studies evaluating cattle with PVD or endometritis found that these diseases will increase days open by 15, decrease the risk of pregnancy by 150 days in milk (DIM) by 31%, and decrease 21-day pregnancy rate by 16%.[10] More recent work using validated case definitions have found a 27% relative reduction in 21-day pregnancy rate, a 32-day increase in median time to pregnancy, and an increase in culling risk.[10]

ROLE OF METABOLIC CHALLENGES AND IMMUNE DYSFUCTION IN THE DEVELOPMENT OF UTERINE DISEASE

The process of parturition is associated with numerous physiologic changes. Many of these changes are associated with the onset of milk production and primarily include hypocalcemia and negative nutrient balance. Milk production peaks 4-6 weeks post-calving in most cattle. Unfortunately, the amount of nutrients taken in with the diet during this period of time are nowhere near the amount needed to maintain body condition. As a result, cattle begin to mobilize fat and protein to maintain themselves so as to spare all available glucose for milk production. Initially, blood glucose decreases and, as a result, insulin levels drop concurrently. These changes trigger fat mobilization. Non-esterified fatty acids (NEFAs) are then transported into the blood by albumin so that they may be used as an energy source. It is important to note that negative nutrient balance is, to some extent, a normal phenomenon and, in many animals, mobilization of adipose tissue is regulated and the release of NEFAs is balanced. Unfortunately, a large proportion of transition cattle experience an excessive negative nutrient balance and develop disease issues.

Recent research has shown that lipid mobilization is associated with infectious diseases in the transition dairy cow.[11,12] Adipocytes and macrophages that reside in adipose tissue are sources of pro-inflammatory mediators. It has been shown that certain pro-inflammatory mediators block downstream insulin signaling and exacerbate insulin resistance. Several studies have shown a link between metabolic disorders in the transition period and the subsequent development of infectious disease. For example, there is an association with the development of retained fetal membranes and an increased risk of mastitis.[12] Also, cattle suffering from ketosis are two times more likely to develop mastitis than healthy cattle.[12] The reasons for this increased risk of infectious diseases in cattle with metabolic disturbances are likely multifactorial. Nevertheless, newer research is beginning to show that NEFAs have significant effects on immune function, some of which are reviewed below.

Along with these physiologic changes come alterations in normal immune function. Some of the earliest studies evaluating immune function in the periparturient period.revealed significant reductions in neutrophil random migration, phagocytosis, superoxide production, chemiluminesence, and iodination.[13] Other work found that periparturient cattle have significantly lower expression of NOD-1 mRNA and NOD-1 protein in neutrophils than cattle in mid-lactation.[14] In addition, NOD-1 deficient neutrophils have reduced phagocytosis and reactive oxygen species (ROS) production compared to controls.[14] In other work, lymphocyte responses to chemical mitogens was investigated and revealed significant impairment of lymphocyte blastogenic activity in response to phytohemagglutnin and concavalin-A.[15] Seminal work evaluating peripheral blood T-cell dynamics demonstrated significant declines in all T cell subsets in the days and weeks prior to calving.[16] These studies demonstrated that virtually all T cell subsets ($CD4^+$, $CD8^+$, γδ) decline prior to parturition and do not return to pre-calving numbers until approximately 2 weeks after calving.[16] More recently, studies have found an increase in the proportion of T cells expressing the CD4 and CD25 receptors on their surface, a phenotype often associated with the presence of regulatory T cells (T_{regs}).[17] While the significance of this finding is unknown, T_{regs} are known to be important in mice and people to prevent rejection of the fetus *in utero*.

From these studies it appears that, during the immediate period before and after calving, significant changes in immune function and immune cells numbers occur and these changes are likely being driven by several factors, mainly production of colostrum for the newborn calf and to some extent the stresses of parturition and lactation. Indeed, one study found that, if the mammary gland is surgically removed, the changes in lymphocyte subsets noted previously no longer occur.[18,19] Other work has clearly demonstrated that the transfer of mononuclear cells from dam to calf in colostrum is important for optimal development of the neonatal immune system.[20-23] When looking at the role of stress, most early investigations focused on the influence of cortisol on immune function. In fact, it has been shown that both dexamethasone and cortisol both down-regulate the expression of L-selectin on bovine neutrophils.[24] In addition, while L-selectin is down regulated on neutrophils, migration into the uterus is not affected. However, dexamethasone downregulates the production of reactive oxygen species (ROS) generated by neutrophils once in the uterus, a factor that may suggest impaired ability to kill microbes.[24] Other work that evaluated gene expression in neutrophils isolated from periparturient dairy cattle found that, when compared to cattle at other stages of production, periparturient cattle have lower expression of genes encoding cytochrome b, and rig/ribosomal protein S15, two genes essential for respiratory metabolism. In addition, the expression of these two genes was correlated to serum progesterone concentrations.[25] Studies that evaluated the effect of parity on neutrophil function found that cows in their 4th or greater lactation have much greater impairment of neutrophil function than younger cattle.[26] Thus, for a long period of time it was believed that hormonal changes occurring around the time of parturition, namely an increase in plasma cortisol and estrogen concentrations and decreased plasma progesterone concentrations, drive certain changes in immune function. Nevertheless, more recent work has called these theories into question as research has shown that ovarian steroids had no effect on inflammatory responses in *ex vivo* endometrial explants.[27]

More recently, the influence of hypocalcemia and negative energy balance has also been intensely studied. Indeed, hypocalcemia is associated with decreased intracellular concentrations of calcium within mononuclear cells.[28] In addition, the response of these cells to various stimuli is impaired greatly when compared to normocalcemic controls. Studies that evaluated the effect of serum non-esterified fatty acids (NEFAs) on functionality of immune cells and found that NEFAs significantly impair both lymphocyte proliferative responses to mitogens and oxidative burst capacity in isolated neutrophils. Two groups have studied the relationship between serum NEFA concentration and neutrophil function and found that animals with uterine health disorders have significantly higher serum NEFA concentrations, lower prepartum dry matter intake (DMI), and greater impairments in neutrophil activity when compared to healthy controls.[29,30] One study also noted that cattle with uterine health disorders have lower neutrophil glycogen content, a factor that may reduce cellular energy stores and contribute to impaired activity.[30] Studies on the effect of NEFAs on inflammatory responses in bovine endothelial cells found that excessive concentrations of NEFAs increase the expression of adhesion molecules and cytokine mRNA.[31] Furthermore, the activity of cyclooxygenase 2 (COX2) is increased by NEFAs in a concentration dependent manner.[31] This same group evaluated the expression of glucose transporters (GLUT) in monocytes of periparturient cattle and found significantly decreased expression of GLUT1, GLUT3, and GLUT4 in all animals studied.[32] In addition, this study found that both lipopolysaccharide (LPS) and TNF-α increase the expression of these same transporters.[32]

DIAGNOSTIC TESTS FOR UTERINE DISEASES

Tests commonly used for diagnosing reproductive tract diseases include cervical diameter, uterine horn size, thickness of the uterine wall, and volume of intrauterine fluid. In addition, evaluation of vaginal discharge score and collection of cervical and uterine swabs to assess numbers and types of inflammatory cells have been studied. Unfortunately, no gold standard diagnostic tests exits so exact cut-points to differentiate between healthy and diseased animals do not exist. When using uterine cytology, it is recommended that the sample be collected with the use of a cytobrush and a threshold of > 6% neutrophils be used to distinguish between healthy and diseased cattle.[33] For uterine lavage, > 10% neutrophils in a sample collected from cattle > 26 days in milk has been suggested to be the threshold to distinguish between healthy and diseased cattle.[33] For PVD, the presence of mucopurulent vaginal discharge or a cervical diameter > 7.5 cm at > 26 days in milk have been shown to be predictive of pregnancy by 120 days in milk.[33] Newer work evaluating the optical density of uterine lavage fluid at 620 nm has been published and shows promise for use in the diagnosis of PVD.[34] Other work evaluating the ability of reagent test strips evaluating leukocyte esterase, protein, and pH showed no advantage when compared to conventional cytology.[35]

TREATMENT OF SUBCLINICAL ENDOMETRITIS AND PURULENT VAGINAL DISCHARGE

There are currently no products labeled for the treatment of cattle with subclinical endometritis or PVD in the United States. In Canada, Europe, Australia, and New Zealand, a product containing 500 mg of cephapirin benzathine is labeled for use as an intrauterine infusion in cattle with these disorders. While this product is not currently available to veterinarians in the United States, intrammamary products containing lower concentrations of cephapirin benzathine are available and have been used and, anecdotally, show promise. Nevertheless, this constitutes extra-label drug use and must follow the guidelines set forth by AMDUCA. Historically, clinicians believed that injections of prostaglandin were beneficial for cattle with these conditions.[36] However, current evidence suggests that, while prostaglandin likely has no untoward effects on cattle health, its use in cattle diagnosed with these conditions is ineffective.[36] More recently, an infusion of 50% dextrose has been evaluated as a potential therapy for cattle with these uterine health disorders. In this study, 240 ml of 50% dextrose was infused into the uterine lumen. Cattle treated with dextrose were significantly more likely to have a clear uterine discharge by day 14 post-infusion.[37] In addition, pregnancy per AI (P/AI) was 29.8% in the treated group compared to 21.1% in the control group.[37] While more work is needed to better characterize this treatment modality, intrauterine infusions of 50% dextrose show promise as a non-antibiotic alternative for cattle with certain uterine health disorders.

REFERENCES

1. LeBlanc SJ. Animal 2014;8 Suppl 1:54.2. LeBlanc SJ. Reprod Domest Anim 2012;47 Suppl 5:18.3. Prunner I, et al. Theriogenology 2014;82:1316.4. Wagener K, et al. Vet J 2014;202:527.5. Wagener K, et al. Vet Microbiol 2015;175:286.6. Sheldon IM, et al. Theriogenology 2006;65:1516.7. Dubuc J, et al. J Dairy Sci 2010;93:5764.8. Overton MW, et al. Dairy Cattle Reproductive Council Convention 2008;39.9. Dubuc J, et al. J Dairy Sci 2011;94:1339.10. LeBlanc SJ, et al. Theriogenology 2011;76:1610.11. Sordillo LM, et al. Anim Health Res Rev 2009;10:53.12. Sordillo LM, et al. Vet Clin North Am Food Anim Pract 2013;29:267.13. Kehrli ME, Jr., et al. Am J Vet Res 1989;50:207.14. Tan X, et al. Vet Immunol Immunopathol 2012;150:133.15. Kehrli ME, Jr., et al. Am J Vet Res 1989;50:215.16. Kimura K, et al. J Dairy Sci 1999;82:315.17. Oliveira LJ, et al. Reproduction 2008;136:481.18. Nonnecke BJ, et al. J Dairy Sci 2003;86:2359.19. Kimura K, et al. J Dairy Sci 2002;85:1437.20. Reber AJ, et al. Vet Immunol Immunopathol 2008;123:305.21. Reber AJ, et al. Vet Immunol Immunopathol 2008;123:186.22. Reber AJ, et al. Am J Vet Res 2005;66:1854.23. Donovan DC, et al. Am J Vet Res 2007;68:778.24. Burton JL, et al. J Leukoc Biol 1995;57:317.25. Madsen SA, et al. Vet Immunol Immunopathol 2002;86:159.26. Gilbert RO, et al. Vet Immunol Immunopathol 1993;36:75.27. Saut JP, et al. Reproduction 2014;148:593.28. Kimura K, et al. J Dairy Sci 2006;89:2588.29. Hammon DS, et al. Vet Immunol Immunopathol 2006;113:21.30. Galvao KN, et al. J Dairy Sci 2010;93:2926.31. Contreras GA, et al. J Dairy Sci 2012;95:5011.32. O'Boyle NJ, et al. J Dairy Sci 2012;95:5709.33. de Boer MW, et al. J Dairy Sci 2014;97:3983.34. Machado VS, et al. J Dairy Sci 2012;95:1349.35. Cheong SH, et al. Theriogenology 2012;77:858.36. Lefebvre RC, et al. Vet Clin North Am Food Anim Pract 2012;28:79.37. Brick TA, et al. J Dairy Sci 2012;95:1894.

HYPERKETONEMIA IN EARLY LACTATION DAIRY CATTLE: EPIDEMIOLOGY

Jessica McArt, DVM, PhD
Ithaca, NY

INTRODUCTION

The ability of dairy cattle to adapt to the natural change of energy balance in early lactation is an important aspect of the transition period, as the demands for milk production cannot be met by feed intake alone. In order to maintain homeorrhesis during this period of negative energy balance, cows break down adipose tissue to produce non-esterified fatty acids (NEFA), which are partially converted to ketone bodies (e.g. β-hydroxybutyrate; BHBA) and can both act as alternate fuel sources. An elevation of blood NEFA and BHBA is normal during this time in order for cows to adapt to the physiologic changes that occur from late gestation through early lactation, however cows that have a poor adaptive response to negative energy balance will produce excessive NEFA and/or BHBA. This excessive production, particularly in cows diagnosed with hyperketonemia (a blood BHBA concentration ≥1.2 mmol/L), has been shown to have detrimental effects on immune function, milk production, and overall health. Given that 85-95% of hyperketonemic cows do not show signs consistent with clinical ketosis[1,2], the health and production consequences of a poor transition into lactation are often unseen.

EPIDEMIOLOGY

Hyperketonemia, by definition, is the elevation of ketone bodies above a physiologically normal range. Although historically the terms "subclinical" and "clinical" ketosis have been used to describe this metabolic state of ketone elevation, classification into these two categories is quite subjective. It is important to remember that expression of signs consistent with ketosis has much individual-animal variation and that manifestation of these signs most likely suggests a more severe hyperketonemia compared with cows that do not show clinical signs. For this reason, rather than refer to different severities of the same disease as subclinical or clinical, I prefer to use the term "hyperketonemia" which encompasses both groups. Recent research has provided solid and repetitive evidence of quantitative blood BHBA concentration cut-points that result in increased risks of subsequent disease development and production loss. This cut-point falls somewhere around 1.2 to 1.4 mmol/L depending on the desired sensitivity and specificity[6-8]; a lower cut-point will result in a higher test sensitivity and lower specificity and a higher cut-point will results in a lower sensitivity and higher specificity.

The incidence of early lactation hyperketonemia (number new cases divided by the number of cows at risk), has been reported in only 2 studies[2,9] as repetitive testing for incidence estimation is time and labor intensive. One of these studies[2] followed 1,717 cows from 3-16 days in milk (DIM) in 4 free-stall, total mixed ration fed herds in New York and Wisconsin. Using a Precision Xtra meter, all cows were monitored for hyperketonemia (BHBA ≥1.2 mmol/L) on a Monday, Wednesday, Friday testing schedule, and each cow was tested 6 times between 3-16 DIM. Peak incidence occurred at 5 DIM (Figure 1a) with 22% of cows testing hyperketonemia positive for the first time; 75% of cows that developed hyperketonemia between 3 and 16 DIM were first positive between 3 and 7 DIM. Peak prevalence of hyperketonemia (number of positive cows divided by the number of cows tested) also occurred at 5 DIM when almost 29% of cows tested were hyperketonemic (Figure 1b). Median time to resolution of hyperketonemia (BHBA < 1.2 mmol/L) was 5 days.

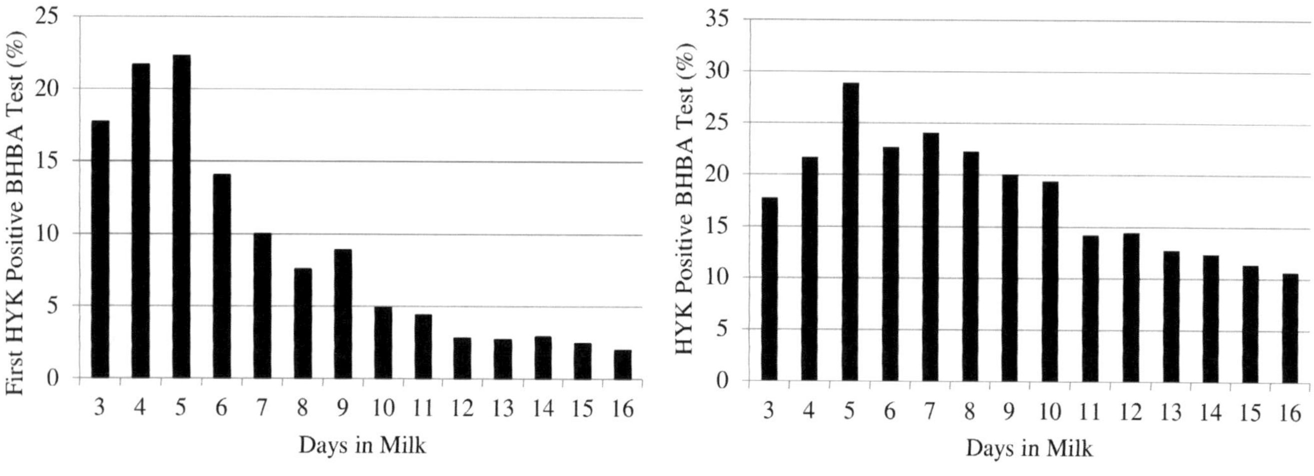

Figure 1. Histogram of a) incidence and b) prevalence of hyperketonemia (HYK) in 1,717 Holstein dairy cows undergoing repeated testing for HYK from 3 to 16 DIM. A positive test was defined as a blood BHBA concentration ≥ 1.2 mmol/L (Adapted from McArt et al., 2012).

Given that the frequency of BHBA testing has an impact on the incidence of hyperketonemia and that the median time to resolution of hyperketonemia is 5 days, BHBA testing should occur at least twice per week in order to appropriately estimate a herd's incidence of hyperketonemia. A herd's prevalence of hyperketonemia is much easier to estimate given the ease of cross-sectional sampling. Studies have found that the incidence of hyperketonemia is approximately 2 times the prevalence[2,9], which allows for extrapolation of a herd's hyperketonemia incidence from prevalence testing. Herd incidence of hyperketonemia has been found to range from 27-57% and herd prevalence of hyperketonemia from 0-71% (Table 1).

Table 1. Summary of recent research on incidence and prevalence of elevated prepartum and postpartum β-hydroxybutyrate concentrations in dairy cows. Cow-level incidence and prevalence refer to the number of positive animals across all herds combined, whereas herd minimum and maximum refer to individual herd-level incidence and prevalence (Adapted from McArt et al., 2013).

Study	No. Herds	No. Cows	Cut-point (mmol/L)	Time	Freq.	Cow-level (%)	Herd Min (%)	Herd Max (%)
Post-partum incidence[a]								
Duffield et al., 1998	25	507	≥1.2	1 to 63 DIM	1/week	59	-	-
	25	507	≥1.4	1 to 63 DIM	1/week	43	-	-
McArt et al., 2012b	4	1,717	≥1.2	3 to 16 DIM	3/week	44	27	57
Post-partum prevalence								
LeBlanc et al., 2005	20	1,063	≥1.2	1 to 7 DIM	-	20	-	-
Duffield et al., 2009	1,010	987	≥1.2	1 to 7 DIM	-	24	-	-
	1,010	941	≥1.2	8 to 14 DIM	-	25	-	-
Ospina et al., 2010a	104	1,318	≥1.2	3 to 14 DIM	-	18	0	71
Seifi et al., 2011	16	849	≥1.0	1 to 7 DIM	-	31	-	-
Chapinal et al., 2012a	45	2,069	≥1.4	1 to 7 DIM	-	12	-	-
	45	2,069	≥1.2	8 to 14 DIM	-	20	-	-
Suthar et al. (2013)[b]	528	5,884	≥1.2	2-15 DIM	-	22	11	37

DIM, days in milk

[a] Cows were classified as positive if they had at least one test above the cut-point.

[b] Prevalence in 10 European countries.

INDIVIDUAL ANIMAL HEALTH EVENTS AND PRODUCTION

Multiple studies have shown that the risk or odds of metritis, displaced abomasum (DA), and early lactation culling are increased for hyperketonemic animals[2,6,8,10]. Measurements of reproductive performance have shown more inconstancies in their association with elevated BHBA in early lactation. In addition, most studies report a decrease in milk production (in the range of 5 lbs/cow/day) in hyperketonemic cows compared to non-hyperketonemic animals, although this effect appears to wane as lactation progresses[1,8,10]. Many of the risks regarding subsequent disease development and milk production loss increase as blood BHBA concentration increases[2,6] (i.e. the more severely hyperketonemic a cow is, the higher her chances of incurring additional disease events as well as a more pronounced drop in milk yield; Figure 2). The time of onset of hyperketonemia also affects the risk of subsequent disease development, reproduction, and milk production. Cows that became hyperketonemic between 3 to 7 DIM were over 6 times more likely to develop a DA, 4.5 times more likely to be removed from the herd, 0.7 times as likely to conceive to first service, and made almost 5 pounds less milk per cow per day for the first 30 DIM than cows first testing SCK positive between 8 and 16 DIM[2]. Thus it is important to identify these hyperketonemic animals early in lactation in order to reduce the risk of subsequent negative events.

Figure 2. Regression plot of mean predicted milk yield per milking for the first 30 DIM by blood BHBA concentration of first positive hyperketonemia test (BHBA ≥ 1.2 mmol/L) for 369 Holstein dairy cows undergoing repeated testing for hyperketonemia from 3 to 16 DIM. The solid line represents the best fit; 95% confidence intervals are shown for each predicted milk yield by BHBA concentration (Adapted from McArt et al. 2012).

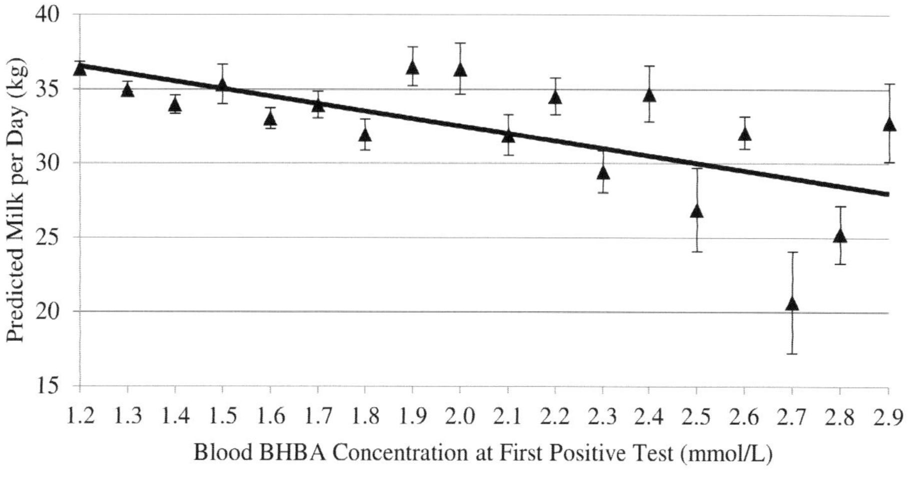

HERD-LEVEL HEALTH EVENTS AND PRODUCTION

Herds with a high prevalence of hyperketonemia are more likely to have a higher incidence of postpartum disease, poorer reproduction, and lower milk production. Research completed in over 150 herds in the United States and Canada suggests that this association between herd hyperketonemia prevalence and disease and production outcomes occurs when ≥ 15-25% of animals have elevated postpartum blood BHBA concentrations[11,12]. Ospina et al. (2010) found that 40% of the 100 herds tested were above the herd alarm level (Figure 3).

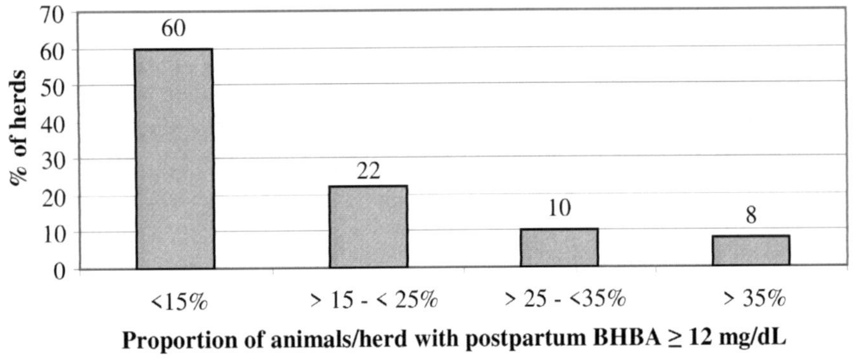

Figure 3. The prevalence of herds in which more than 15% of sampled animals had BHBA concentration ≥12mg/dL (Adapted from Ospina et al. 2010).

PREDICTION OF HYPERKETONEMIA DEVELOPMENT

Studies have also evaluated the risk factors associated with development of hyperketonemia[13,14,15]. McArt et al. (2013) looked at risk factors in dry cows to help predict which cows went on to develop ketosis between 3 and 5 DIM, as most cows first develop SCK during this time and, as mentioned above, these cows are at a higher risk for negative events than cows first developing SCK later in lactation. Cows with pre-calving NEFA ≥ 0.30 mEq/L were almost 2 times more likely to develop ketosis than cows with NEFA < 0.30 mEq/L; similarly, cows birthing male calves were 1.8 times more likely to develop ketosis than cows birthing female calves. In addition, cows with a calving ease ≥ 3 on a scale of 1 to 5 were 2.6 times more likely to develop ketosis than cows with a calving ease < 3, cows that gave birth to a dead calf were 2.2 times more likely to develop ketosis than cows that gave birth to a live calf, and parity ≥ 3 cows were 3 times more likely to develop ketosis than their younger herdmates. Additional work done by Suthar et al. (2013) in over 5,000 cows in 10 European countries identified country, parity ≥3, and summer season as strong risk factors for hyperketonemia development. Similarly, Vanholder et al. (2015) found that cows with a body condition score >3.0 at calving, advanced parity, and calving in the spring and summer increased the risk of developing hyperketonemia within the first 2 weeks of lactation. Interestingly, an increased yield of colostrum at first milking was also associated with an increased risk of hyperketonemia development. Thus for herds that chose to focus their ketosis testing rather than test all fresh cows, special attention should be paid to cows with high pre-calving NEFA, cows with high body condition scores, older cows, and cows that have had difficulty birthing.

CONCLUSIONS

Excessive negative energy balance in early lactation is an issue in many dairy herds. As most cases of hyperketonemia are subclinical in nature, associated disease events, early removal from the herd, and production losses are often unrecognized. Cows that develop hyperketonemia within the first week of lactation are at a greater risk of subsequent disease development and milk loss; this risk increases with higher concentrations of blood BHBA. Individual and herd-level testing should focus on the first week to two weeks of lactation in order to optimize individual animal treatment and herd management practices.

REFERENCES

[1] Duffield TF, et al. J Dairy Sci 2009:92:571. [2] McArt JAA, et al. J Dairy Sci 2012:95:5056. [3] Grummer R, J Dairy Sci 1993:76:3882. [4] Herdt TH, Vet Clin N Am – Food An 2000:16:215. [5] Oetzel GR, Vet Clin N Am – Food An 2004:20:651. [6] LeBlanc SJ, et al. J Dairy Sci 2005:88:159. [7] Ospina PA, et al. J Dairy Sci 2010:93:546. [8] Chapinal N, et al. J Dairy Sci 2011:94:4897. [9] Duffield TF, et al. J Dairy Sci 1998:81:2866. [10] Ospina PA, et al. J Dairy Sci 2010:93:1596. [11] Ospina PA, et al. J Dairy Sci 2010:93:3595. [12] Chapinal N, et al. J Dairy Sci 2012:95:5676. [13] McArt JAA, et al. Vet J 2013:198:560. [14] McArt JAA, et al. J Dairy Sci 2013:96:198. [15] Vanholder T, et al. J Dairy Sci 2015:98:880.

HYPERKETONEMIA IN EARLY LACTATION DAIRY CATTLE: DIAGNOSTICS AND TREATMENT

Jessica McArt, DVM, PhD
Ithaca, NY

INTRODUCTION

Evaluation and monitoring of individual and herd-level negative energy balance in early lactation has revolved around measurement of non-esterified fatty acids (NEFA) and β-hydroxybutyrate (BHBA). Blood NEFA concentrations have been found to be a stronger indicator of disease, milk production, and reproductive performance than blood BHBA concentrations[1,2]; however, current technology does not allow for easy collection and processing of blood samples for NEFA analysis. To obtain samples and accurate results, blood for NEFA analysis should be collected in an EDTA or non-anticoagulant tube, placed on ice immediately after collection, kept at 4°C until processing, and the plasma or serum separated within 24 hours[3]. Samples with a hemolytic index ≥ 300 should be interpreted with caution. Due to a natural rise of NEFA a few days before parturition, collection of samples between 14 and 3 days prior to expected calving is optimal[3]; some samples may need to be discarded if cows then calve outside of this window. There is currently no true "cowside" test for NEFA, and current costs for NEFA analysis range from $11-16 at the New York State Animal Health Diagnostic Center and the Michigan State Diagnostic Center. Hyperketonemia monitoring through blood BHBA measurement is less expensive and more practical than determination of blood NEFA concentrations and will be discussed below.

Given the ease and economics of cowside BHBA testing, knowledge of methods for individual animal identification of hyperketonemia and herd-level monitoring approaches are important tools for bovine veterinarians from both a disease and economic standpoint. Recent research has estimated the financial impact of hyperketonemia on both a component cost per case (i.e. the sum of the cost of diagnostics, therapeutics, labor, death loss, and future milk production, culling, and reproduction losses) and a total cost per case basis (i.e. the component cost plus cost related to hyperketonemia-attributable diseases). The average component cost per case was estimated to be $117; the average total cost per case when accounting for costs attributable to displaced abomasa and metritis was estimated to be $289[4]. Given the high total cost, reported incidences of hyperketonemia at 40-60% highlight the importance of appropriate nutrition and management strategies as well as herd-level monitoring and individual animal treatment plans.

INDIVIDUAL ANIMAL TESTING METHODS

As BHBA is the most stable ketone body in blood[5], the gold standard for detection of hyperketonemia is laboratory analysis of blood BHBA via a kinetic enzymatic assay. However, there are numerous cowside tests varying in accuracy that are available for practical on-farm diagnosis of hyperketonemia. The most common ketone tests use either urine, milk, or blood, although the sensitivity and specificity of each type of test varies tremendously (Table 1). To complicate matters when using a urine test, only 50% of cows can typically be induced to urinate while sampling[6]. Meters for blood BHBA monitoring can be purchased for $20-50 with the largest complaint being that they don't work well in cold temperatures.

Test	Reading	Sensitivity	Specificity	Cost per test	Lab BHBA Reference Conc.
Urine – acetoacetic acid (Ketostix)[6]					
	≥15 mg/dL (small)	78%	96%	$0.20	1.4 mmol/L
	≥40 mg/dL (moderate)	49%	99%	$0.20	1.4 mmol/L
Milk – acetoacetic acid (KetoCheck)[6]					
	Purple color	41%	99%	$0.60	1.4 mmol/L
Milk – BHBA (KetoTest)[6,7]					
	100 µmol/L	73-80%	76-96%	$1.70	1.4 mmol/L
	200 µmol/L	27-59%	90-99%	$1.70	1.4 mmol/L
Milk – BHBA (PortaBHB)[8]					
	100 µmol/L	89%	80%	$2.00	1.4 mmol/L
	200 µmol/L	40%	100%	$2.00	1.4 mmol/L
Blood – BHBA (Precision Xtra)[9,10,11]					
	1.2 mmol/L	85-96%	94-98%	$1.50	1.2 mmol/L
	1.4 mmol/L	90-100%	98-100%	$1.50	1.4 mmol/L
Blood – BHBA (Novamax)[12]					
	1.2 mmol/L	97%	82%	$2.75	1.2 mmol/L

Testing should be focused on cows that are in the first 2 weeks of lactation with even more emphasis placed on animals 3-9 days in milk (DIM), as a cow that is hyperketonemic during this period has the highest risk for subsequent negative health and production effects[13]. Additionally, test sick cows that are off feed, appear depressed, or have had a recent drop in milk production. Whereas these tests have great use as tools for diagnosis of hyperketonemia in individual animals, they can also be incorporated into a program for herd-level monitoring in order to assist with management decisions. It is important to note that the sensitivity and specificity of different tests can have a large impact on individual test results and, and as a consequence, herd diagnosis and treatment[14]. The probability of correctly identifying a herd as positive (i.e. having a prevalence of hyperketonemia above a certain threshold) decreases when a test with low sensitivity is used (Table 2).

	Herd Level Prevalence				
	0.2	0.3	0.4	0.5	0.6
HSe based on individual animal test sensitivity = 41%, specificity = 99%[31] (Ketocheck powder; Great States Animal Health, St. Joseph, MO)					
20[a] animals sampled	0.27	0.49	0.69	0.83	0.91
35[b] animals sampled	0.2	0.49	0.74	0.89	0.96
HSe based on individual animal test sensitivity = 78%, specificity = 96%[29] (Ketostix; Bayer Corp, Elkhart, IN)					
20[a] animals sampled	0.76	0.93	0.98	1	1
35[b] animals sampled	0.82	0.97	1	1	1
HSe based on individual animal test sensitivity = 96%, specificity = 97%[29] (Precision Xtra meter; Abbott Laboratories, Abbott Park, IL)					
20[a] animals sampled	0.84	0.97	1	1	1
35[b] animals sampled	0.91	0.99	1	1	1

[a] Three or more (~ 15%) animals sampled would need to test positive to call the herd positive.[47]
[b] Five or more (~ 15%) animals sampled would need to test positive to call the herd positive.[47]

Table 2. Estimation of herd-level sensitivity (HSe) for tests with different individual animal-level sensitivity and specificity in a herd with 1,000 animals (Adapted from Ospina et al. 2013).

HERD-LEVEL MONITORING

Once you have decided which tool you will use to monitor ketones in your herd, there are two important measurements which will help you evaluate the degree of ketosis on a farm: incidence and prevalence. Incidence is the number of cows that develop ketosis any time during early lactation (typically about the first 2-3 weeks in milk) divided by the total number of cows that have gone through this early lactation time period. In order to calculate incidence, choose a group of cows to follow through the early lactation period; determining the incidence of hyperketonemia requires repeated testing of cows throughout this risk period. Testing must occur twice or more per week in order to accurately assess the incidence of hyperketonemia. This is necessary because the median time for the resolution of hyperketonemia is approximately 5 days[15]. If, for example, 15 cows within a group of 50 cows develop hyperketonemia sometime in early lactation, the incidence is 30% (15/50). Prevalence is a snapshot of the amount of hyperketonemia in the herd at one point in time. It is calculated by dividing the number of hyperketonemic cows by the number of cows sampled at that point in time. For example, if 3 cows were hyperketonemic out of 20 cows tested that day, the prevalence would be 15%. Prevalence is more current and a lot easier to determine than incidence. It is used for general herd monitoring and for gauging the effectiveness of changes in dry or fresh cow management. Although the prevalence of hyperketonemia in a herd is much easier to determine than its incidence, the incidence of hyperketonemia must be known in order to estimate the overall negative impacts of hyperketonemia on herd performance. The repeated testing necessary to determine the incidence of hyperketonemia within a herd is daunting and requires testing a large number of cows twice weekly for about the first two weeks of lactation. Fortunately, the incidence of hyperketonemia can be estimated from its prevalence; the incidence of hyperketonemia has been reported to be approximately 2.2 times the prevalence[13, 16]. The number of animals tested in order to determine herd-level prevalence of hyperketonemia is also important, as an increase in sample size will increase the precision of the estimate (Table 3).

Table 3. Sample size necessary to estimate the prevalence of animals with elevated energy metabolites with various numbers of animals at risk, confidence intervals, and desired precision around the estimate (Adapted from Ospina et al. 2013).

Animals at Risk	Confidence Interval (%)	Sample Size (±10)[a]	Sample Size (±5)[a]
35	75	14	25
35	85	18	28
35	90	20	30
35	95	22	31
35	99	26	33
70	75	17	39
70	85	23	46
70	90	27	50
70	95	33	55
70	99	43	61

[a] Precision around the confidence interval.

INDIVIDUAL ANIMAL TREATMENT

Once a hyperketonemic animal is identified, immediate treatment is warranted to prevent negative subsequent health events and production losses. An excellent systematic review of hyperketonemia treatment in lactating dairy cattle found only 10 publications that reported a clearly defined method of hyperketonemia diagnosis before initiation of treatment, included a control group, had defined outcomes of interest, and used appropriate statistical methodology[17]. Interestingly, no studies have evaluated the effect of intravenous dextrose administration on hyperketonemia although this treatment is commonplace. Treatment of hyperketonemic cows with oral propylene glycol has been shown to improve resolution of hyperketonemia, decrease subsequent negative health events, and increase milk production[15,18]. Propylene glycol acts by increasing blood glucose concentrations through its conversion to propionate as well as its actions in decreasing peripheral tissue glucose demand[19]. A dose of 300 mL of propylene glycol orally once a day for 3-5 days is recommended given the median duration of a hyperketonemic event[15]. Glucocorticoids have been used for years to treat hyperketonemia given their action in stimulating gluconeogenesis and providing glucose precursors, however few studies have scientifically supported the use of glucocorticoids in the treatment of hyperketonemia. Similarly, few studies have described the benefits of B vitamin use in the treatment of hyperketonemia, although two studies have reported a decreasing trend of blood BHBA concentration in hyperketonemic cows treated with a butaphosphan and cyanocobalamin combination product[20,21].

ECONOMICS OF TESTING AND TREATMENT

Figure 1 outlines a suggested approach for monitoring of hyperketonemia and action recommendations in herds following prevalence sampling. This flow chart is based from work that evaluated 4 different testing and treatment strategies at varying herd incidences of hyperketonemia using stochastic and iterative modeling[22]. Results indicate that at herd incidences above 50%, blanket treatment of all fresh cows with 5 days of oral propylene glycol starting at 5 DIM was the most cost-effective strategy. At incidences between 15 and 50%, testing cows that are 3 through 9 DIM two days per week (e.g. Mondays and Thursdays) and treating hyperketonemic cows with 5 days of oral propylene glycol was the most cost-effective strategy, although testing all cows 3 through 16 DIM one day per week (e.g. Mondays) also provided a positive return on investment. For a herd with a 40% incidence of hyperketonemia that freshens 1,000 cows per year, choosing to test cows two days per week and treating the positives will return a benefit of $10,000 to $25,000 per year.

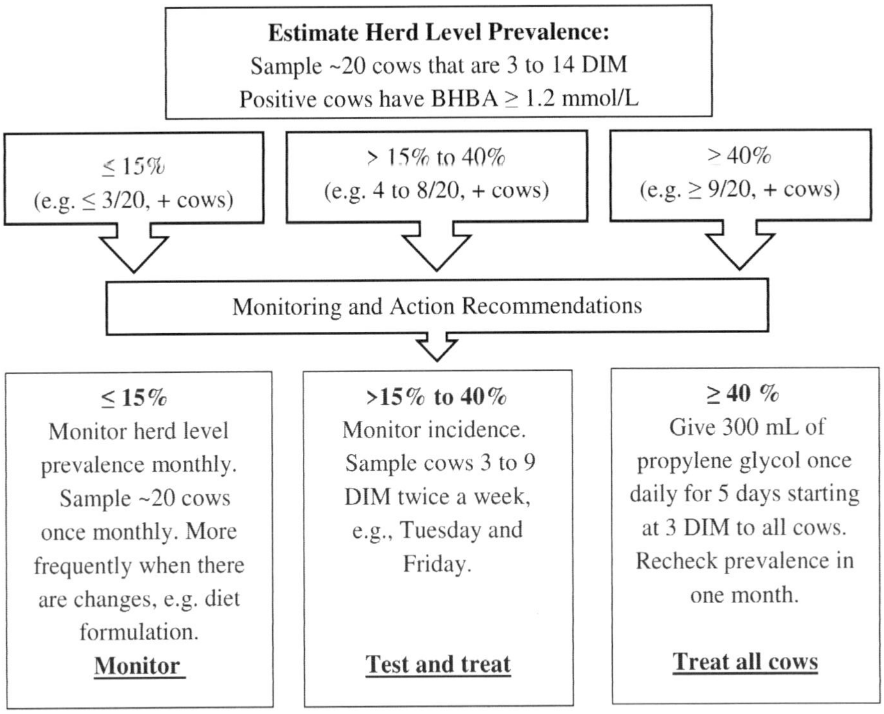

Figure 1. Monitoring recommendations for BHBA testing based on herd-level prevalence of elevated BHBA (Adapted from Ospina et al., 2013).

PUTTING IT ALL TOGETHER

Choose a test with a high sensitivity and specificity. Conduct a hyperketonemia prevalence assessment (sample 15 to 20 cows) in a herd in order to approximate herd incidence and determine the best testing and treatment plan. For herds with an estimated incidence greater than 50%, where blanket treatment with PG is initiated, repeated prevalence testing is necessary after management changes to determine if treating all fresh cows remains the best option. For herds with an incidence of 15 to 50%, either a one or two day per week testing strategy will allow for repeated monitoring of herd incidence and improve individual animal health. Repeated incidence or prevalence testing is recommended in order to evaluate changes in transition cow management and allow appropriate adjustment of farm hyperketonemia testing and treatment protocols. Remember, the goal is to not treat many, if any, cows with propylene glycol, but rather have transition cow management strategies in place such that the prevalence of hyperketonemia is lower than 10%.

REFERENCES

[1] Ospina PA, et al. J Dairy Sci 2010:93:546. [2] Chapinal N, et al. J Dairy Sci 2011:94:4897. [3] Stokol T, et al. J Dairy Sci 2005:88:3139. [4] McArt JAA, et al. J Dairy Sci 2015: http://dx.doi.org/ 10.3168/jds.2014-8740. [5] Tyopponen J, et al. Acta Vet Scand 1980:21:55. [6] Carrier J, et al. J Dairy Sci 2004:87:3725. [7] Geishauser T, et al. J Dairy Sci 2000:83:296. [8] Denis-Robichaud J, et al. Bovine Pract 2011:45:97. [9] Iwersen M, et al. J Dairy Sci 2009:92:2618. [10] Konkol KS, et al. Proc. AABP 2009:290. [11] Voyvoda H, et al. Res Vet Sci 2010:89:344. [12] Mahrt A, et al. Tierärztl Prax 2014:42:5. [13] McArt JAA, et al. J Dairy Sci 2012:95:5056. [14] Ospina PA, et al. Vet Clin N Am – Food An 2013:29:387. [15] McArt JAA, et al. J Dairy Sci 2011:94:6011. [16] Duffield TF, et al. J Dairy Sci 1998:81:2866. [17] Gordon J, et al. Vet Clin N Am – Food An 2013:29:433. [18] McArt JAA, et al. J Dairy Sci 2012:95:2505. [19] Kristensen NB, et al. J Dairy Sci. 2007:90:4707. [20] Lohr B, et al. Tier Umschau 2006:61:187. [21] Gordon J, et al. J Dairy Sci 2012:95:177. [22] McArt JAA, et al. Prev Vet Med 2014:117:170.

UPDATE ON PORCINE EPIDEMIC DIARRHEA VIRUS IN THE U.S.

David C. Van Metre, DVM, DACVIM
Fort Collins, CO

INTRODUCTION

Porcine epidemic diarrhea (PED) was first described in pigs of all ages in the United Kingdom in 1971. The virus (porcine epidemic diarrhea virus, or PEDV) was identified by Belgian investigators in 1978.[1] Subsequently, the disease was detected in scattered outbreaks of relatively limited impact in Europe; however, in Asia and China in particular, multiple outbreaks of PED have occurred since the 1980s, with recent outbreaks associated with severe clinical disease.[2] In mid-2013, at least two closely related strains of PEDV emerged in the United States at approximately the same time and spread rapidly among the national herd. The ongoing US outbreak is characterized by diarrhea and vomiting in pigs of all ages, with high morbidity in all age groups and high mortality (90-100%) in preweaned piglets. The economic impact of PED in the United States is still being tallied. According to one economic model, an annual net cost – including the swine industry and related food and commodity industries, as well as consumer costs from higher pork prices – could range from $900M – 1.8 B USD.[3] This review will provide a broad summary of the taxonomy, history, epidemiology, clinical signs, treatment, and prevention of this recently emerged and costly disease.

The clinical and epidemiologic features of a related, more recently emerged viral pathogen, porcine deltacoronavirus, (PDCoV) will be briefly summarized. This virus was first detected in pigs in Hong Kong in 2012; subsequently, PDCoV was detected in the U.S. in February, 2014, in association with an outbreak of diarrheal disease in a sow herd in Iowa.[2] Co-infection of pig populations with PEDV and PDCoV has been documented in the U.S. and elsewhere. The diseases induced by these viruses are clinically identical, prompting categorization of the separate diseases into one, Swine Enteric Coronavirus Disease (SECD).

TAXOMONY

Porcine epidemic diarrhea virus (PEDV) is a single-stranded, enveloped, positive-sense RNA virus of the genus *Alphacoronavirus*, family Coronaviridae, subfamily Coronavirinae.[4] This subfamily currently consists of 4 genera: *Alphacoronavirus*, *Betacoronavirus*, *Gammacoronavirus*, and *Deltacoronavirus*. Infection of susceptible hosts typically results in respiratory or gastrointestinal disease.[1,4]. Alpha- and betacoronaviruses are considered to have originated from bats, whereas gamma- and deltacoronaviruses have genetic signatures suggesting avian ancestry.[4] A variety of vertebrates, ranging from giraffes to sparrows to humans, can serve as hosts for coronaviruses.[3]

Other members of the genus *Alphacoronavirus* include the feline infectious peritonitis virus, the transmissible gastroenteritis (TGE) virus and its close relative, the porcine respiratory coronavirus. The genus *Betacoronarvirus* includes the severe acute respiratory syndrome (SARS) coronavirus, the Middle Eastern Respiratory Syndrome (MERS) coronavirus, and the bovine and equine coronaviruses. Avian viruses such as the infectious bronchitis virus predominate in the *Gamma-* and *Deltacoronavirus* genera; PDCoV belongs to the latter.[4] Coronaviruses replicate with low genetic fidelity, resulting in high mutation rates and broad genetic diversity among progeny viruses.[5] These features make coronaviruses particularly adept at adaptation to new hosts and ecological niches, as evidenced by the emergence of several novel coronaviruses, some of which are zoonotic, in the last two decades.

HISTORY OF THE U.S. OUTBREAK

In late April of 2013, diagnosticians at the Iowa State University Veterinary Diagnostic Laboratory (ISUVDL) received clinical samples from a private veterinary practitioner who provided services to a 2,400-sow breed-to-wean farm in central Iowa.[6] The farm was a shower-in, shower-out facility with a robust biosecurity program. The most recent addition of breeding gilts into the farm occurred one month prior to the outbreak. The attending practitioner reported watery, fetid diarrhea and vomiting that quickly spread to the entire population of preweaned (suckling) piglets; within 2-3 days, 95% of affected piglets had died. Diarrhea and anorexia, with occasional vomiting, were observed in most (~90%) of the sows in the affected farrowing units and gilts in farrowing or gestation; a smaller percentage of older sows showed clinical signs (15%); no fatalities occurred in adult pigs. A presumptive diagnosis of TGE was made, and feedback of homogenized feces and intestinal contents was initiated for all adult pigs on the premises, while diagnostic samples were submitted for confirmation of the diagnosis.

Within 10 days of the initial outbreak on the first farm, disease outbreaks with remarkably similar clinical features to the first farm occurred on two farms in northeastern Iowa and two breed-to-wean farms within the same production system in Indiana.[6] Diagnostic samples from these farms were submitted to the ISUVDL, also with the presumptive diagnosis of TGE. The two Indiana farms shared a common source of feed and were shower-in, shower-out facilities with rigorous biosecurity practices. The Indiana farms had been closed to introduction of live replacement gilts for several years. Feedback of feces and homogenized intestines was initiated on these farms as well.

At necropsy, affected piglets from all initial farms showed marked dehydration, cutaneous staining with yellow feces, nearly empty stomachs with little curdled milk inside, thin-walled small intestines, and yellow, watery intestinal contents containing flecks of undigested milk curds. Microscopic lesions were limited to the small intestine and consisted of enterocyte sloughing from the sides and tips of villi, with villous blunting. Villous tips were either eroded or covered by cuboidal to squamous epithelial cells.[6] These findings were consistent with malabsorptive diarrhea caused by enterocyte loss, subsequent villous blunting, and proliferation of poorly differentiated crypt cells to cover denuded villi. These findings are commonly found in other viral infections of the small intestine.

Immunohistochemistry and PCR tests were negative for TGE virus, porcine respiratory coronavirus, and rotaviruses; however, electron microscopic examination of feces and intestinal contents showed the presence of coronavirus-like particles.[6] A pan-coronavirus PCR confirmed the presence of a coronavirus in samples from all affected farms. Sequence analysis of these PCR products was conducted using the Basic Local Alignment Search Tool (BLAST, National Center for Biotechnology Information, National Institutes of Health). All PCR products were found to have high genetic homology to PEDV strains identified in China in 2011 - 2012. There was very high sequence identity (99.6-100%) among the PCR products from the different farms in Iowa and Indiana. These farms were separated by >100 miles and shared no personnel, transportation resources, veterinarians, or seedstock suppliers.[6]

CLINICAL SIGNS AND TREATMENT

As seen in the initially infected farms, subsequent outbreaks on other U.S. farms were characterized by high morbidity and mortality in suckling piglets, with case fatality risk in this age group typically ranging from 90-100%.[2] The clinical signs of PED were often initially considered to be caused by TGE, and it is now apparent that the two diseases are clinically indistinguishable. Affected suckling piglets appear to die from dehydration and hypothermia; affected piglets are sometimes found huddling on the sow for warmth.[7] The duration from the onset of clinical signs to death in baby piglets may be several hours to 3 days.

In nursery pigs, grow-finish pigs, and adults, morbidity estimates have varied, possibly owing to environmental factors and the emergence of at least one variant of PEDV that appear to be less virulent.[8] Further, although the effect of the virus on the enterocytes is the same in suckling piglets and adults, villus regeneration is much more rapid in adult pigs (3-4 days) than in younger pigs (6-7+ days) .[9] Morbidity risk estimates in nursery, grower, and adult pigs are clouded by the common practice of intentional exposure (feedback) and intentional co-mingling of affected pigs with healthy pigs upon confirmation of a diagnosis in these older pigs. These measures are enacted in order to establish herd immunity as rapidly as possible, thereby limiting the duration of the outbreak and facilitating cleaning and disinfection of facilities and equipment once a particular unit is vacated.[7] Diarrhea is the most consistent clinical finding, with vomiting and anorexia being common, albeit more variably seen, clinical signs. Clinical signs in older pigs tend to abate within approximately a week. Mortality risk is relatively low (1-3%).[2,7] A delay of 2-3 weeks in reaching market weight can be expected for grow-finish pigs, and an increase in nursery pigs who fail to thrive is expected.[7]

Treatment is non-specific. Provision of supplemental heat and additional bedding are considered vital. Maintaining dry conditions can be challenging, and frequent cleaning and drying of affected pens may be needed. Feeding of palatable gruels, tube feeding, feeding from mats, and provision of bowls of electrolyte solutions have been utilized. At present, there are no published data on treatment efficacy.

TRANSMISSION

Pigs are the sole host species identified to date. Fecal-oral transmission is considered to be the primary route of transmission.[2] Viral replication takes place in the porcine jejunum , ileum, and colon.[10] Fecal shedding of PEDV occurs around 24 hours post infection, with diarrhea appearing concurrently or up to 12 hours later.[10] Infection of the neonatal GI tract generates large quantities of virus. Experimental infection of gnotobiotic baby piglets with a virulent U.S. field strain of PEDV resulted in fecal shedding over 10^{12} genome equivalents of virus/ml), with markedly less viral load in serum ($\sim10^{5}$-10^{8}/ml).[10] Shedding of virus in surviving pigs (typically of weaning age or older) lasts for 1 week in most cases, with prolonged shedding (up to 2 months) in some; importantly, shedding can continue well beyond the resolution of clinical disease.[11] Seroconversion of infected animals occurs as early as 7 days post-infection.[12] At present, the following issues regarding immunity to PEDV remain unclear: 1) The duration of immunity induced in surviving pigs to the emergent U.S. variants of PEDV; 2) the magnitude of cross-protection among PEDV variants conferred by the host's immune response; and 3) the efficacy of two vaccines conditionally licensed for PED control in the U.S.[2] Given the genetic heterogeneity of PEDV isolates to date, development and validation of serologic tests and challenge studies that define immunity may require considerable time.[2]

Recently, PEDV was found to be capable of replicating in porcine alveolar macrophages, but the significance of the respiratory tract in disease pathogenesis and transmission remains unclear.[13] Infectious PEDV was detected in room air of pigs experimentally inoculated with PEDV, and viral RNA could be detected in air samples up to 10 miles away from infected farms.[14] However, air collected under field conditions was not found to be infective. To date, airborne transmission is considered possible for PEDV.[2]

PEDV survives to a variable duration in matrices such as manure and lagoon slurry, with colder temperatures generally favoring prolonged survival.[2] The role of pig feed and feed components such as porcine spray-dried porcine plasma (SDPP) in PEDV transmission remains unclear at present.[2] It is generally considered that contamination of any feed type or feed equipment by fecal matter from infected pigs could result in transmission of PEDV, so production, transport, and storage practices for feeds and feed supplements likely influence transmission risk for these matrices. PEDV genomic material can be demonstrated in certain tissues of geese, buzzards, and feral cats, although it is not clear at present if this represents viral replication in these animals or mechanical carriage.[2] As described below, transportation is considered to play a central role in transmission among farms.

EPIDEMIOLOGY

Although the route of entry of the virus into North America remains unknown, the detection of the virus at multiple locations over a short time frame supports the potential for incursion via a feedstuff or feed ingredient. Notably, testing of archived samples over the summer of 2013 subsequently revealed that the first detection of PEDV in the US occurred in mid-April from samples obtained on an Ohio farm.[15] With the exception of the two Indiana farms, none of the initially affected farms shared a common feed supplier or mill.[6] Given that the clinical signs of PEDV infection are relatively mild in older pigs, it is possible that the virus could have circulated among grow-finish or adult pigs prior to entry into breeding farms, wherein the most demonstrative infections (naïve baby piglets) can occur.

There appears to be little to no cross-protection to PEDV induced by vaccination or natural infection with the TGE virus, so the U.S. (and, ultimately the North American) swine populations were immunologically naïve to PEDV upon its introduction.[2] Despite the perceived high level of biosecurity practiced by most large commercial swine farms, PEDV quickly spread from Midwestern states into pig populations in neighboring and distant states. At the time of preparation of this paper (mid-December 2014), PEDV has been identified in pigs in 27 States.[16] During 2013-2014, PEDV was detected in pigs in Mexico, Canada, the Dominican Republic, Peru, Colombia, and Ecuador.[2] Relatively small-scale outbreaks of PEDV were reported in Germany and Italy in 2014; however, despite the high sequence homology of European PEDV isolates to the U.S. isolates, the clinical disease in European pigs appears to be relatively mild.[2] This disparity may be due in part to pre-existing, cross-protective immunity in some European swine populations imparted by endemic PEDV infections, but this remains unproven.

In retrospect, rapid dissemination of this virus in the U.S. appears to have been facilitated by the central role that transportation plays in this nation's pork industry. Contaminated trucks and trailers used to move pigs to harvest facilities were considered a likely source of infectious virus, given that these vehicles had not been consistently cleaned and disinfected between loads of animals prior to the outbreak.[17] Viral RNA can be present in incoming trucks, outgoing trucks, or in the same truck at arrival and departure from harvest facilities.[17] Rendering trucks appear to pose a particularly high risk, given that these vehicles are used to haul cadavers from potentially infected premises to a central location, with variable cleaning and disinfection practices for the vehicles.[18] Other potential means of mechanical transmission of PEDV include feed delivery trucks, manure hauling and spreading vehicles and machinery, equipment repair or service vehicles, and personnel who move from site to site. At present, the optimal downtime for veterinarians moving from farm to farm remains unclear, but given that the virus is not capable of replication in humans, strict adherence to biosecurity practices are considered likely to be effective in preventing transmission by people.

REGULATORY ISSUES

Neither PEDV nor PDCoV are zoonotic agents, nor are they listed as reportable diseases to the World Organization for Animal Health (OIE). In response to the rapid spread of PEDV during 2013-14 and emergence of PDCoV in the U.S. in early 2014, the USDA mandated in that all new cases of SECD be reported to State or Federal authorities as of June 5, 2014.[19] Subsequent to that order, the USDA reported improvement in the quality of data regarding affected premises location, diagnostic test results, and sharing of information among regulatory, industry, and diagnostic laboratory sources.[19] PEDV is also reportable in Japan, Korea, and the Canadian provinces of Quebec, Manitoba, and Alberta.[2]

PREVENTION

Anecdotal and experimental evidence supporting the role of transportation in PEDV dissemination has prompted revised, very strict biosecurity measures for all types of vehicles and personnel involved in the swine industry. These practices include designation of a line of separation – a physical boundary that is clearly delineated by signage and communication of protocols – that limits crossing of equipment and personnel from the exterior of the farm to the interior. Similar protocols are in place at harvest facilities Protocols for transport of feed, live and dead pigs, and manure have been established by the National Pork Board.[20] Thorough cleaning, disinfection, and thermal-assisted drying procedures are being enacted for vehicles used to haul pigs.

Detailed protocols have been developed for elimination of PEDV from infected premises. Briefly, these protocols include closure of the herd to new introductions, cessation of continuous flow and adaptation of all-in, all-out facilities management, effective feedback exposure to achieve infection and immunity in the entire population, diagnostic testing, cleaning and disinfection, environmental sampling, and use of seronegative sentinel animals to document the lack of infective virus in a previously-positive farm.[7,20] Taken together, these measures are intended to: 1) induce mucosal immunity in the herd – particularly vital in sows and replacement gilts; 2) eliminate the virus from the farm environment; and 3) prevent re-introduction of virus.

Interestingly, many producers and veterinarians see a silver lining in the SEDC outbreak. It has revealed flaws in what were considered highly effective biosecurity programs and practices on US pig farms, transportation systems, waste management, and harvest / rendering facilities. Over the long term, measures to mitigate SEDC spread, particularly by transportation, will enhance preparedness against other, potentially even more serious emerging and foreign animal diseases.[21]

REFERENCES

[1]Huang et al, **MBio** 2013, 4(5):00737-13. doi:10.1128/mBio.00737-13 [2]**EFSA AHAW Panel 2014**, doi.10.2903/j.efsa.2014.3877. 3. Paarlberg, **Working Paper #14-4**, Purdue University. http://ageconsearch.umn.edu/bitstream/174517/2/144.Updated%20Estimated%20Economic%20Welfare%20Impacts%20of%20PEDV.pdf , 2014. [4]Woo et al, **J Virol** 2012, 86 (7):3995. [5]Vlasova et al, **J Gen Virol** 2011, 92: 1369. [6]Stevenson et al, **J Vet Diagn Invest** 2013, 25(5):649. [7]Yeske, **Proc NAVC** Lg An 2014, 28. [8]Wang et al, **Emerg Infect Dis** 2014, 20(5):917. [9]Kelly et al, **J Pediatr** 1972, 80:925. [10]Jung et al, **Emerg Inf Dis** 2014, 20(4):662. [11]Hesse et al, **Proc 23rd IPVS Congress** 2014. [12]Gerber et al, **Vet J** 2014, 202:33-36. [13]Park et al, **Virus Res** 2014, 191:143. [14]Alonso et al, **Vet Res** 2014, 45:73. [15]US DA, http://www.aasv.org/aasv%20website/Resources/Diseases/PED/PEDv_weekly_report_070313.pdf. , 2013 [16]USDA , http://www.aphis.usda.gov/animal_health/animal_dis_spec/swine/downloads/secd_sit_rep_12_11_14.pdf , 2014. [17]Lowe et al, **Emerg Infect Dis** 2014, 20(5):872. [18]Morrison B, http://www.swinecast.com/dr-bob-morrison-porcine-epidemic-diarrhea-virus-pedv-transmission, 2014. [19]USDA, http://www.aphis.usda.gov/publications/animal_health/2014/SA_SECD_progress.pdf , 2014. [20]National Pork Board, http://www.pork.org/pork-checkoff-research/pedv/%E2%80%8E , 2014. [21]Barker J, AMVC Management Services, pers comm, 2014.

THE VETERINARIAN'S ROLE IN MANAGING PATIENTS WITH REGULATORY CONCERNS

David C. Van Metre, DVM, DACVIM
Fort Collins, CO

INTRODUCTION

Livestock health and food safety have long been foundational professional priorities for veterinarians of all nations. However, over the past quarter century, global trade in livestock, livestock products, and feedstuffs has undergone profound expansion and diversification. Concurrently, transportation systems have evolved to enable people, livestock, and food products to cross oceans and continents in hours, rather than days to weeks. These changes have increased the potential for exotic or emerging infectious disease agents to gain entry into even relatively secure livestock production systems. The ongoing outbreak of porcine epidemic diarrhea virus in North America provides the most sobering example – a disease that was once considered to be European, and Asian animal health problem has globalized, with substantial consequences for the American swine population, pig farmers, and consumers.[1,2] Maintaining livestock health and food safety have become extraordinarily complex in a globalized market.

The potential for incursion of novel infectious agents into this nation's livestock population is projected to be progressively more influenced by events that occur outside of our borders. Expansion of the human population and the concurrent expansion of global demand for high biological value protein are projected to drive expansion and intensification of animal production systems within the developing world.[3] These intensive production systems could foster novel interactions among livestock, the environment, and people, creating the potential for emergence and trans-boundary movement of novel or exotic infectious disease agents. If a particular infectious disease spreads among those populations, the "global weight of infection" – essentially, the mass and distribution of the infectious agent - combined with the mobility afforded by enhanced trade and transportation, could overcome the best of farm and ranch biosecurity programs.

Globalization of livestock trade has not occurred in a political vacuum. This is an era of sustained asymmetric conflict, wherein small groups of radicalized, motivated individuals can influence the security of vastly larger nations. Intentional disruption of food supplies or food-related economies has long served as a means of inciting hardship and unrest; these activities fit well into the asymmetric conflict model. As a result, the concept of food security – essentially, development of a resilient, reliable, safe, and defendable food supply - has gained prominence in national defense planning.[4]

Secure movement of food from the farm to the fork food requires veterinary oversight at all stages. Private practitioners are commonly regarded as serving on the front line of food security oversight, primarily because they are well-trained and well-positioned to detect infectious diseases that threaten food safety or trading status. Regulatory veterinary medicine (RVM) is also integral to that oversight. RVM is defined here as the animal health professionals specifically dedicated to development, implementation, and supervision of policies and practices that protect of animal health and food safety at the local, state, and national levels. The efforts of RVM professionals are integral to maintaining stable trade relations with other countries; in so doing, RVM professionals directly support the capability of our producers to access export markets. Given the growing animal health and food safety challenges brought forth by globalization, the relevance of RVM to national and global food security can only expand.

WHAT IS THE INTERNIST'S ROLE?

The evolution of global livestock trade and the integral role of RVM in food security are issues that have not likely gone unnoticed by the large animal internist. However, it is an interesting exercise to pause and ask, "What role(s) does the large animal internist have in RVM?" There are no data to objectively assess this question. Describing a role is inherently a subjective process, and it is likely influenced by factors such as the internist's species of interest and the location of his or her practice. For example, given that vesicular stomatitis, a reportable disease of horses and livestock, has occurred in the Southwestern United States in 7 of the past 10 years, it is possible that internists in affected states may have communicated with RVM professionals more frequently than those in states with a relatively more "quiet" reportable disease history. I do not consider myself to be a content expert on this topic. Therefore, to better define the role of the internist in RVM, in January, 2015, I consulted with two individuals who I considered to be able to provide expert opinion. I posed the question above, and guided responses to include the most vital messages to be imparted to the audience of this forum. The two interviews were conducted separately.

I should add that novel RVM terminology can be potentially confusing. Many interactions among practicing veterinarians occur with RVM professionals who are technically employed by each state; these individuals may have job responsibilities that overlap significantly with federally-based RVM personnel (e.g. the APHIS:VS Area Veterinarian in Charge) located within the same state. The term State Animal Health Official (SAHO) is commonly applied to those RVM professionals who directly interface with practicing veterinarians.

The topics below represent the combined opinions of the individuals whom I contacted.

Decentralization of RVM Personnel Resources

While practitioners have always been the mainstay of surveillance for transboundary infectious diseases, they are playing an expanding role in outbreak response measures. Highly infectious diseases such as foot and mouth disease have the potential to spread rapidly across multiple states, enabled in large part by the central role that efficient transportation plays in movement of livestock and livestock products in this country. Such large-scale outbreaks have high potential to exceed local and state RVM resources, as illustrated by the 2003 outbreak of Exotic Newcastle Disease in the Southwestern United States. In such scenarios, recruitment of additional veterinarians from within and outside of affected states will be essential to generate sufficient manpower to contend with the outbreak. Many states have established emergency response corps of veterinarians, veterinary technicians, animal scientists, and individuals in affiliated animal agriculture industries. The National Animal Health Emergency Response Corps was established in

2001 to serve as a source of qualified veterinarians that are able to be deployed, as paid USDA temporary employees, to assist in animal disease outbreaks or natural disasters that affect animal populations.[5] These response corps provide essential surge capacity for large-scale problems that may exceed dedicated SAHO resources.

SAHOs cannot be expected to be all things to all animals at all times; in other words, they cannot be content experts for all contingencies. The internist's training in gathering historical, physical, and diagnostic test data makes them subject matter experts in a broad array of health issues in a variety of species, with particular capability to recognize clinical signs or transmission patterns that are not characteristic of endemic diseases. Large animal internists would be valued members of state and federal veterinary response teams. Federal accreditation of the internist would be essential for certain response activities, a caveat that was deemed by one expert to be a priority item to communicate to this audience.

For internists employed at veterinary colleges, the students represent a potential source of additional manpower for assisting SAHOs in certain tasks. Veterinary students have been recruited to obtain epidemiologic survey data by telephone from owners of horses diagnosed with West Nile Virus infection.[6] Similar measures were enacted at my institution in October, 2014, to gather epidemiologic data from owners of horses diagnosed with vesicular stomatitis in a large Colorado outbreak. In addition to easing data collection duties from busy SAHOs, these activities enabled veterinary students to directly participate in RVM activities.

Reportable (Listed) Diseases

The infectious diseases of livestock that are required to be reported to SAHOs vary from state to state. The content experts both stated that formal, frequent review of reportable disease lists are essential tasks for large animal internists. Internists frequently participate in residency training programs; those trainees may originate from states or countries far removed from a given internist's practice. As a result, the trainee may not be familiar with the list of reportable diseases for the state(s) in which he or she practices. An example cited was a case of equine herpesvirus-1 infection recently diagnosed at a university teaching hospital over a holiday weekend. The SAHO received notice of the diagnosis from the diagnostic laboratory several days after the positive result was communicated to the internal medicine resident. Such communication failures can delay implementation of appropriate control measures, with potentially serious consequences. This issue is particularly applicable for diseases that require reporting to multiple agencies; for example, a diagnosis of rabies or plague in Colorado requires reporting to the SAHO and the state's public health agency.[7]

Specialty practice can, at times, be somewhat insulated from the regulatory realm, in that referral cases and field interventions focus on ill animals, while interstate commerce, harvest, and importation of livestock involve predominantly healthy livestock. This creates a potential disconnect between the internist's daily practice priorities those considered to be routine within the RVM realm. One solution to this problem would be involvement of SAHOs in clinical rounds or some similar, regularly-scheduled communication program to maintain awareness of RVM issues within the specialty practitioners and staff.

As an exercise, consider the needs for reporting in the following scenarios. The responses in italics are mine.

- While on a field call to examine a lame horse, you see evidence that the pigs on the premises are being fed kitchen waste. *This constitutes garbage feeding to swine, which is illegal in some states and subject to federal oversight and inspection in all states, as dictated by the 1980 Swine Health Protection Act*[8]. *Diseases such as African Swine Fever and Classical Swine Fever can be spread by garbage feeding, so proper cooking of kitchen or restaurant waste is required.*
- You obtain necropsy samples from a caprine abortion outbreak, and the rancher indicates that a family member has fallen ill with flu-like symptoms over the course of the outbreak. *Tests for agents with zoonotic potential should be conducted. The family member should be seen by a physician, and the veterinarian may offer consultation with the physician about potential zoonoses. State regulations would dictate which agent(s) are reportable.*
- Samples from a pig farm experiencing an outbreak of diarrhea with high neonatal mortality are submitted to a veterinary diagnostic laboratory. Infection with porcine deltacoronavirus is confirmed; the laboratory that conducted the diagnostic test is one included in the National Animal Health Laboratory Network (NAHLN). *No further reporting requirements are necessary if the laboratory is part of the NAHLN system, provided that the samples were submitted with the appropriate information regarding the location of the farm, the sample type, date sampled, and type of swine operation (sow / nursery / finisher).*[9]
- A ram with unilateral epididymal enlargement is found to be seropositive to *Brucella ovis*. *The reporting requirements for confirmed cases of* Brucella ovis *infection vary by state.*[10]
- Several cases of stillbirth are reported on multiple sheep and goat farms in your practice area. Necropsy lesions consist of hydranencephaly and arthrogryposis. Tests for endemic infectious disease agents (bluetongue virus, Cache Valley virus, border disease virus / bovine viral diarrhea virus) are negative. *Testing for Schmallenberg virus is indicated, and suspect cases should be reported to the SAHO or APHIS:VS Area Veterinarian in Charge.*[11]

These scenarios are intended to illustrate the broad scope of potential scenarios wherein the large animal internist may manage cases with regulatory implications. When faced with the potential for an animal or livestock operation to include concerns of potential importance to SAHOs, the experts clearly felt that prompt communication with SAHOs is the most prudent, efficient, and effective means by which the internist can serve RVM.

SUMMARY

The complex, expanding global livestock trade carries with it rapidly evolving risks for incursion of novel infectious diseases into a nation's livestock population. Internists are well positioned to serve integral roles in RVM. These include direct involvement in emergency response activities, remaining current on reportable disease requirements, and seeking novel and efficient methods to foster communication between SAHOs, internists, practice staff, and trainees.

REFERENCES

[1] **EFSA AHAW Panel** 2014, doi.10.2903/j.efsa.2014.3877. [2] Paarlberg P, **Working Paper #14-4**, Purdue University. http://ageconsearch.umn.edu/bitstream/174517/2/144.Updated%20Estimated%20Economic%20Welfare%20Impacts%20of%20PEDV.pdf , 2014. [3] Zessin ,K **J Vet Med** 2006; 53:7. [4]Bertini C et al **Foreign Aff** 2009; 88:93. [5]USDA, http://nsu.aphis.usda.gov/animal_health/emergency_management/downloads/brovet.pdf . Accessed 29 January 2015. [6]Salazar P et al, **J Am Vet Med Assoc** 2004;225(2):267. [7]Colorado Department of Agriculture, https://www.colorado.gov/pacific/sites/default/files/Colorado%20Reportable%20Diseases.6.10.14_0.pdf . Accessed 29 January 2015. [8] US Government Publishing Office, http://www.gpo.gov/fdsys/pkg/STATUTE-94/pdf/STATUTE-94-Pg2229.pdf. Accessed 29 January 2015. [9] USDA, http://www.aphis.usda.gov/newsroom/2014/06/pdf/secd_federal_order.pdf . Accessed 29 January 2015. [10]Center for Food Security and Public Health Iowa State University, 2009, http://www.cfsph.iastate.edu/Factsheets/pdfs/brucellosis_ovis.pdf . Accessed 29 January 2015. [11]USDA, http://www.aphis.usda.gov/animal_health/animal_diseases/schmallenberg/downloads/schmallenberg_case_definition_guidance.pdf . Accessed 29 January 2015.

BARTONELLOSIS: A ONE HEALTH APPROACH TO AN EMERGING INFECTIOUS DISEASE

Edward B. Breitschwerdt, DVM, DACVIM (Small Animal Internal Medicine)
Raleigh, NC

INTRODUCTION

Bartonella species are fastidious Gram-negative bacteria that are highly adapted to a mammalian reservoir host and within which the bacteria usually cause a long-lasting intraerythrocytic bacteremia.[1-3] These facts are of particular importance to veterinarians and physicians, as an increasing number of animal reservoir hosts have been identified for various *Bartonella* species. Among numerous other examples, *Bartonella henselae* has co-evolved with cats, *Bartonella vinsonii* subsp. *berkhoffii* has co-evolved with dogs and wild canines, and *Bartonella bovis* has co-evolved with cattle. Importantly, the list of reservoir-adapted *Bartonella* species, including a large number of rodent species that might serve as "pocket pets", continues to grow exponentially, as new *Bartonella* spp. are discovered.[2-3] Prior to 1990, there were only two named *Bartonella* species, whereas there are now at least 32 named and numerous unnamed or candidatus species, based upon deposited GenBank sequences or preliminary reports, respectively.

In the natural reservoir host, chronic bacteremia with a *Bartonella* species can frequently be detected by blood culture or PCR in outwardly healthy individuals. In contrast, the diagnostic detection of a *Bartonella* spp. in a non-reservoir adapted host can be extremely difficult. Most, although not all diseases caused by *Bartonella* spp., occur in accidental hosts and these organisms are being increasingly implicated as a cause of zoonotic infections.[4-8] Until recently, mechanisms that facilitate persistent *Bartonella* bacteremia in mammals were not well understood. Recent reports have identified an intra-endothelial and intra-erythrocytic localization for these bacteria, which represents a unique strategy for bacterial persistence.[2,3] Non-hemolytic intracellular colonization of erythrocytes and endothelial cells would preserve the organisms for efficient vector transmission, protect *Bartonella* from the host immune response, and potentially contribute to decreased antimicrobial efficacy. Other *in vitro* studies indicate that *Bartonella* spp. can infect dendritic cells, microglial cells, pericytes, monocytes and CD34+ bone marrow progenitor cells.

EPIDEMIOLOGY

Bartonella henselae was initially isolated from an HIV-infected human and subsequently from cats, dogs, horses, marine mammals and other small terrestrial mammals. *Bartonella vinsonii* subsp. *berkhoffii* was initially isolated from a dog with endocarditis in our North Carolina laboratory in 1993[9] and subsequently from cats, coyotes and human patients. Retrospectively, long-term administration of immunosuppressive doses of corticosteroids for a presumptive diagnosis of systemic lupus erythematosus with cutaneous vasculitis may have facilitated the isolation of the original type strain of *B. vinsonii (berkhoffii)* from this dog that subsequently developed endocarditis. Due to the relatively recent recognition that dogs can be infected with *B. vinsonii (berkhoffii)*, *B. henselae* and potentially other *Bartonella* spp., seroprevalence data is somewhat limited.[3] Seroprevalence was determined in 1,920 sick dogs from North Carolina or surrounding states that were evaluated at a veterinary teaching hospital. Using a reciprocal titer of >32, only 3.6% of sick dogs had antibodies to *B. vinsonii (berkhoffii)*. Risk factors that could be associated with seroreactivity included: heavy tick exposure (Odds ratio 14.2), cattle exposure (OR 9.3), rural vs. urban environment (OR 7.1) and heavy flea exposure (OR 5.6). Experimental flea transmission of *B. henselae* to dogs has now been confirmed in the laboratory (Lappin and Breitschwerdt, unpublished data). Also, using sera from dogs experimentally infected with *R. rickettsii* or *Ehrlichia canis*, cross reactivity to *Bartonella* antigens was not detected. However, 36% of serum samples derived from dogs naturally infected with *E. canis* were reactive to *B. vinsonii* antigens. As *E. canis* is transmitted by *Rhipicephalus sanguineous*, this tick may be involved in the transmission of *B. vinsonii*. The possibility of tick transmission was further supported by two additional studies involving dogs infected with one or more *Ehrlichia* spp. from the same geographic region, in which seroreactivity to *B. vinsonii (berkhoffii)* antigens was 30% and 89%, respectively. Seroprevalence, using *B. vinsonii (berkhoffii)* antigens, was 10% (4/40 dogs) in dogs with suspected tick-borne illness from Israel and 36% in dogs with fever and thrombocytopenia from Thailand. Using an ELISA assay, 35% of 869 samples, derived from coyotes in California, contained antibodies to *B. vinsonii (berkhoffii)* antigens. Current data indicates that canine and human exposure to *B. henselae* and *B. vinsonii (berkhoffii)* can be found throughout much of the United States and most tropical and subtropical regions of the world.

Studies from Hawaii, the United Kingdom and Japan identified *B. henselae* seroprevalences of 6.5% (2/31dogs), 3.0% (3/100 dogs) and 7.7% (4/52). In our laboratory, *B. henselae* is the most common *Bartonella* species found in sick dogs using the BAPGM enrichment blood culture platform. The pathogenicity of all *Bartonella* spp. in dogs is poorly understood; however, like humans, dogs can be infected with numerous *Bartonella* spp. *B. henselae* was amplified and sequenced from the liver of a dog with peliosis hepatis, a unique pathological lesion also reported in *B. henselae* infected people. *B. henselae* DNA was also amplified from a dog with granulomatous hepatitis, a histopathological lesion reported with some frequency in *B. henselae*-infected children and adults. Similarly, *B. clarridgeae* DNA has been amplified and sequenced from the liver of a Doberman pincher with copper storage disease and from the aortic valve of a dog with vegetative valvular endocarditis. *Bartonella elizabethae*, a species that infects rodents, was found in a dog that had experienced chronic weight loss culminating in sudden unexplained death. Based upon a large seroepidemiological, controlled study from the University of California (Davis), dogs that were seroreactive to either *B. henselae*, *Bartonella clarridgeae* or *B. vinsonii (berkhoffii)* were referred for evaluation of lameness, neutrophilic polyarthritis, nasal discharge, epistaxis and splenomegaly.[10]

In cats, *B. henselae* and *B. clarridgeiae* have been amplified or grown most frequently from cats and the fleas collected from cats. Flea-associated transmission has been well documented amongst cats. In flea endemic areas, *Bartonella* spp. seroprevalence rates in cats can be greater than 90% and bacteremia rates can be greater than 50%. Granulomatous myocarditis has been reported in cats, naturally or experimentally-infected with *B. henselae*. *B. vinsonii* (*berkhoffii*) caused recurrent osteomyelitis in a cat.[3,11]

PATHOGENESIS

Although as yet unproven, *B. vinsonii (berkhoffii)* is presumably transmitted to dogs by the bite of an infected tick. Based upon antidotal evidence, dogs may become infected with *B. henselae* by a cat bite or scratch, analogous to cat scratch disease in people. *B. vinsonii* appears to cause chronic intraerythrocytic and endothelial cell infections in dogs for extended periods of time, potentially resulting in vasoproliferative pathologies. Similar to other highly adapted intracellular vector-transmitted pathogens, the factors that ultimately result in these bacteria causing disease manifestations are yet to be determined. If similar to babesiosis, another intraerythrocytic pathogen, stress, hard work, parturition, concurrent infection with other organisms or therapeutic immunosuppression may contribute to the development of pathology. Following experimental inoculation of SPF dogs with culture grown *B. vinsonii (berkhoffii)*, there was sustained suppression of peripheral blood CD8+ lymphocytes, accompanied by an altered cell surface phenotype and an increase in CD4+ lymphocytes in the peripheral lymph nodes.[11] Therefore, infection with *B. vinsonii (berkhoffii)* appears to induce a degree of chronic immunosuppression that might predispose dogs to other infectious agents, resulting in a wide array of clinical manifestations in naturally-infected dogs. Less is known about pathogenesis of disease in cats but some mechanisms are likely similar. Cats infected with *Bartonella* spp. are commonly co-infected with hemoplasmas and at times, more than one *Bartonella* sp. However, whether co-infections magnify disease manifestations of either genera is unclear and in most studies co-infections did not appear to potentiate illness. From an evolutionary perspective, it is obvious that vectors, vector-borne organisms, and animal and human hosts have developed a highly adapted form of interaction.These factors serve to illustrate the potential difficulty in establishing causation in cats, dogs or people co infected with multiple tick transmitted pathogens. Recently, we proposed an addition to Koch's postulates entitled the Postulate of Comparative Infectious Disease Causation.[11] By satisfying this postulate, *Bartonella* species appear to be able to cause, endocarditis, granulomatous inflammatory diseases, particularly involving heart, liver, lymph nodes, and spleen, persistent intravascular infections and the induction of vasoproliferative tumors.[3,9,12-14]

CLINICAL FINDINGS

The spectrum of disease associated with *Bartonella* infection in dogs and most other animal species is currently unknown. Endocarditis, has been reported in cats, dogs and humans, infected with a spectrum of *Bartonella* spp. In some dogs, intermittent lameness, bone pain, epistasis or fever of unknown origin can precede the diagnosis of endocarditis for several months, whereas other dogs will present with an acute history of cardiopulmonary decompensation.[3,9] Cardiac arrhythmias secondary to myocarditis can be detected in dogs without echocardiographic evidence of endocarditis. Granulomatous lymphadenitis has been associated with *B. vinsonii (berkhoffii)* and *B. henselae* in dogs. *B. vinsonii (berkhoffii)* and other *Bartonella* species appear to contribute to the development of dermatologic lesions indicative of a cutaneous vasculitis, panniculitis, as well as anterior uveitis, polyarthritis, meningoencephalitis and immune-mediated hemolytic anemia.[3,10-15] Additional research efforts, using carefully designed case controlled studies are necessary to establish the frequency and extent to which *Bartonella* spp. contribute to dermatological, ocular, orthopedic, neurological or hematological abnormalities in dogs (and humans).

Clincally, many disease manifestations have also been attributed to *Bartonella* spp. infections in cats.[3] However, it is very difficult to prove disease associations in cats in the field because of the high prevalence rates in non-clinical carriers. In research cats that are infected by exposure to *C. felis*, fever, endocarditis and myocarditis are the most common disease manifestations. As discussed for dogs, additional case controlled studies are needed in cats.

DIAGNOSIS

Thrombocytopenia, anemia, which frequently can be immune-mediated, and neutropenia or neutrophilic leukocytosis are the hematological abnormalities in dogs that are seroreactive or BAPGM enrichment blood culture/PCR positive.[12,15] Thrombocytopenia is found in approximately half, eosinophilia approximately one third of infected dogs and monocytosis frequently occurs in Bartonella endocarditis. Hematological abnormalities have been rarely reported in cats, but similar to dogs, a subset of Bartonella-infected cats are neutropenic. Serum biochemical abnormalities are usually very mild or nonexistent in both cats and dogs. In cats, *Bartonella* spp. antibodies have correlated with polyclonal hyperglobulinemia and hypoglycemia.[16]

As *B. vinsonii (berkhoffii)* antibodies are infrequently detected (<4%) in sick or healthy (<1%) dog populations in endemic regions, detection of *B. vinsonii (berkhoffii)* antibodies in a sick dog provides strong clinical evidence for prior exposure and potentially active infection. For this reason, treatment of seroreactive dogs or dogs from which any *Bartonella* spp. DNA is detected in blood or tissue samples would be recommended.

ISOLATION AND MOLECULAR DETECTION OF *BARTONELLA* SPECIES

Because conventional microbiological isolation techniques lack sensitivity, bartonellosis is usually diagnosed by PCR amplification of organism specific DNA sequences and/or through serological testing. Recently, the development of a more sensitive isolation approach, using BAPGM (*Bartonella* alpha *Proteobacteria* growth medium) followed by real time PCR has greatly facilitated the molecular detection or isolation of *Bartonella* species from the blood of sick or healthy animals, including dogs and human beings.[5,6,12] Obviously, the relative sensitivity of the diagnostic methods used to detect *Bartonella* species infection greatly influences an investigator's ability to establish disease causation or a clinicians ability to initiate appropriate treatment. The use of this optimized microbiological approach has facilitated the recognition of blood-borne *Bartonella* spp. infections in dogs, horses, human beings and porpoises.[17] Diagnostic testing (animals and humans) for *Bartonella* species (serology, PCR and BAPGM Enrichment Blood Culture/PCR) is available through Galaxy Diagnostics, Inc. (contact@galaxydx.com). In cats, serology, PCR or culture combined with serology is recommended and can be procured at Galaxy Diagnostics Inc. and Colorado State University (www.dlab.colostate.edu)

PATHOLOGIC FINDINGS

In dogs, pathologic findings associated with *Bartonella* spp. infection include endocarditis, myocarditis, granulomatous lymphadenitis, granulomatous hepatitis, osteomyelitis, bacillary angiomatosis and peliosis hepatitis.[3,9,12,13] Multifocal areas of severe myocardial inflammation can be found in dogs with *B. vinsonii* (*berkhoffii*) endocarditis. Although not specific for bartonella infections, organisms can be detected in diseased tissues using silver stains, particularly in acute bartonella infections, such as acute regional lymphadenitis (cat scratch disease). During chronic infections, organisms are often too few in number to be detected in tissues by silver staining, unless a fulminate infection is localized to heart valves. The cardiac abnormalities noted in cats to date are similar to those described for dogs.[9] It seems likely that the spleen plays an important immunomodulatory role in controlling persistent *Bartonella* spp. bacteremia in animals and people.[13] The extent to which *Bartonella* spp. induce splenic pathology deserves additional research consideration.

THERAPY

To date, an optimal protocol has not been established for the treatment of bartonella infections in cats, dogs, or people.[3,17] Regardless of the antibiotic(s) that is used for treatment, a long duration of antibiotic administration (4-6 weeks) may be necessary to eliminate the infection. Due to the rapid development of resistance, macrolides (azithromycin) I no longer recommend these antibiotics for treating *Bartonella* infections. Fluoroquinolones in combination with doxycycline are currently being used by the author to treat clinical cases of bartonellosis, while exploring efficacy through experimental infection studies. Doxycycline alone does not appear to eliminate *B. vinsonii (berkhoffii), B. henselae* or *B. clarridgeae* in cats, dogs or other animal species. Serum antibody titers decrease rapidly (3-6 months) and are generally no longer detectable in dogs that recover following antimicrobial therapy. Therefore, post-treatment serology may be a useful adjunct to BAPGM/PCR to determine if therapeutic elimination of bartonella infections has been achieved. Whether there is clinical benefit to follow serologic or molecular assay results in cats has not been widely studied, but most treated cats do not become seronegative in the short term. However, bacteremia can resolve after treatment or resolve spontaneously in some cats, whereas other cats remain bacteremic despite four to six weeks of antibiotic (documented for several antibiotic regimens) administration, and despite resolution of clinical abnormalities (such as lethargy, inappetance and fever).

PREVENTION

Although somewhat circumstantial, there is increasing evidence that *Bartonella* species can be transmitted by fleas and ticks to cats, dogs and human beings.[17] Based upon scientific evidence generated during the past several decades, vector-transmitted pathogens can induce clinical manifestations ranging from acute fatal illness (i.e. Rocky Mountain spotted fever, ehrlichiosis, babesiosis and bartonellosis) to chronic debilitating disease states (ehrlichiosis, babesiosis, borreliosis, and bartonellosis). Therefore, minimizing or eliminating flea and tick exposure is perhaps of greater veterinary and public health importance today, than during any previous time in history. When rigorous flea and tick control measures are instituted, it is highly probable that transmission of *Bartonella* species to pets and their owners will be greatly reduced or eliminated.[18]

PUBLIC AND OCCUPATIONAL HEALTH CONSIDERATIONS

There is increasing evidence to support an important role for *Bartonella* species as a cause of a spectrum of disease manifestations in human patients.[4,17,19-25] Due to extensive contact with a variety of animal species, veterinary professionals appear to be at occupational risk for infection because of frequent exposure to *Bartonella* spp.; therefore, these individuals should exercise increased precautions to avoid arthropod bites, arthropod feces (i.e. fleas and lice), animal bites or scratches and direct contact with bodily fluids from sick animals.[26] As *Bartonella* spp. have been isolated from cat, dog or human blood, cerebrospinal fluid, joint fluid, aqueous fluid, seroma fluid and from pleural, pericardial and abdominal effusions, a substantial number of diagnostic biological samples collected on a daily basis in veterinary practices could contain viable bacteria. The increasing number of defined *Bartonella* spp., in conjunction with the high level of bacteremia found in reservoir-adapted hosts, which represent the veterinary patient population, ensures that all veterinary professionals will experience frequent and repeated exposure to animals harboring these bacteria. Therefore, personal protective equipment, frequent hand washing and avoiding cuts and needle sticks have become more important as our knowledge of this genus has improved and various modes of transmission have been defined. Physicians should be educated as to the large number of *Bartonella* spp. in nature, the extensive spectrum of animal reservoir hosts, and the diversity of confirmed and potential arthropod vectors, current limitations associated with diagnosis and treatment efficacy, and the ecological and the medical complexities induced by these highly evolved intravascular, endotheliotropic bacteria.

REFERENCES

[1]Breitschwerdt EB, et al **Clin Microbiol Rev** 2000;13:428. [2]Chomel BB, et al **Ann N Y Acad Sci** 2003;990:1. [3]Breitschwerdt EB, et al **J Vet Emerg Crit Care** 2010;20:8. [4]Breitschwerdt EB, et al **J Neuroparasitology** 2012;3:1. [5]Mascarelli PE, et al **Parasit Vectors** 2013;6:98. [6]Maggi RG, et al **Parasit Vectors** 2013;6:101. [7]Maggi RG, et al **Parsit Vectors** 2013;6:106. [8]Balakrishnan N, et al **J Med Microbiol** 2013;62:338. [9]Chomel BB, et al **Ann N Y Acad Sci** 2009;1166:120. [10]Henn JB, et al **Am J Vet Res** 2005;66:688. [11]Breitschwerdt EB, et al **J Comp Pathol** 2013;1:1. [12]Perez C, et al **J Vet Intern Med** 2011;25:805. [13]Varanat M, et al **J Vet Intern Med** 2011;25:1284. [14]Beerlage C, et al **Med Microbiol Immunol** 2012: 319. [15]Pérez Veraa C, et al **Comp Immunol Microbiol Infect Dis** 2013;36:481. [16]Whittemore et al **J Vet Intern Med** 2012;26:639. [17]Breitschwerdt EB **ILAR J** 2014;55:46. [18]Lappin MR, et al **Parasit Vectors** 2013;25:26. [19]Breitschwerdt EB, et al **J Clin Microbiol** 2011;49:3415. [20]Cherry NA, et al **Vector Borne Zoonotic Dis** 2011;11:1425. [21]Maggi RG, et al **Diagn Microbiol Infect Dis** 2011;71:43. [22]Breitschwerdt EB, et al **J Clin Microbiol** 2007;45:270. [23]Breitschwerdt EB, et al **J Clin Microbiol** 2010;48:2289. [24]Oliveira AM, et al **J Vet Intern Med** 2010;24:1229. [25]Rossi MA, et al **Vet Dermatol** 2015;26(1):60-e22. [26]Balakrishnan N, et al **J Clin Microbiol** 2014;52:3447.

ALDOSTERONE DEFICIENCY: MORE COMMON THAN YOU THINK

Ellen N. Behrend, VMD, PhD, DACVIM (Small Animal Internal Medicine)
Auburn, AL

INTRODUCTION

Early studies suggest that mitotane (Lysodren) and trilostane (Vetoryl), the mainstays of medical treatment for canine hyperadrenocorticism (HAC), have minimal effect on aldosterone concentration.[1-4] In 6-10% of dogs receiving mitotane, the zona glomerulosa, the site of aldosterone production, is believed to be destroyed.[5,6] The occurrence of complete deficiency is not dose-related, can occur at any time after initiation of therapy and is not predictable.[5] Trilostane may have greater effects than mitotane on aldosterone secretion as its main mechanism of action is inhibition of the enzyme 3-beta-hydroxysteroid dehydrogenase, which is present in the hormone synthesis cascade upstream of aldosterone synthesis. Adrenocortical necrosis can occur to varying degrees in dogs on trilostane therapy.[7,8] It was further believed that aldosterone deficiency would only develop after cortisol deficiency, i.e. if a dog was not cortisol-deficient, it could not be aldosterone-deficient.

Clinical experience, however, suggested that aldosterone deficiency might be present in dogs receiving mitotane or trilostane despite the published information. Only serum electrolytes are routinely monitored to assess for the presence of mineralocorticoid sufficiency during treatment of HAC. Given that aldosterone deficiency has potentially life-threatening consequences such as hyperkalemia, hyponatremia, severe volume depletion, hypotension and death, we sought to further investigate the possibility that aldosterone deficiency existed in dogs on mitotane and trilostane therapy.

CASE PRESENTATION

"Chico", a 9 year old, 3.5 kg, FS Chihuahua, had been receiving maintenance mitotane therapy. Mitotane was discontinued as her basal and ACTH-stimulated cortisol concentrations were both non-detectable. Prednisone (1 mg daily) was prescribed. One month later, her cortisol concentrations were still non-detectable and prednisone was continued. On physical examination after another month, her body temperature was 97.4F and her blood pressure measured by Doppler was 66 mm Hg systolic. On a biochemistry profile, sodium concentration was 144 mmol/L (142-151), chloride was 105 mmol/L (105-117) and potassium was 5.2 mmol/L (3.6-4.9). Cortisol concentrations were now in the ideal range for a dog with pituitary-dependent HAC (PDH) receiving mitotane; basal and ACTH-stimulated concentrations were 34 nmol/L and 40 nmol/L, respectively (ideal 30-150 nmol/L). Serum basal aldosterone concentrations was non-detectable; ACTH-stimulated concentration was 247 pmol/L (197-2,103). Fludrocortisone acetate was prescribed and the dog responded well to therapy.

TESTING FOR ALDOSTERONE DEFICIENCY

Although disease states, e.g. hypoadrenocorticism, as well as therapy for diseases such as HAC may affect aldosterone secretion, the optimal protocol for evaluation of aldosterone has not been determined. The ACTH stimulation test is typically utilized to assess glucocorticoid secretion. The premise of the test is that, in order to assess secretory reserve, a maximally stimulating ACTH dose is administered and a sample is drawn post-ACTH when cortisol concentrations peak. Measurement of peak is paramount; after the peak, other factors such as clearance rate of ACTH can affect test results and do not speak to adrenocortical secretory capability. Although the same test protocol has been repeatedly used to evaluate aldosterone secretory capacity[3,5,9,10], use of the test for such has never been validated. In dogs with spontaneous hypoadrenocorticism, electrolyte concentrations do not correlate with aldosterone concentrations.[11,12] One possible reason for the lack of correlation is a poor test protocol.

In order to try to establish an ACTH stimulation test protocol for aldosterone, we first analyzed samples from a previously published study[13] that established the maximally stimulating dose of ACTH with respect to serum cortisol concentration and the time of peak response. In short, 5 dose-response trials were performed in 10 healthy dogs.[14] Each dog was given 5 doses (1, 0.5, 0.1, 0.05, and 0.01 ug/kg) of cosyntropin (Cortrosyn) IV with a 2-week wash out period between each dose. Blood samples for serum aldosterone concentration determination were obtained before and at 10, 20, 30, 40, 50, 60, 120, 240 minutes after ACTH injection.

Serum aldosterone concentration reliably increased in all dogs after the administration of each dose of cosyntropin. A dose of 1.0 ug/kg cosyntropin stimulated the highest serum aldosterone concentrations. Unfortunately, the dose needed for maximum stimulation was not identified. The time of peak response was at 10 or 20 minutes, not the traditional 60-min sampling time, suggesting that the testing protocol used previously for aldosterone assessment may be suboptimal.

In a follow-up study, 10 additional, privately owned, clinically normal dogs were used.[15] Each dog received 3 doses of cosyntropin (5, 10, and 20 ug/kg) IV in random order with a minimum 2 -day washout period between each dose. Of the 3 doses, 5 ug/kg cosyntropin was the minimum dose that stimulated maximal serum aldosterone concentration. Serum aldosterone concentrations were consistently increased in each dog. The time of peak response was at 30 minutes.

Since our study, another has assessed timing of aldosterone response. If a higher dose of cosyntropin is used (250 ug/dog), the peak is prolonged and samples can be collected at 60-min post-ACTH.[12] However, there is no apparent advantage to using a higher dose and the cost of the cosyntropin could be much more.

Unfortunately, even with an optimized protocol, testing for aldosterone deficiency can be difficult. The test can be used to show that a dog is not deficient. On the other hand, typically in dogs with primary hypoadrenocorticism, for example, basal and ACTH-stimulated aldosterone concentrations are low, but not always.[12] Furthermore, dogs without hypoadrenocorticism can have basal and/or ACTH-stimulated aldosterone concentrations below the detection of the assay.[12] Thus, lack of measurable aldosterone concentrations is not synonymous with aldosterone deficiency. Measurement of plasma renin activity would help in many cases, if such an assay becomes commercially available.

EVIDENCE OF ALDOSTERONE DEFICIENCY BEING COMMON IN DOGS TREATED FOR PDH

Given the knowledge that 30 minutes post-ACTH is the optimal sampling time, we undertook a clinical study using dogs that were receiving mitotane or trilostane therapy for treatment of PDH.[16] Seventy-nine ACTH stimulation tests (5 ug/kg cosyntropin IV) were performed on client-owned dogs with PDH: 42 on mitotane-treated dogs and 37 on trilostane-treated dogs. In the mitotane treatment group, one dog was enrolled four times, eight dogs were enrolled three times, and three dogs were enrolled two times. In the trilostane treatment group, four dogs were enrolled three times. Cortisol concentrations were measured at 0 and 60 minutes and aldosterone concentrations at 0, 30, and 60 minutes. Ten historical clinically healthy controls[15] were included. Basal sodium and potassium concentrations were measured.

The aldosterone concentrations in the mitotane- and trilostane-treated dogs at 30 and 60 minutes post-ACTH were significantly lower than in clinically healthy dogs; no significant difference was detected in aldosterone concentration between 30 and 60 minutes in treated dogs. However, a significantly higher percentage of dogs had a decreased aldosterone secretory reserve at 30 min than 60 min. At 30 minutes, 49% and 78% of trilostane- and mitotane-treated dogs had a decreased secretory reserve, respectively. Thus, decreased aldosterone secretory reserve is common in trilostane- and mitotane-treated dogs. Aldosterone concentration at 30 minutes post-ACTH stimulation identifies more cases of decreased aldosterone secretory reserve than conventional testing at 60 min.

Aldosterone concentrations at 30- and 60-min post-ACTH were significantly lower in dogs receiving mitotane therapy compared with trilostane regardless of the level of control of the PDH. Furthermore, the aldosterone concentrations would not be expected to change during the day in dogs treated with mitotane as compared to trilostane; trilostane therapy suppresses cortisol concentrations for a shorter period of time compared to mitoane.[17] In other words, the aldosterone suppression in trilostane-treated dogs might be present only transiently. Whether we evaluated the dogs at the time of maximal trilostane-induced inhibition of aldosterone secretion is unknown, since the effect of trilostane waxes and wanes and timing is patient-dependent.

A significant relationship was not detected between basal or ACTH-stimulated aldosterone concentrations and serum electrolyte concentrations. Similarly, in dogs with spontaneous hypoadrenocorticism, electrolyte concentrations do not correlate with aldosterone concentrations.[11,12] In our study, for both drugs combined, 67% and 94% of dogs with decreased aldosterone secretory reserve were normonatremic and normokalemic, respectively. Conversely, a number of dogs with a normal reserve were also either hyponatremic (n=12) or hyperkalemic (n=2). Therefore, hyperkalemia and hyponatremia are insensitive markers of aldosterone deficiency. Accordingly, aldosterone concentrations should be measured in patients with normal electrolyte concentrations that have clinical signs suggestive of or are at risk for decreased aldosterone secretory reserve. Detection of decreased reserve can help predict which patients may need mineralocorticoid supplementation routinely or during times that aldosterone secretory reserve is important, e.g. dehydration and hypovolemia.

ALDOSTERONE DEFICIENCY CANNOT PREDICT RECOVERY FROM OVERSUPPRESSION

In dogs made hypocortisolemic with mitotane or trilostane, it is impossible to predict if and when adrenal function will recover. Dogs with damage to the zona glomerulosa with resultant hypoaldosteronism could, in theory, be less likely to recover because the zona glomerulosa contains the progenitor cells for the other layers. Thus, we sought to determine if measuring aldosterone as well as cortisol may be a more efficient means of evaluating adrenal damage and allow for prediction of recovery.

Eighteen dogs with low cortisol concentrations (non-detectable at baseline and <1.4 μg/dL (<40 nmol/L) at 60 minutes) had follow-up ACTH stimulation tests performed.[16] At the time hypocortisolemia was documented, 15 dogs had 60-minute aldosterone concentrations that were undetectable or below the reference range. Nine of the 15 dogs with decreased aldosterone secretory reserve (6 mitotane-treated and 3 trilostane-treated dogs) had follow-up ACTH stimulation tests that documented adrenocortical recovery within 0.5 to 5 months. Six of the 15 dogs (4 mitotane-treated and 2 trilostane-treated) did not have documented adrenocortical recovery during a follow-up period of 1-7 months. At the time hypocortisolemia was initially detected, 3 dogs had 60-minute aldosterone concentrations within the reference range. Of the 3 dogs, two had documented recovery of cortisol secretory ability 2 and 5 months later; however the third dog (on mitotane) remained hypocortisolemic through the end of the follow-up at 21 months. Therefore, the presence of decreased aldosterone secretory reserve cannot be used to predict permanent hypoadrenocorticism, nor can an aldosterone concentration within the reference range be used to predict a rapid recovery from hypocortisolism.

GLUCOCORTICOID THERAPY CAN CAUSE ALDOSTERONE DEFICIENCY

Serum potassium concentrations and the renin-angiotensin system are the main controllers of aldosterone synthesis and release from the zona glomerulosa of the adrenal gland, but ACTH has a role. Thus, we sought to determine if ACTH deficiency secondary to glucocorticoid administration could suppress serum aldosterone concentration.

Five healthy dogs received an anti-inflammatory dose of prednisone (0.55 mg/kg, PO, q 12 h) and 5 received a placebo for 28 days.[18] Basal and ACTH-stimulated cortisol and aldosterone concentrations and serum electrolyte concentrations were measured before (day 0 [baseline]), during (days 7, 14, 21, and 28), and after (days 35 and 42) treatment.

Basal and ACTH-stimulated cortisol concentrations did not change from baseline values in placebo-treated dogs. In prednisone-treated dogs, serum chloride and corrected chloride concentrations were significantly lower on days 7, 14, 21, and 28 and serum bicarbonate concentrations were significantly higher on days 14, 21, and 28, compared with baseline values. Serum cortisol concentrations before and after ACTH stimulation testing were significantly lower than baseline values during prednisone treatment. Serum aldosterone concentration after ACTH stimulation testing was significantly lower than baseline on day 35 (i.e., 1 week after discontinuation of prednisone treatment) but returned to baseline by day 42 in prednisone-treated dogs.

Significant hypochloremia developed in the dogs during prednisone administration. On occasion, dogs may be suspected of having developed naturally occurring hypoadrenocorticism during glucocorticoid treatment. To determine whether complete adrenocortical insufficiency (which is typical in cases of spontaneous hypoadrenocorticism) is present in those dogs, clinicopathologic evaluations for expected electrolyte abnormalities are often performed. However, the results of our study indicate that such practice is inappropriate because marked hypochloremia may be present in the face of normal serum aldosterone concentration and aldosterone secretory capacity in dogs receiving exogenous glucocorticoid treatment. In addition, if a patient is tested for hypoadrenocorticism within one week of cessation of glucocorticoid treatment, decreased ACTH-stimulated serum aldosterone concentration cannot be used as evidence of spontaneous disease. The mechanism and clinical importance of the changes remain to be determined. Lastly, dogs that have signs of a hypoadrenal crisis shortly after cessation of glucocorticoid treatment may benefit from mineralocorticoid administration because their mineralocorticoid secretory reserve may be diminished.

CASES OF ALDOSTERONE DEFICIENCY

Three types of isolated aldosterone deficiency have been documented in the literature. First, a 9-year-old male German shepherd dog was evaluated for clinical and clinicopathological changes that were suggestive of Addison's disease.[19] However, the basal and ACTH-stimulated plasma cortisol concentrations were high and normal, respectively. Plasma renin activity was elevated and serum aldosterone concentration was low. A diagnosis of hyperreninemic hypoaldosteronism was made. Replacement fludrocortisone resulted in complete normalization of the electrolyte and fluid imbalances. Serum cortisol concentrations were similar approximately 5 months later. The cause of the hyperreninemic hypoaldosteronism was not determined, however the dog did have pleural effusion; effusions can cause hyponatremia and hyperkalemia.

Conversely, a case of hyporeninemic hypoaldosteronism was described in a 13-year-old female spayed Yorkshire terrier with hyperkalemia but no associated clinical signs.[20] In human medicine, hyporeninemic hypoaldosteronism is a common cause of persistent, asymptomatic hyperkalemia in older patients with renal disease, diabetic nephropathy, or chronic tubulointerstitial disease. The dog had mild renal azotemia and significant proteinuria. Fludrocortisone was prescribed, but the dog was lost to follow-up.

Last, in a dog with primary hypoadrenocorticism, mineralocorticoid deficiency preceded glucocorticoid deficiency.[21] A 3-year-old castrated male Doberman pinscher presented for head and neck tremors, weight loss, poor appetite, increased thirst and lethargy. Hyponatremia and hypochloremia were present. Basal and ACTH-stimulated cortisol concentrations were within reference ranges. One week later, hyperkalemia was also present; basal and ACTH-stimulated aldosterone concentrations were below the reference range and fludrocortisone therapy was initiated. Six weeks after initial presentation, cortisol deficiency and primary hypothyroidism were diagnosed.

REFERENCES

[1]Goy-Thollot I, Pechereau D, Keroack S, et al. J Small Anim Pract 2002; 43: 489-92. [2]Golden DL, Lothrop CD. J Vet Int Med 1988; 2: 121-5. [3]Sieber-Ruckstuhl NS, Boretti FS, Wenger M, et al. Domest Anim Endocrinol 2006; 31: 63-75. [4]Wenger M, Sieber-Ruckstuhl NS, Müller C, Reusch C. Effect of trilostane on serum concentrations of aldosterone, cortisol, and potassium in dogs with pituitary-dependent hyperadrenocorticism. Am J Vet Res 2004; 65: 1245-50. [5]Kintzer PP, Peterson ME. J Vet Int Med 1991; 5: 182-90. [6]Kintzer PP, Peterson ME. J Am Vet Med Assoc 1994; 225: 54-61. [7]Chapman PS, Kelly DF, Archer J, et al. J Small Anim Pract 2004; 45: 307-10. [8]Reusch CE, Sieber-Ruckstuhl N, Wenger M, et al. Vet Rec 2007; 160: 219-24. [9]Neiger R, Ramsey IK, O'Conner J, et al. Vet Rec 2002; 150: 799-804. [10]Ruckstuhl NS, Nett CS, Reusch C. Am J Vet Res 2002; 63: 506-12. [11]Muller C, Boretti FS, Wenger M, et al. Kleintierpraxis 2007; 52: 216-24. [12]Baumstark ME, Sieber-Ruckstuhl NS, Muller C, et al. J Vet Intern Med 2014; 28: 154-9. [13]Martin LG, Behrend EN, Mealey KL, et al. Am J Vet Res 2007; 68: 555-60. [14]Behrend EN, Martin L, Lee HP, et al. J Vet Int Med 2007; 21: 597. [15]Carlson KJ, Behrend EN, Martin LG, et al. J Vet Intern Med 2010; 24: 685. [16]Reid LE, Behrend EN, Martin LG, et al. J Vet Intern Med 2014; 28: 443-50. [17]Vaughn MA, Feldman EC, Hoar BR, et al. J Am Vet Med Assoc 2008; 232: 1321-8. [18]Corrigan AM, Behrend EN, Martin LG, et al. Am J Vet Res 2010; 71: 649-54. [19]Lobetti RG. J South African Vet Assoc 1998; 69: 33-5. [20]Kreissler JJ, Langston CE. J Vet Intern Med 2011; 25: 944-8. [21]McGonigle KM, Randolph JF, Center SA, et al. J Am Anim Hosp Assoc 2012; 49: 54-7

COMPREHENSIVE REVIEW: VACCINES AND VACCINOLOGY

Michael J. Day, BSc, BVMS (Hons), PhD, DSc, DECVP, FASM, FRCPath, FRCVS
Bristol, United Kingdom

INTRODUCTION

'Vaccinology' may be defined as the science of vaccine development and delivery in the field. Vaccinology incorporates the overlapping fields of microbiology, immunology and clinical medicine and so advances in vaccinology are a reflection of scientific advances in all three of these areas.

ADVANCES IN IMMUNOLOGY

Key advances in understanding the immune response to infectious agents underpin the development of new and improved vaccines that seek to replicate aspects of the natural protective immune response. The most significant advances in immunology related to this concept have been in the recognition of functional subsets of helper T lymphocytes and the immunoregulatory cytokine profiles that align with each of these subsets. The knowledge that Th1 cells producing interferon (IFN)-γ promote cell-mediated immunity (CMI), while Th2 cells producing interleukin (IL)-4, IL-5, IL-9 and IL-13 underpin humoral immunity, allows immunological explanation of the fact that the strongest 'correlates of protection' for particular vaccines are either CMI or humoral immunity. An excellent example of where this new knowledge has impacted on canine vaccine development has been in the production of new vaccines to protect against leishmaniosis. Investigations of the spontaneously arising canine disease have identified genetically-determined 'resistant' and 'susceptible' phenotypes in dogs infected with *Leishmania infantum* by sand flies. Resistant dogs mount a strong Th1 response with IFN-γ production signalling infected macrophages to kill intracellular amastigotes, while susceptible dogs respond to infection by mounting an inappropriate Th2-regulated humoral immune response that leads to clinical signs and ultimately death. *Leishmania* vaccines therefore seek to replicate the natural protective Th1 immune response[1].

ADVANCES IN MICROBIOLOGY

Key advances in microbiology have been the molecular revolution led by polymerase chain reaction diagnostics and the ability to characterize organisms phylogenetically. This science has allowed investigation of strain variation in infectious agents and how such variation arises in the field. In terms of small animal vaccine preventable diseases, such studies have allowed characterization of the evolution of the canine parvovirus (CPV) and recognition that current vaccines can cross-protect against emergent variants (i.e. CPV2c). New vaccines have been introduced which contain a broader spectrum of field-relevant organisms; for example, the introduction of feline calicivirus (FCV) vaccines containing two strains of the virus, and the introduction of canine *Leptospira* vaccines containing three or four serogroups of the bacterium that are tailored to the field prevalence of these serogroups in different geographical areas[2].

Future developments in human vaccinology must keep pace with the changing nature of the human population – with a greater proportion of ageing individuals and changing 'niche' lifestyle requirements. Human vaccinology is progressing by exploiting modern molecular microbiological techniques; for example the use of 'reverse vaccinology' in which the microbial genome is screened for potential vaccine candidates[3].

ADVANCES IN VACCINE DELIVERY

In recent decades we have seen rapid progress in the technology of vaccine production for small animals. Traditional 'infectious' or 'non-infectious' whole organism vaccines have led to products based on antigenic molecules (subunit vaccines), recombinant vector organisms or naked DNA technology. Vaccine delivery has also expanded from traditional subcutaneous injection by needle to intranasal and now oral vaccines and percutaneous delivery using high-pressure needle-free apparatus. In the case of the latter delivery methods, mucosal delivery of vaccine antigens provides a much more relevant local immune protection and percutaneous delivery allows more efficient targeting of epidermal and dermal dendritic cell populations with reduced doses of vaccine[4]. Improving the safety of feline vaccination has been explored by considering a range of anatomical locations for optimum delivery of injectable vaccines; most recently tail vaccination[5].

VACCINATION GUIDELINES

The driver for vaccination guidelines has been a global focus on vaccine safety and investigations of the nature and prevalence of rare vaccine associated adverse events. These adverse events, including the feline injection site sarcoma and canine immune-mediated haemolytic anaemia, have fuelled concerns about vaccine safety and discussions as to how vaccination might be delivered better to companion animals. The response of the veterinary profession has been to establish a number of scientific expert groups, which have provided evidence-based advice on optimum methods of small animal vaccination. These vaccine guidelines have come from the American Academy of Feline Practitioners[6], the European Advisory Board on Cat Diseases[7] (for feline vaccination), the American Animal Hospital Association[8] (for canine vaccination) and the World Small Animal Veterinary Association Vaccination Guidelines Group[9] (for both cats and dogs).

FUNDAMENTAL GUIDELINES CONCEPTS

All of the above guidelines propose a number of fundamental concepts in small companion animal vaccinology that have driven, and continue to drive, change in vaccination practice globally.

Evidence-based Vaccinology

Wherever possible, vaccination guidelines have embraced the concept of evidence-based veterinary medicine. However, for vaccinology this poses a challenge, because much scientific data related to vaccine development, efficacy and adverse events has traditionally been commercially sensitive and not published in the peer-reviewed scientific literature. Although there has now been a culture shift in this practice, guidelines advice must still sometimes be based on empirical 'first principles' science.

Herd Immunity

Small companion animal veterinary practitioners are reminded that the concept of 'herd immunity' applies equally to dogs and cats living in a particular geographical area as it does to the human or production animal population. Optimistic estimations of the percentage of vaccinated dogs and cats, even in developed countries, are of the order of 30 – 50% and so we are far from achieving optimum herd immunity for canine and feline populations.

Infectious Disease Surveillance

Informed decision making on the use of small animal vaccines is often impaired by lack of knowledge about the local presence or prevalence of specific infectious agents. Emerging surveillance schemes offered by manufacturers are commendable and large projects based on data mining from practice-based medical record programmes (e.g. the UK Small Animal Surveillance Network [SAVSNET] project run by the British Small Animal Veterinary Association and the University of Liverpool) should allow further progress in the future.

Minimize Individual Vaccination

A fundamental principle of all guidelines is that the vaccine load on individual animals should be minimized to that providing essential protective immunity. The classification of vaccines as 'core', 'non-core' and 'not recommended' and the introduction of core vaccines with extended minimum duration of immunity has simplified identification of the minimum vaccine requirements for dogs and cats. Selection of non-core vaccines is more challenging and should be based on factors such as local disease prevalence, geographical location and the lifestyle and risk factors for the individual pet.

Serological Testing

Newly available in-practice serological test kits enable the rapid identification of protective immunity for a number of core vaccines. Guidelines promote the use of these tests in order to confirm protective immunity after puppy vaccination, to minimize future vaccination in an animal that has recovered from a vaccine-associated adverse event, and to determine the core revaccination requirements for adult animals.

The Annual Health Check

Vaccination guidelines groups promote the concept of the annual health check rather than the 'annual vaccine booster' as the primary reason for a yearly visit to the veterinary practice. The annual health check addresses vaccination as just one aspect of regular preventive healthcare and is more about marketing of professional services than science. Implementing the health check system is shown to impact positively on the business of veterinary practice.

Vaccination for Public Health

Vaccination guidelines emphasize the importance of companion animal vaccination in the control of zoonotic infectious diseases such as canine rabies virus infection and leishmaniosis. Mass vaccination campaigns in rabies endemic African and Asian countries are proving successful in controlling the disease locally[10], but there is a long road ahead to meet the WSAVA-OIE target of global elimination of this infection by 2030. In Brazil, scientific field studies have also shown the impact of vaccination of dogs against leishmaniosis on the disease in the human population[1].

FUTURE CHALLENGES

The future of companion animal vaccinology is ensured and likely to be accompanied by further scientific advances in products and methods of delivery. The current challenge for the profession is in creating global equality by ensuring that full product ranges, allowing vaccination in accordance with guidelines principles, are made available throughout the world[11].

REFERENCES

[1]Palatnik-de-Sousa CB. **Frontiers Immunol** 2012; 3: 69. [2]Klaasen HLBM et al. **Vet Immunol Immunopathol** 2014; 158:26. [3]Rappuoli R et al. **Nature Revs Immunol** 2011; 11:S65. [4]Li N et al. **Vaccine** 2011; 29:6179. [5]Hendricks CG et al. **J Feline Med Surg** 2014; 16:275. [6]Scherk MA et al. **J Feline Med Surg** 2013; 15:785. [7]Welborn LV et al. **J Am Anim Hosp Assoc** 2011; 47:1. [8]Hosie MJ et al. **J Feline Med Surg** 2013; 15:540. [9]Day MJ et al. **J Small Anim Pract** 2010; 51:338. [10]Lembo T et al. **PLoS Negl Trop Dis** 2010; 4:2. [11]Day MJ et al. **J Small Anim Pract** 2015; in press.

VACCINATION AGAINST *BORDETELLA BRONCHISEPTICA* IN DOGS: A CRITICAL REVIEW OF THE LITERATURE

John Ellis, DVM, PhD
Saskatoon, Saskatchewan, Canada

Presentation extracted from review article "How well do vaccines for *Bordetella bronchiseptica* work in dogs? A critical review of the literature 1977–2014" from The Veterinary Journal. Available at http://dx.doi.org/10.1016/j.tvjl.2015.02.006.

VACCINATION AGAINST CANINE LEPTOSPIROSIS

George E. Moore, DVM, MS, PhD, DACVPM, DACVIM
West Lafayette, IN

INTRODUCTION

Leptospirosis is a globally important zoonotic disease occurring in all continents and in both industrialized and developing countries. The disease can affect humans and dogs in both rural and urban settings, and affects many mammalian food-producing species, e.g. cattle, pigs, sheep, and goats. Because leptospirosis vaccines have been investigated in numerous species, it is valuable to understand some general concepts about the organism and these vaccines in general before considering canine leptospirosis vaccines specifically.

CELL COMPOSITION AND PROTEIN ANTIGENS

Leptospires are spiral or helicoidal (corkscrew-shaped) bacteria with distinctive hooks on one or both ends. Leptospires are obligate aerobic spirochetes with optimal growth at pH 7.2-7.6 and temperature 28-30°C. They derive their major energy and carbon sources through beta-oxidation of long chain (>C15) fatty acids. These essential fatty acids are paradoxically also toxic to the leptospire, but albumin (in the leptospires' environment) absorbs large quantities of fatty acids and slowly releases them at concentrations that are non-toxic to the leptospires.

Leptospires are composed of cytoplasmic membrane and peptidoglycan cell wall, surrounded by an outer membrane. Besides a diversity of outer membrane proteins (OMPs), leptospires contain lipopolysaccharide (LPS) which is absent in the outer membrane of other pathogenic spirochetes such as *Borrelia*. There are variations however in the carbohydrate composition of LPS in different serovars, reflecting antigenic diversity. Fortunately, leptospiral LPS has lower endotoxic activity than enterobacterial LPS,[1] yet LPS presumably provides a number of general benefits that facilitate long-term survival of the organism in the environment after the bacteria are shed in the urine of reservoir hosts.

The leptospiral outer membrane is complex with features that are unique among bacteria. The most abundant surface-exposed OMPs are lipoproteins (Lip) that are integrated into the membrane lipid bilayer. The leptospiral surface undergoes significant changes during key transitions in its life cycle from a free-living state outside the hosts to living in the tissues of carrier host animals. It has long been known that leptospires recovered from kidneys of infected animals were antigenically distinct from cultivated organisms.[2] Thus, culture conditions that mimic the host environment/body strongly affect the gene expression and production levels of a number of OMPs. Lig (leptospiral immunoglobulin-like repeat) proteins, specifically LigA and LigB, are upregulated during infection; and their production is also dramatically enhanced by increasing culture osmolarity to physiologic levels found in mammalian hosts. Temperature and iron concentration can also affect expression of surface proteins.

IMMUNE RESPONSE TO INFECTION

Leptospira causes two distinctly different disease manifestations depending on the mammalian and the infective serovar. In reservoir hosts, infection leads to a chronic, nearly asymptomatic disease state; whereas in incidental hosts it can lead to a mild or acute, life-threatening illness. Innate immunity is stimulated when leptospiral LPS activates macrophages through Toll-like receptors (TLRs). Failure of TLRs to detect leptospiral LPS results in infection with the spirochete. Mice with intact TLR2 and TLR4 receptors are resistant to infection; human TLR4 however fails to recognize leptospiral LPS, making humans susceptible to infection. Other TLRs, e.g. TLR5, can induce macrophage production of IL1-β, IL-6, IL-10, IFN, and TNF-α in virulent infections.

PROTECTIVE IMMUNE RESPONSE

The nature of protective immunity against leptospirosis infection is poorly understood. The chief mechanism of immunity to leptospires is believed to be solely or primarily humoral, as protective immunity has been conveyed by passive transfer of monoclonal antibody alone (antibodies directed against LPS). Other leptospiral protein antigens can also elicit protective immunity.

A role for cell-mediated immunity in protection against leptospirosis has more recently been found in cattle.[3] Peripheral blood mononuclear cells in vaccinated cattle produced IFN-γ after antigenic challenge, and thus protective immunity against serovar Hardjo infection in cattle correlated with establishment of Th1 immunity. It is unknown if this is serovar or host species specific.

VACCINES FOR HUMANS

As early as 1933, a heat-killed whole cell vaccine made from *Leptospira* cultures was reported to protect mine workers in endemic areas of Japan.[4] Current licensed human vaccines against leptospirosis are produced in France, Japan, China, and Cuba.[5] All four vaccines are killed/inactivated whole cells or outer envelope fractions. Antibody (MAT) titers after vaccination are noted to be significantly lower than those developed after natural infection, but efficacy rates of 60-100% are reported. Reported duration of immunity varies from 6-12 months in some studies to at least 3 years in others. To maintain immunity, the French vaccine recommends a booster every 2 years and the Japanese vaccine at least every 5 years.[5] Reported side effects of whole cell vaccines in humans include systemic (fever, chills, malaise) and localized reactions. Concern over reactions to host proteins led to the development of a leptospiral vaccine derived from leptospires grown in a chemically defined protein-free medium,[6] however growth in such media is generally poorer and have not gained widespread use.

VACCINES FOR LIVESTOCK

Vaccination is considered the easiest and often the only practical method of control of leptospirosis in animals. Its use is limited by availability, expense, quality, and appropriateness of the antigens in the vaccine in terms of relevance to the species and country. Efficacy data for some products is poor in terms of microbiological protection data, and often based on inappropriate studies, including unnatural routes of challenge and very short duration of immunity studies.[7] Vaccines have been most widely used in cattle and pigs, and can contain from one to 5 or more serovars.

VACCINES FOR DOGS

Although there are nearly 200 serovars of pathogenic *Leptospira*, the serogroups most frequently reported (based on serology) from dogs in the US are Canicola, Grippotyphosa, Icterohaemorrhagiae, and Pomona [note: these serogroups include several serovars, including serovars that carry the same name as the serogroup itself]. Other serogroups that have been recognized in dogs in other countries include Australis [serovars Australis and Bratislava], Autumnalis [serovar Autumnalis], Bataviae [serovar Bataviae], and Sejroe [serovars Serjoe, Medanensis, and Saxkoebing].

Based on testing of stray and owned dogs in US in the 1950s, seropositivity for exposure/infection to serogroups Canicola (host reservoir – dogs) and/or Icterohaemorrhagiae (host reservoir – rats) was documented in up to 22% of dogs. The first developed canine leptospirosis vaccines in the US were therefore bivalent products directed against these two serogroups. Cross-reaction (on diagnostic tests) or cross-protection (from natural infection or vaccination) between these 2 serogroups was infrequently reported, consistent with known genetic variability between serogroups.

Two additional serogroups Grippotyphosa and Pomona, known as pathogens in cattle and pigs, became increasingly recognized as canine pathogens in the mid- to late-1990s. Documentation of these serogroups in wildlife suggested that urban wildlife may be important or primary reservoir hosts for leptospire spread to dogs. The major veterinary biologics manufacturers in the US now market tetravalent (Canicola, Grippotyphosa, Icterohaemorrhagiae, and Pomona) leptospirosis vaccines for dogs. Leptospiral vaccines for dogs are usually formulated without adjuvants, alum, or aluminum hydroxide, although vaccines for cattle often include adjuvants to increase immunity. Maternally derived immunity in dogs declines by 8-10 weeks of age.

Although MAT seroreactivity to serovars Autumnalis and Bratislava are noted in dogs, failure to culture these specific pathogens from sick dogs has also "failed" to prove to USDA the need for these additional serovars in US canine vaccines. Some cross-reactivity has been noted between these two serovars and Grippotyphosa and Pomona, and it is unknown if some limited vaccinal cross-protection exists. I am not aware of any documented evidence of infection by these two non-vaccinal serovars in vaccinated dogs, yet exposure risk is undoubtedly low.

DURATION OF IMMUNITY

Although post-vaccinal titers are notoriously short-lived, e.g. 3-8 months, in dogs, there has been virtually no epidemiological evidence of increased infection risk before annual boosters. Challenge studies at 12 months post-vaccination have now demonstrated that protection against infection is provided without detectable titers. Due to similarities in bacterin manufacturing, canine leptospirosis vaccines are considered to have 12-14 months duration of immunity if administered in accordance with the manufacturer's instructions. Some owners of valuable dogs in high-risk areas, e.g. water sport or hunting dogs, may desire semi-annual boosters but there is no published evidence that this is necessary with newer tetravalent vaccines.

Infection in a vaccinated dog is most likely due to failure to give initial 2-vaccination series, infection by a non-vaccinal serovar, or failure by the bacterin to invoke initial protective immunity.

CLAIMS OF PROTECTION OR DISEASE PREVENTION

The USDA Center for Veterinary Biologics is the federal regulatory agency regulating leptospirosis vaccines for companion animals and livestock. All leptospirosis vaccines must be shown to protect host animals from a virulent challenge prior to licensure. The potency test is codified in the Code of Federal Regulations and is a hamster potency assay in which ≥80% of vaccinated hamsters must be protected from a virulent challenge by each serovar. This assay was developed in the early 1970's and has been codified since the mid-late 1970's. Some manufacturers are either using or migrating toward in vitro assays (ELISA) for potency testing in lieu of the hamster potency assay. These are monoclonal antibody based assays in which the potency of the vaccine is determined relative to a reference vaccine. These assays exist for leptospiral serovars Canicola, Grippotyphosa, Icterohaemorrhagiae, and Pomona. The vaccine must still be shown to protect host animals against challenge from individual serovars.

There is no requirement to publish results of experimental clinical trials required for licensing of a new veterinary vaccine. A review of 4 published canine leptospirosis vaccine studies (4 different manufacturers) in the last decade revealed that vaccinated and control/placebo groups had 6-10 dogs per group for each serovar tested, approximately 10^7-10^9 *Leptospira* organisms were administered to each dog as a "challenge", and these organisms were administered intraperitoneally (IP) +/- in the conjunctival sac. The high doses and routes are necessary to produce clinical disease in control/placebo dogs and demonstrate vaccine efficacy.

There are different label claims which can be used on a biological. The highest claim is "Prevention of infection", a claim intended to convey the ability of a product to prevent infection upon exposure to a disease agent may be made only for products able to prevent all colonization or replication of the challenge organism in vaccinated and challenged animals. Next(lower) is "Prevention of disease", a claim intended to convey the ability of a product to prevent disease after exposure to a disease agent may be made for products shown to be effective in preventing clinical disease in vaccinated and challenged animals. If the product is highly effective and the entire 95% confidence interval estimate for efficacy is at least 80%, a label statement "for the prevention of disease due to [specific microorganism]" may be used. Next is "Aid in disease prevention", for products preventing disease in a statistically and clinically significant manner, but with a 95% confidence interval less than that required to support a claim of disease prevention. The (lower) claim "Aid in disease control" is intended to convey the ability of a product to control disease after exposure to a disease agent as for products shown to alleviate disease severity or reduce disease duration.[8] The claims are based on experimental studies, and the degree to which they are translatable to natural exposure/challenge in leptospirosis is unknown. Due to the high doses of organism and challenge routes of administration, e.g. IP, in licensing studies one might expect better protection in your practice than the label claim.

ADVERSE EVENTS FOLLOWING VACCINATION

Leptospirosis vaccines were originally manufactured in the US for protecting livestock/cattle, and first generation canine vaccines were produced in a similar fashion. As such, canine leptospirosis vaccines were reputed to commonly cause adverse events, most typically allergic (type I) hypersensitivity reactions. In 2005, separate studies identified risk factors for vaccine-associated allergic reactions as prior exposure to fetal calf serum or bovine serum albumin (BSA) and increased quantitative exposure by multiple concurrent vaccinations. Although BSA has been noted to be an important (nearly obligate) component in leptospiral growth media, it is a large ionized molecule which can be filtered/reduced in final vaccine manufacturing. Due to safety concerns, newer "4-way" generation of leptospirosis vaccines are believed to have less BSA (and other excipients) than former 2-way vaccines due to better manufacturing filtration methods. USDA does not regulate the amount of BSA that can be present in a dose of vaccine, but manufacturers use varied methods of reducing BSA and other proteins in the final product. Even if BSA is reduced, likely lowering the likelihood of a reaction, other bovine-origin excipient proteins may still remain the vaccines. Manufacturers are not required to publicly report end concentrations of excipient proteins.

Although vaccine safety is a reported component of licensing trials, safety data of any vaccine at this point is likely biased (overly optimistic) as study dogs probably have not had repeated exposures to the new vaccine, it is not administered with other concurrent vaccines, and/or high-risk breeds/dogs may not be represented sufficiently in the trial.

FUTURE VACCINES – CONCEPTS AND DIRECTIONS

For nearly a century very little has changed about leptospiral vaccines. All licensed vaccines are whole cell (or primarily whole cell), inactivated bacterins. A major drawback of bacterin vaccines is that elicited immunity is directed mainly against the leptospiral lipopolysaccharide, a T-independent antigen. The LPS structure is very complex and also is not identical/homologous across serovars, thus there is minimal prospect of LPS-based vaccine in the near future.

There are no licensed recombinant leptospirosis vaccines in veterinary or human medicine. Nevertheless the use of recombinant DNA techniques has caused a resurgence of activity to try to identify protective antigens and develop subunit vaccines. The outer membrane lipoprotein LipL32 is the most studied leptospiral protein, as it is the most abundant protein in the cell and outer membrane. Although reported to elicit protective immunity, there are problems with inadequate or weak challenge dose, small numbers of animals, and inappropriate analyses. It is also possible that LipL32 is a protective antigen in some species but not in others. In the case of humans at least, there is a compelling argument that LipL32 does not mediate protection.[9]

Numerous other recombinant proteins (up to 238 recombinant proteins in one report) have been tested with similar problems and a lack of protective immunity. Even if a mechanism of immunity can be identified in a laboratory animal species, e.g. hamster or mouse, the identity of the antigen cannot necessarily be extrapolated from laboratory animals to production animals to companion animals or to humans. Furthermore, antigens identified for one serovar may not be generalizable across other different serovars. The search for one antigen "to rule them all" remains the elusive Holy Grail of leptospiral vaccinology.

REFERENCES

[1]Nahori, et al. **J Immunol** 2005;175:6022. [2]Haake DA, Matsunaga J. **Mol Microbiol** 2010;77:805. [3]Naiman, et al. **Infect Immun** 2002;70:6147. [4]Wani H. **Zeitschrift Immunitätsforschung Experimentelle Therapie** 1933;79:1. [5]Koizumi N, Watanabe H. Leptospirosis. In: Barrett and Stanberry (eds). **Vaccines for Biodefense and Emerging and Neglected Disease**, 2009, Elsevier Inc., 1291. [6]Shenberg E, Torten M. **J Infect Dis** 1975;128:62. [7]Ellis WA. **Curr Top Microbiol Immunol** 2015;387:99. [8]USDA VS Memo 800.202, dtd 2014.10.27. [9]Adler B. **Curr Top Microbiol Immunol** 2015;387:251.
Additional references available on request.

HEPATIC BIOPSY: THE ACQUISITION

Keith Richter, DVM, DACVIM (SAIM)
San Diego, CA

INTRODUCTION

Hepatobiliary diseases can be challenging to diagnose with clinical signs and laboratory abnormalities alone, and often requires hepatic tissue sampling for cytologic or histopathologic evaluation. Recent advances in imaging technology, the use of multiple imaging modalities, and newer biopsy methods have resulted in improvement in the veterinary clinician's ability to safely procure hepatic tissue for evaluation. There are several means of obtaining hepatic samples including fine needle aspiration (FNA), ultrasound-guided biopsy, laparoscopy, and laparotomy. All techniques have both advantages and disadvantages.

PRE-BIOPSY CONSIDERATIONS

Persistent elevation of serum hepatic enzymes, elevated serum bile acids, focal or diffuse hepatic masses or echotextural changes, or the need to assess response to therapy are all indications for hepatic sampling. The primary concern for any hepatic sampling technique is the chance for hemorrhage, although it is unclear what the best predictor of hemorrhage is in patients about to undergo hepatic sampling. Ewe studied 200 human patients and evaluated bleeding laparoscopically, and found no correlation with any *in vitro* coagulation test and "liver bleeding time".[1] These findings have been subsequently observed in other studies.[2,3] Studies have shown that clotting times assessing proteins induced by vitamin K antagonism (PIVKA) are more sensitive in detecting coagulopathies in patients with hepatic disease.[4,5] However, in a pilot study performed at the authors' hospital, hepatic bleeding times assessed via laparoscopy did not correlate with PIVKA time. Thus it appears that peripheral blood coagulation indices are generally unreliable guidelines regarding the risk of hemorrhage following liver biopsy. In most cases of significant hemorrhage, technical errors such as damaging a large vessel are the cause rather than persistent oozing from a needle biopsy site. In one study in normal dogs, biopsies taken from the left lateral hepatic lobe using a biopsy punch, biopsy needle, ligature method, laparoscopic biopsy forceps, and ultrasonically activated scalpel resulted in minimal hemorrhage (<2ml).[6]

FINE NEEDLE ASPIRATES

Fine needle aspiration is typically performed with ultrasound guidance and involves obtaining a small amount of hepatic tissue for cytologic analysis. Using a 22 gauge, 1.5 inch needle without a syringe, the needle is rapidly agitated in and out of the hepatic tissue and simultaneously twisted to obtain a sample. This method relies on capillary action rather than suction to acquire tissue within the needle, which results in less hemodilution. The sample is then expelled onto a glass slide for cytologic examination. Often three to five separate attempts are made to increase the sample size and diversity. Ultrasound imaging allows the sampling of focal lesions or diffuse abnormalities. Aspiration sites must be chosen based on accessibility and limited risk for damaging local structures including the gallbladder and large vessels. The clinician must decide whether the relative risk of sampling such lesions is the best decision. Sedation or anesthesia may be necessary depending on the individual patient's temperament.

Fine needle aspiration is a quick and inexpensive method of hepatic sampling that requires little to no sedation and carries a low risk for hemorrhage. Multiple sites can be easily sampled and the procedure can usually be performed on an outpatient basis. The primary disadvantage of FNA is the questionable accuracy. The sample size often limits the number of available cells to obtain an accurate diagnosis, and hemodilution makes it difficult to assess whether inflammatory cells were present in the liver or peripheral blood. There are several important elements used to interpret pathologic information to make clinical decisions, including lobular architecture, presence and location of inflammation within a lobule, presence and severity of fibrosis, location and amount of metal accumulation, vascular abnormalities, and lobule heterogeneity. These criteria cannot be accurately examined using a cytologic sample obtained using FNA. Several studies have compared FNA cytology to biopsy with histopathology. In one study with a total of 34 cases, there was good correlation in 35% of cases, partial correlation in 35% cases, and no correlation in 30% of cases.[7] Poor correlation was found with a variety of histologic changes including vacuolar change, lipidosis, cholestasis, inflammation, and neoplasia. In a similar study with 97 cases, complete agreement between FNA and histopathology was seen in only 30% of cases of dogs; 25% agreement with inflammation, 14% agreement with neoplasia (mainly carcinoma), and 64% agreement with vacuolar hepatopathy.[8] In cats, there was overall agreement in 51% of cases; 27% agreement with inflammation, 33% agreement with neoplasia (lymphoma), and 64% agreement with vacuolar hepatopathy.[8] Although vacuolar hepatopathy was the most sensitive diagnosis, it was also the most common misdiagnosis. In another study, the best correlation between hepatic cytology and biopsy was seen with lipidosis, lymphoma, and carcinoma whereas the worst performance was seen with inflammatory and fibrotic disorders.[9]

In summary, given its clear limitations, FNA is best used as an adjunctive diagnostic modality in conjunction with other techniques or clinical findings, and certainly does not replace histopathology. The clinician must be aware of its inherent inaccuracy before undertaking FNA and relying on the cytologic findings.

ULTRASOUND-GUIDED BIOPSIES

Ultrasound-guided hepatic biopsy uses a cutting-type needle as a sampling tool. Automated needles are preferred and should be either completely automated or semi-automated. Completely automated needles thrust the inner obturator followed by the outer cutting sheath into the liver in a fraction of a second. The needles can be operated with one hand while the other hand operates an ultrasound transducer to allow precise placement of the biopsy instrument. There is minimal displacement of the liver, a shorter intra-parenchymal phase, and a more reliable yield of tissue. Using the rapid cutting action, the hepatic tissue tends to be less fragmented. Examples of completely automated needles include the Monopty© needle (C.R. Bard, Inc, Tempe AZ), Max-Core needle© (C.R. Bard,

Inc, Tempe AZ) Magnum© needle (C.R. Bard, Inc, Tempe AZ), and Achieve© needle (CareFusion, San Diego CA). Semi-automated needles require manual placement of the internal obturator into the liver, followed by an automatic thrusting of the outer sheath by a spring-loaded mechanism. Examples of semi-automated needles include the Temno© needle (CareFusion, San Diego CA) and Vet-Core© needle (Smiths Medical PM, Inc., Norwell, MA). These needles have the additional advantage of control over the final needle position, since the tip of the needle can be precisely localized before the outer cutting sheath is deployed. More recently a novel end-cut needle has been developed (BioPince™, Angiotech, Vancouver, BC Canada). Rather than the side-notch style of the needles described above, this needle has an end-cut design in which the entire lumen and almost the whole length of advancement of the needle are used to capture the specimen. A coaxial pincer, which slides through a slit at the tip of the device, cuts and encloses the full core specimen. The author generally uses a 16-gauge needle for ultrasound-guided hepatic biopsy.

Local anesthesia and minimal sedation or general anesthesia is used depending on the patient. Care must be taken prior to taking samples to ensure that vessels and other organs are not within the path of the needle. For diffuse lesions, the transducer is typically placed caudal and to the left of the xiphoid, and aimed at the left medial or lateral lobes. In patients with a small liver, it may be difficult to adequately visualize the needle without gastric gas interference. Placing these animals in a 45-degree right lateral oblique position can reduce this interference. If the animal is under general anesthesia, an assistant can compress a re-breathing bag to hold the animal in deep inspiration which serves to move the diaphragm and liver caudally to improve visualization. The area is surgically prepped, and a stab incision is made in the skin at the desired needle insertion site. The ultrasound transducer is covered with sterile wrap and sterile lube is used to enhance skin contact. While one hand maneuvers the transducer, the other hand advances the needle into the liver under direct ultrasound visualization. The image should be optimized to maximize the chance of recognizing the needle within the liver. To allow distinction of the needle from other echogenic structures, the needle can be gently moved in and out with minimal movements (attempting to move the needle within the abdominal cavity rather than the needle within the liver). Occasionally the needle cannot be seen, and indirect evidence of organ penetration must be used such as movement of the liver or visualization of movement at the liver border. The needle is then directed so the trajectory will avoid other structures when it is fired. The needle is fired, and then removed. For most cases, four to five samples are obtained, and are submitted for aerobic/anaerobic culture, histopathologic evaluation, and metal (copper, zinc, and iron) quantification. Careful examination for post-biopsy hemorrhage is then performed. External digital pressure may be used to help control hemorrhage in smaller patients; an abdominal compression wrap is usually ineffective for this.

This method requires minimal sedation in many patients, has a low to moderate cost, and provides adequate samples for histopathology. In one study, 96 percutaneous transabdominal hepatic needle biopsy samples were obtained with no adverse consequences noted.[10] However, this study was performed in normal dogs, and in the authors' experience this sampling method has the highest risk of any sampling method for hemorrhage. Additional disadvantages include the need for sedation or anesthesia in some patients, difficulty of imaging small livers, difficulty of obtaining liver tissue in patients with fibrosis, and samples that carry a questionable representation of the underlying hepatic pathology. This questionable accuracy is in most part due to potential for sampling error. This method still results in a relatively small sample size, possible fragmentation of fibrous tissue, and may not enable sampling of abnormalities located in other lobes (the left medial or lateral lobes are generally sampled due to their ease of imaging). In one study, percutaneous hepatic sampling using core biopsies resulted in 92% diagnostic quality samples, however these were not compared to large wedge biopsy to assess the accuracy of this method.[11] In another study, the diagnostic accuracy of the Tru Cut needle biopsy was compared to the gold standard of surgical wedge biopsy of the liver in 124 patients.[12] The overall discordance between the two methods was 53% in dogs and 50% of cats, with >60% discrepancy occurring with chronic hepatitis or cirrhosis, cholangitis/cholangiohepatitis, portosystemic vascular anomalies, microvascular dysplasia, fibrosis, and miscellaneous disorders. These are the disorders that are the most commonly seen among dogs and cats with hepatobiliary disease. The greatest accuracy was with neoplasia (80% concordance). In this study, needle biopsies commonly had <4 portal triads. Use of a 14-gauge needle versus 18-gauge may reduce this discordance as it raises the number of portal triads sampled from an approximate mean of four to seven. However the larger needle size carries a higher risk of complications, including hemorrhage. One study in human beings concluded that at least 6 to 8 portal triads are necessary for proper evaluation of hepatic biopsy specimens.[13] Another study in normal dogs showed that the use of a 16-gauge biopsy needle resulted in inconsistent procurement of adequate tissue samples (42% of samples had <6 portal triads, 17% were fragmented to the extent that they were unacceptable for evaluation).[14] Another more recent study showed that a 14 ga needle contained a mean of 2.9 portal triads. These authors recommended multiple biopsies be obtained to reduce sampling error, though that comes with increased risk.[15]

In summary, ultrasound-guided hepatic biopsy is relatively easy to perform, but has questionable accuracy. The accuracy may be increased by using a larger gauge needle, but this carries a greater risk of post-biopsy hemorrhage. If the patient is suspected of having inflammatory disease, vascular abnormalities, significant fibrosis, or is at risk for hemorrhage, laparoscopy or laparotomy should be considered.

LAPAROSCOPIC BIOPSIES

Laparoscopy is performed under general anesthesia with the patient in dorsal recumbency tilted 45 degrees to the left. A Veress needle is placed at the level of the umbilicus into the peritoneal space through a stab incision using a #11 blade. After ensuring no obstruction and negative pressure using the infusion and aspiration of saline, the abdomen is insufflated with carbon dioxide gas and maintained at a pressure of approximately 12 mm Hg. A scope port (cannula) is then placed 4 cm right lateral to the Veress needle. The Veress needle is then removed and replaced with an instrument port. Hepatic sampling is achieved using a "spoon" or oval cup biopsy forceps. Multiple samples are obtained under direct visualization, and samples are submitted for aerobic/anaerobic culture, histopathologic evaluation, and metal (copper, zinc, and iron) quantification. Following procurement of all the biopsy specimens, the sites are inspected for hemorrhage. The abdomen is then decompressed, and lidocaine and bupivacaine are infused into the peritoneal cavity through either port. Both the instrument and scope ports are removed, and the port site incisions are closed using either a cruciate or simple interrupted pattern in the body wall, subcutaneous tissue, and skin.

This technique enables gross evaluation of the entire liver, extra-hepatic biliary system, and surrounding structures while obtaining multiple large specimens of liver. The ability to obtain multiple samples decreases the risk for sampling artifact in cases of regional diversity within the liver. Additionally, by directly visualizing the hepatic parenchyma, the clinician can correlate the histopathologic findings and clinical data with the gross appearance of the liver to render the most accurate diagnosis. In one study, "spoon" or oval cup forceps used for hepatic biopsy in normal dogs resulted in consistent diagnostic samples with a mean of 16 to18 portal triads per biopsy specimen.[14] The authors concluded that biopsy samples obtained using this instrument yielded accurate representation of the hepatic parenchyma. Another study in clinical patients revealed diagnostic quality biopsy specimens in all 80 dogs that underwent laparoscopic liver biopsy.[16] They were all considered to contain a sufficient number of portal triads to obtain a histologic diagnosis. The number of portal triads in each sample ranged from 8 to 13, and the mean total number of portal triads submitted per patient was 39. However, disagreements in histologic diagnoses were identified in 7 of 49 (14%) of the dogs for which biopsy samples from multiple sites were available for review.[16] This suggests that multiple samples should always be obtained, ideally from different lobes. These studies were then contradicted by a more recent study that showed only a mean of 3.4 portal triads with a cup biopsy forceps, though these authors also point out that the diagnostic accuracy is increased when multiple biopsies are obtained.[15] Laparoscopy also enables the visualization of smaller masses and irregularities that may not be evident with ultrasonographic imaging.

There is generally minimal bleeding during this procedure, even in patients with *in vitro* coagulopathies. Using a "spoon" or oval cup forceps typically results in a marked decrease in the amount of hemorrhage when compared to needle biopsies. Any hemorrhage can be directly visualized for adequate clot formation. If hemorrhage persists, direct pressure using a blunt probe for five minutes can be used. If the site continues to bleed, electrocautery can be applied to the biopsy site or a topical hemostatic agent (Gel Foam®) can be placed directly on the biopsy site using laparoscopic forceps. In one study, 76 of 80 dogs (95%) survived to hospital discharge after undergoing laparoscopic liver biopsy.[16] Three (4%) of dogs required conversion to laparotomy (none of which were due to hemorrhage). Another 3 dogs (4%) required a blood transfusion, all of which were anemic prior to the biopsy. In this study, the complication rate was too low to identify risk factors that could be used to predict the development of complications.[16] In the author's experience, conversion to laparotomy or significant hemorrhage requiring blood transfusion is extremely rare. The main disadvantages of laparoscopy are the need for general anesthesia, increased cost, and the need for technical expertise and equipment. There is also the potential for inadvertent trauma to adjacent organs during trocar/cannula placement.

SURGICAL BIOPSIES

Wedge biopsy via laparotomy is another potential method for obtaining hepatic biopsies. If a random liver biopsy is needed and a section of liver is protruding, a guillotine suture can be used. A pre-formed encircling ligature of 4-0 monofilament absorbable suture material is placed around the protruding section of liver. The ligature is then tightened until it has crushed the hepatic parenchyma. After completing several throws in the knot, the sample is excised 1-2 mm distal to the ligature using Metzenbaum scissors or a scalpel blade. If a specific area of liver is needed, a sample can be obtained using the transfixation method or a biopsy punch. The transfixation method entails placing a ligature through the liver lobe approximately 8-10 mm from its edge. The ligature is tightened to crush through the hepatic parenchyma of half the desired biopsy specimen. An additional throw is made at a right angle to the first ligature, and this throw is tightened to crush the parenchyma of the second half of the specimen. The sample is removed 1-2 mm distal to the crushed area using a scalpel blade or Metzenbaum scissors. If the desired area does not lie near the edge of a liver lobe, a 6 or 8 mm biopsy punch can be utilized (an 8 mm biopsy punch yields a mean of 12 portal triads).[15] The biopsy punch should be inserted into the liver ensuring not to penetrate the opposite surface. If the biopsy site is close to the hilus, extra caution must be used so that no more than half of the thickness of the liver is penetrated. The biopsy sample is removed from the liver using scissors. Hemorrhage can be controlled by filling the defect with a topical hemostatic agent (Gel Foam®) and applying digital pressure for 3 to 5 minutes. Akin to laparoscopy, this method has similar advantages and disadvantages as listed above. It is more invasive, but it allows easier biopsy of other abdominal organs (such as intestine and mesenteric lymph node) and the ability to perform therapeutic maneuvers (such as hepatic mass removal or biliary diversion).

REFERENCES

1. Ewe K: **Dig Disc Sci** 1981; 26:388. 2. McVay PA, et al: **Am J Clin Pathol** 1990; 94:747. 3. Bigge LA, et al: **J Am Anim Hosp Assoc** 2001; 37:228. 4. Center SA, et al: **J Vet Intern Med** 2000; 14:292. 5. Center S A, et al: **J Vet Intern Med** 1998; 12:214. 6. Vasanjee SC, et al: **Vet Surg** 2006; 35:86. 7. Fondacaro JV, et al: **Abst. Proceedings of the 17th ACVIM Forum** 1999; 719. 8. Wang, KY, et al: **JAVMA** 2004; 224:75. 9. Roth L: **Vet Clin Pathol** 2001; 30:35. 10. Hitt ME, et al: **Am J Vet Res** 1992; 53:785. 11. Barr F: **J Small Anim Pract** 1995; 36:105. 12. Cole TL, et al: **JAVMA** 2002; 220:1483. 13. Bravo AA, et al: **N Engl J Med** 2001; 344:495. 14. Vasanjee SC, et al: **Vet Surg** 2006; 35:86. 15. Kemp SD, et al: **J Vet Intern Med** 2015; 29: 51. 16. Petre SL, et al: **J Am Vet Med Assoc** 2012; 240:181.

HEPATIC BIOPSY: A PATHOLOGISTS PERSPECTIVE

John M. Cullen VMD PhD DACVP FIATP
Raleigh, NC

INTRODUCTION

Liver disease is a frequently encountered problem in small animal practice. The use of diagnostic terminology is a key issue affecting communication between pathologists as well as between pathologists and clinicians. The diagnostic terminology used by pathologists to describe liver diseases in various countries and even different training programs is often inconsistent, leading to confusion for clinicians interpreting pathology reports and those who would like to compare the pathology results from separate studies. One effort to resolve this potential Tower of Babel was undertaken by the World Small Animal Veterinary Association (WSAVA) when it formed a group of experienced clinicians and pathologists to develop a standardized format for diagnostic terminology that, it was hoped, will lead to greater uniformity in diagnoses and better communication between clinicians and pathologists alike.

A robust exchange of information between the clinician and the pathologist is essential to obtain as much value from a biopsy as is possible. Just as a clinician sees history and baseline data for a patient as essential to making a diagnosis, the pathologist needs as much pertinent clinical data as possible to interpret a biopsy. Submissions with "Elevated liver enzymes" as the entire data set are not going to get as useful biopsy report or comment as they might. Similarly, the pathologist has a responsibility to communicate the type and extent of injury and to comment on the additional steps that can be taken to gain greater precision in the diagnosis and to provide an informed comment to clarify what is known and what possible interpretations could be made from the biopsy. In this presentation I will discuss some of the common ancillary tests are discussed and some features of a diagnosis that involve some diagnostic nuance to effectively communicate with clinicians.

WHAT ARE "SPECIAL STAINS?" WHEN SHOULD THEY BE APPLIED AND WHAT INFORMATION DO YOU GET?

Hematoxylin and eosin are the routine stains of diagnostic histopathology and are very valuable in virtually all applications. In hepatic pathology there are several special stains that can be used to more fully characterize various disease processes. Some of these are described below:

Copper and other Pigment Stains

Rhodanine or rubeanic acid stain copper or copper - binding proteins and are useful stains to detect the presence, the lobular location of copper and intensity of staining. Copper staining can be used as an adjunct to copper level determination by other laboratory methods or, using morphometric analysis, hepatic copper can been quantified. Given the contemporary increase in canine and occasionally feline hepatic copper virtually all inflamed livers and many other livers with unexplained enzyme elevations should be stained with copper stains to assess copper concentrations. This particularly important because this in one form of chronic hepatitis that can be treated specifically. Symptomatic treatment alone without copper determination deprives the patient of targeted resolution and can lead to unnecessary morbidity and possibly liver failure. Lobular localization can be important because impaired copper excretion is usually manifest by centrilobular accumulation of copper. Periportal copper accumulation is more often associated with other disorders such as chronic inflammation and possibly cholestasis. Estimation of copper burden can be done by several methods. A subjective grading scale has been used in which copper scores range from I-IV and scores over III are associated with copper elimination disorders. As mentioned earlier, morphometric methods may provide a more precise measurement of hepatic copper. A cautionary note that should be mentioned is that this approach requires lobular architecture to be maintained. Once injured livers develop bridging fibrosis and nodular regeneration, copper distribution is often altered with abundant copper in the collapsed stroma and fibrous septa and the regenerative nodules are often copper-free or have much diminished copper. At this stage it is difficult to get an accurate measurement as the nodules are low in copper content and the septa are much higher. While there is no agreed upon amount of copper to determine when copper is driving liver injury, some histologic landmarks can be helpful. For example, if there is centrilobular injury present and copper is over roughly 800 ppm dw this may be a situation in which to consider chelation. The presence of parenchymal macrophage aggregates that contain foamy macrophages and copper is a landmark that I use to suggest copper-induced injury needs a therapeutic response.

Fig. 1 A and B. Typical hepatic copper staining with Rhodanine (left) and negative (right).

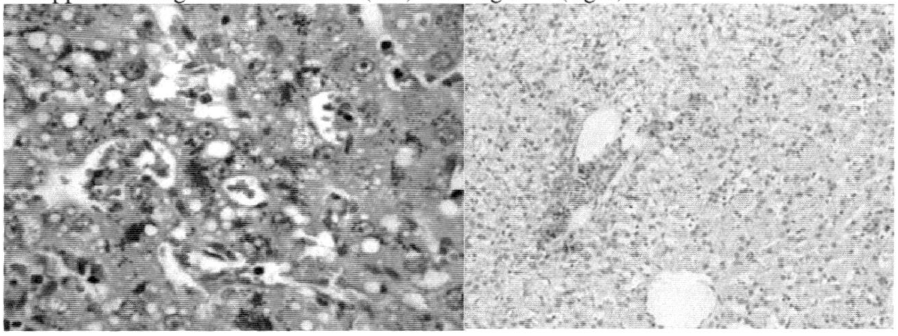

It is helpful to identify all hepatic pigments. First, copper can be eliminated, but abundant lipofuscin (Schmorl's stain or acid fast stains or PAS with diastase) can indicate an excessive oxidative injury to cell membranes. In such cases anti-oxidant therapy may be helpful. Iron in the form of hemosiderin is readily detected using the Prussian Blue stain. Since iron, like copper can participate in the Fenton reaction causing oxidative injury, excessive iron has the potential to be injurious. Certain nondomestic species can suffer from excessive iron causing hemochromatosis, but I am not familiar with chronic iron-driven liver injury in domestic species. Often elevated iron accompanies increased copper so some synergy is possible. Bile can be stained with Hall's stain, but in my experience the bile is already evident on H&E by the time the Hall's stain detects bile. PAS will detect carbohydrates and diastase will remove glycogen. This combination can demonstrate glycogen (by removal with diastase) and non-glycogen carbohydrates seen in some storage disorders or lipofuscin/ceroid remain stained after diastase treatment.

WHY USE CONNECTIVE TISSUE STAINS?

There are two main purposes for the connective tissue stains. Reticulin stains outline the lobules and clearly reveal changes like centrilobular collapse and nodular regeneration that can be subtle changes in H&E-stained sections. The purpose for connective tissue stains such as Sirius Red or Masson's Trichrome is to evaluate the amount and location of fibrosis. As of yet, we do not have an agreed upon classification of hepatic fibrosis and there are several in use for medical pathology. Several systems are used and, simplified, consist of none, minimal, mild, moderate and severe, although with slightly varying criteria. Early fibrosis may resolve without significant impact, but location can be key as subendothelial fibrosis can lead to significant hepatic impairment with only modest amounts of fibrous tissue, as seen in lobular dissecting hepatitis. Other forms of fibrosis-centrilobular or periportal have different causes and possibly prognoses. Certainly more is bad, but as a profession we lack longitudinal studies to assess the "point of no return" when a certain amount of fibrosis is destined to manifest as end-stage fibrosis or cirrhosis. Serial biopsies would help lift the veil of ignorance that obscures our prognostic window currently.

Fig. 2. A and B. Sirius Red Stain for collagen (left) and reticulin stain (right)

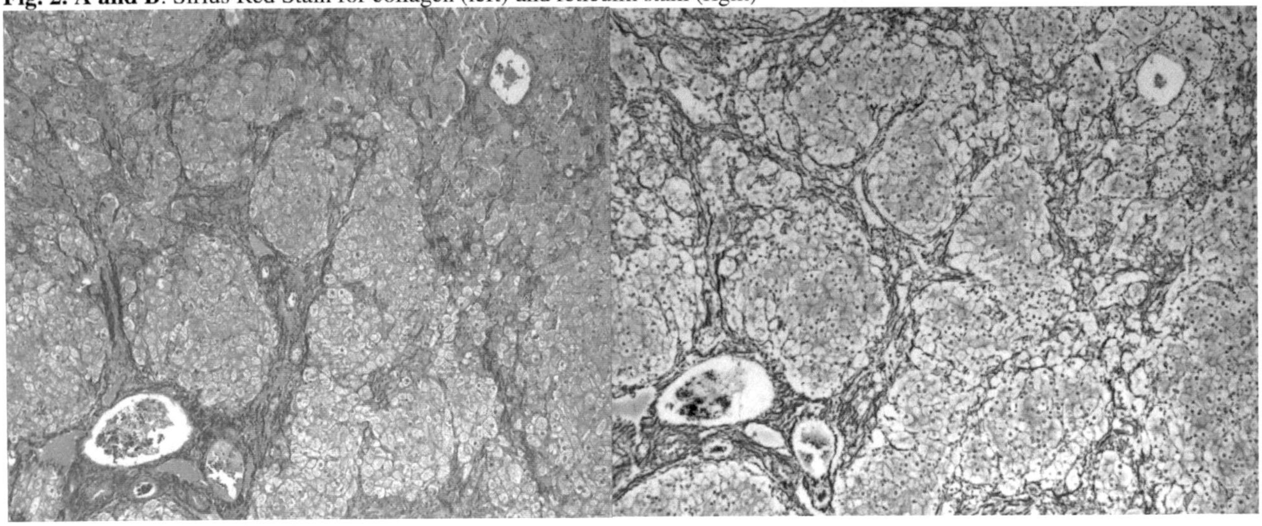

Lipid Stains

Several stains such as Oil Red O or osmium can be used to detect lipid. Osmium is a much more precise stain, but highly toxic and can only be used safely in appropriate laboratory settings. Oil Red O is more like finger painting to the pointillism of the osmium stain. Regardless, in some circumstances it is important to identify lipid versus glycogen to better understand the underlying metabolic issues leading to the vacuole formation.

WHAT IS "VACUOLAR HEPATOPATHY?"

Vacuolar hepatopathy is not a very helpful term and is generally unsatisfactory for pathologists and clinicians alike. This term refers to an excess of vacuoles in the hepatocellular cytoplasm. Not all vacuoles are alike the genesis of each can vary. Some vacuoles, usually a combination of glycogen and lipid are typical of hepatocytes, particularly at the ultrastructural level. Depending on the period of fasting before biopsy acquisition some vacuoles are normal in dogs and cats. It is useful to separate vacuoles into types when possible. Possibilities include: macrovesicular lipid, microvesicular lipid and glycogen, as well as mixed and some abnormal storage products. Phospolipid is also possible in uncommon circumstances, usually involving cationic amphiphilic therapeutic drugs. Some specific forms of hepatic vacuolization are discussed below.

Hepatocellular Steatosis (Lipidosis, Fatty Change)

Hepatic steatosis has two phenotypic forms, but both form distinct, round clear vacuoles. In macrovesicular hepatocellular steatosis the accumulation of lipids forms vacuoles that are larger than the size of the nucleus and that tend to displace the hepatocellular nucleus. These lipid vacuoles most commonly form as a result of dietary excess or starvation or in some early forms of feline hepatic lipidosis and diabetes mellitus. They are usually associated with only modest hepatic injury. Microvesicular steatosis (vacuoles smaller than the size of the nucleus) can be associated with injury to mitochondria and more severe liver dysfunction (diabetic ketoacidosis, small breed hypoglycemia syndrome) than macrovesicular lipidosis. Hepatocellular steatosis should be described by distribution (zonal or diffuse), severity, and size of the cytoplasmic vacuoles (microvesicular, macrovesicular, or mixed).

Hepatocellular glycogen accumulation (Steroid-induced hepatopathy)

This condition is most often caused by an excess of endogenous or exogenous corticosteroids and evident in dogs (and rabbits) primarily. The characteristic lesion is the presence of ballooned hepatocytes with abundant vacuoles and only diaphanous wisps of hepatocellular cytoplasm evident. These vacuoles are ragged and clear. Possible additional causes include an excess of other types of steroid hormones that may be produced by the adrenal cortex, particularly in older dogs and occasionally drugs such as D-penicillamine or progestational compounds. Normal female dogs increase the amount of glycogen in their livers at different states of the estrus cycle which can add to confusion in intact females.

It is worth trying to distinguish lipid from glycogen in dogs with hepatocellular vacuoles, as each tends to have a different cause. Elimination of the known causes based on clinical history can also help-estrus cycle, pseudopregnancy, Cushing's disease, chelation therapy etc. is useful. Drug history should be evaluated. Some therapeutic drugs such as amiodarone can cause phospholipid accumulation in hepatocytes and Kupffer cells which resembles steatosis. After this effort there are still many cases that may have one or the other type of vacuole or even mixed vacuoles.

Fig. 3. Steroid Hepatopathy (left) and microvesicular steatosis(right)

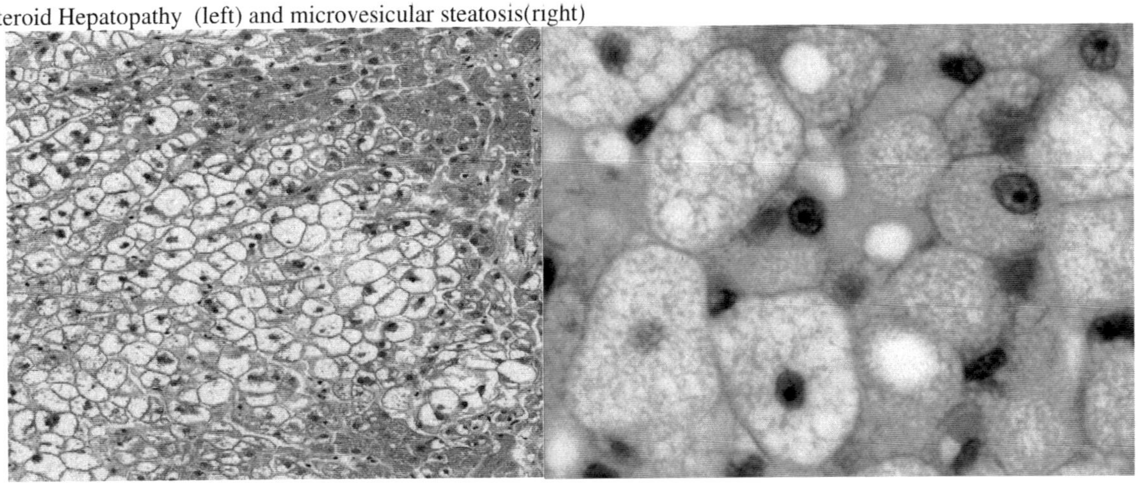

ACUTE AND CHRONIC HEPATITIS

Acute hepatitis is characterized morphologically by a combination of inflammation, hepatocellular apoptosis and necrosis, and, in some instances, regeneration. The proportion and detailed nature of these components vary widely according to the cause, the host response and the passage of time and it is necessary to include in the diagnosis the type, pattern and extent of the necrosis and inflammation as well as the possible etiology. These features are a critical to report to the attending clinician to guide appropriate therapy and prognosis. Acute hepatitis may be self-limiting or the initiation of a chronic process and this is often a function of the etiology. The distinction between acute and chronic hepatitis can sometimes be difficult to distinguish due to overlapping features. Common rule outs for acute hepatitis include viruses, such as adenovirus in dogs, herpes in neonates and these are typically diagnosed by the presence of inclusion bodies. Corona virus infections causing FIP also damage the liver, although this often has a more chronic presentation. Acute hepatitis can also result from septicemic bacterial diseases. These cause multifocal random and confluent necrosis with macrophage proliferation and/or infiltration with neutrophils and in later stages lymphocytes and plasma cells or cause non-specific reactive hepatitis. Enteric bacteria such as *Escherichia coli* and *Salmonella spp* are relatively common, as are *Streptococcus spp., Pasteurella spp.* and *Brucella spp.*, other many other organisms *(C. piliformis* and *Leptospira spp*) are possible. *Toxoplasma gondii* and related protozoa, such as Neospora and sarcocycstis, cause multisystemic disease involving the liver, lung, brain and other organs in the cat and dog.

Chronic Hepatitis

Chronic inflammation of the liver is rare in cats, but a common problem in dogs. The pathogenesis of this process in most cases remains unclear. It is very likely that chronic hepatitis has a multifactorial etiology that may include infectious, autoimmune, drug-induced and copper-associated causes. As in acute hepatitis and it is necessary to include in the diagnosis the type, pattern and extent of the fibrosis and inflammation as well as the possible etiology. The severity of inflammation is an important element of the diagnosis because this shapes the therapeutic plant. The word "hepatitis" can trigger a panoply of therapeutic approaches, but these should vary depending on the extent of inflammation and in some cases no therapy is needed. (see Non-specific reactive hepatitis below) As mentioned above, it is essential to check for copper in dogs with chronic hepatitis is essential because specific chelation therapy or dietary restriction of copper can eliminated the hepatic injury. The type and site of inflammation is also important. For example, chronic neutrophilic cholangitis may have a mix of neutrophils and mononuclear cells, but usually requires antibiotic therapy, while in aged cats, lymphocytic/plasmacytic infiltrates alone targeting the bile ducts is indicative of lymphocytic cholangitis treated with anti-inflammatory therapy. The pathologist should communicate the extent of ongoing inflammation as this is likely to drive continued fibrosis and is a feature that clinical intervention can influence. The extent of fibrosis needs to be clearly discussed to clearly identify the likely compromise of hepatic function at the time of the biopsy. The risk of progression to end stage liver (cirrhosis) can be estimated, but we lack accurate criteria.

Fig. 4. Chronic hepatitis. Mononuclear cell inflammation and fibrosis are evident,(left) and example of fibrosis staging scheme (right).

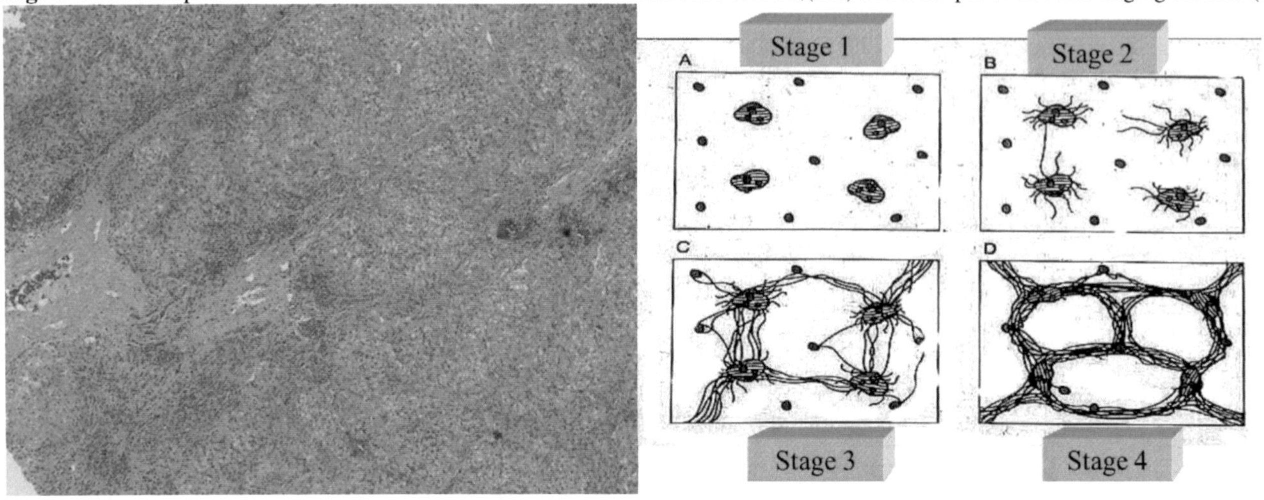

Cirrhosis

Cirrhosis is the end-stage of chronic hepatitis and is a diffuse process of the liver characterized histologically by fibrous septa, shunting of afferent and efferent vessels, and conversion of normal liver architecture into abnormally structured parenchymal nodules. As in chronic hepatitis it is essential to include in the diagnosis the extent of the fibrosis, the activity of the disease and the possible etiology. A unified scheme of classification for fibrosis is still needed, as noted above, but Liver failure is likely to ensue so accurate classification is important. In general, fibrosis is more severe at the periphery of the hepatic lobes so laparascopic biopsies that are taken only from the free margins of the liver may lead to an over estimation of the degree of liver injury.

Cirrhosis is a relatively common condition in dogs often associated with the presence of multiple portosystemic collaterals. Some animals may have compensated cirrhotic disease and show no or minor clinical signs, other animals show decompensation of the liver function and liver failure. Cirrhosis is only rarely seen in cats and, in my experience, as a result of congenital hepatic fibrosis.

Non-specific reactive hepatitis

This is not an ideal morphologic diagnosis as it can be confusing, but nonetheless it is useful term. Non-specific reactive hepatitis is an inflammatory response of the liver to a variety of extrahepatic disease processes, especially febrile illnesses and inflammation somewhere in the splanchnic bed, or is the residuum of previous inflammatory intrahepatic disease. The lesion is characterized by an inflammatory infiltrate in portal areas and in the parenchyma without evident hepatocellular necrosis. In acute extrahepatic diseases there is a slight to moderate infiltrate in the parenchyma or the portal areas composed of neutrophils mainly. The infiltrate varies in intensity between portal areas and some normal portal areas may be found. There is, slight to marked leukocytosis and Kupffer cell proliferation in the sinusoids and some neutrophils in the connective tissue around the hepatic veins. In chronic extrahepatic diseases or in case of residual intrahepatic disease the inflammation is usually mononuclear with plasma cells and lymphocytes and pigmented macrophages in the portal areas and around the hepatic veins as well as some plasma cells and lymphocytes with or without single or small aggregates of pigment laden macrophages in the parenchyma. The key point is that this is a reaction of the liver to injury elsewhere and therapy should be directed at the cause of insult, not the liver, in same way that you do not treat the lymph nodes when you have the flu.

Fig. 5. Nonspecific reactive hepatitis: Note inflammatory cells limited to portal tract.

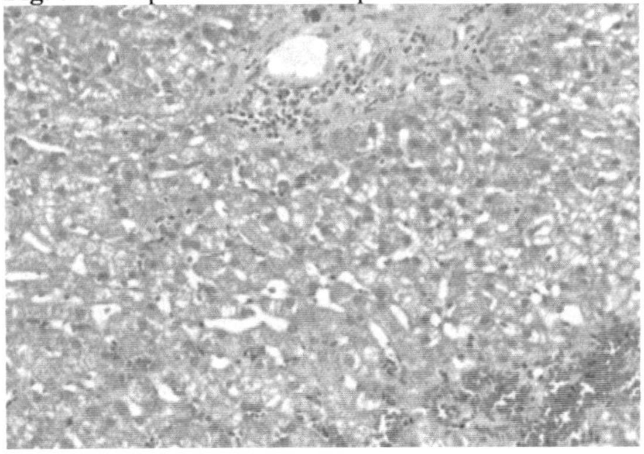

IS THERE "TOXIC HEPATITIS?"

There are many hepatotoxins that affect dogs and cats. The most common feature is injury or death of hepatocytes or biliary epithelium. Manifestations of liver toxicity, one or more of which may occur with each toxin, include: no morphologic abnormalities, hepatocellular swelling, lipidosis, necrosis (usually in a specific pattern), inflammation, and eventually fibrosis if exposure is long term. Most acute drug toxicity produces lesions in the centrilobular region. As part of the host response to tissue necrosis there is an influx of cells-macrophages and others. Pathologists are trained to assign "-itis" to any lesion with inflammatory cells, but in this case it is prudent for the pathologist to "read through" the inflammatory cells send to clean up debris and acknowledge that they are a secondary event in a primary toxic insult. This should be clearly communicated to the clinician.

REFERENCES

Anthony PP, et al. The Morphology of Cirrhosis. Recommendations on Definition, Nomenclature, and Classification by a Working Group Sponsored by the World Health Organization. J Clin Pathol 1978;31:395-414. Armstrong PJ, Blanchard G. Hepatic lipidosis in cats. Vet Clin North Am Small Anim Pract 2009;39:599-616 Boomkens SY, et al. Hepatitis with special reference to dogs. A review on the pathogenesis and infectious etiologies, including unpublished results of recent own studies. Vet Q 2004;26:107–114. Center SA, et al. Digital image analysis of rhodanine-stained liver biopsy specimens for calculation of hepatic copper concentrations in dogs. Am J Vet Res 2013;74:1474-1480) Cullen, J.M. Brown, D. L., Liver, Biliary System and Exocrine Pancreas. In: Pathologic Basis of Veterinary Disease (Chapter 8), pp. 405- 458, 2012. Goodman ZG. Grading and staging systems for inflammation and fibrosis in chronic liver diseases. J Hepatol 2007;47:598-607. Hoffman, G. van den Ingh, TS, Bode, P, and Rothuizen, J., Copper-associated chronic hepatitis in Labrador retrievers. JVIM 20:856-861, 2006. Holt MP, et al. Identification and characterization of infiltrating macrophages in acetaminophen-induced liver injury. J Leukoc Biol 2008;84:1410-1421.Ishak K, et al. Histological grading and staging of chronic hepatitis. J Hepatol 1995;22:696–699Mandigers, PJJ, van den Ingh, TS, Bode, P, and Rothuizen, J., Improvement in liver pathology after 4 months of D-penicillamine in 5 Doberman pinschers with subclinical hepatitis, JVIM 19:40-43, 2005. Rothuizen, J., Buch, S., Charles, J., Cullen, J., Desmet, V., Szatmari, V., Twedt, D., van den Inge, T., Van Winkle, T., Washabau, R., Standards for Clinical and Histological Diagnosis of Canine and Feline Liver Diseases (WSAVA). Saunders Elsevier, 2006. Shih, JL, Keating, JH, Freeman, LM, and Webster, CR, Chronic hepatitis in Labrador retrievers: clinical presentation and prognostic factors. JVIM 21:33-39, 2007.Spee, B, Arends, B, van den Ingh, TS, et al, Copper metabolism and oxidative stress in chronic inflammatory and cholestatic liver diseases in dogs. JVIM 20:1085-1092, 2006.

WHAT TO EXPECT FROM A LIVER BIOPSY: AN INTERNIST'S PERSPECTIVE

Sharon A. Center, DVM, DACVIM
Ithaca, NY

INTRODUCTION

Definitive diagnosis of hepatobiliary disease usually requires liver biopsy. Sampling methods include: fine-needle aspiration (cytology), core needle-biopsy (cutting versus non-cutting; manual vs spring-triggered automated), laparoscopic and surgical collections. Several authors have documented inaccuracy of small needle biopsies (18 G), single samples, and aspiration cytology. Aspiration samples and tiny needle biopsies can confirm severe feline hepatic lipidosis glycogen type vacuolation (dogs), certain forms of neoplasia, and sometimes, septic inflammation. The quality of automated needle core sample varies with operator, hepatic disease distribution, extent of fibrosis, and pathologist expertise. Best 14-16 G reviewed on our pathology service are those collected with a SpeedyBell semiautomatic guillotine needle, placed at full biopsy depth (US imaging) before being triggered. It is useful to acknowledge that a16 g needle biopsy represents ~1/50,000th of the liver; this may not accurately represent multifocal disease or the extent of fibrosis. In cases of fragmented needle samples the rendered diagnosis can become somewhat "creative".

HEPATIC LOBULAR / ACINAR STRUCTURE & LESION DISTRIBUTION

Accurate disease characterization requires samples disclosing acinar/lobular structure. Lobular structure is defined by circulatory distribution from portal triads (Zone 1), across the Intermediate zone (Zone 2), to Hepatic Venules/Veins (Zone 3, outflow to the vena cava). (Acinar Nomenclature: centrilobular regions = Periacinar location). Portal tracts are comprised of: *monads* (single vessel or ductule structure), *dyads* (any pair of vessels or vessel and ductule structure), and *triads* (~80% of portal tracts). Accurate biopsy interpretation requires a minimum of 12-15 complete portal *triads* which also assures adequate sampling of liver lobules. Samples from several different liver lobes are recommended to ensure adequate lesional sampling and lobular representation. This can be difficult with needle biopsy collection in small patients, especially cats where vasovagal responses may lead to respiratory arrest and severe bradycardia.

VASCULATURE

Because arteries, veins, & lymphatics can reflect disease, understanding their origin and structure is relevant.

Hepatic Arteries

~25% of hepatic blood flow, primarily perfuse the *Peribiliary Plexus* surrounding intrahepatic bile ducts.

Cystic Artery

A single branch from the proper hepatic artery perfuses the Gallbladder (GB); injury can result in GB necrosis.

Portal Veins

~ 65-75% of hepatic blood flow; provides essential splanchnic hepatotrophic substances. The first portal vein branch perfuses the caudate lobe, then distributes to the right and left liver lobes. Avoid biopsying the caudate lobe in a dog with suspected Portal Vascular Malformation- as this liver lobe likely will have the most normal portal venous perfusion for that patient.

Hepatic Venules

Surrounded by smooth spiral "Throttling Muscle" in dogs, but not cats. An arteriole vasa vasorum nourishes this muscle, supplied by branches from the Internal Thoracic and Phrenic arteries. Mast cells surround hepatic vein branches in dogs and may orchestrate contraction of the Throttling Muscle (smooth muscle surrounding the hepatic veins) in response to hypovolemic or antigenic shock. Inflammatory cells collect in centrilobular regions in dogs with splanchnic derived antigenemia / bacteremia.

Lymphatics

Portal, sublobular (centrilobular region surrounding hepatic venules), and superficial or subcapsular lymphatics function as the lymphatic drainage system for the liver and provide 25-50% of lymph flowing through the thoracic duct. Initial portal lymphatics have a discontinuous /absent basement membrane, are tubulosaccular, and have numerous valves directing unidirectional flow. Collecting lymphatics, located downstream, serve as drainage conduits; these have mural smooth muscle and valves. Of hepatic lymph (collected into the space of Disse) 80% or more drains into portal lymphatics, the remainder draining to sublobular and capsular lymphatics. An adventitial lymphatic plexus surrounds the sublobular veins, within loosely woven collagen fibers enshrouding the vein. This anatomic orientation provides a pre-lymphatic pathway that can expand with increased hepatic lymph accumulation. Sublobular lymphatics drain to lymphatics running along the inferior vena cava.

ZONAL LOCALIZATION

The essential first step in characterizing hepatobiliary disorders. This cannot be discerned from aspirate /imprint cytology or tiny tru-cut needle biopsies (esp. if fragmented). Lesions involving bile ducts / ductules may include: hyperplasia, dysplasia, necrosis, circumferential fibrosis (reflecting mechanical bile flow stasis or periductal inflammation), intraluminal infiltrates, transmigrating inflammatory cells (crossing duct epithelium), and epitheliotropic lymphocyte targeting of ductal epithelium (e.g., destructive cholangitis / cholangiohepatitis [CCHS]). Parenchymal lesions are characterized as necroinflammatory (NI) or non-necroinflammatory (non-NI). If non-NI the process should be further assessed for: infiltrative [neoplasia, amyloid], vacuolar [lipid or glycogen], degenerative [hydropic degeneration, vacuolar, lipid), or regenerative changes. "Reactive Hepatitis", a catch-all term proposed to represent nonspecific low grade hepatic parenchymal injury (small random lipogranulomas or inflammatory foci) may be misleading or confusing. These changes may be mistakenly described as hepatitis in small biopsy samples or liver aspirate cytology. Reading the morphologic biopsy report is more informative than use of this catch al term. Severe centrilobular injury is often described as "portal-to-portal bridging" or as" portal focused injury". Instead, portal-to-portal bridging may merely reflect that the only remaining identifiable structures are portal tracts / regions (bile duct remnants).

IMPORTANT FEATURES

Adequate biopsies allow inspection for: 1) processes targeting biliary epithelium, 2) assessment of bile ducts to RO destructive cholangitis and ductopenia, 3) detection of bile flow stasis: dilated ± tortuous bile ducts, intraluminal biliary debris (mucinous, cellular, bacterial, mineral)- canalicular cholestasis is relatively uncommon and may when observed may reflect sepsis (compromised organic anion pumps), 4) identification / characterization of portal inflammatory infiltrates, 5) "limiting plate" disruption (interface between portal elements and periportal hepatocytes) by inflammatory infiltrates ± necrotic hepatocytes (Portal Hepatitis), 6) processes impacting perfusion: portal veins, hepatic arteries, central veins, lymphatics, and 7) regional remodelling (zones 1,2, or 3 or random), regenerative of dysplastic nodule formation, fibrosis, or bridging collapse. Adequate liver biopsy samples also allow 8) inspection of centrilobular regions including small branches of hepatic venules and larger sublobular veins. These regions also should be evaluated for perivenular: 9) inflammation, 10) accumulation of pigmented macrophages (copper, hemosiderin, lipofuscin), 11) hepatocellular cytosolic pigment granules (copper, lipofuscin, iron), 12) hepatocellular necrosis or necrotizing parenchymal collapse, 13) sinusoidal circulatory obstruction (Sinusoidal Occlusion Syndrome, SOS) or congestion, 14) centrilobular lymphatic distention, 15) venule thrombi, and 16) centrilobular fibrosis.

OBSERVATIONS ASSISTING DEFINITIVE DIAGNOSES & UNDERSTANDING LESION PATHOPHYSIOLOGY

Lipogranulomas

Single or coalesced foamy macrophages containing membranous debris; these reflect antecedent regional injury and often contain lipofuscin pigment or hemosiderin. Lipofuscin implicates prior oxidative "stress".

Lipofuscin Pigment

Often referred to as 'aging pigment' represents oxidized protein and lipid moieties, phagocytized by macrophages or spontaneously formed in centrilobular hepatocytes. This pigment is routinely observed in centrilobular hepatocytes in domestic cats ≥ 7 yrs of age and can be appreciated on cytologic inspection of hepatocytes collected by aspiration sampling. Diffuse hepatocellular distribution of lipofuscin, its retention in macrophages mantling damaged parenchyma, and lipofuscin associated with copper-laden hepatocytes reflects pathologic processes.

"Copper Granulomas"

Unique lipogranulomas integrated with lymphocytes, pigmented macrophages, variable neutrophils and occasional plasma cells, with intralesional eosinophilic refractile copper-protein aggregates (H&E staining) in macrophages. Adjacent hepatocytes typically display copper-protein cytosolic aggregates. A copper specific stain such as Rhodanine is needed for definitive confirmation.

Immune-Mediated Nonsuppurative Inflammation

Association of lymphocytes with single necrotic hepatocytes (satellitosis), sinusoidal lymphocytosis or variable zonal aggregates <u>with</u> evidence of zonal injury confirmed with special stains.

Lymphatic Distention / Engorgement

Acquired Sinusoidal Hypertension: this develops secondary to architectural remodelling (sinusoidal collagenization in the space of Disse), and increases ultralymph formation leading to lymphatic distention (portal, sublobular veins, subcapsular lymphatics. Remodeling can substantially increase ultralymph formation, for example, lymph flow in a cirrhotic liver increases 30-fold, correlating with acquired sinusoidal hypertension.

Sinusoidal Occlusion Syndrome (SOS): centrilobular congestion & increased ultralymph formation caused by collapse of centrilobular sinusoids. Develops secondary to zone 3 injury that may reflect centrilobular p450 cytochrome activated drug adduct or toxin formation (e.g., NSAID, CCNU, alkaloids in holistic/naturopathic/herbal remedies) or severe hypoxia.

Increased Arterial Flow: dominant arterial flow in portal venous hypoperfusion →HABR (see below) increases ultralymph.

Hepatic Venule Outflow Obstruction: thrombus (*Budd Chiari Syndrome*), pericardial tamponade increases ultralymph formation.

Complete Thoracic Duct Ligation / Obstruction In Cat: increases ultralymph accumulation by restricting extrahepatic drainage.

Perioperative Fluid Administration: can expand portal, sublobular, subcapsular lymphatics by increasing ultralymph formation.

Portal Venous Hypoperfusion

Anatomic features of reduced portal venous perfusion include: changes reflecting *Lobular Atrophy* (diminished exposure to splanchnic derived hepatotrophic substances) and the *Compensatory Hepatic Arterial Buffer Response* (*HABR*). *Features of Lobular Atrophy:* 1) small hepatocytes, 2) increased binucleated hepatocytes, 3) miniaturized portal tracts, 4) disorganized hepatic cords (random ± widened sinusoids], 5) abnormally close apposition of portal tracts and centrilobular regions, and 6) microhepatia. *HABR* increases flow in the *Peribiliary Arterial Plexus*, the dominant recipient of hepatic arterial flow. This vascular plexus is closely intertwined with portal bile ductules- and can appear "tangled" with these structures. In response to increased flow & pressure the small arterials 1) coil and develop 2) neovascular twigs causing 3) increased cross and tangential sections of arterials in portal tracts, with 4) thick muscular walls, 5) development of "orphaned" arterials in hepatic parenchyma distant to other portal tract elements, and 6) increased ultralymph formation (engorged portal and sublobular lymphatics ± subcapsular lymphatics. Dogs with congenital *Portal Venous Hypoperfusion* also may develop lipogranulomas distributed: randomly, within portal tracts, and /or shrouding hepatic venules. In severe cases, hepatic venular compression obstructs circulatory outflow causing a *Veno-occlusive Effect (*outflow obstruction). In dogs with microvascular dysplasia (MVD, microscopic malformation of tertiary portal vein branches), this change may provoke acquired portosystemic shunt (APSS) formation. Finding a "*Fusion Complex*" – conjoined hepatic venule and portal tract usually reflects a congenital vascular malformation. <u>The cause of *Portal Venous Hypoperfusion* cannot usually be discerned only from a liver biopsy-</u> it cannot differentiate PSVA from MVD. This requires CT or radiographic angiography, US bubble study, or isotope imaging or consideration of Protein C activity. *Portal Venous Hypoperfusion* also may reflect acquired changes to splanchnic or hilar portal venous vasculature. Notably, in large breed dogs with visceral abnormalities impacting on splanchnic "portal" circulation (e.g., large splenic mass lesions, splenic torsions, gastric dilatation / volvulus, chronic GI foreign bodies, GI neoplasia, severe IBD leading to portal vasculitis) or any dog with a portal thrombus. The Mass Effect on splanchnic perfusion reflects physiologic response to acquired reduction in *Hepatopetal Circulation*. The portal vein is a low pressure system (5-8 mm Hg) lacking valves that can be considerably impacted by local compressive /occlusive influences. Any lesion impinging or obstructing portal circulation can initiate HABR and subsequent histologic changes of *Portal Venous Hypoperfusion.*

Hepatocellular Vacuolation

Should be characterized as a *Glycogen-type Vacuolar Hepatopathy (VH)*, *Ballooning Degeneration*, *Hydropic Degeneration*, or *Lipid Vacuolation (LV,* microvesicular, macrovesicular, or mixed). *Microvesicular LV* may reflect mitochondrial injury (dysfunctional electron transport chain or fatty acid β-oxidation). If diffuse and associated with necrosis, this may reflect hepatotoxin exposure (e.g., Aflatoxin). Small focal aggregates of hepatocytes with microvesicular *LV* are common in inappetent cats (this does not predict eminent Feline Hepatic Lipidosis) and hyperlipidemic dogs (e.g., Min. Schnauzers, Shelties). Glycogen-type VH and LV are easily differentiated on well fixed, thin liver sections stained with H&E. *Glycogen-type VH* : small homogenous pink-glassy cytosolic "pools" are observed in hepatocytes at the biopsy edge (faster fixation stops glycogen metabolism). In the center of the biopsy, Glycogen vacuoles appear "wispy" without definite margins and cause a "rarefied" cytoplasmic appearance. Affected cells are distended (up to 10-12 fold normal) causing hepatomegaly and liver pallor when diffuse and severe. *Glycogen–type VH* may reflect systemic or splanchnic inflammatory processes, systemic neoplasia, as well as Adrenal Hyperplasia Syndromes. Association of *VH* with "*Dysplastic"* hepatocellular foci raises concern regarding risk for hepatocellular carcinoma and Adrenal Hyperplasia involving sex hormones, as we described in Scottish Terriers (also observed in many other dog breeds, unpublished information, SAC).

Cholecystectomy for GB Disease / Obstruction

GB changes cannot be assessed unless the GB is promptly sectioned before formalin fixation. Immediately after cholecystectomy, samples of bile (if GB mucocele [GBM] biliary amalgam) and mucosal epithelium (scrape with scalpel) should be cytologically investigated for inflammatory infiltrates and bacteria (up to 32%GBM have culturable bacteria). Particulate material in bile has highest yield for bacterial isolation/identification. GB and liver biopsies (from two or more liver lobes) should be histologically evaluated. Liver samples are investigated for typical GBM related *Glycogen-type VH* and short term incomplete mechanical cholestasis (dilated, tortuous, medium sized bile ducts with intraluminal mucinous debris and minimal periductal inflammation) or longer term mechanical obstruction (> 5 days: periductal inflammation [neutrophils, macrophages, fewer lymphocytes, with variable periductal fibrosis). Increasing chronicity of GB obstruction leads to advanced changes: thick laminating periductal fibrosis, marked duct tortuosity, periductal edema with mixed inflammatory infiltrates (esp. neutrophils).

Cholangiohepatitis also may be observed. Histologic features of a "typical GBM" include cystic mucosal hyperplasia; intraluminal laminated biliary conglomerate, and variable wall inflammation depending on duration and tautness of GB distension (there may be no inflammation, so no cholecystitis). Chronic GBM associated with massive GB distention leads to wall thinning ± mucosal ulcerations, ± ischemic necrosis (sometimes associated with overt vascular thrombi). Chronic inflammation may lead to wall thickening (fibrosis and granulation response). Discovered choleliths should be crushed and cultured. Tan choleliths reflect bilirubinate / glycoprotein aggregates, brown reflect bilirubinate-calcium carbonate and are more likely infection stones, and black stone reflect hemobilia antecedent to cholecystectomy or chronic hemolysis/heme turnover.

Feline Destructive / Sclerosing CCHS

Multiple lobes should be biopsied (n=3 is best) because disease activity varies among liver lobes in this disorder. Destructive cholangitis is characterized by: 1) lymphocytic inflammatory infiltrates (67% = T cells), 2) epitheliotropic lymphocyte targeting of ductal epithelium, 3) transmural lymphocyte duct invasion, 4) small- or medium-sized bile duct involution, 5) peripheralization of bile ducts to margins of the portal tract, 6) overt ductopenia, 7) portal tract lipogranulomas marking sites of duct involution (loss), 8) lymphoid follicle formation, and 9) periportal copper accumulation.

Extramedullary Hematopoiesis (EMH)

Commonly associated with NI liver injury, regenerative responses, *Glycogen-type VH*, blood loss anemia, and within extinct parenchyma. EMH reflects differentiation of hepatic pluripotential stem cells to hematopoietic lineage. Stimulation may involve inflammatory cytokines or induction phenomenon and regional hypoxia. Aspiration cytology of liver with EMH aggregates that are primarily granulocytic- may lead to erroneous impression of suppurative inflammation / hepatitis.

Ductal Plate Malformations (DPM)

Congenital malformations involving structure /function of primary cilia interfering with tubulogenesis and bile duct embryogenesis. Affected dogs and cats are typically anicteric and presented for increased liver enzyme activity. Enzyme activity is attributable to variable cholangitis (often bacterial) and sometimes pathologic copper accumulation (esp. dogs). Severe DPM phenotypes display portal tract expansion with malformed biliary ductules and increased extracellular matrix (ECM, type 1 collagen) that bridge portal regions, causing *Presinusoidal Portal Hypertension* and formation of APSS. This severe phenotype is categorized as *Congenital Hepatic Fibrosis*. DPM phenotypes differ markedly but most may be confused with cholangitis / cholangiohepatitis. They differ from acquired lesions in that the ECM deposition is more severe, ductule malformations are unique in severity and morphology, and the presence of a signature histologic feature: direct juxtaposition of biliary ductules and hepatocytes. Some DPM dogs lack a GB, have vascular malformations (portosystemic shunts), and atrophied or atretic left liver lobes.

Nodular Hyperplasia (NH), Dysplastic Foci, Hepatocellular Carcinoma(HCCA) / Hepatic Adenoma (HA)

NH are non-encapsulated, usually $\leq$ 2 cm expansile lesions of hyperplastic hepatocytes. Hepatic cord width is usually normal (1 cell wide) but nodules have a modified lobular architecture with central and portal elements irregularly spaced, and normal reticulin substructure. NH lesions lack marginal parenchymal collapse and fibrosis and thus differ from regenerative nodules. They also have smooth margins and a phenotype that blends with surrounding hepatocytes or may be partially vacuolated (*VH or LV*). NH may associate with increased ALP and are sometimes are detected during US imaging or grossly during exploratory surgery or laparoscopy. Biopsy of adjacent normal tissue helps distinguish NH which may be confused with HCCA/HA if only small needle cores or derm punch biopsies are collected, or when only nodular tissue is sampled. *Dysplastic foci* (*DF*) are potentially pre-neoplastic proliferative hepatocellular foci observed in dogs with *Glycogen-type VH*. These are non-encapsulated, non-compressive, variably sized, expansile nodules of non-vacuolated hepatocytes, with reduced reticulin substructure, and wide (2-cells wide, normal = 1 cell width) disorganized hepatic cords. *DF* display irregular (serrated) margins that interface with adjacent "normal *VH*" affected hepatocytes. These also lack remodeled marginal lesions (fibrosis, parenchymal collapse). Because *DF* associates with canine *Glycogen-type VH*, increased ALP activity is common. *DF* are recognized in unaffected liver lobes in dogs with HCCA (e.g., Scottish

Terriers, also other breeds) and we have predicted HCCA formation in some dogs based on recognition of *DF* lesions. Available data suggest that *DF* associate with increased sex hormones (androgens, progestins). Variable US appearance depends on DF size, number, and distribution. *DF* may be mistaken for nodular regeneration or NH without special stains (reticulin for substructural characterization, Masson's Trichrome characterizing absence of perilesional collagen). *HCCA* is far more common than *HA* in dogs and is a rare diagnosis in the cat. Canine HCCA have numerous phenotypes (several often observed in a single tumor/patient), and are variable in size (may be small, massive, infiltrate multiple lobes, or be multifocal). Mass lesions are compressive causing mechanical cholestasis, impaired perfusion, and necrosis in adjacent non-neoplastic liver. Fed by branches of the hepatic artery, tumor rupture can provoke catastrophic abdominal hemorrhage. Rapid tumor growth can lead to development of necrotic foci that become secondarily infected by enteric opportunists leading to hepatic abscessation (may be emphysematous). Thick hepatic cords (up to 12 cells across) are markedly disorganized (chaotic), lack reticulin substructure, and may display macro-canaliculi. Numerous sections may be needed for diagnosis. Biopsy of non-involved liver is important to investigate for *Glycogen-type VH* and *DF* implicating adrenal hyperplasia syndromes.

Copper Associated Hepatopathy (CuAH)

By far the most common cause of acquired NI liver disease in dogs; now also characterized in cats. Referred to as "*Cu Hepatopathy, CuAH*" because injury may be subtle. Carprofen associated liver injury in Labrador Retrievers is often augmented by co-existent *CuAH* – this combination can be lethal owing to augmented centrilobular necrosis/ collapse. Once *CuAH* is recognized, Cu must be removed (chelated), antioxidants administered and lifelong dietary Cu restriction mandated. Thin H&E stained liver sections usually disclose refractile eosinophilic cytosolic granules in centrilobular hepatocytes and classic "Copper Granulomas" in random or centrilobular distribution. Rhodanine stained sections are mandated to confirm suspected *CuAH.* Cu quantification is now completed using Digitally Scanned Rhodanine Stained Liver Sections; this method has greater accuracy than bench Cu analysis because it considers the entire biopsy and circumvents naïve sampling of non-representative tissue (regenerative nodules, fibrotic regions, extinct parenchyma where Cu is low). Digital scanning for Cu quantification is performed at Cornell (additional minor fee for scanning and technical support) with data reported on a dry weight µg/gm basis. Analysis is offered for samples that have been histologically evaluated and stained by our diagnostic laboratory.

Iron Distribution & Pathologic Significance

Mesenchymal iron (Fe, in macrophages) accumulates where cells have been regionally damaged (heme from cytochromes), secondary to heme release elsewhere (hemolysis, repeated blood transfusion, body cavity or enteric hemorrhage), in response to systemic iron loading (iatrogenic supplements), or because inflammatory cytokines provoke macrophage iron uptake and retention.. Similar to humans, mesenchymal Fe accumulation (hemosiderosis) is common in chronic NI liver disease. Hemosiderosis increases risk for oxidative injury because stored Fe remains in flux with the redox reactive free Fe pool. Like Cu, Fe functions as a transition metal catalyzing the Haber-Weiss and Fenton reactions which generate injurious superoxide and hydroxyl radicals. Fe also activates cells involved with hepatic inflammation and fibrogenesis (e.g., Kupffer cells, Stellate or Ito cells that transform into sinusoidal myofibrocytes). The hepatotoxic and pro-fibrogenic potential of Fe can be dramatic at moderate levels of tissue metal overload when other inciting insults are encountered (e.g., severe acute anemia or splanchnic or systemic hypotension). Thus, finding large amounts of stainable Fe suggests need for chronic antioxidant support and investigation of causal factors that might be remediated. CuAH often associates with increased mesenchymal Fe stores which can be mobilized during the chelation process (proven by post treatment biopsy assessment). In hemochromatosis (hepatocyte iron accumulation), a rare canine or feline diagnosis, intermittent phlebotomy or chronic specific iron chelation is needed. Hemosiderin is an insoluble complex of degraded ferritin and large ferric hydroxide chains that is easily recognized on routine H&E staining. Perls stain identifies membrane-bound lysosomal hemosiderin (Fe^{+2}) while assessment of Fe^{+2} & Fe^{+3} is done with Prussian blue.

MAXIMIZING DIAGNOSTIC YIELD FROM A LIVER BIOPSY

Cytology

Imprint cytology initially stained with Wrights-Giemsa (*Diff Quik®*), then *Gram* stain if bacteria are identified assists with initial antimicrobial selection. Collecting bile for culture and cytology is indicated if cholestatic or GB disease is suspected- aspiration of GB bile for culture in suspected major bile duct obstruction is strongly contraindicated unless done during surgical exploration or alternative temporary drainage of the GB is being orchestrated.

Cultures

Aerobic & anaerobic bacterial, ± fungal (fungal if known pyogranulomatous inflammation) should be routinely submitted. Combining bile & tissue samples reduces expense; culture with dual sample collection increases the chance of identifying disease related bacteria. If aspiration samples only are possible, consider inoculating blood culture media to increase microbial yield.

Formalin Fixed Tissue & Cornell Special Liver Staining Panel

For more than a decade a panel of stains additional to routine H&E staining have been used at Cornell. This has remarkably augmented comprehensive histopathologic evaluation of liver biopsies. After initial examination of H&E stained sections, liver panel stains improve diagnostic accuracy by objectifying characterization of lobular atrophy, parenchymal collapse, lobular remodeling, fibrosis, detection of regenerative nodules, and unequivocal identification of the primary zonal or lobular focus of the disease process.

Reticulin Stain: reticulin fibrils (fine fibrils of Type III collagen & other ECM components) comprise hepatic substructure. Reticulin staining discloses hepatic cord organization and continuity (disorganization, parenchymal collapse) and bridging that may be too subtle for H&E detection. Reticulin staining also helps identify regenerative nodules, *NH, DF*, and *HCCA*. It clarifies lobular atrophy versus stromal collapse (collapse being associated with condensation of reticulin fibers) and ascertains without doubt the zonal location of regional hepatocellular injury (e.g., centrilobular vs portal vs random).

Masson's Trichrome Stain: stains fibrillar collagen (Type 1) blue and smooth muscle magenta. Simultaneous evaluation of Masson's Trichrome and Reticulin objectifies fibrosis and remodeling. Without these stains fibrosis may be erroneously deduced and regenerative nodules overlooked. Together we also use these stains to confirm DF in dogs with VH. Using digital scans of MT stained liver in dogs with DPM we also have estimated portal tract collagen deposition. Masson's trichrome also assists in the localization of hepatic venules in dogs with suspected SOS or veno-occlusive lesions.

Prussian Blue Stain: Stains iron blue (both Fe^{+2} & Fe^{+3}) differentiating brown/tan cytoplasmic inclusions observed in H&E stained macrophages and hepatocytes from ceroid-lipofuscin lipopigment.

Rhodanine Stain: cytosolic Cu-protein aggregates stain orange; can apply to hepatic aspirates (cytology specimens) in dogs in crisis where CuAH suspected. However, not recommended mode of CuAH diagnosis. Interpretation complicated by lack of acinar distribution; samples require estimation of % hepatocytes sampled with Cu^+ granules. Rhodanine staining is the basis for Digital Scanning Quantification of Hepatic Cu Concentration. Cu does not accumulate in dogs secondary to cholestasis as it does in humans and cats.

SPECIAL STAINS- UNIQUE LESIONS, FEATURES & CIRCUMSTANCES

Periodic Acid Schiff (PAS)

With and without diastase (amylase) can confirm storage of cytosolic glycoproteins and glycogen in hepatocytes. Uptake of PAS by ceroid-lipofuscin lipopigment does not resolve with diastase (glycogen is digested) which can be used to confirm lipofuscin accumulation.

Oil Red O Stain

Stains neutral fat globules orange but is ONLY USED ON frozen sections or tissue NOT embedded in paraffin.

Congo Red

Stains amyloid, viewed under polarized light for apple green birefringent. Amyloid may also stain with Masson's Trichrome but in some cases, remains unrecognized without Congo Red. Amyloid accumulates in the space of Disse and may not be notable unless suspected (unexplained hepatic lobe fracture in cats, sinusoidal atrophy).

Halls Stain

Stains bilirubin confirming canalicular bile stasis, macrophage phagocytosis of bile, and bile "lakes", has low utility.

Toludine Blue or Giemsa

Used to confirm mast cells sometimes involved with zone 3 inflammation in dogs with IBD/

Stains for Pyogranulomatous Hepatitis Lesions

Infectious Agent Panel- Gram Stain (gram + & gram - bacteria; Ziehl Neelsen (acid fast bacteria: esp. mycobacteria (TB group and atypical groups), Fite Faraco (acid fast, a more "gentle" acid fast stain), PAS (fungi), GMS (fungi). **PCR** for mycobacteria is sent to College of Veterinary Medicine, Washington Animal Disease Diagnostic Laboratory- requesting testing for typical and atypical mycobacteria.

Special Immunohistochemistry: not all inclusive

a. lymphocyte panels for feline CCHS, less commonly used for canine Chronic Hepatitis (CD3-T cell, BLA36- B cell, others)
b. cytokeratin (CK19) marks biliary epithelium when ductopenia suspected (esp. useful in feline sclerosing/destructive CCHS)
c. α-smooth muscle actin (α-SMA): myofibrocyte spindle cells (useful in suspected Lobular Dissecting Hepatitis)
d. a_1-antitrypsin: positive PAS + globules, unique in some Cocker Spaniel degenerative vacuolar hepatopathy
e. oxidative injury markers: Hydroxynonenal (4-HNE), Nitrotyrosine (NT), 8-OH-Guanosine (8-OH-dG)
g. neoplasia: designation of cell type (e.g. carcinoma vs mesenchymal vs neuroendocrine vs histiocytic)
i. IHC or other Molecular Testing: specific for infectious agents: e.g. toxoplasmosis, Leishmania, many others.
j. Lipoteichoic Acid: gram "+" cell wall IHC
l. Pathogen Associated Molecular Pattern Expression: Toll-like receptor 4

FISH

Fluorescent in situ hybridization

Universal eubacterial primer may detect small numbers of organisms undetected by histopathologic visualization or special stains but cannot speciate (identify organism) without proposed genus. Sections are cut from paraffin embedded formalin fixed tissue. Special treatment is need for detection of mycobacterial-type bacteria.

REFERENCES

Arey LB, et al: The relation of the smooth muscle in the hepatic veins to shock phenomena. Anat Rec 1920;18:219. Center SA, et al: Digital image analysis of rhodanine-stained liver biopsy specimens for calculation of hepatic copper concentrations in dogs. AmJVet Res 2013;74:1474-80. Cole TL, et al: Diagnostic comparison of needle and wedge biopsy specimens of the liver in dogs and cats. J Am Vet Med Assoc. 2002;15;220:1483-90. Cortright CC, et al: Clinical features of progressive vacuolar hepatopathy in Scottish Terriers with and without hepatocellular carcinoma:114 cases (1980-2013). J Am Vet Med Assoc. 2014;245:797-808. Kemp SD, et al: A comparison of liver sampling techniques in dogs. J Vet Intern Med. 2015;29:51-7. Kemp SD: Histopathologic variation between liver lobes in dogs. JVetInternMed. 2015;29:58-62. Lefkowitch JH: Special stains in diagnostic liver pathology. Sem Diag Pathol 2006;23:190-8. Ohtani O, et al. Lymph circulation in the liver. Anat Rec 2008;291:643-52. Tajiri S: The terminal distribution of the hepatic artery. Acta Med Okayama 1960;14:215–25. Takahashi-Iwanaga H, et al: Canine hepatic vein branches associated with subendothelial mast cells and an adventitial lymphatic plexus. Arch Histol Cytol 1990;53 Suppl:189-97.Wang KY, et al: Accuracy of ultrasound-guided fine-needle aspiration of the liver and cytologic findings in dogs and cats: 97 cases (1990-2000). J Am Vet Med Assoc. 2004;224:75-78. Warren A, et al: Histopathologic features, immunophenotyping, clonality, and eubacterial fluorescence in situ hybridization in cats with lymphocytic cholangitis/cholangiohepatitis. Vet Pathol 2011;48:627-41

CLUES FROM APPEARANCE AND POSTURE

Rodney S. Bagley, DVM, DACVIM (Neurology and Small Animal Internal Medicine)
Ames, IA

Through examination and assessment of functions controlled by the nervous system, clues to the presence and location of disease within the nervous system are found. As the nervous system is a part of the entire body, the neurologic examination is an extension of a complete physical examination. It is important to perform a complete physical examination in addition to any specialty assessment to avoid overlooking important abnormalities that might ultimately lead to the clarification of the disease process.

Observing the animal is paramount in determining the presence, location, and extent of any disease process. Patient examination identifies functional and anatomical abnormalities associated with disease. The significance of abnormalities identified by subsequent diagnostic testing can be determined only by an understanding of the associated clinical signs found in the animal. Many clinically observable neurologic abnormalities are indicative of dysfunction of specific functional components within the nervous system. Clinical signs resulting from dysfunction of the nervous system, however, are rarely pathognomonic for a specific disease process. Instead, clinical neurologic abnormalities reflect the location of involvement of the nervous system with disease. Regardless, a significant amount of information can be collected relatively quickly by simply observing how the animal is postured, reacts in the environment, moves, and walks. In some instances, observation of specific movements or postures is extremely helpful in determining where abnormalities exist within the nervous system.

Of critical importance is the actual "label" the observer places ("thinks") based on what they see, hear, feel, and in some cases smell. The label, for example, may be "head tilt" or some other medically learned description. This is an important step in clinical diagnosis as an inappropriate "label" will influence subsequent steps in the diagnostic process, and influence medial communication.

Information Gathered Through Observation of the Animal

What does the examiner observe about the animal when viewing the animal from a distance (a certain perspective)? Astute and accurate observation is one of the most important of all clinical skills during physical and neurologic assessments. The way an animal moves, acts, or looks often provides important clues to the location and type of clinical dysfunction. As nervous system dysfunction is often reflected in abnormalities of appearance, posture, and movement, it is helpful to allow the animal to move freely around the examination room while watching for movement traits and characteristics. An obvious exception to allowing the animal to move freely would apply to the animal that may have a traumatic vertebral fracture or subluxation or is dangerous to itself, the owner, or the health care team such as by being unduly aggressive.

As the veterinary hospital often contains many new and different sights and smells (environmental stimuli), most animals presented for clinical examination should be interpreted as "aware" in this environment. These animals should remain alert and orient themselves to obvious environmental visual, sound, smell, or touch stimuli. Owner questioning is often necessary to clarify the "personality" of the individual animal and obtain information concerning how the animal has acted in similar environments in the past. Owner comments such as, "He's much more "quiet" than he usually is" or "He's always this scared when he comes to the veterinarian" are often very helpful for making a subjective assessment of the animal's overall awareness of various environmental stimuli.

Abnormal postures and movement disorders are usually observed during this portion of the examination. Examples of these types of abnormalities include head tilting, head turning, falling, wandering, circling, limping, or vocalizing as if painful.

Mental Awareness, Consciousness, and Behavior

Clinical signs of alterations in consciousness and behavior range from subtle (e.g., an animal that is just less active than normal) to severe (e.g., comatose and unable to respond to any external stimuli). The "normal" state of an animal's consciousness is dynamic and is dependent on a complex interaction between stimuli and response, which varies from individual to individual and, in some cases, from moment to moment. Most importantly, it is virtually impossible for a clinician to determine what an animal "thinks". Clinical assessments of consciousness, therefore, are often crudely categorized into general groups of responses.

Behavior

Normal behavior is commonly determined for a population of animals, but is also dependent on individual patient characteristics. Alterations in behavior may include aggression; vocalization; excessive or abnormal sexual activity; changes in eating, chewing, drinking, urinating, defecating; as well as changes in personality. Excessive activity and repeated unusual behavior such as tail chasing or fly biting might also occur.

Abnormalities of Posture and Body Position at Rest

Posture is evaluated at rest and as the animal is placed in abnormal positions throughout the examination.

A *head tilt* is an abnormal posture wherein one ear is deviated more ventrally than the opposite. Head tilting is most often associated with disease of the vestibular or cerebellar systems. The head tilt is usually directed toward the side of the lesion in the nervous system especially if the lesion involves the peripheral vestibular apparatus. Occasionally, with lesions involving the caudal cerebellar peduncle between the brain stem and cerebellum or flocculonodular lobe of the cerebellum, the head tilt is directed away from the side of the lesion, the so-called paradoxical vestibular syndrome. This clinical sign is indicative of an "unconscious proprioceptive" problem.

A *head turn* is where the nose remains parallel to the horizon, however, the nose is directed toward the flank or caudal aspect of the body. This has sometimes been referred to as a "yaw". This posture may be a component of the adversive syndrome, which consists of a head turn (sagittal axis of head perpendicular to the ground) and circling. The animal may also intermittently orient the head laterally or toward the caudal aspect of the body. A head turn occurs most often with unilateral supratentorial lesions. The head turn and circling are usually toward the side of the lesion. A similar abnormality, termed "hemi-inattention" or "hemi-neglect," may be seen wherein the animal will only eat from one side of its food bowl. In some situations of lesions of the (caudal) brain stem or cervical region, the head may be turned to be positioned tightly against the neck or trunk.

Abnormalities of Limb Posture

Animals with neurologic dysfunction may stand with their limbs more laterally positioned than normal. This is often referred to as a *wide-based stance.* This appearance is most often associated with an abnormality of proprioceptive function. Lesions of the nervous system commonly associated with a wide-based stance include the vestibular system, cerebellum, and spinal cord. This posture may be an attempt by the animal to prevent itself from falling to one side. Conversely, animals that stand with the limbs very close together and directly under the trunk may have abnormalities of muscle strength, such as with disease of the peripheral nerves, neuromuscular junction, or muscles themselves.

Spasticity is increased tone of the muscles of the limbs. Opisthotonus is a body posture where the neck is held in extension with the top of the head directed toward the mid-back region. In a recumbent animal, spasticity and/or opisthotonus can be seen in numerous situations. These include decerebrate rigidity, decerebellate rigidity, and Schiff-Sherrington syndrome.

Decerebrate rigidity is characterized by opisthotonus and extension of all limbs. Usually the animal is stuporous or comatose. This is due to loss of descending input (from the supratentorial structures) to the brain stem structure responsible for flexion and extension of the limbs. Lesions that interrupt the descending supratentorial influences the brain stem will more predominantly alter the function of the flexor systems compared to the extensor systems. The extensor systems essentially act unopposed to bring about extension of the limbs. Lesions of the (mesencephalon) midbrain result in this postural abnormality.

Lesions of the cerebellum may result in a similar posture. *Decerebellate rigidity,* is characterized by opisthotonus with thoracic limb extension, flexion of the pelvic limbs up under the body due to contraction of the sublumbar muscles. If the lesion involves the ventral aspects of the cerebellum, extensor rigidity of all limbs is seen. Importantly, as the lesion associated with this posture does not affect consciousness, the animal with decerebellate rigidity has normal mental awareness and consciousness. This clinical feature serves to differentiate decerebellate from decerebrate rigidity. Affected animals can assume decerebellate posture episodically resulting in some referring to this as a "cerebellar seizure." This is not a true seizure disorder but rather an episodic postural abnormality.

Schiff-Sherrington posture is characterized by thoracic limb extension and inability to move the pelvic limbs normally. Spinal reflexes in the pelvic limb are usually normal. This results from a lesion in the thoracolumbar spinal segments that interrupts the ascending inhibitory impulses originating in the lumbar grey matter and terminating on cells responsible for extension of the thoracic limbs.

Severe extensor rigidity of the limbs and other muscles is characteristic of *tetanus* . While tetanus usually affects the entire body, localized tetanus involving one limb is occasionally noted. Episodic extensor rigidity of the limbs may be associated with a variety of abnormalities of the central or peripheral nervous systems and muscles. Episodic muscle contraction resulting in a stiff appearance, albeit intermittently, can result in disorders such as *myotonia* (failure of muscle relaxation), *reflex myoclonus, seizure,* and metabolic abnormalities such as *hypoglycemia, electrolyte abnormalities*, and muscle *cramping* syndromes.

An intermittent or consistent flexion of a limb may be associated with soft tissue injury of the limb, muscle fibrosis, and skeletal and joint abnormalities. When walking, this is commonly referred to as a (non-weight bearing) *lameness.* If the limb is not anatomically fused in this position, this posture most often is the result of pain somewhere or around the limb. If the pain originates from a spinal or peripheral nerve abnormality, this is termed a *"nerve root signature"*. Nerve root signature may involve either the thoracic or pelvic limbs.

Abnormalities of Body Trunk Posture

Scoliosis is an abnormal lateral deviation of the vertebral column. This may be due to a problem of the vertebrae, ribs, or associated muscles and ligaments. Abnormal muscle support due to denervation of the paraspinal musculature contributing to asymmetric lateral muscle tension and subsequent vertebral deviation. In dogs, scoliosis often occurs secondary to intramedullary spinal cord lesions such as syringomyelia or hydromyelia. *Kyphosis*, an abnormal flexion primarily of the thoracolumbar spine, may be a postural abnormality associated with pain, or may result from abnormalities of the vertebrae, ligaments, ribs, paraspinal muscles, and spinal and other peripheral nerves. Often, a kyphotic posture is the result of pain originating in the cervical area, thoracolumbar spine, or abdomen.

Lordosis is a ventral deviation of the spine and is uncommon in small animals. This may be a reflection of weakness or impairment of the paraspinal muscles. Occasionally, animals with thoracic vertebral or thoracic cavity disease will tend to flex the thoracic spine ventrally compared to the cervical and lumbar regions. In some instances, the neck and head may secondarily be positioned in a more extended position.

Ventral flexion of the neck (ventral neck flexion) is mostly commonly seen in cats and immature dogs. The animal's head will be directed ventrally toward the ground and usually caudally toward the ventral thoracic region. While this posture may result from cervical pain, it is more often a reflection of weakness of the cervical muscles. Animals with paralysis of the cervical muscles may also have a head and neck that appear to "flop" in a dependent (compared to gravity) direction.

Abnormal Spontaneous and Involuntary Movements

All movement results from muscular action. Abnormal movements commonly result from dysfunction of the nervous system or muscular functions. Movement can be voluntary or involuntary. Abnormalities of voluntary movement are discussed below under Assessment and Abnormalities of Gait in this chapter.

Involuntary movement disorders are less well classified in animals compared to humans. Terms such as tics, twitches, shivering, shuddering, and fasciculation are often used to describe episodic, irregular muscle movements or depolarizations associated with muscle contraction. Some movement disorders are classified as "hyperkinetic" or *dyskinesias* or *myokymias*. Dyskinesias are defined as impairment of the power of voluntary movements resulting in fragmented or incomplete movements. Involuntary movements are usually manifested through abnormal motion of the limbs, trunk, or head. Flexing movements of the limbs may occur. These types of episodes need to be differentiated from partial seizure activity and muscle disorders.

Tremor is a rhythmic, oscillatory, involuntary disorder of movement resulting from alternate or synchronous contraction of reciprocally innervated, antagonistic muscles. Tremor can be localized to one body area or be generalized (whole body). Localized tremor usually involves the head or pelvic limbs.

Myoclonus is a shock-like contraction of a muscle or muscles that tends to occur repeatedly in a rhythmic pattern. It is akin to the rhythmic depolarization and contraction that occurs in the heart with each beat. Myoclonus, if present, often involves the thoracic limb, however, a pelvic limb or the facial muscles including the tongue may also be involved. Myoclonus in dogs is usually the result of distemper infection that establishes a pacemaker-like depolarization of local motor neurons, however, it has been associated with lead toxicity and other forms of nervous system inflammation.

M*yotonia* occurs as a congenital problem in various breeds including the Chow-Chow and cats. Acquired myotonia can result from toxicity. Animals with myotonia have sustained muscle contraction, which is initiated voluntarily or with stimulation, and sustained involuntarily. The excessive muscle contraction is thought to be due to an abnormal muscle cell membrane that supports persistent depolarization. Animals with myotonia may have a short-stilted gait and apparent stiffness or extensor rigidity of the limbs. Affected animals also may have large, bulky muscles on palpation. Muscle dimpling may occur with direct muscle percussion.

REFERENCES

Bagley, R. S., 2005. Fundamentals of Clinical Neurology, Blackwell.

Bagley, R. S., Gavin, P. G., Silver, G. M., et al. 2000. Syringomyelia and hydromyelia in dogs. Compen Cont Ed 22:471-479.

Bagley, R. S., Pluhar, G. E., Alexander, J. E. 1994. Lateral intervertebral disk extrusion causing lameness in a dog. J Amer Vet Med Assoc 205:181-183.

Bagley, R. S., Silver, G. M., Seguin, B., Lincoln, J. D., Britt, L. G. 1997 Scoliosis and associated cystic spinal cord abnormality in a dog. J Am Vet Med Assoc 211:573-575.

Silver, G. M., Bagley, R. S., Moore, M. P. 1997. Myoclonus of the tongue. Prog Vet Neurol 7:137-140.

THE IMPORTANCE OF STEP (INTER-STEP OR CROSSSTEP) DISTANCE AND SEQUENCE AT GAIT

Rodney S. Bagley, DVM, DACVIM (Neurology and Small Animal Internal Medicine)
Ames, IA

Gait is commonly defined as a regularly repeating series of leg movements during walking or running. The nervous system controls the actions of the muscles, bones, joints, and associated connective tissue important for this function. Disease of any or all of these components can result in a gait abnormality. Normal walking is produced by the recruitment of stepping reflexes, which alternate between the extensor and flexor muscles. There is, however, a complex of actions involving multiple nervous system pathways associated with voluntary movement.

The Importance of the Step (inter-step or cross-step) Distance

Clinical evaluation of gait usually involves observation of the animal's movements during walking and, when indicated, running or trotting. This is best accomplished by having a handler walk the animal over a flat, non-slippery area (such as concrete, black top, carpet, or grass). Initially, the step distance (distance between where both thoracic limbs contact the ground, for example) should be determined. The step length is assessed as either normal, increased, or decreased. Step length determination will help in establishing whether a gait abnormality is associated with disease of the nervous system.

Generally, animals with UMN neurological disease have normal or increased step lengths, where dogs with orthopedic or neuromuscular disease have shorten step lengths. Animals with UMN disease may appear irregular and uncoordinated, having a tendency to sway from side to side; some may fall to the side. Their feet may contact the ground with increased force. Doberman pinschers with cervical vertebral malformation/malarticulation, for example, may over-flex or "sink" in the hock joints during weight bearing, presumably the result of motor or sensory weakness. These animals may rebound from this sinking resulting in hypermetria during the flight phase of the subsequent step. Also, many dogs with neurologic disease vary the step distance with subsequent steps, giving the gait an irregular temporal pattern. This results in some steps being longer and some steps being shorter. Irregular step distances is often a clue to an underlying neurologic disease compared to, for example, a musculoskeletal disease where the step distance, while abnormal, tends to be consistently abnormal in a stereotypic pattern. These same animals tend have greater impact with their limbs and the ground. This may reflect an abnormality of proprioception wherein the animal is uncertain when and where the limb will touch the ground.

Dogs with LMN disease may hold the head low (possibly because of neck muscle weakness) and take short, "choppy" steps. Some have a kyphotic posture. They may tire easily and appear unwilling to perform. Animals with orthopedic disease may have a similar gait abnormality. Bilateral cruciate ligament rupture is one such example wherein the pelvic limb step length is short. Normal conscious proprioception should suggest the possibility of an underlying orthopedic or musculoskeletal disease. Poor limb perfusion, such as from a partial bilateral iliac arterial thrombus or from a right-to-left PDA, may also result in a short, "choppy" pelvic limb gait. The clinical signs may worsen with exercise (similar to myasthenia gravis) and differential cyanosis (caudal aspects of the body) may be seen with the latter disease. Dogs with vertebral pain may also have short steps or be reluctant to move.

Animals have varying normal gait characteristics depending upon such factors as their body size, limb conformation, and shape. The degree of diversity even among a single species such as dogs is great, and can result in characteristic stepping movements and body movements for that breed. While seemingly simple, one of the first questions to be addressed when evaluating animals for abnormalities of movement is, "Does this animal have a gait abnormality or is this movement normal for the breed?"

As abnormalities of numerous body components can affect gait, it is important to determine which of these components is the cause of the dysfunction. Some gait abnormalities are more characteristic of nervous system disease, while other abnormalities are more characteristic of musculoskeletal or other system abnormalities.

Descriptive terminology for voluntary movement disorders is not universally agreed upon.

Ataxia is a failure of muscular coordination or an irregularity of muscle action. In some situations, it is defined as "lack of an axis", implying an inability to travel along a straight line. Ataxia is usually associated with a "falling" or "swaying" of the body from side to side. It is also commonly present in association with other forms of abnormal limb movements such as dysmetria.

Ataxia results from a variety of anatomic lesions within the nervous system, most commonly of the cerebellum, vestibular system, and spinal cord sensory pathways. Ataxia, without motor dysfunction, usually implies cerebellar or cerebellar pathway disease ("unconscious" proprioception). Sensory ataxia due to loss of joint position sense is often made worse by blindfolding the animal, effectively preventing visual compensatory mechanisms.

Dysmetria is improper estimation of distance during muscular activity. Dysmetria includes both *hypo-* and *hypermetria.* With hypermetria, voluntary muscular movement overreaches the intended goal and is essentially a failure to limit the degree of flexion. With hypometria, voluntary movement falls short of the intended goal. Hypermetria is more commonly recognized than hypometria. Both of these abnormalities are most often associated with lesions of the cerebellum or cerebellar pathways. In the instance of hypermetria, for example, the loss of cerebellar input, which normally limits the flexion phase of gait, results in the exaggerated movement.

"*Spasticity*" is a state of increased muscle tone. This is reflected as a stiff appearance of the limbs, a lack of normal flexion, or limbs that are maintained more extended during the flight phase of gait. This is occasionally referred to as "floating." Spasticity commonly results from upper motor neuron (UMN) lesions.

Paresis is derived from the Greek word for relaxation and implies weakness of motor function without complete paralysis (some voluntary motion remains). *Paralysis* implies loss of motor function from either a neurologic or muscular abnormality. The term "paresis," however, has different connotations including limb dysfunction from various causes. In many clinical descriptions, paresis and conscious proprioceptive abnormalities are used interchangeably, with the term "paresis" implying a conscious proprioceptive abnormality.

Varying degrees of paresis exist; some animals retain the ability to walk while others are unable to support weight. During gait, paresis may be manifested in a number of ways. This is most obvious when the animal contacts the dorsum of the paws (drags) along the ground. Abnormal toenail wear, especially when the toenails of the two longest toes (P3 and P4) are worn to a greater extent than the shorter (P2 and P5) toes, may suggest underlying paresis. In some animals, when the limb contacts the ground or when the animal's body weight is placed on limb, the paw rotates or twists. In other instances, the limb may overflex when the animal contacts the ground. This is often followed by an overflexion during the subsequent flight phase of the gait. In still other instances, the animal impacts the ground with greater force than normal. This may be a reflection of poor proprioceptive knowledge of where the limb is in space and subsequently when this limb should impact the ground.

Animals that are paretic often will have irregular step distances. These same animals tend to have greater impact with their limbs and the ground. This may reflect an abnormality of proprioception wherein the animal is uncertain when and where the limb will touch the ground. When the foot does impact the ground the animals may swivel or twist, usually with the caudal aspect of the limb moving more laterally and the toes becoming more medially placed (i.e., the front of the foot ends up pointing more medially).

Supratentorial disease will not dramatically affect gait. Paresis at gait is usually evident with lesions in the midbrain caudal. The severity of the gait impairment increases as the lesion occurs progressively more caudally in the central nervous system. For example, supratentorial lesions may result in significant hemiparesis when postural reactions are tested, however, gait remains relatively normal. With brain stem or spinal cord lesions, the associated paresis usually results in obvious gait impairment. Depending upon where within the nervous system the lesion occurs, gait may be altered differently. Wide circling toward the side of a unilateral supratentorial lesion is common. Occasionally, spasticity may be noticed, with the limb (primarily a thoracic limb) appearing stiff, floating, and overreaching.

Lesions of the brain stem and cervical spinal cord will often have dramatic effects on gait, frequently impairing the ability to stand and generate a gait. If the lesion occurs unilaterally, ipsilateral hemiparesis is noted. If pathways are affected bilaterally, tetraparesis will be present. Spinal reflexes should reflect the UMN nature of the lesion.

"Circling" is propulsive walking in circles. This is usually found in animals with disease involving either the supratentorial or vestibular areas. Usually, but not always, the animal circles toward the side of a unilateral supratentorial lesion or vestibular disease lesion. Animals with supratentorial disease will also tend to overstep with the thoracic limb contralateral to the direction of the circling. The paw appears to be "thrown" cranially, wherein the paw is quickly overextended prior to impacting the ground and may result in a "flipping" movement of the paw.

Gait Abnormalities and Lesions of Selective Peripheral Nerves

Lesions of the *femoral nerve* result in an inability to extend the stifle and support weight. The animal may have difficulty bringing the limbs forward and may "bunny-hop" or have shorter step distance in the affected limb. Bunny-hopping is the simultaneous advancement of the pelvic limbs, and has classically been associated with UMN spinal cord disease (spinal dysraphism/myelodysplasia). Animals with orthopedic disease (bilateral hip dysplasia or coxofemoral pain) may have the same gait abnormality.

Lesions of the *obturator nerve* often do not affect gait in dogs and cats. If an animal is placed on a slippery surface, however, the affected pelvic limb(s) may slide laterally with weight bearing.

Lesions of the *sciatic nerve* result in a characteristic gait wherein the distal foot is almost "thrown" forward by the movements of the more proximal muscles of the limb. When the animal is weight bearing, the hock joint may overflex resulting in the dropped-hock appearance. Atrophy of the muscles of the limb is common with LMN disease. Atrophy associated with sciatic disease is often obvious in the cranial tibial muscle group. With lesions of the *cranial gluteal nerve,* the stifle may become abducted during the stance phase of gait.

Lesions of the *radial nerve* affect the animal's ability to extend the carpus. If the lesion occurs proximally in the radial nerve, the elbow may be positioned more ventrally than normal. The limb may be carried off of the ground with the elbow flexed if the musculocutaneous nerve is intact.

With lesions of the *medial, ulnar, musculocutaneous,* and *suprascapular nerves,* gait is usually unaffected. The carpal joint may be overextended with medial and ulnar lesions. With musculocutaneous nerve paralysis, elbow flexion is poor and this joint may be held overextended.

Gait Abnormalities and Musculoskeletal Problems

Certain gait abnormalities are more characteristic of a problem in the musculoskeletal system. *Fibrotic myopathy* of the semimembranous and semitendinous muscles result in one such characteristic gait abnormality. At faster speeds, there is an awkward jerking in a medial direction of the affected limb during the flight phase of limb movement. If the animal is observed from behind, as the affected pelvic limb begins to be lowered following the zenith of the flight phase of gait, the distal foot (from the hock level) is rapidly pulled in a medial direction due to the decreased elasticity of the fibrotic muscle.

With fibrosis of the *infraspinatous* and *supraspinatous muscle,* the thoracic limb at the shoulder lever may be more outwardly rotated, with the foot contacting the ground in a more sideways position.

Animals with *rupture of the gastrocnemius tendon* may sink at the hock joint when weight bearing. A similar phenomenon may occur with abnormal sciatic innervation to the gastrocnemius muscle. If, however, the gastrocnemius tendon ruptures and the superficial digital flexor tendon remains intact, as weight is placed on the limb and the hock sink, the toes will be held in a more flexed position. This characteristic toe position does not occur with neurologic weakness of the gastrocnemius muscle. Affected animals may have a leg-carrying "lameness", and stand straight in the stifles when weight bearing. Animals often have associated muscle atrophy of the limb including the semimembranous, semitendinous, and gastrocnemius muscles. This may sometimes be confused with dancing Doberman syndrome.

Iliopsoas muscle disease may result in pelvic limb lameness. Animals may position the hocks in a more adducted way (cow-hocked).

Gait Abnormalities and Various System Diseases

Stiffness associated with *decreased step length* is commonly seen with diseases of the peripheral neuromuscular apparatus (LMN cell body, nerve roots, peripheral nerve, neuromuscular junction, and muscle), musculoskeletal disease, metabolic weakness, or poor perfusion of the limb.

Dogs with neuromuscular disease may also have a stiff, stilted, choppy gait due primarily to muscle weakness. These abnormalities may be episodic and worsen, as in the case of myasthenia gravis, as the level of exercise increases. A similar appearance may occur in dogs with pain, primarily from orthopedic disease.

Poor perfusion to the limb, as associated with a cardiovascular or local vascular problem, results in decreased limb oxygenation, increased anaerobic metabolism, and subsequent muscle weakness. Systemic metabolic disease may also result in abnormalities of muscle metabolism and subsequent muscle weakness. Differential cyanosis may be presence with some shunting of limb perfusion such as in the instance of a right-to-left PDA.

Lameness (decreased or non-weight bearing on a limb) is usually associated with pain of the limb from musculoskeletal disease. A similar clinical abnormality can also occur with nervous system dysfunction (and presumably pain), referred to as "*nerve root signature*". This abnormality often occurs in a single thoracic or pelvic limb. Often, the affected limb may appear painful upon manipulation, mimicking an orthopedic problem.

Bunny-hopping is the simultaneous advancement of the pelvic limbs, and has classically been associated with UMN spinal cord disease (spinal dysraphism/myelodysplasia). It can also occur with diseases that result in bilateral decreases in step distance. Occasionally, animals with orthopedic disease (bilateral hip dysplasia or coxofemoral pain) or poor perfusion to the limbs will have the same gait abnormality.

REFERENCES

Bagley, R. S. 2005. Fundamentals of Clinical Neurology, Blackwell Publishing.
Brown, C. M. 1986. Dog Locomotion and Gait Analysis. Hoflin Publishing.
Dewey C.W. 2008. A Practical Guide to Canine and Feline Neurology, Wiley Blackwell.
Lyon, M. 1950, Reprinted 2002. The Dog in Action. Orange Judd Publishing Company.

INTERPRETTING EXAGGERATED MOVEMENTS

Sean G. Sanders, DVM, PhD, DACVIM (Neurology)
Seattle, WA

INTRODUCTION

An exaggerated movement is a normal movement that is simply more pronounced. This may be purposeful as in a cat that hisses in order to send a message or involuntary such as a cat with disease of the cerebellum. The aim of this discussion is the focus on exaggerated movements that are involuntary and the interpretation of these movements. Excessive or exaggerated movements are *hyperkinetic*. The forces involved in the motion (kinetics) and the motion of the body parts (kinematics) results in abnormal, asymmetrical movements[1]. A paroxysmal movement disorder refers to hyperkinetic movements such as seizures (epileptic or non epileptic), sleep related movements and behavioral disorders. While it is often easy to recognize something is wrong, determining what is wrong, how to proceed with diagnostic testing and ultimately how to help the patient can be very challenging. Certain clinical pictures will easily come to mind with experience. Pattern recognition is a valuable tool but not always 100% accurate. A logical or systematic approach to syndromically classify the abnormality is the first step in an evidence based medicine approach. Some disorders will be obvious, others will be vague, however, knowing which system is affected and being able to define the abnormality is the first step after recognition. Exaggerated movements usually involve gait or posture, are commonly involuntary and they tend to involve a single system. The correct interpretation of the hyperkinetic movement is the first step after recognizing a problem exists. Taking a careful history from the pet owner, focusing on when the movement started, was or is there a trigger, does it worsen with activity or improve with activity, is it associated with a particular phase of movement, is pain involved and how does it respond to medications, can be very helpful in honing in on what system is affected. Exaggerated movements related to gait or posture include paroxysmal, neuromuscular and those resulting from the central nervous system[2].

PAROXYSMAL

Paroxysmal movements are those movements that happen on a reoccurring basis and are stereotypical. These movements are usually very patterned (stereotypical) and fairy easy to recognize either by historical description, direct observation or video analysis from a pet owner's own recording. A seizure disorder would be the most familiar example. Disorders of normal or abnormal sleep also tend to be paroxysmal. Muscle spasms can be a bit confusing and the difference between myotonia and a "cramp" may be inconsequential.

Seizures

A seizure is a paroxysmal, transient, stereotypic, abnormal electrical discharge of neurons within the brain, associated with either lack of voluntary control of movement or dyscognition[3]. If an enduring substrate occurs in the brain, which could lead to multiple seizures then the seizure disorder is termed *epileptic*. A brain tumor, encephalitis or a stroke are examples of posibel causes of epilepsy. If no enduring substrate in the brain is present, then the seizures are not epileptic. A metabolic or reactive cause of seizures would be an example as once the underlying cause of the seizures (toxin or hepatic encephalopathy, etc.) is eliminated there is no reason for further seizures. Seizures are often confused with muscle weakness, vestibular disease, neck pain and sleep disorders. Generalized seizures are fairly easy to recognize and are commonly referred to as fits, convulsions or attacks by pet owners. Focal seizures on the other hand, maybe more difficult to recognize. These can include episodes of staring off into space, fly biting, lip smacking or other automatisms. Focal seizures may involve loss of consciousness (focal without cognition) or intact consciousness (focal with cognition). An automatism is an action performed without conscious thought or intension. The actions are usually normal movements performed at abnormal times. Automatisms are a frequent event associated with a seizure. Video recordings of the events can be very helpful in determining on a gross level if a seizure is occurring. However, for more discrete events an electroencephalographic recording (EEG) is often necessary to make a final determination. Absence seizures have been poorly documented in the veterinary literature, although the use of awake or ambulatory EEG recordings can be very helpful in making the diagnosis of this and many other seizure disorders[4].

Sleep Disorder

Sleep disorders tend to be fairly easy to recognize and differentiate from epilepsy with the exception of atonic seizures "drop attacks". They occur during sleep in most instances (cataplexy excluded). They may be referred to as *Sleep - related Movement* Disorder and like many seizures exhibit exaggerated movements and behaviors, which tend to be associated with vocalization, such as dreaming ("chasing rabbits") or *Periodic Limb Movements of Sleep* and *Rhythmic Movement Disorder of Sleep*. Parasomnias may be reported more frequently as they tend to appear more significant. Sudden startle attacks (night terrors) associated with episodes of extreme fear, aggression or another mimetic state can be more challenging to differentiate from a seizure. *Nocturnal Frontal Lobe Epilepsy* is a seizure disorder often mistaken for night terrors in humans[5]. An ambulatory EEG can be very helpful with this parasomnia if there is doubt that they are truly a sleep disorder as opposed to epileptic activity. During a night terror, the dog or less commonly cat, will bolt startle awake for no apparent reason, cry, scream and often run around the room in an inconsolable manner. The mimetic or emotional state will often dissipate within a few minutes and they go right back to sleep. Contrasted to most seizures where there is a persistent state of disorientation or ataxia, a dog or cat with a parasomnia will often be "normal" immediately after the event. *Catathrenia* (sleep associated groaning) is also categorized as a parasomnia and pet owners may incidentally report this condition. A hypnic jerk is a sudden startle movement usually associated with the beginning of sleep. These feelings of "falling" as they are often described in people likely occur in animals but tend to be very sporadic and may go unnoticed. Cataplexy on the other hand, while quite rare is easy to recognize in most of its forms. Catapexy is a sudden "sleep attack" characterized by paralysis of movement and sudden loss of muscle tone. The animal (dog) collapses, after a brief and intense emotional response (play or eating food). The episodes last for a few seconds to a few minutes and the dog wakes up suddenly as if nothing had happened. The patient may have their eyes wide open and are conscious or they may immediately enter into REM sleep or mixed stages of sleep associated with the EEG. They "wake up" just as suddenly and often go right back to eating or playing. The disorder in dogs has been genetically linked to certain families of Dobermans, Labrador Retrievers and Dachshunds. Sporadic mutations also exist. Dysfunction of the *orexin* (hypocretin) receptor due to mutation of the type II orexin receptor gene (*OxR2*) has been identified as the cause[6].

Behavioral Disorders

Certain behavioral disorders may exhibit exaggerated movements. Behaviors such as tail chasing, spinning, flank sucking or obsessive compulsive events may be difficult to distinguish from a seizure disorder. At a basic level the described behavior may contain most of the elements used to define an epileptic seizure, however to be truly the result of epileptiform activity, an EEG must be performed during the event. In many cases these disorders are easy to distinguish because they do not posses a post ictal period or disorientation, ataxia or dementia. Furthermore, many behavioral disorders have a "trigger" that either preced the behavior or is associated with the onset of the activity. The behaviors of certain breeds, namely Bull Terriers and their stereotypical tail chasing or spinning behavior, are through to be a focal form of undiagnosed seizure disorder and more recently a potential animal model of Autism Spectral Disorder of humans[7]. As ambulatory EEG becomes more readily available we now have a diagnostic tool that possesses the ability for clinical neurologists to separate a behavioral disorder from a neurological disorder so that appropriate treatment (behavior modification or antiseizure medication) can be pursued.

NEUROMUSCULAR

Involuntary hyperkinetic movements as a result of neuromuscular disorders include tremors, myotonia, myokymia, muscle fasciculation and tetanus. These movements are due to disorders of the peripheral nerves and muscles, two systems, which are intimately involved in the generation of normal movements.

Tremors

A tremor is an involuntary, approximately rhythmic, oscillatory or sometimes sinusoidal movement of a body part. It can exist in normal and abnormal animals. It differs from a myoclonic movement (such as a tic or twitch) in that it possesses the associated rhythmic, oscillatory or sinusoidal (smooth, repetitive oscillation) movement, as opposed to a brief shock like contraction. Tremors are characterized by their frequency in Hertz or cycles per second, their location (head, foot etc.), and the condition through which they occur (resting, standing or action). A tremor may be physiologic or pathologic. Most physiologic tremors are unnoticed (the beating of the heart or *ballistocardiogram* for example). Many occur as a normal aging process such as *essential* tremor or *senile* tremor. Some are associated with "nervous" dogs, usually showing up in the pelvic limbs and worsening with activity or excitement (like getting ready to chase a ball) but then go away with rest. However, an enhanced physiologic tremor may become a physiologic tremor of *pathological* amplitude. These tremors are an exaggeration of the normal physiologic tremor and may be caused by metabolic abnormalities (electrolyte disturbances, hypoglycemia, etc.), stress, exercise, certain medications (e.g., metoclopramide, trazodone diphenhydramine, etc.), and toxins (e.g., lead, organophosphates, etc.). Pathologic tremors originating from the neuromuscular system may include those due to neuropathies, myopathies, and junctionopathies, which are often associated with lower motor neuron signs and *weakness*. Tremors, as a distinct clinical sign are vague. We often need to rely on the development or inclusion of other clinical signs in order to associate a tremor with a particular system or disease. For example tremor along with a metabolic abnormality reflected in routine blood work or a tremor, along with abnormalities of electrodiagnostic testing and muscle biopsy. Finally, a tremor associated with a compressive neuropathy such as that from intervertebral disk disease or nerve root tumors are not common but will usually manifest along with pain associated with movement and other neurological deficits.

Myotonia

Myotonia is characterized by delayed relaxation of skeletal muscles following voluntary contraction, which persists (prolonged contraction / delayed relaxation). The patients appear to have a stiff and very stilted gait and often have trouble getting up or down. When they stand, they have a typical "saw horse" stance. Several forms exist. One of the most common and easily mistaken for other disease is the acquired form associated with untreated or uncontrolled hyperadrenocorticism. This is often referred to as "Cushing's myotonia" or a "pseudomyotonia". Clinically, it may manifest as extreme stiffness and be mistaken for orthopedic disease or compressive spinal cord disease. The patient often has other physical signs and a history suggestive of Cushing's disease. If it is suspected, the patient should be tested for hyperadrenocorticism and treated appropriately. Treatment of the underlying disease in this advanced state is rarely helpful in alleviating the stiffness associated with myotonia. Procainamide or mexiletine may help by stabilizing neuronal cell membranes. The congenital form has been identified in Chow Chow Dogs and Miniature Schnauzers. Genetic testing is available for the Miniature Schnauzer[8]. In some forms the clinical signs are due to abnormal chloride conductance of the myocyte membrane. This leads to hyperpolarization of the myocyte resulting in prolonged contraction. Abnormalities of swallowing and prehension may also accompany the syndrome. Affected muscles will occasionally have a persistent "dimple" after being struck with a plexor. The diagnosis is confirmed (although usually not necessary) through the demonstration of electromyographic changes commonly referred to "dive bomber" sounds. These high frequency, waxing and waning, myoclonic discharges are easily recognized. Muscle biopsy is usually normal. Muscle contractions persist even while under general anesthesia.

Myokymia

Myokymia is probably a manifestation of a more inclusive disorder called *neuromyotonia*. Neuromyotonia refers to episodic muscle contraction secondary to hyperexcitability of motor nerve axons. This pattern of abnormal muscle contraction often looks like undulating muscle contractions of the underlying muscle as though worms are under the surface. Neuronal disease followed by motor unit axonal sprouting in the terminal fields of affected neurons is an indication of denervation. It may suggest a healing process after an injury to neurons or in the case of the Jack Russell Terrier, a genetically transmitted disease also associated with cerebellar ataxia collapse, seizures and episodes of hyperthermia. A missense mutation has been identified in Jack Russell Terriers of the *KCNJ10* gene coding for the inwardly rectifying potassium channel Kir4.1[9].

Muscle Fasciculation

Muscle fasciculations result from disease of lower motor neurons and manifest as synchronous contraction of the all the skeletal muscle under the influence of a single motor neuron unit. They may also arise from irritability of the axon of a motor neuron and they can be completely benign occurrences. Benign fasciculations are usually spontaneous and self-resolving events. They can be exacerbated by activity or excitability. Drug reactions (especially methylxanthines such as theophylline or caffeine), as well as electrolyte disturbances may also result in transient muscle fasciculation. Disease or irritation of the motor neuron or its axon results in more persistent and typically progressive clinical sign. Similar to many disorders of movement, careful observation must look for other clinical signs (present or developing) that help the clinician determine the origin of the abnormality.

Tetany

Tetany, not to be mistaken for the disease *Tetanus* is the intermittent and variable contraction of extensor muscles commonly resulting from hypocalcemia or hyperphosphatemia. Hypocalcemia causes increased cellular permeability of sodium, which results in depolarization of neuronal cell membranes, and a propensity for action potentials to be generated. The most common cause is hyperparathyroidism associated with serum calcium levels below 7.0 mg/dl. Dogs show clinical signs of weakness (inability to rise), excessive and spontaneous extensor muscle contractions and variable degrees of hyperthermia due to excessive muscle contraction. Diagnosis is typically made based on suspicion and routine blood work. Treatment by restoring normal calcium, phosphorous and vitamin D levels is usually curative while the underlying pathology is addressed. Initial relaxation of skeletal muscles using benzodiazepines may also be helpful from a supportive standpoint.

CENTRAL NERVOUS SYSTEM

Disorders of the central nervous system, which may result in hyperkinetic or exaggerated movements can be categorized as those due to a disorder of the cerebellum and those that are non-cerebellar. As stated previously movement disorders if not associated with a very specific pattern of recognition tend to be early vague signs of a disorder. Often the development of other clinical signs, specific to another aspect of the nervous system is necessary for the clinician to make any meaningful presumptions. Probably the most widely recognized pattern of hyerkinetic movements associated with a disorder of the central nervous system are disorders of the cerebellum.

Cerebellum

The cerebellum is responsible for the generation of normal rate and range of movements. Disorders of the cerebellum may result in abnormalities in the rate, range, direction and force of motor movements. Lesions of the cerebellum often cause ipsilateral deficits due to the anatomical pathways from which it influences; the crossed rostral cerebellar peduncle influences movement generated by the crossed corticospinal and rubrospinal tracts. Weakness (paresis or proprioceptive deficits) is not present with pure disease of the cerebellum because proprioceptive tracts do not course through the cerebellum. Disorders of the cerebellum commonly manifest as hypermetria (or less commonly hypometric), ataxia, dysmetria and tremors. Tremors are most often in the form of course (3-5 Hz) action related or *intention* tremors. Ataxia in of itself, is not specific to a disorder of the cerebellum but is often present and symmetrical. A hypermetric ("toy soldier") gait is often associated with cerebellar disease. Vestibular signs and absent menace with preservation of vision may also be found with cerebellar dysfunction. Congenital hyperkinetic movements are usually associated with a variety of cerebellar signs and are difficult to miss. Feline cerebellar hypoplasia associated with gestational infection with the feline panleukopenia virus is one of the most commonly any easily recognized disorders. Like many patients with hyperkinetic movements due to a disorder of the cerebellum, intention tremors, jerky movements (dysmetria), hypermetria and trunkal ataxia with *normal* proprioception are present. Clinical signs may be progressive or static. Many congenitally affected animals will improve with time. Acquired disease of the cerebellum results in similar signs although progression is more rapid with inflammatory or neoplastic causes, slow and potentially stabilizing with degenerative conditions and potentially improving with trauma. Dogs and cats tend to compensate well with disease of the cerebellum (trauma, stroke, post operative, toxin) and may have a very good life if allowed enough time and supportive care to heal.

Tetanus

The disease *Tetanus* and the clinical sign tetany are often intermixed. Tetanus is the sustained contraction of a group or groups of muscles. It is most commonly seen (although rare) in dogs and cats infected with the *Clostridium tetani* bacteria. The bacteria once ingested or more commonly gaining entrance to the body through a wound, turns into the vegetative form which creates the exotoxin, tetanospasm. Tetanospasm binds to peripheral nerve terminals and is then transported through retrograde axonal transport to the neuronal cell body in the spine or brainstem. The toxin blocks the release of the inhibitory neurotransmitters glycine and GABA through the selective cleavage of synaptobrevin II, which is necessary for exocytosis of the inhibitory neurotransmitters.

Clinical signs may vary from mild stiffness to extreme rigidity though interaction with spinal cord and brainstem motor neurons. Opistothonus, respiratory difficulty, trismus, a sardonic or grimacing grin and extensor rigidity are common. Treatment utilizing skeletal muscle relaxants such as benzodiazepines, penicillin or metronidazole and supportive care are essential in simple cases. Ventilator support parenteral feeding may be necessary in severe cases. Resolution of clinical signs may take a few weeks to months.

Tremor

Non cerebellar tremors of the central nervous system must usually be associated with the presence or absence of other localizing signs to help the clinical localize the lesion. Degeneration of the nigrostriatal pathway famous for the cause of Parkinson disease in humans is also seen in horses that ingest the toxic weeds yellow star thistle (*Centaurea solstitialis*) and Russian knapweed (*Centaurea repans*)[10]. The "Parkinsonian" tremor of humans, which affects fine motor muscles such as the hands and fingers, may be similarly represented by the slow, rhythmic, oscillatory movement of a horse's muzzle in those animals clinically affected with Yellow Star Thistle poisoning. This particular type of tremor has not been recognized in dogs or cats. An erratic or uneven tremor with a low frequency (<4.5 Hz), commonly referred to a rubral tremor in humans results from a lesion in the cerebellar outflow tracts to the midbrain or the thalamus. These are not recognized in dogs and cats but could certainly result from a lesion such as a stroke. They can occur with rest, intention and posture. *Mixed* tremors, however, are recognized in veterinary patients. Orthostatic tremor (OT) has been described in young (1–2 years old at onset) adult giant-breed dogs, including Great Danes, Mastiffs, and one Scottish Deerhound. These dogs exhibited tremors of all four limbs while standing, which appeared as a constant "quivering" or "shivering" of the limbs. Fine rippling of the musculature was apparent while the dogs were standing. Dogs with OT tend to shift weight from limb to limb and have difficulty lying down from a standing position. The tremors disappear when affected dogs are recumbent, while walking or running, when leaning against objects, or when picked up (to remove all weight support). When a dog with OT is in lateral recumbency, if an examiner pushes against a paw (causing isometric limb muscle contraction), the tremor can be induced. These dogs do not demonstrate fatigue with exercise, and no deficits are apparent on neurologic examination. This disorder appears to be very slowly progressive in dogs and also may respond to oral phenobarbital therapy. Other suggested treatments for people with OT include gabapentin and clonazepam. Clonazepam is actually the preferred drug choice for people, but it is probably ill-advised for canine use due to the rapid development of drug tolerance to this drug in dogs.

Head Bobbers

Repetitive, intermittent head bobbing (vertical or horizontal) is a poorly understood movement disorder, which is often encountered in dogs, particularly young adult Boxer dogs, English and French Bulldogs, and Doberman Pinschers. This is also a benign condition for which the underlying mechanism remains elusive, although ambulatory EEG has been shown to demonstrate epileptiform activity in some patients[11,12]. Among the possibilities for causes of this disorder is a potentially heritable disorder of the stretch mechanism affecting the neck musculature similar to that described for essential tremor. However, this tremor syndrome differs from essential tremor in that it is paroxysmal (characterized by short, frequent and stereotypic symptoms). As with other postural tremors, the tremors tend to subside if the head is supported (i.e., not supported by neck muscles). However, they may also be present (although often attenuated) when the patient's head is supported (such as on a pillow while lying down). The tremors also tend to stop if the dog is distracted (e.g., calling patient's name, offering a toy or food). Other terms for this condition include head-bobber and postural repetitive myoclonus.

Myoclonus

Myoclonus is a sudden, quick, involuntary muscle contraction. It may be solitary (as in a shock like twitch), repetitive or intermittent. Repetitive myoclonus, typically of a single muscle group in a young dog may be due to distemper myoclonus. It is usually non progressive but can be debilitating if it prevents normal function (sleep, prehension, ambulation, etc.). Any upstream disorder of the nervous system can cause a muscle to exhibit a myoclonic twitch or myoclonus, which is why it is often a very vague clinical sign. Taken in context with the development of other neurological deficits or clinical signs, imaging finding and disease progression over time, the clinician may be able to make a presumptive diagnosis.

REFERENCES

[1]Millis D, et al. **Canine Rehabilitation & Physical Therapy** 2004. Saunders, St. Louis. [2]Podell M. **Vet Clinics North Am**. 2004;34(6):1435. [3]DeLahunta A, et al. **Veterinary Neuroanatomy and Clinical Neurology**. 2009. Saunders, St. Louis. [4]Poma R, et al. **Epileptic Disord** 2010;12(2):138. [5]Derry CP, et al. **Arch Neurol** 2006;63(5):705. [6]Toth LA, et al. **Comp Med** 2013;63(2):91. [7]Tsillioni I, et al. **Transl Psychiatry** 2014;4(epub):e466. doi:10.1038/tp.2014.106. [8]University of Pennsylvania PennGen Laboratories. [9]Gilliam D, et al. **J Vet Int Med** 2014;28(3):871. [10]Sanders SG, et al. **Vet Rad and Ultrasound** 2001;42(4):291. [11]Bush WW, et al. **Proc ACVIM Forum** 2013. [12]James F, et al. **Proc ACVIM Forum** 2014.

INTERPRETTING RESTRICTIVE MOVEMENTS

Sean G. Sanders, DVM, PhD, DACVIM (Neurology)
Seattle, WA

INTRODUCTION

The clinical evaluation of a patient with hypokinetic or restricted movements tends to be less enigmatic than those with hyperkinetic or exaggerated movements. Perhaps this is because we have better clues and in most cases the clinical signs tend to be less vague and more specific. We are more comfortable looking at a patient that is having trouble moving as opposed to a patient who is moving too much. Restricted movements can be looked at as those that are the result of lameness, weakness and disorders of posture.

LAMENESS

When a clinician is presented with a lame patient, identifying the area to examine is a simple process of observation...in most cases. Lameness may be due to pain or a conformational abnormality that results in restricted movement (and therefore usually no pain). Traditionally, we think of lameness or limping as a patient who does not want to walk. In this sense, the lameness is likely due to pain and pain is often associated with self-restriction of activity. Lameness may be looked at from an orthopedic / musculoskeletal perspective or a neurological. When we discuss these aspects, it is important to remember, we mostly talk about *generalizations*. In general, pain can be the source of restricted mobility and restricted mobility may result from a musculoskeletal problem, a neurological problem or a metabolic problem. Eliminating a metabolic problem from your localization is often easily performed by a thorough history. Animals with metabolic disease, tend to look and act *sick*.

Musculoskeletal

Generally speaking, the more severe the lameness (and radiographically normal) the more likely the problem is a neurological form of lameness. Chronic orthopedic disease and mild orthopedic disease tends to cause non-weight bearing lameness and shifting leg lameness but the patients are usually not holding their limb off of the ground. If you see a patient packing a limb, neurological cause of the lameness should be a little more suspected. There are of course many exceptions. A slipped proximal femoral physis in a young dog or other severe orthopedic disease would be one exception. Gait analysis is important to try and pinpoint the affected body part so a more specific examination can be performed[1]. With musculoskeletal disease, there should not be any neurological deficits, although a patient may be reluctant to perform normal proprioception testing (knuckling) they usually will if the painful limb is supported. Specific clinical signs including, swollen or painful joint or joints, pain on palpation or manipulation of a specific body part and history will help the clinical isolate the affected area and develop a list of differentials. *Shifting leg* lameness and pain associated with multiple joints (including spinal pain) or multiple swollen joints is often a clue for polyarthritis. Intermittent lameness or pain with no instability may be associated with mild disease such as an incomplete cruciate ligament tear, osteoarthritis or soft tissue disease. Chronic, intermittent forelimb lameness is common with shoulder disease. The older and larger the patient, the more likely it is shoulder. The younger the patient the more likely it is elbow. Acute onset or acute onset followed by chronic lameness in the forelimbs is commonly found with intervertebral disk disease of the cervical spine. Lameness associated with musculoskeletal disease is often present in all phases of movement or gait. If going up stairs or inclines exacerbates the lameness think pelvic limbs, whereas worsening lameness going down stairs or hills is usually associated with forelimb, shoulder or cervical disease.

Some of the most enigmatic cases include bilateral cruciate rupture, semimembranosus tendinopathy and traumatic myopathies. Bilateral cruciate disease causes a slow, stiff and shuffling gait in both rear limbs with (typically) instability of the stifles. *Bunny hopping* is commonly seen with bilateral cruciate disease but can also be observed with severe coxofemoral disorders, neurological disease and polyarthritis. *Semimembranosus tendinopathy* is almost exclusively seen in young to middle aged German Shepherd dogs presented for a "lameness". This is an excellent example of a conformational abnormality. The diagnosis is made through gait analysis and is characteristic of an internal rotation of the tarsus and external rotation of the stifle causing a *mechanical* lameness in one limb. *Traumatic myopathies*, muscle sprain or tendon strain often result in muscle contracture leading to abnormal conformation commonly visibly on gait analysis or at rest. *Infraspinatus contracture* and rupture of the *superficial digital flexor* (Achilles tendon) are examples of more obvious disorders. Less obvious causes of lameness due to muscle contracture or injury include gastrocnemius, iliopsoas, gemelli and obturator muscle injuries (hip adductors)[2,3].

Neurological

Pinched nerves are painful. Severe forms of lameness from a painful aspect should focus attention on the nerve root signature. The nerve root signature is the clinical sign associated with distinct lameness that results from a compressed nerve either in the spinal canal (e.g. herniated disk), as the nerves exit the foramen (e.g. stenosis), or the nerve itself (e.g. tumor or trauma). In the most simple case a neurological deficit, usually in the form of knuckling or another sign of a proprioceptive deficit, allows the clinician to immediately determine the problem is in the nervous system. A "drunken" or ataxic gait further strengthens the localization to the nervous system. There is no need to check for deep pain if voluntary motor function is present.

Perhaps the most vague of presenting clinical signs are those associated with lumbosacral disk disease or cauda equine syndrome. If reflex changes, pain on lordosis testing, changes in tail carriage or incontinence are not present (which are very localizing signs), often history will help with the suspicion. Pain associated with transitional movements, such as getting up and down, pain or "travelling" while eliminating, or pain going up and down stairs does not help differentiate lumbosacral disease (herniated disk, stenosis or neoplasia) from coxofemoral or stifle disease. However, startle responses, "looking" back at the base of tail or obsessive licking of specific parts of the distal rear limb may lend suspicion to a pinched nerve in the cauda equina or peripheral nerve in the pelvic plexus. In order to differentiate the neurological cause of lameness from the orthopedic or musculoskeletal, the early "vague" signs must be supportive by history, imaging findings and analysis of disease progression over time. As disease progresses the signs tent to become more specific and eventually the disease will "declare itself".

A short and choppy, long and lopey or "two engine" gait is often observed in dogs with caudal cervical disease. In these cases the front limbs demonstrate a short and choppy gait, almost as if the patient is walking on hot coals. The rear limbs have a disconnected long and lopey gait with the pelvis often dropped to the ground. Dropped tarsi in the cat is a not a very specific sign but is most commonly associated with disease of the nervous system or a metabolic cause of weakness. Pattern recognition can be very important.

WEAKNESS

Dogs and cats with weakness *cannot* walk, those with lameness do not *want* to walk. Pure weakness could be a result of a disorder of the nervous system, cardiac disease or a metabolic disorder such as anemia. Weakness affecting just the rear limbs would localize most of the time caudal to T2, however, due to the somatotopic organization of descending spinal cord tracts, the pelvic limbs may be affected first with cervical disease, as well as, thoracic or lumbar disease. Other signs such as obvious cervical pain or decreased reflexes in the thoracic limbs may lead one to suspect cervical disease. If the weakness (assume proprioceptive deficits) exist in the front limbs and the rear limbs are normal, a *central cord* lesion is suspected localized to C1 – T2, depending on thoracic limb reflexes. Thoracic limb reflexes are accurate in localizing a lesion (C1 – C5 or C6 – T2) approximately 65% of the time[4]. Weakness affecting the rear limbs may also be a sign of a right to left shunt. If suspected (and it is best to always check), look for differential cyanosis in the mucus membranes of the perineum, prepuce or vagina. The weakness is usually worse with exercise. If we focus on weakness only involving the nervous system we can narrow down the cause to upper motor neuron disease, lower motor neuron disease, disorders of the neuromuscular junction and muscles. All have the potential to create a "drunken", ataxic, or "weak gait". The patient often shifts from side to side when they walk and in most cases the further they go the weaker they become. With pure cerebellar disease causing ataxia, there should be no weakness. Weakness does not necessarily mean a proprioceptive deficit but most of the time upper motor neuron general proprioceptive gaits usually have a proprioceptive deficit (conscious or unconscious). Weakness of one rear limb can result from thromboembolic disease (pulses should be palpated). Muscles are usually very painful. A saddle thrombus (dog or cat) may result in bilateral rear limb weakness. If suspected (poor pulse quality) consider comparison of forelimb to hindlimb blood pressure.

With upper motor neuron weakness, the patient's muscle tone is usually normal to exaggerated. Reflexes are also normal or exaggerated and the lesion would localize to the spinal cord or the brain depending on other localizing deficits. The bladder is usually normal to difficult to express and the patient may exhibit dysuria. The patient will often be able to support their weight if propped up and in extreme cases may have a "saw horse" appearance. If extreme extensor tonicity is encountered, consider *Tetanus* and myotonia on the list of differentials. Muscle atrophy is slow and mild to moderate due to disuse atrophy in cases of upper motor neuron disease.

With lower motor neuron weakness, muscle tone is reduced and often flaccid. Reflexes are more likely to be abnormal (obviously reduced) and the patient may exhibit incontinence (fecal and/or urinary if a L4 – S1 localization) with a bladder that is flaccid and easy to express. Keep in mind that in a patient with multiple injuries or disorders to the nervous system a lower motor neuron sign will always cover up an upper motor neuron sign in the spine. This is most likely observed in cases of spinal or plexus trauma. These patients cannot support their weight if propped up. If brachial plexus trauma occurs, the patient will usually present with a sudden onset, typically non painful at the time of exam, forelimb weakness which is asymmetrical and often only one side affected. The patient will usually be able to hold the limb up but drags the dorsum of the paw. This is due to damage of the radial nerve, which is most commonly affected with brachial plexus traumatic injuries due to its short length as it exits the spine (C7, C8 and T1 spinal nerve roots). The patient loses function to the triceps (losing the ability to support weight in the affected limb) and the extensor muscles of the antebrachium losing the ability to rotate the paw dorsally to land on the palmar surface of the paw. Muscle atrophy tends to be rapid and severe from denervation atrophy. If diffuse lower motor neurons signs are present (decreased reflexes in all limbs), disorders such as botulism, tick paralysis and polyradiculoneuritis should be suspected.

The most common disorder of the neuromuscular junction is myasthenia gravis. If the patient shows weakness or fatigue with exercise and has *a normal* neurological exam, myasthenia or a myasthenia – like syndrome should be suspected. A tensilon test can be very helpful in a clinical setting to support the diagnosis. Serum should be submitted to look for acetylcholine receptor antibodies prior to treatment with immunosuppressive agents. Acetylcholine esterase inhibitors will not affect the blood test. Congenital forms of myasthenia gravis will not test positive for acetylcholine receptor antibodies[5].

Because the muscle is the end organ of the efferent nervous system any disorder of the nervous system has the potential to affect the muscles. This may be in the form or spasticity, rigidity, flaccidity and ultimately weakness. Primary muscle disease can mimic many other components of the nervous system, however, there are a few clues which would suggest a myopathy as opposed to a neuropathy. Myopathies should not show any hyperesthesia (increased sensitivity) or paresthesia (licking a spot on a limb). Muscles may be painful in cases of myositis. A fever is also more commonly associated with polymyositis and less commonly associated with meningitis. Myoglobinuria (very dark or coke colored urine) may be present. If atrophy and reduced reflexes are present early in the course of disease (days to a few weeks), the problem is likely a neuropathy. Atrophy of the muscles is associated with late stage or chronic disease. Disorders of the muscle are more likely to be symmetrical unless the result of a vascular accident such as a thromboembolism. Weakness may be episodic or associated with extreme activity. Often when the patient is evaluated they appear normal and have a normal neurological exam. Video of the event tends to be helpful in determining if the nervous system is affected or another area needs to be considered.

POSTURE

The pattern recognition of a few postures associated with restrictive disease can be very helpful. The most common postures observed clinically include; kyphosis, head and neck neutral, Schiff-Sherrington, decerebrate and decerebellate postures. An animal with a head turn will usually turn toward the lesion. A *head turn* is characterized by the head remaining level, and the nose deviated away from midline. This usually implies forebrain disease and may be accompanied by circling in the direction of the head turn. A head turn should be differentiated from *torticollis*. Torticollis may result from cervical lesions resulting in the abnormal curving or twisting of the neck. A *head tilt* means that one ear is held lower than the other. Head tilts imply vestibular or cerebellar disease. The head tilt is usually toward the side of the lesion. A head tilt is the result of decreased input from the vestibular apparatus to the vestibular nuclei in the brainstem, or a lesion of the vestibular nuclei. The vestibular nuclei, by default want to extend the muscles of the forearm and neck. When one side loses input (due to a lesion), the other side now has more extension than the lesioned side, which pushes the animal toward the side of the lesion. An exception to the rule is *Paradoxical Vestibular Syndrome,* in this case the head tilt is away from the lesioned side and the weakness is on the same side as the lesion. Paradoxical vestibular disease results from injury of the flocculonodular lobe of the cerebellum or the caudal cerebellar peduncle. Thalamic disease and cerebellar disease can also cause vestibular signs. Central vestibular disease can be easily differentiated from peripheral vestibular disease by the presence or absence of conscious proprioceptive deficits and any other cranial nerve deficits with the exception of cranial nerve VII (Facial nerve) and especially multiple cranial nerve deficits.

Kyphosis is arching of the back and may be due to pain or muscle spasticity. Animals will often arch the spine to relieve pressure on compressed nerve roots. If obvious spinal pain is present, one should question the need to check for spinal pain by palpation. Kyphosis may also be present with upper thoracic and cervical lesions.

Head and neck neutral is a very common and easily recognized sign of cervical pain. Most pet owners can easily recognize this sign as many have experienced it themselves. The dog or cat will walk with the nose pointed to the ground, will turn its whole body to look and often look out of the corner of their eyes (the "guilty" look) if called. This sign may also be present with brainstem compression.

The *Schiff – Sherrington* posture is characterized by extension of the thoracic limbs and paralysis of the pelvic limbs with normal reflexes in the pelvic limbs. It results from a severe lesion to the T3 – L3 spinal cord segments which causes a loss of innervation from fibers (Border cells in the dorsal aspect of the ventral gray matter concentrated between L1 – L4) originating in the lumbar segments and ascending to the cranial thoracic segments, that serve to inhibit the thoracic limb extensors.

Decerebrate rigidity is characterized by opisthotonus (severe extension of the head and neck) and extension of all limbs. Animals in decerebrate rigidity are usually *not conscious*. This results from a severe brainstem lesion that functionally separates the forebrain from the brainstem. Forebrain, descending input to brainstem structures, which are responsible for extension/flexion of the limbs, is inhibited. The pontine reticulospinal system is responsible for limb extension but is independent of forebrain influence. The medullary reticulospinal system is responsible for limb flexion and is *dependent* on supratentorial influence. By cutting off supratentorial influence the two systems are out of balance and extensor function dominates.

Decerebellate rigidity is characterized by opisthotonos, extension of the thoracic limbs, extension or flexion of the pelvic limbs. If the lesion involves the ventral cerebellum, the pelvic limbs will be extended. The main differentiating feature between decerebrate and decerebellate posturing is the *maintenance of consciousness* associated with decerebellate posture. This posture occurs with severe cerebellar lesions.

REFERENCES

[1]Millis D, et al. **Canine Rehabilitation & Physical Therapy** 2004. Saunders, St. Louis. [2]Stahl C, et al. **Vet Rad and Ultrasound** 2010;51(4):380. [3]Laksito MA, et al. **Aust Vet J** 2011;89(4):117. [4]Forterre F, et al. **J Am Vet Med Assoc** 2010;232(4):559. [5]Shelton D, et al. **Proc ACVIM Forum** 2014.

MAGNETIC RESONANCE IMAGING OF INTRACRANIAL MALFORMATIONS IN DOGS AND CATS

Edward MacKillop, DVM, DACVIM (Neurology)
Pittsburgh, Pennsylvania

INTRODUCTION

Intracranial malformations are a common cause of neurologic deficits in young and developing animals. Both genetic and environmental factors can lead to malformation of the brain during its development. Brain injury *in utero* may be caused by infection (often viral), toxicity, hypoxia-ischemia or trauma.[1-3] The gross lesion, as is seen with intracranial imaging, represents a combination of both developmental malformation and atrophy. Macrophage responses predate astroglial repair within the developing central nervous system (CNS) and so resorption of necrotic tissue may occur without evident astrocytosis or glial scarring to suggest prior injury.[1] Hence, there may be no clear pathologic distinction between primary (developmental) and secondary (atrophic) malformation of the brain. Clinical signs of a CNS malformation in dogs and cats are usually present from a young age although some malformations, such as intracranial cysts, may not cause neurologic dysfunction until adulthood.[4] Magnetic resonance (MR) imaging has revolutionized the diagnosis of intracranial malformations in humans[5] and animals.[6]

HYDROCEPHALUS

An increase in the intracranial volume of cerebrospinal fluid (CSF) is termed hydrocephalus. An excess of CSF may accumulate within the ventricular system (internal hydrocephalus) or, less commonly, in the subarachnoid space surrounding the brain (external hydrocephalus)[7-9]. Hydrocephalus may develop due to obstruction within the ventricular system, insufficient CSF resorption or, less commonly, by overproduction of CSF secondary to choroid plexus neoplasia or hyperplasia.[7,8,10] Obstruction to CSF flow within the ventricular system may arise from a space-occupying mass, malformation, meningoencephalitis or hemorrhage. Hydrocephalus may also be seen with age-associated cerebral atrophy[11] or degenerative encephalopathy,[12] although this reflects a compensatory change due to loss of neural parenchyma.

Congenital hydrocephalus is most commonly seen in toy and brachycephalic breeds of dog. Clinical signs of hydrocephalus include altered mentation, blindness, ataxia, propulsive circling, ventrolateral strabismus and seizures[7,8]. If hydrocephalus develops before closure of the cranial sutures, macrocephaly and large, persistent fontanelle(s) may be noted. There is significant variability in the size of lateral ventricles in dogs, especially amongst breeds[13]; however dilation of the ventricular system beyond the lateral ventricles suggests clinically significant hydrocephalus. Progressive enlargement of the lateral ventricles causes thinning of the cortical mantle, such that sometimes only few millimeters of normal cortex remain. The interthalamic adhesion may appear atrophied and this is best appreciated on a midsagittal T2-weighted (T2W) image. Cerebrospinal fluid may dissect along periventricular white matter creating diverticula and clefts that are readily seen on routine MR sequences.[14] Periventricular (interstitial) edema resulting from transependymal flow of CSF is seen as a rim of hyperintensity surrounding the ventricular system on T2W and fluid attenuated inversion recovery (FLAIR) images.[15] With severe hydrocephalus, the septum pellucidum often atrophies or ruptures leading to continuity between the two lateral ventricles.[13]

Cerebrospinal fluid produced within the ventricular system normally drains into the subarachnoid space through the lateral aperture of the fourth ventricle. The narrowest point in the ventricular system is the mesencephalic aqueduct. Stenosis of the aqueduct leads to obstructive hydrocephalus of the lateral and third ventricles with dilation of the cavity of the olfactory bulb and third ventricular recesses. Severe aqueductal stenosis with fusion of the rostral colliculi has been reported in dogs.[2,16] An aqueductal flow void on transverse T2W images may be seen in normal dogs but has been associated with turbulent CSF flow and hydrocephalus in humans.[15,17] Phase-contrast (PC) MRI, 3D heavily T2W sequences and contrast enhanced MR cisternography have proven useful in determining the etiology and treatment of hydrocephalus in humans but are not routinely used in veterinary imaging.[15]

In cases of severe hydrocephalus, the atrophic cerebral cortex is fragile and may be prone to tearing with trauma causing subdural hematoma formation.[18] Intraparenchymal hemorrhage may also develop as high pressure CSF dissects along white matter tracts.[14] The appearance of intracranial hemorrhage on MR imaging is time dependent and will vary based on the natural evolution of a hematoma. T2*-weighted sequences exploit the natural tendency of iron pigment to cause magnetic field inhomogeneity and are useful in confirming intracranial hemorrhage.

HYDRANENCEPHALY/PORENCEPHALY

Hydranencephaly or porencephaly is seen when there is destruction of cerebral parenchyma *in utero* and brain tissue is replaced by a CSF filled cavity.[1-3,6] This process may be localized to one region of the cerebrum or may involve the majority of one or both cerebral hemispheres. Porencephaly is a fluid filled cavity lined by white matter that may communicate within the ventricular system or subarachnoid space.[1] With hydranencephaly, part or all of the cerebrum is replaced by CSF without any residual neural parenchyma. The underlying causes for porencephaly and hydranencephaly are the same with simply less damage occurring in cases of porencephaly. Common etiologies for hydranencephaly/porencephaly are hypoxia-ischemia, toxicity, trauma or infection (usually viral).[1-3,21] Perinatal infection of cats with feline panleukopenia virus is known to cause cerebellar hypoplasia/atrophy and may occasionally cause hydranencephaly.[22]

HOLOPROSENCEPHALY AND HYPOPLASIA/APLASIA OF THE CORPUS CALLOSUM

Holoprosencephaly (HPE) is a failure of the forebrain to divide normally into two discrete cerebral hemispheres and exists in three forms: alobar, semilobar and lobar.[1,23] Lobar HPE is the least severe variant with fusion of only the most rostral and ventral portions of the cerebral hemispheres. Hypoplasia/aplasia of the corpus callosum (CC) is seen as a feature of HPE but may be identified as an isolated malformation. Normal CC anatomy (rostrum, genu, body, splenium) is best seen on a midsagittal MR image. Hypodipsic hypernatremia has been reported with both lobar HPE and hypoplasia/aplasia of the CC in dogs[24,25]. Staffordshire Bull Terriers and Miniature Schnauzers are overrepresented in reports of HPE suggesting a breed predisposition[24,25]. Rostrocerebellar diverticulum and Chiari-like malformation have been reported as concurrent abnormalities with HPE and CC hypoplasia/aplasia in dogs[25].

LISSENCEPHALY

Lissencephaly is a disorder of cortical neuronal migration in which the developing brain to fails to form the normal surface convolutions, giving the cerebral surface a smooth appearance.[1,2] There may be a complete absence of gyri (agyria) or in milder cases abnormally broad folds (pachygyria).[26] Histologically, the cerebral cortex is thickened and lacks the normal laminar pattern of organization, with only four discernible layers rather than the normal six.[1,2] In humans with lissencephaly, the pathologic phenotype varies dependent upon the genotypic mutation; for example, an anterior to posterior gradient (A>P) of pachygyria to agyria (i.e. more severe changes in the posterior lobes) is associated with mutations in LIS1 (also known as PAFAH1B1) or TUBA1A genes.[26] In contrast, classical lissencephaly with a P>A gradient in males is most commonly associated with a mutation in the DCX gene. Descriptions of lissencephaly in dogs and cats are limited but there have been reports in both species with MR diagnosis of lissencephaly in dogs.[2,3,27] Transverse MR images of the brain demonstrate a smooth, thickened cortical mantle with some preservation of gyri/sulci frontally and ventrally.[27] In dogs there is a breed predilection for Lhasa Apso dogs, suggesting a hereditary basis for this malformation.[2,27] Clinical signs of lissencephaly primarily reflect a forebrain disturbance: abnormal mentation, behavioral abnormalities (e.g. problematic house training), delayed postural reactions and epileptic seizures.[27] Lissencephaly with cerebellar hypoplasia (LCH) has been reported in both dogs and cats; this combination is characteristic of mutations in the RELN and TUBA1A genes in people.[2,27] A cat with microencephaly, LCH and hypoplasia of the corpus callosum was not found to have any significant mutation in the TUBA1A gene.[28]

POLPMICROGYRIA

Polymicrogyria is a rare malformation of the cerebrum in domestic animals characterized by an excessive number of small, histologically anomalous gyri.[1,2] Polymicrogyria is a primary brain malformation in both dogs and humans but may also be seen in people following intrauterine cerebral ischemia or *in utero* infection.[1,29] In dogs, polymicrogyria has been reported in a number of Standard Poodles, where it is suspected to have a hereditary basis, and a single Golden Retriever.[29,30] Polymicrogyria may be generalized or lobar in distribution but has been consistently found in the occipital lobes of affected dogs. In Standard Poodles, polymicrogyria has been detected antemortem via MR imaging.[29,30] Polymicrogyria is best appreciated on T2W dorsal plane images but can be seen on routine transverse T2W images. All Standard Poodles with polymicrogyria were found to have thinning of the cortical mantle with variable lateral ventriculomegaly that may be asymmetrical.[29] Neurologic deficits due to polymicrogyria are varied but all affected animals have central blindness or visual deficits, which is consistent with the occipital lobe distribution of the lesion. Naturally occurring polymicrogyria has not been documented in cats.

MENINGOCOELE AND MENINGOENCEPHALOCOELE

Meningoencephalocele is a congenital protrusion of brain tissue and meninges covered by skin through a defect in the skull (cranioschisis).[1,2] If there is protrusion of just meninges without any brain tissue then this is referred to as a meningocele. Most meningocele or meningoencephalocele will be clinically detectable from birth as a fluctuant swelling palpable over the skull.[31] Ethmoidal and basal meningoencephalocele may only be detected with advanced imaging. Ethmoidal meningoencephalocoeles (EME) have been documented with both CT and MR imaging in dogs.[32,33] With EME, there may be unilateral or bilateral herniation of the olfactory bulb into the nasal cavity. Seizures, behavioral abnormalities and inappropriate mentation have been reported with EME.

INTRACRANIAL CYSTS

Arachnoid cysts

Intra-arachnoid accumulations of CSF around the brain or spinal cord have been called arachnoid cysts.[34-36] However, these lesions lack a true epithelial lining and are more appropriately defined as arachnoid pseudocysts or diverticula. The term "arachnoid cyst" is pervasive throughout the human and veterinary literature and so, despite being a misnomer, it is unlikely that these lesions will be reclassified.[6] Intracranial arachnoid cysts (IAC) are usually considered a primary malformation although they may develop secondary to meningoencephalitis, trauma or subarachnoid hemorrhage.[34-36] Intracranial arachnoid cysts in dogs and cats are predominantly found in the caudal fossa with the majority cysts occurring dorsal to the colliculi and rostral to the cerebellum (quadrigeminal cistern in humans).[34,35] The distinction between quadrigeminal cistern arachnoid cyst and diverticulum of the caudodorsal third ventricle or rostral aspect of the fourth ventricle may not be clear with conventional MR imaging. The term rostrocerebellar diverticulum accurately describes all three of these abnormalities.[6] Contrast enhanced MR cisternography (CE-MRC) is a valuable technique for distinguishing hydrocephalus from IAC in people[37].

Clinical signs associated with IAC are variable and many have been detected as an incidental finding.[56,59] Small and brachycephalic breeds of dog, especially the Shih Tzu, are overrepresented in all reports of IAC and age at diagnosis is highly variable (3 months to 10 years).[34,36] In dogs, there are rare reports of arachnoid cysts within the 4th ventricle causing severe obstructive hydrocephalus and syringohydromyelia[38]. All reports of feline IAC have been in young Perisan cats.[35] Arachnoid cysts may not be readily distinguished from other intracranial cysts with CT but have a distinct appearance on MR imaging. On MR, the signal characteristics of fluid within an arachnoid cyst is identical to CSF on all sequences unless there has been hemorrhage within the cyst.[34,36] Most IAC are best appreciated on sagittal T2W images. Enhancement of an arachnoid cyst following intravenous contrast is not expected unless there is concurrent hemorrhage or meningoencephalitis.[6,36]

Epidermoid and Dermoid Cysts

Intracranial epidermoid and dermoid cysts develop following aberrant inclusion of non-neural ectoderm during neural tube closure or invagination of secondary neural vesicles, such as the otic vesicle.[39,40] These cysts are lined by stratified squamous epithelium and gradually enlarge as desquamated keratinocytes, keratinaceous debris and cholesterol accumulate in the lumen.[39,40] Dermoid cysts also contain epidermal adnexal structures such as hair follicles, sebaceous and apocrine glands, and fat.[40] Intracranial epidermoid and dermoid cysts have been reported sporadically in dogs, horses, rats and mice.[2,41-43] They are invariably found in the caudal fossa; most are within the fourth ventricle. Clinical signs may not manifest until late in life and epidermoid cysts are occasionally an incidental finding at necropsy.[41,43] IAC and epidermoid cysts appear as similar homogeneous hypoattenuating masses on CT, although they may be distinguished by MR.[41] Epidermoid and IAC are usually hypointense in T1W images and hyperintense in T2W images, whereas epidermoid cysts remain iso- to hyperintense in FLAIR images while IAC are hypointense.[6] Diffusion weighted imaging (DWI) has also been used to distinguish IAC from epidermoid cysts in humans.

In contrast, dermoid cysts tend to be heterogeneous, but predominantly hyperintense on both T1W and T2W fast spin-echo images due to their fat content.[6,42] Regions of low signal may reflect hair or calcification within the cyst.[42] Dermoid cysts may have low signal on fat suppression sequences, a feature that aids differentiation from other causes of high signal on T1W images, such as methemoglobin or melanin. Enhancement of the rim of an epidermoid or dermoid cyst may be observed after gadolinium administration. Rupture of the wall of a dermoid cyst can release fat droplets into the CSF.[6] Acute meningoencephalitis was described in a dog following rupture of an epidermoid.[4]

Choroid Plexus Cyst

There are two reports of choroid plexus cysts of the fourth ventricle in dogs diagnosed with CT and MR imaging.[6,44,45] On MR, the cyst was hypointense to brain parenchyma on T1W images and hyperintense on T2W images.[45] Unlike other intracranial cysts, there was marked, homogenous contrast enhancement of the cyst wall.

MALFORMATIONS OF THE CEREBELLUM

Generalized hypoplasia of the cerebellum occurs as both a primary malformation and secondary to viral infection.[1,2] In cats, cerebellar hypoplasia and atrophy is well documented following *in utero* or perinatal infection with feline panleukopenia virus.[2,22] Canine parvovirus may rarely cause a similar lesion in dogs.[46] Congenital cerebellar hypoplasia must be distinguished from cerebellar cortical atrophy, in which the cerebellum is normal at birth and subsequently loses parenchyma. On MR, animals with cerebellar cortical abiotrophy will have a normal to slightly small cerebellum with an increase in CSF between the folia. There is often greater conspicuity of cerebellar white matter due to loss of grey matter. This lesion can be subtle on transverse images but is clearly visible on a T2W midsagittal image of the brain. This should be readily distinguishable from animals with cerebellar hypoplasia in which the cerebellum is significantly smaller than normal with a greater volume of CSF surrounding then entire cerebellum.

Complete absence of the cerebellar vermis or its caudal aspect has been described in dogs with or without hypoplasia of the cerebellar hemispheres.[47-49] On MR, the vermis is either absent or hypoplastic and there is a large cystic region with signal characteristics of CSF between the cerebellar hemispheres.[49] In humans, cerebellar vermian aplasia or hypoplasia, cystic dilatation of the fourth ventricle and enlargement of the posterior fossa is the Dandy-Walker malformation (DWM).[46] Dandy-Walker variant is a term used to describe milder cases of DWM or those without posterior fossa enlargement; however, there is no consensus on this terminology in the human medical literature.[46] Concurrent hydrocephalus, agenesis of the corpus callosum and polymicrogyria may also be seen with but are not required for the diagnosis of DWM.[1] Dandy-Walker-like malformation has been described in dogs with antemortem diagnosis using MR imaging.[2,46] Cerebellar vermian hypoplasia and Dandy-Walker-like malformation should be included in the differential diagnosis of cystic lesions of the caudal fossa but should be readily distinguished from other intracranial cysts based on abnormal cerebellar anatomy.

HAMARTOMA

Hamartoma is a disorganized, non-neoplastic proliferation of cells and tissue that normally occur in an organ.[6] Most intracranial hamartoma are either neural or vascular, with clinical signs referable to the location of the lesion.[51-54] Neural hamartomas are rare in both humans and domestic animals. A single case of hypothalamic hamartoma has been described in a dog without antemortem imaging.[52] Hypothalamic hamartoma in people are seen on MR as noncontrast-enhancing, noncalcified masses that are isointense to grey matter on T1W images and iso- to hyperintense on T2W images.[5]

Vascular hamartoma may occur anywhere in the body but tend to have the most serious consequences when present in or around the CNS. Most descriptions of intracranial vascular hamartoma are in older or aged dogs.[50,52,55] There is a single case report of a cerebellar vascular hamartoma in a young cat.[53] The most widely accepted classification system for intracranial vascular hamartomas in people divides these lesions into arteriovenous malformations (AVM), cavernous malformation (CVM), capillary telangiectases and venous malformations.[54,56] Arteriovenous and cavernous malformations are clinically relevant because of their predisposition to acute hemorrhage and/or thrombosis. The MR appearance of both AVM and CVM are well described in humans with occasional reports in dogs and cats.[52-56] Arteriovenous malformation are prone to spontaneous hemorrhage/hematoma that will vary on MR dependent on the chronicity of hemorrhage.[56] Intraventricular and subarachnoid hemorrhage may be seen with AVM in people. The nidus of an AVM is a tangle of interconnected arteries and veins that is seen on MR as a cluster of signal voids due to rapidly flowing blood.[6,54] Gadolinium enhancement is seen in the slower flowing, usually venous channels, within the nidus. Cavernous malformations, also known as cavernomas or cavernous angiomas, are made up of large thin walled sinuses without intervening neural parenchyma.[54,55,56] With MR, CVM are lobulated heterogeneous lesions with a peripheral hypointense ring on T1W and T2W images representing hemosiderin deposition secondary to chronic hemorrhage.[56] Cavernous malformations typically enhance with gadolinium.[56] Gradient echo sequences are useful for confirming the presence of intracranial hemorrhage associated with both AVM and CVM. The primary differentials for vascular hamartoma on MR imaging are spontaneous hematoma and hemorrhagic neoplasm.

REFERENCES

1. Harding BN, Copp AJ. Malformations. In: Graham DI, Lantons PL (eds): Greenfield's Neuropathology. London: Arnold, 2002;357.
2. Summers BA, Cummings JF, de Lahunta A. Malformations of the Central Nervous System. In: Summers BA, Cummings JF, de Lahunta A (eds): Veterinary neuropathology. St. Louis: Mosby, 1995;68-94.
3. de Lahunta A, Glass E. Development of the nervous system: Malformation. In: de Lahunta A, Glass E (eds): Veterinary neuroanatomy and clinical neurology. St. Louis: Saunders, 2009;23.
4. O'Brien DP et al. J Am Anim Hosp Assoc 1990; 26:582.
5. Caruso PA, Roberston R, Setty B, Grant Ellen. Disorders of brain development. In: Atlas, SA (ed): Magnetic resonance imaging of the brain and spine, 4th edition. Philadelphia. Lippincott Williams & Wilkins, 2009;194.
6. MacKillop E. Vet Radiol Ultrasound. 2011;52:S42
7. Harrington ML et al. Vet Clin North Am: Small Anim Pract 1996;26:843.
8. Thomas WB Vet Clin North Am: Small Anim Pract 2010; 40:143.
9. Dewey CW et al. J Am Anim Hosp Assoc 2003;39:567.
10. Summers BA et al. Vet Pathol 1983;20:513.
11. Hasegawa D et al. Vet Radiol Ultrasound 2005;46:452.
12. Asakawa MG et al. Vet Radiol Ultrasound 2010;451:155.
13. De Haan CE et al. Vet Radiol Ultrasound 1994;35:83.
14. Wünschmann A et al. Vet Pathol 2001;38:67.
15. Kartal MG et al. Insights Imaging. 2014;5:531.
16. Ryan CT et al. Vet Radiol Ultrasound. 2014;55:292.
17. Scrivani PV et al. Vet Radiol Ultrasound. 2009;50:269.
18. Nykamp S et al. Vet Radiol Ultrasound 2001;42:511.
19. Schmidt MJ et al. Vet Radiol Ultrasound. 2012;53:142.
20. Davies ES et al. Vet Rec. 2012 Feb;170:179.
21. Govaert P. Semin Fetal Neonatal Med. 2009;14:250.
22. Sharp NJ et al. J Comp Pathol 1999;121:39.
23. Hahn JS et al. 2004;31:79.
24. Shimokawa et al. J Vet Med Sci 2009;71:1387.
25. Gonçalves R, et al. J Vet Intern Med. 2014;28:1275.
26. Fry AE et al. Am J Med Genet C Semin Med Genet. 2014;166C:198.
27. Saito M et al. Vet Radiol Ultrasound. 2002;43:331.
28. Herrmann A et al. Tierarztl Prax Ausg K Kleintiere Heimtiere. 2011;39:116.
29. Van Winkle et al. Prog in Vet Neurol 1995;5:66.
30. Jurney C et al. J Vet Intern Med 2009;23:871.
31. Dewey CW et al. Vet Surg. 2011 Jun;40:473.
32. Martlé VA, et al. Vet Surg 2009;38:515.
33. Jeffery N. J Small Anim Pract 2005;46:89.
34. Matiasek LA et al. J Vet Intern Med 2007;21:1021-1026.
35. Lowrie M et al. J Fel Med Surg 2009;11:711.
36. Vernau KM et al. Vet Radiol Ultrasound 2002;43:449.
37. Algin O et al. AJNR Am J Neuroradiol. 2013;34:14.
38. Bazelle N et al. Vet Radiol Ultrasound. 2014 Nov 11. doi: 10.1111/vru.12218
39. Netsky, MG. Surg Neurol 1988;29:477-483.
40. Osborn AG et al. Radiology 2006;239:650.
41. Platt SR et al. Vet Radiol Ultrasound 1999;40:454.
42. Targett MP et al. Vet Radiol Ultrasound 1999;40:23.
43. Kornegay JN et al. Vet Pathol 1982; 19:646.
44. Galano HR et al. Vet Radiol Ultrasound 2002;43:349.
45. Brewer DM et al. J Small Anim Pract 2010;51:169.
46. Schatzberg SJ et al. J Vet Intern Med 2003;17:538.
47. Kornegay JN. Vet Pathol 1986;23:374.
48. Niesen CE. Semin Pediatr Neurol 2002;9:320.
49. Schmidt MJ et al. Vet Radiol Ultrasound 2008;49:264.
50. Smith SH, et al. Vet Pathol 2001;38:108.
51. Cook RW. Vet Pathol 1977;14:138.
52. Thomas WB et al. Prog Vet Neurol 1995;6:20.
53. Stalin CE et al. J Feline Med Surg 2008;10:206.
54. Zabramski JM et al. Neurosurg Clin N Am 1999;10:395.
55. Schoeman JP et al. J S Afr Vet Assoc 2002;73:207.
56. Atlas SW, Do HM. Intracranial vascular malformations and aneurysms. In: Atlas, SA (ed): Magnetic resonance imaging of the brain and spine, 4th edition. Philadelphia: Lippincott Williams & Wilkins, 2009;695.

BIOMARKERS IN VETERINARY NEUROLOGY

Christopher L. Mariani, DVM, PhD, DACVIM (Neurology)
Raleigh, NC

INTRODUCTION

A biomarker may be as broadly defined as "any observable feature that can be used to inform on the state of the patient", and in this context may include biochemical or molecular species, monitoring parameters (e.g., blood pressure, spO_2), imaging characteristics, and other parameters.[1] The term is most commonly used to refer to biochemical or molecular species analyzed within a biological sample, but this also includes a broad range of substances (e.g., proteins, electrolytes, amino acids, sugars, lipids) assayed from a variety of sources such as biological fluids (blood, urine, cerebrospinal fluid [CSF]) or tissues. This definition is also quite broad and would comprise parameters assessed as part of routine complete blood counts, serum biochemical panels, urinalyses, CSF evaluations or special immunohistochemical markers on tissue biopsies (e.g., vimentin or glial fibrillary acidic protein [GFAP] expression). The focus of this talk will be novel analytes within the blood and CSF of veterinary patients with neurologic disorders that might be used to aid in clinical decision-making.

The information gained from such analyses might take several forms but we will focus here on improvements in diagnosis, therapy and prognostication. Conventional diagnostic tests routinely used in veterinary neurology include diagnostic imaging (e.g., magnetic resonance imaging, computed tomography, myelography, radiography), CSF analysis and electrodiagnostic tests. Although effective in reaching a definitive or presumptive diagnosis in many patients, a substantial number of animals with neurologic disease cannot be diagnosed with such a strategy. Specific challenges include the differentiation of some inflammatory and neoplastic conditions, differentiation of stroke from inflammation, and weighing the relative contribution of a disease process to dysfunction when there are several disorders or comorbidities present. Histopathological examination of a tissue biopsy remains the gold standard for diagnosis but the central nervous system (CNS) provides challenges to tissue procurement not faced with other organ systems. Although biopsy of CNS lesions is becoming more widespread, such procedures are invasive, costly, associated with potential morbidity, require specialized interpretation and are frightening for many owners. Thus a biomarker obtained through less invasive means that could provide information relative to diagnosis would be a welcome addition to the clinician's toolbox.

In addition to providing information on broad categories of disease (e.g., inflammation vs. neoplasia), potential diagnostic uses for biomarkers include differentiation of different neoplasms or inflammatory conditions that appear similar with other tests (e.g., meningioma vs. histiocytic sarcoma; necrotizing meningoencephalitis [NME] vs. granulomatous meningoencephalitis [GME]) or correlation with the grade of a neoplasm (and therefore impacting therapeutic options and prognosis). In terms of treatment, a biomarker might theoretically predict a response to a specific therapeutic intervention or might be used to monitor the response of the patient to the therapy. Finally, a marker that correlated to ultimate prognosis might be quite useful in some scenarios. One example would be a marker that might better predict the response of dogs to decompressive surgery for intervertebral disk herniation when there is a complete spinal cord injury (i.e., absent nociception).

BIOMARKERS IN SPECIFIC DISEASE CONDITIONS

Intervertebral Disk Disease

A variety of CSF biomarkers have been evaluated in dogs with intervertebral disk disease, including glutamate, beta-2-microglobulin, myelin basic protein, creatine kinase (CK), lactate, aspartate aminotransferase (AST), matrix metalloproteinase-9 (MMP-9), cartilage oligomeric matrix protein, clusterin, glial fibrillary acidic protein (GFAP), and tau protein.[2-11] Some studies have shown increases in these markers in dogs with IVDD relative to control dogs, which might provide diagnostic information.[4,7] Somewhat more importantly, other studies have attempted to show an association of these CSF markers with the severity of injury,[3,7,8] or the likelihood of recovery/prognosis.[2,5,6,8,11,12]

Several serum markers have also been evaluated in dogs with intervertebral disk herniation. Sato et al. found that dogs with progressive myelomalacia after acute thoracolumbar intervertebral disk herniation were more likely to have detectable serum GFAP than dogs with disk herniation but without myelomalacia.[12] Nishida et al. assessed phosphorylated neurofilament subunit NF-H in the serum of dogs with acute intervertebral disk herniation and found differences between dogs with different injury severities as well as higher concentrations in dogs that did not regain the ability to ambulate after surgery.[13]

Central Nervous System Inflammation

C-reactive protein (CRP) is an acute phase protein produced within the liver that is upregulated with conditions causing systemic inflammation. Dogs with steroid-responsive meningitis-arteritis (SRMA) have dramatically elevated serum CRP concentrations, which decrease dramatically after effective immunosuppressive therapy.[14,15] Other acute phase proteins show similar changes in dogs with SRMA, including serum amyloid A (SAA) and alpha-1-acid glycoprotein (AGP). With clinical signs consistent with relapse of the disease, CRP and SAA concentrations are again elevated, even when CSF cytology is within normal range.[14,15] Thus, CRP and SAA may be useful biomarkers in supporting a diagnosis of SRMA as well as to detect relapse of disease. Our laboratory (The Comparative Neuroimmunology & Neurooncology Laboratory [CNNL]) has evaluated serum CRP concentrations in dogs with other forms of CNS inflammation, including GME and NME. Concentrations in these dogs appear to be much more variable than in those with SRMA, which is consistent with the limited published results from other groups.[15]

Serum immunoglobulin A (IgA) is also elevated above the reference range in dogs with untreated SRMA. However, unlike some of the acute phase proteins, this analyte does not vary substantially with effective therapy (or at relapse).[16,17] Thus, although having some utility in achieving an initial diagnosis, serum IgA has limited use in monitoring the response to therapy over time.

Miyake et al. found that pugs with NME had increased concentrations of GFAP in their serum.[18] Interestingly, such elevations were not found in other CNS diseases or in other breeds with NME. Thus, serum GFAP may be a useful diagnostic biomarker for NME in the pug.

GFAP within the CSF has also been evaluated as a marker for NME, and dogs with this disease had higher GFAP concentrations than healthy controls, dogs with noninflammatory CNS diseases and dogs with other inflammatory CNS diseases.[19] This research group and others have also shown the diagnostic utility of anti-GFAP autoantibodies within the CSF of dogs with NME.[19-23] However, serum autoantibody concentrations, although detectable in dogs with NME, are much less discriminatory, as they are increased in dogs with other diseases, including disorders outside of the CNS.[24]

Following a proteomic analysis of CSF from dogs with meningoencephalitis of unknown etiology, Nakamura et al. identified neuron-specific enolase (NSE) as a potential biomarker of MUE.[25] Dogs with MUE had increased NSE concentrations in CSF compared to normal controls, and this marker was able to discriminate these two groups of dogs with excellent accuracy. However, dogs with other CNS disease may show elevations in CSF NSE (see below), and this marker requires further characterization before routine clinical use.

The acute phase proteins CRP, SAA, and AGP have been evaluated within the CSF as biomarkers for SRMA and showed apparent increased concentrations. However, these were more variable and less dramatic than the changes seen in serum and were not useful for detecting relapses.[14,15] These results likely reflect the predominant production of these proteins within the liver and release into the serum, with subsequent leakage into the CSF after CNS inflammation. Similar to serum IgA, CSF IgA concentrations are dramatically increased in dogs with SRMA, but do not change substantially with therapy or with relapse of clinical signs.[16,17]

Other analytes that have been evaluated as potential biomarkers in dogs with CNS inflammation include beta-2-microglobulin, interleukin-6, transforming growth factor beta-1, and vascular endothelial growth factor (VEGF).[4,26] Our laboratory has evaluated the expression of VEGF within the CSF of dogs with meningoencephalitis and found increased concentrations when compared to normal controls or dogs with other CNS disorders (excepting neoplastic disease; see below).[27] Matrix metalloproteinase-9 is an enzyme responsible for the degradation of extracellular matrix and is expressed by a number of cells including leukocytes, which utilize MMP-9 to move from the blood into the CNS. Normal canine CSF does not have detectable MMP-9.[28] However, studies in the CNNL have shown MMP-9 in the CSF of substantial numbers of dogs with meningoencephalitis and meningomyelitis. We have also documented increased concentrations of a variety of cytokines and chemokines within the CSF of dogs with inflammatory CNS disease, notably IL-2 and CCL2.[29]

Lactate is being increasingly investigated as a biomarker of disease in dogs. Although first evaluated in blood, CSF lactate is receiving increased attention from the veterinary community and has been evaluated in normal dogs and those with a variety of experimental and naturally occurring diseases.[11,30-35] In human patients CSF lactate can be used to discriminate bacterial from aseptic meningitis and potentially subtypes of multiple sclerosis.[36-38] Studies at the CNNL have shown substantial elevations in CSF lactate in a subset of dogs with inflammatory CNS disease although the underlying reasons for these changes are still under investigation.

Central Nervous System Neoplasia

Relatively fewer studies have been published evaluating novel biomarkers in dogs with CNS neoplasia. However, VEGF is one molecule that has received some attention, due to the widely recognized role of VEGF in promoting angiogenesis, a necessary factor for sustained neoplastic growth. VEGF was detectable in the plasma of a subset of dogs with intracranial tumors and was more likely to be detected in dogs with astrocytomas than those with meningiomas and oligodendrogliomas. In addition, VEGF concentrations were greater in dogs with higher-grade tumors.[39]

In humans, elevated concentrations of VEGF have been documented within the CSF of patients with meningeal metastases,[40,41] CNS leukemia[42] and astrocytomas.[43] Our laboratory has examined CSF VEGF concentrations in dogs with a variety of intracranial neoplasms and found increased concentrations compared to dogs with other CNS diseases.[27] In addition, the CSF from patients with gliomas and choroid plexus tumors had higher concentrations than CSF from patients with other neoplasms.

Work at the CNNL has also examined the expression of MMP-9 within the CSF of dogs with intracranial neoplasms. We found expression of MMP-9 in the CSF of these patients, which was more prevalent in dogs with certain tumors, particularly round cell and choroid plexus tumors.[44] We have also begun investigating CSF lactate in these patients, and have documented increased lactate concentrations in a subset of dogs.

Other Neurologic Conditions

Both clusterin and myelin basic protein have been evaluated as potential biomarkers within the CSF of dogs with degenerative myelopathy, and showed increased concentrations relative to control dogs.[9,45] Vascular endothelial growth factor concentrations within the CSF were significantly increased in dogs with experimentally induced hypothyroidism relative to normal control dogs.[46] Dogs with GM1-gangliosidosis showed elevated CSF concentrations of AST, lactate dehydrogenase, LDH, NSE and myelin basic protein relative to normal control dogs.[47]

CONCLUSIONS

Although biomarkers of disease have been used for decades in veterinary and human medicine, interest in novel biomarkers has intensified in recent years. These markers hold the promise of improving diagnostic capabilities, guiding and monitoring therapy and refining the prognosis for patients with neurological disorders. The main biological sources for such patients are serum and cerebrospinal fluid, although other sources (e.g., urine) are possible. Although such work is encouraging, each biomarker must be carefully validated and tested before routine clinical use. Ultimately, it is likely that panels of different biomarkers will improve upon the diagnostic, therapeutic and prognostic utility of individual analytes for various disorders.

REFERENCES

[1]Carpenter KL, et al. Front Neurol 2015;6:26. [2]Levine GJ, et al. J Vet Intern Med 2010;24:890. [3]Levine JM, et al. Am J Vet Res 2006;67:283. [4]Munana KR, et al. Vet Clin Path 2007;36:173. [5]Nagano S, et al. Res Vet Sci 2012;93:1487. [6]Nagano S, et al. Res Vet Sci 2011;91:482.
[7]Olby NJ, et al. J Neurotrauma 1999;16:1215. [8]Roerig A, et al. Vet J 2013;197:253. [9]Shafie IN, et al. Cell Stress Chaperones 2014;19:311. [10]Tokunaga S, et al. Spine 2010;35:4. [11]Witsberger TH, et al. J Am Vet Med Assoc 2012;240:555. [12]Sato Y, et al. J Vet Med Sci 2013;75:949. [13]Nishida H, et al. Vet Surg 2014;43:289. [14]Lowrie M, et al. Vet J 2009;182:125. [15]Bathen-Noethen A, et al. J Vet Intern Med 2008;22:1149. [16]Lowrie M, et al. J Vet Intern Med 2009;23:862. [17]Maiolini A, et al. Vet J 2012;191:219. [18]Miyake H, et al. J Vet Med Sci 2013;75:1543. [19]Toda Y, et al. Vet Rec 2007;161:261. [20]Shibuya M, et al. J Vet Med Sci 2007;69:241. [21]Uchida K, et al. Vet Pathol 1999;36:301. [22]Matsuki N, et al. J Vet Med Sci 2004;66:295. [23]Matsuki N, et al. J Vet Med Sci 2009;71:99. [24]Fujiwara K, et al. Vet Rec 2008;162:592. [25]Nakamura K, et al. J Vet Med Sci 2012;74:751. [26]Maiolini A, et al. BMC Vet Res 2013;9:23. [27]Niman ZE, et al. J Vet Intern Med 2010;24:697. [28]Bergman RL, et al. Am J Vet Res 2002;63:1359. [29]Mariani CL. Annual Forum of the American College of Veterinary Internal Medicine, Montreal 2009. [30]Bashar AH, et al. Interact Cardiovasc Thorac Surg 2008;7:262. [31]Javaheri S, et al. Am J Respir Crit Care Med 1994;150:78. [32]Pugliese M, et al. Prog Neuropsychopharmacol Biol Psychiatry 2005;29:603. [33]Sugi T, et al. Stroke 1975;6:715. [34]Galan-Rodriguez A, et al. Vet Rec 2013;173:249. [35]Caines D, et al. Can J Vet Res 2013;77:297. [36]Majwala A, et al. Am J Trop Med Hyg 2013;88:127. [37]Kleine TO, et al. Brain Res Bull 2003;61:287. [38]Jongen PJ, et al. J Neurol Neurosurg Psychiatry 1997;63:446. [39]Rossmeisl JH, et al. Am J Vet Res 2007;68:1239. [40]Groves MD, et al. J Neurooncol 2009;94:229. [41]Stockhammer G, et al. Neurology 2000;54:1670. [42]Tang YT, et al. Leuk Res 2013;37:359. [43]Sampath P, et al. Cancer Control 2004;11:174. [44]Mariani CL, et al. Am J Vet Res 2013;74:122. [45]Oji T, et al. Vet Clin Path 2007;36:281. [46]Pancotto T, et al. Vet Clin Path 2010;39:485. [47]Satoh H, et al. Brain Res 2007;1133:200.

TRANSLATING NEXT GENERATION GENOMICS INTO A CLINICAL DIAGNOSIS

Dennis P. O'Brien, DVM, PhD, DACVIM (Neurology)
Columbia, Missouri

Advances in genomic research are rapidly expanding our understanding of disease and clinicians must keep pace. The ready availability of artificial insemination and frozen semen means that a popular sire can be widely used. Unfortunately, he will also be carrying recessive genes that can contribute to disease as well as those for the desirable traits that made him popular. It is not until the F2 and later generations that the detrimental genes are manifest, and by then they may be widely distributed in the breed (Figure 1). Next generation sequencing permit the evaluation of the entire genome of an individual at rapidly decreasing cost. This will allow breeding stock to be screened for such mutations prior to their appearance. It has also become a tool for diagnosing unknown diseases in human medicine. To use this new tool effectively, the clinician must understand the language of genomics and the limitations as well as the strength of various gene discovery strategies and tests.

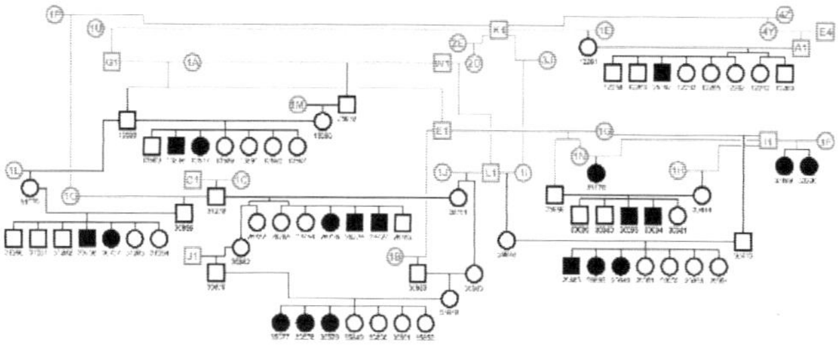

Figure 1: *A popular sire will spread hidden, disease-causing recessive alleles along with the desirable traits that made him popular. It is not until F2 or later generations that the disease becomes apparent (shaded symbols) and by then those alleles can be widely disseminated in the population.*

SPEAKING THE LANGUAGE OF GENOMICS

The terminology used in DNA studies and describing mutations may not be familiar to many clinicians. This section provides an overview of how genes are named and mutations described. Table 1defines some commonly used terms.

The names of genes are expressed as acronyms that are italicized (e.g. *RAB3GAP1*) to differentiate them from the protein (e.g. Rab3GAP1) or other acronyms. Domestic animal genes have followed the convention used for human genes of using all capital letters while rodent gene names use lower case after the first capital (*Rab3gap1*). Like many things in biology and medicine, the names applied to genes evolve over time as we learn more about their biology, which can lead to confusion when the names vary in the literature.

Many genes are named based on the disease phenotype. For example, the neuronal ceroid lipofuscinoses (NCLs) are a group of lysosomal storage diseases characterized by accumulation in neurons of autofluorescent proteinaceous material that resembles ceroid and lipofuscin. The genes involved are named the *CLN* genes (Ceroid Lipofuscinosis, Neuronal). Genes often are numbered consecutively as different phenotypes or different modes of inheritance were recognized in humans. *CLN1, CLN2, CLN3* and *CLN4* are the first 4 phenotypes described and are differentiated by age of onset: infantile, late-infantile, juvenile and adult respectively[1]. In contrast, the genes for the hereditary peripheral neuropathies Charcot-Marie-Tooth disease were divided initially by type of neuropathy with *CMT1* being predominantly axonal and *CMT2* predominantly demyelinating. Both the initial diseases were dominant traits and when autosomal recessive forms were identified the axonal and demyelinating versions were name *CMT3* and *CMT4* respectively[2].

Alternatively, the name of a gene may reflect a biochemical or structural component of the nervous system such as an enzyme or ion channel. In this instance, the acronym will abbreviate the molecule and often will be followed by a number or letter referring to a subunit or subtype. For example, *GRM1* refers to the gene which codes for the glutamate receptor, metabotropic, type 1. Often, as more is learned about a disease and gene, the molecular name replaces the phenotype-based name. For example, the original name for *GRM1* was *SCAR13*, the 13th form of autosomal recessive spinocerebellar ataxia to be recognized in humans[1]. Some genes have whimsical, quirky names such as sonic hedgehog (*SHH*) which was named after the spikey appearance of fruit fly larvae with a mutation in the gene. Knowing the origin provides a context to understand the name, but some clinicians have objected to such names because it can be difficult to discuss with a client a very serious disease when the name of the gene sounds frivolous[3].

Table 1: Definitions of some terms used in genetics[1]

Phenotype: the physical manifestations of a trait which can be influenced by genetics, epigenetics, and environment
Genotype: the genetic makeup of an individual with respect to the gene being considered (the alleles)
Allele: one of alternate forms of a gene or locus
Segregation: the separation of alleles during meiosis which determines which allele is passed to an offspring
Recombination: the two copies of a chromosome may cross over during meiosis which results in exchange of genetic material and a new combination of alleles.
Locus: the location of a gene or genes on a chromosome; often used as a vague proxy for gene in gene mapping studies before the specific gene is identified
Codon: the 3 nucleotide code for an amino acid or the initiation or stop of protein translation.
Variant: variation in the DNA sequence of an individual from the reference sequence. This could be a disease causing mutation or it could be neutral.
Single nucleotide polymorphisms (SNPs); variants where only a single nucleotide is substituted (e.g. A>G); SNPs are the most common type of variant and are for example used to identify differences between individual chromosomes for association studies
Frame shift: deletion or insertion of any number of nucleotides not divisible by 3 shifting the reading frame of the codons which follow. This in turn changes the subsequent amino acid sequence and the location of the stop codon.

Mutations in a gene are named based on the nucleotide or amino acid affected and the change the mutation produces. The term mutation is often used to refer to a disease-causing variant, although geneticists prefer to use the term variant to describe any difference in nucleotide or amino acid sequence, regardless of whether it is common or rare and regardless whether it is functionally neutral or causally involved in a specific phenotype. The term polymorphism is used to refer to a non-disease causing variant or a variant with a high frequency in the population. Such variants are utilized to identify difference between individuals for gene mapping purposes.

The nomenclature for a specific mutation begins with the name of the protein or gene, followed by a "p" or a "c" to indicate whether an amino acid change in the protein or a nucleotide change in the coding sequence respectively are being described (Figure 2). A simple change in the sequence will be described as shown in the figure. A change in a codon that creates a premature stop codon which signals the ribosome to stop translating the RNA sequence into protein will usually lead to a non-functional protein. Such a change is indicated as "X" or "*". An insertion or deletion of nucleic acids (indicated by "_") will shift the reading frame of the sequence and change to subsequent codons often producing a premature stop codon. This is indicated by "fs" for frameshift and then the position where the premature stop would occur.

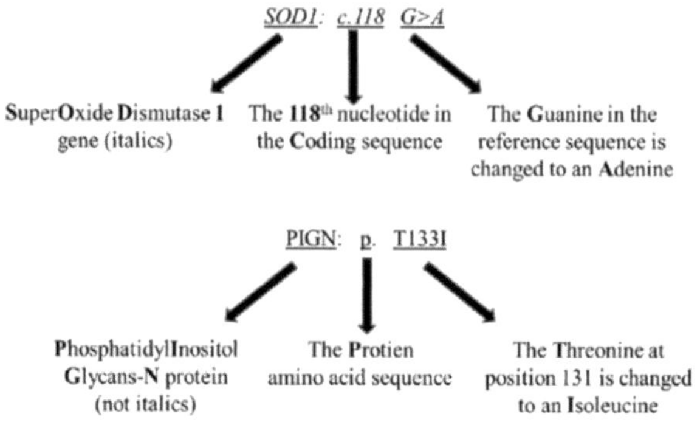

Figure 2: *Examples of the nomenclature used to describe variants affecting a gene (top) or a protein (bottom).*

GENE DISCOVERY STRATEGIES & DNA TESTS

The simplest strategy to identify a mutation causing disease is the candidate gene approach. A gene will be a candidate if it has been associated with the same phenotype in another species or if there is biochemical evidence to point toward a specific enzyme or other protein as the cause. For example, if immunohistochemistry demonstrated that dystrophin was absent in a dog with muscular dystrophy, the dystrophin gene would be a candidate, or if malonic acid was found in the urine, the gene for the enzyme that catabolizes conjugated malonic acid, malonylCoA decarboxylase, becomes a candidate. The coding regions of the candidate would be sequenced and the family being studied evaluated for concordance between any variants identified and the disease phenotype.

Often, however, no gene has been associated with a comparable disease or there are too many genes associated with a phenotype (e.g. seizures) to evaluate all of them this way. Then gene mapping can be used to limit the range of genes to be considered. Current techniques use "SNP Chips" which contain a panel of thousands of SNPs specific for that species. These polymorphic markers are spread throughout all the chromosomes and can be used to identify differences between individuals. If SNPs are near the disease causing mutation on a chromosome, they will segregate in the population along with it. If a group of SNPs is statistically more common in affected dogs than normal dogs, then the disease is said to be "linked" to those markers (Figure 3). Recombinations break that link in some individuals and define the borders of the region where the mutant allele resides, the "locus". Genes within the locus then become candidates to sequence based on their position.

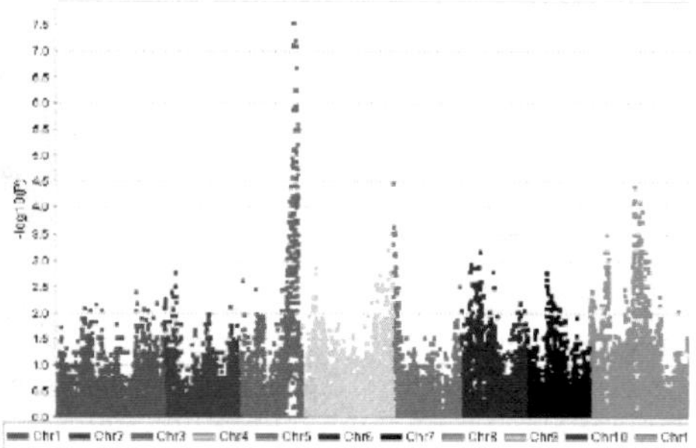

Figure 3: *A "Manhattan plot" shows the odds that a marker is found more commonly in animals with a hereditary disease compared to normal animals. Each dot represents a single SNP being evaluated. In this example there is a significant peak of association on the end of chromosome 3.*

The markers which show the greatest association with the disease can then be used for a "linked marker" DNA test. Such test can help identify carriers of a mutation but have their limitations. Though the marker is associated with the disease, it may be segregating in the population independently as well resulting in false-positive results. Conversely, recombinations can break the link between that marker and the disease. This improves mapping by narrowing the range of the locus but can produce false negative results on a test. Fortunately, sequencing of the positional candidates often identifies the specific mutation permitting subsequent development of "direct test" for that specific mutation. A direct test will be highly sensitive and specific for that individual mutation. Different mutations in the same gene can cause identical phenotypes, however, and a DNA test developed for one mutation will not detect another. For example the *SOD1c:118G>A* mutation is associated with degenerative myelopathy in most breeds, including Bernese Mountain Dogs[4]. In this breed, however, some necropsy confirmed cases were clear of that mutation and a different mutation, *SOD1c:A>T*, was identified in those dogs[5]. Thus both tests would need to be run in this breed.

NEXT GENERATION SEQUENCING

Next generation sequencing (NGS) uses high-throughput technology to generate sequence data on the whole genome or whole exome (coding sequences only) of an individual. Using this approach for gene discovery in hereditary diseases is becoming routine in human medicine and has been successfully applied to veterinary medicine as well[6,7]. NGS can interrogate the entire genome of an individual for disease causing variants but the potential for false negative and false positive findings must be recognized.

The sequence data must first be aligned to the canine reference sequence and variants from the reference identified. The current annotation of the reference sequence is greatly improved, but still will contain errors and omissions which can affect the alignment. Also because the reference sequence is derived largely from one individual, not all of it will be representative of the canine population as a whole. For example, if the Boxer used to generate the reference sequence had been homozygous for the degenerative myelopathy allele, it would have been considered the reference and not identified as abnormal.

Millions of variants from the reference will be identified from the alignment, many of which will simply be normal variations. These must then be filtered through bioinformatics algorithms to identify those likely to cause disease. The algorithm may select variants only within the coding sequence (exome) and adjacent regions where promoter and splice cites reside, codon changes that will alter an amino acid or produce a stop codon, deletions or insertions that produce a frame shift, or alterations to splicing signals. These can be further filtered for being homozygous or heterozygous in the patient depending on the mode of inheritance being modeled. Some mutations, such as large deletions or insertions or copy number variants may not be recognized. This filtering could still identify thousands of potentially disease-causing variants.

The technique used at the University of Missouri to further limit this list is a case-control comparison. We compare the variants identified in the patient against not only the reference sequence, but also against whole genome sequences from > 100 canids either generated at MU or shared with us by other researchers. These dogs suffered from some other disease or were considered clinically normal. Thus the disease causing allele should not be present in that population. Even with these selection criteria, homozygous variants could occur only in the patient by chance producing false positive results and other levels of evidence, as outlined in Table 2, are needed to indicate that the variant causes the disease.

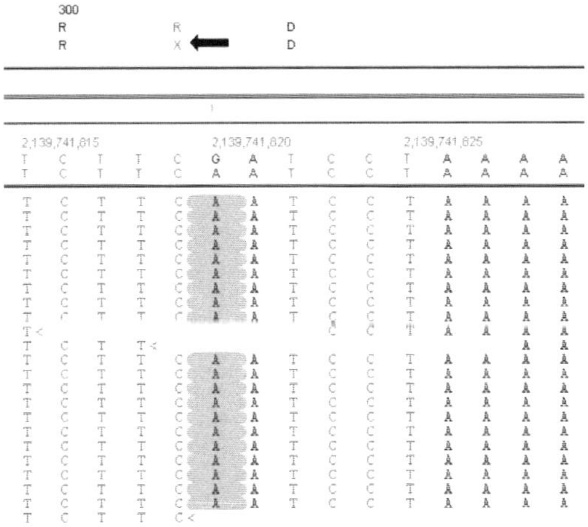

Figure 4: The G>A variant (shaded) in this whole genome sequence predicts a premature stop codon (arrow) in the gene and a non-functional protein. Yet this dog had no signs expected for that loss of function and a mutation in another gene that was responsible for the disease being investigated. Note the anti-sense orientation.

The example in Figure 4 shows a homozygous *G>A* mutation identified by NGS. It produces a premature stop codon that predicted truncation of the protein eliminating domains essential to function. That function, however, did not make sense in terms of the biology of the disease. When other family members were evaluated, the variant was not concordant, that is it did not segregate as predicted if it was the causal mutation while another, less dramatic variant was concordant. Even when a variant is perfectly concordant, when ~20,000 genes have been evaluated, some variants may be concordant by chance in a small sample or may be a tightly linked normal variant that segregates with the causal mutation. Other levels of evidence are needed to suggest causality.

In the case above, the protein affected by the second variant made sense in terms of the biology of the disease and is highly expressed in the brain. Polyphen, a program that evaluates the variant for evolutionary conservation and effects on protein structure, predicted that the mutation would be damaging. Such predictions, however, are based on models and may not be correct. Randomly selected DNA samples from other breeds did not contain the variant suggesting it was rare with one notable exception. The variant was also found in another breed which had a very similar phenotype. Associating the mutation with the same disease independently in two different breeds provided further proof of causality. Ultimate proof of causality would include all the levels of evidence in Table 2, but some such as generation of transgenic mice and rescue by gene replacement are beyond the scope of most studies. Valuable clinical and breeding decisions can still be made, but the strength of the evidence must be recognized.

Next generation sequencing has been called the most powerful diagnostic tool since the roentgenogram[8]. It is already being used in the diagnosis of syndromes in individual children, and advances in technology and bioinformatics will soon make it a consideration in veterinary practice. The vision of these advances is a personalized medicine where the genetic background of an individual are a starting point in diagnostic testing and therapeutic planning. In veterinary medicine, such information can also play an important role in breeding decisions. Studies in large human populations suggest that variability in penetrance and expressivity of even simple Mendelian traits is much greater than expected[9,10]. Thus role of the clinician in interpreting test results in the context of clinical signs will not diminish. Biology is more complex than the genome with environmental and epigenetic factors affecting health. Being able to assess those with a knowledge of the genomic background can only improve our ability to diagnose and treat disease.

Table 2: Types of evidence that support a causal relationship between a variant and a disease (Modified from MacArthur[11])

Level	Type of evidence	Example
Gene or Protein	Biochemical function	Shared with other genes known to be associated with disease or consistent with the biology
	Protein interactions	Product interacts with proteins previously implicated in the disease
	Expression	Protein expressed in the tissue of interest
	Disruption	Protein function or gene expression altered in the patient
Variant	Association	Significantly more common in affected vs controls
	Segregation	Segregates in the family as expected and other pathogenic variants cosegregating are unlikely
	Frequency	Rare in the general population (other breeds)
	Conservation	In an evolutionarily conserved area
	Predicted effect	In a location such as an enzyme active site or protein-protein interaction site
	Protein disruption	Alters level, splicing or function of the protein in patient or model
	Recapitulation	Produces similar phenotype in model system
	Rescue	Phenotype in model rescued by replacing gene

REFERENCES

1. O'Brien DP & Leeb T. JVIM 2014;28:1186-1198. 2. Granger N. Vet J 2011;188:274-285. 3. Maclean K. Nature 2006;439:266. 4. Zeng R, et al. JVIM 2014;28:515-521. 5. Wininger FA, et al. JVIM 2011;25:1166-1170. 6. Guo J, et al. BMC Vet Res 2015;10:960. 7. Gilliam et al. JVIM 2014;28:871-877. 8. Hennekam RC & Biesecker LG. Hum Mut 2012;33:884-886. 9. Xue Y, et al. Am J Hum Genet 2012;91:1022-1032. 10. Norton N, et al. Circ Card Gen 2012;5:167-174. 11. MacArthur DG, et al. Nature 2014;508:469-476.

THORACOLUMBAR AND LUMBAR INTERVERTEBRAL DISK DISEASE IN ENGLISH COCKER SPANIELS

Steven De Decker, DVM, PhD, DECVN, MvetMed
London, Hatfield, United Kingdom

INTRODUCTION

Thoracolumbar and lumbar intervertebral disk extrusion (IVDE) is a common spinal disease in dogs. Although every dog can be affected, chondrodystrophic breeds are predisposed with Dachshunds strongly overrepresented.[1-3] Thoracolumbar IVDE has been therefore best characterized in this particular breed. While the estimated lifetime prevalence of intervertebral disk disease is 2 to 3.5% in the overall dog population, it is estimated to be 19 to 25% in Dachshunds, with a much higher risk of disease occurrence compared to the general dog population.[1,3] Dachshunds most commonly develop clinical signs between 4 and 6 years of age and the majority (55-75%) of IVDE's occur in the region of the thoracolumbar junction with the intervertebral disk spaces between T11 and L2 most often affected.[1-3] The English Cocker Spaniel is among the seven breeds at highest risk of IVDE.[1,2] Although anecdotal reports suggest a different anatomical distribution of IVDE along the vertebral column in English Cocker Spaniels, no studies have yet characterized the clinical characteristics of thoracolumbar and lumbar IVDE in this specific breed.

COMPARSION BETWEEN ENGLISH COCKER SPANIELS AND DACHSHUNDS

As illustrated above, thoracolumbar and lumbar IVDE is best characterized in the Dachshund. To evaluate if breed-specific characteristics exist in English Cocker Spaniels, it could be suggested to compare selected clinical aspects of IVDE in this specific breed with those in Dachshunds.

A recent study compared therefore the clinical presentation and anatomical distribution of IVDE between Dachshunds and English Cocker Spaniels.[4] Affected English Cocker Spaniels were older (although still in the reported age range of IVDE in chondrodystrophic breeds), presented with a different spectrum of clinical signs, had a different distribution of IVDE along the vertebral column, had more often IVDE confined to a single intervertebral disk space, and were less often treated surgically than Dachshunds. The most important result of this study was the confirmation that English Cocker Spaniels demonstrated indeed a different anatomical distribution of IVDE along the vertebral column compared to Dachshunds. As expected, more than 80% of Dachshunds demonstrated IVDE's affecting the thoracolumbar intervertebral disk spaces (T10-L2 intervertebral disk spaces) and less than 2% demonstrated IVDE's affecting the caudal-lumbar intervertebral disk spaces (L5-S1 intervertebral disk spaces). In contrast, only 44% of English Cocker Spaniels demonstrated IVDE's affecting the thoracolumbar intervertebral disk spaces, while more than 25% demonstrated IVDE affecting the caudal-lumbar intervertebral disk spaces. These findings suggest that, although the thoracolumbar intervertebral disk spaces were still most often affected, English Cocker Spaniels seem to be predisposed to develop caudal-lumbar and even lumbosacral IVDE's.

It is well recognized that the anatomical distribution of a spinal lesion determines the type of clinical signs. It was therefore not surprising that English Cocker Spaniels also demonstrated other differences in clinical presentation compared to Dachshunds.

CLINICAL PHENOTYPES IN ENGLISH COCKER SPANIELS

This altered anatomical distribution of IVDE along the vertebral column can have considerable clinical consequences. When performing a more detailed analysis of the above mentioned group of English Cocker Spaniels, two distinct clinical phenotypes could be identified.[5]

The first group compromised English Cocker Spaniels with IVDE affecting the thoracolumbar (44% of the studied population) and mid-lumbar (30% of the studied population) intervertebral disk spaces. This group formed the majority of affected English Cocker Spaniels and presented with a similar clinical presentation to the classical presentation in Dachshunds and other chondrodystrophic dogs. These English Cocker Spaniels demonstrated a per-acute or acute onset of clinical signs and often had neurological deficits with non-ambulatory paraparesis or paraplegia as the main presenting clinical sign. English Cocker Spaniels with thoracolumbar or mid-lumbar IVDE responded well to surgical decompression with all animals discharged within 20 days of hospitalization and more than 80% able to ambulate independently at the time of discharge.

The second group compromised English Cocker Spaniels with caudal-lumbar IVDE (26% of the studied population). The vast majority of these dogs did not demonstrate neurological deficits on presentation and displayed a different spectrum of clinical signs. These dogs presented most often with spinal hyperesthesia or unilateral pelvic limb lameness as the predominant clinical sign. English Cocker Spaniels with caudal-lumbar IVDE also presented with a longer duration of clinical signs and more frequently received medical treatment prior to referral compared to dogs with thoracolumbar and mid-lumbar IVDE. Interestingly, in 30% of English Cocker Spaniels with caudal-lumbar IVDE that underwent magnetic resonance imaging the extruded material could not be readily visualized on T2W sagittal images. In these dogs, extruded material was only noted after careful inspection of the transverse images. English Cocker Spaniels with caudal-lumbar IVDE's also demonstrated more rapid and complete postoperative recoveries compared to dogs with thoracolumbar and mid-lumbar IVDE's. They had shorter hospitalisation times and returned more rapidly to a neurologically normal status.

It is not surprising that ECS with caudal-lumbar IVDE displayed a different spectrum of clinical signs and did not always demonstrate overt neurological deficits. Extruded disk material in this part of the vertebral canal typically results in cauda equina instead of spinal cord compression. Unilateral pelvic limb lameness without obvious neurological deficits is a common clinical presentation of dogs with cauda equina compression resulting from degenerative lumbosacral stenosis.

CLINICAL IMPORTANCE

It is important to realize that more than a quarter of English Cocker Spaniels with thoracolumbar/lumbar IVDE present with extrusions affecting the caudal-lumbar intervertebral disk spaces. The majority of these dogs will not demonstrate neurological deficits and present with unilateral pelvic limb lameness as the predominant clinical sign. The clinical presentation of these dogs is therefore difficult to differentiate from dogs with primary orthopaedic disease, which can result in delayed recognition of a possible neurological disease.

Additionally, the strongly lateralized extrusions are not always readily recognized on routine T2-weighted sagittal magnetic resonance images, which complicate recognizing the neurological nature of clinical signs further. It is therefore possible that caudal-lumbar IVDE in English Cocker Spaniels is currently under-recognized and probably under diagnosed. English Cocker Spaniels with caudal-lumbar IVDE improve remarkably well after surgery and this type of treatment was associated with rapid and complete recoveries. This excellent response to surgery highlights further the importance of reaching an accurate and timely diagnosis.

It is therefore important to consider caudal-lumbar IVDE in an English Cocker Spaniel with unilateral pelvic limb lameness without neurological deficits.

FUTURE PERSPECTIVES

It is currently unclear why English Cocker Spaniels present more often with caudal-lumbar IVDE compared to Dachshunds (and other breeds). The etiology of intervertebral disk disease is considered to be multifactorial with genetic, biomechanical, conformational and anatomical factors involved.[6]

Recent studies have proposed an important role of genetics into the etiology of intervertebral disk disease in Dachshunds. Several genes have been suggested, some of which are associated with the expression of chondrodystrophy.[7] Chondrodystrophy is characterized by disturbed endochondral ossification, resulting in dogs with disproportionally short limbs and relative long spines. This was highlighted in a recent study, which demonstrated that dogs with short limbs and longer backs, the so called 'long and low morphologies', were at increased risk of developing thoracolumbar IVDE.[2] The Dachshund, with its extreme long back and short limbs, is considered the prototype of a chondrodystrophic dog breed. There is however uncertainty about which breeds should be defined as being truly chondrodystrophic and it is therefore currently unclear if the English Cocker Spaniel should be considered as such. English Cocker Spaniels have a different morphology with a less pronounced chondrodystrophic 'phenotype' compared to Dachshunds. Their vertebral column length relative to the height at the withers is much smaller than in Dachshunds, but it is however similar to those in German Shepherd Dogs.[2] It is therefore possible that different etiological factors contribute to a different degree in the development of IVDE in English Cocker Spaniels and Dachshunds. In other words, it is possible that factors strongly predisposing Dachshunds to develop thoracolumbar IVDE (e.g. genetics and chondrodystrophic phenotype) are less important in English Cocker Spaniels and vice versa.

Several studies have evaluated morphological characteristics that could potentially predispose German Shepherd Dogs to develop degenerative lumbosacral stenosis. Several anatomical factors have been identified, including presence of transitional vertebrae, and different congruency, shape and inclination of the caudal-lumbar and lumbosacral articular facets.[8-10] These anatomical variations can cause altered biomechanics of the caudal-lumbar and lumbosacral vertebral column, thereby potentially predisposing this region to accelerated degenerative processes, such as intervertebral disk degeneration. Further studies are therefore indicated to evaluate regional anatomical and biomechanical characteristics of the vertebral column in English Cocker Spaniels.

REFERENCES

[1]Bergknut N, et al. **J Am Vet Med Assoc** 2012;240:1300. [2]Packer RMA, et al. **Plos One** 2013:8;e69650. [3]Goggin JE, et al. **Am J Vet Res** 1970;31:1687. [4]De Decker S, et al. **Proc Annual ECVN-ESVN symposium** 2014. [5]Cardy TJA, et al. **Proc Annual ECVN symposium** 2014. [6]Smolders LA, et al. **Vet J** 2013;195:292. [7]Mogensen MS, et al. **Front Genet** 2012;3:225. [8]Seiler GS, et al. **Am J Vet Res** 2002;63:86. [9]Rossi F, et al. **Vet Radiol Ultrasound** 2004;45:381. [10]Benninger MI, et al. **Am J Vet Res** 2006;67:43.

UPDATE ON DISSEMINATED IDIOPATHIC SKELETAL HYPEROSTOSIS (DISH) IN DOGS

Steven De Decker, DVM, PhD, DECVN, MvetMed
London, Hatfield, United Kingdom

INTRODUCTION

Diffuse idiopathic skeletal hyperostosis (DISH) is a systemic disease characterized by calcification and ossification within ligaments and entheses of both the axial and appendicular skeleton. Although DISH can affect multiple anatomical structures, it affects most often the ventral longitudinal ligament resulting in the formation of contiguous bone ventral to the vertebral column. This typically results in complete bony fusion of consecutive vertebral segments.[1] There is controversy surrounding the definition of DISH[2] and this disorder can easily be confused with severe spondylosis deformans. DISH is relative commonly encountered in people, and is also increasingly recognized in dogs. Although imaging studies of affected dogs often demonstrate spectacular abnormalities, the occurrence of DISH is only rarely associated with clinical signs. Although the clinical relevance is currently unclear, recent studies have reported interesting clinical syndromes in people and dogs with DISH. It is increasingly recognized that DISH can affect the biomechanical properties of the vertebral column, which could then initiate a cascade of potential harmful events in affected and adjacent non-affected vertebral segments.

CONTROVERSIAL DEFINITION

In both human and veterinary medicine, there is debate considering the diagnostic criteria for DISH. This has led to a multitude of terms to refer to this disorder. Two definitions of DISH can be found in the veterinary literature:

The diagnostic criteria most often used in dogs and people are those proposed by Resnick and Niwayama[1]: (1) flowing calcification and ossification along the ventrolateral aspect of at least four contiguous vertebral bodies, (2) preservation of the intervertebral disk width and the absence of overt radiographic changes indicative of degenerative intervertebral disc disease, and (3) the absence of articular process ankylosis, sacroiliac joint erosion, sclerosis or intra-articular osseous fusion.

Although this definition excludes abnormalities affecting the dorsal vertebral column, several cases reports have described dogs with prominent abnormalities of dorsal vertebral structures, including the articular and spinous processes.[3-5] Therefore Morgan and Stavenborn[6] have proposed additional diagnostic criteria: (1) Flowing calcification and ossification along ventral and lateral aspects of three contiguous vertebral bodies leading to segmental bony ankylosis, (2) Relative preservation of disc width within involved areas and absence of extensive radiographic changes of degenerative disc disease such as end-plate sclerosis, nuclear calcification, or localized spondylosis deformans, (3) Peri-articular osteophytes surrounding true vertebral joints, (4) Formation of pseudoarthrosis between the bases of the spinous processes, (5) Peri-articular osteophytes and calcification and ossification of soft tissue attachments in both axial and peripheral skeleton, (6) Peri-articular osteophytes, sclerosis, and ankylosis of sacroiliac joints, (7) Bony ankylosis of the symphysis pubis. Four of the first 5 criteria have to be present to support a diagnosis of DISH. It is obvious that this (rather strict) definition, in contrast to the definition proposed by Resnick and Niwayama, requires dorsal vertebral abnormalities to be present before a diagnosis of DISH can be made. A proportion of dogs with DISH will not demonstrate dorsal vertebral column abnormalities, making this definition again difficult to use.

Although the diagnostic criteria proposed by Resnick and Niwayama are most often used in both veterinary and human medicine, it has been suggested that using these criteria would allow identification of only advanced stages of the disease. In human medicine, there is therefore a variability of definitions with alternative diagnostic criteria including less than 4 affected vertebral segments.[7]

THE DOG AS A TRANSLATIONAL MODEL

DISH is relative commonly diagnosed in people with varying prevalence among different geographical regions. Although the underlying etiopathogenesis is unknown for both people and dogs, suggested risk factors in people include male gender, higher age, obesity, type 2 diabetes mellitus, hypertension and genetics.[8] In agreement with the situation in people, DISH in dogs is more common in older animals and males. The prevalence in the overall dog population is estimated to be 3.8%, with a much higher prevalence of 40.6% in the Boxer breed.[9] Socio-demographic risk factors, such as obesity and diabetes mellitus, are typically not present in Boxers. DISH in the Boxer breed has therefore been suggested to represent a natural occurring translational disease model to investigate the etiopathogenesis, genetics and treatment of DISH in people.[10]

CLINICAL SIGNS IN PEOPLE WITH DISH

Also in people is DISH most often considered an incidental finding and only a minority of people will develop clinical signs. An important difference with dogs is that DISH in people often affects the cervical vertebral column, while this anatomical region is only rarely affected in dogs. This difference in anatomical distribution is also reflected in a different spectrum of clinical signs between the two species.

Mechanical obstruction

Excessive bone formation ventral to the cervical vertebral column can cause dysphagia and airway obstruction due to compression and irritation of the esophagus and upper respiratory structures. Difficulties during intubation for anesthetic procedures have also been reported. Improvement can occur after surgical removal of the bone mass.[8]

Spinal hyperaesthesia

Spinal hyperaesthesia is uncommonly reported in people with DISH and is most likely related to compression or irritation of spinal nerve roots or vertebral canal stenosis. Vertebral canal stenosis due to DISH has been reported occasionally.[8]

Spinal Fractures

Several studies have documented an increased risk of vertebral fractures in people with DISH. Affected people can develop severe and unstable vertebral fractures after relative minor trauma or low-energy impacts.[11] Although the exact reason for this increased fracture risk is currently unknown, it is postulated that increased vertebral stiffness associated with DISH causes decreased flexibility and increased biomechanical stress on the vertebral column. This is expected to interfere with normal energy distributing, making the affected vertebral segments more vulnerable to fracture.[11]

A recent study has demonstrated decreased vertebral bone mineral density in Boxers with DISH compared to Boxers without DISH.[12] The relationship between bone mineral density, bone strength and fracture risk is well documented. Fractures associated with decreased BMD are commonly referred to as fragility fractures. It is therefore possible that decreased vertebral bone mineral density contributes to an increased fracture risk in DISH patients. Although the underlying cause of decreased vertebral bone mineral density in dogs with DISH is currently unknown, this is possibly related to altered biomechanics of the vertebral column. Physical exercise and mechanical stimulation are essential for maintaining bone mass and strength. Loss of biomechanical function results in altered bone metabolism characterised by decreased bone formation and increased bone resorption (referred to as disuse osteoporosis). It can therefore be hypothesized that abnormal mechanical loading of DISH-affected vertebral bodies could result in alterations of bone metabolism with loss of trabecular bone mass.

Involvement of the appendicular skeleton

DISH is a systemic disease and therefore more peripheral structures can also be affected. Periarticular hyperostosis is a well-reported cause of peripheral joint pain in people and should not be confused with arthritis.[7,8]

CLINICAL SYNDROMES IN DOGS

As stated above, the occurrence of DISH is often not associated with clinical signs and this disorder is therefore often considered an incidental finding on imaging studies including the vertebral column. It is therefore important to investigate potential other causes before clinical signs are attributed to DISH. Recent studies have identified new insights into the clinical relevance of DISH in dogs. Clinical signs can be caused by direct impingement of nervous structures by excessive bone or indirectly by biomechanical alterations of the vertebral column.

Spinal hyperaesthesia and 'stiffness'

Affected dogs can present with spinal hyperaesthesia and a stiff gait. DISH typically affects the ventral and ventrolateral aspects of the vertebral column. It is therefore likely that spinal hyperaesthesia is caused by compression and irritation of the spinal nerve roots at the level of the intervertebral foramen. It is suggested that the stiff gait is associated with increased stiffness and decreased flexibility of affected vertebral segments.[9]

Adjacent segment disease

Adjacent segment disease is a well-known complication after surgical treatment of cervical spondylomyelopathy in dogs. Although the exact etiopathogenesis of adjacent segment disease in dogs is currently unknown, it is suggested that spinal fusion can cause increased biomechanical stress as the adjacent non-fused vertebral segment. This can then ultimately result in intervertebral disk degeneration at this adjacent site and recurrence of clinical signs. Similar to this surgical complication, adjacent segment disease has been reported in dogs with DISH. Although intervertebral disks at the level of the affected segments are typically not degenerated, the intervertebral disks at the adjacent non-fused segments seem vulnerable to degeneration and herniation.[13]

Vertebral fractures

Although the association between DISH and an increased risk of spinal fractures is well recognized in people, there is only one case report of an affected dog developing two episodes of spinal fractures after relative minor trauma.[14] As stated above, decreased vertebral bone mineral density has been demonstrated in dogs with DISH, but the association between this finding and an increased risk of spinal fractures is currently unclear.

Dorsal vertebral column changes and lumbosacral syndrome

Dorsal vertebral column abnormalities, including articular process hypertrophy, periarticular new bone formation, pseudoarthrosis between spinous processes, and thickening of the dorsal lamina have been reported in dogs. These changes occur both at segments fused by DISH and adjacent non-affected vertebral segments. Although these abnormalities can occur without development of clinical signs, a proportion of dogs suffered from a unique lumbosacral syndrome. In these dogs, cauda equina compression was caused by extreme enlargement of the articular processes. Affected animals demonstrated excessive DISH formation involving L7, but not the lumbosacral junction.[4] Although it cannot be excluded that articular process hypertrophy represented an unusual expression of DISH, it was hypothesized that the fused lumbar vertebral column caused increased biomechanical stress on the lumbosacral junction where still some mobility persisted. Experimental studies have demonstrated degenerative changes of the articular processes and spinous process hypertrophy following implant assisted caudal lumbar spinal hypomobility.[15,16]

REFERENCES

[1]Resnick D, et al. **Radiology** 1976;119:559. [2]Greatting HH, et al. **Vet Radiol Ultrasound** 2011;52:472. [3]Woodard JC, et al. **Vet Pathol** 1985;22:317. [4]De Decker S, et al. **Vet Rec** 2014;174:632. [5]Ciepluch MF, et al. **Vet Radiol Ultrasound** 2015;56:E5. [6]Morgan JP, et al. **Vet Radiol Ultrasound** 1991;32:65. [7]Mazières B **Joint Bone Spine** 2013;80:466. [8]Mader R, et al. **Nat Rev Rheumatol** 2013;9:741. [9]Kranenburg HC, et al. **Vet J** 2011;190:84. [10]Kranenburg HC, et al. **Eur Spine J** 2010;19:1325. [11]Westerveld LA, et al. **Eur Spine J** 2009;18:145. [12]De Decker S, et al., **Proc Annual Symposium ECVN-ESVN** 2014. [13]Ortega M, et al., **Vet Radiol Ultrasound** 2012;53:128. [14]Kornmayer M, et al. **Vet Comp Orthop Traumatol** 2013;26:76. [15]Little JS, et al. **Spine J** 2004;4:153. [16]Homb NM, et al. **J manipulative Physiol Ther** 2012;35:367.

ETHICAL CONSIDERATIONS IN MAINTAINING PARALYSED PETS

Patrick J. Kenny, BVSc, DACVIM (Neurology), DECVN
London, UK

INTRODUCTION

Veterinarians are commonly presented with paralyzed (paraparetic/plegic or tetraparetic/plegic) pet animals because of spinal cord injury (SCI) secondary to trauma, intervertebral disc herniation & fibrocartilagenous embolism[1], spinal neoplasia[2], degenerative myelopathy[3] or other cause. The treatment of these animals – particularly their maintenance as pets if prognosis for functional recovery is poor, can give rise to ethical dilemmas and induce a degree of "moral distress"[4] in the owner and/or veterinary staff. However, some pets under certain circumstances appear to tolerate paralysis and can have a good quality of life (QOL).[5,6]

Clinicians are accustomed to using structured approaches to solve clinical problems such vomiting, seizures or lameness. A rigorous, structured approach to defining the ethical issues at hand, consideration of ethical concepts alongside patient and owner factors, and application of practical strategies to minimize the impact of the decision may greatly assist the clinician in resolving ethical dilemmas and formulating acceptable treatments plans.

ETHICAL THEORY

Morality has been defined as "an informal public system applying to all rational persons, governing behavior that affects others, and includes what are commonly known as moral rules, ideals & virtues, and has the lessening of evil or harm as its goal."[7] Ethics have been defined as "a group of moral principles or set of values… governing the behavior of an individual or profession."[8] Though some make subtle distinctions between the two terms, in the context of clinical medicine their similarities in meaning greatly exceed any differences, and so are used synonymously by many authors.[4]

Ethical theory has been a subject of scholarship since antiquity.[9] Contemporary ethical theory has at its foundation in two systems: utilitarianism and deontology. Under a utilitarian system, the morality of an act is determined solely by analysis of its consequences. As a given act may have complex consequences, the degree of moral rightness is in direct proportion to the net good or harm produced. A necessary condition of its application is that the possible consequences of a decision must be considered and the costs/benefits weighed. Deontology determines the morality of an act on the basis of the intent, positing that moral choices should be made following certain rules (*deon* from the Greek for "duty"). Most medical/veterinary legislation and professional codes of conduct are rooted in both systems (utilitarianism and deontology), as neither alone is sufficient to encompass and adequately address all clinical-ethical issues.[4] For example, though most clinicians will innately use a utilitarian analysis of the costs & benefits of an intervention before recommending it to the owner, utilitarianism alone does not take into account the rights of the individual – such as their right to self-determination, and the necessity for valid consent. Rigid application of a rules-based system (deontology) may be an impediment to a good outcome, e.g. in the case of "do not cause pain". An analysis of cost (a temporary, small degree of inflicted pain) compared to benefit (accurate diagnosis, appropriate treatment, and subsequent long-term decrease in pain/disability) makes applying the intent of this rule practicable clinically.

Bioethics (the branch of ethic that encompasses clinical ethics) is widely accepted as being based in four essential concepts: that of autonomy, non-maleficence, beneficence and justice.[10] Respect for autonomy implies a respect for decisions made by individuals. For a decision to be autonomous, it must be made intentionally, with understanding, and unconstrained by influence that determines it.[11] Consent therefore can only be a truly autonomous decision if the individual has the capacity to decide, has received adequate information with which to make the decision, and has not been coerced into the decision.[12] The concept of nonmaleficence can be defined by the ancient maxim *primum non nocere* (above all, do no harm).[4] Beneficence is the moral duty to promote good and may be considered the primary ethical foundation of the relationship between clinician and patient. It is articulated in the Hippocratic oath,[13] veterinary oaths[14] and veterinary codes of ethics and conduct internationally.[15,16] The concept of justice in a bioethical context refers to the fair distribution of societal resources according to need, merit or desert.[4] These concepts are useful in clinical practice as they embody important moral ideals, but may come into conflict in certain clinical contexts, such as the tension between respect for autonomy and beneficence in cases where the decisions of the patient/owner are incompatible with adequate care, or promote harm.

ETHICAL PRACTICE

Resolving Ethical Issues

A sound understanding of clinical ethics can help clinicians identify, understand and resolve ethical issues in practice.[17] Most ethical issues can be resolved satisfactorily; competing goods are not usually problematic, competing harms or losses are more challenging, where the best choice may be the "least bad option". The later case may be considered a true "dilemma" – not just a difficult choice, but a choice between two or more undesirable alternatives. A framework for solving ethical problems may distinguish ethical dilemmas from more unremarkable ethical issues.[4] For example, though spinal surgery is an invasive procedure, whether or not to perform surgery on a paralyzed dog with an intervertebral disc herniation is not a dilemma if the prognosis with surgery is good and the owners are willing. Despite the temporary disruption to the dog and its environment, the intervention is likely to benefit it and the owner over the longer term – more so than not intervening or euthanizing the dog. Deciding whether to treat and maintain a paralyzed pet over the long term or to euthanize it has the potential to be a dilemma; extent to which the animal's QOL may be impaired must be balanced against the death of the animal; and decide which, under the particular circumstances, would be preferable.

Many practical approaches to resolving ethical dilemmas have been developed;[17-21] recurring components include:

1. Establishing the medical facts: diagnosis, prognosis and co-morbidities. A frequently cited maxim amongst clinical ethicists asserts "good ethics relies on good facts."[4]
2. Establishing what the ethical issues are, and their relative importance.

3. Establishing the concerns, values and preferences of the patient (or surrogate, or owner), shared decision making and obtaining valid consent.
4. Establishing what practical considerations must be addressed to minimize the impact of the decision.

Welfare Considerations & Quality of Life

The welfare of the animal should be the core consideration in resolving clinical-ethical dilemmas in veterinary medicine, taking primacy over all other concerns.[15,16] Well established principles of animals welfare, such as the "five freedoms"[22] (1. Freedom from thirst, hunger and malnutrition; 2. Freedom from pain, injury & disease; 3. Freedom from fear and distress; 4. Freedom from physical and thermal discomfort; 5. Freedom to perform most normal forms of behavior) may be a useful start, though really at best reflect a minimum standard, and do not account for all welfare considerations that may arise for a clinical patient.

Considering an animal's QOL - both immediate and anticipated - is commonly done when deciding on treatment or non-treatment/euthanasia. Paralysis due to SCI, neurodegenerative disease or other cause results in a dramatic change to the individual and their lifestyle; SCI has been described as "one of the greater calamities that can befall man".[23] Factors that affect the QOL of humans with SCI include loss of mobility, loss of independence, loss of bladder and bowel functions, loss of sexual function, sensory dysfunction, pain, respiratory dysfunction and the development of co-morbidities such as infection and pressure sores.[24] QOL has psychological as well as physical roots,[25,26] being influenced by level of social support, socioeconomic status, as well as clinical status.[27] Clinical factors that may affect the QOL of animals with SCI or disease are similar; motor dysfunction (loss of ambulation), sensory dysfunction, pain, urinary & fecal incontinence, decubital ulceration and skin abrasions.[1] QOL assessment of an animal has traditionally based on subjective assessments made by the owner and clinician, influenced by the clinician's appraisal of the prognosis based on clinical facts, and perception and experience of anticipated outcomes. But how is an acceptable QOL best determined? This may be unclear for both humans and animals, as no agreement exists on the definition of QOL and how it should be measured,[24] and as a concept is multifactorial and subjective.[28]

QOL measurement tools, both general and disease specific, have been used in human medicine for several decades, where QOL has quite reasonably been considered an outcome measure in its own right.[29] While veterinarians are disadvantaged in that our patients can't contribute directly to these measurements, QOL tools have been validated in human pediatric medicine.[30] Use of QOL measurements in humans have been found to have a number of advantages such as ensuring pertinent physical and psychosocial problems are not overlooked, prioritizing more relevant problems over less relevant ones, monitoring treatment effects, facilitating shared decision making, and improving quality of communication between patients, caregivers and healthcare professionals.[24] QOL scores are commonly used in the care of humans with SCI[24,31-35] & ALS[25-27] among other neurological diseases. Though there are an increasing number of studies on the development of QOL tools for animals,[28,36-46] and veterinary studies assessing QOL as an outcome measure,[47-54] there are few to date that have assessed at the impact of SCI on an animal.[38,47] One study[47] found that ambulatory dogs that had suffered SCI had a significantly higher perceived QOL than non-ambulatory dogs with SCI, and that underlying injury or duration of injury was not of significance. While this study is an important first step in attempting to measure QOL in veterinary SCI patients, it included a wide variety of breeds (26 + mixed breeds), and ages (range 0.33-17 years), where the ability of dogs and owners to cope with the consequences of SCI may have differed between groups. A recent systematic review of several different QOL measurement tools for SCI in humans concluded that tool selection should be guided by the specific purpose and context in which it is used (including patient population and setting).[33]

There are many factors that may impact the QOL of an animal (and owner of an animal) with paralysis, including: size, age, breed, species, temperament, underlying disease process, location of lesion (para- or tetraparesis), severity of paresis/paralysis, sensory deficits, urinary & fecal continence, co-morbidies, resultant change in lifestyle and ability to perform normal behaviors, and level & quality of rehabilitation and nursing care. Though many experienced clinicians will have their opinions, the relative impact of these factors on QOL of different populations (large dogs/small dogs, active dogs/sedentary dogs, dogs/cats, etc.) remains to be tested through validated QOL measures.

Decision Making

To make a rational decision (and give valid consent) the owner needs to be educated on the clinical facts at hand (diagnosis and prognosis), realistic expectations of the timeframe of anticipated clinical course, and the impact that managing a paralyzed pet is likely to have on them as a financial and nursing commitment over the short and longer term. Consent is an ongoing process[4] and a decision to continue care (as opposed to euthanasia) may be revised as patient or owner factors change.

"Would you do it if it were your dog?" Owners struggling to make difficult treatment decisions commonly ask veterinarians this question. What is a "life worth living", and when would euthanizing the pet be preferable? The point at which "doing everything" becomes "too much" is an emotive topic for pet owners, veterinary staff and the general public, many of whom come with their own preconceived ideas and opinions. It is sad fact that depression and suicide rates among SCI patients are higher than that of the general population,[35] and that many quadriplegic patients will request removal from life-sustaining ventilation.[55] But it should also be noted that QOL is not a static phenomenon, and that QOL scores change with time and often improve as patients adjust to new circumstances.[24] Non-patients frequently estimate the QOL of people with chronic health conditions to be worse than the patient's own perception. Members of the public often underestimate the QOL of people with paraplegia.[56] Caregivers tend to underestimate the QOL of their patients with ALS.[26] In many situations, euthanizing a paralyzed animal is the most humane and appropriate course of action. Some dogs, however, appear to be able to have a good QOL over a long period of time;[5] in cases such as these euthanasia would have acted against the dog and owners best interests.

A fundamental ethical and legal principle applied when decisions on forgoing life-sustaining treatments of people are made is that the patient's wishes are considered the gold standard directive.[57] If the patient does not have the capacity to decide for themself, and no clear and specific advance directive is available, then a surrogate decision maker (usually a close relative) acts on the patients' behalf. Substituted judgment asks that they attempt to make a decision in the manner that the patient would have, taking into account the patients identity-constituting narrative.[57] Veterinarians are in a similar situation to physicians caring for patients without decision-making capacity. The concept of patient narrative is worth considering when anticipating the impact of paralysis on an individual animal. A toy breed dog rarely out of a handbag or its owners arms may be less impacted by paralysis than a previously highly active working dog; the ability of these dogs to cope with immobility, and therefore the welfare and ethical implications, are likely very different.

Minimizing the impact of the decision.

Strategies must be implemented to mitigate the physical and psychological impact of paralysis. Adequately caring for a paralyzed pet requires considerable medical and nursing support. Urinary tract infections are highly prevalent.[58] Adequate bedding and vigilant skin care are necessary to decrease the occurrence of decubital ulcers.[59] Spinal surgery solely aimed at improve QOL (rather than neurological function) has been reported in a dog with apparently good results.[5] It is important that caregiver issues are also addressed.[27]

Though the evidence base supporting specific modalities of physical therapy in humans with SCI is low,[60] there is strong consensus that physical therapy is important in promoting recovery, mitigating physical comorbidities and improving psychological health.[61] There is similarly a paucity of evidence regarding the use of physical therapy in managing paralyzed animals,[62] though it can facilitate social interaction, maintain the pet-owner bond, improve client education and support therapy at home.[63] Longer survival times reported in dogs with degenerative myelopathy undergoing physiotherapy[6] may in part reflect an improvement of perceived QOL. Physical activity has been shown to improve QOL in humans with SCI;[64] it may be reasonable to assume that the same is true for animals. Use of a wheel-cart enables mobility and the opportunity to perform more normal behaviors.

CONCLUSION

When faced deciding whether or not to maintain a paralyzed pet, a structured approach analyzing the clinical and ethical considerations for the patient at hand, including consideration of the patient narrative may help the owner and veterinarian determine the most desirable course of action. Further studies investigating the QOL of specific populations of paralyzed animals may help clinicians weight up the relative importance of various disease, patient and owner factors, to further improve decision making.

REFERENCES

1. Granger N & Carwardine D **Vet Clin Small Anim** 2014;44:1131–1156.
2. Bagley RS **Vet Clin North Am Small Anim Pract** 2010;40(5):915–927.
3. Coates JR & Wininger FA **Vet Clin North Am Small Anim Pract** 2010;40(5):929–950 .
4. Bernat JL **Ethical Issues in Neurology, 3rd ed.** Lippincott Williams & Wilkins, 2008.
5. Tertuliano Marinho PV, et.al **J Small Anim Pract** 2014 ;55(10):538–541.
6. Kathmann I, et.al **J Vet Intern Med** 2006;20(4):927–932.
7. Gert B **Morality: Its Nature and Justification.** Oxford University Press, 2005:14.
8. Pfaff DW, ed. **Ethical Questions in Brain and Behavior.** Springer Verlag, 1983:1.
9. Carrick P **Medical Ethics in Antiquity.** D Reidel Publishing Co, 1985.
10. Beauchamp TL, Childress JF **Principles of Biomedical Ethics, 5th ed.** Oxford University Press, 2001.
11. Beauchamp TL, Childress JF **Principles of Biomedical Ethics, 5th ed.** Oxford University Press, 2001:120-124.
12. Childress JF **Hastings Cent Rep** 1990;20(1):12-17.
13. Emery AE **J Med Biogr** 2013;21(4):198-9.
14. http://www.wsava.org/sites/default/files/Global%20Veterinary%20Oaths.pdf
15. https://www.avma.org/kb/policies/pages/principles-of-veterinary-medical-ethics-of-the-avma.aspx
16. http://www.rcvs.org.uk/advice-and-guidance/code-of-professional-conduct-for-veterinary-surgeons/
17. Lo B **Resolving Ethical Dilemmas; A Guide for Clinicians 5th ed.** Lippincott Williams & Wilkins, 2013.
18. Fletcher JF, et.al **Introduction to Clinical Ethics.** University Publishing Group, 1995.
19. Siegler M **Arch Intern Med** 1982;142:2178-2179.
20. Snyder L, Leffler C **Ann Intern Med** 2005;142:560-582.
21. Mullen S & Main D **In Practice** 2001;23:394-401.
22. Rogers Brambell FW **Brambell Report.** London, 1965.
23. Guttmann L **Spinal cord injuries: Comprehensive management and research.** Blackwell Scientific Publications; 1973.
24. Dijkers MPJM **J Rehabil Res Dev** 2005;42(3 Suppl 1):87–110.
25. Simmons Z, et.al **Neurology** 2000;55(3):388–392.
26. Trail M, et.al **J Neurol Sciences** 2003;209(1-2):79–85.
27. Chiò A, et.al **J Neurol Neurosurg Psychiatr** 2004;75(11):1597–1601.
28. Wojciechowska JI, et.al **Am J Vet Res** 2005;66(8):1461–1467.
29. Niessen SJM **J Small Anim Pract** 2011;52(9):453-454.
30. Klassen A, et.al **Health Qual Life Outcomes** 2003;1:81.
31. Hitzig SL, et.al **J Spinal Cord Med** 2013;36(6):600–615.
32. Hitzig SL, et.al **Disabil Rehabil** 2013;35(16):1408–1414.
33. Ravenek KE, et.al **Disabil Health J** 2012;5(4):213–223.
34. Boakye M, et.al **Journal of Neurosurgery-Spine** 2012;17(1 Suppl);29–37.
35. Chevalier Z, et.al **Spinal Cord** 2009;47(11):778–782.
36. Wojciechowska JI, et.al **Am J Vet Res** 2005;66(8):1453–1460.
37. Brown DC, et.al **Am J Vet Res** 2007;68(6):631–637.
38. Budke CM, et.al **J Am Vet Med Assoc** 2008;233(6):925-930.
39. Favrot C, et.al **Vet Dermatol** 2010;21(1):63–69.

[40]Niessen SJM, et.al **J Vet Intern Med** 2010;24:1098-1105.
[41]Noli C, et.al **Vet Dermatol** 2011;22(4):335–343.
[42]Noli C, et.al **Vet Dermatol** 2011;22(4):344–351.
[43]Yeates JW, et.al **J Small Anim Pract** 2011;52(9):459–463.
[44]Niessen SJM, et.al **J Vet Intern Med** 2012;26(4):953–961.
[45]Iliopoulou MA, et.al **J Am Vet Med Assoc** 2013;242(12):1679–1687.
[46]Lavan RP **Vet J** 2013;197(3):578–582.
[47]Levine JM, et.al **J Am Vet Med Assoc** 2008;233(6):931-935.
[48]Ahlstrom LA, et.al **J Vet Pharmacol Ther** 2010;33(6):573–582.
[49]Muñana KR, et.al **J Vet Intern Med** 2012;26(2):341–348.
[50]Peddle GD, et.al **J Vet Cardiol** 2012;14(1):253–259.
[51]Rutherford L, et.al **Vet J** 2012;194(3):294–298.
[52]Haggstrom J, et.al **J Vet Intern Med** 2013;27(6):1441–1451.
[53]Wessmann A, et.al **J Vet Intern Med** 2014;28(2):510–514.
[54]Giuffrida MA, Kerrigan SM **J Vet Intern Med** 2014;28(6):1824–1829.
[55]Maynard FM **West J Med** 1991;154(5);617–619.
[56]Ubel PA, et.al **Med Decis Making** 2001;21(3):190–199.
[57]Kuczewski MG **J Law Med Ethics** 1999;27(1):32–36.
[58]Olby NJ, et.al **J Vet Intern Med** 2010;24:1106–11.
[59]Sikes R **Probl Vet Med** 1989;1(3):467-77.
[60]Whiteneck G, et.al **J Spinal Cord Med** 2009;32(3):251–259.
[61]Wing PC, et.al **J Spinal Cord Med** 2008;3(4):408–479.
[62]Drum MG **Vet Clin North Am Small Anim Pract** 2010;40:181–93.
[63]Sims C, et.al **Vet Clin Small Anim** 2015;45:123–143.
[64]Stevens SL, et.al **J Spinal Cord Med** 2008:31(4);373–378.

HIGHS AND LOWS OF MEDICAL MARIJUANA IN THE TREATMENT OF EPILEPSY

Michael Podell, MSc, DVM, DACVIM (Neurology)
Chicago, IL

INTRODUCTION

"We shall, by and by, want a world of hemp more for our own consumption." - John Adams, U.S. President quote on Hemp.

No more prescient thought could have come from one of our forefathers over 200 years ago as the United States (US) and the world enters the 21st century in the midst of controversy on the social, legal and medical use of marijuana. The conglomeration of fact and fantasy through personal experiences, public health concerns, legislative reactivity, and medical anecdotal reports has resulted in a stream of almost indecipherable internet information. To help clear the smoke over the controversy, the purpose of this presentation is to provide an overview of the current state of knowledge on the potential benefit that the active pharmacologic compounds of marijuana, Δ-9 tetrahydrocannabinol (THC) and cannabidiol (CBD), may have in the treatment of epilepsy in general, and more specifically, in canine epilepsy.

WHY CONSIDER MEDICAL MARIJUANA IN THE TREATMENT OF EPILEPSY?

Epilepsy is a serious and prevalent disease in people and dogs. Over 50 million people are estimated to have epilepsy worldwide, [1] with an estimated prevalence range of 28% to 37% of drug-resistant epilepsy (DRE) in these patients[2] as defined by failure to respond to two or more antiepileptic drugs (AED) according to the International League Against Epilepsy definition.[3] Premature death, social stigma, and reduced quality of life is often the outcome for many individuals. Dogs also a have a high incidence of epilepsy, affecting a variety of breeds. The actual prevalence of dogs that develop DRE is unknown, but estimates range from 20 to 30%.[4]

These statistics hold true despite the exponential introduction of AED over the past 20 years, with 24 approved AED for people in the US alone. On a smaller level, a similar increase in drug selection is available for use in dogs, with 4 to 6 drugs available for use on a regular basis. Therefore, the problem at hand is that a strong need to improve seizure control in people and dogs exists while maintaining a high quality lifestyle. As veterinarians, we also face the challenge of balancing the economics and lifestyle of the owner with patient care.

HISTORICAL USE OF CANNABIS

Cannabis is a ubiquitous plant found throughout the world in a variety of habitats and climates. Known use for textile fiber in the Far and Middle East regions dates back over 5000 years.[5] *Cannabis sativa* and *indica* are the two most common flowering plants of the Cannabis *genus*. Through the centuries, cannabis has been used to produce hemp fiber for rope, paper and clothing, supplement for livestock feed, edible seeds and oil for cooking, and even in religious ceremonies. Interspersed was the noted use of *C. sativa* in homeopathic remedies for arthritis, malaria, gastrointestinal disorders, menstrual disorders, and fever in Chinese medicine.[5] The transition to western medicine practice started in the 1800's as a pain medication for headache and other ailments. More recently, cannabis has been used to treat glaucoma, nausea and vomiting (often chemotherapy related), anxiety, spasticity, peripheral painful neuropathy and epilepsy (see below). Nabiximols (Sativex®) is derived from plant extract, contains an equal THC: CBD and is approved for use in spasticity associated with multiple sclerosis. Nabilone (Cesamet®) and dronabinol (Marinol®) are synthetic THC compounds approved for use in the US for chemotherapy induced nausea and vomiting.

REGULATORY ISSUES RELATED TO MEDICAL MARIJUANA

Medical marijuana use is now legal in 23 states and the District of Columbus with an estimated 2,434,192 prescribed uses through the end of 2013. (http://medicalmarijuana.procon.org/) Marijuana, however, is a Schedule I substance and that the federal government considers growing, distribution, or possession of marijuana to be a federal crime regardless of the state laws.
The Department of Justice memo of August, 2013 clarified the government's prosecutorial priorities and stated that the federal government would rely on state and local law enforcement to "address marijuana activity through enforcement of their own narcotics laws." Interesting, the federal legal definition of marijuana " all parts of the plant *Cannabis sativa*, whether growing or not; the seeds thereof; the resin extracted from any part of the plant; and every compound, manufacture, salt, derivative, mixture, or preparation of such plant, its seeds or resin" and does not include *C. indica* as a prohibited plant.

State law currently determines the potential uses for medical marijuana. Physicians and patients must be registered with the state in which the drug is to be used. To date, no state has legislated the use of medical marijuana by licensed veterinarians for any purpose or species.

OVERVIEW OF CANNABIS PHARMACOLOGY

Chemotypes

The term cannabinoids refers to the chemical substances isolated from *C. Sativa*, and to its derivatives and transformation products, of which 80 terpenophenolic compounds have been identified.[6] Phytocannabinoids refer to only those compounds that are plant derived, of which 11 sub-types have been defined. These can be divided into psychotropic forms, which contain THC, or non-psychotropic forms, which do not contain THC.

Psychotropic form

Δ-9 THC is the main psychotropic constituent found in cannabis. Δ-9 THC is a potent compound that binds to two G-protein coupled cell membrane receptors, cannabinoid type 1 (CB_1) and cannabinoid type 2 (CB_2). Endogenous endocannabinoid ligands anandamide and 2-arachidonylglycerol bind to CB_1 and CB_2 receptors, respectively. Table 1 summarizes the mechanisms of action.

Non-psychotropic form

Cannabidiol (CBD) is the primary non-psychotropic constituent in cannabis and is a CB_1 receptor antagonist that down regulates G-protein coupled activation. As such, it can inhibit some of the psychoactive effect of Δ-9 THC when used in combination. Unlike Δ-9 THC, CBD is a multi-target compound with influences not only on different receptors, but also a more widespread effect in the brain. In vitro and in vivo research has demonstrated CBD to have analgesic, anti-inflammatory and anti-oxidant effects.[7]

Table 1. Mechanisms of action of Δ-9 THC and Cannabidiol (CBD)

	Action/Effect	Δ-9 THC	CBD
Cannabinoid type 1 receptor	Psychotropic Neuroprotection	Agonist	Antagonist
Cannabinoid type 2 receptor	Locomotion effect	Agonist	Agonist
5-HT receptor (serotonin)	Increase serotonin	Inactive	Agonist
TRPA 1 cation receptor	Decrease calcium flux	Inactive	Agonist
Glycine receptors	Analgesic	Inactive	Agonist
Adenosine uptake	Anti-inflammatory	Inactive	Inhibit
Fatty acid amide hydrolase	Anti-oxidant	Inactive	Inhibit

Pharmacokinetics and drug metabolism

The comparative pharmacokinetics of CBD is presented in Table 2. Although limited data was found for the dog from a single paper.[8] there are several similarities present with people: CBD has low bioavailability due to high first past effect through the liver, the drug is primarily metabolized through the liver, and there is a high volume of distribution due to the lipophilic nature of the drug.[9] Of interest was the lack of any detectable plasma concentration in 3 or the 6 dogs after oral administration of 180 mg of CBD. While oral CBD is limited in its bioavailability, once in the circulation, there is a rapid distribution through the body with expected prolonged elimination in people. Inconclusive data is available to comment on oral pharmacokinetics in dogs. Drug- drug interaction should be considered, as CBD is a strong cytochrome P450 inhibitor.[10] As such, altered metabolism of other hepatic metabolized AED may be affected.

Table 2. Comparison of CBD pharmacokinetics between human and dog.

	Human	Dog	
	Oral	Oral	Intravenous
Bioavailability	6%	3 dogs: 0 3 dogs: 13-19	6-10
Concentration max (ug/L)	3 +/- 3.1	0 to NR	13.6
T max (hour)	2.8 +/- 1.3	0 to NR	NR
Volume of distribution (L/kg)	32	0 to NR	100
Terminal elimination half-life (hour)	18-32	0 to NR	9
Primary method of metabolism	Cytochrome P450	Cytochrome P450	
Clearance (ml/min)	960-1560	0 to NR	265-288
Protein binding	High	High	

NR: not reported

Safety

The safety profile of cannabis is people is difficult to quantitate due to the high inter-individual variability.[11] Overall, CBD is considered to have a high therapeutic index with a good safety profile due to the antagonistic effect on the CB1 receptor. Δ-9 THC, however, can lead to a variety of direct and indirect psychological and physical adverse effects. Among these possible reactions are loss of coordination, reduced attention span, loss of short term memory, delayed gastric emptying and increased gastric reflux, reduced male fertility, and altered immune-system function.[11]

CBD AND EPILEPSY

Mechanism of action: The exact anticonvulsant mechanism of CBD is not known. One hypothesis focuses on reduction of glutamate mediated neuronal excitation through modulation of calcium flux via G-coupled receptor protein 55, TRP channels, and/or voltage gaited anion –selective channel protein 1.[9] Multiple mechanisms are most likely in place due to the number of receptor activations present.

Animal models: The majority of animal model research has been done in mice and rats using the standardized maximum electroshock (MES), method for seizure induction. Significant anticonvulsant activity was first shown in this acute seizure model in mice.[12] CBD decreased seizure frequency and onset in 3 seizure induction models of MES, Penicillin and PTZ-induced rats at doses ranging from 10 to 100 mg/kg via intraperitoneal injection prior to convulsant challenge.[13] CBD has also been shown to increase the after-discharge threshold and reduce after-discharge amplitude, duration and propagation in electrically kindled limbic seizures in rats.[14]

No scientific studies on the benefit of CBD in spontaneous or induced seizures in dogs was found. One study documented a proconvulsant effect of Δ-9 THC in dogs exposed to marijuana smoke prior to subconvulsant penicillin exposure, suggesting a decrease in seizure threshold,[15] which also seen in rodents.

More recently, epileptic dogs were found to have altered endocannabinoid cerebrospinal fluid concentrations.[17] Dogs with idiopathic epilepsy had significantly higher concentration of anandamide (AEA), an endogenous CB1 receptor ligand, compared to control dogs. More severe seizures and longer duration of epilepsy were also corrected with higher AEA levels. The authors postulated that increased AEA are in response to down regulation of CB1 receptors with progressive loss of AEA effectiveness to reduce neuronal hyperexcitability.

Human studies

Historical: Doctors Russell Reynolds and William Gowers described the first reported use of cannabis in the treatment of epilepsy in a person in 1861.[18] This middle-aged man was refractory to bromide therapy and successfully treated with extracts of *C. indica* three times a day. The "fits ceased at once" without recurrence for 6 months when the patient discontinued treatment, but remitted once treatment was restarted.

Anecdotal: A variety of anecdotal reports of improvement in seizure control with CBD exists in varying degrees of accuracy. In a recent survey of 19 parents whose children suffered from intractable seizures, 42% reported a > 80% reduction in seizures after use of CBD-enriched cannabis. Individual success stories with dramatic results are found in the internet and even the literature.[19]

Open label, randomized study: Three open-label, non-randomized but placebo control clinical CBD epilepsy trials have been reported between 1978 and 1990.[9] Conclusions from these studies are limited due to the small sample cohort, (total of 17 treated and 19 placebo patients), limited duration of treatment (weeks to months), varying dosing schedule, and heterogeneous patient population and poor retention. A total of 4 of the 17 treated patients were reported to have reduction in seizure frequency.

Randomized-blinded study: Only one double-blinded, randomized study performed on DRE patients with temporal lobe epilepsy was identified.[20] This study compared CBD 100 mg t.i.d. treatment groups to placebo for 6 month followed by a cross-over design for another 6 months. No change in seizure frequency between groups was noted.

CAN DOGS BENEFIT FROM AND CONTRIBUTE TO THE STUDY OF CANNABIS IN CHRONIC EPILEPSY?

The basis of providing sound medical treatment advice is the ability to provide predicative outcomes. This predictability for CBD treatment in epilepsy can only be obtained by moving away from autonomous decision making by individuals to evidence-based, randomized double – blinded and placebo controlled studies. Strong-selection bias exists in the current methodology of CBD epilepsy treatment in people whereby individual doctors treat individual patients with varying types of cannabis extracts. The lack of federal funding in this area of research has prohibited the start of larger epilepsy centers to establish the needed studies.

So, how can dogs afflicted with epilepsy be helped and offer help to people? Similar to people, dog have a high incidence of unprovoked (idiopathic) epilepsy, can respond well to many of the same AED drugs, and develop about the same prevalence of DRE over time. Dogs have similar CB1 and CB2 receptors types and endocannabinoid ligands, similar CBD metabolism, and can tolerate chronic exposure. As such, canine epilepsy is a suitable spontaneous model for comparison.

Many hurdles lay ahead in the study of CBD in epilepsy treatment for people and dogs. As the stigma of marijuana is transformed from ignorance to acceptance, as consistent drug compounding can be obtained under strict legal guidelines, and as veterinarians are accepted more into the one health initiative, then we be able to know the answer to question. Maybe John Adams was right: we will need a "world hemp for our own consumption", if not our health.

REFERENCES

1. Kwan P et al. New Eng J Med 2011; 365:919; 2. Tellez-Zenteno JF et al. Epilepsia 2014;55:829; 3. Kwan P et al Epilepsia 2010;51:1069; 4. Podell M Vet Clin NA, 1996; 5. Abel EL NY, NY; Plenum Press, 1980; 6. Mechoulan R, Hamos L .Chem Phys Lipids 2000;180;4; 7. Russo E . Oxford Press, 2014;Chapt 2;23. 8. Samara E et al. Drug Metab Disp 1988;16:469. 9. Devinsky O et al. Epilepsia 2014;55:791; 10. Hawksworth G, McArdle K London, UK, Pharmaceut Press, 2004; 11. Grotenherman F Oxford Press 2014:Chapt 38;674. 12. Karler R, Tukanais SA J. Clin Pharm 1981;21:437S. 13. Jones NA et al. J Pharm Exptl Therap 2010;332:569; 14. Jones NA et al. Seizure 2012;21:344; 15. Labrecque G et al. Can J Physiol Pharmacol 1978;56:87; 16. Stadniki SW et al Life Sci 1974; 14:463; 17. Gesell FK et al. BMC Vet Res 2013;9:262; 18. Reynolds JR. London, UK. John Churchill, 1861. 19. Maa E, Figi P. Epilepsia 2014;55:783. 20. Trembley B, Sherman M. Intl Conf Cannibis and Cannibidiol 1990.

EVOLVING ROLE & SCIENCE OF IMMUNOTHERAPEUTICS IN VETERINARY ONCOLOGY

Philip J. Bergman, DVM, MS, PhD, DACVIM (Oncology)
Bedford Hills, NY
Chad M. Johannes, DVM, DACVIM (SAIM, Oncology)
Kansas City, KS

INTRODUCTION

The term "immunity" is derived from the Latin word immunitas, which refers to the legal protection afforded to Roman senators holding office. While the immune system is normally thought of as providing protection against infectious disease, the immune system's ability to recognize and eliminate cancer is the fundamental rationale for the immunotherapy of cancer. Multiple lines of evidence support a role for the immune system in managing cancer, including: 1) spontaneous remissions in cancer patients without treatment, 2) the presence of tumor-specific cytotoxic T-cells within tumor or draining lymph nodes, 3) the presence of monocytic, lymphocytic and plasmacytic cellular infiltrates in tumors, 4) the increased incidence of some types of cancer in immunosuppressed patients and 5) documentation of cancer remissions with the use of immunomodulators. With the tools of molecular biology and a greater understanding of mechanisms to harness the immune system, effective tumor immunotherapy is becoming a reality. This new class of therapeutics offers a more targeted and therefore precise approach to the treatment of cancer. It is extremely likely that immunotherapy will have a place alongside the classic cancer treatment triad components of surgery, radiation therapy and chemotherapy within the next five to ten years.

I. TUMOR IMMUNOLOGY

A. Cellular Components

The immune system is generally divided into two primary components: the innate immune response, and the highly specific, but more slowly developing adaptive or acquired immune response. Innate immunity is rapidly acting but typically not very specific and includes physico-chemical barriers (e.g. skin and mucosa), blood proteins such as complement, phagocytic cells (macrophages, neutrophils, dendritic cells (DCs), and natural killer (NK) cells) and cytokines which coordinate and regulate the cells involved in innate immunity. Adaptive immunity is thought of as the acquired arm of immunity which allows for exquisite specificity, an ability to remember the previous existence of the pathogen (ie "memory") and differentiate self from non-self, and importantly the ability to respond more vigorously upon repeat exposure to the pathogen. Adaptive immunity consists of T and B lymphocytes. The T cells are further divided into CD8 (cluster of differentiation) & MHC (major histocompatibility complex) Class I cytotoxic helper T cells (CD4 & MHC class II), natural killer (NK) cells and regulatory T cells. B lymphocytes produce antibodies (humoral system) which may activate complement, enhance phagocytosis of opsonized target cells and induce antibody dependent cellular cytotoxicity (ADCC). B-cell responses to tumors are thought by many investigators to be less important than the development of T-cell mediated immunity, but there is little evidence to fully support this notion. The innate and adaptive arms of immunity are not mutually exclusive; they are linked by: 1) the innate response's ability to stimulate and influence the nature of the adaptive response, and 2) the sharing of effector mechanisms between innate and adaptive immune responses.

Immune responses can be further separated by whether they are induced by exposure to a foreign antigen (an "active" response) or if they are transferred through serum or lymphocytes from an immunized individual (a "passive" response). While both approaches have the ability to be extremely specific for an antigen of interest, one important difference is the inability of passive approaches to confer memory. The principal components of the active/adaptive immune system are lymphocytes, antigen-presenting cells and effector cells. Furthermore, responses can be sub-divided by whether they are specific for a certain antigen, or a non-specific response whereby immunity is attempted to be conferred by upregulating the immune system without a specific target. These definitions are helpful as they allow methodologies to be more completely characterized, such as active-specific, passive-nonspecific, etc.

B. Immune Surveillance

The idea that the immune system may actively prevent the development of neoplasia is termed "cancer immunosurveillance". Sound scientific evidence supports some aspects of this hypothesis including: 1) IFN-gamma protects mice against the growth of tumors, 2) mice lacking IFN receptor were more sensitive to chemically-induced sarcomas than normal mice and were more likely to spontaneously develop tumors, 3) mice lacking major components of the adaptive immune response (T and B cells) have a high rate of spontaneous tumors, and 4) mice that lack IFN and B/T cells develop tumors, especially at a young age.

C. Immune Evasion by Tumors

There are significant barriers to the generation of effective antitumor immunity by the host. Many tumors evade surveillance mechanisms and grow in immunocompetent hosts as easily illustrated by the overwhelming numbers of people and animals succumbing to cancer. There are multiple ways in which tumors evade the immune response including: 1) Immunosuppressive cytokine production (e.g. TGF and IL-10), 2) Impaired DC function via inactivation ("anergy") and/or poor DC maturation through changes in IL-6/IL-10/VEGF/GM-CSF, 3) Induction of cells called Regulatory T cells (Treg), which were initially called suppressor T cells (CD4/CD25/CTLA-4/GITR/Foxp3 positive cells which can suppress tumor-specific CD4/CD8+ T cells), 4) MHC I loss through structural defects, changes in B2-microglobulin synthesis, defects in transporter-associated antigen processing or actual MHC I gene loss (ie allelic or locus loss) and 5) MHC I antigen presentation loss through B7-1 attenuation (B7-1 is an important co-stimulatory molecule for CD28 mediated T-cell receptor and MHC engagement) when the MHC system in #4 remains intact.

II. NON-SPECIFIC TUMOR IMMUNOTHERAPY

Dr. William Coley, a New York surgeon in the early 1900's, noted that some cancer patients developing incidental bacterial infections survived longer than those without infection. Coley developed a bacterial "vaccine" (killed cultures of S. marcescens & S. pyogenes; "Coley's toxins") to treat people with sarcomas which provided complete response rates of approximately 15%. Unfortunately, high failure rates and significant side effects lead to discontinuation of this approach. His seminal work laid the foundation for nonspecific modulation of the immune response in the treatment of cancer. There are numerous non-specific tumor immunotherapy approaches ranging from biological response modifiers (BRM) to recombinant cytokines.

A. Biological Response Modifiers

BRM's are molecules that can modify the biological response of cells to changes in its external environment, which in the context of cancer immunotherapy could easily span non-specific and specific immunotherapies. This section will be discussing non-specific BRM's (sometimes termed "immunopotentiators") which are often related to bacteria and/or viruses.

One of the earliest BRM discoveries after Coley's toxin was the use of BCG (Bacillus Calmette-Guérin; interestingly, Guérin was a veterinarian). BCG is the live attenuated strain of Mycobacterium bovis and intra-vesical instillation in the urinary bladder causes a significant local inflammatory response which results in anti-tumor responses. The use of BCG in veterinary patients has been investigated with numerous types of cancers including urinary bladder carcinoma, osteosarcoma, lymphoma, prostatic carcinoma, transmissible venereal tumor, mammary tumors, sarcoids, squamous cell carcinoma and others. LDI-100, a product containing BCG and human chorionic gonadotropin (hCG), was compared to vinblastine in dogs with measurable grade II or III mast cell tumors. Response rates were 28.6% and 11.7%, respectively, and the LDI-100 group had significantly less neutropenia. It is particularly exciting for the field of veterinary cancer immunotherapy to potentially be able to use a BRM product which has greater efficacy and less toxicity than a chemotherapy standard of care. Unfortunately, LDI-100 is not commercially available at present.

Corynebacterium parvum is another BRM which has been investigated for a number of tumors in veterinary medicine including melanoma and mammary carcinoma. Other bacterially-derived BRM's include attenuated Salmonella typhimurium (VNP20009), Mycobacterial cell wall DNA complexes (MCC; abstracts only at present) and bacterial superantigens. Mycobacterial cell walls contain muramyl dipeptide (MDP) which can activate monocytes and tissue macrophages. MTP-PE (muramyl tripeptide-phosphatidylethanolamine) is an analog of MDP. When encapsulated in multilamellar liposomes (L-MTP-PE), monocytes and macrophages uptake MTP leading to activation and subsequent tumoricidal effects through induction of multiple cytokines, including IL–1a, IL-1b, IL-7, IL-8, IL-12, and tumor necrosis factor (TNF). L-MTP-PE has been investigated in numerous tumors in human and veterinary patients, including osteosarcoma, hemangiosarcoma and mammary carcinoma.

Oncolytic viruses have also been used as non-specific anti-cancer BRM's in human and veterinary patients. Adenoviruses have been engineered to transcriptionally target canine osteosarcoma cells and have been tested in vitro and in normal dogs with no major signs of virus-associated side effects. Similarly, canine distemper virus (CDV), the canine equivalent of human measles virus, has been used in vitro to infect canine lymphocyte cell lines and neoplastic lymphocytes from dogs with B and T cell lymphoma, with high infectivity rates, suggesting that CDV may be investigated in the future for treatment of dogs with lymphoma.

Imiquimod (AldaraTM) is a novel BRM that is a TLR7 (toll-like receptor 7) agonist. Imiquimod has been reported as a successful treatment for Bowen's disease (multicentric squamous cell carcinoma in situ) and other skin diseases in humans. Twelve cats with Bowen's like disease were treated topically with imiquimod 5% cream and initial and all subsequent new lesions responded in all cats. An additional cat (with pinnal actinic keratoses and squamous cell carcinoma) and dog with cutaneous melanocytomas has subsequently been reported to have been successfully treated with topical imiquimod 5% cream. It therefore appears that Imiquimod 5% cream is well tolerated, and further studies are warranted to further examine its usefulness in cats and dogs with other skin tumors that are not treatable through standardized means.

B. Recombinant Cytokines, Growth Factors & Hormones

A number of investigations using recombinant cytokines, growth factors or hormones in various fashions for human and veterinary cancer patients have been reported to date. Many have investigated the in vitro and/or in vivo effects of the soluble cytokine (e.g. interferons, IL-2, IL-12, IL-15, etc.), liposome encapsulation of the cytokine (e.g. liposomal IL-2), or use a virus, cell, liposome-DNA complex, plasmid or other mechanism to expresses the cytokine (e.g. recombinant poxvirus expressing IL-2). Important to note, the European Committee for Medicinal Products for Veterinary Use (CVMP) adopted a positive opinion in March, 2013 for the veterinary product Oncept IL-2 (feline Pox virus expressing recombinant feline IL-2) and the USDA gave this product a conditional license in March, 2015. It is to be used in addition to surgery and radiation in cats with large fibrosarcomas without metastasis or lymph node involvement, to reduce the risk of relapse and increase the time to relapse.

III. CANCER VACCINES

The ultimate goal for a cancer vaccine is elicitation of an anti-tumor immune response which results in clinical regression of a tumor and/or its metastases. There are numerous types of tumor vaccines in phase I-III trials across a wide range of tumor types. Responses to cancer vaccines may take several months or more to appear due to the slower speed of induction of the adaptive arm of the immune system.

The immune system detects tumors through specific tumor-associated antigens (TAAs) and/or abnormal disease-associated antigens (DAAs) that are potentially recognized by both CTLs and antibodies. TAAs and/or DAAs may be common to a particular tumor type, be unique to an individual tumor or may arise from mutated gene products such as ras, p53, p21 and/or others. While unique TAAs may be more immunogenic than the other aforementioned shared tumor antigens, they are not practical targets because of their narrow specificity. Most shared tumor antigens are normal cellular antigens that are overexpressed in tumors. The first group to be identified was termed cancer testes antigens due to expression in normal testes, but they are also found in melanoma, and various other solid tumors such as the MAGE/BAGE gene family. These notes will highlight those tumor vaccine approaches which appear to hold particular promise in human clinical trials and many which have been tested to date in veterinary medicine.

A variety of approaches have been taken to date to focus the immune system on the aforementioned targets, including 1) Whole cell, tumor cell lysate and/or subunit vaccines (autologous, or made from a patient's own tumor tissue; allogeneic, or made from individuals within a species bearing the same type of cancer; or whole cell vaccines from irradiated tumor cell lines with or without immunostimulatory cytokines), 2) DNA vaccines which immunize with syngeneic and/or xenogeneic (different species than recipient) plasmid DNA designed to elicit antigen-specific humoral and cellular immunity (to be discussed in more detail later), 3) Viral vector-based methodologies designed to deliver genes encoding TAAs and/or immunostimulatory cytokines, 4) DC or CD40-activated B cell vaccines (which are commonly loaded or transfected with TAAs, DNA or RNA from TAAs, or tumor lysates), 5) adoptive cell transfer (the "transfer" of specific populations of immune effector cells in order to generate a more powerful and focused antitumor immune response) and 6) Antibody approaches such as monoclonal antibodies, anti-idiotype antibodies (an idiotype is an immunoglobulin sequence unique to each B lymphocyte, and therefore antibodies directed against these idiotypes are referred to as anti-idiotype) or conjugated antibodies. The ideal cancer immunotherapy agent would be able to discriminate between cancer and normal cells (ie specificity), be potent enough to kill small or large numbers of tumor cells (ie sensitivity) and lastly be able to prevent recurrence of the tumor (ie durability).

This author (PJB) has developed a xenogeneic DNA vaccine program for melanoma in collaboration with human investigators from Memorial Sloan-Kettering Cancer Center. Preclinical and clinical studies by our laboratory and others have shown that xenogeneic DNA vaccination with tyrosinase family members (e.g. Tyrosinase, GP100, GP75, others) can produce immune responses resulting in tumor rejection or protection and prolongation of survival while syngeneic vaccination with orthologous DNA does not induce immune responses. While tyrosinase may not appear to be a preferred target in amelanotic canine melanoma due to poor expression when assessed by immunohistochemistry (IHC), more appropriate/sensitive PCR-based studies and other IHC-based studies document significant tyrosinase overexpression in melanotic and amelanotic melanomas across species. These studies provided the impetus for development of a xenogeneic tyrosinase (or similar melanosomal glycoproteins) DNA vaccine program in canine malignant melanoma (CMM). Cohorts of dogs received increasing doses of xenogeneic plasmid DNA encoding either human Tyrosinase (huTyr), murine GP75 (muGP75), murine tyrosinase (muTyr), muTyr ± HuGM-CSF (both administered as plasmid DNA) or muTyr "off-study" intramuscularly biweekly for a total of four vaccinations. We and collaborators have investigated the antibody and T-cell responses in dogs vaccinated with huTyr. Antigen-specific (huTyr) IFNγ T-cells were found along with 2- to 5-fold increases in circulating antibodies to huTyr which can cross react to canine Tyrosinase, suggesting the breaking of tolerance. The clinical results with prolongation in survival have been reported previously. The results of these trials demonstrate that xenogeneic DNA vaccination in CMM: (1) is safe, (2) leads to the development of anti-tyrosinase antibodies and T cells, (3) is potentially therapeutic, and (4) is an attractive candidate for further evaluation in an adjuvant, minimal residual disease Phase II setting for CMM. Based on these studies a multi-institutional safety and efficacy trial for USDA licensure in dogs with locally controlled stage II/III oral melanoma was initiated in 2006 with granting of conditional licensure in 2007, which represented the first U.S. governmental regulatory agency approval of a vaccine to treat cancer across species. Results of this licensure trial documented a statistically significant improvement in survival for vaccinates vs controls and a full licensure for the HuTyr-based canine melanoma vaccine from USDA-CVB was received in December, 2009 (OnceptTM, Merial, Inc.).

Tumor immunology and immunotherapy is one of the most exciting and rapidly expanding fields at present. Significant resources are focused on mechanisms to simultaneously maximally stimulate an antitumor immune response while minimizing the immunosuppressive aspects of the tumor microenvironment. The recent elucidation and blockade of immunosuppressive cytokines and/or the negative costimulatory molecule CTLA-4 along with the functional characterization of myeloid derived suppressor cells (MDSC) & T regulatory cells, may dramatically improve cell-mediated immunity to tumors. As investigators more easily generate specific anti-tumor immune responses in patients, we will need to remain vigilant to not push the immune system into pathologic autoimmunity. In addition, immunotherapy is unlikely to become a sole modality in the treatment of cancer as the traditional modalities of surgery, radiation and/or chemotherapy are extremely likely to be used in combination with immunotherapy in the future. Like any form of anti-cancer treatment, immunotherapy appears to work best in a minimal residual disease setting, suggesting its most appropriate use will be in an adjuvant setting with local tumor therapies such as surgery and/or radiation. Similarly, the long held belief that chemotherapy (non-corticosteroid) attenuates immune responses from cancer vaccines is beginning to be disproven through investigations on a variety of levels.

Additional exciting new approaches which appear to confer a survival benefit in human metastatic melanoma include the use of the anti-CTLA-4 antibody, ipilimumab (YervoyTM, Bristol-Myers Squibb) and the selective BRAF inhibitors, vemurafenib (ZelborafTM, Genetech) and dabrafenib (GSK2118436, GlaxoSmithKline), in patients who are BRAF V600 mutation positive. Fifty-three percent of human patients receiving concurrent administration of Ipilimumab and a new antibody directed against human programmed death 1 (PD-1) receptor, nivolumab, had an objective tumor response with tumor reductions of 80% or greater. As a small subset of dogs with malignant melanoma have exon 11 KIT gene mutations, the more routine use of KIT testing by PCR of CMM and subsequent use of c-kit small molecule inhibitors (particularly in dogs with advanced stage disease and/or lack of response to OnceptTM), should be considered. Furthermore, with somatic mutations in NRAS and PTEN being found in canine malignant melanoma similar to human melanoma hotspot sites, these may represent logical druggable targets in the future.

In summary, the future looks extremely bright for immunotherapy. Similarly, the veterinary oncology profession is uniquely able to greatly contribute to the many advances to come in this field. Unfortunately, what works in a mouse will often not reflect the outcome in human cancer patients. Therefore, comparative immunotherapy studies utilizing veterinary patients may be able better "bridge" murine and human studies. To this end, a large number of cancers in dogs and cats appear to be remarkably stronger models for counterpart human tumors than presently available murine model systems. This is likely due to a variety of reasons including, but not limited to, extreme similarities in the biology of the tumors (e.g. chemoresistance, radioresistance, sharing metastatic phenotypes and site selectivity, etc.), spontaneous syngeneic cancer (vs. typically an induced and/or xenogeneic cancer in murine models), and finally that the dogs and cats that are spontaneously developing these tumors are outbred, immune-competent and live in the same environment that humans do. These authors ardently look forward to the time when immunotherapy plays a significant role in the treatment and/or prevention of cancer in human and veterinary patients.

REFERENCES

- Bergman PJ. Biologic Response Modification. In: Rosenthal RC. Veterinary Oncology Secrets. Philadelphia: Hanley & Belfus, Inc.; 2001. 79-82.
- Baxevanis CN, Perez SA, Papamichail M. Cancer immunotherapy. Crit Rev Clin Lab Sci 2009;46:167-189.
- Reilly RT, Emens LA, Jaffee EM. Humoral and cellular immune responses: independent forces or collaborators in the fight against cancer? Curr Opin Investig Drugs 2001;2(1):133-135.
- Smyth MJ, Godfrey DI, Trapani JA. A fresh look at tumor immunosurveillance and immunotherapy. Nat Immunol 2001;2(4):293-299.
- Owen LN, Bostock DE. Proceedings: Tumour therapy in dogs using B.C.G. Br J Cancer 1974;29:95.
- MacEwen EG. An immunologic approach to the treatment of cancer. Vet Clin North Am 1977;7:65-75.
- Theilen GH, Hills D. Comparative aspects of cancer immunotherapy: immunologic methods used for treatment of spontaneous cancer in animals. J Am Vet Med Assoc 1982;181:1134-1141.
- MacEwen EG. Approaches to cancer therapy using biological response modifiers. Vet Clin North Am Small Anim Pract 1985;15:667-688.
- Henry CJ, Downing S, Rosenthal RC et al. Evaluation of a novel immunomodulator composed of human chorionic gonadotropin and bacillus Calmette-Guerin for treatment of canine mast cell tumors in clinically affected dogs. Am J Vet Res 2007;68:1246-1251.
- MacEwen EG, Patnaik AK, Harvey HJ, Hayes AA, Matus R. Canine Oral Melanoma: Comparison of Surgery Versus Surgery Plus Corynebacterium parvum. Cancer Investigation 1986;4(5):397-402.
- Thamm DH, Kurzman ID, King I et al. Systemic administration of an attenuated, tumor-targeting Salmonella typhimurium to dogs with spontaneous neoplasia: phase I evaluation. Clin Cancer Res 2005;11:4827-4834.
- Dow SW, Elmslie RE, Willson AP, Roche L, Gorman C, Potter TA. In vivo tumor transfection with superantigen plus cytokine genes induces tumor regression and prolongs survival in dogs with malignant melanoma. J Clin Invest 1998;101:2406-2414.
- Kurzman ID, MacEwen EG, Rosenthal RC et al. Adjuvant therapy for osteosarcoma in dogs: results of randomized clinical trials using combined liposome-encapsulated muramyl tripeptide and cisplatin. Clin Cancer Res 1995;1:1595-1601.
- Vail DM, MacEwen EG, Kurzman ID et al. Liposome-encapsulated muramyl tripeptide phosphatidylethanolamine adjuvant immunotherapy for splenic hemangiosarcoma in the dog: a randomized multi-institutional clinical trial. Clin Cancer Res 1995;1:1165-1170.
- Gill VL, Bergman PJ, Baer KE, Craft D, Leung C. Use of imiquimod 5% cream (AldaraTM) in cats with multicentric squamous cell carcinoma in situ: 12 cases (2002-2005). Veterinary and Comparative Oncology 2008;6:55-64.
- Dow S, Elmslie R, Kurzman I, MacEwen G, Pericle F, Liggitt D. Phase I study of liposome-DNA complexes encoding the interleukin-2 gene in dogs with osteosarcoma lung metastases. Hum Gene Ther 2005;16:937-946.
- Quintin-Colonna F, Devauchelle P, Fradelizi D et al. Gene therapy of spontaneous canine melanoma and feline fibrosarcoma by intratumoral administration of histoincompatible cells expressing human interleukin-2. Gene Ther 1996;3(12):1104-1112.
- Bergman PJ. Anticancer vaccines. Vet Clin North Am Small Anim Pract 2007;37:1111-1119.
- Alexander AN, Huelsmeyer MK, Mitzey A et al. Development of an allogeneic whole-cell tumor vaccine expressing xenogeneic gp100 and its implementation in a phase II clinical trial in canine patients with malignant melanoma. Cancer Immunol Immunother 2006;55(4):433-442.
- Gyorffy S, Rodriguez-Lecompte JC, Woods JP et al. Bone marrow-derived dendritic cell vaccination of dogs with naturally occurring melanoma by using human gp100 antigen. J Vet Intern Med 2005;19(1):56-63.
- Bergman PJ, Camps-Palau MA, McKnight JA et al. Development of a xenogeneic DNA vaccine program for canine malignant melanoma at the Animal Medical Center. Vaccine 2006;24(21):4582-4585.
- Bergman PJ, McKnight J, Novosad A et al. Long-term survival of dogs with advanced malignant melanoma after DNA vaccination with xenogeneic human tyrosinase: a phase I trial. Clin Cancer Res 2003;9(4):1284-1290.
- Smedley RC, Lamoureux J, Sledge DG, Kiupel M. Immunohistochemical diagnosis of canine oral amelanotic melanocytic neoplasms. Vet Pathol 2011;48(1):32-40.
- Phillips JC, Lembcke LM, Noltenius CE et al. Evaluation of tyrosinase expression in canine and equine melanocytic tumors. Am J Vet Res 2012;73(2):272-278.
- Liao JC, Gregor P, Wolchok JD et al. Vaccination with human tyrosinase DNA induces antibody responses in dogs with advanced melanoma. Cancer Immun 2006;6:8.
- Goubier A, Fuhrmann L, Forest L et al. Superiority of needle-free transdermal plasmid delivery for the induction of antigen-specific IFNgamma T cell responses in the dog. Vaccine 2008;26:2186-2190.
- Grosenbaugh DA, Leard AT, Bergman PJ et al. Safety and efficacy of a xenogeneic DNA vaccine encoding for human tyrosinase as adjunctive treatment for oral malignant melanoma in dogs following surgical excision of the primary tumor. Am J Vet Res 2011;72(12):1631-1638.
- Ottnod JM, Smedley RC, Walshaw R, Hauptman JG, Kiupel M, Obradovich JE. A retrospective analysis of the efficacy of Oncept vaccine for the adjunct treatment of canine oral malignant melanoma. Vet Comp Oncol 2013;11(3):219-229.
- Wolchok JD, Yuan J, Houghton AN et al. Safety and immunogenicity of tyrosinase DNA vaccines in patients with melanoma. Mol Ther 2007;15:2044-2050.
- Manley CA, Leibman NF, Wolchok JD et al. Xenogeneic Murine Tyrosinase DNA Vaccine for Malignant Melanoma of the Digit of Dogs. J Vet Intern Med 2011;25(1):94-99.
- Phillips JC, Blackford JT, Lembcke LM, Grosenbaugh DA, Leard AT. Evaluation of Needle-free Injection Devices for Intramuscular Vaccination in Horses. J Eq Vet Sci 31, 738-743. 2011.
- Wolchok JD, Kluger H, Callahan MK et al. Nivolumab plus ipilimumab in advanced melanoma. N Engl J Med 2013;369(2):122-133.
- Chu PY, Pan SL, Liu CH, Lee J, Yeh LS, Liao AT. KIT gene exon 11 mutations in canine malignant melanoma. Vet J 2012.

CLINICAL APPLICATION OF IMMUNOTHERAPEUTICS IN VETERINARY ONCOLOGY

Chad M. Johannes, DVM, DACVIM (SAIM, Oncology)
Kansas City, KS
Philip J. Bergman, DVM, MS, PhD, DACVIM (Oncology)
Bedford Hills, NY

IMMUNO-ONCOLOGY DEVELOPMENTS FOR CANINE LYMPHOMA

With nearly 20 FDA-approved monoclonal antibodies (MAbs) in the United States for treatment of various malignancies, passive immunotherapy is a key element in therapy guidelines in human oncology. Rituximab, a monoclonal antibody targeting the CD20 antigen on the surface of B-lymphocytes, was approved by the FDA in 1997 and has become a component of standard-of-care therapy for a number of human B-cell disorders. Expression of CD20 has been confirmed in canine B-cell lymphomas by immunohistochemistry using anti-human CD20 polyclonal antibodies that recognize the intracellular domains of CD20.[1] Unfortunately, rituximab does not bind canine CD20 and thus cannot be utilized for passive immunotherapy in dogs.[2] Development of canine monoclonal antibodies for treatment of B-cell and T-cell lymphomas would represent a huge potential step forward in our ability to more effectively treat these aggressive cancers in dogs.

1. **Anti-CD20 Monoclonal Antibody (MAb) for Aid in Treatment of Canine B-cell Lymphoma (Aratana)**
   A caninized anti-CD20 MAb (AT-004) was fully licensed by the USDA in January 2015 as an aid in the treatment of canine B-cell lymphoma. This MAb is commercially marketed by Aratana Therapeutics, Inc. Initial clinical data presented in abstract form at the Veterinary Cancer Society Conference in 2014 showed promising signs of biologic activity.

   The first study was an open-label pilot study evaluating the treatment of canine B-cell lymphoma with doxorubicin with or without the anti-CD20 MAb. Dogs enrolled were chemotherapy-naive with high-grade (WHO stage III–IV) B-cell lymphoma. Dogs were randomized 3:1 to doxorubicin (DOX; 30 mg/m^2 intravenously at week 1) plus anti-CD20 MAb (5 mg/kg intravenously weekly at weeks 2–8) or DOX alone (phase 1). In phase 2, relapsed dogs previously treated with DOX+anti-CD20 MAb could receive DOX (30 mg/m^2 at week 1) plus anti-CD20 MAb (5 mg/kg at weeks 2, 4, 6, 8), with additional courses permitted. Response rate, progression-free survival (PFS), overall survival (OS), and safety were evaluated. Seventy-five percent (9/12) of dogs treated with DOX+anti-CD20 MAb achieved complete remission vs 60% (3/5) receiving DOX alone (phase 1). Median (95% CI) progression-free survival (PFS) with DOX+anti-CD20 MAb was 98 (64–259) days vs 57 (29–77) days with DOX alone (P<0.0014). In phase 2 (8 relapsed dogs), median PFS (95% CI) with DOX+anti-CD20 MAb was 113 (14–163) days. For dogs receiving ≥1 cycle DOX+anti-CD20 MAb, overall survival (OS) was 316 (117–838) days. DOX+anti-CD20 MAb was generally well tolerated, with few additional adverse events (transient infusion reaction, n=4) vs DOX alone.[3]

   The second study was a prospective double-blind, randomized, placebo-controlled study evaluating the treatment of canine B-cell lymphoma with abbreviated L-CHOP chemotherapy and the canine anti-CD20 MAb. Dogs were treated with one 4-week cycle of L-CHOP chemotherapy and then randomized to either receive treatment with the canine MAb or placebo. Once remission was lost, all dogs received one dose of doxorubicin followed by treatment with the canine MAb. Efficacy was assessed by evaluation of measurable lymph nodes, cytology and clinical status. Safety was based on Veterinary Cooperative Oncology Group's common terminology criteria (VCOG-CTCAE v1.1). Dogs (n=27 enrolled, 24 randomized) all had measurable, cytologically-confirmed and flow cytometrically-immunophenotyped B-cell lymphoma. The median progression-free survival (PFS) and overall survival (OS) times in the canine MAb arm of the study were 167 and 325 days, respectively, compared to 93.5 and 177 days for the placebo arm (p<0.0001 for both PFS and OS values). Adverse events were restricted to the L-CHOP cycle.[4]

   Additional clinical studies are underway to further evaluate the best use and timing of this anti-CD20 monoclonal antibody in conjunction with cytotoxic chemotherapy.

2. **Anti-CD52 Monoclonal Antibody for Aid in Treatment of Canine T-cell Lymphoma (Aratana)**
   A caninized anti-CD52 monoclonal antibody (AT-005) was conditionally licensed by the USDA in January 2014 as an aid in the treatment of canine T-cell lymphoma. This monoclonal antibody is commercially marketed by Aratana Therapeutics, Inc. Anti-CD52 is directed against a cell-surface glycoprotein expressed by T-cells but not hematopoietic stem cells. Initial pharmacokinetic data was presented in abstract form at the Veterinary Cancer Society Conference in 2014.

   The bioavailability and safety of intravenously administered anti-CD52 MAb was evaluated in dogs with high-grade (WHO stage IV–V) T-cell lymphoma. Nine dogs (6 – 12 years old) received anti-CD52 MAb (~2.5 mg/kg) twice weekly for 4 weeks, followed by once weekly every other week. Plasma samples were collected before each dose and after the first and last doses; in 4 of these dogs, samples also were collected at 4, 8, 24, and 48 hours after dose 1. Plasma antibody concentrations were measured via ELISA. Pharmacokinetic parameters were calculated, using WinNonlin

Pharmacokinetic software. Plasma anti-CD52 MAb was detected immediately after dose 1, and plasma concentrations increased until doses 4–6, when steady-state was reached. Mean pharmacokinetic parameters were: single-dose: AUC, 8363.87 μg/mL*h; elimination half-life, 123.67 h; maximum plasma concentration (Cmax), 48.65 μg/mL; clearance, 0.45 (mg/kg)/(μg/mL)/h; volume of distribution (Vd), 51.86 (mg/kg)/(μg/mL); multiple-dose: AUC, 9147.3 μg/mL*h; elimination half-life, 158.2 h; Cmax, 40.9 μg/mL; clearance, 0.3 (mg/kg)/(μg/mL)/h; Vd, 64.6 (mg/kg)/(μg/mL). Neither clinically relevant changes in hematology or serum chemistry values, nor any episodes of hypotension or anaphylactoid reactions were noted. Canine anti-CD52 MAb is detectable in plasma after a single dose (initial mean half-life, 5.2 days) and accumulates after multiple doses (mean half-life, 6.6 days). Anti-CD52 MAb is well tolerated in dogs with T-cell lymphoma.[5]

Aratana is currently conducting two prospective multi-center, double-blind, randomized, placebo-controlled studies to further evaluate the anti-CD52 MAb in conjunction with cytotoxic chemotherapy for treatment of canine T-cell lymphoma. The first study will assess the benefit of adding the anti-CD52 MAb to a multi-agent chemotherapy protocol for dogs with intermediate to high grade T-cell lymphoma. All enrolled dogs will receive multi-agent chemotherapy (CHOP protocol), which will be supplemented by either placebo or the monoclonal antibody beginning with the second week of the CHOP protocol. Dogs will be randomized equally (1:1) to receive antibody or placebo during the chemotherapy protocol. This study was fully enrolled (49 dogs) in January 2015.[6]

The second study will assess the benefit of adding the anti-CD52 MAb to a single-agent CCNU chemotherapy protocol for dogs with intermediate to high grade T-cell lymphoma. All enrolled dogs will receive single-agent CCNU chemotherapy on Weeks 1 and 4 of the study. After two cycles of CCNU treatment, dogs that achieve a partial or complete response will be randomized equally (1:1) to receive anti-CD52 MAb or placebo twice a week for the next 4 weeks. Dogs will continue to receive the monoclonal antibody or placebo every other week for another 8 weeks. Sixty dogs will be enrolled in this study, with the goal of randomizing 40 dogs.[6]

These studies, as well as further prospective clinical studies, will help define the best use of the anti-CD52 MAb along with cytotoxic chemotherapy in dogs with T-cell lymphoma.

3. **Anti-CD20 Monoclonal Antibody (Elanco)**
The generation and characterization of a rituximab-like anti-CD20 antibody intended as a candidate treatment for canine B-cell lymphoma has recently been described.[7] A panel of anti-canine CD20 monoclonal antibodies was generated using a mouse hybridoma approach. Mouse monoclonal antibody 1E4 was selected for construction of a canine chimeric molecule based on its rank ordering in a flow cytometry-based affinity assay. 1E4 binds to approximately the same location in the extracellular domain of CD20 as rituximab, and 1E4-based chimeric antibodies co-stain canine B cells in flow cytometric analysis of canine leukocytes using an anti-canine CD21 antibody. Two of the four reported canine IgG subclasses (cIgGB and cIgGC) can bind to canine CD16a, a receptor involved in antibody-dependent cellular cytotoxicity (ADCC). Chimeric monoclonal antibodies were assembled using canine heavy chain constant regions that incorporated the appropriate effector function along with the mouse monoclonal 1E4 anti-canine CD20 variable regions, and expressed in CHO cells. Both 1E4-cIgGB and 1E4-cIgGC were observed to significantly deplete B-cell levels in healthy beagle dogs. The *in vivo* half-life of 1E4-cIgGB in a healthy dog was ~14 days. The antibody 1E4-cIgGB has been selected for further testing and development as an agent for the treatment of canine B-cell lymphoma.[7]

4. **Anti-CD20 Monoclonal Antibody (IDEXX)**
A newly established anti-canine CD20 monoclonal antibody that can be used to identify B-cells by flow cytometry and that stably binds an epitope promoting macrophage-mediated phagocytosis of canine B-cell lymphoma cells *in vitro* has also recently been reported.[8] This novel anti-canine CD20 mAb has been designated as 6C8. It recognized the extracellular domain of canine CD20 and showed high-affinity binding to canine CD20 in solution, as well as in its native conformation on canine B-cells. The 6C8 target was expressed invariably in B-cell lineage cells, but not in T-cells or in myeloid cells. 6C8 promoted phagocytosis of B-cell lymphoma cells by macrophages, but in its current framework, it did not induce direct cytotoxicity or complement dependent cytotoxicity. The 6C8 MAb may be useful as a diagnostic tool to phenotype B-cells, and could potentially be integrated as a tool for passive immunotherapy to treat dogs with B-cell disorders.[8]

5. **ImmuneFx™ Cancer Vaccine (Morphogenesis Inc)**
A multi-indication cancer vaccine, ImmuneFx™, has been described. Pre-clinical studies have indicated that ImmuneFx™ acts by providing a priming action for the immune system. The priming antigen is supplied to the patient's own tumor cells or *in vivo* in the form of a DNA plasmid. The cellular machinery expresses the antigen on the surface of the tumor cells where it alerts the body's scavenging cells (antigen presenting cells) to the fact that these cells are now foreign. Once this priming action has initiated the immune response, the antigen presenting cells attack the tumor cells, digest them and display all the antigens, those specific to the tumor cells as well as the foreign antigen, to T-cells which then initiate a response specific to all the antigens. Preclinical studies have shown that the vaccine can be produced for

and safely delivered to dogs with naturally occurring lymphoma and a study in dogs with B-cell lymphoma is reported to be underway.[9]

6. **Telomerase vaccine (Takis)**[10]
A DNA telomerase therapeutic vaccine has been described. As observed in humans, telomerase reverse transcriptase (TERT) activity is largely confined to tumor tissues and absent in the majority of normal dog tissues. Therefore, dog TERT (dTERT) may constitute a valid target for translational cancer immunotherapy. The ability of adenovirus serotype 6 (Ad6) and DNA electroporation (DNA-EP) to induce immune responses against dTERT in dogs affected by lymphoma was evaluated. The vaccine was combined with standard chemotherapy regimen [cyclophosphamide, vincristine, prednisone (COP)]. dTERT-specific immune response was induced in 13 out of 14 treated animals (93%) and remained detectable and long-lasting with the absence of autoimmunity or other side effects. The survival time of vaccine/Chemo-treated dogs was significantly increased over historic controls of Chemo-treated animals (>97.8 versus 37 weeks, respectively, P = 0.001).[11] This therapeutic vaccine is currently being studied in dogs with lymphoma.[12]

IMMUNO-ONCOLOGY DEVELOPMENTS FOR CANINE OSTEOSARCOMA

A recombinant HER2/neu expressing Listeria therapeutic vaccine (AT-014) has been developed by Aratana and is being studied as an aid in the treatment of canine osteosarcoma. In a study conducted at the University of Pennsylvania, Dr. Nicola Mason administered AT-014 to 18 dogs with appendicular osteosarcoma following amputation and chemotherapy (4 doses of carboplatin). The median survival time of 11 historical control dogs was 316 days; the median survival time for the treated group has not yet been reached (p<0.0001). However, eleven of the 18 treated dogs have surpassed the MST of the control group and 8 were alive as of 03 Dec 2014. Adverse events were mild to moderate and primarily consisted of fever, lethargy, and nausea/vomiting.[13]

ANTI-EGFR MONOCLONAL ANTIBODY

The epidermal growth factor (EGFR) family is becoming an important target for immunotherapy in human oncology. Immunoglobulins such as cetuximab (Erbitux®; Merck) for the treatment of colorectal carcinoma overexpressing EGFR (epidermal growth factor receptor, ErbB-1) or trastuzumab (Herceptin®; Genentech) for metastatic breast cancer overexpressing HER2 (human epidermal growth factor receptor-2, ErbB-2), have become standard components of therapy.

Generation of a canine anti-EGFR (Erb1) antibody for passive immunotherapy in canine cancer patients has been recently described.[14] As the tumor-associated antigen EGFR (ErbB-1) is highly conserved between humans and dogs, and considering the effectiveness of the anti-EGFR antibody cetuximab in human clinical oncology, the development of a caninized version of this antibody, can225IgG, for comparative oncology studies was undertaken. Variable region genes of 225, the murine precursor of cetuximab, were fused with canine constant heavy gamma and kappa chain genes, respectively, and transfected into Chinese hamster ovary (CHO) DUKX-B11 cells. 480 clones were screened and the best clones were selected according to productivity and highest specificity in EGFR-coated ELISA. Upon purification with Protein G, the recombinant cetuximab-like canine IgG was tested for integrity, correct assembly, and functionality. Specific binding to the surface of EGFR-overexpressing cells was assessed by flow cytometry and immunofluorescence; moreover, binding to canine mammary tissue was demonstrated by immunohistochemistry. In cell viability and proliferation assays, incubation with can225IgG led to significant tumor cell growth inhibition. Moreover, this antibody mediated significant tumor cell killing via phagocytosis in vitro. On the basis of its cetuximab-like binding site and the expression of a 91% homologous EGFR molecule in canine cancer, this antibody may be a promising research compound to establish passive immunotherapy in dog patients with cancer.[14]

IMMUNO-ONCOLOGY DEVELOPMENTS FOR CANINE MELANOMA

1. **Human CSPG4 DNA Vaccine**
Due to the many similarities with its human counterpart, canine malignant melanoma (CMM) is a valuable model in which to assess the efficacy of novel therapeutic strategies. The immunogenicity, safety, and therapeutic efficacy of a human chondroitin sulfate proteoglycan-4 (hCSPG4) DNA-based vaccine has recently been evaluated.[15] The homology between hCSPG4 and cCSPG4 amino-acidic sequences stands at more than 80%, providing the rationale for using an hCSPG4 DNA vaccine in the CMM model. Dogs with stage II–III surgically resected CSPG4-positive oral MM were treated with monthly intramuscular plasmid administration, which was followed immediately by electroporation (electrovaccination) for at least 6, and up to 20, months. The immunogenicity, safety, and therapeutic efficacy of the vaccine were evaluated. hCSPG4 electrovaccination caused no clinically relevant local or systemic side effects and resulted in significantly longer overall and disease-free survival times in 14 vaccinated dogs as compared with 13 nonvaccinated controls. All vaccinated dogs developed antibodies against both hCSPG4 and cCSPG4. Seven vaccinated dogs were also tested for a cCSPG4-specific T-cell response and only two gave a detectable interferon (IFN) gamma response.[15]

2. **Canine Melanoma Vaccine, DNA (Oncept®, Merial)**

Preclinical and clinical studies have shown that xenogeneic DNA vaccination with tyrosinase family members (eg, tyrosinase, GP100, and GP75) can produce immune responses resulting in tumor rejection or protection and prolongation of survival whereas syngeneic vaccination with orthologous DNA does not induce immune responses. These studies provided the impetus for development of a xenogeneic tyrosinase (or similar melanosomal glycoproteins) DNA vaccine program in canine malignant melanoma (CMM). Cohorts of dogs received increasing doses of xenogeneic plasmid DNA encoding either human tyrosinase (huTyr), murine GP75, murine tyrosinase (muTyr), muTyr _ human granulocyte-monocyte (HuGM)-CSF (both administered as plasmid DNA), or muTyr off-study intramuscularly biweekly for a total of 4 vaccinations. The antibody and T-cell responses in dogs vaccinated with huTyr have been investigated. Antigen-specific (huTyr) IFN-g T cells were found along with 2- to 5-fold increases in circulating antibodies to huTyr, which can cross-react to canine tyrosinase, suggesting the breaking of tolerance.[16,17]

The clinical results with prolongation in survival have been reported previously.[18,19] The results of these trials demonstrate that xenogeneic DNA vaccination in CMM (1) is safe, (2) leads to the development of antityrosinase antibodies and T cells, (3) is potentially therapeutic, and (4) is an attractive candidate for further evaluation in an adjuvant, minimal residual disease phase II setting for CMM. Based on these studies, a multi-institutional safety and efficacy trial for US Department of Agriculture (USDA) licensure in dogs with locally controlled stage II/III oral melanoma was initiated in 2006 with granting of conditional licensure in 2007, which represented the first US governmental regulatory agency approval of a vaccine to treat cancer across species. Results of this licensure trial documented a statistically significant improvement in survival for vaccinates versus controls and a full licensure for the HuTyr-based canine melanoma vaccine from the USDA Center for Veterinary Biologics was received in December 2009 (Oncept, Merial, Duluth, GA).[20]

Kaser-Hotz and colleagues[21] reported on concurrent use of Oncept and external beam radiation because many dogs with oral malignant melanoma may not be able to undergo surgery for local tumor control. This pilot study determined that concurrent use was well tolerated with no unexpected toxicities. Ottnod and colleagues[22] performed a single-site retrospective study on 30 dogs with stage II to III oral malignant melanoma (15 each with and without use of Oncept). They determined that those dogs receiving Oncept did not achieve a greater progression-free survival, disease-free interval, or median survival time than dogs that did not receive the vaccine. Contrary to the aforementioned prospective USDA 5-site licensure trial,[20] this study had less than 35% of cases treated surgically with margins 1 mm or more, suggesting a significant lack of local tumor control. Furthermore, contrary to the aforementioned prospective USDA 5-site licensure trial, the Ottnod and colleagues study, similar to most noncontrolled retrospective studies, had a wide variety of other treatments used in both the nonvaccinated and vaccinated groups and the cause (in the context of local or distant disease) of death and/or progression of disease was not reported.

REFERENCES

[1]Jubala CM, et al. Vet Pathol 2005;42:468-476. [2]Impellizeri JA, et al. Vet J 2006;171:556-558. [3]Bulman-Fleming J, et al. Abstract, 2014 Veterinary Cancer Society; St. Louis, MO. [4]Ogilvie GK, et al. Abstract, 2014 Veterinary Cancer Society; St. Louis, MO. [5]Rodriguez CR, et al. Abstract, 2014 Veterinary Cancer Society; St. Louis, MO. [6]http://aratana.com/for-veterinarians/clinical-studies. [7]Rue SM, et al. Vet Immunol Immunop 2015; http://dx.doi.org.10.1016/j.vetimm.2015.02.004. [8]Ito D, et al. Leuk Lymphoma 2015;56(1):219-25. [9]http://www.morphogenesis-inc.com/ImmuneFx.html. [10]http://www.takisbiotech.it/index.php/news/30-takis-veterinary-cancer. [11]Peruzzi D, et al. Mol Ther 2010;18(8):1559-67. [12]http://www.vetcancertrials.org/studies/vschv-telomerase-vaccine-for-canine-cancer. [13]Mason N, et al. Abstract, 2014 ACVIM Forum; Nashville, TN. [14]Singer J, et al. Mol Cancer Ther 2014;13(7):1777-90. [15]Riccardo F, et al. Clin Cancer Res 2014; 20(14):3753-62. [16]Liao JC, et al. Cancer Immun 2006;6:8. [17]Goubier A, et al. Vaccine 2008;26:2186-90. [18]Bergman PJ, et al. Vaccine 2006;24(21):4582-5. [19]Bergman PJ, et al. Clin Cancer Res 2003;9(4):1284-90. [20]Grosenbaugh DA, et al. Am J Vet Res 2011;72(12):1631-8. [21]Herzog A et al. Schweiz Arch Tierheilkd 2013;155(2):135-42. [22]Ottnod JM, et al. Vet Comp Oncol 2013;11(3):219-29.

CONTROVERSIES IN ONCOLOGY: LYMPHOMA PHENOTYPING AND STAGING

CA Clifford[1]
PJ Bergman[2]
[1] Malvern, PA
[2] Bedford Hills, NY

Introduction

The goal of this lecture is to provide pros and cons toward the implementation of full staging for dogs with lymphoma and specifically discussing phenotyping of canine lymphoma. Lymphoma (LSA) is the most common tumor of the cat and represents approximately 80-90% of hematopoietic tumors in cats. LSA is the third most common tumor in the dog with an estimated annual incidence of 13-24/100,000 dogs at risk. The mean age of cats diagnosed with LSA over 20 years ago was 2-5 years of age, however, recent reports suggest the mean age of cats diagnosed with LSA is now 8-12 years. The mean age of dogs afflicted with LSA remains stable at 6-9 years of age, however, the range of age in dogs can be as short as weeks to months. The most common site of LSA diagnosis in cats from over 20 years ago was mediastinal and/or multicentric, whereas recent reports suggest the most common site presently is alimentary.

Traditionally, lymphoma therapy was considered a recipe and protocols were commonplace that were considered a "one size fits all" for therapy. With the advent of newer immunologic, molecular and histologic diagnostics, the treatment and diagnostics of lymphoma are changing. Currently, decisions' regarding which chemotherapeutic agents/protocols to be utilized is now based upon tumor grade, location, histologic subtype and phenotype.

Staging

The defined "gold standard" of staging for lymphoma entails

- bloodwork (CBC, chemistry panel),
- 3-view chest radiographs,
- Abdominal ultrasound
- Bone marrow aspirates
- Lymph node aspirates or biopsy
- Phenotyping

The older literature does suggest an improvement in survival for stage I-III vs stage V, however, an integral question remains which is whether staging is truly needed to adequately and successfully treat a patient with lymphoma. Is there a basic amount of staging that is deemed sufficient in order to comfortably treat? The goal of staging is certainly to determine, to best of one's ability, the true extent of the disease within a patient. Many of the staging diagnostics will yield prognostic information that may guide treatment decisions. However, unlike in human medicine, the diagnostics of staging in veterinary oncology are "out of pocket" for owners thus there is an often a finite amount of financial resources for each client. Knowing this an oncologist needs to be pragmatic in choosing which staging diagnostics are deemed most important, especially in a financially limited situation. Meaning, will this test change my treatment protocol or have an impact on this owner's decision to treat? If the answer is yes, then certainly, the test is warranted, If the answer is no, can you strongly argue that it is a necessity in order to treat the patient? In light of this, one could argue as to whether a bone marrow aspirate or abdominal ultrasound are truly required staging tests. Several studies have not documented a statistical difference in outcomes between dogs with stage III vs IV disease. Along those lines, how is a stage IV determined; is this based upon the ultrasonographic description or are all patients having cytology performed of the liver and spleen to fully document involvement?

Phenotyping

On a very basic level, lymphoma is broken down into two main phenotypes, B and T cell. Traditionally 2/3 of dogs with lymphoma are classified as B cell and 1/3 are T cell. A minor percentage (<2%) are deemed "null cell." Phenotyping of lymphoma patients can be achieved through a variety of tests including immunohistochemistry, immunocytochemistry, PARR and flow cytometry. T cell lymphoma is more commonly associated with certain breeds including the boxer, golden retriever, Bassett Hound, Bulldog, Australian shepherd, Asian lap dogs and Siberian husky. T cell LSA is also associated with certain anatomic forms including cutaneous (epitheliotropic, AKA ELSA), mediastinal, hepatic and gastrointestinal. Many studies have documented a worse prognosis for dogs with T-cell lymphoma and for this this reason, many oncologists have begun modifying protocols based upon phenotype. Further support of this was based upon a retrospective study in which the response of T cell LSA to a single dose of doxorubicin was ~50% vs ~100% for dogs with B cell LSA. Some oncologists like to incorporate more alkylating agents into T cell protocols, partially based on the higher response rates noted in dogs with ELSA treated with lomustine (CCNU). Although, only 17% experienced a complete remission (CR), 61% experienced a partial remission (PR). Furthermore, the combination of L-asparaginase, mechlorethamine, vincristine, procarbazine, and prednisone (L-MOPP) was investigated in dogs with T-cell lymphoma and was associated with a complete remission rate of 78%, and overall survival 270 days; however, >20% were alive at >900 days. The challenge with this protocol is cost, availability of some of the agents and difficulty of administration and toxicity. Currently protocols with substitutions of CCNU for doxorubicin and Elspar in each cycle are underway, however, data are lacking. There does exist one report evaluating only T cell phenotype treated with a standard CHOP based protocol and its associated survival is similar to that noted in other studies.

Additional support for the importance of phenotyping comes from human experience with monoclonal antibodies, specifically Rituximab which target human CD20, in combination with standard of care CHOP based protocols (ie "R-CHOP"). Monoclonal antibody (mAb) can be used to specifically bind to target cells or proteins. This may then stimulate the patient's immune system to attack those targeted cells and remove them from the body. In human oncology, monoclonal antibodies have been developed for T and

B cell Lymphoma which enables the immune system to recognize, attack and remove them. Normal lymphocytes cell can be replenished, as stem cells within the bone marrow are not targeted and as such normal cells are replenished but the cancer lymphocytes are not. These have now become standard of care therapy in human oncology. Cancer cell killing is thought to be via three mechanisms:

- Antibody-dependent cellular cytotoxicity (ADCC)
- Complement-mediated cytotoxicity (CMC)
- Induction of apoptosis (natural cell death)

We now have access to two distinctly different veterinary monoclonal antibodies for dogs with T cell (conditional approval by USDA; Aratana Therapuetics: http://www.aratana.com/) and B cell (full approval by USDA; Novartis/Elanco). To date they have been shown to safe and have a reasonable expectation of efficacy. Each are being evaluated in separate large scale trials (www.vetcancertrials.org) along with standard chemotherapy. If similar efficacy to their "human" counterparts is noted, they will become standard of care in veterinary medicine and further pointing toward the necessity to phenotype all LSA patients.

REFERENCES

1. Thalheim L; Williams LE, Borst LB, et al. J Vet Intern Med. 2013;27:1509-16.
2. Regan RC, Kaplan MSW, Bailey DB. VCO. 2013;11:287-95.
3. Modiano JF, Breen M, Burnett RC, et al. Cancer Res 2005;65:5654
4. Lurie DM, Lucroy MD, Griffey SM et al. VCO 2004;2;171–175
5. Risbon R, de Lorimier LP, Skorupski K, et al. JVIM 2006;20:1389–1397
6. Beaver LM, Strottner G, Klein MK. J Am Vet Med Assoc. 2010;237:1052-5.
7. Brodsky EM, Mauldin GN, Lachowicz JL, et al. J Vet Intern Med. 2009;23:578-84
8. Rebhun RB, Kent MS, Borrofka S, et al. VCO 201;9:38–44.
9. Avery PR, Burton J, Bromberek JL, et al. J Vet Intern Med. 2014 Mar;28(2):538-46.
10. Chun R. Top Companion Anim Med. 2009 Aug;24(3):157-62.
11. Avery A. Top Companion Anim Med. 2009 Aug;24(3):144-50.
12. Lana SE, Jackson TL, Burnett RC, et al. J Vet Intern Med 2006;20(2):329-334.
13. Kiupel M, Teske E, Bostock D. Vet Pathol 1999;36:292-300.
14. Dobson JM, Blackwood LB, McInnes EF, et al. J Small Anim Pract 2001;42:377-384.
15. MacEwen EG, Brown NO, Patnaik AK, et al. J Am Vet Med Assoc 1981;178:1178-1181.
16. MacEwen EG, Hayes AA, Matus RE, et al. J Am Vet Med Assoc 1987;190:564-568.

CONTROVERSIES IN ONCOLOGY: MAST CELL TUMOR PANELS

PJ Bergman[1]
CA Clifford[2]
[1] **Bedford, NY**
[2] **Malvern, PA**

Introduction

Mast cell tumors (MCT) are the most common tumor in dogs, comprising close to 25% of all diagnosed skin tumors and the second most common tumor in the cat. These tumors may be seen in any age dog, but are more typically found in middle aged to older patients. Breeds at an increased risk for developing these tumors include boxers, Boston terriers, Labrador retrievers, beagles and schnauzers. The etiology of MCTs is largely unknown, although genetic factors and molecular alterations may play prominent roles.

Eighty-five to ninety percent of dogs and cats with MCT's have solitary lesions. It is important to note that not all dogs or cats with multiple MCT's have metastatic or systemic mastocytosis. Studies suggest that well-differentiated MCT's are slow-growing, usually < 3-4 cm in diameter, without ulceration of overlying skin, variably alopecic and commonly are present for more than 6 months. In contrast, poorly differentiated MCT's are rapidly growing, variably sized (but generally large), with ulceration of the underlying skin and inflammation/edema of surrounding tissues and lastly rarely are present for more than 2-3 months before presentation. Since most MCT's are of moderate-differentiation, signs may be somewhere between these two extremes.

Fine needle aspiration and cytology (FNAC) is the mainstay for diagnosis of MCT prior to surgical removal. Mast cells of MCT's have a characteristic discrete cell cytological appearance with eccentrically placed nuclei and abundant red to purple (ie metachromatic) cytoplasmic granules. Occasional MCT's, predominately undifferentiated MCT's, do not have the classic metachromatic cytoplasmic granules and must be diagnosed via other means (histopathology, special stains, etc.). Once a diagnosis is obtained, staging (looking for disease elsewhere) is routinely recommended, however, the completeness of staging is presently extremely controversial. After an FNAC diagnosis of MCT has been made, staging diagnostics may include full physical examination, bloodwork/urinalysis, FNAC of any local lymph nodes and abdominal ultrasound, thoracic radiography and bone marrow aspiration/cytology. The use of buffy coat cytology and liver/spleen FNAC is presently controversial in the routine staging of dogs with MCT and these authors does not routinely employ these diagnostics for staging of MCT's in dogs. Some oncologists have begun to either not routinely utilize bone marrow aspiration & cytology (BMAC) for MCT staging, or have begun to utilize results of CBC/plt to delineate whether or not to perform a BMAC.

Recently, there has been an increased focus on understanding MCT-related genetic and molecular alterations, including the identification of activating mutations in the growth pathway involving c-kit, the receptor for stem cell factor, which promotes the malignant process within mast cell tumors when mutated. Since this mutation is not present in all dogs, it is likely only a contributing mechanism to a more complex process of neoplastic transformation.

Controversies with MCTs

The pathology and behavior of a MCT depends substantially on the histologic pattern of the tumor, which is associated with the Patnaik tumor grade. However, there is variation within the grading scheme used among pathologists that lends itself to subjectivity in assigning a tumor to a particular category. Traditionally, these categories are grade I (well-differentiated or low grade), grade II (intermediate grade), and grade III (poorly-differentiated or high grade). Subsequently, guiding the appropriate therapy and providing prognostic information based on the traditional grading scheme has become complicated and unpredictable. The introduction of a two tier system (Kiupel; Michigan State) of low and high grade designed to minimize the subjectivity was introduced several years ago and has since been further validated. In a study of 137 surgically resected cutaneous MCTs, the relationship between grade and survival was evaluated. All grade I MCTs were low grade in the Kiupel system, and all grade III were deemed high grade. Among grade II, 71 (85.6%) were low grade, and 12 (14.4%) were high grade, with a 1-year survival probability of 94% and 46%. Per this study, the 2-tier system had a high prognostic value and was able to correctly predict the negative outcomes of some grade II MCTs. Currently pathologists are reporting both grading schemes.

Mitotic index (MI) is defined as the number of mitoses per 10 high power fields and has proven to generally be a useful prognostic factor. An MI >5 is associated with a poor prognosis (median survival time of 2 months), whereas an MI ≤ 5 suggests a more favorable prognosis, with a reported median survival time (MST) of 70 months. In the new two-tier system (Kiupel) a MI of 7 or > is associated with a high grade tumor. Many oncologist will use the MI, at least in part, to help dictate the need for additional therapy, ie. A low grade tumor with a MI of 0-1 that was completely excised generally warrants only local control as approximately 5% of these dogs will have their MCT's behave aggressively.

To assist in more objectively categorizing a grade II MCT, which account for 70-80% of all MCTs, various immunohistochemical and molecular tests of tumor cell proliferation are now available in addition to routine histopathology. These markers of proliferation are readily performed on tissue biopsy samples in the form of MCT panels (www.dcpah.msu.edu/Sections/Immunohistochemistry/FAQ.php#08). These assays provide clinicians with the ability to make more sound recommendations regarding the appropriate adjuvant therapy. This particular panel evaluates argyrophilic staining nucleolar organizing regions (AgNOR), proliferation cell nuclear antigen (PCNA), Ki-67, c-kit pattern assessment, and PCR for c-kit gene mutation. AgNOR frequency is an indirect measure of tumor cell proliferation and may be equally or even more important as the tumor grade in terms of predicting the biologic behavior of the MCT. PCNA and Ki-67 are also indirect measures of tumor cell proliferation and are most useful when interpreted in conjunction with one another. c-kit is the protein receptor for stem cell factor found in many cells including mast cells. A genetic mutation of the c-kit gene, which encodes for the c-kit receptor itself, has been identified in some MCTs. When mutated, the receptor is constitutively active and promotes the malignant process within the cells. Mutations within the c-kit gene are associated with a more aggressive phenotype. With the new two tier Michigan State system used more universally over time, ideally this may decrease the need for further MCT panels, as low grade tumors for the most part are associated with longer term survival.

***When should one use a MCT panel*?**

If one goes on the assumption that the new grading scheme will soon become standard of care, then the data suggest 5% of "low grade" tumors will behave more aggressively. Ideally, an MCT panel, may help differentiate these patients and thereby guide the clinician to either a more aggressive surgery or adjuvant therapy that could benefit that patient. To run MCTs panels on all "low grade" tumors seems overkill to help identify the 5% of atypical cases, it also takes away the validity of grading scheme if it is being suggested to be run in conjunction with grading. Ideally, the MCT panel should be reserved for cases in which the biopsy results do not fit the clinical picture; ie a more aggressive site (muzzle), fast growing tumor, severely ulcerated, large tumor, lymph node involvement, etc.

For high grade tumors, performing portions of the panel may be useful for making treatment decisions. Knowing a kit mutation status in a high grade tumor may change the treatment choice from purely chemotherapy to a combination of chemotherapy and a TKI, whereby a higher response rate is noted in kit mutation positive patients. Currently studies are underway utilizing mutation status to help logically guide treatment decision for a more personalized medicine approach, albeit on a very basic level.

REFERENCES

1. McCaw DL. Tumors of the skin, subcutis, and other soft tissues: Section D - Mast cell tumors. In: Henry CJ, Higginbotham ML, eds. Cancer Management in Small Animal Practice, St. Louis: Elsevier, 2010:317–321.
2. Thamm DH, Mauldin EA, Vail DM. J Vet Intern Med. 1999 13:491-497.
3. Patnaik AK, Ehler WJ, MacEwen EG. Vet Pathol. 1984 21:469-474.
4. Romansik EM, Reilly CM, Kass PH, et al. J Am Vet Med Assoc. 2004;218:1120-1123.
5. Kiupel M, Webster JD, Bailey KL, et al. Vet Pathol. 2011;48:147-55.
6. Sabattini S, ScarpaF, Berlato D, et al. Vet Pathol 2014.
7. Seguin B, Leibman NF, Bregazzi VS, et al. J Am Vet Med Assoc. 2001 Apr 1;218(7):1120-3.
8. Scase TJ, Edwards D, Miller J, et al. J Vet Intern Med. 2006;20(1):151-8.
9. Seguin B, Besancon MF, McCallan JL, et al. J Vet Intern Med 2006;20(4):933-40.
10. Ozaki K, Yamagami T, Nomura K, et al. J Vet Med Sci. 2007 Nov;69(11):1117-21.
11. London CA, Malpas PB, Wood-Follis SL, et al. Clin Cancer Res. 2009 Jun 1;15(11):3856-65.
12. Hahn KA, Legendre AM, Shaw NG et al. Am J Vet Res. 2010 Nov;71(11):1354-61. Erratum in: Am J Vet Res. 2011 Feb;72(2):247.
13. Smrkovski OA, Essick L, Rohrbach BW, et al. Vet Comp Oncol. 2013 Jul 12.

Controversies in Oncology: Metronomic Therapy

CA Clifford[1]
PJ Bergman[2]
[1] Malvern, PA
[2] Bedford Hills, NY

Introduction

Chemotherapy has traditionally been administered at a maximum tolerated dose (MTD) designed to kill a proportion of rapidly dividing cells at the time of administration. Unfortunately MTD dosing is associated with inevitable and potentially life-threatening side effects to tissues such as bone marrow progenitor cells and intestinal epithelium. Despite this, for most cancers in dogs, traditional MTD dosing has yielded little advances in long-term survival in spite of dose intensification/modifications protocols.

A shift in the paradigm of chemotherapy administration is currently underway. Instead of using cycles of potentially toxic MTD chemotherapy with long breaks to allow recovery of normal cells, metronomic chemotherapy utilizes frequent, even daily, administration of chemotherapeutics at a dosage well below the MTD, without prolonged breaks. A key feature of metronomic is compression of the schedule of drug administration with the cumulative dose being significantly less than with MTD-based chemotherapy. The result of which should be a favorable toxicity profile and the potential for long-term treatments without treatment breaks.

Unlike MTD chemotherapy, where the tumor cells are the primary targets of therapy, metronomic therapy appears to target the tumor microenvironment including the endothelial cells and in this context, the chemotherapeutics may act as anti-angiogenic agents. Many chemotherapy agents administered at the MTD result in apoptosis of tumor endothelial cells and this effect is lost in standard protocols which require prolonged recovery time. Lower and more frequent dosages continue to yield endothelial cell apoptosis; however, without the required breaks between treatments, the repair of endothelial cells is prevented. Evidence also suggests that metronomic therapy is capable of responses even in drug resistant cancers via targeting the more genetically stable endothelial cells. There is also suppression on the mobilization of circulating endothelial progenitor cells (CEPs) from the bone marrow (angiogenesis). These CEPs provide a source of endothelial cells in newly formed blood vessels and evidence has shown a decrease in CEP number and viability with metronomic therapy. A third potential mechanism of metronomic therapy involves increasing the production of the body's own natural angiogenesis inhibitors. Several studies have documented an increase in thrombospondin-1 (TSP1), an inhibitor of angiogenesis, in vivo and in vitro to LMD cyclophosphamide. Finally, metronomic therapy may selectively remove/lower T-regulatory lymphocytes decreasing immune tolerance.

Data from human studies suggest a favorable toxicity profile and likely lower cost for metronomic therapy. Thus it may be very appealing for our veterinary patients. Several studies using metronomic therapy in humans have been promising, leading to a number of clinical trials assessing metronomic therapy alone or in combination with vascular targeting agents (nonsteroidal agents (NSAIDs), small molecule inhibitors (tyrosine kinase inhibitors) or monoclonal antibodies). Although there is mounting evidence of metronomic efficacy, optimal treatment regimens are still not defined and an important challenge is the need to evaluate optimal treatments using relevant surrogate markers of metronomic response such as circulating angiogenic growth factors and circulating CEPs. Changes in trial design are necessary when assessing metronomic therapy, since the target of therapy is the vasculature or other cells in the tumor in addition to the malignant cells, objective clinical benefits may take considerable time to develop. Often this is only manifested as disease stabilization.

In veterinary medicine, data exists further documenting the mechanism of action behind the use of metronomic therapy in cancer patients. Cyclophosphamide administered at 15mg/m2 orally on a daily basis did decrease tumor microvessel density based upon serial biopsy at day 14. Although a short term study, decreases in tumor angiogenesis was also noted. Immunomodulation via decreased T regulatory cells in patients with soft tissue sarcomas has been noted in patients receiving Cyclophosphamide and Chlorambucil documented decreased T regulatory cell levels in patients with soft tissue sarcomas.

Thus far several studies evaluating metronomic therapy in dogs and cats have been reported using a variety of protocols containing Cyclophosphamide, Chlorambucil, Etoposide, Lomustine and a tyrosine kinase inhibitor (Toceranib, Zoetis) in conjunction with a nonsteroidal. The side effect profile seems favorable with sterile hemorrhagic cystitis noted in dogs with Cyclophosphamide ranging from 5-12% depending upon study. Anecdotally, the incidence may be higher as more patients are treated and for longer time frames, thus it is imperative that routine evaluation of the urine be performed. Toxicity leading to protocol cessation was noted with Lomustine and more recently chronic bone marrow suppression has been noted with Chlorambucil.

Multiple tumor types have been evaluated, but particular interest is noted with canine soft tissue sarcoma (STS; decreased microvessel density, decrease T regulatory cells and improved disease free interval vs surgery alone), mammary carcinoma, transitional cell carcinoma, hemangiosarcoma (HAS; improved disease free interval for a combination metronomic protocol vs standard doxorubicin dosing) and histiocytic sarcoma among others. This novel approach may be more cost effective, however, prospective, randomized and long-term studies evaluating responses and safety are lacking. With the exception of the HSA and one STS study, the vast majority were looking at bulky disease and often patients with advanced disease. The microscopic disease setting would appear to be the most beneficial time to institute such a protocol as evidenced by the data for STS and HSA. Two new studies, however, using metronomic therapy post standard therapy, but still in the microscopic disease setting did not note a benefit in patients with HSA and osteosarcoma. A recent evaluation also suggested higher toxicity in dogs with OSA receiving MTD chemotherapy along with a metronomic protocol.

Although there is mounting suggestive evidence of efficacy associated with metronomic therapy, optimal treatment regimens are still not defined. An important challenge is the need to evaluate optimal treatments using relevant surrogate markers of response which may include circulating angiogenic growth factors, T-regulatory cell levels and circulating CEPs. Innovation in trial design is also necessary when assessing LMD. Since the target of therapy is the vasculature or other cells in the tumor in addition to the malignant cells, objective clinical benefits may take considerable time to develop and may be initially manifested only as disease stabilization. In veterinary oncology, we are still learning which dosing schedules and histologies are optimal and therefore metronomic therapy should NOT be a "catch all" therapy when standardized therapy is not available or tenable for the client.

REFERENCES

1. Kerbel RS, Kamen BA. Nat Rev Cancer. 2004;4:423-36.
2. Kamen BA, Glod J, Cole PD. J Pediatr Hematol Oncol. 2006;28:325-7.
3. Shaked Y, Emmenegger U, Francia G, et al. Cancer Res. 2005;15:7045-51.
4. Kerbel RS. Science 2006;312:1171-5.
5. Fremder E, ShakedY. Mechanisms of Action of Low-Dose Metronomic Chemotherapy. Metronomic Chemotherapy 2014, pp 23-38
6. Damber JE, Vallbo C, Albertsson P, et al. Cancer Chemother Pharmacol. 2006;58:354-60.
7. Shaked Y, Emmenegger U, Francia G, et al. Cancer Res. 2005;15:7045-51.
8. Kerbel RS, Klement G, Pritchard KI, et al. Ann Oncol. 2002;13:12-5.
9. Young SD, Whissell M, Noble JC, et al. Clin Cancer Res 2006;12:3092-8.
10. Bocci G, Francia G. Metronomic Chemotherapy. Pharmacokinetics and Pharmacogenetics of Metronomic Chemotherapy. 2014, pp 229-246
11. Troy CG, Huckle WR, Rossmeisl JH, et al. J Vet Intern Med. 2006;20:144-50.
12. Allen DK, Waters DJ, Knapp DW, et al. J Vet Intern Med. 1996;10:231-4.
13. Burton JH, Mitchell L, Thamm DH, et al. JVIM 2011;25:920-926.
14. Lana S, U'ren L, Plaza S, et al. JVIM 2007;21:764-769.
15. Mitchell L, Thamm DH and Biller BJ. JVIM2012;26:355-362.
16. Elmslie RE, Glawe P and Dow SW. JVIM 2008;22:1373-1379.
17. Leach TN, Childress MO, Greene SN. VCO 2012;10:102-112.
18. Schrempp DR, Childress MO, Stewart JC, et al. JAVMA 2013;242:1534-1538.
19. Brachal S, Walshaw R, Danton T, et al. JSAP 2014;55:369-374.
20. Biller B. VCNA 2014;44:817-829.
21. Leo C, Stell A, Borrego J, et al. Evaluation of low-dose metronomic (LDM) cyclophosphamide toxicity in cats with malignant neoplasia. J Fel Med and Surgery 2014 (EPUB ahead of print)

CONTROVERSIES IN ONCOLOGY: MELANOMA VACCINE

PJ Bergman[1]
CA Clifford[2]
[1] **Bedford Hills, NY**
[2] **Malvern, PA**

Introduction

Canine malignant melanoma (CMM) of the oral cavity, nail bed, foot pad and mucocutaneous junction is a spontaneously occurring, highly aggressive and frequently metastatic neoplasm. CMM is the most common oral tumor in dogs and commonly occurs in Poodles, Dachshund, Scottish terriers, Golden retrievers, Schanuazer and Rottweiler. The tumor is most commonly located on the gingiva (~50%), buccal/labial (~30%), tongue (~5%) and hard palate (~15%). Biologically, these are deemed aggressive with a high metastatic rate, although recent evidence suggest some locations may carry a better prognosis. There is also reported a "low grade" subtype that may carry a more favorable prognosis. The history for most dogs involves signs such as: halitosis, abnormal chewing, weight loss, hypersalivation, bleeding from the mouth, or the visualization of a mass.

Staging

Dogs with oral melanomas are generally staged via

- Tumor aspiration: Oral melanomas typically exfoliate easily resulting in highly cellular slides. The slides are typically predominated by many round to irregular to spindle shaped cells with a variable about of cytoplasm. The cytoplasm is typically lightly basophilic and will contain either no or many greenish-black granules or globules. Many high-grade malignant melanomas will contain no visible melanin histologically (hence the term amelanotic melanoma), but a small number of cells will contain a slight dusting of melanin that can be seen cytologically. These malignant melanomas, commonly arising from the oral cavity, typically exhibit numerous criteria of malignancy, such as: large round to irregular nuclei with very coarsely stippled nuclear chromatin, multiple prominent nucleoli of variable size and shape, binucleation, moderate to marked anisokaryosis, and abnormal mitotic figures. Macrophages (melanophages) are commonly present and may consist of many large vacuoles full of melanin pigment.
- Routine labwork (CBC, Clinical Chemistry, Urinalysis)
- Radiographs (Three-view Thorax)
- Lymph node aspiration for cytology review: A metastatic lymph node can only be identified by finding non-lymphoid cells (melanocytes) with malignant criteria. In contrast, melanophages are commonly found in benign lymph nodes and their presence does not warrant a diagnosis of metastatic melanoma.
- Advanced imaging (Computed Tomography)

Therapy

CMM and advanced human melanoma (HM) are diseases that are initially treated with aggressive local therapies including surgery and/or fractionated radiation therapy; however, systemic metastatic disease are common sequelae. Based on these similarities, CMM appears to be a good clinical model for evaluating new treatments for advanced HM. Canine patients with advanced disease (WHO stage II, III or IV) have a reported median survival time of 1-5 months with standardized therapies. A combination of hypo-fractionated radiation therapy and chemotherapy have a reported median survival time of one year in stage I oral CMM. Human patients with deep AJCC stage II or stage III disease (locally advanced or regional lymph node involvement) have at least a 50% chance of recurrence after surgical resection; patients with stage IV melanoma (distant metastases) have a median survival of less than ten months and most of these patients eventually die of melanoma. Standard systemic therapy is dacarbazine chemotherapy in HM, and carboplatin chemotherapy in CMM. Unfortunately, response rates to chemotherapy in humans or dogs with advanced melanoma range from 8-28% with little evidence that treatment improves survival. It is easily evident that new approaches to this disease are desperately needed and multiple methodologies have been reported to date.

In cases where the only a marginal resection has been attempted, additional local therapy with either a second more aggressive surgery or radiation therapy (6 weekly treatments to the site and draining lymph node) is recommended.

In cases deemed nonresectable, or if an owner is unwilling to pursue surgery, radiation therapy has a high response rate (70-90%). Most protocols utilized are classified as "hypofractionated", in that they require only a few treatments but each treatment has a high dose per treatment (fraction). Generally these are clinician dependent but are either three to four weekly fractions of 8 Gray (Gy) or six weekly fractions of 6 Gy. Side effects (oral mucositis) are generally mild and self-limiting with hypofractionated protocols and the majority of radiation oncologists will include not only the primary tumor site or residual scar but the draining lymph node bed in the radiation fields. Similar to surgery, patients treated with radiation therapy may achieve local control but generally succumb to metastatic disease and survival times of 7-9 months are common.

Due to the high metastatic rate associated with melanoma systemic therapy is warranted. Responses to chemotherapy (Carboplatin, Cisplatin, Dacarbazine, Melphalan and Doxorubicin) has been noted in the bulky disease setting, but generally these are partial responses and not durable. Carboplatin, specifically was evaluated in dogs with malignant melanoma of various anatomic location in patients with local-regional control and no significant improvement in survival. Using the immune system to target cancer, is not a new therapeutic modality but overcoming tolerance is a challenge. A novel therapeutic DNA vaccine (OnceptTM) has been developed and this xenogeneic model (using DNA from one species to elicit a response against another species) significantly improved survival compared to historical controls in a prospective, 5-site USDA-regulated 110 dog study. This is the first US government licensed therapeutic vaccine for the treatment of cancer in either animals or humans. Importantly, the vaccine is safe and with side effects rarely noted. The vaccine is administered via a new "needleless" transdermal device. The vaccine protocol requires four treatments on an every other week basis and booster vaccines every 6 months. Antigen-specific immune responses have been seen in all species investigated to date including mice, rats, dogs, horses and humans.

A controversy has arisen regarding the use of the vaccine which stems from anecdote, personal opinion and several retrospective studies with small numbers of patients. One retrospective study by Ottnod et al evaluated patients treated with local regional control that received adjuvant melanoma vaccine vs. those that did not. The results did not demonstrate a difference in this subgroup of patients, however the numbers between groups were quite small (15 per arm) and it is impossible to delineate from the paper whether patients died from local recurrence vs distant metastatic disease.

A second retrospective evaluated the addition of the melanoma vaccine in dogs with digit melanoma with local regional control. Traditionally, the literature suggests a MST of 1 year with local regional control alone. The overall MST for dogs treated with loco-regional control and xenogeneic DNA vaccine was 476 days with a 1-year survival rate of 63%. Although, one cannot compare studies, the overall survival times were similar, however, it is important to note that the authors used a new staging scheme in which the majority of the patients were deemed higher stage than that seen in previous digit melanoma studies. It is also worth noting that many patients had delays of 60 days or greater prior to receiving vaccine, which was statistically found to have impacted reduced survival.

A question that has been raised comes from the use of a historical control in the melanoma vaccine data that may not accurately reflect true survival with surgery alone. A large retrospective study evaluated 151 dogs with naturally occurring oral malignant melanomas treated by excision with or without adjuvant therapies. The overall median survival time was 346 days and a clear survival benefit was not evident with any systemic adjuvant therapy ((335 days) and did not (352 days)) receive systemic adjuvant therapy, including melanoma vaccine therapy or chemotherapy. The authors do note, however, the overall number of dogs in these groups receiving adjuvant therapy was quite low (e.g. 12 dogs receiving Oncept). Furthermore, based on its retrospective nature, it was difficult if not impossible to delineate stage-dependent differences in survival making up the 346 day median survival. A second recent retrospective study consisting of 70 dogs, 36 of which a stage could be determined, were evaluated for PFI and ST following curative-intent surgery. Of this group, 51 were completely excised and 29 received adjuvant therapy. The overall Median PFI and ST were 508 and 723 days. Presence of metastasis at diagnosis, higher tumor stage, administration of adjuvant disease and tumor size were all significantly negatively associated with ST and PFI. Although difficult to compare studies, the PFI and ST noted in this study with curative-intent surgery is longer than that which made up the data in the historical control for the melanoma vaccine even though all prior studies of local control in stage II/III oral CMM typically had 3-7 month median survival times.

Clearly, evidence based medicine tells us one cannot truly compare the results of retrospective studies to the same scrutiny of a prospective designed trial. Well-designed prospective studies comparing contemporaneous controls would definitively answer the question as to the true value of the vaccine.

REFERENCES

1. Bergman PJ, et al. In: Withrow and MacEwen's Small Animal Clinical Oncology. 5th ed. 2013:321.
2. Spangler WL, Kass PH: Vet Pathol 43:136, 2006.
3. Williams LE, Packer RA: J Am Vet Med Assoc 222:1234, 2003.
4. Smedley RC, et al. Vet Pathol. 2011;48(1):54.
5. DeNicola DB. Round cells. In Cowell and Tyler's Diagnostic Cytology and Hematology of the Dog and Cat, 4th Ed. Valenciano AC and Cowell RL, eds. Elsevier Mosby, St. Louis, MO, 2014.
6. Busam KJ, Jungbluth AA. Adv Anat Pathol 6: 12-18, 1999.
7. Rassnick KM, et al. J Am Vet Med Assoc. 2001;218:1444.
8. Brockley LK, et al. New Zeal Vet J. 2013;61(1):25
9. Ottnod JM, Smedley RC, Walshaw R, et al. VCO 2013;11:219–229
10. Manley CM, et al. JVIM 2011;25:94–99
11. Henry CJ, Brewer WG, Whitley EM, et al. JVIM 2005;19:720–724.
12. Boston SE, Lu X, Culp WTN, et al. JAVMA 2014;245:401-407
13. Tuohy JL, Selmic LE, Worley DR, et al. JAVMA 2014;245:1266-1273.
14. Bergman PJ, McKnight J, Novosad A, et al. Clin Cancer Res 2003; 9(4):1284-1290.
15. Bergman PJ, Camps-Palau MA, McKnight JA, et al. Vaccine 2006; 24(21):4582-5.
16. Liao JC, Gregor P, Wolchok JD et al. Cancer Immun 2006; 6:8.
17. Goubier A, Fuhrmann L, Forest L et al. Vaccine. 2008 Apr 24;26(18):2186-90.
18. Wolchok JD, Yuan J, Houghton AN et al. Mol Ther. 2007 Nov;15(11):2044-50. Epub 2007 Aug 28.
19. Yuan J, Ku GY, Gallardo HF et al. Cancer Immun. 2009 Jun 5;9:5.
20. Bergman PJ & Wolchok JD. Cancer Therapy 2008, 6:817-826.
21. Grosenbaugh DA, Leard AT, Bergman PJ, et al. Am J Vet Res. 2011 Dec;72(12):1631-8.

PERSONALIZED CHEMOTHERAPY FOR CANINE LYMPHOMAS: ARE WE THERE YET?

Michael O. Childress, DVM, MS, DACVIM (Oncology)
West Lafayette, IN

CHEMOTHERAPY OF CANINE LYMPHOMAS – DOES ONE SIZE FIT ALL?

Lymphomas in dogs are possessed of stunning, and at times vexing, clinical heterogeneity. It has long been known that lymphomas can occur in any organ of the body, and that response to chemotherapy and prognosis vary in accordance with anatomic tumor location. The recently published World Health Organization (WHO) classification system[1,2] recognizes over 30 histologic subtypes of lymphoma in dogs, reaffirming the notion of biological diversity within these tumors. Despite an increasing awareness and understanding of the biological and clinical heterogeneity of canine lymphomas, our ability to diversify therapy based upon this understanding has not kept pace. For example, although it has long been known that tumor immunophenotype has a profound influence upon prognosis for survival following chemotherapy (in that dogs with T-cell lymphomas usually experience inferior survival times when compared to dogs with B-cell lymphomas), it is still unclear how best to modify treatment based upon this trait.[3-5] But the need to diversify therapeutic strategies for dogs with lymphoma extends beyond simply treating various tumor subtypes with different standard treatments. Perhaps the most frustrating manifestation of biological heterogeneity among canine lymphomas is that clinical tumor behavior varies widely even *within* histologically-defined subtypes.

This phenomenon of intra-subtype heterogeneity is strikingly evident in diffuse large B-cell lymphoma (DLBCL), the most common lymphoma subtype, which accounts for 33-50% of all lymphomas in dogs.[1,2,6,7] Doxorubicin-based combination chemotherapy (i.e. CHOP) affords durable cancer remission and extends survival for the majority of dogs with DLBCL. However, the degree of clinical benefit provided by chemotherapy varies significantly from dog to dog; a small proportion will experience unexpectedly prolonged survival,[8] while some dogs have minimally chemoresponsive cancers and die within weeks of diagnosis, regardless of treatment. Therefore, while it is clear that chemotherapy is of demonstrable benefit to the *average* dog with DLBCL, therapeutic gain is not apportioned equally to all dogs. This is unfortunate, as chemotherapy may be expensive and of potentially significant toxicity. A method for identifying *a priori* which individual dogs are most and least likely to benefit from chemotherapy would be of great use to veterinarians and pet owners. Such a method would ideally provide an accurate means to inform the decisions of both parties as to whether the pursuit of chemotherapy would be, in the words that so many pet owners use to frame the choice, "worth it."

THE PREMISE, THE PROMISE, AND THE PITFALLS OF PERSONALIZED CANCER THERAPY

It is clear that therapeutic decisions must be made for dogs with lymphomas in the context of imperfect predictors of therapeutic response. The same conundrum, in truth, permeates the management of all veterinary and human cancers, and this fact speaks to the need to specifically tailor therapy based upon the expected therapeutic response of each individual patient's tumor – that is, to *personalize* cancer therapy. The idea of personalized cancer therapy is not new. Oncologists have modified and tweaked treatment protocols for individual patients based upon pharmacogenomic and pharmacokinetic considerations, as well as clinical judgment and experience, for decades.[9] However, the current notion of *personalized cancer therapy (PCT)* is focused on using drugs to target the mutant genes (or, most commonly, the mutant oncoproteins which they encode) critical to sustaining the growth of an individual patient's cancer.

The modern paradigm of PCT in humans has already yielded some remarkable successes. The first, and still most dramatic, among these was the use of imatinib to achieve sustained cancer remissions in approximately 90% of patients with chronic myelogenous leukemia CML).[10] Imatinib is a relatively specific inhibitor of the BCR-ABL tyrosine kinase fusion protein which forms as a result of a reciprocal translocation between human chromosomes 9 and 22. Crucial to the dramatic success of imatinib was that this translocation is essentially ubiquitous in CML and that its presence is an absolute requirement for the survival of most CML cells.

The situation exemplified by BCR-ABL in CML, in which a single mutation is nearly the sole contributor to cancer progression, is relatively rare in clinical oncology. Indeed, subsequent attempts at PCT directed at a single oncogenic target met with more muted success. In some cases, this could be attributed to an incomplete understanding of the genetic context in which targeted therapy would be effective. For example, while the epidermal growth factor receptor (EGFR) is highly expressed in non-small cell lung cancer (NSCLC), targeted therapy using gefitinib, a small molecule inhibitor of EGFR, in unselected patients with NSCLC afforded a disappointing objective response rate of only 8% and did not extend survival time relative to placebo treatment.[11] However, when gefitinib was given to patients with NSCLC bearing activating mutations in the EGFR, an objective tumor response occurred in over 70% of patients and progression-free survival time was significantly longer than in patients treated with carboplatin-paclitaxel.[12] Another example of the importance of genetic context in targeted cancer therapy involves the use of cetuximab and panitumumab, both therapeutic monoclonal antibodies directed against the EGFR, in the treatment of metastatic colorectal cancer. While both drugs have been shown to provide clinical benefit, this benefit is restricted to patients whose tumors possess a wild-type *K-*

ras gene.[13,14] Colorectal cancers bearing *K-ras* mutations, on the other hand, are refractory to therapy with cetuximab and panitumumab, despite the fact that EGFR is expressed by a large proportion of these tumors. Thus, a lesson drawn from early attempts at PCT is that expression of a putative therapeutic drug target by a cancer is, by itself, not an accurate predictor of drug response if the genetic context in which the target is expressed is unfavorable.

In addition to the limitation imposed by genetic context on the anticipated therapeutic benefit from PCT, other limitations have also recently emerged. First, objective responses by solid tumors to targeted therapies may be dramatic, but are often brief. Drug resistance to single-agent targeted therapy of solid tumors typically develops rapidly. In some cases, knowledge of mutations in the original target protein or an understanding of the parallel cell signaling pathways which confer drug resistance has informed the development of alternative[15] or combination[16] targeted drug therapies which can be employed with renewed success. However, this knowledge is painstaking and time-consuming to acquire, and the time needed to acquire this knowledge is simply not a luxury that patients with advanced cancers can afford. A second limitation to current PCT approaches is that cancer genotyping may simply fail to elucidate a viable therapeutic drug target in many tumors. This problem was evident in the IMPACT (Initiative for Molecular Profiling and Advance Care Therapy) study recently completed at the University of Texas MD Anderson Cancer Center.[17,18] In this study, in which patient tumor samples were analyzed for gene mutations known to affect response to a variety of drugs at the time of enrollment, no targetable molecular alterations could be identified in over 40% of study participants' tumors. These limitations to the current genomics-based approach to PCT – the difficulty in predicting the effect of genetic context on therapeutic response, the inability to respond rapidly to the emergence of resistance to targeted therapy, and the inability to deliver targeted therapy in an informed fashion in the absence of a detectable drug target – point to the need for a novel approach.

GENOTYPIC VS. PHENOTYPIC CANCER PROFILING

The aforementioned specific limitations notwithstanding, a more general, yet inherent, limitation to current PCT approaches is that they rely largely upon the use of genotypic information (e.g. the presence of specific gene mutations) to predict a phenotypic response by the tumor (e.g. tumor regression). A major pitfall to this approach is that cancer genomes are enormously complex and chaotic; distinguishing *driver mutations* – those which are essential to cancer progression – from *passenger mutations* – those which are a product of genomic instability but are of no functional consequence to the tumor – is often difficult.[19] A PCT assay which extracts phenotypic information, such as onset of apoptosis following exposure to a drug, from a tumor sample might be useful for overcoming this problem.

Phenotypically-oriented chemosensitivity assays, have in fact, already been developed, and are conceptually much older than genotypic assays. In 1957, Wright and colleagues first reported on the correlation between *in vivo* response to chemotherapy in human cancer patients and *in vitro* response to drugs in primary cell cultures established from these patients' tumors.[20] Several similar approaches have been employed since then, which have collectively afforded sensitivity of approximately 80-90% and specificity of 60-70% in predicting response to chemotherapy.[21] Several shortcomings, however, have limited the routine implementation of phenotypic chemosensitivity assays in the oncology clinic. For one, clinical trials evaluating the validity of these assays in cancer patients have frequently suffered from design flaws such as small sample size, lack of randomization, lack of control groups, and high patient dropout rates.[22] Additionally, due to their requirement for establishing primary cultures from patient tumor cells prior to assay performance, phenotypic chemosensitivity tests are technically cumbersome, and provide results only after a significant delay (i.e. weeks) which would not be acceptable for guiding the therapy of many patients with advanced cancer. Furthermore, primary cell culture may not adequately recapitulate the biology of a tumor *in situ* – the disaggregation of tumor cells from a biopsy into monolayer or suspension cultures disrupts cell signaling pathways occurring between the cancer cells themselves or between cancer and stromal cells at the tissue microenvironmental level. These tissue-level interactions are known to exert a powerful influence on tumor response to drug therapy,[24] and failure to adequately model them *in vitro* may lead to over- or underestimation of the degree of clinical benefit a drug will afford to a patient *in vivo*. A chemosensitivity test which can be performed rapidly upon minimally disrupted living tissue is the obvious means to overcome these shortcomings to past phenotypic assays.

Biodynamic imaging (BDI), a novel live-tissue imaging technology recently developed at Purdue University, offers significant promise to circumvent the limitations inherent to previously characterized phenotypic chemosensitivity tests. BDI measures temporal variation in cellular and subcellular motion by recording dynamic patterns of short-coherence light scattering through intact, living tissue samples up to 1 mm thick.[25,26] Dynamic motion is an excellent biomarker by which to assess phenotypic changes within cells (and tissues) because *any change in cellular physiology elicited by a microenvironmental cue will ultimately result in altered cell motion*. Cell motility metrics captured by BDI have been used to characterize disparate phenotypic responses to a variety of anticancer drugs applied to tumor tissue samples both *in vitro* and *ex vivo*.[25,26] An *et al.* also demonstrated that BDI data obtained *ex vivo* correlated with objective response to cisplatin chemotherapy *in vivo* in a murine subcutaneous tumor xenograft model of ovarian carcinoma,[25] suggesting a potential application of BDI to PCT. This potential is supported by results from a recent pilot study in dogs with multicentric lymphomas.[27] In this study, tumor biopsy samples from 10 dogs were treated with doxorubicin (10 μM) *ex vivo*, then analyzed by BDI. Motility metrics captured by BDI were correlated with objective response to, and progression-free survival time following, doxorubicin chemotherapy *in vivo*. The sensitivity and specificity of BDI in predicting response to doxorubicin chemotherapy both exceeded 90%. In addition to its predictive accuracy, BDI also provides data in a temporally relevant fashion, in contrast to previous phenotypic chemosensitivity assays. In fact, all BDI data were acquired within *hours* of performing tumor biopsies in the 10 dogs involved in the aforementioned study. BDI therefore represents a potentially highly useful phenotypic assay for PCT, and it is currently undergoing continued investigation in dogs with lymphomas to validate these initial promising results.

COMBINED PROFILING FOR PERSONALIZED CANCER THERAPY – THE BEST OF BOTH WORLDS?

The current approach to PCT based upon genotypic profiling has shortcomings, illustrating the need for validated, clinically practical phenotypic profiling methods. Past attempts at phenotypic profiling; however, were cumbersome and time consuming, which limited their clinical utility. Novel phenotypic profiling techniques, such as BDI, offer new opportunities to address the limitations of genotypic profiling. BDI could be very useful for selecting therapies for cancer patients in whom a genotypically-defined drug target cannot be found. BDI could also be used to rapidly select rescue therapies for patients whose cancers have become resistant to front-line targeted therapy and do not carry a rapidly identifiable genomic determinant of drug resistance.

But BDI, like all PCT techniques, has limitations. At present, BDI cannot specify novel targets for drug therapy. BDI also cannot characterize the molecular mechanisms of resistance to targeted therapies. These limitations, though, *are* addressable with genotypic profiling techniques. In fact, the relative merits of each approach seem to counterbalance the other's relative drawbacks, which argues for a combined genotypic-phenotypic approach in the next iteration of PCT. In addition to improving PCT for individuals, a combined approach would also accelerate cancer drug development for the benefit of all patients. Phenotypic profiling would provide a means for early identification of patient subsets most likely to respond to novel drugs. These patients could quickly become the focus of further studies to define, with genotypic profiling, the molecular determinants of drug response. With the subset of patients with a given cancer most likely to respond to a novel drug identified *a priori*, clinical trials using targeted cancer therapies could be conducted more efficiently and regulatory approval for new drugs could occur more quickly.

Such an expedited approach to drug development would certainly stand to benefit many patients, both human and canine, but would be of particular importance in the advancement of new therapies for lymphomas. In humans, there has been little progress in the development and approval of novel targeted therapies for DLBCL since the introduction of rituximab for B-cell lymphomas in the late 1990s. In dogs with DLBCL and other lymphomas, the forecast for novel therapies is bleaker still – while therapeutic monoclonal antibodies are on the horizon, the impact they will have on disease management is, as yet, uncertain. Also, as is the case in humans, some dogs with DLBCL are likely to have tumors incurable with combined chemoimmunotherapy. Therefore, it seems imperative that we optimize methods for PCT in dogs with DLBCL and other lymphomas now so that the next array of therapeutic challenges posed by this diverse family of cancers can be anticipated, novel treatment strategies can be proactively implemented, and patient outcomes can continue to improve. Combining the merits of current genotypic PCT techniques with novel, but promising, phenotypic techniques like BDI could be an effective way of meeting the therapeutic challenges posed by lymphomas, and is an approach deserving of further consideration.

REFERENCES

1. Valli VE, et al. *Vet Pathol* 2011;48:198-211. 2. Valli VE, et al. *Vet Pathol* 2013;50:738-48. 3. Beaver LM, et al. *J Am Vet Med Assoc* 2010;237:1052-5. 4. Brodsky EM, et al. *J Vet Intern Med* 2009;23:578-584. 5. Rebhun RB, et al. *Vet Comp Oncol* 2010;9:38-44. 6. Ponce F, et al. *Vet Pathol* 2010;47:414-33. 7. Vezzali E, et al. *Vet Comp Oncol* 2009;8:38-49. 8. Marconato L, et al. *J Am Vet Med Assoc* 2011;238:480-5. 9. Longo DL. *Oncologist* 2013;18:644-5. 10. Druker BJ , et al. *N Engl J Med* 2006;355:2408-17. 11. Thatcher N, et al. *Lancet* 2005;366:1527-37. 12. Mok TS, et al. *N Engl J Med* 2009;361:947-57. 13. Karapetis CS, et al. *N Engl J Med* 2008;359:1757-65. 14. Amado RG, et al. *J Clin Oncol* 2008;26:1626-34. 15. Guilhot F, et al. *Blood* 2007;109:4143-50. 16. Long GV, et al. *N Engl J Med* 2014;37:1877-88. 17. Tsimberidou A, et al. *Clin Cancer Res* 2012;18:6373-83. 18. Tsimberidou A, et al. *Clin Cancer Res* 2014;20;4827-36. 19. Chin L, et al. *Nat Med* 2011;17:297-303. 20. Wright JC, et al. *N Engl J Med* 1957;257:1027-11. 21. Nygren P, et al. *Expert Opin Med Diagn* 2008;2:349-360. 22. Schrag D, et al. *J Clin Oncol* 2004;22:3631-38. 24. Junttila MR, et al. *Nature* 2013;501:346-54. 25. An R, et al. *Appl Opt* 2013;52:A300-9. 26. Nolte DD, et al. *J Biomed Opt* 2011;16:087004. 27. Custead M, et al. *Proceedings: Annual Conference of the Veterinary Cancer Society*. St. Louis, MO:2014;p41.

THE ROLE OF CHEMOIMMUNOTHERAPY FOR THE TREATMENT OF CANINE LYMPHOMA

Gregory K. Ogilvie, DVM, DACVIM (SAIM & Oncology), DECVIM-CA (Oncology)
Carlsbad, CA

INTRODUCTION

Lymphoma is the most common hematopoietic tumor in the dog and cat that generally responds well to chemotherapy resulting in an excellent quality of life and happy clients. There are many types of lymphoma, each with a unique biological behavior, response to therapy and thus, prognosis. There are many differences between lymphoma in dogs and cats. The disease is usually not painful in either species, and it responds very well to systemic chemotherapy. Quality of life is generally excellent, especially if the veterinary health care team does everything possible to minimize adverse effects. The purpose of this brief manuscript is to review some key points about canine lymphoma and its treatment, including the amazing new world of chemoimmunotherapy.

Overview

Lymphoma accounts for approximately 90% of canine hematopoietic tumors in the dog. Genetics and the environment have both been cited as factors that can influence the prevalence of this disease in the dog.[1-5] Some believe this disease is increasing in prevalence, with boxers, bulldogs, bull mastiff, golden retrievers and rottweilers being over represented in several studies.[1-6] Boxers may have more T cell lymphomas than the general population.[7] Golden retrievers have been reported to have three times and rottweilers are 2.5 and 6 times more likely to develop lymphoma than other breeds. [8-13] C type virus particles have been observed in cases of canine lymphoma and leukemia although it is unknown if these viruses are innocent passengers or involved in causality.[14,15] Similarly, factors in the immediate environment have been implicated such as environmental factors, industrial paints, solvents, herbicides such as 2, 4-D, exposure to electromagnetic fields, and diseases of the immune system such as immune thrombocytopenia.[14-24]

Lymphoma is uncommon in young dogs; affected dogs are typically middle-aged. One controversial study suggested gender and neutering is a predisposing factor for developing lymphoma, at least in Golden Retrievers. In that study, the authors stated that nearly 10% of male dogs neutered early in their lives were diagnosed with lymphoma, 3 times more than intact males.

Diagnostics[1-24]

When evaluating an animal for treatment of lymphoma, it is important to not only obtain a definitive diagnosis but also assess the general health of the patient by clinical examination and ancillary diagnostics. The clinical stage of a dog with lymphoma is determined in part by performing a physical examination. All lymph nodes should be palpated; accurate measurements should be recorded to help the clinician assess response to therapy. Careful attention should be directed to examine the eyes for any evidence of iritis, uveitis or hyphema. A delayed identification of this finding can result in blindness. Other, extranodal sites should be examined as well. Each dog is clinically staged based on the results of physical examination, clinical laboratory testing (i.e., complete blood count [CBC], biochemical profile, urinalysis, and bone marrow cytology), and imaging procedures (i.e., radiography and ultrasonography). In each case, a biopsy or cytology and flow cytometry of the affected lymphoid tissue is necessary to confirm the diagnosis and determine the cell type (histologic grading and immunophenotype). The purpose of doing these tests is to confirm the diagnosis and extent of lymphoma, the health of the patient, and to determine which therapy is likely to be most effective which should result in the best quality of life. By knowing where the lymphoma begins, it is possible to subsequently reassess the patient to determine the response to therapy. The most important diagnostics include:

- **Cytology:** Samples obtained from lymph node, liver, spleen, skin or other organs that might be altered by lymphoma for cytological review is practical, cost effective, accurate and expeditious. These samples are easily obtained by repeatedly directing an 18-22 gage needle into the site in question without the needle attached to the syringe. The goal is to fill the needle with cells from the tissue in question. The sample is then expressed onto a slide and gently spread across the glass using a number of techniques. The slide is then stained and examined under the microscope. Cytology will always be the most common way of diagnosing lymphoma despite the fact that histopathology and flow cytometry are generally more accurate and precise. This diagnostic test can be performed in house although confirmation by a board certified clinical pathologist is important. Occasionally, it may be difficult to differentiate a low-grade lymphoma from a reactive lymphoid disease. In this situation, the lymphocytes in the malignant disease would be expected to have a monoclonal population, but those in the reactive disease would not. Flow cytometry can help delineate the benign from the malignant disorder.[26]
- **Biopsy:** A histopathologic biopsy is generally considered more accurate than results from cytology. Needle core, wedge, or excisional biopsy samples should be placed into formalin (1 part tissue to 10 parts formalin) and then sent to the pathology laboratory. Special stains can be applied to determine the origin of the cells, however flow cytometry is becoming a preferred diagnostic test.
- **Flow cytometry:** This diagnostic test is performed in the laboratory from cells obtained via needle and syringe and placed in special media to keep the cells alive. The test not only confirms the diagnosis of lymphoma, it determines if the cells are B or T cell in origin, if they express class II major histocompatibility complex (MHC), and the cell size of the lymphoma. The clinician should check with the laboratory performing the test to determine the optimal sample handling technique.
- **Blood and urine tests:** Dogs with lymphoma can have bone marrow infiltration and its associated peripheral blood changes including thrombocytopenia, anemia, and either neutropenia or neutrophilia and/or leukemic cells. Changes in biochemical profile parameters are also commonly seen in dogs with lymphoma including elevated liver enzymes, hypercalcemia with

consequent elevations in ionized calcium, magnesium, blood urea nitrogen (BUN) and creatinine with an isosthenuric urine specific gravity. The presence of hypercalcemia, anemia, leukemic cells in the peripheral blood, presence of pan- or bicytopenia are all negative prognostic indicators. As reported below, flow cytometry is a key part in directing effective chemoimmunotherapy.

- **Radiographs or computerized tomography of the chest:** The presence of hilar or sternal lymphadenopathy is seen fairly commonly, especially among dogs with T cell lymphoma. Some dogs with lymphoma will present with subtle respiratory signs that may be associated with a diffuse interstitial pattern of the lungs.
- **Ultrasound or computerized tomography of the abdomen:** Abdominal lymphadenopathy, hepatomegaly or splenomegaly may not be appreciated without these diagnostics. Ultrasonographic features of splenic and sometimes hepatic involvement includes a honeycombed or "chicken wire" -like appearance to that organ.
- **Bone marrow aspirate:** The common misperception is that a bone marrow is not needed if the complete blood count is within normal limits. This is not true, as infiltration of the bone marrow, and thus a poor prognosis can occur when blood cells are normal or abnormal. Thus, the presence of lymphoma in the bone marrow suggests not only an unfavorable prognosis, but also concern about the exaggerated myelosuppression associated with the use of chemotherapeutic agents.

TREATMENT[1-18]

Treatment for malignant lymphoma must be systemic in nature since it is a multi-system disease. Chemotherapy is most frequently utilized; however, some immunotherapy has been effective. Within the immunotherapy field, monoclonal antibodies have shown the most promise. Untreated, dogs affected with malignant lymphoma live an average of only six weeks once a diagnosis has been made. With chemotherapy, dogs can survive for 6-10 months with an excellent quality of life. Dogs may receive Doxorubicin every 21 days at a dosage of 30 mg /m^2.

Chemotherapy[15-32]

Malignant lymphoma is one of the most responsive forms of cancer presented to the practitioner. With appropriate chemotherapy or chemoimmunotherapy protocols, nearly 90% of animals placed on therapy should reach complete remission. Of those that reach remission, approximately 80-90% will maintain a reasonable (>6 months) time frame of excellent quality life. Cost is not inexpensive; however, it is within reasonable financial reach of many clients, thus allowing therapy to be possible.

Although marginally effective, prednisone is inexpensive and often used in combination with other drugs to treat lymphoma. With prednisone therapy, the average pet lives 2 months. One-third of the dogs and cats treated with prednisone will go into complete remission, one-third will go into partial remission, and one-third will not respond at all.

Doxorubicin is one of the most effective single agent treatments for lymphoma in dogs.[29] Of the dogs treated with Doxorubicin, 81% developed a complete and partial remission. The duration of remission is approximately 9 months. Dogs treated with Doxorubicin and then switched to COP (cyclophosphamide, Oncovin [vincristine], prednisone) had a higher second remission rate compared to those started on COP and then switched to Doxorubicin.

The COP protocol is effective for inducing a remission in 75% of dogs with lymphoma.[30] A median duration of remission of 6 months is commonly seen. Approximately twenty percent of the dogs treated with the COP regimen are in remission at 1 year. The addition of Doxorubicin to the COP regiment (COPA) resulted in longer remission time (7 months versus 6 months). A complete remission was attained in 84% of dogs treated with COPA. An additional 7% achieved a partial remission. Twenty-two percent of dogs treated with the COPA protocol were in remission at one year.

Garrett *et al*[28] compared a maintenance-free chemotherapy protocol based on CHOP to a similar protocol with a maintenance phase for the treatment of canine lymphoma. Fifty-three dogs with multicentric lymphoma were treated with a 6-month modified version of the University of Wisconsin (UW)-Madison chemotherapy protocol. Disease-free interval (DFI) and survival were compared to a historical control group of 55 dogs treated with a similar protocol with a prolonged maintenance phase. Remission rate for the study dogs was 94.2%. The remission and survival between the 2 groups did not differ significantly. Thus, we, and others, believe the 6-month chemotherapy protocol based on CHOP with no maintenance phase provided is equal to a similar protocol with a prolonged maintenance phase.

Chemoimmunotherapy[32-36]

One monoclonal antibody developed for the treatment of cancer in humans or animals is canine lymphoma/monoclonal antibody 231 (CL/MAb 231; Synbiotics Corporation, San Diego, California), which recognizes canine lymphoma cells.[32,33] It mediates antibody-dependent cellular cytotoxicity against a canine lymphoma cell line and has been reported to prolong remission duration when used in combination with chemotherapy in dogs with lymphoma.[33] The median survival time of dogs treated with the monoclonal antibody and chemotherapy was 591 days, as compared to historical controls, which had a median survival time of 189 days. Additional studies are needed to further understand the clinical utility of this treatment modality.

Rituximab, a chimeric murine/human IgG monoclonal antibody that targets the B-cell antigen CD20 is now considered the standard of care for the management of human B-cell lymphoproliferative disorders including diffuse large B-cell lymphoma (DLBCL).[34,35] This aggressive form of DLBCL has many similarities to the most common form of lymphoma in dogs.[336] In both species, DLBCL generally responds well to combination chemotherapy, however the duration of remission is often limited. The addition of the species-specific antibody Rituximab in combination with chemotherapy to treat DLBCL in people is associated with little increase in toxicity, yet enhanced efficacy. The increased frequency in the use of Rituximab in the management of lymphoid malignancies has grown to the point that it presented a huge logistical challenge for the human medical system because the antibody was originally given by slow infusion, a time intensive procedure. Therefore, more cost and time effective procedures have been developed to safely and effectively infuse the antibody.[37] Thus, the availability of a similar antibody that targets the canine CD20 antibody combined with the lessons learned on how to safely and effectively incorporate this type of therapy could have a powerful impact for the management of DLBCL.

The discovery, development, and commercial production of a fully USDA licensed Rituximab-like chimeric murine/canine monoclonal antibody that specifically and precisely targets the B-cell antigen CD20 in dogs is notable.[1] In fact, if this construct is confirmed to be safe and effective, then the potential to benefit dogs with a long list of benign or malignant diseases that are caused by B-cell lymphocytes that express CD20 is high. That is exactly what occurred with the development and subsequent use of Rituximab. A double blind, randomized, placebo controlled study was initiated to be the very first to test the hypothesis that the Rituximab-like chimeric murine/canine monoclonal antibody is safe and effective for the treatment of CD20 positive DLBCL in the dog.[39] In the study, dogs were enrolled in a prospective, randomized, double blind, placebo controlled clinical trial. All dogs were treated with one 4-week cycle of an L-CHOP chemotherapy protocol followed by either the canine lymphoma antibody or a placebo. Once remission was lost, a single dosage of doxorubicin was given followed by the canine lymphoma antibody. Assessment of antitumor efficacy was based on modified Response Evaluation Criteria in Solid Tumors (RECIST) guidelines and toxicity was based on criteria delineated by Veterinary Cooperative Oncology Group's common terminology criteria for adverse events (VCOG-CTCAE).[40] The median proliferation free interval and overall survival time respectively of dogs with DLBCL treated with one cycle of the L-CHOP protocol followed by the antibody or the placebo was much longer in the antibody arm compared to the placebo arm. Toxicities that were observed in this protocol were mild and restricted to the L-CHOP cycle with no significant adverse events in the antibody or placebo arms. On the contrary, there was a significant improvement in client perceived quality of life parameters in the antibody, but not the placebo treated dogs. Future publications will elucidate the details of this work. This antibody was purchased by Aratana during the acquisition of Veterinary Therapeutics. Aratana will likely market the product in 2015.

A second monoclonal antibody has received the first conditional licensure for use as an aid to treating T cell lymphoma in dogs that express CD-52. These CD-52 antigens are expressed on most dogs with T cell lymphoma and thus, a monoclonal antibody directed to this epitope can bind and initiate a specific cytotoxic event. The initiation of remission prior to the administration of the antibody with combination chemotherapy will increase the probability of inducing a sustained response. This antibody was purchased by Aratana during the acquisition of Veterinary Therapeutics. Aratana will likely market the product in 2015. Work done at California Veterinary Specialists Angel Care Cancer Center affirmed its safety and efficacy for the treatment of T cell lymphoma in dogs. Future publications will elucidate the details of this work.

REFERENCES

1. Dorn CR et al: *Am J Vet Res* 28:993–1001, 1967. 2. Schneider R: *J Natl Cancer Inst* 70:971–977, 1983. 3. Teske E: *Vet Q* 16:209–219, 1994. 4. Dobson JMet al: *J Small Anim Pract* 43:240–246, 2002. 5. Clancy E et al: *Proc 19th Ann Conf Vet Cancer Soc* 71. 1999. 6. Priester WA: *J Natl Cancer Inst* 39:833–845, 1967. 7. Lurie DM et al: *Vet Comp Oncol* 2:171–175, 2004. 8. Onions DE: *J Natl Cancer Inst* 72:909–912, 1984. 9. Edwards DS et al: *Vet Comp Oncol* 1:200–206, 2003. 10. Moore AS et al: *J Vet Intern Med* 15:348–354, 2001. 11. Zemann BI et al: *J Vet Intern Med* 12:465–470, 1998. 12. Jagielski D et al: *J Vet Med A Physiol Pathol Clin Med* 49:419–424, 2002. 13.Teske E et al: *Vet Q* 16:134–136, 1994.14. Onions D: *Eur J Cancer* 16:345–350, 1980. 15. Tomley FM et al: *Brit J Cancer* 45:644, 1982. 16. O'Brien DJ et al: *Prev Vet Med* 42:1–15, 1999. 17. Gavazza A et al: *J Vet Intern Med* 15:190–195, 2001. 18. Hayes HM et al: *J Natl Cancer Inst* 83:1226–1231, 1991. 19. Reynolds PM et al: *Cancer Epidemiol Biomarkers Prev* 3:233–237, 1994. 20. Carlo GL et al: *Regul Toxicol Pharmacol* 16:245–252, 1992.21. Kaneene JB et al: *Vet Human Toxicol* 41:164–170, 1999. 22. Hayes HM et al: *Environ Res* 70:119–125, 1995. 23. Reif JS et al: *Am J Epidemiol* 141:352–359, 1995. 24. Keller ET: Immune-mediated diseases as a risk factor for canine lymphoma. *Cancer* 70:2334–2337,1992. 25 Torres de la Riva G et al: PLoS One. 2013;8(2):e55937. 26. Burnett RC et al: *Vet Pathol* 40:32–41, 2003. 27. Morrison-Collister KE et al: *Vet Comp Oncol* 1:180–190, 2003. 28. Garrett LD et al: *J Vet Intern Med* 16:704–709, 2002. 29. Valerius KD et al: *JAVMA* 210:512–516, 1997. 30. Cotter SM et al: *J Am Anim Hosp Assoc* 19:159–165, 1983. 31. Rosales C et al: *Cell Immunol* 115:420–428, 1988. 32. Jeglum KA:. *Vet Clin North Am Small Animal* Pract 26(1): 73-85, 1996. 33. Smith MR: *Oncogene* 22(47):7359-7368, 2003. 34. Vose JM et al: *J Clin Oncol.* 2001 Jan 15;19(2):389-97. 35. Vose JM et al: *Leuk Lymphoma.* 2005 Nov;46(11):1569-73. 36. Marconato L et al: *Hematol Oncol.* 2013 Mar;31(1):1-9. 37. Sehn LH et al: *Blood.* 2007 May 15;109(10):417. 38. Patent US 20110217298, Geneviève Hansen, Monoclonal Antibodies. 39. Ogilvie GK et al: *J Vet Intern Med.* In Press, 2015. 40. Veterinary cooperative oncology group: *Vet Comp Oncol.* 2011 doi: 10.1111/j.1476-5829.2011.00283.x

APHERESIS AND CANINE BONE MARROW TRANSPLANTATION

Steven E. Suter, VMD, PhD, DACVIM (Oncology)
Raleigh, NC

INTRODUCTION

Apheresis, which in Greek means "a taking away", is an extracorporeal therapy that has been an important part of physician-based medicine since the early 1970's. With the development of sophisticated cell separator machines and the commercial availability of human cytokines, apheresis is an essential component in the treatment of a wide variety of human diseases, including cancer. Fortunately, with the availability of used, human cell separator machines to veterinarians employed at both academic institutions and high-end specialty practices, apheresis is beginning to be recognized as a valuable tool in the treatment of a variety of canine diseases. This review will discuss how cell separators (also called apheresis machines) work, briefly discuss physician-based applications of apheresis, and review the current state of companion animal apheresis.

CELL SEPARATOR MACHINES

All cell separator machines, regardless of the manufacturer, are designed to separate specific blood components, most often based on cell density centrifugation. Therefore, an integral component of all machines is a large scale centrifuge that is able to continuously separate blood into various fractions, including WBCs, platelets, and serum using continuous flow centrifugation (CFC). This means the blood can be collected into the machine, centrifuged, and returned to the patient simultaneously. The component of interest is typically pumped into a separate collection bag while the rest of the unused blood components are returned back to the patient. Therefore, patients undergoing apheresis typically have a catheter placed in each arm or a dual-lumen catheter placed in one arm that allows blood to enter into the machine and pump the uncollected portions of the blood back into the patient. Because an anti-coagulant, such as ACD-A, is added to the blood after it leaves the patient and before it enters the machine, apheresis patients actually experience very little loss of blood volume during the procedure. The most popular cell collectors for physician-based medicine are the COBE® Spectra and Spectra Optia® apheresis systems manufactured by TerumoBCT (formerly CaridianBCT) and CS-3000 *Plus* and Amicus™ apheresis systems manufactured by Fresenius Kabi (formally Fenwal, Inc). Other popular machines, due to their small size and portability are the Haemonetics® MCS®+8150 Multicomponent Collection System (red blood cells and plasma only) and the MCS®+9000 Mobile Platelet Collection System.

PHYSICIAN-BASED APPLICATIONS OF APHERESIS

The most widely used application of cell separator machines in physician-based medicine is the collection of platelets (plateletpheresis) for use in cancer patients receiving chemotherapy, radiation treatments, or bone marrow transplants, as well as patients with aplastic anemia and blood coagulation disorders. It takes approximately 5 whole units of blood to harvest the amount of platelets that can be obtained from one 90-minute plateletpheresis. Therefore, one unit of platelets harvested using plateletpheresis can be used to treat 2-3 patients (www.redcrossblood.org/donating-blood/types-donations/platelet-donation). Plateletpheresis is considered absolutely safe with the most common side effect being a tingling sensation in the lips due to transient hypocalcemia that is treated with calcium-containing antacid tablets.

Therapeutic plasma exchange (TPE, a form of plasmapheresis) is a procedure where the cell separator machine separates and removes the patient's plasma which is then replaced with either donated plasma or a protein solution such as 5% human albumin. TPE is typically used to remove disease-causing antibodies that cause a variety of immune-mediated diseases, including Guillain-Barre Syndrome, CIDP (chronic inflammatory demyelinating polyneuropathy), and myasthenia gravis. It can also be used to treat TTP (thrombotic thrombocytopenia purpura) patients in an effort to reduce the thickness of the blood. TPE can also be used to remove large amounts of medication if the need arises.

LDL apheresis removes LDL cholesterol from the blood of patients with hypercholesterolemia using two specific types of apheresis machines approved by the US Food and Drug Administration. The Liposorber uses a filter covered with dextran that binds LDL cholesterol and removes it from the circulation. Another machine, called HELP uses heparin to bind and remove LDL cholesterol. Since hypercholesterolemia is usually caused by the combination of mutations in a variety of genes (especially LDLRAP1) and lifestyle choices, treatment using LDL apheresis is usually life long.

Photopheresis is a very specialized type of apheresis where the harvested white blood cells are exposed to a photoactive drug called methoxsalen which renders them sensitive to ultraviolet light. The cells are then exposed to ultraviolet (*uv*) light inside the machine and returned to the patient. This procedure can improve the symptoms of certain conditions such as Graft Versus Host Disease (GVHD), cutaneous T-cell lymphoma, or organ transplant rejection, although it is currently not known exactly how photophoresis works. The length and number of procedures needed to provide benefit is highly variable.

Leukopheresis is used to remove extra white blood cells from the blood in patients with diseases that can cause very high white blood cell counts, including acute lymphoblastic and acute myeloblastic leukemias. Leukophoresis in these patients can significantly decrease blood viscosity and also decrease the chance of acute tumor lysis syndrome in patients starting induction chemotherapy.

Peripheral blood stem cell collections are used to harvest monocytes from patients after some type of blood cell mobilization, such as G-CSF (Neupogen) administration. A small percentage of the harvested monocytes (~1%-4%) are CD34+ progenitor cells that can lead to complete hematologic reconstitution after some type of myeloablative therapy (such as total body irradiation or high-dose chemotherapy). This procedure is used to collect CD34+ progenitor cells for their use in both autologous and allogeneic bone marrow transplants when treating patients with lymphoma, leukemia, and multiple myeloma. The cells can also be used to treat other conditions including myelodysplastic syndromes and childhood neuroblastoma.

COMPANION ANIMAL APHERESIS

Plasmapheresis (also called therapeutic plasma exchange or TPE-the separation of plasma from blood to be used in primary and supportive disease management) is the most commonly reported apheresis procedure reported in the veterinary literature. In horses, plasmapheresis has been explored using continuous-flow cell separator machines based on data showing that plasma harvested from equine blood using fast erythrocyte sedimentation contains a high number of erythrocytes that could cause recipient sensitization[1,2] and evidence that bacterial contamination can occur using an open collection system.[2] In two studies, Feige *et al* found that automated plasmapheresis using a Blutz separator AS 104 (Fresenius, Stans, Switzerland) was the method of choice when compared to whole blood collection followed by either centrifugation or gravity sedimentation and that the plasmapheresis procedure itself was very well tolerated by the donor horses.[3,4] To this Author's knowledge the use of TPE to treat equine patients with immune-mediated diseases has not been reported.

Canine plasmapheresis has also been reported using both membrane filtration and continuous-flow centrifugation. Bartges *et al* used discontinuous membrane filtration plasmapheresis (Therapore, Therapeutic Apheresis Systems, Sarns/3M, Ann Arbor, MI) twice, in conjunction with corticosteroids, to treat a dog with acquired myasthenia gravis, which dropped the AchR antibody titer by ~70%.[5] Crump et al used continuous membrane filtration (Prisma CRRT and TPE Control Unit, Gambro Inc., Lakewood, CO) to treat a dog with immune-mediated hemolytic anemia.[6] In both reports, plasmapheresis was performed safely with minimal side effects and the dogs responded clinically.

Canine TPE has also been reported, although the majority of studies were performed in the 1980's. Matus *et al*, used a cell separator (IBM 2997 cell separator, IBM Corp., Princetown, NJ) to treat 3 dogs with an IgA monoclonal gammopathy and secondary hyperviscosity.[7] In all 3 dogs, one total plasma volume was removed and replaced with Lactated Ringer's Solution (LRS) and all 3 dogs responded clinically within 24 hrs of treatment. This same group also treated 1 dog with systemic lupus erythematous using a combination of plasmapheresis and immunoabsorption. Plasmapheresis was performed 8 times over a period of 45 days using the same cell separator and the harvested product was then passed over a column loaded with Staphylococcus purified protein A/SSPA, which binds canine IgG classes of antibodies.[8] This dog also responded well clinically. Based on this case report, Matus *et al* then treated 5 dogs with systemic lupus[9] and 2 dogs with immune-mediated hemolytic anemia[10] using plasmapheresis alone, again using the same IBM cell separator. One total plasma volume was removed and replaced with LRS and all dogs responded clinically within 24 hrs of the procedure.

Although, as stated previously, one of the primary uses of cell separator machines in physician-based medicine is plateletpheresis, canine plateletpheresis has been rarely reported and remains an underutilized resource in the clinical veterinary setting. Callan *et al* used a Terumo COBE® Spectra machine to assess the safety and feasibility of plateletpheresis using fourteen normal adult dogs.[11] In all cases, a high-quality, leukoreduced (<1 x 10^6 WBCs) platelet concentrate (mean total yield of 3.3 x 10^{11} platelets) was collected in a mean procedure time of only 105 min. Since citrate-induced hypocalcemia was noted in some dogs, prophylactic IV calcium supplementation was recommended. Appleman *et al* then extended this work by examining the effects of platelet *in vitro* function and *in vivo* survival using two different methods of cryopreservation (6% DMSO or Thrombosol).[12] Although cryopreserved platelets using either method were not able to mimic fresh platelets in terms of *in vivo* platelet recovery, the authors concluded that both cryopreserved products could be considered for dogs with acute, life-threatening bleeding when fresh platelet products are not available.

Dogs have been undergoing leukopheresis in a research setting since 1967.[13,14] However, the collection of large amounts of cells needed for routine clinical use using continuous flow centrifugation was not possible until the cloning and large scale production of recombinant hematopoietic cytokines in the 1990s and the development of highly automated computer-controlled cell separators soon thereafter. Although there is a large body of research literature utilizing dogs as a preclinical model to develop leukopheresis techniques to collect a sufficient number of peripheral blood CD34+ cells to allow complete hematologic reconstitution following myeloablative therapy, the use of leukopheresis in the veterinary clinical setting is still in its infancy.

In 2006, Lupu *et al* reported the use of a COBE® Spectra cell separator to harvest enough peripheral blood CD34+ progenitor cells (>5 x 10^6/kg) for use in an allogeneic bone marrow transplant setting.[15] Their choice of cell separator was based on the experience of the group at the Fred Hutchinson Cancer Research Center who showed that automated apheresis using a COBE® Spectra machine developed for humans could also be used in dogs.[16] Our group at North Carolina State University also found that an adequate number of peripheral blood progenitor cells could be safely harvested from dogs using a Fresenius Kabi CS-3000*Plus* cell separator.[17] We also showed that the CD34+ component of the harvest product was adequate to achieve complete hematopoietic reconstitution following lethal total body irradiation.[18] Finally, since all cell separator machines are designed for adult humans and their use in small children and dogs can be challenging, our group showed, similar to human pediatric data, that apheresis using a COBE® Spectra cell separator can be safely performed in dogs weighing <14 kg if appropriate machine priming techniques are utilized.[19] The smallest dog our group has performed apheresis on weighed 7.3 kg (unpublished data). This combination of older and more recent data clearly shows that canine leukopheresis is safe in a clinical setting and the procedure can harvest an adequate amount of hematopoietic progenitor cells for use in the canine bone marrow transplant setting.

LEUKOPHORESIS FOLLOWED BY BONE MARROW TRANSPLANTATION

Early research, performed mainly at the Fred Hutchinson Research Center, proved that canine peripheral blood CD34+ progenitor cells harvested using human cell separators could lead to complete hematologic reconstitution following lethal myeloablative therapy.[20-22] Further work showed that dogs with lymphoma could be successfully treated in both an autologous[23,24] and allogeneic[15,25] bone marrow transplant setting with long-term survivals longer than chemotherapy alone. More recently, our group at

North Carolina State University used a combination of total body irradiation and autologous peripheral blood CD34+ progenitor cells harvested using either a CS-3000 *Plus* or COBE® Spectra cell separator, to treat dogs with high-grade B- and T-cell lymphoma.[26,27] Five of 15 (33%) dogs with B-cell lymphoma transplanted while in remission for their disease remained alive >2 years (median overall survival, 524 days), while two of 13 (15%) of dogs with T-cell lymphoma transplanted while in remission for their disease remained alive >2 years (median overall survival, 240 days). Importantly, when we transplanted dogs with B-cell lymphoma who were not in clinical remission, their median disease-free intervals were significantly shorter than those who were in remission (52 days vs. 456 days). In dogs with T-cell lymphoma, although a fewer percentage of dogs were alive >2 years when compared with the B-cell lymphoma dogs, the median overall survival from the time of diagnosis to death was 602 days (range 246-928 days), which is significantly longer than when dogs are treated with chemotherapy alone.[28]

Our group has also performed 3 allogeneic bone marrow transplants using DLA-matched donor peripheral blood CD34+ cells[15] to treat two dogs, one with B- and one with T-cell lymphoma (unpublished data). Both dogs underwent complete hematologic reconstitution and achieved full donor engraftment within 6 weeks of transplant. The dog with B-cell lymphoma died of unexplained causes ~6 months post-transplant, while the dog with T-cell lymphoma remains alive ~26 months post-transplant. We also treated a dog with acute large granular lymphocytic leukemia using DLA-matched donor peripheral blood CD34+ cells and this dog remains disease-free at ~34 months post-transplant (manuscript accepted). Both autologous and allogeneic bone marrow transplantation as a treatment of canine lymphoproliferative disorders is currently offered at 4 private practices in the United States (CA and OH).

SUMMARY

Although cell separator machines are used daily in physician-based medicine to treat a wide variety of diseases, their use in veterinary-based medicine remains largely unexplored. Recent studies show that dogs can tolerate apheresis using human cell separator machines and that specific blood components can be harvested in a clinical veterinary setting. Moreover, it is clear that animals with a variety of diseases, including cancer, can achieve positive clinical outcomes when treated with cell separator harvest products. As such, apheresis in veterinary medicine remains a largely untapped resource that should expand in the future as more veterinarians become aware of the technology and more machines become available to our profession.

REFERENCES

1. Becht JL, et al. **Current Therapy in Equine Medicine**1987; 317. 2. Morris DD. **Equine Internal Medicine** 1998; 598. 3. Feige K, et al. 2003. **J. Vet. Med** 2003;50:185. 4. Feige, K, et al. **Vet J** 2005;169:102. 5. Bartges JW, et al. **J Am Vet Med Assoc** 1990;196:1276. 6. Crump KL., Seshadri R. **J Vet Emerg Crit Care** 2009;19:375. 7. Matus RE, et al. **J Am Vet Med Assoc** 1983;183:215. 8. Matus RE, et al. **J Am Vet Med Assoc** 1983;182:499. 9. Matus RE, et al. **J Am Vet Med Assoc** 1985;187:595. 10. Matus RE, et al. **J Am Vet Med Assoc** 1985;186:691. 11. Callan, MB, et al. **Transfusion** 2008;48:2214. 12. Appleman EH, et al. **J Vet Int Med** 2009;23:138. 13. Storb R, et al. **Blood** 1967;30:805. 14. Buckner D, et al. **Blood** 1968;31:653. 15. Lupu M, et al. **J Am Vet Med Assoc** 2006;5;728. 16. Lupu M, et al. **J Vet Int Med** 2008;22:74. 17. Suter SE, **J Vet Int Med** 2011;25:1406. 18. Escobar C, et al. **Vet Path** 2012;49:341. 19. Posner LP, et al. **Vet Anaesth Analg** 2013;40:403. 20. Cavins JA, et al. **Blood** 1964;23:38. 21. Appelbaum FR, **Exp Hematol** 1979;7:7. 22. Bodenberger U, et al. **Exp Hematol** 1980;8:384. 23. Weiden PL, et al. **Blood** 1979;54:1039. 24. Appelbaum FR, et al. **Transplantation** 1986;42:19. 25. Storb R, et al. **Blood** 1967;30:805. 26. Willcox JL, et al. **J Vet Int Med** 2012;26:1155. 27. Warry EE, et al. **J Vet Int Med** 2014;28:529. 28. Rebhun RB, et al. **Vet Comp Oncol** 2010;9:38. 27.

THROUGH THE LOOKING GLASS: PARTNERING WITH INDUSTRY

Michael D. Lucroy, DVM, MS, DACVIM (Oncology)
Greenfield, IN

INTRODUCTION

Over the past two decades, veterinarians in academic and private practice have had increasing opportunities to partner with industry. This has come about due to several factors, including decreasing availability of government-sponsored funding, increasing acceptance of the "One Health" concept of comparative medicine, evolving models of drug discovery and development[1], and the exponential growth of the small biotech industry. Many types of partnerships are possible depending on the needs of each party; and, can range from cross-disciplinary problem-solving ventures, through iterative collaborative relationships.[2] Other partnerships may represent testing arrangements. The fruit of these relationships is the commercialization of technology that provides an economic reward to the participants and benefits society by creating useful applications of the knowledge generated at universities. Challenges may arise with industry partnerships when the needs or motivations of the other parties are not understood or considered when entering the relationship.

THE METRICS

Academics

For junior faculty members, success is often measured in terms of the amount and source of extramural funding, and the number of peer-reviewed publications that are leveraged from this funding. Traditionally, funding from government sources (NIH, NSF, etc.) has been valued the highest. However, funding available from agencies like the NIH have been declining since 2003 (when adjusted for inflation).[3] Diminishing funds leads to increased competition for scarce resources and many junior faculty members are unable to obtain the coveted R01 grant. Increasingly, promotion and tenure committees have begun to view research partnerships with industry in a more favorable light so that junior faculty members can receive due credit for their intellectual contributions. Many universities also have clinical tracks for veterinary faculty members, and collaborations with industry are often well-suited to their applied research interests.

Biotech

Small-to-medium sized biotech companies are generally focused on the commercialization of a pharmaceutical, biologic, or a device asset that may be a discovery or invention of one of the company founders. In other cases, they may have secured a license to develop the asset from a university or other company. The investors or shareholders of these biotech companies measure success by the speed at which the asset is advanced towards commercialization. Because many of these companies are small (sometimes 2 to 10 employees), their goal is typically to add value to the asset (e.g., generating Phase I data), then sell the asset or entire company. Some companies opt to grow as the asset matures and secure funding from venture capitalists or take the company public and issue stock to facilitate progress through Phase II and Phase III clinical trials. Some startup companies are turning to crowdfunding as an alternative to traditional funding paths.[4]

Pharma

Large pharmaceutical companies are for-profit enterprises. As a whole, they are judged by their investors and shareholders on their ability to bring innovation to the marketplace while being profitable. Within a company, various R&D teams represent the face of the company, and are the people entering into partnerships with practicing veterinarians or universities. These teams may be deemed successful by their management when research projects are completed according to timeline and budget predictions.

Practice

In private practice, one measure of success is the veterinarian's economic influence on the practice and may be reflected by gross revenue, caseload growth, client retention, accounts receivable, referral catchment area, etc. A more subjective measure of success may be reputation in the community (which can be defined geographically and professionally).

THE MOTIVATION

Academics

Universities commonly engage in partnerships with industry for several reasons, including the ability to introduce their students to career options, exert an influence on technology development, raise the profile of their faculty members, and create a revenue stream.[5] This revenue may take several forms, including "fee-for-service" with testing or other research agreements, royalties and milestone payments from out-licensing university-created technology, student scholarships, post-doc funding, sponsorships, naming rights, etc.

Biotech

Often biotech companies partner with universities or veterinarians in practice to further the development of their technology assets. As mentioned above, their technology assets may actually be licensed from a university. Universities offer biotech companies access to specialized facilities and expertise that may not be feasible to replicate, and lend credibility. When pursuing preclinical studies – especially when client-owned animals are involved – universities and veterinarians in practice represent a valuable resource for biotech companies in terms of expertise and available clinical cases for study.

Pharma

Pharmaceutical companies often have many university and practicing veterinarian relationships. As the costs for developing drugs continues to rise, many companies are expanding their in-licensing of technology from universities and small biotechs in favor of increasing in-house discovery efforts. Pharma also forms partnerships to conduct client-owned animal studies that support development of human drugs as preclinical work or specifically for developing a veterinary drug. Pharmaceutical companies involved with animal health also seek input from veterinarians to inform product concept development and other marketing activities.

Practice

Veterinarians in practice may choose to partner with industry for a variety of reasons, including the curiosity of observing firsthand how a new therapy works in the field, participating in a well-funded clinical trial to allow a greater number of clients to afford advanced care, and the opportunity to participate in the drug development process. In some instances, the networking from these partnerships may help a veterinarian transition into an industry position, or allow greater participation by becoming a key opinion leader for the company.

THE MARRIAGE

There are several common partnerships between industry, biotechs, universities and practicing veterinarians. The potential issues that may arise as a result of these partnerships are typically addressed in the written agreement (or contract) that defines the relationship. Some of the more common topics that are covered in these agreements include confidentiality, ownership of any new intellectual property, rights to the data generated, publication rights, payment terms, license options, termination provisions, etc.[6] Typically universities, pharma, and biotech companies have their legal and business development teams negotiate the terms in these agreements. Private practice veterinarians entering into an agreement should consider getting legal counsel to ensure they understand their rights and obligations relating to the contract.

Common types of partnerships include non-disclosure agreements, material transfer agreements, licensing agreements, testing (service) agreements, and collaboration or research agreements. These may vary considerably in their complexity and expectations and obligations from the each of participating parties.

THE MADNESS

Despite the best intentions, good rapport, and a carefully crafted agreement, sometimes challenges arise during research collaborations. Some misunderstandings may simply reflect the different priorities among the various parties. Areas of potential friction spots include: importance of time/sense of urgency, timing of payments, data reporting requirements, timing and nature of publications, and the scope of work. Industry partners may change direction or focus based on competing business needs, changes in the competitive landscape, changes in pipeline development priorities or changes in funding (particularly true of small biotech companies). Veterinarians in academic and private practice may face challenges in accruing sufficient cases for a clinical study, or may underestimate the amount of work required to participate in a GCP trial.

Although partnering with industry may seem to be like" going through the looking glass", these types of relationships can be rewarding for all stakeholders. Indeed these partnerships are the mechanism by which new innovations are developed into clinically useful tools.

REFERENCES

1. Tralau-Stewart CJ, *et al. Drug Disc Today*. 2009;14(1-2):95-101.
2. Jacob M, *et al. R&D Mgmt*. 2000;30(3):255-262.
3. Johnson JA. Brief History of NIH Funding: Fact Sheet. CRS R43341 Dec 23, 2013.
4. Brenan J, *Pulse, IEEE*. 2104; 5(1):59-62.
5. Prigge GW, *Industry Higher Ed*. 2005;19(3):221-229.
6. Drozdoff V, et al. *CSH Perspect Med*. 2015; doi: 10.1101/cshperspect.a021014

NOVEL CANCER THERAPIES – YEAR IN REVIEW

Douglas H. Thamm, VMD, DACVIM (Oncology)
Fort Collins, CO

INTRODUCTION

This review will focus on novel animal cancer therapies that are moving forward toward regulatory approval through the 2 main US agencies, the FDA-Center for Veterinary Medicine (FDA-CVM) which regulates drugs, and the US Department of Agriculture (USDA), which regulates biologics. 2014 brought the conditional licensure of 1 novel drug through the FDA-CVM (Paccal Vet), the full licensure of 1 biologic through the USDA (AT-004, Canine Lymphoma Monoclonal Antibody, B-cell), and the conditional licensure of 1 biologic through the USDA (AT-005). In addition, among the 6 therapies with MUMS (minor use / minor species) designation through the FDA-CVM for canine cancer, 2 of these (KPT-335 and rabacfosadine succinate) have meaningful published data regarding safety and efficacy in dogs.

PACCAL VET

The taxanes are a class of antineoplastic drug with potent and broad antitumor activity in humans; however, their broad use in veterinary medicine has been hampered by the high incidence of infusion-related hypersensitivity reactions owing to the emulsifier necessary to deliver the poorly soluble drug paclitaxel,[1] and the prohibitive cost of the more water soluble analog docetaxel. A new formulation of paclitaxel (Paccal Vet, Oasmia Pharmaceutical AB) has been made water-soluble through a mixed micellar preparation utilizing retinoid acid derivatives (XR-17), resulting in improved infusion-related tolerability.

A pilot, open-label clinical trial of Paccal Vet was performed in 32 dogs with neoplasia.[2] In this study, the maximum tolerated dose (MTD) was established at 150 mg/m^2. Early (day 4-5) myelosuppression and gastrointestinal toxicity were the most commonly encountered adverse effects (AE): one dog experienced an infusion-related hypersensitivity reaction. The reported overall response rate (ORR) in evaluable patients was 82% (8 CR, 11 PR, n=23). Responses were observed in mast cell tumor (MCT), squamous cell carcinoma (SCC), mammary carcinoma, lymphoma, histiocytic sarcoma, and fibrosarcoma. However, 7 dogs were not included in response assessment because they "did not receive the protocol-mandated three treatment cycles of Paccal Vet". Their reason for early removal was not provided, but if this was for early disease progression then the true ORR may be less than that reported.

A subsequent European study evaluated the clinical efficacy and adverse effect profile of Paccal Vet in macroscopic canine MCT.[3] In this study, neutropenia was again the most common AE. The reported ORR was 59%, with a median time to progression of 247 days. However, concurrent with this study a prospective, randomized trial comparing Paccal Vet and lomustine for MCT was conducted in the US and EU.[4] In this study, the confirmed ORR at 14 weeks was only 7% (vs 1% for lomustine), and the best ORR was 23% (vs 10% for lomustine). While superiority was demonstrated for Paccal Vet over lomustine, the relatively low ORR in this study may have diminished enthusiasm for pursuing Paccal Vet for canine MCT.

Paccal Vet currently has conditional FDA approval for the treatment of nonresectable stage III, IV or V mammary carcinoma in dogs that have not received previous chemotherapy or radiation therapy, and resectable and nonresectable squamous cell carcinoma (SCC) in dogs that have not received previous chemotherapy or radiation therapy.[5] Approval for mammary carcinoma appears to be based on observed responses in 4 of 7 dogs treated in previous studies (1 CR, 3 PR). Progression free survival (PFS) in responding patients ranged from 56 to >498 days. Approval for SCC appears to be based on observed responses in 5 of 16 dogs treated in previous studies (1 CR, 4 PR). PFS in responding patients ranged from 63 to >388 days.

AT-004 AND AT-005

Monoclonal antibodies directed against the CD20 antigen expressed on a majority of human B cell lymphomas (e.g. rituximab) have changed the standard of care for the treatment of many forms of B-cell malignancy, owing to improvements in outcome when combinations of conventional antineoplastic therapy and anti-CD20 antibody therapy are used.[6] Clinical benefit has been observed in terms of increases in complete response rates (76-86% vs 63-68%) and improvements in long-term outcome (multivariate hazard ratios 0.44-0.58) in humans with diffuse large B cell lymphoma (DLBCL). Surprisingly, after more than 20 years of human use, mechanisms responsible for the antitumor effects of anti-CD20 antibody therapy remain incompletely understood. Binding to the lymphoma cell and activation of host immune cells via the Fc component (so-called antibody-dependent cellular cytotoxicity), complement fixation, and the possible cross-linking of CD20 on the cell surface leading to apoptosis induction are all possible mechanisms.[6]

AT-004 (Canine Lymphoma Monoclonal Antibody, B-Cell, Aratana Therapeutics / Novartis Animal Health) is a caninized monoclonal antibody directed against the canine CD20 antigen. It is given as an intravenous infusion. Although no publications describing safety or efficacy of this antibody have appeared in the peer-reviewed literature at the time of preparation of this manuscript, it has received full approval through the USDA. The results of 2 studies employing this antibody in dogs with B cell lymphoma were presented at the 2014 Veterinary Cancer Society meeting.[7,8] In one study, AT-004 was administered in conjunction with a single dose of doxorubicin (n=12) and outcome compared to dogs receiving a single dose of doxorubicin alone (n=5). Complete response rate was numerically higher in the antibody group (75% vs 40%), and median PFS was statistically improved (98 vs 57 days).[8] In a second study, dogs were treated with a single 4-week cycle of L-CHOP chemotherapy and then randomized to receive AT-004 or placebo (n=24 total evaluable patients). The median PFS was improved in dogs receiving AT-004 (167 vs 93.5 days).[7] These results are certainly encouraging; however, neither of the chemotherapy protocols utilized in the aforementioned studies resemble the standard of care for treating canine lymphoma, and it remains to be seen whether integration of AT-004 with more conventional chemotherapy will result in similar improvements in outcome. It is also important to note that different anti-CD20 antibodies may have different mechanisms of action and different degrees of efficacy, owing to differences in antibody isotype, complement avidity, ability to redistribute CD20 on the cell surface and ability to induce direct cell death.[6] Thus, the results obtained with the human antibody products may not be directly translatable to canine antibody products.

Alemtuzumab (Campath) is a humanized monoclonal antibody directed against the CD52 antigen, which is expressed in humans on B and T lymphocytes, NK cells, monocytes, macrophages, and some dendritic cells. Surveys of CD52 expression on canine hematopoietic cells have not been published. Alemtuzumab has shown significant efficacy when utilized as a single agent and in combination in humans with peripheral T cell lymphoma and cutaneous T cell lymphoma. Reactivation of latent viral infection is one of the most commonly observed adverse effects.[9]

AT-005 (Aratana Therapeutics) is a caninized monoclonal antibody directed against the CD52 antigen, that has conditional licensure from the USDA. While no studies of efficacy have been published, a pharmacokinetic and toxicity study in dogs with T-cell lymphoma was presented at the 2014 Veterinary Cancer Society meeting.[10] Hypotension and anaphylactoid reactions were reported but considered "infrequent" and "rare" respectively, and overall AT-005 was considered well tolerated. Two randomized trials of AT-005 in conjunction with chemotherapy are underway as of January 2015: one is combining AT-005 or placebo with a conventional L-CHOP based protocol, and the other is combining AT-005 or placebo with 2 doses of lomustine.[11]

RABACFOSADINE SUCCINATE

Rabacfosadine (Tanovea, VetDC), previously referred to as GS-9219 or VDC-1101, is a double prodrug which was designed to preferentially deliver and accumulate the guanine nucleotide analog PMEG and its active phosphorylated metabolite, PMEG disphosphate (PMEGpp) in lymphoid cells while avoiding systemic exposure to PMEG. Rabacfosadine has been granted MUMS designation for canine lymphoma treatment by the FDA-CVM but is not currently approved.

Three published studies have evaluated the safety and efficacy of rabacfosadine when administered as monotherapy, according to a variety of doses and schedules, to dogs with lymphoma.[12-14] In the largest study, rabacfosadine was administered to 38 dogs with naïve or refractory lymphoma, according to one of 4 schedules: every 21 days, every 14 days, every 7 days, or daily for 5 days, repeated every 21 days. The ORR was 79% (61% CR, 18% PR) and responses were observed across all treatment cohorts. Median remission duration was 128 days. The ORR was 100% in previously untreated dogs (76% CR) and 62% in pretreated dogs (47% CR). ORR and median remission duration were both higher in dogs with B cell lymphoma. Self-limiting neutropenia and gastrointestinal toxicity were observed in some patients. A unique toxicity was a cumulative dermatopathy characterized by local superficial erythema and pruritus, most often localized to the periauricular region, dorsum and inguinal region. This resolved in all dogs with drug discontinuation. A small number of dogs developed pulmonary fibrosis as a late complication (months after study completion while in CR), so careful monitoring for pulmonary changes has been recommended. Additional recent publications demonstrate antitumor activity in canine multiple myeloma and cutaneous T cell lymphoma.[15-16] Additional studies with rabacosadine in canine lymphoma are ongoing.[17]

KPT-335

KPT-335 (Verdinexor, Karyopharm Therapeutics) is an orally available selective inhibitor of nuclear export (SINE) which has been evaluated *in vitro* against multiple canine tumor types,[18-19] and has shown efficacy in dogs with lymphoma.[18] KPT-335 is an inhibitor of a protein called exportin-1 (XPO-1), which is responsible for the binding and nuclear export of a large variety of important cancer-associated genes. KPT-335 has been granted MUMS designation for canine lymphoma treatment by the FDA-CVM but is not currently approved. *In vitro,* KPT-335 demonstrated particular efficacy against lymphoma cells, and in a clinical trial, objective responses were observed in 2 of 17 dogs in the initial dose-finding study, both of which were dogs with lymphoma. In a small expansion cohort in lymphoma, responses were observed in an additional 2 of 6 dogs, which persisted for 35 and 354 days. Gastrointestinal toxicity, including inappetence, diarrhea, vomiting, and weight loss, were the most commonly observed adverse effects. Information regarding the status of ongoing clinical investigations with KPT-335 was not available at the time of Proceedings submission.

CONCLUSIONS

2014 brought some useful new tools into the veterinary oncology clinician's armamentarium, and it appears that a number of additional novel therapies, both novel cytotoxic agents and targeted therapies, are on the horizon. It is hoped that what we are seeing represents the "tip of the iceberg", and that equally great strides will be made in the coming years.

REFERENCES

[1]Poirier VJ, et. al. **J Vet Intern Med** 2004;18(2):219. [2]von Euler H, et. al. **Vet Comp Oncol** 2013;11(4):243. [3] Rivera P, et. al. **J Sm An Pract** 2013;54(1):20. [4]Vail DM, et. al. **J Vet Intern Med** 2012;26;598. [5]Paccal Vet FOIA summary. Available from: http://www.fda.gov/downloads/AnimalVeterinary/Products/ ApprovedAnimalDrugProducts/FOIADrugSummaries/UCM396889.pdf . [6] Lim SH, et. al. **J Immunol** 2014;193(4):1519. [7]Ogilvie G, et. al. **Proc Vet Cancer Soc** 2014:50. [8]Bulman-Fleming J, et al. **Proc Vet Cancer Soc** 2014:51. [9]Zinzani PL, et al. **Leuk Lymphoma** 2012;53(3):789. [10]Rodroguez C, et al. **Proc Vet Cancer Soc** 2014:123. [11]Aratana Therapeutics / Animal Clinical Investigation websites: www.aratana.investorroom.com, www.animalci.com/trials. [12]Reiser H, et. al. **Clin Cancer Res** 2008;14(9):2824. [13]Vail DM, et. al. **Clin Cancer Res** 2009;15(10):3503. [14]Lawrence J, et. al. **Vet Radiol Ultrasound** 2009;50(6):660. [15]Thamm DH, et. al. **BMC Vet Res** 2014;10:30. [16]Morges MA, et. al. **J Vet Intern Med** 2014;28(5):1569. [17]VCS Clinical Trials Database: www.vetcancertrials.org. [18] London CA, et. al. **PLoS One** 2014;9(2):e87585. [19]Breit MN, et. al. **BMC Vet Res** 2014;10:160.

NANOTHERAPEUTICS IN VETERINARY ONCOLOGY

Sandra Axiak-Bechtel, DVM, DACVIM (Oncology)
Columbia, MO

INTRODUCTION

The use of nanotechnology in the treatment of disease is to improve the pharmaceutical index of the drug they carry. For example, nanoparticles can enable the use of chemotherapeutics that otherwise insoluble without toxic solubilizing agents (such as paclitaxel). Nanoparticles may increase the concentration of the therapeutic agent at its intended site - the tumor- through the enhanced permeability and retention effect resulting in increased tumor toxicity but decreased systemic toxicity. And nanoparticles may have reduced clearance compared to the parent chemotherapy resulting in sustained, controlled release.[1] These characteristics of nanoparticles have led to a steady increase in the rate of research in nanomedicine, with an increased understanding in synthesis, biodistribution, and exciting potential cancer therapies.

ENHANCED PERMEABILITY AND RETENTION EFFECT

Nanoparticles of greater than 8nm in size can passively target tumor tissues through the enhanced permeability and retention effect (EPR). EPR refers to the preferential distribution of large molecules to tumor tissue due to the presence of large fenestrations within tumor blood vessels coupled with poor lymphatic drainage.[2,3]

Tumor angiogenesis stimulates the rapid development of new blood vessels for continued tumor growth, however, these vessels are irregular with a discontinuous epithelium and do not have the basement membrane found in normal blood vessels. The result is leaky tumor vasculature - large interendothelial junctions ranging from 200-2000nm allow the passage of large molecules from the bloodstream into the tumor with little to no resistance. Therefore, nanoparticles will be passively distributed within tumor tissue over normal tissue. Additionally, tumors lack sufficient lymphatic function, resulting in minimal drainage and reuptake of nanoparticles into the interstitial fluid and then lymphatics and enhanced retention of administered nanoparticles.[2-3]

The distribution of nanoparticles within the tumor is affected by the extravasation rate from blood vessels, diffusion through extravascular tissue, and the particle interaction with cellular targets within the tumor environment. The extravasation rate and diffusion are mainly the result of diffusive and convective forces within the tumor, which vary among different tumors and nanoparticle concentration. Other tumor factors affecting EPR include vessel architecture, interstitial fluid composition, extracellular matrix composition, phagocyte infiltration, and necrosis. Nanoparticle size for cancer therapy is considered optimal between 8 and 100nm, as larger particles may be more easily recognized and cleared by the reticuloendothelial system. Other nanoparticle factors that can enhance or reduce the EPR effect are blood circulation time of the particle, charge, shape, and surface characteristics.[2-3]

NANOPARTICLES UTILIZED IN VETERINARY CANCER THERAPY

Ideal therapeutic nanoparticles are biocompatible and therefore ignored by the immune system, in addition to breaking down to form harmless metabolic products. A variety of nanoparticles are currently under investigation for use in cancer diagnosis, imaging, and therapy. Here we will discuss those nanoparticulate therapies successfully used in dogs and cats with cancer to date. Research is ongoing for long term toxicity and improved nanoparticle characteristics, such as longer circulation times, improved localization to tumor tissues, and effective removal.

Liposomes

Liposomal nanoparticles are the most common nanoparticle used in veterinary cancer therapy. Liposomes are vesicles with an aqueous core encapsulated by one or more phospholipid bilayers. This type of nanoparticle can carry both hydrophobic and hydrophilic substances and are biodegradable. Liposomes are biocompatible but are prone to clearance through the reticuloendothelial system. Techniques to increase the half-life of liposomes include PEGylation or lipid cross-linking. Toxicities associated with the liposomes alone are uncommon but hypersensitivity reactions have been reported.[1,4]

Micelles

Micelle nanoparticles have a hydrophobic core with a hydrophilic shell. This hydrophilic shell allows for water solubility and avoidance of reticuloendothelial system clearance. There are 4 types of shells for micelles, phospholipid, pluronic, poly ($_L$-amino acid), and polyester. Miceller nanoparticles also have low toxicity and are easily produced.[1]

Metallic nanoparticles

Metals are also used to prepare nanoparticles, with gold the most commonly investigated. Metallic nanoparticles are easily produced, stable, and can be easily modified and conjugated. Gold nanoparticles can be utilized for cancer imaging, delivery of cancer drugs, radiosenstizers for tumor tissue, and the gold radioisotope can be used for cancer therapy. Toxicity is dependent on the metal utilized, the amount administered, and the route of administration.[1,4]

Magnetic nanoparticles

Most magnetic nanoparticles are composed of iron oxide; one main advantage to magnetic particles is the ability to localize to tumor sites with the use of a high grade magnetic field over tumor sites. The surface of iron oxide nanoparticles can be easily engineered to augment biocompatibility and biodegradability. Furthermore, they degrade via leakage of free iron ions, which become incorporated into hemoglobin and recycled through normal physiologic processes.[1,4]

Nanocrystals

Nanocrystals consist of pure drug; minimal to no surface active agents are present for stabilizing the nanoparticle. Nanoparticle size ranges from 10-1000nm and the preparation technique allows previously insoluble drugs to become a nanosuspension in water, without the need for solubilizing agents.[5]

NANOPARTICULATE CHEMOTHERAPEUTICS

Liposomal doxorubicin (Doxil®)

Doxorubicin is entrapped in liposomes containing a small fraction of polyethylene glycol (PEG), which aids in decreasing rapid clearance by the reticuloendothelial system seen without PEGylation. The liposomal doxorubicin has an increased circulation time and decreased cardiotoxicity and myelosuppression compared to standard doxorubicin formulations.

Reported side effects are hypersensitivity reactions, gastrointestinal side effects, and palmar-plantar erythrodysesthia (PPE). PPE is a unique side effect of liposomal doxorubicin in both dogs and cats, is associated with increased dose intensity, and lesions are characterized by swelling, pain, erythema, and desquamation of the axilla, inguinal region, and footpads. Pyridoxine (vitamin B6) can be used as an oral supplement daily during liposomal doxorubicin therapy, and can delay the onset in addition to reducing the severity of PPE in dogs.[6] Delayed nephrotoxicity is also reported in cats receiving liposomal doxorubicin.[7]

In cats, liposomal doxorubicin has successfully been used in the treatment of bulky injection site sarcomas, although no difference was noted between free doxorubicin and liposomal doxorubicin groups.[7] When used in the treatment of canine splenic hemangiosarcoma post-splenectomy, no difference in survival was noted between free doxorubicin and liposomal doxorubicin groups; similar survival times were also reported for dogs when liposomal doxorubicin was administered intraperitoneally.[8-9] However, further study is indicated in other tumor types, including lymphoma, other sarcomas, and carcinomas in which responses have been reported.[6]

Liposomal clodronate

Clodronate is a bisphosphonate that results in osteoclast apoptosis; additionally, clodronate kills macrophages. Tumor associated macrophages are in part responsible for immune tolerance of tumors, and depletion of tumor associated macrophages has resulted in positive responses in rodent models. The uptake of liposomes by macrophages has been utilized for therapeutic purposes with liposomal clodronate. In dogs, one study revealed that liposomal clodronate resulted in decreased tumor associated macrophages in soft tissue sarcomas. In dogs with histiocytic sarcoma, a tumor of macrophages or dendritic cells, liposomal clodronate resulted in tumor reduction in 2 of 5 dogs. Further study is indicated to take advantage of the known clearance mechanism for liposomes and the resultant target cell population.[10-11]

Liposomal cisplatin

A lipohilic cisplatin derivative (cis-bis-neodecanoato-transe-R,R-1,2-diaminocyclohexane platinum(II)) was formulated in liposomes and used to treat cats with bulky oral squamous cell carcinomas, with the liposomes altering drug pharmacokinetics to increase safety, as standard cisplatin results in fatal pulmonary edema in cats. Dose limiting side effects reported included gastrointestinal upset, neutropenia, and thrombocytopenia. Anaphylaxis was also reported. Liposomal cisplatin was not effective for tumor control in cats with oral squamous cell carcinoma.[12]

A pegylated liposomal formulation of cisplatin was administered to dogs with osteosarcoma and compared to carboplatin chemotherapy following limb amputation in the microscopic disease setting. Although the median disease free interval and survival times were comparable between the 2 groups, the liposomal cisplatin was administered safely without the diuresis protocol normally utilized to prevent cisplatin induced nephrotoxicity. Furthermore, the dose of cisplatin used was 5 times that of free cisplatin that can be safely used in the dog.[13]

Micellar paclitaxel (Paccal® Vet-CA1)

Paclitaxel is a taxane chemotherapeutic commonly used in human medicine, however, paclitaxel is insoluble in water and the solvent utilized results in hypersensitivity reactions in dogs, even with pre-treatment utilizing antihistamines and corticosteroids. Paccal® Vet is a miceller formulation of paclitaxel, and this formulation allows paclitaxel to be administered without the solvent. A dose finding study revealed no hypersensitivity reactions and a variety of tumor responses.[14]

Of importance is the conditional approval status of this particular formulation, meaning that Paccal® Vet can only be used for its labeled indications and cannot be used off-label. Currently, Paccal® Vet is labeled to treat nonresectable stage III, IV, or V mammary carcinoma in dogs that have not received previous chemotherapy or radiotherapy, and resectable and nonresectable squamous cell carcinoma in dogs that have not received previous chemotherapy or radiotherapy.

NANOPARTICULATE IMMUNOTHERAPY

Liposomal muramyl tripeptide

Muramyl dipeptide is a peptidoglycan that is contained in both gram positive and gram negative bacteria, and is recognized by the immune system as a pathogen associated molecular pattern; once recognized by the immune system, it indices both humoral and cellular immune activity. As the immune system plays an important role in fighting cancer and is often blunted by factors secreted by the tumor, immune system activation can lead to increased tumor control and better prognoses. Liposomal muramyl tripeptide phosphatidylethanolamine (L-MTP-PE) is a synthetic, lipophilic, liposomal formulation of muramyl tripeptide which targets macrophages and monocytes for more specific immune stimulation.

L-MTP-PE improved survival times in dogs with early stage oral melanoma following surgery[15], improved overall survival times in dogs with osteosarcoma following amputation and cisplatin[16], and prolonged disease free and overall survival in dogs with splenic hemangiosarcoma following surgery and combined with chemotherapy[17]. L-MTP-PE was not effective compared to empty liposomes following surgery in canine mammary carcinoma[18] or feline mammary carcinoma following surgery[19].

Liposomal interleukin-2

A phase I study of liposomal interleukin (IL)-2 encoding DNA was undertaken in 20 dogs with chemotherapy resistant osteosarcoma pulmonary metastasis. IL-2 expression was detected in all dogs, systemic immune activation was noted (fever, leukogram changes, increased monocyte and natural killer cell activity), and 3 dogs had an objective response.[20]

Inhalational liposomal IL-2 was used to treat 2 dogs with pulmonary metastasis and 2 dogs with primary lung carcinomas. Responses were noted in 2 of 4 dogs with metastatic osteosarcoma (complete regression) and 1 of 2 dogs with primary lung carcinoma (stable disease). Side effects were minimal.[21]

METALLIC NANOPARTICLES

Gold nanoparticles

Due to the above described advantages of nanoparticles, investigation is ongoing on the use of metallic nanoparticles for intralesional cancer therapy. The glycoprotein gum arabic was utilized as a stabilizer for gold nanoparticles and provided a nontoxic coating to prevent nanoparticle agglomeration. Gum arabic coated ^{198}gold nanoparticles are easily produced, stable, safe in the mouse and pig model, and have the potential for better dose distribution with decreased off target effects compared to traditional seed brachytherapy. Intralesional injection of gum arabic coated ^{198}gold nanoparticles was found to have minimal short term side effects and potential efficacy in dogs with prostate cancer; 1 dog died to urethral obstruction and therefore urethral stenting is recommended prior to treatment. Of the 8 dogs evaluated for response, 2 had partial remission and 4 had stable disease.[22] Further study is ongoing.

Iron oxide magnetic nanoparticles

Twenty cats with injection site sarcomas were pre-treated with plasmid coding for feline granulocyte-macrophage colony factor stimulating factor attached to iron oxide magnetic nanoparticles intralesionally. Following injection, a magnetic field was applied to the tumor for one hour with the goal of localizing GM-CSF effects to the tumor region and avoiding systemic side effects. Following treatment, the tumors were surgically excised. At day 360 of the study, 10 cats were recurrence free and no significant treatment toxicities were observed.[23]

CONCLUSION

Nanomedicine is an exciting area of research in cancer therapy with the end goal of better targeting tumor tissue while sparing normal tissue and decreasing side effects. Nanoparticles also have significant potential to aid in cancer diagnosis and imaging in addition to the treatment of other diseases, including infectious and autoimmune diseases. Current research is focused on the design of nanoparticles with better tumor targeting and retention, alternative routes of nanoparticle administration to improve the therapeutic index, and to develop strategies for simplifying nanoparticle production to decrease cost while increasing production.

REFERENCES

[1]Underwood C et al. The Vet J 2012; 193: 12-23; [2]Bertrand N et al. Adv Drug Del Rev 2014; 66: 2-25; [3]Chow EKH et al. Sci Trans Med 2013; 5: 216rv4; [4]Nazir S et al. Nanomedicine: NBM 2014; 10: 19-34; [5]Irache JN et al. Vet Parist 2011; 180: 47-71; [6]Vail DM et al. Cancer Chemother Pharmacol 1997; 39: 410-6; [7]Poirier VJ et al. J Vet Intern Med 2002; 16: 726-31; [8]Teske E et al. Vet Comp Onc 2011; 9: 283-9; [9]Sorenmo K et al. J Vet Intern Med 2007; 21: 1347-54; [10]Hafemean SD et al. Cancer Immunol Immunother 2010; 59: 441-452; [11]Guth AM et al. Vet Comp Onc 2013; 11: 296-305; [12]Fox LE et al. Am J Vet Res 2000; 61: 791-5; 13Vail DM et al. Cancer Chemother Pharmacol 2002; 50: 131-6; [14]vonEuler H et al. Vet Comp Oncol 2013; 11: 243-55; [15]MacEwen et al. Clin Cancer Res 1999; 5: 4249-58; [16]Kurzman ID et al. Clin Cancer Ares 1995; 1: 1595-601; [17]Vail DM et al. Clin Cancer Res 1995; 1: 1165-70; [18]Teske E et al. Anticancer Res 1998; 18: 1015-9; [19]Fox LE. Cancer Biother 1995; 10: 125-30; [20]Dow S et al. Hum Gene Ther 2005; 16: 937-46; [21]Khanna C et al. Cancer 1997; 79: 1409-21; [22]Axiak-Bechtel et al, Int J Nanomedicine 2014; 9: 5001-11; 23 Huttinger et al J Gene Med 2008; 10: 655-67.

PROGRESS IN CANCER IMMUNOTHERAPY: SUCCESS ON THE PATH TO A CURE

Katie L. Anderson, BS and Jaime F. Modiano VMD, PhD
St. Paul, MN

INTRODUCTION

The field of cancer immunotherapy seeks to harness and enhance the ability of the immune system to eliminate cancer. The application of immunotherapy has gained considerable interest in because of recent, unprecedented success in the treatment of human patients with lymphoma and melanoma.[1,2] Immunotherapy holds similar promise for companion animals with cancer. The immune system can uniquely target cancer cells or the tumor environment while minimizing damage to normal tissues, and immune cells can reach locations that are otherwise inaccessible using conventional approaches.[2] Additionally, immunological memory can provide long lasting, durable clinical responses.[2]

Cancer immunotherapy can generate anti-tumor immune responses through a variety of mechanisms. Passive immunotherapy involves the transfer of biological reagents, such as monoclonal antibodies (mAbs) or antigen-specific adaptive immune cells, into the cancer patient.[2] Active immunotherapy seeks to elicit an anti-tumor response from the patient's own immune system, typically through vaccination.[2] Here, we will focus on strategies that utilize adaptive immunity, and specifically discuss recent advances in veterinary cancer immunotherapy emphasizing technologies that originated from – or that are being developed at the University of Minnesota in collaboration with academic and industrial partners.

PASSIVE IMMUNOTHERAPY

Monoclonal Antibodies

The administration of mAbs has been well established in human oncology for several decades.[3] Anti-CD20 (Rituximab), anti-EGFR (Cetuximab), and anti-HER2 (Trastuzumab) consistently increase patient remissions with minimal toxicity in human B-cell lymphoma, lung and other epithelial cancers, and HER2-positive breast cancers, respectively.[1,3] Recently, the consistent expression of CD20, EGFR, and HER2/neu has been confirmed in various canine cancers, and chimeric, caninized versions of these antibodies have been developed.[4-8] Three distinct anti-canine CD20 antibodies have patents filed or awarded, and each has documented diagnostic or clinical potential. For example, the 6C8 anti-canine CD20 mAb showed labeling of 100% of canine B-cell lymphomas tested, it enhanced antibody dependent cellular cytotoxicity, and it promoted macrophage phagocytosis of canine B-cell lymphoma cells *in vitro*.[8] The 1E4 anti-canine CD20 antibody showed significant benefit (reduced tumor burden) when used as monotherapy in a preclinical model of mice harboring canine CLBL1 B cell lymphoma xenografts *in vivo*.[9] And the AT-004 anti-canine CD20 was reported to increase median progression free survival of dogs with B cell lymphoma by 74 days and overall survival by 148 days in a prospective clinical trial.[10] Similarly, an anti-canine EGFR mAb decreased proliferation of canine mammary carcinoma cells by 40-60% and enhanced macrophage phagocytosis and tumor cell killing *in vitro*.[4] Ongoing and future clinical studies will continue to evaluate the efficacy of these reagents in the veterinary clinic.

In addition to Rituximab, antibodies that prevent binding of CD47 to its receptor on macrophages, SIRP-α, also improve innate anti-tumor responses *in vitro* and *in vivo* using pre-clinical xenograft models of human leukemia and lymphoma.[1,11] The interaction between CD47 and SIRP-α inhibits macrophage phagocytosis; essentially it functions as a "don't eat me" signal, and CD47 blockade enhances the capacity of macrophages to attack tumors by removing this inhibition. Preliminary data indicate that these therapeutic properties of anti-CD47 are retained in the setting of canine B-cell lymphoma, and ongoing studies are designed to examine the capacity of CD47 blockade to enhance adaptive (T-cell) anti-tumor immune responses.

ACTIVE IMMUNOTHERAPY

In situ immunization with adenovirus-Fas Ligand

Stimulation of adaptive, T-cell anti-tumor immunity can be achieved through *in situ* immunization strategies that promote inflammation and necrosis at the primary tumor site. This response then primes T cells to locate and destroy tumor cells at distant metastatic sites. When expressed at supraphysiologic levels such as those that can be achieved using adenovirus vectors, and when delivered to tumors that are resistant to Fas-mediated apoptosis, FasL acts as a neutrophil chemoattractant and induces massive suppurative inflammation. This in turn attracts other innate immune cells and creates a favorable environment for antigen presentation to tumor infiltrating T cells, which can then mount a transplantable systemic anti-tumor response.[12-14] Administration of FasL was initially tested in the clinical setting by intratumoral administration of naked plasmid DNA in five dogs with oral malignant melanoma.[15] This study showed that FasL administration was safe and reduced local tumor burden, and in four of five subjects treated, it was associated with durable objective responses. *In situ* immunization with FasL was subsequently tested using a replication-deficient adenovirus platform in a clinical study of 56 dogs with spontaneous osteosarcoma in the neoadjuvant to standard of care setting.[12] Predictably, tumors with reduced Fas expression showed increased levels of local inflammation, necrosis, and lymphocytic infiltration upon treatment with Ad-FasL; dogs with elevated inflammation scores (>2-3) also demonstrated increased median survival times compared to a group that received standard of care therapy alone (359 vs 221 days).[12] Tumors with elevated Fas expression developed lower levels of inflammation (score ≤1), and the median survival time of these dogs did not differ from that of dogs that received only the standard of care.[12] Overall, this study suggests that Ad-FasL gene-based immunotherapy improved outcomes in a subset of dogs with cancer, specifically those harboring tumors that are resistant to Fas-mediated death signals, making the therapy amenable to patient selection through companion diagnostic tests.

Administration of Attenuated Bacteria

Genetically modified anaerobic bacteria, such as *Salmonella typhimurium*, display tropism for hypoxic tumor tissue and can be used as vectors for gene delivery to induce tumor cytotoxicity and to stimulate an anti-tumor immune response.[16] In a phase I dose-finding clinical trial at the University of Wisconsin, an attenuated *S. typhimurium* strain was administered intravenously to forty-one dogs with a variety of spontaneous malignancies. Bacteria were cultured from 42% of tumors. Although one study related death occurred, four dogs achieved a complete response, two dogs achieved a partial response, and 10% of the dogs achieved disease stabilization, demonstrating the therapeutic potential of this strategy.[16] A recently completed clinical trial at the University of Minnesota also utilized an attenuated *S. typhimurim* that was rendered avirulent, but remained highly immunogenic.[17] In these genetically modified bacteria, a gene needed for replication was replaced with the same gene fused to a truncated IL-2 gene, making it so that only those bacteria that produce the cytokine can survive and proliferate.[17] The bacteria were developed into an oral form to reduce systemic toxicity, and the safety and efficacy of this formulation were confirmed in pre-clinical mouse models.[18] When administered orally to 19 dogs in the adjuvant setting, this therapy was safe and it improved disease free interval, particularly in dogs with highly aggressive tumors.[19] Other bacterial vectors such as *Listeria monocytogenes* have also been investigated as vectors to enhance immunotherapy. In a phase I clinical trial, attenuated *L. monocytogenes* engineered to express a chimeric human HER2/neu fusion protein was administered to 17 dogs with appendicular osteosarcoma. This agent significantly delayed metastatic disease and increased overall survival compared to historical controls.[20]

Oncolytic Virotherapy

Oncolytic viruses that preferentially infect, propagate in, and lyse cancer cells have shown considerable promise in several canine clinical trials. Intratumoral injection of a canine adenovirus genetically modified to express CD40L led to complete tumor rejection in 5 of 19 canine melanoma patients with no significant toxicity.[21] Several poxviruses and vaccinia viruses have also demonstrated considerable anti-tumor activity in xenograft models of canine mammary adenocarcinoma and soft-tissue sarcoma and are currently entering clinical trials.[22] Finally, the maximum tolerated dose of a vesicular stomatitis virus engineered to express IFN-β and the sodium-iodide symporter (NIS) has been established in purpose-bred dogs by researchers at the Mayo Clinic and the University of Tennessee[23] and will be used in a future clinical trial for dogs with osteosarcoma. The extent to which these viruses specifically enhance adaptive anti-tumor immunity in companion animals is a subject of contemporary interest.

Anti-cancer Vaccines

Therapeutic cancer vaccines utilize a variety of approaches to induce immune activation, including the injection of whole cell or tumor cell lysates, peptide antigens, plasmid DNA, or activated immune cells primed with tumor antigens. Autologous tumor-cell lysate vaccines supplemented with an immune adjuvant have used to treat dogs with spontaneous meningioma at the University of Minnesota.[24] Each of the 11 dogs vaccinated developed a polyclonal, tumor-reactive antibody response, and all showed infiltration of antibody-producing plasma cells into the tumor, supporting a cause-effect relationship between the therapy and the generation of anti-tumor immunity. Vaccination with autologous immune cells exposed to tumor antigens *ex vivo* also has been used to treat dogs with melanoma and B-cell lymphoma.[25,26] In a recent clinical trial, activated B cells infused with autologous tumor RNA were administered to dogs with B-cell lymphoma. Vaccination stimulated anti-tumor immune responses and improved the rate of durable second remissions in these dogs.[26]

FUTURE DIRECTIONS AND CONCLUDING REMARKS

Various recent advances in cancer immunotherapy have yet to be translated to companion animal species. One of these is adoptive transfer of genetically engineered T cells containing chimeric antigen receptor (CARs) that recognize tumor cells and incorporate signaling motifs that promote lymphocyte activation. CAR-expressing canine T cells with anti-tumor activity against canine osteosarcoma cells *in vitro* have been reported,[27] demonstrating the potential to apply this technology to canine cancers in the near future. Another exciting technology is immune checkpoint blockade using mAbs against CTLA-4 and PD-1 or PD-L1. Reagents to implement this approach for veterinary use have not yet been developed.

Overall, immunotherapy has the potential to revolutionize the field of veterinary oncology. Implementation will require coordinated efforts to optimize dose, schedule, and route of administration for each unique approach.[2] Severe systemic side effects are possible, including autoimmunity and inflammation, liver toxicity, and death. Furthermore, innovative strategies will be needed to avoid or minimize resistance;[3] therefore, it is likely that immunotherapy will initially find its place as an adjuvant to conventional cancer treatments.[28] Despite the challenges, immunotherapy presents opportunities to achieve durable responses with improved quality of life in companion animals with cancer.

REFERENCES

[1]Chao MP, et al. **Cell** 2010;142;699–713. [2]Mellman I, et al. **Nature** 2011;480;480–489. [3]Makkouk A and Weiner GJ. **Cancer Res** 2015;75;5–10. [4]Singer J, et al. **Mole Cancer Ther** 2014;13;1777–1790. [5]Beall MJ. **US Patent** 20110091483 A1. [6]Rue R, et al. **Patent** WO 2013063186 A1. [7]Hansen G. **US Patent** 8337842 B2. [8]Ito D, et al. **Leuk Lymphoma** 2014;1–25. [9]Ito D, et. al. **ASH Meeting on Lymphoma Biology**, 2014; Vol. 1, p 102. [10]Ogilvie G. **Veterinary Cancer Society Annual Meeting**, 2014, St. Louis, MO. [11]Majeti R, et al. **Cell** 2009;138;286–299. [12]Modiano JF, et al. **Mol Ther** 2012;20;2234–2243. [13]Modiano JF, et al. **Cancer Ther** 2004;2;61-570. [14]Modiano JF, et al. **Clin Immunol** 2004;112;54-65. [15]Bianco SR, et al. **Cancer Gene Ther** 2003;10;726-736. [16]Thamm DH, et al. **Clin Cancer Res** 2005;11;4827–4834. **[17]Saltzman, DA.** Expert Opin Biol Ther **2005;4;443-449.** [18]Sorenson BS, et al. **J Pediatr Surg** 2008;43(6), 1153–1158. [19]Fritz SE, et al. **Veterinary Cancer Society Annual Meeting**, 2013, Minneapolis, MN. [20]Gnanandarajah JS, et al. **J Immunother Cancer** 2014;2(Suppl 3):P55 [21]Westberg S, et.al. **J Immunother** 2013;36(6);350-8. [22]Gentschev I, et al. **Viruses** 2014;6;2122–2137. [23]LeBlanc AK, et al. **Hum Gene Ther Clin Dev** 2013;24;174–181. [24]Andersen BM, et al. **Cancer Res** 2013;73;2987–2997. [25]Gyorffy S, et al. **J Vet Intern Med** 2005;19;56–63. [26]Sorenmo KU, et al. PLoS ONE 2011;6;e24167. [27]Mata M, et al. **J Immunother** 2014;37(8);407-15 [28]Bergman PJ. **Vet Clin North Am Small Anim Pract** 2007;37;1111–1119.

CANCER ACID ENVIRONMENT: IMPLICATIONS FOR THERANOSTIC APPROACHES

Enrico P. Spugnini, DVM, PhD., DACVIM (Oncology), DECVIM-CA (Oncology)
Rome, Italy

INTRODUCTION

Tumor cells rely on H^+ exchangers to relieve themselves from the dangerous protons byproduct of cancer metabolism that could trigger a cascade of lytic enzymes that ultimately would lead to self-digestion. Among these mechanisms, one of the most investigated are the vacuolar H^+-ATPases (V-ATPases).[1-4] V-ATPases are ATP dependant H^+ transporters that utilize the energy freed by the hydrolysis of ATP with the active transport of protons from the cytoplasm to the lumen if intracellular compartments or, if located within the cytoplasmic membrane, the extracellular compartment. Structurally speaking, the V-ATPases are composed of a peripheral domain (V_1) that carries out ATP hydrolysis and an integral domain (V_0) responsible for exchanging protons.[5] The regulatory mechanisms of V-ATPases are neutralized in malignant tumor cells that overexpress lysosomal proteins on the cell surface.[6,7] The acidic tumor environment is a consequence of anaerobic glucose metabolism with secondary production of lactates byproducts through the upregulation of hypoxia-inducible factor 1α or can be due to inadequate tumor perfusion, tumor hypoxia or enhanced transmembrane pH regulation.[8,9] These pumps, coupled with other ion exchangers, play a crucial role in the establishment and maintenance of malignant tumor environment and promote the selection of more aggressive cell phenotypes able to survive in this highly selective ambient.

ROLE OF V-ATPASES IN TUMOR SPREAD

V-ATPases play a critical role in the maintenance of an appropriate relatively neutral intracellular pH, an acidic luminal pH, and an acidic extracellular pH by actively pumping protons either through ion exchange mechanisms or by segregating H^+ within cytoplasmic organelles that are subsequently expelled.[10] It is hypothesized that the low extracellular pH of tumors might trigger proteases, leading to the dissolution of extracellular matrix. This phenomenon, as is well known, significantly contributes to tumor invasion and dissemination.[11-12] In fact, tumor invasion depends on tumor acidifying ability that ultimately leads to secretion and activation of several classes of proteases. [13-14] The most commonly triggered enzymes are MMP-2, MMP-9, cathepsin B, and cathepsin L. Low pH can also up-regulate or activate proangiogenic factors such as VEGF-A and IL-8.[15-16]

ROLE OF V-ATPASES IN CHEMORESISTANCE

The acidic microenvironment caused by changes in the pH gradient between the intracellular and the extracellular compartments as well as the pH gradient between the cytoplasm and the intracellular organelles can be significantly involved in the mechanism of drug resistance.[17,18] There are several mechanisms involved in this phenomenon, including decreased uptake or neutralization of weakly basic drugs by the acidic tumor microenvironment (ion trapping) or the sequestration of chemotherapy drugs within lysosomal vesicles.[18-21] An accelerated turn over of acidic vesicles may represent an additional tumor strategy of drug resistance based on counteracting current transportation[20,21]. Several investigators developed new approaches to better characterize tumor pH in animal models mostly through imaging systems in order to identify novel targets. A research consortium has been created to pool together investigators with different skills in order to formalize the potentially most effective approaches to this biological riddle.[20]

RESULTS IN PRECLINICAL STUDIES

Several studies in murine models have shown that the neutralization of the H^+ transporters could result in increased tumor control and extended survival. A recent study showed that blocking V-ATPases through RNA interference led to prevention of cancer metastases in a murine model. As a result, new approaches have been developed to modulate drug efficacy within the low pH tumor milieu including the use of RNA interference, bicarbonates or the induction of metabolic alkalosis.[21,22] Finally, two recently published articles describe the chemosensitizing action of proton pump inhibitors in a murine model of orthotopic cutaneous melanoma, a well known chemo-refractory neoplasm, opening a novel field of investigation.[23,24] Meanwhile, Acridine Orange, was used in mice carrying osteosarcoma as a diagnostic and therapeutic drug coupled with phototherapy. It has various unique biological activities, and has been shown to be a useful fluorescent dye specific for DNA and RNA, a pH indicator, photosensitizer, antitumor and antimalarial drug, and a detector of bacteria and parasites, apoptosis and sperm mobility. This molecule has shown the tendency to concentrate in an acid environment such as tumors and to be more slowly metabolized by tumors than other tissues.[25] Finally, electroporation could be as well a technique to modulate tumor acid environment, thus improving drug uptake and efficacy.[26]

CURRENT CLINICAL APPLICATIONS IN HUMANS AND PETS

Diagnosis/Prognosis

Many studies pointed out that there exists a very interesting association between oral squamous cell carcinoma in human patients and overexpression of *ATP6V1C1* 54, which seems to be the main gene involved in the regulation of V-ATPase enzymes and the acidity of solid oral tumors. [28] Moreover in brush cytology samples it has been shown that *ATP6V1C1* levels were significantly higher in patients with OSCC than in healthy controls, with expression increasing with higher tumor stage. ROC analysis showed that the measurement of *ATP6V1C1*expression levels is a highly sensitive and specific diagnostic method. [29]

Therapy

Surgeons are exploiting the property of Acridine Orange to be concentrated within tumors to show and remove residual disease during marginal surgeries for the therapy of sarcomas.[30,31] Several phase II trials are ongoing in humans with advanced oral carcinoma, lymphoma and melanoma that combine PPIs with conventional chemotherapy. The preliminary results appear very promising, especially for chemoresistant subtypes such as chondroblastic osteosarcoma in humans receiving neoadjuvant chemotherapy for limb sparing procedures, in this cohort pretreatment with PPI resulted in a 50% increase of responders.[32] A recent meta-analysis on patients with head and neck carcinoma receiving antacid therapy (either PPI or histamine 2 blockers) during their therapies had a significant improvement in their overall survival.[33] A phase II trial has been conducted to identify the best drug(s), their dose and schedule in dogs and cats with advanced tumors showing an high percentage of chemoresistance reversal, moreover, a single arm, non-randomized phase II open study, with historical control group suggested that alkalization could greatly improve the efficacy of metronomic chemotherapy with minimal side effects.[34,35] Currently studies are being conducted in pets to devise novel protocols with inhibitors of acidity and chemotherapy and to test new drugs currently in the final development phase..

FUTURE DIRECTIONS

Novel inhibitors of the tumor deprotonating pumps are being made and evaluated in cancer patients, in combination with different chemotherapy agents in order to maximize tumor control. Finally, besides being a potential target of anticancer drugs, it is conceivable that V-ATPases might become a predictive factor of tumor behavior and final outcome through the immunohistochemical evaluation of their expression and cellular distribution in tumor biopsies. Perhaps it is time to consider acidity among the malignant phenotypes of tumors and include this unique feature in our treatment strategies.

REFERENCES

[1]**Spugnini et al.** J Exp Clin Cancer Res **2010; 29: 44.** [2]**Forgac.** Nat Rev Mol Cell Biol **2007 ; 8: 917.** [3]**Cipriano et al.** Biochim Biophys Acta **2008; 1777: 599.** [4]**Jefferies et al.** Arch Biochem Biophys **2008; 476: 33.** [5]**Arai et al.** J Biol Chem **1988; 263: 8796.** [6]**Saitoh et al.** J Biol Chem **1992; 267:5700.** [7]**Glunde et al.** Neoplasia **2003; 5: 533.** [8]**Gatenby et al.** Nat Rev Cancer **2004; 4: 891.** [9]**Fais et al.** Cancer Res **2007; 67: 10627.** [10]**Nishi et al.** Nat Rev Mol Cell Biol **2002; 3: 94.** [11]**Martinez-Zaguilan et al.** Am J Physiol **1993; 265: 1015.** [12]**Martínez-Zaguilán et al.** Clin Exp Metastasis **1996; 14: 176.** [13]**Razaq et al.** Eur Spine J **2003 ; 12: 341.** [14]**Webb et al.** Novartis Found Symp **2001; 240:169.** [15]**Koukourakis et al.** Br J Cancer **2003; 89: 877.** [16]**Rofstad et al.** Cancer Res **2006; 66: 6699.** [17]**Mahoney et al.** Biochem Pharmacol **2003; 66: 1207.** [18]**Simon et al.** Proc Nat Acad Sci USA **1993; 91: 1128.** [19]**Raghunand et al.** Biochem Pharmacol **1999; 57: 1047.** [20]**Spugnini et al.** Biochim Biophys Acta. **2014; in press.,** [21] **Taylor et al.** Drug Resist Updat **2015; in press.** [22]**Raghunand et al.** Biochem Pharmacol **1999; 57: 309.** [23]**Raghunand N.** Methods Mol Med **2006; 124: 347.** [24]**Luciani et al.** J Natl Cancer Inst **2004; 96:1702.** [25]**De Milito et al.** Int J Cancer **2010; 127: 207.**[26] **Kusuzaki et al.** In Vivo **2007; 21: 205.** [27] **Spugnini et al.** Microsc Res Tech **2007; 70: 1041.** [28] **Pérez-Sayáns et al.** Cancer Letters **2012; 295:135.** [29] **Pérez-Sayáns et al. Cancer Biology & Therapy 2010,9:12, 1-8.** [30]**Kusuzaki et al.** Curr Pharm Des **2012;18:1414.** [31]**Matsubara et al.** Clin Orthop Relat Res **Epub.** [32]**Ferrari et al**. J Transl Med **2013;11:268.** [33] **Papagerakis et al.** Cancer Prev Res (Phila) **2014; 7:1258-69.** [34]**Spugnini et al.** J Trasl Med **2011; 9:221.** [35]**Spugnini et al.** J Trasl Med **2014; 12:225.**

ELECTROPORATION IN ONCOLOGY: CURRENT CONCEPTS

Enrico P. Spugnini, DVM, PhD, DACVIM (Oncology), DECVIM-CA (Oncology)
Rome, Italy

INTRODUCTION

Electroporation is a technique that aims at ablating tumor cells through the direct application of current (irreversible electroporation or IRE) or at increasing the delivery of drugs through the cell membrane by applying proper permeabilizing pulses. Chemotherapy drugs are effective provided that they reach their site of action. The primary obstacle for chemotherapy agents is the cytoplasmic membrane, especially for lipophobic agents such as bleomycin. Electrochemotherapy (ECT) couples the administration of an anticancer agent (usually bleomycin or cisplatin), to the delivery of electric pulses having appropriate waveforms. The application of permeabilizing pulses leads to perturbation of the cell membrane, thus resulting in increased uptake of chemotherapeutic drugs, ultimately leading to cell apoptotic death.[3]Clinical trials evaluating humans treated with ECT for cutaneous neoplasms were first reported in the early 90s, and have been followed by many trials that confirm its potential for the treatment or palliation of dermatological tumors.[3-5] Recently, a standard operating procedure has been proposed.[6]

ELECTROCHEMOTHERAPY EQUIPMENTS

Different waveforms have been adopted by investigators: exponentially decaying, square, rectangular or biphasic. In general the number of pulses delivered is set in 8 single pulses applied per cm of tumor area at a voltage of 1300 V/cm (800 V/cm for intraoperative use), with a duration of 100 µs and a frequency of 1 Hz. Treatments are repeated until the whole tumor area is covered. We recently developed a shortened protocol with decreased morbidity and maintained efficacy.[7]

RESULTS IN VETERINARY MEDICINE USING SQUARE PULSES BASED ECT

The first clinical trial in veterinary patients involved the treatment of feline sarcomas recurring after radiation therapy using bleomycin and Interleukin 2 producing cells coupled with square pulses.[8] The authors did not report any clinical response but claimed prolonged control for cats undergoing ECT compared to untreated patients. More recently, CDDP or bleomycin based ECT has been reported to treat cutanous neoplasms in dogs and cats as well as sarcoids in horses.[9-11]

EVALUATION OF BIPHASIC PULSES BASED ECT IN PETS

Experimental design

Our group investigated in the past 12 years the feasibility of electrochemotherapy in companion animals affected by different neoplasms. We adopted trains of biphasic pulses delivered as a burst instead of single impulses. The electric waveform were generated by a *Chemopulse* equipment. This device is built up by a toroidal core transformer generating a roughly rectangular pulse which is split in two halves that are sequentially driven to obtain a biphasic pulse. Pulses are not singularly produced but are created in bursts. The equipment allows choosing electric field intensity, number of pulses and pulse duration. The standard wave train used for this experiment was set to 8 pulses of 50 + 50 µs (biphasic pulse), and the electric field intensity was 1250 V/cm. The pulse repetition frequency is 1 Hz while the frequency of burst repetition is 1 kHz, resulting in a total burst duration of 7.1 ms. Bleomycin was locally injected at the tumor site and in the tumor's surrounding tissues. Five minutes after the injection, trains of biphasic pulses were administered through caliper electrodes as above described. Adherence of the caliper electrodes to the lesions was maximized using an electrophoresis gel. The treatment was repeated after 1 week.[12] More recently, a shortened protocol involving 8 pulses with 10µs interpulse interval and a total duration of 3.1ms has been devised.[7] The adoptions of a dedicated software for the impulse stabilization and optimization resulted in improved control and minimal side effects.[7]

RESULTS

In the first trial, ECT was used to directly attack neoplasms.[13] Overall response rate was 80%, with 40% long lasting remissions (in excess of 1 year). This phase I/II study led to the development of several electrodes specific for the different body districts.[14] On the basis of the preliminary studies, cohorts of dogs and cats affected by melanoma, soft tissue sarcoma mast cell tumor and nasal squamous cell carcinoma were enrolled in phase II studies.

Melanoma Ten patients with malignant melanoma (MM) of the oral cavity were enrolled in a phase II study. The overall response rate was 80% (median time to recurrence 6 months), with a 50% of patients in remission after 1 year.[15] Of interest, most of the long term responders experienced a vitiligo like discoloration at the treatment site, potentially suggestive of recruitment of the immune system after uncovering of deep antigens.

Soft tissue sarcoma A total of 72 cats affected by soft tissue sarcoma were assigned to three different groups: 1) surgery alone, 2) surgery coupled with intraoperative ECT, 3) surgery coupled with post-operative ECT.[16] The median control times were respectively 4, 12 and 19 months. In this study were identified some prognostic factors such as previous treatment and tumor size. A cat with a recurring hemangiopericytoma was also referred at the time of the study, and is still in remission 2.5 years after the completion of the ECT treatment.[17] The only side effects were tumor dehiscence in 2 patients and focal inflammation in 2 patients. Systemic side effects were not reported among the enrolled patients.

In another study, adjuvant ECT was adopted to increase local control in a group of 22 dogs, obtaining a mean time to recurrence of 730 days, with 50% of the dogs still desease free at the time of writing.[18] In one dog with a large, unresectable sarcoma, ECT has been used to decrease tumor size and allow surgical excision.[19] More recently, ECT has been exploited to reintroduce potentially toxic drugs in the oncology protocols. Specifically, a cohort of 64 cats with incompletely excised soft tissue sarcoma have been enrolled in a study involving the use of cisplatin as ECT agent and matched with a control cohort of 14 cats treated with surgery. All the cats tolerated the treated very well, despite the high reported toxicity of the drug when administered systemically. In terms of tumor control mean time to recurrence was 666 days and 180 days, respectively.[20]
A potentially novel application of ECT as suggested by a recent case report, is in the treatment of aponeurotic fibromatosis, this rare neoplasms, albeit benign, poses a significant challenge for its local invasiveness and tumor location.[21]

Mast cell tumor Adjuvant bleomycin-based ECT was used in 28 dogs with incompletely excised mast cell tumors. Overall response rate was 85%. The median time to recurrence was not reached. Estimated time to recurrence was 52.76±6.5 months.[22] One dog had wound dehiscence. In 2009 Kodre et coll. compared square pulses based ECT to surgery alone in the treatment of mast cell tumors, using intralesional cisplatin, and claiming a longer control with ECT.[23] Another study, enrolling 37 dogs, evaluated the efficacy of cisplatin for the treatment of incompletely excised canine mast cell tumors obtaining a mean disease-free interval of 1218 days with 78% of the patients experiencing local control at different times.[24]

Squamous cell carcinoma a small cohort of 9 cats with sun induced squamous cell carcinoma of the nasal planum were treated with intralesional bleomycin coupled with ECT with a rate of complete responses of 77%.[25] A more recent article describes the use of square pulses based ECT for superficial squamous cell carcinoma in cats, reporting a 81% of complete responders.[26]

CURRENT PROTOCOLS

Electrochemotherapy is mostly used as a adjuvant therapy, locally injecting the chemotherapy drugs bleomycin or cisplatin and then following with the application of permeabilizing pulses. For larger tumors drugs can be administered locally or systemically, mostly on the basis of tumor volume, firmness or resistance to insertional electrodes.[12] Caution should be exerted when using electrochemotherapy in patients previously treated with radiation therapy due to the risk of radiation recall.[27]

ELECTROCHEMOTHERAPY IN OTHER SPECIES

ECT has been used successfully in horse to treat sarcoids using intralesional bleomycin and to treat melanomas using cisplatin as chemotherapy agent, either directly or combined with surgery. The treatment has been well tolerated and resulted in high percentage and durable responses. [28-30]

ECT has been used to treat tumors in exotic animal as well. Our group described the usefulness of adjuvant ECT in the treatment of advanced cutaneous tumors in rabbits and in a turtle.[7,31]

ELECTROGENETHERAPY

The first report of electrochemogene therapy in veterinary cancer patients has been presented by Draghia and colleagues in 2002. The authors investigated the possibility to palliate several cancer induced complications such as cachexia, anemia, anorexia, and decreased activity levels that many veterinary patients experienced by delivering a plasmid encoding for insulin-like growth factor-I.[32] More recently, electrochemogene therapy using plasmids encoding for IL-12 has been successfully investigated in small cohorts of dogs with a variety of spontaneous neoplasms.[33] Finally a recent prospective study on canine mast cell tumor recruited eight dogs for a total of 11 lesions that received from one to four sessions of electrochemogene therapy with IL-12.[34] The treatment was well tolerated but the tumor response was extremely variable.

IRREVERSIBLE ELECTROPORATION

This technique aims at inducing focal tumor destruction by using direct tumor ablation through the delivery of electric pulses at higher voltage than RE (used in ECT), alone or combined with chemotherapy. Two reports have been published so far describing the successful treatment of a canine glioma patient and of a canine patient with a large histiocytic sarcoma. [35,36] These two reports are very promising and warrant further investigation to confirm the efficacy of this approach.

CONCLUSION

Electroporation in its different applications (irreversible and reversible) is a safe and efficacious approach to veterinary solid tumors. Its low cost and ease of administration makes it a valuable addition to the currently available oncological therapies. One advantage of this technique is the possibility of repeated treatments in case of local recurrence and the recruitment of the patients' immune system, as suggested by the low incidence of metastatic disease among most treated patients and by the tumor selection observed among some pets with recurring disease.

FUTURE DIRECTIONS

Ongoing studies are evaluating the efficacy of ECT when other drugs than bleomycin are used, in particular anthracyclines and their analogues (mitoxantrone) and platinum compounds such as CDDP and carboplatin. Studies are being performed to characterize the membrane alterations induced by trains of biphasic pulses in order to improve the efficacy of the technique.[37] Novel electrochemotherapy equipments adopting different electrical parameters are currently being developed. Finally, our group is assessing the potentials for electrogenetherapy using electric pulses to drive chemosensitizing plasmids within tumor cells as well as participating to the transfer of this technique to human patients.

REFERENCES

[1] Keet, Gehl and Lee (Eds) Clinical Aspects of Electroporation, Springer Ed, 2011. [2]Gothelf et al. **Cancer Treat Rev** 2003; 29: 371. [3]Belehradek et a. **Cancer** 1993; 72: 3694. [4]Daskalov et al.**IEEE Eng Med Biol** 1999 ; 18 : 62. [5]Gehl et al. **Melanoma Res** 2000; 10: 585. [6]Mir et al. **Eur J Cancer** 2006 ; 4(S) : 14. [7] Spugnini et al. **J Cell Physiol.** 2014;229:1177-81. [8].Mir et al. **Br J Cancer** 1997 ; 76 : 16. [9]Tozon et al. **Anticancer Res** 2001; 21: 2483. [10]Rols et al. **Bioelectrochem** 2002 ; 55 : 101. [11]Tozon et al. **Anticancer Res** 2005 ; 25 : 839. [12] Spugnini et al. **Methods Mol Biol**. 2014;1121:247-56. [13] Spugnini et al. **J Exp Clin Cancer Res** 2003 ; 22 : 571. [14]Spugnini et al. **J Exp Clin Cancer Res** 2005; 24 : 245. [15]Spugnini et al. **Melanoma Res** 2006; 16 : 23. [16]Spugnini et al. **Cancer Chemoter Pharmacol**; 2007; 59 : 375. [17]Spugnini et al. **Vet Rec** 2006; 159 : 598. [18]Spugnini et al. **In Vivo** 2007; 21: 819.[19]Spugnini et al. **Vet Rec** 2008; 162: 186. [20] Spugnini et al. **J Transl Med** 2011; 9; 152. [21]Spugnini et al. **Can Vet J**. 2013; 54:606-9 [22]Spugnini et al. **Anticancer Res** 2006; 26: 4585. [23] Kodre et al**. In Vivo**. 2009; 23:55-62. [24]Spugnini et al. **J Vet Intern Med** 2011; 25: 407.[25] Spugnini et al. **Vet J** 2009; 179: 117. [26]Tozon et al. **J Feline Med Surg**. 2014;16:291-9. [27] Spugnini et al. **In Vivo** 2008; 22: 751-53.. [28]Tamzali et al. **Equine Vet J** 2012; 44:214. [29]Spugnini et al. **J Equine Vet Sci** 2011; 31: 430: [30] Scacco L et al. **J Equine Vet Sci** . 2013; 33: 637-639. [31] Lanza et al. **J Am Vet Med Assoc** in press. [32]Draghia et al**. Mol Ther** 6:830–836. [33]Reed et al . **Cancer Gene Ther** 17:457–464. [34]Pavlin et al. **Radiol Oncol** 45:31–39.[35]Garcia et al. **Technol Cancer Res Treat**. 2011;10:73-83. [367]Neal et al. **J Clin Oncol**. 2011; 29):e372-7.[37]Spugnini et al. **Microsc Res Tech** 2007; 70: 1041.

POLYUNSATURATED FATTY ACIDS AND CANCER: FRONTIERS FOR CURE AND CONTROL?

Gregory K. Ogilvie, DVM, DACVIM (SAIM & Oncology), DECVIM-CA (Oncology)
Carlsbad, CA

Nutritional therapy is a key component for the prevention and treatment of cancer and cancer cachexia. Nutrients are potent substances that not only provide nutrients, but they can and do actually treat medical disorders, including cancer. Nutritional therapy and supplements are in a non regulated area: some take liberty in overstating their benefits. Thus, nutritional therapy and supplements, like all other treatments should be placed into those that work, those that are being evaluated, and those that do not work. The later should be excluded from use. Specific nutrients supplents can be used as powerful tools to reduce toxicity associated with chemotherapy, radiation therapy, and is important to enhance healing subsequent to surgery. There is little doubt that nutritional intervention must begin early and must be followed through aggressively to gain maximum benefit.....long before the patient exhibits evidence of weight loss, debilitation, or anorexia begin which can in turn enhance response to therapy and improve quality of life.

CANCER AND ITS EFFECTS ON THE BODY: TARGETS FOR NUTRITIONAL INTERVENTION

Cancer causes dramatic changes in metabolism resulting in elevated lactate concentrations, especially when the patient is given simple carbohydrates orally or intravenously, alterations and protein metabolism resulting in protein wasting to the benefit of the host, and alterations in lipid metabolism. In fact, fat loss accounts for the majority of weight loss occurring in cancer cachexia. Therefore, it is not surprising that human beings and animals with cancer have dramatic abnormalities in lipid metabolism.[1-10] The decreased lipogenesis and increased lipolysis observed in humans and rodents with cancer cachexia result in increased levels of free fatty acids, very low density lipoproteins, triglycerides, plasma lipoproteins, and hormone dependent lipoprotein lipase activity, while levels of endothelial derived lipoprotein lipase decrease.[1-10] Recently lipid profiles in dogs with a lymphoma were studied.[1-2] It was determined that many of the alterations seen in other species with cancer were also present in dogs. These abnormalities did not normalize when clinical remission is obtained. The clinical significance of these abnormal lipid profiles in dogs with lymphoma is not known, however, abnormalities in lipid metabolism have been linked to a number of clinical problems including immunosuppression.

The clinical impact of the abnormalities in lipid metabolism may be lessened with dietary therapy. In contrast to carbohydrates and proteins, some tumor cells have difficulty utilizing lipid as a fuel source while host tissues continue to oxidize lipids for energy.[1-3] This has led to the hypothesis that diets relatively high in fat may be of benefit to the animal with cancer when compared to a diet high in simple carbohydrates. Further research may reveal that the type of fat, rather than the amount, may be of greater importance. This has resulted in one product, Impact that has been reported to benefit mean nitrogen intake, nitrogen balance, in vitro lymphocyte mitogenesis, time for wound healing, the prevalence of wound complications, and the duration of hospitalization.[1-3] Studies of polyunsaturated fatty acids (PUFA's) of the n-3 series, especially eicosapentaenoic (EPA) and docosahexaenoic acid (DHA), indicate that these fatty acids may prevent the development of carcinogen-induced tumors, the growth of solid tumors, as well as the occurrence of cachexia and metastatic disease in experimental tumor models. Fatty acids of the n-3 series have been shown to normalize elevated blood lactic acid and insulin levels in non-malignant conditions. In contrast, PUFA's of the n-6 series appear to enhance tumor development and metastases. These data, along with the epidemiological findings of an inverse relationship between dietary n-3 fatty acid intake and incidence of some cancer, is the basis of research to evaluate the potential benefit of n-3 fatty acids in the prevention of cancer cachexia and therapy of malignancy in cancer patients.

NUTRITION AND CANCER FOR THE CLINICIAN: THE EVIDENCE

A study was recently completed in dogs with lymphoma.[14] A double blind, randomized study was recently reported to evaluate the hypothesis that polyunsaturated n-3 fatty acids and arginine can improve metabolic parameters, decrease chemical indices of inflammation, enhance quality of life, and extend disease-free interval and survival time in dogs treated for lymphoma. In that study, dogs fed the experimental diet had significantly higher serum levels of polyunsaturated n-3 fatty acids docosahexaenoic acid (C22:6) and eicosapentaenoic acid (C20:5) as well as arginine when compared to controls. Both diets were formulated to be relatively low in simple carbohydrates, with moderate amounts of highly bioavailable proteins. This formulation is designed to enhance the effect of n-3 fatty acids. Higher serum levels of these n-3 fatty acids were associated with lesser plasma lactic acid responses to intravenous glucose and diet tolerance testing. Increasing C22:6 levels was significantly associated with longer disease free interval and survival time for dogs with stage III lymphoma fed the experimental diet. An update is noted above.

Another study was recently completed that was designed to determine the effect of a diet supplemented with n-3 fatty acids and arginine on irradiated skin and oral mucosa, carbohydrate metabolism and quality of life in a group of dogs with nasal tumors (Vet Cancer Soc Proceedings 1998). This study showed that fatty acids of the n-3 series normalize elevated blood lactic acid. In a dose dependent manner, n-3 fatty acids result in decreased histologic evidence of radiation damage to skin and mucosa and improve performance scores in dogs with malignant nasal tumors. This would obviously be of great benefit to the cancer patient receiving radiation therapy. These two studies confirm that diets supplemented with polyunsaturated n-3 fatty acids are of benefit for the cancer patient.

The bottom line is that n-3 fatty acids in moderate amounts appear to benefit the cancer patient. More specifically, a diet relatively high in n-3 fatty acids and relatively now in simple carbohydrates has been shown not only to improve alterations in metabolism associated with cancer, but also improve response to chemotherapy and decrease the adverse effects associated with radiation therapy.

One might wonder about the toxicity of using high dosages of n-3 fatty acids and the cancer patient. Our research has shown that the use of relatively high dosages of n-3 fatty acids does not cause abnormal prolongation of platelet aggregation or delayed wound healing. We recently completed a double blind, randomized study to test the hypothesis that an n-3 supplemented diet would not cause significant alternations in hemograms or select biochemical profile parameters when compared to a placebo diet not supplemented with polyunsaturatedn-3 fatty acids. Thirty-four dogs with lymphoma and 8 dogs with hemangiosarcoma were randomized to receive one of two diets supplemented with menhaden fish oil and arginine (experimental diet) or an otherwise identical diet supplemented with soybean oil (control diet). Diets were fed for 18 weeks before and after remission were attained with up to 5 dosages of doxorubicin in the case of dogs with lymphoma, and after surgery and for 18 weeks during and after doxorubicin in the case of the dogs with hemangoisarcoma. Parameters examined included hemograms and select biochemical parameters (glucose, BUN, creatinine, phosphorus, calcium, total protein, albumin globulin, A/G ratio, cholesterol, t-bilirubin, sodium, potassium, chloride, bicarbonate, anion gap). There were no significant differences in weights or kcal consumption between groups. The only significant difference occurred for serum cholesterol. Dogs that received the n-3 supplemented diet had significantly lower serum cholesterol concentrations than dogs fed the control diet. Dietary polyunsaturated fatty acids have been shown to reduce serum cholesterol concentrations in many species, and may be important in those disease conditions associated with increased serum cholesterol. Changes in serum cholesterol may reflect changes in cell membrane lipid content that can affect tumor cell metastasis or immune cell responsiveness to malignant cells.

PUFA'S, CANCER AND MMP'S

It appears that fatty acids may regulate matrix metalloproteinases. In order to understand this further, a review of matrix metalloproteinases is in order. Matrix metalloproteinases (MMPs) are a family of zinc dependant enzymes that are produced by multiple tissues throughout the body, and are involved in the degradation of extracellular matrix and basement membrane components. There are currently at least seventeen unique MMPs characterized[3] that can be sub-divided into groups according to substrate preferences: collagenases, stromelysins, gelatinases, and membrane bound enzymes. Matrix metalloproteinases are active during normal tissue remodeling processes such as embryogenesis and wound healing, as well as pathologic processes such as rheumatoid arthritis, degenerative joint disease, cancer metastasis and invasion, and angiogenesis. MMP-2 and MMP-9 have been associated with malignant tumor progression and metastasis. For tumor cells to become metastatic, they must cross at least two basement membranes to enter and exit the vasculature at the primary and metastatic sites. MMP's 2 and 9 are thought to be important in this process, as well as facilitating tumor invasion into surrounding normal tissues. These enzymes can be present in a progelatinase and an active gelatinase form.C In several human tumor types, levels of MMP's 2 and 9 have been found to be increased when compared to normal or benign tissue. These include breast, prostate, colon, gastric, and bladder cancers. Studies done with rodent colon, lung and breast tumor models show that the inhibition of MMPs can lead to a more favorable outcome.

This knowledge led us to hypothesis that the presence of MMP 2 and 9 in the tumor, serum and plasma correlated with the presence of canine malignancies and these levels correlate with histologic grade of tumor. In this study tumor, plasma, serum and stromal tissue were harvested from dogs with various histologically confirmed and graded malignancies prior to therapy and banked at -70C and subsequently analyzed by gelatin zymography to detect MMP 2 and 9 activity compared to commercially available MMP 2 and 9 human standards. Dogs with measurable levels of active form MMP 9 were over 3 times more likely to come out of remission than subjects who had zero levels. Dogs with the active form of MMPs present were, in general, significantly more likely to come out of remission by 90 days than subjects with no active form of MMPs present. This was especially true when active form of MMPs were present in dogs with lymphoma. In general, dogs with stage IV lymphoma had the highest levels of MMP's and had the poorest outcome compared to dogs with stage III disease. These data allow some to draw the conclusion that MMP inhibitors may indeed be of value in treating dogs with cancer.

In order to determine the ability of n-3 fatty acids to inhibit matrix metalloproteinases, we performed a blind, randomized study designed to evaluate the hypothesis that short-term supplementation with polyunsaturated n-3 fatty acids can decrease serum, plasma and tumor concentrations of MMP 2 and 9, VEGF, and lactate and and enhance concentrations of tissue inhibitors of MMP's (TIMP-2). Nineteen dogs with osteosarcoma (9) and soft tissue sarcomas (10) were randomized to receive one of two diets identical in all ways except the experimental diet was supplemented with menhaden fish oil while the otherwise control diet was supplemented with soybean oil. Diets were fed prior to biopsy and for 2-3 weeks until the tumor was biopsied again. Dogs fed the experimental diet higher median serum levels of the n-3 fatty acids docosahexaenoic acid (C22:6) and eicosapentaenoic acid (C20:5) and lower concentrations of the n-6 fatty acids, 18:2 and 20:4 when compared to controls. Dogs fed the control diet had higher glucose concentrations and lower plasma concentrations of pro and active MMP-2 and lactate. In addition, the concentration of pro MMP 9 and active MMP 2 decreased in the experimental group and increased in the control group. Concentrations of TIMP-2 and VEGF increased in both groups.

"OTHER" NUTRIENTS AND CANCER

Zinc deficiency is a common finding in human cancer patients, especially those with head and neck cancer. Cell-mediated immunity and natural killer (NK) cell activity are sub-optimal in zinc-deficient patients. Zinc-deficient cancer patients experience increased treatment morbidity, treatment delays, and unplanned hospitalizations. Chromium deficiency is a causative factor in decreased glucose tolerance, a recognized paraneoplastic syndrome. It can lead to hyperinsulinemia and hyperlactatemia, both of which have been observed in human and animal cancer patients. Iron metabolism is altered in the presence of inflammation or neoplasia. Most commonly, metabolism is shifted toward sequestration, which is manifested by a decrease in serum iron concentration and total iron binding capacity, and an increase in serum ferritin concentration. Iron sequestration may be the mechanism for anemia of chronic disease because of difficulty using iron in a storage pool.

Iron transferrin and ferritin have been linked to cancer risk and cancer cell growth.[67] Lung cancer, colon, bladder and esophageal cancer in people has been highly correlated with increased serum iron and increased transferrin saturation.[67] This may be because many tumor cells require iron for growth

One key question comes up often : should antioxidants be used to support the cancer patient. One answer to this question come from the knowledge that polyunsaturated fatty acids (PUFA) increase the sensitivity of breast cancer cells to cytotoxic drugs that generate an oxidative stress and found this effect to be abolished by vitamin E, suggesting the involvement of lipoperoxides in cell death. The most efficient PUFA was docosahexaenoic acid (DHA). Since an enrichment with n-3 PUFA of the diet of breast cancer patients has been reported to lead to an increased level of DHA in breast adipose tissue (*Bagga et al, J Natl Cancer Inst 1997, 89: 1123*), a dietary intervention should provide an effective means to increase DHA availability in tumor tissues and thereby should increase drug efficacy on tumors (*Shao et al, Nutr Cancer, 1997, 28:63*). This circumstantial evidence now require direct demonstration through a clinical trial.

REFERENCES

1. Ogilvie GK, Vail DM. Metabolic alterations and nutritional therapy for the veterinary cancer patient. In Withrow SJ, MacEwen EG, Clinical Veterinary Oncology, Philadelphia, WB Saunders, 1996.
2. Ogilvie GK, Moore AS. Nutritional Support, In: Managing the Veterinary Cancer Patient: A Practice Manual. Veterinary Learning Systems, 124-127, 1995.
3. Ogilvie GK. Paraneoplastic sundromes. In Ettinger S, Feldman E (eds). Textbook of Veterinary Internal Medicine (fourth edition). WB Saunders, Philadelphia, In Press 1998.
4. Vail DM, Ogilvie GK, Wheeler SL, Fettman MJ t al: Alterations in carbohydrate metabolism in canine lymphoma. J of Vet Intern Med 4:8-11, 1990.
5. Vail DM, Ogilvie GK, Fettman MJ, Wheeler SL: Exacerbation of hyperlactatemia by infusion of lactated Ringer's solution in dogs with lymphoma. J Vet Intern Med 4:228-232, 1990.
6. Ogilvie GK, Vail DM, Wheeler SJ. Effect of chemotherapy and remission on carbohydrate metabolism in dogs with lymphoma. Cancer 69:233-238, 1992.
7. Ogilvie GK, Walters LM, Salman MD, et al. Alterations in select aspects of carbohydrate, lipid and amino acid metabolism in dogs with non-hematopoietic malignancies, Am J Vet Res, 8:62-66, 1994.
8. Ogilvie GK, Ford RD, Vail DM et al: Alterations in lipoprotein profiles in dogs with lymphoma. J Vet Intern Med. 1994;8:62-66.
9. OgilvieGK, Fettman MJ, Mallinckrodt CH, Walton JA, Hansen RA, Davenport DJ, Gross KL, Richardson KL, Rogers Q, Hand MS. Effect of fish oil and arginine on remission and survival time in dogs with lymphoma. Cancer, Submitted 1998.
10. Ogilvie GK, Salman MD, Fettman MJ et al. Effect of anesthesia and surgery on energy expenditure determined by indirect calorimetry in dogs with malignant and non-malignant condition. Am J Vet Res, 1996;57:1321-326. 80. 80. Ogilvie GK, Walters LM, Fettman MJ, et al. Energy expenditure in dogs with lymphoma fed two specialized diets. *Cancer* 1993;71:3146-3152
11. Ogilvie GK, Salman MD, Fettman MJ et al. Resting energy expenditure in dogs with nonhematopoietic malignancies before and after excision of tumors. Am J Vet Res, 1996;57:1463-1467.
14. Ogilvie GK, Fettman MJ, Mallinckrodt CH et al. Effect of fish oil, arginine, and doxorubicin chemotherapy on remission and survival time for dogs with lymphoma: a double-blind, randomized placebo-controlled study. Cancer 2000 15;88(8):1916-1928.

HYPERBARIC OXYGEN THERAPY FOR THE VETERINARY CANCER PATIENT

A. Elizabeth Hershey, DVM, DACVIM (Oncology), CVA
Phoenix, AZ

INTRODUCTION

Tissue hypoxia has proven to be a causative factor in cancer progression. Many solid tumors will contain areas of acute or chronic hypoxia[1]. The severity of hypoxia can vary both within and among different tumor types[2]. Tumor cells ability to adapt to the hypoxic microenvironment can allow tumor cells to survive and proliferate in this hostile milieu[3]. Multiple reports have demonstrated that decreased oxygen tension selects for more aggressive cells and induces multiple cellular adaptations which can sustain and foster cancer progression and growth[4]. Hypoxia is reported to result in cellular responses which improve oxygenation and viability through induction of angiogenesis, alteration in metabolism by increased glycolysis and upregulation of genes involved in cell survival and apoptosis[5]. Surprisingly, tumor angiogenesis does not appear to be enhanced in cancerous tissues treated with HBOT. In fact, HBOT was shown to induce an antiangiogenic effect in 2 mammary tumor models[6, 7, 8] and a glioma model[9]. Hypoxia has also been shown to increase genetic instability, activate invasive growth and preserve the undifferentiated cell state[1,3]. Hypoxic conditions can also lead to development of a metastatic phenotype and contribute to chemotherapy or radiation resistance[10].

The delivery of increased oxygen to the tissues with hyperbaric oxygen therapy (HBOT) can potentially be used to overcome hypoxia in tumor tissue. As in normal tissue, pO2 in cancer tissue increases significantly during HBOT[11]. Elevation of the tumor oxygen pressure has been shown to be preserved clinically for approximately 30 minutes after HBOT[12,13]. Thus it is hypothesized that HBOT can improve or overcome chemotherapeutic resistance by increasing both tumor perfusion and cellular sensitivity. In a mammary tumor model, researchers found that the uptake of chemotherapy is increased for the duration and immediately after HBOT[14]. In one study, HBOT increased the tumorocidal action of doxorubicin on rat sarcoma cells in vitro[15]. A lower dose of doxorubicin when combined with HBOT could obtain the same effect as higher dose doxorubicin alone. The chemosensitizing effect of HBOT was reproduced in a metastatic rat lung tumor model. In this model, HBOT plus doxorubicin significantly decreased the number of lung metastases and the lung weight relative to doxorubicin or HBOT alone. Other chemotherapy agents that have shown improved efficacy in vivo when combined with HBOT include paclitaxel, carboplatin and cisplatin[16,17].

HBOT has emerged in human oncology as one of the most effective treatments for late term radiation effects. HBOT has been effectively used to treat radiation-induced damage to the head, neck, chest wall, abdomen and pelvis. HBOT may prevent tooth loss or collapse of the jaw bone in patients previously treated for head or neck cancers[18], promote successful skin grafts or flaps following reconstructive surgery[19], and eliminate radiation hemorrhagic cystitis[20]. The effect of HBOT as a radiation sensitizer is still unclear.

Recent research has demonstrated that T cells "avoid" going into the hypoxic areas of tumors. To overcome this problem in mice with lung tumors, the investigators had animals breathe supplemental oxygen at 60%. Having a higher concentration of oxygen throughout the body improved oxygenation inside the tumors allowing immune cells to enter the tumors and attack them, thereby extending the animals' survival. The possibility exists that the use of hyperbaric oxygen therapy might enhance the efficacy of immunotherapies.

CLINICAL APPLICATION OF HBOT IN VETERINARY ONCOLOGY

HBOT can be a useful adjunct in treating the veterinary cancer patient. The ideal protocol for treating cancer patients is not known and may vary depending on the goal of treatment (pain control, tumor reduction, chemosensitization). Recommended protocols may vary depending on what an owner's schedule and finances will allow. Unlike acute conditions such as pancreatitis or snake envenomation with a planned finite number of treatments, we consider HBOT more of a long term chronic therapy for many cancer patients. In our limited experience, HBOT appears to be safe for long term treatment. We have one patient who has been receiving HBOT 1-2 times weekly for a nasal chondrosarcoma for over 1 year. (at time of writing this patient has received 100 treatments).This patient had previously received stereotactic radiation therapy 2 months prior to starting HBOT. This patient's severe hip arthritis has also improved with HBOT. This patient recently started treatment with the Morphogenesis ImmuneFx vaccine, an autologous cancer vaccine, in combination with HBOT.

We are investigating whether chemotherapy agents may be safely combined with HBOT. Chemotherapy, whether oral or intravenous is administered immediately after completion of HBOT, generally within 5-10 minutes, while tumor pO2 is still theoretically high. There is evidence in humans that combining HBOT with chemotherapy may potentiate the toxicity of the chemotherapy drug. To date we have combined metronomic cyclophosphamide, doxorubicin or carboplatin chemotherapy with HBOT with no complications. When administering chemotherapy drugs (with the exception of metronomic dosing), the dose of a drug is initially administered at a 50% dose reduction. If no side effects are noted, the dose of the chemotherapy is escalated by 25-35% each treatment until the full standard dose is reached or until side effects are noted (gastrointestinal or neutropenia). In 3 dogs treated with doxorubicin, the highest dose that could be administered concurrently with HBOT was 25 mg/m2. We are currently investigating carboplatin administered concurrently with HBOT in several dogs with various cancers.

We have treated several tumor types, including brain tumors (trigeminal nerve sheath tumor and glioma), nasal tumors, splenic hemangiosarcoma, osteosarcoma, metastatic carcinosarcoma and lymphoma. Most patients tolerate the treatment well with no complications. Spinal and brain tumors are treated at 1.5 ATA for 60 minutes. All other tumor types are treated at 2.0 ATA for 60 minutes. Brain tumor patients with severe neurological symptoms (circling, pacing, and falling repeatedly) can pose challenges to avoid self-trauma during the course of treatment. A lymphoma patient with severe cellulitis and tissue necrosis from a doxorubicin extravasation healed successfully with 10 treatments of HBOT.

In summary, HBOT is safe for cancer patients and does not appear to pose a threat of enhanced tumor growth. The potential benefits include palliation of pain, enhancement of chemotherapy efficacy, and possibly tumor reduction or control. The ideal protocol for the treatment of cancer is not known and may vary with treatment goal.

REFERENCES

1. Michieli P. **Cell Cycle** 2009:8:3291.
2. Vaupel P et al. **Cancer Metastasis Rev.** 2007:26:225.
3. Harris AL. **Nat Rev Cancer** 2002: 2: 38.
4. Moen I et al. **Targ Oncol 2012 7:233. (review)**
5. Holmquist L et al. **Adv Exp Med Biol.** 2006: 587:179.
6. Raa A et al. **BMC Cancer.** 2007:7:23.
7. Moen I et al. **PLoS One** 2009:4:e6381.
8. Moen I et al. **BMD Cancer.** 2012:12:21.
9. Stuhr LE et al. **J Neurooncol.** 2007:85:191.
10. Shannon AM et al. **Cancer Treat Rev.** 2003:29:297.
11. Brizel DM et al. **Br J Cancer.** 1995: 72:1120.
12. Kinoshita Y et al. **Br J Cancer.** 2000: 82:88.
13. Beppu T et al. **J Neurooncol.** 2002: 58:47.
14. Moen I. **BMC Cancer.** 2009:9:446.
15. Petre PM et al. **J Thorac Cardiovasc Surg** 2003:125:85.
16. Ohguri T et al. **Int Hyperthermia.** 2009:25: 160.
17. Selvendiran K et al. **Cancer Biol Ther.** 2010:10:386.
18. Bennett HM et al. **Cochrane Database Syst Rev.** 2012.
19. Gupta P et al. **Indian J Otolaryngol Head Neck Surg**. 2013: 65:140.
20. Nolen D et al. **Head Neck.** 2014: 36: 1701.
21. Norkool DM et al. **J Urology.** 1993:150:332.
22. Hatfield SM et al. **Sci Transl Med** 2015: 7:1 (immunotherapy)

HYPERBARIC OXYGEN THERAPY FOR THE VETERINARY INTERNIST

Andrew A. Turkell, DVM, CVA, CCRT
Boca Raton, Florida

There are over 1,000 Human Hyperbaric centers throughout the United States, most of which have multiple chambers. That's thousands of chambers treating people every day. The FDA has only approved 14 uses for hyperbarics that will be reimbursed by insurance and still there are thousands of human chambers around the country but less than 50 veterinary chambers. For this reason veterinarians must look at the science of human hyperbarics for direction. Also, the vast majority of veterinarians understands and can predict the astonishing connection between human and animal health. They can understand why whales get breast cancer, why koala bears get STDs, why pandas develop eating disorders and why flamingos get heart attacks? Veterinarians understand that no medical problem is uniquely human…and neither is any treatment.

The Undersea and Hyperbaric Medical Society (UHMS)-an organization representing physicians, nurses and technicians in the field of hyperbaric medicine-met with the Food and Drug Administration (FDA) and recommended HBOT for 14 specific conditions. These 14 conditions and these ONLY are listed for reimbursement by insurance companies. These conditions are eased or aggravated by reduced oxygen level in body tissue. More than 30 years later, those approved conditions remain much the same along with the problems that has stymied the forward progress of HBOT. Recent research has shown that HBOT can help treat a variety of other non-approved health problems; however, because of insurance regulations, institutional politics and other bureaucracy, many science supported indications remain unapproved. Veterinary medicine may be one key to developing the science of HBOT further and faster.

14 FDA APPROVED USES FOR HYPERBARIC OXYGEN THERAPY (HBOT)

1. HBOT for Air or Gas Embolism

An air or gas embolism is caused by air in the arteries caused by diving or an invasive medical procedure that punctures an artery or lung.

2. HBOT for Carbon Monoxide Poisoning

Carbon monoxide poisoning or CO poisoning complicated by cyanide poisoning. Also includes poisoning from methylene chloride.

3. HBOT for Gas Gangrene

Clostridial myositis and Myonecrosis.

4. HBOT for Crush injury

Acute ischemias, usually caused by heavy equipment

5. HBOT for Decompression Sickness

Decompression sickness is one of the first conditions treated with HBOT. It is brought on when a diver ascends too quickly and does not allow the oxygen in the body to expand at a safe rate.

6. for Arterial Insufficiencies

Including problem wounds; this category includes diabetic foot ulcers, which afflict one out of five people with type 2 diabetes and can lead to amputation. Recently added to this category was central retinal artery occlusion, or "stroke of the eye," from blockages in the arteries of the eye.

7. HBOT for Severe Anemia

Severe Anemia is any acute, severe blood loss, such as from a wound on the battlefield or severe trauma.

8. HBOT for Intracranial Abscess

An accumulation of infected material; these abscesses of the brain are common in patients with abnormal immune systems.

9. HBOT for Necrotizing soft tissue infections

From "flesh-eating bacteria," these severe infections usually progress rapidly.

10. HBOT for Osteomyelitis

Chronic bone infections that resist standard treatment most common in the lower leg after severe trauma.

11. HBOT for Delayed Radiation Injury

Radiation damages blood vessels, and the lack of blood supply eventually can cause wounds to form in soft tissue and bone.

12. HBOT for Compromised Skin Grafts and Flaps

Grafts and flaps of skin and other tissue (cartilage, bone, fat) are used in reconstructive surgery (such as breast reconstruction after a mastectomy). In some cases, blood supply to the graft or flap is compromised, causing complications.

13. HBOT for Thermal burn Injury

From fire or heat.

14. HBOT for Idiopathic Sudden Sensorineural Hearing Loss

Classically defined as a hearing loss of at least 30 dB over at least three contiguous frequencies - occurring within three days. This is the latest approved indication added by UHMS Board of Directors late 2011.

The 14 above indications have been detailed in physiology, medicine, and clinical evidence behind the commonly accepted indications for hyperbaric oxygen therapy (HBOT).

The following indications have been introduced due to a growing body of evidence in support of using HBOT, or where the application of HBOT as a routine treatment is already advocated by some practitioners. For some of these indications, accumulating evidence may continue to support the effectiveness of HBOT, whereas for others, it appears more likely that further evidence will confirm that HBOT does not improve clinical outcome. These indications have summarized clinical evidence known as systematic reviews. Systematic reviews maybe defined as a literature review that focuses on a specific question and that tried to identify, appraise, select and synthesize all high-quality clinical research evidence relevant to that question. These reviews should then present clear conclusions and recommendations consistent with the evidence located and appraised.

If there are no conclusive truly high quality trials, then it is appropriate that a review examine in detail the highest quality evidence available. It is likely, however that such reviews may only be able to draw moderate recommendations for practice.

1. Stroke

   Hyperbaric oxygenation as an adjunct therapy in strokes due to thrombosis.
   A review of 122 patients
   American Stroke Association
   A Division of American Heart Association
   "Discussion…It does suggest that a significant number of our patients with completed stroke may benefit from HBOT".

   There have been more than 30 human studies showing that HBOT can minimize injury and improve outcome if used shortly after a stroke. Other research shows that it can improve function even years after a stroke-particularly in patients who had strokes that don't involve paralysis.

   Strokes in veterinary medicine
   Ischaemic strokes have been associated with many medical conditions in dogs and cats: kidney disease, heart disease, under or over-active thyroid glands, Cushing disease, diabetes and hypertension. Other less common causes of blockage of the blood vessels supplying the brain include clogging by a fragment of tumor, fat, parasites or spinal cartilage.

   Hemorrhagic strokes can be seen with diseases that interfere with blood clotting as seen with some rodent poisons (warfarin-like products), immune-mediated thrombocytopenia, and congenital clotting diseases, disease causing high blood pressure (kidney disease, heart disease, Cushing's or thyroid disease), inflammation of the arteries (vasculitis) or abnormal development of the blood vessel in the brain. Other causes of bleeding in the brain include head trauma, bleeding from a brain tumor or from a tumor spreading to the brain (especially common with tumors of the spleen). Many of these may benefit from Hyperbaric Oxygen Therapy (HBOT).

2. Hyperbaric Oxygen Treatment of Postoperative Neurosurgical Infections.
   39 consecutive patients
   Larsson, Engstrom et al. Stockholm Sweden

   Objective: To evaluate the clinical usefulness of hyperbaric oxygen therapy for neurosurgical infections after craniotomy or laminectomy

   Conclusion: HBOT is a safe, powerful treatment for postoperative cranial and spinal wound infections and should be included in the neurosurgical armamentarium.

3. Hyperbaric Oxygen Treatment for Inflammatory bowel disease: a systemic review analysis
   Rossignol Review 2012.

   Thirteen studies of HBOT in Crohn's disease and 6 studies in ulcerative colitis were identified. In all studies participants had severe disease refractory to standard medical treatments, including corticosteroids, immunomodulators and anti-inflammatory medications. In Crohn's disease 31/40 (78%) had clinical improvements with HBOT, while all 39 patients with ulcerative colitis improved.
   Daniel Rossignol.

   Conclusions: HBOT lowered markers of inflammation and oxidative stress and ameliorated IBD in both human and animal studies. Most treated patients were refractory to standard medical treatments. Additional studies are warranted to investigate the effects of HBOT on biomarkers of oxidative stress and inflammation as well as clinical outcomes in individuals with IBD.

4. Hyperbaric Oxygen Therapy as a treatment for stage-1 avascular necrosis of the femoral head
   In Veterinary Medicine this is Legg Calve Perthes Disease.
   The Journal of Bone and Joint Surgery, N.D.Reis, et.al. Rambam Medical Center, Haifa, Israel.

   Avascular necrosis (AVN) is a potentially crippling disease affecting mainly young adults. This study selected 12 patients suffering from Steinberg stage-1 AVN of the femoral head whose lesions were 4 mm thick or more. 81% of the patients who received HBOT showed a return to normal on MRI compared to 17% in the untreated group.

5. Stem Cell Mobilization by Hyperbaric Oxygen
   Steven Thom, et.al. 2005.

University of Pennsylvania Medical Center, Philadelphia, PA.

In over a course of 20 HBOT treatments, circulating CD-34 cells increased eightfold.
It was concluded that HBOT mobilizes stem/progenitor cells by stimulating nitric oxide synthesis.

6. Potential Effects of Hyperbaric Oxygen Therapy in Acute Pancreatitis
Christine Cuthbertson, et.al. University of Melbourne, Australia.

Object: To extract from published reports the effects of HBOT on inflammatory disease, in particular acute pancreatitis.
Conclusions: Acute pancreatitis impairs the pancreatic and systemic microcirculation and causes acute inflammation (20-30% develop pancreatic necrosis). These processes are potentially improved by HBO therapy.

7. Hyperbaric Oxygen: Its application in cardiology: a historical perspective and personal journey
Ellestad, et.al. Cardiology Review 2009.

Hyperbaric oxygen (HBO) was first used by placing a patient with the "bends" in a pressure chamber and increasing the pressure to 3 atm (atmospheres) with 100% oxygen. It was soon recognized that HBO could also facilitate wound healing in infected ischemic extremities. Before the pump oxygenator was developed, it was also shown to facilitate corrective surgery, especially in congenital heart disease. In 1997, after encouraging animal work, HBO therapy was used in conjunction with thrombolytics to reduce the magnitude of injury in patients with acute myocardial infarction. The benefit is probably due to the decrease in reperfusion injury, which has been shown to be a major cause of myocardial cell death when the culprit artery is opened with a catheter. The most recent development is to infuse saline saturated with HBO into the opened coronary artery immediately after angioplasty. An advantage of this treatment has been demonstrated in experimental animals and in man. Future applications of HBO treatment in cardiovascular disease await more research, but at this time it seems to have considerable promise.

Research has shown that HBOT before or shortly after open heart surgery can reduce by 30% the common postsurgical loss of cognitive function.

Important: HBOT typically does not replace other treatments; it adds healing power to existing therapies. It is safe medical procedure but not without risk. For example, patients with severe congestive heart failure should be screened carefully before treatment HBOT could worsen the condition.

8. Reflections on the neurotherapeutic effects of hyperbaric oxygen
Shai Efrati and Eshel Ben-Jacob, Sackler School of Medicine and Tel-Aviv University 2014.

Traumatic brain injury (TBI) and stroke are the major causes of brain damage and chronic neurological impairments. There is no agreed-upon effective metabolic intervention for TBI and stroke patients with chronic neurological dysfunction. Clinical studies published this year present convincing evidence that hyperbaric oxygen therapy (HBOT) might be the coveted neurotherapeutic method for brain repair. Here we discuss the multi-faceted role of HBOT in neurotherapeutics, in light of recent persuasive evidence for HBOT efficacy in brain repair and the new understanding of brain energy management and response to damage. We discuss optimal timing of treatment, dosage, suitable candidates and promising future directions.

The April 1, 2012 issue of DVM Magazine printed the "Indications for hyperbaric oxygen therapy By dvm360.com staff, DVM360 MAGAZINE.

Hyperbaric oxygen therapy allows a patient to breathe high concentrations of oxygen by placing the patient in a chamber and increasing the air pressure around them. By breathing 100 percent oxygen under increased pressure, veterinarians can increase the amount of oxygen dissolved in the blood by as much as 15 times normal concentrations. Hyperbaric therapy increases pressure thereby condensing the oxygen molecules in the alveolus, making more molecules available at the alveolar-capillary interface for diffusion into the blood.

Indications currently used by veterinarians include
Central Nervous System

- Cranial/spinal cord trauma
- Cerebral/global ischemia
- Compressive chord diseases
- Fibro-cartilaginous emboli
- Cortical blindness
- Tetraparesis
- Peripheral nerve injury

Musculoskeletal
• Athletic injuries
• Tendonitis
• Desmitis
• Periostitis
• Laminitis
• Myositis
• Crush injuries

Infectious Diseases
• Osteomyelitis
• Septic arthritis
• Blastomycosis
• Lyme disease

Cardiovascular
• Hypotension
• Shock, all causes
• Cardiac infarction
• Acute anemia
• Reperfusion disease
• Carbon monoxide/cyanide toxicity
• Smoke inhalation
• Lymphangitis

Respiratory
• Exercise induced pulmonary hemorrhage
• Pleuritis
• Pulmonary edema

Wounds
• Thermal burns
• Compromised grafts/flaps
• Envenomation—spider, snake

Gastrointestinal
• Pancreatitis
• Peritonitis
• Reperfusion

Genito-urinary
• Infertility

Cancer
• Pre- and post-radiation therapy

Source: The University of Tennessee Center for Veterinary Hyperbaric Medicine

The Future of Hyperbaric Oxygen Therapy In Humans

HBOT as an alternative to Post Traumatic Stress Disorder and Traumatic Brain Injury. HBOT has been a game changer for treating military veterans with mild traumatic brain injury or post-traumatic stress disorder. A healthy brain uses 100% of the oxygen the body provides; however, trauma requires additional oxygen in order to heal brain tissue effectively. HBOT provides that oxygen.

The Future of Hyperbaric Oxygen Therapy in Pets

Central Nervous System Conditions

Intracranial disease (stroke, trauma)/spinal cord trauma

Cerebral hypoxia refers to a decrease in blood flow to the brain, which typically occurs during head trauma. After a severe head injury, traumatic brain damage may occur as the brain swells, wedging sensitive tissues against the skull. Hyperbaric oxygen therapy (HBOT) works to treat brain injury by constricting the brain's blood vessels, yet delivering more oxygen. This type of therapy works by increasing oxygen content in plasma, a part of the blood that normally does not carry oxygen throughout the body. HBOT reduces the swelling and increases the blood supply.

Treat at 1.5 ATA for 45 60 minutes. You may repeat the treatment after a 4 hour rest interval. Treat 10 times and reassess. Up to 40 treatments may be commonly used.

Cerebral/Global Ischemia

Hyperbaric oxygen therapy has been used in a number of conditions characterized by global ischemia (as opposed to focal ischemia of stroke), and anoxia leading to impairment of consciousness. Conditions such as coma due to brain injury and anoxia associated with anesthetic accidents, drowning and hanging as a dog might experience if he fell off a grooming table while tethered at the neck.

Treat at 1.5 ATA for 45 60 minutes. You may repeat the treatment after a 4 hour rest interval. Treat 10 times and reassess. Up to 40 treatments may be commonly used.

Compressive cord diseases

Spinal trauma often creates neurological damage as a result of swelling of these tissues within a confined space, loss of blood and oxygen supply, and the sequential effects of these factors on nervous tissue. HBOT reduces the swelling and increases the blood supply. The treatment of spinal cord injury is focused on repair and functional restoration. This involves promoting axonal sprouting, activation of idling and non-functional neurons while promoting neovascularization (new capillary formation) of damaged areas. The extent of neurovascular deterioration can be significantly diminished with early and continued HBOT implementation. Hyperbaric oxygen treatment is not recommended as a "cure" and is not the missing link in spinal cord injury but certainly has a major role as part of the developing answer in the treatment of spinal cord and brain injury.

Treat at 1.5 ATA for 45-60 minutes. You may repeat the treatment after a 4 hour rest interval. Treat 10 times and reassess. Up to 40 treatments may be commonly used.

Fibro-cartilaginous emboli

Is an extraordinarily rare disease in humans but not in dogs. Embolic material from a disrupted disc somehow reaches the arterial or venous circulation of the spinal cord causing an infarction, or an obstruction of the blood supply to a region of the spinal cord, resulting in death of local tissue. Canine cases of fibro-cartilaginous emboli often show rapid improvement when promptly treated by HBOT. It is also especially important to include rehabilitative therapies in these primarily large breed patients to make them ambulatory as soon as possible.

Treat at 1.5 ATA for 45-60 minutes. You may repeat the treatment after a 4 hour rest interval. Treat 10-20 times and reassess.

Paraparesis/Tetraparesis

Paraparesis refers to partial or complete loss of voluntary motor function in the pelvic limbs. Similar involvement of all four limbs is termed tetraparesis. Paraparesis generally results from spinal cord lesions caudal to the second thoracic spinal cord segment, whereas tetraparesis occurs because of lesions cranial to this segment. Intervertebral disc disease, lumbo sacral syndrome and vertebral stenosis are common spinal cord diseases affecting dogs and cats. A study released on March 24th 2010 in the Journal of Neurosurgery has clearly demonstrated the importance of supplemental HBOT following spinal cord injury. In this study, spinal cord injuries were induced in rats and HBOT was applied to the one group of rats. In this controlled setting, the researchers were able to observe that rats had less severe spinal cord injuries when treated with hyperbaric oxygen therapy. More importantly, researchers found that this group had an increase in key growth factors that may lead us to a better understanding of the mechanism of action for the results seen with the application of hyperbaric oxygen following spinal cord injuries.

Treat at 1.5 ATA for 45-60 minutes for 30-40 treatments. You may repeat the treatment on the same day after a 4 hour rest interval.

Cortical blindness

Is the total or partial loss of vision in a normal appearing eye caused by damage to the visual area in the brain's occipital cortex. The affected eye is physically normal and healthy, but damage to the brain results in full or partial vision loss. The pupil of a cortically blind eye still dilates and constricts in response to changes in light, because this reaction is a reflex, and does not rely on the brain. The most common cause of cortical blindness is oxygen starvation to the occipital lobe of the brain. Causes of cortical blindness can be: neoplasia, stroke, toxin-related or primary ischemic in origin. Common toxins include mercury and lead and metabolic causes include cardiac arrest or head trauma. Cortical blindness occurring after an anesthetic accident during routine anesthesia is known to respond well and regain normal vision.

Treat at 1.5 ATA for 45-60 minutes for 30-40 treatments or until symptoms resolve. You may repeat the treatment on the same day after a 4 hour rest interval.

REFERENCES

1. Tissue gas measurements during hyperbaric oxygen exposure, in Smith[ed]. Proc 6th Int. Congress on Hyperbaric Med, Aberdeen Press, 1977, p1182
2. Jain KK, Textbook of Hyperbaric Medidne 5th ed. Cambridge,Ma. Hogrefe &Huber, 2009
3. Bakker, D Cramer, FS [eds].Hyperbaric Surgery: Perioperative Care. FlagstaffAz. Best, 2004
4. Kindwall EP, Whelan HT [eds], Hyperbaric Medicine Practice 2nd ed. Revised. Flagstaff, Az. "Hyperbaric Oxygen Therapy accelerates neurologic recovery after 15 minute complete global cerebral ischemia in dogs". Critical Care Med. 1992; 20[11]:p1588
5. TakahashiM, et al. "Improved reversibility of traumatic midbrain syndrome wit application of hyperbaric oxygen pressure" Acta Neurochir. 1974:30:p247
6. Allan McGavin Sports Medicine Centre, Division of Sports Medicine, University of British Columbia, Vancouver, Canada.
7. Contreras FL et al."The effect of hyperbaric oxygen on glucose utilization in freeze traumatized rat brain" J Neurosurg.1988:65:p615
8. Vlodavsky, E et al."Hyperbaric Therapy reduces neuroinflammation and expression of matrix etalltoproteinase".
9. Traumatic brain injury "Neuropathology" Appl Neurobiology. 2006; 32:1:p40
10. Maeda, N "Experimental studies on the effect of decompression procedures and hyperbaric oxygenation for the treatment of spinal cord injury" J Nat'l Med Assoc. 1965; 16:p429
11. Kelly DL et al, "Effects of hyperbaric oxygenation and tissue oxygen studies in experimental paraplegia" Neurosurg.1972:36:p425

12. Neubauer RA, 1st International symposium for the use of Hyperbaric Oxygenation in the Neurosciences, 2006, p78
13. Kazantseva NV, 1st Int Symposium for the use of Hyperbaric Oxygenation intheNeurosciences,2006 p32
14. Workman W. Hyperbaric Facility Safety: A Practical Guide.Flagstaff, Az. Best,1999. Journal of Neurotrauma. 2010 Jun; 27(6):1121
15. Hyperbaric oxygen therapy for acute smoke inhalation injuries. Meyer GW, Hart GB, Strauss MB Source Department of Baromedicine, Long Beach Memorial Medical Center, CA 90801
16. Hyperbaric oxygen therapy for wound complications after surgery in the irradiated head and neck: A
review of the literature and a report of 15 consecutive patients Erik B. Neovius MD1,*, Magnus G. Lind MD, PhD1, Folke G. Lind MD, PhD2 Article first published online: 7 DEC 199 DOI: 10.1002/(SICI)10970347(199707)19:4<315::AID HED10>3.0.CO;2
17. Hyperbaric oxygen therapy for wound healing and limb salvage: a systematic review. Goldman Source Wound Care and Rehab Medicine, LLC, P.O. Box 18870, Philadelphia, PA 19119, USA. robert.goldman@verizon.net
18. Undersea and Hyperbaric Medical Society, Inc. 10531 Metropolitan Avenue Kensington, Maryland 20895 UHMS Publication Number 30 CR (HBO) 1992
19. The Merck Veterinary Manuel © 2011; Merck Sharp & Dohme Corp., a subsidiary of Merck & Co., Inc.Whitehouse Station, NJ USA. All Rights Reserved.
20. WebMD Veterinary Reference from "Dog Owner's Home Veterinary Handbook"
21. Potential effects of hyperbaric oxygen therapy in acute pancreatitis. Cuthbertson CM, Christophi C.Source Department of Surgery, University of Melbourne, Austin Hospital, Melbourne, Victoria, Australia.c.cuthbertson@pgrad.unimelb.edu.au
21. HVM HBOT Quick Reference Guide, Compiled by Andrew Turkell DVM, CVA, CCRT and Paul Jaffe, DVM
22. Hyperbaric Oxygen for Neurological Disorders, John Zhang, MD 2008, Best Publishing Co.
23. Physiology and Medicine of Hyperbaric Oxygen Therapy, Neuman & Thom. Saunders Elsevier.
24. Textbook of Hyperbaric Medicine, 5th edition K.K. Jain, Hogrefe, 2009

SPONTANEOUS ANIMAL MODELS UNSUNG HEROS IN BONE CANCER PAIN

Dorothy Cimino Brown, MS, DVM, DACVS
Philadelphia, PA

THE PROBLEM

Seventy-five percent to 90% of human patients with advanced cancer will experience significant, life-altering, cancer-induced pain. Severe pain is especially associated with tumors involving bone destruction. The frequency and intensity of the pain tends to increase during the advanced stages, over weeks or months. With increased bone destruction and time, the pain intensifies and can incapacitate the affected individual. Bone pain is the most common pain syndrome encountered in cancer patients, yet the available treatments are still limited. The efficacy of potent opioids with or without NSAIDs is variable and can produce significant cognitive and gastrointestinal side effects. Novel analgesics with greater efficacy are clearly needed. The development of optimal analgesics for bone cancer pain has been hampered by several factors. Of particular importance is the lack of a full set of clinically appropriate animal models for both basic science translational research and clinical trials of promising agents.

Both rat and mouse models of bone cancer pain are described and are induced models. For example, a rat model involves injection of mammary carcinoma cells into the tibia and a mouse model involves the injection of osteolytic mouse sarcoma cells into the intramedullary space of the femur. In the rat model, mechanical allodynia, hyperalgesia, and reduced weight bearing become consistently evident 14 days after injection of tumor cells and by day 20 there is major structural destruction of the bone and displaced fractures. This leaves a 7-day window for studying the neurochemistry of cancer pain and the effectiveness of antinociceptive agents. The mouse model shows variable bone destruction by 21 days after injection, from no signs of lysis in 21% of the animals, to major destruction of the femur in 39%. On day 21 pain induced behaviors are reported in all of the animals that are injected with sarcoma cells. This model also provides only a very narrow window for studying the neurochemistry of cancer pain and the effectiveness of antinociceptive agents

These rodent bone cancer pain models mirror some of the changes seen in human clinical bone cancer and they yield extensive information about basic mechanisms of bone cancer pain. However, these experimental pain states with a 7-day progression of induced disease in rodents cannot adequately represent the weeks/months of complex evolution of clinical bone cancer pain that occurs in patients. Such an abbreviated time does not allow for clinically relevant testing of novel analgesics. In addition, working with induced tumors that generate very rapid disease progression in inbred rodent populations and artificial environments raises some degree of concern over the applicability of information as it relates to naturally occurring clinical cancer pain in people.

In addition, one of the urgent problems associated with translational pain research is a mismatch in pain evaluation tools between preclinical and clinical research. The tools used in animal model preclinical research are primarily for the assessment of stimulus-induced pain, such as thermal hyperalgesia testing, While such investigations add to our understanding of the neurobiology of the sensory system, translation of such information into human clinical applications has often proven to be unsuccessful. With only 21% of drugs beginning phase I trials getting to market, and clinical period costs growing five times as fast as preclinical period costs, most large companies recognize that identifying better drug development models could be their best chance of modernizing the drug development process and preventing clinical trial failures late in development.

A comprehensive assessment of human clinical cancer pain requires going beyond the characteristics and severity of the pain and assessing the ultimate impact the pain has on the patient's quality of life. It is important to determine the degree to which the pain interferes with the patients functioning. This is also true in the case of bone cancer pain in the companion animal. While "self-reporting" pain is obviously not an option when assessing pain in companion dogs, the owners of those dogs, make daily assessments of the degree of pain and quality of life. Once an unacceptable level of pain and dysfunction is present, the caregiver searches out effective treatments. If none are available they opt for euthanasia for the animal. This observer assessment of pain and function by an owner to a companion animal usually occurs daily in dogs with chronic painful conditions, as is the case in people with chronic pain that are incapable of self-reporting (cognitively impaired or pediatric populations). While many aspects of human clinical pain are difficult to reproduce in experimental animals, numerous aspects are exhibited in companion animals with spontaneous painful conditions and would be useful models in translational research.

A SOLUTION

A large animal model of naturally occurring bone cancer, that is observable in the animal's natural environment, and that more closely mirrors the progression of clinical disease in humans, is likely to be a very useful model of human clinical bone cancer pain. Canine bone cancer is such a naturally occurring model. Osteosarcoma is a common bone tumor in dogs that bears striking resemblance to osteosarcoma in humans. The tumors are histologically indistinguishable, have the same biologic behavior and disease progression. The standard of care for palliation of dogs with osteosarcoma of a single long bone is amputation followed by chemotherapy. However, a significant population of dogs does not receive this treatment. Their size, weight, and overall condition prohibits amputation for fear the dog will not be functional on three legs. In other cases, an owner opts not to pursue aggressive or radical procedures in a terminal pet. For these dogs, the standard of care becomes managing the pain and loss of function caused by the bone tumor for as long as humanely possible. It is this group of dogs that can be recruited as an animal model for human bone cancer pain.

Dogs with osteosarcoma, that are not candidates for amputation, have an evolution of bone cancer pain that mirrors the human presentation. The frequency and intensity of the pain tends to increase over weeks or months. This is manifested in the need to give analgesics and increase or change the dose to allow continued weight bearing on the affected limb and improve the activity of the animal. As the disease progresses, weight bearing produces frequent episodes of breakthrough pain which become difficult to control. These dogs often undergo euthanasia within several months of the diagnosis of bone cancer because of uncontrolled bone pain and associated loss of function. This evolution of bone pain over weeks to months better approximates the human condition than the rodent models and allows enough time to evaluate the effectiveness of antinociceptive agents through the evolution of the pain process. This time is still short enough, however, to ensure rapid accrual of data through companion animal clinical trials.

DOGS TO THE RESCUE FOR NEUROTOXIN RESEARCH

There are two recent examples of studies in dogs with bone cancer being instrumental in moving extremely novel analgesic interventions, intrathecal neurotoxins, forward to human clinical trials. Research and development programs for both intrathecal substance-p saporin and intrathecal resiniferatoxin included proof-of-concept studies in dogs with bone cancer. Positive results in these studies propelled human clinical trials forward as well as opened the door to consideration for regulatory approval for veterinary patients as well.

Substance P-saporin (SP-SAP), a chemical conjugate of substance P and a recombinant version of the ribosome-inactivating protein, saporin. When administered intrathecally, SP-SAP acts as a targeted neurotoxin producing selective destruction of superficial neurokinin-1 receptor–bearing cells in the spinal dorsal horn. In seventy companion dogs with bone cancer pain randomized to receive intrathecal substance P-saporin or standard-of-care therapy alone, owner-requested unblinding for inadequate pain control occurred significantly sooner in animals receiving standard-of-care therapy alone, suggesting an analgesic effect of substance P-saporin. A phase 1 study of substance p-saporin in terminal cancer patients with intractable pain is now open at the University of Texas Southwestern Medical Center (ClinicalTrials.gov Identifier: NCT02036281).

Resiniferatoxin is the most potent amongst all known agonists for the transient receptor potential vanilloid 1 (TRPV1) receptor. TRPV1 is a calcium permeable non-selective cation channel expressed on the peripheral and central terminals of small diameter sensory neurons. When applied to the sensory neuron perikarya, the prolonged calcium influx induced by RTX causes cytotoxicity and death of only those sensory neurons that express the TRPV1 ion channel. Thus, intrathecal RTX administration leads to selective targeting and permanent deletion of the TRPV1-expressing C-fiber neuronal cell bodies in the dorsal root ganglia. Loss of these C-fiber neurons interrupts the transmission of pain information from the body to second-order spinal cord neurons, which in turn convey the information to the brain. In seventy two companion dogs with bone cancer pain randomized to receive intrathecal RTX or standard-of-care therapy alone, owner-requested unblinding for inadequate pain control occurred significantly sooner in animals receiving standard-of-care therapy alone, suggesting an analgesic effect of RTX. A phase I study of the intrathecal administration of resiniferatoxin for treating severe refractory pain associated with advanced cancer is now open at the National Institutes of Health Clinical Center (ClinicalTrials.gov Identifier: NCT00804154).

SUMMARY

The development of new compounds to treat chronic pain has risen dramatically in the past decade, and there is a need for more predictive animal models to bridge the gap between discovery of a candidate compound and its introduction into humans. A novel approach to the development of animal models for assessing intervention efficacy is to focus on companion animals that spontaneously develop disease, and consider the outcome measures in that model, which have meaning for animals as well as for people. By focusing some efficacy studies on animals that spontaneously develop and naturally progress through the same diseases of clinical concern, with outcomes designed specifically to represent those of importance in human clinical studies, an animal model that yields efficacy results more predictive of clinical outcome could evolve. Further, the information gained in the testing of a novel therapeutic in these dogs is directly useful as preclinical data for human application development as well as clinical data for veterinary development, which can potentially benefit both humans and animals.

REFERENCES

Brown DC, Agnello K, Iadarola MJ. Intrathecal resiniferatoxin in a dog model: Efficacy in bone cancer pain. In press Pain

Brown DC, Agnello K. Intrathecal substance P-saporin in the dog: efficacy in bone cancer pain. Anesthesiol 119:1178-1185 (2013)

Brown DC, Iadarola MJ, et al. Physiologic and antinociceptive effects of intrathecal resiniferatoxin in a canine bone cancer model. Anesthesiol103:1052-1059 (2005)

Dionne RA, Witter J. NIH-FDA Analgesic Drug Development Workshop: Translating scientific advances into improved pain relief. Clin J Pain 19:139–47 (2003)

Hayashida K. Substance P-saporin for bone cancer pain in dogs: can man's best friend solve the lost in translation problem in analgesic development?. Anesthesiol 119:999-1000 (2013)

Hunter AJ. Have animal models of disease helped or hindered the drug discovery process? Ann N Y Acad Sci 1245:1-2 (2011)

Mantyh, P.W. A mechanism based understanding of cancer pain. Pain 96, 1-2 (2002)

Mao, J. Translational pain research: bridging the gap between basic and clinical research. Pain 97, 183-187 (2002)

Medhurst, S.J., Walker, K., et al.. A rat model of bone cancer pain. Pain 96, 129-140 (2002)

Mogil JS. Animal models of pain: progress and challenges. Nat Rev Neurosci 10:283-294 (2009)

Rice AS, Brown DD, et al.Animal models and the prediction of efficacy in clinical trials of analgesic drugs: a critical appraisal and call for uniform reporting standards. Pain 139:243-247 (2008)

Schwei, M.J., Honore, P., et al. Neurochemical and cellular reorganization of the spinal cord in a murine model of bone cancer pain. J Neurosci 19, 10886-10897 (1999)

Tufts Center for the Study of Drug Development. Post-approval R&D raises total drug development costs to $897 million. Impact Rep 5:1–4 (2003)

Vail, D.M. & MacEwen, E.G. Spontaneously occurring tumors of companion animals as models for human cancer. Cancer Invest 18, 781-792 (2000)

SPONTANEOUS ANIMAL MODELS: UNSUNG HEROES

Jaime F. Modiano, VMD, PhD
St. Paul, MN

INTRODUCTION

Diligent efforts over the past 50 years have allowed veterinarians to adapt protocols devised for human patients to treat companion animals with cancer. This has provided pet owners reasonable options ranging from palliative care to therapies with curative intent. In addition to adapting protocols using conventional modalities (surgery, radiation, and cytotoxic chemotherapy), we have seen targeted drugs and immunotherapy approaches enter the veterinary space.[1,2]

Historically, veterinarians have leveraged information from basic and clinical research in cancer medicine to extend the quantity and improve the quality of life for companion animals. Now, we find ourselves in a position to turn the tables and enable veterinary medicine to make contributions that positively impact the human health care industry. The last decade has seen important milestones in our understanding of cancer causation and therapy in domestic animals. Perhaps most importantly, we are starting to recognize when and where canine (and feline) cancer research can be applied most efficiently to accelerate clinical translation. Here, we will enumerate some important milestones in comparative oncology and use vascular tumors as an example of an area where veterinary cancer research can be transformational for animals and humans alike.

CANINE AND HUMAN CANCERS: ARE THEY THE SAME?

Similarities in the natural history of canine and human cancers have been recognized for several decades.[3] But confirmation of the molecular homology of these diseases had to await the progress in genomic sequencing that came with the molecular biology revolution. Completion of the canine genome sequence was a critical milestone in these efforts.[4] This in turn set the stage to test hypotheses of molecular equivalence between canine and human cancers, with the first definitive answer documenting the presence of orthologous recurrent chromosomal translocations in canine and human chronic myelogenous leukemia, chronic lymphocytic leukemia, and Burkitt lymphoma published in 2008.[5]

Since then, additional work has shown molecular equivalence in a variety of tumors. However, as we have probed more deeply into the pathogenesis of common cancers, it has become clear that these diseases are more heterogeneous than might have been anticipated. The concept that tissue-specific stem cell divisions can explain the risk for most human cancers[6] is fully consistent with the heterogeneity we observe in canine cancers. We have proposed that this is directly proportional to the error rates of DNA polymerases, adding an important stochastic component to the risk that can be attributed to heritable factors.[7]

WHEN CANINE CANCER LEADS THE WAY

The use of dogs to study the most common human cancers presents unique challenges. For example, cancers of the lung, head and neck, bladder, and bone marrow in humans are strongly influenced by the epidemic use of tobacco products. None of these can be adequately modeled in dogs (or cats) in the natural setting, with the possible exception of isolated cases of secondary exposure to tobacco carcinogens.[8-13] There are similarities in tumors of the breast, prostate, and gut between the two species, but the overlap in the spectrum of molecular pathology in these cancers seems to be more limited, so models must be carefully vetted to avoid reaching conclusions that might not only be invalid, but could retard progress in the field and reduce confidence in comparative approaches. Indeed, this is generally true for any model, so data from models should be interpreted narrowly and with due caution.

One example where canine cancer research has been at the forefront of progress for several decades is osteosarcoma.[14] Osteosarcoma in humans is an orphan disease with fewer than 1,000 cases diagnosed annually; this is an obstacle to recruit cases into clinical trials and to obtain material for study. Another impediment to advancements in osteosarcoma is the age distribution of the affected population. This cancer is predominantly a disease of children and adolescents, so the informed consent process to allow their participation in research studies is necessarily more laborious.[15] Still, the survival of patients with osteosarcoma showed dramatic improvement between 1973 and 1984 with the introduction of adjuvant chemotherapy, but it has been static since.[15] There have been modest enhancements in quality of life; yet, even though survival is shorter in elderly patients, the impact of osteosarcoma on adolescents, as well as on their families, caregivers, and the extended community (for example, the sociological impact on their peers) is disproportionate to the numbers affected due both to the significant morbidity and years of life lost.

In contrast to humans, osteosarcoma occurs frequently in dogs (at least 10-fold higher, *i.e.*, more than 10,000 cases per year). The disease arises spontaneously (it is not induced) and has similar patterns of disease progression and metastasis when chronology is considered in the context of lifespan, offering opportunities to understand etiopathogenesis as well as to develop and test new therapies.[16] Our own research has recently shown how we can use the canine disease strategically to reduce the challenges associated with tumor heterogeneity, and potentially improve prediction for this disease.[17-20] In turn, we have extended these findings from pre-clinical studies[19,21] through successful treatment of dogs with spontaneous disease, and further yet to identification of biomarkers for response.[22] Our data have progressed beyond the realm of academic interest to patent applications for diagnostic and prognostic tests, as well as into the process of investigational new drug applications.

We must note, however, that there are important differences between human and canine osteosarcoma. For example, osteosarcoma can occur in dogs of any age, but the disease is most common in middle age and older dogs (>8 years). The epidemiology of canine osteosarcoma also shows a measurable "pediatric" peak between 18-24 months. So an important, contemporary question is whether pediatric osteosarcoma and adult osteosarcoma in dogs and in humans (the latter being mostly associated with long-standing bone diseases such as Paget's) represent unique subsets or a continuum the same disease.

TACKLING VASCULAR SARCOMAS

The diverse cellular ontogeny and heterogeneity of sarcomas has provided an impetus to develop molecular approaches to classify these tumors.[23,24] This has uncovered two categories of sarcomas: one consisting of tumors with simple genetic alterations, and the other comprising sarcomas with complex, chaotic karyotypes.[23,25] Most sarcomas with simple genetic aberrations are translocation-related cancers such as Ewing's sarcoma, alveolar rhabdomyosarcoma, myxoid/round cell liposarcoma, and angiomatoid fibrous histiocytoma.[26] Sarcomas with complex karyotypes include tumors such as malignant fibrous histiocytoma, dedifferentiated liposarcoma, and leiomyosarcoma.[23,25] The remainder of this article will focus on malignant blood vessel tumors of humans (angiosarcomas) and dogs (hemangiosarcomas).

Natural History and Genetics of Angiosarcoma and Hemangiosarcoma

Angiosarcomas are rare, highly aggressive cancers that originate from blood vessel-forming cells. Our understanding of the pathology of human angiosarcomas is rudimentary; these tumors are diagnosed infrequently, accounting for 0.01 – 0.1% of all human cancers, and they display extensive heterogeneity. This has made the development of effective therapies challenging. More than half of the patients with this disease die within the first year of diagnosis and tumor related mortality is extremely high.[27] It is clear that a different paradigm is needed to address the pathology and to develop effective treatments for this disease. Exposure to genotoxic agents such as vinyl chloride, thorium, and arsenic, to radiation therapy, and to viruses (for Kaposi sarcoma) have been implicated in the etiology of some angiosarcomas, but most patients that develop angiosarcoma do not have histories of exposure to these risk factors and the cause of their disease is unknown.[28,29]

The events that drive progression of angiosarcoma are also poorly understood. There is debate regarding the role of common pathways of tumorigenesis in this disease.[27] Angiosarcomas fall into the category of sarcomas with complex karyotypes,[25,30] although recurrent abnormalities of specific genes and pathways have been found. RAS mutations, KDR (kinase insert domain receptor, also known as VEGFR2) mutations, MYC gene amplification, and alterations in the TP53, CDKN2, NF-κB/IL-6, PIK3CA/AKT/mTOR pathways have all been reported in angiosarcoma.[30-36] Yet, the studies represent small case series, RAS mutations appear to be almost exclusively seen in tumors caused by exposure to vinyl chloride, and the significance of other abnormalities is unclear. Thus, driver events for angiosarcoma still must be rigorously confirmed and validated.

Unlike the rare human angiosarcoma, hemangiosarcoma of dogs occurs spontaneously and commonly (tens of thousands of cases in the US each year), providing a fundamental opportunity to use the canine disease as a discovery system to identify causative or pathognomonic events for human angiosarcoma and to develop novel treatments. In the canine disease, topology is associated with different natural histories. Cutaneous hemangiosarcoma can be locally invasive, but tends to be confined regionally and is amenable to treatment with surgery and adjuvant chemotherapy. This cutaneous form will not be discussed further here. Visceral and subcutaneous hemangiosarcomas are aggressive, highly metastatic and extremely chemoresistant tumors. The course of disease is insidious, so the visceral form in particular is usually diagnosed in advanced stages or after death from tumor rupture and catastrophic hemorrhage. Several studies have reported an association between microangiopathic disease (anemia, thrombocytopenia, and acanthocytosis) and hemangiosarcoma,[37,38] suggesting that diligent analysis of laboratory tests obtained as part of routine well-health exams could be useful for timely detection of this disease. However, these changes are not sensitive or specific.[39] Unfortunately, the common clincopathological features of hemangiosarcoma also do not answer the questions as to whether the tumors show site or substrate specificity and whether they originate from a single cell type. An apparent breed-specific predilection for different topological forms of the disease has been noted,[40] suggesting tumors at different sites could have different etiology. On the other hand, these differences may be driven by other microenvironmental factors that are independent from the cell of origin. It is fair to say that development of tests that can be used to detect hemangiosarcoma before it creates a risk for lethal hemorrhage is an important unmet need in veterinary oncology.

We have shown that canine hemangiosarcoma can be categorized into angiogenic, inflammatory, and adipogenic molecular subtypes based on cell-autonomous properties[41] and on their interactions with the microenvironment.[41,42] Despite their complex, chaotic karyotypes,[43] we have identified a spectrum of recurrent inter-chromosomal fusion events using next generation RNA-seq. These seem to be significantly associated with the different molecular subtypes, allowing us to consider development of specific targeted therapies to attack this disease.

Treatment of Angiosarcoma and Hemangiosarcoma

Treatment of angiosarcoma with conventional therapies is unrewarding, and experimental therapies have produced marginal responses. For example, inhibition of the VEGF-VEGFR axis with bevacizumb (anti-VEGF antibody) or sorafenib (VEGFR kinase inhibitor) produced ~15% objective response rates,[44,45] but they were ineffective in the vast majority of angiosarcoma patients.

The pathological features, morphological organization, and clinical behavior of visceral and subcutaneous canine hemangiosarcoma are virtually indistinguishable from human angiosarcoma, and the response to treatment is similarly dismal. The benefit of adjuvant chemotherapy to prolong progression-free survival and overall survival of dogs with subcutaneous and visceral hemangiosarcoma is beyond question,[46-51] and in fact, this is the only modality that is clearly efficacious to treat this disease. Furthermore, the risk of death from hemangiosarcoma is associated with the potential for thrombosis and embolism in major organs, or more commonly from fatal hemorrhage. Thus, this is one disease where surgery and chemotherapy can be indicated even in the face of metastatic disease.[52] Still, the commonly observed survival for dogs with splenic hemangiosarcoma (the most common form) treated with the standard of care ranges from 3 to 6 months.[53] Dogs without gross metastasis at diagnosis survive slightly longer, with up to ~15% expected to survive >12 months. This is consistent with data from 112 cases seen at our hospital in a 30-month period between 2010 and 2012. The poor response of hemangiosarcoma to conventional therapy has led to numerous controlled studies and anecdotal instances of experimental therapy for this disease. Most studies using combinations of cytotoxic drugs,[52,54-64] metronomic dose schedules,[65] autologous vaccines,[66] and sundry immune-based and anti-angiogenic therapies[67] have not shown consistent improvement over the accepted standard of care. Recent reports show conflicting data regarding the potential benefit of adding dacarbazine to conventional doxorubicin-based protocols; resolution of this issue will require further study.[52,68] Perhaps the most promising experimental therapy for this disease was the use of adjuvant L-MTP-PE,[69,70] suggesting that immunotherapy could have a role in the treatment of this disease.

A GLIMPSE OF SUCCESS: ATTACKING THE TUMOR MICROENVIRONMENT

Over the past decade, we have developed a resource of hemangiosarcoma-derived cell lines with representation of anatomical and breed diversity, and we have examined phenotypic and molecular commonalities among these cell lines, defining characteristics of cell surface antigen expression,[71,72] tumor suppressor gene inactivation,[35] and genome-wide gene expression profiles.[41,73,74] An important aspect arising from these data was the observation that hemangiosarcoma cells in isolation (cell culture) have phenotypic properties suggestive of bone marrow ontogeny and specifically of bone marrow derived stem cells.[72] Alone, these data could not distinguish if the tumors themselves arise from a cell in the bone marrow that migrates to a vascular plexus, or if they originate from a bone fide stem cell. Our gene expression profiling data offered some clues, however, as the recurrent signature of these genes was associated with enrichment of angiogenic and pro-inflammatory genes, with no evidence of tissue specificity.[73] More recent data corroborated these findings and also showed that hemangiosarcomas could be stratified according to the expression of pro-inflammatory genes (inflammatory subtype), endothelial cell-matrix interaction genes (vascular or angiogenic subtype), and pro-adipogenic and connective tissue-forming genes (adipogenic subtype).[41] These genes reflected the composition of the tumor microenvironment, and potentially the balance of hypoxia and inflammation during tumor formation, as well as RXR and PPAR activation. Few sarcomas show the metastatic potential and chemoresistance seen in hemangiosarcoma, and recent work using these models shows progress and promise in defining the metabolic mechanisms that account for this resistance.[75-78]

We have demonstrated that hemangiosarcoma cells respond to cues from the microenvironment and adopt various functions as part of this response.[41] These functions include not only the potential to form anatomically distinct structures or tumors, but potentially, also the capacity to direct other cells in their environment to do so. Hemangiosarcoma cells include subpopulations that retain traits associated with "cancer stem cells", including self-renewal, chemoresistance, and increased tumorigenicity *in vivo*.[41,79,80] Each tumor may achieve these properties by independently disabling or enabling molecular networks associated with self-renewal and survival. Inflammation is also a recurrent feature of hemangiosarcoma. The tumor cells express functional receptors for cytokines, chemokines, and other bioactive molecules including IL-8, CXCL12, nicotine, and modified sphingosines, each of which initiates biologically relevant signals upon binding. We documented a direct correlation between IL-8 expression and reactive tumor microenvironments showing active inflammation, fibrosis, and coagulation in tumor tissues and in cell lines, indicating tumor autonomous effects were dominant and suggesting that IL-8 production is part of an adaptive mechanism used by hemangiosarcoma cells to modulate their microenvironment.[42] The importance of IL-8 in the formation of a tumor niche was further illustrated by our observation that IL-8 blockade inhibited tumor cell survival and engraftment in the mouse xenograft model.[42] Combining xenograft models with *in situ* gene signatures from dog "patient" tumors and from cell lines will allow us to test predictions about mechanisms that are operative during the early stages of tumorigenesis in hemangiosarcoma.

By dissecting unique molecular properties of hemangiosarcomas and their interactions with the microenvironment, we also were able to select and test a novel therapeutic approach. This uses a genetically engineered, bispecific ligand targeted toxin (BST) that can kill conventional tumor cells as well as their progenitor or "cancer stem cells" with high specificity.[80] The BST strategy also opens a line to attack their environmental niche, and specifically the blood vessels that feed the tumor.[81] The promising results *in vitro*, the efficacy in laboratory animal models, and documentation of safety in toxicological studies provided a foundation for translation: we designed an adaptive clinical trial (SRCBST) to extend this therapy to dogs with spontaneous splenic hemangiosarcoma in the minimal residual disease setting.[82] The experimental treatment consists of a single cycle (3 doses 2-days apart) of the experimental drug followed by a conventional 5X doxorubicin (30 mg/m^2) chemotherapy protocol. The adaptive design uses an innovative method to assess safety and efficacy simultaneously. Additionally, we have built increments into the trial that will allow us to use positron emission tomography and x-ray computed tomography (PET-CT) to monitor minimal residual disease and relapse. Twenty dogs have been treated as of the date this article was written. Thus far, the data show that the therapy is safe and an efficacy signal is emerging.

Much work remains to be completed, but our data will undoubtedly facilitate translation of this drug to human patients with vascular sarcomas. It is apparent that some or many properties of these tumors are dependent on interactions between tumor cells and their local microenvironment, and so a multi-modal approach will likely be necessary to achieve durable responses in a majority of patients. The interactions appear to depend on metabolic adaptations of the progenitor cells, the microanatomy of the niche, and the capability of the tumor cells to alter the niche by recruitment or reprogramming of local stromal cells.[42] Local or systemic microangiopathy may precede the disease, but it also may be a consequence of the vascular disruption it creates. Platelets and platelet derived factors, inflammation, and hypoxia are probably key drivers of the disease, and all of these will need to be considered in the ongoing process of therapeutic development.

REFERENCES

[1]Bergman PJ. **Vet Clin North Am Small Anim Pract** 2007;37:1111-1119. [2]London CA, et al. **Clin Cancer Res** 2009;15:3856-3865. [3]MacEwen EG. **Cancer Metastasis Rev** 1990;9:125-136. [4]Lindblad-Toh K, et al. **Nature** 2005;438:803-819. [5]Breen M and Modiano JF. **Chromosome Res** 2008;16:145-154. [6]Tomasetti C and Vogelstein B. **Science** 2015;347:78-81. [7]Modiano JF. **Withrow and MacEwen's Small Animal Clinical Oncology (5th Edition)**, Elsevier/Saunders; 2013:1-15. [8]Bertone ER, et al. **Am J Epidemiol** 2002;156:268-273. [9]Bertone ER, et al. **J Vet Intern Med** 2003;17:557-562. [10]Bukowski JA, and Wartenberg D. **Environ Health Perspec** 1997;105:1312-1319. [11]Bukowski JA, et al. **J Toxicol Envir Health Part A** 1998;54:579-591. [12]Reif JS, et al. **Am J Epidemiol** 1998;147:488-492. [13]Reif JS and Cohen D. **Arch Environ Health** 1971;22:136-140. [14]Withrow SJ, et al. **Clin Orthop** 1991:159-168. [15]Mirabello L, et al. **Cancer** 2009;115:1531-1543. [16]Khanna C, et al. **Nat Biotechnol** 2006;24:1065-1066. [17]Sarver AL, et al. Orphanet J Rare Dis 2013;8:7. [18]Scott MC, et al. **Bone** 2011;49:356-367. [19]Thayanithy V, et al.. **PLoS ONE** 2012;7:e43720. [20]Thayanithy V, et al. **Bone** 2012;50:171-181. [21]Cannon CM, et al. **Vet Comp Oncol** 2013 (AOP). [22]Modiano

JF, et al. **Mol Ther** 2012;20:2234-2243. [23]Taylor BS, et al. **Nat Rev Cancer** 2011;11:541-557. [24]. Neuville A, et al. **Am J Surg Pathol** 2013;37:1259-1268. [25]Helman LJ and Meltzer P. **Nat Rev Cancer** 2003;3:685-694. [26]Mertens F, et al. **Semin Oncol** 2009;36:312-323. [27]Antonescu C. **Mod Pathol** 2014;27 Suppl 1:S30-38. [28]Cohen SM, et al. Toxicol Sci 2009;111:4-18. [29]Koch M, et al. **J Surg Oncol** 2008;97:321-329. [30]Italiano A, et al. **Cancer** 2012;118:5878-5887. [31]Antonescu CR, et al. **Cancer Res** 2009;69:7175-7179. [32]Italiano A, et al. **Genes Chromosomes Cancer** 2012;51:569-578. [33]Yang J, et al. **Cancer Res** 2012;72:4682-4695. [34]Tate G, et al. **Cancer Genet Cytogenet** 2007;178:160-162. [35]Dickerson EB, et al. **Vet Pathol** 2005;42:618-632. [36]Weihrauch M, et al. **Environ Mol Mutagen** 2002;40:36-40. [37]Rebar AH, et al. **Vet Pathol** 1980;17:443-454. [38]Hirsch VM, et al. **Canadian Vet J** 1981;22:152-155. [39]Warry E, et al. **Vet Clin Path** 2013;42:465-470. [40]Andersen NJ, et al. **InTech**; 2011:149-174. [41]Gorden BH, et al. **Am J Pathol** 2014;184:985-995. [42]Kim JH, et al. **Exp Cell Res** 2014;232:155-164. [43]Thomas R, et al. **Chromosome Res** 2014;22:305-319. [44]Maki RG, et al. **J Clin Oncol** 2009;27:3133-3140. [45]Park MS, et al. **Curr Opin Oncol** 2010;22:351-355. [46]de Madron E, et al. **J Am Vet Med Assoc** 1987;190:887-891. [47]Ogilvie GK, et al. **Vet Pathol** 1989;195:1580-1583. [48]Hammer AS, et al. **J Vet Intern Med** 1991;5:160-166. [49]Sorenmo KU, et al. **J Vet Intern Med** 1993;7:370-376. [50]Ogilvie GK, et al. **J Vet Intern Med** 1996;10:379-384. [51]Helfand SC. **Cancer Ther** 2008;6:457-462. [52]Alvarez FJ, et al. **Canadian Vet J** 2013;49:370-377. [53]Clifford CA, et al. **J Vet Intern Med** 2000;14:479-485. [54]Sorenmo K, et al. **J Vet Intern Med** 2000;14:395-398. [55]Payne SE, et al. **Vet Comp Oncol** 2003;1:171-179. [56]Sorenmo KU, et al. **J Vet Intern Med** 2004;18:209-213. [57]Spugnini EP, et al. **In vivo** 2006;20:125-127. [58]Kim SE, et al. **Vet Pathol** 2007;231:1550-1557. [59]Sorenmo K, et al. **J Vet Intern Med** 2007;21:1347-1354. [60]Wiley JL, et al. **Vet Comp Oncol** 2010;8:221-233. [61]Dervisis NG, et al. **J Am Anim Hosp Assoc** 2011;47:170-178. [62]Teske E, et al. **Vet Comp Oncol** 2011;9:283-289. [63]Kahn SA, et al. **Canadian Vet J** 2013;54:237-242. [64]Mullin CM, et al. **Vet Comp Oncol** 2014. [65]Lana S, et al. **J Vet Intern Med** 2007;21:764-769. [66]U'Ren LW, et al. **J Vet Intern Med** 2007;21:113-120. [67]Sahora AI, et al. **J Vet Intern Med** 2012;26:1169-1176. [68]Finotello R, et al. **Vet Comp Oncol** 2015 (AOP). [69]MacEwen EG, et al. **J Drug Target** 1994;2:391-396. [70]Vail DM, et al. **Clin Cancer Res** 1995;1:1165-1170. [71]Fosmire SP, et al. **Lab Invest** 2004;84:562-572. [72]Lamerato-Kozicki AR, et al. **Exp Hematol** 2006;34:870-878. [73]Tamburini BA, et al. **BMC Cancer** 2010;10:619. [74]Tamburini BA, et al. **PLoS ONE** 2009;4:e5549. [75]Dickerson EB, et al. **Translat Oncol** 2013;6:158-168. [76]Andersen NJ, et al. **Mol Cancer Ther** 2013;12:1701-1714. [77]Gorden BH, et al. **Vascular Cell** 2014;6:20. [78]Khammanivong A, et al. **Vet Comp Oncol** 2014 (AOP). [79]Frantz AM. **Doctoral Dissertation**. University of Minnesota; 2012. [80]Schappa JT, et al. **Int J Cancer** 2013;133:1936-1944. [81]Oh S, et al. **J Neurosurg** 2011;114:1662-1671. [82]Koopmeiners JS and Modiano JF. **Clin Trials** 2014;11:38-48.

THE HEALTHCARE COMMUNITY'S ROLE IN SAFE AND EFFECTIVE ANIMAL-ASSISTED THERAPY

Lisa M. Freeman, DVM, PhD, DACVN and Christopher W. Olsen, DVM, PhD
North Grafton, MA, and Madison, WI

INTRODUCTION

In recent years, there has been great enthusiasm for animal visitation. It is easy to observe the benefits of an animal in providing comfort and unconditional acceptance to people in diverse settings, such as hospitals, nursing homes, juvenile facilities, and reading programs. In addition, research is now beginning to demonstrate objective benefits of animal visitation and other forms of animal-assisted interventions (AAI). There are two main categories of AAIs: Animal-assisted activities (AAA), in which there are no specific and individualized treatment goals; and, animal-assisted therapy (AAT), which is a goal-directed intervention that is an integral part of the treatment process and which is directed by a health/human service provider. Animal visitation can be highly valuable in both formats.

While there is often great enthusiasm for participation in animal visitation and for developing animal visitation programs, the process should not be undertaken lightly and should be carefully planned. Development to implementation of a new program will involve a series of questions, concerns, and logistical hurdles. Veterinarians, physicians, and other healthcare providers and facility administrators play a critical role in ensuring that animal visitation is conducted in a way that is effective and safe for the individual being visited, their families, the facility and its personnel, the handler, and the animal. This requires a clear understanding of shared goals and responsibilities among all parties.

IMPORTANT CONSIDERATIONS

Individual Participation

One of the first issues for an individual wishing to participate in AAA/AAT is to find an organization that provides training, registration, mentoring, and liability insurance. While it is possible to begin visiting as an individual (e.g., taking one's apparently friendly dog to a nursing home) without being part of any formal program, it is strongly recommended to be part of an organization that provides all of the above. Many facilities (and the general public) assume that all visitation animals have gone through training, testing, and registration; have similar health requirements; and, that they are covered by liability insurance. Unfortunately, this is not the case. Variability among different animal visitation groups (or individuals visiting on their own, outside of any organization) can risk the good work of all visitation programs. Among the options for animal visitation organizations that individuals can join, there are several national organizations and hundreds of local organizations. Animal visitation groups can serve a number of roles, including: training owner/handlers; evaluating (both as regards behavior and health status) and registering animals or teams; identifying and matching facilities with visitation teams; and, education and research. Some do all of these while others are more informal; some have very rigorous standards, while others have minimal health and safety requirements and provide no liability insurance. Programs also can vary in their focus including elder care facilities, hospitals, library/school reading programs, or at-risk youth homes. No matter which group is chosen, a number of important factors (listed below) must be considered. Veterinarians talking to owners interested in visitation should encourage them to contact a national or established and reputable local organization. Veterinarians can familiarize themselves with animal visitation programs in their region and what each program's individual requirements are (e.g., behavioral testing, health screening, training, insurance, length of time commitment, etc.).

Program Development

Animal visitation programs may be initiated by a specific facility desiring animal visitations, or by a group of individuals or an organization interested in providing AAA/AAT. In either case, comprehensive, multi-disciplinary planning is critical to success. Additionally, it should be anticipated that successful AAA/AAT will involve different requirements for different types of facilities, e.g., visitation sites and infection control procedures will vary between a rehabilitation center with generally healthy patients versus a children's hospital with a highly immunocompromised patient population. Issues to anticipate at the facility level include, but are not limited to, location for the visits, logistics of facility personnel involvement, questions regarding infection control and allergy control, post-visit cleaning protocols, and psychological impacts (both positive and potentially negative) for patients/families/facility personnel/animal handlers. Development to implementation of a successful, effective and safe facility-based program *de novo* can be expected to take one to two years.

The Animal

Some organizations, such as Pet Partners®, allow multiple species (e.g., dogs, cats, horses, rabbits, guinea pigs), while others are limited to dogs only (which are generally considered the safest animal for AAA/AAT programs). One should also consider the age of animals allowed in the program. This is often a minimum of one year of age, but can vary among organizations. Some organizations, for example, allow puppies or kittens to participate in visitation, but this is not recommended. Because a bond is required for a good working team approach to visitation, another consideration for some organizations is how long the animal has been working with the owner/handler. Some organizations do not allow a newly purchased or adopted animal to be eligible for visitation until it has worked with the owner/handler for a minimum amount of time. The eligibility of animals with certain disabilities (e.g., amputation, blindness) also varies among visitation programs; disabilities among the animals themselves can be either positive or negative attributes depending on the specific program and target population for therapy. Programs that do allow them require the handlers to work with their veterinarian to determine if the animal is physically able to participate in visitation. They can make excellent visitation animals.

Most organizations do not provide training of the animal but for dogs, general obedience or the American Kennel Club's Canine Good Citizen® program are good starting points. However, obedience-type training is not sufficient for an animal to be a good visitation animal (nor are there obedience classes for all the other species that can be visitation animals), and it is important to be aware that not all animals are suitable for visitation. The animal should be controllable, predictable, and reliable in a variety of situations.

Typical 'Attributes of a Great Therapy Animal' (from the Pet Partners website® and personal experiences):

- Is comfortable being crowded by a group of people
- People-oriented/sociable, friendly and confident; willing to focus on the target patient and not be overly focused on the owner/handler
- Will initiate contact, stay engaged, make eye contact, and allow their behaviors to be re-directed
- Is able to cope with stressful situations, e.g., situations with noise, visual distractions, and multiple moving people
- Knows how to respect personal boundaries; doesn't jump up on people (risks for both physical injury and dislodgement of catheters, medical monitors, etc.)
- Is well-mannered and completely non-aggressive towards other animals (including the other therapy animals if visits occur in groups) and people
- Is comfortable being touched, at times awkwardly or persistently
- Is controllable, predictable and reliable despite distractions
- Ability to be cued from different positions
- Able to disregard food or toys on cue, i.e. with a "leave it" command
- Comfort level around health care equipment

The Handler

It is important to emphasize that to practice animal visitation at the highest level of safety and efficacy, it is not only the animal that should have the right characteristics, but also that the handler must have the right skills, aptitude and appropriate motivation. Therefore, some organizations actually focus on training the owner/handler and evaluating the animal/handler *team*.

Typical 'Attributes of a Great Handler' (from the Pet Partners website® and personal experiences):

- Is friendly – makes eye contact, smiles, etc.
- Is a proactive advocate for their animal, but recognizes that the reason for participating in AAA/AAT is the good of the patient, not their or their dog's self-promotion
- Is a considerate and caring leader
- Develops a relationship with their animal that reflects a loving partnership with subtle but consistent control
- Is consistently and unequivocally committed to preparing themselves and their animal for each visit according the program's requirements – no short-cuts
- Assesses a visit before, during and after – their animal's behavior, their own behavior, and the people and physical aspects in the surroundings
- Knows how to enhance their animal's abilities to be its best with the people it's visiting
- The handler is aware of signs of animal stress, behaviors, and positions, and can reassure their animal with cues, or commands (as needed) to help the animal be successful.

Health Requirements

All animals participating in visitation should be healthy and not taking antibiotics at the time of the visit, free of parasites (internal and external), and have appropriate vaccinations, whether or not these are required by the organization with which they volunteer. Facilities that have animals visit should also ask for documentation of these practices for each animal. Pet Partners® is the only national animal visitation organization that has a policy precluding animals eating raw meat foods from participating in their therapy animal program, but local programs may (and should) prohibit raw meat diets. Raw meat diets are an important risk factor for zoonotic disease, particularly in the populations often visited by therapy animals. An additional consideration that should be addressed for visitation animals is raw dried or freeze-dried pet treats that have bacterial risks similar to those for raw meat diets (e.g., rawhide chews, pig ears, bully sticks, freeze-dried liver treats).

Specific animal health screening programs vary among programs, but may include procedures including periodic: physical examinations; CBC and blood chemistry analyses; haircoat, oropharynx and/or fecal cultures for bacterial zoonotic agents; and, screening by serology or direct pathogen detection methods (PCR, antigen-detection) for other zoonotic agents (e.g., *Leptospira*). Involving DVM or other animal health students in both behavioral evaluation and medical screening of the visitation animals can be an excellent, applied, active-learning modality to build on didactic knowledge and practice clinical skills.

Infection control and hygiene are particularly important issues to consider at the levels of the animal handler, the visiting animal, the target patient, and the visitation facility. Handlers should be trained in routine hand hygiene and how to appropriately cleanup after the (hopefully very unlikely) urine or fecal accident during a visit. Visiting animals may be required to be bathed within a certain time period preceding a visit and to wear a vest or coat that both identifies them as allowable in the facility and to reduce dander release to the environment. Bathing also reduces the saliva on an animal's coat, which is a primary source of allergens to people and also reduces risk for transmission of zoonotic disease. The more rigorous the requirements, the less likely the risks will be and the easier it is to gain approval by facilities. Facilities and health care providers should think through the requirements they will institute for their individual programs to minimize health risks. Finally, facilities may limit visits to a single room with easily disinfected floor and furniture surfaces, and have a specific route that the visiting animals take in/out of the facility to allow consistent disinfection immediately after a visit.

Evaluating Teams

An ideal behavior evaluation should test both skills and aptitude of the animal and handler team. Many visitation programs develop and utilize their own behavior evaluation protocols and procedures. Wherever possible, an evaluating team should include several individuals, including both veterinarians and current program animal handlers. As much as possible it is ideal to perform the behavior evaluation of prospective dogs in a setting similar in appearance and "feel" to where the actual visits will take place.

Re-evaluation of Teams

Depending upon the organization, re-evaluation may or may not be required. Pet Partners® requires re-evaluation of teams every two years. As animals age or develop health problems, their suitability for visitation or the type of visitation that is most appropriate also can change. For example, if dog develops osteoarthritis, his ability to walk on slippery floors or climb stairs at a hospital may decline so the types of visits that are appropriate also can change. Some animals may enjoy visitation less as they age (although many seem to enjoy it more and more!) and may tire more easily. This can be difficult for owner/handlers to see, especially when the visitation is so personally gratifying. Therefore, regular re-evaluation of the team by an objective trained evaluator is important. Finding an organization that requires re-evaluation (or requesting re-evaluation of the animal, even if it is not required), helps to keep the activity safe and non-stressful for the animal and also helps to maintain a safe and respected visitation program overall.

Summary

Anecdotal benefits of animal therapy abound, and some formal evidence for benefits exists in the refereed literature. However, there needs to be more peer-reviewed evidence of the various benefits that animal visitation can provide and on the most effective methods to enhance safety for all those involved. Veterinarians and human healthcare providers can work together to enhance the benefits of animal visitation, and to ensure the safest and most effective programs possible.

RECOMMENDED READING

- Barker SB, Wolen AR. The benefits of human-companion animal interaction: A review. *J Vet Med Educ* 2008; 35:487-495.
- Beetz A, Uvnas-Moberg A, Julius H, et al. Psychosocial and psychophysical effects of human-animal interactions: the possible role of oxytocin. *Front Psychol* 2012;3:234.
- Friedmann E, Son H. The human-companion animal bond: How humans benefit. *Vet Clin Small Anim* 2009;39:293-326.
- Kaminski M, Pallino T, Wish J. Play and pets: the physical and emotional impact of child-life and pet therapy on hospitalized children. *Children's Health Care* 2002;31:321-325.
- Lefebvre SL, Golab GC, Christensen E, et al. Guidelines for animal-assisted interventions in health care facilities. *Am J Infect Control* 2008;36:78-85. Available at: http://www.petpartners.org/document.doc?id=659
- Lefebvre SL, Peregrine AS, Golab GC, et al. A veterinary perspective on the recently published guidelines for animal-assisted interventions in health-care facilities. *J Am Vet Med Assoc* 2008;233:394-402.
- Palley LS, O'Rourke PP, Niemi SM. Mainstreaming animal-assisted therapy. *ILAR J* 2010;53:199-207.
- Pet Pals® program, University of Wisconsin-Madison School of Veterinary Medicine and American Family Children's Hospital. Available at: http://www.vetmed.wisc.edu/giving/petpals/
- Pet Partners®. Pet Partners® Therapy Animal Program. Available at: http://www.petpartners.org/page.aspx?pid=259 (accessed January 7, 2015).
- Pet Partners®. Pet Partners® resource on articles and references. http://www.petpartners.org/aaa-and-aat-resources#Horses (accessed January 7, 2015).
- Schaffer CB. Enhancing human-animal relationships through veterinary medical instruction in animal-assisted therapy and animal-assisted activities. *J Vet Med Educ* 2008;35:503-510.
- Selby A, Smith-Osborne A. A systematic review of effectiveness of complementary and adjunct therapies and interventions involving equines. *Health Psychol* 2013;32:418-432.
- Tufts Paws for People. Frequently asked questions (how facilities can start a safe and effective animal visitation program, getting started in animal-assisted therapy, etc). http://vet.tufts.edu/paws/ (accessed January 7, 2015).

UNDERLYING DISEASE SCREENING IN DOGS WITH IMMUNE-MEDIATED HEMOLYTIC ANEMIA

Linda Kidd, DVM, PhD, DACVIM
Pomona, CA

PATHOPHYSIOLOGY OF LOSS OF IMMUNE-TOLERANCE

Immune mediated diseases are important causes of morbidity and mortality in dogs. Studies suggest that the pathology associated with self-attack by the immune system may result from inappropriate specific targeting of self-antigen, modification of self-antigen so that it is foreign, or from a less specific, "innocent bystander" effect.

Several regulatory mechanisms are in place in order to prevent self-attack by the immune system. Central tolerance mechanisms result in the deletion of auto- reactive B and T cells during development. However, evidence suggests that this process is imperfect. Naïve T and B cells that have escaped central tolerance mechanisms and have the ability to target and react to self-antigen are present in normal individuals. Several peripheral tolerance mechanisms exist to prevent these cells from being activated and/or to down-regulate already activated cells to keep them in check.

A central player in maintaining peripheral tolerance is the dendritic cell (DC).[1] The DC is a professional antigen presenting cell that processes and presents antigen in the context of MHC I and II to naïve T cells. CD4 + T cells that engage the MHCII /Ag complex in the presence of a costimulatory signal differentiate into T helper (Th) cell subtypes (ie Th-1, Th-2, Th-17 and others). The type of Th cell subset that develops depends on the signaling/cytokine milieu that is present at the time of antigen presentation. It is important to emphasize that a costimulatory signal [eg binding of DC CD80/86 (B7.1 and B7.2) with T cell CD28] is required to activate a T cell upon initial receptor engagement. Dendritic cells do not express the requisite costimulatory signal unless they have been activated by certain triggers, such as signaling through pattern recognition receptors (PRRs) by pathogen associated molecular patterns (PAMPs). LPS and RNA sequences from bacteria and viruses, respectively, are examples of PAMPs. Cytokines like TNFα also facilitate DC maturation and expression of costimulatory signals. Without the presence of a costimulatory signal the T cell is rendered anergic or undergoes apoptosis. Thus engagement of the TCR in the absence of a costimulatory signal is an important peripheral tolerance mechanism.

Additional signals result in T cell proliferation and polarization as they interact with DCs. These signals include cytokines like IL-12. Signaling induced by cytokines causes differentiation of T cell subtypes that mediate different responses in the immune system. For example, Th-1 cells enhance cell mediated immunity whereas Th-2 enhance activation of B cells, while Th-17 cells help recruit neutrophils to fight extracellular bacteria. The ability of DC to facilitate the differentiation of a particular T cell subset may affect self-tolerance as activation of particular subsets is associated with immune-mediated diseases.

The DC also facilitates tolerance by dampening the immune response after activation. For example, engagement of the DC surface molecule CD80/86 with the Tcell surface molecule CTLA-4 results in downregulation of T cell activity. DCs also appear to direct the differentiation and proliferation of a subset of T regulatory cells known as inducible T regs. However, most T regs are produced in the thymus and are known as "naturally occurring" T regs. These important regulatory cells secrete inhibitory cytokines and exert direct cytotoxicity and inactivate effector T cells by engagement of Fas Ligand and constitutively expressed CTLA-4. Thus T regs limit the potential of uncontrolled immune responses to cause disease.[2] Many other downregulatory mechanisms are in place to control the immune response. Some that may be important in the pathophysiology of immune-mediated disease include complement regulatory proteins and mechanisms that downregulate immunoglobulin production by plasma cells.

Dysregulation of these and other homeostatic mechanisms of the immune system result in immune-mediated disease. For example, recent evidence suggests that insufficient or dysregulated Treg responses plays a role in the loss of tolerance that occurs in IMHA and ITP.[3,4] In addition, a decrease in complement regulatory proteins that protect host cells from complement mediated destruction has been shown in people with IMHA.[3] Defects in CTLA-4 may also contribute to the pathogenesis of ITP and IMHA, although one study failed to identify anti-CTLA antibodies in dogs with IMHA.[5] Mouse models and studies in people support the concept that an imbalance in the production of cytokines plays a role in the pathogenesis of IMHA. For example IL-17 has been shown to be an important mediator of IMHA in mouse models. IL-17 is also increased in human patients with IMHA and levels are correlated with anti-RBC antibody production, and negatively correlated with hemoglobin.[6] Polyclonal lymphocyte activation or activation of DCs that result in co-stimulatory signaling molecule expression has been suggested as a possible mechanisms of loss of peripheral tolerance for human IMHA as well.[3]

Many factors are hypothesized to contribute to dysregulation of the immune system resulting in immune-mediated disease. These include genetic predisposition, and the presence of various triggers such as drugs, infection and other environmental factors, neoplasia, and concurrent inflammatory/immune-mediated disease. Genetic polymorphisms and epigenetic changes targeting genes involved in the induction and down-regulation of the immune response may play a role. Breed pre-dispositions for IMHA are well established in dogs. Polymorphisms in the genes that code for the MHC complex, which provide the "self-context" for antigen presentation to T cells, have also been associated with IMHA in dogs.[7] Defects in complement regulatory proteins have been associated immune mediated disease in people and animal models.

Despite this and other evidence of genetic influence on development of immune-mediated disease, studies of twins have shown that having a genetic predisposition does not necessarily result in the development of immune-mediated disease. Furthermore, not all patients with immune mediated disease have a recognized genetic predisposition. Environmental factors appear to play a role in disease development by impacting the regulation of the immune response or providing triggers that activate the immune system.

Indeed some immune-mediated diseases have been shown to be seasonal in people and in dogs, supporting the role of environmental influence in the development of these diseases. Evidence in human medicine suggests that climactic environmental factors can influence vitamin D and melatonin levels, which appear to have complex effects on immune regulation. Seasonal variation of prevalence of certain infectious agents has also been hypothesized to explain seasonal variation in immune-mediated disease. Interestingly, seasonal variation in the incidence of canine IMHA has been shown in some, but not other studies.[8,9,10,11,12] Furthermore, the time of year with increased incidence of canine IMHA appears to vary with among hospitals and possibly microclimate.[11] Research is ongoing to determine the effect of environmental influences on immune-mediated disease.

Evidence suggests that infections can cause immune mediated disease by a number of mechanisms. Activation of immune cells that target self-antigen may occur by engagement of PRRs by bacterial or virus PAMPs. This may result in DCs expressing self-antigen to simultaneously express the requisite costimulatory molecule on their surface thus providing the second signal to a self-reactive T cell. Molecular mimicry, where an antigen on a pathogen closely mimics a self-antigen might induce an auto-reactive response by this mechanism. Polyclonal activation by superantigens from infectious agents is also theorized to cause activation of auto-reactive T cells by providing a requisite costimulatory signal. Modification of self-antigen so that it is seen as foreign, and nonspecific damage to self as a result of immune activation may also occur during infections. For example, the immune system may appropriately target *Babesia* antigen on a red blood cell destroying or damaging the cell as an "innocent bystander." Antibodies that specifically target red cell antigens also develop in dogs infected with *Babesia* perhaps through exposure or modification of red cell antigens during this process. Thus in addition to causing immune mediated disease through non-specific damage, *Babesia* appears to induce true autoimmunity.

It is important to note that for most infections associated IMHA in dogs, reports are limited to descriptive case reports or retrospective studies discovering infection at the time of diagnosis of IMHA. Larger studies assessing increased risk associated with infection or mechanistic investigation are lacking. That said, for some agents, such as *Babesia*, demonstration of anti-red cell antibodies in dogs after experimental infection has been documented, along with a clinical diagnosis of IMHA. Further study of the mechanisms of induction of IMHA is required to prove cause and effect for most infectious agents.

Like infections, neoplasia also may result in the disruption of tolerance, peripheral immune dysregulation and alteration of self-antigens. For example, in people, lymphomas have been shown to express the costimulatory signal B7 and act as APC's. Surprisingly, tumors can also support the differentiation of naïve T cells.[13] For example, AIHA and ITP are common complicating autoimmune diseases in people with chronic lymphocytic leukemia (CLL). Evidence suggests that neoplastic leukemia cells can act as APC's of red cell antigens and induce AIHA. Mutations, upregulation of fetal proteins, synthesis of proteins that are normally sequestered, and tumor mediated necrosis and cell damage may result in expression and/or alteration of self-antigen susceptible to auto-immune attack.[13]

More than 100 drugs have been associated with immune mediated hemolytic anemia in people. Drug-dependent (meaning the drug must be present for the antibody to bind in vitro) and drug independent mechanisms (true AIHA) are described. Antibodies can form to the drug alone, the drug plus an endogenous carrier molecule or the carrier molecule alone. The exact mechanism of the induction of drug induced immune mediated destruction of red cells is controversial.[14]

IDENTIFYING POTENTIAL TRIGGERS OF IMHA IN DOGS

Infection

Theoretically, infection with any agent could induce immune mediated disease in a susceptible individual, although based on pathophysiology, some may be more likely to induce a particular type of immune mediated disease than others. Obviously, it is not possible to test for every infectious disease agent. Therefore the minimum database in working up these patients should include a thorough review of history and physical examination findings that might suggest a nidus of infection in a particular body system. CBC, chemistry, heartworm testing, imaging, urinalysis and urine culture can also help localize an infectious process. Consideration should be given to fecal examination as parasitic infection has also been associated with IMHA in dogs. Given the overall case context, additional infectious disease screening for particular agents can be somewhat tailored to the individual patient. Variables that should be taken into consideration in the investigation of possible infectious triggers include the specific clinical abnormalities are present, signalment, and geographic locale. Particular consideration should be given to vector borne disease testing as IMHA has been associated with several of these agents (see below). Infection with parasites, bacteria and parvovirus have also been reported as underlying diseases that preceded or were associated with a diagnosis of canine IMHA.[12,15,16]

The clinical abnormalities associated with several vector-borne diseases are similar. IMHA; thrombocytopenia, polyarthritis, vasculitis, positive ANA testing and protein losing nephropathy are common. However some infections are more commonly associated with a given clinical finding than others. For example, *Babesia sp Anaplasma sp, Bartonella sp, Ehrlichia sp, Dirofilaria immitis*, and hemotropic mycoplasmas are examples of vector-borne disease agents that have been associated with IMHA in dogs whereas *Borrelia burgorferi* and *Rickettsia rickettsii* have not. **Repeated testing and testing using more than one modality (ie both PCR and serology) are often required to detect infection with vector-borne disease agents, depending on the agent involved and the phase of infection. Comprehensive serologic testing combined with comprehensive PCR testing is a good starting point.** [17,18] Repeated testing can be justifiably focused on a given agent depending on the constellation of clinical signs, signalment and geographic locale (see below).

In addition to consideration of specific clinical abnormalities, signalment can also help direct infectious disease testing. For certain breeds it is important to be aggressive about ruling out underlying infection that may cause an immune mediated disease. For example it is very important to aggressively rule out *Babesia gibsoni* in a pit bull with IMHA. (Maggi, Kidd) Greyhounds and other dogs exposed to *Rhipicephalus sanguineus* ticks may be at increased risk for *Ehrlichia canis, Babesia canis, Anaplasma platys* and other pathogens thought to be transmitted by this tick. It is also equally important *not* to assume that immune-mediated disease is idiopathic in predisposed breeds. Theoretically, infection is as likely (or perhaps more likely) to cause IMHA in a Cocker Spaniel as any other breed in a similar environment.

Consideration of geographic locale can also help tailor the diagnostic approach to infectious disease testing in these patients. For example, the prevalence of infection or exposure to vector-borne disease in dogs differs by geographic locale. Vector borne disease prevalence varies with geographic distribution of its tick vectors and more aggressive testing for specific agents should be considered in dogs living in particular regions or with a specific travel histories. That said, the geographic distribution of vectors is expanding, thus comprehensive testing with panels may help detect pathogens that would not otherwise be suspected.

Other environmental factors

Other environmental factors such as bee stings and food have been associated with IMHA in dogs. Information regarding dietary changes or insect stings may provide meaningful information when taking a history in these patients.[19,20]

Neoplasia

Both solid tumors (eg lung, breast, ovarian, kidney) and hemolymphatic neoplasms (eg CLL, lymphomas, chronic lymphocytic leukemia) have been associated with AIHA in people. In fact, a recent review concluded that AIHA has been associated with almost all common and some very rare cancers.[21] It is important to note that in patients with early stage resectable cancer, surgery resulted in resolution of AIHA within a few weeks, some had sustained remissions, and that all of these AIHA were refractory to steroid treatment before surgery. Interestingly, in patients with advanced stage cancers with metastasis, a few cases exhibited a beneficial effect of resection of the primary tumor. Investigation into the prevalence of malignancy associated IMHA in dogs has not been well-reported. In dogs, an association of ITP but not IMHA was found with lymphoma in one study.[22] Other neoplasms specifically mentioned in association with a case of IMHA in dogs include sarcoma, carcinoma pheochromocytoma and mastocytoma.[8,15,23] "Neoplasia" is mentioned nonspecifically as a reason to exclude dogs from several studies of idiopathic IMHA. Consideration of signalment, history and collecting a minimum database as described above can help localize the presence of neoplastic disease in dogs with IMHA.

Other inflammatory processes

Pancreatitis, systemic pyogranulomatous inflammation and Lupus have been associated with IMHA in dogs but whether they and other inflammatory and immune mediated diseases act as triggers is unknown.[8,12,15] Identifying and treating concominant inflammatory disorders may theoretically be important in resolving associated IMHA.

Drugs and vaccination

A recent review of drug induced AIHA in people noted that antimicrobials are most commonly implicated in drug dependent form, followed by NSAIDS and anti-neoplastic agents.[14] Potentiated sulfonamides, penicillin, cephalosporins and carprofen have been reported in association with IMHA in dogs. Obtaining a good history and eliminating exposure to potential initiating factors is important because in people removal of the offending drug can result in disease resolution without immunosuppression if the antibodies are drug dependent. Some laboratories provide services to help confirm the mechanism of drug induced hemolysis.

Some but not other studies have demonstrated a temporal relationship between vaccination and IMHA in dogs. It is important to determine the vaccination history when gathering a history in these patients. If a vaccination was given shortly before (within approximately 1 month) it is reasonable to assume that it may have played a role in disease development if other possible triggers have been eliminated.

SUMMARY

Underlying disease including, but not limited to infection, inflammation, neoplasia may result in immune dysregulation and a loss of self-tolerance that results in IMHA in susceptible patients. Environmental factors may also play a role in disease development. A thorough history, physical examination, and baseline diagnostic testing can provide clues that may help localize an infectious, neoplastic or inflammatory nidus, and identify a potential cause. Consideration of case context such as specific clinical abnormalities, signalment, and geographic locale, can help the clinician formulate a diagnostic testing plan that is appropriate for an individual patient. Failure to aggressively investigate and eliminate the presence of underlying disease likely contributes to treatment failure. Further research identifying disease associations and mechanisms of loss of immune tolerance in dogs with IMHA may help prevent and treat this devastating disease.

REFERENCES

1. Chung et al. Clin Dev Immunol 2013 Article ID 972865 online. 2. Shevach 2014 Immunol Rev 259 88. 3. Barros Transf Med Rev 2010 24(3) 195. 4. Nishimoto et al. Semin. Hematol. 50(1) S 1 2013 S 43. 5. Khatlani J et al. Immunother 2003 26(1) 12. 6. Xu et al Exp Hematol 2012 40 994. 7. Kennedy et al. Tissue Antigens 2005 502.8. Klag et al. JAVMA1993 202 (5) 783. 9. Reimer et al. JAAHA 1999 35(5) 384. 10. Burgess et al. JVIM 2000 14(4) 456. 11 Kidd et al. JVECC 2014 24(3) 311. 12. Carr et al. 2002 JVIM 16 504. 13. Maverekis et al. Clin Rev Allerg Immunol 2012 42 135. 14. Garratty Blood Rev. 2010 24(4-5):143. Jackson et al Can Vet J 1985 26 245.16. Lobetti et al. 2001 J S Afr Vet Assoc. 72(1):52. 17. Maggi et al Parasites and Vect 2013 6 103. 18. Kidd et al. JVIM 2014 28:976 1064.abs. 19. Smith et al JVECC 2006 16(2) S1. 20. Noble et al JAVMA 1999 214(7) 1026. 21. Puthenparambil et al. Wien Klin Woch 2010 122 229. 22. Keller Cancer 1992;70(9):2334. 23. Mellanby JSAP 2004 45 (1) 21. 20.

PERSPECTIVES ON IMMUNE-MEDIATED DISEASE

Jeffrey Toll, VMD, DACVIM (SAIM)
Miami Beach, FL

INTRODUCTION

Immune-mediated diseases commonly observed in small animal veterinary practice include (1) immune-mediated hemolytic anemia – IMHA, (2) immune-mediated thrombocytopenia – ITP, (3) immune-mediated polyarthritis - IMPA, (4) glomerulonephritis - GN, and (5) systemic lupus erythematosus (SLE). Severe morbidity and mortality related to these and other immune-mediated syndromes continue to be observed despite ongoing advances in basic and applied scientific disciplines including immunopathology, genetics, infectious disease detection, and clinical pharmacology. Over the last 25-30 years, diagnostic and therapeutic algorithms for immune-mediated diseases have evolved both in conjunction with and (sometimes) in spite of these ongoing advances and often out of frustration with continuing morbidity and mortality. With references to the existing scientific evidence pertaining to immune-mediated diseases and syndromes, today's discussion uses case examples to explore principles of clinical immunology, clinical decision-making, and possible novel diagnostic and therapeutic strategies. This clinical workshop is intended to stimulate discussion of specific aspects of existing clinical practices and standards of care that may affect diagnostic and therapeutic success, as well as stimulating areas of future scientific study of immune-mediated diseases.

AUTOIMMUNE DISEASE VS IMMUNE-MEDIATED DISEASE

The terms "autoimmune disease" and "immune-mediated disease" are often used somewhat interchangeably in clinical practice. I'd like to start our discussion by suggesting that the synonymous use of these 2 terms may inadvertently bias the way in which we conceptualize these diseases, and therefore it may affect the way we approach the diagnostic, therapeutic, and monitoring process in individual patients. As discussed in a companion lecture in this workshop, there are cases in which a specific immunologic mechanism can be documented to target a specific organ (e.g. anti-platelet antibody can be identified in some patients with ITP). Such mechanisms are the result of complex immunopathology leading to (in this example) autoantibody production, pathology which is not a diagnosis unto itself. In other words, the production of autoantibody (or complement or cytotoxic T cells or other highly specific and targeted mechanisms of immunologic damage) does not prove causation. The clinical manifestation(s) of an antigen- specific injury or attack against the "self" based solely on a genetic defect is no longer believed to be a fully accurate characterization. Even in human SLE, the archetypal autoimmune disease, manifestations of disease are considered the result of the interaction between (often multiple) "susceptibility genes" and environmental factors, alluding to the widely accepted concept of the immunologic "trigger."[1] A trigger is effectively defined as an agent or event or other exposure(s) that is capable of provoking the development of an immune-mediated consequence or disease. The trigger concept is therefore entirely coexistent with any hereditary or genetic predisposition towards immune-mediated disease that an individual may possess. Triggers may then contribute to the loss of tolerance to self antigens, but they may also result in a less immunologically specific "innocent bystander" type of injury, for example, as produced by immune complex deposition (e.g. glomerulonephritis). Detailed description of molecular, humoral, and other aspects of normal and abnormal immune function will be covered in a companion lecture.

CASE EXAMPLE

A case example will help to illustrate some of the aspects of immunologic triggering that could provide insight into a typical challenging clinical patient seen in a primary care or referral practice. A five year old male neutered Labrador Retriever in south Florida was presented for evaluation of persistent (3 months), severe, IMPA. The dog had an uneventful medical history and received yearly exams, vaccines, and heartworm preventative. He initially presented 8 months after his last annual appointment. The primary care veterinarian had previously performed a complete workup, including a detailed history, physical exam, and diagnostic investigation which included CBC, biochemical analysis, urinalysis, fecal flotation, HW Ag, vector-borne disease screening, radiographs (thorax, abdomen, multiple joints), Coomb's/ANA testing (ANA positive at 1:64), and arthrocentesis. Non-septic suppurative inflammation was identified in multiple joints and culture of blood and all body fluids were negative. Response to empirical doxycycline and immunosuppressive dosages of prednisone was partial/incomplete and was unaffected by addition of a second immunosuppressive agent (azathioprine), prompting referral. Physical exam revealed a Cushingoid-appearing dog with a 1/6 early systolic cardiac murmur, moderate cranial organomegaly, and moderate warm painful swelling of multiple large joints. All routine testing except for advanced imaging had been performed prior to referral, so those tests were performed in order to complete the data base. Abdominal ultrasound revealed presumptive steroid hepatopathy. Echocardiography revealed trivial mitral regurgitation and a solitary adult heartworm (subsequent antigen and microfilaria testing was repeatedly negative). Heartworm adulticide therapy successfully eliminated the heartworm, at which time polyarthritis resolved completely and all immunosuppressive medication was withdrawn. The ANA was negative within 6 months of heartworm treatment.

IMMUNOLOGIC TRIGGERS OF IMMUNE-MEDIATED DISEASE

Immunologic triggers must be sufficiently antigenic or otherwise influential to the immune system to initiate the cascade of events leading either to the loss of self-tolerance, the production of antigen-antibody complexes, or other mechanisms of immune-mediated tissue injury. In the human literature, although some associations between cause/trigger and effect/immune-mediated disease have already been made, investigators speculate that a vast array of chemical, bacterial, viral, fungal, and other agents may produce or conspire to produce the proliferation of immune-mediated diseases observed since the 1940's. The strength of the veterinary scientific evidence varies with the type of immune-mediated disease, the general category of the described trigger (infectious, pharmacologic/biological, environmental, dietary, neoplastic), as well as within the proposed trigger categories. For example, infection with *Babesia spp.*, hemotrophic *Mycoplasma spp.*, and other vector-borne pathogens has been shown to result in the production of anti-erythrocyte antibodies.[2] It is therefore not surprising that the search for vector-borne pathogens is a cornerstone of the diagnostic paradigm in many immune-mediated syndromes (and rightfully so in many geographic locales). Over 50% of dogs with infective endocarditis in one study presented with polyarthritis[3] (this was the rationale for performing the echocardiogram in the case under discussion), thus the association between the two is relatively common and diagnostically reliable. Beyond a few selected infectious diseases in selected immune-mediated disease states, the scientific evidence becomes much weaker when considering the multitude of other infectious agents, and dietary, environmental, and other ubiquitous antigens. Although the association between neoplasia and immune-mediated disease has been made in companion animals (for example: thymoma and myasthenia gravis in dogs), immune-mediated diseases do not generally dominate the clinical presentation of canine and feline neoplasia. Using canine immune-mediated hemolytic anemia as a reference point, where the search for a trigger is unsuccessful in a majority of clinical patients and published mortality rates are 30-70%, it is reasonable to speculate that there are a multitude of triggers, known or unknown and overlooked or under-recognized with regard to causation. We can look to our existing understanding of basic immunology for potential explanations for the lack of diagnostic and therapeutic success in veterinary patients.

In the clinical case example of polyarthritis associated with occult heartworm infection, important clinical and immunologic principles may be illustrated. As a common organism in the south Florida community where the patient lives, heartworm is a ***common antigen***, regardless of whether or not it is acting as a pathogen. The dog did not have heartworm disease as typically defined by clinical signs, routine antigen testing, or thoracic radiography. Based on history, physical examination, and routine diagnostics, heartworm disease was not on the differential diagnosis in this case, but heartworm infection was present. Consider the possibility/probability that ***common antigens are the most likely triggers of common immune-mediated diseases, regardless of their ease of identification or the degree/amount of antigen/trigger present.***

This leads us to consider the concept of ***immunologic hypersensitivity*** in the context of immune-mediated disease and the case example. Subacute or chronic hypersensitivity may be a useful construct within which to consider the triggering of the loss of self-tolerance, the development of immune complex deposition, and other mechanisms of immunologic damage. A solitary adult male heartworm had no apparent typical cardiopulmonary consequence in the patient in question, but it was presumably capable of stimulating immune complex disease manifesting as polyarthritis. A solitary parasite was ***sufficiently antigenic to induce a clinically apparent life-threatening immune response but insufficient in quantity to be readily measured*** through conventional antigen testing (immune complexes may interfere with heartworm antigen testing as well). While the concept of anaphylactic/acute hypersensitivity is based on the presence of a relatively minute or undetectable/immeasurable amount of antigenic trigger, and other hypersensitivity-based diseases are relatively common, we do not generally apply this concept to the subacute and chronic immune-mediated syndromes under discussion. Feline allergic bronchitis/asthma is an example of a common hypersensitivity-based syndrome. Also consider *Trichuris vulpis* (whipworm) infection in some dogs, where clinical signs are sometimes severe relative to the worm burden, where detection of parasite ova/diagnostic confirmation is challenging in the face of severe clinical signs. Immunologic hypersensitivity may provide one explanation for the limitations in the existing diagnostic paradigm for immune-mediated diseases, insofar as ubiquitous environmental, infectious, and others agents may be marginalized as potential triggers simply because they are overly common, not classically associated with the observed disease manifestation, or ostensibly non-pathogenic or generally benign.

In the case example, heartworm is an antigenic trigger rather than what is normally considered to be a pathogen. Extending this concept further, it may be applied to other infectious and other microorganisms, as well as to other non-infectious antigen categories. Regarding the role of some microorganisms and infectious agents in the triggering of immune-mediated diseases, ***it may be as important for an infectious agent to be antigenic than to be pathogenic or invasive***. This conceptually expands the list of potential antigenic triggers in some unexpected ways and is supported by scientific evidence regarding microbial, parasitic, and other infections. Consider that the heartworm in this dog is a foreign antigen in blood and then consider risk factors for exposure to other potential foreign antigens that may enter the bloodstream. Heartworm microfilaria must have preceded development of an adult heartworm – repeated/ongoing exposure to microfilariae, even if transient, could be antigenic in a hypersensitive individual. There is evidence in humans that serial exposures/infections are as potentially important as acute or chronic infections in triggering autoimmune disease in genetically predisposed individuals, introducing the concept of the ***total burden of infections over time.*** Infection with heartworm also implies potential or even likely exposure to *Acanthocheilonema reconditum* (formerly *Dipetalonema*), a *common antigen* in heartworm endemic areas. These filariae are not identified by conventional heartworm antigen testing but they were frequently identified through microfilaria (Knott) testing prior to the availability of the heartworm antigen test and are still prevalent in *Dirofilaria*-endemic regions.. It is interesting to speculate whether infection with a so-called non-pathogen like *A. reconditum*, serving as a foreign antigen in blood, could be associated with a commonly idiopathic immune-mediated disease in dogs such as IMHA or ITP. Likewise non-spotted fever group *Rickettsiae*, to which exposure/infection is documented in selected geographic regions,[4] common gastrointestinal parasites and bacteria, and a whole host of other organisms. Gram-negative bacterial cell wall, viruses, and other agents, pathogenic and non-pathogenic, have also been associated with immune-mediated diseases in humans and mice (which may explain prodromal or comorbid vomiting and diarrhea sometimes observed in dogs presenting with IMHA and other immune diseases). Extending the concept of hypersensitivity to environmental, dietary, and other antigens may have additional value.

Lastly, because of the complex relationship between the trigger and the susceptibility genes, it may not be easy to predict which type of immune-mediated sequelae may be produced by a particular trigger. Put another way, ***the association between a particular antigen and ultimate disease expression may seem improbable*** until the scientific body of knowledge is expanded. Consider the fact that heartworm does not infect joints. Consider the data showing that up to 30% of human patients with inflammatory bowel disease (IBD) develop polyarthritis. As a practical matter, the search for triggers/common antigens may be somewhat similar for a whole host of otherwise diverse immune-mediated syndromes.

DIFFERENTIATING BETWEEN DIAGNOSTIC AND THERAPEUTIC FAILURE IN IMMUNE-MEDIATED DISEASE

Referral of the heartworm-infected patient was initiated because of perceived therapeutic failure. The medical work-up performed prior to referral was relatively complete. Whether or not the average practitioner would have performed a different battery of tests is worthy of discussion. However, there was no specific reason to doubt the quality of the workup, the results, or the compliance of the owners.

In modern clinical practice, the diagnostic process for investigation of immune-mediated diseases appears to be algorithmic but non-standardized. It is often based on a number of factors including the specific disease in question, the severity of the disease, infectious disease prevalence, experience of the individual clinician, published materials, and other factors. To some extent, the clinician's comfort and confidence in the diagnostic algorithm and the quality of the acquired data base will determine how ongoing case evolution is interpreted and case management is conducted. Published diagnostic success rates for common immune-mediated syndromes such as IMHA, ITP, and polyarthritis are relatively low. As a result, transient, delayed, or poor clinical responses to therapy are often interpreted as manifestations of therapeutic failure, especially when specific tests for immune-mediated disease such as Coomb's and/or ANA are positive (our heartworm dog was positive for ANA) or a known breed association is concerned. However, consistent interpretation of ongoing disease expression as therapeutic failure is neither logical nor necessarily accurate. As the heartworm-infected dog illustrates, ongoing clinical disease may reflect persistent immunologic activation. While immunosuppressive therapy may diminish the immune response to a variable extent, it may not produce a complete or durable remission. If we again invoke the concept of immunologic hypersensitivity, we may understand why anything less than complete immunoablative therapy or removal of the trigger(s) would fail to produce the desired result. From a clinical perspective, differentiating a diagnostic failure from a therapeutic failure may pose an ongoing challenge. In cases where the response to therapy is rapid, complete, and durable, there is little incentive or need to remain diagnostically vigilant. However, based on the concept of immunologic hypersensitivity, ***the presence of a persistent trigger is most likely in cases where response to therapy is delayed, inadequate, or transient.*** In such cases, ongoing diagnostic testing and creative empiricism based on antigenic prevalence may be worthwhile adjuncts to more traditional therapeutic trials.

REFERENCES

[1]Bach J-F. **New England Journal of Medicine** 2002;347(12):911-920. [2]Farwell GE et.al **J Am Vet Med Assoc** 1982 Mar 1;180(5):507-11. [3]Sykes, JE, et.al J Am Vet Med Assoc 2006 Jun 1;228(11):1735-47. [4]Comer JA et.al J Am Vet Med Assoc 2001 Jun 1;218(11):1780-2.

EVIDENCE-BASED APPROACH TO IMMUNE-MEDIATED DISEASE

Jeffrey Toll, VMD, DACVIM (SAIM)
Miami Beach, FL

INTRODUCTION

Immune-mediated diseases commonly observed in canine and feline patients include (1) immune-mediated hemolytic anemia – IMHA, (2) immune-mediated thrombocytopenia – ITP, (3) immune-mediated polyarthritis, (4) glomerulonephritis - GN, and (5) systemic lupus erythematosus (SLE). Severe morbidity and mortality related to these and other immune-mediated syndromes persists despite ongoing advances in basic and applied scientific disciplines including immunopathology, genetics, infectious disease detection, and clinical pharmacology. Over the last 25-30 years, diagnostic and therapeutic algorithms for immune-mediated diseases have evolved both in conjunction with and (sometimes) in spite of these ongoing advances and often out of frustration with continuing morbidity and mortality. With references to the existing scientific evidence pertaining to immune-mediated diseases and syndromes, today's discussion uses case examples to explore principles of clinical immunology, clinical decision-making, and possible diagnostic and therapeutic strategies. This clinical workshop is intended to stimulate discussion of specific aspects of existing clinical practices and standards of care that may affect diagnostic and therapeutic success. Please refer to the companion lecture in this series, "Perspectives on immune-mediated disease," as the following material is a continuation thereof.

PERSPECTIVES ON A RATIONAL APPROACH TO IMMUNE-MEDIATED DISEASE

In the first half of this discussion, entitled "Perspectives on immune-mediated disease," the following definitions, immunologic, and clinical concepts were explored as they pertain to the approach to immune-mediated disease...

Use of the term "autoimmune" may serve to overemphasize the role of genetics in the development of disease

Existing evidence supports the idea that a combination of susceptibility genes and environmental factors result in the development of disease in most clinical scenarios. Use of the term "immune-mediated" may be more descriptive and objective as it relates to our concept of these diseases, and it may serve to reinforce the necessity to consider acquired causes of disease throughout the evolution of an individual case, especially given that susceptibility genes are not easily quantified in a clinical setting at this time.

The concept of the immunologic "trigger" is central to the diagnostic workup

A trigger is effectively defined as an agent or event or other exposure(s) that is capable of provoking the development of an immune-mediated consequence or disease. In most cases, the trigger may be considered to be an antigen within one of the following categories...infectious/microbial/biological, environmental, chemical, pharmacologic, dietary, and neoplastic.

Common triggers are most likely to be common antigens

Although some cause-and-effect relationships between infectious and (few) other antigens, and specific immune-mediated sequelae have been documented, the scientific evidence suggests that a vast array of common antigens in multiple categories are theoretically capable of triggering a variety of immune-mediated diseases and syndromes for which an etiology has historically remained elusive. It is intuitive that antigens common to the patient's environment would be most likely to come in contact with the patient's immune system, whether they are infectious agents or other common microbial or other antigens.

Immunologic hypersensitivity is a useful lens through which to view common triggers

Immunologic hypersensitivity may provide one explanation for the limitations in the existing diagnostic paradigm for immune-mediated diseases, insofar as ubiquitous environmental, infectious, and others agents may be marginalized as potential triggers simply because they are overly common, not classically associated with the observed disease manifestation, or ostensibly considered non-pathogenic.

"Antigen" may be as important as "pathogen" in triggering disease expression

Acceptance of this possibility expands the differential diagnosis for immune-mediated diseases, not only with regard to infectious and other microorganisms, but also to the inclusion of a whole array of other common antigens.

Hypersensitivity may assist in understanding ongoing disease expression in the face of immunosuppressive therapy

Delayed, inadequate, or transient responses to immunosuppressive therapy suggest ongoing immunologic stimulation as much as they do therapeutic failure. Ongoing diagnostic vigilance may be a necessity in the continued care of the patient with immune-mediated disease unless remission is rapid and sustained. Given the potential array of antigenic triggers, detailed history and creative empiricism may be as important as routine testing, antimicrobial and immunosuppressive drugs, and other therapy.

Diagnostic and therapeutic failures may be clinically indistinguishable

Side-effects or other complications of therapy, such as opportunistic infections, may also create the impression of therapeutic failure. Consider the benefits of additional historical, environmental, and diagnostic information when attempting to manage patients whose disease is refractory, recurrent, or progressive. Couple the tendency to intensify immunosuppressive drug treatments with review, repetition, and expansion of the diagnostic data base.

A host of diverse immune-mediated syndromes share a common immunopathogenesis

With an understanding that genetic and acquired factors may result in a variety of immune-mediated diseases in humans and animals, and in light of the current state-of-the-art regarding the understanding of mechanisms whereby such diseases may occur, it is possible to conclude that a number of common diseases and syndromes observed in canine and feline practice can be approached using the concepts outlined above. These include IMHA, ITP, IMPA, GN, SLE, and others (anterior uveitis, vasculitis, meningoencephalitis, etc.).

Using additional case examples, we can continue to develop additional concepts consistent with the evidence-based approach to immune-mediated disease…

DIFFERENTIATING BETWEEN DIAGNOSTIC AND THERAPEUTIC FAILURE IN IMMUNE-MEDIATED DISEASE (CONTINUED)

Immune-mediated diseases may not be self-perpetuating if the trigger or triggers can be identified and/or eliminated. In the companion lecture in this series, we discussed a canine patient with IMPA secondary to infection with a single adult heartworm. Unremitting polyarthritis and pyrexia were observed while the patient was receiving treatment with two immunosuppressive drugs. We used this case example to illustrate the possibility that immunosuppressive therapies may not be capable of complete suppression of immune-mediated disease if the immunologic trigger for that disease is not eliminated; i.e. the patient is experiencing ongoing hypersensitivity to an ongoing antigenic trigger. The opposite may also be true in some cases…elimination of an immunologic trigger may be sufficient to resolve the disease without the need for specific immunosuppressive therapy. Consider the following patient…

CASE EXAMPLE

A 7 month old male neutered mixed breed puppy in south Florida was referred for evaluation of rapidly progressing anemia and thrombocytopenia. The dog had been adopted from a local animal shelter 10 days prior to referral, where routine immunizations had been administered 14 days prior to referral. The shelter veterinarian prescribed doxycycline the prior day. On presentation, the puppy was febrile and pale, with multifocal petechiations, tachycardia, and splenomegaly. A moderate regenerative anemia (HCT = 20%), moderate leukopenia (WBC = 2,200), and severe thrombocytopenia (17,000) were present. Slide agglutination was positive and the Coomb's test later returned positive as well (1:16). Numerous *Babesia canis* and occasional *Anaplasma platys* organisms were identified on examination of a peripheral blood smear. Radiographs confirmed splenomegaly only. An *Ehrlichia canis* titer was negative on presentation but not repeated. Fecal flotation and heartworm testing were negative. Urinalysis was normal. Treatments included (1) a single transfusion with fresh whole blood, (2) administration of imidocarb diproprionate on day 1 and day 15, (3) continued doxycycline administration, (4) empirical deworming, and (5) fluid and nutritional support. No steroids or other immunosuppressive treatments were administered. Pyrexia and novel petechiations resolved within 48 hours, the complete blood count normalized by Day 8, and the Coomb's test was negative at Week 3.

DISCUSSION

The main purpose of presenting this specific case is to highlight resolution of an immune-mediated hemolytic anemia solely with an infectious disease treatment protocol. Although positivity for anti-erythrocyte antibody in this scenario is further confirmation of an immune-mediate hemolytic process already well-documented in dogs infected with *Babesia canis*, it is not unusual for dogs with acute babesiosis or anaplasmosis to present without easily identified circulating organisms in blood. In such cases, the differential diagnosis becomes relatively lengthy and an empirical therapeutic protocol is generally initiated hours or days before a confirmed diagnosis, if confirmation is indeed ever obtained. Immunosuppressive therapy is often a component of such empirical therapy and may be justifiable in light of clinical circumstances that may include rapid or severe hemolysis, life-threatening spontaneous hemorrhage, lack of access to blood products, and other factors. However, the ability of an infectious disease treatment protocol to interrupt and resolve an aggressive immune-mediated disease process is worthy of reflection. Further, if all patients with immune-mediated disease are reflexively treated with immunosuppressive therapy, we can no longer appreciate the patients for which such treatment is unnecessary or potentially harmful…

CASE EXAMPLE

A nine month old male neutered Yorkshire terrier in south Florida presents to the primary care veterinarian for acute lethargy and discolored urine. Initial assessment reveals the following abnormalities…(1) pyrexia (2) splenomegaly on radiographs; no other radiographic abnormalities (3) regenerative anemia (4) a negative in-house test for HW/E. canis/Anaplasma (5) positive saline agglutination test (6) spherocytes and reticulocytes observed on peripheral blood review by a clinical pathologist (7) hemoglobinuria. Doxycycline, imidocarb, prednisone, and iron supplementation were prescribed. Azathioprine was added on Day 4 when no response to treatment was observed and the puppy was referred for a blood transfusion on Day 8. On admission, large numbers of *Histoplasma capsulatum* were observed within neutrophils and macrophages during examination of a peripheral blood smear and a diagnosis of histoplasmosis was made.

DISCUSSION

The clinical presentation of this puppy was quite similar to the presentation of the puppy with babesiosis/anaplasmosis except that an infectious disease was not identified on the initial exam of a peripheral blood smear. Diagnostic testing did reveal an immune-mediated anemia. Evidence of vector-borne or other infections was not found on routine serologic testing, peripheral blood exam, or in urine, feces, or survey radiographs. Splenomegaly, present in about 40% of dogs with IMHA, was not considered sufficiently discriminating to imply a specific etiology. Since histoplasmosis is considered rare in south Florida, it was not on the differential diagnosis. With the completion of the initial data base, the decision to use immunosuppressive therapy was based on a constellation of clinical findings that would likely satisfy the average clinician's diagnostic requirements for a diagnosis of immune–mediated disease. However, ***ongoing diagnostic efforts in the face of continued disease expression provided the opportunity to make an etiologic diagnosis rather than a diagnosis of idiopathic disease***.

The only difference between the two case examples was that the dog with babesiosis originated in Florida, and the dog with histoplasmosis was obtained from a Florida pet shop that sourced their puppies from Missouri. While ***H. capsulatum*** is more commonly associated with a chronic non-regenerative anemia than an acute immune-mediated anemia, it ***is a common antigen in Missouri*** and therefore must be considered a more likely trigger for an immune-mediated syndrome.

The use of immunosuppressive therapy in this dog was justified on the basis of identification of an immune-mediated anemia and lack of an obvious underlying disease. However, this therapy was believed to be responsible for the more rapid dissemination of the organism, clinical progression of disease in the face of therapy, and the prolonged time to recovery that was subsequently observed. Observation of ongoing disease expression and patient death could have reasonably been expected to occur with continuation of the initial treatment plan. Other potential side-effects of immunosuppressive treatments include adverse reactions, opportunistic infections, and expense of therapy and monitoring.

SCIENTIFIC EVIDENCE REGARDING ON MORBIDITY, MORTALITY, AND TREATMENT EFFICACY, AND FUTURE DIRECTIONS

Consider the concepts we have been discussing when critically evaluating existing and future scientific studies of immune-mediated diseases such as IMHA, ITP, IMPA, GN, and SLE, or when designing future studies.

REFERENCES

[1]Chang YT et al. Postgrad Medicine J. 2010 Jul;86(1017)443-4. doi 10. 1136/pgmj.2010.099465.

THE BEST VETERINARY IMMUNOLOGY PAPERS OF 2014

Linda Kidd, DVM, PhD, DACVIM
Pomona CA

INTRODUCTION

Quality research is the cornerstone of advancement in the prevention, diagnosis and treatment of disease in dogs and cats. In recent years, the quality and quantity of research in canine and feline internal medicine and immunology has increased dramatically, possibly due to advances in basic science and human clinical research, increased opportunities for veterinarians to train in research, the creation of funding opportunities specific to companion animal health, and efforts to make publication criteria more stringent and consistent. The "best" veterinary immunology papers of 2014 were identified as follows.

METHODS

Pubmed.gov of the United States National Library of Medicine National Institutes of Health was searched using the terms (dog OR cat) or (canine OR feline) AND immune*. The publication date filter was set from January 1st 2014-December 31st 2014. e-publications ahead of print, articles that did not investigate immunology questions pertaining to dogs or cats, and review articles were eliminated from the initial list by scanning titles and abstracts. Studies that investigated basic immunology; the immunopathogenesis of infectious disease, immune-mediated disease or neoplastic disease; immunology in the context of diagnostic testing for infectious or immune-mediated disease; and immunology in the context of treatment of immune-mediated, infectious or neoplastic disease were identified. Studies pertinent to diseases commonly seen by internal medicine specialists were given special consideration. Quality of scientific method was evaluated by examining the authors basis for hypothesis generation, the use of appropriate methodology and whether conclusions were justified. To ensure that a diversity of topics of interest to internists were represented, efforts were made to include papers representing several of the above categories in the final list of best papers. However, bias to what the author found most interesting and important was inherent in the selection process.

RESULTS

Using the search terms specified in the methods, 387 articles were initially identified. 247 were eliminated for either being unrelated to cats or dogs or immunology, review articles, or preliminary e-publications. 140 articles were scrutinized further according to the criteria listed in methods. Categorization of the papers with regard to topic were as follows: basic immunology n=13, immunopathology of infectious disease (including vaccine development) n=64, immunopathology of immune-mediated disease n=38, immunopathology of neoplastic disease n=5, immunology in the context of diagnostic testing for immune-mediated disease n=2, immunology in the context of treatment of immune-mediated disease n=4, immunology in the context of treatment of infectious disease n=7, immunology in the context of diagnostic testing for infectious disease n=2, immunology in the context of treatment of neoplastic disease n=5.

Selected papers in alphabetical order by the last name of the first author were as follows:

1. Caviezel LL, Raj K, and Giger U. Comparison of 4 direct Coombs' test methods with polyclonal antiglobulins in anemic and nonanemic dogs for in-clinic or laboratory use. J Vet Intern Med 2014 28 593-591.

2. LeVine DN, Birkenheuer AJ, Brooks MB, Nordone SK, Bellinger DA, Jones SA, Fischer TH, Oglesbee SE, Frey K, Brinson NS, Peters AP, Marr HS, Motsinger-Reif A, Gudbrandsdottir S, Bussel JB, Key NS. A novel canine model of immune thrombocytopenia: has immune thrombocytopenia (ITP) gone to the dogs? British J Haematol, 2014, 167, 110–120.

3. Maeda S, Ohno K, Uchida K, Igarashi H, Goto-Koshino Y, Fujino Y, Tsujimoto H. Intestinal protease-activated receptor-2 and fecal serine protease activity are increased in canine inflammatory bowel disease and may contribute to intestinal cytokine expression. J. Vet Med. Sci. 76(8): 1119-1127.

4. Miller MM, Petty CS, Tompkins MB, Fogle JE. CD4+CD25+ T regulatory cells activated during feline immunodeficiency virus infection convert T helper cells into functional suppressors through a membrane-bound TGFβ / GARP-mediated mechanism. Virology Journal 2014, 11:7.

5. Nafe L A, Dodam JR, Reinero CR. In-vitro immunosuppression of canine T-lymphocyte-specific proliferation with dexamethasone, cyclosporine, and the active metabolites of azathioprine and leflunomide in a flow-cytometric assay. Can J Vet Res 2014;78:168–175.

CONCLUSIONS

Many high quality papers investigating questions in the field of canine and feline immunology were published in the year 2014 making the selection process difficult. Continued efforts to expand funding opportunities and further training of veterinary researchers will encourage additional quality research that advances the health and well-being of cats, dogs and people.

THYROID CARCINOMA: UPDATES IN DIAGNOSIS, SUPPORTIVE CARE, AND DEFINITIVE TREATMENT

Sandra Bechtel, DVM, DACVIM (Oncology) and Anne Corrigan, DVM, DACVIM (SAIM)
Columbia, MO and Grenada, West Indies

INTRODUCTION

Thyroid anatomy/ physiology

The thyroid gland consists of groups of cells arranged in follicles around central thyroglobulin containing colloid. The cells produce and secrete triiodothyronine (T3) and thyroxine (T4) which are created from iodides combined with and stored in the thyroglobulin, and can be removed, transported through the cells and released upon stimulation of thyroid stimulating hormone (TSH), which is necessary for all steps in thyroid hormone synthesis and release. This organ requires a large blood supply which carries the iodides necessary for hormone construction. T3 and T4 are released and are transported through the blood on carrier proteins to the target cells and are absorbed into cells slowly, most all of the T4 is converted to T3 which has a high affinity for intracellular and nuclear receptors and cause increases in the transcription of a large number of proteins and other substances. The ultimate function of thyroid hormones is to increase metabolism, high quantities can increase the basal metabolic rate 60-100%.[1]

Thyroid tumors

Most thyroid tumors arise from the thyroid gland, however, they can also develop in residual thyroid tissue, present from the base of the tongue to the base of the heart. In necropsy studies, benign adenomas are common and can represent up to 50% of thyroid masses, however, these masses are small, noninvasive, silent, and not clinically detectable and are incidental findings. In dogs, approximately 90% of clinically detectable tumors are thyroid carcinomas and 10% are adenomas, whereas in cats only 1-3% of clinically detectable thyroid masses are malignant carcinomas.[2]

CANINE THYROID CARCINOMA

Most canine thyroid carcinomas are of follicular cell origin, which is subdivided histologically into papillary, follicular, compact (solid), mixed or anaplastic. Although these subtypes are prognostic in people, they have not been found to be prognostic in dogs. With immunohistochemistry (IHC), all follicular carcinomas stain positive for thyroglobulin and thyroid transcription factor 1. Parafollicular or medullary (C-cell) carcinomas are less common, representing 36% of thyroid carcinomas, and stain positive for calcitonin, calcitonin gene-related peptide, thyroid transcription factor-1, chromogranin A, and neuron specific enolase; parafollicular thyroid carcinomas are negative for thyroglobulin with IHC. Thyroid carcinomas are most often large, invasive, and poorly encapsulated. The etiology of thyroid neoplasia in dogs is unknown. Thyroid tumors retain thyroid stimulating hormone (TSH) receptors and a study of hypothyroid beagles that did not receive thyroid supplementation revealed increased incidence of thyroid tumors, leading to a suspicion of TSH trophism without normal negative feedback. Thyroid irradiation is associated with an increased incidence of thyroid carcinoma in people, rodents and dogs. And an iodine deficient diet is also hypothesized to increase TSH and potentially increase risk of thyroid tumor development.[2-4]

Signalment, presentation, and clinical signs

The median age at diagnosis is 9 to 11years, with golden retrievers, beagles, boxers, and Siberian huskies overrepresented. Familial medullary thyroid carcinoma has been reported in a family of Alaskan malamutes. There is no sex predilection.

The left and right thyroid glands are affected with equal frequency and up to 60% are bilateral. Ectopic thyroid tumors are found anywhere between the base of the tongue and the base of the heart, with the tongue base, cranial mediastinum, and heart commonly reported. Thyroid carcinomas tend to be large, infiltrative, and fixed ventral cervical masses, although well circumscribed and freely moveable thyroid carcinomas are reported in 30% of dogs.

A palpable cervical mass is the most common presenting complaint. Differential diagnoses include an abscess, granuloma, salivary mucocele, lymphatic metastasis of a tonsillar squamous cell carcinoma, carotid body tumor, or sarcoma. Clinical signs include cough, tachypnea, dyspnea, dysphagia, dysphonia, laryngeal paralysis, Horner's syndrome, and facial edema. Acute hemorrhage can occur if the cervical vasculature is invaded by tumor. Although uncommon, signs of hyperthyroidism may also be present including polyuria, polydipsia, polyphagia, weight loss, and muscle wasting.[2]

Diagnosis and staging

A complete blood count, chemistry panel, and urinalysis do not reveal changes specific to thyroid carcinoma. In 60% of dogs with thyroid carcinoma, a thyroid panel will reveal euthyroidism; approximately 10% are hyperthyroid and 30% are hypothyroid (likely because of destruction of normal thyroid tissue or suppression of TSH). Loco-regional lymph node palpation and cytology are indicated in cases of suspected thyroid carcinoma, in addition to thoracic radiographs. Abdominal ultrasound should be performed if indicated. Fine needle aspiration and cytology should also be performed, noting that non-diagnostic samples are common because of hemodilution, since thyroid carcinomas have a high vascular density. Moreoverdefinitive determination of malignancy may not be possible if the tumor well differentiated, although cells of neuroendocrine origin implicatethyroid carcinoma in a dog with a large cervical mass. Histopathology will allow for definitive diagnosis using biopsy or surgical excision.[2]

Advanced imaging is helpful in determining a definitive treatment plan. Cervical ultrasound can guide fine needle aspiration and confirm thyroid gland involvement. CT scan or MRI is indicated to determine if the mass is surgically resectable and for surgical planning. However, it is important to note that should a client be interested in ^{131}I treatment, the iodinated contrast commonly used for CT scans will interfere with thyroid scans and ^{131}I treatment. Therefore, CT with iodinated contrast should be avoided when ^{131}I treatment is a consideration.

Scintigraphy is a useful tool to determine if ^{131}I is a potential treatment option for canine thyroid carcinomas, and the dog does not need to be hyperthyroid for this treatment to be considered. If a thyroid carcinoma is capable of trapping ^{99m}Tc-pertechnetate or if the tumor can trap and organify ^{123}I or ^{131}I, the primary tumor can be visualized on scintigraphy and ^{131}I is a valid treatment option, even if the tumor cannot complete the remaining steps for synthesis and secretion of functional thyroid hormone.[5-6]

Up to 35-40% of dogs with follicular thyroid carcinoma have detectable metastasis at presentation, and up to 80% will develop metastasis. Despite these high metastatic rates, metastases tend to be slow growing and dogs can enjoy a long term good quality of life when metastases are present. The metastatic potential of thyroid carcinomas increases with increasing tumor size (> $23cm^3$) and the presence of bilateral disease. The most common metastatic sites are regional lymph nodes (including mandibular, retropharyngeal, cranial cervical), lungs, and late in stage of disease, liver.[2]

Treatment

Numerous treatment options are available for, and successful in, treating canine thyroid carcinomas. In one study, 25-50% of thyroid carcinomas were amenable to surgery at time of diagnosis. Those considered not surgically resectable had significant invasion into the trachea, larynx, jugular veins, and carotid sheath. With surgery alone to treat freely movable tumors, the median survival time was 3 years compared to 6 to 12 months, if the thyroid mass was invasive. Potential complications of surgery are laryngeal paralysis, megaesophagus, hypocalcemia (because of removal of parathyroid glands with the thyroid), and upper airway distress.[2,7]

For thyroid tumors determined to be unresectable or those incompletely excised with surgery, external beam radiation therapy and ^{131}I radiotherapy are commonly used and successful treatment options.

Using external beam radiation therapy in the face of bulky tumor, the reported time to maximal tumor reduction is between 8 and 22 months, with a progression free survival 80% at 1 year and 72% at 3 years. Fine fractionation protocols are preferred because of potential of late side effects to larynx, esophagus, and trachea. Early effects of radiation therapy, which resolve with supportive care, include alopecia, erythema, and mucositis. Late effects are uncommon with currently utilized protocols but may include skin fibrosis, alopecia, chronic tracheitis, or dry cough. Median progression free survival and survival times are reported at 24-45 months.[8-10]

Similarly, ^{131}I can be used effectively post-operatively or in the face of bulky disease, if scintigraphy reveals that the thyroid carcinoma is capable of trapping or trapping and organifying ^{99m}Tc-pertechnetate or ^{123}I, respectively. Median survival times are reported at 28-34 months if metastasis is not present, and 12 months with evidence of metastasis. The major side effect of ^{131}I is myelosuppression, and, although asymptomatic in most cases, complete blood counts are recommended weekly for 4-6 weeks following treatment.[2, 5-6]

Chemotherapy can also be considered for dogs with non-resectable tumors. Responses of 30-50% have been reported with cisplatin or doxorubicin chemotherapy. The role of chemotherapy following local disease control is not yet well defined. Furthermore tyrosine kinase inhibitors may play a role in the treatment of non-resectable disease. Canine thyroid carcinomas express VEGFR2, PDGFR, and KIT, which are inhibited by toceranib. One large study of toceranib included 15 dogs with bulky thyroid carcinoma, and reported 26% partial response and 53% stable disease.[2, 11-12]

For dogs with metastasis present at the time of diagnosis, it is important to note that metastases are generally slowly growing. In one study investigating hypofractionated (course fractionation) external beam radiation therapy, the overall median survival time was 1.8 years for 13 dogs, 5 of which had pulmonary metastatic disease.[12] Metastatic disease has also been reported to respond to ^{131}I in some cases, and chemotherapy in others.

FELINE THYYROID CARCINOMA

Bilateral feline thyroid hyperplasia and adenomas are present in ~70% of feline hyperthyroidism cases, and hyperthyroidism is the most common endocrinopathy of cats. Common clinical signs include weight loss, polyphagia, palpable goiter, heart murmurs and tachycardia. Diagnosis is usually made with high normal to elevated total T4 measurements.[13]

Thyroid carcinomas are uncommon, representing 1-3% of hyperthyroid cats, and follicular carcinoma is the most common type diagnosed. Nonfunctional and nonsecretory thyroid tumors are rare in cats. The clinical presentation is similar to cats with hyperthyroidism caused by thyroid adenomas. Nuclear scintigraphy may reveal patchy or irregular uptake within the thyroid, extension or invasion of the thyroid mass, or uptake at metastatic sites. However, scintigraphy is not a sensitive diagnostic tool for determining benign versus malignant disease.

In contrast to most adenomas, carcinomas tend to recur following surgery and not respond to the lower doses of ^{131}I used to treat adenomas. Thyroid carcinomas are firm, fixed, and invasive, and potentially involve multiple masses. Definitive diagnosis requires histopathology. Metastasis is reported in up to 70% of cases, usually to the lymph nodes and lungs. Staging should include a minimum database and thoracic radiographs. The treatment of choice for thyroid carcinomas in cats is surgical resection followed by high dose ^{131}I, with survivals reported to range from 10 months to 41 months. ^{131}I has also been used as the sole treatment with a median survival time of 814 days. Because cats with thyroid carcinomas have larger tumor burdens, and malignant cells may not trap and retain iodine so efficiently, higher doses are recommended. Medical management with methimazole may also be used to control clinical sings in cats where surgery or ^{131}I are not viable treatment options. [14]

REFERENCES

[1] Guyton and Hall Textbook of Medical Physiology 10th ed. 2000, 858-865; [2]Liptak JM Clin Tech Small An Pract 2007; 22: 75-81; [3]Carver JR et al. Vet Surg 24; 1995 315-19; [4]Campos M et al. J Vet Intern Med 2014; 28: 1805-13; [5]Turrell JM et al. J Vet Med Assoc 2006; 229: 542-8; [6]Worth AJ et al. Aust Vet J 2005; 83: 208-14; [7]Radlinsky Vet Clin Small Anim Pract 2007; 37: 798-98; [8]Mayer MN et al Can Vet J 2007; 48: 761-63; [9]Pack L et al. Vet Rad US 2001; 5: 471-74; [10]Brearley MJ et al. J Small Anim Pract 1999; 40: 206-10; [11]Fineman LS et al. J Am Anim Hosp Assoc 1998; 34: 109-12; [12]London C et al Vet Comp Oncol 2012; 10:

194-205; [13] Mooney in Ettinger and Feldman Textbook of Veterinary Internal Medicine 7th ed. 2010 [14]Barber LG Vet Clin North Am Small Anim Pract 2007; 37: 755-73.

APUDOMAS: DIAGNOSIS, SUPPORTIVE CARE, AND DEFINITIVE TREATMENT

Anne Corrigan, DVM, MS, ACVIM (SAIM) and Sandra Bechtel, DVM, ACVIM (Oncology)
Grenada, West Indies and Columbia, MO

INTRODUCTION

APUDomas are tumors arising from amine precursor uptake and decarboxylation (APUD) cells, which synthesize and metabolize biogenic amines. These cells are normally located in the gastrointestinal tract, hypothalamus, pituitary gland, C-cells of the thyroid, parathyroid glands, and placenta and tumors originating from APUD cells are named after their endocrine product. The most common APUDomas in veterinary medicine are insulinoma and pheochromocytoma; other APUDomas reported are gastrinoma, glucagonoma, and carcinoid. Pheochromocytomas will be covered in a separate session.

INSULINOMA

Insulinomas are functional tumors arising from pancreatic beta cells. Insulinomas are generally seen in medium to large breed dogs with Labrador retrievers, golden retrievers, German shepherd dogs, German pointers, Irish setters, boxers, West Highland terriers, and mixed breed dogs overrepresented. The median reported age at diagnosis is 9-10 years, with a range of 3 to 15 years, and there is no sex predilection.[1]

Normal glucose homeostasis

The liver, insulin and glucagon are primarily responsible for glucose maintenance. Ingested glucose stimulates insulin release and most of the absorbed glucose is taken into the liver for storage as glycogen, between meals the insulin and glucagon levels will fluctuate and the liver will release glucose to maintain normoglycemia between meals. Severe or prolonged hypoglycemia can cause epinephrine, cortisol, and growth hormone release which act as anti-insulin hormones, increasing hepatic glucose release and increasing the utilization of other tissues to increase gluconeogenesis and decrease cellular glucose utilization.[2]

Clinical signs

Although insulinomas may produce glucagon, somatostatin, pancreatic polypeptide, growth hormone, insulin like growth factor-1, and gastrin in addition to insulin, clinical signs are those associated with hypoglycemia. The tumor cells secrete insulin independent of glucose, and hyperinsulinemia results in hypoglycemia through suppression of glucose release and production. Signs of neuroglycopenia are weakness, ataxia, collapse, disorientation, behavioral changes, and seizures. Catecholamine release stimulated by hypoglycemia may lead to muscle tremors, shaking, anxiety, and hunger. Clinical signs are usually intermittent and precipitated by fasting, exercise, excitement, eating. Paraneoplastic peripheral neuropathy has also been reported.[1]

Diagnosis

A normal or elevated blood insulin in the face of hypoglycemia (<60 mg/dL) is supportive of a diagnosis of insulinoma. Serum fructosamine and glycosylated hemoglobin can also be measured and are expected to be low in dogs with chronic hypoglycemia. The insulin:glucose ratio and amended insulin: glucose ratio are not sensitive or specific for a diagnosis of insulinoma. Differential diagnoses to consider are extra pancreatic neoplasia resulting in paraneoplastic hypoglycemia (for example, hepatoma, leiomyoma, leiomyosarcoma, lymphoma), pancreatic islet cell hyperplasia, decreased production or increased consumption of glucose (for example hypoadrenocorticism, hepatic insufficiency, glycogen storage disease, sepsis), iatrogenic insulin overdose, or toxic causes of insulin release (xylitol toxicity).[1]

Imaging can be helpful in ruling out other neoplastic causes of increased insulin production and potentially localizing the insulinoma. Abdominal ultrasound is an inexpensive and rapid screening tool. Ultrasound detects less than 50% of insulinomas and may need to be followed up with further imaging. Single-photon emission computed tomography (SPECT) with ^{111}In-DTPA-D-Phe1-octreotide, a radiolabeled somatostatin analogue, has a reported sensitivity of 50% for detecting insulinomas in dogs. Computed tomography (CT) is the most sensitive of imaging modalities investigated for insulinoma in dogs and identified 10 of 14 primary tumors in one study. The most sensitive tool for identifying the primary tumor is intraoperative pancreatic palpation.[3]
Diagnosis is confirmed on histopathology with the use of immunohistochemistry. Stains used to confirm the diagnosis of insulinomas in dogs are insulin, chromagranin A, neuron-specific enolase, and synaptophysin.[1]

Treatment

Following stabilization and staging, surgical resection of the mass will provide both definitive diagnosis and therapy. Most insulinomas are either visible or palpable at time of surgery. Resection of suspected metastatic lesions is also recommended as this will improve glucose homeostasis. Possible post-operative complications reported include pancreatitis, persistent hypoglycemia, transient hyperglycemia, diabetes mellitus, and exocrine pancreatic insufficiency.

For acute hypoglycemia in a dog or cat with suspected insulinoma, a slow bolus of dextrose can be used followed by continuous rate infusion; note that this can stimulate further unregulated insulin release. Additional medical therapies are available for dogs that have continued hypoglycemia post-operatively or for whom surgical resection is not a viable option; these include prednisone, diazoxide, octreotide, a diet high in fat, protein, and complex carbs, and avoiding exercise. Prednisone is insulin-antagonizing, gluconeogenic, and glycogenolytic. Diazoxide is a nondiuretic benzothiadiazine that suppresses insulin release from beta-cells, stimulates hepatic gluconeogenesis and glycogenolysis, and inhibits cellular uptake of glucose; response rates of 70% are reported. Octreotide is a somatostatin receptor ligand that inhibits synthesis and secretion of insulin by pancreatic beta cells; response rates of 50% are reported.[1]

Streptozotocin is an alkylating chemotherapy agent that is similar in structure to the glucose transporter 2 (GLUT 2) transmembrane carrier protein. Since pancreatic beta cells have high concentrations of GLUT 2 transporters, streptozotocin is selectively toxic to these cells. Two different dosing protocols are described: 500mg per m^2 once every 3 weeks intravenously with 7 hours of saline diuresis or the same dosing and administration scheme utilized every 2 weeks. Important to note is that the diuresis protocol is used to minimize necrosis of the renal tubular epithelium, which is both dose related and cumulative. Other side effects are emesis within 24 hours of administration, hepatotoxicity due to macromolecule alkylation resulting in increased alanine aminotransferase, increased creatinine, hypoglycemic collapse, and diabetes mellitus.[4-6]

Biologic behavior and prognosis

Insulinomas are more likely to be malignant in dogs than in people, and histopathologic criteria (such as nuclear atypia, mitotic index, and infiltration) do not reflect the biologic behavior[7]. In dogs, the metastatic rate is 50%, with metastasis most often occurring in the liver and lymph nodes. Prognostic factors include tumor size (< 2cm have a longer disease free interval than masses > 2cm), Ki67 index, which is predictive of disease free interval and survival, post-operative glucose, pre-operative insulin, and the presence of metastasis.

The median survival time reported following partial pancreatectomy is 12 to14 months and the World Health Organization staging scheme is prognostic (Stage I: T1N0M0; Stage II: T1N1M0; Stage III: T1 N0 or 1 M1). In dogs with stage I disease, 50% are free of hypoglycemia at 14 months; in dogs with stage II and III disease, less than 20% are free of hypoglycemia at 1 year. With the use of medical therapy alone, the reported median survival time is 196 days. Dogs with stage II and III disease treated with streptozotocin were normoglycemic for median of 163 days in one study.[5]

GASTRINOMA

Gastrin

Gastrin is produced by G cells in the distal stomach and duodenum; gastrin release is stimulated by protein in the stomach, a high gastric pH. or vagal stimulation. Its primary action is to stimulate acid secretion by the parietal cells, with secondary actions of stimulating gastric motility and growth of gastric epithelium. Gastrin may also be produced by D cells of the pancreas in the fetus and neonate, and is the suspected cell of origin of gastrinomas in the pancreas. The most common location of gastrinomas in the dog and cat is the pancreas, with one report of duodenal gastrinoma in the dog.[8]

Clinical Signs

Gastrinomas result in the excess release of gastric acid, leading to Zollinger-Ellison syndrome, which is the triad of a non-beta cell neuroendocrine tumor in pancreas with hypergastrinemia and multiple gastrointestinal ulcerations. Therefore, common clinical signs are vomiting, weight loss, melena, abdominal pain, anorexia, hematemesis, hematochezia, and diarrhea. Malabsorption also occurs and is a result of acidification of intestinal contents, leading to inactivation of lipase, precipitation of bile salts, interference with chylomicron formation, damage to mucosal cells, and altered water and electrolyte absorption.

Diagnosis

A complete blood count, chemistry panel, and urinalysis reflect protein and blood loss through GI ulcers and electrolyte changes indicative of vomiting. Diagnostic imaging (thoracic radiographs and abdominal ultrasound) is helpful in staging dogs with suspected gastrinoma, although the primary tumor is usually quite small (most are less than 1 cm), and therefore not often visualized on abdominal radiographs or ultrasound. A persistent elevation of serum gastrin concentrations (> 3 times the upper limit of normal) is supportive of gastrinoma if other causes of hypergastrinemia (renal failure, gastric outflow obstruction, achlorhydria, atrophic gastritis, gastric dilation and volvulus, liver disease, etc.) have been ruled out. Also of note are medications that can lead to hypergastrinemia, including antacids, H^2 receptor antagonists, proton pump inhibitors, and glucocorticoids. Other, less commonly utilized diagnostic tests include secretin stimulation, calcium challenge, and gastrin secretory testing. Histopathology and immunohistochemistry for gastrin expression is needed for definitive diagnosis.[8-9]

Treatment

Medical management is aimed at supportive care using either histamine 2 (H2) receptor antagonists to block the stimulatory effect of gastrin on hydrochloric acid or proton pump inhibitors to inhibit the hydrogen-potassium ATPase pump of the parietal cell. In the parietal cell, maximal gastrin secretion requires all 3 parietal receptor sites (histamine, gastrin, and acetylcholine) to be occupied – because H^2 antagonists block one of these receptors, the doses required for adequate control are variable and will increase over time. Proton pump inhibitors are more effective at suppressing gastric acid secretion and may therefore be more effective at controlling clinical signs in dogs and cats with gastrinomas.[8]

Exploratory laparotomy is the treatment of choice for gastrinomas. Even if complete resection is not possible, partial resection will enhance the effectiveness of supportive medical therapy. In addition, excision of deep or perforating ulcers can take place at the time of tumor removal. In dogs and cats, 80% are reported to have GI ulceration, with 20% havingperforating ulcers. The most common tumor location is the right lobe of the pancreas, followed by the body and then left lobe of the pancreas.[8]

When surgery is not a viable option, octreotide, a somatostatin analogue, acts directly on gastrinoma cells to inhibit gastrin release and gastric acid production. In a report of 2 dogs, octreotide was used successfully for 10 to 14 months in addition to other medical therapies.[8-10]

Biologic behavior and prognosis

Gastrinomas in the dog and cat are highly metastatic to the liver, lymph nodes, spleen, peritoneum, SI, omentum, and mesentery. Up to 85% of dogs and cats have metastasis at the time of diagnosis. Therefore, the overall prognosis is guarded to poor. With surgical and/ or medical management, survival times reported range from 1 week to 2 years, with most survival times less than 8 months.[8-10]

GLUCAGONOMA

Glucagon is produced by the alpha cells of the islets of Langerhans in the pancreas. Production and release of glucagon is normally stimulated by hypoglycemia, and glucagon release results in a net of increased blood glucose concentrations through decreased glycogen synthesis, increased glycogenolysis, and increased gluconeogenesis in the liver. Glucagonomas originate from the alpha cells of the pancreatic islets, are not reported in domestic cats, and are rare in dogs.

The most common presenting clinical sign is the paraneoplastic syndrome superficial necrolytic dermatitis, which is also known as diabetic dermatopathy, hepatocutaneous syndrome, or necrolytic migratory erythema. Superficial necrolytic dermatitis (SND) comprises hyperkeratosis, crusting, ulceration and erosion of the footpads, mucocutaneous junctions, external genitalia, distal extremities, pressure points, and ventral abdomen. Additional findings in dogs with glucagonoma are hyperglycemia, overt diabetes mellitus, hypoaminoacidemia, small intestinal diarrhea, anemia, and increased liver enzymes. Abdominal ultrasound or CT scan may reveal a pancreatic mass. A diagnosis of glucagonoma is suspected based on elevated plasma concentrations of glucagon, and definitive diagnosis is based on histopathology and immunohistochemistry for glucagon. While most glucagonomas in dogs are in the pancreas, glucagonoma has also been reported in the liver, spleen, adrenal glands, and mesenteric lymph nodes. A diagnosis of SND can be determined on histopathology (other causes of SND include advanced liver disease, phenobarbital, and in cats pancreatic carcinoma).

The treatment of choice for glucagonoma is surgical resection. The overall prognosis is poor as metastasis is common. Somatostatin analogues and dacarbazine have been reported palliative measures in people with glucagonoma. SND, if present, may improve after surgery or medical therapy; and other reported treatments for SND are amino acid infusions, oral protein supplementation using protein powder of egg yolks, zinc, and essential fatty acids.[11]

CARCINOID

Carcinoids are rare neoplasms originating from enterchromaffin cells, which are dispersed cells of neuroendocrine system found in the GI tract, liver, gallbladder, and pancreas. Carcinoid syndrome is caused by systemic release of 5-hydroxytrtyptamine, bradykinin, prostaglandin, histamine, and substance P. Clinical signs are dependent on tumor location and definitive diagnosis is made using histopathology and immunohistochemistry. Carcinoids stain with chromagranin, synaptophysin (GI endocrine cells), and neuron specific enolase. Grimelius stain (argyrophylic) will stain intracytoplasmic secretory granules. The treatment of choice is surgical resection. Metastasis is common to the lymph nodes, mesentery, peritoneal wall, lung, pancreas, and liver.[12-13] Because of the rarity of this tumor type in dogs, chemotherapy recommendations are not available. One dog with an intestinal carcinoid was treated with surgical resection and carboplatin chemotherapy, and was alive at 18 months with no evidence of recurrence.[13]

Gallbladder carcinoids in the dog may presented with acute emesis, icterus, melena, hematemesis, and hemobilia; cholecystectomy may be considered, if widespread metastasis is not present.[14] Carcinoids have also been reported in cats, with extra-hepatic carcinoids potentially having a good 1 year prognosis, if surgically resected.[15] Octreotide may be considered as a medical therapy if widespread metastasis is present, and will alleviate clinical signs in 80% of people with carcinoids.

REFERENCES

[1]Goutal CM et al. J Am Anim Hosp Assoc 2012; 48: 151-163; [2] Guyton and Hall Textbook of Medical Physiology 10th ed. 2000, 858-865; [3]Robben JH. J Vet Intern Med 2005; 19: 15-22; [4]Moore AS et al. J Am Vet Med Assoc 2002; 221: 811-818; [5]Northrup NC et al. J Vet Intern Med 2013; 27: 483-90; [6]Bell R et al. J Sm Anim Pract 2005; 46: 247-250; [7]Buishand FO et al. The Vet J 2010; 185: 62-67; [8]Hughes SM. N Z Vet J 2006; 54: 242-7; [9]Vergine M et al. The Vet J 2005; 170: 141-3; [10]Altschul M et al. J Small Anim Pract 199; 38: 286-91; [11]Mizuno T et al. Vet Dermatol 2009; 20: 72-9; [12]Albers TM et al. J Vet Diagn Invest 1998; 10: 116-118; [13]Spugnini EP et al. In vivo 2008; 22: 759-62; [14]Lippo NJ et al J Vet Intern Med 2008; 22: 1249-52. [15]Rossmeisl JH et al. J Am Anim Hosp Assoc 2002; 38: 61-66.

FELINE OBESITY: CAUSES, CONSEQUENCES AND MANAGEMENT

Jacquie S. Rand, BVSc, DVM, DACVIM; Delisa J. Appleton BVSc (Hons)
Gatton, Australia

CAUSES OF FELINE OBESITY

Regulation of food intake and body weight involves a complex balance between long-term control of fat mass involving insulin, adrenal steroids and leptin signals to the CNS and short-term, meal-related signals. For example, leptin increases in parallel with fat mass and acts in the brain through neuropeptide Y to decrease eating. In the peripheral tissues, it has a safety valve action, so that when excess energy is ingested, leptin induces partial uncoupling of electron transport involved in forming high energy ATP bonds and deposition of fat, resulting in some of the ingested excess energy being released as heat. Clearly, the capacity of leptin to maintain an ideal body weight is limited. Although free living cats normally limit their food intake to their energy requirements, many pet cats appear unable to regulate energy balance. Despite elevated circulating leptin levels in obese cats associated with increased fat mass, they continue to overeat and gain weight. This paradox of increased leptin concentrations in obesity has been observed in other species and is hypothesized to be a consequence of 'leptin resistance'.

A number of risk factors have been identified for obesity. Physical inactivity, extended sleeping periods and indoor confinement are associated with an increased incidence of obesity. Neutering results in weight gain, predominantly as fat mass in both genders. With neutering, reductions in roaming and in general physical activity occur along with increased food intake and the efficiency of energy utilization. Neutered male cats are more likely than females to gain excessive weight. The incidence of feline obesity increases with age peaking between the ages of six and eight years. Many pure bred cats such as Siamese or Abyssinians are leaner than most mixed breed cats, suggesting a genetic influence on body weight may exist in cats, as it does in humans. Chronic administration of corticosteroids and progestins, particularly megoestrol acetate, induce polyphagia resulting in weight gain. The human-animal-relationship of owners of overweight cats is characterized by a higher intensity of the bond between owner and cat, and by increased humanization compared with the relationship in owners of normal weight cats.

CONSEQUENCES OF FELINE OBESITY

Obesity in companion animals has been shown to predispose to, or cause a number of medical conditions including glucose intolerance and diabetes mellitus, as well as surgical and anaesthetic complications, reproductive compromise, non-allergic skin conditions, lameness, feline lower urinary disease, hepatic lipidosis and immuno-incompetence.

In rodents, dogs, and humans, obesity is linked to increased susceptibility to infection, altered immunity and impaired host defense mechanisms, and in cats, increased bacterial skin infections have been reported. Hypertriglyceridaemia was also present in cats after becoming obese.

In a study where domestic neutered cats were allowed to eat ad libitum for 10 months, average bodyweight increased by 44% and plasma leptin increased three-fold, but cats continued to ingest excessive amounts of energy to their needs, albeit the rate of increase in bodyweight was decreased. Weight gain was associated with hyperinsulinaemia and reduced insulin sensitivity. In fact, the insulin sensitivity index (S_I) as measured by Bergman's Minimal model, was 53% lower in cats after weight gain, indicating that obese cats were significantly less sensitive to insulin compared to when they were of normal-weight. Cats with the highest basal insulin concentrations after weight gain were also the most insulin resistant. Glucose effectiveness (S_G), was also significantly decreased in obese cats indicating that glucose utilization, independent of insulin, was reduced. Other characteristic metabolic features of obesity evident in the obese cats include fasting hyperinsulinaemia and an exaggerated insulin response to glucose during the glucose tolerance test. Approximately 25% of obese pet cats older than 7 years of age have impaired glucose tolerance and are at risk of developing diabetes. Once fasting blood glucose is 135 mg/dL (7.5 mmol/L)(normal $\leq$ 117 mg/dL (6.5 mmol/L) or higher, the risk of diabetes developing within the following 12 months is high, based on a study of diabetic cats in remission. Before gaining weight, cats that developed impaired glucose tolerance tended to have higher insulin concentrations, and lower insulin sensitivity and glucose effectiveness than cats that maintained normal glucose tolerance with weight gain. This suggests that some cats may have an underlying predisposition to develop glucose intolerance, and if these cats become obese and inactive, they may be more at risk of developing overt type 2 diabetes mellitus.

Unlike their ancestors, modern domestic cats no longer hunt for food, have become relatively inactive and tend to overeat resulting in an increased incidence of overweight and obese cats. These changes parallel those that have occurred in the urbanization of many of the world's indigenous populations that have high levels of diabetes. Furthermore cats are predominantly fed commercial diets, which are often energy dense, require no or little energy expenditure on the part of the cat to procure, and typically contain significant levels of highly digestible cereal which cats as obligate carnivores, did not evolve to eat. These factors increase the demand for insulin production from the pancreatic beta cells and lend credence to the hypothesis of 'beta-cell exhaustion' as an important element in the pathogenesis of feline type 2 diabetes mellitus.

In both diabetic cats and humans, amyloid deposition in the pancreatic islets is the most consistent histological finding. Increased secretion of amylin is one factor associated with increased deposition. Amyloid is formed from amylin (IAPP), a hormone which is co-secreted with insulin, and secretion is increased when insulin secretion is increased, for example in obesity. In diabetic cats, a large amount of amyloid is deposited in the islets, replacing the cells and leading to a reduction in overall beta cell mass.

ASSESSMENT AND MANAGEMENT OF FELINE OBESITY

A safe and effective weight reducing regime involves restricting caloric intake to 60%-75% of estimated caloric requirements for maintenance of body weight (for maintenance of stable body weight approximately 60 kcal/kg of ideal body weight per day), with the aim to maintain weight loss at approximately 1.0 to 1.5% per week. Regular weight checks, initially fortnightly then monthly, with appropriate adjustments of intake should be made. There is considerable individual variation in caloric requirements among cats, and food intake needs to be adjusted individually to ensure appropriate weight loss. Moreover methods to increase energy output such as backyard activity enclosures, laser pointers and food cubes should be instituted. A study found that active playing for 10 minutes daily was as effective as calorie restriction in achieving weight loss in obese cats. Several dietary manipulations have been investigated including the addition of chromium or carnitine, vitamin A supplementation, and altering macronutrient sources. Most commercial weight loss diets have added carnitine and fiber, decreased fat, and increased carbohydrate and protein compared to standard maintenance diets.

Carnitine

Carnitine, a co-factor of fatty acid metabolism, has been shown to promote the maintenance of lean body mass during weight reduction. In obese cats, carnitine supplementation during a weight loss program increased the rate of fatty acid oxidation and accelerated the rate of weight loss, possibly via increasing energy utilization. Importantly, it helped protect cats from hepatic lipidosis during weight loss.

Fibre

Traditionally, a high fiber diet has been promoted for weight management in cats, although there is limited scientific evidence to substantiate its effectiveness. Furthermore unwanted side effects such as increased fecal bulk and frequency of defecation may contribute to the lack of compliance of such diets in clinical practice. The inclusion of dietary fermentable fiber however, may be beneficial in the management of obesity. Fermentation of fiber by intestinal bacteria produces short chain fatty acids (SCFA) which in turn, stimulate the release of proglucagon from the intestinal lining cells. Proglucagon is further broken down to glucagon-like peptide-1 (GLP-1), which increases insulin secretion from the pancreatic islets.

Protein

The provision of meals containing adequate amounts of amino acids, particularly essential amino acids, is necessary for the anabolic effect of insulin on muscle protein synthesis. Increasing dietary protein content facilitates loss of fat mass and sparing muscle mass in overweight cats during weight loss. In a study of diabetic cats, a high protein, low carbohydrate diet has been shown to induce satiety as well as reduce the dose of insulin required. In another study in normal cats, compared to a high protein and low carbohydrate diet, a low protein and high carbohydrate diet produced hyperinsulinaemia and decreased nonesterified fatty acids (NEFA) mobilization which may in the long term, lead to weight gain, obesity and beta-cell 'exhaustion' in cats fed ad libitum.

Carbohydrate

The proportion of calories ingested as carbohydrate is one of the main determinants of post-prandial glucose and insulin concentrations. Diets with restricted carbohydrate content (<20% of calories) are appropriate for preventing diabetes in predisposed cats and for managing diabetes. Dietary carbohydrate source is also a consideration when developing diets for both healthy cats as well as for obese, glucose intolerant cats. Novel carbohydrate sources such as cassava (tapioca) and lentils are associated with very little post-prandial glucose and/or insulin responses in cats when compared with rice, which is traditionally used in cat foods. The inclusion of these types of starches in feline diets may improve glucose control in cats, reduce insulin and possibly amylin secretion, thereby minimizing pancreatic amyloid deposition and beta-cell 'exhaustion' However, at the time of writing, few commercial foods have novel low glycemic carbohydrate sources such as cassava, Including low glycemic carbohydrate sources in feline weight loss diets would facilitate reducing fat content of the diet without the adverse effect of increasing postprandial glucose concentrations from increased carbohydrate content. Therefore, at the current time, diets with restricted carbohydrate content are likely important for management of obesity-induced glucose intolerance and diabetes. A new addition to feline weight loss diets is a commercial diet which alters gene expression of obese cats to an expression pattern more similar to that of lean cats; its effect on glucose tolerance is currently unknown.

REFERENCES

Appleton DJ, Rand JS , Sunvold GD. Plasma leptin concentrations in cats: Reference range, effect of weight gain, and relationship with adiposity as measured by dual energy x-ray absorptiometry. *Journal of Feline Medicine and Surgery,* 2000; 191-9

Appleton DJ, Rand JS, Sunvold GD. Insulin sensitivity decreases with obesity, and lean cats with low insulin sensitivity are at greatest risk of glucose intolerance with weight gain. *J Feline Med Surg* 2001;3:211-228.

Appleton DJ, Rand JS , Sunvold GD. Insulin sensitivity decreases with obesity, and lean cats with low insulin sensitivity are at greatest risk of glucose intolerance with weight gain. *Journal of Feline Medicine and Surgery* 2001, 211-28

de-Oliveira LD, Carciofi AC, Oliveira MC, et al. Effects of six carbohydrate sources on cat diet digestibility and postprandial glucose and insulin response. *J Anim Sci* 2008;86:2237-2246.

Kienzle E, Bergler R, Ziegler D et al. Human-animal relationship and overfeeding in cats (Abstr.), in: *Proceedings,* The 2000 Purina Nutririon Forum, 2000; 130.

Giles,R., Gruffydd-Jones, T.J. & Sturgess, C.P. (2003) A preliminary investigation into the effect of different strategies for achieving weight loss in cats (Abstract). Proceedings of the 46th Annual Congress of the British Small Animal Veterinary Association, Birmingham, UK, p. 546.

Zoran DL, Rand JS. The role of diet in the prevention and management of feline diabetes. *Vet Clin North Am Small Anim Pract* 2013;43:233-243.

MANAGING CHALLENGING CASES OF FELINE DIABETES

Linda M. Fleeman, BVSc, PhD, MANZCVS
Melbourne, VIC, Australia

Management of diabetes in cats can be a very rewarding feature of veterinary practice. However, not all diabetic cats will respond predictably to treatment. This presentation will discuss practical guidelines and a logical approach to solving the most common problems encountered when treating diabetic cats.

DIFFICULTY MONITORING GLYCEMIC RESPONSE – RESULTS DO NOT MATCH THE CAT'S CLINICAL SIGNS

Glucose homeostasis is a dynamic process that can change very rapidly. Cats with diabetes have dysfunctional glucose homeostasis and so there is even more pronounced variability of blood glucose concentrations, especially in response to stress or illness. A common error when managing diabetes is to assume that the condition can be 'stabilized' and that measurements taken on one day are representative of diabetic control on other days. A more appropriate approach is to establish an ongoing monitoring regimen that will detect the changing trends in the animal's response to treatment. This is especially important for cats because improvement in glycemic control and resolution of glucose toxicity can result in substantially decreased insulin requirement and diabetic remission. Improved activity, resolution of polydipsia and polyuria, and weight gain are all indicators of improved glycemic control. In cats, it is important to realize that excellent glycemic control (for example, as evidenced by persistent negative glycosuria) can occur at any time once clinical signs have resolved, and may occur within days, weeks, or even months of the last insulin dose adjustment. Once the blood glucose concentration is consistently below the renal threshold, insulin requirement can decrease substantially. However, the rate of diabetic remission in cats appears to be optimized if the insulin treatment is withdrawn gradually, with a dose decrease every 1-2 weeks. In contrast, sudden cessation of insulin can result in recurrence of hyperglycemia, return of glucose toxicity, and increased requirement for exogenous insulin.

Micro-management with frequent insulin dose adjustment should generally be avoided. It is typically recommended that insulin dosage adjustments should not be made any more frequently than once every 1-2 weeks. The exception is following an episode of hypoglycemia, which should always prompt reduction of the insulin dose (unless there was a dosing error). When presented with a patient that has shown a very variable response to insulin and inconsistent glycemic results, it is often helpful to review carefully the history with the owner to determine if there were any periods when there was resolution of clinical signs. It is important to also note the insulin dose administered before any occurrences of hypoglycemia. A reasonable approach is to resume the treatment regimen that the animal was receiving when there was good glycemic control and tomonitor carefully the response for a minimum of 2 weeks. This information can then provide baseline data for an ongoing regimen of monitoring and treatment adjustment.

UNEXPECTED HYPOGLYCEMIA

Recording urine glucose at home provides a useful 'early warning system' for the risk of clinical hypoglycemia. Decreasing the insulin dose within 2 weeks of the onset of negative glycosuria will ensure that hypoglycemia is avoided in the majority.

The owner's insulin administration protocol should be carefully evaluated whenever hypoglycemia occurs to identify dosing errors. For cats that are prone to hypoglycemia, longer-acting insulin preparations such as glargine insulin are recommended as these will likely have a smoother action and so minimize periods of both hyper- and hypo-glycemia. It is strongly recommended that injections are administered at strict 12-hour intervals. Irregular timing of insulin injections can lead to overlap of insulin action with that of the previous dose. If it is not possible to administer an insulin injection on time, then the best approach is to miss that injection and resume insulin administration at the next injection time. Missing a single injection will usually have negligible consequences. In contrast, late administration of insulin can lead to increased insulin action (and therefore over-dose) if the following insulin injection is administered on time. The usual meals can be fed whenever an insulin injection is missed.

It is important to note that feeding in cats will not typically cause a postprandial increase in blood glucose concentration and so is not a reliable treatment for hypoglycemia. In particular, low carbohydrate foods such as those typically recommended for diabetic cats will not be helpful in this situation. However, liberal application of glucose syrup or honey to the oral mucosa will usually effectively control clinical signs although repeated dosing might be required. The glucose syrup or honey can also be administered per rectum, which is usually the safest option in an animal with severe seizures. Once a cat has recovered, glucose syrup or honey can be added to cat food. Cats cannot taste sweet foods and usually appear not to notice honey if it is mixed thoroughly with food. If oral glucose is not successful, the hypoglycemia should be managed initially with an intravenous bolus of 1 mL/kg (0.5 g/kg) 50% dextrose administered slowly. This can be followed with a constant rate infusion of 2.5% dextrose solution at 6 mL/kg/hr with the flow rate adjusted based on subsequent blood glucose measurements.

FINICKY APPETITE AT MEAL TIMES

It is important to clarify first whether the owner is concerned because a cat that typically eats all meals rapidly and completely is now refusing to eat all or part of a meal (indicating illness requiring prompt veterinary assessment), or if there is concern because a finicky cat is not adjusting well to new dietary recommendations for diabetes management. For some diabetic cats after treatment, decreased polyphagia resulting from improved diabetic control causes marked reduction in appetite which might be concerning for owners. An important difference between cats and dogs is that meals do not seem to influence blood glucose concentrations in cats so much as they do in dogs. Because cats do not have a clinically relevant postprandial blood glucose response, it is not so important to match meals with insulin injections in cats and there is less concern about hypoglycemia if a cat does not eat at the time of insulin injections. This is especially the case for the low carbohydrate diets that are typically recommended for these patients. This allows more flexibility with the feeding of diabetic cats.

INSULIN RESISTANCE OR AN UNEXPECTED INCREASE IN INSULIN REQUIREMENT

Most diabetic cats require doses of 5 U or less of insulin administered every 12 hours. Occasionally an individual cat will need 6 or 7 U before good glycemic control is achieved. Insulin resistance in cats can thus be defined as poor diabetic control at insulin doses exceeding 7 U/cat. Insulin doses should always be based on estimated ideal body weight rather than actual body weight in underweight or overweight cats. Once persistent negative glycosuria is achieved and insulin requirement decreases, many cats will go into remission and the remainder will typically require 3 U or less of insulin every 12 hours. (An unexpected increase in insulin requirement is recognized when a cat that has been in long-term diabetic remission needs insulin therapy again, or when a cat that has needed only a low insulin dose for months unexpectedly requires higher doses.)…this sentence needs restructuring

The major differential diagnoses for a diabetic cat to require more than 7 U of insulin every 12 hours or for an unexpected increase in insulin requirement are:

1. Error in insulin handling or administration
2. Obesity
3. Concurrent disease or drug therapy
4. Compensatory hyperglycemia secondary to insulin overdose (Somogyi phenomenon).

1. Error in insulin handling or administration

Insulin can become inactivated if exposed to temperatures >30°C or light for prolonged periods. An expedient method of ruling out the possibility of inactivated insulin when investigating insulin resistance is to change to a new vial of insulin. Insulin suspensions must be thoroughly mixed before administration or doses might vary greatly.

Although experienced owners rarely report difficulty with administration of insulin to their pets, it is important to review their injection techniques for errors whenever insulin resistance is investigated. A wide range of insulin syringes is available and inadvertently changing to a different type of syringe can lead to dosing errors. Dosing errors are less frequent with insulin dosing pens than with needles and syringes. However, insulin dosing pens must be primed before administration of each dose to ensure there is no air in the system. It is also important to check that the dosing dial has returned to the 'zero' position after each dose.

2. Obesity

Obesity is a common reason for a diabetic cat to have increased insulin requirement. Reversible insulin resistance is induced and if weight loss occurs insulin requirement can rapidly return to normal and care must be taken to avoid hypoglycemia.

Weight gain is initially a desirable outcome when treating diabetic animals as it is an indicator of good glycemic control. However, once this is achieved, overweight and obesity-prone diabetic cats should be carefully managed to attain and maintain an ideal body condition. Both body weight and body condition should be monitored long-term, and calorie restriction will sometimes be necessary.

3. Concurrent disease or drug therapy

Every diabetic cat that is not overweight and has increased insulin requirement should be assessed for concurrent disease. Concurrent disease or drug therapy causing insulin resistance may be suspected based on history and physical examination findings and can be further investigated by performing hematology, serum biochemistry, urinalysis, and microbial culture and sensitivity of the urine. Any medical condition can theoretically cause insulin resistance that affects diabetic control, and dental disease, urinary tract infection, hyperthyroidism, and chronic renal failure are commonly recognized. Systemic and topical corticosteroids are the drugs most commonly associated with insulin resistance. If the concurrent disease state can be resolved or drug treatment withdrawn, insulin sensitivity can be expected to improve and there will be increased risk of hypoglycemia unless the insulin dose is decreased.

Acromegaly should be considered if all other diseases are ruled out. The classic textbook signs of acromegaly are not always present in affected diabetic cats, with some showing no sign other than high insulin requirement. Diagnosis is typically based on increased serum IGF-1 concentration. It is important that an IGF-1 assay that has been validated for cats is used. To optimize the chance of detecting increased IGF-1 concentration, cats must first be treated with insulin for at least 4-6 weeks. This is because portal insulin induces IGF-1 production and acromegalic cats can give normal IGF-1 results when they are insulin deficient. It has been suggested that the incidence of acromegaly in diabetic cats might be underestimated.

4. Compensatory hyperglycemia secondary to insulin overdose (Somogyi phenomenon)

The Somogyi effect (hypoglycemia-induced hyperglycemia) is a well-known phenomenon that may occur when diabetic patients are overdosed with insulin. The overdose causes a transient period of hypoglycemia. Counter-regulatory hormones including glucagon, epinephrine, and cortisol are released to prevent further lowering of blood glucose. This may result in a period of hypoglycemia that lasts only minutes which prevents the animal from experiencing the potentially serious consequences caused by low blood glucose. These hormones thus cause life-saving insulin resistance; however, the counter-regulatory response is sometimes excessive and subsequently results in hyperglycemia that may persist for hours to days. Compensatory hyperglycemia secondary to insulin overdose is thus a common cause of insulin resistance in diabetic cats. There is sometimes a period of good glycemic control that is followed by deteriorating glycemic control despite increasing insulin doses. However the period of good glycemic control may be very brief and is sometimes missed, especially if dose adjustment is based only on results of serial blood glucose concentration curves or fructosamine concentration without careful consideration of the clinical signs.

These patients show signs of poor diabetic control such as polydipsia and polyuria. They might present with very unpredictable blood glucose results that vary anywhere from 30-550 mg/dL (1.7-30 mmol/L) despite consistent insulin dosing.

Compensatory hyperglycemia often appears to persist for several days following insulin-induced hypoglycemia and such cases will often roughly follow a 3-day cycle fluctuating between hyper- and hypo-glycemia and might present with intermittent hypoglycemic clinical signs. More commonly, the hypoglycemic events go unnoticed by owners; they may also occur at night (nocturnal hypoglycemia) making these events even less likely to be witnessed. Some owners will report episodes of yowling by their diabetic cat in the early hours of the morning. In insulin-treated diabetic people, hypoglycemic events occur more frequently during the night than during the day and the same is likely true for cats. This portrays the difficulty encountered in practice when daytime curves may be the only information gathered. In some cases, use of a continuous glucose monitoring system or 24 hour serial glucose curves may be of assistance with these difficult to catch hypoglycemic events.

For diabetic cats with poor glycemic control that are receiving insulin doses greater than 1.5 U/kg, where administration/dosing errors and concurrent disease/drugs have been ruled out, it is recommended that the insulin dose be decreased to 0.5 U/kg and the response to this change monitored. If insulin resistance is due to compensatory hyperglycemia secondary to insulin overdose, there is typically gradual clinical improvement over the following weeks. If there is another cause of insulin resistance, clinical signs typically become much worse within a few days and the previous insulin dose can be resumed and the investigation for another cause continued.

USES AND LIMITATIONS OF SERIAL BLOOD GLUCOSE CURVES

The principle reason for performing a glucose curve is to find the lowest blood glucose concentration (nadir) resulting from the current insulin therapy. Serial sampling is required because the timing of the nadir is typically very variable. A glucose curve can thus provide information when making decisions about changing insulin doses that cannot be obtained from any other method of monitoring including history or fructosamine levels. The goal of the glucose curve is to track the changes in blood glucose concentration that occur over a 12 to 24 hour period after insulin administration. The ideal curve reflects what happens to the patient's blood glucose in the home environment. This information is used to assess how the blood glucose concentrations for a given dose of insulin and a set amount of food are likely influencing the cat's clinical signs at home. Serial blood glucose curves can thus assist with decision making when adjustment of the insulin dose is indicated.

It is very important to recognize that blood glucose curves have many limitations and are often unreliable diagnostic tools in cats. Multiple factors can influence blood glucose concentrations on any given day resulting in lower or more commonly higher results than might occur on a 'typical' day. The diabetic state causes dysfunction of all the normal physiological mechanisms that control blood glucose concentration, which means that diabetic cats are prone to widely variable and unpredictable blood glucose concentrations. Therefore a cat that has experienced good diabetic control for a week or more can unexpectedly have very high blood glucose concentrations for hours or days. If a blood glucose curve is performed at such a time, results may suggest poor control in animals that are well controlled. These factors can make interpretation of a curve difficult, especially if assessed in isolation. This highlights the importance of always assessing the glucose curve in light of the animal's history and clinical signs. Significant day to day variation in blood glucose concentrations has been shown to occur in patients with no variation in insulin dosing or food intake. Stress in particular can substantially increase glucose concentrations, especially if the cat struggles before or during blood sampling or has a concurrent illness. For this reason minimizing stress for any patient undergoing a glucose curve is essential. It is important that all sampling is performed with minimal handling and disturbance of the patient.

Glucose curves performed by the owner at home may help to avoid some hospital related stressors. Blood glucose monitoring at home will particularly appeal to those owners who desire more autonomous control over their cats' diabetes. Owners who choose this method of monitoring sometimes need to be advised against over-zealous blood glucose measurement and interpreting the results themselves. Others might be unwilling to add home blood glucose monitoring because they feel it is an added source of stress. Key advantages of performing glucose curves at home are that they are less costly for the owner and, if confusing results are obtained on one day, it is often more practical to repeat a curve at home than for the cat to return to the clinic. It can also be a great advantage for an owner to have the ability to test blood glucose concentration whenever a diabetic cat shows signs consistent with hypoglycemia. The same protocol should be followed for serial blood glucose curves at home as for those performed in the hospital. Spot checks are useful whenever the owner is concerned about the well-being of the cat, but are not a substitute for a full blood glucose curve. A practical approach is to use knowledge of the cat's clinical signs to guide the timing of home-generated blood glucose curves. For example, if there is marked variability of 24-hour water intake and/or urine glucose results, owners can be advised to perform a glucose curve on a day when the cat has negative glycosuria or does not drink much water. This approach would increase the chance of detecting hypoglycemia.

Successful monitoring of diabetic cats must aim to recognize when the clinical signs are resolving and then identify when the insulin dose needs to be decreased to avoid hypoglycemia. No matter how and when blood glucose results are obtained they only provide a small piece of the puzzle of good feline diabetes management. Glucose curves in isolation are of limited use but their clinical value can be greatly improved when combined with a detailed history. Owners of diabetic cats should be encouraged to keep a daily diary or spreadsheet of their cats' progress at home and to provide this for review at each reassessment. Relevant information collected in the home environment will help to minimize over-interpretation of glucose curves and inappropriate insulin dose changes. Successful outcomes are most likely if detailed patient information is reviewed at each reassessment and clients are supported when learning injection and monitoring techniques while establishing a practical routine at home.

BLASTOCYSTIS IN COMPANION ANIMALS
Craig G Ruaux, BVSc PhD, DACVIM (SAIM)
Corvallis, OR

INTRODUCTION AND TERMINOLOGY

The genus *Blastocystis* is a complex grouping of enteric protozoa commonly detected in a broad variety of species, including most mammals, many avian species, reptiles and amphibians. Phylogenetically, *Blastocystis* is a member of the Stramenopiles (phylum Heterokonta), an extremely diverse and complex phylum of predominantly algal genera. This organism is one of the most common gastrointestinal parasites of human beings, with reported prevalence rates ranging from 5% in some western countries to >76% in some parts of the developing world.[1,2] Originally, species were named on the basis of the host organism (i.e. *Blastocystis hominis, B. ratti*), however it is now well recognized that there is no single species that infects only humans (or any other host organism), and the entire group of organisms is now identified as genus *Blastocystis* with numbered subtypes, these subtypes being determined by small subunit ribosomal DNA (SSU-rDNA) gene sequences.[3] At the time of writing there are at least 14 known subtypes of *Blastocystis* identifiable on the basis of SSU-rDNA sequences.

In addition to the remarkable genetic diversity within this genus, there is a high degree of morphologic variation, with vacuolar, "granular", amoeboid, and cystic forms identified. Any subtype from any host organism may be identified in any of the morphologic forms, although the amoeboid form is thought to be closely linked to disease in humans.[4] Transmission is thought to be predominantly fecal-oral; there is also some evidence for water-borne transmission. The high prevalence rates reported from human beings in the developing world likely reflect differences in public sanitary infrastructure, closer exposure to open sewers, and closer exposure to animal waste than in the developed world.

BLASTOCYSTIS AS A HUMAN ENTERIC PATHOGEN

There exists considerable controversy in the human medical literature regarding the pathogenic potential of *Blastocystis* organisms.[5] Many workers feel that there is a link between this organism and irritable bowel syndrome (IBS) in humans,[6-8] and some reports have shown a relatively high risk of carriage in human patients with ulcerative colitis.[7] Interestingly, there appears to be quite marked geographic and subtype differences in the potential pathogenicity of this organism. Workers in the developing world have demonstrated increased risk of *Blastocystis* carriage in human patients with IBS and inflammatory bowel disease,[7,8] while workers in the developed world (particularly Europe) have reported a low prevalence of these organisms in human patients with chronic gastrointestinal disease.[9,10]

When considered a specific disease entity, Blastocystosis in humans is associated with anorexia or hyporexia, diarrhea, bloating, abdominal pain, skin rashes, steatorrhea, and weight loss.[1] The natural history of this disease can be prolonged (months to years), there are no known reliably effective therapies, and spontaneous resolution, rapid re-infection following apparently successful therapy, and resolution of clinical signs without effective clearance of the organism have all been reported.[11]

POSSIBLE MECHANISMS OF PATHOGENICITY

While the actual pathogenic potential of the various *Blastocystis* subtypes for their hosts is an area of ongoing controversy, there are recognized characteristics of these organisms that are consistent with a potential role as a gastrointestinal pathogen. At least some isolates of Subtype 4 (previously *B. ratti*) are able to induce apoptosis, actin rearrangement and loss of barrier function in cultured rat intestinal epithelial cells.[12] This same strain has been shown not to activate toll-like receptor signaling in a human monocyte cell line, suggesting that the parasite may be able to avoid recognition by the innate immune system.[13] In unpublished data from our laboratory, we have observed this same lack of induction of TLR-2 in CACO-1 cells, along with apparent inhibition of the serotonin-reuptake transporter protein.

Several subtypes have been shown to produce a variety of cysteine proteinases, these proteases have been shown to be capable of cleaving secretory IgA, and similar proteases are recognized as virulence factors in other enteric protozoal pathogens, including a variety of *Trichomonas* spp., *Giardia* spp. and *Entamoeba histolytica*.[14,15]

In a recently reported study of fecal microbiome structure and diversity in humans with IBS symptoms, infection with *Blastocystis* was associated with decreased numbers of *Faecalibacterium prausnitzii* and Bifidobacteria,[16] a pattern that is commonly recognized in a variety of chronic enteropathic and dysbiotic diseases. Interestingly, this pattern was also present in asymptomatic male control patients that were positive for *Blastocystis*, but at this time this finding is one of correlation rather than causation.

PREVALANCE IN COMPANION ANIMALS, ZOONOTIC POTENTIAL?

The true prevalence of *Blastocystis* carriage in companion animals is unclear at this time, with most of the parasitology literature in this area focused on the domestic dog. Early data from subtropical and tropical environments suggested quite high carriage rates for *Blastocystis* in dogs, with rates as high as 70% reported for carriage in shelter-resident dogs in a subtropical environment,[17] although this finding has been thrown into some doubt by recent PCR-based studies from the same environment.[18] Carriage rates as high as 100% have been reported in domestic dogs from Thailand, however this was in a limited number of samples (n=3).[19] By comparison, in a cross-sectional study of military working dogs and personnel at a military dog center in Thailand, the prevalence in 189 dogs was 3.7%, while the prevalence in military personnel (n=317) at the same center was 14.5%.[20] This pattern of lower prevalence in dogs than humans is repeated in several other studies, leading some authors to conclude that dogs are unlikely to act as either natural hosts or primary zoonotic reservoirs for infection.[18]

In a study of asymptomatic individuals and pet animals living in the same household as humans symptomatic for Blastocystosis (n=11), 8/8 in-contact animals (5 dogs, 3 cats) were positive for *Blastocystis* via fecal PCR, with 7/8 of the in-contact animals carrying the same strain as the symptomatic in-contact human.[21] Interestingly, none of the in-contact animals or in-contact humans (n=17) in this admittedly small study showed clinical signs consistent with Blastocystosis. These data suggest that transmission between humans and companion animals is possible, however this is most likely a transient phenomenon and it appears unlikely that domestic dogs and cats represent a significant risk for zoonotic transmission.

Data recently reported from our laboratory assessed the prevalence and subtypes of *Blastocystis* in shelter-resident and client owned dogs in the US Pacific Northwest (specifically, Portland, OR and the southern Willamette valley).[22] In this study, we screened 104 shelter-resident dogs, 105 shelter-resident cats, 51 client-owned dogs and 52 client owned cats using a standard nested PCR methodology. In this study 10/103 dogs and 12/105 cats were positive for *Blastocystis*, however only 4/22 identified *Blastocystis* were of Subtype 1 (a potential human pathogen strain), with the remainder being Subtypes 10, 8 and 14, none of which are typically considered human enteric pathogens. All of the animals testing positive for *Blastocystis* in this study were shelter-resident, only one client-owned dog showed a faint positive result on PCR, this finding could not be replicated and thus the sample was considered negative. Shelter-resident animals (both cats and dogs) were significantly more likely to test positive for *Blastocystis* than client-owned animals, this is consistent with several prior observations that shelter-resident animals are more prone to enteric parasite infection.[23,24]

IS BLASTOCYSTIS A PATHOGEN IN COMPANION ANIMALS?

Within the veterinary and parasitology literature there is a remarkable dearth of data regarding any link between *Blastocystis* carriage and clinical signs of disease in companion animals. A solitary case report describes a significant parasite burden in a mixed-breed dog with vomiting, diarrhea, and weight loss. This animal was definitively diagnosed with exocrine pancreatic insufficiency and marked hypocobalaminemia, which is sufficient to explain the observed clinical signs.[25]

In our recent study of *Blastocystis* prevalence in companion animals in the Pacific Northwest we assessed the relationship between the presence of a parasite burden and signs of gastrointestinal disease of any kind. While records of gastrointestinal disease signs, typically diarrhea, were common in the medical records of the shelter-resident cats (26/105, 24.7%) and less so in dogs (9/103, 8.7%), there was no significant relationship seen between carriage of *Blastocystis* and the presence of gastrointestinal signs in the shelter-resident animals. Four of 52 (7.7%) client owned cats showed owner-reported gastrointestinal signs in the week prior to collection, while 2/51 (3.9%) of client-owned dogs showed gastrointestinal signs in the week prior to collection. None of the client-owned animals showing gastrointestinal signs tested positive for *Blastocystis*. Shelter-resident cats were significantly more likely to have gastrointestinal disease signs overall than client-owned cats (P=0.0099, Relative Risk 3.21, Fischer's Exact Test), while there was no significant relationship between shelter-residence and presence of gastrointestinal signs in dogs (P=0.339, Fischer's Exact Test). Similar data regarding the frequency of diarrhea in shelter-resident dogs and cats have been reported previously.[26] From the data reported in this study, it appears unlikely that the presence of *Blastocystis* infection is a significant contributor to diarrhea in shelter-resident cats in the Pacific Northwest of the USA.

CONCLUSIONS

At this time there is a distinct lack of evidence for a relationship between *Blastocystis* carriage and the presence of gastrointestinal disease in companion animals, however the rate of infection and subtypes present in animals with chronic enteropathies is currently unknown. There is at least some concern that animals in contact with humans with Blastocystosis may act as secondary reservoirs for infection, but there is little evidence at this time that domestic animals represent a significant source of zoonotic spread of these organisms to uninfected human beings. Given the widely varying apparent pathogenicity of differing subtypes, and apparent geographic variation in common subtypes in other species, it would be interesting to determine if similar differences in prevalence and common subtypes are present in domestic animals in differing geographic areas.

REFERENCES

[1]Tan KSW. **Clin Micro Rev**. 2008 Oct 1;21(4):639. [2]Amin O. **Res J Parasitol**. 2006;1(1):1. [3]Stensvold CR et al. **Trends Parasitol**. 2007 Mar;23(3):93. [4]Tan TC & Suresh KG. **Parasitol Res**. 2005;98(3):189. [5]Sekar U et al. **Trop Parasitol**. 2013 Jan;3(1):35. [6]Poirier Pet al. **PLoS Pathog**. 2012;8(3):e1002545. [7]Cekin AH, et al. **BMC Gastroenterol**. 2012;12(1):122. [8]Jimenez-Gonzalez DE, et al. **Parasitol Res**. 2012 Mar;110(3):1269. [9]Rossen NG, et al. **Eur J Clin Microbiol Infect Dis**. 2015 Feb:1. [10]Petersen AM & Stensvold CR. **Scand J Gastroenterol** 2013; 48(5):638. [11]Traub RJ, et al. **J Clin Gastroenterol**. 2010; 44(2):85. [12]Puthia MK, et al. **Inf Immun**. 2006;74(7):4114. [13]Teo JD, et al. **PLoS ONE**. 2014;9(2):e89036. [14]Mirza H & Tan KSW. **Parasitol Res**. 2009;104(2):355. [15]Sio SWS, et al. **Parasitol Res**. 2006; 99(2):126. [16]Nourrisson C, et al. **PLoS ONE**. 2014; 9(11):e111868. [17]Duda A, et al. **Vet Parasitol**. 1998;76(1-2):9. [18]Wang W, et al. **Parasites & Vectors**. 2013; 6(1):1. [19]Parkar U, et al. **Parasitology**. 2007;134(03):359. [20]Leelayoova S, et al. **J Med Assoc Thai**. 2009;92 Suppl 1:S53. [21]Nagel R, et al. **Intern Med J**. 2012;42(11):1187. [22]Ruaux C & Stang BV. **PLoS ONE**; 2014;9(9):e107496. [23]Hill SL, et al. **J Am Vet Med Assoc**. 2000;216(5):687. [24]Scaramozzino P, et al. **Vet J**. 2009;182(2):231. [25]Chapman S, et al. **Vet Clin Path**. 2009;38(1):59. [26]Bybee SN, et al. **J Vet Intern Med**. 2011;25(4):856.

CANINE AND FELINE HISTOPLASMOSIS

Andrew Hanzlicek, DVM, MS, DACVIM (SAIM)
Stillwater, OK

INTRODUCTION

Histoplasma capsulatum, the causative agent of histoplasmosis, is a thermally dimorphic soil-borne fungus found in temperate and subtropical regions throughout the world. In the U.S. histoplasmosis is most commonly found in the Ohio, Missouri, and Mississippi river valleys and in other Midwestern states such as Oklahoma.[1-3] Histoplasmosis has been diagnosed in many species including the dog, cat, and human.

Traditionally *Histoplasma capsulatum* has been divided into 3 varieties based on geographical distribution and phenotypic and pathogenic characteristics. *Histoplasma capsulatum* var. *capsulatum* is the cause of histoplasmosis throughout most of the world including in dogs and cats in North America. *Histoplasma capsulatum* var. *duboisii* is the cause of African histoplasmosis in humans in Africa and Japan. African histoplasmosis has not been reported in dogs or cats. *Histoplasma capsulatum* var. *farciminosum* is the cause of epizootic lymphangitis in horses and donkeys in Africa and disseminated histoplasmosis in dogs in Japan.[4] More recently phylogenetic analyses of *H. capsulatum* from around the world described 8 genetically distinct clades with 7 distinct phylogenetic species. The 3 varieties of *H. capsulatum* are distributed across the 8 phylogenetic clades indicating that the traditional variety classifications are phylogenetically meaningless.[5]

In the 1980s and 1990s histoplasmosis became an increasingly important disease in humans.[6] This was due to an increase in prevalence of immune system compromise associated with solid organ and hematopoietic cell transplants, immunosuppressive drug treatment for chronic inflammatory diseases, and most importantly the HIV/AIDS pandemic. Due in part to more effective antiretroviral therapy, over the past decade, in some areas the incidence of histoplasmosis has decreased in humans with HIV/AIDS and has increased in humans with immune system compromise due to other causes.[7] Histoplasmosis has been reported to be the second most common endemic mycosis in cats and in one study was the most common systemic mycosis in dogs.[8,9]

PATHOGENESIS

In the environment *H. capsulatum* is found as a free living mycelium. Mycelia produce small spores or microconidia (2-5μm) and large spores or macroconidia (5-13 μm). *H. capsulatum* grows best in warm, humid, acidic conditions and has an environmental niche for nitrogen rich soil. *H. capsulatum* is found most abundantly in the top 2 cm of soil and mycelia rarely sporulate in deeper soil. *H. capsulatum* can tolerate cold environmental temperatures with an adequate amount of moisture but cannot survive persistent temperatures greater than 40C. *H. capsulatum* has been found in the feces of multiple species of birds and bats. Outbreaks commonly involve the disruption and aerosolization of soil or dust contaminated with bird or bat feces.

Inhalation and less commonly ingestion or direct cutaneous inoculation of microconidia or hyphal fragments is the source of infection. Due to the small size, once inhaled microconidia are able to penetrate deeply into the lungs. Here the microconidia bind to specific surface receptors and are phagocytized by macrophages, neutrophils, and dendritic cells. Once within the cells the microconidia convert to yeast. This transformation is in large part stimulated by heat stress, the uncoupling of oxidative phosphorylation, and resultant extensive changes in gene expression. Pro-inflammatory carbohydrates in the microconidia cell wall are replaced with carbohydrates in the yeast cell wall that aide in evasion of the immune system and survival within the macrophage. It is here that the yeast reproduces by budding.

With an appropriate immune response many infections are eliminated without lasting signs of severe disease. Transient flu-like symptoms are the most commonly reported in humans with acute pulmonary histoplasmosis. Without an appropriate immune response, or with an overwhelmingly large inoculum, *H. capsulatum* organisms may persist. Disease may be localized to the pulmonary tree or disseminated when yeast are transported via the lymphatics and hematogenously throughout the body. In humans disseminated disease is uncommon in immunocompetent individuals and chronic respiratory conditions predispose to chronic pulmonary histoplasmosis. In some immunocompetent humans and dogs, *H. capsulatum* may lie dormant, only to be reactivated at a later time, maybe years later.[10] Although not well documented this may also occur in cats.

Based on seroprevalence and positive fungal culture in apparently healthy dogs and cats from endemic areas, it is believed that, like humans, many dogs and cats have transient or sub-clinical disease.[11,12] In one study over half the dogs sampled had *H. capsulatum* cultured from tissues 2 ½ years after exposure.[10] These findings suggest that persistence of the organism may be common, even in dogs without clinical signs. It remains unknown how the size of inoculum or the immune system is associated with disease in dogs and cats. Identification of a cause of immunosuppression is uncommon. Intuitively because dogs and cats are lower to the ground and possibly more likely to disturb contaminated soil or dust (sniffing, digging, etc.), as compared with humans, large inoculums may play a more important role in the epidemiology in these species. There are reports of multiple pets or pets and pet-owners in the same household concurrently developing histoplasmosis. These outbreaks are suggestive of a common large environmental source.

A seasonal association with the incidence of histoplasmosis has been reported. An increase of incidence from January to April has been reported in cats.[8] A bimodal increase of incidence from February to April and September to November has been reported in dogs.[9] In French Guiana the incidence of disseminated histoplasmosis in humans is highest with a short rainy season and long dry season.[13] This is thought to be due to the dry weather being conducive to outdoor activity and the aeration of spores. Interestingly, cats restricted to the indoors are commonly affected. In one study 34% of cats diagnosed with histoplasmosis were considered indoor only.[14] It has been postulated that the presence of indoor plants and potting soil or an unfinished basement could be the source of infection.[14]

CLINICAL FINDINGS

Clinical disease varies in severity and is a result of the fungal burden and the organ system(s) involved. At least three clinical forms of histoplasmosis are recognized in dogs and cats – pulmonary, disseminated, and gastrointestinal. Histoplasmosis isolated to the GI tract is relatively less common in cats as compared with dogs. A primary cutaneous form is uncommon but has been described in a cat.[15]

In cats disseminated disease is the most common (68-95%) followed the pulmonary form (5-38%).[16-18] One study found disease isolated to the gastrointestinal tract in 14% of cats.[16] Disseminated disease also appears to be the most commonly diagnosed form in dogs. Organ systems commonly affected in disseminated disease, in addition to the respiratory system, include the liver, spleen, lymph nodes and gastrointestinal tract in the dog and the liver, spleen, and lymph nodes in the cat. Bone, joint, bone marrow, cutaneous/subcutaneous, oral/mucocutaneous, renal, adrenal, ocular, and CNS infiltration have also been seen in both species.

Signalment

Any age, breed, or either sex may be affected. Dogs from 2y to 7y are at increased risk with an average age of 3.6y.[9] The Brittany, Weimaraner, and Pointer in addition to dogs in the American Kennel Club working, sporting, and terrier groups may be predisposed.[9] Data in cats is mixed, with an early study showing a mean age of 3.9y and more recent studies showing a median age of 6.2y, 7.7y, and 9y, respectively.[8,14,16,19] In one retrospective study Persians were predisposed. There is no apparent sex predilection in either dogs or cats.

Clinical Signs

After exposure the incubation period is reported to be 12-16 days in humans. Clinical signs began 3-5 days after experimental intratracheal inoculation of *H. capsulatum* yeast in dogs.[20] Disease is commonly insidious and clinical signs are generally non-specific. Clinical signs common in both species include lethargy or weakness, weight loss, anorexia, and fever unresponsive to antibiotics. With pulmonary involvement dogs commonly cough whereas cats are more likely to be dyspneic and/or tachypnic without a cough. Mixed bowel diarrhea with mucous, blood, and tenesmus is common in dogs. With ocular involvement blindness may be the first or only sign noticed by the pet-owner. Occasionally abdominal distension may be noted and is caused by ascites and/or organomegaly. The pet-owner may notice lameness with bone and/or joint involvement. Rarely cutaneous and/or sub-cutaneous nodules are the first or only outward signs of disease.

Physical Examination

Physical examination findings correlate with clinical signs and the body system(s) involved. In many cases physical examination findings are nonspecific. Common findings include poor body condition, adventitial lung sounds, hepatomegaly, splenomegaly, lymphadenopathy, and mucosal pallor. In some cases jaundice and ascites may also be present. Ulcerative oral and nodular or ulcerative cutaneous or mucocutaneous lesions may be found. Cutaneous fragility has been described in a cat with histoplasmosis.[21] In this case the dermis and subcutis were infiltrated. Lameness and joint effusion, most commonly of the carpal and tarsal joints, can be present with osseous or joint involvement. The soft tissues overlying the bone and/or joint are often swollen. Ulcerative skin lesions may be found directly overlying the affected joints. Vision loss may be due to retinal detachment and may be seen with conjunctivitis, blepharitis, optic neuritis, and chorioretinitis. Rarely signs of neurologic disease are seen with CNS involvement.

Laboratory Findings

The most common laboratory abnormality in both dogs and cats is anemia which is commonly non-regenerative. Hypochromasia and microcytosis may be seen with iron deficiency secondary to GI blood loss. Leukopenia and/or thrombocytopenia may be seen with bone marrow involvement. Pancytopenia is more common in cats as compared with dogs. Other less commonly reported findings include a leukocytosis with or without a left shift, monocytosis, eosinophilia, and lymphopenia. Hypoalbuminemia is the most common serum biochemistry abnormality. Globulin concentrations are usually normal or increased. With liver involvement, liver enzyme elevations are common and may be accompanied by hyperbilirubinemia. Hypercalcemia secondary to granulomatous inflammation has been reported but is uncommon. In one study hypocalcemia was found in 43% of cats.[16] In this study ionized calcium was not measured and the hypocalcemia was thought to be due to hypoalbuminemia and a decrease in the protein bound fraction of calcium.[16] Urinalysis is commonly normal but may reveal proteinuria and/or bilirubinuria.[14] Hematuria has been reported with urinary bladder involvement in a cat.[22]

Imaging

Normal radiographic or ultrasonographic appearance does not rule out infiltration of an organ. With successful treatment abnormalities seen on imaging studies resolve or greatly improve. Permanent tissue architectural changes (scarring) may lead to minimal persistent abnormalities. Pulmonary involvement may be accompanied by any radiographic lung pattern which may be seen with lymphadenopathy and on occasion pleural effusion. An interstitial pattern is most common and has been described as unstructured, fine, miliary, nodular, or linear. Both bronchial and alveolar lung patterns have also been described but are less common. On occasion a large fungal granuloma may cause a pulmonary mass like lesion.

With abdominal involvement radiographs may show hepatomegaly, splenomegaly, or a loss of serosal detail due to emaciation or free abdominal fluid. A recent study of splenic infiltration in cats showed that the spleen was commonly large and hypoechoic.[19] In many cases the spleen had a mottled echotexture and an irregular serosal margin. Abdominal ultrasound may also show free abdominal fluid; hepatomegaly with or without hyperechogenicity; lymphadenopathy; or intestinal or gastric wall thickening with or without a loss of normal layering.

With osseous involvement radiographs are consistent with osteomyelitis and joint effusion. Osteolysis with some periosteal bone proliferation of the metaphysis is commonly the most striking feature. Pathologic fractures and overlying soft tissue swelling may also be present.

Diagnosis

Traditionally a positive fungal culture from affected tissue or fluid sample has been considered definitive for the diagnosis of histoplasmosis.[23] Culture has the disadvantages of lacking sensitivity, requiring a specialized laboratory, posing risk to laboratory personnel, and having a slow turn-around time (2-4 weeks). In dogs and cats the diagnosis of histoplasmosis is most commonly made by finding the yeast phase of *H. capsulatum* in affected tissue or fluid samples. Yeasts are oval, 2-4μm in diameter, with narrow budding. Yeast are commonly found within macrophages and neutrophils on cytology or histopathology of lung, liver, spleen, lymph node, skin, and bone marrow aspirates or biopsies, respectively. In dogs organisms are commonly found on cytology from rectal scrape. On occasion organism may be found on blood smear in circulating neutrophils, monocytes, or eosinophils. Organisms may also be found in pleural or abdominal fluid, lung lavage fluid, synovial fluid, or rarely in CSF.

Antigen testing

Recently a commercially available antigen enzyme immunoassay (EIA; MiraVista Diagnostics® Indianapolis, IN) has shown promise in dogs and cats. The antigen, galactomannan, is a component of the fungal cell wall. Common test samples include serum, urine, and BAL fluid. In humans the sensitivity is dependent upon the form and severity of disease and immune status of the patient. The sensitivity (when both urine and serum are tested) is approximately 90% for disseminated histoplasmosis, 83% for acute histoplasmosis, 88% for chronic pulmonary histoplasmosis, and only 30% for subacute histoplasmosis.[24,25] The specificity is geographically dependent as there is cross-reactivity with other endemic mycoses. This includes *Blastomyces dermatitidis, Coccidioides immitis, Paracoccidioides basiliensis, and Penicillium marneffei.*[26] In the absence of other mycoses the specificity is very high.[27]

In cats the antigen EIA has been shown to have a sensitivity of 94.4% (95% confidence interval 74.2-99.0%) when only urine was tested.[28] Based on results from 19 dogs with confirmed histoplasmosis and 41 dogs without histoplasmosis, the antigen EIA when only urine is tested has a sensitivity of 89.5% and a specificity of 100% (Unpublished data). This study was performed in an area where other mycoses are uncommon. As in humans, in dogs and cats the performance of the antigen EIA is likely dependent upon the severity and form of disease. Interestingly, the only false negatives seen in the two aforementioned studies were in animals with clinical signs isolated to the GI tract. The author has documented negative antigen tests in cats with other localized disease including the eye and lymph node. For the diagnosis of acute pulmonary histoplasmosis in humans the highest sensitivity is achieved with testing both serum and urine for antigen combined with serology.[29] In the author's experience antigenuria in the absence of antigenemia is more common than the opposite scenario. For maximum sensitivity urine and serum should both be submitted for antigen testing.

Serology

There is a paucity of information regarding the utility of serology for the diagnosis of histoplasmosis in dogs and cats. In general, disadvantages of serology include the inability to distinguish between active infection and past exposure and the time lapse for seroconversion in acute disease. Methods for *H. Capsulatum* antibody detection include immunodiffusion and complement fixation. For the diagnosis of acute pulmonary histoplasmosis in humans complement fixation had a sensitivity of 64.3% and immunodiffusion had a sensitivity of 17.2%.[29] At this time it is unknown how including serology with antigen testing may affect the overall sensitivity and specificity for the diagnosis of histoplasmosis in dogs and cats.

TREATMENT

The goals of treatment are to resolve clinical signs, eliminate the organism, and minimize the chance of disease relapse. Current clinical challenges include the high cost of antifungal therapy and treatment monitoring, long term medication administration (especially important in cats), and determination of when to discontinue antifungal therapy and high relapse rates. These challenges are not mutually exclusive.

Antifungal therapy

Based on studies in humans and mice, itraconazole has traditionally been considered the oral antifungal of choice.[30,31] In one small study, itraconazole was effective and well tolerated in all 8 cats treated, although 2 cats required a second course of treatment.[18] Fluconazole has been recommended with ocular or CNS disease or when itraconazole is not tolerated by the patient. Recently this dogma has been challenged. A retrospective study in cats failed to demonstrate a significant difference in mortality or disease relapse rates in cats treated with itraconazole as compared to those treated with fluconazole.[14] Cats treated with itraconazole or fluconazole had a 38% and 29% mortality rate, respectfully. Even with generics of both drugs being available itraconazole is considerably more expensive. In the author's hospital itraconazole is 2.5-5 times as costly as fluconazole. For this reason the author commonly uses fluconazole in both dogs and cats with success. Gastrointestinal upset and hepatotoxicity are possible side-effects and liver enzymes should be monitored during treatment with both drugs. Itraconazole is fat soluble and is better absorbed when administered with a fatty meal. Antacids may decrease the absorption of itraconazole. Fluconazole can be given on an empty stomach and is not affected by concurrent administration of antacids.

The Infectious Diseases Society of America (IDSA) guidelines recommend the use of liposomal amphotericin B in more severe disease such as severe acute pulmonary and progressive disseminated histoplasmosis in humans.[32] Amphotericin B decreased fungal burden more rapidly than itraconazole in a rodent model and in humans with disseminated disease.[31,33,34] There is some evidence that fluconazole may antagonize the antifungal effects of amphotericin B.[31] In severe or life-threatening infection in dogs and cats liposomal amphotericin B followed by itraconazole should be considered. Disadvantages of liposomal amphotericin B include the need for intravenous administration, expense of drug, and possible nephrotoxicity. As compared with the older deoxycholate formulation, liposomal formulations increase delivery to the site of infection and increase the tolerable dose thus improving efficacy while decreasing nephrotoxicity.[35-38]

Second-line oral antifungals include voriconazole and posaconazole. Information in regards to the treatment of histoplasmosis with either drug remains limited. There are published reports of both drugs being used successfully to treat histoplasmosis in humans. Posaconazole was effective as a salvage therapy in six humans with severe histoplasmosis who had all failed traditional therapy.[39] Posaconazole was shown to be highly effective when compared with itraconazole and amphotericin B in experimentally infected immunocompromised mice. Being closely related to fluconazole, voriconazole may not be effective in patients who have failed fluconazole therapy.

Corticosteroids

The IDSA recommends corticosteroid therapy in acute and chronic pulmonary, pericardial, mediastinal lymphadenitis, and rheumatologic forms of histoplasmosis in humans.[32] In one study, the administration of corticosteroids with or without concurrent administration of antifungal drug was effective at relieving airway obstruction in dogs with hilar lymphadenopathy.[40] Information is lacking with regards to the use of corticosteroids in other forms of histoplasmosis in dogs and cats. One rationale for using corticosteroids with active infection is the concern for an inflammatory response to dying organisms with antifungal treatment. Corticosteroids therapy should be considered when signs of dyspnea or hypoxemia are present. Except for cases of granulomatous lymphadenitis with inactive disease, an anti-inflammatory dose is recommended. In general the duration of treatment is short, 7-14 days and should be given with concurrent antifungal therapy. With active infection prolonged or high dose corticosteroid therapy should be avoided.

THERAPEUTIC MONITORING

Since the availability of itraconazole and fluconazole, 2 studies have reported disease relapse rates of 25% and 40% of cats treated, respectively.[14,18] Similar published data is not available for dogs. Inappropriate treatment duration may be one cause of disease relapse. Treatment may be costly and stressful to the pet-owner and animal. Optimizing treatment duration to minimize disease relapse, cost of treatment, and stress is desirable. There is no consensus on appropriate antifungal treatment duration as published recommendations include a duration of 60 days to 6 months or 30 days to 2 months past resolution of clinical signs.[1,41,42]

Antigen levels may provide an objective and non-invasive tool in guiding treatment duration. Antigen levels, as quantified by the antigen EIA, decrease with antifungal treatment and increase with disease relapse in humans.[34,43] The IDSA clinical guidelines recommends measuring antigen levels at baseline, 2 weeks and 1 month after initiating antifungal drug therapy, every 3 months during therapy, and at least 6 months after the discontinuation of antifungal therapy.[32] Urine and serum antigen levels decrease during antifungal treatment and increase in disease relapse in both dogs and cats (Unpublished data). How antigen clearance correlates with clinical remission and how antigen levels can be used for therapeutic monitoring in dogs and cats is currently being investigated. A conservative approach would be to treat until all three of the following criteria are met: 1) antigen clearance, 2) at least 1 month past resolution of clinical signs, and 3) at least 1 month past resolution of imaging abnormalities. Treatment should rarely be shorter than 6 months in duration. Although more research is needed, monitoring antigen levels after the discontinuation of antifungal therapy may provide an early means of detection of disease relapse.

PROGNOSIS

Published prognosticators are lacking in both dogs and cats. Mortality rates before the availability of itraconazole and fluconazole were 80-90%.[2,3,17,44] Current data is lacking in dogs but two recent publications in cats report a mortality rate of 33% and 45%, respectively, with the latter study including cats treated with ketoconazole.[14,16] Research aimed at identifying early prognosticators that can help determine which animal may benefit from more aggressive and costly therapy is needed.

REFERENCES

1. Kerl ME. Update on canine and feline fungal diseases. Vet Clin North Am Small Anim Pract 2003;33:721-747. 2. Clinkenbeard KD, et al. J Am Vet Med Assoc 1988;193:1443-1447. 3. Clinkenbeard KD, et al. J Am Vet Med Assoc 1987;190:1445-1448. 4. Murata Y, et al. Med Myc 2007;45:233-247. 5. Kasuga T, et al. Mol Ecol 2003;12:3383-3401. 6. Adenis AA,et al. Curr Trop Med Rep 2014;1:119-128. 7. Myint T, et al. J of the Int Assoc Provid AIDS Care 2014;13:100-105. 8. Davies C, et al. J Am Anim Hosp Assoc 1996;32:380-391. 9. Selby LA, et al. Am J Epidemiol 1981;113:133-139. 10. Smith CD, et al. Am J Epidemiol 1976;104:299-305. 11. Turner C, et al. Am J Vet Res 1972;33:137-141. 12. Rowley DA, et al. J Infect Dis 1954;95:98-108. 13. Hanf M, et al. J AIDS Clinic Res 2012;3:175. 14. Reinhart JM, et al. J Feline Med Surg 2012;14:841-848. 15. Fischer NM, et al. Vet Dermatol 2013;24:635-638, e158. 16. Aulakh HK,et al. J Am Anim Hosp Assoc 2012;48:182-187. 17. Wolf AM, et al. J Am Anim Hosp Assoc 1983;20:995-998. 18. Hodges RD, et al. J Vet Intern Med 1994;8:409-413. 19. Atiee G, et al. Vet Radiol Ultrasound 2014;55:310-314. 20. Farrell RL, et al. Proc Soc Exp Biology Med 1953;84:51-54. 21. Tamulevicus AM, et al. J Am Anim Hosp Assoc 2011;47:e36-41. 22. Taylor AR, et al. J Am Anim Hosp Assoc 2012;48:203-208. 23. Kauffman CA. Curr Opin Infect Dis 2008;21:421-425. 24. Wheat LJ, et al. Diagn Microbiol Infect Dis 2002;43:29-37. 25. Hage CA, et al. Clin Infect Dis 2011;53:448-454. 26. Wheat J, et al. Clin Infect Dis 1997;24:1169-1171. 27. Abbate M, et al. J Am Soc Nephrol 2006;17:2974-2984. 28. Cook AK, et al. J Feline Med Surg 2012;14:512-515. 29. Swartzentruber S, et al. Clin Infect Dis 2009;49:1878-1882. 30. McKinsey DS, et al. Clin Infect Dis 1996;23:996-1001. 31. LeMonte AM, et al. J Infect Dis 2000;182:545-550. 32. Wheat LJ, et al. Clin Infect Dis 2007;45:807-825.33. Wheat LJ, et al. Antimicrob Agents Chemother 2001;45:2354-2357. 34. Hage CA, et al. Clin Vaccine Immunol 2011;18:661-666. 35. Janoff AS, et al. J Liposome Res 1998;3:451-471. 36. Swenson CE, et al. Antimicrob Agents Chemother 1998;42:767-771. 37. Barrett JP, et al. Clin Ther 2003;25:1295-1320. 38. Bekersky I, et al. Pharm Res 1999;16:1694-1701. 39. Restrepo A, et al. J Infect 2007;54:319-327. 40. Schulman RL, et al. J Am Vet Med

Assoc 1999;214:1345-1348. 41. Bromel C, et al. Clin Tech Small Anim Pract 2005;20:227-232. 42. Greene CE. Histoplasmosis. In: Greene CE, ed. Infectious Diseases of the Dog and Cat., Third edSaunders; 2006. 43. Wheat LJ, et al. Ann Intern Med 1991;115:936-941. 44. Mitchell M, et al. Can Vet J 1980;21:95-100.

RATIONAL USE OF ANTACID THERAPY IN THE DOG AND CAT

Stanley L. Marks BVSc, PhD, DACVIM (Small Animal Internal Medicine, Oncology), DACVN
Davis, CA

The gastric mucosa is frequently exposed to HCl, pepsin, bile acids, lysolecithin, non-steroidal anti-inflammatory drugs, and other noxious factors.[1,2] Despite this exposure, the gastric mucosa is able to maintain its structural integrity in healthy people and animals and resist injury by 0.1 mol/L HCl and pepsin present in the gastric lumen. This remarkable feat is due to an impressive array of defense mechanisms that include: 1) a pre-epithelial mucus-bicarbonate-phospholipid "barrier" and 2) a continuous layer of surface epithelial cells connected by tight and gap junctions, which generates bicarbonate, mucus, phospholipids, trefoil peptides, prostaglandins, and heat shock proteins. Other protective mechanisms that contribute to the 3rd line of defense include continuous epithelial cell renewal, continuous blood flow through mucosal microvessels, an endothelial microvascular "barrier", the generation of nitric oxide and hydrogen sulfide, and sensory innervation.[1] Mucosal defense is also regulated in part by the central nervous system (vagal stimulation, corticotrophin releasing factor, thyrotropin releasing factor, melatonin), and hormones (ghrelin, CCK, gastrin, adrenal corticosteroids, and growth factors such as EGF).[1,2]

Gastric acid secretion is under nervous and hormonal control. This physiologic process is controlled by a number of redundant second messenger pathways activated as a result of he binding of gastrin, acetylcholine, histamine, and prostaglandins to specific receptors on the basolateral surface of parietal cells. An increase in cytosolic calcium mediates the stimulatory effect of acetylcholine and gastrin, whereas the activation of adenylate cyclase with generation of cyclic AMP (cAMP) mediates the stimulatory effect of histamine. Strong potentiation between histamine and either gastrin or acetylcholine reflects post-receptor interaction between the pathways as well as the ability of acetylcholine and gastrin to release histamine from mucosal enterochromaffin-like (ECL) cells.[3] Although gastrin is the major circulating stimulus of acid secretion, it is likely that gastrin does not stimulate the parietal cells directly, but mobilizes histamine from the ECL cells in the oxyntic mucosa. Gastrin acts on CCK2 receptors to control the synthesis of ECL cell histamine, accelerating the formation of the histamine-forming enzyme, histidine decarboxylase (HDC). Stimulation of the proton pump (H^+, K^+-ATPase) to secrete hydrogen ions into the gastric lumen in exchange for potassium ions is the ultimate factor in acid secretion. A feedback system terminates gastric acid secretion after sufficient acid secretion has occurred. A decrease of intragastric pH stimulates somatostatin (SST) release from antral D cells. This peptide inhibits acid secretion and also blunts gastrin release and its stimulatory effect on CCK2 receptors located on parietal and ECL cells. In addition, acidification of the duodenum triggers secretin release, which also inhibits gastric acid secretion.[4] The inhibitory effects of SST and endogenous prostaglandins (EPs) on acid secretion are mediated by receptors (SST 2 and EP 3, respectively) coupled by guanine nucleotide-binding proteins to inhibition of adenylate cyclase activity. All the pathways (both stimulatory and inhibitory) converge on and modulate the activity of the luminal enzyme (i.e. H^+, K^+ -ATPase).[5]

Chemically, all the available proton pump inhibitors (PPIs) consist of a benzimidazole ring and a pyridine ring, but vary in the specific side ring substitution.[6] They are all weak protonatable pyridines, with a pKa of about 4.0 for omeprazole and lansoprazole, about 3.9 for pantoprazole, and about 5.0 for rabeprazole. As a result, they accumulate specifically and selectively in the secretory canaliculus, the highly acidic space of the parietal cell. The selectivity of PPIs for the parietal cell proton pump stems from the fact that H^+, K^+-ATPase is the only pump in the body that generates a sufficiently steep proton gradient of more than 1:1,000,000, corresponding to pH 0.8–1.0 inside the canaliculus of a secreting parietal cell. PPIs are the most potent inhibitors of gastric acid secretion available and are most effective when the parietal cell is stimulated to secrete acid postprandially, a relationship that has important clinical implications for timing of administration. Because the amount of H^+, K^+ -ATPase present in the parietal cell is greatest after a prolonged fast, PPIs should be administered before the first meal of the day.

Clinical Application of PPIs and H_2RAs in People

PPIs are effective for treatment of all acid-related disorders, i.e. peptic ulcer (PU) and eradication of *Helicobacter pylori*, Zollinger-Ellison syndrome (ZES), gastroesophageal reflux disease (GERD) and its complications, nonsteroidal anti-inflammatory drugs (NSAID)-associated gastroduodenal ulcers, as well as PU bleeding. In a landmark study, Burget et al. found a highly significant correlation between ulcer healing and the degree of acid suppression, the duration of acid suppression, and the length of therapy.[7] The shape of the contour expression of these relationships showed that healing rate increased as the duration of suppression increased and as gastric pH increased. Being more potent and longer lasting than H_2RAs, PPIs are therefore more effective in healing both gastric and duodenal ulcers and this observation has been confirmed in several meta-analyses.[8,9]

Once-daily PPI dosing inhibits maximal acid output by about 66% after 5 days in people. Thus, the occasional use of a PPI taken on an 'as needed' basis would not be expected to reliably provide adequate acid inhibition and would be unlikely to produce a consistent or satisfactory clinical response (in contrast to the H_2RAs, which have a more rapid onset of action).[10] Because of the delay in optimal acid inhibition, the initial use of twice-daily dosing (for the first 2–3 days) may be helpful in achieving more rapid inhibition of gastric acid secretion in people. Clear instructions in human practice concerning empirical therapy of PPIs are lacking. That PPIs are often taken inappropriately is revealed by a US survey that showed that only 27% of GERD patients dosed their PPI correctly (i.e. up to 60 min before any meal of the day).[11]

H_2RAs inhibit acid secretion by competitively blocking histamine type-2 receptors on the parietal cell, thus reducing basal and stimulated gastric acid secretion. Pepsin secretion is also reduced, which results in decreased peptic activity. The use of H_2RAs can suppress gastric acid production by 37 to 68% over 24 hours in people and there does appear to be a dose-response relationship between H_2RAs and gastric ulcer prevention and healing. Both a meta-analysis and a Cochrane review found that standard doses of H_2RAs were effective at reducing the risk of duodenal but not gastric NSAID-associated ulcers.[12,13] The same Cochrane review concluded that high dose (double-dose) H_2RAs were effective at preventing chronic NSAID-related duodenal and gastric ulcers and also significantly reduced symptoms of abdominal pain compared with placebo.[13] There are currently three combination products aimed at decreasing the risk of NSAID-associated GI complications that are approved for human use in the US. Each of these combinations – diclofenac and misoprostol, naproxen and esomeprazole, or ibuprofen and famotidine have been shown to lower the incidence of GI toxicity compared to NSAID use alone.[14]

Clinical Applications of PPIs and H_2RAs in Dogs

The effects of H2-receptor antagonists and proton pump inhibitors have been investigated in several studies in healthy dogs. Unfortunately, critical assessment of antacids in experimental animal models or dogs with spontaneous disease are sparse, and most of the studies evaluating the efficacy of antacids have been completed in healthy animals. In people, optimal gastroduodenal ulcer healing and treatment of gastroesophageal reflux disease occur at an intragastric pH of ≥ 3 for approximately 75% of the day, and a pH of ≥ 4 for approximately 67% of the day, respectively.[15] In dogs, effective clinical doses of anti-secretory drugs have not been established, but clinical experience suggests that the recommended oral omeprazole dose (0.7-1.0 mg/kg q 24h) may not suppress acid secretion adequately. This impression is underscored by the findings of Bersenas et al. who determined the degree of gastric acid suppression associated with the administration of 4 gastric acid suppressants (ranitidine 2 mg/kg IV q 12h; famotidine 0.5 mg/kg IV q 12h; pantoprazole 1 mg/kg IV q 12h; and omeprazole 1 mg/kg PO q 24h or 1 mg/kg PO q 12h) compared to saline solution in 12 healthy Beagles.[16] Famotidine, pantoprazole, and omeprazole significantly suppressed gastric acid secretion compared to saline (placebo); however, ranitidine did not. In addition, administration of famotidine q 8h instead of q 12h did not affect gastric pH significantly, suggesting that increased frequency of administration of this H_2RA is of no benefit. Omeprazole administered q 12h instead of q 24h had a significant effect on gastric acid secretion and was the only regimen tested that approached the potential therapeutic efficacy for acid-related disease when assessed by criteria used for human patients.

The ability to accurately and noninvasively measure intragastric pH using catheterless radiotelemetric pH monitoring devices in dogs and cats since 2002 has resulted in renewed understanding of the effects of various antacids on intragastric pH in animals. Several important studies have been published utilizing the Bravo system for assessment of intragastric pH in healthy dogs and cats following administration of PPIs and H_2RAs. Comparison of the gastric acid suppressing effects of an equine omeprazole paste (Gastrogard®) with omeprazole tablets and famotidine tablets (1.0 – 1.3 mg/kg q 12h) was performed in 6 healthy adult dogs.[17] The dose of omeprazole administered (1.5-2.6 mg/kg q 24h) was slightly higher than the recommended dose, and both omeprazole formulations provided superior acid suppression in dogs compared to famotidine or placebo. Mean ± SD % of time the intragastric pH was ≥ 3 was 62.5 ± 6.1% for omeprazole tablet, 54.1 ± 8.3% for omeprazole paste, 21.5 ± 13.6% for famotidine, and 6.1 ± 16.8% for placebo. Famotidine and placebo were not significantly different from each other in this study.

The potential benefits of co-administering a more rapidly acting H_2RA (famotidine) with a PPI (pantoprazole) intravenously versus pantoprazole alone in 12 healthy dogs has also been investigated.[18] Many veterinarians administer the rapidly acting famotidine in the first few days of treatment in combination with the slower acting, but more potent PPIs to critically ill dogs or animals at high risk for GI bleeding. The potential concern with this practice however is that the simultaneous treatment with an H_2RA might decrease the efficacy of PPIs because of the requirement for PPI activation and trapping in the acidic canalicular environment of the parietal cell. There were no significant differences in the mean % time intragastric pH was ≥ 3 and 4 between the two groups, and pantoprazole administered alone achieved pH goals established for humans with acid-related disorders. These results should be interpreted cautiously because there is evidence for the benefits of combining an H_2RA at night with PPI therapy for people with acid reflux experiencing nocturnal acid breakthrough.[19]

The efficacy of omeprazole (0.85 mg/kg q 24h) and standard dose famotidine (1mg/kg q 24h) and high-dose famotidine (2 mg/kg q 12h) administered orally was evaluated in racing Alaskan sled dogs that have been shown to be at high risk for gastric erosions and ulcerations.[20,21] The three most important findings in the studies were that a) administration of famotidine at the standard dose of 1mg/kg q 24h was not as effective in lessening the severity of gastric lesions in dogs running 330 miles as it was in dogs running 100 miles; b) administration of high-dose famotidine (2 mg/kg q 12h) was roughly similar in its effects in lessening the severity of gastric lesions as was the standard dose of famotidine; and c) administration of omeprazole at the recommended dose was superior to both doses of famotidine in lessening the severity of gastric lesions.

The benefits of PPIs in people with moderate to severe gastroesophageal reflux disease (GERD) have been well documented and several studies have investigated the effects of PPIs and H_2RAs on esophageal pH in dogs in an effort to manage or prevent reflux esophagitis and esophageal stricture formation associated with general anesthesia.[22] Administration of esomeprazole to 22 dogs undergoing elective orthopedic procedures was associated with a significant increase in gastric and esophageal pH throughout the surgery procedure but did not have an impact on the number of reflux events. Nevertheless, the maintenance of an esophageal pH > 4 in all of the dogs treated probably minimized the likelihood of reflux esophagitis and stricture formation in the 36% of dogs that had $\geq$ 1 reflux event because the proteolytic pH range for the conversion of pepsinogen to pepsin is between 1.5 and 3.5,[23] and esophageal acid clearance does not occur during sleep or anesthesia. There is compelling in vivo evidence in experimental animal models of esophagitis induced in rabbits, dogs, and cats that the dripping or infusion of an acidic solution of hydrochloric acid alone (pH 1.6–2.0) on the esophagus over a period of 30–60 minutes is associated with minimal to no damage to the esophagus, but the addition of pepsin to the hydrochloric acid is associated with both macroscopic (eg, edema, hyperemia, hemorrhage) and microscopic (eg, epithelial loss, erosions, inflammation, mucosal, and submucosal hemorrhage) evidence of acute esophagitis.[24] Further studies are warranted to determine the benefits of pre-anesthetic administration of PPIs to prevent esophagitis and stricture formation in dogs.

Misoprostol, a synthetic prostaglandin E1 analog, has been shown to significantly reduce endoscopically detectable aspirin-induced gastroduodenal mucosal lesions in healthy dogs compared to a placebo group.[25] Misoprostol was administered orally at 3 μg/kg q 8h and aspirin was administered at 35 mg/kg q 8h and lesions were scored on days 5, 14, and 30 of the study. In contrast to the previous study, misoprostol administered at 2 μg/kg q 8h had no effect in preventing gastric lesions in dogs with acute intervertebral disc disease following administration of a single dose of dexamethasone at 2 mg/kg followed by a tapering regime of prednisolone (2 mg/kg to 0.5 mg/kg over the course of 5-6 days).[26]

Clinical Applications of PPIs and H_2RAs in Cats

There are fewer studies critically evaluating the efficacy of PPIs and H_2RAs in cats compared to dogs. Dosages for antacids in cats have been extrapolated from results of studies in dogs. The Bravo system was used by Parkinson et al. for the wireless measurement of intragrastric pH following twice daily administration of omeprazole (fractionated tablet or reformulated paste) versus standard doses of famotidine administered q 12h.[27] The mean percentage time ± SD that intragastric pH was ≥ 3 was 68.4 ± 35.0% for the fractionated omeprazole tablet, 73.9 ± 23.2% for the reformulated omeprazole paste, 42.8 ± 18.6% for famotidine, and 16.0 ± 14.2% for placebo. Importantly, there were no significant differences in the plasma concentrations between the 2 omeprazole formulations and both were in a range associated with acid suppression. This is the first study in small animals to show that fractionated enteric-coated omeprazole is an effective acid suppressant despite disruption of the enteric coating. In addition, the acid suppressive effects of famotidine were significantly weaker than that of both omeprazole formulations. The study did not evaluate the effects of omeprazole administered q 24h and further studies are warranted in cats comparing once daily versus twice daily administration of PPIs.

Adverse Effects of Chronic PPI and H_2RA Administration

The long-term effects of PPI and H2RA administration have not been well studied in dogs and cats to date and effects have been extrapolated from human studies. The most commonly reported adverse effects associated with chronic PPI therapy include vitamin B12 deficiency, diarrhea, enteric infections, particularly *Clostridium difficile* infection, increased risk of hospital acquired pneumonia, and the potential increased risk of fractures in geriatric people.[28] The effect of omeprazole administration on the gastric and duodenal microbiota was evaluated in 8 healthy dogs administered omeprazole at 1.1 mg/kg q 12h for 15 days.[29] Gastric fluid pH was significantly increased at the end of the treatment period (median: 7.1; range: 4.2-8.1); however, omeprazole did not induce major changes in the microbiome in either the stomach or the duodenum. It should be emphasized that the approach used allowed only the assessment of relative changes in the proportions of bacterial groups within the total population and significant changes in the total number of bacteria in response to omeprazole treatment may not have been discovered.

In conclusion, there is a strong need for veterinarians to better define the clinical applications of PPIs and H_2RAs in small animals in an effort to minimize the injudicious administration of these drugs. Many animals are administered omeprazole, famotidine and sucralfate concurrently when there is little to no benefit documenting the additive effects of co-administering these drugs. The long-term effects of these drugs must be better investigated for animals requiring chronic antacid therapy, and the potential concerns of H_2RA tolerance must be studied. The indications for once daily versus twice daily administration of PPIs needs to be better defined, particularly in cats, and studies should strive to evaluate the effects of these drugs in clinically affected animals when feasible.

REFERENCES

[1]Tarnawski A, et al. **Curr Pharmaceut Design** 2013;19:126-132; [2]Laine L, et al. **Gastroenterol** 2008;135:41-60; [3]Hersey SJ, et al. **Physiol Rev** 1995;75:155-189; [4]Chey WY, et al. **J Physiol Pharmacol** 2003;54(suppl 4):1005-112; [5]Aihara T, et al. **Pharmacol Ther** 2003; 98:109–127; [6]Stedman CAM, et al. **Aliment Pharmacol Ther** 2000;14:963-978; [7]Burget DW, et al. **Gastroenterol** 1990;99:345–351; [8]Poynard T, et al. **Eur J Gastroenterol Hepatol** 1995;7:661-665; [9]Salas M, et al. **BMC Gastroenterol** 2002;2:17-23; [10]Wolfe MM, et al. **Gastroenterol** 2000;118(suppl 1):S9-S31; [11]Pezanoski J, et al. Gastroenterol 2003;124(suppl):A228; [12]Koch M, et al. **Arch Intern Med** 1996;156:2321-2332; [13]Rostom A, et al. **Cochrane Database Syst Rev** 2002;4:CD002296; [14]Schiff M, et al. **Expert Rev Gastroenterol Hepatol** 2012;6:25-35; [15]Bell NJ, et al. Digestion 1992;51(suppl 1):59-67; [16]Bersenas AM, et al. **Am J Vet Res** 2005;66:425-431; [17]Tolbert K, et al. **J Vet Int Med** 2011;25:47-54; [18]Tolbert K, et al. **J Vet Int Med** 2015;29:556-560; [19]Mainie I, et al. **J Clin Gastroenterol** 2008;42:676-9; [20]Williamson KK, et al. **J Vet Int Med** 2007;21:924-927; [21]Williamson KK, et al. **J Vet Int Med** 2010;24:285-288; [22]Zacuto A, et al. **J Vet Int Med** 2012;26:518-525; [23]Hirschowitz BI. **Yale J Biol Med** 1999;72:133–143; [24]Lanas A, et al. **Gastroenterol** 1999;116:97–107; [25]Johnston SA, et al**. J Vet Int Med** 1995;9:32-38; [26]Neiger R, et al. **J Vet Int Med** 2000;14:33-36; [27]Parkinson S, et al. **J Vet Int Med** 2015;29:104-12; [28]Sheen E, et al. **Dig Dis Sci** 2011;56:931-950; [29]Garcia-Mazcorro JF, et al. **FEMS Microbiol Ecol** 2012;80:624-636.

INTRODUCTION TO THE EMERGING PIPRANT THERAPEUTIC CLASS: PROSTANOID RECEPTOR ANTAGONISTS

Kristin Kirkby Shaw, DVM, MS, PhD, DACVS, DACVSMR
Seattle, WA

INTRODUCTION

Prostanoids are metabolites of arachidonic acid and have been the target of medicinal therapies for centuries. However until recently, preparations aimed at inhibiting the action of prostanoids have, with varying degrees of specificity, prevented their production and thus impacted both pathologic and homeostatic activity. This presentation and accompanying manuscript will review the generation and physiologic activity of prostanoids, current drugs used to inhibit their action, and introduce a new class of therapeutics that may provide targeted therapy by antagonizing prostanoid receptors.

Arachidonic Acid Cascade Revisited

Arachidonic acid (AA) is an omega 6 poly-unsaturated fat that is an integral part of the cell membrane and is liberated from the cell membrane by the enzyme phospholipase A_2 (PLA_2) as part of normal cellular metabolism and following cell injury. The freed AA molecule is then shuttled into one of three pathways that ultimately generate three groups of molecules: prostanoids, leukotrienes and epoxides. This manuscript will focus on the prostanoid group.

The enzyme cyclooxygenase 1 (COX-1) converts AA to the intermediate product prostaglandin (PG) H2 and ultimately five active prostanoid molecules: Thromboxane (TXA), PGD_2, PGI_2 (prostacyclin), $PGF_2\alpha$, and PGE_2.[1,2] The enzyme COX-2 converts AA to PGI_2 and PGE_2. The COX-2 enzyme is inducible, meaning it is up-regulated during injury and inflammation; however, COX-2 is also an important mediator of gastric and renal homeostasis.

In the mid-1990s prostanoid receptors were identified and continue to be characterized.[1] These receptors have varying distributions throughout the body and mediate the activities of each prostanoid. That is, each prostanoid interacts with at least one distinct receptor to result in a physiologic effect, and varying physiologic effects are mediated by different receptors (Table 1).

Inhibiting Prostanoid Production: The Good and the Bad

Prostanoids are responsible for a wide range of homeostatic functions including regulation of renal hemodynamics and ion transport, gastrointestinal cytoprotection and motility, vascular and bronchial smooth muscle activity, immune function and platelet aggregation. However pain, inflammation, and acute phase allergic reaction are also mediated by prostanoids, in particular PGE_2 and PGD_2, and thus therapeutics are often used to decrease the activity of these molecules in various disease states. Steroids and cyclooxygenase inhibiting drugs have until now been the primary options for treating pain and inflammation, and because of their mechanism of action, these approaches have some draw backs that are discussed below.

Corticosteroids inhibit the enzyme PLA_2, acting up-stream of the COX enzymes in preventing the metabolism of AA.[3] Corticosteroids prevent the generation of prostanoids, leukotrienes, and epoxides, thus having broad anti-inflammatory action but also inhibiting the production of many homeostatic AA metabolites. Short-acting corticosteroids can be important therapeutics in the management of acute allergic reactions, and while they may provide pain relief and reduce the symptoms of allergy with chronic administration, the untoward side effects make them a poor choice for long term treatment of pain and/or atopy in veterinary patients.

Salicylic acid is a natural product derived from plants such as willow bark and meadowsweet, and its medicinal properties of pain relief and fever reduction have been utilized for centuries.[4] Aspirin, or acetylsalicylic acid, is a synthetic drug made from salicylic acid. Aspirin irreversibly binds to the COX enzymes and inhibits the formation of all prostanoids.[2,5] While aspirin is effective at reducing fever, pain and inflammation, significant adverse effects are well-recognized in humans and animals, in particular gastrointestinal irritation/ ulceration. Furthermore, irreversible inhibition of platelet TXA by aspirin leads to reduced aggregation that lasts up to 7- 10 days (the life span of the affected platelets). This anti-thrombotic property can be useful therapeutically but should also be recognized as a potentially significant unwanted side effect of aspirin administration.

Cyclooxygenase inhibiting non-steroidal anti-inflammatory drugs (COX inhibiting NSAIDs) decrease inflammation through inhibition of COX enzymes and subsequent inhibition of prostaglandin production.[2] This class includes aspirin, as described above, as well as other non-selective COX inhibitors such as ketoprofen and indomethacin and COX-2 preferential and selective inhibitors, such as carprofen, meloxicam, deracoxib, firocoxib, and robenacoxib.

The two cyclooxygenase isoenzymes, COX-1 and COX-2, were isolated in the 1980s and shortly thereafter it was discovered that COX-2 is up-regulated in response to tissue injury and is associated with inflammation and pain.[2] Drug development then concentrated on molecules that selectively inhibited this enzyme in hopes of sparing the so-called "house-keeping" prostaglandins produced by COX-1. The coxib class of drugs, or selective COX-2 inhibitors, was first marketed in human medicine in 1999. In humans, COX-2 selective products such as rofecoxib (Vioxx®) and celecoxib (Celebrex®) did show improved GI tolerability; however, they were linked to significant cardiovascular side effects and thus, all but one product have been removed from the US market.[6] The underlying cause for these adverse cardiovascular events may be attributed to COX-2 inhibition of PGI_2 (prostacyclin). PGI_2 is a potent vasodilator and has anti-thrombotic properties (an endogenous antagonist of TXA), thus decreased production of PGI_2 by coxib NSAIDs may lead to increased thrombosis and vasoconstriction, two methods of potentiating cardiovascular disease.[6] The severe cardiovascular adverse events associated with coxib NSAIDs in humans have not been observed in dogs taking COX-2 selective NSAIDs. However, severe gastrointestinal, renal and hepatic adverse events have been reported in dogs taking these products.[2]

Both COX-1 and COX-2 activity are important in maintaining gastrointestinal integrity and renal hemodynamics in dogs, primarily through the actions of PGE_2 and PGI_2.[2] COX-2 is constitutively expressed in the canine kidney and involved in regulation of vessel tone and salt and water balance. In the stomach and intestinal tract, COX-1 is the primary mediator of gastro-protective PG (PGE_2 and PGI_2) production in the healthy animal. But, COX-2 is up-regulated in areas of gastrointestinal irritation and it is important in promoting the healing of gastrointestinal ulcerations.[2] This suggests that administration of a selective COX-2 inhibitor in the face of gastrointestinal irritation, such as due to stress or administration of aspirin, corticosteroids or a non-selective cyclooxygenase inhibiting NSAID, may reduce healing and potentiate ulceration and perforation.[7] Another potential mechanism for gastrointestinal injury secondary to cyclooxygenase inhibiting NSAID administration is the increased production of inflammatory leukotrienes. Inhibition of the COX pathways may result in increased AA that must be metabolized by the LOX pathway.

The four FDA approved canine COX inhibitor NSAIDs have all demonstrated efficacy for managing osteoarthritic pain and unless contra-indicated, are considered to be effective treatments for the pain associated with OA. However, these COX-inhibiting NSAIDs, as a class, carry the potential for adverse effects including gastrointestinal ulceration and perforation and renal insufficiency.[2] While COX-2 selective and preferential NSAIDs are associated with fewer side effects compared to aspirin and older, non-selective COX inhibitor NSAIDs, many dogs still do not tolerate these drugs, and alternative pharmacologic pain relief is indicated. Currently, no other class of pharmaceuticals is approved by the FDA for the treatment of pain and inflammation associated with OA in dogs. Thus, it is evident that an FDA approved product that demonstrates targeted pain relief with improved tolerability is needed.

THE PIPRANT CLASS: PROSTAGLANDIN RECEPTOR ANTAGONISTS

After the identification of prostanoid receptors in the 1990s, significant research ensued into methods of targeting each receptor with synthetic agonists and antagonists.[1] In October 2013 the World Health Organization defined a newly recognized class of prostaglandin receptors antagonists as the **piprant class.**[8] There are two receptors that currently appear to be particularly promising as therapeutic targets for the piprant class: the EP4 and DP2 receptors.[2]

EP4 Receptor Antagonists

Of the five prostanoids produced from AA, PGE_2 is the one primarily involved in causing pain and inflammation; however, it also plays an important role in gastroprotection and renal function.[9-14] PGE_2 has at least four distinct receptors (EP1, EP2, EP3, EP4) distributed within different tissues, mediating diverse and sometimes overlapping processes (Table 1).[2,9] Over a decade of research using experimental models (including canine) has led to the discovery that while all four EP receptors may mediate various types of pain, the EP4 receptor is most prominently involved in inflammation and pain associated with OA, including central sensitization. For example, Lin *et al.* reported that in rodents, the EP4 receptor is expressed by sensory neurons, and the level of EP4 receptors increased following peripheral inflammation.[10] Administration of an EP4 antagonist drug decreased pain and hypersensitivity in rodents. Other investigators have described similar anti-inflammatory and analgesic effects using selective EP4 receptor antagonists in rodent models of acute and chronic inflammation.[11,14,15] The EP4 receptors in humans, dogs and rodents are very similar and therefore it is believed that EP4 antagonists will provide relief from arthritic pain in humans and veterinary clinical patients, without affecting the synthesis and widespread activity of PGE_2 and other prostanoids.

Grapiprant is a new therapeutic in the piprant class that has demonstrated selective blockade of the EP4 receptor in rodents, humans, and dogs.[16] Grapiprant has undergone numerous experimental and pilot studies in dogs, and a multi-site, placebo controlled, randomized field trial of grapiprant for the control of pain and inflammation associated with OA in dogs was recently completed and will be submitted to the FDA in support of approval.[17] The results of these initial studies in dogs with OA were promising suggesting that this therapeutic provides control of the pain and inflammation associated with OA. Safety studies in laboratory dogs have demonstrated an excellent safety profile, and this safety has been confirmed in two large field effectiveness studies in dogs with OA.[18] The relative lack of toxic effects with grapiprant compared to those that have been reported in dogs treated with COX-inhibiting NSAIDs is consistent with grapiprant's unique mechanism of action. Grapiprant may offer a more targeted, and potentially more tolerated, method of pain management in dogs with OA.

DP2 Receptor Antagonists

PGD_2 is primarily produced by IgE-activated mast cells and thus plays a central and significant role in the allergic response.[1,19] There are at least two receptors that mediate the biological effects of PGD_2: DP1 and DP2 (also known as CRTH2).[1,19-21] Binding of PGD2 to the DP1 receptor results in smooth muscle relaxation, vasodilation, and bronchodilation. The DP2 receptor is expressed on TH2 lymphocytes, eosinophils and basophils, and binding PGD_2 leads to chemotaxis and activation of these cells—early events in an allergic airway or skin reaction.[19]

In vitro and experimental models have confirmed that the DP2 receptor is integrally involved in airway and dermal allergic response, and administration of DP2 agonists in rodent models results in characteristic signs of allergy.[20,21] Furthermore, genetically DP2 deficient mice fail to demonstrate dermal infiltration of lymphocytes and eosinophils, decreased nasal mucosal inflammation, and decreased production of IgE in response to allergens.[20,21] Based on these findings, DP2 receptor antagonists are currently being developed as pharmaceutical methods of treating atopy, allergic rhinitis, and asthma in veterinary and human patients.[21]

The contribution of each prostanoid receptor to the indicated physiologic role varies in significance. Some receptors are the sole or primary mediator of a physiologic function, while in other cases the receptor may play a minor role in the physiologic function but is included for completeness. The full extent of these prostanoid receptor's activities are still being investigated.

Table 1: Physiologic roles and receptors of the prostanoids.[1,9,19]

Prostanoid + Receptor	Physiologic Role of Prostanoid-Receptor Interaction	Potential Therapeutic Application of Receptor Antagonist
PGE_2—EP1	PAIN: Mediates acid-induced visceral pain IMMUNE: Mediates stress responses (ACTH secretion, stress behavior) GI: Stomach: Inc. HCO_3 (rat), potentiates mucus secretion Intestine: Inc. motility KIDNEY: Ion regulation in CD CANCER: Promotes chemical carcinogenesis	Cystitis, Cancer (colon, breast, skin), Stroke, Hypertension
PGE_2—EP2	PAIN: Facilitates pain transmission and joint inflammation IMMUNE: Decreases dendritic cell differentiation, increases neutrophil recruitment, inhibit T cell proliferation GI: Decrease intestinal motility KIDNEY: Vascular smooth mm relaxation; Inc. renin release REPRO: Mediates COX-2 induced mammary hyperplasia Facilitates ovulation and fertilization LUNG: Bronchodilation; Dec. airway inflammation CANCER: Mediates angiogenesis	None in development; no current rationale or commercial incentive to develop.
PGE_2—EP3	PAIN: Fever, endotoxin and viral induced pain IMMUNE: Suppress Type I allergic reaction GI: Stomach: Dec. HCL stomach; Intestine: Inc. peristalsis; Inc. HCO_3 secretion duodenum, cytoprotection (rats) KIDNEY: Regulate ion transport CANCER: Mediates tumor and chronic inflammation-associated angiogenesis	Decrease fever, endotoxin and post-herpetic related pain, rheumatoid arthritis, atherothrombotic disease, allergy, cough, lung infection, overactive bladder, cancer
PGE_2—EP4	PAIN: Mediate joint inflammation, hyperalgesia/ allodynia/ central sensitization associated with OA; protect against NMDA-induced excitotoxicity IMMUNE: Increases dendritic cell motility but decreases release of cytokines GI: Stomach: Inc. HCL secretion, HCO3 and mucus secretion; Intestine: Inc. HCO3 duodenum (human) Dec. hypermotility, smooth mm relaxation; cytoprotection (rats) Colon: Inc. mucus secretion, protect against colitis, angiogenesis and promote mucosal healing; up regulated during colitis KIDNEY: Afferent arteriole vascular tone (regulate GFR), regulate renin excretion VASCULAR: Close Ductus Arteriosus in utero, vascular relaxation; angiogenesis BONE: Influences bone formation and resorption CANCER: Promote proliferation, cell survival, invasiveness, angiogenesis, cell migration, metastasis	Decrease Inflammation, hyperalgesia, allodynia, particularly related to joints/ osteoarthritis; cancer (colon), macular degeneration
PGD_2—DP1	VASCULAR: Vasodilation (allergic erythema and edema), inhibition of platelet aggregation LUNG: Bronchodilation IMMUNE: Suppress cytokine production by dendritic cells NEURO: Regulates sleep, pain recognition	Decreased allergy-related vasodilation/ facial flushing/ nasal congestion
PGD_2—DP2 (aka CRTH2)	IMMUNE: Recruitment, activation of TH2 lymphocytes, eosinophils, basophils	Decrease allergic airway disease (atopic eosinophilic asthma, rhinitis) Decreased conjunctivitis Atopic dermatitis
$PGF_2\alpha$--FP	REPRO: Regulate reproduction, pregnancy	Prevent pre-term labor, cardiovascular disease,

	NEURO: Endogenous anti-convulsant PAIN: Anti-allodynia KIDNEY: Renal water and ion transport EYE: Regulate intra-ocular pressure	traumatic brain injury *FP agonists are currently used to treat glaucoma
PGI_2—IP	PAIN: Sensitize pain receptors GI: GI protection KIDNEY/HEMODYNAMIC: vasodilation (afferent arteriole), renin release, inhibit platelet aggregation	Pain, inflammation, over-active bladder *IP agonists are currently used to treat pulmonary hypertension and critical limb ischemia in humans
TXA—TP *PGD_2 also ligand for TP	VASCULAR: Platelet aggregation, vasoconstriction LUNG: Bronchoconstriction CANCER: regulate cell proliferation, migration, angiogenesis, metastasis	Anti-thrombotic, allergy, cough, asthma, diabetic retinopathy

REFERENCES

[1]Woodward DF, et al. Pharmacol Rev 2011;63(3):471. [2]KuKanich B, et al. Vet Anesth Analgesia 2012;39:69. [3]Stahn C, et al. Mol cell Endocrinol 2007;275:71. [4]Vane JR. Nat New Biol 1971;231:232. [5]Flower R. BMJ 2003;327(7415):572. [6]Harirforoosh S, et al. J Pharm Pharm Sci 2013;16(5):821. [7]Lascelles BDX, et al. J Am Vet Med Assoc 2005;227:1112. [8]World Health Organization. Pre-stems: suffixes used in the selection of INN. October 2013. [9]Sugimoto Y, et al. J Biol Chem. 2007;282:11613. [10]Lin CR, et al. J Pharmacol Exp Therapeutics 2006;319(3): 1096. [11]Kawabata A. Biol Pharm Bull. 2011;34:1170. [12]Dey I, et al. Br J Pharmacol. 2006 ;149:611. [13]Breyer MD. Am J Physiol Renal Physiol. 2000;279:F12. [14]Konya V, et al. Pharmacol Ther. 2013;138:485. [15]Nakao K, et al. J Pharmacol Exp Therapeutics 2007;322(2):686. [16]Okumura T, et al. J Pharm Pharmacol. 2008;60:723. [17]www.aratana.com [18]Rausch-Derra L, et al. ACVIM Forum, Nashville, TN, 2014. [19]Pettifur R, et al. Nature Reviews 2007;6:313. [20]Townley RG, et al. Ann Allergy Asthma Immunol 2012;109:365. [21]Pettifur R, et al. J Pharmacol Exp Therapeutics 2012;340:473.

THORACIC RADIOGRAPHY VERSUS COMPUTED TOMOGRAPHY: OPTIMIZING IMAGING FOR RESPIRATORY CASE MANAGEMENT

Carol Reinero, DVM, DACVIM, PhD
Columbia, MO

OVERVIEW

Diseases of the respiratory tract are a common reason why dogs and cats seek veterinary medical care. This lecture will focus on intrathoracic respiratory diseases which may manifest with clinical signs of cough, wheeze, exercise intolerance, tachypnea, and/or respiratory distress. There are a wide variety of diseases which affect the airways and pulmonary parenchymal which include infectious disorders (bacterial, fungal, viral, protozoal, parasitic), inflammatory processes (e.g., chronic bronchitis, asthma, eosinophilic bronchitis, aspiration pneumonia, foreign body pneumonia, eosinophilic pneumonia, etc.), neoplasia (primary or metastatic), pulmonary edema (non-cardiogenic or cardiogenic), pulmonary hemorrhage, interstitial lung diseases, structural or functional disorders (hypoplastic trachea, collapsing trachea or mainstem bronchi, bronchomalacia, bronchiectasis, bronchopulmonary dysplasia, lung lobe torsion, bullae or cavitary lesions). Obtaining a thorough history and completing both a distance and hands on physical examination can provide important clues to help narrow the list of differentials, but ultimately other diagnostic tests need to be performed.

Traditionally, thoracic radiography has been critical to help discriminate cardiac versus pulmonary disease as a cause of the clinical signs. Thoracic radiography has widespread availability, is cost effective and provides rapid information about cardiac and pulmonary vessel size as well as information about the airways and pulmonary parenchyma. Radiography can also help determine if disease is focal, multifocal or diffuse and can give a general idea of anatomic severity (i.e., how much of the pulmonary parenchyma appears to be involved and if the changes are mild, moderate or severe). However, there are important limitations to thoracic radiography which makes computed tomography appealing. Computed tomography allows for three dimensional reconstruction of anatomic structures with very good discrimination of tissue types.

This lecture will use a series of clinical cases of respiratory disease in dogs and cats to highlight optimal use of each imaging modality.

THORACIC RADIOGRAPHY

Thoracic radiography is one of the most important diagnostic screening tools for canine and feline respiratory disorders. It allows for a rapid assessment of the thoracic cavity without the need for general anesthesia and often even without sedation. Cardiac and respiratory abnormalities may be visualized in a two-dimensional "snapshot". Multiple views (right and left lateral, dorsoventral and ventrodorsal) allow complimentary displays of different regions of the heart and lungs. Location of lesions is a critical aspect of radiologic diagnosis and may help generate an appropriate differential list. While not able to capture dynamic lesions as well as fluoroscopy, some dynamic changes may be appreciated from radiographs taken on inspiration and compared with those on exhalation.

When evaluating respiratory diseases, the four common patterns described are a bronchial pattern (seen as "donuts" or "tram tracks" and being reflective of airway-oriented disease), an interstitial pattern (hazy, lacy or reticulated pattern reflecting disease of the interstitium), an alveolar pattern (noted by air bronchogram with black, air filled airways against a white background where the alveoli have been flooded with fluid, cells or matrix, reflective of disease in the gas exchange units/alveoli) and a vascular pattern (prominent, large and potentially tortuous vessels). Unfortunately, lung patterns are rarely clear cut and many if not most cases have mixtures of these patterns. Interpretation of patterns appears to be highly subjective between individuals, even board certified radiologists. Additionally, there is controversy about the utility of the pattern to be useful to generate a differential list of diagnoses.[1] Radiographic diagnoses that suggest bronchial, interstitial or alveolar locations of pathology do not always correlate with histologic lesions which makes more advanced imaging modalities such as computed tomography, appealing.

Despite being commonly used and relatively inexpensive, there are important limitations to thoracic radiography. Early lesions may be missed or appear less severe than the clinical picture (e.g., aspiration pneumonia). It is recommended that serial monitoring be performed if the initial set of thoracic radiographs does not fit the severity of respiratory signs. Along similar lines, radiographic resolution of disease may lag behind clinical improvement. Another limitation of thoracic radiography is that small lesions (e.g., nodules less than 3-5 mm) are occult radiographically. Thus micrometastases or other infectious/inflammatory nodules may be missed in the earliest stages of disease. In dogs or cats with severe respiratory distress, positioning and the need for restraint while capturing images could lead to decompensation of their clinical status. Quality of radiographs are affected by poor technique or hypoinflation due to poor inspiration or obesity. It must also be remembered that pulmonary parenchymal patterns are generally non-specific and other diagnostics are likely to be necessary. Overall, the sensitivity and specificity of thoracic radiography for the diagnosis of lower respiratory tract disease is low.[2] Thus, a more sensitive and ideally more specific imaging modality would be highly useful for respiratory diseases of dogs and cats.

COMPUTED TOMOGRAPHY

In humans, computed tomography (CT) has revolutionized imaging of the respiratory system. Computed tomography has been shown to provide very detailed information on important structures in the thorax, including lungs, mediastinum, pleural cavity and chest wall which is enhanced by knowledge of cross-sectional anatomy.[3] Once veterinarians gain more experience correlating the clinical picture, CT images and cytologic/histologic results for different disease processes in dogs and cats, we should be able to achieve many of the same benefits on the human side. This will require a multidisciplinary approach with input from clinicians, radiologists and pathologists.

There are multiple advantages of CT scans over radiography. Computed tomographic images can be acquired rapidly. For example, at the author's institution, 1 mm thickness images of an entire feline thorax can be acquired in less than 5 seconds using a 64-row detector CT scanner. Although personnel must leave the room (or stand behind a protective lead sheet) during image acquisition, there is the ability to use a plexiglass chamber known as a VetMouse Trap® to "restrain" an awake and unsedated animal in respiratory distress. Supplemental oxygen can ben plumbed into this chamber during the procedure if needed. It may be possible to use heavy sedation in place of general anesthesia to acquire scans in larger dogs. If general anesthesia is to be used, atelectasis can occur secondary to the anesthesia in the gravity dependent regions of the lung, especially if there is a long delay between induction of anesthesia and acquisition of the scan and if the animal is placed in one recumbency (eg right lateral recumbency) for that duration. The author's preference for the highest quality scans which provide information on dynamic changes is to use a ventilator capable of generating breath holds to help acquire the scans. In human medicine, people are instructed to take a deep breath and hold it while an inspiratory series of images are acquired. Similarly, they may be asked to exhale maximally and hold for CT acquisition for expiratory scans. Since dogs and cats are not capable of breath holds on command, the ventilator can provide an inspiratory breath hold and an expiratory breath hold. Alternatively, if a ventilator with breath hold capabilities is not available, hyperventilation immediately prior to scanning can be attempted to decrease ventilatory drive. While this may induce the animal to not breathe for the period of the scan, which is very helpful to eliminate motion artifact from the lungs inflating or deflating during CT acquisition, the animals rarely will be fully inflated during this "induced" breath hold.

One of the most useful features of CT is the ability to acquire three planes (e.g. sagittal, transverse and dorsal (coronal)) which allows for three dimensional volume rendering. This provides far more detail than the two dimensional images obtained with thoracic radiography. Another advantage of the three dimensional CT is that there is a lack of superimposition of anatomic structures. Finally, high resolution algorithms also provide unprecedented diagnostic quality images allowing more precise localization and extent of lesions compared to radiography.

To fully understand the advantages of CT, some background information on image acquisition and interpretation is helpful. Unlike conventional radiography in which a broad x-ray beam is emitted from a stationary tube and passes through the dog or cat to render a two dimensional image, the CT's x-ray tube rotates 360 degrees and emits a narrow fan-shaped x-ray around the patient as the patient is moved in small increments (axial scan) or the patient is imaged continuously (helical scan).[4] Thus, each CT slice or image represents a thin section of the patient which can be "stacked up" to render a three dimensional image. Thus, transverse, sagittal, and dorsal planes can be viewed. In looking at each image, the amount of attenuation can be assessed by a number called the Hounsfield unit (HU). The Hounsfield units range from -1000 for air, 0 for water, and +3000 for compact cortical bone. CT can distinguish differences between various soft tissues which have different HUs.

CONSIDERATIONS FOR SELECTING RADIOGRAPHY, COMPUTED TOMOGRAPHY OR BOTH

Initial choice of imaging modality will be dependent on multiple factors including but not limited to availability, convenience, owner finances, and need for sedation/anesthesia. Dogs or cats in respiratory distress that will fit into a VetMouse Trap benefit from a hands-off imaging approach in which supplemental oxygen can simultaneously be provided during CT imaging. Even with an oxygen mask or flow-by oxygen, taking thoracic radiographs of dogs or cats in respiratory distress may put them at risk for decompensation; thus, unrestrained CT would be preferable in this situation. For quick and comparatively inexpensive screening of the thoracic cavity in stable patients, radiography is likely to remain the initial diagnostic modality of choice. However, if results do not correlate with the clinical picture, or if there is high suspicion of a lesion not seen on radiography, then CT is an excellent next step. If additional information is needed based on thoracic radiography (e.g., suspicion, but not definitive evidence of a lung lobe torsion) then CT is again appropriate. CT is also useful to guide biopsy or to perform surgical planning.

In this lecture clinical case of dogs and cats with different types of respiratory disease will be used to illustrate optimal use of each modality. A discussion of why each imaging modality was chosen and what the advantages and disadvantages are will be undertaken.

REFERENCES

1. Nykamp SG et al. *Compend Contin Educ Pract Vet* 2002:25-36.
2. Gadbois J et al. *JAVMA* 2009;234:367-375.
3. De Rycke LM, et al. *AJVR* 2005;66:512-524.
4. Whatmough C, et al. *Compend Contin Educ Pract Vet* 2006:789-800.

UPDATE ON RESPIRATORY THERAPEUTICS: LESSONS FROM HUMAN MEDICINE?

Laura A. Nafe, DVM, MS, DACVIM (SAIM)
Stillwater, Oklahoma

INTRODUCTION

Respiratory diseases are a common presenting complaint to small animal veterinarians. Many respiratory conditions in dogs and cats have a similar presentation and pathophysiology to human respiratory diseases. Specific conditions in dogs and cats for which the human equivalent may provide insight into novel therapeutics include inflammatory airway disease (e.g.; asthma, chronic bronchitis), chronic cough, and gastroesophageal reflux disease (GERD). Management of these conditions can be frustrating for both the owner of the pet and the clinician. In general, small animal respiratory medicine has few therapies that we utilize on a routine basis. Commonly administered therapeutics include anti-inflammatory medications (e.g., glucocorticoids), bronchodilator agents, antimicrobials, and anti-tussive agents. Although many of these drug classes are utilized in human medicine, there are other specific therapeutics utilized in human medicine that may prove to be beneficial in dogs and cats with respiratory disease.

INFLAMMATORY AIRWAY DISEASE

Inflammatory airway diseases are among the more common respiratory conditions diagnosed, with feline asthma and canine chronic bronchitis being among the most common. Cats develop a clinical syndrome that closely resembles human asthma, and just like humans, there are variations in severity and response to standard therapies. Standard management of feline asthma includes a combination of environmental alterations (e.g.; minimizing allergen exposure), a glucocorticoid (oral or inhaled) and a bronchodilator. Management of human asthma is similar in the majority of cases. However, there is small population of human asthmatics (approximately 5-10%) that, despite appropriate therapy, continue to have multiple severe respiratory exacerbations per year.[1] These patients are categorized as having "severe refractory asthma," and a variety of adjunctive therapies are being investigated for use in these patients. There is likely a similar population of refractory feline asthmatic patients that may benefit from alternative therapeutics.

NOVEL ANTI-INFLAMMATORY THERAPEUTICS

Novel therapeutics showing promise in severe human asthmatics include nebulized lidocaine, phosphodiesterase-4 (PDE-4) inhibitors, chronic macrolide administration, and vitamin D supplementation.[1-3] Other management strategies that deserve further investigation include chemokine receptor CXCR2 antagonists, heliox-driven nebulization, and magnesium sulfate (IV and inhaled).[4] The anti-inflammatory and bronchodilatory effects of nebulized lidocaine have been evaluated in cats with experimentally induced asthma.[2] Results indicated that nebulized lidocaine is effective as a bronchodilator (based on reduced airway resistance), although all cats had continued airway eosinophilia.[2] While lidocaine may be beneficial from a bronchodilatory perspective, it cannot be recommended as monotherapy for treatment of feline asthma.

The administration of chronic macrolide therapy in human asthma and chronic bronchitis/COPD patients is controversial. Although the addition of azithromycin in patients with severe asthma reduced the frequency of severe exacerbations, recent guidelines on the management of severe asthma do not recommend routine use of macrolide antibiotics unless recurrent respiratory infections are documented. This is likely due to the concern for chronic therapy inducing bacterial resistance in patient population at risk for secondary respiratory infections. From a veterinary perspective, this same concern is also true.

Vitamin D3, a naturally occurring steroid hormone, plays many important roles in health. In addition to regulation of calcium homeostasis, vitamin D3 regulates the function of the immune system and is thought to have anti-inflammatory properties. Vitamin D3 deficiency has been implicated as a contributing factor to asthma, as countries furthest from the equator have been shown to have an increased prevalence of this respiratory disease.[3] Asthmatic children being treated with steroids and having concurrent vitamin D3 deficiency have been shown to have more frequent exacerbations.[3] There are multiple studies that provide evidence to support vitamin D3 deficiency as a potential cause or contributing factor in human asthma and chronic rhinosinusitis. However, we must consider that these findings may merely be an "effect" of the disease condition rather than a "cause." Patients with upper and lower airway disease are less likely to be active outdoors, and therefore, exposure to sunlight may be reduced in these patients. Vitamin D3 has not been evaluated in dogs and cats with respiratory disease and warrants further investigation as a potential diagnostic and therapeutic option in refractory cases of inflammatory airway disease.

CHRONIC COUGH

Cough is an important natural defense mechanism, helping to prevent accumulation of bacteria and any substances that do not below in the respiratory tract. Chronic cough is defined as a cough of at least 2 months duration.[5] If severe, chronic cough can result in disturbance of sleep, syncope, urinary stress incontinence, and potentially rib fracture in more severe cases. Chronic cough is a common condition in dogs and may occur secondary to tracheomalacia and/or bronchomalacia, chronic bronchitis, pulmonary neoplasia, severe left atrial enlargement, and perihilar lymphadenopathy. In dogs with tracheomalacia resulting in tracheal collapse, suppression of cough is the mainstay of therapy to aid in stopping the cycle of cough that will result in tracheitis followed by more coughing which will worsen the tracheitis. Canine cough is most commonly managed with hydrocodone, butorphanol, and/or dextromethorphan. Hydrocodone is the most reliable anti-tussive in my clinical experience. Unfortunately, hydrocodone can be difficult to obtain (controlled), cause sedation and constipation, and result in tolerance over time. Lomotil and tramadol have been advocated as alternative anti-tussive agents in veterinary medicine, but these medications have not been evaluated for this purpose in human medicine.

Neuropathic cough

Cough occurring secondary to airway inflammation and remodeling is common and likely the most common cause of cough in small animals. Neuropathic cough is also thought to occur in humans and may play a role in the cough stimulus. Evidence for this theory is supported by increased concentrations of substance P and neurokinin A in sputum samples from asthmatic coughers and increased neuropeptide calcitonin-gene-related peptide (CGRP) in airway lavage from chronic coughing patients.[5] Based on the neurogenic cough mechanism theory, treatment of human cough with amitriptyline and gabapentin has been evaluated with some success.[5,6] Coughing patients with symptoms of central sensitization have been found to have an increased response to gabapentin. The use of these medications for cough suggests that some patients with refractory, chronic cough (regardless of the initial trigger) may benefit from an adjunctive therapy that reduces cough via an alternative mechanism.

GASTROESOPHAGEAL REFLUX DISEASE (GERD)

Gastroesophageal reflux disease is the third most common cause of cough in humans, second to asthma and post-nasal drip. In addition to being a primary cause of cough, GERD serves as an important cause of refractory cough in patients with asthma and chronic bronchitis.[7] Although difficult to prove in our veterinary patients with cough, there are many canine cases of cough of which GERD may be the cause or a contributing factor to their ongoing cough (especially if refractory to standard therapy). Two major mechanisms for GERD associated cough have been suggested: 1) vagally mediated esophageal transbronchial reflex mechanism;" 2) microaspiration of refluxed contents into the airways.[7] The first mechanism indicates that gastric acid or distention of the caudal esophagus can stimulate a cough via a neuropathic mechanism. The presence of acid in the esophagus of patients with asthma and chronic cough stimulates the release of substance P and neurokinin A into the lungs, resulting in airway microvascular leakage and bronchoconstriction.[8] Microaspriation of gastric contents results in small volume refluxate causing laryngeal, tracheal and bronchial inflammation. In humans, this condition is best diagnosed based on a combination of clinical signs (e.g.; heartburn), laryngeal inflammation on oral exam, voice hoarseness, and a 24-hour ambulatory esophageal pH monitoring. Some clinicians may collect a bronchoalveolar lavage sample and evaluate for the presence of pepsin or bile acids to confirm aspiration.[8] Although these mechanisms make sense pathophysiologically as to how GERD causes chronic cough, there are some that believe GERD is merely a contributor and never a sole cause of cough. There have been no studies evaluating GERD-related cough in dogs or cats.

Management of Gastroesophageal Reflux Disease (GERD)

GERD is often treated with a combination of weight reduction, diet change (high protein, low fat), acid suppressant therapy, prokinetic therapy, and potentially surgery. In several controlled human clinical trials, proton-pump inhibitor (PPI) therapy has shown the most promise in reducing chronic cough in patients with suspected GERD.[8] Prokinetic agents would appear to be a common addition to PPI therapy in patients with suspected GERD-associated chronic cough; however, in human medicine prokinetic agents are not usually a first-line therapy. Azithromycin has been evaluated for treatment of concurrent GERD in lung transplant patients (70% have increased esophageal acid exposure) with results supporting a prokinetic effect of this therapy.[8] From a veterinary perspective, many dogs with chronic cough likely have a component of GERD contributing to their clinical signs. Although we do not have clinical veterinary studies to support this theory, the combination of consistent clinical signs along with inflamed upper airway tissue (laryngeal, pharyngeal, nasopharyngeal), proximal tracheal hyperemia and/or increased vascularity with or without signs of lower airway inflammation is indicative of GERD. I will also often concurrently perform a brief esophagoscopy in these patients to evaluate for evidence of esophagitis. Clinical response to PPI therapy with or without prokinetics is currently the best method we have to diagnosis GERD related cough in our veterinary patients. It is possible that some of these patients may also respond to medications that may reduce neuropathic mechanisms of cough (e.g., gabapentin), although this medication may only control the cough and would not address the GERD.

REFERENCES

[1]Kane et al. *Chronic Resp Dis* 2015; 12: 69 [2]Nafe et al. *J Fel Med and Surg* 2013; 15: 712. [3]Yawn J et al. *J Steroid Biochemistry & Mol Bio* 2015. [4]Rodrigo GJ. *co-pulmonary medicine* 2015; 21: 22. [5]Chung KF. *J of Thoracic* Dis 2014; 6: S699. [6]Ryan et al. *The Lancet* 2012; 380: 1583. [7]Kahrilas et al. *Lung* 2014; 192: 39. [8]Pauwels et al. *Pul Pharm and Therapeutics* 2009; 22: 135.

UNDERSTANDING FELINE CKD: PATHOGENESIS AND PROGRESSION

Rosanne E. Jepson, BVSc, MVetMed, PhD, DECVIM-CA, DACVIM, MRCVS
London, UK

INTRODUCTION

Chronic kidney disease (CKD) is a common condition in the ageing feline population defined as an alteration in structure or function of the kidneys that has been present for typically 2-3 months. Early studies report that between 20-30% of cats over the age of 12-15 years will have evidence of CKD (Lulich et al., 1992, Bartlett et al., 2010) and that whilst up to 50% of cases may have a specific underlying renal pathology (e.g. polycystic kidney disease, renal dysplasia, renal amyloidosis) the remainder have evidence of tubulointerstitial nephritis (Dibartola et al., 1987, Lucke, 1968, Minkus et al., 1994). More recently, studies specifically evaluating a first opinion population of cats would support that the prevalence of primary renal pathology is somewhat lower (~16%) (Chakrabarti et al., 2012a) with the majority of cats demonstrating non-specific chronic tubulointerstitial nephritis. The degree of tubular degeneration, interstitial inflammation, fibrosis, glomerulosclerosis correlates, as may be anticipated, with International Renal Interest Society (IRIS) staging (McLeland et al., 2014) and the severity of lesions identified in IRIS stage 3 and 4 supports that, in order to modify the development and progression of CKD, we need to target therapies at a much earlier time.

AETIOLOGY OF FELINE CHRONIC KIDNEY DISEASE

The underlying aetiology of feline CKD is poorly understood. There is evidence to support that by the time azotaemic CKD is diagnosed there is already substantial renal damage present (McLeland et al., 2014). Indeed evidence of tubulointerstitial inflammation can be identified in cats that are considerably younger than the typical CKD demographic (Lawler et al., 2006) raising the theory that the changes identified are part of a natural ageing process in the cat. Furthering this concept of the 'ageing feline kidney' recent work has supported telomere shortening in proximal and distal tubular epithelial cells from cats with CKD when compared with both healthy geriatric and young cats (Quimby et al., 2013). Telomeres represent the protective caps covering the end of chromosomes providing genome stability. A number of factors including insufficient telomerase activity, oxidative stress and deletions can contribute to telomere shortening and loss, which could ultimately result in DNA damage, cell cycle arrest, apoptosis and cell death. Comparable findings were not identified when telomere length was compared in either the skin or liver suggesting that these were renal specific findings. When telomeres reach a critical level of shortening, they are detected as DNA damage and result in a form of cell cycle arrest referred to as cellular senescence. A greater degree of cellular sensesence was also identified in the kidneys of geriatric cats with CKD then in healthy cats (Quimby et al., 2013).

Other factors such as diet (Hughes et al., 2002, Dibartola et al., 1993), feline immunodeficiency virus (White et al., 2010, Baxter et al., 2012), morbillivirus infection (Woo et al., 2012) and vaccination (Lappin et al., 2006, Lappin et al., 2005, Whittemore et al., 2010) have been investigated as potential contributing factors, although remain largely unsubstantiated. Similarly, other key events in the course of an individual cat's life may be contributory e.g. exposure to nephrotoxic drugs, episodes of acute kidney injury, episodes of hypoperfusion/hypotension and ureterolithiasis but their relative importance on the development of CKD in the feline population is unknown.

Clinical studies have evaluated biochemical risk factors for the development and progression of azotaemia and also the survival of cats with CKD. In particular, urine protein to creatinine ratio (UP/C) and creatinine concentration have been associated with the development of azotaemia, (Jepson et al., 2009), UP/C and phosphate predicted a progressive phenotype of CKD (Chakrabarti et al., 2012b) and serum phosphorus concentration (Boyd et al., 2008), IRIS stage and creatinine (Boyd et al., 2008, Kuwahara et al., 2006, King et al., 2007, Syme et al., 2006), age and UP/C (Syme et al., 2006).

PROGRESSION OF RENAL DISEASE

The course of disease for cats with naturally occurring CKD can be very variable. However, irrespective of the inciting cause, CKD is ultimately considered a progressive condition. Early studies proposed that haemoadaptive mechanisms occurring as a consequence of nephron loss resulted in the development of glomerular hypertrophy, hypertension and hyperfiltration in order to maintain single nephron GFR (Brenner, 1985, Hostetter et al., 1981) and that, although initially beneficial in terms of maintaining GFR, ultimately these adaptations were detrimental resulting in the loss of further nephrons and progression of disease. Evidence supporting this occurrence in the cat has been demonstrated experimentally in renal ablation models (Brown and Brown, 1995).

Proteinuria both as a consequence of glomerular hypertension and hyperfiltration is proposed to contribute to the development of renal inflammation and fibrosis. Filtered protein is reabsorbed by megalin and cubulin mediated endocytosis within the proximal tubular cells and this processing of protein within the tubules may be a stimulus for upregulation of both pro-inflammatory and pro-fibrotic mediators including TGF-β, monocyte chemotractant protein-1 (Abbate et al., 2002, Eddy, 2004, Perico et al., 2005). Polar release of these mediators to the basolateral membrane of tubular cells provides a mechanism by which tubulointerstitial inflammation may be perpetuated. In cats proteinuria has been significantly associated with the development of azotaemia (Jepson et al., 2009), having a progressive phenotype of CKD (Chakrabarti et al., 2012b), renal fibrosis score at post-mortem examination (Chakrabarti et al., 2012a) and survival of cats with both CKD (Syme et al., 2006) and hypertension (Jepson et al., 2007). However, although both angiotensin converting enzyme inhibitors and angiotensin receptor blockers significantly reduce magnitude of proteinuria in cats, the benefit in terms of slowing progression of CKD or improving survival in cats with CKD has yet to be demonstrated with anti-proteinuric therapies (King et al., 2006).

The renin-angiotensin-aldosterone system is widely accepted to be upregulated in patients with CKD and RAAS activation has been reported in cats with experimentally induced renal disease (Watanabe and Mishina, 2007). However, plasma concentrations must be interpreted with caution as they may not correlate with tissue RAAS (Kobori et al., 2007). Mitani and colleagues investigated the intra-renal RAAS in both cats and dogs (Mitani et al., 2012). The study found no association between expression of renin and angiotensin II with histopathological evidence of fibrosis in the cat although a positive association was demonstrated in the dog. RAAS has a direct role in altering renal haemodynamics and mediating glomerular hypertension and hyperfiltration. However, angiotensin II is also known to mediate alteration in glomerular permiselectivity (Benigni et al., 2004) and to be one of the strongest stimuli for TGFβ production and mediators of renal fibrosis such as connective tissue growth factor and endothelin (Remuzzi et al., 2005). Although beneficial effects of RAAS inhibition have been identified in human patients with a variety of aetiologies of renal disease, a beneficial effect in terms of progression of renal disease or improving survival has not been identified to date in cats.

Hypoxia has also been proposed as mechanisms intrinsically linked to renal inflammation, fibrosis and the progression of CKD. Despite receiving ~20% of cardiac output, the kidney and in particular the medullary interstitium operates in a state of hypoxia. A number of mechanisms contribute to the development of hypoxia. The development of glomerulosclerosis can limit peritubular capillary blood flow and RAAS activation and the vasoconstrictive effects of angiotensin II on the efferent arteriole exacerbate this further. Loss of peritubular capillaries due to inflammation and fibrosis within the interstitial space can further reduce blood supply to the tubules and the physical presence of both inflammation and fibrosis increases the diffusion distance for oxygen supply to the tubules. Hypoxia induces production of hypoxia inducible factors (HIF), which in turn promote the expression of target genes such as vascular endothelial growth factor (VEGF) in order to try and promote angiogenesis and improve blood supply. However, overall hypoxia has a negative impact tubular cells and particularly those within the proximal tubule which are considered more susceptible to hypoxia due to their high metabolic activity and dependence on aerobic metabolism (Kawakami et al., 2014).

Renal oxidative stress occurs when there is an imbalance between the production of reactive oxygen species (ROS) and anti-oxidant defense mechanisms. ROS are highly reactive molecules, which can cause damage to DNA, lipid, protein, carbohydrate resulting ultimately in structural and functional cellular damage that leads to apoptosis and necrosis, stimulating inflammation and fibrosis. Renal cells, particularly tubular cells, have a high metabolic activity and therefore have a high production rate of ROS. Oxidative stress occurs in CKD when hyperfiltration and hyperfunctioning of remaining nephrons leads to increased production of ROS. Other factors such as age, proteinuria, RAAS activation and angiotensin II, hyperphosphataemia, inflammation and regions of ischaemia and hypoxia can also contribute to the generation of ROS. There is limited information about oxidative stress (Keegan and Webb, 2010, Krofič Žel et al., 2014) both supporting some degree of anti-oxidant mechanism activation in cats with CKD.

Finally hyperphosphataemia has been associated with progression of renal disease. The proposed pathogenesis is through renal mineralization promoting inflammation and fibrosis. Alternative proposed mechanisms include an association between hyperphosphataemia and vascular calcification and endothelial dysfunction leading to areas of ischaemia and hence stimulating pro-fibrotic mechanisms (Cozzolino et al., 2005). In cats plasma phosphate concentrations have been associated with interstitial fibrosis but not with tubular mineralization (Chakrabarti et al., 2012a). However, phosphate restricted renal diets have been demonstrated to reduce histopathological changes in cats with experimentally induced renal damage and to improve survival in naturally occurring CKD (Ross et al., 1982, Ross et al., 2006, Elliott et al., 2000).

The kidney has a limited capacity for expression of pathology and tubulointerstitial fibrosis is recognized as the final common pathway in the progression of all renal disease whereby gradual expansion of fibrosis destroys normal tissue architecture. Commensurate with this, renal fibrosis has been reported, both in humans and in cats, to be the histopathological finding which correlates best with renal function and plasma creatinine concentration (Chakrabarti et al., 2012a, Farris et al., 2011). If preventing the initiation of CKD is not possible, it becomes important to understand the pathophysiology and factors influencing the progression of tubulointerstitial inflammation and fibrosis such that future therapeutics can target these pathways with the goal of slowing progression of disease.

RENAL FIBROSIS: THE FINAL COMMON PATHWAY

Healthy renal interstitium is composed of sparse cells (fibroblasts and dendritic cells) embedded in an extracellular matrix composed of collagen (I, III, VII), fibronectin and extracellular matrix glycoproteins e.g. tenascin. Renal fibrosis usually begins with focal areas of inflammation secondary to an inciting injury, activation of mesenchymal cells progressing to generate expanding areas of fibrosis and scarring (Prunotto et al., 2011). Intially after injury, the recruitemtn of activated fibroblasts, termed myofibroblasts, is a beneficial response and forms part of the normal repair mechanism. However, in CKD, an excessive or maladaptive response occurs leading to the continuation of mechanisms that promote renal fibrosis (Kramann et al., 2013). Renal fibrosis is characterized by excessive accumulation of matrix proteins, e.g. collagen IV and laminin and myofybroblasts. Four major cell types are important for the genesis of progressive renal fibrosis; myofibroblasts responsible for the accumulation of extracellular matrix, inflammatory cells such as macrophages, microvascular endothelial cells and tubular epithelial cells (Eddy, 2014).

The origin of myofibroblasts in the kidney has been the source of considerable controversy. Based on recent studies it is most likely that the primary sources are renal fibroblasts and microvascular pericytes which undergo a process of transformation (Eddy, 2013). Other minor contributing sources particularly in the later stages of CKD can include fibrocytes (from bone marrow) and local renal cells including tubular epithelium and endothelium (Eddy, 2014, Kramann et al., 2013). Myofibroblasts have the characteristics of both fibroblasts and smooth muscle cells. Key features include their expression of alpha smooth muscle actin (α-SMA) and vimentin, which are not expressed in normal fibroblasts and their ability to secrete pericellular matrix containing collagen and glycosaminoglycans.

The transdifferentiation of fibroblasts and pericytes to myofibroblasts can be activated by a number of cell signaling mechanisms including platelet derived growth factor (PDGF), vascular endothelial growth factor (VEGF), TGF-β, connective tissue growth factor (CTGF), WNT/LRP62, Ephrin B and Hedgehog pathways. Detachment of pericytes may be related to metalloproteinases such as the ADAM and ADAMTS families. Inflammatory cells, particularly macrophages may also play an important role in inflammation, tissue repair and fibrosis. Sub-populations of macrophages may be responsible for production of particular inflammatory mediators, for example IL1β, TNF-α, chemokines (MCP-1) potentiating inflammation versus the production of TGF-β and PDGF promoting myofibroblast activation (Kramann et al., 2013).

Tubular epithelial cells have a variable role in renal fibrosis. Early in the course of the disease injured tubular epithelial cells may be a source of inflammatory mediators, e.g. cytokines, growth factors (TGFβ, PDGF, FGF) and reactive oxygen species secreted either via the paracellular route or to the basolateral membrane to the tubulointerstitium (Eddy, 2014). In the tubulointerstitial compartment, such inflammatory mediators can promote inflammation, immune response and the activation, transformation and proliferation of myofibroblasts. Urinary complement factors and cytokines may also activate release of these inflammatory substances from tubular epithelial cells and protein handling by the tubular cells has also been implicated. Later in the disease process, lack of regeneration and loss of tubular epithelial cells may contribute to progression of disease and loss of nephrons.

Loss of the interstitial capillary network otherwise known as peritubular capillary rarefaction plays a major role in the development of hypoxia, oxidative stress, further tubular cellular injury and fibrosis (Eddy, 2014, Mimura and Nangaku, 2010). It is hypothesized that pericytes are important for vascular stabilization stabilizing the endothelial cells of the renal peritubular capillary bed. Detachment of pericytes and their transformation to myofibroblasts may lead to instability of the capillary bed and ultimately capillary rarefaction (Kramann et al., 2013). In addition, increase in permeability of microvasculature and leakage of plasma proteins such as fibrinogen and albumin into the interstitium may trigger an inflammatory response (Yamaguchi et al., 2012).

Finally, direct interaction with environmental stimuli such as hypoxia, proteinuria and hyperglycaemia may play a role in myofibroblast induction and proliferation.

Knowledge to date in feline medicine

To date two studies have evaluated myofibroblast presence within feline kidneys. The study by Sawashima identified that α-SMA index was significantly higher in cats with tubulointerstitial nephritis particularly in the periglomerular and peritubular regions than in cats without evidence of CKD and that α-SMA index was significantly correlated with fibronectin deposition and creatinine concentration (Sawashima et al., 2000). A further study demonstrated similar associations between α-SMA, tubular vimentin expression, fibrosis and creatinine concentration (Yabuki et al., 2010).

Based on experimental studies, TGF-β is an important mediator in the process of renal inflammation and fibrosis. Production of TGF-β can be induced by a number of stimuli including activation of the renin-angiotensin-aldosterone system (RAAS), hypoxia, oxidative stress, modulation of renal haemodynamics and proteinuria. TGF-β concentrations have been evaluated in cats with CKD. Two studies have identified that urinary TGF-β to creatinine ratios were significantly higher in cats with CKD than in healthy cats (Habenicht et al., 2012, Arata et al., 2005).

The study by Habenicht and colleagues also investigated urinary concentrations of IL8, monocyte chemottractant protein-1 (MCP-1) and VEGF as ratios to urine creatinine as potential markers of renal inflammation and injury (Habenicht et al., 2012). No significant difference in MCP-1:creatinine ratio could be detected although IL8: creatinine ratio were significantly higher in cats with CKD than control cats. Conflicting information has been published with relation to urinary VEGF: creatinine ratios which may be considered a marker of hypoxia. The study by Habenicht demonstrated significantly lower urinary VEGF:creatinine ratios in cats with CKD than in healthy controls whilst in the study by Chakrabarti and colleagues which examined urinary VEGF:creatinine ratio as a marker of hypoxia identified a positive significant association with progression of azotaemia and reduced survival (Chakrabarti et al., 1012). A shortfall of these studies is that histopathology was not routinely available although in other species studies support that the correlation between urinary cytokine levels and renal inflammation is strong (Bobkova et al., 2006).

Transglutaminase 2 (TG-2) has recently been investigated in cats. TG-2 is a calcium dependent cross-linking enzyme from the transglutaminase family, which plays an important role in stabilizing extracellular matrix thereby promoting extracellular matrix deposition and resistance to degradation. TG-2 can also cross-link TGF-β with the extracellular matrix contributing to renal fibrosis by activation of TGF-β. There is a strong relationship between TG-2 expression and renal fibrosis in humans with CKD and also rodent models (Johnson et al., 2003, Johnson et al., 1999). This has recently been extrapolated to the cat where TG-2 activity was positively correlated with renal histopathology scoring, plasma creatinine, phosphate and urea concentrations (Sánchez-Lara et al., 2014).

Although we are only beginning to appreciate the mechanisms involved in the genesis and progression of renal fibrosis in the cat, such mechanisms are very likely to be common to all species. Renal fibrosis is a key target currently for novel therapeutic agents that may slow the progression of renal disease and it is anticipated that interventions, which modulate fibrosis may therefore be beneficial both in terms of slowing disease and improving survival of patients.

REFERENCES

ABBATE, M., ZOJA, C., ROTTOLI, D., CORNA, D., TOMASON, S. & REMUZZI, G. 2002. *Kidney Int,* 61, 2066-2077. ARATA, S., OHMI, A., MIZUKOSHI, F. BABA, K., OHNO, K., SETOGUCHI, A. & TSUJIMOTO, H. 2005. *J Vet Med Sci,* 67, 1253-5. BARTLETT, P. C., VAN BUREN, J. W., NETERER, M. & ZHOU, C. 2010. *Prev Vet Med,* 94, 264-71. BAXTER, K. J., LEVY, J. K. EDINBORO, C. H., VADEN, S. L. & TOMPKINS, M. B. 2012. *Journal of Veterinary Internal Medicine,* 26, 238-243. BENIGNI, A., GAGLIARDINI, E. & REMUZZI, A. 2004. *Semin Nephrol,* 24, 131-140. BOBKOVA, I. N., CHEBOTAREVA, N. V., KOZLOVSKAIA, L. V., VARSHAVSKII, V. A. & GOLITSYNA, E. P. 2006. *Ter Arkh,* 78, 9-14. BOYD, L. M., LANGSTON, C., THOMPSON, K., ZIVIN, K. & IMANISHI, M. 2008. *Journal of Veterinary Internal Medicine,* 22, 1111-1117. BRENNER, B. M. 1985. *Am J Physiol* 249, F324-F337. BROWN, S. A. & BROWN, C. A. 1995. *American Journal of Physiology - Regulatory, Integrative and Comparative Physiology,* 269, R1002-R1008. CHAKRABARTI, S., SYME, H. & ELLIOTT, J. 1012. *Journal of Veterinary Internal Medicine,* 26, 1524. CHAKRABARTI, S., SYME, H. M., BROWN, C. A. & ELLIOTT, J. 2012a. *Vet Pathol.* CHAKRABARTI, S., SYME, H. M. & ELLIOTT, J. 2012b. *Journal of Veterinary Internal Medicine,* 26, 275-281. COZZOLINO, M., BRANCACCIO, D., GALLIENI, M. & SLATOPOLSKY, E. 2005. *Kidney Int,* 68, 429-436. DIBARTOLA, S. P., BUFFINGTON, C. A., CHEW, D. J., MCLOUGHLIN, M. A. & SPARKS, R. A. 1993. *Journal of American Veterinary Medical Association,* 202, 744-751. DIBARTOLA, S. P., RUTGERS, H. C. ZACK, P. M. & TARR, M. J. 1987. *J Am Vet Med Assoc,* 190, 1196-1202. EDDY, A. A. 2013. *Nat Med,* 19, 964-966. EDDY, A. A. 2014. *Kidney Int Suppl (2011),* 4, 2-8. EDDY, A. L. 2004. *Nephrol Dial Transplant,* 19, 277-281. ELLIOTT, J., RAWLINGS, J. M., MARKWELL, P. J. & BARBER, P. J. 2000. *Journal of Small Animal Practice,* 41, 235-242. FARRIS, A. B., ADAMS, C. D., BROUSAIDES, N., DELLA PELLE, P. A., COLLINS, A. B., MORADI, E., SMITH, R. N., GRIMM, P. C. &

COLVIN, R. B. 2011. *Journal of the American Society of Nephrology,* 22, 176-186. HABENICHT, L. M., WEBB, T. L., CLAUSS, L. A., DOW, S. W. & QUIMBY, J. M. 2012. *Journal of Feline Medicine and Surgery.* HOSTETTER, T. H., OLSON, J. L., RENNKE, H. G., VENKATACHALAM, M. A. & BRENNER, B. M. 1981. *Am J Physiol,* 241, F85-F92. HUGHES, K. L., SLATER, M. R., GELLER, S., BURKHOLDER, W. J. FITZGERALD, C. 2002. *Prev Vet Med,* 55, 1-15. JEPSON, R. E., BRODBELT, D., ELLIOTT, J. & SYME, H. M. 2007. *J Vet Intern Med,* 21, 402-409. JEPSON, R. E., BRODBELT, D., VALLANCE, C., SYME, H. M. & ELLIOTT, J. 2009. *J Vet Intern Med,* 23, 806-13. JOHNSON, T. S., EL-KORAIE, A. F., SKILL, N. J., BADDOUR, N. M., EL NAHAS, A. M., NJLOMA, M., ADAM, A. G. & GRIFFIN, M. 2003. *J Am Soc Nephrol,* 14, 2052-62. JOHNSON, T. S., SKILL, N. J., EL NAHAS, A. M., OLDROYD, S. D., THOMAS, G. L. DOUTHWAITE, J. A., HAYLOR, J. L. & GRIFFIN, M. 1999. *J Am Soc Nephrol,* 10, 2146-57. KAWAKAMI, T., MIMURA, I., SHOJI, K., TANAKA, T. & NANGAKU, M. 2014. *Kidney Int Suppl (2011),* 4, 107-112. KEEGAN, R. F. & WEBB, C. B. 2010. *Journal of Veterinary Internal Medicine,* 24, 514-519. KING, J. N., GUNN-MOORE, D. A., TASKER, S., GLEADHILL, A. & STREHLAU, G. 2006. *J Vet Intern Med,* 20, 1054-64. KING, J. N., TASKER, S., GUNN-MOORE, D. A. & STREHLAU, G. 2007. *J Vet Intern Med,* 21, 906-16. KOBORI, H., NANGAKU, M., NAVAR, L. G. & NISHIYAMA, A. 2007. *Pharmacol Rev,* 59, 251-87. KRAMANN, R., DIROCCO, D. P., MAAROUF, O. H. & HUMPHREYS, B. D. 2013. *Curr Pathobiol Rep,* 1. KROFIČ ŽEL, M., TOZON, N. & NEMEC SVETE, A. 2014. *Journal of Veterinary Internal Medicine,* 28, 130-136. KUWAHARA, Y., OHBA, Y., KITOH, K., KUHWAHARA, N. & KITAGAWA, H. 2006. *Journal of Small Animal Practice,* 47, 446-450. LAPPIN, M. R., BASARABA, R. J. & JENSEN, W. A. 2006. *J Feline Med Surg,* 8, 353-6. LAPPIN, M. R., JENSEN, W. A., JENSEN, T. D., BASARABA, R. J., BROWN, C. A., RADECKI, S. V. & HAWLEY, J. R. 2005. *American Journal of Veterinary Research,* 66, 506-511. LAWLER, D. F., EVANS, R. H., CHASE, K., ELLERSIECK, M., LI, Q., LARSON, B. T., SATYARAJ, E. & HEININGER, K. 2006. *J Feline Med Surg,* 8, 363-71. LUCKE, V. M. 1968. *J Pathol Bacteriol,* 95, 67-91. LULICH, J. P., OSBORNE, C. A., O'BRIEN, T. D. & POLZIN, D. J. 1992. *Compend Cont Ed Pract Vet,* 14, 127-152. MCLELAND, S. M., CIANCIOLO, R. E., DUNCAN, C. G. & QUIMBY, J. M. 2014. *Vet Pathol.* MIMURA, I. NANGAKU, M. 2010. *Nat Rev Nephrol,* 6, 667-78. MINKUS, G., REUSCH, C., HÖRAUF, A., BREUER, W., DARBÈS, J., KRAFT, W. & HERMANNS, W. 1994. *Journal of Small Animal Practice,* 35, 465-472. MITANI, S., YABUKI, A., TANIGUCHI, K. & YAMATO, O. 2012. *Journal of Veterinary Medical Science,* advpub. PERICO, N., CODREANU, E., SCHIEPPATI, A. & REMUZZI, G. 2005. *Kidney Int,* 67, S79-S82. PRUNOTTO, M., GHIGGERI, G., BRUSCHI, M., GABBIANI, G., LESCUYER, P., HOCHER, B., CHAYKOVSKA, L., BERRERA, M. & MOLL, S. 2011. *Journal of Proteomics,* 74, 1855-1870. QUIMBY, J. M., MARANON, D. G., BATTAGLIA, C. L., MCLELAND, S. M., BROCK, W. T. & BAILEY, S. M. 2013. *Am J Physiol Renal Physiol,* 305, F295-303. REMUZZI, G., PERICO, N., MACIA, M. & RUGGENENTI, P. 2005. *Kidney Int,* 68, S57-S65. ROSS, L. A., FINCO, D. R. & CROWELL, W. A. 1982. *Am J Vet Res,* 43, 1023-1026. ROSS, S. J., OSBORNE, C. A., KIRK, C. A., LOWRY, S. R., KOEHLER, L. A. & POLZIN, D. J. 2006. *J Am Vet Med Assoc,* 229, 949-957. SÁNCHEZ-LARA, A. C., ELLIOTT, J., SYME, H. M., BROWN, C. A. & HAYLOR, J. L. 2014. *Veterinary Pathology Online.* SAWASHIMA, K., MIZUNO, S., MIZUNO-HORIKAWA, Y., SHIMADA, A., KUDO, T. & KUROSAWA, T. 2000. *Am J Vet Res,* 61, 1080-6. SYME, H. M., MARKWELL, P. J., PFEIFFER, D. & ELLIOTT, J. 2006. *J Vet Intern Med,* 20, 528-535. WATANABE, T. & MISHINA, M. 2007. *Journal of Veterinary Medical Science,* 69, 1015-1023. WHITE, J. D., MALIK, R., NORRIS, J. M. & MALIKIDES, N. 2010. *Journal of the American Veterinary Medical Association,* 236, 424-429. WHITTEMORE, J. C., HAWLEY, J. R., JENSEN, W. A. & LAPPIN, M. R. 2010. *Journal of Veterinary Internal Medicine,* 24, 306-313. WOO, P. C. Y., LAU, S. K. P., WONG, B. H. L., FAN, R. Y. Y., WONG, A. Y. P., ZHANG, A. J. X., WU, Y., CHOI, G. K. Y., LI, K. S. M., HUI, J., WANG, M., ZHENG, B.-J., CHAN, K. H. & YUEN, K.-Y. 2012. *Proceedings of the National Academy of Sciences,* 109, 5435-5440. YABUKI, A., MITANI, S., FUJIKI, M., MISUMI, K., ENDO, Y., MIYOSHI, N. & YAMATO, O. 2010. *Res Vet Sci,* 88, 294-9.YAMAGUCHI, I., TCHAO, B. N., BURGER, M. L., YAMADA, M., HYODO, T., GIAMPIETRO, C. & EDDY, A. A. 2012*Nephron Exp Nephrol,* 120, e20-31.

UNDERSTANDING FELINE CKD: MINERAL AND BONE DISORDER

Rosanne E. Jepson, BVSc, MVetMed, PhD, DACVIM, DECVIM-CA, MRCVS
London, UK

INTRODUCTION

Chronic kidney disease (CKD) is a common condition in both feline and canine medicine and the concept of secondary renal hyperparathyroidism (SRHP) is well recognised. However, phosphorus homeostasis and the pathophysiology of SRHP has been, and remains, incompletely understood. In recent years, the discovery of phosphotonins (a factor responsible for inhibition of renal tubular reabsorption of phosphorus) and in particular Fibroblast Growth Factor 23 (FGF23) has revolutionised our understanding of the control of phosphorus and the changes that occur in phosphorus homeostasis during CKD. Based on the realisation that PTH is not the only player in the regulation of phosphorus and calcium homeostasis in CKD the term secondary renal hyperparathyroidism, in human medicine, has largely been superceded by the term mineral and bone disorder. In the last 2-3 years, studies have begun to investigate the role that FGF23 may play in our veterinary patients.

Regulation of phosphorus is intrinsically linked to the regulation of calcium with both electrolytes being under the influence of parathyroid hormone (PTH) and calcitriol (1, 25 dihydroxycholecalciferol). Calcitriol increases calcium and phosphorus absorption from the gastrointestinal tract whilst PTH increases calcium and phosphorus resorption from bone, increases calcium reabsorption and decreases phosphorus reabsorption from the tubular filtrate in the kidney. For many decades we have widely accepted the 'trade off hypothesis'. Reduction in the number of functioning nephrons leads to a reduced capacity to excrete various solutes in the urine including phosphorus with resulting hyperphosphatemia. This retention of phosphorus, by the law of mass action, results in a relative decline in free calcium concentration. A combination of reduced free calcium and increased phosphorus concentration contribute to stimulating production of PTH (Figure 1). Simlutaneously, increasing phosphorus concentrations have an inhibitory action on the production of calcitriol, which also has a stimulatory action on PTH production. As PTH inhibits reabsorption of phosphorus in the proximal tubules, increased PTH production serves to try and increase phosphorus excretion. At the same time increased PTH concentrations will also stimulate release of calcium and phosphorus from bone and increase production of calcitriol, which will increase absorption of both calcium and phosphorus from the small intestine. However, with continually declining renal function phosphorus retention continues to occur such that the body becomes phosphorus loaded and the SRHP develops. The 'trade-off' for increasing phosphorus excretion by remaining functioning nephrons is therefore the gradual but persistent increase in PTH concentration.

Chronically increased PTH production can ultimately result in bone demineralization (renal osteodystrophy) and complex of phosphorus and calcium can result in dystrophic mineralisation of soft tissues. When this mineralisation occurs within the kidney concern is raised that it can contribute to progressive renal injury.

Figure 1: The 'trade-off hypothesis' in the regulation of phosphorus in chronic kidney disease

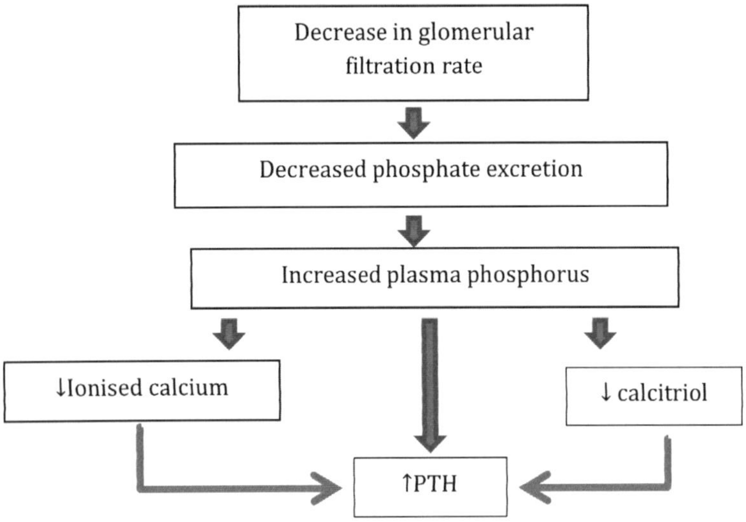

The prevalence of SRHP increases with advancing stage of CKD. An early study performed by Barber and Elliott in 80 cats indicated that 20% of cats with compensated CKD (mean [creatinine] 2.6 mg/dL), 49% of cats with uremic CKD (mean [creatinine] 3.6 mg/dL), and 100% of cats with end stage CKD (mean [creatinine] 10.3 mg/dL) demonstrated hyperphosphtemia whilst 47%, 87% and 100% of cats in those respective groups showed hyperparathyroidism.[1] Calcitriol concentrations were assessed in 31 cats and were found to be below the reference interval in 35%. The higher prevalence of hyperparathyroidism than hyperphosphatemia in cats adds support to the theory that increase in PTH concentrations occurs before the onset of hyperphosphatemia. This has also been seen in a group of non-azotemic cats >9 years which were followed longitudinally until the development of azotemia. Those cats that went on to develop azotemia in a 12 month period had significantly increased PTH concentrations at entry to the study than cats which remained non-azotemic over the same time period. [2] Hyperphosphatemia has been associated with reduced survival in cats. [3,4]

Evidence from naturally occurring feline CKD also indicates that hyperphosphatemia is associated with progression of disease.[5] The management of hyperphosphatemia and SRHP with dietary manipulation and use of low phosphorus diets can therefore have substantial impact on our patients with CKD.[6-10] Evidence supports that maintaining plasma phosphorus concentrations within the IRIS targets for each stage of CKD improves survival and can improve the clinical manifestations of SRHP.

DISCOVERY OF FIBROBLAST GROWTH FACTOR 23 (FGF23)

In 1959 Prader and colleagues published a case report of an 11 year old girl with a rib mass and acquired hypophosphatemia, renal phosphate wasting and rickets which resolved after surgical excision of the mass. At that time it was recognised that the mass lesion must have been secreting what was referred to as a 'rachitogenic substance' likely to be causing the electrolyte abnormalities. Today this condition would be referred to as tumor-induced osteomalacia (TIO) as a consequence of excess production of FGF23 by the neoplastic lesion. However, it was not until 2000 that the importance of FGF23 in phosphate homeostasis was recognised and implicated in multiple hypo- and hyper-phosphatemic disorders.[11]

FGF23, WHAT IS IT AND WHAT DOES IT DO?

FGF23 belongs to and shares structural similarity with other members of the fibroblast growth factor family. The protein has two functional domains; an N-terminal, which is common to all fibroblast growth factor proteins, and a C-terminal, which confers the specialised activity of FGF23.[11] FGF23 is predominantly produced by osteocytes and osteoblasts in the skeleton although transcripts have also been identified in human heart, liver and thyroid/parathyroid tissues.[12,13] Within the circulation FGF23 can either be detected as an intact form (iFGF23) or broken into N or C-terminal fragments but it is widely accepted that iFGF23 is responsible for the majority of biological action.

FGF23 interacts with cell surface membrane bound receptors (FGFR1, 3c, 4) but also requires interaction with a co-receptor, α-Klotho.[14,15] FGF23 binds more avidly to its specific receptors in the presence of α-Klotho. Although FGFR are widely distributed throughout the body, the limited distribution of α-Klotho confers tissue specificity to FGF23, focussing its actions within the parathyroid, kidney and pituitary.[16] The primary physiological functions of FGF23 are to stimulate renal excretion of phosphate, to decrease production of calcitriol by inhibiting 1- α hydroxylase and stimulating the catabolic 24-hydroxylase and to inhibit PTH secretion. Given the known function of FGF23 in the regulation of phosphorus, the primary site of action is believed to be the proximal tubular cells, where it regulates the transcription and translation of the NaPi-2a sodium phosphate cotransporter.[17]

FGF23 production is regulated by both phosphorus concentrations and calcitriol. In health, high phosphorus intake results in increasing FGF23 concentrations and an increase in renal phosphorus excretion, although the mechanism by which phosphorus sensing occurs has not yet been determined.[18] Calcitriol stimulates FGF23 secretion and in a typical negative feedback loop FGF23 reduces calcitriol.[19,20] This sensing is believed to occur via the vitamin D receptor. FGF23 inhibits production of PTH and PTH stimulates FGF23 both directly and indirectly through PTH mediated increases in calcitriol. An interesting current area of research is the role of iron and hypoxia inducible factor (HIF) in the regulation of FGF23.[21,22]

Circulating FGF23 concentrations are reported to increase with reducing renal function in human patients. iFGF23 appears to be the major form of detectable FGF23 in such patients and, although this is attributed to increased bone production, reduced renal excretion may also be contributing.[23,24] With advancing CKD, increasing FGF23 concentrations have been associated with higher serum phosphate, increased fractional excretion of phosphate, lower eGFR and reduced calcitriol concentrations[25].[26,27] The reduction in calcitriol concentrations is independent of reduction in GFR suggesting that reduced calcitriol is predominantly the consequence of inhibition by FGF23, rather than reduction in renal mass. Studies suggest that FGF23 concentrations begin to increase early in the course of renal disease prior to evidence of hyperphosphatemia and SRHP and that therefore FGF23 may act as a sensitive screening test to identify patients, which are developing mineral metabolism disorders in early CKD.[28] It has been proposed that FGF23 increases early in the course of CKD either as a consequence of phosphate retention due to reduced renal excretion or could be the consequence of decreased α-klotho expression in the kidney. Others suggest that early in the course of CKD there may be a primary FGF23 excess, perhaps as a consequence of change in bone metabolism, reducing plasma phosphorus, increasing phosphorus renal excretion and decreasing α-klotho expression.[29] Nevertheless, based on the discovery of FGF23 an updated 'trade-off' hypothesis can be proposed (Figure 2a & b).

Figure 2a: A new 'trade-off' hypothesis for early CKD

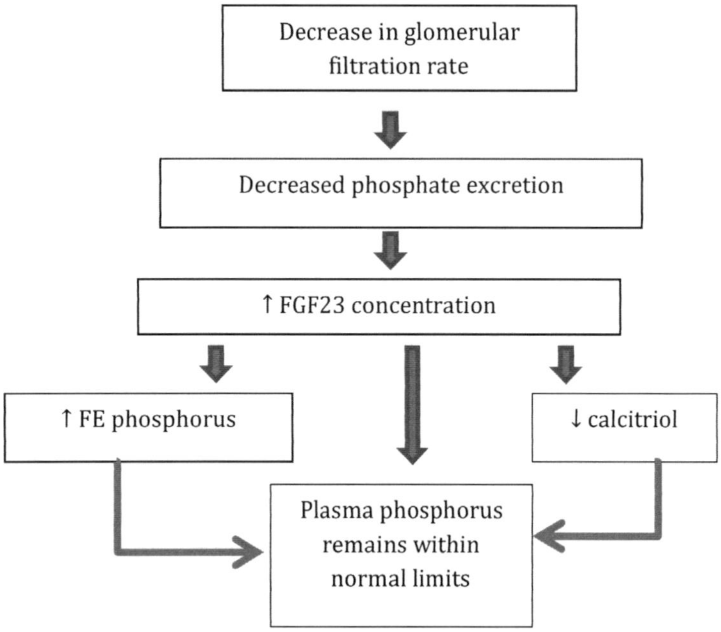

FE; Fractional excretion, FGF23; Fibroblast Growth Factor 23

In early CKD an increase in phosphate retention will drive increase in FGF23 production. Increased FGF23 will inhibit transcription and translation of the NaPi-2a cotransporter in the proximal tubule of remaining functioning nephrons, limiting reabsorption of phosphorus and facilitating phosphorus excretion. Increased FGF23 in combination with increased phosphorus will inhibit 1α-hydroxylase activity thereby decreasing calcitriol production and absorption of phosphorus and calcium from the gastrointestinal tract. In the parathyroid gland increased FGF23 will directly inhibit the production of PTH thereby limiting release of calcium and phosphorus from bone as a consequence of the action of PTH.

FGF23 AS A RISK FACTOR IN HUMANS

FGF23 has been associated with mortality in incident end stage renal disease (ESRD), in kidney transplant patients and in patients with pre-dialysis CKD.[30,31,32] Studies have also associated FGF23 with progression of CKD and as a cardiovascular risk factor although whether it is acting purely as a marker or is directly implicated in disease progression remains to be fully determined. Preliminary data indicates that dietary manipulation of phosphate intake will aid in modulating FGF23 concentrations both in healthy individuals and in those with CKD and that FGF23 concentrations are also lowered with administration of phosphate binders.[18,33-40]

EVALUATION OF FGF23 IN CATS

In the past 3 years preliminary studies have been performed investigating the role of FGF23 in phosphorus homeostasis in cats with CKD. A commercially available human ELISA has been validated for quantification of FGF23 in feline EDTA plasma. In a study by Geddes and colleagues FGF23 concentrations were found to significantly increase with advancing IRIS stage of CKD. Cats in IRIS stage 2 and 3 were subdivided according to whether they met the IRIS targets for phosphate regulation. Those cats that had plasma phosphate concentrations greater than the IRIS targets were found to have significantly higher FGF23 concentrations than those cats where phosphorus concentrations were within target.[41] A further study by Finch and colleagues evaluated FGF23 concentrations in non-azotemic cats > 9 years. These cats were followed longitudinally until the development of azotemia. Those cats that developed azotemia within a 12 month period, at baseline, had significantly higher FGF23 concentrations than those cats which remained non-azotemic suggesting that as in humans plasma FGF23 concentrations may be an early biomarker for alteration in renal phosphorus homeostasis and early non-azotemic CKD.[2] Finally, by Williams and colleagues has evaluated both PTH and FGF23 in cats with hyperthyroidism. This study aimed to determine whether either of these factors were associated with the development of azotemia after treatment of hyperthyroidism or survival in cats with treatement of hyperthyroidism. Unfortunatly, neither were predictive of the development of azotemia in this setting but FGF23 concentrations were significantly associated with all cause mortality.[42] Future study is warranted to investigate the role that dietary and phosphate binder manipulation has on FGF23 concentrations in the cat and whether modulation of FGF23 is beneficial in terms of survival of cats with CKD.

Figure 2b: A new 'trade-off' hypothesis for late CKD

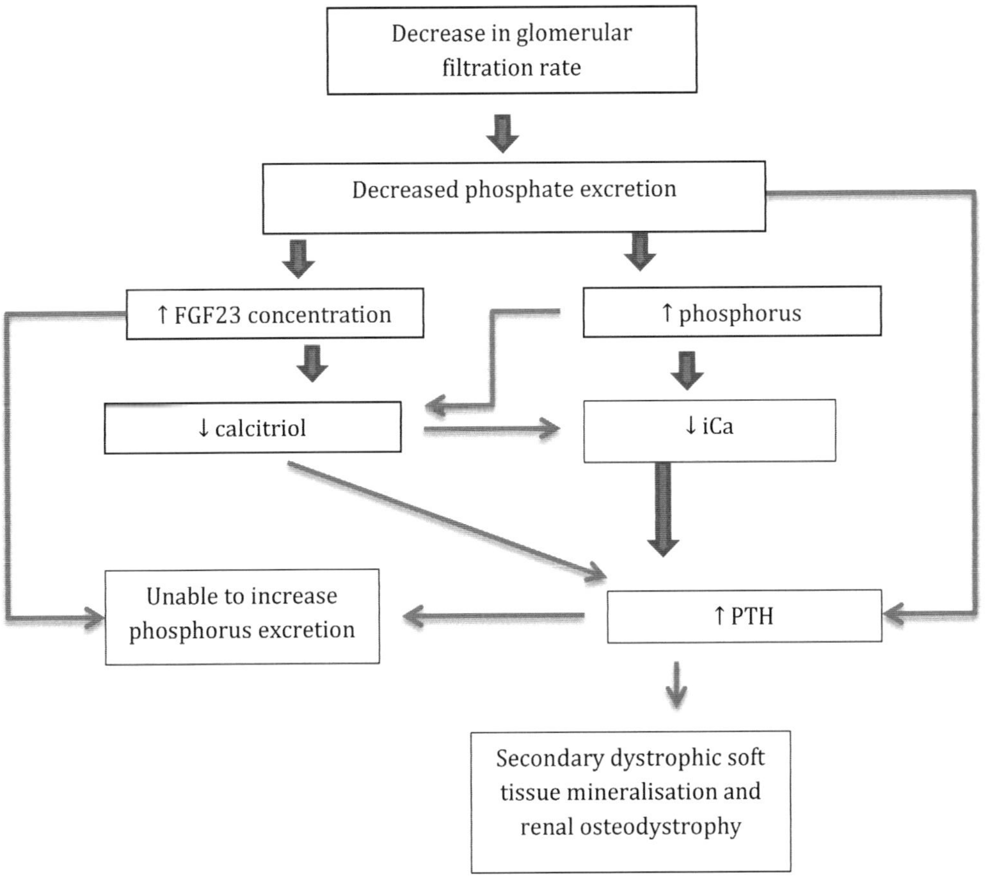

Later in the course of CKD, FGF23 concentrations will continue to increase as a consequence of declining GFR and increased production from osteocytes and osteoblasts. Plasma phosphorus concentrations will begin to increase with declining GFR and will further stimulate FGF23 production. End organ resistance to FGF23 develops due to reduced expression of the FGFR and α-klotho. This can result in FGF23 losing its direct inhibitory action on PTH production. Decreased calcitriol concentrations will reduce calcium absorption from the gastrointestinal tract, stimulating PTH production and will also increase PTH production due to loss of calcitriol-mediated inhibition at the parathyroid gland. PTH production will also be stimulated by increased plasma phosphorus concentrations. Therefore although a combination of increased PTH and FGF23 initially compensate and increase phosphorus excretion overt hyperphosphatemia ultimately develops with declining renal function.

REFERENCES

1. Barber PJ, Elliott J. J Small Anim Pract 1998;39:108-116. 2. Finch NC, Syme HM, Elliott J. Journal of the American Veterinary Medical Association 2012;241:1326-1335. 3. Boyd LM, Langston C, Thompson K, et al. Journal of Veterinary Internal Medicine 2008;22:1111-1117. 4. King JN, Tasker S, Gunn-Moore DA, et al. J Vet Intern Med 2007;21:906-916. 5. Chakrabarti S, Syme HM, Elliott J. Journal of Veterinary Internal Medicine 2012;26:275-281. 6. Ross LA, Finco DR, Crowell WA. Am J Vet Res 1982;43:1023-1026. 7. Ross SJ, Osborne CA, Kirk CA, et al. J Am Vet Med Assoc 2006;229:949-957. 8. Barber PJ, Rawlings JM, Markwell PJ, et al. Journal of Small Animal Practice 1999;40:62-70. 9. Plantinga EA, Everts H, Kastelein J, et al. Veterinary Record 2005;157:185-187. 10. Elliott J, Rawlings JM, Markwell PJ, et al. J Small Anim Pract 2000;41:235-242. 11. Bhattacharyya N, Chong WH, Gafni RI, et al. Trends in Endocrinology & Metabolism 2012;23:610-618. 12. Riminucci M, Collins MT, Fedarko NS, et al. The Journal of Clinical Investigation 2003;112:683-692. 13. White KE, Evans WE, O'Riordan JLH, et al. Nat Genet 2000;26:345-348. 14. Kurosu H, Ogawa Y, Miyoshi M, et al. Journal of Biological Chemistry 2006;281:6120-6123. 15. Urakawa I, Yamazaki Y, Shimada T, et al. Nature 2006;444:770-774. 16. Ben-Dov IZ, Galitzer H, Lavi-Moshayoff V, et al. The Journal of Clinical Investigation 2007;117:4003-4008. 17. Baum M, Schiavi S, Dwarakanath V, et al. Kidney Int 2005;68:1148-1153. 18. Antoniucci DM, Yamashita T, Portale AA. Journal of Clinical Endocrinology & Metabolism 2006;91:3144-3149. 19. Yu X, Sabbagh Y, Davis SI, et al. Bone 2005;36:971-977. 20. Collins MT, Lindsay JR, Jain A, et al. Journal of Bone and Mineral Research 2005;20:1944-1950. 21. Rankin Erinn B, Wu C, Khatri R, et al. Cell 2012;149:63-74. 22. Deger S, Erten Y, Pasaoglu O, et al. Clinical and experimental nephrology 2013;17:416-423. 23. Shimada T, Urakawa I, Isakova T, et al. Journal of Clinical Endocrinology & Metabolism 2010;95:578-585. 24. Stubbs JR, He N, Idiculla A, et al. Journal of Bone and Mineral Research 2012;27:38-46. 25. Gutierrez O, Isakova T, Rhee E, et al. Journal of the American Society of Nephrology 2005;16:2205-2215. 26. Shigematsu T, Kazama JJ, Yamashita T, et al. American Journal of Kidney Diseases 2004;44:250-256. 27. Larsson T, Nisbeth U, Ljunggren O, et al. Kidney Int 2003;64:2272-2279. 28. Isakova T, Wahl P, Vargas GS, et al. Kidney Int 2011;79:1370-1378. 29. Wolf M. Kidney Int 2012;82:737-747. 30. Isakova T XHYW, et al. JAMA 2011;305:2432-2439. 31. Wolf M, Molnar MZ, Amaral AP, et al. Journal of the American Society of Nephrology 2011;22:956-966. 32. Gutiérrez OM, Mannstadt M, Isakova T, et al. New England Journal of Medicine 2008;359:584-592. 33. Ferrari SL, Bonjour J-P, Rizzoli R. Journal of Clinical Endocrinology & Metabolism 2005;90:1519-1524. 34. Burnett S-AM, Gunawardene SC, Bringhurst FR, et al. Journal of Bone and Mineral Research 2006;21:1187-1196. 35. Isakova T, Gutiérrez OM, Smith K, et al. Nephrology Dialysis Transplantation 2011;26:584-591. 36. Moe SM, Zidehsarai MP, Chambers MA, et al. Clinical Journal of the American Society of Nephrology 2011;6:257-264. 37. Koiwa F, Kazama JJ, Tokumoto A, et al. Therapeutic Apheresis and Dialysis 2005;9:336-339. 38. Oliveira RB, Cancela ALE, Graciolli FG, et al. Clinical Journal of the American Society of Nephrology 2010;5:286-291. 39. Gonzalez-Parra E, Gonzalez-Casaus ML, Galán A, et al. Nephrology Dialysis Transplantation 2011;26:2567-2571. 40. Yilmaz MI, Sonmez A, Saglam M, et al. American Journal of Kidney Diseases

2012;59:177-185. 41. Geddes RF, Finch NC, Elliott J, et al. Journal of Veterinary Internal Medicine 2013;27:234-241. 42. Williams TL, Elliott J, Syme HM. Journal of Small Animal Practice 2012;53:561-571.

METABOLIC CONSEQUENCES OF GUT DYSBIOSIS AND INFLAMMATION

Jan S. Suchodolski, DMV, AGAF, PhD, DACVM (SAIM)
College Station, TX

INTRODUCTION

Molecular sequencing studies have shown the importance of the microbiome for host health and have also revealed microbial changes in the composition of the microbiota (i.e., microbial dysbiosis) in gastrointestinal (GI) disease. The intestinal microbiota is defined as the collection of all living microorganisms (i.e., bacteria, fungi, protozoa, and viruses) that inhabit the GI tract. Recent studies have described the gastrointestinal microbiota as highly complex, consisting of several thousand bacterial phylotypes.[1,2] The mammalian GI tract harbors an estimated 10^{10}-10^{14} microbial cells, approximately 10 times more than the number of host cells. This highly complex microbial-host ecosystem plays a crucial role in health and disease.

Many ingested nutrients, but also drugs and other xenobiotics are metabolized by intestinal microbes and then absorbed by the host. Therefore, it is obvious that a balanced composition of the intestinal microbiota will play an important part in intestinal homeostasis, and any changes in the microbiota may directly or indirectly influence metabolic host pathways. Especially convincing evidence has been gathered associating alterations in the composition of the intestinal microbiota with acute and chronic inflammation of dogs and cats.[1,3] However, the focus of the last few years was on describing phylogenetic changes within the microbiome in GI disease, but little knowledge existed about the functional alterations associated with this microbial dysbiosis. The field is now moving more into assessing functional changes due to intestinal inflammation and concurring microbial dysbiosis. Novel metabolomics approaches using multiple mass spectrometry platforms allow assessing changes in metabolite profiles, either produced by the host or microbiota, and therefore both together yield a better understanding of the pathophysiology of GI disease.

MICROBIOME IN HEALTH AND DISEASE

Molecular sequencing methods based on 16S rRNA genes are now the standard in assessing intestinal microbiota. It has been shown that each individual animal harbors very unique microbial communities, with compositional differences along the GI tract.[7] The canine microbiome harbors on average 10 different bacterial phyla, with *Proteobacteria, Firmicutes, Actinobacteria, Bacteroidetes, Spirochaetes,* and *Fusobacteria* being predominant. Additionally, the GI tract harbors various fungi, archaea, protozoa, and viruses.[4]

A balanced microbial ecosystem is of benefit to the host through many mechanisms. The intestinal microbiota protects the host from invasion by potentially pathogenic organisms. This is achieved through competition for oxygen and nutrients, competition for mucosal adhesion sites, creation of a physiologically restrictive environment for non-resident bacterial species (e.g., production of substances that are toxic to other bacteria, changes in pH and redox potential, or hydrogen sulfide production), and secretion of antimicrobial substances (e.g., bacteriocins). Bacterial constituents within the intestine produce, for example, vitamins (e.g., folate) and immunomodulatory compounds such as indole and short chain fatty acids (SCFA; such as butyrate, propionate, and acetate). These SCFA are important for mucosal growth and epithelial cell proliferation. Lactate, produced by microbial fermentation, serves directly as an energy source for the host, or can be metabolized by bacteria into SCFA. Of interest is that this is an area with many unknowns, as many of these metabolic exchanges between host and microbes have not been well examined. While some studies have evaluated changes in SCFA in response to dietary changes, very little is known how these important metabolites are altered in GI inflammation. This is an emerging area, as more and more it becomes recognized that there are highly complex interactions between host metabolites and microbe-derived metabolites and these may be disrupted in GI disease, as in cases of GI inflammation metabolites may either be not absorbed across the epithelium or they may leak into the gut lumen.

Over the past few years strong evidence has been gathered demonstrating that the resident gut microbiota is involved to some part in the pathogenesis of inflammatory bowel diseases (IBD) in humans, dogs, and cats.[4] There are many similarities in the observed microbial changes across humans and veterinary species. Often, a reduction in *Clostridium* clusters XIVa and IV (i.e., *Faecalibacterium, Lachnospiraceae* and *C. coccoides* subgroups) with concurrent increases in Enterobacteriaceae (e.g., *E. coli*) are associated with IBD.

Several studies have recently demonstrated alterations in microbiota in dogs with IBD. In the small intestine, the proportions of Fusobacteria, Bacteroidaceae, Prevotellaceae, and Clostridiales were significantly increased in healthy dogs. In contrast, specific bacterial genera within Proteobacteria, including Diaphorobacter and Acinetobacter, were either more abundant or more frequently identified in IBD dogs.[5] In another study, fecal samples from dogs with acute hemorrhagic diarrhea had significant decreases in Blautia, Ruminococcaceae including Faecalibacterium, and Turicibacter spp., and significant increases in genus *Sutterella* and *Clostridium perfringens* compared to healthy dogs.[6]

METABOLIC CHANGES IN GASTROINTESTINAL DISEASE

Due to the important role of microbial-derived metabolites for host health, it is important to understand the metabolic consequences of GI inflammation and dysbiosis. Several studies have been reported that evaluate selected metabolites that are altered in GI disease. Due to complex interactions between intestinal absorption and microbial metabolism, the exact cause for changes in serum concentrations of serum metabolites is often unknown, but a better understanding of the physiological pathways is helpful to potentially pinpoint specific diseases.

An example for this approach is the measurement of serum concentrations of cobalamin and folate, two important markers for GI disease. The uptake of cobalamin (Vitamin B12) and folate from the small intestine is depended on several factors and can therefore be utilized as an indirect marker for the assessment of gastrointestinal disease. Disorders that may affect serum cobalamin and/or folate concentrations include small intestinal inflammation, exocrine pancreatic insufficiency (EPI), and small intestinal dysbiosis (SIBO/ARD). Due to different sites of absorption, specific changes in serum cobalamin and folate concentrations can yield information on the localization and potentially the severity of intestinal disease. Cobalamin absorption is complex and requires a functioning digestive system. Dietary cobalamin is bound to carrier protein, which is digested in the stomach. Then cobalamin is bound to R-protein, which is degraded by pancreatic enzymes in the small intestine, and the free cobalamin is bound by intrinsic factor. Intrinsic factor/ cobalamin complexes are then absorbed by specific mucosal receptors located in the ileum. Major disorders that interfere with cobalamin uptake are exocrine pancreatic insufficiency (EPI), distal or diffuse small intestinal inflammatory disease (damage of receptors), and excess bacterial utilization of cobalamin in bacterial dysbiosis. Decreased absorption ultimately leads to metabolic consequences on a cellular level, with an accumulation of serum methylmalonic acid (MMA) concentration which is often dramatically increased in patients with cobalamin deficiency. Dietary folate is typically present in the poorly absorbable polyglutamate form. In the proximal small intestine, the brush border enzyme folate deconjugase removes all but one glutamate residue from the molecule. The uptake of folate monoglutamate occurs through specific carriers located in the proximal small intestine. In dogs increased serum folate concentrations can be indicative of small intestinal dysbiosis (SIBO/ARD), as resident bacteria in the distal small intestine (i.e., ileum) and in the large intestine can produce large quantities of folate. Because the carriers responsible for folate uptake are located exclusively in the proximal small intestine, folate produced in distal sections of the intestine cannot be absorbed. However, if folate producing bacteria proliferate in the proximal small intestine, the bacterial folate can be absorbed by the host resulting in increased serum folate concentrations. In contrast, diffuse inflammation in the proximal GI tract may lead to damage of receptors and decreased serum folate concentrations.

Another important pathway that has been shown to be altered in GI disease is bile acid metabolism. Changes in intestinal bile acid metabolism has been implicated as an important factor in intestinal inflammation in human IBD patients as well as *C. difficile* infection, and the therapeutical correction of altered bile acid concentrations leads to improvement of intestinal inflammation.[7] Unconjugated bile acids are toxic to epithelial cells which may result in increased intestinal permeability. Furthermore, altered bile acid profiles lead to changes in fat absorption from the small intestine. Secondary bile acids are produced exclusively by the intestinal microflora. Bacteria, such as *Clostridium hiranonis* exhibit 7α-dehydroxylating activity, resulting in the 7α-dehydroxylation and biotransformation of the primary bile acids, cholic and chenodeoxycholic acid, into the secondary bile acids, deoxycholic and lithocholic acid, respectively. Altered serum unconjugated bile acids (SUCA) have also been associated with altered small intestinal microbiota in dogs.[8]

A recent study has shown an increased serum D-lactate concentration in cats with various gastrointestinal diseases.[9] The D-enantiomer of lactic acid is not normally found in any appreciable quantities in serum from mammals. The increase in serum D-lactate in cats with gastrointestinal disease is possibly due to disarrangements of the intestinal microbiota and increased bacterial production of D-lactate. D-lactate has been shown to lead to neurological signs in some cats.[10]

The above selected examples show that altered microbiota lead to systemic metabolic consequences. However, more in-depth studies evaluating additional metabolites are needed. More recent studies have utilized an untargeted metabolomics approach and measured several hundred metabolites in serum or feces of dogs with GI disease. One study evaluated the serum metabolite profile of dogs with idiopathic IBD before and after treatment. Serum and fecal samples were collected from healthy dogs (n=10), and from dogs with IBD (n=12) before and after 21 days of standard medical therapy (e.g., elimination diet and administration of anti-inflammatory drugs). A total of 359 serum metabolites were profiled using a GC-TOF/MS platform.[11] Before treatment, the abundance of *Blautia, Faecalibacterium*, and *Turicibacter* was significantly decreased in dogs with IBD relative to healthy dogs. While the CIBDAI significantly decreased, no significant differences in the proportions of bacterial groups were found before and after treatment. Serum concentrations of 3-hydroxybutyrate, hexuronic acid, ribose, and gluconic acid lactone were significantly more abundant in dogs with IBD. These metabolites are associated with the pentose phosphate pathway and indicate the presence of oxidative stress in dogs with IBD. Of interest is that the altered metabolic profiles did not normalize after 3 weeks of therapy, suggesting that the intestinal inflammation persisted despite improvement in clinical severity indices of these dogs.

Another recent study in our laboratory evaluated the fecal metabolome in dogs with IBD. A total of 797 metabolites were analyzed using various mass spectrometry platforms. Significant alterations in various metabolic pathways were noticed. Dogs with IBD had altered SCFA metabolism (decreased propionate), altered bile acid metabolism (primary bile acids were generally increased in dogs with disease, while secondary bile acids were generally decreased in dogs with IBD), altered glycolysis pathways and altered tryphtophan-indole pathways. Furthermore, markers of oxidative stress were also highly altered in fecal samples of dogs with IBD. Among the redox components acting in the GI lumen, cystine, cysteine-s-sulfate, and cysteine were increased in disease. Though glutathione disulfide and glutathione were not detected, 5-oxoproline was increased along with several gamma-glutamyl amino acids. Changes in components of the gamma-glutamyl cycle may suggest sequestration of intracellular glutathione by either enterocytes or luminal bacteria under oxidative stress driven by abundance of leukocytes and production of reactive oxygen species. Abnormalities in the redox equilibrium were further supported by increases in the cystine/cysteine redox pair.

Another study evaluated serum and urine metabolites and fecal microbiome changes in dogs with acute diarrhea. Significant alterations were observed in the fecal microbiome (decrease in *Faecalibacterium, Blautia, Clostridiales)* and these microbiome changes were associated with altered tryptophan metabolism and SCFA. The relative serum concentrations of kynurenic acid and urine concentrations of 1H-indole, 2-methyl- and 5-methoxyindole-3-carbaldehyde were significantly decreased in acute diarrhea.

CONCLUSIONS

The above studies indicate that acute and chronic GI disease is associated with microbial dysbiosis and altered systemic metabolic states. Better understanding of the interactions of microbial-derived metabolites and the host will yield insights into the pathophysiology of GI diseases. Furthermore, some of these metabolic pathways can be targeted to develop better treatment approaches.

REFERENCES

1. Frank DN, Amand ALS, Feldman RA, et al. Molecular-phylogenetic characterization of microbial community imbalances in human inflammatory bowel diseases. *Proceedings of the National Academy of Sciences of the United States of America* 2007;104:13780-13785.
2. Suchodolski JS, Dowd SE, Westermarck E, et al. The effect of the macrolide antibiotic tylosin on microbial diversity in the canine small intestine as demonstrated by massive parallel 16S rRNA gene sequencing. *BMC Microbiol* 2009;9:210.
3. Suchodolski JS. Microbes and gastrointestinal health of dogs and cats. *J Anim Sci* 2010:1-34.
4. Honneffer JB, Minamoto Y, Suchodolski JS. Microbiota alterations in acute and chronic gastrointestinal inflammation of cats and dogs. *World J Gastroenterol* 2014;20:16489-16497.
5. Suchodolski JS, Dowd SE, Wilke V, et al. 16S rRNA gene pyrosequencing reveals bacterial dysbiosis in the duodenum of dogs with idiopathic inflammatory bowel disease. *Plos ONE* 2012;7:e39333.
6. Suchodolski JS, Markel ME, Garcia-Mazcorro JF, et al. The fecal microbiome in dogs with acute diarrhea and idiopathic inflammatory bowel disease. *Plos ONE* 2012;7:e51907.
7. Duboc H, Rajca S, Rainteau D, et al. Connecting dysbiosis, bile-acid dysmetabolism and gut inflammation in inflammatory bowel diseases. *Gut* 2013,62.531-539.
8. Melgarejo T, Williams DA, O'Connell N, et al. Serum unconjugated bile acids in dogs with small intestinal bacterial overgrowth. *Gastroenterology* 1997.
9. Packer RA, Moore GE, Chang CY, et al. Serum d-lactate concentrations in cats with gastrointestinal disease. *J Vet Intern Med* 2012;26:905-910.
10. Packer RA, Cohn LA, Wohlstadter DR, et al. D-lactic acidosis secondary to exocrine pancreatic insufficiency in a cat. *Journal of Veterinary Internal Medicine* 2005;19:106-110.
11. Minamoto Y, Otoni CC, Steelman SM, et al. Alteration of the fecal microbiota and serum metabolite profiles in dogs with idiopathic inflammatory bowel disease. *Gut Microbes* 2014:1-15.

HEMOLYTIC BLOOD TRANSFUSION REACTIONS IN DOGS: RECOGNITION AND PREVENTION

Marie-Claude, Blais, DMV, DACVIM
Montréal, QC, Canada

INTRODUCTION

What can be done when a transfusion reaction occurs? From a meticulous blood donor selection to the blood product administration itself, every step is important to offer safe and effective transfusion therapy. This comprehensive review will present practical recommendations on how to recognize, investigate and most importantly, how to prevent hemolytic transfusion reactions, emphasizing blood compatibility.

CLASSIFICATION OF TRANSFUSION REACTIONS

Hemolytic transfusion reactions mostly occur during or shortly after a transfusion (acute hemolytic transfusion reaction (AHTR); within $\leq$ 24 hours), but later onset may also occur (delayed hemolytic transfusion reaction). These reactions can be immunologic or non-immunologic. It may not be easy to differentiate between the two, but the cause can be ascertained by careful evaluation of blood bank management, patient's clinical history and signs, and laboratory investigations.

Non-immune hemolytic transfusion reactions occur when red blood cells (RBCs) are damaged before the transfusion or during its administration. They may result in hemoglobinemia and hemoglobinuria. Although the clinical signs are usually milder, acute, life-threatening or fatal, non-immune hemolytic transfusion reactions have recently been reported by *Patterson et al* in 4 dogs most likely secondary to improper RBCs storage[17].

Acute immune hemolytic transfusion reactions are generally the most severe transfusion reactions and are caused by blood type incompatibilities. Because dogs do not possess clinically significant naturally-occurring alloantibodies, such reactions generally are only observed in previously transfused dogs ($\geq$4 days) receiving RBC products of mismatched blood type.

Delayed immune hemolytic transfusion reactions are associated with the production of RBC antibodies within days of a transfusion, which result in a significant decline in the patient's post-transfusion hematocrit. Because they are not associated with acute clinical signs, they may be clinically unnoticed.

RECOGNITION OF TRANSFUSION REACTIONS

It is critical to monitor patients throughout the administration of blood products. The recognition of transfusion reactions may be improved by using a standardized transfusion monitoring form. In addition to containing information concerning the blood unit administered (unit identification, blood type of donor and recipient, volume administered, crossmatch status), baseline clinical data should be recorded including: temperature, heart rate, pulse quality, respiration rate/effort, mucous membrane color/capillary refill time, blood pressure, mentation as well as any other pertinent observations concerning the patient. It is necessary to monitor the patient closely and to treat any manifestations (e.g. skin rashes, tachycardia, tachypnea, sudden aggressive behavior) that develop during a transfusion as a potential transfusion reaction.

Different acute transfusion reactions share common manifestations (Table 1). In addition, some features can be obscured by the patient's underlying condition or treatments. For instance, fever is one of the most common presentations in an acute hemolytic transfusion reaction, but can also be caused by a febrile non-hemolytic transfusion reaction (patient's reaction to donor white blood cells or to biological response modifiers in the component) or a septic reaction (bacterial contamination).

In dogs sensitized by previous blood transfusions, the binding of antibodies (anti-DEA 1 or others) to donor RBCs leads to RBC lysis, activation of complement and coagulation pathways and cytokine release. Complement activation induces hemolysis, while activation of the coagulation cascade and thrombin generation predisposes the patient to DIC. Consequences of intravascular hemolysis are hemoglobinemia associated with a rapid decline in PCV and a positive Coombs test, and hemoglobinuria. Hypotension may also be present. Severe anxiety is often described in human medicine ("feeling of impending doom"). If this reaction is not recognized, the patient can progress to renal ischemia, acute renal failure, DIC, and death.

Table 1: Clinical signs of acute transfusion reaction in dogs

Clinical signs	Type of transfusion reaction	Comments
Fever	FNHTR, AHTR, TRALI, bacterial contamination, or unrelated to the blood transfusion.	Defined as a temperature >39.5°C or a rise of 1-2 °C from baseline. Can coexist with other symptoms such as chills, rigors, myalgia, nausea or vomiting, dyspnea, hypotension (≥30mmHg below baseline) and tachycardia (HR > 40 bpm above baseline)
Urticaria, hives or pruritus Angioedema	Allergic transfusion reaction or anaphylaxis	Can be mild and localized or more severe with generalized urticarial
Dyspnea or hypoxia	TRALI, TACO or severe allergic transfusion reaction	
Pulmonary oedema	TRALI or TACO	
Hypotension	AHTR, severe allergic transfusion reaction, anaphylaxis, bacterial contamination, TRALI, TACO or unrelated to the transfusion	
Severe anxiety	AHTR or bacterial contamination	
Acute onset of bleeding	AHTR or bacterial contamination	Caused by disseminated intravascular coagulation

FNHTR: febrile non-hemolytic transfusion reaction; AHTR: acute hemolytic transfusion reaction; TRALI: transfusion-related acute lung injury; TACO: transfusion-associated circulatory overload.

Table adapted from Al-Riyami AZ et al. SQU Medical Journal. August 2014[1]

INITIAL CLINICAL ASSESSMENT AND MANAGEMENT

The first step to approaching any acute transfusion reaction is to stop the transfusion. While maintaining the venous access with an isotonic solution such as 0.9% NaCl, the patient's airway, breathing and circulation (ABC) need to be assessed next. The blood unit should be examined for the presence of any unusual clumps, particulate matter or discoloration suggestive of bacterial contamination. Finally, although it is unlikely in veterinary medicine that a clerical error leads to the wrong patient receiving the wrong blood product, it remains essential to verify that the identification of the blood donor is available for eventual transfusion reaction investigations.

Following the successful management of mild FNHTR or a mild allergic transfusion reaction (e.g. antihistamines and/or steroids), the transfusion may be resumed in patients with no or limited changes in their vital signs and who are lacking other clinical signs, but at a slower rate and under direct observation. In patients with all other types of transfusion reactions, the transfusion should be discontinued. Any suspicion of an acute hemolytic transfusion reaction or bacterial contamination upon visual assessment of the unit or by the patient's clinical signs, should prompt further investigation. If an AHTR is suspected, any hypotension should be promptly addressed in order to maintain an adequate renal blood flow.

LABORATORY INVESTIGATIONS

Reports of moderate or severe transfusion reactions should prompt testing to rule out AHTR and bacterial contamination. Post-transfusion blood (EDTA and serum) and urine samples should be collected, and the implicated unit should be put aside. In addition to performing/repeating compatibility testing (see below), a complete blood count, Coombs test, coagulation panel, and renal and liver function tests should be considered. The urine should be inspected for the presence of hemoglobinuria. If there is any suspicion of bacterial contamination, blood cultures should be obtained from the patient and the implicated unit.

If the cause of the acute hemolytic reaction is not obvious (ie immunologic or non-immunologic), a thorough evaluation of the hospital transfusion practices should be done including inspection of in-house transfusion material (collection bags, pumps, transfusion sets, fluids), blood collection techniques (ex.: negative pressure used during collection if suction is used), blood storage conditions and administration techniques. As a preventive measure, the color and presence of gross hemolysis (centrifuged blood sample of the bag) should be evaluated for all blood bags before administration.

As previously mentioned, compatibility testing between the donor and the recipient should be performed/repeated. Pre and post-transfusion crossmatch should be performed to assess the serologic compatibility or incompatibility between donor and recipient. An incompatible major crossmatch result supports an acute immune hemolytic anemia. A standardized tube crossmatching is an appropriate technique, which requires very little material but requires some expertise. Even with experience, the grading of the agglutination reactions (+/- hemolysis) is subjective. Gel column techniques (DiaMed and DMS gel tube assay) have been evaluated and found to be simple, sensitive, and standardized methods to crossmatch dogs. Following the discontinuation of DiaMed's veterinary products in North America, our clinical laboratory (Université de Montréal) has continued to perform all crossmatch tests using the gel column technology but with human saline cards provided by Ortho Clinical. The sensitivity of the gel column technique and the objectivity of the results make it an ideal tool when investigating potential blood incompatibility cases.

Obviously, a pre-transfusion compatible crossmatch does not prevent sensitization of the patient against donor cells within the following weeks. In the same regard, it does not give information about the blood type of the dog. Therefore, previously transfused dogs should always be crossmatched. A minimum of 4 days following a transfusion is generally required for a dog to produce alloantibodies against a blood type it does not possess, which may last in circulation for months to years; this is supported by our preliminary data obtained following the sensitization of two *Dal*-negative dogs.

Because of the strong antigenicity of DEA 1 and its prevalence in the canine population (grossly 50% of dogs are DEA 1 positive, with geographical and breed variation), typing of donors and recipients for DEA 1 is strongly recommended. Indeed, DEA 1 invariably elicits a strong alloantibody response after sensitization of a DEA 1- dog with DEA 1+ blood, and may be responsible for an acute hemolytic transfusion reaction in previously transfused dogs. Standardized DEA 1 blood typing is readily available and may be performed using monoclonal antibodies based on agglutination reactions (DMS/RapidVet-H) or on chromatographic techniques (Alvedia).

Transfusion reactions against DEA 4, *Dal*, and another unspecified common RBC antigen have been described, and other clinically important blood types may be found in the future. If DEA 1 incompatibility does not explain a blood incompatibility, extended blood typing is recommended. Unfortunately, typing for all recognized DEA might be difficult as some reagents are no longer commercially available (polyclonal antibodies from sensitized dogs are used for typing). Little has been published on the prevalence of blood type according to dog breed, which may eventually help narrow the search for compatible blood. A limited survey revealed that Doberman pinschers and Dalmatians are at increased risk of blood incompatibility, as >1/3 of Doberman pinschers and >1/4 of Dalmatians are *Dal*-negative while all non-Doberman blood donors tested to date (>130 dogs) are *Dal*-positive.[11] *Dal* blood typing is now available at the Blood Bank of the University of Montreal and at the Transfusion Laboratory at the University of Pennsylvania's School of Veterinary Medicine (PennGen). The administration of "test volumes" of donor blood to assess blood-type compatibilities is highly inappropriate, as small amounts (< 1 ml) of incompatible blood can result in life-threatening reactions.

CONCLUSION

The prompt recognition, management and reporting of transfusion reactions are important aspects in ensuring patient safety during and after transfusions. Most transfusion reactions can be avoided by careful selection of healthy donors, appropriate collection, storage, and administration techniques, adequate pre-transfusion compatibility testing (blood typing +/- crossmatching), and the transfusion of only required blood components. As transfusion reactions may have overlapping manifestations, only thorough investigation can identify the cause, which may be crucial to improve hospital transfusion practice and the safety of patients. Finally, blood incompatibility investigations require time and expertise, but remain essential to improve our knowledge in the constantly growing field of transfusion medicine.

REFERENCES

[1]Acierno MA et al. **J Vet Intern Med** 2014; 28: 592-598. [2]Al-Riyami AZ et al. **SQU Medical Journal** 2014; 14(3): 306-318. [3]Blais et al. **J Vet Intern Med** 2007; 21: 281-286. [4]Blais et al. **J Vet Intern Med** 2009; 23: 462-465. [5]Blois SL et al. **J Vet Emerg Crit Care** 2013; 23: 340–343. [6]Conversy B et al. **J Am Anim Hosp Asso** 2013; 49 (5) : 301-307. [7]Ergul Ekiz E et al. **Vet Clin Pathol** 2011: 40: 518–523. [8]Ferreira RR et al, **Vet Clin Pathol 40**: 198–201. [9]Giger U et al. **Am J Vet Res** 2005; 66:1386–1390. [10]Giger U.: In: **Kirk's CVT** 2014. [11]Goulet S et al. Research Abstract . **J Vet Intern Med** 2014; 28(3); 976-1134. [13]Hohenhaus AE **Transfus Med Rev** 2004; 18:117–122. [14]Iazbik MC et al. **Vet Clin Patho** 2010; 39:433-435. [15]Kessler RJ et al. **Vet Clin Patho** 2010; 39:306–316. [16]Kohn B et al. **J Vet Diagn Invest** 2012; 24:539–545. [17]Patterson J et al. **J Vet Intern Med** 2011; 25:927–933. [18]Seth M et al. **Am J Vet Res** 2012; 73:213–219. [19]Tinegate H et al. **British Journal of Haematology** 2012; 159: 143-153. [20]Zimring JC and Spitalnik SL. **Annu Rev Pathol Mech Dis** 2015; 10: 83-110.

FLOW CYTOMETRY FOR THE INTERNIST: SOMETHING TO KNOW ABOUT?

Oliver A Garden, BSc, BVetMed, PhD, FHEA, MSB, MRCVS, DACVIM, DECVIM-CA
London, UK

INTRODUCTION

Since its invention in the late 1960s and the introduction of commercial machines in the early 1970s, the flow cytometer has revolutionized biological research and the human clinical diagnostic laboratory[1]. With applications in fields ranging from hematology to oncology and theriogenology, this technology is now beginning to gain traction in veterinary diagnostic work[2-6]. This presentation will introduce some technical concepts that are key to an understanding of flow cytometry, before reviewing current and future veterinary applications of this exciting technology.

PRINCIPLES OF THE FLOW CYTOMETER AND FLUORESCENCE

Flow cytometry allows the phenotypic characterization of cells after their labelling with fluorescently-tagged antibodies (Abs) against specific surface or intracellular antigens[1,7]. The method requires the cells to pass in single-file through a laser beam. Photons of light energy are 'absorbed' by the fluorescent probes (excitation), which then emit light that is detected by a photomultiplier tube (PMT) and digitized (emission); this signal, proportional to the amount of Ab binding to each cell, is then represented in various formats for analysis. The key elements of the process – fluidics, optics, light and fluorochromes – are summarized below.

Fluidics and optics

The fluidics system comprises a central channel through which the sample of cells is injected, enclosed by an outer sheath of faster-flowing fluid. The drag of the sheath fluid on the central channel has the effect of focusing the passage of cells into single file (hydrodynamic focusing), thus preventing blockage of the instrument nozzle – typically 70 μm in diameter – and allowing the interrogation of individual cells. After hydrodynamic focusing, the cells pass through two or more beams of laser light. Light scattering and fluorescence emission provides information on the physical properties and phenotype of the cells. Light scattered in the forward direction (FSC), typically up to 20° from the laser's beam axis, conveys information about cell size, while that scattered at 90° to the beam axis (side scatter; SSC) conveys information about cellular granularity[7]. The combination of FSC and SSC permits the distinction of different populations of cells in heterogeneous samples such as blood. Separate fluorescence (FL-) channels detect the light emitted from fluorochromes, the specificity of detection controlled by optical filters that transmit (pass) light of only certain wavelengths. To permit the simultaneous detection of multiple signals, the choice and order of optical filters is carefully configured (Figure 1).

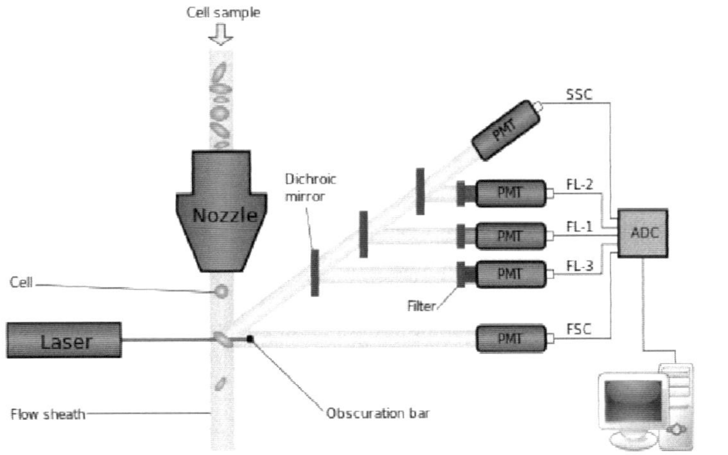

Figure 1: Basic design of a flow cytometer (figure not to scale; from Kierano, Wikipedia)
Abbreviations: SSC=side scatter; PMT=photomultiplier tube; FL=fluorescence channel; FSC=forward scatter; ADC=analog-to-digital converter

Some flow cytometers have the capability of charging the fluid stream as it exits the nozzle of the fluidics system according to pre-set selection criteria, thus allowing electrostatic sorting of cells of a particular phenotype (fluorescence-activated cell sorting; FACS™)[1].

Light and fluorochromes

Light is a form of electromagnetic energy that has both a frequency and wavelength, the latter of which determines its color. Fluorochromes are dyes that absorb and emit light energy, the emitted light having a longer wavelength (less energy) than the absorbed light owing to a phenomenon called the Stokes' shift. Each fluorochrome has a maximal absorbance and a maximal emission wavelength, the former defining the laser line used for excitation and the latter defining the FL channel (detector) most optimally used to measure the emitted fluorescence. Over 30 different fluorochromes are currently commercially available, exceeding the number of detectors in even the most advanced flow cytometers[8]. Single dyes include those that have been available for 30 years or more (e.g. fluorescein isothiocyanate; FITC) and newer generation dyes with increased photostability and fluorescence (e.g. Alexa Fluor® dyes). Tandem dyes, in which one fluorochrome transfers light energy to a second fluorochrome by the phenomenon of fluorescence resonance energy transfer (FRET), provide a means of increasing the number of colors that can be analyzed from a single laser wavelength. A new class of inorganic dyes, semiconductor nanocrystals called quantum dots, have broad excitation but narrow emission spectra, further increasing spectral versatility. Overlap of the emission spectra of the organic dyes in any one FL channel requires 'compensation' to be undertaken by the machine, in which the overlapping signals are deducted from the total signal recorded by any one detector to give an accurate representation of the signal attributed to the fluorochrome being measured by that detector[9].

Data analysis

Flow cytometry data may be represented in one dimension by means of a single-parameter histogram, or two dimensions by means of a dot plot, density plot or contour plot[7]. The phenotypic properties of specific populations may be interrogated by gating them for further analysis, each *gate* comprising one or more *regions* or areas drawn on a plot of flow cytometric data.

CLINICAL APPLICATIONS

Though still a specialized diagnostic modality, flow cytometry has found utility in a variety of arenas in small animal internal medicine and oncology. Current and future applications are summarized below.

Automated hematology analyzers

Several automated hematology analyzers incorporate the technology of flow cytometry, using multi-angle light scatter to enumerate red and white blood cells (WBCs) and platelets, as well as to convey information on red cell and platelet size (e.g. CELL-DYN 3700, Abbott Laboratories; ADVIA 120® System, Siemens; Sysmex XT-2000iV™, SYSMEX; LaserCyte Dx™, IDEXX). The ADVIA 120 System also incorporates automated peroxidase cytochemistry to improve lymphoid and myeloid cell distinction, while the SYSMEX machine captures side fluorescence emitted from a polymethine dye to yield information on cellular DNA and RNA content. Taking analytical capability to the next level, the CELL-DYN Sapphire machine incorporates automated three-color flow cytometry to document the expression of CD61, CD3, CD4 and CD8 after treatment of the cells with the relevant monoclonal Abs, but veterinary software for this machine is currently not available and the CELL-DYN 3700 is due for retirement in 2017. Despite these impressive technological advances, review of a blood smear by a clinical pathologist is still essential for an accurate WBC differential count, since monocytes, eosinophils and basophils are not infallibly distinguished by these analyzers; furthermore, information on both RBC and WBC size and morphological appearance is also diagnostically important.

Immunophenotyping lymphoma and leukemia

Lineage determination of hemolymphoid neoplasia currently remains the most widespread clinical application of flow cytometry beyond its use in hematology analyzers. This technique is well established in canine oncology[10-15] and is now also becoming available in feline oncology[16]. There is increasing recognition of the prognostic value of immunophenotypic information, including B *versus* T cell origin[17,18], MHC class II expression[19,20], and the presence of absence of minimal residual disease[21,22], knowledge of which may fashion specific treatment regimens[23,24]. Our own work has suggested that the proportion of regulatory T cells in affected lymph nodes is also of prognostic importance[25]. Indeed, international consortia have been convened to discuss best practice in diagnostic flow cytometry for lymphoid neoplasms[26]. A panel of monoclonal Abs is applied to the patient sample, usually a fine needle aspirate of an affected lymph node or a blood sample, before analysis of the expression pattern of key antigens, including CD45 (all leukocytes), CD34 (hematopoietic progenitor cells), CD14 (monocytes, macrophages), CD11b (granulocytes, monocytes), CD11c (dendritic cells), CD3 or CD5 (T cells), CD4 (T helper cells), CD8 (cytotoxic T cells), CD79a or CD79b (all B cell developmental stages), CD21 (mature B cells), MHC class II (B cells, monocytes, macrophages and activated T cells), IgM (B cells) and IgG (B cells after class switching)[3,5,6,15,27]. Other samples that may be immunophenotyped in appropriate cases include bone marrow, cerebrospinal fluid (CSF), body cavity fluids, and bronchoalveolar lavage fluid, though in all cases cell counts are usually undertaken first to confirm that the harvested population is sufficiently numerous to allow reliable analysis. Clear B and T cell distinctions are usually possible in the case of lymphoid neoplasms, but aberrant patterns have been documented, including the simultaneous expression of B and T cell markers or lymphoid and myeloid cell markers[28,29]. In all cases, cytological and, or histopathological review of samples is undertaken first to establish the underlying diagnosis; flow cytometry cannot replace the need for fundamental morphological characterization of the cells.

Determination of DNA-ploidy and proliferative index

DNA-aneuploidy and increased fractions of cells in the S-phase of the cell cycle are features of neoplasia that may be detected by DNA dyes such as propidium iodide and 7-aminoactinomycin-D. Data on the prognostic value of these analyses are mixed, some studies finding no correlation between aneuploidy or high S-phase fraction and prognosis[30,31], and others reporting an association of these phenomena with a poorer prognosis[32-34]. Proliferation of cells in the tumor may also be gauged by labeling for Ki67, a non-histone nuclear protein that is expressed in the G1, S, G2 and M phases of the cell cycle. Measurement of Ki67 expression by flow cytometry has allowed the grade of malignancy to be determined in canine lymphoma[35].

Immune-mediated hematological disease

A number of studies have explored the utility of flow cytometry in the context of immune-mediated hemolytic anemia (IMHA), immune-mediated thrombocytopenia (ITP) and immune-mediated neutropenia (IMN), though it is uncommonly undertaken for these indications in practice. Flow cytometry offers greater sensitivity than the direct antiglobulin (Coombs') test for the diagnosis of IMHA[36] and is associated with high specificity[37]. It may also yield prognostic information, cases with two or more immunoglobulin isotypes suffering more severe anemia than those with a single isotype[38]. Platelet-bound antibodies may be documented by flow cytometry as an aid to the diagnosis of ITP in both dogs[39,40] and cats[41]. Platelet activation may also be detected by measuring the expression of P-selectin (CD62)[42-44], though modern hematology analyzers automatically yield this information in the form of the mean platelet component concentration (MPC) and MPC distribution width. As a measure of active thrombopoiesis, reticulated platelets (young platelets containing residual RNA) may be measured by flow cytometry[45], but these data are similarly captured by modern analyzers. Flow cytometric assays for neutrophil-bound immunoglobulin[46,47] and neutrophil function assays, including those for phagocytosis and oxidative burst[48], have also been described, but are all rarely done in practice.

Other applications and the future

Flow cytometry has been sporadically applied in a variety of additional veterinary contexts, including the documentation of primary[49,50] and secondary immunodeficiencies[2], particularly to monitor the decrease in CD4:CD8 T cell ratio associated with feline immunodeficiency virus infection[51]. Lymphocyte subset analysis has also been used in the monitoring of systemic lupus erythematosus, in which active disease is associated with an increased CD4:CD8 ratio[52], and in tracking immunosenescence, in which advancing age is associated with a decreased CD4:CD8 ratio[53-56]. Flow cytometry has found utility in canine blood typing[57]; assessment of the immune response in canine trypanasomiasis[58], leishmaniosis[59,60], arthritis[61] and steroid-responsive meningitis-arteritis[62]; phenotypic profiling of canine intestinal intraepithelial lymphocytes in health[63] and disease[64]; and documentation of autoantibodies in dogs with a variety of tumors[65], measurement of the expression of platelet glycoprotein complex IIb-III in canine Glanzmann thrombasthenia[66], and assessment of the phenotype of tumor-infiltrating lymphocytes in canine mammary carcinoma[67]. A number of studies have also used this technology in the assessment of canine and feline semen quality[68,69].

The future of flow cytometry in veterinary internal medicine and oncology is bright, encouraged by the decreasing footprint and price, yet improving user-friendliness, of modern machines, the increasing realization of the diagnostic value of currently available assays, and the potential for future development of a number of additional assays, some of which may be offered in a multiplexed format to maximize what can be done with scarce clinical samples.

REFERENCES

1. Herzenberg LA, et al **Clin Chem** 2002;48:1819; 2. Chabanne L, et al **Methods Cell Sci** 2000;22:199; 3. Culmsee K, Nolte I **Methods Cell Sci** 2002;24:49; 4. Tarrant JM **Vet J** 2005;170:278; 5. Weiss DJ **Vet Clin Pathol** 2002;31:72; 6. Wilkerson MJ **Vet Clin North America Small Anim Prac** 2012;42:53; 7. Cram LS **Methods Cell Sci** 2002;24:1; 8. Chattopadhyay PK, et al **Immunol** 2008;125:441; 9. Maecker HT, et al **Cytometry Part A** 2004;62:169; 10. Avery AC, Avery PR **Vet Clin North America Small Anim Prac** 2007;37:267; 11. Burkhard MJ, Bienzle D **Vet Clin North America Small Anim Prac** 2013;43:1331; 12. Comazzi S, Gelain ME **Vet J** 2011;188:149; 13. Papakonstantinou S, et al **Irish Vet J** 2013;66:6; 14. Sozmen M, et al J **Vet Diag Investig** 2005;17:323; 15. Weiss DJ **Vet Pathol** 2001;38:512; 16. Guzera M, et al **Vet Comp Oncol** 2014, doi 10.1111/vco.12098; 17. Comazzi S, et al **J Vet Intern Med** 2011;25:100; 18. Williams MJ, et al **J Vet Intern Med** 2008;22:596; 19. Avery PR, et al **J Vet Intern Med** 2014;28:538; 20. Rao S, et al **J Vet Intern Med** 2011;25:1097; 21. Aresu L, et al **Vet J** 2014;200:318; 22. Marconato L, et al **Vet J** 2013;197:776; 23. Regan RC, et al **Vet Comp Oncol** 2013;11:287; 24. Valli VE, et al **Vet Pathol** 2013;50:738; 25. Pinheiro D, et al **PLOS One** 2014;9:e105027; 26. Comazzi S, et al **Hematol Oncol** 2014;32:68; 27. Culmsee K, et al **J Vet Med** 2001;48:199; 28. Gelain ME, et al **Vet Immunol Immunopathol** 2008;121:179; 29. Wilkerson MJ, et al **Vet Immunol Immunopathol** 2005;106:179; 30. Clemo FA, et al **Vet Pathol** 1994;31:207; 31. Teske E, et al **Exp Hematol** 1993;21:579; 32. Ayl RD, et al **Vet Pathol** 1992;29:386; 33. Hellmen E, et al **Vet Pathol** 1993;30:20; 34. Hellmen E, Svensson S **J Comp Pathol** 1995;113:327; 35. Poggi A, et al **Vet Comp Oncol** 2013, doi 10.1111/vco.12078; 36. Quigley KA, et al **J Vet Diag Investig** 2001;13:297; 37. Morley P, et al **J Vet Intern Med** 2008;22:886; 38. Harkin KR, et al **J Am Vet Med Assoc** 2012;241:227; 39. Cortese L, et al **Vet Rec** 2009;164:778; 40. Dircks BH, et al **J Am Vet Med Assoc** 2009;235:960; 41. Kohn B, et al **J Fel Med Surg** 2006;8:254; 42. Moritz A, et al **Am J Vet Res** 2005;66:325; 43. Ridyard AE, et al **J Small Anim Prac** 2010;51:296; 44. Weiss DJ, Brazzell JL **J Vet Intern Med** 2006;20:682; 45. Maruyama H, et al **J Vet Med Sci** 2009;71:195; 46. Weiss DJ **J Vet Intern Med** 2007;21:440; 47. Weiss DJ, Henson M **Vet Clin Pathol** 2007;36:373; 48. Eickhoff S, et al **Vet Immunol Immunopathol** 2004;101:109; 49. Bauer TR, et al **Pediatric Res** 2004;55:363; 50. Felsburg PJ, et al **Vet Immunol Immunopathol** 1999;69:127; 51. Tompkins MB, Tompkins WA **Vet Immunol Immunopathol** 2008;123:45; 52. Chabanne L, et al **Autoimmunity** 1995;22:1; 53. Faldyna M, et al **Vet Immunol Immunopathol** 2001;82:23; 54. Heaton PR, et al **J Nutr** 2002;132:1655s; 55. Heaton PR, et al **J Nutr** 2002;132:1607s; 56. Watabe A, et al **Vet Immunol Immunopathol** 2011;142:189; 57. Acierno MM, et al **J Vet Intern Med** 2014;28:592; 58. Shadomy SV, et al **Clin Diag Lab Immunol** 2004;11:313; 59. de Almeida Leal GG, et al **Vet Parasitol** 2014;205:472; 60. Rosypal AC, et al **Vet Parasitol** 2005;131:45; 61. Muir P, et al **Vet Surg** 2011;40:753; 62. Maiolini A, et al **J Neuroinflamm** 2012;9:226; 63. Luckschander N, et al **Vet Res** 2009;40:58; 64. Haas E, et al **J Vet Intern Med** 2014;28:1708; 65. Rosner K, et al **Res Vet Sci** 2007;82:27; 66. Boudreaux MK, Lipscomb DL **Vet Pathol** 2001;38:249; 67. Estrela-Lima A, et al **BMC Cancer** 2010;10:256; 68. Nizanski W, et al **Reprod Domest Anim** 2012;47 Suppl 6:215; 69. Siemieniuch MJ, Woclawek-Potocka I **Reprod Domest Anim** 2008;43:633.

FLOW CYTOMETRY: APPLICATIONS IN SMALL ANIMAL MEDICINE

Todd M. Archer, DVM, MS, DACVIM (Small Animal Internal Medicine)
Mississippi State, MS

INTRODUCTION

Flow cytometry is a technology which allows measurement of one or several characteristics of single cells or particles as they flow through a beam of light in a fluid stream. The information obtained can be both qualitative and quantitative. Cells in solid tissue must first be separated or disaggregated and suspended in a fluid before analysis. Samples are then prepared for analysis, often by labeling the cells of interest with antibodies or DNA specific dyes which the flow cytometer can detect. Fluorescence labels will help realize the full potential of the flow cytometer. Current state of the art flow cytometers can analyze up to 18 or more parameters of individual cells. After entering the flow cytometer, the processed samples are passed within the middle of a fluid stream to the laser, with each cell emitting scattered and/or fluorescence light as it passes through the laser beam. Detectors then collect these signals. The detectors produce electronic signals based on the light scattering and fluorescent properties, producing information about the subpopulations within the sample. This technology is an integral component to veterinary research, particularly in the fields of hematology and immunology.

MECHANICS OF FLOW CYTOMETRY

There are three main systems on a flow cytometer. These include the fluidics system, optics and detection, and signal processing.[1] The fluidics system transports the particles of interest to the laser. It accomplishes this by transporting the sample in the center of a fluid stream, with only one cell moving through the laser at a given moment. Sheath fluid within the flow cytometer serves as the fluid stream. In a process called hydrodynamic focusing, the sample core remains in the center of the fluid stream due to laminar flow and by maintaining the sample pressure and sheath fluid pressure different from each other.[2,3] There is a setting on the flow cytometer to select the sample pressure. By increasing the sample pressure, the flow rate of the sample is increased because the width of the sample core is increased. This might allow some cells to pass though the laser off center and the data may be less resolved. A lower flow rate will decrease the width of the of the sample core and allow the majority of cells to pass through the center of the laser, allowing greater resolution.[1]

Light scattering occurs when light is deflected from cells. This occurs largely based on the physical properties of the cell such as the size and internal complexity of the cell. Forward-scattered light (FSC) is roughly, but not exactly, proportional the size of the cell, as it is determined when the light is passed in the forward direction.[1,4] Side-scattered light (SSC) is proportional to the internal complexity (or cell granularity). It is determined by detectors arranged at approximately 90° to the laser beam, picking up light which has passed into the cell and then been reflected from within the cell.[1,4] By analyzing the FSC and SSC simultaneously, differentiation of cell types can be achieved within certain types of samples, like leukocytes.

A fluorescent compound absorbs light energy, subsequently allowing electrons to be raised to a higher energy level and then decaying back to their ground state, causing photons of light to be emitted.[1] The range of emitted wavelengths is considered a particular fluorescent compound's emission spectrum. Lasers are used in flow cytometry to create a specific wavelength of light which allows excitation of more than one fluorochrome. Two commonly used fluorochromes are fluorescein and phycoerythrin (PE). When multiple fluorochromes are used to label different items of interest, the peak emission wavelength for each must be far enough apart that separate detectors can resolve each signal. Fluorochromes can be conjugated to antibodies such that the antibodies can label specific antigens or biomarkers on either the surface or the interior of cells, allowing differentiation of the targets of interest. This can allow for identification of different cell populations and count their relative percentages as well as label biomarkers or antigens and quantify their presence in the population of cells investigated.

The major components of the optical system include the excitation optics as well as the collection optics. The laser and focusing lenses make up the excitation optics. The alignment of these allow for the laser beam to consistently intercept the sample stream. The individual cells are interrogated by the laser, with scattered light as well as a fluorescent signal emitted. The collection optics consists of the detectors that collect the signals after the laser beam has passed through each cell. As each cell passes through the laser beam, signals for SSC, FSC, and fluorescence are emitted. The signals are routed to the appropriate detectors by a system of mirrors and optical filters. With each signal, a voltage pulse is generated and is converted to a digital number and displayed as a position on a data plot.

The flow cytometer is able to store the data for later retrieval and analysis. Cell populations can be displayed in differing formats. This can include single parameters being displayed as a single histogram or multiple parameters displayed simultaneously. Gating is often utilized to isolate a specific population of cells for analysis. Data collected from cells within a specific gate can then be displayed in subsequent plots. Some flow cytometers also allow for sorting to occur after the cells have passed through the laser beam. This allows for the capture of cells of interest for further analysis.

CONSIDERATIONS FOR ANALYZING SAMPLES

When proposing a flow cytometric assay or research experiment, the antibodies to be used as well as how the target cell population will be identified are important details to be considered. Antibodies must be determined based on the species studied and the target antigen. The ideal antibody is species specific, but due to the limited availability of antibodies in veterinary medicine, antibodies known to be cross-reactive to the target species are also often used.[4] The target antigen can be a surface antigen or an intracellular antigen. As a first step, the cells are often washed or incubated in a buffer. The cells are then divided into aliquots for staining of positive and negative controls as well as the samples of interest incubated with appropriate antibodies. Negative controls are often cells without antibody added, cells with isotype control antibodies added, or fluorescence minus one (FMO) controls.[5] These negative control sample without antibodies (or unstained control) serves to detect autofluorescence or background staining of the cells of interest. The isotype control sample is used to confirm that the primary antibody binding is specific and not a result of non-specific binding. The isotype control antibody is an antibody which has the same immunoglobulin isotype as the antibody of interest for the experiment, but which is not expected to react to the cell population of interest. The cells with only one antibody (one color) act as compensation controls for each fluorochrome which are used to allow proper setting of the flow cytometer. The cells which contain only secondary antibody are used to check for non-specific binding of secondary antibody to treated cells.

Antibodies can be applied for direct staining or indirect staining.[1] In direct staining, antibodies are directly conjugated to fluorochromes which target the cell marker of interest. In indirect staining, a 2-step staining process in used which includes the application of the direct antibody to the target cell marker and the subsequent application of a secondary fluorescent antibody which is detected by the flow cytometer. By incorporating multiple fluorochrome dyes that emit light at different wavelengths, multiple antigens can be identified on the cell population of interest. The intensity of emitted light is proportional to amount of bound antibody.[4]

APPLICATIONS IN SMALL ANIMAL MEDICINE

Flow cytometry can be used for a wide range of research and clinical purposes. These can include evaluation of ploidy and cell cycle analysis, immunophenotyping, functional cell assays, and determination of specific cell antigens under specified conditions. Many of these applications are often used under research conditions because of the need to process and analyze samples within hours of collection, reducing the clinical practicality of this methodology. However, many important research investigations have been completed in small animal medicine because of the utility of flow cytometry. Depending on the study of interest, flow cytometry can be used to evaluate erythrocytes, platelets, as well as leukocytes. A few examples may include the following categories.

Evaluation of the Immune System

Flow cytometry plays an important role in the identification and evaluation of cells of the immune system. As an example, canine peripheral blood leukocyte subsets can be stained, identified, and sorted by flow cytometry, providing total numbers as well as percentage of leukocyte populations in a sample. These can include CD4+ and CD8+ T cells, CD21+ B cells, natural killer cells, CD14+ monocytes, macrophages, dendritic cells, as well as CD4+CD25+FOXP3+ regulatory T-cells.[4,5]

Pharmacodynamic Evaluation in Dogs

Our research group has completed work surrounding investigations into pharmacodynamic monitoring and the use of cyclosporine in dogs. Pharmacodynamic assays look at the effects of drugs on the target cells instead of the disposition of the drug within the body. These assays are used in human medicine to help determine the adequacy of therapy for improving treatment outcomes in clinical transplant patients. We have used flow cytometry to develop assays to measure the production of different intracellular cytokines and the response of these cytokines after exposure to cyclosporine. In dogs, we have shown a decrease in production of the intracellular cytokines IL-2 and IFN-γ both *in-vitro* and *in-vivo*.[6,7]

Diagnosis of Hematological Malignancy

Fluorescent labeled antibodies can be used for immunophenotyping of lymphoma patients.[4,8,9] Once a diagnosis of lymphoma has been made via cytological or histopathological evaluation, samples aspirated from lymph nodes or masses can be stained with appropriate antibodies to identify the cell lineage. Using flow cytometry, the populations of cells can be identified as T-Cells or B-Cells in origin, allowing very useful diagnostic and prognostic information. In one study, lymph node aspirates from normal dogs were compared to lymph node aspirates from dogs with lymphoma, with all samples evaluated by flow cytometry.[10] The lymph node aspirates from normal dogs identified a mixture of T and B lymphocytes, with the proportion of CD3, CD4, CD8, and CD21 cells in the lymph node aspirates very similar to those found in the peripheral blood of normal dogs. The B-cell lymphoma patients had > 90% expression of CD21 and lacked expression of CD3, CD4, and CD8. The T cell lymphoma patients had a high expression of positive staining for both CD3 and CD4.

REFERENCES

[1]Wilkerson MJ **Vet Clin Small Anim** 2012; 42:53-71. [2]Tarrant JM **Vet Jour** 2005; 170:278-288. [3]Brown M, et.al **Clin Chem** 2000; 46:1221-1229. [4]Thomason J, et.al **J Vet Med Res** 1(1):1004. [5]Weiss DJ **Vet Clin Pathol** 2002; 31:72-82. [6]Archer TM, et.al. **J Vet Intern Med** 2011, 25:1391-1397. [7]Fellman CL, et.al. **Vet Immunol Immunopathol** 2011; 140:175-180. [8]Reggeti F, et.al **Vet Path** 2011; 48(1):223-235. [9]Burkhard MJ, et.al **Vet Clin Small Anim** 2013; 43:1331-1347. [10]Gibson D, et.al **J Vet Intern Med** 2004; 18:710-717.

PROTEIN-LOSING ENTEROPATHIES: HYPOALBUMINEMIA IS NOT ALWAYS BAD NEWS

Michael Willard, DVM, DACVIM (SAIM)
College Station, Texas

HYPOALBUMINEMIA

When concerned with low protein from any cause, one should measure serum albumin concentration, not the serum total protein concentration. Do not use human clinical pathology laboratories; their technology sometimes does not detect canine albumin with the result that they routinely report serum albumin concentrations of < 1.5 gm/dl in clinically normal dogs. If the patient has substantial hypoalbuminemia, the first step is to examine the skin for obvious lesions which can cause protein loss. Cutaneous lesions sufficient cause such hypoalbuminemia will be obvious; you should be able to just look at the patient and know if this is the problem. Next, hepatic function testing (e.g., resting and post-prandial serum bile acid concentrations) and a urinalysis should be requested. If there is any doubt about the amount of protein being lost in the urine, then a urine protein:creatinine ratio will quantify the magnitude of urinary protein loss. Severe hypoalbuminemia (i.e., < 2 gm/dl) in an animal with diarrhea suggests a protein-losing enteropathy (PLE); however, diarrhea (even when severe) in no way is sufficient to eliminate hepatic disease as the cause of the hypoalbuminemia. Furthermore, a very substantial number of dogs and cats with PLE do not have vomiting or diarrhea. Some only have ascites, and some are found fortuitously on routine blood work. This may be especially true of dogs with primary intestinal lymphangiectasia.

In general, once severe, exudative cutaneous disease, protein-losing nephropathy, and hepatic insufficiency are eliminated, then PLE becomes a diagnosis of exclusion in patients with a serum albumin < 2.0 gm/dl. Fecal concentrations of alpha-1 protease inhibitor can be used as a means of confirming PLE if there is confusion because of concurrent hepatic or renal disease. The major use for this test in clinical medicine seems to be the hypoalbuminemic patient in which you strongly suspect PLE (e.g., based upon it having severe diarrhea or having hypocholesterolemia), but which also has PLN and/or hepatic disease. However, there are several nuances to this test, especially in sample collection which makes it potentially difficult to interpret. We seldom need this test in clinical practice. Finally, it is important to note that PLE may be associated with a low, normal or increased serum globulin concentration.

Perhaps the most important point of this discussion is that, while hypoalbuminemia has repeatedly been reported to be a poor prognostic sign in patients with chronic GI disease, there are some patients that respond well to appropriate therapy.[1] Therefore, diagnosing PLE is not necessarily a cause for despair. That being said, because many dogs and cats with PLE have severe alimentary tract disease that needs to be diagnosed promptly to maximize the chance for successful therapy, aggressive diagnostics are typically appropriate. Although therapeutic trials can be chosen in place of classic diagnostic tests in many of the more common alimentary tract diseases (e.g., dietary-responsive diarrhea, antibiotic-responsive enteropathy, parasites), such an approach is generally ill-advised if the serum albumin concentration is less than 2.0 g/dl. This is true because it may be necessary to perform an antibiotic and/or dietary therapeutic trial for 3-6 weeks in order to ascertain if it is being effective, and a patient with severe PLE can become markedly worse in that time, especially if the serum albumin concentration is falling rapidly.

Any GI disease can cause PLE if it is severe enough. Many acute GI diseases cause PLE (e.g., parvoviral enteritis); however, these diseases are comparatively easier to treat than chronic GI diseases causing PLE. Therefore, the focus in this lecture is PLE in animals with chronic GI disease. The major causes of PLE in adult dogs tend to be intestinal lymphangiectasia, inflammatory bowel disease (IBD), alimentary tract lymphoma (LSA), and fungal infections (i.e., histoplasmosis and pythiosis). Other causes include alimentary tract ulceration/erosion, severe disease of intestinal crypts, antibiotic-responsive enteropathy, and parasites. The major causes of PLE in juvenile dogs tend to be parasites and chronic intussusception. Cats with PLE usually have IBD or alimentary tract lymphoma.

Many dogs with PLE have hypocholesterolemia. Pets with protein-losing nephropathies usually have hypercholesterolemia, while those with hepatic insufficiency often have hypocholesterolemia. Fecal examinations for parasites are appropriate. Although parasites are an uncommon cause of PLE in adult animals, pets in select environments (e.g., confined areas where patients can reinfect themselves) may incur substantial parasitic loads.

DIAGNOSTIC APPROACH TO THE PLE PATIENT

Once PLE has been diagnosed, intestinal biopsy is usually the ultimate means of establishing a diagnosis. Biopsy can be done via laparotomy, laparoscopy, or endoscopy. Feeding a small, fatty meal (use canned food, not dry, and add in cream or corn oil) the night before the procedure might make it easier to diagnose lymphangiectasia. Flexible endoscopy, when done by someone who is trained in how to take diagnostic tissue samples and submit them is usually more than adequate to obtain diagnostic samples. However, if endoscopy will be used to biopsy the small intestines, one should first ultrasound the abdomen to make sure that there are no focal infiltrates that are out of reach of the endoscope, or which might be more easily diagnosed by ultrasound-guided fine needle aspiration. Furthermore, there are ultrasonographic changes that can be found in the small intestine (e.g., streaks in the submucosa) that can be diagnostic. Radiographs and barium series are seldom as sensitive as ultrasound. If flexible endoscopy will be done, one should biopsy both the duodenum and ileum. There have been numerous cases in which lymphangiectasia, IBD or LSA were obvious in the ileum but not in the duodenum. It is not necessary to enter the ileum with the endoscope to obtain a good tissue sample of the ileal mucosa.

Laparotomy and laparoscopy are good means of obtaining diagnostic samples, but it is surprisingly easy to procure non-diagnostic samples with these techniques (i.e., "full-thickness sample" is not synonymous with "diagnostic sample"). Endoscopy does have the advantage of allowing one to visualize mucosal lesions that are "invisible" when looking at the serosa. In some cases, the diagnosis can only be obtained by biopsying these focal lesions. If full-thickness biopsies are obtained in severely hypoalbuminemic animals, then serosal patch grafting will minimize the risk of suture line leakage. A nonabsorbable or a poorly absorbable suture (PDS) should also be used.

INTESTINAL LYMPHANGIECTASIA

Intestinal lymphangiectasia seems particularly common in Yorkshire terriers[2] and Soft-Coated Wheaten terriers, but may occur in any breed. Sometimes these dogs have distinct ultrasonographic findings: "steaks" in the mucosa that represent dilated lymphatics.[3,4] While histopathology is obviously the desired means of diagnosis, one can sometimes make a definitive diagnosis based upon grossly visible endoscopic findings (i.e., numerous, erratic, grossly engorged lacteals seen as large white blebs on the mucosa).[1] These lesions are "fragile" and apparently may be destroyed by biopsying them (both endoscopically and surgically) if the endoscopist or surgeon is not careful. It is important to note that lymphangiectasia can be a relatively localized disease in the intestines, being present in only the ileum or only the jejunum or only the duodenum; therefore, it is important to biopsy as much of the intestinal tract as possible. Furthermore, if one biopsies the intestines and cannot find a cause of PLE, sometimes lymphangiectasia can be tentatively diagnosed by elimination (i.e., by eliminating IBD, lymphoma, parasites, intussusception, fungal infections, etc).

Diagnosis by means of endoscopic biopsy is certainly possible if the endoscopist is trained in taking high quality tissue samples. However, recent work has demonstrated that poor quality mucosal biopsies (e.g., primarily villus tips or substantial "squash" artifact) makes is much more difficult or even impossible to find the lesions. If one is taking high quality tissue samples (i.e., total length of the villi plus subvillus mucosa down to the border of the mucosa and muscularis mucosa), it typically takes about 6-7 tissue samples to have 90-99% confidence in finding lymphangiectasia. However, it can take 5-7 times as many tissue samples to have the same assurance if you are obtaining poor quality tissue samples that primarily consist of villus tips.[5]

When doing endoscopy, it is important that ileal biopsies be taken in addition to the typical duodenal biopsies. We are finding that ileal biopsies often reveal lesions not found on duodenal biopsies. This is true for lymphangiectasia as well as lymphoma and other lesions. With basic training, an endoscopist should be able to obtain ileal biopsies endoscopically at least 85% of the time. Typically, ileal biopsies are often of higher quality than duodenal biopsies.

Therapy for intestinal lymphangiectasia revolves around an ultra-low fat diet, preferably with anti-inflammatory therapy designed to alleviate the lipogranuloma formation that commonly occurs within the intestinal wall and/or mesentery. Supplementation with medium chain triglyceride oil (MCT) used to be recommended because MCT oil supposedly bypasses intestinal lymphatics thus preventing further rupturing of the lacteals. Pancreatic enzymes were often added to the diet to ensure digestion of the medium chain triglyceride oil. MCT oil is seldom used anymore, probably because appropriate dietary therapy is usually more than sufficient. Feeding homemade diets that are highly digestible and ultra-low in fat (e.g., white turkey meat plus potato or rice) is often very helpful in these patients. Commercial low fat diets can be used very successfully, but they need to have the lowest possible fat content. Such a diet can be so successful that it might occasionally be appropriate to use it as a therapeutic trial. Dogs with lymphangiectasia often show a marked increase in serum albumin concentration within 7-14 days of starting such a diet.

The importance of lipogranulomas in the intestinal wall and mesentery is uncertain.[6] However, it is hypothesized that some patients fail to respond to appropriate dietary therapy because of formation of very large or excessive numbers of lipogranulomas that so completely obstruct the intestinal lymphatics that even an ultra-low fat diet cannot prevent lacteal rupture. Therefore, once a diagnosis of lymphangiectasia is made (either by histology, grossly at endoscopy, or tentatively by response to an ultra-low fat diet), it seems to be appropriate to use anti-inflammatory therapy designed to prevent granuloma formation/enlargement. Prednisolone, azathioprine, and/or cyclosporine are commonly used for this purpose. I do not like prednisolone, simply because of all the side effects it has in these patients. Be aware that if you use cyclosporine, it is critical that you measure blood levels of the drug. Not only is there a major difference between patients in how much they absorb, but the bioavailability of the same product may change as the intestine heals.

If the serum albumin is very low (e.g., $\leq$ 1.3 gm/dl), one is often tempted to administer a plasma transfusion while waiting to see what effect the diet will have. However, it is exceedingly difficult to increase the serum albumin concentration by transfusing PLE patients with plasma because so much of the albumin is quickly lost out the gut. You would probably have to give at least two units of plasma to a 15 lb dog in order to raise the serum albumin from 1.0 gm/dl to 1.8 gm/dl, and sometimes you would have to give 3 or 4 units. If it is critical to raise the plasma oncotic pressure, then administering hetastarch may be preferred because it costs less than plasma and it stays in the intravascular compartment longer than albumin.

These patients may be a increased risk for hypomagnesemia which may potentiate the problem of hypocalcemia.[7] At this time, we do not know how important it is to supplement magnesium to patients, but severe hypomagnesemia can be resolved by a constant rate infusion of magnesium sulfate.

Anecdotally, octreotide might help some patients that do not respond to more conventional therapy.[8,9]

CRYPT LESIONS

Lesions of the intestinal crypts have been recognized as being associated with PLE in dogs. We have identified two different lesions of the small intestinal crypts that can cause PLE. One type is characterized by crypts (usually duodenal) that are filled and somewhat distended with proteinaceous fluid and necrotic inflammatory cells.[10] While such dilated crypts can be found in many animals, including clinically normal dogs, finding large numbers of them in multiple tissue samples seems to be consistently associated with PLE. We do not know if this is a cause-and-effect relationship, or if the dilated crypts are simply a marker for some other process but are not causing the protein loss themselves. Several of these patients have responded to therapy with elemental diets, total parenteral nutrition, prednisolone, azathioprine, and/or metronidazole. We have seen this lesion associated with IBD as well as lymphangiectasia (especially in Yorkshire terriers).

A second type of crypt lesion that appears to be less common than the first is characterized by focal accumulations of mucus causing massive distention of the intestinal crypts.[11] This has been reported once before, and we have seen two such cases. The most important aspect of diagnosis seems to be the fact that the lesion may be very focal, almost appearing as ulcers when looking at the intestinal mucosa through an endoscope. Therapy similar to that used on animals with the other form of crypt lesion may be helpful. We have used cyclosporine, but do not know if it is helpful, or if the clinical response is due to the other drugs that the patient is receiving.

These lesions have not been commonly reported. Recent work has shown that these lesions are extremely easy to miss if poor quality endoscopic biopsies are performed. While 7-12 high quality tissue samples (i.e., full length of villi plus subvillus mucosa down to the level of the muscularis mucosa) will find these lesions 90-99% of the time, about 7 times as many tissue samples will be needed if poor quality samples primarily consisting of villus tips are submitted.[5]

REFERENCES

[1] Peterson PB, et al. Protein-losing Enteropathies. **Vet Clin N Am** 2003;33:1061. [2]Kimmel SE, et al. Hypomagnesemia and hypocalcemia associated with protein-losing enteropathy in Yorkshire terriers: five cases (1992-1998). **J Am Vet Med Assoc** 2000;217:703. [3] Sutherland-Smith J, et al. Ultrasonographic intestinal hyperechoic mucosal striations in dogs are associated with lacteal dilation. **Vet Radiol Ultra** 2007;48:51. [4]Kull PA, et al. Clinical, clinicopathologic, radiographic, and ultrasonographic characteristics of intestinal lymphangiectasia in dogs: 17 cases (1996-1998). **J Am Vet Med Assoc** 2001;219:197. [5]Willard MD, et al. Effect of sample quality on the sensitivity of endoscopic biopsy for detecting gastric and duodenal lesions in dogs and cats. **J Vet Int Med** 2008;22:1084. [6]Romanucci M, et al. Xanthogranulomatous inflammation of the small bowel in a dog. **Vet Path** 2008;45:207. [7]Bush WW, et al. Secondary hypoparathyroidism attributed to hypomagnesemia in a dog with protein-losing enteropathy. **J Am Vet Med Assoc** 2001;219:1732. [8]Markham KM, et al. Octreotide in the treatment of thoracic duct injuries. **Amer Surg** 2000;66:1165. [9]Filik L. et al. A case of intestinal lymphangiectasia successfully treated with slow-release octreotide. **Dig Liver Dis** 2004;36:687. [10]Willard MD, et al. Intestinal crypt lesions associated with protein-losing enteropathy in the dog. **J Vet Int Med** 2000;14:298. [11]Willard MD, et al. Protein-losing enteropathy associated with cystic mucoid changes in the intestinal crypts of two dogs. **J Am Anim Hosp Assoc** 2003;39:187.

TRADITIONAL & NON-TRADITIONAL THERAPIES FOR IBD: EVIDENCE-BASED APPROACH

Albert E. Jergens, DVM, PhD, DACVIM (Small Animal Internal Medicine)
Ames, IA

INTRODUCTION

Idiopathic inflammatory bowel disease (IBD) is a common cause of chronic gastrointestinal (GI) disease in dogs and cats. Accumulating evidence in human IBD and experimental colitis in animal models suggests that the resident microbiota contribute to the pathogenesis of chronic immunologically-mediated intestinal inflammation. Dogs with IBD have distinctly different duodenal microbial communities compared to healthy dogs. Additionally, canine histiocytic ulcerative colitis now is recognized to be associated with adherent and invasive *E. coli* (AIEC), and affected dogs frequently respond to fluoroquinolone antimicrobial therapy. Gastrointestinal signs in cats with IBD have been linked to altered microbial composition with subsets of intestinal bacteria and histopathologic inflammation. Thus, there appears to be a sound rationale for evaluating IBD treatment protocols that target enteric bacteria and the harmful immune responses that they might evoke.

DESCRIPTION OF THERAPIES FOR CANINE AND FELINE IBD

Overview

Therapeutic intervention for IBD is aimed at counteracting inflammation and dysbiosis and combating nutritional (cobalamin) deficiencies. There is good clinical evidence that dietary constituents contribute to gut inflammation and that elimination diets are of value in reducing disease severity. Drug treatment recommendations are determined by disease severity and clinical course (e.g., presence of mild IBD versus more serious gut inflammation leading to PLE), the involved segments of the alimentary tract, clinician experience, size of the patient impacting drug costs, and the potential side-effects associated with use of some medications. Non-traditional modalities include treatment with n-3 fatty acids, prebiotics, and/or probiotics. Current guidelines for IBD therapy are largely empirical as published randomized controlled trials (RCT) are sparse. Most clinicians prefer a sequential step-up approach for lymphocytic-plasmacytic IBD (most common), using diet → antibiotics → steroids → various immunosuppressive drugs. The optimal diet or drug combination, as well as duration of therapy for induction and maintenance of remission of clinical signs, have not been determined and are based on therapeutic trials. There is little to no published information on pre/probiotic therapy for chronic enteropathy in dogs and cats.

Treatment/Management/Prevention

• ***Dietary modification*** is an important primary or secondary therapy in dogs and cats with IBD. In general, a highly digestible diet is fed to reduce antigenic stimulation and decrease intestinal inflammation. Ideally, use a diet that has low fat levels, moderate protein levels, and moderate to decreased carbohydrate levels. An elimination diet using a novel intact protein or hydrolyzed proteins (lower molecular weight) will reduce antigenic stimulation, and are particularly useful in animals with dietary sensitivity.

• ***Antibiotic treatment*** is premised on the belief that bacterial antigens initiate and drive host responses leading to chronic intestinal inflammation. Generally they are used for an initial 2 week course of therapy in conjunction with diet to induce clinical remission; they are also advocated by some clinicians as a useful single drug agent in dogs and cats having mild IBD.

- Metronidazole may work by multiple mechanisms including anti-bacterial, anti-protozoal, and immunomodulatory actions.
- Tylosin is more commonly administered for clinical signs of colitis in dogs.

Scientific data substantiating the anti-inflammatory effects of either antibiotic in dogs or cats with IBD are lacking.

• ***Immunosuppressive drugs*** - different clinical trials and case series describing IBD therapy attest to the efficacy of immunosuppressives in inducing short-term (weeks to months) clinical remission in dogs and cats.

FOR CANINE IBD

- One standard therapy with glucocorticoids uses **prednisone** at a starting dose of 1-2 mg/kg PO daily for 2 weeks as induction therapy in small and medium sized dogs. Dosage reductions at 25% q 2 weeks are then recommended if the desired clinical response is noted – 50% reduction in baseline CIBDAI and/or maintenance of serum albumin > 2g/dL.
- Dogs > than 25 kg body weight are best treated with prednisone on the basis of body surface area (BSA) expressed as M^2. Higher steroid doses in these dogs will often cause adverse effects including PU/PD, polyphagia, and excessive panting. Guidelines for prednisone therapy using BSA are empirical, and published guidelines are unavailable.
- Use **budesonide** (1-3 mg/day) in diabetic dogs or those with a history of adverse reactions to prednisone/prednisolone. Consider budesonide as a first-choice glucocorticoid in large breed dogs (> 25 kg) to avoid adverse effects which are commonly observed when prednisone is used at traditional immunosuppressive (1-2 mg/kg/day) doses. Budesonide is also useful as a <u>maintenance agent</u> in dogs requiring long-term immunosuppression due to severe intestinal disease or recurring clinical signs (flares). In my experience, there is a 10-14 day "lag time" before clinical response to budesonide is realized.
- **Azathioprine** may be used as a maintenance agent in canine IBD – its use in human IBD is well recognized; however, there are surprisingly few studies in the veterinary literature attesting to its efficacy in the treatment of canine IBD. **Chlorambucil** is an alternative maintenance agent to be considered and is the adjunct treatment of choice along with glucocorticoids in canine PLE.
- **Cyclosporine** may be used as a first line immunosuppressive in dogs with severe IBD; however, drug costs may limit its use in mid-sized and larger dogs. It is also reported to be efficacious in the therapy of steroid-refractory IBD based on a single published study.

FOR FELINE IBD

- One standard therapy with glucocorticoids uses **prednisolone** at a starting dose of 1-2 mg/kg PO daily for 2 weeks as induction therapy. Dosage reductions at 25% q 2 weeks are then recommended if the desired clinical response is noted.
- Glucocorticoids should be used with caution in cats to avoid aggravation of cardiac disease and promotion of diabetes mellitus.
- Chlorambucil may be used with steroids in cats having severe IBD or when biopsy results cannot conclusively discriminate between IBD versus alimentary LSA.

• ***Cobalamin*** is supplemented in dogs and cats, as needed, with severe enteropathy. Failure to diagnose and treat hypocobalaminemia may delay clinical recovery in spite of appropriate dietary and pharmacologic therapies.

• ***Probiotics and prebiotics*** may benefit dogs and cats with gastrointestinal disease, including IBD.

- Probiotics are live bacterial cultures (e.g., *Lactobacilli, Bifidobacteria* spp) that confer health benefits to the host. Multiple mechanisms of action have been suggested to explain the protective effects of probiotics in intestinal inflammation, including (1) inhibition of pathogenic enteric bacteria, (2) improved epithelial/mucosal barrier function, and (3) altered immunoregulation via down-regulated pro-inflammatory cytokine secretion. These products must be administered continuously for their effects to be realized.
- Prebiotics are complex carbohydrates, such as fructo-oligosaccharides (FOS), that promote the growth of beneficial bacterial species. Prebiotics promote the establishment of beneficial bacteria (probiotics). Probiotic + prebiotic = synbiotic. Pro/prebiotic trials for treatment of canine or feline IBD are sparse.

PREVIOUS PUBLISHED DATA ON MEDICAL THERAPY FOR IBD

Drug Therapy for IBD

Study	Trial Design	Drug Therapy	Result
Canine IBD			
Jergens (1992)	RS	Pred, MTZ, 5-ASA	> 80% response
Jergens (2003)	PT	Pred, MTZ, 5-ASA	> 80% response
Craven (2004)	RS	Pred, MTZ, 5-ASA, tylosin	Confounding variables; ↓ albumin = negative outcome
Westermarck (2005)	PT	Tylosin	↓↓ diarrhea in all dogs
Munster (2006)	PT	Pred, MTZ, 5-ASA, AZA	↓↓ diarrhea with immunosuppressive drugs
Allenspach (2006)	PT	Pred	> 75% response
Allenspach (2006)	PT	Cyclosporine	> 80% response in steroid refractory IBD
Luckschander (2006)	PT	Pred	> 50% response
Garcia-Sancho (2007)	PT	Pred + MTZ	> 75% response
Allenspach (2007)	PT	Pred, cyclosporine	> 60% response
Schreiner (2008)	PT	Pred	↓ albumin = negative outcome
Mansfield (2009)	PT	Enrofloxacin	> 85% response in HUC
Jergens (2009)	RCT	Pred vs Pred+MTZ	> 80% response in both groups
Feline IBD			
Jergens (1992)	RS	Pred, MTZ	> 80% response
Dennis (1994)	RS	Pred, tylosin, 5-ASA	70% response
Dennis (1994)	RS	Pred	> 50% response
Hart (1994)	RS	Pred, MTZ	> 75% response
Jergens (2010)	PT	Pred	100% response

CONTEMPORARY DATA

Dye (2013)	RCT	budesonide	80% remission
Volkmann (2013)	RCT	AST-120	↓ CIBDAI
Rossi (2014)	RCT	VSL#3/pred	↓ CIBDAI
Jergens (2014)	RCT	VSL#3/pred	↓ CIBDAI

Nutritional Therapy for IBD

Study	Species (No.)	Diet	Primary/ Adjunct Role	Response
Nelson (1988)	Dog (13)	Controlled	Primary	Signs resolved
Dennis (1993)	Cat (28)	Controlled	Adjunct	50% respond
Guilford (1991)	Cat (60)	Controlled	Adjunct	80% respond
Hart (1994)	Cat (55)	Elimination	Primary	50% respond
Jergens (2003)	Dog (58)	Elimination	Adjunct	80% respond
Allenspach (2007)	Dogs (70)	Elimination	Adjunct	60% respond
Jergens (2009)	Dogs (54)	Elimination	Adjunct	80% respond
Jergens (2010)	Cat (17)	Elimination	Adjunct	100% respond

CONTEMPORARY DATA

Mandigers (2010)	hydrolyzed protein	LT remission
Okanishi (2014)	ultralowfat diet	↓ CIBDAI
Rossi (2014)	elimination diet	↓ CIBDAI
Jergens (2014)	elimination diet	↓ CIBDAI

• ***Omega-3 fatty acids*** may produce eicosanoids that are less inflammatory than those produced from arachidonic acid. Taken together, almost all studies in animal models and humans suggest some beneficial effects of n-3 PUFAs in IBD but the mechanism remains controversial. Trials demonstrating the efficacy of n-3 PUFAs in canine or feline IBD have not been published.

SUGGESTED READINGS

Craven M, Simpson JW, Ridyard AE, et al. Canine inflammatory bowel disease: retrospective analysis of diagnosis and outcome in 80 cases (1995-2002). J Small Anim Pract 2004; 45:336-342.

Jergens AE, Schreiner CA, Frank DE, *et al.* A scoring index for disease activity in canine inflammatory bowel disease. J Vet Intern Med 2003; 17:291-297.

Allenspach K, Wieland B, Grone A, et al. Chronic enteropathies in dogs: evaluation of risk factors for negative outcome. J Vet Intern Med 2007; 21:700-708.

Allenspach K, Rufenacht S, Sauter S, et al. Pharmacokinetics and clinical efficacy of cyclosporine treatment of dogs with steroid-refractory inflammatory bowel disease. J Vet Intern Med 2006; 20:239-244.

Dye TL, Diehl KJ, Wheeler SL, et al. Randomized, controlled trial of budesonide and prednisone for the treatment of idiopathic inflammatory bowel disease in dogs. J Vet Intern Med. 2013; 27:1385-91.

Simpson K, Jergens A. Pitfalls and progress in the diagnosis and management of canine inflammatory bowel disease. The Veterinary clinics of North America Small animal practice 2011;41:381-398.

Trepanier L. Idiopathic inflammatory bowel disease in cats. Rational treatment selection. Journal of feline medicine and surgery 2009;11:32-38.

Shell L, Rothrock K. Inflammatory bowel disease – Associate database – VIN; http:/www.vin.com/Members/Associate/Associate.plx?

FOOD RESPONSIVE CHRONIC DIARRHEA
THE VIEWS OF THE INTERNIST AND THE NUTRITIONIST

Vincent Biourge, DVM, PhD, DACVN
Aimargues, France
Olivier Dossin, DVM, PhD, DECVIM-CA (Internal Medicine)
Toulouse, France

Food responsive diarrhea (FRD) is the most common form of idiopathic chronic enteropathy (ICE) in dogs [1-3]. The definition encompasses all forms of chronic diarrhea associated with intestinal inflammation without any objective cause that respond clinically to a dietary change. The other types of ICE are antibiotic responsive diarrhea (ARD) and steroid/immunosuppressive drug responsive diarrhea (SRD). This presentation will be limited to canine FRD.

WHAT CAUSES FRD?

The exact mechanism of FRD is unknown but the condition has been associated with intestinal immune-mediated inflammation with possibly genetic and environmental factors such as the diet or the intestinal microbiome. It is very likely that a spectrum of adverse food reactions, ranging from intolerance to true allergy, is involved as the main component. However, intestinal dysbiosis may also be a key factor with disturbances of the gut microbiote as consequences of the FRD or as the initiating event through associated mucosal damages exposing immune sensitive cells of the mucosa to alimentary-derived antigens.

On the inflammatory side, nuclear factor kappa B expression is upregulated in the duodenum of dogs with FRD and SRD, [4] and the inflammatory cytokine profile is similar in both conditions [5]. Regarding the immune cell populations of the duodenal mucosa, dogs with FRD express IgA plasma cells and CD4+ cells while dogs with SRD have an increase in IgM plasma cells and a balanced CD4 and CD3+ profile [6]. Dogs with FRD and SRD have an increase in serum IgE and IgG against a wide spectrum of food antigens but there is a lot of overlap between the two groups [7]. Among the functional changes occurring in ICE, the gene expression of trefoil factors (essential proteins involved in epithelial reconstruction) is upregulated in the duodenum and downregulated in the colon. Moreover, TFF 3 is more upregulated in the duodenum of dogs with FRD when compared to SRD dogs, suggesting that epithelial repair might be more active in FRD [8].

The role of intestinal bacteria in FRD has been recently highlighted by studies showing the Pattern Recognition Receptors of the TLR family have different expression in the mucosa of affected dogs when compared to healthy controls [9-11]. Moreover, in German Shepherd dogs with ICE, including dogs with FRD, there is a polymorphism of the gene encoding for TLR 4 and 5 [12]. These findings suggest that dogs with FRD and other form of ICE might have specific inflammatory/immune-mediated reaction triggered by commensal or occasional bacteria of the intestinal microbiome.

Dysbiosis is also a probable component of FRD. Several changes of microbial communities have been reported in canine ICE, including FRD (review in [13]). These changes are variable from one study to another, but the most common ones are an increase in bacteria of the Proteobacteria phyla and a decrease in the phyla Firmicutes and Bacteroidetes. The mucosal changes associated with FRD range from mild to severe but are most of the time in a mild to moderate range. Some dogs have increased eosinophils in the lamina propria, and eosinophils density decrease after clinical improvement with dietary manipulation as well as the density of mononuclear cells [14]. Some ultrastructural changes have recently been reported in FRD such as increased microvillous spacing, decreased microvillous height and mitochondrial lesions. Some of these changes, but not all, normalize after treatment and clinical improvement [14].

A recent study has suggested that inflammation is more severe in dogs with SBD when compared to dogs with FRD [15]. It is possibly an explanation why the first ones do require steroids while the others do respond to food trial only. This suggests that FRD and SRD, and also possibly some forms of ARD, are the same spectrum of a disease condition with different severity of immune/inflammatory reaction due possibly to different host genetic patterns but also different changes in the digestive ecosystem.

What is the prevalence of true food allergy among FRD cases? In a study including 70 cases, 80% of the dogs could be fed their original diet after 14 weeks of hypoallergenic diet without any recurrence of clinical signs over the study period. Only 20% of the cases relapsed after switching to their original diet, and 2 dogs (5%) were considered as true food allergy cases after test protein challenge[1]. However, in another study with a different diet the proportion of dogs with FRD that relapsed after being replaced on the initial diet was 70%[3].

CLINICAL PICTURES OF FOOD RESPONSIVE DIARRHEA

Dogs with FRD are usually younger than dogs with SRD but there is lot of age overlap [2, 10]. Labrador retrievers and boxers might be overrepresented among FRD dogs [14]. Classical clinical signs include diarrhea (small large or mixed bowel), vomiting, and inappetence. Affected dogs can have hematemesis or melena, bouts of abdominal pain, borborygmus and flatulence. Poor hair coat condition is sometimes observed, and some dogs develop pruritic dermatitis and also anal pruritus. They usually have long lasting clinical signs with a history of various treatments often resulting in minor improvement with relapse or failure. Clinicopathological changes are usually non-specific and may include mildly decrease albumin and calcium concentration [10, 14, 16]. Some dogs may have eosinophilia on CBC.

Do not rule out FRD in dogs that have had an incomplete food trial. The dietary history of a dog with chronic diarrhea should be very thorough, and the aim of the dietary changes to rule out FRD is to provide a protein source that the dog has never eaten in his life. This can be achieved with hydrolyzed protein but also with original sources of fish or meat protein such as pork, rabbit or kangaroo in an homemade or commercial diet.

DIAGNOSTIC WORK UP OF FRD

The diagnostic work up of such cases aims at ruling out extra-intestinal diseases such as liver, renal diseases, hypoadrenocorticism (atypical) and pancreatic diseases. Then, infectious causes of chronic diarrhea must be ruled out with fecal examination or systematic deworming with a broad-spectrum drug such as fenbendazole, with or without an anti-coccidian medication. Serum cobalamin is sometimes profoundly decreased in dogs with FRD. It is very important to measure serum cobalamin and to start parenteral cobalamin in selected cases because low cobalamin might be a reason for treatment failure in FRD. In severe cases, C reactive protein may help to follow response to treatment in FRD.

Ultrasonography is an important step in the work up of a dog with ICE. Its main objective is to look for changes that would point in the direction of another disease condition such as neoplasia or granulomatous diseases (eg fungal disease). If there are more pronounced ultrasonographic changes in an area that could not be reached by endoscopy it might also be an indication to perform full thickness laparoscopic/celiotomy biopsies. There are no specific changes that point in the direction of FRD but the small intestinal wall echostructure is more likely to be unremarkable in FRD than in SRD [17]. However, jejunal lymph nodes might be enlarged in both conditions [17].

Do we need to scope dogs with FRD? The decision to biopsy is a case by case one and depend on the clinical picture and the likelihood that the dog has FRD vs other intestinal conditions such as granulomatous inflammation or neoplasia. Except in a clinical research setting, a case with a high probability of FRD does not always require endoscopy and intestinal biopsy before engaging in dietary trial. However in case of dietary failure, endoscopy and biopsy are necessary.

Finally before embarking on dietary treatment clinical cases of FRD must be scored according to the CIBDAI or CCECAI clinical score because they have proven very useful for objective assessment of response to treatment [1, 18].

WHAT TYPE OF DIET SHOULD BE USED?

The diet is a key factor in the management of patients with FRD and, as already mentioned above, a thorough dietary history is necessary to choose the best diet for the patient. Novel protein and hydrolyzed protein based diets have been used with success in FRD. An issue with novel protein diets is that their availability might be limited. This is not the case for hydrolysate-based diet.

Patients suffering from chronic diarrhea are commonly in poor body condition. Except if contraindicated, they should thus be fed a diet with high level of dietary fat (> 16% fat in a dry diet). Dividing the daily intake in 3 to 4 meals, will optimize the digestion of the nutrients. It is commonly recommended to feed a diet enriched in fiber in patients showing mostly signs of large bowel diarrhea. The fibers should be a mix of soluble and insoluble fiber (i.e.: Cellulose, beet pulp). Of course the diet must be complete and balanced, but this is the case for all the diets.

CLINICAL RESPONSE AND FOLLOW UP

Clinical response in FRD dogs usually occurs within the first 2 weeks of treatment, especially concerning GI signs such as vomiting and diarrhea[14]. Dogs with FRD can sometimes be fed with their original diet after a few months, but some of them must be maintained on the therapeutic diet for a long time. In case of failure, probiotics can be tried because they have positive effects in some patients with FRD. The next steps in case of failure are antibiotic trial to exclude ARD and immunosuppressive drugs in case of SRD, provided that other intestinal conditions such as neoplasia have been ruled out. Overall, the outcome of FRD is significantly better than for other forms of ICE at short (weeks) and long term (months) [2-3].

REFERENCES

1.Allenspach, K., et al., *Chronic enteropathies in dogs: evaluation of risk factors for negative outcome.* J Vet Intern Med, 2007. **21**(4): p. 700-8.
2.Culverwell, C., D.L. Chan, and K. Allenspach, *Short term and long term outcome in dogs diagnosed with chronic enteropathies.* J Vet Intern Med, 2014. **28**(3): p. 1038.
3.Mandigers, P.J., et al., *A randomized, open-label, positively-controlled field trial of a hydrolyzed protein diet in dogs with chronic small bowel enteropathy.* J Vet Intern Med, 2010. **24**(6): p. 1350-7.
4.Luckschander, N., et al., *Activation of nuclear factor-kappaB in dogs with chronic enteropathies.* Vet Immunol Immunopathol, 2010. **133**(2-4): p. 228-36.
5.Allenspach, K., *Clinical immunology and immunopathology of the canine and feline intestine.* Vet Clin North Am Small Anim Pract, 2011. **41**(2): p. 345-60.
6.German, A.J., E.J. Hall, and M.J. Day, *Immune cell populations within the duodenal mucosa of dogs with enteropathies.* J Vet Intern Med, 2001. **15**(1): p. 14-25.
7.Foster, A.P., et al., *Serum IgE and IgG responses to food antigens in normal and atopic dogs, and dogs with gastrointestinal disease.* Vet Immunol Immunopathol, 2003. **92**(3-4): p. 113-24.
8.Schmitz, S., et al., *Expression of trefoil factor genes in the duodenum and colon of dogs with inflammatory bowel disease and healthy dogs.* Vet Immunol Immunopathol, 2013. **151**(1-2): p. 168-72.
9.Allenspach, K., et al., *Evaluation of mucosal bacteria and histopathology, clinical disease activity and expression of Toll-like receptors in German shepherd dogs with chronic enteropathies.* Vet Microbiol, 2010. **146**(3-4): p. 326-35.
10.Burgener, I.A., et al., *Upregulation of toll-like receptors in chronic enteropathies in dogs.* J Vet Intern Med, 2008. **22**(3): p. 553-60.
11.McMahon, L.A., et al., *Expression of Toll-like receptor 2 in duodenal biopsies from dogs with inflammatory bowel disease is associated with severity of disease.* Vet Immunol Immunopathol, 2010. **135**(1-2): p. 158-63.
12.Kathrani, A., et al., *Polymorphisms in the TLR4 and TLR5 gene are significantly associated with inflammatory bowel disease in German shepherd dogs.* PLoS One, 2010. **5**(12): p. e15740.
13.Honneffer, J.B., Y. Minamoto, and J.S. Suchodolski, *Microbiota alterations in acute and chronic gastrointestinal inflammation of cats and dogs.* World J Gastroenterol, 2014. **20**(44): p. 16489-97.
14.Walker, D., et al., *A comprehensive pathological survey of duodenal biopsies from dogs with diet-responsive chronic enteropathy.* J Vet Intern Med, 2013. **27**(4): p. 862-74.
15.Im Hof, M., et al., *Urinary leukotriene E4 concentrations as a potential marker of inflammation in dogs with inflammatory bowel disease.* J Vet Intern Med, 2012. **26**(2): p. 269-74.
16.Equilino, M., et al., *Evaluation of serum biochemical marker concentrations and survival time in dogs with protein-losing enteropathy.* J Am Vet Med Assoc, 2015. **246**(1): p. 91-9.

17.Gaschen, L., et al., *Comparison of ultrasonographic findings with clinical activity index (CIBDAI) and diagnosis in dogs with chronic enteropathies.* Vet Radiol Ultrasound, 2008. **49**(1): p. 56-64.
18.Jergens, A.E., et al., *A scoring index for disease activity in canine inflammatory bowel disease.* J Vet Intern Med, 2003. **17**(3): p. 291-7.

REASSESSMENT OF "NORMAL VALUES" IN DOGS AND CATS WITH CHRONIC KIDNEY DISEASE

Gregory F. Grauer, DVM, MS, DACVIM (Small Animal Internal Medicine)
Manhattan, Kansas

Chronic kidney disease (CKD) is a major cause of morbidity and mortality in dogs and cats. The prevalence of CKD has been estimated to be 0.5-1.0% in dogs and 1-3% in cats,[1,2] but it increases with age, especially in cats. It has been estimated that as many as 30-50% of cats 15 years of age or older have CKD.[3-5] Nephron damage associated with CKD is usually irreversible and often progressive. In dogs the progressive loss of renal function tends to be common, linear, and relatively rapid compared with cats. Cats may have stable renal for months to years and be relatively unaffected by the CKD or they may have slowly progressive disease over several years. In some cases, cats will be stable for months to years but then experience an abrupt, unpredictable decline in renal function. Soft tissue mineralization, systemic hypertension, intraglomerular hypertension, and proteinuria have been associated with progression of CKD. It's usually not possible to improve renal dysfunction caused by CKD. Although not proven by prospective, controlled clinical trials, it's logical to assume that early diagnosis of CKD can improve clinical outcomes for dogs and cats. There is firm evidence for dietary treatment and increasing evidence that anti-proteinuric treatments can slow the progressive nature of CKD.

EARLY DIAGNOSIS OF CKD: SERUM CREATININE CONCENTRATION

Early, non-azotemic CKD (IRIS Stage 1) can be diagnosed in dogs and cats with abnormal renal palpation or renal imaging findings, persistent renal proteinuria, or urine concentrating deficits due to renal disease. In most cases however, CKD is diagnosed on the basis of persistent azotemia superimposed on an inability to form hypersthenuric urine (although some cats with CKD retain the ability to concentrate urine). Serum creatinine concentrations are commonly used as a marker of glomerular filtration rate (GFR) in dogs and cats. Creatinine is produced from the non-enzymatic degradation of creatine and creatine phosphate in skeletal muscle and therefore serum creatinine concentrations reflect the patient's muscle mass as well as GFR. Interpretation of serum creatinine concentrations can also be influenced by the method of analysis (Jaffe's reaction vs. enzymatic and bench-top vs. reference laboratory). One of the most disconcerting aspects of interpretation of serum creatinine is the relatively large variation in reference intervals between laboratories which can lead to false-positive and false-negative azotemia.[6,7] Reference ranges need to be individualized to each laboratory but many veterinary nephrologists have suggested that renal azotemia may begin with serum creatinine concentrations lower than most reference ranges (i.e., 1.4 and 1.6 mg/dl in dogs and cats, respectively).[8]

Serum Creatinine Concentrations and IRIS CKD Stages for Dogs and Cats

Serum Creatinine Concentrations (mg/dl)	**Stage 1** (Non-azotemic CKD)	**Stage 2** (Mild renal azotemia)	**Stage 3** (Moderate renal azotemia)	**Stage 4** (Severe renal azotemia)
Cats	<1.6	1.6-2.8	2.9-5.0	>5.0
Dogs	<1.4	1.4-2.0	2.1-5.0	>5.0

Serum creatinine concentrations must always be interpreted in light of the patient's muscle mass, urine specific gravity, and physical examination findings in order to rule out pre- and post-renal causes of azotemia. In dogs there can be a large variation in muscle mass (e.g., miniature poodle vs. greyhound) that will tend to increase the breadth of reference ranges.[9] Despite these potential confounding issues, longitudinal assessment of serum creatinine concentrations (analyzed by consistent methodology), remains an excellent tool to assess renal function and diagnose early CKD. For example, a serum creatinine concentration that increases from 0.7 to 1.4 mg/dl over several years, without evidence of dehydration or an increase in muscle mass, could indicate at least a 50% reduction in GFR. The reduction in renal function may be > 50% due to compensatory hypertrophy of remaining nephrons increasing renal functional capacity.

Serum symmetrical dimethylarginine (SDMA) is derived from intranuclear methylation of L-arginine by protein-arginine methyltransferases and released into the circulation after proteolysis. SDMA is eliminated primarily by renal clearance and represents a potential biomarker for diagnosing and monitoring CKD.[10,11] In two recent longitudinal studies, one in dogs and one in cats that developed CKD, SDMA concentrations increased above normal approximately 17 months prior to serum creatinine concentration increasing above reference range (> 1.8 mg/dl in dogs and > 2.1 mg/dl in cats).[10,12] Interestingly, if a serum creatinine concentration of > 1.6 mg/dl had been considered abnormal in the feline study,[10] both serum creatinine and SDMA would have identified renal azotemia in these cats at nearly the same time. It should also be noted that the 21 healthy geriatric control cats in this study had serum creatinine concentrations between 0.7 and 1.6 mg/dl with a mean of 1.2 mg/dl.[10]

EARLY DIAGNOSIS OF CKD: URINE PROTEIN/CREATININE RATIO

Proteinuria in dogs and cats with chronic kidney disease (CKD) can occur due to glomerular and/or tubular lesions. Glomerular proteinuria can be caused by loss of integrity of or damage to the capillary wall (e.g., immune complex disease and x-linked hereditary nephropathy). It is also likely that increases in glomerular capillary pressure increases the amount of filtered plasma protein. Intraglomerular hypertension may result from loss of nephrons (loss of autoregulation) and from systemic hypertension being transmitted into glomerular capillaries. Structural glomerular disease and CKD are often accompanied by systemic hypertension which can exacerbate intraglomerular hypertension and glomerular proteinuria. Tubular proteinuria occurs when tubular reabsorption of protein from the glomerular filtrate is compromised. Tubular proteinuria is typically of lesser magnitude compared with glomerular proteinuria. Reduced tubular reabsorption of protein in dogs and cats with CKD can occur with chronic tubulointerstial injury. Whether caused by capillary wall lesions, tubular lesions, or intraglomerular hypertension, excessive quantities of protein in the glomerular filtrate may contribute to additional glomerular and tubulointerstitial lesions leading to loss of additional nephrons. Proteinuric renal disease and systemic hypertension are often co-existent and therefore, it is difficult at times to separate the effects of high systemic and intraglomerular pressures and proteinuria.

Diagnosis of renal proteinuria in cats and dogs with CKD should be accomplished in a step-wise fashion.[13] In health and in disease, albumin is the major urine protein in dogs and cats. The specificity of the dipstick screening test for albuminuria is poor (especially in cats) and therefore confirmation of traditional dipstick positive proteinuria should be confirmed with a more specific follow-up test such as the sulfosalicylic acid turbidimetric test, UPC, or species specific albuminuria assays.[14] The second step in assessment of proteinuria is to determine its origin (physiologic or benign proteinuria and pre- and post-renal proteinuria need to be ruled out). Renal proteinuria is persistent and associated with a benign or inactive urine sediment (hyaline casts may be observed in the urine sediment in cases of renal proteinuria). Persistent proteinuria is defined as at least two positive tests at two week intervals. Subsequently, via serial monitoring of the UPC, it should be determined if the proteinuria is stable, increasing, or decreasing over time.

IRIS Classification of Renal Proteinuria in Dogs and Cats

Urine Protein/Creatinine Ratio (UPC)	Classification
< 0.2	Non-Proteinuric
0.2-0.4 (Cats); 0.2-0.5 (Dogs)	Borderline Proteinuric
> 0.4 (Cats); > 0.5 (Dogs)	Proteinuric

Current recommendations suggest that persistent proteinuria of renal origin of a magnitude ≥ UP/C of 0.4 in cats and ≥ 0.5 in dogs with azotemic CKD should be treated with a renal diet and an angiotensin converting-enzyme inhibitor (ACEI).[13] Borderline proteinuria is defined as a UP/C between 0.2 and 0.5 in dogs and 0.2 and 0.4 in cats and warrants increased monitoring. It's interesting to note however that borderline and even "normal" levels of proteinuria in cats have been associated with poor outcomes.

In cats with naturally occurring CKD, relatively mild proteinuria (UPC 0.2-0.4) increased the risk for death or euthanasia 2.9 fold compared with cats with UPCs < 0.2.[15] In a prospective, longitudinal cohort study of non-azotemic cats ≥ 9 years, 95 cats (median age = 13) were followed for 12 months or until death or azotemia developed. Azotemia was defined as a serum creatinine concentration > 2.0 mg/dl and 29/95 (30.5%) cats developed azotemia. Proteinuria at presentation (median UPC of 0.19 vs. 0.14) was significantly associated with development of azotemia in these geriatric cats.[16] Finally, when client-owned cats with stable CKD (n=112) were compared with client-owned cats with progressive CKD (n=101), median UPCs in the progressive group were higher when compared with the stable group (0.27 vs. 0.14).[17] A 0.1 increase in UPC was associated with a 24% increase in risk of progression of CKD.[17]

MANAGEMENT OF CKD: SYSTOLIC BLOOD PRESSURE

Current recommendations state that blood pressure should be measured in a quiet area prior to examining the patient, typically in the presence of the owner and after a 5-10 minute period of acclimation. The ACVIM Panel on Hypertension suggests discarding the first measurement, then obtaining a minimum of 3, preferably 5-7, consecutive measurements with less than 10-20% variability in systolic blood pressure.[18] The animal's disposition, body position, and heart rate, the cuff size and measurement site as well as all measured values should be recorded in the medical record. Many clinicians suggest that hypertension be documented on more than one occasion before accepting the diagnosis (unless ocular lesions compatible with systemic hypertension already exist).

IRIS Classification of Systolic Blood Pressure in Dogs and Cats

Systolic Blood Pressure (mm Hg)	Risk of Target Organ Damage	Arterial Pressure (AP) Category
<150	Minimal	AP0
150-159	Low	AP1
160-179	Moderate	AP2
≥180	High	AP3

IRIS blood pressure sub-staging for dogs and cats with CKD is based on risk of target organ (ocular, neurologic, cardiac, and renal) damage. Not long ago indirect systolic blood pressure measurements > 170-180 mmHg were considered the threshold for hypertension. Despite the inherent difficulties with indirect blood pressure measurement in dogs and cats, it may be appropriate to consider systolic hypertension to be present at lower pressures (e.g., ≥ 160 mm Hg).

In a recent study, 45 dogs with naturally occurring CKD were divided into three groups based on initial systolic blood pressure were followed for up to 2 years. The blood pressure groups were defined as: High (161-201 mmHg), n = 14; Intermediate (144-160 mmHg), n = 15, and Low (107-143 mmHg), n = 16. The initial high systolic blood pressure group had increased risk of uremic crisis and death when compared with the low pressure group (median survival of < 200 days vs. > 400 days).[19] In cats with a remnant kidney-wrap model of CKD, systolic hypertension (mean pressure of 168 vs. 113 mmHg) was associated with reduced GFR (1.34 vs. 3.55 ml/min/kg), increased UPC (1.2 vs. 0.1), and increased glomerulosclerosis.[20] Similarly, in dogs with the remnant kidney-wrap model of CKD, systolic hypertension (> 160 mm Hg) was associated with reduced GFR, increased UP/C ratios, and increased mesangial matrix accumulation, tubular lesions, fibrosis, and cellular infiltrates.[21] Cats with progressive CKD had higher systolic blood pressures than did cats with stable CKD (155 vs. 147 mm Hg).[17] Finally, in 69 cats with naturally occurring CKD, high time-averaged systolic blood pressure (159 vs. 136 mm Hg), was correlated with glomerulosclerosis and hyperplastic arteriolosclerosis.[22]

MANAGEMENT OF CKD: SERUM PHOSPHORUS CONCENTRATIONS

Similar to serum creatinine reference ranges, there is variability in serum phosphorus reference ranges between laboratories. This variability is likely, at least in part, due to higher serum phosphorus concentrations observed in healthy, young, growing puppies and kittens. Soft tissue mineralization of the kidney causes irreversible nephron damage and is associated with CKD progression in dogs and cats. Cats with stable CKD had lower serum phosphorus concentrations than did those cats with progressive CKD (4.4 vs. 5.1 mg/dl).[17] Serum phosphorus concentration is a predictor of CKD progression in cats, with a 41% increase in the risk of progression for every 1.0 mg/dl increase in serum phosphorus concentration.[17] In 80 client-owned cats with CKD, serum phosphorus concentrations were correlated with renal interstitial fibrosis.[22] In dogs with CKD, a serum calcium times phosphorus product > 70 mg/dl was a poor prognostic indicator.[23] These findings have prompted closer scrutiny of therapeutic targets for serum phosphorus in dogs and cats with CKD. IRIS treatment guidelines suggest the following targets which are typically well within laboratory reference ranges.

IRIS CKD Stage	Target Serum Phosphorus (mg/dl)	Management Options
1	2.7 – 4.5	Renal diet or normal diet with enteric binder
2	2.7 – 4.5	Renal diet +/- enteric binder
3	2.7 – 5.0	Renal diet with enteric binder
4	2.7 – 6.0	Renal diet with enteric binder

SUMMARY

Closer monitoring of serum creatinine and UPC may facilitate early diagnosis of CKD in dogs and cats. Longitudinal assessment of these parameters will almost always provide better data than will one-time evaluations. No laboratory test is perfect; trending laboratory data, with the same test methodology, will tend to improve diagnostic sensitivity. It should also be noted that typically there was overlap in UPC, systolic blood pressure, and serum phosphorus concentrations in the studies cited here and broad application of study findings to individual animals vs. groups may be problematic. Once CKD has been diagnosed, standard of care renoprotective treatment includes a renal diet +/- an enteric phosphate binder for hyperphosphatemia and angiotensin-converting enzyme inhibitors and/or calcium channel blockers for proteinuria and hypertension. Tighter control of hyperphosphatemia, renal proteinuria, and systolic hypertension may improve treatment outcome.

REFERENCES

1. Lund EM, et al. *J Am Vet Med Assoc* 1999; 214:1336-1341. 2. Brown SA. In August JR (ed): *Consultations in Feline Internal Medicine*. St. Louis, Elsevier Saunders, 2010, pp 475-482. 3. Polzin DJ, et al. In: Kirk RW and Bonagura JD (eds): *Kirk's Current Veterinary Therapy XI*. Philadelphia: Saunders, 1992, pp 848-853. 4. Ross SJ, et al. In August JR (ed): *Consultations in Feline Internal Medicine*. St Louis: Elsevier Saunders, 2005, pp 389-398. 5. Lulich JP, et al. *Compend Cont Educ Pract Vet* 1992; 14:127-153. 6. Ulleberg T, et al. *Acta Vet Scand* 2011; 53:2538. 7. Boozer L, et al. *J Vet Intern Med* 2002; 16:354 (abstract). 8. www.IRIS-Kidney.com (CKD Staging guidelines for dogs and cats). 9. Miyagawa Y, et al. *J Vet Med Sci* 2010; 79:1129-1136. 10. Hall JE, et al. *J Vet Intern Med* 2014; 28:1676-1683. 11. Braff J, et al. *J Vet Intern Med* 2014; 28:1699-1701. 12. Yeramilli M, et al. *J Vet Intern Med* 2014; 28:1084-1085 (abstract). 13. Lees GE, et al. *J Vet Intern Med* 2005; 19:377-385. 14. Lyon SD, et al. *J Am Vet Med Assoc* 2010; 236:874-879. 15. Syme HM, et al. *J Vet Intern Med* 2006; 20:528-535. 16. Jepson RE, et al. *J Vet Intern Med* 2009; 23:806-813. 17. Chakrabarti S, et al. *J Vet Intern Med* 2012; 26:275-281. 18. Brown S, et al. *J Vet Intern Med* 2007; 21:542-558. 19. Jacob F, et al. *J Am Vet Med Assoc* 2003; 222:322-329. 20. Mathur S, et al. *Am J Vet Res* 2004; 65:1006-1013. 21. Finco DR, et al. *J Vet Intern Med* 2003; 17:406-407 (abstract). 22. Chakrabarti S, et al. *Vet Pathol* 2013; 50:147-155. 23. Lippi I, et al. *J Am Vet Med Assoc* 2014; 245:1135-1140.

SDMA FOR EARLY DETECTION OF CHRONIC KIDNEY DISEASE

Jean A. Hall, DVM, PhD, DACVIM (Small Animal Internal Medicine)
Corvallis, OR

INTRODUCTION

Chronic kidney disease (CKD) is a major cause of morbidity in dogs and cats, especially as they age. CKD represents a progression of decreasing renal function leading to kidney failure. Early recognition of CKD is desirable in order to know when to initiate renoprotective interventions that may slow the progression of CKD. Early indicators of kidney damage include a reduction in glomerular filtration rate (GFR) and impaired urine concentrating ability. Later indicators include azotemia, with elevated serum creatinine (sCr) and urea nitrogen concentrations. Once dogs and cats are diagnosed with CKD, they are staged accorded to guidelines developed by the International Renal Interest Society (IRIS) and accepted by the American and European Societies of Veterinary Nephrology and Urology. The four tier IRIS CKD staging system is based on sCr concentration, the magnitude of proteinuria as measured by the urine protein:creatinine ratio, and blood pressure.[1]

Renal function in dogs and cats with CKD should be regularly monitored. Determination of GFR is a valuable tool for estimating renal function and staging kidney disease.[2] Measuring iohexol plasma clearance appears to be the simplest and most accurate method for determining GFR in clinical practice. Yet measuring GFR is technically cumbersome and expensive, such that sCr remains the standard surrogate for GFR. However, sCr has limitations as a marker of kidney function, most notably insensitivity because it remains in the normal reference interval until GFR is reduced approximately 75%.[3] In addition, other non-renal factors can influence sCr, including endogenous production by muscle such that age, breed, and body weight may influence sCr concentration.[4] Although serial evaluations of sCr within the same animal increase its sensitivity for detecting progressive changes in GFR, there is a need for better biomarkers for monitoring CKD. Novel renal function biomarkers are being evaluated to find more sensitive indicators of altered renal function. Serial evaluation of biomarkers provides a means of monitoring response to diet and other therapies.

Symmetric dimethylarginine has been shown to be an accurate and precise biomarker for calculating estimated GFR in humans with coronary artery disease,[5] as well as a more sensitive biomarker than sCr for assessing renal dysfunction in humans.[6,7] A meta-analysis of 18 studies involving human patients showed that 1/SDMA concentrations correlated highly with inulin clearance (r = 0.85), and SDMA concentrations correlated highly with sCr concentrations (r = 0.75).[8] In one study with human subjects, plasma SDMA concentrations correlated significantly with both GFR (measured by an iodothalamate clearance technique; r = -0.84, P < 0.01) and sCr (r = 0.89, P < 0.01). In fact, the correlation coefficient for SDMA and GFR was almost identical to that of sCr and GFR (r = -0.85; P < 0.01).[9]

In brief, methylated arginine is a byproduct of protein methylation. There are three main species of methylated arginine: monomethylarginine, asymmetric dimethylarginine (ADMA), and SDMA.[10] Subsequent protein degradation of methylated proteins yields individual methylated arginine amino acids. Whereas ADMA is metabolized enzymatically by dimethylarginine dimethylaminohydrolase, SDMA is excreted primarily (≥ 90%) by renal clearance in humans.[11,12] Because SDMA is eliminated by the kidneys, serum concentrations are affected by changes in GFR. Prolonged oral supplementation with L-arginine does not influence serum concentrations of SDMA[13] because serum SDMA derives predominantly from methylated nuclear proteins.[10]

SDMA ALLOWS EARLY DETECTION OF CKD IN CATS[14]

Cats with CKD came from a colony of over 400 DSH cats. Cats in the colony ranged in age from 1 to 19 years, with approximately 25% of cats >10 years. The cats with CKD included cats that were persistently azotemic over an extended period, typically for > 3 months (n=15). CKD was diagnosed in nonazotemic cats (with stable sCr concentrations) based on GFR testing by iohexol clearance (n=4). Nonazotemic cats with calcium oxalate kidney stones were also considered as having CKD (n=2). SDMA concentrations were determined using LC-MS, and shown to correlate with GFR (r = -0.79). The median (range) of GFR measurements in healthy geriatric cats was 1.94 mL/min/kg (1.34 to 3.79 mL/min/kg). The lower 2.5 percentile, which was used to establish the lower limit of normal, was determined to be 1.26 mL/min/kg. All cats with GFR <1.36 mL/min/kg (which corresponds to a >30% decrease below the median GFR) were considered to have CKD whereas cats about this threshold were considered to have normal renal function. Serum SDMA concentrations were shown to increase above the normal reference interval before sCr concentrations increased above the normal reference interval in 17/21 cats by a mean of 17 months (range, 1.5 to 48 months).

Using the relationship between SDMA concentration and GFR, and setting the upper limit of the normal reference interval for SDMA at 14 µg/dL corresponded to a GFR of 1.47 mL/min/kg, which represents an approximately 24% decrease from median GFR of healthy cats. Setting the upper limit of the normal reference interval for sCr at 2.1 mg/dL corresponded to a GFR of 0.78 mL/min/kg, which represents an approximately 60% decrease from median GFR of healthy cats.

SDMA ALLOW EARLY DETECTION OF CKD IN DOGS

Dogs with CKD came from a colony of over 420 dogs. Dogs in the colony ranged in age from 1 to 14 years, with approximately 24% of dogs >10 years. All dogs included in the study had necropsy confirmed CKD. Serum SDMA concentrations were shown to increase above the normal reference interval before sCr concentrations increased above the normal reference interval in the approximately 20 dogs that met inclusion criteria, by a mean of approximately 11 months (range, 1 to 31 months).

PROSPECTIVE HEALTHY AGING STUDIES IN CLIENT-OWNED PETS AND COLONY POPULATIONS

SDMA and sCr concentrations were assessed at baseline and after 6 months in client-owned geriatric dogs and cats entered into prospective healthy aging studies. All cats (n= 80) were ≥ 9 years, spayed or neutered, and free of any chronic disease based on

PE, CBC, serum chemistry profile, TT4, negative FIV/FeLV, and UA results at the start of the study. Dogs (n=211) were included based on body weight and age: small sized dogs (<11.4 kg) and medium sized dogs (11.5 to 22.7 kg) had to be ≥ 9 years, whereas dogs >22.7 kg had to be ≥ 7 years at the start of the study. All dogs were spayed or neutered and free of any chronic disease based on PE, CBC, serum chemistry profile, TT4, negative HW test, and UA. Using normal reference intervals from regional diagnostic laboratories, the upper reference interval for SDMA was 14 µg/dL and for sCr was 2.0 mg/dL (dogs) and 2.5 mg/dL (cats). Analysis of baseline and 6 month serum samples showed the percentage of cats with elevated SDMA was 14%; the percentage of cats with elevated sCr was 0%. Analysis of baseline and 6 month serum samples in dogs showed the percentage with elevated SDMA was 2.6%; the percentage of dogs with elevated sCr was 0%.

REFERENCES

1. International Renal Interest Society. IRIS 2009 Staging of CKD. http://www.iris-kidney.com. Accessed July 8, 2013.
2. Von Hendy-Willson VE, Pressler BM. An overview of glomerular filtration rate testing in dogs and cats. Vet J 2011;188:156-165.
3. Finco DR, Brown SA, Vaden SL, et al. Relationship between plasma creatinine concentration and glomerular filtration rate in dogs. J Vet Pharmacol Ther 1995;18:418-421.
4. Braun JP, Lefebvre HP. Kidney function and damage. In: Kaneko JJH, Harvey JW, Bruss ML, eds. Clinical Biochemistry of Domestic Animals, 6th ed. London: Elsevier; 2008:485–528.
5. Bode-Boger SM, Scalera F, Kielstein JT, et al. Symmetrical dimethylarginine: a new combined parameter for renal function and extent of coronary artery disease. J Am Soc Nephrol 2006;17:1128-1134.
6. Dixon JJ, Lane K, Dalton RN, et al. Symmetrical dimethylarginine is a more sensitive biomarker of renal dysfunction than creatinine. Critical Care 2013;17:P423.
7. Payto D, El-Khoury JM, Bunch D, et al. SDMA outperforms serum creatinine-based equations in estimating kidney function compared with measured GFR. Clin Chem 2014;60:S26.
8. Kielstein JT, Salpeter SR, Bode-Boeger SM, et al. Symmetric dimethylarginine (SDMA) as endogenous marker of renal function--a meta-analysis. Nephrol Dial Transplant 2006;21:2446-2451.
9. Fliser D, Kronenberg F, Kielstein JT, et al. Asymmetric dimethylarginine and progression of chronic kidney disease: The mild to moderate kidney disease study. J Am Soc Nephrol 2005;16:2456-2461.
10. Bedford MT, Richard S. Arginine methylation an emerging regulator of protein function. Mol Cell 2005;18:263-272.
11. Schwedhelm E, Boger RH. The role of asymmetric and symmetric dimethylarginines in renal disease. Nat Rev Nephrol 2011;7:275-285.
12. Kielstein JT, Boger RH, Bode-Boger SM, et al. Marked increase of asymmetric dimethylarginine in patients with incipient primary chronic renal disease. J Am Soc Nephrol 2002;13:170-176.
13. Rytlewski K, Olszanecki R, Korbut R, et al. Effects of prolonged oral supplementation with L-arginine on blood pressure and nitric oxide synthesis in preeclampsia. Eur J Clin Invest 2005;35:32-37.
14. Hall JA, Yerramilli M, Obare E, et al. Comparison of serum concentrations of symmetric dimethylarginine and creatinine as kidney function biomarkers in cats with chronic kidney disease. J Vet Intern Med 2014;28:1676-1683.

INAPPETENCE IN FELINE CHRONIC KIDNEY DISEASE: EXPLORING ETIOLOGY AND MANAGEMENT

Jessica Quimby, DVM, PhD, DACVIM
Fort Collins, CO

THE UNIQUE FELINE

Cats have a higher requirement for protein and amino acids than other species. When nutrition is inadequate, energy is derived from mobilization of amino acids from muscles stores as opposed to fat.[1] Elderly cats are also unique in comparison to other species as they have stable to increased level of metabolism, as opposed to decreased metabolism.[2] A reduced ability to digest protein and fat has been documented in elderly cats. These combined factors make the utilization of high quality, easily digestible food product critical for these patients.[3] Cats are also particularly sensitive to changes in environmental factors, such as the timing and location of feeding, as well as the food type; smell and "mouth feel" may also play into their willingness to eat.[4]

NUTRITION AND DIET IN CKD

Several studies have documented the therapeutic value of specially formulated diet in the management of CKD, including decreased incidence of uremic crisis and increased survival.[5, 6] These diets typically contain conservative amounts of high quality protein, adequate non-protein calories, and are restricted in phosphorus. The failure of the patient to eat the diet negates the benefit of dietary management, and therefore a key therapeutic target for these patients is the maintenance of appetite and food intake. Cats with CKD often suffer from poor appetite, and although renal diets contain adequate dietary protein, the animal will be protein deficient if not eating its caloric requirement. Management of long term hyporexia is important. Poor body condition is associated with decreased prognosis in several species and has a negative effect on immune function, wound healing and strength.[1] Additionally, poor appetite is perceived by owners as a significant quality of life concern and anorexia in companion animals can cause emotional distress to owners.[7]

DOES "UREMIC GASTRITIS" EXIST IN CATS?

Cats with CKD have been shown to have elevated concentrations of gastrin that increase with the severity of renal failure,[8] but the relationship between gastrin, gastric acid secretion, and gastric pathology has not been investigated. The presence of gastric acid results in negative feedback to decrease the secretion of gastrin. In humans and dogs, gastrin is excreted by the kidneys, and it is hypothesized that as renal function declines, hypergastrinemia develops, resulting in gastric hyperacidity.[8] Cats that have gastrin-secreting tumors with levels of hypergastrinemia similar to those found in cats with CKD have significant gastric pathology; however no study has shown this to be the case in cats with CKD.[9] Even in humans, the development of hyperacidity in association with CKD appears to be inconsistent, and may be related to the presence of *Helicobacter spp.* Infection.[10] Thus there is very little available evidence on which to base recommendations for the use of acid-reducing medications such as H_2 blockers, proton pump inhibitors or sucralfate in cats with uremia.[11-14]

A recent study evaluated the type and prevalence of histopathologic lesions in the stomach of cats with CKD.[15] Cats with CKD appear more likely to have gastric fibrosis and mineralization, rather than uremic gastropathy lesions previously described in dogs and humans. Therefore, the administration of gastric protectants such as sucralfate may not be justified, unless obvious clinical evidence of gastrointestinal hemorrhage such as melena or hematemesis is appreciated. The notable frequency of gastric mineralization, presumably as a consequence of metastatic mineralization, may highlight the need for more aggressive control of hyperphosphatemia and renal secondary hyperparathyroidism in cats with CKD. Gastrointestinal symptoms in these animals may not necessarily be the result of gastric lesions, but perhaps the consequence of circulating uremic toxins interacting with the chemoreceptor trigger zone in the brain. Medical management of gastrointestinal symptoms with anti-emetic and anti-nausea drugs may therefore be more appropriate. The exact role of hypergastrinemia in contributing to gastric hyperacidity and/or gastric lesions in cats with CKD is still unclear.

THERAPIES

Medical management of underlying disease

Whenever possible it is obviously ideal to address the underlying disease condition. However in some chronic diseases, i.e. kidney disease, this is not possible and therefore medical management of complications of chronic disease are ideal. Dehydration and anemia can both potentially play a role in inappetence and therefore should be addressed when applicable.

Environmental factors and food choices

As cats are particularly sensitive to environmental factors, suggestions for appetite enhancement include: quiet environment with exclusive access without interference from bothersome household members, novel food type in cases of possible aversion, similar food type in case of food exclusivity, warming the food particularly if olfaction is an issue, alternatively chilling the food if aromas appear to result in nausea, social interaction while eating, feeding small frequent meals as premature satiety is associated with many disease states.

Pain management

For both acute and chronic medical conditions, appropriate management of pain is critical to improving appetite. Opioids are perhaps most commonly utilized. Buprenorphine has a significantly longer half-life and may be useful for in home palliation of chronic pain. Butorphanol may also have some visceral analgesic properties. Opioids should be used with caution in patients with ileus. Maropitant may also have some visceral analgesic properties.[16]

Anti-nausea Therapy

Cats with CKD likely suffer from nausea, vomiting and inappetence as a result of uremia – a buildup of toxins in blood– that affects the chemoreceptor trigger zone in brain. In addition, uremia may have effects on the intestinal tract that lead to further unwillingness to eat. Several anti-nausea therapies have become recently available. These include maropitant, ondansetron and dolasetron. These drugs work at nausea center in brain as well as in gut and can be given as an injection. Ondansteron has been documented to be helpful in human patients suffering from uremia.[17] However recent pharmacokinetic studies in cats have demonstrated that oral bioavailability of ondansetron is poor in cats (~35%) and the half -life is very short (approximately 1 hour) making it a q 8hr medication.[18] Subcutaneous ondansetron has a slightly longer half-life of 3 hours. In addition to its appetite-stimulating properties, mirtazapine demonstrates anti-nausea properties as it acts at the 5HT3 receptor similarly to ondansetron and dolasetron.[19, 20] Limiting gastric acidity with the use of H_2 blockers or proton pump inhibitors such as famotidine or omeprazole respectively anecdotally appears to palliate inappetence in some patients, however as previously mentioned both the degree of hyperacidity present in CKD and the efficacy of these medications for management of cats with CKD remains unproven. Recent studies of the effect of omeprazole on the gastric pH in cats indicates that it is superior to famotidine in its ability to inhibit acid production and twice daily dosing may be necessary.[21]

Cerenia Clinical Trial: Cerenia is commonly used for acute vomiting. A pharmacokinetic and toxicity study in cats indicated that longer-term usage appears safe.[22] A recent study assessed the efficacy of Cerenia for management of chronic vomiting and inappetence associated with feline CKD.[23] When given daily for two weeks, Cerenia was demonstrated to palliate vomiting associated with chronic kidney disease, however did not appear to significantly improve appetite or result in weight gain in cats with Stage II and III CKD within the timeframe of the study.

Appetite Stimulant Therapy

Cyproheptadine has been used for some time as an appetite stimulant and has anecdotal efficacy in many patients, however its efficacy has never been scientifically evaluated. Twice daily administration is necessary in many cases and this can prove a challenge for owners, particularly longterm. Sedation is a common side effect. Mirtazapine has become more commonly used and recent exploration of its pharmacodynamics and pharmacokinetics have provided information for more effective use in cats.[24, 25] Pharmacodynamic studies have illustrated that it can be a potent appetite stimulant, but higher doses are more commonly associated with side effects (hyperexcitability, vocalization, tremors). Smaller, more frequent doses are recommended. The half-life is short enough that it could be administered daily in normal cats. Renal disease delays clearance and in these patients, every other day administration is recommended.[24] Owners should be aware mirtazapine and cyproheptadine cannot not be administered concurrently; cyproheptadine is in fact used as an antidote for serotonin effects of mirtazapine overdose.

Mirtazapine Clinical Trial: A recent placebo-controlled, double-masked crossover clinical trial was performed to evaluate the effects of mirtazapine on body weight, appetite, and vomiting in cats with CKD.[20] Mirtazapine is an effective appetite stimulant in cats with CKD and resulted in significantly increased appetite and weight. Mirtazapine also appears to have anti-emetic properties and resulted in significantly decreased vomiting in cats with CKD. This drug could be a useful adjunct to the nutritional management of cats with CKD.

Table 1: Commonly used appetite stimulant and anti-nausea medications

Drug	Receptor	Location of action	Dosage
Butorphanol	ENK	Cerebral cortex and CRTZ	0.2 – 0.4 mg/kg SQ, IV q 6-8 hrs
Buprenorphine	*mu*	Various	0.005 – 0.01 mg/kg SQ, buccal mucosa
Metoclopramide	D2 (weak)	CRTZ (weak)	0.2 – 0.4 mg/kg SQ, IV q 6-8hrs
Ondansetron	5HT3	CRTZ and GI afferent	0.5 – 1.0 mg/kg IV, SQ, PO q 8 hrs
Dolasetron	5HT3	CRTZ and GI afferent	0.6 – 1.0 mg/kg IV, SQ, PO q 24hrs
Maropitant	NK-1	Emetic center, CRTZ, GI	1 mg/kg SQ, PO q 24hrs
Mirtazapine	5HT3	CRTZ and GI afferent	1.87mg q 24 hrs in normal cats, q 48 hrs in kidney disease.
Cyproheptadine	H1	Various	2-4 mg per cat q 12-24 hrs

LEARNED FOOD AVERSION

Care should be taken to select the appropriate patients for appetite enhancement as learned food aversion is thought by most to be prevalent in cats.[4] Learned food aversion occurs when the patient associates nausea, pain or other physical manifestations of disease with the act of eating or the sight or scent of food. Even after the underlying illness is resolved, this aversion may remain. Therefore it is critical that cats that are overtly nauseous – drooling, gagging, turning away from food - particularly in hospital or in acute illness are not forced to eat lest food aversion be created.[4]

If cats are too nauseous or critical to even consider oral feeding, or have not responded to appetite encouragement after 3-5 days, placement of an enteral feeding tube should be considered. Nasoesphageal, esophageal or gastrotomy tube can be chosen depending on the type and duration of feeding desired.[1] Esophageal feeding tubes can be a valuable tool for long-term management of CKD patients as food, medications and water can be easily given without stressing the patient. Parenteral feeding should be considered in cats that cannot tolerate enteral feeding. Additionally many clinicians feel that prescription diets (i.e. renal) should not be fed in hospital during a crisis lest an aversion be created to the diet desired for long term management. The best candidates for pharmacological enhancement of appetite are cats leaving the hospital with their acute crisis resolved, and cats with chronic disease in the home environment.

REFERENCES

[1]Chan DL. **J Feline Med Surg** 2009; 11: 925-33. [2]Laflamme DP. **Vet Clin North Am Small Anim Pract** 2005; 35: 713-42. [3]Sparkes AH. **Top Companion Anim Med** 2011; 26: 37-42. [4]Michel KE. **J Feline Med Surg** 2001; 3: 3-8. [5]Geddes RF, et al. **J Vet Intern Med** 2013; 27: 1354-61. [6]Ross SJ, et al. **J Am Vet Med Assoc** 2006; 229: 949-57. [7]Reynolds CA, et al. **J Vet Intern Med** 2010; 24: 1421-6. [8]Goldstein RE, et al. **J Am Vet Med Assoc** 1998; 213: 826-8. [9]Liptak JM, et al. **J Feline Med Surg** 2002; 4: 27-42. [10]El Ghonaimy E, et al. **Nephron** 1985; 39: 86-94. [11]Bartges JW. **Vet Clin North Am Small Anim Pract** 2012; 42: 669-92, vi. [12]Polzin DJ. **Vet Clin North Am Small Anim Pract** 2011; 41: 15-30. [13]Polzin DJ, et al. In: Bonagura JD, Twedt DC, editors. Kirk's Current Veterinary Therapy XIV. 2009: 872-8. [14]Roudebush P, et al. **J Feline Med Surg** 2009; 11: 195-210. [15]McLeland SM, et al. **J Vet Intern Med** 2014: [16]Boscan P, et al. **Am J Vet Res** 2011; 72: 1576-9. [17]Ljutic D, et al. **Kidney Blood Press Res** 2002; 25: 61-4. [18]Quimby JM, et al. **J Vet Pharmacol Ther** 2014; 37:348-353. [19]Riechelmann RP, et al. **Am J Hosp Palliat Care** 2010; 27: 106-10. [20]Quimby JM, et al. **Vet J** 2013; 197: 651-5. [21]Parkinson S, et al. **J Vet Intern Med** 2015; 29: 104-12. [22]Hickman MA, et al. **J Vet Pharmacol Ther** 2008; 31: 220-9. [23]Quimby JM, et al. **J Feline Med Surg** 2014 [24]Quimby JM, et al. **J Vet Intern Med** 2011; 25: 985-9. [25]Quimby JM, et al. **J Vet Pharmacol Ther** 2011; 34: 388-96.

Exploring Options for Recurrent UTI

Shelly Vaden, DVM, PhD, DACVIM (Small Animal Internal Medicine)
Raleigh, NC

INTRODUCTION

It is estimated that 14% of female dogs will develop a urinary tract infection (UTI) within their lifetime.[1] Although fewer male dogs are diagnosed with UTI than are female dogs, the percentage of male dogs with UTI increases with advancing age. Bacterial UTI is less common in cats, accounting for only 1-3% of all cases of feline lower urinary tract disease in North America. As cats age they are more commonly diagnosed with a UTI; the percentage of older cats (i.e., > 10 years of age) with UTI may be as high as 50%. Thankfully, recurrent UTI, defined as >2 UTI in a 6 month period or >3 UTI in a year, is encountered less frequently than simple, uncomplicated UTI. While the management of dogs with acute, uncomplicated UTI is fairly straightforward, the management of dogs with chronic UTI, due to either recurrence or persistence, can be anything but straightforward.

Antibiotic therapy is standard treatment for UTI but overuse leads to changes in the normal host microbiome, with selection for resistant bacterial strains. Although severe adverse effects of antibiotic therapy for UTI may be infrequent, mild adverse effects are probably not and chronic antimicrobial administration is expensive for the client. There has been a shift of focus to the prevention of UTI in at risk people. This discussion will focus on different ways that recurrent UTI might be prevented, including strengthening of the host defense mechanisms, prophylactic antibiotics, and novel therapeutic approaches.

PATHOGENS

In recent decades resistant pathogens are more commonly encountered in many areas of the world, further complicating the management of animals with UTI. Of particular concern are the uropathogenic *Escherichia coli* (UPEC). UPEC is the most common etiologic agent for UTI in most mammalian species. Investigators in Taiwan reported "an alarming rate of antimicrobial resistance among E. coli isolates from dogs with UTI."[2] However, in Sweden the resistance to UPEC appears to be stable or even improving.[3] It is possible that these differences reflect different pattern of antibiotic usage. As a profession, we need to investigate methods to reduce antibiotic usage in animals at risk for recurrent UTI in hopes of reducing antimicrobial resistance in UPEC.

UPEC present a particular challenge because of their ability of UPEC to establish intracellular residence, thereby avoiding normal host defense mechanisms and makes them more difficult to eradicate with traditional antibiotic regimens.[4] When UPEC take up intracellular residence, there are 3 potential fates. They can be recycled to the cell surface where they are exocytosed, live a quiescent nonreplicative existence, or live a highly replicative phase in the cell cytoplasm.[5] These latter two fates allow UPEC to establish a reservoir in the urinary tract. When the epithelial cells are exfoliated, the UPEC become free to attach to other epithelial cells and cause another UTI.

Biofilms formation is another mechanism by which E. coli can avoid the host immune defenses. A biofilm is an accumulation of microorganisms and their extracellular products forming a structured community on a surface.[4] It can be difficult to eliminate a biofilm with antibiotics, in part because bacteria living in a biofilm behave differently than they would in culture. Bacteria in biofilms grow more slowly, making them less susceptible to the mechanisms of action of many antimicrobial agents. Furthermore, these bacteria might not express antimicrobial binding proteins to the same extent as they would if grown in culture. Antimicrobial agents may not be able to penetrate the full depth of a biofilm. In fact, bacteria in a biofilm can survive antimicrobial agents at concentrations that are 1000-1500 times higher than what would be needed to kill the same bacteria in culture.[4] Biofilm formation is associated with fluoroquinolone resistance in UPEC isolated from dogs.[6] It might be possible to prevent bacterial from forming biofilms if a UTI is treated early but this can also present a problem for veterinarians because early signs of a UTI might be missed in dogs and cats.

HOST DEFENSE MECHANISMS

Normal host immune mechanisms, both systemic and local (or mucosal) are crucial for the prevention of bacterial colonization and UTI in normal animals. These immune mechanisms are highly complex and may be incompletely understood but it is clear that interruption of normal host defense mechanisms can lead to repeat infections. Proper identification and management of disorders known to lead to immunosuppression (e.g., diabetes mellitus, hyperadrenocorticism) is essential in the prevention of UTI in patients affected by these disorders. Likewise, animals being administered drugs that are known to disrupt the immune system (e.g., corticosteroids, chemotherapeutic agents) should be appropriately monitored for the development of UTI.

Normal urine composition does not favor bacterial growth and is, therefore, a natural host defense against bacterial invasion. Although urine pH has long been considered to be an important factor inhibiting bacterial growth, the reality is that it is not possible to use this inhibitory characteristic in a clinical setting. In people bacterial growth occurs freely when urine pH is 5.0-7.0.[7] However, when urine pH is <4.5 or >9.0, there is effectively no growth of any organism. On the other hand, high urine osmolality, as measured by urine specific gravity, also prevents bacterial growth. Animals with recurrent UTI be thoroughly evaluated for all correctable causes of polyuria and polydipsia.[8] Physicians have long advocated forced fluid intake in people with UTI theorizing this would flush the urinary tract of pathogens, reduce bacterial growth and division in the bladder and allow the bladder mucosal defenses to operate more effectively. However, high urine flow rates may select for bacteria that are able to strongly adhere to the bladder mucosa. Given the protective effect of highly concentrated urine and the fact that "flushing the bladder" does not appear to be effective, this recommendation for lower UTI seems unwarranted. Contrary, there may be some benefit of forcing fluids in prevention or treatment of pyelonephritis. Increased flow rates of filtrate through the kidney might help promote a more isotonic environment in the renal medulla that could promote more effective granulocyte mobilization into the areas of infection.

Normal anatomy is also important in UTI prevention. Congenital abnormalities of the urinary tract should be identified and corrected in young animals with recurrent UTI. In recent years it has become apparent that hooded vulva (aka recessed vulva, excessive perivulvar skin folds) is a clinical relevant abnormality in female dogs with recurrent UTI.[9,10] Hooded vulva is frequently associated with dermatitis, creating ideal conditions for bacterial growth. Inwardly folding perivulvar hairs might wick bacteria into the vestibule. The sometimes-proposed management strategy of repeat cleaning of the folds can be onerous for the owners and does not always appear to be effective. Episioplasty has been reported to be effective in the resolution of the clinical signs associated with UTI in 84-100% of dogs with hooded vulva. We generally recommend episioplasty in dogs with >50% coverage of the vulva by surrounding folds AND a history of recurrent UTI. Vestibulovaginal septal remnants may be incidental findings in some dogs whereas, in other dogs, the might be associated with recurrent UTI. One study demonstrated a significant reduction in the rate of UTI after ablation of the septal remnants.[11]

Normal micturition is another important factor in normal defenses against UTI. In theory, normal voiding should eliminate bacteria from the urinary tract. However, in actuality voiding may not mechanically free bladder of adherent bacteria that are in a biofilm and clearance from the bladder may be more closely associated with antibacterial activity of mucosal surface than with voiding. By mechanisms that seem to be counterintuitive, UPEC adherence may actually be strengthened by sheer stress induced during urine flow. Nonetheless, bladder dysfunction and urine retention have clearly been shown to be risk factors for UTI in dogs. Frequent voiding with complete emptying may be an important factor in prevention of UTI in dogs and cats that are at risk of recurrent UTI. Drugs (e.g., bethanecol, cisapride) or intermittent urethral catheterization can be used to facilitate more complete emptying of the bladder in animals with documented detrusor dysfunction (e.g., spinal cord injury). Likewise, urinary incontinence is associated with a break in the barrier against ascending UTI. Effective management of urinary incontinence may aid in the prevention of UTI in some dogs.

PROPHYLACTIC ANTIMICROBIAL THERAPY

The use of prophylactic antibiotics has long been advocated in dogs with chronic UTI, often initiated after potential underlying causes are corrected and UTI has been eliminated. Once daily, bedtime dosing is used to eliminate ascending bacteria before UTI can be established. Pulse therapy (i.e., short period of antibiotic administration given at regular intervals such as one week out of every month) has also been suggested but may be more likely to induce bacterial resistance than bedtime dosing. In general, antibiotics that are relatively safe and inexpensive should be selected (e.g., amoxicillin, amoxicillin-clavulanic acid, nitrofurantoin). First and second generation cephalosporins generally should not be used for this purpose because of their ability to induce global antimicrobial resistance.

Methenamine mandelate may be effective in situations where bedtime antibiotics have failed to prevent UTI. This agent is converted to formaldehyde in acidic urine; therefore, concurrent administration of urinary acidifiers may be required for the drug to be effective. In 2011, the National Toxicology Program declared that formaldehyde was carcinogenic. Therefore, methenamine carries a theoretical risk of causing bladder cancer. It is currently recommended that methenamine therapy in people be limited to one week in duration, making it difficult to recommend this drug for long-term administration in dogs.[12]

The question remains whether or not prophylactic antimicrobial therapy is effective in dogs at all. Large meta-analyses of human studies suggest that any benefit is likely to be small and of limited clinical significance, with recurrent UTI returning after cessation of prophylactic therapy.[13] Clinical experience suggests that the same may be true in dogs. Antibiotic prophylaxis may also be associated with more adverse drug effects and increased antimicrobial resistance of both the bacteria causing the UTI and those residing in the host microbiome.

TOPICAL THERAPY

Because of the problems associated with systemic antibiotic administration and the creation of resistance, there has been increased interest in topical therapy for the treatment of recurrent UTI. Protocols are available for the use of intravesical antibiotics, including gentamicin, tobramycin, vancomycin, and piperacillin but very few studies support their efficacy or safety. Absorption of agents instilled in the bladder is presumed to be minimal but could be increased when mucosal inflammation is present. Animals given an aminoglycoside via intravesical instillation should have periodic serum aminoglycoside concentrations measured. Tris-EDTA or dilute chlorhexidine solution, instilled through indwelling or intermittently placed catheters, may be safer and carry a reduced risk of creating antibacterial resistance when compared with using antibiotics.

A defective glycosaminoglycan (GAG) layer in the bladder has been linked to the development of recurrent UTI. This has led people to instill hyaluronic acid or chondroitin sulfate in the bladder. Because heparin makes up 15% of the GAG layer and heparin is relatively inexpensive, heparin has also been instilled in bladders of people with recurrent UTI. In one study of people with difficult to treat recurrent UTI, heparin given intravesicular once weekly for 6 weeks was associated with a reduction in UTI from 4.4 infections in the 6 months before treatment to 1.1 infections in the 6 months after treatment.[14] Available data has not demonstrated that oral GAG administration will replenish the GAG layer in the bladder.

CRANBERRY AND D-MANNOSE

Cranberries have long been suggested for the prevention of UTI.[15] Cranberries contain organic acids, fructose, glucose and phytochemicals. Certain phytochemicals, such as A-type proanthocyanidins (PACs), inhibit the adhesion of p-fimbriated UPEC to uroepithelial cells. It is this inhibitory property, and not mild acidifying property, of cranberries that is effective in the prevention of UTI. Cranberry products might also reduce the symptoms of UTI by suppressing the inflammatory response.

Human clinical study results have been somewhat contradictory. Many cranberry products are available, varying from purified PACs to cranberry extracts to cranberry juice. Some cranberry products may actually be devoid of PACs. Differences in dose, dosing frequency and cranberry products as well as the low bioavailability of PACs and variable patient compliance (especially those studies using cranberry juice) probably contribute to the varying results. However, a meta-analysis concluded that there is statistical evidence that cranberry extract reduces the recurrence of UTIs, particularly in young women.[16] A more recent meta-analysis concluded that more evidence is needed to support the use of cranberry over other preventative strategies, but these other strategies often use intravaginal preparations which can be difficult to apply to veterinary practice.[17] The incidence of UTI can be reduced by as much as 35-40% in women consuming cranberry products.

D-mannose has also been shown to inhibit bacterial adhesion to the bladder wall via the same mechanisms as cranberry.[18] Some cranberry products have added D-mannose; however, there is not a clear benefit of these products. Whenever possible, cranberry products that are as pure as is possible should be used because of concerns about unknown adverse effects for any other ingredients given at high doses for a prolonged duration.

PROBIOTICS

Because most UTIs in dogs are believed to be ascending infections, maintenance of a healthy microbial population of the vestibule and vagina may be important in female dogs. In most women, the vaginal microbiome is remarkably similar, with lactobacilli being the most dominant colonizer.[19] Lactic acid producing bacteria (LAB) decrease the vaginal pH and inhibit the growth of other types of bacteria, including the pathogens E. coli, P. Mirabilis and Staphylococcus sp. Women with recurrent UTI have been reported to have a high incidence of bacterial vaginosis, which is an altered vaginal microbiome with reduced numbers of LAB. Because of this, probiotics have been used in women with recurrent UTI in an attempt to reestablish a healthy vaginal microbiome. Probiotic lactobacilli are believed to improve urogenital health in women through immune modulation, pathogen displacement, and creation of a niche less favorable for pathogenic bacterial proliferation. Probiotics may help replenish the normal microbes, displacing pathogens and leading to a return to the lactobacilli-dominated state found in healthy women, preventing ascending UTI. However, studies have failed to demonstrate a clear benefit of probiotics in women with recurrent UTI with different LAB strains yielding different results and some requiring intravaginal administration.[12]

Although often recommended, even less is known about the use of probiotics in dogs with UTI or the effect these agents have on the canine vaginal microbiome. The normal canine vaginal microbiome consists of a mixed population of anaerobic and aerobic bacteria, with greater numbers of bacteria present in the vestibule than in the cranial vagina. Common isolates include E. coli, staphylococci, enterococci, Bacteroides sp. and mycoplasma. Less is know about LAB in the canine vaginal microbiome because many published studies did not include culture enrichment for LAB species. The vaginal pH of dogs is higher (6-7.5) than is the vaginal pH of women (<4.5), which might suggest that there are fewer LAB colonies in the canine vagina. In a study of healthy estrous dogs, healthy spayed dogs, and spayed dogs with recurrent UTI, LAB were isolated from fewer than 22% of dogs and did not differ between groups.[20] It is not even clear that bacterial vaginosis occurs in dogs or that it is a risk factor for the development of recurrent UTI. The vaginal microbiota of spayed female dogs with and without a history of recurrent UTI were not significantly different. E. coli and S. psuedintermedius were the most common pathogens isolated and were more commonly isolated than LAB species.[21] When normal dogs without a history of recurrent UTI were given an oral lactobacillus containing probiotic, the prevalence of vaginal LAB did not increase.[22] This raises questions about the ability of orally-administered probiotics to alter the vaginal microbiome in dogs.

ESTROGENS

Estrogens may have many positive effects on the defense mechanisms of the urinary tract; low estrogen might induce structural and chemical changes that facilitate UTI.[23] Estrogens promote more effective bladder contractility; low estrogen in women is associated with a higher residual volume post-voiding, a risk factor for recurrent UTI. Estrogens promote the growth of LAB; low estrogen has been associated with bacterial vaginosis. Estrogens also enhance the production of antimicrobial peptides, β-defensin 2 and α-defensin 2, and promote bladder epithelial cell proliferation, which in turn provides better protection to the underlying tissues. The use of intravaginal estrogen products is strongly associated with lower UTI frequencies. However, oral administration of estrogen is only weakly associated with lower frequency of UTI.

IMMUNOMODULATION

The immunostimulant OM-89 is an extract of 18 different serotypes of heat-killed UPEC. Oral administration of OM-89 was associated with an approximate 50% reduction in recurrent UTI when compared with placebo.[12]

ACUPUNCTURE

In a study of 5 management strategies for recurrent UTI, acupuncture was the second most effective preventive method with the most effective being daily antibiotic prophylaxis.[24] Although the mechanism is not understood, acupuncture has been associated with a reduced rate of UTI recurrence when compared with untreated or sham treated women.[25,26]

REFERENCES

[1]Bartges JW. **Vet Clin North Am Small Anim Pract** 2004;34:923. [2]Chang SK et al. **J Vet Med Sci.** In press. [3]Windahl U, et al. **BMC Veterinary Research**. 2014,10:217. [4]Tenke P, et al. **World J Urol.** 2012;30:51. [5]Sivick KE, et al. **Infect Immun.** 2010;78:568. [6]Oliveira M et al. **BMC Research Notes.** 2014;7:499. [7]Yeaw RC. **J Urol.** 1940;44:699. [8]Beetz R. **European J Clin Nutr.** 2003;57:S52. [9]Hammel SP, et al. **JAAHA.** 2002;38:79. [10]Lightner BA, et al. **JAVMA.** 2001;219:1577. [11]Burdick et al. **JAVMA.** 2014;244:944. [12]Geerlings SE, et al. **Infect Dis Clin N Am.** 2014;28:135. [13]Williams G, et al. **Curr Opin Inf Dis.** 2009;22:72. [14]Ablove T, et al. **Ther Adv Uro.** 2013:5:303. [15]Vasileiou I, et al. **Nutrition Res.** 2013;33:595. [16]Jepson RG, et al. **Cochrane Database Syst Rev.** 2009;4. [17]Jepson RG, et al. **Cochrane Database Syst Rev.** 2012;10. [18]Kranjcec B. et al. **World J Urol** 2014;32:79. [19]Witkin SS, et. al. **Best Pract Res Clin Obst Gyn.** 2007;21:347. [20]Bailey CS, et. al. **Clin Therio.** 2012;4:405 [21]Hutchins RG, et al. **JVIM.** 2014;28:300. 22Hutchins RG, et al. **JVIM.** 2013;27:1368. [23]Luthje P, et al. **Maturitas.** 2014;77:32. [24]Eells SJ, et al. **Clin Infect Dis**. 2014;58:147. [25]Alraek T, et al. **Am J Pub Health**. 2002;92:1609. [26]Aune A, et al. **Scand J Prim Health Care**. 1998;16:37.

PULMONARY THROMBOEMBOLISM: RISK FACTORS AND PREVENTION

John M. Thomason, DVM, MS, DACVIM
Starkville, MS

INTRODUCTION

One of the most serious and deadly respiratory conditions in dogs is a pulmonary thromboembolism (PTE). Similar to human medicine, the true incidence of pulmonary thromboembolism in dogs is unknown.[1,2] Several studies in humans have tried to identify the true incidence, risk factors, and characteristics of PTE; however, these retrospective studies demonstrated that only 30% of patients with a PTE were properly diagnosed prior to post-mortem.[3] Due to the difficulty in diagnosing this disease, it is estimated that between 200,000 and 600,000 humans are affected by pulmonary thromboembolism each year.[2,4,5] There are no veterinary studies that provide an accurate estimation of the number of cases affected with a PTE. Without this vital information, the true number of cases in veterinary medicine is probably greatly under-estimated.[3,4]

PULMONARY THROMBOEMBOLISM

Pulmonary thromboembolism is the obstruction of a pulmonary arterial bed that adversely affects the respiratory and cardiovascular systems.[6] This obstruction can be caused by a thrombus that forms in the vascular bed or with the development of a thrombus at a distant location and travels through the venous circulation, and obstructs the pulmonary vascular bed. An embolism can be created by a variety of structures that include platelets, parasites, fat, air, and neoplastic cells.[4] The exact mechanisms of thrombus formation is not completely understood,[7] but typically requires the appropriate vascular conditions of a primary disease.[3,4,6] Some of the factors/vascular conditions include a hypercoagulable state, blood stasis, and vascular endothelial cell damage.[3,6]

There are numerous primary diseases in dogs that have been identified as a predisposing cause for venous thrombus formation and pulmonary thromboembolism. These predisposing diseases include immune-mediated hemolytic anemia (IMHA), cardiac disease (heartworm disease), disseminated intravascular coagulation (DIC), pancreatitis, protein-losing nephrophathy, hyperadrenocorticism, sepsis, neoplasia, trauma, and surgery.[3,4,6,8] Immune-mediated hemolytic anemia is one of the most common underlying condition that leads to PTE formation. However, no matter the primary disease or the method of thrombus formation, the development of a pulmonary thromboembolism will cause similar respiratory and cardiovascular abnormalities.

The pulmonary vasculature not only supplies un-oxygenated blood to the lungs, but it also acts as a filter of venous blood to prevent emboli from reaching the arterial system. Most pulmonary thromboemboli are small and do not create respiratory or cardiac dysfunction.[1] Massive pulmonary emboli are uncommon when compared to the total amount of emboli,[1] but can create a life-threatening condition if the thrombus is large enough to occlude a majority of the pulmonary artery or main stem branches. This occlusion will generate an acute increase in pulmonary vascular resistance and pulmonary arterial pressure.[1] Small thromboemboli typically do not cause clinical signs because the pulmonary system has a large reserve of lung capacity. Approximately half of the pulmonary circulation in humans can be obstructed without an increase in the pulmonary artery pressure; while dogs can lose up to 60% of their pulmonary vasculature before an increase in pulmonary arterial pressure.[1]

Pulmonary thromboembolism has a devastating impact on the lungs. The exact mechanism is not completely understood, but some of these proposed mechanisms include vasoconstriction, peripheral airway constriction, decreased surfactant production, pulmonary parenchymal infarction, and pulmonary edema.[1,12] However, no matter the mechanism, the principle cause of death with a PTE is due to the abrupt rise in pulmonary vascular resistance.[12]

PRESENTATION

There is no breed, age or sex that is predisposed to development of a PTE; however, this condition is more common with middle to older aged dogs.[3,4,6,8] The severity of clinical signs is proportionate to the degree of vascular obstruction. Most dogs demonstrate a variety of clinical signs that include dyspnea, tachypnea, and depression. Even though dyspnea and tachypnea are some of the most common clinical signs, only about 60 – 90% of patients demonstrate these clinical signs.[3] Johnson et al. reported that about 40% of dogs with a pulmonary thromboembolism had an abnormal neurologic status.[3] Possible explanations for these abnormalities include cerebral emboli, transient hypoxemia, or cerebral ischemia.[3] Other clinical signs include weakness, cyanosis, and sudden death. Physical examination findings are typically unremarkable and non-specific. Auscultation of the lungs may demonstrate crackles or diminished lungs sounds if pulmonary edema and pleural effusion is present, respectively. However, normal lung sounds could also be present during auscultation. Pain and tachycardia are also nonspecific physical examination findings that may be present. Unfortunately, one of the most reliable clinical finding is a sudden onset of dyspnea with no known respiratory disease.[6] Approximately 80% of dogs with a PTE had an intravenous catheter.[3]

TREATMENT/PREVENTION

Treatment for PTE can be a very difficult and challenging. Initial therapy beings with identifying the potential risk factors and treating the primary disease.[4] Supportive care, fluid administration and oxygen supplementation are necessary. Oxygen supplementation usually does not improve the tachypnea/dyspnea, but it may help ventilate all viable lung fields. Early and aggressive thrombolytic treatment may help dissolve the emboli. However, one of the primary concerns with these medications is uncontrollable hemorrhage.[6] Removal of the emboli can be performed with specialized equipment and surgical thrombectomy, but this procedure is rarely performed in dogs.[4] Due to the severe nature of these patients, a majority of these patients will die before effective therapy can be administered.

To prevent thrombus formation, prophylactic medications are often administered to inhibit either platelet function or the clotting cascade. Anticoagulant therapy with unfractionated heparin, enoxaparin, and warfarin can be administered to prevent additional thrombus and emboli.[4,6] Many of these preventative medications can only be given by injection and/or are cost prohibitive for long term administration, leaving oral anti-platelet therapy as the most affordable option for prophylactic therapy. In fact, the use of low-dose aspirin has been shown to improve survival rates in dogs diagnosed with IMHA.[10]

Unfractionated Heparin

Naturally occurring unfractionated heparin (UFH) is mixture of glycosaminoglycans that is heterogeneous with respect to molecular size, anticoagulant activity and pharmacokinetic properties.[11] Heparin binds antithrombin and increases its rate of activity by as much as 1000 fold. The heparin:antithrombin complex acts by inhibiting the coagulation factors IIa (thrombin), IXa, Xa, XIa and XIIa. It is estimated that only about one third of naturally occurring UFH contain the binding site for antithrombin and the remaining two thirds have minimal anticoagulant activity.[12] Thrombin and factor Xa are most sensitive to the activities of the heparin:antithrombin complex. Heparin must bind both to thrombin and antithrombin in order to inhibit thrombin's activity. By inactivating thrombin, heparin not only prevents fibrin formation but also inhibits thrombin induced platelet activation and continued activation of coagulation factors V and VIII.[13,14,15] The biologic effect of UFH is also heterogeneous because only one third of the molecules will bind antithrombin. Because of this variability in anticoagulant activity, UFH therapy must be monitored closely and doses adjusted to avoid bleeding complications. The laboratory test used to measure UFH effect is the activated partial thromboplastin time (aPTT). The accepted target range for UFH therapy has been an increase in aPTT to 1.5 to 2.5 times the normal control value.

Low Molecular Weight Heparin

Low molecular weight heparins (LMWH) are derived from UFH and have been developed for use in response to some of the difficulties associated with UFH dosage and monitoring. In general LMWHs have reduced anti-factor IIa activity relative to anti-factor Xa activity and have better pharmacokinetic properties. LMWH heparin fractions derived from standard UFH have progressively smaller effects on aPTT as the mean molecular weight of the fraction decreases while the activity against factor Xa is preserved.[16]

The various LMWHs preparations can create some variability in their pharmacokinetic properties and anticoagulant profiles thus they are not clinically interchangeable.[11] The low molecular weight fragments derived from UFH have a reduced affinity for binding to plasma proteins or cells affording the LMWHs a much more predictable dose response.[17] The mechanism of action of LMWH is similar to UFH in that they produce anticoagulation by binding to and catalyzing the activity of antithrombin. The LMWH preparations show superior pharmacokinetic properties over UFH in that they are nearly 100% bioavailability following subcutaneous administration and peak anti-factor Xa activity consistently occurs between 3 and 5 hours following administration with a much more predictable dose response relationship.[18] Because of the predictability of the dose response relationship for LMWHs, they are typically administered at a fixed dose and monitoring is not generally necessary

Aspirin

Aspirin (acetylsalicylic acid) inhibits platelet function by irreversibly acetylating a hydroxyl group on the COX enzyme, preventing arachidonic acid from reaching the active site of the enzyme.[19,20] By acetylating the COX enzyme, aspirin prevents platelet thromboxane A_2 synthesis, and inhibits platelet activation and function.[19,21] Since platelets do not contain a nucleus, and are unable to synthesize new and functioning COX enzymes, aspirin-induced platelet dysfunction will persist for the entire lifespan of the platelet, and function will not return until the drug is discontinued and new platelets, not exposed to aspirin, are released from the bone marrow. The actions of COX-2 at the vascular endothelium, which is less sensitive than COX-1 to inhibition by aspirin at anti-platelet doses, allows for continued COX-2 dependent PGI_2 synthesis of anti-platelet doses, in spite of the early suppression of COX-1.[22]

The aspirin-induced inhibition of the COX enzyme, not only inhibits the prostaglandins needed for platelet function, but also inhibits all prostaglandins generated through the arachidonic acid cascade. These prostaglandins play key roles in mediating inflammation, tissue homeostasis, renal blood flow, and gastrointestinal wall protection. Aspirin's ability to inhibit the COX enzyme will also inhibit pro-inflammatory prostaglandin synthesis, making this medication an effective anti-inflammatory therapy. Unfortunately, by inhibiting the COX enzyme and decreasing prostaglandin synthesis, aspirin can cause several life-threatening side effects that include gastrointestinal hemorrhage and renal failure.[23] An anti-inflammatory dose, or high dose, of aspirin (10 mg/kg twice daily), is very effective at consistently inhibiting platelet function, but can lead to multiple gastrointestinal and renal side effects, which makes this aspirin dose an undesirable preventative therapy.

A high dose of aspirin permanently inhibits COX function in all cells that express this enzyme, including vascular endothelial cells. However, unlike the higher, anti-inflammatory dose, a low dose of aspirin (0.5-1 mg/kg once daily) will inhibit platelet thromboxane synthesis and function without permanently inhibiting COX function of the vascular endothelium, allowing prostacyclin synthesis and vasodilation to continue.[24] Lower doses will also eliminate the potentially devastating side effects that occur with high dose aspirin. Even when administered with concurrent high doses of prednisone, there does not appear to be a significant increase in the incidence of gastrointestinal lesions during low dose aspirin therapy. However, the concurrent administration of prednisone and low-dose aspirin may increase the frequency of mild, self-limiting diarrhea.[25] Low dose aspirin, both in humans and dogs, is a beneficial treatment for the prevention of thromboembolic complications. In fact, the administration of low doses of aspirin has been associated with an improved survival rate in dogs with immune-mediated hemolytic anemia (IMHA).[26]

In people, orally dosed aspirin is absorbed rapidly from the stomach and upper intestine reaching peak plasma levels approximately 30 to 40 minutes after dosing. The bioavailability of ingested aspirin is 40 to 50% over a very wide range of doses.[27] In people, the platelet COX-1 enzyme is acetylated in the pre-systemic circulation prior to metabolism in the liver and therefore the antiplatelet effect is independent of systemic bioavailability.[27] There remains considerable debate around several issues regarding aspirin therapy in people. Questions still remain as to the optimal dose of aspirin, possible antithrombotic effects of aspirin unrelated to the inhibition of platelet TXA_2 production and the possibility that some patients are "resistant" to the anti-platelet effects of aspirin.

Clopidogrel

Clopidogrel is the most commonly used ADP receptor antagonist in human medicine.[28] Clopidogrel is an irreversible ADP receptor antagonist that permanently inhibits the function of circulating platelets. In humans, clopidogrel, administered with and without aspirin, significantly reduces the incidence of thrombotic events following myocardial infarction, ischemic stroke, vascular death, and restenosis of intracoronary stents, and is a recommended therapy for patients at risk for thrombosis.[28,29]

Similar to humans, clopidogrel in dogs requires hepatic conversion into the active metabolite. Measured by whole blood aggregation, platelet function is inhibited within 3 hours of clopidogrel administration in most dogs. Platelet function, as measured by optical platelet aggregometry, remains inhibited for 24 hours after the first dose of clopidogrel. In most dogs, platelet function returns to pre-clopridogrel levels within 4 days of the drug being discontinued. Some dogs, however, still had a significantly decreased response to ADP activation up to 5 days after drug discontinuation, suggesting an individual variation in the duration of platelet inhibition with clopidogrel.[30]

Recently, a small prospective study evaluated the safety of clopidogrel (single loading dose of 10 mg/kg PO, followed by a maintenance dose of 2-3 mg/kg q 24 h) in canine IMHA patients, and compared the clinical efficacy of single agent clopidogrel to single agent low dose aspirin (0.5 mg/kg PO, q 24 h) or a combination of low dose aspirin and clopidogrel. When compared to the low dose aspirin, clopidogrel appeared to be a safe treatment option, with no difference in the incidence of gastrointestinal side effects and no evidence of excessive hemorrhage. Two dogs, one treated with clopidogrel and one treated with low dose aspirin, did not survive to discharge due to suspected thromboembolism. Thromboembolic events were not suspected in any dogs treated with both clopidogrel and low dose aspirin. However, while there was no significant difference in survival rate, duration of hospitalization, transfusion requirements, thromboembolism, or rate of relapse between the three treatment groups, the study had insufficient power (only 8 cases per group) to truly determine whether each treatment option actually provided a comparable outcome.[31]

Clopidogrel has also been used for the prevention and treatment of canine aortic thromboembolism. In one case report, clopidogrel was used as the sole therapy for 6 months in one dog with ATE, but the patient showed progressive ambulatory dysfunction.[32] In a second small case series, three dogs with ATE were treated with clopidogrel and aspirin, and clinical signs improved in one dog, slowly progressed in the second, and acutely reoccurred in the third. None of these 3 dogs were treated with clopidogrel as the sole anti-platelet therapy.[33] Given the small number of cases published to date, a true determination of the value of clopidogrel for the treatment and prevention of canine ATE cannot be determined.

REFERENCES

1. Dennis JS. *Compendium* 1991;13(12):1811. 2. LaRue MJ, et al. *J Am Vet Med Assoc* 1990;197(10):1368. 3. Johnson LR, et al. *J Vet Intern Med* 1999;13(4):338. 4. MacDonald KA, et al. *Textbook of Veterinary Internal Medicine*, ed 6. 2005;1284. 5. Sobieszczyk P, et al. *Circulation 2008;*2002(106):1748. 6. Dennis JS. *Compendium 1993;*15(12):1595. 7. Weiss DJ, et al. *J Vet Intern Med* 2006;20(3):682. 8. Norris CR, et al. *J Am Vet Med Assoc* 1999;215(11):1650. 9. Elliott CG. *Chest* 2008;101:163. 10. Weinkle T, et al. *J Am Vet Med Assoc* 2005; 226:1869. 11. Hirsch J, et al. *Chest*. 2004;126:188S. 12. Lam LH, et al. *Biochem Biophys Res Commun* 1976;69:570. 13. Ofosu FA, et al. *Biochem J* 1987;243:579. 14. Ofosu FA, et al. *Biochem J* 1989;257:143. 15. Beguin S, et al. *Thromb Haemost* 1988;60:457. 16. Harenberg J. *Semin Thromb Hemost* 1990;16:12. 17. Hirsh J, et al. *Blood* 1992;79:1. 18. Handeland GF, et al. *Eur J Clin Pharmacol* 1990;39:107. 19. Bjorkman D. *Am J Med* 1998;105(1B):8S. 20. Guthikonda S, et al. *J Thromb Haemost* 2007;5:490. 21. Patrono C, et al. *Chest* 2001;119:39S. 22. Clarke R, et al. *N Engl J Med* 1991;325:1137. 23. Catella F, et al. *Proc Natl Acad Sci USA* 1986;83:5861. 24. Rackear D, et al. *J Am Anim Hosp Assoc* 1988;24(1):23. 25. Graham, et al. *J Vet Intern Med* 2009;23:482. 26. Weinkle T, et al. *J Am Vet Med Assoc* 2005; 226:1869. 27. Pedersen A, et al. *N Engl J Med* 1984;311:1206. 28. McNicol A, et al. *J Pharmacol Sci* 2003;93:381. 29.Wijeyeratne Y, et al. *Br J Clin Pharmacol* 2011;72(4):647. 30. Brainard B, et al. *Am J Vet Res* 2010;71:822. 31. Mellett A, et al. *J Vet Intern Med* 2011;25:71. 32. Winter R, et al. *J Vet Cardiol* 2012;14:333. 33. Lake-Bakaar G, et al. *J Am Vet Med Assoc* 2012;241:910.

DIAGNOSIS OF IDIOPATHIC HYPERCALCEMIA IN CATS: HOW MUCH IS ENOUGH TO BE SURE?

Dennis J. Chew, DVM, DACVIM
Joao Felipe deBrito Galvao*, MV, MS, DACVIM
Valerie J. Parker, DVM, DACVIM, DACVN
Columbus, OH
***Downers Grove, IL**

HOW COMMON IS HYPERCALCEMIA IN CATS?

Detection of hypercalcemia based on measurement of serum total calcium appears to be less common in populations of sick cats than that encountered in dogs. However, the frequency of finding hypercalcemia in cats has dramatically increased in many regions of the world over the past 20 years mostly due to the diagnosis of idiopathic hypercalcemia (IHC).[1-7] Hypercalcemia is most often initially defined in primary care practice by the finding of increased serum total calcium on routine serum biochemistry. Mild hypercalcemia based on serum total calcium is often overlooked during analysis of serum biochemical profiles, so the frequency of hypercalcemia is likely to be more common than generally appreciated. Mild serum total hypercalcemia is frequently attributed to hemoconcentration from dehydration.

Total serum calcium cannot be reliably used to predict the metabolically active ionized calcium fraction in cats.[8] There was an overall diagnostic discordance of 40% during evaluation of 434 feline serum samples using total calcium to predict ionized calcium in cats of one study. Ionized hypercalcemia and normocalcemia were underestimated and ionized hypocalcemia was overestimated. There was still substantial discordance when serum total calcium was "adjusted" based on serum albumin or serum total protein, an effect more pronounced in cats with CKD.[8] Hypercalcemia based on total calcium was identified in 7% and in 17% of these cats when based on serum ionized calcium. Of the 75 cats with ionized hypercalcemia, only 35% also had a parallel increase in the serum total calcium; 64% had normal serum total calcium and 1% had low serum total calcium. Of the 240 samples with normal ionized calcium concentration, 1.7% had increased serum total calcium, 53.3% had normal serum total calcium, and 45% had low serum total calcium. Based on these findings, the extent of hypercalcemia known to exist when ionized calcium is measured is under-diagnosed when serum total calcium is used as the screening test for calcium disturbances. Thus, screening cats for calcium metabolic disorders is better served by the measurement of ionized calcium.

Ionized calcium concentrations can be measured using in-house analyzers or by outside laboratories. Reference ranges for ionized calcium vary substantially between serum or plasma samples and whether the analyzer is hand-held or a bench top machine. Reference ranges should be established on site for each practice when using in-house devices. Ideally, a sample should be collected anaerobically. Exposure to air can alter the pH of blood or serum, which alters ionized calcium usually as a consequence of loss of carbon dioxide (CO_2). Loss of CO_2 may result in a more alkaline pH, which promotes more calcium binding to albumin and subsequently may result in a lower ionized calcium measurement. Conversely, an acidic pH may result in a higher ionized calcium measurement. Whole blood or heparinized plasma are also acceptable specimens for analysis of ionized calcium; however, ionized calcium and pH are more stable in serum than the other sample types. It is important not to measure ionized calcium from EDTA plasma samples or to contaminate serum tubes with EDTA, as EDTA will bind calcium and result in falsely low ionized calcium measurements.[1] When ionized calcium is measured from anti-coagulated plasma samples, it is important to standardize the technique so that the same amount of blood and anti-coagulant are added to the tubes in each instance. Otherwise differences in dilution and calcium binding will affect the reported results.

CHARACTERIZATION OF HYPERCALCEMIA

Once ionized hypercalcemia has been identified, the next step is to determine if the process is PTH-dependent (high PTH from failure to suppress abnormal parathyroid glands) or PTH-independent (PTH is appropriately suppressed as the response of normal parathyroid glands). Compared to dogs, cats have a higher frequency of PTH-independent hypercalcemia.[3,9] In a study of 322 cats, ionized hypercalcemia was parathyroid independent in 82%, equivocal in 10%, and parathyroid-dependent in 8%.[9]

Details of PTH physiology and its measurement in cats are provided in a recent review.[10] PTH must always be interpreted along with ionized calcium in the same sample in order to determine appropriateness of the PTH concentration. Normal parathyroid glands (PTG) are exquisitely sensitive to fluctuations in ionized calcium, especially when ionized calcium is low. When ionized calcium is low, PTH secretion will dramatically increase if the parathyroid glands are responding appropriately; when ionized calcium is high, PTH secretion will decrease but detection below the reference range is often not possible due to assay limitations. Overlap between the reference range and the assay sensitivity reduces the ability of any assay to detect PTH concentration that has been suppressed below normal. This ability to detect suppression of PTH is crucial in conditions in which a low PTH is the key finding for diagnosis (e.g., differentiating primary hyperparathyroidism from IHC). Currently, it appears that the third generation "whole" PTH assay[11] that uses a primary antibody directed at the C-terminal region together with a secondary antibody directed at amino acids 1-4 on the N-terminal region works best in cats due to greater affinity of this assay for feline PTH. Values measured by the whole PTH assay are more in line with "true" concentration of PTH in other species and they are more likely to reflect the concentrations of the biologically relevant compound since the 1-4 region is most crucial for activation of the PTH receptor. The lower limit of detection (sensitivity) overlaps with the lower limit of the calculated reference interval, which means that accurate assessment as to whether PTH is adequately suppressed or not during states of ionized hypercalcemia is difficult to know with certainty.[10] The degree of PTH suppression in cats with various causes of hypercalcemia has not been reported.

In those with parathyroid-independent hypercalcemia, malignancy-associated hypercalcemia (MAH) needs to be excluded. MAH most often results from humoral mechanisms as the tumor secretes calcemic substances such as PTHrP into the circulation; local osteolytic hypercalcemia is far less common. Approximately 10% of cats evaluated for causes of hypercalcemia had PTHrP greater than the reference range in one report. Carcinomas were most commonly diagnosed in cats with increased PTHrP; lymphoma was the cause in 1 cat.[9] When PTHrP is reported to be high, the presence of malignancy is likely. A low or undetectable PTHrP does not exclude malignancy as the cause for hypercalcemia since other cytokines that cause calcemia can be elaborated by the tumor instead of PTHrP on occasion. The sensitivity for a positive PTHrP test result in malignancy with or without hypercalcemia is not known in cats.

If the diagnostic evaluation does not reveal malignancy as the cause for parathyroid-independent hypercalcemia, (PTHrP and body cavity imaging) evaluation of circulating vitamin D metabolite may be useful in determining the underlying cause or mechanism for the hypercalcemia. Hypervitaminosis D is classically characterized by increased concentrations of circulating 25(OH)-vitamin D (calcidiol) following excess ergo/cholecalfierol exposure from food[12,13] or from cholecalciferol-containing rat-bait.[14,15] Increased circulating calcitriol has been reported in cats with granulomatous disease and hypercalcemia, likely the result of unregulated conversion of calcidiol to calcitriol by activated macrophages.[16-18] Exposure to the vitamin D metabolite calcipotriene used in anti-psoriasis creams is not detected by assays for calcidiol or calcitriol. During states of ionized hypercalcemia, renal synthesis of calcitriol should be suppressed if normal homeostatic systems are operative. Calcitriol is a paraneoplastic substance contributing to the hypercalcemia of malignancy in some dogs[19] but circulating concentrations of calcitriol have not been systematically investigated in cats with a variety of underlying causes of hypercalcemia. We recommend submitting samples to a DEQAS (Vitamin D External Quality Assessment Scheme) certified laboratory to ensure analytical reliability of assays used to measure calcidiol and calcitriol. An additional benefit would be to submit samples to a laboratory that also measures 24,25(OH)$_2$-vitamin D as part of a vitamin D metabolite panel (e.g. Heartland Assays, Ames Iowa).

WHAT ARE THE CAUSES OF HYPERCALCEMIA IN CATS?

The frequency for the occurrence of total serum hypercalcemia from biochemical panels from sick or well cats is not known. The only large survey of the causes of hypercalcemia in cats was reported based on measurement of serum total calcium in 2000.[20] Ionized hypercalcemia concentration has been sporadically reported in cats with specific diseases, but not in a series of cats with varying causes of hypercalcemia. Idiopathic hypercalcemia, CKD (discussed further below) and neoplasia are the most common and important differential diagnoses to exclude as the cause for parathyroid independent hypercalcemia. Overt hypervitaminosis D, granulomatous disease, and hypoadrenocorticism are other far less common causes of hypercalcemia in cats. Calcium oxalate urolithiasis was reported to be associated with hypercalcemia in cats; however, it is likely that hypercalcemia preceded the formation of stones rather than the urolithiasis acting as a stimulus for the formation of hypercalcemia.[20,21] IHC was not considered as a diagnostic category in one large study of cats with hypercalcemia,[20] but in another study the occurrence of IHC in 20 cats was published that same year.[21] Primary hyperparathyroidism was infrequently diagnosed as the cause of the hypercalcemia at a teaching hospital (4 of 71 cats)[20] but this diagnosis is far more frequently made by veterinary endocrine referral laboratories.[1,9]

Based on the number of consultations by veterinary internists and endocrinologists, as well as sample submissions to endocrine laboratories, idiopathic hypercalcemia (IHC) is currently the most-common cause of hypercalcemia in cats in North American and likely so in other parts of the world.[1,2,5-7] IHC in cats was first diagnosed in the late 1990's. IHC is not likely to be a diagnosis that was missed prior to this time due to our group's longstanding interest in and surveillance for disorders of calcium metabolism in cats.[22] It is possible that some of the increase in the diagnosis of IHC can be attributed to greater veterinary care for cats, including more widespread use of biochemical screening in primary care practices. The expected lifespan for cats has increased over this time and many changes in animal husbandry that include food and the environment have occurred that in some way could be operative in the development of IHC.[5,6]

While MAH is the number one cause of pathological hypercalcemia in the dog,[23] it occurs far less frequently in the cat. Based on serum total calcium and how the data is parsed, MAH is 3rd in frequency behind IHC and CKD in cats with hypercalcemia.[20] In dogs the overwhelming cause of MAH is lymphoma with occasional carcinoma as the diagnosis[23] whereas in cats lymphoma and carcinomas each account for about 1/3 of the cases.[20] Patients with MAH are usually "sick" as it takes a reasonably large tumor burden to synthesize the messengers that result in hypercalcemia. MAH is less likely to be the diagnosis if the hypercalcemia persists for a long period of time without the cat showing more clinical signs. The less sick the cat is in the face of persistent ionized hypercalcemia, the more the likelihood for the diagnosis to be that of IHC or primary hyperparathyroidism.

Ingestion of the house plant day-blooming jessamine (*Cestrum diurnum*) that contains calcitriol-glycosides is cited as a cause of hypercalcemia in cats,[24] but there are no detailed reports. Jasmine is sometimes confused with day-blooming jessamine, but contains no active vitamin D metabolites.[25] There are anecdotes that cats can develop hypercalcemia following the licking of anti-psoriasis cream off human skin, but there are no reports proving that this actually occurs either. Comments on veterinary list-serves have noted an association of hypercalcemia with the use of lactulose to treat constipation in cats but this has not been verified. A study in normal adult men showed that lactulose increased intestinal absorption of calcium and increased urinary calcium excretion but circulating calcium concentration did not change.[26] It is important to fully know the calcium status of constipated/obstipated cats before any treatment, as ionized hypercalcemia can occur before treatment as a contributing mechanism or cause of the constipation. Hypoadrenocorticism is associated with hypercalcemia in sick cats based on measurement of serum total calcium at the time of diagnosis but the hypercalcemia quickly abates when treatment is started.[27] Ionized hypercalcemia was noted in one cat with iatrogenic hypoadrenocorticism.[28]

SIGNALMENT AND CLINICAL SIGNS OF IHC CATS

In a report from 427 cats with IHC evaluated at an endocrinology laboratory, the age at diagnosis ranged from 0.5 to 20 years (mean 9.8 ± 4.6 yr). Males and females were equally represented in this study. Long-haired cats were noted to be overrepresented at

27% of the cases in this report,[29] but not in a recent case-control epidemiological study (data analyzed post Todd Green Master's 2008).

No clinical signs were noted in 46% of IHC cats. Other clinical signs were largely related to gastrointestinal signs, including mild weight loss (18%), chronic constipation (5%), vomiting and decreased appetite. IBD was diagnosed in 6% of the IHC cats of this study. Lower urinary tract signs may be observed, especially if urolithiasis is present. Uroliths or renoliths were observed in 15%, and calcium oxalate stones were specifically noted in 10% of cases. Polyuria/polydipsia has not been frequently reported in cats with IHC.[29]

In many instances, hypercalcemia based on measurement of total serum calcium is fortuitously discovered following submission of serum samples from wellness examinations, pre-anestheic evaluation of seemingly healthy individuals, those with routine medical conditions, and those from cats forming calcium-oxalate stones. Hypercalcemia is also sometimes discovered following submission of samples from cats with seemingly trivial clinical complaints like intermittent vomiting of hairballs. Though many cats with IHC do not have obvious clinical signs at first look, a more careful review of the history and physical examination often discloses some abnormality that could be explained by persistence of chronic ionized hypercalcemia. This includes low-grade weight loss, loss of muscle mass, and lethargy. Intermittent vomiting and constipation are also possibly due to adverse effects of ionized hypercalcemia on gut motility. Chronic ionized hypercalcemia is a risk factor for the genesis of calcium oxalate urolithiasis and for the development of chronic renal injury resulting in CKD that may take months to years to develop.

HOW IS THE DIAGNOSIS OF IHC ESTABLISHED?

The diagnosis of IHC is one of exclusion after initially confirming that the ionized calcium is increased. All the known causes of hypercalcemia should ideally be eliminated – this kind of workup can be exhaustive and expensive. The increase in circulating ionized calcium in IHC can be mild, moderate, or severe, as it can also be with other causes of hypercalcemia. Often mild increases in total or ionized calcium that are discovered fortuitously tend to increase over time, but to a varying magnitude. We have observed the ionized calcium concentration to fluctuate into and above the reference range, especially when the hypercalcemia is marginal in magnitude. The repeatability of increased serum total or ionized calcium in cats with IHC over the same day or between days has not been systematically studied. Hypercalcemia has often been assumed to exist at a constant concentration in dogs and cats with a variety of underlying diseases, but wide fluctuations in circulating total calcium have been reported at times in humans with primary hyperparathyroidism.[30] We have observed large fluctuations in total and ionized calcium concentrations on occasion in some cats with IHC and those with primary hyperparathyroidism. The effect of fasting or feeding prior to collection of blood samples to measure ionized calcium have not been investigated in cats with IHC; relatively small increases in ionized calcium occur following meals in normal cats.[1]

In order to exclude other causes of hypercalcemia, a minimum database, including a CBC, biochemistry profile and urinalysis, should be performed. Additionally, analysis of PTH and 25-hydroxyvitamin D are necessary to rule out hyperparathyroidism and hypervitaminosis D as the cause of the hypercalcemia (see sections above). The typical pattern for calcium regulatory hormones in IHC would be for the PTH concentration to be within the reference range (often lower end), the PTHrP concentration to be undetectable, and to have a normal serum ionized magnesium concentration.[29] Most 25-hydroxyvitamin D and calcitriol concentrations are usually within the reference range but a few cats with IHC have been noted to have values increased above the reference range.[21,29] PTHrP was undetectable in 9 of 11 cats with IHC but was increased in 2 cats without evidence for neoplasia.[21] Detection of an increased level of PTHrP increases the suspicion for hypercalcemia of malignancy but a normal or low PTHrP concentration does not definitively exclude neoplasia.

Chest radiographs are useful to rule out metastatic pulmonary nodules and mediastinal lymphoma that may be associated with hypercalcemia, though unlike in dogs, mediastinal lymphoma is not common in cats. A combination of abdominal radiographs and ultrasonography can be useful to determine the presence of urolithasis (kidney, ureter, bladder, urethra), obstructive nephropathy from the stones, or the presence of inflammatory/infiltratitve masses that could be associated with the genesis of the hypercalcemia. Treatment recommendations and prognosis may change with the presence of stones and their location.

Many cats with IHC can still effectively concentrate their urine > 1.030,[21] which differs from dogs with hypercalcemia. Dogs often present with polyuria and dilute urine in the hyposthenuric range due to inhibitory effects of calcium on ADH and the collecting tubule. In cats without CKD, a surprising preservation of ability to concentrate urine in the face of hypercalcemia has been noted in two reports.[20,21] In one study, including cats with CKD, the mean urine specific gravity was 1.032, with a range of 1.012 to 1.060.[21] In a study of 5 normal female cats, diuresis was not induced following oral prednisolone treatment.[31]

CAN A CAT WITH CKD ALSO HAVE IHC?

It is critically important to measure ionized calcium in CKD patients since hypercalcemia based on total serum calcium can exist with or without increases in ionized calcium. PTH is often increased or high reference range (renal secondary hyperparathyroidism) in routine CKD patients; PTH is normal to suppressed in cats with CKD and IHC. Approximately 20% of 102 cats[8] and 490 dogs[32] with azotemic CKD were reported to have hypercalcemia based on assessment of serum total calcium. When these same animals were evaluated following measurement of ionized calcium, cats (29%) with ionized hypercalcemia were detected about 3 times as frequently as in dogs (9%). Serum total calcium underestimates the number of cats with CKD and ionized hypercalcemia while overestimating the number of dogs with ionized hypercalcemia.[8,32] These animals were not characterized further as to any relationship to PTH or calcium regulatory hormones. In another study, ionized hypercalcemia occurred in 10 of 40 (25%) azotemic CKD cats; only 5 (12.5%) of these same cats had hypercalcemia based on serum total calcium measured at the same time. Most of the CKD cats with ionized hypercalcemia were found in those characterized as IRIS stage 2 CKD (6 of 12), whereas 4 of 22 cats were in IRIS stage 3 and 0 of 6 cats in IRIS stage 4 were observed with ionized hypercalcemia.[33] Though not reported as such in the publication, further communication with the primary author revealed that only 2 of the 10 cats with ionized hypercalcemia had increased serum total calcium at the same time. PTH was within the reference range in 6 of 10 cats with ionized hypercalcemia. PTH was suppressed to the lower 1/3 of the reference range in 3 of these 6 cats with ionized hypercalcemia but PTH was within the upper half of the reference range in 2 of the 6 cats. Increased PTH was found in the remaining 4 CKD cats with ionized hypercalcemia, possibly documenting tertiary hyperparathyroidism as the cause for the hypercalcemia in this population. It is also possible that failure to fully suppress PTH secretion was responsible for some of the increase in ionized calcium for those 6 cats that had PTH concentrations within the reference range.[33] At least some of the cats with reference range PTH would be considered to have IHC. Serum phosphorus was > 6.1 mg/dl in 2 of these 10 CKD cats with ionized hypercalcemia and was > 4.1 mg/dl in 10 of 10 cats. In tertiary hyperparathyroidism, PTH will be obviously increased whereas in IHC, PTH will be normal to suppressed.[34] In a report of 20 cats with IHC, 3 had azotemic CKD for 6 to 10 months before ionized hypercalcemia developed. Five cats had azotemic CKD and ionized hypercalcemia discovered at the same time. In the remaining 12 cats, ionized hypercalcemia was discovered in the absence of azotemia. Azotemia did develop in 3 of these 12 cats ranging from18 to 40 months later. Cause and effect from the ionized hypercalcemia to create CKD could not be established.[21]

FUTURE INVESTIGATION OF IHC

It is likely that multiple underlying cause(s) or mechanism(s) will be discovered for what we presently call IHC in cats. Up-regulating and down-regulating genetic predispositions as well as epigenetic changes that alter calcium metabolism potentially will be identified. There is value in the measurement of calcidiol, calcitriol, and 24,25(OH)2-vitamin D in some human patients with hypercalcemia.[35] Measurement of calcitonin may also be informative in cats with IHC, as a subgroup of seemingly normal experimental cats have been identified that do not secrete calcitonin commensurate with the degree of hypercalcemia.[36] The usefulness for measurement of urinary fractional excretion of electrolytes has been questioned in cats,[37] but a recent report suggests it might have clinical utility for calcium metabolic problems in cats.[38] The phosphatonin FGF-23 along with Klotho plays an important role in the regulation of phosphorus metabolism but it also has effects on calcium metabolism. FGF-23 inhibits synthesis of calcitriol by decreasing the activity of 1-alpha-hydroxylase. FGF-23 also increases the activity of 24,25-hydroyxlase, which enhances degradation of calcidiol and calcitriol.[39,40] Increased FGF-23 can be found in some forms of MAH in humans but hypercalcemia develops in some models of FGF-23 knockout mice.[40,41] Studies of the FGF-23-Klotho-vitamin D metabolite-axis could provide important clues to the underlying pathophysiology of IHC.

REFERENCES

[1]Schenck PA, et. al. **Fluid, electrolyte, and acid-base disorders in small animal practice** 2011:120-194. [2]Schenck PA, et. al. **Vet Clin North Am Small Anim Pract** 2008;38:449-453, viii. [3]Refsal KR, et. al. **Vet Clin North Am Small Anim Pract** 2001;31:1043-1062. [4]Peterson M. **ACVIM Forum** 2013. [5]Peterson ME. **Insights into Veterinary Endocrinology** 2012. [6]Peterson ME. Insights into Veterinary Endocrinology 2012. [7]de Brito Galvao JF, et. al. **Veterinary Focus** 2011;21:27-34. [8]Schenck PA, et. al. **Can J Vet Res** 2010;74:209-213.[9]Bolliger AP, et. al. **Vet Clin Pathol** 2002;31:3-8. [10]Parker VJ, et. al. **J Fel Med Surg** 2015. [11]Pineda C, et. al. **Domest Anim Endocrinol** 2012;42:256-264. [12]Wehner A, et. al. **J Feline Med Surg** 2013;15:730-736. [13]Morris JG, et. al. **J Nutr** 1999;129:909-912. [14]Peterson EN, et. al. **J Am Vet Med Assoc** 1991;199:904-906. [15]Moore FM, et. al. **J Am Vet Med Assoc** 1988;193:1099-1100. [16]Mealey KL, et. al. **J Am Vet Med Assoc** 1999;215:959-962, 946. [17]Sharma OP. **Curr Opin Pulm Med** 2000;6:442-447. [18]Stern JA, et. al. **J Am Anim Hosp Assoc** 2011;47:e116-120. [19]Rosol TJ, et. al. **Endocrinology** 1992;131:1157-1164. [20]Savary KC, et. al. **J Vet Intern Med** 2000;14:184-189. [21]Midkiff AM, et. al. **J Vet Intern Med** 2000;14:619-626. [22]Chew DJ, et. al. **J Natl Cancer Inst** 1975;11:46-52. [23]Messinger JS, et. al. **J Vet Intern Med** 2009;23:514-519. [24]Drazner FH. **J Am Vet Med Assoc** 1981;178:1252-1256. [25]Brownie CF. **J Am Vet Med Assoc** 1987;191:613-614. [26]Seki N, et. al. **J Nutr Sci Vitaminol (Tokyo)** 2007;53:5-12. [27]Peterson ME, et. al. **J Vet Intern Med** 1989;3:55-58. [28]Smith SA, et. al. **J Am Anim Hosp Assoc** 2002;38:41-44. [29]Schenck PA, et. al. **J Vet Intern Med (Abstract)** 2004;442. [30]Wood GM, et. al. **Postgrad Med J** 1987;63:741-744. [31]Geyer N, et. al. **Vet Ther** 2007;8:239-246. [32]Schenck PA, et. al. **Am J Vet Res** 2005;66:1330-1336. [33]Giovaninni LH, et. al. **Pesquisa Veterinária Brasileira** 2013;33:229-235. [34]Jamal SA, et. al. **J Clin Densitom** 2013;16:64-68. [35]Jacobs TP, et. al. **J Clin Endocrinol Metab** 2014;99:708-712. [36]Pineda C, et. al. **Am J Vet Res** 2013;74:1514-1521. [37]Finco DR, et. al. **Am J Vet Res** 1997;58:1184-1187. [38]Pimenta MM, et. al. **J Fel Med Surg** 2013;15:825. [39]de Brito Galvao JF, et. al. **J Vet Emerg Crit Care** 2013;23:134-162. [40]Wolf M. **Kidney Int** 2012;82:737-747. [41]Shimada T, et. al. **J Clin Invest** 2004;113:561-568.

NUTRITIONAL MANAGEMENT OF IDIOPATHIC HYPERCALCEMIA (IHC) IN CATS

Valerie J. Parker, DVM, DACVIM, DACVN
Joao Felipe deBrito Galvao*, MV, MS, DACVIM
Dennis J. Chew, DVM, DACVIM
Columbus, OH
***Downers Grove, IL**

This is the 2nd seminar in a series of 3 on IHC by the same authors at this Forum. The minimum criteria for the diagnosis of IHC includes documentation of ionized hypercalcemia in association with suppressed PTH, and evidence for absence of other known causes of hypercalcemia. Further details about diagnosis are provided in the 1st session. Management of IHC usually begins with a dietary recommendation to attempt to restore normocalcemia. Reports of treatment outcome following dietary change are quite limited, so diet recommendations are largely based on expert opinion and uncontrolled case studies in small numbers of cats. We have observed decreased circulating ionized calcium in some cats following dietary change, but the magnitude and duration of this decrement can be quite variable. Future studies comparing test and control diets are needed to determine the effects, if any, of altering intake of nutrient(s) on concentrations of the calcium regulatory hormones PTH, calcidiol, calcitriol, and 24,25(OH)2-vitamin D in addition to that for ionized calcium. Treatment with bisphosphonates and glucocorticosteroids usually are reserved for cats with IHC that fail dietary treatment; this will be reviewed in the 3rd session.

DOES DIET CREATE IHC?

Diet is not likely to be a direct cause of IHC, otherwise most cats eating commercial diets would develop hypercalcemia. It seems more likely that hypercalcemia develops only in a "sensitive cat in a provocative environment" – in this case, diet would be the provocateur. It is possible that ingestion of sufficient amounts of some dietary nutrient(s) "unmasks" an underlying metabolic problem that results in the development of hypercalcemia. It is also possible that some dietary constituents could influence expression of genes involved in calcium metabolism. In these instances, hypercalcemia would not develop if the "sensitive" cat were not exposed to the specific nutrient(s), or to some threshold intake of them.

IS IHC A NUTRIENT-SENSITIVE DISEASE?

This may be true for some cats in which dietary modification results in the restoration of normocalcemia, at least transiently. This will be discussed further below. We have encountered some cats in which dietary modification appeared to effectively restore normocalcemia, as well as cats with IHC in which there was no decrease in ionized calcium following dietary modification. Given our present level of understanding, we cannot predict with certainty just what change(s) in nutrient intake result in restoration of normocalcemia, or if what we observe is a placebo-related response in some cases.

IS THERE ONE SPECIFIC DIETARY NUTRIENT ON WHICH WE SHOULD FOCUS THAT WILL CONSISTENTLY DECREASE CIRCULATING IONIZED CALCIUM?

Regulation of the circulating calcium concentration is dynamic and complex. It has not been determined how much of the hypercalcemia in IHC cats results from too much dietary calcium intestinal absorption, increased bone resorption, reduced renal excretion of calcium, or combinations of these processes. Many of the nutrients in the diet interact with each in ways that affect dietary calcium absorption and not all calcium in the diet is biologically available for absorption.[1] Vitamin D is one obvious dietary nutrient that can affect intestinal absorption of calcium and it also has effects on osteoclastic bone resorption that can contribute to the degree of calcemia.[2] Vitamin A has effects on the osteoclast that can work in concert with vitamin D to increase bone resorption.[3] Other dietary factors affecting intestinal calcium absorption and bone resorption include concentrations of vitamin K[4], phytates[1], fiber[1] (discussed further below), and specific carbohydrates.[5] The contribution of carbohydrates in the diet to genesis of hypercalcemia is not usually given much attention, but hypercalcemia and calcium oxalate crystalluria were induced in a rat model when 30% sucrose was added to diet, effects that did not happen when 30% lactose was added to the diet.[5] While one of the simplest nutrients to target is dietary calcium, it is important to remember that simply prescribing a lower calcium diet may be too simplistic as ultimately other factors influence intestinal calcium absorption and calcium homeostasis. Although manipulation of single nutrients may be possible with the use of specially-made experimental diets, it is quite challenging to study the effects of a commercially-available feline diets on calcium metabolism when multiple nutrients vary.

WHAT DO WE KNOW ABOUT DIETARY CALCIUM CONTENT IN THE MANAGEMENT OF IHC?

Some veterinary nutritionists recommend diets to treat IHC based on a decreased calcium content on a g calcium/1000 kcal (Mcal) energy basis.[6] Minimal and maximal nutrient recommendations for cat food are provided by the Association of American Feed Control Officials (AAFCO) and the National Research Council (NRC). Most diets sold over-the-counter should meet AAFCO requirements; however, veterinary therapeutic diets may be specifically modified in order to provide certain nutrients at concentrations less than AAFCO minimums. The average calcium content of grocery store foods in the USA is approximately 2.0 to 3.0 g calcium per Mcal (200-300 mg/100 kcal), though some contain up to 6.0 g calcium per Mcal (600 mg per 100 kcal).[7] Some of the highest calcium diets are "high-fiber" diets; thus one must carefully weigh the pros and cons of recommending a high-fiber diet for dietary management of IHC when there is some evidence that reducing dietary calcium may be effective in restoring normocalcemia. Nutrient concentrations of diets can be found either in product guides or by contacting the diet manufacturer, but this information is not readily available from the routine diet label. Nutrient profiles are constantly evolving and this information may change up to every 6-12 months. For feline adult maintenance, the NRC recommended allowance (RA) is 0.72 g calcium per Mcal [8] and the AAFCO minimum is 1.5 g calcium per Mcal.[9]

In order to rationally choose a diet that is lower in calcium intake, a complete diet history must be obtained; this includes the cat's primary diet, treats and any supplements. Once this starting dietary calcium concentration has been determined, the veterinarian can determine which diets would provide less calcium, considering potential comorbid conditions. A veterinary therapeutic renal diet low in calcium (and phosphorus) content may be appropriate for a cat with hypercalcemia and CKD, but it may not be appropriate to feed a young, otherwise healthy cat a reduced phosphorus, reduced protein diet. A home-prepared diet providing approximately 0.60 g calcium per Mcal has been used successfully by one group of nutritionists to restore normocalcemia in some cats with IHC within one month of feeding.[6]

Feeding of a high protein and low carbohydrate food similar to what cats would eat in the wild (i.e., 40-60% of calories from protein; 30-50% of calories from fat, and <15% of calories from carbohydrates) has been recommended to effectively lower serum calcium concentration in some cats with IHC, especially those with low magnitude hypercalcemia.[10,11] This nutrient profile is what would be expected from veterinary therapeutic diets designed for cats with diabetes mellitus and also many over-the-counter canned feline diets. In reviewing these types of diets however, it should be noted that calcium content varies from about 1.5 to 5.5 g per Mcal.

WHAT DO WE KNOW ABOUT DIETARY VITAMIN D CONTENT IN THE MANAGEMENT OF IHC?

IHC is not the result of obvious excess dietary vitamin D intake since serum concentrations of 25(OH)-vitamin D have been within the reference range in most cats with IHC. However, the minimal requirement for vitamin D in cats is debatable since reference ranges have been established in cats fed vitamin D–supplemented diets. Normal concentrations of 25(OH)-vitamin D could still potentially be associated with IHC in cats if there are up-regulating mutations in the VDR (vitamin D receptor). These possibilities have not yet been investigated.

Although the amount of vitamin D supplied in most diets is not listed on the diet label, this information should be obtainable from the pet food manufacturer.[6] Cats may be affected by hypervitaminosis D even when companies claim a certain expected concentration of vitamin D, possibly due to errors in diet formulation.[12] Vitamin D content in commercial diets is the sum of what is provided naturally by animal and plant ingredients as well as that added by the vitamin pre-mix. It has been recommended by some to avoid a diet that contains organ meat or oily marine fish as sources that increase vitamin D when formulating a home-prepared diet.[6]

For adult cats, the NRC-RA for dietary vitamin D3 (cholecalciferol) is 70 IU per Mcal. The safe upper limit (SUL) is listed as 7,520 IU per Mcal.[8] AAFCO minimum and maximum recommendations for feline adult maintenance are 125 and 2,500 IU per Mcal, respectively.[9] Clearly, there is a wide range of acceptable dietary vitamin D in commercial cat foods. Feeding a diet formulated to be low in vitamin D content at < 200 IU per Mcal has been recommended in dietary treatment of cats with IHC.[10,11]

The vitamin D requirement specifically for feline adult maintenance has not been studied; thus, recommendations for kitten growth has been extended to the adult cat population.[8] Determination of toxicity of dietary vitamin D has mostly used the endpoints of hypercalcemia, tissue calcification, and the presence or absence of renal pathology.[8] It has been concluded that cats are more resistant to dietary vitamin D toxicity than other species based on these endpoints since the SUL for cats appears to be 9 times higher than that for growing dogs.[8,13] It should be noted that these endpoints for vitamin D toxicity are for advanced changes of overt hypervitaminosis D. Studies are lacking that determine early end points that mark the development of hypervitaminosis D in cats. Hypercalciuria is a standard early marker of excess vitamin D supplementation in human medicine.[14,15] Hypercalciuria occurs as an early homeostatic mechanism that initially prevents or delays the onset of hypercalcemia. Dietary studies that determine urinary calcium excretion at various levels of dietary intake of vitamin D are sorely needed to further understand the pathophysiology and treatment of IHC.

WHAT DO WE KNOW ABOUT DIETARY VITAMIN A CONTENT IN THE MANAGEMENT OF IHC?

Commercial cat foods often contain relatively high concentrations of vitamin A. In combination with high vitamin D concentrations, this may potentially work in concert to affect calcium metabolism in ways that could contribute to the development of IHC in a susceptible population of cats. Most focus has been on dietary vitamin D for its potential contribution to the development of IHC, but vitamin A deserves more attention for its possible contribution to the development and maintenance of IHC in cats. For feline adult maintenance, the NRC-RA for vitamin A is 833 IU per Mcal and the SUL is 83,325 IU per Mcal. AAFCO minimum and maximum dietary recommendations vitamin A are 1,250 IU per Mcal and 187,500 IU per Mcal, respectively. Similarly to vitamin D nutrient requirements, there is a considerably large range of acceptable vitamin A concentrations for feline diets.

HOW HELPFUL ARE LESS ACIDIFYING OR RENAL DIETS IN MANAGEMENT OF IHC?

Anecdotally, the feeding of veterinary therapeutic renal diets has been associated with restoring normocalcemia in some cats with IHC; the specific mechanisms as to how this benefit is achieved are not known. Veterinary therapeutic renal diets are less acidifying than maintenance diets and are lower in calcium and phosphorus concentrations. The decreased intake of dietary calcium could lead to a decrease in the amount of intestinal calcium absorption. However, feeding a reduced phosphorus diet could enhance renal calcitriol synthesis (effects on renal 1-alpha-hydroyxlase), thus potentially offsetting the advantage of decreased intestinal calcium absorption from lower dietary calcium intake. In one study, 2 of 15 cats with CKD developed ionized hypercalcemia after being fed a low phosphorus, low protein diet. Ionized calcium normalized after discontinuing dietary phosphorus and protein restriction.[16] Chronic dietary acidification increases the availability of the ionized form of calcium in the intestinal lumen facilitating absorption. Chronic acidification can also increase bone resorption as a buffering response to chronic acidosis, which then contributes to an increase in circulating ionized calcium in normal cats[17] and so could potentially contribute to hypercalcemia in disease. Feeding a non-acidifying diet may decrease the release of calcium from bone and decrease calcium filtered through the glomerulus.[18]

HOW HELPFUL ARE HIGH FIBER DIETS IN RESTORATION OF NORMOCALCEMIA IN CATS WITH IHC?

Higher fiber diets were associated with the restoration of normocalcemia in 5 of 5 cats with calcium oxalate stones and a likely diagnosis of IHC (high ionized calcium concentration). Four of these cats were switched to a veterinary therapeutic diet high in insoluble fiber (Hill's Prescription Diet w/d). One of the cats was fed an over-the-counter feline maintenance diet with added psyllium, a soluble fiber supplement. In the cats with reported follow-up, hypercalcemia resolved based on measurement of serum total calcium but none of the cats had ionized calcium concentration monitored. Another study showed that a dietary change to higher fiber in some IHC cats failed to restore normocalcemia; these cats were not specifically selected to have calcium oxalate urinary stones.[19]

The effects of fiber on intestinal absorption of calcium are complex and depend on the type and amount of fiber in the diet and the interactions with other nutrients in the diet. It has been theorized that supplemental fiber may lead to increased binding of intestinal calcium, preventing its absorption, and also to decreased intestinal transit time through the small intestine, reducing calcium absorption.[20,21] The salutary effect of a higher fiber diet, if any, is not simply due to the binding of calcium to fiber. It appears to be common practice for most manufacturers to increase the concentration of calcium in high-fiber diets to offset the potential for decreased absorption. The concept of choosing a "high-fiber" diet can be challenging given that most companies only provide the crude fiber concentration, which is not a good indicator of total dietary fiber.[22] Other nutrient alterations found in higher fibers diets such as vitamin A and vitamin D concentrations could also affect treatment outcomes of "high fiber" diets. Studies investigating the effects of soluble and insoluble fiber on IHC are needed.

HOW HELPFUL ARE HIGHER SALT DIETS IN MANAGEMENT OF IHC?

The addition of sodium chloride is potentially useful in cats with IHC as long as the added salt enhances renal calcium excretion without an increased risk of calcium oxalate uroliths. Increased dietary sodium chloride increased urine volume but did not increase calcium oxalate relative supersaturation (RSS) in a small number of young healthy cats.[23] However, an increase in urinary excretion of calcium does not always correlate to the development of calcium urolithiasis, since the concentration of calcium in the urine also depends on the degree of water excreted at the same time.[24] The feeding of increased dietary salt is variably effective in creating diuresis that is reflected by decreases in urinary specific gravity and increased water consumption. The USG collected from random urine samples significantly decreased in normal older cats fed a higher salt diet from a baseline of 1.047 to 1.034 by 3 months in one study.[25]It appears that USG determined in samples from 24-hour collections more consistently demonstrates decreased USG during high dietary salt intake. USG decreased from 1.051 to 1.045 without an effect on the urinary concentration of calcium in one study of cats fed a higher salt diet.[26] The effects of additional dietary sodium chloride have yet to be reported in cats with IHC.

Treatment with higher salt content diets have not been studied in IHC cats, with or without calcium oxalate stones. Higher salt intake potentially could promote increased water intake, volume expansion, and a dilution effect that would decrease circulating ionized calcium to some degree. Increased water turnover would then create a more dilute urine that should help prevent calcium oxalate stone growth by reducing RSS. Increasing salt intake up to 3.7 g per Mcal has been reported to be safe without detection of deleterious effects on renal function, cardiovascular function, and systemic blood pressure when studied in normal cats, geriatric cats, and cats with surgically reduced renal mass.[23,25,27-29] Future studies of higher dietary salt intake for treatment of cats with IHC are warranted.

HOW HELPFUL ARE UROLITHIASIS DIETS IN MANAGEMENT OF IHC?

Decreased dietary calcium is not recommended alone in non-hypercalcemic calcium oxalate stone formers, as lower calcium diets (without also decreasing oxalate intake in parallel) have been reported as a risk factor for the genesis of calcium oxalate stones [21,30-33] Recent literature in humans and dogs indicates that there might be an advantage of increased dietary Ca intake, which is thought to be related to interactions between Ca and oxalate in the intestinal lumen.[34,35]

Ionized hypercalcemia resolved after feeding a canned formulation calcium oxalate–prevention diet that is no longer commercially available (Hill's Prescription Diet c/d-oxl; later called Hills Prescription Diet x/d) in two of three cats with IHC and calcium oxalate uroliths. The third cat had a reduction in the magnitude of hypercalcemia but did not achieve normocalcemia. Hypercalciuria decreased in these cats that had previously formed a calcium oxalate urinary stone but none were hypercalcemic based on total calcium. Both urinary calcium excretion and concentration significantly decreased with the calcium oxalate-prevention diet.[36]

Cats with calcium oxalate stones were enrolled in a study to determine the recurrence rate between two different diets. The test diet was designed to be higher in antioxidant and n-3 fatty acid concentrations but lower in calcium and magnesium concentrations. This diet was compared to a diet similar to popular grocery store foods. Twelve of 36 cats developed hypercalcemia based on measurement of either total or ionized calcium during the two years of this study. Cats fed the test diet developed fewer episodes of hypercalcemia. RSS and the concentration of calcium in the urine were lower in cats fed the test diet. The occurrence of hypercalcemia in this population of cats forming calcium oxalate stones was noted to be higher than generally recognized.[37] The percentage of cats with calcium oxalate urolithiasis and hypercalcemia at the time of diagnosis in an older report by the same investigator was estimated to be 35%.[38] Calcium oxalate calculi have also been noted in 10% of cats with presumed IHC.[39]

HOW SHOULD CATS WITH IHC BE EVALUATED AFTER A DIET CHANGE HAS BEEN IMPLEMENTED?

Ionized calcium should be measured following 4 to 6 weeks of feeding the new diet. In some cats, there will be a dramatic decrease in ionized calcium to within the reference range or closer to it. Continued feeding of the newly recommended diet should be continued for another 2 to 3 months to see if any observed benefits are sustained. It is possible that feeding of a new diet will require 2 to 3 months to see maximal benefits, if any. If ionized calcium remains above the reference range and the IHC cat has few clinical signs, recommending a change to another diet with a different nutrient profile (calcium, vitamin D, vitamin A, sodium) is one path that can be followed. A systematic approach to changing diets for cats that fail in restoration of normocalcemia following the initial diet change recommendation has not yet been developed. Consideration for treatment with bisphosphonates or prednisolone should be considered for those with high levels of ionized hypercalcemia and/or clinical signs associated with hypercalcemia not responding to dietary change.

REFERENCES

[1]Camara-Martos F, et. al. **Biol Trace Elem Res.** 2002;89:43-52. [2]Cline J. **Top Companion Anim Med** 2012;27:159-164. [3]Johansson S, et. al. **J Bone Miner Res** 2001;16:1899-1905. [4]Beulens JW, et. al. **Br J Nutr** 2013;110:1357-1368. [5]Hennequin C, et. al. **Urol Res** 1998;26:57-63.[6]Fascetti AJ, et. al. **Applied Veterinary Clinical Nutrition** 2012:289-300. [7]Parker V. **Personal communication**, 2013. [8]**National Research Council** 2006;366-367. [9]**AFFCO** 2012;149-150.[10]Peterson M. **ACVIM Forum** 2013.[11]de Brito Galvao JF, et. al. **August's Consultations in Feline Internal Medicine**, 2015.[12]Wehner A, et. al. **J Feline Med Surg** 2013;15:730-736.[13]Sih TR, et. al. **Am J Vet Res** 2001;62:1500-1506. [14]Cranney A, et. al. **Am J Clin Nutr** 2008;88:513S-519S. [15]Parfitt AM, et. al. **Am J Clin Nutr** 1982;36:1014-1031. [16]Barber PJ, et. al. **J Small Anim Pract** 1999;40:62-70.[17]Ching SV, et. al. **J Nutr** 1989;119:902-915. [18]Rosol TJ, et. al. **Vet Clin Pathol** 1995;24:49-63. [19]Midkiff AM, et. al. **J Vet Intern Med** 2000;14:619-626. [20]McClain HM, et. al. **J Am Anim Hosp Assoc** 1999;35:297-301. [21]Parivar F, et. al. **J Urol** 1996;155:432-440. [22]de-Oliveira LD, et. al. **J Anim Physiol Anim Nutr (Berl)** 2012;96:895-906. [23]Biourge VC, et. al. **J Vet Intern Med** 2001;15:301. [24]Xu H, et. al. **J Vet Intern Med** 2006;20:738. [25]Reynolds BS, et. al. **J Vet Intern Med** 2013;27:507-515. [26]Hawthorne AJ, et. al. **J Nutr** 2004;134:2128S-2129S. [27]Chetboul V, et. al. **PloS one** 2014;9:e97862. [28]Luckschander N, et. al. **J Vet Intern Med** 2004;18:463-467.[29]Cowgill LD, et. al. **J Vet Intern Med** 2004;18:415. [30]Lekcharoensuk C, et. al. **J Am Vet Med Assoc** 2001;219:1228-1237. [31]Dijcker JC, et. al. **Veterinary Record** 2012;171:46. [32]Dijcker JC, et. al. **Vet Journal** 2012;194:235-239. [33]Dijcker JC, et. al. **Nutr Res Rev.** 2011;24:96-110. [34]Stevenson AE, et. al. **Vet Ther** 2004;5:218-231.[35]Penniston KL, et. al. **Urology** 2009;73:484-489. [36]Lulich JP, et. al. **J Am Anim Hosp Assoc** 2004;40:185-191. [37]Lulich J, et. al. **ACVIM Forum** 2014. [38]Osborne CA, et al. **Vet Clin North Am Small Anim Pract** 1996;26:217-232.[39]Schenck PA, et. al. **J Vet Intern Med (Abstract)** 2004;442.

ORAL ALENDRONATE IN THE TREATMENT OF IDIOPATHIC HYPERCALCEMIA IN CATS

Joao Felipe de Brito Galvao*, MV, MS, DACVIM
Dennis J. Chew, DVM, DACVIM
Valerie J. Parker, DVM, DACVIM, DACVN
***Downers Grove, IL**
Columbus, OH

SHOULD ALL CATS WITH IDIOPATHIC HYPERCALCEMIA (IHC) RECEIVE TREATMENT?

Cats with minimal increases in circulating calcium concentrations are often ignored in clinical practice since many of these cats have mild or no apparent clinical signs. Even though obvious clinical signs are often not apparent, subtle clinical signs often exist as reviewed in the first session in this series. Excess calcium can be toxic to cells, exerting either physiological or structural effects particularly in the central nervous system, gastrointestinal tract, heart, and kidneys. Mineralization of soft tissues is an important potential complication related to the presence of ionized hypercalcemia that is in part determined by the concomitant concentration of serum phosphorus, but this does not develop in all IHC cats. When the calcium (mg/dl) times phosphorus (mg/dl) product is greater than 60, soft-tissue mineralization can be severe.[1] Cats with IHC usually have serum phosphorus concentration below 5.5 mg/d, which likely limits mineralization of soft tissues early in the process in the absence of CKD. The clinical outcome for cats with IHC that have not been treated has not been established following the initial diagnosis. An argument can be made to withhold treatment when an IHC cat has no recognizable signs, no identified risk factors for urolithiasis or CKD, and the increase in ionized calcium is minimal. A stronger argument can be made to treat IHC cats in which the ionized calcium concentration continues to escalate. When the ionized calcium increases to 1.0 mg/dl or more above the reference range, we recommend starting treatment even if the cat has no clinical signs. The strongest argument to start treatment exists for cats that have ongoing weight loss, depression, vomiting, constipation, urinary stones, emergence of CKD and or development of sub-maximally concentrated urine. The consequences of long-standing hypercalcemia can be devastating in those that develop calcium oxalate urolithiasis from the associated hypercalciuria, especially when renal or ureteral stones form that obstruct urine outflow. Recurrent calcium oxalate stone formation is more likely in IHC cats in which normocalcemia has not been restored. Persistence of ionized hypercalcemia provides ongoing risk to injure renal tissue already suffering from CKD (see discussion of CKD and IHC in the first session of this series).

SHOULD IHC CATS BE TREATED WITH DIET BEFORE DRUG THERAPY?

As detailed in the second session in this series, some cats do respond to a dietary change based on calcium, vitamin D, and vitamin A content of the diet addition to other nutrient concentration that can affect calcium metabolism. If normocalcemia is to be restored with dietary management, it usually occurs within 6 to 8 weeks of feeding the new diet. If normocalcemia has not been restored at this point, another diet with a different nutrient profile can be chosen, but drug therapy is more commonly chosen when the first diet change fails. Historically, treatment with glucocorticosteroids had been chosen as the next treatment but treatment with bisphosphonates is another option that can be considered. We do not recommend starting drug therapy immediately after the diagnosis of IHC since dietary treatment is effective in restoration of normocalcemia in some cats.

GLUCOCORTICOSTEROID TREATMENT OF IHC

Treatment with glucocorticoids restores normocalcemia or dramatically reduces the ionized calcium concentration in most cats with IHC, at least initially. A maximal decline in calcium to within the reference range often requires dose escalation and the beneficial effect may be transient. Approximately 80% of cats with IHC become normocalcemic with 1.5 to 2.0 mg/kg/day prednisone per day, but some may require increasing doses to remain normocalcemic over time.[2] It is important to not prescribe glucocorticosteriods before the diagnosis of the hypercalcemia has been established with some certainty, otherwise cytolytic effects in LSA and myeloproliferative disorders will make definitive diagnosis difficult or impossible. A mild calcium-lowering effect can be exerted by use of glucocortocosteroids in other forms of malignancy-associated hypercalcemia and in those with primary hyperparathyroidism. It is also preferred to have biopsy-proven IBD before the start of glucocorticosteroids.

The administration of glucocorticosteroids decreases serum calcium concentration by some combination of decreased intestinal absorption of calcium, decreased renal tubular calcium resorption, and decreased skeletal mobilization of calcium. There is some concern that glucocorticosteroids may increase urinary excretion of calcium, contributing to calcium oxalate urolith formation as occurs in some species, but little is known regarding the effects of glucocorticosteroids on filtration and tubular reabsorption of calcium in the cat. Daily prednisolone (oral liquid at average 2.2 mg/kg) treatment did not result in diuresis or an increase in calcium excretion in 5 healthy female cats.[3] This was a two-week study in cats with normocalcemia. Urinary calcium safety results as seen in this report may not be the same in cats with hypercalcemia undergoing prolonged treatment. Cats usually do not exhibit the obvious side effects from glucocorticosteroid treatment seen in dogs, such as severe polyuria and polydipsia and panting. Cats require higher doses to become immunosuppressed. Long-term treatment with prednisolone does contribute to muscle wasting[4-6] and possible induction of diabetes mellitus in some cats.

Oral prednisolone achieves greater maximal concentration in the circulation than does oral prednisone in the cat, possibly due to greater GI absorption of prednisolone or less hepatic conversion of prednisone to prednisolone.[7] Prednisolone is given orally at 5 – 10 mg/cat/day for 1 month before reevaluation. If the ionized calcium concentration is normal, this dose is continued for several months. If serum ionized calcium concentration is still increased, the dose is increased by 5 mg/cat/day. Some cats may require as much as 15 to 20 mg of prednisolone per day to restore normocalcemia.

Though prednisolone can be effective in restoration of normocalcemia in IHC cats, we now usually consider prednisolone as treatment after oral bisphosphonate treatment has failed to restore normocalcemia. In these instances, prednisolone is prescribed in addition to the oral bisphosphonate, but much lower doses of prednisolone may now be effective during combination drug therapy.

BISPHOSPHONATE TREATMENT FOR IHC CATS?

Historically, oral bisphosphonates have been recommended to treat IHC cats when dietary modification and prednisolone treatment have been unsuccessful in restoration of normocalcemia. Oral alendronate has become our preferred option to treat IHC cats after dietary modification has failed to restore normocalcemia.[8] Even though not extensively reported, we now consider bisphosphonate therapy a safer alternative to glucocorticosteroid use in cats that failed dietary intervention. Treatment with bisphosphonates may be useful to decrease the magnitude of hypercalcemia in cats with IHC by altering osteoclastic bone resorption. Even if increased bone resorption is not the main mechanism generating hypercalcemia, it is possible that benefits to decrease circulating calcium could still occur. Bisphosphonates reduce the activity and number of osteoclasts following binding to hydroxyapatite. Alendronate has been shown to lower urinary calcium excretion and reduction in calcium oxalate and calcium hydrogen phosphate supersaturation in rats with genetic hypercalciuric.[9] An alendronate-induced decrease in urine supersaturation should be beneficial in preventing stone formation in IHC cats.

IV treatment with bisphosphonates is almost never needed in IHC since the hypercalcemia is chronic and the cats are usually not in an acute crisis. The safety and efficacy of pamidronate given intravenously to three cats with hypercalcemia have been reported.[10,11] Mild hypocalcemia[11] and severe hypocalcemia and hypophosphatemia[10] was observed in two of the three cats described. Adequate hydration is essential when treating with bisphosphonates since these drugs may cause nephrotoxicity, especially at higher doses given intravenously. The administration of IV pamidronate every 1 to 2 months could be considered for an IHC cat with obvious clinical signs and high magnitude hypercalcemia when the owners are not able to administer oral medication.

HOW SAFE ARE BISPHOSPHONATES?

The long-term safety and efficacy of oral alendronate therapy has not been reported in cats. The safety and efficacy of oral alendronate treatment given once weekly for 6 months was reported in 12 cats with IHC.[12] Two of the 12 cats developed mild ionized hypocalcemia at 6 months of treatment. These 2 cats were not clinical for the hypocalcemia, and none of the 12 cats showed clinical side-effects from the alendronate (doses ranged from 5 to 20 mg/cat per week).[12] In humans, hypocalcemia, hypophosphatemia, and hypomagnesemia are well known side effects of bisphosphonate treatment.[13,14] The package insert for alendronate insert includes hypocalcemia and hypophosphatemia as possible adverse effects. Mild hypocalcemia is sometimes encountered during oral alendronate treatment of IHC in cats, but overt clinical signs are not usually encountered.[12] One cat with IHC that was treated with IV pamidronate developed moderate and then severe ionized hypocalcemia 5 and 6 days respectively after treatment; this cat also developed severe hypophosphatemia.[10] One cat that failed enrollment criteria developed acute malaise and high fever within a few hours after its first and only dose of oral alendronate,[12] a finding that has been noted to rarely occur in humans following acute cytokine release.[15,16] This cat was normal the next day. In another report, a single cat diagnosed with IHC was initially treated with IV pamidronate and subsequently given 5 to 10 mg/cat of alendronate per week to successfully control circulating calcium for 15 months.[10] We have followed some IHC cats undergoing alendronate treatment for over 2 years without reported clinical side effects.[2] The safety of oral alendronate treatment for cats with IHC and CKD has not been specifically studied, but we have not observed any documented decreases in renal function that we could attribute directly to the alendronate. Two of the 12 cats with IHC of one report had CKD at the time of initial diagnosis and no decrement in renal function was observed over the 6 months of that study.[12] Bisphosphonate treatment is sometimes used in people with CKD to attenuate arterial mineralization.[17]

Drug-induced esophageal damage (erosive esophagitis and esophageal stricture) and gastritis are of concern in humans taking oral bisphosphonates.[18-21] We have not observed the development of these lesions, nor have they been reported by others, following oral alendronate treatment in IHC cats. Studies in dogs suggest a direct adverse effect of alendronate on the esophageal mucosa rather than a simple mechanical effect from the pill. In one study in dogs, the presence of alendronate in the esophagus for 1 hour caused minor mucosal damage, but esophagitis was exacerbated when gastric juice containing alendronate was refluxed into esophagus.[20] It is thought that alendronate sodium is converted to free acid in the presence of gastric juice. No esophageal lesions and trivial gastric lesions (deemed not clinically significant by the pathologist) were reported from necropsy in cats undergoing relatively high doses (9 mg/kg twice a week in tuna juice) of oral alendronate for 49 weeks.[22]

Osteonecrosis of the jaw (ONJ) develops in small number of human patients receiving long-term bisphosphonate treatment.[23] ONJ has not been observed by us or reported by others in cats treated with bisphosphonates. ONJ is mostly reported in humans given high doses of IV bisphosphonates to treat cancer; it is far less frequently encountered following oral bisphosphonate treatment for postmenopausal osteoporosis.[13] The alveolar bone region of the mandible has a high basal turnover rate for bone and so may be especially prone to necrosis of bone matrix following exposure to bisphosphonates. It is not clear if bisphosphonate exposure creates bone necrosis directly or if this is secondary to reduced bone turnover.[13] ONJ was not described in cats with dental lesions treated for 49 weeks with oral alendronate.[22] Daily oral alendronate at 0.2 mg/kg or 1.0 mg/kg given to skeletally mature female beagles for 3 years resulted in matrix necrosis of the mandible in 25% of the dogs treated with the lower dose and in 33% of the dogs treated with the higher dose. No dogs displayed exposed bone lesions in this study.[24] IV zolendronate given to dogs at 0.06 mg/kg daily for 6 months resulted in more osteonecrosis of the mandible than that encountered with oral alendronate.[25]

An increased risk for bone fracture has been reported in humans on long-term bisphosphonate treatment presumably because of the increased brittleness of bone due to bisphosphonate therapy.[26] Bisphosphonate treatment in humans generally does not exceed 3 years due to concerns that acquired bone pathology outweighs previous benefits.[13] The biological half-life of bisphosphonates within bone is 200 days, 3 years, and 12 years in rats, dogs, and humans respectively, but this is not known for cats.[27] We have become aware of two cats that developed pathologic fractures following 9 and 5 years of treatment with weekly oral alendronate. Healthy male mongrel dogs treated with 5 mg daily oral alendronate for 23 weeks (average of 1 mg/kg/week) did not result in bone weakness or changes in structural and mechanical properties of bone.[28]

WHAT IS OUR TYPICAL TREATMENT PROTOCOL FOR ORAL ALENDRONATE TO TREAT IHC?

Any food in the stomach can drastically reduce the absorption of alendronate to near zero – bisphosphonates are poorly absorbed at best. The oral bioavailability of alendronate when administered in water to cats is reported to be about 3.0%.[22] This decreased by approximately ten-fold when alendronate was formulated in tuna juice.[22] To maximize intestinal absorption of alendronate, we recommend fasting cats overnight for 12 hours prior to the administration of medication, giving the pills in nothing other than tap water, and then feeding the cat two hours later. Though not specifically studied, an 18-hour fast prior and 4-hour fast post-pill might be a better protocol to achieve the highest possible intestinal absorption.[22] We do not recommend the administration of alendronate in pill pockets due to concern about decreased intestinal absorption that could occur. For the same reason, we do not recommend alendronate that has been formulated by compounding pharmacies in flavored solution or suspension. It is possible that some cats with IBD and IHC will not absorb alendronate as well as IHC cats without obvious GI disease. Prednisolone therapy may be needed in these cats initially to allow greater alendronate treatment later.

Given the risk of esophagitis and stricture associated with oral bisphosphonate treatment in humans, we advise extra caution to prevent esophageal tissue damage following oral alendronate administration in cats. The starting dose is usually 10 mg/cat (NOT per kg) per week initially. We recommend administration of whole tablets only, as cut tablets may increase exposure of the esophagus and stomach to adverse effects. We recommend "buttering" the cat's lips/nose as this has been shown to increase salivation and swallowing which contributes to decreased transit time and less time for mucosal contact from the pill.[29] The effect of butter on intestinal absorption of alendronate has not been specifically studied, but use of this as part of our treatment protocol has effectively restored normocalcemia in many cats. Five to 6 ml of tap water is administered via syringe to provide an additional measure to prevent the pills from getting caught in the esophagus.[30] Using these preventative measures, we have not yet observed any signs of esophagitis in cats treated with alendronate.

Ionized calcium should be measured 3-4 weeks after starting the 10 mg weekly dose of alendronate. If the ionized calcium has declined substantially but is not yet in the normal range, we recommend waiting another 3 to 4 weeks to see if a further reduction in circulating ionized calcium will occur. Some cats return to normocalcemia on 10 mg oral alendronate per week, whereas other cats require dose escalation to do so. If the ionized calcium remains above the reference range at this next visit, increase the dose to 20 mg once each week, or alternate giving 10 mg one week followed by 20 mg the next week to provide an average of 15 mg per week. As mentioned previously, we do not advise splitting alendronate tablets due to increased possibilities for toxic injury to the esophagus. Once the ionized calcium enters the reference range, we recommend reevaluation in 1, 3, and 4 to 6 months if the ionized calcium remains stable within the reference range. Over 90% of IHC cats had normal ionized calcium concentration at some point during treatment at an average alendronate dose of 15 mg/cat per week (10-40 mg/cat/week).[2] Many IHC cats return to normocalcemia following a 10 mg once weekly dose of oral alendronate, whereas some IHC cats will require 20 mg weekly to achieve normocalcemia. Rarely, 30 or 40 mg/cat/week oral alendronate will be needed to restore normocalcemia. Alendronate dose reduction should be prescribed for cats that achieve very low reference range ionized calcium in order to prevent the development of overt hypocalcemia. For cats that develop overt hypocalcemia, alendronate treatment should be discontinued, at least temporarily.

WHAT DO WE KNOW ABOUT ALENDRONATE FORMULATION?

We do not recommend any kind of alendronate that has been formulated by compounding pharmacies in flavored solution or suspension due to the severe decreases in intestinal absorption. A commercial alendronate suspension is available for human use. Use of this preparation would require administration of a large volume due to its low alendronate concentration. Additionally, the commercial suspension is flavored, which is likely to be unpleasant for cats. All of our initial treatment studies for IHC used the brand name of oral alendronate (Fossamax®), so we do not have experience regarding the efficacy of generic alendronate. Though transdermal alendronate is supplied by some compounding pharmacies, the effectiveness of these preparations has not been reported. It requires special methods of chemical application, patches, or gel-based iontophoresis to allow adequate transcutaneous absorption of alendronate in experimental animals or humans.[31-34]

WHEN SHOULD BISPHOSPHONATE TREATMENT BE STOPPED FOR IHC CATS?

The fasting protocol should be reviewed in detail with the owner to make sure that lack of compliance is not an issue for failure to restore normocalcemia despite dose escalation of oral alendronate. As with all oral treatments in cats, some owners are not capable of administering pills on a regular basis. Alendronate treatment should be stopped in IHC cats that fail to regain normocalcemia despite 30 to 40 mg weekly doses. Alternatively, prednisolone can be added on top of alendronate to see if a beneficial effect can be gained to lower circulating calcium during combination therapy.

It is not known how long oral alendronate treatment should be continued in those IHC cats that have regained normocalcemia for long periods of time. It is possible that the salutary effects to keep circulating calcium concentrations within the reference range may last long after alendronate is discontinued due to its long half-life in bone, but this has not been specifically studied.

Though bisphosphonate treatment is very often effective in restoration of normocalcemia in IHC cats, it would be far preferable to find the underlying cause(s) of IHC so that drug therapy would no longer be needed. Guidelines as to how long bisphosphonate treatment can safely be given to cats with any disease have yet to be established. We are concerned that some cats are now receiving bisphosphonate therapy for years that may be detrimental to the cat's long-term bone health (based on emerging reports of pathological fractures in some cats). It may not be enough to just monitor calcium and renal function status in IHC cats during treatment interventions. The measurement of calcium regulatory hormones (PTH, calcitonin, calcidiol, calcitriol, $24,25(OH)_2$-vitamin D, FGF-23 Klotho) before and after treatment interventions will likely reveal important components for the pathophysiology of IHC in cats and may provide targets to be altered during therapy, and also information to ensure long-term safety. Our new recommendation is to include baseline long bone radiographs for all IHC cats being treated with oral bisphosphonates for more than one year, and then yearly thereafter to more readily detect early bone injury that may be developing. Long-term safety studies in cats treated with oral alendronate are sorely needed.

REFERENCES

[1]O'Neill WC. **Kidney Int** 2007;72:792-796. [2]de Brito Galvao JF. **unpublished observations**, 2013.[3]Geyer N, et. al. **Vet Ther** 2007;8:239-246.[4]Peterson M. **ACVIM Forum** 2013. [5]Peterson ME. **Insights into Veterinary Endocrinology,** 2012. [6]Peterson ME. **Insights into Veterinary Endocrinology**, 2012. [7]Graham-Mize CA, et. al. **Advances in Veterinary Dermatology** 2005;5:152-158. [8]de Brito Galvao JF, et. al. **August's Consultations in Feline Internal Medicine**, 2015. [9]Bushinsky DA, et. al. **Kidney Int** 1999;55:234-243. [10]Whitney JL, et. al. **J Feline Med Surg** 2011;13:129-134.[11]Hostutler RA, et. al. **J Vet Intern Med** 2005;19:29-33. [12]Hardy BT, et. al. **J Vet Intern Med** 2015;29:200-206. [13]Papapetrou PD. **Hormones** 2009;8:96-110. [14]Tanvetyanon T, et. al. **Ann Oncol** 2006;17:897-907. [15]Hewitt RE, et. al. **Clin Exp Immunol** 2005;139:101-111. [16]Recker RR, et. al. **Am J Med** 2009;122:S22-32. [17]Vo TM, et. al. **World J Cardiol** 2014;6:216-226. [18]Abraham SC, et. al. **Mod Pathol** 1999;12:1152-1157. [19]Aibar Arregui MA, et. al. **Rev Esp Enferm Dig** 2011;103:338-339. [20]Peter CP, et. al. **Dig Dis Sci** 1998;43:1998-2002. [21]Ueda K, et. al. **Gastrointest Endosc** 2011;73:1037-1038. [22]Mohn KL, et. al. **J Vet Dent** 2009;26:74-81. [23]Stepaniuk K. **J Vet Dent** 2011;28:277-281. [24]Allen MR, et. al. **J Oral Maxillofac Surg** 2008;66:987-994. [25]Burr DB, et. al. **Orthod Craniofac Res** 2009;12:221-228. [26]Isaacs JD, et. al. **Clin Orthop Relat Res** 2010;468:3384-3392. [27]Lin JH, et. al. **Drug Metab Dispos** 1991;19:926-932. [28]Fischer KJ, et. al. **J Biomed Mater Res B Appl Biomater** 2006;76:143-148. [29]Griffin B, et. al. **J Vet Intern Med** 2003;17:445. [30]Westfall DS, et. al. **J Vet Intern Med** 2001;15:467-470. [31]Whang J, et. al. **Drug Dev Ind Pharm** 2011;37:491-497. [32]Kusamori K, et. al. **J Bone Miner Res** 2010;25:2582-2591. [33]Katsumi H, et. al. **J Pharm Sci** 2012;101:3230-3238. [34]Choi A, et. al. **Drug Deliv** 2010;17:249-254.

UPDATE ON INSULIN TREATMENTS FOR DOGS AND CATS: INSULIN DOSING PENS AND MORE

Ann L Thompson, BVSc, MANZCVS, MS, DACVIM; Patty A Lathan, VMD, MS, DACVIM; Linda M Fleeman, BVSc PhD, MANZCVS,
Gatton, Queensland, Australia

Introduction

Insulin therapy is the cornerstone of management of diabetic dogs and cats, and many insulin types have been shown to be effective. The wide range and variety of insulin products that are currently available might seem overwhelming to clinicians preparing to treat a diabetic dog or cat. They could therefore be tempted to use a traditional, familiar product rather than consider a new option which might provide clinical benefit for their patient or practical benefit for the owner. This paper will allow veterinarians to make informed choices about the most appropriate and practical treatment option for individual cases. Traditionally, insulin has been administered using an insulin syringe, but insulin pens have become readily available over the past few years. Insulin dosing pens are designed to be used by people with no formal medical training with the goal of making the task of measuring and administering insulin doses easier, less painful, and more accurate and precise. They can provide the same benefit for diabetic animals and their owners. An important advantage of insulin dosing pens is that much of the confusion and difficulty of measuring insulin using syringes is eliminated.

What insulin products are currently available?

Type of insulin	Conc	Brand name(s)	Manufacturer	Multi-use vials	Pen vials	Injection pens: Disposable prefilled	Injection pens: Reusable
			VETERINARY INSULIN PREPARATIONS				
insulin zinc suspension (lente)	40 U/mL	Vetsulin; Caninsulin	MSD Animal Health	2.5 & 10 mL	2.7 mL	none	Vetpen [a,b]
protamine zinc insulin (PZI)	40 U/mL	ProZinc	Boehringer Ingleheim	10 mL	none	none	none
			TRADITIONALLY AVAILABLE HUMAN INSULIN PREPARATIONS				
regular insulin	100 (& 500) U/mL	Humulin R; Humulin S	Eli Lilly	3 & 10 mL (500u/mL 20mL)	3 mL Humulin S	none	HumaPens [a,b]
regular insulin	100 U/mL	ActRapid	Novo Nordisk	10 mL	3 mL	none	NovoPens [a,b]
regular insulin	100U/mL	Insuman Rapid	Sanofi Aventis	none	3 mL	none	Autopen[b] ClickSTAR[b] ALLSTAR [b] JuniorSTAR[a]
isophane (NPH) insulin	100 U/mL	Humulin N; Humulin I	Eli Lilly	10 mL & 3mL	3 mL	Humulin N & Humulin I KwikPen[a]	HumaPens [a,b]
isophane (NPH) insulin	100 U/mL	Protophane	Novo Nordisk	10 mL	3 mL	Innolet[a]	NovoPens [a,b]
isophane (NPH) insulin	100 U/mL	Insuman Basal	Sanofi Aventis	5 mL & 10 mL	3 mL	SoloSTAR[b]	Autopen[b] ClickSTAR[b] ALLSTAR [b] JuniorSTAR[a]
50% regular & 50% isophane (NPH)	100 U/mL	Mixtard 50/50	Novo Nordisk	none	3 mL	Innolet[a]	NovoPens [a,b]
50% regular & 50% isophane (NPH)	100 U/mL	Insuman Comb 50	Sanofi Aventis	none	3 mL	none	Autopen[b] ClickSTAR[b] ALLSTAR [b] JuniorSTAR[a]
30% regular & 70% isophane (NPH)	100 U/mL	Mixtard 30/70	Novo Nordisk	none	3 mL	Innolet[a]	NovoPens [a,b]
30% regular & 70% isophane (NPH)	100 U/mL	Humulin 70/30; Humulin M3	Eli Lilly	10 mL & 3 mL	3 mL	Humulin 70/30; Humulin M3 KwikPen[b]	HumaPens [a,b]
25% regular & 75% isophane (NPH)	100 U/mL	Insuman Comb 25	Sanofi Aventis	5 mL	3 mL	SoloSTAR[b]	Autopen[b] ClickSTAR[b] ALLSTAR [b] JuniorSTAR[a]
			NEW HUMAN INSULIN PREPARATIONS				
insulin lispro	100 U/mL	Humalog	Eli Lilly	3 & 10 mL	3 mL	Humalog KwikPen[b]	HumaPens [a] [b] Autopen [b]
insulin aspart	100 U/mL	NovoLog; NovoRapid	Novo Nordisk	10 mL	3 mL	Novolog FlexPen[b]	NovoPens [a,b]
insulin glulisine	100 U/mL	Apidra	Sanofi Aventis	10 mL	3 mL	Apidra SoloSTAR[b]	Autopen[b] ClickSTAR [b] ALLSTAR [b] JuniorSTAR[a]
50% lispro & 50% lispro protamine	100 U/mL	Humalog Mix 50/50	Eli Lilly	10 mL	none	Humalog Mix50 KwikPen[b]	none
30% aspart & 70% aspart protamine	100 U/mL	NovoLog Mix 70/30; NovoMix 30	Novo Nordisk	10 mL	3 mL NovoMix 30	Novolog Mix 70/30 FlexPen[b]	NovoPen [a,b]
25% lispro & 75% lispro protamine	100 U/mL	Humalog Mix 75/25	Eli Lilly	10 mL	none	Humalog mix25 KwikPen[b]	none
insulin glargine	100 U/mL	Lantus	Sanofi Aventis	10 mL	3 mL	Lantus SolarSTAR[b]	Autopen[b] ClickSTAR [b] ALLSTAR [b] JuniorSTAR[a]

insulin glargine	100 U/mL	Abasria	Eli Lilly	none	3 mL	KwikPen[b]	HumaPens [a b]
insulin detemir	100 U/mL	Levemir	Novo Nordisk	10 mL	3 ml	Levemir FlexPen[b] Levemir FlexTouch[b]	NovoPens [ab]

[a] 0.5 U increments   [b] 1 U increments

Note that availability of cartridges, insulin types and pens will vary depending on country.

The insulin products currently registered for veterinary use are Caninsulin/Vetsulin®, and ProZinc®. Caninsulin/Vetsulin® is porcine lente insulin registered for use in both dogs and cats that has been available worldwide for many years. Both veterinary insulin types are less concentrated than human insulin at 40 U/mL. To avoid dosing errors, it is recommended to always administer 40 U/mL insulin with either U-40 syringes or the manufacturer's insulin dosing pen. All other insulin products used in veterinary practice are prescribed off label. Almost all of these insulin preparations have a concentration of 100 U/mL (exception is 500 U/mL Humulin R®). Several human insulin types have been used extensively in veterinary medicine, including the traditionally available insulin preparations such as regular (crystalline), neutral protamine Hagedorn (NPH or isophane), and NPH insulin mixes. More recently some of the newer human preparations, including glargine and detemir, have gained favor in managing pets with diabetes.

Insulin treatment recommended for diabetic dogs – What's the evidence?

Historically, the most universally used maintenance insulin types in dogs have been NPH and lente insulin. Over the past several years, however, additional insulin options have been explored, partially due to the availability of newer insulin analogues for human use (such as detemir and glargine), and partially due to intermittent availability of lente insulin. The most recent canine studies have evaluated the use of NPH, porcine lente, glargine, detemir, and recombinant human PZI insulin preparations. Unfortunately, for logistical reasons, large, controlled studies comparing the efficacies of different types of insulin in diabetic dogs have not been published. Although NPH insulin is the insulin of choice for some clinicians, the duration of activity is too short for twice daily dosing in some dogs. The recommended starting dose is 0.25-0.50 U/kg q12h. In one study including well-controlled diabetic dogs, the mean dose of NPH insulin was 0.63 U/kg (range, 0.40 – 0.97 U/kg) q12h.[1] Since porcine insulin has the same amino acid sequence as canine insulin, porcine lente insulin zinc suspension (Vetsulin® in the United States; Caninsulin® everywhere else) is a rational choice for use in dogs. It contains 65% crystalline zinc insulin and 35% amorphous zinc insulin (although the proportions of the two components are still described in some countries by the conventional product label of 70% and 30%, respectively), resulting in two peaks in insulin concentration at 4 and 11 hours, and lasts approximately 16 hours in dogs. [2,3] Thus, twice daily dosing is necessary in most dogs, with a recommended starting dose of 0.25 – 0.50 U/kg q12h. Due to the suspension of two types of insulin, the manufacturer recommends that porcine lente insulin zinc suspension be shaken vigorously prior to administration.

Detemir (Levemir®) is a long-acting, recombinant insulin analogue. Its prolonged duration of activity is due to its binding to and slow-release from albumin. Little data are available on the use of detemir in diabetic dogs, but two small studies reported effective control of hyperglycemia, but hypoglycemia (or clinical signs that may be associated with hypoglycemia) occurred in 4/10 dogs in one of the studies. [4,5] Detemir appears to be a promising option for use in dogs in which the duration of action of NPH or porcine lente insulin is too short, but close monitoring of blood glucose concentration by owners at home is necessary to avoid hypoglycemia. Detemir is more potent in dogs than human insulin. The modifications made to human insulin (to allow detemir to bind to albumin) cause detemir to bind to the human insulin receptor only 25% as avidly as human insulin. Thus, the marketed formulation was produced to have four times as much insulin per unit as other insulin preparations (24 nmol, as opposed to 6 nmol). However, detemir appears to bind to canine insulin receptors with more affinity than it does to human receptors.[6] Thus, the recommended starting dose for diabetic dogs is 0.1 U/kg q12h. For this reason, the use of detemir in smaller dogs (<10 kg) is problematic and often not practical.

Glargine (Lantus®) is a long-acting, recombinant insulin analogue that forms subcutaneous microprecipitates, allowing for slow absorption and prolonged duration of activity. In humans, glargine is said to be "peakless". Two published studies have evaluated its use in dogs with naturally-occurring diabetes mellitus. One study identified an unpredictable nadir in almost all of the glucose curves, whereas the other study found no difference between the mean minimum and maximum blood glucose concentrations (peakless).[4,7] Dogs in the Hess study were also placed on a diet high in soluble fiber at the time of enrollment, which might have contributed to different findings. Both studies concluded that glargine can be used to obtain good diabetic control in dogs. The median dose at the end of the studies was between 0.5-0.6 U/kg q12h, and their recommended starting dose was 0.3 U/kg q12h; however, these authors find that a range of 0.25-0.50 U/kg q12h, similar to other insulin types (except detemir), is clinically appropriate.

Insulin treatment recommended for diabetic cats – What's the evidence?

Currently, the most frequently recommended insulin preparations for use in cats are glargine, rhPZI, and porcine lente zinc suspension; detemir has also been evaluated. Although NPH has been used in the past, the duration of activity in cats (less than 8 hours) is too short for twice daily dosing.[8] Glargine, rhPZI, porcine lente, and detemir insulin have all been demonstrated to achieve glycemic control in cats with twice daily dosing. [9-12] Recommended starting dose is typically 0.25 – 0.50 U/kg q12h (rounded down to the nearest unit, and generally ≤2U/cat q12h). It is worth noting that due to species differences in metabolism detemir is not as potent in cats as in dogs hence standard insulin doses are recommended in cats.

In addition to availability, ease of use, and expense, a major factor influencing insulin choice for diabetic cats is remission rate. Remission has been documented using each of these insulin types. One small study showed that the use of glargine insulin resulted in a higher remission rate than the use of either PZI or porcine lente insulin.[13] However, a larger study verifying these results has not been published, and the remaining studies evaluating remission rates in diabetic cats have been variably controlled for predisposing and concurrent conditions, duration of diabetes, diet, monitoring frequency, and other factors, making it extremely difficult to compare the effect of insulin type. For example, Roomp and Rand showed a 64% remission rate in cats fed a low carbohydrate diet when owners checked blood glucose concentration at least three times daily and adjusted glargine dose according to an algorithm, and a 67% remission rate using a similar protocol with detemir.[11] Bennett et al achieved a similar remission rate of 68% in cats fed a low carbohydrate diet that were monitored using clinical parameters and fructosamine concentrations, with most cats receiving PZI.[14] In a retrospective study, Zini et al reported a 56% remission rate in cats using porcine lente insulin and a 72% remission rate in cats using glargine, but diet and monitoring procedure was not controlled due to the retrospective nature. [15] Despite the inconsistent control, these studies suggest that glargine, and likely detemir, facilitate the highest remission rate in diabetic cats, particularly when a low carbohydrate diet and intensive monitoring are employed. However, both PZI and porcine lente insulin can result in remission rates >50% as well.

In many countries, rhPZI is now commercially available as ProZinc® (Boehringer Ingelheim), but commercial production of PZI has been intermittent in the past. As a result, many compounding pharmacies offer their own formulations. Since they are compounded, these products are not approved by the USA FDA. However, the United States Pharmacopeia (USP), an independent organization that sets standards for the pharmaceutical industry, sets specifications for PZI. A recent study evaluated various USP specifications in vials of PZI from 12 different compounding pharmacies, using 3 bottles of each product at 4 different time points. Specifications evaluated included endotoxin content, pH, zinc content, insulin content, and other factors that could influence insulin pharmacokinetics. The study found that only 1 of the products met all USP specifications in all vials tested, and that the use of these compounded insulin products could result in inconsistent glycemic control.[16] Thus, the use of compounded insulin is not recommended.

Insulin dosing pens - why use them?

Pens have greater dose accuracy compared to insulin syringes, especially with low doses of insulin. In all accuracy studies syringes tend to overdose low volumes of insulin. Casella et el demonstrated that pediatric nurses when attempting to deliver a 0.5 U syringe dose on average drew up a 95% overdose. [17,18] In a study where over thirteen hundred diabetics compared pens to insulin syringes, 90% of people reported no or minimal pain with the pen injection compared to 50% using standard syringes.[19] In the same study 92% of pen users thought the pens were easy to use compared to 39% of syringe users and 91% would recommend pens to others.

What pens are available?

Once an appropriate insulin type has been selected based on the specific requirements of the diabetic patient then syringes and vials or pen devices and cartridges can be chosen. Caninsulin/Vetsulin® is the only veterinary insulin product that also has an insulin pen for veterinary patients. Two pens are available, a maximum 8 U/dose VetPen® (0.5 -8.0 U) with dosing increments of 0.5 U, and a maxium16 U/dose VetPen® (1-16 U) with dosing increments of 1 U. The pens are plastic, but reusable, and have been tested up to 3000 uses, which would be over 4 years in a diabetic receiving twice daily insulin. Insulin cartridges should be discarded after 6 weeks of use. It is recommended that the needle is kept in the skin for 5 seconds during dosing with VetPen®. There are many pens available for human patients with features that may provide benefit for dogs and cats such as 0.5 U increments, memory, and large dials for people with visual or dexterity impairment. Refer to the table for the reusable and disposable pens available to use with human insulin products.

Pen needles are usually between 29 and 32 gauge and come in differing lengths ranging between 4 mm and 12.7 mm. The current recommendation in human medicine favors the shorter pen needles; however the longer 10-12.7 mm needles are recommended for veterinary patients. The technique used for subcutaneous injection is different for dogs and cats compared with people and so, if the needle is too short for veterinary patients, there is risk of intradermal insulin delivery. Once insulin vials are loaded into a pen or a pre-filled pen is in use they should generally be stored out of the refrigerator at room temperature less than 25C or 77F and away from light. However, manufacturer's instructions vary and should always be followed. It is recommended to discard most vials after 28 days or earlier if they become discolored. It is important that all users refer to individual product inserts for specific recommendations, as some insulin preparations recommend vials be discarded after as little as 10 days or up to 42 days. [20]

Tips and tricks for improving compliance

Insulin dosing pens all function in a similar manner, although there are minor differences. Manufacturers provide useful written instructions that can be used as a guide during the consultation and later referred to by owners at home. The following check list may also be used when instructing owners on use of the pen.

1. Familiarize with the components of the pen, including the operation of the dose selector dial and the release trigger.
2. Practice one-handed operation of the pen. For most people, the most comfortable approach is to hold the pen against the palm with the four fingers and then use the thumb to press the trigger.
3. For re-useable pens, load an insulin cartridge into the pen. This step is not necessary with pre-filled insulin dosing pens.
4. Always mix the insulin before each injection, typically by rolling the pen 10 times and then inverting the pen 10 times. This is crucial for insulin suspensions such as Caninsulin/Vetsulin® (MSD Animal Health), but also advisable for solutions such as Lantus® (glargine) insulin (Sanofi).
5. Prime the insulin dosing pen. Attach the needle onto the pen, use the dose selector to dial the priming dose of either 1 or 2 U according to the manufacturer's instructions, hold the pen vertically with the needle upwards, and push the release button. The goal is for a stream of insulin to be ejected from the needle tip. If a stream is not seen, the pen must be primed again until this is achieved. Pens might need to be primed several times when an insulin cartridge is first inserted. This process ensures there is no air in the system and provides reassurance that the pen is working properly.
6. Practice operating the pen with the dose that will be prescribed for the pet. Use the dose selector to dial the recommended dose and then use a one–handed grip of the pen to push the release trigger. It can be helpful to advise the owner to note the time taken for the pen to deliver the dose as this provides a practical guide for how long the needle must be left in the animal's skin during injection.

7. Finally, observe the owner's technique as they administer their first injection with the insulin dosing pen. Counsel against rushing the process by recommending that four successive steps are used: 1. Gently pinch up the pet's skin, 2. Slide the needle through the skin, 3. Press the release trigger, 4. Withdraw the needle.
8. Check that the dose selector dial has returned to the '0' position, which confirms that the full insulin dose was delivered. Note that it is normal for a drop of insulin to be seen on the needle tip after removing the needle from the injection site. There is no need to be concerned that the full dose was not delivered if the drop is seen.

REFERENCES

[1]Palm CA, et al, **J Vet Intern Med** 2009;23:50, [2]Graham PA, et al, **J Small Anim Pract** 1997;38:434, [3]Fleeman LM, et al, **Vet Rec** 2009;164:232. [4]Fracassi FH, et al, **ECVIM Congress** 2012. [5]Ford SLR, et al; **ACVIM Forum** 2010. [6]European Medicines Agency. Initial Scientific Discussion for the Approval of Levemir [internet] http://www.ema.europa.eu/docs/en_GB/document_library/EPAR_-_Scientific_Discussion/human/000528/WC500036658.pdf. 2004 [cited Dec 14, 2014]. [7]Hess RS & Drobatz KJ, **J Am Vet Med Assoc**.2013;243:1154. [8]Wallace MS, et al, **Dom An Endocr** 1990;7:509. [9]Roomp K & Rand J, **J Fel Med Surg** 2012;14:566. [10]Nelson RW, et al, **J Vet Intern Med** 2009;23:787. [11]Roomp K & Rand J, **J Fel Med Surg** 2009;11:668. [12]Michiels L, et al, **J Fel Med Surg** 2008;10:439. [13]Marshall RD, et al, **J Fel Med Surg** 2009;11:683. [14]Bennett N, et al, **J Fel Med Surg** 2006;8:73. [15]Zini E, al, **J Vet Intern Med** 2010;24:1314. [16]Scott-Moncrieff JCR, et al, **J Am Vet Med Assoc** 2012;240:600. [17]Casella SJ, et al, **Pediatrics** 1993;91:1155. [18]Gnanalingham M, et al, **Archives of Disease in Childhood** 1998;79:59. [19]Graff MR, et al, **Clin Therap** 1998;20:486. [20]Pearson TL, **J Diab Sci Tech** 2010;4:522.

SUBCLINICAL TO MILD HYPERTHYROIDISM IN CATS: ISSUES IN DIAGNOSIS AND TREATMENT

Mark E. Peterson, DVM, DACVIM
New York, NY

INTRODUCTION

Hyperthyroidism is one of the commonly diagnosed disorders in older cats, present in approximately 10% of senior and geriatric cats (>10 years) in the USA today.[1] This feline disease is being diagnosed at a much earlier stage than it was a decade or more ago, in part because of the annual geriatric screening programs for hyperthyroidism employed by many veterinarians. Because many cats are currently diagnosed with milder degrees of hyperthyroidism than in the past, they tend to have fewer (and less severe) clinical signs and abnormalities detected on physical examination. Many cats diagnosed today have no clinical issues discerned by their owners and are only first recognized as being hyperthyroid on the basis of serum thyroid hormone test results (i.e., slightly high total or free T4 concentrations).[2]

This paper will review what is currently known about mild and subclinical hyperthyroidism in cats. My aim is to clarify common questions, which include the following: (1) how common is mild and subclinical hyperthyroidism? (2) how should we diagnose mild and subclinical hyperthyroidism? (3) how do we best treat these conditions? (4) and finally, do we even need to treat mild and subclinical hyperthyroidism or is monitoring the best way of managing these cats?

WHAT IS SUBCLINICAL HYPERTHYROIDISM?

The term "subclinical" denotes the presence of disease without obvious clinical signs or symptoms, and generally indicates that evolution of the disease is at an early stage. In man, the availability of sensitive assays for thyroid-stimulating hormone (TSH) allows for identification of patients who have low or undetectable serum TSH concentrations (< 0.5 µU/mL) with normal serum concentrations of total T4, free T4 and T3, a constellation of biochemical findings defined by the diagnostic term, subclinical hyperthyroidism.[3-5] The highly-sensitive third-generation TSH assays currently used in human patients are capable of detecting levels as low as 0.01 to 0.02 µU/mL.[6,7] Subclinical hyperthyroidism can be divided into two categories: low but detectable TSH levels (0.1 to 0.4 µU/mL; Grade I or mild), and suppressed TSH levels (< 0.1 µU/mL; Grade II or severe).[8,9] Again, since the term subclinical denotes disease without obvious signs, both subclinical hyperthyroidism (Grade I and II) is a biochemical definition, not a clinical one. By contrast, patients with "overt" hyperthyroidism have more severe thyroid derangements, with completely suppressed serum TSH levels, clearly high serum thyroid hormone concentrations, and the presence of clinical signs of thyrotoxicosis.[2]

Does subclinical hyperthyroidism occur in cats? It would certainly make sense that all hyperthyroid cats would go through a preclinical stage of disease before overt hyperthyroidism and clinical signs develop. In support of this, the studies by Wakeling et al report very good evidence for the existence of a subclinical phase of hyperthyroid disease in cats.[10-12] In the most recent prospective study, the finding of an undetectable TSH concentration (< 0.03 ng/mL) in euthyroid cats was associated with an increased risk for the subsequent diagnosis of hyperthyroidism over the following 14 or more months.[12] In that study, the cats with subclinical hyperthyroidism had none of the signs commonly associated with overt hyperthyroidism, such as weight loss, tachycardia, hypertension or presence of a murmur; all initially had normal serum T4 and free T4 concentrations that eventually rose into the hyperthyroid range, as overt hyperthyroidism developed.[12]

A feline-specific TSH assay has not yet been developed for commercial use; however, assays for measuring canine TSH (cTSH) are widely available, and it has been reported that these cTSH assays can be used to help diagnose or exclude hyperthyroidism in cats.[1,2,10-12] As noted above, the term subclinical hyperthyroidism denotes the absence of obvious clinical signs and is defined biochemically as normal serum concentrations of total and free T4 with low to undetectable serum TSH levels.[3-5] However, the commonly used cTSH assay has a suboptimal sensitivity when used for the measurement of feline TSH. In other words, this assay cannot accurately distinguish between low-normal and truly low values in cats, so that both euthyroid and hyperthyroid cats can have TSH concentrations that are undetectable (i.e., below the limit of quantification of the assay or <0.03 ng/mL).[1,2,10-12] Therefore, diagnosis of subclinical hyperthyroidism is difficult to make in cats and can only be achieved with use of thyroid scintigraphy[13-16] or dynamic thyroid testing.[17-19] In any case, since these cats are not overtly hyperthyroid and most have no clinical signs or physical exam abnormalities, it is unclear if we should treat these cats at this pre-clinical phase of their disease anyway (see below).

DEFINING MILD AND OCCULT HYPERTHYROIDISM IN CATS

Mild or occult hyperthyroidism accounts for almost 30% of the cats that are currently referred to our practice. Biochemically, we can define mild hyperthyroidism on the basis of finding a serum T4 concentration in the high-normal to only slightly high range (>2.5 to < 6 µg/dl; reference range, 0.8-4.0 µg/dl) in a cat with only mild clinical features of thyroid disease.[20,21] Of cats with mild disease, serum T4 and T3 concentrations remain within the reference range limits in ≈20% and 80% respectively (Table 1). Almost all of these cats have a high free T4 value, but some have free T4 concentrations that fluctuate in and out of the reference range. Serum TSH concentrations are undetectable (i.e., below the limit of quantification of the assay or <0.03 ng/mL) in most, but not all of these cats with mild thyroid disease.[1,2,12,21] In cats in which mild hyperthyroidism is suspected, I do recommend that a complete thyroid panel be repeated, with determination of total T4, free T4 by dialysis, and cTSH measurement. The finding of high serum free T4 with suppressed TSH concentrations is consistent with mild hyperthyroidism, whereas normal free T4 values or measureable TSH concentrations go against this diagnosis (Table 1).[21]

We routinely perform thyroid scintigraphy in all of our cats referred for suspected hyperthyroidism, which is used to confirm the presence of one or more hot thyroid nodules (increased thyroid uptake and high thyroid/salivary ratio).[16] In cats with mild or occult hyperthyroidism, scintigraphy findings also include a high thyroid-to-salivary ratio, a high thyroid-to-background ratio, or a high percentage of uptake of the administered pertechnetate,[13-16] all of which confirm the hyperthyroid state (Table 1).

ISSUES AND PROBLEMS IN DIAGNOSIS OF SUBCLINICAL TO MILD HYPERTHYROIDISM

Cats with mild or early hyperthyroidism can be difficult to diagnosis since total T4 concentrations may remain within the reference range limits in cats, leading to the term "occult" hyperthyroidism for this syndrome.[17-19] Some of these cats also suffer from concurrent nonthyroidal illness (e.g., CKD or gastrointestinal disease), which leads to suppression of high T4 concentrations down to within reference range limits, further complicating the diagnosis.[20,22,23]

Because of these problems in diagnosis of occult hyperthyroidism, use of dynamic testing (T3 suppression test or TRH stimulation test) can be recommended and can be very helpful.[17-19] However, these testing procedures are fraught with difficulties in owner compliance and equivocal test results. Therefore, such dynamic testing is rarely done today. What most veterinarians currently do today when normal T4 values are found in cats with suspected "occult" hyperthyroidism is to "add-on" a serum free T4. The use of free T4 determinations are indeed helpful in diagnosis of these cats, since free T4 values will be high in over 98% of hyperthyroid cats.[20] Although free T4 is a more sensitive diagnostic test than total T4, the specificity is relatively poor, with up to 20% of sick (and some clinically normal) euthyroid cats having false-positive free T4 results.[20,21,23,24]

When selecting a diagnostic test, it is useful to consider the test sensitivity and specificity of the individual test in the population tested. When cats with occult or mild hyperthyroidism are tested and compared to euthyroid cats referred for suspected hyperthyroidism but are subsequently found not to have the disease, the results are revealing (Table 1). The total T4 test, with its relatively high sensitivity and specificity (and low cost), remains the initial screening test of choice for hyperthyroidism in cats; however, the test sensitivity in cats with mild disease is only slightly more than 80%, with many positive results still being only borderline high (Table 1). Serum T3 measurements are not useful for diagnosis of mild thyroid disease due to the fact that almost 80% of cats with mild hyperthyroidism maintain normal T3 levels (Table 1). The free T4 by dialysis test has very relatively high sensitivity (90%) in cats with mild disease, but is also associated with a relative high rate of false-positive results in euthyroid cats (≈20% in our practice). The clinical issues associated with the low test specificity of the free T4 test are highlighted by the relatively high number of euthyroid cats that are referred for radioiodine treatment to my practice on the basis of having a high serum free T4 value alone; most of these cats are relatively asymptomatic, none have a palpable thyroid nodule, and all have normal results on thyroid scintigraphy, thus confirming that they are not hyperthyroid.

Table 1 - Diagnostic sensitivity and specificity for serum hormone concentrations (T4, T3, free T4, TSH) and scintigraphic findings in cats with occult or mild hyperthyroidism, euthyroid cats suspected of having hyperthyroidism but without the disease, and clinically normal cats

Thyroid Hormone or Thyroid Scintigraphic Ratio	Test Sensitivity	Test Specificity
Total T_4 concentration	82%	93%
Total T_3 concentration	20%	97%
Free T_4 concentration	90%	80%
Serum TSH concentration	95%	65%
Thyroid-to-salivary ratio	96%	99%
Thyroid-to-background ratio	85%	99%

Many veterinarians mistakenly believe that the finding of a high free T4 concentration alone, especially when the total T4 concentration is within the upper half of the reference interval, confirms a diagnosis of hyperthyroidism in cats. Our findings that ≈20% of these cats turned out to be euthyroid makes it clear that the free T4 test should *never* be used alone to confirm a diagnosis of hyperthyroidism; overreliance on free T4 testing will lead to misdiagnosis and inappropriate treatment of many euthyroid cats.

Despite the limitations of the serum cTSH test, with its inability to differentiate low-normal results from truly suppressed or undetectable values, we believe that measurement of serum TSH concentration has diagnostic value in cats with early or mild hyperthyroidism, especially when measured in conjunction with total and free T4 concentrations. The diagnostic test sensitivity is a bit higher than either the total or free T4 concentration, but the specificity is low (Table 1). However, the finding of a total T4 within the upper third of the reference interval combined with both a high free T4 and a suppressed TSH concentration is consistent with a diagnosis of early hyperthyroidism. On the other hand, the finding of high-normal T4 with a normal free T4 or measurable TSH concentration makes hyperthyroidism much less likely. In this latter case, one should reevaluate the diagnosis, withhold treatment, and continue to monitor unless other evidence for hyperthyroidism is present (e.g., compatible clinical signs, palpable thyroid nodule, positive thyroid scan results).

ISSUES AND PROBLEMS IN TREATMENT OF SUBCLINICAL TO MILD HYPERTHYROIDISM

In human endocrine practice, the necessity of treatment for asymptomatic patients with subclinical hyperthyroidism is an open question, with much debate regarding the physiological effects of subclinical thyroid disease and the indications for treatment.[3-5,6,15] The criteria for treatment of subclinical thyroid disease are controversial, and individualized judgment is mandatory in order to evaluate the grade and clinical consequences of the disorder in any given patient. This is less of an issue for cats, in that subclinical hyperthyroidism is extremely difficult to diagnose with certainty. In cats in which subclinical hyperthyroidism is suspected based on the findings of undetectable serum TSH, normal total and free T4, no overt clinical signs, but thyroid autonomy based on dynamic pituitary-thyroid testing or thyroid scintigraphy, monitoring for progression of disease is certainly preferred over any treatment.

But what about cats with very mild or occult hyperthyroidism (i.e., cats with high-normal T4 concentrations, slightly high free T4 values, and undetectable TSH concentrations)? Should these cats automatically be treated or is close monitoring the best and safest option? Although the answer to this is not known for certain, we have no evidence that milder degrees of hyperthyroidism causes significant harm to a cat, at least on a short-term basis.

If the cat is not symptomatic, no thyroid nodule is palpated, and the diagnosis is based only on the finding of a high serum free T4 and/or low TSH concentration, it is wisest not to start treatment. In these symptomatic cats, I recommend close observation and monitoring in 1 month and every 3-6 months thereafter with a thorough repeat physical examination (body weight, heart rate, and thyroid palpation) and complete thyroid panel (total T4, free T4, and TSH determinations). Once we have better evidence for hyperthyroidism, such as compatible clinical signs, palpable thyroid nodule, higher total T4 concentration, or positive thyroid scan results, treatment can then be reconsidered.

If the cat with occult hyperthyroidism does have clinical signs (e.g. weight loss, muscle wasting, tachycardia) and a thyroid nodule is either palpated or documented with thyroid imaging, treatment for the mild hyperthyroidism should be considered once significant nonthyroidal illness is excluded. But which is the best treatment in these cats?

Antithyroid drugs, such as methimazole, are effective at reversing the biochemical abnormalities of mild hyperthyroidism, providing owner and cat compliance is good (the cat can be successfully treated with oral or transdermal medication). The risks of side effects, mostly mild but some serious, are relatively common.[25,26] Again, if the cat has been diagnosed biochemically with mild or occult hyperthyroidism but no clinical signs are present, one could argue that the risks of adverse reactions might outweigh the benefits of antithyroid drug treatment, since we are not "curing" the underlying thyroid tumor with this treatment. In cats with known or suspected nonthyroidal illness, a trial of antithyroid drug therapy might be warranted to find out whether any clinical signs are improved once the total and free T4 concentrations have normalized. If no improvement in clinical signs is observed, the diagnosis of hyperthyroidism should be questioned and the cat reevaluated.

In general, the daily doses used in cats with mild or occult hyperthyroidism should be less than the average or typical dose employed, since their goiter is smaller and the initial thyroid values are lower. To that end, we usually start treatment with 1.25-2.5 mg of methimazole per day, either once daily or divided, with dose increments given if needed. If long-term treatment is given and the disease progresses (remember that methimazole does not slow thyroid tumor growth),[27] higher doses of methimazole would be expected.

Surgical thyroidectomy is an extremely effective treatment for hyperthyroidism that is simple, quick, curative (thyroid tumor is removed). In practice, it is often considered the treatment of choice particularly if radioactive iodine is unavailable. However, surgery comes with its own list of disadvantages, including iatrogenic hypoparathyroidism and hypothyroidism. Especially in cats with mild or occult hyperthyroidism and only slight goiter, it is frequently difficult to differentiate unilateral from bilateral thyroid tumor involvement.[25,28] If only one thyroid lobe is removed in a cat with bilateral thyroid disease, relapse of hyperthyroidism is common within a few months.

Radioactive iodine treatment has been used for the treatment of cats for 35 years and has many advantages over other treatment methods. Radioiodine is safe, well tolerated, and effective at correcting the biochemical abnormalities of mild hyperthyroidism.[25,29] A single administration of radioiodine restores euthyroidism in most (95%) hyperthyroid cats.

However, use of the "standard" I-131 doses commonly used to treat cats (e.g., 4.0 mCi) is not recommended, since those high doses will invariably render the great majority of cats with mild or occult hyperthyroidism hypothyroid and dependent on lifelong L-T4 treatment. We have previous reported that ultra-low doses of radioiodine ($\leq$ 2.0 mCi) are highly effective in restoring euthyroidism without inducing hypothyroidism in most cats with milder forms of hyperthyroidism.[30]

REFERENCES:

[1]Peterson ME. **J Endocrinol** 2014;223:T97. [2]Peterson ME. **J Feline Med Surg** 2013;15:765. [3]Bahn RS, et al. **Thyroid** 2011;21:593. [4]Cooper DS, et al. **Lancet** 2012;379:11422. [5]Garg A, et al. **Br Med Bull** 2013;107:101. [6]Kasagi K, et al. **Clin Endocrinol (Oxf)** 1999;50:185. [7]Ross DS. **Endocrinol Metab Clin North Am** 2001;30:245. [8]Hollowell JG, et al. **J Clin Endocrinol Metab** 2002;87:489. [9]Mitchell AL, et al. **Clin Endocrinol (Oxf)** 2010;72:292. [10] Wakeling J, et al. **J Vet Intern Med** 2006;20:726. [11]Wakeling J, et al. **Thyroid** 2007;17:1201. [12] Wakeling J, et al. **J Vet Intern Med** 2011;25:1057. [13]Intenzo C, et al. **Clin Nucl Med** 2011;36:e107. [14]Canbaz F, et al. **Hell J Nucl Med** 2004;7:203. [15]Smith TA, et al. **J Vet Intern Med** 1996;10:185. [16]Peterson ME, et al. **Vet Radiol Ultrasound** 2015;56:84. [17]Peterson ME, et al. **J Vet Intern Med** 1990;4:233. [18]Peterson ME, et al. **J Vet Intern Med** 1994;8:279. [19]Peterson ME. **August's Consultations in Feline Internal Medicine** 2001:145. [20]Peterson ME, et al. **J Am Vet Med Assoc** 2001;218:529. [21]Peterson ME, et al. **J Vet Intern Med** 2015; submitted. [22]McLoughlin MA, et al. J Am Anim Hosp Assoc 1993;29:227. [23]Wakeling J, et al. **J Small Anim Pract** 2008;49:287. [24]Mooney CT, et al. **J Am Vet Med Assoc** 1996;208:2004. [25]Mooney CT & Peterson ME. **Manual of Canine and Feline Endocrinology** 2012;199. [26]Daminet S, et al. **J Small Anim Pract** 2014;55:4. [27]Peterson ME, et al. **J Feline Med Surgery** 2015:in press. [28]Birchard SJ. **Clin Tech Small Anim Pract** 2006;21:29. [29]Peterson ME, et al. **Kirk's Current Veterinary Therapy XV**. 2014: e112-e122. [30]Peterson ME, et al. **J Vet Intern Med** 2014;28:1031.

PRIMARY, SECONDARY, ATYPICAL, AND PSEUDO HYPOADRENOCORTICISM: WHAT'S NEW AND DIFFERENT?

Rhett Nichols, DVM, DACVIM (SAIM)
Smithtown, NY

INTRODUCTION

The term hypoadrenocorticism (HA) includes all conditions in which adrenal steroid hormones fall below the requirement of animal. The two major forms of this disorder are **primary hypoadrenocorticism (PH)** due to lesions or disease processes that affect the entire adrenal cortex, and **secondary hypoadrenocorticism (SH)**, due to insufficient ACTH release from the pituitary gland. The majority of dogs suffer from PH, which is also known as Addison's disease. Usually, PH results from progressive destruction of all layers of the adrenal cortex; the end result is an absolute deficiency of glucocorticoid and mineralocorticoid secretion. Typically dogs with PH have laboratory abnormalities consistent with mineralocorticoid deficiency such as hyponatremia and hyperkalemia. However, 10 to 30% of dogs with PH have normal sodium and potassium concentrations; these dogs are classified as having **atypical hypoadrenocorticism (AH)** (1,2,3,4,5). And lastly, in rare cases, dogs may present with clinical signs and laboratory abnormalities that mimic PH such as vomiting, diarrhea, weakness, lethargy, hyponatremia, and hyperkalemia, or some combination of these, yet have normal adrenal function. This latter condition is termed **pseudo-Addison's disease** and is associated with Tricuris vulpis (whipworm) infestation (6,7). Because there are so many variants of HA there is often confusion over which diagnostic test to perform and how to interpret the results, especially when asking the question "does the dog have PH, AH, or SH?" A definitive diagnosis is critical when employing a treatment strategy that includes a mineralocorticoid, because, at the very least, mineralocorticoid therapy is *unnecessary* in approximately 20% of the dogs diagnosed with HA, not to mention the adverse effects and cost issues over the lifespan of the dog.

SCREENING TESTS FOR HYPOADRENOCORTICISM

Na:K ratio: Hyponatremia and hyperkalemia with a sodium-to-potassium ratio < 27:1 is considered a characteristic feature of PH (8). In various studies evaluating disorders with these electrolyte abnormalities, HA was the most common explanation for the presence of concurrent hyperkalemia and hyponatremia (9). As a result of this observation, many practitioners place great emphasis on the Na:K ratio as useful test to aid in the diagnosis of HA. Unfortunately the Na:K ratio lacks sensitivity and specificity. For example, in one study, only 28% of the dogs with a Na:K ratio < 25:1 had HA (10). In the same study when the Na:K ratio was lowered to < 20:1, nearly 1/3 of the dogs *did not* have the disorder (10*). And lastly, in dogs with AH and SH, which may account for up to 20% or more of the cases of HA, the Na:K ratio would not be expected to be abnormal.*

Resting cortisol concentrations: In situations where the diagnosis of HA is in question, and the pet owner has money concerns, a resting cortisol is a practical and inexpensive screening test to help determine if a confimatory test (ie, the ACTH response test) is indicated. A resting cortisol concentration > 2 ug/dl (55 nmol/L) rules out the disorder in the vast majority of cases, while a resting cortisol < 2 ug/dl is not diagnostic and the ACTH stimulation test is recommended if HA is suspected (11).

TESTS TO CONFIRM THE DIAGNOSIS OF HYPOADRENOCORTICISM

The ACTH response test. The gold standard for confirming the diagnosis of HA is the ACTH response test. Dogs with HA have an absent or blunted response to ACTH. In a group of 407 dogs with PH, all dogs showed post-ACTH cortisol levels less than or equal to 2 mcg/dl (12). In a separate group, 21 of 23 dogs with SH showed post-ACTH cortisol levels less than or equal to 2 mcg/dl (12). However, two dogs in this group showed post-ACTH cortisol concentrations > 2 mcg/dl but their responses were still quite blunted and remained < 5 mcg/dl (139.5 nmol/L) (12).

Plasma renin and eACTH levels and the aldosterone-to-renin and cortisol-to-ACTH ratio. As mentioned above, the diagnosis of HA is usually made by measuring plasma or serum cortisol before and after stimulation with ACTH. Subnormal responses to ACTH indicate decreased adrenal reserve and establish the diagnosis of adrenal insufficiency as far as cortisol is concerned. This approach ignores the role of aldosterone, although it is widely assumed that hyponatremia and/or hyperkalemia coupled with cortisol deficiency are surrogates for low aldosterone levels. Furthermore, subnormal cortisol responses to ACTH do not permit differentiation of AH from SH.

An alternative approach for diagnosing hypocortisolism and hypoaldosteronism is to determine changes in the interrelationships of relevant endogenous hormones (13,14,15). The combination of a low plasma cortisol concentration and high plasma ACTH concentration is regarded as the hallmark of PH although there is some overlap between healthy dogs and dogs with PH (13,14). This overlap disappears when the cortisol-to-ACTH ratio (CAR) is calculated. With aldosterone deficiency, the inability to retain sodium leads to hypovolemia, which, in turn, stimulates renin release. Thus, PH is also characterized by low plasma aldosterone concentration and high PRA, which is contrary to what occurs with SH, whereby plasma aldosterone does not decrease and PRA does not increase. As with plasma cortisol and eACTH levels, there is some overlap with aldosterone and renin levels between healthy dogs and dogs with PH, however this overlap disappears when the aldosterone-to-renin ratio (ARR) is calculated (14,15).

Summary impact point. Dogs with PH, which include the subset of dogs with AH, would be expected to have a significantly lower ARR and CAR compared to healthy dogs, while dogs with SH would be expected to have a normal ARR coupled with a low-normal to undetectable eACTH concentration (14).

TESTS TO CONFIRM THE DIAGNOSIS OF SECONDARY HYPOADRENOCORTICISM

Secondary hypoadrenocorticism is a variant of HA confirmed by decreased cortisol concentrations before and after ACTH administration, low to low-normal eACTH levels, normal electrolyte concentrations, and a demonstrated need for chronic low dose glucocorticoid (eg, prednisolone dosed at 0.1-0.2 mg/kg/day) but not mineralocorticoid therapy (12). The spontaneous form of this disease may be caused by a pituitary tumor or cranial-cerebral trauma, however, the iatrogenic form of SH from chronic glucocorticoid therapy is much more common than the spontaneous disease. In a series of 205 dogs with HA, 18 (11.4%) met the above criteria for SH (12). Ten dogs had undetectable eACTH levels and 8 dogs had low-normal levels. All 18 dogs responded clinically to glucocorticoid administration. Interestingly, aldosterone concentrations which would be expected to be normal with SH were low in 10 of 14 dogs studied. Low aldosterone levels have also been reported in humans with SH and may be the result of chronic ACTH deficiency secondary to long-standing glucocorticoid therapy (15).

Summary impact point: To differentiate PH from SH, further studies are often required, including basal eACTH and aldosterone levels and a corticotrophin releasing hormone (CRH) stimulation test which is rarely performed in veterinary medicine (16). In dogs with PH, the basal eACTH levels are high and there is an exaggerated response of eACTH levels to CRH, and resting aldosterone levels are low to low-normal (14,16). In dogs with SH, the basal eACTH levels are low to low-normal, there is no increase in eACTH concentrations following the administration of CRH, and aldosterone levels are often normal, but low levels do not rule out this condition (14,16).

TREATMENT AND MONITORING STRATEGIES FOR PRIMARY HYPOADREDNOCORTICISM

Traditional treatment and monitoring approach with fludrocortisone acetate (FC) and desoxycorticosterone pivilate (DOCP). The treatment of PH consists of lifelong hormone replacement (17). Mineralocorticoids usually are replaced by either FC or DOCP. Fludrocortisone acetate is administered PO and is the most commonly used treatment for humans with primary PH and has been recommended for dogs with PH for many years (18). This preparation possesses potent mineralocorticoid activity and some glucocorticoid activity. The starting dose of FC is 0.01-0.02 mg/kg/day PO divided twice daily. Traditionally, the efficacy of FC treatment is assessed by regularly monitoring clinical signs, renal function, and serum electrolyte concentrations. Dose adjustments are made in increments of 0.025-0.1 mg/dog depending on the size of the dog.

DOCP is a parentally administered long-acting mineralocorticoid alternative. DOCP has no glucocorticoid activity, and additional glucocorticoid supplementation is recommended in all cases, although surprisingly not necessary in some patients. The starting dose and dosing interval is 2.2 mg/kg IM or SC q25d. Because the recommended starting dose is often excessive (19), and the dosing interval can be extended in many dogs, and the cost of this product is quite considerable, the following dose adjustments can be made. If the electrolyte concentrations (ie, sodium and potassium) are normal at 14 days after injection, the DOCP dose is decreased by 5-10% at the next injection (20). If the electrolyte concentrations are normal at the time of the next injection, the injection interval is increased by 1 day (20). These dose and dose interval guidelines can be repeated monthly for several months until hyponatremia or hyperkalemia occurs. Many dogs can be dosed with as little as 1 mg/kg DOCP following the above protocol for a stepwise dose reduction, and a few dogs have been reported to remain clinically normal and maintain normal electrolytes with a dosing interval as long as 3 months.

Glucocorticoids: It is recommended that all dogs with HA be treated with a glucocorticoid such as prednisone or prednisolone. In general, the goal is to reach a prednisolone dosage of no more than 0.1 to 0.2 mg/kg PO q24h. In dogs with signs of glucocorticoid overdose (eg, polyuria, polydipsia, and weight gain) the glucocorticoid dose is decreased and given daily or every other day. And in dogs treated with FC, the glucocorticoid is stopped if possible. Alternatively, the synthetic steroid cortisone acetate can be substituted for prednisolone (21). Its shorter half-life and lower overall activity means it is less likely to cause polyuria and polydipsia. Generally a daily dose of approximately 0.5 to 1 mg/kg provides adequate glucocorticoid coverage. Methyprednisolone (Medrol), the 6-methyl derivative of prednisolone, is also associated with less polyuria than that seen with prednisone or prednisolone. The daily maintenance dosage for this glucocorticoid is similar to prednisolone (0.2 mg/kg).

Monitoring approach for the 21st century: In human medicine, standard monitoring of patients with PH consists of monitoring serum electrolyte concentrations, blood pressure, and plasma renin activity (PRA) (15,22). PRA is increased in human patients with PH and decreases during treatment and the same is true in dogs (14). Lack of aldosterone in patients with PH is accompanied by high concentrations of plasma angiotensin II that can be estimated indirectly by species-specific renin concentrations or PRA, whereby angiotensin is generated by a plasma sample during timed incubation.

Recent clinical studies at Zurich University showed that PRA in dogs with newly diagnosed PH was significantly higher before treatment when compared to normal dogs or in dogs with disorders that mimic PH (23). After the start of mineralocorticoid treatment PRA activity decreased significantly. However, the PRA response was different for FC versus DOCP treatment. The PRA of dogs treated with FC remained very high and unchanged while the PRA of dogs treated with DOCP decreased into the low end of the reference range. In addition, significantly more dogs treated with DOCP had serum sodium or potassium values within the reference range than did dogs treated with FC.

Summary impact points (23)

1). PRA can be used to titrate mineralocorticoid treatment in dogs with PH.

2). PRA should be within the reference range for optimal treatment of PH with electrolyte abnormalities.
3). DOCP lowers PRA and normalizes serum electrolyte values in a more predictable way than does FC.
4). PRA is a promising tool for establishing the optimal dosage for mineralocorticoid replacement therapy and monitoring mineralocorticoid treatment in dogs with PH. Moreover, DOCP treatment more effectively suppresses PRA compared to FC treatment and therefore is the preferred mineralocorticoid treatment of choice.

TREATMENT AND MONITORING STRATEGIES FOR SH AND AH

Dogs with SH and AH share certain laboratory findings – both disorders have an undetectable or blunted post-ACTH cortisol concentration and normal electrolytes. However, there are specific differences which include the following: 1) with AH, eACTH and renin levels are high and aldosterone concentrations are low to low-normal resulting in a low ARR (14). With SH, eACTH levels are low to low-normal, renin concentrations are normal, and aldosterone levels are often normal but may be low, however the ARR is normal (14). And lastly, one could argue for prognostic and treatment purposes, a MRI or CT scan should be recommended for all dogs diagnosed with SH to rule out a pituitary tumor (16).

From a treatment perspective, it is recommended the all dogs with AH and SH be treated with only a glucocorticoid (see previous section on glucocorticoids). However, some dogs with AH will develop hyperkalemia and/or hyponatremia within weeks to months after the original diagnosis necessitating mineralocorticoid replacement therapy, and in other dogs with AH, these electrolyte changes never occur. Unfortunately, the need for mineralocorticoids cannot be predicted. One proactive monitoring approach is to determine sodium and potassium concentrations every 3 to 4 months to assess the possible need for mineralocorticoid replacement. Another suggestion, although somewhat controversial and costly, is to treat all AH dogs with a low dose of DOCP (1 mg/kg) "just in case" the need for mineralocorticoid therapy arises.

REFERENCES

1.Thompson AL, et al: J Am Vet Med Assoc 2007;230:1190-1194 2.Rogers W, et al: J Am Vet Med Assoc 1981;179:155-158 3. Lifton SJ, et al: J Am Vet Med Assoc 1996;209:2076-2081 4. Sadek D, Schaer M: J An Anim Hosp Assoc 1996;32:159-163 5.Baumstark ME, et al. J Vet Intern Med 2014;28:154-159 6.DiBartola SP, et al. J Amer Vet Med Assoc 1985;187:60-63 7. Graves TK, et al. J Vet Intern Med 1994;8:287-289 8. Peterson ME, Kintzer PP, J Am Vet Med Assoc 1996; 208(1):85 9. Roth L and Tyler R D. J Vet Diagn Invest 1999 11:60 10. Neiger R and Gunderson HC. BSAVA Congress Scientific Proceedings 2003:538 11.Lennon EM, et al. J AmVet Med Assoc 2007;231:413-416 12. Feldman EC, Nelson RW. Canine and Feline Endocrinology and Reproduction. 3rd ed. St. Louis WB Saunders; 2004. pp. 394–439. 13. Lathan P, et al. J Vet Intern Med 2014;28:1546-1550 14. Javadi S, et al. J Vet Intern Med 2006;20:556-561 15. Oelkers W, et al. J Clin Endo Metab 1992;75:259-264 16. Rijnberk A, Koostra HS.Clinical Endocrinology of Dogs and Cats. 2nd ed, Hannover, Germany, Schlutersche 2010. Pp 101-110 17. Klein SC, Peterson ME. Can Vet J 2010;51:63-69 18. Kintzer PP. Peterson ME. J Vet Intern Med 1997;11:43-49 19. Bates JA, et al. Aust Vet J 2013;91:77-82 20. Klein SC, Peterson ME. Can Vet J 2010;51:179-184 21. Church DB. BSAVA Manuel of Canine and Feline Endocrinilogy. 3rd ed, BSAVA, Gloucester UK 2004, pp. 178-179 22. Reisch N, Arlt W. Endocrinol Metab Clin North Am 2009;38:407-418 23. Baumstark ME, et al. J Vet Intern Med 2014;28:1471-1478

ADVANCES IN MAXIMIZING DIABETIC REMISSION

Jacquie Rand, BVSc, DVSc, MACVS, DACVIM
Gatton, Australia

INTRODUCTION

The primary goal of management of a newly diagnosed diabetic cat has changed from just controlling clinical signs and avoiding clinical hypoglycaemia, to one where the primary goal is to achieve diabetic remission, and in some cats, a cure for diabetes. Preliminary data show that about 25-30% of cats that achieve remission, have no evidence of glucose intolerance on a glucose tolerance test, meaning essentially that they are cured. The remaining 70% have various levels of impaired glucose intolerance, and hence should be classified as pre-diabetic.

With appropriate therapy, remission is achievable in > 80% of newly-diagnosed diabetic cats, that is, cats are able to maintain euglycaemia without the need for insulin therapy. However, remission rates are substantially lower in cats with long-standing diabetes where hyperglycaemia has been inadequately controlled because of inappropriate insulin type or dose adjustments, or because underlying disease such as acromegaly precludes achieving excellent glycaemic control. Typically, long-term, poorly controlled cats referred for management of their diabetes, have remission rates of 30% or less, because there are inadequate numbers of functioning beta cells remaining to maintain euglycaemia without exogenous insulin therapy.

Whatever the underlying cause of diabetes in an individual cat, chronic hyperglycaemia causes glucotoxic and lipotoxic damage to beta cells which initially is reversible but later irreversible. Therefore, therapy needs to be aimed at rigorously controlling blood glucose and should be instituted as soon as possible after diagnosis. Early institution of rigorous glycaemic control results in remission rates of > 80%, whereas waiting 6 months or longer to institute the same rigorous control results in substantially lower remission rates, which are in the order of 30%. The key factors in achieving high remission rates in newly-diagnosed diabetic cats are firstly, use of long-acting insulin, secondly, feeding a low carbohydrate diet and thirdly, frequent monitoring, ideally home monitoring, together with appropriate adjustment of insulin dose to achieve normal or near-normal blood glucose concentrations. For those cats not achieving remission, the goals are to resolve clinical signs and avoid clinical hypoglycaemia.

INSULIN THERAPY

The long-acting insulins glargine and detemir remain the preferred treatment for diabetes mellitus in cats to maximize the probability of remission. Their effectiveness and safety is enhanced when combined with a low-carbohydrate diet. Achieving good glycaemic control and hence remission with intermediate-acting potent insulins such as NPH and lente is often difficult, and increases the risk of clinical hypoglycaemia. Published data suggests that the long-acting insulins glargine and detemir provide better glycaemic control and reduced risk of clinical hypoglycaemia when given twice daily and combined with a low-carbohydrate diet. More importantly, their use results in a significantly higher probability for remission.

Cats presented with diabetic ketoacidosis can be treated with subcutaneous glargine during stabilisation. When used intramuscularly or intravenously, glargine acts like regular insulin and can be used for initial stabilisation in place of regular insulin.

Glargine can be safely instituted at 0.5 IU/kg q12h and serial blood glucose curves should be obtained daily for 3 days either in hospital or at home. When evaluating the blood glucose curve using glargine, it is often more useful to assess pre-insulin glucose concentration rather than the nadir glucose. We have found that in some cats, it takes 3--5 days for a good glucose-lowering effect to be seen in the glucose curves, possibly because of the long duration of action and carry-over effect of glargine. Almost all cats will need to have their initial dose reduced within 2 weeks and many will achieve remission within 4-6 weeks.

In contrast to glargine, detemir is a newer synthetic insulin analogue with long duration of action through modification of the insulin molecule via addition of an acylated fatty acid chain. This modification facilitates reversible binding to plasma proteins, particularly albumin, from where it is released slowly into plasma. The modification also prolongs self-association in the injection depot, which prolongs absorption from subcutaneous tissue at the injection site and contributes to the long duration of action.

Detemir results in similar remission rates and time to remission as glargine. The same dosing prococol as for glargine can be used. Although, the median maximum dose reported from one study (1.75 IU/cat q12h) was approximately 30% less than with glargine (2.5 IU/cat q12h) for a similarly designed study, this might be because the cats on detemir were older and had higher rates of concurrent renal disease, which might have decreased the renal clearance of detemir. The final dose required for management of a long-term diabetic cat when stable may not be that much different from glargine in many cats. It is important to be aware that cats can exhibit an initial increased sensitivity to detemir (and sometimes glargine) which is transient and typically lasts for 24-48hrs. If changing a diabetic cat from another insulin such as lente insulin to detemir, particularly if on a high dose of insulin (>3 U/injection), start with about half the dose of glargine and increase dose within 48hrs if insufficient glucose lowering is occurring. Keep increasing it every 3-7 days until blood glucose is controlled., it is important to note that in dogs, the dose of detemir is approximately ¼ of that for glargine. The advantages of detemir is that it can be diluted with a special diluting medium supplied by the manufacturer (although it appears to be difficult to obtain from the manufacturer in USA), saline or sterile water and that the duration of action might be longer and more predictable than with glargine. The advantage of glargine is that when administered intramuscularly or intravenously, it has a similar action profile to regular insulin and can be used in the treatment of cats with diabetic ketoacidosis.

Monitoring and adjusting insulin dose when using glargine or detemir should be based on a number of parameters, including pre-insulin and nadir glucose concentration (Table 1), water intake, urine glucose concentration and clinical assessment. Cats treated with glargine or detemir should have a negative, 1+ urine glucose (scale 0--4+) and a value of 3+ or 4+ indicates that a dose increase is required or a change to a longer-acting insulin. However, adjusting dose based on clinical signs or urine glucose is suboptimual and will result in lower rates and delayed onset of remission. Owners should be encouraged to do home glucose monitoring to facilitate tight glycemic control. Blood glucose frequently doubles during the drive to the veterinary clinic and takes several hours to return to normal, which compounds evaluation of response to insulin and increases the probability of clinical hypoglycaemia at home or too conservative insulin therapy which decreases the probability of remission. Meters designed for use with feline blood are advantageous in achieving remission because the accuracy is greater than meters designed for human blood which tend to read lower than the blood glucose actual value. Meters selected should use a low blood volume (eg 0.3uL) to facilitate owners successfully getting a blood glucose measurement at home. Owners performing home monitoring still need to return frequently for veterinary visits, ideally once a week, until remission is achieved, so the home log of glucose concentrations and insulin dose can be reviewed and dose adjusted as required.

New developments in management of human diabetes include GLP-1 agonists such as exenatide, which now exists in a once-a-week form. This drug stimulates insulin secretion but only in the presence of glucose, so is not associated with severe clinical hypoglycaemia in the way that insulin therapy is. In addition it stimulates beta cell regeneration. Another new development is a once-a-week insulin for human use, which potentially could provide a once-a-day or once-every-other-day therapy for cats.

FEEDING: LOW CARBOHYDRATE, RESTRICTED PHOSPHORUS DIET

Diets low in carbohydrate reduce postprandial hyperglycaemia and insulin concentrations in healthy cats and are associated with increased remission rates in diabetic cats. Data from diabetic cats indicate that low-carbohydrate diets result in better clinical control, reduced insulin requirements and increased rates of diabetic remission One study demonstrated a significantly higher remission rate using a diet with approximately 12% of energy from carbohydrate (Hills m/d) compared to 26% carbohydrate (Hills w/d); the protein contents were similar. In the first few days of treatment, if cats refuse these low-carbohydrate diets, they should be offered any palatable food. However, a commercial low-carbohydrate diet should be used in diabetic cats, unless contraindicated by other disease. When required, dietary management of renal disease using a restricted-protein diet should take precedence over dietary management of diabetes; acarbose can be used to reduce glucose absorption from the gastrointestinal tract. However, acarbose works well only in cats that consume their food within a short time after it is offered and is much less effective in cats which eat multiple small meals during the day, which is the feeding pattern more often observed in cats with IRIS stage 3 or 4 renal disease. Acarbose substantially decreases postprandial blood glucose in cats fed a high carbohydrate diet, but has no real benefit in cats eating a low carbohydrate diet.

Obesity in cats markedly reduces insulin sensitivity, and hence obese diabetic cats should be fed restricted energy so they lose 1--2% bodyweight per week. Weight loss improves insulin sensitivity, and may reduce insulin requirements. Although many diabetic cats achieve remission without substantial weight loss, achieving an ideal bodyweight is critical for maintenance of long-term remission. Due to the decreased postprandial hyperglycaemia with a low-carbohydrate diet, diets with less than 20% of energy from carbohydrate should be used for obese diabetic cats during the energy-restriction phase. In contrast to feline diabetic diets, most feline weight loss diets are low-fat, high-carbohydrate diets and are not recommended for newly diagnosed diabetic cats being managed to achieve remission.

There is increasing data now available linking renal disease to diabetes in cats. Therefore, given the frequency of renal disease in diabetic cats (17% in one study in 5-10 year old cats), attention should be paid to the phosphate content of the diet. Many grocery-line low-carbohydrate--high-protein feline maintenance diets are high in phosphate. Cats in stage 1 or 2 renal failure which do not require protein or phosphate restriction should be placed on a low-carbohydrate diet formulated for feline diabetics that is low in phosphate; high-phosphate, ultra low-carbohydrate diets are best avoided.

Continued home monitoring is important in diabetic cats in remission, because those with more severe prediabetes (fasting blood glucose $\geq$ 135 mg/dL (7.5 mmol/L) and 3 hour blood glucose concentration after 1 g/kg glucose IV of > 252 mg/dL (14 mmol/L)) have an 88% probability of being overtly diabetic within 9 months, unless other therapies such as low dose insulin or insulin sensitizing drug therapy is instituted.

In conclusion, the principal goal of therapy has changed from controlling clinical signs to that of aiming for remission, and in some cases, cure of their diabetes. Key factors in achieving these goals are the use of the long-acting insulins glargine and detemir, low carbohydrate, restricted-phosphorous diets, and close monitoring, ideally with home monitoring, and appropriate dose adjustments. New therapies are emerging from human studies and they may have a future role in improving management of feline diabetes and reducing frequency of therapy.

REFERENCES

Bloom CA[1], Rand J Feline diabetes mellitus: clinical use of long-acting glargine and detemir J Feline Med Surg. 2014 Mar;16(3):205-15.

Gottlieb S, Rand JS. Remission in cats: including predictors and risk factors. *Vet Clin North Am Small Anim Pract* 2013;43:245-249.

Rand JS. Feline Diabetes Mellitus. In: Rand JS (ed). Clinical Endocrinology of Companion Animals. Ames, Iowa, USA: Wiley-Blackwell; 2013, pp 169-90.

Roomp K Rand JS. Management of Diabetic Cats with Long-acting Insulin. In Veterinary Clinics of North America: *Small animal practice: Feline Diabetes* (Ed.) Rand JS, Saunders/Elsevier 2013; 43(2): 251-266

Singh R, Rand JS, Coradini M, Morton JM. Effect of acarbose on postprandial blood glucose concentrations in cats fed low and high-carbohydrate diets, on line JFMS Jan 2014.

Zoran, D, Rand JS. The Role of Diet in the Prevention and Management of Feline Diabetes. In Veterinary Clinics of North America: *Small animal practice: Feline Diabetes* (Ed.) Rand JS, Saunders/Elsevier 2013; 43(2): 234-244

Table 1. Parameters for changing insulin dosage when using glargine or detemir in diabetic cats and monitoring once a week in the initial stabilization phase

Parameter used for dosage adjustment	**Change in dose**
Begin with 0.5 U/kg of ideal body weight BID if blood glucose ≥ 20 mmol/L (≥360 mg/dL) or 0.25/kg BID if blood glucose is lower. Do not exceed a total dose of 3U/cat/injection regardless of body weight. Do not increase in first week unless minimum response to insulin occurs, but decrease if necessary. If no blood glucose monitoring is occurring in first week, begin with 1 U/cat BID.	
If pre-insulin blood glucose concentration >12 mmol/L (>216 mg/dL) provided nadir is not in hypoglycemic range **or** If nadir blood glucose concentration > 10 mmol/L (>180mg/dL)	Increase by 0.25-1U depending on magnitude of dose (greater or less than 3 U/cat) and how close blood glucose is to 180mg/dL
If pre-insulin blood glucose concentration 10 - 12 mmol/L (180 - 216 mg/dL) provided nadir is not in hypoglycemic range **and/or** Nadir blood glucose concentration in normal range, that is, 3.5-6.5 mmol/L (65-117mg/dL) at home, or 4-9mmol/L (72-160mg/dL) on day of admission to clinic	Same dose
If nadir glucose concentration is 3.5 -4 mmol/L (65-72 mg/dL)	Use nadir glucose, water drunk, urine glucose and next pre-insulin glucose concentration to determine if insulin dose is decreased or maintained.
If pre-insulin blood glucose concentration <10 mmol/L (<180mg/dL) **and/or** If nadir blood glucose concentration <3.5 mmol/l (<65 mg/dL)	Reduce by 0.25- 1U depending on magnitude of dose (greater or less than 3 U/cat) and how close blood glucose is to 180mg/dL **or** if pre-insulin blood glucose concentration <10 mmol/L (180mg/dL) and total dose is 0.5 – 1 U once a day, stop insulin and check for diabetic remission
If clinical signs of hypoglycemia are observed	Reduce dose by 50%

CASE-BASED APPLICATIONS OF THORACIC FAST (TFAST)

Gregory R. Lisciandro, DVM, DABVP, DACVECC
San Antonio, TX

Text Focused Ultrasound Techniques for the Small Animal Practitioner, Wiley © 2014

Introduction

The clinical utility of TFAST^3 and its use for the rapid diagnosis of pleural and pericardial effusion, and pneumothorax (and the lung point concept) in trauma, triage (non-trauma) and tracking (monitoring) cases in the emergent and critical care settings will be reviewed (Lisciandro et al., JVECC 2008; Lisciandro, JVECC 2011; Boysen and Lisciandro, VCNA 2013; McMurray et al., JVECC 2014). The T^3 designation encompasses these 3 patient subsets and avoids the onslaught of confusing acronyms in human medicine in which similar abbreviated formats are given different acronyms when applied to different subsets of human patients. Thus, TFAST^3 becomes a universal term that has exact clarity in the 5-acoustic windows used by the veterinary TFAST^3 sonographer (Lisciandro, JVECC 2011; Boysen and Lisciandro, VCNA 2013).

The TFAST^3 ultrasound format has greater potential to positively guide clinical course and improve patient outcome by detecting conditions and complications otherwise occult based on traditional means of physical examination, laboratory and thoracic radiographic findings. For example, the TFAST^3 Diaphragmatico-Hepatic (DH) View, should always be used as a window into the pleural and pericardial spaces (Lisciandro, JVECC 2015); and the TFAST^3 DH View complements the TFAST^3 Pericardial (PCS) Site Views. By applying TFAST^3 with this mindset, the misdiagnosis of fluid-filled heart chambers for pleural or pericardial effusion is less likely to occur; and thereby, avoid the potential for catastrophic mistakes such as performing centesis on a heart chamber (Lisciandro, JVECC 2015). Carrying the example further regarding the diagnosis of pericardial effusion: 1) ultrasound is the gold standard for the diagnosis or pericardial effusion, 2) both physical examination and thoracic radiography have poor and unreliable sensitivity, and 3) many cases are inappropriately treated because pericardial effusion is not suspected and the diagnosis is missed.

Finally, TFAST^3 findings are made more clinically relevant for the clinician, client, and referring veterinarian, by using standardized ultrasound format (TFAST^3), and standardized goal-directed templates for medical records (see below) (Lisciandro et al., JVECC 2008; Lisciandro, JVECC 2011; Boysen and Lisciandro, VCNA 2013).

Terminology

In the human literature, there has been a confusing onslaught of multiple acronyms for similar ultrasound examinations by non-radiologists, i.e. FFAST, EFAST, HHFAST, BOAST, INBU, FEEL, FALLS, FATE, RUSH, etc. Veterinarians would be best served by using AFAST^3, TFAST^3 (thoracic) and Vet BLUE (lung exam) (Lisciandro, JVECC 2011; Boysen and Lisciandro, VCNA 2013). When all 3 formats are used together the exam is referred to a Global FAST^3 or GFAST^3. Because any of the three GFAST^3 ultrasound format (AFAST^3, TFAST^3 and Vet BLUE) may be diagnostic, the author suggests that comprehensive exams of the abdomen and thorax be designated as "complete abdominal ultrasound" and "complete echocardiography" respectively, similar to the human literature (formal used instead of complete). Finally, terms such as "Focused" and "Targeted" and "COAST^3" (cage-side organ assessment with sonography for trauma, triage and tracking) have been used for specific sonographic interrogation of organs in both human and veterinary medicine.

Patient Positioning and Preparation

TFAST^3 may be performed immediately following AFAST^3 when AFAST^3 is performed in right lateral recumbency because the depth is essentially the same moving from the abdomen to the TFAST^3 left and right Pericardial (PCS) Site Views; and the right PCS view is optimized in lateral recumbency with the heart moving against the thoracic wall displacing air-filled lung allowing for the left ventricular short-axis "mushroom" volume status view. However, in respiratory distressed or compromised patients, the TFAST^3 format may be performed in sternal or standing position because it is generally less stressful and safer for an unstable patient. In cases in which only TFAST^3 is needed initially, sternal and standing positioning is often adequate to answer the clinical questions regarding pleural and pericardial effusion and the presence or absence of pneumothorax (and its degree by the "Lung Point").

Fur is NOT shaved but rather parted and alcohol and/or gel applied. Alcohol should not be used if electrical defibrillation is anticipated.

A Schematic of the TFAST^3 Exam

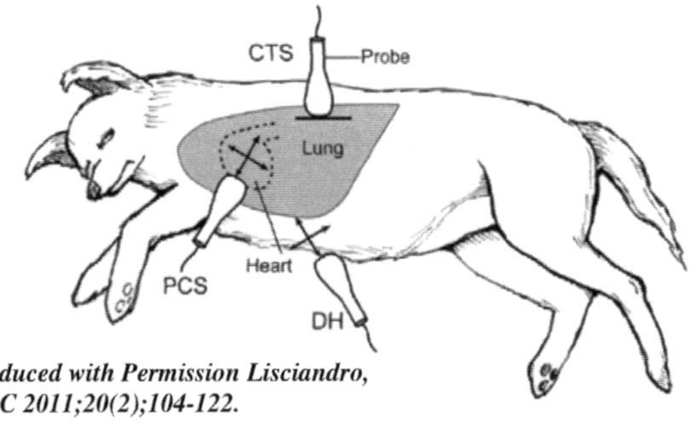

Reproduced with Permission Lisciandro, JVECC 2011;20(2);104-122.

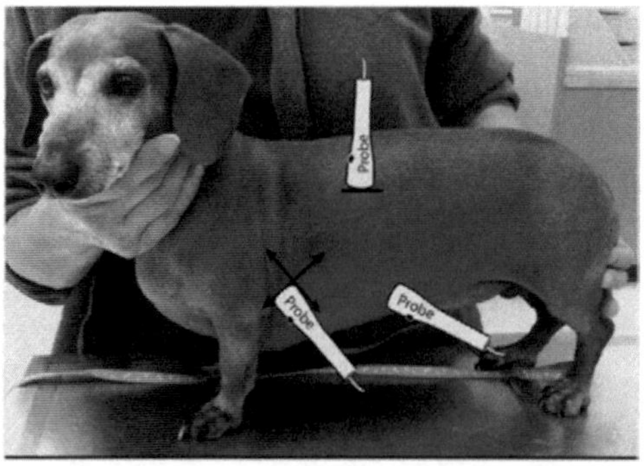

This material is reproduced with permission of John Wiley & Sons, Inc, Focused Ultrasound Techniques for the Small Animal Practitioner, Wiley ©2014 and FASTvet © 2014

The TFAST^{3} exam illustrated on a dog in right lateral recumbency and in standing position. Standing and sternal recumbency are safer for respiratory compromised dogs and cats. TFAST^{3} consists of 5-views named as follows: the ***stationary horizontally probe-positioned*** Chest Tube Site (CTS) view; ***and the dynamically spotlighted*** Pericardial Site (PCS) view, and ***the dynamically spotlighted*** DH view (used in both TFAST^{3} and AFAST^{3}). CTS and PCS views are bilaterally applied; however, the DH view is a single view.

Strengths and Weaknesses of TFAST^{3} Views

The **CTS view** is best used to rule out pneumothorax (PTX) and survey for lung pathology (the wet lung vs. dry lung concept [see Vet BLUE Proceedings]). The CTS view is the highest point on the thoracic wall where lung may be visualized against the thoracic wall along an intercostal space where a cap of air would rise in the event that PTX was present. To avoid false positives, the search for the "Lung Point" next takes place. Its concept is explained below.

The **PCS view** is best used to screen for the presence of pleural or pericardial fluid. It also may be used for volume status assessment (left ventricular short-axis "mushroom" view) and for the "quick peak" left atrial to aortic ratio (LA:Ao) to rule out left-sided heart failure. However, Vet BLUE can rapidly rule out any clinically-relevant left-sided heart failure by the finding of "dry lungs all fields" (see Vet BLUE Proceedings).

The **DH View** may be superior for the detection of pericardial (and pleural effusion) fluid because of less air interference from the lung at the PCS views; and the acoustic window provided by the liver and gallbladder (Lisciandro, JVECC 2015). The DH strategy is to image near the solid muscular apex of the heart where a heart chamber is unlikely to be misinterpreted for pleural or pericardial effusion (Lisciandro, JVECC 2015). The finding has been described and referred to as the "Racetrack Sign" by the author (Lisciandro, Focused Ultrasound Techniques for the Small Animal Practitioner; Lisciandro, JVECC 2015).

*By using multiple views (the DH View and PCS views) abiding by the sage axiom "1 view is no view", the probability of misinterpreting a fluid-filled heart chamber for pericardial and pleural effusions is lessened.

Basic Lung Findings at CTS View- Glide Sign, Pneumothorax, Lung Rockets & Step Sign

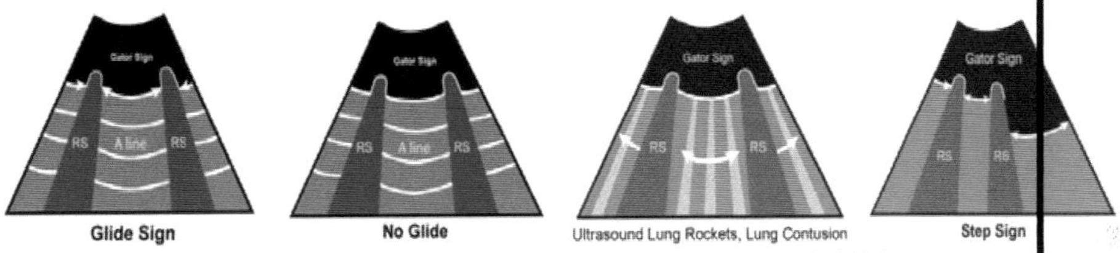

Lisciandro JVECC 2011; Lisciandro et al. JVECC 2008

Reproduced with Permission Lisciandro et al. JVECC 2008 18(3):258-269 and Lisciandro 2011; 20(2):104-122.

The illustrations depict in order from left to right **1)** Normal dry lung (**Glide Sign** with A-lines) **2)** Pneumothorax (PTX) (**No Glide** with A-lines) **3) Ultrasound Lung Rockets** (ULRs, also called B-lines) (hyperechoic laser-like streaks that do not fade extending to the far-field obliterating A-lines and oscillating in synchronization with inspiration and expiration) ruling out PTX at that point on the thorax and further indicating that the lungs are wet (interstitial edema or lung contusion). In trauma, ULRs represent lung contusions until proven otherwise. By counting the number of ULRs at each intercostal space, semi-quantification of degree of interstitial edema may be made (see Vet BLUE). **4) Step Sign** representing deviation from the expected linear to and fro movement of the pulmonary-pleural interface suggesting chest wall (i.e. intercostals tear(s), fractured rib(s), subpleural hematoma) and pleural space disease or conditions (i.e. effusions, diaphragmatic hernia, masses) in trauma; however, in non-trauma there are subsets of the **Step Sign** referred to as the **Shred Sign, Wedge of V Sign (PTE), Tissue Sign, and Nodule Sign** that represent forms of lung consolidation/infiltration (see Vet BLUE Proceedings). These lung ultrasound signs are well described in the textbook **Focused Ultrasound Techniques for the Small Animal Practitioner ©Wiley 2014.**

Key Points

Ultrasound does not transmit through air and thus deeper more centrally located lung pathology will be missed by TFAST^{3} and Vet BLUE necessitating thoracic radiography or other advanced imaging.

Use of TFAST^{3} for the "Lung Point" – The Degree of PTX

Cross-sectional canine thoraces (sternal recumbency) depicting the **quantification of the degree of pneumothorax** (PTX) as **partial or massive** by searching for the **Lung Point** defined as the level at which lung re-contacts the thoracic wall. In the absence of the **Glide Sign** and absence of **Ultrasound Lung Rockets (ULRs),** the probe is moved sequentially in a ventral manner as numerically labeled from dorsal to ventral. **A)** Normal thorax in which PTX has been excluded **B)** PTX has been determined at position 1 and the **Lung Point** is found at position 2 and thus PTX is determined to be partial. **C)** PTX has been determined and a lung point is nonexistent at any of the 3 probe positions indicating massive PTX (CTS, chest tube site; PTX, pneumothorax) or considered indeterminate based on clinical signs.

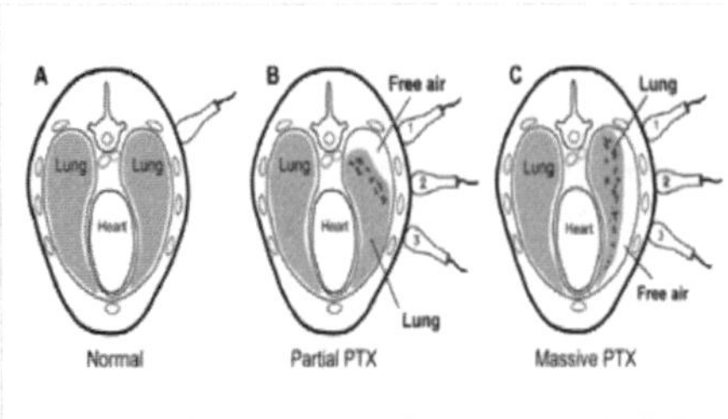

Reproduced with Permission Lisciandro, JVECC 2011; 20(2):104-122

Key Points

By dividing the thorax into thirds when searching for the lung point a subjective assessment of partial vs. massive PTX may be made. Moreover, when PTX is suspected at the CTS view, by locating the Lung Point sensitivity is increased and smaller pneumothoraces may be monitored by using/recording the distance from the CTS to the lung point. <u>Important:</u> Use the distance of the Lung Point from the Chest Tube (CTS) view to track (monitor) the worsening or resolution of PTX including post-invasive procedures (in place of thoracic radiography) including thoracotomy, thoracoscopy, chest tube placement/removal, thoracocentesis, lung lobe aspirate, etc.

Use of Lung Ultrasound Respiratory Distress - The "Wet Lung" vs. "Dry Lung" Pattern-based Approach

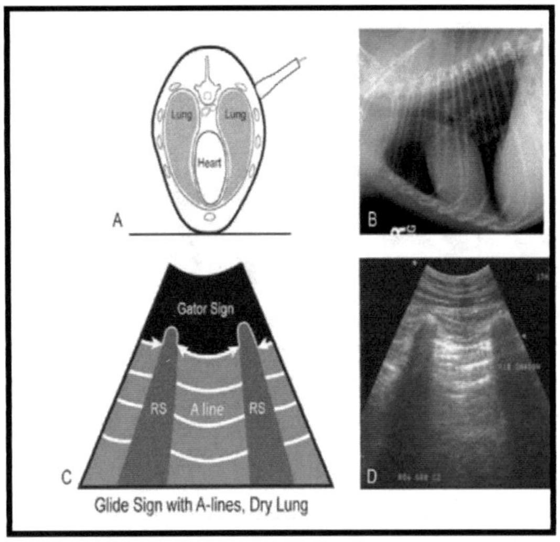

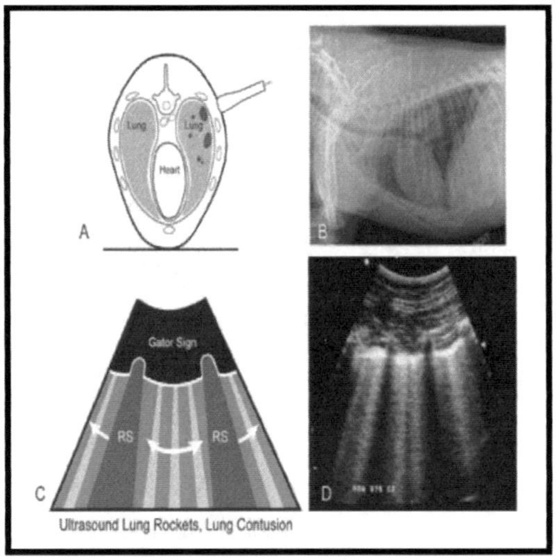

This material is reproduced with permission of John Wiley & Sons, Inc, <u>Focused Ultrasound Techniques for the Small Animal Practitioner</u>, Wiley ©2014 and FASTVet.com © 2014

Because lung pathology was detected via the TFAST[3] Chest Tube Site (CTS) view, the Vet BLUE lung scan was more recently developed and is currently being evaluated for its usefullness in respiratory distressed dogs and cats (Lisciandro et al. Vet Radiol & US 2014). **Basic lung ultrasound findings premise themselves** on **the "Wet Lung"** vs. **"Dry Lung" concept.** The **figure on the left** shows the ultrasound probe placement on the thoracic wall with line drawings and an actual ultrasound B-mode still image along with a corresponding lateral thoracic radiograph.

Since a **Glide Sign** exists with A-lines (reverberation artifact) the **lung at the periphery is considered "dry."** In contrast, the figure on the right shows **ultrasound lung rockets (ULRs) and is considered "wet**." A pattern-based distribution of wet vs. dry lung by named ***lung regions*** (caudo-dorsal, perihilar, middle, and cranial lung regions) may be used to rule in and rule out various lung conditions (see Vet BLUE Proceedings). More advanced lung ultrasound findings and more detail regarding lung ultrasound uses are published in the textbook **<u>Focused Ultrasound Techniques for the Small Animal Practitioner</u>** © Wiley 2014.

Basic Imaging Principles of the TFAST PCS view – Global Heart View with Its Pericardium

A basic echocardiography skill for all FAST sonographer is to be able to perform the left ventricular short-axis "mushroom" view. As far-fetched as it seems, the novice or hasty sonographer by not identifying the heart chambers, left and right ventricles, can easily mistake the right ventricle for pericardial or pleural effusion by not performing the orientation the same way every time, and by not identifying the right ventricle and its papillary muscle.

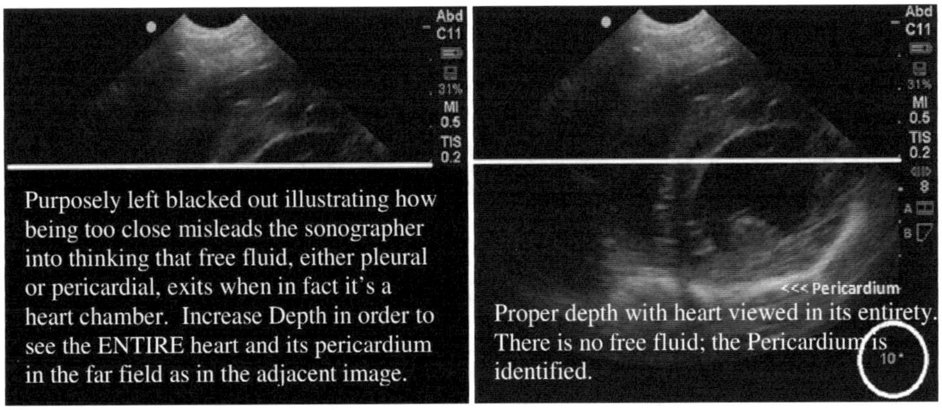

Pitfall of the PCS View: Shows how, by too shallow depth (**image on the left**), you can easily mistake heart chambers, here the right ventricle, for pericardial (or pleural effusion). To prevent this mistake, always increase depth (**image on the right**). By this recommended practice, the heart is seen in its entirety, confirmed by adequately imaging the pericardium as your landmark in the far field, as the brightest white line (labeled in image above). In differentiating between pleural and pericardial fluid, multiple views (right and left PCS and the DH view) prevent mistakes of misidentifying one for the other; or an enlarged right ventricle, or other heart chambers, for pleural or pericardial effusion; or the papillary muscle of the right ventricle for a mass.

Pericardial (on left) **vs. Pleural Effusion** (on right) **using the TFAST PCS and DH Views**

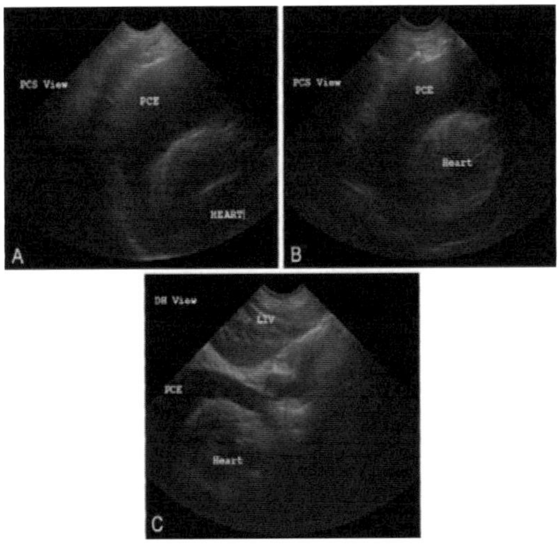

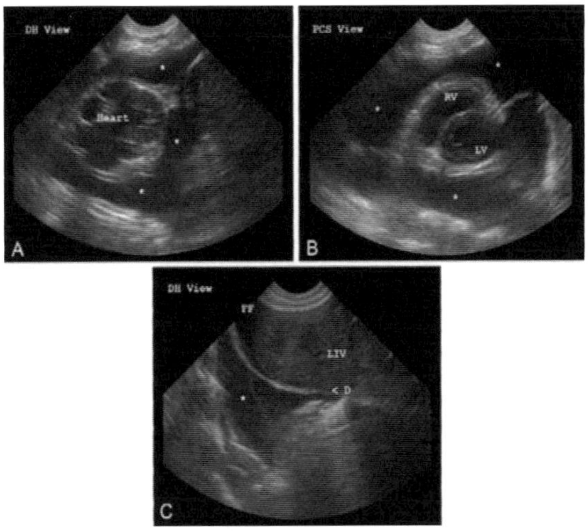

The FAST³ Diaphragmatico-Hepatic (DH) View

The classic DH view nicknamed the "**D**esignated **H**itter" since the DH is part of AFAST³ and TFAST³. In dogs, the DH view begins with imaging the **gallbladder "kissing" the diaphragm.** Once this classic view is appreciated, fanning through the liver lobes, gallbladder, and the diaphragm should then take place while directing the **marker toward the patient's head staying longitudinal** throughout the view (do not rotate the probe, it's unnecessary). Because the DH view advantageously (less lung [air] interference) serves as an acoustic window (via the liver and gallbladder) into the thorax. The orientation of the gallbladder immediately against the diaphragm is less reliable in cats.

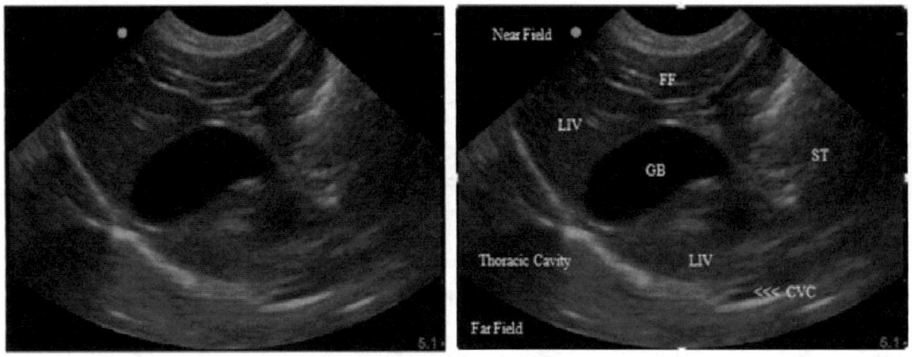

FASTvet.com © 2014

ALWAYS look into the thorax for the rapid detection of **pleural and pericardial effusion** (Lisciandro, JVECC 2015). Once these basic structures are proficiently imaged by the sonographer, the **caudal vena cava** should also become part of the DH view. Depth should be adjusted on your machine so that the pleural and pericardial spaces (25-33% of the far-field) are routinely examined (may not be possible in larger dogs). Questionable findings within the thorax using the DH view should be confirmed via TFAST^3 (PCS view) or Vet BLUE or both or by serial exams (repeating TFAST^3 and Vet BLUE 4-hours later).

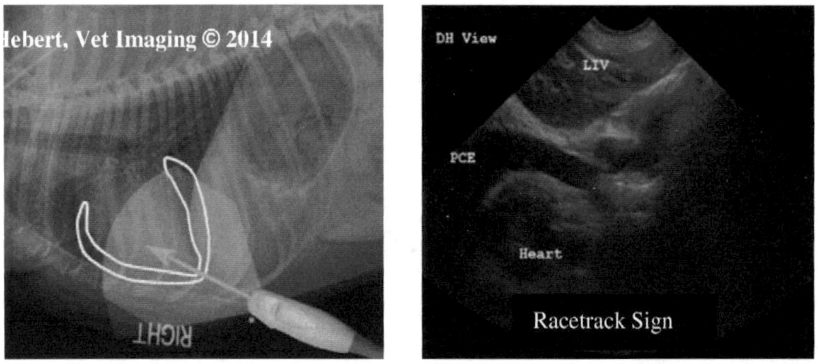

This material is reproduced with permission of John Wiley & Sons, Inc, Focused Ultrasound Techniques for the Small Animal Practitioner, Wiley ©2014

Key Points: DH use for Pericardial and **Pleural effusion.** Since the canine and feline heart do not consistently rest against the diaphragm (as in humans), the heart and its pericardial sac may not be ultrasonographically visualized because of air interference (interposing lung) between the heart and the diaphragm. However, in most cases of clinically relevant pericardial effusion in dogs, the diagnosis is made via the DH view (Lisciandro, JVECC 2015) because the distended (with fluid) pericardial sac moves against the diaphragm displacing lung (now an acoustic window). Moreover, the DH view helps avoid mistaking heart chambers (most catastrophic of mistakes if centesis is performed on the heart) for pericardial or pleural effusions because **the solid muscular apex of the heart is closest to the xiphoid** (rather than fluid-filled heart chambers at the PCS views). We refer to the conspicuous rounding rim of anechoic (black) fluid the **"Racetrack Sign."** It is important to note that in cats, the most common cause of pericardial effusion (generally < 5 mm) is heart failure; and pericardiocentesis is rarely indicated in felines.

Clinical Indications/Applications of AFAST^3

The use of TFAST^3 should be simply stated as an "extension of the physical exam" for nearly ALL dogs and cats that are abnormal. FAST^3 formats (AFAST^3 and its applied fluid scoring system [AFS], TFAST^3 and Vet BLUE) should be adopted as BASIC "screening tests" just as we have been trained to perform minimally basic blood tests (so-called Quick Assessment Tests [PCV/TS, serum character, BUN, Creatinine, Blood Glucose, and ALT]).

- Blunt trauma
- Penetrating trauma
- Collapse, apparent collapse
- Undifferentiated hypotension
- Acutely decompensated cardiac case, detection of left atrial tears
- Respiratory distress (since there are non-respiratory look-a-likes [hemoabdomen, cardiac tamponade, anaphylaxis, high fever, and others])
- Post-interventional at-risk bleeding (surgery, percutaneous procedures (lung lobe aspirates), Thoracoscopy, chest tube)
- Post-interventional at-risk pneumothorax (surgery, percutaneous procedures (lung aspirates, tracheal wash), thoracoscopy, chest tube)

Monitoring PTX, pleural and pericardial effusions
Detecting and monitoring cardiogenic and non-cardiogenic pulmonary edema (Vet BLUE better)
Patient monitoring during fluid resuscitation and during hospitalized care (used in combination with AFAST[3] and Vet BLUE called Global FAST [GFAST[3]])

Goal-directed Template for TFAST[3]
*Right and left sides are listed in templates for the CTS and PCS views

***CTS glide sign:**	**Present** (normal) -- no Pneumothorax **or** **Absent** – Pneumothorax **Indeterminate**
***CTS lung rockets**:	**Present** (no PTX) – interstitial lung fluid (edema, hemorrhage) **or** **Absent** – no interstitial lung fluid (edema, hemorrhage) **Indeterminate**
***CTS step sign:**	**Present** – concurrent thoracic wall trauma (rib fractures, hematoma, intercostal muscle tear) or pleural space disease is suspected **or** **Absent** - no concurrent thoracic wall trauma or pleural space disease is suspected **Indeterminate**
***PCS view:**	**Absent**- no pleural or pericardial fluid **Present** - pleural or pericardial fluid or both (mild, moderate, or severe) **Indeterminate**
LV filling (short-axis):	**Adequate** suggesting normovolemia **or** **Inadequate** suggesting hypovolemia **or** **Indeterminate**

DH View:	**Pleural effusion:**	absent, present (mild, moderate, severe) or indeterminate
	Pericardial effusion:	absent, present (mild, moderate, severe) or indeterminate
	Hepatic venous distension:	present, absent or indeterminate
	Caudal vena cava characterization:	FAT, flat or bounce or indeterminate

Cardiac tamponade: **Present or Absent or Indeterminate**

Comments: ____

KEY: CTS = chest tube site; **PCS** = pericardial sac; **LV** = left ventricle, **PTX** = pneumothorax
(Note: The TFAST[3] exam is a rapid ultrasound procedure used to help detect major chest wall, lung, and pleural and pericardial space problems as a screening test in order to better direct resuscitation efforts or manage hospitalized critically ill patients. TFAST[3] exam is not intended to replace thoracic radiographs, or complete echocardiography.)

REFERENCES

1. Lisciandro GR, Lagutchik MS, Mann KA, et al. Evaluation of a thoracic focused assessment with sonography for trauma (TFAST) protocol to detect pneumothorax and concurrent thoracic injury in 145 traumatized dogs. *J Vet Emerg Crit Care* 2008; 18(3):258-269.
2. Boysen SR, Rozanski EA, Tidwell AS, et al. Evaluation of a focused assessment with sonography for trauma protocol to detect free abdominal fluid in dogs involved in motor vehicle accidents. *J Am Vet Med Assoc* 2004; 225(8):1198-1204.
3. Lisciandro GR. Focused abdominal (AFAST) and thoracic (TFAST) focused assessment with sonography for trauma, triage and monitoring in small animals. *J Vet Emerg Crit Care* 2011;20(2):104-122 .
4. Boysen SR, Lisciandro GR. The use of Ultrasound in the Emergency Room (AFAST and TFAST). *Vet Clin North Am Small Anim Pract* 2013;43(4):773-97.
5. Lisciandro GR, Fosgate GT, Fulton RM. Frequency of ultrasound lung rockets using a regionally-based lung ultrasound examination named veterinary bedside lung ultrasound exam (Vet BLUE) in 98 dogs with normal thoracic radiographical lung findings. *Vet Rad & US* 2014; 55(3):315-22.
6. Lisciandro GR. Chapter 9: The Thoracic (TFAST) Exam; Chapter 10: The Vet BLUE Lung Scan; *In* Focused Ultrasound for the Small Animal Practitioner, Editor, Lisciandro GR. Wiley Blackwell: Ames IA 2014.
7. Lisciandro GR and Armenise A. Chapter 16: Focused or COAST[3] - CPR, Global FAST and FAST ABCDE. *In* Focused Ultrasound for the Small Animal Practitioner, Editor, Lisciandro GR. Wiley Blackwell: Ames IA 2014.
8. Lisciandro GR. Chapter 55: Ultrasound in Animals. *In* Critical Care Ultrasound (human textbook), Editors Lumb and Karakitsos. Elsevier: St. Louis, MO 2014.
9. Lisciandro GR. Evaluation of initial and serial combination focused assessment with sonography for trauma (CFAST) examination of the thorax (TFAST) and abdomen (AFAST) with the application of an abdominal fluid scoring system in 49 traumatized cats. Abstract. *J Vet Emerg Crit Care* 2012;22(2):S11.
10. Lisciandro GR. The detection of pericardial effusion in 24 dogs using the diaphragmatico-hepatic (DH) view of FAST exams. *J Vet Emerg Crit Care, accepted 2014.*

11. McMurray J, Boysen S, Chalhoub S. Focused Assessment with Sonography in Non-trauma dogs and cats in the emergency and critical care setting. Abstract. *J Vet Emerg Crit Care,* 2014; 24(S1):S28.

THE GLOBAL FAST (GFAST) TRIAD FOR PATIENT MONITORING AND CPR

Gregory R. Lisciandro, DVM, DABVP, DACVECC
San Antonio, TX
***FASTvet*™ and *FAST Saves Lives!*™**
Text Focused Ultrasound Techniques for the Small Animal Practitioner, Wiley © 2014

Introduction

The evolution of veterinary abbreviated ultrasound formats has extended beyond the abdominal format and a simple fluid positive or fluid negative approach since the landmark publication by Boysen and colleagues (JAVMA 2004). In 2008, the thoracic FAST format was developed and referred to as TFAST for the rapid diagnosis of pneumothorax and other thorax-related injury (Lisciandro et al., JVECC 2008). In 2009, the abdominal FAST referred to as AFAST was studied with a target-organ approach rather than naming of external sites so that sonographer would be more aware anatomically about the actual organs and structures at each of the AFAST views (Lisciandro et al., JVECC 2009). In the same study, the AFAST-applied fluid scoring system lent more meaning to what a positive meant by semi-quantitating the degree of hemorrhage and advocating for the use for serial AFAST exams with abdominal fluid scoring. In the study, it was found that dogs could be subdivided into small volume bleeders, unlikely to develop anemic, and large or big volume bleeders that predictably developed anemic. Moreover, it was found that of the high-scoring dogs, approximately 33% required blood transfusion.

Most recently, a rapid, abbreviated, lung ultrasound screening format has been developed called Vet BLUE (blue for cyanosis and BLUE for bedside lung ultrasound exam) (Lisciandro et al., VRUS 2014). The Vet BLUE is an extension from the TFAST chest tube site. The Vet BLUE is a pattern-based approach premised on 5 basic lung ultrasound findings including dry lung (A-lines with glide), wet lung (ultrasound lung rockets also called B-lines), Shred Sign, Tissue Sign, and Nodule Sign (see Lung Ultrasound in Animals Proceedings).

In 2011 it was proposed that veterinary ultrasound terminology be adjusted to prevent the onslaught of confusing acronyms in human medicine for abbreviated ultrasound formats many of which have similar objectives. Thus, Abdominal FAST (AFAST), Thoracic FAST (TFAST) are helpful with non-trauma subsets of patients including those triaged, and tracked post-interventionally. Thus, by adding a T3 to the acronym there is no need at this time to create additional acronyms. Hence, AFAST^3 and TFAST^3 formats apply to subsets of dogs and cats including trauma, triage (collapse, weakness), and tracking (post-interventionally, monitoring during hospitalized care). Vet BLUE should also be applied with the T3 mindset. When performing all 3 formats the author uses the term Global FAST^3 or GFAST^3.

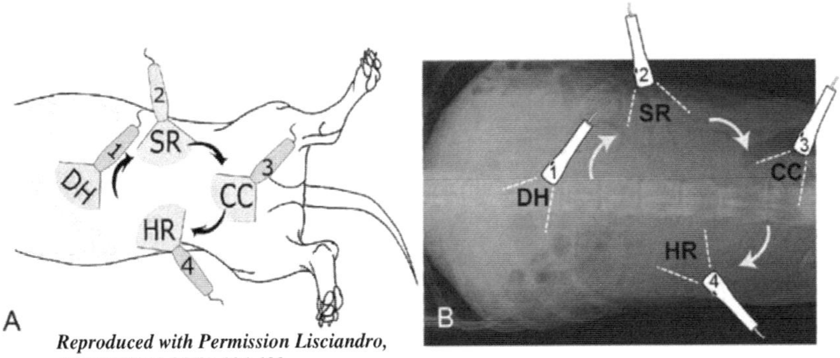

Reproduced with Permission Lisciandro, JVECC 2011;20(2);104-122.

The Abdominal FAST^3 (AFAST^3) format pictorially and on a abdominal radiograph of a dog. The target-organ approach and systematic order of the exam is shown.

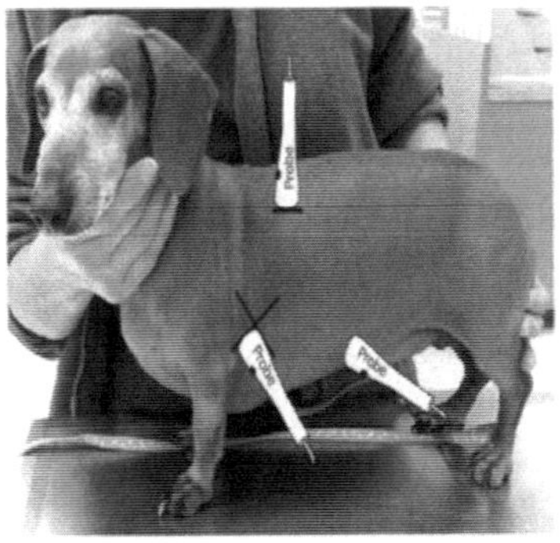

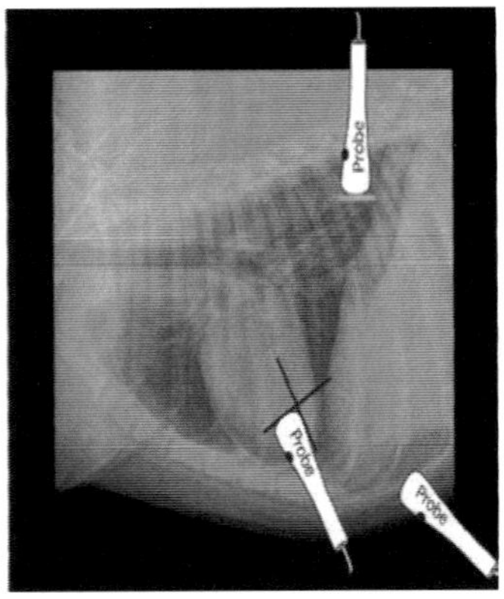

The Thoracic FAST^3 (TFAST^3) format on a dog and on a lateral thoracic radiograph of a dog showing the Diaphragmatico-Hepatic (DH) View, the bilaterally applied Pericardial Site (PCS) and Chest Tube Site (CTS) views for detection pleural, pericardial fluid and pneumothorax.

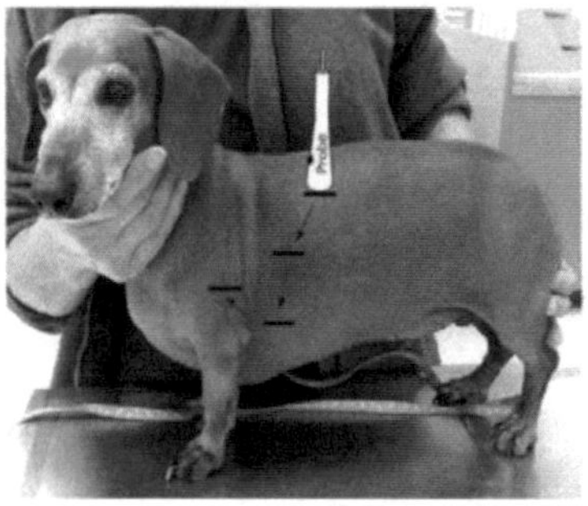

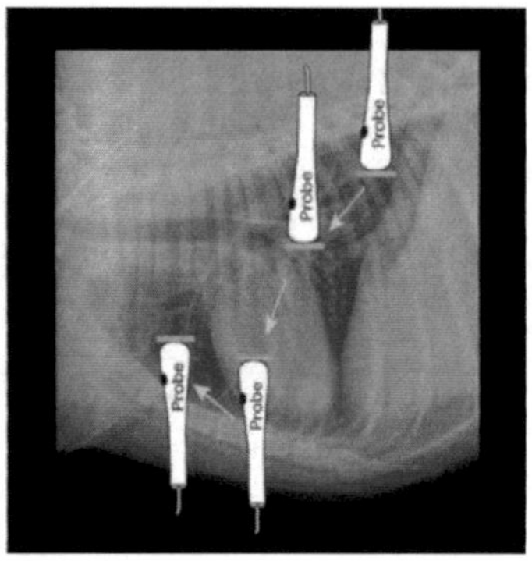

The Vet BLUE lung exam, BLUE for cyanosis, and Bedside Lung Ultrasound Exam is rapid; and an extension from the Chest Tube Site of TFAST for a regionally-based respiratory assessment. Bilaterally applied views from caudo-dorsal to perihilar to middle and ending at the cranial lung lobe ***regions.***

Recently, measurement of central venous pressure via central venous venous catheters for guiding fluid therapy was debunked (Marik et al., 2012); thus, non-invasive ultrasound use for volume assessment has become an even more popular topic. By taking advantage of sonographic findings during abbreviated FAST-like protocols, e.g., the RUSH exam (Rapid Ultrasound for Shock), emergency and critical care physicians are paying attention to cardiac filling and contractility and the dynamic changes of the inferior vena cava. Another approach involving lung, called the FALLS approach (Fluid Assessment Limited by Lung Sonography, Lichtenstein and Karakitsos, 2012), is also being used. Lung ultrasound is less challenging and readily teachable (dry lung vs. wet lung artifacts) when compared to echocardiography. A recent comparative study compared transthoracic echocardiographic and inferior caval technically challenging measurements to subjective assessment (the eyeball approach) regarding patient volume status during resuscitation. It was found that volume status was consistent between the two methodologies in most human patients. By adding in the FALLS-like strategy (Vet BLUE), a comprehensive format similar to ultrasound-based volume assessment strategies published in human medicine is possible during Global FAST^3.

Lastly, the use of ultrasound during CPR has been shown to be an effective monitoring tool and means to rule out potentially treatable causes of CPR. The FEER and FEEL protocols (Focused Echocardiography Evaluation during Resuscitation, Focused Echocardiography Evaluation during Life support) documented how ultrasound exceeds physical examination and electrocardiography during CPR-advanced life support. The studies documented the misdiagnoses of asystole and pulseless electrical activity (PEA) based on physical examination and electrocardiogram findings. Moreover, theses studies documented that ambient light did not significantly affect ultrasound imaging. Other CPR-related studies have shown how ultrasound can rapidly document endotracheal tube placement by both direct visualization of the endotracheal [ET] tube in the cervical area or indirectly by observation of the glide sign; and other formats such as FAST and RUSH exams for the rapid detection of treatable causes of CPR that are missed or delayed by laboratory testing and imaging. The American Heart Association guidelines advocate knowing your H's and T's for rapid assessment for causes of cardio-pulmonary arrest that may be treatable. The H's (acidosis [H+], Hyper-, hypokalemia, Hypoglycemia, Hypotension [bleeds, AX by TFAST^3, AFAST^3]) are generally covered by a venous blood gas; and the T's by using point-of-care Global FAST^3, including pericardial effusion/cardiac Tamponade (TFAST), Tension pneumothorax (TFAST), Thrombo-embolism [PTE] (Vet BLUE), and Toxins (author adds anaphylaxis [AX] here) (FAST^3 DH View for Gallbladder Halo Sign).

Use of Global FAST^3 for Small Animal Patient Monitoring

By taking a cerebral approach to the views during Global FAST^3, fluid resuscitation and patient monitoring of small animals during hospitalized care may be optimized in minutes and point-of-care. The approach is 3-tiered, including the FAST^3 DH view, the TFAST^3 right Pericardial View, and the use of Vet BLUE. Fur is not clipped but rather parted and acoustic coupling medium is applied (alcohol should not be used if defibrillation is potentially used).

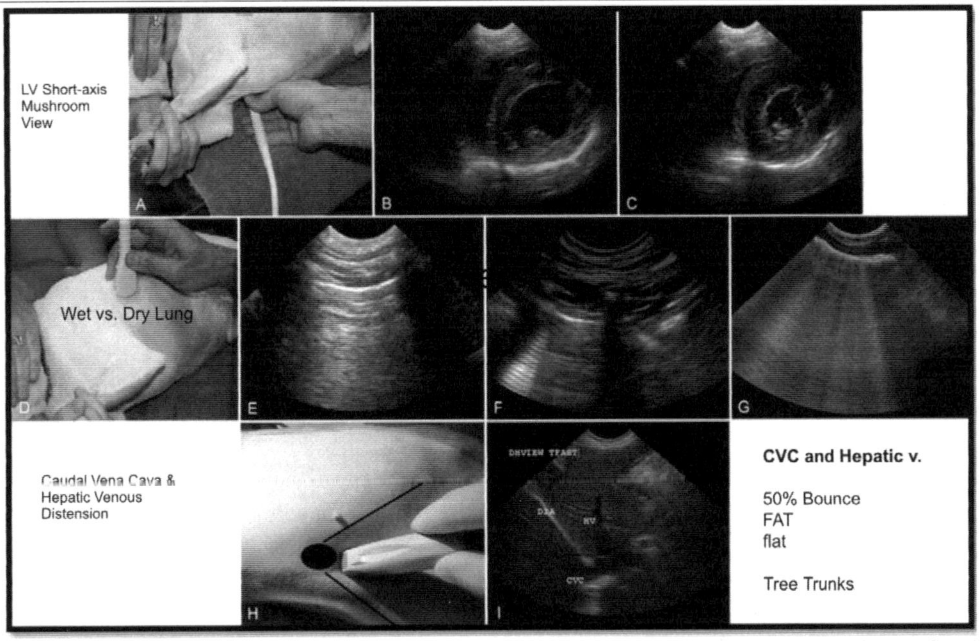

This material is reproduced with permission of John Wiley & Sons, Inc, Focused Ultrasound Techniques for the Small Animal Practitioner, Wiley ©2014

The Use of the FAST[3] DH – Caudal Vena Caval and Hepatic Venous Distension
Preload and Right-sided Cardiac Status

The caudal vena cava (CVC) at the level of the diaphragm may be rapidly characterized as FAT (distended), flat, or having a bounce representing volume overload, hypovolemia, and somewhere near normalcy, respectively. The normal "bounce" should be 50% change in diameter during inspiration and expiration in spontaneously ventilating dogs and cats. Measuring the CVC using M-mode can be challenging without more advanced ultrasound training; however, by visually characterizing the CVC at the FAST[3] DH view and correlating with clinical impression and other findings (blood pressure, physical exam findings, blood lactate), the sonographer gains information regarding volume status (preload).

When the imaging of the CVC is problematic, fanning through the FAST DH[3] View still gains indirect information since the hepatic veins are not normally conspicuous. Distension of hepatic veins due to back flow from the intrathoracic CVC, are readily appreciated likened to tree trunks (referred to as the Tree trunk Sign by the author). Anytime hepatic venous distension is sonographically visualized investigation of the right-heart (pericardial effusion, right atrial and ventricular dilatation), and consideration of volume overload (excessive preload) should be considered.

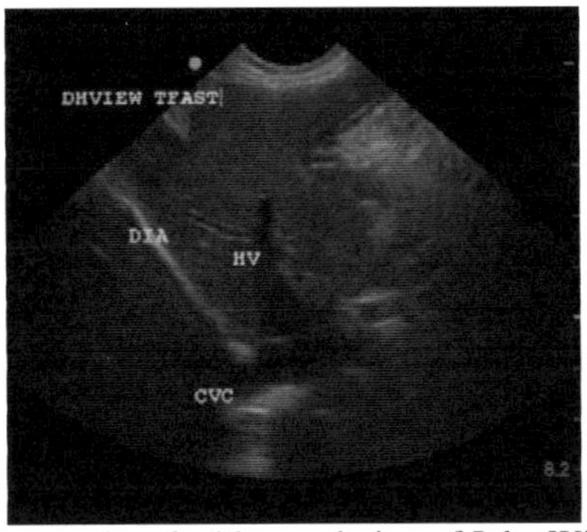

The FAST[3] Diaphragamtico-Hepatic (DH) View, part of AFAST[3] and TFAST.[3]
The caudal vena cava (CVC) passes through the diaphragm (DIA) with a distended FAT cava and tree trunk appearing hepatic veins (HV) from backflow of blood (Tree trunk Sign).

This material is reproduced with permission of John Wiley & Sons, Inc, Focused Ultrasound Techniques for the Small Animal Practitioner, Wiley ©2014

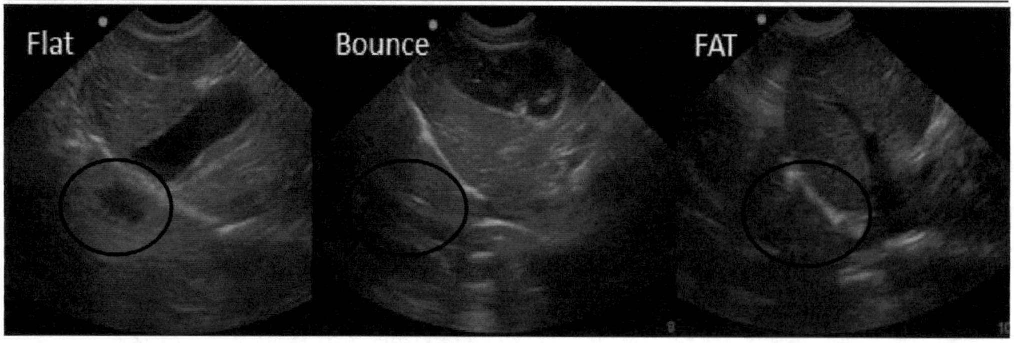

FASTVet.com © 2015

The Use of the TFAST^3 Right Pericardial View - Left Ventricular "Mushroom" View
Contractility (Cardiac Function) and Volume ("Mushroom')

The short-axis left ventricular "mushroom" view is easily learned by the non-radiologist/ cardiologist veterinarian. The view is helpful to re-enforce the clinical impression of volume status assessed by characterization of the CVC (FAT, flat or bounce) at the FAST^3 DH View. Moreover, contractility may be subjectively assessed subjectively as hyperdynamic, hypodynamic or normal.

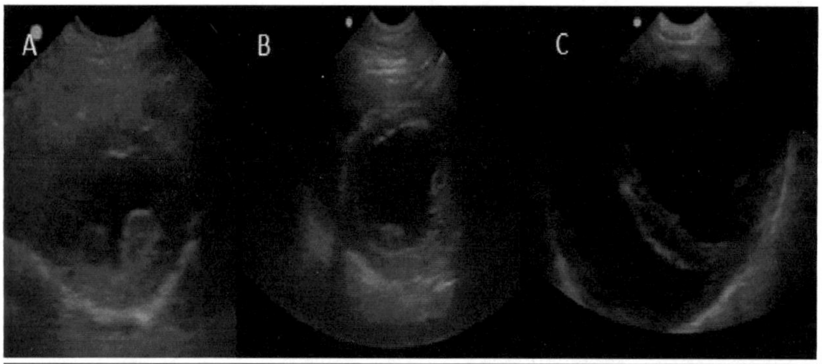

Left ventricular short-axis "mushroom" view, A) Poor B) Normal C) Dilated (Dilated Cardiomyopathy). Each is obvious in real-time imaging and the view easily teachable.

FASTVet.com © 2015

The Use of Vet BLUE – "Wet Lung" vs. "Dry Lung"
Left-sided Cardiac Status

Because the "wet lung" vs. "dry lung" concept is easily recognized using lung ultrasound (Vet BLUE), the presence or absence of ultrasound lung rockets [ULRs] also called B-lines provides important clinical information regarding left-sided cardiac status and left-sided volume overload. The author recommends acquiring a baseline Vet BLUE on all hospitalized dogs and cats prior to intravenous fluid therapy (better a baseline Global FAST^3 [GFAST^3]) and then used serially as clinical course dictates (or minimally once daily on rounds) for the detection of complications or a diagnosis that has become obvious in time. Normal and clinically ill dogs and cats have had baseline frequency of lung ultrasound artifacts compared to normal radiographical lungs; and it was found that ULRs (also called B-lines) are uncommon when compared to radiographically normal lungs (Lisciandro et al. VRUS 2014). In dogs and cats with dry lungs that are becoming wet (development of ULRs), the attending clinician should re-evaluate fluid therapy and investigate left-sided heart status, e.g. is left-sided heart failure occurring?

ULRs have been shown to correlate with extravascular lung water, and thus are sentinels for worsening respiratory status. If treatment strategy is not adjusted, the interstitial-alveolar edema may progress to aleveolar flooding which is much more difficult to treat. Using the pattern-based approach, other causes of wet lung artifacts, such as pneumonia, can often be discriminated using the Vet BLUE pattern-based approach. Moreover, the use of lung ultrasound potentially triggers additional testing and imaging.

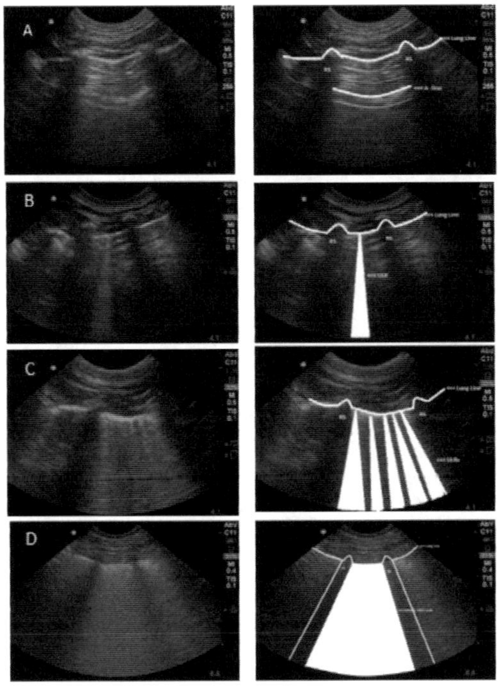

FASTVet.com © 2015

Basic Lung Ultrasound is the **"wet lung"** vs. **"dry lung"** concept that is easily teachable.

Ultrasound Lung Rockets (ULRs) also called B-lines are counted and recorded over a single intercostal space.

A) Dry Lung, Glide Sign and A-lines

B) Single ULR

C) >3 when ULRs are still recognized individually

D) Infinity when the ULRs are confluent and cannot be recognized individually

Lisciandro et al. Frequency of ULRS in Dogs with Normal Radiographic Lung, *Vet Radiol & Ultrasound* 2014

The Use of Global FAST during CPR

Being point-of-care and rapid in properly trained sonographers, Global FAST[3] is clinically helpful ultrasound format that may be used for monitoring (presence of organized heart beat (ROSC), more accurately diagnosing asystole, pulseless electrical activity (PEA), and determining proper placement of the endotracheal tube. Furthermore, GFAST[3] may be used diagnostically for the rapid detection of potentially treatable conditions such as pericardial effusion/tamponade (ultrasound is the gold standard), tension pneumothorax (TFAST[3]; A-lines without Glide Sign), PTE (TFAST[3]/Vet BLUE; RV to LV ratio, Wedge or V Sign), internal hemorrhage (AFAST[3]: abdominal cavity, retro-peritoneal space; TFAST[3]; pericardial sac, pleural cavity; Vet BLUE; lung ultrasound findings), and anaphylaxis (FAST[3] DH VIEW; gallbladder halo sign [other rule outs should be considered, see Gallbladder Halo Sign Proceedings]).

Endotracheal Tube Placement Confirmation Using Ultrasound

Direct confirmation of proper endotracheal tube placement by sonographically imaging the cervical region and finding the appearance of a single echogenic tube (proper placement); or the sonographic finding of 2 echogenic tubes (improper placement; 1 tube is the trachea, the other within the esophagus).

Indirect confirmation of proper endotracheal tube placement by sonographically observing the to and fro movement of the Glide Sign or ULRs with breaths given during positive pressure ventilation. If neither of these is observed, then either the endotracheal tube has been placed in the esophagus or pneumothorax is present.

Contractility, the Presence of Organized Contractions

It was shown in the FEEL/FEER study that asystole and pulseless electrical activity (PEA) were misdiagnosed by auscultation and electrocardiogram. In a significant percentage of humans diagnosed with asystole or PEA, ultrasound clearly showed coordinated cardiac movement; thus, CPR is stopped when the possibility for survival exists (Breitkreutz et al., 2007).

Knowing Your H's and T's – and PEA often caused by Treatable Conditions

The American Heart Association CPR Guidelines advocate ruling out the H's and T's is during CPR-advanced life support to maximize survival since the patient with cardio-pulmonary arrest (CPA) may in fact have a treatable conditions if the condition is expediently determined.

The H's include the following (means to diagnose/or conditions causing each in parentheses): hypovolemia (hemorrhage in abdominal cavity, retroperitoneal space [AFAST[3]], pleural cavity and pericardial sac [TFAST[3]] and lungs [Vet BLUE]) , Hypoxia (TFAST[3], Vet BLUE), Hydrogen ion (acidosis [blood gas]), Hyper- hypokalemia (blood gas), Hypothermia, and unofficially Hypoglycemia (not official but an important rule out).

The T's include the following with means to diagnose/or conditions causing each in parentheses: Toxins (FAST[3] DH View for gallbladder halo sign for anaphylaxis in dogs only), Tension pneumothorax (TFAST[3]), Thrombosis (TFAST[3] and Vet BLUE for RV dilatation and lung ultrasound [Wedge or V Sign]), Tamponade (TFAST[3]), and Trauma (pneumothorax, internal bleed-see hypotension [TFAST[3], AFAST[3]]). Certainly other conditions leading to CPA can also be ruled out such as pleural space disease (effusion, hernia). By performing Global FAST^{3} and a blood gas analysis, a very comprehensive survey has been performed point-of-care when minutes count for proper evaluation and diagnosis leading to patient survival.

How to Rapidly and Efficiently Perform Global FAST^{3}

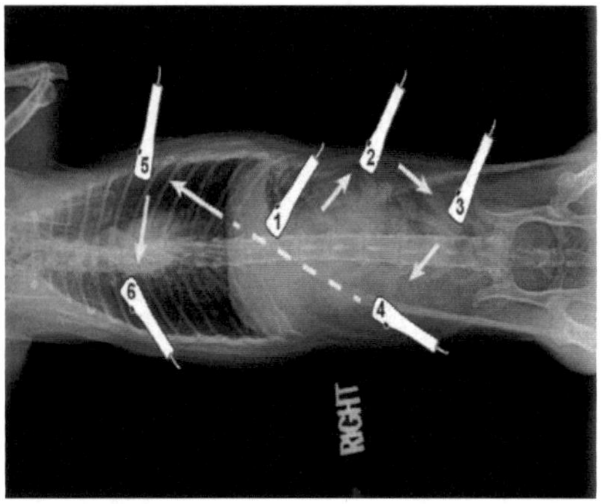

The manner in which to minimize time and maximize efficiency when performing Global FAST^{3} number 1-6 beginning at the DH view, then to SR, CC and ending at the HR view. GFAST^{3} is performed on the abdominal preset, use of a microconvex probe, with minimal changes in depth. Depth may then be decreased to finish with Vet BLUE.

The Use of FAST-ABCDE for CPR

An extended approach beyond Global FAST^{3} has also been evaluated called FAST-ABCDE (A for Airway [upper airway exam], B for Breathing [lower airway exam], C for Circulation (abdominal and thoracic), D for Disability [ocular nerve sheath diameter and fractures], and E for Exposure [serial exams]) (Armenise et al., JVECC 2013). The FAST-ABCDE approach is more technically challenging and requires probe changes utilizing linear, microconvex, and sector or phase arrays probes. Because of these probe changes and additional regions of scanning, the FAST-ABCDE format is more time-consuming than Global FAST^{3}.

Breaking down the FAST-ABCDE components, the A for Airway upper airway component is technically challenging (evaluation or the larynx and trachea as well as cervical structures); the B for Breathing is more extensive than the 8-point Vet BLUE regional approach by searching through all intercostal spaces through upper, middle and lower thirds. The C component represents Circulation (both the abdomen and thorax) and is similar to AFAST^{3} and TFAST^{3} (without using the abdominal fluid scoring system). The D component is for Disability, which includes the measurement of optic nerve sheath diameter (ONSD) as well as screens for intraocular injury (lens luxation, intra-ocular bleeds, retinal detachments). Reference values have been reported in a small population of dogs (Armenise et al., JVECC 2013). The use of the ONSD as a monitoring tool for intracranial hypertension is intriguing especially for use in unconscious, heavily sedated or anesthetized dogs and cats. However, before using ultrasound the eye, the sonographer should be aware of taking appropriate safety measures regarding acceptable types of acoustic coupling gel and technique. Finally, the E Component represents Exposure and the use of frequent serial FAST-ABCDE exams to detect changes/developments. The FAST-ABCDE format's use for CPR is described in more detail in the textbook Focused Ultrasound Techniques for the Small Animal Practitioner, Wiley ©2014 (Lisciandro and Armenise).

The Use of Ultrasound for Fracture Detection

The rapid point-of-care detection of fractures, Focused Musculoskeletal, can be important in certain circumstance including screening for rib fractures, long bone fractures, and skull fractures. Efficacy has been shown in children and foals. In the example below, a dog has a large hematoma and neurological signs. The clinical question as to whether the dog has a skull fracture is answered point-of-care.

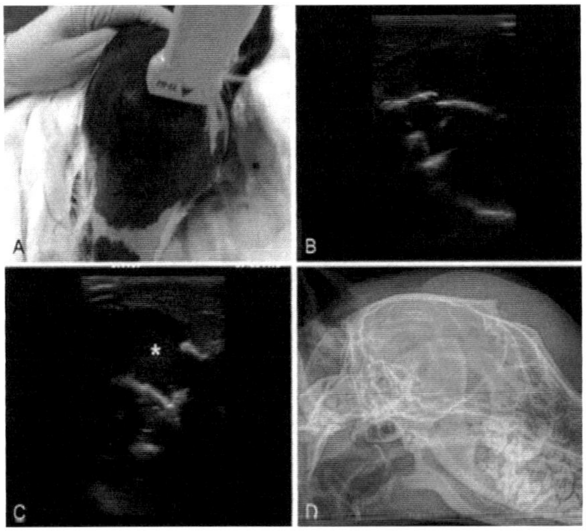

Fracture detection using ultrasound. The dog shown had blunt force trauma to its head and has a large hematoma. The skull fracture is diagnosed point-of-care through the large hematoma. The skull fracture is apparent by the presence of the incongruity of the cortical bone referred to as a "step" shown most clearly on the ultrasound image B) along with the radiographic confirmation D).

In summary, use of the Global FAST[3] monitoring status and CPR, and the use of FAST-ABCDE, and the use of ultrasound for fracture detection, Focused Musculoskeletal, present some interesting and creative manners in which to use ultrasound to potentially improve small animal patient care and outcome as has been shown/suggested in human medicine.

References:

1. Lisciandro GR. Focused abdominal (AFAST) and thoracic (TFAST) focused assessment with sonography for trauma, triage and monitoring in small animals. *J Vet Emerg Crit Care* 2011;20(2):104-122 .
2. Lisciandro GR, Fosgate GT, Fulton RM. Frequency of ultrasound lung rockets using a regionally-based lung ultrasound examination named veterinary bedside lung ultrasound exam (Vet BLUE) in 98 dogs with normal thoracic radiographical lung findings. *Vet Rad & US* 2014; 55(3):315-22.
3. Armenise A, Neri L, Storti E, et al. Evaluation of a FAST-ABCDE protocol (Focused Assessment with Sonography for Trauma- Airway, Breathing, Circulation, Disability, and Exposure) to detect multiple injuries in canine trauma patients: Preliminary data. Abstract. *J Vet Emerg Crit Care* 22(S2):S20.
4. Armenise A, Ricciardi M, Giannuzzi AP, et al. 2012. Ultrasonographic measure of the optic nerve sheath in canine patients with signs of raised intracranial pressure. A preliminary study. Abstract. *J Vet Emerg Crit Care* 22(S2):S20.
5. Nelson NC, Drost WT, Lerche P, et al. 2010. Noninvasive estimation of central venous pressure in anesthetized dogs by measurement of hepatic venous blood flow velocity and abdominal venous diameter. *Vet Radiol & Ultrasound* 51(3):313–323.
6. Lisciandro GR and Armenise A. Chapter 16: Focused or COAST[3] - CPR, Global FAST and FAST ABCDE. *In* Focused Ultrasound for the Small Animal Practitioner, Editor, Lisciandro GR. Wiley Blackwell: Ames IA 2014.
7. Lisciandro GR. The detection of pericardial effusion in 24 dogs using the diaphragmatico-hepatic (DH) view of FAST exams. *J Vet Emerg Crit Care, accepted 2014.*
8. McMurray J, Boysen S, Chalhoub S. Focused Assessment with Sonography in Non-trauma dogs and cats in the emergency and critical care setting. Abstract. *J Vet Emerg Crit Care,* 2014; 24(S1):S28.
9. Boller M, Fletcher DJ. 2012. RECOVER evidence and knowledge gap analysis on veterinary CPR. Part 1: Evidence analysis and consensus process: collaborative path toward small animal CPR guidelines. *J Vet Emerg Crit Care* 22 (S1): S4–S12.
10. Breitkreutz R, Price S, Steiger HV, et al. 2010. Focused echocardiographic evaluation in life support and peri-resuscitation of emergency patients: A prospective trial. *Resuscitation* 81: 1527-33.
11. Breitkreutz R, Walcher F, Seeger FH, et al. 2007. Focused echocardiographic evaluation in resuscitation management (FEER): Concept of an advanced life support–conformed algorithm. *Crit* Care *Med* 35(S5):1527-33.
12. Lichtenstein D. 2012. Fluid administration limited by lung sonography: the place of lung ultrasound in assessment of acute circulatory failure (the FALLS-protocol). *Expert Rev Respir Med* 6(2):155-62.
13. Ferrada P, Evans D, Wolfe L et al. Findings of a randomized controlled trial using limited transthoracic echocardiogram (LTTE) as a hemodynamic monitoring tool in the trauma bay. J Trauma Acute Care Surg 2013;76(1):31-38.
14. Marik PE, Callazo R. Does Central Venous Pressure Reflect Fluid Responsiveness? An Updated Meta-analysis and a Plea for Some Common Sense. Crit Care Med 2013;41(7):1774-81.
15. Neumar RW, Otto CW, Link MS, et al. 2010. Part 8: Adult advanced cardiovascular life support: 2010 American Heart Association guidelines for cardiopulmonary resuscitation and emergency cardiovascular care. *Circulation* 122:S729-S767.

LUNG ULTRASOUND IN SMALL ANIMALS: THE VET BLUE

Gregory R. Lisciandro, DVM, DABVP, DACVECC
San Antonio, TX
***FASTvet*™ and *FAST Saves Lives!*™**
Text **Focused Ultrasound Techniques for the Small Animal Practitioner, Wiley © 2014**

INTRODUCTION AND HISTORY

The use of lung ultrasound dates back into the late 1980s, notably with the ultrasound diagnosis in 1987 of pneumothorax (PTX) by a veterinarian, an equine practitioner, Dr Norman Rantanen. Within a year, the ultrasound diagnosis of PTX was similarly described in human medicine. In 1988, Dr Roy Philly dubbed the ultrasound probe the "modern stethoscope", a remarkable foresight made unbelievably 27 years ago.

More recently, the ultrasound probe has been dubbed the "visual stethoscope" (Moore 2011) because lung ultrasound artifacts are objectively and clearly discernible independent of patient or ambient noise. Moreover, lung ultrasound has been definitively shown to exceed traditional means of chest auscultation and supine chest radiography in humans with common respiratory conditions (Volpicelli 2012, Lichtenstein 2008). As long ago as 1997, the lung ultrasound finding, then referred to as comet tails, representing forms of interstitial edema was documented in humans by Lichtenstein and colleagues.

After that time, it seemed that the focus of lung ultrasound over the next several years changed from the pursuit of lung pathology to further characterization of the ultrasound diagnosis of pneumothorax (PTX). Several comparative studies clearly showed that PTX could be accurately diagnosed, rapidly and point-of-care, by lung ultrasound; and that lung ultrasound exceeded the accuracy, sensitivity and specificity of supine chest radiography. Furthermore, ultrasound-diagnosed PTX was shown to compare quite favorably with computerized tomography (CT), considered the gold standard for the diagnosis of PTX. During this same time, a structured lung ultrasound format was developed called "EFAST" for "extended FAST" by Kirkpatrick and colleagues (2004). EFAST was named as such since it was an additional FAST scan that extended from the FAST abdominal views.

Because PTX was a real-time finding during B-mode use, other ultrasound modalities were created to document PTX in medical records including the "Power Slide" using power Doppler (Kirkpatrick) and the "seashore sign" and "stratosphere sign" using M-mode (Lichtenstein and colleagues), representing dry lung and PTX, respectively. Again, power Doppler and M-mode were primarily modes for documenting PTX and not diagnosing PTX. The author has found in small animal medicine there often is too much movement to effectively use these documenting modalities. In addition to clearly demonstrating the ultrasound diagnosis of PTX, Lichtenstein and colleagues showed how the search for the "lung point", where lung re-contacts the chest wall, not only helped determined the degree of PTX, but also increased the sensitivity of diagnosing PTX using ultrasound (2000). The "lung point" debunked the myth that the ultrasound diagnosis of PTX was an "all or none" diagnosis by showing that partial vs. massive PTX could be determined using lung ultrasound. Another off-shoot of the use of lung ultrasound in trauma, was the finding that lung contusions, also referred to as lung blast, could be easily recognized by non-radiologist sonographers (Soldati 2006; Ball 2009).

In 2004 Jambrik and colleagues refocused lung sonographers on the pursuit of lung pathology in non-trauma subsets of human patients. In 2006 Volpicelli and colleagues re-enforced a scanning format and additionally showed that the counting of ultrasound lung rockets (also called B-lines) correlated with the degree of lung edema found on computerized tomography. In Chest 2008, Lichtenstein and colleagues published a clinical paper in which they showed that a pattern-based, regional approach, called the BLUE Protocol, could diagnose the most common presenting causes of respiratory disease in human patients with high sensitivity, specificity and accuracy including asthma, COPD, lung edema, PTE, and pneumonia. The BLUE protocol had a remarkable overall accuracy of 90.5%; and the BLUE exam only took minutes helping direct clinical course and diagnostics without the insensitivities of traditional means of physical examination and chest auscultation; and without the delays of waiting for chest radiography and other testing.

In another Lichtenstein publication (2009), they found a correlation between pulmonary capillary wedge pressure (invasive) and the presence of lung edema (B-lines or ultrasound lung rockets); and conversely they found that in the absence of lung edema (B-lines or ultrasound lung rockets) that clinically relevant left-sided heart failure could be rapidly ruled out (minutes) with high sensitivity and specificity, point-of-care, and within minutes of patient presentation. A remarkably powerful capability for such a simple, safe, radiation-sparing, point-of-care portable imaging modality, called lung ultrasound. In 2012, an international lung ultrasound consensus statement was made by a group of international lung ultrasound experts. In an evidence-based document, statements were developed regarding the efficacy and clinical utility of lung ultrasound use for various respiratory conditions, re-affirming the positive diagnostic and monitoring potential of lung ultrasound.

However, the use of terms such as lung sliding (glide sign in veterinary medicine), and A-lines, and B-lines, C-lines and PLAP continue to thwart the evolution and widespread use of lung ultrasound. These terms are confusing and difficult to grasp in contrast to analogous terms proposed by the author and still appearing in the human literature including glide sign (veterinary term, same as lung sliding in human medicine), A-lines (same, air reverberation artifact), ultrasound lung rockets (B-lines in human medicine), and Shred Sign, Tissue Sign, and Nodule Sign for lung consolidation/infiltration (called C-lines, PLAP, in human literature). These terms have been proposed in the veterinary literature in a clinical review (Lisciandro 2011), textbook (Focused Ultrasound Techniques for the Small Animal Practitioner, Wiley 2014), and in peer-reviewed clinical study by Lisciandro and colleagues (2014).

USE OF LUNG ULTRASOUND FORMATS IN SMALL ANIMALS

The reluctance to pro-actively apply lung ultrasound to small animals with respiratory distress is irrational in many respects. The overriding belief that air-filled lung creates insurmountable obstacles, and the continued belief in small animal medicine that imaging lung is difficult to perform leading to mistakes, perpetuate lung ultrasound's delayed use in small animals (dogs and cats). Thoracic FAST called TFAST (Lisciandro et al. 2008) was the first standardized abbreviated ultrasound exam of the thorax that included the Chest Tube Site (CTS) for lung surveillance for detection of PTX. Because of the finding of lung pathology found during TFAST^3, the author extended lung surveillance from the TFAST^3 CTS with the addition of 6 more lung views. The name of this novel regionally-based lung ultrasound exam is Vet BLUE ("Vet" for veterinary and "BLUE" blue for cyanosis and bedside lung ultrasound exam) (Lisciandro et al. 2014). The Vet BLUE regional sites include the caudodorsal lung lobe region (cdll), the perihilar lung lobe region (phll), the middle lung lobe region (mdll), and the cranial lung lobe region (crll). Each is named as a region because the naming do not directly correlate with anatomical names of lung lobes.

The Thoracic FAST^3 (TFAST^3) Exam

Reproduced with Permission (left figure) Lisciandro, JVECC 2011;20(2):1104-122.

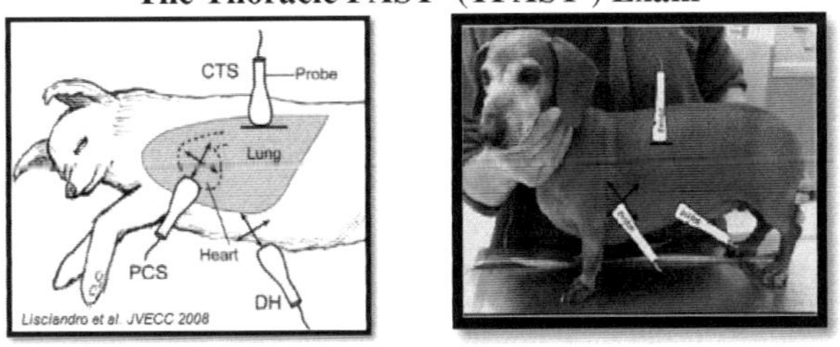

This material (right figure) is reproduced with permission of John Wiley & Sons, Inc, Focused Ultrasound Techniques for the Small Animal Practitioner, Wiley ©2014

Patient Preparation

NO SITES are SHAVED! All images shown by the author are unshaved sites at which the fur is parted and alcohol is applied to the skin and a small amount of acoustic gel to the probe head (most ultrasound manufacturers warn against placing alcohol on the probe head because of alcohol's damaging effects-check with your ultrasound machine manufacturer). **No images from cases in this talk were shaved.**

Patient Positioning

TFAST^3 and **Vet BLUE** are ***performed in sternal recumbency or standing in respiratory distressed or compromised small animals.*** TFAST^3 consists of 5-points as follows: the **stationary** horizontally probe-positioned ***left and right*** Chest Tube Site (CTS) view; the dynamically spotlighted ***left and right*** Pericardial Site (PCS) view; and the ***newer 5th point DH view*** (Lisciandro 2011), for the rapid detection of pleural and pericardial effusion and in some instances lung pathology (Lisciandro 2014).

The "Gator Sign" – Basic Lung Ultrasound Orientation

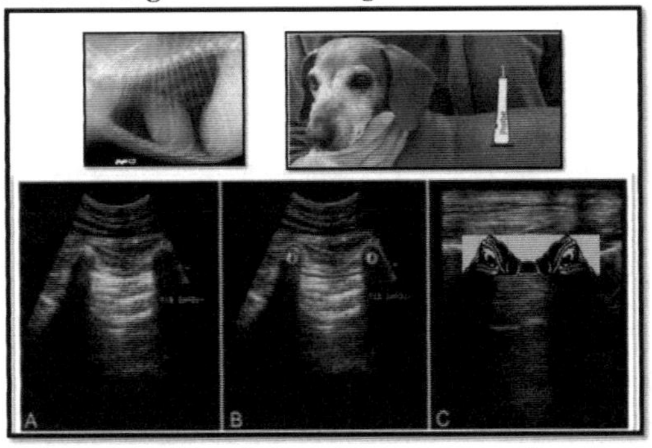

Reproduced with Permission Lisciandro, JVECC 2011;20(2):1104-122.

This material is reproduced with permission of John Wiley & Sons, Inc, Focused Ultrasound Techniques for the Small Animal Practitioner, Wiley ©2014

Probe Orientation

Lung ultrasound orientation **is ALWAYS the same** with the visualization of the **Gator Sign** to **properly identify the pulmonary-pleural interface** or **the "Lung Line"** (the surface of the lung). The **probe is held** perpendicular to the long-axis of the ribs; **Depth** is generally set between **4-6 cm; Frequency** is generally set between **5-10 MHz;** and **a microconvex probe** is preferred over a linear probe because the probe is acceptable for AFAST^3, TFAST^3 and Vet BLUE (Global FAST^3). A phase-array or sector probe is not recommended because its focal point is too small. A liner probe may be used; however, it is not ideal for the AFAST^3 and TFAST^3 portions of Global FAST^3 (GFAST^3). The rounded rib heads are likened to the eyes, and the pulmonary-pleural (PP-line) interface to the bridge of its nose, as a partially submerged gator (alligator) peers at the sonographer. The proximal white line is the focus of ALL lung ultrasound. The major orientation error is looking beyond the PP-line (or "lung line") and mistaking A-line artifacts for the PP-line or "lung line."

Search for the "Lung Point" – The Degree of Pneumothorax

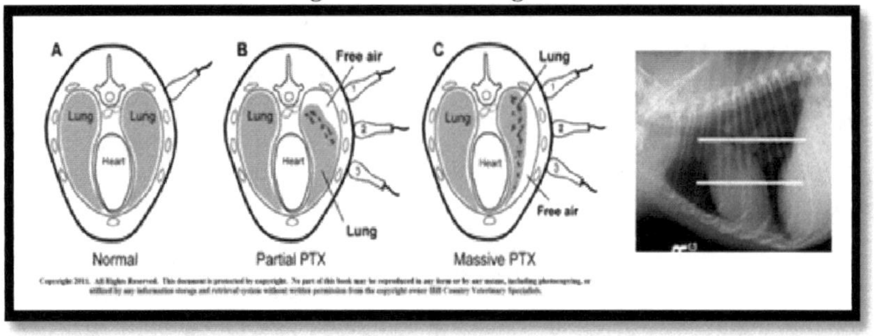

Reproduced with Permission Lisciandro, JVECC 2011 20(2):104-122.

The Lung Point

The search for the "Lung Point" is extremely important concept to understand. The Lung Point is the position along the thorax where collapsed lung recontacts the chest wall. The finding of the Lung Point increases the sensitivity of PTX. The Lung Point is determined by finding either a glide sign with A-lines or ultrasound lung rockets or other lung ultrasound findings supporting lung against the thoracic wall. The **"Lung Pulse"** has been described in people and its frequency is unknown in veterinary medicine. The Lung Pulse if the finding of lung against the thoracic wall, however, because of severe collapse, the glide sign does not move with inspiration and expiration but rather with the heart beat. **A**) Thorax in which pneumothorax (PTX) has been excluded. **B**) PTX has been determined at position 1 and the Lung Point is found at position 2 (PTX is partial). **C**) PTX has been determined and a Lung Point is nonexistent at any of the 3 probe positions (PTX is massive).

Key Point

By dividing the thorax into thirds when searching for the Lung Point a subjective assessment of partial vs. massive PTX may be made. Do not move in small increments when searching for the Lung Point, rather drop the probe down to the middle third then lower third then you're your way back toward the CTS to determine the Lung Point position relative to the CTS. Recording the distance is a way to monitor PTX (worsening, improving, and resolution). Moreover, when PTX is suspected at the CTS view, by locating the Lung Point sensitivity is increased and smaller pneumothoraces may be monitored by using/recording the distance from the CTS to the lung point.

Vet BLUE Examination

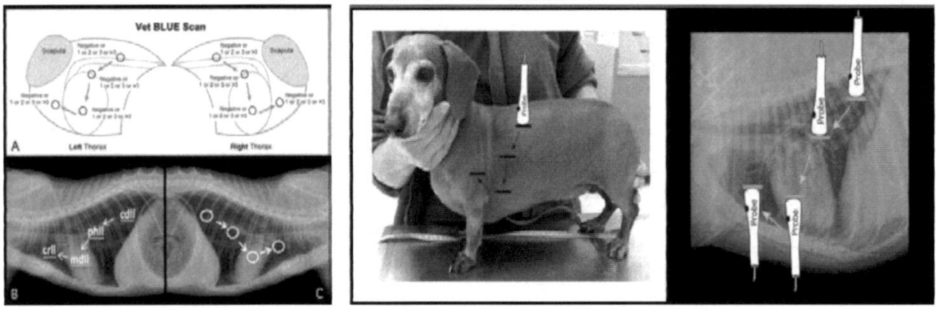

This material is reproduced with permission of John Wiley & Sons, Inc, Focused Ultrasound Techniques for the Small Animal Practitioner, Wiley ©2014 and FASTVet.com © 2014

The **Vet BLUE lung examination** is a **screening test** performed identically as the probe is positioned at the CTS view of TFAST[3]. The probe is then moved through regional locations that are bilaterally applied as follows: **caudodorsal lung lobe region** (cdll - same as the TFAST[3] CTS view, upper third, 8-9Th intercostal space)), **perihilar lung lobe region** (phll – 6-7th intercostal space, middle third), **middle lung lobe region** (mdll – 4-5th intercostal space, lower thrid), and **cranial lung lobe region** (crll – 2nd-3rd intercostal space, lower third). The **maximum number of ULRs** over the respective single intercostal space at each view is recorded. The counting system is as follows: **1, 2, 3, >3** (when ULRs are still recognized as individuals), and **infinity** ∞ (when the ULRs blend into one another becoming confluent [also called white lung]).

Key Point

Perform the Vet BLUE the same way every time. We suggest that you begin on the LEFT and go from dorsal to ventral, move to the right side and do the same, dorsal to ventral. This allows you to think about the pattern in the same manner every time and *helps you remember the findings at each site. Also, by completing the Vet BLUE at the right cranial lung lobe region (crll) region increase your depth, and do your right TFAST pericartdial view and proceed with the increased depth to AFAST and Global FAST (GFAST) is finished in < 4 minutes by the appropriately trained sonographer!*

Vet BLUE for Respiratory Distress – 5 Basic Lung Ultrasound Findings
"Wet Lung" vs. "Dry Lung" - Basic Lung Ultrasound 101
Shred Sign, Tissue Sign, Nodule Sign - Advanced Lung Ultrasound 202

Wet vs. Dry Lung

Basic easily recognizable lung ultrasound findings are categoriazed into **the Wet Lung** vs. **Dry Lung concept** (Lisciandro, JVECC 2011). A Glide Sign with A-lines (reverberation artifact) at the lung line is considered **"Dry Lung"** only to be confounded with PTX (A-lines and No Glide Sign). However, many patients in which the probability of PTX is very low, then spending additional time finding the Glide Sign becomes less important and A-lines alone suffice. **Ultrasound Lung Rockets (ULRs)** are considered **"Wet Lung"** and oscillate to and fro with inspiration and expiration and must extend to the far field obliterating A-lines (Lisciandro, JVECC 2011).

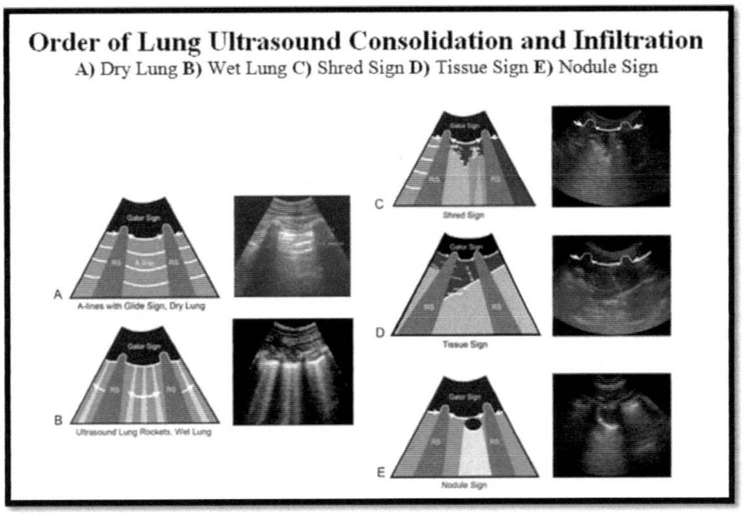

This material is reproduced with permission of John Wiley & Sons, Inc, <u>Focused Ultrasound Techniques for the Small Animal Practitioner</u>, Wiley ©2014 and FASTVet.com © 2014

Regionally-based Respiratory Pattern Approach Using Vet BLUE

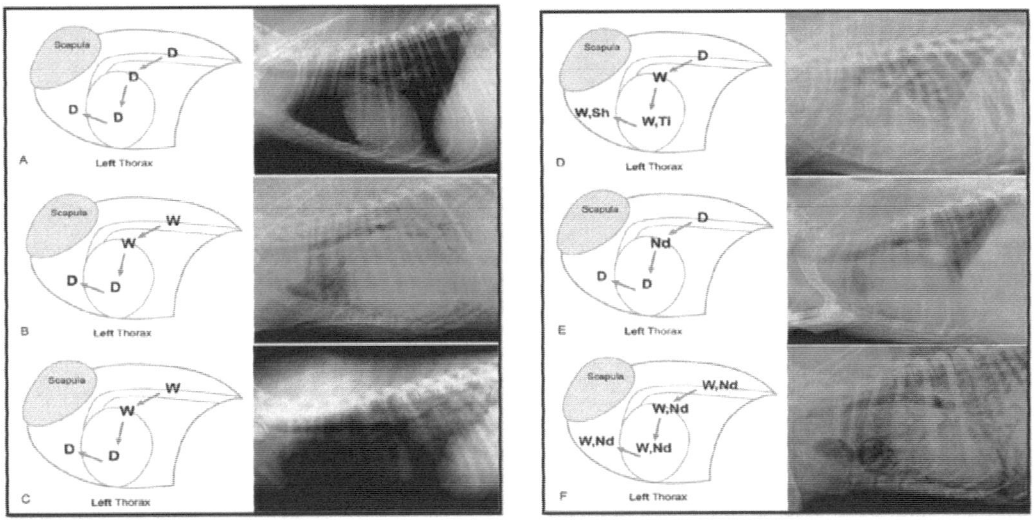

This material is reproduced with permission of John Wiley & Sons, Inc, <u>Focused Ultrasound Techniques for the Small Animal Practitioner</u>, Wiley ©2014 and FASTVet.com © 2014

Clinical Cases

Examples of Vet BLUE regionally-based patterns. **A)** Dry Lung all fields rules out clinically relevant left-sided heart failure, suggests upper airway obstruction, feline asthma, COPD, PTE and non-respiratory look-a-likes. **B)** Wet Lung in dorsal, perihilar, and middle lung lobe regions suggests cardiogenic lung edema (left-sided heart failure, volume overload from intravenous fluids). **C)** Wet Lung in dorsal lung lobe regions suggests forms of non-cardiogenic lung edema. **D)** Wet Lungs in ventral fields with or without signs of consolidation (Shred Sign/Tissue Sign), suggest pneumonia. **E)** Solitary nodule. **F)** Multiple nodules suggest metastatic disease or granulomatous disease.

Vet BLUE Diagnostic Algorithm for Respiratory Distress and Conditions

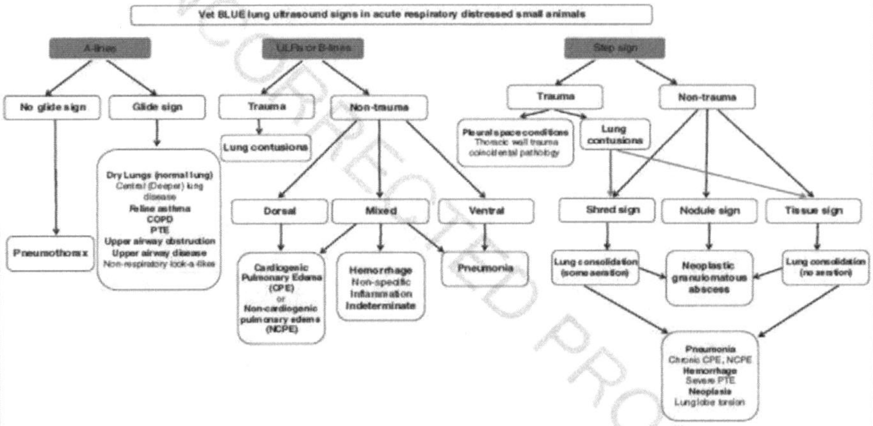

Figure 10.17. Summary of lung ultrasound signs and differential diagnoses. © Gregory Lisciandro

The Global FAST Triad for Volume Status

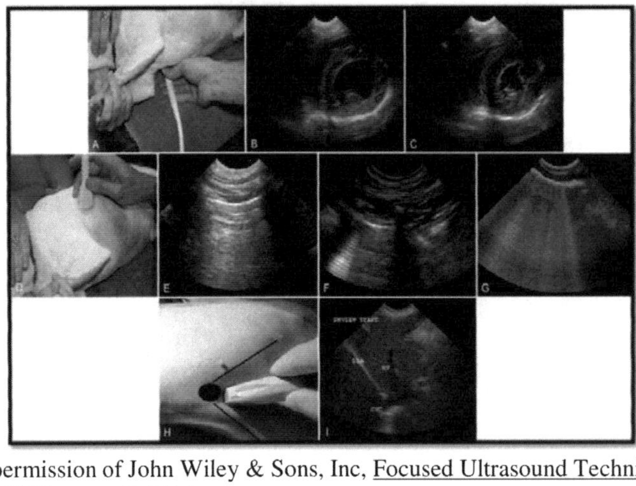

Global FAST^3 should be used for rapid evaluation of patient volume status pre-, during, and post- fluid resuscitation by using the "GFAST^3 Triad" (similar to Ferrada et al. 2013). The use of CVP via central lines for fluid resuscitation and its teaching "should be abandoned" (Marik et al. 2013). **1) Top Row** - the left ventricular short-axis "mushroom" view (TFAST^3 right PCS view) for volume status [~ preload] and contractility) **2) Middle Row -** Dry Lung vs. Wet Lung help determine the absence or presence of clinically relevant non-cardiogenic and cardiogenic pulmonary edema (Lichtenstein and Karakitsos 2012); and numbers of ULRs correlate with degree of interstitial-alveolar edema. **3) Bottom Row** – caudal vena cava at the diaphragm (using the characterization of FAT, flat or bounce) and hepatic vein (HV) distention or "tree trunks" (TFAST^3/AFAST^3 DH view for right-sided cardiac status and volume [preload]); normally, the hepatic veins ***are not*** readily apparent as the drain into the caudal vena cava (CVC) (Lisciandro 2014).

REFERENCES

1. Lisciandro GR, et al. Frequency and number of ultrasound lung rockets (B-lines) using a regionally based lung ultrasound examination named vet blue (veterinary bedside lung ultrasound exam) in dogs with radiographically normal lung findings. *Vet Radiol and Ultrasound* 2014;55(3):315-22.
2. Lisciandro GR, et al. Frequency and number of ultrasound lung rockets (B-lines) using a regionally based lung ultrasound examination named vet blue (veterinary bedside lung ultrasound exam) in cats with radiographically normal lung findings. *In Review.*
3. Lisciandro GR, et al. Accuracy of Focused Assessment with Sonography for Trauma (FAST) to detect pneumothorax in 145 dogs with blunt and penetrating trauma. *J Vet Emerg Crit Care* 2008; 18(3):258-269.
4. Lisciandro GR. Abdominal (AFAST) and thoracic (TFAST) focused assessment with sonography for trauma, triage, and tracking (monitoring) in small animal emergency and critical care. *J Vet Emerg Crit Care* 2011; 21(2):104-119.
5. Boysen SR, Lisciandro GR. The use of Ultrasound in the Emergency Room (AFAST and TFAST). *Vet Clin North Am Small Anim Pract* 2013;43(4):773-97.
6. Lisciandro GR. Chapter 55: Ultrasound in Animals. *In* Critical Care Ultrasound (human textbook), Editors Lumb and Karakitsos. Elsevier: St. Louis, MO 2014.
7. Lisciandro GR. Chapter 9: The Thoracic (TFAST) Exam; Chapter 10: The Vet BLUE Lung Scan. *In* Focused Ultrasound for the Small Animal Practitioner, Editor, Lisciandro GR. Wiley Blackwell: Ames IA 2014.
8. Lisciandro GR and Armenise A. Chapter 16: Focused or COAST[3] - CPR, Global FAST and FAST ABCDE. *In* Focused Ultrasound for the Small Animal Practitioner, Editor, Lisciandro GR, Wiley Blackwell: Ames IA 2014

THE GALLBLADDER HALO SIGN: MORE THAN ANAPHYLAXIS

Gregory R. Lisciandro, DVM, DABVP, DACVECC
San Antonio, TX
FASTvet^{TM} **and *FAST Saves Lives!***^{TM}
Text Focused Ultrasound Techniques for the Small Animal Practitioner, Wiley © 2014

Introduction

The target-organ approach of AFAST^3 and the AFAST^3-applied fluid scoring system, the FAST^3 DH view, and the TFAST^3 format that also assesses heart and lung, encourages the sonographer using these abbreviated formats to look beyond the simplified objective of fluid-positive or fluid-negative FAST approaches. The Global FAST^3 approach using these 3 techniques (AFAST^3, TFAST^3 and Vet BLUE) is much more cerebral mindset by taking advantage of structures viewed during GFAST^3 that provide important clinical information helping favorably direct patient care including volume status and parenchymal changes.

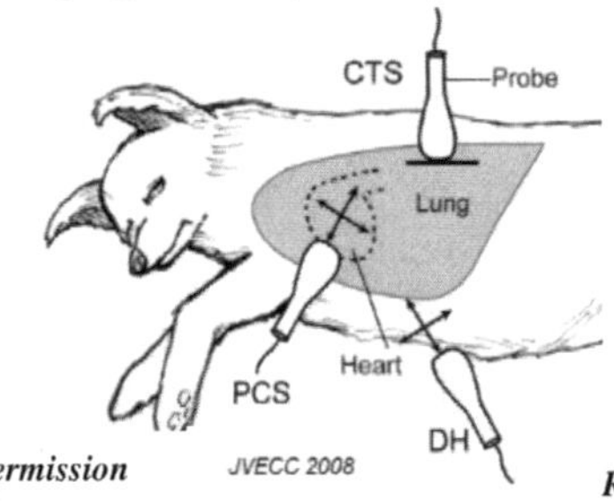

Reproduced with Permission Lisciandro JVECC 2011;20(2):104-122.

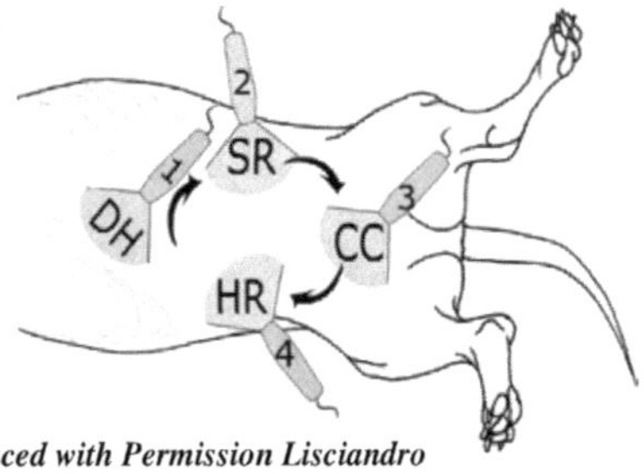

Reproduced with Permission Lisciandro et al. JVECC 2009;19(5):426-437.

A classic example is the observation of the gallbladder at the FAST^3 DH view (part of both the AFAST^3 and TFAST^3 formats). The gallbladder is fluid-filled and thus easily recognized by the non-radiologist since ultrasound images best through fluid; and is reliably imaged in dogs and cats at their FAST^3 DH view when placed in right lateral recumbency. In fact, the gallbladder is the starting point for the FAST^{3}DH view. There are 3 major features to consider when viewing the gallbladder including its shape, lumen contents and its wall.

The Diaphragmatico-Hepatic (DH) View

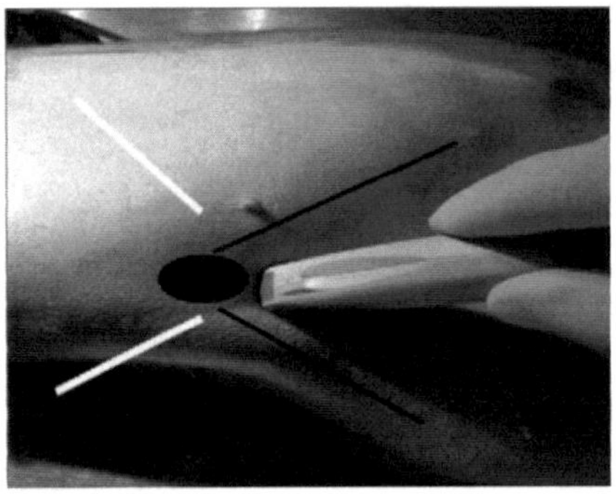

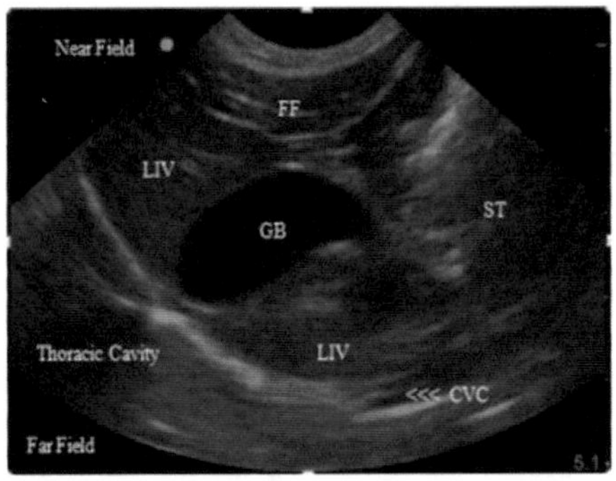

Classic FAST^3 DH View showing probe position at the xiphoid and the associated ultrasound image with the gallbladder (normal wall) immediately against the diaphragm. LIV: liver; GB: gallbladder; CVC: caudal vena cava; FF: falciform fat. FASTVet.com © 2014

In normalcy, the gallbladder sonographically is generally oval in longitudinal (sagittal) orientation with a lumen that is homogeneously anechoic (black); however, exceptions exist, and dogs and more commonly cats may have bi-lobed (oddly-shaped) gallbladders; and in dogs the gallbladder lumen may have echogenic (gray) material, referred to as sludge, that may also be considered normal when placed into clinical context. On the other hand, the gallbladder wall is quite reliably a thin hyperechoic (white) line in both the canine and the feline species (despite being reported to be considered as normal to a thickness of 2-3 mm). In summary, the sonographic features of the canine and feline gallbladder are easy to appreciate by non-radiologist sonographers during both AFAST^3 and TFAST^3 at the DH view.

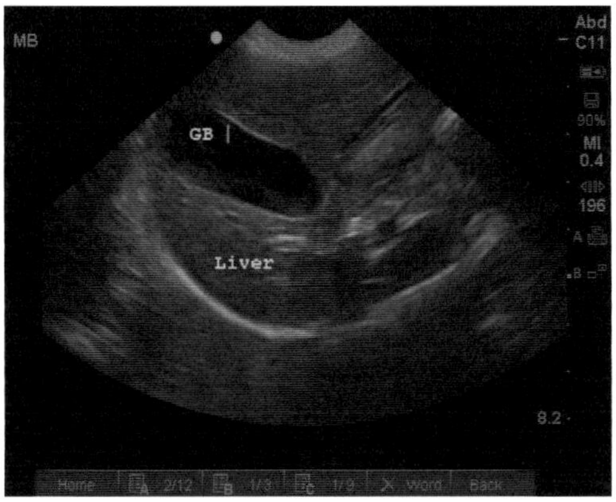

Image showing the normal gallbladder wall as a thin hyperechoic (white) line at the FAST^{3} DH View.

Compare to gallbladder wall edema referred to as the "Gallbladder Halo Sign" in subsequent images.

In 2009, Quantz et al. performed a landmark study that correlated the presence of a thickened gallbladder wall, referred to as the gallbladder halo sign, with anaphylaxis (AX) in dogs because their shock organ, is the gastro-intestinal tract and liver. In contrast, the shock organ of cats and humans is the lung (thus neither humans nor cats get an AX-induced gallbladder halo sign). In previous AX research using canine research models, the sonographic finding of the gallbladder wall thickening was noted but had never been applied clinically. What Quantz et al. found was that when dogs experience AX, the massive release of histamine causes the hepatic sphincters to tighten resulting in massive hepatic venous congestion that causes gallbladder wall edema within minutes. The gallbladder wall edema is easily recognized sonographically by its hyperechoic (white) inner and outer walls with a sonolucent line (typically anechoic [black]) in between. In other words, the gallbladder wall becomes layered as white, black, and white; and thus dubbed the "gallbladder halo sign." Moreover, Quantz et al. found that the gallbladder halo sign was not only sensitive and specific for AX, but also an immediate finding, in contrast to the delay of the spike of the historical marker of serum alanine transaminase (ALT), which may take 2-4 hours.

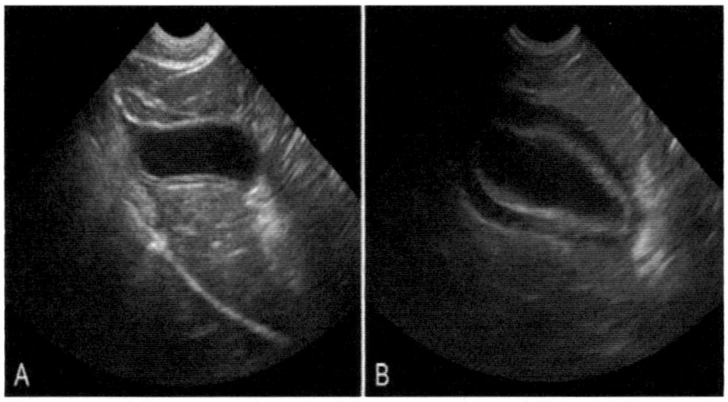

Varying degrees of gallbladder wall edema referred to as the "Gallbladder Halo Sign" in different dogs with anaphylaxis at the FAST^{3} DH View. A) mild B) severe.

Note the sonographic layering of white, black and white of lumen wall, its sonolucent center, and outer wall respectively.

Compare to normal thin hyperechoic (white) line in previous image.

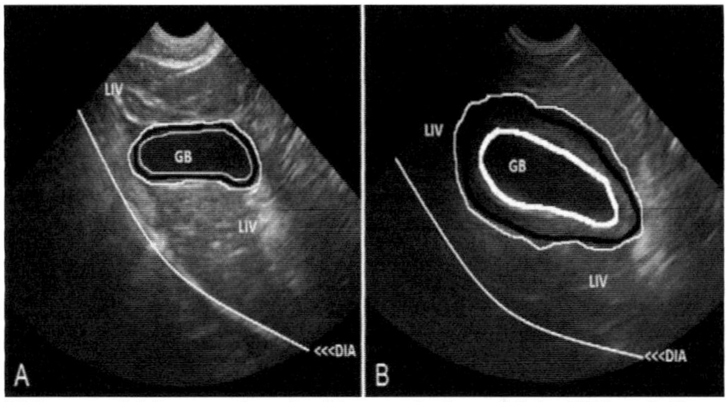

Modified from Focused Ultrasound Techniques for the Small Animal Practitioner, Wiley ©2014 with the sonographic layering placed as white, black and white. Compare to previous unlabeled image.

GB:gallbladder; LIV: liver; DIA: diaphragm

Importantly, it should be noted that the diagnosis of AX in human patients is problematic because there is no good rapid test for diagnosis, and serum markers (tryptase, PAF, histamine) are not immediately available when clinical decisions, e.g., the administration of epinephrine (EPI), are being made by attending physicians. As a result, it has been reported that up to 57% of AX cases are misdiagnosed in the emergency department. Furthermore, one study found that only 14% of human with anaphylaxis deaths were treated with EPI prior to cardiopulmonary arrest; and only 62% received EPI overall (EPI given at time of CPR). The reluctance in giving EPI lies in the myth that EPI can cause fatal reactions in and of itself; so both physicians (fail to administer up to 80% of the time) and veterinarians (unknown % of time) are over-cautious in EPI administration. However, it has been discussed more recently in human medicine that the levels of EPI given through administration of standard doses are 1/20th the toxic levels and thus benefit greatly outweighs risk.

Veterinarians should be made aware of the use of the gallbladder halo sign as supportive evidence for the diagnosis of AX in dogs but also be aware of the other causes for gallbladder wall edema. Hasty diagnosis of AX and treatment with EPI, large volumes of crystalloids, and histamine (H1 and H2) blockers and gluococorticoids may not be in the best interest the patient if misdiagnosed. Other potential causes of the gallbladder halo sign include right-sided heart failure, pericardial effusion and tamponade, volume overload, 3rd spacing from hypoalbuminemia and vasculitis, primary gallbladder disease and pancreatitis.

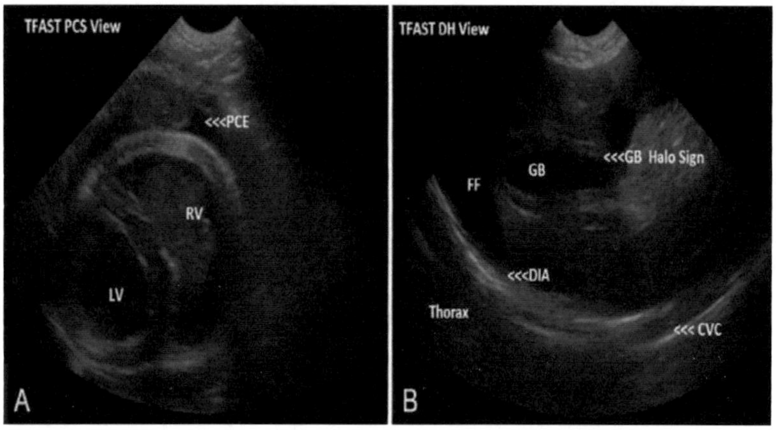

The Gallbladder Halo Sign and FAT (distended) CVC in a dog with pericardial effusion (PCE)

PCE should be ruled out in collapsed, weak dogs suspected of having AX.

LV: left ventricle; RV: right ventricle; PCE: pericardial effusion; DIA: diaphragm; GB: gallbladder; CVC: caudal vena cava; FF: free abdominal fluid.

Regarding right-sided heart failure/conditions including right-sided heart disease (valvular, functional, obstructive [PHT, PTE]), pericardial effusion with and without cardiac tamponade, and right-sided volume overload (overzealous fluid administration), resultant backflow causes venous congestion of the caudal vena cava (FAT) and the hepatic venous system (Tree trunk Sign [Lisciandro, Focused Ultrasound Techniques for the Small Animal Practitioner, Wiley © 2014]). The venous congestion of both the caudal vena cava and the hepatic venous system is obvious sonographically; and gallbladder wall edema is a secondary consequence (Nelson et al., VRUS 2010). The phenomenon occurs in humans (and dogs) and has been termed the "cardiac gallbladder" (Lisciandro, unpublished). To the author's knowledge, the characterization of the caudal vena cava and hepatic venous distension holds true in cats as in dogs and humans: however, the "cardiac gallbladder" phenomenon does not occur in cats (has never been documented) or is extremely rare.

In normalcy, the hepatic veins of dogs and cats are not readily apparent when fanning through the FAST^3 DH view. When hepatic venous congestion is suspected (the Tree trunk Sign) by the sonographer, then right-sided heart problems (look at the heart via the FAST^3 DH view, and add-on TFAST Pericardial Views) should be investigated (rule out pericardial effusion, right ventricular distension) as well as consider volume overload in dogs and cats on fluid therapy. Moreover, the use of the dynamic characterization of caudal vena cava (FAT, flat and bounce) is invaluable for sorting out dogs with weakness and collapse. Consider a dog (or cat) with AX, the caudal vena cava will be "flat" due to lack of venous return from profound hypovolemic/distributive shock. In contrast, consider a dog collapsed from pericardial effusion, the caudal vena cava will be "FAT" from the obstruction of blood flow from the liver to the right atrium. The caudal vena cava is diametrically opposed in these to conditions that can have very similar clinical presentations (the collapsed or weak dog).

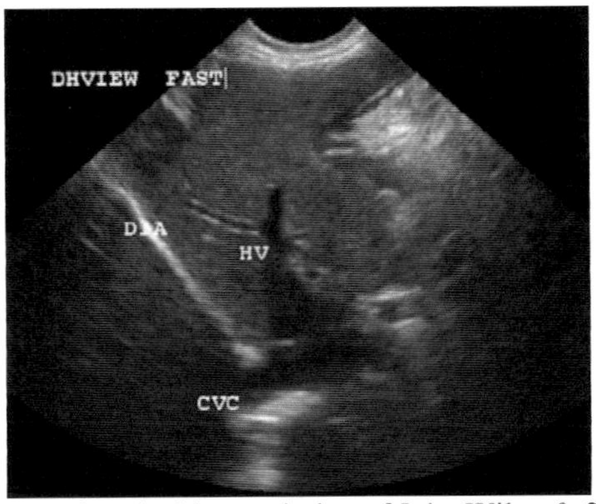

Showing the FAT (distended) caudal vena cava (CVC) as it passes through the diaphragm (DIA) and the distended hepatic veins appearing as tree trunks thus referred to as the Tree trunk Sign. The character of the CVC is completely different in AX (flat) and PCE (FAT) or right-sided heart failure.

Lastly, dogs with anaphylaxis not uncommonly develop degrees of coagulopathic hemoabdomen, typically abdominal fluid scores (AFS) of 1 or 2, but may have higher scores of AFS 3 and 4. Treatment consists of correcting the coagulopathy with transfusion, e.g., fresh frozen plasma and very rarely red blood cells. Monitoring of the degree of hemorrhage, its worsening, and its resolution, can be done with serial AFS. Most lower-scoring AFS 1 and 2 dogs have normal clotting times and their AFS will return to 0 (negative) within 24-hours. Even dogs with high-scoring AFS 3 and 4 can self resolve without fresh frozen plasma. It's important to recognize the complication, and not over-react in stable patients with normal clotting times since many will self resolve (and even some with abnormal coagulation profiles will still resolve without transfusion products); and not unknowingly take an AX-coagulopathic hemoabdomen to surgery, which could be catastrophic for the dog (Lisciandro, unpublished).

REFERENCES

1. Quantz JE, Miles MS, Reed AL, et al. 2009. Elevation of alanine transaminase and gallbladder wall abnormalities as biomarkers of anaphylaxis in canine hypersensitivity patients. *J Vet Emerg Crit Care* 2009;19(6):536–544.
2. Lisciandro GR. Focused abdominal (AFAST) and thoracic (TFAST) focused assessment with sonography for trauma, triage and monitoring in small animals. *J Vet Emerg Crit Care* 2011;20(2):104-122 .
3. Lisciandro GR. Chapters 2: The Abdominal (AFAST) Exam. *In* Focused Ultrasound for the Small Animal Practitioner, Editor, Lisciandro GR. Wiley Blackwell: Ames IA 2014.
4. Lisciandro SC. Chapter 4: Focused or COAST Spleen. *In* Focused Ultrasound for the Small Animal Practitioner, Editor, Lisciandro GR. Wiley Blackwell: Ames IA 2014.
5. Hittmair KM, Vielgrader HD, Loupal G. Ultrasonographic gallbladder wall thickness in cats. *Vet Radiol & Ultrasound* 2001; 42:149-155.
6. Spaulding KA. 1993. Ultrasound corner: Gallbladder wall thickness. *Vet Radiol & Ultrasound* 34:270-272.
7. Lisciandro GR, Lagutchik MS, Mann KA, et al. Evaluation of an abdominal fluid scoring system determined using abdominal focused assessment with sonography for trauma (AFAST) in 101 dogs with motor vehicle trauma. *J Vet Emerg Crit Care* 2009; 19(5):426-437.
8. Nelson NC, Drost WT, Lerche P, et al. Noninvasive estimation of central venous pressure in anesthetized dogs by measurement of hepatic venous blood flow velocity and abdominal venous diameter. *Vet Radiol & Ultrasound* 2010; 51(3):313–323.
9. Lisciandro, GR. Alanine aminotransferase level (ALT) as a marker for hemoabdomen detected by abdominal FAST (AFAST) in dogs with automobile trauma. Abstract. *J Vet Emerg Crit Care,* 2014; 24(S1):S11.
10. Pumphrey RS. Lessons for management of anaphylaxis from a study of fatal reactions. Abstract. *Cin Exp Allergy* 2000;30(11):1144-50.
11. Chad L, Ben-Shoshan M, Asaia Y. A majority of parents of chioeldfren with peanut allergy fear using the epineprhrine auto-injector. Abstract. *Allergy* 2013;68:1605-09.
12. Lisciandro GR. Case series of dogs with the gallbladder halo sign associated with pericardial effusion and right-sided heart failure during FAST exams. Unpublished, 2014.
13. Lisciandro GR et al. Case series of 11 dogs with anaphylaxis and hemoabdomen. Unpublished, 2014.

Hospital Acquired Infection I– Risk Factors and Pathogens

Dennis Spann, DVM, DACVIM
Roseville, California

INTRODUCTION

Hospital acquired infections and antimicrobial resistant infections are problems of increasing magnitude and complexity. Hospital acquired infections are defined as infections not present at admission, but acquired in the healthcare setting. The World Health Organization calls it a problem of sufficient magnitude that it "threatens the achievements of modern medicine." The true prevalence of hospital acquired infections in veterinary species is unknown because of the lack of an infrastructure and mechanisms to monitor the problem adequately. Hospital acquired infections are problematic because they jeopardize patients and staff, increase cost, have the potential to undermine the trust of our clients, and contribute to increased antibiotic use and global bacterial resistance. Sources of infection include endogenous (skin, respiratory tract, gastrointestinal tract) or exogenous (care providers, visitors, equipment, or contaminated environment) sources of microorganisms. Bacterial infections are not a new scourge to medicine. Bacteria have always been potent killers. Similarly, infections and healthcare associated infections have always been a risk factor of diseased and traumatized patients, and to people seeking medical care or surgical interventions. Prior to the advent of the antibiotic in the middle of the 20th century, infections frequently were life-threatening and specific pharmacotherapy was scant. With the miracle of the antibiotics, infections became more manageable and antibiotics seemed to usher in an era where, with enough energy and scientific application, infections could be controlled if not altogether conquered. This has proven to be hubris and cracks in the antibiotic armor appeared early. Organisms resistant to penicillin were identified in 1940; 3 years prior to widespread use of penicillin. In fact Alexander Fleming and Howard Flory discussed the threats of antibiotic resistance in their 1945 Nobel Prize acceptance speeches.

Any use of antibiotics, whether in human and veterinary medicine, creates selection for antibiotic resistance. It is estimated that up to 50% of antibiotic prescriptions are not needed or are improperly prescribed. This creates a strong selection for resistance. It is likely that veterinary antibiotic prescriptions have similar shortcomings. Of particular concern is the use of antibiotics for growth promotion in agriculture which creates a vast selective pressure for antibiotic resistance and the high potential for cross contamination of feces into feed and ultimately into food stuff if there is a breach in slaughter house or food prep protocols. At the very least, feedlot antibiotics and health care environments create low level antibiotic presence thought to stimulate the uptake of DNA from the environment and the development of resistance. Antibiotics to promote growth are not requisite to the production of a robust food supply and the FDA has proposed a non-binding pathway towards ending their use outside of treatment of specific bacterial illnesses.

Bacteria are showing resistance to antibiotics at an increasing rate and the rate of development of new antibiotics has slowed. Estimates vary but conservatively, in the United States, infection complicates 10% of hospitalizations and 75% of the hospital-acquired infections are resistant. Of the approximately 2 million people affected, 23,000 deaths result each year. Just one organism, methicillin-resistant *Staphylococcus aureus* (MRSA), kills more Americans every year than emphysema, HIV/AIDS, Parkinson's disease and homicide combined. Additional years of illness and debility result as well as lost production. A British study commissioned by Prime Minister David Cameron concluded that failure to battle drug-resistant infections and their causes could result in 10 million extra deaths a year by 2050 at a cost of $100 trillion to the global economy. Clearly, this is not an insignificant problem. Resistance to antibiotics has followed the same general course and developed within a few years of each antibiotic's introduction. Currently the rate that bacteria develop antibiotic resistance is outstripping the development of new antibiotics. Medicine has belatedly come to appreciate, or at least address, the power of vast numbers of reproducing organisms in the face of evolutionary pressure. Bacteria develop spontaneous mutations that, in a population of billions, result in occasional antibiotic resistance. Those clones with appropriate resistance are selected for, and their DNA conveys advantages to the bacteria that contained them. Bacterial resistance occurs by 4 basic of mechanisms: inactivation or modification of the antibiotic, alteration of a binding site so the antibiotic is less or ineffective, modification or use of an alternative biochemical pathway, or decreased penetration or increased efflux of the antibiotic. Bacteria can develop resistance by mutations that they pass vertically or by obtaining DNA horizontally from other organisms of the same or different species. DNA conferring resistance can be contained on plasmids, in transposons, and in DNA fragments from the environment. DNA can be obtained via conjugation, transfection, or environmental sampling. Low dose antibiotics in the environment potentiate the uptake of resistance DNA. Bacteria with resistance to multiple drugs are of particular concern. There are many examples of antibiotic resistant bacteria that have been implicated in infections in people. Two larger categories of gram negative bacteria with antibiotic resistance are the carbepenem resistant enterobacteriacea (CRE) and organisms containing extended spectrum beta lactamases (ESBL). Examples of gram-positive bacteria with antibiotic resistance include methicillin resistant *Staphylococcus* spp. Other important species exhibiting resistance include *K. pneumoniae, A. baumannii,*, *P. aeruginosa*, Salmonella and *Enterococcus* spp.

This problem is a growing one. A Pub Med search for antibiotic resistance in 1990 yielded 180 citations, in 2013 it returned 1550. In the absence of effective antibiotics, we face a return to a pre-antibiotic era associated with high morbidity and mortality for pediatric, geriatric, and critically ill patients as well as patients undergoing minor procedures suffering from relatively minor injuries. To visualize the challenges associated with antibiotic resistance, we need look no further than low and middle-income countries where newer generation antibiotic availability is often limited. Due to the development of resistance to first line antibiotics, 70% of neonatal infections cannot be treated according to the recommendations of the World Health Organization. Without antibiotics, 40 – 50% of people getting hip replacement will develop infection and up to 30% of those patients will die. In 2010, in Thailand, resistant infections secondary to *E.coli*, *Klebsiella pneumoniae*, *Acinetobacter baumannii*, methicillin resistant *Staphylococcus spp.* and multiple drug resistant *Pseudomonas aeruginosa* resulted in 35,000 deaths.

RISK FACTORS

Risk factors for infection in human and veterinary medicine are similar and include medical or surgical, facility and staff and host factors. Some well established risk factors include antibiotic administration, critical illness, complex surgical procedures, prolonged hospitalization, immune suppression, mechanical ventilation and invasive medical instrumentation. Mechanical ventilation is an especially important risk factor as that intervention combines several individual risk factors for hospital-acquired infection. These patients are critically ill, often have extended hospital stays, they usually have indwelling urinary and intravenous catheters and they have cleanliness, and mobility issues that don't plague more ambulatory patients. Facility factors that increase risk are those facilities with heavy patient loads and low staff to patient ratios. Facilities that have difficult to clean surfaces are at high risk for environmental contamination. ICUs are a particular threat because they house critically ill patients who may be immobile and have increased risk of exposure to vomit, feces and urine. These patients are highly instrumented and frequently have breaches in the cutaneous defense system due to intravenous or arterial catheterization. ICU patients may have hypoalbuminemia which predisposes them to edema and poor circulation and, because of prolonged recumbency they could develop pressure sores and ulcers. Sicker patients develop atelectasis which may predispose them to pneumonia. Surgical interventions increase infections because they breach the natural immune barrier provided by the skin, they disrupt tissue, vascular and lymphatic supply, and they can be associated with foreign material implantation which is a potential substrate for biofilm formation and infections.

PATHOGENS

In a collection of hospital reporting to National Health Care Safety Network on hospital acquired infections, eight pathogen groups accounted for about 80% of reported pathogens: Staphylococcus aureus (16%), Enterococcus spp. (14%), Escherichia coli (12%), coagulase-negative staphylococci (11%), Candida spp. (9%), Klebsiella pneumoniae (and Klebsiella oxytoca; 8%), Pseudomonas aeruginosa (8%), and Enterobacter spp. (5%). An additional medically important organism is *Acinetobacter baumannii*. Gram negative infections with enterbacteriaecea, *Acinetobacter ssp.*and *Pseudomonas ssp.* are the most worrisome because they are becoming resistant to most available antibiotics. The same is true to a lesser extent with gram positive strains like staphylococcus and enterococcus. In the human arena, the CDC considers carbepenem resistant enterobacteriaceae to be an urgent threat while MRSA, multiply drug resistant (MDE) acinetobacter and pseudomonas, vancomycin resistant enterococcus and extended spectrum enterobacteriaceae are all serious threats.

- *E. coli and Klebsialla ssp.* are the most commonly encountered carbepenem resistant enterobacteriaceae (CRE) in people. They are resistant to the penem class of antibiotics (imipenem, carbepenem, meropenem etc.). These are the last line of defense antibiotics for many of the more resistant enterobacteriaceae. These patients are at increased risk of death from these infections. They are associated with wound infections, pneumonia, skin infections and urinary tract infections. Types of CRE are sometimes known as *Klebsiella pneumoniae* carbapenemase (KPC), and New Delhi metallo-beta-lactamase (NDM). KPC and NDM are enzymes that breakdown carbapenems and make them ineffective. Both of these enzymes, as well as the enzyme Verona Integron-Mediated Metallo- β -lactamase), have also been reported in Pseudomonas. Healthy patients are not at great risk for these infections. Patients with extensive invasive devices like catheters, ventilators patients and patients on prolonged courses of antibiotics are at increased risk. CRE infections are very dangerous and because they are so difficult to treat are associated with mortality rates as high as 50%. If identified, personal protection including gloves, gowns and masks are essential. Monitoring of contact with patients is also warranted.

- Extended spectrum beta lactamase enterbacteriaceae are resistant to penicillin and cephalosporins. They represent approximately 19% of enterobacteriaceae. They are still sensitive to the carbepenems, but that requirement increases the use of last resort antibiotics and provides selective pressure for carbepenem resistance. Human patients with blood stream infections of ESBL enterobacteriaceae have a 53% mortality rate. Management of these cases adds $40,000 per case per year in additional health care costs.

- *Clostridium difficile* is not a common pathogen in veterinary medicine. In studies it was frequently isolated in veterinary hospital settings, but it can be isolated in healthy and diseased animals and humans. It is clearly a pathogen of significance in people and has also been reported in a hospital infection outbreak in a veterinary teaching hospital. It is not currently highly resistant, but is common in patients who are treated with other antibiotics (250,000 infections, 14,000 deaths and $1 billion in extra medical costs). In humans it is considered a pathogen of urgent concern. Clostridium difficile infection can range from asymptomatic to fatal hemorrhagic diarrhea. Signs can be large bowel, small bowel or both. Chronic diarrhea is also a possibility.

- *Acinetobacter* is a gram negative coccobacillus. It is ubiquitous in nature and can be isolated from water, soil and animals. It can become multiple drug resistant. This means resistant to 3 or more classes of drugs. These organisms are uncommonly isolated causes of hospital acquired infection but are a particular threat due to their ability to colonize inanimate surfaces like medical equipment, furniture, fixtures and form biofilms. *Acinetobacter spp.* infections are commonly associated with pneumonia, blood stream infections, urinary tract, wound and ventilator associated infections. In the human health care experience 63% of *Acinetobacter spp.* isolates are multidrug resistant. A unique strength of *Acinetobacter* is its ability to survive prolonged periods of desiccation. Outbreaks have been reported secondary to improperly sterilized pressure transducers and bedding. Infections are common on catheters and endotracheal tubes. Often they are part of mixed infections. Most hospital isolates are MDR and limited options exist. Treatment should be governed by culture results. Most are extensively resistant and resistance comes from target modification, increased efflux of antibiotics. They obtain resistance through contact with the DNA from a resistant organism. In some cases it was transferred from a *Pseudomonas.*

- *Pseudomonas aeruginosa,* a gram negative rod is not an emerging pathogen. It accounts for approximately 8% of healthcare associated infections. 13% of *Pseudomonas spp.* are multiple drug resistant. It has long been a complication of hospitalization and surgical procedures and is a species that naturally and frequently multiplies drug resistance. Strains have been reported that are resistant to fluorquinolones, aminoglycosides, cephalosporines and carbopenems. It can colonize patients and inanimate objects and is capable of forming biofilms. *Pseudomonas* is ~~a~~ frequently a complicating infection in lungs, urinary tract and wounds. *Pseudomonas* has multiple factors that make it a difficult pathogen in the healthcare environment. Inappropriate use of antibiotics can induce a variety of resistance factors in *Pseudomonas spp.*. Carbapenems (e.g., imipenem) induce the formation of thicker and more robust biofilms. Fluoroquinolones and aminoglycosides induce biofilm formation. Ciprofloxacin has been demonstrated to enhance the frequency of mutation to carbapenem resistance. Tetracycline is not very effective against this organism, but is known to induce the type III secretion system and consequently enhance cytotoxicity of *P. aeruginosa* in vivo.

- Methicillin Resistant *Staphylococcus pseudintermedius* and *Staphylococcus aureus* are a frequent cause of HAI in veterinary and human medicine respectively. Apparent community acquired infections occur as well. MRSP and MRSA are significant risks for wound and surgical infections, blood stream infection and urinary tract infection. These organisms are frequently resistant to most available antibiotics in a class. Human medicine reported over 80,000 infections and 11,300 deaths. Numbers are not available for veterinary medicine. Asymptomatic carriage has been demonstrated in nasal swabs in veterinary professionals. The role of health care giver carriage in transmission is still undefined.

- *Enterococcus spp.* cause a variety of urinary tract, bloodstream and surgical infections. In human medicine approximately 1/3 of enterococcus infections are vancomycin resistant. Vancomycin resistant enterococcus are difficult to treat with conventional antibiotic approaches. Ampicillin resistant enterococcus isolates are also problematic and appear to express a distinct genetic repertoire than the *Enterococcus* species that colonize the GI tract in people and dogs. *Enterococcus spp.* can form biofilms that enhance their resistance to antibiotics and host defenses. Dogs leaving an ICU after 2-9 day courses of antibiotics had a gut flora enriched for a less biodiverse and more resistant strain of *Enterococcus.* A suggestion was made to consider limited or protected human contact to prevent cross infection from patient to owners or other pets.

- *Salmonella* is a gram negative rod. *Salmonella ssp.* infections affect approximately 1.4 million people per year. 15,000 are hospitalized and approximately 500 deaths are attributable to *Salmonella* infections. *Salmonella* are not a common source of hospital acquired infections in veterinary medicine but there are reports of outbreaks in large and small hospitals. These infections can originate as food borne infections from raw or contaminated food, from exposure to healthy or symptomatic infected patients or from fecal exposure. Signs can vary from asymptomatic carrier, to GI signs of varying severity. Bloodstream infections, sepsis and death are also possible. Antibiotic usage plays two roles in Salmonella transmission. Antibiotic usage lowers the necessary infective dose by removing protective commensal bacteria from the GI tract and antibiotics select for more resistant strains of *Salmonella.* Epidemiologic studies show that infected animals represent a risk to owners, veterinary staff, other hospitalized patients, and if improperly managed are a serious risk for environmental contamination.

Hospital acquired infections are a menace to our patients, and threaten to undermine the trust of our clients. The magnitude of the veterinary problem is unknown because there is an underdeveloped infrastructure to monitor the prevalence of hospital acquired infections. There is a self interest not to emphasize hospital acquisition of infections. Further there is no central or regional data collection center like the National Healthcare Safety Network in veterinary medicine to track trends in infection and microbial resistance patterns. There is frequently negative financial pressure from owners to obtain cultures in adequate numbers. Despite these shortcomings, we will be facing increasing risks of hospital acquired, and particularly, antibiotic resistant infections. As a profession, we need to be more involved in recognizing the problem locally and globally. This is a common problem in human and veterinary medicine and we need to participate in serious efforts to define and solve the problem together. The clear task ahead of our profession is to improve stewardship of antibiotics. We will have to educate ourselves and our clients about the importance of culturing and judicious use of antibiotics. We are also responsible for educating ourselves and our staffs of the risk factors for hospital acquired infections, and implementing programs to identify and minimize the risk factors for our patients. The second lecture in this series will be a discussion on approaches to address risk factors for and techniques to lower the rate of hospital acquired infections.

REFERENCES

WHO - Pub: © World Health Organization 2014 WHO/HSE/PED/AIP/2014.2, U.S. Outpatient Antibiotic Prescribing, 2010N Engl J Med 2013; 368:1461-1462, J. Antimicrob. Chemother. Antibiotic Resistance — Problems, Progress, and Prospects. NEJM 371;19 2014, Carl Nathan, M.D., and Otto Cars, M.D., Ph.D. "Does the use of antibiotics in food animals pose a risk to human health? A critical review of published data." Journal of Antimicrobial Chemotherapy. (2004) 53 (1): 28-52. Wegener, Henrik. "Antibiotics in animal feed and their role in resistance development." Current Opinion in Microbiology. 6. no. 5 (2003): 439-45. Jeters RT,et al. Tetracycline-associated transcriptional regulation of transfer genes of the Bacteroides conjugative transposon CTnDOT. *J. Bacteriol.* 191, 6374–6382 (2009) Infectious Diseases Society of America, Spellberg, et al. (2011) Combating antimicrobial resistance: policy recommendations to save lives. Clin Infect Dis 52(Suppl. 5): S397–S428#209 Guidance for IndustryThe Judicious Use of Medically Important Antimicrobial Drugs in Food-Producing Animals, Antibiotic resistant threats in the United States 2013. CDC publication 508. Sievert et al. *Infection Control and Hospital Epidemiology Vol.* 34, No. 1 (January 2013) , pp. 1-14, . Addressing the Emergence and Impact of Multidrug-Resistant Gram-Negative Organisms: A Critical Focus for the Next Decade Ebbing Lautenbach, MD,

MPH, MSCE;1 Eli N. Perencevich, MD, MS2. Infection control and hospital epidemiology April 2014, vol. 35, no. 4, Biofilm formation in *Acinetobacter baumannii* Francesca Longi et al. New Microbiologica, 37, 119-127, 2014, . The Role of *Acinetobacter baumannii* as a Nosocomial Pathogen for Dogs and Cats in an Intensive Care Unit. Francey Thierry et al, J *Vet Intern Med* 2000;14:177–183, Yuji Morita et al. Responses of *Pseudomonas aeruginosa* to antimicrobials. Frontiers in Microbiology, Jan 8, 2014 J.S. Weese et al.Outbreak of *Clostridium difficile*–Associated Disease in a SmallAnimal Veterinary Teaching Hospital *J Vet Intern Med* 2003;17:813–816, Spellberg, B. et al.Trends in antimicrobial drug development: Implications for the future. Clin.Infect.Dis.2004, 38,1279–1286, Upsala Journal of Medical Sciences. 2014; 119: 209–214. Otto Carrs. Securing access to effective antibiotics for current and future generations. Whose responsibility?, J. Scott Weese et al. Bacterial Enteritis in Dogs and Cats: Diagnosis, Therapy, and Zoonotic Potential. Vet Clin Small Anim 41 (2011) 287–309, John F. Prescott, W.J. Brad Hanna, Richard Reid-Smith, Kelli Drost, Antimicrobial drug use and resistance in dogs. Can Vet J 2002;43:107–116, Anuradha Ghosh et al Dogs Leaving the ICU Carry a Very Large Multi-Drug Resistant Enterococcal Population with Capacity for Biofilm Formation and Horizontal Gene Transfer. PLoS One. 2011; 6(7): e22451.Published online 2011 Jul 19. PMCID: PMC3139645, Cherry B et al. Salmonella Typhimurium outbreak associated with veterinary clinic. Emerg Infect Dis 2004;10:2249–2251, Wright JG, et al. Multidrug-resistant Salmonella Typhimurium in four animal facilities. Emerg Infect Dis 2005;11:1235–1241, . Jennifer G. Wright et al. Multidrug-resistant Salmonella Typhimurium inFour Animal Facilities. Emerging Infectious Diseases • www.cdc.gov/eid • Vol. 11, No. 8, August 2005

Hospital Acquired Infection II: Preventative Strategies

Dennis Spann, DVM, DACVIM
Roseville, CA

INTRODUCTION

Hospital or healthcare acquired infections (HAI's) are emerging as a major health care risk in human medicine and are a serious concern in veterinary medicine as well. Coupled with the increasing frequency of multiple resistant pathogens, multiple governmental bodies have sounded an alarm to increase awareness and mobilize clinicians to adapt strategies that minimize the development of antibiotic resistance and the development of HAIs. HAI's are frequently associated with microbial resistance because organisms in the hospital setting have a relatively strong selective pressure from frequent use of antibiotics in hospitalized patients. The magnitude of the problem is unknown in veterinary medicine, but it is clearly not insignificant. On the human side, the World Health Organization has called bacterial resistance a problem of sufficient magnitude to "reverse the advances of modern human medicine." Prevention of hospital acquired infections and prevention of resistant infections are complimentary tasks. Several strategies need to be employed to best prevent occurrence of HAIs. There are several key components of an infection control program: infection surveillance (passive or active depending on facility size and resources), minimization of exposure through hand hygiene and contact control, environmental contamination control, patient risk assessment, education, and antibiotic stewardship.

ESTABLISHMENT OF AN INFECTION CONTROL POLICY

The first step to establishing an infection control policy is to identify an infection control officer. This will be the lead person in development of a written policy and will be the go-to person for case management and troubleshooting. This person should also take the lead in coordinating the development of an infection control policy. This policy should be a written document that is readily available and disseminated to all staff. It will be most useful if it is practical, terse and readable. A bulleted reference version can also help make the policy more useful to post and implement. The protocol should address specific procedures from handling patient inquiries on the phone through discharge, case transfer or autopsy and all the contacts and procedures between admission and discharge.

- Hand hygiene is the act of disinfecting hands with soap and water or utilizing an antiseptic sanitizer to decrease the bacterial load on health-care workers' hands. Pathogens can survive an hour or more on clinicians' hands and the degree of contamination is proportional to the time spent with clinical cases without effective washing. Despite this knowledge, clinicians routinely have compliance less than 50% in observational studies with reports as low as 17%. Washing should last 20 seconds and contact all surfaces and up the forearms. The time to sing "happy birthday to you" twice approximates 20 seconds. Washing should occur before and after contact with a patient, before and after contact with the patient's environment (cages, blankets, etc.), bodily fluids, and before putting on or taking off gloves. Wearing gloves does not replace hand washing, and wearing the same pair of gloves through multiple cases is not helpful.

- Personal protective equipment is the second component to prevent us from becoming fomites. Street clothes should be covered by smocks or lab coats in the clinic. Scrubs and lab coats shouldn't be worn outside of the clinic. Lab coats, gowns, scrubs and smocks should be laundered frequently (daily would be best), and changed with first gross contamination or after handling any patient with known or suspected infectious disease. Gloves should be worn when handling bodily fluids, when performing laundry or when cleaning environmental surfaces. Face protection should be used if splashes or other aerosols are expected, particularly during dentistry. Foot covers should be worn when in isolation or when cleaning grossly contaminated laundry or working with patients with infections. Many newer recommendations are suggesting clinicians should be bare below the elbows, free of jewelry and watches, and ties. Stethoscopes and other shared equipment should be cleaned and disinfected between patients.

- Environmental risk reduction includes assuring high levels of patient cleanliness, identification of patients at risk of soiling themselves and the environment, removal of organic debris and sterilization of the environment (cages, floors, exam tables, sinks, stethoscopes, thermometers, clippers and other commonly used equipment), and appropriate use of patient spacing or isolation. Soiling with any bodily discharges should be rapidly cleaned and disinfected. Linens (towels, blankets, etc.) should be changed at least daily. Cleaning and disinfection need to be separate steps. Cleaning to remove gross organic material mechanically, usually with soap and water; while disinfection helps remove microscopic pathogens with disinfectant solutions. Mop buckets should be changed frequently throughout the day. A review of disinfectants is beyond the scope of this talk, but the choice of disinfectant depends on the surface, degree of penetration required, the pathogens being destroyed and toxicity considerations. Choices often involve alcohols, biguanides, halides, peroxides, phenolics and quaternary ammoniums. Care should be given to assure hygienic handling of all bodily fluids and discharge. These biological wastes should be treated and disposed as such. Bandages should be kept clean, dry and free of strike through. Areas in the hospital and lab that are contacted by biologics should be cleaned and disinfected. Waste should be categorized as routine versus biological waste. Biomedical waste includes: sharps, blood or bodily fluid soaked material including bandages, tissues and dead animals. Elimination areas should be separate for patients shedding infectious organisms. Ideally, a cleanable surface should be chosen for elimination for those patients. Clients should generally be restricted from visiting animals with infectious diseases. Raw food should never be fed to hospitalized patients.

- Surgical patients are at risk for hospital-acquired infection. First, surgery creates a breach in the skin, which is the natural barrier for infectious agents. Surgery creates some degree of vascular and lymphatic compromise and tissue damage. Environmental cleanliness, surgical scrub and gown cleanliness and hand hygiene are paramount. Sterilization of equipment is an obvious necessity and indicators of sterilization are essential. Anesthetic equipment should be kept very clean since hands touch patients and then the equipment. Endotracheal tubes should be meticulously clean and ideally sterilized. Preoperative clipping should be as close to the surgical time as possible. Dirty patients may require a bath prior to surgery. Different clipper blades should be used for contaminated and clean procedures and both should be regularly cleaned and replaced. Perioperative antibiotics, when indicated, should be given so that they are at therapeutic levels at the time of first incision. Post-operatively, incisions should not be touched with bare hands to avoid introducing bacteria. Other important possible sources of infection could be failure to change disinfection scrub pots or inappropriate cleaning of multidose vials.

- Wounds are important avenues of infection. When handling wounds, sterile gloves should be used. Glove changes should occur as necessary as the wound prep becomes cleaner. Bandages must be carefully monitored as part of specifically timed patient orders and any strike-through warrants a bandage change. Soiled bandages should be treated as infectious material. Personal protection is needed if the wound has known multi-resistant bacteria and these patients should have some form of isolation.

- Patient factors must also be considered. Identify at-risk patients. These are critically ill patients, patients that have gastroenteritis, immune compromised patients (from immune suppressants, chemotherapy or sepsis), patients with difficult to heal wounds or wounds that produce copious discharge. Animals with coughing and or oculonasal discharge. Animals with these problems should have hygiene practiced aggressively to prevent infection and to prevent spread of that infection should it occur. Immobile, debilitated, heavily instrumented animals are all at risk for developing and potentially spreading ordinary and resistant infections. Known infectious animals should be isolated to prevent exposure to other patients. Suspected infectious patients should be separated from healthy and at-risk patients.

- Tracking infections is an important aspect of infection control. Records of patients' infections and particularly records of resistant infections should be maintained. Ideally, the hospital location of every patient should be recorded where possible. A binder of positive cultures can be maintained and resistance patterns monitored. Maintaining and reviewing records avoids conjecture and vague anecdotal assessments. One of the significant limitations in veterinary medicine is the lack of centralized collection of data from multiple hospitals. This handicaps the recognition of evolving resistance patterns in our patients and hospitals. Reference labs could serve as a valuable resource for recognition of local and regional resistance patterns. Knowledge of local susceptibility patterns would allow for educated empirical choices of antibiotics while waiting for culture results. A collaborative effort between academia and reference labs could be very productive in this effort.

- Antibiotic stewardship is integral to preventing the increase of antibiotic resistance and hospital acquired infections. Antibiotics should be delayed long enough to obtain appropriate cultures, then administered aggressively. Antibiotic choice should be based on culture results when available and de-escalation should occur as indicated once cultures are final. We should have a decision process and create a criterion for use of higher tier antibiotics. When empirical choices are required, antibiotics should be chosen based on the likely organisms for the organ system affected. Antibiotics should be chosen on pharmacologically sound principles. They should be able to reach therapeutic levels in the affected organ. They should be used for an appropriate length of time for the disease process. Owners should be educated about the importance of antibiotic compliance to avoid treatment failure and contributing to antibiotic resistance.

Summary

Prevention and limitation of hospital acquired infections and prevention of antibiotic resistance requires an organized multifaceted approach. Veterinary hospitals of every size should have formalized infection control policies. We currently lack some of the centralized information collection needed to track infection and resistance patterns. Despite this limitation, institution of well organized infection control policies can provide guidelines for optimal management of our routine and at-risk patients. Improved infectious disease control is going to require the development of centralized information database about infection patterns and resistance. This will require greater efforts individually to obtain appropriate cultures, coupled with academic and industry collaboration to develop a clearing house for larger patterns of infection and resistance than can be appreciated by individual clinicians or practices.

REFERENCES

Infection Prevention and Control Best Practices for Small Animal Veterinary Clinics – Canadian Committee on Antibiotic Resistance. Healthcare Personnel Attire, Recommendations to help prevent healthcare-associated infections transmitted through clothing. Society for Healthcare Epidemiology of AmericaStudy contact: Gonzalo Bearman MD, MPH. Ther Adv Drug Saf 2014, Vol 5(6) 229 – 241. B. Spellberg et al. Trends in Antimicrobial Drug Development: Implications for the Future Clin Infect Dis. (2004) 38 (9): 1279-1286. Chang-Ro Lee et al. Strategies to Minimize Antibiotic Resistance Int. J. Environ. Res. Public Health 2013, 10, 4274-4305, Sabino et al. Can Vet J 2006;47:779–782. Jumaa et al. Hand hygiene: Simple and complex. International Journal of Infectious Diseases (2005) 9, 3—14

PETS & PUBLIC HEALTH: CONVEYING THE BALANCE OF RISKS AND BENEFITS

Maureen E.C. Anderson, DVM, DVSc, PhD, DACVIM (LAIM), Jason W. Stull, VMD, MPVM, PhD, DACVPM
Guelph, Ontario, Canada

INTRODUCTION

Veterinarians play an important role in promoting public health through education of clients regarding zoonotic disease risks. Conveying the balance of the risks and benefits of animal contact can be difficult given time and privacy constraints with clients, as well as stigma or misinformation regarding some animal-related topics or human disease conditions. Understanding important knowledge gaps, attitudes, and high disease-risk practices of the public allows for targeted, effective discussions with clients on this topic.

PET OWNERSHIP AND THE GENERAL PUBLIC

In a recent survey of the general population of Ontario, Canada, 64% of participants currently had a pet in their household.[1] Pet ownership remained high (55%) for households with individuals at higher risk for zoonotic infections (i.e. < 5 years, ≥ 65 years, immunocompromised),[1] including households with children diagnosed with cancer (63%).[2] Species owned appear to be similar between higher- and lower-risk households and included dogs (68%), cats (48%), fish (13%), exotic mammals (7%) such as hamsters, and reptiles and birds (each 6%). Most (77%) households with a child previously diagnosed with cancer that obtained a new pet, acquired a high-risk pet. Some animal species are considered to be "high-risk" as data on pathogen carriage and/or zoonotic transmission has resulted in established guidelines recommending their exclusion or cautioning ownership in households with high disease-risk individuals. This list includes dogs/cats less than six months of age, reptiles, amphibians, rodents, baby poultry and exotic species.[3] Although general pet ownership is typically not discouraged for those at high risk for infectious disease, in most situations high-risk pets should not be acquired or, if present, should be temporally housed elsewhere until the risk of infectious disease is lower. The placement of high-risk pets into high-risk households should be discouraged and is an important educational topic for clients. For the delivery of this message to be most effective, veterinary staff must be aware of the current general disease risk of their clients' households (e.g., are there members or frequent visitors that are very young, elderly, immunocompromised, or pregnant); see "Advising clients from high-risk households." However, it is still important to remember that lower-risk households are not no-risk, and the status of a household can change either rapidly or gradually. All pet owners can therefore benefit from education about zoonotic disease risks and simple, every day ways to mitigate them.

What does the public know about the potential risk of pet-associated zoonotic disease?

Recent research suggests the answer to this question is "not very much." When given a list of 11 infectious pathogens, respondents from the general population were only able to correctly classify just over half based on their potential to be transmitted from pets to people. Perhaps this is not overly surprising, as a minority of individuals (36%) recalled having ever received information regarding pet-associated disease risks. Responses were similar regardless of whether there were high-risk household members (and thus one would hope knowledge level would be higher than the general public).[1,2] Perhaps most concerning is that most respondents were comfortable with their current level of knowledge, and thus perhaps unlikely to seek additional information or implement suggested changes.

Despite a number of readily available brochures and posters on pet-associated zoonotic disease offered from reliable sources (e.g., the Centers for Disease Control and Prevention Healthy Pets - Healthy People, Pets are Wonderful Support, Worms and Germs Blog – Ontario Veterinary College), it appears many veterinary clinics do not utilize such resources (in one study,[4] only 57% of veterinary practices had such materials available to clients). Making these resources available to clients (e.g. in waiting rooms, on websites, or provided when indicated by a topic of discussion) may be helpful in educating clients and promoting more discussion. Additionally, techniques that utilize active formats (e.g., veterinary staff initiating discussion) may be more effective, due to individuals' aforementioned comfort with an often-deficient knowledge level.

What pet husbandry practices occur in households that may increase disease risk?

The way in which we interact with our pets can play an important role in disease risk. In one survey,[5] dogs were reported to lick a household child's face at least several times per week in approximately 24% of households. As pets harbour many commensal organisms in their oral cavity that can result in significant disease in high-risk people (e.g. *Capnocytophaga canimorsus*), and that can cause infection even in healthy individuals (e.g. *Pasteurella multocida*), and particularly in the case of coprophagic animals that could potentially harbour enteropathogens in their mouths, this practice should be discouraged. Household children were often involved in cleaning up pet fecal material (25-63%).[5] Additionally, practices that increase zoonotic disease risk were frequently identified in many households; some fed high-risk foods (i.e. raw eggs, raw meat, or raw animal product treats) to their dogs (28%) or cats (3%); 14% of reptile-owning households allowed the pet to roam through the kitchen or washed it in the kitchen sink.[5] These practices are important routes for pathogen transmission that could be easily avoided if clients are made aware that lower-risk alternatives exist. Another, often controversial, practice is allowing pets to sleep on a person's bed. Many people (and many animals!) find sleeping with their pet (owner) very comforting, so there are benefits. From an infection control standpoint, in addition to factors such as the type of pet and the immune status and age of the owner, a key consideration is what constitutes "on the bed." The lowest risk situation is likely a pet that sleeps at the foot of the bed, ideally on an additional blanket that can be washed regularly. The opposite of this is the pet that sits and sleeps on a person's pillow (either when the person is there or not) as there is obviously a high degree of face/mouth/mucous membrane contact with the pillow, and anything the animal may have shed on it. The key is not necessarily to tell all clients what practices should and should not be done, but to make them aware of the risks and the alternatives so they can make the decision themselves.

Advising clients from high-risk households

Knowledge, attitudes and practices surrounding pet-associated zoonotic diseases appear to be similar for households with and without high-risk members. It is particularly important not only for individuals in high-risk households to be given information about mitigating infectious disease risks, but also to help them understand the information and how to implement safer practices. The key appears to be to identify high-risk households so that effective, targeted discussion can occur. In one study,[2] pet-owning respondents with children previously diagnosed with cancer that visited a veterinarian recalled that veterinarians rarely (14%) inquired about the immune status of household members. All who were asked about their child's medical condition told the veterinary staff about their child's condition, while only approximately half of those not asked provided the staff with this information. Thus, if staff ask, chances are clients will supply information about household-disease risk.[2] However, when veterinary staff were informed of the child's condition, in few cases (30%) did the veterinarian make any specific comments about the pet in relation to the child's medical condition.[2] If veterinary staff ask about household-disease risk, it is important to be prepared to provide targeted education and/or direct clients toward resources to address questions. Specific ownership and husbandry recommendations should be considered with high-risk households; there are a number of additional resources on this topic.[6-9]

THERAPY PETS

Most companion animals have a limited ability to spread infection in terms of the number and types of people with which they have contact. However, "therapy pets," such as those used in hospital/patient visitation programs and other types of animal-assisted interventions or activities (AAAs), have contact with a much larger number of people than the average household pet. Furthermore, these people, being patients in hospitals or residents of long-term care centers or similar facilities, are at higher risk for infections and complications thereof. It is therefore understandable why the health status, carrier status (in terms of zoonotic pathogens), and the type of human-animal interactions of therapy pets should be carefully scrutinized.

A cross-sectional study of 102 therapy dogs in Ontario, Canada reported that at least one zoonotic agent was isolated from 80% of the animals involved.[10] Toxigenic strains of *Clostridium difficile* were the most common, being isolated from the feces of 41 (40%) dogs, and another 17 (17%) dogs were shedding non-toxigenic strains of *C. difficile*. Other potentially zoonotic agents identified in small numbers of animals from this group included multidrug-resistant *Escherichia coli*, *Salmonella*, *Giardia*, roundworms and hookworms.[10] It has been shown that dogs involved in AAAs in healthcare facilities are at greater risk of acquiring methicillin-resistant *Staphylococcus aureus* (MRSA) and *C. difficile* than dogs involved in non-healthcare related AAAs. Licking patients and accepting treats were significant risk factors for acquiring these pathogens.[11] Furthermore, it is has been shown that a dog's haircoat can be transiently contaminated with MRSA, *C. difficile* and potentially other pathogens (both zoonotic and non-zoonotic) that they may encounter in a healthcare or long-term care facility.[12] Lefebvre and Weese[12] showed that MRSA contamination of a therapy pet's haircoat could be transferred to a person's hand after one minute of petting. Another study implicated a resident cat in transmission of MRSA between patients at a long-term care facility.[28] Reasonable precautions should be taken to reduce the potential risk of pathogen transmission between patients and therapy pets.[13,14]

Every time a person has contact with an animal, there is a risk of transmission of any number of infectious pathogens (the same is also true every time a person has contact with another person). Factors that can increase the risk of infection include decreased immunity in the person, increased contact with the animal (making it more likely for transmission to occur, or resulting in transmission of a larger number of infectious organisms), and increased virulence of the pathogen.

In a hospital or long-term care facility setting, it is not possible to control the health status of individuals, as their compromised state is often the very reason they are there. However, it is prudent to restrict or at least modify the kind of animal contact that occurs with the most severely ill and susceptible patients. For example, avoid any licking of the face, hands or broken skin (especially around invasive medical devices such as catheters), avoid patients kissing the animal, do not allow patients to feed the animal treats, and avoid direct contact with the patient's bedding by placing the animal on a separate towel/sheet that is removed at the end of the visit (assuming it is necessary to place the animal on the patient's bed at all).[13] Hand hygiene both before and after contact with the therapy pet is also critical.[13]

The virulence and number of organisms being shed by therapy pets cannot be directly controlled (or even predicted) in many cases, but there are a few measures that can be taken to help reduce this risk. Animals showing overt signs of illness (including systemic, gastrointestinal, respiratory or dermatologic disease) should not enter health or long-term care facilities. A sick animal is more likely to be shedding pathogenic organisms, and may be more prone to infection by organisms it encounters in the environment or through patient contact. In healthy animals, regular preventative healthcare and routine deworming and flea control are certainly important, but will only help to reduce or eliminate a small number of the organisms of concern. Potential risk factors that have been identified for carriage of various zoonotic pathogens in cats and dogs include age and source: young animals (less than one year of age) and animals kept in high-density populations (e.g. shelters, kennels/catteries) may be more likely to shed parasites such as *Giardia*, *Cryptosporidium*, *Toxoplasma*, roundworms and hookworms. Likewise, cats that are allowed to roam outdoors and hunt are more likely to shed some of these organisms, as well as *Salmonella*. It can also be difficult to predict the temperaments of very young animals (less than one year of age), and bites and scratches may occur more commonly from aggression, fear, or play. Therefore, in order to reduce the risk to patients, it is reasonable (and logical) to exclude these types of animals from being therapy pets. Therapy pets also need to be clean and well groomed, as this reduces contamination of the hair coat (minute amounts of fecal matter are of particular concern), as well as shedding.

As with pet ownership, many individuals may stand to benefit from AAAs, but the benefits must be weighed against the potential risks for each patient and each animal. Giving careful consideration to which patients can be visited, what animals should be allowed to visit, and reasonable precautions for reducing potential pathogen transmission during patient-animal interactions can help to reduce (though never eliminate) the infectious disease risks inherent to AAAs.

PETTING ZOOS

Petting zoos and other animal exhibits facilitate contact of the public with live animals as part of agricultural fairs and other seasonal or permanent attractions. They are entertaining, but also serve an important educational function for individuals of all ages regarding animals and animal husbandry, and also help to increase compassion for animals of different kinds. However, as with any type of human-animal contact, there is an associated risk of infectious disease transmission, which varies according to the animals involved, the individuals in contact with them and the degree of contact that occurs. Risks are higher for young children, with whom petting zoos and similar events are often particularly popular, but elderly and immunocompromised individuals may also visit such exhibits. This is yet another area in which veterinarians, even if they normally deal exclusively with companion animals, are uniquely suited to offer information and advice to members of the public for the benefit of public health.

Numerous disease outbreaks have been associated with animal contact at public venues. The most commonly reported pathogen in these outbreaks is *Escherichia coli* O157:H7, but other pathogens include *Salmonella*, *Cryptosporidium*, *Giardia*, *Coxiella burnetti* (Q fever) and dermatophytes (ringworm). There is also potential risk for the transmission of other zoonotic pathogens including *Campylobacter* and rabies. Several organizations have published recommendations for management of live animal exhibits such as petting zoos, intended to help decrease the risk of infectious disease transmission from animals to the public,[15-18] but implementation of these recommendations is still poor in many facilities.[19,20]

According to the National Association of State Public Health Veterinarians, "the recommendation to wash hands is the most important prevention step for reducing the risk for disease transmission associated with animals in public settings."[16] However, hand hygiene compliance at petting zoos and similar public exhibits has been reported to be poor,[20-21] and failure to perform hand hygiene has been identified as a significant risk factor for disease in numerous outbreaks associated with such events.[22-26] Children have been found to have better compliance with hand hygiene than adults at petting zoos in two studies.[21,27] This could potentially be due to increased awareness of the need for hand hygiene among children as a result of educational programs in schools, or due to parents instructing children to perform hand hygiene after visiting the exhibit but failing to do so themselves. While healthy adults are at lower risk of infection than children, it is still equally important for parents and guardians to perform hand hygiene, as indirect transmission may results from contamination of their own hands followed by close contact with their children.

The risk of infectious disease transmission between animals and people in any setting can never be completely eliminated. However, by increasing adherence to management and safety recommendations and increasing hand hygiene compliance among visitors at public animal exhibits the risk can be reduced. This will allow these exhibits to continue to play their role in educating the public about animals and animal husbandry and promoting compassion towards different animal species.

REFERENCES

[1]Stull JW, et al. **BMC Public Health** 2012;12:553. [2]Stull, et al. **J Pediatr** 2014;165(2):348-55. [3]CDC. **MMWR Recomm Rep** 2009;58(RR-11):1-166. [4]Lipton BA, et al. **J Am Vet Med Assoc** 2008;233(8):1242-9. [5]Stull JW, et al. **BMC Public Health** 2013; 20(3):485-7. [6]Stull JW, et al. **Vet Clin North Am Small Anim Pract** 2014;Epub ahead of print. [7]Weese JS, et al. Companion animal zoonoses. Toronto: Wiley-Blackwell; 2011. [8]Rabinowitz PM, et al. Human-animal medicine. Missouri: Saunders; 2010. [9]Worms and Germs Blog. Available at: http://www.wormsandgermsblog.com. [10]Lefebvre SL, et al. **J Hosp Infect**. 2006;62(4):458–66. [11]Lefebvre SL, et al. **J Am Vet Med Assoc**. 2009;234(11):1404–17. [12]Lefebvre SL, et al. **J Hosp Infect**. 2009 Jul 1;72(3):268–9. [13]Lefebvre SL, et al. **Am J Infect Control**. 2008;36(2):78–85. [14]Lefebvre SL, et al. **J Am Vet Med Assoc**. 2008;233(3):394–402. [15]Ministry of Health Long-Term Care Onatrio. Recommendations to prevent disease and injury associated with petting zoos in Ontario. 2009:1–28. [16]National Association of State Public Health Veterinarians Animal Contact Compendium Committee 2013. **J Am Vet Med Assoc** 2013;243:1270–1288. [17]Environmental Health Directorate. Petting Zoo Guidelines. Western Australia, 2007:1–18. [18]Bender JB, et al. **J Am Vet Med Assoc**. 2004;224(7):1105–9. [19]Heuvelink AE, et al. **Epidemiol Infect**. 2007;135(7):1174–83. [20]Weese JS, et al. **Clin Infect Dis**. 2007;45(1):10–5. [21]McMillian M, et al. **J Am Vet Med Assoc**. 2007;231(7):1036–8. [22]Steinmuller N, et al. **Clin Infect Dis**. 2006;43(12):1596–602. [23]Crump JA, et al. **N Engl J Med**. 2002;347(8):555–60. [24]CDC. **MMWR Morb Mortal Wkly Rep**. 2001;50(15):293–7. [25]Friedman CR, et al. **J Pediatr**. 1998;132(5):802–7. [26]Smith KE, et al. **Pediatr Infect Dis J**. 2004;23(12):1098–104. [27]Anderson MEC, et al. **Epidemiol Infect**. 2012;140(1):182–90. [28]Scott GM, Thomson R, Malone-Lee J, et al. Cross-infection between animals and man: possible feline transmission of Staphylococcus aureus infection in humans? **J Hosp Infect** 1988;12:29–34.

FELINE IMMUNODEFICIENCY VIRUS: AN EVIDENCE-BASED CLINICAL REVIEW OF NATURALLY-ACQUIRED INFECTION

Annette L. Litster BVSc, PhD, FANZCVSc (Feline Medicine), MMedSci (Clinical Epidemiology)
West Lafayette, IN

PREVALENCE AND RISK FACTORS

Worldwide prevalence of feline immunodeficiency virus (FIV) varies from 1 to 31% in healthy cats and is generally higher in sick cats, but the exact prevalence depends on the study design used. In a US-based study, Levy et al. surveyed 345 veterinary clinics (FIV-positives - 3.1% of 9970 cats) and 145 animal shelters (FIV-positives - 1.7% of 8068 cats; Levy, 2006). Prevalence figures for Canada and Japan are as follows: Canada – 4.3% of 11,144 cats tested in 2007 (Little, 2011); Japan – 23.2% of 1770 outdoor cats (Nakamura, 2010). The major risk factors for FIV infection are as follows: Age – adult; Gender – male (MN 4.3%, MI 3.3%) and intact status; Lifestyle – free-roaming/outdoor access; and Health status – Current illness (6.1%; Levy, 2006). Of shelter cats studied by Levy et al. (2006), 1.4% were owner relinquished, 1.6% were stray and 3.9% were feral.

TRANSMISSION

By far the most common mode of transmission is via bite wounds from infected cats. Also documented, but *much* less common, is vertical transmission from an infected mother to kittens during pregnancy, birth or lactation; although this seems to occur more commonly with laboratory viral strains rather than field strains. FIV transmission can also be achieved via blood donation from an infected cat. Mucosal transmission via oral, rectal or vaginal mucosa has only been demonstrated under laboratory conditions and mucosal infection requires up to 10,000x more virus than other routes (Dandekar 1992). Fomite transmission is not an important route for FIV as FIV quickly loses infectivity outside the host and is susceptible to all commonly used disinfectants. There can be viral strain differences in the relative ease of transmission via various routes. In one recently published study of a large 'mixed' household of FIV-positive and FIV-negative cohabiting cats, FIV transmission did not occur over years of exposure and vertical transmission of FIV was not demonstrated in 19 kittens born to 5 FIV-positive queens (Litster, 2014).

PATHOGENESIS

Acute infection

Initial infection is often clinically silent, although there can be mild fever and/or lymphadenopathy. During this phase there are large amounts of virus circulating and CD4 (helper) and CD8 (cytotoxic) T-lymphocytes decline. The host response to initial infection results in production of anti-FIV antibodies, reducing the circulating viral load. There is also an increased CD8 T-lymphocyte count, so that CD4:CD8 ratio is reduced.

Long asymptomatic period

After initial infection, there is an extended asymptomatic period, during which progressive dysfunction of immune system can occur. The CD4 T-lymphocyte count generally declines, resulting in reduced CD4:CD8 ratio, but does not always cause clinical signs. Non-regenerative anemia, lymphopenia and neutropenia can also occur. Because cell-mediated immunity is reduced, antibody-mediated immunity can be stimulated, with an accompanying increased serum globulin concentration. FIV-positive cats respond adequately to vaccination, unless advanced disease is present.

AIDS-related complex phase and terminal phase

These stages do not always/usually occur and many FIV-infected cats can live long and healthy lives, especially if regular veterinary attention is provided. Clinical signs, if they occur, might take years to develop and clinical signs are generally seen in older cats. These can include chronic inflammation of the oral cavity and skin; secondary infections (viral, bacterial, fungal, protozoal); neoplasia, especially lymphoma; signs of neurological or renal disease; and slow, progressive weight loss.

DIAGNOSIS

Diagnosis – Antibody tests

Most cats produce antibodies within 60 days of exposure, but it may take much longer (12 months) if viral exposure is low. Point-of-care testing is available to be performed on serum, plasma or whole blood. It is highly sensitive and specific and detects antibodies to FIV core proteins. The Western blot test is a laboratory-based test which has previously been used to confirm an ELISA-positive result, but might not be as sensitive or specific as the original SNAP test (Levy, 2004). Positive FIV antibody test results can sometimes occur in uninfected cats as antibody tests cannot distinguish between antibodies that are produced in response to a natural infection and those that are produced in response to FIV vaccination. Vaccine-induced antibodies can persist for at least 1 year after vaccination and perhaps over 9 years (Crawford and Levy, 2007). Positive antibody test results can also be produced in uninfected kittens (<6 months old) when their mother is FIV-positive (infected or vaccinated).

Diagnosis – Antigen tests

These tests detect FIV viral protein. A commercially available FIV PCR test can potentially distinguish cats that are vaccinated but FIV-uninfected from FIV-infected cats. This test relies on adequate amounts of certain amino acid sequences from field strains of FIV being 'recognized' by the test. Strain information is also provided. Virus isolation testing can be performed at reference laboratories and is a 'reference standard' method that takes at least 28 days to perform.

Diagnosis – Which cats to test

The following cats should be tested for FIV - sick cats; cats and kittens that will be group housed; cats and kittens at adoption, and a minimum of 60 days later if negative; cats with recent exposure to FIV-positive cats or cats of unknown FIV status, especially if there is a bite wound, and a minimum of 60 days later if negative. Cats living with FIV-infected cats should be tested annually. High-risk cats, such as outdoor, free-roaming cats, and cats with bite wounds should also be considered for FIV testing. Testing should always be performed before considering vaccination against FIV and in blood donor cats before blood collection.

TREATMENTS FOR FIV

There have been a few published reports of treatments targeting the immune system or the virus in FIV-infected cats. In one study, an 8-week course of recombinant feline interferon was given to 7 naturally infected FIV-positive cats (3 healthy, 4 unhealthy) and 5 untreated FIV-positive cats were enrolled as controls (Doménech, 2011). In the FIV-infected group, the healthy/mildly unhealthy cats remained stable (4 cats) and the unhealthy cats had improved clinical scores (3 cats). Recombinant feline interferon is not available in the USA. Another study used oral human interferon as a low-dose oral treatment (Pedretti, 2006) in 30 naturally-infected unhealthy FIV-positive cats (24 treated, 6 placebo) over a 14 month treatment period. Investigators documented clinical improvement in the first 2 months and treated cats had significantly longer survival than placebo-treated cats, but there was no change in CD4:CD8 ratio or other hematological parameters. Hartmann, et al., (1992) used zidovudine (AZT) to treat FIV-infected cats in a placebo-controlled study with a 3-week treatment period. Stomatitis and CD4:CD8 improved, but AZT can cause dose-dependent anemia which often resolves in the first 3 weeks of treatment and so is not suitable for cats with signs of bone marrow suppression. Additionally, AZT-resistant strains of FIV could potentially arise. Nine-(2-phosphonylmethoxyethyl) adenine (PMEA) was also used by the same group and was associated with clinical improvement in a placebo-controlled study conducted over a 3-week treatment period. PMEA caused more severe anemia than AZT-treated cats.

VACCINATION AGAINST FIV

FIV vaccination is classified as a non-core vaccine by the AAFP Vaccine Guidelines (AAFP, 2013), only to be administered to cats in specific risk categories, such as outdoor cats that fight and cats living with FIV-positive cats in unstable relationships. It should be remembered that FIV antibody tests cannot distinguish between vaccinated and infected cats, so FIV vaccination should be performed in conjunction with microchipping so that cats are properly identified. The currently available FIV vaccine is an inactivated (killed) vaccine against subtypes A and D, which also cross-protect against subtype B. Published challenge studies have shown a 20-100% 'preventable fraction' (proportion protected by vaccination in excess of proportion that is naturally resistant; Yamamoto et. al., 2007). The AAFP Vaccine Guidelines do not recommend FIV vaccination for shelter use and it is widely agreed that resources are better used elsewhere, such as spay/neuter/rabies vaccination programs. FIV vaccination requires 3 doses to be effective and protection is strain-dependent (Yamamoto et. al., 2007). Additionally, reduced aggression in spayed/neutered cats makes FIV transmission less likely. Free-roaming cats are more likely to be presented as strays at veterinary hospitals and shelters where it might be assumed that they are FIV-infected which could result in a negative outcome for the FIV-positive testing yet uninfected cat.

FACTORS ASSOCIATED WITH DISEASE PROGRESSION

Clinical staging for FIV is not well characterized or widely adopted. CD4 T-lymphocyte count and CD4:CD8 decline in the terminal stages of disease, accompanied by decreased IL-2 and increased TNF-α. During the course of infection, errors also occur during viral replication, resulting in 'evolution' of the virus over time. During this process, 'natural selection' of viral variants that resist the host immune response and lead to progression of disease can potentially occur. One study evaluated 33 naturally infected cats that were divided into high and low viral load groups at enrollment (Goto, 2002). The survival of the high viral load group was significantly reduced compared to the low viral load group over the next 4 years and viral loads increased just prior to death.

PROGNOSIS

There have been a number of retrospective studies of cats naturally infected with FIV. One study described a closed household with endemic FIV, FeLV and feline coronavirus observed over 10 years (Addie, 2000). Nine of 26 cats were initially infected with FIV; 6 additional cats became infected with FIV by the end of the 10-year study period. FIV infection did not adversely affect life expectancy in this household. In a retrospective Canadian study (Ravi, 2010), 39 FIV-positive cats were compared with 22 FIV-negative cats over approximately 8 years. The survival time of FIV-positive cats after diagnosis was not different from FIV-negative cats.

REFERENCES

Addie D **Vet Rec** 2000;146:419. Bandecchi P **Vet Rec** 2006;158(16):555. Crawford PC, and Levy JK **Vet Clin N Am Small An Pract** 2007;37(2):335. Doménech A, et al **Vet Immunol Immunopathol** 2011;143:301. Goto Y, et al **J Virol** 2002;76(19):10079. Hartmann K, et al **Vet Immunol Immunopathol** 1992 35(1-2):167. Levy JK, et al **JAVMA** 2004;225(10):1558. Levy JK, et al **JAVMA** 2006;228(3):371. Litster, A **Vet J** 2014;201(2):184. Little S **Vet Immunol Immunopathol** 2011;143(3-4):243. Nakamura Y, et al **J Vet Med Sci** 2010;72(8):1051. Pedretti E, et al **Vet Immunol Immunopathol** 2006;109(3-4):245. Ravi M, et al **Can Vet J** 2010;51(3):271. Yamamoto, et al **Am J Vet Res** 1988;49:1246. Yamamoto et. al **AIDS** 2007;21:547.

ADDITIONAL READING

Burkhard MJ and Dean GA **Curr HIV Res** 2003;1(1):15. Dandekar S **J Virology** 1992;66:4044. Medeiros S, et al **Virol J** 2012;9: 99. Pu R, et al **AIDS** 1995;9: 235. Rogers AB and Hoover EA **J Inf Dis** 1998;178:960. Sellon R, et al **J Virol 1994;68:3380.** Shelton GH **J Am An Hosp Assoc 1989;25:7.** Shelton GH 1990. **J Acq Imm Def Syn** 1990;3:623. Ueland K and Neese LL **Vet Immunol Immunopathol** 1992;33:301.

UNDERSTANDING THE ADAPTIVE IMMUNE RESPONSE: INTRACELLULAR PATHOGENS

Jonathan E. Fogle, DVM, PhD, DACVIM (Small Animal Internal Medicine)
Raleigh, NC

INTRODUCTION

The adaptive immune response to pathogens is a complex process and can follow many different pathways. In general, the cytotoxic T lymphocyte (CTL) response, also known as the $CD8^+$ T cell response, is directed toward intracellular pathogens including viruses, intracellular bacteria, and some protozoa and fungi. Organisms that infect macrophages are a special case and will be discussed below. First we will discuss viral infection as the prototypical intracellular infection and highlight differences with other types of intracellular infection. Viral entry into host cells signals the cells to produce Type I interferons (IFNs), including IFN-α and IFN-β, which act in an autocrine and paracrine fashion. At the same time, plasmacytoid dendritic cells (DCs) also begin to make massive amounts of Type I IFNs. These early events help to shape the entire downstream immune response, directing the response toward elimination of intracellular organisms. Type I interferons induce resistance to viral replication, promote display of cell surface markers recognized by natural killer (NK) cells, and activate NK cells which act as an early defense against intracellular infections. As the name implies, NK cells recognize virus-infected cells by alterations in surface receptors such as down-regulation of non-classical MHC-I molecules, which inhibit NK cell killing, and upregulation of NK stimulatory molecules such as MIC and CD48 to induce killing of the target cell.[1] Natural killer cells produce large amounts of interferon-γ (IFN-γ, a Type II interferon) which activates antigen presenting cells such as macrophages and dendritic cells, thus initiating the adaptive immune response. In the local lymph tissue, antigen presenting cells (APCs), such as dendritic cells, begin presenting viral antigen via MHC I and MHC II which interact with the T cell receptor (TCR) on the $CD8^+$ T cell surface and $CD4^+$ T cell surface, respectively. Engagement of the TCR along with co-stimulatory signals from the APC leads to activation of virus specific $CD8^+$ CTLs and $CD4^+$ T helper (Th) cells. At the same time, B cells internalize antigen via their B cell receptor, activating the B cell to present antigen to Th cells via MHC II and to differentiate into antibody producing plasma cells.

Once activated, the CTL is able to then kill virus infected cells, one after another, utilizing a variety of mechanisms. First, the virus-infected cells must present viral antigen via MHC I. The CTL then engages the virus-infected cell through its TCR, recognizing the viral antigen / MHC I complex. This interaction creates an "immunologic synapse" and the activated CTL reorganizes its cytoskeleton to direct cytotoxic contents toward the target cell. CTLs contain cytotoxic granules that store cytotoxic proteins including perforin, granzyme, and granulysin. Perforin secreted by the CTL acts by forming a pore in the target cell membrane. This allows the toxic proteins granzyme and granulysin to enter the cell and induce apoptosis, without damaging neighboring, uninfected cells. CTLs are also able to activate cell surface receptors on the virus infected target cell that induce target cell death. Finally, activated CTLs secrete a variety of potent antiviral cytokines such as IFN-γ and lymphotoxin that stimulate other lymphocytes and induce an "anti-viral state" in neighboring uninfected cells. At the same time, plasma cells have differentiated to produce antibody that neutralizes virus as virus particles are released from infected cells. The combination of a robust CTL response combined with the production of antibody leads to elimination of the infectious organism, or if the organism escapes the acute phase, establishment of a persistent infection.

CHRONIC VIRAL INFECTION

Feline immunodeficiency virus (FIV) is an elegant example of how a virus is able to subvert a vigorous acute immune response and establish a chronic infection. All of the immune mechanisms discussed above occur during acute FIV infection and contribute to the development of early, nonspecific clinical signs such as fever and malaise. In experimental infections, peak plasma viremia occurs 2-3 weeks post infection and is accompanied by profound lympholysis resulting in generalized B and T lymphocyte depletion. In FIV, an early rebound in the $CD8^+$ cell count and the quality of the CTL response corresponds with a decline in plasma viremia. This early increase in the peripheral blood $CD8^+$ cell count combined with the decreased $CD4^+$ cell count, leads to the inverted $CD4^+$ to $CD8^+$ ratio that is associated with lentiviral infections. Following generalized lymphocyte depletion a rebound lymphocytosis may also be evident on CBC evaluation and is associated with the lymphadenopathy sometimes noted during physical examination of the acute FIV patient.

$CD4^+CD25^+$ T regulatory (Treg) cells are cells of the immune system that control the immune responses of B cells and other T cells (discussed in detail in part 3 of this series). Our laboratory has shown that FIV infects Treg cells and that Treg cells support viral replication. These cells are long-lived and once activated they potently suppress the immune function of B cells and T cells which hampers the elimination of the virus, allowing a chronic infection to become established and they contribute to the progressive immune dysfunction that is well described in FIV infections.[2]

One of the more puzzling aspects that occurs during many chronic viral infection (particularly AIDS lentivirus infection) is the development of chronic, generalized immune activation in the face of obvious immune dysfunction. During the course of FIV infection, $CD4^+$ and $CD8^+$ lymphocytes progressively up-regulate surface markers that define them as activated T lymphocytes. The progressive replacement of naïve T lymphocytes with activated lymphocytes suggests chronic immune activation / stimulation. A constant cycle of immune activation and subsequent apoptosis helps to explain, in part, the eventual immune exhaustion and lymphopenia that defines the development of AIDS.

OTHER INTRACELLULAR ORGANISMS

As with viral infection, DCs, present antigen via MHC I and MHC II which interact with the T cell receptor (TCR) on the $CD8^+$ T cell surface and $CD4^+$ T cell surface, respectively. Engagement of the TCR along with co-stimulatory signals from the DC leads to activation of $CD4^+$ T helper (Th) cells and $CD8^+$ T cells that target infected cells. $CD4^+$ Th cells can also be activated in an antigen non-specific manner through TLRs and other pattern recognition receptors. Activation of $CD4^+$ Th cells is particularly important for elimination of complex intracellular organisms. For example, $CD4^+$ T cells are essential for clearance of Salmonella, which is related to the amount of IFN-γ, IL2, and other stimulatory cytokines they produce which, in turn, heightens macrophage and $CD8^+$ responses.[4] Likewise, in mouse models of *Cryptosporidium* sp infection, $CD4^+$ T cells are essential for elimination of infection Similarly, HIV/AIDS patients with decreased $CD4^+$ Th counts infected with *C. parvum* exhibited protracted enteritis and difficulty eliminating the infection, when compared to HIV/AIDS patients with higher $CD4^+$ T cell counts.[5] Although the role of $CD4^+$ Th cells in eliminating these infections is still not completely understood, Th cells may be directly responsible for pathogen elimination and it is clear that they are integral to coordinating and heightening both $CD8^+$ T cell and macrophage responses.

ORGANIMS THAT INFECT MACROPHAGES: A SPECIAL CASE

Certain pathogens, such as FIPV and *Mycobacteria sp*., have evolved strategies to survive within the macrophage. Macrophages begin to engulf bacteria in infected tissue soon after pathogen invasion or are targeted by the pathogen for entry. At the same time, dendritic cells located in lymphoid tissue have begun processing and presenting microbial antigen to Th cells via MHC II on their surface. The local cytokine environment, dominated by IFN-γ and IL-12, favors T helper type 1 (Th1) polarization. The Th1 cells then leave the lymphoid tissue and travel to the site of infection. As Th1 cells begin to accumulate at the site of infection, they interact with the macrophage. The macrophages are able to present microbial antigen to the Th1 cell. The Th1 cell then sends two specific signals to the macrophage to heighten its anti-pathogen function. First, the Th1 cell secretes IFN-γ which stimulates the macrophage to upregulate CD40 on its surface.[1] The Th1 cell also has CD40-ligand on its surface, which then stimulates the macrophage CD40 as the second signal. Once the macrophage is activated, the phagosomes containing the microorganisms fuse more efficiently with lysosomes, facilitating the death of the intracellular organism. The macrophage also has heightened secretion of antimicrobial substances and it increases MHC II expression to facilitate enhanced mycobacterial antigen processing and presentation to more Th1 cells. If the Th response is insufficient or is misdirected, the organism maintains a persistent infection within the macrophage leading to granuloma formation and chronic inflammation. In summary, the Th1 response is polarized by a cytokine environment consisting of IFN-γ and IL-12, Th1 cells secrete IL-2 and IFN-γ, and the major function of Th1 cells is activation of macrophages to facilitate killing of intracellular bacteria.

REFERENCES

[1] Murphy K, **Janeway's Immunobiology**, 2011. [2] Fogle J, et al **Retrovirology** 2010; 26(2):201-16. [3]Perryman LE, et al **J Virol.** 1994; 68(3):1459-67. [4]Yrlid U, et al **Infect Immun.** 2001; 69(9):5726-35. [5] Flanigan T, et al **Ann Intern Med.** 1992; 116(10):840-2.

UNDERSTANDING THE ADAPTIVE IMMUNE RESPONSE: EXTRACELLULAR PATHOGENS

Jonathan E. Fogle, DVM, PhD, DACVIM (Small Animal Internal Medicine)
Raleigh, NC

INTRODUCTION

The adaptive immune response has two basic pathways directed against pathogens. Simplistically put, either the pathogen is inside the cell or the pathogen is outside the cell. We have examined the immune response to intracellular pathogens in the previous section. The response to viruses is predominantly by antigen presenting cells (APCs) and $CD8^+$ T cells, while the immune response to more complex intracellular organisms requires coordination of APC, $CD8^+$ T cell and $CD4^+$ T helper (Th) responses. In contrast, B cells and $CD4^+$ Th cells are central to the development of the adaptive immune response to extracellular organisms.

INITIATION OF THE ADAPTIVE IMMUNE RESPONSE

Following pathogen entry, cells of the innate immune system such as neutrophils, macrophages, and tissue dendritic cells (DCs) very quickly recognize and respond to the organism. As in the immune response to intracellular organisms, the early response by the cells of the innate immune system polarizes and drives the subsequent adaptive immune response. Neutrophils, macrophages and dendritic cells express a wide range of receptors that recognize pathogen-associated molecular patterns, such as toll-like receptors (TLRs). By activating these receptors, neutrophils bind and engulf pathogens and secrete inflammatory cytokines. Similarly macrophages bind and engulf pathogens and secrete cytokines such as IL-1 and TNF-α which promote fever, inflammation, and begin to recruit immune cells to the site. Very quickly, lymphocytes are trapped in local lymph nodes, in response to these early signals and initiate the adaptive immune response. Extracellular pathogens, helminthes, and allergens generally are poor stimulators of IFN-γ and promote a local immune environment that favors an IL-4 dominated response.

It is important to remember that both the cellular arm (T cell) and humoral arm (antibody production) of the adaptive immune response are activated in response to these early signals. Dendritic cells in the local environment engulf pathogen material, cellular debris, and travel to the local lymph node to process and present antigen. Extracellular antigen is generally presented on MHC II molecules as 15-25 amino acid length peptides. The MHC II / peptide complex interacts with the T cell receptor (TCR) to stimulate $CD4^+$ Th cell activation. Upon activation, these T cells grow, proliferate and differentiate into effector and memory populations. B cells require two signals for activation. The first signal is cross-linking of the B cell receptor which initiates the activation signal. The B cell receptor (BCR) is simply the antibody molecule anchored to the B cell membrane and the BCR complex also contains other molecules such as CD79a and CD79b that cluster together. B cell co-receptors, such as CD19 and CD81 also cluster with the BCR complex to initiate the second signal and activate the B cell.[1] B cell activation results in B cell proliferation, growth, division, and differentiation into plasma cells and memory cells.

The germinal center of the lymph node contains a B cell rich zone, dendritic cells, and is generally found adjacent to the T cell rich zone. This is important because this allows for the interaction between B cells and T cells. When the BCR is activated, along with growth and differentiation, the B cell also engulfs antigenic material. In the same process mentioned above for dendritic cells, B cells are able to present antigen to Th cells via MHC II. If this process occurs, the B cell and T cell undergo powerful co-stimulation, with each stimulating the other to rapidly grow, differentiate, and proliferate.[2] Although this phenomenon doesn't occur for every antigen, linking the humoral and cellular immune responses, greatly magnifies the overall response to the pathogen. Proliferating B cells and T cells migrate to other lymphoid tissues and continue development and within a few days almost all lymphoid tissue throughout the body contains the antigen-specific B cells and T cells.

ANTIBODY KINETICS

The initial time period in which little or no antibody is produced is known as the lag phase, and generally lasts about seven days. Initially IgM is produced by the first wave of activated plasma cells. Very quickly, over the next seven to fourteen days, B cells undergo "isotype switching" and begin to produce other types of antibodies. In general, most extracellular organisms promote IgG production (helminth parasites being an exception, with the production of both IgG and IgE). About twenty one days after the initial infection, IgG concentrations peak in the serum and will decline over the next fourteen days, as the infection is resolved.

OPSONIZATION

Antibody molecules are distributed throughout the fluids and tissues of the body to neutralize pathogens. High affinity IgG can bind and inactivate exotoxins produced by bacteria. The coating of a pathogenic organism with antibody initiates the process of opsonization. The definition of opsonization is to coat an antigen with any substance that promotes ingestion by phagocytic cells. The most important opsonins are antibody (particularly IgM and IgG) and complement. Following differentiation to a plasma cell, the plasma cell makes antibody that is able to bind and coat the extracellular pathogen. The binding by multiple antibody molecules promotes the activation and binding of complement components, particularly C3b. Components of the complement system can be activated to disrupt the membrane of the organism (C5-C8), sometimes known as the membrane attack complex (MAC). Phagocytic cells have cell surface receptors that recognize complement (complement receptors) and antibody bound to antigen (Fc receptors, FcR). The FcR is denoted by which antibody class it binds. For example, the Fc-gamma receptors (FcγRs) bind to IgG (gamma, the Greek letter to denote G).[2] In general, FcRs recognize antibody / antigen complexes. FcγRs and complement receptors on the surface of phagocytic cells induce rapid phagocytosis and killing of antibody and complement coated structures, thus linking the innate and humoral immune response to efficiently eliminate pathogens.

MEMORY RESPONSES

As mentioned above, each time a B cell is activated, about half of these cells become effector cells, eventually differentiating into antibody producing plasma cells. Importantly, the remaining population of B cells differentiates into long-lived memory B cells. Although the mechanism differs somewhat with species and age, several rounds of subsequent exposure to the pathogen lead to the induction of a larger and broader memory B population. Evidence suggests that after several rounds of differentiation, the memory B cell population is maintained in an antigen-independent manner, probably through cytokine signals that promote survival.[4]

REFERENCES

[1] Murphy K, **Janeway's Immunobiology** 2011. [2] Sibley et al, **Vet Immunol Immunopathol**. 2013; 15;151(3-4):229-34. [3]Talinton D, **Nat Rev Immunol**. 2006; (10):785-90. [4]Maruyama M et al **Nature** 2000; (407) 636–642.

T REGULATORY CELLS IN AUTOIMMUNITY, INFECTIOUS DISEASE, AND NEOPLASIA

Jonathan E. Fogle, DVM, PhD, DipACVIM (Small Animal Internal Medicine)
Raleigh, NC

INTRODUCTION

$CD4^+CD25^+$ T regulatory (Treg) cells are cells of the immune system that control the immune responses of B cells and other T cells. Treg cell lineages have not been fully defined; however, there are at least two major populations of Treg cells. The first major population, which was the first to be discovered, is known as natural Treg cells (nTregs), and they originate in the thymus. Natural Tregs are an important link between central and peripheral tolerance to self-antigen and are superb in their ability to control autoimmune phenomenon. Once activated, nTregs then act in a non-specific manner, inhibiting autoimmune T cells from responding to self-antigen. Later studies demonstrated a second, peripherally induced Treg population, known as pathogen-induced Treg cells (sometimes called adaptive Tregs or "aTregs") that modulate immune responses to pathogens and tumors. Adaptive $CD4^+CD25^+$ Treg cells appear to be an integral component of the immune response to pathogens, controlling the magnitude and duration of the protective immune response and the collateral damage associated with unchecked inflammation. Treg cells suppress other T cell and B cell responses through a variety of mechanisms including the secretion of TGF-β and IL-10, and in our research model, FIV, through contact-mediated inhibition via membrane-bound TGF-β targeting TGF-β receptor II on activated lymphocytes. Certain tumors, viruses and parasites have taken advantage of Treg activation to escape elimination by the immune response.

AUTOIMMUNITY

As different murine models of autoimmune disease were developed and studied, it became evident to many researchers that autoimmune T cells were escaping the thymus and could be found in peripheral tissues. These autoimmune T cells could remain silent in the periphery for long periods of time, or could be activated by various means to produce autoimmune disease. Different groups theorized that there were subpopulations of T cells, which were suppressing these auto-reactive T cells. These groups of researchers systematically began depleting different T cell subsets in an effort to produce autoimmune disease in different murine models. One particular subset was defined by constitutive expression of CD25, the alpha chain of the IL-2 receptor. Sakaguchi et al demonstrated that $CD4^+CD25^+$ T cells were able to prevent the development of autoimmunity.[1] Shortly after that, several investigations from Shevach et al also showed that once these $CD4^+CD25^+$ T cells were activated they went on to inhibit many other activated T helper cells by suppressing the production of interleukin 2 (IL-2), a major lymphocyte growth factor. They also showed that these $CD4^+CD25^+$ cells were anergic, meaning that these cells did not grow and divide when stimulated through their T cell receptor like other T cells. However, stimulation promoted potent suppressor function in the $CD4^+CD25^+$ subset.[2] In 2001, researchers identified mutation in the FOXP3 gene as the cause for scurfy in mice and IPEX in humans. Scurfy and IPEX are characterized by uncontrolled proliferation of lymphocytes leading to autoimmune disease. FOXP3 was eventually shown to be essential to the development of $CD4^+CD25^+$ Treg cells. Experiments demonstrated that retroviral transduction of the FOXP3 gene into FOXP3 defective mice converted the transduced T cells into Treg cells, which prevented autoimmune disease associated with scurfy. In the span of a few short years, several of the major defining characteristics of T regulatory cells were discovered: the $CD4^+CD25^+$ phenotype, FOXP3 expression, the ability to inhibit T cell activation (particularly in autoimmune models), the inhibition of IL-2 production by T helper cells, and the anergic response to stimulation.

INFECTIOUS DISEASE

A compelling argument for Treg cells as major modulators of immune responses to infectious agents derives from the observation that escape of malaria parasites from the host's protective immune response and establishment of a chronic, ultimately fatal infection requires $CD4^+CD25^+$ Treg cells.[3] During *Leishmania* infection, Belkaid et al. reported that $CD4^+CD25^+$ Treg cells accumulate and are activated (pathogen-induced) at the site of infection, where they suppress the protective immune response, thereby allowing persistence of the parasite. Depletion of the $CD4^+CD25^+$ T cells enhanced the protective immune responses, resulting in complete eradication of the parasite. However, complete eradication also made the mice vulnerable to re-infection with the parasite, while a low-level, persistent infection conferred protection from subsequent *Leishmania* challenge.[4] Accumulation of Treg cells has been identified in canine leishmaniasis and it is likely that they play an important role in modulating inflammation and T cell responses to the parasite.

Investigations have demonstrated that both primary and memory immune responses to chronic viral infections such as herpes simplex virus type 1 (HSV-1) and HSV-2 are regulated by $CD4^+CD25^+$ Treg cells. Suvas et al reported that in vivo antibody depletion of $CD25^+$ T cells prior to infection increased both primary and memory $CD8^+$ cytotoxic T cell responses to HSV-1 challenge.[5] Their results suggested that immunosuppressive Treg cells are activated during acute HSV-1 infection, and that elimination of Treg cells during the acute state of infection could result in a more robust protective $CD8^+$ immune response and more rapid clearance of the virus. Subsequently the same investigators demonstrated that depletion of $CD25^+$ T cells prior to challenge with HSV-1 increased the severity of the ocular inflammatory lesions that correlated with an increase in HSV-specific $CD4^+$ T cell responses. While the HSV-1 keratitis model may be an exception, the important issue that arises from these studies is that depletion of Treg cells allows for a more robust protective T cell response to pathogens, yet the more aggressive $CD4^+$ and $CD8^+$ anti-viral immune response may incite immunopathology in the target tissue. Thus, $CD4^+CD25^+$ Treg cells appear to be an integral component of the immune response to pathogens, controlling the magnitude and duration of the protective immune response and the collateral immunopathology associated with unchecked inflammation.

One of the more puzzling aspects unique to lentiviral infections is the development of chronic, generalized immune activation in the face of obvious immune dysfunction. During the course of both HIV and FIV infection, $CD4^+$ and $CD8^+$ lymphocytes progressively up-regulate surface markers that define them as activated T lymphocytes. The progressive replacement of naïve T lymphocytes with activated lymphocytes suggests chronic immune activation / stimulation. The suppressive activity of Treg cells during acute and chronic FIV on activated $CD4^+$ and $CD8^+$ cells helps explain the seemingly paradoxical observation of immune activation in the face of obvious immune dysfunction. FIV infection also results in increased $CD4^+$ and $CD8^+$ lymphocyte apoptosis. A constant cycle of immune activation and subsequent apoptosis helps to explain, in part, the eventual immune exhaustion and lymphopenia that defines the development of AIDS.

NEOPLASIA

The role of Treg cells in tumor pathogenesis is complex. It is likely that Treg cells are able to suppress $CD4^+$ and $CD8^+$ T cell anti-tumor responses. Further evidence suggests that Treg cells may also influence infiltrating macrophage phenotype and may also contribute to tumor angiogenesis and metastasis. Bridging the gap between infectious disease and cancer, approximately 20-50% of patients with Hodgkin's Lymphoma (HL) have the oncongenic herpes virus EBV present in the lymphoma cells. Using peripheral blood chemotaxis assays, investigations have shown that EBV is able to recruit T regulatory cells to tumor sites using this mechanism, which may contribute to both tumor and viral persistence. Treg cells have been identified as contributing to pathogenesis in a wide variety of human neoplasms including metastatic melanoma, metastatic breast cancer, and various carcinomas.[6] Recently, $Foxp3^+$ T regulatory cells have been identified in the blood and local lymph nodes in a variety of canine neoplasms such as lymphoma, melanoma and osteosarcoma and it is likely that the presence of Treg cells is associated with the development of metastasis in carcinomas and other tumors.[7] In canine osteosarcoma, a decreased peripheral $CD8^+$ T cell to peripheral Treg cell ratio is associated with a significantly shorter survival time.[8] Low-dose, metronomic cyclophosphamide has been shown to decrease Treg cell numbers and angiogenesis in dogs with sarcomas and ongoing studies suggest that metronomic therapy with other T cell modulatory drugs may reduce Treg cell numbers in dogs with neoplasia. Collectively, these results demonstrate that Treg cells are important targets for therapeutic intervention in a variety of canine neoplasias.

REFERENCES

[1]Sakaguchi S et al, **J Immunol** 1995; 155(3):1151-64. [2]Thornton AM , et al 1998**; J Exp Med** 188(2): 287-96. [3]Hisaeda H, et al 2004; **Nat Med** 10(1):29-30. [4]Belkaid Yet al, **Nature** 2002 5;420(6915):502-7. [5]Suvas S et al, **J Immunol.** 2004; 172(6):3719-24. [6]Facciabene A et al, Caner Res. 2012 72(9):2162-71[7]Horiuchi et al, Microbiol Immunol 2009; 3(8):468-74. [8]Biller BJ et al, J Vet Intern Med. 2010 24(5):1118-23.

IMMUNE MECHANISMS UNDERLYING TYPE I-IV HYPERSENSITIVITY REACTIONS

Jonathan E. Fogle, DVM, PhD, DACVIM (Small Animal Internal Medicine)
Raleigh, NC

TYPE I HYPERSENSITIVTY REACTIONS

Type I hypersensitivity reactions (HST I) range from mild forms of atopy to life threatening anaphylaxis. The underlying immune mechanisms are the same and differ only in magnitude and location. There are three major components involved in the development of HST I antibody: soluble antigen, antibody generated against the soluble antigen, and the subsequent stimulation of immune effector cells. Soluble antigens known be involved in HST I include food antigens, plant antigens or insect antigens and are transported across mucosal surfaces by M cells and probably by other cells of the GI epithelium. These antigens are usually present in low concentrations and M cells transfer antigen from the intestinal lumen to dendritic cells (antigen presenting cells) within the mucosal associated lymphoid tissue (MALT).[1] After processing antigen, antigen presenting cells (APCs) such as dendritic cells (DCs) present antigen to $CD4^+$ T helper (Th) cells via MHC II. The processing and presentation of soluble antigen combined with the local cytokine environment polarizes the T cell response, leading to $CD4^+$ Th cells which signal isotype switching of B cells to produce mainly IgE. Atopy is a genetic predisposition to produce IgE in response to antigens that normally induce the production of other classes of antibody and is thought to be the underlying cause for many HST I conditions. Following internalization of its cognate antigen, the B cell becomes activated leading to growth, division, and differentiation into IgE producing plasma cells. The B cell is also capable of presenting antigen to Th cells via MHC II, leading to co-stimulation of both the B cell and T cell, thus greatly magnifying the immune response.[2] Following their interactions with the immune cells in the MALT, B cells and T cells that recognize soluble antigens then migrate through mucosal lymphatics and into the blood where they achieve broad distribution to distant sites such as the intestines, airways, and skin, which are known as the effector sites.[1] At the effector site, the Th cells are able to stimulate the differentiation of B cells into plasma cells for the production of IgE, while Th cells and other immune cells also secrete cytokines that attract mast cells and eosinophils. Plasma cells produce IgE that binds to the surface of mast cells via a specialized Fc receptor the Fc-epsilon receptor. Once a mast cell is coated by IgE, soluble antigen from food, plants, or insects can cross-link IgE on the mast cell surface leading to mast cell degranulation. Products of mast cell degranulation such as histamine, serotonin, eotaxin and other cytokines lead to inflammation, smooth muscle constriction, fluid secretion into intestine and lung, and further attraction of other inflammatory cells such as eosinophils and basophils.[3] Eosinophils that migrate into the area may become activated, resulting in degranulation with the release of substances such as major basic protein and phospholipases that cause tissue injury and further inflammation. The mast cell is central to this phenomenon with degranulation leading to the clinical signs associated with HST I reactions such as pulmonary and GI edema that is often associated with anaphylaxis, chronic pruritis which is often present in cases of flea allergy, atopy and food allergies; and chronic intestinal inflammation resulting from food allergies. These events also contribute to the histopathological findings noted on skin and intestinal biopsies such as lymphocytic infiltration, increase in mast cells, and eosinophilia with varying degrees of edema and inflammation.

TYPE II HYPERSENSITIVITY REACTIONS

Type II hypersensitivity reactions (HST II) can involve immune reactions to foreign or self antigen. Examples include autoimmune hemolytic anemia (AIMHA) and myasthenia gravis. There are three major components involved in development of HST II: cell surface antigen or extracellular antigen (such as extracellular matrix protein), IgG directed against these antigens, and immune effector cells. Immunoglobulin displayed on the B cell surface is known as the B cell receptor (BCR). B cells are constantly circulating through lymphoid tissue in search of antigen that will bind to their specific BCR. As mentioned above, when a B cell recognizes antigen via its BCR, the antigen is internalized and can be re-displayed via MHC II. The B cell can then migrate to T cell rich areas of lymphoid tissue where a $CD4^+$ T helper (Th) cell may recognize the antigen / MHC II complex on the B cell surface. If this occurs, both the B cell and the T cell communicate and activate one another through a complex set of activation signals.[2] It is unclear how B cells initially encounter and internalize self antigens but it may occur during the breakdown of aged or damaged tissues. Upon activation by the Th cell, the B cell will grow, divide, and differentiate into memory B cells and plasma cells which begin producing IgG directed against the antigen. Once IgG binds the antigen in sufficient quantities, activation of the complement cascade occurs. Complement can facilitate direct damage to the cell through the formation of the membrane attack complex and through the attraction of inflammatory cells to the site. Neutrophils and macrophages have cell surface receptors that recognize IgG (Fc receptors) and cell surface receptors that recognize complement. Clinical and laboratory findings are typically associated with the specific anatomic location(s) where the HST II occurs. For example, anemia, icterus and spherocytosis are often noted in AIMHA resulting from intravascular and extravascular hemolysis, while generation of antibodies against acetylcholine receptors at the neuromuscular junction results in the development of myasthenia gravis.

TYPE III HYPERSENSITIVITY REACTIONS

Type III hypersensitivity reactions (HST III) can involve immune reactions against foreign antigen or immune reactions against self antigens. Clinical examples of HST III include certain forms of polyarthritis and glomerulonephritis. There are three major components involved in development of HST III: soluble antigen, IgG directed against these antigens, and immune effector cells. As with HST II above, when a B cell recognizes antigen via its BCR, the antigen is internalized and re-displayed on the B cell surface via MHC II. The B cell can migrate to T cell rich areas of lymphoid tissue where a $CD4^+$ T helper (Th) cell may recognize the antigen / MHC II complex on the B cell surface and the B cell and the T cell communicate and activate one another through a complex set of activation signals. For HST III reactions to cause immunopathology there must be a chronic source of soluble antigen in slight excess of the IgG produced. Once sufficient quantities of antigen / antibody complexes have accumulated, activation of the complement cascade occurs. Complement proteins facilitate the attraction of inflammatory cells to the site. Neutrophils and macrophages have cell surface receptors that recognize the Fc region of IgG and cell surface receptors that recognize complement. Because the phagocytes are unable to engulf the accumulated immune complexes, they release their contents, resulting in tissue damage and inflammation at the site of immune complex deposition.[2] In HST III reactions the accumulation of immune complexes tends to occur in areas rich in blood flow with a large network of branching capillaries. These areas are often in locations that modify plasma to produce specialized fluids such as aqueous humor, cerebrospinal fluid, joint fluid, and urine. Clinical and laboratory findings are typically associated with the specific anatomic location(s) where immune complexes accumulate. For example, immune complex deposition in the joints and kidneys result in polyarthritis and glomerulonephritis.

TYPE IV HYPERSENSITIVITY REACTIONS

Type IV hypersensitivity reactions (HST IV) are the most diverse in pathogenesis. In type I through type III hypersensitivity reactions, immunoglobulin is involved in the initial encounter with antigen and disease pathogenesis, however, in HST IV reactions, either $CD4^+$ Th cells or $CD8^+$ T cells are involved in the initial encounter with antigen. Examples of HST IV reactions include contact dermatitis and tuberculin skin reactions. Antigens that stimulate HST IV can either be cell associated or soluble antigens. Cell associated antigens typically stimulate a $CD8^+$ cytotoxic T lymphocyte response which leads to the destruction of the cell. Soluble antigens that promote macrophage activation with $CD4^+$ Th activation are implicated in contact dermatitis reactions. Soluble antigens that lead to $CD4^+$ Th production and IgE production stimulate an influx of eosinophils and mast cells that produce a HST IV that is quite often seen along with a HST I (see previous section).

REFERENCES

[1]Fogle et al **Compend Contin Educ Vet.** 2007; 29(5):290-302. [2] Murphy K, **Janeway's Immunobiology** 2011. [3]Shmuel DL and Cortes Y, **J Vet Emerg Crit Care** 2013; 23(4):377-94.

GLUCAGON IN DIABETES

Tamara S. Hannon, MD, MS (Pediatric Endocrinology)
Indianapolis, IN

THE PANCREAS[1]

Exocrine Function

The exocrine pancreas is physically located around the periphery of the pancreatic islets. Exocrine glands are organized as acinar and duct cells which secrete pancreatic enzymes important to digestion directly into the small intestine via the pancreatic duct. The exocrine pancreas is not affected in type 1 diabetes. Chronic pancreatitis, which sometimes accompanies diabetes, is associated with decline in exocrine pancreas functioning.

Endocrine Function

The endocrine component of the pancreas consists of islet cells, which are surrounded by the exocrine acinar and duct cells. The endocrine pancreas includes α-cells which secrete glucagon, β-cells which secrete insulin and amylin, δ-cells which secrete somatostatin, and F cells which secrete pancreatic polypeptide directly into the bloodstream. The two hormones central to the pathophysiology of diabetes are secreted by α and β-cells.

Paracrine Function

Paracrine function is defined as a hormone's action on an adjacent or nearby cell. In the case of the pancreatic islet, the cells rely on paracrine functioning to stimulate and inhibit the secretion of glucose-regulating hormones. Specifically, insulin secretion by islet β-cells directly inhibits the secretion of glucagon from adjacent α-cells. It is not falling blood sugar which signals the secretion of glucagon, but rather falling or absent local insulin concentrations.

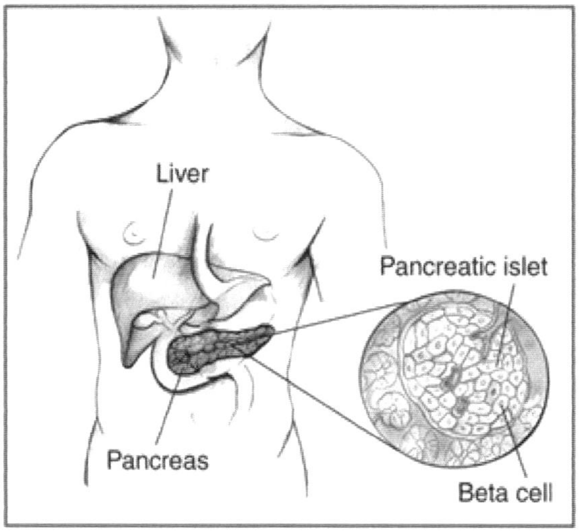

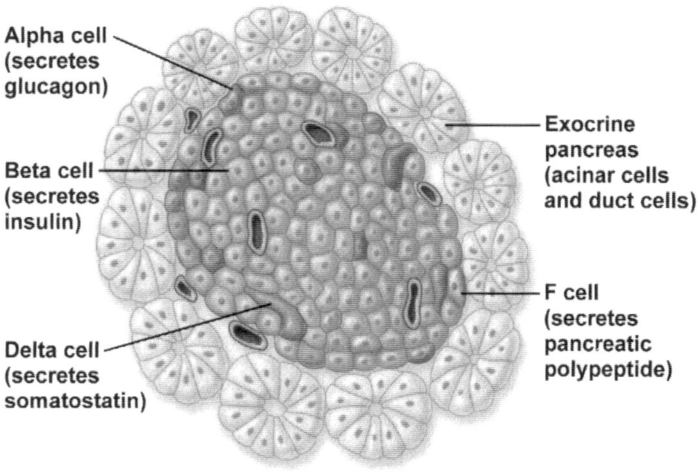

Pearson Education, 2011

PATHOPHYSIOLOGY OF DIABETES

Insulin Deficiency

Universal features of diabetes mellitus are insulin resistance (especially in type 2 diabetes, but also present in type 1 diabetes), and insulin deficiency resulting from reduced or absent β-cell function (as is the case with autoimmune-mediated type 1 diabetes). As a result of lack of insulin action, there is post-prandial hyperglycemia due to lack of insulin action on insulin sensitive tissue to take up glucose, and increased circulating free fatty acids due to the failure to inhibit liposis. There is also fasting hyperglycemia from lack of suppression of hepatic glucose production and lack of insulin paracrine action to suppress glucagon secretion. Glucagon excess is thus, part of the aberrant pathophysiology which occurs in diabetes mellitus.

Glucagon Excess

Glucagon is counter-regulatory to insulin. The primary actions of glucagon are increased hepatic glucose production via glycogenolysis and gluconeogenesis, and ketogenesis, to protect against hypoglycemia. The currently accepted model of the pathophysiology is that glucose-responsive β-cells exhibit paracrine regulation of the proximal α-cells: intraislet insulin suppresses glucagon secretion.[2] It is established that there are increased circulating glucagon concentrations in diabetes mellitus.[3] Inappropriately high circulating concentrations of glucagon contribute to hyperglycemia, and hyperglycemia also promotes continued secretion of glucagon. Hyperglucagonemia is present in every form of poorly-controlled diabetes. Interestingly, in animal models in which glucagon is absent or non-functional, the metabolic effects of insulin deficiency (hyperglycemia, ketosis, acidosis) are greatly reduced.[4-7]

INSULIN AND GLUCAGON EFFECTS IN DIFFERENT ORGANS[8]

	Insulin Effects	Glucagon Effects
Liver	Inhibits glycogenolysis Inhibits glyconeogenesis Stimulates glycolysis Stimulates glycogenesis Inhibits ketogenesis	Stimulates glycogenolysis Stimulates glyconeogenesis Inhibits glycolysis Inhibits glycogenesis Stimulates ketogenesis
Adipose	Facilitates fat storage Inhibits lipolysis	
GI Tract		Stimulates ghrelin secretion in the stomach Decreases gastrointestinal mobility
Heart	Inotropic effects Increases glucose uptake, myocardial efficiency	Increases heart rate and contractility Improves AV conduction
Brain	Stimulates feeding	Suppresses feeding
Kidney	Increases sodium reabsorption	Promotes sodium excretion Stimulates K excretion Stimulates bicarbonate secretion Increases glomerular filtration rate
B-cell	Decreases glucose-stimulated insulin secretion	Promotes glucose-stimulated insulin secretion
α-cell		Stimulates α-cell exocytosis

IMPLICATION FOR THE TREATMENT OF DIABETES

Reducing Hyperglucagonemia

In patients with insulin-dependent diabetes, there are therapeutic maneuvers that can help improve hyperglucagonemia. These include:

- Giving insulin ahead of the meal. When insulin is given late, there is further hyperglycemia and glucagon excess.
- Giving sufficient insulin doses with all meals and snacks. When food is eaten without insulin (or without enough insulin), this promotes hyperglycemia and glucagon excess.
- Giving sufficient basal insulin

There are few additional pharmacologic options for reducing hyperglucagonemia which occurs in insulin-deficient diabetes. Amylin is co-secreted with insulin from the β-cells (100:1 ratio), and so amylin production absent in the case of immune-mediated type 1 diabetes and in type 2 diabetes where there is poor β-cell functioning. Amylin functions include slowing gastric emptying (reducing the appearance of glucose in the blood), promotion of satiety, and inhibition of glucagon. Amylin slows the rise in blood sugar when eating and reduces hyperglucagonemia, contributing to glycemic control.[9] Amylin is available in injectable form for patients with diabetes. It is associated with improved glycemic control, nausea and weight loss.

Another potential suppressor of hyperglucagonemia is the adipose-secreted hormone, leptin. Continuous infusion of leptin has been found to meaningfully reduce glucagon concentrations and promote glycemic control. This hormone is under further investigation.[9]

Hypoglycemia

Insulin-induced hypoglycemia is a common, acute, complication of diabetes that can lead to mild cognitive effects that are reversed with ingestion of glucose, or more severe effects including hypoglycemic seizure, coma, and death. With insulin-induced hypoglycemia where sufficient concentrations of insulin are inhibiting endogenous glucagon secretion, glucagon is not appropriately counter-regulatory as glucagon secretion is decreased or absent. In this case, simple sugars are required to be ingested if the individual is capable of doing so. An injection of a pharmacologic dose of glucagon can be given to immediately promote hepatic glycogenolysis and increased blood glucose concentrations.

For patients who are dependent on insulin for treatment of diabetes, exogenous insulin has effects on endogenous glucagon concentration. Because the islet concentrations of exogenously-given insulin are smaller than locally secreted insulin, glucagon secretion is incompletely suppressed, contributing to post-prandial hyperglycemia. This often leads to increasing doses of insulin to control the hyperglycemia. At some point, the concentration of insulin realized in the islet are sufficient to fully suppress glucagon. This can then lead to precipitous hypoglycemia. These extremes of hyperglycemia and hypoglycemia are often characterized as "brittle diabetes", a term referring to patients who have much difficulty obtaining acceptable glycemic control. However, it is very likely that these patients have no endogenous insulin production, no endogenous suppression of glucagon, and require a threshold concentration of insulin which then "pushes them over the edge" resulting in hypoglycemia that is not adequately countered due to suppression of endogenous glucagon. In these cases, shot of exogenous pharmacologic doses of glucagon are necessary to reverse the hypoglycemia.

REFERENCES

1. *Endocrinology: An Integrated Approach.* Oxford: BIOS Scientific Publishers; 2001.
2. Unger RH, Cherrington AD. Glucagonocentric restructuring of diabetes: a pathophysiologic and therapeutic makeover. *The Journal of clinical investigation.* Jan 3 2012;122(1):4-12.
3. Edgerton DS, Cherrington AD. Glucagon as a critical factor in the pathology of diabetes. *Diabetes.* Feb 2011;60(2):377-380.

4. Lee Y, Berglund ED, Wang MY, et al. Metabolic manifestations of insulin deficiency do not occur without glucagon action. *Proceedings of the National Academy of Sciences of the United States of America.* Sep 11 2012;109(37):14972-14976.
5. Lee Y, Wang MY, Du XQ, Charron MJ, Unger RH. Glucagon receptor knockout prevents insulin-deficient type 1 diabetes in mice. *Diabetes.* Feb 2011;60(2):391-397.
6. Conarello SL, Jiang G, Mu J, et al. Glucagon receptor knockout mice are resistant to diet-induced obesity and streptozotocin-mediated beta cell loss and hyperglycaemia. *Diabetologia.* Jan 2007;50(1):142-150.
7. Qureshi SA, Rios Candelore M, Xie D, et al. A novel glucagon receptor antagonist inhibits glucagon-mediated biological effects. *Diabetes.* Dec 2004;53(12):3267-3273.
8. Jiang G, Zhang BB. Glucagon and regulation of glucose metabolism. *American journal of physiology. Endocrinology and metabolism.* Apr 2003;284(4):E671-678.
9. Edgerton DS, Johnson KM, Cherrington AD. Current strategies for the inhibition of hepatic glucose production in type 2 diabetes. *Front Biosci (Landmark Ed).* 2009;14:1169-1181.

DIABETIC KETOACIDOSIS IN CHILDREN

Tamara S. Hannon, MD, MS (Pediatric Endocrinology)
Indianapolis, IN

DEFINITIONS

Diabetic ketoacidosis (DKA) is a serious and emergent, acute complication of diabetes, caused by insulin deficiency. The lack of insulin results in a catabolic state, with endogenous glucose production, proteolysis, and lipolysis.[1,2] Metabolic acidosis develops as a result of lipolysis and the formation of ketone bodies in the absence of glucose availability in insulin dependent tissues. DKA is the leading cause of morbidity and mortality in youth with type 1 diabetes. It can occur in youth with type 2 diabetes as well, although this is less common, as a result of an absolute or relative deficiency of insulin. The following represent the biochemical criteria for the diagnosis of DKA:

- Hyperglycemia, blood glucose of >200 mg/dL (11 mmol/L) **AND**
- Metabolic acidosis, defined as a venous pH <7.3 or plasma bicarbonate <15 mEq/L (15 mmol/L) **AND**
- Ketosis, determined by the presence of ketones in the blood or urine.

Hyperosmolar hyperglycemic state (HHS)[3] is a serious and emergent, acute complication of diabetes which is distinguished from DKA by marked hyperglycemia (plasma glucose >600 mg/dL [>33.3 mmol/L]) but with minimal or only mild acidosis (serum bicarbonate >15 mmol/L, venous pH >7.25 or arterial pH >7.30). There is an absence of ketosis, and only small ketonuria. The extreme dehydration and hyperosmolar state (effective serum osmolality >320 mOsm/L) leads to symptoms and altered consciousness. HHS is reported mainly among black adolescents with type 2 diabetes, but has occurred in whites and other race/ethnicity groups as well.

EPIDEMIOLOGY

Despite polyuria, polydipsia, and weight loss being the universal key features of new-onset, or uncontrolled diabetes, DKA is present at the initial diagnosis in approximately 30% of cases of new-onset type 1 diabetes in youth.[4-6] Risk factors for presenting with DKA at initial diagnosis of type 1 diabetes include: young age (less than 5 years), lower socioeconomic status and parental education, and missed initial diagnosis (i.e. being given another diagnosis, such as gastroenteritis without a proper evaluation).[7,8]

The majority of cases of DKA, however, occur in individuals who have previously been diagnosed but their diabetes is poorly controlled. In fact, the risk of developing DKA in youth with established type 1 diabetes is up to 10 percent per patient per year. Recurrent episodes of DKA with established type 1 diabetes mellitus are primarily the result of poor adherence to diabetes treatment and missed insulin injections (or insulin pump failure). Intercurrent illness makes it more difficult to control the diabetes due to both increased insulin resistance and not being able to eat normally. Increased stress occurring during intercurrent illness, infection, extreme emotional stress, or for other reasons, increases the secretion of catecholamines, cortisol, and glucagon, which promote insulin resistance, endogenous glucose production, lipolysis leading to ketoacid production. Being treated with medications which increase insulin resistance (glucocorticoids for example) can also precipitate DKA if adjustments in insulin doses are not made. Specific risk factors associated with DKA in youth with established type 1 diabetes include: poor metabolic control (high hemoglobin A1C), dehydration associated with gastroenteritis, psychiatric disorders, family conflict, lack of regular access to medical care and/or diabetes treatment medicines and supplies, insulin omission/insulin pump failure. As in most chronic health conditions, a small percentage of patients require a large percentage of healthcare resources.

Ketosis and DKA can occur in children with type 2 diabetes mellitus secondary to absolute or relative insulin deficiency, particularly in African-American children. The incidence of DKA in youth with type 2 diabetes is unclear, but it occurs less commonly than in type 1 diabetes.[9-11]

PATHOPHYSIOLOGY

Actions of Insulin (ANABOLIC)		Insulin Deficiency (CATABOLIC)
Glucose uptake in insulin-sensitive tissues	→	Hyperglycemia, increased serum osmolality; cellular starvation
Enhance amino acid uptake	→	Decreased protein synthesis
Enhance protein synthesis	→	Decreased protein synthesis
Promote glycogen synthesis	→	Decreased glycogen stores
Promote fat deposition	→	Increased lipolysis, increased circulating FFA
Suppress lipolysis	→	Increased lipolysis, increased circulating FFA
Suppress ketogenesis	→	Decreased ketone utilization, ketosis
Suppress glucagon & glycogenolysis	→	Increased glucagon & glycogenolysis
Suppress gluconeogenesis	→	Increased gluconeogenesis (glycerol and amino acids)

Insulin deficiency can be absolute, as in the case of new onset type 1 diabetes, or relative. States of relative insulin deficiency include intercurrent illness, infection, emotional/surgical/physical stress, the use of glucocorticoids, and puberty (decreased insulin sensitivity). The hyperglycemia resulting from insulin deficiency results in hyperosmolality, leading to an osmotic diuresis and volume depletion. The combination of hypovolemia and continued ketosis leads to metabolic acidosis. If there is sufficient insulin to suppress ketogenesis and ketosis, this leads to a pure hyperosmolar state (HHS) with hypovolemia. Prostaglandins and prostacyclins released from adipose tissue mediate decreased vascular resistance, nausea, vomiting, and abdominal pain. Decreased perfusion and decreased glomerular filtration rate associated with severe dehydration can result in lactic academia and severe metabolic acidosis. The metabolic perturbations, hyperosmolality, and severe osmotic diuresis result in electrolyte disturbances, including: hyponatremia (must be corrected for hyperglycemia), hypokalemia, hypophosphatemia, hypocalcemia, and hypomagnesemia. [12]

PRESENTATION[1,2]

Symptoms	Signs
Polyuria	Flushed cheeks; warm, dry skin
Polydipsia	Fruity breath (ketosis)
Nausea	Hyperventilation (Kussmaul breathing)
Vomiting	Dehydration
Abdominal pain	Tachycardia
Weakness	Hypotension
Blurred vision	Decreased level of consciousness
Sleepiness	Coma
Weight loss	

EVALUATION

Laboratory evaluation begins with making the diagnosis: glucose (hyperglycemia), pH and bicarbonate (acidosis), ketosis (serum hydroxybutyrate, urine ketones). Once the diagnosis is established, electrolytes (including Na, K, Cl, Ca, Phos, Mg), BUN, Cr are obtained. Other examinations include neurologic evaluation, laboratory evaluation for intercurrent infection if suspected, and cardiorespiratory monitoring.

Typical laboratory characteristics of DKA and HHS[13,14]

	DKA			HHS
	Mild	Moderate	Severe	
Plasma glucose (mg/dL)	>250	>250	>250	>600
(mmol/L)	>13.9	>13.9	>13.9	>33.3
Arterial pH	7.25 - 7.30	7.00 - 7.24	<7.00	>7.30
Serum bicarbonate (mEq/L)	15 - 18	10 - <15	<10	>18
Urine ketones*	Positive	Positive	Positive	Small
Serum ketones - Nitroprusside reaction	Positive	Positive	Positive	≤ Small
Serum ketones - Enzymatic assay of beta hydroxybutyrate (normal range <0.6 mmol/L)[Δ]	3 to 4 mmol/L	4 to 8 mmol/L	>8 mmol/L	<0.6 mmol/L
Effective serum osmolality (mOsm/kg)[◊]	Variable	Variable	Variable	>320
Anion gap[§]	>10	>12	>12	Variable
Alteration in sensoria or mental obtundation	Alert	Alert/drowsy	Stupor/coma	Stupor/coma

DKA: diabetic ketoacidosis; HHS: hyperosmolar hyperglycemic state.

TREATMENT[15-18]

Diabetic Ketoacidosis/Hyperosmolar Hyperglycemic Crisis

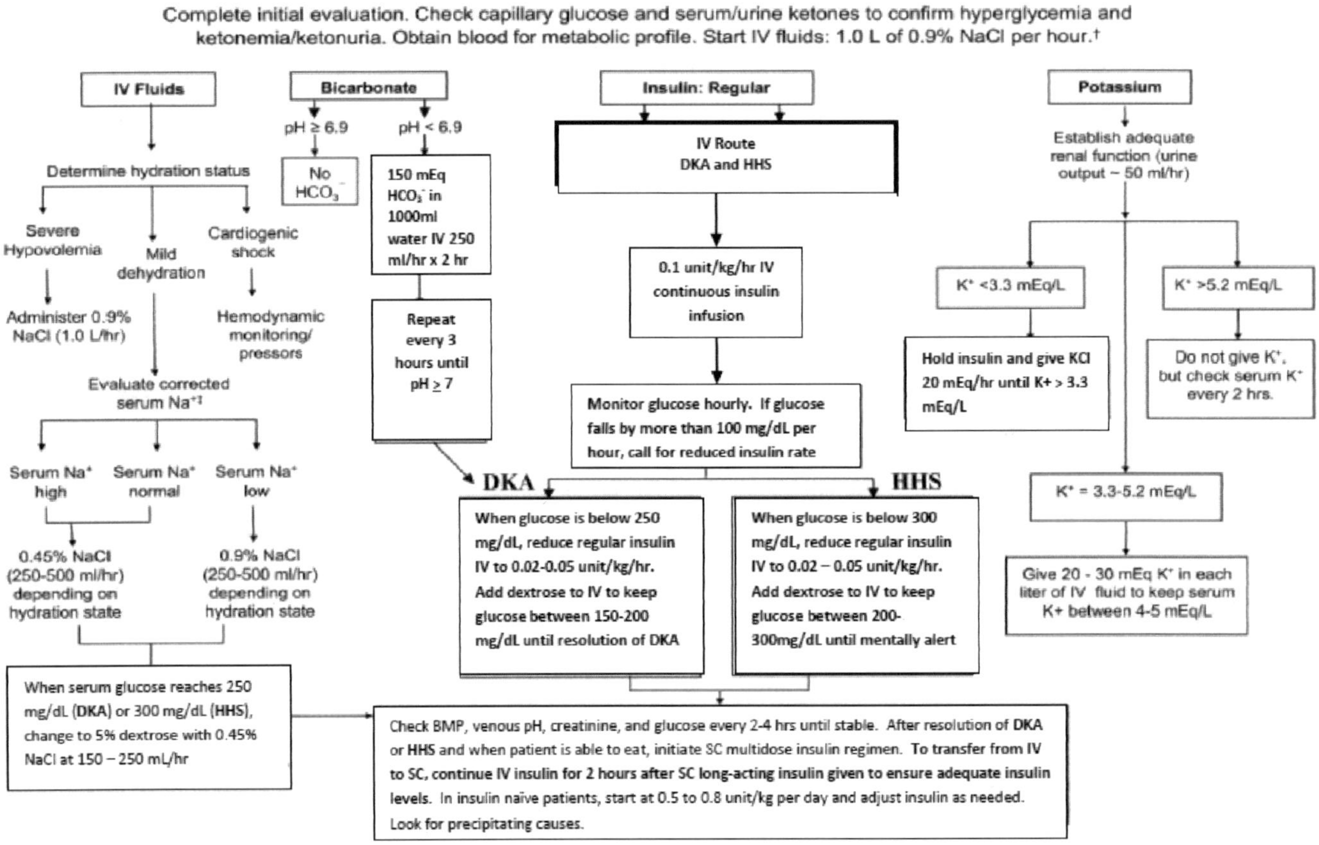

COMPLICATIONS

Cerebral Edema[19] **-** Reported mortality rates for DKA range from 0.15 to 0.51 percent. Cerebral edema accounts for the majority of deaths (60 to 90 percent). Detectable cerebral edema occurs in 0.3 to 1 percent of children with DKA and has a high mortality rate of 21 to 24 percent. Although cerebral edema is felt to be exacerbated by rapid fluid administration and correction of metabolic abnormalities, up to 20 percent of cases of cerebral edema occur before the initiation of therapy.[20-22] Children with DKA are monitored hourly for signs of cerebral edema (intensifying headache, altered consciousness, vomiting, irritability, increasing lethargy) throughout the course of treatment. If cerebral edema is clinically suspected (not on the basis of CT scan which is not recommended as it delays therapy), it should be addressed without delay by the administration of mannitol and/or 3% saline.[23]

Venous Thrombosis - Deep venous thrombosis may be related to a prothrombotic state and the use of femoral central venous catheters in DKA.[24]

Aspiration - Altered state of consciousness combined with vomiting can lead to aspiration and diffuse pulmonary pneumonitis. Patient with an altered state of consciousness should be intubated to protect the airway.

Cardiac arrhythmia - Cardiac arrhythmias may be seen with either hypokalemia or hyperkalemia, and other electrolyte abnormalities.[25]

Sample Treatment Flow Sheet for Monitoring DKA / HHS[26]

Patient's name: ______________________

Weight: Initial: __________ **After 24 hours:** __________

Date: Hour:	0	1	2	3	4	5	6
General information							
Mental status*							
Temperature							
Pulse							
Respiration/depth†							
Blood pressure							
Serum glucose (mg/dL)							
Serum ketones							
Urinary ketones							
Electrolytes							
Serum sodium (mEq/L)							
Serum potassium (mEq/L)							
Serum chloride (mEq/L)							
Serum bicarbonate (mEq/L)							
Serum blood urea nitrogen (mg/dL)							
Effective osmolality: 2 (measured serum sodium [mEq/L]) + glucose (mg/dL)/18							
Anion gap (mEq/L)							
Arterial blood gases							
pH: venous (V); arterial (A)							
Pa_{O_2}							
Pa_{CO_2}							
O_2 saturation							
Insulin							
Units in past hour							
Route							
Intake of fluids/metabolites							
0.45% saline (mL) in past hour							
0.9% saline (mL) in past hour							
5% dextrose (mL) in past hour							
Potassium chloride (mEq) in past hour							
Phosphate (mmol) in past hour							
Other							
Output							
Urine (mL)							
Other							

*—*A = alert; D = drowsy; S = stuporous; C = comatose.*
†—*D = deep; S = shallow; N = normal.*

REFERENCES

1. Kecskes SA. Diabetic ketoacidosis. *Pediatric clinics of North America.* Apr 1993;40(2):355-363.
2. Casteels K, Mathieu C. Diabetic ketoacidosis. *Reviews in endocrine & metabolic disorders.* May 2003;4(2):159-166.
3. Zeitler P, Haqq A, Rosenbloom A, Glaser N. Hyperglycemic hyperosmolar syndrome in children: pathophysiological considerations and suggested guidelines for treatment. *The Journal of pediatrics.* Jan 2011;158(1):9-14, 14 e11-12.
4. Elding Larsson H, Vehik K, Bell R, et al. Reduced prevalence of diabetic ketoacidosis at diagnosis of type 1 diabetes in young children participating in longitudinal follow-up. *Diabetes care.* Nov 2011;34(11):2347-2352.
5. Fritsch M, Schober E, Rami-Merhar B, Hofer S, Frohlich-Reiterer E, Waldhoer T. Diabetic ketoacidosis at diagnosis in Austrian children: a population-based analysis, 1989-2011. *The Journal of pediatrics.* Nov 2013;163(5):1484-1488 e1481.
6. Klingensmith GJ, Tamborlane WV, Wood J, et al. Diabetic ketoacidosis at diabetes onset: still an all too common threat in youth. *The Journal of pediatrics.* Feb 2013;162(2):330-334 e331.
7. Cengiz E, Xing D, Wong JC, et al. Severe hypoglycemia and diabetic ketoacidosis among youth with type 1 diabetes in the T1D Exchange clinic registry. *Pediatric diabetes.* Sep 2013;14(6):447-454.
8. White K, Kolman ML, Wexler P, Polin G, Winter RJ. Unstable diabetes and unstable families: a psychosocial evaluation of diabetic children with recurrent ketoacidosis. *Pediatrics.* Jun 1984;73(6):749-755.
9. Pinhas-Hamiel O, Dolan LM, Zeitler PS. Diabetic ketoacidosis among obese African-American adolescents with NIDDM. *Diabetes care.* Apr 1997;20(4):484-486.
10. Umpierrez GE, Casals MM, Gebhart SP, Mixon PS, Clark WS, Phillips LS. Diabetic ketoacidosis in obese African-Americans. *Diabetes.* Jul 1995;44(7):790-795.
11. Musey VC, Lee JK, Crawford R, Klatka MA, McAdams D, Phillips LS. Diabetes in urban African-Americans. I. Cessation of insulin therapy is the major precipitating cause of diabetic ketoacidosis. *Diabetes care.* Apr 1995;18(4):483-489.

12. Davis SM, Maddux AB, Alonso GT, Okada CR, Mourani PM, Maahs DM. Profound hypokalemia associated with severe diabetic ketoacidosis. *Pediatric diabetes.* Nov 27 2014.
13. Kitabchi AE, Umpierrez GE, Miles JM, Fisher JN. Hyperglycemic crises in adult patients with diabetes. *Diabetes care.* Jul 2009;32(7):1335-1343.
14. Kitabchi AE, Umpierrez GE, Murphy MB, Kreisberg RA. Hyperglycemic crises in adult patients with diabetes: a consensus statement from the American Diabetes Association. *Diabetes care.* Dec 2006;29(12):2739-2748.
15. White PC, Dickson BA. Low morbidity and mortality in children with diabetic ketoacidosis treated with isotonic fluids. *The Journal of pediatrics.* Sep 2013;163(3):761-766.
16. Basnet S, Venepalli PK, Andoh J, Verhulst S, Koirala J. Effect of normal saline and half normal saline on serum electrolytes during recovery phase of diabetic ketoacidosis. *Journal of intensive care medicine.* Jan-Feb 2014;29(1):38-42.
17. Koves IH, Leu MG, Spencer S, et al. Improving care for pediatric diabetic ketoacidosis. *Pediatrics.* Sep 2014;134(3):e848-856.
18. White NH. Management of diabetic ketoacidosis. *Reviews in endocrine & metabolic disorders.* Dec 2003;4(4):343-353.
19. Rosenbloom AL. Cerebral edema in diabetic ketoacidosis and other acute devastating complications: recent observations. *Pediatric diabetes.* Mar 2005;6(1):41-49.
20. Glaser NS, Ghetti S, Casper TC, Dean JM, Kuppermann N. Pediatric diabetic ketoacidosis, fluid therapy, and cerebral injury: the design of a factorial randomized controlled trial. *Pediatric diabetes.* Sep 2013;14(6):435-446.
21. Glaser NS, Tancredi DJ, Marcin JP, et al. Cerebral hyperemia measured with near infrared spectroscopy during treatment of diabetic ketoacidosis in children. *The Journal of pediatrics.* Oct 2013;163(4):1111-1116.
22. Glaser NS, Wootton-Gorges SL, Buonocore MH, et al. Subclinical cerebral edema in children with diabetic ketoacidosis randomized to 2 different rehydration protocols. *Pediatrics.* Jan 2013;131(1):e73-80.
23. Decourcey DD, Steil GM, Wypij D, Agus MS. Increasing use of hypertonic saline over mannitol in the treatment of symptomatic cerebral edema in pediatric diabetic ketoacidosis: an 11-year retrospective analysis of mortality*. *Pediatric critical care medicine : a journal of the Society of Critical Care Medicine and the World Federation of Pediatric Intensive and Critical Care Societies.* Sep 2013;14(7):694-700.
24. Gutierrez JA, Bagatell R, Samson MP, Theodorou AA, Berg RA. Femoral central venous catheter-associated deep venous thrombosis in children with diabetic ketoacidosis. *Critical care medicine.* Jan 2003;31(1):80-83.
25. Kuppermann N, Park J, Glatter K, Marcin JP, Glaser NS. Prolonged QT interval corrected for heart rate during diabetic ketoacidosis in children. *Archives of pediatrics & adolescent medicine.* Jun 2008;162(6):544-549.
26. Kitabchi AE, Wall BM. Management of diabetic ketoacidosis. *American family physician.* Aug 1999;60(2):455-464.

FELINE HYPERALDOSTERONISM

Hans S. Kooistra, DVM, PhD, DECVIM-CA (Internal Medicine)
Utrecht, The Netherlands

SUMMARY

Primary hyperaldosteronism is the most common disorder of the adrenal gland in cats. Cats affected with primary hyperaldosteronism are typically presented with muscle weakness due to hypokalemic myopathy and/or complications of arterial hypertension, such as acute blindness. The plasma concentration of aldosterone is relatively high and plasma renin activity is low in cats with primary hyperaldosteronism. Unilateral adrenalectomy is the treatment of choice for confirmed unilateral primary hyperaldosteronism. If surgery is not possible or if the adrenocortical disease is bilateral, cats can be treated with the mineralocorticoid-receptor antagonist spironolactone.

INTRODUCTION

Primary hyperaldosteronism, also referred to as primary aldosteronism, low-renin hyperaldosteronism and Conn's syndrome, is an adrenocortical disorder which is characterized by autonomous aldosterone secretion originating from either neoplastic or hyperplastic zona glomerulosa tissue.[1] Primary hyperaldosteronism was first described in a cat in 1983, and was initially regarded a rare disease in felines. However, the number of case reports has risen considerably in the past 15 years. Increased awareness of the disease will probably lead to a further increase in detected cases.

The inappropriately high aldosterone secretion in primary hyperaldosteronism has a number of serious implications. The most acute repercussions are associated with the typical mineralocorticoid actions of aldosterone on the kidneys. Excessive mineralocorticoid receptor activation in the epithelial cells of the distal nephron results in potassium depletion and sodium and water retention, and thereby leads to hypokalemia, systemic arterial hypertension, or both. Hence, cats affected with primary hyperaldosteronism are typically presented with muscle weakness due to hypokalemic myopathy and/or complications of arterial hypertension, such as acute blindness associated with retinal detachment and/or intraocular hemorrhage.[2] In addition, primary hyperaldosteronism has been associated with progressive renal dysfunction in cats.[3]

DIAGNOSIS

In primary mineralocorticoid excess, the plasma concentration of aldosterone is characteristically high and plasma renin activity (PRA) is immeasurably low. In hyperaldosteronism due to adrenocortical tumor plasma aldosterone concentration (PAC) is usually highly elevated. In idiopathic hyperaldosteronism PAC is usually only slightly elevated or within the upper limit of the reference range. As hypokalemia is a predominant factor in lowering PAC, in the presence of hypokalemia moderately elevated aldosterone values can be regarded as inappropriately high. The PRA must also be taken into account. The combination of a high-normal or elevated PAC and low PRA indicates persistent aldosterone synthesis in the presence of little or no stimulation by the renin-angiotensin system. In humans, the PAC:PRA ratio (ARR) is considered to be a very useful aid in diagnosing primary hyperaldosteronism. This also seems to be true for cats with idiopathic hyperaldosteronism.

The ARR is elevated in about 10% of human patients with arterial hypertension and most of these have excess aldosterone production from both adrenal cortices. The diagnostic value of the ARR is principally determined by the sensitivity of the renin assay and interpretation should rest upon comparison with an appropriate control population. The ARR is currently regarded as the most reliable means of detecting primary hyperaldosteronism, but the measurements should be repeated if the initial result is inconclusive or difficult to interpret because of suboptimal sampling conditions.

An alternative diagnostic approach may be measurement of the urinary aldosterone:creatinine ratio (UACR). Cats excrete smaller quantities of aldosterone and its 18-glucuronidated metabolite in urine than do humans or dogs, but nevertheless the UACR can be determined. This would allow the development of a dynamic test, such as employing a suppressive agent that would reduce the UACR in healthy individuals but have little or no effect in those with primary hyperaldosteronism. In 42 healthy cats the upper limit for the UACR was 46.5×10^{-9}. The administration of sodium chloride did not significantly lower the UACR but administration of fludrocortisone (0.05 mg/kg body weight) reduced it by 44-97% (median 78%). In a cat with an aldosterone-producing adrenocortical carcinoma the UACR was within the reference range and was not lowered by fludrocortisone administration.[4] This test may prove to be a practical noninvasive diagnostic tool, but further evaluation is required, particularly with regard to its discriminatory power in diagnosing idiopathic hyperaldosteronism.[5]

Subtype classification - differentiating between tumorous and non-tumorous mineralocorticoid excess -requires diagnostic imaging. Ultrasonography and computed tomography have been used in cats to identify and characterize adrenal tumors.[6] As in humans, the findings are not always immediately conclusive. The visualization of a small aldosteronoma may pose problems while nodular hyperplasia might be interpreted as microadenoma.

TREATMENT

Unilateral adrenalectomy is the treatment of choice for confirmed unilateral primary hyperaldosteronism. Preoperatively and perioperatively hypokalemia should be controlled as well as possible, by oral and/or intravenous supplementation. Analogous to the postoperative management of hypercortisolemia due to adrenocortical tumor, temporary fludrocortisone therapy could be considered. However, in the reported cases such postsurgical measures have not been necessary and their omission does not seem to have had deleterious effects.

If surgery is not possible or if the adrenocortical disease is bilateral, medical treatment is possible with the mineralocorticoid-receptor antagonist spironolactone and oral supplementation with potassium gluconate. The initial doses are 2 mg spironolactone/kg given twice daily and 0.5 mmol potassium gluconate/kg given twice daily. Persistent arterial hypertension can be treated with the calcium blocker amlodipine (0.1-0.2 mg/kg). In cases of adrenocortical tumor medical treatment may lead to resolution of symptoms and signs such as the myopathy in cats, but complete normalization may not be achieved. Particularly plasma potassium tends to remain below the reference range, despite increasing doses of both spironolactone and potassium.

Doses of spironolactone >4 mg/kg may cause anorexia, diarrhea, and vomiting. These side effects may be due to interference by spironolactone with aldosterone action on transepithelial electrolyte transport in the distal colon.

Experience is very limited, but medical treatment appears to be preferable in cats with hyperaldosteronism due to bilateral adrenocortical hyperplasia. The hyperaldosteronism is usually somewhat milder than in cases due to tumor and normokalemia may be maintained for a long period with spironolactone alone or together with low doses of potassium.

PROGNOSIS

After complete removal of a unilateral nonmetastasized mineralocorticoid-producing tumor, the prognosis can be excellent, without any medication. In both forms the disease may be associated with renal insufficiency. Successful removal of the tumor will probably prevent further progression of aldosterone-induced arteriolar sclerosis and interstitial fibrosis in the kidneys. The prognosis may not be as favorable in cats with idiopathic hyperaldosteronism treated with spironolactone, for this treatment will not abolish the mineralocorticoid excess as definitely as surgery may do.

REFERENCES

[1]Galac S, et al. In: Clinical Endocrinology of Dogs and Cats, 2nd ed. 2010:93. [2]Djajadiningrat-Laanen SC, et al. **J Feline Med Surg** 2011;13:641.[3]Javadi S, et al. **Domest Anim Endocrinol** 2005;28:85. [4]Djajadiningrat-Laanen SC, et al. **J Vet Intern Med** 2008;22:1283. [5]Djajadiningrat-Laanen SC, et al. **J Vet Intern Med** 2013;27:1493. [6]Ash RA, et al. **J Feline Med Surg** 2005;7:173.

FELINE THYROID STORM

Cynthia R. Ward, VMD, PhD, DACVIM
Athens, GA

Thyroid storm is a syndrome described in human medicine to define a multi systemic disorder resulting from organ exposure to excessive levels of thyroid hormone. This form of acute thyrotoxicosis can be life-threatening and is a significant cause of mortality in human emergency rooms. Thyrotoxicosis is a term used to describe any condition in which there is an excessive amount of circulating thyroid hormone whether from excess production and secretion from an overactive thyroid gland, leakage from a damaged thyroid gland, or from an exogenous source. In contrast, hyperthyroidism describes thyroid gland hyperfunction. Therefore, acute thyrotoxicosis may occur in the absence of thyroid gland hyperfunction, although this is rare in veterinary medicine. In humans, thyroid storm can occur at any age. It can be present in euthyroid patients as well as treated and partially-treated hyperthyroid patients. Although this is a well recognized syndrome in human medicine, it has not been described as a clinical entity in veterinary medicine. Hyperthyroidism is a common disease in older feline patients. It is also rarely seen in dogs with thyroid carcinoma. Hyperthyroid cats that present with acute exacerbation of thyrotoxicosis may be described to have thyroid storm. This chapter will discuss the human syndrome and define a similar syndrome in hyperthyroid veterinary patients. The clinical signs of and treatment modalities for feline thyroid storm will also be presented.

PATHOGENESIS

Although the exact pathogenesis of thyroid storm remains to be elucidated, several factors appear to be involved.

High levels of circulating thyroid hormones

Although, one would expect circulating thyroid hormones to be increased in patients with thyroid storm, there is no difference between serum thyroid hormone levels in these patients and in more stable hyperthyroid patients in human medicine. This results in the diagnosis being made primarily based on clinical signs.[1]

Rapid, acute increases in circulating thyroid hormones

The magnitude of change in the serum thyroid hormone level may be more important than the actual level itself. This would explain the occurrence of thyroid storm following radioactive iodine therapy and thyroid surgery that damage the thyroid gland causing release of hormone and also, following abrupt cessation of anti-thyroid medication resulting in the rapid rise of serum thyroid hormone levels.[2] Certainly, non-thyroidal illness, known to be a precipitating factor in human medicine, has been shown to alter binding of thyroid hormones to their carriers and could be responsible for rapid alteration of circulating thyroid hormone as well.

Hyperactivity of the sympathetic nervous system

Activation of the sympathetic nervous system has been implicated in the onset of thyroid storm.[3] Many of the clinical signs and physiological symptoms seen in thyroid storm are similar to those seen during catecholamine excess. In humans, it has been shown that serum and urine catecholamine levels are within normal limits during thyroid storm. However, it is known that thyroid hormones can alter tissue sensitivity to catecholamines, at the cell surface receptor as well as the intracellular signaling levels, and this increased sensitivity may result in the clinical signs seen during thyroid storm. In addition, many of these clinical signs are controlled by β adrenergic blockade.

Increased cellular response to thyroid hormones.

This effect has been implicated in the cause of thyroid storm resulting from infection, sepsis, hypoxemia, hypovolemia and lactic or ketoacidosis.[4]

PRECIPITATING EVENTS

In most cases of thyroid storm in humans, a precipitating event can be identified, although no known causes are found in up to 2% of cases.[5] The most common events are infection, thyroidal and nonthyroidal surgery, radioactive iodine therapy, administration of iodinated contrast dyes, administration of stable iodine, withdrawal of antithyroid medication, amiodarone therapy, ingestion of excessive amounts of exogenous thyroid hormone, vigorous palpation of the thyroid, severe emotional stress, and a variety of acute nonthyroidal illnesses. Common events that may precipitate thyroid storm in feline hyperthyroid patients may include: radioactive iodine therapy, abrupt withdrawal of antithyroid medication, thyroid surgery, vigorous thyroid palpation, stress, administration of stable iodine compounds as well as any of the other causes found as precipitating factors in humans.

CLINICAL SIGNS

Thyroid storm is the acute exacerbation of clinical signs of thyrotoxicosis; however, the diagnosis of thyroid storm in human medicine is based upon the prevalence of 4 major clinical signs. These include: fever, CNS effects from mild agitation to seizures or coma, GI-hepatic dysfunction ranging from vomiting/diarrhea and abdominal pain to unexplained jaundice, and cardiovascular effects including sinus tachycardia, atrial fibrillation and congestive heart failure. The combination of these clinical signs along with identification of a precipitating event allows for the diagnosis of thyroid storm.[5] In cats presenting with presumed thyroid storm, many of these clinical signs also occur. Auscultation may reveal cardiac murmur or arrhythmia, most often a gallop rhythm. Crackles or dullness in the lung fields indicating pulmonary edema or pleural effusion, respectively, associated with congestive heart failure may also be ausculted.[6] Additional clinical signs that may be associated with the event in cats include mild to severe hypertension.[7]

Retinopathies, including hemorrhage, edema, degeneration or even retinal detachment may be found especially in hypertensive thyrotoxic cats.[8] Tachypnea and hypothermia may be present, and absent limb motor function may be detected as a result of thromboembolic disease occurring from acute thyrotoxicosis.[9] Sudden death may also occur. Severe, acute muscle weakness and ventroflexion of the neck may be seen in acutely thyrotoxic cats, often associated with hypokalemia.

DIAGNOSIS

The diagnosis of thyroid storm is based upon identification of the presence of thyrotoxicosis, appropriate clinical signs, and evidence of a precipitating event.[5] Thyrotoxicosis in hyperthyroid cats is demonstrated by an elevated total T_4 level or a total T_4 level in the high normal range combined with an elevated free T_4. In some cases the total T_4 may be in the normal range in a hyperthyroid cat, but it is expected that in cases of thyroid storm, the total T_4 and free T_4 will be above the normal range. In human medicine, thyroid storm is diagnosed based upon a point system assigned to each of the main clinical components: fever, CNS signs, GI signs, and cardiovascular signs as well as presence or absence of a precipitating event.[5] In hyperthyroid feline patients, thyroid storm may be diagnosed based upon the presence of clinical signs of acute thyrotoxicosis. In some cases a precipitating event may be identified.

LABORATORY ABNORMALITIES

Laboratory abnormalities are those seen resulting from uncomplicated thyrotoxicosis;[10] there is no distinguishing laboratory value(s) in cases of feline thyroid storm. In the hyperthyroid cat, hematology abnormalities may include a mild erythrocytosis, macrocytosis, and Heinz body formation. In humans with thyroid storm, a leukocytosis with left shift in the absence of active infection/inflammation has been identified.[11] In hyperthyroid cats a mature neutrophilia, lymphopenia, eosinopenia are more commonly identified as a stress response. Biochemical abnormalities seen in people with thyroid storm include a mild hyperglycemia and hypercalcemia. Elevated liver enzymes are seen as well and hyperbilirubinemia may be seen in severe cases; this finding carries a poor prognosis. In hyperthyroid cats, elevated liver enzymes, a mild hyperglycemia and severe hypokalemia may be seen in acute thyrotoxicosis. A decreased sodium:potassium ratio may be seen in thyrotoxic cats who present in heart failure with pleural effusions.[12] Radiographs may reveal an enlarged heart or evidence of congestive heart failure.

TREATMENT

Treatment of thyroid storm is aimed at controlling the four major problematic areas: 1) to reduce the production/secretion of thyroid hormones, 2) to counteract the peripheral effects of thyroid hormones, 3) to provide systemic support, and 4) to identify and eliminate the precipitating factor.[13]

Reduction in the production or secretion of new thyroid hormones

The thioimidazole compound, methimazole, inhibits iodine incorporation into tyrosyl residues of thyroglobulin and thus prevents the synthesis of active thyroid hormone. In this way, methimazole should be the first line of defense against thyroid storm. However, it does not prevent the secretion of already formed thyroid hormones. Methimazole may be given orally, transdermally, or even rectally in cats. The dose should be at the high end in cats that have normal renal function at 5 mg twice/daily. If there is suspected renal insufficiency or failure, the dose of methimazole should be reduced by half.

Methimazole will block the formation of new active thyroid hormone, but other therapy must be instituted in order to prevent further secretion of formed hormone which is stored in high concentrations in the thyroid gland. This can be done by treatment with stable iodine compounds such as potassium iodide. These compounds, in large doses, can also decrease the synthesis rate of thyroid hormone. They must be given 1 hour after methimazole administration since a large load of iodine will initially stimulate thyroid hormone production. Potassium iodide may be given at 25 mg every 8 hours. Instead of potassium iodide, lipid-soluble radiographic contrast agents such as, iopanoic acid, may be given at 100 mg bid. Although iopanoic acid is available in parenteral form, oral dosing is safer since it is a very hyperosmolar agent. This compound has the additional advantages of blocking peripheral conversion of T_4 to T_3, blocking T_3 binding to its receptor, and inhibiting thyroid hormone synthesis.

Inhibition of peripheral effects of thyroid hormone

The most rapid relief of signs caused by thyroid storm is by medications that block the β adrenergic receptors such as propranolol and atenolol. The nonselective β blocker, propranolol most commonly used as a sympatholytic in human medicine, is inherently difficult to use in cats due to its poor oral bioavailability and short half life, requiring dosing q 8 hours. Its use has been largely superseded by that of atenolol because of its selectivity and the once daily dosing regime. However, propranolol has been shown to inhibit the peripheral conversion of T_4 to T_3, although this effect happens slowly.[14] Therefore, its use may be advantageous is severely thyrotoxic cats. Propranolol should be used toward the high end of the dose to ensure β adrenergic blockade at 5 mg PO q 8 hrs or 0.02 mg/kg IV over 1 minute. Alternatively, atenolol, a selective β1 adrenergic blocker that has better oral bioavailability, may be used at 1 mg/kg q 12-24 hour. In acute situations, the short acting β1 blocker, esmolol, may be used intravenously at a loading dose of 0.5 mg/kg IV over 1 minute; followed by a constant rate IV infusion of 10-200 mcg/kg/minute.

An extreme method to fight the peripheral actions of excess thyroid hormones is to reduce the systemic levels present. Peritoneal dialysis, plasmapheresis, and hemodialysis have been used in human medicine as well as the use of cholestyramine, which inhibits enterohepatic circulation of thyroid hormones by binding to the GI tract. These methods are rarely utilized in human patients and probably have limited use in veterinary thyroid storm patients.

Systemic Support

The third arm of treatment for thyroid storm involves reversing the effects of thyroid hormones on the body. Fever should be treated by the judicious use of ice packs and cooled parenteral fluid administration. Volume depletion is another common systemic effect of thyroid storm and this should be treated with crystalloid fluid replacement. Colloid therapy is generally not indicated unless severe GI disease or other syndrome resulting in low oncotic pressure is present. Potassium supplementation should be added as necessary remembering that some patients with thyroid storm become acutely hypokalemic. Dextrose supplementation of 5-10% should be considered as well as B vitamin supplementation to combat potential thiamine deficiency in hyperthyroid cats.

Cardiac disturbances are common with thyroid storm in people, and it is not uncommon for cats with thyroid storm to present in cardiac failure that must be managed. β adrenergic blockade therapy as described above may be also be helpful to manage cardiac failure due to its effects in reducing the elevated heart rate caused by thyrotoxicosis. Furosemide (1-4 mg/kg IV/IM; 0.5-2 mg/kg PO q.d.), ACE inhibitors (enalapril or benazapril at 0.5-2 mg/kg PO bid), isosorbide dinitrate (0.5-2 mg/kg PO q 8-12 hr), nitroglycerine (0.5-1.5 in q 8-12 hr cutaneously), or hydralazine (0.5-1 mg/kg IV; 0.5-2 mg/kg PO q 12 hr) may be useful to manage feline heart failure, but must be used with care in patients with renal compromise. In all cases, medications should be started at the lowest levels and titrated up to effect, and blood pressure must be carefully monitored. Supraventricular arrhythmias are also common in human thyroid storm, with the most common disturbance being atrial fibrillation. In thyrotoxic feline patients atrial fibrillation can also occur. β adrenergic receptor blockade as described above is a first line defense in treating these arrhythmias. A sequela in feline patients with heart failure or atrial fibrillation is thromboembolic disease. Anticoagulation therapy should be considered to include low dose aspirin (5 mg/cat q 72 h), heparin (200-400 U/kg SQ q hr until aPTT is 1.5-2 times prolonged), and low molecular weight heparin (100 U/kg SQ q 6 hr). Hypertension is often a complication of thyroid storm in cats. Blood pressure in these cats should be checked and anti-hypertensive therapy instituted as appropriate to include: β blockade as discussed above or amlodipine (0.625-1.25 mg/cat q.d.). In acute cases of hypertension, nitroprusside may be used as a constant rate infusion at 0.5-5 mcg/kg/min

In humans with thyroid storm, a relative adrenal insufficiency can be found due to increased cortisol clearance. This prompts the use of glucocorticoid therapy. No such studies have been done in feline patients and the use of glucocorticoid therapy in patients with thyroid storm is controversial.

Eradication of the precipitating factor

In human thyroid storm, a precipitating factor is one of the criteria that defines the disease. Presence of a precipitating factor should be thoroughly investigated in cats presenting with thyroid storm. A full work up including full hematology, biochemical analysis, urinalysis, retroviral testing, blood pressure measurement, and imaging studies should be performed on these cats. Abnormal findings should be further examined by specialized testing. If another abnormality is identified, it should be treated in order to prevent recurrence of thyroid storm.

OUTCOME

Although thyroid storm is an uncommon presentation in human emergency rooms, there is a significant rate of mortality in those patients with this syndrome. Rapid recognition of the problem as well as aggressive treatment are necessary for a successful outcome. Thyroid storm is not as well a defined syndrome in feline medicine, although acute manifestations of thyrotoxicosis result in a syndrome that can be considered feline thyroid storm. Veterinary recognition of this syndrome may be lacking so it is unknown what the true incidence and mortality from thyroid storm may be in cats. However, it is certainly recognized that death may arise from untreated, acute thyrotoxicosis. As in humans, it is anticipated that early recognition and aggressive treatment of feline thyroid storm will improve veterinary patient survival.

REFERENCES

1.Tietgens ST, Leinung MC. *Med Clin North Am* 1995;79:169-84., 2.Maussier ML, D'Errico G, Putignano P, et al. *Rays* 1999;24:263-72., 3.Silva JE. Catecholamines and the sympathoadrenal system in thyrotoxicosis In: L. E. Braverman and U. R.D., eds. *Werner and Ingbar's The Thyroid: A Fundamental and Clinical Text*. Philadelphia, PA: Lippinkott, Williams and Wilkins, 2000;642-651., 4.Boelaert K, Franklyn JA. *J Endocrinol* 2005;187:1-15., 5.Burch HB, Wartofsky L. *Endocrinol Metab Clin North Am* 1993;22:263-77., 6.Kienle RD, Bruyette D, Pion PD. *Vet Clin North Am Small Anim Pract* 1994;24:495-507., 7.Elliott J, Barber PJ, Syme HM, et al. *J Small Anim Pract* 2001;42:122-9., 8.Maggio F, DeFrancesco TC, Atkins CE, et al.. *J Am Vet Med Assoc* 2000;217:695-702., 9.Smith SA, Tobias AH, Jacob KA, et al. Arterial thromboembolism in cats: acute crisis in 127 cases (1992-2001) and long-term management with low-dose aspirin in 24 cases. 17:73-83., 10.Mooney CT. Hyperthyroidism In: S. J. Ettinger and E. C. Feldman, eds. *Textbook of Veterinary Internal Medicine*. 6 ed. St. Louis: Elsevier Saunders, 2005;1544-1560., 11.Pimentel L, Hansen KN. *J Emerg Med* 2005;28:201-9. 12. Bell R, Mellor DJ, Ramsey I, et al. *Vet Clin Pathol* 2005;34:110-4., 13.Sarlis NJ, Gourgiotis L. *Rev Endocr Metab Disord* 2003;4:129-36., 14.Geffner DL, Hershman JM. *Am J Med* 1992;93:61-8.

PLEUROPNEUMONIA: APPROACH TO PREVENTION, DIAGNOSIS, AND TREATMENT IN LARGE ANIMALS

William Gilsenan, VMD, DACVIM (LAIM)
Blacksburg, VA
Rose Nolen-Walston, DVM, DACVIM (LAIM)
Kennett Square, PA

INTRODUCTION

Bacterial infection of the lower respiratory tract is a common disease that affects all large animals. While extension of infection into the pleural cavity and resultant accumulation of pleural effusion are commonly observed in horses, bacterial respiratory disease typically results in bronchopneumonia in ruminants. The development of bacterial pneumonia in any large animal seldom occurs without a predisposing factor. Pleuropneumonia in horses is unusual without a history of long-distance transport. Bronchopneumonia in cattle similarly occurs most frequently in beef cattle after long-distance transport ("shipping fever") or in intensively-managed dairy calves ("enzootic calf pneumonia"). Management of pneumonia in any large animal can be a challenging undertaking in both field and hospital settings. Factors that contribute to this challenge include the frequently protracted course of therapy, cost of therapy, and the development of antimicrobial resistance leading to treatment failure.

There are several species-specific features of the pulmonary anatomy that the large animal practitioner should consider. Most mammalian species exhibit interalveolar pores of Kohn, a microanatomical feature that allows air to travel between alveoli, preventing rupture hyperinflated alveoli to vent air into adjoining alveoli, thus protecting the lung from volutrauma. These pores of Kohn are virtually absent in cattle, however, and over-inflated alveoli burst during extreme respiratory effort leading to pulmonary emphysema, bullae, and pneumothorax that can escape the parietal pleura and cause subcutaneous gas accumulation leading to crepitus over the withers and back. Another difference seen within mammals are the presence of pulmonary intravascular macrophages (PIMS). These unique inflammatory cells are present in equids and ruminants but not in rats, humans, or dogs. PIMS confer an increased susceptibility to the effects of the Gram-negative cell wall toxin, endotoxin. Animal species without PIMs are able to tolerate high levels of both inhaled and hematogenous lipopolysaccaride without showing significant pulmonary vascular responses, inflammation or edema.

In ruminants, viral respiratory pathogens (bovine herpesvirus, bovine viral diarrhea virus, bovine respiratory syncytial virus, parainfluenza type 3 virus) probably incite colonization of the lower respiratory tract with bacteria by impairing defense mechanisms such as the mucociliary escalator and by inducing immunosuppression. These bacterial pathogens, particularly *Pasteurella multocida* and *Mannheimia haemolytica*, are normal inhabitants of the upper respiratory tract. Bovine coronavirus and *Mycoplasma bovis* are two pathogens whose role in the development of BRDC is not quite clear.[1] A recent study indicated, though, that nasopharyngeal carriage of *M. bovis* was associated with an increased odds of clinical signs, lung consolidation, and reduced average daily gain.[2] Chronic bronchopneumonia typically results in anaerobic bacterial colonization by *Trueperella pyogenes*. Because equine pleuropneumonia also results from impairment of the mucociliary escalator, normal bacterial inhabitants of the upper respiratory tract, such as *Streptococcus equi* ssp. *zooepidemicus*, are the primary agents causing disease in initial stages of pneumonia. In later stages of disease, anaerobic bacterial colonization, most commonly by *Bacteroides fragilis*, occurs.

DIAGNOSIS

Appropriate diagnosis of pneumonia in the large animal begins with a thorough physical examination. Severity and chronicity of disease both heavily influence the clinical signs that are observed upon initial examination. In all large animal species, dullness, tachypnea, dyspnea, spontaneous coughing, and mild to moderate fever are the most commonly observed physical examination findings. Absence of any of these signs, however, should not eliminate clinical suspicion of respiratory disease. Bilateral nasal discharge might be observed and will likely be mucopurulent or purulent in character. Nasal discharge that has a fetid odor is strongly suggestive of anaerobic bacterial infection; the presence of anaerobic bacteria is indicative of longstanding infection. The trachea should be palpated and squeezed in any large animal to assess for the presence of an inducible cough; this finding is occasionally the once clinical sign of pneumonia in the juvenile ruminant.

Careful auscultation of the thorax and trachea in a quiet environment is helpful in the characterization of lower respiratory disease. The thoracic fields vary substantially in size and shape between horses and ruminants. The eleventh rib marks the caudodorsal extent of the thoracic field in ruminants. Airflow at rest will be more easily auscultated in ruminants compared to horses. Referred upper airway noise is common in ruminants so thorough auscultation of the trachea should always be performed to differentiate this finding from actual bronchovesicular sounds. The limitations of thoracic auscultation of the ruminant must be considered during evaluation of a patient. Buczinski et al. determined that auscultation of the thorax has a sensitivity of 5.6% when compared to identification of consolidated lung via ultrasonographic examination.[3] The ability of thoracic auscultation to predict lung pathology is similarly poor in sheep.[4] Rebreathing examination is generally necessary to appropriately assess lung sounds in horses.

During rebreathing examination, the authors prefer to assess four categories during the procedure: the presence of airflow and any abnormal lung sounds in all thoracic fields, the ability of the animal to tolerate rebreathing examination without becoming distressed, the ability of the animal to recover following rebreathing examination, and the presence of cough at any time during rebreathing examination. When considering the first category, airflow should be present in all thoracic fields and the absence of lung sounds is always abnormal. In horses, the absence of lung sounds typically occurs in ventral lung fields due to the accumulation of pleural fluid. The point along a dorsal-ventral axis at which lung sounds can be heard is helpful in quantification of pleural fluid; this point should be described in reference to its position compared to the patient's point-of-the-shoulder. Again, the accumulation of pleural fluid is rare in pneumonic ruminants. The absence of lung sounds dorsally occurs in cases of pneumothorax. Pneumothorax is an unusual sequela to pneumonia in large animals but if anything occurs more frequently in ruminants compared to horses.

Pneumothorax in adult cattle typically occurs secondary to chronic bronchopneumonia.[5] The patient's body condition should always be considered when airflow is absent or difficult to identify. A healthy horse should be able to tolerate the presence of a rebreathing bag for several minutes without becoming distressed and their respiratory effort should return to baseline within 5-6 breaths after removal of the rebreathing bag. Ruminants will generally be less tolerant of rebreathing examination but should not experience distress when taking maximal breaths. Respiratory effort should return to baseline within 30 seconds of the end of rebreathing examination. Coughing during or after rebreathing examination is never normal. Due to its ease and cost, rebreathing examination should be performed on any large animal in which respiratory disease is suspected. Serial rebreathing examinations are quite helpful in assessing progress or lack thereof in a patient that requires long-term management.

With respect to information gathered from clinical laboratory work, white blood cell count and fibrinogen concentration are probably the most valuable indices when assessing severity of bacterial pneumonia. Because white blood cell count tends to normalize in large animals with chronic infection, it is the authors' opinion that fibrinogen concentration is more helpful than the former in monitoring a response to treatment following baseline evaluation. Normal fibrinogen concentration (<375 mg/dL in horses; <800 mg/dL in cattle) is rarely observed in cases of bacterial pneumonia; this finding should prompt the clinician to consider other causes of respiratory disease such as interstitial pneumonia. Severe bacterial pneumonia is typically accompanied by a profound hyperfibrinogenemia, particularly in horses with pleuropneumonia that have developed pulmonary abscessation. Biochemistry panel findings are variable in large animals; abnormalities typically reflect concurrent disease.

Thoracic ultrasonography of both hemithoraces can be performed rapidly and should be employed in any case of suspected pneumonia whenever equipment is available. Given the inability of ultrasonographic waves to travel through air, ultrasonography is only capable of evaluating the pleural surface and non-aerated (consolidated) lung. Ultrasonography is useful in determining the volume and location of fluid within the pleural cavity and also in determining appropriate sites for thoracocentesis and pleural drainage. The usage of thoracic ultrasonography in cattle, particularly calves, with BRDC is becoming increasingly common. Considering the poor sensitivity of thoracic auscultation, thoracic ultrasonography has great potential to assist in the earlier identification of lung consolidation in subclinically and clinically affected cattle alike.[3] Thoracic radiography is utilized less commonly in the adult large animal and is probably most useful in the identification of pulmonary abscesses in horses with pleuropneumonia. The presence of pulmonary abscesses is particularly important to determine because this finding substantially increases the duration of therapy and the cost of therapy. In cases of pleuropneumonia, performing radiography prior to pleural drainage is not helpful. Thoracic radiography is useful for assessing severity of disease in foals as radiographic appearance of pneumonia depending on its etiologic agent (*Rhodococcus equi* vs. *Streptococcus equi* ssp. *zooepidemicus* vs. mixed infection secondary to aspiration). Thoracic radiography may be helpful in assessing severity of disease in calves with BRDC but is unlikely to change initial choice of antimicrobials and overall initial treatment plan.

Lastly, thorough evaluation of a large animal with pneumonia includes the sterile collection of samples prior to initiation of therapy for bacterial culture and sensitivity. In the horse, sampling of the airways is best accomplished via transtracheal wash or endoscopically-guided endotracheal wash. In cases of pleuropneumonia, samples of pleural fluid can and should be collected. Samples should be submitted for aerobic and anaerobic bacterial culture and sensitivity; cytology using Diff-Quik and Gram staining techniques might aid in preliminary determination of etiologic agent prior to institution of empiric antimicrobial therapy. Financial constraints notwithstanding, collection of sterile samples is probably less important in ruminants with bacterial pneumonia for two primary reasons. First, antimicrobial susceptibility patterns do not differ vastly among major BRDC pathogens. Second, antimicrobial selection in food animals in the United States is limited as it is and susceptibility patterns might not help in the overall management of the animal. Regardless, collection of sterile samples should be pursued in herds in which antimicrobial resistance is suspected and in individual animals that do not respond to initial empiric therapy. In cases or herds in which involvement of *Mycoplasma bovis* is suspected, samples should be tested for *M. bovis* via polymerase chain reaction. Notably, *M. bovis* is difficult to culture and diagnosis of *M. bovis* via culture is not recommended. It is the opinion of the authors that the orientation of the ruminant trachea and carina make collection of fluid via transtracheal or endotracheal wash somewhat more difficult than performing the procedure in the horse.

TREATMENT

The primary components of therapy of bacterial pneumonia in large animals are antimicrobial therapy, pain management, and supportive care. Pleural drainage should be considered in horses with substantial pleural effusion, tachypnea, or dyspnea. Placement of an indwelling thoracic catheter facilitates continuous drainage of the pleural cavity and should be considered in horses with severe pleuropneumonia.

Consideration of the major causes of bacterial respiratory disease in large animals is necessary when formulating an empiric treatment plan. In cases of equine pleuropneumonia, it is the authors' preference to institute broad spectrum antimicrobial therapy with sufficient anaerobic bacterial coverage. In horses demonstrating clinical signs of endotoxemia or sepsis, intravenous antimicrobial therapy with potassium penicillin (22,000 IU/kg, IV, q6h) and gentamicin (6.6-8.8 mg/kg, IV, q24h) should be instituted. Importantly, penicillin's limited efficacy against *Bacteroides fragilis* demands the inclusion of metronidazole (15-20 mg/kg, PO, q6-8h; not exceeding 60 mg/kg/d). In cases in which intravenous antimicrobial therapy cannot be used, chloramphenicol (33-50 mg/kg, PO, q6-8h) or a combination of trimethoprim-sulfamethoxazole (30 mg/kg, PO, q12h) and metronidazole (15-20 mg/kg, PO, q6-8h; not exceeding 60 mg/kg/d) may be sufficient. Because complete resolution of equine pleuropneumonia and pulmonary abscessation may require months of antimicrobial therapy, the development of antimicrobial resistance is not uncommon and should be anticipated. Response to therapy can be assessed objectively by close monitoring of rectal temperature, serial blood fibrinogen concentrations and serial thoracic radiography. Worsening of disease during the treatment period warrants re-evaluation of the patient via collection of sterile samples. Adjustment of an antimicrobial therapeutic regimen may become necessary. For cases with severe fibrinous pleuritis, standing thoracotomy with rib resection and pleural lavage may be useful as a salvage procedure, and some horses may even return to athletic function.[6] Anti-fibrinolytic therapy (such as intrapleural tissue plasminogen activator) has been reported to be beneficial in case studies but no data supporting its efficacy have been published.[7]

The approach to initial antimicrobial therapy in the ruminant with pneumonia is slightly different. The two major bacterial BRDC pathogens (*P. multocida* and *M. haemolytica*) are both Gram-negative coccobacilli and antimicrobial susceptibility patterns likely vary more by strain that by species. Initial selection of an antimicrobial might therefore seem arbitrary without knowledge of susceptibility of BRDC pathogens in a herd. Many antibiotics are labeled for the treatment of BRDC in cattle. In the last 5 years, two macrolide antibiotics (tildipirosin and gamithromycin) received FDA approval for the treatment of BRDC, joining tulathromycin, also a macrolide, on the market. Tilmicosin, an older macrolide, is being used less frequently in the United States due to its risk to human health and due to the efficacy of newer macrolides. Other antimicrobials that may be used in the treatment of BRDC include florfenicol, ceftiofur, enrofloxacin, danofloxacin, and oxytetracycline. A recent meta-analysis employed mixed treatment comparison to assess efficacy of available antimicrobials used to treat cases of undifferentiated BRDC in cattle.[8] Identification of a single superior antimicrobial was unsurprisingly not possible but notably oxytetracycline and ceftiofur were routinely outperformed by other available antimicrobial choices. In cases of undifferentiated BRDC, initial antimicrobial choice should be influenced by safety, availability, cost and clinical impression of efficacy within the herd in question. Extra-label use of enrofloxacin is strictly prohibited in the United States and its use as a first-line antimicrobial cannot be justified. It is important to remember that the lack of cell wall that is characteristic of *Mycoplasma* spp., rendering β-lactam antibiotics (penicillins, cephalosporins) ineffective.

PREVENTION

Prevention of pneumonia in horses is probably best addressed through examination of risk factors. In multiple studies, long distance road transport is the primary associated factor in predisposing horses to pneumonia, with general anesthesia and post-esophageal obstruction also increasing likelihood. There is considerable consensus that "shipping fever" in horses is secondary to prolonged confinement with the head tied in an elevated position, which results in a reduction in tracheal mucociliary clearance, tracheal inflammation and bacterial colonization of the airways.[9-11] Transport-associated pneumonia is best prevented by altering transport techniques to include longer rest periods, loose boxes rather than tie stalls, and better air quality within the vehicle. Horses being shipped unrestrained have normal BAL fluid that does not contain higher numbers of bacteria, but over distances above 1000 miles, signs of lower respiratory infection were seen in more than a third of horses despite being transported in loose boxes. Pharmacologic interventions to prevent transport-associated pneumonia have focused on specific risk factors. As the main etiologic agent, *Streptococcus equi* ssp. *zooepidemicus* is almost invariably sensitive to penicillin, the prophylactic use of this antimicrobial agent was investigated using a long-distance transport model (head elevation for 48 hours), but had no significant effect on airway contamination.[12] The data on enrofloxacin as a prophylactic therapy are mixed; though one very large, well-conducted study found that its use did not reduce shipping fever,[13] another study suggested that both enrofloxacin and marbofloxacin outperformed a saline control.[14] However, based on the prevalence of Gram-positive organisms as etiologic agents, the use of enrofloxacin (which is ineffective against most streptococcal species) should probably be avoided. Clenbuterol, an oral β-2 agonist that has been shown to increase tracheal mucociliary clearance in the horse, was also evaluated using a 48 hour head elevation protocol which found a trend towards faster tracheal mucociliary clearance in horses receiving clenbuterol, and also a small reduction (0% versus 33%) in tracheal β-hemolytic *Streptococcus* spp. numbers after treatment with clenbuterol versus placebo.[15]

Approach to prevention of enzootic calf pneumonia in dairy calves and of shipping fever in beef cattle varies because of the differences in risk factors and management that precipitate disease in each setting. Vaccination against the major viral respiratory pathogens and management of the environment are the primary components of prevention of BRDC in all cattle. In a recent meta-analysis, vaccination against these pathogens was demonstrated to decrease BRDC morbidity risk in naturally and experimentally exposed beef cattle.[16] Bacterins against *P. multocida, M. haemolytica,* and *H. somni* are available but have a poorer efficacy when compared to vaccines against viral agents.[17] Due to the risks associated with transport and processing of beef cattle, timing of these events with respect to timing of vaccination complicates the development of programs for the prevention of shipping fever. Preconditioning has been employed as one strategy that may minimize coinciding stressors and, in turn, decrease rates of BRDC.[18] Metaphylaxis, or control, of BRDC, involves mass medication of livestock and is a common practice, being used in nearly 60% of American feedlots.[19] Despite the variety of strategies that can be employed when timing vaccination and processing in beef cattle, effectively preventing shipping fever has proven to be a challenge. This challenge has manifested as an actual increase in mortality rates of finishing cattle and as an increase in overall incidence of BRDC over the last 10-15 years.[18,20] Concerns regarding the cost of antimicrobial drug use and the apparent increase in treatment failure are justified. More recent research efforts have been directed at utilizing genetic selection for cattle that are less susceptible to BRDC. Evidence thus far indicates a sufficient level of heritability to allow for exploitable genetic variation.[21] Common genomic regions associated with BRDC susceptibility have been identified.[22] Future challenges include enhancing our understanding of these regions and determining how to implement a breeding selection program that will use this information to impact overall BRDC resistance.

REFERENCES

[1]Griffin D. **Vet Clin Food Anim.** 2010 26(2):381; [2]Francoz D et al. **J Vet Intern Med.** 2015 29(1):381; [3]Buczinski S et al. **J Vet Intern Med.** 2004 28(1):234; [4]Scott P et al. **Vet J.** 2010 186(1);53; [5]Slack JA et al. **J Am Vet Med Assoc.** 2004 225(5):732; [6]Hilton H, et al. **Vet Surg.** 2010 Oct;39(7):847; [7]Rendle DI, et al. **Aust Vet J.** 2012;90(9):358; [8]O'Connor AM et al. **Prev Vet Med.** 2013 110(2):77; [9]Racklyeft DJ et al. **Aust Vet J** 2000 78(8):549; [10]Racklyeft DJ et al. **Aust Vet J.** 1990 Nov;67(11):402; [11]Stull CL et al. **Equine Vet J** 2002 34(6):550; [12]Raidal SL et al. **Aust Vet J.** 1997 Feb;75(2):126; [13]Davis E. et al, **Vet Ther.** 2006;7(3):232; [14]Endo Y, et al. **J Vet Med Sci.** 2014 [epub]; [15]Norton JL, et al. **J Vet Intern Med.** 2013;27(6):1523; [16]Theurer ME et al. **J Am Vet Med Assoc.** 2015 246(1):126; [17]Griffin D et al. **Vet Clin Food Anim.** 2010 26(1):57; [18]Hilton WM. **Anim Health Res Rev.** 2014 15(2):120; [19]Neibergs HL et al. **Beef Improv Fed Ann Meeting**, Lincoln, NE. 2014; [20]Engler M et al. **Anim Health Res Rev.** 2014 15(2):126; [21]Berry DP. **Anim Health Res Rev.** 2014 15(2):151; [22]Niebergs HL et al. **BMC Genomics.** 2014 15:1164;

COMPARATIVE ASPECTS OF INFECTIOUS PULMONARY DISEASE

Daniela Bedenice, DMV, DACVIM (LAIM), DACVECC and Elizabeth Rozanski, DVM, DACVIM (SAIM), DACVECC
North Grafton, MA

This hour will focus on comparing and contrasting the clinical signs, diagnosis, etiology, treatment and outcome of infectious pulmonary diseases. The diseases covered will include transport and shelter associated respiratory disease in dogs and cats, and Aspergillus infection.

Transport and rescue associated respiratory disease

Animals are commonly moved around the country for purposes of rehoming, or for use in agricultural pursuits. Shipping is associated with physiological stresses, as well as exposure to a variety of infectious organisms. Complications associated with respiratory illness associated with transport can be mild and self-limited or fatal. Pleuropneumonia, which is predominantly a large animal disease was covered in the previous hour. Dogs and cats are considered to rarely develop pleuropleumonia, although some cases of pyothorax may represent extension of pulmonary infection.

Cats

Due to the large number of homeless cats throughout the country, it is far less common for cats and kittens to be transported. However, cat rescues are commonplace, and viral respiratory disease, particularly rhinotracheitis (herpes) is a frequent complication. Many cats are sub-clinically infected with herpes virus, and the virus recrudesces when they are stressed. Viral infections are a particular concern in kittens for several reasons, including the potential for secondary bacterial pneumonia and even death, and from rehoming perspective, kittens may lose an eye due to severe ocular damage or may require so long to recover from their infection that they are no longer "cute" and subsequently adoptable. Techniques for prevention of respiratory infection in cats include strict isolation of affected cats and vaccination on entry to shelter environments. Recent research studies have evaluated different approaches for control of signs. Ballin et al evaluated the use of recombinant interferon omega in naturally occurring viral respiratory infection, and found no benefit in clinical signs, but perhaps a decrease in calici viral load. Commercially available hyperimmune serum was found to result in a more rapid improvement in clinical signs, although all kittens were improved within a week. (Friedl et al). Litster and colleagues evaluated the impact of a single dose of famciclovir at admit on clinical signs and viral shedding and found no advantage.

Dogs

Respiratory infection in dogs being transported to new homes is very common. Bacterial infection, most commonly Bordetella bronchiseptica and Mycoplasma is readily transmitted from dog to dog, particularly in enclosed environments associated with transport. Vaccinations may be effective in reducing the incidence or severity of disease, but is not effective in all cases. Canine influenza has emerged as a new infection and been implicated in some canine respiratory disease in the United States. Other bacteria such as *E. coli, Klebsiella* and *Pseudomonas* may also colonize airways after an initial insult with a viral disease. Occasionally, it may be difficult to distinguish kennel cough complex from canine distemper virus. Infectious tracheobronchitis in adult dogs, with the potential exception of the emerging canine influenza, tends to be mild and more commonly an irritating cough rather than critical illness. Specific breeds of puppies, including bulldogs and toy breeds (eg. Yorkies) appear to be at an increased risk of infection and even death.

Clinical signs include a cough, lethargy, and difficulty breathing. Puppy pneumonia is a classical example of "community-acquired" pneumonia. This suggest aerosol transmission and predicts a high likelihood of effective first-line antibiotics. Physical examination is occasionally unremarkable, although many puppies have mild fevers and seem depressed. Some puppies will have other congenital defects such as hernias, undescended testicles or congenital heart disease. Usually, tracheal palpation will trigger a paroxysm of coughing. Routine laboratory testing is usually normal for a pediatric patient although an inflammatory leukogram may be present and characteristic serum chemistry profile results of puppies are observed. Puppies, particularly toy breeds, may be hypoglycemic if they have been anorexic and often appear to be outwardly ill. The crucial decision at this step is if the puppy is sick enough to hospitalize. This decision will reflect the degree of illness of the puppy, as well as practical constraints, such as a hospital staffing, isolation facilities and owner finances. Owners should be advised as to the unpredictable nature of pet stores in helping either finance hospitalization or return the puppy's purchase price. Most (all?) clients grow very fond of individual puppies very rapidly, and are often reluctant to return them to sub-optimal environments. However, the hospitalization costs for such puppies can be significant, particularly if oxygen or intensive nursing care is required. The primary therapy focuses on supportive care, including fluids (subcutaneously or intravenously), antibiotics, and monitoring appetite and blood glucose levels. Nebulization and gentle coupage may be very helpful in clearing infection. Antibiotic choices are usually empiric, even if a tracheal wash is performed, it is several days until bacterial culture and sensitivity data are available. Antimicrobials with good activity against *Bordetella* include aminoglycosides, fluoroquinolones, chloramphenicol, trimethoprim-sulfa and tetracyclines. Systemically ill puppies should receive parenteral antibiotics. The overall prognosis for puppy pneumonia is typically excellent, although some rare dogs have persistent cough and certainly the *Bordetella* organisms may be remarkably persistent!

In shelter environments, significant research has surrounded methods of treating and preventing respiratory infections; however there is limited evidence for any form of prevention short of isolation of affected dogs, and preferential therapy with doxycycline for affected dogs. It is also important to recall the role of heartworm infection in respiratory signs in dogs, particularly in non-endemic areas.

Pyothorax/Pleuropneumonia

As outlined in last hour's lecture, infected pleural effusion as an extension of pneumonia is common in large animals. In dogs and cats, infected pleural effusion is most often an infection primarily of the pleural space, although there may be extension from the lungs in the form of ruptured pulmonary abscesses.diagnosis is made by cytological identification of bacterial organisms in pleural effusion, and ideally by positive bacterial culture and sensitivity results. Cytological examination should also document suppurative inflammation with toxic change and intracellular bacteria. In dogs and cats, most commonly pyothorax is a mixed population of bacteria, and the fluid is clearly purulent in appearance. In people, as empyema is often less apparently infected, and complementary fluid analysis, including evaluation such as glucose, LDH or amylase are occasionally employed. Amylase is used to look for salivary leakage (eg. esophageal perforation); dogs and cats lack salivary amylase so this is not beneficial in small animals.

Treatment for pyothorax may be medical or surgical (thoracotomy). Medical therapy includes placing thoracostomy tubes and draining the effusion, +/- lavage; while surgical therapy includes debriding infected tissues. Cats, and less commonly dogs are occasionally treated as outpatients, with varying degrees of success. There is some evidence to suggest that thoracotomy is indicated in dogs, particularly where Actinomyces is a concern (Rooney); however, in cats there is little evidence for either form of treatment (in-patient medical versus surgery) is preferred. In all cases, broad spectrum antibiotics, including coverage for anaerobic organisms is warranted.

Aspergillus infection (Aspergillosis)

Dogs

Aspergillus is a world-wide fungi that has hundreds of species. Most aspergillus are non-pathogenic, however, some species may cause infection either in the immunocompent or immunocompromised. Aspergillus infection (aspergillosis) in dogs is most often localized to the nasal cavity and sinuses. Diagnosis is based upon identification of fungal hyphae as plaques, and typically destruction of the nasal turbinates. Histopathology is useful as well. Treatment with anti-fungal agents is warranted, with a fair to good prognosis. Interestingly, canine sinonasal aspergillus is not typically associated with immunosuppression. Fungal culture for species and ideally sensitivity is recommended to guide treatment. Disseminated aspergillosis or fungal diskospondylitis has also been reported, particularly in German Shepherds. Aspergillus infection localized to the pulmonary parenchyma has also been sporadically described in dogs, particularly shepherds. Specifically, cavitary pulmonary aspergillosis has been previously documented in a German Shepherd Dog and was associated with a solitary lesion within the left cranial lung lobe that was associated with cough and hemoptysis. Surgical resection was palliative, but the disease ultimately recurred, and the dog was euthanized (Kulendra 2010). Three other cases with pulmonary aspergillus have been reported, with good outcomes following surgical removal of affected tissues (Whitley 2010,Adamama-Moraitou 2011). Finally, in 2012, Trempala and Herold described a mixed-breed dog with spontaneous pneumothorax secondary to aspergillus, which resolved following surgical resection of a lesion. Thus, while aspergillus in dogs is primarily sinonasal, it is possible to have systemic disease, or localized pulmonary disease. In disseminated disease, urinary hyphae may be visualized but this is not true in isolated pulmonary disease. There is also a hypersensitivity disorder, termed allergic bronchopulmonary aspergillosis (ABPA), which is described in people, primarily those with asthma and cystic fibrosis and has been alluded to in dogs, but not reported yet.

Cats

Cats have not commonly been thought have aspergillus, but recent evidence has supported that they may develop infection associated with *Apergillus felis,* specifically a sino-orbital infection. Cats may also develop sinonasal aspergillosis, which is most often associated with Aspergillus fumigatus and Aspergillus niger infection. Molecular methods are useful to correctly identify the fungi. Prognosis is dependent on the type of infection, with nasal infections better than sino-orbital infection.

Horses

While guttural pouch mycosis is well-recognized in horses, there is also a recent report by Headley and colleagues from southern Brazil describing two horses with widespread "metastatic" aspergillosis, suspected to originate from pulmonary bronchopneumonia. Additionally, almost 30 years ago, Slocumbe and colleagues described pulmonary aspergillosis in 19 horses associated with acute enteritis (primarily Salmonella). In general, pulmonary aspergillosis is considered a rare disease in horses as well, with more common evidence of immunocompromise. Aspergillus and other molds are intermittently implicated in inflammatory airway disease, and contamination of hay with molds is considered a trigger in some horses.

Alpaca

Cavitary pulmonary aspergillosis was recently described in an adult alpaca with type II diabetes, showing large, rounded, cavitated lesions on thoracic radiographs. Transtracheal aspirate and intralesional, ultrasound-guided fluid aspiration both identified neutrophilic inflammation with evidence of cellular necrosis and fungal hyphae. The numbers and location of these hyphae in abnormal cystic structures within the pulmonary parenchyma was consistent with a primary fungal pathogen. Subsequent culture confirmed Aspergillus fumigatus infection. The alpaca failed to respond to both systemic (Voriconazole, 250 mg IV once daily) and intralesional (Amphotericin B) antifungal therapy long term. Post mortem examination confirmed marked, multifocal and locally extensive, suppurative, interstitial pneumonia with extensive cavitary pulmonary aspergillosis or chronic necrotizing aspergillosis (CNA). Chronic necrotizing aspergillosis (CNA) is an indolent, cavitary, and infectious process of the lung parenchyma secondary to local invasion by Aspergillus species, which is usually slowly progressive over weeks to months. This syndrome is rare in both humans and animals and the available literature is based on individual case reports and small case series. In humans, CNA most commonly affects middle-aged and elderly patients with altered local defenses, secondary to diabetes mellitus, alcoholism, chronic liver disease, low-dose corticosteroid therapy, malnutrition, and connective tissue diseases such as rheumatoid arthritis. Chronic necrotizing aspergillosis is generally associated with underlying chronic lung diseases such as COPD, previous pulmonary tuberculosis, thoracic surgery, radiation therapy, pneumoconiosis, cystic fibrosis, lung infarction, or (less commonly) sarcoidosis (Zmeili 2007). Risk factors in the affected alpaca may have included age, chronic type II diabetes and underlying pulmonary parenchymal disease.

Other animals

Aspergillosis is widely described a variety of other species, include rhea, chickens, and turkeys. Other mammalian species, such as sheep and goats are also occasionally affected. Aspergillosis should be considered a potential primary pathogen or complicating pathogen in a variety of diseases. Of specific interest in the pulmonary world is primary pulmonary aspergillosis and complications of allergic airway disease associated with secondary infection with aspergillus.

REFERENCES

"Shipping fever" Ballin et al Vet J. 2014 Dec;202(3):466-70, Friedl et al Vet J. 2014 Sep;201(3):316-21, Litster et al Vet J 2014 Nov 25.(14)00458-4.*Maxwell E Vet Parasitol2014206(1-2)71-7, Schulz BS et al, Vet J 2014201(3)356-359*

Pyothorax *Rooney MB et al* Am Vet Med Assoc. 2002 Jul 1;221(1):86-92.

Aspergillus Hartmann K J Feline Med and Surgery 2013 15(7) 605-610, Barrs VR Vet Clin North Am Small Anim Pract 2014, 44(1) 51-73. Cohn LA Vet Clin North Am Small Anim Pract 2014(44) 75-89. Trempala and Herold, JVECC 2013 23(6) 624-630.

Adamama-Moraitou KK et al. JAAHA 2011 47(2) e13-18. Whitely N et al JSAP 2010 51(10) 561, Kulendra E et al. JSAP 2010 51(5) 271-274.Slocumbe RF and Slauson DO. Vet Patho 1988 25(4) 277-281. Headley et al. Mycopathologia 2014 177 129-135.

COMPARATIVE ASPECTS OF NON-INFECTIOUS PULMONARY DISEASE

Daniela Bedenice, DMV, DACVIM (LAIM), DACVECC, Rose Nolen-Walston DVM, DACVIM (LAIM) and Elizabeth Rozanski, DVM, DACVIM (SAIM), DACVECC
North Grafton, MA

This hour will focus on comparing and contrasting the clinical signs, diagnosis, etiology, treatment and outcome of non-infectious pulmonary diseases. The diseases covered will include pulmonary fibrosis (PF) and Inflammatory Airway Disease/Feline asthma.

Pulmonary fibrosis

Pulmonary fibrosis results in a restrictive pathology, with a short and shallow respiratory pattern, with exercise intolerance and ultimately death from hypoxemia if the disease is allowed to run its natural course. The etiology of PF in people is unknown; genetic factors are responsible for some cases, while smoking may result in the development of other cases. Pulmonary fibrosis has been considered a disease of aberrant healing, with epithelial damage from any insult progressing to a fibrotic change. Pulmonary fibrosis is considered especially deadly because by the time that clinical signs are apparent, a large amount of lung tissue has been replaced by scar tissue, and is not amenable to therapies. Survival time in people is typically less than 3 years from diagnosis. In people, lung transplant is the only effective form of therapy, but is accompanied by a high long-term complication rate and median survival at 4.5 years (Kistler).

Dogs

Pulmonary fibrosis is a chronic progressive syndrome associated with a grave prognosis in all species. In people, PF is a subset of interstitial lung disease (ILD), with more specific characterization based upon computed CT scanning and lung biopsy. While multiple forms of interstitial lung disease likely exist in dogs, idiopathic pulmonary fibrosis (IPF) is considered the most common based upon characteristic clinical signs and imaging findings, with histopathology being less commonly pursued. This feature may be important, as earlier reports of IPF in people were occasionally sprinkled with individuals responding well to anti-inflammatory therapy, presumably due to a more inflammatory condition. It is well-recognized in people with true IPF that prednisone does not provide a survival benefit. In 2014, two drugs were introduced and approved that resulted in improved survival times and stabilization of lung function, but not clear gains in lung function. These drugs includes pirfenidone and nintedanib (Staitieh et al). No formal evaluation of these drugs have occurred in dogs with naturally-developed PF, although pharmacokinetic data exists from the drug development phase.

Diagnosis of ILD is typically suspected based upon appropriate clinical signs (shortness of breath, and tachypnea), and radiographic evidence of interstitial infiltrates. Confirmation of a disease in dogs requires biopsy at this point in time although clinical suspicion is reasonable given the potential morbidity associated with surgical biopsy. Computed tomography is being used with increasing frequency in dogs, and given the rapidly of current-day scanners, a thoracic CT can be accomplished with minimal morbidity. Computed tomography scans are useful for excluding other diseases processes (e.g. bronchitis) and in evaluating the severity of the infiltrates/fibrotic changes. Classic IPF abnormalities include honeycombing, ground glass densities, subpleural foci/reticulation, and traction bronchiectasis.

The cause of PF remained unknown in dogs. There has been no confirmation of viral causes. A variety of causes have been associated with different forms of interstitial lung disease in dogs, including cases associated with exposure to waterproofing spray, or to infectious disease (distemper virus and Leishmania). In West Highland White Terriers, there is a familiar nature to PF although PF is also seen in other breeds of dogs, most frequently other terriers, Chihuahuas, and poodles.

Cats

Pulmonary fibrosis in cats is considered far less common that in dogs. There is no recognized breed predilection; however, most pet cats are of unknown parentage, with unknown sibling status. Clinical signs in cats include tachypnea and respiratory distress without evidence of another underlying cause such as congestive heart failure. One report exists of a cat who developed PF associated with treatment with CCNU. PF in cats appears to more commonly be associated with pulmonary neoplastic lesions. From a clinical perspective, it is wise to recall that an isolated lung mass should not result in tachypnea, and that surgical intervention for a lung mass should typically be reserved for cats that otherwise feel systemically well. Due to the rarity of the disease, there have been limited studies in cats, and no formal evaluation of therapies.

Horses

Since 2008, equine gammaherpesvirus EHV-5 has been associated with the nodular form of interstitial pneumonia, termed equine multinodular pulmonary fibrosis (EMPF). Although a broad variety of etiologies may induce interstitial pneumonia, the lung has a limited range of morphologic responses to injury. The initial insult generally results in characteristic injury to the alveolar wall (alveolar epithelium, capillary endothelium and basement membrane) resulting in acute alveolitis. During the exudative phase, proteinaceous fluid leaks into the alveolar space, and inflammatory cells accumulate within the alveolus and alveolar interstitium, resulting in a loss of functional alveolar-capillary gas exchange units (Wong 2008).This event is followed by a proliferative phase (subacute to chronic phase) characterized by hyperplasia of type II pneumocytes, accumulation of fibroblasts and inflammatory cells, and changes in the extracellular matrix of the lung parenchyma. Ultimately, the response to the chronic injury will progress to pulmonary fibrosis in the terminal phase (Wong 2008). Two distinct gross patterns of lesions are present at necropsy, with the more common pattern presenting as multifocal coalescing nodules of fibrosis within the lung parenchyma. Grossly, nodules are firm and vary in size from < 1 cm to 5 cm and are pale tan to white. A less common pattern involves multiple discrete nodules, varying in size up to 8 to 10 cm, separated by larger areas of grossly unaffected lung (Wong 2008).

The exact temporal sequence and mechanism by which gamma herpes virus induces pulmonary fibrosis in horses still remains controversial, and has been more extensively studied in murine models. Fibrogenesis is likely initiated by chronic viral injury to the alveolar epithelial cells, leading to production of the pro-fibrotic cytokine TGF beta. TGF beta leads to increased production of the chemokine CCL-2 and subsequent activation and proliferation of fibroblasts with progressive collagen deposition. Furthermore, latent gamma herpes virus (HV) infection may cause the respiratory system to become susceptible to opportunistic infectious agents, due to an altered immune response. Williams et al noted that gamma herpes viruses have been previously shown to encode viral homologues of cytokine and chemokine receptors, which may alter the host immune response if expressed. In fact EHV-5 possess a gene which encodes for a homologue of IL-10, which is thought to suppress the host immune response by way of down-regulation of MHC Class II expression. Fibrosis may also be initiated due to recurrent inflammatory responses related to repeated viral reactivation from its latent state in B cells and alveolar epithelial cells. Latently infected alveolar epithelial cells are known to produce higher levels of TGF beta in vitro, which can initiate a pro-fibrotic state.

Several case reports and case series have described EMPF and its association with viral infection to date. More specifically, a larger case study by Williams et al (2007) describes the pathologic and virologic findings from 24 horses with progressive nodular fibrotic lung disease associated with EHV-5 infection. Gross lesions consisted of multiple nodules of fibrosis throughout the lungs. Histologically, there was marked interstitial fibrosis, often with preservation of an "alveolar like" architecture, lined by cuboidal epithelial cells. The airways contained primarily neutrophils and macrophages. Rare macrophages contained large eosinophilic intranuclear viral inclusion bodies; similar inclusion bodies were also found cytologically. The inclusions were identified as herpesviral-like particles by transmission electron microscopy in a single horse. In situ hybridization was used to detect EHV-5 nucleic acids within occasional macrophage nuclei. With polymerase chain reaction (PCR), the herpesviral DNA polymerase gene was detected in 19/24 (79.2%) of affected horses and 2/23 (8.7%) of the control horses. Virus genera-specific PCR was used to detect EHV-5 in all of the affected horses and none of the control horses. EHV-2 was detected in 8/24 (33.3%) of affected horses and 1/9 (11.1%) of the control horses. The authors proposed that EMPF develops in association with infection by the equine gamma-herpesvirus EHV-5 (Williams 2007). Recently (2013) Williams et al experimentally reproduced EMPF in EHV-5 inoculated horses. The authors concluded that gamma-herpesvirus EHV-5 can induce lung fibrosis in the horse, and hypothesize that induction of fibrosis occurs while the virus is latent within the lung. The latter study is the first example of a gamma-herpesvirus inducing lung fibrosis in the natural host (Williams 2013).

There is limited information on the prevalence of EHV-5 infection within the lower respiratory tract of horses. However, a recent study by Bell et al (2006) detected EHV-5 in 64% of nasal samples of young racehorses, and Wang et al detected EHV-5 in 48% of nasal swabs from 5- to 9-month-old Thoroughbred foals. Williams et al reported an age range of 4-28 years and a mean age of 14.5 years in horses with EMPT. These data may therefore suggest a prolonged lag period between exposure to the virus and development of EMPF in affected animals.

From the clinical perspective, the manifestation of EMPF can be diverse during the early phase of disease, and may mimic any number of lung conditions (bacterial pneumonia, fungal pneumonia, IAD). Affected horses will ultimately show evidence of fever with significant hyperfibrinogenemia on hematologic screening. Classis radiographic changes may range from a diffuse miliary lung pattern to the presence of focal or diffuse coalescing nodules. Similarly, thoracic ultrasound of the lung may reveal multifocal hypoechoic nodular to diffuse densities. Trans-tracheal aspirate findings are often diverse with a definitive diagnosis being established on lung biopsy and associated quantitative EHV-5 PCR analysis (TTA, BAL, biopsy).

Combined anti-viral and systemic corticosteroid treatment is most commonly recommended. However, cost may limit the long-term use of oral valacyclovir, famciclovir or valganciclovir, even though pharmacokinetic analyses suggest adequate systemic bioavailability in horses. In contrast, oral Acyclovir is poorly bioavailable following oral administration in horses.

Airway disease

Asthma affects an increasing number of people and is characterized by bronchoconstriction, airway smooth muscle hypertrophy and excessive mucus production. While uncommonly fatal, the disease is associated with a significant impact on quality of life, and work and school attendance. Treatment of people with asthma includes avoidance of aeroallergens, inhaled and oral glucocorticoids, and bronchodilators.

Cats

While airway disease is common in dogs, only cats have true asthma, or reactive airway disease. Some cats have more chronic bronchitis syndrome, characterized by a cough, wheeze and excessive mucus production, while other cats have a more classical bronchoconstrictive disease, with acute onset of respiratory distress and severe expiratory flow limitation. From a physiological standpoint, normal lung function is dependent upon adequate lung mechanics and gas exchange. Lung mechanics are defined as the mechanical properties of the pulmonary system and are comprised of ratios of airflow rate, airway pressures and lung volumes. Compliance represents the distensibility of the lungs, and primarily reflects the lung parenchyma; low compliance is associated with clinical entities such as pulmonary contusions and edema. Lung compliance is not typically affected by lower airway disease in cats. Lung resistance primarily represents ease of airflow; most of the airway resistance (due to branching airway generations) is in the larger airways. Lower airway disease should increase airway resistance, induce bronchoconstriction, increased airway mucus, and/or airway smooth muscle hypertrophy. Gas exchange may be impaired by flow limitation and would be expected to result in hypoxemia; however, arterial blood gases are rarely measured in cats with respiratory distress.

The underlying cause surrounding the development of lower airway disease remains undetermined in most cats. Viral respiratory infections are a common cause of asthma in genetically susceptible infants and children. While viral respiratory infections are common in kittens, the role that respiratory infections may have in the development of later lower airway disease in cats is unclear. Lower airway disease is suspected to represent allergy to some inhaled allergen in most cases, but how or why a specific cat becomes sensitized remains unclear. Some cats appear to worsen seasonally, while others appear to have year round clinical signs. Certain cats appear to react to more dusty kitty litters, or the presence of indoor pollution, including environmental tobacco smoke or potpourri scent. (Figure 1) Some cats with lower airway disease also appear to have chronic rhinitis, although this syndrome has not been formally evaluated as it has been in people. Allergy testing is rarely performed in cats with suspected asthma, but one pilot study in cats showed a high rate of positive responses to skin testing. Serum immunoglobulin E assays have variable reliability in cats, and care should be taken in interpreting the results. Experimentally, cats may be sensitized to various antigens, such as Bermuda grass or *Ascaris suum* and a clinical syndrome similar to lower airway disease may be stimulated with re-exposure. While experimental models are useful to elucidate some mechanistic aspects of airway disease, validation of any results in naturally occurring cases living in a home environment is essential before making therapeutic conclusions. In contrast to horses, cats tolerate long-term therapy with prednisolone quite well; however inhaled glucocorticoids may be substituted effectively. Additional novel therapies that are being evaluated include maropitant, tyrosine-kinase inhibitors, stem cells and allergen-specific immunotherapy.

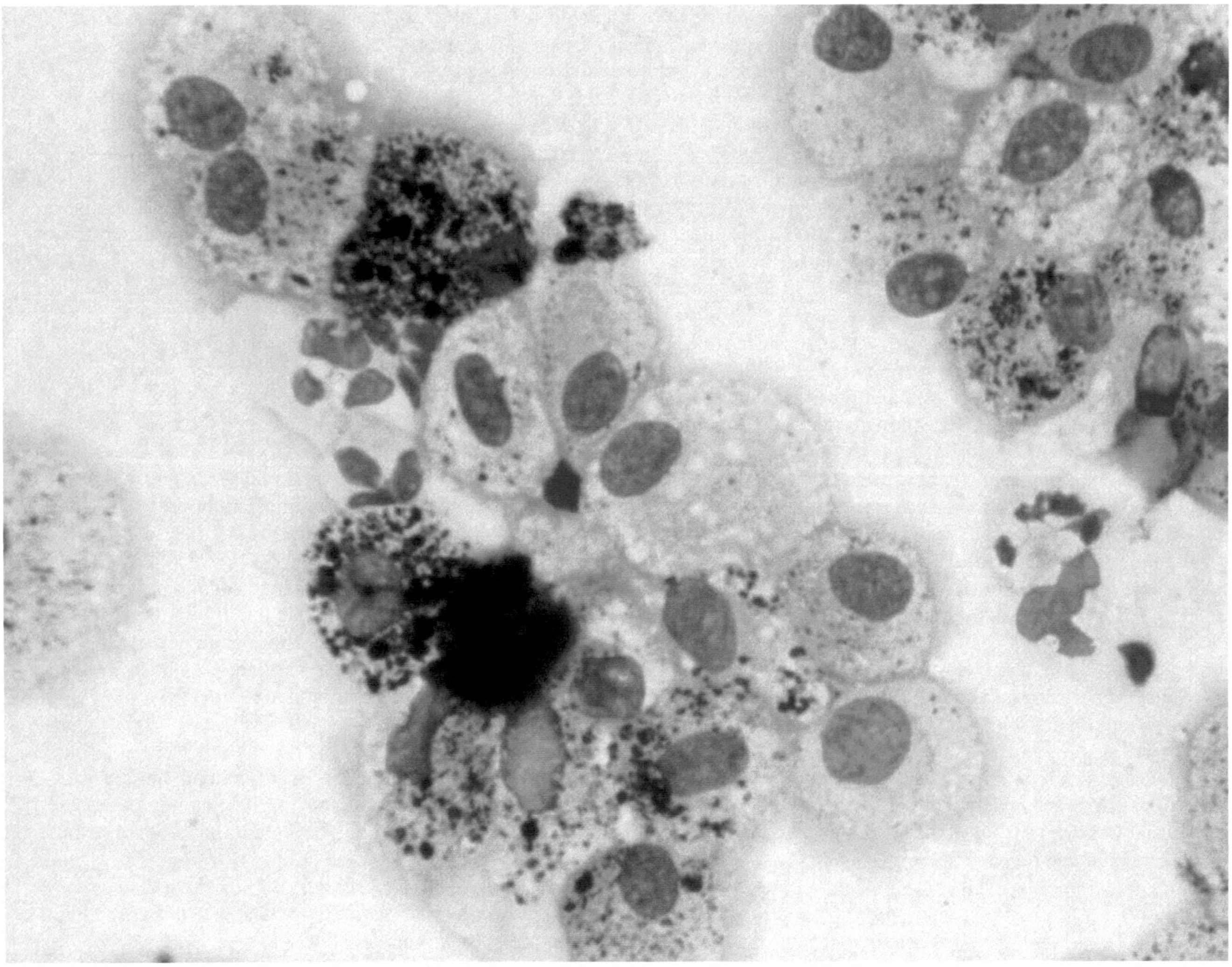

Figure 1. Bronchoalveolar lavage demonstrating carbon particles from a cat from a smoking home.

Horses

There are two non-infectious inflammatory small airway diseases in horses: Recurrent Airway Obstruction (RAO; "heaves") and Inflammatory Airway Disease (IAD). The more asthma-like of the two diseases, RAO, is characterized by reversible episodic dyspnea that resembles asthma attacks in humans and cats. IAD is milder and horses show no clinical signs at rest, but may have cough, nasal discharge, tracheal mucus, and exercise limitations characterized by exercising hypoxemia (Couetil). Summer pasture–associated obstructive pulmonary disease has the same signs as RAO but is seen in horses in the Southeastern US that are turned out at pasture in the warmer months,(Costa) whereas most standard RAO horses undergo exacerbations that are brought on by exposure to barn dust or hay dust. Both diseases involve airway obstruction, airway inflammation, and hyperreactivity of the smooth muscle of the small airways to non-specific stimuli.

In equids, inflammatory processes of the small airways are characterized by an increased percentage of lymphocytes, monocytes, neutrophils, mast cells, eosinophils or both in BALF; RAO is characterized by airway neutrophilia in BALF usually of >25% neutrophils, whereas IAD, which has a similar presumed etiopathology but does not exhibit episodic dyspnea at rest, although the specific criteria for the diagnosis of IAD by use of BALF focus on mild neutrophilia, eosinophilia, mastocytosis, or both. (Couetil) Horses with neutrophilic BALF are more likely to be older, have a cough, and have abnormally low exercising arterial oxygen tension, whereas those with pulmonary eosinophilia or mastocytosis are usually racehorses, younger, and are examined for poor performance rather than cough, and show increased baseline respiratory reactance and resistance, as well as hyperreactivity to histamine bronchoprovocation.(Hare, Hoffman) Asthma, the most common non-infectious inflammatory airway disease of humans, has traditionally been viewed as an allergic response to inhaled triggers such as pollens or other aeroallergens, resulting in a predominantly eosinophilic inflammation. Recently, several asthma subtypes are increasingly being associated with neutrophilic rather than eosinophilic inflammation of the airways (Pierce). These include occupational asthma, which occurs in workers exposed to inhaled particulates ranging from industrial dust and chemicals to urban air pollution, especially ozone and diesel fumes. Additionally, recent data suggest that up to half of all cases of asthma in human patients are characterized as having neutrophils as the predominant inflammatory cell type. This "occupational asthma" pathophysiology may have considerable relevance to equine medicine considering the high loads of dust and endotoxin experienced in the breathing zone of stabled horses. In horses as well as humans, there is still no clear consensus whether lower airway inflammation results mainly from aeroallergens versus a hypersensitivity to inhaled toxins, or whether in fact there are two distinct disease processes with different causes, treatments, and outcomes. Genetic and viral factor may also contribute to the pathogenesis of IAD and RAO.

Basic pathophysiology would suggest that differing inflammatory cell profiles may well be the result of different etiologies. The classic allergic, or atopic, response of type I hypersensitivity involves a T-helper cell 2 response characterized by IgE-mediated expression of IL-4, IL-5 and IL-13 that are responsible for eosinophil and mast cell recruitment, whereas neutrophilic infiltrates may be consistent with a non-allergic T-helper cell 1 response to airborne particles, viral disease, or other inhaled stimuli, or simply a non-specific response to IL-8.(Lavoie,Hughes) Patients with IAD have been shown to have significant increases in pro-inflammatory cytokines as well as differences in gene expression associated with airway neutrophilia versus mastocytosis. In horses, no association between season and cytologic profile of BAL fluid, nor major effects of IAD subtype on pulmonary gas exchange at exercise was seen in this population of horses (Nolen-Walston).

Therapy

Avoidance of inhaled triggers is usually successful in managing the disease when it is logistically possible. However, this means that horses that are triggered by hay and barn dust must be kept outside at pasture 100% of the time, which is rarely acceptable either logistically or emotionally for many owners and especially racehorses and high-level sporthorses. Other environmental modifications include using wood shavings instead of straw for bedding, feeding soaked hay or dust-free hay preparations, and maintaining a low-dust barn, especially focusing on removing the horse from the stall during cleaning and for the two hours afterwards. For the majority of cases, though, pharmacologic therapy is essential. As both IAD and RAO are inflammatory diseases, glucocorticoids form the mainstay of therapy. In most cases of RAO and many of IAD, remission is best achieved using systemic corticosteroids such as dexamethasone or prednisolone, and remission is often able to be maintained with inhaled medications such as fluticasone or beclomethasone. Many horses can also benefit from addition of a bronchodilator, especially during exacerbations of RAO or during early therapy. Clenbuterol is the only FDA-approved medication for horses with reversible bronchospasm. A β2-adrenoceptor agonist, clenbuterol is administered as an oral preparation with a wide dosage range (0.8 to 3.2 μg/kg, q 12 h for up to 30 days). Clenbuterol has a dose-dependent effect, but it is not effective in all horses. At the commonly administered dose of 0.8 μg/kg, only 25% of horses with RAO have a clinical response, which increases to 75% at an administered dose of 3.2 μg/kg.(Erichsen) However, in asthmatic humans as well as in other mammals with experimentally induced bronchoconstriction, chronic treatment with β2-receptor agonists results in tolerance or tachyphylaxis (desensitization) to the effects of the drug. This can be a positive attribute in that adverse effects (e.g. sweating, tachycardia, and agitation) attributable to administration of β2-receptor agonists decrease over the first few weeks of use; however, it also is associated with a decrease in efficacy of the bronchodilatory effects of the treatment and a worsening of asthma control (Cockcroft). In horses, tachyphylaxis to clenbuterol occurs within 14 days of twice daily administration resulting in a clinically relevant reduction in its bronchodilatory effects. This may be ameliorated by co-administration with corticosteroids as it is in humans, but this has not been investigated in the horses. As a result, courses longer than 14 days are not recommended for horses – animals with heaves and IAD should not use clenbuterol as a daily maintenance drug (Read).

REFERENCES

Kistler KD, Nalysnyk L, Rotella P, Esser D. BMC Pulm Med. 2014 Aug 16;14:139. doi: 10.1186/1471-2466-14-139. Walsh SLF and Hansell DM. Eur Radio 2010 20: 1859-1867; Staiteh BS et al. BMC Pulm Med. 2014 Aug 16;14:139; Willams Vet Pathol 2014 51(2) 372-384. Roman J et al Ann Am Thorac Soc. 2013 Dec;10(6):S224-9; Heikkilä-Laurila HP and Rajamäki MM. Vet Clin North Am Small Anim Pract. 2014 Jan;44(1):129-42 Couëtil LL, et al. J Vet Intern Med 2007;21:356; Costa LR, et al. Am J Vet Res 2000;61(2):167; Hoffman AM, et al. Am J Vet Res 1998;59:176; Hare JE et al. J Vet Intern Med 1998;12:163; Pearce N et al. Thorax 2000;55(5):424; Lavoie JP, et al. J Vet Intern Med 2011;25(2):322; Hughes KJ, et al. Vet Immunol Immunopathol 2011;140(1-2):82; Nolen-Walston RD, et al. J Am Vet Med Assoc. 2013 15;242(8):1138; Erichsen DF et al. Equine Vet J 1994;26:331; Cockcroft DW, et al. Lancet 1993;342:833; Read JR, et al. Am J Vet Res. 2012 Jan;73(1):140.

FOOD AND FIBER ANIMAL RESPIRATORY CASE STUDIES

William F. Gilsenan, VMD, DACVIM (LAIM)
Blacksburg, VA

SCENARIO

This clinical case scenario evaluates a small dairy farm in southwest Virginia. This dairy raises Holstein and Jersey heifers from birth until adulthood on its premises. One major problem that has challenged this dairy is its incidence of enzootic calf pneumonia and the incidence of treatment rates and antimicrobials use. Despite aggressive antimicrobial therapy, an unacceptable percentage of calves develop chronic bronchopneumonia; in severe cases, enzootic calf pneumonia has led to stunted growth, emaciation, generalized poor production, and death.

Several characteristics of this dairy (group housing and automatic feeding) make it unique from most American dairies. However, the purpose of this scenario is to emphasize the importance of management factors in the prevention of enzootic calf pneumonia. Challenges and benefits of group dairy heifer housing will be addressed. Additionally, this clinical case scenario will investigate the role of colostral management, antimicrobial resistance, and management and monitoring of the individually affected calf in the approach to solving this herd problem.

VITAMIN D IN PULMONARY DISEASE

Claire R. Sharp, BSc, BVMS (Hons), MS, DACVECC
North Grafton, MA

INTRODUCTION

In human and veterinary medicine, vitamin D has classically been thought of in the context of bone health and calcium metabolism.[1] In recent years, however, there has been a growing recognition of the extra-skeletal roles of vitamin D. It is now known that most tissues and cells in the body have a vitamin D receptor (VDR) and that many have the enzymatic machinery to convert 25-hydroxyvitamin D to the active 1,25-dihydroxyvitamin D. Either directly or indirectly, 1,25-dihydroxyvitamin D controls the transcription of more than 200 genes, and is thereby associated with functions such as cellular proliferation, differentiation, apoptosis and angiogenesis.[2] In human medicine, vitamin D deficiency and/or insufficiency has been associated with a wide variety of chronic diseases including cancer, infectious and inflammatory diseases, autoimmune disease, renal disease, and cardiovascular disease. This lecture, as part of the veterinary comparative respiratory society lectures at the ACVIM Forum 2015 will focus on the potential role of vitamin D in pulmonary disease. Although there is little work in veterinary medicine to date, vitamin D has been evaluated in pulmonary diseases in human medicine such as asthma, respiratory infections, chronic obstructive pulmonary disease, cystic fibrosis and lung cancer. Despite a plethora of research in the last decade, the question of whether vitamin D is a marker or mediator of these diseases remains largely unanswered.

SYNTHESIS AND METABOLISM OF VITAMIN D

The main source of vitamin D for people is exposure to sunlight, specifically UVB radiation. Upon UVB exposure, 7-dehydrocholesterol in the skin is converted to pre-vitamin D_3, which is immediately converted to vitamin D_3 in a heat-dependent process. Dietary intake is a less significant source of vitamin D in people but the predominant source of vitamin D in dogs and cats. Dietary vitamin D_2 and D_3 are incorporated into chylomicrons and transported by the lymphatics into circulation. In circulation vitamin D binds to vitamin D binding protein (VDBP), and to a lesser extent albumin, which transports it to the liver. Hepatic vitamin-D-25-hydroxylase converts vitamin D to 25-hydroxyvitamin D [25(OH)D]. Although biologically inactive, 25(OH)D is the major circulating form of vitamin D, and is used clinically to determine vitamin D status. Subsequently 25(OH)D is converted by 1-α-hydroxylase in the kidney to 1,25-dihydroxyvitamin D [1,25(OH)D]. 1,25[OH]D is the active form of vitamin D that has functions in calcium homeostasis including promoting small intestinal calcium absorption and calcium reabsorption from bone.

VITAMIN D DEFICIENCY VS. INSUFFICIENCY

In human medicine, the ideal serum concentration of 25(OH)D is controversial, and serum concentrations in apparently healthy people vary widely. Vitamin D deficiency is most commonly defined as 25(OH)D < 20 ng/mL (<50 nmol/L). This cut-off is based on the Institute of Medicine of the National Academies (in the United States), 2011 guidelines. However, these guidelines are based on the levels required to maintain adequate bone health and don't take into consideration other potential roles of vitamin D in health and disease. It is known that 25(OH)D concentrations are inversely correlated with PTH concentrations until 25(OH)D concentrations reach 30-40 ng/mL, at which point PTH concentrations plateau, suggesting that the PTH plateau may be useful as a marker of vitamin D adequacy.[3] Many now refer to 25(OH)D concentrations of 21-29 ng/mL as relative vitamin D insufficiency. In people, toxicosis is observed with serum concentrations of 25(OH)D >150 ng/mL (>374 nmol/L). Vitamin D deficiency is a global public health problem due to high and increasing prevalence, and associated morbidity.

Even less is understood about vitamin D sufficiency vs. deficiency and insufficiency in small animal veterinary medicine. At many diagnostic laboratories, the 'normal' reference interval is extremely wide. For example at the Michigan State University Diagnostic Center for Population and Animal Health, the reported reference interval for 25(OH)D in cats is 65-170 nmol/L. A recent study using PTH plateau to assess 25(OH)D adequacy in dogs found that variation in PTH plateaued at concentrations of 100 – 120 ng/mL;[4] a range higher and more narrow than reported 'normal' reference intervals at diagnostic laboratories. Not surprisingly, 25(OH)D is affected markedly by dietary intake in our small animal veterinary patients.

VITAMIN D DEFICIENCY AND PULMONARY DISEASE – WHAT IS THE EVIDENCE?

Vitamin D deficiency has been epidemiologically associated with many respiratory diseases of significance in human health. This lecture will focus on asthma and respiratory infections, as these are two areas in which disease is similar across species and preliminary data in veterinary medicine exists.

Asthma

The vitamin D hypothesis postulates that lower vitamin D concentrations are causally associated with increased asthma risk and severity in children and adults. There are numerous studies showing that lower serum 25(OH)D concentrations are associated with worse asthma control (eg. reduced steroid responsiveness, increased asthma exacerbations), increased airway hyperresponsiveness, and lower lung function (eg. lower FEV_1). *In vitro* studies also suggest a role for vitamin D in airway remodeling and modulating T-cell driven immune and inflammatory responses. Unfortunately, while epidemiologic studies show association and *in vitro* studies suggest mechanisms of disease modulation they cannot confirm a causative role of vitamin D deficiency in asthma. This is a theme of the vitamin D and respiratory disease literature; ie. association $\neq$ causation. Numerous authors have pointed out the challenges with these associations; for example that vitamin D deficiency is often an indirect marker of other confounding factors such as physical activity (ie. more time outdoors) which itself modulates the development and progression of respiratory disease.[6] Additionally, the association with disease severity may be that patients that are sicker due to their asthma spend more time indoors and hence have lower serum 25(OH)D concentrations (ie. reverse causation).[5]

To help better understand the role of vitamin D in asthma pathogenesis, a logical step is to perform studies investigating the effects of vitamin D supplementation in clinical patients. A large clinical trial was recently published that evaluated the effect of Vitamin D_3 on asthma treatment failures in adults with symptomatic asthma and lower vitamin D levels – the VIDA randomized, double-blind, parallel, placebo controlled, clinical trial.[7] The objective of this study was to determine if vitamin D supplementation would improve the clinical efficacy of inhaled corticosteroids (ciclesonide) in patients with symptomatic asthma and lower vitamin D levels (<30 ng/mL). The treatment group (n=201) received oral vitamin D3 (100,000 IU once, then 4,000 IU/day for 28 weeks) vs. placebo (n=207). In this study vitamin D_3 therapy did not reduce the rate of first treatment failure or exacerbation, the overall treatment failure rate, or the overall exacerbation rate. The only statistically significant outcome was a small difference in the overall dose of ciclesonide required to maintain asthma control (111.3 ug/d) in the VD_3 group vs. placebo (126.2 ug/d). The authors concluded that their findings do not support vitamin D_3 supplementation in patients with symptomatic asthma. Interestingly though, only about 80% of patients in the VD_3 group achieved target 25(OH)D concentrations of $\geq$ 30 ng/mL; suggesting a variable response to VD_3 therapy. The increase in 25(OH)D from baseline to 12 weeks was associated with a reduction in the rate of treatment failures and exacerbations. Although this trial was negative for the primary outcomes, additional clinical trials are needed, perhaps with individually adjusted VD_3 therapy to achieve target serum concentrations. An opinion piece published after the VIDA trial discussed some of the unique challenges of conducting vitamin D trials, and suggested potential ways to address them.[5] In addition to the aforementioned challenge of fixed dosing schedules not achieving sufficiency in some patients (which is likely due at least in part to genetic variation in VDBP, VDR etc.), other challenges include variability in sun exposure during the trial, high variability in vitamin D potency in different drug preparations, and off-protocol vitamin D consumption and/or UV-B exposure.[5]

Although asthma, or inflammatory airway diseases with similar phenotype to asthma, is also a common respiratory disease in cats and horses, little is known about the role of vitamin D in disease pathogenesis in these species. One small study presented in abstract form found no difference in serum 25(OH)D concentrations in cats with asthma (n=12; 114 +/- 36 nmol/L) vs. healthy control cats (n=9; 110 +/- 39 nmol/L). Interestingly, two cats with more severe signs had particularly low 25(OH)D concentrations (65 and 69 nmol/L). More studies are needed to better understand the role of vitamin D deficiency in asthma in cats and horses.

Respiratory infections

Evidence for a role of vitamin D in respiratory infections includes temporal associations, experiments establishing biological plausibility, and epidemiologic associations. Perhaps the first suggestion that vitamin D deficiency could be associated with respiratory infections came from the clear <u>temporal association</u>; both serum vitamin D concentrations and viral upper respiratory tract infections show seasonal oscillations that are worse in the winter months.[13] Additionally, an association between vitamin D deficiency and respiratory infections has <u>biological plausibility</u> since it is now well recognized that vitamin D is involved in numerous aspects of the immune response to pathogens. As part of the innate immune response, vitamin D is involved in the induction of antimicrobial peptides (eg. cathelcidin and β-defensin) that can kill microbes and act as chemoattractants to recruit other immune / inflammatory cells. This has been demonstrated specifically in the context of respiratory infections such as *Mycobacterium tuberculosis*.[9] As with the asthma story, a plethora of studies have also suggested <u>epidemiologic associations</u>. For example, in a prospective cohort study of adults in Connecticut, 25(OH)D > 95 nmol/L was associated with a 2-fold reduction in the number of upper respiratory infections over a single winter.[10]

When it comes to clinical trials of vitamin D supplementation in URTI, the results are unfortunately inconclusive. For example, a recent RCT in New Zealand, failed to show a benefit of monthly administration of 100,000 IU of vitamin D (after an initial dose of 200,000 IU) on the incidence or severity of URTIs in healthy adults (n=161 placebo, 161 treatment) over an 18 month period, despite effectively raising serum 25(OH)D. Nonetheless a recent systemic review and metanalysis of RCTs concluded that their results indicate that vitamin D has a protective effect against respiratory tract infections, and that once-daily dosing seems to be the most effective dosing regimen. Our research group has conducted studies of feline upper respiratory tract infections in shelter cats; data analysis is pending.

Vitamin D status has also been investigated in pneumonia patients. A recent retrospective cohort study evaluated vitamin D levels and the risk of developing community-acquired pneumonia and sepsis in a population of adults in Utah and Idaho. Cases and controls had to have a serum 25(OH)D concentration in their electronic medical record 3-15 months prior to their hospital admission. Of 11,455 patients diagnosed with pneumonia, 132 had recent vitamin D levels record and were designated case. Controls were randomly selected from a pool of patients without CAP admitted in the same time period and matched 1:1 with cases by age, sex, race and season of 25(OH)D measurement. In this study, 25(OH)D concentrations < 37nmol/L were strongly associated with increased odds of hospitalization for community acquired pneumonia (OR 2.57; 95% CI 1.08-6.08). In another study, 25(OH)D deficiency was associated with increased mortality in patients admitted to hospital with community acquired pneumonia during winter.[14] Patients with 25(OH)D < 30 nmol/L had a higher 30-day mortality compared with patients with sufficient 25(OH)D (>50 nmol/L) (OR 12.7; 95% CI 2.2-73.3, p=0.004). These associations were not explained by differences in age, comorbidities, or the severity of the acute illness. Our group has also conducted studies of serum vitamin D concentrations in puppies with community-acquired pneumonia; data analysis is pending.

CONCLUSIONS

The epidemiology and biologic plausibility of an association between vitamin D deficiency and various respiratory diseases is tantalizing, however in general, more work needs to be done before we know whether vitamin D supplementation to people with low vitamin D levels might change their disease course. I look forward to more work in veterinary medicine to investigate the epidemiology of vitamin D deficiency in our patients and potential pulmonary disease associations.

REFERENCES

[1] Dittmer KE and Thompson KG. Vet Pathol 2011;48(2):389-407. [2]Holick MF NEJM 2007;357(3):266-281. [3]Saliba Am J Medicine 2011;124(12):1165-1170. [4]Selting K et al. Vet Comp Oncol 014 Jul 8. doi: 10.1111/vco.12101. [Epub ahead of print]. [5]Brehm JM. Curr Allergy Asthma Rep 2014;14:461. [6]Foong RE and Zosky GR. Nutrients 2013;5:2880-2900. [7]Castro M et al. JAMA 2014;311(20):2083-2091. [8]Rozanski EA et al. J Vet Int Med 2011; 25(3):743. [9]Liu PT et al. Science 2006 311:1770-1773. [10]Sabetta JR et al. PLoS One 2010 e11088. doi: 10.1371/journal.pone.0011088. [11]Murdoch DR. JAMA 2012;308(13):1333-9. [12]Bergman P et al. PLoS One 2013; 8(6): e65835. [13]Maguire JL. BMC Pediatrics. 2014(14):37. [14]Leow L et al. Respirology 2011;16(4):611-6.

MAKING FRIENDS WITH YOUR PULMONARY PATHOLOGIST: HOW HISTOPATHOLOGY CAN HELP

Samuel Jennings, DVM DACVP
Tufts University, North Grafton, MA

Pathology can provide vital clues and in some cases, the final answer, surrounding the cause for a patient's respiratory dysfunction. There are several ways to collect samples for diagnosis with varying degrees of invasiveness. Several (and generally the most commonly performed) are with the realm of the clincial pathologist, so we will only discuss those briefly, largely in the context of when they might be indictated.

Options for sampling for cytology:

Transtracheal wash- Helpful for trachea and bronchial cytology (eg, infection, eosinophils); less useful for interstitial disease
BAL- Better for interstitial diseases, but may get get adequate airway sampling as well.
Fine needle aspirate- useful for direct sampling of a focal lesion with ultrasound/CT guidance, iatrogenic pneumothorax is possible
Effusion cytology – useful for diseases involving the pleural space when sufficient pleural effusion is present

As in all cases, sample handing can influence outcome. If sample evaulation will be delayed, remember to make slides of TTW or BAL sample. Additionally, lube on an endotracheal tube can contaminate a sample and possible prevent bacterial growth. For many cases, such as neoplasia, eosiniphilic pneumopathy, or pneumonia, cytological evalution may be adequate for a final (or near final) diagnosis. However, in many cases, a tissue sample is required to confirm a diagnosis.

Options for sampling for histopathology:

1) Open lung biopsy- Provides plenty of tissue for evaluation, but may be associated with signficant morbidity or even mortality, particularly in severely compromised patients. Caudal lobe lesions may be biopsied through the diaphragm, which may be an easier recovery than a thoracotomy.
2) Thoroscopic biospy- Less morbidity, potentially smaller samples. May be limited in location of sampling. Good for experienced operators. Procedure-induced atelectasis can be signficant.
3) Tru-cut biopsy- Significant risk of pneumothorax in small animals, helpful in mass lesions or areas of consolidation.
4) Bronchoscopic biopsy- OK for airways, but samples are small and unlikely to reach deeper tissues samples. Some clinicians are comfortable aspirating masses near the carina trans-bronchially.
5) Complete autopsy- While unable to directly benefit to the patient itself, but a full postmortem exam is the best way to fully evaulate the lungs and address lesions in other organ systems. Critical to furthering our knowledge of interstitial lung diseases in veterinary species.
6) Postmortem "lung-opsy"- While some clients may be resistant to complete autopsy, for animals with unclear disease, a postmortem sample can often help the family as well as help clinicians and subsquently future dogs to understand the disease affecting a specific pet. Lacks "full context" of a complete autopsy, but still may yield valuable information.

How to get good samples:

1) Diffusely affected lung- 1 cm+ sections from several diffuerent locations is preferred. If sample is atelectatic, tissues can be gently, partially inflated by instilling samll amounts of formalin into the parenchyma via a fine gauge needle and syringe. Samples should be fixed in 10 parts formalin to 1 part tissue. Consider subsamples for culture and cytologic impression smears. Consult with your pathologist AHEAD to brainstorm ideas of unusual diagnoses or additional diagnostics that might require fresh tissues.
2) Nodules- Try to get several, if possible. Larger nodules often have central necrosis. Sampling multiple lesions or parts of a single lesion that look or feel different (similar to dermatology) can maximize the liklihood of a diagnostic samples.
3) Mixed patterns- Sometimes, more than one process is occuring. For example, it is not uncommon for cats with pulmonary fibrosis to also have pulmonary neoplasia. Multiple samples from different areas will help catch these situations.
4) If you have the option for an entire lobe (postmortem most commonly), try to fix inflated (eg, run a catheter into the bronchi and infuse with formalin to about 15-20 cmh20 (so inflated). We often run a red rubber catheter with a stopcock, purse-string the catheter in place, inflate the lung, then close the stopcock so the lung is inflated under pressure. This allows for excellent fixation and allows for more accurate evaluation of alveolar septa. This can create some mild artifacts, so let the pathologist know you did this.

Does history matter to a pathologist?

YES- the more specific you can get with your history, the more helpful your results will be. Just as internists use history and signalment so do pathologists. We don't like being "in the dark" any more than you do with mystery cases, and enjoy being able to be useful in confirming a suspected diagnosis. Occasionally people worry that they will bias us by providing information, but this risk is much lower than the risk of getting a more generic interpretation. Knowing about diseases elsewhere in the animal (e.g. history of oral melanoma or renal failure), imaging findings (diffuse vs focal/multifocal disease, are there nodules?), and clinical pathology (including cytologic findings) can make a big difference in some cases.

What about EM?

EM can be expensive, requires special fixatives, and requires expertise in interpretation. Discussion with a pathologist prior to sampling is highly recommended. With regards to the respiratory system, this is most commonly considered in the work up of primary ciliary dyskinesia. The most recent list of places to obtain EM includes (from the SAIM-list serve, others may be available).

a) University of Pennsylvania Section of Medical Genetics, VHUP Room 4033 3850 Spruce Street, Philadelphia, PA 19104-6010, 215-898-0029
b) Va-Md Regional College of Vet Med (VT) is 540-231-4621 and ask for Kathy Baker. She can help you arrange EM for dyskinesia.
c) Dr. Matti Kiupel at Michigan State University 517-432-2670 is his office and 517-353-1683 is the Pathology Department Kiupel@dcpah.msu.edu

As always, these options may change. Your local pathologist will be happy to help you find a location and understand sample processing.

What are special stains? Do I want them?

Additional stains that can be performed come in 2 general flavors: histochemical and immunohistochemical.

Histochemical stains (like Gram, Giemsa, PAS, GMS, trichrome) can be helpful in identifying infectious agents or evaluating for changes in connective tissue content. Most labs will generally run the stains they feel will be helpful without additional charges.

Immunohistiochemical stains are generally used to help evaluate for specific infectious agents (e.g., herpesviruses, canine distemper, toxoplasma, etc) or characterize cell populations by highlighting particular cytoskeletal components (e.g. cytokeratins, smooth muscle actin, vimentin), adhesion molecules (e.g., CD18), trascription factors (TTF-1), or other features. These are generally more expensive tests to run and most labs require approval and additional charges to run them. The pathologist should be happy to recommend particular stains and discuss the rational behind each. In many cases, a panel of stains may be necessary. Turnaround times vary from lab to lab. If you have a high suspicion for particular diseases, pre-approval of IHC can speed up the process.

How come there is not an end-diagnosis, and just a bunch of words?

While some diagnoses are relatively straight-forward (such as well differentiated carcinoma or pneumonia), in many cases, the answer may not be so clear cut. In some cases, this uncertainty can be addressed with carefully chosen histochemical or immunohistochemical stains (particularly with neoplasia). In some cases, tissue changes can be accounted for by multiple types of insults and a list of possibilities need to be considered. In other cases, we may be limited by the state of veterinary knowledge. This is often the case with chronic interstitial lung diseases. Human classification schemes of idiopathic interstitial lung disease are complex and have been developed and refined by the radiological and clinicopathological correlation of hundreds of cases. We simply haven't fully addressed the variety of pathologic appearances chronic interstitial diseases in veterinary medicine, so we are often left providing descriptive, morphologic diagnoses than may not end with a known "entity" on the bottom line. This doesn't mean you haven't gained anything from the biopsy as the description and morphologic diagnosis can still provide information on the presence/character/severity of inflammation, presence/severity of fibrosis, what compartments are involved (e.g., bronchi, bronchioles, alveoli, pleura), how extensive the lesions are, etc.

Please call your pathologist for advice on what your biopsy means if you have questions; we are eager to help you and your patients.

Picking a specific pathologist

For neoplasia and many classic entities, most pathologists are up to the task. Just as internists commonly have areas of specific focus and interest, so do pathologists. The subspecialty of dermatohistopathology is perhaps the most established, however, other subspecialties like hepatobiliary pathology and bone pathology are continuing to grow. Certainly pulmonary pathology is of greater interest to certain pathologists, so in uncommon cases a second opinion may be warranted.

The rest of this hour will focus on cases, coupled with a review of the clinical presentation and imaging findings.

CHOOSING THORACIC IMAGING: CONUNDRUMS AND CONCLUSIONS

Trisha J Oura, DVM, DACVR
North Grafton, MA

INTRODUCTION

Diagnostic imaging of the thorax is essential for the diagnosis and management of most pulmonary diseases. Options include thoracic radiography, fluoroscopy, computed tomography (CT), magnetic resonance imaging (MRI), and ultrasonography. Each modality had specific strength, indications and limitations. Working closely with a radiologist is very useful for planning and performing imaging, and similar to internal medicine, input *prior* to completing a procedure is particularly useful. When deciding on the best imaging modality, it is important to consider what you are looking for; for example, a dynamic process such as airway collapse or obstruction would be best imaged by fluoroscopy, while a more static lesion such as a lung tumor or diffuse interstitial disease may be adequately documented with radiographs or CT. In patients with a concern for pulmonary thromboembolism (PTE), CT angiography may be required to make a diagnosis. Such advanced studies require more skill on the part of the imaging technician and interpreting specialist, and consistent protocols should be established ahead of time. Imaging can also confirm disease, but is not a replacement for histopathology.

THE USE OF CT

For internists, although many cases have had multiple radiographs taken prior to arrival, it may be prudent to start with repeat radiographs if there has been a change in patient status, a small suspicious lesion on initial imaging, or inadequate patient positioning. However, one common question concerning thoracic radiographs is "Should we just CT instead of re-radiograph?" The answer to this is most often "Yes, probably, but we won't know until we look." As shown in several studies, CT is often (though not always) more sensitive than radiographs for detecting pulmonary nodules[1-3]. However, using CT as a gold standard may be flawed without histopathology confirming true metastatic lesions rather than granulomas. Also, most information available for survival times are based on metastatic disease identified on radiographs, not CT, so this may make it harder to provide clients with accurate information. In addition, availability and practicality may limit the use of CT for repeat metastatic checks, though this may change over the coming years as CT scans become more widely available and scan times are limited, permitting both awake scans or those under sedation alone. In addition to identifying pulmonary nodules, CT often provides new information regarding intra-thoracic pathology that is present, the location and extent of disease, involvement of mediastinal structures, and may even result in a change in differential diagnoses when compared to thoracic radiographs[4].

If used, anesthetic protocols are exceeding important if evaluation of the lung parenchyma is the primary goal of the CT study. Despite the tomographic nature and increased sensitivity of CT, anesthesia-induced atelectasis may limit complete evaluation of lung parenchyma, resulting in missed lesions. To limit atelectasis, studies are typically performed in sternal recumbency, and in most studies, maximal inspiratory breath-hold (positive end expiratory pressure, PEEP) is the best for highlighting lesions (Figure 1). In cases of suspected air trapping, an expiratory scan at zero end expiratory pressure (ZEEP) might also be useful and to better document air trapping. Motion artifacts can also decrease image quality and limit confidence in identifying causative lesions. For critically ill patients, awake or sedated CT studies can also be considered, though increased motion artifact should be expected.

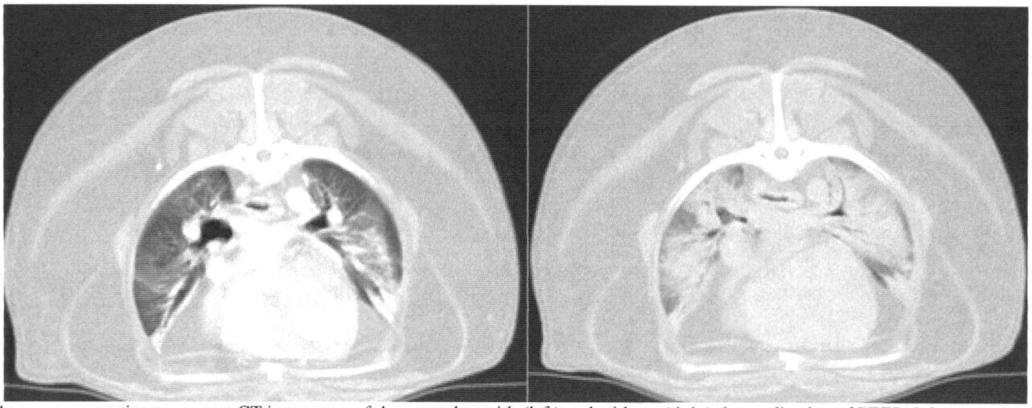

Figure 1. These representative transverse CT images are of the same dog with (left) and without (right) the application of PEEP. It is easy to see the difference in the appearance of the lung parenchyma. The image on the left was also acquired post intravenous contrast medium administration.

Common indications for thoracic CT include pulmonary masses, spontaneous pneumothorax, and interstitial lung disease. Pulmonary masses may be incidental findings or may result in cough, shortness of breath or pleural effusion. For over 20 years, CT has been shown to useful in veterinary medicine to identify lung masses and assist in classification of tumor types[5,6]. CT can be especially good in establishing the extent of the mass or presence of enlarged lymph nodes and may be used in conjunction with ultrasound-guided tissue sampling of more peripheral lesions (Figure 2). However, CT identification of tumor extent and metastasis may not change the surgical approach compared to radiographs alone, and some surgeons may prefer to proceed directly to exploratory thoracotomy rather than CT.

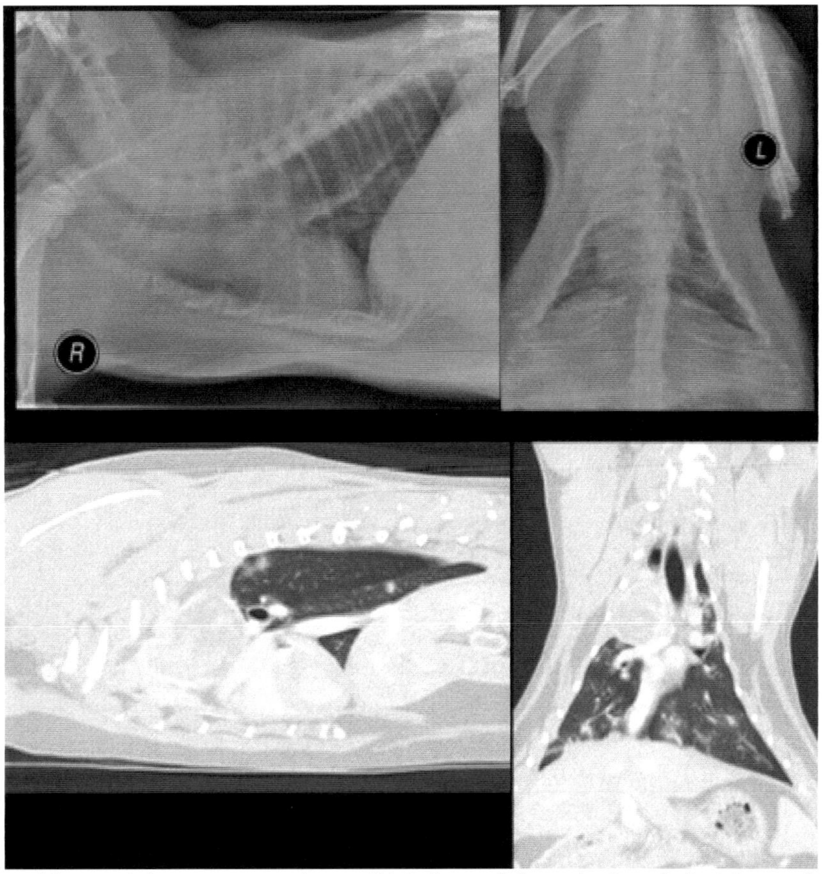

Figure 2. This is a 12 yo cat presented for a cough that was nonresponsive to antibiotic therapy. Orthogonal radiographs (top) show a soft tissue opacity within the right cranial thorax partially superimposed on the intrathoracic trachea and silhouetting with the cranial aspect of the cardiac silhouette. Sagittal and dorsal planar reformatted thoracic CT images (bottom) revealed that this opacity was a well-demarcated mass, likely of pulmonary origin.

Another condition that presents a diagnostic challenge is spontaneous pneumothorax as it is a complicated condition with underlying bulla or bleb often being challenging to identify. While CT has been shown to be more sensitive than thoracic radiographs for identification of pulmonary bullae, a recent student indicated that CT had low sensitivity and low positive predictive value for identifying the causative lesion in canine spontaneous pneumothorax[7]. While this study suggests CT prior to exploratory thoracotomy may be unnecessary, CT should still be considered in animals with potential mass or those with a suspicious history, physical examination or pre-operative thoracic radiographs.

CT angiography has replaced traditional angiography in human medicine for diagnosis of pulmonary thromboembolism and has replaced nuclear scintigraphy and traditional angiography for identification of emboli in veterinary patients. Intravascular filling defects are often easily appreciated and documented and timely anticoagulant therapy may be instituted. However, performing a CT angiogram requires special timing of contrast medium administration and image acquisition to highlight the defect and may necessitate additional technician training and use of a power injector. Although CT angiography is traditionally performed under general anesthesia, sedated studies can also be considered in critically ill patients (Figure 3).

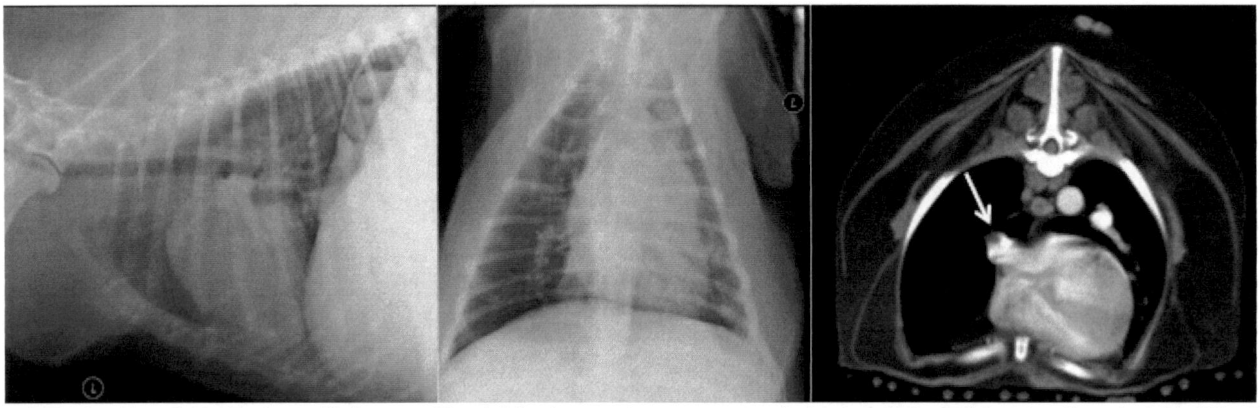

Figure 3. This is a 10 yo Golden Retriever with shortness of breath following a craniotomy to remove a brain tumor. Orthogonal thoracic radiographs (left/middle) did not reveal a cause for the clinical signs. On a transverse thoracic CT image following administration of intravenous iodinated contrast medium (right), a round, well-defined filling defect is noted within a right pulmonary artery (white arrow) consistent with pulmonary thromboembolism.

Finally, interstitial lung disease is an emerging area of pulmonology and CT scans are essential in order to highlight the abnormalities associated with these uncommon clinical entities. Although CT is the gold standard for identifying and categorizing interstitial lung disease in humans, the use of CT in veterinary medicine is often limited by a lack of histopathology confirming specific diseases and a lack of follow up with clinical cases. These are frustrating cases for internists and radiologists alike! CT may be useful for specific interstitial diseases such as acute respiratory distress syndrome/acute lung injury (ARDS/ALI). In dogs already under anesthesia for intermittent positive pressure ventilation, CT can detect lung heterogenicity and changes in disease distribution during ventilation, which may be prognostic[8].

CONCLUSIONS

Radiographs are the first line of defense for imaging the pulmonary system. They are easily obtained and provide vital information concerning thoracic pathology. However, if a patient is not responding to conventional therapy, it may be prudent to pursue other forms of imaging. Fluoroscopy is available at most universities and many specialty hospitals and can be very useful for dynamic lesions. However, many clinicians may opt for thoracic CT when pursuing additional imaging modalities. CT has been shown to be more sensitive for identifying some lesions, but CT can be frustrating in veterinary medicine without patient follow up or histologically confirmed diagnoses for processes such as interstitial lung disease. Looking to the future, better CT/histopathology correlates are needed if we are to optimize use of this modality in our patients.

REFERENCES

[1]Nemiac S, et al. **J Vet Int Med** 2006;20:508, [2]Alexander K, et al. **Vet Radiol Ultrasound** 2012;53(3):258, [3]Armbrust L, et al. **J Am Vet Med Assoc** 2012;240(9):1088, [4]Prather A, et al. **Vet Radiol Ultrasound** 2005;46(2):114, [5]Spann D, et al. **Vet Radiol Ultrasound** 1998;39:532, [6]Barrett L, et al. **Vet Radiol Ultrasound** 2014;55(5):480, [7]Reetz JA et al. **J Am Vet Med Assoc** 2013 Jul 15;243(2):244-51, [8]Oura T, et al. **Vet Radiol Ultrasound** Abstracts from 2012 ACVR scientific conference 2012;53(6):673.

HIGH FLOW OXYGEN DELIVERY/HELIOX THERAPY

CHRISTINE GUENTHER DVM, DACVECC and ELIZABETH ROZANSKI DVM, DACVECC, DACVIM (SAIM)

INTRODUCTION

Oxygen administration is frequently used in veterinary medicine to treat hypoxia. Non-invasive or conventional oxygen administration methods include nasal cannula, oxygen cage, hood oxygen and flow by/mask administration. With conventional oxygen therapy (COT), FIO2 is limited to 30-75%. In addition, other factors including desiccation and irritation of the upper airway with nasal oxygen, limited hands on access to patients in an oxygen cage and patient tolerance with flow by oxygen can limit the efficacy of these therapies. In patients with severe hypoxia (paO2 < 60 mmHg) who fail COT more aggressive treatment is warranted.

ADVANCED OXYGEN DELIVERY METHODS

In veterinary medicine, mechanical ventilation (MV) is typically recommended to improve oxygenation and reduce the work of breathing in dogs and cats failing COT. While MV is effective at reversing hypoxemia, it is associated with complications such as lung injury, infection, prolonged immobility and sedation and high associated labor and costs. Similar challenges with MV have been identified in human medicine and other non-invasive strategies for improving oxygenation have been developed including continuous positive airway pressure (CPAP) and a relatively new technology called high flow oxygen therapy (HFOT). Several veterinary studies have been performed evaluating CPAP. While use of CPAP in dogs and cats showed improvement in oxygenation, patients had difficulty being fitted for a mask due to variable facial conformation. They also required heavy sedation approaching anesthetic levels requiring intubation for airway protection thus negating any potential clinical benefit.

HIGH FLOW OXYGEN THERAPY (HFOT)

High flow oxygen therapy is a nasal oxygen delivery system that accurately delivers humidified and heated oxygen to the patient. Membrane technology is used to transfer heated water vapor into a gas stream at 99.9% relative humidity. A water jacket delivery tube is used to protect the humidified gas from heat loss. This greatly improves patient comfort and tolerance allowing for higher flow rates to be used (up to 60L/min). HFOT can also be blended with room air to adjust FiO2 allowing the patient to be accurately weaned from the therapy as the patients' pulmonary function improves as well as decreasing the associated risks of oxygen toxicity.

USE OF HFOT IN PEOPLE

HFOT has been used in human medicine for over 10 years in patients with pulmonary dysfunction who have failed COT. In addition to allowing for higher flow rates and therefore a higher FiO_2, research has shown an additional benefit of a mild positive end expiratory pressure (PEEP) of 1cmH_2O for every 10L/min of flow. Complications with HFOT seen in people include patient compliance and risk of persistent airway leak attributable to the PEEP associated with this therapy. The use of HFOT in people does not require heavy sedation or constant bedside monitoring. People receiving HFOT can eat, drink and talk without interruption of oxygen delivery. HFOT units are portable allowing for the patients to be ambulatory.

HFOT VETERINARY STUDIES

A pre-clinical study was done in six healthy sedated dogs comparing the effects of HFOT to COT. The data showed that HFOT at 20L/min and 30L/min significantly increased PaO2 compared to standard nasal oxygen administered at 100 ml/kg/min. There was no significant difference between paCO2, pH and transpulmonary pressures measured by esophageal balloon catheter between any group. Gastric distension as measure by radiographs was noted post therapy in 1/6 dogs.

A retrospective pilot clinical study (in press, JVECCS) was done looking at 6 dogs with severe hypoxia failing COT where MV was recommended but declined by owners. HFOT resulted in a significant increase in PaO2 when compared to COT. Hypoxemia completely resolved resulting in weaning from oxygen therapy in 4/6 dogs with 3/6 patients surviving to discharge. Complications documented included lack of patient tolerance requiring mild sedation in one dog and persistence of a pre-existing pneumothorax in one dog. All dog were monitored by ICU technicians in a manner similar to dogs requiring COT i.e. they did not require a constant bedside presence.

HFOT EQUIPMENT AND LABOR

There are several different high flow oxygen delivery systems available in the human market. The The cost of the system we purchased was around $3500 (Vapotherm Precision Flo®). In order to adjust FIO2 an air compressor is needed which we purchased for around $2500. Other required components include calibrated filters, sterile distilled water for humidification and specialized nasal cannulas calibrated to the system

In our hospital, all patients on supplemental oxygen therapy are housed in the ICU. Labor caring for dogs receiving HFOT is about the same as that for dogs receiving COT. The system is easy to set up and use. As compared to MV, there is little specialty expertise needed. Other gas delivery (heliox and nitric oxide) are available with separately purchased units.

HFOT CLINICAL IMPRESSIONS

Our hospital has been using HFOT in dogs failing COT for about 2 years. Our general impression is that owners are more likely to pursue HFOT as compared to MV. This is likely due to a combination of lower overall costs to the owner as well as a guarded prognosis associated with ventilation in pulmonary diseases. We have found that owners seem more likely to continue HFOT on a longer basis as compared to patients undergoing MV. Besides financial and prognostic reasons, it may be that because dogs getting HFOT don't need heavy sedation, owners may be more apt to continue treatment when they can see their pet awake and interactive. HFOT is not a substitute for MV and dogs in our hospital have failed HFOT requiring more aggressive oxygen delivery with MV. The benefit of PEEP with HFOT has not been shown in dogs and would be negated in dogs that pant. In pulmonary diseases that are highly PEEP responsive such as congestive heart failure, MV may be a superior treatment option to HFOT. However, in some dogs with severe pulmonary edema where owners have declined MV, HFOT has resolved hypoxia. Other uses for HFOT in our hospital have included recovery from anesthesia, especially in brachycephalic dogs and for oxygen supplementation in dogs and cats undergoing bronchoalveolar lavage or tracheal wash. A multicenter clinical trial is currently underway comparing outcomes in dogs failing COT in facilities where HFOT is available vs. facilities where HFOT is not available.

HFOT CONCLUSIONS

Preliminary data suggests the use of HFOT is a clinically viable and cost effective initial alternative to mechanical ventilation in dogs with severe hypoxemia failing conventional oxygen therapies.

HELIOX THERAPY

HELIOX THERAPY INTRODUCTION

While oxygen therapy is well-established in critical care medicine, the use of other gas mixtures has been less well-defined. Room air is a mixture of approximately 21% oxygen and 79% nitrogen, with supplemental oxygen most frequently provided at 35-40% via an oxygen cage, nasal oxygen or via a mechanical ventilator. Oxygen therapy is primarily indicated for treatment of respiratory distress, from lower airway or parenchymal disease, with pleural space disease treated best with drainage of air or fluid, and upper airway disease treated with sedation/anxiolytics or tracheostomy. However, supplemental oxygen makes sense when addressing parenchymal lung disease, with alterations in the capillary-alveolar membrane. Increasing concentrations of oxygen may (should) subsequently increase the likelihood of maintaining adequate oxygen delivery to the tissues. Excessive oxygen (> 60% inspired oxygen concentration) has been associated with lung injury in healthy animals and is also associated with atelectasis due to absorption (eg. absorption atelectasis). Therefore, for upper and lower airway diseases, which are characterized primarily by airway obstruction, there is interest in reducing the resistance to airflow, which would reduce the work of breathing, and subsequently limit the need for supplemental oxygen. Resistance is defined as the change in pressure/change in air flow, and reported in the units cmH20/L/sec. The bulk of airway resistance is appreciated in the upper and conducting airways. This is in contrast to diseases primarily affecting lung compliance, which is defined as change in volume/change in pressure (ml/cmH20). Resistance is proportional to diameter of the airway as well as if the flow is laminar or turbulent. Laminar or turbulent flow is dependent on several characteristics, including the viscosity of the gas itself. This has led to the idea of using helium in exchange for nitrogen in order to decrease airway resistance/work of breathing. Helium, which is the second most abundant noble gas in the atmosphere, is far less dense than nitrogen. The point at which gas flow transitions from turbulent to laminar will impact airway resistance. This transition is determined several factors including the branching pattern of the airways and the diameter and wall characteristics of the airways. In people, laminar flow is typically present by the second generation of bronchi, however with disease, turbulent flow may progress far into the lower airways. Recall that turbulent flow INCREASES the work of breathing, and may result in respiratory failure. The flow of gas through an obstructed airway is turbulent and dependent upon the density of inspired gas. The type of flow through an airway can be predicted by the Reynolds number, a dimensionless number given by the following equation:

$$Re = \rho v L / \mu$$

Where ρ is density, v is kinematic velocity, L is length and μ is viscosity. A *Re* of <2000 favors the formation of laminar flow. By decreasing the density (ρ) of inspired gas, the Re decreases as well. The densities of oxygen and nitrogen at standard barometric pressure are 1.429g/L and 1.251g/L, respectively. The density of helium however, is considerably lower (0.179g/L). By simply removing the non-physiologically active portion of inspired gas (nitrogen) and replacing it with a lower density gas (helium), both the gas density and *Re* are dramatically reduced, favoring formation of laminar flow through obstructed airways. Helium is an inert, non-toxic molecule and there are no reported contraindications to its use. Heliox is produced commercially as a medical-grade homogenized mixture of helium and oxygen. It is produced in various concentrations, including 78:22, 80:20, 70:30 and 60:40 ratios of helium: oxygen. Oxygen concentrations greater than 40% increase the overall gas density and should be avoided due to decreased efficacy. Typically, heliox is administered through a heliox-specific regulator and flow meter. If another flow meter is used, a conversion factor must be employed.

USES IN PEOPLE

Reported uses of heliox therapy in human medicine include asthma, upper airway obstruction, laryngotracheobronchitis ("Croup"), and bronchiolitis. Studies provide conflicting information as to clear benefit from Heliox, and all studies reiterate that Heliox is only a form of supportive care, not a specific therapy and that treatment (eg. bronchodilators) should be aggressively pursued even if heliox is used.

Asthma is a very common disease in children and adults. Most patients respond to bronchodilators and supplemental oxygen. However due to narrowed airways, the use of Heliox has been considered a reasonable additional therapy, particularly if used to try to avoid intubation. Similar to veterinary medicine, if intubation may be avoided, that is certainly preferred. Multiple studies in children have had somewhat conflicting results, with the major conclusion being that having access to heliox may help individuals, it unlikely to benefit most patients about the standard of care. Upper airway obstruction in common in children since they have smaller airways, similar to our dog and cat patients. For any cause, Heliox will improve gas flow and subsequently oxygenation, decrease work of breathing, and typically prevent the need for intubation. Premature babies in particular appear to benefit from heliox, as similar to puppies and kittens, they have relatively large tongues and airway obstruction is frequent. In children with burns, prolonged intubation may lead to post-extubation stridor, and heliox was also shown to be useful in these cases. Children often get viral laryngitis, this can cause narrowing of the laryngeal lumen. However, treatment with nebulized racemic epinephrine and heliox were equally effective for children with croup. Bronchiolitis (inflammation of the bronchioles) is a common infectious disease in children, and while heliox has used extensively with somewhat mixed results, with the suggestion of benefit in affected children. Heliox is used less commonly in adults, but may be used in patients with COPD, especially as a method to avoid intubation. Intubation is particularly to be avoided in this group of patients due to the subsequent difficultly in extubation. Occasional efforts have been made to use heliox in patients receiving mechanical ventilation, but difficulty adapting the specialized ventilator equipment for use with heliox have limited ease of experimental studies. Limited studies for the use of heliox in experimental ARDS in rats have suggested that the use of heliox may limit airway inflammation and prevent progression of disease.

Heliox is also widely used in diving to prevent the "bends" which is due to formation of gas bubbles in tissues due to abrupt changes in pressurization.

VETERINARY USES

Reports of heliox use in veterinary species are currently sparse. The first report of its use in companion animals was an experimental model of upper airway obstruction in dogs. Pressure-volume loops were nearly identical between dogs breathing heliox with obstruction and breathing room air without obstruction. This study demonstrated a significant decrease in work of breathing and airway pressure in dogs with obstructed airways receiving heliox 70:30.

Proposed/theoretical areas for the use of heliox in animals might include the following.

Upper airway obstruction

1) Brachycephalic disease: Heliox may be of benefit in brachycephalic dogs experiencing an acute exacerbation of chronic airway obstruction in the emergency and/or post-operative settings. Typically, these dogs are treated with sedation, which may diminish their ability to protect the airway, increasing the risk of aspiration pneumonia. In more severe cases, general anesthesia and intubation may be necessary, which may also increase the risk of an aspiration event. Brachycephalic dogs have a decreased ability to clear pulmonary infections due to their airway conformation. Therefore, an effort should be made to avoid the morbidity associated with the currently accepted treatment standards. With heliox therapy, it is possible that sedation and general anesthesia may be avoided if partial airway obstruction is overcome by an increase in flow. A small pilot study (Figure 1) in our laboratory showed that dogs with BC conformation had improved flow rates when breathing heliox as measured by TBVFL.
2) Laryngeal paralysis (LP). LP is very common in older large breed dogs, and affected dogs often present with severe inspiratory stridor. Similar to brachycephalic dogs, sedation may be relatively contra-indicated due to risks of aspiration, although may be required to permit adequate oxygenation and reduce of patient anxiety.
3) Tracheal collapse is also a common LARGE Airway disease, and could benefit from the use of heliox. This is particularly appealing as the other options for managing tracheal collapse, if oxygen and sedation are not adequate, includes intubation. Extubation of these dogs may be particularly challenging.

Lower airway disease

4) Feline asthma may be a reasonable condition to use heliox, as airway resistance in severely affected patients will be high. It is possible that delivery of heliox via an incubator of enclosed cage will improve outcome. The actual benefit of heliox over simple oxygen, bronchodilators and glucocorticoids is unknown.
5) Chronic bronchitis with acute exacerbations may also be associated with airway narrowing. Recall that dogs rarely get true bronchospasm/constriction.

HOW TO PROVIDE HELIOX

Administration of heliox requires a tank of premixed gas, which should be available from any local gas distributor. Heliox can be made with various concentrations, although most often, it is provided as 79% helium and 21% oxygen.

It essential that it be delivered by tight fitting facemask or within a cage as even small leaks significantly decrease any benefit for heliox therapy. Helium is very difficult/impossible to humidify, but may also be delivered through nasal cannula. Heliox may also result in increased heat loss so the patient's comfort should be closely monitored. Heliox tanks are much more expensive than oxygen tanks, and this cost should be factored into any treatment. (roughly $300). Further research is needed to identify which cases in veterinary medicine will benefit from heliox therapy, but it is novel treatment that should be considered.

REFERENCES

HFOT Reference List

1) Darin J, et al. Resp Care. 1982; 27(1): 41-50
2) Ward J. Resp Care. 2013; 58(1): 98-122.
3) Dunphy E, et al. JVECC. 2002; 12(4): 245-251.
4) Clare M, Hopper K. Compendium. 2005;March: 195-208.
5) Hopper K,et al. JAVMA. 2007; 230: 64-75.
6) Briganti A et al. JVECC. 2010; 20(5): 503-508.
7) Brown J, et al. JVECC. 2009; 19(5): 416-425.
8) Parke R, et al. Resp Care. 2013; 58(10): 1621-1624
9) Hedge S. et al. Pediatrics. 2013; 131: e939-e944.
10) Daly J et al, JVECCS 2014 24(S1): S5

Heliox Reference List

1) Beurskens C et al PLoS One. 2013 Oct 18;8(10).
2) Rodrigo GJAnn Allergy Asthma Immunol. 2014 Jan;112(1):29-34
3) Staffieri F, Equine Vet J. 2009 Nov;41(8):747-52.
4) Moraa ICochrane Database Syst Rev. 2013 Dec 7;12
5) Martinón-Torres F. J Crit Care. 2012 Apr;27(2):220.e1-9
6) Byers C, JVECC 2008 (18(6) 586-593.

RECOGNIZING AND MANAGING ACUTE KIDNEY INJURY

Marie E. Kerl DVM, MPH, DACVIM (SAIM), DACVECC
Columbia, MO

Acute, intrinsic renal failure (ARF) is the abrupt, sustained decline in glomerular filtration resulting in azotemia. The kidneys are uniquely susceptible to acute damage due to the large volume of blood filtered for each cardiac cycle, regional differences in blood supply within the kidney, and a high rate of metabolic activity of renal tubular cells. [1] In critical illness, ARF occurs from ischemic events, nephrotoxicity, hypovolemia, drugs, or secondary to the systemic inflammatory response. Common events predisposing to ARF are listed in Table 1.

Development of acute kidney injury occurs through a series of steps that include initiation, extension, maintenance, and recovery. The initiation phase occurs with exposure to the toxicant or the ischemic event.[2] Extension occurs during the response to injury and the subsequent renal damage that exacerbates the response to the initial insult. Recognizing the possibility of renal injury and administering appropriate therapy during the initiation phase prevents the establishment of ARF or lessens the severity of injury. The maintenance phase consists of established tubular lesions and nephron dysfunction. The recovery phase occurs with nephron repair and return to function, with compensatory hypertrophy of the remaining functional nephrons. In general, renal recovery occurs over 6-8 weeks following establishment of renal failure.[3,4,5]

Table 1: Common Causes of Canine and Feline Acute Renal Failure[2,3,4]

Common Nephrotoxic Agents	Common Ischemic Events	Other Events
Therapeutics	Shock	Leptospirosis
Aminoglycosides	Hypovolemic	Pyelonephritis
Cephalosporins	Cardiogenic	Glomerulonephritis
Tetracycline	Distributive	Hypercalcemia
Amphotericin B	Obstructive	Ureteral obstruction
Cisplatin	Decreased cardiac output	Urethral obstruction
Acetaminophen	Trauma	Systemic immune- mediated disease
Aspirin	Hypotension	
Ibuprofen	Hyperthermia/hypothermia	
Radiographic contrast agents	Extensive cutaneous burns	
Nontherapeutics	Renal vessel thrombosis	
Heavy metals	Hyperviscosity syndromes	
Ethylene glycol	Nonsteroidal anti-inflammatory agents inhibiting renal prostaglandins	
Pesticides		
Herbicides		
Envenomnation		
Hemoglobin/Myoglobin		
Raisins / Grapes		
Easter lily		

Diagnosis of Acute Renal Failure

Evaluating specific gravity prior to fluid administration is the best method to determine renal concentrating ability in the dehydrated animal. When pre-treatment urine is not available, urine obtained post fluid administration can be used to evaluate for cellular casts indicative of tubular injury, crystals from renal intoxicants, or inflammatory cells. Although isosthenuria will persist for several weeks following the onset of acute renal failure, urine sediment parameters can be useful to monitor for hospital acquired infection from an indwelling catheter, or for renal injury when nephrotoxic drugs (e.g. amphotericin B, cisplatin) are utilized for therapy. Acute ethylene glycol intoxication can cause an overabundance of calcium oxalate monohydrate crystals following metabolism of the intoxicant. Granular casts are created by tubular debris following acute tubular necrosis, and are seen in acute tubular injury. Urinary casts are fragile and deteriorate within 15-20 minutes of passage; therefore evaluation of fresh urine sediment in-house is necessary to improve the chances of detection.

Evaluation of the serum biochemical profile is the mainstay for diagnosing renal failure.[2] Urea nitrogen and creatinine provide an indication of renal function; however the complete profile also allows for evaluation of electrolyte and protein changes that occur as a result of kidney disease, and sometimes provides clues to the underlying cause of disease. Urea nitrogen is a useful marker of changes in renal filtration. This small molecule (MW 60) is produced in the liver as a byproduct of protein metabolism, freely filtered at the level of the glomerulus, and is about 50% reabsorbed in the renal tubules under normal tubular flow conditions. Creatinine is the other common renal marker on the biochemical profile. Creatinine (MW 115) is a byproduct of muscle breakdown that is freely filtered at the glomerulus and neither reabsorbed nor secreted by the tubules. Therefore, it is viewed as a more sensitive marker for renal disease, although is subject to pre-renal and post-renal elevation similar to BUN. Even if serial creatinine values remains within a laboratory reference range, an elevation of creatinine compared to a previous sample most likely represents a decline in glomerular filtration rate (GFR). Likewise, relatively small incremental increases in creatinine represent significant decreases in GFR in the individual.

Electrolytes that are important for evaluation of the patient with renal failure include phosphorous, potassium, sodium, chloride, bicarbonate, and calcium. Phosphorous is managed similarly to BUN by the kidneys, and an increase indicates a reduction in glomerular filtration. Since phosphorous is an intracellular anion, severe muscle trauma, muscle reperfusion, or tumor lysis syndrome can cause an increase that is not due to renal failure. Phosphorous elevation is not useful in determining whether the azotemia is acute or chronic since either situation can cause hyperphosphatemia. Potassium can be decreased from potassium-wasting renal disease (more commonly seen in cats), or elevated with acute oliguric/ anuric or end-stage chronic renal failure. Hyperkalemia can be life-threatening and require emergency treatment to decrease potassium. Identification of hyponatremia and hypochloremia can be consistent with hypoadrenocorticism. Likewise, hyponatremia can occur in animals with polyuric acute renal failure if there are severe fluid losses from the kidneys. Metabolic acidosis frequently occurs with acute renal failure. Mild hypercalcemia or hypocalcemia can also occur with various presentations of renal failure.[2]

TREATMENT FOR ACUTE RENAL FAILURE

Treatment strategies for acute renal failure are aimed at correcting, hypovolemia, restoring hydration, quantifying urine output to appropriately calculate fluid needs, and managing uremia. Renal recovery following establishment of ARF will continue for several weeks: therefore, animals experiencing acute renal failure should be protected against dehydration or hypovolemia during and after hospitalization. Fluid therapy should be initiated with an isotonic sodium-based, electrolyte solution (replacement crystalloid solution). Fluid dose should be calculated to include maintenance fluid needs, dehydration, and contemporary losses from vomiting or diarrhea. Maintenance fluid requirement can be calculated at a dose of 55-65ml/kg/day. Maintenance fluid requirement combines insensible losses (water used for metabolic reactions, normal fecal losses, ventilation, and mucous membrane evaporation) of 20 ml/kg/day, and estimated urine losses of 35-45 ml/kg/day. Maintenance needs should be adjusted daily, based upon response of the individual patient, and hydration status. Maintenance fluid dose should be given over 24 hours.

Volume of fluid needed for correction of dehydration is based upon the estimate of percent dehydration from clinical parameters. The following formula is used to translate calculated dehydration into fluid requirement: Dehydration Dose = (% estimated dehydration) (Body weight in kg) (1000ml/kg). Rate of administration of the dehydration dose may be divided over 6-24 hours, in addition to the maintenance fluid dose. Excessively rapid volume administration may result in hypervolemia that may precipitate third-spacing of fluid into extravascular spaces, leading to pulmonary edema or pleural effusion. Contemporary losses from vomiting or diarrhea should be estimated, and added to the calculated fluid requirement. Fluid should be administered via constant rate infusion pump if available. Timely evaluation of hydration status is imperative to avoid overhydration.[6]

Acute renal failure can cause anuria, oliguria, or polyuria. Assessment of urine output is essential in tailoring fluid therapy to the individual. The polyuric acute renal failure patient is easier to manage that the patient with oliguria or anuria since hyperkalemia and third spacing of fluids are less likely to occur. Indwelling urinary collection systems provide the most accurate measure of urine output. Drawbacks of urinary catheterization include iatrogenic ascending urinary infection, and urethral trauma causing urethritis.[7] If oliguria or anuria is confirmed, continuation of fluid therapy at a standard maintenance rate will result in hypervolemia and third spacing of fluid.[6] Common clinical manifestations of third spacing of fluid include peripheral edema, pulmonary edema, or pleural effusion. If urine output is inappropriately low, fluid therapy should be altered to an "ins and outs" plan.[2] This type of fluid dosing implies that the amount of fluid administered would equal the volume of fluid being lost in the urine, plus an additional quantity to account for "insensible losses" such as metabolic water needs, respiratory losses, and gastrointestinal losses. Before initiating "ins and outs" therapy, the patient should be euvolemic and hydrated. If urine output is not known but anuria is suspected, the first time interval treatment may be administered with little or no replacement for urine loss. Insensible losses are 20 ml/kg/day. Urinary output is measured over a set period of time, and fluid rate for the subsequent time period would include that volume of urine output administered plus insensible losses for that time period. Time intervals can be set at the clinician's discretion, but are typically between one to six hours.

Potassium is regulated by renal excretion, and oliguria or anuria causes failure of excretion and hyperkalemia. Hyperkalemia causes life-threatening cardiac arrhythmias, culminating in cardiac arrest if left untreated. Ultimately, hyperkalemia must be resolved by improving glomerular filtration rate to promote excretion. Strategies to treat hyperkalemia include drug therapy to shift potassium intracellularly, reducing myocardial sensitivity to potassium, and promoting urinary output. Treatment to lower serum potassium by causing an intracellular shift of potassium includes administration of intravenous regular insulin at ¼ U/kg, followed by intravenous bolus of dextrose at 1 gram for each unit of insulin, and treatment with 5% dextrose for 4-6 hours. Alternatively, intravenous sodium bicarbonate therapy will cause a metabolic alkalosis causing potassium to shift intracellularly.[8] To protect myocardial function while initiating other therapy to lower potassium, calcium gluconate may be given slowly intravenously. Approximate volume of calcium gluconate is 0.3-0.4 ml/kg intravenously. This treatment does not lower serum potassium, but makes the myocardium less sensitive to the effects of hyperkalemia until other therapy is initiated.[8] If polyuria develops, hypokalemia may occur from renal potassium wasting. Therefore, regular potassium monitoring is essential in treatment of acute renal failure.

If oliguria or anuria persists despite restoration of vascular volume and resolution of dehydration, drug therapy may be initiated to promote urine output. Traditional choices for this therapy include mannitol, furosemide, and low-dose dopamine.[2] In human medicine, low-dose dopamine has been shown to increaser renal workload and energy demand and is no longer recommended. Prior to initiating any of these therapies, clinical assessment of hydration status, confirmation of appropriate weight gain, and assessment of central venous pressure are indicated. Administration of diuretic agents while a renal failure patient is dehydrated will exacerbate renal injury. Mannitol produces an osmotic diuresis to cause more urine to remain in the tubules. Typically, one bolus dose is administered at 250-500 mg/kg intravenously. If urine output does not improve within 1 to 2 hours, therapy may be repeated one time with caution. The risk of repeated administration of mannitol, or of administration in the hypervolemic patient, includes hypervolemia and subsequent third-spacing into the pulmonary space.

Furosemide causes loss of activity of the Na-K-2Cl transport pump in the ascending limb of the loop of Henle.[2] Recommended dose for initiation of therapy is 2 mg/kg as an IV bolus. This dose may be incrementally increased to 20 mg/kg by repeat dosing every 2-4 hours. Alternatively, furosemide may be administered via constant rate infusion at 0.25-1.0 mg/kg/hr. Constant rate infusion therapy requires administration by a constant rate infusion pump, but may be somewhat more effective by causing vasodilation in addition to the diuretic effect. Controlled studies comparing the two methods of administration in veterinary patients are lacking. Furosemide administration in cases of acute renal failure from aminoglycoside toxicosis should be avoided since furosemide exacerbates accumulation of aminoglycoside antibiotics in the proximal renal tubule. If drug therapy is successful in promoting urine flow, fluid rates must be adjusted accordingly to avoid dehydration and repeated renal injury. If the patient remains oliguric despite this therapy, eventually life-threatening hyperkalemia and third-spacing of fluid into the pulmonary space will result. The only options for ongoing therapy include peritoneal dialysis or hemodialysis.[9] Dialysis therapy reduces uremic waste products by promoting passage of uremic waste products across a semipermeable membrane (an artificial dialyzer for hemodialysis, or the peritoneal surface for peritoneal dialysis). Peritoneal dialysis is technically difficult and labor-intensive, however is more widely available than hemodialysis. Hemodialysis is currently available at a few universities and specialty referral centers in the United States. Dialysis does not cure kidney failure but can "buy time" to avoid life-threatening consequences of azotemia while waiting for renal recovery to occur.[9] Depending upon the underlying cause of renal failure, when the patient is stabilized, renal transplant may be an option to restore renal function.

Serum biochemical values should be monitored at 24-48 hours to evaluate azotemia and electrolyte abnormalities. When azotemia has resolved or is no longer changing, fluid therapy should be gradually reduced. The patient should be eating and drinking freely during fluid taper so that water consumption will increase to meet physiologic needs. Renal recovery usually takes 4-8 weeks following establishment of renal failure. However if the patient compensates for polyuria by increased drinking, the patient may be able to maintain and recover at home. Repeat episodes of dehydration should be strictly avoided during this time.

MANAGEMENT OF UREMIA

Uremia causes injury to various organ systems such as the gastrointestinal tract, pulmonary system, hematologic system, cardiovascular system, and central nervous system. Managing azotemia is the best way to limit damage of other organ systems. While azotemia is severe, a variety of ancillary therapies can be used to limit nausea and protect the GI tract. Nutritional support is important; however sick patients with acute renal failure will be unlikely to accept a diet change to a "kidney-type" diet. If there is no appetite, a feeding tube may be placed. If phosphorous remains elevated following rehydration and the patient is eating, oral phosphate binders (aluminum hydroxide, calcium carbonate) may be administered to bind intestinal phosphorous. Oral ulcers should be rinsed daily with an antiseptic solution. If renal failure is the result of bacterial infection (pyelonephritis, leptospirosis), appropriate antibiotic therapy is indicated.

REFERENCES

[1]Racusen LC. **Acute Renal Failure: A Companion to Brenner & Rector's The Kidney**. Philadelphia: Saunders, 2001;1. [2]Ross L. **Vet Clin North Am Small Anim Pract** 2011;41(1):1. [3]Behrend EN, et al. **J Am Vet Med Assoc** 1996;208(4):537. [4]Stokes JE, et al. **Compend Contin Educ Pract Vet** 2006;36(5):387. [5]Anderson RJ. **Acute Renal Failure: A Companion to Brenner & Rector's The Kidney**. Philadelphia: Saunders, 2001;157. [6]Prowle JR, et al. Nat Rev Nephrol 2010;6(1);107. [7]Smarick SD, et al. **J Am Med Assoc** 2004;224(12):1936.[8]Thomovsky EJ. **Vet Med** 2011;106:352. [9]Langston CE. **Compend Contin Educ Pract Vet** 2002;24(7):540.

COMPLICATED FELINE URINARY OBSTRUCTION

Marie E. Kerl DVM, MPH, DACVIM (SAIM), DACVECC
Columbia, MO

PRESENTING COMPLAINTS

Feline urethral obstruction (FUO) represents a common disease presentation for veterinarians in emergency practice. Urinary obstruction occurs secondary to feline lower urinary tract disease (FLUTD). Whereas male and female cats can develop FLUTD, male cats more commonly develop FUO due to differences in urethral anatomy. While cats with complete urinary obstruction rarely present a diagnostic challenge, treatment can be complicated by the comorbidities caused by urine retention, bladder rupture resulting in uroabdomen, and risk of repeat urinary obstruction.[1]

Historical complaints consistent with FUO include agitation, increased vocalization, frequent trips to the litter box and attempts to void, indiscriminate urination, pollakiuria, hematuria, lethargy, increased urine odor, and soiling of the perineum. Vomiting can accompany episodes of straining, and owners sometimes mistake straining to urinate for constipation. If the obstructive phase was unrecognized by the owner and electrolyte changes or bladder rupture have occurred, the owner may find the cat mentally dull, stuporous, or comatose. Physical examination findings for cats that have recently obstructed include a large or turgid bladder that is painful on palpation. Depending on the duration of obstruction, other examination findings include dehydration, mental dullness, stupor, bradycardia, and arrhythmia.[2] If the metabolic status has deteriorated to shock, additional findings include pallor, hypothermia, cool extremities, and poorly palpable peripheral pulses. If the urinary bladder has ruptured, it may not be palpable and a fluid wave may be present.

INITIAL UNBLOCKING AND MANAGEMENT

Relief of urinary obstruction requires sedation for virtually all cats, and sometimes requires general anesthesia for severe obstruction. Following sedation, the perineum is surgically prepared, and the cat is positioned in dorsal recumbency with hind legs advanced cranially to expose the perineal area. The penis is exteriorized by applying digital pressure on the ventral surface of the prepuce. Once the penis is visualized, the prepuce can be held tightly around the penis to keep the penis exteriorized. Maintaining sterility, urethral catheterization is performed. A variety of catheters can be used for initial unblocking. Care must be taken to avoid excessive urethral trauma. Attaching an extension set to the distal end of the catheter is helpful to flush the urethra with saline while attempting catheterization. This action will sometimes help to dislodge distal obstructions.[2] Following initial unblocking with a temporary catheter, a more flexible catheter can be placed and attached to the prepuce with a tape butterfly and suture or using the attachments on the catheter itself. Recent investigation into factors that increase the risk of repeat obstruction following unblocking identified the use of 5-Fr, as opposed to 3.5-Fr, red rubber catheters as increasing repeat urinary obstruction in cats.[3] Urethral catheters should be attached to a closed urinary collection system, and managed with appropriate hygiene to minimize the risk of ascending infection. There is no set time to leave an indwelling urinary catheter in place following unblocking.[4] Depending on the severity of the original obstruction, catheters can stay in place for 24-48 hours, and then removed to assess urethral patency.

The most life-threatening electrolyte abnormality in the urinary obstruction cat is hyperkalemia. Cats that present with dehydration, dull or stuporous mentation, bradycardia or other cardiac arrhythmias in association with urinary obstruction should receive immediate intravenous catheterization for management and an EKG. Cardiac rhythm abnormalities associated with hyperkalemia include tall, tented T waves, diminished or absent P waves, and wide, bizarre QRS complexes. Cardiac changes are typically only seen with hyperkalemia exceeding 7mmol/L, and both the magnitude and the rapidity of elevation contribute to the severity of clinical signs. If hyperkalemia is left untreated, cardiac asystole occurs causing cardiopulmonary arrest. If EKG changes are present, treatment for hyperkalemia should be initiated even if potassium cannot be measured immediately.[2] Treatments for hyperkalemia are effective either by shifting potassium intracellularly, or by restoring the electronegative potential of cardiac conducting cells. In addition, unblocking will provide the route by which potassium is excreted, and intravenous fluid therapy will increase renal potassium excretion. Intravenous catheter should be placed immediately in order to give emergency treatments to lower potassium and to administer fluid treatment and other therapies. If the cat is comatose, sedation may not be needed. Analgesic therapy is indicated prior to unblocking. Two treatment strategies have been used to shift potassium into the intracellular space: regular insulin and dextrose (or dextrose alone), and sodium bicarbonate. Regular insulin stimulates the Na^+/K^+ ATPase to cause intracellular movement of potassium. Dextrose is given following insulin administration to prevent hypoglycemia. If dextrose is given alone, insulin release is stimulated from the pancreatic beta cells. The dose of insulin is 1 unit per cat intravenously, followed by 0.5 g/kg of 50% dextrose diluted for intravenous administration by bolus, and then administration for 2.5% dextrose in fluids for 4-6 hours to prevent hypoglycemia. This treatment will take 20-30 minutes to be effective. Sodium bicarbonate administration induces a metabolic alkalosis which in turn causes intracellular shifting of potassium. Bicarbonate administration also will cause complexing of calcium that can result in ionized hypocalcemia, so this treatment is not recommended as a first-line therapy. Bicarbonate dose is 1 mEq/kg once, given slowly, and effectiveness takes 20-30 minutes. For moderate to severe hyperkalemia and electrolyte changes, 10% calcium gluconate should be administered to restore the normal difference between the resting membrane potential and the threshold potential. The dose is 3 mL/cat (1-1.5 mg/kg) given IV slowly over 3-5 minutes. The EKG changes should resolve if this treatment is effective; dosing can be repeated one time if necessary.[2]

Fluid therapy should be calculated initially for maintenance and rehydration. Urine output should be monitored carefully, both to insure patency of the catheter and to insure that polyuria does not cause the urine output volume to exceed volume of fluid administered. Once the cat is rehydrated and fluid balance is restored, fluid therapy can be adjusted to slightly exceed the urine output volume. Cats with moderate to severe urinary obstruction are expected to have a post-obstructive diuresis for 3-5 days following unblocking. The kidneys are unresponsive to normal physiologic mechanisms to concentrate urine due to the quantity of uremic waste products being excreted and the back-pressure on the renal tubules causing transient tubular dysfunction. During this phase, fluid treatment can be transitioned to "ins and outs" therapy, which is calculated as "insensible losses" of 20 mL/kg/day, and additional fluid volume to match hourly urine output volume from the previous 4-6 hours. Normal hydration must be established prior to starting this therapy since there is no rehydration replacement in this method of fluid calculation.

MANAGEMENT POST-UNBLOCKING AND COMPLICATIONS

In addition to appropriate fluid therapy and maintaining urethral patency, analgesic therapy should be administered to maintain patient comfort. Intermittent opioid therapy with buprenorphine provides sufficient maintenance of analgesia for most cats once the urethral obstruction is relieved. Antibiotics are not indicated in most cats since bacterial urinary infection occurs in less than 2% of FUO cats that are first-time offenders.

There is no ideal length of time for maintaining an indwelling urethral catheter following unblocking for feline urinary obstruction (FUO). Reasonable catheter duration is from 0-48 hours, depending on the severity of the obstruction, presence of physical obstruction (e.g. urethral calculi) and need to closely monitor urine output volume. If the urethral obstruction is very distal and resolves easily, maintaining an indwelling catheterization might not be necessary. Eisenberg et al identified that cats with urethral catheters in for a shorter duration of time were at greater risk for reobstruction than those with catheters in for longer periods; however, median time in the short duration group was 24.5 hours and in the long duration group was 26 hours.[4]

If an indwelling catheter is maintained, it must be attached to a closed urine collection system (UCS) to maintain sterility of the urinary system. An Elizabethan collar should be utilized to prevent the cat from removing the catheter prematurely, or chewing the catheter in two and creating a urinary foreign body. Cats should also be hospitalized in a facility with 24-hour monitoring to prevent repeat obstruction of the catheter or the UCS from patient movement or debris in the urinary bladder. The urinary catheter can be gently flushed with 2-3 mL of sterile saline every 4-6 hours if there is debris in the bladder to maintain catheter patency. Once the catheter is removed, the cat should be able to void completely and empty the bladder.

There are 2 reasons for failure to empty the bladder following unblocking and catheter removal: repeat urethral obstruction, and detrusor atony. Causes for repeat obstruction include reestablishment of urethral plug from debris in the bladder, urethritis, urethral injury, or urethral calculus or stricture. Manual expression and evaluation of urine stream can be attempted; however, excessive manual pressure can cause bladder injury and urine reflux into the ureters and renal pelvises. If the bladder does not empty easily, repeat sedation and urinary catheterization should be performed. If the catheter passes easily and no obstruction is found, intermittent catheterization to empty the bladder might allow urethral inflammation to subside. If a physical obstruction is identified, the catheter should remain in place. If radiographic imaging was not performed initially, these should be performed at repeat catheterization to evaluate for urethral calculi, being careful to include the entire urethra in the study. If urethral tear has occurred, the catheter should remain in place for 5-7 days.

Clinical signs associated with detrusor atony include large bladder with no attempts to void and urinary incontinence due to overflow. Depending on the duration of distention, detrusor atony can be partial or complete. Medical management can be instituted with prazosin, a selective alpha-1 antagonist at 0.25-0.5 mg per cat orally every 12-24 hours to relax the internal urethral sphincter, and bethanechol, 1.25-2.5 mg per cat orally every 8-12 hours. Failure to respond following an appropriate duration of medical management is consistent with permanent detrusor atony.

Following serious urinary obstruction, cats undergo an obligate phase of diuresis termed "postobstructive diuresis". This occurs due to the need to excrete a large quantity of uremic waste following obstruction, and kidney injury caused by excessive back-pressure on the renal tubules that disrupts absorptive capabilities. The duration and severity of the diuresis phase is directly associated with the severity of obstruction. Cats with a simple urinary obstruction that are not systemically ill at the time of presentation have minimal to no obligate diuresis phase. Urine output exceeds 2 mL/kg/hr, and may be dramatically higher.[5] During the obligate diuresis phase, fluid must be dosed to prevent dehydration. Body weight and hydration status should be carefully assessed twice daily to minimize the risk of dehydration and overhydration. Following 2-5 days of this treatment plan, if the cat is eating and drinking and azotemia has resolved fluid taper can begin. If the cat becomes dehydrated during the taper, a longer period of intravenous therapy is required.

Resolving urethral obstruction is extremely difficult in some cats. Possible causes for difficulty unblocking include urethral calculi, anatomical abnormality, urethral injury from trauma or prior catheterization attempts, and urethral stricture. Excessive back-pressure from an overdistended urinary bladder can also make catheterization more difficult. If urethral catheterization attempts are unsuccessful, cystocentesis can be attempted to relieve back-pressure.[6] The bladder is more prone to injury from needle insertion at this time, so attempts at cystocentesis should not be undertaken lightly. A 22 or 23 gauge needle attached to an extension line or butterfly catheter can be inserted at the base of the bladder while the bladder is manually isolated and held to the body wall. Following needle insertion, an assistant should withdraw as much urine as possible via syringe attached to the extension set to relieve back-pressure, and also to empty the bladder in case urethral catheterization is not possible. Avoid multiple needle insertions into the urinary bladder to limit risk of obstruction.

Obligate diuresis following unblocking can result in potassium-wasting which leads to hypokalemia. Cats that become hypokalemic exhibit a variety of clinical signs including muscle weakness, cervical ventroflexion, cardiac arrhythmias, and ventilatory failure. Following resolution of the urinary obstruction and hyperkalemia, intravenous fluids can be supplemented with potassium chloride (KCl) at a rate not to exceed 0.5 mEq/kg/hr to manage hypokalemia. Typically, if patient serum potassium is between 3.5-5.0 mEq/l, addition of 20 mEq KCl per liter of isotonic crystalloid fluid with administration at a rate calculated for that patient will maintain potassium in a normal range. Be careful with addition of potassium chloride since doses of greater than 0.5 mEq/kg/hr can cause cardiac standstill and death. Hypokalemia is typically short-lived and resolves with return to normal urinary function; however, oral supplementation of potassium in the form of potassium gluconate is useful for long-term supplementation.

There is no set time point at which perineal urethrostomy is indicated. Idiopathic feline lower urinary disease can be self-limiting, and normal urinary function can be restored following obstructive episodes even if the episode is severe. Certainly, if an obstruction is permanent (e.g. urethral stricture), perineal urethrostomy is indicated. Repeat urinary obstructions can cause death if not addressed rapidly, and some clients may elect perineal urethrostomy to prevent a future life-threatening obstruction.[7] Finally, veterinarians may provide surgical treatment as a way to limit future cost of unblocking and hospitalization for cats that have failed appropriate medical management when clients have financial limitations that would force a decision of euthanasia with repeat unblocking episodes.

REFERENCES

[1]Gunn-Moore DA. **J Fel Med Surg** 2003;5:133. [2]Thomovsky EJ. **Vet Med** 2011;106:352. [3]Hetrick PF, et al. **J Am Med Assoc** 2013;243(4):512. [4]Eisenberg BW, et al., et al. **J Am Med Assoc** 2013;243(8):1140. [5]Francis BJ, et al. **J Fel Med Surg** 2010;12(8):606. [6]Cooper ES, et al. **J Am Med Assoc** 2010;237(11):1261. [7]Ruda L, et al. **J Sm Anim Pract** 2012;53:693.

NON-REGENERATIVE ANEMIAS IN DOGS
Amy Koenigshof, DVM, MS, DACVECC
East Lansing, MI

Anemias are common in critically ill dogs and can be classified based on the underlying mechanism of the anemia. Hemolytic anemias and blood loss anemias usually have a robust regenerative response, however it can take 3-4 days after the initial insult for the regenerative response to occur. Often these dogs present with acute onset of their clinical signs. Non-regenerative anemias typically have a longer history of less severe clinical signs. The anemia is usually slowly progressive due to the relatively long lifespan of the dog red blood cell of around 104 days.

Diagnostic approach

In any dog presenting with anemia just a few simple tests can help differentiate amongst the major cause of anemia: hemorrhage, hemolysis and decreased production of red blood cells. A packed cell volume (PCV) and total solids (TS) should always be performed. Most commonly, except in peracute hemorrhage, if bleeding is the cause of the anemia the PCV and TS will both be decreased. With hemolysis and decreased production the PCV is decreased, but the TS is often normal to slightly increased. A blood smear can help differentiate between a regenerative anemia (likely to be from hemorrhage or hemolysis) and a non-regenerative anemia (decreased production, or a pre-regenerative anemia). Evidence of regeneration on a blood smear can include polychromasia, and anisocytosis. Spherocytes will often be seen in cases of immune-mediated hemolytic anemia. To definitively identify a regenerative response, quantification of the reticulocyte count is necessary. Typically a reticulocyte count greater than 60,000 reticulocytes/μL is considered regenerative. A CBC will allow for quantification of the reticulocyte count as well as evaluation of the other cell lines.

Once the dog's anemia has been classified as non-regenerative based on a more chronic history (ruling out a pre-regnerative state), the CBC and reticulocyte count a full-work up to identify the cause is necessary. A chemistry and urinalysis allow for identification of systemic illnesses such as chronic kidney disease, severe liver disease, or hypoadrenocorticism as the cause of the anemia. If the chemistry and physical exam are supportive of hypoadrenocoritcism or hypothyroidism endocrine testing (ACTH stimulation test or a thyroid profile) should be performed. Chest radiographs and abdominal radiographs or ultrasound are performed to look for neoplasia. Depending on geographic location testing for tick borne illness, particularly *Ehrlichiosis*, may be indicated. Less common causes of non-regenerative anemias including cobalamin deficiency may be suspected if there are other clinical signs. Anemia. Cobalamin and folate levels can be measured if a deficiency is suspected. Other critical illnesses and chronic illnesses can also lead to a non-regenerative and are usually identified with the diagnostics discussed above. If a systemic cause of the anemia is not found, a bone marrow aspirate and biopsy are needed to identify the cause.

Systemic illness

Chronic renal failure is a common cause of non- regenerative anemia. The main cause of anemia in these patients decreased erythropoietin production, however chronic gastrointestinal blood loss can also contribute to the anemia. In patients with liver disease inflammation, abnormal nutrient metabolism and defective iron handling likely all contribute to anemia. Dogs with hypoadrenocorticism often have a mild to moderate anemia. This may be masked on initial exam due to severe dehydration. Contributing factors to anemia in these patients include lack of stimulation of erythropoiesis by glucocorticoids and potentially chronic gastrointestinal blood loss. Hypothyrodisim can also lead to anemia through decreased erythropoiesis. Contributing factors likely include decreases erythropoietin, decreased responsiveness to erythropoietin, and lack of stimulation of the hematopoietic stem cells by thyroid hormone. In dogs with neoplasia both treatment and disease can contribute to the anemia. Neoplasia can lead to increased apoptosis and inflammation which decreases red blood cell production. While acute infection with *Ehrlichiosis* tends to cause a regnereative aneima, chronic *Ehrlichiosis* can lead to a non-regnerative anemia. Though the mechanism of this anemia is not fully understood, infection of red blood cell precursors has been documented. Systemic inflammation during chronic or critical illness can lead to a non-regenerative anemia primarily through iron sequestration. Iron sequestration is thought to be a defense mechanism against bacteria, making this micronutrient scarce and unavailable for use by bacteria. Blood loss can also contribute to the anemia seen with critical illness. Most commonly this is from gastrointestinal ulceration, however frequent blood sampling, especially in small patients can significantly contribute to the anemia.

BONE MARROW DISEASE

Toxicity

Exposure to toxins such as estrogen or cytotoxic drugs used to treat neoplasia can lead to non-regenerative anemias. In addition estrogen secreting tumors such as Sertoli cell tumors or Lydig cell tumors can cause anemia. Whole or half body radiation commonly leads to pancytopenia. Removal of the source and supportive care (antibiotics in neutropenic patients, transfusions as needed) usually leads to resolution of the anemia.

Bone Marrow Necrosis

Bone marrow necrosis from occlusion of small blood vessels can be caused by numerous infectious, immunologic and neoplastic diseases. Idiopathic bone marrow necrosis has also been reported. In cases of systemic illness, treatment of the underlying cause and supportive care can lead to recovery of the bone marrow. In cases of idiopathic bone marrow necrosis, the disease can resolve with just supportive care.

Neoplasia

Bone marrow neoplasia leads to anemia by crowding out the normal cells in the bone marrow. Most commonly neoplasia will lead to a pancytopenia because of crowding of all precursor cells. Blastic cells can be seen on bone marrow analysis.

Immune-mediated disease

Immune-mediated destruction of red blood cell precursors in the bone marrow is becoming a more well recognized cause non-regenerative anemias in dogs. Historically these patients have been diagnosed with "non- regenerative immune mediated anemia," however with new research showing frequent phagocytosis of red blood cell precursors a new term: Precursor-targeted immune mediated anemia (PIMA) has been proposed. A recent retrospective study involving dogs with PIMA identified a similar predilection for females as is seen in dogs with immune mediated hemolytic anemia (IMHA), however no breed predilection was identified. Dogs with PIMA usually have a longer history of mild clinical signs prior to presentation when compared to dogs with IMHA due to the relatively long life span of red blood cells and lack of systemic illness from hemolysis. A diagnosis of PIMA is made based on the above diagnostic approach combined with a bone marrow aspirate and biopsy. On bone marrow exam, rubriphagocytosis is commonly seen as well as myelofibrosis. Maturation arrest with expansion of red blood cell precursors prior to the phase being phagocytosed is also commonly seen. The cornerstone of treatment for PIMA is packed red blood cell transfusions as needed and immunosuppression. Prednisone (2 mg/kg/day) is the main immunosuppressant used. The author commonly combines prednisone with another immunomodulating agent such as azathioprine or mycophenolate. It typically takes 1-2 months (or even longer in some cases) for dogs to become regenerative once they are started on immunosuppressive therapy. During this time period dogs commonly need transfusions as supportive care. Similarly to dogs with IMHA, thromboembolic disease has been documented in dogs with PIMA. Unlike dogs with IMHA where thromboembolism is most common in the first week of treatment, dogs with PIMA can develop thromboembolism later during the course of treatment. Thromboprophylaxis should be considered for these patients.

REFERENCES

Gravelyn T, Koenigshof A, Jutkowitz A, Scott M. Prevalence of thromboembolic events in dogs with precursor-targeted immune-mediated anemia. *J Vet Emerg*. 2014; 24(S1): S15.

Gravelyn T, Koenigshof A, Scott M, Lucindi C, Jutkowitz A. Clinical features of precursor-targeted immune-mediated anemia in dogs: 65 cases (2004-2013). *J Vet Emerg*. 2014; 24(S1): S7.

Stokol T, Blue JT, French TW. Idiopathic pure red cell aplasia and non-regenerative immune-mediated anemia in dogs: 42 cases (1988-1999). *J Am Vet Med Assoc*. 2000; (9) 1429-36.

Weiss DJ. Bone marrow pathology in dogs and cats with non-regenerative immune-mediated hemolytic anemia and pure red cell aplasia. *J Comp Pathol*. 2008; (1): 46-53.

APPROACH TO THE BLEEDING PATIENT:

Amy Koenigshof, DVM, MS, DACVECC
East Lansing, MI

Hemorrhage is a common condition in the emergency setting. Patients may present with external hemorrhage or they can present with signs of lethargy, weakness or collapse. The location of hemorrhage can help differentiate defects in primary hemostasis (thrombocytopenia/thrombocytopathia) from disorders of secondary hemostasis (coagulation factors). Patients with disorders of primary hemostasis often bleed from mucosal surfaces. They may present with epistaxsis, petechial and ecchymotic hemorrhages, GI bleeding, or bleeding from the gums. Patients with disorders of secondary hemostasis usually bleed into body cavities such as the abdomen, thorax or pericardium. They may also bleed into tissues surround the upper airway and trachea, or joints. While these guidelines hold true for most patients, evaluation of a patient for hemorrhage should always include assessment of both primary and secondary hemostasis.

Evaluation of primary hemostasis

Disorders of primary hemostasis may be caused by thrombocytopenia or thrombocytopathia. In a bleeding patient, automated platelet counts that reveal thrombocytopenia should always be confirmed by a manual platelet count. Platelet clumping can falsely lower the automated platelet count. Evaluation of platelet function is more challenging. The buccal mucosal bleed time (BMBT) is used as a screening test for thrombocytopathias. The BMBT is performed using a standardized bleeding time device and filter paper to wick away the blood from the incision, care must be used to not disturb a forming clot. To perform the BMBT, the patient is placed in lateral recumbency and the upper lip is everted using a gauze strip loosely tied around the maxilla. The standardized BMBT device is placed on the buccal mucosa and triggered, creating a standard incision that is 1 mm in width and 5 mm in length. Filter paper is used to blot the blood distal to the incision, using care not to touch the incision or a forming clot. The timing is stopped when a clot forms. Normal BMBT in dogs is less than 4 minutes. BMBT should not be performed in dogs with significant thrombocytopenias (<100,000 platlets/ µL) as thrombocytopenia will cause a prolonged BMBT. Anemia also prolongs the BMBT. The BMBT is a good screening test for platelet function, however it is not sensitive and the results are often inconsistent between users. Other assessments of platelet function often require specialized equipment such as the platelet function analyzer (PFA), platelet aggregometry or flow cytometry.

Evaluation of secondary hemostasis

Traditional coagulation tests such as the activated clotting time (ACT), prothrombin time (PT) and activated partial thromboplastin time (APTT) provide assessment of secondary hemostasis. The ACT is a bed-side test that evaluates the intrinsic and common pathways. The conventional method for assessment of ACT involves placing 2 mL of whole blood into a prewarmed (37°C) ACT vacuum tube which contains diatomaceous earth. The diatomaceous earth causes contact activation (activation of the intrinsic pathway). After the blood is placed in the tube, the tube is inverted 5 times to mix the blood and then incubated at 37°C for one minute. After the incubation period, the tube is inverted every 5-10 seconds to evaluate for clot formation. Normal ACT is 60-110 seconds in the dog and 50-75 seconds in the cat. Severe thrombocytopenia (<10,000/µL) may prolong the ACT. Many automated instruments are also available for ACT measurement. The aPTT also evaluates the intrinsic and common pathways. It is performed on citrated plasma, therefore it is not affected by thrombocytopenia. The PT evaluates the extrinsic and common pathways. Like the aPTT it is performed on citrated plasma.

Viscoelastic Testing

Viscoelastic testing allows for evaluation of both primary and secondary hemostasis. Whole blood is used which allows for evaluation of the interaction of coagulation factors, platelets, red blood cells and white blood cells. Typically citrated whole blood is use for the sample. Calcium is added back to the blood just prior to starting the test. Viscoelastic testing result can be greatly affected by preanalytical variations in sample handling. To minimize the effects of preanalytical variations, other activators such as tissue factor or kaolin can be added to the sample. Viscoelastic testing also offers the advantage of being able to identify a prothrombic state in the patient. However, viscoelastic testing results have not been correlated to clinical bleeding or risk for thrombus development and therefore must be interpreted with caution.

COMMON CAUSES OF BLEEDING

Thrombocytopenia

Clinical bleeding from thrombocytopenia usually does not occur until the platelet count decreases below 50,000 platelets/µL. The most common cause of severe thrombocytopenia (<30,000/µL) is immune mediated thrombocytopenia (IMT). During the diagnostic work-up of a patient suspected of having IMT, other causes of thrombocytopenia should be ruled-out including a history of recent vaccination or medications (sulfa drugs, etc.), infectious agents (tick borne disease, FeLV), neoplasia (chest radiographs, and abdominal radiographs or ultrasound), or bone marrow disorders (presence of other cytopenias). The treatment of IMT involves immunosuppression with prednisone (2 mg/kg/day). Many patients can be treated with prednisone alone, however adjuvant drugs such as azathioprine (2 mg/kg/day for 7 days then 2 mg/kg every 48 hours) are occasionally needed. Vincristine (0.02 mg/kg IV) causes release of platelets from megakarocytes in the bone marrow and has been shown to reduce the length of hospitalization of dogs with IMT. Platelet transfusions are not cost effective or indicated in patients with IMT. Transfused platelets are destroyed by the patient within hours of administration. Patients should be slowly weaned off steroids over 4-6 months, rechecking the platelet count after each reduction in prednisone dose.

Thrombocytopathia

Von Willebrand disease (vWD) is the most common hereditary disorder of platelet function in dogs. There are three different types of vWD, classified by plasma vWF concentration and vWF composition.

Type	Plasma vWF concentration (vWF:Ag)	vWF composition	Clinical bleeding
Type 1	Low	Normal distribution	Mild to moderate
Type 2	Variable	Disproportionate loss of high molecular weight multimers	Moderate to severe
Type 3	Complete lack of vWF	None	Severe

Type 1 is the most common form of vWD. Patients with vWD have a normal platelet count and normal coagulation times, the BMBT is often prolonged. Definitive diagnosis is made by assessing plasma vWF:ag levels. Bleeding patients suspected of having vWD (or known to have vWD) should be treated with cryoprecipitate or fresh frozen plasma. Transfusion of packed red blood cells (pRBCs) may also be needed if the patient has bleed to the point of a clinically significant anemia. Desmopressin acetate (DDAVP) can be administered subcutaneously (1 μg/kg) to patients with type 1 vWD to reduce the chances of severe hemorrhage when invasive procedures such as surgery are necessary.

Many other inherited disorders have been reported in dogs and cats but are beyond the scope of this discussion. Drugs such as synthetic colloids and NSAIDS can also cause platelet dysfunction.

Acquired coagulopathies

The most common acquired coagulopathy in dogs is ingestion of anticoagulant rodenticides. Anticoagulant rodenticides inhibit vitamin K epoxide reductase, the enzyme required to activate vitamin K which is required to carboxylate (and therefore activate) the vitamin K dependent coagulation factors. Coagulation factors II, VII, IX and X as well as the anticoagulant proteins C and S require vitamin K for activation. Diagnosis of anticoagulant rodenticide toxicity as the cause of bleeding in a patient can be made from a history of ingestion and/or assessment of coagulation times. Patients who have ingested an anticoagulant rodenticide will have a prolonged ACT. Their PT and aPTT will be prolonged, but the PT will be more prolonged than the aPTT due to factor VII (in the extrinisic pathway) having the shortest half-life. Measurement of blood levels of anticoagulant rodenticides can confirm the diagnosis; however the patient must be treated before the results of this test are available. Patients bleeding from anticoagulant rodenticide toxicity require replacement of coagulation factors with either fresh frozen plasma or frozen plasma (10-20 mL/kg). Vitamin K_1 is started as soon as possible (2.5-5 mg/kg orally) and continued for 30 days. The PT should be rechecked 48 hours after the last dose of Vitamin K_1 to ensure that the length of therapy was sufficient.

Inherited coagulopathies

Hemophilia A (Factor VIII deficiency) is the most common congenital coagulation abnormality. Patients with hemophilia A tend to bleed into body cavities, form hematomas following relatively small traumas, or bleed into joints. It is an X-linked recessive trait, so males are most commonly affected. Patients with hemophilia A will have a prolonged aPTT and a normal PT. Definitive diagnosis is made via specific factor assay.

Hemophilia B (Factor IX deficiency) has also been described in dogs and cats. Clinical signs are similar to patients with hemophilia A. The aPTT will be prolonged and the PT normal in patients with hemophilia B. Definitive diagnosis is made via measurement of factor IX levels.

Trauma

Trauma is a common cause of bleeding. History and physical exam usually lead to a suspicion of trauma. Hemoabdomen and hemothorax are common in trauma patients. Treatment of bleeding from trauma involves supportive care including volume resuscitation. Occasionally pRBC transfusions are required if the patient remains clinical for its anemia after volume expansion. Patients with severe trauma and bleeding may even require plasma transfusion; this should be guided by evaluation of coagulation times. Rarely, patients with trauma will require surgery to stop the bleeding. Surgery should be considered in patients who cannot be stabilized medically (following fluid expansion and transfusion of blood products). In the case of a traumatic hemoabdomen that cannot be medically stabilized, the cause of bleeding is often fracture of the spleen or liver. Fractured bones may also lacerate arteries, necessitating emergency surgery.

Bleeding masses

Spontaneous hemoabdomens are commonly caused by bleeding masses on the liver or spleen. Most of these masses are either on the spleen or liver. The most common cause of a bleeding mass is neoplasia, with up to 75-80% of bleeding splenic masses being cancerous. Hematomas can also form and rupture leading to a hemoabdomen (20-25% of bleeding splenic masses). Assessment of coagulation (PT and aPTT) should be performed to rule out rodenticide as the cause of bleeding. A CBC and serum chemistry are performed to evaluate platelet count, severity of anemia and organ function. Chest radiographs are performed to evaluate for metastatic spread of cancer to the chest. Abdominal ultrasound may be useful to identify which organ the mass is originating from, however malignant masses cannot be differentiated from benign masses using ultrasound. Ultrasound delays the time to definitive treatment (surgery) so the author does not perform ultrasound unless it is needed to help an owner decide if they want to pursue surgery. Treatment of these patients involves volume resuscitation with crystalloids and transfusion of pRBCs if necessary followed by surgery to remove the bleeding mass.

REFERENCES

Aronsohn MG, Dubiel B, Roberts B, Powers BE. Prognosis for acute nontraumatic hemoperitoneum in the dog: a retrospective analysis of 60 cases (2003-2006). J Am Anim Hosp Assoc. 2009, 45(2): 72-77.

Callan MB, Giger U. Effect of desmopressin acetate administration on primary hemostasis in Doberman Pinschers with type-1 von Willebrand disease as assessed by a point-of-care instrument. Am J Vet Res 2002, 63(12):1700-1706.

Carr AP, Nibblett BM, Panciera DL. Von Willebrand's disease and other hereditary coagulopathies. In: Bonagura JD and Twedt DC. *Kirk's Current Veterinary Therapy XIV.* St. Louis. Saunders. 2009.

DeClementi C, Sobczak BR. Common rodenticide toxicoses in small animals. Vet Clin North Am Small Anim Pract. 2012, 42(2): 349-360.

Murphy MJ. Rodenticide Toxicoses. In: Bonagura JD and Twedt DC. *Kirk's Current Veterinary Therapy XIV.* St. Louis. Saunders. 2009.

Rozanski EA, Callan MB, et al. Comparison of platelet count recovery with use of vincristine and prednisone or prednisone alone for treatment for severe immune-mediated thrombocytopenia in dogs. J Am Vet Med Assoc. 2002, 220(4): 477-481.

IMMUNE MEDIATED HEMOLYTIC ANEMIA: BEYOND STEROID THERAPY

Lenore Bacek, DVM, MS, DACVECC
Auburn, AL

INTRODUCTION

Immune-mediated hemolytic anemia (IMHA) is an often fatal disease with mortality rates reported as high as 50-70% with significantly higher rates initially in the first two weeks of treatment.[1-3] It is the most common cause of hemolytic anemia in dogs and is most often a primary or idiopathic process, accounting for 60-75% of cases. IMHA can also be secondary to any number of infectious, neoplastic, drug-related or other conditions. The destruction of the red blood cells is a result of a type II (antibody-mediated) hypersensitivity reaction in which cells are lysed after the binding of immunoglobulins and complement, or both.[4] IgG or IgM antibodies are directed against the red blood cell membrane. Intravascular hemolysis is a result of IgM-mediated intravascular complement activation in contrast to IgG-mediated extravascular hemolysis by macrophages within the mononuclear phagocytic system, primarily in the liver and spleen.[2] The destruction by the mononuclear phagocytic system is mediated by Fc receptors on the surface of the macrophage. The Fc receptor binds to the Fc portion of the IgG molecule leading to phagocytosis of the red blood cell.[4] On the other hand, with complement activation, red blood cell membranes can become damaged allowing water leakage into the cytoplasm and subsequent swelling and lysis.[5]

Treatment of IMHA is largely based on clinical experience with a lack of prospective studies evaluating an optimal therapeutic course. Although the mainstay for treatment is still glucocorticoids, multiple treatments are currently used including other immunosuppressive therapies such as azathioprine, cyclosporine, mycophenolate mofetil, leflunomide, liposomal clodronate, human immunoglobulin, splenectomy and therapeutic plasmapheresis. There is no evidence for increased survival with the addition of multiple immunosuppressive therapies. However, side effects of glucocorticoids may be significantly detrimental and the addition of another immunosuppressive agent may allow for dose-reductions or complete discontinuation. Although there are no definitive parameters for adding more immunosuppressive therapy, a lack of response to steroid therapy, intravascular hemolysis, or continued agglutination may be reasonable indications .[4]

MYCOPHENOLATE MOFETIL (MMF)

Mycophenolate mofetil (MMF) is metabolized in the liver into mycophenolic acid, which acts by inhibiting the enzyme inosine 5'-monophosphtae dehydrogenase. This enzyme is necessary for de novo purine synthesis and therefore, essential for lymphocyte proliferation.[7] Because of this, MPA inhibits T and B lymphocyte production leading to suppressed cell-mediated and humoral immune responses.[7,8] Mycophenolate mofetil is used widely in human medicine for the prevention of renal allograft rejection and treatment of immune mediated diseases.[9] Its use in small animal medicine has been similar.[10,11]

A starting dose of 10-20 mg/kg orally or intravenously divided into two doses is recommended. However, canine pharmacokinetic studies show rapid but variable oral absorption with significant enterohepatic recirculation with variation in clearance that may necessitate alterations in dose and interval.[12,13] Common dose-limiting side effects include diarrhea, vomiting, weight loss, lethargy, and decreased appetite. A pilot study looking at a dose of 10-15 mg/kg orally every 8 hours found this dosing regimen was not recommended due to the severity of gastrointestinal signs. In two dogs, a necropsy was performed and revealed ulcerative enterocolitis.[14] The colonic lesions are similar to those found in people treated with mycophenolate.[15] In contrast, a retrospective study of 30 dogs comparing immunosuppressive protocol found that the combination of glucocorticoids and mycophenolate at a dose of 20 mg/kg/day produced less side effects than other protocols. They also found no difference in short-term outcome in dogs treated with glucocorticoids plus mycophenolate versus another immunosuppressive combined with steroids.[16] Some advantages of using mycophenolate include the availability of an intravenous formulation in dogs that cannot tolerate oral medications, lack of myelosuppresion, cost, and relatively rapid onset of action.[17] A 500 mg tablet currently costs about $7, the IV formulation (500 mg/50 mls) is $118 and 500 mg of the liquid suspension is $9 at our institution.

LEFLUNOMIDE

Leflunomide is an isoxazole immunomodulatory drug whose mechanism of action is not completely understood. Leflunomide is converted in the liver and intestinal mucosa to its primary metabolite, a malononitriloamide (A771726 or teriflunomide) that inhibits dihydro-orotate dehydrogenase. Lymphocytes require this enzyme for de novo synthesis of pyrimidine for activation and proliferation. Both T and B canine lymphocytes appear sensitive to the anti-proliferative effects of the malononitriolamides.[19] At higher concentrations, there is inhibition of protein cytokine and growth factor receptor-associated tyrosine kinase activity.[18,19] It has been used in both humans and dogs for rheumatoid arthritis and other immune mediated diseases.[16] Reported dosages range from 2 to 4 mg/kg/day orally with 12-hour serum trough level monitoring recommended.[20,21] A pharmacokinetic study performed indicated dosing once daily at a dose of 3-4 mg/kg is appropriate based on the half-life.[22] Side effects are rare and include self-limiting hematemesis and/or hematochezia, decreased appetite, mild anemia and lethargy.[23] There are limited studies documenting the use of leflunomide. Prospective studies are needed for more information, however there are some retrospective studies. A diabetic dog treated for Evans' Syndrome with leflunomide and intravenous immunoglobulin was in remission with no relapse over approximately 2 years of follow-up.[21] A recent retrospective case series of 14 dogs with immune-mediated polyarthritis treated with leflunomide documented a safe and effective alternative with minimal side effects.[20] A 20 mg tablet currently costs $6.50 at our institution.

HUMAN INTRAVENOUS IMMUNOGLOBULIN

Intravenous immunoglobulin (IVIG) is used extensively in human medicine as a treatment for a variety of diseases including immune-mediated anemia.[25] Human IVIG is the product of pooled plasma from thousands of donors and is comprised of highly purified immunoglobulin G. There are several, not completely understood, mechanisms by which IVIG acts on the immune system including inhibition of B and T cells and complement, reduced immunoglobulin production and perhaps most importantly, blockage of the Fc receptors on mononuclear cells by the IVIG. It is this mechanism that is though to decrease red blood cell phagocytosis immediately during the acute phase of the disease.[26] Human IVIG has been used as an adjunctive treatment for canine IMHA, ITP and for multiple dermatologic conditions.[27-29] However, IMHA studies have failed to show a clear benefit. A retrospective study of 10 dogs demonstrated a potential benefit for short-term stabilization but no long-term benefit was demonstrated.[27] A recent prospective clinical trial did not demonstrate a difference in initial response, transfusion requirements or hospitalization time between dogs that received glucocorticoids alone or glucocorticoids and hIVIG. There is a wide range of doses reported with 0.5 g/kg daily over 6-12 hours seeming to be the most common starting dose. IgG can be administered peripherally with a filter through a dedicated line. Repeated treatments are not recommended due to immediate or delayed reactions.[28] hIVIG may be cost limiting (approximately $250 for 2.5 grams) for veterinary patients in some circumstances but may be of value in dogs that do not respond glucocorticoid therapy within 1 week.[29,30] Adverse effects include hypersensitivity, anaphylaxis, hypercoagulability, fluid overload, and hypotension. Larger prospective studies are needed for more information.

SPLENECTOMY

In the human literature, there is support for splenectomy as a primary treatment for IMHA with survival rates of close to 100%.[31]Part of the underlying pathophysiology of IMHA involves antibody-mediated destruction of red blood cells by the spleen. Hence, removing the spleen would potentially be beneficial. Many human patients show clinical improvement following a splenectomy after failure of a steroid trial.[31] The concept of splenectomy as a treatment for canine IMHA is decades old but has yet to become commonplace. A study from 1985 demonstrated survival in 3 of 3 dogs with IMHA treated with splenectomy.[32] A retrospective study in 2009 evaluated 10 dogs that had a splenectomy as part of their early treatment for IMHA. In this study, there was a decrease in transfusion requirements and an increase in hematocrit. Potential complications include increased susceptibility to infectious organisms, bleeding, thromboembolism, and general risk associated with surgery and anesthesia. No dogs experienced any complications related to the surgery.[33] A prospective study is needed for more information.

THERAPEUTIC PLASMAPHERESIS

Therapeutic plasmapheresis is used as an adjunct therapy for a variety of diseases in human medicine including autoimmune disorders such as myasthenia gravis, Goodpasture's syndrome, IMHA and thrombotic thrombocytopenic purpura.[34] The process of plasmapheresis involves separation of plasma from whole blood. At that point, the plasma can be returned to the patient, replaced or discarded. Essentially, therapeutic plasmapheresis refers to the process where plasma, which is thought to contain components that are causing disease, is removed from circulation. These disease causing components are most commonly pathogenic autoantibodies, immunoglobulins or paraproteins but can also be inflammatory mediators such as endotoxin.[34-36] In cases of severe, acute immune mediated hemolytic anemia, plasmapheresis may be beneficial in rapid removal of antibodies, immune complexes and complement components allowing time for traditional immunosuppressive therapies to work while decreasing transfusion requirements.[35] Plasmapheresis may either be performed using centrifugation or, more recently, membrane filtration.[37] With membrane filtration, the blood components are separated based on size. Plasma passes readily through the filter while cells are retained and returned to the patient. This technique involves a microporous membrane allowing more complete separation of the cellular components from plasma resulting in a decrease loss in red blood cells. Hemodialysis equipment, such as the Gambro PrismaFlex machines, is capable of plasmapheresis. The filter used in plasmapheresis has a much larger pore size to allow much larger molecules such as immunoglobulins to pass through versus filters used in hemodialysis.[37] The distribution of IgG and IgM is key to the effectiveness of plasmapheresis; 45% of IgG is intravascular compared to 70% of IgM.[34] Potentially, this may be a more useful modality for cases of intravascular hemolysis. A replacement fluid is necessary to maintain intravascular volume and colloid support. In humans, typically albumin is chosen while in veterinary medicine, fresh frozen plasma seems to be reasonable.[37]

In 2009, a case report reviewing the usage of therapeutic plasmapheresis for the treatment of canine IMHA was published with promising results. This case report was also the first to utilize the membrane filtration technique. The decreases in both IgG and IgM levels are transient and may make multiple plasmapheresis treatments necessary. The dog in the case report only required one treatment and went on to do well clinically with a corresponding decrease in hemolysis.[37] At the author's institution, therapeutic plasmapheresis using membrane filtration was instituted in a similar case of refractory immune-mediated hemolytic anemia. Unfortunately, the dog continued to be transfusion dependent and was ultimately euthanized demonstrating that multiple treatments or more aggressive concurrent immunosuppressive therapy may be necessary.

Complications of plasmapheresis include hemorrhage from the dialysis catheter site, hypotension, hypothermia, coagulation abnormalities, and citrate-induced hypocalcemia if using sodium citrate as the anticoagulant. Therapeutic plasmapheresis requires specialized equipment and training and is quite expensive. The cost is variable depending on the size of the patient and replacement fluid volume needed. The filter alone for one treatment is $500 at our institution.

CONCLUSION

Immune-mediated hemolytic anemia is a very common and often very severe disease in our small animal patients. Unfortunately, there is a subset of patients that do not respond adequately to glucocorticoid therapy alone and require additional immunosuppressive agents or therapies. At this time, there is no clear evidence to support any single therapy over another. Multiple prospective trials are necessary.

REFERENCES

1. Reimer M et al J Am Anim Hosp Assoc 1999;35:384.
2. Piek C et al J Vet Intern Med 2008; 22:366.
3. Weinkle T et al J Am Vet Med Assoc 2005; 226:1869.
4. McCullough S Vet Clin North Am Small Anim Pract 2003;33:1295.
5. Stewart A et al Compen Contin Educ Pract Vet 1993;15(3):372.
6. Piek C et al BMC Vet Research 2011;7(15).
7. Braun et al Transplantation Proceedings 2002;34:1745.
8. Howard et al British Journal of Haematology 2002;117:712.
9. Danovitch G Transplantation 2005;80:S272.
10. Broaddus K, et al Veterinary Surgery 2006;32:125.
11. Yuki M et al Australian Veterinary Journal 2007;85:495.
12. Lupu M et al Biology of Blood and Marrow Transplantation 2006;12:1352.
13. Langman L et all Transplant 1996;61(1):87.
14. West L et al J Vet Emerg Crit Care 2014;24(2):226.
15. Selbst M et al Modern Path 2009;22:737.
16. Wang A et al Journ Small Animal Pract 2013;54:399.
17. Whitley N et al Jour Small Anim Pract 2011;52:70.
18. Gregory C et al Transplant Proc 1998;30(8):4143.
19. Gregory C et al Transplant Proc 1998;30(4):1047.
20. Colopy S et al J Am Vet Med Assoc 2010;236(3):312.
21. Bianco D et al J Am Anim Hosp Assoc 2009;45:147.
22. Boothe D et al ACVIM Proceedings 2014
23. Carr A et al J Vet Intern Med 2002;16(5):504.
24. Ulutas B et al J Vet Emerg Crit Care 2006;16:141.
25. Flores G et al Am J Hematol 1993;44(4):237.
26. Kazatchkine M et al N Engl J Med 2001;345(10):747.
27. Scott-Moncrieff J et al J Am Vet Med Assoc 1997;210(11):1623.
28. Murphy E et al J Infus Nurs 2005;28(4):265.
29. Kellerman D et al J Vet Intern Med 1997;11(6):327.
30. Whelan M et al J Vet Emerg Crit Care 2009;19(2):158.
31. Akpek G et al Am J Hematol 1999;61:98.
32. Feldman B et al J Am Vet Med Assoc 1985;187:617.
33. Horgan J et al J Vet Emerg Crit Care 2009;19(3):254.
34. Arslan O et al Turk J Hematol 2003;20:7.
35. Barges J et al Sem Vet Med Surg 1997;12:170.
36. Miller E J Vet Intern Med 1991;6(4):206.
37. Crump K et al J Vet Emerg Crit Care 2009;19(4):375.

RECOGNITION AND TREATMENT OF THE HYPERCOAGULABLE CANINE PATIENT

Lenore Bacek, DVM, MS, DACVECC
Auburn, AL

INTRODUCTION

One of the leading causes of mortality in immune-mediated hemolytic anemia is thromboembolic disease, with documented lesions in approximately 80% of dogs at necropsy.[1-5] The etiology is complex and there is a lack of evidence supporting one anticoagulant treatment protocol over another.

WHY DOES IT HAPPEN?

In immune-mediated hemolytic anemia, the hypercoagulable, prothrombotic state is multifactorial and not completely understood but likely includes a pro-inflammatory state, platelet hyper-reactivity, altered blood viscosity, hemolysis, decreased anticoagulant factors and treatment itself.[1,3,5,7] In human medicine, studies have clearly demonstrated a link between inflammation and the initiation of coagulation.[8] In times of health, tissue factor (TF) is not expressed on endothelial cells or in the vascular space. This localizes the initiation of coagulation to TF-bearing surfaces secondary to endothelial injury or activation of cells.[9] However, during pro-inflammatory states, cytokines induce expression of tissue factor on the surface of cells including monocytes and endothelial cells. Because of this, pathologic intravascular coagulation occurs.[10] A recent study documented upregulation of tissue factor gene expression in dogs with IMHA along with a proinflammatory state characterized by a leukocytosis, left shift and monocytosis.[11] This is similar to the previous finding that a moderate to marked leukocytosis correlated with the finding of ischemic necrosis at necropsy. This necrosis was thought to be secondary to thromboembolic disease or anemic hypoxia.[5] Increased levels of other inflammatory mediators such as C-reactive protein have also been documented. This study also documented a correlation between CRP and white blood cell counts at certain time points.[12] Another study evaluating the acute phase reaction found higher CRP and α1-acid glycoprotein in IMHA dogs compared to healthy control dogs.[13]

Microparticles also play an important role in the hypercoagulability documented with IMHA. Microparticles are present in both health and disease states; they are formed from the cell membrane of their cell of origin and express the same cell surface molecules as their parent cell. These cells include platelets, lymphocytes, erythrocytes, endothelial cells and monocytes.[14] When cells, such as platelets or damaged red blood cells, become activated, they release phosphatidylserine (PS) exposing microparticles.[15] Phosphatidylserine provides a procoagulant surface allowing the initiation of the coagulation cascade.[16] In human medicine, certain disease processes causing hemolysis, such as sickle cell anemia, will increase the expression of PS on the red blood cell.[17] Other cell types, such as endothelial cells and monocytes, generate microparticles that express tissue factor, leading to the initiation of coagulation and formation of thrombin. The formation of thrombin leads to continued platelet activation.[18]

Platelet hyper-reactivity has been demonstrated in both human and veterinary medicine using flow cytometry.[7] In one study, P-selectin, which is transported to the surface during platelet activation, was measured. Dogs with IMHA had P-selection expression eight times higher than control dogs. A significant proportion (7/20) dogs developed clinical evidence of thromboembolic disease.[18] Another study showed evidence of platelet activation in dogs with IMHA, specifically with simultaneous severe thrombocytopenia.[7]

Some of the treatments used in IMHA may also contribute to overall hypercoagulability. Corticosteroids have been shown to induce a hypercoagulable state demonstrated by thromboelastography.[19] Another study looked at the effects of steroids with and without low-dose aspirin on coagulation factors. The dogs that received prednisone were found to have evidence of hypercoagulability as demonstrated by changes in thromboelastography (increased MA) as well as increased fibrinogen and decreased antithrombin. The concurrent administration of ULDA did not affect the TEG values.[20] Other medications such as cyclosporine may also cause a hypercoagulable effect.[21] Intravenous immunoglobulin is sometimes used as a treatment for cases that are refractory to corticosteroid administration. IVIG was administered to healthy dogs and was found to cause a pro-inflammatory and hypercoagulable state.[22] Other treatment variables that may influence coagulation include placement of central catheters and large numbers of blood transfusions.[23]

HOW CAN WE DIAGNOSIS IT?

Traditional coagulation testing readily identifies a hypocoagulable state but identifying a hypercoagulable state can be more challenging. Evidence of a hypercoagulable state often includes a low platelet count, increase in fibrinogen, increase in d-dimers, decrease in antithrombin and changes in thromboelastography. Multiple studies have documented thrombocytopenia as a risk factor for thromboembolism and/or mortality.[6,24] In IMHA, antithrombin is most likely low due to a consumptive process. One study prospectively evaluating dogs with immune-mediated hemolytic anemia documented an increase in fibrinogen in 85% of dogs, decrease in antithrombin in 50% of dogs and increased d-dimers in 80% of patients although mortality was not associated with any specific finding.[1] Antithrombin is crucial to the inhibition of coagulation and decreased levels correlate with an increased risk of thromboembolic disease. A retrospective study looking at antithrombin in a variety of disease populations demonstrated with 76% of dogs with IMHA had a low antithrombin and, in this subset of patients, was a prognostic marker.[25]

Thromboelastography (TEG) is a helpful diagnostic tool to assess the entire hemostatic system and identify a hypercoagulable state. A retrospective study of 39 dogs with IMHA determined that most of the dogs were hypercoagulable (~85%) based on thromboelastography parameters. Interestingly, they also found a 100% mortality rate in the dogs with a normal coagulation index.[26] Another study evaluated dogs before treatment was started. They found that all of the dogs had high fibrinogen, the majority had high d-dimers and a hypercoagulable TEG tracing compared to healthy dogs.[2] Both of the previously mentioned studies only evaluated TEG tracings at a single time point. A recent prospective study serially evaluated the tracings over a 5-day period. At admission, IMHA dogs were significantly hypercoagulable compared to control dogs.[4]

HOW CAN WE TRY TO PREVENT IT?

In human patients, the selection of thromboprophylactic therapy is based on the underlying pathophysiology of the disease. Based on necropsy findings, IMHA can cause both venous and arterial thrombosis suggesting that both an anti-platelet and anticoagulant therapeutic may be helpful. A variety of drug therapies have been used including unfractionated heparin, low molecular weight heparin, low dose aspirin, and clopidogrel.[27-31]

Unfractionated heparin potentiates the effect of antithrombin by inhibiting factors Xa, IIa, VIIa, XIa and XIIa to attempt to decrease the development of venous thromboembolic events. Close monitoring must be performed to avoid bleeding. Typically, the activated partial thromboplastin time is used to monitor the dose of heparin with a goal of 1.5 to 2 times baseline. Unfortunately, high fibrinogen levels can affect PTT, which is often seen in IMHA patients.[1] Monitoring of anti-factor Xa activity can also be used but this test is not as widely available. At this time, there is no standard dosing and most dosing is based on studies in healthy patients.[32] A prospective study using a set-dose at 300 u/kg subcutaneously every 6 hours was not reliable in achieving the anti-Xa activity desired. The therapeutic range desired was 0.35 to 0.7 U/mL.[27] Helmond *et al* compared a fixed dose to an individually daily-adjusted dose of heparin based on anti-Xa activity. There was a higher incidence of thromboembolic events in the fixed dose group compared to the adjusted group (5 vs. 2) as well as a higher mortality rate at 180 days in the fixed group (88%) compared to the adjusted group (13%).[28] Thromboelastography is also a potential way to monitor the effects of unfractionated heparin. In healthy dogs, there were progressive TEG changes after a 200 U/kg SQ dose of heparin. Unfortunately, the TEG may be too sensitive to use clinically.

Low molecular weight heparins, such as enoxaparin or dalteparin, are used more often in human medicine due to more predictable bioavailability, similar efficacy, less frequent dosing and potential decreased need for intensive monitoring. However, pharmacokinetics vary between species. A pharmacokinetic study of enoxaparin in healthy dogs revealed that a dose of 0.8 mg/kg subcutaneously every 6 hours was able to maintain target levels of anti-Xa activity. There were no bleeding complications in this study.[34] A retrospective study looking at Enoxaparin use in 8 dogs with IMHA revealed a long-term survival rate of 75% with minimal evidence of hemorrhage. This study used a median dosage of 0.89 mg/kg every 6 hours in 7 dogs and every 8 hours in 1 dog.[29] Dalteparin is another low molecular weight heparin that is used in both human and veterinary medicine. One study compared the effects on anti-Xa activity by low-dose heparin (300 u/kg/d), high-dose heparin (900 u/kg/day after a 100 u/kg bolus) and dalteparin at 100 u/kg every 12 hours subcutaneously. They determined that only the high-dose heparin increased the anti-Xa to the therapeutic range in 2 of 6 dogs, with hemorrhage noted in some dogs.[35] Thromboelastography may also be an important tool for monitoring treatment with low molecular weight heparin.[36,37]

Aspirin inhibits COX-1 and therefore, the production of thromboxane A_2, a potent platelet agonist. Ultra low does aspirin is able to inhibit the production of thromboxane A_2 while still allowing the production of prostacyclin. An exact dose has yet to be determined in dogs. A common dose of 0.5 mg/kg/day inhibits aggregation in some normal dogs, while a dose of 0.5 mg/kg every 12 hours was more efficacious.[38,39] Many studies have evaluated the effects of low dose aspirin in normal dogs. One study compared several doses (0.5, 2 and 10 mg/kg) of aspirin using TEG Platelet Mapping versus whole blood impedance aggregometry. They determined that an antiplatelet effect was seen at 0.5 and 10 mg/kg with platelet mapping but only at 10 mg/kg with aggregometry.[40] Another study documented to significant effect on platelet function at a dose of 0.5 mg/kg/day using either the PFA-100 or optical aggregometry.[41] A recent study demonstrated the expression of both COX 1 and 2 by platelets, with aspirin increasing the expression of COX-1 but still inhibiting platelet function.[42] Platelet function with low-dose aspirin was also assessed by aggregometry (turbidimetric and impedance) and PFA-100 closure time. The results indicated the majority of dogs were aspirin responders with turbidimetric aggregometry being the preferred technique.[43] Cyclosporine has also been shown to decrease the effectiveness of low-dose aspirin based on PFA-100 and turbidimetric aggregometry.[44]

Clopidogrel inhibits platelet aggregation in dogs by irreversibly inhibiting the ADP receptor.[45] One study demonstrated a dose of at least 2 mg/kg seems necessary for platelet inhibition.[46] A recent prospective study evaluated the effect of aspirin versus clopidogrel versus both with respect to survival at discharge and 90 days. The results suggest that clopidogrel is both safe and has similar survival rates to aspirin alone.[30]

There have been no large-scale prospective studies comparing the efficacy of anticoagulation protocols. In one retrospective study, treatment groups were evaluated in 151 dogs. Treatment groups included: no anticoagulant, ultra low-dose aspirin, ULDA plus heparin and heparin alone. Results suggested a significant improvement in both short and long-term survival in the group treated with ULDA.[47] Newer drugs, such as factor X inhibitors, may prove to be beneficial with further research.

CONCLUSION

Overall, large, prospective studies are necessary to determine dogs at the highest risk for thromboembolic disease as well the most appropriate anticoagulation protocol for immune-mediated hemolytic anemia patients.

REFERENCES

1. Scott-Moncrieff J et al J Amer Anim Hosp Assoc 2001;37:220.
2. Fenty R et al J Amer Vet Med Assoc 2011;238(4):463.
3. Sinnott V et al J Vet Emerg Crit Care 2009;19(5):484.
4. Goggs R et al Vet J 2011;191:347.
5. McManus P et al J am Vet Med Assoc 2001;218:1308.
6. Carr A et al J Vet Intern Med 2002;16:504.
7. Ridyard A et al J Sm Anim Pract 2010;51:296.
8. Esmon C Trends Immunol 2004;25(10):536.
9. Smith S J Vet Emerg Crit Care 2009;19(1):3.
10. Mackman N et al Anesth Analg 2009;108(5):1447.

11. Piek C et al Vet Immun and Immunopath 2011;144:346.
12. Griebsch C et al Vet Clin Path 2009;38(4):421.
13. Mitchell K et al J Vet Inter Med 2009;23:585.
14. Herring et al J Vet Intern Med 2013;27:1020.
15. Owens A et al Thromb Haemost 2010;104(3): 432.
16. Kidd L et al J Vet Emerg Crit Care 2013;23(1):3.
17. Zwaal R et al Cell Mol Life Sci 2005;62(9):971.
18. Weiss D et al J Vet Intern Med 2006;20:682.
19. Flint S et al Am J Vet Res 2011;72:1325.
20. O'Kell A et al Am J Vet Res 2012;73(10):1569.
21. Thomason J et al J Vet Intern Med 2012;26:1389.
22. Tsuchiya R et al J Vet Intern Med 2009;23;1164.
23. Klein M et al J Am Vet Med Assoc 1989;195;246.
24. Piek C et al J Vet Intern Med 2008;22:366.
25. Kuzi S et al J Vet Intern Med 2010;25:587.
26. Sinnott V et al J Vet Emerg Crit Care 2009;19(5):484.
27. Breuhl E et al J Am Anim Hosp Assoc 2009;45:125.
28. Helmond S et al J Vet Intern Med 2010;24:597.
29. Bianco et al ECVIM Congress 2011
30. Mellett A et al J Vet Intern Med 2011;25:71.
31. Stiller A et al ACVIM Congress 2013.
32. Kellerman D et al J Vet Intern Med 1995;9:187.
33. Pittman J et al J Vet Emerg Crit Care 2010;20(2):216.
34. Lunsford K et al J Am An Hosp Assoc 2009;45:261.
35. Scott K et al J Vet Emerg Crit Care 2009;19(1):74.
36. Brainard B et al Amer J Vet Res 2012;73:1577.
37. Jessen L et al Vet Clin Path 2008;37(4):363.
38. Sharpe K et al Am J Vet Res 2010;71(11):1294.
39. Rackear D et al J Am Anim Hosp Assoc 1988;24(1):23.
40. Puglia G et al IVECCS 2010.
41. Thomason J et al ACVIM 2010.
42. Shearer L et al ACVIM 2009.
43. Haines J et al ACVIM 2014.
44. Thomason J et al ACVIM 2014.
45. Brainard B et al Am J Vet Res 2010;71(7):822.
46. Biermann K et al ECVIM 2013.
47. Weinkle T et al J Am Vet Med Assoc 2005;226(11):1869.

REVISITING PLASMA TRANSFUSIONS: INDICATIONS AND CONTROVERSIES ASSOCIATED WITH FRESH FROZEN PLASMA

Adesola Odunayo, DVM, MS, DACVECC
Knoxville, Tennessee

INTRODUCTION

Transfusion of fresh frozen plasma is commonly performed in veterinary and human medicine for a variety of indications. The major indications for the use of plasma in human patients are to prevent bleeding and stop bleeding. [1] Indications for use in veterinary medicine includes the replacement of coagulation factors during active bleeding, hypoalbuminemia, pancreatitis, treatment of prolonged clotting times without clinical bleeding and parvovirus enteritis. In one veterinary study, hypoalbuminemia was the most common indication for plasma transfusion.[2] Other indications in that study included replacement of coagulation factors (33%), provision of immunoglobulins (11%) and replacement of alpha-macroglobulin (9%).

Fresh frozen plasma (FFP) is the most common plasma product utilized in human patients with other products like frozen plasma (FP) or thawed plasma, used less frequently. FFP is a single donor plasma preparation, produced either by whole blood fractionation or plasmapheresis.[3] FFP units are separated from the red blood cells and platelets of whole blood donations and frozen (usually at -30 degrees C) within 8 hours of collection. Proteins and coagulation factors in FFP remain viable for at least 12 months when stored appropriately. Frozen plasma (FP) is plasma that is separated from whole blood but not frozen within 8 hours, or FFP that has been frozen but stored between 1 and 4 years. The main difference between FFP and FP is the loss of the heat labile coagulation factors, V and VIII.[3]

Critically ill patients are among the highest users of FFP, with up to a third of critically ill human patients receiving FFP in one study.[4] However, human studies on FFP have cited that it is often misused.[1] Reasons cited for the misuse of FFP include limited knowledge of its efficacy in specific situations and ignorance of its use and its increased availability. Other than the misuse of FFP, there are significant adverse effects associated with the use of FFP in human patients, including transfusion related volume overload, transfusion related acute lung injury and transmission of infectious disease. The rational use of FFP in human patients is being better defined through the use of published guidelines establishing recommended uses of FFP.[3,5] This session will evaluate the evidence for the use of FFP for various indications in human and veterinary patients.

Hypoalbuminemia

Hypoalbuminemia is a common finding in critically ill patients. Causes may include one or the combination of the following reasons- increased loss, decreased synthesis, redistribution of albumin to extravascular compartments or dilution of albumin within the vascular space.[6,7] Replenishment of albumin in the critically ill patients may be helpful in restoring some of the important functions of albumin, including maintenance of colloid oncotic pressure. Consequences of hypoalbuminemia may be significant in critically ill patients and hypoalbuminemia has been shown to be a poor predictor of outcome in several human studies, although this link is still being debated.[7-9] FFP is not routinely used in human patients for the replacement of albumin in hypoalbuminemic patients. This is primarily due to the low concentration of albumin present in one unit of FFP. The advent of human serum albumin solutions (HSA) and synthetic colloids have virtually eliminated the routine use of hypoalbuminemia for intravascular support. Recently, the FDA has placed a black box warning on the use of synthetic colloids in critically ill human patients. [10,11] FFP is still routinely used in veterinary medicine for the treatment of hypoalbuminemia although it is thought to be a very inefficient way of raising serum albumin concentrations. In one veterinary study, 63% of the dogs in the study were given FFP transfusions to correct low serum albumin concentrations.[2] This study did not evaluate post transfusion serum albumin concentrations. It is currently accepted that it takes 22mls/kg of FFP to raise albumin concentrations by 0.5g/dL. This volume of FFP may predispose critically ill patients to volume overload and pulmonary edema

Coagulation Disorders

The prophylactic administration of FFP for the management of coagulopathy is a common reason for the administration of FFP. In one veterinary study, 80% of FFP use over a two year period was used to treat patients with coagulopathies.[12] Another study found that 47% of FFP transfusions, given over a three month time frame, were administered to dogs to provide coagulation factors.[2] When available, specific coagulation factor concentrates are preferred over the use of FFP in cases of single coagulation factor deficiencies but these concentrates have limited availability in veterinary medicine. Coagulopathies, defined as any disorder of blood coagulation, may present with patients with active bleeding or no clinical signs. Coagulopathies are usually identified based on bench top screening tests available in human and veterinary hospitals. Coagulation screening tests play a major role in determining what patients are going to be transfused with FFP. Standard first-line coagulation tests include the prothrombin time (PT), activated partial thromboplastin time (aPTT), international normalized ratio (INR), thromboelastography (TEG) and the activated clotting time (ACT). The PT, aPTT and ACT were developed to investigate coagulation factor deficiencies in patients with a history of bleeding.[3] Similarly, the INR was developed as a means to standardize the PT across various testing locations. These coagulation screening tests were not designed to assess risk of procedural bleeding and cannot be used confidently in that regards.[3,13] While these tests are run frequently, there is a lot of debate about their applied value in patients who do not have any evidence of clinical bleeding.

Limitations of the first-line screening tests include the normal biological variation. Technically, 2.5% of healthy individuals have abnormally prolonged coagulation times. Another limitation is that inherited bleeding disorders have low prevalence and indiscriminate screening for these disorders results in a low positive predictive value and a high number of false positives. These tests can also be falsely affected by sampling error (heparin contamination, inappropriate collection and storage and inappropriate dilution of the sample with citrate). The aPTT, a test of the intrinsic and common pathway of coagulation, can detect factor deficiencies not associated with clinical bleeding (i.e. factor XII or Hageman deficiency in cats). The PT, though more standardized than the aPTT can be affected by varying composition of the tissue factor reagents, resulting in variable sensitivities.[13] In human patients, clinicians are encouraged not to make a decision to administer FFP based on changes in the coagulation screening tests because of the inherent problems associated with interpreting the test results and the lack of evidence demonstrating improved outcome after FFP administration. In the 2012 Surviving Sepsis guidelines, FFP was only recommended when there was active bleeding or a planned invasive procedure.[14] Several human studies have also shown that the administration of FFP does not normalize pre-transfusion coagulation screening tests.[15,16] There are many conditions that may cause coagulopathies in veterinary patients including gastric dilatation and volvulus, neoplasia, pancreatitis, hepatic dysfunction, gastrointestinal disease, renal disease, hyperadrenocorticism and cardiac disease. There are no known studies evaluating the effect of plasma transfusions on outcomes in veterinary patients with coagulopathies. With the very limited information available, the role of FFP in veterinary patients with coagulopathies is still unknown at this point.

Pancreatitis

FFP has been used to treat pancreatitis in human and veterinary patients. Early deaths from pancreatitis are usually due to massive inflammatory responses which results in multiple organ failure. While the exact mechanisms which trigger the immune response are not completely understood, it is thought that auto digestion and activated leukocytes play a major role in the pathogenesis of acute pancreatitis. FFP is thought to be beneficial in these patients by replenishing circulating proteins, especially the natural occurring antiproteases.[17] The plasma antiproteases are the third largest group of plasma proteins, after albumin and immunoglobulins. They remain active after plasma is frozen and α1 antiproteases and α2 macroglobulin account for more than 90% of antiprotease capacity in human beings. α2 macroglobulin concentrations have been demonstrated to fall during acute pancreatitis and administration of FFP has reliably increased plasma α2 macroglobulin concentrations in human patients.[17] However, studies have failed to demonstrate an improved outcome in human patients that were treated with FFP for pancreatitis.[17-19] In a set of guidelines published, FFP transfusion was not recommended as a standard of care for human patients with pancreatitis.[20] In one rodent pancreatitis model, administration of high volume FFP was shown to have a beneficial effect on survival but the same authors found no benefit when high volume FFP was given to human patients with pancreatitis.[18,21] A few veterinary studies have evaluated the use of FFP in pancreatitis. One study found that 10% of FFP transfusions given were administered to dogs with pancreatitis, with most of those dogs receiving FFP for replenishing α2 macroglobulin concentrations.[12] A more specific study compared dogs with pancreatitis treated with FFP to a control group that was not treated with FFP.[22] This study failed to show any survival benefit from plasma administration. Instead dogs that received plasma had a significantly higher mortality rate. The authors of this study tried to ensure both groups were balanced by looking at the number of dogs that had evidence of SIRS and there was no difference between both groups. It is possible that dogs with more severe form of pancreatitis got plasma (which might explain the increased mortality) but this study did not show any morbidity or mortality benefit from FFP transfusions.

Disseminated Intravascular Coagulation (DIC)

DIC is a syndrome that occurs acutely in the course of severe underlying disease.[23] Clinical conditions associated with DIC include trauma, sepsis, malignancies, pancreatitis, renal disease, heat stroke and immune mediated diseases. DIC is characterized by widespread activation of coagulation, which leads to formation of fibrin and thrombotic occlusion of small and midsize vessels. This ultimately leads to compromise of blood supply to end organs, which along with hemodynamic and metabolic derangements, contributes to multiple organ failure.[23] The use and depletion of platelets and coagulation proteins during DIC may induce bleeding, which may be severe in some cases. DIC, though difficult to definitively diagnose, may be characterized by prolonged (or shortened) PT, aPTT and INR, elevated fibrin degradation products (FDPs), elevated D-dimers, thrombocytopenia and low levels of coagulation inhibitors, such as antithrombin. DIC is treated by effective management of the underlying disease. Completely addressing the underlying disease may not be possible or may take some time and other supportive therapy is usually advocated to disrupt the cycle of thrombosis and hemorrhage. Supportive therapy generally advocated includes treatment with whole blood (or packed red blood cell transfusions), platelet transfusions, FFP, unfractionated (UFH) and low molecular weight heparin (LMWH), synthetized protease inhibitors (SPI) and activated protein C (APC).

FFP and platelet substitutions have historically been important in the treatment of DIC in human and veterinary medicine. There used to be concerns about factor replacement because of the fear that administration of more coagulation factors will "feed the fire" and worsen thrombosis. This fear has not been confirmed by clinical evidence and replacement of coagulation factors in patients in DIC with active bleeding is still being performed by some physicians. There are very limited randomized controlled trials evaluating the use of FFP in DIC in human patients. One study evaluated the effect of platelet and FFP transfusions in neonates with DIC.[24] In this study, resolution of DIC and survival were not different in the 3 treatment groups. Outcome of DIC was dependent on the success of treatment of the underlying disease process and aggressive supportive care, including restoration of blood pressure. A consensus statement released by the British Society for Haematology had similar recommendations for the use of FFP in DIC.[25] In this evidence based guidelines, it was recommended that FFP transfusions should not be initiated based on prolonged coagulation tests but should be considered in patients with active bleeding and those requiring an invasive procedure. Initial doses of 15mls/kg were suggested although it was noted that doses as high as 30mls/kg may be required to provide complete resolution of coagulation factor deficiencies. Specific clotting factor concentrates may also be considered for volume overloaded patients requiring FFP transfusions. The use of FFP was not advocated in chronic DIC or in the absence of hemorrhage. The use of FFP as part of treatment of DIC is common in veterinary medicine. However there is a paucity of studies available looking at the effect of treatment with FFP on various outcomes on veterinary patients with DIC. One study evaluated the efficacy of FFP on the coagulogram in dogs with DIC. [26]The results showed that FFP given at a dose of 6-10mls/kg intravenously every 12 hours normalized coagulation parameters. It was unclear how long the dogs were treated for and what the final volume of plasma used was. There was no control group utilized and the final outcome of the cases was undefined in this small group of dogs (6). To the authors' knowledge, there are no other studies evaluating the use of plasma on outcome on veterinary patients with DIC.

Hepatic Disease

Coagulation disorders are common in patients with hepatic disease as the physiology of blood coagulation is closely linked to liver function.[27,28] The mechanisms underlying the coagulopathy are many. The liver synthesizes most coagulation factors, their inhibitors and fibrinolytic proteins. In addition, the liver is involved in the regulation of coagulation by facilitating the clearance of activated clotting and fibrinolytic factors and activation complexes. Liver dysfunction can be associated with impaired synthesis of coagulation factors, production of abnormally functioning coagulation factors, vitamin K deficiency, impaired posttranslational modification of vitamin dependent clotting factors, thrombocytopenia, thrombocytopathia, dysfibrinogenemia, increased fibrinolysis, consumptive coagulopathy (DIC) and impaired clearance of activated complexes. In human patients, overt bleeding tends to be unusual in acute hepatitis (except for fulminant hepatitis or massive hepatic necrosis).[27] Increased bleeding tendencies are seen in human patients with cirrhotic disease and patients undergoing invasive procedures or surgery.[27] This is slightly different from what has been suggested in dogs where dogs with acute liver disease are thought to be more likely to have spontaneous bleeding compared to dogs with chronic hepatitis.[29,30] Hepatic disease associated coagulopathy has been reported in veterinary medicine. Ninety three percent of dogs with naturally occurring hepatic disease had at least one abnormal coagulation test result and dogs with congenital portosystemic shunts have been shown to have alterations in their coagulation factors.[28,31] In a study, 57% of dogs with hepatic dysfunction had one or more coagulation abnormalities. The coagulopathies were more severe in dogs with cirrhosis and was thought to be secondary to reduced synthesis instead of increased consumption of coagulation factors.[28] Coagulation abnormalities have also been reported in cats with hepatic disease, where one study documented at least one coagulation abnormality in 82% of the cats evaluated.[32]

Studies evaluating FFP administration to prevent hemorrhage in human patients with hepatic disease induced coagulopathy have been published. Although these studies administered FFP at a dose of 6-15mls/kg, no significant differences in bleeding events or transfusion requirements were observed between FFP treated and control patients.[33,34] There are very few veterinary studies evaluating the use of FFP in correcting coagulation abnormalities in canine and feline patients with liver disease. In one feline study, six cats required FFP transfusions for coagulopathies primarily associated with liver disease.[35] This study did not describe the changes in coagulation times after the FFP transfusions to see if there was a beneficial effect. In another canine study, a proportion of dogs with hepatic dysfunction were treated with FFP to replace coagulation factors. This study also did not evaluate the effect of the transfusion on coagulation times abnormalities.[2] More studies are needed to see if FFP transfusions in veterinary patients with hepatic dysfunction affect standard coagulation parameters, outcome and prognosis.

Parvoviral Enteritis

FFP has historically been recommended for use in veterinary patients with parvo viral enteritis. The rationale behind the use of FFP is the provision of immunoglobulin which may hasten recovery and improve outcome in patients with parvovirus infection. The infused antibodies are thought to neutralize the free virus in plasma and suppress the release of new virus from infected cells. FFP also provides albumin, coagulation factors and serum protein inhibitors, which may be helpful in this group of patients. CPV-immune plasma is produced from dogs that have been naturally infected with parvovirus, who presumably have higher concentrations of circulating antibodies than naturally infected dogs. CPV-immune plasma and FFP have been anecdotally used in the management of dogs with naturally occurring parvovirus infections. A study evaluated the efficacy of a single dose of CPV-immune plasma for the treatment of dogs with parvovirus infection.[36] Administration of a single 12-mL dose of immune plasma did not ameliorate clinical signs, reduce viremia or hasten hematologic recovery compared to dogs in the control group. This study had a few limitations including small sample size and a potential low dose of CPV-immune plasma used. However, there is no other evidence to suggest or refute the fact that CPV-immune plasma or FFP should be routinely used in patients with parvovirus infection.

Consequences Of Inappropriate FFP Use

FFP is thought to be one of the most hazardous blood component transfusion although the overall level of risk is low in human patients. Potential complications of FFP transfusions include acute allergic and anaphylactic reactions, transfusion-related acute lung injury (TRALI), Transfusion associated circulatory overload (TACO) and transmission of infectious diseases. It is therefore very important that the clinician evaluates the risk benefit ratio before administering a FFP transfusion.

REFERENCES

1. Kor DJ, Stubbs JR, Gajic O. Perioperative coagulation management–fresh frozen plasma. *Best Practice & Research Clinical Anaesthesiology.* 2010;24(1):51-64.
2. Logan JC, Callan MB, Drew K, et al. Clinical indications for use of fresh frozen plasma in dogs: 74 dogs (October through December 1999). *Journal of the American Veterinary Medical Association.* 2001;218(9):1449-1455.
3. Stanworth SJ. The evidence-based use of FFP and cryoprecipitate for abnormalities of coagulation tests and clinical coagulopathy. *ASH Education Program Book.* 2007;2007(1):179-186.
4. Holland LL, Brooks JP. Toward rational fresh frozen plasma transfusion the effect of plasma transfusion on coagulation test results. *American journal of clinical pathology.* 2006;126(1):133-139.
5. Stanworth SJ, Grant-Casey J, Lowe D, et al. The use of fresh-frozen plasma in England: high levels of inappropriate use in adults and children. *Transfusion.* 2010;51(1):62-70.
6. Throop JL, Kerl ME, Cohn LA. Albumin in health and disease: causes and treatment of hypoalbuminemia. *Compend Contin Educ Pract Vet.* 2004;26(12):940-949.
7. Dubois MJ, Vincent A. Use of albumin in the intensive care unit. *Transfusion Alternatives in Transfusion Medicine.* 2002;4(3):80-84.
8. Goldwasser P, Feldman J. Association of serum albumin and mortality risk. *Journal of clinical epidemiology.* 1997;50(6):693.
9. Mangialardi RJ, Martin GS, Bernard GR, et al. Hypoproteinemia predicts acute respiratory distress syndrome development, weight gain, and death in patients with sepsis. *Critical care medicine.* 2000;28(9):3137-3145.
10. Aubron C, Bellomo R. Infusion of hydroxyethyl starch-containing fluids. *Minerva Anestesiol.* 2013;79:1088-1092.
11. Nolan J, Mythen M. I. Hydroxyethyl starch: here today, gone tomorrow. *British journal of anaesthesia.* 2013;111(3):321-324.
12. Snow SJ, Ari Jutkowitz L, Brown AJ. Retrospective Study: Trends in plasma transfusion at a veterinary teaching hospital: 308 patients (1996–1998 and 2006–2008). *Journal of Veterinary Emergency and Critical Care.* 2010;20(4):441-445.
13. Chee Y-L, Greaves M. Role of coagulation testing in predicting bleeding risk. *The Hematology Journal.* 2003;4(6):373-378.

14. Dellinger RP, Levy MM, Rhodes A, et al. Surviving Sepsis Campaign: international guidelines for management of severe sepsis and septic shock, 2012. *Intensive Care Med.* 2013;39(2):165-228.
15. Dzik W. Predicting hemorrhage using preoperative coagulation screening assays. *Current hematology reports.* 2004;3(5):324-330.
16. Abdel-Wahab OI, Healy B, Dzik WH. Effect of fresh-frozen plasma transfusion on prothrombin time and bleeding in patients with mild coagulation abnormalities. *Transfusion.* 2006;46(8):1279-1285.
17. Chen C-C, Wang S-S, Lee F-Y. Action of antiproteases on the inflammatory response in acute pancreatitis. *Jop.* 2007;8(4 Suppl):488-494.
18. Leese T, Holliday M, Watkins M, et al. A multicentre controlled clinical trial of high-volume fresh frozen plasma therapy in prognostically severe acute pancreatitis. *Annals of the Royal College of Surgeons of England.* 1991;73(4):207.
19. Leese T, Holliday M, Heath D, Hall A, Bell P. Multicentre clinical trial of low volume fresh frozen plasma therapy in acute pancreatitis. *British journal of surgery.* 1987;74(10):907-911.
20. Wyncoll D. The management of severe acute necrotising pancreatitis: an evidence-based review of the literature. *Intensive Care Med.* 1999;25(2):146-156.
21. Leese T, West KP, Morton DB, Bell PR. Fresh frozen plasma therapy in acute pancreatitis: an experimental study. *International journal of pancreatology.* 1988;3(6):437-447.
22. Weatherton LK, Streeter EM. Evaluation of fresh frozen plasma administration in dogs with pancreatitis: 77 cases (1995–2005). *Journal of Veterinary Emergency and Critical Care.* 2009;19(6):617-622.
23. Levi M, Ten Cate H. Disseminated intravascular coagulation. *New England Journal of Medicine.* 1999;341(8):586-592.
24. Gross SJ, Filston HC, Anderson JC. Controlled study of treatment for disseminated intravascular coagulation in the neonate. *The Journal of pediatrics.* 1982;100(3):445-448.
25. Levi M, Toh C, Thachil J, Watson H. Guidelines for the diagnosis and management of disseminated intravascular coagulation. *British journal of haematology.* 2009;145(1):24-33.
26. Bruchim Y, Aroch I, Saragusty J, Waner T. Disseminated intravascular coagulation. *Compendium.* 2008;30(10).
27. Kaul V, Munoz SJ. Coagulopathy of liver disease. *Current treatment options in gastroenterology.* 2000;3(6):433-437.
28. Prins M, Schellens C, Van Leeuwen M, Rothuizen J, Teske E. Coagulation disorders in dogs with hepatic disease. *The Veterinary Journal.* 2010;185(2):163-168.
29. Dereszynski DM, Center SA, Randolph JF, et al. Clinical and clinicopathologic features of dogs that consumed foodborne hepatotoxic aflatoxins: 72 cases (2005–2006). *Journal of the American Veterinary Medical Association.* 2008;232(9):1329-1337.
30. Dunayer EK, Gwaltney-Brant SM. Acute hepatic failure and coagulopathy associated with xylitol ingestion in eight dogs. *Journal of the American Veterinary Medical Association.* 2006;229(7):1113-1117.
31. Badylak SF. Coagulation disorders and liver disease. *Veterinary Clinics of North America: Small Animal Practice.* 1988;18(1):87-93.
32. Lisciandro SC, Hohenhaus A, Brooks M. Coagulation abnormalities in 22 cats with naturally occurring liver disease. *Journal of Veterinary Internal Medicine.* 1998;12(2):71-75.
33. Gazzard B, Henderson J, Williams R. Early changes in coagulation following a paracetamol overdose and a controlled trial of fresh frozen plasma therapy. *Gut.* 1975;16(8):617-620.
34. Lerner R, Nelson J, Sorcia E, et al. Evaluation of Solvent/Detergent-Treated Plasma in Patients with a Prolonged Prothrombin Time. *Vox sanguinis.* 2000;79(3):161-167.
35. Couto CG, Gray TL. Clinical use of blood products in cats: a retrospective study (1997–2000). *Journal of veterinary internal medicine.* 2004;18(4):529-532.
36. Bragg RF, Duffy AL, DeCecco FA, et al. Clinical evaluation of a single dose of immune plasma for treatment of canine parvovirus infection. *Journal of the American Veterinary Medical Association.* 2012;240(6):700-704.

REVIEW OF TOP 5 TRANSFUSION RELATED JOURNAL ARTICLES IN THE LAST YEAR

Adesola Odunayo, DVM, MS, DACVECC
Knoxville, Tennessee

INTRODUCTION

Last year was a busy year for transfusion medicine related publications in veterinary medicine. There were over 15 transfusion related prospective and retrospective peer reviewed small animal articles published across different veterinary journals. While they were all equally important, this session will review the "top 5" articles published in veterinary transfusion medicine in 2014. Though somewhat subjective, the articles were picked based on the number of times they were cited by other articles, viewed and downloaded from the website of the publishing journal. Articles that would appeal to a wide variety of people in the audience were also ranked higher. Review articles and case studies were not considered for this session. A list of other transfusion medicine related articles published in 2014 are available in the reference section. [1-12]

Incidence of Acute Lung Injury in Dogs Receiving Transfusions

This was a prospective observational study performed at the University of Wisconsin School of Veterinary Medicine.[13] The study was supported by an intramural grant provided by the University of Wisconsin. The purpose of the study was to document that Veterinary Acute Lung Injury (VetALI) occurs in dogs receiving transfusions of blood products within 24-48 hours after the transfusion and to determine the incidence of VetALI within this population. Their hypothesis was that VetALI occurs in animals receiving transfusion and the incidence will be similar to TRALI in people.

TRALI is defined as Transfusion Related Acute Lung Injury.[14] It is defined as new acute lung injury that occurs within 6 hours of transfusion of one or more blood products.[14,15] It is the leading cause of transfusion associated morbidity and mortality in human medicine. It is a clinical syndrome that results in hypoxemia and non-cardiogenic pulmonary edema during or after a blood transfusion. It is reported by the FDA that between 2005 and 2010, 48% of all transfusion related mortality was secondary to TRALI.[14,15] Critically ill human patients tend to have the highest incidence of TRALI. While the pathogenesis of TRALI is still not completely clear, it is thought that alterations of the immune system play a primary role in its development. Simply stated, TRALI is thought to occur when antibodies directed against human leukocyte antigens are transfused to a recipient. The recipients usually have primed neutrophils in their pulmonary endothelium as a result of their inflammatory state. The transfused antibodies stimulate the primed neutrophils to release proteases, which leads to endothelial damage, capillary leak and pulmonary edema. [14,16] Multiparous female blood donors are thought to have a higher incidence of having HLA antibodies, possibly due to exposure to fetal blood during pregnancy. Hence many programs have moved towards utilizing only male donors for fresh frozen plasma (FFP). [16,17] TRALI is still under recognized and underreported in human medicine because the clinical signs are nonspecific and variable.[14]

In this study, VetALI was defined based on the 2007 veterinary ARDS/ALI consensus definition.[18] Sixty-six dogs requiring transfusion of blood or plasma containing units that had never received a transfusion before were enrolled in the study. Twelve dogs were excluded either because they died or were euthanized, medical records were incomplete or mixed venous samples were obtained instead of arterial samples. All dogs included in the study had thoracic radiographs and an arterial blood gas (or pulse oximetry) before the transfusion was given. Red blood cells or plasma products were administered over 1-6 hours, based on the directives of the attending clinician. Thoracic radiographs and a blood gas (or pulse oximetry) were repeated 24-48 hours after the transfusion. All radiographs were interpreted by a board certified veterinary radiologist. The Chi-squared analysis was used to determine if the incidence of VetALI in this study population was significantly different from a 10% incidence of ARDS in a population of ill dogs that did not receive a transfusion.

Their results showed that two dogs had a PaO2:FIO2 ratio of less than 300 after transfusion with radiographic changes that could be consistent with VetALI. None of the two dogs underwent an echocardiogram. The incidence of VetALI in the study was 3.7%, which was significantly less than what is reported in people (the study referenced an incidence of 25% in human medicine). Consistent with a common challenge in diagnosing TRALI in people, the study could not absolutely prove that the two dogs developed VetALI secondary to the transfusion (TRALI), instead of secondary to their primary illnesses. The main limitation of this study was that most of the dogs received red cell transfusion and only 18 out of 54 dogs (33%) received a plasma transfusion. TRALI in human medicine is most commonly seen with FFP transfusions and both dogs that developed radiographic changes after transfusion got a FFP transfusion. The second limitation is that the reproductive status of the donors was not described in the paper. Due to a confirmed association between multiparous females and HLA antibodies, it would be interesting to see if plasma from dogs that have had at least one litter of puppies are more likely to trigger VetALI in dogs.

Effect of Duration of Packed Red Blood Cell Storage on Morbidity and Mortality in Dogs After Transfusion: 3,095 cases (2001–2010)

This study was a much needed additional to veterinary literature related to transfusion medicine.[19] Over at least the last decade, a lot of focus has been placed in human medicine on the age of transfused red cells. [20-23] In human medicine, red cells stored longer than 14 days have been correlated with untowed consequences in traumatized, critically ill and perioperative patients.[23] Biochemical, metabolic, mechanical and morphologic lesions tend to be seen in packed red blood cells (pRBC) after about 2 weeks of storage. These in vitro changes are commonly known as "storage lesions". The most commonly described storage lesions include impaired nitrous oxide metabolism, depletion of cellular 2,3 DPG, disruption of the membrane sodium-potassium pump, generation of inflammatory mediators and oxidative damage. [24] Several studies in human patients have described the association between the age of red blood cells and clinically significant adverse effects, including infection, organ failure, increased oxygen stay and death.[24-26] However, some studies did not identify an increased risk of infection, ICU hospitalization or death in patients that got aged red blood cells.[27] At this time, there is no strong consensus suggestion a change in clinical practice in the administration of aged cells in human patients. [27] There have been veterinary studies evaluating storage lesions associated with pRBC but no study has evaluated the effect of aged red cells on clinical patients.[28,29]

This was a retrospective study performed at the Matthew J Ryan Veterinary Hospital of the University of Pennsylvania. Records were reviewed to identify dogs who received a pRBC transfusion between January 2001 and December 2010. The guideline for maximal storage of pRBC according to the blood bank (Penn Animal Blood Bank) was 28 days but could be extended to 33 days based on demand. Information was obtained from medical records on parameters determined before the transfusion, including complete blood count, serum biochemistry analysis and coagulation results. Animals were recorded as having specific preexisting system/organ dysfunction based on criteria described in the paper. Animals were monitored during the transfusion and it was noted if a transfusion reaction occurred, based on specified criteria.

Three thousand and ninety five dogs were included in the study. There was a distribution of pRBC storage time from 0-40 days, with the median of 23 days for the oldest pRBC administered. Approximately 20% of the dogs received pRBC stored less than 14 days. There was an association between the length of pRBC storage and development of coagulation and thromboembolic disease after administration of the oldest unit of pRBC. A larger volume of pRBC administered was also associated with an increased incidence for the development of multiple organ dysfunction, individual organ failure, sepsis, DIC and thromboembolic disease. There was no difference in outcome or 30 day survival when comparing dogs who received only pRBC stored for less than 14 days to those who received pRBC stored for greater than 14 days. However, for dogs with hemolysis, a longer duration of pRBC storage was associated with non-survival to discharge. For every 7 day increase in storage, there was a 0.79 lesser odds of 30 day survival (95% CI, 0.64–0.97; P = .024).

The biggest limitation to this study is that most of the dogs (about 80%) received pRBC transfusions that had been stored for greater than 14 days. This could have potentially prevented the detection of differences between the two groups. The retrospective nature of this study also serves as a limitation as it makes it difficult to completely document adverse effects associated with transfusions. Prospective case controlled studies are definitely warranted evaluating the effects of the age of pRBC in veterinary patients.

An Ex Vivo Evaluation of Efficacy of Refrigerated Canine Plasma

This prospective laboratory based study was performed at Tufts Cummings School of Veterinary Medicine.[30] Fresh frozen plasma (FFP) is best stored frozen as it is the best method to preserve clotting factors, including the heat labile clotting factors V and VIII. This article highlights the fact that thawing FFP may be time consuming and the period between the FFP is thawed could lead to major blood loss in a critically ill patient. The article also goes ahead to describe how plasma is stored as refrigerated liquid plasma (RP) and frozen plasma (FP) in high volume trauma hospitals in the human field. RP is available for immediate transfusion when needed. A few human studies demonstrated that coagulation factors are held at levels adequate for hemostasis when FFP is stored in the refrigerator for 28-35 days. [31-33] The other purported advantage of storing plasma frozen is to prevent bacterial growth. The goals of the study were to determine the length of time one unit of FFP thaws in a water bath. They also set out to evaluate coagulation stability of canine plasma during a two week period. Finally, they wanted to evaluate the RP for evidence of bacterial growth during storage in the refrigerator.

For the plasma thaw analysis portion of the study, six units of canine plasma from the blood bank were individually thawed in a 37 degrees centigrade water bath. The units were considered thawed when ice particles could not be determined from manual palpation and visual inspection of the bags. Whole blood was then collected from 9 canine healthy donors, fractionated into plasma and aliquoted into two fractions. One fraction was stored as RP while the other fraction was stored in the freezer as FFP. 4 ml aliquots were obtained from RP units on days 1,5,7 and 14 for bacterial culture. A 4ml aliquot for bacterial culture was obtained from the FFP units on day 14 after they were thawed. Coagulation factor analysis was performed for factors V, VII, VIII, IX and X as well as PT and PTT.

Results showed that RP showed a significant decrease in coagulation factor concentration and a significant increase in clotting times, however no values obtained were outside established reference ranges. It took at least 30 minutes for the FFP units to thaw and there was no evidence of bacterial growth in all units, including the RP plasma units. Surprisingly, there was no significant change in factor V activity over time in both the RP and FP units since it is a heat labile factor. While there was a statistically significant change in factor VIII activity for RP, there was still benefit in using it clinically in the patient with hemophilia A as levels after two weeks of storage were still within reference range.

While most veterinary hospitals may not transfuse enough plasma to justify keeping RP, the clinical utility of this study lies in FFP units that are thawed and subsequently discarded after patient death/euthanasia or change in the medical plan of the patient. This study may support storage of thawed FFP units in the refrigerator although this needs to be specifically evaluated by a different study.

Assessment of Clinical and Laboratory Variables as a Guide to Packed Red Blood Cell Transfusion of Euvolemic Anemic Dogs

For many years, intensivists in human hospitals have debated on transfusion triggers for critically ill patients.[34-36] There is currently no consensus in human and veterinary patients regarding if patients will benefit from a blood transfusion. A critical hemoglobin concentration, which would serve as a trigger for providing a transfusion, has also not been determined in human and veterinary patients. This prospective study was performed at the University of Glasgow Small Animal Hospital, School of Veterinary Medicine. The hypothesis of the study was that physiologic variables including, hemoglobin concentration, hematocrit, CvO2, lactate concentration, and standardized clinical assessment of anemic dogs, would improve after a red cell transfusion. The goal was to identify levels of these variables common to the dogs included in the study, which could be extrapolated to other anemic dogs as guidelines for when to provide a red cell transfusion.

Anemic dogs admitted to the hospital that were determined to require a blood transfusion were included in the study (2010-2012). An Anemic Dog Clinical Assessment Score (ADCAS) was designed for the purpose of the study and evaluated five parameters, including mucous membrane color, pulse quality, heart rate, respiratory rate and mentation/exercise tolerance. Dogs were blood typed (DEA-1.1 positive or negative) before transfusion and type compatible pRBC units were administered. The volume of blood administered was at the discretion of the attending clinician. A venous blood sample was taken before and within

1 hour after transfusion for PCV measurement. Lactate concentration and oxygenation values (venous partial pressure of oxygen, PvO2, and venous oxygen saturation of hemoglobin, SvO2), hemoglobin concentration and hematocrit, from blood gas analysis were assessed within 5 minutes of sample collection from surplus blood.

Twenty-four dogs were included, with a total of 30 transfusion cases. Transfusion was deemed by the attending clinician, independently of the ADCAS, lactate concentration, and oxygenation values, to be successful in 21 dogs (27/30 transfusions) based on improvement in clinical examination findings and increase in hematocrit. All variables in the ADCAS, including the total score, decreased significantly with transfusion ($P < .001$). Hemoglobin concentration, hematocrit, and CvO2 increased significantly with transfusion ($P < .001$), and lactate concentration decreased significantly ($P = .006$), whereas PvO2 and SvO2 did not change significantly. With the significant improvement in clinical and laboratory values, this study was able to identify variables that can be utilized by practicing veterinarians to guide the decision to provide a transfusion. The limitations of this study include the small sample size utilized in the study. It would also have been ideal if the SvO2 were measured rather than calculated. There was also the bias introduced due to the fact that most of the cases that received the transfusion were referred and likely had significant clinical signs associated with their severe anemia. This study has practical clinical relevance and can be used not only to determine when a patient requires a transfusion, but may also be used to evaluate the clinical response to the transfusion.

Influence of cross-match on post transfusion packed cell volume in feline packed red blood cell transfusion

Type specific blood products have been advocated in use in cats due to the well described evidence of naturally occurring alloantibodies. However there is recent evidence to suggest that there is a high prevalence of naturally occurring alloantibodies in the *Mik*-negative recipient cats against the *Mik* antigen expressed in donor cats. The presence of these alloantibodies may lead to transfusion reactions and decreased efficacy of packed red cell transfusion. Pretransfusion compatibility testing may help identify incompatibilities and reduce the risks for transfusion reactions or decreased efficacy of transfused red cells. The hypothesis of this study was that type-specific, c ross-match compatible transfusions of pRBCs would yield significantly higher PCV increases post transfusion per volume of pRBCs transfused than type specific, non cross-matched transfusions.

This was a retrospective analysis carried out from 2000-2010 at the Cornell University College of Veterinary Medicine. Two hundred nine cats receiving pRBC transfusions were identified and pertinent information from their medical records were recorded. Blood type was identified by a laboratory personnel and 36 cats had a cross-match performed for 43 type-specific, cross-match compatible transfusions. The mean dose of pRBCs administered, as well as the volume of IV fluids administered, between cats that had cross-matches and cats that did not was not significantly different. Cats that had a cross-match performed had a significantly larger mean change in PCV scaled to dose of pRBCs administered than cats that did not have a cross-match performed.

This was an interesting study being that all cats received type-specific pRBC transfusions yet their results showed that cats receiving pRBC units deemed compatible by major cross-match showed a significantly greater increase in the PCV scaled by dose or pRBCs administered. This study did not report a rate of transfusion reactions in their study population as the information was not well documented in the medical records but a potential reason for the lower than expected increase in PCV in the non cross-matched cats is premature destruction of the transfused red cells.

While this study had multiple limitations, including its retrospective nature and the inability to compare blood typing methods, they raise an important question regarding the standard of only performing cross-matches in feline patients that have received a prior blood transfusion. Their study identified at least one incompatible unit of pRBC in 16.5% of the cross matches performed, which also raises an important point of considering cross-matches whenever possible in all cats, regardless of their prior transfusion status.

REFERENCES

1. Acierno M, Raj K, Giger U. DEA 1 Expression on Dog Erythrocytes Analyzed by Immunochromatographic and Flow Cytometric Techniques. *Journal of Veterinary Internal Medicine.* 2014;28(2):592-598.
2. Corsi R, McMichael MA, Smith SA, et al. Cytokine concentration in stored canine erythrocyte concentrates. *Journal of Veterinary Emergency and Critical Care.* 2014;24(3):259-263.
3. Heikes BW, Ruaux CG. Effect of syringe and aggregate filter administration on survival of transfused autologous fresh feline red blood cells. *Journal of Veterinary Emergency and Critical Care.* 2014;24(2):162-167.
4. Hoareau GL, Jandrey KE, Burges J, Bremer D, Tablin F. Comparison of the platelet-rich plasma and buffy coat protocols for preparation of canine platelet concentrates. *Veterinary Clinical Pathology.* 2014;43(4):513-518.
5. Holowaychuk MK, Leader JL, Monteith G. Risk factors for transfusion-associated complications and nonsurvival in dogs receiving packed red blood cell transfusions: 211 cases (2008–2011). *Journal of the American Veterinary Medical Association.* 2014;244(4):431-437.
6. Kisielewicz C, Self I, Bell R. Assessment of Clinical and Laboratory Variables as a Guide to Packed Red Blood Cell Transfusion of Euvolemic Anemic Dogs. *Journal of Veterinary Internal Medicine.* 2014;28(2):576-582.
7. Lacerda LA, Hlavac NR, Terra SR, Back FP, Jane Wardrop K, González FH. Effects of four additive solutions on canine leukoreduced red cell concentrate quality during storage. *Veterinary Clinical Pathology.* 2014;43(3):362-370.
8. Reed N, Espadas I, Lalor SM, Kisielewicz C. Assessment of five formulae to predict post-transfusion packed cell volume in cats. *Journal of feline medicine and surgery.* 2014:1098612X13517254.
9. Spada E, Miglio A, Proverbio D, et al. Signalment and blood types in cats being evaluated as blood donors at two Italian university blood banks. *Veterinary medicine international.* 2014;2014.
10. Tasker S, Barker E, Day M, Helps C. Feline blood genotyping versus phenotyping, and detection of non-AB blood type incompatibilities in UK cats. *Journal of Small Animal Practice.* 2014;55(4):185-189.
11. Weltman JG, Fletcher DJ, Rogers C. Influence of cross-match on posttransfusion packed cell volume in feline packed red blood cell transfusion. *Journal of Veterinary Emergency and Critical Care.* 2014;24(4):429-436.
12. Spada E, Proverbio D, Martino PA, Baggiani L, Perego R, Roggero N. Ammonia concentration and bacterial evaluation of feline whole blood and packed red blood cell units stored for transfusion. *International Journal of Health, Animal Science and Food Safety.* 2014;1(2).
13. Thomovsky EJ, Bach J. Incidence of acute lung injury in dogs receiving transfusions. *Journal of the American Veterinary Medical Association.* 2014/01/15 2013;244(2):170-174.
14. Berdai M, Labib S, Harandou M, Berdai MA, Labib S, Harandou M. TRALI in the perioperative period. *Vascular Health and Risk Management.* 2015;11:9-23.

15. Sridhar G, Menis M, Selvam N, et al. Transfusion-related acute lung injury (TRALI) occurrence, risk factors, and outcome: a nested case-control study. *Internet J Hematol.* 2013;9.
16. Bux J, Sachs UJH. The pathogenesis of transfusion-related acute lung injury (TRALI). *British Journal of Haematology.* 2007;136(6):788-799.
17. Muller M, Van Stein D, Binnekade J, Van Rhenen D, Vlaar A. Male Only Fresh Frozen Plasma Donation Policy And The Impact On The Onset Of Transfusion Related Acute Lung Injury; A Meta-Analysis. *Am J Respir Crit Care Med.* 2014;189:A5003.
18. Wilkins PA, Otto CM, Baumgardner JE, et al. Acute lung injury and acute respiratory distress syndromes in veterinary medicine: consensus definitions: The Dorothy Russell Havemeyer Working Group on ALI and ARDS in Veterinary Medicine. *Journal of Veterinary Emergency and Critical Care.* 2007;17(4):333-339.
19. Hann L, Brown DC, King LG, Callan MB. Effect of Duration of Packed Red Blood Cell Storage on Morbidity and Mortality in Dogs After Transfusion: 3,095 cases (2001–2010). *Journal of Veterinary Internal Medicine.* 2014;28(6):1830-1837.
20. Biffl WL, Moore EE, Offner PJ, Ciesla DJ, Gonzalez RJ, Silliman CC. Plasma from aged stored red blood cells delays neutrophil apoptosis and primes for cytotoxicity: abrogation by poststorage washing but not prestorage leukoreduction. *Journal of Trauma and Acute Care Surgery.* 2001;50(3):426-432.
21. Offner PJ. Age of blood: does it make a difference? *CRITICAL CARE-LONDON-.* 2004;8:S24-S26.
22. Cohn SM, DeRosa M, Kumar A, et al. Impact of the age of transfused red blood cells in the trauma population: a feasibility study. *Injury.* 2014;45(3):605-611.
23. Daly A, Raval JS, Waters JH, Yazer MH, Kameneva MV. Effect of blood bank storage on the rheological properties of male and female donor red blood cells. *Clinical hemorheology and microcirculation.* 2014;56(4):337-345.
24. Gregg R, Brunskill SJ, Wilkinson KL, Doree C, Trivella M, Stanworth S. Transfusion of fresher versus older red blood cells for all conditions. *The Cochrane Library.* 2014.
25. Offner PJ, Moore EE, Biffl WL, Johnson JL, Silliman CC. Increased rate of infection associated with transfusion of old blood after severe injury. *Archives of Surgery.* 2002;137(6):711-717.
26. Tinmouth A, Fergusson D, Yee IC, Hébert PC. Clinical consequences of red cell storage in the critically ill. *Transfusion.* 2006;46(11):2014-2027.
27. Aubron C, Nichol A, Cooper DJ, Bellomo R. Age of red blood cells and transfusion in critically ill patients. *Ann Intensive Care.* 2013;3(1):2.
28. McMichael M, Smith S, Galligan A, Swanson K, Fan T. Effect of leukoreduction on transfusion-induced inflammation in dogs. *Journal of Veterinary Internal Medicine.* 2010;24(5):1131-1137.
29. Herring JM, Smith SA, McMichael MA, et al. Microparticles in stored canine RBC concentrates. *Veterinary Clinical Pathology.* 2013;42(2):163-169.
30. Grochowsky AR, Rozanski EA, Laforcade AM, et al. An ex vivo evaluation of efficacy of refrigerated canine plasma. *Journal of Veterinary Emergency and Critical Care.* 2014;24(4):388-397.
31. Lamboo M, Poland D, Eikenboom J, et al. Coagulation parameters of thawed fresh-frozen plasma during storage at different temperatures. *Transfusion Medicine.* 2007;17(3):182-186.
32. Thiele T, Kellner S, Hron G, et al. Storage of thawed plasma for a liquid plasma bank: impact of temperature and methylene blue pathogen inactivation. *Transfusion.* 2012;52(3):529-536.
33. Nilsson L, Hedner U, Nilsson I, Robertson B. Shelf-life of bank blood and stored plasma with special reference to coagulation factors. *Transfusion.* 1983;23(5):377-381.
34. Corwin HL, Gettinger A, Pearl RG, et al. The CRIT Study: Anemia and blood transfusion in the critically ill—Current clinical practice in the United States*. *Critical care medicine.* 2004;32(1):39-52.
35. Vincent JL, Baron J-F, Reinhart K, et al. Anemia and blood transfusion in critically ill patients. *Jama.* 2002;288(12):1499-1507.
36. Hébert PC, Fergusson DA, Stather D, et al. Revisiting transfusion practices in critically ill patients*. *Critical care medicine.* 2005;33(1):7-12.

A BREAK DOWN OF CARDIOVASCULAR PHYSIOLOGY: THE WIGGERS DIAGRAM

Hally E. Niehaus, BS, RVT, VTS (Cardiology)
Akron, OH

INTRODUCTION

Cardiovascular physiology can be very intimidating, but the Wiggers diagram can help us understand its complexity and lessen that intimidation. By breaking down the cardiac cycle into a graph format, one can appreciate the pressures and volumes within the heart along with their timing relative to the ECG and the coinciding heart sounds.

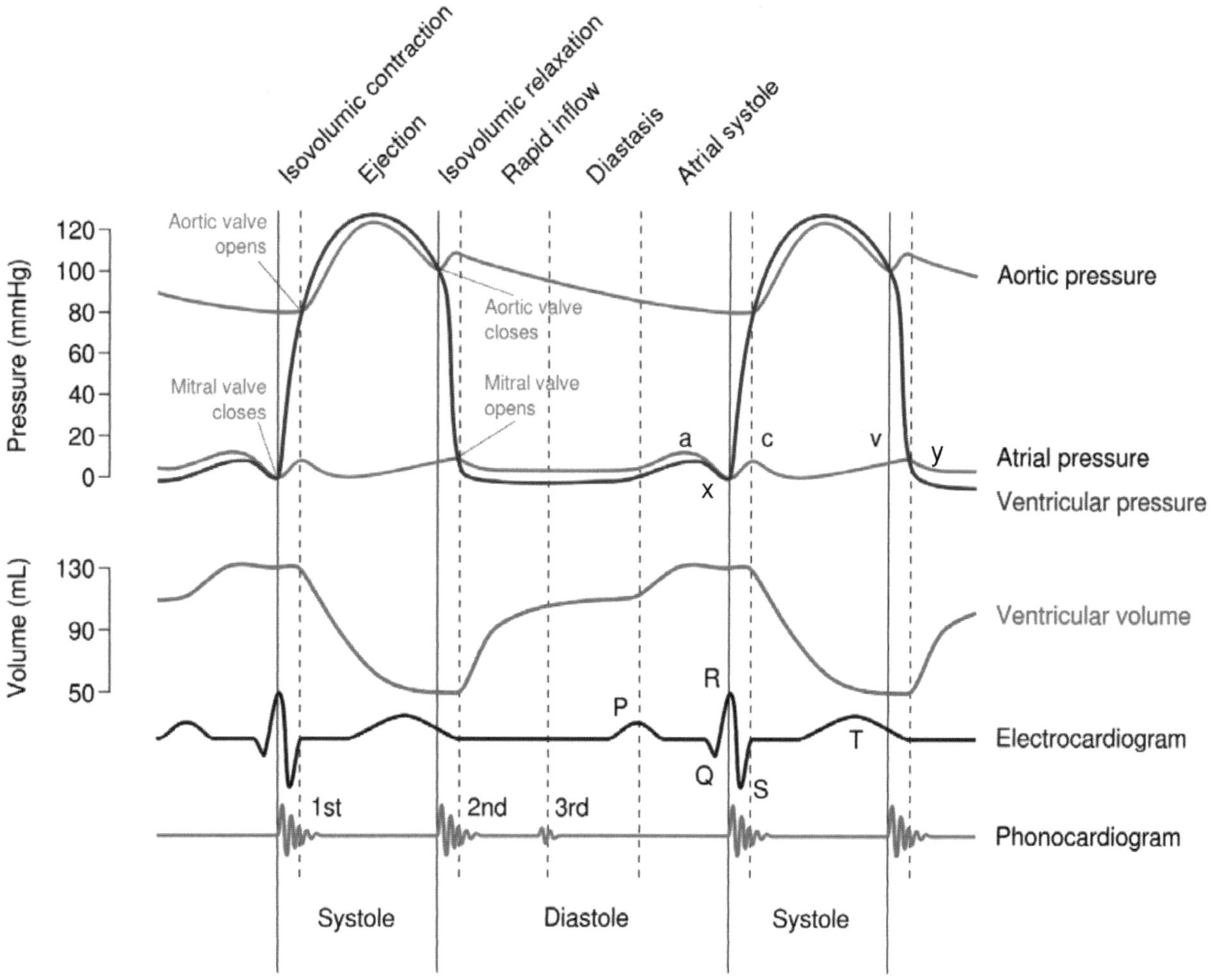

BASIC CARDIAC ANATOMY

There are four chambers in the heart: the right atrium and ventricle and the left atrium and ventricle. Along with the four heart chambers, there are four major heart valves. The mitral valve separates the left atrium and ventricle, whereas the tricuspid valve separates the right atrium and ventricle. There are also the aortic valve which is located between the left ventricular outflow tract and the aorta, and the pulmonic valve which is located between the right ventricular outflow tract and the pulmonary artery. Blood travels from the lungs through the pulmonary veins into the left atrium, across the mitral valve into the left ventricle exiting from the left ventricular outflow tract across the aortic valve into the aorta and out to the body. Blood flow returns from the body to the right atrium via the vena cavae and crosses the tricuspid valve to the right ventricle, where blood is then pumped through the right ventricular outflow tract, across the pulmonic valve into the pulmonary artery, and out to the lungs.

CARDIAC CYCLE

The heart pumps in a cyclical fashion continually pumping blood through the body. Each cardiac cycle refers to a full heart beat from the beginning of the SA node initiating depolarization and contraction of the heart to the ventricular repolarization. The ventricle contracting, or pumping blood, is known as systole and the period of relaxation and filling is diastole. Each cardiac cycle encompasses all of systole and diastole.

Systole

Systole occurs between the first and second heart sounds and has two phases. The start of ventricular contraction causes a rise in pressures within the ventricle resulting in closure of the AV valves (mitral and tricuspid) and marks the beginning of systole. The end of diastole and beginning of systole occur at the same time. During the first phase there is a brief moment, approximately 30-50msec, after the atrioventricular (AV) valves close where both the AV valves and the aortic and pulmonic valves are simultaneously closed. This period is termed isovolumic or isovolumetric contraction because there is no change in volume within the heart at this time. The second and final phase in systole is known as ejection and can be divided into rapid and reduced ejections. The aortic and pulmonic valves open at the beginning of this phase of systole.

Diastole

Diastole is the relaxation of the ventricles which allows the aortic and pulmonic valves to close and the AV valves to open due to pressure changes in the heart. As in systole, diastole also starts with a very brief period of time when all the valves are closed simultaneously, isovolumic or isovolumetric relaxation. This is the first of four phases of diastole. Early or rapid filling (inflow) is the second phase of diastole and is at the time when the AV valves are opening. Diastasis (slow filling) follows early filling and the length of this phase is dependent on the patient's heart rate. Those with faster heart rates may lose diastasis all together. The fourth and final phase of diastole is the atrial kick or atrial systole. Following the atrial kick marks the beginning of another systole. Again, the end of diastole and beginning of systole occur at the same time.

HEART SOUNDS

Cardiac auscultation is the art of listening to heart sounds. The first heart sound (S1) or "lub" is heard at the beginning (or initiation) of systole and coincides with the AV valves closing. The second heart sound (S2) or "dub" is heard at end systole when the aortic and pulmonic valves are closing, initiating the diastolic or filling phase of the cardiac cycle. These sounds are heard from vibrations of the blood pool not the valves closing in and of themselves. A third heart sound (S3) and fourth heart sound (S4) or gallop can be heard during diastole in some patients with cardiac disease. S3 can be heard during early filling and S4 can be heard at the time of atrial systole.

A murmur is an abnormal sound heard during auscultation of the heart when there is abnormal blood flow or turbulence across one of the valves. Murmurs can also be heard for non-valvular reasons such as a patent ductus arteriosis, an atrial septal defect or a ventricular septal defect.

NORMAL ELECTRICAL CONDUCTION OF THE HEART

The sinoatrial (SA) node, located in the right atrial wall just ventral to the opening of the cranial vena cava, initiates depolarization and contraction of the heart. Immediately following the SA node depolarization, the atria depolarize resulting in the P wave on the ECG. The impulse travels to the AV node, located within the atrioventricular septum, where there is a slight pause in the electrical transfer allowing the ventricles to fill (PR interval). The electrical impulse then travels through the His-Purkinje system (bundle of His, bundle branches, and Purkinje network) and cardiac myocytes resulting in the QRS complex representing electrical activation of the ventricles. Electrical activation is typically followed by ventricular contraction (mechanical systole). Ventricular repolarization is seen with the T wave.

CARDIAC PRESSURES

The atrial pressure waveforms consist of a, c, and v waves, and x and y descents. The a and v waves are the two major positive waves and are each followed by the negative x and y descents, respectively. The c wave may be present at the onset of ventricular systole on the right and left atrial pressure pulses. The a is atrial systole and follows the P wave on the ECG. Ventricular pressures rise above atrial pressures at the beginning of mechanical systole resulting in the AV valves closing with the x descent following. The v wave is inscribed when the atria fill from venous return and peaks at the opening of the AV valves. The y descent is after the AV valves open and the atria empty into the ventricles (as the ventricular pressure falls below atrial pressures). Pressure is measured at the peak of the a and v waves with the mean pressure being determined electronically. The left heart pressures are greater than right heart pressures.

There are two phases (systole and diastole) within the ventricular pressure waveforms. There is a small rise in ventricular pressure with each step up of ventricular volume, and then a marked rise during mechanical systole. The AV valves close when the ventricular pressure rises above the atrial pressures. The aortic and pulmonic valves open when the ventricular pressures rise above the aortic/pulmonic pressures. The aortic and pulmonic valves close when the ventricular pressures fall below the aortic/pulmonic pressures as systole ends. Finally, the ventricles begin to relax and pressure falls. The AV valves open when the ventricular pressures fall below the atrial pressures.

Table 1. Normal intracardiac pressures (mmHg)

Site:	Systolic	Diastolic	Mean
Right atrium	4-6	0-4	2-5
Right ventricle	15-30	<5	-
Pulmonary artery	15-30	5-15	8-20
Left atrium	5-12	<8	<10
Left ventricle	95-150	<10	-
Aorta	95-150	70-100	80-110

VENTRICULAR VOLUME

The volume in the ventricles depends on the timing within the cardiac cycle. During diastole, the ventricle fills to its maximal volume known as end-diastolic volume (EDV). EDV is reached when the AV valves close at the end of diastole/start of systole. There is rapid passive ventricular filling in early diastole followed by a slowing of ventricular filling or diastasis. The length of diastasis is determined by the heart rate and does not occur in patients with a tachycardia. Atrial systole, or the atrial kick, occurs towards the end of diastole after the P wave of the ECG and adds approximately 20% of the filling volume in a normal animal. Based on pressures within the heart, the aortic and pulmonic valves open and lead to the onset of ventricular ejection lasting until the end of systole, when the aortic and pulmonic valves close. At this point the ventricle has the smallest volume of blood for that cycle, known as end-systolic volume (ESV). The EDV minus the ESV is the total stroke volume (TSV) of the ventricle, or the amount of blood ejected during systole.

REFERNCES

Katz; *Physiology of the Heart*; 2011: 305-312. Kittleson, et al; *Small Animal Cardiovascular Medicine*; 1998: 12-14, 124-125. Pasquini, et al; *Anatomy of Domestic Animals*; 1997: 382-404.

CARDIAC DISEASE AND ANESTHESIA

Christopher Stauthammer, DVM, DACVIM (Cardiology)
Saint Paul, MN

Anesthetic management of patients with underlying heart disease can be very challenging as most anesthetic drugs depress cardiac function, alter vascular tone and heart rate as well as their potential for arrhythmogenesis. The subsequent reduction in cardiac output and arterial blood pressure decrease tissue perfusion with potential organ damage and patient death. Limited cardiac reserve as a consequence of the underlying heart disease enhances the cardio-depressant effects and restricts common methods of patient stabilization.

Patient Evaluation

Proper preanesthetic evaluation of the patient helps minimize deleterious effects of anesthesia, improving the overall patient safety. The anesthetist should review the patient's history in search for evidence of clinical signs indicating deterioration of patient status and current medication regiment. The physical examination should also focus on the clinical status of the patient's heart disease including hemodynamic assessment with blood pressure, hydration status, respiratory function and the presence of pleural/abdominal effusion. Blood urea nitrogen, creatinine and electrolytes need to be evaluated due to the high incidence of concurrent renal disease and diuretic induced electrolyte abnormalities. Thoracic radiographs are essential for assessment of cardiac size and the presence of pulmonary edema. ECG and echocardiography provide useful information but are not necessary in all cases. Anesthesia is contraindicated in patients with decompensated heart disease (ASA Catergory IV & V) until stabilization of cardiac complications.

Anesthetic Drug Selection

Due to the vast array of cardiovascular diseases affecting dogs and cats, one anesthetic protocol is not appropriate for all patients. Appropriate drug selection is based on an understanding of the cardiovascular pathophysiology and the cardiovascular effects of the anesthetic drugs. A multimodal approach using reduced dosages of drugs from multiple pharmacologic categories is preferred to minimize extreme cardiovascular effects of any one drug.

Medications to Avoid

While there are circumstances when the below drugs could be used in patients with heart disease, it is best to avoid them. Most of these medications are no longer in common use anyhow.

- Alpha-2 Adrenergic Drugs
    - Intense vasoconstriction and depression of myocardial contractility
    - Bradycardia
- Barbiturates
    - Potent cardiovascular depressant effects
    - Arrhythmogenic
- Halothane
    - Greatest decrease on contractility of inhalants, arrhythmogenic

Premedications

The benefits of premedicating cardiac patients with a combination of a sedative and analgesic include reducing preoperative anxiety and stress, preemptive analgesia, lowering the amounts of induction agents and inhalants and a smoother recovery. Opioids do not affect contractility or stimulate arrhythmias. They may cause a mild decrease in blood pressure secondary to vasodilation and vagally mediated bradycardia responsive to anticholinergics. Most opioids do not possess profound sedative effects when administered alone in dogs and may trigger hyperexcitability in cats. Opioids used in combination with benzodiazepines for neuroleptanalgesia generally result in adequate sedation. The benzodiazepines also produce minimal cardiovascular effects with the potential for a mild vasodilation and myocardial depression.

Acepromazine may be administered for sedation in patients with cardiovascular disease. However, it causes vasodilation secondary to alpha adrenergic blockade. Pending the disease, low dose acepromazine is beneficial with the decreased peripheral resistance or afterload increasing cardiac output. The sedative and vasodilatory effects of acepromazine are of long duration (4-6 hours) and nonreversible.

Anticholinergics increase heart rate and myocardial oxygen consumption, increase the possibility of cardiac arrhythmias and decrease the threshold for ventricular fibrillation. Some patients may not tolerate excessive increases in heart rate. Therefore, anticholinergics should not be a routine component of the anesthetic protocol but reserved for treatment of bradycardia that limits cardiac output.

Induction drugs

The use of inhalant gases for induction by mask or chamber is not recommended for patients with cardiovascular disease due to associated stress, respiratory depression and potential for hypotension.

Propofol causes a dose dependent decreased contractility and hypotension similar to that of the thiobarbituates. Use of preanesthetic medication greatly reduces dose of required propofol necessary for induction.

Etomidate is the preferred drug for severe cardiac disease due to its minimal effects on myocardial contractility, heart rate and cardiac output. Etomidate may trigger vasodilation decreasing blood pressure. Etomidate has minimal effects on arrhythmia development. It may produce a rough induction with patient excitement, muscle twitching and vocalization. Effects are minimized by the administration of a benzodiazepine or acepromazine prior to administration.

Ketamine is a commonly used induction agent in heart disease but never to be used alone. Ketamine indirectly increases heart rate, blood pressure and contractility by stimulation of the sympathetic nervous system. Such effects are beneficial in diseases with decreased contractility and low heart rates but can be detrimental in patients with hypertrophic cardiomyopathy, hypertension, hyperthyroidism and arrhythmias. Ketamine should also be used with caution in patients with end stage heart failure as a dose dependent decrease in myocardial contractility may occur secondary to catecholamine exhaustion.

Maintenance Drugs

Inhalant anesthetics all cause dose dependent cardiovascular depression. Isoflurane and sevoflurane have similar effects of hypotension with a dose dependent peripheral vasodilation and minimal effects on rhythm and contractility. The heart rate usually remains constant.

A drug with less cardio-depressant effects may be administered as a CRI during the maintenance phase to further reduce or even eliminate the dose of inhalant necessary for maintenance of a surgical plane. A CRI of an opioid, ketamine or a combination are often used in patients with severe disease to minimize the inhalant. In dogs, lidocaine may also be added for arrhythmia suppression.

Fluids

Most patients will require fluid support for maintenance of blood pressure. Ideally, a sodium restricted crystalloid fluid (0.45% NaCl/2.5% dextrose) should be used in the face of heart disease. The rate of fluid therapy however, is far more important than fluid type. The standard anesthetic fluid rate of 10 ml/kg/hr would likely result in fluid overload in patients with moderate disease. A general rate of 3-5 ml/kg/hr for patients with mild to moderate disease and 2-3 ml/kg/hr for patients with severe disease is recommended. Left atrial size is useful in determining fluid rate. Fluid rate is of lesser concern for those patients with strictly right sided heart disease. Extreme caution should be exercised with colloid administration.

Ancillary Therapy

A CRI of a positive inotropic or pressor drug is often necessary. However, positive inotropic therapy should be used with caution in patients with a pressure overload state (aortic/pulmonic stenosis) and those with diastolic dysfunction. Dobutamine directly increases myocardial contractility and heart rate with a subsequent increase in cardiac output and blood pressure. Dobutamine triggers a mild vasodilation thereby decreasing ventricular afterload. Patients with diminished contractility are generally started on a CRI (5-15 mcg/kg/min) prior to induction. The dose is adjusted based on blood pressure and heart rate. Dobutamine is arrhythmogenic and may result in significant tachycardia especially at higher dosing. May trigger seizures in cats. Dopamine is also commonly administered as a CRI for pressor support. Dopamine triggers vasoconstriction at higher doses increasing ventricular afterload and mitral regurgitation.

Specific Heart Diseases

Mitral Valve Disease

The predominate concern in patients with mitral valve disease is minimizing vasoconstriction. In small breeds contractility is generally well preserved, however marked deterioration of ventricular contractility is common in large breeds of dogs.

Vasoconstriction increases the afterload to the left ventricle with a subsequent increase in mitral regurgitation and potential development of pulmonary edema and arrhythmias. Vasodilation and increasing contractility will directly reduce mitral regurgitation. Low dose acepromazine may be used to decrease mitral regurgitation. Include opioids for reduction of inhalant drug concentrations with the choice of opioid determined by level of expected pain. Etomidate, profol or ketamine/diazepam are all suitable induction drugs. Isoflurane or sevoflurane are appropriate for maintenance, and vasodilatory effects of the inhalants will further reduce mitral regurgitation.

Dilated Cardiomyopathy (DCM)

The predominate feature of DCM is severe myocardial contractile dysfunction with subsequent chamber dilation, congestive heart failure and arrhythmias. Atrial fibrillation is quite common in all affected breeds. Fatal ventricular arrhythmias are common in Dobermans and Boxer dogs. Areas of concern include maintaining contractility and prevention of arrhythmias. Intotropic support should be addressed throughout the anesthetic period. Propofol should be avoided due to its myocardial depressant effects. Induction with etomidate, ketamine/diazepam or fentanyl is recommended. A CRI of lidocaine is recommended for dogs with ventricular arrhythmias. Lidocaine may be combined with morphine/ketamine or fentanyl CRI for additional reduction of inhalant concentration.

Hypertrophic Cardiomyopathy

The predominate feature of HCM is diastolic dysfunction secondary to decreased myocardial relaxation and ventricular compliance. Ventricular systolic function is generally normal to enhanced. Perfusion abnormalities are common secondary to increased wall thickness and coronary artery abnormalities. The main area of concern is preventing tachycardia. Increasing the heart rate diminishes diastolic filling with a subsequent increase in atrial pressures and development of heart failure. Myocardial perfusion is also reduced with an increase in myocardial metabolism. The anesthetist should avoid medications that may increase heart rate including anticholinergics and ketamine.

Congenital Heart Disease

Congenital heart diseases predominately affect the heart in one of two ways, volume overload secondary to shunting (septal defect, patent ductus arteriosus) or a pressure overload secondary to a stenotic valve/artery (aortic and pulmonic stenosis). Diseases resulting in a volume overload should be handled within a similar manner as mitral valve disease. Systemic vasodilation is also beneficial for left to right shunting as it reduces the pressure gradient and subsequently the amount of shunting through the defect. Generally, contractility is preserved in patients with a left to right shunt. Patients with a right to left shunt either have severe pulmonary hypertension or pulmonic stenosis and an ASA classification of 4 or 5. The predominate concern with a pressure overload is a limited ability to compensate for vasodilation due to a restricted cardiac output. Patients with severe wall thickening will likely have diastolic dysfunction and myocardial perfusion abnormalities similar to HCM. Contractility is generally normal except in patients with advanced disease. Patients with a pressure overload should be managed similar to HCM with careful attention in maintaining heart rate and limiting vasodilation. The development of left sided congestive heart failure is rare with aortic stenosis, and hence a larger rate of fluid administration is generally safe as long as the left atrium is of normal size.

HYPERTHYROID CATS HAVE HEARTS TOO!
UNDERSTANDING THE SECONDARY EFFECTS OF HYPERTHYROIDISM ON THE HEART

Kristie Garcia LVT,VTS (Cardiology)

Overview

Hyperthyroidism in cats has only been recognized recently. There are secondary effects on the rest of the body that are well documented and are expected. The symptoms that the patient presents with are easily recognized, and the veterinarian will run the appropriate testing to officially diagnose the hyperthyroidism. Some of the symptoms, usually in geriatric cats, include: weight loss, ravenous appetite, poor coat and tachycardia.

The veterinarian will typically order radiographs, ECG, and blood work, including CBC, Chemistry, T4, fT4 and TSH. The CBC ensures that the patient isn't anemic, as this is a possibility. The chemistry makes sure the liver and kidney values are within normal ranges. Hyperthyroid cats seem to have elevated liver values but histologically they are unremarkable. Kidney values are a little more difficult to properly assess. The increase in blood pressure secondary to the hyperthyroidism may diurese the kidneys, making the numbers look to be within normal limits. This may not be discovered unless an owner elects to stop therapy and / or pursue I-131 therapy.

Treatment

There are a few accepted treatments for hyperthyroidism in the cat. They are all well known and include: diet control, methimazole, and I-131 radioactive therapy.

When a course of therapy has been chosen by the owner and the veterinarian, the follow up for the patient includes recheck bloodwork and radiographs.

The bloodwork upon recheck may show a decrease in the thyroid values. This doesn't reflect the reality of the secondary effects on the heart, unless the therapy chosen was I-131.

Cardiac-Specific effects on the heart

The treatment selected by the owner and veterinarian determines the severity of the secondary effects on the heart. The diets and the medications are all used most commonly, however the heart still is under stress from the thyroid. The I-131 therapy , in fact, cures the hyperthyroidism and therefore the heart benefits much more than with other therapies.. When the patient is diagnosed, the hyperthyroidism may have already been stressing the heart. They may present with tachycardia, arrhythmias, gallop rhythm upon auscultation, and perhaps the disease has advanced enough to cause pleural effusion. The thyrotoxic heart looks like a hypertrophic cardiomyopathy upon echocardiogram by a cardiologist. The question becomes: is it primary HCM or secondary changes from the hyperthyroidism? The answer is the response to I-131 therapy. If the changes are secondary to hyperthyroidism, the damaging changes to the myocardium will reverse and improve.

Conclusion

Hyperthyroidism has effects on the heart that seem to be uncontrolled even when the blood work shows that the thyroid should be under control. The hyperthyroid cardiac patient will need follow up cardiac care for the rest of its life. As technicians, our job is to try and understand as much as we can about the effects of acquired disease on the whole body of our patients. In this case, the hyperthyroid patient will indeed become a cardiac patient and understanding this disease process is crucial to helping them.

DILATED CARDIOMYOPATHY, THE TROUBLE WITH HAVING A BIG HEART

Michelle St John, LVT, VTS Cardiology
Latham, NY

INTRODUCTION

Dilated cardiomyopathy (DCM) is a disease that affects the heart muscle causing it to become thin and weak resulting in poor contractility, eventually leading to congestive heart failure. There is no definitive cause of dilated cardiomyopathy (DCM) but there appears to be a genetic relationship in certain pure breed dogs such as Boxers, Doberman Pinschers, Great Danes, and Irish Wolfhounds. DCM is predominately a disease that affects middle aged, giant breed dogs, but smaller breeds such as Cocker Spaniels may be affected. Cats are rarely diagnosed with DCM. Knowing that DCM has a familiar component, it is important to have dogs of particular breeds screened for DCM while they are in an active breeding program. Various diagnostic tests are used to screen pets for DCM, such as echocardiography, Holter monitoring, and genetic DNA testing to name a few. It should also be noted that there is no curative treatment for DCM.

SCREENING

As stated, genetics are assumed to play a role in patients developing DCM, so breeders of commonly affected breeds are encouraged to screen their dogs yearly with echocardiography and Holter monitoring (24 hour ECG recording) while they are actively being used for breeding. It is also helpful to have a DNA genetic test performed once for each dog in the breeding program. Owners are encouraged to screen their dogs yearly if there is any family history of DCM whether they are intended for breeding or not.

Diagnostic tests used to screen patients for DCM are echocardiography, electrocardiogram (ECG), Holter monitoring (24 hour ECG recording), and DNA genetic testing. Echocardiography is the gold standard for DCM diagnosis. Echocardiography is the only way to evaluate myocardial contractility. Documenting how well the ventricles are contracting in systole and diastole are achieved by calculating the fractional shortening, ejection fraction, and EPSS. The fractional shortening is calculated from the M-Mode and the ejection fraction is calculated from a 2-D right short axis, right four chamber long axis, and left apical image. Both measure the percentage of change in the ventricle from diastole to systole. The EPSS (E point to septal separation) is measured from the M-mode at the mitral valve. The smaller the EPSS, (the point of the mitral valve to the septum), the better. Positive diagnosis is made with a fractional shortening < 25% in asymptomatic patients and <15% in symptomatic patients. Normal EPSS should be 6mm or less, and left ventricular dimensions in diastole should be <46mm in the unaffected patient. One complicating factor is that Dobermans can have reduced myocardial function, looking as though their ventricle is not contracting well and measurements fall close to the abnormal category but in fact, these dogs are still normal. This is when genetic testing and Holter monitoring can all come into play to help determine whether these patients are at risk of developing DCM.

Genetic testing is still a working diagnosis but there have been a few genes isolated for Dobermans and Boxers that can help predict if a patient has the potential for developing DCM. Genetic testing is done via cheek swab or EDTA blood sample.

ECGs and Holter monitors are used to evaluate a patient for ventricular arrhythmias. Few breeds, such as the Boxer and Doberman, frequently encounter fatal arrhythmias prior to being symptomatic for DCM. If a Boxer or Doberman has any VPCs noted on a 5 minute resting ECG, it is concerning that this patient will develop DCM. Any patient that has a Holter showing more than 50 VPCs in a 24 hour time frame, they are also of concern of developing DCM and more so if there are documented couplets or triplets (2 and 3 VPCs in a row). .

Generally speaking, chest radiographs and routine lab work, such as a biochemistry profile and complete blood count, will all be within normal limits in the asymptomatic patient. When patients present with symptoms, signs of congestion and heart enlargement are noted on chest films but still little changes in lab work other than mild liver or renal elevations due to congestion and poor profusion.

SECONDARY CAUSES OF DCM

Patients that are diagnosed with DCM should also be screened for other health factors that can cause changes to the heart muscle mimicking DCM. Low thyroid level, deficiencies in amino acids (taurine and carnitine), prolonged rapid arrhythmias, viral or infectious diseases (myocarditis) and ischemia can cause the heart muscle to have diminished contractility. Certain chemotherapy drugs, such as doxorubicin, can be cardio-toxic leading to decreased myocardial function. Patients that are intended to be placed on doxorubicin, should have a pretreatment echocardiogram and ECG done, and again as their treatment continues. As long as the first echo and ECG were normal, and the patient appears to be handling chemotherapy well, the echo and ECG is repeated prior to the fifth dose of doxorubicin.

Prolonged tachycardia, such as SVT (supraventricular tachycardia) can over time cause the heart to weaken and have decreased contractility. This condition has also been referred to as TICM (Tachycardia-Induced Cardiomyopathy). Once these arrhythmias are treated and under control, the hearts appearance can reverse over time and regain some normalcy to its contractility.

Deficiencies in essential amino acids can also cause the heart to mimic DCM. Patients that are fed vegetarian home cooked diets or diets that are low in taurine (such as lamb based diets) should have a taurine level checked if suspicious of having DCM. If patients are found be deficient, their diet can be supplemented and after 3-6 months, their echocardiogram rechecked to see if any improvement is noted. The same is true for hypothyroidism. Any large breed dog, especially Dobermans, if their echocardiogram measurements are questionable, a thyroid panel should be checked. If found to be hypothyroid, after proper supplementation, their echo parameters may improve.

Unfortunately changes that have arisen from ischemia and other infiltrative processes (neoplasia) are going to be permanent. Even when treated with proper medications, the changes noted are more than likely not reversible. For these patients, medications will be used to help improve/ maintain cardiac output and control arrhythmias.

OCCULT VS OVERT DCM

DCM can be broken down into two stages, occult DCM and overt DCM. Occult DCM is the pre-clinical phase and can be very hard to diagnose since the patient is not yet symptomatic with clinical signs of DCM. Overt DCM is when the patient has clinical signs of DCM.

Occult DCM, as previously stated, is the pre-clinical phase. Patients are not showing symptoms of being affected by DCM. At home, they are not exercise intolerant, appetite is normal, and they are eupneic with no signs of distressed breathing when exerting themselves or at rest. Physical exam is almost always normal. Again, certain breeds, such as Boxers and Dobermans, are prone to ventricular arrhythmias with DCM and premature beats may be noted during the physical exam. Some patients may have a soft apical systolic murmur, and many have a notable gallop (S3 gallop) sound. In fact, a gallop sound is one of the most common physical exam findings with patients having occult DCM. Patients are diagnosed with occult DCM when they are asymptomatic but have decreased contractility on their echocardiogram. There can be a prolonged pre-clinical phase, with no documented time frame as to when a patient diagnosed with occult DCM will progress to overt DCM.

A patient with overt DCM will be showing clinical signs of having DCM. Exercise intolerance, weakness, lethargy, cough, tachypnea or dyspnea, weight loss, poor appetite, ascites and syncope can be common findings with overt DCM. Physical exam findings may include: soft systolic murmur, S3 gallop, muffled heart sounds, irregular heart rate and rhythm if atrial fibrillation develops, increased breath sounds, audible crackles, tachypnea, poor femoral pulses, pulse deficits if ventricular arrhythmia present, jugular venous distension, liver and spleen enlargement upon palpation, an abdominal fluid wave if ascites present. Depending on the degree of myocardial dysfunction, the patient may be entering or already in congestive heart failure. Once patients develop congestive heart failure, the prognosis is guarded and limited to an average of 6 months or less. Although not commonly diagnosed, the same is true for cats. A cat's physical exam and clinical signs will be similar to the dog. Poor appetite, lethargy, difficulty breathing, abnormal heart rate and rhythm, soft murmur, and jugular pulses may all be present with affected cats. Echocardiography will definitively diagnose a cat with DCM.

Again the same diagnostic tests are used to evaluate the degree of myocardial dysfunction with the addition of thoracic radiographs. Depending on the degree of heart failure, radiographs will show generalized cardiomegaly, distended pulmonary veins, pulmonary edema and pleural effusion with left sided heart failure. A distended caudal vena cava, ascites and hepatomegaly are noted with right heart failure. ECGs will be used to diagnose atrial fibrillation and ventricular arrhythmias, and monitor the effects of medications used to control these dysrhythmias. A Holter monitor may or may not be applied based on the severity of notable arrhythmia documented on the ECG. Echocardiography is still the most useful in diagnosing overt DCM. The heart will have a rounded, globoid shape with dilated cardiac chambers, dramatic loss of ventricular wall motion, regurgitant leaks along the mitral valve and tricuspid valve. EPSS will be increased, the FS% will be reduced, and the ejection fractioning will be <40%.

MEDICAL MANAGEMENT

Treatments for patients with overt DCM are the same as for any other patient in heart failure, with the addition of medications that help control heart rate and dysrhythmias. The goals of medications used are to help control congestion, improve cardiac output, and control arrhythmias. Furosemide is the most commonly used diuretic for control of pulmonary edema and pleural effusion. Pimobendan, a positive inotropic agent, helps increase ventricular contractility and also causes systemic and pulmonary vasodilation. Despite increasing contractility, it does not increase myocardial oxygen demand. Patients that have developed atrial fibrillation, may be placed on digoxin to help slow the heart rate. Digoxin also has a positive inotropic effect by increasing the force of contractions but also slows the conduction through the A-V node, hence slowing the overall heart rate. ACE inhibitors act as mixed vasodilators and help minimize sodium and water retention.

After patients have been placed on heart failure medications, it is important to follow up with lab work checking electrolyte and renal parameters about 7-10 days post treatment. A digoxin level should be checked 7-10 days after being placed on digoxin and again after any dose adjustment. The frequency of rechecking lab work and other tests (such as chest films) depends on the clinician but usually the first recheck is about one week after starting medications and then after any medication adjustments are made.

CONCLUSION

DCM is an acquired myocardial disease that affects middle aged, giant breed dogs, and can go undetected in many patients since the only way to positively diagnose DCM is with echocardiography. Many patients are not screened with echocardiography since they are not intended for breeding or there is no known family history. Many patients with DCM are asymptomatic for a period of time, and many are not evaluated by a cardiologist until they are in overt DCM and/or congestive heart failure. Once patients are symptomatic, their prognosis is poor with average of 6 months to live.

Veterinary technicians have an important role in client education. From the unexpected breed screening exam that turns into the diagnosis of DCM, to the syncopal Doberman in congestive heart failure. Owners need to know that DCM is not a curable disease, but there are medications available to try and provide a good quality of life. Clients should be educated on how to calculate resting respiratory rates, what the signs of heart failure are (tachypnea, increased cough, and lethargy), signs of digoxin toxicity for those placed on digoxin (lack of appetite, vomiting, and/or diarrhea). Also, it is important for owners to realize that those patients with ventricular arrhythmias are at risk of syncope and/or sudden death. It has been documented that once a patient with DCM goes into congestive heart failure, they have about an average of 6 months to live. By communicating regularly with the owners and making medications adjustments as needed, we can help support the owners and patient until the end of life decision has been made.

REFERENCES

1. Ware W, et al. Cardiovascular Disease 2011;280
2. Tilley LP, et al. Manual of Canine and Feline Cardio 2001;III;137

BREEDING SOUNDNESS EXAMINATION IN THE STALLION FOR CREDENTIALED VETERINARY TECHNICIANS

Shirley Sandoval, BAS, LVT, VTS-LAIM
Palouse, WA

INTRODUCTION

The breeding soundness examination (BSE) is an important aspect of any breeding program. It provides the veterinarian with information regarding the health and breeding potential of the stallion.[1,2,3] These results allow the veterinarian to provide recommendations for the management of individual stallions to optimize their breeding season.

Veterinary technicians and technologist can be instrumental in assisting in these exams by preparing equipment and supplies, preparing and handling animals, and collecting and processing samples. Being knowledgeable of the BSE process and potential diseases associate with breeding stallions will assist the technician or technologist in anticipating the needs and safety of all personnel participating in the examination.

WHY PERFORM A BSE?

Ideally, a BSE should be performed at the beginning of the breeding season for each stallion. [1,2,3] In addition, these exams may be part of a pre-purchase exam of breeding stallions, be requested as part of a stallions insurance exam, or be included on a certificate of veterinary inspection (health certificate). Breeding soundness exams can also be implemented as part of the training of a novice stallion for collection; or prior to collecting a stallion to ship cooled semen or to cryopreserve semen. The BSE provides a view into the reproductive health and efficiency for the stallion at the time of the exam.

It is the responsibility of the veterinary health care team to educate breeders on the importance of annual BSE for all breeding stock as the reproductive potential of individuals can change over time, environmental influences and trauma. Commonly, breeders request a BSE when there is a question of a stallions' fertility.

WHAT IS INCLUDED IN A BSE?

History

Each BSE begins with the patients' history (Signalment. What is the stallion used for currently? Has he had any medical problems? Has he sired any offspring? If so, how many? Etc...). Always ask what the owner/breeders goals are for the stallion.

Physical Exam

A thorough physical examination of the stallion, including external genitalia, is performed making note of any anatomical abnormalities or gait difficulties. The testicles are palpated and each measured for volume using the height, width and length.[1,2,3] Depending upon the situation, the physical exam (part or all) may be best performed after the semen collection; especially if the stallion becomes easily agitated or requires sedation.[3] Avoid the use of phenothiazine tranquilizers as they can result in the inability of the stallion to retract his penis (paraphimosis).[3]

Teaser Mare

Prior to semen collection, a teaser mare is prepared. This may be an intact mare in estrus or an ovariecomized mare pre-treated hormonally to express estrus. The mares' tail is wrapped and the perineal area cleaned with mild soapy water, rinsed and dried. Breeding hobbles are employed on some mares to prevent injury to the stallion. The mare handler may also require a twitch to assist in restraining the mare during the procedure. When using a breeding phantom, the phantom should be cleaned and disinfected prior to the procedure and between collections of individual stallions. The teaser mare may be restrained at or near the head of the phantom to provide stimulation for the stallion during the collection.

Stallion Preparation

Exposure to the teaser mare can assist in stimulating the stallion to drop his penis. If the stallion is not clean, his prepuce and penis should be washed with warm water and mild soap. A sterile towel is used to dry off the penis prior to collection.

Bacterial Sampling

Bacteriological swab samples are taken of the surface of the penis, urethral fossa and prepucial fold **prior** to washing. Then pre and post ejaculate urethral swabs are also taken.[1,3] If testing for contagious equine metritis (CEM) swabs should be taken at this time. All swabs for CEM should be inoculated in Ames media with charcoal then submitted to an approved laboratory. CEM is a reportable disease and some states require a negative test for importation of semen.[5]

Artificial Vagina

Artificial vaginas (AV) used to collect stallions come in a variety of styles and sizes.[4] They all employ rubber liners that are filled with hot water (42- 45 ºC+); with each stallion dictating the preferred temperature and presser of the AV (desired temperature and pressures should be documented in the patients' record for future collections). In addition, sterile plastic liners may be employed to provide additional biosecurity and +/- filtration systems for the AV's. Unfortunately, some stallions dislike the feel of the plastic and may refuse this type of lined AV. Lubricate the AV liners with sterile, non-spermicidal lubricant.[4] Collection bottles should be covered to protect the semen from light and cold shock.[3]

Special care is important for the use, longevity and biosecurity of the rubber liners. Follow the manufactures recommendations on cleaning and storing.

Collection

All stallions should be handled with caution and appropriate personal protective attire should be worn.[4] Additional care should be taken when working with novice stallions as they can be unpredictable. Every practice should have a protocol in place for who will handle the stallion during collection (Owner, stallions' handler or hospital staff). Ask if the owner/breeder/handler has any specific signal they use to alert the stallion he is going to be breeding or collected. The stallions' libido is assessed during the collection process. Each stallion is collected twice, one hour apart.[1] Each ejaculate is evaluated separately.[1]

The collection equipment and most semen evaluation supplies have strict temperature requirements to optimize the quality of the spermatozoa being evaluated.[4]

Semen Evaluation Supplies

Heated microscope stage, although not required, is useful in keeping the samples at a constant temperature (37°C) during evaluation.[4] A slide warmer or incubator may be employed to keep the supplies that come in contact with semen warm (slides, pipettes, stain, coverslips, sample vials, etc…).

Semen Evaluation

Initial evaluation for each ejaculate includes the semen color, quantity and concentration. Concentration can be calculated using a hemocytometer, spectrophotometer or a computer assisted sperm analysis (CASA) system. Motility is evaluated using a drop of extended semen on a pre-warmed slide with a pre-warmed coverslip.[4] Motility decreases as the slide cools, therefore it is imperative to evaluate the semen immediately. A warm stage is advantageous; however rewarming the slide should improve the motility. Morphology slides are made with semen and eosin-nigrosin stain (Live Dead Stain) on warmed slides.[4] Care must be taken to prevent iatrogenic defects on morphology slides. Slides which may be too cold or too warm can cause mid-piece reflection, while hypo-osmotic shock may result in coiled tails. Detached heads may be a result of excessive pressure during slide preparation.

Evaluation of 100-200 cells is performed under oil immersion (x1000), allowing that more than one defect may be present on a single spermatozoon. Along with normal spermatozoa, evaluation of defects with the head, mid-piece and tail are counted. Detached heads are evaluated, however headless mid-pieces are not.[4]

Longevity Testing

Sperm longevity testing is performed by placing extended semen in a shipping container and chilled to 5°C. The sample is evaluated at 6, 12, 24 and 48 hours.[4]

ADDITIONAL DIAGNOSTIC TESTING

Venereal Diseases

Contagious Equine Metritis (CEM) is a sexually transmitted disease caused by the bacteria *Taylorella equigenitalis*. Although it is considered a foreign animal disease, sporadic outbreaks of CEM have been diagnosed in the USA since the early 1970's. This has resulted in the costly quarantine and treatment of the affected animals.

Breeding stallions can be tested at the time of collection for contagious equine metritis (CEM) by culturing the penis, prepuce, urethral fossa and sinus, pre and post ejaculate distal urethra, and semen. All samples should be placed in Ames media with charcoal and shipped on ice. These samples should be taken concurrently with the bacteria swabs previously mentioned. Samples must be submitted to a USDA approved laboratory for analysis as CEM is a reportable disease and some states require a negative test for importation of semen.

Equine Viral Arteritis (EVA) is an androgen dependent viral disease. Once infected, stallions can become carriers and spread the virus in their ejaculate. A modified live virus vaccine is available, but requires vaccinated horses to be quarantined during the post vaccination period to prevent potential spread of infection to naive horses. The EVA vaccination program requirements vary from state to state; contact your state health official to ensure compliance with the agency policies. Annual boosters are recommended and should be administered at least 4 weeks prior to the breeding season. An official certificate of EVA vaccination should be provided for each vaccinated animal. All animals should test negative for EVA prior to the initial vaccination. Note: some countries will not allow importation of vaccinated animals regardless of their negative pre-vaccination status.

COMMON COMPLAINTS

Abnormal Ejaculate

Asthenozoospermia occurs when the motility of a majority of the spermatazoa within an ejaculate is less than normal.

Hemospermia is the presence of red blood cells in the ejaculate, which depending upon the severity may cause infertility. Lesions that lead to this condition may be found externally on the penis or internally along the reproductive tract. These lesions may be acute or chronic. A long flexible endoscope can aid in the diagnosis of internal lesions located between the distal urethra and bladder.

Oligozoospermia occurs when there are low numbers of spermatozoa present in an ejaculate. Depending upon the cause, treatment or management of this condition may be possible.

Pyospermia is the presence of pus in the ejaculate. It is indicative of an infection within the reproductive tract such as a seminal vasculitis. Treatment depends upon the location of the lesion and organism(s) involved.

Teratospermia is the abnormal morphology of the spermatozoa. This occurs when the maturation process is not completed appropriately and the spermatozoon develops abnormally.

Urospermia occurs when the ejaculate is contaminated with urine. This complaint may stem from a physical bladder sphincter anomaly or it may be a behavioral problem.

ADDITIONAL DIAGNOSTIC PROCEDURES

When abnormalities are noted, additional diagnostic tests may be indicated. Transrectal palpation and ultrasound are indicated when stallions BSE results are less than satisfactory and abnormalities of the accessory sex glands are being considered. In addition, endoscopy may be indicated to rule out disease processes within the urinary tract of stallions.

Externally, ultrasonography of the scrotum and testicles provides additional information not necessarily noted on palpation; it is also used to guide testicular biopsies.

USA SEMEN SHIPPING REQUIREMENTS

Most states require a certificate of veterinary inspection (CVI) or health certificate from the stallion. With respect to frozen semen, the common requirement is a CVI from the date of collection. A negative Equine Infectious Anemia (EIA) or Coggins test may also be required. Some states require negative EVA or CEM testing from the date of collection. A few states require a semen import permit number. As these requirements change often, calling the state veterinarian from the state of import should occur for each shipment[5].

REFERENCES

[1]Turner R, Clin Tech Eq Prac 2005; 4: 257-268, [2]Griffin PG, J Eq Vet Sci 2000;20(3):168-171, [3]Crabtree J, In Prac 2010; 32:22-28, [4]Crabtree J, In Prac 2010;232:58-63 2010, [5]Dennis S, J Eq Vet Sci 2014; 34(7): 897-902

SOLUTION FOR IV SOLUTIONS: AGRICULTURAL ANIMAL IV FLUIDS

Shirley Sandoval, BAS, LVT, VTS-LAIM
Palouse, WA

INTRODUCTION

It is well known the medical community is currently experiencing a shortage of intravenous fluids. This can directly affect the ability of the large animal veterinary health care team to provide IV fluid therapy in their patients. The shortage is also creating inflated prices with the limited available commercial fluids, resulting in a substantial increase in the cost of veterinary services.[1] With respect to the food and fiber animals, there is added concern over food safety as we are called to treat these patients.[3]

FLUID THERAPY OPTIONS

When working with agricultural animals, the veterinary healthcare team must include the importance of food safety in all treatment protocols, including fluid therapy.

Many times fluid loss in ruminants can be replaced by an oral drench, however animals which are greater than 8% dehydrated may benefit from intravenous (IV) fluid therapy or a combination of oral and IV.[2]In animals that present with diseases that have potential for endotoxemia or hypovolemic shock, IV fluid therapy is prudent.[2]

Working with adult cattle the volume of fluid necessary to treat dehydration and electrolyte imbalance can become quite large; obtaining this volume of fluids may be difficult and economically challenging. As such, the sterilization of IV fluids prepared in house may become a viable option if the current trend continues.

Preparation of materials

When preparing fluids for IV use, it is important to choose the highest quality supplies available. When access to commercial fluids is available, they should be the first choice. If this route is unavailable, alternatives must be considered.

Containers for fluid therapy should be autoclavable, reusable, and easy to clean. Carboys work well and come in a variety of sizes. The limitation for container size will be determined by the available autoclave. Use of a sterile delivery system, IV tubing, and catheter is essential. Aseptic technique while handling the components is equally important.

Tap water can contain additives and contaminates that may not be eliminated through the sterilization process.[2]These contaminants, including endotoxins and pyrogens, can be avoided by the use of sterile distilled water.[2] Reagent grade salts can also be sterilized for use in IV fluids, however reagent grade sodium bicarbonate must be filtered prior to use.

A laminar flow hood provides an aseptic area to work when adding the ingredients to the fluids. The operator should wear sterile gloves and use sterile measuring devices for this process.

Regulations

The regulations governing veterinary medicine vary from state to state. It is important that veterinary personnel are familiar with the rules and laws governing veterinary medicine; and in this case, the specific rules pertaining to the care and treatment of food and fiber animals.

REFERENCES

[1]James, K http://veterinarynews.dev360.com/saline-shortage-impacting-fluid-supply-prices?rel=canonical [2]4Roussel, AJ, Vet Clin Food Anim 30(2014)429-439

MRSA IN HORSES AND EQUINE PERSONNEL

Maureen E.C. Anderson, DVM, DVSc, PhD, DACVIM (LAIM)
Guelph, Ontario, Canada

INTRODUCTION

Methicillin-resistant *Staphylococcus aureus* (MRSA) is one of the most important hospital-associated pathogens in human medicine, and is also recognized as a community-associated and zoonotic pathogen.[1-6] While capable of infecting (i.e. causing clinical disease) or colonizing (i.e. producing a subclinical carrier state) numerous species, including companion animals, MRSA has been isolated with particular frequency in certain livestock species, including horses,[5-9] swine and cattle.[10-12] Personnel who work with these species need to be aware of the potential for transmission of MRSA to and from their patients and what they can do to prevent it, in order to avert wider dissemination of this important pathogen in both human and animal populations.

BIOLOGY OF MRSA

Staphylococcus aureus is a commensal bacterium which colonizes the anterior nares of approximately 30% of people,[13-15] and has been found to colonize the same site in approximately 7% of horses.(Anderson MEC. Unpublished data. January 2006.) This bacterial species has a wide variety of virulence factors at its disposal, making it an excellent opportunistic pathogen capable of causing a broad range of clinical disease, from superficial skin and soft tissue infections to necrotizing pneumonia.[16,17] The acquisition of the *mecA* gene by *S. aureus* results in the production of a unique bacterial cell wall protein called PBP2a, which renders the organism resistant to not only methicillin, but all beta-lactam antimicrobials. An MRSA isolate is therefore inherently resistant to all penicillins, cephalosporins, carbapenems and monobactams, but may also carry a variety of resistance genes to other antimicrobial classes, depending on the specific strain. Oxacillin-resistant *S. aureus* is synonymous to MRSA. Colonization with MRSA of people in the general population has been estimated between 0.2-3.5%.[14,15,18,19]

Molecular typing methods have shown that the MRSA isolates obtained most commonly from horses are not the same types as those most commonly found in humans. One particular type of MRSA (designated CMRSA-5 or USA500 depending on the region) and a few closely related strains used to account for the majority of equine MRSA isolates.[5,20,21] This MRSA type is relatively uncommon in people,[22] which suggests that it may be somehow adapted to survive in horses, although a recent study suggests this is not the result of increased affinity for equine corenocytes.[23] The Panton-Valentine leukocidin (PVL) genes, which are associated with many highly virulent community-associated MRSA infections in humans and the USA300 community-associated MRSA clone,[24,25] are general absent from MRSA isolates found in horses. More recently, sequence type 398 (ST398) MRSA has become the most common MRSA clone isolated from horses in Europe[26] (having supplanted CC8 in Germany[27]), and has been involved in outbreaks in equine hospital facilities in several countries.[28-31] This clone was first associated with food animals (especially swine) and has spread from there back to the human population.[10,12,32] A very recent study further identified an epidemic sub-clone (clade C) of ST398 which appeared to have spread within and between equine hospitals, among both patients and staff, but which is otherwise rare in the human population,[33] similar to the USA500.

MRSA IN HORSES

Methicilin-resistant *S. aureus* can cause a wide variety of infections in horses. The "niche" for *S. aureus* is skin and soft tissue infections (SSTIs), particularly surgical site or wound infections, but reported MRSA infections in horses include pleuropneumonia, osteomyelitis, arthritis, omphalophlebitis, sinusitis, dermatitis and IV catheter site infections.[6,7,20,34] Survival in one retrospective review of 115 cases of MRSA infection in hospitalized horses was 83%, with MRSA being a contributing factor to non-survival in approximately 64% of those cases that died or were euthanized.[34]

Clinically there is no way to differentiate an MRSA infection from an infection caused by a methicllin-susceptible strain of *S. aureus*. Culture and antimicrobial susceptibility testing are therefore required to detect MRSA and determine an appropriate course of treatment, while also considering specific patient and infection factors (e.g. tissue affected, mixed infection, etc.). Studies have shown that most MRSA isolates from horses are still susceptible to common antimicrobials outside the beta-lactam class, although they are frequently resistant to more than just this one class, particularly to tetracycline and gentamicin.[5-9,20] The use of drugs such as vancomycin, which are of major importance in the treatment of MRSA infections in human medicine, is therefore unnecessary in most of these cases, and may be dangerous in terms of promoting the development of vancomycin resistance in numerous commensal bacterial species. It is also important to consider the correlation between *in vitro* test results and treatment response *in vivo*. For example, MRSA isolates can rapidly develop resistance to fluoroqinolones *in vivo*, even when *in vitro* testing initially indicates susceptibility to agents in this class.[35] Fluoroquinolones are therefore not recommended for treatment of human MRSA infections, and it is reasonable to extend this recommendation to horses as well.

If it is suspected that a horse has an MRSA infection based on isolation of this pathogen from other animals with which the horse has had direct or indirect contact, a culture specimen must still be collected for submission prior to commencing any antimicrobial therapy, in order to confirm the diagnosis and to monitor the outbreak isolate for changes in antimicrobial susceptibility over time. In a highly suspected case, first line antimicrobial therapy should be based on culture and susceptibility results from other isolates from the same outbreak. Empirical therapy for MRSA in horses is not recommended – an infection should not be treated as MRSA unless the pathogen has been isolated from the horse or from another animal which may have directly or indirectly transmitted it to the horse, in which case antimicrobial susceptibility results should be available.

Colonization of horses is usually detected by means of culture of a swab inserted 8-10 cm into one naris and withdrawn in contact with the nasal mucosa. Colonization of other body sites may occur as well as it does in humans, but the nares are the most sensitive screening site.[36,37] It appears that the majority of horses are not persistently colonized, and will clear the MRSA on their own if re-exposure can be prevented.[36,38] Drug therapy is therefore not advocated unless clinical signs of infection are present. Risk factors for MRSA infection and colonization have not been extensively investigated in horses as they have in humans. Studies have identified colonization as a risk factor for infection in horses.[39] Administration of antimicrobials (ceftiofur or aminoglycosides) in hospital,[39] administration of antimicrobials within 30 days of hospital admission, admission to the neonatal intensive care unit or a non-surgical service,[40] and residing on a farm with more than 20 horses[5] or on any farm on which a horse was previously identified as colonized with MRSA have been identified as risk factors for colonization in horses.[40] Identification of MRSA (both infection and colonization) in horses is important in the community setting and in veterinary hospitals in order to implement effective infectious disease control protocols.

MRSA IN PERSONNEL WHO WORK WITH HORSES

A study of veterinary personnel attending a veterinary internal medicine conference identified working with large animals (including horses) as a significant risk factor for MRSA colonization, with 15.6% of large animal personnel being positive for MRSA on a single nasal swab, compared to only 4.4% of small animal personnel.[41] Furthermore, 13/15 (87%) of isolates from the large animal personnel were consistent with CMRSA-5 based on pulsed-field gel electrophoresis (PFGE), which supported the hypothesis that the high prevalence was a result of contact with horses. Another study of veterinarians and veterinary students in the Netherlands found a 4.6% prevalence of MRSA colonization among those who had contact with livestock, compared to 0% prevalence among students with no livestock contact and a 1.0% prevalence among the general population in that country.[42] It is worth noting, however, that this study did not evaluate contact with horses specifically, and that a previous survey of healthy horses in the Netherlands failed to demonstrate the presence of MRSA at that time.[43] In a study performed at the 2006 Convention of the American Association of Equine Practitioners, colonization with MRSA was found in 26/257 (10.1%) of study participants,[44] which is a higher prevalence than that found in studies of the general population.[14,15,18,19] Independent risk factors for colonization were having been diagnosed with MRSA colonization or infection in the last year (odds ratio (OR) 8.41, $p = 0.015$) or having treated a horse diagnosed with MRSA in the last year (OR 3.27, $p = 0.039$). Similar studies performed at a variety of other veterinary conferences have found the prevalence of MRSA among equine veterinary personnel to be 11.8-21.4%.[41,45,46] One other significant risk factor for MRSA colonization of equine personnel identified in a previous study was regular contact with more than 20 horses.[5]

Clinical MRSA infection of personnel associated with direct contact with colonized horses has also been reported. One incident involved infection of a tattoo site of a veterinary resident with a CMRSA-5 strain that was indistinguishable from that isolated from two horses that were recently under the resident's care.[6] Evidence of horse-to-human transmission was also described in an outbreak of MRSA skin infections in three neonatal foal intensive care unit personnel following close contact with a foal that was positive for MRSA on a nasal swab collected on admission, despite contact precautions including gowns and gloves.[47] Colonization of ten other personnel in the hospital, all of whom had had contact with the affected foal, was also detected. All of the MRSA isolates from the infected and colonized individuals as well as the foal and its mare were indistinguishable from one another and were consistent with CMRSA-5.[47] There is also one report of an MRSA ST398 wound infection in a teenage girl that was suspected to have been the result of contact with a foal colonized with an indistinguishable strain.[48] During an outbreak involving a rare MRSA strain in an equine hospital in Israel, colonization of 11.5% of hospital personnel with the same strain was reported.[49]

PREVENTING SPREAD OF MRSA BETWEEN HORSES AND HUMANS

Large horse farms and boarding stables can be compared to human long-term care facilities in terms of the potential for disease transmission, because animals are housed in close contact with a relatively small group of common caretakers, frequent veterinary contact and frequent antimicrobial use. Isolation or cohorting of infected and colonized horses should help to prevent widespread dissemination of MRSA in the subclinical form,[38] and therefore also the number of potential infections. Quarantine and screening of horses entering a hospital or returning to a barn from a hospital can also be considered in situations considered higher risk. Although the role of the environment in the spread of MRSA between horses and between horses and humans is still unclear, nasal carriage by horses has been shown to be positively correlated to environmental contamination with MRSA, both within and outside of the hospital setting.[28,50,51] It is important to pay particular attention to the use, cleaning and disinfection of certain kinds of equipment that are likely to have a high degree of contact with a horse's nose (e.g. twitches, muzzles, feed/water buckets).[51] Non-patient areas can also become easily contaminated,[28] likely by colonized staff and/or by staff transiently carrying MRSA, particularly on their hands if basic infection control measures such as hand hygiene are not adequately performed.[52]

In contrast to veterinary medicine, hand hygiene in the human healthcare setting has been studied extensively, and it is considered to be one of the most important infection control tools for control of MRSA and other hospital-associated pathogens.[53,54] Hand washing with soap and water between cases considered potentially infectious (OR 0.27, $p = 0.009$), and between farms (OR 0.35, $p = 0.047$) were found to be protective against nasal colonization with MRSA among equine veterinary personnel in one study;[44] this was the first study to demonstrate a statistically significant association between hand hygiene practices and a measurable clinical outcome in veterinary medicine. Alcohol-based products consistently demonstrate superior ability to reduce bacterial counts on the hands of human healthcare workers,[54] and are considerably more convenient for use in ambulatory veterinary practice, where sinks and running water are not always readily available. Under experimental conditions, Traub-Dargatz et al. demonstrated an equal to superior effect of use of alcohol-based hand sanitizers compared to antibacterial soap and water following routine physical examination of horses.[55] These alcohol-based products are typically combined with emollients to help prevent drying and damage to the skin, as often occurs with frequent hand washing.[54] However, alcohol-based hand sanitizers are only meant to be used when hands are not visibly soiled.[55] In large animal veterinary practice, visible dirt and debris are commonly present on the hands, which may necessitate mechanical removal using soap and water.

SUMMARY

Methicillin-resistant *Staphylococcus aureus* (MRSA) is an important zoonotic and equine pathogen worldwide. Equine veterinary personnel need to be aware that MRSA is present in the equine population so that further dissemination of this pathogen can be curbed. Veterinary personnel who work with horses are at increased risk of MRSA colonization, but close attention to routine infection control measures such as hand hygiene may help decrease this risk.

REFERENCES

[1]Cosgrove SE, et al. **Clin Infect Dis** 2003, 36:53–59. [2]Cosgrove SE, et al. **Infect Control Hosp Epidemiol** 2005, 26:166–174. [3]Gorak EJ, et al. **Clin Infect Dis** 1999, 29:797–800. [4]Herold BC, et al. **JAMA** 1998, 279:593–598. [5]Weese JS, et al. **J Am Vet Med Assoc** 2005, 226:580–583. [6]Weese JS, et al. **Emerg Infect Dis** 2005, 11:430–435. [7]Baptiste KE, et al. **Emerg Infect Dis** 2005, 11:1942–1944. [8]O'Mahony R, et al. **Vet Microbiol** 2005, 109:285–296. [9]Seguin JC, et al**. J Clin Microbiol** 1999, 37:1459–1463. [10]Huijsdens XW, et al. **Ann Clin Microbiol Antimicrob** 2006, 5:26. [11]de Neeling AJ, et al. **Vet Microbiol** 2007, 122:366–372. [12]Feßler AT, et al. **Vet Microbiol** 2012, 160:77–84. [13]Choi CS, et al. **J Microbiol Immunol Infect** 2006, 39:458–464. [14]Kuehnert MJ, et al. **J Infect Dis** 2006, 193:172–179. [15]Salgado CD, et al. **Clin Infect Dis** 2003, 36:131–139. [16]CDC. **MMWR Morb Mortal Wkly Rep** 2007, 56:325–329. [17]Cunha BA. **Clin Microbiol Infect** 2005, 11 Suppl 4:33–42. [18]Graham PL, et al. **Ann Intern Med** 2006, 144:318–325. [19]Lu P-L, et al. **J Clin Microbiol** 2005, 43:132–139. [20]Cuny C, et al. **Euro Surveill** 2006, 11:44–47. [21]Moodley A, et al. **J Antimicrob Chemother** 2006, 58:1118–1123. [22]Christianson S, et al. **J Clin Microbiol** 2007, 45:1904–1911. [23]Akridge HD, et al. **Vet Dermatol** 2013, 24:525–e124. [24]Vandenesch F, et al. **Emerg Infect Dis** 2003, 9:978–984. [25]Morgan MS. **Int J Antimicrob Agents** 2007, 30:289–296. [26]Bergström K, et al. **Acta Veterinaria Scandinavica** 2012, 54:11. [27]Vincze S, et al. **PLoS ONE** 2014, 9:e85656. [28]van Duijkeren E, et al. **Vet Microbiol** 2010, 141:96–102. [29]Van den Eede A, et al. **Vet Microbiol** 2009, 133:138–144. [30]Bergström K, et al. **Acta Veterinaria Scandinavica** 2012, 54:14. [31]Sieber S, et al. **Microb Drug Resist** 2011, 17:471–478. [32]van Cleef BAGL, et al. **Emerg Infect Dis** 2011, 17:502–505. [33]Abdelbary MMH, et al. **PLoS ONE** 2014, 9:e88083. [34]Anderson M, et al. **Equine Vet J** 2009, 41:401–405. [35]Gorwitz RJ, et al. Strategies for Clinical Management of MRSA in the Community: Summary of an Experts' Meeting Convened by the Centers for Disease Control and Prevention. 2006:1–24. [36]Bergström K, et al. **Vet Microbiol** 2013. [37]Van den Eede A, et al. **Vet J** 2012, 193:408–411. [38]Weese JS, et al. **Equine Vet J** 2005, 37:510–514. [39]Weese JS, et al. **J Vet Intern Med** 2006, 20:182–186. [40]Weese JS, et al. **Can Vet J** 2007, 48:921–926. [41]Hanselman BA, et al. **Emerg Infect Dis** 2006, 12:1933–1938. [42]Wulf M, et al. Emerg Infect Dis 2006, 12:1939–1941. [43]Busscher JF, et al. Vet Microbiol 2006, 113:131–136. [44]Anderson MEC, et al. **Vet Microbiol** 2008, 129:410–417. [45]Jordan D, et al. **Aust Vet J** 2011, 89:152–159. [46]Burstiner LC, et al. **Vet Surg** 2010, 39:150–157. [47]Weese JS, et al. **Vet Microbiol** 2006, 114:160–164. [48]van Duijkeren E, et al. **Emerg Infect Dis** 2011, 17:1137–1139. [49]Schwaber MJ, et al. **Vet Microbiol** 2013, 162:907–911. [50]Peterson AE, et al. **Vet Microbiol** 2012, 160:539–543. [51]Weese JS, et al. **J Vet Intern Med** 2004, 18:468–470. [52]van Balen J, et al. **Vet Res** 2014, 45:31. [53]Muto CA, et al. **Infect Control Hosp Epidemiol** 2003, 24:362–386. [54]Boyce JM, et al. **Am J Infect Control** 2002, 30:S1–46. [55]Traub-Dargatz JL, et al. **Can Vet J** 2006, 47:671–676.

EQUINE SKIN DISEASES: COMMON AND NOT SO COMMON

Susan L. White DVM, MS, DACVIM (LAIM)
Mountain Rest, SC

Skin diseases often cause unsightly and disfiguring lesions in horses. In some cases the disease process may affect the ability to use the horse and thus are a common reason owners seek veterinary care. A careful history and physical examination will allow one to formulate a rule out list, and may render a diagnosis. Pruritus is arguably the most common reason horses are presented for skin disease.[1] Pruritus often leads to self-inflicted trauma, alopecia, and moderate to severe secondary skin lesions. Because of the many possible diagnosis of a pruritic skin disease the history, especially the environment, seasonality, age of onset and progress of the disease is very important in making a diagnosis. Time spent obtaining a detailed, complete history is rewarding and often equally as important as the examination and diagnostic tests.

A large percentage of horses affected with seasonal pruritic dermatitis are **hypersensitive to the bites of insects**. Any biting insect, including *Culicoides spp.* (No-see-ums, midges, punkies), stable flies (*Stomoxys calcitrans*), horn flies (*Hematobia sp*), black flies or "buffalo gnats" (*Simulian spp.)*, horse and deer flies (*Tabanidae* and *Chrysops spp.*), mosquitoes (primarily *Culex* and *Aedes spp.*), can cause insect hypersensitivity dermatitis. "Chiggers" (Trombiculids) and ticks can also lead to hypersensitivity reactions. Insect bite induced hypersensitivities are characterized by intense pruritus which often leads to excoriation, extensive hair loss, secondary infections and chronically to hyperkeratosis and lichenification.[2] Several different clinical syndromes have been associated with hypersensitivities to different insects. Queensland or Sweet Itch, initiated by the bites of hematophagous *Culicoides* species, is classically characterized by a diffuse dorsal distribution with severe involvement of the mane and tail area. However, there are several species of *Culicoides* in different geographic areas and different species have different preferred areas of feeding. There are, for example, over 1000 species of Culicoides worldwide and as many as 30 species active in any one area.[3] Work by Greiner in Florida has demonstrated multiple species feed on horses in different regions of Florida.[4] These different species have variable feeding patterns with some species having a predilection for biting the ventral aspect of the body instead of the dorsal.[3] The only region generally not affected by *Culicoides* is the flank.[5] An hereditary predisposition to develop hypersensitivity to *Culicoides spp.* has been shown in some breeds of horses and clinical evidence supports a familial tendency for hypersensitivity in many, if not all, breeds.[6,7]

Culicoides gnats transmit the filarial nematode *Onchocerca cervicalis*. The adult nematode lives in the nuchal ligament, microfilaria migrate through connective tissue to the skin. The microfilaira are found in the highest number in the skin of the ventral abdomen and chest, although they may be found in the head, neck, withers and eye. Hypersensitivity to the micro filaria produces pruritus and clinical signs similar to gnats other biting insects. Avermectin anthelmentics are effective against *Onchocerca*, thus hypersensitivity to this parasite is less common than before. However, with the increase in strategic worming to resistance among nematodes, Onchocerca hypersensitivity may be increasing.

Bites of horn flies have also been implicated in **Ventral Midline Dermatitis**, which is characterized by focal lesions which often ooze serum. Horn flies are usually found on ventral regions of horses when in the sun, but may be anywhere when solar radiation is not severe. Bites of *Simulian spp.* are most commonly associated with lesions on the head and within the inner pinnae, but in "swarm attacks" may bite any region of the horse. A generalized ventral distribution of lesions may also be caused by black flies. *Simulian* flies are blood suckers and bites are covered with small accumulations of dried blood. In mass attacks urticaria and angioedema may occur and can be associated with systemic signs of weakness with an increased pulse and respiratory rate, particularly in small young animals. Stable flies are vicious biters (as opposed to house flies *(Musca spp)* that feed on body secretions). Initial lesions may consist of erythema, wheals, and raised nodules with a central crust. Exudative dermatitis of the legs due to reactions to extensive stable fly bites and/or urticaria and large nodules that are pruritic characterize hypersensitivities to stable flies.

Mosquito bites produce wheals and papules that are variably pruritic and are often painful. The lesions do not have a central crust as the bites of other insects. Most individual lesions resolve in 1-2 days, however larger persistent nodules may develop in certain individuals.

Chiggers and harvest mites (trombiculids) are free living and are found in undeveloped or overgrown fields and wooded areas. Adults are plant parasites, but the 6 legged larval form requires blood or tissue fluids for further development. Mites are non-host specific and normally feed on small rodents. Infestations of horses may occur if horses are pastured or ridden in contaminated areas and are most prevalent in the summer and fall. Feeding larvae cause papules with 1-2mm crusts and can be intensely pruritic. Larvae remain on the host for 2-7 days before dropping off to molt and are often absent by the time the horse is presented for examination. The larvae are 0.2-0.4 mm long and yellow-orange to red in color and, if found, confirm the diagnosis.[8]

Horses may develop hypersensitivities to ticks, either larvae ("seed ticks") or adults. Bites may induce individual large nodules with a central crust and/or ulcer or multiple bites may induce generalized edema, multiple nodules or ulcerated crusted areas. Very small larvae are easily overlooked even in massive infestations. Pruritus may be extreme, resulting in severe secondary lesions. Ticks are most commonly found on the legs, tail, head and ears.(SLWhite, unpublished data)

In cool weather **lice** and **chorioptes mange** are common conditions exhibiting pruritus. Chorioptes mites are surface inhabitors that feed on epidermal debris. Their 3 week life cycle is complete on the host, however they can survive up to 70 days off host, and can be transmitted by direct or indirect contact. Mite infestations are more common in cool, wet conditions and less common in hot dry conditions. Mites may persist and may only be found around coronet when clinical signs decrease in summer. Draft horses and others with feathered fetlocks (Andalusians) are most commonly affected, especially in winter. Pruritus may be intense or absent; the tail may be involved. Occasionally horses with widespread lesions are seen. Mites are easily found on skin scraping in cool weather. Mites are very active and fast and will crawl off the slide before examination, thus an insecticide should be added to the oil used for skin scrapings.[8]

Lice are highly host specific obligate parasites that spend their entire life cycle (20-40 days) on the host. Biting lice (Bovicola equi, formerly Damalinia spp.) prefer the dorsal trunk, sucking lice (Haematopinus and Linognathus spp) favor the mane, neck, poll, tail and fetlocks. Clinical signs are dependent on number of lice and site of infestation. Scaling, rough haircoat, alopecia, erosions and/or ulcers from self trauma due to pruritus may be seen. Nits (eggs), 1-2 mm long, are attached to hairs by a clear adhesive secretion by female lice. Under favorable environmental conditions lice can live 2-3 weeks off the host, but < 7 days is typical. Transmission is by direct and indirect contact. Lice populations are larger in the winter and therefore clinical signs are usually seen in the winter. Other predisposing factors are long hair coat, overcrowding, poor nutrition and possibly other stresses associated with cold weather. Although sucking lice are controlled with avermectins, since biting lice and chorioptic mites feed on epithelial tissue debris they are poorly controlled by parenterally administered avermectins. These parasites are most effectively treated with topical agents such as lyme sulfur or permethrin products. Persistence of parasite populations occur in low numbers often on asymptomatic animals, thus all in contact animals and common tack or grooming equipment should be treated.

Dermatophytosis (ringworm) is variably pruritic. Dermatophytosis is a superficial contagious and zoonotic fungal skin disease spread by direct contact and fomites. It can be spread from host to host, soil to host and host to man to host. *Trichophyton equinum* is the most common cause of dermatophytosis in horses. Other dermatophytes are *T. mentagrophytes, T. verrucosum, Microsporum equinum (M. canis) and M. gypseum. M. gypseum* is the only dermatophyte that is normally a soil inhabitant (the others are zoophilic and have adapted to animals) and produces more severe cases of dermatophytosis.[9]

Transmission can be by direct contact but is often transmitted by contaminated objects. In many clinical situations, there will be no evidence of transmission to other animals in contact with the affected one. Young animals are most susceptible. Lesions usually begin as focal lesions with initial urticaria. Central alopecia, scaling and crusting appears; pruritus is variable. Some infestations may be accompanied by a marked response to fungal products and have a more generalized edema, particularly on the legs which may be painful and/or pruritic. The usual distribution is in the bridle, saddle and girth area especially in large training stables where more than one horse affected, as the tack spreads the disease. Atypical forms can include extensive crusting and scaling with little alopecia, or in the case of *M. gypseum* severe inflammatory reaction with suppuration crusts and ulceration and, if on the lower limbs, edema. Lesions may be confined to the posterior aspect of the pastern and heel and present as pastern dermatitis, or more rarely be confined to the coronary band. Lesions may wax and wane dependent on environmental conditions and are often self limiting.

Spontaneous regression of dermatophytosis occurs in 3 weeks to 3 months. Therapy may shorten the course and reduce spread and environmental contamination. Sunlight, good nutrition (including adequate vitamin A) and decontamination of environment, including all objects (tack, brushes, etc) as they are often the source of infection help speed resolution of the disease. Although repeated bathing with iodine or chorhexidine shampoos is discouraged (they have no residual action and physical act of bathing may result in further hair loss and disperse infective spores in the coat) the author believes an initial gentle shampoo to remove surface dirt and crustsis beneficial. Shampoos containing ketaconazole may also be used. After that rinses may be applied with a very soft brush daily for 5-7 days. The act of **gently** brushing on rinses encourages the medication to get close to the opening of hair follicles, where the infection resides.

Symptomatic treatment for pruritus In all cases of pruritic dermatitis managerial procedures to decrease exposure to exciting agents as well as symptomatic therapy to reduce pruritus is warranted. All avenues available to decrease exposure to insects should be fully implemented. Heat, humidity and solar radiation exacerbate pruritus thus the provision of shade and wind currents by fans can provide relief. Simple feeds and whole grains are better than mixed multiple grain sweet feeds. Cold water rinses and shampoos can rehydrate dry skin and reduce the amount of topical allergens on the skin. Colloidal oatmeal, pramoxine (a topical anthesthetic) and 1% hydrocortisone shampoos (hydrocortisone is not absorbed) may reduce pruritus and minimize or reduce the amount of systemic anti-pruritics needed. A complete and detailed investigation into the etiology of the disease should occur simultaneously with general symptomatic care. The client must understand, in the case of allergies, that the condition is life long and the horse will need continuous management and or therapy. Often a patient may be symptom free with low exposure to inciting antigens and symptomatic as the antigen "load" increases. Antigen exposure is additive, thus comprehensive management is needed to best control clinical disease.[2]

Dermatophilosis is a common superficial crusting and pustular dermatitis which is not pruritic. It is caused by *Dermatophilus congolensis*, an actinomycete (gram positive), branching filamentous bacterium, which forms spores. Synonyms used by horsemen for this condition are rain scald, rain rot, cutaneous streptothricosis and winter fungus. It occurs worldwide and there is a high association with rain, poor grooming, poor nutrition, ticks and lack of sunlight. Consequently the disease is usually a fall and winter disease but can occur when significant prolonged wet conditions prevail. The organism may be maintained on carrier animals, both domestic and wild, and zoospores may survive for prolonged periods in the environment. Certain farms, even under good management, have a high incidence of dermatophilosis in resident horses. Conversely, certain horses are more susceptible to dermatophilosis than others, often exhibiting severe disease where herd mates kept in identical conditions do not. To establish an infection there must be skin damage, either from maceration of the epidermis due to prolonged wetting, bites of insects and arthropods, or minor skin injuries from vegetation (briars and other brush).[10] In hot humid environments horses may develop dermatophilosis under the saddle area due to repeated sweating and lack of cleanliness.

In the most common presentation of the disease lesions are found over the dorsum of the back as scab-like, exudative crusts which coagulate, causing the hair to stick together. The hair tufts stand erect in a paint-brush effect. In early infections the lesions may be felt better than seen by running your hand over the hair, the underlying scabs will be evident. In more advanced infections the hair tuffs stand up. By pulling out the hair tufts, the underlying skin is moist, raw, and may be hemorrhagic. The hair roots extend through the scab in the characteristic "paint brush" effect. Untreated severe lesions may result in deep ulcers that extend into the subcutaneous tissues or, in extreme cases, deeper. In the active stage, the lesions are painful. Dermatophilosis may also present with extensive epidermal thickening and alopecia. Lesions may be found on the lower legs if horses are kept in moist pasture conditions, particularly in deep grass and are often a component of pastern dermatitis. Foals may have lesions on their legs, ventral abdomen and sides of the trunk and hindquarters from lying in wet pastures. Regardless of the form, this disease is the only exudative, painful, non-pruritic common disease in clinical practice.

Diagnosis is often made on clinical presentation and can be confirmed by cytology, biopsy and culture. Cytological preparations of crusts soaked in saline and minced may be stained with Diff Quik®, Gram's stain of new methylene blue. If exudate is present on the underside of the crusts simple impression smears can be used for cytological examination. *D. congolensis* appears as gram positive branching multi-septate hyphae with cuboidal packets of coccoid cells in parallel rows (railroad track appearance).

In most cases the disease can be treated with crust removal and sufficient housing to keep the horse dry. In early superficial infections the crusts can be removed by grooming. The zoospores in the crusts are resilient and thus should be disposed of rather than left in the environment of the horse. Likewise any grooming equipment, tack and blankets should be thoroughly cleaned and confined to individual horses. Crusts can also be removed by bathing. Any appropriate equine shampoo is adequate, although most references suggest biocidal shampoos or rinses (2% - 4% chlorohexidine). Crust removal may take days to weeks, depending on the severity of the disease and the degree of pain exhibited by the patient. Warm water soaks prior to active scrubbing help soften crusts and ease removal.

Horses may be treated with parenteral penicillin or oral trimethoprim potentiated sulfonamides, particularly when the lesions are deep and generalized. Future cases can be prevented by optimizing nutrition, cleanliness of the horse and keeping the horse dry. In some situations (pasture maintained horses) provision of a run in shed may be sufficient to avoid further infections

Bacterial folliculitis may present as a superficial **pyoderma** to deep ulcerated lesions. The causative agent is usually a coagulase positive *Staphylococcus aureus* or *S. intermedius*, although *Corynebacterium pseudotuberculosis* and occasionally other organisms may also cause folliculitis. Mild lesions may be characterized by alopecia and exfoliation which may be regional or generalized. The lesions may be circular with crusts and epidermal collarettes that must be differentiated from dermatophytosis. The lesions may present as a few papules with crusts to a more generalized miliary dermatitis. Lesions are often found associated with tack and other equipment (blankets, fly masks), particularly on young horses, or horses kept in unclean conditions, or horses with suboptimal nutrition, and/or immuno-compromised horses.[9] Staphylococcal dermatitis is often alone or one of multiple inciting agents in pastern dermatitis. The lesions may begin with multiple encrusted papules that enlarge and coalesce. Lesions generally begin on the plantar surface but may spread to encircle much of the leg and migrate proximally to include the fetlock and lower cannon bone. Staphylococcal granulomas or botryomycosis are most frequently associated with improper castration technique, but may be found anywhere on the body and resemble fungal granulomas. Lesions may be variably pruritic and/or painful. Lesions associated with tack and pastern dermatitis are most often painful.

Pastern dermatitis (scratches, grease heel) is a cutaneous reaction pattern that may have one or more causative factors. The lesions are often more severe on the plantar surface of the hind legs than the forelegs. Lesions may vary from a few papular or crusty lesions to severe, deep, exudative and painful lesions extending from the coronary band to the fetlock. Lesions are often associated with moist and/or unsanitary conditions. Horses with white hair and non-pigmented skin on their lower legs may have a component of solar irritation as part of the syndrome.[11] Management should include keeping the area dry and clean. Gentle washing should be performed with nonirritating shampoos; the area thoroughly dried and, in some cases, protective dry warps. If management alone does not improve the condition then specific diagnostic tests should be performed. Etiologic agents include dermatophytes, bacteria, autoimmune processes, ultraviolet light reaction, and in some cases be a reflection of systemic disease.

In draft horses pastern lesions of Chronic Progressive Lymphedema must be differentiated from the more typical cause. Draft horse breeds, including Shires, Clydesdales, Belgian draft horses, Gypsy Vanners, English Cobs, several German draft horse breeds, and Friesians may be affected by this condition. In these horses there is a defect in the elastin of the connective tissue which results in an accumulation of lymph and gradual swelling of the lower limbs. Eventually the swelling results in multiple folds of the skin and ulceration with possible secondary infections. The lesions are usually found around the circumference of the limbs, which helps differentiate it from other causes of pastern dermatitis. The condition is multifactorial with a genetic component and is not curable. Management with pressure bandages early in the course of the disease may slow its progression but not halt it.[12]

Urticaria is a cutaneous reaction pattern that may occur due to a variety of causes, both immunologic and non-immunologic. Urticaria is usually a flat topped steep walled lesion that is symmetrically distributed on the body. The lesions pit with digital pressure and may be nonpainful or painful and variably pruritic.[13] Many occurrences of uritcaria are short lived and resolve without specific treatment. An example of such occurrences is the appearance of urticaria when a horse is attending a show or when initially housed in a new boarding stable. Often the etiolgy is not determined. In other cases urticaria may persist for weeks or reoccur multiple times. In these cases a detailed and in depth examination is needed to determine the cause. Immunologic causes of urticarial include allergies to insect bites, foods or plants such as in atopic individuals or drugs.[13] Less common, autoimmune diseases such as pemphigus may also present with urticaria. Infections (bacteria, viral, fungal, parasitic) may induce urticaria. Any drug that a horse is receiving when urticarial develops should be discontinued.

REFERENCES

1.Knottenbelt DC Vet Clinics N Am: Eq Prac 2012:28(1):131 2.Marsella R Vet Clinics N Am: Eq Prac 2013;29(3):551 3. Fadock VJ et al Eq Vet J 1990;22:236 4. Greiner EC. Vet Clinic N Am: Eq Pract 1995;11(1):29 5. Greiner EC et al Med Vet Entomol 1990;4:375 6. Steinman et al Vet Rec 2003;152:748 7. Anderson LS et al Immunogenetics 2012;64:201 8. Bergall K Clin Tech in Eq Prac 2005;4(4):296 9. Weese JS et al Vet Clinics N Am:Eq Prac 2013;29(3):559 10. White SD Clin Tech Eq Prac 2005;4(4):302 11. Psalla DI Eq Vet J. 2013;198(2):524 12. Affolter VK Vet Clinic N Am: Eq Pract 2013;29(3):589 13. Fadock VA *Veterinary Allergy* Wiley 2014:338

CHEMOTHERAPY SAFETY AND ADMINISTRATION

Kathi L. Smith, RVT, VTS (Oncology)
Portland, ME

INTRODUCTION

An increasing number of clients are choosing to treat their pets that have been diagnosed with cancer. The safety of the health care members must be made a priority due to the potential health risks associated with the preparation and administration of antineoplastic drugs being used in veterinary medicine.

BACKGROUND

Antineoplastic drugs have been in clinical use for more than 50 years in human medicine.[1] There are numerous studies in animals and human patients regarding the carcinogenic and toxic properties of these drugs. In animal studies, it has been shown that antineoplastic drugs are known to be:

1) *Carcinogenic*: cancer causing substances or agents,
2) *Teratogenic*: causing malformation of an embryo or fetus,
3) *Mutagenic*: causing gene or chromosomal damage, may also (but not necessarily) act as a carcinogen, or cause
4) *Serious organ damage* or other toxic manifestation i.e. cardiotoxicity (doxorubicin), hemorrhagic cystitis (cyclophosphamide), thrombocytopenia (lomustine).[2,3,4]

It wasn't until 1979 that studies were instituted looking at the occupational risks associated with low dose exposure of these drugs to the health care workers handling them. The lack of existing exposure records for health care workers along with the fairly recent interest in occupational exposure to antineoplastic drugs, have made studies to their long-term side effects difficult.[3,4]

The risk to health care workers using antineoplastic drugs is difficult to monitor due to the inherent toxicity of the drugs, individual susceptibility, concurrent exposure to known carcinogens such as cigarette smoke, and the level (occasional vs. constant) or type (drug or material) of exposure.[3,4]

There have been studies in oncology nurses that show adverse reproductive outcomes with occupational exposure to antineoplastic drugs. Birth defects, fetal losses, infertility, ectopic pregnancies, low birth weights, and menstrual dysfunctions have been reported.[3,4]

Work place studies revealed that the chemotherapeutic drug doxorubicin remained stable on various surfaces for up to seven weeks and other studies examined airflow exposure in HEPA filters from cyclophosphamide contamination.[3,4]

Studies have been done that show that health care workers who wear protective gear (gloves, gowns and facial shields) are in fact at a lower risk of exposure than those who do not.[2,3,4]

In 1986 the Occupational Safety & Health Administration (OSHA) published guidelines for health care workers exposed to low-dose occupational doses of cytotoxic drugs. This was in response to a series of Technical Assistance Bulletins that were developed by the American Society of Hospital Pharmacists and recommendations from the Federal Occupational Safety and Health Administration, the National Institutes of Health, the American Medical Association.[2,3,4]

These guidelines are readily available and veterinary hospitals that offer chemotherapy to their clients should have formal procedures instituted into their practice. With proper training and protection, chemotherapy preparation and administration can be performed safely.[2,3,4]

METHODS OF EXPOSURE

There are several ways that technicians can be inadvertently exposed to antineoplastic drugs. *Inhalation exposure* occurs during the mixture and/or administration of cytotoxic drugs. This is due to the aerosolizing of the drug. Studies have been done, measuring air levels of cytotoxic drugs (cyclophosphamide) when biological safety cabinets (BSC) were not in use. Increased levels of the drugs were reported in the HEPA air filter. This is also an opportunity for respiratory exposure to those employees changing the air filters. *Absorption exposure* occurs during the absorption of the drug through the skin and mucous membranes. There are case reports (McDiarmid & Egan, 1988)[2,3,4] of health care workers that developed severe acute allergic reactions from accidental cytotoxic drug spills. *Ingestion exposure* occurs through contact with contaminates such as food, cosmetics or cigarettes. *Inoculation exposure* occurs while recapping needles after drug administration. Recapping needles should be discouraged for this reason. *Clinical exposure* is probably the most common source of exposure and can be associated with needle withdrawal from a pressurized vial, drug transfers from different containers, opening of glass ampules, expulsion of air from drug filled syringes, equipment failure or improperly set up equipment, exposure to excreta from patients that have received cytotoxic drugs, and crushing or breaking of tablets or capsules of cytotoxic drugs.[2,3,4]

EQUIPMENT FOR PREPARING AND ADMINISTERING CHEMOTHERAPEUTICS

A class II vertical laminar flow biologic safety cabinet (BSC) should be part of any facility that administers antineoplastic drugs. These devices insure that aerosolized particles are properly vented away from the veterinary technician preparing the drug, and that containment is made easier in the event of an accidental spill. Biological safety cabinets can be found relatively inexpensively.[2,3,4]

Closed double-membrane systems for drug transfer such as Equashield® or PhaSeal® are excellent for use with multi-dose drug vials. They insure that drug aerosolization does not occur and that the drug vial maintains the correct amount of pressurization. They also prevent the possibility of absorption exposures.[3,4,5]

Absorbent mats with a leak-free plastic backing prevent contamination of the work surface in the event of an accidental spill from the injection port. Disposable diapers can be used if chemo safety mats are unavailable. [3,4,5]

Using re-sealable plastic bags for used chemotherapy administration materials will help eliminate possible aerosolization or contamination of the drugs after they are disposed of in the hard waste chemotherapy containers.[3,4,5]

Personal protection should always include the use of face shields that cover the entire face if a BSC is unavailable. Dust particle facemasks and safety goggles should only be used as a last resort due to the potential risk of exposure. Surgical masks should never be used because they do not protect against the breathing of aerosolized drugs.[3,4,5]

Powder-free chemo gloves should be used instead of latex exam gloves as they are made of a thicker material.[3,4,5]

Disposable, solid front, moisture barrier gowns with cuffs should be worn to prevent accidental exposure to skin or clothing. Gloves should be worn both under the cuff and over the cuff to ensure that no bare skin of the wrist or forearm is exposed. Gowns can be worn several times before disposing if proper preparation techniques are maintained. Those staff members that are assisting during chemotherapy administration should also wear protective gear.[3,4,5]

Leur-lok® syringes should be utilized instead of slip syringes. This will help eliminate accidental separation of the needle from the syringe thus preventing a chemo spill situation.[3,4,5]

Gauze 4x4's soaked with alcohol are used for covering the injection port, which will help prevent aerosolization at the injection port. Porous white tape is better for anchoring catheters in place because it's not as sticky as some of the other brands and is easier to remove.[3,4,5]

Due to the regulated nature of disposal of cytotoxic waste, proper waste containers are required. There must be one for all soft waste items, which include gloves, gown, tape, gauze, and absorbent mats. Waste such as syringes, bottles, vials, needles, indwelling IV catheters, butterfly catheters, IV drips sets and fluid bags need to be in a yellow chemotherapy hard waste container.[3,4,5]

CHEMOTHERAPY PREPARATION

Preparing chemotherapeutic agents should ideally be done away from the general flow of traffic, preferably in a separate, isolated room with a biologic safety cabinet on a counter at waist level. Unfortunately, most hospitals are unable to accommodate all of the above recommendations. However, steps can be taken to ensure the safety of the veterinary technician preparing cytotoxic drugs and other staff members in the immediate area.[3,4,5]

Antineoplastic drugs should be prepared in a low traffic area that is away from windows, doorway or drafts. This will help eliminate secondary exposure to the rest of the staff.[3,4,5]

Syringes should not be filled more than two-thirds full to prevent the plunger from separating from the syringe barrel and prime IV drip sets prior to adding antineoplastic drug to fluid bags. This can be prevented with the use of closed systems such as Equashield or PhaSeal.

Veterinary technicians who prepare the cytotoxic drugs should NEVER eat, drink, apply make-up or smoke in any of the areas where these drugs are prepared.[3,4,5]

Any drawers or cabinets that contain chemotherapeutic agents must be labeled properly. All antineoplastic drugs that require refrigeration must be kept in their own refrigerator. *No other drugs, food, beverages or other products are allowed in the chemotherapy refrigerator.*[3,4,5]

CHEMOTHERAPY ADMINISTRATION

Prior to treating with a patient with a chemotherapeutic agent, there are protocols that should be in place to prevent problems during the administration:[5]

1) *Identify the patient to be treated.* All patients should have ID bands to prevent mix-ups
2) *Check for the current weight* (treatment day weight, not last weeks)
3) *Choose drug to be administered*
4) *Calculate dose and ask another person to cross check*
5) *Choose appropriate administration materials* (butterfly vs. indwelling catheter)

Veterinary technicians should always wash their hands prior to preparing and administering chemotherapeutic drugs. This helps prevent absorption of any drug that may have inadvertently contaminated the skin.[3,4,5]

Choosing the peripheral vein used for the administration is a personal choice. The veins typically used are the cephalic, and lateral (dogs) or medial (cats) saphenous. Other sites may be used but these are the most common. Some facilities are still using central in-dwelling catheters that are surgically placed into the jugular vein for long-term use.

After choosing the appropriate vein, shave the injection site (remember that hair grows much slower in cancer patients) and prepare aseptically with soft, cotton gauze material. Contact time of the cleaning agent on the skin is more effective as an antimicrobial agent than the physical scrubbing. This will also help prevent irritation to the injection site tissue. Patients are also less likely to lick the area after administration if it is not irritated. Dogs can form lick granulomas and care must be taken to prevent secondary bacterial infections from occurring. Using the lateral saphenous instead of the cephalic will usually help in these patients. Place the absorbent mat under the patient's leg to absorb any cytotoxic drugs that might accidentally spill.

Proper restraint is critical when administering antineoplastic drugs. There should always be another technician or veterinary tech assistant to help restrain the patient. If necessary, sedation should be used to prevent any possibility of perivascular injection of the drug(s).[5]

Extravasation of certain chemotherapeutic drugs will cause severe tissue necrosis, sometimes requiring amputation of the affected limb. It is for this reason that administration should always be via a "clean stick". Second attempts in the same vein can be done only if it is proximal to the first attempt. There are several "schools of thought" on how often the same vein can be used. Some follow the "use it till you lose it" approach, while others rotate veins with each treatment. Once again, this is a personal choice situation.[5]

Following the placement of the butterfly or indwelling catheter, flush thoroughly with ≈12 mls of 0.9% saline to ensure the catheter is properly situated. Heparinized saline should not be used due to the fact that heparin can precipitate with certain chemotherapeutic agents (doxorubicin).[5]

Just prior to the initiation of the chemotherapy, place an alcohol soaked 4x4 or cotton ball over the injection port. As discussed earlier, this will help eliminate aerosolization of the drug.

Administer the chemotherapeutic agent(s) at the appropriate time interval for those particular drugs. Certain drugs, such as vincristine or vinblastine can be given as a bolus. Other drugs, such as doxorubicin must be given slowly over a longer period of time.

Following the chemotherapeutic administration, flush with another 0.9% saline injection to eliminate the escape of any of the drug perivascularly after the needle or catheter has been removed. Place a pressure wrap to prevent hematoma formation.

Place all contaminated hard supplies (syringe, needles, catheters, bottles) into a re-sealable plastic bag and place into the appropriate hard waste yellow chemotherapy container. Gloves and absorbable mats go into the soft chemotherapy waste containers.

Always wash your hand after completion of any chemotherapeutic administration.[2,3,4,5]

ACCIDENTAL CHEMOTHERAPY SPILLS

In the event of a chemotherapy spill, there are very important steps that must be followed to ensure the safety of the veterinary health care staff.[2,3,4,5]

1) Restrict the contaminated area from the rest of the staff (only the person performing the spill clean-up should be in the immediate area)
2) The clean-up person should wear protective gear (gown, gloves, face shield or mask & goggles)
3) Mop up all liquid spills with absorbent disposable towels or pads
4) Collect powder spills with damp (water or alcohol soaked) disposable towels or pads
5) Rinse the contaminated area with water, then wipe area again
6) Small spill (<3 mls) wipe with 70% alcohol 4x4's
7) Large spill (> 3 mls) clean with detergent --NOT BLEACH!!!!
8) Place in appropriate chemo soft waste container
9) Wash your hands following spill clean

CONCLUSION

Chemotherapy safety is everyone's responsibility. Establishing standard operating procedures will ensure the safety of your veterinary health care staff and your patients.

REFERENCES

[1]Morrison W.B. J Vet Intern Med 2010; 24:1249-1262. [2]Arrington DM, et al. Am J Hosp Pharm 1993;50:1170-1174.[3]NIOSH List of Antineoplastic and Other Hazardous Drugs in Healthcare Setting 2010. [4]CDC Workplace Safety and Health NIOSH ALERT 2004. [5]Moore A, et al. Oncology for Veterinary Technicians and Nurses, Wiley Blackwell; 2010:77-84.

MAKING SENSE OF LYMPHOMA DIAGNOSTICS

Joanne L Intile, MS, DVM, DACVIM (Oncology)
Riverhead, NY

INTRODUCTION

Lymphoma is a hematopoietic neoplasm of lymphocytes. It is one of the most commonly diagnosed and treated cancers in dogs and cats. The term "lymphoma" actually refers to a diverse group of blood cancers, rather than a single disease process, and many different subtypes are described in both people and animals.

Terminology surrounding the categorization of different forms of lymphoma in veterinary patients is confusing, mainly resulting from lack of standardization of clinicopathologic reports and lack of utilization of a consistent classification scheme.

To be clinically useful, a cancer classification system should provide information about response to therapy, duration of maintenance of remission, and survival. Examples of classification schemes for lymphoma include:

Anatomic Site (multicentric, alimentary, mediastinal, hepato-splenic, extra-nodal)
Cellular Morphology (small cell lymphocytic, large cell lymphoblastic, plasmacytoid, cleaved)
Cellular Grade (high, intermediate, low)
Growth Pattern (follicular, diffuse)
Phenotype (B cell, T cell, T cell rich B cell, non-B non-T cell)

Considering features of the above-mentioned classification schemes, the majority of canine lymphomas are high- (65%) or intermediate- (20%) grade, multicentric, B-cell while feline lymphomas are most often high-grade, intestinal, T-cell. Low-grade lymphomas occur less frequently in both species, and are characterized by a slow rate of growth and an indolent biological behavior. Biopsy and histopathology are necessary to confirm a diagnosis of indolent lymphoma.

DIAGNOSING LYMPHOMA

Achieving a confirmative diagnosis of lymphoma can be surprisingly challenging in some cases and confusion often arises from inconsistencies among clinicopathological and histopathological reports.

The following diagnostic tests are recommended in cases where lymphoma is suspected:

1. Cytology: For dogs presenting with typical clinical signs of lymphoma (specifically, dogs with generalized lymph node enlargement), diagnosing lymphoma based on cytology of fine needle aspirates interpreted by a trained cytopathologist is appropriate.

   Fine-needle aspiration of enlarged peripheral lymph nodes, or affected visceral organs or nodes, usually provides specimens of adequate cellular content and detail to make a definitive diagnosis. Cytologically, lymph node or tissue aspirates identify a monomorphic population of lymphoid cells of large (lymphoblastic), intermediate, or small (lymphocytic) size.

   Despite ease of diagnosis, conventional cytology is limited for differentiating and categorizing the heterogeneous spectrum of lymphomas with regard to morphologic subtype (diffuse versus follicular, cleaved versus non-cleaved, etc.) and histologic grade (high versus low). However, at minimum, a cytology report should include information related to the size, shape, and complexity of the observed neoplastic cells so an inference can be made as to whether the lymphoma is of a lymphoblastic or lymphocytic origin. Immunophenotype (see below) can be determined using cytology samples, though this is not routinely offered.

2. Histopathology: Because of the inherent limitations associated with cytology, histopathologic tissue evaluation remains the gold standard for diagnosing lymphoma. Histopathology provides the additional morphologic information required for definitive classification, as well as for determining grade. Tissue can also be used for determining immunophenotype (see below). These are essential pieces of data necessary for determining a patient's prognosis as well as their optimal treatment plan.

   Tru-Cut®, needle core, or incisional lymph node biopsies are adequate for most patients. For dogs with solitary (stage I) or regional lymph node (stage II) enlargement, surgical excision of an entire node is preferable. Low-grade histology is often found in dogs with stage I or II disease, and is most accurately diagnosed and classified based on examining the distribution of neoplastic cells within the nodal architecture, as well as cellular morphology.

   Histopathology reports should include information as to whether or not the lymphoma is high-grade (diffuse large cell, centroblastic, and immunoblastic) or low-grade (small lymphocytic or centrocytic lymphoma), and pathologists should comment on regional or diffuse involvement of the architecture of the node when applicable.

   In situations where cytology and/or histopathology fails to confirm lymphoma, but disease is still suspected, advanced molecular techniques are available for definitive diagnosis, including flow cytometry for specific cell surface markers called cluster of differentiation (CD) antigens, and clonality testing using PCR for antigen receptor rearrangement (PARR.)

3. Flow cytometry: This is a laser-based, biophysical technology employed in cell counting, cell sorting, biomarker detection and protein engineering. Flow cytometry works by suspending live cells, labeled with antibodies that bind to proteins expressed on the cell surface, in a stream of fluid, and passing them by an electronic detection apparatus. With regard to lymphoma, different types of lymphocytes express different cell surface proteins (for example T cells express the protein CD3, and B cells express the protein CD21). The cells are analyzed in the flow cytometer, which provides enumeration of how many cells of each type are present within a given sample. The information allows us to determine the lineage of the cells present, and whether they are homogeneous (more consistent with neoplasia) or heterogeneous (more consistent with a reactive process.) Flow cytometry can be performed on blood or bone marrow samples, thoracic or abdominal fluid samples, and fine needle aspirates of tissues when samples are stored in a proper mixture of saline/serum. Viability of the cells is of major importance; therefore careful sample handling during all portions of analysis is of utmost importance.

4. PARR: Clonality testing using PARR attempts to differentiate reactive expansions of cell lines from neoplastic proliferations by examining for different gene rearrangements. Because most neoplastic outgrowths result from the clonal expansion of one malignantly transformed cell, PCR techniques can differentiate if a lymphocyte expansion within the sample occurs as a consequence of cancer (e.g. lymphoma) versus inflammation (e.g. reactive or hyperplastic lymphocytosis).

   Specifically, a polymerase chain reaction (PCR) assay is used to detect rearrangements in the B-cell receptor (surface Ig) or T-cell (TCR) genes. Individual B and T-lymphocytes have unique cell surface antigen-binding receptors that arise from rearrangement events between V, D, and J gene segments.

   During PARR, DNA is extracted from cells and gene sequences for the T-cell or B-cell receptors are amplified. If the population of lymphocytes is clonal, as seen in neoplastic conditions such as lymphoma, a single receptor will be amplified, and results will show a single band on the PCR assay. If the population of lymphocytes is polyclonal, as seen in reactive conditions, multiple receptors will be amplified, and results will show multiple bands on the PCR assay. False positive single bands can be obtained in cases of infections diseases (e.g. Ehrlichia canis and Borrelia). Clonality is not useful for differentiating an acute leukemia from a stage V lymphoma. PCR can be performed on fixed or fresh tissue samples.

IMMUNOPHENOTYPING

Immunophenotyping involves the detection of cell-surface antigens using antibodies against lineage-specific markers. Each hematopoietic cell will express unique markers during differentiation and maturation. Antibodies generated against these markers that can be used to identify the lineage of a particular cell. For lymphoma, this translates into determining whether the cell of origin is either a B-lymphocyte or T-lymphocyte. Immunophenotyping is most routinely performed using cytochemistry or immunohistochemistry, however by nature of their mechanism of action, both flow cytometry and PARR provide information as to immunophenotype.

Cytochemistry involves detecting the presence or absence of intracellular enzymes via the application of substrates that react with the enzyme to produce color changes that are detected using light microscopy. Cytochemistry is performed on cytology smears from blood, bone marrow, or tissue samples. As myeloid cells contain more enzymes than lymphoid cells, cytochemical stains are considered more valuable for identifying cells of myeloid origin, whereas lymphoid cells are typically negative for all cytochemical markers. Common cytochemical stains used in veterinary medicine include peroxidase for identifying granulocytes and monocytes, chloroacetate esterase for identifying granulocytes, non-specific esterase for identifying monocytes and megakaryocytes, and alkaline phosphatase for identifying monoblasts.

Immunohistochemistry: Immunohistochemistry involved the application of antibodies to fixed (acetone for cytology or formalin for histology) or fresh tissue (for flow cytometry – see above) samples. In veterinary medicine, there are many more antibodies available for identifying lymphoid cells than for myeloid cells. Common antigens used for confirming a B-cell lineage are CD3, CD21, pax-5, and BLA-36. Common antigens for confirming T-cell lineage are CD3, CD4, and CD8. CD34 is generally considered a marker for hematoipoietic stem cells and therefore, can be useful for differentiating an acute leukemia (which arises within the bone marrow) from lymphoma (which arises in extramedullary sites). Neoplastic cells from dogs with acute leuekmias will generally be positive for CD34, where as neoplastic lymphocytes from dogs with lymphoma are typically negative.

BIOMARKERS

Biomarkers are measurable indicators of particular biological states or conditions and can be used to detect, screen, diagnose, treat, and monitor disease. With regard to lymphoma testing in dogs and cats, there are several commercially available tests evaluating examining different biomarkers. Most frequently, assays measure serum levels of thymidine kinase (TK) and C-reactive protein (CRP). The utility of these markers is not well established, but emphasis is often placed on their ability to detect minimal residual disease (MRD).

TK is a protein involved in the synthesis of DNA precursors. TK is expressed during the S-G2 portions of the cell cycle in dividing cells, and levels increase with increased rate of cellular proliferation. TK levels correlate to the proliferative activity of lymphoid cells (and less likely with proliferation of other kinds of tumor cells.) Elevated TK levels are also associated with viral infections and inflammatory conditions.

Serum TK levels tend to be higher in dogs with cancer than in healthy dogs, however there is a great amount of overlap in levels measured from healthy dogs, dogs with cancer, and dogs with other diseases. Additionally, dogs with lymphoma can have normal serum TK levels.

TK levels have also been measured in cats and a reference interval was established from clinically healthy cats and cats diagnosed with lymphoma and in inflammatory gastrointestinal disease. Cats with lymphoma had significantly higher serum thymidine kinase activity than healthy cats or cats with inflammatory disease and cats with non-hematopoietic neoplasia.

CRP is the major acute phase protein produced in response to inflammation and cytokine release. Serum CRP levels correlate to the duration and severity of inflammatory response. Causes of inflammation are varied, and include infection, autoimmune disease, and cancer. Therefore, CRP is considered a sensitive marker for inflammation, but unfortunately, it is a relatively non-specific as to the nature of the inflammation it represents.

In dogs, CRP is elevated in at least some kinds of cancer, and serum levels are generally elevated in dogs with cancer compared to healthy dogs. As with TK, there is significant overlap between these two groups, and some dogs with cancer have normal serum CRP and some healthy patients have elevated serum CRP. Dogs with lymphoma who are in remission, with only microscopically detectable cancer cells in their bodies, generally have lower CRP than dogs with measurable lymphoma. This places potential value on serum CRP levels as a marker for remission status and relapse of disease.

Clinical cases will be presented as correlates.

REFERENCES

Aresu et al. **Vet J.** 2014;200(2):318.

Burkhard MJ and Beinzle D. **Vet Clin North Am Small Anim Pract.** 2013;43(6):1331

Cowell RL et al. **Vet Clin North Am Small** Anim Pract. 2003;33(1):47.

Lana et al. **J Vet Intern Med.** 2006;20(2):329.

Nielsen L et al. **J Vet Intern. Med.** 2007;21(6):1231.

Sato H, et al. **J Vet Med Sci.** 2014;76(6):807.

Selting KA et. al **Vet Comp Oncol.** 2013;epub ahead of print.

Thalheim L et al. L **Vet Intern Med.** 2013;(27(6):1509.

Taylor SS, et. al **J Feline Med Surg.** 2013;15(2):142.

vonEuler H et al. **J Vet Intern Med.** 2004;18(5):696.

RADIATION ONCOLOGY FOR VETERINARY TECHNICIANS: BIOLOGY, PHYSICS AND THERAPY

Jessica Lawrence, DVM, DACVIM (Oncology), DACVR (Radiaton Oncology), MRCVS
Edinburgh, UK

INTRODUCTION

Cancer is the leading natural cause of death in cats and dogs and the prevalence of cancer in pet animals is rising. Cancer is not simply one disease, but many different diseases with various biologic behaviors, standards of care and prognoses. There are many misconceptions about treatment options and the quality of life of companion animals living with cancer but it is our role as veterinarians and veterinary technicians to be compassionate and knowledgeable resources for our clients/patients.

Background

Radiation therapy (RT) began in the 19th century, following Roentgen's description of x-rays in 1895 and the discovery of radium by Marie & Pierre Curie in 1898.[1-3] Radiation oncology now refers to the medical use of ionizing radiation (IR) as an integral part of cancer treatment by killing or controlling malignant cells. The primary purpose of RT has remained to provide local and/or loco-regional treatment, although half body or total body irradiation is occasionally administered. Depending on the clinical situation, the primary intent of treatment is either palliative or definitive (potentially curative) and is commonly used in conjunction with other therapies such as surgery and/or systemic therapy. RT has been available in a number of veterinary facilities for over 30 years and the number of units and degree of sophistication continually grows.[4-6]

Considerations When Choosing Veterinary Patients for Radiation Therapy

Dogs and cats (and occasionally other species such as rabbits) should be evaluated carefully prior to offering radiation therapy in order to ensure that they are good candidates. It is important to understand the tumor type, biologic behavior and extent of disease to determine if radiation therapy is indicated and likely to aid in control. Radiation therapy alone or in combination with surgery may be a very appropriate treatment modality to manage patients with local or loco-regional disease; adjuvant chemotherapy can also be administered concurrently. Prospective planning often involves a concerted effort between surgeons, medical oncologists, and radiation oncologists to determine the best approach to managing a tumor and what is a good option for one patient may not be suitable for another patient with a similar tumor.

Staging, or a determination of the extent of cancer, is important in order to determine if localized disease is present. Diagnostic tests typically relate to the primary tumor type and may include complete blood count, biochemical profile, urinalysis, aspiration cytology of regional lymph nodes, thoracic and/or abdominal imaging, and advanced imaging. Computed tomography (CT) is one of the most useful tools for tumor imaging as it is critical for computer-assisted radiation treatment planning. Based on CT images, subjective assessments can be made initially regarding extent of disease, proximity of critical tissues near or in the treatment field (such as the eyes in the case of dogs with nasal tumors), and prognosis.

RADIATION PHYSICS: HOW DOES RADIATION INTERACT WITH TISSUE?

All matter is composed of atoms. An atom consists of a nucleus with one or more protons and neutrons held together by strong intranuclear forces, forming the positive charge of the atom. There is a surrounding cloud of electrons held in place by electrostatic forces, which forms the negative charge. Radiation interacts with biological material by removing an orbiting electron, a process termed ionization. Ionizing radiation consists of both x-rays and γ-rays: x-rays are man-made and arise from a linear accelerator while γ-rays occur from radioactive decay (^{60}Cobalt). X-rays and γ-rays are photons used in the treatment of RT, but photons also include radio waves, visible light, and ultraviolet rays. Simplistically, a photon is any packet or wave of energy traveling through space at the speed of light; the energy of the photon dictates the type of physical interaction with tissue. With low energy photons (diagnostic), photons are absorbed differentially in tissue, which is dependent on the density of the tissue. This photoelectric effect is what provides the differentiation between bone, soft tissue, and fat on a diagnostic radiograph. With high-energy electrons (therapeutic), physical interactions with tissue occur predominantly via Compton scattering, which minimizes the differential absorption across various tissues, providing a more uniform dose of radiation across a variety of tissues. With Compton scattering, a photon interacts with a loosely bound electron, giving up all of its energy to the atom. This electron is then ejected from the atom along with a second photon emitted in a different direction; both the electron and secondary photon can interact with additional tissue.

RADIATION BIOLOGY: HOW DOES RADIATION INDUCE DAMAGE?

High-energy photons collide with orbiting electrons of biologic molecules in tissue, which leads to the ejection of electrons. These electrons can directly damage biologic targets (such as DNA) or, more commonly, interact with water (cells are made up of 80% water) to produce free radicals that damage the target (indirect damage). Photons are considered "skin-sparing", as there is a buildup region of tissue needed for maximum radiation effect. Radiation using electrons (electron therapy) "skips" a step, as electrons from a linear accelerator act like the secondary fast electrons that are created by x-rays following interaction with tissue. DNA within tumor cells is generally considered to be the principal target of therapeutic photons, although damage to cell membranes and blood vessels may also contribute to tumor control.[1-3] Oxygen essentially "fixes" radiation damage to DNA and prevents the cell from easily repairing the damage.[1] Cells that are in certain phases of the cell cycle are also more sensitive to radiation as DNA damage is more likely to be lethal to the cell. [1] Both the oxygen effect and the reassortment of the cell into a sensitive phase of the cell cycle provide a basis for fractionating radiation (dividing up the overall dose into many doses) to allow time for reoxygentation and reassortment; it also supports the use of radiation for smaller tumor volumes.

Several consequences can occur in the cell after DNA damage – damage can be classified as lethal, sublethal, or potentially lethal.[1] Cells can repair DNA damage if enough time is present after radiation delivery, whereas other cells (lymphocytes, mast cells) die instantly from receiving a "hit" of radiation; the most common consequence is a disruption in mitosis (cell division).[1] Clinical consequences are dependent on a number of factors, including the number of cells present, the intrinsic radio-sensitivity of the tumor cells, the tumor microenvironment, and cell cycle kinetics. Radiation is generally more effective on microscopic disease compared to macroscopic tumors although there are exceptions depending on the tumor type. As most cells sustain "reproductive" damage, measurable tumors often do not shrink immediately but may respond after several weeks of RT.

RADIATION THERAPY

Radiation Terminology

Radiation terminology can be confusing and some terms (SRS and SRT) are inaccurately used as synonyms. Common terminology and abbreviations are provided below to improve the reader's understanding of various commonly used terms.[1-5]

Dose – amount of radiation absorbed by the patient (Gray = Gy = 1 Joule/kg; the older term "rad" is not used any longer).

Fraction – individual administration of radiation.

Fractionation – radiation dose is divided into fractions over the course of several days to weeks depending on the intent of treatment.

Protocol – particular dose and number of fractions to be given to a particular tumor.

External beam radiation – radiation delivered from outside the body by a machine that focuses high-energy photons at a tumor.

3-D Conformal RT (3D-CRT) – conformal radiation therapy that is image-based and improves radiation dose distribution across the tumor volume, which can limit the normal tissue involved.

Intensity-modulated RT (IMRT) – improved radiation planning and delivery to change the intensity of the radiation beams aimed at the target volume to improve conformity of the radiation dose distribution to tumor with a rapid fall-off in radiation dose to adjacent normal structures.

Image-guided RT (IGRT) – allows the detection of patient and tumor positioning errors through the use of advanced images acquired prior to each treatment (usually IMRT treatments). IGRT provides a "quality assurance" that the radiation beams are directed precisely to the target volume (tumor).

Stereotactic radiosurgery (SRS) is a single treatment involving a large fraction size (15-20 Gy), usually to treat small well-delineated tumors.

Stereotactic radiation therapy (SRT) is a small number of treatments (3-8) involving a large fraction size (10-20 Gy) used to treat small well-defined lesions.

Plesiotherapy – "surface" radiation with limited penetration (2-3 mm), usually delivered by a Strontium-90 (^{90}Sr) probe.

Brachytherapy – short distance radiation (commonly internal radiation), usually delivered through use of radioactive seeds.

Clinical Radiation Therapy

The primary goal of radiation therapy is to precisely target the tumor volume with cytotoxic doses of radiation while minimizing dose to surrounding normal structures. This involves multiple steps in order to ensure the tumor volume is adequately treated. The overall dose and fractionation protocol is largely based on the particular tumor type and grade, adjacent normal tissue, the inherent radiosensitivity of the tumor cells, and clinical evidence (and sometimes clinical experience). Advanced imaging (often with CT alone but sometimes a combination or registration of CT and MRI and/or PET-CT) aids in the delineation of tumor volumes and target structures. Typically a radiation oncologist outlines 3 main tumor volumes as well as normal tissues (organs at risk) within the imaging. Tumor volumes include: gross tumor volume (GTV), clinical target volume (CTV) to account for microscopic disease, and planning target volume (PTV) to account for intrinsic and extrinsic setup error. The radiation prescription is typically prescribed to the PTV. Developments in computerized treatment planning have vastly improved the ability of the radiation oncologist to target the tumor and reduce the radiation dose to normal structures; image-based, 3D-CRT and IMRT have both evolved from software development and both are dependent upon proper patient positioning and verification of positioning with imaging (on-board imaging) for each treatment. Because IMRT changes the intensity of the beams and allows buildup of radiation dose within the target but a rapid decrease in radiation dose to surrounding normal tissue, IMRT drastically reduces normal tissue toxicity and improves quality-of-life for both the patient and the owner.[4,7]

Animal owners, veterinarians and veterinary technicians are correct to be concerned about the development of toxicity and it is important that a thorough discussion occurs between all parties prior to starting radiation therapy. Treatment toxicity depends on the fractionation schedule, the time over which radiation occurs, and the normal tissues that are in the radiation field. As radiation can damage any cell, both tumor and normal tissue react similarly. Therapeutic radiation attempts to exploit the difference in repair capabilities between normal and tumor tissue.[1-3] Tumor cells have impaired repair mechanisms and more cells in the cell cycle by nature of their increased proliferative rate compared to normal tissue. Cell killing follows exponential kinetics; therefore for any given dose of radiation, the same fraction of cells in the population will be killed.[1-5] Thus, large tumors require higher doses of radiation to induce a complete response compared to smaller tumors.

Normal Tissue Toxicity

Normal tissue in the radiation field is "dose-limiting" with respect to the maximum dose that can be safely administered. The dose per fraction, treatment schedule, and total dose affect the tolerance of the normal tissues. Fractionation refers to the delivery of smaller doses of radiation a number of times in order to reduce the toxicity to normal tissues and often to improve the overall response of a tumor. Giving radiation once or twice a day allows normal cells to repair but does not provide adequate repair of tumor cell damage. The probability of tumor control is always weighted against the probability of complications. At lower doses of radiation, the probability of complications is low but with moderate chances of tumor control – increasing the dose may gain a higher chance of tumor control but at the price of significantly higher complication risk.[1-5]

Acute radiation toxicity is typically reversible and self-limiting and occurs during or shortly after therapy; effects are confined to the irradiated area. Rapidly proliferating tissues such as epithelium, mucosa, and tumor cells are affected most in the acute setting. Discomfort is expected in the short-term and early effects can be unpleasant for both the owners and the pet. Acute toxicities are predominantly dependent on the total dose of radiation administered (the higher the dose, the greater the toxicity) but also on the duration and frequency of treatment (shorter overall treatment time increases toxicity) and the fraction size (the higher the fraction size, the greater the toxicity).[1-3]

Late radiation side effects are typically adverse events that occur 6 months or greater after completion of radiation therapy and are related to the structures within the irradiated field. Late effects typically develop in non-renewing tissues such as nerve, bone, and muscle. Consequential late effects can occur which are directly related to unexpected, severe acute side effects that do not recover. These non-renewing tissues are dose limiting; dose to tumor is typically prescribed such that the probability of the occurrence of clinically-relevant late effects is < 5%. Late tissue damage is progressive and irreparable in most cases and results from vascular damage, fibrosis, and loss of cells. The likelihood of late radiation effects is heavily dependent on the size of the fraction (the greater the fraction size, the greater probability of late toxicity), although the overall dose prescribed is also important. Table 1 outlines a variety of acute and late adverse events that can be seen depending upon the site irradiated.

Table 1: Table of radiation-induced toxicities that can occur in the dog and cat

Skin	Radiation acute side effects	Radiation late side effects
Skin & Subcutis	Moist desquamation Alopecia	Fibrosis and contraction Non-healing ulcer Leukotrichia Alopecia Secondary tumor (sarcoma)
Extremity	Nail or pad slough	Neuropathy Muscle fibrosis and contraction
Spinal Cord	Delayed "acute" transient myelopathy (rare)	Myelopathy Infarction Radiculopathy (lower motor neuron syndrome)
Cervical Region	Pharyngitis Esophagitis Tracheitis	Hypothyroidism Esophageal stricture
Oral Cavity	Mucositis	Periodontal disease Xerostomia (dry mouth)
Nasal cavity	Mucositis Nasal discharge	Chronic rhinitis (nasal discharge)
GI Tract	Gastritis, Enteritis or Gastroenteritis Colitis / Anusitis	Stricture
Eye	Blepharitis/ Blepharospasm Conjunctivitis Corneal ulceration Keratitis Uveitis	Cataract Keratoconjunctivitis Sicca (KCS)
Brain	Encephalopathy (rare) Edema (rare) Systemic effects: lethargy, nausea, vomiting (rare)	Encephalopathy Infarction / Hemorrhage
Bone		Osteoradionecrosis Secondary tumor formation (sarcoma)
Kidney	Acute nephropathy	Fibrosis leading to nephropathy

Factors that dictate the likelihood of late toxicity support the rationale behind choosing definitive versus palliative radiation protocols for a particular case. Definitive radiation protocols are typically administered daily (Monday through Friday) for 2-4 weeks in a row depending on the underlying tumor type. In the author's practice, definitive protocols are recommended for pets with a good prognosis and that have a high likelihood of survival beyond 6 months. The dose given each day in this type of protocol is lower in order to "protect" normal tissue from developing significant late toxicities that could negatively impact quality of life. Palliative protocols are typically reserved for animals with a poor prognosis. The goals of palliative therapy are to alleviate pain associated with the tumor, decrease the size of any mass causing obstruction (a laryngeal mass), and to otherwise alleviate debilitating signs of cancer (bleeding from a large ulcerated oral mass). Palliative radiation therapy is not intended to prolong survival (although indirectly often prolong survival) but to improve the quality of life for the time remaining. There are many different protocols for palliative radiation however they often involve radiation administered once weekly or twice weekly for 3-4 doses or for 4-5 total doses administered daily. Because the dose of radiation is higher for each fraction, if pets do well (> 6 months) with treatment, there is a higher likelihood of clinically relevant late toxicity.

Most tumors are ideally treated in the microscopic disease setting; therefore postoperative therapy is most common, when surgical margins are incomplete or narrow. However, there are situations in which preoperative radiation therapy is indicated. Cats with injection site sarcomas often have extremely large tumors (the "tip of the iceberg" effect) that are best treated with preoperative radiation followed by aggressive surgical resection. Occasionally dogs with unresectable tumors can be managed as cats with injection site sarcomas. Many tumors are responsive to radiation, however the overall response tends to be variable and dependent on the signalment, tumor histology, location, grade, and/or stage of disease. It is always important to remember that normal critical structures adjacent to the radiation field limit the amount of radiation that can be administered; occasionally doses must be modified to avoid significant injury. A general guide for tumor types that are responsive to radiation is provided in Table 2.

Table 2: Common tumor types in dogs and cats and respective radiation responses[4,5]

Tumor	Response to radiation
Ameloblastoma (Acanthomatous epulis)	Excellent
Oral Fibrosarcoma	Poor to fair
Oral Melanoma	Fair to good local control
Oral Squamous Cell Carcinoma (dog)	Good
Oral Squamous Cell Carcinoma (cat)	Poor
Nasal Adenocarcinoma	Good
Nasal Sarcoma	Good
Nasal Poorly Differentiated Carcinoma or Squamous Cell Carcinoma	Poor to fair
Mast Cell Tumor	Good to excellent
Lymphoma (localized)	Excellent
Perianal Gland Adenoma	Excellent
Anal Sac Apocrine Gland Adenocarcinoma	Good to excellent
Perianal Gland Adenocarcinoma	Good
Salivary Gland Adenocarcinoma	Good to excellent
Squamous Cell Carcinoma (non-oral)	Poor to fair
Thyroid Carcinoma	Good
Multilobular Osteochondrosarcoma (MLO)	Unknown – fair?
Brain Tumors	Poor to good
Soft Tissue Sarcoma	Good to excellent
Osteosarcoma	Good (palliative)
Prostatic Adenocarcinoma	Poor to Fair
Transitional Cell Carcinoma	Poor to fair

Summary

Radiation therapy is appropriate for a number of different scenarios including: 1) potentially any single site disease, 2) in the pre-surgical or post-surgical setting, 3) radioresponsive tumors in the gross (measurable) disease setting 4) palliation of pain or temporary local control to improve function. Generally speaking, tumors will be more responsive to radiation in the microscopic disease setting, thus surgery is often indicated first. It is important to differentiate the goals of radiation treatment and to be prepared for relevant acute and late toxicities.

REFERENCES

[1]Hall EJ. **Radiobiology for the Radiologist**. 7th ed. 2011. [2]Halperin EC et al (eds). **Perez & Brady's Principles and Practice of Radiation Oncology**. 6th ed. 2013; Lippincott Williams & Wilkins, Philadelphia; [3]Morgan MA, et al. Radiation Therapy. In: **DeVita, Hellman, and Rosenberg's Cancer: Principles and Practice of Oncology**. 9th ed. 2011; Lippincott Williams & Wilkins, Philadelphia: 289. [4]LaRue SM, Gordon IK. Radiation Therapy. In: **Small Animal Clinical Oncology**. 5th ed. 2013; Elsevier, St Louis: 180. [5]McEntee MC. **J Am Anim Hosp Assoc** 2006;42:94. [6]Farrelly J, McEntee MC. **Vet Radiol Ultrasound** 2014;55:638. [7]LaRue SM et al. **Vet Clin North Am Small Anim Pract** 2014;44:909.

Respiratory Tumors in Dogs and Cats: Update on Current Options & Outcomes

Jessica Lawrence (Rausch), DVM, DACVIM (Oncology), DACVR (Radiation Oncology), MRCVS
Edinburgh, UK

INTRODUCTION

Respiratory tumors are relatively uncommon in both the dog and cat and can include tumors affecting the nasal planum, nasal and/or sinus cavities, larynx, trachea, and lungs. Nasal or sino-nasal tumors are the most common tumor to affect both dogs and cats and are the primary focus of the lecture, although tumors of the nasal planum and lungs are not uncommon in referral hospitals.

NASAL PLANUM SQUAMOUS CELL CARCINOMA

Incidence & Biologic Behavior

Squamous cell carcinoma (SCC) of the nasal planum is uncommon in dogs but relatively common in cats. SCC has been correlated with ultraviolet light exposure, unpigmented tissue, and possibly papillomavirus.[1-3] SCC is more common in older, white or lightly pigmented cats particularly in geographic areas with abundant sunlight. It tends to be a locally proliferative disease and can range from carcinoma in situ (very shallow lesion) to a deeply infiltrative tumor.[1] Most cats have a chronic history; owners often report an initial crust-like lesion that progresses to a superficial erosion and to a deep, often painful ulcer. Most feline tumors are easily visible on the external aspect of the planum while canine tumors develop on the inner aspect of the planum.[1]

Diagnosis & Staging

Biopsy is required for definitive diagnosis, as cytology is often consistent with inflammation. Bleeding is common following biopsy but is typically transient. Advanced imaging with computed tomography (CT) or magnetic resonance imaging (MRI) may help define the extent of disease, particularly in the dog, and can be done prior to biopsy or following diagnosis.[1] Regional lymph node aspirates and three-view thoracic radiographs (CXR) are often performed for full staging but rarely demonstrate metastasis. Complete blood count (CBC), serum chemistry profile and urinalysis are primarily used to identify co-morbidities.

Therapy & Outcome

For cats with small, superficial nasal planum SCC, a variety of treatments are possible including surgical resection ("nosectomy"), strontium radiation therapy (plesiotherapy), cryosurgery, laser ablation, photodynamic therapy and intralesional chemotherapy.[1,4-6] The primary benefit of surgical resection is that adequate evaluation of margins status is possible. Small tumors carry an excellent prognosis although cats may be at risk for SCC in other locations. Deeper tumors can be more difficult to control and nosectomy is likely the most appropriate treatment for larger, invasive lesions. Most cats heal well following surgery with good cosmetic outcome. For dogs, nosectomy or "nasal planectomy" is typically the treatment of choice with the extent of surgery being dependent on the caudal extent of the tumor.[1,7] Cosmetic outcome in the dog is good following postoperative healing although support staff and owners may need to be prepared for the postoperative appearance. Prognosis for both cats and dogs is generally good to excellent following surgery with good margins. For dogs and cats with larger lesions, prognosis can still be fair to good following surgery or palliative measures.

NASAL (SINO-NASAL) TUMORS

Incidence & Biologic Behavior

Nasal tumors are rare in dogs and cats, accounting for approximately 1% of all tumors in dogs and < 1% in cats.[1] Despite the relatively low incidence reported, nasal tumors comprise a large fraction of the types of tumors that are treated by radiation oncologists.[8] Most dogs are older (10 years) and medium to large breed dogs tend to be over-represented. Risk factors may include dolicocephalic conformity, living in an urban environment (filtering of pollutants), exposure to fossil fuel combustion products, and possibly tobacco smoke exposure.[9-10] Carcinoma (adenocarcinoma, squamous cell carcinoma, and undifferentiated or poorly differentiated carcinoma) is the most common tumor type, comprising approximately 2/3 of all nasal tumors in dogs; sarcoma is the second most common (fibrosarcoma, osteosarcoma, chondrosarcoma, and undifferentiated sarcoma), comprising almost 1/3 of tumors.[11] Most nasal tumors are characterized by progressive local invasion and late onset of metastasis.[12] There is a low rate of metastatic disease at the time of diagnosis, however it may be as high as 45% at the time of death.[13] The most common metastatic sites include the local lymph nodes and lung although other reported sites include the brain, kidneys, liver, skin, and bones.[1,13]

Sinonasal cancer in the cat is less common than in the dog. Most cats are older (10 years) and there are no obvious breed or sex predilections. The most common nasal tumors in cats include lymphoma and carcinoma, although a variety of other tumor types can occur. Almost all cats with nasal lymphoma are FeLV negative.[1] Sinonasal tumors in cats display similar biologic behavior to those in dogs in that they are generally locally invasive and associated with a low metastatic rate at the time of diagnosis.[1,14]

Most dogs and cats present with a history of clinical signs attributable to nasal disease, including nasal discharge, congestion, epistaxis, sneezing, reverse sneezing, coughing, nasal deformity, pawing at the face, stertor, ocular discharge or protrusion, or neurological signs (rare). Epistaxis has many differentials and is not pathognomonic for cancer. Idiopathic rhinitis, fungal rhinitis and lymphoplasmacytic rhinitis can often cause severe nasal signs. Dogs with localized nasal disease may be more likely to have unilateral disease than bilateral disease; however over 50% of dogs presenting for epistaxis associated with systemic disorders have unilateral epistaxis.[15] Chronic nasal signs (>1-3 months) may be more likely attributed to intranasal disease rather than systemic disease.[1,15]

Diagnosis & Staging

Definitive diagnosis of sinonasal neoplasia requires a tissue biopsy although other information may point to a presumptive diagnosis of cancer. Advanced imaging with CT or MRI is superior to nasal radiograph series and may aid in obtaining an adequate tissue sample. CT provides excellent anatomic detail of the tumor and normal adjacent structures, extent of disease for accurate staging, assessment of the integrity of the cribriform plate, and evaluation of presence or absence of extension into the cranial vault, and can later be used for radiation therapy (RT) planning.[16] MRI can determine the extent of disease and may be helpful to determine extent of invasion into the olfactory lobe or farther within the brain, however there is a decreased ability to assess bone changes.[16] Biopsy techniques are variable and often dependent on the clinician's comfort and experience; techniques such as vigorous nasal flushing, blind trans-nasal biopsy, FNA or biopsy of facial deformities, and rhinoscopic-guided biopsies may all yield a diagnosis. Because secondary bacterial overgrowth and inflammatory changes often accompany neoplastic changes within the nasal or oral cavity, cytology or small superficial biopsy samples are discouraged. Although bleeding is common following biopsy, most nasal biopsy sites stop bleeding within 5 minutes although intermittent oozing for several hours to days is not unusual. In cases with severe hemorrhage, ligation of the ipsilateral carotid artery can help control epistaxis.[1]

As most tumors are locally invasive, full staging for nasal tumors is straightforward in dogs and includes CBC, serum chemistry profile, urinalysis, CXR, CT, and fine needle aspiration cytology (FNA) of loco-regional lymph nodes. In cats diagnosed with nasal lymphoma, additional staging includes FeLV/FIV test, abdominal ultrasound and bone marrow aspirate to ensure that lymphoma is truly localized to the nasal cavity. If systemic lymphoma is identified, therapeutic options will change.[17]

Therapy & Outcomes

Definitive therapy is directed primarily at control of advanced local disease. While surgery is typically first-line treatment for most localized tumors, radiation therapy (RT) is considered the standard-of-care treatment for nasal tumors in both dogs and cats. Surgery alone is considered a poor choice for treatment of malignant nasal disease as it is difficult to obtain margins and therefore local recurrence is common early after therapy.[18] In dogs, various alternative treatments to RT and surgery have been described, including a combination of surgery and RT, targeted therapy, chemotherapy, COX-inhibitors, brachytherapy, cryotherapy, photodynamic therapy, immunotherapy, and carotid artery ligation.[1] Without treatment, the prognosis is poor with survival times ranging from 1-3 months upwards of 9 months; dogs with nasal carcinoma that present without epistaxis were more likely to survive an average of 7-9 months compared to dogs with epistaxis (3 months).[1,19] Outcomes following RT vary with technique and radiation prescription parameters but median survival is typically close to 16-19 months with definitive-intent RT. Routine monitoring with CT is recommended to evaluate response to treatment. Re-irradiation is possible in most cases at the time of recurrence although there is a higher risk for significant late effects.[20,21] In cats with nasal carcinoma or sarcoma, there is limited information regarding outcome but average survival time is typically over 12-18 months.[22] In cats with primary nasal lymphoma, RT or RT and chemotherapy yield good outcomes with survival ranging from 6 months to 2.7 years.[1,17,23] The inclusion of RT in a treatment protocol appears important for long-term control. The timing of chemotherapy and its absolute role in initial treatment is unclear, as fewer than 20% of cases develop systemic lymphoma and failure is often month to years following treatment. The role of chemotherapy for most non-lymphoma nasal tumors is unclear although may provide palliation of clinical signs. Targeted therapy with tyrosine kinase inhibitors is of interest and may be useful in conjunction with standard therapy although prospective studies are needed in order to evaluate its utility.

RT allows treatment of the entire nasal cavity and a margin for microscopic disease; with recent advances in treatment planning software, imaging verification of patient positioning prior to treatment, and improved machine capabilities, acute and chronic radiation-induced toxicity can be reduced and better controlled.[1,24,25] Sinonasal tumors pose a challenge to treat with radiation as they are positioned in close proximity to critical structures such as the eyes and brain, thus causing complex target volumes. Intensity modulated radiation therapy (IMRT) and stereotactic radiation therapy (SRT) approaches aid in the delivery of more uniform radiation dose distribution throughout the target while minimizing dose to normal structures. IMRT refers to the delivery of radiation through a large number of fields of non-uniform intensity that are modulated spatially and temporally to allow a "build-up" radiation dose in the target while reducing the dose immediately adjacent to the target. Because there is a steep drop off in radiation dose from the target to adjacent tissue, image-guided setup verification is necessary with either daily portal radiographs or daily CT images immediately prior to each fraction of radiation. In a study evaluating ocular toxicity in dogs receiving definitive intent RT for sinonasal tumors, IMRT significantly decreased the incidence of both acute and late toxicity compared to historical controls treated with a cobalt unit.[24] SRT involves a precise 3-D mapping technique to guide the delivery of large doses of radiation over a small number of treatments (2-8) to a precise target while sparing adjacent normal tissues with a steep dose drop-off in dose. Preliminary investigation have suggested that SRT is well tolerated with good response durations but further studies are needed.[1,25] The increased availability of these treatment units and software upgrades will hopefully permit dose escalation (hopefully yielding improved tumor control) yet spare normal tissue to maintain an excellent quality-of-life for veterinary patients.

PRIMARY LUNG TUMORS

Incidence & Biologic Behavior

While one of the most common causes of cancer-related death in humans, primary lung cancer is relatively uncommon in both dogs and cats. Most dogs are older and breeds such as Boxers, Dobermans, Irish Setters, and Bernese Mountain dogs are possibly predisposed to developing primary lung tumors.[1] Cats tend to be older and Persians may have a slight breed predisposition.[1] A clear underlying cause has not been identified in either dogs or cats although second-hand smoke and urban pollutants may play a role.[1] Most primary lung tumors in the dog are bronchoalveolar carcinomas although other carcinomas (adenocarcinoma and squamous cell carcinoma) comprise approximately 15%.[1,26] In the cat, adenocarcinoma is the most common primary lung tumor.[27]

Most primary lung tumors in the dog are solitary but metastasis can occur to lymph nodes, other parts of the lung, and to extrathoracic sites (digits, bones). Some histologic subtypes (squamous cell carcinoma and anaplastic carcinoma) have much higher rates of metastasis than the more common bronchoalveolar carcinomas. Metastasis is more common in the cat and occurs in approximately 75% of cases.[27] The "lung-digit" syndrome is a well-described phenomenon in cats, where cats with primary lung tumors present with digital metastases.[28]

Approximately 25-50% of dogs may be diagnosed with an incidental primary lung tumor during a routine screening for other purposes.[1,29] Clinical signs are variable and can include respiratory signs such as chronic cough and exercise intolerance to non-specific signs such as lethargy, inappetence, and weight loss; cats are more likely to develop gastrointestinal signs compared to dogs.[1] Pleural effusion is uncommon in the dog but occurs with some regularity in the cat; if present, pleural fluid can be evaluated cytologically; the presence of neoplastic effusion tends to be associated with a poorer prognosis compared to cases without effusion.[1]

Staging & Diagnosis

Three view thoracic radiographs are standard for an initial workup of lung tumors. Primary lung tumors occur more commonly in the caudal lung fields; tracheobronchial lymphadenopathy can be difficult to assess on radiographs unless there is marked lymphadenomegaly. Thoracic CT is more accurate in the assessment of the tracheobronchial lymph nodes and more sensitive in the detection of pulmonary nodules compared to radiographs and thus a pre-surgical CT is often performed.[30] Positron emission tomography (PET) or PET-CT techniques have drastically altered management of lung tumors in humans; PET and PET/CT units are increasingly available for companion animals and it is likely that functional imaging will play a role in the future for sentinel lymph node mapping or to guide further investigations. Complete blood count, chemistry profile, and urinalysis complete routine staging but are predominantly used to identify co-morbid diseases. Clinicians vary on their opinions regarding the utility of trans-thoracic FNA for peripherally located tumors. Samples can be obtained via CT-guidance or ultrasound-guidance but the diagnostic utility of obtaining a sample before surgery is questionable.

Therapy & Outcome

Surgery is the standard-of-care treatment for primary pulmonary tumors; thoracotomy or sternotomy or thoracoscopy are common techniques used for partial or total lung lobectomy. Tracheobronchial lymph nodes should be evaluated histologically following extirpation or biopsy to allow for full staging, although depending on the surgical approach, it may not be feasible.[30] Histologic assessment of both the primary lung mass and the tracheobronchial lymph nodes can aid in prognosticating along with CT characteristics. Dogs with well differentiated, low grade, small (< 5 cm) masses with normal lymph nodes have an excellent prognosis with most surviving > 1 year following surgery.[1,29,30] In cats, histologic differentiation is an important prognostic factor, with well differentiated tumors carrying the best prognosis.[1,31] Cats with lung-digit syndrome carry a poor prognosis.[28]

For dogs with incompletely resected tumors and/or with evidence of metastatic disease to lymph nodes or other lungs, chemotherapy is often utilized following adequate surgical resection. In humans, cisplatin-based chemotherapy is typically recommended following surgery or RT however few studies have objectively evaluated chemotherapy in dogs and cats with lung tumors. Cisplatin, carboplatin, or vinorelbine are often chosen based on previous observations of response in the dog.[1] Vinorelbine is attractive as it is well-tolerated and there is a high concentration of drug in the lungs following administration in humans.[32] In the author's experience, dogs with metastatic disease remain candidates for surgery, particularly if they are clinically affected. In 5 dogs with metastatic lung carcinoma, surgical resection for cytoreduction followed by vinorelbine has yielded favorable results in 5 dogs, with 3 dogs still alive > 12 months following diagnosis. There is little information on postoperative chemotherapy in the cat and prognosis can be difficult to predict. Inhalational chemotherapy, tyrosine kinase inhibitors, intra-cavitary chemotherapy, interventional radiologic techniques, and RT have not been well studied in dogs and cats to date. Novel RT machines capable of delivering IMRT with image-guidance (with or without respiratory gating) play an integral role in the management of lung tumors in humans, and it is likely that these improved delivery techniques will find a place in veterinary patient management.

REFERENCES

[1]Withrow SJ, et al. Tumors of the Respiratory System. In: **Small Animal Clinical Oncology**, 5th ed. 2013; Elsevier, St Louis: 432. [2]Munday et al. **N Z Vet J** 2009;57:248. [3]Hargis AM. **Comp Cont Ed Pract Vet** 1981;3:287. [4]Lana et al. **J Am Anim Hosp Assoc** 1997;33:329. [5]Bexfield NH et al. **J Vet Intern Med** 2008;22:1385. [6]Hammond G et al. **J Am Vet Med Assoc** 2007;231:736. [7]Lascelles BD et al. **J Am Anim Hosp Assoc** 2004;40:137. [8]McEntee M. **Vet Radiol Ultrasound** 2004;45:476–479. [9]Reif JS et al. **Arch Environ Health** 1971;22:136. [10]Bukowski JA et al. **J Toxicol Environ Health** 1998;54:579. [11]Madewell BR et al. Am J Vet Res 1976;37:851. [12]Henry CJ et al. **J Vet Intern Med** 1998;12:436. [13]Patnaik AK. **J Am Anim Hosp Assoc** 1989;25:103. [14]Mukaratirwa S et al. **J Feline Med Surg** 2001;3:235. [15]Bissett SA et al. **J Am Vet Med Assoc** 2007;231:1843-1850. [16]Drees et al. **J Small Anim Pract** 2009;50:334. [17]Haney SM et al. **J Vet Intern Med** 2009;23:287. [18]Degner D. **Clin Tech Small Anim Pract** 2006;21:64 [19]Rassnick KM et al. **J Am Vet Med Assoc** 2006;229:401. [20]Bommarito DA et al. **Vet Radiol Ultrasound** 2011;52:207. [21]Gieger T et al. J Am Anim Hosp Assoc 2013;49:318. [22]Theon et al J Am Vet Med Assoc 1994;204:78. [23]Taylor SS et al. **J Small Anim Pract** 2009;50:584. [24]Lawrence JA et al. **Vet Radiol Ultrasound** 2010;51:561. [25]LaRue SM, Gordon IK. Radiation Therapy. In: **Small Animal Clinical Oncology**. 5th ed. 2013; Elsevier, St Louis: 180. [26]Griffey SM et al. **Carcinogenesis** 1998;19:959. [27]Hahn KA et al. **J Am Vet Med Assoc** 1997;211:1257. [28]van der Linde-Sipman JS et al. **Vet Q** 2000;22:141. [29]McNiel EA et al. **J Am Vet Med Assoc** 1997;211:1422. [30]Paoloni MC et al. **J Am Vet Med Assoc** 2006;228:1718.[31]Hahn KA et al. **Vet Surg** 1998;27:307. [32]Chabner B et al. **Cancer Chemotherapy and biotherapy: principles and practice**, 5th ed. 2006;Lippincott Williams & Wilkins, Philadelphia.

ALTERNATIVE AND RAW DIETS – ENLIGHTENING CLIENTS TO BENEFIT PATIENTS

Ed Carlson, CVT, VTS (Nutrition)
Woburn, Massachusetts

INTRODUCTION

All-natural, organic, corn-free, grain-free, gluten-free, raw, Biologically Appropriate Raw Food (BARF), kibble, canned, semi-moist, freeze-dried, frozen; there are so many types of pet foods available today that the list seems endless! How do you know which to recommend to your clients or to feed your own pets? This can be quite a controversial topic! Advocates of many diet types often adamantly believe their choice of diet is the absolute best, and some will passionately defend their beliefs, whether based on facts, anecdotal information, or misinformation. Let's take a look at some diet option types, pros/cons, facts and misconceptions.

HOME COOKED DIETS

Some owners enjoy cooking for their pets; others feel that they are providing superior nutrition, while some believe their pet will not eat or will not enjoy commercial foods, while others prepare limited ingredient recipes. Preparing home cooked diets for pets is often more time consuming and expensive than feeding commercially available options. There are countless books, websites, articles, and blogs with recipes for home cooking dog and cat foods. Unfortunately very few of these recipes are actually complete and balanced. Clients that wish to prepare their pets food at home should be strongly encouraged to consult with a veterinary nutritionist to ensure that the diet is nutritionally complete. Clients should also be counseled to follow the recipe exactly as formulated and not substitute or eliminate ingredients.

Limited ingredient home cooked recipes designed for elimination trials can be very useful to determine if a patient is sensitive to a particular ingredient. These diets are intended to only be used on a short term basis under the supervision of a veterinarian as they are often not nutritional balanced. Once the ingredient the pet is sensitive to is identified the veterinarian will often recommend a complete and balanced commercially available diet or consult with a veterinary nutritionist to develop a complete and balanced home cooked recipe. Owners preparing these elimination diets must not make any changes to ingredients unless directed to do so by a veterinarian. These patients should also not receive any treats or foods other than those in the elimination diet. It is extremely important to stress to clients they must follow the recipe exactly as formulated and not to substitute or eliminate ingredients.

ORGANIC AND ALL NATURAL DIETS

The term "organic" refers to the handling and processing of ingredients and products; it does not describe the quality of a product. In order to use this term on the label, companies that produce pet foods and treats must comply with the USDA's National Organic Program (NOP) regulations. These regulations include ingredient sourcing, ingredient handling, manufacturing, labeling and certification of products using the term "organic" on the product label. When "natural" is listed on a food label, the food should not contain any chemically synthesized ingredients. If "natural with added vitamins and minerals" appears on the label, the vitamins and minerals may be chemically synthesized. "Holistic", "human grade", "premium" and "gourmet" are purely marketing terms and have no legal definition.

CORN FREE DIETS

Corn has also been labeled by some owners/breeders as an inexpensive filler that causes allergies and is not well digested by dogs and cats. Actually corn provides a highly available source of complex carbohydrates and is an excellent source of linoleic acid (an essential fatty acid which is important for healthy skin), essential amino acids and fiber. Despite what some owners believe documented allergies to corn are quite rare. The ingredients that are most commonly associated to adverse reactions in dogs are beef, dairy products, and wheat; to lesser extent lamb, chicken egg, chicken, and soy. In cats beef, dairy products, and fish are the most common ingredients reported to cause adverse food reactions. Actual food allergies are estimated to be responsible for only 1% of skin disease. Food sensitivities often improve from a diet change, but this is not due to a food allergy. Rather other factors such as an increase in digestibility of the new diet, fat, or fiber content.

GRAIN FREE DIETS

There have been no studies indicating that grain free diets are superior or healthier for dogs and cats. While some grain free pet foods provide excellent nutrition, others may not. Potatoes and tapioca, which have a lower nutritive value than grains, are often used in place of grain in these diets. In fact, potatoes and tapioca have less protein and more sugar than corn or oats. Some owners mistakenly believe grains are common causes of food allergies, however as discussed above this is incorrect. Another common misconception is that dogs and cats eating foods containing grains are at increased risk of developing diabetes; there is no evidence to support this belief.

GLUTEN FREE DIETS

Corn gluten, which is found in some pet foods, is often blamed for GI issues by some owners and breeders as associated with what they perceive as celiac disease. Celiac disease is rare in dogs and has been reported primarily in Irish Setters. In reality, pets with celiac disease react to the protein in wheat, rye, and barley not corn gluten. Gluten is the concentrated protein from grain after all the starch is removed. Corn gluten is actually an excellent source of protein containing greater than 60% protein and is highly digestible.

RAW, BARF, FREEZE DRIED AND FROZEN DIETS

Commercial raw diets are available in many forms with frozen and freeze-dried being the most common types. These are often marketed as complete; others are intended to be fed with additional supplements. The Internet is also full of recipes for Bones and Raw Food or Biologically Appropriate Raw Food (BARF) homemade diets. Many of these advocate a formula of 60% raw meaty bones with the remainder of the diet consisting of green vegetables, eggs, milk, brewer's yeast, muscle meat and organ meats, yogurt, and sometimes with some grain and legumes. Meals are not expected to be balanced, but rather (in theory) to be balanced overall by varying the ingredients from one meal to the other. Articles explaining "how to feed raw" quite often advocate feeding muscle meats, organ meats, bones, and fish, with or without vegetables, fruits, grains, herbs, vitamins, and supplements. While some dedicated pet owners will follow these instructions in order to provide this "over time balanced approach" many others, however well intended, will not. Many owners will begin to make substitutions or not continue to dedicate the time and effort this method of feeding may require.

Studies on bacterial contamination of raw foods found that 20% to 35% of raw poultry and 80% of raw food diets for dogs tested positive for Salmonella spp, and 30% of stool samples from dogs fed these diets were positive for Salmonella spp. Raw pet food diets have also tested positive for Esherichia coli and Yersinis enterocolitica. Healthy pets may not show clinical signs of illness when ingesting these bacteria however immunocompromised, the very young and the very old are at significate risk. The pathogens found in raw meat also pose a significate health risk to young children, the elderly and immunocompromised people. It is therefore extremely important for anyone preparing these diets to practice proper hand washing and that all utensils, bowls, and surfaces used for the preparation of raw pet food diets are properly cleaned and disinfected. Raw meat may also contain parasites such as Echinococcus spp, Neospora caninum, Sarcocystis spp, Toxocara canis, Toxoplasma gondii, and Taenia spp. It is also important to note that the feces of pets consuming raw diets can contaminate the environment that may also pose health risks to people and other pets.

CLIENT COMMUNICATION AND EDUCATION

What if the diet a client is feeding isn't complete and balanced? Or if you do not agree with the type of diet a client has chosen to feed? How can you explain your concerns and educate a client that their perceptions are incorrect?

- Listen to the client's reasons for feeding the diet in question and their concerns with other diet options.
- Ask open-ended questions to ensure you understand their viewpoint. Be careful not to sound or appear judgmental.
- Avoid making disapproving comments, using a negative tone of voice, facial expressions, and body language. If you are perceived as judgmental, condescending, closed minded, or unwilling to listen and consider their point of view or beliefs, clients are likely to become defensive or simply unwilling to listen to your advice.
- Gently point out any misconceptions or false information offering references to the correct information.
- Some owners turn to alternative diets out of concern that their pet will not eat more conventional diet options. These owners may express concern that their pet not eating as the cause for not wanting to change diets.
- Some owners will not be interested in hearing your advice. Others may listen; perhaps discuss or debate the issue with you, but will not be willing to make any change to their feeding practices. Be respectful and do not take it personally; making a nutritional recommendation does not ensure the owner will follow your recommendation.
- As with any medical advice, document the patient's nutritional history and your nutritional recommendation in the patient record.

NUTRITIONAL RECOMMENDATION

Remember to make a nutritional recommendation for every patient every time they present to your hospital. Consider the patients age, weight, body condition and muscle condition scores, life stage, life style, and any health issues. Recommend the diet you believe is the best for that particular patient but remember you may also need to consider the owners perception of what is the best diet for their pet. When you have done your best to educate your client but they insist on feeding a diet you do not think is best, consider if there is a way to improve the patient's diet. Perhaps the client would agree to have a veterinary nutritionist evaluate the diet and if not nutritionally balanced suggest improvements. If they are feeding a grain-free diet that does not contain highly digestible ingredients perhaps you can help them find one that does. If corn free is all the client will consider, research the available options and recommend the one you believe is best suited for your patient. Suggest or provide a referral to a veterinary nutritionist for the client that wants to prepare a home cooked diet or a raw diet.

Remember the goal of a nutritional recommendation is to provide the patient with the best possible diet. If the client is not willing to follow your recommendation, what recommendations can you make to improve the patient's diet? Be a nutritional advocate for your patients!

REFERENCES

Hand, et. Al, Small Animal Clinical Nutrition, 5th edition

L. Freeman, et. al, Grain-free diets: An alternative option, but don't dismiss the grains, 2012

Case LP, et al.,eds. Canine and Feline Nutrition 3rd edition, 2011

Freeman LM, et. al, Current knowledge about the risks and benefits of raw meat-based diets for dogs and cats. JAVMA 243:1549-1558, 2013

A NUTRITIONAL APPROACH TO DIFICULT DIABETIC FELINES

By Ed Carlson, CVT, VTS (Nutrition)
Woburn, MA

INTRODUCTION

Nutrition plays a vital role in the heath of cats. This is particularly important for cats suffering from multiple diseases, perhaps none more important than cats diagnosed with diabetes mellitus. Experts agree that the most common type of feline diabetes mellitus (DM) (assumed to affect 80 to 90% of diabetic cats) is a type-2-like DM that is similar to type-2 diabetes in humans. Over the past several years the goal of treatment of diabetic cats has changed from simply controlling the disease to achieving non-insulin dependence (remission).

DIET TYPE

One popular nutritional recommendation for diabetic cats is a diet with increased fiber that is high in carbohydrates. The increased dietary fiber is beneficial to slowing gastric emptying which slows gastrointestinal glucose absorption, increasing insulin sensitivity, and improving the control of nutrient metabolism. This approach can be effective in the control of diabetes and the clinical signs of the disease in some cats.

Another popular nutritional recommendation for diabetic cats is a high protein/low carbohydrate diet. The theory behind this type of formulation is based on the concept that cats, being obligate carnivores, have a digestive system that is better designed to process protein than carbohydrates and starch.

A study comparing two canned diabetic diets, one high protein/low carbohydrate and the other with high fiber/moderate carbohydrates showed the rate of diabetic remission to be higher in the high protein diet compared to the higher fiber diet, >60% versus approximately 40% (Bennett et al, 2006). Other studies have since shown a low carbohydrate/high protein diet to be more effective in achieving and maintaining remission in diabetic cats. "In general, it is much easier to maintain near normal glycemia in insulin treated, diabetic cats when they are fed a high protein, low carbohydrate diet. Postprandial hyperglycemia is almost absent, and the average blood glucose level is reduced." (Encyclopedia of Feline Clinical Nutrition) One study of the use of a high-fiber diet showed improved glycemic control in cats compared with low-fiber diets. (Nelson RW, Scott-Moncrieff C, Feldman EC, et al.) However, many of the high-fiber diets formulated for cats also contain high levels of carbohydrates. It is therefore important to select a high fiber diet with moderate carbohydrates when using this method.

NON-INSULIN DEPENDENCE (REMISSION)

Research shows cats newly diagnosed with diabetes mellitus, when managed with a combination of appropriate insulin therapy <u>and appropriate dietary management,</u> can achieve remission, no longer requiring insulin therapy. Estimates of how often remission occurs vary greatly with some studies reporting remission in only 20% of felines, with other studies reporting as high as 80% to 90%. Remission is mostly likely to occur in the first year after diagnosis, however diabetic cats already receiving exogenous insulin therapy may benefit from and may achieve remission with nutritional intervention.

Diabetic cats already receiving insulin therapy often require a reduction in the amount of insulin they require when switched to a low carbohydrate/high protein diet. Cats already receiving insulin therapy, when changing from a high carbohydrate to a low carbohydrate diet, should have the initial insulin dose reduced by 30% to 50% to avoid hypoglycemia. It is therefore important to monitor these patients to ensure their blood glucose does not drop to a dangerous level.

Multiple studies have shown approximately 25% of diabetic cats that achieved remission reverted to overt diabetes again requiring exogenous insulin to control the disease. It is important to continue feeding cats that are in remission a low carbohydrate/high protein diet to increase their chances of remaining in remission.

OBESITY

Obese cats are four times more likely to develop diabetes than cats of normal body weight and about 60% of obese cats will become diabetic. (Hoenig, 2006, Encyclopedia of Feline Clinical Nutrition, Pibot, Biourge, Elliott) It is therefore important to begin educating owners of kittens and cats on the health risks linked to obesity, including the predisposition to developing diabetes mellitus, from their very first visit to the veterinary clinic. Body weight needs to be monitored at each visit. Developing a nutritional recommendation for every feline patient and remembering to educate owners to reduce the caloric intake at the time of neutering is important to reducing the risk of the animal developing DM.

Once a cat is diagnosed with diabetes mellitus, weight loss is more important than ever. Insulin resistance induced by obesity can be reversible once the animal returns to normal body weight. A low carbohydrate/high protein diet is the best option for many obese cats, not only to facilitate weight loss and maintain muscle mass, but to also provide them the best chance of achieving remission. Gradual weight loss in obese cats is crucial; loss of 0.5 to 1% of initial body weight per week is considered safe to avoid hepatic lipidosis. Many diabetic cats have poor body condition scores; high protein diets may assist in the prevention of hepatic lipidosis during weight loss, aid in replacement of muscle loss, and help to increase metabolism.

NUTRITIONAL RECCOMMENDATION

While a low carbohydrate/high protein diet may work very well with many diabetic cats, there are times when this type of diet is contraindicated. One common example of this is a diabetic cat with advanced renal dysfunction which may require protein and phosphorus restriction.

Avoid beginning to feed a veterinary diabetic diet to sick, hospitalized, or anorexic cats to avoid the chance of a food aversion. These patients should be offered a food they are likely to eat as encouraging them to eat is far more important than attempting to get them to eat a diet designed for cats with diabetes mellitus. Once healthy and eating readily, a change in diet should be made very gradually over 10 to 14 days.

There is a wide variety of veterinary formulas designed for diabetic cats available in both canned and dry form. Many veterinary health care professionals prefer to recommend canned diets for diabetic felines believing canned diets to be lower in carbohydrates then the dry formulas. The extrusion process used in many current veterinary diets has been improved in recent years allowing for dry diets with high protein and low carbohydrate content to be available. At this time, the author is unaware of any studies indicating canned veterinary diabetic diets to be superior to the dry versions. Although having had several feline patients achieve remission while eating canned diets, the author recommends the canned version for cats that will eat it.

Many diabetic cats do well being fed twice a day with insulin being injected just after meals. In fact, this method is commonly recommended by many veterinarians who instruct owners to give only half of their cats' normal insulin dose if the animal does not eat or only eats a minimal amount of food. However, many cats fed on an adlib basis prior to diagnosis may resist changing to being fed twice daily. Postprandial glucose levels in cats fed a high protein diet appear to increase only slightly as compared to cats fed a high carbohydrate diet. One unpublished study indicates that the timing of insulin injection had little effect on metabolic control leading some to believe the composition of the diet to be more important than the timing of meals. However, food must be available once insulin is administered to prevent life-threatening hypoglycemia. The author prefers the timed feeding method, if accepted by the patient. It is important to remember that some cats may do better when free fed. Whichever method is used, the patient weight should be closely monitored to avoid obesity.

CLIENT EDUCATION

In a survey of 610 owners of diabetic cats, 63.8% reported diet was NOT prescribed! 93% of cat owners, participating in this survey reported changing their cat's diet, however only 29.2% were feeding a diet prescribed by their veterinarian. After the cat was diagnosed with diabetes mellitus, 37% of the owners in this survey reported getting information on diabetes from the Internet, 34% from their veterinarian, and 11% from veterinary technicians. (Owner Experiences in Treating Dogs and Cats Diagnosed with Diabetes Mellitus in the United States. JAAHA, 2014).

Veterinary technicians provide client education on a variety of subjects; in fact veterinary technicians often spend more time with clients than the veterinarian. Veterinary technicians are ideally suited to educate clients on the importance of providing the proper nutrition to diabetic cats. Developing a hospital feline diabetic nutritional recommendation that all the veterinarians and technicians at the veterinary hospital agree on ensures that clients receive a consistent message from all team members. It will also allow the staff to develop client handouts explaining the hospital's medical and nutritional recommendations. Remember clients often only retain a small fraction of what is told to them which makes consistent messaging and written materials they can refer to at home extremely important.

In communicating information to cat owners be sure to allow ample time to explain all aspects of diabetes mellitus to owners of newly diagnosed cats. Provide owners with detailed discharge instructions and client handouts. Important points to cover orally and in writing include:

- How the disease affects cats
- The negative effect obesity has on glycemic control
- How to store, handle, and administer insulin
- The warning signs of hypoglycemia
- The important role nutrition plays in control and possible remission of the disease
- Emphasize that the role of the veterinary technician is to answer questions and provide support! Encourage owners to call or email you with any questions or concerns that they might have.

Call and/or email owners of newly diagnosed diabetic patients two days after discharge, again in a week, again in two weeks and continue frequent follow-up and communication. Be supportive and continue to educate. Remember owners will have many questions; if you are not answering their questions they might turn to questionable sources of information such as Doctor Google!

REFERENCES

Small Animal Clinical Nutrition, 5th Edition, Hand, Tharcher, Remillard, Roudebush, Novomy
Encyclopedia of Feline Clinical Nutrition, Pibot, Biourge, Elliott
AAHA Diabetes Management Guidelines for Dogs and Cats
Feline Diabetes Mellitus: Diagnosis, Treatment, and Monitoring, Compendum 12/2008, Lori Rios, DVM, PhD, DACVIM, Cynthia Ward, VMD, PhD, DACVIM
Nelson RW, Scott-Moncrieff C, Feldman EC, et al. Effect of dietary insoluble fiber on control of glycemia in cats with naturally acquired diabetes mellitus. JAVMA 2000;216(7):1082-1088.
Tanaka A, Inoue A,Takeguchi A, et al. Comparison of expression of glucokinase gene and activities of enzymes related to glucose metabolism in livers
Debra L. Zoran, DVM, PhDa, Jacquie S. Rand, BVSc, DVSc, MANZVS, DACVIM. Vet Clin Small Anim 43 (2013) 233–243
Owner Experiences in Treating Dogs and Cats Diagnosed with Diabetes Mellitus in the United States. 2014JAAHA. Karina P. Aptekmann, DVM, MS, PhD etal

NUTRITION AND ACID-BASE BALANCE

Ann Wortinger BIS, LVT, VTS (ECC, SAIM, Nutrition)
Allen Park, MI

Blood pH is a measurement of the hydrogen ion concentration, and is expressed as a –log and has no units of measure. An increase in hydrogen ions results in a decrease in pH, and a decrease in hydrogen ions, causes an increase in pH. (1) Buffering of hydrogen ions is done through an intricate interaction between the lungs, liver and kidneys. The primary buffering agents are blood proteins, bicarbonate/carbonic acid and phosphates.

Acid-base disturbances never are a primary problem, but always related to other disturbances within the body. The types of disturbances seen will depend on the primary cause and the body's ability to compensate.

Slight deviations in blood pH in either direction, whether it is acidosis or alkalosis, can denature body proteins leading to metabolic mayhem. When pH is outside the ideal range, enzymes cannot catalyze their reactions, hemoglobin cannot carry oxygen and functional proteins cannot function. Any metabolic process that utilizes enzymes or oxygen is affected. (1)

Production of Hydrogen Ions

Normal energy metabolism is the primary source of hydrogen ions in the body. All energy enters the body through the diet that is fed to the animal. (1) Hydrogen is not the only product of energy metabolism, other acids are also produced that must be neutralized by the body. We are familiar with a number of these, such as the short chain fatty acids produced with fiber digestion in the colon, acetoacetic acid, butyric acid and acetic acid. With fatty acid digestion we get the production of ketoacids that can be used for energy. When excess ketoacids are produced due to increase fat catabolism, our patients can become ketotic.

Hydrogen ion concentration can also determine the structure and function of proteins within the body, permeability of cell membranes, the distribution of electrolytes and structure of connective tissues. (2)

Carbohydrates

Carbohydrates have the general chemical structure of CH_20, repeating depending on the individual carbohydrate. Monosaccharides are composed of a single carbohydrate unit, disaccharides are composed of 2 units, and polysaccharides are composed of many units. Hydrogen and oxygen are found in the same ratio as water, and lead to the name of this class of nutrients, hydrogenated carbons or carbohydrates. (3)

The primary purpose of carbohydrates in the diet to provide materials for formation of other compounds, provide energy to the body and to serve as a source of stored energy in the form of fat. (3) Carbohydrates provide the carbon skeletons for formation of non-essential amino acids, when metabolized for energy production; they yield carbon dioxide and water, and through metabolic reactions they provide a source of heat for the body. (3) Carbon dioxide formed during energy metabolism dissolves in water found in the blood to form carbonic acid. Carbonic acid further dissociates to form hydrogen ions and bicarbonate. (1)

Proteins

Amino acids are structurally composed of carbon, hydrogen, oxygen, nitrogen and sometimes sulfur and phosphorus. The presence of the nitrogen side branch (amine group) is what makes a protein a protein. This amine group is important for synthesis of nonessential amino acids and other nitrogen containing compounds. (5)

All enzymes involved in metabolism and digestion are composed of proteins, as are the hormones that control the homeostatic process in the body and all structural components such as muscle and organs.

Blood cells and albumin that are responsible for transporting substances such as oxygen, carbon dioxide, drug and protein-bound substances through the blood are composed of protein. Blood proteins are important buffers in the acid-base system and are synthesized primarily in the liver from digested amino acids. (5)

Fats

Fats produce primarily carbon and hydrogen ions when digested; the number released is determined by the relative saturation of the carbon molecules with hydrogen. Saturated fats contain the highest number of hydrogen molecules (the carbon is saturated with hydrogen). Unsaturated fats contain double bonds, either singly or in multiples, this lowers the number of hydrogen molecules available for release during digestion. (4)

The primary purpose of fats in the diet is to provide energy, improve diet palatability and serve as a source for fat-soluble vitamins and the essential fatty acids in the omega 3 and omega 6 families.

Acid and Base Production

Generation of acids and bases during digestion is determined by oxidation of sulfur containing amino acids and organic acid anions of various alkali salts. (8) The intestines are directly involved in generation of food-derived acids and bases due to the differences in intestinal absorption rates for different nutrients. Carbohydrates are quickly absorbed through the intestinal tract, while proteins and minerals are more slowly absorbed. (8)

Bioavailability of specific nutrients can also affect renal acid loads and dietary intake. Studies in dogs have demonstrated that diets high in egg proteins may be more acidifying than diets utilizing vegetable based proteins. (6) Eggs are high in sulfur-containing amino acids such as methionine and cysteine. When sulfur is found in an amino acid, typically an oxygen molecule has been replaced with sulfur. This exchange lowers the electronegativity of the molecule making it less reactive and less hydrophobic. This in turn produces acids with higher pHs. Taurine, an essential amino acid for cats, is also a sulfur-containing amino acid. In dogs taurine is viewed as nonessential as it can be synthesized from methionine. In dogs and cats, taurine is the most abundant free amino acid in animal tissues. (7)

Lower protein diets may result in less acid production during digestion, especially if they contain reduced amount of sulfur-containing amino acids. (6) Veterinary therapeutic renal diets are usually formulated to be mildly alkalinizing through the addition of salts, specifically potassium citrate that are metabolized to bicarbonate. (6)

Metabolic acidosis increases catabolism and degradation of skeletal muscle proteins, disrupts intracellular metabolism and promotes dissolution of bone minerals. These cellular disruptions can exacerbate azotemia, contribute to loss of lean body mass and further promote renal osteodystrophy. Metabolic acidosis also increases the likelihood that hypokalemia will occur or can exacerbate preexisting hypokalemia as potassium moves out of the cells in response to metabolic acidosis. As this potassium reaches the renal threshold, it is permanently removed from the system by the kidneys. (9)

Buffering

Bicarbonate (an alkali) and carbonic acid (an acid) in the body fluids help to protect the body against changes in acidity by acting as buffers for other acids produced. Chemical reactions of the buffer system are under primary control of the lungs and kidneys, compromising the respiratory and metabolic compensation systems. (1)

The lungs control the concentration of carbonic acid by increasing or decreasing the respiratory rate. (1) With a decreased respiratory rate, carbon dioxide is retained in the system, causing a decrease in the pH. With increased respiratory rate, the animal is said to be 'blowing off CO_2', causing a decrease in carbon dioxide levels in the blood, increasing pH levels.

The kidneys are essential to regulation of acid-base balance. They have a key role in excretion of metabolically derived nonvolatile acids, such as sulfates, and hydrogen ions. (9) The kidneys also control the concentration of bicarbonate by either absorbing or excreting it, dependent whether the pH needs to be increased or decreased. The kidney's goal is to maintain the body's total acid burden at a nearly constant level. (1)

The liver is the source of the buffer proteins as well as the phosphate buffers. It cannot adjust carbon dioxide levels as the lungs can, or produce bicarbonate and dispose of hydrogen ions as the kidneys can, but nonetheless it is an integral part of the acid-base buffering system as protein is the largest volume of buffers found.

The skeleton represents a large but not endless alkaline reservoir due to the present of calcium and phosphorus salts. Even mild forms of long-term, low grade metabolic acidosis can impair skeletal architecture and structural stability. (2)

Dietary Disease Management

Diseases that affect any one or more of these systems pose the potential to deregulate the delicate balance that exists within the body. Loss of intestinal contents through vomiting or diarrhea can produce hypokalemia, with or without acidosis. Hypokalemia can result in metabolic alkalosis. (10) A highly digestible diet composed of good quality proteins can provide an acidifying effect to the body. (6)

Metabolic acidosis is usually evident with stage III and IV renal failure. Dietary alkalization therapy will usually improve the clinical signs of anorexia, lethargy, nausea, vomiting, muscle weakness and weight loss. (9) Many commercial pet foods in the United States are formulated to be acidifying. These diets should be discontinued in favor of therapeutic veterinary diets designed to help moderate the acidifying effects of the disease. (6)

Rarely is it appropriate to administer bicarbonate to adjust acidosis in the blood. Fluid resuscitation will re-expand the vascular space, improve peripheral perfusion and help alleviate lactic acidosis. If a metabolic alkalosis is present, Lactated ringer's solution would not be an appropriate fluid resuscitation choice, as it can further exacerbate the pH balance. (10)

Conclusion

Understanding the role of nutrition in disease management can have a very powerful effect in managing many of our more commonly encountered diseases. It is infinitely easier to manipulate a diet than to administer more medications to our patients ☺

Being able to anticipate what changes will occur to the acid-base balance due to disease processes, and adapt the diet to counter these is a constant and interesting challenge.

REFERENCES

1) Whitney E, Rolfes SR. Water and the Major Minerals. In Understanding Nutrition, 11th ed. Thomson Wadsworth, Belmont CA 2008. Pg 406-408
2) Vormann J, Remer T. Dietary, Metabolic, Physiologic and Disease-related aspects of Acid-Base Balance. In The Journal of Nutrition. 2006 pg 413S-414S. downloaded 8/17/14
3) Wortinger A. Protein and Amino Acids. In Nutrition for Veterinary Technicians and Nurses. Wiley-Blackwell, Ames IO 2007. Pg 25-28
4) Wortinger A. Carbohydrates. In Nutrition for Veterinary Technicians and Nurses. Wiley-Blackwell, Ames IO 2007. Pg 15-18
5) Wortinger A. Fats. In Nutrition for Veterinary Technicians and Nurses. Wiley-Blackwell, Ames IO 2007. Pg 19-20
6) Buffington T, Holloway C, Abood S. Clinical Dietetics. In Manual of Veterinary Dietetics. Elsevier, St Louis MO. 2004. Pg 104-105
7) Brosnan J, Brosnan M. The sulfur-containing Amino Acids: an overview. In Journal of Nutrition 2006. Pg 1636S-1640S. downloaded 8/29/14
8) Remer T. Influence of nutrition on acid-base balance- metabolic aspects. In European Journal of Nutrition. 2001 Oct;40(5): 214-220. Downloaded 8/29/14
9) Eliott D. Nutritional Management of Kidney Disease. In Applied Veterinary Clinical Nutrition. Eds Fascetti A, Delaney S. Wiley-Blackwell, Ames IO. 2012. Pg 256-257
10) Nelson RW, Couto CG. General Therapeutic Principles. In Small Animal Internal Medicine, 4th ed. Mosby Elsevier St Louis, MO. 2009. Pg 395-396

NUTRITIONAL MANAGEMENT OF PATIENTS WITH VOMITING AND DIARRHEA

Ann Wortinger BIS, LVT, VTS (ECC, SAIM, Nutrition)
Allen Park, MI

Introduction

One of the most important things to remember when dealing with a patient with vomiting or diarrhea, is that neither of these are in themselves diseases, but rather signs of disease.

Causes of Vomiting

Vomiting can be caused by any number of things, not all of which are related to the stomach. Vomiting can be induced with motion sickness, ingestion of emetogenic substances, gastrointestinal tract obstruction, abdominal infection, inflammation or extra-gastrointestinal diseases that stimulate the chemoreceptor trigger zone (2). The most important clinical sign of chronic gastric disease is vomiting (2).

Causes of Diarrhea

Acute diarrhea is most typically caused by dietary indiscretion or intolerance, intestinal parasites and infectious diseases such as *Salmonella* or *E.coli* (3). Chronic diarrhea is most commonly caused by intestinal disease, though other systemic diseases may affect intestinal function and can induce secondary hypersecretion or intestinal malabsorption (3). The most common cause of chronic vomiting and diarrhea in dogs and cats is caused by the constellation of diseases found under the heading of inflammatory bowel disease (4).

Acute diarrhea can often be treated symptomatically and will resolve within 2-3 days with minimal treatment. Chronic diarrhea is rarely self-limiting, with treatment being based on definitive diagnosis (2).

To Withhold Food or Not

For years our standard treatment for acute vomiting or diarrhea has been to withhold food for 12-48 hours. The belief has been that by removing food from the already upset GI tract we could allow it to clear itself of luminal contents that may be causing the problem, to help prevent mucosal cell abrasion by the GI contents, deprive the opportunistic luminal bacteria of nutrients preventing proliferation, preventing absorption of dietary antigens that may be present due to maldigestion and to permit restabilization of the brush border enzyme function (4, 5).The problem with this approach is that in almost all cases of enteritis, there is a decreased motility with delayed gastric emptying and reduced segmental contractions. Therefore fasting would not immediately provide any physical rest or clearing for the GI tract(1).

Evidence suggests that the prokinetic effect of feeding may decrease the vomiting response in some patients (1). We know that oral fasting alone can induce an intestinal insult, with mucosal atrophy occurring within 12 hours of a fasting state (1, 3).

Treatment Options

The primary goal for management of vomiting or diarrhea is to maintain delivery of nutrition to the GI tract, prevent nutrient deficiency and malnutrition (5).The most common first step is introduction of a highly digestible, low fat diet (2, 4). Many times we will recommend that the patient be started on small amounts of a bland diet. There is little evidence to support this recommendation, and the term 'bland' is poorly defined and vague. Diet selection should be based on the specific disease being treated, the area of the GI tract that is affected and the ability of the diet to promote tissue healing and maintain remission of clinical signs (5).

Our long-term goal in management of vomiting and diarrhea is to help repair the damaged GI tract lining, restore normal GI bacterial populations, to promote normal GI motility and function, to support the immune system and decrease GI tract inflammation (5).

Adverse Food Reactions and Food Responsive Diarrhea

Food allergies in dogs and cats are seen as being uncommon, though some clinicians believe that they are more likely to occur with diets that are consumed during periodsof acute gastritis (4). If an adverse food reaction is suspected, the current recommendation is to feed a highly digestible, single source novel protein diet. The ideal diagnostic and long-term management of suspected cases of food allergies is based on using novel proteins or protein hydrolysates and monitoring patient response (1).

Dietary management of these conditions can be difficult in practice because they are slow to respond, and the animals remain at risk for relapse if they gain access to the problem ingredient. A minimum trial of 3-4 weeks is recommended, though some clinicians will go for 3-4 months before making any diet changes (2).

Gluten, specifically wheat gluten has received a lot of press lately with the advent of new, 'grain-free' and 'gluten-free' diets for pets. Gluten is a plant protein that provides an excellent source of amino acids. On average, gluten contains 80-82% protein, with low dietary fiber and a digestibility of approximately 99%. Irish setters are the only breed of dog or cat in which gluten sensitivity has been documented. For these dogs a gluten-free diet would be recommended. Rice and corn do not contain gluten, where wheat, rye, barley and oats all contain gluten (2).

Protein

A protein source that is of high quality, easily digested and assimilated, and contains all the essential amino acids in the correct proportions would be recommended. Quality of a protein is measured by digestibility, as determined by feeding trials and by the biologic value of the protein. The protein source with the highest biologic value is eggs with a biologic value of 100.

Proteins in the diet are essential for secretion of hormones and enzymes required for digestion. These include the release of the pancreatic enzymes as well as the hormones insulin, gastrin, cholecystokinin (1, 3). Minimum dietary protein levels of 85-88% dry matter (DM) are recommended with a protein digestibility of > 92% (3).

Fats

While fats are the most calorically dense nutrient and inclusion in the diets would benefit those patients who are malnourished, absorption of fats through the intestinal lymphatic's may be impaired and could contribute to a postprandial increase of fluid influx into the intestinal tract (1, 5).The content is recommended to be ~ 11-15% DM or less of the diet (5).

The presence of the Ω-3 class of fatty acids has been found to be helpful in management of inflammatory responses in the body, including those occurring in the GI tract (5).

Carbohydrates and Fibers

Similar to proteins, carbohydrates should be easily digested and assimilated by the intestinal tract. The carbohydrate with the highest digestibility is cooked and blended white rice (5). Fibers are useful in the management of GI disease, as well as maintenance of GI health. The type of fiber used as well as the level found in the diet is important (5). Those fibers that result in production of the short-chain fatty acid (SCFA) butyrate are preferred. Butyrate is the preferred energy source for the colonocytes, obtaining ~ 70% of their energy from luminally derived SCFA (5). The addition of highly fermentable fibers that produce high levels of SCFAs can cause diarrhea and flatulence and may interfere with the digestion and absorption of other nutrients (5).

Conclusion

Diet selection will be ultimately determined by the cause of the signs as well as individual patient response. There are many commercially available therapeutic diets that offer options for management of intestinal diseases including low fat, novel protein and modified fiber diets. As there are no physical test finding, lab test results or historical facts that are predictive of which feeding method will be most successful, individual response and follow up with the veterinary staff is necessary for optimal success.

REFERENCES

1) Cave N. (2012) Nutritional Management of Gastrointestinal Diseases in Applied Veterinary Clinical Nutrition (eds Fascetti AJ, Delaney SJ). Wiley-Blackwell Ames, IO. pp 175-204

2) German A, Zentek J. (2006) The most common digestive diseases: the role of nutrition in Encyclopedia of Canine Clinical Nutrition (eds Pibot P, Biourge V, Elliott D). Royal Canin, Aimargues France EU. pp 103-123

3) Zentek J, Freiche V. (2008) Digestive diseases in cats: the role of nutrition in Encyclopedia of Feline Clinical Nutrition (eds Pibot P, Biourge V, Elliott D). Royal Canin, Aimargues France EU. pp 85-92

4) Buffington CA, Holloway C, Abood SK. (2004) Clinical Dietetics in Manual of Veterinary Dietetics. Elsevier Saunders, St. Louis MO. pp 82-89

5) Case L, Daristotle L, Hayek M, Raasch M. (2011) Nutritional management of gastrointestinal disease in Canine and Feline Nutrition 3rd ed. Mosby Elsevier, Maryland Heights MO. pp 455-472

EQUINE METABOLIC SYNDROME AND NUTRITION

Kara M. Burns, MS, MEd, LVT, VTS (Nutrition)
Wamego, KS

Equine metabolic syndrome (EMS) is one of the most common endocrine disorders affecting horses. EMS is a collection of endocrine and metabolic abnormalities associated with the development of laminitis and consists of three major components:

1. increased adiposity in specific locations (regional adiposity) or generally (obesity)
2. insulin resistance (IR) with hyperinsulinemia
3. predisposition to development of laminitis.

It is important to note that laminitis is a consequence of EMS and if management changes are made, laminitis can be avoided. EMS focuses on hyperinsulinemia. EMS also includes insulin resistance, increased fat depositis, hyperleptinemia, and hypertriglyceridemia. EMS has also been referred to as insulin resistance syndrome, peripheral Cushing's syndrome, and prelaminitic metabolic syndrome. Historically, clinical signs of EMS were thought to be a result of hypothyroidism. However, the low resting thyroid hormone concentrations noticed in some horses with EMS are considered a consequence of obesity.

EMS has been found to develop in young adult and middle-aged horses. Many horses with EMS exhibit a predisposition toward obesity which is first seen upon adulthood and is often intensified by environmental conditions that promote weight gain.

The potential for equine metabolic syndrome appears in all domesticated breeds of horses. Although higher prevalence is seen in ponies, Morgan Horses, Paso Finos, and Norwegian Fjord breeds the syndrome is recognized in many other breeds as well. Horses are mature when EMS is first recognized. However the onset of laminitis is typically a result of the conditions under which the horse has been kept. It is important that healthcare team members are aware that genetically predisposed horses allowed to become obese and graze on pasture with abundant grass, may develop laminitis at a young age. Conversely, other susceptible horses which are managed appropriately can avoid laminitis altogether.

It is imperative that veterinary technicians familiarize themselves with the physical characteristics of EMS: generalized obesity, regional adiposity, or both. Regional adiposity presents as cresty neck in horses. A negative correlation between neck circumference and insulin sensitivity has been found. Other manifestations of regional adiposity include abnormal adipose tissue deposits close to the tailhead, within the prepuce, or in indiscriminate areas beneath the skin. Many horses are found to be pasture fed when laminitis first develops. Also, laminitis episodes are more prevalent in the spring and fall. Other presenting issues/signs in EMS horses are infertility problems. In obese horses, pedunculated lipomas may lead to colic and should be cause for concern.

Pathophysiology

Nutritional management and exercise are two important modifying factors that can affect expression of the phenotype and should be considered in horses that are genetically predisposed to EMS. The healthcare team should recognize that an overfed, genetically predisposed horse is expected to express the EMS phenotype. Conversely, the same horse is a reduced risk for EMS if kept at an ideal body condition and weight, fed appropriately, and exercised regularly. Obesity is an important modifying factor because it can induce IR, and this raises insulin concentrations. The concept of pathologic fat must also be considered because adipose tissues secrete proinflammatory cytokines as obesity develops. This is accompanied by a shift in adipokine production, with increased leptin secretion and decreased adiponectin production. Detection of high leptin concentrations confirms that changes in fat metabolism have occurred, and hyperleptinemia is associated with insulin dysregulation. Hyperinsulinemia and hyperleptinemia might contribute to the so-called easy keeper metabolic state recognized in horses with EMS.

A key component of EMS is insulin dysregulation, and postprandial hyperinsulinemia may offer an explanation as to why affected horses grazing on pasture develop laminitis.[1] Hyperinsulinemia-induced IR is also a potential mechanism for insulin-induced laminitis. Insulin resistance within endothelial cells would be expected to promote vasoconstriction, alter normal blood flow dynamics, and reduce nutrient delivery to laminar tissues.[1]

Hyperinsulinemia can develop as a result of increased pancreatic beta cell secretion or as a result of slowed clearance from the blood. Incretin hormones may experience alterations in regulation from increased insulin secretion following a meal. Incretin hormones stimulate insulin secretion and slow gastric emptying as glucose concentrations rise after feeding, thus minimizing postprandial hyperglycemia. Alterations in incretin hormones may lead to the the development of postprandial hyperinsulinemia and gives reason to laminitis in horses grazing on pastures that are rich in simple sugars, starch, and protein.

Diagnostic testing recommendations have shifted with greater importance being placed on oral glucose testing. As stated, postprandial hyperinsulinemia is a chief concern, particularly when horses are consuming grass on pasture. Horses which may be genetically prone to postprandial hyperinsulinemia may also be inclined to become obese and have an increased risk for laminitis. Previous recommendations focused on fasting insulin concentrations and the combined glucose-insulin test (CGIT). Both remain useful tests and are still recommended for diagnosing IR, but they do not assess the insulin response to ingested sugars, which may be the first manifestation of insulin dysregulation in equids.[1]

In an attempt to detect postprandial hyperinsulinemia and identify horses with a greater risk for laminitis, the oral sugar test (OST) was introduced. The OST is performed through the administration of corn syrup by mouth thus providing a measured dose of glucose that induces transient hyperinsulinemia. This test is easier to administer as corn syrup is readily available from grocery stores and is easily administered with a dose syringe. The OST provides the team and the patient with relatively easy testing as well as small amounts of glucose needing to be administered. Owners may be reluctant to administer sugars to horses with suspected insulin dysregulation but the healthcare team should remind owners that horses do not develop laminitis as a result of the OST. In these cases, a two-step approach is recommended. The first step is to measure fasting insulin concentrations because horses with moderate or severe insulin dysregulation would be more likely to develop laminitis, and these animals have high fasting insulin concentrations. If fasting insulin concentrations are within reference range (<20 μU/mL by radioimmunoassay), the OST is performed. This approach is useful considering the heightened awareness of endocrinopathic laminitis. However, it is highly recommended that the second step be executed to identify mildly affected horses. The healthcare team can also assess postprandial hyperinsulinemia by collecting two blood samples 30 minutes apart within the 1.5- to 2.5-hour period after feeding.

Hypertriglyceridemia is recognized as a predictor of laminitis risk in ponies. The cutoff values of 57 and 94 mg/dL have been established from two studies including the same ponies. A lower cutoff value of 27 mg/dL has been proposed for horses on the basis of the current EMS genetics study, and breed-specific reference ranges are anticipated to be coming in the near future. Leptin can also be measured, and hyperleptinemia (>4 ng/mL) indicates abnormal fat tissues which secrete excessive amounts of leptin. Hyperleptinemia is associated with insulin dysregulation. Affected horses should be assessed by performing an OST, even if fasting insulin concentrations are within reference range.

Obesity

One of the most important parts of managing EMS is through obesity management. It is recognized that obesity is an exacerbating factor for insulin dysregulation in horses. Obesity is a disease condition which causes additional health risks, including increased production of inflammatory cytokines, equine hyperlipemia, infertility, and pedunculated lipomas. A weight loss program should be implemented in obese horses by limiting access to pasture, removing grain from the diet, and feeding hay in amounts equivalent to 1.5% of current body weight (15 lb hay for a 1000-lb horse) per day. Owners should be advised to provide a vitamin-mineral supplement. Those horses on hay-only diets may benefit from a ration balancer containing additional protein. If there is no improvement in body condition after 4 weeks, the amount of hay should be lowered to 1.5% of the ideal body weight per day, and after another 4 weeks without progress, the amount may be lowered to 1.25% of ideal body weight. In obese ponies, it may be necessary to reduce the amount of hay fed to 1% of body weight, but this is the minimal level recommended. It is preferable to provide feed in smaller amounts more frequently throughout the day. An exercise regimen should also be developed and instituted based on the individual horse.

Horses diagnosed with EMS should have limited access to pasture as oftentimes the obese body condition will be maintained unless grass intake is limited. Recommended strategies for limiting grass consumption include housing the obese horse with a companion in a small grass paddock that is one third to one half of an acre (~ square with 120- to 150-ft dimensions). Grazing muzzles have also become more prevalent in weight management programs and can be placed on the horse during turnout. Other approaches to limiting grass intake in horses:

- short (1-hour) turnout periods
- strip grazing with an electric fence
- housing in a dirt paddock.

Each patient should be individually assessed before formulating a management plan. A mildly affected horse with EMS that is obese as a result of overfeeding will generally respond well to energy restriction and can be returned to pasture with a muzzle in place. Grazing in the early morning is likely to be safer for horses with insulin dysregulation, except after a hard frost when grasses rapidly accumulate sugars. In contrast, a severely hyperinsulinemic horse that has recently developed laminitis should remain off pasture until insulin concentrations improve. Responses to management changes should then be reassessed and housing conditions adjusted over time.

Horses held off pasture should receive 1000 IU vitamin E per day as a supplement because access to green grass has been restricted. Patients that are laminitic should not be exercised until hoof structures have stabilized, but horses that are sound can be exercised regularly. Exercise recommendations must be adjusted to the individual horse and owner, but the goals are to increase energy consumption and improve insulin sensitivity. A greater amount of energy is expended when the exercise regimen is strenuous and sustained, so owners are encouraged to commit time to exercising their horse as frequently as possible. Daily exercise for a minimum of 1 hour, with trotting and cantering, work on hillsides, or time on a treadmill, is a goal for weight reduction programs. Even if the horse cannot be exercised at this level, turnout with other horses and walking in hand are recommended starting points.

Key Nutritional Factors

If OST insulin concentrations are significantly increased and postprandial hyperinsulinemia is a concern, hay with low nonstructural carbohydrate (NSC) content should be utilized. NSC include simple sugars, starch, and fructans, and the NSC content of the hay is calculated by taking the sum of water-soluble carbohydrate and starch percentages on a dry matter basis. Calculating NSC with this method includes fructans, which are less likely to contribute to postprandial insulinemic responses; therefore a different approach is to take the sum of the starch and ethanol-soluble carbohydrate percentages. When using this approach, hay with NSC content of less than 10% is often recommended. If the owner is able to acquire hay with marginally higher NSC content, it is recommended to soak the hay for 30 to 60 minutes in cold water. This will aid in lowering the sugar content before feeding. However, the amount of water-soluble carbohydrate lost through soaking will vary considerably among different batches of hay. Owners should be reminded that complete feeds and bagged forages are also available for horses with EMS.

Feeds that are lower in starches and sugar should be selected, and treats such as sugar cubes should be avoided altogether. Some horses with EMS have a normal or lean body condition and require additional calories. A diet consisting of hay, pelleted low-NSC feed, balanced vitamin and mineral supplement, and 0.5 cup vegetable oil (equal to 125 mL; contains approximately 100 g fat) twice daily can be recommended in these cases. If the owner cannot afford a commercial low-NSC feed, molasses-free beet pulp can be fed as an alternative, although it should be soaked before feeding to remove simple sugars and lower the risk for esophageal obstruction.

Recommendations for reducing the incidence of pasture-associated laminitis in genetically susceptible horses include limiting time on pasture or use of a grazing muzzle. These recommendations are particularly important when the grass is green and growing rapidly, first drying out at the beginning of a summer drought, rapidly growing after a heavy rain, or entering the dormant phase in winter. Pasture access should be limited when the grass is in any of these dynamic phases.

It is theorized that chronic obesity and insulin dysregulation increase the risk for pituitary pars intermedia dysfunction (PPID), so veterinary technicians should closely monitor these patients. Owners should look for signs such as haircoat changes, loss of muscle mass, and shifts in metabolism. Pituitary pars intermedia dysfunction should be investigated if a horse that previously battled obesity begins looking thinner and requires more energy for maintenance. Skeletal muscle atrophy and generalized hypertrichosis may not be evident initially, but shedding is sometimes delayed for a few weeks, and regional hypertrichosis can be observed. The development of PPID in a horse with underlying insulin dysregulation is significant because hyperadrenocorticism can induce IR and increase insulin concentrations. Higher insulin concentrations increase the risk for laminitis, and this problem is observed in middle-aged horses that develop PPID in addition to EMS. It is therefore important to diagnose and medically manage PPID in horses with a previous diagnosis of EMS.

Veterinary technicians should assist the veterinarian and the horse owner in maintaining the proper weight in horses. This is of utmost importance to the overall well-being of the animal, especially in reducing the risk of EMS and laminitis.

REFERENCES

1. Frank N. Equine Metabolic Syndrome. In *Robinson's Current Therapy in Equine Medicine, 7th ed.* Sprayberry KA and Robinson NE, eds. 2015. Elsevier, St. Louis
2. Frank N. Insulin Resistance and Equine Metabolic Syndrome. In Equine Internal Medicine, 3rd ed. Reed SM, Bayly WM, Sellon DC, eds. 2010. Elsevier, St. Louis.
3. Burns, KM. Nutrition. *Large Animal Internal Medicine*. Lien L, Loly S., Ferguson S., eds. 2014. Wiley-Blackwell. Ames, Iowa
4. P.J. Johnson: The equine metabolic syndrome peripheral Cushing's syndrome,. Vet Clin North Am Equine Pract. 18, 2002, 271–293.
5. N. Frank, S.B. Elliott, L.E. Brandt, et al.: Physical characteristics, blood hormone concentrations, and plasma lipid concentrations in obese horses with insulin resistance. J Am Vet Med Assoc. 228, 2006, 1383–1390.
6. K.H. Treiber, D.S. Kronfeld, T.M. Hess, et al.: Evaluation of genetic and metabolic predispositions and nutritional risk factors for pasture-associated laminitis in ponies. J Am Vet Med Assoc. 228, 2006, 1538–1545.
7. Vervuert I, Coenen M, Dahlhoff S, et al: Fructan concentrations in grass, silages and hay. In Proceedings of the nineteenth Equine Science Society symposium :309-310, 2005.
8. M. Slawik, A.J. Vidal-Puig: Lipotoxicity, overnutrition and energy metabolism in aging. Ageing Res Rev. 5, 2006, 144–164.
9. M.M. Vick, B.A. Murphy, D.R. Sessions, et al.: Effects of systemic inflammation on insulin sensitivity in horses and inflammatory cytokine expression in adipose tissue. Am J Vet Res. 69, 2008, 130–139.
10. C.F. Kearns, K.H. McKeever, V. Roegner, et al.: Adiponectin and leptin are related to fat mass in horses. Vet J. 172, 2006, 460–465.
11. K.E. Asplin, M.N. Sillence, C.C. Pollitt, et al.: Induction of laminitis by prolonged hyperinsulinaemia in clinically normal ponies. Vet J. 174, 2007, 530–535.
12. Carter RA, Treiber KH, Geor RJ, et al.: Prediction of incipient pasture-associated laminitis from hyperinsulinemia, hyperleptinemia, and generalized and localized obesity in a cohort of ponies. 7. 41, 2009, 171–178.
13. S.R. Bailey, N.J. Menzies-Gow, P.A. Harris, et al.: Effect of dietary fructans and dexamethasone administration on the insulin response of ponies predisposed to laminitis. J Am Vet Med Assoc. 231, 2007, 1365–1373.
14. Frank N. Equine metabolic Syndrome. In Large Animal Internal Medicine, 5th edition. Smith BP, ed. 2015. Elsevier, St. Louis.

NUTRITIONAL MANAGEMENT OF EQUINE OBESITY

Kara M. Burns, MS, MEd, LVT, VTS (Nutrition)
Wamego, KS

As is being seen in epidemic proportions in humans, cats, and dogs; obesity is also being diagnosed extensively in companion-animal horses. Because of this increase in prevalence in horses, the veterinary healthcare team has an even greater interest in weight loss and weight management. As with other species there are numerous risk factors that are associated with obesity. Obesity in horses has been associated with insulin resistance. Furthermore, obesity and insulin resistance have been linked to increased risk of laminitis (particularly the pasture-associated form of this disease). Also, overweight and obese horses have an increased risk for developmental orthopedic disease, osteoarthritis, and other skeletal and joint related conditions.

When implementing a weight management program, nutrition should be assessed and a nutritional recommendation made. Nutrition is the cornerstone to a weight management program and combined with exercise and owner education, the veterinary technician plays a key role in executing and managing the weight program designed for the obese horse.

OBESITY ASSESSMENT

Obesity can be defined as an expanded mass of adipose tissue in the body. Obesity is not just a cosmetic issue; it is a health issue.

The most common cause of obesity in horses is also the easiest to correct – overfeeding. The healthcare team is responsible for assessing the nutritional practices of the horse owner. First time horse owners, single-horse owners, and pony owners commonly feed their animals too much. Ponies seem to be particularly susceptible to obesity, perhaps because their size renders them more easily overfed.

The most widely utilized method for assessing overweight and obesity in horses is the evaluation of body condition using a 9-point scale. Body-condition scoring (BCS) in horses is a subjective method of estimating a horses' fat stores and is important tool to help assess the nutritional status of the patient. BCS measures subcutaneous fat deposition in six areas:

1. crest of the neck
2. withers
3. behind the shoulder
4. over the ribs
5. along the back
6. around the tailhead

Fat deposits may be asymmetric in distribution, thus the need to evaluate all six areas of the body on both sides. A score between 1 and 9 is assigned, with 1 indicating emaciation and 9 indicating extreme obesity.

The veterinary healthcare team should perform BCS and body weight evaluation regularly. Likewise, the owner should be educated as to how to perform BCS on their horse. Additional weight in any animal increases the risk for a variety of disease conditions and can exacerbate existing disease conditions. Body weight is measured by use of a calibrated large-animal scale. However, weighing by scale is not always possible. In these instances weight can be assessed by utilizing weight tapes. Weight tapes estimate the horses' weight using accurate body measurements of girth circumference and body length.

$$\text{Lb/body weight} = [\text{heart girth(in)}]^2 \times \text{length (in)}/330$$
$$\text{Kg/body weight} = [\text{heart girth(cm)}]^2 \times \text{length (cm)}/11800$$

The formula above is considered most accurate when used on mature horses with a BCS in the range of 5 to 7. The healthcare team should also be advised that weight estimates based on heart-girth measurements do not account for variation caused by hydration state, pregnancy, and the weight of intestinal contents.

Definition and Prevalence of Obesity

Although a general definition or body condition score of obesity in horses and ponies does not truly exist, it has been accepted that horses with a BCS of 8 (fat) or 9 (extremely fat) are defined as obese. Animals with a BCS of 7 may be considered overweight if not obese.

Obesity in horses and ponies as in other species is caused simply by too many calories in; not enough calories expended. Horses today are likely to be kept in stalls or small pens for much of the day, and may not require more than maintenance energy intake. Still, these horses may be fed more than the maintenance energy requirement, especially if grains, sweet feeds, and other high caloric density foods are added to the diet without calculating amount to feed for the individual horse. Obesity also may be a problematic in pasture fed horses and ponies given unlimited access to pasture. This is especially true in the spring and fall seasons, when pasture forage is plentiful and rich in energy.

Equine Metabolic Syndrome

Obesity in horses is considered a risk factor for laminitis. One theory linking obesity with laminitis attributes the increased weight of the horse to increased load on the feet. Nevertheless, the increased risk of laminitis in obese horses is undoubtedly related to the attendant insulin resistance. Associations among obesity, insulin resistance, and laminitis have been observed in horses, and has been termed equine metabolic syndrome. Other effects related to obesity include:

- impaired thermoregulation in hot weather

- reduced athletic performance
- increased risk of joint injuries

Obesity and Inflammation

Overall obesity is considered to be a chronic inflammatory state, in part mediated by increased macrophage infiltration of adipose tissue. Recent studies in horses have also associated obesity with increased expression of inflammatory cytokines. Adipokines secreted by adipocytes and macrophages in adipose tissue include tumor necrosis factor-α (TNF-α), interleukins, leptin, adiponectin, resistin, and plasminogen-activator inhibitor type 1. Leptin provides information to the brain regarding the availability of fat stores in the body, promoting satiety and reduction in food intake when energy balance is positive or fat stores are plentiful. In horses, leptin concentrations have been found to vary with season, level of fitness, and gender. Also, a positive association between BCS and leptin has been documented.

Weight Management Programs With Horses

Regardless of the species, weight management is a simple approach - eat less and exercise more. The healthcare team, specifically the veterinary technician, plays an integral part in the weight management of horses and in developing a weight management program with the veterinarian and the owner. Weight management in horses involves the following:

- Nutritional Evaluation –
    - Evaluation of the current feeding program
    - Evaluation of housing
    - Evaluation of what feed is being provided (including supplementary feed, hay, pasture quality, treats, time allowed for grazing)
    - Evaluation of the amount of feed being given

- Assessment of exercise:
    - How many hours per week is involved in physical activity?
    - Information on the activity level will form the basis for development of recommendations for physical activity.

- Setting realistic weight loss goals and regularly monitoring progress. An effective weight loss regimen should result in the loss of around 1 unit of BCS (approximately 20 to 25 kg for a light-breed horse) over a 4- to 6-week period. Body weight (via scale or weight tape) and BCS should be assessed regularly, at least every 2 to 4 weeks, during the weight reduction program to insure progress is monitored and the program amended as required.

- Once the target weight and BCS have been reached a weight management program should be implemented. The healthcare team should assess body weight and BCS monthly to ensure that the feeding program is appropriate to the current level of physical activity and other environmental influences (such as ambient conditions) on energy requirements.

Key Nutritional Factors

The first step in the managing overweight or obese horses is calorie restriction. Trial and reevaluation will be necessary to achieve the goal weight and BCS in individual horses. The amount of food being fed and the make-up of the food are two important points for the healthcare team to consider. In those pasture fed horses, it may be necessary to move the horse from pasture feeding to stalls to monitor dietary intake.

Restrictive grazing may be allowed. The following techniques allow the horse to graze while minimizing forage intake:

- grazing muzzles (attached to a breakaway halter)
- strip grazing behind other horses
- mowing the pasture and removing clippings before allowing access
- putting a deep layer of wood chips over a small paddock

Some horses will not tolerate a grazing muzzle and for those that do tolerate the muzzle it is imperative that the horse has the ability to consume water.

Food for overweight and obese horses should be higher in fiber and lower in nonstructural carbohydrate (NSC) (starches, sugars, and fructans). Horses at maintenance require approximately 2.0% to 2.5% of their body weight as forage or forage plus supplement to meet daily nutrient requirements. To help restrict calories and encourage weight loss, grain and other concentrated sources of calories (such as commercial sweet feeds and feeds containing added fats) in the diet should be reduced in overweight horses or totally eliminated in obese horses.

The most crucial nutrient in any species is water. All animals should have access to fresh, clean water at all times. A 10% loss in total body water causes serious illness, while a 15% loss may result in death! Horses on average will drink 6 gallons of water per day. In conditions of extreme heat or stress this number increases to 25 gallons per day. Technicians should remind owners that the more grain their horse eats the more water the horse will need.

Treats, whether carrots, apples, or commercially purchased treats, should also be eliminated or the amount calculated into the daily caloric allowance. Forage, in the form of hay or hay substitute (chop, chaff, or haylage), should be the primary or sole energy-providing component of the ration. In some areas, forage-based, low-calorie feeds complete with vitamins and minerals are available commercially; this type of diet offers convenience for the owner and may be used as a substitute for hay or fed as a component of the ration along with hay. (Geor, 2010)

The veterinary technician should be fully involved in the weight management plan and educate owners on the plan and the risks to the horse should the weight management program not work. During the first month, hay should be provided at approximately 2% of current body weight per day with clients instructed to weigh the ration. These steps alone or with an increase in physical activity may result in some weight loss. However, with others further reductions to hay at 1.5% of current body weight, and then 1.0 to 1.5% of target body weight, maybe required. Continuous assessment and reassessment are crucial throughout the weight management program. It is not recommended to decrease forage provision below 1.0% of body weight as this may result in unwanted behaviors such as wood chewing, ingestion of bedding, and coprophagy. It is also recommended that the daily food amount calculated be divided into multiple feedings (three to four) per day. (Geor and Harris, 2009)

Grass hay that is mature (visible seed heads and a high stem-to-leaf ratio), is higher in fiber and lower in energy and NSC; and is appropriate forage for obese horses. Alfalfa hay or legumes are less desirable due to the higher energy and NSC content. Haylage is generally lower in NSC content when compared with hay, but is more palatable. The increase in palatability may cause higher total NSC intake. An NSC content less than 10% is recommended. Otherwise, the water-soluble carbohydrate content (sugars and fructans) can be considerably reduced by soaking the hay in clean water for 20 to 30 minutes before feeding. It is recommended to avoid feeding significant amounts of poorly digestible, highly silicated forages as it has been suggested that this may increase the risk of impaction colic.

When feeding for weight loss, forage-only diets do not provide adequate protein, minerals, or vitamins. Consequently this lack of protein may result in loss of muscle mass rather than loss of fat. For those horses it is recommended to supplement the forage with a low calorie commercial ration balancer product containing sources of high-quality protein and a mixture of vitamins and minerals to balance the low vitamin E, vitamin, copper, zinc, selenium, and other minerals typically found in mature grass hays. It is important to remind the owners that products such as these are designed to be fed in small quantities (e.g., 0.5 to 1.0 kg daily). If needed this supplement can be mixed with chaff (hay chop) to increase the size of the meal and extend feeding time, which may alleviate boredom in animals provided a restricted diet.

When discussing weight management/weight reduction programs with horse owners, several additional factors should be considered as approaches to extend feeding time and relieve boredom in the face of limited feed provision, the need for individual rather than group feeding, potential need for a change in the type of bedding, and potential use of pharmacologic agents or nutritional supplements that are purported to enhance weight loss or mitigate comorbidities such as insulin resistance.

Hay nets with small openings or double hay nets may be used to extend feeding time in horses. Mixing the feed with chaff or chopped straw is also a way to increase the size of the meal and encourage the horse to take longer to ingest the meal. For horses that may be housed with other horses, the obese horse may need to be separated to watch the intake of the obese horse. As mentioned above those horses in stalls may turn to ingesting their bedding when their daily calories are restricted. This behavior substantially increases daily caloric intake and potentially increases the risk of colic. In this situation, use of shavings, paper, or other nonstraw alternatives for bedding is recommended.

Veterinary technicians should assist the veterinarian and the horse owner in maintaining the proper weight in horses. This is of utmost importance to the overall well-being of the animal. To make this happen a weight management program should be designed and implemented for the individual horse. An effective weight management program incorporated by the veterinary healthcare team should include the following:

- Performing a nutritional history
- Calculating amount to feed based on weight and BCS and target weight
- Begin a feeding regimen focused on the key nutritional factors above for weight management
- Restricting access to pasture, particularly during the growing seasons.
- Increasing physical activity to promote energy expenditure in horses without musculoskeletal injury and lameness.
- Ensuring a gradual transition to new diet
- Follow up and monitor the horse's body weight and body condition on a regular basis.

REFERENCES/SUGGESTED READING

Becvarova I, Pleasant RS, Thatcher CD. Clinical assessment of nutritional status and feeding programs in horses. *Veterinary Clinics of North America, Equine Practice. Clinical Nutrition*. Vol. 25, No. 1, April, 2009.

Burns, KM. Nutrition. *Large Animal Internal Medicine*. Lien L, Loly S., Ferguson S., eds. 2014. Wiley-Blackwell. Ames, Iowa

Carter RA, Treiber KH, Geor RJ, et al.: Prediction of incipient pasture-associated laminitis from hyperinsulinemia, hyperleptinemia, and generalized and localized obesity in a cohort of ponies. *7*. 41, 2009, 171–178.

Foreman JH. Clinical approach to commonly encountered problems. *Equine Internal Medicine, 3rd Edition*. Reed, et al, eds. W.B. Saunders Company, 2010.

Geor RJ, Harris P. Dietary Management of the Obese Horse. In *Current Therapy in Equine Medicine*, 6th Edition. Robinson, Sprayberry, eds. W.B. Saunders Company, 2009.

Geor RJ. Aspects of clinical nutrition. *Equine Internal Medicine, 3rd Edition*. Reed, et al, eds. W.B. Saunders Company, 2010.

Henneke DR, Potter GD, Krieder JL, et al.: Relationship between body condition score, physical measurements and body fat percentage in mares. *Equine Veterinary Journal*. 15, 1983, 371–372.

Treiber KH, Kronfeld DS, Hess TM, et al.: Evaluation of genetic and metabolic predispositions and nutritional risk factors for pasture-associated laminitis in ponies. J Am Vet Med Assoc. 228, 2006, 1538–1545.

FORAGE QUALITY & BALANCING FORAGE RATIONS FOR HORSES

Megan Shepherd, DVM, PhD, DACVN
Blacksburg, VA

INTRODUCTION

The American College of Veterinary Nutrition (ACVN)'s circle of nutrition highlights the three main components of a nutritional assessment: patient assessment, diet/ration assessment, and feeding management assessment. To the latter point, how you feed is as important as what you feed. For this talk I will touch on the patient and the feeding management, but will focus on diet/ration (what you feed), specifically forage.

The two main parts of patient assessment are body weight and body condition score. Body condition is important for determining whether the current body weight is appropriate or not and thus will influence what and how you feed. You also want to determine the patient's physiologic status (signalment, life-stage) and if a clinical abnormality is present. Some clinical abnormalities are diet responsive (i.e. obesity), some aren't. For the purpose of the current topic, I will focus on feeding the healthy adult horses and will touch on growth as a representation of a more demanding life-stage.

IMPORTANCE OF FORAGE

Horses *generally* consume water, forage, concentrates, and supplements (minerals, vitamins, nutraceuticals, etc.). Water is always the most important ingredient in a horse's ration. Forage is the next most important and makes up 50-100% of a horse's diet/ration. The horse evolved consuming forage and still thrive on a forage-based diet thanks to their hindgut (cecum & colon) microbes (bacteria, fungi). Forage is a good source of energy and several essential nutrients (i.e. protein, calcium, phosphorus, and vitamin E). Forage is an important source of carbohydrates, especially fiber. Cellulose, an insoluble fiber, is the most common carbohydrate in forage and is slowly fermented by equine hindgut microbes. Slow fermentation helps to keep hindgut lumen pH near neutral, which maintains a symbiotic microbial environment. Products from microbial fermentation of carbohydrates, namely volatile fatty acids, include acetate, propionate, and butyrate. Butyrate is important for colonocyte health and thus hindgut health. Furthermore, there are physical effects of fiber that benefit the gastrointestinal tract.

FORAGES & FORAGE QUALITY

Forage includes both grasses and legumes. Forage comes fresh, dried, and processed forms, which all serve a unique purpose. I am not going to provide an in-depth review of forage types. However, for the context of my background, the common grasses fed in my region (Southwest Virginia) are cool-season grasses and include Tall Fescue (*Festuca arundinacea*), Orchard Grass (*Dactylis glomerata*) and Timothy (*Phleum pratense*). Tall Fescue is a dominant forage across the country; the latter two are more popular to grow and feed as hay. Bermudagrass (*Cynodon dactylon*) is a popular warm-season grass and is fed in the coastal regions of Virginia. The common legumes fed in Virginia are cool-season legumes and include Alfalfa (*Medicago sativa*), as hay, and clover (*Trifolium repens* [white]/*pretense* [red]), in pasture or mixed hay.

There are *general* differences between grasses and legumes with respect to nutritional quality. Legumes are *generally* more energy dense, higher in protein, and higher in calcium. However, it is important to consider the stage of plant maturity at which the pasture forage is consumed on pasture or cut to make hay. An immature grass could have a higher energy density and higher protein content than a more mature legume.

Stage of plant maturity is the single most influential factor on forage quality. There are numerous other factors (i.e. sun, rain, temperature, soil fertility) that also influence forage growth and thus quality. Forage quality can be estimated using olfactory, visual and tactile senses such as assessing smell (i.e. moldy, fresh), leaf:stem ration, presence of seed heads, color and texture. A hay with a low leaf:stem ratio and presence of seed heads is more mature and likely be of poorer quality. Conversely, hay with a high leaf to stem ratio will be softer and have greater nutritional value (i.e. protein, increased digestibility). Hay should also be free of mold (black spots) and foreign debris (sticks, leaves). Hay should be green in color and have a fresh/pleasant aroma (not moldy).

The most accurate way to determine forage quality is to send a representative sample off for chemical nutrient analysis (at the end of the document). When assessing forage quality, you should consider the parts of the plant cell, the cell wall and the cell contents. The cell wall serves as the skeleton for the plant and as forage matures, the cell wall to cell content ratio increases. As plant biomass increases, quality decreases. A forage quality analysis should include analyses of crude protein (CP), neutral detergent fiber (NDF), and acid detergent fiber (ADF). Neutral detergent fiber and ADF generally represent the cell wall component of the forage. More specifically, NDF represents cellulose, lignin, and hemicellulose; this fraction is used as a positive indicator of forage bulk. Therefore, the higher the NDF the more bulky the hay and the less a horse is willing to consume. Acid detergent fiber (ADF) represents cellulose and lignin; this fraction is used as a negative indicator of digestibility (total tract digestibility, including microbial fermentation of fiber). Lignin is not digested/fermented by microbes and negatively impacts the digestibility/fermentability of the forage; therefore, higher ADF forage is less digestible than low ADF forage. As forage matures, CP generally decreases while NDF and ADF increase. *General* guidelines for dietary protein are 8% CP in the total ration on a dry matter basis (DM/DMB). I generally like to see NDF less than 70% DMB for horses. I generally like to see ADF less than 40% DMB for horses.

Energy density of grass hays vary, in *GENERAL*, from 0.8-1.0 Mcal per pound on an as fed basis (AF) [6]; therefore, if fed at equal weight (i.e. 2% BW per day in total dry matter [DM = 100% of a feed - % moisture]) a 1100lb horse fed 24 pounds AF of hay per day could consume anywhere from 17.6 to 22.0 Mcal (17,600 – 22,000 kcal) per day! Furthermore, voluntary dry matter intake of various forages by adult horses ranges from 1.4-5.4 % BW per day with consumption typically greatest for legume-based forages and lowest for high NDF forages [1-5]. Therefore, adult maintenance horses (i.e. not growing, breeding, pregnant, lactating, or working) could quite easily over consume calories relative to requirements when fed a moderate to good quality forage free choice. Conversely,

a forage of poor quality (> 70% NDF DMB) could be offered free choice, but even an adult horse at maintenance may become satiated before he/she meets his/her energy requirements.
If a horse is to consume all forage in the form of fresh pasture, ensure that the pasture contains enough forage (grass +/- legume), meaning that leaf height is no less than 3-4 inches and there is at least a ½ acre to graze per horse. When pasture forage is not adequate (especially in the winter months), the hay fed should be of adequate quality for that horse's physiologic status.

FORAGE RATIONS

The focus of a horse's ration must be on what forage is available and the quality of that forage. The rest of the ration should be built around the forage. How do you know if the forage you are feeding is appropriate and appropriately fed? Check body condition score monthly, at least spanning any diet change period including seasonal changes in pasture.

Up to 50% of a horse's diet may include concentrates (i.e. grains and commercial feeds) if the forage does not meet the horse's energy and protein requirements alone. In *general*, if horses are receiving fortified concentrates (i.e. commercial concentrates that contain vitamins and minerals), they may not need an additional source of trace minerals. *General* rule of thumb is that horses would need to consume at least 0.5% of body weight per day (i.e. 5 pounds of concentrate per day for a 1,000lb horse) in a fortified concentrate in order to not need further trace mineral supplementation. HOWEVER, keep in mind, this is a *GENERAL* rule of thumb and will not apply across all concentrates since different feeds have different concentrations of vitamins and minerals.

In addition to calories, it is important to ensure daily protein requirements are being met [3] (Table 1). If a forage is used exclusively to meet the horse's energy and protein requirements, then the forage should contain adequate protein; see guidelines in Table 1. In *general*, it is advisable to feed forages/feeds with at least 8% crude protein DMB. If you chose to feed a forage that does not meet your protein goal (Table 1), you will need to add a more concentrated source of protein to the diet (i.e. alfalfa, balancer pellet, life-stage feed).

Table 1: Total dry matter intake (DMI) *provided* and CP content *required* in the forage when feeding to meet energy[a] and protein[b] requirements for an adult horse.

Mcal/lb DM	DMI (%BW)	% Crude P DMB
0.8	1.7	7.4
0.9	1.5	8.4
1.0	1.4	9.0

[a] Minimum DE (digestible energy) requirements[3] = 30.3 kcal/BW_{kg}; for example, a 1,100lb horses would need 15,150 kcal/day (15 Mcal/day).
[b] Average CP (crude protein)[3] = 1.26g/ BW_{kg} ; for example, a 1,100lb horse would need 630 g CP/day.

When energy and protein requirements are met with forage alone, I like to balance the forage-based ration with a calorie-free vitamin-mineral supplement (~22% NaCl, ~16ppm Se). White and red (i.e. trace) salt blocks are a good source of sodium chloride (95-99% NaCl) and may be fed free choice near a clean water source. A horse would have to over consume sodium chloride in order to consume adequate copper, selenium, and zinc from these trace mineral blocks. Furthermore, not all red salt blocks contain selenium. If you are in a selenium deficient area, like Virginia, mineral sources should contain selenium. Vitamin-mineral supplements come in granular and block forms; granular forms are easier to quantify. When energy, but not protein requirements are being met by forage alone, I like to compliment the forage with a commercial ration balancer pellet. Ration balancer pellets are a concentrated source of protein, vitamins, and minerals, are meant to be fed in small quantities, and would replace the need for a vitamin-mineral supplement.

For more demanding life-stages (i.e. growth, late gestation, lactation, work), the reliance on concentrates increases. Don't forget, however, that forage quality determines how much concentrate is needed.

I will present a ration example for adult maintenance and for growth to highlight how to balance forage rations for horses.

When in doubt, contact our Clinical Nutrition Consultation Service
vetnutrition@vt.edu; 540-231-4621; 540-231-6448 (fax)
http://www.vetmed.vt.edu/vth/nutrition.asp

SUBMITTING A FORAGE SAMPLE FOR NUTRIENT ANALYSIS

To take a sample, sample for forage, follow instructions on the "Taking a Hay Sample" form http://www.equi-analytical.com/TakingASample/TakingAHaySample.pdf. Note, that each BATCH of forage (i.e. first cutting orchard grass or second cutting orchard grass, or alfalfa, etc.) must be sampled and nutritionally analyzed separately. Dried forage (hay) is most commonly sampled, but fresh forage (i.e. grass pasture) can be sampled and analyzed as well. For hay (grass or alfalfa hay), 10 representative bales (or 10%) from EACH BATCH should be sampled with a hay corer. You can either borrow a hay corer from your local extension office or ask to have an extension agent come to your farm to sample the hay. To contact your local extension agent, search online your state with "ext". For example, to get a list of Virginia extension offices, you would search "Virginia ext", select the "Virginia Cooperative Extension" site, then select "Local Offices" to search offices by county. If you would like for us to contact your local extension agent, we are happy to do so.

To submit the sample for analysis, complete the "Sample Information" form http://www.equi-analytical.com/Services/ForageInfoEqui.pdf. I prefer the "TRAINER" analysis ($54). Include pertinent information on the "Sample

Information" sheet to properly identify the sample. Include whether it is hay or pasture, which cutting, and what type of forage as specific as possible (i.e. timothy versus cool-season grass). Send samples to Equi-Analytical.

Equi-analytical Laboratories
730 Warren Road
Ithaca, New York 14850
1-877-819-4110
service@equi-analytical.com
http://www.equi-analytical.com/default.htm

There are other forage labs out there, but Equi-analytical is the one I am most familiar with.

REFERENCES

[1]Crozier et al., **J Anim Sci** 1997 75:1651; [2]Edouard et al., **Animal** 2008 2:1526; [3]NRC, 2007; [4]Smith et al., **J Agr Sci** 2007 145:273; [5]Staniar et al., **J Anim Sci** 2010 88:3296; [6]Equi-Analytical 2004

CENTRAL LINES: PLACEMENT, INDICATIONS, AND USES

Jordin Karalunas LVT, VTS (ECC)

Central lines have become a necessity in the majority of hospitalized patients in the critical care setting. Despite the many uses they provide, there is an increase in patient compliance and comfort when used appropriately. Central lines are not just for fluid administration; they can provide us with additional patient parameters which affect patient care and treatment plans.

Indications

Total Parenteral Nutrition, or TPN, must have a dedicated route for administration. Due to the risk for bacterial contamination, and vaso-irritation/phlebitis due to its' high osmolality, TPN should always be administered through a central line. Partial/Peripheral Parenteral Nutrition, or PPN such as Procalamine, should also be administered via central line. PPN is also in jeopardy of bacterial contamination, in addition to vaso-iritation due to its high osmolality. In addition to nutrition, multi-lumen central lines allow incompatible fluids, additives, blood components and drugs to be given simultaneously without interference. Patients who require multiple blood samplings, such as a DKA (diabetic ketoacidosis), will also benefit from a central line, as blood sampling can be performed using a three syringe technique. This decreases frequency and discomfort from multiple peripheral venipunctures. Central venous pressures, or CVP, can be obtained through a central line using a manometer. Central venous pressures give us an estimate of right atrial blood pressure. This tells us how well patients are tolerating fluid therapy through CVP trends, and if there is an increased risk for fluid overload. Cardiac patients, trauma patients, septic and acute renal failure patients, and those requiring high volume diuresis are examples of candidates for CVP monitoring. Also, patients who are fractious often benefit from central lines. Central lines allow access to the line without being in a compromised position, such as at the animals' feet. Instead it allows catheter management and blood sampling without restraint, and the discomfort of phlebotomy.

Contraindications

It is contraindicated to place central lines in patients who are coagulopathic, are hypercoagulopathic (such as Immune Mediated Hemolytic Anemia), or where skin infection is evident at the site of insertion. It is important to note which disease processes cause an increased risk of thrombosis prior to placing a central line. Jugular central lines are contraindicated in patients who have suffered head or neck trauma, such as A-A luxation or traumatic brain injury. Avoid jugular central lines in patients who are in respiratory distress, have laryngeal paralysis, cerebral edema or cranial disease. It may be beneficial in a select group of these patients to place a central line via the femoral vein into the caudal vena cava, instead of the jugular.

Products

There are a few different methods and products available for placing central lines.

Peel-Away Sheath Technique: This technique requires the peel-away sheath introducer/catheter. It is an over the needle sheath that is gently peeled away, after inserting the catheter through the sheath. (ex: MILA cath)

Guide Wire Seldinger Technique: This method incorporates the use of a needle or peripheral catheter, guide wire, vasodilator, and central line catheter. It involves the use of a guide wire to pass the vasodilator and central line into the vessel. (ex: Aarrow)

Through the Needle Catheter: Requires a through the needle catheter unit. After placing the needle in the vessel, the catheter is fed through the needle and the needle is backed out and covered with a guard. (ex: Intracath)

The silicone/polyurethane material that constructs central lines will allow long-term placement and are less thrombogenic, as opposed to peripheral catheters which are made up of polypropylene/polyethylene/polyvinylchloride.

Placement

Regardless of placement technique, it is imperative the process be treated in a sterile manner. This requires sterile gloves, a sterile central line kit, hair removal, three scrub prep, and patient draping.

Measurement: Because central lines are available in a variety of sizes and lengths, measurement is crucial when placing a central line. For jugular central lines, accurate measurement can be achieved by measuring from the desired site of insertion to the caudal aspect of the shoulder. Ideal placement is in the cranial vena cava within the thoracic inlet, cranial to the right atrium. This allows accurate CVP monitoring. Femoral placed central lines should be long enough to enter the caudal vena cava to allow CVP monitoring.

Anesthesia: Most critical patients will not require chemical intervention for central line placement. However, more alert and stable patients may require some form of injectable sedation, or combination of sedation and pain medication (Neuroleptanalgesia). Commonly used medications for placement are: Propofol, Fentanyl, torbugesic, butorphanol, midazolam, and valium. Please keep in mind these are titrated to effect. An assistant should be administering drugs, flow by oxygen, as well as monitoring the patient throughout the procedure. Flow by oxygen is recommended during the procedure due to the respiratory depressant effects of opiods, as well as the compromised state of critically ill patients.

Supplies: Be sure to have all supplies ready prior to placement. Clippers, scrub and alcohol scrub, assistant, monitoring equipment, central line kit, bandage materials, patients own bag of saline flush and heparin/saline flush (500u heparin/250cc 0.9% NaCl), sterile gloves, oxygen, and sedative if needed.

Placement: Locate desired point of insertion by having an assistant hold off the vein. Prepare site by clipping a large area of hair, roughly 4"x4". Using a three scrub technique, alternating a chlorhexadine solution and alcohol, clean site thoroughly and let air dry. (Dirtier patients may require more cleaning.) Open pack and, wearing sterile gloves, familiarize yourself with the contents of the central line kit. Once familiarized, apply drape to patient, and if indicated administer sedation. Ask assistant to hold off again, locate desired point of insertion, and begin placement of central line. (Techniques may vary). After central line placement, be sure to obtain blood back in all lumens of the central line, to clear air from the lines. Then flush all lines with saline, clamp, and cap. Be sure to suture in place using 3 points of contact.

Bandaging: Keeping the central line clean is a necessity. Applying a bandage to prevent nosocomial infections while hospitalized is mandatory. Using gauze squares around the insertion site can not only act as a barrier, but it also provides cushion between the patient and the port of the central line. After applying gauze squares, cast padding can be wrapped around neck/leg 2-3 times, followed by 2-3 rows of roll gauze. This not only protects the central line, it also keeps the central line stable. A few rows of vet wrap will help hold everything together. Then secure all lumens with tape, and be sure to label all lumens for IVF, blood sampling, or TPN. (It is highly recommended a lumen be preserved for TPN only.) Lastly, be sure 2-3 fingers can easily fit under the bandage, to avoid constricting the patient's airway. Remember, bandaging is an art!

Confirmation: Be sure to verify accurate central line placement with a radiograph. This will allow the determination of accurate CVP's. It will also prevent arrhythmias if the central line requires backing out and the use of a spacer. Often, measurement is inaccurate, and the central line can end up in the right atrium, leading to arrhythmias.

Maintenance

Now that the central line is in place and ready for use, it is crucial maintenance is not overlooked. Bandage changes daily, using aseptic technique while wearing gloves, is recommended. During bandage changes monitor for fever, swelling, phlebitis, thrombosis formation, leaking, and redness. If noted, central line removal may be indicated. Monitor all lumens and lines for remnants of blood, as this can serve as a bacterial medium. It is also important to change out caps, t-sets, and lines daily. Routine alcohol swabbing of injection ports, personal fluid bag ports, and IV line connections is also warranted. Due to the requirements of central line cleanliness, patients should have their own bag of plain saline, heparin/saline flush and sterile water (if reconstituted medications are ordered). It is also recommended to fit canine patients with a harness, and avoid the use of neck leads while a central line is in place.

Central lines allow ease of treatment, patient compliance and comfort, as well as diagnostic capabilities. They allow advanced care, and require a higher standard of skill and patient management.

MYTHS AND FACTS OF TRANSFUSION MEDICINE

Kenichiro Yagi, BS, RVT, VTS (ECC, SAIM)
Los Altos, CA

INTRODUCTION

Veterinary transfusion medicine and blood banking has received increased interest from the veterinary community over the past several years as blood products have become increasingly available and the knowledge of their use being more widely disseminated. While there are risks to using transfusion products in practice, the benefits of these products have far outweighed them, and the demand attempted to be met by blood banks typically leave them in a shortage in a consistent basis. Despite the increase in use of blood products and transfusion therapy becoming more commonplace, there are many questions surrounding their use. Many of these questions are starting to be answered through a combination of veterinary and human evidence emerging. Some of the evidence further clarifies already known concepts while others confirm long time suspicions, and others yet teach us something new.

Can RBCs Be Given Through an Infusion Pump?

"Can blood be given through an infusion pump?" is a common and highly debated topic in both human and veterinary care. Delivering RBC products through an infusion pump provides accurate control over the transfusion rate and volume but has been theorized to damage RBCs reducing the effectiveness of the transfusion.

The Truth: Two veterinary studies have been recently published evaluating in vivo RBC survival time with common transfusion methods in the dog and cat. The canine study (McDevitt 2011) observed:

- 100% chance of survival to 49 days post transfusion with gravity delivery (no pump)
- 50% chance of survival to 49 days with a volumetric pump (typical peristaltic IV pump)
- 0% chance of survival beyond 24 hours with a syringe pump

The infusions with syringe pumps were performed through an 18μm pore size filter instead of a 170-260μm administration set, and this is suspected to be the largest contributing factor to the low survival rate. The feline study observed no significant difference in survival time up to 42 days post transfusion between gravity delivery and syringe pump infusions despite the filter pore size difference (Heikes 2014).

Those convinced use gravity delivery as their method for canine transfusions with careful monitoring of transfusion rate through frequent drip rate checks as the current best evidence-based practice. Others await further evidence before changing their practices as experience argues effectiveness of transfusions delivered through mechanical infusion devices. The good news: nobody is debating over transfusion methods for cats.

PRBC Has an Expiration Date of 42 Days(?)

Current practices in blood banking involve the usage of APS and additive nutrient solution which are labeled for 42 days of storage. More recent evidence gathered over the past decade indicates stored red blood cells to have impaired RBC survival, reduced efficacy as an oxygen carrier, and even incite adverse effects in the recipient causing mortality and morbidity. These changes are seen as early as 7 to 14 days into storage, and involve a collection of biochemical, biomechanical, and oxidative changes to the RBC and storage solution, all collectively referred to as "storage lesions".

Mature RBCs lack mitochondria and rely on glycolysis for ATP production, leading to a lowered pH. ATP production is reduced by the acidic environment, combined with depletion, leads to decreased RBC membrane integrity. Lowered pH also affects 2,3 diphosphoglycerate (2,3 DPG) level reducing hemoglobin's effectiveness as oxygen carriers, though this effect is reversible and not significant in cats. Hemoglobin in longer stored RBC products contain free hemoglobin and microparticles that scavenge nitrous oxide (NO) upon transfusion and cause a vasoconstrictive effect impairing blood flow, stimulate coagulation, induce oxidative damage, and cause proinflammatory effects. Microparticles, which are vesicles that have budded off of cellular components, induce proinflammatory and procoagulant effects. Stored RBCs show morphologic changes to echinocytes and spheroechinocytes leading to a loss of deformability and impairment in normal flow through capillaries. Oxidative damage leads to increased hemolysis and methemoglobin formation decreasing viable RBC count and oxygen carrying capacity.

There are many complicated mechanisms in play during RBC storage. To summarize the effects, storages lesions can lead to impaired RBC survival, reduce the efficacy of RBCs as oxygen carriers, and induce adverse effects such as arrhythmias, thrombosis, systemic inflammation, transfusion-related acute lung injury (TRALI), acute respiratory distress syndrome (ARDS), hypotension, and multiple organ dysfunctions. These changes occur as early as 7-14 days into storage, making supplying our patients with safe transfusion products a realistic challenge. Storage lesions and their clinical impact is a topic of ongoing investigation while blood banks strive to balance provision of fresher products and minimizing wasting of blood.

First Transfusions Are "Free"?

Compatibility testing for canine blood transfusions has traditionally been omitted in the interest of swift transfusions and financial considerations. This comes from the a widespread notion that the "first transfusions are free for dogs", intended to state that canine RBC transfusions can be given without blood type matching (without typing the donor or recipient) or cross matching yet be performed without signs of immunologic complications, namely acute hemolytic transfusion reactions or anaphylaxis. This statement is made with the understanding that the most clinically significant dog erythrocyte antigen (DEA) is DEA 1, responsible for inciting acute hemolytic transfusion reactions when preexisting alloantibodies for the antigen is present. In 98% of the population, these antibodies are not present, so the first mismatched transfusion will only result in sensitization of the immune system to the antigen, leading to the development of antibodies over a course of approximately 4 days. This leads to a delayed hemolytic transfusion reaction, often asymptomatic as long as the patient has overcome the initial incident of anemia, or clinical symptoms of anemia as well as bilirubinemia and bilirubinuria may arise.

Given the asymptomatic or mild nature of clinical signs, many have accepted this reason to forgo compatibility testing. However, the sensitization will lead to an acute hemolytic transfusion reaction in subsequent mismatched transfusions, resulting in hemolysis of transfused cells and likeliness of anaphylaxis. By omitting compatibility testing, we run the risk of priming a patient for such reaction in the next transfusion which may be handled similarly if the patient's transfusion status is not noticed. A medical team may be placed in a situation where the transfusion status of the patient may be unknown (pet brought in by pet sitter who thinks there had been no transfusions, or adopted dogs who "probably" has not had a transfusion). In the case a patient presents with risk of imminent death from anemia, this practice may be justified with the knowledge of the risk. Blood typing of all blood donors and stocking of DEA 1 negative blood is highly recommended for use in these situations to avoid sensitization of the patient to DEA 1. If there is any uncertainty in the transfusion history of the patient, cross matching is appropriate as erythrocyte antigens aside from DEA 1 exist with limited knowledge on consequences from patients sensitized for these miscellaneous antigens (some reports of AHTR exist). Transfusions of canine RBCs without compatibility testing are not "free", and certainly have the hidden costs of DHTR and sensitization. Cats possess alloantibodies for the RBC antigens foreign to them (aside from the very rare type AB cats), leading to hemolytic transfusion reaction even with first exposure.

DEA 1 Negative is the Universal Blood Type?

The concept of "universal" blood type indicates a blood type that can be given to any member of the same species without expectation of an immunologic reaction related to blood type mismatches. Because DEA 1 is the one antigen we know most about and leads to AHTR when mismatched for the second time, blood from DEA 1 negative dogs can be given without sensitization of DEA 1 negative and DEA 1 positive recipients, and often is considered as "universal". There are, however, other RBC antigens such as DEA 3 through 8, *dal*, and other less known antigens confirmed to exist which can lead to sensitization when mismatched transfusions occur. Therefore, a donor should be tested negative for every RBC antigen we are capable of testing in order to truly considering it "universal". This creates a challenge as 98% of the canine population is positive for DEA 4, and a donor negative in DEA 4 is virtually impossible to find. Fortunately this is not a clinical issue since the recipient is likely DEA 4 positive as well, allowing the blood types to match. Another challenge lies in our current inability to routinely test for DEA other than 1, 4, and 7 through a reference lab due to a lack of testing anti-sera (and anything aside from DEA 1 not available as in-house kits), preventing complete typing of our donors and timely testing of our recipients. Given our knowledge of additional RBC antigens, we should consider DEA 1 negative, 4 positive, 7 negative blood type as the "least antigenic", and type our donors for all DEAs we are capable of, given finances permit it. DEA 1 negative can be considered safe blood to use from anecdotal evidence as reports of hemolytic transfusion reactions are rare, and cross matches should detect incompatibility issues arising from repeated exposure to the less known erythrocyte antigens. Cats have no universal donors, though type AB cats may receive transfusions from both type A and B donors.

Are Blood Transfusions between Different Species Possible?

Common knowledge dictates that blood product transfusions should be between members of the same species in order to prevent immunologic consequences. Despite this "common" fact, one may be surprised to hear an ongoing research to test interspecies transfusions, or xenotransfusions. Early experiments in blood transfusion in the 1600's document a human patient receiving sheep blood, and showing no signs of reaction (at least on first exposure). Porcine red blood cells with modified antigens have been a topic of research in compatibility as human blood substitute. In the veterinary field, feline blood is consistently in short supply, especially for patients with the rare blood type of B. Type B cats can only be transfused with type B blood as introduction of a small volume of type A blood will result in an acute hemolytic reaction and anaphylaxis. In addition, even for type A patients, blood supply may be short causing delays or inability to obtain blood products in a timely manner as the patient suffers life-threatening anemia. In these situations, veterinarians have attempted to use canine blood as a source of blood as it is more readily available, and can easily tolerate the small volume donations.

There is limited amount of evidence available from a few studies conducted on canine to feline transfusions. The results of the studies concluded felines do not possess naturally occurring alloantibodies against canine erythrocytes. Compatibility testing methods such as slide-agglutination test and cross-matching only revealed agglutination on the minor-crossmatch. Of the total of 62 transfusions performed between the various studies, 5 cats showed signs of mild reactions, with tachypnea and pyrexia within 24 hours of the start of transfusion. Development of antibodies against canine RBCs were seen 4 to 7 days after the transfusion, indicating the transfusion led to sensitization of the immune system to the foreign antigens. Because of this, the life span of the transfused RBCs was approximately 4 days due to delayed hemolytic transfusion reactions while feline to feline transfusions allow RBCs to last 30 days. Subsequent transfusions resulted in anaphylaxis and were fatal in 66% of documented cases.

While transfusion of dog blood to a feline patient is not the best solution to supplementing oxygen carrying capacity, it may be justifiable when faced with imminent death of the feline patient and without blood. A responsible medical team would discourage dog to cat transfusion and consider the method for situations where the patient 1) has no source of compatible cat blood (Type B cat with no stocked blood, donor, or nearby hospital with stock, for example) or hemoglobin based oxygen carrier solutions, 2) is imminently going to pass away without a transfusion or compatible blood will not be obtained soon enough (truly dying animal), 3) is expected to benefit from a short term oxygen carrying capacity gain, and 4) the owner understands risks and consequences. The method of xenotransfusion should not become a common practice and effort for practices to maintain a good source of cat blood should always be pursued without considering canine blood as "backup".

Premedicating Reduce Chances of Reactions?

Premedication, or administration of antihistamines, glucocorticoids, or antipyretics in anticipation of immunologic complications to counter histamine and inflammatory mediators and suppress the effects, have been a traditional practice in transfusion medicine. While there is no veterinary evidence available addressing the efficacy of this practice, there are a number of human studies observing no difference in incidence of type I hypersensitivity reactions (allergic reaction) or febrile non-hemolytic transfusion reactions (FNHTR). Some clinicians reason that administration of premedication potentially masks early symptoms of immunologic complications delaying required interventions for treatment, advocating against it. Evaluation of the difference in severity between recipients with premedication or without premedication has not been performed, and remains a question whether this reasoning is valid. Human evidence is unfortunately not always directly translatable into veterinary practice, though expectations of similar physiological mechanisms exist. Studies evaluating effects of premedication and efficacy in prevention of hemolytic transfusion reactions are not apparently available, and the theoretical benefit is no justification for forgoing proper compatibility testing.

Is Warming of Blood Products Necessary?

Warming of blood products in the interest of prevention hypothermia in the recipient is a consideration during blood product administration. Concerns for hemolysis of erythrocytes when warming during transfusion exist, and studies point towards little to no difference in markers for hemolysis *in vitro* when blood is warmed to typical body temperature. However, at non emergent administration rates, blood reaching the patient through the line placed in a room temperature environment is easily at room temperature upon reaching the patient, and will not contribute to a significant decrease in body temperature. In the case of rapid transfusions of large volumes into small patients, warming of the blood may be indicated with care taken to be evenly warmed to 35-37°C and not exceed 42°C close to the patient to minimize loss of heat. Hypothermia is also a documented complications related to massive transfusions. Aside from these situations, in many cases warming effort directed at the patient is most effective in treating hypothermia without causing risk of damage to the RBCs.

Is Plasma Indicated for Use in Hypoproteinemia? Parvoviral Enteritis?

Plasma contains many proteins of interest, namely hemostatic proteins, albumin, and immunoglobulins. Hypoproteinemia, specifically hypoalbuminemia, occurs in many critically ill patients with protein-losing disorders including protein-losing enteropathies, protein-losing nephropathies, liver failure, trauma, burn wounds, etc. This leads to a loss of intravascular colloid osmotic pressure (COP), and subsequent consequences. Administration of plasma products (fresh frozen plasma, frozen plasma, or cryosupernatant) have been used as a method in supplementing albumin for COP.

However, the amount of plasma required to raise the patient's albumin level by 1g/dL is approximately 40-50mL/kg. This is equivalent to 1.1L of plasma (9.5 units) for a 50# patient. The amount of plasma required to make a significant difference in the measurable level of albumin is both cost prohibitive and pose a large immunologic risk to the patient. Synthetic colloids provide higher COP for the volume transfused, without the immunologic risks associated with biological products. In addition, whether increasing the albumin level to a normal value (>2g/dL) will lead to increased chances of a positive outcome is still unclear, and difficult to advocate.

Similar concepts can be applied to the usage of plasma products derived from survivors of parvovirus infection. Clinicians have theorized that transfusion of plasma containing antibodies against canine parvovirus (CPV) will aid in recovery from CPV infections. A study evaluating use of a single dose of plasma containing CPV antibodies in its efficacy versus saline placebo saw no significant difference in reducing clinical signs, viremia, or speeding recovery. The volume used in this study (12mL) may be a limitation to the efficacy of the compared treatment, though the amount of plasma required for an adequate dosage of antibodies is unknown, and is likely to be at similar or higher levels of dosage for albumin supplementation. Thus, same concerns prevent use of plasma in this manner.

Summary

As we can see, there are both myths, truths, and everything in between regarding transfusion medicine, which continues to be an active area of research to develop better knowledge and more accurate standards in order to optimize our practice leading to the best possible outcomes.

REFERENCES

[1]Bragg RE. J Amer Vet Med Assoc 2012;240(6):700-704. [2]Chiang V. J Vet Emerg Crit Care 2011;21(6):639-647. [3]Heikes BW. J Vet Emerg Crit Care 2014;24(2):162-167. [4]Kisielewicz C. Vet Anaesth Analg 2014;41(3):233-42. [5]Marti-Carvaial AJ. Cochrane Database Syst Rev 2010;(6):CD007539. [6]McDevitt RI. J Vet Emerg Crit Care 2011; 21(3):209-216.

MANAGEMENT OF CANINE IMMUNE-MEDIATED HEMOLYTIC ANEMIA (IMHA)

Laura A. Nafe, DVM, MS, DACVIM (SAIM)
Stillwater, Oklahoma

INTRODUCTION

Immune-mediated destruction of red blood cells is an example of a type II hypersensitivity reaction that is caused by binding of anti-RBC antibodies (IgG, IgM) or components of the complement system.[1] Immune-mediated hemolytic anemia can be subdivided based on the inciting cause (primary or secondary) or the mechanism of red blood cell destruction (intravascular and extravascular hemolysis). Although pathophysiologically IMHA can be divided into extravascular and intravascular hemolysis, many patients may have a component of both (although one may predominate). Unfortunately, canine IMHA has a mortality rate reported to range from 30-70% and veterinary studies evaluating the most effective treatment are lacking. IMHA may occur as a single clinical entity or in combination with immune-mediated thrombocytopenia (ITP). When both IMHA and ITP are present this condition is known as "Evans Syndrome."

Although the focus of this manuscript and presentation will be on management of canine IMHA, understanding features of this potentially fatal disease is important in understanding the various treatments available.

CATEGORIES OF IMHA

Approximately 60-75% of canine IMHA cases are classified as primary, meaning the immune system is attacking presumably normal RBCs. [1] Autoantibodies target self-antigens on the surface of the RBC membrane. Although there are some breed and sex predispositions (eg; Cocker Spaniels, female) that have been identified, the underlying cause of primary IMHA is multifactorial and poorly understood. Secondary IMHA results from antibody or complement mediated destruction of RBC that have been "altered." The antigen alteration that occurs may be due to underlying cancer, RBC parasites (*Mycoplasma spp., Babesia spp.*), infectious diseases (although we commonly test for tick-borne diseases this association is weak), drugs (specifically antibiotics) and vaccination (weak association). The exact alteration that triggers an immune response is not usually known.

Extravascular hemolysis occurs when RBCs that have been targeted for destruction via the binding of anti-erythrocyte antibodies are identified and phagocytized by macrophages in the spleen and liver. Extravascular hemolysis often results in partial erythrophagocytosis, resulting in spherocyte formation (RBC morphology change) and hyperbilirubinemia and bilirubinuria. In general, extravascular is more common than intravascular hemolysis. Intravascular hemolysis occurs when antibodies (IgM) bound to the red blood cell surface activates complement that generates a membrane attack complex (MAC). This MAC attaches to the RBC membrane and punches holes in the membrane, allowing an influx of extracellular components (electrolytes, water), leading to RBC swelling and eventual lysis. The outcome is a release of free hemoglobin into the bloodstream, resulting in hemoglobinemia and hemoglobinuria. In general, intravascular hemolysis is associated with more severe disease and carries a worse prognosis.

CLINICAL PRESENTATION AND DIAGNOSIS OF IMHA

Dogs with IMHA can have a variety of clinical signs. These signs often vary depending on the severity and cause of red blood cell destruction. Most patients are presented for evaluation of acute, non-specific signs, including lethargy, weakness, anorexia and other signs gastrointestinal signs. These signs are usually the result of moderate to severe anemia and associated systemic inflammation. Diagnosis is based on clinicopathological abnormalities consistent with IMHA. This includes a moderate macrocytic, hypochromic regenerative anemia with evidence of spherocytosis and/or autoagglutination and hyperbilirubinemia. Dogs with IMHA will often have a mild to moderate thrombocytopenia and increased white blood cell count that may be markedly increased with an increased number of band neutrophils. An effort should be made to rule-out secondary causes of IMHA. It is recommended that most patients have imaging performed (to help rule-out cancer) and vector-borne disease testing.

MANAGEMENT OF IMHA

Management of IMHA includes immunosuppression, thromboprophylaxis, and supportive care measures to help prevent and treat concurrent conditions. Management strategies may vary among clinicians and on a case-by-case basis. It is important to realize that the many of the treatment protocols employed are based heavily on clinical experience and anecdotal opinion, rather than strong evidence-based research.

Immunosuppressive and Immunomodulatory Therapy

Treatment with immunosuppressive medications is the mainstay of canine IMHA management. The goal is to introduce medications that will reduce immune destruction of red blood cells and allow the number of new red blood cells being produced to exceed the number being destroyed. Glucocorticoids (e.g., prednisone, dexamethasone) are administered to all canine IMHA patients with or without an additional immunosuppressive agent. When an additional agent is added to the treatment protocol, which second-line medication is chosen is based on clinician preference, cost, tablet/capsule size, and potential adverse effects. Potential second or third line immunosuppressive agents commonly administered to dogs with IMHA include azathioprine, cyclosporine, mycophenolate, and leflunomide.[2] There is minimal evidence to support the use of one of these agents over another. One retrospective evaluation indicated that azathioprine was second line immunosuppressive agent of choice for management of canine IMHA; however, the inherent bias of a retrospective study when evaluating treatment of this disease underscores the value of these results.[3] Azathioprine and cyclosporine are most commonly used, with mycophenolate gaining popularity in recent years. Leflunomide is used more commonly for immune-mediated polyarthritis.

Human intravenous immunoglobulin (hIVIg) infusions are occasionally administered to patients with IMHA as an alternative means of slowing immune-destruction of red blood cells. IVIg is a pool of human immunoglobulins (primarily IgG) that result in multiple immunomodulatory properties.[4] While IVIg reduces immune destruction of red blood cells via multiple mechanisms, the primary mechanism is macrophage Fc receptor blockade. In doing so, the macrophage Fc receptors are occupied preventing antibody bound red blood cells from binding.[4] While IVIg may be a beneficial adjunctive therapy in some canine IMHA patients, the current expense ($1000-3000/infusion) and lack of definitive evidence in the treatment of IMHA limit the use of this immunomodulatory agent.

Drug	Mechanism of Action	Dose	Potential adverse effects	Other considerations
Glucocorticoid	Alters gene transcription. Down regulate Fc receptor expression, decrease antigen processing, suppress T cell function	Prednisone: 2 mg/kg/day Dexamethasone 0.2-0.3 mg/kg/day	Increased thirst, increased urination, increased appetite, panting, muscle wasting, weight gain, opportunistic infections, GI ulceration	Inexpensive, many formulations available
Azathioprine	Thiopurine analog	1-2 mg/kg/day PO x 7-14 days; then decrease to 1-2 mg/kg PO every other day	GI upset, hepatotoxicity, bone marrowsuppression, pancreatitis	Inexpensive Single size (50 mg tablet)
Cyclosporine	Calcineurin inhibitor	4-5 mg/kg PO q 12 hr	GI upset, gingival hyperplasia, opportunistic infections, hepatotoxicity?	Expensive Capsules and liquid formulation
Mycophenolate	Purine synthesis inhibitor	6-10 mg/kg PO q 12 hr	GI upset, hepatotoxicity, bone marrow suppression	
Leflunomide	Pyrimidine synthesis inhibitor	1-4 mg/kg PO q 24 hr	GI upset, bone marrow suppression, hepatotoxicity?	Primarily used for management of IMPA
hIVIg	Fc receptor blockade	0.5-1.5 g/kg IV over 6-12 hours	Anaphylactoid reaction, increased risk of thromboembolism, fluid overload	Expensive Do not repeat infusion

THROMBOPROPHYLAXIS

The most common cause of natural death in canine IMHA patients is related to thromboembolism, specifically pulmonary thromboembolism (PTE). Predisposition to thromboembolic events is due to vascular stasis (hospitalization, IV catheter), endothelial damage (hypoxia) and hypercoagulability (free hemoglobin, blood transfusion, glucocorticoid therapy). Although the treatment of IMHA with blood transfusion(s) and glucocorticoids contributes to the hypercoagulable state, recent evidence suggests that canine IMHA patients are hypercoagulable (based on thromboelastography) prior to treatment.[5]

Prevention of thromboembolic events is managed with a platelet inhibitor medication (low dose aspirin or clopidogrel) and/or heparin (unfractionated or low molecular weight). Warfarin is not commonly used in veterinary medicine due to a narrow therapeutic index and required frequent monitoring. Most canine IMHA patients are managed with a platelet inhibitor medication, with clopidogrel gaining popularity in recent years. The addition of heparin therapy is clinician and case dependent. There is minimal evidence to support administration of one platelet inhibitor over the other for the prevention of thromboembolic events in patients with IMHA. A retrospective study found that ultra low-dose aspirin therapy improved short and long-term survival in dogs with IMHA, but this study was published prior to the use of clopidogrel in veterinary medicine.[3]

Unfractionated heparin is most cost effective (although still expensive) and can be administered via intermittent subcutaneous injections or as a constant rate infusion (CRI). Efficacy of unfractionated heparin is thought to be more predictable when administered via CRI; however, this requires more intensive care and a CRI is not practical for outpatient care. Low molecular weight heparin has a more predictable dose range and efficacy; however, expense often precludes use in most of our veterinary patients. One small study evaluated individually adjusted unfractioned heparin in canine IMHA and found that those who had their heparin dosages adjusted based on efficacy were more likely to survive.[6]

SUPPORTIVE CARE

Intravenous fluid therapy

IV fluid therapy is important to maintain hydration and support kidney function in patients with IMHA. The presence of pigmenturia (hemoglobinuria, bilirubinuria) can result in renal tubular injury. In addition, these patients are often dehydrated with decreased appetite, requiring IV fluids to maintain hydration and electrolyte balance.

Red blood cell transfusion(s)

Many canine IMHA patients will require at least one blood transfusion, with severe or refractory cases potentially requiring multiple transfusions in the early phase of their disease. When available, packed red blood cells are the preferred blood product, as to minimize the risk of transfusion reaction and hypercoagulability. Dogs must be "matched" to a donor if they are receiving an additional blood transfusion 3-5 days after the previous transfusion. Monitoring for transfusion reactions is an important aspect of patient management.

Gastroprotectants and Anti-emetics

Dogs with IMHA may have concurrent signs of GI upset and GI erosion/ulceration may develop due to decreased perfusion and/or oxygen delivery to the GI tract. In addition, high doses of glucocorticoids can predispose to GI ulceration. Development of GI bleeding in these patients may contribute to continued anemia and confuse management strategies. As a result, treatment with an acid suppressant +/- GI coating agents (e.g.; sucralfate) is often indicated. If the patient is vomiting and/or anorexic, anti-emetics may also be indicated.

REFERENCES

[1] Swann et al. *J Vet Intern Med* 2015; 29: 7. [2] Viviano KR. *Vet Clin Small Anim* 2013; 43: 1149. [3] Weinkle et al. *JAVMA* 2005; 226: 1869. [4] Whelan MF et al. *J Vet Emerg Crit Care* 2009; 19: 158. [5] Goggs R et al. *Vet J* 2012; 191: 347. [6] Helmond et al. *J Vet Intern Med* 2010; 24: 597.

Video Analysis of Gait Disorders: Technician Program
Karen Dyer Inzana, DVM, PhD, DACVIM (Neurology)
Blacksburg, VA

Organization of the Nervous System

The ability to walk depends on the coordinated function of the axial skeleton, joints, tendons, skeletal muscles and the nervous system. From a neurological perspective, a complex system of motor neurons as well as a variety of sensory input from skin, joints and muscles must work together to produce a normal gait. The motor system is most easily divided into 3 components; lower motor neuron systems, which are controlled by upper motor neuron centers which are, in turn, controlled by higher centers including the cerebellum and forebrain.

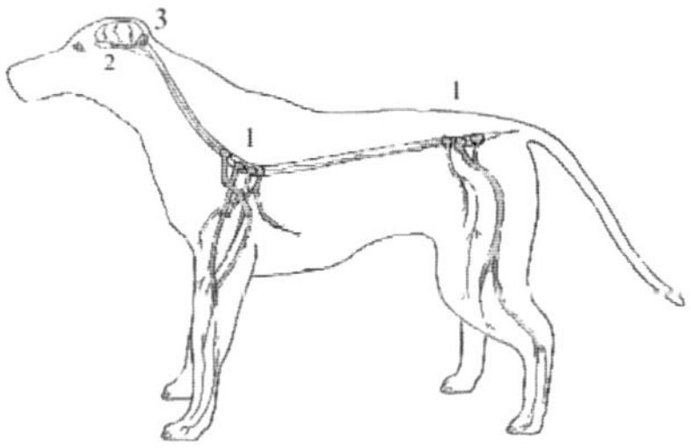

(1) Lower motor neurons have their origin in gray matter of the spinal cord (or brainstem for Cranial Nerves) and extend out of the nervous system to supply peripheral muscles. (2) Upper motor neurons originate primarily in the brainstem (a few in the cerebral cortex) and extend down white matter tracts in the spinal cord to regulate the firing of lower motor neurons. (3) Upper motor neurons are regulated by a series of feedback circuits through forebrain and cerebellum that make movement voluntary and coordinated.

The sensory system originates in skin, joints, and muscles and provides constant input to all components of the motor systems to provide information about where the muscles, joints, and limbs are in space, a physiological function termed proprioception.

Normal Gaits

When considering normal dog gaits, it is best to divide them into symmetrical or asymmetrical gaits. With symmetric gaits movement on one side of the body mirrors movement on the opposite side with near even placement of the feet with each movement. Examples of symmetric gaits are the walk, trot, and pace. In Asymmetric gaits such as a gallop, both sides move independently and stride length varies among limbs. A good example of normal gaits may be viewed at this website: http://vanat.cvm.umn.edu/gaits/info.html The walk is a four-beat gait, in which each limb steps (lift/swing/support/thrust) sequentially. Either two or three limbs are on the ground and the alternate limbs on the front and rear advance in order. The head and neck come up slightly during the support phases of the front limbs and down during the swing phases. Clearly, the characteristics of the gait vary with different breeds and limb length, but the important observation is that movement is coordinated and stride length is equal in each limb. The trot is a two beat gait where the right and left diagonal limbs move together to support weight. The head and neck should remain level. The canter and the gallop are three or 4 beat gaits depending on the speed with the front inside limb leading. The normal sequence is the forelimb followed by the opposite hind limb and then the diagonal. As the speed increases to a gallop there is more suspension between the final diagonal phase making it a 4 beat gait. The normal sequence is leading forelimb, opposite hind limb, diagonal with the rear landing before the forelimb with a period of suspension between.

Clinical Evaluation

Neurological examinations should always begin with observation of at a distance, preferably with the animal moving freely off leash in a space such as an examination room taking note of any abnormal behaviors, body posture, or movements. It is best to evaluate an animals gait on a non-slick surface such as carpet, mowed grass, or other textured surface. The symmetric gaits of walk and trot are easiest to evaluate. Circling, moving over objects such as blocks, curbs, steps can also help detect subtle deficits. Three distinctive abnormal gaits maybe observed; Ataxia, Paresis, and Lameness. Each may be seen in animals with neurological diseases.

1) Ataxia is difficulty performing normal coordinated movements. There are three types of ataxia.
   a) Cerebellar ataxia is caused by injury to portions of the cerebellum. The rate of movement as well as the range of movement is often affected creating a hypermetric or high-stepping gait. There is generally marked trunkal sway and goal directed tremor of the head and neck (intention tremor).
   b) Vestibular ataxia is created by an imbalance of input from normal and diseased vestibular systems. There generally is a head tilt and tendency to fall or roll to the side of the injury. Abnormal eye movements (nystagmus and strabismus) are frequently seen with vestibular ataxia.
   c) Somatosensory or Proprioceptive ataxia is caused by injury to sensory nerves, ascending sensory pathways in the spinal cord, brain stem and forebrain. It is characterized by a loss of sensory recognition of where the limb and trunk are in relations to other parts. This gait appears clumsy and uncoordinated with a base-wide stance and the limbs placed away from the center of gravity.

In the spinal cord, motor systems run in close proximity to sensory fibers, so it is difficult to have a pure sensory ataxia without some loss of motor control as well. Animals with upper motor neuron weakness often have a delayed onset of protraction (swing phase) increased stride length, and the limb is placed in an abnormal position when it lands. Muscle tone is preserved in UMN paresis so the gait often appears spastic due to poorly regulated or exaggerated reflex activity in the limbs.

2) Paresis or weakness is caused by injury to motor units (peripheral motor nerves, skeletal muscle, or the neuromuscular junction). Animals with these conditions appear weak, have a hard time supporting weight for any length of time and tend to sit or lay down frequently. Often the head and neck are ventroflexed due to weakness of the neck and back muscles. Paretic gait is manifested by shortened strides and a tendency to move the rear limbs closer to the center of gravity. Hock weakness characterized by either excessive movement (over flexion) or plantagrade stance may be seen. When asked to stand paretic animals may have a fine generalized tremor simply due to weakness.
3) Lameness generally appears as if the limb was painful. Animals may hold the limb up, be reluctant to bear weight on it and therefore have a shortened stride length in the affected limb. The contralateral limb will be used to bear most of the weight so it will have a longer stance phase and longer stride length. If an animal has a bilateral orthopedic disease (for example bilateral cruciate disease) the gait of both affected limbs are shortened and the animal will tend to shift weight to the font limbs. As a rule, limbs tend to be carried with orthopedic diseases while neurological diseases causes the limb to drag or scuff the nails as the animal walks. The primary exception to this rule is if the animal has nerve root irritation causing shooting limb pain in which case the animal may carry that limb. Nerve root irritation is one example where neurological diseases may be difficult to differentiate from orthopedic lameness.

Abnormal Movements

Abnormal movements may occur constantly, or only be intermittent. They may be generalized, occurring throughout the head, neck and body or may be localized. These are best classified into the following categories.

1) Myotonia: results from a stimulus induced persistent, repetitive muscle contraction without relaxation. Abnormal movements appear during sudden movement especially is the animal is stressed or startled. There is limb stiffness that causes a short stiff appearing gait and occasionally collapse (or fainting). Affected animals may appear more heavily muscled
2) Myokymia is similar to myotonia. It results in repetitive contractions of small bands of muscle fibers (continuous muscle fiber activity) that causes episodic stiffness of the limbs often leading to collapse with the limbs remaining in a tetanic stiffness. Undulating involuntary movements may also be observed in the skin due to contraction of underlying bands of muscles. Myokymia may be exacerbated during excitement and episodes may last a few minutes to hours. This syndrome is most likely caused by abnormalities in the nerves supplying the muscle fibers so the term neuromyokymia may be more correct.
3) Tetanus or Tetany is caused by persistent muscle (tetanus) or intermittent (tetany) contraction most prominent in extensor muscles. It is most commonly seen in animals that have lost the ability to control extensor muscle tone due to loss of inhibitory neurotransmitters. The most common form is caused by Clostridium tetani infection but occasionally congenital abnormalities or immune mediated damage to receptors (stiff person syndrome) may be seen.
4) Myoclonus is a movement characterized by the sudden contraction of a group of muscles followed by relaxation. A sporatic form of myoclonus would look like a twitch of a group of muscles that is intermittent in origin. This can be caused by structural diseases in the forebrain and represent a form of epilepsy. It can also be idiopathic in origin. Repetitive myoclonus results in more consistant twitching of a group of muscles. This may be constant (constant repetitive myoclonus) and occurs during rest as well as during movement, or may be only action related (action-related repetitive myoclonus) affecting groups of skeletal muscles that are being activated and disappearing during rest. This type of myoclonus is often classified as tremors because of their appearance.

    a) Intention tremors as seen with cerebellar disorders is a tremor that is generally confined to the head and neck and occurs during periods of behavioral focus such as with eating or sniffing something.
    b) Whole body action tremors occur with myelin disorders, intoxications, electrolyte abnormalities, or abnormalities in neurotransmitter function.
    c) Postural repetitive myoclonus is a type of tremor that occurs only in muscles required to support weight. It may be episodic and confined to the head and neck in some breeds of dogs such as Doberman pinchers, Bulldogs, Boxer dogs were it is often referred to as head bobbing. The other form of postural repetitive myoclonus referred to as orthostatic tremors occurs in limb muscles, especially those needed to support weight. A benign form occurs in the pelvic limbs of aged dogs and disappears when the animal starts to move. A more severe form has been reported in Great Dane and Scottish deerhounds. This syndrome is most severe when the dog attempts to lie down or postures to eat, drink or eliminate. It disappears with rest and during normal ambulation. Phenobarbital and gabapentin have been beneficial in ameliorating clinical signs.
    d) Dyskinesia is a general term used to describe movement disorders characterized by involuntary contractions of groups of muscles in a conscious animal. The movements may take the form of fast contractions of groups of muscles creating a flailing appearance (Ballism and Chorea) or more prolonged contraction of muscles (dystonia or athetosis). Most reported cases of these type of movements tend to be familial and poorly characterized. Examples are Scottie cramps, Chinook seizures, Spike disease in Border terriers, Dyskinesia in Bichon frise, and Collapsing Cavalier King Charles spaniels.

During this session, video's of each of these abnormalities will be presented. This will be followed by a series of unknowns where the audience will be asked to provide their opinion. The overall goal is to recognize different forms of movement abnormalities in dogs.

REFERENCES

Datt VL and Fletcher TF. Gait Foot-Fall Patterns 2012 Supported by University of Minnesota College of Veterinary Medicine available online at http://vanat.cvm.umn.edu/gaits/info.html; Caprette H and Dyce J. 2002. Canine Lameness Learning Module available online at http://accad.osu.edu/~hcaprett/COTA_741_sp04/CanineLO_090504.html; de Lahunta A, Glass EN, Kent M. Classifying involuntary muscle contractions. Comp Cont Educ Pract 2006;28:516–530.; de Lahunta A, Glass E. Veterinary neuroanatomy and clinical neurology. 3rd ed. St. Louis: W.B. Saunders; 2009; Deuschl G, Elble R. Essential tremor—neurodegenerative or nondegenerative disease towards a working definition of ET. Mov Disord 2009;24:2033–2041; Donaldson I, CD Marsden, Schneider S, and Bhatia K. Marsden's Book of Movement Disorders. Oxford University Press, Oxford 2012; Thomas WB. Evaluation of veterinary patients with brain disease. Vet Clin Small Anim 2010; 40: 1–19; Vite CH. Myotonia And Disorders Of Altered Muscle Cell Membrane Excitability. Vet Clin Small Anim. 2002; 32: 169-187; Gillette RL, Angle TC. Recent developments in canine locomotor analysis: A review. The Veterinary Journal 2008,178:165–176' Nunamaker DM, Balauner PD. Normal and Abnormal Gait in Newton CD and Nunamaker DM eds. Textbook of Small Animal Orthopaedics Philadelphia J.B. Lippincott Company 1985 available online at http://cal.vet.upenn.edu/projects/saortho/chapter_91/91mast.htm

MANAGING SPINAL CORD INJURY: REHABILITATION AND ETHICAL DILEMMAS

Patrick J. Kenny, BVSc, DACVIM (Neurology), DECVN
London, UK

Veterinarians, veterinary technicians and rehabilitation staff commonly encounter paralyzed (paraparetic/plegic or tetraparetic/plegic) pets due to spinal cord injury (SCI) secondary to trauma, intervertebral disc herniation & fibrocartilagenous embolism[1], spinal neoplasia[2], degenerative myelopathy[3] or other cause. The treatment of these animals – particularly their maintenance as pets if prognosis for functional recovery is poor, can give rise to ethical dilemmas and cause a degree of "moral distress"[4] in the owner and/or veterinary staff. However, some pets under certain circumstances appear to tolerate paralysis and can have a good quality of life (QOL).[5,6]

RESOLVING CLINICAL-ETHICAL ISSUES

Most ethical issues can be resolved satisfactorily; competing goods are not usually problematic, competing harms or losses are more challenging, where the best choice may be the "least bad option". The later case may be considered a true "dilemma" – not just a difficult choice, but a choice between two or more undesirable alternatives. A framework for solving ethical problems may distinguish ethical dilemmas from more unremarkable ethical issues.[4] For example, though spinal surgery is an invasive procedure, whether or not to perform surgery on a paralyzed dog with an intervertebral disc herniation is not a dilemma if the prognosis with surgery is good and the owners are willing. Despite the temporary disruption to the dog and its environment, the intervention is likely to benefit it and the owner over the longer term – more so than not intervening or euthanizing the dog. Deciding whether to treat and maintain a paralyzed pet over the long term or to euthanize it has the potential to be a dilemma; extent to which the animal's QOL may be impaired must be balanced against the death of the animal; and decide which, under the particular circumstances, would be preferable.

Many practical approaches to resolving ethical dilemmas have been developed;[7-11] recurring components include:

1. Establishing the medical facts: diagnosis, prognosis and co-morbidities. A frequently cited maxim amongst clinical ethicists asserts "good ethics relies on good facts."[4]
2. Establishing what the ethical issues are, and their relative importance.
3. Establishing the concerns, values and preferences of the patient (or surrogate, or owner), shared decision making and obtaining valid consent.
4. Establishing what practical considerations must be addressed to minimize the impact of the decision.

WELFARE CONSIDERATIONS OF PARALYSED PETS

The welfare of the animal should be the core consideration in resolving clinical-ethical dilemmas in veterinary medicine, taking primacy over all other concerns.[12,13] Well established principles of animals welfare, such as the "five freedoms"[14] (1. Freedom from thirst, hunger and malnutrition; 2. Freedom from pain, injury & disease; 3. Freedom from fear and distress; 4. Freedom from physical and thermal discomfort; 5. Freedom to perform most normal forms of behavior) may be a useful start, though really at best reflect a minimum standard, and do not account for all welfare considerations that may arise for a clinical patient.

Considering an animal's QOL is commonly done when on deciding treatment or non-treatment/euthanasia. Paralysis due to SCI, neurodegenerative disease or other cause results in a dramatic change to the individual and their lifestyle; SCI has been described as "one of the greater calamities that can befall man".[15] Factors that affect the QOL of humans with SCI include loss of mobility, loss of independence, loss of bladder and bowel functions, loss of sexual function, sensory dysfunction, pain, respiratory dysfunction and the development of co-morbidities such as infection and pressure sores.[16] QOL has psychological as well as physical roots,[17,18] being influenced by level of social support, socioeconomic status, as well as clinical status.[19] Clinical factors that may affect the QOL of animals with SCI or disease are similar; motor dysfunction (loss of ambulation), sensory dysfunction, pain, urinary & fecal incontinence, decubital ulceration and skin abrasions.[1] QOL assessment of an animal has traditionally based on subjective assessments made by the owner and clinical staff, influenced by the clinician's appraisal of the prognosis based on clinical facts, and perception and experience of anticipated outcomes. But how is an acceptable QOL best determined? This may be unclear for both humans and animals, as no agreement exists on the definition of QOL and how it should be measured.[16] QOL scores are commonly used in the care of humans with SCI[16,20-24] & ALS[17-19] among other neurological diseases. There are few studies that have assessed at the impact of SCI on an animal.[25,26] One study[26] found that ambulatory dogs that had suffered SCI had a significantly higher perceived QOL than non-ambulatory dogs with SCI, and that underlying injury or duration of injury was not of significance. While this study is an important first step in attempting to measure QOL in veterinary SCI patients, it included a wide variety of breeds (26 + mixed breeds), and ages (range 0.33-17 years), where the ability of dogs and owners to cope with the consequences of SCI may have differed between groups. A recent systematic review of several different QOL measurement tools for SCI in humans concluded that tool selection should be guided by the specific patient population in which it is used.[22] There are many factors that may impact the QOL of an animal (and owner of an animal) with paralysis, including: size, age, breed, species, temperament, underlying disease process, location of lesion (para- or tetraparesis), severity of paresis/paralysis, sensory deficits, urinary & fecal continence, co-morbidies, resultant change in lifestyle and ability to perform normal behaviors, and level & quality of rehabilitation and nursing care. The relative impact of these factors on QOL of different populations (large dogs/small dogs, active dogs/sedentary dogs, dogs/cats, etc.) remains to be tested through validated QOL measures.

"Would you do it if it were your dog?" Owners struggling to make difficult treatment decisions commonly ask veterinarians this question. What is a "life worth living", and when would euthanizing the pet be preferable? The point at which "doing everything" becomes "too much" is an emotive topic for pet owners, veterinary staff and the general public, many of whom come with their own preconceived ideas and opinions. It is sad fact that depression and suicide rates among SCI patients are higher than that of the general population,[24] and that many quadriplegic patients will request removal from life-sustaining ventilation.[27] QOL, however, is not a static phenomenon, and that QOL scores change with time and often improve as patients adjust to new circumstances.[16] Non-patients are not good judges of a patient's QOL; members of the public often underestimate the QOL of people with paraplegia,[28] and caregivers tend to underestimate the QOL of their patients with ALS.[18] In many situations, euthanizing a paralyzed animal is the most humane and appropriate course of action. Some dogs, however, appear to be able to have a good QOL over a long period of time;[5] in cases such as these euthanasia would have acted against the dog and owners best interests. The concept of patient narrative[29] (i.e. its previous life experience) is worth considering when anticipating the impact of paralysis on an individual animal and deciding whether treatment or euthanasia is most humane. A toy breed dog rarely out of a handbag or its owners arms may be less impacted by paralysis than a previously highly active working dog; the ability of these dogs to cope with immobility, and therefore the welfare and ethical implications, are likely very different.

REHABILITATION

Owners need to be educated on the clinical facts at hand (diagnosis and prognosis), and be given a realistic expectation of the timeframe for improvement if recovery is possible. Adequately caring for a paralyzed pet requires considerable medical and nursing support. Urinary tract infections are highly prevalent.[30] Adequate bedding and vigilant skin care are necessary to decrease the occurrence of decubital ulcers.[31] Consent is an ongoing process[4] and a decision to continue care (as opposed to euthanasia) may be revised as patient or owner factors change.

For humans with SCI, treatment is ideally managed in integrated centers by multidisciplinary teams of health care professionals.[32,33] A successful encounter between patient and rehabilitation staff is considered crucial for a satisfactory outcome of the rehabilitation process.[34] Though the evidence base supporting specific modalities of physical therapy in humans with SCI is low,[35] there is strong consensus that physical therapy is important in promoting recovery, mitigating physical comorbidities and improving psychological health.[36] There is similarly a paucity of evidence regarding the use of physical therapy in managing paralyzed animals,[37] though it can facilitate social interaction, maintain the pet-owner bond, improve client education and support therapy at home.[38] Longer survival times reported in dogs with degenerative myelopathy undergoing physiotherapy[6] may in part reflect an improvement of perceived QOL. Physical activity has been shown to improve QOL in humans with SCI;[39] it may be reasonable to assume that the same is true for animals.

REFERENCES

[1]Granger N & Carwardine D **Vet Clin Small Anim** 2014;44:1131–1156.
[2]Bagley RS **Vet Clin North Am Small Anim Pract** 2010;40(5):915–927.
[3]Coates JR & Wininger FA **Vet Clin North Am Small Anim Pract** 2010;40(5):929–950 .
[4]Bernat JL **Ethical Issues in Neurology, 3rd ed.** Lippincott Williams & Wilkins, 2008.
[5]Tertuliano Marinho PV, et.al **J Small Anim Pract** 2014 ;55(10):538–541.
[6]Kathmann I, et.al **J Vet Intern Med** 2006;20(4):927–932.
[7]Lo B **Resolving Ethical Dilemmas; A Guide for Clinicians 5th ed.** Lippincott Williams & Wilkins, 2013.
[8]Fletcher JF, et.al **Introduction to Clinical Ethics.** University Publishing Group, 1995.
[9]Siegler M **Arch Intern Med** 1982;142:2178-2179.
[10]Snyder L, Leffler C **Ann Intern Med** 2005;142:560-582.
[11]Mullen S & Main D **In Practice** 2001;23:394-401.
[12]https://www.avma.org/kb/policies/pages/principles-of-veterinary-medical-ethics-of-the-avma.aspx
[13]http://www.rcvs.org.uk/advice-and-guidance/code-of-professional-conduct-for-veterinary-surgeons/
[14]Rogers Brambell FW **Brambell Report.** London, 1965.
[15]Guttmann L **Spinal cord injuries: Comprehensive management and research.** Blackwell Scientific Publications; 1973.
[16]Dijkers MPJM **J Rehabil Res Dev** 2005;42(3 Suppl 1):87–110.
[17]Simmons Z, et.al **Neurology** 2000;55(3):388–392.
[18]Trail M, et.al **J Neurol Sciences** 2003;209(1-2):79–85.
[19]Chiò A, et.al **J Neurol Neurosurg Psychiatr** 2004;75(11):1597–1601.
[20]Hitzig SL, et.al **J Spinal Cord Med** 2013;36(6):600–615.
[21]Hitzig SL, et.al **Disabil Rehabil** 2013;35(16):1408–1414.
[22]Ravenek KE, et.al **Disabil Health J** 2012;5(4):213–223.
[23]Boakye M, et.al **Journal of Neurosurgery-Spine** 2012;17(1 Suppl);29–37.
[24]Chevalier Z, et.al **Spinal Cord** 2009;47(11):778–782.
[25]Budke CM, et.al **J Am Vet Med Assoc** 2008;233(6):925-930.
[26]Levine JM, et.al **J Am Vet Med Assoc** 2008;233(6):931-935.
[27]Maynard FM **West J Med** 1991;154(5):617–619.
[28]Ubel PA, et.al **Med Decis Making** 2001;21(3):190–199.
[29]Kuczewski MG **J Law Med Ethics** 1999;27(1):32–36.
[30]Olby NJ, et.al **J Vet Intern Med** 2010;24:1106–11.
[31]Sikes R **Probl Vet Med** 1989;1(3):467-77.
[32]Guttmann L **Paraplegia** 1979;17:96–121.
[33]Donovan WH et.al **Paraplegia** 1984;22(5):282–290
[34]Ohry A **Spinal Cord** 1987;25(3):293–295
[35]Whiteneck G, et.al **J Spinal Cord Med** 2009;32(3):251–259.
[36]Wing PC, et.al **J Spinal Cord Med** 2008;3(4):408–479.
[37]Drum MG **Vet Clin North Am Small Anim Pract** 2010;40:181–93.
[38]Sims C, et.al **Vet Clin Small Anim** 2015;45:123–143.
[39]Stevens SL, et.al **J Spinal Cord Med** 2008:31(4);373–378.

A COLLABERATIVE APPROACH TO NEUROLOGICAL DISORDERS

Edward MacKillop, DVM, DACVIM (Neurology) and Meredith Wille, CVT
Pittsburgh, Pennsylvania

INTRODUCTION

Complicated medical problems in dogs and cats require the veterinarian and veterinary technician to work as a team in order to diagnose and resolve the problem. Thorough communication between staff members not only improves patient care but also makes each member of the team feel involved in the resolution of the pet's problem. Disorders of the brain or intracranial disease can often seem confusing or overwhelming; however, a team based approach can help facilitate the diagnosis and treatment, whether it is medical or surgical.

CASE HISTORY

A thorough clinical history or anamnesis is essential in determining the cause of a patient's problem. The onset of signs and progression should be established. The owner should be questioned about any medical therapy (drug, dosage, frequency and length of therapy) and the patient's response to therapy. Frequently after a technician takes a history, the owner may report slightly different information to the doctor having had more time to think about things. Open ended questions ("How did it all start?") facilitate communication as a dialogue and give the client an opportunity to recall more information. Closed ended questions ("What day did the problem start?") limit the client to a list of answers or yes/no, often stalling the conversation.

NEUROANATOMIC LOCALIZATION

The goal of a focused neurologic examination is to evaluate a patient for abnormalities referable to the brain, spinal cord, peripheral/cranial nerves or neuromuscular system. A lesion can be localized to one of the following areas: prosencephalon (forebrain), brainstem, cerebellum, C1-C5, C6-T2, T3-L3 or L4-S3 spinal cord segments, neuromuscular system or a specific peripheral/cranial nerve. In more simplified terms, a lesion can be localized to rostral to the foramen magnum (intracranial), caudal to the foramen magnum (spinal) or neuromuscular (peripheral nerve, muscle or the neuromuscular junction).

Clinical signs of intracranial disease depend upon the region of the brain that is affected.[1] A forebrain lesion usually does not affect gait generation but will cause a postural deficit (knuckling or hopping) on the opposite side of the body (contralateral). Vision or sensation may also be impaired contralateral to a forebrain lesion. While the gait may be normal, there may be compulsive circling or a tendency to orient in a single direction. Personality changes or alteration in level of consciousness are suggestive of forebrain disease. Epileptic seizures may be the only sign of an intracranial lesion, especially in dogs.[2] Vague clinical signs such as lethargy or an increased tendency to hide are often seen in cats with forebrain disease.[3] Intracranial disease affecting the brainstem typically causes an abnormal gait seen in the patient as generalized or vestibular ataxia. A lesion in the cerebellum causes hypermetria or a bursting, exaggerated quality to the gait. Postural reactions are abnormal with brainstem or cerebellar disease and deficits are typically ipsilateral. Facial paralysis, dysphagia, pupillary asymmetry or other cranial nerve deficits can be seen with disorders of the brainstem. Stupor or coma is associated with severe brainstem disease causing disruption of the ascending reticular activating system (ARAS).

Patients suspected to have intracranial disease should be monitored carefully as signs can progress rapidly. The patient should be kept in a part of the hospital where they are readily visible and mentation should be checked regularly as this is often the first thing to change.

DIFFERENTIAL DIAGNOSIS OF INTACRANIAL DISEASE

Based on localization of the lesion, signalment and progression of neurologic signs, a short list of potential causes (the differential diagnosis) is formulated. The DAMNIT V or VITAMN D mnemonic is helpful in formulating a differential diagnosis. The following is an example of a differential diagnosis for a 6 yo Mc domestic short haired cat with progressive intracranial signs (primary differentials are in **bold)**

- Degenerative – lysosomal storage disease
- Anomaly – hydrocephalus
- Metabolic – portosystemic shunt, thiamine deficiency
- **Neoplasia – meningioma, lymphoma,** glioma, cholesterinic granuloma
- **Inflammatory – FIP virus**, Toxoplama gondii, Cryptococcus sp., **chronic bacterial infection or abscess**
- Trauma/Toxicity - trauma is not considered likely with progressive nature, chronic lead intoxication
- Vascular – vascular anomaly (hamartoma), subdural hematoma

BRAIN IMAGING

Imaging of the brain may provide a presumptive diagnosis or "clues" as to the underlying cause. Radiographs of the head are not usually helpful unless there is skull trauma, a large tumor of bone (i.e. multilobular osteochondrosarcoma) or pneumocephalus secondary to a recent craniotomy. Computed tomography (CT) and/or magnetic resonance imaging (MRI) are often available for brain imaging at veterinary referral hospitals.

Computed tomography (CT) uses x-rays to image the brain in thin, typically, transverse slices. Those images can be reconstructed with post processing software to generate sagittal and dorsal plane images. CT imaging is useful in detecting skull trauma, intracranial hemorrhage and mass lesions; however, ischemic stroke, inflammatory disease or subtle lesions may be missed or only vaguely defined. Intravenous contrast typically helps to define mass lesions and the accuracy for predicting tumor type is approximately 80%.[4]

Magnetic resonance imaging (MRI) takes advantage of the tendency for hydrogen protons within water to align along a magnetic field. After the protons have aligned, an excitation radiofrequency is applied causing misalignment of the hydrogen protons. As the protons return to their original energy state and realign, radiofrequency energy is emitted, creating a signal that is registered by a receiver coil.[5] Different MRI sequences can be used to detect pathology within the brain and determine the nature of a lesion. T2-weighted (T2W) sequences cause water to appear white and tend to be best for picking pathology. T1-weighted sequences are better for anatomy and are usually performed before and after contrast to determine if a lesion enhances or "lights up." Tumors and inflammatory lesions frequently enhance with contrast whereas degenerative diseases and acute ischemic stroke do not. FLAIR is a T2-W sequence that suppresses normal CSF and can help to pick up disease around the ventricle or determine the nature of an intracranial cyst.[6] Hemo or T2*W sequences look for hemorrhage (acute or chronic) within the brain. Diffusion imaging is used in the diagnosis of acute stroke. Normal MR imaging of the brain does not exclude the presence of intracranial disease. One study of dogs with inflammatory CSF found that 24% (6/25) had MR images that were interpreted as normal.[7]

INTRACRANIAL PRESSURE AND BRAIN HERNIATION

The Monro-Kelly doctrine states that the cranial compartment is rigid and holds a limited volume made up of brain tissue, blood and cerebrospinal fluid (CSF). A change in one of these constituents will inevitably cause a change in the others.[8] Intracranial pressure (ICP) is the pressure exerted between intracranial tissue/fluid and the skull.[9] For ICP to stay the same, an increase in one of the intracranial components is compensated by a decrease in the others (intracranial compliance). Compliance is, however, limited and eventually an increase in tissue, fluid or blood within the cranial cavity leads to an increase in ICP. Elevated ICP reduces cerebral blood flow causing brain tissue hypoxia, dysfunction and eventually death. Cerebral perfusion pressure (CPP) is the difference between the pressure pushing blood into the brain (mean arterial pressure or MAP) and the pressure keeping blood out (ICP):[9]

$$CPP = MAP - ICP$$

An increase in ICP will cause a decrease in CPP due to a greater force keeping blood from flowing into the cranial vault:

$$\downarrow CPP = MAP - \uparrow ICP$$

In this circumstance, brain perfusion and cerebral blood flow is maintained by a relative increase in systemic blood pressure:

$$\text{Normal } CPP = \uparrow MAP - \uparrow ICP$$

A significant increase in systemic arterial pressure leads to reflex bradycardia via baroreceptors. This has been called the Cushing's reflex after the neurosurgeon Harvey Cushing.

A space-occupying mass, inflammation, hemorrhage or edema within the cranial vault will initially lead to displacement of CSF and, to a lesser degree, blood. Brain tissue will eventually begin to shift causing brain herniation within the cranial cavity, or into the foramen magnum. If there is a breach in the cranial vault (skull fracture, craniectomy or fontanelle), brain tissue may herniate through this opening. Severe mental depression or obtundation is an early sign of elevated ICP and brain herniation. If the cerebral hemisphere herniates under the tentorium cerebelli (caudal transtentorial brain herniation), there may be compression of the oculomotor nerve and leading to ipsilateral pupillary dilation.[9] Severe herniation of the cerebrum will push the cerebellum into or through the foramen magnum causing compression of the medulla oblongata. In the patient, this causes signs of central vestibular disease or opisthotonos with extensor rigidity.

TREATMENT FOR INTRACRANIAL DISEASE

Therapy for intracranial disease is directed at reducing intracranial pressure and normalizing brain perfusion. Intracranial pressure may be reduced in the following ways:[9]

- Elevating the head to approximately 30 degree above the heart to facilitate venous drainage.
- Mannitol is a monosaccharide osmotic diuretic that has been demonstrated to be effective in decreasing vasogenic, cytotoxic and interstitial brain edema and consequently decreasing ICP.[10] Mannitol should be administered over 10-20 minutes at a dosage of 0.25-1 g/kg. An easy "rule of thumb" for mannitol is 1 mL per pound of body weight of a 20% solution. Dosing may be repeated every 6-8 hours.
- Furosemide (0.6 – 1 mg/kg IV) can be given to lower ICP or potentiate the effects of mannitol
- Corticosteroids have been proven to decrease cerebral vascular permeability, reduce CSF production, and scavenge free radicals in experimental models of brain edema. These effects are well proven in treating peritumoral brain edema. Dexamethasone sodium phosphate is administered at dose of 0.15-0.25 mg/kg IV q 8-12 hrs. The benefits of corticosteroids (low and high dose methyprednisolone and dexamethasone) for head trauma have been evaluated in numerous large clinical trials evaluating clinical outcome, ICP or both and found no substantial benefit.[11]

- Hyperventillation to decrease $PaCO_2$ causes cerebral vasoconstriction and subsequently reduces ICP. Vasoconstriction within the normal brain shunts blood to abnormal regions thus improving perfusion to ischemic brain tissue.
- Removal of CSF can improve intracranial pressure if obstructive hydrocephalus is present. A ventriculoperitoneal shunt can be placed in the lateral or 4th ventricle if there is congenital hydrocephalus or obstructive hydrocephalus secondary to a brain tumor.
- Removal of a portion of the skull with durotomy can improve ICP if traditional therapy has failed. Craniectomy is most commonly performed to remove or cytoreduce a space occupying mass.

ANESTHETIC CONSIDERATIONS FOR INTRACRANIAL DISEASE AND CRANIOTOMY

When making an anesthetic plan for a patient with intracranial disease, anesthetic drug choice and careful monitoring should be in the forefront of the plan. It is of primary importance to support cardiovascular and respiratory function in order to maintain CPP and appropriate ICP.[12] A good understanding of the disease process for which the patient is undergoing anesthesia will also aid in developing an anesthetic plan. As much preparatory work as possible should be completed before the patient undergoes general anesthesia. This includes presurgical shaving, placing a second IV catheter, and an arterial catheter for direct blood pressure monitoring. Induction drugs are carefully chosen based on the specific needs of the patient. General goals of anesthesia are to maintain MAP (mean arterial blood pressure) between 70-80mmHg, end tidal carbon dioxide level ($ETCO_2$) in the range of 25-35mmHg (with a preference of close to 30), and a heart rate of 120-160 beats per minute (bpm) for cats and 60-110 bpm for dogs. A dopamine constant rate infusion (5-10mg/kg/min) can be used to improve blood pressure. Fluid overload may exacerbate vasogenic cerebral edema, so fluids should be used judiciously. Hyperthermia can cause an increase in ICP, therefore vigilant monitoring of core body temperature will aid in decisions of how to maintain appropriate body temperature.

Premedication should consider whether a patient with intracranial disease has a history of epileptic seizures. For most patients, methadone (0.3mg/kg IV) along with phenobarbital (4mg/kg IV) or levetiracetam (20 mg/kg IV) will sedate the patient and allow for smooth induction. Propofol is the most commonly used induction drug for patients with intracranial disease. Many patients will become apneic after induction with propofol, increasing $ETCO_2$ +/- ICP, and so patients should be intubated immediately and placed on a mechanical ventilator. Ketamine, at anesthetic doses, may increase ICP and should be avoided in patients with intracranial disease. Etomidate may seem like a good choice due to the limited effects it has on the cardiovascular function; however it has been associated with cerebral hypoxia and ischemic injury.

Correct choice of endotracheal tube size is important when anesthetizing a patient with intracranial disease because coughing and gagging at intubation can contribute to elevating ICP. When intubating cats, lidocaine should be used to prevent laryngospasm. Most of the common gases used today are very safe for use in patients with intracranial disease. Desflurane has been associated with higher ICP and should be avoided. Isoflurane or sevoflurane have a fairly equivalent desirability with regard to their use as a maintenance anesthetic during surgery. Sevoflurane may be more desirable due to its slightly lower blood: gas partition coefficient resulting in a more rapid induction and recovery from anesthesia; the advantages of this characteristic may, and likely will be diminished with the use of various premedications and adjunct analgesics. The use of adjunct analgesia intra-op will help reduce the MAC (minimum alveolar concentration) needed to maintain the patient with inhalation anesthetics during surgery; higher MAC levels depress the cardiovascular and respiratory systems resulting in lower MAP and higher $ETCO_2$ levels. Opioids are typically used for intra-op analgesia and can be continued post-operatively; fentanyl is administered at a constant rate infusion (CRI) at 3-10mcg/kg/hr or morphine at 0.1-0.5mg/kg/hr. Adding lidocaine to run between 0.25-1 mg/kg/hr can aid in a lowering MAC. The use of remifentanil, in substitution for fentanyl, has become popular in recent years at some institutions as it is quickly metabolized in the plasma resulting in faster recovery from respiratory and mental depression secondary to opioid administration.[13]

A patient with elevated ICP may benefit total intravenous anesthesia (TIVA) as most inhalant anesthetics cause vasodilation which can increase ICP. A common protocol for TIVA involves fentanyl (0.3 - 0.7 ug/kg/min IV) or remifentanil in combination with a propofol (0.1 - 0.5 mg/kg/min IV). With TIVA, recovery is typically delayed and so plan for post-op ventilation should be in place. In our experience, TIVA is a viable option for dogs undergoing craniotomy; however most cats require some inhalant to minimize propofol associated hemolysis and post-operative anemia.

CRANIOTOMY POST-OPERATIVE CARE

Post-operative care following craniotomy follows the same principles as for managing the patient under anesthesia. A continuous monitor (ECG or pulse oximetry) should be placed on the patient for at least the first 24 hours following surgery. Blood pressure (direct or indirect) should be checked regularly to monitor for Cushing's reflex. Occlusion of the jugular veins and jugular venipuncture should be avoided since this may increase ICP. The head should be kept elevated and staff/owners are instructed not to pet the patient on the head. We typically maintain our patients on a fentanyl/lidocaine CRI running fentanyl at 1.5-3mcg/kg/hr and lidocaine at 0.25-1mg/kg/hr. Sedation with midazolam or diazepam (0.1-0.2 mg/kg IV) may be used keep the patient calm. Cefotaxime (50 mg/kg IV TID) or clindamycin (10-15mg/kg IV or PO BID) are used perioperative or post-operatively for antibiotic prophylaxis because of good intracranial penetration.

The first night following surgery, the patient's mentation is checked hourly. A lubricating ophthalmic ointment is applied every 4-6 hours since many patients do not have a good blink reflex or are unable to fully close their eyes due to sutures in the skin of the head. Antiepileptic drugs (AED) are continued post-operatively if there have been presurgical seizures but are not typically given prophylactically. Additional mannitol may be given as needed. Aspiration pneumonia is a major potential complication for dogs following craniotomy.[14] Maropitant (1 mg/kg SQ or IV slowly over 10 minutes) and/or metoclopramide (1-2 mg/kg/d as a CRI) should be administered to minimize post-operative regurgitation and vomiting. Famotidine (0.5-1 mg/kg IV q12-24 hours) helps to lower gastric pH which may minimize lung injury in cases that do aspirate. Electrolyte levels are monitored on the first and second day following surgery since mannitol can cause wasting of potassium. Most cats will need a supplementation of potassium chloride in their maintenance fluids. Hemocrit is monitored daily since anemia, potentially requiring a blood transfusion, is a common occurrence in cats following craniotomy.[3]

Once the patient is interested in food, which is usually first or second day post-op, they are transitioned to oral medications. For pain management, tramadol (2-5mg/kg PO BID-TID) is typically used in dogs and buprenorphine (0.01 mg/kg PO TID) in cats. Gabapentin (5-10 mg/kg PO BID-TID) can be used for adjunct pain management in dogs or cats. If a patient is non-ambulatory, due to disease or drugs, following craniotomy we will perform passive range of motion exercises, massaging all 4 limbs, turning the patient every 4 hours along with reflex stimulation. If the patient is able to get up and walk on their own then physiotherapy and proprioceptive exercises are instituted early to hasten recovery.

REFERENCES

1. Leigh EJ et al. Vet Radiol Ultrasound. 2008;49:11.
2. Bagley et al. J Am Vet Med Assoc. 1999 Sep 15;215:818.
3. Sessums K et al. Compend Contin Educ Vet. 2009 Jul;31:330.
4. Turrel JM et al. J Am Vet Med Assoc. 1986;188:851
5. Hecht S et al. Vet Clin North Am Small Anim Pract. 2010;40:21.
6. Benigni L et al. Vet Radiol Ultrasound. 2005;46:287.
7. Lamb CR et al. Vet Radiol Ultrasound. 2005;46:17.
8. Mokri, B. Neurology. 2001;56:1746.
9. Bagley RS. Vet Clin North Am Small Anim Pract. 1996;26:711.
10. Brain Trauma Foundation, Inc, American Association of Neurological Surgeons. Part 1: guidelines for the management of severe traumatic brain injury. New York (NY): Brain Trauma Foundation, Inc.; 2000.
11. Edward P et al. Lancet. 2005;365:1957.
12. Armitage-Chan EA et al. J Vet Emergency and Crit Care. 2007;17: 5.
13. Warne LN et al. Can Vet J. 2014 Nov;55:1083.
14. Fransson BA et al. Vet Surg. 2001;30:432.

RECURRENT UTIS: CAN ANYTHING REALLY BE DONE TO STOP THEM

Shelly Vaden, DVM, PhD, DACVIM (SAIM)
Raleigh, NC

INTRODUCTION

It is estimated that 14% of female dogs will develop a urinary tract infection (UTI) within their lifetime.[1] Although fewer male dogs are diagnosed with UTI than are female dogs, the percentage of male dogs with UTI increases with advancing age. Bacterial UTI is less common in cats, accounting for only 1-3% of all cases of feline lower urinary tract disease in North America. As cats age they are more commonly diagnosed with a UTI; the percentage of older cats (i.e., > 10 years of age) with UTI may be as high as 50%. Thankfully, recurrent UTI, defined as >2 UTI in a 6 month period or >3 UTI in a year, is encountered less frequently than simple, uncomplicated UTI. While the management of dogs with acute, uncomplicated UTI is fairly straightforward, the management of dogs with chronic UTI, due to either recurrence or persistence, can be anything but straightforward.

In recent decades resistant pathogens are more commonly encountered, further complicating the management of animals with UTI. Of particular concern are the uropathogenic *Escherichia coli* (UPEC) and *Enterococcus sp*. UPEC is the most common etiologic agent for UTI in most mammalian species. The ability of UPEC to establish intracellular residence and form biofilms allows the UPEC to avoid normal host defense mechanisms and makes them more difficult to eradicate with traditional antibiotic regimens.[2]

Antibiotic therapy is standard treatment for UTI but overuse leads to changes in the normal host microbiota and selection for resistant bacterial strains. Although severe adverse effects of antibiotic therapy for UTI may be infrequent, mild adverse effects are probably not and chronic antimicrobial administration is expensive for the client. There has been a shift of focus to the prevention of UTI in at risk people. This discussion will focus on different ways that recurrent UTI might be prevented, including strengthening of the host defense mechanisms, prophylactic antibiotics, and novel therapeutic approaches.

HOST IMMUNE SYSTEM

Normal host immune mechanisms, both systemic and local (or mucosal) are crucial for the prevention of bacterial colonization and UTI in normal animals. These immune mechanisms are highly complex and may be incompletely understood but it is clear that interruption of normal host defense mechanisms can lead to repeat infections. Proper identification and management of disorders known to lead to immunosuppression (e.g., diabetes mellitus, hyperadrenocorticism) is essential in the prevention of UTI in patients affected by these disorders. Likewise, animals being administered drugs that are known to disrupt the immune system (e.g., corticosteroids, chemotherapeutic agents) should be appropriately monitored for the development of UTI. UPEC can avoid local immunity by establishing an intracellular residence. There are 3 potential fates for intracellular UPEC. They can be recycled to the cell surface where they are exocytosed, live a quiescent nonreplicative existence, or live a highly replicative phase in the cell cytoplasm.[3] These latter two fates allow UPEC to establish a reservoir in the urinary tract. When the epithelial cells are exfoliated, the UPEC become free to attach to other epithelial cells and cause a UTI.

URINE COMPOSITION

Normal urine composition does not favor bacterial growth and is, therefore, a natural host defense against bacterial invasion. Although urine pH has long been considered to be an important factor inhibiting bacterial growth, the reality is that it is not possible to use this inhibitory characteristic in a clinical setting. In people bacterial growth occurs freely when urine pH is 5.0-7.0.[4] However, when urine pH is <4.5 or >9.0, there is effectively no growth of any organism. It would be difficult and unwise to therapeutically manipulate an animal to produce urine that has a pH of <4.5 or >9, rendering the use of urine pH in the prevention of UTI impractical. On the other hand, high urine osmolality, as measured by urine specific gravity, also prevents bacterial growth. Animals with reduced urine concentrating abilities should be thoroughly evaluated for all correctable causes of polyuria and polydipsia.[5] Physicians have long advocated forced fluid intake in people with UTI theorizing that this would flush the urinary tract of pathogens, reduce bacterial growth and division in the bladder and allow the bladder mucosal defenses to operate more effectively because they would not be overwhelmed by larger numbers of bacteria. The reality is the high flow rates through the bladder may select for bacteria that are able to strongly adhere to the bladder mucosa. Given the protective effect of highly concentrated urine and the fact that "flushing the bladder" by increasing urine flow does not appear to be effective, this recommendation for lower UTI seems unwarranted. However, there may be some benefit of forcing fluids in prevention or treatment of pyelonephritis. Increased flow rates of filtrate through the kidney might help promote a more isotonic environment in the renal medulla that could promote more effective granulocyte mobilization into the areas of infection.

ANATOMY

Normal anatomy is also important in UTI prevention. Congenital abnormalities of the urinary tract should be identified and corrected in young animals with recurrent UTI. In recent years it has become apparent that hooded vulva (also called recessed vulva or excessive perivulvar skin folds) is a clinical relevant abnormality in female dogs with recurrent UTI.[6,7] Hooded vulva is frequently associated with dermatitis, creating ideal conditions for bacterial growth. Animals with hooded vulvas may have perivulvar hairs that fold inward, wicking bacteria into vestibule. Hooded vulvas are often found in young dogs with infantile vulvas but also in obese dogs and in older dogs with reduced skin elasticity. The sometimes-proposed management strategy of repeat cleaning of the folds can be onerous for the owners and does not always appear to be effective. Episioplasty has been reported to be effective in the resolution of the clinical signs associated with UTI in 84-100% of dogs with hooded vulva. We generally recommend episioplasty in dogs with >50% coverage of the vulva by surrounding folds AND a history of recurrent UTI.

Vestibulovaginal septal remnants are bands found in the vestibule, at the opening of the vagina. These bands vary from being very thin, like an imperforate hymen, to being thick enough to give the illusion of vaginal duplication. In some dogs these are incidental findings whereas, in other dogs, the might be associated with recurrent UTI. Bands are identified during vaginoscopy. Thin bands may be digitally broken whereas thicker bands may require surgical or laser ablation. One study demonstrated a significant reduction in the rate of UTI after ablation of the septal remnants.[8]

MICTURITION

Normal micturition is another important factor in normal defenses against UTI. In theory, normal voiding should eliminate bacteria from the urinary tract. However, in actuality voiding may not mechanically free bladder of adherent bacteria that are in a biofilm and clearance from the bladder may be more closely associated with antibacterial activity of mucosal surface than with voiding. By mechanisms that seem to be counterintuitive, UPEC adherence may actually be strengthened by sheer stress induced during urine flow. Nonetheless, bladder dysfunction and urine retention have clearly been shown to be risk factors for UTI in dogs. Frequent voiding with complete emptying may be an important factor in prevention of UTI in dogs and cats that are at risk of recurrent UTI. Likewise, urinary incontinence is associated with a break in the barrier against ascending UTI. Effective management of urinary incontinence may aid in the prevention of UTI in some dogs.

Dogs with intervertebral disc disease have a higher incidence of urinary tract disease when compared with the general population. Dogs with neurologic dysfunction often create a unique challenge by having chronic urinary incontinence or retention that cannot be effectively managed. Judicious cleansing of the perineum, as well as keeping the area dry, is required in dogs with urinary incontinence, in hopes to prevent urine scald and bacterial colonization of the perianal area. Perineal hair should be clipped in longhaired dogs. Urinary retention is even more common than urinary incontinence in this population of dogs. The bladder can be difficult to manually express and the effectiveness of manual expression in emptying the bladder is highly variable. The effectiveness of expression can be assessed by intermittent bladder ultrasound. Drugs (e.g., bethanecol, cisapride) or intermittent urethral catheterization can be used to facilitate more complete emptying of the bladder. Indwelling urethral catheterization has a high risk of causing UTI and should be avoided. When it is absolutely necessary to maintain an indwelling urethral catheter for a prolonged period of time, topical therapy as discussed below should be considered.

PROPHYLACTIC ANTIMICROBIAL THERAPY

The use of prophylactic antibiotics has long been advocated in dogs with chronic UTI. It has been suggested that prophylactic antibiotic administration be initiated after potential underlying causes are corrected and UTI has been eliminated. Once daily, bedtime dosing (i.e., antibiotics given at bedtime, after voiding) is used to eliminate ascending bacteria before UTI can be established. Pulse therapy (i.e., short period of antibiotic administration given at regular intervals such as one week out of every month) has also been suggested but may be more likely to induce bacterial resistance than bedtime dosing. In general, antibiotics that are relatively safe and inexpensive should be selected (e.g., amoxicillin, amoxicillin-clavulanic acid, nitrofurantoin). First and second generation cephalosporins generally should not be used for this purpose because of their ability to induce global antimicrobial resistance.

Methenamine mandelate may be effective in situations where bedtime antibiotics have failed to prevent UTI. This agent is converted to formaldehyde in acidic urine; therefore, concurrent administration of urinary acidifiers may be required for the drug to be effective. In 2011, the National Toxicology Program declared that formaldehyde was carcinogenic. Therefore, methenamine carries a theoretical risk of causing bladder cancer. It is currently recommended that methenamine therapy in people be limited to one week in duration, making it difficult to recommend this drug for long-term administration in dogs.[9]

The question remains whether or not prophylactic antimicrobial therapy is effective in dogs at all. Large meta-analyses of human studies suggest that any benefit is likely to be small and of limited clinical significance, with recurrent UTI returning after cessation of prophylactic therapy.[10] Clinical experience suggests that the same may be true in dogs. Antibiotic prophylaxis may also be associated with more adverse drug effects and increased antimicrobial resistance of both the bacteria causing the UTI and those residing in the host microbiome.

TOPICAL THERAPY

Because of the problems associated with systemic antibiotic administration and the creation of resistance, there has been increased interest in topical therapy for the treatment of recurrent UTI. Protocols are available for the use of intravesical antibiotics, including gentamicin, tobramycin, vancomycin, and piperacillin but very few studies support their efficacy or safety. Absorption of agents instilled in the bladder is presumed to be minimal but could be increased when mucosal inflammation is present. Any animal given an aminoglycoside via intravesical instillation should have periodic serum aminoglycoside concentrations measured. Protocols are also available for topical therapy with Tris-EDTA or dilute chlorhexidine solution, generally instilled through indwelling or intermittently placed catheters. These agents may be safer and carry a reduced risk of creating antibacterial resistance when compared with using antibiotics.

A defective glycosaminoglycan layer in the bladder has been linked to the development of recurrent UTI. This has led people to instill hyaluronic acid or chondroitin sulfate in the bladder. Because heparin makes up 15% of the glycosaminoglycan layer and heparin is relatively inexpensive, heparin has also been instilled in bladders of people with recurrent UTI. In one study of people with difficult to treat recurrent UTI, heparin was given intravesicular once weekly for 6 weeks. This was associated with a reduction in UTI from 4.4 infections in the 6 months before treatment to 1.1 infections in the 6 months after treatment.[11] All of these treatments warrant further investigation in dogs.

CRANBERRY

Cranberries have long been suggested for the prevention of UTI.[12] Cranberries contain organic acids, fructose, glucose and phytochemicals. Certain phytochemicals, such as A-type proanthocyanidins (PACs), inhibit the adhesion of p-fimbriated UPEC to uroepithelial cells. It is this inhibitory property, and not mild acidifying property, of cranberries that is effective in the prevention of UTI. Cranberry products might also reduce the symptoms of UTI by suppressing the inflammatory response.

Although initial studies in people were somewhat contradictory, a meta-analysis concluded that there is statistical evidence that cranberry extract reduces the recurrence of UTIs, particularly in young women. The incidence of UTI is reduced by approximately 35-40% in women consuming cranberry products. There is evidence to suggest that inhibition of bacterial adherence to bladder epithelial cells is dose-dependent and lasts for approximately 10 hours. It is possible that an insufficient dose or dosing frequency was used in some of the trials in which a beneficial effect was not found. Many cranberry products are available – for both people and animals. Select products are extracts of purified PACs. Purified PACs have been shown to reduce bacterial adhesion in a bacterial adhesion model in dogs. D-mannose has also been shown to inhibit bacterial adhesion to the bladder wall via the same mechanisms as cranberry. Some cranberry products have added D-mannose; however, there is not a clear benefit of these products. Whenever possible, cranberry products that are as pure as is possible should be used because of concerns about unknown adverse effects for any other ingredients given at high doses for a prolonged duration.

PROBIOTICS

Because most UTIs in dogs are believed to be ascending infections, maintenance of a healthy microbial population of the vestibule and vagina may be important in female dogs. In most women, the vaginal microbiome is remarkably similar, with lactobacilli being the most dominant colonizer.[13] Lactic acid producing bacteria (LAB) decrease the vaginal pH and inhibit the growth of other types of bacteria, including the pathogens E. coli, P. Mirabilis and Staphylococcus sp. Women with recurrent UTI have been reported to have a high incidence of altered vaginal microbiome with reduced numbers of lactic acid producing bacteria, a condition called bacterial vaginosis. Because of this, probiotics have been used in women with recurrent UTI in an attempt to reestablish a healthy vaginal microbiome. Probiotic lactobacilli are believed to improve urogenital health in women through immune modulation, pathogen displacement, and creation of a niche less favorable for pathogenic bacterial proliferation. Probiotics may help replenish the normal microbes, displacing pathogens and leading to a return to the lactobacilli-dominated state found in healthy women, preventing ascending UTI. However, studies have failed to demonstrate a clear benefit of probiotics in women with recurrent UTI.

Although often recommended, even less is known about the use of probiotics in dogs with UTI or the effect these agents have on the canine vaginal microbiome. The normal canine vaginal microbiome consists of a mixed population of anaerobic and aerobic bacteria, with greater numbers of bacteria present in the vestibule than in the cranial vagina. Because bacteria in the vagina originates in the gastrointestinal tract and skin, common isolates include E. coli, staphylococci, enterococci, Bacteroides sp. and mycoplasma. Less is know about LAB in the canine vaginal microbiome because many published studies did not include culture enrichment for LAB species. The vaginal pH of dogs is higher (6-7.5) than is the vaginal pH of women (<4.5), which might suggest that there are fewer LAB colonies in the canine vagina. In a study of healthy estrous dogs, healthy spayed dogs, and spayed dogs with recurrent UTI, LAB were isolated from fewer than 22% of dogs and did not differ between groups.[14]

Before considering the use of probiotics in the treatment of canine recurrent UTI, it would seem reasonable to determine if bacterial vaginosis occurs in dogs and, if so, to determine if it is a risk factor for recurrent UTI in dogs. The vaginal microbiota of spayed female dogs with a history of recurrent UTI and without a history of recurrent UTI were not significantly different. E. coli and S. psuedintermedius were the most common pathogens isolated and were isolated more commonly than LAB species.[15] When normal dogs without a history of recurrent UTI were given an oral lactobacillus containing probiotic, the prevalence of vaginal LAB did not increase.[16] This raises questions about the ability of orally-administered probiotics to alter the vaginal microbiome in dogs.

REFERENCES

1. Bartges JW. **Vet Clin North Am Small Anim Pract** 2004;34:923. 2. Tenke P, et al. **World J Urol** 2012;30:51. 3. Sivick KE, et al. **Infect Immun** 2010;78:568. 4. Yeaw RC. **J Urol** 1940;44:699. 5. Beetz R. **European J Clin Nutr** 2003;57:S52. 6. Hammel SP, et al. **JAAHA** 2002;38:79. 7. Lightner BA, et al. **JAVMA** 2001;219:1577. 8. Burdick et al. **JAVMA** 2014;244:944. 9. Geerlings SE, et al. **Infect Dis Clin N Am** 2014;28:135. 10. Williams G, et al. **Curr Opin Inf Dis** 2009;22:72. 11. Ablove T, et al. **Ther Adv Uro** 2013:5:303. 12. Vasileiou I, et al. **Nutrition Res** 2013;33:595. 13. Witkin SS, et. al. **Best Pract Res Clin Obst Gyn** 2007;21:347. 14. Bailey CS, et. al. **Clin Therio** 2012;4:405 15. Hutchins RG, et al. **JVIM** 2014;28:300. 16. Hutchins RG, et al. **JVIM** 2013;27:1368.

CYTOLOGICAL LOOK AT COMMONLY PRESENTING EFFUSIONS

Amy M. Oliver, LVT, VTS (ECC)
Auburn, AL

INTRODUCTION

Effusion is defined as an accumulation of excessive fluid within a body cavity. These include the pericardial, pleural, and peritoneal cavities. Fluid analysis and cytology from these cavities is the most effective and inexpensive way to start diagnosing a problem.

OBTAINING A SAMPLE

Collecting a sample requires skill, but can be easily mastered with proper equipment and practice. Thoracocentesis is performed to obtain a sample of pleural effusion. The art of this procedure is in identifying where the fluid lies. Fluid will accumulate in the ventral aspect of any cavity when the patient is standing or toward the down side if lying in lateral recumbency. Radiographs or ultrasound can be used as a helpful guide to direct the clinician or technician to a fluid pocket for sample collection. Performing a thoracic focused assessment with sonography for trauma (TFAST) can highlight pockets of fluid present in the thoracic cavity. Five specific sites are observed for evidence of a pleural effusion, pericardial effusion, pulmonary disease, or signs consistent with a pneumothorax. These include the chest tube site on both sides of the chest, the pericardial site on both sides of the chest, and a view of the diaphragm [Figure 1]. If an ultrasound is not available, a shaved field should be sterilely prepped between the 7th and 9th rib to ensure the needle is caudal to the heart. Though complications are minimal if proper technique is used, they include laceration of underlying lung tissue, causing pneumothorax, hemothorax, or pulmonary hemorrhage. This procedure can be not only diagnostic, but therapeutic in patients experiencing respiratory distress due to accumulation of large volumes of fluids.

Another cavity in the thorax that may accumulate fluid is the pericardial cavity, defined as the space between the pericardial sac and the heart. Pericardiocentesis requires skill and is only performed in emergent cases where the patient is experiencing cardiac tamponade. Cardiac tamponade is a serious condition in which there is increased pressure exerted on the heart due to the accumulation of fluid in the pericardial sac. Increasing pressure on the outside of the heart results in a decrease in the filling capacity of the heart. If the filling capacity of the heart is decreased, then cardiac output is impaired. This condition is life-threatening and can cause sudden death. If the fluid accumulation is slow, the pericardium stretches and the patient may lack obvious clinical signs. In the absence of clinical signs, pericardiocentesis is rarely performed. Possible complications include ventricular premature contractions, laceration of the coronary artery, exsanguination by aspirating a chamber of the heart, and sudden death. This procedure should never be performed without the accompaniment of an electrocardiogram. Hair should be clipped and the site sterilely prepped to obtain a sample. Pericardiocentesis is a classic example of a procedure that is definitely therapeutic (often saving the patient) and may also be potentially diagnostic.

Abdominocentesis is the collection of abdominal or peritoneal fluid that has accumulated. Locating fluid pockets can be done ultrasonographically to aid in collection. Fluid tends to be found in one of four areas in the abdomen. An abdominal focused assessment with sonography for trauma (AFAST) utilizes a four quadrant approach. This assesses the caudal abdomen where fluid encircles the bladder, the lateral abdomen (left and right) around each kidney, and the cranial abdomen around the diaphragm and liver [Figure 1]. If an ultrasound is unavailable, radiographs can show accumulation and guide a "blind stick" to obtain a diagnostic sample. Preferentially, hair should be shaved over the site and prepped sterilely prior to sample collection.

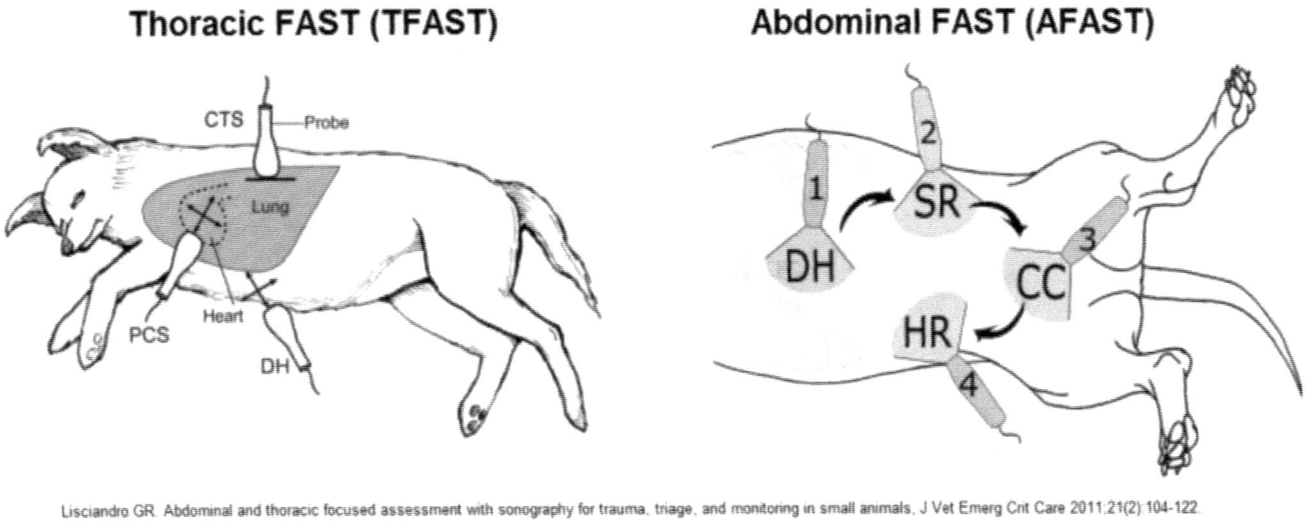

Figure 1. Diagnostic use TFAST and AFAST for effusion.

WHAT SHOULD I DO ONCE I'VE COLLECTED FLUID?

The amount of fluid needed to perform an adequate analysis depends on the cellularity of the fluid. The more transparent the fluid is, the more fluid needed for cytologic preparations to identify cells. In most instances, approximately 5 to 10 ml of aspirated fluid should be saved for cytology and cultures with antimicrobial susceptibility analysis. Placing samples of the fluid in an EDTA tube should be submitted for total nucleated cell count, total protein levels, specific gravity, and cytologic examination. In addition, fluids should be placed in a tube without any type of anticoagulant, including lithium heparin, calcium citrate, and EDTA, for chemical analysis. This is important for samples such as chylous effusion in the thorax or suspected urine in the abdomen. Samples for culture and susceptibility should be placed in a separate tube by sterile technique; performed by changing the needle on the sample syringe, swabbing the top of a red top tube with alcohol, allowing it to air dry, then inserting the new needle into the cleaned red top tube allowing the vacuum suction to obtain a small aliquot of fluid to be plated for microbial testing.

Once samples have been allocated into their respective tubes, they should be submitted as soon as possible for analysis. If this is not possible, they can be stored under refrigeration for up to 12 hours for aseptic samples. If blood is present, there is evidence that the longer the sample is stored, there is an increased likelihood of cell lysis. There is no specific scientific data regarding stability of cells in different types of fluids.

APPEARANCE OF FLUID

In most cases of effusion, the appearance of the sample can give some indication on the cellularity. Fluids with a high cell or lipid concentration have a cloudy appearance. Fluids appearing to be clear can be expected to have a low cell count. The color can indicate bile, urine, or blood presence. Examination of the supernatant after centrifugation can identify cell types and aid in diagnoses [Table 1].

Table 1. Appearance of different types of fluids.

Type of fluid	*Appearance*
Hemorrhagic	Red, pink, or yellow; similar to blood in consistency; does not clot
Chylous or pseudochylous	Opaque; milky; white or pink
Transudate	Translucent, colorless, similar to water
Modified transudate	Clear, cloudy, bloody, or serous
Exudate	Viscous; straw-colored; serosanguinous; seropurulent
Bilious	Turbid, yellow, green, brown

PREPARATION

Fluid analysis and cytology are the most important part of diagnosing the nature of effusion. Nucleated cell counts are obtained by following the usual technique for white blood cell counts. An automatic cell count should not be used if there is fibrin strands, clots, or cell aggregates. Chemistry profiles are utilized to identify triglyceride, glucose, lactate, and creatinine levels. Cytology by manual slide preparation is the most informative in most cases.

Cytological preparations should be obtained as soon as possible after sample collection to maintain cellular integrity. The sample is subjected to centrifugation, then the concentration of cells, or sediment, is fixed to a microscope slide. Markedly turbid, viscous, or samples containing a lot of debris should be prepared by doing a "crush" smear. This is done by putting the sediment at the end of one slide and using the flat side of a second slide to smear the sample across the length of the slide. This helps separate the cells for identification. Hemorrhagic, sanguinous fluids, and fluids of less cellularity should be prepared similar to a blood smear by placing a drop of sample on the end of a slide and using the edge of a second slide to smear the drop to a feathered edge. Once preparations have been obtained, they can either be stained or fixed to await microscopic review.

CELLS TYPES IN EFFUSIONS

There are various types of cells that can be present in effusions of any body cavity. Mesothelial cells, neutrophils, lymphocytes, red blood cells, eosinophils, macrophages, and neoplastic cells can all be observed microscopically. The absence of tumor cells in cytology does not rule out the presence of neoplasia as the cause of the effusion. It is very difficult to differentiate between tumor cells and activated mesothelial cells, or macrophages. Cytological review requires knowledge and experience. Analyses of effusive fluids from cats can differ slightly from results in dogs [Table 2].

Table 2. Analysis of effusion types.

Type of effusion	*TP (g/dL)*	*WBC (#/uL)*	*NC (#/uL)*	*Specific Gravity*	*Predominant cell type*	*Other possible findings*
Hemorrhagic	>3.0	>1,000	Variable	>1.015	Erythrocytes, leukocytes	Macrophages with evidence of erythrophagocytosis
Chylous	>3.0	Variable	<10,000	>1.018	Acute: small lymphocytes Chronic: non-degenerate neutrophils, mixed population	Lymphoblasts, fat droplets may be present
Transudate *Cat	<2.5 *<3.5	<1,000	<1,500	<1.017	Mesothelial, macrophages	Lymphocytes, non-degenerate neutrophils
Modified transudate *Cat	2.5-7.5	500-10,000	1,000-7,000 *<5,900	1.017-1.025	Mixed population	
Exudate *Cat	>3.0 *>3.5	>5,000	>7,000 *>5,900	>1.025	Neutrophils; non-degenerative (non-septic) or degenerative (septic)	Macrophages, mesothelial, lymphocytes, eosinophils, neoplastic cells, bacteria/fungi, urinary crystals
Bilious (exudative)	>3.0	>5,000	>7,000	>1.025	Mixed population with green-black or yellow material phagocytized by macrophages	Biliary crystals

TP: total protein; NC: nucleated cells

The number one cause of pericardial effusion in dogs in neoplasia, especially if the patient is more than 7 years old. Cytology of pericardial effusion is almost always hemorrhagic and non-diagnostic. Reactive mesothelial cells can appear to be neoplastic, when in fact they are not. The most common types of neoplasia are hemangiosarcomas and chemodectomas. Neoplastic cells are rarely found because neither type of these neoplasia are highly exfoliative. On rare occasion, lymphoma can be diagnosed by the presence of affected lymphocytes. Since most effusion of the pericardium yields high levels of erythrocytes, a packed cell volume should be obtained and compared to the peripheral sample. A coagulation panel should also be performed to rule out a coagulopathy in the case of hemorrhagic pericardial fluid.

Pleural effusion can present as a chylothorax, hemothorax, or pyothorax, in addition to transudative effusion. Chylous fluid has a pink or white appearance and is usually high in triglycerides that do not dissipate when centrifuged. This fatty fluid is easily identified. Blood or sanguinous fluid present in the pleural space is indicative of an active or chronic bleed. There will be a high red cell count on review of the cytology. A coagulation panel should be reviewed to rule out a coagulopathy in these cases. Pyothorax indicates sepsis within the thoracic cavity. Intracellular bacteria will be present, as well as degenerate neutrophils. A translucent, or clear fluid can be expected to have a low cellularity and usually contains only mesothelial cells.

Abdominal, or peritoneal effusion can vary immensely in cell type. Ruptured bladders and urethral tears yield urine in the abdomen, while, bilious effusion come from a ruptured gall bladder. Septic fluids can come from any infectious or inflammatory agent. If urine is suspected, analyzing a creatinine level on the fluid could be indicated to determine the presence of uroabdomen. In the case of bile peritonitis, the presence of bilirubin in the fluid can be a definitive indicator. Septic peritonitis could be determined by comparing lactate and glucose levels in the abdominal fluid to that of the peripheral blood sample. Lactate levels of abdominal fluid that are 2.0mmol/L or greater than the blood lactate level is indicative of septic peritonitis in dogs. Cats have not proven to follow this trend. Glucose levels with a difference of 20 mg/dL in effusion compared to peripheral blood has also been a positive indicator for septic peritonitis. Transudative fluids can be present as well.

CAUSATIVE AGENTS OF EFFUSIONS

There are a multitude of causes of effusive fluids in the various cavities of patients. From trauma to coagulopathies and heart failure to neoplasia, effusions are a common occurrence that can help diagnose a disease process. Knowledge of the type of fluid present can lead to a potential understanding of the pathology that is associated with the effusion [Table 3].

Table 3. Examples of indicated pathologies associated with effusion types.

Types of Effusion	*Mechanism of effusion production*	*Potential pathologies*
Hemorrhagic	Leakage from blood vessels	Abdominal or thoracic trauma Neoplasia (hemangiosarcomas) Alteration in secondary hemostasis
Chylous	Leakage from lymphatic vessels, usually obstructive	Compression or obstruction of thoracic duct (mediastinal neoplasia, granulomas, thrombus) Congestive heart failure Cardiomyopathies, heartworm disease, pericardial effusions Diaphragmatic hernia Lung lobe torsion
Exudate	Increased vascular permeability due to inflammation	Bacteria, fungi, viruses, protozoa, parasites Neoplasia Sterile foreign body Bile peritonitis Uroperitoneum Necrotizing pancreatitis Gastrointestinal perforation
Transudate	Excessive water diffusion from vascular space due to increasing capillary hydrostatic pressure and decreasing oncotic pressure Lymphatic drainage impairment	Hepatic portal hypertension Hypoalbuminemia Hepatic vein obstruction Caudal vena cava obstruction Right heart congestive failure Cardiac tamponade Cranial mediastinal masses Diaphragmatic hernia Lung lobe torsion

REFERENCES

1. Silverstein, DC, et al. Pleural Space Disease. *Small Animal Critical Care Medicine*. 2nd Ed. St. Louis, MO: Saunders Elsevier; 2014:151
2. Radlinsky, MG. Pleural Effusion. *Mechanisms of Disease in Small Animal Surgery*, 3rd Ed. Ithaca: International Veterinary Information Service (www.ivis.org), 2014; Document No. A5657.0314.
3. De Laforcade AM, et al. Biochemical analysis of pericardial fluid and whole blood in dogs with PE. *J Vet Intern Med*. 2005; 19 (6): 833-6
4. Silverstein, DC, et al. Pericardial Disease. *Small Animal Critical Care Medicine*.2nd Ed. St. Louis, MO: Saunders Elsevier; 2014: 239
5. Cortellini, S, et al. Plasma lactate concentrations in septic peritonitis: A retrospective study of 83 dogs (2007–2012). *J Vet Emer Crit Care*.2014; online
6. Silverstein DC, et al. Peritonitis. *Small Animal Critical Care Medicine*. 2nd Ed. St. Louis, MO; Saunders Elsevier; 2014. 643
7. Lisciandro, GR, Abdominal and Thoracic focused assessment with sonography for trauma, triage, and monitoring in small animals. *J Vet Emer Crit Care. 2011; 21(2); 104*

Mast Cell Tumours: New Grading, New Treatments

Antony S. Moore, BVSc, MVSc, MANZCVSc, DACVIM (Oncology)
Wauchope NSW, Australia

INTRODUCTION

In dogs, mast cell tumours (MCTs) are most commonly found in the cutaneous tissue; and this is the most common malignancy of the skin in dogs. In most dogs, tumours are solitary, but in about 10 - 15%, the tumours are multiple. Tumours usually occur in older dogs (mean age 9 years) with no sex predilection. Boxers, Rhodesian ridgebacks, pugs, Boston terriers, pit-bull terriers and Weimaraners are at high risk (4 to 8 times more than the general population) for developing MCTs. Shar-Peis, particularly young dogs, are predisposed to developing MCT, and these tumours are often poorly differentiated and biologically aggressive.

It is uncommon to diagnose MCTs without skin involvement in dogs. Mast cell tumours vary greatly in appearance, and no estimate of their malignancy or prediction of their behavior can be made on clinical appearance alone. Some MCTs may be present for months to years before rapidly disseminating; others act aggressively from the beginning. Occasionally, mechanical manipulation during examination of this tumour causes degranulation of mast cells, producing erythema and wheal formations. This phenomenon has been observed in both dogs and cats (the "Darier's sign") and is considered of diagnostic significance. Owners may report that the tumour enlarges rapidly and then diminishes in size over a period of about 24 hours. Such a history should increase the clinician's suspicion of mast cell tumour.

The clinical appearance of MCTs in dogs may vary widely but diagnosis is relatively easy using aspiration cytology. Presurgical aspiration of these tumours provides a cytology specimen characterized by *round cells* that may have well-stained and large cytoplasmic granules (well-differentiated) or that may be more anaplastic with small, poorly staining cytoplasmic granules. Eosinophils are often seen in aspirates of MCTs because of eosinophil chemotaxis to histamine release.

GRADING

While diagnosis of MCT often can be made by fine needle aspiration cytology; excisional biopsy is required for accurate histologic grading of the tumour. Histopathologic grading of the tumour has been correlated with both recurrence and survival, with Patnaik grade 1 MCT essentially having benign behaviour, and grade 3 being more likely to show malignant behaviour. Grade 2 MCT have an "intermediate" behaviour.[1]

One study found that mitotic index was strongly prognostic for survival with grade 2 MCT with a mitotic index (per 10 high power fields) of < or =5 having a survival time of 70 months, compared to 5 months if the mitotic index was >5.[2] Another study looking at mitotic index suggested 3 levels of prognostic significance, with dogs that had tumours with mitotic index <1 having very long survivals; with mitotic index 1 – 7 having median survival of 18 months; and dogs with mitotic index >7 having median survival of 3 months. Our clinical experience has been that we have occasionally seen some MCT behave aggressively when the mitotic index was less than 5. Usually these have a mitotic index of 2 or more; but have we seen lymph node metastasis with MCT with a mitotic index of 0 or 1.

A recent pathology revaluation of the grading system for MCT proposed a new two-tiered scheme that identified dogs at high risk regardless of their Patnaik grade.[3] In that study a Kiupel high grade MCT was characterized by: at least 7 mitotic figures in 10 hpf; or at least 3 multinucleated (3 or more nuclei) cells in 10 hpf; or at least 3 bizarre nuclei (highly atypical with marked indentations, segmentation, and irregular shape) in 10 hpf; or karyomegaly (nuclear diameters of at least 10% of neoplastic mast cells vary by at least two-fold). One advantage in this system is its simplicity for pathologists in applying, eliminating some of the subjectivity inherent in the Patnaik system. The study showed a median survival time of <4 months for high grade tumours and >2 years for low grade tumours. The report showed that all dogs with Kiupel high grade MCT had a uniformly negative outcome with their MCT; however not all dogs with low grade tumours had uniformly positive outcomes. The difficulties in relying on the Kiupel system are that it is less informative for dogs with low grade tumours than the Patnaik system, and that little clinical / therapeutic data has been published using this system – the majority is based on the Patnaik system and mitotic index. Therefore, clinically, while the Kiupel system is being used more and more; we still use the Patnaik system plus mitotic index to a large extent, and it is very helpful when the pathologist provides all 3 pieces; as recently suggested by the VCS Oncology-Pathology Working Group on MCT.

STAGING

All dogs with MCTs should be staged to determine the extent of their disease. This is especially important for dogs being considered for aggressive surgery such as amputation, or radiation therapy. To establish the stage of a dog with a cutaneous MCT, the following information is needed:

1. Complete Blood Count (CBC), Serum Chemistry Profile and Urinalysis

2. Lymph Node Aspirates

The clinician should perform fine-needle aspiration of the regional lymph node if the node is enlarged. The presence of clusters of mast cells (and eosinophils) is an indication that the MCT *may* no longer be confined to the primary site.[4] Mast cells may also infiltrate a regional lymph node in a dog with a MCT as an inflammatory response to the tumour; therefore a suspicious cytology result should be confirmed by biopsy.

3. Radiographs and Ultrasonography

Splenomegaly or hepatomegaly may indicate spread of MCTs systemically. Pulmonary metastasis of MCTs is rare. Ultrasonography is most useful for staging a dog with MCT if used in conjunction with histopathology or cytology.

4. Bone Marrow Aspirates

The presence of >1% mast cells indicates systemic spread of the neoplasm. In a recent report 4.5% of dogs with cutaneous MCT had either >1% or abnormal mast cells in a bone marrow aspirate.

5. Miscellaneous Tests

Faecal occult blood tests may be useful in evaluating patients with mast cell disease. In many cases, faeces may contain small amounts of blood that are insufficient to produce melena. Evidence of gastrointestinal bleeding in a patient with a MCT should prompt the clinician to treat with medications that block the effects of mast cell hyperhistaminaemia (i.e., H_2 blockers, such as famotidine).

6. Buffy Coat Smears

Care must be exercised in interpreting buffy coat smears, because mastocythaemia has been reported in a variety of canine acute inflammatory diseases. For this reason, most oncologists now feel that this test has limited applicability to staging of dogs with MCT.

PROGNOSTIC FACTORS

Recent clinical research has led to the identification of prognostic factors for dogs undergoing therapy with surgery and radiation therapy, and this same research has led to changes in thinking about other previously well accepted prognostic factors.

Major Prognostic Factors

Tumour Grade: Recent studies using aggressive surgical technique, and histology to examine margins (rather than the surgeon's clinical impression), has shown that dogs with grade 2 MCT have a lower rate of local recurrence and longer survival rates than was previously believed. Nonetheless grading is still an important prognostic factor. Grade 3 tumours are more likely to be incompletely excised, more likely to metastasize and nearly 4 times more likely to result in death than tumours of lower grades.

Mitotic index: As discussed above, mitotic index is clearly and important prognostic factor, regardless of which grading system is used. Depending on the study, different cutoffs for the mitotic index predict different outcomes, but in all studies and systems, the lower the mitotic index the better. Mitotic index is always counted in the area of the tumour that has the highest mitotic activity, and is always reported as number of mitoses per 10 high power fields (hpf).

Surgical Margins: The completeness of excision (i.e., whether the surgical margins are "dirty") is an important prognostic factor and also important in determining if further surgery or adjunctive radiation therapy is needed. There is often disparity between the surgeon's expectation of margins and those assessed by histopathology. In one study 22 of 59 tumours thought to have been excised widely had either questionable (10) or incomplete (12) excision based on histological examination. However, not all incompletely resected MCT will recur. Nonetheless, dogs with incomplete excisions are more likely to develop metastatic disease. Because of this finding, and because tumour recurrence is more common following incomplete excision, the clinician is counseled to obtain clear surgical margins wherever possible, and not to rely on a marginal excision.

Tumour stage: Although it is strictly considered stage 3, the presence of multiple MCTs does not necessarily confer a worse prognosis.[5] Dogs with lymph node metastases are nearly 8 times as likely to die of MCT as those without. As noted above, the presence of small numbers of mast cells on cytology of a lymph node is suspicious for metastasis but not conclusive; so this is an indication for biopsy for confirmation. A potential problem with staging is that small numbers of mast cells may be found in the circulation, spleen, liver and bone marrow, so the significance of such a finding is unclear. In one study the presence of small numbers of mast cells in these locations did not seem to influence survival in dogs with grade 2 MCTs.

Minor Prognostic Factors

c-kit (CD117) Expression Pattern: All MCT express c-kit but the pattern of expression within cells has been found to be prognostic, with 3 different patterns identified. Pattern 1 (perimembrane) staining is favourable, with only 2.4% of 42 dogs dying of MCT; pattern 3 (diffuse cytoplasmic) staining is unfavourable, with 38.5% of 13 dogs dying of MCT, and pattern 2 (focal cytoplasmic) staining is intermediate, with 25.6% of 43 dogs dying of MCT. It should be noted that staining pattern is not the same as c-kit mutation status, so it is not predictive of response to tyrosine kinase inhibitors (e.g. toceranib, masitinib).

Ki-67 Expression: Ki-67 is a proliferative index that has been shown to have some prognostic value. In Patnaik grade 2 tumours with <1.8% positivity, 77 to 95% of dogs lived 3 years; whereas dogs with tumours that had >1.8% positivity had 3-year survival of 21 – 33%.

TREATMENT

Control of canine MCTs involves the use of surgery, chemotherapy, or radiation therapy, either individually or in combination.

Surgery

Surgical excision is indicated if the tumour is solitary and evidence of lymph node involvement or systemic spread is lacking. Excision should be wide and deep to a minimum margin of 2-3 cm around the perceived borders of the tumour and one fascial plane below. With this approach, recurrence of grade 1 and grade 2 MCTs is very low. One study examined the completeness of surgical excision at margins 1cm, 2cm and 3cm from the edges of grade 1 and 2 MCT. All grade 1 tumours were excised 1 cm from the tumour borders, while only 75% of grade 2 tumours were completely excised at the same distance.[6]

Surgical excision should be aggressive. All excised tissue should be examined histologically for completeness of tumour excision. Extension of the tumour beyond the surgical borders or a report of "close" margins should prompt wider excision if this is possible.[7;8] A second excision should include the previous excision site plus lateral margins of 2 cm and additional deep tissue.[9] If the tumour cannot be completely excised due to tumour location or other factors, or if it is a grade 3 MCT, further therapy is indicated.[10]

The dog should be evaluated for radiation therapy, if available. Chemotherapy should be considered if staging disclosed metastases, or if the MCT is Patnaik grade 3, Kiupel high grade, or if the mitotic index is high (see above).

Radiation Therapy

Mast cell tumours are quite sensitive to the effects of radiation therapy, even at moderate doses. Dogs that have no measurable evidence of disease after surgical removal of MCTs but which had incomplete excision on histologic examination of excised tissues has significantly longer tumour control and survival following radiation therapy than other dogs.[11-13] Thus, post-surgical radiation therapy for incompletely excised tumours seems beneficial. Mast cell tumours of the extremities often present the greatest challenge for complete surgical excision. For well- or moderately differentiated tumours in these locations, combined modalities of less aggressive surgical "debulking" followed by radiation therapy may be a more acceptable treatment—both functionally and cosmetically.

Palliative radiation therapy may relieve symptoms of extensive or systemic disease. When the tumour is poorly differentiated or metastasis is already confirmed, high-dose intermittent radiation treatments may improve the quality of life by stopping bleeding or reducing the size of a bulky or irritating tumour. In these cases, a fully fractionated course would be costly and reduce the amount of time spent by owners with the dog. A short, coarsely fractionated series of treatments may provide relief from symptoms, although it will not increase life span. Systemic therapy, as outlined later, should also be considered.

Systemic Therapy

Metastatic disease occurs more frequently in dogs that had Patnaik grade 3 / Kiupel high grade MCT, in dogs with MCT with high mitotic index, and in dogs that had incomplete surgical excision of their cutaneous tumour. When MCTs have metastasized or spread systemically, localized therapies, such as surgery or radiation, are appropriate only as palliation of discomfort or mechanical obstruction. For these dogs, systemic therapy is required.

Corticosteroids are primarily palliative, but some long-term responses do occur. Oral prednisone is given as long as the tumour does not progress. Anecdotally, dogs that are tumour-free after six months have a lower incidence of recurrence; therefore, therapy is usually discontinued at this time.

Vinblastine and prednisolone were used to treat MCTs in one study. The response rate in dogs with measurable disease was 47%; there were 5 complete responses and 2 partial responses. The median response duration was 5 months (1 to >22 months). Dogs with lower grade tumours seemed to respond better. Vinblastine is myelosuppressive, and a CBC should be performed weekly and prior to administering the drug.

CCNU (Lomustine) was given to 19 dogs. One had a complete response for 15 months, and 7 had a partial response for an average of 3 months (1 to 9 months) (42% overall response rate).[14] Another study, using a lower dosage level reported a lower response rate of 1%.[15] CCNU can be combined with prednisolone. Concurrent SAMe has been shown to reduce the risk and severity of hepatic toxicity. If using CCNU, monitor CBC (especially platelet count), creatinine and liver enzymes, and discontinue if thrombocytopenia or increased liver enzymes or creatinine occur.

Vinblastine and CCNU in an alternating combination protocol have also been evaluated. In one study,[16] a 57% response rate was seen in dogs with macroscopic disease. Dogs with macroscopic disease had a median progression free survival time (PFST) of 30 weeks and a median overall survival time (OST) of 35 weeks. Dogs with microscopic disease had a median PFST of 35 weeks and a median OST of 48 weeks. In another study,[17] response rate in dogs with nonresectable MCTs was 65%. Overall median progression-free survival (PFS) time in dogs treated in the adjuvant setting was 16 months.

Vincristine was found to be an inactive agent for the treatment of MCT and often causes severe gastrointestinal toxicity. It is not recommended as a first-line chemotherapy agent for the treatment of MCT.

Mutations in the proto-oncogene tyrosine kinase receptor c-kit were shown to lead to constitutive phosphorylation of the gene product, and are believed to be important in the development and progression of at least some canine MCT. There is no evidence that such mutations are breed associated. Recently there has been an increasing interest in c-kit mutation status due to the release of tyrosine kinase inhibitors (TKIs) specifically for use in veterinary patients. Two TKIs have undergone registrational studies in dogs with mast cell tumours.

Masitinib (Masivet / Kinavet; AB Science) primarily targets c-kit. A placebo-controlled randomized double blind phase III clinical trial in 202 dogs with non-metastatic grade 2 or 3 MCTs,[18] showed a 12-month complete response (CR) rate of 11.1% (vs. 2.9% with placebo (not significantly different)), and "controlled disease" (overall response rate (ORR) plus stable disease (SD)) of 23.1% (vs. 5.9% with placebo (significantly different)). Dogs with Kit mutations had 27.3% CR (vs. 0% with placebo (not significant), and 31.8% "controlled disease" (vs. 0% with placebo (not significant). The study did report significantly prolonged time to progression (TTP) of 118 days (vs. 75 days with placebo), which was greater in dogs with Kit mutations (230 days vs. 42 days with placebo). A more recent study reported improved survival in dogs with nonresectable MCTs compared with results for the placebo, with 59 of 95 (62.1%) and 9 of 25 (36.0%) dogs alive at 12 months and 33 of 83 (39.8%) and 3 of 20 (15.0%) dogs alive at 24 months, respectively; and median overall survival times of 20.5 months and 10.7 months. Masivet is approved in Europe and the UK for dogs with recurrent or non-resectable grade 2 or 3 MCT regardless of c-kit mutation status, and as Kinavet in the US.

Toceranib (Palladia; Pfizer Animal Health) is similar to sunitinib: a human TKI, in that it is active against several members of the split-kinase RTK family (VEGFR, PDGFR and Kit), and has both anti-angiogenic and anti-tumour activity. In a trial in 57 dogs with various cancers (carcinomas, sarcomas, MCTs, melanomas, and lymphomas) the overall response (complete response plus partial response) rate was 28%. In a placebo-controlled randomized study in dogs with non-resectable grade 2 and 3 MCTs,[19] 86 Palladia treated dogs had an overall response rate (ORR, complete response (CR) plus partial response (PR)) of 37.2% (7 CR, 25 PR) compared to 7.9% (5 PR) with placebo (n=63). Following placebo escape, the total ORR for all 145 dogs receiving Palladia was

42.8% (21 CR, 41 PR); dogs with Kit mutations had an ORR of 60% (vs. 31.3% in dogs without mutations). Dogs without lymph node metastasis had higher response rate (48%, vs. 33% in dogs with metastasis). Grade 3 or 4 adverse events were seen in 20.7% of the dogs treated with toceranib. Combination protocols of Palladia with other chemotherapy drugs are being developed.[20]

Palliation of Paraneoplastic Symptoms

Ancillary drug therapy is important with canine MCTs.

Animals with mastocytosis or any bulky mast cell disease should receive H_2 antagonists, as rapid degranulation of neoplastic mast cells may follow surgery or chemotherapy. Elevated systemic histamine levels may also be seen with recurrent disease. The objective of the therapy is to prevent gastrointestinal ulceration associated with elevated levels of histamine and to treat ulcers already present. This is most likely to occur in dogs with larger, bulky disease, with recurrence of cutaneous disease, or with systemic spread of MCT. H_2 antagonists such as famotidine reduce gastric acid production by competitive inhibition of the action of histamine on H_2 receptors of the gastric parietal cells. Omeprazole, which inhibits gastric acid production by the gastric parietal cells through proton pump inhibition, may also be used. Dogs with clinical evidence of gastrointestinal ulceration and bleeding may benefit from sucralfate therapy. Sucralfate reacts with stomach acid to form a highly condensed, viscous, adherent, paste-like substance that binds to the surface of both gastric and duodenal ulcers. The barrier formed protects the ulcer from potential ulcerogenic properties of pepsin, acid, and bile, allowing the ulcer to heal.

H_1 antagonists should be considered for use in addition, before and after surgical removal of canine MCTs to help prevent the negative effects of local histamine release on fibroplasia and wound healing. Loratadine has been shown to be very effective at inhibiting histamine release by blocking degranulation from normal canine mast cells, and therefore may be a good choice for palliation of dogs with MCT.

REFERENCES

1.Murphy S, et al. **Vet Rec** 2004;154(24):743. 2.Romansik EM, et al. **Vet Pathol** 2007;44(3):335. 3. Kiupel M, et al. **Vet Pathol** 2011;48(1):147. 4.Krick EL, et al. **Vet Comp Oncol** 2009;7(2):130. 5. Murphy S, et al. **Vet Rec** 2006;158(9):287. 6.Fulcher RP, et al. **JAVMA** 2006;228(2):210. 7.Weisse C, et al. **JAAHA** 2002;38:71. 8.Séguin B, et al. **JAVMA** 2001;218:1120. 9.Kry KL, et al. **Vet Surg** 2014;43(2):182. 10.Schultheiss PC, et al. **JAVMA** 2011;238(11):1464. 11. LaDue T, et al. **Vet Radiol & Ultrasound** 1998;39:57. 12.Frimberger AE, et al. **JAAHA** 1997;33(4):320-324. 13.Al Sarraf R, et al. **JVIM** 1996;10(6):376. 14.Rassnick KM, et al. **JVIM** 1999;13(6):601. 15.Vail DM, et al. **JVIM** 2012;26(3):598. 16.Cooper M, et al. **Vet Comp Oncol** 2009;7(3):196. 17. Rassnick KM, et al. **Vet Comp Oncol** 2010;8(2):138. 18.Hahn KA, et al. **Am J Vet Res** 2010;71(11):1354. 19.London CA, et al. **Clin Cancer Res** 2009;15(11):3856. 20.Robat C, et al. **Vet Comp Oncol** 2012;10(3):174.

MULTIDISCIPLINARY APPROACH TO LYMPHADENOMEGALY

Anne Corrigan, DVM, DACVIM (SAIM) and Sandra Axiak-Bechtel, DVM, DACVIM (Oncology)
Grenada, West Indies and Columbia, MO

INTRODUCTION

Lymph nodes (LNs) are bean shaped filters that function as a station for macrophages, dendritic cells and lymphocytes - all cells required for innate and acquired immune responses. LNs also act as filters to evaluate infectious and inflammatory agents and triggers, allowing the cells of the immune system to evaluate and respond to the individual threats. There are many lymph nodes and collections of lymphoid tissue in the body, and the most commonly palpable ones in dogs and cats are the mandibular, prescapular and popliteal; the less commonly identified (or identified if enlarged) LNs include the axillary, inguinal, retropharyngeal, and sublumbar (palpated if enlarged via rectal palpation).[1] Internal lymph nodes can be evaluated with imaging techniques, including the sternal and mesenteric lymph nodes.[2] LNs are placed strategically throughout the body to drain and protect different areas, near areas where infectious agents commonly enter. The overall function of the LNs allow them to be sentinels for local or systemic disease by draining regions of the body they can help identify or localize many different disease processes.[3]

Anatomy

LNs consist of a capsule, subcapsular sinus, cortex (predominantly B cells in follicles), paracortex (predominantly T-cells and dendritic cells) and medulla (containing draining sinuses). The cells are arranged in cords containing plasma cells, phagocytic antigen presenting cells (APC) and T cells; the cords allow lymph to surround the cells to ensure adequate filtration and contact with the immune cells. Afferent and efferent lymphatics bring and remove lymph to the LNs and allow the sinuses and tissues to filter the fluid. Dendritic cells and macrophages are closely packed next to the fluid channels to allow them to identify and trap antigens. Germinal centers develop in the follicles when B cells are activated and begin dividing.[4]

Physiology

Lymph flow occurs though afferent lymphatics into the lymph nodes through the medullary sinuses and then into the efferent vessels, which ultimately empty into the venous system. These nodes are lined with white blood cells (WBC) allowing recognition and trapping of infectious agents (such as bacteria, fungi, protozoa and viruses) or inflammatory mediators from the surrounding tissues. Macrophages can, using the example of a bacterium, phagocytize the bug and prevent it from continuing on throughout the body in addition to triggering an inflammatory response to prepare the body for infection and tissue injury. Cytokines and lymphokines are released which will cause vasodilation and increased capillary permeability, stimulating more WBC to come to the area. The result is collection of fluid and proteins in the lymphatics, causing edema, and additional swelling of the region.[3,4]

LYMPHADENOMEGALY

The unique anatomy and physiology of LNs not only allows the use of LNs to help assess the patient as a whole, but also to help localize disease processes. A single enlarged lymph node is consistent with local disease such as trauma, infection, localized metastatic neoplasia, or abscessation associated with the region that lymph node drains. Generalized lymphadenomegaly is more commonly associated with systemic diseases or infections (especially fungal infections), immune mediated diseases and systemic neoplasia. Classifications of lymphadenomegaly include hyperplasia, reactive lymphadenopathy, neoplastic disease, immune mediated disease and extramedullary hematopoiesis.

History and Presenting Complaints

Many owners will notice obvious clinical signs including: a mass or multiple masses, tissue swelling, lameness, or their dog or cat paying more attention to a particular area of the body. If systemic disease is occurring nonspecific signs such as general decrease in well-being, lethargy, depression or a change in behavior. Often mild to moderate lymphadenomegaly will be an incidental finding upon physical examination.

Physical Examination

Careful palpation of the lymph nodes during a complete physical examination is necessary. Some lymphomas will have severe generalized lymph node enlargement but many will be not quite as obvious. If a single node is enlarged close attention should be paid to the local area that the regional lymph node drains. Often evaluation will identify a reason, such as a cat with an abscess, or a wound or foreign body in a dog's footpad. Evaluate dental health in dogs and cats with enlarged mandibular lymph nodes. Identify masses or swelling in the area of the nodes, and evaluate for mechanical obstruction of the drainage pathway and associated tissue/limb edema. If swelling is severe, the animal may present with abnormal function of the region, for example dysphagia, lameness, respiratory distress (especially with retropharyngeal LN enlargement), coughing (especially with perihilar LN enlargement), or dyschezia.

Diagnostic Tests

If the patient is stable with no obvious contraindications and consideration of coagulation status, the first logical step in evaluating an enlarged LN is fine needle aspiration with cytology. Cytology is inexpensive, has low complication rates, is easy to perform and quick. If multiple LNs are enlarged multiple sites should be sampled. In-house slide evaluation will allow for preliminary results and assurance of a diagnostic sample prior to pathologist evaluation.

A normal lymph node is small and surrounded by perinodal fat; therefore, cytology of a normal lymph node may contain perinodal fat and small amounts of small mature lymphocytes. Multiple aspirates may be needed for a diagnostic sample. In contrast, enlarged LNs are easier to palpate and aspirate. Since lymph nodes are heterogeneous in nature as described above, sampling multiple areas of the LN is helpful. For example, once the needle is inserted into the lymph node, repeatedly advance, partially withdraw, and redirect in multiple directions. Most lymph node aspirates do not require using suction with a syringe, as this may cause some fragile lymphocytes to rupture and interfere with cytologic interpretation.

The first step in cytological evaluation is to determine if a diagnostic sample is present by determining the presence of expected cells (lymphocytes), inflammation (if present), and neoplastic cells or infectious agents if present. If neoplasia is identified or suspected in any patient, the mass and local LN should both be aspirated to evaluate for diagnosis and metastasis. Differential staining or gram staining may assist in differentiating causes of lymphadenitis. If the cellular population is not clearly indicative of the cause, LN biopsy (core, wedge, or whole LN excisional options) with histopathology may have to follow for confirmation of the diagnosis.[1,5]

If systemic disease is suspected, a complete blood count with differential, serum chemistry panel, and urinalysis should be performed as a minimum data base. Imaging may be necessary such as chest radiographs to assess the sternal and perihilar LNs and abdominal radiographs and ultrasound to evaluate the abdominal and mesenteric LNs. Imaging will also help assess for signs of neoplasia or sites of infection or inflammation.

Normal Lymph Nodes

Normal lymph nodes are soft and symmetrical. Cytologically, they are composed of small, mature lymphocytes (75-85%) with round nuclei approximately 1 to 1.5 times the size of a red blood cell; the total size is similar to a neutrophil. Normal lymphocytes have a scant amount of basophilic staining cytoplasm surrounding the nucleus. Intermediate to large sized lymphocytes (lymphoblasts) are also present, although low in number (less than 10-15%) with nuclei approximately 1.5 to 3 times the size of a red blood cell and total size ranges from that of a neutrophil to 4 times the size of a red blood cell. Additionally, intermediate lymphocytes and lymphoblasts will have more cytoplasm than small mature lymphocytes. Plasma cells, characterized by deep blue cytoplasm and a prominent clear Golgi zone are present in low numbers; small numbers of macrophages, eosinophils, neutrophils, and mast cells may also be noted and should be less than 1% of total cells.[6]

Hyperplasia and Reactive Lymph Nodes

Hyperplasia of the lymph nodes is proliferation of lymphocytes and plasma cells due to antigenic stimulation. This is particularly common with any infectious agents including bacterial antigenic stimulation or inflammatory mediators from local tissue injury or inflammation. This is not to be confused with infection or inflammation of the node itself. A hyperplastic LN has increased lymphocytes, but no difference in the % population of the different sub-types of lymphocytes or with a very mild increase in lymphoblasts or plasma cells. In both hyperplastic and reactive LNs, small mature lymphocytes predominate, plasma cells may represent 5-10% of the cell population, and increased macrophages may be present. When the proportion of lymphoblasts or plasma cells increases (~10%-50% of the nodal cells), it is labeled a reactive lymph node. The clinical causes are not different between the two terms (hyperplasia and reactive LN).[7]

Lymphadenitis

Lymphadenitis, or infection/inflammation of the LN, occurs when WBCs - neutrophils, eosinophils, or macrophages- migrate into the lymph node. This can occur with certain particular infectious agents, allergic, or immune mediated disease.[3] Lymphadenitis usually has an overrepresentation of one type of inflammatory cell identified in the cytology sample, such as eosinophilic lymphadenitis due to allergic skin diseases, or neutrophilic/pyogenic or granulomatous lymphadenitis which can occur with certain infections including *Neorickettsia* sp, *Mycobacterium* sp, and some fungal agents.[5] A combination of WBC infiltration may also occur with an immune mediated disease associated with lymphadenomegaly and a secondary bacterial infection causing further inflammation.[3]

Neoplastic Lymph Nodes

Lymphadenomegaly due to neoplasia can be due to primary lymphoid neoplasia (lymphoma) or secondary neoplasia from metastatic disease. Lymphoma in the dog most commonly results an overabundance of lymphoblasts, greater than 50% and usually 80-90% of the lymph node population. Small cell lymphoma occurs less commonly in dogs and is generally not diagnosed on cytology alone, as the predominant cell population consists of small mature lymphocytes. Multicentric lymphoma in cats is uncommon; common sites of lymphoma are the gastrointestinal tract and mediastinum.

Metastatic disease to the lymph nodes is also quite common from tumors that metastasize via lymphatics, including carcinomas, melanomas, and round cell tumors (particularly mast cell tumors). The presence of cells not normally found in the lymph node (such as epithelial cells) or an increase in numbers (such as large aggregates of mast cells) suggest metastatic neoplasia. It is important to note that metastatic cancer cells can be present in a lymph node that palpates normally on examination.[5,7]

DIFFERENTIAL DIAGNOSES FOR LYMPHADENOMEGALY

Single lymph node enlargement

Enlargement of a single lymph node may be indicative of localized infection or inflammation, or localized neoplasia. A variety of agents can be responsible for LN enlargement and examples of localized infection include, fungi, protozoa, metazoa, or algae. Localized inflammation can be due to significant dental disease/ oral infection or localized abscess formation. Neoplasia can include metastatic neoplasia (some examples are oral melanoma, mast cell tumor, carcinoma) or localized lymphoma.

Generalized lymph node enlargement

Generalized lymph node enlargement may be indicative of systemic infection or inflammation, immune mediated disease, or systemic neoplasia. Systemic infections include bacteria, fungi, or viruses. Systemic inflammation from immune mediated disease and systemic neoplasia such as lymphoma also cause generalized lymphadenomegaly.

Cytology can differentiate between many of the broad categories and allow guidance for further diagnostics tests, treatment, and prognosis. Important to note in the diagnosis of lymphoma is that the use of glucocorticoids prior to obtaining a diagnosis can lead to difficulty in diagnosing lymphoma at a later time.

CONCLUSION

Enlarged lymph nodes may be caused by a variety of etiologies. Differentiating between various causes of enlarged lymph nodes commonly utilizes cytology, which often provides an easy, inexpensive, and rapid differentiation between infectious, inflammatory, or neoplastic lymph nodes. Lymph node histopathology can also be used should cytology be non-diagnostic.

REFERENCES

[1]Cowell, et al, Vet Clin Small Anim 33 (2003) 47-67; [2]Dyce, Sack and Wensing; Textbook of Veterinary Anatomy 4th ed 2009 ; Guyton and Hall Textbook of Medical Physiology 10th ed. 2000, 697-704; [4]Tizard Veterinary Immunolgy 9th ed. 2013; [5] Seidlecki in Ettinger and Feldman Textbook of Veterinary Internal Medicine 7th ed. 2010 393-395; [6]Cowell et al. Diagnostic Cytology and Hematology of the Dog and Cat 2008; 3rd ed: 180-91; [7]Stockham and Scott Fundamentals of Veterinary Clinical Pathology 2008; 2nd ed: 363-65; [8]Greene's Infectious Diseases of the Dog and Cat. 4th ed.2012.

IMMUNOSUPPRESSIVE TREATMENT FOR MENINGOENCEPHALITIS

Meredith Wille, CVT
Pittsburgh, Pennsylvania

INTRODUCTION

Inflammation the brain or meningoencephalitis may develop secondary to an infectious etiology (virus, parasite, protozoa, bacteria, fungal organism) or as an idiopathic inflammatory condition. In dogs, a number of idiopathic encephalitides have been reported and a definitive diagnosis is made based on histopathology. The term meningoencephalitis of unknown etiology (MUE) has been proposed as a collective term for granulomatous meningoencephalomyelitis (GME), necrotizing meningoencephalitis (NME) and necrotizing leukoencephalitis (NLE) and other noninfectious encephalitides when a histopathologic diagnosis has not been made. The pathogenesis for MUE is suspected to be autoimmune or immune mediated and many patients will improve with immunosuppression. The diagnosis is presumptive after the exclusion of infectious causes or based on a high index of suspicion due to breed, imaging findings and/or cerebrospinal fluid analysis. Without aggressive treatment, prognosis and median survival is poor. Immunosuppressive therapy appear to offer the patient the longest survival time; however treatment for MUE is still subjective and more research is needed to determine the optimal medication regimen.

MENINGOENCEPHALITIS OF UNKNOWN ETIOLLGY

Granulomatous meningoencephalitis

GME is an inflammatory CNS disorder of unknown cause. It can be found in any breed; however it is most common in young to middle aged toy and terrier breeds. Female dogs may have a predisposition to the disease[1]. GME generally affects the brain and/or spinal cord. It comes in three forms: disseminated (mutlifocal), focal and ocular. Disseminated GME is the most common form and presents with acute, rapidly progressive neurologic deficits.[10] Focal GME can be acute or slowly progressive with signs referral to single, mass-like granuloma. The ocular manifestation causes acute visual impairment or blindness and may be seen in combination with disseminated or focal GME. Median survival times for different forms are focal-114 days, multifocal-8 days, and focal (other)-59 days[2].

Necrotizing Meningoencephalitis

NME has previously been known as Pug dog encephalitis; however it can be found in many different breeds of dogs. It is characterized by cavitary necrosis in the neuroparenchyma. Most patients present for seizures or forebrain signs (circling, confusion, postural deficits). On magnetic resonance imaging (MRI), lesions are multifocal and often asymmetric within the prosencephalon affecting both grey and white matter. There is generally contrast enhancement on T1 weighted images.

Necrotizinge Leukoencephalitis

NLE is historically called Yorkie encephalitis; however it has been found in many different breeds and is no longer considered a "breed specific" disease. It may be seen in older dogs and typically carries a more guarded prognosis due to loss of brain tissue. MRI images show multiple, asymmetric lesions within the within the white matte of the forebrain and/or brainstem. Lesions are hyperintense on T2 weighted and FLAIR images and represents cystic regions of necrosis.

THERAPY

Once a presumptive diagnosis has been made, there should be a discussion with the pet's owners regarding a treatment plan. This is a decision not taken lightly and should be thought out with all options explained. There are no guarantees for response to therapy and many patients are not able to leave the hospital despite aggressive treatment. Research has shown that without the aggressive treatment median survival is days to weeks, so whatever the choice of treatment it is the best chance at survival. Treatment may be lifelong and owners should be prepared for that. Some of the treatments can be expensive; however, there are some options to that, so hopefully that is not the deciding factor. Euthanasia is sometimes the best option, for the pet and the family.

Corticosteroids

Corticosteroids are at the center of therapy for MUE. Corticosteroids suppress the inflammatory process in the body. They achieve this by interacting with specific intracellular receptor proteins and alter the expression of corticosteroid-responsive genes. Steroid treatment as the only treatment has not been thoroughly investigated, the majority of the research and survival times is based on multimodal therapy. It may resolve signs in some dogs, but insufficiently. Also long-term therapy with high dose corticosteroids comes with many side effects such as polyuria-polydipsia, polyphagia, weight gain, hepatotoxicity, gastrointestinal ulceration, pancreatitis, and iatrogenic hyperadrenocorticism[9]. The following prednisone protocol has been proposed as therapy for MUE: 1.5 mg/kg twice a day for three weeks, 1.0 mg/kg twice a day for six weeks, 0.5 mg/kg twice a day for three weeks, 0.5 mg/kg once a day for three weeks, 0·5 mg/kg every other day indefinitely (may reduce to 0·25 mg/kg every other day). Alternatively, prednisone is started at 1-2mg/kg PO q 12hrs then tapered every 10-14 days to the lowest effective dose (LED). Recent research has shown that adjunct therapy that will combine the use of a long corticosteroid treatment along with adding another immunosuppressive medication. There are several different medications that will fall in to this category and aid in the treatment of MUE.[9]

Cytosine arabinoside

Cytosine arabinoside (CA) is a synthetic nucleoside analogue that crosses the BBB (blood to brain barrier). It competitively inhibits DNA polymerase and prevents DNA repair. CA is metabolized via liver, plasma, and gastrointestinal tract. Side effects can be myelosuppression, vomiting, diarrhea, and hair loss and are usually dose dependent. There are options in administering the medication and more studies need to be done on which dosing is more effective. CA can be administered subcutaneously at 50mg/m2 every 12 hours for 2 consecutive days and repeated every three to six weeks[3]. Alternatively CA is given at 200 mg/m2 diluted in sodium chloride and administered intravenously over 24 hours. A CBC is assessed 10-14 days after first treatment then a CBC and chemistry profile is checked before each treatment.[6] When given in combination with prednisone therapy survival ranges 46 to 1025 days. As a chemotherapeutic agent, CA should be treated as such and clients with a suppressed immune system should not handle the medication or any of the excrement from the patient for 3-5 days following treatment. If there are accidents in the house they can be cleaned with a dilute bleach solution.

Procarbazine

Procarbazine is and antineoplastic alkylating agent. In human medicine it is used to treat Hodgkin's lymphoma. It is lipid soluble and crosses the BBB. By inhibiting insertion of essential DNA precursors it also disrupts RNA and protein synthesis. Side effects include myelosuppression, nausea, vomiting, hepatic dysfunction, and neurotoxicity. Dosing is recommended at 25 to 50mg/m2/day PO, as an adjunct to corticosteroids[8]. Lab work that needs to be run are a CBC once weekly for the first month of therapy and monthly while on the medication. The greatest concerns are the WBCs and platelets. Median survival on this treatment plan is 15 months.

Cyclosporine

Cyclosporine is an immunosuppressive agent that acts by directly suppressing T lymphocyte activation and proliferation. It will also prevent synthesis of several cytokines. This drug has poor BBB permeability, however it does concentrate effectively in the cerebral endothelial cells. With meningoencephalitis, there is disruption of the BBB allowing therapeutic concentration of cyclosporin within the brain. T-cell responses are also initiated in peripheral lymph nodes and so the peripheral effects of cyclosporin may help to control intracranial disease. One of the benefits of cyclosporine is that it works rapidly and reaches a steady state blood level within 24-48 hours. Side effects include diarrhea, anorexia, vomiting, gingival hyperplasia, excessive shedding, and insulin resistance. At high doses nephrotoxicity has been seen. The dose range for cyclosporine is broad: 3-15mg/kg PO q 12-24 hours. The most commonly used dose is 6-10mg/kg PO q 24 hours as adjunct medication with corticosteroids and/or a chemotherapeutic agent. Ketoconazole can be added at 8mg/kg PO q 24 hours[7]. One of the goals of adding cyclosporine to the medication regiment is to be able to taper the corticosteroids. CBC and chemistry monitoring is not necessary with cyclosporine although cyclosporine levels may be checked to help direct therapy. Cyclosporin is best administered on an empty stomach to improve absorption. Median survival with a combination of cyclosporine and prednisone is 930 days.[9]

CCNU

CCNU is an antineoplastic agent with potent immunosuppressive properties that relate to its toxic effect on lymphocytes. It is highly soluble and crosses BBB alkylating both DNA and RNA. Central nervous system penetration is determined from the plasma/CSF ratio. This medication is generally used to treat tumors of the CNS and lymphoma. Side effects commonly seen are vomiting, diarrhea, bone marrow suppression, and hepatotoxicity. Thrombocytopenia can be seen as a cumulative effect. It is typically used in conjunction with corticosteroids and the dose administered is 30-60 mg/m2 PO every 4-6 weeks.[9] It is recommended to run a CBC 10-14 days after the first few doses to monitor for neutropenia and CBC and serum chemistry before each dose. At our clinic, patients with severe MUE are typically induced with cytosine arabinoside and then transitioned to CCNU orally every 4-6 weeks in combination with prednisone and cyclosporine. Further evaluation of this course of treatment is needed. Recent studies show that with the addition of this medication median survival times are getting longer.

Mycophenolate mofetil

MMF is a lymphocyte specific drug that decreases the recruitment of inflammatory cells, as well as suppressing lymphocyte proliferation. This is a relatively new course of treatment and the efficacy needs to be evaluated more in depth. So far results are encouraging. The most common side effect is hemorrhagic diarrhea. As an adjunct medication dosed at 20mg/kg PO q 12 hours for one month then tapered to 10mg/kg PO q 12 hours.[9] There are no exact numbers on median survival with this treatment course.

Leflunomide

Leflunomide is known to be very effective in autoimmune diseases. It inhibits T- and B- cell proliferation, suppresses immunoglobulin production, and interferes with cell adhesion. It may also have some antiviral properties. Side effects are quite rare when administered to dogs.[9]There are some reports of thrombocytopenia and hemorrhagic colitis. Dosing is 1.5mg-4mg/kg PO q 24 hours.[9] Patients treated for MUE with leflunomide has been done due to a poor response with corticosteroid therapy. Prognoses of this treatment have not been evaluated and the overall treatment plan is in great need of further studies.

NURSING CARE

From a technician standpoint, treatment of a patient suspected to have MUE, can be very challenging and there are several things to keep in mind. Having a "fresh" and second IV catheter to administer a chemotherapeutic agent, as well as insuring a balance of fluids, supplements, antibiotics, and anti-epileptic medications as needed. If chemotherapeutic medications are being used wearing gloves is a good plan, some of that is to protect the care giver however most of the medications being used have a very low level remaining in the any excrement. A much larger purpose of wearing gloves is to protect the patient from any other bacterial infections being transferred to them by contact from other patients or other areas around the clinic while be housed there. Inappetence, nausea, and diarrhea are common with virtually all the medications in any of the treatment options. Caution in what is fed, as well as significant encouragement in eating is also very helpful for a better recovery.

Depending on how mobile the patient is good padding is always a good idea. Frequently these patients are not able to get up and move around on their own. If that is the case, physical therapy should also be added to the regimen of care. Passive range of motion, limb massage, and reflex stimulation are all good modalities to add in due to the fact that the patient can have those treatments without being disconnected from any IV chemo medications. Continuing physical therapy is also important for the owner to do. As the patient progresses they can add cavelletis, a wobble board, and balance donuts all for help in coordination exercises. These are easy and inexpensive for the owner to do at home and help them feel involved in the treatment plan for their own pet.

Once a patient is discharged from the hospital supporting and encouraging the owner to continue with the intensive nursing will only better the chances and lengthen survival. Some owners are extremely overwhelmed at the number of medications, the frequency they are given, and the scary terms of a "chemo" medication. Taking the time to fully explain to an owner what the purpose of each medication and what the desired result/reason of choice for them can be a comforting conversation to have. Making suggestion such as a calendar of medication schedules so that the owner can visualize when to taper medications. Also having pill containers to help know which ones should be given with or without food, given in the morning and/or evening etc. Remember that some clients have never had to take any medications for themselves and it can be very hard to know or understand why all these drugs are so important for their pet to take.

SUMMARY

The value of this lecture is that the technician will have a better understanding of how to deal with a patient in this condition and how to communicate with clients that have pets being treated for a disease that can go on for many years of the pet's life. The prognosis of MUE has improved in recent years with more studies of the variety of treatment options. It is very clear that there is a great need for a more basic understanding of the disease so as to be able to target specific therapies. All of this in the hopes of improving survival times of a still mysterious disease process.

REFERENCES

1 Braund KG. Granulomatous meningoencephalomyelitis. J Am Vet Med Assoc. 1985 Jan 15;186(2):138-41.

2 Muñana KR, Luttgen PJ. Prognostic factors for dogs with granulomatous meningoencephalomyelitis: 42 cases (1982-1996). J Am Vet Med Assoc. 1998 Jun 15;212(12):1902-6.

3 Nuhsbaum MT, Powell CC, Gionfriddo JR, Cuddon PA.Treatment of granulomatous meningoencephalomyelitis in a dog. Vet Ophthalmol. 2002 Mar;5(1):29-33.

4 Adamo FP, O'Brien RT. Use of cyclosporine to treat granulomatous meningoencephalitis in three dogs. J Am Vet Med Assoc. 2004 Oct 15;225(8):1211-6, 1196.

5 Gnirs K. Ciclosporin treatment of suspected granulomatous meningoencephalomyelitis in three dogs. J Small Anim Pract. 2006 Apr;47(4):201-6.

6 Zarfoss M, Schatzberg S, Venator K, Cutter-Schatzberg K, Cuddon P, Pintar J, Weinkle T, Scarlett J, Delahunta A. Combined cytosine arabinoside and prednisone therapy for meningoencephalitis of unknown aetiology in 10 dogs. J Small Anim Pract. 2006 Oct;47(10):588-95.

7 Adamo PF, Rylander H, Adams WM. Ciclosporin use in multi-drug therapy for meningoencephalomyelitis of unknown aetiology in dogs. J Small Anim Pract. 2007 Sep;48(9):486-96.

8 Coates JR, Barone G, Dewey CW, Vitale CL, Holloway-Azene NM, Sessions JK. Procarbazine as adjunctive therapy for treatment of dogs with presumptive antemortem diagnosis of granulomatous meningoencephalomyelitis: 21 cases (1998-2004). J Vet Intern Med. 2007 Jan-Feb;21(1):100-6.

9 LAUREN R. TALARICO AND SCOTT J. SCHATZBERG Idiopathic granulomatous and necrotising infl ammatory disorders of the canine central nervous system: a review and future perspectives *Journal of Small Animal Practice* (2010) 51, 138–149 DOI: 10.1111/j.1748-5827.2009.00823.x

10 Susan N. Fitzmaurice, Small Animal Neurology. 2010. Pg 66-68

INTRODUCTION TO ELECTRODIAGNOSTIC TESTING FOR VETERINARY TECHNICIANS

Edward MacKillop, DVM, DACVIM (Neurology)
Pittsburgh, Pennsylvania

INTRODUCTION

Neuromuscular disorders cause weakness and/or muscle atrophy due to dysfunction of the lower motor neuron (LMN), nerve root, peripheral nerve, muscle or neuromuscular junction. Abnormalities may affect the whole body or be localized to one region (a single limb, nerve or set of muscles). Generalized neuromuscular weakness affects the whole body and is characterized by a short-strided, shuffling gait and fatigue with exercise[1]. Muscle tone and segmental spinal reflexes are generally decreased though in some conditions they may be preserved. Ataxia may be seen with diseases of the brain or spinal cord but is not a feature of neuromuscular disease. With LMN disorders or peripheral nerve disease (neuropathy), there is rapid muscle atrophy (denervation atrophy) often within 7-10 days of clinical signs. Cranial nerve abnormalities such as facial paralysis or fatigue of the blink reflex are seen with certain neuromuscular disorders such as myasthenia gravis[2]. Dysphagia or megaesophagus may develop due to dysfunction of skeletal muscle in the upper gastrointestinal tract (pharynx or esophagus). Severe cases of generalized neuromuscular disease will progress to nonambulatory tetraparesis or tetraplegia. In these cases, the patient should be monitored carefully for respiratory distress due to hypoventilation or aspiration pneumonia.

The differential diagnosis of neuromuscular disease can be difficult; many neuromuscular diseases look clinically similar but have different underlying causes and treatment. The differential list can best generated by looking at the onset of signs (acute vs. chronic) and the distribution of problem (localized vs. whole body). Myasthenia gravis, polymyositis, muscle necrosis, polyradiculoneuritis (Coon hound paralysis), tick paralysis and botulism all cause acute, generalized weakness[2-5]. Slowly progressive generalized neuromuscular dysfunction is seen with degenerative muscle diseases such as muscular dystrophy and Labrador myopathy, peripheral neuropathy/laryngeal paralysis, metabolic neuropathy (hypothyroid or diabetic neuropathy), and paraneoplastic neuropathy[6-8]. Localized deficits are seen with malignant peripheral nerve sheath tumor, idiopathic facial paralysis, masticatory muscle myositis and hypothyroid neuropathy[4,6]. Patchy or mulitifocal neuropathy may be seen with lymphoma as a paraneoplastic syndrome or secondary to lymphomatous infiltration of peripheral or cranial nerves[9,10].

ELECTRODIAGNOSTIC TESTING

Electrophysiologic testing confirms the presence of neuromuscular disease and localizes the problem within the neuromuscular system (LMN, nerve or nerve root, muscle or the neuromuscular junction). Electrodiagnostic evaluation may provide a presumptive diagnosis in diseases such as Coon hound paralysis[3] or may direct the clinician to pursue nerve or muscle biopsy. Veterinary patients are typically anesthetized for electrodiagnostic testing since spontaneous movements or poor patient compliance confounds the evaluation. A typical electrodiagnostic evaluation for neuromuscular disease in dogs and cats includes electromyography (EMG), nerve conduction study (NCS) and repetitive nerve stimulation (RNS).

ELECTROMYOGRAPHY

Electromyography is the study of the electrical activity of muscle. Normal resting muscle is electrically quiet, appearing on the monitor as a flat line. When the connection between a nerve and muscle is disrupted, the muscle is considered denervated. Denervated, inflamed or degenerating muscle is prone to spontaneous activity on EMG[11,12]. Loss of muscle mass because of chronic orthopedic disease or poor mobility does not cause spontaneous activity and so EMG is useful in separating disuse muscle atrophy from neurogenic or myopathic atrophy.

Concentric or monopolar needles are used for EMG in animals[12]. Surface electrodes are not used because of the poor contact with skin due to hair coat. A concentric needle is a hollow needle, similar to a hypodermic needle, which compares the electric potential between the stainless steel shaft and the center of the needle. Monopolar needles are used to compare the electric potential between two needles to determine voltage. Both types are sold for single usage but may be cleaned with alcohol and reused unless the needle tip has become blunt. When the EMG needle initially enters a muscle there is a short burst of electrical noise called insertional activity representing transient mechanical injury to muscle fibers. Insertional activity is brief and should not continue once the needle stops moving. Other electrical activities that may be found in normal muscle are miniature end-plate potentials (MEPPs), end-plate spikes and motor unit action potentials (MUAPs). MEPPs represent a localized change in the muscle fiber electrical potential due to random release of small amounts of neurotransmitter from the nerve terminal. MEPPs are seen as a low amplitude change from the baseline and sound similar to the noise heard when listening to a sea shell. End-plate spikes are often seen with MEPPs and represent single muscle fibers that are contracting due spontaneous release of neurotransmitter. MUAP are seen with mild voluntary contraction of muscle by the patient[11,12].

Fibrillation potentials, positive sharp waves, complex repetitive discharges and myotonic potentials are the four common types of abnormal spontaneous activity seen on EMG[12] (Figure 1). Fibrillation potentials and positive sharp waves both stem from the spontaneous firing of single muscle fibers due to ether nerve or muscle disease. Fibrillation potentials usually appear as a sharp downward spike of variable height with a subsequent upward deflection. They produce a sound on the loudspeaker reminiscent of eggs frying in a pan or the wrinkling of tissue paper[11]. Positive sharp waves are a single downward spike (confusingly down is positive when evaluating EMG) followed by a slower negative wave that is smaller in amplitude and longer in duration[11,12].

Myotonic discharges are most commonly associated with myotonia congenita. This is a disorder of excessive muscle stiffness, especially with stimulation. Look up "fainting goats" on Youtube.com to see a comical example of myotonia congenita.

Myotonic discharges naturally wax and wane without movement of the needle. They produce a very distinct sound similar to a dive bomber or an accelerating/decelerating motorcycle.[11,13]

Complex repetitive discharges (CRD) appear on the monitor as the name suggests: there is a complex waveform and it repeats over and over again. They start and stop suddenly, unlike myotonic discharges, and produce a sound on the loudspeaker that is similar to a machine gun[11]. CRD are classically associated with myopathy due to chronic corticosteroid administration or hyperadrenocorticism (Cushing's myopathy)[7]; however, other muscles diseases, especially muscular dystrophy, and chronic nerve denervation are also associated with CRD.[11]

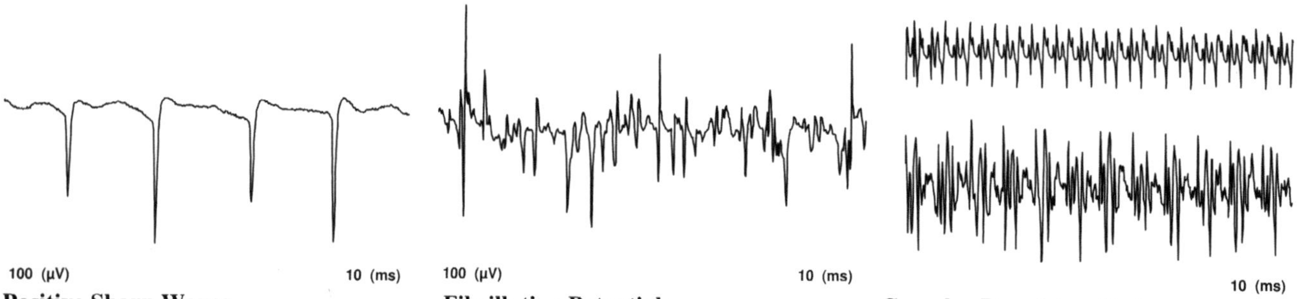

Positive Sharp Waves **Fibrillation Potentials** **Complex Repetitive Discharges**

Figure 1. EMG Abnormalities

NERVE CONDUCTION STUDY

Motor nerve conduction testing involves stimulating a peripheral nerve at two or more sites and recording a compound muscle action potential (CMAP) from a muscle innervated by the nerve.[11,12] A nerve can be thought of as a biological power cord made up of two basic components: axons and myelin. The axon is equivalent to a copper wire and myelin is the insulation around a wire that speeds up conduction. Axonal diameter also contributes to the speed of conduction; larger diameter axons conduct impulses faster than smaller fibers. Nerve disease may affect axons, myelin or both.

Once recording needles (the same needles used in EMG) have been placed in a muscle and the nerve is stimulated, a CMAP waveform is displayed on the monitor. The waveform is evaluated for when it starts (onset latency or o), height (peak or p), duration and appearance, all of which may be altered by nerve disease. Physiologic temporal dispersion is a normal phenomenon that can be seen when stimulating proximally (high up) and distally (low down) on a limb.[11] With proximal stimulation there is a greater distance for the nerve impulse to travel and slower conducting fibers will trail behind faster fibers. This leads to a CMAP that is slight longer duration and lower amplitude than that seen with distal stimulation.

The nerve is stimulated at two points and the time it takes for the CMAP to start is measured by the machine. These two numbers are subtracted from each other to determine the difference in time (Figure 2).

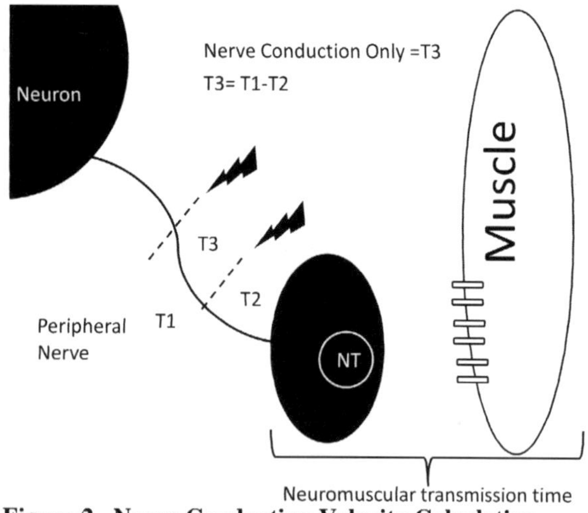

Figure 2. Nerve Conduction Velocity Calculation

The distance between the two points is then measured manually with a tape measure. Speed or velocity is distance divided by time (how far you travel over how long it takes to get to your destination) and so this information determines nerve conduction velocity (NCV). Nerve diseases will often cause a decrease in NCV, especially if there is disruption of myelin. You may wonder why the nerve can't simply be stimulated at one point and then measure the time and distance. When stimulating a nerve in one place and measuring the time it takes for muscle contraction, you are measuring not only the time it takes for nerve conduction but also the time it takes for release of neurotransmitter (NT) and binding of NT to receptors on the muscle (neuromuscular transition time, see RNS section). Imagine trying to figure out how fast you are driving on a highway on your way to work without using a speedometer. If you time yourself from the point you start driving on the highway to when you get to work, you are also including the time you travel on the off ramp, on smaller roads and any traffic lights on the way. If you time yourself starting at two different points on the highway and then subtract those numbers you will only be looking at speed on the highway but not after exiting. Normal NCV in dogs and cats is usually >50-60 m/s[12]; however, published normal values should be reviewed when testing a patient since NCV will vary with different nerves and species. Age and body temperature may also alter NCV.

Abnormalities on NCS can be broadly categorized as follows: reduced amplitude of the CMAP +/- delayed onset, delayed onset (increased latency) with normal amplitude, absent response.[11] Typically, a decrease in CMAP amplitude with or without a decrease in onset latency occurs due to axonal disruption. Delayed onset can be a sign of demyelination or axonal disease affecting the fastest conducting fibers. An absent response may be due to complete disruption of a nerve (axonotmesis or neurotmesis). Polyphasia, defined as greater than 4 phases (up or down waves relative to the baseline), may also be seen with nerve disease due to asynchronous arrival of electrical impulse at the muscle (Figure 3).[14]. Slowed conduction velocity also leads to abnormal temporal dispersion, seen as an increase in the CMAP peak duration.[14]

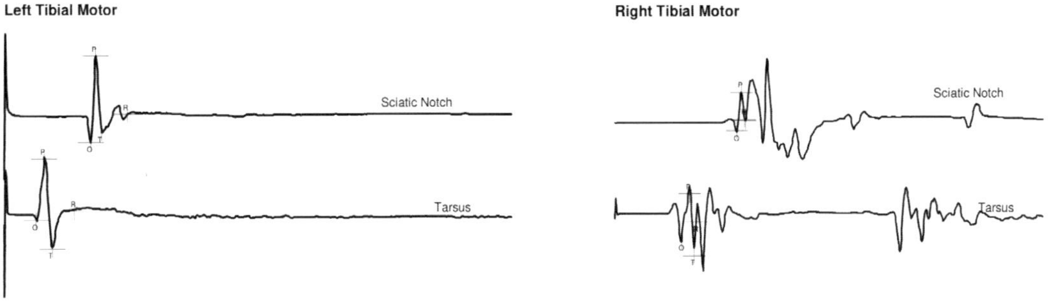

Normal CMAP **Abnormal CMAP – Polyphasia with Temporal Dispersion**

Figure 3. Nerve Conduction Study

REPETITIVE NERVE STIMULATION

Repetitive nerve stimulation (RNS) tests the function of the neuromuscular junction[15]. Stimulation of a peripheral motor nerve cause release of the neurotransmitter acetylcholine (Ach) at the neurotransmitter junction. Acetylcholine then binds to nicotinic Ach receptors on the muscle causing muscle depolarization and contraction. The enzyme, acetylcholinesterase, then cleaves Ach into acetate and choline, terminating the signal.[15] Within the nerve terminus, Ach is packaged into synaptic vesicles available as a readily releasable store. A second pool of reserve vesicles exists to replenish the readily releasable vesicles. With low frequency stimulation there is an initial drop in release of Ach but this is eventually balanced out by the reserve vesicles replenishing the readily releasable pool. Once the replenishment of vesicles equals released vesicles, the amount of Ach in the junction reaches equilibrium and a recorded CMAP maintains a level height. Thus, in normal animals with low frequency stimulation (<3 Hz) the CMAP waveform may decrease a little initially but should not decrease >10%. In animal with myasthenia gravis, no abnormalities are noted on EMG or nerve conduction study; however, with supramaximal repetitive nerve stimulation there is a >10% decrease in CMAP height (Figure 4).[15]

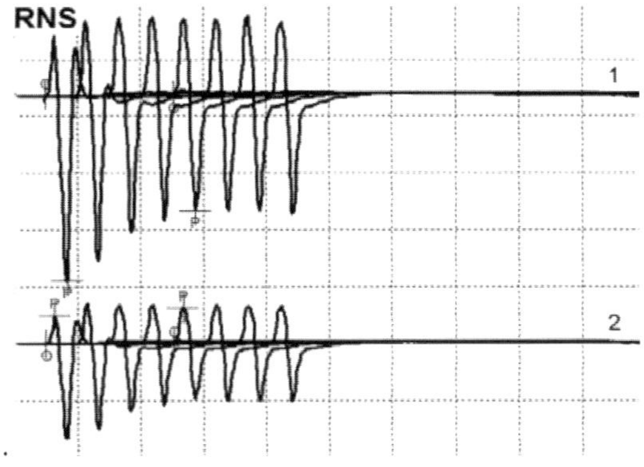

Figure 4. Decremental Response in a Cat with Myasthenia Gravis

REFERENCES

1. Glass EN et al. Vet Clin North Am Small Anim Pract. 2002;32:1.
2. Dewey et al. J Vet Intern Med. 1997; 11:50-7.
3. Cuddon PA. J Vet Intern Med 1998; 12:294-303.

4. Podell M. Vet Clin North Am Small Anim Pract. 2002;32:14.
5. Shelton GD. Vet Clin North Am Small Anim Pract. 2004;34:1469.
6. Cuddon PA. Vet Clin North Am Small Anim Pract. 2002; 32:207.
7. Platt SR. Vet Clin North Am Small Anim Pract. 2002;32:125.
8. Shelton GD et al. J Vet Intern Medicine 1987; 1:36.
9. Cavana P et al. J Feline Med Surg. 2009;11:869.
10. Pfaff AM et al. J Am Anim Hosp Assoc. 2000;36:57.
11. Kimura J. Electrodiagnosis in Diseases of Nerve and Muscle: Principles and Practice, ed 4. Philadelphia, FA Davis, 1989.
12. Cuddon PA. Vet Clin North Am Small Anim Practice 2002; 32:31
13. Vite CH. Vet Clin North Am Small Anim Pract. 2002;32:169.
14. Bromberg MB. Phys Med Rehabil Clin N Am. 2013; 24:153.
15. Meriggioli, MN, Howard Jr., JF, Haper, CM Neuromuscular Junction Disorders: Diagnosis and Treatment (Neurological Disease and Therapy), Boca Raton, Taylor and Francis Group, 2003: 1.
16. Indrieri RJ et al. J Am Vet Med Assoc. 1983; 182:57.

CLINICAL ASSESSMENT OF ACUTELY SPINAL INJURED PATIENTS

Laurie Cook, DVM, DACVIM (Neurology)

INTRODUCTION

Spinal injuries are one of the more common neurologic problems seen in clinical practice in dogs and cats. The most common cause of spinal cord injuries in dogs is acute inter-vertebral disc extrusions seen in dachshunds and other chondrodystrophic breeds. Other causes of spinal cord injuries include vascular events such as a fibrocartilaginous embolism , exogenous causes of trauma, and others. Assessment of neurologic function, bladder function and other body systems is important for establishing prognosis and providing optimal patient care.

NEUROLOGIC EXAM OF SPINAL INJURED PATIENTS

Proper assessment of the neurologic function in spinal injured patients is key to determining prognosis and for providing proper nursing care and physical therapy. Ideally neurologic assessment should be performed when the animal is systemically stable and without the influence of sedatives or pain medications which can alter the neurologic exam. Prognosis can differ depending on the type of spinal injury. A discussion of the major types of spinal injury follows this section.

Assessment of gait, motor function and proprioception

The gait exam is very important to assessing the degree of neurologic injury present. Ambulatory patients can be assessed by placing them on the floor with a leash or loose in a small room and watching them walk. The gait should be evaluated for ataxia, weakness or lameness. Non-ambulatory patients should be placed on the floor and supported either with a hand under the hind end, with a belly sling or supported by the tail and allowed to walk to see if they are able to move their legs. Non-ambulatory patients may be paretic, meaning they are weak but still have motor (movement) in their limbs, or plegic, meaning they have no motor present. (Table 1) Dogs that have lost motor function may also lose pain sensation. (1)

Table 1: The modified Frankel Scale is derived from the human system to evaluate the extent of spinal cord injuries: (3)

Grade 0	Paraplegia with absent deep pain sensation
Grade 1	Paraplegia with absent superficial pain sensation
Grade 2	Paraplegia with intact pain sensation
Grade 3a	Non-ambulatory paraparesis, inability to bear weight
Grade 3b	Non-ambulatory paraparesis, ability to bear weight
Grade 4	Ambulatory paraparesis
Grade 5	Normal gait with spinal pain

Proprioceptive placing can help to distinguish spinal cord injuries from orthopedic problems. Often dogs with spinal cord injuries will have proprioceptive deficits caudal to the site of the injury. Dogs with orthopedic problems will have normal proprioceptive placing.

Extensor tone and spinal reflexes may help to identify patients that have lower motor neuron (LMN) injuries v.s. upper motor neuron (UMN) injuries. LMN injuries are those that affect the peripheral nerves that directly supply the limbs (cervical or lumbosacral intumescence). UMN injuries affect the spinal cord above the region of the intumescence and interfere with the spinal cord tracts that influence the peripheral nerves. Patients with UMN injuries will have normal to increased muscle tone and reflexes. Patients with LMN injuries often have reduced to absent muscle tone and reflexes and atrophy within a few days of injury. Some LMN injuries have a worse prognosis for recovery.

Assessment of pain sensation

Pain sensation is the conscious recognition of a noxious stimulus applied to the toe caudal to the site of the spinal lesion. Assessment of pain sensation is important to assessing prognosis in dogs with spinal injuries that are non-ambulatory paraplegic. This assessment requires application of a noxious/painful stimulus to the toes of the pelvic limb and watching for a behavioral demonstration that the dog feels the stimulus. A toe or toe web can be pinched with fingers but for some dogs that are stressed, a finger pinch may not be enough to elicit a response even if pain sensation is present. If the patient doesn't react to a finger pinch, hemostats can be used to pinch the toe to provide a stronger stimulus. Usually dogs that have pain sensation will cry out or turn around and look at the toe pinch, try to bite, try to get away, or show other signs of pain. Commonly assessment of pain sensation can be confused with withdrawal reflex of the limb being tested. A withdrawal reflex occurs when a noxious stimulus is applied to toes and the patient pulls the leg back. This is a local spinal reflex arc occurring in the L7-S2 spinal cord segments and does not require pain sensation to be present. Therefore, if a dog pulls its leg back but does not show a behavioral response to the stimulus, it does not indicate presence of pain sensation. Dogs with thoracolumbar spinal cord injuries above the L7-S2 spinal cord segments can still have a withdrawal reflex with absent pain sensation. (1) Further discussion of the impact of loss of pain sensation is discussed under clinical conditions following this section.

Assessment of bladder function

Non-ambulatory paraplegic patients, regardless of whether they have pain sensation, may lose the ability to urinate. Patients with injuries above the L7 spinal cord segments can have an UMN bladder while patients with injuries affecting the sacral segments or nerve roots can have a lower motor neuron bladder.

UMN bladders

With lesions above the L7 spinal segments, they may have an UMN bladder due to interruption of the tracts that control urination from the brainstem. Usually this happens with injuries severe enough to cause a loss of motor (movement) function in the limbs. These patients aren't able to initiate urination and bladders can become over distended with urine. The bladder typically feels firm and turgid. They can have overflow incontinence, as the bladder becomes overfull, and will leak urine frequently. Urethral sphincter tone is normal to increased and can impede bladder expression. UMN bladders typically have a good prognosis for recovery as long as the spinal cord injury recovers. Usually they regain ability to urinate as they regain motor function.

LMN bladders

With lesions affecting the sacral spinal cord segments, patients may have a LMN due to interruption of the pelvic nerve. These patients also cannot initiate urination and the bladder becomes over distended with urine. However, since the pelvic nerve directly innervates the detrusor muscle, the bladder will feel flaccid and floppy (analogous to a Ziploc® baggie filled with fluid). Also, the pudendal nerve is commonly affected which means the external urethral sphincter tone is lost. Therefore they often dribble urine constantly. With manual expression, it's fairly easy to get urine to spill out, but difficult to empty the bladder due to loss of the bladder muscle tone. LMN bladders often have a more guarded prognosis for recovery since direct pelvic nerve injury is less able to recover than spinal cord injuries.

Bladder management

In either case, the bladder needs to be kept small to prevent development of recurrent urinary tract infections, urine scald due to overflow incontinence, or detrusor muscle atony. UMN bladders can usually be managed by manual expression. If urethral sphincter tone is high, they can be more difficult to express. Addition of prazosin or phenoxybenzamine, both α adrenergic antagonists, may help to reduce urethral sphincter tone and make bladder expression easier. LMN bladders can be harder to keep empty with manual bladder expression due to loss of detrusor muscle tone. Urinary catheterization or cystostomy tubes can be used to manage these bladders if needed. Failure to keep the bladder small often results in overfilling of the bladder and subsequent overflow incontinence. Frequent leaking or dribbling of urine often indicates overflow incontinence. If bladders remain overdistended for long periods of time, the detrusor muscle can become stretched out and atonic, and then may not be functional even if the neurologic injury can recover. (1)

Anal Sphincter tone

Injuries to the sacral segments of the spinal cord or nerve roots can cause damage to the pudendal nerve which innervates the anal spincter and perineal region. With loss of pudendal nerve function, the anal sphincter may appear open and flaccid. Tone can be assessed also by rectal examination. The perineal reflex and presence of pain sensation around the anus can help assess the degree of injury. Loss of anal sphincter tone is often associated with fecal incontinence.

Spinal shock

Spinal shock may be seen in acute to peracute spinal cord injuries from any cause (IVDD, FCE, Trauma). Interruption of ascending and descending spinal tracts results in flaccid paralysis below the site of the injury. Therefore, dogs with UMN spinal cord injury may appear to have a LMN injury. Spinal shock may be present for 12-48 hours but then usually resolves.

ASSESSMENT OF CLINICAL CONDITIONS

Intervertebral disc extrusions (IVDD)

Intervertebral disc extrusions usually occur due to degenerative disc disease. The degenerated disc material can extrude into the spinal canal and cause contusion and compression of the spinal cord and nerve roots. (2) Less commonly disc extrusions occur secondary to trauma. In both cases, the extruded disc leads to signs including spinal pain, weakness, ataxia, paralysis and loss of ability to urinate.

Assessment of the patient's neurological status is important for prognosis and for deciding on the appropriate therapy i.e. medical therapy v.s. surgery. Mildly affected patients may be ambulatory with just spinal pain without neurologic deficits or only mild deficits such as mild ataxia or paresis. Often these patients can be managed with restricted activity and medications for pain control.

With more severe or progressive injuries, patients may become non-ambulatory paraparetic or paraplegic. Pain sensation is one of the most important prognostic factors in dogs that are paraplegic due to a disc extrusion. Presence of pain sensation in thoracolumbar disc patients indicates that the prognosis is still good to excellent – 90-95% chance of recovering ability to walk with surgery. Absence of pain sensation indicates a more complete spinal cord injury and poorer prognosis for recovery of ability to walk. If surgically decompressed within 24 hours of losing pain sensation, patients have a 50-60% chance of recovering ability to walk. If they have lost pain sensation for longer than 24 hours, particularly for multiple days, then chances for recovery drop significantly. (1,2)

Fibrocartilaginous embolism (FCE)

Dogs with FCE present with variable neurologic deficits indicative of a spinal cord injury. The injury is a result of loss of blood supply to a region of the spinal cord due to embolization of a spinal cord blood vessel by a fragment of fibrocartilage believed to originate from an intervertebral disc. These patients may cry out when they initially go down, but are usually non-painful at the time they present to the veterinarian, though occasionally spinal pain may be detected. Often their neurologic deficits are asymmetrical with 1 side affected much worse than the other side, though some are more symmetrically affected. Hemi or tetra paresis or plegia may occur with cervical FCE's. Para or mono paresis or plegia may occur with thoracolumbar/lumbosacral FCE's. Prognosis is fair to good as long as motor and pain perception is present. As with other spinal injuries, patients with lower motor deficits may have a more guarded prognosis for recovery. Loss of pain sensation also indicates a guarded to poor prognosis for recovery. (1)

Exogenous causes of spinal trauma

Exogenous causes of spinal trauma include being hit by a car, falls, being kicked or stepped on by a horse or cow, owner inflicted injuries, gun shot wounds, and others. Often these are accompanied by fractures or luxation/subluxation of the spine. Primary injury to the spinal cord is caused by contusion, shearing, laceration and traction, etc. Because these stresses are different than with compressive spinal cord injuries (contusion and compression), these injuries may carry a more guarded prognosis than compressive spinal cord injuries depending on the severity of the injury. (3,4)

First and foremost, trauma patients should be evaluated systemically to ensure that they are stable. Basic assessments of ABC and vital parameters should be first priority as well as assessments of more immediately critical injuries i.e. hemorrhage, thoracic or abdominal injuries, concurrent head injuries. Maintenance of systemic blood pressure and oxygenation may help to prevent further spinal cord injury by maintaining perfusion to the spinal cord. If a spinal injury is suspected, spinal manipulation should be kept to a minimum. The patient can be secured to a back board to minimize spinal manipulation. (3)

Once the patient is hemodynamically stable, a neurologic examination can be performed. Patients that are ambulatory with mild deficits usually have a good prognosis for recovery. Patients that are non-ambulatory with or without motor function, if they maintain pain sensation, may recover function, but risk of permanent injury and therefore incomplete recoveries is higher than with compressive spinal cord injuries. When these patients lose pain sensation, the prognosis for recovery of ambulation is considered poor. (1,2)

REFERENCES

1) Dewey, CW. A Practical Guide to Canine and Feline Neurology. 2nd ed. 2010.
2) Jeffrey ND et al. JVIM 2013; 27: 1318-1333.
3) Park EH et al. J Vet Emerg Crit Care 2012;22(2):160-178.
4) Webb et al. Can Vet J. 2010;51:598-604

MOVEMENT DISORDERS

Dennis P. O'Brien, DVM, PhD, DACVIM (Neurology)
Columbia, Missouri

Movement disorders are a group of condition characterized by involuntary movements. Seizures would technically fall under this definition, but they are classified separately. Seizures are characterized by abnormal electrical activity in the cerebral cortex seen on EEG, and are typically accompanied by loss of consciousness and post-ictal behavior changes. Focal seizures, however, may not affect consciousness, and some movement disorders are also episodic. Thus the border between episodic movement disorders and seizure disorders are often unclear. Diseases that cause visible contractions of muscle fibers, such as fasciculations, but do not cause movement of the body part are not classified as movement disorders. This session will focus on recognizing the different types of movement disorders.

TREMORS

Tremor is defined as a rhythmic, oscillatory movement of body parts. They are further broken down by the frequency of the oscillations and when the tremors occur. Feedback systems have a natural tendency to oscillate and in muscles such feedback systems produce a physiologic tremor. Under most conditions, physiologic tremor is barely perceptible, but things like stress, fear, fatigue, weakness, and drugs such as caffeine can produce an exaggerated physiologic tremor. These tremors are typically fast and small amplitude and tend to occur when standing still, diminishing with movement. Disease processes such peripheral nerve disease, pheochromocytoma, hypoglycemia will also exaggerate physiologic tremors. They can also be normal in aged dogs.

Orthostatic tremors occur when standing and affect primarily the hind limbs. They tend to be a little bit slower and coarser than physiologic tremor. The dogs often hesitant to sit or stand, and thus appear weak. When the dog walks, however, the movements show no evidence of weakness and the tremors resolve. They also resolve when the dog is sitting or lying down. The cause is not known but they have been reported most commonly in Great Danes[1]. Bulldogs and Doberman Pincers show benign head tremors that appear to be hereditary[2,3]. These may be a variation on orthostatic tremors since the head must also be supported against gravity. Both appear to be benign conditions.

Intention tremors are a hallmark of cerebellar disease of any cause. The term intention tremor may not be appropriate for animals since we cannot know what they intend to do, but it captures the key feature of the tremor so the term is still used. These tremors only occur when the animal is making a goal directed movement. They disappear completely at rest. They are most apparent in the head but can involve the entire body or limbs. The head tremors can become dramatic when the animal eats and the limb tremors when they try to walk down stairs. The tremor is very slow and coarse.

MYOCLONUS

Myoclonus is a very brief, sharp contraction of a muscle, group of muscles or the entire body. It can be single or repeated rhythmically. Rhythmic myoclonus lacks the smooth, oscillatory character of a tremor. Focal myoclonus is a common sequelae to canine distemper infections. Generalized myoclonus is often a sign of a more generalized neurodegenerative disease[4,5]. Sometimes myoclonus can be elicited by stimulation. A hereditary myoclonus in Irish Wolfhounds called startle disease or hyperekplexia is caused by a dysfunction of the inhibitory neurotransmitter glycine[6]. Without this inhibition, the normal startle response is exaggerated. In other conditions the jerks occur spontaneously. The latter is sometimes call myoclonic epilepsy but is often not clear if the cortical epileptic discharges that would classify it as a seizure disorder are present or not. Negative myoclonus looks similar clinically, but is actually caused by brief loss of tone in a muscle followed by a jerk as the animal catches itself.

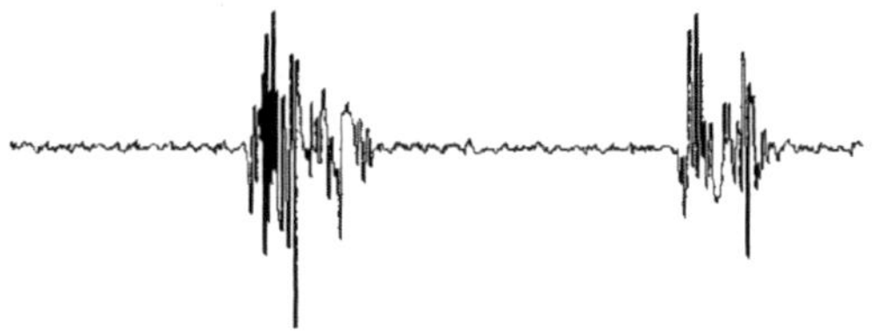

An EMG recording of muscle activity shows the repetitive, brief, "lightning-like" burst of firing that characterizes myoclonus.

Opsoclonus-myoclonus is an unusual form of myoclonus affecting the eyes as well as the body. The body appears more like a generalized tremor, but the eyes make very rapid, uncontrollable darting movements. It is thought to be an autoimmune disease.

CEREBELLAR ATAXIA

Cerebellar ataxia is the most common movement disorder diagnosed in veterinary practice. Cerebellum is from the Greek meaning "the little brain" and is tucked behind the cerebral cortex at the base of the brain. The cerebellum is responsible for the fine-tuning of movement. When an animal is learning to walk or perform any motor task, the cerebellum monitors commands coming from the forebrain motor systems and receives feedback from the vestibular system and proprioception sensors in muscles about balance and execution of the movement respectively. Motor learning occurs in the cerebellum as synapses are strengthened or weakened to perfect the movement. Coton de Tulears with Bandera's neonatal ataxia have a mutation in the neurotransmitter receptor that mediates the weakening of synaptic connections. As a result, motor learning is abolished and they can never

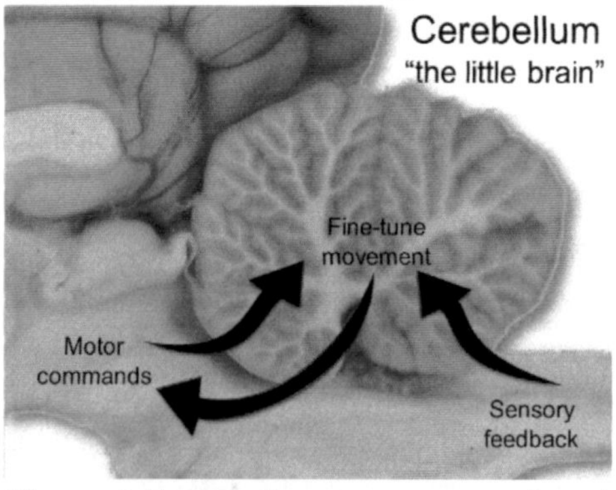

The cerebellum monitors motor commands and vestibular and proprioception feedback to fine-tune movement.

walk[7]. Interestingly, the menace response is also a learned motor response and not a reflex. Thus it does not develop until after weaning age and it is often lost in cerebellar disease. Once the movement has been learned, the cerebellum utilizes what has been learned to adjust the motor program in real time in response to sensory feedback so that movements are smooth.

In addition to intention tremor, cerebellar lesions produce cerebellar ataxia. Animals with cerebellar ataxia are not weak and have no proprioception deficits. Instead the range and force of movement are not regulated properly, the result is an exaggerated, goose-stepping gait. This is most apparent when the animal is moving quickly or trying to do a more skilled movement like walking down stairs. In severe cases, the animal may have all four feet off the ground or flip over backwards. Because the cerebellum works intimately with the vestibular system, affected animals may lose their balance with sudden movements or develop nystagmus which may be vertical. Occasionally, affected dogs will fall and adopt a contorted posture which is sometimes mistaken for a seizure.

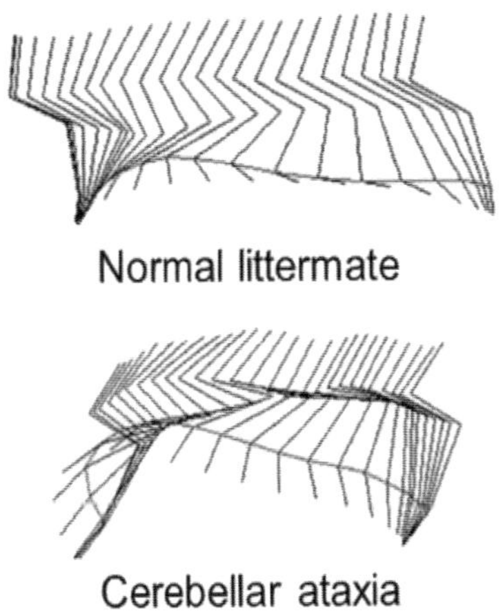

The exaggerated flexion of the elbow and "Goose-stepping" is seen in this stick-figure representation of the gait in cerebellar ataxia compared to normal.

Many acquired diseases including infections such as *Neospora caninum* encephalitis, tumors, and strokes commonly affect the cerebellum and produce cerebellar ataxia and intention tremors. An idiopathic tremor syndrome called idiopathic tremors or idiopathic cerebellitis has been recognized. It is thought to be an immune mediated disease. Affected dogs develop an acute onset of severe generalized tremors and cerebellar ataxia. Most dogs improve with anti-inflammatory drugs and diazepam. Maltese and other white dogs are most commonly affected and the condition was originally called "little white shakers", but we now know it can affect any breed.

Developing the circuitry of the cerebellum requires interaction of a variety of developmental signals and the process of regulating movement in real time once the cerebellum has developed is very complex. Thus mutations in a variety of genes can disrupt cerebellar development or function, and hereditary cerebellar ataxias are common. Developmental disorders are typically apparent a neonates. For example, in Eurasier dogs, mutations in a gene that is part of the signaling pathway that regulates brain development cause Dandy-Walker syndrome, a failure of the midline of the cerebellum to develop, and cerebellar ataxia[8]. Functional disruptions will also cause cerebellar ataxia. Potassium channels are critical for regulating neuronal excitability. Jack Russell Terriers with mutations in one of these channels that is found in high density in the cerebellum have severe spinocerebellar ataxia because the increased excitability disrupts the timing of impulses essential for coordinating movement. The dogs show other signs of excessive excitability of neurons such as seizures and myokymia, uncontrollable, rippling contractions of the muscles at rest[9].

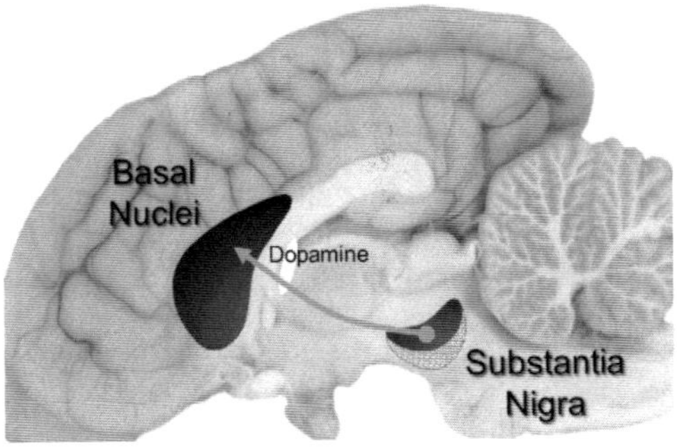

Dopamine neurons from the substantia nigra provide a go/no-go signal to motor programs from the basal nuclei.

PARKINSONISM

Parkinson's disease is a common neurodegenerative disease of older people. It is caused by degeneration of an area of the brainstem, the substantia nigra. Neurons in this area that use dopamine as a neurotransmitter project forward to synapse on the basal nuclei, clusters of neurons below the cortex that are a key part of the motor system. This system is the gate-keeper for movement. Dopamine is an important component of the "reward system" in the brain. Based on past experience, this system gives either a "go" or a "no-go" signal to the motor programs. Without the dopamine input, the system is stuck in "no-go" and movement cannot be initiated. Although the resting tremor is the sign of Parkinson's disease that everyone recognizes, it is the inability to initiate movement that devastates patients with Parkinson's' disease. These signs of difficu lty initiating movement and resting tremors are called parkinsonism and can occur in other hereditary, toxic, and degenerative diseases in people.

Parkinson's disease per se has not be identified in dogs or cats, but the signs of parkinsonism are seen in some hereditary movement disorders of dogs. Chinese Crested dogs and Kerry Blue Terriers have a juvenile-onset, hereditary disease called canine multiple system degeneration[10,11]. As the name implies, multiple parts of the motor system in the brain degenerate beginning about 4 months of age. The cerebellum is the first to degenerate and the dogs have classic cerebellar ataxia. Affected dogs have tremors, but they are intention tremors associated with movement in contrast to the tremors of parkinsonism which occur at rest. Except for primates, none of the experimental or hereditary forms of parkinsonism in animals show a resting tremor. Beginning at about 6-8 months of age, affected dogs begin having difficulty initiating movement. They shift their weight forward until they finally propel themselves forward in a gait called

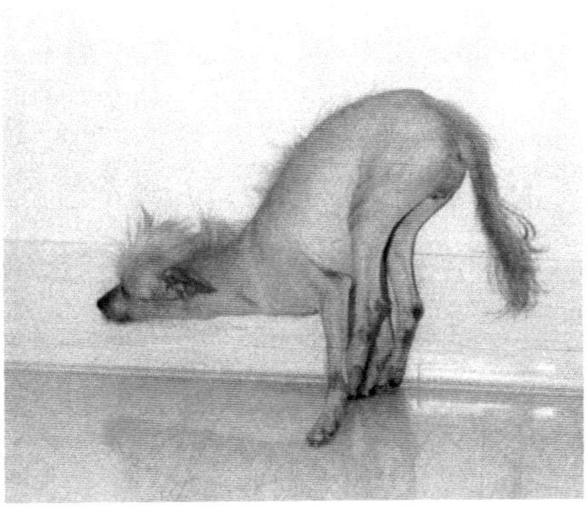

Parkinsonism is characterized by an inability to initiate movement. Affected dogs adopt a hunched posture, lean forward, but are unable to step forward.

festination. With further progression of the disease, they are unable to initiate voluntary movements. They adopt a hunched-up posture when they try to walk and rock until they lose their balance and fall. In addition to degeneration of the cerebellum, affected dog have degeneration of the substantia nigra and basal nuclei.

DYSKINESIA AND DYSTONIA

Parkinsonism is considered a hypokinetic disease of the basal nuclei because there is difficulty initiating movement. Another group of diseases are called hyperkinetic movement disorders because the basal nuclei are unable to stop movements and involuntary movements, (dyskinesia) or abnormal muscle tone (dystonia) occurs. Dyskinesias can be focal or generalized. Focal dyskinesia can involve the facial muscles or single limbs. In the limbs, proximal or distal muscles may be preferentially affected. The movements can either be slow or rapid and they may be continuous or episodic (paroxysmal). In humans, dyskinesias are divided into three general categories; athetosis, chorea, and ballism. Athetosis is a slow, continuous, writhing movement involving primarily the distal arm. The movements may be somewhat purposeful. Chorea (from the Greek "to dance") also affects predominantly distal limb muscles, but the movements are totally involuntary. They are more rapid and fragmented. Ballism is also a rapid movement, but it affects predominantly proximal muscles. The result is a flailing or flinging movement. Dystonia is a sustained contraction of muscles. The axial muscle or limb muscles can be affected. The limbs can either be held in flexion or extension and sometimes a tremor of the affected muscles accompanies dystonia. As with dyskinesias, the episodes can last from minutes to hours but tend to be more infrequent. Often a mix of different dyskinesias and dystonia will occur in the same patient. Acquired dyskinesias can be drug induce or secondary to basal nuclei lesions from infections or stroke.

Chinook dogs have a paroxysmal dyskinesia consisting of primarily dystonic episodes with flexion of the limbs and twisting of the torso. The episodes may last for up to an hour. The dogs remain conscious during the episodes and show no post-ictal behavior.

The best characterized dyskinesias in veterinary medicine are paroxysmal and appear to be hereditary. Like seizures, the animals are normal between episodes. In contrast to generalized seizures, animals remain conscious during paroxysmal dyskinesias and they do not have the life threatening complications sometimes seen in seizures. Some dyskinesias have clear triggers such as stress or excitement while others do not. The episodes can occur infrequently to multiple times per day and they can last minutes to hours. The episodes can vary from simple flexion of one limb while walking which looks like a lameness, to frantic, alternating flexion and extension of multiple limbs. The paroxysmal dyskinesias will sometimes respond to antiepileptic drugs or the carbonic anhydrase inhibitor acetazolamide.

Chinook dogs have a paroxysmal dystonia that appears to be inherited as an autosomal recessive trait[12]. Affected dogs have episodes of dystonia and tremors lasting up to an hour. They are conscious during the episodes and appear perfectly normal on recovery. Occasional flailing movements of a limb can occur also. In Cavalier King Charles Spaniels, a mutation has been identified in a gene involved in connections between cells and fiber guidance in the brain development in dogs with episodic falling syndrome[13]. Although the dogs do collapse during these episodes, they do not lose consciousness and they have increased tone in the limbs resulting in them adopting a "deer stalking" posture. Canine epileptoid cramping syndrome or Spike's disease is a movement disorder that affects Border Terriers. As the name implies, the episodes resemble seizures, but the dog remains conscious throughout. Most episode last from 2-30 minutes but may last for hours. The episodes consist primarily of tremors and dystonia affecting the head and body. Excessive intestinal noise and vomiting or diarrhea may accompany the episodes leading some owners to suspect a dietary link[14].

Conclusion

Movement disorders are being increasingly recognized. In the case of hereditary diseases, current gene discovery techniques permits us to identify the mutations responsible and eliminate the disease from affected breeds. Understanding the basis of these hereditary diseases can suggest new pathways to explore for effective therapies for acquired conditions. What we can learn from our patients can also shed light on the comparable human diseases.

REFERENCES

1. Garosi LS, et al. Journal of Veterinary Internal Medicine 2005;19:606-609. 2. Guevar J, et al. Movement disorders : 2014;29:191-194. 3. Wolf M, et al. Disorders 2011;26:2381-2386. 4. Lohi H, et al.Science 2005;307:81. 5. Awano T, et al. Molecular Genetics & Metabolism 2006;89:254-260. 6. Gill JL et al. NeurobiolDis 2011;43:184-189. 7. Zeng R et al. Journal of Veterinary Internal Medicine 2011;25:267-272. 8. Gerber M, et al. PLoS One 2015;10:e0108917. 9. Gilliam D, et al. Journal of veterinary internal medicine / American College of Veterinary Internal Medicine 2014;28:871-877. 10. O'Brien DP, et al. Journal of Heredity 2005;96:727-734. 11. deLahunta A, Averill DR. Journal American Veterinary Medical Association 1976;168:1119-1124. 12. Packer RA, et al. Journal of Veterinary Internal Medicine 2010:24:1305-1313. 13. Gill JL, et al. NeurobiolDis 2012;45:130-136. 14. Black V, et al. The Journal of small animal practice 2013.

DEALING WITH A DIFFICULT DOCTOR (SUPPORT STAFF'S GUIDE TO SANITY IN THE WORKPLACE)

Jordin Karalunas LVT, VTS (ECC)
Oakland, CA

The veterinary hospital is full of difficult people including clients, support staff, receptionists, and veterinarians. How to appropriately deal with difficult support staff and clients has been discussed in staff meetings everywhere. Dealing with the difficult veterinarian is a more delicate topic that is not routinely brought up for discussion. Personality types, self assessment, conflict management, and resolution will be explored.

Personality types vary, and bring constant challenges to the workplace. The road to becoming a veterinarian requires dedication, hard work, and sacrifice. To achieve these traits requires a specific personality type. Most commonly, a "type A" personality is encountered. "Type A" personalities tend to be perfectionists, , impatient, , and have an abnormal sense of urgency. Frequently, this personality type can leave others feeling resentful, intimidated, , overwhelmed, and frustrated. These attributes must be delicately handled when dealing with a difficult doctor. Typically that doctor is the employer. If the doctor is not the employer, they are still head of the veterinary team, which puts them in a leadership position. It is important to remember that being a veterinarian does not mean having good leadership or communication skills. Each individual use the tools they possess to communicate effectively, which is the key to conflict prevention.

The difficult doctor may be one displaying unprofessional behavior. These behaviors can include, but are not limited to, gossiping, , hostility, and vulgarity.. It is important to remember that these behaviors should not be tolerated, and should be addressed in a professional manner. When confronted with unprofessional behavior, do not take it personally. It may be advisable to remove oneself from the situation, to prevent acting instinctively rather than rationally. We respond to conflict based on our perceptions. Talking about what took place is healthy to gain an outsiders' prospective. However, it is unprofessional to discuss it in the workplace. Gossip does not resolve conflict; it feeds it. Discussing the problem with a friend or family member is preferred. Once the conflict has been thought through, it is important to forge a resolution. What was the conflict about? What can be done to prevent it from occurring in the future? How should the conflict be resolved with the doctor? It is necessary to have potential solutions ready for discussion to avoid further conflict. Take time to process the conflict in a healthy manner; however, it is important to act promptly to resolve conflict effectively. When meeting with the doctor, take into consideration their weaknesses, management and communication style. Do they need all of the details, or just bullet points? Is the doctor very sensitive? Doesthe management need to be present for the meeting? Be sure to go into the meeting with poise, confidence, and a positive attitude. Be respectful of the doctor to encourage the same response.. It is important to only convey the issue at hand. Do not generalize, or bring up past problems. If clarification of the conflict is needed, ask for clarification. When an agreement has been reached, reiterate the salient points to confirm the consensus Prepare to apologize if you are in the wrong. Willingness to admit wrongdoing will inspire open mindedness. Show respect for others' opinions. When communicating the conflict resolution, take note of your body language. Body language is over 50% of how we communicate. Being gentle and calm can aid in successful resolution. Using the right words can quickly resolve conflict without blame. Words such as "we", "our", "the challenge" and "goal" are preferred over "I", "you", or "problem". Remember that conflict is an opportunity for growth. In the event a one-on-one conflict resolution is ineffective, meeting with the management should be the next step. It is important to follow human resources' protocols when handling this type of challenge in the workplace.

It is important to know the doctor's expectations of the support staff. Often the demands are very basic in nature. The support staff is expected to listen, understand the task, complete the task efficiently and proficiently, and treat the doctor with respect. In addition to the doctor's needs, it is important to convey certain qualities to avoid interstaff conflict. Characteristics such as , empathy, , fairness, and sound judgment can diffuse conflict with the difficult doctor. These behaviors command respect. Building rapport with the difficult doctor can also prevent conflict. It allows the support staff to see the veterinarian as a person, which encourages empathy. Avoid assumption and moral judgment in regards to the difficult doctor. Remember that we work in a place ofextreme suffering at times. To expect that it does not affect the doctor and staff is unreasonable. Compassion fatigue and burnout, as well as personal problems such as health and family problems, encompass us all. Taking this into consideration may help us understand the difficult doctor a little better.

The support staff should also take a look internally. Are theirfeelings being managed appropriately? Are healthy boundaries in place? Is life balanced between work, and personal time? Is the staff maintaining a healthy diet and exercise? Has introspection been encouraged to recognized personality traits and flaws? Answering these questions can help us handle conflict in a healthy and productive manner. Some tips for stress relief in the work environment are listed below:

-Take breaks
-Say "no"
-Share the workload
-Find humor in the day and laugh
-Ask for help
-Give yourself credit
-Give others credit
-Breathe deeply
-Forgive and let go
-Pick your battles
-Reinforce good behavior
-Ignore bad behavior ("We should re-visit this issue at a later date, after we have had time to reflect")
-Don't respect the bad behavior, respect the person

-You are in control. Your only problem is the task at hand.

In the event conflict with the difficult doctor is attributed to malpractice, animal abuse or mishandling, harassment of any kind, or drug abuse, it is recommended that management be consulted prior to meeting for potential conflict resolution.

To remain sane amidst a difficult doctor requires the support staff to tend to their own emotional needs. Remembering that you can't change the person, but instead how you react to them, has a significant impact on your quality of life. Though it is hard to show respect during yelling, swearing, and immaturity, reaction to the situation means giving up control and allowing the discord to escalate. The real art is to say the right thing in the right moment, and not to say the wrong thing at the most tempting time.

COMMUNICATING DIFFICULT DIAGNOSTIC TESTING AND MEDICALCASES TO PET OWNERS

Saundra E Willis DVMA, ACVIM (Small Animal Internal Medicine)
Mukilteo, WA

INTRODUCTION

Complicated medicine disease and diagnostic testing can be challenging for pet owners. It can be frustrating for owners when they don't understand the VALUE of testing. They will often decline an option that is not understood. This is particularly true if the veterinarian has not clearly and concisely outlined what they think the problem or problems might be, the potential disease(s) causing the problem(s) and the diagnostic tests needed to rule-in or out those potential diseases.

"Enough tests already just treat" owners say. Or one may hear from an owner: "They are just going to do a bunch of tests" when a friend suggests a visit to the veterinarian for an ill pet. The veterinarian may or may not do a "bunch of tests" but how does the veterinarian know how to treat the patient if they don't know what's wrong? The most expedient way to achieve lasting relief and return to health is by uncovering what the problem really is by diagnosing the disease, so that treatment can be definitive and not simply empirical. Diagnosing disease also makes financial sense. Money is not wasted on treatments for diseases the pet does not have.

As part of the online survey of the Bayer Veterinary Care Usage Study, pet owners were asked under what circumstances they would take their pet to the veterinarian more often. Four attributes that ranked highest were: If I knew I could prevent problems and expensive treatment later, if I was convinced it would help my pet live longer, if each visit was less expensive and if I rally believed my pet needed examinations more often[1]. These findings suggested that pet owners need to fully understand the health implications for their pets and the economic benefits for themselves before they will increase their visits to the veterinarian.

As a small animal internal medicine consultant for a local veterinary diagnostic laboratory, I speak to veterinarians from all different walks of the profession regarding the medical aspects of cases but we also discuss client communication. Creating a healthcare team with the client and veterinary staff the benefit of the pet is key to the management of a complicated case in medicine. The veterinary team cannot do it alone without the understanding and participation of the pet owner, from initial history to final dose of the treatment medications.

I am a passionate believer that we need to do what is best for the pet. That is why owners come to us, their family veterinarian. We are their source of information. Our clients may "find out" about additional options for their pet's illness through the internet and/or a neighbor and ask us about these but I firmly believe we need to be the primary source of this information. We are the experts that listen to their concerns, examine their pets, formulate an idea of what may be going on, provide testing recommendations to help make a diagnosis, and initiate a treatment plan. We will modify our recommendations depending on many factors as discussed below but we should not prejudge our clients' interest in their pet, willingness to treat, ability to afford care and not offer what we think is best. We should not withhold information from clients but educate and inform them so that can make the bet decision for their pet. I remember walking with a cardiologist who was speaking to a close friend on the phone. Their older dog, which had been seeing their family veterinarian for foreleg lameness, was just diagnosed with bone cancer on radiographs taken for the first time one year after the dog originally presented for the problem. The owner wanted to know why it had taken so long for their veterinarian to take radiographs. That was not an easy question to answer.

After nearly 20 years in diagnostic laboratory work consulting with small animal veterinarians, I have developed some methods for communicating diagnostic testing and complicated medical disease. Most ideas I have gathered from listening to veterinarians, hearing what they are saying and through them, what their clients are saying. Some I have learned directly from clients whom I have either spoken to directly when they have called the laboratory or advised on WebMD as an AVMA expert. I am sensitive to the client concern regarding the rising costs of veterinary medicine and the perception that these increasing costs are solely attributed to the business of veterinary practice and our salaries. The truth is veterinary medicine has advanced and we have many more options in diagnostics and treatments to offer the owner for their pet. That being said, a thorough work-up does not mean running every test possible and going out on tangents. Sometimes we need to focus our time and finances on supportive care and empirical treatment. It is all about quality of life.

POMR AND DIFFERENTIAL DIAGNOSES

In the initial office call, the veterinarian will take a detailed history and perform a thorough physical exam. The veterinarian may know at that time what the problem is and what treatment needs to be done. But generally more information is needed to make a diagnosis. The veterinarian will create a problem list and a list of diseases that might cause each problem, termed a differential list. This systematic method of diagnosing disease, termed the problem oriented medical record (POMR) helps to organize a potentially complicated process. Diagnostic testing, imaging, etc are then performed to rule-in or out the differentials on the list generally starting with the most likely diagnosis on down to the least. While awaiting test results, symptomatic therapy which can include fluids, appetite stimulants, intestinal protectants, and pain medications is often provided for symptomatic relief. Not only will the pet generally start feeling better, but the owner will feel less angst at their discomfort. This should foster trust in the veterinarian and a more favorable inclination toward additional recommendations. It is difficult for the client to await treatment until test results are returned and generally not necessary. To the client this likely feels like we are doing nothing for the pet during that time. That is why early symptomatic care is important and honestly, most often does not interfere with diagnosis.

Diagnostic testing generally starts with noninvasive screening tests such as the general cbc/chemistry/UA panel and progresses to more complicated and invasive testing such as imaging, endoscopy, and/or biopsy. During this process, the differential list is narrowed and testing is continued until: 1) the diagnosis is made; 2) the pet gets better with symptomatic and/or empirical therapy; 3) the owner's finances and/or desire for more testing limits further work.

As mentioned above, avoid unnecessary testing. With every test there needs to be a reason to run them and a plan for using the results. Often the results help the veterinarian determine the next step, whether diagnostic or therapeutic. Sometimes testing simply provides piece of mind, i.e. a diagnosis of cancer when the owner is not considering chemotherapy but needed to know why their pet was ill. Tests should not be run if the results would not provide necessary information or piece of mind. Owners nearly always have at least some financial restrictions and the veterinarian needs to work within their budget.

Veterinary staff is integral to this process. Veterinarians start the conversation with pet owners but it is often the receptionist, technician or assistant whom the owner goes to for understanding and clarification. Communication to the pet owner is a team effort. Staff often pull results off the fax machine, access them online and/or record them in the record and their understanding of these tests and the medical diseases will make their job more rewarding and the staff member a better resource for the practice.

COMMUNICATION

Veterinary staff needs to consider issues beyond the pet in communicating diagnostic testing and medical disease. What are the client's needs? The profession has only recently begun to focus on the need for communication skills. We can be the most intelligent, skilled veterinarian in the community but if we cannot communicate with our clients, educate, explain and empathize, we will not succeed. People learn differently. Some learn visually, some are auditory learners. While diagramming a diagnostic workup on a white board might work well for a visual learner, others would understand better by talking about the diagnostic tests and hearing the options. Some clients just want whatever is best for their pet and don't need to spend time on the details. They will quickly sign off on a treatment plan for their pet. Others need to discuss each step of a financially to support each and every diagnostic test before they are done. Veterinary staff needs to be aware that clients have different needs and communication styles that factor into their information gathering.

The veterinarian should make their best recommendation and pause allowing the client to speak next and accept or suggest alterations to the recommendation. In one study the odds for adhering to a dental or surgical recommendation in companion animal practice were seven times greater when the recommendation was clear compared to one that was vague and ambiguous[2]. After making a recommendation the veterinarian should ask "What questions do you have?" not "Do you have any questions?" These are two different questions that will result in very different answers. Clients nearly always have questions. If dialogue is encouraged, the veterinarian may learn where there is a need to communicate more information, what misconceptions the client might have, and what limitations including financial that might be in play. We become and resource and we create a health care partnership with the owner focused on their pet.

The veterinarian needs to consider expectations of the pet owner regarding their pet's care. Veterinarians generally don't ask but may assume that they want us to find out why their pet is ill and treat until cure. But sometimes their expectations are not to diagnose but to maintain quality of life through symptomatic care. I remember a veterinarian telling me that the owners of an older large breed dog just wanted "nature to run its course". The pet had a two week history of anorexia and weight loss for which a physical exam and initial blood work were normal. The veterinarian planned to do nothing further. Literally taken, the client's comment would be interpreted to do nothing. Figuratively taken, as I interpreted it, the owner didn't want advanced diagnostics and treatment but I suspect they wanted some symptomatic treatment for their dog to make it feel better. I encouraged the veterinarian to clarify the owner's wishes and suggest supportive treatment for the pet. As one young veterinarian once said to me, "the answer is not to do nothing".

There may be other barriers to care and they should be explored. Are the clients away most of the day and can't medicate the pet? Is the family a multi-pet household which makes feeding a special diet challenging? Does the client have personal including medical limitations that make treating a pet difficult? What experiences and/or perceptions does the client have regarding the disease? In a study looking at client expectations when seeking oncology care at a referral hospital, health care provider awareness of clients' understanding and experiences with cancer facilitated client action and patient intervention as well as enhancing the clients' psychological well being[3]. In the same study, the manner in which information was communicated provided necessary information but, more importantly, created a humanistic environment to provide psychological support to help the client successfully cope with their pet's life threatening cancer[4]. What about the pet? Can anyone including professional staff actually pill that cat? Lastly, cost concerns. There are now more options with pet insurance and payment plans. These financial choices should ideally be discussed with the client before the pet becomes ill. Clients need to be reassured that they are still doing the best for their pet even though they have cost limitations the diagnostics and treatments are carefully selected.

THE HEALTHCARE OR TREATMENT PLAN

The estimate for costs should be a natural continuation of the diagnostic plan based on the differential list arising from the problem list. The veterinarian will create a healthcare or treatment plan (estimates are for cars sales) that provides a range of costs, from a minimal work up to the full recommendation. It is difficult to anticipate every cost as additional testing and treatments will vary based on the results of preliminary work. Furthermore we will often have several options for the work up: plan A, B or C depending on how completely the client wants to pursue the pet's illness. For example, does the client wish to do Plan A which includes all recommended diagnostic and treatment plans including a potential for additional testing such as ultrasound and biopsy? Do they want some diagnostics but not all? Do they want limited diagnostics and supportive care? It is important to give the client all the possible scenarios and costs up front. As the work up progresses, the client can follow the treatment plan and see where the work up is going and will best understand any adjustments in the plan that need to be made. People do like surprises but not when they involve costs. The treatment plan should make sure that the client's finances are used most effectively, i.e. avoiding using up all the client's financial resources on testing leaving no monies available for treatment.

REFERENCES

1. Volk J, et al JAVMA 2011;238(10)1275.
2. Kanji C et al JAVMA 2012;240(4)427.
3. Stoewen D et al JAVMA 2014;245(7)773.
4. Stoewen D et al JAVMA 2014;245(7)785.

SECRET TRICKS THEY NEVER TAUGHT YOU IN CVT SCHOOL

Amy E. DeClue, DVM, MS, DACVIM (SAIM) and Dennis R. Spann DVM, DACVIM (SAIM)
Columbia, MO and Roseville, CA

In this lecture we will discuss and show videos of ways to improve or simplify various technical procedures. We will review simple changes that can streamline procedures and promote success every time. Some examples of topics we will cover include: How to organize supplies to optimize convenience and clean up; tricks to improve success with phlebotomy; tips on peripheral and central venous catheter placement; keeping animals warm during anesthesia; how to be successful at Doppler blood pressure measurement; improving yield from nasal flushes and blind BALs; how to use the "threading" technique for intranasal clotrimasole therapy; ways to make bone marrow aspiration easier; methods to secure continuous glucose monitor probes, fool-proof female urinary catheter placement; vascular access for puppies and kittens; how to secure PEG tubes, tips for esophageal and gastric foreign body removal. If you want to learn the tricks, you have to come to the lecture!

ARTERIAL BLOOD GASES

Kenichiro Yagi, BS, RVT, VTS (ECC, SAIM)
Los Altos, CA

INTRODUCTION

Evaluation of the respiratory status of the patient can be accomplished in numerous ways. One of the most accessible methods is visual assessment of respiratory effort and respiratory rate, as well as auditory evaluation through auscultation of the chest cavity. Both of these methods allows one to detect a patient who is having difficulty breathing, as well as giving some hints to the type of respiratory disorder. When evaluating the severity of respiratory compromise, visual assessment and auscultation is not always sufficient. A more objective look at respiratory status can be made through more sophisticated, and sometimes invasive methods to determine the patient's blood gas values. There are many gases dissolved and carried by the blood in various forms. Two that are of the most interest for respiratory function are oxygen and carbon dioxide. Oxygen is required for efficient energy production through oxidative phosphorylation through aerobic respiration, and CO_2 level is controlled by the body in order to maintain a normal pH value suitable for life. Arterial blood gas values and ranges displayed in Table 1.

Oxygenation

The importance of maintaining adequate delivery of oxygen (DO2) lies in the difference of the amount of adenosine triphosphate (ATP) produced in the presence and absence of oxygen. ATP is considered the "currency of cellular energy", providing energy for cellular processes required to maintain life as phosphate groups are cleaved. This results in energy being released and the formation of adenosine diphosphate (ADP) or adenosine monophosphate (AMP). ATP is involved in cellular signaling, DNA and RNA synthesis, muscle contraction, cytoskeletal maintenance, active transporting, and many other cellular functions. There is a finite amount of ATP available within a body, and a constant recycling of ADP and AMP into ATP is required to keep up with energy demands. In the presence of oxygen, 38 ATP molecules are generated from metabolism of a single glucose molecule undergoing oxidative phosphorylation occurring in the mitochondria. In contrast, a single glucose molecule yields two ATP molecules through anaerobic metabolism. The presence of oxygen is imperative in efficient energy generation.

Provided there is adequate intravascular volume and tissue perfusion, DO_2 is dependent on the oxygen level contained in the blood (arterial oxygen content, CaO_2) and how quickly the body can circulate the blood to the tissues (cardiac output, CO). The resulting mathematical expression of DO_2 is: $DO_2 = CaO_2 \times CO$. Oxygen contained within blood exists in two forms; dissolved in the plasma and bound to hemoglobin. The amount of oxygen dissolved in plasma depends on the partial pressure of oxygen (PaO_2), with 1 mmHg creating enough tension to result in 0.0031mL of dissolved O2 per dL of plasma. Each gram of hemoglobin is able to theoretically carry 1.39mL of O_2 when fully bound with oxygen, making up a significant portion of oxygen content of blood. In reality, there are portions of dysfunctional hemoglobin lowering this to approximately 1.34mL. In addition, not every hemoglobin molecule will be fully bound to oxygen in every situation (SaO_2, or arterial oxyhemoglobin saturation), adding some variability. With all of these considerations in mind, the resultant formula to quantify DO_2 is the following, expressing the impact lowered hemoglobin concentration and saturation of the hemoglobin will have on overall delivery of oxygen: $DO_2 = [(1.34 \times Hgb \times SaO_2) + (0.0031 \times PaO_2)] \times CO$. The respiratory system is responsible for providing SaO_2 and PaO_2.

The respiratory system ensures adequate oxygenation of the blood to provide sufficient DO_2 to meet the body's oxygen consumption by bringing air with oxygen (~21% in the atmosphere) to the blood-gas barrier and allowing diffusion of O_2 into the plasma. The diffused oxygen is then taken up by hemoglobin in the red cells, and carried to the rest of the body. Oxygenation can be decreased by reduced oxygen content of breathed in air (high altitudes or rebreathing of expired air), increased thickness of the blood-gas barrier (pulmonary edema, pneumonia, pulmonary fibrosis), reduction of the amount of air moved in and out of the lungs (*hypoventilation* caused by pleural space disease, respiratory muscle dysfunction, respiratory suppression), ventilation of alveoli without proper perfusion (*dead space ventilation* caused by pulmonary thromboembolism, pulmonary vascular injury or hypotension), and areas of the lungs with dysfunction in ventilation even though receiving proper perfusion (*shunting* such as in pulmonary edema, pneumonia, acute respiratory distress syndrome, atelectasis, and airway obstruction).

A physical sign seen in patients with severe hypoxemia is cyanosis, or a blue color to the mucous membranes. Cyanosis becomes apparent when there is more than 5 g/dL of deoxyhemoglobin present in the blood. An average hemoglobin level in dogs is approximately 13-17 g/dL, and in cats is approximately 10-14 g/dL. This means the oxygen saturation of hemoglobin will be at a significantly decreased level, on average of 61-70% for dogs and 50-64% for a cat, before cyanosis is seen. Patients presenting with cyanosis are severely compromised in their DO_2 and require immediate attention.

Oxygenation can be better gauged through measurement of PaO_2, serving as an indicator of pulmonary function. PaO2 can be measured through blood gas analysis of arterial blood, requiring arterial blood sampling and a blood gas analyzer. A patient with normal respiratory function breathing room air will have a PaO_2 of 80-100mmHg. PaO_2 of less than 80mmHg qualifies as hypoxemia, and 60mmHg is considered severe hypoxemia.

Pulse oximetry allows non-invasive measurements of the percentage of oxygenated functional hemoglobin in the arterial bloodstream, utilizing the concept of light absorption. The saturation of oxygen measured by pulse oximetry (SpO_2) closely reflects SaO_2 and can be used to estimate the PaO_2 level. The oxygen-hemoglobin dissociation curve expresses the relationship between SaO_2 and PaO_2. A SaO_2 of 95-98% corresponds to a PaO_2 of 80-100mmHg. A SaO_2 below 90% indicates a PaO_2 of less than 60mmHg. Pulse oximetry has its limitations, including falsely reading in the presence of significant levels of dysfunctional hemoglobin species (methemoglobin, carboxyhemoglobin), inconsistent readings with movement, poor perfusion, anemia, and pigmented skin.

Interpretation of oxygenation and pulmonary function based on hemoglobin saturation is especially difficult when the patient is provided inspired oxygen level (FiO_2) higher than that of room air (21%, or 0.21). Simplistically speaking, the PaO_2 can be expected to be about 4-5x the FiO_2 expressed as a percentage in patients with normal pulmonary function. Patients on room air (FiO_2 21%) will be expected to have a PaO_2 of 80-100mmHg, with 95-98% oxygen saturation of hemoglobin. Patients on oxygen therapy receiving 40%, 60%, 80%, and 100% for example, will be expected to have a PaO_2 of 160-200mmHg, 240-300mmHg, 320-400mmHg, and 400-500mmHg, respectively. Oxygen saturation of hemoglobin with PaO_2 above 100mmHg will consistently read 98-100%, and therefore will not allow for estimation of PaO_2 beyond 100mmHg. Because of this limitation, an actual PaO_2 measurement through blood gas analysis is necessary in order to gauge the efficiency of gas exchange.

By determination of PaO_2, a value called PaO_2:FiO_2 Ratio (PF ratio) can be calculated. For example, a patient breathing air that is 21% oxygen and having a PaO_2 of 100mmHg has a PF ratio of 100 divided by 0.21 (decimal expression of 21%), which equals 476. A PF ratio of 400-500 indicates normal pulmonary function, 200-300 indicates a degree of pulmonary dysfunction seen with acute lung injury, and anything less than 200 indicate severe pulmonary dysfunction often seen with acute respiratory distress syndrome. The PF ratio is important in making patient progress in recovery from their pulmonary dysfunction quantifiable. A patient with a PaO_2 of 200mmHg on 80% FiO_2 on day 1, having a PaO_2 of 140mmHg on 40% FiO_2 on day 2, has increased the PF ratio from 250 to 350, indicating improved pulmonary function between the two measurements even though the PaO_2 has decreased in value. In long term oxygen therapy, delivering oxygen supplementation allowing the PaO_2 to be between 80-100mmHg for any given pulmonary function is most effective in providing adequate DO_2 and minimizing effects of oxidative damage (which is more likely to occur with higher oxygen concentration).

Another calculated value called the A-a or alveolar-arterial gradient, is the difference between alveolar and arterial concentration of O_2, indicating the efficiency of gas exchange. At room air, the A-a gradient is normally around 10-15mmHg, but can be larger with lung pathology causing more significant ventilation/perfusion mismatching.

Carbon Dioxide Control

CO_2 is produced by tissues as a metabolic byproduct of energy production. The body normally maintains control of CO_2 levels in order to control the pH level of the body. An accumulation of CO_2 causes an increase in levels of carbonic acid, leading to higher levels of dissociated hydrogen ions, leading to a more acidic environment (lower pH). A reduction in CO_2 level will lead to a decrease in hydrogen ions, leading to a more basic environment. This effect is called respiratory acidosis and respiratory alkalosis, respectively. Maintenance of the pH in a normal range of 7.35-7.45 is important, as there are various negative effects caused by abnormal pH. Acidosis (pH < 7.35) for example, can cause cardiovascular effects (decreased contractility, arrhythmias), central nervous system dysfunction (CNS depression, increased intracranial pressure), respiratory dysfunction (decreased oxygen uptake by hemoglobin), and enzyme dysfunction. Alkalosis (pH >7.45) can cause similar cardiovascular effects, decreased cerebral blood flow, decreased oxygen unloading by hemoglobin, and enzyme dysfunction.

The amount of CO_2 eliminated by the body largely depends on the movement of air in and out of the alveoli to perform gas exchange, or ventilation. Room air contains about 0.04% CO_2 (0.3mmHg), and the replacement of gas within the alveoli with fresh room air will promote diffusion of CO_2 out of the blood stream into the gas within the alveoli, which in turn gets expired out of the lungs and airway. A normal CO_2 levels within the blood is approximately 35-45mmHg in dogs, and 30-40mmHg in cats, and can be measured by blood gas analysis (PaCO2 if arterial or PvCO2 if venous). The difference in PCO_2 in the pulmonary capillaries and alveoli create a pressure gradient required for gas exchange (high to low; high in the capillary, low in the alveoli).

PCO_2 is largely influenced by the amount of air that can be moved in and out of the alveoli, or alveolar ventilation. The value will increase with hypoventilation leading to hypercapnia (>45mmHg) in cases of respiratory depression (suppression of respiration due to drugs, neuromuscular disease, CNS disease), inability to expand the lungs (pleural space disease, compromise to chest walls), or increased resistance to breathing (narrowed airway). Hyperventilation and subsequent hypocapnia (low $PaCO_2$) can be seen in patients with increased RR due to anemia and hypoxia. When the patient experiences metabolic acidosis, compensatory increase in respiratory effort and hyperventilation is often seen, countering the metabolic acidosis effect with respiratory alkalosis. This occurs because the presence of hydrogen ions will stimulate the respiratory center of the brain to increase respiratory efforts.

Table 1. Arterial Blood Gas Values and Ranges

	Human	Dog	Cat
pH	7.35-7.45	7.35-7.46	7.31-7.46
$PaCO_2$ (mmHg)	35-45	32-43	26-36
Base Deficit (mmol/L)	-2 to +2	+1 to -5	-2 to -8
HCO3 (mmol/L)	22-26	18-26	14-22
PaO_2 (mmHg), sea level	80-105	80-105	95-115

Source: Silverstein DC, Hopper K. *Small Animal Critical Care Medicine, 2e*

The $PaCO_2$ can be estimated by measurement of End-tidal CO_2 ($ETCO_2$). The CO_2 content in the gas present at the probe at the end of expiration is measured to obtain this value. The $ETCO_2$ in a normal cardiovascular and respiratory situation, is within

5mmHg of the $PaCO_2$. The $ETCO_2$ is most easily measured when an endotracheal tube is placed in a patient (anesthetic procedure or mechanically ventilated patients, for example). There are nasal tubes and masks available which allow for less invasive $ETCO_2$ measurement.

Arterial Blood Gas Analysis

Arterial blood samples need to be collected in order to perform blood gas analysis for respiratory assessment. Sampling can be performed periodically through a syringe and needle, with common sampling sites including the dorsal pedal and femoral arteries. Radial, auricular, and lingual arteries are less commonly, but are good alternatives when access to dorsal and femoral arteries are limited or have been attempted and exhausted. Proper knowledge of anatomy and experience with palpation of each vessel is required to accurately puncture the artery. Successful sampling with minimal trauma and stimulation to the vessel is recommended, since stimulation will cause constriction of the artery and make sampling more difficult. If frequent or repeated sampling is needed, an arterial catheter becomes more desirable and useful, especially with patients on mechanical ventilation and with severe respiratory compromise. The placement of an arterial catheter also gives the ability to measure direct arterial pressure, which is beneficial in many critical care patients. Complications of arterial catheters include thrombophlebitis and embolism.

The collected sample should be analyzed immediately to prevent mixing with room air, resulting in inaccurate results. Prolonged exposure to room air will result in a reduced $PaCO_2$ and an inaccurate PaO2 (direction depends on what the sample contains, heading towards 150mmHg). Even if the sample is kept from being exposed to air, delay in analysis will allow the metabolic processes of RBCs to alter blood gas results. Because of this, placing the sample on ice to maintain 4°C is recommended if the sample needs to wait for analysis. This will keep the sample results stable for up to 6 hours. The heparin used beyond 10% of the sample volume will result in significantly diluted results. Coating of the syringe with heparin by pulling heparin into the syringe and then expelling it is recommended to minimize the dilution effect. Self-venting arterial sampling syringe sets are commercially available preloaded with dry heparin pellets, which minimizes dilution and facilitates sampling, as well.

There is a controversy that exists in blood gas analysis involving temperature correction of the values. The sample is analyzed at 37°C in the analyzer. The patient's body temperature may be different from 37°C, which changes the blood gas values. The body temperature can also affect the affinity for oxygen with hemoglobin, and alter metabolic rates and gas solubility. Because all of the effects of hypothermia and hyperthermia are not completely understood, the necessity of temperature correction in calculated blood gas values is also unknown.. If temperature correction is applied in clinical practice, it must be done consistently with all samples.

Summary

Arterial blood gas analysis, while requiring extensive training and comfort in sampling techniques, is a very valuable tool in evaluating the patient's respiratory status. Both oxygenation and ventilation status, as well as pulmonary function, is most accurately assessed through blood gas analysis, providing a better idea of patient progress through treatment. While blood gas analysis is a powerful tool, a veterinary technician can utilize a combination of low-tech patient assessment skills and high-tech assessment tools to grasp the entire picture of the patient's respiratory status.

REFERENCES

[1]Burkitt JM, Davis H, Advanced Monitoring and Procedures for Small Animal Emergency and Critical Care 2012. [2]Crawford AH, Halfacree ZJ, et al, J Feline Med Surg 2011;13(10):762-767. [3]Johns SM, Nelson OL, Gay JM, J Vet Intern Med 2012;26(5):1134-1139. [4]Silverstein DC, Hopper K, Small Animal Critical Care Medicine 2009;67-144. [5]Smart L, Jandrey KE, J Vet Emerg Crit Care 2008;19(40):393-398. [6]Thunberg B, Lantz GC, J Am Anim Hosp Assoc 2010;46(6):418-24.

CARDIAC AUSCULTATION – HOW TO INTERPRET WHAT YOU HEAR
A Primer for Veterinary Technicians
William Rausch, DVM, DACVIM (Cardiology)
Portland, Ore

THE FIRST HEART SOUND

Cardiovascular sounds in the dog and cat can be characterized as either transient heart sounds or murmurs. Transient heart sounds are of relatively short duration and include the normal first and second heart sounds. The first heart sound (S1) is initiated at the onset of ventricular systole. Under normal circumstances S1 is louder, longer and lower pitched than the second heart sound (S2). S1 is temporally associated with closure of mitral and tricuspid valves and is heard best in the region of these valves (often in the 5th intercostal space). The sound is not necessarily created by atrioventricular (AV) valve leaflets meeting per se. Rather, it is thought to be primarily due to the sudden acceleration and deceleration of blood flow and tensing of the AV valvular apparatus associated with valve closure. There is a correlation between the intensity of the first heart sound and the degree of sympathetic tone, so any disease process which increases sympathetic tone – or even simply anxiety or fear, can increase the intensity of S1. First heart sounds can also be louder in animals who are thin, anemic or tachycardic. The first heart sound is often of diminished intensity in animals who have pathology between the heart and stethoscope placement. This could include body conformity, obesity, pleural effusion, any pericardial disease or severe pulmonary disease. Other conditions sometimes associated with a softer S1 sound include dilated cardiomyopathy, shock and first-degree AV block.

THE SECOND HEART SOUND

The second heart sound is of slightly shorter duration and higher frequency relative to the first heart sound. With a normal heart rate and rhythm, there will be a longer pause after the second heart sound than there is between the first and second sound and this relatively longer period can help properly identify the two sounds. This relative pause following S2 is easier to identify the slower the overall heart rate. During sinus tachycardia in a dog however, or even at upper-normal heart rates in felines in a clinical setting, the duration of diastole shortens to a greater degree than systole and the pause between the sounds becomes more similar. In this case, it helps to remember that S2 is the higher pitched sound and that S1 is timed more closely with a palpable arterial pulse.

The second heart sound is generally heard best in the region of the aortic and pulmonic valves (left hemithorax in the 3rd and 4th intercostal space). S2 occurs at the end of ventricular systole, near the end of the T wave on the ECG and is temporally associated with closure of the pulmonic and aortic valves. Under normal circumstances aortic valve closure precedes pulmonic valve closure by a time interval too small to be perceived by the human ear, and thus this composite sound (S2) is heard as a single sound.

SPLIT HEART SOUNDS

If physiologic or pathologic conditions allow the mitral and tricuspid valves to close asynchronously, S1 can be divided into two distinct component sounds. Generally the time between AV valve closures must be at least 20 milliseconds in order for the distinction to be readily audible, but a phonocardiograph can be more sensitive for this finding. It is often an incidental innocent finding in healthy large breed dogs. A split first heart sound can also be a rate dependent phenomenon and when so is usualy more easily heard at lower heart rates. Split first heart sounds can also be caused by pathologic conditions that cause asynchronous AV valve closure such as bundle branch block, certain arrhythmias and congenital AV valve stenosis. In practice, a split first heart sound in a healthy large breed adult dog with no murmur and a normal ECG is usually not of clinical significance. It must be differentiated in its timing from a gallop sound or systolic click and examples will be given in lecture.

The second heart sound is also a composite heart sound - with contributions from both aortic and pulmonic valve closure. Splitting of the second heart sound can be recognized if the normal time interval between semilunar valve closures is accentuated by conditions that delay pulmonic valve closure or markedly delay aortic valve closure. Rarely, a split S2 can be recognized as a normal finding in healthy large breed dogs during inspiration at low heart rates. In these rare cases, intrathoracic pressure becomes more negative during inspiration which promotes right ventricular filling. The resultant increase in RV filling leads to a slightly longer RV ejection period and thus delayed pulmonic valve closure. Pathologic splitting of S2 can be caused by certain arrhythmias or congenital heart diseases such as atrial septal defects, right-to-left shunting PDAs and aortic stenosis. The most common and clinically important cause for a split S2 sound is delayed closure of the pulmonic valve as a result of pulmonary hypertension. The pulmonary hypertension can be caused by heartworm disease, chronic pulmonary disease or less commonly congenital heart disease. If a split S2 is ausculted in a clinical situation where chronic pulmonary disease is supected, pulmonary hypertension should be suspected and further evaluation with echocardiography and thoracic radiographs may be indicated. A split S2 is generally the only auscultation abnormality noted in a young dog with a right-to-left shunting patent ductus arteriosus and this should be considered a strong differential in a young dog with a split S2 with exercise intolerance or pelvic limb weakness.

GALLOP SOUNDS

The third and fourth heart sounds are not normally auscultable in dogs and cats. The third heart sound (S3) is a low frequency sound thought to be associated with abnormally sudden deceleration of ventricular inflow during early diastole. The fourth heart sound (S4) is associated with atrial systole and is thus heard late in diastole just prior to S1 when present. An auscultable S4 in a dog or cat generally indicates that an atrium is contracting to fill a ventricle that is already distended or otherwise non-compliant (such as due to a feline cardiomyopathy). As S3 and S4 are both low frequency sounds, they are best appreciated using the bell of the stethoscope with only gentle pressure against the skin. They are named gallop sounds because of the rhythmic similarity to the sound of a galloping horse. The gallop sound that is created when these extra sounds are present can be described as protodiastolic (S3 only), presystolic (S4 only) or a summation of S3 and S4. The distinction between these three types of gallop sounds can be difficult at the high heart rates that these patients commonly present with. However, keep in mind that it is far more important that a gallop rhythm is recognized than categorized, as gallop sounds are often the first or only finding suggesting myocardial dysfunction.

Gallop sounds in canine patients are relatively uncommon, but when heard are most often associated with severe myocardial dysfunction and excessive ventricular volume – most commonly as a result of dilated cardiomyopathy, end-stage AV valvular disease or left-to-right shunting congenital heart disease. They need to be differentiated from split heart sounds, which tend to occur in different canine patient populations as described above. Gallop sounds are not uncommon in feline patients with heart disease. They are often heard in cats with conditions associated with diastolic dysfunction such as hypertrophic cardiomyopathy, hyperthyroidism, systemic hypertension and restrictive and dilated cardiomyopathy. Ventricular compliance decreases with age in humans, and occasionally geriatric cats will have new gallop sounds in the absence of remarkable echocardiographic changes. The presence of a new gallop sound in a cat is a strong indication for further evaluation by echocardiography, blood pressure, laboratory work and/or radiographs as appropriate. New gallop sounds are also occasionally heard in cats after they receive IV fluid treatment for an unrelated disease. These cats may have had some degree of diastolic dysfunction due to occult heart disease that was revealed by the sudden increase in vascular volume.

HEART MURMURS

Heart murmurs are sustained sounds associated with abnormal turbulent blood flow within the heart or proximal great vessels. This turbulence can be caused by benign anatomic or physiologic variation, valvular insufficiencies, outflow tract disturbances, abnormal communication between chambers, anemia or neoplasia. The murmurs caused by anemia are due in large part to the drop in blood viscosity associated with anemia and thus are more likely to occur with more severe anemia. Many of these same patients will have increased flow velocities associated with the increase in cardiac output triggered by the anemia. The propensity for these changes in fluid viscosity and velocity to cause non-laminar flow in a tube was described by Osborne Reynolds in 1883. The Reynold's number (R) is a dimensionless number that can help to predict the likelihood of turbulent flow under given conditions of flow velocity, blood viscosity and vessel size.

$$R = \frac{\text{Radius x velocity x density}}{\text{Viscosity}}$$

Cardiac murmurs can be described by characteristics such as their timing, volume, location and sound quality. These descriptions allow for initial clinical deduction as well as clear communication between clinicians. The timing of the murmur is described relative to its place within the cardiac cycle. Systolic murmurs often begin with the first heart sound (S1), but can begin any time before the second heart sound (S2). Systolic murmurs end with or prior to S2. Diastolic murmurs begin with or following S2 and end with or prior to the next S1. Therefore it is first necessary to identify S1 and S2 in order to place the timing of a murmur. Continuous murmurs can be heard throughout the entirety of the cardiac cycle. The timing of a murmur can be further described with modifiers such as holo- and pan- (as in holosystolic) or in their relative duration by prefixes such as pre-, early-, mid- or late. The intensity of a murmur is most often described using the six grades scale. As we will discuss, although there is a loose correlation between grade of murmur and disease severity in some conditions, this is far from true in many cases and imaging studies are required to answer questions of disease type and severity with adequate specificity and sensitivity.

Other descriptive characteristics of murmurs include location, radiation and shape. The location, or point of maximum intensity (PMI), is simply the point where it can be heard the loudest with the stethoscope. If there is a palpable thrill associated the murmur, the PMI is usually at the location of the thrill. If a loud left-basilar murmur can be heard with similar character but slightly diminished intensity at the right cardiac base, it is said to radiate to the right base. Location and radiation help in constructing a list of differentials for the murmur in a particular case. The shape of a murmur refers to any variation in the intensity of the murmur during the cardiac cycle. These are more easily recognized on a phonocardiogram and examples will be shown, but with practice and patience these can be appreciated on exam. Plateau murmurs are of nearly equal intensity throughout the murmur. Although there are numerous exceptions, this is often the sound quality created by AV valve insufficiencies. Crescendo-decrescendo murmurs rise to a peak intensity and then soften during the course of a murmur. Again, exceptions abound, but this sound quality is often heard in congenital outflow tract disturbances such as sub-aortic stenosis and pulmonic stenosis.

With this information as background the lecture will focus on how various abnormal heart sounds might be interpreted in a variety of common clinical scenarios. We will discuss how these auscultory findings allow us to predict the presence or absence of a particular heart disease for an individual patient. We will also discuss the heart diseases that often do not create any physical exam abnormalities and how the diagnostic approach to these patients may differ. Case examples will be used with a focus on acquired and congenital heart disease as well as "innocent" ausculatory abnormalities.

CARDIAC EMBRYOGENESIS: NORMAL DEVELOPMENT

Harris Edward Durham Jr, CVT LATG, VTS (Cardiology)
Columbia, MO

INTRODUCTION

Any study of the heart needs to begin with an understanding of how the heart develops in the embryo. This knowledge is extremely valuable, especially when trying to evaluate patient with congenital heart defects. If the heart does not develop properly or at least close to properly the fetus will not survive to birth. This presentation will present an overview of a complex series of events that represents a field of study unto itself. During the formation of the heart *in-utero*, the orientation in the embryo changes several times which can add to the challenge of understanding the development. The terms cranial and caudal are used to indicate the orientation of the heart related to the embryo forebrain.

EARLY DEVELOPMENT

The development of the heart begins with the formation of the *cardiac tube.* The cardiac tube is situated in the pericardial cavity below the head of the embryo and begins as two tubes that fuse together creating a single lumen. It is anchored at the cranial end by *ventricular aorta* and the caudal end by the *omphalomesenteric veins.* It is about this time that the heart begins to beat moving blood from caudal to its cranial aspect. Since the cardiac tube in fixed at both ends, as the heart grows, it convolutes, folding over itself to form the convoluted *cardiac loop* at about 23 days. This rudimentary structure contains the *cardiac jelly.* The cardiac jelly is an acellular mass of glycoproteins and glyosaminoglycans that function as a basic valve in this early heart and will go on to become important in the development of the cardiac valves. Blood is circulated from the venous side at the caudal pole through the arterial side at the cranial pole. During this stage of development the cardiac tube widens at the caudal aspect as the cranial begins to move caudal pushing the bottom to the side convoluting the structure into what eventually becomes the ventricles and begins the development of the aortic arches. As the cardiac loop grows, the venous portion gets pushed toward the cranial portion of the embryonic heart as the ventricular portion drapes down. The *sinus venosus* gets pushed cranial and has an atrial like structure in the cranial dorsal portion that collects blood returning to the heart, and a ventricular like portion at the caudoventral region with the bulbus cordis extending up to outflow tract which eventually becomes the *truncus arteriosus.* (Fig 1) From the ventral view, the two primitive atria can be seen projecting on either side of the truncus. The *atrioventricular sulcus* develops along the sagittal plane and the familiar structure of the heart begins appear, with multiple veins entering posteriorly allowing blood to drop caudally into the primitive ventricles to be ejected from the aortic arch at the cranial anterior. The narrowing between the primitive ventricle and atria created by bulboventriculer sulcus creates the *atrioventricular canal* (AV canal) with the cardiac jelly acting as a unidirectional valve.

SEPTATION

At about 27 days the heart begins to septate. There are three septa that develop simultaneously; the ventricular, the atrial and spiral septation of the truncus arteriosus into the aorta and the main pulmonary artery. At the same time the endocardial cushions develop from each side of the AV canal and in the center from the cardiac jelly. Ventricular septation begins at the apex with the muscular septum growing toward the atrioventricular junction. It will finally fuse with the central endocardial cushion (Fig 2) and leave on opening from the ventricles to the truncus arteriosus. An interventricular foramen persists until the complete septation of the truncus arteriosus. The beginning of the atrial septum also grows from the outer wall of the common septum toward the AV canal. This *septum primum,* nearly completely separates the two atria. The *ostium primum* is an opening at the distal end of this developing septum which allows blood to shunt from the right to left atrial sides. As it closes, a perforation (*ostium secundum)* develops higher up in the septum primum, creating the *fossa ovalis* allowing for the continued shunting of blood in the fetal heart. A second atrial septum then develops, the *septum secundum* overlapping creating the *foramen ovale* creating a functional valve over the foramen ovale allowing blood to only pass from the right atrium to the left in the fetal heart. (Fig 2) A spiraling septum develops in the truncus arteriosus that separates into the two great vessels exiting the heart, the pulmonary artery and the aorta. The spiral septum starts as two ridges of tissue that spiral down the long axis of the truncus arteriosus toward the ventricles. As the spiral septum grows, the truncus twists and divides into the two vessels, accounting for the crossing of pulmonary artery over the aorta in the adult heart. The interventricular foramen is ultimately obliterated by masses of tissue from the spiral septum, endocardial cushion tissue, and the muscular ventricular septum. The last phase of development is the formation of the cardiac valves. The AV valves are derived from folding of tissue in the ventricular walls and the endocardial cushions. The mesenchymal cells of the endocardial cushions transform into endocardial tissue forming the AV valves, which perforate along their cusps, then coalesce and form the chordae tendineae and papillary muscles. The semilunar valves are formed from masses of cardiac jelly in the aortic and pulmonary artery lumens seeded with mesenchymal cells covered by endocardial tissue.

FORMATION OF MAJOR VESSELS

Both the pulmonary artery and the aorta derive from the differentiating aortas of the cardiac loop in early cardiac development, and exit the cardiac loop from a single orifice which then divides to form a "V". This "V" then develops six pairs of aortic arches supplying different portions of the embryo. (Fig 3)) As the circulation system matures, the first second and fifth pairs regress completely. The left fourth aortic arch forms the aorta and the sixth forms the pulmonary arteries. The right fourth aortic arch forms the base of the subclavian artery. The third pair develops into the internal carotid arteries. The distal portion of the left sixth aortic arch remains connected to the aorta to create the *ductus arteriosus.* Pulmonary veins grow from the left atrium to connect with the pulmonary venous drainage as the lungs are formed from the primitive foregut. The development of the lungs is completely independent of the heart.

FETAL CIRCULATION

The general arrangement of the adult circulatory system occurs as two circuits in series, that is two separate circulatory paths in which the end of one feeds into the beginning of the other. In the fetus, circulation occurs as two circuits in parallel, or two circulatory paths that cross over each other at strategic areas to incorporate placental blood flow and bypass the unused lungs. (Fig 4) As highly oxygenated blood enters the fetus from the placental artery, it passes to the liver via the ductus venosus where mixes with the deoxygenated blood of the lower body. From there it travels to the right atrium and is shunted almost directly into the left atrium. There is very little mixing at this point with deoxygenated blood from the head and forelimbs due to the anatomic proximity of the posterior vena cava to the fossa ovalis which directs the blood almost directly into the left atrium. This blood is still highly oxygenated and goes to the left ventricle and out the aorta. At this point another mixing occurs with some highly oxygenated blood going to the head and some mixing with deoxygenated blood from the right ventricle via the pulmonary artery and ductus arteriosus. This mixed blood then goes to the lower body and to the placenta to be remixed into the maternal circulation. This mixed blood still carries enough oxygen to supply the lower extremities with all the nutrition they need. One the other end the highly oxygenated blood from the left ventricle goes directly to the head and forelimbs to ensure a highly oxygenated blood supply to the developing brain. Deoxygenated blood from the head and forelimbs then return to the right atrium of the fetal heart out the pulmonary artery to mix as described earlier. The circulation remains like this until birth.

As mentioned earlier, during the last phases of the cardiac development, as the atrial septum develops, the fossa ovalis transforms into the foramen ovale. The foramen ovale develops as a one way valve through the intra-arterial septum that allows oxygenated blood to pass from the right atrium to the left, similar to the fossa ovalis, yet will close later to keep blood from shunting from the left atrium to the right in the adult.

It is at parturition and shortly thereafter that two systems in series finally develop. This process starts with the first breath, which drops pulmonary vascular resistance dramatically. Simultaneously the placental circulation is removed, which increases systemic vascular resistance and reverses the blood flow through the ductus arteriosus to move blood from the aorta to the pulmonary artery. The tissue of the ductus arteriosus is highly sensitive to oxygen. When exposed to the increased oxygen content of the arterial blood in the aorta, the musculature of the ductus arteriosus constricts and closes the communication. The net effect of these changes is to decrease the volume of blood flowing through the right side of the heart, and increase it in the left. This changes cause the pressure in the left atrium to increase which functionally closes the foramen ovale and finally separates the two circulations.

In the fetus the size of the individual ventricles is equal and their relationship stay the same for about ten days. Over the first few weeks of life, the systemic blood pressure raises and pulmonary pressures remain static, and this increased workload changes the size of the left ventricle. By about two weeks of age, the proportions of the two ventricles are that of the adult with the right ventricle being about one third as thick as the right. During the next four weeks cardiac growth converts from *hyperplasia* to *hypertrophy,* growth in cardiac mass by myocyte enlargement rather than division. From this stage on, essentially all increases in cardiac mass are related to hypertrophy.

All Figures from: From *Textbook of Canine and Feline Cardiology*; Fox, Sisson, Moise 1999 Saunders

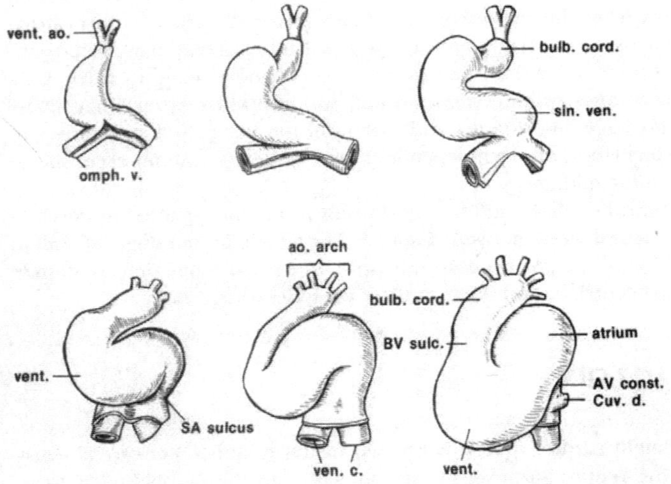

Figure 1

Ventral view of the cardiac tube developing into the convoluted cardiac loop.

ao.arch-aortic arches; AV const-atrioventricular constriction; bulb. cord.-bulbus cordis; BV sulcus-bulboventricular sulcus; Cuv. D-Cuvier's duct; omph. v.-omphalomesenteric veins; SA sulcus-sinoatrial sulcus; sin. ven.-sinus venosus; ven. c.-venae cavae;vent-ventricle; vent. ao.-ventricular aortic roots.

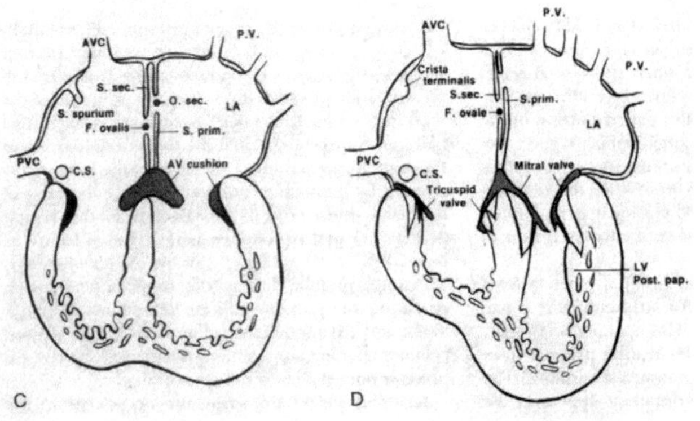

Figure 2

Development of the atrial septum showing fossa ovalis (diagram C) and foramen ovale (diagram D).
AVC- anterior vena cava; C.S.- coronary sinus; F. ovalis-fossa ovalis; F. ovale-foramen ovale; LA-left atrium; LV Post Pap-left ventricular posterior papillary muscle; O. sec.-ostium secundum; O. prim.-ostium primum; PV-pulmonary veins; PVC-posterior vena cava; RA-right atrium; S. sec.-septum secundum; S. prim.-septum primum; S. spurium-septum spurium

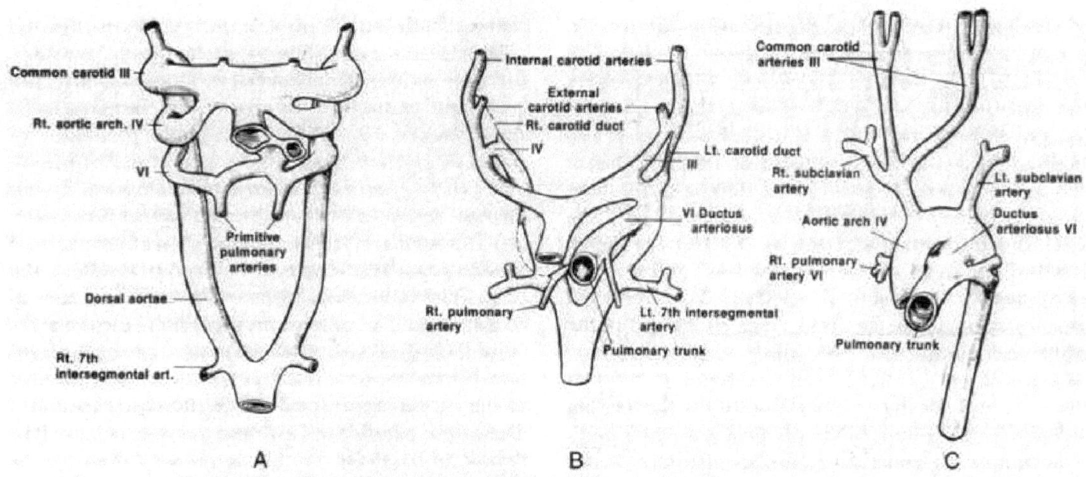

Figure 3

Development of the aortic arches in mammals

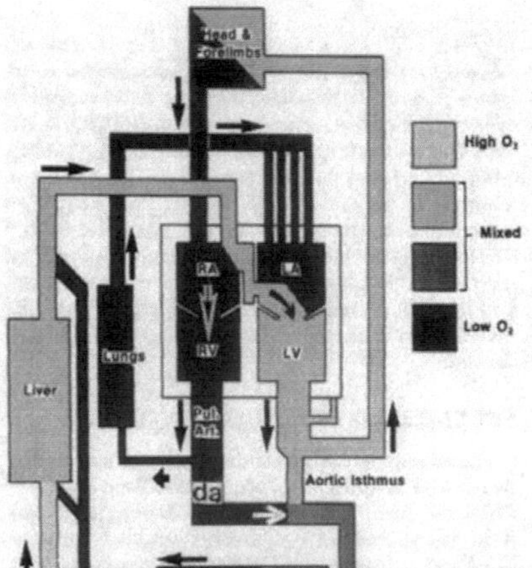

Figure 4

A diagram of fetal blood flow. Note 3 shunts; the foramen ovale, the ductus arteriosus and the ductus venosus

SO YOUR PUPPY HAS A MURMUR ....
TOP THREE CONGENITAL DEFECTS IN CANINES

Kimberly Sederquist, BS, RVT, VTS (Cardiology)
West Lafayette, IN

INTRODUCTION:

Congenital heart diseases (CHD) occur in puppies that are born with heart defects, usually due to the abnormal development of tubes or tissues during the embryologic stage. The three most common canine CHD are: patent ductus arteriosus, subaortic stenosis, and pulmonic stenosis.

PULMONIC STENOSIS

Description

Pulmonic stenosis (PS) is the third most common diagnosed CHD.[1] The pulmonic valve is located between the right ventricle (RV) and the main pulmonary artery (PA). As the right ventricle contracts, under normal physiologic conditions, it builds pressure to open the pulmonic valve allowing blood to pass from the right ventricle to the pulmonary arteries and finally to the lungs. Pulmonic stenosis is a narrowing of the right ventricular outflow tract (RVOT) and can be subvalvular, valvular or supravalvular. This narrowing results in resistance to blood flow from the right ventricle to the pulmonary artery. The right ventricle develops concentric hypertrophy due to the pressure overload. Other sequelae include post stenotic dilatation of the pulmonary artery, ischemic damage to the myocardium, arrhythmias, and right sided heart failure.

The most common breeds affected by pulmonic stenosis include the Bulldog, Miniature Schnauzer, Mastiff, Samoyed, Beagle, Spaniel breeds, Chihuahua, West Highland White Terrier, Keeshond, and Labrador Retriever.[2] The most common form of PS involves a valvular abnormality. There are two types of valvular PS: valvular dysplasia involving malformation of the valve leaflets with or without hypoplasia of the valve annulus and fusion of the peripheral edges of the leaflets with doming of the valve. Bulldogs and other brachycephalic breeds may have a variation of PS in which an anomalous coronary artery wraps around the RVOT below the valve causing a subvalvular stenosis.

Pulmonic stenosis is commonly an incidental finding during a puppy visit. Clinical signs related to this disease are dependent on the severity of the stenosis. Mild stenosis may have no clinical signs. Severe stenosis can result in exercise intolerance, weakness, syncope, ascites (right sided heart failure) and typically manifest when the dog is an adult.

Physical Exam Findings

On physical examination, a systolic ejection type murmur is ausculted due to turbulent blood flow originating at the level of the stenosis. The murmur is typically a loud murmur (grade IV-VI/VI) with point of maximum intensity (PMI) at the left heart base and radiation to the right side. Because PS involves the right side of the heart, the peripheral pulses are typically normal unless arrhythmias are present. Mucous membrane color is usually within normal limits. If right sided heart failure is present, jugular venous distension and pulsation may be noted as well as abdominal ascites.

Diagnostic Testing

The presence of a heart base murmur should raise the suspicion for PS but diagnostic tests are necessary to differentiate it from other congenital heart defects. Thoracic radiographs in moderate to severe cases may demonstrate increased sternal contact on lateral view suggestive of a prominent RV and an enlarged PA may be noted on lateral and dorsoventral/ventrodorsal views. Electrocardiogram (ECG) findings may reveal criteria for RV enlargement and arrhythmias such as ventricular premature complexes. Definitive diagnosis of PS is by echocardiography. Echocardiographic findings include RV concentric hypertrophy, high pulmonic outflow velocities, and dilation of the PA. The type of stenosis can often be determined and the use of spectral flow Doppler can measure velocity of blood flow across the stenotic area. Using the measured velocity, the pressure gradient between the RVOT and PA can be determined. Severity of disease depends on this pressure gradient.

Treatment Options

Treatment options depend on the severity of the disease and the dog's presence of clinical signs. In cases of mild PS, dogs usually do not require interventional therapy. Medical treatment with antiarrhythmic drugs, such as beta blockers can decrease the heart rate, myocardial oxygen demand, and the risk of arrhythmias but are typically reserved for more severely affected animals. In moderate to severe cases, there are open thoracic surgeries or transvenous procedures available. Thoracic surgical options include a patch graft in cases of valvular dysplasia with subvalvular muscle hypertrophy[3] or placement of a conduit in dogs that have an anomalous single right coronary artery causing subvalvular PS.[4] Interventional therapy is most commonly performed via transvenous balloon valvuloplasty. Using fluoroscopic guidance, a catheter is passed across the RVOT and gently pulled back to measure the pressures in the PA, RVOT, RV, and the RA so that the pressure gradients can be measured invasively. An angiography catheter is then placed in the RV to perform an angiogram that allows measurement of the size of the stenosis and pulmonic valve annulus. The size of the balloon catheter will depend on these measurements. The balloon catheter is then placed across the stenosis and rapidly inflated and deflated to stretch and expand the stenotic area. The pressure gradient is measured in comparison with the pre balloon pressures.

Prognosis

Balloon valvuloplasty is not curative. However, most dogs benefit by 50% decrease in pressure gradient of the stenotic area. Post- operative complications are rare with the most common being tricuspid valve insufficiency and pulmonic insufficiency.

SUBAORTIC STENOSIS

Description

Subaortic stenosis (SAS) is the second most commonly diagnosed congenital heart disease in canines.[1] The aortic valve is located between the left ventricle (LV) and aorta in the left ventricular outflow tract (LVOT) As the LV contracts, it builds up pressure to open the aortic valve allowing blood to pass from the LV to the aorta and finally to the body. Subaortic stenosis is a narrowing of the LVOT below the aortic valve. This obstruction is caused by fibrous or fibromuscular tissue forming a ridge or band that may encircle the LVOT or a raised nodule on the ventricular septum of the LVOT. Dynamic SAS occurs when the hypertrophied LV septum bulges inward during systole and creates an obstruction. Systolic anterior motion of the mitral valve may also be a factor. The LV develops concentric hypertrophy in response to the increased afterload. The severely hypertrophied LV leads to an increased myocardial oxygen demand. Abnormalities in diastolic function, as well as coronary blood flow changes, can lead to a hypoxic state. These factors most commonly result in development of life threatening arrhythmias in severely affected dogs.[5]

The breeds most commonly affected by SAS are the Newfoundland, Golden Retriever, Rottweiler, German Shepherd, Boxer, Bull Terrier, Samoyed, Great Dane, and German Short Hair Pointer.[2] The lesions may not be present at birth and can develop during the first 3-8 weeks of age. Some puppies do not develop murmurs until after three weeks of age. The SAS lesions can continue to progress in severity to one year of age. Dogs that are evaluated for breeding purposes cannot be considered clear of SAS unless they are evaluated at one year of age or older.

Clinical signs are dependent on the severity of the stenotic lesion. Clinical signs for the severe form of SAS may include exercise intolerance, syncope, and sudden death.

Physical Exam Findings

Physical exam findings may include an elevated heart rate and a systolic murmur with PMI at the left heart base. The grade of murmur may correlate with severity of disease with grade I-III/VI suggesting milder SAS and IV-V/VI suggesting a more moderate or severe SAS.[5] SAS patients may also develop a to and fro murmur caused by the outflow tract obstruction and the development of aortic insufficiency caused by the high velocity turbulent jet damaging the aortic valve. The murmur can also be ausculted radiating up the carotid arteries and over an open fontanel on the cranium. The murmur can increase in intensity during exercise because of the increase in heart rate. Slow rising and weak pulses may be palpated in severe obstructions. Arrhythmias may be ausculted.

Diagnostic testing

Diagnostic tests include thoracic radiographs, ECG, and an echocardiogram. Radiographs maybe normal or may show left sided cardiomegaly with a loss of the cranial waist on the lateral view and widened mediastinum suggesting dilation of the ascending aorta. An ECG may be normal or show ST-segment depression, left axis deviation, an increase in R wave amplitude in Lead II, and ventricular arrhythmias. In moderate to severe cases an echocardiogram will demonstrate LV concentric hypertrophy, post- stenotic dilatation of the ascending aorta, and a thickened aortic valve. A subvalvular lesion with a hyperechoic fibrous ring below the valve can be visualized. Using color flow Doppler, the turbulent flow caused by the SAS can be isolated and measured using spectral flow Doppler. The peak velocity of blood flow determines the pressure gradient between the LV and the aorta. The pressure gradient determines the severity of the disease and ultimately the prognosis in each case.

Treatment

Dogs with mild SAS do not require treatment. Beta blockers maybe prescribed in moderate to severe cases to reduce the heart rate, help prevent ischemic events, and reduce the risk of fatal arrhythmias. Surgical intervention may also be considered in moderate to severe cases.

Open heart surgery to excise the fibromuscular tissue and balloon valvuloplasty have been investigated, however, they are high risk and there currently is no evidence the outcome would be different than medical treatment alone. The balloon valvuloplasty procedure is similar to that which was previously described for PS. The hemodynamic measurements, angiography, and balloon dilation are focused on the LV and LVOT. The goal is to relieve 50% of the obstruction of flow across the stenotic lesion. Complications of the procedure may include aortic insufficiency and arrhythmias.

Prognosis

Mild cases of SAS will likely live a normal lifespan. Moderate and severe cases have a higher incidence of sudden death. In severe cases, most dogs die within the first three years of life with sudden death representing the outcome in 70 % of those dogs[6]. All dogs with SAS should be considered at increased risk of endocarditis due to the trauma to the aortic valve.

PATENT DUCTUS ARTERIOSUS

Description

Patent ductus arteriosus (PDA) is the most commonly diagnosed congenital heart diseases in dogs.[1] During embryonic development, the distal portion of left sixth aortic arch forms the ductus arteriosus[7]. In the normal fetus, the ductus arteriosus shunts blood from the PA to the aorta because the lungs are collapsed and do not exchange gases. Shortly after birth, the ductus functionally closes due to increased oxygen tension, inhibition of prostaglandins, and the contraction of smooth muscle. The ductus arteriosus becomes the ligamentum arteriosum.

In the case of a PDA, the shunt fails to close. This failure may be due to the PDA being composed of more elastic smooth muscle. The flow of blood through the shunt is from the aorta to the PA, into the lungs, back to the left atrium via the pulmonary veins and into the LV and out the aorta again. This is considered a left to right shunt. This creates a volume overload in the pulmonary veins and left side of the heart. Over time, the volume overload causes eccentric hypertrophy of the LV. Left atrial enlargement and mitral regurgitation may also be present with chronicity. This process can lead to left sided congestive heart failure if left untreated.

The most commonly affected breeds are the Poodle, Maltese, German Shepherd, Collie, Pomeranian, Yorkshire Terrier, English Springer Spaniel, and Shetland Sheepdog.[2] There is a sex predilection for females. If not in left sided congestive heart failure, the puppies are asymptomatic. Detecting a murmur is usually the first indication of a problem with most of these patients. If a puppy presents in left sided congestive heart failure, the signs will include dyspnea, tachypnea and coughing.

Physical Exam Findings

On physical examination a continuous murmur of varying grade (usually grade IV to VI/VI) is located over the left cranial thorax in the axillary region. There may be a palpable thrill felt over the area of the murmur. Pulses are bounding due to a greater difference in systolic and diastolic pressure. Mucous membranes are typically within normal limits.

Diagnostic Tests

Diagnostic tests should include thoracic radiographs, an ECG, and an echocardiogram. Thoracic radiographic findings include generalized cardiomegaly, left atrial enlargement, left ventricular enlargement, a bulge in the aorta, dilation of the main PA (ductus bump) and may include pulmonary edema if the dog is in left sided congestive heart failure. An ECG can be suggestive of LV enlargement indicated by an increased amplitude of the R wave in Leads II, III, avF and lower chest leads and left atrial enlargement represented by a widened p wave. An echocardiogram, using color flow Doppler will allow visualization of the continuous turbulent flow across the PDA and retrograde flow into the PA. Spectral flow Doppler measures the velocity of the continuous PDA flow. The velocity is usually 5m/second which calculates to 100 mm/Hg pressure gradient between the PA (20mmHG) and aorta (120mmHG) Ventricular fractional shortening may decrease in some dogs due to volume overload. The left ventricular eccentric hypertrophy and left atrial enlargement will be present dependent on the size of the ductus and the age of the animal.

Treatment

There is no medical treatment for PDA. Treatment consists of PDA occlusion via direct surgical ligation or interventional procedure. With surgical ligation, a thoracotomy is performed and careful dissection and debridement of the tissue is performed to allow direct visualization of the shunting vessel. The PDA is then ligated using suture. For the interventional closure, a catheter is placed in the femoral artery and advanced proximal to the PDA. An angiogram is performed to determine the size and shape of the PDA. From the angiogram, the occlusion device can be chosen to best fit the PDA shape and the correct size is calculated by measuring at the minimal ductal diameter. The chosen occlusion device is then delivered through the jugular vein or via femoral artery into the PDA. The three occlusion devices commonly used for a canine PDA are: embolization coils, an Amplatzer® Vascular Plug, or an Amplatzer® Canine Ductal Occluder. The embolization coils are free released but require only a 3French system to deploy. The two Amplatzer® devices are attached to a cable, have a controlled release, and require a 4-7 French system depending on the device. Once the device is in place, a post angiogram is performed to confirm occlusion of the PDA. Complications can include hemorrhage, migration, or infection of the device during the interventional procedure or failure to fully occlude the PDA in both the interventional and surgical procedure.

Prognosis

Successful surgical ligation or interventional occlusion procedure is curative. If corrected prior to the onset of clinical signs the prognosis is excellent. If left untreated, 65- 90% of puppies with a PDA will die of left sided congested heart failure within their first year.

Conclusion

Canine congenital heart disease are common and dogs are often asymptomatic at diagnosis. Early recognition, diagnosis, and treatment (if available) generally carries a better prognosis and quality of life for the dog. Because these diseases are hereditary, breeding of affected dogs is discouraged.

REFERENCES

[1]Buchanan JW.Textbook of Canine and Feline Cardiology.1999:457.[2]Bonagura J D, Lehmkuhl, LB. Textbook of Canine and Feline Cardiology. 1999: 471.[3]Tanaka, R, et al, Aust Vet J. 2009; 87: 88.[4]Fujiwara, et al. Journal of Small Animal Practice. 2012; 53: 89. [5]Kittleson M D, Kienle R D. *Small Animal Cardiovascular Medicine.* 1998; 260.6Kienle RD, et al. J Vet Intern Med. 1994; 8:423.[7]Bishop S P. Textbook of. Canine and Feline Cardiology. 1999:7.

ESSENTIALS OF BLOOD PRESSURE MONITORING

Harris Edward Durham Jr, CVT, LATG, VTS (Cardiology)
Columbia, MO

INTRODUCTION

Traditionally, systemic blood pressure (BP) measurement has not been practiced widely in veterinary medicine. In addition to checking temperature, pulse and respiration, BP measurement and pain assessment should be routine parameters obtained on physical examination. BP should **always** be assessed in the following patients: renal failure, vision loss, geriatric patients, and patients receiving cardiovascular medication, general anesthesia, major trauma, and any patient with non-specific signs of weakness or lethargy.

In dogs and cats hypertension is typically secondary to some other systemic disease. This is in contrast to humans in which hypertension is most commonly idiopathic (known as essential hypertension). Hypotension is a serious complication in critical ill patients (e.g. trauma, sepsis) and during anesthesia.

Typically, a patient with hypertension will show no overt signs initially and thus can be hypertensive for a prolonged period before any clinical signs become apparent. Untreated systemic hypertension can lead to serious and irreversible damage to a number of major organ systems. This so-called end-organ damage most often affects the brain, eyes, heart, and kidneys. Hypotension is also a serious condition that if left untreated can lead to shock, organ failure, and death.

Knowledge of the normal regulation of BP is critical to understanding the assessment of hyper- and hypotension. Blood pressure is the product of cardiac output (CO) and the resistance of the vasculature to blood flow. Systemic vascular resistance (SVR) is produced mainly from the precapillary arterioles. CO is the product of heart rate (HR) and stroke volume (SV) which is the amount of blood ejected during one cardiac cycle. Stroke volume is affected by preload, afterload, and contractility. The interactions between blood flow, pressure, and SVR are shown in the equation: $BP = (HR \times SV) \times SVR$

The body continually balances the CO and SVR to maintain homeostasis. Blood volume and its regulation are important aspects of BP control. The systems that respond to changes in BP can be divided into two categories: 1) rapid-responding systems that are mediated predominantly through changes in HR and SVR and are controlled by the autonomic nervous system; and 2) slower-responding, longer duration systems which act primarily via changes in blood volume, and are regulated by the kidney.

The strength of pulse quality on palpation is a reflection of the difference between the systolic and diastolic pressure; it is not a reliable indicator of BP. Normal pulses do not necessarily mean the patient is normotensive and weak pulses are not always an indication of hypotension. Additionally, body condition score has a major influence on the quality of pulses palpated. Most importantly, palpation will not identify hypertension because hypertensive patients often have normal pulses. Therefore the diagnosis of hypo- or hypertension can only be made by using one of the below measuring techniques.

BLOOD PRESSURE MEASUREMENT

Direct BP measurement involves placing catheter into the lumen of an artery. The catheter is then connected to a pressure transducer. The transducer converts the mechanical fluctuations in the fluid into an electrical signal that can be displayed on an oscilloscope or on paper. Direct monitoring is considered the gold standard of BP measurement. Generally this method requires sedation or anesthesia for catheter placement. Complications of direct BP monitoring include infection of the catheter, thromboembolism, and serious hemorrhage if the catheter becomes dislodged. Other limiting factors include the technical difficulty of arterial catheterization, and the careful attention needed to keep the catheter patent.

The three most popular systems for indirect measure of BP are Doppler sphygmomanometery; oscillometry; and photoplethysmography. The Doppler method and the photoplethysmographic method both rely on the Ravi-Ricci principle of detecting flow past a pressurized cuff attached to a manometer to estimate BP. The oscillometric method records the oscillations of the artery under a pressurized cuff.

The Doppler system has been shown to be effective at measuring SBP in dogs and cats. In this method, ultrasound gel is placed on the contact surface of a Doppler crystal which is then placed over the artery distal to the cuff. The characteristic pulsatile flow signal should be audible, but as the cuff is inflated to a pressure greater than that of the BP, the flow signal is lost. As the pressure in the cuff is slowly released, the flow signal returns. The first audible instance of the flow signal returning corresponds with the SBP.

The Doppler method has been shown to be a very poor estimator of DBP, and when diastolic flow sounds can be heard, the results are not considered reliable.[1-4] Additionally, with SBP readings less than 100 mmHg, Doppler methodology underestimates SBP markedly but has good correlation with mean arterial pressure (MAP).[5] Doppler recordings of BP less than 100 mmHg should therefore be considered a MAP, not the true SBP. In our facility, we usually rely on the SBP pressure alone to manage most clinical cases.

A potential challenge of the Doppler method can arise in attempting to locate the flow signal. Weakly audible signals can often be attributed to insufficient coupling gel, hair interfering with the signal, or the crystal not being optimally placed over the artery. Signal is improved by clipping the site at which the Doppler crystal is placed and using liberal amounts of coupling gel. Once the signal has been located, gentle manipulations of the crystal can be used to optimize the signal.

Profound arrhythmias such as atrial fibrillation or frequent atrial or ventricular premature contractions can also confound measurement of BP. These rapid rhythms do not allow for normal filling of the ventricle during diastole, and consequently the SV of individual beats can vary widely. As a result, some beats are transmitted at higher pressures than others and can be detected earlier as cuff pressure is released. These "breakthrough" beats can be as much as 60 mmHg greater than the MAP and should not be used as the determinate of BP. It is more appropriate to make clinical decisions based on the pressure at which most beats are heard consistently.

Oscillometric systems can measure the SBP and DBP and calculate the MAP. Most oscillometric systems display the HR as well. This method generally provides an accurate assessment of the DBP and MAP but somewhat underestimates the actual SBP. Some of these devices are automated once the cuff is placed and secured; a machine inflates the cuff and performs the readings, making oscillometry a convenient monitoring tool during surgery. The average of five oscillometric BP values provides the most accurate reading. Oscillometric methods are susceptible to profound arrhythmias and motion artifact (e.g., shivering) making readings unreliable in awake animals. It can be difficult to obtain oscillometric readings in cats, perhaps because of the small PP in the distal arteries. The HR measured by the device should always be compared with that achieved by direct auscultation; if the HR matches, then the BP readings are more likely to be accurate. Oscillometric methods are less accurate than Doppler in patients in hypotensive states.[5] Other limitations to oscillometric readings include an increased margin of error as the BP increases and time constraints (it takes several minutes per measurement). These combine to make the oscillometric system less efficient for outpatient use when compared with the Doppler method

The photoplethysmographic method works by measuring the attenuation of an infrared beam through tissue and a pressurized cuff to control arterial volume to estimate MAP. This system has the advantage of giving a continuous BP reading and has been shown to be quite accurate in recording "beat-to-beat" variation. However, these systems are not often used in veterinary medicine. The cuffs for the photoplethysmographic system are designed for the human finger. Consequently, the size of patient size is limited to those less than 5 kg, and the cuffs do not always fit well. The device requires up to a minute to obtain a signal, making it less than ideal for conscious patients. There are also reports of difficulty obtaining readings in cats with darkly pigmented skin.

CUFF SELECTION AND PLACEMENT

A fundamental principle of noninvasive BP measurement is that the pressure being measured is inside the cuff, not in the vessel. Thus, if the cuff is placed too loosely on the limb, extra pressure is required to occlude flow through the artery, creating a falsely high reading. A cuff that has been placed too proximal on the limb may slide distal, loosen, and give the same error. If the cuff is placed over a joint, the artery may not be compressed at all, preventing disappearance of the Doppler flow signal. This may be erroneously as a BP greater than 300 mmHg.

Cuff selection is paramount to good technique. The most common errors in measuring BP due to incorrect cuff selection. A cuff that is too large will underestimate BP, and a cuff that is too narrow will overestimate it. For dogs, the correct cuff width should be 40% of limb circumference.[1,6] For cats, the cuff width should be 30% of limb circumference.[7] The circumference of the limb should be measured in centimeters with a pliable measuring tape (tailor's tape), and if measurement falls between two available cuffs, the wider of the two should be selected[8].

There are several sites that are suitable for cuff placement. The forelimbs, the tail, and the hindlimbs may all be used. A cylindrical portion of the limb with minimal bone works best. The areas just proximal to the carpus or just proximal to the tarsus work well. In patients such as dachshunds and basset hounds, the tail is the most suitable location. Cats often tolerate a tail application of the cuff better than a limb. In Doppler or photoplethysmographic methods, the crystal or probe should be placed distal to the cuff.

OTHER FACTORS INFLUENCEING READINGS

Because all forms of noninvasive measurement underestimate true BP, consistency is crucial. Trends can be more useful than absolute values of BP. When a BP is recorded that is abnormal, it is useful to ask several questions. The first question should be "Is my technique good?" In many cases, BP outside of the normal limits can be attributed to poor technique. If technique is adequate, then one should consider the patient's anxiety level. Is the patient showing signs of anxiety or fear? BP measurement should be done in a quiet room free of distractions (e.g., other animals, busy activity). Tachycardia can be suggestive of an increased level of stress. A phenomenon known as white coat hypertension has been reported in humans in whom BP is elevated simply because they are having it measured. This phenomenon has been reported in cats as well. In such cases, the recording should be repeated after the patient has been in the clinic for some time to allow for acclimation. Repeating measurements until a repeatable pressure (within 10 mmHg) is obtained will increase the likelihood of determining the real value.

If the technique is appropriate and the patient reasonably calm but BP is still elevated, it is appropriate to repeat the pressure measurement on another limb or the tail. If both readings are reasonably close, then the recorded pressure is probably real. If the readings are widely different, 40 to 60 mmHg, then it is likely that an error was made in one of the recordings. A single recording of an elevated BP is not necessarily diagnostic of hypertension.

Finally, it is important to ask "what is the patient's cardiac rhythm?" In patients with atrial fibrillation, supraventricular tachycardia, or ventricular tachycardia, interpretation of BP measurements can be complicated by breakthrough beats.

NORMAL ARTERIAL DEFINED

Normal blood pressures have been published for a wide range of species, however the author warns of extreme variability because of environmental factors. Given the spectrum of environmental influences on a given BP measurement, it may be more useful to characterize BP as low, high, or "reasonable." The definition of "normal" BP in dogs and cats is not exact, but studies have investigated the effects of hypertension and its links with systemic disease. BP is variable from moment to moment and should be viewed within a range. The methods of BP measurement (direct versus indirect), temperament and age of the patient, intra- and inter-patient variation, setting variation, sedation, and concurrent diseases all contribute to variability in BP.

In general, normal BP in cats ranges from 120 to 140 mmHg systolic and 70 to 90 mmHg diastolic, with a mean of about 110 mmHg. Dogs show some breed variation with an average BP of 133 mmHg systolic and 75 mmHg diastolic, and a mean of 95 mmHg.

Hypertension has been defined as systolic pressures from 140 mmHg to 180 mmHg or higher, with 160 mmHg most often defining the lower limits of hypertension. One epidemiologic study of BP in 203 cats showed that ocular changes owing to hypertension occur at a systolic pressure of 177.3 (± 28.7) mmHg.

Hypotension or low BP may be considered serious when the mean BP is less than 60 mmHg, because this is the pressure needed for glomerular filtration in the kidneys. Many factors influence BP, and individual measurements need to be interpreted in the proper clinical context.

In 2002, the Veterinary Blood Pressure Society announced some general guidelines for classifying hypertension in animals into mild moderate and severe categories. It must be kept in mind that these are guidelines and that patients in the mild or even moderate categories may only need treatment of the underlying disease. Patients in the severe category require prompt antihypertensive therapy to prevent or treat end-organ damage.

CONCLUSION

Traditionally, BP measurement has not been widely practiced, but many private veterinary practices now have BP monitoring equipment. With the advent of newer recording devices, BP is becoming part of the minimum database for many veterinary patients. With an understanding of the mechanisms of BP regulation and the advantages and limitations of each of the measuring systems, incorporating BP measurement should be a routine part of every veterinary practice.

REFERENCES

1. Binns S, Sisson D, Buoscio DA, Scheffer DJ: Doppler Ultrasonographic, Oscillometric Sphygmomanometric, and Photoplethysmographic Techniques for Noninvasive Blood Pressure Measurement in Anesthetized Cats. *J Vet Intern Med* 9:405-414, 1995.
2. McLeish I: Doppler Ultrasonic Arterial Pressure Measurement in the Cat. *Vet.Rec.* 100:290-291, 1977.
3. Chalifoux A, Dallaire A, Blais D, Larivie're N, Pelletier N: Evaluation of the Arterial Blood Pressure of Dogs by Two Noninvasive Methods. *Can J Comp Med* 49:419-423, 1985.
4. Chalifoux A, Dallaire A, Blais D, Larivie're N, Pelletier N: Evaluation of the Arterial Blood Pressure of Dogs by Two Noninvasive Methods. *Can J Comp Med* 49:419-423, 1985.
5. Caulkett NA, Cantwell SL, Houston DM: A Comparison of Indirect Blood Pressure Monitoring Techniques in the Anesthetized Cat. *Veterinary Anesthesia* 27:370-377, 1998.
6. Egner B, Carr AJ, Brown S, Kraft W, Erhardt W, Henke J, Off W, Ungemach FR, Bagley RS, Maggio F, Davidson M, Stepien RL: *Essential Facts of Blood Pressure in Dogs and Cats.* Germany, BE Vet Verlag, 2003.
7. Grandy JL, Dunlop CI, Hodgson DS, Curtis CR, Chapman PL: Evaluation of the Doppler Ultrasonic Method of Measuring Systolic Arterial Blood Pressure in Cats. *Am J Vet Res* 53:1166-1169, 1992.
8. Valtonen MH, Eriksson LM: The Effect of Cuff Width on Accuracy of Indirect Measurement of Blood Pressure in Dogs. *Res.Vet.Sci.* 11:358-362, 1970.

LITHOTRIPSY FOR CANINE NEPHROLITHS/URETEROLITHS

Julie B. Commons, RVT, VTS (SAIM)
Lafayette, IN

INTRODUCTION

Extracorporeal shock wave lithotripsy (ESWL) is a safe and effective minimally invasive therapy for removal of canine nephroliths and ureteroliths that involves the use of repeated high energy acoustic pulses (shock waves) generated outside the body that readily travel through water and soft tissue until reaching the hard acoustic surface of the urolith where fragmentation occurs. Small fragments pass through the excretory system and void spontaneously or through voiding urohydropropulsion. ESWL is the most common and preferred treatment of nephroliths in people and open surgical procedures on the kidney are rare, as they have been replaced with minimally invasive alternatives that preserve renal function. This procedure was first used in the United States in 1984 at Methodist Hospital in Indianapolis, IN using a Dornier HM3 Lithotriptor (Dornier, Marietta, GA). Larry G. Adams, DVM, DACVIM treated his first canine patient in 1992 using this original HM3 and to date has treated 189 patients. ESWL treatment in small animals is currently available at Purdue University Veterinary Teaching Hospital (West Lafayette, IN) and Animal Medical Center (New York, NY).

NEPHROLITHS and URETEROLITHS

Urolithiasis is the result of multiple interacting underlying abnormalities and risk factors influencing urolith formation in oversaturated urine. This includes breed, gender, age, existing urinary tract infection (UTI), diet, decreased water intake, altered urine pH, lack of inhibitors or presence of promoters in urine, functional urinary tract abnormalities, and abnormalities of metabolism.[1] Upper urinary tract stones account for less then 1-4% of uroliths and are often an incidental finding.[2] The term inactive nephroliths is used when referring to small nephroliths in the renal pelvis not causing any clinical problems. Inactive nephroliths require monitoring for potential obstruction or enlargement. The majority of canine ureteroliths and nephroliths are composed of calcium oxalate (CaOx) followed by struvite.[1] CaOx uroliths are not amendable to medical dissolution and require removal if indicated. Infection-induced struvite nephroliths secondary to pyelonephritis can be medically dissolved, which is preferred with very large non-obstructive nephroliths. Strict adherence to a dissolution diet with appropriate antimicrobial therapy for the entire dissolution period is required for successful dissolution.

Obstruction

Nephroliths and/or ureteroliths can have serious clinical consequences when partial or complete obstruction occurs resulting in progressive renal injury, uremia, and death. Signs of upper urinary tract obstruction in dogs may include hematuria, abdominal pain, fever or hypothermia, vomiting, anorexia, depression, lethargy, emaciation, polyuria, stranguria, and pollakiuria[3]. Bloodwork may reveal azotemia but degree of azotemia may be dependent on hydration status, extent of ureteral obstruction, pre-existing chronic kidney disease (CKD), and function of the contralateral kidney. Urinalysis may have evidence of crystalluria, hematuria, and UTI. Evidence of complete obstruction on ultrasound or worsening azotemia requires immediate intervention, as timing is critical for restoring renal function. The longer the obstruction, the more significant the damage and is often irreversible with a permanent reduction in glomerular filtration rate.[3] Resolution of even partial obstructions can improve overall outcome with greater return of renal function.[4] Multiple intermittent obstructions could result in renal damage and fibrosis that may occur as small nephroliths in the renal pelvis intermittently move into the proximal ureter.[5]

TREATMENT of UPPER URINARY TRACT UROLITHS

Indications for ESWL treatment and removal of upper urinary tract calculi include large or progressively enlarging nephroliths, progression of CKD, recurrent UTI, presence of symptomatic disease, partial or complete obstruction of renal pelvis or ureter resulting in hydronephrosis, and a solitary functional kidney.[6] Bilateral nephroliths can be treated at the same time or staggered 2-4 weeks apart and is dependent on nephrolith size, renal function, and pre-placement of ureteral stents.[2] Many canine patients with ureteral obstruction have a concurrent UTI that must be treated appropriately prior to ESWL being performed. Ideal candidates are dogs with nephroliths <1-2 cm diameter with an indication for treatment, able to tolerate general anesthesia and fluid diuresis post procedure, and any concurrent problems such as CKD, hypertension, and hyperadrenocorticism should be effectively treated.[6]

Clients should be made aware of ESWL treatment cost and be prepared to travel with a 4-7d hospitalization stay for their pet along with the potential of multiple procedures and also re-treatment if indicated, the risk of an obstructive fragment, and follow-up over several months with radiographs and ultrasound. The incidence of stone recurrence is high especially with CaOx and small residual fragments remaining after any intervention can act as nidus for new urolith formation. Recurrence should be considered when deciding a treatment plan because less invasive options have significant advantages over surgical methods. Even after treatment for upper urinary tract obstruction, many dogs remain azotemic so preservation of remaining renal function is extremely important and should include serial monitoring for progression of CKD, high blood pressure, any signs of obstruction, and UTI.[4] Urolithiasis is a frustrating disease that requires monitoring at appropriate intervals to identify recurrences early and minimize risk factors.

URETERAL STENTING

Transient ureteral obstruction during passage of stone fragments causing partial obstruction is the most common complication of ESWL in approximately 10% of patients and requires intervention if occlusion is > 24-48 hours and is more likely to occur in small dogs with large stone burdens or nephroliths >1-2 cm diameter.[2] This risk in canines more susceptible to obstruction may be minimized with pre-placement of ureteral stents that divert urine from the renal pelvis into the urinary bladder bypassing obstruction. Stent-induced passive ureteral dilation occurs in over 98% of canine patients and aids in ESWL for large obstructive ureteroliths or nephroliths by allowing urine and stone fragment passage around the stent.[4] Indwelling multi-fenestrated double-pigtail ureteral stents are made of a soft polyurethane material and range in size from 2.5 Fr to 6 F and are generally well tolerated by small animal patients.

Placement

The greatest risk in ureteral stenting is often the placement itself and should be performed by a veterinary specialist with training and experience. The distal ureter creates a 'J-shaped' hook at the ureterovesicular junction (UVJ) as the ureter enters at the bladder neck and passage of a floppy tipped guidewire retrograde through the working channel of a cystoscope into the ureteral orifice can be challenging resulting in a blind passage and trauma at the UVJ. The guidewire must be advanced up the ureter using fluoroscopic guidance carefully to avoid ureteral perforation/tear. An open-ended ureteral catheter may be placed over the guidewire in order to preform a retrograde ureteropyelogram. A guidewire is then placed through the open-ended ureteral catheter and coiled in the renal pelvis before removing the catheter. The stent is placed retrograde over the guidewire and the blunt end of the positioner catheter is inserted behind the stent in order to advance the stent to the renal pelvis. Under fluoroscopic guidance the stent is positioned in the renal pelvis and the guidewire is withdrawn allowing for the double pigtail to curl in the renal pelvis first and bladder neck secondly. Then the positioner catheter and guidewire are removed.

A limitation to placement using the retrograde approach is small male dogs not able to accommodate the size of a flexible male cystourethroscope. A percutaneous or surgical approach is performed to place the stent retrograde as described above for female dogs. Stents should remain in place for a minimum of 3 weeks to allow adequate ureteral dilation and are removed once evidence that the majority of fragments have passed which may take 4-8 weeks.[2] Alternatively, ureteral stents may be left in place and monitored for any problems such as stent encrustation. Stents are easily removed during a cystoscopic procedure by grasping the distal end in the bladder and providing gentle traction to withdraw the stent through the urethra.

EXTRACORPOREAL SHOCK WAVE LITHOTRIPTORS

There are four components of a shock wave lithotripter: energy source to generate the shock wave, focusing device to concentrate the shock wave energy into a small target zone (focal point), coupling medium to transmit shock waves (SW) and allow entrance into the patient's body, and an imaging system for locating and positioning the urolith into the focal zone.[2] ESWL with the Dornier HM3 produces SW by passing a high-voltage electrical current across a spark-gap electrode located within a water-filled container. The electrohydraulic focusing system utilizes a metal ellipsoid reflector to direct the energy that the spark-gap electrode creates. The transmission of high-energy pressure waves requires a coupling mechanism so energy does not dissipate at the skin surface. The HM3 utilizes a "wet" coupling by placing the patient in a water bath that provides air-free contact with the skin then localizes and monitors uroliths during treatment using biplanar fluoroscopy.

Newer second or third generation lithotriptors typically use an electromagnetic generator and employ a "dry" treatment head at the patient interface through use of a water filled cushion. This cushion must carefully be coupled to the patient's skin using conductive gel to avoid trapping of air bubbles at the cushion-patient interface which would attenuate SW and reduce energy of the pulse.[7] Newer generation lithotriptors have smaller focal zones that cause less peripheral damage but there is less positional flexibility to account for urolith movement.[7] Their lower peak pressures require more overall energy be delivered to the stone for fragmentation. Stone free rates are not as high and re-treatment rates are higher compared to the HM3 although new technologies are continually being developed with an increase in treatment efficacy. The Dornier HM3 is still considered one of the most effective lithotriptors and the standard to which other devices are compared. However, it is no longer in production as dry lithotriptors became favored for user convenience.

Mechanism of Action

A destructive force is generated when SW hit a substance with different properties with two mechanisms of action that result in stone fragmentation; cavitation and stress waves act together to ensure effective and successful fragmentation.[8] Urolith breakage is gradual and repetitive stress waves (positive pressure waves) dominant in initial fragmentation by exerting a circular pressure outside the stone that results in a squeezing mechanism and compression zone inside the stone. Compression-induced microcracks lengthen and expand along pre-existing defects in the urolith. Cavitation primarily takes over after urolith size is decreased taking advantage of cracks and microcracks created by the stress waves. Cavitation is the rapid formation of water-vapor bubbles in the urine surrounding the urolith. Bubbles cluster and as they collapse an explosive secondary shockwave is generated through high-speed fluid microjets that grind away on the surface of the urolith.[8]

Treatment Strategy

Ramping or voltage stepping involves beginning at lower energy settings and increasing slowly to higher energy, which has resulted in improved fragmentation and a protective effect on the kidney due to shock wave induced vasoconstriction that occurs in both the ipsilateral and contralateral kidney.[9] Starting at lower energy settings also aids in patient tolerance under anesthesia by making only small adjustments to their plane of anesthesia during voltage stepping. The pulse repetition frequency or rate should be kept as low as possible to allow for better fragmentation and less renal injury. Increased rates affect cavitation by disrupting bubble cluster activity.[7] The HM3 is designed to trigger SW based on the ECG to avoid shock wave induced arrhythmias secondary to SW and can be set to fire at every beat or every other beat.

The dose (power and number of shocks) will vary with urolith location and be determined by a trained board-certified specialist. The treatment strategy for dogs using the HM3 based on clinical experience and research include: rate of ~60-90 shocks/min, voltage stepping from 13 to 16 kV for nephroliths and 15 to 18 kV for mid to distal ureteroliths, and a dose of 1,400-1,500 shocks per kidney or 2,000-3,000 shocks per mid to distal ureteroliths or 2,000 maximum shocks per proximal ureteroliths and nephroliths that are included in same treatment focal spot.[2] The higher doses are possible when the kidney is not within the focal zone.

Patient Care

ESWL requires general anesthesia in dogs to facilitate positioning, provide analgesia, and maintain a regular respiratory rhythm. The natural movement of the kidney during respiration can move the stone in and out of the focal zone in addition to any patient movement. It is important to keep them at an adequate plane of anesthesia and provide pain control and prevent any unnecessary movement. Hair must be clipped over the entry point to prevent trapping of air bubbles. Patients are placed in a modified lithotripsy hydraulic gantry in dorsal recumbency with the rear legs positioned lower than the front legs and then partially submerged in a warm water bath. The patient is then turned at an oblique angle in order for SW to enter through the lumbar muscles overlying the targeted kidney. Careful monitoring of patient is required to ensure that their airway and cephalic IV catheter remains above the water line during positioning. To prevent pulmonary contusions, the lungs and heart should be shielded from the SW. Fluoroscopic imaging is performed to monitor stone fragmentation and verify positioning within the focal zone. IV contrast can be administered during treatment to enhance visualization of ureteroliths.

Post-Treatment

Post treatment diuresis (90 ml/kg/day) for 2-4 days promotes movement of the stone fragments.[2] The passage of stone fragments may be rapid or take several months. Analgesics are provided for 24 hours and may be extended if fragments are actively moving or clinical signs of pain are observed. Follow up radiographs and urinary tract US are performed 1 day post treatment to assess change in shape and monitor passage of fragments down the ureters. Repeat radiographs and US are recommended every 2-4 weeks and monthly thereafter until all fragments pass.[2] Most patients will retain some small stone fragments in the recesses of the renal pelvis that may serve as nidus for re-formation. Kidney values are monitored on serum chemistry panels and urinalysis following treatment for any evidence of transient or progressive decrease in renal function. Amylase and lipase are monitored for pancreatic injury secondary to SW. Acute pancreatitis confirmed by clinical signs and abdominal ultrasound is rare and only seen in approximately 2-3% of canines.[10]

Urination should be monitored post ESWL treatment. Transient hematuria (<24 hours) is expected and an incidental finding contributed to the direct injury of small blood vessels within the renal parenchyma by shock waves.[11] Urine is collected and strained for fragments. Fragments not voided spontaneously should be removed by voiding urohydropropulsion or stone extraction both of which involve brief general anesthesia. If stone fragments remain in the kidney or ureter after 2-3 months, repeat ESWL may be indicated.[2] Fragments should be sent for quantitative urolith analysis to assist with prevention/recurrence.

Success of ESWL

Approximately 85% of cases undergo successful fragmentation of CaOx nephroliths and 30% require second treatment to gain sufficient fragmentation especially with larger stones compared to a nearly 80% success rate with ureteroliths and roughly 50% requiring two or more treatments with the Dornier HM3.[2] The re-treatment rate for ureteroliths has decreased over the years with the pre-placement of ureteral stents prior to ESWL treatment. There are several factors that contribute to the higher re-treatment rate for ureteroliths: imaging and focusing are more difficult due to their size and interference of overlying tissues, the confined space means less fluid interface compared to the renal pelvis and also less movement which limits fragmentation, and fragments may be imbedded in the ureteral wall preventing passage.[2]

CONCLUSION

Training and specialized equipment is a high cost investment and access to lithotriptors for small animals patients is a major barrier especially when the need to relieve obstruction is urgent. With recent advances in Veterinary Medicine the trend towards less invasive and safer options for management of upper urinary tract stones is increasing for small animal patients and includes options such as ESWL, ureteral stenting, ureteroscopy, surgical-assisted endoscopic nephrolithotomy (SENL) (similar to percutaneous nephrolithotomy) for large stones >15mm, and subcutaneous ureteral bypass (SUB™) systems (Norfolk Vet Products Inc., Skokie, IL). Owners often seek out minimally invasive procedures in order to avoid repeated invasive procedures especially with the high recurrence of CaOx urolithiasis. The causes of urolithiasis are continually being studied in order to reduce, avoid, or prevent recurrence.

REFERENCES

[1]Osborne CA, et al. *Vet Clin North Am Small Anim Pract* 2008;39:183. [2]Adams LG, Goldman CK. *Nephrology and Urology of Small Animals* 2011:340. [3]Hardie EM, Kyles AE. *Vet Clin North Am Small Anim Pract* 2004;34:989. [4]Berent AC. *J Vet Emerg Crit Care* 2011;21:86. [5]Dalby AM. *J Am Vet Med Assoc* 2006;229:1118. [6]Adams LG, Senior DF. *Vet Clin North Am Small Anim Pract* 1999;29:293. [7]Lingeman JE. *Nat Rev Urol* 2009;6:660. [8]Cleveland RO, McAteer JA. *Smith's Textbook on Endourology* 2007:317. [9]Willis LR, et al. *J Endurol* 2005;19:90. [10]Daugherty MA, et al. *J Vet Intern Med* 2004;18:441. [11]McAteer JA, Evan AP. Semin Nephrol 2008;28:200.

DIAGNOSTIC TESTS OF GASTROINTESTINAL DISEASES
Tracy Hill, DVM, DACVIM, PhD, MRCVS, DECVIM-CA
Edinburgh, United Kingdom

INTRODUCTION

Dogs and cats are frequently presented to specialist referral centres for assessment of diseases of the gastrointestinal system. Many complementary diagnostic tests are performed in the assessment of these patients. From the point of view of the specialist veterinary technician, understanding the indications and implications of these diagnostic tests may facilitate communication with both clients and veterinary colleagues. What follows is not an exhaustive list of gastrointestinal diagnostics but instead a sampling of the most commonly performed tests.

MINIMUM DATABASE

COMPLETE BLOOD COUNT

Any chronic inflammatory gastrointestinal disease may result in an inflammatory leukogram (mild to moderate neutrophilia, lymphopenia, monocytosis, and eosinopenia). With intestinal perforations causing septic peritonitis, bile peritonitis, or severe pancreatitis, neutropenia may be present in dogs and cats with systemic inflammatory response syndrome (SIRS) or sepsis. Eosinophilia may be present with intestinal parasitism or some forms of inflammatory bowel disease (IBD).

Anaemia of chronic inflammatory disease (normocytic, normochromic, non-regenerative anaemia) may be present with chronic inflammatory conditions of the gut. Iron deficiency anaemia (microcytic, hypochromic, non-regenerative anemia) can indicate chronic GI blood loss as may occur with gastrointestinal ulceration, intestinal neoplasia, intestinal parasitism, or inflammatory bowel disease. In dehydrated patients with any disease, a relative polycythemia can be seen.

Diseases of the gastrointestinal tract less commonly affect platelet number. Mild to moderate thrombocytopenia can be seen with chronic GI blood loss.

BIOCHEMISTRY

Several components of the biochemistry panel are helpful in assessment of gastrointestinal disease. Albumin can be increased with dehydration. If albumin is decreased, potential causes include decreased production with liver dysfunction, renal loss, or loss through disrupted gastrointestinal mucosa as can occur with protein losing enteropathies, IBD, food or antibiotic-responsive diarrhea, viral enteropathy (parvovirus), GI neoplasia or GI ulceration. It is important in all hypoalbuminemic patients to obtain a urinalysis to rule out protein-losing nephropathy. Assessment of globulins can be helpful in differentiating causes of decreased albumin as globulins are typically reduced in cases of intestinal protein loss but remain unaffected with liver or renal disease. Cholesterol may be decreased with intestinal protein loss or with liver dysfunction.

Urea and creatinine may be elevated in any dehydrated patient. If urea is disproprortionately elevated over creatinine, GI blood loss should be considered.

Gastrointestinal diseases can result in elevation of alkaline phosphatase with functional or physical obstruction of the entrance of the bile duct into the duodenum. This is seen with IBD and intestinal neoplasias.

Though most dogs and cats with diarrhea have isotonic fluid losses, patients with excess free water loss due to diarrhea may be hypernatremic. Hyperchloremia is suggestive of metabolic acidosis with vomiting and diarrhea, whereas hypochloremia occurs with pyloric outflow obstruction and is associated with a metabolic alkalosis. Chronic anorexia, especially in cats, can cause hypokalemia. Relatively hyponatremia with hyperkalemia has been described with certain large bowel diseases, such as whipworm. Hypocalcaemia can develop with any chronic enteropathy; total hypocalcemia occurs due to reduction in albumin though ionized hypocalcemia can also be seen, perhaps due to altered vitamin D metabolism.

URINALYSIS

Urine specific gravity over 1.035 in dogs and 1.040 in cats occurs with dehydration. Assessment of urine protein concentration via dipstick or urine protein:creatinine ratio rules out kidneys as potential source of protein loss.

FECAL ANALYSIS

DIRECT SMEAR

A sample of feces directly microscopically examined can demonstrate Cryptosporidium, Campylobacter, Clostridia, and occasionally Giardia and Tritrichomonas. Care should be taken on interpretation as presence of these organisms indicates presence, not necessarily causality. Histoplasma may be detected with a rectal scraping and direct smear.

FLOAT

Fecal floatation (ideally with centrifugation) detects ova of roundworms, hookworms, whipworms, and occasionally Tritrichomonas, Giardia, Physaloptera and Coccidiosis. Because ova and protozoa may be intermittently shed, repeated fecal floats are necessary to achieve adequate sensitivity of the test.

CULTURE

Although multiple cultures are commercially available including Salmonella spp., Clostridium spp., Camplyobacter spp., Yersinia, spp. and Tritrichomonas, positive cultures uncommonly are the primary cause of disease and are usually secondary infections or normal GI flora that do not need specific treatment so as such are not generally part of a standard GI workup. Fecal culture is indicated in higher risk groups with compatible clinical signs such as young dogs and cats, especially in group-housing situations, or in dogs or cats fed a raw food diet.

FECAL OCCULT BLOOD

Because chronic slow GI blood loss can be clinically undetectable, fecal occult blood testing can confirm the presence or absence of intestinal blood loss. Fecal occult blood testing is indicated in dogs and cats with microcytic hypochromic anemias of unknown etiology or disproportionately increased urea compared to creatinine. Dogs and cats should be fed a vegetarian diet for 3 days before sampling to reduce false positive results.

FECAL α-1 PROTEINASE INHIBITOR

Fecal α-1 proteinase inhibitor is similar in size/mass to albumin, so it is loss at a similar rate in PLE dogs and cats. Because it is a proteinase inhibitor, it resists intestinal proteolytic degradation so its rate of fecal excretion is a relatively stable indication of protein loss. Three pooled samples of feces over 24-48 hour should be used to reduce false negatives. (Steiner p53-54)

ANTIGEN/ANTIBODY TESTS

There are a number of specific antigen or antibody test for pathologic agents, including Clostridial perfringens enterotoxin, Giardia, Cryptosporidium, and Parvovirus. Although not necessarily the most sensitive test available, they have the advantage of high sensitivity combined with ease of use/testing, especially in the case of Giardia and Parvovirus.

SUPPLEMENTARY BLOOD TESTING

TRYPSIN-LIKE IMMUNOREACTIVITY (TLI)

TLI is often performed in dogs and cats with chronic weight loss +/- diarrhea, to rule out exocrine pancreatic insufficiency, especially in overrepresented breeds (German Shepherd dogs, Rough Coated Collies) or in patients with chronic pancreatitis.

PANCREATIC LIPASE IMMUNOREACTIVITY

As many cases of pancreatitis can present with similar clinical signs to many intestinal disorders, a PLI is often indicated to diagnose pancreatitis.

COBALAMIN

Cobalamin, when bound to intrinsic factor, is absorbed in the ileum. Hypocobalaminemia occurs when inflammation in the ileum decreases absorption or there is decreased intrinsic factor released by the pancreas, as in cases of chronic pancreatitis or exocrine pancreatic insufficiency.

FOLATE

The majority of folate is absorbed in the proximal small intestine. Decreased folate can occur with chronic duodenitis. Because certain intestinal bacteria produce folate, elevated serum folate may indicate deregulation of intestinal microflora, as occurs with antibiotic-responsive diarrhea.

POST-PRANDIAL BILE ACIDS

As many hepatopathies can present with primarily gastrointestinal signs (vomiting, diarrhea, anorexia), post-prandial bile acids are often useful to rule out liver disease. Chronic enteropathies can cause mild increases in post-prandial bile acids.

VENOUS BLOOD GAS

Venous blood gas analysis can be helpful in cases of vomiting to differentiate between vomiting of intestinal contents (metabolic acidosis) versus vomiting secondary to pyloric outflow obstruction (metabolic alkalosis). If not included on biochemistry or if the patient presents after hours, venous blood gas analysis will provide electrolyte information that can be evaluated as discussed previously.

BASELINE CORTISOL/ACTH STIMULATION TEST

Atypical hypoadrenocorticism presents with acute or chronic onset of gastrointestinal signs. Baseline cortisol should be measured in suspect patients to rule out glucocorticoid deficient hypoadrenocorticisim. If baseline cortisol is less than 2mg/dl, hypoadrenocorticism cannot be excluded and an ACTH stimulation test should be performed.[1]

IMAGING MODALITIES

THORACIC RADIOGRAPHS

Thoracic radiographs, including the cervical esophagus and pharynx, should be completed in any dog or cat that is presented with signs of dysphagia or regurgitation to evaluate the pharynx and esophagus for mass lesions, foreign bodies, and focal or diffuse megaesophagus. In geriatric animals, thoracic radiographs are often recommended as a screening test for metastatic disease.

ABDOMINAL RADIOGRAPHS

Abdominal radiographs may show intestinal obstruction (2 populations of small bowel or obviously distended portion of the small bowel. Often, orthogonal views are sufficient, however, three-view radiographs can help discern colon from small bowel if that is a concrn. Radiograph may show a radioopaque foreign body or intestinal mass. Decreased detail may be present with peritoneal effusion secondary to other diseases. Pneumoperitoneum indicates gastrointestinal perforation.

ABDOMINAL ULTRASOUND

Ultrasonography can be helpful for detection of intestinal obstructions, mass lesions, and foreign bodies. Chronic enteropathies, such as IBD, may show loss of layering or thickening of the intestinal wall. The location of the thickening within the bowel (duodenum, jejunum, ileum) as well as determining the affected layer of gut are helpful in determining whether endoscopic or surgical biopsies are most appropriate. Abdominal ultrasonography aids in the diagnosis of other differentials such as pancreatitis and liver disease.

ESOPHAGRAM

Esophagrams, or barium swallow, evaluates pharyngeal and esophageal function. Esophagrams should include at least 3-4 consistencies of food: gruel (barium mixed with a bit of a/d), slurry (a/d mixed with a bit of barium), paste (more solid canned food mixed with a bit of barium) and kibble (dry food coated with barium). The use of different consistencies of food increases the likelihood of detecting abnormal function as esophageal or pharyngeal dysfunction is sometimes only seen with one phase. Esophagram should not be performed in cases of confirmed megaesophagus or if the patient has confirmed episodes of aspiration as this can increase the risk of barium aspiration pneumonia.

ABDOMINAL/THORACIC CT

Abdominal computed tomography is occasionally useful in assessment of GI disease. In large, deep-chested dogs, full assessment of the gastric body, pylorus, and proximal duodenum is sometimes difficult; in these cases, abdominal CT can provide an advantage over ultrasound. If mass lesions are detected, abdominal CT can more definitively detect the extent of disease (vascular invasion, origin of mass lesion, etc). If abdominal CT is performed in dogs and cats with mass lesions, thoracic CT may increase detection of metastatic lesions.

NUCLEAR SCINTIGRAPHY

This imaging technique involves administration of radiolabeled tags whose passage through the intestinal tract can be tracked, allowing for assessment of intestinal motility and gastric emptying time. Although helpful in diagnosis of intestinal transit times, it is uncommonly used due to the difficulties associated with handling radioactive materials and patients, as well as the duration of motility that would need to be assessed (up to 36 hours for colonic transit). Nuclear scintigraphy may diagnose portosystemic shunts, an important differential for chronic GI signs in younger dogs, though as CT is more readily available and more definitive it is usually the preferred method of diagnosis.

ENDOSCOPY

Esophageal-gastroduodenoscopy and colonoileoscopy are frequently performed as part of the assessment of dogs and cats with gastrointestinal disease. It allows for definitive diagnosis of many chronic enteropathies, though caution should be taken in interpretation of results as histologic results from endoscopic biopsy samples do not always correlate with full thickness biopsy, especially in cats.[2] Ileoscopy is recommended along with upper gastrointestinal endoscopy for assessment of both vomiting and diarrhea as the ileum may the most severely affected segment of small bowel.[3,4] Endoscopy also allows for assessment and biopsy of any mass lesions seen in the lumen of the esophagus, stomach, duodenum, colon, or distal ileum. Esophageal and gastric foreign bodies may be definitively treated.

SURGICAL EXPLORATORY OR LAPAROSCOPY

Exploratory laparotomy or laparoscopy allow for assessment of the full length of the gastrointestinal tract, including jejunum, which is inaccessible with traditional endoscopy. Full thickness biopsies are the gold standard for diagnosis of chronic enteropathies. In dogs where disease appears localized to the jejunum or is mostly submucosal, full thickness biopsies are advised. In cats, full thickness biopsies should be strongly considered for diagnosis of chronic enteropathy to allow differentiation of IBD from small cell GI lymphoma.[2] Appropriate patient selection is important, as markedly hypoalbuminemic patients may be at higher risk of complications.[5] Exploratory laparotomy can provide definitive treatment for intestinal masses and foreign bodies, gastric dilatation-volvulus, and a number of other gastrointestinal diseases.

ON THE HORIZON

IMPLANTABLE ESOPHAGEAL MONITORS

Some referral centers implant esophageal monitors, which measure pH and impedence. These devices are used to detect gastroesophageal reflux[6], which is likely an underdiagnosed syndrome in dogs. They are also used experimentally in the stomach for assessment of efficacy of various acid-suppressants.[7,8]

BALLOON ENTEROSCOPY

With this technique, one or two balloons over the endoscope are alternately inflated and withdrawn during endoscopy, telescoping the bowel over the endoscope and allowing for assessment of the entire GI tract. It can be utilized for difficult to access lesions in patients that are not good candidates for surgical explore.[9]

CAPSULE ENDOSCOPY: MOTILITY AND IMAGING

There are two forms of capsule endoscopy available in human medicine: imaging capsule endoscopy and capsules that monitor motility and pH. Dogs have been used as models in several studies of these devices and they could prove useful in diagnosis of motility disorders and in cases of intestinal hemorrhage where a mass lesion is not otherwise detected. Their expense likely limits widespread use currently.[8]

REFERENCES

1 Lennon EM et al, JAVMA 2007. 231(3):413-6.
2 Evans SE et al, JAVMA 2006. 229(9):1447-50.
3 Scott JD et al, JVIM 2011. 25(6):1253-7.
4 Procoli F et al, JVIM 2013. 27(2):268-74.
5 Ralphs SC et al, JAVMA 2003. 223(1): 73-7.
6 Zacuto AC et al, JVIM 2012. 26(3): 518-25.
7 Tolbert K et al, JVIM 2011. 25(1): 47-54.
8 Parkinson S et al, JVIM 2015. 29(1): 104-12.
9 Sarria R et al, Vet J. 2013. 195(3):331-6.
10 Boillat CS et al. AJVR 2010. 71(8):898-908.

POLYCYTHEMIA

Meri F. Hall RVT, CVT, LVT, LATG, VTS (SAIM)
Palouse, Washington

Polycythemia is a disease that has been associated with an increase of erythrocytes in the circulation system. In actuality, polycythemia means an increase in numbers of all the circulating cells. The correct terminology is erythrocytosis. Clinically, this is demonstrated by an increase in the packed cell volume (PCV), hematocrit (HCT), hemoglobin concentration or red blood cell (RBC) count. If the only increase hematologically is the RBC parameters, then erythrocytosis is a more appropriate label. Dogs that live at high altitudes and certain dog breeds that naturally have HCTs above the normal reference ranges. Greyhounds for example can have hematocrits (HCT) of 70%. If there is an increase in all cell lines, it is more appropriate to label this is as polycythemia.

Erythrocytosis can be further classified as relative or absolute. Relative erythrocytosis is typically due to the redistribution of RBCs in circulation or an intracompartmental fluid shift. It may be due to the hemoconcentration seen with dehydration. With relative erythrocytosis, the PCV and the total serum protein concentration is also increased. In dogs with hemorrhagic gastroenteritis, there may be a normal total serum protein with an increased PCV. After rehydration with fluid therapy, one would expect the PCV to begin to normalize.

An actual increase in the numbers of RBC would be the cause for absolute erythrocytosis. Absolute erythrocytosis is further delineated by the cause of the erythrocytosis. If the cause is due to a myeloproliferative disorder of the bone marrow, the disease is classified as primary erythrocytosis. This condition is commonly called polycythemia vera. When the bone marrow increases the production of the RBC independent of erythropoietin, all other blood cells are normal in number. The erythropoietin levels are low to normal level when tested.

If there is an increase in the production of erythropoietin in the kidney (orthopotic) or from other sites (heterotopic), the condition is classified as secondary erythrocytosis. This condition occurs due to another pathological condition such as kidney disease. Chronic lung disease, heart disease and congenital defects may predispose an animal to secondary polycythemia. A decrease in the partial pressure of oxygen (PaO2) as can be seen with hypoxia or hypoxemia may result in a compensatory response. The decreased PaO2, results in an increase in the RBCs produced and secondary erythrocytosis occurs. Animal living at high altitudes, renal masses and some breeds have a natural erythrocytosis.

Erythrocytosis may occur acutely and clinical signs may present themselves as abnormalities of the central nervous system. Animals may present with behavior changes, motor or sensory changes and seizures. Cardiopulmonary signs may be present due to the increased blood viscosity. In some cases, the changes may be gradual and most animals do not show clinical signs until the PCV becomes very high. A PCV of 70-80% is common in patients with absolute erythrocytosis. These patients may present with red mucous membranes, polyuria or polydipsia.

To diagnose erythrocytosis the hydration status of the patient must be taken into consideration. If no dehydration is noted in the presence of increased PCV or HCT, then a PaO2 level and potentially an erythropoietin level should be considered. Many hospitals do not have the ability to perform erythropoietin levels. Relative erythrocytosis patients will have normal PaO2 and erythropoietin levels. For secondary erythrocytosis, the PaO2 levels would be decreased with an increase in the erythropoietin level. In polycythemia vera, the PaO2 levels would be within the reference range and the erythropoietin levels may be normal or decreased.

	Packed Cell Volume	Total Protein
Relative erythrocytosis	Elevated	Elevated
Absolute erythrocytosis	Elevated	Normal

To treat relative erythrocytosis the dehydration status should be corrected. If after fluid therapy, the PCV remains elevated performing therapeutic phlebotomy would be indicated. This is accomplished by reducing the circulating red blood cells (approximately 20ml/kg) via jugular vein. This may be repeated gradually with removal of additional blood (5ml/kg) as needed. To prevent hypotension that can result due to the sudden decrease in the blood volume, an equivalent volume of isotonic crystalloid fluid should be administered via a peripheral catheter.

In patients with secondary erythrocytosis due to hypoxia or hypoxemia, it may be impossible to determine the blood gas status of the patient until after therapeutic phlebotomy has been performed. With erythrocytosis it is imperative to diagnose and treat the congenital abnormality or underlying disease. If polycythemia vera is suspected, a veterinary oncologist should be consulted.

If it is determined that the animal has polycythemia vera, hydroxyurea is administered for 7-10 days. Phlebotomy should be repeated if necessary based on clinical signs of the patient. Patients diagnosed with polycythemia vera have been reported to have survival times of greater than 2 years. Complete blood counts should be performed every 4 to 8 weeks for patients treated with hydroxyurea because of the myelosuppressive complications with the drug. Prognosis for animals with secondary erythrocytosis is entirely dependent on the primary disease.

REFERENCES

Merrill, L et al, *Small Animal Internal Medicine for Veterinary Technicians and Nurses, 2012,* Wiley Blackwell
Nelson, Cuoto, *Small Animal Internal Medicine, Fifth edition,* 2014, Elseiver
Ettinger, Feldman, Et.al,*Textbook of Veterinary Internal Medicine, 7th edition,* 2010, Saunders

NEOSPORA FOR THE VETERINARY TECHNICIAN

Meri Hall, RVT, LATG, LVT, CVT, VTS (SAIM)
Palouse, Washington

Animals are infected by a multitude of viral, bacterial, fungal and protozoal agents. *Nespora caninum* is a common protozoa that infects companion and agricultural animals alike. Like *Toxoplasma gondii* , *Neospora caninum* requires an intermediate host to complete the lifecycle. Prior to 1988, many dogs were falsely diagnosed with Toxoplasmosis because of the similarities of the oocysts of *Neospora caninum.* Neosporosis affects canines, equine, cattle, goats and deer. There have been no reports of domestic cats infected with *N. caninum*, but has been reported in wild felids. *N. caninum* has been reported in every continent and major countries including the United States, Germany, Australia, New Zealand, Belgium and Canada.

Neospora canium can affect animals via horizontal transmission or vertical transmission. Intermediate hosts include cattle, pigs and chickens. Canines and canine species are the definitive host for the parasite. There has been no documented cases of domestic cats with natural cases of Neosporosis, but wild felids have had documented infections.

Dogs become infected by the ingestion of bradyzoites, but can be transmitted transplacentally. Bradyzoites are found in the placental tissue and aborted fetus of infected cattle. The protozoa undergo sexual reproduction in the gastrointestinal tract of dogs and the oocysts are shed in the feces within 24 hours of ingestion. Dogs will shed the oocyst for months. Once shed, the oocyst must have exposure to oxygen, high temperature and humidity. This causes the oocyst to sporulate. The sporulated oocyst can survive for months to years in the environment. Environmental transmission occurs through water or on plants. When the sporulated oocyst is ingested by an intermediate host, becomes a tachyzoite and forms cysts where they become bradyzoites. When the bradyzoites are ingested by the definitive host, they mature and oocysts are produced. Bradyzoites can be found for the life of the host.

Identification of oocysts in dogs with diarrhea has been reported. The intracellular replication of the tachyzoite causes cellular necrosis and can spread to skeletal muscle, spinal column, brains and heart tissue. In cattle, abortions are seen from 3 months of gestation to full term. There are an increased number of still births, abortions and resorption of fetuses reported. Up to 90% of some herds can be affected. If a calf survives, they may be poor doers and show neurologic signs. Adult cattle may have ataxia, depression, anorexia, or fever depending on the location of the bradyzoite.

The clinical signs in dogs are primarily neuromuscular. *N. caninum* is most common in pups less than six months of age who present with ascending paralysis, hyperextension of the hind limbs and muscle atrophy. Pups can show clinical signs at 3-9 weeks of age and death is common. In dogs over six months of age, the clinical manifestation can be regional or generalized. Behavioral changes, seizures, head tilt, blindness, fever, Horner's syndrome, cough, dyspnea, myocarditis, megaesophagus and vomiting can be seen. Non neuromuscular signs may include myocarditis, pneumonia, coughing, lethargy, pyrexia and dermatitis. In adult dogs, animals can present with ataxia, paresis, cranial nerve deficits and seizures. Dogs may present with a myositis with weakness, stiff gait, tetraparesis, muscle wasting or lower motor neuron signs.

Diagnostics include hematology, serum chemistry and cerebrospinal fluid analysis. Hematology may reveal a non-regenerative anemia and neutropenia. Increased creatine kinase (CK) and aspartate aminotransferase (AST) are seen on serum chemistry. The cerebrospinal fluid analysis may show an increase in the protein level along with inflammatory cells. The organism may be seen on microscopic examination of the cerebrospinal fluid. It is rare to see bradyzoites in the cerebrospinal fluid. The presence of oocysts in the feces is not conclusive because there are other parasites with oocysts that are similar in appearance. Titers can be performed on serum and cerebrospinal fluid for IgG and IgM. In adult dogs, magnetic resonance imaging (MRI) may show an inflammatory lesion within the brain or spinal column depending on the location of the lesion.

Treatment is not curative, the clinical signs may reappear months to years after treatment. The treatment for cattle includes embryo transfer and culling of affected cattle. Antibiotics and supportive care is indicated for dogs that are infected. Any littermates of pups less than 6 months of age should be treated as well. Breeding of females that have had a litter with neosporosis should not occur since the bradyzoites will infect future litters. The administration of glucocorticoids may activate encysted bradyzoites and lead to clinical signs in affected dogs.

REFERENCES

Merrill, L et al, *Small Animal Internal Medicine for Veterinary Technicians and Nurses, 2012,* Wiley Blackwell

Nelson, Cuoto, *Small Animal Internal Medicine, Fifth edition,* 2014, Elseiver

Ettinger, Feldman, Et.al,*Textbook of Veterinary Internal Medicine, 7th edition,* 2010, Saunders

BLOODWORK MONITORING DURING CHEMOTHERAPY

Jackie M. Wypij, DVM, MS, DACVIM (Oncology)
Urbana, IL

INTRODUCTION

Serial evaluation of hematologic and biochemical parameters in veterinary chemotherapy patients is a crucial aspect of patient monitoring. Changes may be reflective of the primary or metastatic cancer, paraneoplastic syndromes, chronic disease/inflammation, and/or the effects of systemic chemotherapy. Veterinary patients have lower incidences of adverse side effects compared to human chemotherapy patients. Prevention and early detection of abnormal parameters are crucial to minimizing clinical problems. The veterinary technician is most often the first person to see bloodwork results, and a critical "triage" of the bloodwork can maximize the outcome for the patient. Nearly 90% of in-clinic sample processing and more than three-fourths of necessary equipment maintenance are performed by veterinary technicians and assistants.[1] Approximately 90% of clinics perform routine quality assurance.[1] The accuracy of the information from these in-clinic results are vital to patient management.

Evaluation of bloodwork involves complex data interpretation, as an occasional "abnormal" is expected in healthy animals. Because of the method by which reference ranges are conventionally determined, 5% of normal animals may have values outside the defined reference range. In general, most adverse chemotherapy side effects in veterinary patients are mild and transient and should be graded according to the VCOG scale.[2] Chemotherapy dose adjustments and treatment delays are then implemented if needed clinically based on type of cancer and treatment protocol as well as monitoring parameters, individual patient characteristics, and concurrent medications.

Changes in hematologic and biochemical parameters are complicated by both the effects of chemotherapy agents as well as the impact of the cancer on multiple organ systems. For example, chemotherapy can cause gastrointestinal mucosal damage, altered bacterial flora, and myelosuppression resulting in neutropenia that could lead to sepsis. In addition to the gastrointestinal tract, other common portals of entry (opportunistic bacteria) include the genitourinary tract, respiratory tract, and skin. At risk patients for sepsis include small animals and dogs treated for lymphoma. Organ dysfunction (e.g. kidneys, liver) from chronic degenerative disease, poly-pharmacy, or cancer infiltration can inhibit metabolism and excretion of chemotherapy agents leading to increased toxicity to the patient even at standard doses.

Venous blood sampling in feline and canine chemotherapy patients should generally be performed via jugular venipuncture unless there are underlying patient concerns such as known thrombocytopenia or coagulopathy. This allows peripheral veins to be utilized for treatment purposes and reduces the blood turbulence and hemolysis that can lead to sample artifacts. Common sampling issues that can effect hematologic and biochemical results include lipemia, hemolysis, patient stress, and clot formation in EDTA/heparin samples. Red top tubes should be allowed to clot for 30-60 minutes prior to serum separation to minimize artifactual changes.[3]

HEMATOLOGY

Chemotherapy drugs are nearly universal in causing some degree of myelosuppression (decreased bone marrow activity) resulting in altered production of red blood cells, white blood cells, and platelets. Approximately 4-7 days are required for the bone marrow to generate a response to cell loss. Chemotherapy-induced neutropenia is most common. Primary cancers rarely cause neutropenia although diseases such as feline retroviruses can also lead to neutropenia. In dogs, most conventional chemotherapy agents result in a neutrophil nadir (low point) at ~7-10 days, although some drugs such as carboplatin may have a later nadir often at 10-14 days. Thus, for traditional chemotherapy agents a post-chemotherapy CBC is generally recommended at 7-10 days. In contrast, newer receptor tyrosine kinase inhibitors (RTKIs) while also myelosuppressive are monitored on a different schedule. In general, for RTKIs, complete blood counts (along with serum biochemistry profiles and urinalysis) is suggested at baseline, after two weeks of therapy, and then monthly thereafter. Dose reductions and/or drug holidays may be indicated. Depending on the individual animal, treatment protocol, and drug used, most conventional chemotherapy agents can be administered with a baseline neutrophil count of 2000/μL. When evaluating a post-chemotherapy CBC, there is a low risk of infection with neutrophil counts of 2000/μL or higher. In general, feline chemotherapy nadirs can be harder to predict with some drugs having delayed or double nadirs.

Thrombocytopenia is common in veterinary cancer patients, with up to 40% of animals presenting with thrombocytopenia and nearly one-third of dogs presented with thrombocytopenia eventually diagnosed with cancer.[4] Thrombocytopenia may be induced by chemotherapy or cancer via immune on non-immune mediated destruction, sequestration (hepatic/splenic tumors), macroscopic or microscopic bleeding, increased consumption/disseminated intravascular coagulation, and myelophthisis (displacement of bone marrow tissue via tumor infiltration or fibrosis). Common cancers causing thrombocytopenia are leukemia, lymphoma, multiple myeloma, histiocytic sarcoma, disseminated mast cell disease, and hemangiosarcoma.[4] The most common chemotherapy drug causing thrombocytopenia is CCNU (lomustine) particularly with cumulative dosing. In addition to platelet numbers, platelet function may be adverse affected by cancer or cancer therapy, and many geriatric, cancer-bearing dogs are on NSAID therapy either as part of the cancer treatment or to treat degenerative joint disease which may impact their platelet function. Normal platelets lifespan is approximately 5-7 days. Thrombocytopenia should always be verified with a manual count, particularly in cats as feline blood samples are particularly prone to clumping leading to pseudo thrombocytopenia.

Anemia is also common in veterinary cancer patients, with one-third of anemic dogs eventually diagnosed with cancer. Anemia can also be a prognostic factor; for example dogs with multicentric lymphoma that are anemic at diagnosis have a worse prognosis. The common cancers causing anemia are similar to those causing thrombocytopenia, and include leukemia, lymphoma, multiple myeloma, histiocytic sarcoma, disseminated mast cell disease, and hemangiosarcoma.[4] Anemia is often a chronic problem and may be induced by chemotherapy or the primary cancer via immune on non-immune mediated destruction, macroscopic or microscopic bleeding, anemia of chronic/inflammatory disease (cytokine-mediated), nutritional/iron deficiency, neoplastic kidney infiltration (decreased erythropoietin), and myelophthisis (displacement of bone marrow tissue via tumor infiltration or fibrosis). Given that the normal red blood cell life-span is ~70 days in the cat and ~120 days in the dog, anemia may not be immediately apparent until the circulating population is depleted. Feline red blood cells are also more sensitive to oxidative damage resulting in anemia.

A concurrent manual blood smear should be evaluated for all chemotherapy patients, particularly focusing on red cell morphology, platelet count, leukocyte differential, and presence of atypical cells such as blast cells. EDTA blood is ideal for blood smear preparation to preserve cell morphology and minimize cell clumping. In most studies of normal dogs automated CBC machines appear to be accurate. However, automated counters are unable to accurately differentiate and count abnormal cells (e.g. blast cells, cell fragments, band neutrophils) leading to error codes or "flags" which should be followed up. Up to 20% of dogs undergoing chemotherapy have at least one abnormality on a blood smear with the potential to change the treatment plan. When evaluating the smear, look for platelet clumps, microfilariae, and large cells in the feathered edge and look for rouleau formation and RBC agglutination in the body of the slide. Estimate platelet and WBC counts and examine cell morphology in the monolayer zone. To estimate a platelet count, count platelets in 100x (oil). Each platelet is equivalent to ~10-20K platelets. A general rule (per 100x) is 8-10 (minimum), 10-15 (standard), and 35-40 (high end).

SERUM BIOCHEMISTRY

Kidney parameters

Azotemia is the elevation in blood urea nitrogen (BUN) and creatinine and may be caused by pre-renal (dehydration), renal (kidney disease), or post-renal factors (urinary obstruction). Inappropriately elevated BUN without a comparable increase in creatinine in a patient should suggest looking for an underlying cause of an increased nitrogenous load, most commonly gastric bleeding from treatment (e.g. NSAIDs or chemotherapy) or a primary tumor (e.g. gastric cancer or histamine release from mast cell tumor degranulation resulting in gastric ulceration). Some chemotherapy agents, most notably the platinum agents cisplatin and carboplatin, are nephrotoxic. Doxorubicin is considered nephrotoxic in cats and should also be used with caution in dogs with underlying kidney insufficiency. When evaluating azotemia, it is crucial to evaluate a concurrent urinalysis focusing on urine specific gravity and sediment changes including urine protein. Some forms of cancer (e.g. bladder or prostate cancer) or anatomical functional changes as a result of cancer therapy (e.g. hind-end paralysis, fecal incontinence, etc.) may predispose to susceptibility to urinary tract infections, which in the setting of chemotherapy-induced neutropenia can lead to bacterial translocation of opportunistic pathogens resulting in sepsis. Therefore, routine urinalysis +/- culture is recommended in symptomatic or at-risk patients.

Hepatic parameters

The major hepatic parameters include 1) liver enzymes, 2) bilirubin and 3) pseudo-function tests. Alanine aminotransferase (ALT) is a cytosolic enzyme that may be elevated in both acute and chronic liver disease. ALT elevation is a result of hepatocellular damage (transient or chronic) with maximal elevation ~2 days from acute injury. The severity is comparable to the number of leaky hepatocytes with a marked increase more common in diffuse disease vs. focal lesions. However, ALT elevation may not correlate with severity of overall hepatocellular damage particularly in chronic disease. ALT half-life is much longer in the dog than the cat.

In comparison, alkaline phosphatase (ALP) is a membrane-bound enzyme and is more suggestive of cholestasis and biliary obstruction than hepatocellular damage. ALP is not liver-specific, and can also be produced by other organs (e.g. bone) and is routinely elevated in the setting of exogenous (e.g. prednisone) or endogenous glucocorticoids (e.g. hyperadrenocorticism). In-house analyzers are unable to differentiate total vs. corticosteroid-induced ALP. Dogs often have mild, non-specific elevations in ALP however any elevation in ALP in the cat is suggestive of hepatic disease/ cholestasis. Gamma glutamyl transferase (GGT) is another membrane bound enzyme and an additional indicator of cholestasis. GGT may be more useful than ALP as it is not significantly induced by glucocorticoids and drugs however most drugs that induce ALP also cause hepatocellular swelling which secondarily causes intrahepatic cholestasis and elevated GGT. In cats, increased GGT and ALP supports cholestasis with ALP changes often more subtle. A relatively greater increase in ALP than GGT is suggestive of hepatic lipidosis in cats. Aspartate aminotransferase is not routinely used by most laboratories because of its nonspecific nature.

The liver is important for metabolism and excretion of common chemotherapy agents, most notably doxorubicin and the vinca alkaloids vincristine and vinblastine. In patients with primary and metastatic liver cancer, biochemical changes are common and can include elevations in ALT, GGT, and bilirubin. Common chemotherapy drugs with biliary excretion such as doxorubicin and the vinca alkaloids are contraindicated in animals with elevated bilirubin attributed to cholestasis. Interpretation of bilirubin must also include evaluation of sample hemolysis in light of clinical vs. artifactual hemolysis. Hepatobiliary excretion of these drugs is dependent on the drug transporter P-glycoprotein, which can be functionally deficient particularly in some herding breeds (Collie, long haired whippets, Australian shepherds, others). Evaluation of the mutation in affected breeds (ABCB1-1Delta status) is imperative prior to initiation of these chemotherapy agents and the test can be performed on saliva or blood samples. Dogs with p-glycoprotein deficiency are increased risk of life-threatening severe adverse effects even at reduced doses

Dogs with primary (vs. metastatic) liver tumors often have hypoglycemia, hypoalbuminemia, hypergammaglobulinemia, and elevated bile acids. Altered pseudo-function tests (low glucose, low BUN, low albumin, low cholesterol) may be suggestive of hepatic insufficiency. Advanced liver function tests include bile acids and ammonia levels however these are not a part of routine screening. In addition to decreased hepatic production, albumin may also be lost through the intestines (protein losing enteropathy or PLE) or the kidneys (protein losing nephropathy or PLN). Both cancer and some chemotherapy agents can cause PLN, most notably some receptor tyrosine kinase inhibitors. If there is no evidence of hepatic dysfunction or PLE, a urinalysis including sediment, urine culture, and urine protein:creatinine ratio may be indicated.

The most common chemotherapy drug to induce hepatotoxicity in canine patients is CCNU (lomustine) although clinically significant hepatic injury is seemingly uncommon in cats.[5] CCNU is routinely used for lymphoma, histiocytic sarcoma, mast cell tumors, and brain tumors. CCNU can cause delayed, cumulative dose-related, chronic hepatotoxicity that can be fatal. Up to 85% of dogs receiving CCNU treatment will develop elevated ALT usually within 1-3 doses.[6-8] Approximately half of the cases of severe elevations are abrupt although serial monitoring may identify an increasing trend in ALT in the other half of cases.[8] Reported risk factors for CCNU hepatotoxicity include ≤5 years of age.[8] Possible mechanisms include glutathione depletion and lipid peroxidation resulting in hepatotoxicity. Denamarin ® (S-Adenosylmethionine or SAMe and silybin) has been shown to both minimize and rescue ALT elevations in CCNU-treated dogs.[7]

Electrolytes

Severe electrolyte abnormalities are most commonly a sequelae of acute tumor lysis syndrome (ATLS) occurring within 1-2 days of initial chemotherapy treatment. Rapid destruction of lymphoma cells leads to release of intracellular contents resulting in hyperphosphatemia, hyperkalemia, hypocalcemia, and metabolic acidosis. These electrolyte imbalances can overwhelm the kidneys' ability to compensate and can result in renal failure, DIC, and death. Suspected factors predisposing for ATLS include underlying renal dysfunction, large tumor burden, stage IV or V lymphoma, high pre-treatment serum alkaline phosphatase activity, rapid tumor regression, and dehydration.[9,10]

REFERENCES

[1]Bell R, et al. Vet Clin Pathol. 2014 Jun;43(2):185-92. [2]Veterinary cooperative oncology group - Vet Comp Oncol. 2011 Jul 20. [3]Gilor S, Gilor C. Top Companion Anim Med. 2011 May;26(2):109-18. [4]Childress MO. Vet Clin North Am Small Anim Pract. 2012 Jan;42(1):123-55. [5]Musser ML, et al. J Feline Med Surg. 2012 Dec;14(12):871-5. [6]Heading KL, et al. Aust Vet J. 2011 Apr;89(4):109-16. [7]Skorupski KA, et al. J Vet Intern Med. 2011 Jul-Aug;25(4):838-45. [8]Hosoya K, et al. Vet Comp Oncol. 2009 Dec;7(4):244-55 [9]Mylonakis ME, et al. Aust Vet J. 2007 May;85(5):206-8. [10]Vickery KR, Thamm DH. J Vet Intern Med. 2007 Nov-Dec;21(6):1401-4.

RESCUE PROTOCOLS FOR CANINE LYMPHOMA

Jackie M. Wypij, DVM, MS, DACVIM (Oncology)
Urbana, IL

INTRODUCTION

Relapsed vs. Refractory/resistant lymphoma

Standard treatment of canine multicentric lymphoma most commonly utilizes a "CHOP"-style protocol, and variations on this theme have been explored in the setting of drug naïve lymphoma. These protocols incorporate (C) cyclophosphamide, (H) hydroxydaunorubicin (doxorubicin), (O) Oncovin® (vincristine) and (P) prednisone. There is an approximately 80-95% overall response rate (ORR) with a median survival of approximately one year, and up to 25% of dogs achieving 2+ years survival. However lymphoma is considered a terminal disease as nearly all dogs will eventually come out of remission. Upon relapse (recurrence of clinical signs attributable to lymphoma), re-induction with the same protocol is often beneficial especially if the first remission was durable upon discontinuation of treatment. A rescue or alternative chemotherapy protocol should be used in the setting of refractory or resistant lymphoma, or if completion of chemotherapy was recent (~4weeks). Rescue protocols can also be used in the setting of relapsed lymphoma without a re-induction with the same initial induction protocol. In interpreting clinical studies of rescue protocols it is important to note that data from relapsed and resistant lymphoma patients are often aggregated, despite the likely difference in their response to future therapies. In addition, many patients considered relapsed or refractory have been treated with a variety of non-standard chemotherapy protocols, sometimes without doxorubicin which has traditionally been considered one of the most effective drugs for lymphoma.

Reasons for relapse may include inadequate dosing or frequency of chemotherapy (e.g. treatment delays), inadequate concentration of drug in affected tissues (e.g. CNS), multidrug resistance leading to the development of resistant clones or tumor stem cells, decreased intracellular drug accumulation (decreased drug uptake e.g. altered pH; increased drug efflux e.g. ABC transporters), decreased activation of prodrugs, increased systemic clearance of drug, increased intracellular clearance of drug (e.g. cytosolic inactivation via glutathione), altered drug target, increased expression of cell repair proteins (e.g. nuclear excision repair), upregulation of inhibitors of apoptosis (e.g. XIAPs), and downregulation of pro-apoptotic proteins (e.g. p53).

Goals and monitoring for rescue protocols

In the setting of a first-line protocol for naïve lymphoma, the goals are to achieve a durable, complete remission (CR) using a defined course of chemotherapy. If using a re-induction protocol, the overall response rate (ORR) and duration of remission (DOR) are expected to be approximately 50% of initial treatment protocol. Dogs achieving CR vs. partial remission (PR) will have improved DOR and survival. Given the improved outcome for dogs achieving a CR, it is important to maximize intensity while maintaining quality of life. Although assessment is not standardized across studies, it is generally accepted that if a dog achieves CR, it should be maintainable for 3-4wks. Historically, the ORR of most canine lymphoma rescue chemotherapy protocols is ~30-50% with a progression free survival (PFS) of 1.5-2.5 months. Although a course of rescue chemotherapy may initially be prescribed inside a defined timeframe, the limited sustainability of most rescue protocols eventually leads to continuous chemotherapy administration using sequential protocols upon repeat clinical failure. There are diminishing returns in terms of expected response and outcome with sequential protocols. At this point, durable partial remission or stable disease are reasonable options as long as quality of life is maintained. Although partial remission may not be statistically beneficial for survival, the patient's quality of life is maximized and this may be an acceptable outcome for the client. Complete remission (CR) is commonly defined as complete clinical resolution of tumor burden. In most studies, partial remission (PR) is defined as <100 but >50% reduction in tumor burden and stable disease as <50% reduction or <25% increase in tumor burden with no new disease. Progressive disease (PD) is most commonly defined as >25% increase in tumor burden. In some studies, alternative measurements are used to define response to therapy.

The "best" rescue protocol

When discussing the "best" treatment protocol with a client, there are many factors which must be considered. When using a multi-drug rescue protocol, consider previous use of drug (or drug class) in the patient, potential overlapping toxicity, and different/complementary mechanisms of action. Client factors include: toxicity of initial protocol, financial impact, acceptable cost-benefit given diminishing returns (effectiveness vs. likelihood of side effects), time/travel impact, balancing achievable time with quality of life, willingness to commit to life-long continuous chemotherapy (vs. initial defined protocol), and anticipatory and repeated grief. Patient factors include: patient tolerability of drug treatment, cumulative myelosuppression with long term chemotherapy, concomitant disease conditions (e.g. liver, kidney), maximum cumulative dose, stress to patient with long-term repeated blood sampling and catheterization, and scarring of peripheral veins. Cancer factors include: impact of multidrug resistance, cross resistance in or across classes of drugs, and protocol intensity. Practitioner factors include: familiarity with drugs, window of safety (i.e. if this protocol doesn't work when will I safely be able to try something else?), drug availability, and staff/facility limitations (e.g. chemo hood).

Multidrug Rescue Protocols

Re-induction with the same CHOP-style initial protocol is generally considered the best medical option. In one small study, 10/10 dogs that achieved CR with a UW Madison protocol were successfully re-induced with the same protocol.[1] In a larger study this response was confirmed with and ORR 100% in 95 dogs; 78% CR with a DOR of 159 days. A long first remission (≥289 days) was a positive prognostic factor.[2]

A commonly used rescue protocol is the combination of mechlorethamine (Mustargen, alkylating agent), vincristine (vinca alkaloid), procarbazine (alkylating agent), and prednisone; (MOPP). This protocol can be expensive and time-intensive and may not be feasible in a general practice given safety issues of mechlorethamine. In one study of relapsed lymphoma in dogs previous treated with a doxorubicin-based protocol, ORR was 65% with a PFS of 61 days.[3] In dogs achieving CR (31%), PFS was 63 days.[3] Hematologic toxicity is generally mild. Of note, 12% of dogs had a DOR greater than that achieved with their initial protocol, including 20% of dogs with SD.[3] In most cases when considering a MOPP protocol, the patient has previously received vincristine and therefore this drug may have little added benefit to the protocol in these patients. In a modified protocol with dose-intensified mechlorethamine and no vincristine (MPP), ORR was 34% with a PFS of 56 days.[4] A subset of dogs achieving CR (17%) had an extended median PFS of 238 days.[4] Related protocols BOPP and LOPP substitute other alkylating agents BCNU (carmustine) and CCNU (lomustine), respectively, for the mechlorethamine. In two studies, response between protocols was similar, with ORR of 50% for BOPP and 52-61% for LOPP in previously CHOP-treated dogs with a PFS up to 129 days.[5,6] Approximately one-third of dogs achieve a CR with a PFS of up to 130 days.[5,6] In the first study, dose intensity was higher, however there was significant hematologic toxicity and an unacceptable 13% treatment-related mortality.[5] In the related study evaluating a modified LOPP with a lower dose intensity, the toxicity was much lower.[6] Another variation of the MOPP protocol is MOMP (mechlorethamine, vincristine, melphalan and prednisone) which replaces procarbazine with melphalan. In a recent study, dogs treated with MOMP had an ORR of 51% (56 days), with a CR of 12% (81 days) with similar response rates for B and T cell immunophenotype.[7]

Anthracycline-type combinations have also proven useful as rescues for canine lymphoma. The combination of dexamethasone, melphalan, actinomycin-D, and cytarabine arabinoside is referred to as DMAC. In one study in relapsed dogs, there was a 72% ORR with a PFS of 61 days; a subset of dogs (44%) achieved a CR with an extended PFS of 112 days.[8] However in that study only 22% of patients had received a previous chemotherapy protocol incorporating an anthracycline. A reduced response was seen in a related study of dogs that had previously received CHOP protocols with an ORR of 43% and PFS of 24 days. In the subset of dogs that achieved CR (16%) PFS was 63 days.[9] These studies identify thrombocytopenia as a major concern, with up to 34% grade III-IV hematologic toxicity.[8-9]

The anthracyclines have also been used in combination with intravenous DTIC (dacarbazine, an atypical alkylating agent that acts as a temozolomide prodrug) or oral temozolomide. In a small study of 15 dogs relapsed from doxorubicin, the combination of doxorubicin and DTIC afforded a 46% ORR (all CR) with a median survival time (MST) of 105 days.[10] In a group of previous CHOP-treated dogs, doxorubicin or actinomycin-D in combination with either dacarbazine or temozolomide afforded a 71% ORR (58% CR) with PFS of 75 days.[11]

Alkylating agent combinations are also reported as rescue protocols. DTIC in combination with CCNU in CHOP-resistant patients afforded a 35% ORR (23% CR) with PFS of 62 and 83 days, respectively.[12] In this combination, toxicity is of concern; approximately 25% of animals had grade II-IV hematologic or hepatic toxicity.[12] CCNU +/- prednisone has also been used in combination with L-asparaginase in two studies of dogs previously treated with CHOP therapy, with an overall response rate of 77-87% and a PFS of 63-70 days. More than half of dogs achieve CR (52-65%) with a PFS of 90-111 days.[13,14] Myelosuppression, thrombocytopenia (cumulative) and hepatotoxicity are significant concerns with CCNU therapy.

Single agent protocols

Single agent protocols are attractive because of their often reduced cost and time commitment. In dogs not previously receiving doxorubicin-based chemotherapy, single-agent doxorubicin is likely to be the most effective single agent drug. In one study including both doxorubicin-naïve and previously doxorubicin-treated dogs there was an ORR of 75% (CR 42%) with a DOR of approximately five months.[15] Most CHOP-style protocols include four doses of doxorubicin (120mg/m2 cumulative dose) and therefore even in a previous CHOP-treated dog, additional doses of doxorubicin can be used. Caution should be used as you approach the maximum cumulative lifetime dose, including appropriate monitoring for cardiac dysfunction. Another drug in the same class used as a single agent is actinomycin-D. Reported response to actinomycin is variable. In one study, there were no responders while in another, there was an ORR of 41% with all of those dogs having CR, with an extended PFS of 129 days.[16,17] Mitoxantrone is a related anthracenedione with similar mechanism of action. As a rescue agent, ORR to single agent mitoxantrone is 21-47%, with PFS of 84-126 days in dogs achieving CR.[18-20]

The alkylating agent CCNU is commonly used as a single agent and is an easily administered oral formulation. In two studies of CCNU alone or with prednisone, ORR was 27-59%, with a PFS of 15-86 days.[21-22] The non-traditional alkylating agent DTIC has also been used as a single agent in previously CHOP-treated dogs, with ORR 35% and DOR of 43d. Complete responders were rare, with a similar duration of response.[23] Thrombocytopenia is a common side effect. DTIC is given over a 4-5 hour CRI and pretreatment with anti-emetics is necessary.

Vinblastine has also been reported as a single agent. In a small study of CHOP-resistant dogs, there was an ORR of 20% for 21-48 days.[24] Anecdotally, L-asparaginase has also been used as a rescue agent particularly as it is well-tolerated, with low risk of myelosuppression. If a patient is refractory to the current chemotherapy agent and is myelosuppressed, L-asparaginase can provide short-term effect while the neutrophil count is recovering to allow an alternative protocol. As a foreign protein, hypersensitivity reactions can occur and the risk increases with consecutive doses. Pretreat with diphenhydramine at 2mg/kg SQ/IM is recommended to minimize this complication. Resistance to L-asparaginase appears to develop quickly.

Comparison of protocols

It is important to note that the various published protocols are not truly comparable. The majority of studies are retrospective rather than prospective studies, leading to inherent biases. Staging of dogs is variable between studies, and may be impacted by stage migration. Inclusion/exclusion criteria, response criteria, reporting of results, statistical power, clinical parameters, and pretreatment are variable between studies.

Future potential

Modifications of current protocols may yield slightly improved responses or limit toxicity, perhaps using alternative anthracyclines such as epirubicin or oral anti-mitotic agents. Recent work in canine lymphoma has focused on novel therapeutics such as procaspase activators, nucleotide prodrugs, heat shock protein peptide chaperones, bone marrow transplantation, monoclonal antibody treatment, receptor tyrosine kinase inhibition, radiation therapy, and activated B cell therapy. These and other treatments may offer additional survival advantages in a first-line or rescue setting for canine lymphoma and/or minimize toxicity to maintain quality of life.

Table: Chemotherapy protocols for relapsed canine multicentric lymphoma

	Protocol	ORR(CR), %	DOR/PFS Overall (CR), days
Multidrug protocols			
Re-induction, CHOP type	CHOP type[1,2]	100 (78-100)	NR, 159
MOPP-type	MOPP[3]	65 (31)	61 (63)
	MPP[4]	34 (17)	56 (238)
	BOPP/LOPP[5,6]	50-61 (27-36)	98-129 (112-130)
	MOMP[7]	51 (12)	56 (81)
Anthracycline-type	DMAC[8,9]	43-72 (16-44)	24-61 (63-112)
	Anthracycline/non-traditional alkylating agent[10,11]	46-71 (46-58)	NR, 75
Alkylating combo	DTIC/CCNU[12]	35 (23)	62 (83)
	CCNU/L-asparaginase +/- prednisone[13,14]	77-87 (52-65)	63-70 (90-111)
Single-agent			
Anthracycline-type	Doxorubicin[15]	75 (42)	140 (NR)
	Actinomycin-D[16,17]	0-41 (0-41)	0-129 (0-129)
	Mitoxantrone[18-20]	21-47 (26-47)	21, NR (84-126)
Alkylating agent	CCNU[21,22]	27-59 (7, NR)	15-86 (110)
	DTIC[23]	35 (3)	43 (44)
Vinca alkaloid	Vinblastine[24]	20 (6)	21-48d

REFERENCES

[1]Garrett LD, et al. J Vet Intern Med. 2002 Nov-Dec;16(6):704-9. [2]Flory AB, et al. J Am Vet Med Assoc. 2011 Feb 15;238(4):501-6. [3]Rassnick KM, et al. J Vet Intern Med. 2002 Sep-Oct;16(5):576-80. [4]Northrup NC, et al. Vet Comp Oncol. 2009 Mar;7(1):38-44. [5]LeBlanc AK, et al. Vet Comp Oncol. 2006 Mar;4(1):21-32. [6]Fahey CE, et al. J Am Vet Med Assoc. 2011 Jul 15;239(2):209-15. [7]Back AR, et al. Vet Comp Oncol. 2013 Aug 5. [Epub ahead of print] [8]Alvarez FJ, et al. J Vet Intern Med. 2006 Sep-Oct;20(5):1178-83. [9]Parsons-Doherty M, et al. Can Vet J. 2014 Feb;55(2):175-80. [10]Van Vechten M, et al. J Vet Intern Med. 1990 Jul-Aug;4(4):187-91. [11]Dervisis NG, et al. J Am Vet Med Assoc. 2007 Aug 15;231(4):563-9. [12]Flory AB, et al. J Vet Intern Med. 2008 Jan-Feb;22(1):164-71. [13] Saba CF, et al. J Vet Intern Med. 2007 Jan-Feb;21(1):127-32. [14] Saba CF, et al. J Vet Intern Med. 2009 Sep-Oct;23(5):1058-63. [15] Calvert CA, Leifer CE. J Am Vet Med Assoc. 1981 Nov 15;179(10):1011-2.[16]Moore AS, et al. J Vet Intern Med. 1994 Sep-Oct;8(5):343-4. [17]Bannink EO, et al. J Am Vet Med Assoc. 2008 Aug 1;233(3):446-51. [18] Lucroy MD, et al. J Vet Intern Med. 1998 Sep-Oct;12(5):325-9. [19]Moore AS, et al. J Am Vet Med Assoc. 1994 Jun 15;204(12):1903-5. [20] Ogilvie GK, et al. J Am Vet Med Assoc. 1991 May 1;198(9):1618-21. [21] Rusk A, et al. ClinCancer Res. 2006 Dec 15;12(24):7456-64. [22]Moore AS, et al. J Vet Intern Med. 1999 Sep-Oct;13(5):395-8. [23]Griessmayr PC, et al. J Vet Intern Med. 2009Nov-Dec;23(6):1227-31. [24]Bailey DB, et al. J Vet Intern Med. 2008 Nov-Dec;22(6):1397-402.

PARANEOPLASTIC SYNDROMES

Jackie M. Wypij, DVM, MS, DACVIM (Oncology)
Urbana, IL

INTRODUCTION

Paraneoplastic syndrome (PNS): Systemic disorder with clinical signs at a distant site from the primary tumor or metastases. Caused by: 1) tumor-produced factors (hormones, cytokines, peptides) with signaling activity, 2) tumor-depletion of normal substances or 3) host immune response to tumor antigens. In human oncology, PNS are very common with ~50-70% incidence, and are often the first clinical sign of cancer. In veterinary oncology, PNS are less common but are of significant importance because of their impact on morbidity and mortality.[1] In addition, they may be useful as a "first sign" of internal cancer and may be useful for monitoring cancer recurrence or progression.

HYPERCALCEMIA

Hypercalcemia is perhaps the most common PNS recognized in veterinary oncology and is more common in dogs than cats. Calcium regulation is complex, with the parathyroid gland, intestines, kidneys, and bone all involved in regulation. Total calcium consists of ~50% ionized, ~40% protein-bound and 10% complexed. Although hypophosphatemia is a normal physiologic response to hypercalcemia, the negative effects of calcium on the kidneys can result in a vicious cycle of elevated calcium/phosphorus and kidney damage. A calcium-phosphorus product of >70 is of concern. Concurrent evaluation of azotemia and urine specific gravity is needed. Mechanisms of kidney damage can include calcium deposition in tissues as well as arteriole vasoconstriction, altered glomerular permeability, increased renal vascular resistance, and acute tubular necrosis. In addition to renal injury/failure, hypercalcemia can result in neurologic and cardiac compromise in later stages.

Clinically, the most common sign is polyuria/polydipsia in dogs and decreased appetite in cats.[2,3] Cancer-associated hypercalcemia occurs via two primary mechanisms. The more common mechanism is humoral hypercalcemia of malignancy associated with parathyroid-related peptide (PTHrp) production or other soluble factors. Hypercalcemia affects 10-30% of human cancer patients with >80% being PTHrp-mediated. PTHrp is found in low levels in some normal tissues and necessary in fetal development and lactation. PTHrp-producing tumors include lymphoma (most commonly T cell), AGASACA, and multiple myeloma. The second mechanism is osteolytic disease from local infiltration of tumor cells into bone causing paracrine osteolysis. This is associated with metastatic or locally invasive carcinomas (prostate, mammary, squamous cell), hematological tumors (lymphoma, multiple myeloma) and rarely osteosarcoma or other sarcomas. Paracrine factors with osteoclast activating activity include growth factors, interleukins, prostaglandins, and RANKL. Approximately two-thirds of dogs and one-third of cats with hypercalcemia are diagnosed with cancer. In dogs, >75% of tumors are lymphoma with ~15% AGASACA and other carcinomas.[2] Other reported tumors include multiple myeloma, thyroid carcinoma, squamous cell carcinoma, mammary gland carcinoma, melanoma, primary lung tumors, chronic lymphocytic leukemia, renal tumors, thymoma, acanthomatous ameloblastoma, and parathyroid gland tumors. In cats, carcinomas (bronchogenic pulmonary CA, squamous cell CA) and hematological tumors (lymphoma and multiple myeloma) each account for about 1/3 of cancer-related hypercalcemia.[3]

A thorough physical exam should be performed with special emphasis on lymph node palpation, anal sac palpation, fundic examination, abdomen palpation, and thoracic compression. Useful diagnostic tests include a minimum database (CBC, serum biochemistry, urinalysis), ionized calcium, thoracic radiographs, abdominal ultrasound, +/- cytology of lymph nodes or masses. Total calcium is poorly predictive of ionized calcium levels.[4,5] Additional diagnostic tests might include ultrasound of the cervical area, PTH and PTHrp levels, bone survey radiographs, and bone marrow aspirate.

Treatment: Beyond treatment of the underlying cause, the primary mechanism of therapy is to promote renal calciuresis and inhibit bone reabsorption. Glucocorticosteroids favorably alter calcium balance (intestinal and renal effects) in mild cases. A diagnosis should be obtained prior to starting glucocorticosteroids. In emergent cases saline diuresis and furosemide (in euhydrated patients) are effective. For severe/refractory cases, bisphosphonates (pamidronate or zoledronate) are considered standard of care. Other potential treatments such as salmon calcitonin, mitramycin, and gallium nitrate are limited by toxicity, efficacy, and cost. Pamidronate is often most feasible due to cost at 1-2mg/kg intravenously with minimal side effects.[6,7]

ANEMIA

Anemia is one of the most common clinical problems related to veterinary cancer and can lead to significant morbidity however true paraneoplastic anemia is rarely the cause. Paraneoplastic anemia is usually an immune-mediated anemia with tumor production of antigens mimicking or binding reticulocytes. The most common cancer resulting in paraneoplastic anemia is lymphoma. Other tumors including sarcomas (hemangiosarcoma, histiocytic sarcoma) or other hematologic cancers may have an immune mediated component. A rare PNS-associated anemia results from hyperestrogenism from estrogen-producing ovarian or testicular neoplasms (e.g. Sertoli cell tumor) resulting in myelotoxicity including a nonregenerative anemia. Hemophagocytic syndrome can be a indirect effect via stimulation of non-malignant histiocytes or a paraneoplastic syndrome (usually histiocytic sarcoma).

More commonly anemia is cancer-related but not truly paraneoplastic. Examples include disordered iron storage and metabolism, shortened RBC lifespan, decreased marrow response, gastrointestinal or internal bleeding, disseminated intravascular coagulation, microangiopathic anemia, myelophthisis (crowding of bone marrow by cancer cells), and chemotherapy result in anemia in veterinary cancer patients. Hemophagocytic histiocytic sarcoma is most common, however other tumor types with direct erythrophagocytic ability include mast cell tumor, plasma cell tumors, lymphoma, hemangiosarcoma, and osteosarcoma.[8] While many cancer-associated causes of anemia also result in thrombocytopenia, myelophthisis usually causes pancytopenia. Anemia secondary to chemotherapy is usually mild in veterinary patients.

Treatment of paraneoplastic immune-mediated anemia focuses on treating/removing underlying cause, treating with immunosuppressive corticosteroids, and supportive care with blood products. Limited clinical information is available on other potential bone marrow stimulants such as erythropoietin/darbopoeitin, androgen therapy, lithium carbonate, or GHRH plasmid-based[9] therapy in veterinary cancer patients.

THROMBOCYTOPENIA

Thrombocytopenia is also common in veterinary and human cancer patients and can lead to potentially fatal complications. Similarly, true paraneoplastic thrombocytopenia is less common than other causes of decreased platelets. Similar to anemia (and often concurrent with), the most common cancer resulting in paraneoplastic thrombocytopenia is lymphoma. Histiocytic sarcoma and hemangiosarcoma, thymoma, and other hematologic cancers may also have an immune-mediated component however the anemia and thrombocytopenia is often multi-factorial. Less commonly, lymphoma can cause indirect, paraneoplastic hemophagocytic syndrome and rarely, hyperestrogenism from estrogen-producing ovarian or testicular neoplasms (e.g. Sertoli cell tumor) can result in myelotoxicity including thrombocytopenia. More commonly, thrombocytopenia is cancer-associated but non-paraneoplastic, from causes such as consumption (e.g. gastrointestinal or internal bleeding, DIC), decreased marrow response, myelophthisis, sequestration, or hemophagocytic syndrome (usually hemophagocytic histiocytic sarcoma). Some chemotherapy drugs preferentially affect megakaryocytic lineage (e.g. lomustine [CCNU], carboplatin) but are often reversible.

Treatment of paraneoplastic immune-mediated thrombocytopenia focuses on treating/removing underlying cause and treating with immunosuppressive corticosteroids. Blood products (fresh whole blood, platelet-rich plasma, platelet concentrate) are of limited benefit. There is limited clinical information available on other potential bone marrow stimulants such as oprelvekin (rHIL-11) or thrombopoietin in veterinary cancer patients. In addition, both erythropoietin and thrombopoietin receptors have been identified on tumor cells suggesting clinical implications of stimulation.[10,11]

CACHEXIA

Cachexia (loss of lean body mass) is present in ~50% of human oncology patients. Veterinary cancer patients are commonly overweight, with more than 55% with a body condition score (BCS) of 6/9 or greater.[12] In dogs while <5% of patients are cachectic, 15-35% have evidence of clinical muscle wasting and 37% had recently lost ≥5% of body weight.[12] Similar results are seen in feline patients, with 60% having reduced fat mass and 91% reduced muscle mass.[13] Two studies have demonstrated that weight loss prior to and/or during chemotherapy is a poor prognostic factor in cats.[14,15] It is known that lean body mass loss occurs before substantial weight loss can be detected and thus cachexia is likely a larger concern than is commonly thought.

The mechanisms leading to cancer cachexia are complex. First, inflammatory cytokines and leptin can affect the hypothalamic-gastrointestinal axis resulting in anorexia leading to inadequate nutrient intake. However the clinical syndrome may be characterized by weight loss in the face of potentially adequate calorie/nutrient intake. Cancer-associated mediators (e.g. inflammatory cytokines, catecholamines, cortisol, insulin, glucagon) also alter nutrient metabolism. This limits ability to utilize fat leading to amino acid consumption as the primary energy source leading to catabolization of lean body mass. This results in a negative energy balance; in advanced stages all body components are utilized and the catabolic state may be irreversible. In human cancer patients, loss of lean body mass has direct negative effects on strength, immune function, wound healing, and survival. Similar derangements in immunological and hematological parameters have been identified in dogs with cachexia.[16] The effects of cachexia (e.g. weakness, anorexia, weight loss, and perceived poor quality of life) are major contributing factors to euthanasia decisions. Although there is no apparent difference in energy requirements in cancer-bearing dogs,[17] the role of gastrointestinal alterations, altered nutrient absorption, or the balance between orexigenic vs. anorexigenic factors has not been described. Inflammatory cytokines (TNF, interleukin-1β, and interleukin-6) may be primary causative factors in cachexia because they increase energy metabolism and accelerate loss of lean body mass.

Treatment consists of early and aggressive nutritional support. Dietary modification, assisted feeding, alternative feeding strategies (e.g. esophagostomy tube), and appetite stimulants (e.g. mirtazapine, cyproheptadine, dronabinol) often are useful in maximizing food intake and quality of life. In addition to palatability, incorporating low simple carbohydrates, higher levels of fat, moderately high quality protein, and Ω-3 fatty acids may be ideal.

HYPERTROPHIC OSTEOPATHY

Hypertrophic osteopathy (HO) is an uncommon PNS reported in dogs and less commonly in cats usually associated with large intrathoracic tumors in dogs. Intra-abdominal tumors as well as non-neoplastic processes (e.g. thoracic infectious processes) may also cause HO. The syndrome presents as a shifting leg lameness with painful long bones, warmth, and swelling particularly distally. Lethargy, ocular discharge, and episcleral injection are often present.[18] Perpendicular periosteal new bone proliferation starts distally and may progress proximally. Concurrent hematologic and biochemical changes include anemia, neutrophilia, and elevated ALP.[18] The underlying pathophysiology remains undetermined at molecular level however proposed mechanisms including pulmonary vascular shunting, vagal abnormalities, and aberrant angiogenesis via VEGF, PDGF, or other humoral factors.[19] Treatment options include treat/remove underlying cancer/disease, vagal nerve transection, and palliative support including NSAIDs or corticosteroids, other analgesics, and bisphosphonates.

CUTANEOUS MANIFESTATIONS

A variety of PNS-related cutaneous manifestations have been reported. Most common are skin changes related to hypercortisolism. Necrolytic migratory erythema (NME) or superficial necrolytic dermatitis in dogs is a syndrome most often associated with certain chronic liver diseases (hepatocutaneous syndrome) or pancreatic glucogonomas (rarely insulinoma).[20,21] This crusting, exudative, alopecia is often perioral, periocular, and/or involving the genitals and distal extremities with concurrent hyperkeratosis of the footpads. Animals are often diabetic/hyperglycemic, although serum glucagon levels may not be elevated. Treatment includes octreotide (somatostatin analogue) and surgical resection. Paraneoplastic pemphigus is another rare form of PNS-skin disease in dogs. Clinical signs are similar to other pemphigus lesions with exfoliative dermatitis and oral ulcers; reported with thymoma, cutaneous lymphoma, and splenic sarcoma.

In cats, the two most common PNS-related skins disorders are thymoma-related exfoliative dermatitis and carcinoma-related alopecia (usually pancreatic, also hepato-biliary carcinoma). Feline thymoma-related exfoliative dermatitis is thought to be T-cell mediated, with abnormal antigen presentation of neoplastic thymic epithelial cells resulting in immature auto reactive T cells that attack keratinocytes. This presents as a nonpruiritic, erythematous dermatitis with exfoliative scaling primarily on the head, cervical region, ears, chest and trunk. A primary differential is non-thymoma associated exfoliative dermatitis, reported as a suspected immune-mediated disease in cats without evidence of neoplasia.[22] Both diseases have similar histopathology characteristics. Feline carcinoma-related alopecia is characteristically glistening, non-fragile alopecia that starts on the ventrum, then legs, and eventually the face, often sparing the pinna and top of head. Histopathologically, this disease appears quite distinct from thymoma-related exfoliative dermatitis. Differentials include skin fragility disorders related to hepatic disease and hyperadrenocorticism.

OTHER

Various other paraneoplastic syndromes are reported, including hypoglycemia, polycythemia, anaphylaxis, fever, myasthenia gravis, hyperviscosity syndrome, and neurologic deficiencies.

REFERENCES

[1]Belshaw Z. Abstract, 18th ECVIM-CA Congress, 2008 [2]Messinger JS, et al. J Vet Intern Med. 2009 May-Jun;23(3):514-9. [3]Savary KC, et al. J Vet Intern Med. 2000 Mar-Apr;14(2):184-9. [4]Schenk PA, Chew DJ. Can J Vet Res. 2010 Jul;74(3):209-13. [5]Schenck PA, Chew DJ. Am J Vet Res. 2005 Aug;66(8):1330-6. [6]Hostutler RA, et al. J Vet Intern Med. 2005 Jan-Feb;19(1):29-33. [7]Wypij JM, Heller DA. Vet Med Int. 2014;2014:675172. [8]Barger AM, et al. Vet Clin Pathol. 2012 Dec;41(4):587-9. [9]Bodles-Brakhop AM, et al. Mol Ther. 2008 May;16(5):862-70. [10]Rohrer Bley C, et al. Eur J Cancer. 2007 Mar;43(5):963-7. [11]Sottnik JL, Thamm DH. Vet Comp Oncol. 2010 Jun;8(2):96-102. [12]Michel KE, et al. J Vet Intern Med. 2004 Sep-Oct;18(5):692-5. [13]Baez JL, et al. J Feline Med Surg. 2007 Oct;9(5):411-7. Epub 2007 Apr 23. [14] Krick EL, et al. J Feline Med Surg. 2011 Dec;13(12):976-83. [15]Hadden AG, et al. J Vet Intern Med 2008; 22: 153–7. [16]Freeman LM. J Vet Intern Med 26.1 (2012): 3-17. [17]Ogilvie GK, et al. Am J Vet Res 1996;57:1463–1467. [18]Withers SS, et al. Vet Comp Oncol. 2013 Mar 14. [19]Atkinson S1, Fox SB.J Pathol. 2004 Jun;203(2):721-8. [20]Isidoro-Ayza M, et al. Vet Pathol. 2014 Jul;51(4):805-8. [21]Oberkirchner U, et al. Vet Dermatol. 2010 Oct;21(5):510-6. [22]Linek M, et al. Vet Dermatol. 2015 Feb;26(1):40-5, e12-3.

AUTHOR INDEX

AUTHOR INDEX

AUTHOR INDEX

AUTHOR INDEX

AUTHOR INDEX

American College of Veterinary Internal Medicine
1997 Wadsworth Boulevard Suite A
Lakewood CO 80214-5293

CURRAN ASSOCIATES INC
proceedings.com